To convert from	To	Multiply by
Mass Flow Rate		
lb_m/h	kg/s	1.26×10^{-4}
lb_m/s	kg/s	0.4536
Mass Flux, Mass Velocity		
$lb_m/h \cdot ft^2$	$kg/s \cdot m^2$	1.356×10^{-3}
Power		
$ft \cdot lb_f/h$	$W = J/s = N \cdot m/s$	3.766×10^{-4}
$ft \cdot lb_f/s$	$W = J/s = N \cdot m/s$	1.356
hp	$W = J/s = N \cdot m/s$	745.7
Btu (IT)/h	$W = J/s = N \cdot m/s$	0.2931
Pressure		
lb_f/ft^2	$Pa = N/m^2$	47.88
lb_f/in^2	$Pa = N/m^2$	6895
atm	$Pa = N/m^2$	1.013×10^5
bar	$Pa = N/m^2$	1×10^5
torr = mmHg	$Pa = N/m^2$	133.3
in. Hg	$Pa = N/m^2$	3386
in. H_2O	$Pa = N/m^2$	249.1
Specific Heat		
Btu (IT)/$lb_m \cdot °F$	$J/kg \cdot K = N \cdot m/kg \cdot K$	4187
$cal/g \cdot °C$	$J/kg \cdot K = N \cdot m/kg \cdot K$	4187
Surface Tension		
lb_f/ft	N/m	14.59
dyne/cm	N/m	0.001
erg/cm^2	N/m	0.001
Thermal Conductivity		
Btu (I/T) $\cdot ft/h \cdot ft^2 \cdot °F$	$W/m \cdot K = J/s \cdot m \cdot K$	1.731
cal (IT) $\cdot cm/s \cdot cm^2 \cdot °C$	$W/m \cdot K = J/s \cdot m \cdot K$	418.7
Velocity		
ft/h	m/s	8.467×10^{-5}
ft/s	m/s	0.3048
Viscosity		
$lb_m/ft \cdot s$	$kg/m \cdot s$	1.488
$lb_m/ft \cdot h$	$kg/m \cdot s$	4.134×10^{-4}
cP	$kg/m \cdot s$	0.001
Volume		
ft^3	m^3	0.02832
L	m^3	1×10^{-3}
gal (US)	m^3	3.785×10^{-3}

SEPARATION PROCESS PRINCIPLES

SEPARATION PROCESS PRINCIPLES

SECOND EDITION

J. D. Seader

Department of Chemical Engineering
University of Utah

Ernest J. Henley

Department of Chemical Engineering
University of Houston

John Wiley & Sons, Inc.

ACQUISITIONS EDITOR   Jennifer Welter

SENIOR PRODUCTION EDITOR   Patricia McFadden

OUTSIDE PRODUCTION MANAGEMENT   Ingrao Associates

MARKETING MANAGER   Frank Lyman

SENIOR DESIGNER   Kevin Murphy

PROGRAM ASSISTANT   Mary Moran-McGee

MEDIA EDITOR   Thomas Kulesa

FRONT COVER: Designed by Stephanie Santé using pictures with permission of Vendome Copper & Brass Works, Inc. and Sulzer Chemtech AG.

This book was set in 10/12 Times Roman by Interactive Composition Corporation and printed and bound by Courier/Westford. The cover was printed by Phoenix Color.

This book is printed on acid free paper. ∞

To order books or for customer service please, call 1-800-CALL WILEY (225-5945).

 ISBN-13 978- 0-471-46480-8

Printed in the United States of America

10 9 8 7

About the Authors

J. D. Seader is Professor Emeritus of Chemical Engineering at the University of Utah. He received B.S. and M.S. degrees from the University of California at Berkeley and a Ph.D. from the University of Wisconsin. From 1952 to 1959, Seader designed processes for Chevron Research in Richmond, California, and from 1959 to 1965, he conducted rocket engine research for Rocketdyne in Canoga Park, California. Before joining the faculty at the University of Utah, where he served for 37 years, he was a professor at the University of Idaho. Combined, he has authored or coauthored 110 technical articles, eight books, and four patents, and also coauthored the section on distillation in the sixth and seventh editions of *Perry's Chemical Engineers' Handbook*. Seader was a trustee of CACHE for 33 years, serving as Executive Officer from 1980 to 1984. For 20 years he directed the use and distribution of Monsanto's FLOWTRAN process simulation computer program for various universities. Seader also served as a director of AIChE from 1983 to 1985. In 1983, he presented the 35th Annual Institute Lecture of AIChE; in 1988 he received the Computing in Chemical Engineering Award of the CAST Division of AIChE; in 2004 he received the CACHE Award for Excellence in Chemical Engineering Education from the ASEE; and in 2004 he was a co-recipient of the Warren K. Lewis Award for Chemical Engineering Education of the AIChE. For 12 years he served as an Associate Editor for the journal, *Industrial and Engineering Chemistry Research*.

Ernest J. Henley is Professor of Chemical Engineering at the University of Houston. He received his B.S. degree from the University of Delaware and his Dr. Eng. Sci. from Columbia University, where he served as a professor from 1953 to 1959. Henley also has held professorships at the Stevens Institute of Technology, the University of Brazil, Stanford University, Cambridge University, and the City University of New York. He has authored or coauthored 72 technical articles and 12 books, the most recent one being *Probabilistic Risk Management for Scientists and Engineers*. For 17 years, he was a trustee of CACHE, serving as President from 1975 to 1976 and directing the efforts that produced the seven-volume set of "Computer Programs for Chemical Engineering Education" and the five-volume set, "AIChE Modular Instruction." An active consultant, Henley holds nine patents, and served on the Board of Directors of Maxxim Medical, Inc., Procedyne, Inc., Lasermedics, Inc., and Nanodyne, Inc. In 1998 he received the McGraw-Hill Company Award for "Outstanding Personal Achievement in Chemical Engineering," and in 2002, he received the CACHE Award of the ASEE for "recognition of his contribution to the use of computers in chemical engineering education." He is President of the Henley Foundation.

Preface to the Second Edition

NEW TO THIS EDITION

"Time and tide wait for no man" and most certainly not for engineering textbooks. The seven years since publication of the first edition of "Separation Process Principles" have witnessed: (1) advances in the fundamentals of mass, heat, and momentum transport and wide availability of computer programs to facilitate the application of complex transport mathematical models; (2) changes in the practice of chemical engineering design; and (3) restructuring of the chemical engineering curriculum. In response to what we have noted and what has been pointed out in strong reviews solicited by the publishers, we have included the following revisions and additions to this second edition:

- A new section on dimensions and units to facilitate the use of the SI, AE, and CGS systems, which permeate applications to separation processes.
- The addition to each chapter of a list of instructional objectives.
- Increased emphasis on the many ways used to express the composition of chemical mixtures.
- New material on the thermodynamics of difficult mixtures, including electrolytes, polymer solutions, and mixtures of light gases and polar organic compounds.
- Tables of typical diffusivity values.
- Table of formulae and meanings of dimensionless groups.
- A subsection on the recent theoretical analogy of Churchill and Zajic.
- New sections on hybrid systems and membrane cascades.
- Discussions of the fourth generation of random packings and high-capacity trays.
- A brief discussion of the rate-based multicell model.
- New section on optimal control as a third mode of operation for batch distillation.
- New discussion on concentration polarization and fouling.
- New sections on ultrafiltration and microfiltration.
- New subsection on Continuous, Countercurrent Adsorption Systems.
- Revision of the subsection on the McCabe–Thiele Method for Bulk Separation by adsorption.
- New subsection on Simulated (and True) Moving Bed Systems for Adsorption.

The following three chapters were not in the first edition of the book, but were available in hard copy, as supplemental chapters, to instructors. They are now included in the second edition:

Chapter 16 on Leaching and Washing, with an added subsection on the espresso machine.

Chapter 17 on Crystallization, Desublimation, and Evaporation.

Chapter 18 on Drying of Solids, including Psychrometry.

In the first edition, each topic was illustrated by at least one detailed example and was accompanied by at least three homework exercises. This continues to be true for most of the added topics and chapters. There are now 214 examples and 649 homework exercises. In addition, 839 references are cited.

TOPICAL ORGANIZATION

As with the first edition, the second edition covers equilibrium-based and rate-based models. The book is organized and divided into four parts. Part 1, which consists of five chapters, presents fundamental concepts. Chapter 1 introduces the many methods used to separate chemical mixtures, particularly in industrial applications. Chapter 2 is a review of solution thermodynamics as applied to separation problems. Chapter 3 covers the basic principles of diffusion and mass transfer, which are applied to rate-based models. The use of phase equilibrium and mass-balance equations for single equilibrium-stage models is presented in Chapter 4, while Chapter 5 treats cascades of equilibrium stages and hybrid separation systems.

The remaining three parts of the book are organized according to the method of separation. Part 2, consisting of Chapters 6–12, covers separations achieved by phase addition or creation. Chapters 6 through 8 cover absorption and stripping of dilute solutions, binary distillation, and ternary liquid–liquid extraction, with an emphasis on graphical methods of solution. Chapters 9 through 11 present computer-based methods widely used in process simulation programs, such as ASPEN PLUS, HYSYS, PRO-II, CHEMCAD, BATCH PLUS, and SUPERPRO DESIGNER, for multicomponent, equilibrium-based models of vapor–liquid and liquid–liquid separations. Chapter 12 treats multicomponent, rate-based models, while Chapter 13 focuses on binary and multicomponent batch distillation.

Part 3, consisting of Chapters 14 and 15, treats separations using barriers and solid agents, which have found increasing applications in industrial and laboratory operations. Chapter 14 covers rate-based models for membrane separations, while Chapter 15 is concerned with equilibrium-based and rate-based models of adsorption, ion exchange, and chromatography, which use solid or solid-like sorbents.

Separations involving a solid phase that undergoes a change in chemical composition are covered in Part 4, which consists of the final three chapters. Chapter 16 treats the selective leaching of material from a solid into a liquid solvent. Crystallization from a liquid and desublimation from a vapor are discussed in Chapter 17, which also includes evaporation. Chapter 18 is concerned with the drying of solids and includes a section on psychrometry.

Chapters 6, 7, 8, 14, 15, 16, 17, and 18 each begin with a detailed description of an industrial application of the subject separation process to help orient the student to actual equipment. Where appropriate, the development of theory is accompanied by appropriate historical content.

WEBSITES

Throughout the book, websites are occasionally cited that present useful, supplemental material. Students and instructors are also encouraged to use a search engine, such as Google, to locate additional information on old or new developments. Consider two examples: (1) McCabe–Thiele diagrams, which were first presented 80 years ago and (2) new developments in tower packings. A search on the former gives more than 700 websites listed in the order of the number of times accessed. A search on the latter references several websites.

Websites that have proven useful to students at the University of Houston include:

1. www.chemspy.com—Finds terms, definitions, synonyms, acronyms, and abbreviations; and provides links to tutorials and the latest news in chemical engineering and chemistry. Also assists in finding safety information, scientific publications, and worldwide patents.

2. webbook.nist.gov—Contains thermochemical data for more than 7000 compounds and thermophysical data for 34 fluids.

3. www.chemistry.about.com/od/chemicalengineerin1/index.htm—Includes articles and links to websites concerning topics in chemical engineering, including modeling, mass transfer, membrane separations, and thermodynamics.

4. www.matche.com—Provides capital cost data for many types of chemical processing equipment.

RESOURCES FOR INSTRUCTORS

Several resources for instructors may be found at the website: *www.wiley.com/college/seader.* Included are:

(1) Instructor's Solutions Manual giving detailed solutions to all homework exercises

(2) errata to various printings of the book

(3) a copy of a Preliminary Examination used successfully for a number of years by one of the authors to establish and ensure the preparedness of students for taking a course in separations, equilibrium-stage operations, and/or mass transfer. This closed-book, 50-minute examination, which has been given on the second day of the course, consists of 10 straightforward problems that cover the following topics studied by students in previous courses on fundamental principles of chemical engineering and heat transfer: Temperature conversions, pressure conversions, mass-to-moles conversion, mass balance around a generic separator, conversion of flow quantities, average residence time in a vessel, ideal gas law, heat duty for a heat exchanger, heat conduction, and heat convection. Students must retake the examination until all 10 problems are solved correctly.

These resources are password-protected, and are available only to instructors who adopt the text for their course. Visit the instructor section of the book website to register for a password.

RESOURCES FOR STUDENTS

Resources for students are also presented at the website: *www.wiley.com/college/seader.* Included are:

(1) an extensive list of Study Questions for each chapter to encourage students to read the book by helping them to confirm a suitable understanding of the descriptive material,

(2) Answers to Selected Odd-numbered Homework Exercises,

(3) a discussion of Problem-solving Techniques,

(4) Suggestions for Completing Homework Exercises, and

(5) Errata.

SUGGESTED COURSE OUTLINES

Undergraduate instruction on separation processes is incorporated in the chemical engineering curriculum following courses on fundamental principles of chemical engineering, chemical and engineering thermodynamics, fluid mechanics, and heat transfer. Courses that cover separation processes may be titled: Separations, Equilibrium-Stage Operations, Mass Transfer and Rate-Based Operations, or Unit Operations. This book contains sufficient material to be used in courses based on any of these four titles according to the following suggested outlines, depending on the number of semester credit hours.

SEPARATIONS OR UNIT OPERATIONS

3 Credit Hours: Chapters 1, 3, 4, 5, 6, 7, 8, (14, 15, or 17).
4 Credit Hours: Chapters 1, 3, 4, 5, 6, 7, 8, 9, 14, 15, 17
5 Credit Hours: Chapters 1, 3, 4, 5, 6, 7, 8, 9, 10, 13, 14, 15, 16, 17, 18

EQUILIBRIUM-STAGE OPERATIONS

3 Credit Hours: Chapters 1, 4, 5, 6, 7, 8, 9, 10
4 Credit Hours: Chapters 1, 4, 5, 6, 7, 8, 9, 10, 11, 13

MASS TRANSFER AND RATE-BASED OPERATIONS

3 Credit Hours: Chapters 1, 3, 6, 7, 8, 12, 14, 15
4 Credit Hours: Chapters 1, 3, 6, 7, 8, 12, 14, 15, 16, 17, 18

ACKNOWLEDGMENTS

In the preparation of the first and second editions of this book, a number of instructors provided valuable comments and suggestions. Among these, we wish to acknowledge Professors Richard G. Akins of Kansas State University, Paul Bienkowski of the University of Tennessee, C. P. Chen of the University of Alabama in Huntsville, William L. Conger of Virginia Polytechnic Institute and State University, Kenneth Cox of Rice University, R. Bruce Eldridge of the University of Texas at Austin, Rafiqul Gani of the Institut for Kemiteknik, Ram B. Gupta of Auburn University, Shamsuddin Ilias of North Carolina A&T State University, William A. Heenan of Texas A&M University–Kingsville, Kenneth R. Jolls of Iowa State University of Science and Technology, Alan M. Lane of the University of Alabama, Richard L. Long of New Mexico State University, Jerry Meldon of Tufts University, John Oscarson of Brigham Young University, Timothy D. Placek of Tufts University, Randel M. Price of Christian Brothers University, Michael E. Prudich of Ohio University, Daniel E. Rosner of Yale University, Ralph Schefflan of Stevens Institute of Technology, Ross Taylor of Clarkson University, and Vincent Van Brunt of the University of South Carolina.

J. D. Seader
Ernest J. Henley

Contents

Chapter 8   **Liquid–Liquid Extraction with Ternary Systems   295**

Chapter 9   **Approximate Methods for Multicomponent, Multistage Separations   344**

Chapter 18   **Drying of Solids   695**

Nomenclature

Latin Capital and Lowercase Letters

A constant in equations of state; constant in Margules equation; area for mass transfer; area for heat transfer; area; coefficient in Freundlich equation; absorption factor $= L/KV$; total area of a tray; frequency factor

A_a active area of a sieve tray

A_b active bubbling area of a tray

A_d downcomer cross-sectional area of a tray

A_{da} area for liquid flow under downcomer

A_h hole area of a sieve tray

A_{ij} binary interaction parameter in van Laar equation

$\bar{A}_{ij}$ binary interaction parameter in Margules two-constant equation

A_j, B_j, C_j, D_j material-balance parameters defined by (10-7) to (10-11)

A_M membrane surface area

A_p pre-exponential (frequency) factor

A_w specific surface area of a particle

a activity; constants in the ideal-gas heat capacity equation; constant in equations of state; interfacial area per unit volume; surface area; characteristic dimension of a solid particle; equivalents exchanged in ion exchange; interfacial area per stage

$\bar{a}$ interfacial area per unit volume of equivalent clear liquid on a tray

a_h specific hydraulic area of packing

a_{mk} group interaction parameter in UNIFAC method

a_v surface area per unit volume

B constant in equations of state, bottoms flow rate; number of binary azeotropes

B^0 rate of nucleation per unit volume of solution

b molar availability function $= h - T_0 s$; constant in equations of state; component flow rate in bottoms; surface perimeter

C general composition variable such as concentration, mass fraction, mole fraction, or volume fraction; number of components; constant; capacity parameter in (6-40); constant in tray liquid holdup expression given by (6-50); rate of production of crystals

C_1 constant in (6-126)

C_2 constant in (6-127)

C_D drag coefficient

C_F entrainment flooding factor in Figure 6.24 and (6-42)

C_L constant in (6-132) and Table 6.8

C_V constant in (6-133) and Table 6.8

C_h packing constant in Table 6.8

C_o orifice coefficient

C_P, C_p specific heat at constant pressure; packing constant in Table 6.8

$C_{P_V}^o$ ideal gas heat capacity at constant pressure

c molar concentration; constant in the BET equation; speed of light

c^* liquid concentration in equilibrium with gas at its bulk partial pressure

c' concentration in liquid adjacent to a membrane surface

c_m metastable limiting solubility of crystals

c_s humid heat; normal solubility of crystals

c_t total molar concentration

Δc_{limit} limiting supersaturation

$D, \mathcal{D}$ diffusivity; distillate flow rate; amount of distillate; desorbent (purge) flow rate; discrepancy functions in inside-out method of Chapter 10.

D_B bubble diameter

D_E eddy diffusion coefficient in (6-36)

D_e, D_{eff} effective diffusivity [see (3-49)]

D_H diameter of perforation of a sieve tray

D_i impeller diameter

D_{ij} mutual diffusion coefficient of i in j

D_K Knudsen diffusivity

D_L longitudinal eddy diffusivity

$\bar{D}_N$ arithmetic-mean diameter

D_o diffusion constant in (3-57)

D_P, D_p effective packing diameter; particle diameter

$\bar{D}_p$ average of apertures of two successive screen sizes

D_s surface diffusivity

$\bar{D}_S$ surface (Sauter) mean diameter

D_T tower or vessel diameter

$\bar{D}_V$ volume-mean diameter

$\bar{D}_W$ mass-mean diameter

d component flow rate in distillate

d_e equivalent drop diameter; pore diameter

d_H hydraulic diameter $= 4r_H$

d_m molecule diameter

d_p droplet or particle diameter; pore diameter

d_{vs} Sauter mean diameter defined by (8-35)

E — activation energy; dimensionless concentration change defined in (3-80); extraction factor defined in (4-24); amount or flow rate of extract; turbulent diffusion coefficient; voltage; wave energy; evaporation rate

E^0 — standard electrical potential

E_b — radiant energy emitted by a black body

E_D — activation energy of diffusion in a polymer

$E_{i,j}$ — residual of equilibrium equation (10-2)

E_{MD} — fractional Murphree dispersed-phase efficiency

E_{MV} — fractional Murphree vapor efficiency

E_{OV} — fractional Murphree vapor point efficiency

E_o — fractional overall stage (tray) efficiency

E_p — activation energy

$E_{\lambda,b}$ — radiant energy of a given wavelength emitted by a black body

$E\{t\}dt$ — fraction of effluent with a residence time between t and $t + dt$

$\mathscr{E}$ — number of independent equations in Gibbs phase rule

ΔE^{vap} — molar internal energy of vaporization

e — entrainment rate; heat transfer rate across a phase boundary

F — Faraday's constant $= 96,490$ coulomb/g-equivalent; feed flow rate; force; F-factor defined below (6-67)

F_b — buoyancy force

F_d — drag force

F_F — foaming factor in (6-42)

F_g — gravitational force

F_{HA} — hole-area factor in (6-42)

F_{LV}, F_{LG} — kinetic-energy ratio defined in Figure 6.24

F_P — Packing factor in Table 6.8

F_{ST} — surface tension factor in (6-42)

F_V — solids volumetric velocity in volume per unit cross-sectional area per unit time

$F\{t\}$ — fraction of eddies with a contact time less than t

$\mathscr{F}$ — number of degrees of freedom

f — pure-component fugacity; Fanning friction factor; function; component flow rate in feed; residual

f_f — fraction of flooding velocity

f_i — fugacity of component i in a mixture

f_v — volume shape factor

$\bar{f}$ — partial fugacity

f_ω — function of the acentric factor in the S-R-K and P-R equations

G — Gibbs free energy; mass velocity; volumetric holdup on a tray; rate of growth of crystal size

G_{ij} — binary interaction parameter in NRTL equation

g — molar Gibbs free energy; acceleration due to gravity

g_c — universal constant $= 32.174\ \text{lbm} \cdot \text{ft/lbf} \cdot \text{s}^2$

g_{ij} — energy of interaction in NRTL equation

H — Henry's law coefficient defined in Table 2.3; Henry's law constant defined in (3-50); height or length of vessel; molar enthalpy

$\bar{H}$ — partial molar enthalpy

H' — Henry's law coefficient defined by (6-121)

H_j — residual of energy balance equation (10-5)

ΔH_{ads} — heat of adsorption

ΔH_{cond} — heat of condensation

ΔH_{crys} — heat of crystallization

ΔH_{dil} — heat of dilution

ΔH_{sol}^{sat} — integral heat of solution at saturation

ΔH_{sol}^{∞} — heat of solution at infinite dilution

ΔH^{vap} — molar enthalpy of vaporization

H_G — height of a transfer unit for the gas phase $= l_T/N_G$

H_i — distance of impeller above tank bottom

H_L — height of a transfer unit for the liquid phase $= l_T/N_L$

H_{OG} — height of an overall transfer unit based on the gas phase $= l_T/N_{OG}$

H_{OL} — height of an overall transfer unit based on the liquid phase $= l_T/N_{OL}$

$\mathcal{H}$ — humidity

$\mathcal{H}'_m$ — molal humidity

$\mathcal{H}_P$ — percentage humidity

$\mathcal{H}_R$ — relative humidity

$\mathcal{H}_s$ — saturation humidity

$\mathcal{H}_w$ — saturation humidity at temperature T_w

HETP — height equivalent to a theoretical plate

HETS — height equivalent to a theoretical stage (same as HETP)

HTU — height of a transfer unit

h — molar enthalpy; heat-transfer coefficient; specific enthalpy; liquid molar enthalpy; height of a channel; height; Planck's constant $= 6.626 \times 10^{-34}\ \text{J} \cdot \text{s/molecule}$

h_d — dry tray pressure drop as head of liquid

h_{da} — head loss for liquid flow under downcomer

h_{dc} — clear liquid head in downcomer

h_{df} — height of froth in downcomer

h_f — height of froth on tray

h_l — equivalent head of clear liquid on tray

h_L — specific liquid holdup in a packed column

h_t — total tray pressure drop as head of liquid

h_w — weir height

h_σ — pressure drop due to surface tension as head of liquid

I — electrical current

i — current density

J_i — molar flux of i by ordinary molecular diffusion relative to the molar-average velocity of the mixture

j_D — Chilton–Colburn j-factor for mass transfer $\equiv N_{St_M}(N_{Sc})^{2/3}$

j_H — Chilton–Colburn j-factor for heat transfer $\equiv N_{St}(N_{Pr})^{2/3}$

j_M — Chilton–Colburn j-factor for momentum transfer $\equiv f/2$

j_i mass flux of i by ordinary molecular diffusion relative to the mass-average velocity of the mixture.

K equilibrium ratio for vapor–liquid equilibria; equilibrium partition coefficient in (3-53) and for a component distributed between a fluid and a membrane; overall mass-transfer coefficient; adsorption equilibrium constant

K' overall mass-transfer coefficient for UM diffusion

K_a chemical equilibrium constant based on activities

K_c solubility product; overall mass-transfer coefficient for crystallization

K_D equilibrium ratio for liquid–liquid equilibria

K'_D equilibrium ratio in mole- or mass-ratio compositions for liquid–liquid equilibria

K_G overall mass-transfer coefficient based on the gas phase with a partial pressure driving force

K_{ij} molar selectivity coefficient in ion exchange

K_L overall mass-transfer coefficient based on the liquid phase with a concentration driving force

K_s capacity parameter defined by (6-53)

K_W wall factor given by (6-111)

K_X overall mass-transfer coefficient based on the liquid phase with a mole ratio driving force

K_x overall mass-transfer coefficient based on the liquid phase with a mole-fraction driving force

K_Y overall mass-transfer coefficient based on the gas phase with a mole ratio driving force

K_y overall mass-transfer coefficient based on the gas phase with a mole-fraction driving force

K_r restrictive factor for diffusion in a pore

k thermal conductivity; mass-transfer coefficient in the absence of the bulk-flow effect

k' mass-transfer coefficient that takes into account the bulk-flow effect as in (3-229) and (3-230)

k_c mass-transfer coefficient based on a concentration, c, driving force; thermal conductivity of crystal layer

k_{ij} binary interaction parameter

k_i mass-transfer coefficient for integration into crystal lattice

k_N constant

k'_N constant

k_p mass-transfer coefficient for the gas phase based on a partial pressure, p, driving force

k_x mass-transfer coefficient for the liquid phase based on a mole-fraction driving force

k_y mass-transfer coefficient for the gas phase based on a mole-fraction driving force

$\bar{L}$ liquid molar flow rate in stripping section

L liquid; length; height; liquid flow rate; underflow flow rate; crystal size

L' solute-free liquid molar flow rate; liquid molar flow rate in an intermediate section of a column

L_B length of adsorption bed

L_e entry length

L_{pd} predominant crystal size

L_S liquid molar flow rate of sidestream

LES length of equilibrium (spent) section of adsorption bed

LUB length of unused bed in adsorption

L_w weir length

l constant in UNIQUAC and UNIFAC equations; component flow rate in liquid; length

l_{ij} binary interaction parameter

l_M membrane thickness

I_T packed height

M molecular weight; mixing-point amount or flow rate, molar liquid holdup

M_i moles of i in batch still

$M_{i,j}$ residual of component material-balance equation (10-1)

M_T mass of crystals per unit volume of magma

M_t total mass

m slope of equilibrium curve; mass flow rate; mass

m_c mass of crystals per unit volume of mother liquor

$\bar{m}_i$ molality of i in solution

m_p mass of adsorbent or particle

m_s mass of solid on a dry basis; solids flow rate

m_v mass evaporated; rate of evaporation

m_x tangent to the vapor–liquid equilibrium line in the region of liquid-film mole fractions as in Figure 3.22

m_y tangent to the vapor–liquid equilibrium line in the region of gas-film mole fractions as in Figure 3.22

MTZ length of mass-transfer zone in adsorption bed

N number of phases; number of moles; molar flux $= n/A$; number of equilibrium (theoretical, perfect) stages; rate of rotation; number of transfer units; cumulative number of crystals of size, L, and smaller; number of stable nodes; molar flow rate

N_A number of additional variables; Avogadro's number $= 6.022 \times 10^{23}$ molecules/mol

N_a number of actual trays

N_{Bi} Biot number for heat transfer

N_{Bi_M} Biot number for mass transfer

N_D number of degrees of freedom

N_E number of independent equations

N_{Eo} Eotvos number defined by (8-49)

N_{Fo} Fourier number for heat transfer $= \alpha t/a^2 =$ dimensionless time

N_{Fo_M} Fourier number for mass transfer $= Dt/a^2 =$ dimensionless time

N_{Fr} Froude number $=$ inertial force/gravitational force

N_G number of gas-phase transfer units defined in Table 6.7

N_L number of liquid-phase transfer units defined in Table 6.7

N_{Le} Lewis number $= N_{Sc}/N_{Pr}$

N_{Lu} Luikov number $= 1/N_{Le}$

N_{min} minimum number of stages for specified split

N_{Nu} Nusselt number $= dh/k =$ temperature gradient at wall or interface/temperature gradient across fluid ($d =$ characteristic length)

N_{OG} number of overall gas-phase transfer units defined in Table 6.7

N_{OL} number of overall liquid-phase transfer units defined in Table 6.7

N_{Pe} Peclet number for heat transfer $= N_{Re}N_{Pr} =$ convective transport to molecular transfer

N_{Pe_M} Peclet number for mass transfer $= N_{Re}N_{Sc} =$ convective transport to molecular transfer

N_{Po} Power number defined in (8-21)

N_{Pr} Prandtl number $= C_P\mu/k =$ momentum diffusivity/thermal diffusivity

N_R number of redundant equations

N_{Re} Reynolds number $= du\rho/\mu =$ inertial force/viscous force ($d =$ characteristic length)

NRX   number of reactions

N_{Sc} Schmidt number $= \mu/\rho D =$ momentum diffusivity/mass diffusivity

N_{Sh} Sherwood number $= dk_c/D =$ concentration gradient at wall or interface/concentration gradient across fluid ($d =$ characteristic length)

N_{St} Stanton number for heat transfer $= h/GC_P$

N_{St_M} Stanton number for mass transfer $= k_c\rho/G$

NTU   number of transfer units

N_T total number of crystals per unit volume of mother liquor; number of transfer units for heat transfer

N_t number of equilibrium (theoretical) stages

N_V number of variables

N_{We} Weber number defined by (8-37)

$\mathcal{N}$ number of moles

n molar flow rate; moles; constant in Freundlich equation; number of pores per cross-sectional area of membrane; number of crystals per unit size per unit volume

n_c number of crystals per unit volume of mother liquor

n^0 initial value for number of crystals per unit size per unit volume

n_+, n_- valences of cation and anion, respectively

P pressure; power; electrical power

P', P'' difference points

$\mathcal{P}$ parachor; number of phases in Gibbs phase rule

P_c critical pressure

P_M permeability

$\bar{P}_M$ permeance

P_r reduced pressure, P/P_c

P^s vapor pressure

P_P^s vapor pressure in a pore

P_0 adsorbate vapor pressure at test conditions

p partial pressure

p^* partial pressure in equilibrium with liquid at its bulk concentration

p_j, q_j material-balance parameters for Thomas algorithm in Chapter 10

Q rate of heat transfer; volume of liquid; volumetric flow rate

Q_C rate of heat transfer from condenser

Q_L volumetric liquid flow rate

Q_{ML} volumetric flow rate of mother liquor

Q_R rate of heat transfer to reboiler

Q_k area parameter for functional group k in UNIFAC method

q relative surface area of a molecule in UNIQUAC and UNIFAC equations; heat flux; loading or concentration of adsorbate on adsorbent; feed condition in distillation defined as the ratio of increase in liquid molar flow rate across feed stage to molar feed rate

$\bar{q}$ volume-average adsorbate loading defined for a spherical particle by (15-103)

q^e surface excess in liquid adsorption

q_L liquid flow rate across a tray

R universal gas constant:

1.987 cal/mol · K or Btu/lbmol · °F
8315 J/kmol · K or Pa · m³/kmol · K
82.06 atm · cm³/mol · K
0.7302 atm · ft³/lbmol · °R
10.73 psia · ft³/lbmol · °R;

molecule radius; amount or flow rate of raffinate; ratio of solvent to insoluble solids; reflux ratio; drying-rate flux; inverted binary mass-transfer coefficients defined by (12-31) and (12-32)

R' drying-rate per unit mass of bone-dry solid

R_c drying-rate flux in the constant-rate period

R_f drying-rate flux in the falling-rate period

R_k volume parameter for functional group k in UNIFAC method

R_L liquid-phase withdrawal factor in (10-80)

R_{min} minimum reflux ratio for specified split

R_p particle radius

R_V vapor-rate withdrawal factor in (10-81)

r relative number of segments per molecule in UNIQUAC and UNIFAC equations; radius; ratio of permeate to feed pressure for a membrane; distance in direction of diffusion; reaction rate; fraction of a stream exiting a stage that is removed as a sidestream; molar rate of mass transfer per unit volume of packed bed

r_c radius at reaction interface

r_H hydraulic radius $=$ flow cross section/wetted perimeter

r_p	pore radius	u_{mf}	minimum fluidization velocity
r_s	radius at surface of particle	u_o	hole velocity for sieve tray; superficial gas velocity in a packed column
r_w	radius at tube wall	u_s	superficial velocity
S	solid; rate of entropy; total entropy; solubility equal to H in (3-50); cross-sectional area for flow; solvent flow rate; mass of adsorbent; stripping factor $= KV/L$; surface area; inert solid flow rate; flow rate of crystals; supersaturation; belt speed; number of saddles	u_V	gas velocity
		u_0	characteristic rise velocity of a droplet
		V	vapor; volume; vapor flow rate; overflow flow rate
S_{ij}	separation factor in ion exchange	V'	vapor molar flow rate in an intermediate section of a column; solute-free molar vapor rate
S_g	surface area per unit volume of a porous particle	V_B	boilup ratio
S_x	residual of liquid-phase mole-fraction summation equation (10-3)	V_H	holdup as a fraction of dryer volume
		V_{LH}	volumetric liquid holdup
S_y	residual of vapor-phase mole-fraction summation equation (10-4)	V_{ML}	volume of mother liquor in magma
s	molar entropy; fractional rate of surface renewal; relative supersaturation	V_p	pore volume per unit mass of particle
		V_V	volume of a vessel
s_p	particle external surface area	$\bar{V}$	vapor molar flow rate in stripping section
SF	split fraction defined by (1-2)	$\mathcal{V}$	number of variables in Gibbs phase rule
SP	separation power or relative split ratio defined by (1-4); salt passage defined by (14-70)	v	molar volume; velocity; component flow rate in vapor; volume of gas adsorbed
SR	split ratio defined by (1-3)	$\bar{v}$	average molecule velocity
T	temperature	v_i	species velocity relative to stationary coordinates
T_c	critical temperature		
T_g	glass-transition temperature for a polymer	v_{i_D}	species diffusion velocity relative to the molar average velocity of the mixture
T_{ij}	binary interaction parameter in UNIQUAC and UNIFAC equations	v_c	critical molar volume
		v_H	humid volume
T_m	melting temperature for a polymer	v_M	molar average velocity of a mixture
T_0	datum temperature for enthalpy; reference temperature; infinite source or sink temperature	v_p	particle volume
		v_r	reduced molar volume, v/v_c
T_r	reduced temperature $= T/T_c$	v_s	molar volume of crystals
T_s	source or sink temperature	v_0	superficial velocity
T_v	moisture evaporation temperature	Σ_V	summation of atomic and structural diffusion volumes in (3-36)
t	time; residence time		
$\bar{t}$	average residence time	W	rate of work; width of film; bottoms flow rate; amount of adsorbate; washing factor in leaching $= S/RF_A$; baffle width; moles of liquid in a batch still; moisture content on a wet basis; vapor sidestream molar flow rate; weir length
t_b	time to breakthrough in adsorption		
t_c	contact time in the penetration theory		
t_E	elution time in chromatography		
t_F	feed pulse time in chromatography	W_{min}	minimum work of separation
t_L	contact time of liquid in penetration theory; residence time of crystals to reach size L	WES	weight of equilibrium (spent) section of adsorption bed
t_{res}	residence time	WUB	weight of unused adsorption bed
U	superficial velocity; overall heat-transfer coefficient; liquid sidestream molar flow rate; reciprocal of extraction factor	W_s	rate of shaft work
		w	mass fraction; width of a channel; weighting function in (10-90)
U_a	superficial vapor velocity based on tray active bubbling area	X	mole or mass ratio; mass ratio of soluble material to solvent in underflow; moisture content on a dry basis; general variable; parameter in (9-34)
U_f	flooding velocity		
u	velocity; interstitial velocity		
$\bar{u}$	bulk-average velocity; flow-average velocity	X^*	equilibrium moisture on a dry basis
$\bar{u}_r$	relative or slip velocity	X_B	bound moisture content on a dry basis
u_{all}	allowable velocity	X_c	critical free moisture content on a dry basis
u_c	velocity of concentration wave in adsorption	X_T	total moisture content on a dry basis
u_{ij}	energy of interaction in UNIQUAC equation	X_i	mass of solute per volume of solid
u_L	superficial liquid velocity		

X_m	mole fraction of functional group m in UNIFAC method	y	mole fraction in vapor phase; distance; mass fraction in extract; mass fraction in overflow
x	mole fraction in liquid phase; mole fraction in any phase; distance; mass fraction in raffinate; mass fraction in underflow; mass fraction of particles	$\mathbf{y}$	vector of mole fractions in vapor phase
		Z	compressibility factor $= Pv/RT$; total mass; height
x'	normalized mole fraction $= x_i / \sum_{j=1}^{C} x_j 1$	Z_f	froth height on a tray
		Z_L	length of liquid flow path across a tray
$\mathbf{x}$	vector of mole fractions in liquid phase	$\bar{Z}$	lattice coordination number in UNIQUAC and UNIFAC equations
x_n	fraction of crystals of size smaller than L	z	mole fraction in any phase; overall mole fraction in combined phases; distance; overall mole fraction in feed; dimensionless crystal size; length of liquid flow path across tray
Y	mole or mass ratio; mass ratio of soluble material to solvent in overflow; pressure-drop factor for packed columns defined by (6-102); concentration of solute in solvent; parameter in (9-34)	$\mathbf{z}$	vector of mole fractions in overall mixture

Greek Letters

α	thermal diffusivity, $k/\rho C_P$; relative volatility; surface area per adsorbed molecule	Λ_{ij}	binary interaction parameter in Wilson equation
α^*	ideal separation factor for a membrane	λ	mV/L; radiation wavelength
α_{ij}	relative volatility of component i with respect to component j for vapor-liquid equilibria; parameter in NRTL equation	λ_+, λ_-	limiting ionic conductances of cation and anion, respectively
		λ_{ij}	energy of interaction in Wilson equation
$\alpha_j, \beta_j \gamma_j$	energy-balance parameters defined by (10-23) to (10-26)	μ	chemical potential or partial molar Gibbs free energy; viscosity
β_{ij}	relative selectivity of component i with respect to component j for liquid–liquid equilibria	ν	momentum diffusivity (kinematic viscosity), μ/ρ; wave frequency; stoichiometric coefficient
Γ	film flow rate/unit width of film; thermodynamic function defined by (12-37)	$v_k^{(i)}$	number of functional groups of kind k in molecule i in UNIFAC method
Γ_k	residual activity coefficient of functional group k in UNIFAC equation	ξ	fractional current efficiency; dimensionless distance in adsorption defined by (15-115); dimensionless warped time in (11-2)
γ	specific heat ratio; activity coefficient		
Δ	change (final − initial)	π	osmotic pressure; product of ionic concentrations
δ	solubility parameter; film thickness; velocity boundary layer thickness; thickness of the laminar sublayer in the Prandtl analogy	ρ	mass density
		ρ_b	bulk density
δ_c	concentration boundary layer thickness	ρ_M	crystal density
δ_{ij}	Kronecker delta	ρ_p	particle density
ϵ	exponent parameter in (3-40); fractional porosity; allowable error; tolerance in (10-31)	ρ_s	true (crystalline) solid density
		σ	surface tension; interfacial tension; Stefan-Boltzmann constant $= 5.671 \times 10^{-8}\,\text{W/m}^2 \cdot \text{K}^4$
ϵ_b	bed porosity (external void fraction)		
ϵ_D	eddy diffusivity for diffusion (mass transfer)	σ_I	interfacial tension
ϵ_H	eddy diffusivity for heat transfer	$\sigma_{s,L}$	interfacial tension between crystal and solution
ϵ_M	eddy diffusivity for momentum transfer	τ	tortuosity; shear stress; dimensionless time in adsorption defined by (15-116); retention time of mother liquor in crystallizer; convergence criterion in (10-32)
ϵ_p	particle porosity (internal void fraction)		
η	Murphree vapor-phase plate efficiency in (10-73)		
θ	area fraction in UNIQUAC and UNIFAC equations; dimensionless concentration change defined in (3-80); correction factor in Edmister group method; cut equal to permeate flow rate to feed flow rate for a membrane; contact angle; fractional coverage in Langmuir equation; solids residence time in a dryer; root of the Underwood equation, (9-28)	τ_{ij}	binary interaction parameter in NRTL equation
		τ_w	shear stress at wall
		v	number of ions per molecule
		Φ, Φ'	volume fraction; parameter in Underwood equations (9-24) and (9-25)
		$\bar{\Phi}$	local volume fraction in the Wilson equation
θ_L	average liquid residence time on a tray	$\phi\{t\}$	probability function in the surface renewal theory
κ	Maxwell-Stefan mass-transfer coefficient in a binary mixture	ϕ	pure-species fugacity coefficient; association factor in the Wilke-Chang equation; recovery

factor in absorption and stripping; volume fraction; concentration ratio defined by (15-125)

$\bar{\phi}$	partial fugacity coefficient
ϕ_{df}	froth density
ϕ_e	effective relative density of froth defined by (6-48)
ϕ_s	particle sphericity
Ψ	segment fraction in UNIQUAC equation; V/F in flash calculations; E/F in liquid–liquid

equilibria calculations for single-stage extraction; sphericity defined before Example 15.7

Ψ_o	dry-packing resistance coefficient given by (6-113)
ψ	fractional entrainment; loading ratio defined by (15-126); sphericity
ω	acentric factor defined by (2-45); segment fraction in UNIFAC method

Subscripts

A	solute
a,ads	adsorption
avg	average
B	bottoms
b	bulk conditions; buoyancy
bubble	bubble-point condition
C	condenser; carrier; continuous phase
c	critical; convection; constant-rate period
cum	cumulative
D	distillate, dispersed phase; displacement
d	drag; desorption
d,db	dry bulb
des	desorption
dew	dew-point condition
ds	dry solid
E	enriching (absorption) section
e	effective; element
eff	effective
F	feed
f	flooding; feed; falling-rate period
G	gas phase
GM	geometric mean of two values, A and B = square root of A times B
g	gravity
gi	gas in
go	gas out
H,h	heat transfer
I, I	interface condition
i	particular species or component
in	entering
irr	irreversible
j	stage number
k	particular separator; key component

L	liquid phase; leaching stage
LM	log mean of two values, A and B = $(A - B)/\ln(A/B)$
LP	low pressure
M	mass transfer; mixing-point condition; mixture
m	mixture; maximum
max	maximum
min	minimum
N	stage
n	stage
O	overall
o,0	reference condition; initial condition
out	leaving
OV	overhead vapor
P	permeate
R	reboiler; rectification section; retentate
r	reduced; reference component; radiation
res	residence time
S	solid; stripping section; sidestream; solvent; stage; salt
s	source or sink; surface condition; solute; saturation
T	total
t	turbulent contribution
V	vapor
W	batch still
w	wet solid-gas interface
w,wb	wet bulb
ws	wet solid
X	exhausting (stripping) section
x,y,z	directions
δ	at the edge of the laminar sublayer
0	surroundings; initial
∞	infinite dilution; pinch-point zone

Superscripts

E	excess; extract phase
F	feed
ID	ideal mixture
(k)	iteration index

LF	liquid feed
o	pure species; standard state; reference condition
p	particular phase

R raffinate phase

s saturation condition

VF vapor feed

¯   partial quantity; average value

∞ infinite dilution

(1), (2)   denotes which liquid phase

I, II   denotes which liquid phase

*   at equilibrium

Abbreviations

Angstrom 1×10^{-10} m

ARD   asymmetric rotating disk contactor

atm   atmosphere

avg   average

BET   Brunauer-Emmett-Teller

BP   bubble-point method

B-W-R   Benedict-Webb-Rubin equation of state

bar   0.9869 atmosphere or 100 kPa

barrer   membrane permeability unit, 1 barrer = 10^{-10} cm^3 (STP) $\cdot$ cm/(cm$^2 \cdot$ s $\cdot$ cm Hg)

bbl   barrel

Btu   British thermal unit

C_i paraffin with i carbon atoms

$C_i^=$ olefin with i carbon atoms

C-S   Chao-Seader equation

°C   degrees Celsius, K-273.2

cal   calorie

cfh   cubic feet per hour

cfm   cubic feet per minute

cfs   cubic feet per second

cm   centimeter

cmHg   pressure in centimeters head of mercury

cP   centipoise

cw   cooling water

EMD   equimolar counter diffusion

EOS   equation of state

ESA   energy separating agent

ESS   error sum of squares

eq   equivalents

°F   degrees Fahrenheit, °R 459.7

FUG   Fenske-Underwood-Gilliland

ft   feet

GLC-EOS   group-contribution equation of state

GP   gas permeation

g   gram

gmol   gram-mole

gpd   gallons per day

gph   gallons per hour

gpm   gallons per minute

gps   gallons per second

H   high boiler

HHK   heavier than heavy key component

HK   heavy-key component

hp   horsepower

h   hour

I   intermediate boiler

in.   inch

J   joule

K   degrees Kelvin

kg   kilogram

kmol   kilogram-mole

L   liter; low boiler

LHS   left-hand side of an equation

LK   light-key component

LLK   lighter than light key component

L-K-P   Lee-Kessler-Plöcker equation of state

LM   log mean

LW   lost work

lb   pound

lb$_f$ pound-force

lb$_m$ pound-mass

lbmol   pound-mole

ln   logarithm to the base e

log   logarithm to the base 10

M   molar

MSMPR   mixed-suspension, mixed-product removal

MSC   molecular-sieve carbon

MSA   mass separating agent

MW   megawatts

m   meter

meq   milliequivalents

mg   milligram

min   minute

mm   millimeter

mmHg   pressure in mm head of mercury

mmol   millimole (0.001 mole)

mol   gram-mole

mole   gram-mole

N   newton; normal

NLE   nonlinear equation

NRTL   nonrandom, two-liquid theory

nbp   normal boiling point

ODE   ordinary differential equation

PDE   partial differential equation

POD   Podbielniak extractor

P-R   Peng–Robinson equation of state

ppm   parts per million (usually by weight)

PSA   pressure-swing adsorption

psi   pounds force per square inch

psia   pounds force per square inch absolute

PV   pervaporation

RDC rotating-disk contactor

RHS right-hand side of an equation

R-K Redlich-Kwong equation of state

R-K-S Redlich-Kwong-Soave equation of state (same as S-R-K)

RO reverse osmosis

RTL raining-bucket contactor

°R degrees Rankine

SC simultaneous-correction method

SG silica gel

S.G. specific gravity

SR stiffness ratio; sum-rates method

S-R-K Soave-Redlich-Kwong equation of state

STP standard conditions of temperature and pressure (usually 1 atm and either 0°C or 60°F)

s second

scf standard cubic feet

scfd standard cubic feet per day

scfh standard cubic feet per hour

scfm standard cubic feet per minute

stm steam

TSA temperature-swing adsorption

UMD unimolecular diffusion

UNIFAC UNIQUAC functional group activity coefficients

UNIQUAC universal quasi-chemical theory

VOC volatile organic compound

VPE vibrating-plate extractor

vs versus

VSA vacuum-swing adsorption

wt weight

y year

yr year

μm micron = micrometer

Mathematical Symbols

d differential

e exponential function

erf(x) error function of $x = \frac{1}{\sqrt{\pi}} \int_0^x \exp(-\eta^2) d\eta$

erfc(x) complementary error function of $x = 1 - \mathrm{erf}(x)$

exp exponential function

f function

i imaginary part of a complex value

ln natural logarithm

log logarithm to the base 10

∂ partial differential

{ } delimiters for a function

|| delimiters for absolute value

$\sum$ sum

π product; pi = 3.1416

Dimensions and Units

Chemical engineers must be proficient in the use of three systems of units: (1) the International System of Units, *SI System* (Systeme Internationale d'Unites), which was established in 1960 by the 11th General Conference on Weights and Measures and has been widely adopted; (2) the *AE* (American Engineering) *System,* which is based largely upon an English system of units adopted when the Magna Carta was signed in 1215 and is the preferred system in the United States; and (3) the *CGS* (centimeter-gram-second) *System,* which was devised in 1790 by the National Assembly of France, and served as the basis for the development of the SI System. A useful index to units and systems of units is given on the website at http://www.sizes.com/units/index.htm

Engineers must deal with *dimensions* and *units* to express the dimensions in terms of numerical *values*. Thus, for 10 gallons of gasoline, the dimension is volume, the unit is gallons, and the value is 10. As detailed in NIST (National Institute of Standards and Technology) Special Publication 811 (1995 edition), which is available at the website http://physics.nist.gov/cuu/pdf/sp811.pdf, units are *base* or *derived*.

BASE UNITS

The base units are those that cannot be subdivided, are independent, and are accurately defined. The base units are for dimensions of length, mass, time, temperature, molar amount, electrical current, and luminous intensity, all of which can be measured independently. Derived units are expressed in terms of base units or other derived units and include dimensions of volume, velocity, density, force, and energy. In this book we deal with the first five of the base dimensions. For these, the base units are:

Base Dimension	SI Unit	AE Unit	CGS Unit
Length	meter, m	foot, ft	centimeter, cm
Mass	kilogram, kg	pound, lb_m	gram, g
Time	second, s	hour, h	second, s
Temperature	kelvin, K	Fahrenheit, °F	Celsius, °C
Molar amount	gram-mole, mol	pound-mole, lbmol	gram-mole, mol

DERIVED UNITS

Many derived dimensions and units are used in chemical engineering. Several are listed in the following table:

Derived Dimension	SI Unit	AE Unit	CGS Unit
Area = $Length^2$	m^2	ft^2	cm^2
Volume = $Length^3$	m^3	ft^3	cm^3
Mass flow rate = Mass/Time	kg/s	lb_m/h	g/s
Molar flow rate = Molar amount/Time	mol/s	lbmol/h	mol/s
Velocity = Length/Time	m/s	ft/h	cm/s
Acceleration = Velocity/Time	m/s^2	ft/h^2	cm/s^2
Force = Mass · Acceleration	newton, N = $1 \text{ kg} \cdot m/s^2$	lb_f	dyne = $1 \text{ g} \cdot cm/s^2$

(Continued)

Derived Dimension	SI Unit	AE Unit	CGS Unit
Pressure = Force/Area	pascal, Pa = $1\ N/m^2$ = $1\ kg/m \cdot s^2$	lb_f/in^2	atm
Energy = Force $\cdot$ Length	joule, J = $1\ N \cdot m$ = $1\ kg \cdot m^2/s^2$	ft $\cdot$ lb_f, Btu	erg = 1 dyne $\cdot$ cm = $1\ g \cdot cm^2/s^2$, cal
Power = Energy/Time = Work/Time	Watt, W = $1\ J/s = 1\ N \cdot m/s$ = $1\ kg \cdot m^2/s^3$	hp	erg/s
Density = Mass/Volume	kg/m^3	lb_m/ft^3	g/cm^3

OTHER UNITS ACCEPTABLE FOR USE WITH THE SI SYSTEM

A major advantage of the SI System is the consistency of the derived units with the base units. However, some acceptable deviations from this consistency and some other acceptable base units are given in the following table:

Dimension	Base or Derived SI Unit	Acceptable SI Unit
Time	s	minute (min), hour (h), day (d), year (y)
Volume	m^3	liter (L) = $10^{-3}\ m^3$
Mass	kg	metric ton or tonne (t) = 10^3 kg
Pressure	Pa	bar = 10^5 Pa

PREFIXES

Also acceptable for use with the SI System are decimal multiples and submultiples of SI units formed by prefixes. The following table lists the more commonly used prefixes:

Prefix	Factor	Symbol
giga	10^9	G
mega	10^6	M
kilo	10^3	k
deci	10^{-1}	d
centi	10^{-2}	c
milli	10^{-3}	m
micro	10^{-6}	μ
nano	10^{-9}	n
pico	10^{-12}	p

USING THE AE SYSTEM OF UNITS

The AE System is more difficult to use than the SI System because of the units used with force, energy, and power. In the AE System, the force unit is the pound-force, lb_f, which is defined to be numerically equal to the pound-mass, lb_m, at sea-level of the Earth. Accordingly, Newton's second law of motion is written,

$$F = m\frac{g}{g_c}$$

where F = force in lb_f, m = mass in lb_m, g = acceleration due to gravity in ft/s^2, and to complete the definition, $g_c = 32.174\ lb_m \cdot ft/lb_f \cdot s^2$, where $32.174\ ft/s^2$ is the acceleration due to gravity at sea-level of the Earth. The constant, g_c, is not used with the SI System or the CGS System because the former does not define a kg_f and the CGS System does not use a g_f.

Thus, when using AE units in an equation that includes force and mass, incorporate g_c to adjust the units.

Example

A 5.000-pound-mass weight, m, is held at a height, h, of 4.000 feet above sea-level. Calculate its potential energy above sea-level, P.E. $= mgh$, using each of the three systems of units. Factors for converting units are given on the inside back cover of this book.

SI System:

$$m = 5.000\,\text{lb}_\text{m} = 5.000(0.4536) = 2.268\,\text{kg}$$
$$g = 9.807\,\text{m/s}^2$$
$$h = 4.000\,\text{ft} = 4.000(0.3048) = 1.219\,\text{m}$$
$$\text{P.E.} = 2.268(9.807)(1.219) = 27.11\,\text{kg}\cdot\text{m}^2/\text{s}^2 = 27.11\,\text{J}$$

CGS System:

$$m = 5.000\,\text{lb}_\text{m} = 5.000(453.6) = 2268\,\text{g}$$
$$g = 980.7\,\text{cm/s}^2$$
$$h = 4.000\,\text{ft} = 4.000(30.48) = 121.9\,\text{cm}$$
$$\text{P.E.} = 2268(980.7)(121.9) = 2.711 \times 10^8\,\text{g}\cdot\text{cm}^2/\text{s}^2$$
$$= 2.711 \times 10^8\,\text{erg}$$

AE System:

$$m = 5.000\,\text{lb}_\text{m}$$
$$g = 32.174\,\text{ft/s}^2$$
$$h = 4.000\,\text{ft}$$
$$\text{P.E.} = 5.000(32.174)(4.000) = 643.5\,\text{lb}_\text{m}\cdot\text{ft}^2/\text{s}^2$$

However, the accepted unit of energy for the AE System is ft · lb$_\text{f}$, which is obtained by dividing by g_c. Therefore, P.E. $= 643.5/32.174 = 20.00$ ft · lb$_\text{f}$.

Another difficulty with the AE System is the differentiation between energy as work and energy as heat. As seen in the preceding table, the work unit is ft · lb$_\text{f}$, while the heat unit is Btu. A similar situation exists in the CGS System with corresponding units of erg and calorie (cal). In older textbooks, the conversion factor between work and heat is often incorporated into an equation with the symbol J, called Joule's constant or the mechanical equivalent of heat, where,

$$J = 778.2\,\text{ft}\cdot\text{lb}_\text{f}/\text{Btu} = 4.184 \times 10^7\,\text{erg/cal}$$

Thus, in the previous example, the heat equivalents are

AE System:

$$20.00/778.2 = 0.02570\,\text{Btu}$$

CGS System:

$$2.711 \times 10^8/4.184 \times 10^7 = 6.479\,\text{cal}$$

In the SI System, the prefix M, mega, stands for million. However, in the natural gas and petroleum industries of the United States, when using the AE System, M stands for thousand and MM stands for million. Thus, MBtu stands for thousands of Btu, while MM Btu stands for millions of Btu.

It should be noted that the common pressure and power units in use for the AE System are not consistent with the base units. Thus, for pressure, pounds per square inch, psi or lb$_\text{f}$/in.^{2}, is used rather than lb$_\text{f}$/ft^2. For power, hp is used instead of ft · lb$_\text{f}$/h, where, the conversion factor is

$$1\,\text{hp} = 1.980 \times 10^6\,\text{ft}\cdot\text{lb}_\text{f}/\text{h}$$

CONVERSION FACTORS

Physical constants may be found on the inside front cover of this book. Conversion factors are given on the inside back cover. These factors permit direct conversion of AE and CGS values to SI values. The following is an example of such a conversion together with the reverse conversion.

Example

Convert 50 psia (lb_f/in^2 absolute) to kPa:
The conversion factor for lb_f/in^2 to Pa is 6895, which results in
$$50(6895) = 345000 \text{ Pa or } 345 \text{ kPa}$$

Convert 250 kPa to atm:
250 kPa = 250000 Pa. The conversion factor for atm to Pa is
1.013×10^5. Therefore, dividing by the conversion factor,
$$250000/1.013 \times 10^5 = 2.47 \text{ atm}$$

Three of the units [gallons (gal), calories (cal), and British thermal unit (Btu)] in the list of conversion factors have two or more definitions. The gallons unit cited here is the U.S. gallon, which is 83.3% of the Imperial gallon. The cal and Btu units used here are international (IT). Also in common use are the thermochemical cal and Btu, which are 99.964% of the international cal and Btu.

FORMAT FOR EXERCISES IN THIS BOOK

In numerical exercises throughout this book, the system of units to be used to solve the problem is stated. Then when given values are substituted into equations, units are not appended to the values. Instead, the conversion of a given value to units in the above tables of base and derived units is done prior to substitution in the equation or carried out directly in the equation as in the following example.

Example

Using conversion factors on the inside back cover of this book, calculate a Reynolds number, $N_{Re} = Dv\rho/\mu$, given $D = 4.0$ ft, $v = 4.5$ ft/s, $\rho = 60$ lb_m/ft^3, and $\mu = 2.0$ cP (i.e., centipoise).

Using the SI System (kg-m-s),
$$N_{Re} = \frac{Dv\rho}{\mu} = \frac{[(4.00)(0.3048)][(4.5)(0.3048)][(60)(16.02)]}{[(2.0)(0.001)]} = 804,000$$

Using the CGS System (g-cm-s),
$$N_{Re} = \frac{Dv\rho}{\mu} = \frac{[(4.00)(30.48)][(4.5)(30.48)][(60)(0.01602)]}{[(0.02)]} = 804,000$$

Using the AE System (lb_m-ft-h)
Convert the viscosity of 0.02 g/cm · s to $lb_m/ft \cdot h$:
$$N_{Re} = \frac{Dv\rho}{\mu} = \frac{(4.00)[(4.5)(3600)](60)}{[(0.02)(241.9)]} = 804,000$$

SEPARATION PROCESS PRINCIPLES

Part 1

Fundamental Concepts

In the first five chapters, fundamental concepts are presented that apply to processes for the separation of chemical mixtures. Emphasis is on industrial processes, but many of the concepts apply to small-scale separations as well. In Chapter 1, the role of separation operations in chemical processes is illustrated. Five general separation techniques are enumerated, each being driven by energy and/or the addition of mass to alter properties important to separation. For each technique, equipment types are briefly described. Various ways of specifying separation operations are discussed, including component recovery and product purity, and the use of these specifications in making mass balances is illustrated. The selection of feasible equipment for a particular separation problem is briefly covered.

The degree to which a separation can be achieved depends on differing rates of mass transfer of the individual components of the mixture, with limits dictated by thermodynamic phase equilibrium. Chapter 2 is a review of thermodynamics applicable to separation operations, particularly those involving fluid phases. Chapter 3 is an extensive discussion of mass transfer of individual components in binary mixtures under stagnant, laminar-flow, and turbulent-flow conditions, by analogy to conductive and convective heat transfer wherever possible.

Many separation operations are designed on the basis of the limit of attaining thermodynamic phase equilibrium. Chapter 4 covers mass-balance calculations for phase equilibrium in a single contacting stage that may include vapor, liquid, and/or solid phases. Often the degree of separation can be greatly improved by using multiple contacting stages, with each stage approaching equilibrium, in a cascade and/or by using a sequence of two or more different types of separation methods in a hybrid system. These are of great importance to industrial separation processes and are briefly described in Chapter 5, before proceeding to subsequent chapters in this book, each focusing on detailed descriptions and calculations for a particular separation operation. Included in Chapter 5 is a detailed discussion of degrees-of-freedom analysis, which determines the number of allowable specifications for cascades and hybrid systems. This type of analysis is used throughout this book, and is widely used in process simulators such as ASPEN PLUS, CHEMCAD, and HYSYS.

Chapter 1

Separation Processes

The separation of chemical mixtures into their constituents has been practiced, as an art, for millennia. Early civilizations developed techniques to (1) extract metals from ores, perfumes from flowers, dyes from plants, and potash from the ashes of burnt plants, (2) evaporate sea water to obtain salt, (3) refine rock asphalt, and (4) distill liquors. The human body could not function for long if it had no kidney, a membrane that selectively removes water and waste products of metabolism from blood.

Separations, including enrichment, concentration, purification, refining, and isolation, are important to chemists and chemical engineers. The former use *analytical separation methods,* such as chromatography, to determine compositions of complex mixtures quantitatively. Chemists also use small-scale *preparative separation techniques,* often similar to analytical separation methods, to recover and purify chemicals. Chemical engineers are more concerned with the manufacture of chemicals using economical, large-scale separation methods, which may differ considerably from laboratory techniques. For example, in a laboratory, chemists separate and analyze light-hydrocarbon mixtures by gas–liquid chromatography, while in a large manufacturing plant a chemical engineer uses distillation to separate the same hydrocarbon mixtures.

This book presents the principles of large-scale component separation operations, with emphasis on methods applied by chemical engineers to produce useful chemical products economically. Included are treatments of classical separation methods, such as distillation, absorption, liquid–liquid extraction, leaching, drying, and crystallization, as well as newer methods, such as adsorption and membrane separation. Separation operations for gas, liquid, and solid phases are covered. Using the principles of separation operations, chemical engineers can successfully develop, design, and operate industrial processes.

Increasingly, chemical engineers are being called upon to deal with industrial separation problems on a smaller scale, e.g., manufacture of specialty chemicals by batch processing, recovery of biological solutes, crystal growth of semiconductors, recovery of valuable chemicals from wastes, and the development of products (such as the artificial kidney) that involve the separation of chemical mixtures. Many of the separation principles for these smaller-scale problems are covered in this book and illustrated in examples and homework exercises.

1.0 INSTRUCTIONAL OBJECTIVES

After completing this chapter, you should be able to:

- Explain the role of separation operations in an industrial chemical process.
- Explain what constitutes the separation of a chemical mixture and enumerate the five general separation techniques.
- Explain the use of an energy-separating agent (ESA) and/or a mass-separating agent (MSA) in a separation operation.
- Explain how separations are made by phase creation or phase addition and list the many separation operations that use these two techniques.
- Explain how separations are made by introducing selective barriers and list several separation operations that utilize membranes.
- Explain how separations are made by introducing solid agents and list the three major separation operations that utilize this technique.
- Explain the use of external fields to separate chemical mixtures.
- Calculate component material balances around a separation operation based on specifications of component recovery (split ratios or split fractions) and/or product purity.

- Use the concepts of key components and separation power to measure the degree of separation between two key components.
- Make a selection of feasible separation operations based on factors involving the feed, products, property differences among chemical components, and characteristics of different separation operations.

1.1   INDUSTRIAL CHEMICAL PROCESSES

The chemical industry manufactures products that differ in chemical content from process feeds, which can be (1) naturally occurring raw materials, (2) plant or animal matter, (3) chemical intermediates, (4) chemicals of commerce, or (5) waste products. Especially common are oil refineries [1], which, as indicated in Figure 1.1, produce a variety of useful products. The relative amounts of these products produced from, say, 150,000 bbl/day of crude oil depend on the constituents of the crude oil and the types of refinery processes. Processes include distillation to separate crude oil into various boiling-point fractions or cuts, alkylation to combine small hydrocarbon molecules into larger molecules, catalytic reforming to change the structure of medium-size hydrocarbon molecules, fluid catalytic cracking to break apart large hydrocarbon molecules, hydrocracking to break apart even larger molecules, and other processes to convert the crude-oil residue to coke and lighter fractions.

A chemical process is conducted in either a *batchwise, continuous,* or *semicontinuous* manner. The operations may be classified as *key operations,* which are unique to chemical engineering because they involve changes in chemical composition, or *auxiliary operations,* which are necessary to the success of the key operations but may be designed by mechanical engineers as well because the auxiliary operations do not involve changes in chemical composition. The key operations are (1) chemical reactions and (2) separation of chemical mixtures. The auxiliary operations include phase separation, heat addition or removal (to change temperature or phase condition), shaft-work addition or removal (to change pressure), mixing or dividing of streams or batches of material, solids agglomeration, size reduction of solids, and separation of solids by size.

The key operations for the separation of chemical mixtures into new mixtures and/or essentially pure components are of central importance. Most of the equipment in the average chemical plant is there to purify raw materials, intermediates, and products by the separation techniques described briefly in this chapter and discussed in detail in subsequent chapters.

Block-flow diagrams are used to represent chemical processes. They indicate, by square or rectangular blocks, chemical reaction and separation steps and, by connecting lines, the major process streams that flow from one processing step to another. Considerably more detail is shown in *process-flow diagrams,* which also include auxiliary operations and utilize symbols that depict more realistically the type of equipment employed. The block-flow diagram of a continuous process for manufacturing hydrogen chloride gas from evaporated chlorine and electrolytic hydrogen [2] is shown in Figure 1.2. The heart of the process is a chemical reactor, where the high-temperature gas-phase combustion reaction, $H_2 + Cl_2 \rightarrow 2HCl$, occurs. The only auxiliary equipment required consists of pumps and compressors to deliver feeds to the reactor and product to storage, and a heat exchanger to cool the product. For this process, no separation operations are necessary because complete conversion of chlorine occurs in the reactor. A slight excess of hydrogen is used, and the product, consisting of 99% HCl and small amounts of H_2, N_2, H_2O, CO, and CO_2, requires no purification. Such simple commercial processes that require no separation of chemical species are very rare.

Some industrial chemical processes involve no chemical reactions, but only operations for separating chemicals and phases, together with auxiliary equipment. A block-flow diagram for such a process is shown in Figure 1.3, where wet natural gas is continuously separated into six light-paraffin

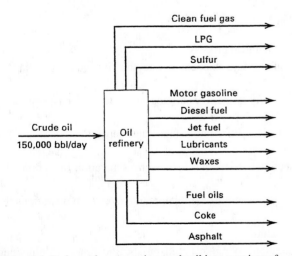

Figure 1.1 Refinery for converting crude oil into a variety of marketable products.

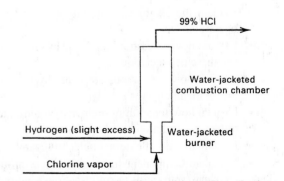

Figure 1.2 Synthetic process for anhydrous HCl production.

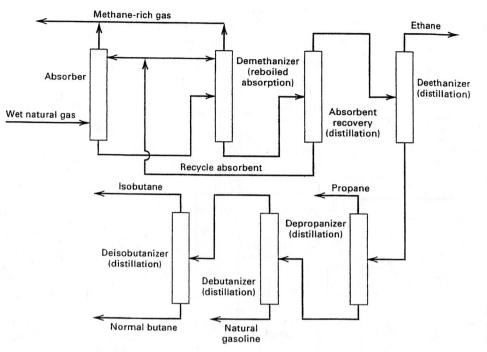

Figure 1.3 Process for recovery of light hydrocarbons from casinghead gas.

hydrocarbon components and mixtures by a train of separators [3]. A train or sequence of separators is used because it is often impossible to produce more than two products with a single piece of separation equipment.

Many industrial chemical processes involve at least one chemical reactor accompanied by one or more separation trains. An example is the continuous, direct hydration of ethylene to ethyl alcohol [4]. The heart of the process is a reactor packed with solid-catalyst particles, operating at 572 K and 6.72 MPa (570°F and 975 psia), in which the hydration reaction, $C_2H_4 + H_2O \rightarrow C_2H_5OH$, takes place. Because of thermodynamic equilibrium limitations, the conversion of ethylene is only 5% per pass through the reactor. The unreacted ethylene is recovered in a separation step and recycled back to the reactor. By this recycle technique, which is common to many industrial processes, essentially complete conversion of the ethylene fed to the process is achieved. If pure ethylene were available as a feedstock and no side reactions

occurred, the relatively simple process in Figure 1.4 could be constructed, in which two by-products (light ends and waste water) are also produced. This process uses a reactor, a partial condenser for ethylene recovery, and distillation to produce aqueous ethyl alcohol of near-azeotropic composition (93 wt%). Unfortunately, a number of factors frequently combine to increase the complexity of the process, particularly with respect to separation-equipment requirements. These factors include impurities in the ethylene feed, and side reactions involving both ethylene and feed impurities such as propylene. Consequently, the separation system must also deal with diethyl ether, isopropyl alcohol, acetaldehyde, and other chemicals. The resulting industrial process, shown in Figure 1.5, is much more complicated. After the hydration reaction, a partial condenser and high-pressure water absorber recover unreacted ethylene for recycling. The pressure of the liquid from the bottom of the absorber is reduced, causing partial vaporization. Vapor is separated from the

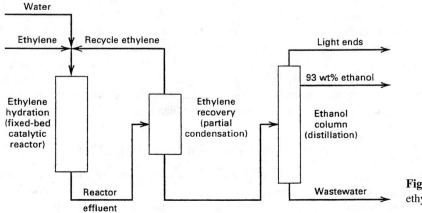
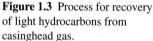
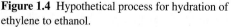

Figure 1.4 Hypothetical process for hydration of ethylene to ethanol.

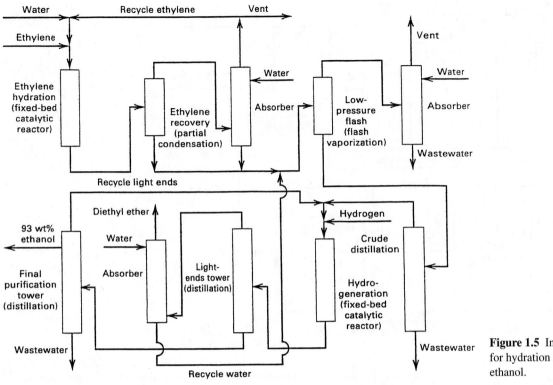

Figure 1.5 Industrial process for hydration of ethylene to ethanol.

remaining liquid in the low-pressure flash drum. Vapor from the low-pressure flash is scrubbed with water in an absorber to remove alcohol and prevent its loss to the vent gas. Crude, concentrated ethanol containing diethyl ether and acetaldehyde is distilled overhead in the crude-distillation (stripper) column and then catalytically hydrogenated in the vapor phase to convert acetaldehyde to ethanol. Diethyl ether is removed by distillation in the light-ends tower and scrubbed with water in an absorption tower. The final product is prepared by distillation in the final-purification tower, where 93 wt% aqueous ethanol product is withdrawn several trays below the top tray, light ends are concentrated in the so-called pasteurization-tray section above the product-withdrawal tray and recycled to the catalytic-hydrogenation reactor, and wastewater is removed from the bottom of the tower. Besides the separation equipment shown, additional separation steps may be necessary to concentrate the ethylene feed to the process and remove impurities that poison the catalysts. In the development of a new process from the laboratory stage through the pilot-plant stage, experience shows that more separation steps than originally anticipated are usually needed.

The above examples serve to illustrate the importance of separation operations in industrial chemical processes. Such operations are employed not only to separate a feed mixture into other mixtures and relatively pure components, to recover solvents for recycle, and to remove wastes, but also, when used in conjunction with chemical reactors, to purify reactor feeds, recover reactants from reactor effluents for recycle, recover by-products, and recover and purify products

to meet required specifications. Sometimes a separation operation, such as absorption of SO_2 by limestone slurry, may be accompanied by a chemical reaction that serves to facilitate the separation. In this book, emphasis is on separation operations that do not rely on concurrent chemical reactions; however, reactive distillation is discussed in Chapter 11.

Chemical engineers also design products. A significant product that involves the separation of chemicals is the espresso machine for making a cup of coffee that is superior to that made in a filter-drip machine. The goal in coffee making is to leach from the coffee beans the best oils, leaving behind ingredients responsible for acidity and bitterness in the cup of coffee. The espresso machine accomplishes this by conducting the leaching operation rapidly in 20–30 seconds with water at high temperature and pressure. If the operation is carefully controlled, the resulting cup of espresso, if immediately consumed, has: (1) a topping of creamy foam that traps the extracted chemicals, (2) a fullness of body due to emulsification, and (3) a richness of aroma. Typically, 25% of the coffee bean is extracted and the espresso contains less caffeine than filtered coffee. Cussler and Moggridge [13] and Seider, Seader, and Lewin [14] discuss many other examples of products designed by chemical engineers, some of which involve the separation of chemicals.

1.2 MECHANISM OF SEPARATION

Mixing of chemicals is a spontaneous, natural process that is accompanied by an increase in entropy or randomness. The inverse process, the separation of that mixture into its

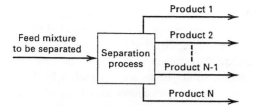

Figure 1.6 General separation process.

constituent chemical species, is not a spontaneous process; it requires an expenditure of energy. A mixture to be separated usually originates as a single, homogeneous phase (solid, liquid, or gas). If it exists as two or more immiscible phases, it is often best to first use a mechanical means based on gravity, centrifugal force, pressure reduction, or an electric and/or magnetic field to separate the phases. Then, appropriate separation techniques are applied to each phase.

A schematic diagram of a general separation process is shown in Figure 1.6. The feed mixture can be vapor, liquid, or solid, while the two or more products may differ in composition from each other and the feed, and may differ in phase state from each other and/or from the feed. The separation is accomplished by forcing the different chemical species in the feed into different spatial locations by any of five general separation techniques, or combinations thereof, as shown in Figure 1.7. The most common industrial technique, Figure 1.7a, involves the creation of a second phase (vapor, liquid, or solid) that is immiscible with the feed phase. The creation is accomplished by energy (heat and/or shaft-work) transfer to or from the process or by pressure

reduction. A second technique, Figure 1.7b, is to introduce the second phase into the system in the form of a solvent that selectively dissolves some of the species in the feed. Less common, but of growing importance, is the use of a barrier, Figure 1.7c, which restricts and/or enhances the movement of certain chemical species with respect to other species. Also of growing importance are techniques that involve the addition of solid particles, Figure 1.7d, which act directly or as inert carriers for other substances so as to cause separation. Finally, external fields, Figure 1.7e, of various types are sometimes applied for specialized separations.

For all the techniques of Figure 1.7, separations are achieved by enhancing the rate of mass transfer by diffusion of certain species relative to mass transfer of all species by bulk movement within a particular phase. The driving force and direction of mass transfer by diffusion is governed by thermodynamics, with the usual limitations of equilibrium. Thus, both transport and thermodynamic considerations are crucial in separation operations. The rate of separation is governed by *mass transfer,* while the extent of separation is limited by *thermodynamic equilibrium*. These two topics are treated in Chapters 2, 3, and 4. Fluid mechanics and heat transfer also play important roles, and applicable principles are included in appropriate chapters, particularly with respect to phase separation, phase change, pressure drop, temperature change, and entrainment.

The extent of separation achieved between or among the product phases for each of the chemical species present in the feed depends on the exploitation of differences in molecular, thermodynamic, and transport properties of the

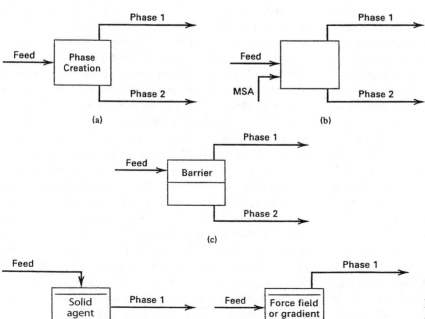

Figure 1.7 General separation techniques: (a) separation by phase creation; (b) separation by phase addition; (c) separation by barrier; (d) separation by solid agent; (e) separation by force field or gradient.

species in the different phases present. Some properties of importance are:

1. **Molecular properties**

Molecular weight	Polarizability
van der Waals volume	Dielectric constant
van der Waals area	Electric charge
Molecular shape (acentric factor)	Radius of gyration
Dipole moment	

2. **Thermodynamic and transport properties**

Vapor pressure	Adsorptivity
Solubility	Diffusivity

Values of these properties for many substances are available in handbooks, specialized reference books, and journals. Many of these properties can also be estimated using computer-aided, process-simulation programs. When they are not available, these properties must be estimated or determined experimentally if a successful application of the appropriate separation operation(s) is to be achieved.

1.3 SEPARATION BY PHASE ADDITION OR CREATION

If the feed mixture is a homogeneous, single-phase solution (gas, liquid, or solid), a second immiscible phase must often be developed or added before separation of chemical species can be achieved. This second phase is created by an *energy-separating agent* (ESA) and/or added as a *mass-separating agent* (MSA). Application of an ESA involves heat transfer and/or transfer of shaft work to or from the mixture to be separated. Alternatively, vapor may be created from a liquid phase by reducing the pressure. An MSA may be partially immiscible with one or more of the species in the mixture. In this case, the MSA frequently remains the constituent of highest concentration in the added phase. Alternatively, the MSA may be completely miscible with a liquid mixture to be separated, but may selectively alter the partitioning of species between liquid and vapor phases. This facilitates a more complete separation when used in conjunction with an ESA, as in extractive distillation.

Although separations that use an ESA are generally preferred, an MSA can make possible a separation that is not feasible with an ESA. Disadvantages of the use of an MSA are: (1) need for an additional separator to recover the MSA for recycle, (2) need for MSA makeup, (3) possible contamination of the product with the MSA, and (4) more difficult design procedures.

When two immiscible fluid phases are contacted, intimate mixing of the two phases is important in enhancing mass-transfer rates so that the thermodynamic-maximum degree-of-partitioning of species can be approached more rapidly. After phase contact, the separation operation is completed by employing gravity and/or an enhanced technique, such as centrifugal force, to disengage the two phases. Table 1.1 is a compilation of the more common industrial-

separation operations based on interphase mass transfer between two phases, one of which is created by an ESA or added as an MSA. Graphic symbols that are suitable for process-flow diagrams are included in the table. Vapor and liquid and/or solid phases are designated by *V*, *L*, and *S*, respectively. Design procedures have become fairly routine for the operations prefixed by an asterisk (*) in the first column of Table 1.1. Such procedures have been incorporated as mathematical models into widely used commercial computer-aided, chemical-process simulation and design (CAPD) programs for continuous, steady-state operations and are treated in considerable detail in subsequent chapters of this book. Batchwise modes of these operations are also treated in this book when appropriate.

When the feed mixture includes species that differ widely in their tendency to vaporize and condense, *partial condensation* or *partial vaporization,* Separation Operation (1) in Table 1.1 may be adequate to achieve the desired separation or recovery of a particular component. A vapor feed is partially condensed by removing heat, and a liquid feed is partially vaporized by adding heat. Alternatively, partial vaporization can be caused by *flash vaporization,* Operation (2), by reducing the pressure of the feed with a valve. In both of these operations, after partitioning of species by interphase mass transfer has occurred, the resulting vapor phase is enriched with respect to the species that are most volatile (most easily vaporized), while the liquid phase is enriched with respect to the least volatile species. After this single contact, the two phases, which, except near the critical region, are of considerably different density, are separated by gravity.

Often, the degree of species separation achieved by a single, partial vaporization or partial condensation step is inadequate because the volatility differences among species in the feed are not sufficiently large. In that case, it may still be possible to achieve a desired separation of the species in the feed mixture, without introducing an MSA, by employing *distillation,* Operation (3) in Table 1.1, the most widely utilized industrial separation method. Distillation involves multiple contacts between countercurrently flowing liquid and vapor phases. Each contact consists of mixing the two phases to promote rapid partitioning of species by mass transfer, followed by phase separation. The contacts are often made on horizontal trays (referred to as *stages*) arranged in a vertical column as shown in the symbol for distillation in Table 1.1. Vapor, while flowing up the column, is increasingly enriched with respect to the more volatile species. Correspondingly, liquid flowing down the column is increasingly enriched with respect to the less-volatile species. Feed to the distillation column enters on a tray somewhere between the top and bottom trays, and often near the middle of the column. The portion of the column above the feed entry is called the *enriching* or *rectification section,* and that below is the *stripping section.* Feed vapor starts up the column; feed liquid starts down. Liquid is required for making contacts with vapor above the feed tray, and vapor is required for making contacts with liquid below the feed tray.

Table 1.1 Separation Operations Based on Phase Creation or Addition

Separation Operation	Symbol[a]	Initial or Feed Phase	Created or Added Phase	Separating Agent(s)	Industrial Example[b]
Partial condensation or vaporization* (1)		Vapor and/or liquid	Liquid or vapor	Heat transfer (ESA)	Recovery of H_2 and N_2 from ammonia by partial condensation and high-pressure phase separation (Vol. 2, pp. 494–496)
Flash vaporization* (2)		Liquid	Vapor	Pressure reduction	Recovery of water from sea water (Vol. 24, pp. 343–348)
Distillation* (3)		Vapor and/or liquid	Vapor and liquid	Heat transfer (ESA) and sometimes work transfer	Purification of styrene (Vol. 21, pp. 785–786)
Extractive distillation* (4)		Vapor and/or liquid	Vapor and liquid	Liquid solvent (MSA) and heat transfer (ESA)	Separation of acetone and methanol (Suppl. Vol., pp. 153–155)
Reboiled absorption* (5)		Vapor and/or liquid	Vapor and liquid	Liquid absorbent (MSA) and heat transfer (ESA)	Removal of ethane and lower molecular weight hydrocarbons for LPG production (Vol. 14, pp. 384–385)
Absorption* (6)		Vapor	Liquid	Liquid absorbent (MSA)	Separation of carbon dioxide from combustion products by absorption with aqueous solutions of an ethanolamine (Vol. 4, pp. 730–735)
Stripping* (7)		Liquid	Vapor	Stripping vapor (MSA)	Stream stripping of naphtha, kerosene, and gas oil side cuts from crude distillation unit to remove light ends (Vol. 17, pp. 199–201)

(*Continued*)

Table 1.1 (*Continued*)

Separation Operation	Symbol[a]	Initial or Feed Phase	Created or Added Phase	Separating Agent(s)	Industrial Example[b]
Refluxed stripping (steam distillation)* (8)	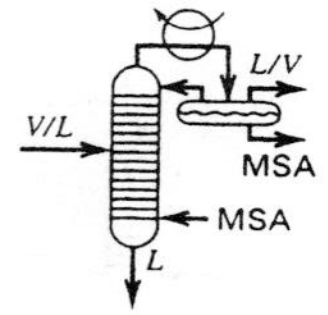	Vapor and/or liquid	Vapor and liquid	Stripping vapor (MSA) and heat transfer (ESA)	Separation of products from delayed coking (Vol. 17, pp. 210–215)
Reboiled stripping* (9)	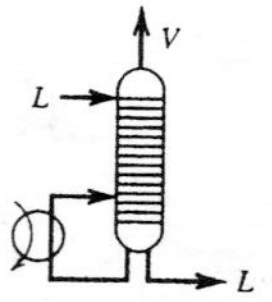	Liquid	Vapor	Heat transfer (ESA)	Recovery of amine absorbent (Vol. 17, pp. 229–232)
Azeotropic distillation* (10)	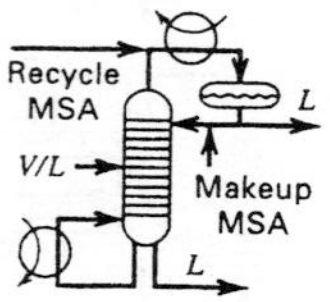	Vapor and/or liquid	Vapor and liquid	Liquid entrainer (MSA) and heat transfer (ESA)	Separation of acetic acid from water using n-butyl acetate as an entrainer to form an azeotrope with water (Vol. 3, pp. 365–368)
Liquid–liquid extraction* (11)	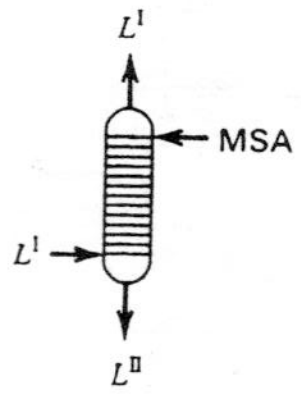	Liquid	Liquid	Liquid solvent (MSA)	Recovery of aromatics (Vol. 9, pp. 707–709)
Liquid–liquid extraction (two-solvent)* (12)	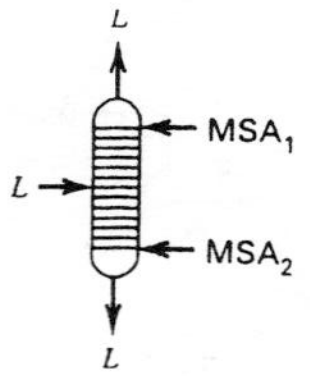	Liquid	Liquid	Two liquid solvents (MSA$_1$ and MSA$_2$)	Use of propane and cresylic acid as solvents to separate paraffins from aromatics and naphthenes (Vol. 17, pp. 223–224)
Drying (13)	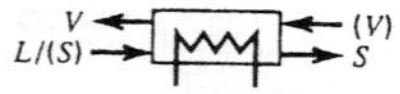	Liquid and often solid	Vapor	Gas (MSA) and/or heat transfer (ESA)	Removal of water from polyvinylchloride with hot air in a fluid-bed dryer (Vol. 23, pp. 901–904)

(*Continued*)

Table 1.1 (*Continued*)

Separation Operation	Symbol[a]	Initial or Feed Phase	Created or Added Phase	Separating Agent(s)	Industrial Example[b]
Evaporation (14)		Liquid	Vapor	Heat transfer (ESA)	Evaporation of water from a solution of urea and water (Vol. 23, pp. 555–558)
Crystallization (15)		Liquid	Solid (and vapor)	Heat transfer (ESA)	Crystallization of p-xylene from a mixture with m-xylene (Vol. 24, pp. 718–723)
Desublimation (16)		Vapor	Solid	Heat transfer (ESA)	Recovery of phthalic anhydride from noncondensible gas (Vol. 17, pp. 741–742)
Leaching (liquid–solid extraction) (17)		Solid	Liquid	Liquid solvent	Extraction of sucrose from sugar beets with hot water (Vol. 21, pp. 907–908)
Foam fractionation (18)		Liquid	Gas	Gas bubbles (MSA)	Recovery of detergents from waste solutions (Vol. 10, pp. 544–545)

*Design procedures are fairly well accepted.

[a]Trays are shown for columns, but alternatively packing can be used. Multiple feeds and side streams are often used and may be added to the symbol.

[b]Citations refer to volume and page(s) of *Kirk-Othmer Encyclopedia of Chemical Technology*, 3rd ed., John Wiley and Sons, New York (1978–1984).

Often, vapor from the top of the column is condensed in a condenser by cooling water or a refrigerant to provide contacting liquid, called *reflux*. Similarly, liquid at the bottom of the column passes through a reboiler, where it is heated by condensing steam or some other heating medium to provide contacting vapor, called *boilup*.

When volatility differences between species to be separated are so small as to necessitate more than about 100 trays in a distillation operation, *extractive distillation*, Operation (4), is often considered. Here, an MSA, acting as a solvent, is used to increase volatility differences between selected species of the feed, thereby reducing the number of required trays. Generally, the MSA, which must be completely miscible with the liquid phase throughout the column, is less volatile than any of the species in the feed mixture and is introduced to a stage near the top of the column. Reflux to the top tray is utilized to minimize MSA content in the top product. A subsequent separation operation, usually distillation, is used to recover the MSA for recycling back to the extractive distillation column.

If condensation of vapor leaving the top of a distillation column is not easily accomplished by heat transfer to cooling water or a refrigerant, a liquid MSA called an *absorbent* may be introduced to the top tray in place of reflux. The resulting separation operation is called *reboiled absorption*, (5). If the feed is all vapor and the stripping section of the column is not needed to achieve the desired separation, the operation is referred to as *absorption*, (6). This operation may not require an ESA and is frequently conducted at ambient temperature and high pressure. Constituents of the vapor feed dissolve in the absorbent to varying extents depending on their solubilities. Vaporization of a small fraction of the absorbent also generally occurs.

The inverse of absorption is *stripping*, Operation (7) in Table 1.1. Here, a liquid mixture is separated, generally at elevated temperature and ambient pressure, by contacting liquid feed with a stripping agent. This MSA eliminates the need to reboil the liquid at the bottom of the column, which may be important if the liquid is not thermally stable. If contacting trays are also needed above the feed tray in order to achieve the desired separation, a *refluxed stripper*, (8), may be employed. If the bottoms product from a stripper is thermally stable, it may be reboiled without using an MSA. In that case, the column is called a *reboiled stripper*, (9). Additional separation operations are required to recover, for recycling, MSAs used in absorption and stripping operations.

The formation of minimum-boiling azeotropic mixtures makes *azeotropic distillation*, (10), another useful tool where separation by distillation is not feasible. In the example cited in Table 1.1, the MSA, *n*-butyl acetate, which forms a heterogeneous (i.e., two liquid phases present), minimum-boiling azeotrope with water, is used as an entrainer to facilitate the separation of acetic acid from water. The azeotrope is taken overhead and then condensed and separated into acetate and water layers. The MSA is recirculated, and the distillate water layer and bottoms acetic acid are removed as products.

Liquid–liquid extraction, (11) and (12), using one or two solvents, respectively, is widely used when distillation is impractical, especially when the mixture to be separated is temperature-sensitive and/or more than about 100 distillation stages would be required. When one solvent is used, it selectively dissolves only one or a fraction of the components in the feed mixture. In a two-solvent extraction system, each solvent has its own specific selectivity for dissolving the components of the feed mixture. Thus, if a feed mixture consists of species A and B, solvent C might preferentially dissolve A, while solvent D dissolves B. As with extractive distillation, additional separation operations are generally required to recover, for recycling, solvent from streams leaving the extraction operation.

A variation of liquid–liquid extraction is *supercritical-fluid extraction*, where the extraction temperature and pressure are slightly above the critical point of the solvent. In this region, solute solubility in the supercritical fluid changes drastically with small changes in temperature and pressure. Following extraction, the pressure of the extract can be reduced to release the solvent, which is then recycled. For the processing of foodstuffs, the supercritical fluid is an inert substance such as CO_2, which will not contaminate the product.

Since many chemicals are processed wet but sold as dry solids, one of the more common manufacturing steps is *drying*, (13), which involves removal of a liquid from a solid by vaporization of the liquid. Although the only basic requirement in drying is that the vapor pressure of the liquid to be evaporated be higher than its partial pressure in the gas stream, the design and operation of dryers represents a complex problem in heat transfer, fluid mechanics, and mass transfer. In addition to the effect of such external conditions as temperature, humidity, air flow rate, and degree of solid subdivision on drying rate; the effect of internal conditions of liquid and vapor diffusion, capillary flow, equilibrium moisture content, and heat sensitivity in the solid must be considered. Although drying is a multiphase mass-transfer process, equipment-design procedures differ from those of any of the other processes discussed in this chapter because the thermodynamic concepts of equilibrium are difficult to apply to typical drying situations, where the concentration of vapor in the gas is so far from saturation, and concentration gradients in the solid are such that mass-transfer driving forces are undefined. Also, heat transfer rather than mass transfer may well be the limiting rate process. Therefore, the typical dryer design procedure is for the process engineer to send a few tons of representative, wet sample material for pilot-plant tests by one or two reliable dryer manufacturers and to purchase the equipment that produces a satisfactorily dried product at the lowest cost. The types of commercial dryers are discussed in detail in *Perry's Chemical Engineers' Handbook* [5] and Chapter 18.

Evaporation, Operation (14) in Table 1.1, is generally defined as the transfer of volatile components of a liquid into a gas by volatilization caused by heat transfer. Humidification and evaporation are synonymous in the scientific sense; however, *humidification* or *dehumidification* implies that one is intentionally adding vapor to or removing vapor from a gas. Major applications of evaporation are humidification, conditioning of air, cooling of water, and the concentration of aqueous solutions.

Crystallization, (15), is carried out in many organic, and almost all inorganic, chemical manufacturing plants where the desired product is a finely divided solid. Since crystallization is essentially a purification step, the conditions in the crystallizer must be such that impurities do not precipitate with the desired product. In *solution crystallization,* the mixture, which includes a solvent, is cooled and/or the solvent is evaporated to cause crystallization. In *melt crystallization,* two or more soluble species, in the absence of a solvent, are separated by partial freezing. A particularly versatile melt crystallization technique is *zone melting* or *refining,* which relies on selective distribution of impurity solutes between a liquid and a solid phase to achieve a separation. Many metals are refined by this technique, which, in its simplest form, involves moving a molten zone slowly through an ingot by moving the heater or drawing the ingot past the heater. The manufacture of single crystals has been a vital development in the semiconductor industry in recent years. Typically single crystals of very high purity silicon are produced worldwide by the Czochralski technique, wherein a single crystal is pulled from a melt. Typical crystal dimensions, after shaping into a uniform rod with diamond grinding machines, are 150-mm diameter $\times$ 1-m long, from which wafers of 675-micron thickness are sawed.

Sublimation is the transfer of a substance from the solid to the gaseous state without formation of an intermediate liquid phase, usually at a relatively high vacuum. Major applications have been in the removal of a volatile component from an essentially nonvolatile one. Examples are separation of sulfur from impurities, purification of benzoic acid, and freeze-drying of foods. The reverse process, *desublimation,* (16), is also practiced, for example, in the recovery of phthalic anhydride from gaseous reactor effluent. The most common application of sublimation in everyday life occurs in the use of Dry Ice as a refrigerant for storing ice cream, vegetables, and other perishables. The sublimed gas, unlike water, does not puddle and spoil the frozen materials.

Liquid–solid extraction, often referred to as *leaching,* (17), is widely used in the metallurgical, natural product, and food industries under batch, semicontinuous, or continuous operating conditions. The major problem in leaching is to promote diffusion of the solute out of the solid and into the liquid solvent. The most effective way of doing this is to reduce the dimensions of the solid to the smallest feasible particle size. For large-scale applications, in the metallurgical industries in particular, large, open tanks are used in countercurrent operation. The major difference between solid–liquid and liquid–liquid systems centers about the difficulty of transporting the solid, or the solid slurry, from stage to stage. For this reason, the solid may be left in the same tank, with only the liquid transferred from tank to tank. In the pharmaceutical, food, and natural-product industries, countercurrent solid transport is provided by complicated

mechanical devices. A supercritical fluid is sometimes used as the solvent in leaching.

In adsorptive-bubble separation methods, surface-active material collects at solution interfaces, establishing a concentration gradient between a solute in the bulk and in the surface layer. If the (very thin) surface layer can be collected, partial solute removal from the solution will have been achieved. The major application of this phenomenon is in ore flotation processes, where solid particles migrate to and attach themselves to rising gas bubbles and literally float out of the solution. This is essentially a three-phase system. *Foam fractionation,* (18), a two-phase adsorptive-bubble separation method, is a process where natural or chelate-induced surface activity causes a solute to migrate to rising bubbles and is, thus, removed as a foam. This method is not covered in this book.

Each equipment symbol shown in Table 1.1 corresponds to the simplest configuration for the operation represented. More complex versions are possible and frequently desirable. For example, a more complex version of the reboiled absorber, Separation Operation (5) in Table 1.1, is shown in Figure 1.8. This reboiled absorber has two feeds, an intercooler, a side stream, and both an interreboiler and a bottoms reboiler. Acceptable design procedures must handle such complex situations. It is also possible to conduct chemical reactions simultaneously with separation operations in a single column. Siirola [6] describes the evolution of an advanced commercial process for producing methyl acetate by the esterification of methanol and acetic acid. The process is conducted in a single column in an integrated process that involves three reaction zones and three separation zones.

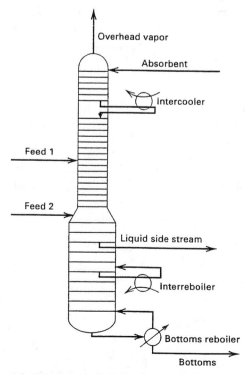

Figure 1.8 Complex reboiled absorber.

1.4 SEPARATION BY BARRIER

The use of microporous and nonporous membranes as semipermeable barriers for difficult and highly selective separations is rapidly gaining adherents. Membranes are fabricated from natural fibers, synthetic polymers, ceramics, or metals, but may also consist of liquid films. Solid membranes are fabricated into flat sheets, tubes, hollow fibers, or spiral-wound sheets, which are incorporated into commercial modules or cartridges, generally available only in certain sizes. For microporous membranes, separation is effected by differing rates of diffusion through the pores; while for nonporous membranes, separation occurs because of differences in both solubility in the membrane and rate of diffusion through the membrane. The most complex and selective membranes are found in the trillions of cells in the human body.

Table 1.2 lists the more widely used membrane separation operations. *Osmosis,* Operation (1) in Table 1.2, involves transfer, by a concentration gradient, of a solvent through a membrane into a mixture of solute and solvent. The membrane is almost nonpermeable to the solute. In *reverse osmosis,* (2), transport of solvent in the opposite direction is effected by imposing a pressure, higher than the osmotic pressure, on the feed side. Using a nonporous membrane, reverse osmosis successfully desalts water. *Dialysis,* (3), is the transport, by a concentration gradient, of small solute molecules, sometimes called crystalloids, through a porous membrane. The molecules unable to pass through the membrane are small, insoluble, nondiffusible particles, sometimes referred to as colloids.

Microporous membranes can be used in a manner similar to reverse osmosis to selectively allow small solute molecules and/or solvents to pass through the membrane and to prevent large dissolved molecules and suspended solids from passing through. *Microfiltration,* (4), refers to the retention of molecules typically in the size range from 0.02 to 10 μm. *Ultrafiltration,* (5), refers to the range from 1 to 20 nm. To retain molecules down to 0.1 nm, reverse osmosis, sometimes called *hyperfiltration,* can be used.

Although reverse osmosis can be used to separate organic and aqueous-organic liquid mixtures, high pressures are required. Alternatively, *pervaporation,* (6), in which the species being absorbed by and transported through the nonporous membrane are evaporated, can be used. This method, which uses much lower pressures than reverse osmosis, but

Table 1.2 Separation Operations Based on a Barrier

Separation Operation	Symbol[a]	Initial or Feed Phase	Separating Agent	Industrial Example[b]
Osmosis (1)		Liquid	Nonporous membrane	—
Reverse osmosis[*] (2)		Liquid	Nonporous membrane with pressure gradient	Desalinization of sea water (Vol. 24, pp. 349–353)
Dialysis (3)		Liquid	Porous membrane with pressure gradient	Recovery of caustic from hemicellulose (Vol. 7, p. 572)
Microfiltration (4)		Liquid	Microporous membrane with pressure gradient	Removal of bacteria from drinking water (Vol. 15, p. 115)
Ultrafiltration (5)		Liquid	Microporous membrane with pressure gradient	Separation of whey from cheese (Vol. 15, pp. 562–564)
Pervaporation[*] (6)		Liquid	Nonporous membrane with pressure gradient	Separation of azeotropic mixtures (Vol. 15, pp. 116–117)
Gas permeation[*] (7)		Vapor	Nonporous membrane with pressure gradient	Hydrogen enrichment (Vol. 20, pp. 709–710)
Liquid membrane (8)		Vapor and/or liquid	Liquid membrane with pressure gradient	Removal of hydrogen sulfide (Vol. 15, p. 119)

[*]Design procedures are fairly well accepted.

[a]Single units are shown. Multiple units can be cascaded.

[b]Citations refer to volume and page(s) of *Kirk–Othmer Encyclopedia of Chemical Technology*, 3rd ed., John Wiley and Sons, New York (1978–1984).

where the heat of vaporization must be supplied, is used to separate azeotropic mixtures.

The separation of gas mixtures by selective *gas permeation,* (7), through membranes, using pressure as the driving force, is a relatively simple process, first used in the 1940s with porous fluorocarbon barriers to separate $^{235}UF_6$ and $^{238}UF_6$ at great expense because it required enormous amounts of electric power. More recently, nonporous polymer membranes are used commercially to enrich gas mixtures containing hydrogen, recover hydrocarbons from gas streams, and produce nitrogen-enriched and oxygen-enriched air.

Liquid membranes, (8), of only a few molecules in thickness can be formed from surfactant-containing mixtures that locate at the interface between two fluid phases. With such a membrane, aromatic hydrocarbons can be separated from paraffinic hydrocarbons. Alternatively, the membrane can be formed by imbibing the micropores with liquids that are doped with additives to facilitate transport of certain solutes, such as CO_2 and H_2S.

1.5 SEPARATION BY SOLID AGENT

Separation operations that use solid mass-separating agents are listed in Table 1.3. The solid, usually in the form of a granular material or packing, acts as an inert support for a thin layer of absorbent or enters directly into the separation operation by selective adsorption of, or chemical reaction with, certain species in the feed mixture. Adsorption is confined to the sur-

face of the solid adsorbent, unlike absorption, which occurs throughout the bulk of the absorbent. In all cases, the active separating agent eventually becomes saturated with solute and must be regenerated or replaced periodically. Such separations are often conducted batchwise or semicontinuously. However, equipment is available to simulate continuous operation.

Adsorption, Separation Operation (1) in Table 1.3, is used to remove components present in low concentrations in nonadsorbing solvents or gases and to separate the components in gas or liquid mixtures by selective adsorption on solids, followed by desorption to regenerate the adsorbents, which include activated carbon, aluminum oxide, silica gel, and synthetic sodium or calcium aluminosilicate zeolite adsorbents (molecular sieves). The sieves differ from the other adsorbents in that they are crystalline and have pore openings of fixed dimensions, making them very selective. A simple adsorption device consists of a cylindrical vessel packed with a bed of solid adsorbent particles through which the gas or liquid flows. Regeneration of the adsorbent is conducted periodically, so two or more vessels are used, one vessel desorbing while the other(s) adsorb(s). If the vessel is arranged vertically, it is usually advantageous to employ downward flow of a gas. With upward flow, jiggling of the bed can cause particle attrition and a resulting increase in pressure drop and loss of material. However, for liquid flow, better distribution is achieved by upward flow. Regeneration is accomplished by one of four methods: (1) vaporizing the adsorbate with a hot purge gas (*thermal-swing adsorption*), (2) reducing the

Table 1.3 Separation Operations Based on a Solid Agent

Separation Operation	Symbol[a]	Initial or Feed Phase	Separating Agent	Industrial Example[b]
Adsorption* (1)		Vapor or liquid	Solid adsorbent	Purification of *p*-xylene (Vol. 24, pp. 723–725)
Chromatography* (2)		Vapor or liquid	Solid adsorbent or liquid adsorbent on a solid support	Separation of xylene isomers and ethylbenzene (Vol. 24, pp. 726–727)
Ion exchange* (3)		Liquid	Resin with ion-active sites	Demineralization of water (Vol. 13, pp. 700–701)

*Design procedures are fairly well accepted.

[a]Single units are shown. Multiple units can be cascaded.

[b]Citations refer to volume and page(s) of *Kirk–Othmer Encyclopedia of Chemical Technology,* 3rd ed., John Wiley and Sons, New York (1978–1984).

pressure to vaporize the adsorbate (*pressure-swing adsorption*), (3) inert purge stripping without change in temperature or pressure, and (4) displacement desorption by a fluid containing a more strongly adsorbed species.

Chromatography, Separation Operation (2) in Table 1.3, is a method for separating the components of a feed gas or liquid mixture by passing the feed through a bed of packing. The feed may be volatilized into a carrier gas, and the bed may be a solid adsorbent (gas-solid chromatography) or a solid-inert support that is coated with a very viscous liquid that acts as an absorbent (gas-liquid chromatography). Because of selective adsorption on the solid adsorbent surface or absorption into liquid absorbent, followed by desorption, different components of the feed mixture move through the bed at different rates, thus effecting the separation. In *affinity chromatography,* a macromolecule (called a ligate) is selectively adsorbed by a ligand (e.g., an ammonia molecule in a coordination compound) that is covalently bonded to a solid-support particle. Ligand–ligate pairs include inhibitors–enzymes, antigens–antibodies, and antibodies–proteins. Chromatography in its various forms is finding use in bioseparations.

Ion exchange, (3), resembles adsorption in that solid particles are used and regeneration is necessary. However, a chemical reaction is involved. In water softening, a typical ion-exchange application, an organic or inorganic polymer in its sodium form removes calcium ions by exchanging calcium for sodium. After prolonged use, the (spent) polymer, which becomes saturated with calcium, is regenerated by contact with a concentrated salt solution.

1.6 SEPARATION BY EXTERNAL FIELD OR GRADIENT

External fields can be used to take advantage of differing degrees of response of molecules and ions to forces and gradients. Table 1.4 lists common techniques, with combinations of these techniques with each other and with previously described separation methods also being possible.

Centrifugation, Operation (1) in Table 1.4, establishes a pressure field that separates fluid mixtures according to molecular weight. This technique is used to separate $^{235}UF_6$ from $^{238}UF_6$, and large polymer molecules according to molecular weight.

If a rather large temperature gradient is applied to a homogeneous solution, concentration gradients can be established and *thermal diffusion,* (2), is induced. It has been used to enhance the separation of uranium isotopes in gas permeation processes.

Natural water contains 0.000149 atom fraction of deuterium. When water is decomposed by *electrolysis,* (3), into hydrogen at the cathode and oxygen at the anode, the deuterium concentration in the hydrogen produced is lower than that in the water. Until 1953, this process was the only commercial source of heavy water (D_2O). In *electrodialysis,* (4), cation- and anion-permeable membranes carry a fixed charge, preventing the migration of species of like charge. This operation can be used to desalinize (remove salts from) sea water. A somewhat related process is *electrophoresis,* (5), which exploits the different migration velocities of charged colloidal or suspended species in an electric field. Positively charged species, such as dyes, hydroxide sols, and colloids, migrate to the cathode; while most small, suspended, negatively charged particles are attracted to the anode. By changing the solvent from an acidic to a basic condition, migration direction can sometimes be changed, particularly for proteins. Electrophoresis is a highly versatile method for separating biochemicals.

Another separation technique for biochemicals and difficult-to-separate heterogeneous mixtures of micromolecular and colloidal materials is *field-flow fractionation,* (6). For the mixture to be separated, an electrical field, magnetic field, or thermal gradient is established in a direction perpendicular to a laminar-flow field. Components of the mixture are driven to different locations in the stream; thus, they travel in the flow direction at different velocities, so a separation is achieved.

Table 1.4 Separation Operations by Applied Field or Gradient

Separation Operation	Initial or Feed Phase	Force Field or Gradient	Industrial Example[a]
Centrifugation (1)	Vapor	Centrifugal force field	Separation of uranium isotopes (Vol. 23, pp. 531–532)
Thermal diffusion (2)	Vapor or liquid	Thermal gradient	Separation of chlorine isotopes (Vol. 7, p. 684)
Electrolysis (3)	Liquid	Electrical force field	Concentration of heavy water (Vol. 7, p. 550)
Electrodialysis (4)	Liquid	Electrical force field and membrane	Desalinization of sea water (Vol. 24, pp. 353–359)
Electrophoresis (5)	Liquid	Electrical force field	Recovery of hemicelluloses (Vol. 4, p. 551)
Field-flow fractionation (6)	Liquid	Laminar flow in force field	

[a]Citations refer to volume and page(s) of *Kirk–Othmer Encyclopedia of Chemical Technology,* 3rd ed., John Wiley and Sons, New York (1978–1984).

1.7 COMPONENT RECOVERIES AND PRODUCT PURITIES

Separation operations are subject to the conservation of mass. Accordingly, if no chemical reactions occur and the process operates in a continuous, steady-state fashion, then for each component, i, in a mixture of C components, the molar (or mass) flow rate in the feed, $n_i^{(F)}$, is equal to the sum of the product molar (or mass) flow rates, $n_i^{(p)}$, for that component in the N product phases, p. Thus, referring to Figure 1.6,

$$n_i^{(F)} = \sum_{p=1}^{N} n_i^{(p)} = n_i^{(1)} + n_i^{(2)} + \cdots + n_i^{(N-1)} + n_i^{(N)} \quad (1\text{-}1)$$

To solve (1-1) for values of $n_i^{(p)}$, from specified values of $n_i^{(F)}$, we need an additional $N-1$ independent expressions involving $n_i^{(p)}$. This gives a total of NC equations in NC unknowns. For example, if a feed mixture containing C components is separated into N product phases, $C(N-1)$ additional expressions are needed. General forms of these expressions, which deal with the extent of separation, are considered in this and the next section. If more than one stream is fed to the separation process, $n_i^{(F)}$ is the summation for all feeds.

Equipment for separating components of a mixture is designed and operated to meet desired or required specifications, which are typically given as *component recoveries* and/or *product purities*. In Figure 1.9, the block-flow diagram for a hydrocarbon separation system, the feed is the bottoms product from a reboiled absorber used to deethanize (i.e., remove ethane and components of smaller molecular weight) a mixture of refinery gases and liquids. The separation process of choice in this example is a sequence of three multistage distillation columns. The composition of the feed to the process is included in Figure 1.9, where components are rank-listed by decreasing volatility, and hydrocarbons heavier (i.e., of greater molecular weight) than normal pentane and in the hexane (C_6)-to-undecane (C_{11}) range are lumped together in a so-called C_6^+ fraction. The three distillation columns of Figure 1.9 separate the deethanized feed into four products: a C_5^+-rich bottoms, a C_3-rich distillate, an iC_4-rich distillate, and an nC_4-rich bottoms. For each column, each component in the feed is partitioned between the overhead and the bottoms, according to a unique *split fraction* or *split ratio* that depends on (1) the component thermodynamic and transport properties in the vapor and liquid phases, (2) the number of contacting stages, and (3) the relative vapor and liquid flows through the column. The *split fraction*, SF, for component i in separator k is the fraction of that component found in the first product:

$$SF_{i,k} = \frac{n_{i,k}^{(1)}}{n_{i,k}^{(F)}} \quad (1\text{-}2)$$

where $n^{(1)}$ and $n^{(F)}$ refer to component molar flow rates in the first product and the feed, respectively. Alternatively, a *split ratio*, SR, between two products, may be defined as

$$SR_{i,k} = \frac{n_{i,k}^{(1)}}{n_{i,k}^{(2)}} = \frac{SF_{i,k}}{(1 - SF_{i,k})} \quad (1\text{-}3)$$

where $n^{(2)}$ refers to a component molar flow rate in the second product. Alternatively, SF and SR can be defined in terms of component mass flow rates.

If the process shown in Figure 1.9 is part of an operating plant with the measured material balance of Table 1.5, the split fractions and split ratios in Table 1.6 are determined from (1-2) and (1-3). In Table 1.5, it is seen that only two of the four products are relatively pure: C_3 overhead from the second column and iC_4 overhead from the third column. The molar purity of C_3 in the C_3 overhead is (54.80/56.00) or

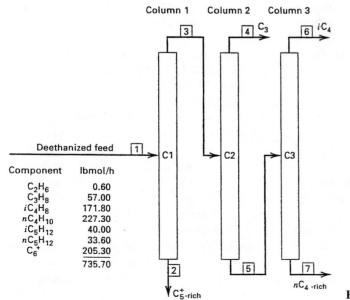

Figure 1.9 Hydrocarbon recovery process.

Table 1.5 Operating Material Balance for Hydrocarbon Recovery Process

	lbmol/h in Stream						
	1	2	3	4	5	6	7
Component	Feed to C1	C_5^+-rich	Feed to C2	C_3	Feed to C3	iC_4	nC_4-rich
C_2H_6	0.60	0.00	0.60	0.60	0.00	0.00	0.00
C_3H_8	57.00	0.00	57.00	54.80	2.20	2.20	0.00
iC_4H_{10}	171.80	0.10	171.70	0.60	171.10	162.50	8.60
nC_4H_{10}	227.30	0.70	226.60	0.00	226.60	10.80	215.80
iC_5H_{12}	40.00	11.90	28.10	0.00	28.10	0.00	28.10
nC_5H_{12}	33.60	16.10	17.50	0.00	17.50	0.00	17.50
C_6^+	205.30	205.30	0.00	0.00	0.00	0.00	0.00
Total	735.60	234.10	501.50	56.00	445.50	175.50	270.00

Table 1.6 Split Fractions and Split Ratios for Hydrocarbon Recovery Process

	Column 1		Column 2		Column 3		Overall Percent Recovery
Component	SF	SR	SF	SR	SF	SR	
C_2H_6	1.00	Large	1.00	Large	—	—	100
C_3H_8	1.00	Large	0.9614	24.91	1.00	Large	96.14
iC_4H_{10}	0.9994	1,717	0.0035	0.0035	0.9497	18.90	94.59
nC_4H_{10}	0.9969	323.7	0.00	0.00	0.0477	0.0501	94.94
iC_5H_{12}	0.7025	2.361	0.00	0.00	0.00	0.00	29.75
nC_5H_{12}	0.5208	1.087	0.00	0.00	0.00	0.00	47.92
C_6^+	0.00	Small	—	—	—	—	100

97.86%, while the iC_4 overhead purity is (162.50/175.50) or 92.59% iC_4. The nC_4-rich bottoms from Column 3 has an nC_4 purity of only (215.80/270.00) or 79.93%. Each of the three columns is designed to make a split between two adjacent components (called the *key components*) in the list of components ordered in decreasing volatility. As seen by the horizontal dividing lines in Table 1.6, the three key splits are nC_4H_{10}/iC_5H_{12}, C_3H_8/iC_4H_{10}, and iC_4H_{10}/nC_4H_{10} for Columns 1, 2, and 3, respectively. From the split fractions listed in Table 1.6, it is seen that all splits are relatively sharp (SF > 0.95 for the light key and SF < 0.05 for the heavy key), except for column 1, where the split ratio for the heavy key (iC_5H_{12}) is not sharp at all, and ultimately causes the nC_4-rich bottoms to be relatively impure in nC_4, even though the split between the two key components in the third column is relatively sharp.

In Table 1.6, for each column, values of SF and SR decrease in the order of the ranked component list. It is also noted in Table 1.6 that SF may be a better quantitative measure of degree of separation than SR because SF is bounded between 0 and 1, while SR can range from 0 to a very large value.

Two other common measures of extent of separation can be applied to each column, or to the separation system as a whole. One measure is the *percent recovery* in a designated system product of each component in the feed to the system.

These values, as computed from the data of Table 1.5, are listed in the last column of Table 1.6. As shown, the component recoveries are all relatively high (>95%) except for the two pentane isomers. The other measure of extent of separation is *product purity*. These purities for the main component were computed for all except the C_5^+-rich product, which is [(11.90 + 16.10 + 205.30)/234.10] or 99.66% pure with respect to the pentanes and heavier. Such a product is a *multicomponent product*. One of the most common multicomponent products is gasoline.

Product impurity levels and a designation of the impurities are included in product specifications for chemicals of commerce. The product purity with respect to each component in each of the three final products for the hydrocarbon recovery process, as computed from the process operating data of Table 1.5, is given in Table 1.7, where the values are also extremely important because maximum allowable percentages of impurities are compared to the product specifications. The C_5^+ fraction is not included because it is an intermediate that is sent to an isomerization process. From the comparison in Table 1.7, it is seen that two products easily meet their specifications, while the iC_4 product barely meets its specification. If the process is equipped with effective controllers, it might be possible to reduce the energy input to the process and still meet C_3 and nC_4-rich product

Table 1.7   Comparison of Measured Product Purities with Specifications

| | mol% in Product | | | | | |
| | Propane | | Isobutane | | Normal Butane | |
Component	Data	Spec	Data	Spec	Data	Spec
C_2H_6	1.07	5 max	0	—	0	—
C_3H_8	97.86	93 min	1.25	3 max	0	1 max
iC_4H_{10}	1.07	2 min	92.60	92 min	{ 83.11	{ 80 min
nC_4H_{10}	0	—	6.15	7 max		
C_5^+	0	—	0	—	16.89	20 max
Total	100.00		100.00		100.00	

specifications. Although the product purities in Table 1.7 are given in mol%, this designation is usually restricted to gas mixtures for which purities in vol% are equivalent to mol%. For liquid mixtures, purities are often specified in wt%. To meet environmental regulations, maximum amounts of impurities in gas, liquid, and solids streams are typically specified in *ppm* (parts per million) or *ppb* (parts per billion), usually by volume (same as moles) for gases and by weight (mass) for liquids and solids. For aqueous solutions, especially those containing acids and bases, common designations for composition are *molarity* or molar concentration (moles of solute per liter of solution, M), *normality* (number of equivalent weights of solute per liter of solution, N), and *molality* (moles of solute per kilogram of solvent). For some chemical products, an attribute, such as color, may be used in place of a purity in terms of composition.

1.8   SEPARATION POWER

Some separations in Table 1.1 are often inadequate for making a sharp split between two key components of a feed mixture, and can only effect the desired recovery of a single key component. Examples are Operations 1, 2, 6, 7, 8, 9, 11, 13, 14, 15, 16, and 17 in Table 1.1. For these, either a single separation stage is utilized as in Operations 1, 2, 13, 14, 15, 16, and 17 or the feed enters at one end (not near the middle) of a multistage separator as in Separation Operations 6, 7, 8, 9, and 11. The split ratio, SR, split fraction, SF, recovery, or purity that can be achieved for the key component depends on a number of factors. For the simplest case of a single separation stage, the factors that influence SR and SF values include: (1) the relative molar amounts of the two phases leaving the separator and (2) thermodynamic, mass transport, and other properties of the key components. For multistage separators, an additional factor must be added, namely, (3) the number of stages and their configuration. The quantitative relationships involving these factors are unique to each type of separator. Therefore, detailed discussion of these relationships is deferred to subsequent chapters, where individual separation operations are discussed in detail. A general but brief discussion of some of the important property factors is given in the next section.

When multistage separators are utilized and the feed mixture enters somewhere near the middle of the separator, such that the separator consists of two sections of stages, one on either side of the feed stage, it is often possible to achieve a relatively sharp separation between two key components. One section acts to remove one key component, while the other section acts to remove the other key component. Examples are Separation Operations 3, 4, 5, 10, and 12 in Table 1.1. For these operations, a convenient measure of the relative degree of separation between two components, i and j, is the *separation power* (also referred to as the relative split ratio and the separation factor), SP, of the separation equipment, defined in terms of the component splits achieved, as measured by the compositions of the two products, (1) and (2):

$$SP_{i,j} = \frac{C_i^{(1)}/C_i^{(2)}}{C_j^{(1)}/C_j^{(2)}} \qquad (1\text{-}4)$$

where C is some measure of composition such as mole fraction, mass fraction, or concentration in moles or mass per unit volume. Most commonly, mole fractions or concentrations are used, but in any case, the separation power is readily converted to the following forms in terms of split fractions or split ratios:

$$SP_{i,j} = \frac{SR_i}{SR_j} \qquad (1\text{-}5)$$

$$SP_{i,j} = \frac{SF_i/SF_j}{(1 - SF_i)/(1 - SF_j)} \qquad (1\text{-}6)$$

Achievable values of SP depend on the number of stages and the relative thermodynamic and mass transport properties of components i and j. In general, when applied to the two key components, components i and j and products 1 and 2 are selected so that $SP_{i,j} > 1.0$. Then, a large value corresponds to a relatively high degree of separation or high separation

Table 1.8 Main Separation Factors for Hydrocarbon Recovery Process

Key-Component Split	Column	Separation Factor, SP
nC_4H_{10}/iC_5H_{12}	C1	137.1
C_3H_8/iC_4H_{10}	C2	7103
iC_4H_{10}/nC_4H_{10}	C3	377.6

power; a small value larger than but close to 1.0 corresponds to a low degree of separation power. For example, if SP = 10,000 and $SR_i = 1/SR_j$, then, from (1-5), $SR_i = 100$ and $SR_j = 0.01$, corresponding to a sharp separation. However, if SP = 9 and $SR_i = 1/SR_j$, then $SR_i = 3$ and $SR_j = \frac{1}{3}$, corresponding to a nonsharp separation.

For the hydrocarbon recovery process of Figure 1.9, the values of SP in Table 1.8 are computed from the data in Tables 1.5 or 1.6 for the main split in each of the three separators. The separation factor in Column C1 is relatively small because the split for the heavy key, iC_5H_{12}, is not sharp. The largest separation factor occurs in column C2, where the separation is relatively easy because of the fairly large volatility difference between the two keys. Much more difficult is the butane-isomer split in Column C3, where only a moderately sharp split is achieved.

When applying the conservation of mass principle to separation operations using (1-1), component specifications in terms of component recoveries are easily applied, while those in terms of split ratios and, particularly, purities are more difficult, as shown in the following example.

EXAMPLE 1.1

A feed, F, of 100 kmol/h of air containing 21 mol% O_2 and 79 mol% N_2 is to be partially separated by a membrane unit according to each of the following four sets of specifications. For each case, compute the amounts, in kmol/h, and compositions, in mol%, of the two products (retentate, R, and permeate, P). The membrane is more permeable to the oxygen.

Case 1: 50% recovery of O_2 to the permeate and 87.5% recovery of N_2 to the retentate.

Case 2: 50% recovery of O_2 to the permeate and 50 mol% purity of O_2 in the permeate.

Case 3: 85 mol% purity of N_2 in the retentate and 50 mol% purity of O_2 in the permeate.

Case 4: 85 mol% purity of N_2 in the retentate and a split ratio of O_2 in the permeate to the retentate equal to 1.1.

SOLUTION

The feed is

$$n_{O_2}^{(F)} = 0.21(100) = 21 \text{ kmol/h}$$

$$n_{N_2}^{(F)} = 0.79(100) = 79 \text{ kmol/h}$$

Case 1: This is the simplest case to calculate because two recoveries are given:

$$n_{O_2}^{(P)} = 0.50(21) = 10.5 \text{ kmol/h}$$

$$n_{N_2}^{(R)} = 0.875(79) = 69.1 \text{ kmol/h}$$

$$n_{O_2}^{(R)} = 21 - 10.5 = 10.5 \text{ kmol/h}$$

$$n_{N_2}^{(P)} = 79 - 69.1 = 9.9 \text{ kmol/h}$$

Case 2: With the recovery for O_2 given, calculate its distribution into the two products:

$$n_{O_2}^{(P)} = 0.50(21) = 10.5 \text{ kmol/h}$$

$$n_{O_2}^{(R)} = 21 - 10.5 = 10.5 \text{ kmol/h}$$

Using the purity of O_2 in the permeate, the total permeate is

$$n^{(P)} = 10.5/0.5 = 21 \text{ kmol/h}$$

By a total permeate material balance,

$$n_{N_2}^{(P)} = 21 - 10.5 = 10.5 \text{ kmol/h}$$

By an overall N_2 material balance,

$$n_{N_2}^{(R)} = 79 - 10.5 = 68.5 \text{ kmol/h}$$

Case 3: With two purities given, write two simultaneous material-balance equations, one for each component, in terms of the total retentate and total permeate.

For nitrogen, with a fractional purity of $1.00 - 0.50 = 0.50$ in the permeate,

$$n_{N_2} = 0.85n^{(R)} + 0.50n^{(P)} = 79 \text{ kmol/h} \tag{1}$$

For oxygen, with a fractional purity of $1.00 - 0.85 = 0.15$ in the retentate,

$$n_{O_2} = 0.50n^{(P)} + 0.15n^{(R)} = 21 \text{ kmol/h} \tag{2}$$

Solving (1) and (2) simultaneously for the total products gives

$$n^{(P)} = 17.1 \text{ kmol/h} \qquad n^{(R)} = 82.9 \text{ kmol/h}$$

Therefore, the component flow rates are

$$n_{N_2}^{(R)} = 0.85(82.9) = 70.5 \text{ kmol/h}$$

$$n_{O_2}^{(R)} = 82.9 - 70.5 = 12.4 \text{ kmol/h}$$

$$n_{O_2}^{(P)} = 0.50(17.1) = 8.6 \text{ kmol/h}$$

$$n_{N_2}^{(P)} = 17.1 - 8.6 = 8.5 \text{ kmol/h}$$

Case 4: First compute the O_2 flow rates using the split ratio and an overall O_2 material balance,

$$\frac{n_{O_2}^{(P)}}{n_{O_2}^{(R)}} = 1.1 \qquad 21 = n_{O_2}^{(P)} + n_{O_2}^{(R)}$$

Solving these two equations simultaneously gives

$$n_{O_2}^{(R)} = 10 \text{ kmol/h} \qquad n_{O_2}^{(P)} = 21 - 10 = 11 \text{ kmol/h}$$

Since the retentate contains 85 mol% N_2 and, therefore, 15 mol% O_2, the flow rates for the N_2 are

$$n_{N_2}^{(R)} = \frac{85}{15}(10) = 56.7 \text{ kmol/h} \qquad n_{N_2}^{(P)} = 79 - 56.7 = 22.3 \text{ kmol/h}$$

1.9   SELECTION OF FEASIBLE SEPARATION PROCESSES

Selection of a best separation process must be made from among a number of feasible candidates. When the feed mixture is to be separated into more than two products, a combination of two or more operations may be best. Even when only two products are to be produced, a hybrid process of two or more different types of operations may be most economical. Only an introduction to the selection of a separation process is given here. A detailed treatment is given in Chapter 7 of Seider, Seader, and Lewin [14].

Important factors in the selection of feasible separation operations are listed in Table 1.9. These factors have to do with feed and product conditions, property differences that can be exploited, and certain characteristics of the candidate separation operations. The most important feed conditions are composition and flow rate, because the other conditions (temperature, pressure, and phase condition) can be altered by pumps, compressors, and heat exchangers to fit a particular candidate separation operation. In general, however, the vaporization of a liquid feed that has a high heat of vaporization, the condensation of a vapor feed with a refrigerant, and/or the compression of a vapor feed can add significantly to the cost. Some separation operations, such as those based on the use of barriers or solid agents, perform best on feeds that are dilute in the species to be recovered. The most important product conditions are the required purities because, again, the other conditions listed can be altered by energy transfer after the separation is achieved.

Table 1.9   Factors That Influence the Selection of Feasible Separation Operations

---

A. Feed conditions
  1. Composition, particularly concentration of species to be recovered or separated
  2. Flow rate
  3. Temperature
  4. Pressure
  5. Phase state (solid, liquid, and/or gas)
B. Product conditions
  1. Required purities
  2. Temperatures
  3. Pressures
  4. Phase states
C. Property differences that may be exploited
  1. Molecular
  2. Thermodynamic
  3. Transport
D. Characteristics of separation operation
  1. Ease of scale-up
  2. Ease of staging
  3. Temperature, pressure, and phase-state requirements
  4. Physical size limitations
  5. Energy requirements

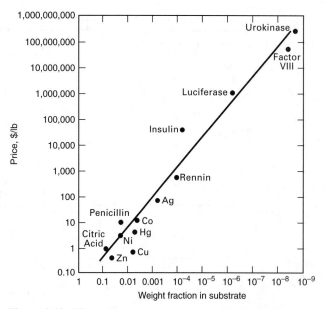

Figure 1.10   Effect of concentration of product in feed material on price [9].

In general, as demonstrated by Sherwood, Pigford, and Wilke [7] and updated recently, using additional data for biological materials from Dwyer [8], by Keller [9], the cost of recovering and purifying a chemical contained in a mixture can depend strongly on the concentration of that chemical in the feed mixture. Keller's correlation is given in Figure 1.10, where it is seen that the more dilute the chemical is in the mixture, the higher is its sales price.

When very pure products are required, either large differences in certain properties must exist or significant numbers of stages must be provided. It is important to consider both molecular and bulk thermodynamic and transport properties, some of which are listed near the end of Section 1.2. Data and estimation methods for many bulk properties are given by Poling, Prausnitz, and O'Connell [10] and for both molecular and bulk properties by Daubert and Danner [11].

Some separation operations are well understood and can be readily designed from a mathematical model and/or scaled up to a commercial size from laboratory data. The results of a survey by Keller [9], shown in Figure 1.11, show that the degree to which a separation operation is technologically mature correlates well with its commercial use. Operations based on a barrier are more expensive to stage than those based on use of a solid agent or the creation or addition of a second phase. Some separation equipment is limited to a maximum size. For capacities requiring a larger size, parallel units must be provided. The choice of single or parallel units must be given careful consideration. Except for size constraints or fabrication problems, the capacity of a single unit can be doubled for an additional investment cost of only about 50%. If two parallel units are installed, the additional investment is 100%. Table 1.10 is a list of the more common separation operations ranked according to ease of scale-up. Those operations ranked near the top are frequently designed without the need for any laboratory data or

Table 1.10   Ease of Scale-up of the Most Common Separation Operations

Operation in Decreasing Ease of Scale-up	Ease of Staging	Need for Parallel Units
Distillation	Easy	No need
Absorption	Easy	No need
Extractive and azeotropic distillation	Easy	No need
Liquid–liquid extraction	Easy	Sometimes
Membranes	Repressurization required between stages	Almost always
Adsorption	Easy	Only for regeneration cycle
Crystallization	Not easy	Sometimes
Drying	Not convenient	Sometimes

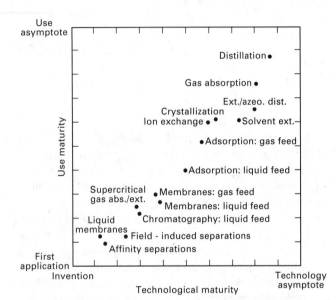

Figure 1.11 Technological and use maturities of separation processes [9].

pilot-plant tests. Operations near the middle usually require laboratory data, while operations near the bottom require pilot-plant tests on actual feed mixtures. Also included in the table is an indication of the ease of providing multiple stages and to what extent parallel units may be required to handle high capacities. A detailed discussion of the selection of alternative techniques for the separation of components from both homogeneous and heterogeneous phases, with many examples, is given by Woods [12]. Ultimately, the process having the lowest operating, maintenance, and capital costs is selected.

EXAMPLE 1.2

Propylene and propane are among the light hydrocarbons produced by thermal and catalytic cracking of heavy petroleum fractions. Propane is valuable as a fuel by itself and in liquefied natural gas (LPG), and as a feedstock for producing propylene and ethylene. Propylene is used to make acrylonitrile monomer for synthetic rubber, isopropyl alcohol, cumene, propylene oxide, and polypropylene.

Although propylene and propane have close boiling points, they are traditionally separated by distillation. Representative conditions are shown in Figure 1.12, where it is seen that a large number of stages is needed and the reflux and boilup flow rates compared to the feed flow rate are also large. Accordingly, considerable attention has been given to the possible replacement of distillation with a more economical and less energy-intensive separation operation. Based on the factors in Table 1.9, the characteristics in Table 1.10, and the list of species properties that might be exploited, given at the end of Section 1.2, propose some feasible alternatives to distillation to produce products from the feed in Figure 1.12.

SOLUTION

First, note that the component feed and product flow rates in Figure 1.12 satisfy (1-1), the conservation of mass. Table 1.11 compares properties of the two species, taken mainly from Daubert and Danner [11], where it is seen that the only listed property that might be exploited is the dipole moment. Because of the asymmetric location of the double bond in propylene, its dipole moment is significantly greater than that of propane, making propylene a polar compound, although weakly so (some define a polar compound as one with a dipole moment greater than 1 debye). Separation operations that can exploit this difference are:

1. Extractive distillation with a polar solvent such as furfural or an aliphatic nitrile that will reduce the volatility of propylene (Ref.: U.S. Patent 2,588,056, March 4, 1952).

Table 1.11   Comparison of Properties for Example 1.2

Property	Propylene	Propane
Molecular weight	42.081	44.096
van der Waals volume, m^3/kmol	0.03408	0.03757
van der Waals area, m^2/kmol $\times 10^{-8}$	5.060	5.590
Acentric factor	0.142	0.152
Dipole moment, debyes	0.4	0.0
Radius of gyration, m $\times 10^{10}$	2.254	2.431
Normal melting point, K	87.9	85.5
Normal boiling point, K	225.4	231.1
Critical temperature, K	364.8	369.8
Critical pressure, MPa	4.61	4.25

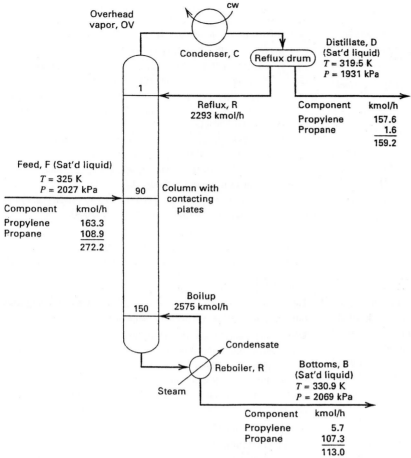

Figure 1.12 Distillation of a propylene–propane mixture.

2. Adsorption with silica gel or a zeolite that will selectively adsorb propylene [Ref.: *J. Am. Chem. Soc.,* **72,** 1153–1157 (1950)].

3. Facilitated transport membranes using impregnated silver nitrate to carry propylene selectively through the membrane [Ref.: *Recent Developments in Separation Science,* Vol. IX, 173–195 (1986)].

SUMMARY

1. Almost all industrial chemical processes include equipment for separating chemicals contained in the process feed(s) and/or produced in reactors within the process.

2. More than 25 different separation operations are commercially important.

3. The extent of separation achievable by a particular separation operation depends on exploitation of the differences in certain properties of the species.

4. The more widely used separation operations involve transfer of species between two phases, one of which is created by energy transfer or the reduction of pressure, or by introduction as a MSA.

5. Less commonly used separation operations are based on the use of a barrier, a solid agent, or a force field to cause species being separated to diffuse at different rates and/or to be selectively absorbed or adsorbed.

6. Separation operations are subject to the conservation of mass. The degree of separation of a component in a separator is indicated by a split fraction, SF, given by (1-2), and/or by a split ratio, SR, given by (1-3).

7. For a sequence, system, or train of separators, overall component recoveries and product purities are of prime importance and are related by material balances to individual SF and/or SR values for the separators in the system.

8. Some separation operations, such as absorption, are capable of only a specified degree of separation for a single species. Other separation operations, such as distillation, can effect a sharp split between two so-called key components.

9. The degree of separation between two key components by a particular separation operation can be indicated by a separation power (separation factor), SP, given by (1-4) and related to SF and SR values by (1-5) and (1-6).

10. For given feed(s) and product specifications, the best separation process must frequently be selected from among a number of feasible candidates. The choice may depend on factors listed in Table 1.9. The cost of recovering and purifying a chemical depends on its concentration in the feed mixture. The extent of industrial use of a separation operation depends on the technological maturity of the operation.

REFERENCES

1. *Kirk-Othmer Encyclopedia of Chemical Technology,* 3rd ed., John Wiley and Sons, New York, Vol. 17, pp. 183–256 (1982).

2. MAUDE, A.H., *Trans. AIChE,* **38,** 865–882 (1942).

3. CONSIDINE, D.M., Ed., *Chemical and Process Technology Encyclopedia,* McGraw-Hill, New York, pp. 760–763 (1974).

4. CARLE, T.C., and D.M. STEWART, *Chem. Ind.* (London), May 12, 1962, 830–839.

5. PERRY, R.H., and D.W. GREEN, Eds., *Perry's Chemical Engineers' Handbook,* 7th ed., McGraw-Hill, New York (1997).

6. SIIROLA, J.J., *AIChE Symp. Ser.,* **91**(304), 222–233 (1995).

7. SHERWOOD, T.K., R.L. PIGFORD, and C.R. WILKE, *Mass Transfer,* McGraw-Hill, New York (1975).

8. DWYER, J.L., *Biotechnology,* **1,** 957 (Nov. 1984).

9. KELLER, G.E., II, *AIChE Monogr. Ser.,* **83**(17) (1987).

10. POLING, B.E., J.M. PRAUSNITZ, and J.P. O'CONNELL, *The Properties of Gases and Liquids,* 5th ed., McGraw-Hill, New York (2001).

11. DAUBERT, T.E., and R.P. DANNER, *Physical and Thermodynamic Properties of Pure Chemicals—Data Compilation,* DIPPR, AIChE, Hemisphere, New York (1989).

12. WOODS, D.R., *Process Design and Engineering Practice,* Prentice-Hall, Englewood Cliffs, NJ (1995).

13. CUSSLER, E.L., and G.D. MOGGRIDGE, *Chemical Product Design,* Cambridge University Press, Cambridge, UK (2001).

14. SEIDER, W.D., J.D. SEADER, and D.R. LEWIN, *Product & Process Design Principles,* John Wiley & Sons, New York (2004).

EXERCISES

Section 1.1

1.1 The book, *Chemical Process Industries,* 4th edition, by R. Norris Shreve and J. A. Brink, Jr. (McGraw-Hill, New York, 1984), contains process descriptions, process flow diagrams, and technical data for processes used commercially in 38 chemical industries. For each of the following processes, draw a block-flow diagram of just the reaction and separation steps and describe the process in terms of just those steps, giving careful attention to the particular chemicals being formed in the reactor and separated in each of the separation operations:

(a) Coal chemicals, pp. 72–74

(b) Natural gas purification, pp. 84–86

(c) Acetylene, pp. 115–117

(d) Magnesium compounds, pp. 174–177

(e) Chlorine and caustic soda, pp. 214–219

(f) Potassium chloride, pp. 269–270

(g) Ammonia, pp. 278–282

(h) Sulfuric acid, pp. 299–310

(i) Fluorocarbons, pp. 321–323

(j) Uranium, pp. 338–340

(k) Titanium dioxide, pp. 388–390

(l) Cottonseed oil, pp. 468–471

(m) Glycerin, pp. 502–503

(n) Industrial alcohol, pp. 530–534

(o) Polyethylene, pp. 587–588

(p) Formaldehyde, pp. 596–598

(q) Styrene, pp. 630–635

(r) Natural-gas liquids, pp. 660–661

Section 1.2

1.2 Explain in detail, using thermodynamic principles, why the mixing of pure chemicals to form a homogeneous mixture is a so-called spontaneous process, while the separation of that mixture into its pure (or nearly pure) species is not.

1.3 Explain in detail, using the first and second laws of thermodynamics, why the separation of a mixture into essentially pure species or other mixtures of differing compositions requires the transfer of energy to the mixture or a degradation of its energy.

Section 1.3

1.4 Compare the advantages and disadvantages of making separations using an ESA versus using an MSA.

1.5 Every other year, the magazine *Hydrocarbon Processing* publishes a petroleum-refining handbook, which gives process-flow diagrams and data for more than 75 commercial processes. For each of the following processes in the November 1990 handbook, list the separation operations of the type given in Table 1.1 and indicate what chemical(s) is(are) being separated:

(a) Hydrotreating (Chevron), p. 114

(b) Ethers (Phillips), p. 128

(c) Alkylation (Exxon), p. 130

(d) Treating of BTX cut (GKT), p. 136

1.6 Every other year, the magazine *Hydrocarbon Processing* publishes a petrochemical handbook, which gives process-flow diagrams and data for more than 50 commercial processes. For each of the following processes in the March 1991 handbook, list the separation operations of the type given in Table 1.1 and indicate what chemical(s) is(are) being separated:

(a) Linear alkylbenzene (UOP), p. 130

(b) Methyl amines (Acid-Amine Technologies), p. 133

(c) Butene-2 (Phillips), p. 144

(d) Caprolactam (SNIA), p. 150

(e) Ethylene glycols (Scientific Design), p. 156

(f) Styrene (Monsanto), p. 188

Section 1.4

1.7 Explain why osmosis is not used as a separation operation.

1.8 The osmotic pressure, π, of sea water is given approximately by the expression $\pi = RTc/M$, where c is the concentration of the dissolved salts (solutes) in g/cm^3 and M is the average molecular weight of the solutes as ions. If pure water is to be recovered from sea water at 298 K and containing 0.035 g of salts/cm^3 of sea water and $M = 31.5$, what is the minimum required pressure difference across the membrane in kPa?

1.9 It has been shown that a liquid membrane of aqueous ferrous ethylenediaminetetraacetic acid, maintained between two sets of microporous, hydrophobic, hollow fibers that are packed in a permeator cell, can selectively and continuously remove sulfur dioxide and nitrogen oxides from the flue gas of power-generating plants. Prepare a detailed drawing of a possible device to carry out such a separation. Show all locations of inlet and outlet streams, the arrangement of the hollow fibers, and a method for handling the membrane liquid. Should the membrane liquid be left in the cell or circulated? Would a sweep fluid be needed to remove the oxides?

Section 1.5

1.10 Explain the differences, if any, between adsorption and gas–solid chromatography.

1.11 In gas–liquid chromatography, is it essential that the gas flow through the packed tube in plug flow? Discuss in detail.

Section 1.6

1.12 In electrophoresis, explain why most small suspended particles are negatively charged.

1.13 In field-flow fractionation, could a turbulent-flow field be used? Why or why not?

Section 1.7

1.14 The feed to Column C3 of the distillation sequence in Figure 1.9 is given in Table 1.5. However, the separation is to be altered so as to produce a distillate that is 95 mol% pure isobutane with a recovery of isobutane in the distillate (SF) of 96%. Because of the relatively sharp separation in Column C3 between iC_4 and nC_4, assume all propane goes to the distillate and all C_5s goes to the bottoms.

(a) Compute the flow rates in lbmol/h of each component in each of the two products leaving Column C3.

(b) What is the percent purity of the normal butane in the bottoms product?

(c) If the purity of the isobutane in the distillate is fixed at 95%, what percent recovery of isobutane in the distillate will maximize the percent purity of normal butane in the bottoms product?

1.15 Five hundred kmol/h of a liquid mixture of light alcohols containing, by moles, 40% methanol (M), 35% ethanol (E), 15% isopropanol (IP), and 10% normal propanol (NP) is distilled in a sequence of two distillation columns. The distillate from the first column is 98% pure M with a 96% recovery of M. The distillate from the second column is 92% pure E with a 95% recovery of E from the process feed. Assume no propanols in the distillate from Column C1, no M in the bottoms from Column C2, and no normal propanol in the distillate from Column C2.

(a) By material balances, assuming negligible propanols in the distillate from the first column, compute the flow rates in kmol/h of each component in each feed, distillate, and bottoms. Draw a labeled block-flow diagram like Figure 1.9. Include the results of the material balances in a table like Table 1.5 and place the table below your block-flow diagram.

(b) Compute the mole-percent purity of the propanol mixture leaving as bottoms from the second column in the sequence.

(c) If the recovery of ethanol is fixed at 95%, what is the maximum purity that can be achieved for the ethanol in the distillate from the second column?

(d) If instead, the purity of the ethanol is fixed at 92%, what is the maximum recovery of ethanol (based on the process feed) that can be achieved?

1.16 A mixture of ethanol and benzene is separated in a network of distillation and membrane separation steps. In one intermediate step, a near-azeotropic liquid mixture of 8,000 kg/h of 23 wt% ethanol in benzene is fed to a pervaporation membrane consisting of a thin ionomeric film of perfluorosulfonic acid polymer cast on a porous Teflon support. The membrane is selective for ethanol such that the vapor permeate contains 60 wt% ethanol, while the non-permeate liquid contains 90 wt% benzene.

(a) Draw a flow diagram of the pervaporation step using the appropriate symbol from Table 1.2 and include on the diagram all of the given information.

(b) Compute the component flow rates in kg/h in the feed stream and in the two product streams and enter these results on the diagram.

(c) What separation operation could be used to further purify the vapor permeate?

Section 1.8

1.17 The Prism gas permeation process developed by the Monsanto Company is highly selective for hydrogen when using hollow-fiber membranes of materials such as silicone-coated polysulphone. In a typical application, a gas at 16.7 MPa and 40°C, containing the following components in kmol/h: 42.4 H_2, 7.0 CH_4, and 0.5 N_2, is separated into a nonpermeate gas at 16.2 MPa and a permeate gas at 4.56 MPa.

(a) If the membrane is nonpermeable to nitrogen, the Prism membrane separation index, on a mole basis (SP) for hydrogen relative to methane is 34.13, and the split fraction (SF) for hydrogen to the permeate gas is 0.6038, calculate the kmol/h of each component and the total flow in kmol/h of both the nonpermeate gas and the permeate gas.

(b) Compute the percent purity of the hydrogen in the permeate gas.

(c) Using an average heat capacity ratio, γ, of 1.4, estimate the outlet temperatures of the two exiting gas streams by assuming the ideal gas law and reversible expansions for each gas and no heat transfer between the two exiting gas streams.

(d) Draw a process-flow diagram of the membrane process and indicate on the diagram for each stream the pressure, temperature, and component flow rates.

1.18 Nitrogen gas can be injected into oil wells to increase the recovery of crude oil (enhanced oil recovery). Usually, natural gas is produced with the oil and it is desirable to recover the nitrogen from the gas for reinjection into the well. Furthermore, the natural gas must not contain more than 3 mol% nitrogen if the natural gas is to be put into a pipeline. A total of 170,000 SCFH (based on 60°F and 14.7 psia) of natural gas containing 18% N_2, 75% CH_4, and 7% C_2H_6 at 100°F and 800 psia is to be processed for N_2 removal. A two-step separation process has been proposed consisting of (1) membrane separation with a nonporous glassy polyimide membrane, followed by (2) pressure-swing adsorption using molecular sieves to which the permeate gas is fed. The membrane separator is highly selective for N_2 ($SP_{N_2,CH_4} = 16$) and completely impermeable to ethane. The pressure-swing adsorption step selectively adsorbs methane, giving 97% pure methane product in the adsorbate, with an 85% recovery of CH_4 fed to the adsorber. The nonpermeate (retentate) gas from the membrane step and adsorbate from the pressure-swing adsorption step are combined to give a methane stream that contains 3.0% N_2. The pressure drop across the membrane is 760 psia. The permeate at 20°F is compressed in two stages to 275 psia and cooled to 100°F before entering the adsorption step. The adsorbate gas, which exits the adsorber during regeneration at 100°F and 15 psia, is compressed in three stages to 800 psia and cooled to 100°F before being combined with nonpermeate gas to give the final pipeline natural gas.

(a) Draw a process-flow diagram of the separation process using appropriate symbols from Tables 1.2 and 1.3. Include the gas compressors and heat exchangers. Label the diagram with all of the data given above, and number all process streams.

(b) Compute by material balances, using the data above, the component flow rates of N_2, CH_4, and C_2H_6 in lbmol/h for all process streams entering and exiting the two separation operations. Place the results in a material-balance table similar to Table 1.5.

Section 1.9

1.19 A mixture of ethylbenzene (EB) and the three isomers (ortho, meta, and para) of xylene is widely available in petroleum refineries.

(a) Based on differences in normal boiling points, verify that the separation between *meta*-xylene (MX) and *para*-xylene (PX) by distillation is far more difficult than the separations between EB and PX, and MX and *ortho*-xylene (OX).

(b) Prepare a list of properties for MX and PX similar to Table 1.11. From that list, which property differences might be the best ones to exploit to separate a mixture of these two xylenes?

(c) Explain why melt crystallization and adsorption are used commercially to separate MX and PX.

1.20 When a mixture of ethanol and water is distilled at ambient pressure, the products are a distillate of ethanol and water of near-azeotrope composition (89.4 mol% ethanol) and a bottoms product of nearly pure water. Based on differences in certain properties of ethanol and water, explain how the following separation operations might be able to recover almost pure ethanol from the distillate:

(a) Extractive distillation

(b) Azeotropic distillation

(c) Liquid–liquid extraction

(d) Crystallization

(e) Pervaporation membrane

(f) Adsorption

1.21 A stream containing 7,000 kmol/h of water and 3,000 parts per million (ppm) by weight of ammonia at 350 K and 1 bar is to be processed to remove 90% of the ammonia. What type of separation operation would you use? If it involves a mass-separating agent, propose one.

1.22 A light-hydrocarbon feed stream contains 45.4 kmol/h of propane, 136.1 kmol/h of isobutane, 226.8 kmol/h of *n*-butane, 181.4 kmol/h of isopentane, and 317.4 kmol/h of *n*-pentane. This stream is to be separated in a sequence of three distillation columns, similar to that in Figure 1.9, into four products: (1) propane-rich,

(2) isobutane-rich, (3) *n*-butane-rich, and (4) combined pentanes-rich. However, the distillate from the first column is to be the propane-rich product; the distillate from Column 2 is to be the isobutane-rich product; and the distillate from Column 3 is to be the *n*-butane-rich product, with the combined pentanes being the bottoms from Column 3. The recovery of each main component in each product is to be 98%. For example, 98% of the propane in the feed stream is to appear in the propane-rich product, and 98% of the combined pentanes in the feed stream is to appear in the bottoms product from Column 3.

(a) Draw a process-flow diagram, similar to Figure 1.9.

(b) Complete a material balance for each column and summarize the results in a table similar to Table 1.5. To complete the material balance, you will have to make some assumptions about the flow rates of: (1) isobutane in the distillates for Columns 1 and 3 and (2) *n*-butane in the distillates for Columns 1 and 2, consistent with the specified recoveries. Assume that propane will not be found in the distillate from Column 3 and pentanes will not be found in the distillate from Column 2.

(c) Calculate the mol% purities of each of the products and summarize your results in a table similar to Table 1.7, but without the specifications, which are not given here.

1.23 The need to remove organic pollutants from wastewater is common to many industrial processes. Separation methods that may be considered are: (1) adsorption, (2) distillation, (3) liquid–liquid extraction, (4) membrane separation, (5) stripping with air, and (6) stripping with steam. Discuss the advantages and disadvantages of each method for this application. Be sure to consider the fate of the organic material.

1.24 Many waste gas streams in processing plants contain volatile organic compounds (VOCs), which must be removed. Recovery of the VOCs may be accomplished by several separation methods, including: (1) absorption, (2) adsorption, (3) condensation, (4) freezing, and (5) membrane separation. Discuss the advantages and disadvantages of each method, paying particular attention to the fate of the VOC. For the case of a stream containing 3 mol% acetone in air, draw a flow diagram for a process based on absorption. Choose a reasonable absorbent and include in your process a means to recover the acetone and recycle the absorbent.

1.25 Describe three methods suitable for the separation of air into nitrogen and oxygen.

1.26 What separation methods can be used to separate azeotropic mixtures of water and an organic chemical such as ethanol?

1.27 An aqueous stream contains 5% by weight of magnesium sulfate. Devise a process, complete with a process-flow diagram, for the production of nearly pure magnesium sulfate from this stream.

1.28 Explain why the separation of a stream containing 10 wt% acetic acid in water might be more economical by liquid–liquid extraction with ethyl acetate than by distillation.

Chapter 2

Thermodynamics of Separation Operations

Thermodynamic properties and equations play a major role in separation operations, particularly with respect to energy requirements, phase equilibria, and sizing equipment. This chapter discusses applied thermodynamics for separation processes. Equations for energy balances, entropy and availability balances, and for determining phase densities and phase compositions at equilibrium are developed. These involve thermodynamic properties, including specific volume

or density, enthalpy, entropy, availability, and fugacities and activities together with their coefficients, all as functions of temperature, pressure, and phase composition. Methods for estimating properties for ideal and nonideal mixtures are summarized. However, this chapter is not a substitute for any of the excellent textbooks on chemical engineering thermodynamics. Furthermore, emphasis here is on fluid phases, with little consideration of solid phases.

2.0 INSTRUCTIONAL OBJECTIVES

After completing this chapter, you should be able to:

- Make energy, entropy, and availability balances around a separation process using the first and second laws of thermodynamics.
- Calculate lost work and second-law efficiency of a separation process.
- Explain the concept of phase equilibria in terms of Gibbs free energy, chemical potential, fugacity, fugacity coefficients, activity, and activity coefficients.
- Understand the concept and usefulness of the equilibrium ratio (*K*-value) for problems involving liquid and/or vapor phases at equilibrium.
- Derive expressions for *K*-values in terms of fugacity coefficients and activity coefficients.
- Write vapor–liquid *K*-value expressions for Raoult's law (ideal), a modified Raoult's law, and Henry's law.
- Calculate density, enthalpy, and entropy of ideal mixtures.
- Utilize graphical correlations to obtain thermodynamic properties of ideal and near-ideal mixtures.
- Use nomographs to estimate vapor–liquid *K*-values of nonideal hydrocarbon and light–gas mixtures.
- Explain how computer programs use equations of state (e.g., Soave-Redlich-Kwong or Peng-Robinson) to compute thermodynamic properties of vapor and liquid mixtures, including *K*-values.
- Explain how computer programs use liquid-phase activity-coefficient correlations (e.g., Wilson, NRTL, UNIQUAC, or UNIFAC) to compute thermodynamic properties, including *K*-values, for nonideal vapor and liquid mixtures at equilibrium.

2.1 ENERGY, ENTROPY, AND AVAILABILITY BALANCES

Most industrial separation operations utilize large quantities of energy in the form of heat and/or shaft work. A study by Mix et al. [1] reports that two quads (1 quad $= 10^{15}$ Btu) of energy were consumed by distillation separations in petroleum, chemical, and natural-gas processing plants in the United States in 1976. This amount of energy was 2.7% of the total U.S. energy consumption of 74.5 quads and is equivalent to the energy obtained from approximately 1 million bbl of crude oil per day over a one-year period. This amount of oil can be compared to 13 million bbl/day, the

average amount of crude oil processed by petroleum refineries in the United States in early 1991. At a crude oil price of approximately \$40/bbl, the energy consumption by distillation in the United States is approximately \$20 trillion per year. Thus, it is of considerable interest to know the extent of energy consumption in a separation process, and to what degree energy requirements might be reduced. Such energy estimates can be made by applying the first and second laws of thermodynamics.

Consider the continuous, steady-state, flow system for a general separation process in Figure 2.1. One or more feed streams flowing into the system are separated into two or

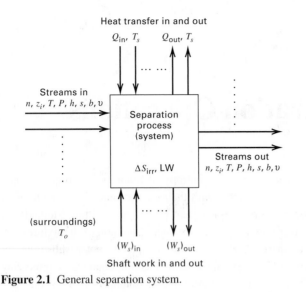

Figure 2.1 General separation system.

more product streams that flow out of the system. For all these streams, we denote the molar flow rates by n, the component mole fractions by z_i, the temperature by T, the pressure by P, the molar enthalpies by h, the molar entropies by s, and the molar availabilities by b. If chemical reactions occur in the process, enthalpies and entropies are referred to the elements, as discussed by Felder and Rousseau [2]; otherwise they can be referred to the compounds. Heat flows in

Table 2.1 Universal Thermodynamic Laws for a Continuous, Steady-State, Flow System

---

Energy balance:

(1) $\sum\limits_{\substack{\text{out of} \\ \text{system}}} (nh + Q + W_s) - \sum\limits_{\substack{\text{in to} \\ \text{system}}} (nh + Q + W_s) = 0$

Entropy balance:

(2) $\sum\limits_{\substack{\text{out of} \\ \text{system}}} \left(ns + \dfrac{Q}{T_s}\right) - \sum\limits_{\substack{\text{in to} \\ \text{system}}} \left(ns + \dfrac{Q}{T_s}\right) = \Delta S_{\text{irr}}$

Availability balance:

(3) $\sum\limits_{\substack{\text{in to} \\ \text{system}}} \left[nb + Q\left(1 - \dfrac{T_0}{T_s}\right) + W_s\right]$

$\quad - \sum\limits_{\substack{\text{out of} \\ \text{system}}} \left[nb + Q\left(1 - \dfrac{T_0}{T_s}\right) + W_s\right] = \text{LW}$

Minimum work of separation:

(4) $W_{\text{min}} = \sum\limits_{\substack{\text{out of} \\ \text{system}}} nb - \sum\limits_{\substack{\text{in to} \\ \text{system}}} nb$

Second-law efficiency:

(5) $\eta = \dfrac{W_{\text{min}}}{\text{LW} + W_{\text{min}}}$

where $b = h - T_0 s =$ availability function

$\quad \text{LW} = T_0 \Delta S_{\text{irr}} = \text{lost work}$

---

or out of the system are denoted by Q, and shaft work crossing the boundary of the system is denoted by W_s. At steady state, if kinetic, potential, and surface energy changes are neglected, the first law of thermodynamics (also referred to as the conservation of energy or the energy balance), states that the sum of all forms of energy flowing into the system equals the sum of the energy flows leaving the system:

(stream enthalpy flows + heat transfer
$\quad$ + shaft work)$_{\text{leaving system}}$ − (stream enthalpy flows
$\quad$ + heat transfer + shaft work)$_{\text{entering system}}$
$\quad = 0$

In terms of symbols, the energy balance is given by Eq. (1) in Table 2.1, where all flow rate, heat transfer, and shaft work terms are positive. Molar enthalpies may be positive or negative depending on the reference state.

All separation processes must satisfy the energy balance. Inefficient separation processes require large transfers of heat and/or shaft work both into and out of the process; efficient processes require smaller levels of heat transfer and/or shaft work. The first law of thermodynamics provides no information on energy efficiency, but the second law of thermodynamics (also referred to as the entropy balance), given by Eq. (2) in Table 2.1, does. In words, the steady-state entropy balance is

(Stream entropy flows
$\quad$ + entropy flows by heat transfer)$_{\text{leaving system}}$
$\quad$ − (stream entropy flows
$\quad$ + entropy flows by heat transfer)$_{\text{entering system}}$
$\quad = $ production of entropy by the process

In the entropy balance equation, the heat sources and sinks in Figure 2.1 are at absolute temperatures T_s. For example, if condensing steam at 150°C supplies heat, Q, to the reboiler of a distillation column, $T_s = 150 + 273 = 423$ K. If cooling water at an average temperature of 30°C removes heat, Q, in a condenser, $T_s = 30 + 273 = 303$ K. Unlike the energy balance, which states that energy is conserved, the entropy balance predicts the production of entropy, ΔS_{irr}, which is the irreversible increase in the entropy of the universe. This term, which must be a positive quantity, is a quantitative measure of the thermodynamic inefficiency of a process. In the limit, as a reversible process is approached, ΔS_{irr} tends to zero. Note that the entropy balance contains no terms related to shaft work.

Although ΔS_{irr} is a measure of energy inefficiency, it is difficult to relate to this measure because it does not have the units of energy/time (power). A more useful measure or process inefficiency can be derived by combining (1) and (2) in Table 2.1 to obtain a combined statement of the first and second laws of thermodynamics, which is given as (3) in Table 2.1. To perform this derivation, it is first necessary to define an infinite source of or sink for heat transfer at the absolute temperature, $T_s = T_0$, of the surroundings. This temperature is typically about 300 K and represents the largest source of coolant associated with the processing plant being analyzed. This might be the average temperature of cooling water, air, or a nearby river, lake, or ocean. Heat transfer associated with this surrounding coolant and transferred from (or to) the process is termed Q_0. Thus, in both (1) and (2) in Table 2.1, the Q and Q/T_s terms include contributions from Q_0 and Q_0/T_0, respectively.

The derivation of (3) in Table 2.1 can be made, as shown by de Nevers and Seader [3], by combining (1) and (2) to eliminate Q_0. The resulting equation is referred to as an *availability* (or *exergy*) balance, where the term availability means "available for complete conversion to shaft work." The stream availability function, b, as defined by

$$b = h - T_0 s \qquad (2\text{-}1)$$

is a measure of the maximum amount of stream energy that can be converted into shaft work if the stream is taken to the reference state. It is similar to Gibbs free energy, $g = h - Ts$, but differs in that the infinite surroundings temperature, T_0, appears in the equation instead of the stream temperature, T. Terms in (3) in Table 2.1 containing Q are multiplied by $(1 - T_0/T_s)$, which, as shown in Figure 2.2, is the reversible Carnot heat-engine cycle efficiency, representing the maximum amount of shaft work that can be produced from Q at T_s, where the residual amount of energy $(Q - W_s)$ is transferred as heat to a sink at T_0. Shaft work, W_s, remains at its full value in (3). Thus, although Q and W_s have the same thermodynamic worth in (1) of Table 2.1, heat transfer has less worth in (3). This is because shaft work can be converted completely to heat (by friction), but heat cannot be converted completely to shaft work unless the heat is available at an infinite temperature.

Availability, like entropy, is not conserved in a real, irreversible process. The total availability (i.e., ability to produce shaft work) passing into a system is always greater than the total availability leaving a process. Thus (3) in Table 2.1 is written with the "into system" terms first. The difference

is the *lost work*, LW, which is also called the loss of availability (or exergy), and is defined by

$$LW = T_0 \Delta S_{\text{irr}} \qquad (2\text{-}2)$$

Lost work is always a positive quantity. The greater its value, the greater is the energy inefficiency. In the lower limit, as a reversible process is approached, lost work tends to zero. The lost work has the same units as energy, thus making it easy to attach significance to its numerical value. In words, the steady-state availability balance is

(Stream availability flows + availability of heat
$\qquad$ + shaft work)$_{\text{entering system}}$ − (stream availability flows
$\qquad$ + availability of heat + shaft work)$_{\text{leaving system}}$
$\quad$ = loss of availability (lost work)

For any separation process, lost work can be computed from (3) in Table 2.1. Its magnitude depends on the extent of process irreversibilities, which include fluid friction, heat transfer due to finite temperature-driving forces, mass transfer due to finite concentration or activity-driving forces, chemical reactions proceeding at finite displacements from chemical equilibrium, mixing of streams at differing conditions of temperature, pressure, and/or composition, and so on. Thus, to reduce the lost work, driving forces for momentum transfer, heat transfer, mass transfer, and chemical reaction must be reduced. Practical limits to this reduction exist because, as driving forces are decreased, equipment sizes increase, tending to infinitely large sizes as driving forces approach zero.

For a separation process that occurs without chemical reaction, the summation of the stream availability functions leaving the process is usually greater than the same summation for streams entering the process. In the limit for a reversible process (LW = 0), (3) of Table 2.1 reduces to (4), where W_{min} is the minimum shaft work required to conduct the separation and is equivalent to the difference in the heat transfer and shaft work terms in (3). This minimum work is independent of the nature (or path) of the separation process. The work of separation for an actual irreversible process is always greater than the minimum value computed from (4).

From (3) of Table 2.1, it is seen that as a separation process becomes more irreversible, and thus more energy inefficient, the increasing LW causes the required equivalent work of separation to increase by the same amount. Thus, the equivalent work of separation for an irreversible process is given by the sum of lost work and minimum work of separation. The *second-law efficiency*, therefore, can be defined as

(fractional second-law efficiency)
$$= \left(\frac{\text{minimum work of separation}}{\text{equivalent actual work of separation}} \right)$$

In terms of symbols, the efficiency is given by (5) in Table 2.1.

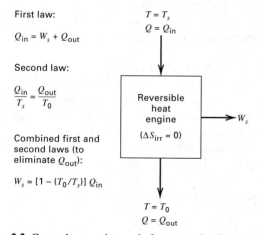

First law:

$$Q_{\text{in}} = W_s + Q_{\text{out}}$$

Second law:

$$\frac{Q_{\text{in}}}{T_s} = \frac{Q_{\text{out}}}{T_0}$$

Combined first and second laws (to eliminate Q_{out}):

$$W_s = [1 - (T_0/T_s)]\, Q_{\text{in}}$$

$T = T_s$
$Q = Q_{\text{in}}$

Reversible heat engine ($\Delta S_{\text{irr}} = 0$) → W_s

$T = T_0$
$Q = Q_{\text{out}}$

Figure 2.2 Carnot heat engine cycle for converting heat to shaft work.

EXAMPLE 2.1

For the propylene–propane separation of Figure 1.12, using the following thermodynamic properties for certain streams, as estimated from the Soave–Redlich–Kwong equation of state discussed in Section 2.5, and the relations given in Table 2.1, compute in SI units:

(a) The condenser duty, Q_C

(b) The reboiler duty, Q_R

(c) The irreversible entropy production, assuming 303 K for the condenser cooling-water sink and 378 K for the reboiler steam source

(d) The lost work, assuming $T_0 = 303$ K

(e) The minimum work of separation

(f) The second-law efficiency

Stream	Phase Condition	Enthalpy (h), kJ/kmol	Entropy (s), kJ/kmol-K
Feed (F)	Liquid	13,338	−4.1683
Overhead vapor (OV)	Vapor	24,400	24.2609
Distillate (D) and reflux (R)	Liquid	12,243	−13.8068
Bottoms (B)	Liquid	14,687	−2.3886

SOLUTION

Place the condenser (C) cooling water and the reboiler (R) steam outside the distillation system. Thus, Q_C and Q_R cross the boundary of the system. The following calculations are made using the stream flow rates in Figure 1.12 and properties above.

(a) Compute condenser duty from an energy balance around the condenser. From (1), Table 2.1, noting that the overhead-vapor molar flow rate is given by $n_{OV} = n_R + n_D$ and $h_R = h_D$, the condenser duty is

$$Q_C = n_{OV}(h_{OV} - h_R)$$
$$= (2,293 + 159.2)(24,400 - 12,243)$$
$$= 29,811,000 \text{ kJ/h}$$

(b) Compute reboiler duty from an energy balance around the entire distillation operation. (An energy balance around the reboiler cannot be made because data are not given for the boilup rate.) From (1), Table 2.1,

$$Q_R = n_D h_D + n_B h_B + Q_C - n_F h_F$$
$$= 159.2(12,243) + 113(14,687)$$
$$+ 29,811,000 - 272.2(13,338)$$
$$= 29,789,000 \text{ kJ/h}$$

(c) Compute the production of entropy from an entropy balance around the entire distillation system. From Eq. (2), Table 2.1,

$$\Delta S_{irr} = n_D s_D + n_B s_B + Q_C/T_C - n_F s_F - Q_R/T_R$$
$$= 159.2(-13.8068) + 113(-2.3886) + 29,811,000/303$$
$$- 272.2(-4.1683) - 29,789,000/378$$
$$= 18,246 \text{ kJ/h-K}$$

(d) Compute lost work from its definition at the bottom of Table 2.1:

$$LW = T_0 \Delta S_{irr}$$
$$= 303(18,246) = 5,529,000 \text{ kJ/h}$$

Alternatively, compute lost work from an availability balance around the entire distillation system. From (3), Table 2.1, where the availability function, b, is defined near the bottom of Table 2.1,

$$LW = n_F b_F + Q_R(1 - T_0/T_R)$$
$$- n_D b_D - n_B b_B - Q_C(1 - T_0/T_C)$$
$$= 272.2[13,338 - (303)(-4.1683)]$$
$$+ 29,789,000(1 - 303/378)$$
$$- 159.2[12,243 - (303)(-13.8068)]$$
$$- 113[14,687 - (303)(-2.3886)]$$
$$- 29,811,000(1 - 303/303)$$
$$= 5,529,000 \text{ kJ/h} \quad \text{(same result)}$$

(e) Compute the minimum work of separation for the entire distillation system. From (4), Table 2.1,

$$W_{min} = n_D b_D + n_B b_B - n_F b_F$$
$$= 159.2[12,243 - (303)(-13.8068)]$$
$$+ 113[14,687 - (303)(-2.3886)]$$
$$- 272.2[13,338 - (303)(-4.1683)]$$
$$= 382,100 \text{ kJ/h}$$

(f) Compute the second-law efficiency for the entire distillation system. From (5), Table 2.1,

$$\eta = \frac{W_{min}}{LW + W_{min}}$$
$$= \frac{382,100}{5,529,000 + 382,100}$$
$$= 0.0646 \quad \text{or} \quad 6.46\%$$

This low second-law efficiency is typical of a difficult distillation separation, which in this case requires 150 theoretical stages with a reflux ratio of almost 15 times the distillate rate.

2.2 PHASE EQUILIBRIA

Analysis of separations equipment frequently involves the assumption of phase equilibria as expressed in terms of Gibbs free energy, chemical potentials, fugacities, or activities. For each phase in a multiphase, multicomponent system, the total Gibbs free energy is

$$G = G(T, P, N_1, N_2, \ldots, N_C)$$

where $N_i =$ moles of species i. At equilibrium, the total G for all phases is a minimum, and methods for determining this minimum are referred to as *free-energy minimization techniques*. Gibbs free energy is also the starting point for the derivation of commonly used equations for expressing phase equilibria. From classical thermodynamics, the total differential of G is given by

$$dG = -S\, dT + V\, dP + \sum_{i=1}^{C} \mu_i dN_i \qquad (2\text{-}3)$$

where μ_i is the chemical potential or partial molar Gibbs free energy of species i. When (2-3) is applied to a closed system consisting of two or more phases in equilibrium at uniform temperature and pressure, where each phase is an open system capable of mass transfer with another phase, then

$$dG_{system} = \sum_{p=1}^{N}\left[\sum_{i=1}^{C}\mu_i^{(p)}dN_i^{(p)}\right]_{P,T} \qquad (2\text{-}4)$$

where the superscript (p) refers to each of N phases in equilibrium. Conservation of moles of each species, in the absence of chemical reaction, requires that

$$dN_i^{(1)} = -\sum_{p=2}^{N}dN_i^{(p)} \qquad (2\text{-}5)$$

which, upon substitution into (2-4), gives

$$\sum_{p=2}^{N}\left[\sum_{i=1}^{C}\left(\mu_i^{(p)} - \mu_i^{(1)}\right)dN_i^{(p)}\right] = 0 \qquad (2\text{-}6)$$

With $dN_i^{(1)}$ eliminated in (2-6), each $dN_i^{(p)}$ term can be varied independently of any other $dN_i^{(p)}$ term. But this requires that each coefficient of $dN_i^{(p)}$ in (2-6) be zero. Therefore,

$$\mu_i^{(1)} = \mu_i^{(2)} = \mu_i^{(3)} = \cdots = \mu_i^{(N)} \qquad (2\text{-}7)$$

Thus, the chemical potential of a particular species in a multicomponent system is identical in all phases at physical equilibrium.

Fugacities and Activity Coefficients

Chemical potential cannot be expressed as an absolute quantity, and the numerical values of chemical potential are difficult to relate to more easily understood physical quantities. Furthermore, the chemical potential approaches an infinite negative value as pressure approaches zero. For these reasons, the chemical potential is not favored for phase equilibria calculations. Instead, fugacity, invented by G. N. Lewis in 1901, is employed as a surrogate.

The partial fugacity of species i in a mixture is like a pseudo-pressure, defined in terms of the chemical potential by

$$\bar{f}_i = C\exp\left(\frac{\mu_i}{RT}\right) \qquad (2\text{-}8)$$

where C is a temperature-dependent constant. Regardless of the value of C, it is shown by Prausnitz, Lichtenthaler, and Azevedo [4] that (2-7) can be replaced with

$$\bar{f}_i^{(1)} = \bar{f}_i^{(2)} = \bar{f}_i^{(3)} = \cdots = \bar{f}_i^{(N)} \qquad (2\text{-}9)$$

Thus, at equilibrium, a given species has the same partial fugacity in each existing phase. This equality, together with

equality of phase temperatures and pressures,

$$T^{(1)} = T^{(2)} = T^{(3)} = \cdots = T^{(N)} \qquad (2\text{-}10)$$

and

$$P^{(1)} = P^{(2)} = P^{(3)} = \cdots = P^{(N)} \qquad (2\text{-}11)$$

constitutes the required conditions for phase equilibria. For a pure component, the partial fugacity, $\bar{f}_i$, becomes the pure-component fugacity, f_i. For a pure, ideal gas, fugacity is equal to the pressure, and for a component in an ideal-gas mixture, the partial fugacity is equal to its partial pressure, $p_i = y_i P$. Because of the close relationship between fugacity and pressure, it is convenient to define their ratio for a pure substance as

$$\phi_i = \frac{f_i}{P} \qquad (2\text{-}12)$$

where ϕ_i is the pure-species fugacity coefficient, which has a value of 1.0 for an ideal gas. For a mixture, partial fugacity coefficients are defined by

$$\bar{\phi}_{iV} \equiv \frac{\bar{f}_{iV}}{y_i P} \qquad (2\text{-}13)$$

$$\bar{\phi}_{iL} \equiv \frac{\bar{f}_{iL}}{x_i P} \qquad (2\text{-}14)$$

such that as ideal-gas behavior is approached, $\bar{\phi}_{iV} \rightarrow 1.0$ and $\bar{\phi}_{iL} \rightarrow P_i^s/P$, where P_i^s = vapor (saturation) pressure.

At a given temperature, the ratio of the partial fugacity of a component to its fugacity in some defined standard state is termed the *activity*. If the standard state is selected as the pure species at the same pressure and phase condition as the mixture, then

$$a_i \equiv \frac{\bar{f}_i}{f_i^o} \qquad (2\text{-}15)$$

Since at phase equilibrium, the value of f_i^o is the same for each phase, substitution of (2-15) into (2-9) gives another alternative condition for phase equilibria,

$$a_i^{(1)} = a_i^{(2)} = a_i^{(3)} = \cdots = a_i^{(N)} \qquad (2\text{-}16)$$

For an ideal solution, $a_{iV} = y_i$ and $a_{iL} = x_i$.

To represent departure of activities from mole fractions when solutions are nonideal, *activity coefficients* based on concentrations in mole fractions are defined by

$$\gamma_{iV} \equiv \frac{a_{iV}}{y_i} \qquad (2\text{-}17)$$

$$\gamma_{iL} \equiv \frac{a_{iL}}{x_i} \qquad (2\text{-}18)$$

For ideal solutions, $\gamma_{iV} = 1.0$ and $\gamma_{iL} = 1.0$.

For convenient reference, thermodynamic quantities that are useful in phase equilibria calculations are summarized in Table 2.2.

Table 2.2 Thermodynamic Quantities for Phase Equilibria

Thermodynamic Quantity	Definition	Physical Significance	Limiting Value for Ideal Gas and Ideal Solution
Chemical potential	$\mu_i \equiv \left(\dfrac{\partial G}{\partial N_i}\right)_{P,T,N_j}$	Partial molar free energy, $\bar{g}_i$	$\mu_i = \bar{g}_i$
Partial fugacity	$\bar{f}_i \equiv C\,\exp\left(\dfrac{\mu_i}{RT}\right)$	Thermodynamic pressure	$\bar{f}_{iV} = y_i P$ $\bar{f}_{iL} = x_i P_i^s$
Fugacity coefficient of a pure species	$\phi_i \equiv \dfrac{f_i}{P}$	Deviation to fugacity due to pressure	$\phi_{iV} = 1.0$ $\phi_{iL} = \dfrac{P_i^s}{P}$
Partial fugacity coefficient of a species in a mixture	$\bar{\phi}_{iV} \equiv \dfrac{\bar{f}_{iV}}{y_i P}$ $\bar{\phi}_{iL} \equiv \dfrac{\bar{f}_{iL}}{x_i P}$	Deviations to fugacity due to pressure and composition	$\bar{\phi}_{iV} = 1.0$ $\bar{\phi}_{iL} = \dfrac{P_i^s}{P}$
Activity	$a_i \equiv \dfrac{\bar{f}_i}{f_i^o}$	Relative thermodynamic pressure	$a_{iV} = y_i$ $a_{iL} = x_i$
Activity coefficient	$\gamma_{iV} \equiv \dfrac{a_{iV}}{y_i}$ $\gamma_{iL} \equiv \dfrac{a_{iL}}{x_i}$	Deviation to fugacity due to composition	$\gamma_{iV} = 1.0$ $\gamma_{iL} = 1.0$

K-Values

A *phase-equilibrium ratio* is the ratio of mole fractions of a species present in two phases at equilibrium. For the vapor–liquid case, the constant is referred to as the *K-value* (vapor–liquid equilibrium ratio or *K*-factor):

$$K_i \equiv \frac{y_i}{x_i} \qquad (2\text{-}19)$$

For the liquid–liquid case, the ratio is referred to as the distribution coefficient or liquid–liquid equilibrium ratio:

$$K_{D_i} \equiv \frac{x_i^{(1)}}{x_i^{(2)}} \qquad (2\text{-}20)$$

For equilibrium-stage calculations involving the separation of two or more components, separation factors, like (1-4), are defined by forming ratios of equilibrium ratios. For the vapor–liquid case, *relative volatility* is defined by

$$\alpha_{ij} \equiv \frac{K_i}{K_j} \qquad (2\text{-}21)$$

For the liquid–liquid case, the *relative selectivity* is

$$\beta_{ij} \equiv \frac{K_{D_i}}{K_{D_j}} \qquad (2\text{-}22)$$

Equilibrium ratios can be expressed by the quantities in Table 2.2 in a variety of rigorous formulations. However, the only ones of practical interest are developed as follows. For vapor–liquid equilibrium, (2-9) becomes, for each component,

$$\bar{f}_{iV} = \bar{f}_{iL}$$

To form an equilibrium ratio, these partial fugacities are commonly replaced by expressions involving mole fractions as derived from the definitions in Table 2.2:

$$\bar{f}_{iL} = \gamma_{iL} x_i f_{iL}^o \qquad (2\text{-}23)$$

or

$$\bar{f}_{iL} = \bar{\phi}_{iL} x_i P \qquad (2\text{-}24)$$

and

$$\bar{f}_{iV} = \bar{\phi}_{iV} y_i P \qquad (2\text{-}25)$$

If (2-24) and (2-25) are used with (2-19), a so-called *equation-of-state form* of the *K*-value is obtained:

$$K_i = \frac{\bar{\phi}_{iL}}{\bar{\phi}_{iV}} \qquad (2\text{-}26)$$

This expression has received considerable attention, with applications of importance being the Starling modification of the Benedict, Webb, and Rubin (B–W–R–S) equation of state [5], the Soave modification of the Redlich–Kwong (S–R–K or R–K–S) equation of state [6], the Peng–Robinson (P–R) equation of state [7], and the Plöcker et al. modification of the Lee–Kesler (L–K–P) equation of state [8].

If (2-23) and (2-25) are used, a so-called *activity coefficient form* of the *K*-value is obtained:

$$K_i = \frac{\gamma_{iL} f_{iL}^o}{\bar{\phi}_{iV} P} = \frac{\gamma_{iL} \phi_{iL}}{\bar{\phi}_{iV}} \qquad (2\text{-}27)$$

Table 2.3 Useful Expressions for Estimating K-Values for Vapor–Liquid Equilibria ($K_i \equiv y_i/x_i$)

	Equation	Recommended Application
Rigorous forms:		
(1) Equation-of-state	$K_i = \dfrac{\bar{\phi}_{iL}}{\bar{\phi}_{iV}}$	Hydrocarbon and light gas mixtures from cryogenic temperatures to the critical region
(2) Activity coefficient	$K_i = \dfrac{\gamma_{iL}\phi_{iL}}{\bar{\phi}_{iV}}$	All mixtures from ambient to near-critical temperature
Approximate forms:		
(3) Raoult's law (ideal)	$K_i = \dfrac{P_i^s}{P}$	Ideal solutions at near-ambient pressure
(4) Modified Raoult's law	$K_i = \dfrac{\gamma_{iL} P_i^s}{P}$	Nonideal liquid solutions at near-ambient pressure
(5) Poynting correction	$K_i = \gamma_{iL}\phi_{iV}^s\left(\dfrac{P_i^s}{P}\right)\exp\left(\dfrac{1}{RT}\int_{P_i^s}^{P} v_{iL}\,dP\right)$	Nonideal liquid solutions at moderate pressure and below the critical temperature
(6) Henry's law	$K_i = \dfrac{H_i}{P}$	Low-to-moderate pressures for species at supercritical temperature

Since 1960, (2-27) has received considerable attention with applications to important industrial systems presented by Chao and Seader (C–S) [9], with a modification by Grayson and Streed [10].

Table 2.3 is a summary of useful formulations for estimating K-values for vapor–liquid equilibrium. Included are the two rigorous expressions given by (2-26) and (2-27), from which the other approximate formulations are derived. The so-called Raoult's law or ideal K-value is obtained from (2-27) by substituting from Table 2.2, for an ideal gas and ideal gas and liquid solutions, $\gamma_{iL} = 1.0$, $\phi_{iL} = P_i^s/P$, and $\bar{\phi}_{iV} = 1.0$. The modified Raoult's law relaxes the assumption of an ideal liquid solution by including the liquid-phase activity coefficient. The Poynting-correction form for moderate pressures is obtained by approximating the pure-component liquid fugacity coefficient in (2-27) by the expression

$$\phi_{iL} = \phi_{iV}^s \frac{P_i^s}{P}\exp\left(\frac{1}{RT}\int_{P_i^s}^{P} v_{iL}\,dP\right) \qquad (2\text{-}28)$$

where the exponential term is the Poynting factor or correction. If the liquid molar volume is reasonably constant over the pressure range, the integral in (2-28) becomes $v_{iL}(P - P_i^s)$. For a light gas species, whose critical temperature is less than the system temperature, the Henry's law form for the K-value is convenient provided that a value of H_i, the empirical Henry's law coefficient, is available. This constant for a particular species, i, depends on liquid-phase composition, temperature, and pressure. As pointed out in other chapters, other forms of Henry's law are used besides the one in Table 2.3. Included in Table 2.3 are recommendations for the application of each of the vapor–liquid K-value expressions.

Regardless of which thermodynamic formulation is used for estimating K-values, the accuracy depends on the particular correlations used for the thermodynamic properties required (i.e., vapor pressure, activity coefficient, and fugacity coefficients). For practical applications, the choice of K-value formulation is a compromise among considerations of accuracy, complexity, convenience, and past experience.

For liquid–liquid equilibria, (2-9) becomes

$$\bar{f}_{iL}^{(1)} = \bar{f}_{iL}^{(2)} \qquad (2\text{-}29)$$

where superscripts (1) and (2) refer to the two immiscible liquid phases. A rigorous formulation for the distribution coefficient is obtained by combining (2-23) with (2-20) to obtain an expression involving only activity coefficients:

$$K_{D_i} = \frac{x_i^{(1)}}{x_i^{(2)}} = \frac{\gamma_{iL}^{(2)} f_{iL}^{o(2)}}{\gamma_{iL}^{(1)} f_{iL}^{o(1)}} = \frac{\gamma_{iL}^{(2)}}{\gamma_{iL}^{(1)}} \qquad (2\text{-}30)$$

For vapor–solid equilibria, a useful formulation can be derived if the solid phase consists of just one of the components of the vapor phase. In that case, the combination of (2-9) and (2-25) gives

$$f_{iS} = \bar{\phi}_{iV}\,y_i\,P \qquad (2\text{-}31)$$

At low pressure, $\bar{\phi}_{iV} = 1.0$ and the solid fugacity can be approximated by the vapor pressure of the solid to give for the vapor-phase mole fraction of the component forming the solid phase:

$$y_i = \frac{(P_i^s)_{\text{solid}}}{P} \qquad (2\text{-}32)$$

For liquid–solid equilibria, a similar useful formulation can be derived if again the solid phase is a pure component. Then the combination of (2-9) and (2-23) gives

$$f_{iS} = \gamma_{iL}x_i f_{iL}^o \qquad (2\text{-}33)$$

At low pressure, the solid fugacity can be approximated by vapor pressure to give, for the component in the solid phase,

$$x_i = \frac{(P_i^s)_{\text{solid}}}{\gamma_{iL}(P_i^s)_{\text{liquid}}} \qquad (2\text{-}34)$$

EXAMPLE 2.2

Estimate the K-values of a vapor-liquid mixture of water and methane at 2 atm total pressure for temperatures of 20 and 80°C.

SOLUTION

At the conditions of temperature and pressure, water will exist mainly in the liquid phase and will follow Raoult's law, as given in Table 2.3. Because methane has a critical temperature of -82.5°C, well below the temperatures of interest, it will exist mainly in the vapor phase and follow Henry's law, in the form given in Table 2.3. From *Perry's Chemical Engineers' Handbook*, 6th ed., pp. 3-237 and 3-103, the following vapor pressure data for water and Henry's law coefficients for CH_4 are obtained:

T, °C	P^s for H_2O, atm	H for CH_4, atm
20	0.02307	3.76×10^4
80	0.4673	6.82×10^4

K-values for water and methane are estimated from (3) and (6), respectively, in Table 2.3, using $P = 2$ atm, with the following results:

T, °C	K_{H_2O}	K_{CH_4}
20	0.01154	18,800
80	0.2337	34,100

The above K-values confirm the assumptions of the phase distribution of the two species. The K-values for H_2O are low, but increase rapidly with increasing temperature. The K-values for methane are extremely high and do not change rapidly with temperature for this example.

2.3 IDEAL-GAS, IDEAL-LIQUID-SOLUTION MODEL

Design procedures for separation equipment require numerical values for phase enthalpies, entropies, densities, and phase–equilibrium ratios. Classical thermodynamics provides a means for obtaining these quantities in a consistent manner from P–v–T relationships, which are usually referred to as *equation-of-state models*. The simplest model applies when both liquid and vapor phases are ideal solutions (all activity coefficients equal 1.0) and the vapor is an ideal gas. Then the thermodynamic properties can be computed from unary constants for each of the species in the mixture in a relatively straightforward manner using the equations given in Table 2.4. In general, these ideal equations apply only at near-ambient pressure, up to about 50 psia (345 kPa), for mixtures of isomers or components of similar molecular structure.

For the vapor, the molar volume and mass density are computed from (1), the ideal-gas law in Table 2.4, which involves the molecular weight, M, of the mixture and the universal gas constant, R. For a mixture, the ideal-gas law assumes that both Dalton's law of additive partial pressures and Amagat's law of additive pure-species volumes apply.

The molar vapor enthalpy is computed from (2) by integrating, for each species, an equation in temperature for the zero-pressure heat capacity at constant pressure, $C_{P_V}^o$, starting from a reference (datum) temperature, T_0, to the temperature of interest, and then summing the resulting species vapor enthalpies on a mole-fraction basis. Typically, T_0 is taken as 0 K or 25°C. Although the reference pressure is zero, pressure has no effect on the enthalpy of an ideal gas. A common empirical representation of the effect of

Table 2.4   Thermodynamic Properties for Ideal Mixtures

Ideal gas and ideal-gas solution:

$$(1)\quad v_V = \frac{V}{\sum\limits_{i=1}^{C} N_i} = \frac{M}{\rho_V} = \frac{RT}{P}, \qquad M = \sum_{i=1}^{C} y_i M_i$$

$$(2)\quad h_V = \sum_{i=1}^{C} y_i \int_{T_0}^{T} (C_P^o)_{iV}\, dT = \sum_{i=1}^{C} y_i h_{iV}^o$$

$$(3)\quad s_V = \sum_{i=1}^{C} y_i \int_{T_0}^{T} \frac{(C_P^o)_{iV}}{T}\, dT - R \ln\left(\frac{P}{P_0}\right) - R \sum_{i=1}^{C} y_i \ln y_i$$

where the first term is s_V^o

Ideal-liquid solution:

$$(4)\quad v_L = \frac{V}{\sum\limits_{i=1}^{C} N_i} = \frac{M}{\rho_L} = \sum_{i=1}^{C} x_i v_{iL}, \qquad M = \sum_{i=1}^{C} x_i M_i$$

$$(5)\quad h_L = \sum_{i=1}^{C} x_i \left(h_{iV}^o - \Delta H_i^{\text{vap}}\right)$$

$$(6)\quad s_L = \sum_{i=1}^{C} x_i \left[\int_{T_0}^{T} \frac{(C_P^o)_{iV}}{T}\, dT - \frac{\Delta H_i^{\text{vap}}}{T}\right]$$
$$- R \ln\left(\frac{P}{P_0}\right) - R \sum_{i=1}^{C} x_i \ln x_i$$

Vapor–liquid equilibria:

$$(7)\quad K_i = \frac{P_i^s}{P}$$

Reference conditions (datum): h, ideal gas at T_0 and zero pressure; s, ideal gas at T_0 and 1 atm pressure.

Refer to elements if chemical reactions occur; otherwise refer to components.

temperature on the zero-pressure vapor heat capacity of a pure component is the following fourth-degree polynomial equation:

$$C_{P_V}^o = [a_0 + a_1T + a_2T^2 + a_3T^3 + a_4T^4]R \quad (2\text{-}35)$$

where the constants a_k depend on the species. Values of the constants for hundreds of compounds, with T in K, are tabulated by Poling, Prausnitz, and O'Connell [11]. Because $C_P = dh/dT$, (2-35) can be integrated for each species to give the ideal-gas species molar enthalpy:

$$h_V^o = \int_{T_0}^T C_{P_V}^o \, dT = \sum_{k=1}^5 \frac{a_{k-1}(T^k - T_0^k)R}{k} \quad (2\text{-}36)$$

The vapor molar entropy is computed from (3) in Table 2.4 by integrating $C_{P_V}^o/T$ from T_0 to T for each species, summing on a mole-fraction basis, adding a term for the effect of pressure referenced to a datum pressure, P_0, which is generally taken to be 1 atm (101.3 kPa), and adding a term for the entropy change of mixing. Unlike the ideal vapor enthalpy, the ideal vapor entropy includes terms for the effects of pressure and mixing. The reference pressure is not taken to be zero, because the entropy is infinity at zero pressure. If (2-35) is used for the heat capacity,

$$\int_{T_0}^T \left(\frac{C_{P_V}^o}{T}\right) dT = \left[a_0 \ln\left(\frac{T}{T_0}\right) + \sum_{k=1}^4 \frac{a_k(T^k - T_0^k)}{k}\right] R$$

$$(2\text{-}37)$$

The liquid molar volume and mass density are computed from the pure species molar volumes using (4) in Table 2.4 and the assumption of additive volumes (not densities). The effect of temperature on pure-component liquid density from the freezing point to the critical region at saturation pressure is correlated well by the empirical two-constant equation of Rackett [12]:

$$\rho_L = AB^{-(1-T/T_c)^{2/7}} \quad (2\text{-}38)$$

where values of the constants A, B, and the critical temperature, T_c, are tabulated for approximately 700 organic compounds by Yaws et al. [13].

The vapor pressure of a pure liquid species is well represented over a wide range of temperature from below the normal boiling point to the critical region by an empirical extended Antoine equation:

$$\ln P^s = k_1 + k_2/(k_3 + T) + k_4T + k_5 \ln T + k_6 T^{k_7}$$

$$(2\text{-}39)$$

where the constants k_k depend on the species. Values of the constants for hundreds of compounds are built into the physical-property libraries of all computer-aided process simulation and design programs. Constants for other

empirical vapor-pressure equations are tabulated for hundreds of compounds by Poling et al. [11]. At low pressures, the enthalpy of vaporization is given in terms of vapor pressure by classical thermodynamics:

$$\Delta H^{\text{vap}} = RT^2\left(\frac{d\ln P^s}{dT}\right) \quad (2\text{-}40)$$

If (2-39) is used for the vapor pressure, (2-40) becomes

$$\Delta H^{\text{vap}} = RT^2\left[-\frac{k_2}{(k_3 + T)^2} + k_4 + \frac{k_5}{T} + k_7 k_6 T^{k_7 - 1}\right]$$

$$(2\text{-}41)$$

The enthalpy of an ideal-liquid mixture is obtained by subtracting the molar enthalpy of vaporization from the ideal vapor molar enthalpy for each species, as given by (2-36), and summing these, as shown by (5) in Table 2.4. The entropy of the ideal-liquid mixture, given by (6), is obtained in a similar manner from the ideal-gas entropy by subtracting the molar entropy of vaporization, given by $\Delta H^{\text{vap}}/T$.

The final equation in Table 2.4 gives the expression for the ideal K-value, previously included in Table 2.3. Although it is usually referred to as the Raoult's law K-value, where Raoult's law is given by

$$p_i = x_i P_i^s \quad (2\text{-}42)$$

the assumption of Dalton's law is also required:

$$p_i = y_i P \quad (2\text{-}43)$$

Combination of (2-42) and (2-43) gives the Raoult's law K-value:

$$K_i \equiv \frac{y_i}{x_i} = \frac{P_i^s}{P} \quad (2\text{-}44)$$

$$y_i = x_i\left(\frac{P_i^s}{P}\right)$$

The extended Antoine equation, (2-39) (or some other suitable expression), can be used to estimate vapor pressure. Note that the ideal K-value is independent of phase compositions, but is exponentially dependent on temperature, because of the vapor pressure, and inversely proportional to pressure. From (2-21), the relative volatility using (2-44) is independent of pressure.

EXAMPLE 2.3

Styrene is manufactured by catalytic dehydrogenation of ethylbenzene, followed by vacuum distillation to separate styrene from unreacted ethylbenzene [14]. Typical conditions for the feed to an industrial distillation unit are 77.5°C (350.6 K) and 100 torr (13.33 kPa) with the following vapor and liquid flows at equilibrium:

	n, kmol/h	
Component	**Vapor**	**Liquid**
Ethylbenzene (EB)	76.51	27.31
Styrene (S)	61.12	29.03

Based on the property constants given below, and assuming that the ideal-gas, ideal-liquid-solution model of Table 2.4 is suitable at this low pressure, estimate values of v_V, ρ_V, h_V, s_V, v_L, ρ_L, h_L, and s_L in SI units, and the K-values and relative volatility.

Property Constants for (2-35), (2-38), (2-39)
(In all cases, *T* is in K)

	Ethylbenzene	Styrene
M, kg/kmol	106.168	104.152
$C_{P_V}^o$, J/kmol-K:		
$a_0 R$	−43,098.9	−28,248.3
$a_1 R$	707.151	615.878
$a_2 R$	−0.481063	−0.40231
$a_3 R$	1.30084×10^{-4}	9.93528×10^{-5}
$a_4 R$	0	0
P^s, Pa:		
k_1	86.5008	130.542
k_2	−7,440.61	−9,141.07
k_3	0	0
k_4	0.00623121	0.0143369
k_5	−9.87052	−17.0918
k_6	4.13065×10^{-18}	1.8375×10^{-18}
k_7	6	6
ρ_L, kg/m³:		
A	289.8	299.2
B	0.268	0.264
T_c, K	617.9	617.1
$R = 8.314$ kJ/kmol-K or kPa-m³/kmol-K $= 8,314$ J/kmol-K		

SOLUTION

Phase mole-fraction compositions and average molecular weights: From $y_i = (n_{iV})/n_V$, $x_i = (n_{iL})/n_L$,

	Ethylbenzene	Styrene
y	0.5559	0.4441
x	0.4848	0.5152

From (1), Table 2.4,

$$M_V = (0.5559)(106.168) + (0.4441)(104.152) = 105.27$$
$$M_L = (0.4848)(106.168) + (0.5152)(104.152) = 105.13$$

Vapor molar volume and density: From (1), Table 2.4,

$$v_V = \frac{RT}{P} = \frac{(8.314)(350.65)}{(13.332)} = 219.2 \text{ m}^3/\text{kmol}$$
$$\rho_V = \frac{M_V}{v_V} = \frac{105.27}{219.2} = 0.4802 \text{ kg/m}^3$$

Vapor molar enthalpy (datum = ideal gas at 298.15 K and 0 kPa): From (2-36) for ethylbenzene,

$$h_{EB_V}^o = -43098.9(350.65 - 298.15)$$
$$+ \left(\frac{707.151}{2}\right)(350.65^2 - 298.15^2)$$
$$- \left(\frac{0.481063}{3}\right)(350.65^3 - 298.15^3)$$
$$+ \left(\frac{1.30084 \times 10^{-4}}{4}\right)(350.65^4 - 298.15^4)$$
$$= 7,351,900 \text{ J/kmol}$$

Similarly,

$$h_{S_V}^o = 6,957,700 \text{ J/kmol}$$

From (2), Table 2.4, for the mixture,

$$h_V = \sum y_i h_{iV}^o = (0.5559)(7,351,900)$$
$$+ (0.4441)(6,957,100) = 7,176,800 \text{ J/kmol}$$

Vapor molar entropy (datum = pure components as vapor at 298.15 K, 101.3 kPa): From (2-37), for each component,

$$\int_{T_0}^{T} \left(\frac{C_{P_V}^o}{T}\right) dT = 22,662 \text{ J/kmol-K} \quad \text{for ethylbenzene}$$
$$\text{and} \quad 21,450 \text{ J/kmol-K} \quad \text{for styrene}$$

From (3), Table 2.4, for the mixture,

$$s_V = [(0.5559)(22,662.4) + (0.4441)(21,450.3)]$$
$$- 8,314 \ln\left(\frac{13.332}{101.3}\right) - 8,314[(0.5559)\ln(0.5559)$$
$$+ (0.4441)\ln(0.4441)] = 44,695 \text{ J/kmol-K}$$

Note that the terms for the pressure effect and the mixing effect are significant for this problem.

Liquid molar volume and density. From (2-38), for ethylbenzene,

$$\rho_{EB_L} = (289.8)(0.268)^{-(1-350.65/617.9)^{2/7}} = 816.9 \text{ kg/m}^3$$
$$v_{EB_L} = \frac{M_{EB}}{\rho_{EB_L}} = 0.1300 \text{ m}^3/\text{kmol}$$

Similarly,

$$\rho_{S_L} = 853.0 \text{ kg/m}^3$$
$$v_{S_L} = 0.1221 \text{ m}^3/\text{kmol}$$

From (4), Table 2.4, for the mixture,

$$v_L = (0.4848)(0.1300) + (0.5152)(0.1221) = 0.1259 \text{ m}^3/\text{kmol}$$
$$\rho_L = \frac{M_L}{v_L} = \frac{105.13}{0.1259} = 835.0 \text{ kg/m}^3$$

Liquid molar enthalpy (datum = ideal gas at 298.15 K): Use (5) in Table 2.4 for the mixture. For the enthalpy of vaporization of ethylbenzene, from (2-41),

$$\Delta H_{EB}^{vap} = 8,314(350.65)^2 \left[\frac{-(-7,440.61)}{(0+350.65)^2} + 0.00623121 \right.$$
$$\left. + \frac{-(9.87052)}{(350.65)} + 6(4.13065 \times 10^{-18})(350.65)^5 \right]$$
$$= 39,589,800 \text{ J/kmol}$$

Similarly,

$$\Delta H_{S}^{vap} = 40,886,700 \text{ J/kmol}$$

Then, applying (5), Table 2.4, using $h_{EB_V}^o$ and $h_{S_V}^o$ from above,

$$h_L = [(0.4848)(7,351,900 - 39,589,800)$$
$$+ (0.5152)(6,957,700 - 40,886,700)]$$
$$= -33,109,000 \text{ J/kmol}$$

Liquid molar entropy (datum = pure components as vapor at 298.15 K and 101.3 kPa): From (6), Table 2.4 for the mixture, using values for $\int_{T_0}^{T}(C_{P_V}^o / T)\, dT$ and ΔH^{vap} of EB and S from above,

$$s_L = (0.4848)\left(22,662 - \frac{39,589,800}{350.65}\right)$$
$$+ (0.5152)\left(21,450 - \frac{40,886,700}{350.65}\right)$$
$$- 8.314 \ln\left(\frac{13.332}{101.3}\right)$$
$$- 8,314[0.4848 \ln(0.4848) + 0.5152 \ln(0.5152)]$$
$$= -70,150 \text{ J/kmol-K}$$

K-values: Because (7), Table 2.4 will be used to compute the K-values, first estimate the vapor pressures using (2-39). For ethylbenzene,

$$\ln P_{EB}^s = 86.5008 + \left(\frac{-7,440.61}{(0+350.65}\right)$$
$$+ 0.00623121(350.65) + (-9.87052)\ln(350.65)$$
$$+ 4.13065 \times 10^{-18}(350.65)^6$$
$$= 9.63481$$
$$P_{EB}^s = \exp(9.63481) = 15,288 \text{ Pa} = 15.288 \text{ kPa}$$

Similarly,

$$P_{S}^s = 11.492 \text{ kPa}$$

From (7), Table 2.4,

$$K_{EB} = \frac{15.288}{13.332} = 1.147$$

$$K_S = \frac{11.492}{13.332} = 0.862$$

Relative volatility: From (2-21),

$$\alpha_{EB,S} = \frac{K_{EB}}{K_S} = \frac{1.147}{0.862} = 1.331$$

2.4 GRAPHICAL CORRELATIONS OF THERMODYNAMIC PROPERTIES

Calculations of estimated thermodynamic and other physical properties for the design of separation operations are most commonly carried out with computer-aided, process design

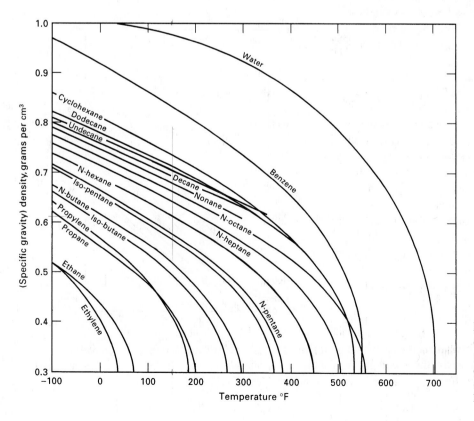

Figure 2.3 Hydrocarbon fluid densities.

[Adapted from G.G. Brown, D.L. Katz, G.G. Oberfell, and R.C. Alden, *Natural Gasoline and the Volatile Hydrocarbons*, Nat'l Gas Assoc. Amer., Tulsa, OK (1948).]

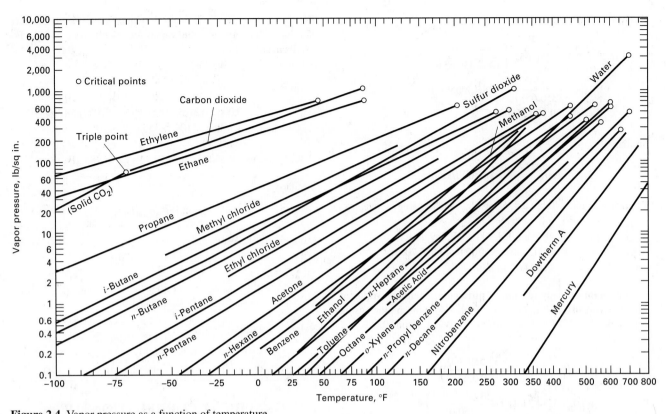

Figure 2.4 Vapor pressure as a function of temperature.

[Adapted from A.S. Faust, L.A. Wenzel, C.W. Clump, L. Maus, and L.B. Andersen, *Principles of Unit Operations,* John Wiley and Sons, New York (1960).]

and simulation programs, such as Aspen Plus, HYSYS, ChemCad, and Pro/II. However, plots of properties can best show effects of temperature and pressure. Some representative plots, which are readily generated by simulation programs, are shown in this section.

Saturated liquid densities as a function of temperature are plotted for some hydrocarbons in Figure 2.3. The density decreases rapidly as the critical temperature is approached until it becomes equal to the density of the vapor phase at the critical point. The liquid density curves are well correlated by the modified Rackett equation (2-38).

Figure 2.4 is a plot of liquid-state vapor pressures for some common chemicals, covering a wide range of temperature from below the normal boiling point to the critical temperature, where the vapor pressure terminates at the critical pressure. In general, the curves are found to fit the extended Antoine equation (2-39) reasonably well. This plot is useful for determining the phase state (liquid or vapor) of a pure substance and for estimating Raoult's law K-values from (2-44) [or (3) in Table 2.3].

Curves of ideal-gas, zero-pressure enthalpy over a wide range of temperature are given in Figure 2.5 for light-paraffin hydrocarbons. The datum is the liquid phase at 0°C, at which the enthalpy is zero. The derivatives of these curves fit the fourth-degree polynomial (2-35) for the ideal-gas heat

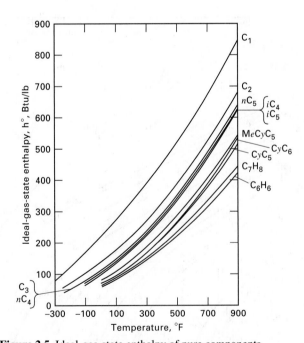

Figure 2.5 Ideal-gas-state enthalpy of pure components.

[Adapted from *Engineering Data Book,* 9th ed., Gas Processors Suppliers Association, Tulsa (1972).]

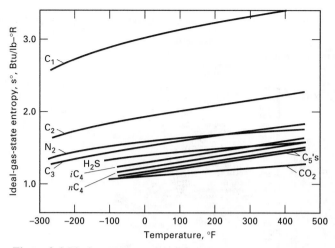

Figure 2.6 Ideal-gas-state entropy of pure components.
[Adapted from *Engineering Data Book,* 9th ed., Gas Processors Suppliers Association, Tulsa (1972).]

capacity reasonably well. Curves of ideal-gas entropy of several light gases, over a wide range of temperature, are given in Figure 2.6.

Enthalpies (heats) of vaporization are plotted as a function of saturation temperature in Figure 2.7 for light-paraffin hydrocarbons. These values are independent of pressure and decrease to zero at the critical point, where vapor and liquid phases become indistinguishable.

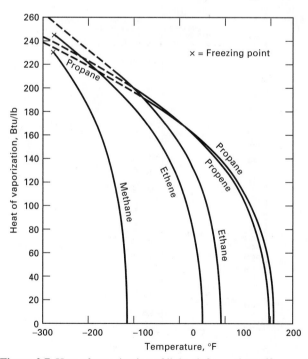

Figure 2.7 Heat of vaporization of light olefins and paraffins.
[Adapted from American Petroleum Institute, Technical Data Book, Washington, DC (Aug. 1963).]

Nomographs for determining the effects of temperature and pressure on the K-values of hydrocarbons and light gases are presented in Figures 2.8 and 2.9, which are taken from Hadden and Grayson [15]. In both charts, all K-values collapse to 1.0 at a pressure of 5,000 psia (34.5 MPa). This pressure, called the *convergence pressure,* depends on the boiling range of the components in the mixture. For example, in Figure 2.10 the components of the mixture (N_2 to nC_{10}) cover a very wide boiling-point range, resulting in a convergence pressure of close to 2,500 psia. For narrow-boiling mixtures, such as a mixture of ethane and propane, the convergence pressure is generally less than 1,000 psia. The K-value charts of Figures 2.8 and 2.9 apply strictly to a convergence pressure of 5,000 psia. A detailed procedure for correcting for the convergence pressure is given by Hadden and Grayson [15]. Use of the nomographs is illustrated below in Exercise 2.4.

No simple charts are available for estimating liquid–liquid equilibrium constants (distribution coefficients) because of the pronounced effect of composition. However, for ternary systems that are dilute in the solute and involve almost immiscible solvents, an extensive tabulation of distribution coefficients for the solute is given by Robbins [16].

EXAMPLE 2.4

Petroleum refining begins with the distillation, at near-atmospheric pressure, of crude oil into fractions of different boiling ranges. The fraction boiling from 0 to 100°C, the light naphtha, is a blending stock for gasoline. The fraction boiling from 100 to 200°C, the heavy naphtha, undergoes subsequent chemical processing into more useful products. One such process is steam cracking to produce a gas containing ethylene, propylene, and a number of other compounds, including benzene and toluene. This gas is then sent to a distillation train to separate the mixture into a dozen or more products. In the first column, hydrogen and methane are removed by cryogenic distillation at 3.2 MPa (464 psia). At a tray in the distillation column where the temperature is 40°F, use the appropriate K-value nomograph to estimate K-values for H_2, CH_4, C_2H_4, and C_3H_6.

SOLUTION

At 40°F, Figure 2.8 applies. The K-value of hydrogen depends on the other compounds in the mixture. Because appreciable amounts of benzene and toluene are present, locate a point (call it A) midway between the points for "H_2 in benzene" and "H_2 in toluene." Next, locate a point (call it B) at 40°F and 464 psia on the T–P grid. Connect points A and B with a straight line and read a value of $K = 100$ where the line intersects the K scale.

In a similar way, with the same location for point B, read $K = 11$ for methane. For ethylene (ethene) and propylene (propene), the point A is located on the normal boiling-point scale and the same point is used for B. Resulting K-values are 1.5 and 0.32, respectively.

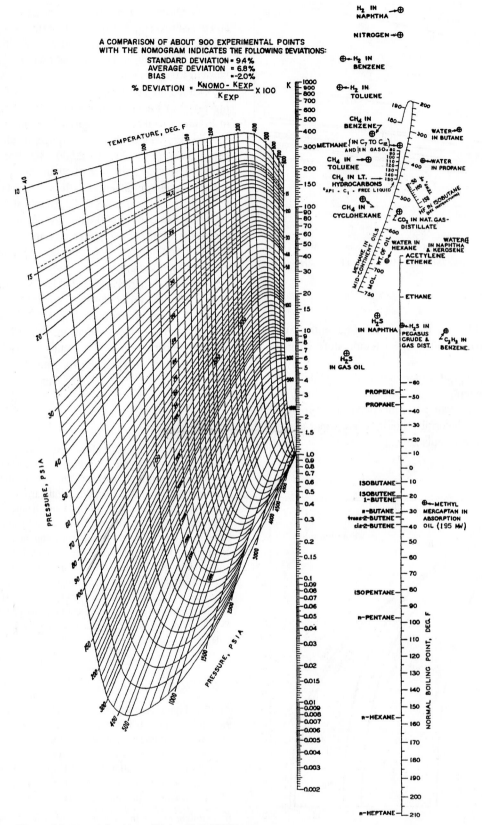

Figure 2.8 Vapor–liquid equilibria, 40 to 800°F.

[From S.T. Hadden and H.G. Grayson, *Hydrocarbon Proc. and Petrol. Refiner,* **40,** 207 (Sept. 1961), with permission.]

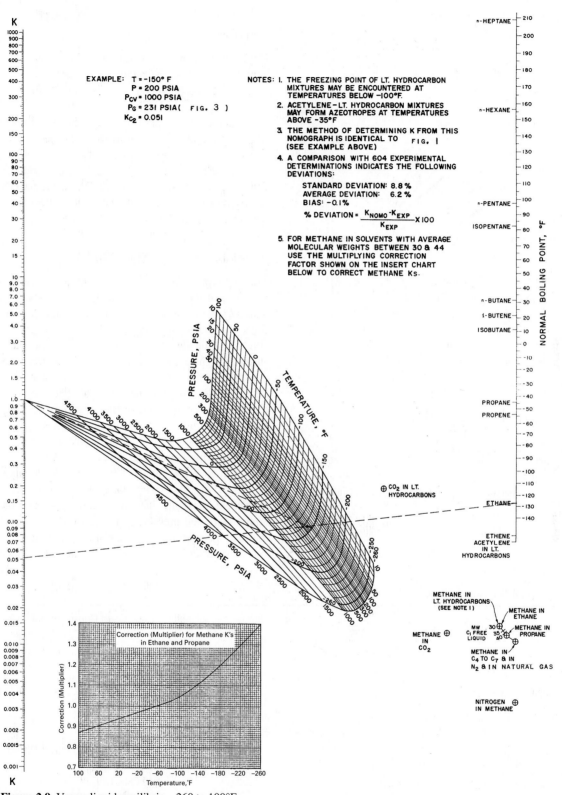

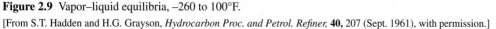

Figure 2.9 Vapor–liquid equilibria, –260 to 100°F.

[From S.T. Hadden and H.G. Grayson, *Hydrocarbon Proc. and Petrol. Refiner,* **40,** 207 (Sept. 1961), with permission.]

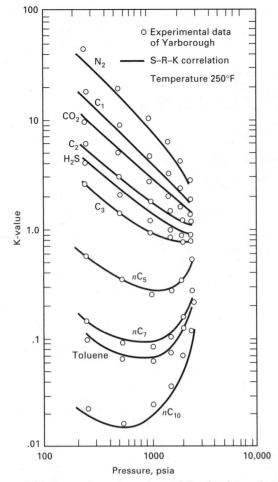

Figure 2.10 Comparison of experimental K-value data and S–R–K correlation.

2.5 NONIDEAL THERMODYNAMIC PROPERTY MODELS

Unlike the equations of Table 2.1, which are universally applicable to all pure substances and mixtures, whether ideal or nonideal, no universal equations are available for computing, for nonideal mixtures, values of thermodynamic properties such as density, enthalpy, entropy, fugacities, and activity coefficients as functions of temperature, pressure, and phase composition. Instead, two types of models are used: (1) P–v–T equation-of-state models and (2) activity coefficient or free-energy models. These are based on *constitutive equations* because they depend on the constitution or nature of the components in the mixture.

P–v–T Equation-of-State Models

The first type of model is a relationship between molar volume (or density), temperature, and pressure, usually referred to as a P–v–T equation of state. A large number of such equations have been proposed, mostly for the vapor phase. The simplest is the ideal-gas law, which applies only at low pressures or high temperatures because it neglects the volume occupied by the molecules and intermolecular forces among the molecules. All other equations of state attempt to correct for these two deficiencies. The equations of state that are most widely used by chemical engineers are listed in Table 2.5. These and other equations of state are discussed in some detail by Poling et al. [11].

Not included in Table 2.5 is the van der Waals equation, $P = RT/(v - b) - a/v^2$, where a and b are species-dependent constants that can be estimated from the critical temperature and pressure. The van der Waals equation was the first successful approach to the formulation of an equation of state for a nonideal gas. It is rarely used by chemical

Table 2.5 Useful Equations of State

Name	Equation	Equation Constants and Functions
(1) Ideal gas law	$P = \dfrac{RT}{v}$	None
(2) Generalized	$P = \dfrac{ZRT}{v}$	$Z = Z\{P_r, T_r, Z_c \text{ or } \omega\}$ as derived from data
(3) Redlich–Kwong (R–K)	$P = \dfrac{RT}{v - b} - \dfrac{a}{v^2 + bv}$	$b = 0.08664 RT_c/P_c$ $a = 0.42748 R^2 T_c^{2.5}/P_c T^{0.5}$
(4) Soave–Redlich–Kwong (S–R–K or R–K–S)	$P = \dfrac{RT}{v - b} - \dfrac{a}{v^2 + bv}$	$b = 0.08664 RT_c/P_c$ $a = 0.42748 R^2 T_c^2 \left[1 + f_\omega \left(1 - T_r^{0.5}\right)\right]^2 /P_c$ $f_\omega = 0.48 + 1.574\omega - 0.176\omega^2$
(5) Peng–Robinson (P–R)	$P = \dfrac{RT}{v - b} - \dfrac{a}{v^2 + 2bv - b^2}$	$b = 0.07780 RT_c/P_c$ $a = 0.45724 R^2 T_c^2 \left[1 + f_\omega \left(1 - T_r^{0.5}\right)\right]^2 /P_c$ $f_\omega = 0.37464 + 1.54226\omega - 0.26992\omega^2$

engineers because its range of application is too narrow. However, its development did suggest that all species might have equal reduced molar volumes, $v_r = v/v_c$, at the same reduced temperature, $T_r = T/T_c$, and reduced pressure, $P_r = P/P_c$. This finding, referred to as the *law* (*principle* or *theorem*) *of corresponding states,* was utilized to develop the generalized equation of state given as (2) in Table 2.5. That equation defines the *compressibility factor, Z,* which is a function of P_r, T_r, and the critical compressibility factor, Z_c, or the *acentric factor,* ω, which is determined from experimental P–v–T data. The acentric factor, introduced by Pitzer et al. [17], accounts for differences in molecular shape and is determined from the vapor pressure curve:

$$\omega = \left[-\log\left(\frac{P^s}{P_c}\right)_{T_r=0.7}\right] - 1.000 \qquad (2\text{-}45)$$

This definition results in a value for ω of zero for symmetric molecules. Some typical values of ω are 0.264, 0.490, and 0.649 for toluene, *n*-decane, and ethyl alcohol, respectively, as taken from the extensive tabulation of Poling et al. [11].

In 1949, Redlich and Kwong [18] published an equation of state that, like the van der Waals equation, contains only two constants, both of which can be determined from T_c and P_c, by applying the critical conditions

$$\left(\frac{\partial P}{\partial v}\right)_{T_c} = 0 \quad \text{and} \quad \left(\frac{\partial^2 P}{\partial v^2}\right)_{T_c} = 0$$

However, the R–K equation, given as (3) in Table 2.5, is a considerable improvement over the van der Waals equation. A study by Shah and Thodos [19] showed that the simple R–K equation, when applied to nonpolar compounds, has an accuracy that compares quite favorably with equations containing many more constants. Furthermore, the R–K equation can approximate the liquid-phase region.

If the R–K equation is expanded to obtain a common denominator, a cubic equation in v results. Alternatively, (2) and (3) in Table 2.5 can be combined to eliminate v to give the compressibility factor, Z, form of the R–K equation:

$$Z^3 - Z^2 + (A - B - B^2)Z - AB = 0 \qquad (2\text{-}46)$$

where

$$A = \frac{aP}{R^2 T^2} \qquad (2\text{-}47)$$

$$B = \frac{bP}{RT} \qquad (2\text{-}48)$$

Equation (2-46), which is cubic in Z, can be solved analytically for three roots (e.g., see *Perry's Handbook,* 7th ed., p. 4-20). In general, at supercritical temperatures, where only one phase can exist, one real root and a complex conjugate pair of roots are obtained. Below the critical temperature, where vapor and/or liquid phases can exist, three real roots are obtained, with the largest value of Z (largest v)

corresponding to the vapor phase—that is, Z_V—and the smallest Z (smallest v) corresponding to the liquid phase—that is, Z_L. The intermediate value of Z is of no practical use.

To apply the R–K equation to mixtures, *mixing rules* are used to average the constants a and b for each component in the mixture. The recommended rules for vapor mixtures of C components are

$$a = \sum_{i=1}^{C}\left[\sum_{j=1}^{C} y_i y_j (a_i a_j)^{0.5}\right] \qquad (2\text{-}49)$$

$$b = \sum_{i=1}^{C} y_i b_i \qquad (2\text{-}50)$$

EXAMPLE 2.5

Glanville, Sage, and Lacey [20] measured specific volumes of vapor and liquid mixtures of propane and benzene over wide ranges of temperature and pressure. Use the R–K equation to estimate specific volume of a vapor mixture containing 26.92 wt% propane at 400°F (477.6 K) and a saturation pressure of 410.3 psia (2,829 kPa). Compare the estimated and experimental values.

SOLUTION

Let propane be denoted by P and benzene by B. The mole fractions are

$$y_P = \frac{0.2692/44.097}{(0.2692/44.097) + (0.7308/78.114)} = 0.3949$$

$$y_B = 1 - 0.3949 = 0.6051$$

The critical constants for propane and benzene are given by Poling et al. [11]:

	Propane	Benzene
T_c, K	369.8	562.2
P_c, kPa	4,250	4,890

From the equations for the constants b and a in Table 2.5 for the R–K equation, using SI units,

$$b_P = \frac{0.08664(8.3144)(369.8)}{4,250} = 0.06268 \text{ m}^3/\text{kmol}$$

$$a_P = \frac{0.42748(8.3144)^2(369.8)^{2.5}}{(4,250)(477.59)^{0.5}}$$

$$= 836.7 \text{ kPa-m}^6/\text{kmol}^2$$

Similarly,

$$b_B = 0.08263 \text{ m}^3/\text{kmol}$$

$$a_B = 2,072 \text{ kPa-m}^6/\text{kmol}^2$$

From (2-50),

$$b = (0.3949)(0.06268) + (0.6051)(0.08263) = 0.07475 \text{ m}^3/\text{kmol}$$

From (2-49),

$$a = y_P^2 a_P + 2y_P y_B (a_P a_B)^{0.5} + y_B^2 a_B$$
$$= (0.3949)^2(836.7) + 2(0.3949)(0.6051)[(836.7)(2,072)]^{0.5}$$
$$+ (0.6051)^2(2,072) = 1,518 \text{ kPa-m}^6/\text{kmol}^2$$

From (2-47) and (2-48) using SI units,

$$A = \frac{(1,518)(2,829)}{(8.314)^2(477.59)^2} = 0.2724$$

$$B = \frac{(0.07475)(2,829)}{(8.314)(477.59)} = 0.05326$$

From (2-46), we obtain the cubic Z form of the R–K equation:

$$Z^3 - Z^2 + 0.2163Z - 0.01451 = 0$$

Solving this equation gives one real root and a conjugate pair of complex roots:

$$Z = 0.7314, \quad 0.1314 + 0.04243i, \quad 0.1314 - 0.04243i$$

The one real root is assumed to be that for the vapor phase.

From (2) of Table 2.5, the molar volume is

$$v = \frac{ZRT}{P} = \frac{(0.7314)(8.314)(477.59)}{2,829} = 1.027 \text{ m}^3/\text{kmol}$$

The average molecular weight of the mixture is computed to 64.68 kg/kmol. The specific volume is

$$\frac{v}{M} = \frac{1.027}{64.68} = 0.01588 \text{ m}^3/\text{kg} = 0.2543 \text{ ft}^3/\text{lb}$$

Glanville et al. report experimental values of $Z = 0.7128$ and $v/M = 0.2478$ ft³/lb, which are within 3% of the above estimated values.

Following the success of earlier work by Wilson [21], Soave [6] added a third parameter, the acentric factor, ω, defined by (2-45), to the R–K equation. The resulting, so-called Soave–Redlich–Kwong (S–R–K) or Redlich–Kwong–Soave (R–K–S) equation of state, given as (4) in Table 2.5, was immediately accepted for application to mixtures containing hydrocarbons and/or light gases because of its simplicity and accuracy. The main improvement was to make the parameter a a function of the acentric factor and temperature so as to achieve a good fit to vapor pressure data of hydrocarbons and thereby greatly improve the ability of the equation to predict properties of the liquid phase.

Four years after the introduction of the S–R–K equation, Peng and Robinson [7] presented a further modification of the R–K and S–R–K equations in an attempt to achieve improved agreement with experimental data in the critical region and for liquid molar volume. The Peng–Robinson (P–R) equation of state is listed as (5) in Table 2.5. The S–R–K and P–R equations of state are widely applied in process calculations, particularly for saturated vapors and liquids. When applied to mixtures of hydrocarbons and/or light gases, the mixing rules are given by (2-49) and (2-50), except that (2-49) is often modified to include a binary interaction coefficient, k_{ij}:

$$a = \sum_{i=1}^{C} \left[\sum_{j=1}^{C} y_i y_j (a_i a_j)^{0.5}(1 - k_{ij}) \right] \quad (2\text{-}51)$$

Values of k_{ij}, back-calculated from experimental data, have been published for both the S–R–K and P–R equations. Knapp et al. [22] present an extensive tabulation. Generally, k_{ij} is taken as zero for hydrocarbons paired with hydrogen or other hydrocarbons.

Although the S–R–K and P–R equations were not intended to be applied to mixtures containing polar organic compounds, they are finding increasing use in such applications by employing large values of k_{ij}, in the vicinity of 0.5, as back-calculated from experimental data. However, a preferred procedure for mixtures containing polar organic compounds is to use a more theoretically based mixing rule such as that of Wong and Sandler, which is discussed in detail in Chapter 11 and which bridges the gap between a cubic equation of state and an activity-coefficient equation.

Another theoretical basis for polar and nonpolar substances is the virial equation of state due to Thiesen [23] and Onnes [24]. A common representation of the virial equation, which can be derived from the statistical mechanics of the forces between the molecules, is a power series in $1/v$ for Z:

$$Z = 1 + \frac{B}{v} + \frac{C}{v^2} + \cdots \quad (2\text{-}52)$$

An empirical modification of the virial equation is the Starling form [5] of the Benedict–Webb–Rubin (B–W–R) equation of state for hydrocarbons and light gases in both the gas and liquid phases. Walas [25] presents an extensive discussion of B–W–R-type equations, which because of the large number of terms and species constants (at least 8), is not widely used except for pure substances at cryogenic temperatures. A more useful modification of the B–W–R equation is a generalized corresponding-states form developed by Lee and Kesler [26] with an important extension to mixtures by Plöcker et al. [8]. All of the constants in the L–K–P equation are given in terms of the acentric factor and reduced temperature and pressure, as developed from P–v–T data for three simple fluids ($\omega = 0$), methane, argon, and krypton, and a reference fluid ($\omega = 0.398$), n-octane. The equations, constants, and mixing rules in terms of pseudo-critical properties are given by Walas [25]. The Lee–Kesler–Plöcker (L–K–P) equation of state describes vapor and liquid mixtures of hydrocarbons and/or light gases over wide ranges of temperature and pressure.

Derived Thermodynamic Properties from P–v–T Models

In the previous subsection, several useful P–v–T equations of state for the estimation of the molar volume (or density) or pure substances and mixtures in either the vapor or liquid phase were presented. If a temperature-dependent, ideal-gas heat capacity or enthalpy equation, such as (2-35) or (2-36), is also available, all other vapor- and liquid-phase properties can be derived in a consistent manner by applying the classical integral equations of thermodynamics given in Table 2.6. These equations, in the form of departure (from the ideal gas) equations of Table 2.4, and often referred to as residuals, are applicable to vapor or liquid phases.

Table 2.6 Classical Integral Departure Equations of Thermodynamics

At a given temperature and composition, the following equations give the effect of pressure above that for an ideal gas.

Mixture enthalpy:

(1) $\left(h - h_V^o\right) = Pv - RT - \int_\infty^v \left[P - T \left(\frac{\partial P}{\partial T}\right)_v \right] dv$

Mixture entropy:

(2) $\left(s - s_V^o\right) = \int_\infty^v \left(\frac{\partial P}{\partial T}\right)_v dv - \int_\infty^v \frac{R}{v} dv$

Pure-component fugacity coefficient:

(3) $\phi_{iV} = \exp\left[\frac{1}{RT} \int_0^P \left(v - \frac{RT}{P} \right) dP \right]$

$= \exp\left[\frac{1}{RT} \int_v^\infty \left(P - \frac{RT}{v} \right) dv - \ln Z + (Z - 1) \right]$

Partial fugacity coefficient:

(4) $\bar{\phi}_{iV} = \exp\left\{ \frac{1}{RT} \int_V^\infty \left[\left(\frac{\partial P}{\partial N_i}\right)_{T,V,N_j} - \frac{RT}{V} \right] dV - \ln Z \right\}$

where $V = v \sum_{i=1}^C N_i$

When the ideal-gas law, $P = RT/v$, is substituted into (1) to (4) of Table 2.6, the results for the vapor, as expected, are

$\left(h - h^o\right) = 0 \qquad \phi = 1$

$\left(s - s^o\right) = 0 \qquad \bar{\phi} = 1$

However, when the R–K equation is substituted into the equations of Table 2.6, the following results for the vapor phase are obtained after a rather tedious exercise in calculus:

$h_V = \sum_{i=1}^C (y_i h_{iV}^o) + RT\left[Z_V - 1 - \frac{3A}{2B} \ln\left(1 + \frac{B}{Z_V}\right) \right]$ (2-53)

$s_V = \sum_{i=1}^C (y_i s_{iV}^o) - R \ln\left(\frac{P}{P^o}\right)$ (2-54)

$- R \sum_{i=1}^C (y_i \ln y_i) + R \ln(Z_V - B)$

$\phi_V = \exp\left[Z_V - 1 - \ln(Z_V - B) - \frac{A}{B} \ln\left(1 + \frac{B}{Z_V}\right) \right]$ (2-55)

$\bar{\phi}_{iV} = \exp\left[(Z_V - 1)\frac{B_i}{B} - \ln(Z_V - B) \right.$

$\left. - \frac{A}{B}\left(2\sqrt{\frac{A_i}{A}} - \frac{B_i}{B} \right) \ln\left(1 + \frac{B}{Z_V}\right) \right]$ (2-56)

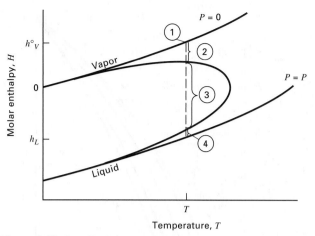

Figure 2.11 Contributions to enthalpy.

The results for the liquid phase are identical if y_i and Z_V (but not h_{iV}^o) are replaced by x_i and Z_L, respectively. It may be surprising that the liquid-phase forms of (2-53) and (2-54) account for the enthalpy and entropy of vaporization, respectively. This is because the R–K equation of state, as well as the S–R–K and P–R equations, are continuous functions in passing between the vapor and liquid regions, as shown for enthalpy in Figure 2.11. Thus, the liquid enthalpy is determined by accounting for the following four effects for a pure species at a temperature below the critical. From (1), Table 2.6, the four contributions to enthalpy in Figure 2.11 are as follows:

$h_L = h_V^o + Pv - RT - \int_\infty^v \left[P - T\left(\frac{\partial P}{\partial T}\right)_v \right] dv$

$= \underbrace{h_V^o}_{\text{(1) Vapor at zero pressure}}$

$\underbrace{+ (Pv)_{V_s} - RT - \int_\infty^{v_{V_s}} \left[P - T\left(\frac{\partial P}{\partial T}\right)_v \right] dv}_{\text{(2) Pressure correction for vapor to saturation pressure}}$

$\underbrace{- T\left(\frac{\partial P}{\partial T}\right)_s (v_{V_s} - v_{L_s})}_{\text{(3) Latent heat of vaporization}}$

$\underbrace{+ [(Pv)_L - (Pv)_{L_s}] - \int_{v_{L_s}}^{v_L} \left[P - T\left(\frac{\partial P}{\partial T}\right)_v \right] dv}_{\text{(4) Correction to liquid for pressure in excess of saturation pressure}}$ (2-57)

where the subscript s refers to the saturation pressure.

The fugacity coefficient, ϕ, of a pure species at temperature T and pressure P from the R–K equation, as given by (2-55), applies to the vapor for $P < P_i^s$. For $P > P_i^s$, ϕ is the fugacity coefficient of the liquid. Saturation pressure corresponds to the condition of $\phi_V = \phi_L$. Thus, at a temperature $T < T_c$, the saturation pressure (vapor pressure), P^s, can be estimated from the R–K equation of state by setting (2-55) for the vapor equal to (2-55) for the liquid and solving, by an iterative procedure, for P, which then equals P^s.

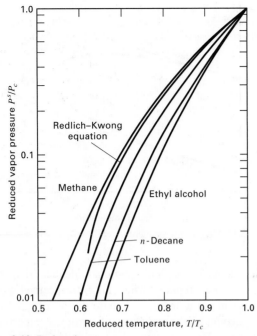

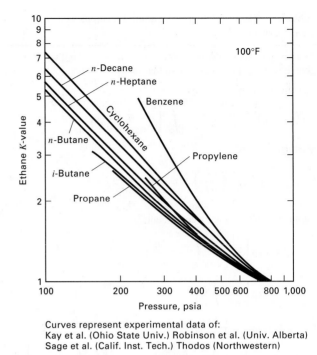

Figure 2.12 Reduced vapor pressure.

Curves represent experimental data of:
Kay et al. (Ohio State Univ.) Robinson et al. (Univ. Alberta)
Sage et al. (Calif. Inst. Tech.) Thodos (Northwestern)

Figure 2.13 K-values of ethane in binary hydrocarbon mixtures at 100°F.

The results, as given by Edmister [27], are plotted in reduced form in Figure 2.12. The R–K vapor-pressure curve does not satisfactorily represent data for a wide range of molecular shapes, as witnessed by the experimental curves for methane, toluene, n-decane, and ethyl alcohol on the same plot. This failure represents one of the major shortcomings of the R–K equation and is the main reason why Soave [6] modified the R–K equation by introducing the acentric factor in such a way as to greatly improve agreement with experimental vapor-pressure data. Thus, while the critical constants, T_c and P_c alone are insufficient to generalize thermodynamic behavior, a substantial improvement is made by incorporating into the $P-v-T$ equation a third parameter that represents the generic differences in the reduced-vapor-pressure curves.

As seen in (2-56), partial fugacity coefficients depend on pure-species properties, A_i and B_i, and mixture properties, A and B. Once $\bar{\phi}_{iV}$ and $\bar{\phi}_{iL}$ are computed from (2-56), a K-value can be estimated from (2-26).

The most widely used $P-v-T$ equations of state for separation calculations involving vapor and liquid phases are the S–R–K, P–R, and L–K–P relations. These equations are combined with the integral departure equations of Table 2.6 to obtain useful equations for estimating the enthalpy, entropy, fugacity coefficients, partial fugacity coefficients of vapor and liquid phases, and K-values. The results of the integrations are even more complex than (2-53) to (2-56) and are unsuitable for manual calculations. However, computer programs for making calculations with these equations are rapid, accurate, and readily available. Such programs are incorporated into widely used steady-state, computer-aided process design and simulation programs, such as Aspen Plus, HYSYS, ChemCad, and Pro/II.

Ideal K-values as determined from Eq. (7) in Table 2.4, depend only on temperature and pressure, and not on composition. Most frequently, ideal K-values are applied to mixtures of nonpolar compounds, particularly hydrocarbons such as paraffins and olefins. Figure 2.13 shows experimental K-value curves for a light hydrocarbon, ethane, in various binary mixtures with other, less volatile hydrocarbons at 100°F (310.93 K) for pressures from 100 psia (689.5 kPa) to *convergence pressures* between 720 and 780 psia (4.964 MPa to 5.378 MPa). At the convergence pressure, separation by operations involving vapor–liquid equilibrium becomes impossible because all K-values become 1.0. The temperature of 100°F is close to the critical temperature of 550°R (305.6 K) for ethane. Figure 2.13 shows that ethane does not form ideal solutions at 100°F with all the other components because the K-values depend on the other component, even for paraffin homologs. For example, at 300 psia, the K-value of ethane in benzene is 80% higher than in propane.

The ability of equations of state, such as S–R–K, P–R, and L–K–P equations, to predict the effect of composition as well as the effect of temperature and pressure on K-values of multicomponent mixtures of hydrocarbons and light gases is shown in Figure 2.10. The mixture contains 10 species ranging in volatility from nitrogen to n-decane. The experimental data points, covering almost a 10-fold range of pressure at 250°F, are those of Yarborough [28]. Agreement with the S–R–K equation is very good.

EXAMPLE 2.6

In the high-pressure, high-temperature, thermal hydrodealkylation of toluene to benzene ($C_7H_8 + H_2 \rightarrow C_6H_6 + CH_4$), excess hydrogen is used to minimize cracking of aromatics to light gases. In practice, conversion of toluene per pass through the reactor is only 70%. To separate and recycle hydrogen, hot reactor effluent vapor of 5,597 kmol/h at 500 psia (3,448 kPa) and 275°F (408.2 K) is partially condensed to 120°F (322 K), with product phases separated in a flash drum. If the composition of the reactor effluent is as follows, and the flash drum pressure is 485 psia (3,344 kPa), calculate equilibrium compositions and flow rates of vapor and liquid leaving the flash drum and the amount of heat that must be transferred using a computer-aided, steady-state, simulation program with each of the equation-of-state models discussed above. Compare the results, including flash-drum K-values and enthalpy and entropy changes.

Component	Mole Fraction
Hydrogen (H)	0.3177
Methane (M)	0.5894
Benzene (B)	0.0715
Toluene (T)	0.0214
	1.0000

SOLUTION

The computations were made with a computer-aided, process-simulation program, using the S–R–K, P–R, and L–K–P equations of state. The results at 120°F and 485 psia are as follows:

	Equation of State		
	S–R–K	P–R	L–K–P
Vapor flows, kmol/h:			
Hydrogen	1,777.1	1,774.9	1,777.8
Methane	3,271.0	3,278.5	3,281.4
Benzene	55.1	61.9	56.0
Toluene	6.4	7.4	7.0
Total	5,109.6	5,122.7	5,122.2
Liquid flows, kmol/h:			
Hydrogen	1.0	3.3	0.4
Methane	27.9	20.4	17.5
Benzene	345.1	338.2	344.1
Toluene	113.4	112.4	112.8
Total	487.4	474.3	474.8
K-values:			
Hydrogen	164.95	50.50	466.45
Methane	11.19	14.88	17.40
Benzene	0.01524	0.01695	0.01507
Toluene	0.00537	0.00610	0.00575
Enthalpy change, GJ/h	35.267	34.592	35.173
Entropy change, MJ/h-K	−95.2559	−93.4262	−95.0287
Percent of benzene and toluene condensed	88.2	86.7	87.9

Because the reactor effluent is mostly hydrogen and methane, the effluent at 275°F and 500 psia, and the equilibrium vapor at 120°F and 485 psia are nearly ideal gases ($0.98 < Z < 1.00$), despite the moderately high pressures. Thus, the enthalpy and entropy changes are dominated by vapor heat capacity and latent heat effects, which are largely independent of which equation of state is used. Consequently, the enthalpy and entropy changes among the three equations of state differ by less than 2%.

Significant differences exist for the K-values of H_2 and CH_4. However, because the values are in all cases large, the effect on the amount of equilibrium vapor is very small. Reasonable K-values for H_2 and CH_4, based on experimental data, are 100 and 13, respectively. K-values for benzene and toluene differ among the three equations of state by as much as 11% and 14%, respectively, which, however, causes less than a 2% difference in the percentage of benzene and toluene condensed. Raoult's law K-values for benzene and toluene, based on vapor-pressure data, are 0.01032 and 0.00350, which are considerably lower than the values computed from each of the three equations of state because deviations to fugacities due to pressure are important in the liquid phase and, particularly, in the vapor phase.

Note that the material balances are precisely satisfied for each equation of state. However, the user of a computer-aided design and simulation program should never take this as an indication that the results are correct.

2.6 ACTIVITY-COEFFICIENT MODELS FOR THE LIQUID PHASE

In Sections 2.3 and 2.5, methods based on equations of state are presented for predicting thermodynamic properties of vapor and liquid mixtures. In this section, predictions of liquid properties based on *Gibbs free-energy models* for predicting liquid-phase activity coefficients and other excess functions such as volume and enthalpy of mixing are developed. Regular-solution theory, which can be applied to mixtures of nonpolar compounds using only constants for the pure components, is the first model presented. This is followed by a discussion of several models that can be applied to mixtures containing polar compounds, provided that experimental data are available to determine the *binary interaction parameters* in these models. If not, group-contribution methods, which have been extensively developed, can be used to make estimates. All models discussed can be applied to predict vapor–liquid phase equilibria; and some can estimate liquid–liquid equilibria, and even solid–liquid and polymer–liquid equilibria.

Except at high pressures, dependency of K-values on composition is due primarily to nonideal solution behavior in the liquid phase. Prausnitz, Edmister, and Chao [29] showed that the relatively simple *regular-solution theory* of Scatchard and Hildebrand [30] can be used to estimate deviations due to nonideal behavior of hydrocarbon–liquid mixtures. They expressed K-values in terms of (2-27), $K_i = \gamma_{iL}\phi_{iL}/\bar{\phi}_{iV}$. Chao and Seader [9] simplified and extended application of this equation to a general correlation for hydrocarbons and some light gases in the form of a compact set of equations especially suitable for use with a digital computer, which was widely used before the availability of the S–R–K and P–R equations.

Simple models for the liquid-phase activity coefficient, γ_{iL}, based only on properties of pure species, are not generally accurate. However, for hydrocarbon mixtures, regular-solution theory is convenient and widely applied. The theory is based on the premise that nonideality is due to differences in van der Waals forces of attraction among the different molecules present. Regular solutions have an endothermic heat of mixing, and all activity coefficients are greater than one. These solutions are regular in the sense that molecules are assumed to be randomly dispersed. Unequal attractive forces between like and unlike molecule pairs tend to cause segregation of molecules. However, for regular solutions the species concentrations on a molecular level are identical to overall solution concentrations. Therefore, excess entropy due to segregation is zero and entropy of regular solutions is identical to that of ideal solutions, in which the molecules are randomly dispersed.

Activity Coefficients from Gibbs Free Energy

Activity-coefficient equations often have their basis in Gibbs free-energy models. For a nonideal solution, the molar Gibbs free energy, g, is the sum of the molar free energy of an ideal solution and an excess molar free energy g^E for nonideal effects. For a liquid solution,

$$
\begin{aligned}
g &= \sum_{i=1}^{C} x_i g_i + RT \sum_{i=1}^{C} x_i \ln x_i + g^E \\
&= \sum_{i=1}^{C} x_i \left(g_i + RT \ln x_i + \bar{g}_i^E \right)
\end{aligned}
\tag{2-58}
$$

where $g \equiv h - Ts$ and excess molar free energy is the sum of the partial excess molar free energies. The partial excess molar free energy is related by classical thermodynamics to the liquid-phase activity coefficient by

$$
\begin{aligned}
\frac{\bar{g}_i^E}{RT} = \ln \gamma_i &= \left[\frac{\partial (N_t g^E / RT)}{\partial N_i} \right]_{P,T,N_j} \\
&= \frac{g^E}{RT} - \sum_k x_k \left[\frac{\partial (g^E / RT)}{\partial x_k} \right]_{P,T,x_r}
\end{aligned}
\tag{2-59}
$$

where $j \neq i$, $r \neq k$, $k \neq i$, and $r \neq i$.

The relationship between excess molar free energy and excess molar enthalpy and entropy is

$$
g^E = h^E - Ts^E = \sum_{i=1}^{C} x_i \left(\bar{h}_i^E - T\bar{s}_i^E \right)
\tag{2-60}
$$

Regular-Solution Model

For a multicomponent, regular liquid solution, the excess molar free energy is based on nonideality due to differences in molecular size and intermolecular forces. The former are

expressed in terms of liquid molar volume and the latter in terms of the enthalpy of vaporization. The resulting model is

$$
g^E = \sum_{i=1}^{C} (x_i v_{iL}) \left[\frac{1}{2} \sum_{i=1}^{C} \sum_{j=1}^{C} \Phi_i \Phi_j (\delta_i - \delta_j)^2 \right]
\tag{2-61}
$$

where Φ is the volume fraction assuming additive molar volumes, as given by

$$
\Phi_i = \frac{x_i v_{iL}}{\sum_{j=1}^{C} x_j v_{jL}} = \frac{x_i v_{iL}}{v_L}
\tag{2-62}
$$

and δ is the solubility parameter, which is defined in terms of the volumetric internal energy of vaporization as

$$
\delta_i = \left(\frac{\Delta E_i^{\text{vap}}}{v_{iL}} \right)^{1/2}
\tag{2-63}
$$

Values of the solubility parameter for many components can be obtained from process simulation programs.

Applying (2-59) to (2-61) gives an expression for the activity coefficient in a regular solution:

$$
\ln \gamma_{iL} = \frac{v_{iL} \left(\delta_i - \sum_{j=1}^{C} \Phi_j \delta_j \right)^2}{RT}
\tag{2-64}
$$

Because $\ln \gamma_{iL}$ varies almost inversely with absolute temperature, $v_i L$ and δ_j are frequently taken as constants at some convenient reference temperature, such as 25°C. Thus, the estimation of γ_L by regular-solution theory requires only the pure-species constants v_L and δ. The latter parameter is often treated as an empirical constant determined by back-calculation from experimental data. For species with a critical temperature below 25°C, v_L and δ at 25°C are hypothetical. However, they can be evaluated by back-calculation from phase-equilibria data.

When molecular-size differences, as reflected by liquid molar volumes, are appreciable, the following Flory–Huggins size correction can be added to the regular-solution free-energy contribution:

$$
g^E = RT \sum_{i=1}^{C} x_i \ln \left(\frac{\Phi_i}{x_i} \right)
\tag{2-65}
$$

Substitution of (2-65) into (2-59) gives

$$
\ln \gamma_{iL} = \ln \left(\frac{v_{iL}}{v_L} \right) + 1 - \left(\frac{v_{iL}}{v_L} \right)
\tag{2-66}
$$

The complete expression for the activity coefficient of a species in a regular solution, including the Flory–Huggins correction, is

$$
\gamma_{iL} = \exp \left[\frac{v_{iL} \left(\delta_i - \sum_{j=1}^{C} \Phi_j \delta_j \right)^2}{RT} + \ln \left(\frac{v_{iL}}{v_L} \right) + 1 - \frac{v_{iL}}{v_L} \right]
\tag{2-67}
$$

EXAMPLE 2.7

Yerazunis, Plowright, and Smola [31] measured liquid-phase activity coefficients for the *n*-heptane/toluene system over the entire concentration range at 1 atm (101.3 kPa). Estimate activity coefficients for the range of conditions using regular-solution theory both with and without the Flory–Huggins correction. Compare estimated values with experimental data.

SOLUTION

Experimental liquid-phase compositions and temperatures for 7 of 19 points are as follows, where H denotes heptane and T denotes toluene:

T, °C	x_H	x_T
98.41	1.0000	0.0000
98.70	0.9154	0.0846
99.58	0.7479	0.2521
101.47	0.5096	0.4904
104.52	0.2681	0.7319
107.57	0.1087	0.8913
110.60	0.0000	1.0000

At 25°C, liquid molar volumes are $v_{H_L} = 147.5$ cm^3/mol and $v_{T_L} = 106.8$ cm^3/mol. Solubility parameters are 7.43 and 8.914 (cal/cm^3)$^{1/2}$, respectively, for H and T. As an example, consider mole fractions in the above table for 104.52°C. From (2-62), volume fractions are

$$\Phi_H = \frac{0.2681(147.5)}{0.2681(147.5) + 0.7319(106.8)} = 0.3359$$

$$\Phi_T = 1 - \Phi_H = 1 - 0.3359 = 0.6641$$

Substitution of these values, together with the solubility parameters, into (2-64) gives

$$\gamma_H = \exp\left\{\frac{147.5[7.430 - 0.3359(7.430) - 0.6641(8.914)]^2}{1.987(377.67)}\right\}$$

$$= 1.212$$

Values of γ_H and γ_T computed in this manner for all seven liquid-phase conditions are plotted in Figure 2.14.

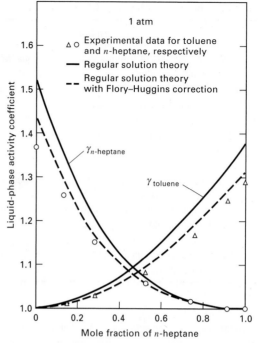

Figure 2.14 Liquid-phase activity coefficients for *n*-heptane/toluene system at 1 atm.

Applying (2-67), with the Flory–Huggins correction, to the same data point gives

$$\gamma_H = \exp\left[0.1923 + \ln\left(\frac{147.5}{117.73}\right) + 1 - \left(\frac{147.5}{117.73}\right)\right] = 1.179$$

Values of γ_H and γ_T computed in this manner are included in Figure 2.14. Deviations from experiment are not greater than 12% for regular-solution theory and not greater than 6% when the Flory–Huggins correction is included. Unfortunately, such good agreement is not always obtained with nonpolar hydrocarbon solutions, as shown, for example, by Hermsen and Prausnitz [32], who studied the cyclopentane/benzene system.

Nonideal Liquid Solutions

When liquids contain dissimilar polar species, particularly those that can form or break hydrogen bonds, the ideal-liquid solution assumption is almost always invalid and the regular-solution theory is not applicable. Ewell, Harrison, and Berg [33] provide a very useful classification of molecules based on the potential for association or solvation due to hydrogen-bond formation. If a molecule contains a hydrogen atom attached to a donor atom (O, N, F, and in certain cases C), the active hydrogen atom can form a bond with another molecule containing a donor atom. The classification in Table 2.7 permits qualitative estimates of deviations from Raoult's law for binary pairs when used in conjunction with Table 2.8. Positive deviations correspond to values of $\gamma_{iL} > 1$. Nonideality results in a variety of variations of γ_{iL} with composition, as shown in Figure 2.15 for several binary systems, where the

Roman numerals refer to classification groups in Tables 2.7 and 2.8. Starting with Figure 2.15a and taking the other plots in order, we offer the following explanations for the nonidealities. Normal heptane (V) breaks ethanol (II) hydrogen bonds, causing strong positive deviations. In Figure 2.15b, similar but less positive deviations occur when acetone (III) is added to formamide (I). Hydrogen bonds are broken and formed with chloroform (IV) and methanol (II) in Figure 2.15c, resulting in an unusual positive deviation curve for chloroform that passes through a maximum. In Figure 2.15d, chloroform (IV) provides active hydrogen atoms that can form hydrogen bonds with oxygen atoms of acetone (III), thus causing negative deviations. For water (I) and *n*-butanol (II) in Figure 2.15e, hydrogen bonds of both molecules are broken, and nonideality is sufficiently strong to cause formation of two immiscible liquid phases (*phase splitting*) over a wide region of overall composition.

Table 2.7   Classification of Molecules Based on Potential for Forming Hydrogen Bonds

Class	Description	Example
I	Molecules capable of forming three-dimensional networks of strong H-bonds	Water, glycols, glycerol, amino alcohols, hydroxylamines, hydroxyacids, polyphenols, and amides
II	Other molecules containing both active hydrogen atoms and donor atoms (O, N, and F)	Alcohols, acids, phenols, primary and secondary amines, oximes, nitro and nitrile compounds with α-hydrogen atoms, ammonia, hydrazine, hydrogen fluoride, and hydrogen cyanide
III	Molecules containing donor atoms but no active hydrogen atoms	Ethers, ketones, aldehydes, esters, tertiary amines (including pyridine type), and nitro and nitrile compounds without α-hydrogen atoms
IV	Molecules containing active hydrogen atoms but no donor atoms that have two or three chlorine atoms on the same carbon atom as a hydrogen or one chlorine on the carbon atom and one or more chlorine atoms on adjacent carbon atoms	$CHCl_3$, CH_2Cl_2, CH_3CHCl_2, CH_2ClCH_2Cl, $CH_2ClCHClCH_2Cl$, and $CH_2ClCHCl_2$
V	All other molecules having neither active hydrogen atoms nor donor atoms	Hydrocarbons, carbon disulfide, sulfides, mercaptans, and halohydrocarbons not in class IV

Nonideal-solution effects can be incorporated into K-value formulations in two different ways. We have already described the use of $\bar{\phi}_i$, the partial fugacity coefficient, in conjunction with an equation of state and adequate mixing rules. This is the method most frequently used for handling nonidealities in the vapor phase. However, $\bar{\phi}_{iV}$ reflects the combined effects of a nonideal gas and a nonideal-gas solution. At low pressures, both effects are negligible. At moderate pressures, a vapor solution may still be ideal even though the gas mixture does not follow the ideal-gas law. Nonidealities in the liquid phase, however, can be severe even at low pressures. Earlier in this section, $\bar{\phi}_{iL}$ was used to express liquid-phase nonidealities for nonpolar species.

When polar species are present, mixing rules can be modified to include binary interaction parameters, k_{ij}, as in (2-51).

The other technique for handling solution nonidealities is to retain $\bar{\phi}_{iV}$ in the K-value formulation, but replace $\bar{\phi}_{iL}$ by the product of γ_{iL} and ϕ_{iL}, where the former quantity accounts for deviations from nonideal solutions. Equation (2-26) then becomes

$$K_i = \frac{\gamma_{iL}\phi_{iL}}{\bar{\phi}_{iV}} \qquad (2\text{-}68)$$

which was derived previously as (2-27). At low pressures, from Table 2.2, $\phi_{iL} = P_i^s/P$ and $\bar{\phi}_{iV} = 1.0$, so (2-68)

Table 2.8   Molecule Interactions Causing Deviations from Raoult's Law

Type of Deviation	Classes	Effect on Hydrogen Bonding
Always negative	III + IV	H-bonds formed only
Quasi-ideal; always positive or ideal	III + III III + V IV + IV IV + V V + V	No H-bonds involved
Usually positive, but some negative	I + I I + II I + III II + II II + III	H-bonds broken and formed
Always positive	I + IV (frequently limited solubility) II + IV	H-bonds broken and formed, but dissociation of Class I or II is more important effect
Always positive	I + V II + V	H-bonds broken only

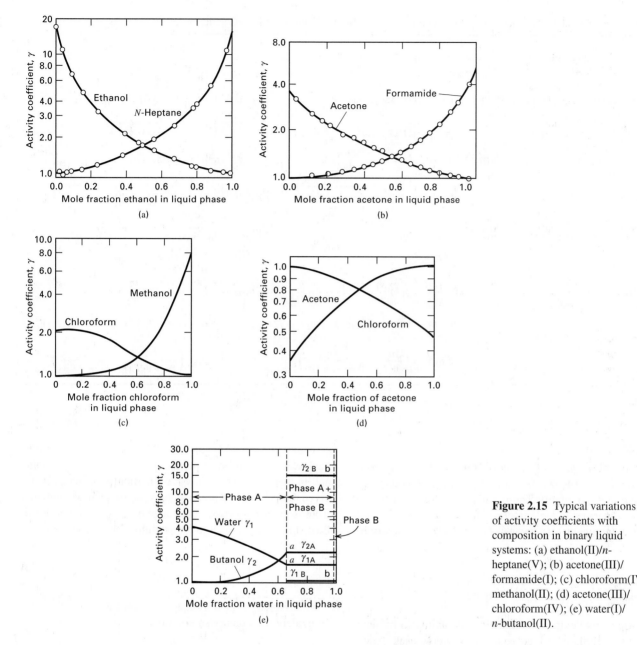

Figure 2.15 Typical variations of activity coefficients with composition in binary liquid systems: (a) ethanol(II)/n-heptane(V); (b) acetone(III)/formamide(I); (c) chloroform(IV)/methanol(II); (d) acetone(III)/chloroform(IV); (e) water(I)/n-butanol(II).

reduces to a modified Raoult's law K-value, which differs from (2-44) only in the γ_{iL} term:

$$K_i = \frac{\gamma_{iL} P_i^s}{P} \qquad (2\text{-}69)$$

At moderate pressures, (5) of Table 2.3 is preferred over (2-69).

Regular-solution theory is useful only for estimating values of γ_{iL} for mixtures of nonpolar species. However, many empirical and semitheoretical equations exist for estimating activity coefficients of binary mixtures containing polar and/or nonpolar species. These equations contain binary interaction parameters, which are back-calculated from experimental data. Some of the more useful equations are listed in Table 2.9 in binary-pair form. For a given activity-

coefficient correlation, the equations of Table 2.10 can be used to determine excess volume, excess enthalpy, and excess entropy. However, unless the dependency on pressure of the parameters and properties used in the equations for activity coefficient is known, excess liquid volumes cannot be determined directly from (1) of Table 2.10. Fortunately, the contribution of excess volume to total mixture volume is generally small for solutions of nonelectrolytes. For example, consider a 50 mol% solution of ethanol in n-heptane at 25°C. From Figure 2.15a, this is a highly nonideal, but miscible, liquid mixture. From the data of Van Ness, Soczek, and Kochar [34], excess volume is only 0.465 cm^3/mol, compared to an estimated ideal-solution molar volume of 106.3 cm^3/mol. Once the partial molar excess functions are estimated for each species, the excess functions are computed from the mole fraction sums.

Table 2.9 Empirical and Semitheoretical Equations for Correlating Liquid-Phase Activity Coefficients of Binary Pairs

Name	Equation for Species 1	Equation for Species 2
(1) Margules	$\log \gamma_1 = A x_2^2$	$\log \gamma_2 = A x_1^2$
(2) Margules (two-constant)	$\log \gamma_1 = x_2^2 [\bar{A}_{12} + 2x_1(\bar{A}_{21} - \bar{A}_{12})]$	$\log \gamma_2 = x_1^2 [\bar{A}_{21} + 2x_2(\bar{A}_{12} - \bar{A}_{21})]$
(3) van Laar (two-constant)	$\ln \gamma_1 = \dfrac{A_{12}}{[1 + (x_1 A_{12})/(x_2 A_{21})]^2}$	$\ln \gamma_2 = \dfrac{A_{21}}{[1 + (x_2 A_{21})/(x_1 A_{12})]^2}$
(4) Wilson (two-constant)	$\ln \gamma_1 = -\ln(x_1 + \Lambda_{12} x_2)$ $+ x_2 \left(\dfrac{\Lambda_{12}}{x_1 + \Lambda_{12} x_2} - \dfrac{\Lambda_{21}}{x_2 + \Lambda_{21} x_1} \right)$	$\ln \gamma_2 = -\ln(x_2 + \Lambda_{21} x_1)$ $- x_1 \left(\dfrac{\Lambda_{12}}{x_1 + \Lambda_{12} x_2} - \dfrac{\Lambda_{21}}{x_2 + \Lambda_{21} x_1} \right)$
(5) NRTL (three-constant)	$\ln \gamma_1 = \dfrac{x_2^2 \tau_{21} G_{21}^2}{(x_1 + x_2 G_{21})^2} + \dfrac{x_1^2 \tau_{12} G_{12}}{(x_2 + x_1 G_{12})^2}$ $G_{ij} = \exp(-\alpha_{ij} \tau_{ij})$	$\ln \gamma_2 = \dfrac{x_1^2 \tau_{12} G_{12}^2}{(x_2 + x_1 G_{12})^2} + \dfrac{x_2^2 \tau_{21} G_{21}}{(x_1 + x_2 G_{21})^2}$ $G_{ij} = \exp(-\alpha_{ij} \tau_{ij})$
(6) UNIQUAC (two-constant)	$\ln \gamma_1 = \ln \dfrac{\Psi_1}{x_1} + \dfrac{\bar{Z}}{2} q_1 \ln \dfrac{\theta_1}{\Psi_1}$ $+ \Psi_2 \left(l_1 - \dfrac{r_1}{r_2} l_2 \right) - q_1 \ln(\theta_1 + \theta_2 T_{21})$ $+ \theta_2 q_1 \left(\dfrac{T_{21}}{\theta_1 + \theta_2 T_{21}} - \dfrac{T_{12}}{\theta_2 + \theta_1 T_{12}} \right)$	$\ln \gamma_2 = \ln \dfrac{\Psi_2}{x_2} + \dfrac{\bar{Z}}{2} q_2 \ln \dfrac{\theta_2}{\Psi_2}$ $+ \Psi_1 \left(l_2 - \dfrac{r_2}{r_1} l_1 \right) - q_2 \ln(\theta_2 + \theta_1 T_{12})$ $+ \theta_1 q_2 \left(\dfrac{T_{12}}{\theta_2 + \theta_1 T_{12}} - \dfrac{T_{21}}{\theta_1 + \theta_2 T_{21}} \right)$

Margules Equations

The Margules equations (1) and (2) in Table 2.9 date back to 1895, and the two-constant form is still in common use because of its simplicity. These equations result from power-series expansions in mole fractions for $\bar{g}_i^E$ and conversion to activity coefficients by means of (2-59). The one-constant form is equivalent to symmetrical activity-coefficient curves, which are rarely observed experimentally.

van Laar Equation

Because of its flexibility, simplicity, and ability to fit many systems well, the van Laar equation is widely used. It was derived from the van der Waals equation of state, but the

Table 2.10 Classical Partial Molar Excess Functions of Thermodynamics

Excess volume:

$(1) \; (\bar{v}_{iL} - \bar{v}_{iL}^{ID}) \equiv \bar{v}_{iL}^{E} = RT \left(\dfrac{\partial \ln \gamma_{iL}}{\partial P} \right)_{T,x}$

Excess enthalpy:

$(2) \; (\bar{h}_{iL} - \bar{h}_{iL}^{ID}) \equiv \bar{h}_{iL}^{E} = -RT^2 \left(\dfrac{\partial \ln \gamma_{iL}}{\partial T} \right)_{P,x}$

Excess entropy:

$(3) \; (\bar{s}_{iL} - \bar{s}_{iL}^{ID}) \equiv \bar{s}_{iL}^{E} = -R \left[T \left(\dfrac{\partial \ln \gamma_{iL}}{\partial T} \right)_{P,x} + \ln \gamma_{iL} \right]$

ID = ideal mixture; E = excess because of nonideality.

constants, shown as A_{12} and A_{21} in (3) of Table 2.9, are best back-calculated from experimental data. These constants are, in theory, constant only for a particular binary pair at a given temperature. In practice, they are frequently computed from isobaric data covering a range of temperature. The van Laar theory expresses the temperature dependence of A_{ij} as

$$A_{ij} = \frac{A'_{ij}}{RT} \qquad (2\text{-}70)$$

Regular-solution theory and the van Laar equation are equivalent for a binary solution if

$$A_{ij} = \frac{v_{iL}}{RT} (\delta_i - \delta_j)^2 \qquad (2\text{-}71)$$

The van Laar equation can fit activity coefficient–composition curves corresponding to both positive and negative deviations from Raoult's law, but cannot fit curves that exhibit minima or maxima such as those in Figure 2.15c.

When data are isothermal, or isobaric over only a narrow range of temperature, determination of van Laar constants is conducted in a straightforward manner. The most accurate procedure is a nonlinear regression to obtain the best fit to the data over the entire range of binary composition, subject to minimization of some objective function. A less accurate, but extremely rapid, manual-calculation procedure can be used when experimental data can be extrapolated to infinite-dilution conditions. Modern experimental techniques are available for accurately and rapidly determining activity coefficients at infinite dilution. Applying (3) of Table 2.9 to the conditions $x_i = 0$ and then $x_j = 0$, we have

$$A_{ij} = \ln \gamma_i^\infty, \quad x_i = 0$$

and

$$A_{ji} = \ln \gamma_j^\infty, \quad x_j = 0 \qquad (2\text{-}72)$$

For practical applications, it is important that the van Laar equation predicts azeotrope formation correctly, where $x_i = y_i$ and $K_i = 1.0$. If activity coefficients are known or can be computed at the azeotropic composition—say, from (2-69), ($\gamma_{iL} = P/P_i^s$, since $K_i = 1.0$)—these coefficients can be used to determine the van Laar constants directly from the following equations obtained by solving simultaneously for A_{12} and A_{21}:

$$A_{12} = \ln \gamma_1 \left(1 + \frac{x_2 \ln \gamma_2}{x_1 \ln \gamma_1} \right)^2 \qquad (2\text{-}73)$$

$$A_{21} = \ln \gamma_2 \left(1 + \frac{x_1 \ln \gamma_1}{x_2 \ln \gamma_2} \right)^2 \qquad (2\text{-}74)$$

These equations are applicable to activity-coefficient data obtained at any single composition.

Mixtures of self-associated polar molecules (class II in Table 2.7) with nonpolar molecules such as hydrocarbons (class V) can exhibit the strong nonideality of the positive-deviation type shown in Figure 2.15a. Figure 2.16 shows experimental data of Sinor and Weber [35] for ethanol

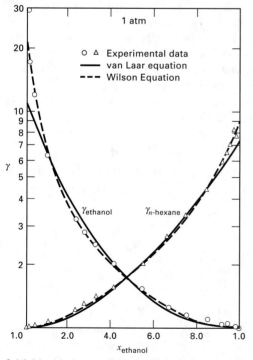

Figure 2.16 Liquid-phase activity coefficients for ethanol/ n-hexane system.

[Data from J.E. Sinor and J.H. Weber, *J. Chem. Eng. Data*, **5**, 243–247 (1960).]

(1)/n-hexane (2), a system of this type, at 101.3 kPa. These data were correlated with the van Laar equation by Orye and Prausnitz [36] to give $A_{12} = 2.409$ and $A_{21} = 1.970$. From $x_1 = 0.1$ to 0.9, the fit of the data to the van Laar equation is reasonably good; in the dilute regions, however, deviations are quite severe and the predicted activity coefficients for ethanol are low. An even more serious problem with these highly nonideal mixtures is that the van Laar equation may erroneously predict formation of two liquid phases (phase splitting) when values of activity coefficients exceed approximately 7.

Local-Composition Concept and the Wilson Model

Since its introduction in 1964, the Wilson equation [37], shown in binary form in Table 2.9 as (4), has received wide attention because of its ability to fit strongly nonideal, but miscible, systems. As shown in Figure 2.16, the Wilson equation, with binary interaction parameters of $\Lambda_{12} = 0.0952$ and $\Lambda_{21} = 0.2713$ determined by Orye and Prausnitz [36], fits experimental data well even in dilute regions where the variation of γ_1 becomes exponential. Corresponding infinite-dilution activity coefficients computed from the Wilson equation are $\gamma_1^\infty = 21.72$ and $\gamma_2^\infty = 9.104$.

The Wilson equation accounts for effects of differences both in molecular size and intermolecular forces, consistent with a semitheoretical interpretation based on the Flory–Huggins relation (2-65). Overall solution-volume fractions ($\Phi_i = x_i v_{iL}/v_L$) are replaced by local-volume fractions, $\bar{\Phi}_i$, which are related to local-molecule segregations caused by differing energies of interaction between pairs of molecules. The concept of local compositions that differ from overall compositions is shown schematically for an overall, equimolar, binary solution in Figure 2.17, which is taken from Cukor and Prausnitz [38]. About a central molecule of type 1, the local mole fraction of molecules of type 2 is shown as $\frac{5}{8}$, while the overall composition is $\frac{1}{2}$.

For local-volume fraction, Wilson proposed

$$\bar{\Phi}_i = \frac{v_{iL} x_i \exp(-\lambda_{ii}/RT)}{\displaystyle\sum_{j=1}^{C} v_{jL} x_j \exp(-\lambda_{ij}/RT)} \qquad (2\text{-}75)$$

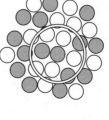

○ 15 of type 1
○ 15 of type 2

Overall mole fractions: $x_1 = x_2 = \frac{1}{2}$
Local mole fractions:

$$x_{21} = \frac{\text{Molecules of 2 about a central molecule 1}}{\text{Total molecules about a central molecule 1}}$$

$x_{21} + x_{11} = 1$, as shown
$x_{12} + x_{22} = 1$
$x_{11} \sim 3/8$
$x_{21} \sim 5/8$

Figure 2.17 The concept of local compositions.

[From P.M. Cukor and J.M. Prausnitz, *Int. Chem. Eng. Symp. Ser. No. 32*, **3**, 88 (1969).]

where energies of interaction $\lambda_{ij} = \lambda_{ji}$, but $\lambda_{ii} \neq \lambda_{jj}$. Following the treatment by Orye and Prausnitz [36], substitution of the binary form of (2-75) into (2-65) and defining the binary interaction parameters as

$$\Lambda_{12} = \frac{v_{2L}}{v_{1L}} \exp\left[-\frac{(\lambda_{12} - \lambda_{11})}{RT}\right] \qquad (2\text{-}76)$$

$$\Lambda_{21} = \frac{v_{1L}}{v_{2L}} \exp\left[-\frac{(\lambda_{12} - \lambda_{22})}{RT}\right] \qquad (2\text{-}77)$$

leads to the following equation for a binary system:

$$\frac{g^E}{RT} = -x_1 \ln(x_1 + \Lambda_{12}x_2) - x_2 \ln(x_2 + \Lambda_{21}x_1) \qquad (2\text{-}78)$$

The Wilson equation is very effective for dilute compositions where entropy effects dominate over enthalpy effects. The Orye–Prausnitz form of the Wilson equation for the activity coefficient, as given in Table 2.9, follows from combining (2-59) with (2-78). Values of $\Lambda_{ij} < 1$ correspond to positive deviations from Raoult's law, while values > 1 correspond to negative deviations. Ideal solutions result from $\Lambda_{ij} = 1$. Studies indicate that λ_{ii} and λ_{ij} are temperature-dependent. Values of v_{iL}/v_{jL} depend on temperature also, but the variation may be small compared to temperature effects on the exponential terms in (2-76) and (2-77).

The Wilson equation is readily extended to multicomponent mixtures by neglecting ternary and higher molecular interactions and assuming a pseudo-binary mixture. The following multicomponent Wilson equation involves only binary interaction constants:

$$\ln \gamma_k = 1 - \ln\left(\sum_{j=1}^{C} x_j \Lambda_{kj}\right) - \sum_{i=1}^{C}\left(\frac{x_i \Lambda_{ik}}{\sum_{j=1}^{C} x_j \Lambda_{ij}}\right) \qquad (2\text{-}79)$$

where $\Lambda_{ii} = \Lambda_{jj} = \Lambda_{kk} = 1$.

As mixtures become highly nonideal, but still miscible, the Wilson equation becomes markedly superior to the Margules and van Laar equations. The Wilson equation is consistently superior for multicomponent solutions. Values of the constants in the Wilson equation for many binary systems are tabulated in the DECHEMA collection of Gmehling and Onken [39]. Two limitations of the Wilson equation are its inability to predict immiscibility, as in Figure 2.15e, and maxima and minima in the activity coefficient–mole fraction relationships, as shown in Figure 2.15c.

When insufficient experimental data are available to determine binary Wilson parameters from a best fit of activity coefficients over the entire range of composition, infinite-dilution or single-point values can be used. At infinite dilution, the Wilson equation in Table 2.9 becomes

$$\ln \gamma_1^\infty = 1 - \ln \Lambda_{12} - \Lambda_{21} \qquad (2\text{-}80)$$

$$\ln \gamma_2^\infty = 1 - \ln \Lambda_{21} - \Lambda_{12} \qquad (2\text{-}81)$$

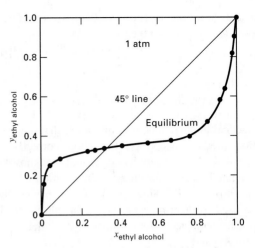

Figure 2.18 Equilibrium curve for n-hexane/ethanol system.

An iterative procedure is required to obtain Λ_{12} and Λ_{21} from these nonlinear equations. If temperatures corresponding to γ_1^∞ and γ_2^∞ are not close or equal, (2-76) and (2-77) should be substituted into (2-80) and (2-81) with values of $(\lambda_{12} - \lambda_{11})$ and $(\lambda_{12} - \lambda_{22})$ determined from estimates of pure-component liquid molar volumes.

When the experimental data of Sinor and Weber [35] for n-hexane/ethanol, shown in Figure 2.16, are plotted as a y–x diagram in ethanol (Figure 2.18), the equilibrium curve crosses the 45° line at an ethanol mole fraction of $x = 0.332$. The measured temperature corresponding to this composition is 58°C. Ethanol has a normal boiling point of 78.33°C, which is higher than the normal boiling point of 68.75°C for n-hexane. Nevertheless, ethanol is more volatile than n-hexane up to an ethanol mole fraction of $x = 0.322$, the minimum-boiling azeotrope. This occurs because of the relatively close boiling points of the two species and the high activity coefficients for ethanol at low concentrations. At the azeotropic composition, $y_i = x_i$; therefore, $K_i = 1.0$. Applying (2-69) to both species,

$$\gamma_1 P_1^s = \gamma_2 P_2^s \qquad (2\text{-}82)$$

If species 2 is more volatile in the pure state ($P_2^s > P_1^s$), the criteria for formation of a minimum-boiling azeotrope are

$$\gamma_1 \geq 1 \qquad (2\text{-}83)$$

$$\gamma_2 \geq 1 \qquad (2\text{-}84)$$

and

$$\frac{\gamma_1}{\gamma_2} < \frac{P_2^s}{P_1^s} \qquad (2\text{-}85)$$

for x_1 less than the azeotropic composition. These critieria are most readily applied at $x_1 = 0$. For example, for the n-hexane (2)/ethanol (1) system at 1 atm (101.3 kPa), when the liquid-phase mole fraction of ethanol approaches zero, temperature approaches 68.75°C (155.75°F), the boiling point of pure n-hexane. At this temperature, $P_1^s = 10$ psia

(68.9 kPa) and $P_2^s = 14.7$ psia (101.3 kPa). Also from Figure 2.16, $\gamma_1^{\infty} = 21.72$ when $\gamma_2 = 1.0$. Thus, $\gamma_1^{\infty}/\gamma_2 = 21.72$, but $P_2^s/P_1^s = 1.47$. Therefore, a minimum-boiling azeotrope will occur.

Maximum-boiling azeotropes are less common. They occur for relatively close-boiling mixtures when negative deviations from Raoult's law arise such that $\gamma_i < 1.0$. Criteria for their formation are derived in a manner similar to that for minimum-boiling azeotropes. At $x_1 = 1$, where species 2 is more volatile,

$$\gamma_1 = 1.0 \tag{2-86}$$

$$\gamma_2^{\infty} < 1.0 \tag{2-87}$$

and

$$\frac{\gamma_2^{\infty}}{\gamma_1} < \frac{P_1^s}{P_2^s} \tag{2-88}$$

For an azeotropic binary system, the two binary interaction parameters Λ_{12} and Λ_{21} can be determined by solving (4) of Table 2.9 at the azeotropic composition, as shown in the following example.

EXAMPLE 2.8

From measurements by Sinor and Weber [35] of the azeotropic condition for the ethanol/n-hexane system at 1 atm (101.3 kPa, 14.696 psia), calculate Λ_{12} and Λ_{21}.

SOLUTION

Let E denote ethanol and H denote n-hexane. The azeotrope occurs at $x_E = 0.332$, $x_H = 0.668$, and $T = 58°C$ (331.15 K). At 1 atm, (2-69) can be used to approximate K-values. Thus, at azeotropic conditions, $\gamma_i = P/P_i^s$. The vapor pressures at 58°C are $P_E^s = 6.26$ psia and $P_H^s = 10.28$ psia. Therefore,

$$\gamma_E = \frac{14.696}{6.26} = 2.348$$

$$\gamma_H = \frac{14.696}{10.28} = 1.430$$

Substituting these values together with the above corresponding values of x_i into the binary form of the Wilson equation in Table 2.9 gives

$$\ln 2.348 = -\ln(0.332 + 0.668\Lambda_{EH})$$
$$+ 0.668 \left(\frac{\Lambda_{EH}}{0.332 + 0.668\Lambda_{EH}} - \frac{\Lambda_{HE}}{0.332\Lambda_{HE} + 0.668} \right)$$

$$\ln 1.430 = -\ln(0.668 + 0.332\Lambda_{HE})$$
$$- 0.332 \left(\frac{\Lambda_{EH}}{0.332 + 0.668\Lambda_{EH}} - \frac{\Lambda_{HE}}{0.332\Lambda_{HE} + 0.668} \right)$$

Solving these two nonlinear equations simultaneously by an iterative procedure, we obtain $\Lambda_{EH} = 0.041$ and $\Lambda_{HE} = 0.281$. From these constants, the activity-coefficient curves can be predicted if the temperature variations of Λ_{EH} and Λ_{HE} are ignored. The results are plotted in Figure 2.19. The fit of experimental data is good except, perhaps, for near-infinite-dilution conditions, where $\gamma_E^{\infty} = 49.82$ and $\gamma_H^{\infty} = 9.28$. The former value is considerably

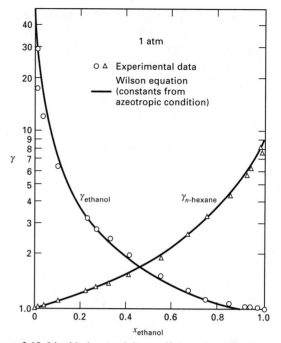

Figure 2.19 Liquid-phase activity coefficients for ethanol/n-hexane system.

greater than the value of 21.72 obtained by Orye and Prausnitz [36] from a fit of all experimental data points. However, if Figures 2.16 and 2.19 are compared, it is seen that widely differing γ_E^{∞} values have little effect on γ in the composition region $x_E = 0.15$ to 1.00, where the two sets of Wilson curves are almost identical. For accuracy over the entire composition range, commensurate with the ability of the Wilson equation, data for at least three well-spaced liquid compositions per binary are preferred.

The Wilson equation can be extended to liquid–liquid or vapor–liquid–liquid systems by multiplying the right-hand side of (2-78) by a third binary-pair constant evaluated from experimental data [37]. However, for multicomponent systems of three or more species, the third binary-pair constants must be the same for all constituent binary pairs. Furthermore, as shown by Hiranuma [40], representation of ternary systems involving only one partially miscible binary pair can be extremely sensitive to the third binary-pair Wilson constant. For these reasons, application of the Wilson equation to liquid–liquid systems has not been widespread. Rather, the success of the Wilson equation for prediction of activity coefficients for miscible liquid systems greatly stimulated further development of the local-composition concept of Wilson in an effort to obtain more universal expressions for liquid-phase activity coefficients.

NRTL Model

The nonrandom, two-liquid (NRTL) equation developed by Renon and Prausnitz [41,42] as listed in Table 2.9, represents an accepted extension of Wilson's concept. The NRTL equation is applicable to multicomponent vapor–liquid,

liquid–liquid, and vapor–liquid–liquid systems. For multi-component vapor–liquid systems, only binary-pair constants from the corresponding binary-pair experimental data are required. For a multicomponent system, the NRTL expression for the activity coefficient is

$$
\ln \gamma_i = \frac{\displaystyle\sum_{j=1}^{C} \tau_{ji} G_{ji} x_j}{\displaystyle\sum_{k=1}^{C} G_{ki} x_k} + \sum_{j=1}^{C} \left[\frac{x_j G_{ij}}{\displaystyle\sum_{k=1}^{C} G_{kj} x_k} \left(\tau_{ij} - \frac{\displaystyle\sum_{k=1}^{C} x_k \tau_{kj} G_{kj}}{\displaystyle\sum_{k=1}^{C} G_{kj} x_k} \right) \right]
$$

$$(2\text{-}89)$$

where

$$ G_{ji} = \exp(-\alpha_{ji}\tau_{ji}) \qquad (2\text{-}90) $$

The coefficients τ are given by

$$ \tau_{ij} = \frac{g_{ij} - g_{jj}}{RT} \qquad (2\text{-}91) $$

$$ \tau_{ji} = \frac{g_{ji} - g_{ii}}{RT} \qquad (2\text{-}92) $$

where g_{ij}, g_{jj}, and so on are energies of interaction between molecule pairs. In the above equations, $G_{ji} \neq G_{ij}, \tau_{ij} \neq \tau_{ji}$, $G_{ii} = G_{jj} = 1$, and $\tau_{ii} = \tau_{jj} = 0$. Often $(g_{ij} - g_{jj})$ and other constants are linear in temperature. For ideal solutions, $\tau_{ji} = 0$.

The parameter α_{ji} characterizes the tendency of species j and species i to be distributed in a nonrandom fashion. When $\alpha_{ji} = 0$, local mole fractions are equal to overall solution mole fractions. Generally α_{ji} is independent of temperature and depends on molecule properties in a manner similar to the classifications in Tables 2.7 and 2.8. Values of α_{ji} usually lie between 0.2 and 0.47. When $\alpha_{ji} < 0.426$, phase immiscibility is predicted. Although α_{ji} can be treated as an adjustable parameter, to be determined from experimental binary-pair data, more commonly α_{ji} is set according to the following rules, which are occasionally ambiguous:

1. $\alpha_{ji} = 0.20$ for mixtures of saturated hydrocarbons and polar, nonassociated species (e.g., *n*-heptane/acetone).

2. $\alpha_{ji} = 0.30$ for mixtures of nonpolar compounds (e.g., benzene/*n*-heptane), except fluorocarbons and paraffins; mixtures of nonpolar and polar, nonassociated species (e.g., benzene/acetone); mixtures of polar species that exhibit negative deviations from Raoult's law (e.g., acetone/chloroform) and moderate positive deviations (e.g., ethanol/water); mixtures of water and polar nonassociated species (e.g., water/acetone).

3. $\alpha_{ji} = 0.40$ for mixtures of saturated hydrocarbons and homolog perfluorocarbons (e.g., *n*-hexane/perfluoro-*n*-hexane).

4. $\alpha_{ji} = 0.47$ for mixtures of an alcohol or other strongly self-associated species with nonpolar species (e.g., ethanol/benzene); mixtures of carbon tetrachloride with either acetonitrile or nitromethane; mixtures of water with either butyl glycol or pyridine.

UNIQUAC Model

In an attempt to place calculations of liquid-phase activity coefficients on a simple, yet more theoretical basis, Abrams and Prausnitz [43] used statistical mechanics to derive an expression for excess free energy. Their model, called UNIQUAC (universal quasichemical), generalizes a previous analysis by Guggenheim and extends it to mixtures of molecules that differ appreciably in size and shape. As in the Wilson and NRTL equations, local concentrations are used. However, rather than local volume fractions or local mole fractions, UNIQUAC uses the local area fraction θ_{ij} as the primary concentration variable.

The local area fraction is determined by representing a molecule by a set of bonded segments. Each molecule is characterized by two structural parameters that are determined relative to a standard segment taken as an equivalent sphere of a unit of a linear, infinite-length, polymethylene molecule. The two structural parameters are the relative number of segments per molecule, r (volume parameter), and the relative surface area of the molecule, q (surface parameter). Values of these parameters computed from bond angles and bond distances are given by Abrams and Prausnitz [43] and Gmehling and Onken [39] for a number of species. For other compounds, values can be estimated by the group-contribution method of Fredenslund et al. [46].

For a multicomponent liquid mixture, the UNIQUAC model gives the excess free energy as

$$
\frac{g^E}{RT} = \sum_{i=1}^{C} x_i \ln\left(\frac{\Psi_i}{x_i}\right) + \frac{\bar{Z}}{2}\sum_{i=1}^{C} q_i x_i \ln\left(\frac{\theta_i}{\Psi_i}\right) - \sum_{i=1}^{C} q_i x_i \ln\left(\sum_{j=1}^{C} \theta_i T_{ji}\right)
$$

$$(2\text{-}93)$$

The first two terms on the right-hand side account for *combinatorial* effects due to differences in molecule size and shape; the last term provides a *residual* contribution due to differences in intermolecular forces, where

$$ \Psi_i = \frac{x_i r_i}{\displaystyle\sum_{i=1}^{C} x_i r_i} = \text{segment fraction} \qquad (2\text{-}94) $$

$$ \theta = \frac{x_i q_i}{\displaystyle\sum_{i=1}^{C} x_i q_i} = \text{area fraction} \qquad (2\text{-}95) $$

where $\bar{Z} =$ lattice coordination number set equal to 10, and

$$ T_{ji} = \exp\left(\frac{u_{ji} - u_{ii}}{RT}\right) \qquad (2\text{-}96) $$

Equation (2-93) contains only two adjustable parameters for each binary pair, $(u_{ji} - u_{ii})$ and $(u_{ij} - u_{jj})$. Abrams and Prausnitz show that $u_{ji} = u_{ij}$ and $T_{ii} = T_{jj} = 1$. In general, $(u_{ji} - u_{ii})$ and $(u_{ij} - u_{jj})$ are linear functions of temperature.

If (2-59) is combined with (2-93), an equation for the liquid-phase activity coefficient for a species in a multicomponent mixture is obtained:

$$\ln \gamma_i = \ln \gamma_i^C + \ln \gamma_i^R$$

$$= \underbrace{\ln(\Psi_i/x_i) + (\bar{Z}/2) \, q_i \ln(\theta_i/\Psi_i) + l_i - (\Psi_i/x_i) \sum_{j=1}^{C} x_j l_j}_{C, \text{ combinatorial}}$$

$$\underbrace{+ q_i \left[1 - \ln \left(\sum_{j=1}^{C} \theta_j T_{ji} \right) - \sum_{j=1}^{C} \left(\frac{\theta_j T_{ij}}{\sum\limits_{k=1}^{C} \theta_k T_{kj}} \right) \right]}_{R, \text{ residual}}$$

$$(2\text{-}97)$$

where

$$l_j = \left(\frac{\bar{Z}}{2} \right)(r_j - q_j) - (r_j - 1) \qquad (2\text{-}98)$$

For a binary mixture of species 1 and 2, (2-97) reduces to (6) in Table 2.9 for $\bar{Z} = 10$.

UNIFAC Model

Liquid-phase activity coefficients must be estimated for nonideal mixtures even when experimental phase equilibria data are not available and when the assumption of regular solutions is not valid because polar compounds are present. For such predictions, Wilson and Deal [47] and then Derr and Deal [48], in the 1960s, presented methods based on treating a solution as a mixture of functional groups instead of molecules. For example, in a solution of toluene and acetone, the contributions might be 5 aromatic CH groups, 1 aromatic C group, and 1 CH_3 group from toluene; and 2 CH_3 groups plus 1 CO carbonyl group from acetone. Alternatively, larger groups might be employed to give 5 aromatic CH groups and 1 CCH_3 group from toluene; and 1 CH_3 group and 1 CH_3CO group from acetone. As larger and larger functional groups are used, the accuracy of molecular representation increases, but the advantage of the group-contribution method decreases because a larger number of groups is required. In practice, about 50 functional groups are used to represent literally thousands of multicomponent liquid mixtures.

To estimate the partial molar excess free energies, $\bar{g}_i^E$, and then the activity coefficients, size parameters for each functional group and binary interaction parameters for each pair of functional groups are required. Size parameters can be calculated from theory. Interaction parameters are back-calculated from existing phase-equilibria data and then used with the size parameters to predict phase-equilibria properties of mixtures for which no data are available.

The UNIFAC (UNIQUAC Functional-group Activity Coefficients) group-contribution method, first presented by Fredenslund, Jones, and Prausnitz [49] and further developed for use in practice by Fredenslund, Gmehling, and Rasmussen [50], Gmehling, Rasmussen, and Fredenslund [51], and Larsen, Rasmussen, and Fredenslund [52], has several advantages over other group-contribution methods: (1) It is theoretically based on the UNIQUAC method; (2) the parameters are essentially independent of temperature; (3) size and binary interaction parameters are available for a wide range of types of functional groups; (4) predictions can be made over a temperature range of 275–425 K and for pressures up to a few atmospheres; and (5) extensive comparisons with experimental data are available. All components in the mixture must be condensable.

The UNIFAC method for predicting liquid-phase activity coefficients is based on the UNIQUAC equation (2-97), wherein the molecular volume and area parameters in the combinatorial terms are replaced by

$$r_i = \sum_k v_k^{(i)} R_k \qquad (2\text{-}99)$$

$$q_i = \sum_k v_k^{(i)} Q_k \qquad (2\text{-}100)$$

where $v_k^{(i)}$ is the number of functional groups of type k in molecule i, and R_k and Q_k are the volume and area parameters, respectively, for the type-k functional group.

The residual term in (2-97), which is represented by $\ln \gamma_i^R$, is replaced by the expression

$$\ln \gamma_i^R = \underbrace{\sum_k v_k^{(i)} \left(\ln \Gamma_k - \ln \Gamma_k^{(i)} \right)}_{\text{all functional groups in mixture}} \qquad (2\text{-}101)$$

where Γ_k is the residual activity coefficient of the functional group k in the actual mixture, and $\Gamma_k^{(i)}$ is the same quantity but in a reference mixture that contains only molecules of type i. The latter quantity is required so that $\gamma_i^R \rightarrow 1.0$ as $x_i \rightarrow 1.0$. Both Γ_k and $\Gamma_k^{(i)}$ have the same form as the residual term in (2-97). Thus,

$$\ln \Gamma_k = Q_k \left[1 - \ln \left(\sum_m \theta_m T_{mk} \right) - \sum_m \frac{\theta_m T_{mk}}{\sum_n \theta_n T_{nm}} \right]$$

$$(2\text{-}102)$$

where θ_m is the area fraction of group m, given by an equation similar to (2-95),

$$\theta_m = \frac{X_m Q_m}{\sum_n X_n Q_m} \qquad (2\text{-}103)$$

where X_m is the mole fraction of group m in the solution,

$$X_m = \frac{\sum_j v_m^{(j)} x_j}{\sum_j \sum_n (v_n^{(j)} x_j)} \qquad (2\text{-}104)$$

and T_{mk} is a group interaction parameter given by an equation similar to (2-96),

$$T_{mk} = \exp \left(-\frac{a_{mk}}{T} \right) \qquad (2\text{-}105)$$

where $a_{mk} \neq a_{km}$. When $m = k$, then $a_{mk} = 0$ and $T_{mk} = 1.0$. For $\Gamma_k^{(i)}$, (2-102) also applies, where θ terms correspond to the pure component i. Although values of R_k and Q_k are different for each functional group, values of a_{mk} are equal for all subgroups within a main group. For example, main group CH_2 consists of subgroups CH_3, CH_2, CH, and C. Accordingly,

$$a_{CH_3,CHO} = a_{CH_2,CHO} = a_{CH,CHO} = a_{C,CHO}$$

Thus, the amount of experimental data required to obtain values of a_{mk} and a_{km} and the size of the corresponding bank of data for these parameters is not as large as might be expected.

The ability of a group-contribution method to predict liquid-phase activity coefficients has been further improved by introduction of a modified UNIFAC method by Gmehling [51], referred to as UNIFAC (Dortmund). To correlate data for mixtures having a wide range of molecular size, they modified the combinatorial part of (2-97). To handle temperature dependence more accurately, they replaced (2-105) with a three-coefficient equation. The resulting modification permits reasonably reliable predictions of liquid-phase activity coefficients (including applications to dilute solutions and multiple liquid phases), heats of mixing, and azeotropic compositions. Values of the UNIFAC (Dortmund) parameters for 51 groups are available in a series of publications starting in 1993 with Gmehling, Li, and Schiller [53] and more recently with Wittig, Lohmann, and Gmehling [54].

Liquid–Liquid Equilibria

When species are notably dissimilar and activity coefficients are large, two and even more liquid phases may coexist at equilibrium. For example, consider the binary system of methanol (1) and cyclohexane (2) at 25°C. From measurements of Takeuchi, Nitta, and Katayama [55], van Laar constants are $A_{12} = 2.61$ and $A_{21} = 2.34$, corresponding, respectively, to infinite-dilution activity coefficients of 13.6 and 10.4 obtained using (2-72). These values of A_{12} and A_{21} can be used to construct an equilibrium plot of y_1 against x_1 assuming an isothermal condition. By combining (2-69), where $K_i = y_i/x_i$, with

$$P = \sum_{i=1}^{C} x_i \gamma_{iL} P_i^s \qquad (2\text{-}106)$$

one obtains the following relation for computing y_i from x_i:

$$y_1 = \frac{x_1 \gamma_1 P_1^s}{x_1 \gamma_1 P_1^s + x_2 \gamma_2 P_2^s} \qquad (2\text{-}107)$$

Vapor pressures at 25°C are $P_1^s = 2.452$ psia (16.9 kPa) and $P_2^s = 1.886$ psia (13.0 kPa). Activity coefficients can be computed from the van Laar equation in Table 2.9. The resulting equilibrium plot is shown in Figure 2.20, where it is observed that over much of the liquid-phase region, three values of y_1 exist. This indicates phase instability. Experimentally, single liquid phases can exist only for cyclohexane-rich mixtures of

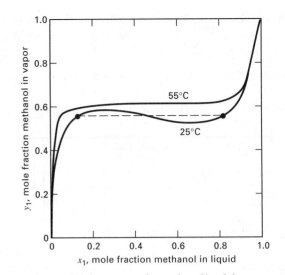

Figure 2.20 Equilibrium curves for methanol/cyclohexane systems.

[Data from K. Strubl, V. Svoboda, R. Holub, and J. Pick, *Collect. Czech. Chem. Commun.*, **35**, 3004–3019 (1970).]

$x_1 = 0.8248$ to 1.0 and for methanol-rich mixtures of $x_1 = 0.0$ to 0.1291. Because a coexisting vapor phase exhibits only a single composition, two coexisting liquid phases prevail at opposite ends of the dashed line in Figure 2.20. The liquid phases represent solubility limits of methanol in cyclohexane and cyclohexane in methanol.

For two coexisting equilibrium liquid phases, the relation $\gamma_{iL}^{(1)} x_i^{(1)} = \gamma_{iL}^{(2)} x_i^{(2)}$ must hold. This permits determination of the two-phase region in Figure 2.20 from the van Laar or other suitable activity-coefficient equation for which the constants are known. Also shown in Figure 2.20 is an equilibrium curve for the same binary system at 55°C based on data of Strubl et al. [56]. At this higher temperature, methanol and cyclohexane are completely miscible. The data of Kiser, Johnson, and Shetlar [57] show that phase instability ceases to exist at 45.75°C, the critical solution temperature. Rigorous thermodynamic methods for determining phase instability and, thus, existence of two equilibrium liquid phases are generally based on free-energy calculations, as discussed by Prausnitz et al. [4]. Most of the empirical and semitheoretical equations for the liquid-phase activity coefficient listed in Table 2.9 apply to liquid–liquid systems. The Wilson equation is a notable exception.

2.7 DIFFICULT MIXTURES

The equation-of-state and activity-coefficient models presented in Sections 2.5 and 2.6, respectively, are inadequate for estimating K-values of mixtures containing: (1) both polar and supercritical (light-gas) components, (2) electrolytes, and (3) both polymers and solvents. For these difficult mixtures, special models have been developed, some of which are briefly described in the following subsections. More detailed discussions of the following three topics are given by Prausnitz, Lichtenthaler, and de Azevedo [4].

Predictive Soave–Redlich–Kwong (PSRK) Model

Equation-of-state models, such as S–R–K and P–R, describe mixtures of nonpolar and slightly polar compounds. Gibbs free-energy activity-coefficient models are formulated for subcritical nonpolar and polar compounds. When a mixture contains both polar compounds and supercritical (light-gas) components (e.g., a mixture of hydrogen, carbon monoxide, methane, methyl acetate, and ethanol), neither method applies. To estimate vapor–liquid phase equilibria for such mixtures, a number of more theoretically based mixing rules for use with the S–R–K and P–R equations of state have been developed. In a different approach, Holderbaum and Gmehling [58] formulated a group-contribution equation of state referred to as the predictive Soave–Redlich–Kwong (PSRK) model, which combines a modified S–R–K equation of state with the UNIFAC model. To improve the ability of the S–R–K equation to predict vapor pressure of polar compounds, they provide an improved temperature dependence for the pure-component parameter, a, in Table 2.5. To handle mixtures of nonpolar, polar, and supercritical components, they use a mixing rule for a, which includes the UNIFAC model for handling nonideal effects more accurately. Additional and revised pure-component and group interaction parameters for use in the PSRK model are provided by Fischer and Gmehling [59]. In particular, [58] and [59] provide parameters for nine light gases (Ar, CO, CO_2, CH_4, H_2, H_2S, N_2, NH_3, and O_2) in addition to UNIFAC parameters for 50 groups.

Electrolyte Solution Models

Solutions of weak and/or strong electrolytes are common in chemical processes. For example, sour water, found in many petroleum plants, may consist of solvent (water) and five dissolved gases: CO, CO_2, CH_4, H_2S, and NH_3. The apparent composition of the solution is based on these six molecules. However, because of dissociation, which in this case is weak, the true composition of the aqueous solution includes ionic as well as molecular species. For sour water, the ionic species present at chemical equilibrium include H^+, OH^-, HCO_3^-, $CO_3^=$, HS^-, $S^=$, NH_4^+, and NH_2COO^-, with the total numbers of positive and negative ions subject to electroneutrality. For example, while the apparent concentration of NH_4 in the solution might be 2.46 moles per kg of water, when dissociation is taken into account, the molality is only 0.97, with NH_4^+ having a molality of 1.49. All eight ionic species are nonvolatile, while all six molecular species are volatile to some extent. Accurate calculations of vapor–liquid equilibrium for multicomponent electrolyte solutions must consider both chemical and physical equilibrium, both of which involve liquid-phase activity coefficients.

A number of models have been developed for predicting activity coefficients in multicomponent systems of electrolytes. Of particular note are the models of Pitzer [60] and Chen and associates [61, 62, and 63], both of which are included in simulation programs. Both models can handle dilute to concentrated solutions, but only the model of Chen and associates, which is a substantial modification of the NRTL model (see Section 2.6), can handle mixed-solvent systems, such as those containing water and alcohols.

Polymer Solution Models

Polymer processing often involves solutions of solvent, monomer, and an amorphous (noncrystalline) polymer, requiring vapor–liquid and, sometimes, liquid–liquid phase-equilibria calculations, for which estimation of activity coefficients of all components in the mixture is needed. In general, the polymer is nonvolatile, but the solvent and monomer are volatile. When the solution is dilute in the polymer, activity-coefficient methods of Section 2.6, such as the NRTL method, can be used. Of more interest are solutions with appreciable concentrations of polymer, for which the methods of Sections 2.5 and 2.6 are inadequate. Consequently, special-purpose empirical and theoretical models have been developed. One method, which is available in simulation programs, is the modified NRTL model of Chen [64], which combines a modification of the Flory-Huggins equation (12-65) for widely differing molecular size with the NRTL concept of local composition. Chen represents the polymer with segments. Thus, solvent–solvent, solvent–segment, and segment–segment binary interaction parameters are required, which are often available from the literature and may be assumed independent of temperature, polymer chain length, and polymer concentration, making the model quite flexible.

2.8 SELECTING AN APPROPRIATE MODEL

The three previous sections of this chapter have discussed the more widely used models for estimating fugacities, activity coefficients, and K-values for components in mixtures. These models and others are included in computer-aided, process-simulation programs. To solve a particular separations problem, it is necessary to select an appropriate model. This section presents recommendations for making at least a preliminary selection.

The selection procedure includes a few models not covered in this chapter, but for which a literature reference is given. The procedure begins by characterizing the mixture by chemical types present: Light gases (LG), Hydrocarbons (HC), Polar organic compounds (PC), and Aqueous solutions (A), with or without Electrolytes (E).

If the mixture is (A) with no (PC), then if electrolytes are present, select the modified NRTL equation. Otherwise, select a special model, such as one for sour water (containing NH_3, H_2S, CO_2, etc.) or aqueous amine solutions.

If the mixture contains (HC), with or without (LG), covering a wide boiling range, choose the corresponding-states method of Lee–Kesler–Plöcker [8, 65]. If the boiling range of a mixture of (HC) is not wide boiling, the selection depends on the pressure and temperature. For all temperatures and pressures, the Peng–Robinson equation is suitable. For all pressures and noncryogenic temperatures, the Soave–Redlich–Kwong equation is suitable. For all temperatures, but not pressures in the critical region, the Benedict–Webb–Rubin–Starling [5, 66, 67] method is suitable.

If the mixture contains (PC), the selection depends on whether (LG) are present. If they are, the PSRK method is recommended. If not, then a suitable liquid-phase activity-coefficient method is selected as follows. If the binary interaction coefficients are not available, select the UNIFAC method, which should be considered as only a first approximation. If the binary interaction coefficients are available and splitting in two liquid phases will not occur, select the Wilson or NRTL equation. Otherwise, if phase splitting is probable, select the NRTL or UNIQUAC equation.

SUMMARY

1. Separation processes are often energy-intensive. Energy requirements are determined by applying the first law of thermodynamics. Estimates of minimum energy needs can be made by applying the second law of thermodynamics with an entropy balance or an availability balance.

2. Phase equilibrium is expressed in terms of vapor–liquid and liquid–liquid K-values, which are formulated in terms of fugacity and activity coefficients.

3. For separation systems involving an ideal-gas mixture and an ideal-liquid solution, all necessary thermodynamic properties can be estimated from the ideal-gas law, a vapor heat-capacity equation, a vapor-pressure equation, and an equation for the liquid density as a function of temperature.

4. Graphical correlations of pure-component thermodynamic properties are widely available and useful for making rapid, manual calculations at near-ambient pressure for an ideal solution.

5. For nonideal vapor and liquid mixtures containing nonpolar components, certain P–v–T equation-of-state models such as S–R–K, P–R, and L–K–P can be used to estimate density, enthalpy, entropy, fugacity coefficients, and K-values.

6. For nonideal liquid solutions containing nonpolar and/or polar components, certain free-energy models such as Margules, van Laar, Wilson, NRTL, UNIQUAC, and UNIFAC can be used to estimate activity coefficients, volume and enthalpy of mixing, excess entropy of mixing, and K-values.

7. Special models are available for polymer solutions, electrolyte solutions, and mixtures of polar and supercritical components.

REFERENCES

1. MIX, T.W., J.S. DWECK, M. WEINBERG, and R.C. ARMSTRONG, *AIChE Symp. Ser., No. 192,* Vol. 76, 15–23 (1980).

2. FELDER, R.M., and R.W. ROUSSEAU, *Elementary Principles of Chemical Processes,* 3rd ed., John Wiley & Sons, New York (2000).

3. DE NEVERS, N., and J.D. SEADER, *Latin Am. J. Heat and Mass Transfer,* **8,** 77–105 (1984).

4. PRAUSNITZ, J.M., R.N. LICHTENTHALER, and E.G. DE AZEVEDO, *Molecular Thermodynamics of Fluid-Phase Equilibria,* 3rd ed., Prentice-Hall, Upper Saddle River, NJ (1999).

5. STARLING, K.E., *Fluid Thermodynamic Properties for Light Petroleum Systems,* Gulf Publishing, Houston, TX (1973).

6. SOAVE, G., *Chem. Eng. Sci.,* **27,** 1197–1203 (1972).

7. PENG, D.Y., and D.B. ROBINSON, *Ind. Eng. Chem. Fundam.,* **15,** 59–64 (1976).

8. PLÖCKER, U., H. KNAPP, and J.M. PRAUSNITZ, *Ind. Eng. Chem. Process Des. Dev.,* **17,** 324–332 (1978).

9. CHAO, K.C., and J.D. SEADER, *AIChE J.,* **7,** 598–605 (1961).

10. GRAYSON, H.G., and C.W. STREED, Paper 20-P07, Sixth World Petroleum Conference, Frankfurt, June 1963.

11. POLING, B.E., J.M. PRAUSNITZ, and J.P. O'CONNELL, *The Properties of Gases and Liquids,* 5th ed., McGraw-Hill, New York (2001).

12. RACKETT, H.G., *J. Chem. Eng. Data,* **15,** 514–517 (1970).

13. YAWS, C.L., H.-C. YANG, J.R. HOPPER, and W.A. CAWLEY, *Hydrocarbon Processing,* **71** (1), 103–106 (1991).

14. FRANK, J.C., G.R. GEYER, and H. KEHDE, *Chem. Eng. Prog.,* **65** (2), 79–86 (1969).

15. HADDEN, S.T., and H.G. GRAYSON, *Hydrocarbon Process., Petrol. Refiner,* **40** (9), 207–218 (1961).

16. ROBBINS, L.A., Section 15, "Liquid-Liquid Extraction Operations and Equipment," in R.H. Perry, D. Green, and J.O. Maloney, Eds., *Perry's Chemical Engineers' Handbook,* 7th ed., McGraw-Hill, New York (1997).

17. PITZER, K.S., D.Z. LIPPMAN, R.F. CURL, Jr., C.M. HUGGINS, and D.E. PETERSEN, *J. Am. Chem. Soc.,* **77,** 3433–3440 (1955).

18. REDLICH, O., and J.N.S. KWONG, *Chem. Rev.,* **44,** 233–244 (1949).

19. SHAH, K.K., and G. THODOS, *Ind. Eng. Chem.,* **57** (3), 30–37 (1965).

20. GLANVILLE, J.W., B.H. SAGE, and W.N. LACEY, *Ind. Eng. Chem.,* **42,** 508–513 (1950).

21. WILSON, G.M., *Adv. Cryogenic Eng.,* **11,** 392–400 (1966).

22. KNAPP, H., R. DORING, L. OELLRICH, U. PLÖCKER, and J.M. PRAUSNITZ, *Vapor-Liquid Equilibria for Mixtures of Low Boiling Substances,* Chem. Data. Ser., Vol. VI, DECHEMA (1982).

23. THIESEN, M., *Ann. Phys.,* **24,** 467–492 (1885).

24. ONNES, K., *Konink. Akad. Wetens,* p. 633 (1912).

25. WALAS, S.M., *Phase Equilibria in Chemical Engineering,* Butterworth, Boston (1985).

26. LEE, B.I., and M.G. KESSLER, *AIChE J.,* **21,** 510–527 (1975).

27. EDMISTER, W.C., *Hydrocarbon Processing,* **47** (9), 239–244 (1968).

28. YARBOROUGH, L., *J. Chem. Eng. Data,* **17,** 129–133 (1972).

29. PRAUSNITZ, J.M., W.C. EDMISTER, and K.C. CHAO, *AIChE J.,* **6,** 214–219 (1960).

30. HILDEBRAND, J.H., J.M. PRAUSNITZ, and R.L. SCOTT, *Regular and Related Solutions,* Van Nostrand Reinhold, New York (1970).

31. YERAZUNIS, S., J.D. PLOWRIGHT, and F.M. SMOLA, *AIChE J.,* **10,** 660–665 (1964).

32. HERMSEN, R.W., and J.M. PRAUSNITZ, *Chem. Eng. Sci.,* **18,** 485–494 (1963).

33. EWELL, R.H., J.M. HARRISON, and L. BERG, *Ind. Eng. Chem.,* **36,** 871–875 (1944).

34. VAN NESS, H.C., C.A. SOCZEK, and N.K. KOCHAR, *J. Chem. Eng. Data,* **12,** 346–351 (1967).

35. SINOR, J.E., and J.H. WEBER, *J. Chem. Eng. Data,* **5,** 243–247 (1960).

36. ORYE, R.V., and J.M. PRAUSNITZ, *Ind. Eng. Chem.,* **57** (5), 18–26 (1965).

37. WILSON, G.M., *J. Am. Chem. Soc.,* **86,** 127–130 (1964).

38. CUKOR, P.M., and J.M. PRAUSNITZ, *Inst. Chem. Eng. Symp. Ser. No. 32,* **3,** 88 (1969).

39. GMEHLING, J., and U. ONKEN, *Vapor-Liquid Equilibrium Data Collection,* DECHEMA Chem. Data Ser., 1–8, (1977–1984).

40. HIRANUMA, M., *J. Chem. Eng. Japan,* **8,** 162–163 (1957).

41. RENON, H., and J.M. PRAUSNITZ, *AIChE J.,* **14,** 135–144 (1968).

42. RENON, H., and J.M. PRAUSNITZ, *Ind. Eng. Chem. Process Des. Dev.,* **8,** 413–419 (1969).

43. ABRAMS, D.S., and J.M. PRAUSNITZ, *AIChE J.,* **21,** 116–128 (1975).

44. ABRAMS, D.S., Ph.D. thesis in chemical engineering, University of California, Berkeley, 1974.

45. PRAUSNITZ, J.M., T.F. ANDERSON, E.A. GRENS, C.A. ECKERT, R. HSIEH, and J.P. O'CONNELL, *Computer Calculations for Multicomponent Vapor-Liquid and Liquid-Liquid Equilibria,* Prentice-Hall, Englewood Cliffs, NJ (1980).

46. FREDENSLUND, A., J. GMEHLING, M.L. MICHELSEN, P. RASMUSSEN, and J.M. PRAUSNITZ, *Ind. Eng. Chem. Process Des. Dev.,* **16,** 450–462 (1977).

47. WILSON, G.M., and C.H. DEAL, *Ind. Eng. Chem. Fundam.,* **1,** 20–23 (1962).

48. DERR, E.L., and C.H. DEAL, *Inst. Chem. Eng. Symp. Ser. No. 32,* **3,** 40–51 (1969).

49. FREDENSLUND, A., R.L. JONES, and J.M. PRAUSNITZ, *AIChE J.,* **21,** 1086–1099 (1975).

50. FREDENSLUND, A., J. GMEHLING, and P. RASMUSSEN, *Vapor-Liquid Equilibria Using UNIFAC, A Group Contribution Method,* Elsevier, Amsterdam (1977).

51. GMEHLING, J., P. RASMUSSEN, and A. FREDENSLUND, *Ind. Eng. Chem. Process Des. Dev.,* **21,** 118–127 (1982).

52. LARSEN, B.L., P. RASMUSSEN, and A. FREDENSLUND, *Ind. Eng. Chem. Res.,* **26,** 2274–2286 (1987).

53. GMEHLING, J., J. LI, and M. SCHILLER, *Ind. Eng. Chem. Res.,* **32,** 178–193 (1993).

54. WITTIG, R., J. LOHMANN, and J. GMEHLING, *Ind. Eng. Chem. Res.,* **42,** 183–188 (2003).

55. TAKEUCHI, S., T. NITTA, and T. KATAYAMA, *J. Chem. Eng. Japan,* **8,** 248–250 (1975).

56. STRUBL, K., V. SVOBODA, R. HOLUB, and J. PICK, *Collect. Czech. Chem. Commun.,* **35,** 3004–3019 (1970).

57. KISER, R.W., G.D. JOHNSON, and M.D. SHETLAR, *J. Chem. Eng. Data,* **6,** 338–341 (1961).

58. HOLDERBAUM, T., and J. GMEHLING, *Fluid Phase Equilibria,* **70,** 251–265 (1991).

59. FISCHER, K., and J. GMEHLING, *Fluid Phase Equilibria,* **121,** 185–206 (1996).

60. PITZER, K.S., *J. Phys. Chem.,* **77,** No. 2, 268–277 (1973).

61. CHEN, C.-C., H.I. BRITT, J.F. BOSTON, and L.B. EVANS, *AIChE Journal,* **28,** 588–596 (1982).

62. CHEN, C.-C., and L.B. EVANS, *AIChE Journal,* **32,** 444–459 (1986).

63. MOCK, B., L.B. EVANS, and C.-C. CHEN, *AIChE Journal,* **28,** 1655–1664 (1986).

64. CHEN, C.-C., *Fluid Phase Equilibria,* **83,** 301–312 (1993).

65. LEE, B.I., and M.G. KESLER, *AIChE Journal,* **21,** 510–527 (1975).

66. BENEDICT, M., G.B. WEBB, and L.C. RUBIN, *Chem. Eng. Progress,* **47** (8), 419 (1951).

67. BENEDICT, M., G.B. WEBB, and L.C. RUBIN, *Chem. Eng. Progress,* **47** (9), 449 (1951).

EXERCISES

Section 2.1

2.1 A hydrocarbon stream in a petroleum refinery is to be separated at 1,500 kPa into two products under the conditions shown below. Using the data given, compute the minimum work of separation, W_{min}, in kJ/h for $T_0 = 298.15$ K.

	kmol/h	
Component	Feed	Product 1
Ethane	30	30
Propane	200	192
n-Butane	370	4
n-Pentane	350	0
n-Hexane	50	0

	Feed	Product 1	Product 2
Phase condition	Liquid	Vapor	Liquid
Temperature, K	364	313	394
Enthalpy, kJ/kmol	19,480	25,040	25,640
Entropy, kJ/kmol-K	36.64	33.13	54.84

2.2 In petroleum refineries, a mixture of paraffins and cycloparaffins is commonly reformed in a fixed-bed catalytic reactor to produce blending stocks for gasoline and aromatic precursors for making petrochemicals. A typical multicomponent product from catalytic reforming is a mixture of ethylbenzene with the three xylene isomers. If this mixture is separated, these four chemicals can then be subsequently processed to make styrene, phthalic anhydride, isophthalic acid, and terephthalic acid. Compute, using the following data, the minimum work of separation in Btu/h for $T_0 = 560°$R if the mixture below is separated at 20 psia into three products.

		Split Fraction (SF)		
Component	Feed, lbmol/h	Product 1	Product 2	Product 3
Ethylbenzene	150	0.96	0.04	0.000
p-Xylene	190	0.005	0.99	0.005
m-Xylene	430	0.004	0.99	0.006
o-Xylene	230	0.00	0.015	0.985

	Feed	Product 1	Product 2	Product 3
Phase condition	Liquid	Liquid	Liquid	Liquid
Temperature, °F	305	299	304	314
Enthalpy, Btu/lbmol	29,290	29,750	29,550	28,320
Entropy, Btu/lbmol-°R	15.32	12.47	13.60	14.68

2.3 Distillation column C3 in Figure 1.9 separates stream 5 into streams 6 and 7, according to the material balance in Table 1.5. A suitable column for the separation, if carried out at 700 kPa, contains 70 plates with a condenser duty of 27,300,000 kJ/h. Using the following data and an infinite surroundings temperature, T_0, of 298.15 K, compute:

(a) The duty of the reboiler in kJ/h

(b) The irreversible production of entropy in kJ/h-K, assuming the use of cooling water at a nominal temperature of 25°C for the condenser and saturated steam at 100°C for the reboiler

(c) The lost work in kJ/h

(d) The minimum work of separation in kJ/h

(e) The second-law efficiency

Assume the shaft work of the reflux pump is negligible.

	Feed (Stream 5)	Distillate (Stream 6)	Bottoms (Stream 7)
Phase condition	Liquid	Liquid	Liquid
Temperature, K	348	323	343
Pressure, kPa	1,950	700	730
Enthalpy, kJ/kmol	17,000	13,420	15,840
Entropy, kJ/kmol-K	25.05	5.87	21.22

2.4 A spiral-wound, nonporous cellulose acetate membrane separator is to be used to separate a gas containing H_2, CH_4, and C_2H_6. The permeate will be 95 mol% pure H_2 and will contain no ethane. The relative split ratio (separation power), SP, for H_2 relative to methane will be 47. Using the following data and an infinite surroundings temperature of 80°F, compute:

(a) The irreversible production of entropy in Btu/h-R

(b) The lost work in Btu/h

(c) The minimum work of separation in Btu/h. Why is it negative? What other method(s) might be used to make the separation?

Feed flow rates, lbmol/h

H_2	3,000
CH_4	884
C_2H_6	120

Stream properties:

	Feed	Permeate	Retentate
Phase condition	Vapor	Vapor	Vapor
Temperature, °F	80	80	80
Pressure, psia	365	50	365
Enthalpy, Btu/lbmol	8,550	8,380	8,890
Entropy, Btu/lbmol-R	1.520	4.222	2.742

Section 2.2

2.5 Which of the following K-value expressions, if any, is (are) rigorous? For those expressions that are not rigorous, cite the assumptions involved.

(a) $K_i = \bar{\phi}_{iL}/\bar{\phi}_{iV}$

(b) $K_i = \phi_{iL}/\phi_{iL}$

(c) $K_i = \phi_{iL}$

(d) $K_i = \gamma_{iL}\phi_{iL}/\bar{\phi}_{iV}$

(e) $K_i = P_i^s/P$

(f) $K_i = \gamma_{iL}\phi_{iL}/\gamma_{iV}\phi_{iV}$

(g) $K_i = \gamma_{iL}P_i^s/P$

2.6 Experimental measurements of Vaughan and Collins [*Ind. Eng. Chem.,* **34,** 885 (1942)] for the propane–isopentane system at 167°F and 147 psia show for propane a liquid-phase mole fraction of 0.2900 in equilibrium with a vapor-phase mole fraction of 0.6650. Calculate:

(a) The K-values for C_3 and iC_5 from the experimental data.

(b) Estimates of the K-values of C_3 and iC_5 from Raoult's law assuming vapor pressures at 167°F of 409.6 and 58.6 psia, respectively.

Compare the results of (a) and (b). Assuming the experimental values are correct, how could better estimates of the K-values be achieved? To respond to this question, compare the rigorous expression $K_i = \gamma_{iL}\phi_{iL}/\bar{\phi}_{iV}$ to the Raoult's law expression $K_i = P_i^s/P$.

2.7 Mutual solubility data for the isooctane (1)/furfural (2) system at 25°C are [*Chem. Eng. Sci.,* **6,** 116 (1957)]

	Liquid Phase I	Liquid Phase II
x_1	0.0431	0.9461

Compute:

(a) The distribution coefficients for isooctane and furfural

(b) The relative selectivity for isooctane relative to furfural

(c) The activity coefficient of isooctane in liquid phase 1 and the activity coefficient of furfural in liquid phase 2 assuming $\gamma_2^{(1)} = 1.0$ and $\gamma_1^{(2)} = 1.0$.

2.8 In petroleum refineries, streams rich in alkylbenzenes and alkylnaphthalenes result from catalytic cracking operations. Such streams can be hydrodealkylated to more valuable products such as benzene and naphthalene. At 25°C, solid naphthalene (normal melting point = 80.3°C) has the following solubilities in various liquid solvents [*Naphthalene,* API Publication 707, Washington, DC (Oct. 1978)], including benzene:

Solvent	Mole Fraction Naphthalene
Benzene	0.2946
Cyclohexane	0.1487
Carbon tetrachloride	0.2591
n-Hexane	0.1168
Water	0.18×10^{-5}

For each solvent, compute the activity coefficient of naphthalene in the liquid solvent phase using the following equations for the vapor pressure in torr of solid and liquid naphthalene:

$$\ln P_{solid}^s = 26.708 - 8,712/T$$

$$\ln P_{liquid}^s = 16.1426 - 3992.01/(T - 71.29)$$

where T is in K.

Section 2.3

2.9 A binary ideal-gas mixture of A and B undergoes an isothermal, isobaric separation at T_0, the infinite surroundings temperature. Starting with Eq. (4), Table 2.1, derive an equation for the minimum work of separation, W_{min}, in terms of mole fractions of the feed and the two products. Use your equation to prepare a plot of the dimensionless group, $W_{min}/RT_0 n_F$, as a function of mole fraction of A in the feed for:

(a) A perfect separation
(b) A separation with $SF_A = 0.98$, $SF_B = 0.02$
(c) A separation with $SR_A = 9.0$ and $SR_B = \frac{1}{9}$
(d) A separation with $SF = 0.95$ for A and $SP_{A,B} = 361$

How sensitive is W_{min} to product purities? Does W_{min} depend on the particular separation operation used?

Prove, by calculus, that the largest value of W_{min} occurs for a feed with equimolar quantities of A and B.

2.10 The separation of isopentane from n-pentane by distillation is difficult (approximately 100 trays are required), but is commonly practiced in industry. Using the extended Antoine vapor pressure equation, (2-39), with the constants below and in conjunction with Raoult's law, calculate relative volatilities for the isopentane/n-pentane system and compare the values on a plot with the following smoothed experimental values [*J. Chem. Eng. Data,* **8,** 504 (1963)]:

Temperature, °F	α_{iC_5,nC_5}
125	1.26
150	1.23
175	1.21
200	1.18
225	1.16
250	1.14

What do you conclude about the applicability of Raoult's law in this temperature range for this binary system?

Vapor pressure constants for (2-39) with vapor pressure in kPa and T in K are

	iC_5	nC_5
k_1	13.6106	13.9778
k_2	−2,345.09	−2,554.60
k_3	−40.2128	−36.2529
k_4, k_5, k_6	0	0

2.11 Operating conditions at the top of a vacuum distillation column for the separation of ethylbenzene from styrene are given below, where the overhead vapor is condensed in an air-cooled condenser to give subcooled reflux and distillate. Using the property constants in Example 2.3, estimate the heat transfer rate (duty) for the condenser in kJ/h, assuming an ideal gas and ideal gas and liquid solutions.

	Overhead Vapor	Reflux	Distillate
Phase condition	Vapor	Liquid	Liquid
Temperature, K	331	325	325
Pressure, kPa	6.69	6.40	6.40
Component flow rates, kg/h:			
Ethylbenzene	77,500	66,960	10,540
Styrene	2,500	2,160	340

2.12 Toluene can be hydrodealkylated to benzene, but the conversion per pass through the reactor is only about 70%. Consequently, the toluene must be recovered and recycled. Typical conditions for the feed to a commercial distillation unit are 100°F, 20 psia, 415 lbmol/h of benzene, and 131 lbmol/h of toluene. Based on the property constants below, and assuming that the ideal gas, ideal liquid solution model of Table 2.4 applies at this low pressure, prove that the mixture is a liquid and estimate v_L and ρ_L in American engineering units.

Property constants for (2-38) and (2-39), where in all cases, T is in K, are

	Benzene	Toluene
M, kg/kmol	78.114	92.141
P^s, torr:		
k_1	15.9008	16.0137
k_2	−2,788.51	−3,096.52
k_3	−52.36	−53.67
k_4, k_5, k_6	0	0
ρ_L, kg/m³:		
A	304.1	290.6
B	0.269	0.265
T_c	562.0	593.1

Section 2.4

2.13 Measured conditions for the bottoms from a depropanizer distillation unit in a small refinery are given below. Using the data in Figure 2.3 and assuming an ideal liquid solution (volume of mixing = 0), compute the liquid density in lb/ft³, lb/gal, lb/bbl (42 gal), and kg/m³.

Phase Condition	Liquid
Temperature, °F	229
Pressure, psia	282
Flow rates, lbmol/h:	
C_3	2.2
iC_4	171.1
nC_4	226.6
iC_5	28.1
nC_5	17.5

2.14 Isopropanol, containing 13 wt% water, can be dehydrated to obtain almost pure isopropanol at a 90% recovery by azeotropic distillation with benzene. When condensed, the overhead vapor from the column splits into two immiscible liquid phases. Use the relations in Table 2.4 with data in Perry's Handbook and the operating conditions below to compute the rate of heat transfer in Btu/h and kJ/h for the condenser.

	Overhead	Water-Rich Phase	Organic-Rich Phase
Phase	Vapor	Liquid	Liquid
Temperature, °C	76	40	40
Pressure, bar	1.4	1.4	1.4
Flow rate, kg/h:			
Isopropanol	6,800	5,870	930
Water	2,350	1,790	560
Benzene	24,600	30	24,570

2.15 A hydrocarbon vapor–liquid mixture at 250°F and 500 psia contains N_2, H_2S, CO_2, and all the normal paraffins from methane to heptane. Use Figure 2.8 to estimate the K-value of each component in the mixture. Which components will have a tendency to be present to a greater extent in the equilibrium vapor?

2.16 Acetone, a valuable solvent, can be recovered from air by absorption in water or by adsorption on activated carbon. If absorption is used, the conditions for the streams entering and leaving are as listed below. If the absorber operates adiabatically, estimate the temperature of the exiting liquid phase using a simulation program.

	Feed Gas	Absorbent	Gas Out	Liquid Out
Flow rate, lbmol/h:				
Air	687	0	687	0
Acetone	15	0	0.1	14.9
Water	0	1,733	22	1,711
Temperature, °F	78	90	80	—
Pressure, psia	15	15	14	15
Phase	Vapor	Liquid	Vapor	Liquid

Some concern has been expressed about the possible explosion hazard associated with the feed gas. The lower and upper flammability limits for acetone in air are 2.5 and 13 mol%, respectively. Is the mixture within the explosive range? If so, what can be done to remedy the situation?

Section 2.5

2.17 Subquality natural gas contains an intolerable amount of nitrogen impurity. Separation processes that can be used to remove nitrogen include cryogenic distillation, membrane separation, and pressure-swing adsorption. For the latter process, a set of typical feed and product conditions is given below. Assume a 90% removal of N_2 and a 97% methane natural-gas product. Using the R–K equation of state with the constants listed below, compute the flow rate in thousands of actual cubic feet per hour for each of the three streams.

	N_2	CH_4
Feed flow rate, lbmol/h:	176	704
T_c, K	126.2	190.4
P_c, bar	33.9	46.0

Stream conditions are

	Feed (Subquality Natural Gas)	Product (Natural Gas)	Waste Gas
Temperature, °F	70	100	70
Pressure, psia	800	790	280

2.18 Use the R–K equation of state to estimate the partial fugacity coefficients of propane and benzene in the vapor mixture of Example 2.5.

2.19 Use a computer-aided, steady-state simulation program to estimate the K-values, using the P–R and S–R–K equations of state, of an equimolar mixture of the two butane isomers and the four butene isomers at 220°F and 276.5 psia. Compare these values with the following experimental results [*J. Chem. Eng. Data*, **7**, 331 (1962)]:

Component	K-value
Isobutane	1.067
Isobutene	1.024
n-Butane	0.922
1-Butene	1.024
trans-2-Butene	0.952
cis-2-Butene	0.876

2.20 The disproportionation of toluene to benzene and xylenes is carried out in a catalytic reactor at 500 psia and 950°F. The reactor effluent is cooled in a series of heat exchangers for heat recovery until a temperature of 235°F is reached at a pressure of 490 psia. The effluent is then further cooled and partially condensed by the transfer of heat to cooling water in a final exchanger. The resulting two-phase equilibrium mixture at 100°F and 485 psia is then separated in a flash drum. For the reactor effluent composition given below, use a computer-aided, steady-state simulation program with the S–R–K and P–R equations of state to compute the component flow rates in lbmol/h in both the resulting vapor and liquid streams, the component K-values for the equilibrium mixture, and the rate of heat transfer to the cooling water. Compare the results.

Component	Reactor Effluent, lbmol/h
H_2	1,900
CH_4	215
C_2H_6	17
Benzene	577
Toluene	1,349
p-Xylene	508

Section 2.6

2.21 For an ambient separation process where the feed and products are all nonideal liquid solutions at the infinite surroundings temperature, T_0, (4) of Table 2.1 for the minimum work of separation reduces to

$$\frac{W_{min}}{RT_0} = \sum_{out} n \left[\sum_i x_i \ln(\gamma_i x_i) \right] - \sum_{in} n \left[\sum_i x_i \ln(\gamma_i x_i) \right]$$

For the liquid-phase separation at ambient conditions (298 K, 101.3 kPa) of a 35 mol% mixture of acetone (1) in water (2) into 99 mol% acetone and 98 mol% water products, calculate the minimum work in kJ/kmol of feed. Liquid-phase activity coefficients at ambient conditions are correlated reasonably well by the van Laar equations with $A_{12} = 2.0$ and $A_{21} = 1.7$. What would the minimum rate of work be if acetone and water formed an ideal liquid solution?

2.22 The sharp separation of benzene and cyclohexane by distillation at ambient pressure is impossible because of the formation of an azeotrope at 77.6°C. K.C. Chao [Ph.D. thesis, University of Wisconsin (1956)] obtained the following vapor–liquid equilibrium data for the benzene (B)/cyclohexane (CH) system at 1 atm:

T, °C	x_B	y_B	γ_B	γ_{CH}
79.7	0.088	0.113	1.300	1.003
79.1	0.156	0.190	1.256	1.008
78.5	0.231	0.268	1.219	1.019
78.0	0.308	0.343	1.189	1.032
77.7	0.400	0.422	1.136	1.056
77.6	0.470	0.482	1.108	1.075
77.6	0.545	0.544	1.079	1.102
77.6	0.625	0.612	1.058	1.138
77.8	0.701	0.678	1.039	1.178
78.0	0.757	0.727	1.025	1.221
78.3	0.822	0.791	1.018	1.263
78.9	0.891	0.863	1.005	1.328
79.5	0.953	0.938	1.003	1.369

Vapor pressure is given by (2-39), where constants for benzene are in Exercise 2.12 and constants for cyclohexane are $k_1 = 15.7527$, $k_2 = -2766.63$, and $k_3 = -50.50$.

(a) Use the data to calculate and plot the relative volatility of benzene with respect to cyclohexane versus benzene composition in the liquid phase. What happens to the relative volatility in the vicinity of the azeotrope?

(b) From the azeotropic composition for the benzene/cyclohexane system, calculate the constants in the van Laar equation. With these constants, use the van Laar equation to compute the activity coefficients over the entire range of composition and compare them, in a plot like Figure 2.16, with the above experimental data. How well does the van Laar equation predict the activity coefficients?

2.23 Benzene can be used to break the ethanol/water azeotrope so as to produce nearly pure ethanol. The Wilson constants for the ethanol(1)/benzene(2) system at 45°C are $\Lambda_{12} = 0.124$ and $\Lambda_{21} = 0.523$. Use these constants with the Wilson equation to predict the liquid-phase activity coefficients for this system over the entire range of composition and compare them, in a plot like Figure 2.16, with the following experimental results [*Austral. J. Chem.*, **7**, 264 (1954)]:

x_1	$\ln \gamma_1$	$\ln \gamma_2$
0.0374	2.0937	0.0220
0.0972	1.6153	0.0519

x_1	$\ln \gamma_1$	$\ln \gamma_2$
0.3141	0.7090	0.2599
0.5199	0.3136	0.5392
0.7087	0.1079	0.8645
0.9193	0.0002	1.3177
0.9591	-0.0077	1.3999

2.24 For the binary system ethanol(1)/isooctane(2) at 50°C, the infinite-dilution, liquid-phase activity coefficients are $\gamma_1^\infty = 21.17$ and $\gamma_2^\infty = 9.84$.

(a) Calculate the constants A_{12} and A_{21} in the van Laar equations.

(b) Calculate the constants Λ_{12} and Λ_{21} in the Wilson equations.

(c) Using the constants from (a) and (b), calculate γ_1 and γ_2 over the entire composition range and plot the calculated points as log γ versus x_1.

(d) How well do the van Laar and Wilson predictions agree with the azeotropic point where $x_1 = 0.5941$, $\gamma_1 = 1.44$, and $\gamma_2 = 2.18$?

(e) Show that the van Laar equation erroneously predicts separation into two liquid phases over a portion of the composition range by calculating and plotting a y–x diagram like Figure 2.20.

Chapter 3

Mass Transfer and Diffusion

Mass transfer is the net movement of a component in a mixture from one location to another where the component exists at a different concentration. In many separation operations, the transfer takes place between two phases across an interface. Thus, the absorption by a solvent liquid of a solute from a carrier gas involves mass transfer of the solute through the gas to the gas–liquid interface, across the interface, and into the liquid. Mass-transfer models describe this and other processes such as passage of a species through a gas to the outer surface of a porous, adsorbent particle and into the adsorbent pores, where the species is adsorbed on the porous surface. Mass transfer also governs selective permeation through a nonporous, polymeric material of a component of a gas mixture. Mass transfer, as used here, does not refer to the flow of a fluid through a pipe. However, mass transfer might be superimposed on that flow. Mass transfer is not the flow of solids on a conveyor belt.

Mass transfer occurs by two basic mechanisms: (1) *molecular diffusion* by random and spontaneous microscopic movement of individual molecules in a gas, liquid, or solid as a result of thermal motion; and (2) *eddy* (turbulent) *diffusion* by random, macroscopic fluid motion. Both molecular and/or eddy diffusion frequently involve the movement of different species in opposing directions. When a net flow occurs in one of these directions, the total rate of mass transfer of individual species is increased or decreased by this bulk flow or *convection effect*, which may be considered a third mechanism of mass transfer. Molecular diffusion is extremely slow, whereas eddy diffusion is orders of magnitude more rapid. Therefore, if industrial separation processes are to be conducted in equipment of reasonable size, fluids must be agitated and interfacial areas maximized. If mass transfer in solids is involved, using small particles to decrease the distance in the direction of diffusion will increase the rate.

When separations involve two or more phases, the extent of the separation is limited by phase equilibrium, because, with time, the phases in contact tend to equilibrate by mass transfer between phases. When mass transfer is rapid, equilibration is approached in seconds or minutes, and design of separation equipment may be based on phase equilibrium, not mass transfer. For separations involving barriers, such as membranes, differing species mass-transfer rates through the membrane govern equipment design.

In a binary mixture, molecular diffusion of component A with respect to B occurs because of different potentials or driving forces, which include differences (gradients) of concentration (ordinary diffusion), pressure (pressure diffusion), temperature (thermal diffusion), and external force fields (forced diffusion) that act unequally on the different chemical species present. Pressure diffusion requires a large pressure gradient, which is achieved for gas mixtures with a centrifuge. Thermal diffusion columns or cascades can be employed to separate liquid and gas mixtures by establishing a temperature gradient. More widely applied is forced diffusion in an electrical field, to cause ions of different charges to move in different directions at different speeds.

In this chapter, only molecular diffusion caused by concentration gradients is considered, because this is the most common type of molecular diffusion in separation processes. Furthermore, emphasis is on binary systems, for which molecular-diffusion theory is relatively simple and applications are relatively straightforward. Multicomponent molecular diffusion, which is important in many applications, is considered briefly in Chapter 12. Diffusion in multicomponent systems is much more complex than diffusion in binary systems, and is a more appropriate topic for advanced study using a text such as Taylor and Krishna [1].

Molecular diffusion occurs in solids and in fluids that are stagnant or in laminar or turbulent motion. Eddy diffusion occurs in fluids in turbulent motion. When both molecular diffusion and eddy diffusion occur, they take place in parallel and are additive. Furthermore, they take place because of the same concentration difference (gradient). When mass transfer occurs under turbulent-flow conditions, but across an interface or to a solid surface, conditions may be laminar or nearly stagnant near the interface or solid surface. Thus, even though eddy diffusion may be the dominant mechanism in the bulk of the fluid, the overall rate of mass transfer may be controlled by molecular diffusion because the eddy-diffusion mechanism is damped or even eliminated as the interface or solid surface is approached.

Mass transfer of one or more species results in a total net rate of bulk flow or flux in one direction relative to a fixed plane or stationary coordinate system. When a net flux occurs, it carries all species present. Thus, the molar flux of an individual species is the sum of all three mechanisms. If N_i is the molar flux of species i with mole fraction x_i, and N is the total molar flux, with both fluxes in moles per unit time per unit area in a direction perpendicular to a stationary plane across which mass transfer occurs, then

$$N_i = x_i N + \text{molecular diffusion flux of } i$$
$$+ \text{ eddy diffusion flux of } i \qquad (3\text{-}1)$$

where $x_i N$ is the bulk-flow flux. Each term in (3-1) is positive or negative depending on the direction of the flux relative to

the direction selected as positive. When the molecular and eddy-diffusion fluxes are in one direction and N is in the opposite direction, even though a concentration difference or gradient of i exists, the net mass-transfer flux, N_i, of i can be zero.

In this chapter, the subject of mass transfer and diffusion is divided into seven areas: (1) steady-state diffusion in stagnant media, (2) estimation of diffusion coefficients, (3) unsteady-state diffusion in stagnant media, (4) mass transfer in laminar flow, (5) mass transfer in turbulent flow, (6) mass transfer at fluid–fluid interfaces, and (7) mass transfer across fluid–fluid interfaces.

3.0 INSTRUCTIONAL OBJECTIVES

After completing this chapter, you should be able to:

- Explain the relationship between mass transfer and phase equilibrium.
- Explain why separation models for mass transfer and phase equilibrium are useful.
- Discuss mechanisms of mass transfer, including the effect of bulk flow.
- State, in detail, Fick's law of diffusion for a binary mixture and discuss its analogy to Fourier's law of heat conduction in one dimension.
- Modify Fick's law of diffusion to include the bulk flow effect.
- Calculate mass-transfer rates and composition gradients under conditions of equimolar, countercurrent diffusion and unimolecular diffusion.
- Estimate, in the absence of data, diffusivities (diffusion coefficients) in gas and liquid mixtures, and know of some sources of data for diffusion in solids.
- Calculate multidimensional, unsteady-state, molecular diffusion by analogy to heat conduction.
- Calculate rates of mass transfer by molecular diffusion in laminar flow for three common cases: (1) falling liquid film, (2) boundary-layer flow past a flat plate, and (3) fully developed flow in a straight, circular tube.
- Define a mass-transfer coefficient and explain its analogy to the heat-transfer coefficient and its usefulness, as an alternative to Fick's law, in solving mass-transfer problems.
- Understand the common dimensionless groups (Reynolds, Sherwood, Schmidt, and Peclet number for mass transfer) used in correlations of mass-transfer coefficients.
- Use analogies, particularly that of Chilton and Colburn, and more theoretically based equations, such as those of Churchill et al., to calculate rates of mass transfer in turbulent flow.
- Calculate rates of mass transfer across fluid–fluid interfaces using the two-film theory and the penetration theory.

3.1 STEADY-STATE, ORDINARY MOLECULAR DIFFUSION

Suppose a cylindrical glass vessel is partly filled with water containing a soluble red dye. Clear water is carefully added on top so that the dyed solution on the bottom is undisturbed. At first, a sharp boundary exists between the two layers, but after a time the upper layer becomes colored, while the layer below becomes less colored. The upper layer is more colored near the original interface between the two layers and less colored in the region near the top of the upper layer. During this color change, the motion of each dye molecule is random, undergoing collisions mainly with water molecules and sometimes with other dye molecules, moving first in one

direction and then in another, with no one direction preferred. This type of motion is sometimes referred to as a *random-walk process,* which yields a mean-square distance of travel for a given interval of time, but not a direction of travel. Thus, at a given horizontal plane through the solution in the cylinder, it is not possible to determine whether, in a given time interval, a given molecule will cross the plane or not. However, on the average, a fraction of all molecules in the solution below the plane will cross over into the region above and the same fraction will cross over in the opposite direction. Therefore, if the concentration of dye molecules in the lower region is greater than in the upper region, a net rate of mass transfer of dye molecules will take place from the lower to the upper region. After a long time, a dynamic equilibrium will be achieved and the concentration of dye will be uniform throughout the solution. Based on these observations, it is clear that:

1. Mass transfer by ordinary molecular diffusion occurs because of a concentration, difference or gradient; that is, a species diffuses in the direction of decreasing concentration.

2. The mass-transfer rate is proportional to the area normal to the direction of mass transfer and not to the volume of the mixture. Thus, the rate can be expressed as a flux.

3. Net mass transfer stops when concentrations are uniform.

Fick's Law of Diffusion

The above observations were quantified by Fick in 1855, who proposed an extension of Fourier's 1822 heat-conduction theory. Fourier's first law of heat conduction is

$$q_z = -k\frac{dT}{dz} \qquad (3\text{-}2)$$

where q_z is the heat flux by conduction in the positive z-direction, k is the thermal conductivity of the medium, and dT/dz is the temperature gradient, which is negative in the direction of heat conduction. Fick's first law of molecular diffusion also features a proportionality between a flux and a gradient. For a binary mixture of A and B,

$$J_{A_z} = -D_{AB}\frac{dc_A}{dz} \qquad (3\text{-}3a)$$

and

$$J_{B_z} = -D_{BA}\frac{dc_B}{dz} \qquad (3\text{-}3b)$$

where, in (3-3a), J_{A_z} is the molar flux of A by ordinary molecular diffusion relative to the molar-average velocity of the mixture in the positive z direction, D_{AB} is the *mutual diffusion coefficient* of A in B, discussed in the next section, c_A is the molar concentration of A, and dc_A/dz is the concentration gradient of A, which is negative in the direction of ordinary molecular diffusion. Similar definitions apply to (3-3b). The molar fluxes of A and B are in opposite directions. If the gas, liquid, or solid mixture through which diffusion occurs is isotropic, then values of k and D_{AB} are independent of direction. Nonisotropic (anisotropic) materials include fibrous and laminated solids as well as single, noncubic crystals. The diffusion coefficient is also referred to as the *diffusivity* and the mass diffusivity (to distinguish it from thermal and momentum diffusivities).

Many alternative forms of (3-3a) and (3-3b) are used, depending on the choice of driving force or potential in the gradient. For example, we can express (3-3a) as

$$J_A = -cD_{AB}\frac{dx_A}{dz} \qquad (3\text{-}4)$$

where, for convenience, the z subscript on J has been dropped, $c =$ total molar concentration or molar density ($c = 1/v = \rho/M$), and $x_A =$ mole fraction of species A.

Equation (3-4) can also be written in the following equivalent mass form, where j_A is the mass flux of A by ordinary molecular diffusion relative to the mass-average velocity of the mixture in the positive z-direction, ρ is the mass density, and w_A is the mass fraction of A:

$$j_A = -\rho D_{AB}\frac{dw_A}{dz} \qquad (3\text{-}5)$$

Velocities in Mass Transfer

It is useful to formulate expressions for velocities of chemical species in the mixture. If these velocities are based on the molar flux, N, and the molar diffusion flux, J, the molar average velocity of the mixture, v_M, relative to stationary coordinates is given for a binary mixture as

$$v_M = \frac{N}{c} = \frac{N_A + N_B}{c} \qquad (3\text{-}6)$$

Similarly, the velocity of species i, defined in terms of N_i, is relative to stationary coordinates:

$$v_i = \frac{N_i}{c_i} \qquad (3\text{-}7)$$

Combining (3-6) and (3-7) with $x_i = c_i/c$ gives

$$v_M = x_A v_A + x_B v_B \qquad (3\text{-}8)$$

Alternatively, species diffusion velocities, v_{i_D}, defined in terms of J_i, are relative to the molar-average velocity and are defined as the difference between the species velocity and the molar-average velocity for the mixture:

$$v_{i_D} = \frac{J_i}{c_i} = v_i - v_M \qquad (3\text{-}9)$$

When solving mass-transfer problems involving net movement of the mixture, it is not convenient to use fluxes and flow rates based on v_M as the frame of reference. Rather, it is preferred to use mass-transfer fluxes referred to stationary coordinates with the observer fixed in space. Thus, from (3-9), the total species velocity is

$$v_i = v_M + v_{i_D} \qquad (3\text{-}10)$$

Combining (3-7) and (3-10),

$$N_i = c_i v_M + c_i v_{i_D} \qquad (3\text{-}11)$$

Combining (3-11) with (3-4), (3-6), and (3-7),

$$N_A = \frac{n_A}{A} = x_A N - cD_{AB}\left(\frac{dx_A}{dz}\right) \qquad (3\text{-}12)$$

and

$$N_B = \frac{n_B}{A} = x_B N - cD_{BA}\left(\frac{dx_B}{dz}\right) \qquad (3\text{-}13)$$

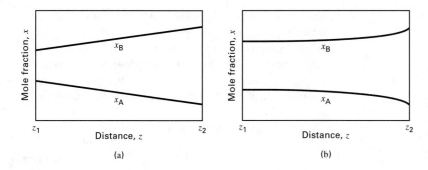

Figure 3.1 Concentration profiles for limiting cases of ordinary molecular diffusion in binary mixtures across a stagnant film: (a) equimolar counterdiffusion (EMD); (b) unimolecular diffusion (UMD).

where in (3-12) and (3-13), n_i is the molar flow rate in moles per unit time, A is the mass-transfer area, the first terms on the right-hand sides are the fluxes resulting from bulk flow, and the second terms on the right-hand sides are the ordinary molecular diffusion fluxes. Two limiting cases are important:

1. Equimolar counterdiffusion (EMD)
2. Unimolecular diffusion (UMD)

Equimolar Counterdiffusion

In equimolar counterdiffusion (EMD), the molar fluxes of A and B in (3-12) and (3-13) are equal but opposite in direction; thus,

$$N = N_A + N_B = 0 \tag{3-14}$$

Thus, from (3-12) and (3-13), the diffusion fluxes are also equal but opposite in direction:

$$J_A = -J_B \tag{3-15}$$

This idealization is closely approached in distillation. From (3-12) and (3-13), we see that in the absence of fluxes other than molecular diffusion,

$$N_A = J_A = -cD_{AB}\left(\frac{dx_A}{dz}\right) \tag{3-16}$$

and

$$N_B = J_B = -cD_{BA}\left(\frac{dx_B}{dz}\right) \tag{3-17}$$

If the total concentration, pressure, and temperature are constant and the mole fractions are maintained constant (but different) at two sides of a stagnant film between z_1 and z_2, then (3-16) and (3-17) can be integrated from z_1 to any z between z_1 and z_2 to give

$$J_A = \frac{cD_{AB}}{z - z_1}(x_{A_1} - x_A) \tag{3-18}$$

and

$$J_B = \frac{cD_{BA}}{z - z_1}(x_{B_1} - x_B) \tag{3-19}$$

Thus, in the steady state, the mole fractions are linear in distance, as shown in Figure 3.1a. Furthermore, because c is constant through the film, where

$$c = c_A + c_B \tag{3-20}$$

by differentiation,

$$dc = 0 = dc_A + dc_B \tag{3-21}$$

Thus,

$$dc_A = -dc_B \tag{3-22}$$

From (3-3a), (3-3b), (3-15), and (3-22),

$$\frac{D_{AB}}{dz} = \frac{D_{BA}}{dz} \tag{3-23}$$

Therefore, $D_{AB} = D_{BA}$.

This equality of diffusion coefficients is always true in a binary system of constant molar density.

EXAMPLE 3.1

Two bulbs are connected by a straight tube, 0.001 m in diameter and 0.15 m in length. Initially the bulb at end 1 contains N_2 and the bulb at end 2 contains H_2. The pressure and temperature are maintained constant at 25°C and 1 atm. At a certain time after allowing diffusion to occur between the two bulbs, the nitrogen content of the gas at end 1 of the tube is 80 mol% and at end 2 is 25 mol%. If the binary diffusion coefficient is 0.784 cm^2/s, determine:

(a) The rates and directions of mass transfer of hydrogen and nitrogen in mol/s

(b) The species velocities relative to stationary coordinates, in cm/s

SOLUTION

(a) Because the gas system is closed and at constant pressure and temperature, mass transfer in the connecting tube is equimolar counterdiffusion by molecular diffusion.

The area for mass transfer through the tube, in cm^2, is $A = 3.14(0.1)^2/4 = 7.85 \times 10^{-3}$ cm^2. The total gas concentration (molar density) is $c = \frac{P}{RT} = \frac{1}{(82.06)(298)} = 4.09 \times 10^{-5}$ mol/cm^3. Take the reference plane at end 1 of the connecting tube. Applying (3-18) to

N_2 over the length of the tube,

$$n_{N_2} = \frac{c D_{N_2, H_2}}{z_2 - z_1}[(x_{N_2})_1 - (x_{N_2})_2]A$$

$$= \frac{(4.09 \times 10^{-5})(0.784)(0.80 - 0.25)}{15}(7.85 \times 10^{-3})$$

$$= 9.23 \times 10^{-9} \text{ mol/s} \quad \text{in the positive } z\text{-direction}$$

$$n_{H_2} = 9.23 \times 10^{-9} \text{ mol/s} \quad \text{in the negative } z\text{-direction}$$

(b) For equimolar counterdiffusion, the molar-average velocity of the mixture, v_M, is 0. Therefore, from (3-9), species velocities are equal to species diffusion velocities. Thus,

$$v_{N_2} = (v_{N_2})_D = \frac{J_{N_2}}{c_{N_2}} = \frac{n_{N_2}}{A c x_{N_2}}$$

$$= \frac{9.23 \times 10^{-9}}{[(7.85 \times 10^{-3})(4.09 \times 10^{-5}) x_{N_2}]}$$

$$= \frac{0.0287}{x_{N_2}} \quad \text{in the positive } z\text{-direction}$$

Similarly,

$$v_{H_2} = \frac{0.0287}{x_{H_2}} \quad \text{in the negative } z\text{-direction}$$

Thus, species velocities depend on species mole fractions, as follows:

z, cm	x_{N_2}	x_{H_2}	v_{N_2}, cm/s	v_{H_2}, cm/s
0 (end 1)	0.800	0.200	0.0351	−0.1435
5	0.617	0.383	0.0465	−0.0749
10	0.433	0.567	0.0663	−0.0506
15 (end 2)	0.250	0.750	0.1148	−0.0383

Note that species velocities vary across the length of the connecting tube, but at any location, z, $v_M = 0$. For example, at $z = 10$ cm, from (3-8),

$$v_M = (0.433)(0.0663) + (0.567)(-0.0506) = 0$$

Unimolecular Diffusion

In unimolecular diffusion (UMD), mass transfer of component A occurs through stagnant (nonmoving) component B. Thus,

$$N_B = 0 \tag{3-24}$$

and

$$N = N_A \tag{3-25}$$

Therefore, from (3-12),

$$N_A = x_A N_A - c D_{AB} \frac{dx_A}{dz} \tag{3-26}$$

which can be rearranged to a Fick's-law form,

$$N_A = -\frac{c D_{AB}}{(1 - x_A)}\frac{dx_A}{dz} = -\frac{c D_{AB}}{x_B}\frac{dx_A}{dz} \tag{3-27}$$

The factor $(1 - x_A)$ accounts for the bulk-flow effect. For a mixture dilute in A, the bulk-flow effect is negligible or small. In mixtures more concentrated in A, the bulk-flow effect can be appreciable. For example, in an equimolar mixture of A and B, $(1 - x_A) = 0.5$ and the molar mass-transfer flux of A is twice the ordinary molecular-diffusion flux.

For the stagnant component, B, (3-13) becomes

$$0 = x_B N_A - c D_{BA}\frac{dx_B}{dz} \tag{3-28}$$

or

$$x_B N_A = c D_{BA}\frac{dx_B}{dz} \tag{3-29}$$

Thus, the bulk-flow flux of B is equal but opposite to its diffusion flux.

At quasi-steady-state conditions, that is, with no accumulation, and with constant molar density, (3-27) becomes in integral form:

$$\int_{z_1}^{z} dz = -\frac{c D_{AB}}{N_A}\int_{x_{A_1}}^{x_A}\frac{dx_A}{1 - x_A} \tag{3-30}$$

which upon integration yields

$$N_A = \frac{c D_{AB}}{z - z_1}\ln\left(\frac{1 - x_A}{1 - x_{A_1}}\right) \tag{3-31}$$

Rearrangement to give the mole-fraction variation as a function of z yields

$$x_A = 1 - (1 - x_{A_1})\exp\left[\frac{N_A(z - z_1)}{c D_{AB}}\right] \tag{3-32}$$

Thus, as shown in Figure 3.1b, the mole fractions are nonlinear in distance.

An alternative and more useful form of (3-31) can be derived from the definition of the log mean. When $z = z_2$, (3-31) becomes

$$N_A = \frac{c D_{AB}}{z_2 - z_1}\ln\left(\frac{1 - x_{A_2}}{1 - x_{A_1}}\right) \tag{3-33}$$

The log mean (LM) of $(1 - x_A)$ at the two ends of the stagnant layer is

$$(1 - x_A)_{LM} = \frac{(1 - x_{A_2}) - (1 - x_{A_1})}{\ln[(1 - x_{A_2})/(1 - x_{A_1})]}$$

$$= \frac{x_{A_1} - x_{A_2}}{\ln[(1 - x_{A_2})/(1 - x_{A_1})]} \tag{3-34}$$

Combining (3-33) with (3-34) gives

$$N_A = \frac{c D_{AB}}{z_2 - z_1}\frac{(x_{A_1} - x_{A_2})}{(1 - x_A)_{LM}} = \frac{c D_{AB}}{(1 - x_A)_{LM}}\frac{(-\Delta x_A)}{\Delta z}$$

$$= \frac{c D_{AB}}{(x_B)_{LM}}\frac{(-\Delta x_A)}{\Delta z} \tag{3-35}$$

EXAMPLE 3.2

As shown in Figure 3.2, an open beaker, 6 cm in height, is filled with liquid benzene at 25°C to within 0.5 cm of the top. A gentle breeze of dry air at 25°C and 1 atm is blown by a fan across the mouth of the beaker so that evaporated benzene is carried away by convection after it transfers through a stagnant air layer in the beaker. The vapor pressure of benzene at 25°C is 0.131 atm. The mutual diffusion coefficient for benzene in air at 25°C and 1 atm is 0.0905 cm²/s. Compute:

(a) The initial rate of evaporation of benzene as a molar flux in mol/cm²-s

(b) The initial mole-fraction profiles in the stagnant air layer

(c) The initial fractions of the mass-transfer fluxes due to molecular diffusion

(d) The initial diffusion velocities, and the species velocities (relative to stationary coordinates) in the stagnant layer

(e) The time in hours for the benzene level in the beaker to drop 2 cm from the initial level, if the specific gravity of liquid benzene is 0.874. Neglect the accumulation of benzene and air in the stagnant layer as it increases in height

SOLUTION

Let A = benzene, B = air.

$$c = \frac{P}{RT} = \frac{1}{(82.06)(298)} = 4.09 \times 10^{-5} \text{ mol/cm}^3$$

(a) Take $z_1 = 0$. Then $z_2 - z_1 = \Delta z = 0.5$ cm. From Dalton's law, assuming equilibrium at the liquid benzene–air interface,

$$x_{A_1} = \frac{p_{A_1}}{P} = \frac{0.131}{1} = 0.131 \qquad x_{A_2} = 0$$

$$(1 - x_A)_{LM} = \frac{0.131}{\ln[(1-0)/(1-0.131)]} = 0.933 = (x_B)_{LM}$$

From (3-35),

$$N_A = \frac{(4.09 \times 10^{-5})(0.0905)}{0.5}\left(\frac{0.131}{0.933}\right) = 1.04 \times 10^{-6} \text{ mol/cm}^2\text{-s}$$

(b) $\dfrac{N_A(z - z_1)}{cD_{AB}} = \dfrac{(1.04 \times 10^{-6})(z - 0)}{(4.09 \times 10^{-5})(0.0905)} = 0.281\, z$

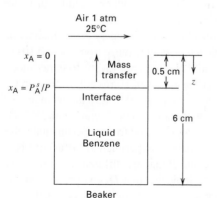

Air 1 atm
25°C

$x_A = 0$

Mass transfer

$x_A = P_A^s/P$

Interface

0.5 cm

z

6 cm

Liquid Benzene

Beaker

Figure 3.2 Evaporation of benzene from a beaker—Example 3.2.

From (3-32),

$$x_A = 1 - 0.869\,\exp(0.281\,z) \tag{1}$$

Using (1), the following results are obtained:

z, cm	x_A	x_B
0.0	0.1310	0.8690
0.1	0.1060	0.8940
0.2	0.0808	0.9192
0.3	0.0546	0.9454
0.4	0.0276	0.9724
0.5	0.0000	1.0000

These profiles are only slightly curved.

(c) From (3-27) and (3-29), we can compute the bulk flow terms, $x_A N$ and $x_B N_A$, from which the molecular diffusion terms are obtained.

	$x_i N$ Bulk-Flow Flux, mol/cm²-s × 10⁶		J_i Molecular-Diffusion Flux, mol/cm²-s × 10⁶	
z, cm	A	B	A	B
0.0	0.1360	0.9040	0.9040	−0.9040
0.1	0.1100	0.9300	0.9300	−0.9300
0.2	0.0840	0.9560	0.9560	−0.9560
0.3	0.0568	0.9832	0.9832	−0.9832
0.4	0.0287	1.0113	1.0113	−1.0113
0.5	0.0000	1.0400	1.0400	−1.0400

Note that the molecular-diffusion fluxes are equal but opposite, and the bulk-flow flux of B is equal but opposite to its molecular-diffusion flux, so that its molar flux, N_B, is zero, making B (air) stagnant.

(d) From (3-6),

$$v_M = \frac{N}{c} = \frac{N_A}{c} = \frac{1.04 \times 10^{-6}}{4.09 \times 10^{-5}} = 0.0254 \text{ cm/s} \tag{2}$$

From (3-9), the diffusion velocities are given by

$$v_{i_d} = \frac{J_i}{c_i} = \frac{J_i}{x_i c} \tag{3}$$

From (3-10), the species velocities relative to stationary coordinates are

$$v_i = v_{i_d} + v_M \tag{4}$$

Using (2) to (4), we obtain

	v_{i_d} Molecular-Diffusion Velocity, cm/s		J_i Species Velocity, cm/s	
z, cm	A	B	A	B
0.0	0.1687	−0.0254	0.1941	0
0.1	0.2145	−0.0254	0.2171	0
0.2	0.2893	−0.0254	0.3147	0
0.3	0.4403	−0.0254	0.4657	0
0.4	0.8959	−0.0254	0.9213	0
0.5	∞	−0.0254	∞	0

Note that v_B is zero everywhere, because its molecular-diffusion velocity is negated by the molar-mean velocity.

(e) The mass-transfer flux for benzene evaporation can be equated to the rate of decrease in the moles of liquid benzene per unit cross section of the beaker. Letting z = distance down from the mouth of the beaker and using (3-35) with $\Delta z = z$,

$$N_A = \frac{cD_{AB}}{z} \frac{(-\Delta x_A)}{(1 - x_A)_{LM}} = \frac{\rho_L}{M_L} \frac{dz}{dt} \tag{5}$$

Separating variables and integrating,

$$\int_0^t dt = t = \frac{\rho_L(1 - x_A)_{LM}}{M_L c D_{AB}(-\Delta x_A)} \int_{z_1}^{z_2} z\, dz \tag{6}$$

The coefficient of the integral on the right-hand side of (6) is constant at

$$\frac{0.874(0.933)}{78.11(4.09 \times 10^{-5})(0.0905)(0.131)} = 21{,}530 \text{ s/cm}^2$$

$$\int_{z_1}^{z_2} z\, dz = \int_{0.5}^{2.5} z\, dz = 3 \text{ cm}^2$$

From (6), $t = 21{,}530(3) = 64{,}590$ s or 17.94 h, which is a long time because of the absence of turbulence.

3.2 DIFFUSION COEFFICIENTS

Diffusivities or diffusion coefficients are defined for a binary mixture by (3-3) to (3-5). Measurement of diffusion coefficients must involve a correction for any bulk flow using (3-12) and (3-13) with the reference plane being such that there is no net molar bulk flow.

The binary diffusivities, D_{AB} and D_{BA}, are mutual or binary diffusion coefficients. Other coefficients include D_{i_M}, the diffusivity of i in a multicomponent mixture; D_{ii}, the self-diffusion coefficient; and the tracer or interdiffusion coefficient. In this chapter, and throughout this book, the focus is on the mutual diffusion coefficient, which will be referred to as the diffusivity or diffusion coefficient.

Diffusivity in Gas Mixtures

As discussed by Poling, Prausnitz, and O'Connell [2], a number of theoretical and empirical equations are available for estimating the value of $D_{AB} = D_{BA}$ in gases at low to moderate pressures. The theoretical equations, based on Boltzmann's kinetic theory of gases, the theorem of corresponding states, and a suitable intermolecular energy-potential function, as developed by Chapman and Enskog, predict D_{AB} to be inversely proportional to pressure and almost independent of composition, with a significant increase for increasing temperature. Of greater accuracy and ease of use is the following empirical equation of Fuller, Schettler, and Giddings [3], which retains the form of the Chapman–Enskog theory but utilizes empirical constants

Table 3.1 Diffusion Volumes from Fuller, Ensley, and Giddings [*J. Phys. Chem*, **73**, 3679–3685 (1969)] for Estimating Binary Gas Diffusivity by the Method of Fuller et al. [3]

Atomic Diffusion Volumes Atomic and Structural Diffusion-Volume Increments			
C	15.9	F	14.7
H	2.31	Cl	21.0
O	6.11	Br	21.9
N	4.54	I	29.8
Aromatic ring	−18.3	S	22.9
Heterocyclic ring	−18.3		

Diffusion Volumes of Simple Molecules			
He	2.67	CO	18.0
Ne	5.98	CO_2	26.7
Ar	16.2	N_2O	35.9
Kr	24.5	NH_3	20.7
Xe	32.7	H_2O	13.1
H_2	6.12	SF_6	71.3
D_2	6.84	Cl_2	38.4
N_2	18.5	Br_2	69.0
O_2	16.3	SO_2	41.8
Air	19.7		

derived from experimental data:

$$D_{AB} = D_{BA} = \frac{0.00143\, T^{1.75}}{P M_{AB}^{1/2} [(\sum_V)_A^{1/3} + (\sum_V)_B^{1/3}]^2} \tag{3-36}$$

where D_{AB} is in cm^2/s, P is in atm, T is in K,

$$M_{AB} = \frac{2}{(1/M_A) + (1/M_B)} \tag{3-37}$$

and $\sum_V$ = summation of atomic and structural diffusion volumes from Table 3.1, which includes diffusion volumes of some simple molecules.

Experimental values of binary gas diffusivity at 1 atm and near-ambient temperature range from about 0.10 to 10.0 cm^2/s. Poling, et al. [2] compared (3-36) to experimental data for 51 different binary gas mixtures at low pressures over a temperature range of 195–1,068 K. The average deviation was only 5.4%, with a maximum deviation of 25%. Only 9 of 69 estimated values deviated from experimental values by more than 10%. When an experimental diffusivity is available at values of T and P that are different from the desired conditions, (3-36) indicates that D_{AB} is proportional to $T^{1.75}/P$, which can be used to obtain the desired value. Some representative experimental values of binary gas diffusivity are given in Table 3.2.

Table 3.2 Experimental Binary Diffusivities of Some Gas Pairs at 1 atm

Gas pair, A–B	Temperature, K	D_{AB}, cm²/s
Air—carbon dioxide	317.2	0.177
Air—ethanol	313	0.145
Air—helium	317.2	0.765
Air—n-hexane	328	0.093
Air—water	313	0.288
Argon—ammonia	333	0.253
Argon—hydrogen	242.2	0.562
Argon—hydrogen	806	4.86
Argon—methane	298	0.202
Carbon dioxide—nitrogen	298	0.167
Carbon dioxide—oxygen	293.2	0.153
Carbon dioxide—water	307.2	0.198
Carbon monoxide—nitrogen	373	0.318
Helium—benzene	423	0.610
Helium—methane	298	0.675
Helium—methanol	423	1.032
Helium—water	307.1	0.902
Hydrogen—ammonia	298	0.783
Hydrogen—ammonia	533	2.149
Hydrogen—cyclohexane	288.6	0.319
Hydrogen—methane	288	0.694
Hydrogen—nitrogen	298	0.784
Nitrogen—benzene	311.3	0.102
Nitrogen—cyclohexane	288.6	0.0731
Nitrogen—sulfur dioxide	263	0.104
Nitrogen—water	352.1	0.256
Oxygen—benzene	311.3	0.101
Oxygen—carbon tetrachloride	296	0.0749
Oxygen—cyclohexane	288.6	0.0746
Oxygen—water	352.3	0.352

From Marrero, T. R., and E. A. Mason, *J. Phys. Chem. Ref. Data,* **1,** 3–118 (1972).

EXAMPLE 3.3

Estimate the diffusion coefficient for the system oxygen (A)/benzene (B) at 38°C and 2 atm using the method of Fuller et al.

SOLUTION

From (3-37),

$$M_{AB} = \frac{2}{(1/32) + (1/78.11)} = 45.4$$

From Table 3.1, $(\sum_V)_A = 16.3$ and $(\sum_V)_B = 6(15.9) + 6(2.31) - 18.3 = 90.96$

From (3-36), at 2 atm and 311.2 K,

$$D_{AB} = D_{BA} = \frac{0.00143(311.2)^{1.75}}{(2)(45.4)^{1/2}[16.3^{1/3} + 90.96^{1/3}]^2} = 0.0495 \text{ cm}^2/\text{s}$$

At 1 atm, the predicted diffusivity is 0.0990 cm²/s, which is about 2% below the experimental value of 0.101 cm²/s in Table 3.2. The experimental value for 38°C can be extrapolated by the temperature dependency of (3-36) to give the following prediction at 200°C:

$$D_{AB} \text{ at } 200°C \text{ and } 1 \text{ atm} = 0.102 \left(\frac{200 + 273.2}{38 + 273.2} \right)^{1.75}$$
$$= 0.212 \text{ cm}^2/\text{s}$$

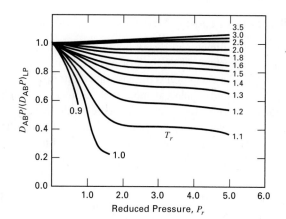

Figure 3.3 Takahashi [4] correlation for effect of high pressure on binary gas diffusivity.

For binary mixtures of light gases, at pressures to about 10 atm, the pressure dependence on diffusivity is adequately predicted by the simple inverse relation (3-36), that is, $PD_{AB} =$ a constant for a given temperature and gas mixture. At higher pressures, deviations from this relation are handled in a manner somewhat similar to the modification of the ideal-gas law by the compressibility factor based on the theorem of corresponding states. Although few reliable experimental data are available at high pressure, Takahasi [4] has published a tentative corresponding-states correlation, shown in Figure 3.3, patterned after an earlier correlation for self-diffusivities by Slattery [5]. In the Takahashi plot, $D_{AB}P/(D_{AB}P)_{LP}$ is given as a function of reduced temperature and pressure, where $(D_{AB}P)_{LP}$ is at low pressure where (3-36) applies. Mixture-critical temperature and pressure are molar-average values. Thus, a finite effect of composition is predicted at high pressure. The effect of high pressure on diffusivity is important in supercritical extraction, discussed in Chapter 11.

EXAMPLE 3.4

Estimate the diffusion coefficient for a 25/75 molar mixture of argon and xenon at 200 atm and 378 K. At this temperature and 1 atm, the diffusion coefficient is 0.180 cm²/s. Critical constants are

	T_c, K	P_c, atm
Argon	151.0	48.0
Xenon	289.8	58.0

SOLUTION

Calculate reduced conditions:
$T_c = 0.25(151) + 0.75(289.8) = 255.1$ K;
$T_r = T/T_c = 378/255.1 = 1.48$
$P_c = 0.25(48) + 0.75(58) = 55.5$;
$P_r = P/P_c = 200/55.5 = 3.6$

From Figure 3.3, $\dfrac{D_{AB}P}{(D_{AB}P)_{LP}} = 0.82$

$$D_{AB} = \frac{(D_{AB}P)_{LP}}{P} \left[\frac{D_{AB}P}{(D_{AB}P)_{LP}} \right] = \frac{(0.180)(1)}{200}(0.82)$$
$$= 7.38 \times 10^{-4} \text{ cm/s}$$

Diffusivity in Liquid Mixtures

Diffusion coefficients in binary liquid mixtures are difficult to estimate because of the lack of a rigorous model for the liquid state. An exception is the case of a dilute solute (A) of very large, rigid, spherical molecules diffusing through a stationary solvent (B) of small molecules with no slip of the solvent at the surface of the solute molecules. The resulting relation, based on the hydrodynamics of creeping flow to describe drag, is the Stokes–Einstein equation:

$$(D_{AB})_\infty = \frac{RT}{6\pi\mu_B R_A N_A} \qquad (3\text{-}38)$$

where R_A is the radius of the solute molecule and N_A is Avagadro's number. Although (3-38) is very limited in its application to liquid mixtures, it has long served as a starting point for more widely applicable empirical correlations for the diffusivity of solute (A) in solvent (B), where both A and B are of the same approximate molecular size. Unfortunately, unlike the situation in binary gas mixtures, $D_{AB} = D_{BA}$ in binary liquid mixtures can vary greatly with composition as shown in Example 3.7. Because the Stokes–Einstein equation does not provide a basis for extending

dilute conditions to more concentrated conditions, extensions of (3-38) have been restricted to binary liquid mixtures dilute in A, up to 5 and perhaps 10 mol%.

One such extension, which gives reasonably good predictions for small solute molecules, is the empirical Wilke–Chang [6] equation:

$$(D_{AB})_\infty = \frac{7.4 \times 10^{-8}(\phi_B M_B)^{1/2}T}{\mu_B v_A^{0.6}} \qquad (3\text{-}39)$$

where the units are cm^2/s for D_{AB}; cP (centipoises) for the solvent viscosity, μ_B; K for T; and cm^3/mol for v_A, the liquid molar volume of the solute at its normal boiling point. The parameter ϕ_B is an association factor for the solvent, which is 2.6 for water, 1.9 for methanol, 1.5 for ethanol, and 1.0 for unassociated solvents such as hydrocarbons. Note that the effects of temperature and viscosity are identical to the prediction of the Stokes–Einstein equation, while the effect of the radius of the solute molecule is replaced by v_A, which can be estimated by summing the atomic contributions in Table 3.3, which also lists values of v_A for dissolved light gases. Some representative experimental values of diffusivity in dilute binary liquid solutions are given in Table 3.4.

Table 3.3 Molecular Volumes of Dissolved Light Gases and Atomic Contributions for Other Molecules at the Normal Boiling Point

	Atomic Volume $(m^3/kmol) \times 10^3$		Atomic Volume $(m^3/kmol) \times 10^3$
C	14.8	Ring	
H	3.7	Three-membered, as in	−6
O (except as below)	7.4	ethylene oxide	
Doubly bonded as carbonyl	7.4	Four-membered	−8.5
Coupled to two other elements:		Five-membered	−11.5
In aldehydes, ketones	7.4	Six-membered	−15
In methyl esters	9.1	Naphthalene ring	−30
In methyl ethers	9.9	Anthracene ring	−47.5
In ethyl esters	9.9		
In ethyl ethers	9.9		Molecular Volume $(m^3/kmol) \times 10^3$
In higher esters	11.0		
In higher ethers	11.0	Air	29.9
In acids (—OH)	12.0	O_2	25.6
Joined to S, P, N	8.3	N_2	31.2
N		Br_2	53.2
Doubly bonded	15.6	Cl_2	48.4
In primary amines	10.5	CO	30.7
In secondary amines	12.0	CO_2	34.0
Br	27.0	H_2	14.3
Cl in RCHClR′	24.6	H_2O	18.8
Cl in RCl (terminal)	21.6	H_2S	32.9
F	8.7	NH_3	25.8
I	37.0	NO	23.6
S	25.6	N_2O	36.4
P	27.0	SO_2	44.8

Source: G. Le Bas, *The Molecular Volumes of Liquid Chemical Compounds,* David McKay, New York (1915).

Table 3.4 Experimental Binary Liquid Diffusivities for Solutes, A, at Low Concentrations in Solvents, B

Solvent, B	Solute, A	Temperature, K	Diffusivity, D_{AB}, $cm^2/s \times 10^5$
Water	Acetic acid	293	1.19
	Aniline	293	0.92
	Carbon dioxide	298	2.00
	Ethanol	288	1.00
	Methanol	288	1.26
Ethanol	Allyl alcohol	293	0.98
	Benzene	298	1.81
	Oxygen	303	2.64
	Pyridine	293	1.10
	Water	298	1.24
Benzene	Acetic acid	298	2.09
	Cyclohexane	298	2.09
	Ethanol	288	2.25
	n-Heptane	298	2.10
	Toluene	298	1.85
n-Hexane	Carbon tetrachloride	298	3.70
	Methyl ethyl ketone	303	3.74
	Propane	298	4.87
	Toluene	298	4.21
Acetone	Acetic acid	288	2.92
	Formic acid	298	3.77
	Nitrobenzene	293	2.94
	Water	298	4.56

From Poling et al. [2].

EXAMPLE 3.5

Use the Wilke–Chang equation to estimate the diffusivity of aniline (A) in a 0.5 mol% aqueous solution at 20°C. At this temperature, the solubility of aniline in water is about 4 g/100 g of water or 0.77 mol% aniline. The experimental diffusivity value for an infinitely dilute mixture is 0.92×10^{-5} cm^2/s.

SOLUTION

$\mu_B = \mu_{H_2O} = 1.01$ cP at 20°C

$v_A = $ liquid molar volume of aniline at its normal boiling point of 457.6 K = 107 cm^3/mol

$\phi_B = 2.6$ for water $M_B = 18$ for water $T = 293$ K

From (3-39),

$$D_{AB} = \frac{(7.4 \times 10^{-8})[2.6(18)]^{0.5}(293)}{1.01(107)^{0.6}} = 0.89 \times 10^{-5} \ cm^2/s$$

This value is about 3% less than the experimental value for an infinitely dilute solution of aniline in water.

More recent liquid diffusivity correlations due to Hayduk and Minhas [7] give better agreement than the Wilke–Chang

equation with experimental values for nonaqueous solutions. For a dilute solution of one normal paraffin (C_5 to C_{32}) in another (C_5 to C_{16}),

$$(D_{AB})_\infty = 13.3 \times 10^{-8} \frac{T^{1.47}\mu_B^\epsilon}{v_A^{0.71}} \qquad (3\text{-}40)$$

where

$$\epsilon = \frac{10.2}{v_A} - 0.791 \qquad (3\text{-}41)$$

and the other variables have the same units as in (3-39).

For general nonaqueous solutions,

$$(D_{AB})_\infty = 1.55 \times 10^{-8} \frac{T^{1.29}\left(\mathcal{P}_B^{0.5}/\mathcal{P}_A^{0.42}\right)}{\mu_B^{0.92}v_B^{0.23}} \qquad (3\text{-}42)$$

where $\mathcal{P}$ is the parachor, which is defined as

$$\mathcal{P} = v\sigma^{1/4} \qquad (3\text{-}43)$$

When the units of the liquid molar volume, v, are cm^3/mol and the surface tension, σ, are g/s^2 (dynes/cm), then the units of the parachor are $cm^3 \text{-} g^{1/4}/s^{1/2}\text{-}mol$. Normally, at near-ambient conditions, $\mathcal{P}$ is treated as a constant, for which an extensive tabulation is available from Quayle [8], who also provides a group-contribution method for estimating parachors for compounds not listed. Table 3.5 gives values of parachors for a number of compounds, while Table 3.6 contains structural contributions for predicting the parachor in the absence of data.

The following restrictions apply to (3-42):

1. Solvent viscosity should not exceed 30 cP.

2. For organic acid solutes and solvents other than water, methanol, and butanols, the acid should be treated as a dimer by doubling the values of $\mathcal{P}_A$ and v_A.

3. For a nonpolar solute in monohydroxy alcohols, values of v_B and $\mathcal{P}_B$ should be multiplied by $8\mu_B$, where the viscosity is in centipoise.

Liquid diffusion coefficients for a solute in a dilute binary system range from about 10^{-6} to 10^{-4} cm^2/s for solutes of molecular weight up to about 200 and solvents with viscosity up to about 10 cP. Thus, liquid diffusivities are five orders of magnitude less than diffusivities for binary gas mixtures at 1 atm. However, diffusion rates in liquids are not necessarily five orders of magnitude lower than in gases because, as seen in (3-5), the product of the concentration (molar density) and the diffusivity determines the rate of diffusion for a given concentration gradient in mole fraction. At 1 atm, the molar density of a liquid is three times that of a gas and, thus, the diffusion rate in liquids is only two orders of magnitude lower than in gases at 1 atm.

Table 3.5 Parachors for Representative Compounds

	Parachor, $cm^3\text{-}g^{1/4}/s^{1/2}\text{-mol}$		Parachor, $cm^3\text{-}g^{1/4}/s^{1/2}\text{-mol}$		Parachor, $cm^3\text{-}g^{1/4}/s^{1/2}\text{-mol}$
Acetic acid	131.2	Chlorobenzene	244.5	Methyl amine	95.9
Acetone	161.5	Diphenyl	380.0	Methyl formate	138.6
Acetonitrile	122	Ethane	110.8	Naphthalene	312.5
Acetylene	88.6	Ethylene	99.5	n-Octane	350.3
Aniline	234.4	Ethyl butyrate	295.1	1-Pentene	218.2
Benzene	205.3	Ethyl ether	211.7	1-Pentyne	207.0
Benzonitrile	258	Ethyl mercaptan	162.9	Phenol	221.3
n-Butyric acid	209.1	Formic acid	93.7	n-Propanol	165.4
Carbon disulfide	143.6	Isobutyl benzene	365.4	Toluene	245.5
Cyclohexane	239.3	Methanol	88.8	Triethyl amine	297.8

Source: Meissner, *Chem. Eng. Prog.,* **45,** 149–153 (1949).

Table 3.6 Structural Contributions for Estimating the Parachor

Carbon–hydrogen:		R—[—CO—]—R′(ketone)	
C	9.0	R + R′ = 2	51.3
H	15.5	R + R′ = 3	49.0
CH₃	55.5	R + R′ = 4	47.5
CH₂ in —(CH₂)ₙ		R + R′ = 5	46.3
$n < 12$	40.0	R + R′ = 6	45.3
$n > 12$	40.3	R + R′ = 7	44.1
		—CHO	66
Alkyl groups			
1-Methylethyl	133.3	O (not noted above)	20
1-Methylpropyl	171.9	N (not noted above)	17.5
1-Methylbutyl	211.7	S	49.1
2-Methylpropyl	173.3	P	40.5
1-Ethylpropyl	209.5	F	26.1
1,1-Dimethylethyl	170.4	Cl	55.2
1,1-Dimethylpropyl	207.5	Br	68.0
1,2-Dimethylpropyl	207.9	I	90.3
1,1,2-Trimethylpropyl	243.5	Ethylenic bonds:	
C₆H₅	189.6	Terminal	19.1
		2,3-position	17.7
Special groups:		3,4-position	16.3
—COO—	63.8		
—COOH	73.8	Triple bond	40.6
—OH	29.8		
—NH₂	42.5	Ring closure:	
—O—	20.0	Three-membered	12
—NO₂	74	Four-membered	6.0
—NO₃ (nitrate)	93	Five-membered	3.0
—CO(NH₂)	91.7	Six-membered	0.8

Source: Quale [8].

EXAMPLE 3.6

Estimate the diffusivity of formic acid (A) in benzene (B) at 25°C and infinite dilution, using the appropriate correlation of Hayduk and Minhas [7]. The experimental value is 2.28×10^{-5} cm²/s.

SOLUTION

Equation (3-42) applies, with $T = 298$ K

$\mathscr{P}_A = 93.7$ cm³-g$^{1/4}$/s$^{1/2}$-mol $\mathscr{P}_B = 205.3$ cm³-g$^{1/4}$/s$^{1/2}$-mol

$\mu_B = 0.6$ cP at 25°C $v_B = 96$ cm³/mol at 80°C

However, because formic acid is an organic acid, $\mathscr{P}_A$ is doubled to 187.4.

From (3-42),

$$(D_{AB})_\infty = 1.55 \times 10^{-8} \left[\frac{298^{1.29}(205.3^{0.5}/187.4^{0.42})}{0.6^{0.92}96^{0.23}} \right]$$

$$= 2.15 \times 10^{-5} \text{ cm}^2/\text{s}$$

which is within 6% of the the experimental value.

The Stokes–Einstein and Wilke–Chang equations predict an inverse dependence of liquid diffusivity with viscosity. The Hayduk–Minhas equations predict a somewhat smaller dependence on viscosity. From data covering several orders of magnitude variation of viscosity, the liquid diffusivity is found to vary inversely with the viscosity raised to an exponent closer to 0.5 than to 1.0. The Stokes–Einstein and Wilke–Chang equations also predict that $D_{AB}\mu_B/T$ is a constant over a narrow temperature range. Because μ_B decreases exponentially with temperature, D_{AB} is predicted to increase exponentially with temperature. For example, for a dilute solution of water in ethanol, the diffusivity of water increases by a factor of almost 20 when the absolute temperature is increased 50%. Over a wide temperature range, it is preferable to express the effect of temperature on D_{AB} by an Arrhenius-type expression,

$$(D_{AB})_\infty = A \exp\left(\frac{-E}{RT}\right) \qquad (3\text{-}44)$$

where, typically the activation energy for liquid diffusion, E, is no greater than 6,000 cal/mol.

Equations (3-39), (3-40), and (3-42) for estimating diffusivity in binary liquid mixtures only apply to the solute, A, in a dilute solution of the solvent, B. Unlike binary gas mixtures in which the diffusivity is almost independent of composition, the effect of composition on liquid diffusivity is complex, sometimes showing strong positive or negative deviations from linearity with mole fraction.

Based on a nonideal form of Fick's law, Vignes [9] has shown that, except for strongly associated binary mixtures such as chloroform/acetone, which exhibit a rare negative deviation from Raoult's law, infinite-dilution binary diffusivities, $(D)_\infty$, can be combined with mixture activity-coefficient data or correlations thereof to predict liquid binary diffusion coefficients D_{AB} and D_{BA} over the entire composition range. The Vignes equations are:

$$D_{AB} = (D_{AB})_\infty^{x_B}(D_{BA})_\infty^{x_A}\left(1 + \frac{\partial \ln \gamma_A}{\partial \ln x_A}\right)_{T,P} \qquad (3\text{-}45)$$

$$D_{BA} = (D_{BA})_\infty^{x_A}(D_{AB})_\infty^{x_B}\left(1 + \frac{\partial \ln \gamma_B}{\partial \ln x_B}\right)_{T,P} \qquad (3\text{-}46)$$

EXAMPLE 3.7

At 298 K and 1 atm, infinite-dilution diffusion coefficients for the methanol (A)/water (B) system are 1.5×10^{-5} cm²/s and 1.75×10^{-5} cm²/s for AB and BA, respectively.

Activity-coefficient data for the same conditions as estimated from the UNIFAC method are as follows:

x_A	γ_A	x_B	γ_B
0.0	2.245	1.0	1.000
0.1	1.748	0.9	1.013
0.2	1.470	0.8	1.044
0.3	1.300	0.7	1.087
0.4	1.189	0.6	1.140
0.5	1.116	0.5	1.201
0.6	1.066	0.4	1.269
0.7	1.034	0.3	1.343
0.8	1.014	0.2	1.424
0.9	1.003	0.1	1.511
1.0	1.000	0.0	1.605

Use the Vignes equations to estimate diffusion coefficients over the entire composition range.

SOLUTION

Using a spreadsheet to compute the derivatives in (3-45) and (3-46), which are found to be essentially equal at any composition, and the diffusivities from the same equations, the following results are obtained with $D_{AB} = D_{BA}$ at each composition. The calculations show a minimum diffusivity at a methanol mole fraction of 0.30.

x_A	D_{AB}, cm²/s	D_{BA}, cm²/s
0.20	1.10×10^{-5}	1.10×10^{-5}
0.30	1.08×10^{-5}	1.08×10^{-5}
0.40	1.12×10^{-5}	1.12×10^{-5}
0.50	1.18×10^{-5}	1.18×10^{-5}
0.60	1.28×10^{-5}	1.28×10^{-5}
0.70	1.38×10^{-5}	1.38×10^{-5}
0.80	1.50×10^{-5}	1.50×10^{-5}

If the diffusivity is assumed linear with mole fraction, the value at $x_A = 0.50$ is 1.625×10^{-5}, which is almost 40% higher than the predicted value of 1.18×10^{-5}.

Diffusivities of Electrolytes

In an electrolyte solute, the diffusion coefficient of the dissolved salt, acid, or base depends on the ions, since they are the diffusing entities. However, in the absence of an electric potential, only the molecular diffusion of the electrolyte molecule is of interest. The infinite-dilution diffusivity of a single salt in an aqueous solution in cm²/s can be estimated from the Nernst–Haskell equation:

$$(D_{AB})_\infty = \frac{RT[(1/n_+) + (1/n_-)]}{F^2[(1/\lambda_+) + (1/\lambda_-)]} \qquad (3\text{-}47)$$

where

n_+ and n_- = valences of the cation and anion, respectively

λ_+ and λ_- = limiting ionic conductances in (A/cm²) (V/cm)(g-equiv/cm³), where A = amps and V = volts

F = Faraday's constant
= 96,500 coulombs/g-equiv

T = temperature, K

R = gas constant = 8.314 J/mol-K

Values of λ_+ and λ_- at 25°C are listed in Table 3.7. At other temperatures, these values are multiplied by $T/334\mu_B$,

Table 3.7   Limiting Ionic Conductances in Water at 25°C, in $(A/cm^2)(V/cm)(g\text{-}equiv/cm^3)$

Anion	λ_-	Cation	λ_+
OH^-	197.6	H^+	349.8
Cl^-	76.3	Li^+	38.7
Br^-	78.3	Na^+	50.1
I^-	76.8	K^+	73.5
NO_3^-	71.4	NH_4^+	73.4
ClO_4^-	68.0	Ag^+	61.9
HCO_3^-	44.5	Tl^+	74.7
HCO_2^-	54.6	$(\frac{1}{2})Mg^{2+}$	53.1
$CH_3CO_2^-$	40.9	$(\frac{1}{2})Ca^{2+}$	59.5
$ClCH_2CO_2^-$	39.8	$(\frac{1}{2})Sr^{2+}$	50.5
$CNCH_2CO_2^-$	41.8	$(\frac{1}{2})Ba^{2+}$	63.6
$CH_3CH_2CO_2^-$	35.8	$(\frac{1}{2})Cu^{2+}$	54
$CH_3(CH_2)_2CO_2^-$	32.6	$(\frac{1}{2})Zn^{2+}$	53
$C_6H_5CO_2^-$	32.3	$(\frac{1}{3})La^{3+}$	69.5
$HC_2O_4^-$	40.2	$(\frac{1}{3})Co(NH_3)_6^{3+}$	102
$(\frac{1}{2})C_2O_4^{2-}$	74.2		
$(\frac{1}{2})SO_4^{2-}$	80		
$(\frac{1}{3})Fe(CN)_6^{3-}$	101		
$(\frac{1}{4})Fe(CN)_6^{4-}$	111		

Source: Poling, Prausnitz, and O'Connell [2].

where T and μ_B are in kelvins and centipoise, respectively. As the concentration of the electrolyte increases, the diffusivity at first decreases rapidly by about 10% to 20% and then rises to values at a concentration of 2 N (normal) that approximate the infinite-dilution value. Some representative experimental values from Volume V of the International Critical Tables are given in Table 3.8.

Table 3.8   Experimental Diffusivities of Electrolytes in Aqueous Solutions

Solute	Concentration, Mol/L	Temperature, °C	Diffusivity, D_{AB}, $cm^2/s \times 10^5$
HCl	0.1	12	2.29
HNO_3	0.05	20	2.62
	0.25	20	2.59
H_2SO_4	0.25	20	1.63
KOH	0.01	18	2.20
	0.1	18	2.15
	1.8	18	2.19
NaOH	0.05	15	1.49
NaCl	0.4	18	1.17
	0.8	18	1.19
	2.0	18	1.23
KCl	0.4	18	1.46
	0.8	18	1.49
	2.0	18	1.58
$MgSO_4$	0.4	10	0.39
$Ca(NO_3)_2$	0.14	14	0.85

EXAMPLE 3.8

Estimate the diffusivity of KCl in a dilute solution of water at 18.5°C. The experimental value is 1.7×10^{-5} cm²/s. At concentrations up to 2N, this value varies only from 1.5×10^{-5} to 1.75×10^{-5} cm²/s.

SOLUTION

At 18.5°C, $T/334\mu = 291.7/[(334)(1.05)] = 0.832$. Using Table 3.7, at 25°C, the corrected limiting ionic conductances are

$$\lambda_+ = 73.5(0.832) = 61.2 \quad \text{and} \quad \lambda_- = 76.3(0.832) = 63.5$$

From (3-47),

$$D_\infty = \frac{(8.314)(291.7)[(1/1) + (1/1)]}{96,500^2[(1/61.2) + (1/63.5)]} = 1.62 \times 10^{-5} cm^2/s$$

which is close to the experimental value.

Diffusivity of Biological Solutes in Liquids

For dilute, aqueous, nonelectrolyte solutions, the Wilke–Chang equation (3-39) can be used for small solute molecules of liquid molar volumes up to 500 cm³/mol, which corresponds to molecular weights to almost 600. In biological applications, diffusivities of water-soluble protein macromolecules having molecular weights greater than 1,000 are of interest. In general, molecules with molecular weights to 500,000 have diffusivities at 25°C that range from 1×10^{-7} to 8×10^{-7} cm²/s, which is two orders of magnitude smaller than values of diffusivity for molecules with molecular weights less than 1,000. Data for many globular and fibrous protein macromolecules are tabulated by Sorber [10] with a few diffusivities given in Table 3.9. In the absence of data, the following semiempirical equation given by Geankoplis [11] and patterned after the Stokes–Einstein equation can be used:

$$D_{AB} = \frac{9.4 \times 10^{-15} T}{\mu_B(M_A)^{1/3}} \tag{3-48}$$

where the units are those of (3-39).

Also of interest in biological applications are diffusivities of small, nonelectrolyte molecules in aqueous gels containing up to 10 wt% of molecules such as certain polysaccharides (agar), which have a great tendency to swell. Diffusivities are given by Friedman and Kraemer [12]. In general, the diffusivities of small solute molecules in gels are not less than 50% of the values for the diffusivity of the solute in water, with values decreasing with increasing weight percent of gel.

Diffusivity in Solids

Diffusion in solids takes place by different mechanisms depending on the diffusing atom, molecule, or ion; the nature of the solid structure, whether it be porous or nonporous,

Table 3.9 Experimental Diffusivities of Large Biological Proteins in Aqueous Solutions

Protein	MW	Configuration	Temperature, °C	Diffusivity, D_{AB}, $cm^2/s \times 10^5$
Bovine serum albumin	67,500	globular	25	0.0681
γ–Globulin, human	153,100	globular	20	0.0400
Soybean protein	361,800	globular	20	0.0291
Urease	482,700	globular	25	0.0401
Fibrinogen, human	339,700	fibrous	20	0.0198
Lipoxidase	97,440	fibrous	20	0.0559

crystalline, or amorphous; and the type of solid material, whether it be metallic, ceramic, polymeric, biological, or cellular. Crystalline materials may be further classified according to the type of bonding, as molecular, covalent, ionic, or metallic, with most inorganic solids being ionic. However, ceramic materials can be ionic, covalent, or most often a combination of the two. Molecular solids have relatively weak forces of attraction among the atoms or molecules. In covalent solids, such as quartz silica, two atoms share two or more electrons equally. In ionic solids, such as inorganic salts, one atom loses one or more of its electrons by transfer to one or more other atoms, thus forming ions. In metals, positively charged ions are bonded through a field of electrons that are free to move. Unlike diffusion coefficients in gases and low-molecular-weight liquids, which each cover a range of only one or two orders of magnitude, diffusion coefficients in solids cover a range of many orders of magnitude. Despite the great complexity of diffusion in solids, Fick's first law can be used to describe diffusion if a measured diffusivity is available. However, when the diffusing solute is a gas, its solubility in the solid must also be known. If the gas dissociates upon dissolution in the solid, the concentration of the dissociated species must be used in Fick's law. In this section, many of the mechanisms of diffusion in solids are mentioned, but because they are exceedingly complex to quantify, the mechanisms are considered only qualitatively. Examples of diffusion in solids are considered, together with measured diffusion coefficients that can be used with Fick's first law. Emphasis is on diffusion of gas and liquid solutes through or into the solid, but movement of the atoms, molecules, or ions of the solid through itself is also considered.

Porous Solids

When solids are porous, predictions of the diffusivity of gaseous and liquid solute species in the pores can be made. These methods are considered only briefly here, with details deferred to Chapters 14, 15, and 16, where applications are made to membrane separations, adsorption, and leaching. This type of diffusion is also of great importance in the analysis and design of reactors using porous solid catalysts. It is sufficient to mention here that any of the following four mass-transfer mechanisms or combinations thereof may take place:

1. Ordinary molecular diffusion through pores, which present tortuous paths and hinder the movement of large molecules when their diameter is more than 10% of the pore diameter

2. Knudsen diffusion, which involves collisions of diffusing gaseous molecules with the pore walls when the pore diameter and pressure are such that the molecular mean free path is large compared to the pore diameter

3. Surface diffusion involving the jumping of molecules, adsorbed on the pore walls, from one adsorption site to another based on a surface concentration-driving force

4. Bulk flow through or into the pores

When treating diffusion of solutes in porous materials where diffusion is considered to occur only in the fluid in the pores, it is common to refer to an effective diffusivity, D_{eff}, which is based on (1) the total cross-sectional area of the porous solid rather than the cross-sectional area of the pore and (2) on a straight path, rather than the pore path, which may be tortuous. If pore diffusion occurs only by ordinary molecular diffusion, Fick's law (3-3) can be used with an effective diffusivity. The effective diffusivity for a binary mixture can be expressed in terms of the ordinary diffusion coefficient, D_{AB}, by

$$D_{eff} = \frac{D_{AB}\epsilon}{\tau} \qquad (3\text{-}49)$$

where ϵ is the fractional porosity (typically 0.5) of the solid and τ is the pore-path tortuosity (typically 2 to 3), which is the ratio of the pore length to the length if the pore were straight in the direction of diffusion. The effective diffusivity is either determined experimentally, without knowledge of the porosity or tortuosity, or predicted from (3-49) based on measurement of the porosity and tortuosity and use of the predictive methods for ordinary molecular diffusivity. As an example of the former, Boucher, Brier, and Osburn [13] measured effective diffusivities for the leaching of processed soybean oil (viscosity = 20.1 cP at 120°F) from 1/16-in.-thick porous clay plates with liquid tetrachloroethylene solvent. The rate of extraction was controlled by the rate of diffusion of the soybean oil in the clay plates. The measured value of

D_{eff} was 1.0×10^{-6} cm^2/s. As might be expected from the effects of porosity and tortuosity, the effective value is about one order of magnitude less than the expected ordinary molecular diffusivity, D, of oil in the solvent.

Crystalline Solids

Diffusion through nonporous crystalline solids depends markedly on the crystal lattice structure and the diffusing entity. As discussed in Chapter 17 on crystallization, only seven different lattice structures are possible. For the cubic lattice (simple, body-centered, and face-centered), the diffusivity is the same in all directions (isotropic). In the six other lattice structures (including hexagonal and tetragonal), the diffusivity can be different in different directions (anisotropic). Many metals, including Ag, Al, Au, Cu, Ni, Pb, and Pt, crystallize into the face-centered cubic lattice structure. Others, including Be, Mg, Ti, and Zn, form anisotropic, hexagonal structures. The mechanisms of diffusion in crystalline solids include:

1. Direct exchange of lattice position by two atoms or ions, probably by a ring rotation involving three or more atoms or ions

2. Migration by small solutes through interlattice spaces called interstitial sites

3. Migration to a vacant site in the lattice

4. Migration along lattice imperfections (dislocations), or gain boundaries (crystal interfaces)

Diffusion coefficients associated with the first three mechanisms can vary widely and are almost always at least one order of magnitude smaller than diffusion coefficients in low-viscosity liquids. As might be expected, diffusion by the fourth mechanism can be faster than by the other three mechanisms. Typical experimental diffusivity values, taken mainly from Barrer [14], are given in Table 3.10. The diffusivities cover gaseous, ionic, and metallic solutes. The values cover an enormous 26-fold range. Temperature effects can be extremely large.

Metals

Important practical applications exist for diffusion of light gases through metals. To diffuse through a metal, a gas must first dissolve in the metal. As discussed by Barrer [14], all light gases do not dissolve in all metals. For example, hydrogen dissolves in such metals as Cu, Al, Ti, Ta, Cr, W, Fe, Ni, Pt, and Pd, but not in Au, Zn, Sb, and Rh. Nitrogen dissolves in Zr, but not in Cu, Ag, or Au. The noble gases do not dissolve in any of the common metals. When H_2, N_2, and O_2 dissolve in metals, they dissociate and may react to form hydrides, nitrides, and oxides, respectively. More complex molecules such as ammonia, carbon dioxide, carbon monoxide, and sulfur dioxide also dissociate. The following example illustrates how pressurized hydrogen gas can slowly leak through the wall of a small, thin pressure vessel.

Table 3.10 Diffusivities of Solutes in Crystalline Metals and Salts

Metal/Salt	Solute	T, °C	D, cm^2/s
Ag	Au	760	3.6×10^{-10}
	Sb	20	3.5×10^{-21}
	Sb	760	1.4×10^{-9}
Al	Fe	359	6.2×10^{-14}
	Zn	500	2×10^{-9}
	Ag	50	1.2×10^{-9}
Cu	Al	20	1.3×10^{-30}
	Al	850	2.2×10^{-9}
	Au	750	2.1×10^{-11}
Fe	H_2	10	1.66×10^{-9}
	H_2	100	1.24×10^{-7}
	C	800	1.5×10^{-8}
Ni	H_2	85	1.16×10^{-8}
	H_2	165	1.05×10^{-7}
	CO	950	4×10^{-8}
W	U	1727	1.3×10^{-11}
AgCl	Ag^+	150	2.5×10^{-14}
	Ag^+	350	7.1×10^{-8}
	Cl^-	350	3.2×10^{-16}
KBr	H_2	600	5.5×10^{-4}
	Br_2	600	2.64×10^{-4}

EXAMPLE 3.9

Gaseous hydrogen at 200 psia and 300°C is stored in a small, 10-cm-diameter, steel pressure vessel having a wall thickness of 0.125 in. The solubility of hydrogen in steel, which is proportional to the square root of the hydrogen partial pressure in the gas, is equal to 3.8×10^{-6} mol/cm^3 at 14.7 psia and 300°C. The diffusivity of hydrogen in steel at 300°C is 5×10^{-6} cm^2/s. If the inner surface of the vessel wall remains saturated at the existing hydrogen partial pressure and the hydrogen partial pressure at the outer surface is zero, estimate the time, in hours, for the pressure in the vessel to decrease to 100 psia because of hydrogen loss by dissolving in and diffusing through the metal wall.

SOLUTION

Integrating Fick's first law, (3-3), where A is H_2 and B is the metal, assuming a linear concentration gradient, and equating the flux to the loss of hydrogen in the vessel,

$$-\frac{dn_A}{dt} = \frac{D_{AB} A \Delta c_A}{\Delta z} \qquad (1)$$

Because $p_A = 0$ outside the vessel, $\Delta c_A = c_A =$ solubility of A at the inside wall surface in mol/cm^3 and $c_A = 3.8 \times 10^{-6} \left(\frac{p_A}{14.7}\right)^{0.5}$, where p_A is the pressure of A in psia inside the vessel. Let p_{A_o} and n_{A_o} be the initial pressure and moles of A, respectively, in the vessel. Assuming the ideal-gas law and isothermal conditions,

$$n_A = n_{A_o} p_A / p_{A_o} \qquad (2)$$

Differentiating (2) with respect to time,

$$\frac{dn_A}{dt} = \frac{n_{A_o}}{p_{A_o}}\frac{dp_A}{dt} \tag{3}$$

Combining (1) and (3),

$$\frac{dp_A}{dt} = -\frac{D_A A (3.8 \times 10^{-6}) p_A^{0.5} p_{A_o}}{n_{A_o} \Delta z (14.7)^{0.5}} \tag{4}$$

Integrating and solving for t,

$$t = \frac{2 n_{A_o} \Delta z (14.7)^{0.5}}{3.8 \times 10^{-6} D_A A p_{A_o}} \left(p_{A_o}^{0.5} - p_A^{0.5} \right)$$

Assuming the ideal-gas law,

$$n_{A_o} = \frac{(200/14.7)[(3.14 \times 10^3)/6)]}{82.05(300+273)} = 0.1515 \text{ mol}$$

The mean-spherical shell area for mass transfer, A, is

$$A = \frac{3.14}{2}[(10)^2 + (10.635)^2] = 336 \text{ cm}^2$$

The time for the pressure to drop to 100 psia is

$$t = \frac{2(0.1515)(0.125 \times 2.54)(14.7)^{0.5}}{3.8 \times 10^{-6}(5 \times 10^{-6})(336)(200)}(200^{0.5} - 100^{0.5})$$

$$= 1.2 \times 10^6 \text{ s} = 332 \text{ h}$$

Silica and Glass

Another area of great interest is the diffusion of light gases through various forms of silica, whose two elements, Si and O, make up about 60% of the earth's crust. Solid silica can exist in three principal crystalline forms (quartz, tridymite, and cristobalite) and in various stable amorphous forms, including vitreous silica (a noncrystalline silicate glass or fused quartz). Table 3.11 includes diffusivities, D, and solubilities as Henry's law constants, H, at 1 atm for helium and hydrogen in fused quartz as calculated from correlations of experimental data by Swets, Lee, and Frank [15] and Lee [16], respectively. The product of the diffusivity and the solubility is called the permeability, P_M. Thus,

$$P_M = DH \tag{3-50}$$

Unlike metals, where hydrogen usually diffuses as the atom, hydrogen apparently diffuses as a molecule in glass.

Table 3.11 Diffusivities and Solubilities of Gases in Amorphous Silica at 1 atm

Gas	Temp, C	Diffusivity, cm²/s	Solubility mol/cm³-atm
He	24	2.39×10^{-8}	1.04×10^{-7}
	300	2.26×10^{-6}	1.82×10^{-7}
	500	9.99×10^{-6}	9.9×10^{-8}
	1,000	5.42×10^{-5}	1.34×10^{-7}
H₂	300	6.11×10^{-8}	3.2×10^{-14}
	500	6.49×10^{-7}	2.48×10^{-13}
	1,000	9.26×10^{-6}	2.49×10^{-12}
O₂	1,000	6.25×10^{-9} (molecular)	
	1,000	9.43×10^{-15} (network)	

For both hydrogen and helium, diffusivities increase rapidly with increasing temperature. At ambient temperature the diffusivities are three orders of magnitude lower than in liquids. At elevated temperatures the diffusivities approach those observed in liquids. Solubilities vary only slowly with temperature. Hydrogen is orders of magnitude less soluble in glass than helium. For hydrogen, the diffusivity is somewhat lower than in metals. Diffusivities for oxygen are also included in Table 3.11 from studies by Williams [17] and Sucov [18]. At 1000°C, the two values differ widely because, as discussed by Kingery, Bowen, and Uhlmann [19], in the former case, transport occurs by molecular diffusion; while in the latter case, transport is by slower network diffusion as oxygen jumps from one position in the silicate network to another. The activation energy for the latter is much larger than for the former (71,000 cal/mol versus 27,000 cal/mol). The choice of glass can be very critical in high-vacuum operations because of the wide range of diffusivity.

Ceramics

Diffusion rates of light gases and elements in crystalline ceramics are very important because diffusion must precede chemical reactions and causes changes in the microstructure. Therefore, diffusion in ceramics has been the subject of numerous studies, many of which are summarized in Figure 3.4, taken from Kingery et al. [19], where diffusivity is plotted as a function of the inverse of temperature in the high-temperature range. In this form, the slopes of the curves are proportional to the activation energy for diffusion, E, where

$$D = D_o \exp\left(-\frac{E}{RT}\right) \tag{3-51}$$

An insert at the middle-right region of Figure 3.4 relates the slopes of the curves to activation energy. The diffusivity curves cover a ninefold range from 10^{-6} to 10^{-15} cm²/s, with the largest values corresponding to the diffusion of potassium in β-Al_2O_3 and one of the smallest values for carbon in graphite. In general, the lower the diffusivity, the higher is the activation energy. As discussed in detail by Kingery et al. [19], diffusion in crystalline oxides depends not only on temperature but also on whether the oxide is stoichiometric or not (e.g., FeO and $Fe_{0.95}O$) and on impurities. Diffusion through vacant sites of nonstoichiometric oxides is often classified as metal-deficient or oxygen-deficient. Impurities can hinder diffusion by filling vacant lattice or interstitial sites.

Polymers

Thin, dense, nonporous polymer membranes are widely used to separate gas and liquid mixtures. As discussed in detail in Chapter 14, diffusion of gas and liquid species through polymers is highly dependent on the type of polymer, whether it be crystalline or amorphous and, if the latter, glassy or rubbery. Commercial crystalline polymers are

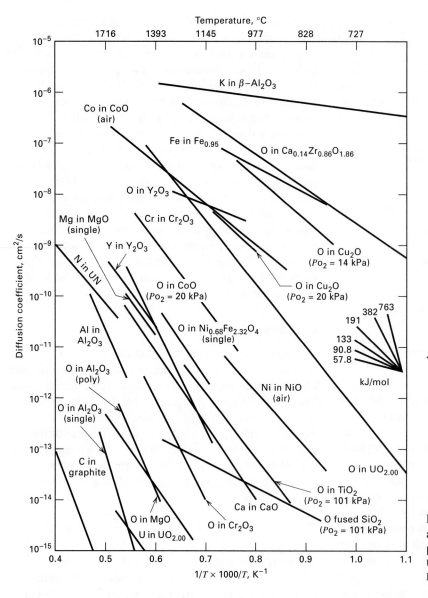

Figure 3.4 Diffusion coefficients for single- and polycrystalline ceramics.

[From W.D. Kingery, H.K. Bowen, and D.R. Uhlmann, *Introduction to Ceramics,* 2nd ed., Wiley Interscience, New York (1976) with permission.]

about 20% amorphous. It is mainly through the amorphous regions that diffusion occurs. As with the transport of gases through metals, transport of gaseous species through polymer membranes is usually characterized by the solution-diffusion mechanism of (3-50). Fick's first law, in the following integrated forms, is then applied to compute the mass transfer flux.

Gas species:

$$N_i = \frac{H_i D_i}{z_2 - z_1}(p_{i_1} - p_{i_2}) = \frac{P_{M_i}}{z_2 - z_1}(p_{i_1} - p_{i_2}) \quad (3\text{-}52)$$

where p_i is the partial pressure of the gas species at a polymer surface.

Liquid species:

$$N_i = \frac{K_i D_i}{z_2 - z_1}(c_{i_1} - c_{i_2}) \quad (3\text{-}53)$$

where K_i, the equilibrium partition coefficient, is equal to the ratio of the concentration in the polymer to the concentration, c_i, in the liquid adjacent to the polymer surface. The product $K_i D_i$ is the liquid permeability.

Values of diffusivity for light gases in four polymers, given in Table 14.6, range from 1.3×10^{-9} to 1.6×10^{-6} cm²/s, which is orders of magnitude less than for diffusion of the same species in a gas.

Diffusivities of liquids in rubbery polymers have been studied extensively as a means of determining viscoelastic parameters. In Table 3.12, taken from Ferry [20], diffusivities are given for different solutes in seven different rubber polymers at near-ambient conditions. The values cover a sixfold range, with the largest diffusivity being that for *n*-hexadecane in polydimethylsiloxane. The smallest diffusivities correspond to the case where the temperature is approaching the glass-transition temperature, where the polymer becomes glassy in structure. This more rigid structure hinders diffusion. In general, as would be expected,

Table 3.12 Diffusivities of Solutes in Rubbery Polymers

Polymer	Solute	Temperature, K	Diffusivity, cm^2/s
Polyisobutylene	n-Butane	298	1.19×10^{-9}
	i-Butane	298	5.3×10^{-10}
	n-Pentane	298	1.08×10^{-9}
	n-Hexadecane	298	6.08×10^{-10}
Hevea rubber	n-Butane	303	2.3×10^{-7}
	i-Butane	303	1.52×10^{-7}
	n-Pentane	303	2.3×10^{-7}
	n-Hexadecane	298	7.66×10^{-8}
Polymethylacrylate	Ethyl alcohol	323	2.18×10^{-10}
Polyvinylacetate	n-Propyl alcohol	313	1.11×10^{-12}
	n-Propyl chloride	313	1.34×10^{-12}
	Ethyl chloride	343	2.01×10^{-9}
	Ethyl bromide	343	1.11×10^{-9}
Polydimethylsiloxane	n-Hexadecane	298	1.6×10^{-6}
1,4-Polybutadiene	n-Hexadecane	298	2.21×10^{-7}
Styrene-butadiene rubber	n-Hexadecane	298	2.66×10^{-8}

smaller molecules have higher diffusivities. A more detailed study of the diffusivity of n-hexadecane in random styrene/butadiene copolymers at 25°C by Rhee and Ferry [21] shows a large effect on diffusivity of fractional free volume in the polymer.

Diffusion and permeability in crystalline polymers depend on the degree of crystallinity. Polymers that are 100% crystalline permit little or no diffusion of gases and liquids. For example, the diffusivity of methane at 25°C in polyoxyethylene oxyisophthaloyl decreases from 0.30×10^{-9} to 0.13×10^{-9} cm^2/s when the degree of crystallinity increases from 0 (totally amorphous) to 40% [22]. A measure of crystallinity is the polymer density. The diffusivity of methane at 25°C in polyethylene decreases from 0.193×10^{-6} to 0.057×10^{-6} cm^2/s when the specific gravity increases from 0.914 (low density) to 0.964 (high density) [22]. A plasticizer can cause the diffusivity to increase. For example, when polyvinylchloride is plasticized with 40% tricresyl triphosphate, the diffusivity of CO at 27°C increases from 0.23×10^{-8} to 2.9×10^{-8} cm^2/s [22].

EXAMPLE 3.10

Hydrogen diffuses through a nonporous polyvinyltrimethylsilane membrane at 25°C. The pressures on the sides of the membrane are 3.5 MPa and 200 kPa. Diffusivity and solubility data are given in Table 14.9. If the hydrogen flux is to be 0.64 kmol/m^2-h, how thick in micrometers should the membrane be?

SOLUTION

Equation (3-52) applies. From Table 14.9,

$$D = 160 \times 10^{-11} \text{ m}^2/\text{s} \quad H = S = 0.54 \times 10^{-4} \text{ mol/m}^3\text{-Pa}$$

From (3-50),

$$P_M = DH = (160 \times 10^{-11})(0.54 \times 10^{-4})$$
$$= 86.4 \times 10^{-15} \text{ mol/m-s-Pa}$$
$$p_1 = 3.5 \times 10^6 \text{ Pa} \quad p_2 = 0.2 \times 10^6 \text{ Pa}$$

Membrane thickness $= z_2 - z_1 = \Delta z = P_M(p_1 - p_2)/N$

$$\Delta z = \frac{86.4 \times 10^{-15}(3.5 \times 10^6 - 0.2 \times 10^6)}{[0.64(1000)/3600]}$$
$$= 1.6 \times 10^{-6} \text{ m} = 1.6 \text{ μm}$$

As discussed in Chapter 14, polymer membranes must be very thin to achieve reasonable gas permeation rates.

Cellular Solids and Wood

As discussed by Gibson and Ashby [23], cellular solids consist of solid struts or plates that form edges and faces of cells, which are compartments or enclosed spaces. Cellular solids such as wood, cork, sponge, and coral exist in nature. Synthetic cellular structures include honeycombs, and foams (some with open cells) made from polymers, metals, ceramics, and glass. The word *cellulose* means "full of little cells."

A widely used cellular solid is wood, whose annual world production of the order of 10^{12} kg is comparable to the production of iron and steel. Chemically, wood consists of lignin, cellulose, hemicellulose, and minor amounts of organic chemicals and elements. The latter are extractable, and the former three, which are all polymers, give wood its structure. Green wood also contains up to 25 wt% moisture in the cell walls and cell cavities. Adsorption or desorption of moisture in wood causes anisotropic swelling and shrinkage.

The structure of wood, which often consists of (1) highly elongated hexagonal or rectangular cells, called tracheids in softwood (coniferous species, e.g., spruce, pine, and fir) and fibers in hardwood (deciduous or broad-leaf species, e.g., oak, birch, and walnut); (2) radial arrays of rectangular-like cells, called rays, which are narrow and short in softwoods but wide and long in hardwoods; and (3) enlarged cells with large pore spaces and thin walls, called sap channels because they conduct fluids up the tree. The sap channels are less than 3 vol% of softwood, but as much as 55 vol% of hardwood.

Because the structure of wood is directional, many of its properties are anisotropic. For example, stiffness and strength are 2 to 20 times greater in the axial direction of the tracheids or fibers than in the radial and tangential directions of the trunk from which the wood is cut. This anisotropy extends to permeability and diffusivity of wood penetrants, such as moisture and preservatives. According to Stamm [24], the permeability of wood to liquids in the axial direction can be up to 10 times greater than in the transverse direction.

Movement of liquids and gases through wood and wood products takes time during drying and treatment with preservatives, fire retardants, and other chemicals. This movement takes place by capillarity, pressure permeability, and diffusion. Nevertheless, wood is not highly permeable because the cell voids are largely discrete and lack direct interconnections. Instead, communication among cells is through circular openings spanned by thin membranes with submicrometer-sized pores, called pits, and to a smaller extent, across the cell walls. Rays give wood some permeability in the radial direction. Sap channels do not contribute to permeability. All three mechanisms of movement of gases and liquids in wood are considered by Stamm [24]. Only diffusion is discussed here.

The simplest form of diffusion is that of a water-soluble solute through wood saturated with water, such that no dimensional changes occur. For the diffusion of urea, glycerine, and lactic acid into hardwood, Stamm [24] lists diffusivities in the axial direction that are about 50% of ordinary liquid diffusivities. In the radial direction, diffusivities are about 10% of the values in the axial direction. For example, at 26.7°C the diffusivity of zinc sulfate in water is 5×10^{-6} cm^2/s. If loblolly pine sapwood is impregnated with zinc sulfate in the radial direction, the diffusivity is found to be 0.18×10^{-6} cm^2/s [24].

The diffusion of water in wood is more complex. Moisture content determines the degree of swelling or shrinkage. Water is held in the wood in different ways: It may be physically adsorbed on cell walls in monomolecular layers, condensed in preexisting or transient cell capillaries, or absorbed in cell walls to form a solid solution.

Because of the practical importance of lumber drying rates, most diffusion coefficients are measured under drying conditions in the radial direction across the fibers. Results depend on temperature and swollen-volume specific gravity.

Typical results are given by Sherwood [25] and Stamm [24]. For example, for beech with a swollen specific gravity of 0.4, the diffusivity increases from a value of about 1×10^{-6} cm^2/s at 10°C to 10×10^{-6} cm^2/s at 60°C.

3.3 ONE-DIMENSIONAL, STEADY-STATE AND UNSTEADY-STATE, MOLECULAR DIFFUSION THROUGH STATIONARY MEDIA

For conductive heat transfer in stationary media, Fourier's law is applied to derive equations for the rate of heat transfer for steady-state and unsteady-state conditions in shapes such as slabs, cylinders, and spheres. Many of the results are plotted in generalized charts. Analogous equations can be derived for mass transfer, using simplifying assumptions.

In one dimension, the molar rate of mass transfer of A in a binary mixture with B is given by a modification of (3-12), which includes bulk flow and diffusion:

$$n_A = x_A(n_A + n_B) - cD_{AB}A\left(\frac{dx_A}{dz}\right) \qquad (3\text{-}54)$$

If A is a dissolved solute undergoing mass transfer, but B is stationary, $n_B = 0$. It is common to assume that c is a constant and x_A is small. The bulk-flow term is then eliminated and (3-54) accounts for diffusion only, becoming Fick's first law:

$$n_A = -cD_{AB}A\left(\frac{dx_A}{dz}\right) \qquad (3\text{-}55)$$

Alternatively, (3-55) can be written in terms of concentration gradient:

$$n_A = -D_{AB}A\left(\frac{dc_A}{dz}\right) \qquad (3\text{-}56)$$

This equation is analogous to Fourier's law for the rate of heat conduction, Q:

$$Q = -kA\left(\frac{dT}{dz}\right) \qquad (3\text{-}57)$$

Steady State

For steady-state, one-dimensional diffusion, with constant D_{AB}, (3-56) can be integrated for various geometries, the most common results being analogous to heat conduction.

1. Plane wall with a thickness, $z_2 - z_1$:

$$n_A = D_{AB}A\left(\frac{c_{A_1} - c_{A_2}}{z_2 - z_1}\right) \qquad (3\text{-}58)$$

2. Hollow cylinder of inner radius r_1 and outer radius r_2, with diffusion in the radial direction outward:

$$n_A = 2\pi L \frac{D_{AB}(c_{A_1} - c_{A_2})}{\ln(r_2/r_1)} \qquad (3\text{-}59)$$

or

$$n_A = D_{AB}A_{LM}\left(\frac{c_{A_1} - c_{A_2}}{r_2 - r_1}\right) \qquad (3\text{-}60)$$

where

$$A_{LM} = \text{log mean of the areas } 2\pi r L \text{ at } r_1 \text{ and } r_2$$

$$L = \text{length of the hollow cylinder}$$

3. Spherical shell of inner radius r_1 and outer radius r_2, with diffusion in the radial direction outward:

$$n_A = \frac{4\pi r_1 r_2 D_{AB}(c_{A_1} - c_{A_2})}{r_2 - r_1} \qquad (3\text{-}61)$$

or

$$n_A = D_{AB} A_{GM} \left(\frac{c_{A_1} - c_{A_2}}{r_2 - r_1} \right) \qquad (3\text{-}62)$$

where A_{GM} = geometric mean of the areas $4\pi r^2$ at r_1 and r_2.

When $r_1/r_2 < 2$, the arithmetic mean area is no more than 4% greater than the log mean area. When $r_1/r_2 < 1.33$, the arithmetic mean area is no more than 4% greater than the geometric mean area.

Unsteady State

Equation (3-56) is applied to unsteady-state molecular diffusion by considering the accumulation or depletion of a species with time in a unit volume through which the species is diffusing. Consider the one-dimensional diffusion of species A in B through a differential control volume with diffusion in the z-direction only, as shown in Figure 3.5. Assume constant total concentration, $c = c_A + c_B$, constant diffusivity, and negligible bulk flow. The molar flow rate of species A by diffusion at the plane $z = z$ is given by (3-56):

$$n_{A_z} = -D_{AB} A \left(\frac{\partial c_A}{\partial z} \right)_z \qquad (3\text{-}63)$$

At the plane, $z = z + \Delta z$, the diffusion rate is

$$n_{A_{z+\Delta z}} = -D_{AB} A \left(\frac{\partial c_A}{\partial z} \right)_{z+\Delta z} \qquad (3\text{-}64)$$

The accumulation of species A in the control volume is

$$A \frac{\partial c_A}{\partial t} \Delta z \qquad (3\text{-}65)$$

Since rate in − rate out = accumulation,

$$-D_{AB} A \left(\frac{\partial c_A}{\partial z} \right)_z + D_{AB} A \left(\frac{\partial c_A}{\partial z} \right)_{z+\Delta z} = A \left(\frac{\partial c_A}{\partial t} \right) \Delta z \qquad (3\text{-}66)$$

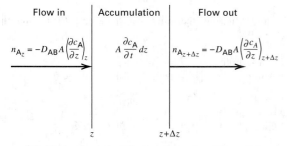

Figure 3.5 Unsteady-state diffusion through a differential volume $A \, dz$.

Rearranging and simplifying,

$$D_{AB} \left[\frac{(\partial c_A/\partial z)_{z+\Delta z} - (\partial c_A/\partial z)_z}{\Delta z} \right] = \frac{\partial c_A}{\partial t} \qquad (3\text{-}67)$$

In the limit, as $\Delta z \to 0$,

$$\frac{\partial c_A}{\partial t} = D_{AB} \frac{\partial^2 c_A}{\partial z^2} \qquad (3\text{-}68)$$

Equation (3-68) is Fick's second law for one-dimensional diffusion. The more general form, for three-dimensional rectangular coordinates, is

$$\frac{\partial c_A}{\partial t} = D_{AB} \left(\frac{\partial^2 c_A}{\partial x^2} + \frac{\partial^2 c_A}{\partial y^2} + \frac{\partial^2 c_A}{\partial z^2} \right) \qquad (3\text{-}69)$$

For one-dimensional diffusion in the radial direction only, for cylindrical and spherical coordinates, Fick's second law becomes, respectively,

$$\frac{\partial c_A}{\partial t} = \frac{D_{AB}}{r} \frac{\partial}{\partial r} \left(r \frac{\partial c_A}{\partial r} \right) \qquad (3\text{-}70)$$

and

$$\frac{\partial c_A}{\partial t} = \frac{D_{AB}}{r^2} \frac{\partial}{\partial r} \left(r^2 \frac{\partial c_A}{\partial r} \right) \qquad (3\text{-}71)$$

Equations (3-68) to (3-71) are analogous to Fourier's second law of heat conduction where c_A is replaced by temperature, T, and diffusivity, D_{AB}, is replaced by thermal diffusivity, $\alpha = k/\rho C_P$. The analogous three equations for heat conduction for constant, isotropic properties are, respectively:

$$\frac{\partial T}{\partial t} = \alpha \left(\frac{\partial^2 T}{\partial x^2} + \frac{\partial^2 T}{\partial y^2} + \frac{\partial^2 T}{\partial z^2} \right) \qquad (3\text{-}72)$$

$$\frac{\partial T}{\partial t} = \frac{\alpha}{r} \frac{\partial}{\partial r} \left(r \frac{\partial T}{\partial r} \right) \qquad (3\text{-}73)$$

$$\frac{\partial T}{\partial t} = \frac{\alpha}{r^2} \frac{\partial}{\partial r} \left(r^2 \frac{\partial T}{\partial r} \right) \qquad (3\text{-}74)$$

Analytical solutions to these partial differential equations in either Fick's law or Fourier's law form are available for a variety of boundary conditions. Many of these solutions are derived and discussed by Carslaw and Jaeger [26] and Crank [27]. Only a few of the more useful solutions are presented here.

Semi-infinite Medium

Consider the semi-infinite medium shown in Figure 3.6, which extends in the z-direction from $z = 0$ to $z = \infty$. The x and y coordinates extend from $-\infty$ to $+\infty$, but are not of interest because diffusion takes place only in the z-direction. Thus, (3-68) applies to the region $z \geq 0$. At time $t \leq 0$, the concentration is c_{A_o} for $z \geq 0$. At $t = 0$, the surface of the semi-infinite medium at $z = 0$ is instantaneously brought to the concentration $c_{A_s} > c_{A_o}$ and held there for $t > 0$. Therefore, diffusion into the medium occurs. However, because

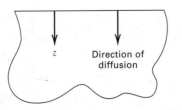

Figure 3.6 One-dimensional diffusion into a semi-infinite medium.

the medium is infinite in the z-direction, diffusion cannot extend to $z = \infty$ and, therefore, as $z \to \infty$, $c_A = c_{A_o}$ for all $t \geq 0$. Because the partial differential equation (3-68) and its one boundary (initial) condition in time and two boundary conditions in distance are linear in the dependent variable, c_A, an exact solution can be obtained. Either the method of combination of variables [28] or the Laplace transform method [29] is applicable. The result, in terms of the fractional accomplished concentration change, is

$$\theta = \frac{c_A - c_{A_o}}{c_{A_s} - c_{A_o}} = \text{erfc}\left(\frac{z}{2\sqrt{D_{AB}t}}\right) \quad (3\text{-}75)$$

where the complementary error function, erfc, is related to the error function, erf, by

$$\text{erfc}(x) = 1 - \text{erf}(x) = 1 - \frac{2}{\sqrt{\pi}}\int_0^x e^{-\eta^2}d\eta \quad (3\text{-}76)$$

The error function is included in most spreadsheet programs and handbooks, such as *Handbook of Mathematical Functions* [30]. The variation of erf(x) and erfc(x) is as follows:

x	**erf(x)**	**erfc(x)**
0	0.0000	1.0000
0.5	0.5205	0.4795
1.0	0.8427	0.1573
1.5	0.9661	0.0339
2.0	0.9953	0.0047
∞	1.0000	0.0000

Equation (3-75) is used to compute the concentration in the semi-infinite medium, as a function of time and distance from the surface, assuming no bulk flow. Thus, it applies most rigorously to diffusion in solids, and also to stagnant liquids and gases when the medium is dilute in the diffusing solute. In (3-75), when $(z/2\sqrt{D_{AB}t}) = 2$, the complementary error function is only 0.0047, which represents less than a 1% change in the ratio of the concentration change at $z = z$ to the change at $z = 0$. Thus, it is common to refer to $z = 4\sqrt{D_{AB}t}$ as the penetration depth and to apply (3-75) to media of finite thickness as long as the thickness is greater than the penetration depth.

The instantaneous rate of mass transfer across the surface of the medium at $z = 0$ can be obtained by taking the derivative of (3-75) with respect to distance and substituting it into Fick's first law applied at the surface of the medium.

Thus, using the Leibnitz rule for differentiating the integral of (3-76), with $x = z/2\sqrt{D_{AB}t}$,

$$
\begin{aligned}
n_A &= -D_{AB}A\left(\frac{\partial c_A}{\partial z}\right)_{z=0} \\
&= D_{AB}A\left(\frac{c_{A_s} - c_{A_o}}{\sqrt{\pi D_{AB}t}}\right)\exp\left(-\frac{z^2}{4D_{AB}t}\right)\bigg|_{z=0}
\end{aligned} \quad (3\text{-}77)
$$

Thus,

$$n_A|_{z=0} = \sqrt{\frac{D_{AB}}{\pi t}}A(c_{A_s} - c_{A_o}) \quad (3\text{-}78)$$

We can also determine the total number of moles of solute, $\mathcal{N}_A$, transferred into the semi-infinite medium by integrating (3-78) with respect to time:

$$
\begin{aligned}
\mathcal{N}_A &= \int_o^t n_A|_{z=0}\,dt = \sqrt{\frac{D_{AB}}{\pi}}A(c_{A_s} - c_{A_o})\int_o^t \frac{dt}{\sqrt{t}} \\
&= 2A(c_{A_s} - c_{A_o})\sqrt{\frac{D_{AB}t}{\pi}}
\end{aligned} \quad (3\text{-}79)
$$

EXAMPLE 3.11

Determine how long it will take for the dimensionless concentration change, $\theta = (c_A - c_{A_o})/(c_{A_s} - c_{A_o})$, to reach 0.01 at a depth $z = 1$ m in a semi-infinite medium, which is initially at a solute concentration c_{A_o}, after the surface concentration at $z = 0$ increases to c_{A_s}, for diffusivities representative of a solute diffusing through a stagnant gas, a stagnant liquid, and a solid.

SOLUTION

For a gas, assume $D_{AB} = 0.1$ cm^2/s. We know that $z = 1$ m $= 100$ cm. From (3-75) and (3-76),

$$\theta = 0.01 = 1 - \text{erf}\left(\frac{z}{2\sqrt{D_{AB}t}}\right)$$

Therefore,

$$\text{erf}\left(\frac{z}{2\sqrt{D_{AB}t}}\right) = 0.99$$

From tables of the error function,

$$\left(\frac{z}{2\sqrt{D_{AB}t}}\right) = 1.8214$$

Solving,

$$t = \left[\frac{100}{1.8214(2)}\right]^2\frac{1}{0.10} = 7{,}540\text{ s} = 2.09\text{ h}$$

In a similar manner, the times for typical gas, liquid, and solid media are:

Semi-infinite Medium	D_{AB}, cm^2/s	Time for $\theta = 0.01$ at 1 m
Gas	0.10	2.09 h
Liquid	1×10^{-5}	2.39 year
Solid	1×10^{-9}	239 centuries

These results show that molecular diffusion is very slow, especially in liquids and solids. In liquids and gases, the rate of mass transfer can be greatly increased by agitation to induce turbulent motion. For solids, it is best to reduce the diffusion path to as small a dimension as possible by reducing the size of the solid.

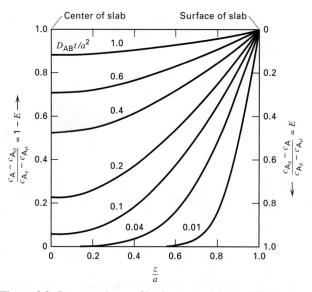

Figure 3.8 Concentration profiles for unsteady-state diffusion in a slab.

[Adapted from H.S. Carslaw and J.C. Jaeger, *Conduction of Heat in Solids,* 2nd ed., Oxford University Press, London (1959).]

Medium of Finite Thickness with Sealed Edges

Consider a rectangular, parallelepiped medium of finite thickness $2a$ in the z-direction, and either infinitely long dimensions in the y- and x-directions or finite lengths of $2b$ and $2c$, respectively, in those directions. Assume that in Figure 3.7a the edges parallel to the z-direction are sealed, so diffusion occurs only in the z-direction and initially the concentration of the solute in the medium is uniform at c_{A_o}. At time $t = 0$, the two unsealed surfaces of the medium at $z = \pm a$ are brought to and held at concentration $c_{A_s} > c_{A_o}$. Because of symmetry, $\partial c_A/\partial z = 0$ at $z = 0$. Assume constant D_{AB}. Again (3-68) applies, and an exact solution can be obtained because both (3-68) and the boundary conditions are linear in c_A. By the method of separation of variables [28] or the Laplace transform method [29], the result from Carslaw and Jaeger [26], in terms of the fractional, unaccomplished concentration change, E, is

$$E = 1 - \theta = \frac{c_{A_s} - c_A}{c_{A_s} - c_{A_o}} = \frac{4}{\pi} \sum_{n=0}^{\infty} \frac{(-1)^n}{(2n+1)}$$
$$\times \exp[-D_{AB}(2n+1)^2 \pi^2 t/4a^2] \cos \frac{(2n+1)\pi z}{2a} \tag{3-80}$$

or, in terms of the complementary error function,

$$E = 1 - \theta = \frac{c_{A_s} - c_A}{c_{A_s} - c_{A_o}} = \sum_{n=0}^{\infty} (-1)^n$$
$$\times \left[\operatorname{erfc}\frac{(2n+1)a - z}{2\sqrt{D_{AB}t}} + \operatorname{erfc}\frac{(2n+1)a + z}{2\sqrt{D_{AB}t}} \right] \tag{3-81}$$

For large values of $D_{AB}t/a^2$, which is referred to as the Fourier number for mass transfer, the infinite series solutions of (3-80) and (3-81) converge rapidly, but for small values

(e.g., short times), they do not. However, in the latter case, the solution for the semi-infinite medium applies for $D_{AB}t/a^2 < \frac{1}{16}$. A convenient plot of the exact solution is given in Figure 3.8.

The instantaneous rate of mass transfer across the surface of either unsealed face of the medium (i.e., at $z = \pm a$) is obtained by differentiating (3-80) with respect to z, evaluating the result at $z = a$, followed by substitution into Fick's first law to give

$$n_A|_{z=a} = \frac{2D_{AB}(c_{A_s} - c_{A_o})A}{a} \times$$
$$\sum_{n=0}^{\infty} \exp\left[-\frac{D_{AB}(2n+1)^2 \pi^2 t}{4a^2} \right] \tag{3-82}$$

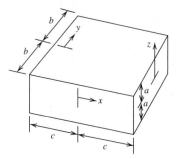

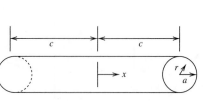

(a) Slab. Edges at $x = +c$ and $-c$ and at $y = +b$ and $-b$ are sealed.

(b) Cylinder. Two circular ends at $x = +c$ and $-c$ are sealed.

(c) Sphere

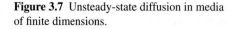

Figure 3.7 Unsteady-state diffusion in media of finite dimensions.

We can also determine the total number of moles transferred across either unsealed face by integrating (3-82) with respect to time. Thus,

$$\mathcal{N}_A = \int_o^t n_A|_{z=a}\, dt = \frac{8(c_{A_s} - c_{A_o})Aa}{\pi^2}$$

$$\times \sum_{n=0}^{\infty} \frac{1}{(2n+1)^2}\left\{1 - \exp\left[-\frac{D_{AB}(2n+1)^2\pi^2 t}{4a^2}\right]\right\}$$

$$(3\text{-}83)$$

In addition, the average concentration of the solute in the medium, $c_{A_{avg}}$, as a function of time, can be obtained in the case of a slab from:

$$\frac{c_{A_s} - c_{A_{avg}}}{c_{A_s} - c_{A_o}} = \frac{\int_o^a (1-\theta)\,dz}{a} \qquad (3\text{-}84)$$

Substitution of (3-80) into (3-84) followed by integration gives

$$E_{avg_{slab}} = (1 - \theta_{ave})_{slab} = \frac{c_{A_s} - c_{A_{avg}}}{c_{A_s} - c_{A_o}}$$

$$= \frac{8}{\pi^2}\sum_{n=0}^{\infty} \frac{1}{(2n+1)^2}\exp\left[-\frac{D_{AB}(2n+1)^2\pi^2 t}{4a^2}\right]$$

$$(3\text{-}85)$$

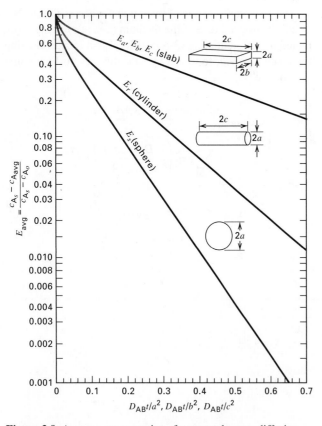

Figure 3.9 Average concentrations for unsteady-state diffusion.

[Adapted from R.E. Treybal, *Mass-Transfer Operations,* 3rd ed., McGraw-Hill, New York (1980).]

This equation is plotted in Figure 3.9. It is important to note that concentrations are in mass of solute per mass of dry solid or mass of solute/volume. This assumes that during diffusion the solid does not shrink or expand so that the mass of dry solid per unit volume of wet solid will remain constant. Then, we can substitute a concentration in terms of mass or moles of solute per mass of dry solid, i.e., the moisture content on the dry basis.

When the edges of the slab in Figure 3.7a are not sealed, the method of Newman [31] can be used with (3-69) to determine concentration changes within the slab. In this method, E or E_{avg} is given in terms of the E values from the solution of (3-68) for each of the coordinate directions by

$$E = E_x E_y E_z \qquad (3\text{-}86)$$

Corresponding solutions for infinitely long, circular cylinders and spheres are available in Carslaw and Jaeger [26] and are plotted in Figures 3.9, 3.10, and 3.11, respectively. For a short cylinder, where the ends are not sealed, E or E_{ave} is given by the method of Newman as

$$E = E_r E_x \qquad (3\text{-}87)$$

Some materials such as crystals and wood, have thermal conductivities and diffusivities that vary markedly with direction. For these anisotropic materials, Fick's second law in the form of (3-69) does not hold. Although the general anisotropic case is exceedingly complex, as shown in the following example, the mathematical treatment is relatively simple when the principal axes of diffusivity coincide with the coordinate system.

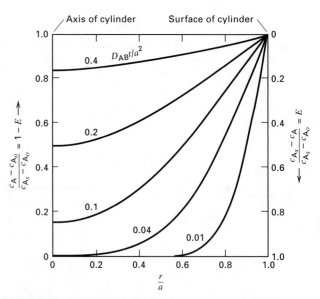

Figure 3.10 Concentration profiles for unsteady-state diffusion in a cylinder.

[Adapted from H.S. Carslaw and J.C. Jaeger, *Conduction of Heat in Solids,* 2nd ed., Oxford University Press, London (1959).]

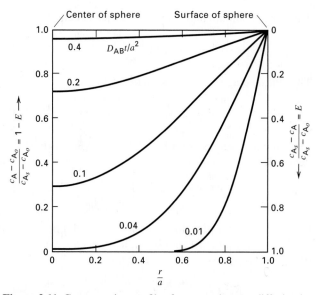

Figure 3.11 Concentration profiles for unsteady-state diffusion in a sphere.

[Adapted from H.S. Carslaw and J.C. Jaeger, *Conduction of Heat in Solids,* 2nd ed., Oxford University Press, London (1959).]

EXAMPLE 3.12

A piece of lumber, measuring $5 \times 10 \times 20$ cm, initially contains 20 wt% moisture. At time 0, all six faces are brought to an equilibrium moisture content of 2 wt%. Diffusivities for moisture at 25°C are 2×10^{-5} cm²/s in the axial (z) direction along the fibers and 4×10^{-6} cm²/s in the two directions perpendicular to the fibers. Calculate the time in hours for the average moisture content to drop to 5 wt% at 25°C. At that time, determine the moisture content at the center of the piece of lumber. All moisture contents are on a dry basis.

SOLUTION

In this case, the solid is anisotropic, with $D_x = D_y = 4 \times 10^{-6}$ cm²/s and $D_z = 2 \times 10^{-5}$ cm²/s, where dimensions $2c$, $2b$, and $2a$ in the x, y, and z directions are 5, 10, and 20 cm, respectively. Fick's second law for an isotropic medium, (3-69), must be rewritten for this anisotropic material as

$$\frac{\partial c_A}{\partial t} = D_x \left[\frac{\partial^2 c_A}{\partial x^2} + \frac{\partial^2 c_A}{\partial y^2} \right] + D_z \frac{\partial^2 c_A}{\partial z^2} \qquad (1)$$

as discussed by Carslaw and Jaeger [26].

To transform (1) into the form of (3-69), let

$$x_1 = x \sqrt{\frac{D}{D_x}} \qquad y_1 = y \sqrt{\frac{D}{D_x}} \qquad z_1 = z \sqrt{\frac{D}{D_z}} \qquad (2)$$

where D is chosen arbitrarily. With these changes in variables, (1) becomes

$$\frac{\partial c_A}{\partial t} = D \left(\frac{\partial^2 c_A}{\partial x_1^2} + \frac{\partial^2 c_A}{\partial y_1^2} + \frac{\partial^2 c_A}{\partial z_1^2} \right) \qquad (3)$$

Since this is the same form as (3-69) and since the boundary conditions do not involve diffusivities, we can apply Newman's method, using Figure 3.9, where concentration, c_A, is replaced by weight-percent moisture on a dry basis.

From (3-86) and (3-85),

$$E_{ave_{slab}} = E_{avg_x} E_{avg_y} E_{avg_z} = \frac{c_{A_{ave}} - c_{A_s}}{c_{A_o} - c_{A_s}} = \frac{5-2}{20-2} = 0.167$$

Let $D = 1 \times 10^{-5}$ cm²/s.

z_1 Direction (axial):

$$a_1 = a \left(\frac{D}{D_z} \right)^{1/2} = \frac{20}{2} \left(\frac{1 \times 10^{-5}}{2 \times 10^{-5}} \right)^{1/2} = 7.07 \text{ cm}$$

$$\frac{Dt}{a_1^2} = \frac{1 \times 10^{-5} t}{7.07^2} = 2.0 \times 10^{-7} t, \text{ s}$$

y_1 Direction:

$$b_1 = b \left(\frac{D}{D_y} \right)^{1/2} = \frac{10}{2} \left(\frac{1 \times 10^{-5}}{4 \times 10^{-6}} \right)^{1/2} = 7.906 \text{ cm}$$

$$\frac{Dt}{b_1^2} = \frac{1 \times 10^{-5} t}{7.906^2} = 1.6 \times 10^{-7} t, \text{ s}$$

x_1-Direction:

$$c_1 = c \left(\frac{D}{D_x} \right)^{1/2} = \frac{5}{2} \left(\frac{1 \times 10^{-5}}{4 \times 10^{-6}} \right)^{1/2} = 3.953 \text{ cm}$$

$$\frac{Dt}{c_1^2} = \frac{1 \times 10^{-5} t}{3.953^2} = 6.4 \times 10^{-7} t, \text{ s}$$

Use Figure 3.9 iteratively with assumed values of time in seconds to obtain values of E_{avg} for each of the three coordinates until (3-86) equals 0.167.

t, h	t, s	$E_{avg_{z_1}}$	$E_{avg_{y_1}}$	$E_{avg_{x_1}}$	E_{avg}
100	360,000	0.70	0.73	0.46	0.235
120	432,000	0.67	0.70	0.41	0.193
135	486,000	0.65	0.68	0.37	0.164

Therefore, it takes approximately 136 h.

For 136 h = 490,000 s, the Fourier numbers for mass transfer are

$$\frac{Dt}{a_1^2} = \frac{(1 \times 10^{-5})(490,000)}{7.07^2} = 0.0980$$

$$\frac{Dt}{b_1^2} = \frac{(1 \times 10^{-5})(490,000)}{7.906^2} = 0.0784$$

$$\frac{Dt}{c_1^2} = \frac{(1 \times 10^{-5})(490,000)}{3.953^2} = 0.3136$$

From Figure 3.8, at the center of the slab,

$$E_{center} = E_{z_1} E_{y_1} E_{x_1} = (0.945)(0.956)(0.605) = 0.547$$

$$= \frac{c_{A_s} - c_{A_{center}}}{c_{A_s} - c_{A_o}} = \frac{2 - c_{A_{center}}}{2 - 20} = 0.547$$

Solving,

$$c_A \text{ at the center} = 11.8 \text{ wt% moisture}$$

3.4 MOLECULAR DIFFUSION IN LAMINAR FLOW

Many mass-transfer operations involve diffusion in fluids in laminar flow. The fluid may be a film flowing slowly down a vertical or inclined surface, a laminar boundary layer that forms as the fluid flows slowly past a thin plate, or the fluid may flow through a small tube or slowly through a large pipe or duct. Mass transfer may occur between a gas and a liquid film, between a solid surface and a fluid, or between a fluid and a membrane surface.

Falling Liquid Film

Consider a thin liquid film, of a mixture of volatile A and nonvolatile B, falling in laminar flow at steady state down one side of a vertical surface and exposed to pure gas, A, on the other side of the film, as shown in Figure 3.12. The surface is infinitely wide in the x-direction (normal to the page). In the absence of mass transfer of A into the liquid film, the liquid velocity in the z-direction, u_z, is zero. In the absence of end effects, the equation of motion for the liquid film in fully developed laminar flow in the downward y-direction is

$$\mu \frac{d^2 u_y}{dz^2} + \rho g = 0 \qquad (3\text{-}88)$$

Usually, fully developed flow, where u_y is independent of the distance y, is established quickly. If δ is the thickness of the film and the boundary conditions are $u_y = 0$ at $z = \delta$ (no-slip condition at the solid surface) and $du_y/dz = 0$ at $z = 0$ (no drag at the liquid–gas interface), (3-88) is readily integrated to give a parabolic velocity profile:

$$u_y = \frac{\rho g \delta^2}{2\mu} \left[1 - \left(\frac{z}{\delta} \right)^2 \right] \qquad (3\text{-}89)$$

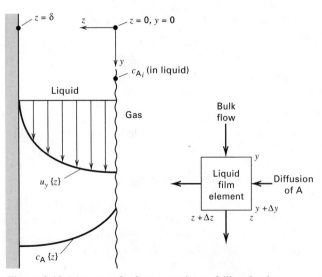

Figure 3.12 Mass transfer from a gas into a falling, laminar liquid film.

Thus, the maximum liquid velocity, which occurs at $z = 0$, is

$$(u_y)_{\max} = \frac{\rho g \delta^2}{2\mu} \qquad (3\text{-}90)$$

The bulk-average velocity in the liquid film is

$$\bar{u}_y = \frac{\int_0^\delta u_y \, dz}{\delta} = \frac{\rho g \delta^2}{3\mu} \qquad (3\text{-}91)$$

Thus, the film thickness for fully developed flow is independent of location y and is

$$\delta = \left(\frac{3 \bar{u}_y \mu}{\rho g} \right)^{1/2} = \left(\frac{3 \mu \Gamma}{\rho^2 g} \right)^{1/3} \qquad (3\text{-}92)$$

where Γ = liquid film flow rate per unit width of film, W.

For film flow, the Reynolds number, which is the ratio of the inertial force to the viscous force, is

$$N_{Re} = \frac{4 r_H \bar{u}_y \rho}{\mu} = \frac{4 \delta \bar{u}_y \rho}{\mu} = \frac{4 \Gamma}{\mu} \qquad (3\text{-}93)$$

where r_H = hydraulic radius = (flow cross section)/(wetted perimeter) = $(W\delta)/W = \delta$ and, by the equation of continuity, $\Gamma = \bar{u}_y \rho \delta$.

As reported by Grimley [32], for $N_{Re} < 8$ to 25, depending on the surface tension and viscosity, the flow in the film is laminar and the interface between the liquid film and the gas is flat. The value of 25 is obtained with water. For 8 to $25 < N_{Re} < 1{,}200$, the flow is still laminar, but ripples and waves may appear at the interface unless suppressed by the addition of wetting agents to the liquid.

For a flat liquid–gas interface and a small rate of mass transfer of A into the liquid film, (3-88) to (3-93) hold and the film velocity profile is given by (3-89). Now consider a mole balance on A for an incremental volume of liquid film of constant density, as shown in Figure 3.12. Neglect bulk flow in the z-direction and axial diffusion in the y-direction. Then, at steady state, neglecting accumulation or depletion of A in the incremental volume,

$$-D_{AB}(\Delta y)(\Delta x)\left(\frac{\partial c_A}{\partial z} \right)_z + u_y c_A|_y (\Delta z)(\Delta x)$$

$$= -D_{AB}(\Delta y)(\Delta x)\left(\frac{\partial c_A}{\partial z} \right)_{z+\Delta z} + u_y c_A|_{y+\Delta y} (\Delta z)(\Delta x) \qquad (3\text{-}94)$$

Rearranging and simplifying (3-94),

$$\left[\frac{u_y c_A|_{y+\Delta y} - u_y c_A|_y}{\Delta y} \right] = D_{AB} \left[\frac{(\partial c_A/\partial z)_{z+\Delta z} - (\partial c_A/\partial z)_z}{\Delta z} \right] \qquad (3\text{-}95)$$

In the limit, as $\Delta z \to 0$ and $\Delta y \to 0$,

$$u_y \frac{\partial c_A}{\partial y} = D_{AB} \frac{\partial^2 c_A}{\partial z^2} \qquad (3\text{-}96)$$

Substituting (3-89) into (3-96),

$$\frac{\rho g \delta^2}{2\mu} \left[1 - \left(\frac{z}{\delta} \right)^2 \right] \frac{\partial c_A}{\partial y} = D_{AB} \frac{\partial^2 c_A}{\partial z^2} \qquad (3\text{-}97)$$

This equation was solved by Johnstone and Pigford [33] and later by Olbrich and Wild [34], for the following boundary conditions:

$$c_A = c_{A_i} \quad \text{at } z = 0 \quad \text{for } y > 0$$
$$c_A = c_{A_0} \quad \text{at } y = 0 \quad \text{for } 0 < z < \delta$$
$$\partial c_A / \partial z = 0 \quad \text{at } z = \delta \quad \text{for } 0 < y < L$$

where L = height of the vertical surface. The solution of Olbrich and Wild is in the form of an infinite series, giving c_A as a function of z and y. However, of more interest is the average concentration at $y = L$, which, by integration, is

$$\bar{c}_{A_y} = \frac{1}{\bar{u}_y \delta} \int_0^\delta u_y c_{A_y} \, dz \tag{3-98}$$

For the condition $y = L$, the result is

$$\frac{c_{A_i} - \bar{c}_{A_L}}{c_{A_i} - c_{A_0}} = 0.7857 e^{-5.1213\eta} + 0.09726 e^{-39.661\eta} + 0.036093^{-106.25\eta} + \cdots \tag{3-99}$$

where

$$\eta = \frac{2 D_{AB} L}{3 \delta^2 \bar{u}_y} = \frac{8/3}{N_{Re} N_{Sc}(\delta/L)} = \frac{8/3}{(\delta/L) N_{Pe_M}} \tag{3-100}$$

$$N_{Sc} = \text{Schmidt number} = \frac{\mu}{\rho D_{AB}}$$
$$= \frac{\text{momentum diffusivity}, \mu/\rho}{\text{mass diffusivity}, D_{AB}} \tag{3-101}$$

$$N_{Pe_M} = N_{Re} N_{Sc} = \text{Peclet number for mass transfer}$$
$$= \frac{4 \delta \bar{u}_y}{D_{AB}} \tag{3-102}$$

The Schmidt number is analogous to the Prandtl number, used in heat transfer:

$$N_{Pr} = \frac{C_P \mu}{k} = \frac{(\mu/\rho)}{(k/\rho C_P)} = \frac{\text{momentum diffusivity}}{\text{thermal diffusivity}}$$

The Peclet number for mass transfer is analogous to the Peclet number for heat transfer:

$$N_{Pe_H} = N_{Re} N_{Pr} = \frac{4 \delta \bar{u}_y C_P \rho}{k}$$

Both Peclet numbers are ratios of convective transport to molecular transport.

The total rate of absorption of A from the gas into the liquid film for height L and width W is

$$n_A = \bar{u}_y \delta W (\bar{c}_{A_L} - c_{A_0}) \tag{3-103}$$

Mass-Transfer Coefficients

Mass-transfer problems involving fluids are most often solved using mass-transfer coefficients, analogous to heat-transfer coefficients. For the latter, *Newton's law of cooling* defines a heat-transfer coefficient, h:

$$Q = h A \, \Delta T \tag{3-104}$$

where

Q = rate of heat transfer

A = area for heat transfer (normal to the direction of heat transfer)

ΔT = temperature-driving force for heat transfer

For mass transfer, a composition driving force replaces ΔT. As discussed later in this chapter, because composition can be expressed in a number of ways, different mass-transfer coefficients are defined. If we select Δc_A as the driving force for mass transfer, we can write

$$n_A = k_c A \, \Delta c_A \tag{3-105}$$

which defines a mass-transfer coefficient, k_c, in mol/time-area-driving force, for a concentration driving force. Unfortunately, no name is in general use for (3-105).

For the falling laminar film, we take $\Delta c_A = c_{A_i} - \bar{c}_A$, which varies with vertical location, y, because even though c_{A_i} is independent of y, the average film concentration, $\bar{c}_A$, increases with y. To derive an expression for k_c, we equate (3-105) to Fick's first law at the gas–liquid interface:

$$k_c A (c_{A_i} - \bar{c}_A) = -D_{AB} A \left(\frac{\partial c_A}{\partial z} \right)_{z=0} \tag{3-106}$$

Although this is the most widely used approach for defining a mass-transfer coefficient, in this case of a falling film it fails because $(\partial c_A / \partial z)$ at $z = 0$ is not defined. Therefore, for this case we use another approach as follows. For an incremental height, we can write for film width W,

$$n_A = \bar{u}_y \delta W \, d\bar{c}_A = k_c (c_{A_i} - \bar{c}_A) W \, dy \tag{3-107}$$

This defines a local value of k_c, which varies with distance y because $\bar{c}_A$ varies with y. An average value of k_c, over a height L, can be defined by separating variables and integrating (3-107):

$$k_{c_{avg}} = \frac{\int_0^L k_c \, dy}{L} = \frac{\bar{u}_y \delta \int_{c_{A_0}}^{c_{A_L}} [d\bar{c}_A/(c_{A_i} - \bar{c}_A)]}{L}$$
$$= \frac{\bar{u}_y \delta}{L} \ln \frac{c_{A_i} - c_{A_0}}{c_{A_i} - \bar{c}_{A_L}} \tag{3-108}$$

In general, the argument of the natural logarithm in (3-108) is obtained from the reciprocal of (3-99). For values of η in (3-100) greater than 0.1, only the first term in (3-99) is significant (error is less than 0.5%). In that case,

$$k_{c_{avg}} = \frac{\bar{u}_y \delta}{L} \ln \frac{e^{5.1213\eta}}{0.7857} \tag{3-109}$$

Since $\ln e^x = x$,

$$k_{c_{avg}} = \frac{\bar{u}_y \delta}{L} (0.241 + 5.1213\eta) \tag{3-110}$$

In the limit, for large η, using (3-100) and (3-102), (3-110) becomes

$$k_{c_{avg}} = 3.414 \frac{D_{AB}}{\delta} \tag{3-111}$$

In a manner suggested by the Nusselt number, $N_{Nu} = h\delta/k$ for heat transfer, where δ = a characteristic length, we define a Sherwood number for mass transfer, which for a falling film of characteristic length δ is

$$N_{Sh_{avg}} = \frac{k_{c_{avg}} \delta}{D_{AB}} \tag{3-112}$$

From (3-111), $N_{Sh_{avg}} = 3.414$, which is the smallest value that the Sherwood number can have for a falling liquid film.

The average mass-transfer flux of A is given by

$$N_{A_{avg}} = \frac{n_{A_{avg}}}{A} = k_{c_{avg}}(c_{A_i} - \bar{c}_A)_{mean} \qquad (3\text{-}113)$$

For values $\eta < 0.001$ in (3-100), when the liquid-film flow regime is still laminar without ripples, the time of contact of the gas with the liquid is short and mass transfer is confined to the vicinity of the gas–liquid interface. Thus, the film acts as if it were infinite in thickness. In this limiting case, the downward velocity of the liquid film in the region of mass transfer is just $u_{y_{max}}$, and (3-96) becomes

$$u_{y_{max}} \frac{\partial c_A}{\partial y} = D_{AB} \frac{\partial^2 c_A}{\partial z^2} \qquad (3\text{-}114)$$

Since from (3-90) and (3-91), $u_{y_{max}} = 3\bar{u}_y/2$, (3-114) can be rewritten as

$$\frac{\partial c_A}{\partial y} = \left(\frac{2D_{AB}}{3\bar{u}_y}\right) \frac{\partial^2 c_A}{\partial z^2} \qquad (3\text{-}115)$$

where the boundary conditions are

$$\begin{aligned}
c_A &= c_{A_0} \quad \text{for } z > 0 \quad \text{and } y > 0 \\
c_A &= c_{A_i} \quad \text{for } z = 0 \quad \text{and } y > 0 \\
c_A &= c_{A_0} \quad \text{for large } z \quad \text{and } y > 0
\end{aligned}$$

Equation (3-115) and the boundary conditions are equivalent to the case of the semi-infinite medium, as developed above. Thus, by analogy to (3-68), (3-75), and (3-76) the solution is

$$E = 1 - \theta = \frac{c_{A_i} - c_A}{c_{A_i} - c_{A_0}} = \text{erf}\left(\frac{z}{2\sqrt{2D_{AB}y/3\bar{u}_y}}\right) \qquad (3\text{-}116)$$

Assuming that the driving force for mass transfer in the film is $c_{A_i} - c_{A_0}$, we can use Fick's first law at the gas–liquid interface to define a mass-transfer coefficient:

$$N_A = -D_{AB} \left.\frac{\partial c_A}{\partial z}\right|_{z=0} = k_c(c_{A_i} - c_{A_0}) \qquad (3\text{-}117)$$

The error function is defined as

$$\text{erf } z = \frac{2}{\sqrt{\pi}} \int_0^z e^{-t^2} dt \qquad (3\text{-}118)$$

Using the Leibnitz rule with (3-116) to differentiate this integral function,

$$\left.\frac{\partial c_A}{\partial z}\right|_{z=0} = -(c_{A_i} - c_{A_0})\sqrt{\frac{3\bar{u}_y}{2\pi D_{AB} y}} \qquad (3\text{-}119)$$

Substituting (3-119) into (3-117) and introducing the Peclet number for mass transfer from (3-102), we obtain an expression for the local mass-transfer coefficient as a function of distance down from the top of the wall:

$$k_c = \sqrt{\frac{3D_{AB}^2 N_{Pe_M}}{8\pi y \delta}} = \sqrt{\frac{3D_{AB}\Gamma}{2\pi y \delta \rho}} \qquad (3\text{-}120)$$

The average value of k_c over the height of the film, L, is obtained by integrating (3-120) with respect to y, giving

$$k_{c_{avg}} = \sqrt{\frac{6D_{AB}\Gamma}{\pi \delta \rho L}} = \sqrt{\frac{3D_{AB}^2}{2\pi \delta L} N_{Pe_M}} \qquad (3\text{-}121)$$

Combining (3-121) with (3-112) and (3-102),

$$N_{Sh_{avg}} = \sqrt{\frac{3\delta}{2\pi L} N_{Pe_M}} = \sqrt{\frac{4}{\pi \eta}} \qquad (3\text{-}122)$$

where, by (3-108), the proper mean to use with $k_{c_{avg}}$ is the log mean. Thus,

$$\begin{aligned}
(c_{A_i} - \bar{c}_A)_{mean} &= (c_{A_i} - \bar{c}_A)_{LM} \\
&= \frac{(c_{A_i} - c_{A_0}) - (c_{A_i} - c_{A_L})}{\ln[(c_{A_i} - c_{A_0})/(c_{A_i} - \bar{c}_{A_L})]}
\end{aligned} \qquad (3\text{-}123)$$

When ripples are present, values of $k_{c_{avg}}$ and $N_{Sh_{avg}}$ can be considerably larger than predicted by these equations.

In the above development, asymptotic, closed-form solutions are obtained with relative ease for large and small values of η, defined by (3-100). These limits, in terms of the average Sherwood number, are shown in Figure 3.13. The

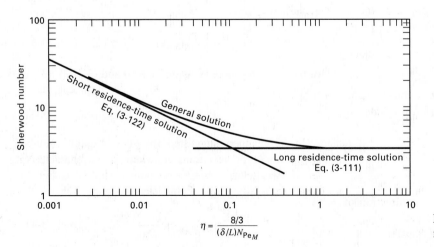

Figure 3.13 Limiting and general solutions for mass transfer to a falling, laminar liquid film.

general solution for intermediate values of η is not available in closed form. Similar limiting solutions for large and small values of appropriate parameters, usually dimensionless groups, have been obtained for a large variety of transport and kinetic phenomena, as discussed by Churchill [35]. Often the two limiting cases can be patched together to provide a reasonable estimate of the intermediate solution, if a single intermediate value is available from experiment or the general numerical solution. The procedure is discussed by Churchill and Usagi [36]. The general solution of Emmert and Pigford [37] to the falling, laminar liquid film problem is included in Figure 3.13.

EXAMPLE 3.13

Water (B) at 25°C, in contact with pure CO_2 (A) at 1 atm, flows as a film down a vertical wall 1 m wide and 3 m high at a Reynolds number of 25. Using the following properties, estimate the rate of adsorption of CO_2 into water in kmol/s:

$$D_{AB} = 1.96 \times 10^{-5} \text{ cm}^2/\text{s}; \qquad \rho = 1.0 \text{ g/cm}^3;$$
$$\mu_L = 0.89 \text{ cP} = 0.00089 \text{ kg/m-s}$$

Solubility of CO_2 in water at 1 atm and 25°C $= 3.4 \times 10^{-5} \text{ mol/cm}^3$.

SOLUTION

From (3-93),

$$\Gamma = \frac{N_{Re}\mu}{4} = \frac{25(0.89)(0.001)}{4} = 0.00556 \frac{\text{kg}}{\text{m} - \text{s}}$$

From (3-101),

$$N_{Sc} = \frac{\mu}{\rho D_{AB}} = \frac{(0.89)(0.001)}{(1.0)(1,000)(1.96 \times 10^{-5})(10^{-4})} = 454$$

From (3-92),

$$\delta = \left[\frac{3(0.89)(0.001)(0.00556)}{1.0^2(1,000)^2(9.807)} \right]^{1/3} = 1.15 \times 10^{-4} \text{ m}$$

From (3-90) and (3-91), $\bar{u}_y = (2/3)u_{y_{max}}$. Therefore,

$$\bar{u}_y = \frac{2}{3} \left[\frac{(1.0)(1,000)(9.807)(1.15 \times 10^{-4})^2}{2(0.89)(0.001)} \right] = 0.0486 \text{ m/s}$$

From (3-100),

$$\eta = \frac{8/3}{(25)(454)[(1.15 \times 10^{-4})/3]} = 6.13$$

Therefore, (3-111) applies, giving

$$k_{c_{avg}} = \frac{3.41(1.96 \times 10^{-5})(10^{-4})}{1.15 \times 10^{-4}} = 5.81 \times 10^{-5} \text{ m/s}$$

To determine the rate of absorption, $\bar{c}_{A_L}$ must be determined. From (3-103) and (3-113),

$$n_A = \bar{u}_y \delta W (\bar{c}_{A_L} - c_{A_0}) = k_{c_{avg}} A \frac{(\bar{c}_{A_L} - c_{A_0})}{\ln[(c_{A_i} - c_{A_0})/(c_{A_i} - \bar{c}_{A_L})]}$$

Thus,

$$\ln \left(\frac{c_{A_i} - c_{A_0}}{c_{A_i} - \bar{c}_{A_L}} \right) = \frac{k_{c_{avg}} A}{\bar{u}_y \delta W}$$

Solving for $\bar{c}_{A_L}$,

$$\bar{c}_{A_L} = c_{A_i} - (c_{A_i} - c_{A_0}) \exp\left(-\frac{k_{c_{avg}} A}{\bar{u}_y \delta W} \right)$$

$$L = 3 \text{ m}, \quad W = 1 \text{ m}, \quad A = WL = (1)(3) = 3 \text{ m}^2$$
$$c_{A_0} = 0, \quad c_{A_i} = 3.4 \times 10^{-5} \text{ mol/cm}^3 = 3.4 \times 10^{-2} \text{ kmol/m}^3$$

$$\bar{c}_{A_L} = 3.4 \times 10^{-2} \left\{ 1 - \exp\left[-\frac{(5.81 \times 10^{-5})(3)}{(0.0486)(1.15 \times 10^{-4})(1)} \right] \right\}$$

$$= 3.4 \times 10^{-2} \text{ kmol/m}^3$$

Thus, the exiting liquid film is saturated with CO_2, which implies equilibrium at the gas–liquid interface. From (3-103),

$$n_A = 0.0486(1.15 \times 10^{-4})(3.4 \times 10^{-2}) = 1.9 \times 10^{-7} \text{ kmol/s}$$

Boundary-Layer Flow on a Flat Plate

Consider the flow of a fluid (B) over a thin, flat plate parallel with the direction of flow of the fluid upstream of the plate, as shown in Figure 3.14. A number of possibilities for mass transfer of another species, A, into B exist: (1) The plate might consist of material A, which is slightly soluble in B. (2) Component A might be held in the pores of an inert solid plate, from which it evaporates or dissolves into B. (3) The plate might be an inert, dense polymeric membrane, through which species A can pass into fluid B. Let the fluid velocity profile upstream of the plate be uniform at a free-system velocity of u_o. As the fluid passes over the plate, the velocity u_x in the direction of flow is reduced to zero at the wall, which establishes a velocity profile due to drag. At a certain distance z, normal to and out from the solid surface, the fluid velocity is 99% of u_o. This distance, which increases with increasing distance x from the leading edge of the plate, is arbitrarily defined as the velocity boundary-layer thickness, δ. Essentially all flow retardation occurs in the boundary layer, as first suggested by Prandtl [38]. The buildup of this layer, the velocity profile in the layer, and the drag force can be determined for laminar flow by solving the equations of continuity and motion (Navier–Stokes equations) for the x-direction. For a Newtonian fluid of constant density and viscosity, in the absence of pressure gradients in the x- and

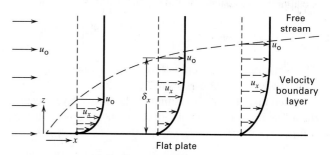

Figure 3.14 Laminar boundary-layer development for flow across a flat plate.

y- (normal to the x–z plane) directions, these equations for the region of the boundary layer are

$$\frac{\partial u_x}{\partial x} + \frac{\partial u_z}{\partial z} = 0 \tag{3-124}$$

$$u_x \frac{\partial u_x}{\partial x} + u_z \frac{\partial u_x}{\partial z} = \frac{\mu}{\rho}\left(\frac{\partial^2 u_x}{\partial z^2}\right) \tag{3-125}$$

The boundary conditions are

$u_x = u_0$ at $x = 0$ for $z > 0$ $u_x = 0$ at $z = 0$ for $x > 0$
$u_x = u_0$ at $z = \infty$ for $x > 0$ $u_z = 0$ at $z = 0$ for $x > 0$

The solution of (3-124) and (3-125) in the absence of heat and mass transfer, subject to these boundary conditions, was first obtained by Blasius [39] and is described in detail by Schlichting [40]. The result in terms of a local friction factor, f_x, a local shear stress at the wall, τ_{w_x}, and a local drag coefficient at the wall, C_{D_x}, is

$$\frac{C_{D_x}}{2} = \frac{f_x}{2} = \frac{\tau_{w_x}}{\rho u_0^2} = \frac{0.322}{N_{Re_x}^{0.5}} \tag{3-126}$$

where

$$N_{Re_x} = \frac{x u_0 \rho}{\mu} \tag{3-127}$$

Thus, the drag is greatest at the leading edge of the plate, where the Reynolds number is smallest. Average values of the drag coefficient are obtained by integrating (3-126) from $x = 0$ to L, giving

$$\frac{C_{D_{avg}}}{2} = \frac{f_{avg}}{2} = \frac{0.664}{(N_{Re_L})^{0.5}} \tag{3-128}$$

The thickness of the velocity boundary layer increases with distance along the plate:

$$\frac{\delta}{x} = \frac{4.96}{N_{Re_x}^{0.5}} \tag{3-129}$$

A reasonably accurate expression for the velocity profile was obtained by Pohlhausen [41], who assumed the empirical form $u_x = C_1 z + C_2 z^3$.

If the boundary conditions,

$u_x = 0$ at $z = 0$ $u_x = u_0$ at $z = \delta$ $\partial u_x/\partial z = 0$ at $z = \delta$

are applied to evaluate C_1 and C_2, the result is

$$\frac{u_x}{u_0} = 1.5\left(\frac{z}{\delta}\right) - 0.5\left(\frac{z}{\delta}\right)^3 \tag{3-130}$$

This solution is valid only for a laminar boundary layer, which by experiment persists to $N_{Re_x} = 5 \times 10^5$.

When mass transfer of A into the boundary layer occurs, the following species continuity equation applies at constant diffusivity:

$$u_x \frac{\partial c_A}{\partial x} + u_z \frac{\partial c_A}{\partial z} = D_{AB}\left(\frac{\partial^2 c_A}{\partial x^2}\right) \tag{3-131}$$

If mass transfer begins at the leading edge of the plate and if the concentration in the fluid at the solid–fluid interface is constant, the additional boundary conditions are

$c_A = c_{A_0}$ at $x = 0$ for $z > 0$,
$c_A = c_{A_i}$ at $z = 0$ for $x > 0$,
and $c_A = c_{A_0}$ at $z = \infty$ for $x > 0$

If the rate of mass transfer is low, the velocity profiles are undisturbed. The solution to the analogous problem in heat transfer was first obtained by Pohlhausen [42] for $N_{Pr} > 0.5$, as described in detail by Schlichting [40]. The results for mass transfer are

$$\frac{N_{Sh_x}}{N_{Re_x} N_{Sc}^{1/3}} = \frac{0.332}{N_{Re_x}^{0.5}} \tag{3-132}$$

where

$$N_{Sh_x} = \frac{x k_{c_x}}{D_{AB}} \tag{3-133}$$

and the driving force for mass transfer is $c_{A_i} - c_{A_0}$.

The concentration boundary layer, where essentially all of the resistance to mass transfer resides, is defined by

$$\frac{c_{A_i} - c_A}{c_{A_i} - c_{A_0}} = 0.99 \tag{3-134}$$

and the ratio of the concentration boundary-layer thickness, δ_c, to the velocity boundary thickness, δ, is

$$\delta_c/\delta = 1/N_{Sc}^{1/3} \tag{3-135}$$

Thus, for a liquid boundary layer, where $N_{Sc} > 1$, the concentration boundary layer builds up more slowly than the velocity boundary layer. For a gas boundary layer, where $N_{SC} \approx 1$, the two boundary layers build up at about the same rate. By analogy to (3-130), the concentration profile is given by

$$\frac{c_{A_i} - c_A}{c_{A_i} - c_{A_0}} = 1.5\left(\frac{z}{\delta_c}\right) - 0.5\left(\frac{z}{\delta_c}\right)^3 \tag{3-136}$$

Equation (3-132) gives the local Sherwood number. If this expression is integrated over the length of the plate, L, the average Sherwood number is found to be

$$N_{Sh_{avg}} = 0.664\, N_{Re_L}^{1/2} N_{Sc}^{1/3} \tag{3-137}$$

where

$$N_{Sh_{avg}} = \frac{L k_{c_{avg}}}{D_{AB}} \tag{3-138}$$

EXAMPLE 3.14

Air at 100°C, 1 atm, and a free-stream velocity of 5 m/s flows over a 3-m-long, thin, flat plate of naphthalene, causing it to sublime.

(a) Determine the length over which a laminar boundary layer persists.

(b) For that length, determine the rate of mass transfer of naphthalene into air.

(c) At the point of transition of the boundary layer to turbulent flow, determine the thicknesses of the velocity and concentration boundary layers.

Assume the following values for physical properties:

Vapor pressure of napthalene = 10 torr
Viscosity of air = 0.0215 cP
Molar density of air = 0.0327 kmol/m^3
Diffusivity of napthalene in air = 0.94×10^{-5} m^2/s

SOLUTION

(a) $N_{Re_x} = 5 \times 10^5$ for transition. From (3-127),

$$x = L = \frac{\mu N_{Re_x}}{u_o \rho} = \frac{[(0.0215)(0.001)](5 \times 10^5)}{(5)[(0.0327)(29)]} = 2.27 \text{ m}$$

at which transition to turbulent flow begins.

(b) $c_{A_o} = 0$ $c_{A_i} = \frac{10(0.0327)}{760} = 4.3 \times 10^{-4}$ kmol/m^3

From (3-101),

$$N_{Sc} = \frac{\mu}{\rho D_{AB}} = \frac{[(0.0215)(0.001)]}{[(0.0327)(29)](0.94 \times 10^{-5})} = 2.41$$

From (3-137),

$$N_{Sh_{avg}} = 0.664(5 \times 10^5)^{1/2}(2.41)^{1/3} = 630$$

From (3-138),

$$k_{c_{avg}} = \frac{630(0.94 \times 10^{-5})}{2.27} = 2.61 \times 10^{-3} \text{m/s}$$

For a width of 1 m,

$A = 2.27$ m^2

$n_A = k_{c_{avg}} A(c_{A_i} - c_{A_o}) = 2.61 \times 10^{-3}(2.27)(4.3 \times 10^{-4})$

$= 2.55 \times 10^{-6}$ kmol/s

(c) From (3-129), at $x = L = 2.27$ m,

$$\delta = \frac{3.46(2.27)}{(5 \times 10^5)^{0.5}} = 0.0111 \text{ m}$$

From (3-135),

$$\delta_c = \frac{0.0111}{(2.41)^{1/3}} = 0.0083 \text{ m}$$

Fully Developed Flow in a Straight, Circular Tube

Figure 3.15 shows the formation and buildup of a laminar velocity boundary layer when a fluid flows from a vessel into a straight, circular tube. At the entrance, plane a, the velocity profile is flat. A velocity boundary layer then begins to build up as shown at planes b, c, and d. In this region, the central core outside the boundary layer has a flat velocity profile where the flow is accelerated over the entrance velocity. Finally, at plane e, the boundary layer fills the tube. From here the velocity profile is fixed and the flow is said to be fully developed. The distance from the plane a to plane e is the entry region.

For fully developed laminar flow in a straight, circular tube, by experiment, the Reynolds number, $N_{Re} = D\bar{u}_x \rho/\mu$, where $\bar{u}_x$ is the flow-average velocity in the axial direction, x, and D is the inside diameter of the tube, must be less than 2,100. For this condition, the equation of motion in the axial direction for horizontal flow and constant properties is

$$\frac{\mu}{r} \frac{\partial}{\partial r} \left(r \frac{\partial u_x}{\partial r} \right) - \frac{dP}{dx} = 0 \qquad (3\text{-}139)$$

where the boundary conditions are

$r = 0$ (axis of the tube), $\partial u_x/\partial r = 0$

and $r = r_w$ (tube wall), $u_x = 0$

Equation (3-139) was integrated by Hagen in 1839 and Poiseuille in 1841. The resulting equation for the velocity profile, expressed in terms of the flow-average velocity, is

$$u_x = 2\bar{u}_x \left[1 - \left(\frac{r}{r_w} \right)^2 \right] \qquad (3\text{-}140)$$

or, in terms of the maximum velocity at the tube axis,

$$u_x = u_{x_{max}} \left[1 - \left(\frac{r}{r_w} \right)^2 \right] \qquad (3\text{-}141)$$

From the form of (3-141), the velocity profile is parabolic in nature.

The shear stress, pressure drop, and Fanning friction factor are obtained from solutions to (3-139):

$$\tau_w = -\mu \left(\frac{\partial u_x}{\partial r} \right)\bigg|_{r=r_w} = \frac{4\mu\bar{u}_x}{r_w} \qquad (3\text{-}142)$$

$$-\frac{dP}{dx} = \frac{32\mu\bar{u}_x}{D^2} = \frac{2f\rho\bar{u}_x^2}{D} \qquad (3\text{-}143)$$

with

$$f = \frac{16}{N_{Re}} \qquad (3\text{-}144)$$

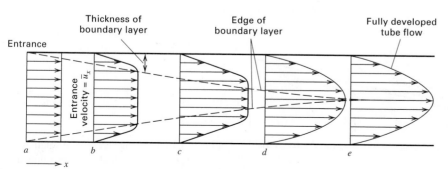

Thickness of boundary layer   Edge of boundary layer   Fully developed tube flow

Entrance

Entrance velocity = $\bar{u}_x$

a b c d e

$\longrightarrow x$

Figure 3.15 Buildup of a laminar velocity boundary layer for flow in a straight, circular tube.

The entry length to achieve fully developed flow is defined as the axial distance, L_e, from the entrance to the point at which the centerline velocity is 99% of the fully developed flow value. From the analysis of Langhaar [43] for the entry region,

$$\frac{L_e}{D} = 0.0575 N_{\text{Re}} \qquad (3\text{-}145)$$

Thus, at the upper limit of laminar flow, $N_{\text{Re}} = 2,100$, $L_e/D = 121$, a rather large ratio. For $N_{\text{Re}} = 100$, the ratio is only 5.75. In the entry region, Langhaar's analysis shows the friction factor is considerably higher than the fully developed flow value given by (3-144). At $x = 0$, f is infinity, but then decreases exponentially with x, approaching the fully developed flow value at L_e. For example, for $N_{\text{Re}} = 1,000$, (3-144) gives $f = 0.016$, with $L_e/D = 57.5$. In the region from $x = 0$ to $x/D = 5.35$, the average friction factor from Langhaar is 0.0487, which is about three times higher than the fully developed value.

In 1885, Graetz [44] obtained a theoretical solution to the problem of convective heat transfer between the wall of a circular tube, held at a constant temperature, and a fluid flowing through the tube in fully developed laminar flow. Assuming constant properties and negligible conduction in the axial direction, the energy equation, after substituting (3-140) for u_x, is

$$2\bar{u}_x \left[1 - \left(\frac{r}{r_w} \right)^2 \right] \frac{\partial T}{\partial x} = \frac{k}{\rho C_P} \left[\frac{1}{r} \frac{\partial}{\partial r} \left(r \frac{\partial T}{\partial r} \right) \right] \qquad (3\text{-}146)$$

The boundary conditions are

$x = 0$ (where heat transfer begins), $T = T_0$, for all r

$x > 0$, $r = r_w$, $T = T_i$ $x > 0$, $r = 0$, $\partial T / \partial r = 0$

The analogous species continuity equation for mass transfer, neglecting bulk flow in the radial direction and diffusion in the axial direction, is

$$2\bar{u}_x \left[1 - \left(\frac{r}{r_w} \right)^2 \right] \frac{\partial c_A}{\partial x} = D_{AB} \left[\frac{1}{r} \frac{\partial}{\partial r} \left(r \frac{\partial c_A}{\partial r} \right) \right] \qquad (3\text{-}147)$$

with analogous boundary conditions.

The Graetz solution of (3-147) for the temperature profile or the concentration profile is in the form of an infinite series, and can be obtained from (3-146) by the method of separation of variables using the method of Frobenius. A detailed solution is given by Sellars, Tribus, and Klein [45]. From the concentration profile, expressions for the mass-transfer coefficient and the Sherwood number are obtained. When x is large, the concentration profile is fully developed and the local Sherwood number, N_{Sh_x}, approaches a limiting value of 3.656. At the other extreme, when x is small such that the concentration boundary layer is very thin and confined to a region where the fully developed velocity profile is linear, the local Sherwood number is obtained from the classic Leveque [46] solution, presented by Knudsen and Katz [47]:

$$N_{\text{Sh}_x} = \frac{k_{c_x} D}{D_{AB}} = 1.077 \left[\frac{N_{\text{Pe}_M}}{(x/D)} \right]^{1/3} \qquad (3\text{-}148)$$

where

$$N_{\text{Pe}_M} = \frac{D \bar{u}_x}{D_{AB}} \qquad (3\text{-}149)$$

The limiting solutions, together with the general Graetz solution, are shown in Figure 3.16, where it is seen that $N_{\text{Sh}_x} = 3.656$ is valid for $N_{\text{Pe}_M}/(x/D) < 4$ and (3-148) is valid for $N_{\text{Pe}_M}/(x/D) > 100$. The two limiting solutions can be patched together if one point of the general solution is available where the two solutions intersect.

Over a length of tube where mass transfer occurs, an average Sherwood number can be derived by integrating the general expression for the local Sherwood number. An empirical representation for that average, proposed by Hausen [48], is

$$N_{\text{Sh}_{\text{avg}}} = 3.66 + \frac{0.0668[N_{\text{Pe}_M}/(x/D)]}{1 + 0.04[N_{\text{Pe}_M}/(x/D)]^{2/3}} \qquad (3\text{-}150)$$

which is based on a log-mean concentration driving force.

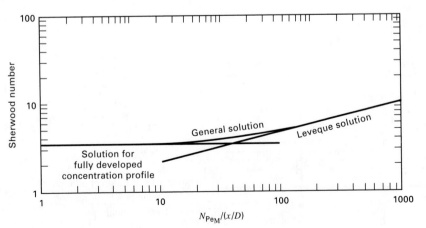

Figure 3.16 Limiting and general solutions for mass transfer to a fluid in laminar flow in a straight, circular tube.

EXAMPLE 3.15

Linton and Sherwood [49] conducted experiments on the dissolution of cast tubes of benzoic acid (A) into water (B) flowing through the tubes in laminar flow. They obtained good agreement with predictions based on the Graetz and Leveque equations. Consider a 5.23-cm-inside-diameter by 32-cm-long tube of benzoic acid, preceded by 400 cm of straight metal pipe of the same inside diameter where a fully developed velocity profile is established. Pure water enters the system at 25°C at a velocity corresponding to a Reynolds number of 100. Based on the following property data at 25°C, estimate the average concentration of benzoic acid in the water leaving the cast tube before a significant increase in the inside diameter of the benzoic acid tube occurs because of dissolution.

Solubility of benzoic acid in water $= 0.0034$ g/cm^3

Viscosity of water $= 0.89$ cP $= 0.0089$ g/cm-s

Diffusivity of benzoic acid in water at infinite dilution
$= 9.18 \times 10^{-6}$ cm^2/s

SOLUTION

$$N_{Sc} = \frac{0.0089}{(1.0)(9.18 \times 10^{-6})} = 970$$

$$N_{Re} = \frac{D\bar{u}_x \rho}{\mu} = 100$$

from which

$$\bar{u}_x = \frac{(100)(0.0089)}{(5.23)(1.0)} = 0.170 \text{ cm/s}$$

From (3-149),

$$N_{Pe_M} = \frac{(5.23)(0.170)}{9.18 \times 10^{-6}} = 9.69 \times 10^4$$

$$\frac{x}{D} = \frac{32}{5.23} = 6.12$$

$$\frac{N_{Pe_M}}{(x/D)} = \frac{9.69 \times 10^4}{6.12} = 1.58 \times 10^4$$

From (3-150),

$$N_{Sh_{avg}} = 3.66 + \frac{0.0668(1.58 \times 10^4)}{1 + 0.04(1.58 \times 10^4)^{2/3}} = 44$$

$$k_{c_{avg}} = N_{Sh_{avg}}\left(\frac{D_{AB}}{D}\right) = 44\frac{(9.18 \times 10^{-6})}{5.23} = 7.7 \times 10^{-5} \text{ cm/s}$$

Using a log-mean driving force,

$$n_A = \bar{u}_x S(\bar{c}_{A_x} - c_{A_0}) = k_{c_{avg}} A \frac{[(c_{A_i} - c_{A_0}) - (c_{A_i} - \bar{c}_{A_x})]}{\ln[(c_{A_i} - c_{A_0})/(c_{A_i} - \bar{c}_{A_x})]}$$

where S is the cross-sectional area for flow. Simplifying,

$$\ln\left(\frac{c_{A_i} - c_{A_0}}{c_{A_i} - \bar{c}_{A_x}}\right) = \frac{k_{c_{avg}} A}{\bar{u}_x S}$$

$$c_{A_0} = 0 \quad \text{and} \quad c_{A_i} = 0.0034 \text{ g/cm}^3$$

$$S = \frac{\pi D^2}{4} = \frac{(3.14)(5.23)^2}{4} = 21.5 \text{ cm}^2 \quad \text{and}$$

$$A = \pi D x = (3.14)(5.23)(32) = 526 \text{ cm}^2$$

$$\ln\left(\frac{0.0034}{0.0034 - \bar{c}_{A_x}}\right) = \frac{(7.7 \times 10^{-5})(526)}{(0.170)(21.5)}$$

$$= 0.0111$$

$$\bar{c}_{A_x} = 0.0034 - \frac{0.0034}{e^{0.0111}} = 0.000038 \text{ g/cm}^3$$

Thus, the concentration of benzoic acid in the water leaving the cast tube is far from saturation.

3.5 MASS TRANSFER IN TURBULENT FLOW

In the two previous sections, diffusion in stagnant media and in laminar flow were considered. For both cases, Fick's law can be applied to obtain rates of mass transfer. A more common occurrence in engineering is turbulent flow, which is accompanied by much higher transport rates, but for which theory is still under development and the estimation of mass-transfer rates relies more on empirical correlations of experimental data and analogies with heat and momentum transfer. A summary of the dimensionless groups used in these correlations and the analogies is given in Table 3.13.

As shown by the famous dye experiment of Osborne Reynolds [50] in 1883, a fluid in laminar flow moves parallel to the solid boundaries in streamline patterns. Every particle of fluid moves with the same velocity along a streamline and there are no fluid velocity components normal to these streamlines. For a Newtonian fluid in laminar flow, the momentum transfer, heat transfer, and mass transfer are by molecular transport, governed by Newton's law of viscosity, Fourier's law of heat conduction, and Fick's law of molecular diffusion, respectively.

In turbulent flow, the rates of momentum, heat, and mass transfer are orders of magnitude greater than for molecular transport. This occurs because streamlines no longer exist and particles or eddies of fluid, which are large compared to the mean free path of the molecules in the fluid, mix with each other by moving from one region to another in fluctuating motion. This eddy mixing by velocity fluctuations occurs not only in the direction of flow but also in directions normal to flow, with the latter being of more interest. Momentum, heat, and mass transfer now occur by two parallel mechanisms: (1) molecular motion, which is slow; and (2) turbulent or eddy motion, which is rapid except near a solid surface, where the flow velocity accompanying turbulence decreases to zero. Mass transfer by bulk flow may also occur as given by (3-1).

In 1877, Boussinesq [51] modified Newton's law of viscosity to account for eddy motion. Analogous expressions were subsequently developed for turbulent-flow heat and mass transfer. For flow in the x-direction and transport in the z-direction normal to flow, these expressions are written in the following forms in the absence of bulk flow in the z-direction:

$$\tau_{zx} = -(\mu + \mu_t)\frac{du_x}{dz} \tag{3-151}$$

$$q_z = -(k + k_t)\frac{dT}{dz} \tag{3-152}$$

$$N_{A_z} = -(D_{AB} + D_t)\frac{dc_A}{dz} \tag{3-153}$$

where the double subscript, zx, on the shear stress, τ, stands for x-momentum in the z-direction. The molecular contributions, μ, k, and D_{AB}, are molecular properties of the fluid and depend on chemical composition, temperature, and pressure; the turbulent contributions, μ_t, k_t, and D_t, depend on the

Table 3.13 Some Useful Dimensionless Groups

Name	Formula	Meaning	Analogy
	Fluid Mechanics		
Drag Coefficient	$C_D = \dfrac{2F_D}{Au^2\rho}$	$\dfrac{\text{Drag force}}{\text{Projected area} \times \text{Velocity head}}$	
Fanning Friction Factor	$f = \dfrac{\Delta P}{L}\dfrac{D}{2\bar{u}^2\rho}$	$\dfrac{\text{Pipe wall shear stress}}{\text{Velocity head}}$	
Froude Number	$N_{Fr} = \dfrac{\bar{u}^2}{gL}$	$\dfrac{\text{Inertial force}}{\text{Gravitational force}}$	
Reynolds Number	$N_{Re} = \dfrac{L\bar{u}\rho}{\mu} = \dfrac{L\bar{u}}{\nu} = \dfrac{LG}{\mu}$	$\dfrac{\text{Inertial force}}{\text{Viscous force}}$	
Weber Number	$N_{We} = \dfrac{\bar{u}^2\rho L}{\sigma}$	$\dfrac{\text{Inertial force}}{\text{Surface-tension force}}$	
	Heat Transfer		
j-Factor for Heat Transfer	$j_H = N_{St_H}(N_{Pr})^{2/3}$		j_M
Nusselt Number	$N_{Nu} = \dfrac{hL}{k}$	$\dfrac{\text{Convective heat transfer}}{\text{Conductive heat transfer}}$	N_{Sh}
Peclet Number for Heat Transfer	$N_{Pe_H} = N_{Re}N_{Pr} = \dfrac{L\bar{u}\rho C_P}{k}$	$\dfrac{\text{Bulk transfer of heat}}{\text{Conductive heat transfer}}$	N_{Pe_M}
Prandtl Number	$N_{Pr} = \dfrac{C_P\mu}{k} = \dfrac{\nu}{\alpha}$	$\dfrac{\text{Momentum diffusivity}}{\text{Thermal diffusivity}}$	N_{Sc}
Stanton Number for Heat Transfer	$N_{St_H} = \dfrac{N_{Nu}}{N_{Re}N_{Pr}} = \dfrac{h}{C_P G}$	$\dfrac{\text{Heat transfer}}{\text{Thermal capacity}}$	N_{St_M}
	Mass Transfer		
j-Factor for Mass Transfer	$j_M = N_{St_M}(N_{Sc})^{2/3}$		j_H
Lewis Number	$N_{Le} = \dfrac{N_{Sc}}{N_{Pr}} = \dfrac{k}{\rho C_P D_{AB}} = \dfrac{\alpha}{D_{AB}}$	$\dfrac{\text{Thermal diffusivity}}{\text{Mass diffusivity}}$	
Peclet Number for Mass Transfer	$N_{Pe_M} = N_{Re}N_{Sc} = \dfrac{L\bar{u}}{D_{AB}}$	$\dfrac{\text{Bulk transfer of mass}}{\text{Molecular diffusion}}$	N_{Pe_H}
Schmidt Number	$N_{Sc} = \dfrac{\mu}{\rho D_{AB}} = \dfrac{\nu}{D_{AB}}$	$\dfrac{\text{Momentum diffusivity}}{\text{Mass diffusivity}}$	N_{Pr}
Sherwood Number	$N_{Sh} = \dfrac{k_c L}{D_{AB}}$	$\dfrac{\text{Convective mass transfer}}{\text{Molecular diffusion}}$	N_{Nu}
Stanton Number for Mass Transfer	$N_{St_M} = \dfrac{N_{Sh}}{N_{Re}N_{Sc}} = \dfrac{k_c}{\bar{u}}$	$\dfrac{\text{Mass transfer}}{\text{Mass capacity}}$	N_{St_H}

L = characteristic length G = mass velocity = $\bar{u}\rho$

Subscripts: M = mass transfer   H = heat transfer

mean fluid velocity in the direction of flow and on position in the fluid with respect to the solid boundaries.

In 1925, in an attempt to quantify turbulent transport, Prandtl [52] developed an expression for μ_t in terms of an eddy mixing length, l, which is a function of position. The eddy mixing length is a measure of the average distance that an eddy travels before it loses its identity and mingles with other eddies. The mixing length is analogous to the mean free path of gas molecules, which is the average distance a molecule travels before it collides with another molecule.

By analogy, the same mixing length is valid for turbulent-flow heat transfer and mass transfer. To use this analogy, (3-151) to (3-153) are rewritten in diffusivity form:

$$\frac{\tau_{zx}}{\rho} = -(\nu + \epsilon_M)\frac{du_x}{dz} \qquad (3\text{-}154)$$

$$\frac{q_z}{C_P\rho} = -(\alpha + \epsilon_H)\frac{dT}{dz} \qquad (3\text{-}155)$$

$$N_{A_z} = -(D_{AB} + \epsilon_D)\frac{dc_A}{dz} \qquad (3\text{-}156)$$

where ϵ_M, ϵ_H, are ϵ_D are momentum, heat, and mass eddy diffusivities, respectively; v is the momentum diffusivity (kinematic viscosity), μ/ρ; and α is the thermal diffusivity, $k/\rho C_P$. As a first approximation, the three eddy diffusivities may be assumed equal. This assumption is reasonably valid for ϵ_H and ϵ_D, but experimental data indicate that $\epsilon_M/\epsilon_H = \epsilon_M/\epsilon_D$ is sometimes less than 1.0 and as low as 0.5 for turbulence in a free jet.

Reynolds Analogy

If (3-154) to (3-156) are applied at a solid boundary, they can be used to determine transport fluxes based on transport coefficients, with driving forces from the wall, i, at $z = 0$, to the bulk fluid, designated with an overbar, $\bar{\ }$:

$$\frac{\tau_{zx}}{\bar{u}_x} = -(v + \epsilon_M)\left.\frac{d(\rho u_x/\bar{u}_x)}{dz}\right|_{z=0} = \frac{f\rho}{2}\bar{u}_x \qquad (3\text{-}157)$$

$$q_z = -(\alpha + \epsilon_H)\left.\frac{d(\rho C_P T)}{dz}\right|_{z=0} = h(T_i - \bar{T}) \qquad (3\text{-}158)$$

$$N_{A_z} = -(D_{AB} + \epsilon_D)\left.\frac{dc_A}{dz}\right|_{z=0} = k_c(c_{A_i} - \bar{c}_A) \qquad (3\text{-}159)$$

We define dimensionless velocity, temperature, and solute concentration by

$$\theta = \frac{u_x}{\bar{u}_x} = \frac{T_i - T}{T_i - \bar{T}} = \frac{c_{A_i} - c_A}{c_{A_i} - \bar{c}_A} \qquad (3\text{-}160)$$

If (3-160) is substituted into (3-157) to (3-159),

$$\left.\frac{\partial\theta}{\partial z}\right|_{z=0} = \frac{f\bar{u}_x}{2(v + \epsilon_M)} = \frac{h}{\rho C_P(\alpha + \epsilon_H)}$$
$$= \frac{k_c}{(D_{AB} + \epsilon_D)} \qquad (3\text{-}161)$$

This equation defines the analogies among momentum, heat, and mass transfer. Assuming that the three eddy diffusivities are equal and that the molecular diffusivities are either everywhere negligible or equal,

$$\frac{f}{2} = \frac{h}{\rho C_P \bar{u}_x} = \frac{k_c}{\bar{u}_x} \qquad (3\text{-}162)$$

Equation (3-162) defines the Stanton number for heat transfer,

$$N_{St_H} = \frac{h}{\rho C_P \bar{u}_x} = \frac{h}{G C_P} \qquad (3\text{-}163)$$

where G = mass velocity = $\bar{u}_x\rho$, and the Stanton number for mass transfer,

$$N_{St_M} = \frac{k_c}{\bar{u}_x} = \frac{k_c\rho}{G} \qquad (3\text{-}164)$$

both of which are included in Table 3.13.

Equation (3-162) is referred to as the Reynolds analogy. It can be used to estimate values of heat and mass transfer coefficients from experimental measurements of the Fanning friction factor for turbulent flow, but only when

$N_{Pr} = N_{Sc} = 1$. Thus, the Reynolds analogy has limited practical value and is rarely applied in practice. Reynolds postulated the existence of the analogy in 1874 [53] and derived it in 1883 [50].

Chilton–Colburn Analogy

A widely used extension of the Reynolds analogy to Prandtl and Schmidt numbers other than 1 was presented by Colburn [54] for heat transfer and by Chilton and Colburn [55] for mass transfer. They showed that the Reynolds analogy for turbulent flow could be corrected for differences in velocity, temperature, and concentration distributions by incorporating N_{Pr} and N_{Sc} into (3-162) to define the following three Chilton–Colburn j-factors, included in Table 3.13.

$$j_M \equiv \frac{f}{2} = j_H \equiv \frac{h}{G C_P}(N_{Pr})^{2/3}$$
$$= j_D \equiv \frac{k_c\rho}{G}(N_{Sc})^{2/3} \qquad (3\text{-}165)$$

Equation (3-165) is the Chilton–Colburn analogy or the Colburn analogy for estimating average transport coefficients for turbulent flow. When $N_{Pr} = N_{Sc} = 1$, (3-165) reduces to (3-162).

In general, j-factors are uniquely determined by the geometric configuration and the Reynolds number. Based on the analysis, over many years, of experimental data on momentum, heat, and mass transfer, the following representative correlations have been developed for turbulent transport to or from smooth surfaces. Other correlations are presented in other chapters. In general, these correlations are reasonably accurate for N_{Pr} and N_{Sc} in the range of 0.5 to 10, but should be used with caution outside this range.

1. Flow through a straight, circular tube of inside diameter D:
$$j_M = j_H = j_D = 0.023(N_{Re})^{-0.2} \qquad (3\text{-}166)$$
$$\text{for}\quad 10{,}000 < N_{Re} = DG/\mu < 1{,}000{,}000$$

2. Average transport coefficients for flow across a flat plate of length L:
$$j_M = j_H = j_D = 0.037(N_{Re})^{-0.2} \qquad (3\text{-}167)$$
$$\text{for}\quad 5 \times 10^5 < N_{Re} = Lu_0\rho/\mu < 5 \times 10^8$$

3. Average transport coefficients for flow normal to a long, circular cylinder of diameter D, where the drag coefficient includes both form drag and skin friction, but only the skin friction contribution applies to the analogy:
$$(j_M)_{\text{skin friction}} = j_H = j_D = 0.193(N_{Re})^{-0.382}$$
$$\text{for}\quad 4{,}000 < N_{Re} < 40{,}000 \qquad (3\text{-}168)$$
$$(j_M)_{\text{skin friction}} = j_H = j_D = 0.0266(N_{Re})^{-0.195}$$
$$\text{for}\quad 40{,}000 < N_{Re} < 250{,}000 \qquad (3\text{-}169)$$

$$\text{with} \qquad N_{Re} = \frac{DG}{\mu}$$

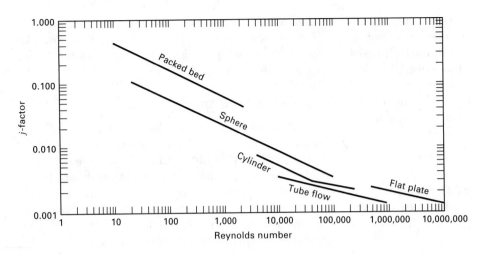

Figure 3.17   Chilton–Colburn j-factor correlations.

4. Average transport coefficients for flow past a single sphere of diameter D:

$$(j_M)_{\text{skin friction}} = j_H = j_D = 0.37(N_{\text{Re}})^{-0.4} \quad (3\text{-}170)$$

$$\text{for} \quad 20 < N_{\text{Re}} = \frac{DG}{\mu} < 100,000$$

5. Average transport coefficients for flow through beds packed with spherical particles of uniform size D_P:

$$j_H = j_D = 1.17(N_{\text{Re}})^{-0.415}$$

$$\text{for} \quad 10 < N_{\text{Re}} = \frac{D_P G}{\mu} < 2,500 \quad (3\text{-}171)$$

The above correlations are plotted in Figure 3.17, where the curves do not coincide because of the differing definitions of the Reynolds number. However, the curves are not widely separated. When using the correlations in the presence of appreciable temperature and/or composition differences, Chilton and Colburn recommend that N_{Pr} and N_{Sc} be evaluated at the average conditions from the surface to the bulk stream.

Other Analogies

New turbulence theories have led to improvements and extensions of the Reynolds analogy, resulting in expressions for the Fanning friction factor and the Stanton numbers for heat and mass transfer that are less empirical than the Chilton–Colburn analogy. The first major improvement was by Prandtl [56] in 1910, who divided the flow into two regions: (1) a thin laminar-flow sublayer of thickness δ next to the wall boundary, where only molecular transport occurs; and (2) a turbulent region dominated by eddy transport, with $\epsilon_M = \epsilon_H = \epsilon_D$.

Further important theoretical improvements to the Reynolds analogy were made by von Karman, Martinelli, and Deissler, as discussed in detail by Knudsen and Katz [47]. The first two investigators inserted a buffer zone between the laminar sublayer and turbulent core. Deissler gradually reduced the eddy diffusivities as the wall was approached.

Other improvements were made by van Driest [64], who used a modified form of the Prandtl mixing length, Reichardt [65], who eliminated the zone concept by allowing the eddy diffusivities to decrease continuously from a maximum to zero at the wall, and Friend and Metzner [57], who modified the approach of Reichardt to obtain improved accuracy at very high Prandtl and Schmidt numbers (to 3,000). Their results for turbulent flow through a straight, circular tube are

$$N_{\text{St}_H} = \frac{f/2}{1.20 + 11.8\sqrt{f/2}(N_{\text{Pr}} - 1)(N_{\text{Pr}})^{-1/3}} \quad (3\text{-}172)$$

$$N_{\text{St}_M} = \frac{f/2}{1.20 + 11.8\sqrt{f/2}(N_{\text{Sc}} - 1)(N_{\text{Sc}})^{-1/3}} \quad (3\text{-}173)$$

Over a wide range of Reynolds number (10,000–10,000,000), the Fanning friction factor is estimated from the explicit empirical correlation of Drew, Koo, and McAdams [66],

$$f = 0.00140 + 0.125(N_{\text{Re}})^{-0.32} \quad (3\text{-}174)$$

which is in excellent agreement with the experimental data of Nikuradse [67] and is preferred over (3-165) with (3-166), which is valid only to $N_{\text{Re}} = 1,000,000$. For two- and three-dimensional turbulent-flow problems, some success has been achieved with the κ (kinetic energy of turbulence)–ϵ (rate of dissipation) model of Launder and Spalding [68], which is widely used in computational fluid dynamics (CFD) computer programs.

Theoretical Analogy of Churchill and Zajic

An alternative to (3-151) to (3-153) or the equivalent diffusivity forms of (3-154) to (3-156) for the development of transport equations for turbulent flow is to start with the time-averaged equations of Newton, Fourier, and Fick. For example, let us derive a form of Newton's law of viscosity for molecular and turbulent transport of momentum in parallel. In a turbulent-flow field in the axial x-direction, instantaneous velocity components, u_x and u_z, are

$$u_x = \bar{u}_x + u'_x$$

$$u_z = u'_z$$

where the "overbarred" component is the time-averaged (mean) local velocity and the primed component is the local fluctuating component that denotes instantaneous deviation from the local mean value. The mean velocity in the perpendicular z-direction is zero. The mean local velocity in the x-direction over a long period Θ of time θ is given by

$$\bar{u}_x = \frac{1}{\Theta} \int_0^\Theta u_x d\theta = \frac{1}{\Theta} \int_0^\Theta (\bar{u}_x + u'_x)\, d\theta \qquad (3\text{-}175)$$

The time-averaged fluctuating components u'_x and u'_z equal zero.

The local instantaneous rate of momentum transfer by turbulence in the z-direction of x-direction turbulent momentum per unit area at constant density is

$$\rho u'_z(\bar{u}_x + u'_x) \qquad (3\text{-}176)$$

The time-average of this turbulent momentum transfer is equal to the turbulent component of the shear stress, τ_{zx_t},

$$\begin{aligned}
\tau_{zx_t} &= \frac{\rho}{\Theta} \int_0^\Theta u'_z(\bar{u}_x + u'_x)\, d\theta \\
&= \frac{\rho}{\Theta}\left[\int_0^\Theta u'_z(\bar{u}_x)\, d\theta + \int_0^\Theta u'_z(u'_x)\, d\theta \right]
\end{aligned} \qquad (3\text{-}177)$$

Because the time-average of the first term is zero, (3-177) reduces to

$$\tau_{zx_t} = \rho\, (\overline{u'_z u'_x}) \qquad (3\text{-}178)$$

which is referred to as a Reynolds stress. Combining (3-178) with the molecular component of momentum transfer gives the turbulent-flow form of Newton's law of viscosity,

$$\tau_{zx} = -\mu \frac{du_x}{dz} + \rho\, (\overline{u'_z u'_x}) \qquad (3\text{-}179)$$

If (3-179) is compared to (3-151), it is seen that an alternative approach to turbulence is to develop a correlating equation for the Reynolds stress, $(\overline{u'_z u'_x})$, first introduced by Churchill and Chan [73], rather than an expression for a turbulent viscosity μ_t. This stress, which is a complex function of position and rate of flow, has been correlated quite accurately for fully developed turbulent flow in a straight, circular tube by Heng, Chan, and Churchill [69]. In generalized form, with a the radius of the tube and $y = (a - z)$ the distance from the inside wall to the center of the tube, their equation is

$$\begin{aligned}
(\overline{u'_z u'_x})^{++} = &\left(\left[0.7 \left(\frac{y^+}{10} \right)^3 \right]^{-8/7} + \left| \exp\left\{ \frac{-1}{0.436 y^+} \right\} \right. \right. \\
&\left. \left. - \frac{1}{0.436 a^+} \left(1 + \frac{6.95 y^+}{a^+} \right) \right|^{-8/7} \right)^{-7/8}
\end{aligned} \qquad (3\text{-}180)$$

where

$$(\overline{u'_z u'_x})^{++} = -\rho \overline{u'_z u'_x}/\tau$$
$$a^+ = a(\tau_w \rho)^{1/2}/\mu$$
$$y^+ = y(\tau_w \rho)^{1/2}/\mu$$

Equation (3-180) is a highly accurate quantitative representation of turbulent flow because it is based on experimental data and numerical simulations described by Churchill and Zajic [70] and in considerable detail by Churchill [71]. From (3-142) and (3-143), the shear stress at the wall, τ_w, is related to the Fanning friction factor by

$$f = \frac{2\tau_w}{\rho \bar{u}_x^2} \qquad (3\text{-}181)$$

where $\bar{u}_x$ is the flow-average velocity in the axial direction. Combining (3-179) with (3-181) and performing the required integrations, both numerically and analytically, lead to the following implicit equation for the Fanning friction factor as a function of the Reynolds number, $N_{Re} = 2a\bar{u}_x\rho/\mu$:

$$\begin{aligned}
\left(\frac{2}{f} \right)^{1/2} = &\, 3.2 - 227 \frac{\left(\dfrac{2}{f} \right)^{1/2}}{\dfrac{N_{Re}}{2}} + 2500 \left[\frac{\left(\dfrac{2}{f} \right)^{1/2}}{\dfrac{N_{Re}}{2}} \right]^2 \\
&+ \frac{1}{0.436} \ln \left[\frac{\dfrac{N_{Re}}{2}}{\left(\dfrac{2}{f} \right)^{1/2}} \right]
\end{aligned} \qquad (3\text{-}182)$$

This equation is in excellent agreement with experimental data over a Reynolds number range of 4,000–3,000,000 and can probably be used to a Reynolds number of 100,000,000. Table 3.14 presents a comparison of the Churchill–Zajic equation, (3-182), with (3-174) of Drew et al. and (3-166) of Chilton and Colburn. Equation (3-174) gives satisfactory agreement for Reynolds numbers from 10,000 to 10,000,000, while (3-166) is useful only from 100,000 to 1,000,000.

Churchill and Zajic [70] show that if the equation for the conservation of energy is time averaged, a turbulent-flow form of Fourier's law of conduction can be obtained with the fluctuation term $(\overline{u'_z T'})$. Similar time averaging leads to a turbulent-flow form of Fick's law of diffusion with $(\overline{u'_z c'_A})$. To extend (3-180) and (3-182) to obtain an expression for the Nusselt number for turbulent-flow convective heat transfer in a straight, circular tube, Churchill and Zajic employ an analogy that is free of empircism, but not exact. The result

Table 3.14 Comparison of Fanning Friction Factors for Fully Developed Turbulent Flow in a Smooth, Straight Circular Tube

N_{Re}	f, Drew et al. (3-174)	f, Chilton–Colburn (3-166)	f, Churchill–Zajic (3-182)
10,000	0.007960	0.007291	0.008087
100,000	0.004540	0.004600	0.004559
1,000,000	0.002903	0.002902	0.002998
10,000,000	0.002119	0.001831	0.002119
100,000,000	0.001744	0.001155	0.001573

for Prandtl numbers greater than 1 is

$$N_{Nu} = \frac{1}{\left(\dfrac{N_{Pr_t}}{N_{Pr}}\right)\dfrac{1}{N_{Nu_1}} + \left[1 - \left(\dfrac{N_{Pr_t}}{N_{Pr}}\right)^{2/3}\right]\dfrac{1}{N_{Nu_\infty}}} \quad (3\text{-}183)$$

where, from Yu, Ozoe, and Churchill [72],

$$N_{Pr_t} = \text{turbulent Prandtl number} = 0.85 + \frac{0.015}{N_{Pr}} \quad (3\text{-}184)$$

which replaces $\overline{(u_z' T')}$, as introduced by Churchill [74],

N_{Nu_1} = Nusselt number for $(N_{Pr} = N_{Pr_t})$

$$= \frac{N_{Re}\left(\dfrac{f}{2}\right)}{1 + 145\left(\dfrac{2}{f}\right)^{-5/4}} \quad (3\text{-}185)$$

N_{Nu_∞} = Nusselt number for $(N_{Pr} = \infty)$

$$= 0.07343\left(\frac{N_{Pr}}{N_{Pr_t}}\right)^{1/3} N_{Re}\left(\frac{f}{2}\right)^{1/2} \quad (3\text{-}186)$$

The accuracy of (3-183) is due to (3-185) and (3-186), which are known from theoretical considerations. Although (3-184) is somewhat uncertain, its effect is negligible.

A comparison of the Churchill et al. correlation of (3-183) with the Nusselt forms of (3-172) of Friend and Metzner and (3-166) of Chilton and Colburn, where from Table 3.13, $N_{Nu} = N_{St}N_{Re}N_{Pr}$, is given in Table 3.15 for a wide range of Reynolds number and Prandtl numbers of 1 and 1,000.

In Table 3.15, at a Prandtl number of 1, which is typical of low-viscosity liquids and close to that of most gases, the Chilton–Colburn correlation, which is widely used, is within 10% of the more theoretically based Churchill–Zajic equation for Reynolds numbers up to 1,000,000. However, beyond that, serious deviations occur (25% at $N_{Re} = 10,000,000$ and almost 50% at $N_{Re} = 100,000,000$). Deviations of the Friend–Metzner correlation from the Churchill–Zajic equation vary from about 15% to 30% over the entire range of Reynolds number in Table 3.15. At all Reynolds numbers, the Churchill–Zajic equation predicts higher Nusselt numbers and, therefore, higher heat-transfer coefficients.

At a Prandtl number of 1,000, which is typical of high-viscosity liquids, the Friend–Metzner correlation is in fairly close agreement with the Churchill–Zajic equation, predicting values from 6 to 13% higher. The Chilton–Colburn correlation is seriously in error over the entire range of Reynolds number, predicting values ranging from 74 to 27% of those from the Churchill–Zajic equation as the Reynolds number increases. It is clear that the Chilton–Colburn correlation should not be used at high Prandtl numbers for heat transfer or (by analogy) at high Schmidt numbers for mass transfer.

The Churchill–Zajic equation for predicting the Nusselt number provides an effective power dependence on the Reynolds number as the Reynolds number increases. This is in contrast to the typically cited constant exponent of 0.8, as in the Chilton–Colburn correlation. For the Churchill–Zajic equation, at a Prandtl number of 1, the exponent increases with Reynolds number from 0.79 to 0.88; at a Prandtl number of 1,000, the exponent increases from 0.87 to 0.93.

Extension of the Churchill–Zajic equation to low Prandtl numbers, typical of molten metals, and to other geometries, such as parallel plates, is discussed by Churchill [71], who also considers the important effect of boundary conditions

Table 3.15   Comparison of Nusselt Numbers for Fully Developed Turbulent Flow in a Smooth, Straight Circular Tube

	Prandtl number, $N_{Pr} = 1$		
N_{Re}	N_{Nu}, Friend–Metzner (3-172)	N_{Nu}, Chilton–Colburn (3-166)	N_{Nu}, Churchill–Zajic (3-183)
10,000	33.2	36.5	37.8
100,000	189	230	232
1,000,000	1210	1450	1580
10,000,000	8830	9160	11400
100,000,000	72700	57800	86000

	Prandtl number, $N_{Pr} = 1000$		
N_{Re}	N_{Nu}, Friend–Metzner (3-172)	N_{Nu}, Chilton–Colburn (3-166)	N_{Nu}, Churchill–Zajic (3-183)
10,000	527	365	491
100,000	3960	2300	3680
1,000,000	31500	14500	29800
10,000,000	267800	91600	249000
100,000,000	2420000	578000	2140000

(e.g., constant wall temperature and uniform heat flux) at low-to-moderate Prandtl numbers.

For calculation of convective mass-transfer coefficients, k_c, for turbulent flow of gases and liquids in straight, smooth, circular tubes, it is recommended that the Churchill–Zajic equation be employed by applying the analogy between heat and mass transfer. Thus, as illustrated in the following example, in (3-183) to (3-186), from Table 3.13, the Sherwood number, N_{Sh}, is substituted for the Nusselt number, N_{Nu}; and the Schmidt number, N_{Sc}, is substituted for the Prandtl number, N_{Pr}.

EXAMPLE 3.16

Linton and Sherwood [49] conducted experiments on the dissolving of cast tubes of cinnamic acid (A) into water (B) flowing through the tubes in turbulent flow. In one run, with a 5.23-cm-i.d. tube, $N_{Re} = 35{,}800$, and $N_{Sc} = 1{,}450$, they measured a Stanton number for mass transfer, N_{St_M}, of 0.0000351. Compare this experimental value with predictions by the Reynolds, Chilton–Colburn, and Friend–Metzner analogies, and by the more theoretically-based Churchill–Zajic equation.

SOLUTION

From either (3-174) or (3-182), the Fanning friction factor is 0.00576.

Reynolds analogy:

From (3-162), $N_{St_M} = \frac{f}{2} = 0.00576/2 = 0.00288$, which, as expected, is in poor agreement with the experimental value because the effect of Schmidt number is ignored.

Chilton–Colburn analogy:

From (3-165),

$$N_{St_M} = \left(\frac{f}{2}\right) \Big/ (N_{Sc})^{2/3} = \left(\frac{0.00576}{2}\right) \Big/ (1450)^{2/3} = 0.0000225,$$

which is 64% of the experimental value.

Friend–Metzner analogy:

From (3-173), $N_{St_M} = 0.0000350$, which is almost identical to the experimental value.

Churchill–Zajic equation:

Using mass-transfer analogs,

$$(3\text{-}184) \text{ gives } N_{Sc_t} = 0.850$$
$$(3\text{-}185) \text{ gives } N_{Sh_1} = 94$$
$$(3\text{-}186) \text{ gives } N_{Sh_\infty} = 1686$$
$$(3\text{-}183) \text{ gives } N_{Sh} = 1680$$

From Table 3.13,

$$N_{St_M} = \frac{N_{Sh}}{N_{Re}N_{Sc}} = \frac{1680}{(35800)(1450)} = 0.0000324,$$

which is an acceptable 92% of the experimental value.

3.6 MODELS FOR MASS TRANSFER AT A FLUID–FLUID INTERFACE

In the three previous sections, diffusion and mass transfer within solids and fluids were considered, where the interface was a smooth solid surface. Of greater interest in separation processes is mass transfer across an interface between a gas and a liquid or between two liquid phases. Such interfaces exist in absorption, distillation, extraction, and stripping. At fluid–fluid interfaces, turbulence may persist to the interface. The following theoretical models have been developed to describe mass transfer between a fluid and such an interface.

Film Theory

A simple theoretical model for turbulent mass transfer to or from a fluid-phase boundary was suggested in 1904 by Nernst [58], who postulated that the entire resistance to mass transfer in a given turbulent phase is in a thin, stagnant region of that phase at the interface, called a film. This film is similar to the laminar sublayer that forms when a fluid flows in the turbulent regime parallel to a flat plate. This is shown schematically in Figure 3.18a for the case of a gas–liquid interface, where the gas is pure component A, which diffuses into nonvolatile liquid B. Thus, a process of absorption of A into liquid B takes place, without desorption of B into

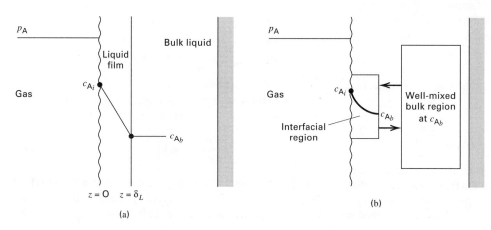

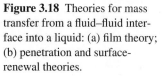

Figure 3.18 Theories for mass transfer from a fluid–fluid interface into a liquid: (a) film theory; (b) penetration and surface-renewal theories.

gaseous A. Because the gas is pure A at total pressure $P = p_A$, there is no resistance to mass transfer in the gas phase. At the gas–liquid interface, phase equilibrium is assumed so the concentration of A, c_{A_i}, is related to the partial pressure of A, p_A, by some form of Henry's law, for example, $c_{A_i} = H_A p_A$. In the thin, stagnant liquid film of thickness δ, molecular diffusion only occurs with a driving force of $c_{A_i} - c_{A_b}$. Since the film is assumed to be very thin, all of the diffusing A passes through the film and into the bulk liquid. If, in addition, bulk flow of A is neglected, the concentration gradient is linear as in Figure 3.18a. Accordingly, Fick's first law, (3-3a), for the diffusion flux integrates to

$$J_A = \frac{D_{AB}}{\delta}(c_{A_i} - c_{A_b}) = \frac{cD_{AB}}{\delta}(x_{A_i} - x_{A_b}) \quad (3\text{-}187)$$

If the liquid phase is dilute in A, the bulk-flow effect can be neglected and (3-187) applies to the total flux:

$$N_A = \frac{D_{AB}}{\delta}(c_{A_i} - c_{A_b}) = \frac{cD_{AB}}{\delta}(x_{A_i} - x_{A_b}) \quad (3\text{-}188)$$

If the bulk-flow effect is not negligible, then, from (3-31),

$$N_A = \frac{cD_{AB}}{\delta} \ln\left[\frac{1 - x_{A_b}}{1 - x_{A_i}}\right] = \frac{cD_{AB}}{\delta(1 - x_A)_{LM}}(x_{A_i} - x_{A_b}) \quad (3\text{-}189)$$

where

$$(1 - x_A)_{LM} = \frac{x_{A_i} - x_{A_b}}{\ln[(1 - x_{A_b})/(1 - x_{A_i})]} = (x_B)_{LM} \quad (3\text{-}190)$$

In practice, the ratios D_{AB}/δ in (3-188) and $D_{AB}/\delta(1 - x_A)_{LM}$ in (3-189) are replaced by mass transfer coefficients k_c and k_c', respectively, because the film thickness, δ, which depends on the flow conditions, is not known and the subscript, c, refers to a concentration driving force.

The film theory, which is easy to understand and apply, is often criticized because it appears to predict that the rate of mass transfer is directly proportional to the molecular diffusivity. This dependency is at odds with experimental data, which indicate a dependency of D^n, where n ranges from about 0.5 to 0.75. However, if D_{AB}/δ is replaced with k_c, which is then estimated from the Chilton–Colburn analogy, Eq. (3-165), we obtain k_c proportional to $D_{AB}^{2/3}$, which is in better agreement with experimental data. In effect, δ depends on D_{AB} (or N_{Sc}). Regardless of whether the criticism of the film theory is valid, the theory has been and continues to be widely used in the design of mass-transfer separation equipment.

EXAMPLE 3.17

Sulfur dioxide is absorbed from air into water in a packed absorption tower. At a certain location in the tower, the mass-transfer flux is 0.0270 kmol SO_2/m²-h and the liquid-phase mole fractions are 0.0025 and 0.0003, respectively, at the two-phase interface and in

the bulk liquid. If the diffusivity of SO_2 in water is 1.7×10^{-5} cm²/s, determine the mass-transfer coefficient, k_c, and the film thickness, neglecting the bulk-flow effect.

SOLUTION

$$N_{SO_2} = \frac{0.027(1,000)}{(3,600)(100)^2} = 7.5 \times 10^{-7} \frac{\text{mol}}{\text{cm}^2\text{-s}}$$

For dilute conditions, the concentration of water is

$$c = \frac{1}{18.02} = 5.55 \times 10^{-2} \text{ mol/cm}^3$$

From (3-188),

$$k_c = \frac{D_{AB}}{\delta} = \frac{N_A}{c(x_{A_i} - x_{A_b})}$$

$$= \frac{7.5 \times 10^{-7}}{5.55 \times 10^{-2}(0.0025 - 0.0003)} = 6.14 \times 10^{-3} \text{ cm/s}$$

Therefore,

$$\delta = \frac{D_{AB}}{k_c} = \frac{1.7 \times 10^{-5}}{6.14 \times 10^{-3}} = 0.0028 \text{ cm}$$

which is very small and typical of turbulent-flow mass-transfer processes.

Penetration Theory

A more realistic physical model of mass transfer from a fluid–fluid interface into a bulk liquid stream is provided by the penetration theory of Higbie [59], shown schematically in Figure 3.18b. The stagnant-film concept is replaced by Boussinesq eddies that, during a cycle, (1) move from the bulk to the interface; (2) stay at the interface for a short, fixed period of time during which they remain static so that molecular diffusion takes place in a direction normal to the interface; and (3) leave the interface to mix with the bulk stream. When an eddy moves to the interface, it replaces another static eddy. Thus, the eddies are alternately static and moving. Turbulence extends to the interface.

In the penetration theory, unsteady-state diffusion takes place at the interface during the time the eddy is static. This process is governed by Fick's second law, (3-68), with boundary conditions

$$c_A = c_{A_b} \quad \text{at} \quad t = 0 \quad \text{for} \quad 0 \leq z \leq \infty;$$
$$c_A = c_{A_i} \quad \text{at} \quad z = 0 \quad \text{for} \quad t > 0; \quad \text{and}$$
$$c_A = c_{A_b} \quad \text{at} \quad z = \infty \quad \text{for} \quad t > 0$$

These are the same boundary conditions as in unsteady-state diffusion in a semi-infinite medium. Thus, the solution can be written by a rearrangement of (3-75):

$$\frac{c_{A_i} - c_A}{c_{A_i} - c_{A_b}} = \text{erf}\left(\frac{z}{2\sqrt{D_{AB}t_c}}\right) \quad (3\text{-}191)$$

where t_c = "contact time" of the static eddy at the interface during one cycle. The corresponding average mass-transfer flux of A in the absence of bulk flow is given by the following form of (3-79):

$$N_A = 2\sqrt{\frac{D_{AB}}{\pi t_c}}(c_{A_i} - c_{A_b}) \qquad (3\text{-}192)$$

or

$$N_A = k_c(c_{A_i} - c_{A_b}) \qquad (3\text{-}193)$$

Thus, the penetration theory gives

$$k_c = 2\sqrt{\frac{D_{AB}}{\pi t_c}} \qquad (3\text{-}194)$$

which predicts that k_c is proportional to the square root of the molecular diffusivity, which is at the lower limit of experimental data.

The penetration theory is most useful when mass transfer involves bubbles or droplets, or flow over random packing. For bubbles, the contact time, t_c, of the liquid surrounding the bubble is taken as the ratio of bubble diameter to bubble-rise velocity. For example, an air bubble of 0.4-cm diameter rises through water at a velocity of about 20 cm/s. Thus, the estimated contact time, t_c, is $0.4/20 = 0.02$ s. For a liquid spray, where no circulation of liquid occurs inside the droplets, the contact time is the total time for the droplets to fall through the gas. For a packed tower, where the liquid flows as a film over particles of random packing, mixing can be assumed to occur each time the liquid film passes from one piece of packing to another. Resulting contact times are of the order of about 1 s. In the absence of any method of estimating the contact time, the liquid-phase mass-transfer coefficient is sometimes correlated by an empirical expression consistent with the 0.5 exponent on D_{AB}, given by (3-194) with the contact time replaced by a function of geometry and the liquid velocity, density, and viscosity.

EXAMPLE 3.18

For the conditions of Example 3.17, estimate the contact time for Higbie's penetration theory.

SOLUTION

From Example 3.17, $k_c = 6.14 \times 10^{-3}$ cm/s and $D_{AB} = 1.7 \times 10^{-5}$ cm²/s. From a rearrangement of (3-194),

$$t_c = \frac{4D_{AB}}{\pi k_c^2} = \frac{4(1.7 \times 10^{-5})}{3.14(6.14 \times 10^{-3})^2} = 0.57 \text{ s}$$

Surface-Renewal Theory

The penetration theory is not satisfying because the assumption of a constant contact time for all eddies that temporarily reside at the surface is not reasonable, especially for stirred tanks, contactors with random packings, and bubble and spray columns where the bubbles and droplets cover a wide range of sizes. In 1951, Danckwerts [60] suggested an improvement to the penetration theory that involves the replacement of the constant eddy contact time with the assumption of a residence-time distribution, wherein the probability of an eddy at the surface being replaced by a fresh eddy is independent of the age of the surface eddy.

Following the Levenspiel [61] treatment of residence-time distribution, let $F(t)$ be the fraction of eddies with a contact time of less than t. For $t = 0$, $F\{t\} = 0$, and $F\{t\}$ approaches 1 as t goes to infinity. A plot of $F\{t\}$ versus t, as shown in Figure 3.19, is referred to as a residence-time or age distribution. If $F\{t\}$ is differentiated with respect to t, we obtain another function:

$$\phi\{t\} = dF\{t\}/dt$$

where $\phi\{t\}dt = $ the probability that a given surface eddy will have a residence time t. The sum of probabilities is

$$\int_0^\infty \phi\{t\}\, dt = 1 \qquad (3\text{-}195)$$

Typical plots of $F\{t\}$ and $\phi\{t\}$ are shown in Figure 3.19, where it is seen that $\phi\{t\}$ is similar to a normal probability curve.

For steady-state flow in and out of a well-mixed vessel, Levenspiel shows that

$$F\{t\} = 1 - e^{-t/\bar{t}} \qquad (3\text{-}196)$$

where $\bar{t}$ is the average residence time. This function forms the basis, in reaction engineering, of the ideal model of a continuous, stirred-tank reactor (CSTR). Danckwerts selected the same model for his surface-renewal theory, using the corresponding $\phi\{t\}$ function:

$$\phi\{t\} = se^{-st} \qquad (3\text{-}197)$$

where $s = 1/\bar{t} = $ fractional rate of surface renewal. As shown in Example 3.19 below, plots of (3-196) and (3-197) are much different from those in Figure 3.19.

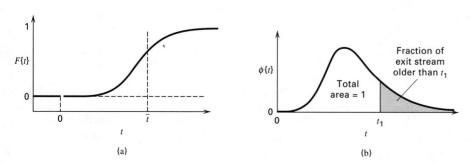

Figure 3.19 Residence-time distribution plots: (a) typical F curve; (b) typical age distribution.

[Adapted from O. Levenspiel, *Chemical Reaction Engineering*, 2nd ed., John Wiley and Sons, New York (1972).]

The instantaneous mass-transfer rate for an eddy with an age t is given by (3-192) for the penetration theory in flux form as

$$N_{A_t} = \sqrt{\frac{D_{AB}}{\pi t}} (c_{A_i} - c_{A_b}) \qquad (3\text{-}198)$$

The integrated average rate is

$$(N_A)_{avg} = \int_0^{\infty} \phi\{t\} N_{A_t} \, dt \qquad (3\text{-}199)$$

Combining (3-197), (3-198), and (3-199), and integrating:

$$(N_A)_{avg} = \sqrt{D_{AB}s} (c_{A_i} - c_{A_b}) \qquad (3\text{-}200)$$

Thus,

$$k_c = \sqrt{D_{AB}s} \qquad (3\text{-}201)$$

The more reasonable surface-renewal theory predicts the same dependency of the mass-transfer coefficient on molecular diffusivity as the penetration theory. Unfortunately, s, the fractional rate of surface renewal, is as elusive a parameter as the constant contact time, t_c.

EXAMPLE 3.19

For the conditions of Example 3.17, estimate the fractional rate of surface renewal, s, for Danckwert's theory and determine the residence time and probability distributions.

SOLUTION

From Example 3.17,

$$k_c = 6.14 \times 10^{-3} \text{ cm/s} \quad \text{and} \quad D_{AB} = 1.7 \times 10^{-5} \text{ cm}^2/\text{s}$$

From (3-201),

$$s = \frac{k_c^2}{D_{AB}} = \frac{(6.14 \times 10^{-3})^2}{1.7 \times 10^{-5}} = 2.22 \text{ s}^{-1}$$

Thus, the average residence time of an eddy at the surface is $1/2.22 = 0.45$ s.
From (3-197),

$$\phi\{t\} = 2.22 e^{-2.22t} \qquad (1)$$

From (3-196), the residence-time distribution is given by

$$F\{t\} = 1 - e^{-t/0.45}, \qquad (2)$$

where t is in seconds. Equations (1) and (2) are plotted in Figure 3.20. These curves are much different from the curves of Figure 3.19.

Film-Penetration Theory

Toor and Marchello [62], in 1958, combined features of the film, penetration, and surface-renewal theories to develop a film-penetration model, which predicts a dependency of the mass-transfer coefficient k_c, on the diffusivity, that varies from $\sqrt{D_{AB}}$ to D_{AB}. Their theory assumes that the entire resistance to mass transfer resides in a film of fixed thickness δ. Eddies move to and from the bulk fluid and this film. Age distributions for time spent in the film are of the Higbie or Danckwerts type.

Fick's second law, (3-68), still applies, but the boundary conditions are now

$$c_A = c_{A_b} \quad \text{at} \quad t = 0 \quad \text{for} \quad 0 \le z \le \infty,$$
$$c_A = c_{A_i} \quad \text{at} \quad z = 0 \quad \text{for} \quad t > 0; \quad \text{and}$$
$$c_A = c_{A_b} \quad \text{at} \quad z = \delta \quad \text{for} \quad t > 0$$

Infinite-series solutions are obtained by the method of Laplace transforms. The rate of mass transfer is then obtained in the usual manner by applying Fick's first law (3-117) at the fluid–fluid interface. For small t, the solution, given as

$$N_{A_t} = (c_{A_i} - c_{A_b}) \left(\frac{D_{AB}}{\pi t}\right)^{1/2} \left[1 + 2\sum_{n=1}^{\infty} \exp\left(-\frac{n^2\delta^2}{D_{AB}t}\right)\right] \qquad (3\text{-}202)$$

converges rapidly. For large t,

$$N_{A_t} = (c_{A_i} - c_{A_b}) \left(\frac{D_{AB}}{\delta}\right)$$
$$\times \left[1 + 2\sum_{n=1}^{\infty} \exp\left(-n^2\pi^2 \frac{D_{AB}t}{\delta^2}\right)\right] \qquad (3\text{-}203)$$

Equation (3-199) with $\phi\{t\}$ from (3-197) can then be used to obtain average rates of mass transfer. Again, we can write two equivalent series solutions, which converge

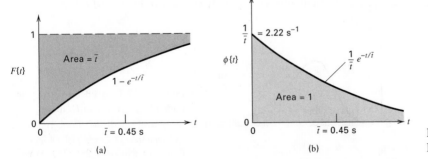

Figure 3.20 Age distribution curves for Example 3.19: (a) F curve; (b) $\phi\{t\}$ curve.

at different rates. Equations (3-202) and (3-203) become, respectively,

$$N_{A_{avg}} = k_c(c_{A_i} - c_{A_b}) = (c_{A_i} - c_{A_b})(s D_{AB})^{1/2}$$
$$\times \left[1 + 2 \sum_{n=1}^{\infty} \exp\left(-2n\delta \sqrt{\frac{s}{D_{AB}}} \right) \right] \quad (3\text{-}204)$$

$$N_{A_{avg}} = k_c(c_{A_i} - c_{A_b}) = (c_{A_i} - c_{A_b}) \left(\frac{D_{AB}}{\delta} \right)$$
$$\times \left[1 + 2 \sum_{n=1}^{\infty} \frac{1}{1 + n^2\pi^2 \dfrac{D_{AB}}{s\delta^2}} \right] \quad (3\text{-}205)$$

In the limit, for a high rate of surface renewal, $s\delta^2/D_{AB}$, (3-204) reduces to the surface-renewal theory, (3-200). For low rates of renewal, (3-205) reduces to the film theory, (3-188). At conditions in between, k_c is proportional to D_{AB}^n, where n is in the range of 0.5–1.0. The application of the film-penetration theory is difficult because of lack of data for δ and s, but the predicted effect of molecular diffusivity brackets experimental data.

3.7 TWO-FILM THEORY AND OVERALL MASS-TRANSFER COEFFICIENTS

Separation processes that involve contacting two fluid phases require consideration of mass-transfer resistances in both phases. In 1923, Whitman [63] suggested an extension of the film theory to two fluid films in series. Each film presents a resistance to mass transfer, but concentrations in the two fluids at the interface are assumed to be in phase equilibrium. That is, there is no additional interfacial resistance to mass transfer. This concept has found extensive application in modeling of steady-state, gas–liquid, and liquid–liquid separation processes.

The assumption of phase equilibrium at the phase interface, while widely used, may not be valid when gradients of interfacial tension are established during mass transfer between two fluids. These gradients give rise to interfacial turbulence resulting, most often, in considerably increased mass-transfer coefficients. This phenomenon, referred to as

the *Marangoni effect,* is discussed in some detail by Bird, Stewart, and Lightfoot [28], who cite additional references. The effect can occur at both vapor–liquid and liquid–liquid interfaces, with the latter receiving the most attention. By adding surfactants, which tend to concentrate at the interface, the Marangoni effect may be reduced because of stabilization of the interface, even to the extent that an interfacial mass-transfer resistance may result, causing the overall mass-transfer coefficient to be reduced. In this book, unless otherwise indicated, the Marangoni effect will be ignored and phase equilibrium will always be assumed at the phase interface.

Gas–Liquid Case

Consider the steady-state mass transfer of A from a gas phase, across an interface, into a liquid phase. It could be postulated, as shown in Figure 3.21a, that a thin gas film exists on one side of the interface and a thin liquid film exists on the other side, with the controlling factors being molecular diffusion through each film. However, this postulation is not necessary, because instead of writing

$$N_A = \frac{(D_{AB})_G}{\delta_G}(c_{A_b} - c_{A_i})_G = \frac{(D_{AB})_L}{\delta_L}(c_{A_i} - c_{A_b})_L$$
$$(3\text{-}206)$$

we can express the rate of mass transfer in terms of mass-transfer coefficients determined from any suitable theory, with the concentration gradients visualized more realistically as in Figure 3.21b. In addition, we can use any number of different mass-transfer coefficients, depending on the selection of the driving force for mass transfer. For the gas phase, under dilute or equimolar counter diffusion (EMD) conditions, we write the mass-transfer rate in terms of partial pressures:

$$N_A = k_p(p_{A_b} - p_{A_i}) \quad (3\text{-}207)$$

where k_p is a gas-phase mass-transfer coefficient based on a partial-pressure driving force.

For the liquid phase, we use molar concentrations:

$$N_A = k_c(c_{A_i} - c_{A_b}) \quad (3\text{-}208)$$

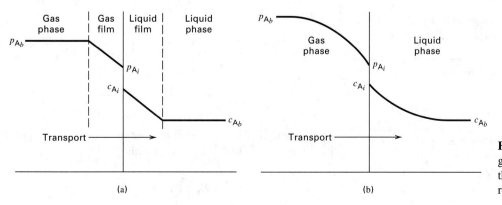

Figure 3.21 Concentration gradients for two-resistance theory: (a) film theory; (b) more realistic gradients.

At the phase interface, c_{A_i} and p_{A_i} are assumed to be in phase equilibrium. Applying a version of Henry's law different from that in Table 2.3,[1]

$$c_{A_i} = H_A p_{A_i} \qquad (3\text{-}209)$$

Equations (3-207) to (3-209) are a commonly used combination for vapor–liquid mass transfer. Computations of mass-transfer rates are generally made from a knowledge of bulk concentrations, which in this case are p_{A_b} and c_{A_b}. To obtain an expression for N_A in terms of an overall driving force for mass transfer, (3-207) to (3-209) are combined in the following manner to eliminate the interfacial concentrations, c_{A_i} and p_{A_i}. Solve (3-207) for p_{A_i}:

$$p_{A_i} = p_{A_b} - \frac{N_A}{k_p} \qquad (3\text{-}210)$$

Solve (3-208) for c_{A_i}:

$$c_{A_i} = c_{A_b} + \frac{N_A}{k_c} \qquad (3\text{-}211)$$

Combine (3-211) with (3-209) to eliminate c_{A_i} and combine the result with (3-210) to eliminate p_{A_i} to give

$$N_A = \frac{p_{A_b} H_A - c_{A_b}}{(H_A/k_p) + (1/k_c)} \qquad (3\text{-}212)$$

It is customary to define: (1) a fictitious liquid-phase concentration $c_A^* = p_{A_b} H_A$, which is the concentration that would be in equilibrium with the partial pressure in the bulk gas; and (2) an overall mass-transfer coefficient, K_L. Thus, (3-212) is rewritten as

$$N_A = K_L(c_A^* - c_{A_b}) = \frac{(c_A^* - c_{A_b})}{(H_A/k_p) + (1/k_c)} \qquad (3\text{-}213)$$

where

$$\frac{1}{K_L} = \frac{H_A}{k_p} + \frac{1}{k_c} \qquad (3\text{-}214)$$

in which K_L is the overall mass-transfer coefficient based on the liquid phase. The quantities H_A/k_p and $1/k_c$ are measures of the mass-transfer resistances of the gas phase and the liquid phase, respectively. When $1/k_c \gg H_A/k_p$, (3-214) becomes

$$N_A = k_c(c_A^* - c_{A_b}) \qquad (3\text{-}215)$$

Since resistance in the gas phase is then negligible, the gas-phase driving force is $p_{A_b} - p_{A_i} \approx 0$ and $p_{A_b} \approx p_{A_i}$.

---

[1]Many different forms of Henry's law are found in the literature. They include

$$p_A = H_A x_A, \quad p_A = \frac{c_A}{H_A}, \quad \text{and} \quad y_A = H_A x_A$$

When a Henry's-law constant, H_A, is given without citing the equation that defines it, the defining equation can be determined from the units of the constant. For example, if the constant has the units of atm or atm/mole fraction, Henry's law is given by $p_A = H_A x_A$. If the units are mol/L-mmHg, Henry's law is $p_A = \frac{c_A}{H_A}$.

Alternatively, (3-207) to (3-209) can be combined to define an overall mass-transfer coefficient, K_G, based on the gas phase. The result is

$$N_A = \frac{p_{A_b} - c_{A_b}/H_A}{(1/k_p) + (1/H_A k_c)} \qquad (3\text{-}216)$$

In this case, it is customary to define: (1) a fictitious gas-phase partial pressure $p_A^* = c_{A_b}/H_A$, which is the partial pressure that would be in equilibrium with the bulk liquid; and (2) an overall mass-transfer coefficient for the gas phase, K_G, based on a partial-pressure driving force. Thus, (3-216) can be rewritten as

$$N_A = K_G(p_{A_b} - p_A^*) = \frac{(p_{A_b} - p_A^*)}{(1/k_p) + (1/H_A k_c)} \qquad (3\text{-}217)$$

where

$$\frac{1}{K_G} = \frac{1}{k_p} + \frac{1}{H_A k_c} \qquad (3\text{-}218)$$

In this, the resistances are $1/k_p$ and $1/(H_A k_c)$. When $1/k_p \gg 1/H_A k_c$,

$$N_A = k_p(p_{A_b} - p_A^*) \qquad (3\text{-}219)$$

Since the resistance in the liquid phase is then negligible, the liquid-phase driving force is $c_{A_i} - c_{A_b} \approx 0$ and $c_{A_i} \approx c_{A_b}$.

The choice between using (3-213) or (3-217) is arbitrary, but is usually made on the basis of which phase has the largest mass-transfer resistance; if the liquid, use (3-213); if the gas, use (3-217). Another common combination for vapor–liquid mass transfer uses mole fraction-driving forces, which define another set of mass-transfer coefficients:

$$N_A = k_y(y_{A_b} - y_{A_i}) = k_x(x_{A_i} - x_{A_b}) \qquad (3\text{-}220)$$

In this case, phase equilibrium at the interface can be expressed in terms of the K-value for vapor–liquid equilibrium. Thus,

$$K_A = y_{A_i}/x_{A_i} \qquad (3\text{-}221)$$

Combining (3-220) and (3-221) to eliminate y_{A_i} and x_{A_i},

$$N_A = \frac{y_{A_b} - x_{A_b}}{(1/K_A k_y) + (1/k_x)} \qquad (3\text{-}222)$$

This time we define fictitious concentration quantities and overall mass-transfer coefficients for mole-fraction driving forces. Thus, $x_A^* = y_{A_b}/K_A$ and $y_A^* = K_A x_{A_b}$. If the two values of K_A are equal, we obtain

$$N_A = K_x(x_A^* - x_{A_b}) = \frac{x_A^* - x_{A_b}}{(1/K_A k_y) + (1/k_x)} \qquad (3\text{-}223)$$

and

$$N_A = K_y(y_{A_b} - y_A^*) = \frac{y_{A_b} - y_A^*}{(1/k_y) + (K_A/k_x)} \qquad (3\text{-}224)$$

where K_x and K_y are overall mass-transfer coefficients based on mole-fraction driving forces with

$$\frac{1}{K_x} = \frac{1}{K_A k_y} + \frac{1}{k_x} \qquad (3\text{-}225)$$

and

$$\frac{1}{K_y} = \frac{1}{k_y} + \frac{K_A}{k_x} \qquad (3\text{-}226)$$

When using correlations to estimate mass-transfer coefficients for use in the above equations, it is important to determine which coefficient (k_p, k_c, k_y, or k_x) is correlated. This can usually be done by checking the units or the form of the Sherwood or Stanton numbers. Coefficients correlated by the Chilton–Colburn analogy are k_c for either the liquid or gas phase. The different coefficients are related by the following expressions, which are summarized in Table 3.16.

Liquid phase:

$$k_x = k_c c = k_c \left(\frac{\rho_L}{M}\right) \qquad (3\text{-}227)$$

Ideal-gas phase:

$$k_y = k_p P = (k_c)_g \frac{P}{RT} = (k_c)_g c = (k_c)_g \left(\frac{\rho_G}{M}\right) \quad (3\text{-}228)$$

Typical units are

	SI	**American Engineering**
k_c	m/s	ft/h
k_p	kmol/s-m²-kPa	lbmol/h-ft²-atm
k_y, k_x	kmol/s-m²	lbmol/h-ft²

When unimolecular diffusion (UMD) occurs under nondilute conditions, the effect of bulk flow must be included in the above equations. For binary mixtures, one method for doing this is to define modified mass-transfer coefficients, designated with a prime, as follows.

Table 3.16 Relationships among Mass-Transfer Coefficients

Equimolar Counterdiffusion (EMD):

Gases: $N_A = k_y \Delta y_A = k_c \Delta c_A = k_p \Delta p_A$

$\qquad k_y = k_c \dfrac{P}{RT} = k_p P$ if ideal gas

Liquids: $N_A = k_x \Delta x_A = k_c \Delta c_A$

$\qquad k_x = k_c c$, where $c =$ total molar concentration (A + B)

Unimolecular Diffusion (UMD):

Gases: Same equations as for EMD with k replaced by $k' = \dfrac{k}{(y_B)_{LM}}$

Liquids: Same equations as for EMD with k replaced by $k' = \dfrac{k}{(x_B)_{LM}}$

When using concentration units for both phases, it is convenient to use:

$\qquad k_G(\Delta c_G) = k_c(\Delta c)$ for the gas phase
$\qquad k_L(\Delta c_L) = k_c(\Delta c)$ for the liquid phase

For the liquid phase, using k_c or k_x,

$$k' = \frac{k}{(1 - x_A)_{LM}} = \frac{k}{(x_B)_{LM}} \qquad (3\text{-}229)$$

For the gas phase, using k_p, k_y, or k_c,

$$k' = \frac{k}{(1 - y_A)_{LM}} = \frac{k}{(y_B)_{LM}} \qquad (3\text{-}230)$$

The expressions for k' are most readily used when the mass-transfer rate is controlled mainly by one of the two resistances. Experimental mass-transfer coefficient data reported in the literature are generally correlated in terms of k rather than k'. Mass-transfer coefficients estimated from the Chilton–Colburn analogy [e.g., equations (3-166) to (3-171)] are k_c, not k'_c.

Liquid–Liquid Case

For mass transfer across two liquid phases, equilibrium is again assumed at the interface. Denoting the two phases by $L^{(1)}$ and $L^{(2)}$, (3-223) and (3-224) can be rewritten as

$$N_A = K_x^{(2)}\left(x_A^{(2)*} - x_{A_b}^{(2)}\right) = \frac{x_A^{(2)*} - x_{A_b}^{(2)}}{(1/K_{D_A}k_x^{(1)}) + (1/k_x^{(2)})} \quad (3\text{-}231)$$

and

$$N_A = K_x^{(1)}\left(x_{A_b}^{(1)} - x_A^{(1)*}\right) = \frac{x_{A_b}^{(1)} - x_A^{(1)*}}{(1/k_x^{(1)}) + (K_{D_A}/k_x^{(2)})} \quad (3\text{-}232)$$

where

$$K_{D_A} = \frac{x_{A_i}^{(1)}}{x_{A_i}^{(2)}} \qquad (3\text{-}233)$$

Case of Large Driving Forces for Mass Transfer

When large driving forces exist for mass transfer, phase equilibria ratios such as H_A, K_A, and K_{D_A} may not be constant across the two phases. This occurs particularly when one or both phases are not dilute with respect to the diffusing solute, A. In that case, expressions for the mass-transfer flux must be revised.

For example, if mole-fraction driving forces are used, we write, from (3-220) and (3-224),

$$N_A = k_y(y_{A_b} - y_{A_i}) = K_y(y_{A_b} - y_A^*) \qquad (3\text{-}234)$$

Thus,

$$\frac{1}{K_y} = \frac{y_{A_b} - y_A^*}{k_y(y_{A_b} - y_{A_i})} \qquad (3\text{-}235)$$

or

$$\frac{1}{K_y} = \frac{(y_{A_b} - y_{A_i}) + (y_{A_i} - y_A^*)}{k_y(y_{A_b} - y_{A_i})} = \frac{1}{k_y} + \frac{1}{k_y}\left(\frac{y_{A_i} - y_A^*}{y_{A_b} - y_{A_i}}\right) \qquad (3\text{-}236)$$

From (3-220),

$$\frac{k_x}{k_y} = \frac{(y_{A_b} - y_{A_i})}{(x_{A_i} - x_{A_b})} \qquad (3\text{-}237)$$

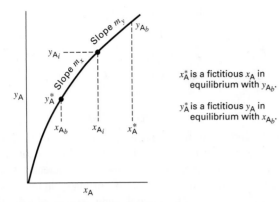

Figure 3.22 Curved equilibrium line.

x_A^* is a fictitious x_A in equilibrium with y_{A_b}.

y_A^* is a fictitious y_A in equilibrium with x_{A_b}.

Combining (3-234) and (3-237),

$$\frac{1}{K_y} = \frac{1}{k_y} + \frac{1}{k_x}\left(\frac{y_{A_i} - y_A^*}{x_{A_i} - x_{A_b}}\right) \quad (3\text{-}238)$$

In a similar manner,

$$\frac{1}{K_x} = \frac{1}{k_x} + \frac{1}{k_y}\left(\frac{x_A^* - x_{A_i}}{y_{A_b} - y_{A_i}}\right) \quad (3\text{-}239)$$

A typical curved equilibrium line is shown in Figure 3.22 with representative values of y_{A_b}, y_{A_i}, y_A^*, x_A^*, x_{A_i}, and x_{A_b} indicated. Because the line is curved, the vapor–liquid equilibrium ratio, $K_A = y_A/x_A$, is not constant across the two phases. As shown, the slope of the curve and thus, K_A, decreases with increasing concentration of A. Denote two slopes of the equilibrium line by

$$m_x = \left(\frac{y_{A_i} - y_A^*}{x_{A_i} - x_{A_b}}\right) \quad (3\text{-}240)$$

and

$$m_y = \left(\frac{y_{A_b} - y_{A_i}}{x_A^* - x_{A_i}}\right) \quad (3\text{-}241)$$

Substituting (3-240) and (3-241) into (3-238) and (3-239), respectively, gives

$$\frac{1}{K_y} = \frac{1}{k_y} + \frac{m_x}{k_x} \quad (3\text{-}242)$$

and

$$\frac{1}{K_x} = \frac{1}{k_x} + \frac{1}{m_y k_y} \quad (3\text{-}243)$$

EXAMPLE 3.20

Sulfur dioxide (A) is absorbed into water in a packed column. At a certain location, the bulk conditions are 50°C, 2 atm, $y_{A_b} = 0.085$, and $x_{A_b} = 0.001$. Equilibrium data for SO_2 between air and water at 50°C are

p_{SO_2}, atm	c_{SO_2}, lbmol/ft^3
0.0382	0.00193
0.0606	0.00290
0.1092	0.00483
0.1700	0.00676

Experimental values of the mass transfer coefficients are as follows.

Liquid phase: $k_c = 0.18$ m/h

Gas phase: $k_p = 0.040\dfrac{\text{kmol}}{\text{h-m}^2\text{-kPa}}$

Using mole-fraction driving forces, compute the mass-transfer flux by:

(a) Assuming an average Henry's-law constant and a negligible bulk-flow effect.

(b) Utilizing the actual curved equilibrium line and assuming a negligible bulk-flow effect.

(c) Utilizing the actual curved equilibrium line and taking into account the bulk-flow effect.

In addition,

(d) Determine the relative magnitude of the two resistances and the values of the mole fractions at the interface from the results of part (c).

SOLUTION

The equilibrium data are converted to mole fractions by assuming Dalton's law, $y_A = p_A/P$, for the gas and using $x_A = c_A/c$ for the liquid. The concentration of the liquid is close to that of pure water or 3.43 lbmol/ft^3 or 55.0 kmol/m^3. Thus, the mole fractions at equilibrium are:

y_{SO_2}	x_{SO_2}
0.0191	0.000563
0.0303	0.000846
0.0546	0.001408
0.0850	0.001971

These data are fitted with average and maximum absolute deviations of 0.91% and 1.16%, respectively, by the quadratic equation

$$y_{SO_2} = 29.74 x_{SO_2} + 6{,}733 x_{SO_2}^2 \quad (1)$$

Thus, differentiating, the slope of the equilibrium curve is given by

$$m = \frac{dy}{dx} = 29.74 + 13{,}466 x_{SO_2} \quad (2)$$

The given mass-transfer coefficients can be converted to k_x and k_y by (3-227) and (3-228):

$$k_x = k_c c = 0.18(55.0) = 9.9\frac{\text{kmol}}{\text{h-m}^2}$$

$$k_y = k_p P = 0.040(2)(101.3) = 8.1\frac{\text{kmol}}{\text{h-m}^2}.$$

(a) From (1) for $x_{A_b} = 0.001$, $y_A^* = 29.74(0.001) + 6{,}733(0.001)^2$ $= 0.0365$. From (1), for $y_{A_b} = 0.085$, we solve the quadratic equation to obtain $x_A^* = 0.001975$.

The average slope in this range is

$$m = \frac{0.085 - 0.0365}{0.001975 - 0.001} = 49.7 .$$

From an examination of (3-242) and (3-243), the liquid-phase resistance is controlling because the term in k_x is much larger than the term in k_y. Therefore, from (3-243), using $m = m_x$,

$$\frac{1}{K_x} = \frac{1}{9.9} + \frac{1}{49.7(8.1)} = 0.1010 + 0.0025 = 0.1035$$

$$\text{or} \quad K_x = 9.66 \frac{\text{kmol}}{\text{h-m}^2}$$

From (3-223),

$$N_A = 9.66(0.001975 - 0.001) = 0.00942 \frac{\text{kmol}}{\text{h-m}^2} .$$

(b) From part (a), the gas-phase resistance is almost negligible. Therefore, $y_{A_i} \approx y_{A_b}$ and $x_{A_i} \approx x_A^*$.

From (3-241), the slope m_y must, therefore, be taken at the point $y_{A_b} = 0.085$ and $x_A^* = 0.001975$ on the equilibrium line. From (2), $m_y = 29.74 + 13,466(0.001975) = 56.3$. From (3-243),

$$K_x = \frac{1}{(1/9.9) + [1/(56.3)(8.1)]} = 9.69 \frac{\text{kmol}}{\text{h-m}^2} ,$$

giving $N_A = 0.00945$ kmol/h-m^2. This is only a slight change from part (a).

(c) We now correct for bulk flow. From the results of parts (a) and (b), we have

$$y_{A_b} = 0.085, \ y_{A_i} = 0.085, \ x_{A_i} = 0.1975, \ x_{A_b} = 0.001$$

$$(y_B)_{LM} = 1.0 - 0.085 = 0.915 \text{ and } (x_B)_{LM} \approx 0.9986$$

From (3-229),

$$k_x' = \frac{9.9}{0.9986} = 9.9 \frac{\text{kmol}}{\text{h-m}^2} \text{ and } k_y' = \frac{8.1}{0.915} = 8.85 \frac{\text{kmol}}{\text{h-m}^2}$$

From (3-243),

$$K_x = \frac{1}{(1/9.9) + [1/56.3(8.85)]} = 9.71 \frac{\text{kmol}}{\text{h-m}^2}$$

From (3-223),

$$N_A = 9.71(0.001975 - 0.001) = 0.00947 \frac{\text{kmol}}{\text{h-m}^2}$$

which is only a very slight change from parts (a) and (b), where the bulk-flow effect was ignored. The effect is very small because here it is important only in the gas phase; but the liquid-phase resistance is controlling.

(d) The relative magnitude of the mass-transfer resistances can be written as

$$\frac{1/m_y k_y'}{1/k_x'} = \frac{1/(56.3)(8.85)}{1/9.9} = 0.02$$

Thus, the gas-phase resistance is only 2% of the liquid-phase resistance. The interface vapor mole fraction can be obtained from (3-223), after accounting for the bulk-flow effect:

$$y_{A_i} = y_{A_b} - \frac{N_A}{k_y'} = 0.085 - \frac{0.00947}{8.85} = 0.084$$

Similarly,

$$x_{A_i} = \frac{N_A}{k_x'} + x_{A_b} = \frac{0.00947}{9.9} + 0.001 = 0.00196$$

SUMMARY

1. Mass transfer is the net movement of a component in a mixture from one region to another region of different concentration, often between two phases across an interface. Mass transfer occurs by molecular diffusion, eddy diffusion, and bulk flow. Molecular diffusion occurs by a number of different driving forces, including concentration (the most important), pressure, temperature, and external force fields.

2. Fick's first law for steady-state conditions states that the mass-transfer flux by ordinary molecular diffusion is equal to the product of the diffusion coefficient (diffusivity) and the negative of the concentration gradient.

3. Two limiting cases of mass transfer are equimolar counterdiffusion (EMD) and unimolecular diffusion (UMD). The former is also a good approximation for dilute conditions. The latter must include the bulk-flow effect.

4. When experimental data are not available, diffusivities in gas and liquid mixtures can be estimated. Diffusivities in solids, including porous solids, crystalline solids, metals, glass, ceramics, polymers, and cellular solids, are best measured. For some solids—for example, wood—diffusivity is an anisotropic property.

5. Diffusivity values vary by orders of magnitude. Typical values are 0.10, 1×10^{-5}, and 1×10^{-9} cm^2/s for ordinary molecular diffusion of a solute in a gas, liquid, and solid, respectively.

6. Fick's second law for unsteady-state diffusion is readily applied to semi-infinite and finite stagnant media, including certain anisotropic materials.

7. Molecular diffusion under laminar-flow conditions can be determined from Fick's first and second laws, provided that velocity profiles are available. Common cases include falling liquid-film flow, boundary-layer flow on a flat plate, and fully developed flow in a straight, circular tube. Results are often expressed in terms of a mass-transfer coefficient embedded in a dimensionless group called the Sherwood number. The mass-transfer flux is given by the product of the mass-transfer coefficient and a concentration driving force.

8. Mass transfer in turbulent flow is often predicted by analogy to heat transfer. Of particular importance is the Chilton–Colburn analogy, which utilizes empirical j-factor correlations and the dimensionless Stanton number for mass transfer. A more accurate equation by Churchill and Zajic should be used for flow in tubes, particularly at high Schmidt and Reynolds numbers.

9. A number of models have been developed for mass transfer across a two-fluid interface and into a liquid. These include the film theory, penetration theory, surface-renewal theory, and the film-penetration theory. These theories predict mass-transfer coefficients that are proportional to the diffusivity raised to an exponent that varies from 0.5 to 1.0. Most experimental data provide exponents ranging from 0.5 to 0.75.

10. The two-film theory of Whitman (more properly referred to as a two-resistance theory) is widely used to predict the mass-transfer flux from one fluid phase, across an interface, and into another fluid phase, assuming equilibrium at the interface. One resistance is often controlling. The theory defines an overall mass-transfer coefficient that is determined from the separate coefficients for each of the two phases and the equilibrium relationship at the interface.

REFERENCES

1. TAYLOR, R., and R. KRISHNA, *Multicomponent Mass Transfer,* John Wiley and Sons, New York (1993).

2. POLING, B.E., J.M. PRAUSNITZ, and J.P. O'CONNELL, *The Properties of Liquids and Gases,* 5th ed., McGraw-Hill, New York (2001).

3. FULLER, E.N., P.D. SCHETTLER, and J.C. GIDDINGS, *Ind. Eng. Chem.,* **58** (5), 18–27 (1966).

4. TAKAHASHI, S., *J. Chem. Eng. Jpn.*, **7**, 417–420 (1974).

5. SLATTERY, J.C., M.S. thesis, University of Wisconsin, Madison (1955).

6. WILKE, C.R., and P. CHANG, *AIChE J.,* **1**, 264–270 (1955).

7. HAYDUK, W., and B.S. MINHAS, *Can. J. Chem. Eng.,* **60,** 295–299 (1982).

8. QUAYLE, O.R., *Chem. Rev.,* **53,** 439–589 (1953).

9. VIGNES, A., *Ind. Eng. Chem. Fundam.,* **5,** 189–199 (1966).

10. SORBER, H.A., *Handbook of Biochemistry, Selected Data for Molecular Biology,* 2nd ed., Chemical Rubber Co., Cleveland, OH (1970).

11. GEANKOPLIS, C.J., *Transport Processes and Separation Process Principles,* 4th ed., Prentice-Hall, Upper Saddle River, NJ (2003).

12. FRIEDMAN, L., and E.O. KRAEMER, *J. Am. Chem. Soc.,* **52,** 1298–1304, 1305–1310, 1311–1314 (1930).

13. BOUCHER, D.F., J.C. BRIER, and J.O. OSBURN, *Trans. AIChE,* **38,** 967–993 (1942).

14. BARRER, R.M., *Diffusion in and through Solids,* Oxford University Press, London (1951).

15. SWETS, D.E., R.W. LEE, and R.C. FRANK, *J. Chem. Phys.,* **34,** 17–22 (1961).

16. LEE, R. W., *J. Chem, Phys.,* **38,** 44–455 (1963).

17. WILLIAMS, E.L., *J. Am. Ceram. Soc.,* **48,** 190–194 (1965).

18. SUCOV, E.W., *J. Am. Ceram. Soc.,* **46,** 14–20 (1963).

19. KINGERY, W.D., H.K. BOWEN, and D.R. UHLMANN, *Introduction to Ceramics,* 2nd ed., John Wiley and Sons, New York (1976).

20. FERRY, J.D., *Viscoelastic Properties of Polymers,* John Wiley and Sons, New York (1980).

21. RHEE, C.K., and J.D. FERRY, *J. Appl. Polym. Sci.,* **21,** 467–476 (1977).

22. BRANDRUP, J., and E.H. IMMERGUT, Eds., *Polymer Handbook,* 3rd ed., John Wiley and Sons, New York (1989).

23. GIBSON, L.J., and M.F. ASHBY, *Cellular Solids, Structure and Properties,* Pergamon Press, Elmsford, NY (1988).

24. STAMM, A.J., *Wood and Cellulose Science,* Ronald Press, New York (1964).

25. SHERWOOD, T.K., *Ind. Eng. Chem.,* **21,** 12–16 (1929).

26. CARSLAW, H.S., and J.C. JAEGER, *Heat Conduction in Solids,* 2nd ed., Oxford University Press, London (1959).

27. CRANK, J., *The Mathematics of Diffusion,* Oxford University Press, London (1956).

28. BIRD, R.B., W.E. STEWART, and E.N. LIGHTFOOT, *Transport Phenomena,* 2nd ed., John Wiley and Sons, New York (2002).

29. CHURCHILL, R.V., *Operational Mathematics,* 2nd ed., McGraw-Hill, New York (1958).

30. ABRAMOWITZ, M., and I.A. STEGUN, Eds., *Handbook of Mathematical Functions,* National Bureau of Standards, Applied Mathematics Series 55, Washington, DC (1964).

31. NEWMAN, A.B., *Trans. AIChE,* **27,** 310–333 (1931).

32. GRIMLEY, S.S., *Trans. Inst. Chem. Eng.* (London), **23,** 228–235 (1948).

33. JOHNSTONE, H.F., and R.L. PIGFORD, *Trans. AIChE,* **38,** 25–51 (1942).

34. OLBRICH, W.E., and J.D. WILD, *Chem. Eng. Sci.,* **24,** 25–32 (1969).

35. CHURCHILL, S.W., *The Interpretation and Use of Rate Data: The Rate Concept,* McGraw-Hill, New York (1974).

36. CHURCHILL, S.W., and R. USAGI, *AIChE J.,* **18,** 1121–1128 (1972).

37. EMMERT, R.E., and R.L. PIGFORD, *Chem. Eng. Prog.,* **50,** 87–93 (1954).

38. PRANDTL, L., *Proc. 3rd Int. Math. Congress,* Heidelberg (1904); reprinted in *NACA Tech. Memo 452* (1928).

39. BLASIUS, H., *Z. Math Phys.,* **56,** 1–37 (1908); reprinted in *NACA Tech. Memo 1256.*

40. SCHLICHTING, H., *Boundary Layer Theory,* 4th ed., McGraw-Hill, New York (1960).

41. POHLHAUSEN, E., *Z. Angew. Math Mech.,* **1,** 252 (1921).

42. POHLHAUSEN, E., *Z. Angew. Math Mech.,* **1,** 115–121 (1921).

43. LANGHAAR, H.L., *Trans. ASME,* **64,** A-55 (1942).

44. GRAETZ, L., *Ann. d. Physik,* **25,** 337–357 (1885).

45. SELLARS, J.R., M. TRIBUS, and J.S. KLEIN, *Trans. ASME,* **78,** 441–448 (1956).

46. LEVEQUE, J., *Ann. Mines,* [12], **13,** 201, 305, 381 (1928).

47. KNUDSEN, J.G., and D.L. KATZ, *Fluid Dynamics and Heat Transfer,* McGraw-Hill, New York (1958).

48. HAUSEN, H., *Verfahrenstechnik Beih. z. Ver. Deut. Ing.,* **4,** 91 (1943).

49. LINTON, W.H. Jr., and T.K. SHERWOOD, *Chem. Eng. Prog.,* **46,** 258–264 (1950).

50. REYNOLDS, O., *Trans. Roy. Soc.* (London), **174A,** 935–982 (1883).

51. BOUSSINESQ, J., *Mem. Pre. Par. Div. Sav.,* XXIII, Paris (1877).

52. PRANDTL, L., *Z. Angew, Math Mech.,* **5,** 136 (1925); reprinted in *NACA Tech. Memo 1231* (1949).

53. REYNOLDS, O., *Proc. Manchester Lit. Phil. Soc.,* **14,** 7 (1874).

54. COLBURN, A.P., *Trans. AIChE,* **29,** 174–210 (1933).

55. CHILTON, T.H., and A.P. COLBURN, *Ind. Eng. Chem.,* **26,** 1183–1187 (1934).

56. PRANDTL, L., *Physik. Z.,* **11,** 1072 (1910).

57. FRIEND, W.L., and A.B. METZNER, *AIChE J.*, **4**, 393–402 (1958).

58. NERNST, W., *Z. Phys. Chem.*, **47**, 52 (1904).

59. HIGBIE, R., *Trans. AIChE*, **31**, 365–389 (1935).

60. DANCKWERTS, P.V., *Ind. Eng. Chem.*, **43**, 1460–1467 (1951).

61. LEVENSPIEL, O., *Chemical Reaction Engineering*, 3rd ed., John Wiley and Sons, New York (1999).

62. TOOR, H.L., and J.M. MARCHELLO, *AIChE J.*, **4**, 97–101 (1958).

63. WHITMAN, W.G., *Chem. Met. Eng.*, **29**, 146–148 (1923).

64. VAN DRIEST, E.R., *J. Aero Sci.*, 1007–1011 and 1036 (1956).

65. REICHARDT, H., *Fundamentals of Turbulent Heat Transfer*, NACA Report TM-1408 (1957).

66. DREW, T.B., E.C. KOO, and W.H. MCADAMS, *Trans. Am. Inst. Chem. Engrs.*, **28**, 56 (1933).

67. NIKURADSE, J., *VDI-Forschungsheft*, p. 361 (1933).

68. LAUNDER, B.E., and D.B. SPALDING, *Lectures in Mathematical Models of Turbulence,* Academic Press, New York (1972).

69. HENG, L., C. CHAN, and S.W. CHURCHILL, *Chem. Eng. J.*, **71**, 163 (1998).

70. CHURCHILL, S.W., and S.C. ZAJIC, *AIChE J.*, **48**, 927–940 (2002).

71. CHURCHILL, S.W., "Turbulent Flow and Convection: The Prediction of Turbulent Flow and Convection in a Round Tube," in *Advances in Heat Transfer,* J.P. Hartnett, and T.F. Irvine, Jr., Ser. Eds., Academic Press, New York, **34**, 255–361 (2001).

72. YU, B., H. OZOE, and S.W. CHURCHILL, *Chem. Eng. Sci.*, **56**, 1781 (2001).

73. CHURCHILL, S.W., and C. CHAN, *Ind. Eng. Chem. Res.*, **34**, 1332 (1995).

74. CHURCHILL, S.W., *AIChE J.*, **43**, 1125 (1997).

EXERCISES

Section 3.1

3.1   A beaker filled with an equimolar liquid mixture of ethyl alcohol and ethyl acetate evaporates at 0°C into still air at 101 kPa (1 atm) total pressure. Assuming Raoult's law applies, what will be the composition of the liquid remaining when half the original ethyl alcohol has evaporated, assuming that each component evaporates independently of the other? Also assume that the liquid is always well mixed. The following data are available:

	Vapor Pressure, kPa at 0°C	Diffusivity in Air m²/s
Ethyl acetate (AC)	3.23	6.45×10^{-6}
Ethyl alcohol (AL)	1.62	9.29×10^{-6}

3.2   An open tank, 10 ft in diameter and containing benzene at 25°C, is exposed to air in such a manner that the surface of the liquid is covered with a stagnant air film estimated to be 0.2 in. thick. If the total pressure is 1 atm and the air temperature is 25°C, what loss of material in pounds per day occurs from this tank? The specific gravity of benzene at 60°F is 0.877. The concentration of benzene at the outside of the film is so low that it may be neglected. For benzene, the vapor pressure at 25°C is 100 torr, and the diffusivity in air is 0.08 cm²/s.

3.3   An insulated glass tube and condenser are mounted on a reboiler containing benzene and toluene. The condenser returns liquid reflux so that it runs down the wall of the tube. At one point in the tube the temperature is 170°F, the vapor contains 30 mol% toluene, and the liquid reflux contains 40 mol% toluene. The effective thickness of the stagnant vapor film is estimated to be 0.1 in. The molar latent heats of benzene and toluene are equal. Calculate the rate at which toluene and benzene are being interchanged by equimolar countercurrent diffusion at this point in the tube in lbmol/h-ft².

Diffusivity of toluene in benzene = 0.2 ft²/h.

Pressure = 1 atm total pressure (in the tube).

Vapor pressure of toluene at 170°F = 400 torr.

3.4   Air at 25°C with a dew-point temperature of 0°C flows past the open end of a vertical tube filled with liquid water maintained at 25°C. The tube has an inside diameter of 0.83 in., and the liquid level was originally 0.5 in. below the top of the tube. The diffusivity of water in air at 25°C is 0.256 cm²/s.

(a) How long will it take for the liquid level in the tube to drop 3 in.?

(b) Make a plot of the liquid level in the tube as a function of time for this period.

3.5   Two bulbs are connected by a tube, 0.002 m in diameter and 0.20 m in length. Initially bulb 1 contains argon, and bulb 2 contains xenon. The pressure and temperature are maintained at 1 atm and 105°C, at which the diffusivity is 0.180 cm²/s. At time $t = 0$, diffusion is allowed to occur between the two bulbs. At a later time, the argon mole fraction in the gas at end 1 of the tube is 0.75, and 0.20 at the other end. Determine at the later time:

(a) The rates and directions of mass transfer of argon and xenon

(b) The transport velocity of each species

(c) The molar average velocity of the mixture

Section 3.2

3.6   The diffusivity of toluene in air was determined experimentally by allowing liquid toluene to vaporize isothermally into air from a partially filled vertical tube 3 mm in diameter. At a temperature of 39.4°C, it took 96×10^4 s for the level of the toluene to drop from 1.9 cm below the top of the open tube to a level of 7.9 cm below the top. The density of toluene is 0.852 g/cm³, and the vapor pressure is 57.3 torr at 39.4°C. The barometer reading was 1 atm. Calculate the diffusivity and compare it with the value predicted from (3-36). Neglect the counterdiffusion of air.

3.7   An open tube, 1 mm in diameter and 6 in. long, has pure hydrogen blowing across one end and pure nitrogen blowing across the other. The temperature is 75°C.

(a) For equimolar counterdiffusion, what will be the rate of transfer of hydrogen into the nitrogen stream (mol/s)? Estimate the diffusivity from (3-36).

(b) For part (a), plot the mole fraction of hydrogen against distance from the end of the tube past which nitrogen is blown.

3.8   Some HCl gas diffuses across a film of air 0.1 in. thick at 20°C. The partial pressure of HCl on one side of the film is 0.08 atm and it is zero on the other. Estimate the rate of diffusion, as mol HCl/s-cm², if the total pressure is (a) 10 atm, (b) 1 atm, (c) 0.1 atm. The diffusivity of HCl in air at 20°C and 1 atm is 0.145 cm²/s.

3.9 Estimate the diffusion coefficient for the gaseous binary system nitrogen (A)/toluene (B) at 25°C and 3 atm using the method of Fuller et al.

3.10 For the mixture of Example 3.3, estimate the diffusion coefficient if the pressure is increased to 100 atm using the method of Takahashi.

3.11 Estimate the diffusivity of carbon tetrachloride at 25°C in a dilute solution of: (a) Methanol, (b) Ethanol, (c) Benzene, and (d) n-Hexane by the method of Wilke–Chang and Hayduk–Minhas. Compare the estimated values with the following experimental observations:

Solvent	Experimental D_{AB}, cm²/s
Methanol	1.69×10^{-5} cm²/s at 15°C
Ethanol	1.50×10^{-5} cm²/s at 25°C
Benzene	1.92×10^{-5} cm²/s at 25°C
n-Hexane	3.70×10^{-5} cm²/s at 25°C

3.12 Estimate the liquid diffusivity of benzene (A) in formic acid (B) at 25°C and infinite dilution. Compare the estimated value to that of Example 3.6 for formic acid at infinite dilution in benzene.

3.13 Estimate the liquid diffusivity of acetic acid at 25°C in a dilute solution of: (a) Benzene, (b) Acetone, (c) Ethyl acetate, and (d) Water by an appropriate method. Compare the estimated values with the following experimental values:

Solvent	Experimental D_{AB}, cm²/s
Benzene	2.09×10^{-5} cm²/s at 25°C
Acetone	2.92×10^{-5} cm²/s at 25°C
Ethyl acetate	2.18×10^{-5} cm²/s at 25°C
Water	1.19×10^{-5} cm²/s at 20°C

3.14 Water in an open dish exposed to dry air at 25°C is found to vaporize at a constant rate of 0.04 g/h-cm². Assuming the water surface to be at the wet-bulb temperature of 11.0°C, calculate the effective gas-film thickness (i.e., the thickness of a stagnant air film that would offer the same resistance to vapor diffusion as is actually encountered at the water surface).

3.15 Isopropyl alcohol is undergoing mass transfer at 35°C and 2 atm under dilute conditions through water, across a phase boundary, and then through nitrogen. Based on the date given below, estimate for isopropyl alcohol:

(a) The diffusivity in water using the Wilke–Chang equation

(b) The diffusivity in nitrogen using the Fuller et al. equation

(c) The product, $D_{AB}\rho_M$, in water

(d) The product, $D_{AB}\rho_M$, in air

where ρ_M is the molar density of the mixture.

Using the above results, compare:

(e) The diffusivities in parts (a) and (b)

(f) The diffusivity-molar density products in Parts (c) and (d)

Lastly:

(g) What conclusions can you come to about molecular diffusion in the liquid phase versus the gaseous phase?

Data:

Component	T_c, °R	P_c, psia	Z_c	v_L, cm³/mol
Nitrogen	227.3	492.9	0.289	—
Isopropyl alcohol	915	691	0.249	76.5

3.16 Experimental liquid-phase activity-coefficient data are given in Exercise 2.23 for the ethanol/benzene system at 45°C. Estimate and plot diffusion coefficients for both ethanol and benzene over the entire composition range.

3.17 Estimate the diffusion coefficient of NaOH in a 1-M aqueous solution at 25°C.

3.18 Estimate the diffusion coefficient of NaCl in a 2-M aqueous solution at 18°C. Compare your estimate with the experimental value of 1.28×10^{-5} cm²/s.

3.19 Estimate the diffusivity of N_2 in H_2 in the pores of a catalyst at 300°C and 20 atm if the porosity is 0.45 and the tortuosity is 2.5. Assume ordinary molecular diffusion in the pores.

3.20 Gaseous hydrogen at 150 psia and 80°F is stored in a small, spherical, steel pressure vessel having an inside diameter of 4 in. and a wall thickness of 0.125 in. At these conditions, the solubility of hydrogen in steel is 0.094 lbmol/ft³ and the diffusivity of hydrogen in steel is 3.0×10^{-9} cm²/s. If the inner surface of the vessel remains saturated at the existing hydrogen pressure and the hydrogen partial pressure at the outer surface is assumed to be zero, estimate:

(a) The initial rate of mass transfer of hydrogen through the metal wall

(b) The initial rate of pressure decrease inside the vessel

(c) The time in hours for the pressure to decrease to 50 psia, assuming the temperature stays constant at 80°F

3.21 A polyisoprene membrane of 0.8-μm thickness is to be used to separate a mixture of methane and H_2. Using the data in Table 14.9 and the following compositions, estimate the mass-transfer flux of each of the two species.

	Partial Pressures, MPa	
	Membrane Side 1	Membrane Side 2
Methane	2.5	0.05
Hydrogen	2.0	0.20

Section 3.3

3.22 A 3-ft depth of stagnant water at 25°C lies on top of a 0.10-in. thickness of NaCl. At time < 0, the water is pure. At time = 0, the salt begins to dissolve and diffuse into the water. If the concentration of salt in the water at the solid–liquid interface is maintained at saturation (36 g NaCl/100 g H_2O) and the diffusivity of NaCl in water is 1.2×10^{-5} cm²/s, independent of concentration, estimate, by assuming the water to act as a semi-infinite medium, the time and the concentration profile of salt in the water when

(a) 10% of the salt has dissolved

(b) 50% of the salt has dissolved

(c) 90% of the salt has dissolved

3.23 A slab of dry wood of 4-in. thickness and sealed edges is exposed to air of 40% relative humidity. Assuming that the two unsealed faces of the wood immediately jump to an equilibrium moisture content of 10 lb H_2O per 100 lb of dry wood, determine the time for the moisture to penetrate to the center of the slab (2 in. from either face). Assume a diffusivity of water in the wood as 8.3×10^{-6} cm²/s.

3.24 A wet, clay brick measuring $2 \times 4 \times 6$ in. has an initial uniform moisture content of 12 wt%. At time = 0, the brick is exposed on all sides to air such that the surface moisture content is maintained at 2 wt%. After 5 h, the average moisture content is 8 wt%. Estimate:

(a) The diffusivity of water in the clay in cm^2/s.

(b) The additional time for the average moisture content to reach 4 wt%. All moisture contents are on a dry basis.

3.25 A spherical ball of clay, 2 in. in diameter, has an initial moisture content of 10 wt%. The diffusivity of water in the clay is 5×10^{-6} cm^2/s. At time $t = 0$, the surface of the clay is brought into contact with air such that the moisture content at the surface is maintained at 3 wt%. Estimate the time for the average moisture content in the sphere to drop to 5 wt%. All moisture contents are on a dry basis.

Section 3.4

3.26 Estimate the rate of absorption of pure oxygen at 10 atm and 25°C into water flowing as a film down a vertical wall 1 m high and 6 cm in width at a Reynolds number of 50 without surface ripples. Assume the diffusivity of oxygen in water is 2.5×10^{-5} cm^2/s and that the mole fraction of oxygen in water at saturation for the above temperature and pressure is 2.3×10^{-4}.

3.27 For the conditions of Example 3.13, determine at what height from the top the average concentration of CO_2 would correspond to 50% of saturation.

3.28 Air at 1 atm flows at 2 m/s across the surface of a 2-in.-long surface that is covered with a thin film of water. If the air and water are maintained at 25°C, and the diffusivity of water in air at these conditions is 0.25 cm^2/s, estimate the mass flux for the evaporation of water at the middle of the surface assuming laminar boundary-layer flow. Is this assumption reasonable?

3.29 Air at 1 atm and 100°C flows across a thin, flat plate of naphthalene that is 1 m long, causing the plate to sublime. The Reynolds number at the trailing edge of the plate is at the upper limit for a laminar boundary layer. Estimate:

(a) The average rate of sublimation in $kmol/s-m^2$

(b) The local rate of sublimation at a distance of 0.5 m from the leading edge of the plate

Physical properties are given in Example 3.14.

3.30 Air at 1 atm and 100°C flows through a straight, 5-cm-diameter circular tube, cast from naphthalene, at a Reynolds number of 1,500. Air entering the tube has an established laminar-flow velocity profile. Properties are given in Example 3.14. If pressure drop through the tube is negligible, calculate the length of tube needed for the average mole fraction of naphthalene in the exiting air to be 0.005.

3.31 A spherical water drop is suspended from a fine thread in still, dry air. Show:

(a) That the Sherwood number for mass transfer from the surface of the drop into the surroundings has a value of 2 if the characteristic length is the diameter of the drop.

If the initial drop diameter is 1 mm, the air temperature is 38°C, the drop temperature is 14.4°C, and the pressure is 1 atm, calculate:

(b) The initial mass of the drop in grams.

(c) The initial rate of evaporation in grams per second.

(d) The time in seconds for the drop diameter to be reduced to 0.2 mm.

(e) The initial rate of heat transfer to the drop. If the Nusselt number is also 2, is the rate of heat transfer sufficient to supply the heat of vaporization and sensible heat of the evaporated water? If not, what will happen?

Section 3.5

3.32 Water at 25°C flows at 5 ft/s through a straight, cylindrical tube cast from benzoic acid, of 2-in. inside diameter. If the tube is 10 ft long, and fully developed, turbulent flow is assumed, estimate the average concentration of benzoic acid in the water leaving the tube. Physical properties are given in Example 3.15.

3.33 Air at 1 atm flows at a Reynolds number of 50,000 normal to a long, circular, 1-in.-diameter cylinder made of naphthalene. Using the physical properties of Example 3.14 for a temperature of 100°C, calculate the average sublimation flux in $kmol/s-m^2$.

3.34 For the conditions of Exercise 3.33, calculate the initial average rate of sublimation in $kmol/s-m^2$ for a spherical particle of 1-in. initial diameter. Compare this result to that for a bed packed with naphthalene spheres with a void fraction of 0.5.

Section 3.6

3.35 Carbon dioxide is stripped from water by air in a wetted-wall tube. At a certain location, where the pressure is 10 atm and the temperature is 25°C, the mass-transfer flux of CO_2 is 1.62 $lbmol/h-ft^2$. The partial pressures of CO_2 are 8.2 atm at the interface and 0.1 atm in the bulk gas. The diffusivity of CO_2 in air at these conditions is 1.6×10^{-2} cm^2/s. Assuming turbulent flow of the gas, calculate by the film theory, the mass-transfer coefficient k_c for the gas phase and the film thickness.

3.36 Water is used to remove CO_2 from air by absorption in a column packed with Pall rings. At a certain region of the column where the partial pressure of CO_2 at the interface is 150 psia and the concentration in the bulk liquid is negligible, the absorption rate is 0.017 $lbmol/h-ft^2$. The diffusivity of CO_2 in water is 2.0×10^{-5} cm^2/s. Henry's law for CO_2 is $p = Hx$, where $H = 9,000$ psia. Calculate:

(a) The liquid-phase mass-transfer coefficient and the film thickness

(b) Contact time for the penetration theory

(c) Average eddy residence time and the probability distribution for the surface-renewal theory

3.37 Determine the diffusivity of H_2S in water, using the penetration theory, from the following data for the absorption of H_2S into a laminar jet of water at 20°C.

Jet diameter = 1 cm, Jet length = 7 cm, and Solubility of H_2S in water = 100 mol/m^3

The average rate of absorption varies with the flow rate of the jet as follows:

Jet Flow Rate, cm^3/s	Rate of Absorption, mol/s $\times 10^6$
0.143	1.5
0.568	3.0
1.278	4.25
2.372	6.15
3.571	7.20
5.142	8.75

Section 3.7

3.38 In a test on the vaporization of H_2O into air in a wetted-wall column, the following data were obtained:

Tube diameter, 1.46 cm, Wetted-tube length, 82.7 cm
Air rate to tube at 24°C and 1 atm, 720 cm^3/s

Temperature of inlet water, 25.15°C, Temperature of outlet water, 25.35°C

Partial pressure of water in inlet air, 6.27 torr, and in outlet air, 20.1 torr

The value for the diffusivity of water vapor in air is 0.22 cm²/s at 0°C and 1 atm. The mass velocity of air is taken relative to the pipe wall. Calculate:

(a) Rate of mass transfer of water into the air

(b) K_G for the wetted-wall column

3.39 The following data were obtained by Chamber and Sherwood [*Ind. Eng. Chem.*, **29**, 1415 (1937)] on the absorption of ammonia from an ammonia-air system by a strong acid in a wetted-wall column 0.575 in. in diameter and 32.5 in. long:

Inlet acid (2-N H$_2$SO$_4$) temperature, °F	76
Outlet acid temperature, °F	81
Inlet air temperature, °F	77
Outlet air temperature, °F	84
Total pressure, atm	1.00
Partial pressure NH$_3$ in inlet gas, atm	0.0807
Partial pressure NH$_3$ in outlet gas, atm	0.0205
Air rate, lbmol/h	0.260

The operation was countercurrent, with the gas entering at the bottom of the vertical tower and the acid passing down in a thin film on the inner wall. The change in acid strength was inappreciable, and the vapor pressure of ammonia over the liquid may be assumed to have been negligible because of the use of a strong acid for absorption. Calculate the mass-transfer coefficient, k_p, from the data.

3.40 A new type of cooling-tower packing is being tested in a laboratory column. At two points in the column, 0.7 ft apart, the following data have been taken. Calculate the overall volumetric mass-transfer coefficient $K_y a$ that can be used to design a large, packed-bed cooling tower, where a is the mass-transfer area, A, per unit volume, V, of tower.

	Bottom	**Top**
Water temperature, °F	120	126
Water vapor pressure, psia	1.69	1.995
Mole fraction H$_2$O in air	0.001609	0.0882
Total pressure, psia	14.1	14.3
Air rate, lbmol/h	0.401	0.401
Column area, ft²	0.5	0.5
Water rate, lbmol/h (approximate)	20	20

Chapter 4

Single Equilibrium Stages and Flash Calculations

The simplest separation process is one in which two phases in contact are brought to physical equilibrium, followed by phase separation. If the separation factor between two species in the two phases is very large, a single contacting stage may be sufficient to achieve a desired separation between them; if not, multiple stages are required. For example, if a vapor phase is in equilibrium with a liquid phase, the separation factor is the relative volatility, α, of a volatile component called the light key, LK, with respect to a less-volatile component called the heavy key, HK, where $\alpha_{LK,HK} = K_{LK}/K_{HK}$. If the separation factor is 10,000, an almost perfect separation is achieved in a single stage. If the separation factor is only 1.10, an almost perfect separation requires hundreds of stages. In this chapter, only a single equilibrium stage is considered, but a wide spectrum of separation operations is described. In all cases, the calculations are made by combining material balances with phase equilibria relations. When a phase change such as vaporization occurs, or when heat of mixing effects are large, an energy balance must be added. In the next chapter, arrangements of multiple stages, called cascades, are described.

4.0 INSTRUCTIONAL OBJECTIVES

After completing this chapter, you should be able to:

- Explain what an equilibrium stage is and why it may not be sufficient to achieve a desired separation.
- Use the Gibbs phase rule to determine the number of intensive variables that must be specified to fix the remaining intensive variables for a system at equilibrium.
- Extend Gibbs phase rule to include extensive variables so that the number of degrees of freedom (number of variables minus the number of independent relations among the variables) can be determined for a continuous separation process.
- Explain and utilize ways that binary vapor–liquid equilibrium data are presented.
- Define relative volatility between two components of a vapor-liquid mixture.
- Use T–y–x and y–x diagrams of binary mixtures, with the concept of the q-line, to determine equilibrium phase compositions.
- Understand the difference between minimum- and maximum-boiling azeotropes and how they form.
- Use component material-balance equations with K-values to calculate bubble-point, dew-point, and equilibrium-flash conditions for multicomponent mixtures.
- Use triangular phase diagrams for ternary systems with component material balances to determine equilibrium compositions of liquid–liquid mixtures.
- Use distribution coefficients, usually determined from activity coefficients, with component material-balance equations to calculate liquid–liquid phase equilibria for multicomponent systems.
- Use equilibrium diagrams with component material balances to determine equilibrium-phase amounts and compositions for solid–fluid systems (leaching, crystallization, sublimation, desublimation, and adsorption) and for light gas–liquid systems (absorption).
- Calculate phase amounts and compositions for multicomponent vapor–liquid–liquid systems.

4.1 THE GIBBS PHASE RULE AND DEGREES OF FREEDOM

The description of a single-stage system at physical equilibrium involves *intensive variables,* which are independent of the size of the system, and *extensive variables,* which do not depend on system size. Intensive variables are temperature, pressure, and phase compositions (mole fractions, mass fractions, concentrations, etc.). Extensive variables include mass or moles and energy for a batch system, and mass or molar flow rates and energy transfer rates for a flow system.

Regardless of whether only intensive variables or both intensive and extensive variables are considered, only a few of the variables are independent; when these are specified, all other variables become fixed. The number of independent variables is referred to as the *variance* or the number of *degrees of freedom*, $\mathcal{F}$, for the system.

The phase rule of J. Willard Gibbs, which applies only to the intensive variables at equilibrium, is used to determine $\mathcal{F}$. The rule states that

$$\mathcal{F} = C - \mathcal{P} + 2 \tag{4-1}$$

where C is the number of components and $\mathcal{P}$ is the number of phases at equilibrium. Equation (4-1) is derived by counting, at physical equilibrium, the number of intensive variables and the number of independent equations that relate these variables. The number of intensive variables, $\mathcal{V}$, is

$$\mathcal{V} = C\mathcal{P} + 2 \tag{4-2}$$

where the 2 refers to the equilibrium temperature and pressure, while the term $C\mathcal{P}$ is the total number of composition variables (e.g., mole fractions) for components distributed among $\mathcal{P}$ equilibrium phases. The number of independent equations, $\mathcal{E}$, relating the intensive variables is

$$\mathcal{E} = \mathcal{P} + C(\mathcal{P} - 1) \tag{4-3}$$

where the first term, $\mathcal{P}$, refers to the requirement that mole or mass fractions sum to one for each phase and the second term, $C(\mathcal{P} - 1)$, refers to the number of independent K-value equations of the general form

$$K_i = \frac{\text{mole fraction of } i \text{ in phase (1)}}{\text{mole fraction of } i \text{ in phase (2)}}$$

where (1) and (2) refer to equilibrium phases. For two phases, there are C independent expressions of this type; for three phases, $2C$; for four phases, $3C$; and so on. For example, for three phases ($V, L^{(1)}, L^{(2)}$), we can write $3C$ different K-value equations:

$$K_i^{(1)} = y_i / x_i^{(1)} \quad i = 1 \text{ to } C$$
$$K_i^{(2)} = y_i / x_i^{(2)} \quad i = 1 \text{ to } C$$
$$K_{D_i} = x_i^{(1)} / x_i^{(2)} \quad i = 1 \text{ to } C$$

However, only $2C$ of these equations are independent, because

$$K_{D_i} = K_i^{(2)} / K_i^{(1)}$$

Thus, the term for the number of independent K-value equations is $C(\mathcal{P} - 1)$, not $C\mathcal{P}$.

Degrees-of-Freedom Analysis

The degrees of freedom is the number of intensive variables, $\mathcal{V}$, less the number of equations, $\mathcal{E}$. Thus, from (4-2) and (4-3),

$$\mathcal{F} = \mathcal{V} - \mathcal{E} = (C\mathcal{P} + 2) - [\mathcal{P} + C(\mathcal{P} - 1)] = C - \mathcal{P} + 2$$

which completes the derivation of (4-1). When the number, $\mathcal{F}$, of intensive variables is specified, the remaining $\mathcal{P} + C(\mathcal{P} - 1)$ intensive variables are determined from the $\mathcal{P} + C(\mathcal{P} - 1)$ equations.

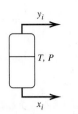

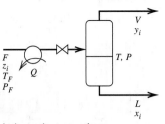

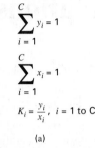

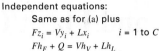

Independent equations:

$$\sum_{i=1}^{C} y_i = 1$$

$$\sum_{i=1}^{C} x_i = 1$$

$$K_i = \frac{y_i}{x_i}, \quad i = 1 \text{ to } C$$

(a)

Independent equations:
Same as for (a) plus
$$Fz_i = Vy_i + Lx_i \qquad i = 1 \text{ to } C$$
$$Fh_F + Q = Vh_V + Lh_L$$

(b)

Figure 4.1 Different treatments of degrees of freedom for vapor–liquid phase equilibria: (a) Gibbs phase rule (considers equilibrium intensive variables only); (b) general analysis (considers all intensive and extensive variables).

As an example, consider the vapor–liquid equilibrium ($\mathcal{P} = 2$) shown in Figure 4.1a, where the equilibrium intensive variables are labels on the sketch located above the list of independent equations relating these variables. Suppose there are $C = 3$ components. From (4-1), $\mathcal{F} = 3 - 2 + 2 = 3$. The equilibrium intensive variables are $T, P, x_1, x_2, x_3, y_1, y_2$, and y_3. If values are specified for T, P, and one of the mole fractions, the remaining five mole fractions are fixed and can be computed from the five independent equations listed in Figure 4.1a. Irrational specifications lead to infeasible results. For example, if the components are H_2O, N_2, and O_2, and $T = 100°F$ and $P = 15$ psia are specified, a specification of $x_{N_2} = 0.90$ is not feasible because nitrogen is not nearly this soluble in water.

In using the Gibbs phase rule, it should be noted that the K-values are not variables, but are thermodynamic functions that depend on the intensive variables discussed in Chapter 2.

The Gibbs phase rule is limited because it does not deal with feed streams sent to the equilibrium stage nor with extensive variables used when designing or analyzing separation operations. However, the phase rule can be extended for process applications, by adding the feed stream and extensive variables, and additional independent equations relating feed variables, extensive variables, and the intensive variables already considered by the rule.

Consider the single-stage, vapor–liquid ($\mathcal{P} = 2$) equilibrium separation process shown in Figure 4.1b. By comparison with Figure 4.1a, the additional variables are z_i, T_F, P_F, F, Q, V, and L, or $C + 6$ variables, all of which are indicated in the diagram. In general, for $\mathcal{P}$ phases, the additional variables number $C + \mathcal{P} + 4$. The additional independent equations, listed below the diagram, are the C component material balances and the energy balance, or $C + 1$ equations. Note that, like K-values, stream enthalpies are not counted as variables because they are thermodynamic functions that depend on intensive variables.

For the general degrees-of-freedom analysis for phase equilibrium, with C components, $\mathcal{P}$ phases, and a single feed phase, (4-2) and (4-3) are extended by adding the number of additional variables and equations, respectively:

$$\mathcal{V} = (C\mathcal{P} + 2) + (C + \mathcal{P} + 4) = \mathcal{P} + C\mathcal{P} + C + 6$$
$$\mathcal{E} = [\mathcal{P} + C(\mathcal{P} - 1)] + (C + 1) = \mathcal{P} + C\mathcal{P} + 1 \qquad (4\text{-}4)$$
$$\mathcal{F} = \mathcal{V} - \mathcal{E} = C + 5$$

For example, if the $C + 5$ degrees of freedom are used to specify all z_i and the five variables F, T_F, P_F, T, and P, the remaining variables are computed from the equations shown in Figure 4.1.[1] To apply the Gibbs phase rule, (4-1), the number of phases must be known. When applying (4-4), the determination of the number of equilibrium phases, $\mathcal{P}$, is implicit in the computational procedure as illustrated in later sections of this chapter.

In the following sections, the Gibbs phase rule, (4-1), and the equation for the number of degrees of freedom of a flow system, (4-4), are applied to (1) tabular equilibrium data, (2) graphical equilibrium data, or (3) thermodynamic equations for K-values and enthalpies for vapor–liquid, liquid–liquid, solid–liquid, gas–liquid, gas–solid, vapor–liquid–solid, and vapor–liquid–liquid systems at equilibrium.

4.2   BINARY VAPOR–LIQUID SYSTEMS

Experimental vapor–liquid equilibrium data for systems containing two components, A and B, are widely available. Sources include *Perry's Handbook* [1] and Gmehling and Onken [2]. Because $y_B = 1 - y_A$ and $x_B = 1 - x_A$, the data are presented in terms of just four intensive variables: T, P, y_A, and x_A. Most commonly T, y_A, and x_A are tabulated for a fixed P for ranges of y_A and x_A from 0 to 1, where A is the more-volatile component ($y_A > x_A$). However, if an azeotrope (see Section 4.3) forms, B becomes the more volatile component on one side of the azeotropic point. By the Gibbs phase rule, (4-1), $\mathcal{F} = 2 - 2 + 2 = 2$. Thus, with pressure fixed, phase compositions are completely defined if temperature is also fixed, and the separation factor, that is, the relative volatility in the case of vapor–liquid equilibria,

$$\alpha_{A,B} = \frac{K_A}{K_B} = \frac{(y_A/x_A)}{(y_B/x_B)} = \frac{(y_A/x_A)}{(1 - y_A)/(1 - x_A)} \qquad (4\text{-}5)$$

is also fixed.

Vapor–liquid equilibria data of the form T–y_A–x_A for 1 atm pressure of three binary systems of industrial importance are given in Table 4.1. Included are values of relative volatility computed from (4-5). As discussed in Chapter 2,

---

[1]The development of (4-4) assumes that the sum of the mole fractions in the feed will equal one. Alternatively, the equation $\sum_{i=1}^{C} z_i = 1$ can be added to the number of independent equations (thus forcing the feed mole fractions to sum to one). Then, the degrees of freedom becomes one less or $C + 4$.

Table 4.1   Vapor–Liquid Equilibrium Data for Three Common Binary Systems at 1 atm Pressure

a. Water (A)–Glycerol (B) System
   $P = 101.3$ kPa
   Data of Chen and Thompson, *J. Chem. Eng. Data,* **15,** 471 (1970)

Temperature, °C	y_A	x_A	$\alpha_{A,B}$
100.0	1.0000	1.0000	
101.7	0.9998	0.9534	250
104.6	0.9996	0.8846	333
109.8	0.9991	0.7731	332
122.0	0.9982	0.5610	440
128.8	0.9980	0.4742	544
140.4	0.9974	0.3622	683
148.2	0.9964	0.3077	627
158.6	0.9940	0.2460	507
175.2	0.9898	0.1756	456
207.0	0.9804	0.0945	481
244.5	0.9341	0.0491	275
282.5	0.8308	0.0250	191
290.0	0.0000	0.0000	

b. Methanol (A)–Water (B) System
   $P = 101.3$ kPa
   Data of J.G. Dunlop, M.S. thesis, Brooklyn Polytechnic Institute (1948)

Temperature, °C	y_A	x_A	$\alpha_{A,B}$
64.5	1.000	1.000	
65.0	0.979	0.950	2.45
66.0	0.958	0.900	2.53
67.5	0.915	0.800	2.69
69.3	0.870	0.700	2.87
71.2	0.825	0.600	3.14
73.1	0.779	0.500	3.52
75.3	0.729	0.400	4.04
78.0	0.665	0.300	4.63
81.7	0.579	0.200	5.50
84.4	0.517	0.150	6.07
87.7	0.418	0.100	6.46
89.3	0.365	0.080	6.61
91.2	0.304	0.060	6.84
93.5	0.230	0.040	7.17
96.4	0.134	0.020	7.58
100.0	0.000	0.000	

c. *Para*-xylene (A)–*Meta*-xylene (B) System
   $P = 101.3$ kPa
   Data of Kato, Sato, and Hirata, *J. Chem. Eng. Jpn.,* **4,** 305 (1970)

Temperature, °C	y_A	x_A	$\alpha_{A,B}$
138.335	1.0000	1.0000	
138.414	0.9019	0.9000	1.0021
138.491	0.8033	0.8000	1.0041
138.568	0.7043	0.7000	1.0061
138.644	0.6049	0.6000	1.0082
138.720	0.5051	0.5000	1.0102
138.795	0.4049	0.4000	1.0123
138.869	0.3042	0.3000	1.0140
138.943	0.2032	0.2000	1.0160
139.016	0.1018	0.1000	1.0180
139.088	0.0000	0.0000	

$\alpha_{A,B}$ depends on T, P, and the compositions of the equilibrium vapor and liquid. At 1 atm, where $\alpha_{A,B}$ is approximated well by $\gamma_A P_A^s / \gamma_B P_B^s$, $\alpha_{A,B}$ depends only on T and x_A, since vapor-phase nonidealities are small. Because of the dependency on x_A, $\alpha_{A,B}$ is not a constant, but varies from point to point. For the three binary systems in Table 4.1, the vapor and liquid phases become richer in the less-volatile component, B, as temperature increases. For $x_A = 1$, the temperature is the normal boiling point of A; for $x_A = 0$, the temperature is the normal boiling point of B. For the three systems, all other data points are at temperatures between the two boiling points. Except for the pure components ($x_A = 1$ or 0), $y_A > x_A$ and $\alpha_{A,B} > 1$.

For the water–glycerol system, the difference in normal boiling points is 190°C. Therefore, relative volatility values are very high, making it possible to achieve a reasonably good separation in a single equilibrium stage. Industrially, the separation is often conducted in an evaporator, which produces a nearly pure water vapor and a glycerol-rich liquid. For example, from Table 4.1, at 207°C, a vapor of 98 mol% water is in equilibrium with a liquid phase containing more than 90 mol% glycerol.

For the methanol–water system, the difference in normal boiling points is 35.5°C. As a result, the relative volatility is an order of magnitude lower than for the water–glycerol system. A sharp separation cannot be made with a single stage. About 30 trays are required in a distillation operation to obtain a 99 mol% methanol distillate and a 98 mol% water bottoms, an acceptable industrial separation.

For the aromatic paraxylene-metaxylene isomer system, the normal boiling-point difference is only 0.8°C. Thus, the relative volatility is very close to 1.0, making the separation by distillation impractical because about 1,000 trays are required to produce nearly pure products. Instead, crystallization and adsorption, which have much higher separation factors, are used commercially to make the separation.

Experimental vapor–liquid equilibrium data for the methanol–water system are given in Table 4.2 in the form of $P-y_A-x_A$ for fixed temperatures of 50, 150, and 250°C. The three sets of data cover a pressure range of 1.789 to 1,234 psia, with the higher pressures corresponding to the higher temperatures. At 50°C, relative volatilities are moderately high at an average value of 4.94 over the composition range. At 150°C, the average relative volatility is only 3.22; and at 250°C, it decreases to 1.75. Thus, as the temperature and pressure increase, the relative volatility decreases significantly. In Table 4.2, for the data set at 250°C, it is seen that as the compositions become richer in methanol, a point is reached in the neighborhood of 1,219 psia, at a methanol mole fraction of 0.772, where the relative volatility is 1.0 and no separation by distillation is possible because the compositions of the vapor and liquid are identical and the two phases become one phase. This is the critical point of a mixture of this composition. It is intermediate between the critical points of pure methanol

Table 4.2   Vapor–Liquid Equilibrium Data for the Methanol–Water System at Temperatures of 50, 150, and 250°C

a. Methanol (A)–Water (B) System
$T = 50°C$
Data of McGlashan and Williamson, *J. Chem. Eng. Data*, **21**, 196 (1976)

Pressure, psia	y_A	x_A	$\alpha_{A,B}$
1.789	0.0000	0.0000	
2.373	0.2661	0.0453	7.64
2.838	0.4057	0.0863	7.23
3.369	0.5227	0.1387	6.80
3.764	0.5898	0.1854	6.32
4.641	0.7087	0.3137	5.32
5.163	0.7684	0.4177	4.63
5.771	0.8212	0.5411	3.90
6.122	0.8520	0.6166	3.58
6.811	0.9090	0.7598	3.16
7.280	0.9455	0.8525	3.00
7.800	0.9817	0.9514	2.74
8.072	1.0000	0.0000	

b. Methanol (A)–Water (B) System
$T = 150°C$
Data of Griswold and Wong, *Chem. Eng. Prog. Symp. Ser.*, **48** (3), 18 (1952)

Pressure, psia	y_A	x_A	$\alpha_{A,B}$
73.3	0.060	0.009	7.03
79.0	0.135	0.022	6.94
85.7	0.213	0.044	5.88
93.9	0.286	0.079	4.67
114.9	0.459	0.186	3.71
139.7	0.610	0.374	2.62
148.6	0.662	0.459	2.31
160.4	0.731	0.578	1.98
177.4	0.832	0.748	1.67
193.5	0.929	0.893	1.57
194.5	0.943	0.913	1.58
196.5	0.960	0.936	1.64
197.7	0.972	0.953	1.71
199.2	0.982	0.969	1.75

c. Methanol (A)–Water (B) System
$T = 250°C$
Data of Griswold and Wong, *Chem. Eng. Prog. Symp. Ser.*, **48** (3), 18 (1952)

Pressure, psia	y_A	x_A	$\alpha_{A,B}$
681	0.163	0.066	2.76
764	0.280	0.132	2.56
818	0.344	0.180	2.39
889	0.423	0.254	2.15
949	0.487	0.331	1.92
994	0.542	0.404	1.75
1049	0.596	0.483	1.58
1099	0.643	0.553	1.46
1159	0.698	0.631	1.35
1204	0.756	0.732	1.13
1219	0.772	0.772	1.00
1234	0.797	0.797	1.00

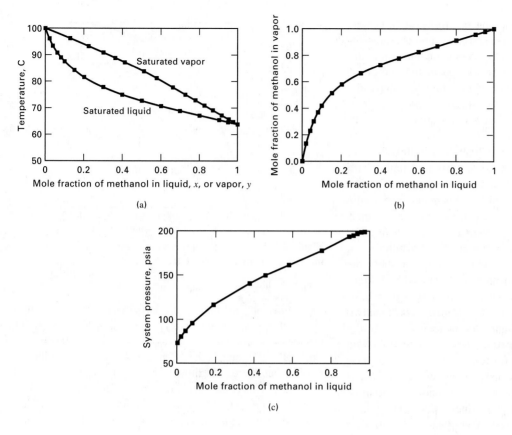

(a)

(b)

(c)

Figure 4.2 Vapor–liquid equilibrium conditions for the methanol–water system: (a) T–y–x diagram for 1 atm pressure; (b) y–x diagram for 1 atm pressure; (c) P–x diagram for 150°C.

and pure water:

$y_A = x_A$	T_c, °C	P_c, psia
0.000	374.1	3,208
0.772	250	1,219
1.000	240	1,154

A set of critical conditions exists for each binary-mixture composition. In industry, distillation columns operate at pressures well below the critical pressure of the mixture to be separated to avoid relative volatilities that approach a value of 1.

The data of Tables 4.1 and 4.2 for the methanol–water system are plotted in three different ways in Figure 4.2: (a) T versus y_A or x_A at $P = 1$ atm; (b) y_A versus x_A at $P = 1$ atm; and (c) P versus x_A at $T = 150$°C. These three plots all satisfy the requirement of the Gibbs phase rule that when two intensive variables are fixed, all other variables are fixed by the governing equilibrium equations and mole-fraction-summation constraints. Of the three diagrams in Figure 4.2, only (a) contains the complete data; (b) does not contain temperatures; and (c) does not contain vapor-phase mole fractions. Although mass fractions could be used in place of mole fractions, the latter are preferred because theoretical phase-equilibrium relations are based on molar properties.

Plots like Figure 4.2a are useful for determining phase states, phase-transition temperatures, equilibrium-phase compositions, and equilibrium-phase amounts for a given feed of known composition. Consider the T–y–x plot in Figure 4.3 for the normal hexane (H)–normal octane (O)

system at 101.3 kPa. Because normal hexane is the more volatile species, the mole fractions are for that component. The upper curve, labeled "saturated vapor," gives the dependency on the dew-point temperature of the vapor mole fraction y_H; the lower curve, labeled "saturated liquid," gives the dependency of the bubble-point temperature on the liquid-phase mole fraction, x_H. The two curves converge at $x_H = 0$, the normal boiling point of pure normal octane (258.2°F),

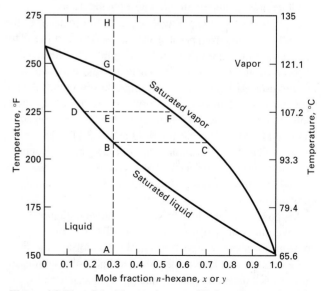

Figure 4.3 Use of the T–y–x phase equilibrium diagram for the normal hexane–normal octane system at 1 atm.

and at $x_H = 1$, the normal boiling point of normal hexane (155.7°F). In order for two phases to exist, a point representing the overall composition of the two-phase binary mixture at a given temperature must be located in the two-phase region between the two curves. If the point lies above the saturated-vapor curve, only a superheated vapor is present; if the point lies below the saturated-liquid curve, only a subcooled liquid exists.

Suppose we have a mixture of 30 mol% H at 150°F. From Figure 4.3, at point A we have a subcooled liquid with $x_H = 0.3 (x_O = 0.7)$. When this mixture is heated at a constant pressure of 1 atm, the liquid state is maintained until a temperature of 210°F is reached, which corresponds to point B on the saturated-liquid curve. Point B is the *bubble point* because the first bubble of vapor appears. This bubble is a saturated vapor in equilibrium with the liquid at the same temperature. Thus, its composition is determined by following a *tie line*, BC from $x_H = 0.3$ to $y_H = 0.7 (y_O = 0.3)$. The tie line is horizontal because the temperatures of the two equilibrium phases are the same. As the temperature of the two-phase mixture is increased to point E, on horizontal tie line DEF at 225°F, the mole fraction of H in the liquid phase decreases to $x_H = 0.17$ (because it is more volatile than O and preferentially vaporizes) and correspondingly the mole fraction of H in the vapor phase increases to $y_H = 0.55$. Throughout the two-phase region, the vapor is at its dew point, while the liquid is at its bubble point. The overall composition of the two phases remains at a mole fraction of 0.30 for hexane. At point E, the relative molar amounts of the two equilibrium phases is determined by the inverse lever-arm rule based on the lengths of the line segments DE and EF. Thus, referring to Figures 4.1b and 4.3, $V/L = DE/EF$ or $V/F = DE/DEF$. When the temperature is increased to 245°F, point G, the dew point for $y_H = 0.3$ is reached, where only one droplet of equilibrium liquid remains with a composition from the tie line FG at point F of $x_H = 0.06$. A further increase in temperature—say, to point H at 275°F—gives a superheated vapor with $y_H = 0.30$. The steps are reversible starting from point H and moving down to point A.

Constant-pressure $x–y$ plots like Figure 4.2b are also useful because the equilibrium-vapor-and-liquid compositions are represented by points on the equilibrium curve. However, no phase–temperature information is included. Such plots usually include a 45° reference line, $y = x$. Consider the $y–x$ plot in Figure 4.4 for H–O at 101.3 kPa. This plot is convenient for determining equilibrium-phase compositions for various values of mole-percent vaporization of a feed mixture of a given composition by geometric constructions.

Suppose we have a feed mixture, F, shown in Figure 4.1b, of overall composition $z_H = 0.6$. To determine the equilibrium-phase compositions if, say, 60 mol% of the feed is vaporized, we develop the dashed-line construction in Figure 4.4. Point A on the 45° line represents z_H. Point B on the equilibrium curve is reached by extending a line, called

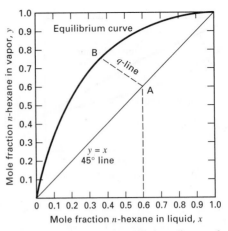

Figure 4.4 Use of the $y–x$ phase equilibrium diagram for the normal hexane–normal octane system at 1 atm.

the *q-line*, upward and to the left toward the equilibrium curve at a slope equal to $[(V/F) - 1]/(V/F)$. Thus, for 60 mol% vaporization, the slope $= (0.6 - 1)/0.6 = -\frac{2}{3}$. Point B at the intersection of line AB with the equilibrium curve gives the equilibrium composition as $y_H = 0.76$ and $x_H = 0.37$. This computation requires a trial-and-error placement of a horizontal line if we use Figure 4.3. The derivation of the slope of the q-line in Figure 4.4 follows by combining the mole-balance equation of Figure 4.1a,

$$F z_H = V y_H + L x_H$$

with the total mole balance,

$$F = V + L$$

to eliminate L, giving the equation for the q-line:

$$y_H = \left[\frac{(V/F) - 1}{(V/F)} \right] x_H + \left[\frac{1}{(V/F)} \right] z_H$$

Thus, the slope of the q-line passing through the equilibrium point (y_H, x_H) is $[(V/F) - 1]/(V/F)$.

Figure 4.2c is the least used of the three plots in Figure 4.2. However, such a plot does illustrate, for a fixed temperature, the extent to which the binary mixture deviates from an ideal solution. If Raoult's law applies, the total pressure above the liquid is

$$\begin{aligned} P &= P_A^s x_A + P_B^s x_B \\ &= P_A^s x_A + P_B^s (1 - x_A) \\ &= P_B^s + x_A (P_A^s - P_B^s) \end{aligned} \tag{4-6}$$

Thus, a plot of P versus x_A is a straight line with intersections at the vapor pressure of B for $x_A = 0$ and the vapor pressure of A for $x_B = 0 (x_A = 1)$. The greater the departure from a straight line, the greater is the deviation from the assumptions of an ideal gas and/or an ideal-liquid solution.

If the pressures are sufficiently low that the equilibrium-vapor phase is ideal and the curve is convex, deviations from Raoult's law are positive, and species liquid-phase activity coefficients are greater than 1; if the curve is concave, deviations are negative and activity coefficients are less than 1. In either case, the total pressure is given by

$$P = \gamma_A P_A^s x_A + \gamma_B P_B^s x_B \qquad (4\text{-}7)$$

If the vapor does not obey the ideal-gas law, (4-7) does not apply. In Figure 4.2c, system pressures are sufficiently high that some deviation from the ideal-gas law occurs. However, the convexity is due mainly to activity coefficients that are greater than 1.

For relatively close (narrow)-boiling binary mixtures that exhibit ideal or nearly ideal behavior, the relative volatility, $\alpha_{A,B}$, varies little with pressure. If $\alpha_{A,B}$ is assumed constant over the entire composition range, the y–x phase-equilibrium curve can be determined and plotted from a rearrangement of (4-5):

$$y_A = \frac{\alpha_{A,B} x_A}{1 + x_A(\alpha_{A,B} - 1)} \qquad (4\text{-}8)$$

For an ideal solution, $\alpha_{A,B}$ can be approximated with Raoult's law to give

$$\alpha_{A,B} = \frac{K_A}{K_B} = \frac{P_A^s/P}{P_B^s/P} = \frac{P_A^s}{P_B^s} \qquad (4\text{-}9)$$

Thus, from a knowledge of just the vapor pressures of the two components at a temperature, say, midway between the two boiling points at the given pressure, a y–x phase-equilibrium curve can be approximated using only one value of $\alpha_{A,B}$. Families of curves, as shown in Figure 4.5, can be used for preliminary calculations in the absence of detailed experimental data. The use of (4-8) and (4-9) is not recommended for wide-boiling or nonideal mixtures.

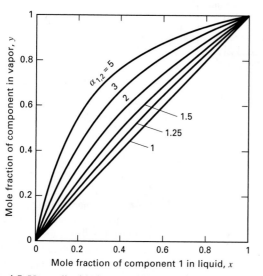

Figure 4.5 Vapor–liquid phase equilibrium curves for constant values of relative volatility.

4.3 AZEOTROPIC SYSTEMS

Departures from Raoult's law frequently manifest themselves in the formation of *azeotropes*, particularly for mixtures of close-boiling species of different chemical types whose liquid solutions are nonideal. Azeotropes are formed by liquid mixtures exhibiting maximum- or minimum-boiling points. These represent, respectively, negative or positive deviations from Raoult's law. Vapor and liquid compositions are identical at the azeotropic composition; thus, all K-values are 1 and no separation of species can take place.

If only one liquid phase exists, the mixture forms a *homogeneous* azeotrope; if more than one liquid phase is present, the azeotrope is *heterogeneous*. In accordance with the Gibbs phase rule, at constant pressure in a two-component system, the vapor can coexist with no more than two liquid phases, while in a ternary mixture up to three liquid phases can coexist with the vapor.

Figures 4.6, 4.7, and 4.8 show three types of azeotropes that are commonly encountered with binary mixtures. The most common type by far is the minimum-boiling homogeneous azeotrope, illustrated in Figure 4.6 for the isopropyl ether–isopropyl alcohol system. In Figure 4.6a, for a temperature of 70°C, the maximum total pressure is greater than the vapor pressure of either component because activity coefficients are greater than 1. The y–x diagram in Figure 4.6b shows that for a pressure of 1 atm the azeotropic mixture occurs at 78 mol% ether. Figure 4.6c is a T–x diagram for a pressure of 101 kPa, where the azeotrope is seen to boil at 66°C. In Figure 4.6a, for 70°C, the azeotrope, at 123 kPa (923 torr), is 72 mol% ether. Thus, the azeotropic composition shifts with pressure. In distillation, the minimum-boiling azeotropic mixture is the overhead product.

For the maximum-boiling homogeneous azeotropic acetone–chloroform system in Figure 4.7a, the minimum total pressure is below the vapor pressures of the pure components because activity coefficients are less than 1. The azeotrope concentrates in the bottoms in a distillation operation.

Heterogeneous azeotropes are always minimum-boiling mixtures because activity coefficients must be significantly greater than 1 to cause splitting into two liquid phases. The region a–b in Figure 4.8a for the water–normal butanol system is a two-phase region where total and partial pressures remain constant as the relative amounts of the two phases change, but the phase compositions do not. The y–x diagram in Figure 4.8b shows a horizontal line over the immiscible region, and the phase diagram of Figure 4.8c shows a minimum constant temperature.

Azeotropes limit the separation achievable by ordinary distillation. It is possible to shift the equilibrium by changing the pressure sufficiently to "break" the azeotrope, or move it away from the region where the required separation must be made. For example, ethyl alcohol and water form a homogeneous minimum-boiling azeotrope of 95.6 wt% alcohol at

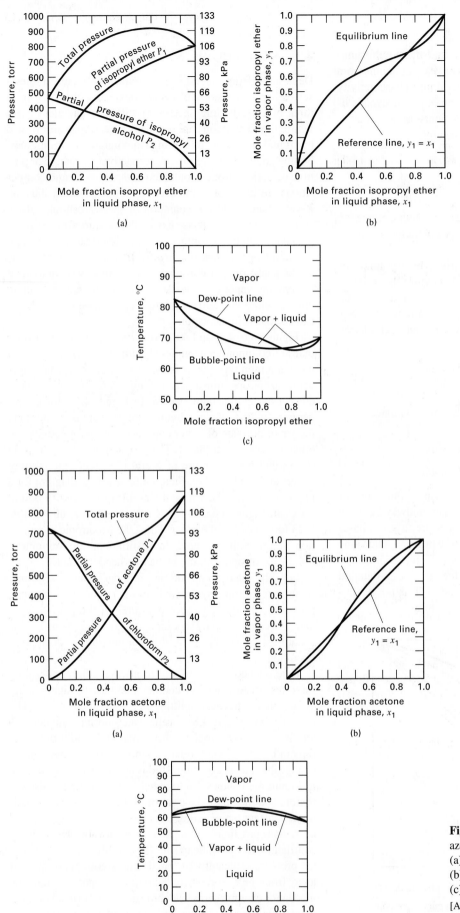

Figure 4.6 Minimum-boiling-point azeotrope, isopropyl ether–isopropyl alcohol system: (a) partial and total pressures at 70°C; (b) vapor–liquid equilibria at 101 kPa; (c) phase diagram at 101 kPa.

[Adapted from O.A. Hougen, K.M. Watson, and R.A. Ragatz, *Chemical Process Principles. Part II,* 2nd ed., John Wiley and Sons, New York (1959).]

Figure 4.7 Maximum-boiling-point azeotrope, acetone–chloroform system: (a) partial and total pressures at 60°C; (b) vapor–liquid equilibria at 101 kPa; (c) phase diagram at 101 kPa pressure.

[Adapted from O.A. Hougen, K.M. Watson, and R.A. Ragatz, *Chemical Process Principles. Part II,* 2nd ed., John Wiley and Sons, New York (1959).]

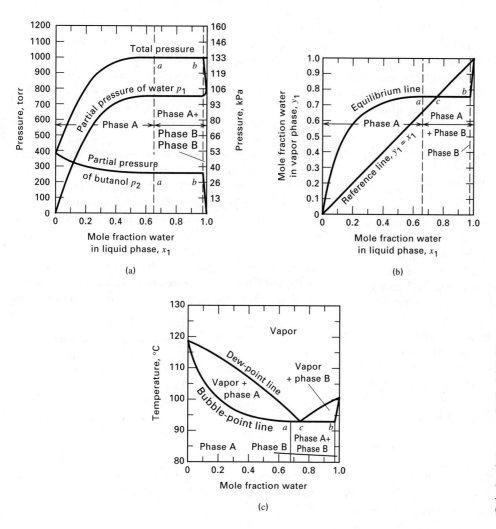

(a)

(b)

(c)

Figure 4.8 Minimum-boiling-point (two liquid phases) water/*n*-butanol system: (a) partial and total pressures at 100°C; (b) vapor–liquid equilibria at 101 kPa; (c) phase diagram at 101 kPa pressure.

[Adapted from O.A. Hougen, K.M. Watson, and R.A. Ragatz, *Chemical Process Principles. Part II*, 2nd ed., John Wiley and Sons, New York (1959).]

78.15°C and 101.3 kPa. However, at vacuums of less than 9.3 kPa, no azeotrope is formed. Ternary azeotropes also occur, and these offer the same barrier to complete separation as do binary azeotropes.

Azeotrope formation in general, and heterogeneous azeotropes in particular, can be employed to achieve difficult separations. As discussed in Chapter 1, an entrainer is added for the purpose of combining with one or more of the components in the feed to form a minimum-boiling azeotrope, which is then recovered as the distillate.

Figure 4.9 shows the Keyes process [3] for making pure ethyl alcohol by *heterogeneous azeotropic distillation*. Water and ethyl alcohol form a binary, minimum-boiling azeotrope containing 95.6 wt% alcohol and boiling at 78.15°C at 101.3 kPa. Thus, it is impossible to obtain pure alcohol (boiling point = 78.40°C) by ordinary distillation at 1 atm. The addition of benzene to an alcohol–water mixture results in the formation of a minimum-boiling, heterogeneous ternary, azeotrope containing, by weight, 18.5% alcohol, 74.1% benzene, and 7.4% water, boiling at 64.85°C. Upon condensation, the ternary azeotrope separates into two liquid layers: a top layer containing 14.5% alcohol, 84.5% benzene, and 1% water, and a bottoms layer of

53% alcohol, 11% benzene, and 36% water, all by weight. The benzene-rich layer is returned as reflux. The other layer is sent to a second distillation column for recovery and recycling of alcohol and benzene. Absolute alcohol, which has a boiling point above that of the ternary azeotrope, is removed at the bottom of the column.

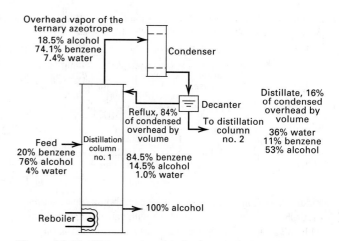

Figure 4.9 The Keyes process for absolute alcohol.

In *extractive distillation,* as discussed in Chapter 1, a solvent is added, usually near the top of the column, to selectively alter the activity coefficients in order to increase the relative volatility between the two species to be separated. The solvent is generally a relatively polar, high-boiling constituent, such as phenol, aniline, or furfural, which concentrates at the bottom of the column.

4.4 MULTICOMPONENT FLASH, BUBBLE-POINT, AND DEW-POINT CALCULATIONS

A *flash* is a single-equilibrium-stage distillation in which a feed is partially vaporized to give a vapor richer in the more-volatile components than the remaining liquid. In Figure 4.10a, a liquid feed is heated under pressure and flashed adiabatically across a valve to a lower pressure, resulting in the creation of a vapor phase that is separated from the remaining liquid in a flash drum. If the valve is omitted, a low-pressure liquid can be partially vaporized in the heater and then separated into two phases in the flash drum. Alternatively, a vapor feed can be cooled and partially condensed, with phase separation in a flash drum, as in Figure 4.10b, to give a liquid that is richer in the less-volatile components. In both cases, if the equipment is properly designed, the vapor and liquid leaving the drum are in equilibrium [4].

Unless the relative volatility is very large, the degree of separation achievable between two components in a single equilibrium stage is poor. Therefore, flashing (partial vaporization) or partial condensation are usually auxiliary operations used to prepare streams for further processing.

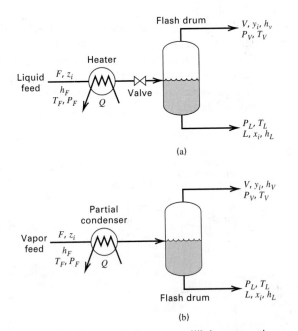

(a)

(b)

Figure 4.10 Continuous, single-stage equilibrium separation: (a) flash vaporization (adiabatic flash with valve, isothermal flash without valve when T_V is specified); (b) partial condensation (analogous to isothermal flash when T_V is specified).

Typically, the vapor phase is sent to a vapor separation system, while the liquid phase is sent to a liquid separation system. Computational methods for a single-stage flash calculation are of fundamental importance. Such calculations are used not only for the operations in Figure 4.10, but also to determine, anywhere in a process, the phase condition of a stream or batch of known composition, temperature, and pressure.

For the single-stage equilibrium operation with one feed stream and two product streams, shown in Figure 4.10, the $2C + 5$ equations listed in Table 4.3 apply. (In Figure 4.10, T and P are given separately for the vapor and liquid products to emphasize the subsequent need to assume mechanical and thermal equilibrium.) They relate the $3C + 10$ variables $(F, V, L, z_i, y_i, x_i, T_F, T_V, T_L, P_F, P_V, P_L, Q)$ and leave $C + 5$ degrees of freedom. Assuming that $C + 3$ feed variables F, $T_F, P_F,$ and C values of z_i are known, two additional variables can be specified. The most common sets of specifications are

T_V, P_V	Isothermal flash
$V/F = 0, P_L$	Bubble-point temperature
$V/F = 1, P_V$	Dew-point temperature
$T_L, V/F = 0$	Bubble-point pressure
$T_V, V/F = 1$	Dew-point pressure
$Q = 0, P_V$	Adiabatic flash
Q, P_V	Nonadiabatic flash
$V/F, P_V$	Percent vaporization flash

Calculation procedures, described in the following for all these cases, are well known and widely used. They all assume that specified values of feed mole fractions, z_i, sum to one.

Isothermal Flash

If the equilibrium temperature T_V (or T_L) and the equilibrium pressure P_V (or P_L) of a multicomponent mixture are specified, values of the remaining $2C + 5$ variables are determined from the same number of equations in Table 4.3.

Table 4.3 Equations for Single-Stage Flash Vaporization and Partial Condensation Operations

Equation		Number of Equations
(1) $P_V = P_L$	(mechanical equilibrium)	1
(2) $T_V = T_L$	(thermal equilibrium)	1
(3) $y_i = K_i x_i$	(phase equilibrium)	C
(4) $F z_i = V y_i + L x_i$	(component material balance)	C
(5) $F = V + L$	(total material balance)	1
(6) $h_F F + Q = h_V V + h_L L$	(energy balance)	1
(7) $\sum_i y_i - \sum_i x_i = 0$	(summations)	1
		$\mathcal{E} = 2C + 5$

$$K_i = K_i\{T_V, P_V, \mathbf{y}, \mathbf{x}\} \qquad h_F = h_F\{T_F, P_F, \mathbf{z}\}$$
$$h_V = h_V\{T_V, P_V, \mathbf{y}\} \qquad h_L = h_L\{T_L, P_L, \mathbf{x}\}$$

Table 4.4 Rachford–Rice Procedure for Isothermal-Flash Calculations When K-Values Are Independent of Composition

Specified variables: $F, T_F, P_F, z_1, z_2, \ldots, z_C, T_V, P_V$

Steps

(1) $T_L = T_V$

(2) $P_L = P_V$

(3) Solve

$$f\{\Psi\} = \sum_{i=1}^{C} \frac{z_i(1 - K_i)}{1 + \Psi(K_i - 1)} = 0$$

for $\Psi = V/F$, where $K_i = K_i\{T_V, P_V\}$.

(4) $V = F\Psi$

(5) $x_i = \dfrac{z_i}{1 + \Psi(K_i - 1)}$

(6) $y_i = \dfrac{z_i K_i}{1 + \Psi(K_i - 1)} = x_i K_i$

(7) $L = F - V$

(8) $Q = h_V V + h_L L - h_F F$

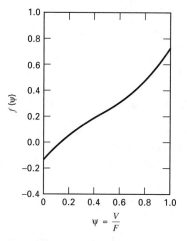

Figure 4.11 Rachford–Rice function for Example 4.1.

The computational procedure, referred to as the *isothermal-flash calculation,* is not straightforward because Eq. (4) in Table 4.3 is a nonlinear equation in the unknowns V, L, y_i, and x_i. Many solution strategies have been developed, but the generally preferred procedure, as given in Table 4.4, is that of Rachford and Rice [5] when K-values are independent (or nearly independent) of equilibrium-phase compositions.

Equations containing only a single unknown are solved first. Thus, Eqs. (1) and (2) in Table 4.3 are solved, respectively, for P_L and T_L. The unknown Q appears only in Eq. (6), so Q is computed only after all other equations have been solved. This leaves Eqs. (3), (4), (5), and (7) in Table 4.3 to be solved for V, L, and all values of y and x. These equations can be partitioned so as to solve for the unknowns in a sequential manner by substituting Eq. (5) into Eq. (4) to eliminate L and combining the result with Eq. (3) to obtain Eqs. (5) and (6) in Table 4.4. Here (5) is in x_i, but not y_i, and (6) is in y_i but not x_i. Summing these two equations and combining them with $\sum y_i - \sum x_i = 0$ to eliminate y_i and x_i gives Eq. (3) in Table 4.4; a nonlinear equation in V (or $\Psi = V/F$) only. Upon solving this equation numerically in an iterative manner for Ψ and then V, from Eq. (4) of Table 4.4, one can obtain the remaining unknowns directly from Eqs. (5) through (8) in Table 4.4. When T_F and/or P_F are not specified, Eq. (6) of Table 4.3 is not solved for Q. By this isothermal-flash procedure, the equilibrium-phase condition of a mixture at a known temperature ($T_V = T_L$) and pressure ($P_V = P_L$) is determined.

Equation (3) of Table 4.4 can be solved iteratively by guessing values of Ψ between 0 and 1 until the function $f\{\Psi\} = 0$. A typical form of the function, as will be computed in Example 4.1, is shown in Figure 4.11. The most widely employed numerical method for solving Eq. (3) of Table 4.4 is Newton's method [6]. A predicted value of the Ψ

root for iteration $k + 1$ is computed from the recursive relation

$$\Psi^{(k+1)} = \Psi^{(k)} - \frac{f\{\Psi^{(k)}\}}{f'\{\Psi^{(k)}\}} \qquad (4\text{-}10)$$

where the superscript is the iteration index, and the derivative of $f\{\Psi\}$, from Eq. (3) in Table 4.4, with respect to Ψ is

$$f'\{\Psi^{(k)}\} = \sum_{i=1}^{C} \frac{z_i(1 - K_i)^2}{[1 + \Psi^{(k)}(K_i - 1)]^2} \qquad (4\text{-}11)$$

The iteration can be initiated by assuming $\Psi^{(1)} = 0.5$. Sufficient accuracy will be achieved by terminating the iterations when $|\Psi^{(k+1)} - \Psi^{(k)}|/\Psi^{(k)} < 0.0001$.

One should check the existence of a valid root ($0 \leq \Psi \leq 1$), before employing the procedure of Table 4.4, by checking to see if the equilibrium condition corresponds to subcooled liquid or superheated vapor rather than partial vaporization or partial condensation. A first estimate of whether a multicomponent feed gives a two-phase equilibrium mixture when flashed at a given temperature and pressure can be made by inspecting the K-values. If all K-values are greater than 1, the exit phase is superheated vapor above the dew point. If all K-values are less than 1, the single exit phase is a subcooled liquid below the bubble point. If one or more K-values are greater than 1 and one or more K-values are less than 1, the check is made as follows. First, $f\{\Psi\}$ is computed from Eq. (3) for $\Psi = 0$. If the resulting $f\{0\} > 0$, the mixture is below its bubble point (subcooled liquid). Alternatively, if $f\{1\} < 0$, the mixture is above the dew point (superheated vapor).

EXAMPLE 4.1

A 100-kmol/h feed consisting of 10, 20, 30, and 40 mol% of propane (3), *n*-butane (4), *n*-pentane (5), and *n*-hexane (6), respectively, enters a distillation column at 100 psia (689.5 kPa) and

200°F (366.5°K). Assuming equilibrium, what mole fraction of the feed enters as liquid, and what are the liquid and vapor compositions?

SOLUTION

At flash conditions, from Figure 2.8, $K_3 = 4.2$, $K_4 = 1.75$, $K_5 = 0.74$, $K_6 = 0.34$, independent of compositions. Because some K-values are greater than 1 and some are less than 1, it is first necessary to compute values of $f\{0\}$ and $f\{1\}$ for Eq. (3) in Table 4.4 to determine if the mixture is between the bubble point and the dew point.

$$f\{0\} = \frac{0.1(1-4.2)}{1} + \frac{0.2(1-1.75)}{1}$$
$$+ \frac{0.3(1-0.74)}{1} + \frac{0.4(1-0.34)}{1} = -0.128$$

Since $f\{0\}$ is not greater than zero, the mixture is above the bubble point.

$$f\{1\} = \frac{0.1(1-4.2)}{1+(4.2-1)} + \frac{0.2(1-1.75)}{1+(1.75-1)}$$
$$+ \frac{0.3(1-0.74)}{1+(0.74-1)} + \frac{0.4(1-0.34)}{1+(0.34-1)} = 0.720$$

Since $f\{1\}$ is not less than zero, the mixture is below the dew point. Therefore, the mixture is part vapor and substitution of z_i and K_i values in Eq. (3) of Table 4.4 gives

$$0 = \frac{0.1(1-4.2)}{1+\Psi(4.2-1)} + \frac{0.2(1-1.75)}{1+\Psi(1.75-1)}$$
$$+ \frac{0.3(1-0.74)}{1+\Psi(0.74-1)} + \frac{0.4(1-0.34)}{1+\Psi(0.34-1)}$$

Solution of this equation by Newton's method using an initial guess for Ψ of 0.50 gives the following iteration history:

k	$\Psi^{(k)}$	$f\{\Psi^{(k)}\}$	$f'\{\Psi^{(k)}\}$	$\Psi^{(k+1)}$	$\left\lvert\dfrac{\Psi^{(k+1)} - \Psi^{(k)}}{\Psi^{(k)}}\right\rvert$
1	0.5000	0.2515	0.6259	0.0982	0.8037
2	0.0982	-0.0209	0.9111	0.1211	0.2335
3	0.1211	-0.0007	0.8539	0.1219	0.0065
4	0.1219	0.0000	0.8521	0.1219	0.0000

For this example, convergence is very rapid, giving $\Psi = V/F = 0.1219$. From Eq. (4) of Table 4.4, the equilibrium-vapor flow rate is $0.1219(100) = 12.19$ kmol/h, and the equilibrium-liquid flow rate from Eq. (7) is $(100 - 12.19) = 87.81$ kmol/h. The liquid and vapor compositions computed from Eqs. (5) and (6) are

	x	y
Propane	0.0719	0.3021
n-Butane	0.1833	0.3207
n-Pentane	0.3098	0.2293
n-Hexane	0.4350	0.1479
	1.0000	1.0000

A plot of $f\{\Psi\}$ as a function of Ψ is shown in Figure 4.11.

Bubble and Dew Points

Often, it is desirable to bring a mixture to the bubble point or the dew point. At the bubble point, $\Psi = 0$ and $f\{0\} = 0$. Therefore, from Eq. (3), Table 4.4,

$$f\{0\} = \sum_i z_i(1 - K_i) = \sum z_i - \sum z_i K_i = 0$$

However, $\sum z_i = 1$. Therefore, the bubble-point equation is

$$\sum_i z_i K_i = 1 \qquad (4\text{-}12)$$

At the dew point, $\Psi = 1$ and $f\{1\} = 0$. Therefore, from Eq. (3), Table 4.4,

$$f\{1\} = \sum_i \frac{z_i(1 - K_i)}{K_i} = \sum \frac{z_i}{K_i} - \sum z_i = 0$$

Therefore, the dew-point equation is

$$\sum_i \frac{z_i}{K_i} = 1 \qquad (4\text{-}13)$$

For a given feed composition, z_i, (4-12) or (4-13) can be used to find T for a specified P or to find P for a specified T.

Because of the K-values, the bubble- and dew-point equations are generally highly nonlinear in temperature, but only moderately nonlinear in pressure, except in the region of the *convergence pressure*, where K-values of very light or very heavy species change radically with pressure, as in Figure 2.10. Therefore, iterative procedures are required to solve for bubble- and dew-point conditions. One exception is where Raoult's law K-values are applicable. Substitution of $K_i = P_i^s/P$ into (4-12) leads to an equation for the direct calculation of bubble-point pressure:

$$P_{\text{bubble}} = \sum_{i=1}^{C} z_i P_i^s \qquad (4\text{-}14)$$

where P_i^s is the temperature-dependent vapor pressure of species i. Similarly, from (4-13), the dew-point pressure is

$$P_{\text{dew}} = \left(\sum_{i=1}^{C} \frac{z_i}{P_i^s} \right)^{-1} \qquad (4\text{-}15)$$

Another useful exception occurs for mixtures at the bubble point when K-values can be expressed by the modified Raoult's law, $K_i = \gamma_i P_i^s/P$. Substituting this equation into (4-12),

$$P_{\text{bubble}} = \sum_{i=1}^{C} \gamma_i z_i P_i^s \qquad (4\text{-}16)$$

Liquid-phase activity coefficients can be computed for a known temperature and composition, since $x_i = z_i$ at the bubble point.

Bubble- and dew-point calculations are used to determine saturation conditions for liquid and vapor streams, respectively. It is important to note that when vapor–liquid equilibrium is established, the vapor is at its dew point and the liquid is at its bubble point.

EXAMPLE 4.2

In Figure 1.9, the nC_4-rich bottoms product from column C3 has the composition given in Table 1.5. If the pressure at the bottom of the distillation column is 100 psia (689 kPa), estimate the temperature of the mixture.

SOLUTION

The bottoms product will be a liquid at its bubble point with the following composition:

Component	kmol/h	$z_i = x_i$
i-Butane	8.60	0.0319
n-Butane	215.80	0.7992
i-Pentane	28.10	0.1041
n-Pentane	17.50	0.0648
	270.00	1.0000

The bubble-point temperature can be estimated by finding the temperature that will satisfy (4-12), using K-values from Figure 2.8. Because the bottoms product is rich in nC_4, assume that the K-value of nC_4 is 1. From Figure 2.8, for 100 psia, $T = 150°F$. For this temperature, using Figure 2.8 to obtain the K-values of the other three hydrocarbons and substituting these values and the z-values into (4-12),

$$\sum z_i K_i = 0.0319(1.3) + 0.7992(1.0) + 0.1041(0.47) + 0.0648(0.38)$$
$$= 0.042 + 0.799 + 0.049 + 0.025 = 0.915$$

Because the sum is not 1.0, another temperature must be assumed and the summation repeated. To increase the sum, the K-values must be greater and, thus, the temperature must be higher. Because the sum is dominated by nC_4, assume that its K-value must be $1.000(1.00/0.915) = 1.09$. This corresponds to a temperature of $160°F$, which results in a summation of 1.01. By linear interpolation, $T = 159°F$.

EXAMPLE 4.3

Cyclopentane is to be separated from cyclohexane by liquid–liquid extraction with methanol at 25°C. In extraction it is important that the liquid mixtures be maintained at pressures greater than the bubble-point pressure. Calculate the bubble-point pressure using the following equilibrium liquid-phase compositions, activity coefficients, and vapor pressures:

	Methanol	Cyclohexane	Cyclopentane
Vapor pressure, psia	2.45	1.89	6.14
Methanol-rich layer:			
x	0.7615	0.1499	0.0886
γ	1.118	4.773	3.467
Cyclohexane-rich layer:			
x	0.1737	0.5402	0.2861
γ	4.901	1.324	1.074

SOLUTION

Because the bubble-point pressure is likely to be below ambient pressure, the modified Raoult's law in the form of (4-16) applies for either liquid phase. If the methanol-rich layer data are used:

$$P_{bubble} = 1.118(0.7615)(2.45) + 4.773(0.1499)(1.89)$$
$$+ 3.467(0.0886)(6.14)$$
$$= 5.32 \text{ psia } (36.7 \text{ kPa})$$

A similar calculation based on the cyclohexane-rich layer gives an identical result because the data are consistent with phase equilibrium theory such that $\gamma_{iL}^{(1)} x_i^{(1)} = \gamma_{iL}^{(2)} x_i^{(2)}$. A pressure higher than 5.32 psia will prevent formation of vapor at this location in the extraction process. Thus, operation at atmospheric pressure is a good choice.

EXAMPLE 4.4

Propylene (P) is to be separated from 1-butene (B) by distillation into a vapor distillate containing 90 mol% propylene. Calculate the column operating pressure assuming the exit temperature from the partial condenser is 100°F (37.8°C), the minimum attainable temperature with cooling water. Determine the composition of the liquid reflux. In Figure 4.12, K-values estimated from Eq. (5), Table 2.3, using the Redlich–Kwong equation of state for the vapor fugacity, are plotted and compared to experimental data [7] and Raoult's law K-values.

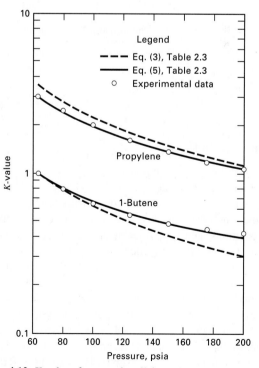

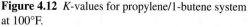

Figure 4.12 K-values for propylene/1-butene system at 100°F.

SOLUTION

The operating pressure corresponds to a dew-point condition for the vapor-distillate composition. The composition of the reflux corresponds to the liquid in equilibrium with the vapor distillate at its dew point. The method of false position [8] can be used to perform the iterative calculations by rewriting (4-13) in the form

$$f\{P\} = \sum_{i=1}^{C} \frac{z_i}{K_i} - 1 \qquad (1)$$

The recursion relationship for the method of false position is based on the assumption that $f\{P\}$ is linear in P such that

$$P^{(k+2)} = P^{(k+1)} - f\{P^{(k+1)}\} \left[\frac{P^{(k+1)} - P^{(k)}}{f\{P^{(k+1)}\} - f\{P^{(k)}\}} \right] \qquad (2)$$

This assumption is reasonable because, at low pressures, K-values in (2) are almost inversely proportional to pressure. Two values of P are required to initialize this formula. Choose 100 psia and 190 psia. At 100 psia, reading the K-values from the solid lines in Figure 4.12,

$$f\{P\} = \frac{0.90}{2.0} + \frac{0.10}{0.68} - 1.0 = -0.40$$

Subsequent iterations give

k	$P^{(k)}$, psia	K_P	K_B	$f\{P^{(k)}\}$
1	100	2.0	0.68	−0.40
2	190	1.15	0.42	+0.02
3	186	1.18	0.425	−0.0020

Iterations are terminated when $|P^{(k+2)} - P^{(k+1)}|/P^{(k+1)} < 0.005$.

An operating pressure of 186 psia (1,282 kPa) at the partial condenser outlet is indicated. The composition of the liquid reflux is obtained from $x_i = z_i/K_i$ with the result

Component	Equilibrium Mole Fraction	
	Vapor Distillate	Liquid Reflux
Propylene	0.90	0.76
1-Butene	0.10	0.24
	1.00	1.00

Adiabatic Flash

When the pressure of a liquid stream of known composition, flow rate, and temperature (or enthalpy) is reduced adiabatically across a valve as in Figure 4.10a, an *adiabatic-flash* calculation is made to determine the resulting temperature, compositions, and flow rates of the equilibrium liquid and vapor streams for a specified pressure downstream of the valve. For an adiabatic flash, the isothermal-flash calculation procedure can be applied in the following iterative manner. A guess is made of the flash temperature, T_V. Then Ψ, V, x, y, and L are determined, as for an isothermal flash, from steps 3 through 7 in Table 4.4. The guessed value of T_V (equal to T_L) is next checked by an energy balance obtained by combining Eqs. (7) and (8) of Table 4.4 with $Q = 0$ to give

$$f\{T_V\} = \frac{\Psi h_V + (1 - \Psi)h_L - h_F}{1,000} = 0 \qquad (4-17)$$

where the division by 1,000 is to make the terms of the order of 1. If the computed value of $f\{T_V\}$ is not zero, the entire procedure is repeated for two or more guesses of T_V. A plot of $f\{T_V\}$ versus T_V is interpolated to determine the correct value of T_V. The procedure is tedious because it involves inner-loop iteration on Ψ and outer-loop iteration on T_V.

Outer-loop iteration on T_V is very successful when Eq. (3) of Table 4.4 is not sensitive to the guess of T_V. This is the case for wide-boiling mixtures. For close-boiling mixtures (e.g., isomers), the algorithm may fail because of extreme sensitivity to the value of T_V. In this case, it is preferable to do the outer-loop iteration on Ψ and solve Eq. (3) of Table 4.4 for T_V in the inner loop, using a guessed value for Ψ to initiate the process:

$$f\{T_V\} = \sum_{i=1}^{C} \frac{z_i(1 - K_i)}{1 + \Psi(K_i - 1)} = 0 \qquad (4-18)$$

Then, Eqs. (5) and (6) of Table 4.4 are solved for x and y, respectively. Equation (4-17) is then solved directly for Ψ, since

$$f\{\Psi\} = \frac{\Psi h_V + (1 - \Psi)h_L - h_F}{1,000} = 0 \qquad (4-19)$$

from which

$$\Psi = \frac{h_F - h_L}{h_V - h_L} \qquad (4-20)$$

If Ψ from (4-20) is not equal to the value of Ψ guessed to solve (4-18), the new value of Ψ is used to repeat the outer loop starting with (4-18).

Multicomponent, isothermal-flash, bubble-point, dew-point, and adiabatic-flash calculations can be very tedious because of their iterative nature. They are unsuitable for manual calculations for nonideal vapor and liquid mixtures because of the complexity of the expressions for the thermodynamic properties, K, h_V, and h_L. However, robust algorithms for making such calculations are incorporated into widely used steady-state simulation computer programs such as ASPEN PLUS, CHEMCAD, HYSYS, and PRO/II.

EXAMPLE 4.5

The equilibrium liquid from the flash drum at 120°F and 485 psia in Example 2.6 is fed to a distillation tower to remove the remaining hydrogen and methane. A tower for this purpose is often referred to as a *stabilizer*. Pressure at the feed plate of the stabilizer is 165 psia (1,138 kPa). Calculate the percent vaporization of the feed if the pressure is decreased adiabatically from 485 to 165 psia by valve and pipeline pressure drop.

SOLUTION

This problem is most conveniently solved by using a steady-state simulation program. If the CHEMCAD program is used with

K-values and enthalpies estimated from the P-R equation of state, the following results are obtained:

	kmol/h		
Component	**Feed 120°F 485 psia**	**Vapor 112°F 165 psia**	**Liquid 112°F 165 psia**
Hydrogen	1.0	0.7	0.3
Methane	27.9	15.2	12.7
Benzene	345.1	0.4	344.7
Toluene	113.4	0.04	113.36
Total	487.4	16.34	471.06
Enthalpy, kJ/h	−1,089,000	362,000	−1,451,000

This case involves a wide-boiling feed, so the procedure involving (4-17) is the best choice. The above results show that only a small amount of vapor ($\Psi = 0.0035$), predominantly H_2 and CH_4, is produced by the adiabatic flash. The computed flash temperature of 112°F is 8°F below the feed temperature. The enthalpy of the feed is equal to the sum of the vapor and liquid product enthalpies for this adiabatic operation.

4.5 TERNARY LIQUID–LIQUID SYSTEMS

Ternary mixtures that undergo phase splitting to form two separate liquid phases can differ as to the extent of solubility of the three components in each of the two liquid phases. The simplest case is shown in Figure 4.13a, where only the solute, component B, has any appreciable solubility in either the carrier, A, or the solvent, C, both of which have negligible (although never zero) solubility in each other. In this case, the equations can be derived for a single equilibrium stage, using the variables F, S, $L^{(1)}$, and $L^{(2)}$ to refer, respectively, to the flow rates (or amounts) of the feed, solvent, exiting extract, and exiting raffinate. By definition, the extract is the exiting liquid phase that contains the solvent and the extracted solute; the raffinate is the exiting liquid phase that contains the carrier, A, of the feed and the portion of the solute, B, that is not extracted. Although the extract is shown in Figure 4.13a as leaving from the top of the stage, this will only be so if the extract is the lighter (lower-density) exiting phase. Assuming that the entering solvent contains no solute, B, it is convenient to write material balance and phase-equilibrium equations for the solute, B. These two equations may be written in terms of molar or mass flow rates. To obtain the simplest result, it is preferable to express compositions of the solute as mass or mole ratios instead of mass or mole fractions.

Let:

F_A = feed rate of carrier A

S = flow rate of solvent C

X_B = ratio of mass (or moles) of solute B, to mass (or moles) of the other component in the feed (F), raffinate (R), or extract (E)

Then, the solute material balance is

$$X_B^{(F)} F_A = X_B^{(E)} S + X_B^{(R)} F_A \qquad (4\text{-}21)$$

and the distribution of solute at equilibrium is given by

$$X_B^{(E)} = K'_{D_B} X_B^{(R)} \qquad (4\text{-}22)$$

where K'_{D_B} is the distribution coefficient defined in terms of mass or mole ratios. Substituting (4-22) into (4-21) to eliminate $X_B^{(E)}$ gives

$$X_B^{(R)} = \frac{X_B^{(F)} F_A}{F_A + K'_{D_B} S} \qquad (4\text{-}23)$$

It is convenient to define an extraction factor, E_B, for the solute B:

$$E_B = K'_{D_B} S / F_A \qquad (4\text{-}24)$$

The larger the value of E, the greater the extent to which the solute is extracted. Large values of E result from large values of the distribution coefficient, K'_{D_B}, or large ratios of solvent to carrier. Substituting (4-24) in (4-23) gives the fraction of B that is not extracted as

$$X_B^{(R)} / X_B^{(F)} = \frac{1}{1 + E_B} \qquad (4\text{-}25)$$

where it is clear that the larger the extraction factor, the smaller the fraction of B not extracted.

Values of mass (mole) ratios, X, are related to mass (mole) fractions, x, by

$$X_i = x_i / (1 - x_i) \qquad (4\text{-}26)$$

Values of the distribution coefficient, K'_D, in terms of ratios, are related to K_D in terms of fractions as given in (2-20) by

$$K'_{D_i} = \frac{x_i^{(1)} / (1 - x_i^{(1)})}{x_i^{(2)} / (1 - x_i^{(2)})} = K_{D_i} \left(\frac{1 - x_i^{(2)}}{1 - x_i^{(1)}} \right) \qquad (4\text{-}27)$$

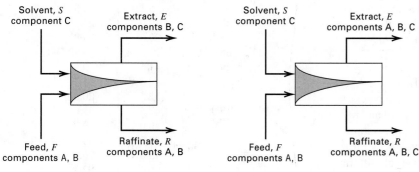

Figure 4.13 Phase splitting of ternary mixtures: (a) components A and C mutually insoluble; (b) components A and C partially soluble.

When values of x_i are small, K'_D approaches K_D. As discussed in Chapter 2, the distribution coefficient, K_{D_i}, which can be determined from activity coefficients using the expression $K_{D_B} = \gamma_B^{(2)}/\gamma_B^{(1)}$ when mole fractions are used, is a strong function of equilibrium-phase compositions and temperature. However, when the raffinate and extract are both dilute in the solute, activity coefficients of the solute can be approximated by the values at infinite dilution so that K_{D_B} can be taken as a constant at a given temperature. An extensive listing of such K_{D_B} values in mass fraction units for various ternary systems is given in *Perry's Handbook* [9]. If values for F_B, $X_B^{(F)}$, S, and K_{D_B} are given, (4-25) can be solved for $X_B^{(R)}$.

EXAMPLE 4.6

A feed of 13,500 kg/h consists of 8 wt% acetic acid (B) in water (A). The removal of the acetic acid is to be accomplished by liquid–liquid extraction at 25°C with methyl isobutyl ketone solvent (C), because distillation of the feed would require vaporization of large amounts of water. If the raffinate is to contain only 1 wt% acetic acid, estimate the kilograms per hour of solvent required if a single equilibrium stage is used.

SOLUTION

Assume that the carrier (water) and the solvent are immiscible. From *Perry's Handbook*, take $K_D = 0.657$ in mass-fraction units for this system. For the relatively low concentrations of acetic acid in this problem, assume that $K'_D = K_D$.

$$F_A = (0.92)(13,500) = 12,420 \text{ kg/h}$$

$$X_B^{(F)} = (13,500 - 12,420)/12,420 = 0.087$$

The raffinate is to contain 1 wt% B. Therefore,

$$X_B^{(R)} = 0.01/(1 - 0.01) = 0.0101$$

From (4-25), solving for E_B,

$$E_B = \frac{X_B^{(F)}}{X_B^{(R)}} - 1 = (0.087/0.0101) - 1 = 7.61$$

From (4-24), the definition of the extraction factor,

$$S = \frac{E_B F_A}{K'_D} = 7.61(12,420/0.657) = 144,000 \text{ kg/h}$$

This is a very large solvent flow rate compared to the feed rate— more than a factor of 10! Multiple stages should be used to reduce the solvent rate or a solvent with a larger distribution coefficient should be sought. For l-butanol as the solvent, $K_D = 1.613$.

In the ternary liquid–liquid system, shown in Figure 4.13b, components A and C are partially soluble in each other and component B again distributes between the extract and raffinate phases. Both of these exiting phases contain all components present in the feed and solvent. This case is by far the most commonly encountered, and a number of different phase diagrams and computational techniques have been

devised to determine the equilibrium compositions. Examples of phase diagrams are shown in Figure 4.14 for the water (A)–ethylene glycol (B)–furfural (C) system at 25°C and a pressure of 101 kPa, which is above the bubble-point pressure, so no vapor phase exists. Experimental data for this system were obtained by Conway and Norton [18]. The pairs water–ethylene glycol and furfural–ethylene glycol are each completely miscible. The only partially miscible pair is furfural–water. Furfural might be used as a solvent to remove the solute, ethylene glycol, from water; the furfural-rich phase is the extract, and the water-rich phase is the raffinate.

Figure 4.14a, an equilateral-triangular diagram, is the most common display of ternary liquid–liquid equilibrium data in the chemical literature. Any point located within or on an edge of the triangle represents a mixture composition. Such a diagram has the property that the sum of the lengths of the perpendicular lines drawn from any interior point to the sides equals the altitude of the triangle. Thus, if each of the three altitudes is scaled from 0 to 100, the percent of, say, furfural, at any point such as M, is simply the length of the line perpendicular to the base opposite the pure furfural apex, which represents 100% furfural. Figure 4.14a is constructed for compositions based on mass fractions (mole fractions and volume fractions are also sometimes used). Thus, the point M in Figure 4.14a represents a mixture of feed and solvent (before phase separation) containing 18.9 wt% water, 20 wt% ethylene glycol, and 61.1 wt% furfural.

The miscibility limits for the furfural–water binary system are at D and G. The miscibility boundary (saturation or binodal curve) DEPRG can be obtained experimentally by a *cloud-point titration;* water, for example, is added to a (clear) 50 wt% solution of furfural and glycol, and it is noted that the onset of cloudiness due to the formation of a second phase occurs when the mixture is 11% water, 44.5% furfural, and 44.5% glycol by weight. Other miscibility data are given in Table 4.5, from which the miscibility curve in Figure 4.14a was drawn.

Table 4.5 Equilibrium Miscibility Data in Weight Percent for the Furfural–Ethylene Glycol–Water System at 25°C and 101 kPa

Furfural	Ethylene Glycol	Water
95.0	0.0	5.0
90.3	5.2	4.5
86.1	10.0	3.9
75.1	20.0	4.9
66.7	27.5	5.8
49.0	41.5	9.5
34.3	50.5	15.2
27.5	52.5	20.0
13.9	47.5	38.6
11.0	40.0	49.0
9.7	30.0	60.3
8.4	15.0	76.6
7.7	0.0	92.3

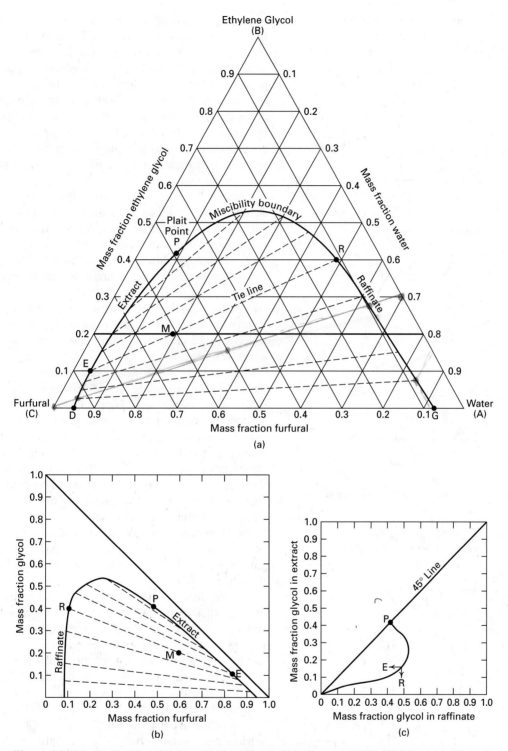

Figure 4.14 Liquid–liquid equilibrium, ethylene glycol–furfural–water, 25°C, 101 kPa: (a) equilateral triangular diagram; (b) right triangular diagram; (c) equilibrium solute diagram in mass fractions (*continues*).

Tie lines, shown as dashed lines below the miscibility boundary in Figure 4.14a, are used to connect points on the miscibility boundary, DEPRG, that represent equilibrium-phase compositions. To obtain data to construct tie lines, such as ER, it is necessary to make a mixture such as M (20% glycol, 18.9% water, and 61.1% furfural),

equilibrate it, and then chemically analyze the resulting equilibrium extract and raffinate phases E and R (in this case, 10% glycol, 3.9% water, and 86.1% furfural; and 40% glycol, 49% water, and 11% furfural, respectively). At point P, the *plait point,* the two liquid phases have identical compositions. Therefore, the tie lines converge

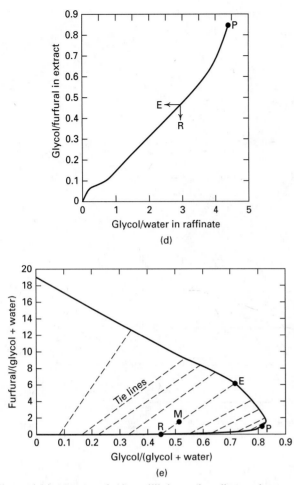

Figure 4.14 (*Continued*) (d) equilibrium solute diagram in mass ratios; (e) Janecke diagram.

Table 4.6 Mutual Equilibrium (Tie Line) Data for the Furfural–Ethylene Glycol–Water System at 25°C and 101 kPa

Glycol in Water Layer, wt%	Glycol in Furfural Layer, wt%
41.5	41.5
50.5	32.5
52.5	27.5
51.5	20.0
47.5	15.0
40.0	10.0
30.0	7.5
20.0	6.2
15.0	5.2
7.3	2.5

to a point and the two phases become one phase. Tie-line data for this system are given in Table 4.6, in terms of glycol composition.

When there is mutual solubility between two phases, the thermodynamic variables necessary to define the equilib-

rium system are temperature, pressure, and the concentrations of the components in each phase. According to the phase rule, (4-1), for a three-component, two-liquid-phase system, there are three degrees of freedom. At constant temperature and pressure, specification of the concentration of one component in either of the phases suffices to completely define the state of the system. Thus, as shown in Figure 4.14a, one value for glycol weight percent on the miscibility boundary curve fixes the composition of the corresponding phase and, by means of the tie line, the composition of the other equilibrium phase.

Figure 4.14b is a representation of the same system on a right-triangular diagram. Here the concentrations in weight percent of any two of the three components (normally the solute and solvent) are given, the concentration of the third being obtained by difference from 100 wt%. Diagrams like this are easier to construct and read than equilateral triangular diagrams.

However, equilateral triangular diagrams are conveniently constructed with the computer program, CSpace, which can be downloaded from the web site at www. ugr.es/~cspace.

Figures 4.14c and 4.14d are representations of the same ternary system in terms of weight fraction and weight ratios of the solute, respectively. Figure 4.14c is simply a plot of the equilibrium (tie-line) data of Table 4.6 in terms of solute mass fraction. In Figure 4.14d, mass ratios of solute (ethylene glycol) to furfural and water for the extract and raffinate phases, respectively, are used. Such curves can be used to interpolate tie lines, since only a limited number of tie lines are shown on triangular graphs. Because of this, such diagrams are often referred to as *distribution diagrams*. When mole (rather than mass) fractions are used in a diagram like Figure 4.14c, a nearly straight line is often evident near the origin, where the slope is the distribution coefficient, K_D, for the solute at infinite dilution.

In 1906, Janecke [10] suggested the equilibrium data display shown as Figure 4.14e. Here, the mass of solvent per unit mass of solvent-free material, furfural/(water + glycol), is plotted as the ordinate versus the mass ratio, on a solvent-free basis, of glycol/(water + glycol) as abscissa. The ordinate and abscissa apply to both phases. Equilibrium conditions are related by tie lines. Mole ratios can be used also to construct Janecke diagrams.

Any of the five diagrams in Figure 4.14 can be used for solving problems involving material balances subject to liquid–liquid equilibrium constraints, as is demonstrated in the following example.

EXAMPLE 4.7

Determine the composition of the equilibrium extract and raffinate phases produced when a 45% by weight glycol (B)–55% water (A) solution is contacted with twice its weight of pure furfural solvent (C) at 25°C and 101 kPa. Use each of the five diagrams in Figure 4.14, if possible.

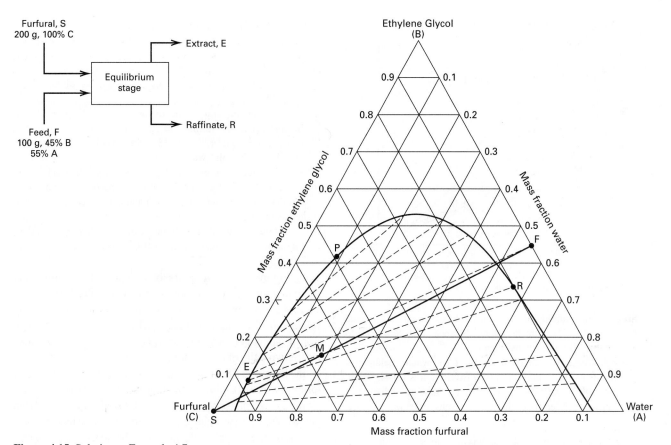

Figure 4.15 Solution to Example 4.7a.

SOLUTION

Assume a basis of 100 g of 45% glycol–water feed. Thus, in Figure 4.13b, the feed (F) is 55 g of A and 45 g of B. The solvent (S) is 200 g of C. Let E = the extract, and R = the raffinate.

(a) By an equilateral-triangular diagram, Figure 4.15:

Step 1. Locate the feed and solvent compositions at points F and S, respectively.
Step 2. Define M, the mixing point, as $M = F + S = E + R$
Step 3. Apply the inverse-lever-arm rule to the equilateral-triangular phase-equilibrium diagram. Let $w_i^{(1)}$ be the mass fraction of species i in the extract, $w_i^{(2)}$ be the mass fraction of species i in the raffinate, and $w_i^{(M)}$ be the mass fraction of species i in the combined feed and solvent phases.
From a balance on the solvent, C: $(F + S)w_C^{(M)} = Fw_C^{(F)} + Sw_C^{(S)}$.

Therefore,

$$\frac{F}{S} = \frac{w_C^{(S)} - w_C^{(M)}}{w_C^{(M)} - w_C^{(F)}} \qquad (1)$$

Thus, points S, M, and F lie on a straight line, and, by the inverse lever arm rule,

$$\frac{F}{S} = \frac{SM}{MF} = \frac{1}{2}$$

The composition at point M is 18.3% A, 15.0% B, and 66.7% C.
Step 4. Since M lies in the two-phase region, the mixture must separate along an interpolated dash-dot tie line into the extract

phase at point E (8.5% B, 4.5% A, and 87.0% C) and the raffinate at point R (34.0% B, 56.0% A, and 10.0% C).
Step 5. The inverse-lever-arm rule applies to points E, M, and R, so $E = M(\overline{RM}/\overline{ER})$. Because $M = 100 + 200 = 300$ g, and from measurements of the line segments, $E = 300(147/200) = 220$ g and $R = M - E = 300 - 220 = 80$ g.

(b) By a right-triangular diagram, Figure 4.16:

Step 1. Locate the points F and S for the two feed streams.
Step 2. Define the mixing point $M = F + S$.
Step 3. The inverse-lever-arm rule also applies to right-triangular diagrams, so MF/MS = $\frac{1}{2}$.
Step 4. Points R and E are on the ends of the interpolated dash-dot tie line passing through point M.

The numerical results of part (b) are identical to those of part (a).

(c) By an equilibrium solute diagram, Figure 4.14c. A material balance on glycol, B,

$$Fw_B^{(F)} + Sw_B^{(S)} = 45 = Ew_B^{(E)} + Rw_B^{(R)} \qquad (2)$$

must be solved simultaneously with a phase-equilibrium relationship. It is not possible to do this graphically using Figure 4.14c in any straightforward manner unless the solvent (C) and carrier (A) are mutually insoluble. The outlet-stream composition can be found, however, by the following iterative procedure.

Step 1. Guess a value for $w_B^{(E)}$ and read the equilibrium value, $w_B^{(R)}$, from Figure 4.14c.

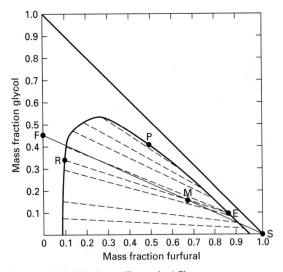

Figure 4.16 Solution to Example 4.7b.

Step 2. Substitute these two values into the equation obtained by combining (2) with the overall balance, $E + R = 300$, to eliminate R. Solve for E and then R.

Step 3. Check to see if the furfural (or water) balance is satisfied using the equilibrium data from Figures 4.14a, 4.14b, or 4.14e. If not, repeat steps 1 to 3 with a new guess for $w_B^{(E)}$. This procedure leads to the same results obtained in parts (a) and (b).

(d) By an equilibrium solute diagram in mass fractions, Figure 4.14d: This plot suffers from the same limitations as Figure 4.13c in that a solution must be achieved by an iterative procedure.

(e) By a Janecke diagram, Figure 4.17:

Step 1. The feed mixture is located at point F. With the addition of 200 g of pure furfural solvent, $M = F + S$ is located as shown, since the ratio of glycol to (glycol + water) remains the same.

Step 2. The mixture at point M separates into the two phases at points E and R, using the interpolated dash-dot tie line, with the coordinates (7.1, 0.67) at E and (0.10, 0.37) at R.

Step 3. Let Z^E and Z^R equal the total mass of components A and B in the extract and raffinate, respectively. Then, the following

balances apply:

Furfural: $7.1Z^E + 0.10Z^R = 200$

Glycol: $0.67Z^E + 0.37Z^R = 45$

Solving these two simultaneous equations, we obtain $Z^E = 27$ g, $Z^R = 73$ g.

Thus, the furfural in the extract $= (7.1)(27 \text{ g}) = 192$ g, the furfural in the raffinate $= 200 - 192 = 8$ g, the glycol in the extract $= (0.67)(27 \text{ g}) = 18$ g, the glycol in the raffinate $= 45 - 18 = 27$ g, the water in the raffinate $= 73 - 27 = 46$ g, and the water in the extract $= 27 - 18 = 9$ g. The total extract is $192 + 27 = 219$ g, which is close to the results obtained in part (a). The raffinate composition and amount can be obtained just as readily.

It should be noted on the Janecke diagram that $\overline{ME}/\overline{MR}$ does not equal R/E; it equals the ratio of R/E on a solvent-free basis.

In Figure 4.14, two pairs of components are mutually soluble, while one pair is only partially soluble. Ternary systems where two pairs and even all three pairs are only partially soluble are also common. Figure 4.18 shows examples, taken from Francis [11] and Findlay [12], of four different cases where two pairs of components are only partially soluble.

In Figure 4.18a, two separate two-phase regions are formed, while in Figure 4.18c, in addition to the two-phase regions, a three-phase region, RST, is formed. In Figure 4.18b, the two separate two-phase regions merge. For a ternary mixture, as temperature is reduced, phase behavior may progress from Figure 4.18a to 4.18b to 4.18c. In Figures 4.18a, 4.18b, and 4.18c, all tie lines slope in the same direction. In some systems of importance, *solutropy,* a reversal of tie-line slopes, occurs.

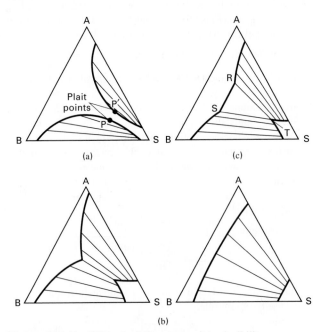

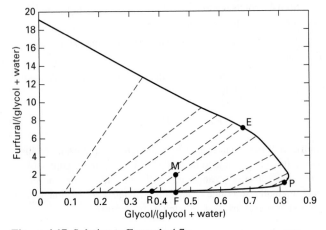

Figure 4.17 Solution to Example 4.7e.

Figure 4.18 Equilibria for 3/2 systems: (a) miscibility boundaries are separate; (b) miscibility boundaries and tie-line equilibria merge; (c) tie lines do not merge and the three-phase region RST is formed.

4.6 MULTICOMPONENT LIQUID–LIQUID SYSTEMS

Quarternary and higher multicomponent mixtures are often encountered in liquid–liquid extraction processes, particularly when two solvents are used for liquid–liquid extraction. As discussed in Chapter 2, multicomponent liquid–liquid equilibria are very complex and there is no compact graphical way of representing experimental phase equilibria. Accordingly, the computation of the equilibrium-phase compositions is best made by a computer-assisted algorithm using activity coefficient equations from Chapter 2 that account for the effect of composition (e.g., NRTL, UNIQUAC, or UNIFAC). One such algorithm is a modification of the Rachford–Rice algorithm for vapor–liquid equilibrium, given in Tables 4.3 and 4.4. To apply these tables to multicomponent, liquid–liquid equilibria, the following symbol transformations are made, where all flow rates and compositions are in moles:

Vapor–Liquid Equilibria
(Tables 4.3, 4.4) | **Liquid–Liquid Equilibria**

Feed, F | Feed, F, + solvent, S

Equilibrium vapor, V | Extract, E ($L^{(1)}$)

Equilibrium liquid, L | Raffinate, R ($L^{(2)}$)

Feed mole fractions, z_i | Mole fractions of combined F and S

Vapor mole fractions, y_i | Extract mole fractions, $x_i^{(1)}$

Liquid mole fractions, x_i | Raffinate mole fractions, $x_i^{(2)}$

K-value, K_i | Distribution coefficient, K_{D_i}

$\Psi = V/F$ | $\Psi = E/F$

Most liquid–liquid equilibria are achieved under adiabatic conditions, thus necessitating consideration of an energy balance. However, if both feed and solvent enter the stage at identical temperatures, the only energy effect is the heat of mixing, which is often sufficiently small that only a very small temperature change occurs. Accordingly, the calculations are often made isothermally.

The modified Rachford–Rice algorithm is shown in the flow chart of Figure 4.19. This algorithm is applicable for either an isothermal vapor–liquid (V–L) or liquid–liquid (L–L) equilibrium-stage calculation when K-values depend strongly on phase compositions. For the L–L case, the algorithm assumes that the feed and solvent flow rates and compositions are fixed. The equilibrium pressure and temperature are also specified. An initial estimate is made of the phase compositions, $x_i^{(1)}$ and $x_i^{(2)}$, and corresponding estimates of the distribution coefficients are made from liquid-phase activity coefficients, using (2-30) with, for example, the NRTL or UNIQUAC equations discussed in Chapter 2. Equation 3 of Table 4.4 is then solved iteratively for $\Psi = E/(F + S)$, from which values of $x_i^{(2)}$ and $x_i^{(1)}$ are computed from Eqs. (5) and (6), respectively, of Table 4.4. The resulting values of $x_i^{(1)}$ and $x_i^{(2)}$ will not usually sum, respectively, to 1 for each liquid phase and are therefore normalized. The normalized values are obtained from equations of the form $x_i' = x_i / \Sigma x_j$, where x_i' are the normalized values that force $\Sigma x_j'$ to equal 1. The normalized values replace the values computed from Eqs. (5) and (6). The iterative

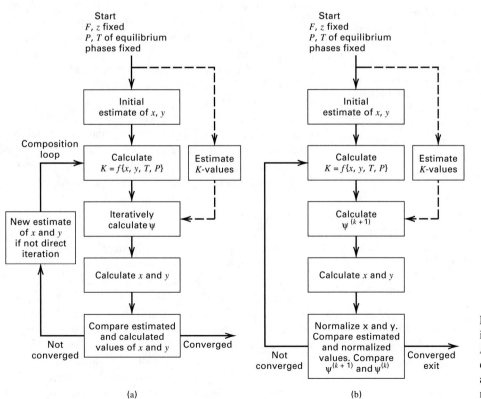

Figure 4.19 Algorithm for isothermal-flash calculation when K-values are composition-dependent: (a) separate nested iterations on Ψ and (x, y); (b) simultaneous iteration on Ψ and (x, y).

procedure is repeated until the compositions $x_i^{(1)}$ and $x_i^{(2)}$, to say three or four significant digits, no longer change from one iteration to the next. Multicomponent liquid–liquid equilibrium calculations are best carried out with a steady-state simulation computer program.

EXAMPLE 4.8

An azeotropic mixture of isopropanol, acetone, and water is being dehydrated with ethyl acetate in a distillation system of two columns. Benzene was previously used as the dehydrating agent, but recent legislation has made the use of benzene undesirable because it is carcinogenic. Ethyl acetate is far less toxic. The overhead vapor from the first column, with the composition given below, at a pressure of 20 psia and a temperature of 80°C is condensed and cooled to 35°C, without significant pressure drop, causing the formation of two liquid phases in equilibrium. Estimate the amounts of the two phases in kilograms per hour and the phase compositions in weight percent.

Component	kg/h
Isopropanol	4,250
Acetone	850
Water	2,300
Ethyl acetate	43,700

Note that the specification of this problem satisfies the degrees of freedom from (4-4), which for $C = 4$ is 9.

SOLUTION

This example was solved with the ChemCAD program using the UNIFAC group contribution method to estimate liquid-phase activity coefficients. The results are as follows:

	Weight Fraction	
Component	Organic-Rich Phase	Water-Rich Phase
Isopropanol	0.0843	0.0615
Acetone	0.0169	0.0115
Water	0.0019	0.8888
Ethyl acetate	0.8969	0.0382
	1.0000	1.0000
Flow rate, kg/h	48,617	2,483

It is of interest to compare the distribution coefficients computed from the above results based on the UNIFAC method to experimental values given in *Perry's Handbook* [1]:

	Distribution Coefficient (wt% Basis)	
Component	UNIFAC	*Perry's Handbook*
Isopropanol	1.37	1.205 (20°C)
Acetone	1.47	1.50 (30°C)
Water	0.0021	—
Ethyl acetate	23.5	—

Results for isopropanol and acetone are in reasonably good agreement at these relatively dilute conditions, considering that no temperature corrections were made.

4.7 SOLID–LIQUID SYSTEMS

Solid–liquid separation operations include leaching, crystallization, and adsorption. In a leaching operation (solid–liquid extraction), a multicomponent solid mixture is separated by contacting the solid with a solvent that selectively dissolves some, but not all, components in the solid. Although this operation is quite similar to liquid–liquid extraction, two aspects of leaching make it a much more difficult separation operation in practice. Diffusion in solids is very slow compared to diffusion in liquids, thus making it difficult to achieve equilibrium. Also, it is virtually impossible to completely separate a solid phase from a liquid phase. A clear liquid phase can be obtained, but the solids will be accompanied by some liquid. In comparison, the separation of two liquid phases is fairly easy to accomplish.

A second solid–liquid system involves the crystallization of one or more, but not all, components from a liquid mixture. This operation is analogous to distillation. However, although equilibrium can be achieved, a sharp phase separation is again virtually impossible.

A third application of solid–liquid systems, adsorption, involves the use of a porous solid agent, which does not undergo phase change or composition change. The solid selectively adsorbs, on its exterior and interior surface, certain components of the liquid mixture. The adsorbed species are then desorbed and the solid adsorbing agent is regenerated for repeated use. Variations of adsorption include ion exchange and chromatography. A solid–liquid system is also utilized in membrane-separation operations, where the solid is a membrane that selectively absorbs and transports certain species, thus effecting a separation.

Solid–liquid separation processes, such as leaching and crystallization, almost always involve phase-separation operations such as gravity sedimentation, filtration, and centrifugation. These operations are not covered in this textbook, but are discussed in Section 18 of *Perry's Handbook* [1].

Leaching

A leaching stage for a ternary system is shown in Figure 4.20. The solid mixture to be separated consists of particles containing insoluble A and solute B. The solvent, C, selectively dissolves B. The overflow from the stage is a solids-free liquid of solvent C and dissolved B. The underflow is a wet solid or slurry of liquid and solid A. In an ideal, equilibrium leaching stage, all of the solute is dissolved by the solvent; none of the solid A is dissolved. In addition, the composition of the retained liquid phase in the underflow is identical to the composition of the liquid overflow, and the

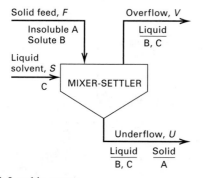

Figure 4.20 Leaching stage.

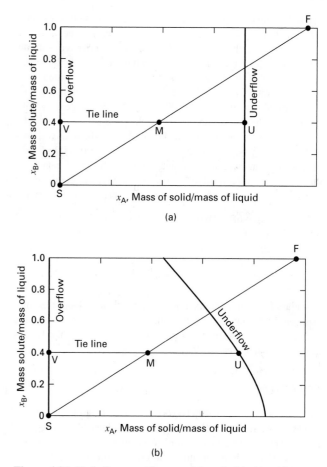

Figure 4.21 Underflow–overflow conditions for ideal leaching: (a) constant solution underflow; (b) variable solution underflow.

overflow is free of solids. The mass ratio of solid to liquid in the underflow depends on the properties of the two phases and the type of equipment used, and is best determined from experience or tests with prototype equipment. In general, if the viscosity of the liquid phases increases with increasing solute concentration, the mass ratio of solid to liquid in the underflow will decrease because the solid will retain more liquid.

Ideal leaching calculations can be carried out algebraically or graphically, with diagrams like those shown in Figure 4.21, using the following nomenclature in mass units:

F = total mass flow rate of feed to be leached

S = total mass flow rate of entering solvent

U = total mass flow rate of the underflow, including solids

V = total mass flow rate of the overflow

X_A = mass ratio of insoluble solid A to (solute B + solvent C) in the feed flow, F, or underflow, U

Y_A = mass ratio of insoluble solid A to (solute B + solvent C) in the entering solvent flow, S, or overflow, V

X_B = mass ratio of solute B to (solute B + solvent C) in the feed flow, F, or underflow, U

Y_B = mass ratio of solute B to (solute B + solvent C) in the solvent flow, S, or overflow, V

Figure 4.21a depicts ideal leaching conditions when, in the underflow, the mass ratio of insoluble solid to liquid, X_A, is a constant, independent of the concentration, X_B, of solute in the solids-free liquid. The resulting tie line is vertical. This case is referred to as *constant-solution underflow*. Figure 4.21b depicts ideal leaching conditions when X_A varies with X_B. This case is referred to as *variable-solution underflow*. In both ideal cases, we assume (1) an entering feed, F, free of solvent such that $X_B = 1$; (2), a solids-free and solute-free solvent, S, such that $Y_A = 0$ and $Y_B = 0$; (3) equilibrium between the exiting liquid solutions in the underflow, U, and the overflow, V, such that $X_B = Y_B$; and (4) a solids-free overflow, V, such that $Y_A = 0$.

As with ternary, liquid–liquid extraction calculations, discussed in Section 4.5, a mixing point, M, can be defined for

$(F + S)$, equal to that for the sum of the two products of the leaching stage, $(U + V)$. Typical mixing points and inlet and outlet compositions are included in Figures 4.21a and b. In both cases, as shown in the next example, the inverse-lever-arm rule can be applied to the line UMV to obtain the flow rates of the underflow, U, and overflow, V.

EXAMPLE 4.9

Soybeans are the predominant oilseed crop in the world, followed by cottonseed, peanuts, and sunflower seed. While soybeans are not generally consumed directly by humans; they can be processed to produce valuable products. Large-scale production of soybeans in the United States began after World War II, increasing in recent years to more than 140 billion pounds per year. Most of the soybeans are processed to obtain soy oil and vitamins like niacin and lecithin for human consumption, and a defatted meal for livestock feed. Compared to other vegetable oils, soy oil is more economical, more stable, and healthier. Typically, 100 pounds of soybeans yields 18 pounds of soy oil and 79 pounds of defatted meal.

To recover oil, soybeans are first cleaned, cracked to loosen the seeds from the hulls, dehulled, and dried to 10–11% moisture. They are then leached with a hexane solution to remove the oil. However, before leaching, the soybeans are flaked to reduce the time required for mass transfer of the oil out of the bean and into hexane. Following leaching, the hexane in the overflow is separated from the soy oil and recovered for recycle by evaporation, while the underflow is treated to remove residual hexane and toasted with high-temperature air to produce defatted meal. Modern soybean-extraction plants crush up to 3,000 tons of soybeans per day. This example is concerned with just the leaching step.

Oil is to be leached from 100,000 kg/h of soybean flakes, containing 19% by weight oil, in a single stage by 100,000 kg/h of a hexane solvent. Experimental data indicate that the oil content of the flakes will be reduced to 0.5 wt%. For the type of equipment to be used, the expected contents of the underflows is as follows:

β, Mass fraction of solids in underflow	0.68	0.67	0.65	0.62	0.58	0.53
Mass ratio of solute in underflow liquid, X_B	0.0	0.2	0.4	0.6	0.8	1.0

Calculate by both a graphical and an analytical method, the compositions and flow rates of the underflow and overflow, assuming an ideal leaching stage. What percentage of the oil in the feed is recovered in the overflow?

SOLUTION

The flakes contain $(0.19)(100,000) = 19,000$ kg/h of oil and $(100,000 - 19,000) = 81,000$ kg/h of insolubles. However, all of the oil is not leached. For convenience in the calculations, lump the unleached oil with the insolubles to give an effective A. The flow rate of unleached oil $= (81,000)(0.5/99.5) = 407$ kg/h. Therefore, take the flow rate of A as $(81,000 + 407) = 81,407$ kg/h and the oil in the feed as just the amount that is leached or $(19,000 - 407) = 18,593$ kg/h of B. Therefore, in the feed, F,

$$Y_A = (81,407/18,593) = 4.38$$
$$X_B = 1.0$$

The sum of the liquid solutions in the underflow and overflow includes 100,000 kg/h of hexane and 18,593 kg/h of leached oil. Therefore, for the underflow and overflow,

$$X_B = Y_B = [18,593/(100,000 + 18,593)] = 0.157$$

This is a case of variable-solution underflow. Using data in the above table, convert values of β, the mass fraction of solids in the underflow, to values of X_A, the mass ratio of insolubles to liquid in the underflow, which by material balance is

$$X_A = \frac{\text{kg/h A}}{\text{kg/h (B + C)}} = \frac{\beta U}{(1-\beta)U} = \frac{\beta}{(1-\beta)} \qquad (1)$$

Using (1), the following values of X_A are computed from the previous table:

X_A	2.13	2.03	1.86	1.63	1.38	1.13
X_B	0.0	0.2	0.4	0.6	0.8	1.0

Graphical Method

Figure 4.22 is a plot of X_A as a function of X_B. Data for the underflow line are obtained from the preceding table. Because no solids

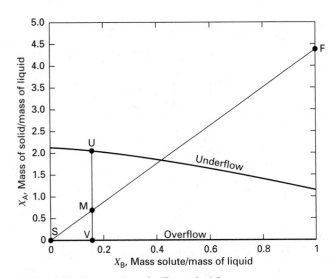

Figure 4.22 Constructions for Example 4.9.

leave in the overflow, that line is drawn horizontally at $X_A = 0$. On this plot are composition points for the feed flakes, F, and the entering solvent hexane, S, with a straight line drawn between them. A point for the overflow, V, is plotted at $X_A = 0$ and, from above, $X_B = 0.157$. Since $Y_B = X_B = 0.157$, the value of X_A in the underflow is obtained at the intersection of a vertical line drawn from the overflow point, V, to the underflow line. This value is $X_A = 2.05$. The two lines $\overline{FS}$ and $\overline{UV}$ intersect at the mixing point, M.

We now compute the compositions of the underflow and overflow. In the overflow, from $X_B = 0.157$, the mass fractions of solute B and solvent C are, respectively, 0.157 and $(1 - 0.157) = 0.843$. In the underflow, using $X_A = 2.05$ and $X_B = 0.157$, the mass fractions of solids, B, and C, are, respectively, $[2.05/(1 + 2.05)] = 0.672$, $0.157(1 - 0.672) = 0.0515$, and $(1 - 0.672 - 0.0515) = 0.2765$.

The inverse-lever-arm rule can be used to compute the amounts of underflow and overflow. Here, the rule applies only to the liquid phases in the two exiting streams because the coordinates of Figure 4.22 are compositions on a solids-free basis. The mass ratio of liquid flow rate in the underflow to liquid flow rate in the overflow is given by the ratio of line $\overline{MV}$ to line $\overline{MU}$. With M located at $X_A = 0.69$, this ratio $= (0.69 - 0.0)/(2.05 - 0.69) = 0.51$. Thus, the liquid flow rate in the underflow $= (100,000 + 18,593)(0.51)/(1 + 0.51) = 40,054$ kg/h. Adding to this the flow rates of carrier and unextracted oil, computed above, gives $U = 40,054 + 81,407 = 121,461$ kg/h or say 121,000 kg/h. The overflow rate $= V = 200,000 - 121,000 = 79,000$ kg/h.

The oil flow rate in the feed is 19,000 kg/h. The oil flow rate in the overflow $= Y_B V = 0.157(79,000) = 12,400$ kg/h. Thus, the percentage of the oil in the feed that is recovered in the overflow $= 12,400/19,000 = 0.653$ or 65.3%. Adding washing stages, as described in Section 5.2, can increase the oil recovery.

Algebraic Method

Instead of using the inverse-lever-arm rule with Figure 4.22, mass-balance equations can be applied. As with the graphical method, $X_B = 0.157$, giving a value from the previous table of $X_A = 2.05$. Then, since the flow rate of solids in the underflow $= 81,407$ kg/h, the flow rate of liquid in the underflow $= 81,407/2.05 = 39,711$ kg/h.

The total flow rate of underflow is $U = 81,407 + 39,711 = 121,118$ kg/h. By mass balance, the flow rate of overflow = $200,000 - 121,118 = 78,882$ kg/h. These values are close to those obtained by the graphical method. The percentage recovery of oil, and compositions of the underflow and overflow, are computed in the same manner as in the graphical method.

Crystallization

Crystallization may take place from aqueous or nonaqueous solutions. The simplest case is for a binary mixture of two organic chemicals such as naphthalene and benzene, whose solubility or solid–liquid phase-equilibrium diagram for a pressure of 1 atm is shown in Figure 4.23. Points A and B are the melting (freezing) points of pure benzene (5.5°C) and pure naphthalene (80.2°C), respectively. When benzene is dissolved in liquid naphthalene or naphthalene is dissolved in liquid benzene, the freezing point of the solvent is depressed. Point E is the *eutectic point*, corresponding to a eutectic temperature (−3°C) and eutectic composition (80 wt% benzene). The word "eutectic" is derived from a Greek word that means "easily fused," and in Figure 4.23 it represents the binary mixture of naphthalene and benzene, as separate solid phases, with the lowest freezing (melting) point.

Temperature–composition points located above the curve AEB correspond to a homogeneous liquid phase. Curve AE is the solubility curve for benzene in naphthalene. For example, at 0°C the solubility is very high, 87 wt% benzene or 6.7 kg benzene/kg naphthalene. Curve EB is the solubility curve for naphthalene. At 25°C the solubility is 41 wt% naphthalene or 0.7 kg naphthalene/kg benzene. At 50°C, the solubility of naphthalene is much higher, 1.9 kg naphthalene/kg benzene. For this mixture, as with most mixtures, solubility increases with increasing temperature.

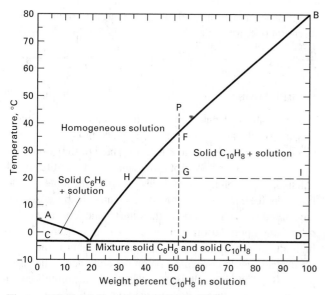

Figure 4.23 Solubility of naphthalene in benzene.
[Adapted from O.A. Hougen, K.M. Watson, and R.A. Ragatz, *Chemical Process Principles. Part I,* 2nd ed., John Wiley and Sons, New York (1954).]

If a liquid solution of composition and temperature represented by point P is cooled along the vertical, dashed line, it will remain a liquid until the line intersects the solubility curve at point F. If the temperature is lowered further, crystals of naphthalene form and the remaining liquid, called the *mother liquor,* becomes richer in benzene. For example, when point G is reached, pure naphthalene crystals and a mother liquor, given by point H on solubility curve EB, coexist at equilibrium, with the composition of the solution being 37 wt% naphthalene. This is in agreement with the Gibbs phase rule (4-1), because with $C = 2$ and $\mathcal{P} = 2$, $\mathcal{F} = 2$ and for fixed T and P, the phase compositions are fixed. The fraction of the solution crystallized can be determined by applying the inverse-lever-arm rule. Thus, in Figure 4.23, the fraction is kilograms naphthalene crystals/kilograms original solution = length of line GH/length of line HI = $(52 - 37)/(100 - 37) = 0.238$.

As the temperature is lowered further until line CED, corresponding to the eutectic temperature, is reached at point J, the two-phase system consists of naphthalene crystals and a mother liquor of the eutectic composition given by point E. Any further removal of heat causes the eutectic solution to solidify.

EXAMPLE 4.10

A total of 8,000 kg/h of a liquid solution of 80 wt% naphthalene and 20 wt% benzene at 70°C is cooled to 30°C to form naphthalene crystals. Assuming that equilibrium is achieved, determine the amount of crystals formed and the composition of the equilibrium mother liquor.

SOLUTION

From Figure 4.23, at 30°C, the solubility of naphthalene is 45 wt% naphthalene. By the inverse-lever-arm rule, for an original 80 wt% solution,

$$\frac{\text{kg naphthalene crystals}}{\text{kg original mixture}} = \frac{(80 - 45)}{(100 - 45)} = 0.636$$

The flow rate of crystals = 0.636 (8,000) = 5,090 kg/h.

The composition of the remaining 2,910 kg/h of mother liquor is 55 wt% benzene and 45 wt% naphthalene.

Crystallization of a salt from an aqueous solution is frequently complicated by the formation of hydrates of the salt with water in certain definite molar proportions. These hydrates can be stable solid compounds within certain ranges of temperature as given in the solid–liquid phase equilibrium diagram. A rather extreme, but common, case is that of $MgSO_4$, which can form the stable hydrates $MgSO_4 \cdot 12H_2O$, $MgSO_4 \cdot 7H_2O$, $MgSO_4 \cdot 6H_2O$, and $MgSO_4 \cdot H_2O$. The high hydrate is stable at low temperatures, while the low hydrate is the stable form at higher temperatures.

A simpler example is that of Na_2SO_4 in mixtures with water. As seen in the phase diagram of Figure 4.24, only one

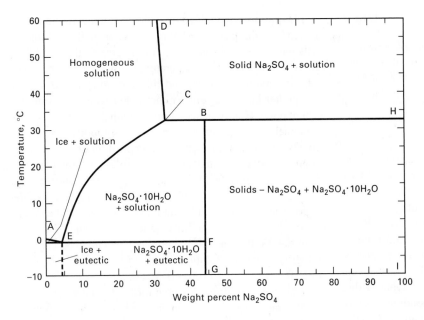

Figure 4.24 Solubility of sodium sulfate in water.

[Adapted from O.A. Hougen, K.M. Watson, and R.A. Ragatz, *Chemical Process Principles. Part I*, 2nd ed., John Wiley and Sons, New York (1954).]

stable hydrate is formed, $Na_2SO_4 \cdot 10H_2O$, commonly known as Glauber's salt. Not shown in Figure 4.24 is the metastable hydrate $Na_2SO_4 \cdot 7H_2O$. Since the molecular weights are 142.05 for Na_2SO_4 and 18.016 for H_2O, the weight percent Na_2SO_4 in the decahydrate is 44.1, which corresponds to the vertical line BFG.

The freezing point of water, 0°C, is at A in Figure 4.24, but the melting point of Na_2SO_4, 884°C, is not shown because the temperature scale is terminated at 60°C. The decahydrate melts at 32.4°C, point B, to form solid Na_2SO_4 and a mother liquor, point C, of 32.5 wt% Na_2SO_4. As Na_2SO_4 is dissolved in water, the freezing point is depressed slightly along curve AE until the eutectic, point E, is reached.

Curves EC and CD represent the solubilities of the decahydrate crystals and anhydrous sodium sulfate, respectively, in water. Note that the solubility of Na_2SO_4 decreases slightly with increasing temperature.

For each region, the coexisting phases are indicated. For example, in the region below GFBHI, a solid solution of the anhydrous and decahydrate forms exists. The amounts of the coexisting phases can be determined by the inverse-lever-arm rule.

EXAMPLE 4.11

A 30 wt% aqueous Na_2SO_4 solution of 5,000 lb/h enters a cooling-type crystallizer at 50°C. At what temperature will crystallization begin? Will the crystals be the decahydrate or anhydrous form? To what temperature will the mixture have to be cooled to crystallize 50% of the Na_2SO_4?

SOLUTION

From Figure 4.24, the original solution of 30 wt% Na_2SO_4 at 50°C corresponds to a point in the homogeneous liquid solution region. If a vertical line is dropped from that point, it intersects the solubility

curve EC at 31°C. Below this temperature, the crystals formed are the decahydrate.

The feed contains $(0.30)(5,000) = 1,500$ lb/h of Na_2SO_4 and $(5,000 - 1,500) = 3,500$ lb/h of H_2O. Thus, $(0.5)(1,500) = 750$ lb/h are to be crystallized. The decahydrate crystals include water of hydration in an amount given by ratioing molecular weights or

$$750 \left[\frac{(10)(18.016)}{(142.05)} \right] = 950 \text{ lb/h}$$

Thus, the total amount of decahydrate is $750 + 950 = 1,700$ lb/h. The water remaining in the mother liquor is $3,500 - 950 = 2,550$ lb/h. The composition of the mother liquor is $750/(2,550 + 750)$ $(100\%) = 22.7$ wt% Na_4SO_4. From Figure 4.24, the temperature corresponding to 22.7 wt% Na_2SO_4 on the solubility curve EC is 26°C.

The amount of crystals can be verified by applying the inverse-lever-arm rule, which gives $5,000 \left[(30 - 22.7)/(44.1 - 22.7) \right] = 1,700$ lb/h.

Liquid Adsorption

When a liquid mixture is brought into contact with a microporous solid, adsorption of certain components in the mixture takes place on the internal surface of the solid. The maximum extent of adsorption occurs when equilibrium is reached. The solid, which is essentially insoluble in the liquid, is the *adsorbent*. The component(s) being adsorbed are called *solutes* when in the liquid and constitute the *adsorbate* upon adsorption on the solid. In general, the higher the concentration of the solute, the higher is the equilibrium adsorbate concentration on the adsorbent. The component(s) of the liquid mixture other than the solute(s), that is, the solvent (carrier), are assumed not to adsorb.

No theory for predicting adsorption-equilibrium curves, based on molecular properties of the solute and solid, is universally embraced. Instead, laboratory experiments must

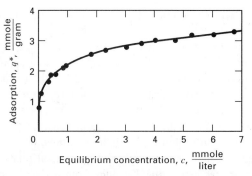

Figure 4.25 Adsorption isotherm for phenol from an aqueous solution in the presence of activated carbon at 20°C.

be performed at a fixed temperature for each liquid mixture and adsorbent to provide data for plotting curves, called *adsorption isotherms*. Figure 4.25, taken from the data of Fritz and Schuluender [13], is an isotherm for the adsorption of phenol from an aqueous solution onto activated carbon at 20°C. Activated, powdered, or granular carbon is a microcrystalline, nongraphitic form of carbon that has a microporous structure to give it a very high internal surface area per unit mass of carbon, and therefore a high capacity for adsorption. Activated carbon preferentially adsorbs organic compounds rather than water when contacted with an aqueous phase containing dissolved organics. As shown in Figure 4.25, as the concentration of phenol in the aqueous phase is increased, the extent of adsorption increases very rapidly at first, followed by a much-slower increase. When the concentration of phenol is 1.0 mmol/L (0.001 mol/L of aqueous solution or 0.000001 mol/g of aqueous solution), the concentration of phenol on the activated carbon is somewhat more than 2.16 mmol/g (0.00216 mol/g of carbon or 0.203 g phenol/g of carbon). Thus, the affinity of this adsorbent for phenol is extremely high. The extent of adsorption depends markedly on the nature of the process used to produce the activated carbon. Adsorption isotherms like Figure 4.25 can be used to determine the amount of adsorbent required to selectively remove a given amount of solute from a liquid.

Consider the ideal, single-stage adsorption process of Figure 4.26, where A is the carrier liquid, B is the solute, and C is the solid adsorbent. Let

c_B = concentration of solute in the carrier liquid, mol/unit volume

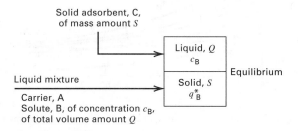

Figure 4.26 Equilibrium stage for liquid adsorption.

q_B = concentration of adsorbate, mol/unit mass of adsorbent

Q = volume of liquid (assumed to remain constant during adsorption)

S = mass of adsorbent (solute-free basis)

A material balance on the solute, assuming that the entering adsorbent is free of solute and that adsorption equilibrium is achieved, as designated by the asterisk superscript on q, gives

$$c_B^{(F)} Q = c_B Q = q_B^* S \qquad (4\text{-}28)$$

This equation can be rearranged to the form of a straight line that can be plotted on the graph of an adsorption isotherm of the type in Figure 4.25, to obtain a graphical solution at equilibrium for c_B and q_B^*. Thus, solving (4-28) for q_B^*,

$$q_B^* = -\frac{Q}{S} c_B + c_B^{(F)} \frac{Q}{S} \qquad (4\text{-}29)$$

The intercept on the c_B axis is $c_B^{(F)} Q/S$, and slope is $-(Q/S)$. The intersection of (4-29) with the adsorption isotherm is the equilibrium condition, c_B and q_B^*.

Alternatively, an algebraic solution can be obtained. Adsorption isotherms for equilibrium-liquid adsorption of a species i can frequently be fitted with the empirical Freundlich equation, discussed in Chapter 15:

$$q_i^* = A c_i^{(1/n)} \qquad (4\text{-}30)$$

where A and n depend on the solute, carrier, and particular adsorbent. The constant, n, is greater than 1, and A is a function of temperature. Freundlich developed his equation from experimental data on the adsorption on charcoal of organic solutes from aqueous solutions. Substitution of (4-30) into (4-29) gives

$$A c_B^{(1/n)} = -\frac{Q}{S} c_B + c_B^{(F)} \frac{Q}{S} \qquad (4\text{-}31)$$

which is a nonlinear equation in c_B that can be solved numerically by an iterative method, as illustrated in the following example.

EXAMPLE 4.12

One liter of an aqueous solution containing 0.010 mol of phenol is brought to equilibrium at 20°C with 5 g of activated carbon having the adsorption isotherm shown in Figure 4.25. Determine the percent adsorption of the phenol and the equilibrium concentrations of phenol on carbon by:

(a) A graphical method

(b) A numerical algebraic method

For the latter case, the curve of Figure 4.25 is fitted quite well with the Freundlich equation (4-30), giving

$$q_B^* = 2.16 c_B^{(1/4.35)} \qquad (1)$$

SOLUTION

From the data given, $c^{(F)} = 10$ mmol/L, $Q = 1$ L, and $S = 5$ g.

(a) *Graphical method.* From (4-29), $q_B^* = -\left(\frac{1}{5}\right)c_B + 10\left(\frac{1}{5}\right) = -0.2c_B + 2$

This equation, with a slope of -0.2 and an intercept of 2, when plotted on Figure 4.25, yields an intersection with the equilibrium curve at $q_B^* = 1.9$ mmol/g and $c_B = 0.57$ mmol/liter. Thus, the percent adsorption of phenol is

$$\frac{c_B^{(F)} - c_B}{c_B^{(F)}} = \frac{10 - 0.57}{10} = 0.94 \quad \text{or} \quad 94\%$$

(b) *Numerical algebraic method.* Applying Eq. (1) from the problem statement and (4-31),

$$2.16c_B^{0.23} = -0.2c_B + 2 \tag{2}$$

or

$$f\{c_B\} = 2.16c_B^{0.23} + 0.2c_B - 2 = 0 \tag{3}$$

This nonlinear equation for c_B can be solved by any of a number of iterative numerical techniques. For example, Newton's method [14] can be applied to Eq. (3) by using the iteration rule:

$$c_B^{(k+1)} = c_B^{(k)} - f^{(k)}\{c_B\}/f'^{(k)}\{c_B\} \tag{4}$$

where k is the iteration index. For this example, $f\{c_B\}$ is given by Eq. (3) and $f'\{c_B\}$ is obtained by differentiating Eq. (3) with respect to c_B to give

$$f'^{(k)}\{c_B\} = 0.497c_B^{-0.77} + 0.2$$

A convenient initial guess for c_B can be made by assuming almost 100% adsorption of phenol to give $q_B^* = 2$ mmol/g. Then, from (4-30),

$$c_B^{(0)} = (q_B^*/A)^n = (2/2.16)^{4.35} = 0.72 \text{ mmol/L}$$

where the (0) superscript designates the starting guess. The Newton iteration rule of Eq. (4) can now be applied, giving the following results:

k	$c_B^{(k)}$	$f^{(k)}\{c_B\}$	$f'^{(k)}\{c_B\}$	$c_B^{(k+1)}$
0	0.72	0.1468	0.8400	0.545
1	0.545	−0.0122	0.9928	0.558
2	0.558	−0.00009	0.9793	0.558

These results indicate convergence to $f\{c_B\} = 0$ for a value of $c_B = 0.558$ after only three iterations. From Eq. (1),

$$q_B^* = 2.16(0.558)^{(1/4.35)} = 1.89 \text{ mmol/g}$$

The result of the numerical method is within the accuracy of the graphical method.

4.8 GAS–LIQUID SYSTEMS

Vapor–liquid systems were covered in Sections 4.2, 4.3, and 4.4. There, the vapor was a mixture of species, most or all of which were condensable. Although the terms *vapor* and *gas* are often used interchangeably, the term *gas* is used to designate a mixture for which the temperature is above the critical temperatures of most or all of the species in the mixture. Thus, the components of a gas mixture are not easily condensed to a liquid. In this section, the physical equilibrium of gas–liquid mixtures is considered.

Even though components of a gas mixture are at a temperature above critical, they can dissolve in an appropriate liquid solvent to an extent that depends on the temperature and their partial pressure in the gas mixture. With good mixing, equilibrium between the two phases can be achieved in a short time unless the liquid is very viscous.

Unlike equilibrium vapor–liquid mixtures, where, as discussed in Chapter 2, a number of theoretical relationships are in use for estimating K-values from molecular properties, no widely accepted theory exists for gas–liquid mixtures. Instead, experimental data, plots of experimental data, or empirical correlations are used.

Experimental solubility data for 13 common gases dissolved in water are plotted over a range of temperature from 0 to as high as 100°C in Figure 4.27. The ordinate is the

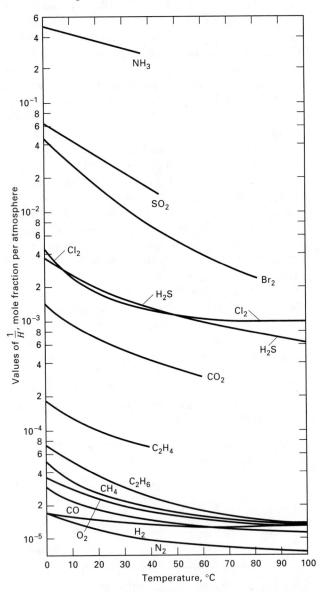

Figure 4.27 Henry's law constant for solubility of gases in water.
[Adapted from O.A. Hougen, K.M. Watson, and R.A. Ragatz, *Chemical Process Principles. Part I*, 2nd ed., John Wiley and Sons, New York (1954).]

equilibrium mole fraction of the gas (solute) in the liquid when the pressure of the gas is 1 atm. The curves of Figure 4.27 can be used to estimate the solubility in water at other pressures and for mixtures of gases by applying Henry's law with the partial pressure of the solute, provided that mole-fraction solubilities are low and no chemical reactions occur among the gas species or with water. Henry's law, discussed briefly in Chapter 2 and given in Table 2.3, is rewritten for use with Figure 4.27 as

$$x_i = \left(\frac{1}{H_i}\right) y_i P \qquad (4\text{-}32)$$

where H_i = Henry's law constant, atm.

For gases with a high solubility, such as ammonia, Henry's law may not be applicable, even at low partial pressures. In that case, experimental data for the actual conditions of pressure and temperature are necessary as in Example 4.14. In either case, calculations of equilibrium conditions are made in the manner illustrated in previous sections of this chapter by combining material balances with equilibrium relationships or data. The following two examples illustrate single-stage, gas-liquid equilibria calculation methods.

EXAMPLE 4.13

An ammonia plant, located at the base of a 300-ft (91.44-m)-high mountain, employs a unique absorption system for disposing of by-product CO_2. The CO_2 is absorbed in water at a CO_2 partial pressure of 10 psi (68.8 kPa) above that required to lift water to the top of the mountain. The CO_2 is then vented to the atmosphere at the top of the mountain, the water being recirculated as shown in Figure 4.28. At 25°C, calculate the amount of water required to dispose of 1,000 ft^3 (28.31 m^3)(STP) of CO_2.

SOLUTION

Basis: 1,000 ft^3 (28.31 m^3) of CO_2 at 0°C and 1 atm (STP). From Figure 4.27, the reciprocal of the Henry's law constant for CO_2 at 25°C is 6×10^{-4} mole fraction/atm. The CO_2 pressure in the absorber (at the foot of the mountain) is

$$p_{CO_2} = \frac{10}{14.7} + \frac{300 \text{ ft } H_2O}{34 \text{ ft } H_2O/\text{atm}} = 9.50 \text{ atm} = 960 \text{ kPa}$$

At this partial pressure, the equilibrium concentration of CO_2 in the water is

$$x_{CO_2} = 9.50(6 \times 10^{-4}) = 5.7 \times 10^{-3} \text{ mole fraction } CO_2 \text{ in water}$$

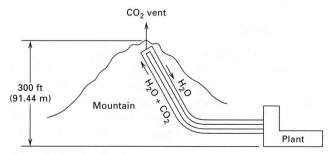

Figure 4.28 Flowsheet for Example 4.13.

The corresponding ratio of dissolved CO_2 to water is

$$\frac{5.7 \times 10^{-3}}{1 - 5.7 \times 10^{-3}} = 5.73 \times 10^{-3} \text{ mol } CO_2/\text{mol } H_2O$$

The total number of moles of CO_2 to be absorbed is

$$\frac{1,000 \text{ ft}^3}{359 \text{ ft}^3/\text{lbmol (at STP)}} = \frac{1,000}{359} = 2.79 \text{ lbmol}$$

or $(2.79)(44)(0.454) = 55.73$ kg.

Assuming all the absorbed CO_2 is vented at the mountain top, the number of moles of water required is $2.79/(5.73 \times 10^{-3}) = 458$ lbmol $= 8,730$ lb $= 3,963$ kg.

If one corrects for the fact that the pressure on top of the mountain is 101 kPa, so that not all of the CO_2 is vented, 4,446 kg (9,810 lb) of water are required.

EXAMPLE 4.14

The partial pressure of ammonia (A) in air–ammonia mixtures in equilibrium with their aqueous solutions at 20°C is given in Table 4.7. Using these data, and neglecting the vapor pressure of water and the solubility of air in water, construct an equilibrium diagram at 101 kPa using mole ratios Y_A = mol NH_3/mol air, and X_A = mol NH_3/mol H_2O as coordinates. Henceforth, the subscript A is dropped. If 10 mol of gas, of composition $Y = 0.3$, are contacted with 10 mol of a solution of composition $X = 0.1$, what are the compositions of the resulting phases at equilibrium? The process is assumed to be isothermal and at atmospheric pressure.

SOLUTION

The equilibrium data given in Table 4.7 are recalculated in terms of mole ratios in Table 4.8 and plotted in Figure 4.29.

Mol NH_3 in entering gas $= 10[Y/(1 + Y)] = 10(0.3/1.3) = 2.3$

Mol NH_3 in entering liquid $= 10[X/(1 + X)] = 10(0.1/1.1) = 0.91$

Table 4.7 Partial Pressure of Ammonia over Ammonia–Water Solutions at 20°C

NH$_3$ Partial Pressure, kPa	g NH$_3$/g H$_2$O
4.23	0.05
9.28	0.10
15.2	0.15
22.1	0.20
30.3	0.25

Table 4.8 Y–X Data for Ammonia–Water, 20°C

Y, mol NH$_3$/mol Air	X, mol NH$_3$/mol H$_2$O
0.044	0.053
0.101	0.106
0.176	0.159
0.279	0.212
0.426	0.265

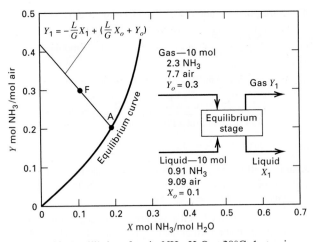

Figure 4.29 Equilibrium for air–NH₃–H₂O at 20°C, 1 atm, in Example 4.14.

A molar material balance for ammonia about the equilibrium stage is

$$GY_0 + LX_0 = GY_1 + LX_1 \qquad (1)$$

where G = moles of air and L = moles of H_2O. Then $G = 10 - 2.3 = 7.7$ mol and $L = 10 - 0.91 = 9.09$ mol. Solving for Y_1 from Eq. (1),

$$Y_1 = -\frac{L}{G}X_1 + \left(\frac{L}{G}X_0 + Y_0\right) \qquad (2)$$

This material-balance relationship is an equation of a straight line of slope $(L/G) = -9.09/7.7 = -1.19$, with an intercept of $(L/G)(X_0) + Y_0 = 0.42$.

The intersection of this material-balance line with the equilibrium curve, as shown in Figure 4.29, gives the ammonia composition of the gas and liquid phases leaving the stage as $Y_1 = 0.195$ and $X_1 = 0.19$. This result can be checked by an NH_3 balance, since the amount of NH_3 leaving is $(0.195)(7.70) + (0.19)(9.09) = 3.21$, which equals the total moles of NH_3 entering.

It is of importance to recognize that Eq. (2), the material balance line, called an *operating line* and discussed in great detail in Chapters 5 to 8, is the locus of all passing stream pairs; thus, X_0, Y_0 (point F) also lies on this operating line.

4.9 GAS–SOLID SYSTEMS

Systems consisting of gas and solid phases that tend to equilibrium are involved in sublimation, desublimation, and adsorption separation operations.

Sublimation and Desublimation

In sublimation, a solid vaporizes into a gas phase without passing through a liquid state. In desublimation, one or more components (solutes) in the gas phase are condensed to a solid phase without passing through a liquid state. At low pressure both sublimation and desublimation are governed by the solid vapor pressure of the solute. Sublimation of the solid takes place when the partial pressure of the solute in the gas phase is less than the vapor pressure of the solid at

the system temperature. When the partial pressure of the solute in the gas phase exceeds the vapor pressure of the solid, desublimation occurs. At equilibrium, the vapor pressure of the species as a solid is equal to the partial pressure of the species as a solute in the gas phase. This is illustrated in the following example.

EXAMPLE 4.15

Ortho-xylene is partially oxidized in the vapor phase with air to produce phthalic anhydride, PA, in a catalytic reactor (fixed bed or fluidized bed) operating at about 370°C and 780 torr. However, a very large excess of air must be used to keep the xylene content of the reactor feed below 1 mol% to avoid an explosive mixture. In a typical plant, 8,000 lbmol/h of reactor effluent gas, containing 67 lbmol/h of PA and other amounts of N_2, O_2, CO, CO_2, and water vapor are cooled to separate the PA by desublimation to a solid at a total pressure of 770 torr. If the gas is cooled to 206°F, where the vapor pressure of solid PA is 1 torr, calculate the number of pounds of PA condensed per hour as a solid, and the percent recovery of PA from the gas if equilibrium is achieved. Assume that the xylene is converted completely to PA.

SOLUTION

At these conditions, only the PA condenses. At equilibrium, the partial pressure of PA is equal to the vapor pressure of solid PA, or 1 torr. Thus, the amount of PA in the cooled gas is given by Dalton's law of partial pressures:

$$(n_{PA})_G = \frac{p_{PA}}{P}n_G \qquad (1)$$

where

$$n_G = (8,000 - 67) + (n_{PA})_G \qquad (2)$$

and n = lbmol/h. Combining Eqs. (1) and (2),

$$(n_{PA})_G = \frac{p_{PA}}{P}[(8,000 - 67) + (n_{PA})_G]$$
$$= \frac{1}{770}[(8,000 - 67) + (n_{PA})_G] \qquad (3)$$

Solving this linear equation gives

$$(n_{PA})_G = 10.3 \text{ lbmol/h of PA}$$

The amount of PA desublimed is $67 - 10.3 = 56.7$ lbmol/h. The percent recovery of PA is $56.7/67 = 0.846$ or 84.6%. The amount of PA remaining in the gas is a very large quantity. In a typical plant, the gas is cooled to a much lower temperature, perhaps 140°F, where the vapor pressure of PA is less than 0.1 torr, bringing the recovery of PA to almost 99%.

Gas Adsorption

As with liquid mixtures, one or more components of a gas mixture can be adsorbed on the surface of a solid adsorbent. Data for a single solute can be represented by an adsorption isotherm of the type shown in Figure 4.25, or in similar diagrams, where the partial pressure of the solute in the gas is used in place of the concentration. However, when two components of a gas mixture are adsorbed and the purpose

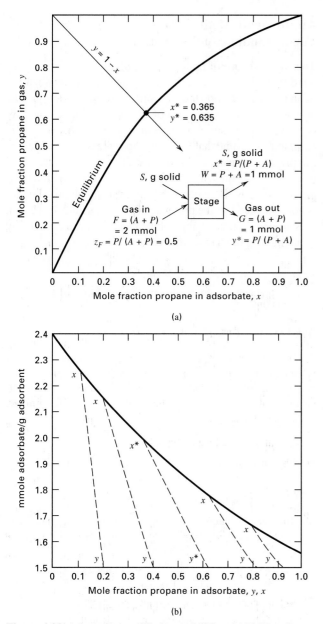

(a)

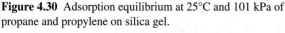

(b)

Figure 4.30 Adsorption equilibrium at 25°C and 101 kPa of propane and propylene on silica gel.

[Adapted from W.K. Lewis, E.R. Gilliland, B. Chertow, and W. H. Hoffman, *J. Am. Chem. Soc.*, **72**, 1153 (1950).]

of adsorption is to separate these two components, other methods of representing the experimental data may be preferred. One such representation is shown in Figure 4.30, from the data of Lewis et al. [15], for the adsorption of a propane (P)–propylene (A) gas mixture with silica gel at 25°C and 101 kPa. At 25°C, a pressure of at least 1,000 kPa is required to initiate condensation (dew point) of a mixture of propylene and propane. However, in the presence of silica gel, significant amounts of the gas are adsorbed at 101 kPa.

Figure 4.30a is similar to a binary vapor–liquid equilibrium plot of the type discussed in Section 4.2. For adsorption equilibria, the liquid-phase mole fraction is replaced by the mole fraction in the adsorbate. For the propylene–propane mixture, propylene is adsorbed more strongly. For example, for an equimolar mixture in the gas phase, the adsorbate contains only 27 mol% propane. Figure 4.30b combines the data for the equilibrium mole fractions in the gas and adsorbate with the amount of adsorbate per unit of adsorbent. The mole fractions are obtained by reading the abscissa at the two ends of a tie line. For example, for equilibrium with $y_P = y^* = 0.50$, Figure 4.30b gives $x_P = x^* = 0.27$ and 2.08 mmol of adsorbate/g adsorbent. Therefore, $y_A = 0.50$, and $x_A = 0.73$. The separation factor analogous to the relative volatility for distillation is

$$(0.50/0.27)/(0.50/0.73) = 2.7$$

This value is much higher than the α-value for distillation, which, from Figure 2.8, at 25°C and 1,100 kPa is only 1.13. Accordingly, the separation of propylene and propane by adsorption has received some attention. Equilibrium calculations using data such as that shown in Figure 4.30 are made in the usual manner by combining such data with material-balance equations, as illustrated in the following example.

EXAMPLE 4.16

Propylene (A) and propane (P), are to be separated by preferential adsorption on silica gel (S) at 25°C and 101 kPa.

Two millimoles of a gas containing 50 mol% P and 50 mol% A is equilibrated with silica gel at 25°C and 101 kPa. Manometric measurements show that 1 mmol of gas is adsorbed. If the data of Figure 4.30 apply, what is the mole fraction of propane in the equilibrium gas and adsorbate, and how many grams of silica gel are used?

SOLUTION

A pictorial representation of the process is included in Figure 4.30a, where W = millimoles of adsorbate, G = millimoles of gas leaving, and z_F = mole fraction of propane in the feed.

The propane mole balance is

$$Fz_F = Wx^* + Gy^* \qquad (1)$$

With $F = 2$, $z_F = 0.5$, $W = 1$, and $G = F - W = 1$, Eq. (1) becomes $1 = x^* + y^*$.

The operating (material-balance) line $y^* = 1 - x^*$ is the locus of all solutions of the material-balance equation, and is shown in Figure 4.30a. It intersects the equilibrium curve at $x^* = 0.365$, $y^* = 0.635$. From Figure 4.30b, at the point x^*, there must be 2.0 mmol adsorbate/g adsorbent; therefore there are $1.0/2 = 0.50$ g of silica gel in the system.

4.10   MULTIPHASE SYSTEMS

In previous sections of this chapter, only two phases were considered to be in equilibrium. In some applications of multiphase systems, three or more phases coexist. Figure 4.31

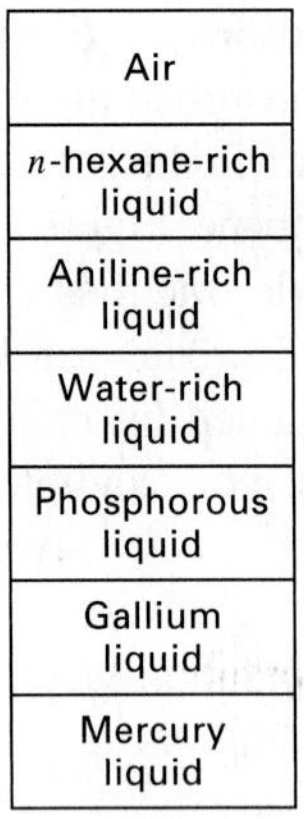

Figure 4.31 Seven phases in equilibrium.

is a schematic diagram of a photograph of a laboratory curiosity taken from Hildebrand [16], which shows seven phases in equilibrium at near-ambient temperature. The phase on top is air, followed by six liquid phases in order of increasing density: hexane-rich, aniline-rich, water-rich, phosphorous, gallium, and mercury. Each phase contains all components in the seven-phase mixture, but the mole fractions in many cases are extremely small. For example, the aniline-rich phase contains on the order of 10 mol% n-hexane, 20 mol% water, but much less than 1 mol% each of dissolved air, phosphorous, gallium, and mercury. Note that even though the hexane-rich phase is not in direct contact with the water-rich phase, an equilibrium amount of water (approximately 0.06 mol%) is present in the hexane-rich phase because each phase is in equilibrium with each of the other phases, as attested by the equality of component fugacities:

$$f_i^{(1)} = f_i^{(2)} = f_i^{(3)} = f_i^{(4)} = f_i^{(5)} = f_i^{(6)} = f_i^{(7)}$$

More practical multiphase systems include the vapor–liquid–solid systems present in evaporative crystallization and pervaporation, and the vapor–liquid–liquid systems that occur when distilling certain mixtures of water and hydrocarbons or other organic chemicals having a limited solubility in water. Actually, all of the two-phase systems considered in the previous sections of this chapter involve a third phase, the containing vessel. However, the material of the container is selected on the basis of its inertness to and lack of solubility in the phases it contains, and therefore the material of the container does not normally enter into phase-equilibria calculations.

Although calculations of multiphase equilibrium are based on the same principles as for two-phase systems (material balances, energy balances, and phase-equilibria criteria such as equality of fugacity), the computations can be quite complex unless simplifying assumptions are made, in which case approximate results are obtained. Rigorous calculations are best made with a computer algorithm. In this section both types of calculations are illustrated.

Approximate Method for a Vapor–Liquid–Solid System

The simplest case of multiphase equilibrium is that encountered in an evaporative crystallizer involving crystallization of an inorganic compound, B, from its aqueous solution at its bubble point in the presence of its vapor. Assume that only two components are present, B and water. In that case, it is common to assume that B has no vapor pressure and water is not present in the solid phase. Thus, the vapor is pure water (steam), the liquid is a mixture of water and B, and the solid phase is pure B. Then, the solubility of B in the liquid phase is not influenced by the presence of the vapor, and the system pressure at a given temperature can be approximated by applying Raoult's law to the water in the liquid phase:

$$P = P_{H_2O}^s x_{H_2O} \qquad (4-33)$$

where x_{H_2O} can be obtained from the solubility of B.

EXAMPLE 4.17

A 5,000-lb batch of 20 wt% aqueous $MgSO_4$ solution is fed to a vacuum, evaporative crystallizer operating at 160°F. At this temperature, the stable solid phase is the monohydrate, with a $MgSO_4$ solubility of 36 wt%. If 75% of the water is evaporated, calculate:

(a) Pounds of water evaporated

(b) Pounds of monohydrate crystals, $MgSO_4 \cdot H_2O$

(c) Crystallizer pressure

SOLUTION

(a) The feed solution is $0.20(5,000) = 1,000$ lb $MgSO_4$, and $5,000 - 1,000 = 4,000$ lb H_2O. The amount of water evaporated is $0.75(4,000) = 3,000$ lb H_2O.

(b) Let W = amount of $MgSO_4$ remaining in solution. Then $MgSO_4$ in the crystals = $1,000 - W$.

MW of H_2O = 18 and MW of $MgSO_4$ = 120.4.
Water of crystallization for the monohydrate
  = $(1,000 - W)(18/120.4) = 0.15(1,000 - W)$.
Water remaining in solution
  = $4,000 - 3,000 - 0.15(1,000 - W) = 850 + 0.15W$.
Total amount of solution remaining
  = $850 + 0.15W + W = 850 + 1.15W$.
From the solubility of $MgSO_4$,

$$0.36 = \frac{W}{850 + 1.15W}$$

Solving: $W = 522$ pounds of dissolved $MgSO_4$.
$MgSO_4$ crystallized = $1,000 - 522 = 478$ lb.
Water of crystallization = $0.15(1,000 - W)$
  = $0.15(1,000 - 522) = 72$ lb.
Total monohydrate crystals = $478 + 72 = 550$ lb.

(c) Crystallizer pressure is given by (4-33). At 160°F, the vapor pressure of H_2O is 4.74 psia. Then water remaining in solution = $(850 + 0.15W)/18 = 51.6$ lbmol.

$MgSO_4$ remaining in solution = $522/120.4 = 4.3$ lbmol.

Hence

$$x_{H_2O} = 51.6/(51.6 + 4.3) = 0.923$$

By Raoult's law, $p_{H_2O} = P = 4.74(0.923) = 4.38$ psia

Approximate Method for a Vapor–Liquid–Liquid System

Another case suitable for an approximate method is that of a mixture containing water and hydrocarbons (HCs), at conditions such that a vapor phase and two liquid phases, HC-rich (1) and water-rich (2) coexist. Often the solubilities of water in the liquid HC phase and HCs in the water phase are less than 0.1 mol% and may be neglected. In that case, if the liquid HC phase obeys Raoult's law, the total pressure of the system is given by the sum of the pressures exhibited by the separate phases:

$$P = P_{H_2O}^s + \sum_{HCs} P_i^s x_i^{(1)} \tag{4-34}$$

For more general cases, at low pressures where the vapor phase is ideal but the liquid HC phase may be nonideal,

$$P = P_{H_2O}^s + P \sum_{HCs} K_i x_i^{(1)} \tag{4-35}$$

which can be rearranged to

$$P = \frac{P_{H_2O}^s}{1 - \sum\limits_{HCs} K_i x_i^{(1)}} \tag{4-36}$$

Equations (4-34) and (4-36) can be used directly to estimate the pressure for a given temperature and liquid-phase composition or iteratively to estimate the temperature for a given pressure. An important aspect of the calculation is the determination of the particular phases present from all six possible cases, namely, V, $V-L^{(1)}$, $V-L^{(1)}-L^{(2)}$, $V-L^{(2)}$, $L^{(1)}-L^{(2)}$, and L. It is not always obvious how many and which phases may be present. Indeed, if a $V-L^{(1)}-L^{(2)}$ solution to a problem exists, almost always $V-L^{(1)}$ and $V-L^{(2)}$ solutions also exist. In that case, the three-phase solution is the correct one. It is important, therefore, to seek the three-phase solution first.

EXAMPLE 4.18

A mixture of 1,000 kmol of 75 mol% water and 25 mol% n-octane is cooled under equilibrium conditions at a constant pressure 133.3 kPa (1,000 torr) from an initial temperature of 136°C to a final temperature of 25°C. Determine:

(a) The initial phase condition

(b) The temperature, phase amounts, and compositions when each phase change occurs

Assume that water and n-octane are immiscible liquids. The vapor pressure of octane is included in Figure 2.4.

SOLUTION

(a) Initial phase conditions are $T = 136°C = 276.8°F$ and $P = 133.3$ kPa $= 19.34$ psia. Vapor pressures at 276.8°F and $P_{H_2O}^s = 46.7$ psia and $P_{nC_8}^s = 19.5$ psia. Because the initial pressure is less than the vapor pressure of each component, the initial phase condition is all vapor, with partial pressures

$$p_{H_2O} = y_{H_2O}P = 0.75(19.34) = 14.5 \text{ psia}$$
$$p_{nC_8} = y_{nC_8}P = 0.25(19.34) = 4.8 \text{ psia}$$

(b) As the temperature is decreased, the first phase change occurs when a temperature is reached where either $P_{H_2O}^s = p_{H_2O} = 14.5$ psia or $P_{nC_8}^s = p_{nC_8} = 4.8$ psia. The corresponding temperatures where these vapor pressures occur are 211°F for H_2O and 194°F for nC_8. The highest temperature applies. Therefore, water condenses first when the temperature reaches 211°F. This is the dew-point temperature of the initial mixture at the system pressure. As the temperature is further reduced, the number of moles of water in the vapor decreases, causing the partial pressure of water to decrease below 14.5 psia and the partial pressure of nC_8 to increase above 4.8 psia. Thus, nC_8 begins to condense, forming a second liquid phase, at a temperature higher than 194°F but lower than 211°F. This temperature, referred to as the *secondary dew point*, must be determined iteratively. The calculation is simplified if the bubble point of the mixture is computed first.

From (4-34),

$$P = 19.34 \text{ psi} = P_{H_2O}^s + P_{nC_8}^s \tag{1}$$

Thus, a temperature is sought, as follows, to cause Eq. (1) to be satisfied:

T, °F	$P_{H_2O}^s$, psia	$P_{nC_8}^s$, psia	P, psia
194	10.17	4.8	14.97
202	12.01	5.6	17.61
206	13.03	6.1	19.13
207	13.30	6.2	19.50

By linear interpolation, $T = 206.7°F$ for $P = 19.34$ psia. Below this temperature, the vapor phase disappears and only two immiscible liquid phases are present.

To determine the temperature at which one of the liquid phases disappears, which is the same condition as when the second liquid phase begins to appear (secondary dew point), it is noted for this case, with only pure water and a pure HC present, that vaporization starting from the bubble point is at a constant temperature until one of the two liquid phases is completely vaporized. Thus, the secondary dew-point temperature is the same as the bubble-point temperature or 206.7°F. At the secondary dew point, the partial pressures are $p_{H_2O} = 13.20$ psia and $p_{nC_8} = 6.14$ psia, with all of the nC_8 in the vapor phase. Therefore, the phase amounts and compositions are

Component	Vapor		H₂O-Rich Liquid
	kmol	y	kmol
H_2O	53.9	0.683	21.1
nC_8	25.0	0.317	0.0
	78.9	1.000	21.1

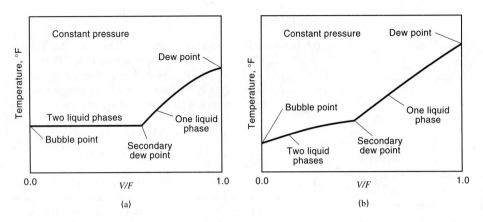

Figure 4.32 Typical flash curves for immiscible liquid mixtures of water and hydrocarbons at constant pressure: (a) only one hydrocarbon species present; (b) more than one hydrocarbon species present.

If desired, additional flash calculations can be made for conditions between the dew point and secondary dew point. The resulting flash curve is Figure 4.32a. If more than one HC species is present, the liquid HC phase does not evaporate at a constant composition and the secondary dew-point temperature is higher than the bubble-point temperature. In that case, the flash is described by Figure 4.32b.

Rigorous Method for a Vapor–Liquid–Liquid System

The rigorous method for treating a vapor–liquid–liquid system at a given temperature and pressure is called a *three-phase isothermal flash*. As first presented by Henley and Rosen [17], it is analogous to the isothermal two-phase flash algorithm developed in Section 4.4. The system is shown schematically in Figure 4.33. The usual material balances and phase-equilibrium relations apply for each component:

$$Fz_i = Vy_i + L^{(1)}x_i^{(1)} + L^{(2)}x_i^{(2)} \tag{4-37}$$

$$K_i^{(1)} = y_i/x_i^{(1)} \tag{4-38}$$

$$K_i^{(2)} = y_i/x_i^{(2)} \tag{4-39}$$

Alternatively, the following relation can be substituted for (4-38) and (4-39):

$$K_{D_i} = x_i^{(1)}/x_i^{(2)} \tag{4-40}$$

These equations can be solved by a modification of the Rachford–Rice procedure if we let $\Psi = V/F$ and $\xi = L^{(1)}/(L^{(1)} + L^{(2)})$, where $0 \le \Psi \le 1$ and $0 \le \xi \le 1$. By combining

(4-37), (4-38), and (4-39) with

$$\sum x_i^{(1)} - \sum y_i = 0 \tag{4-41}$$

and

$$\sum x_i^{(1)} - \sum x_i^{(2)} = 0 \tag{4-42}$$

to eliminate y_i, $x_i^{(1)}$, and $x_i^{(2)}$, two simultaneous equations in Ψ and ξ are obtained:

$$\sum_i \frac{z_i(1 - K_i^{(1)})}{\xi(1 - \Psi) + (1 - \Psi)(1 - \xi)K_i^{(1)}/K_i^{(2)} + \Psi K_i^{(1)}} = 0 \tag{4-43}$$

and

$$\sum_i \frac{z_i(1 - K_i^{(1)}/K_i^{(2)})}{\xi(1 - \Psi) + (1 - \Psi)(1 - \xi)K_i^{(1)}/K_i^{(2)} + \Psi K_i^{(1)}} = 0 \tag{4-44}$$

Values of Ψ and ξ are computed by solving the nonlinear equations (4-43) and (4-44) simultaneously by an appropriate numerical method such as that of Newton. Then the amounts and compositions of the three phases are determined from

$$V = \Psi F \tag{4-45}$$

$$L^{(1)} = \xi(F-V) \tag{4-46}$$

$$L^{(2)} = F-V-L^{(1)} \tag{4-47}$$

$$y_i = \frac{z_i}{\xi(1 - \Psi)/K_i^{(1)} + (1 - \Psi)(1 - \xi)/K_i^{(2)} + \Psi} \tag{4-48}$$

$$x_i^{(1)} = \frac{z_i}{\xi(1 - \Psi) + (1 - \Psi)(1 - \xi)(K_i^{(1)}/K_i^{(2)}) + \Psi K_i^{(1)}} \tag{4-49}$$

$$x_i^{(2)} = \frac{z_i}{\xi(1 - \Psi)(K_i^{(2)}/K_i^{(1)}) + (1 - \Psi)(1 - \xi) + \Psi K_i^{(2)}} \tag{4-50}$$

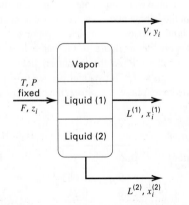

Figure 4.33 Conditions for a three-phase isothermal flash.

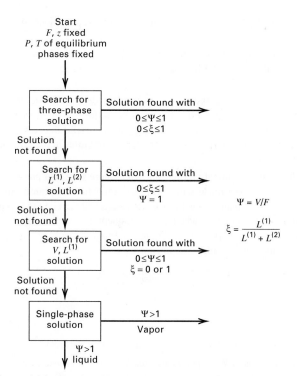

Figure 4.34 Algorithm for an isothermal three-phase flash.

Calculations for an isothermal three-phase flash are difficult and tedious because of the strong dependency of K-values on liquid-phase compositions when two immiscible liquid phases are present. In addition, it is usually not obvious that three phases will be present, and calculations may be necessary for other combinations of phases. A typical algorithm for determining the phase conditions is shown in Figure 4.34. Because of the complexity of the isothermal three-phase flash algorithm, calculations are best made with a steady-state, process-simulation computer program. Such programs can also perform adiabatic or nonadiabatic three-phase flashes by iterating on temperature until the enthalpy balance,

$$h_F F + Q = h_V V + h_{L^{(1)}} L^{(1)} + h_{L^{(2)}} L^{(2)} = 0 \quad (4\text{-}51)$$

is satisfied.

EXAMPLE 4.19

In a process for producing styrene from toluene and methanol, the gaseous reactor effluent is as follows:

Component	kmol/h
Hydrogen	350
Methanol	107
Water	491
Toluene	107
Ethylbenzene	141
Styrene	350

If this stream is brought to equilibrium at 38°C and 300 kPa, compute the amounts and compositions of the phases present.

SOLUTION

Because water, hydrocarbons, and a light gas are present in the mixture, the possibility exists that a vapor and two liquid phases may be present, with the methanol being distributed among all three phases. The isothermal three-phase flash module of the ChemCAD simulation program was used with Henry's law for hydrogen and the UNIFAC method for estimating liquid-phase activity coefficients for the other components, to obtain the following results:

Component	kmol/h		
	V	$L^{(1)}$	$L^{(2)}$
Hydrogen	349.96	0.02	0.02
Methanol	9.54	14.28	83.18
Water	7.25	8.12	475.63
Toluene	1.50	105.44	0.06
Ethylbenzene	0.76	140.20	0.04
Styrene	1.22	348.64	0.14
Totals	370.23	616.70	559.07

As would be expected, little of the hydrogen is dissolved in either of the two liquid phases. Little of the other components is left uncondensed. The water-rich liquid phase contains little of the hydrocarbons, but much of the methanol. The organic-rich phase contains most of the hydrocarbons and small amounts of water and methanol.

Additional calculations at 300 kPa indicate that the organic phase condenses first with dew point = 143°C and secondary dew point = 106°C.

SUMMARY

1. The phase rule of Gibbs, which applies to intensive variables at equilibrium, determines the number of independent variables that can be specified. This rule can be extended to the more general determination of the degrees of freedom (number of allowable specifications) for a flow system, including consideration of extensive variables. The intensive and extensive variables are related by material and energy balance equations together with phase equilibrium data in the form of equations, tables, and/or graphs.

2. Vapor–liquid equilibrium conditions for binary systems are conveniently represented and determined with T–y–x, y–x, and P–x diagrams. The relative volatility for a binary system tends to 1.0 as the critical point is approached.

3. Minimum- or maximum-boiling azeotropes, which are formed by close-boiling, nonideal liquid mixtures, are conveniently represented by the same types of diagrams used for nonazeotropic (zeotropic) binary mixtures. Highly nonideal liquid mixtures can form heterogeneous azeotropes involving two liquid phases.

4. For multicomponent mixtures, vapor–liquid equilibrium-phase compositions and amounts can be determined by isothermal-flash, adiabatic-flash, and bubble- and dew-point calculations. When the mixtures are nonideal, the computations are best done with process-simulation computer programs.

5. Liquid–liquid equilibrium conditions for ternary mixtures are best determined graphically from triangular and other equilibrium diagrams, unless only one of the three components (called the solute) is soluble in the two liquid phases and the system is dilute in the solute. In that case, the conditions can be readily determined algebraically using phase-distribution ratios for the solute.

6. Liquid–liquid equilibrium conditions for multicomponent mixtures of four or more components are best determined with process-simulation computer programs, particularly when the system is not dilute with respect to the solute(s).

7. Solid–liquid equilibrium commonly occurs in leaching, crystallization, and adsorption. Leaching calculations commonly assume that the solute is completely dissolved in the solvent and that the remaining solid leaving in the underflow is accompanied by a known fraction of liquid. Crystallization calculations are best made with a solid–liquid phase equilibrium diagram. For crystallization of inorganic salts from an aqueous solution, formation of hydrates must be considered. Equilibrium adsorption

can be represented algebraically or graphically by adsorption isotherms.

8. Solubility of gases that are only sparingly soluble in a liquid are well represented by a Henry's law constant that depends on temperature.

9. Solid vapor pressure can be used to determine equilibrium sublimation and desublimation conditions for gas–solid systems. Adsorption isotherms and y–x diagrams are useful in determining adsorption-equilibrium conditions for gas mixtures in the presence of a solid adsorbent.

10. Calculations of equilibrium when more than two phases are present are best made with computer simulation programs. However, approximate manual procedures are readily applied to vapor–liquid–solid systems when no component is found in all three phases and for vapor–liquid–liquid systems when only one component distributes in all three phases.

REFERENCES

1. PERRY, R.H., D.W. GREEN, and J.O. MALONEY, Eds., *Perry's Chemical Engineers' Handbook,* 7th ed., McGraw-Hill, New York, Section 13 (1997).

2. GMEHLING, J., and U. ONKEN, *Vapor-Liquid Equilibrium Data Collection,* DECHEMA Chemistry Data Series, **1-8** (1977–1984).

3. KEYES, D.B., *Ind. Eng. Chem.,* **21,** 998–1001 (1929).

4. HUGHES, R.R., H.D. EVANS, and C.V. STERNLING, *Chem. Eng. Progr.,* **49,** 78–87 (1953).

5. RACHFORD, H.H., JR., and J.D. RICE, *J. Pet. Tech.,* **4** (10), Section 1, p. 19, and Section 2, p. 3 (Oct. 1952).

6. PRESS, W.H., S.A. TEUKOLSKY, W.T. VETTERLING, and B.P. FLANNERY, *Numerical Recipes in FORTRAN,* 2nd ed., Cambridge University Press, Cambridge, chap. 9 (1992).

7. GOFF, G.H., P.S. FARRINGTON, and B.H. SAGE, *Ind. Eng. Chem.,* **42,** 735–743 (1950).

8. CONSTANTINIDES, A., and N. MOSTOUFI, *Numerical Methods for Chemical Engineers with MATLAB Applications,* Prentice Hall PTR, Upper Saddle River, NJ (1999).

9. ROBBINS, L.A., in R.H. PERRY, D.H. GREEN, and J.O. MALONEY, Eds., *Perry's Chemical Engineers' Handbook,* 7th ed., McGraw-Hill, New York, pp. 15–10 to 15–15 (1997).

10. JANECKE, E., *Z. Anorg. Allg. Chem.,* **51,** 132–157 (1906).

11. FRANCIS, A.W., *Liquid-Liquid Equilibriums,* Interscience, New York (1963).

12. FINDLAY, A., *Phase Rule,* Dover, New York (1951).

13. FRITZ, W., and E.-U. SCHULUENDER, *Chem. Eng. Sci.,* **29,** 1279–1282 (1974).

14. FELDER, R.M., and R.W. ROUSSEAU, *Elementary Principles of Chemical Processes,* 3rd ed., John Wiley and Sons, New York, pp. 613–616 (1986).

15. LEWIS, W.K., E.R. GILLILAND, B. CHERTON, and W.H. HOFFMAN, *J. Am. Chem. Soc.,* **72,** 1153–1157 (1950).

16. HILDEBRAND, J.H., *Principles of Chemistry,* 4th ed., Macmillan, New York (1940).

17. HENLEY, E.J., and E.M. ROSEN, *Material and Energy Balance Computations,* John Wiley and Sons, New York, pp. 351–353 (1969).

18. CONWAY, J.B., and J.J. NORTON, *Ind. Eng. Chem.,* **43,** 1433–1435 (1951).

EXERCISES

Section 4.1

4.1 Consider the equilibrium stage shown in Figure 4.35. Conduct a degrees-of-freedom analysis by performing the following steps:

(a) List and count the variables.

(b) Write and count the equations relating the variables.

(c) Calculate the degrees of freedom.

(d) List a reasonable set of design variables.

4.2 Can the following problems be solved uniquely?

(a) The feed streams to an adiabatic equilibrium stage consist of liquid and vapor streams of known composition, flow rate, temperature, and pressure. Given the stage (outlet) temperature and pressure, calculate the composition and amounts of equilibrium vapor and liquid leaving the stage.

(b) The same as part (a), except that the stage is not adiabatic.

(c) A multicomponent vapor of known temperature, pressure, and composition is to be partially condensed in a condenser. The outlet pressure of the condenser and the inlet cooling water temperature are fixed. Calculate the cooling water required.

4.3 Consider an adiabatic equilibrium flash. The variables are all as indicated in Figure 4.36.

(a) Determine the number of variables.

(b) Write all the independent equations that relate the variables.

(c) Determine the number of equations.

(d) Determine the number of degrees of freedom.

(e) What variables would you prefer to specify in order to solve a typical adiabatic flash problem?

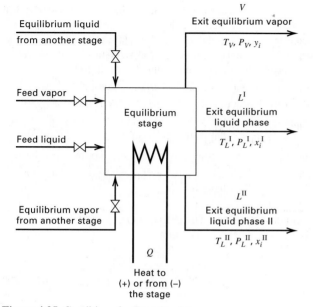

Figure 4.35 Conditions for Exercise 4.1.

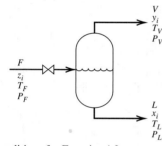

Figure 4.36 Conditions for Exercise 4.3.

4.4 Determine the number of degrees of freedom for a non-adiabatic equilibrium flash for one liquid feed, one vapor stream product, and two immiscible liquid stream products as shown in Figure 4.33.

4.5 Consider the seven-phase equilibrium system shown in Figure 4.31. Assume that air consists of N_2, O_2, and argon. How many degrees of freedom are computed by the Gibbs phase rule? What variables might be specified to fix the system?

Section 4.2

4.6 A liquid mixture containing 25 mol% benzene and 75 mol% ethyl alcohol, in which components are miscible in all proportions, is heated at a constant pressure of 1 atm (101.3 kPa, 760 torr) from a temperature of 60°C to 90°C. Using the following $T–x–y$ experimental data, perform calculations to determine the answers to parts (a) through (f).

EXPERIMENTAL $T–x–y$ DATA FOR BENZENE–ETHYL ALCOHOL AT 1 ATM

Temperature, °C:
78.4 77.5 75 72.5 70 68.5 67.7 68.5 72.5 75 77.5 80.1

Mole percent benzene in vapor:
0 7.5 28 42 54 60 68 73 82 88 95 100

Mole percent benzene in liquid:
0 1.5 5 12 22 31 68 81 91 95 98 100

(a) At what temperature does vaporization begin?

(b) What is the composition of the first bubble of equilibrium vapor formed?

(c) What is the composition of the residual liquid when 25 mol% has evaporated? Assume that all vapor formed is retained within the apparatus and that it is completely mixed and in equilibrium with the residual liquid.

(d) Repeat part (c) for 90 mol% vaporized.

(e) Repeat part (d) if, after 25 mol% is vaporized as in part (c), the vapor formed is removed and an additional 35 mol% is vaporized by the same technique used in part (c).

(f) Plot the temperature versus the percent vaporized for parts (c) and (e).

(g) Use the following vapor pressure data in conjunction with Raoult's and Dalton's laws to construct a $T–x–y$ diagram, and compare it for the answers obtained in parts (a) and (f) with those obtained using the experimental $T–x–y$ data. What do you conclude about the applicability of Raoult's law to this binary system?

VAPOR PRESSURE DATA

Vapor pressure, torr:

20	40	60	100	200	400	760
Ethanol, °C:						
8	19.0	26.0	34.9	48.4	63.5	78.4
Benzene, °C:						
−2.6	7.6	15.4	26.1	42.2	60.6	80.1

4.7 Stearic acid is to be steam distilled at 200°C in a direct-fired still, heat-jacketed to prevent condensation. Steam is introduced into the molten acid in small bubbles, and the acid in the vapor leaving the still has a partial pressure equal to 70% of the vapor pressure of pure stearic acid at 200°C. Plot the kilograms of acid distilled per kilogram of steam added as a function of total pressure from 101.3 kPa down to 3.3 kPa at 200°C. The vapor pressure of stearic acid at 200°C is 0.40 kPa.

4.8 The relative volatility, α, of benzene to toluene at 1 atm is 2.5. Construct an $x–y$ diagram for this system at 1 atm. Repeat the construction using vapor pressure data for benzene from Exercise 4.6 and for toluene from the following table in conjunction with Raoult's and Dalton's laws. Also construct a $T–x–y$ diagram.

(a) A liquid containing 70 mol% benzene and 30 mol% toluene is heated in a container at 1 atm until 25 mol% of the original liquid is evaporated. Determine the temperature. The phases are then separated mechanically, and the vapors condensed. Determine the composition of the condensed vapor and the liquid residue.

(b) Calculate and plot the K-values as a function of temperature at 1 atm.

VAPOR PRESSURE OF TOLUENE

Vapor pressure, torr:

20	40	60	100	200	400	760	1,520
Temperature, °C:							
18.4	31.8	40.3	51.9	69.5	89.5	110.6	136

4.9 The vapor pressure of toluene is given in Exercise 4.8, and that of n-heptane is given in the accompanying table.

VAPOR PRESSURE OF *n*-HEPTANE

Vapor pressure, torr:

20	40	60	100	200	400	760	1,520

Temperature, °C:

9.5	22.3	30.6	41.8	58.7	78.0	98.4	124

(a) Plot an *x–y* equilibrium diagram for this system at 1 atm by using Raoult's and Dalton's laws.

(b) Plot the *T–x* bubble-point curve at 1 atm.

(c) Plot α and *K*-values versus temperature.

(d) Repeat part (a) using the arithmetic average value of α, calculated from the two extreme values.

(e) Compare your *x–y* and *T–x–y* diagrams with the following experimental data of Steinhauser and White [*Ind. Eng. Chem.*, **41**, 2912 (1949)].

VAPOR–LIQUID EQUILIBRIUM DATA FOR *n*-HEPTANE/TOLUENE AT 1 ATM

$x_{n\text{-heptane}}$	$y_{n\text{-heptane}}$	*T*, °C
0.025	0.048	110.75
0.129	0.205	106.80
0.250	0.349	104.50
0.354	0.454	102.95
0.497	0.577	101.35
0.692	0.742	99.73
0.843	0.864	98.90
0.940	0.948	98.50
0.994	0.993	98.35

4.10 Saturated-liquid feed, of $F = 40$ mol/h, containing 50 mol% A and B is supplied continuously to the apparatus shown in Figure 4.37. The condensate from the condenser is split so that half of it is returned to the still pot.

(a) If heat is supplied at such a rate that $W = 30$ mol/h and $\alpha = 2$, as subsequently defined, what will be the composition of the overhead and the bottoms product?

(b) If the operation is changed so that no condensate is returned to the still pot and $W = 3D$ as before, what will be the composition of the products?

$$\alpha = \frac{P_A^s}{P_B^s} = \frac{y_A x_B}{y_B x_A}$$

4.11 It is required to design a fractionation tower to operate at 101.3 kPa to obtain a distillate consisting of 95 mol% acetone (A) and 5 mol% water, and a residue containing 1 mol% A. The feed

liquid is at 125°C and 687 kPa and contains 57 mol% A. The feed is introduced to the column through an expansion valve so that it enters the column partially vaporized at 60°C. From the data below, determine the molar ratio of liquid to vapor in the partially vaporized feed. Enthalpy and equilibrium data are as follows:

Molar latent heat of A = 29,750 kJ/kmol (constant)

Molar latent heat of H_2O = 42,430 kJ/kmol (constant)

Molar specific heat of A = 134 kJ/kmol-K (constant)

Molar specific heat of H_2O = 75.3 kJ/kmol-K (constant)

Enthalpy of high-pressure, hot feed before adiabatic expansion = 0

Enthalpies of feed phases after expansion: $h_V = 27,200$ kJ/kmol, $h_L = -5,270$ kJ/kmol

VAPOR–LIQUID EQUILIBRIUM DATA FOR ACETONE-H_2O AT 101.3 kPA

	T, °C						
	56.7	57.1	60.0	61.0	63.0	71.7	100
Mol% A in liquid:	100	92.0	50.0	33.0	17.6	6.8	0
Mol% A in vapor:	100	94.4	85.0	83.7	80.5	69.2	0

4.12 Using vapor pressure data from Exercises 4.6 and 4.8 and the enthalpy data provided below:

(a) Construct an *h–x–y* diagram for the benzene–toluene system at 1 atm (101.3 kPa) based on the use of Raoult's and Dalton's laws.

	Saturated Enthalpy, kJ/kg			
	Benzene		Toluene	
T, °C	h_L	h_V	h_L	h_V
60	79	487	77	471
80	116	511	114	495
100	153	537	151	521

(b) Calculate the energy required for 50 mol% vaporization of a 30 mol% liquid solution of benzene in toluene, initially at saturation temperature. If the vapor is then condensed, what is the heat load on the condenser in kJ/kg of solution if the condensate is saturated and if it is subcooled by 10°C?

Section 4.3

4.13 Vapor–liquid equilibrium data at 101.3 kPa are given for the chloroform–methanol system on p. 13-11 of *Perry's Chemical Engineers' Handbook*, 6th ed. From these data, prepare plots like Figures 4.6b and 4.6c. From the plots, determine the azeotropic composition and temperature at 101.3 kPa. Is the azeotrope of the minimum- or maximum-boiling type?

4.14 Vapor–liquid equilibrium data at 101.3 kPa are given for the water–formic acid system on p. 13-14 of *Perry's Chemical Engineers' Handbook*, 6th ed. From these data, prepare plots like Figures 4.7b and 4.7c. From the plots, determine the azeotropic composition and temperature at 101.3 kPa. Is the azeotrope of the minimum- or maximum-boiling type?

4.15 Vapor–liquid equilibrium data for mixtures of water and isopropanol at 1 atm (101.3 kPa, 760 torr) are given below.

(a) Prepare *T–x–y* and *x–y* diagrams.

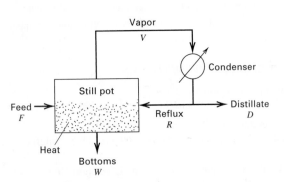

Figure 4.37 Conditions for Exercise 4.10.

(b) When a solution containing 40 mol% isopropanol is slowly vaporized, what will be the composition of the initial vapor formed?

(c) If this same 40% mixture is heated under equilibrium conditions until 75 mol% has been vaporized, what will be the compositions of the vapor and liquid produced?

VAPOR–LIQUID EQUILIBRIUM FOR ISOPROPANOL AND WATER AT 1 ATM

T, °C	Mol% Isopropanol	
	Liquid	**Vapor**
93.00	1.18	21.95
89.75	3.22	32.41
84.02	8.41	46.20
83.85	9.10	47.06
82.12	19.78	52.42
81.64	28.68	53.44
81.25	34.96	55.16
80.62	45.25	59.26
80.32	60.30	64.22
80.16	67.94	68.21
80.21	68.10	68.26
80.28	76.93	74.21
80.66	85.67	82.70
81.51	94.42	91.60

Notes:

Composition of the azeotrope: $x = y = 68.54\%$.

Boiling point of azeotrope: 80.22°C.

Boiling point of pure isopropanol: 82.5°C.

(d) Calculate K-values and values of α at 80°C and 89°C.

(e) Compare your answers in parts (a), (b), and (c) to those obtained from T–x–y and x–y diagrams based on the following vapor pressure data and Raoult's and Dalton's laws.

What do you conclude about the applicability of Raoult's law to this system?

Vapor Pressures of Isopropanol and Water

Vapor pressure, torr	200	400	760
Isopropanol, °C	53.0	67.8	82.5
Water, °C	66.5	83	100

Section 4.4

4.16 Using the y–x and T–y–x diagrams in Figures 4.3 and 4.4, determine the temperature, amounts, and compositions of the equilibrium vapor and liquid phases at 101 kPa for the following conditions with a 100-kmol mixture of nC_6 (H) and nC_8 (C).

(a) $z_H = 0.5$, $\Psi = V/F = 0.2$

(b) $z_H = 0.4$, $y_H = 0.6$

(c) $z_H = 0.6$, $x_C = 0.7$

(d) $z_H = 0.5$, $\Psi = 0$

(e) $z_H = 0.5$, $\Psi = 1.0$

(f) $z_H = 0.5$, $T = 200°F$

4.17 For a binary mixture of components 1 and 2, show that the equilibrium phase compositions and amounts can be computed directly from the following reduced forms of Eqs. (5), (6), and (3)

of Table 4.4.

$$x_1 = (1 - K_2)/(K_1 - K_2)$$

$$x_2 = 1 - x_1$$

$$y_1 = (K_1 K_2 - K_1)/(K_2 - K_1)$$

$$y_2 = 1 - y_1$$

$$\Psi = \frac{V}{F} = \frac{z_1[(K_1 - K_2)/(1 - K_2)] - 1}{K_1 - 1}$$

4.18 Consider the Rachford–Rice form of the flash equation,

$$\sum_{i=1}^{C} \frac{z_i(1 - K_i)}{1 + (V/F)(K_i - 1)} = 0$$

Under what conditions can this equation be satisfied?

4.19 A liquid containing 60 mol% toluene and 40 mol% benzene is continuously distilled in a single-equilibrium-stage unit at atmospheric pressure. What percent of benzene in the feed leaves in the vapor if 90% of the toluene entering in the feed leaves in the liquid? Assume a relative volatility of 2.3 and obtain the solution graphically.

4.20 Solve Exercise 4.19 by assuming an ideal solution and using vapor pressure data from Figure 2.4. Also determine the temperature.

4.21 A seven-component mixture is flashed at a specified temperature and pressure.

(a) Using the K-values and feed composition given below, make a plot of the Rachford–Rice flash function

$$f\{\Psi\} = \sum_{i=1}^{C} \frac{z_i(1 - K_i)}{1 + \Psi(K_i - 1)}$$

at intervals of Ψ of 0.1, and from the plot estimate the correct root of Ψ.

(b) An alternative form of the flash function is

$$f\{\Psi\} = \sum_{i=1}^{C} \frac{z_i K_i}{1 + \Psi(K_i - 1)} - 1$$

Make a plot of this equation also at intervals of Ψ of 0.1 and explain why the Rachford–Rice function is preferred.

Component	z_i	K_i
1	0.0079	16.2
2	0.1321	5.2
3	0.0849	2.6
4	0.2690	1.98
5	0.0589	0.91
6	0.1321	0.72
7	0.3151	0.28

4.22 One hundred kilomoles of a feed composed of 25 mol% n-butane, 40 mol% n-pentane, and 35 mol% n-hexane are flashed at steady-state conditions. If 80% of the hexane is to be recovered in the liquid at 240°F, what pressure is required, and what are the liquid and vapor compositions? Obtain K-values from Figure 2.8.

4.23 An equimolar mixture of ethane, propane, n-butane, and n-pentane is subjected to a flash vaporization at 150°F and 205 psia. What are the expected amounts and compositions of the liquid and vapor products? Is it possible to recover 70% of the ethane in the

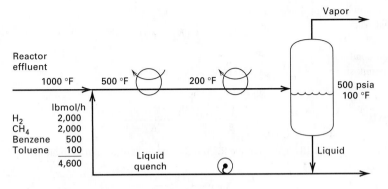

Figure 4.38 Conditions for Exercise 4.24.

vapor by a single-stage flash at other conditions without losing more than 5% of nC_4 to the vapor? Obtain K-values from Figure 2.8.

4.24   The system shown in Figure 4.38 is used to cool the reactor effluent and separate the light gases from the heavier hydrocarbons. K-values for the components at 500 psia and 100°F are

Component	K_i
H_2	80
CH_4	10
Benzene	0.010
Toluene	0.004

(a) Calculate the composition and flow rate of the vapor leaving the flash drum.

(b) Does the flow rate of liquid quench influence the result? Prove your answer analytically.

4.25   The mixture shown in Figure 4.39 is partially condensed and separated into two phases. Calculate the amounts and compositions of the equilibrium phases, V and L.

4.26   The following stream is at 200 psia and 200°F. Determine whether it is a subcooled liquid or a superheated vapor, or whether it is partially vaporized, without making a flash calculation.

Component	lbmol/h	K-value
C_3	125	2.056
nC_4	200	0.925
nC_5	175	0.520

4.27   The overhead system for a distillation column is shown in Figure 4.40. The composition of the total distillates is indicated, with 10 mol% of it being taken as vapor. Determine the pressure in the reflux drum, if the temperature is 100°F. Use the following K-values by assuming that K is inversely proportional to pressure.

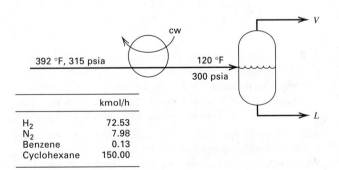

Figure 4.39 Conditions for Exercise 4.25.

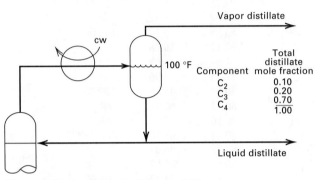

Figure 4.40 Conditions for Exercise 4.27.

Component	K at 100°F, 200 psia
C_2	2.7
C_3	0.95
C_4	0.34

4.28   Determine the phase condition of a stream having the following composition at 7.2°C and 2,620 kPa.

Component	kmol/h
N_2	1.0
C_1	124.0
C_2	87.6
C_3	161.6
nC_4	176.2
nC_5	58.5
nC_6	33.7

Perform the calculations with a computer simulation program using at least three different options for K-values. Does the choice of K-value method influence the results?

4.29   A liquid mixture consisting of 100 kmol of 60 mol% benzene, 25 mol% toluene, and 15 mol% o-xylene is flashed at 1 atm and 100°C.

(a) Compute the amounts of liquid and vapor products and their composition.

(b) Repeat the calculation at 100°C and 2 atm.

(c) Repeat the calculation at 105°C and 0.1 atm.

(d) Repeat the calculation at 150°C and 1 atm.

Assume ideal solutions and use the vapor pressure curves of Figure 2.4 for benzene and toluene. For o-xylene, draw a vapor pressure line that goes through the points (100.2°C, 200 torr) and (144°C, 760 torr).

4.30 Prove that the vapor leaving an equilibrium flash is at its dew point and that the liquid leaving an equilibrium flash is at its bubble point.

4.31 The following mixture is introduced into a distillation column as saturated liquid at 1.72 MPa. Calculate the bubble-point temperature using the K-values of Figure 2.8.

Compound	kmol/h
Ethane	1.5
Propane	10.0
n-Butane	18.5
n-Pentane	17.5
n-Hexane	3.5

4.32 An equimolar solution of benzene and toluene is totally evaporated at a constant temperature of 90°C. What are the pressures at the beginning and end of the vaporization process? Assume an ideal solution and use the vapor pressure curves of Figure 2.4.

4.33 The following equations are given by Sebastiani and Lacquaniti [*Chem. Eng. Sci.*, **22**, 1155 (1967)] for the liquid-phase activity coefficients of the water (W)–acetic acid (A) system.

$$\log \gamma_W = x_A^2[A + B(4x_W - 1) + C(x_W - x_A)(6x_W - 1)]$$
$$\log \gamma_A = x_W^2[A + B(4x_W - 3) + C(x_W - x_A)(6x_W - 5)]$$
$$A = 0.1182 + \frac{64.24}{T(K)}$$
$$B = 0.1735 - \frac{43.27}{T(K)}$$
$$C = 0.1081$$

Find the dew point and bubble point of a mixture of composition $x_W = 0.5$, $x_A = 0.5$ at 1 atm. Flash the mixture at a temperature halfway between the dew point and the bubble point.

4.34 Find the bubble-point and dew-point temperatures of a mixture of 0.4 mole fraction toluene (1) and 0.6 mole fraction n-butanol (2) at 101.3 kPa. The K-values can be calculated from (2-72), the modified Raoult's law, using vapor-pressure data, and γ_1 and γ_2 from the van Laar equation of Table 2.9 with $A_{12} = 0.855$ and $A_{21} = 1.306$. If the same mixture is flashed at a temperature midway between the bubble point and dew point, and 101.3 kPa, what fraction is vaporized, and what are the compositions of the two phases?

4.35 (a) For a liquid solution having a molar composition of ethyl acetate (A) of 80% and ethyl alcohol (E) of 20%, calculate the bubble-point temperature at 101.3 kPa and the composition of the corresponding vapor using (2-72) with vapor pressure data and the van Laar equation of Table 2.9 with $A_{AE} = 0.855$, $A_{EA} = 0.753$.

(b) Find the dew point of the mixture.

(c) Does the mixture form an azeotrope? If so, predict the temperature and composition.

4.36 A binary solution at 107°C contains 50 mol% water (W) and 50 mol% formic acid (F). Using (2-72) with vapor pressure data and the van Laar equation of Table 2.9 with $A_{WF} = -0.2935$ and $A_{FW} = -0.2757$, compute:

(a) The bubble-point pressure.

(b) The dew-point pressure.

Also determine whether the mixture forms a maximum- or minimum-boiling azeotrope. If so, predict the azeotropic pressure at 107°C and the azeotropic composition.

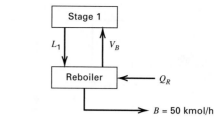

Figure 4.41 Conditions for Exercise 4.38.

4.37 For a mixture consisting of 45 mol% n-hexane, 25 mol% n-heptane, and 30 mol% n-octane at 1 atm, use a simulation computer program to:

(a) Find the bubble- and dew-point temperatures.

(b) Find the flash temperature, and the compositions and relative amounts of the liquid and vapor products if the mixture is subjected to a flash distillation at 1 atm so that 50 mol% of the feed is vaporized.

(c) Find how much of the octane is taken off as vapor if 90% of the hexane is taken off as vapor.

Repeat parts (a) and (b) at 5 atm and 0.5 atm.

4.38 In Figure 4.41, 150 kmol/h of a saturated liquid, L_1, at 758 kPa, of molar composition, propane 10%, n-butane 40%, and n-pentane 50%, enters the reboiler from stage 1. What are the compositions and amounts of V_B and B? What is Q_R, the reboiler duty? Use a simulation computer program to find the answers.

4.39 (a) Find the bubble-point temperature of the following mixture at 50 psia, using K-values from Figure 2.8 or Figure 2.9.

Component	z_i
Methane	0.005
Ethane	0.595
n-Butane	0.400

(b) Find the temperature that results in 25% vaporization at this pressure. Determine the corresponding liquid and vapor compositions.

4.40 As shown in Figure 4.42, a hydrocarbon mixture is heated and expanded before entering a distillation column. Calculate, using a simulation computer program, the mole percent vapor phase and vapor and liquid phase mole fractions at each of the three locations indicated by a pressure specification.

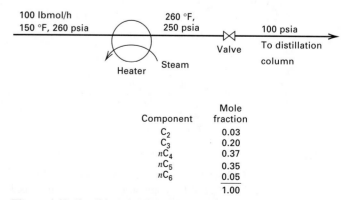

Component	Mole fraction
C_2	0.03
C_3	0.20
nC_4	0.37
nC_5	0.35
nC_6	0.05
	1.00

Figure 4.42 Conditions for Exercise 4.40.

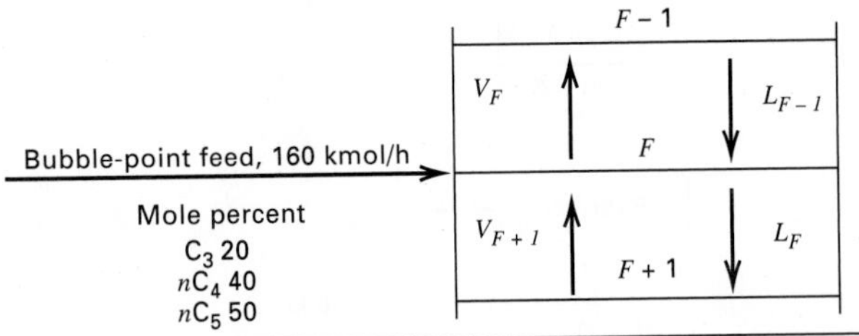

Figure 4.43 Conditions for Exercise 4.41.

Stream	Total flow rate kmol/h	C_3	nC_4	nC_5
L_{F-1}	100	15	45	40
V_{F+1}	196	30	50	20

4.41 Streams entering stage F of a distillation column are shown in Figure 4.43. What is the temperature of stage F and the compositions and amounts of streams V_F and L_F if the pressure is 785 kPa for all streams? Use a simulation computer program to obtain the answers.

4.42 Flash adiabatically, across a valve, a stream composed of the six hydrocarbons given below. The feed upstream of the valve is at 250°F and 500 psia. The pressure downstream of the valve is 300 psia.

Component	z_i
C_2H_4	0.02
C_2H_6	0.03
C_3H_6	0.05
C_3H_8	0.10
iC_4	0.20
nC_4	0.60

Compute using a simulation computer program:

(a) The phase condition upstream of the valve.

(b) The temperature downstream of the valve.

(c) The molar fraction vaporized downstream of the valve.

(d) The mole fraction compositions of the vapor and liquid phases downstream of the valve.

4.43 Propose a detailed algorithm like Figure 4.19a and Table 4.4 for a flash where the percent vaporized and the flash pressure are to be specified.

4.44 Determine algorithms for carrying out the following flash calculations, assuming that expressions for K-values and enthalpies are available.

Given	Find
h_F, P	Ψ, T
h_F, T	Ψ, P
h_F, Ψ	T, P
Ψ, T	h_F, P
Ψ, P	h_F, T
T, P	h_F, Ψ

Section 4.5

4.45 A feed of 13,500 kg/h consists of 8 wt% acetic acid (B) in water (A). The removal of the acetic acid is to be accomplished by liquid–liquid extraction at 25°C. The raffinate is to contain only 1 wt% acetic acid. The following four solvents, with accompanying distribution coefficients in mass-fraction units, are being considered. Water and each solvent (C) can be considered immiscible. For each solvent, estimate the kilograms required per hour if a single equilibrium stage is used.

Solvent	K_D
Methyl acetate	1.273
Isopropyl ether	0.429
Heptadecanol	0.312
Chloroform	0.178

4.46 Forty-five kilograms of a solution containing 30 wt% ethylene glycol in water is to be extracted with furfural. Using Figures 4.14a and 4.14e, calculate:

(a) The minimum quantity of solvent.

(b) The maximum quantity of solvent.

(c) The weights of solvent-free extract and raffinate for 45 kg solvent, and the percent glycol extracted.

(d) The maximum possible purity of glycol in the finished extract and the maximum purity of water in the raffinate for one equilibrium stage.

4.47 Prove that, in a triangular diagram, where each vertex represents a pure component, the composition of the system at any point inside the triangle is proportional to the length of the respective perpendicular drawn from the point to the side of the triangle opposite the vertex in question. It is not necessary to assume a special case (i.e., a right or equilateral triangle).

4.48 A mixture of chloroform ($CHCl_3$) and acetic acid at 18°C and 1 atm (101.3 kPa) is to be extracted with water to recover the acid.

(a) Forty-five kilograms of a mixture containing 35 wt% $CHCl_3$ and 65 wt% acid is treated with 22.75 kg of water at 18°C in a simple one-stage batch extraction. What are the compositions and weights of the raffinate and extract layers produced?

(b) If the raffinate layer from the above treatment is extracted again with one-half its weight of water, what will be the compositions and weights of the new layers?

(c) If all the water is removed from this final raffinate layer, what will its composition be?

Solve this exercise using the following equilibrium data to construct one or more of the types of diagrams in Figure 4.14.

LIQUID–LIQUID EQUILIBRIUM DATA FOR $CHCl_3$-H_2O-CH_3COOH AT 18°C AND 1 ATM

Heavy Phase (wt%)			Light Phase (wt%)		
$CHCl_3$	H_2O	CH_3COOH	$CHCl_3$	H_2O	CH_3COOH
99.01	0.99	0.00	0.84	99.16	0.00
91.85	1.38	6.77	1.21	73.69	25.10
80.00	2.28	17.72	7.30	48.58	44.12
70.13	4.12	25.75	15.11	34.71	50.18
67.15	5.20	27.65	18.33	31.11	50.56
59.99	7.93	32.08	25.20	25.39	49.41
55.81	9.58	34.61	28.85	23.28	47.87

4.49 Isopropyl ether (E) is used to separate acetic acid (A) from water (W). The liquid–liquid equilibrium data at 25°C and 1 atm

(101.3 kPa) are presented below.

(a) One hundred kilograms of a 30 wt% A–W solution is contacted with 120 kg of ether in an equilibrium stage. What are the compositions and weights of the resulting extract and raffinate? What would be the concentration of acid in the (ether-rich) extract if all the ether were removed?

(b) A mixture containing 52 kg A and 48 kg W is contacted with 40 kg of E. What are the extract and raffinate compositions and quantities?

LIQUID–LIQUID EQUILIBRIUM DATA FOR ACETIC ACID (A), WATER (W), AND ISOPROPANOL ETHER (E) AT 25°C AND 1 ATM

Water-Rich Layer			Ether-Rich Layer		
Wt% A	Wt% W	Wt% E	Wt% A	Wt% W	Wt% E
1.41	97.1	1.49	0.37	0.73	98.9
2.89	95.5	1.61	0.79	0.81	98.4
6.42	91.7	1.88	1.93	0.97	97.1
13.30	84.4	2.3	4.82	1.88	93.3
25.50	71.1	3.4	11.4	3.9	84.7
36.70	58.9	4.4	21.6	6.9	71.5
45.30	45.1	9.6	31.1	10.8	58.1
46.40	37.1	16.5	36.2	15.1	48.7

Section 4.6

4.50 Diethylene glycol (DEG) is used as a solvent in the UDEX liquid–liquid extraction process [H.W. Grote, *Chem Eng. Progr.,* **54** (8), 43 (1958)] to separate paraffins from aromatics. If 280 lbmol/h of 42.86 mol% n-hexane, 28.57 mol% n-heptane, 17.86 mol% benzene, and 10.71 mol% toluene is contacted with 500 lbmol/h of 90 mol% aqueous DEG at 325°F and 300 psia, calculate, using a simulation computer program and the UNIFAC L/L method for estimating liquid-phase activity coefficients, the flow rates and molar compositions of the resulting two liquid phases. Is DEG more selective for the paraffins or the aromatics?

4.51 A feed of 110 lbmol/h includes 5, 3, and 2 lbmol/h, respectively, of formic acid, acetic acid, and propionic acid in water. If the acids are extracted in a single equilibrium stage with 100 lbmol/h of ethyl acetate (EA), calculate with a simulation computer program using the UNIFAC method, flow rates and molar compositions of the resulting two liquid phases. What is the order of selectivity of EA for the three organic acids?

Section 4.7

4.52 Repeat Example 4.9 for 200,000 kg/h of hexane.

4.53 Water is to be used in a single equilibrium stage to dissolve 1,350 kg/h of Na_2CO_3 from 3,750 kg/h of a solid, where the balance is an insoluble oxide. If 4,000 kg/h of water is used and the underflow from the stage is 40 wt% solvent on a solute-free basis, compute the flow rates and compositions of the overflow and the underflow.

4.54 Repeat Exercise 4.53 if the residence time is only sufficient to leach 80% of the carbonate.

4.55 A total of 6,000 lb/h of a liquid solution of 40 wt% benzene in naphthalene at 50°C is cooled to 15°C. Assuming that equilibrium is achieved, use Figure 4.23 to determine the amount of crystals formed, and the flow rate and composition of the mother liquor. Are the crystals benzene or naphthalene?

4.56 Repeat Example 4.10, except determine the temperature necessary to crystallize 80% of the naphthalene.

4.57 A total of 10,000 kg/h of a 10 wt% liquid solution of naphthalene in benzene is cooled from 30°C to 0°C. Assuming that equilibrium is achieved, determine the amount of crystals formed and the composition and flow rate of the mother liquid. Are the crystals benzene or naphthalene? Use Figure 4.23.

4.58 Repeat Example 4.11, except let the original solution be 20 wt% Na_2SO_4.

4.59 At 20°C, 1,000 kg of a mixture of 50 wt% $Na_2SO_4 \cdot 10H_2O$ and 50 wt% Na_2SO_4 crystals exists. How many kilograms of water must be added to just completely dissolve the crystals if the temperature is kept at 20°C and equilibrium is maintained? Use Figure 4.24.

4.60 Repeat Example 4.12, except determine the grams of activated carbon to achieve:

(a) 75% adsorption of phenol.

(b) 90% adsorption of phenol.

(c) 98% adsorption of phenol.

4.61 A colored substance (B) is to be removed from a mineral oil by adsorption with clay particles at 25°C. The original oil has a color index of 200 units/100 kg oil, while the decolorized oil must have an index of only 20 units/100 kg oil. The following experimental adsorption equilibrium data have been measured in a laboratory:

c_B, color units/ 100 kg oil	200	100	60	40	10
q_B, color units/ 100 kg clay	10	7.0	5.4	4.4	2.2

(a) Fit the data to the Freundlich equation.

(b) Compute the kilograms of clay needed to treat 500 kg of oil if one equilibrium contact is used.

Section 4.8

4.62 Vapor–liquid equilibrium data in mole fractions for the system acetone–air–water at 1 atm (101.3 kPa) are as follows:

y, acetone in air:	0.004	0.008	0.014	0.017	0.019	0.020
x, acetone in water:	0.002	0.004	0.006	0.008	0.010	0.012

(a) Plot the data as (1) a graph of moles acetone per mole air versus moles acetone per mole water, (2) partial pressure of acetone versus g acetone per g water, and (3) y versus x.

(b) If 20 moles of gas containing 0.015 mole fraction acetone is brought into contact with 15 moles of water in an equilibrium stage, what would be the composition of the discharge streams? Solve graphically. For both parts, neglect partitioning of water and air.

4.63 It has been proposed that oxygen be separated from nitrogen by absorbing and desorbing air in water. Pressures from 101.3 to 10,130 kPa and temperatures between 0 and 100°C are to be used.

(a) Devise a workable scheme for doing the separation assuming the air is 79 mol% N_2 and 21 mol% O_2.

(b) Henry's law constants for O_2 and N_2 are given in Figure 4.27. How many batch absorption steps would be necessary to make 90 mol% pure oxygen? What yield of oxygen (based on total amount of oxygen feed) would be obtained?

4.64 A vapor mixture having equal volumes of NH_3 and N_2 is to be contacted at 20°C and 1 atm (760 torr) with water to absorb a portion of the NH_3. If 14 m^3 of this mixture is brought into contact with 10 m^3 of water and if equilibrium is attained, calculate the percent of the ammonia originally in the gas that will be absorbed. Both temperature and total pressure will be maintained constant during the absorption. The partial pressure of NH_3 over water at 20°C is as follows:

Partial Pressure of NH_3 in Air, torr	Grams of Dissolved NH_3/100 g of H_2O
470	40
298	30
227	25
166	20
114	15
69.6	10
50.0	7.5
31.7	5.0
24.9	4.0
18.2	3.0
15.0	2.5
12.0	2.0

Section 4.9

4.65 Repeat Example 4.15 for temperatures corresponding to the following vapor pressures for solid PA:

(a) 0.7 torr

(b) 0.4 torr

(c) 0.1 torr

Plot the percent recovery of PA versus the solid vapor pressure for the range from 0.1 torr to 1.0 torr.

4.66 Nitrogen at 760 torr and 300°C contains 10 mol% anthraquinone (A). If this gas is cooled to 200°C, calculate the percent desublimation of A. Vapor pressure data for solid A are as follows:

T, °C:	190.0	234.2	264.3	285.0
Vapor pressure, torr:	1	10	40	100

These data can be fitted to the Antoine equation (2-39) using the first three constants.

4.67 At 25°C and 101 kPa, 2 mol of a gas containing 35 mol% propylene in propane is equilibrated with 0.1 kg of silica gel adsorbent. Using the equilibrium data of Figure 4.30, calculate the moles and composition of the gas adsorbed and the equilibrium composition of the gas not adsorbed.

4.68 A gas containing 50 mol% propylene in propane is to be separated with silica gel having the equilibrium properties shown in Figure 4.30. The final products are to be 90 mol% propylene and 75 mol% propane. If 1,000 lb of silica gel/lbmol of feed gas or less is used, can the desired separation be made in one equilibrium stage? If not, what separation can be achieved?

Section 4.10

4.69 Repeat Example 4.17 for 90% evaporation of the water.

4.70 A 5,000-kg/h aqueous solution of 20 wt% Na_2SO_4 is fed to an evaporative crystallizer operating at 60°C. Equilibrium data are given in Figure 4.24. If 80% of the Na_2SO_4 is to be crystallized, calculate:

(a) The kilograms of water that must be evaporated per hour

(b) The crystallizer pressure in torr

4.71 Calculate the dew-point pressure, secondary dew-point pressure, and bubble-point pressure of the following mixtures at 50°C, assuming that the liquid aromatics and water are mutually insoluble:

(a) 50 mol% benzene and 50 mol% water.

(b) 50 mol% toluene and 50 mol% water.

(c) 40 mol% benzene, 40 mol% toluene, and 20 mol% water.

4.72 Repeat Exercise 4.71, except compute temperatures for a pressure of 2 atm.

4.73 A liquid containing 30 mol% toluene, 40 mol% ethylbenzene, and 30 mol% water is subjected to a continuous, flash distillation at a total pressure of 0.5 atm. Assuming that mixtures of ethylbenzene and toluene obey Raoult's law and that the hydrocarbons are completely immiscible in water and vice versa, calculate the temperature and composition of the vapor phase at the bubble-point temperature.

4.74 As shown in Figure 4.8, water (W) and *n*-butanol (B) can form a three-phase system at 101 kPa. For a mixture of overall composition of 60 mol% W and 40 mol% B, use a simulation computer program and the UNIFAC method to estimate:

(a) Dew-point temperature and composition of the first drop of liquid.

(b) Bubble-point temperature and composition of the first bubble of vapor.

(c) Compositions and relative amounts of all three phases for 50 mol% vaporization.

4.75 Repeat Example 4.19 for a temperature of 25°C. Are the changes significant?

Chapter 5

Cascades and Hybrid Systems

In the previous chapter, a single equilibrium stage was utilized to separate a mixture. In practice, a single stage is rarely sufficient to perform the desired separation. This chapter introduces separation *cascades,* which are collections of contacting stages. Cascades are used in industrial processes to (1) accomplish separations that cannot be achieved in a single stage, and/or (2) reduce the required amount of the mass- or energy-separating agent.

A typical cascade is shown in Figure 5.1, where, in each stage, an attempt is made to bring two or more process streams of different phase state and composition into intimate contact to promote rapid mass and heat transfer, so as to approach physical equilibrium. The resulting phases, whose compositions and temperatures are now closer to, or at, equilibrium, are then separated and each is sent to another stage in the cascade, or withdrawn as a product. Although equilibrium conditions may not be achieved in each stage, it is common to design and analyze cascades using equilibrium-stage models. Alternatively, in the case of membrane separations, where

phase equilibrium is not a consideration and mass-transfer rates through the membrane determine the separation, cascades of membranes can enable separations that cannot be achieved by contact of the feed mixture with a single-membrane separator.

Cascades are prevalent in *unit operations,* such as distillation, absorption, stripping, and liquid–liquid extraction. In cases where the extent of separation by a single-unit operation is limited or the energy required is excessive, it is worthwhile to consider a *hybrid system* of two different unit operations, such as the combination of distillation and pervaporation, which is used to separate azeotropic mixtures. In the last decade, with increased awareness of the need for conserving energy, much attention is being given to hybrid systems. This chapter introduces both cascades and hybrid systems. To illustrate the benefits of cascades, the calculations are based on simple models. Rigorous models, best implemented by computer calculations, are deferred to Chapters 10–12.

5.0 INSTRUCTIONAL OBJECTIVES

After completing this chapter, you should be able to:

- Explain how multi-equilibrium-stage cascades with countercurrent flow can achieve a significantly better separation than a single equilibrium stage.
- Explain the difference between a single-section cascade and a two-section cascade and the limits of what each type can achieve.
- Estimate the recovery of a key component in countercurrent leaching and washing cascades.
- Estimate recovery of a key component in each of three types of liquid–liquid extraction cascades.
- Define and explain the significance of absorption and stripping factors.
- Estimate the recoveries of all components in a single-section, countercurrent cascade using the Kremser method.
- Estimate recoveries of all components in a two-section, countercurrent cascade using the Edmister extension of the Kremser method.
- Configure a membrane cascade to improve a membrane separation.
- Explain the merits and give examples of hybrid separation systems.
- Determine degrees of freedom and a set of specifications for a separation process or any element included in the process.

5.1 CASCADE CONFIGURATIONS

Cascades can be configured in many ways, as shown by the examples in Figure 5.2, where stages are represented by either boxes, as in Figure 5.1, or as horizontal lines in Figure 5.2d,e.

Depending on the mechanical design of the stages, cascades may be arranged vertically or horizontally. The feed to be separated is designated by F; the mass-separating agent, if used, is designated by S; and products are designated by P_i.

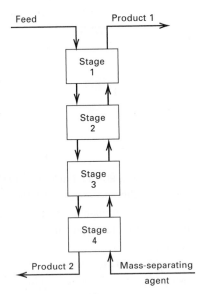

Figure 5.1 Cascade of contacting stages.

In the *countercurrent cascade,* shown in Figures 5.1 and 5.2a, the two phases flow countercurrently to each other between stages. As will be shown in examples, this configuration is very efficient and is widely used for absorption, stripping, liquid–liquid extraction, leaching, and washing. The *crosscurrent cascade,* shown in Figure 5.2b, is, in most cases, not as efficient as the countercurrent cascade, but it is

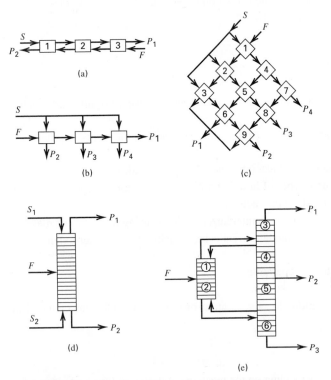

Figure 5.2 Examples of cascade configurations: (a) countercurrent cascade; (b) crosscurrent cascade; (c) two-dimensional, diamond cascade; (d) two-section, countercurrent cascade; (e) interlinked system of countercurrent cascades.

easier to apply in a batchwise manner. It differs from the countercurrent cascade in that the solvent is divided into portions fed individually to each stage.

A complex diamond variation of the crosscurrent cascade is shown in Figure 5.2c. Unlike the two former cascades, which are linear or one-dimensional, the diamond configuration is two-dimensional. One application is to batch crystallization. Feed F is separated in stage 1 into crystals, which pass to stage 2, and mother liquor, which passes to stage 4. In each of the other stages, partial crystallization or recrystallization occurs by processing crystals, mother liquor, or combinations of the two. Final products are purified crystals and impurity-bearing mother liquors.

The first three cascades in Figure 5.2 consist of *single sections* with streams entering and leaving only from the ends. Such cascades are used to recover components from a feed stream and are not generally useful for making a sharp separation between two selected feed components, called *key components.* To do this, it is best to provide a cascade consisting of *two sections.* The countercurrent cascade of Figure 5.2d is often used. It consists of one section above the feed and one below. If two solvents are used, where S_1 selectively dissolves certain components of the feed, while S_2 is more selective for the other components, the process, referred to as *fractional liquid–liquid extraction,* achieves a sharp separation. If S_1 is a liquid absorbent and S_2 is a vapor stripping agent, added to the cascade, as shown, or produced internally by condensation heat transfer at the top to give liquid reflux, and boiling heat transfer at the bottom to give vapor boilup, the process is simple *distillation,* for which a sharp split between two key components can be achieved if a reasonably high relative volatility exists between the two key components and if reflux, boilup, and the number of stages are sufficient.

Figure 5.2e shows an interlinked system of two distillation columns containing six countercurrent cascade sections. Reflux and boilup for the first column are provided by the second column. This system is capable of taking a ternary (three-component) feed, F, and producing three relatively pure products, P_1, P_2, and P_3.

In this chapter, algebraic equations are developed for modeling idealized cascades to illustrate, quantitatively, their capabilities and advantages. First, a simple countercurrent, single-section cascade for a solid–liquid leaching and/ or washing process is considered. Then, cocurrent, crosscurrent, and countercurrent single-section cascades, based on simplified component distribution coefficients, are compared for a liquid–liquid extraction process. A two-section, countercurrent cascade is subsequently developed for a vapor–liquid distillation operation. Finally, membrane cascades are described. In the first three cases, a set of linear algebraic equations is reduced to a single relation for estimating the extent of separation as a function of the number of stages in the cascade, the separation factor, and the flow ratio of the mass- or energy-separating agent to the feed. More rigorous models for design and analysis purposes are developed in subsequent chapters. As will be seen, single relations cannot be obtained from rigorous models because of the nonlinear nature of rigorous models, making computer calculations a necessity.

5.2 SOLID–LIQUID CASCADES

Consider the N-stage, countercurrent leaching–washing process shown in Figure 5.3. This cascade is an extension of the single-stage systems discussed in Section 4.7. The solid feed, entering stage 1, consists of two components A and B, of mass flow rates F_A and F_B. Pure liquid solvent, C, which can dissolve B completely, but not A at all, enters stage N at a mass flow rate S. Thus, A passes through the cascade as an insoluble solid. It is convenient to express liquid–phase concentrations of B, the solute, in terms of mass ratios of solute to solvent. The liquid *overflow* from each stage, j, contains Y_j mass of soluble material per mass of solute-free solvent, and no insoluble material. The *underflow* from each stage is a slurry consisting of a mass flow F_A of insoluble solids, a constant ratio of mass of solvent/mass of insoluble solids equal to R, and X_j mass of soluble material/mass of solute-free solvent. For a given solid feed, a relationship between the exiting underflow concentration of the soluble component, X_N, the solvent feed rate, and the number of stages is derived as follows.

If equilibrium is achieved at each stage, the overflow solute concentration, Y_j, equals the underflow solute concentration in the liquid, X_j, which refers to liquid held by the solid in the underflow. Assume that all soluble material, A_j is dissolved or leached in stage 1 and all other stages are then washing stages for reducing the amount of soluble material lost in the underflow leaving the last stage, N, and thereby increasing the amount of soluble material leaving in the overflow from stage 1. By solvent material balances, it is readily shown that for constant R, the flow rate of solvent leaving in the overflow from stages 2 to N is S. The flow rate of solvent leaving in the underflow from stages 1 to N is RF_A. Therefore, the flow rate of solvent leaving in the overflow from stage 1 is $S - RF_A$.

A material balance for the soluble material around any interior stage n from $n = 2$ to $N - 1$ is given by

$$Y_{n+1}S + X_{n-1}RF_A = Y_nS + X_nRF_A \tag{5-1}$$

For terminal stages 1 and N, the material balances on the soluble material are, respectively,

$$Y_2S + F_B = Y_1(S - RF_A) + X_1RF_A \tag{5-2}$$

$$X_{N-1}RF_A = Y_NS + X_NRF_A \tag{5-3}$$

Assuming equilibrium, the concentration of soluble material in the overflow from each stage is equal to the concentration of soluble material in the liquid part of the underflow from the same stage. Thus,

$$X_n = Y_n \tag{5-4}$$

In addition, it is convenient to define a *washing factor*, W, as

$$W = \frac{S}{RF_A} \tag{5-5}$$

If (5-1), (5-2), and (5-3) are each combined with (5-4) to eliminate Y, and the resulting equations are rearranged to allow the substitution of (5-5), the following equations result:

$$X_1 - X_2 = \left(\frac{F_B}{S}\right) \tag{5-6}$$

$$\left(\frac{1}{W}\right)X_{n-1} - \left(\frac{1+W}{W}\right)X_n + X_{n+1} = 0, \tag{5-7}$$
$$n = 2 \text{ to } N - 1$$

$$\left(\frac{1}{W}\right)X_{N-1} - \left(\frac{1+W}{W}\right)X_N = 0 \tag{5-8}$$

Equations (5-6) to (5-8) constitute a set of N linear algebraic equations in N unknowns, $X_n(n = 1 \text{ to } N)$. The equations are of a tridiagonal, sparse-matrix form, which—for example, with $N = 5$—is given by

$$\begin{bmatrix} 1 & -1 & 0 & 0 & 0 \\ \left(\frac{1}{W}\right) & -\left(\frac{1+W}{W}\right) & 1 & 0 & 0 \\ 0 & \left(\frac{1}{W}\right) & -\left(\frac{1+W}{W}\right) & 1 & 0 \\ 0 & 0 & \left(\frac{1}{W}\right) & -\left(\frac{1+W}{W}\right) & 1 \\ 0 & 0 & 0 & \left(\frac{1}{W}\right) & -\left(\frac{1+W}{W}\right) \end{bmatrix}$$
$$\times \begin{bmatrix} X_1 \\ X_2 \\ X_3 \\ X_4 \\ X_5 \end{bmatrix} = \begin{bmatrix} \left(\frac{F_B}{S}\right) \\ 0 \\ 0 \\ 0 \\ 0 \end{bmatrix} \tag{5-9}$$

Equations of type (5-9) can be solved by Gaussian elimination by starting from the top and eliminating unknowns X_1, X_2, etc. in order to obtain

$$X_N = \left(\frac{F_B}{S}\right)\left(\frac{1}{W^{N-1}}\right) \tag{5-10}$$

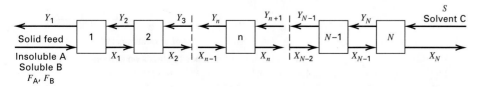

Figure 5.3 Countercurrent leaching or washing system.

By back-substitution, interstage values of X are given by

$$X_n = \left(\frac{F_B}{S}\right)\left(\frac{\sum_{k=0}^{N-n} W^k}{W^{N-1}}\right) \qquad (5\text{-}11)$$

For example, with $N = 5$,

$$X_1 = Y_1 = \left(\frac{F_B}{S}\right)\left(\frac{1 + W + W^2 + W^3 + W^4}{W^4}\right)$$

The purpose of the cascade, for any given S, is to maximize Y_1, the amount of soluble solids dissolved in the solvent leaving in the overflow from stage 1, and to minimize X_N, the amount of soluble solid dissolved in the solvent leaving the underflow with the insoluble material from stage N. Equation (5-10) indicates that this can be achieved for a given soluble-solids feed rate, F_B, by specifying a large

solvent feed rate, S, a large number of stages, N, and/or by employing a large washing factor, which can be achieved by minimizing the amount of liquid underflow compared to overflow. It should be noted that the minimum amount of solvent required corresponds to zero overflow from stage 1, or

$$S_{\min} = RF_A \qquad (5\text{-}12)$$

For this minimum value, $W = 1$ from (5-5) and all soluble solids leave in the underflow from the last stage, N, regardless of the number of stages. Therefore, it is best to specify a value of S significantly greater than $S_{\min}$. Equations (5-10) and (5-5) show that the value of X_N is reduced exponentially by increasing the number of stages, N. Thus, the countercurrent cascade can be very effective. For two or more stages, X_N is also reduced exponentially by increasing the solvent rate, S. For three or more stages, the value of X_N is reduced exponentially by decreasing the underflow ratio R.

EXAMPLE 5.1

Pure water is to be used to dissolve 1,350 kg/h of Na_2CO_3 from 3,750 kg/h of a solid, where the balance is an insoluble oxide. If 4,000 kg/h of water is used as the solvent for the carbonate and the total underflow from each stage is 40 wt% solvent on a solute-free basis, compute and plot the percent recovery of the carbonate in the overflow product for one stage and for two to five countercurrent stages, as in Figure 5.3.

SOLUTION

Soluble solids feed rate $= F_B = 1,350$ kg/h

Insoluble solids feed rate $= F_A = 3,750 - 1,350 = 2,400$ kg/h

Solvent feed rate $= S = 4,000$ kg/h

Underflow ratio $R = 40/60 = 2/3$

Washing factor $W = S/RF_A = 4,000/[(2/3)(2,400)] = 2.50$

Overall fractional recovery of soluble solids $= Y_1(S - RF_A)/F_B$

By overall material balance on soluble solids for N stages,

$$F_B = Y_1(S - RF_A) + X_N RF_A$$

Solving for Y_1 and using (5-5) to introduce the washing factor,

$$Y_1 = \frac{(F_B/S) - (1/W)X_N}{(1 - 1/W)}$$

From the given data,

$$Y_1 = \frac{(1,350/4,000) - (1/2.50)X_N}{(1 - 1/2.50)} \quad \text{or} \qquad (1)$$

$$Y_1 = 0.5625 - 0.6667X_N$$

where, from (5-10),

$$X_N = \left(\frac{1,350}{4,000}\right)\frac{1}{2.50^{N-1}} = \frac{0.3375}{2.50^{N-1}} \qquad (2)$$

The percent recovery of soluble material is

$$Y_1(S - RF_A)/F_B = Y_1[4,000 - (2/3)(2,400)]/1,350 \times 100\%$$

$$= 177.8Y_1 \qquad (3)$$

Results for one to five stages, as computed from (1) to (3), are

No. of Stages in Cascade, N	X_N	Y_1	Percent Recovery of Soluble Solids
1	0.3375	0.3375	60.0
2	0.1350	0.4725	84.0
3	0.0540	0.5265	93.6
4	0.0216	0.5481	97.4
5	0.00864	0.5567	99.0

A plot of the percent recovery of soluble solids as a function of the number of stages is shown in Figure 5.4. Although only a 60% recovery is obtained with one stage, 99% recovery is achieved for five stages. To achieve 99% recovery with one stage, a water rate of 160,000 kg/h is required, which is 40 times that required for five stages. Thus, the use of multiple stages in a countercurrent cascade to increase recovery of soluble material can be much more effective than increased use of a mass-separating agent with a single stage.

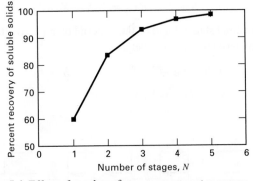

Figure 5.4 Effect of number of stages on percent recovery in Example 5.1.

5.3 SINGLE-SECTION, LIQUID–LIQUID EXTRACTION CASCADES

Three possible two-stage, single-section, liquid–liquid extraction cascades are the cocurrent, crosscurrent, and countercurrent arrangements in Figure 5.5. The countercurrent arrangement is generally preferred because, as will be shown in this section, that arrangement results in a higher degree of extraction for a given amount of solvent and number of equilibrium stages.

In Section 4.5, (4-25), for the fraction of solute, B, that is not extracted, was derived for a single liquid–liquid equilibrium extraction stage, assuming the use of pure solvent, and a constant value for the distribution coefficient, K'_{D_B}, for the solute, B, dissolved in components A and C, which are mutually insoluble. That equation is now extended to multiple stages for each type of cascade shown in Figure 5.5.

Cocurrent Cascade

If additional stages are added in the cocurrent arrangement in Figure 5.5a, the equation for the first stage is that of a single stage. That is, from (4-25) in mass ratio units,

$$X_B^{(1)}/X_B^{(F)} = \frac{1}{1+E} \tag{5-13}$$

where E is the extraction factor, given by

$$E = K'_{D_B}S/F_A \tag{5-14}$$

Since $Y_B^{(1)}$ is in equilibrium with $X_B^{(1)}$ according to

$$K'_{D_B} = Y_B^{(1)}/X_B^{(1)} \tag{5-15}$$

the combination of (5-15) with (5-13) gives

$$Y_B^{(1)}/X_B^{(F)} = K'_{D_B}/(1+E) \tag{5-16}$$

For the second stage, a material balance for B gives

$$X_B^{(1)}F_A + Y_B^{(1)}S = X_B^{(2)}F_A + Y_B^{(2)}S \tag{5-17}$$

with

$$K'_{D_B} = Y_B^{(2)}/X_B^{(2)} \tag{5-18}$$

Combining (5-17) with (5-18), (5-13), and (5-16) to eliminate $X_B^{(1)}$, $Y_B^{(1)}$, and $Y_B^{(2)}$ gives

$$\frac{X_B^{(2)}}{X_B^{(F)}} = \frac{1}{1+E} = \frac{X_B^{(N)}}{X_B^{(F)}} \tag{5-19}$$

Comparison of (5-19) with (5-13) shows that $X_B^{(2)} = X_B^{(1)}$. Thus, no additional extraction takes place in the second stage. This is as expected because the two streams leaving the first stage are at equilibrium and when they are recontacted in stage 2, no additional net mass transfer of B occurs. Accordingly, a cocurrent cascade of equilibrium stages has no merit other than to provide increased residence time.

Crosscurrent Cascade

For the crosscurrent cascade shown in Figure 5.5b, the feed progresses through each stage, starting with stage 1 and finishing with stage N. The solvent flow rate, S, is divided into portions that are sent to each stage. If the portions are equal, the following mass ratios are obtained by application of (5-13), where S is replaced by S/N, so that E is replaced by E/N:

$$X_B^{(1)}/X_B^{(F)} = 1/(1+E/N)$$
$$X_B^{(2)}/X_B^{(1)} = 1/(1+E/N)$$
$$\vdots$$
$$X_B^{(N)}/X_B^{(N-1)} = 1/(1+E/N) \tag{5-20}$$

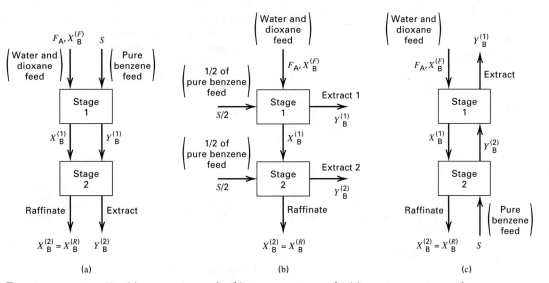

Figure 5.5 Two-stage arrangements: (a) cocurrent cascade; (b) crosscurrent cascade; (c) countercurrent cascade.

Combining the equations in (5-20) to eliminate all intermediate interstage variables, $X_B^{(n)}$, the final raffinate mass ratio is given by

$$X_B^{(N)}/X_B^{(F)} = X_B^{(R)}/X_B^{(F)} = \frac{1}{(1 + E/N)^N} \qquad (5\text{-}21)$$

Interstage values of $X_B^{(n)}$ are obtained similarly from

$$X_B^{(n)}/X_B^{(F)} = \frac{1}{(1 + E/N)^n} \qquad (5\text{-}22)$$

Thus, unlike the cocurrent cascade, the value of X_B decreases in each successive stage. For an infinite number of equilibrium stages, (5-21) becomes

$$X_B^{(\infty)}/X_B^{(F)} = 1/\exp(E) \qquad (5\text{-}23)$$

Thus, even for an infinite number of stages, $X_B^{(R)} = X_B^{(\infty)}$ cannot be reduced to zero.

Countercurrent Cascade

In the countercurrent arrangement for two stages in Figure 5.5c, the feed liquid passes through the cascade countercurrently to the solvent. For a two-stage system, the material balance and equilibrium equations for solute, B, for each stage are as follows.

Stage 1:

$$X_B^{(F)} F_A + Y_B^{(2)} S = X_B^{(1)} F_A + Y_B^{(1)} S \qquad (5\text{-}24)$$

$$K_{D_B}' = \frac{Y_B^{(1)}}{X_B^{(1)}} \qquad (5\text{-}25)$$

Stage 2:

$$X_B^{(1)} F_A = X_B^{(2)} F_A + Y_B^{(2)} S \qquad (5\text{-}26)$$

$$K_{D_B}' = \frac{Y_B^{(2)}}{X_B^{(2)}} \qquad (5\text{-}27)$$

Combining (5-24) to (5-27) with (5-14) to eliminate $Y_B^{(1)}$, $Y_B^{(2)}$, and $X_B^{(1)}$ gives

$$X_B^{(2)}/X_B^{(F)} = X_B^{(R)}/X_B^{(F)} = \frac{1}{1 + E + E^2} \qquad (5\text{-}28)$$

If the number of countercurrent stages is extended to N stages, the result is

$$X_B^{(N)}/X_B^{(F)} = X_B^{(R)}/X_B^{(F)} = 1 \bigg/ \sum_{n=0}^{N} E^n \qquad (5\text{-}29)$$

Interstage values of $X_B^{(n)}$ are obtained in a similar fashion, giving

$$X_B^{(n)}/X_B^{(F)} = \sum_{k=0}^{N-n} E^k \bigg/ \sum_{k=0}^{N} E^k \qquad (5\text{-}30)$$

As with the crosscurrent arrangement, the value of X_B decreases in each successive stage. The amount of decrease for

the countercurrent arrangement is greater than for the crosscurrent arrangement, and the difference increases exponentially with increasing extraction factor, E. Therefore, the countercurrent cascade is the most efficient of the three linear cascades.

For an infinite number of equilibrium stages, the limit of (5-28) gives two results:

$$X_B^{(\infty)}/X_B^{(F)} = 0, \qquad 1 \le E \le \infty \qquad (5\text{-}31)$$

$$X_B^{(\infty)}/X_B^{(F)} = (1 - E), \qquad E \le 1 \qquad (5\text{-}32)$$

Thus, complete extraction can be achieved in a countercurrent cascade, but only for an extraction factor, E, greater than 1.

EXAMPLE 5.2

Ethylene glycol can be catalytically dehydrated completely to *p*-dioxane (a cyclic diether) by the reaction $2\text{HOCH}_2\text{CH}_2\text{HO} \rightarrow \text{H}_2\text{CCH}_2\text{OCH}_2\text{CH}_2\text{O} + 2\text{H}_2\text{O}$. Water and *p*-dioxane have normal boiling points of 100°C and 101.1°C, respectively, and cannot be separated economically by distillation. However, liquid–liquid extraction at 25°C (298.15 K), using benzene as a solvent, is reasonably effective. Assume that 4,536 kg/h (10,000 lb/h) of a 25 wt% solution of *p*-dioxane in water is to be separated continuously by using 6,804 kg/h (15,000 lb/h) of pure benzene. Assuming that benzene and water are mutually insoluble, determine the effect of the number and arrangement of stages on the percent extraction of *p*-dioxane. The flowsheet is shown in Figure 5.6.

SOLUTION

Three different arrangements of stages will be examined: (a) cocurrent cascade, (b) crosscurrent cascade, and (c) countercurrent cascade. Because water and benzene are almost mutually insoluble, (5-13), (5-21), and (5-29) can be used, respectively, to estimate $X_B^{(R)}/X_B^{(F)}$, the fraction of *p*-dioxane not extracted, as a function of the number of stages. From the equilibrium data of Berdt and Lynch [1], the distribution coefficient for *p*-dioxane, $K_{D_B}' = Y_B/X_B$, where Y refers to the benzene phase and X refers to the water phase, varies from 1.0 to 1.4 over the concentration range of interest. For this example, assume a constant value of 1.2. From the given data, $S = 6,804$ kg/h of benzene, $F_A = 4,536(0.75) = 3,402$ kg/h of water, and $X_B^{(F)} = 0.25/0.75 = 1/3$. From (5-14),

$$E = 1.2(6,804)/3,402 = 2.4$$

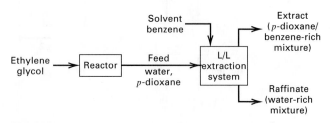

Figure 5.6 Flowsheet for Example 5.2.

Single equilibrium stage

All three arrangements give identical results for a single stage. From (5-13),

$$X_B^{(1)}/X_B^{(F)} = 1/(1 + 2.4) = 0.294$$

The corresponding fractional extraction is

$$1 - X_B^{(1)}/X_B^{(F)} = 1 - 0.294 = 0.706 \quad \text{or} \quad 70.6\%$$

More than one equilibrium stage

(a) Cocurrent cascade. For any number of stages, the percent extraction is the same as for one stage, 70.6%.

(b) Crosscurrent cascade. For any number of stages, (5-21) applies. For example, for two stages, assuming equal flow rates of solvent to each stage,

$$X_B^{(2)}/X_B^{(F)} = 1/(1 + E/2)^2 = 1/(1 + 2.4/2)^2 = 0.207$$

and the percent extraction is 79.3%. Results for other numbers of stages are obtained in the same manner.

(c) Countercurrent cascade. For any number of stages, (5-29) applies. For example, for two stages,

$$X_B^{(2)}/X_B^{(F)} = 1/(1 + E + E^2) = 1/(1 + 2.4 + 2.4^2) = 0.109$$

and the percent extraction is 89.1%. Results for other numbers of stages are obtained in the same manner.

A plot of percent extraction as a function of the number of equilibrium stages for up to five stages is shown in Figure 5.7 for each of the three arrangements. The probability-scale ordinate is convenient because for the countercurrent arrangement, with $E > 1$, 100% extraction is approached as the number of stages approaches infinity. For the crosscurrent arrangement, a maximum percent extraction of 90.9% is computed from (5-23). For five stages, Figure 5.7 shows that the countercurrent cascade has already achieved 99% extraction.

5.4 MULTICOMPONENT VAPOR–LIQUID CASCADES

Countercurrent cascades are used extensively for vapor–liquid separation operations, including absorption, stripping, and distillation. For absorption and stripping, a *single-section cascade* is used to recover one selected component from the feed. For distillation, a *two-section cascade* is effective in achieving a separation between two selected components referred to as the key components. For both cases, approximate calculation procedures relate compositions of multicomponent vapor and liquid streams entering and exiting the cascade to the number of equilibrium stages required. These approximate procedures are called *group methods* because they provide only an overall treatment of the group of stages in the cascade, without considering detailed changes in temperature, phase compositions, and flows from stage to stage.

Single-Section Cascades by Group Methods

Kremser [2] originated the group method by deriving an equation for the fractional absorption of a species from a gas into a liquid absorbent for a multistage countercurrent absorber. His method also applies to strippers. The treatment presented here is similar to that of Edmister [3] for general application to vapor–liquid separation operations. An alternative treatment is given by Smith and Brinkley [4].

Consider first the countercurrent cascade of N adiabatic, equilibrium stages used, as shown in Figure 5.8a, to absorb species present in the entering vapor. Assume that these species are absent in the entering liquid. Stages are numbered from top to bottom. It is convenient to express stream compositions in terms of component molar flow rates, v_i and l_i, in the vapor and liquid phases, respectively. However, in the following derivation, the subscript i is dropped. A material balance around the top of the absorber, including stages 1 through $N - 1$, for any absorbed species gives

$$v_N = v_1 + l_{N-1} \tag{5-33}$$

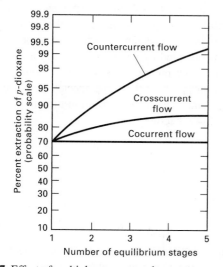

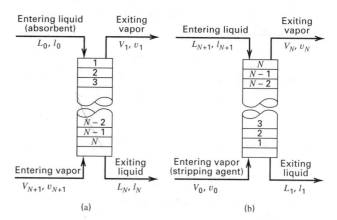

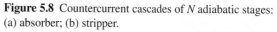

Figure 5.7 Effect of multiple-stage cascade arrangement on extraction efficiency.

Figure 5.8 Countercurrent cascades of N adiabatic stages: (a) absorber; (b) stripper.

where

$$v = yV \tag{5-34}$$

$$l = xL \tag{5-35}$$

and $l_0 = 0$. From equilibrium considerations for stage N, the definition of the vapor–liquid equilibrium ratio or K-value can be employed to give

$$y_N = K_N x_N \tag{5-36}$$

Combining (5-34), (5-35), and (5-36), v_N becomes

$$v_N = \frac{l_N}{L_N/(K_N V_N)} \tag{5-37}$$

An absorption factor A, analogous to the extraction factor, E, for a given stage and component is defined by

$$A = \frac{L}{KV} \tag{5-38}$$

Combining (5-37) and (5-38),

$$v_N = \frac{l_N}{A_N} \tag{5-39}$$

Substituting (5-39) into (5-33),

$$l_N = (l_{N-1} + v_1)A_N \tag{5-40}$$

The internal flow rate, l_{N-1}, is eliminated by successive substitution using material balances around successively smaller sections of the top of the cascade. For stages 1 through $N - 2$,

$$l_{N-1} = (l_{N-2} + v_1)A_{N-1} \tag{5-41}$$

Substituting (5-41) into (5-40),

$$l_N = l_{N-2}A_{N-1}A_N + v_1(A_N + A_{N-1}A_N) \tag{5-42}$$

Continuing this process to the top stage, where $l_1 = v_1 A_1$, ultimately converts (5-42) into

$$l_N = v_1(A_1 A_2 A_3 \ldots A_N + A_2 A_3 \ldots A_N \\ + A_3 \ldots A_N + \cdots + A_N) \tag{5-43}$$

A more useful form is obtained by combining (5-43) with the overall component balance

$$l_N = v_{N+1} - v_1 \tag{5-44}$$

to give an equation for the exiting vapor in terms of the entering vapor and a recovery fraction:

$$v_1 = v_{N+1}\phi_A \tag{5-45}$$

where, by definition, the recovery fraction is

$$\phi_A = \frac{1}{A_1 A_2 A_3 \ldots A_N + A_2 A_3 \ldots A_N + A_3 \ldots A_N + \cdots + A_N + 1}$$
$$= \text{fraction of species in entering vapor that is not} \tag{5-46}$$
absorbed

In the group method, an average effective absorption factor, A_e, replaces the separate absorption factors for each

stage. Equation (5-46) now becomes

$$\phi_A = \frac{1}{A_e^N + A_e^{N-1} + A_e^{N-2} + \cdots + A_e + 1} \tag{5-47}$$

When multiplied and divided by $(A_e - 1)$, (5-47) reduces to

$$\phi_A = \frac{A_e - 1}{A_e^{N+1} - 1} \tag{5-48}$$

Note that each component has a different A_e and, therefore, a different value of ϕ_A. Figure 5.9 from Edmister [3] is a plot of (5-48) with a probability scale for ϕ_A, a logarithmic scale for A_e, and N as a parameter. This plot, in linear coordinates, was first developed by Kremser [2].

Consider next the countercurrent stripper shown in Figure 5.8b. Assume that the components stripped from the liquid are absent in the entering vapor, and ignore condensation or absorption of the stripping agent. In this case, stages are numbered from bottom to top to facilitate the derivation. The pertinent stripping equations follow in a manner analogous to the absorber equations. The results are

$$l_1 = l_{N+1}\phi_S \tag{5-49}$$

where

$$\phi_S = \frac{S_e - 1}{S_e^{N+1} - 1} = \begin{array}{l} \text{fraction of species in entering} \\ \text{liquid that is not stripped} \end{array} \tag{5-50}$$

$$S = \frac{KV}{L} = \frac{1}{A} = \text{stripping factor} \tag{5-51}$$

Figure 5.9 also applies to (5-50). As shown in Figure 5.10, absorbers are frequently coupled with strippers or distillation columns to permit regeneration and recycle of absorbent. Since stripping action is not perfect, recycled absorbent entering the absorber contains species present in the vapor entering the absorber. Vapor passing up through the absorber can strip these as well as the absorbed species introduced in the makeup absorbent. A general absorber equation is obtained by combining (5-45) for absorption of species from the entering vapor with a modified form of (5-49) for stripping of the same species from the entering liquid. For stages numbered from top to bottom, as in Figure 5.8a, (5-49) becomes

$$l_N = l_0\phi_S \tag{5-52}$$

or, since

$$l_0 = v_1 + l_N \\ v_1 = l_0(1 - \phi_S) \tag{5-53}$$

The total balance in the absorber for a component appearing in both entering vapor and entering liquid is obtained by adding (5-45) and (5-53) to give

$$v_1 = v_{N+1}\phi_A + l_0(1 - \phi_S) \tag{5-54}$$

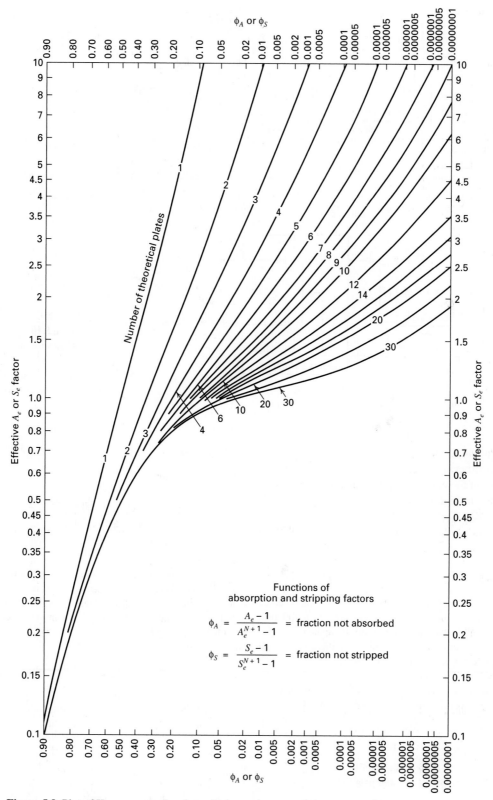

ϕ_A or ϕ_S

Effective A_e or S_e factor

Number of theoretical plates

Functions of
absorption and stripping factors

$$\phi_A = \frac{A_e - 1}{A_e^{N+1} - 1} = \text{fraction not absorbed}$$

$$\phi_S = \frac{S_e - 1}{S_e^{N+1} - 1} = \text{fraction not stripped}$$

ϕ_A or ϕ_S

Figure 5.9 Plot of Kremser equation for a single-section countercurrent cascade.
[From W. C. Edmister, *AIChE J.*, **3**, 165–171 (1957).]

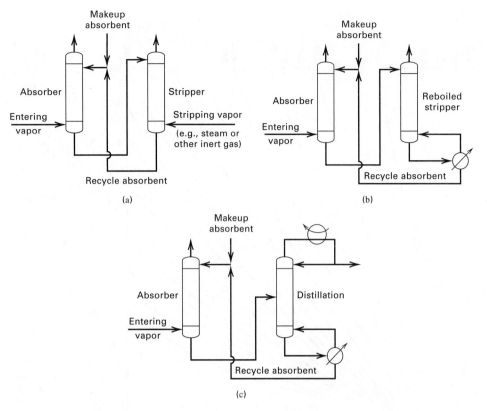

Figure 5.10 Various coupling schemes for absorbent recovery: (a) use of steam or inert gas stripper; (b) use of reboiled stripper; (c) use of distillation.

which is generally applied to each component in the vapor entering the absorber. Equation (5-52) is used for species that appear only in the entering liquid. The analogous equation for a stripper in Figure 5.8b is

$$l_1 = l_{N+1}\phi_S + v_0(1 - \phi_A) \qquad (5\text{-}55)$$

EXAMPLE 5.3

In Figure 5.11, the heavier components in a slightly superheated hydrocarbon gas are to be removed by absorption at 400 psia (2,760 kPa) with a high-molecular-weight oil. Estimate exit vapor and exit liquid flow rates and compositions by the approximate group method of Kremser. Assume that effective absorption and stripping factors for each component can be estimated from the entering values of L, V, and the component K-values, as listed below based on an average entering temperature of $(90 + 105)/2 = 97.5°F$.

SOLUTION

From (5-38) and (5-51),

$$A_i = L/K_i V = 165/[K_i(800)] = 0.206/K_i$$
$$S_i = 1/A_i = 4.85K_i$$
$$N = 6 \text{ stages}$$

Values of ϕ_A and ϕ_S are obtained from (5-48) and (5-50) or Figure 5.9. Values of $(v_i)_1$, the component flow rates in the exit vapor, are computed from (5-54). Values of $(l_i)_6$, the component flow rates in the exit liquid, are computed from an overall component material balance using Figure 5.8a:

$$(l_i)_6 = (l_i)_0 + (v_i)_7 - (v_i)_1 \qquad (1)$$

The computations, which are best made in tabular fashion with a spreadsheet computer program, give the following results:

Component	$K@97.5°F$, 400 psia	A	S	ϕ_A	ϕ_S	v_1	l_6
C_1	6.65	0.0310	—	0.969	—	155.0	5.0
C_2	1.64	0.126	—	0.874	—	323.5	46.5
C_3	0.584	0.353	—	0.647	—	155.4	84.6
nC_4	0.195	1.06	0.946	0.119	0.168	3.02	22.03
nC_5	0.0713	2.89	0.346	0.00112	0.654	0.28	5.5
Oil	0.0001	—	0.0005	—	0.9995	0.075	164.095
						637.275	327.725

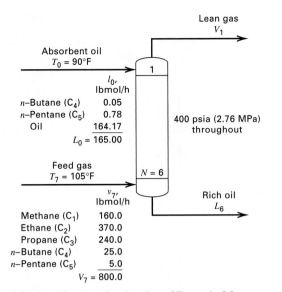

Figure 5.11 Specifications for absorber of Example 5.3.

The above results indicate that approximately 20% of the gas is absorbed. Less than 0.1% of the absorbent oil is stripped.

Two-Section Cascades

A single-stage flash distillation produces a vapor that is somewhat richer in the lower-boiling constituents than the feed. Further enrichment can be achieved by a series of flash distillations in which the vapor from each stage is condensed, then reflashed. In principle, any desired product purity can be obtained by a multistage flash technique, provided a suitable volatility difference exists and a suitable number of stages is employed. In practice, however, the recovery of product is small, heating and cooling requirements are high, and relatively large quantities of various liquid products are produced.

As an example, consider Figure 5.12a, where n-hexane (H) is separated from n-octane by a series of three flashes at 1 atm (pressure drop and pump needs are ignored). The feed to the first flash stage is an equimolar bubble-point liquid at a flow rate of 100 lbmol/h. A bubble-point temperature calculation yields 192.3°F. If the vapor rate leaving stage 1 is set equal to the amount of n-hexane in the feed to stage 1, the calculated equilibrium exit phases are as shown. The vapor V_1 is enriched to a hexane mole fraction of 0.690. The heating requirement is 751,000 Btu/h. Equilibrium vapor from stage 1 is condensed to bubble-point liquid with a cooling duty of 734,000 Btu/h. Repeated flash calculations for stages 2 and 3 give the results shown. For each stage, the leaving molar vapor rate is set equal to the moles of hexane in the feed to the stage. The purity of n-hexane is increased from 50 mol% in the feed to 86.6 mol% in the final condensed vapor product, but the recovery of hexane is only 27.7(0.866)/50 or 48%. Total heating requirement is 1,614,000 Btu/h and liquid products total 72.3 lbmol/h.

In comparing feed and liquid products from two contiguous stages, we note that liquid from the later stage and the feed to the earlier stage are both leaner in hexane, the more volatile species, than the feed to the later stage. Thus, if intermediate streams are recycled, intermediate recovery of hexane is improved. This processing scheme is depicted in Figure 5.12b, where again the molar fraction vaporized in each stage equals the mole fraction of hexane in the combined feeds to the stage. The mole fraction of hexane in the final condensed vapor product is 0.853, just slightly less than that achieved by successive flashes without recycle. However, the use of recycle increases recovery of hexane from 48% to 61.6%. As shown in Figure 5.12b, increased recovery of hexane is accompanied by approximately 28% increased heating and cooling requirements. If the same degree of heating and cooling is used for the no-recycle scheme in Figure 5.12a as in Figure 5.12b, the final hexane

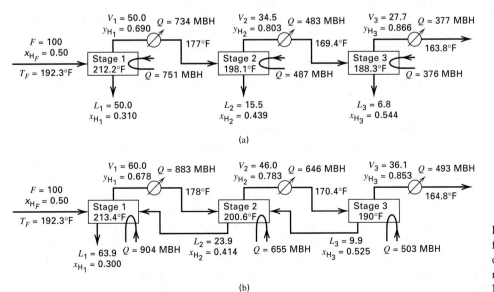

Figure 5.12 Successive flashes for recovering hexane from octane: (a) no recycle; (b) with recycle. Flow rates in lbmol/h. MBH = 1,000 Btu/h.

mole fraction is reduced from 0.866 to 0.815, but hexane recovery is increased to 36.1(0.815)/50 or 58.8%.

Both of the successive flash arrangements in Figure 5.12 involve a considerable number of heat exchangers and pumps. Except for stage 1, the heaters in Figure 5.12a can be eliminated if the two intermediate total condensers are converted to partial condensers with duties of $734 - 487 = 247$ MBH (MBH = 1,000 Btu/h) and $483 - 376 = 107$ MBH. Total heating duty is now only 751,000 Btu/h, and total cooling duty is 731,000 Btu/h. Similarly, if heaters for stages 2 and 3 in Figure 5.12b are removed by converting the two total condensers to partial condensers, total heating duty is 904,000 Btu/h (20% greater than the no-recycle case), and cooling duty is 864,000 Btu/h (18% greater than the no-recycle case).

A considerable simplification of the successive flash technique with recycle is shown in Figure 5.13a. The total heating duty is provided by a feed boiler ahead of stage 1. The total cooling duty is utilized at the opposite end to condense totally the vapor leaving stage 3. Condensate in excess of distillate is returned as reflux to the top stage, from which it passes successively from stage to stage countercurrently to vapor flow. Vertically arranged adiabatic stages eliminate the need for interstage pumps, and all stages are contained within a single piece of less expensive equipment. The set of stages is called a *rectifying section*. As discussed in Chapter 2, such an arrangement is thermodynamically inefficient, however, because heat is added at the highest temperature level and removed at the lowest temperature level.

The number of degrees of freedom for the arrangement in Figure 5.13a is determined by the method of Chapter 4 to be $(C + 2N + 10)$. If all independent feed conditions, number of stages (3), and all stage pressures (1 atm), bubble-point liquid leaving the condenser, and adiabatic stages are specified, two degrees of freedom remain. These are specified to be a heating duty for the boiler and a distillate rate equal to that of Figure 5.12b. Calculations result in a mole fraction of 0.872 for hexane in the distillate. This is somewhat greater than that shown in Figure 5.12b.

The same principles by which we have concluded that the adiabatic, multistage, countercurrent-flow arrangement is advantageous for concentrating a light component in an overhead product can be applied to the concentration of a heavy component in a bottoms product, as in Figure 5.13b. Such a set of stages is called a *stripping section*.

Figure 5.13c, a combination of Figures 5.13a and 5.13b with a liquid feed, is a complete column for rectifying and stripping a feed to effect a sharper separation between a selected more volatile component, called the *light key,* and a less volatile component, called the *heavy key component,* than is possible with either a stripping or an enriching section alone. Adiabatic flash stages are placed above and below the feed. Recycled liquid reflux, L_R, is produced in the condenser and vapor boilup, V_1, in the reboiler. The reflux ratios are L_R/V_N and L_2/V_1 at the top and bottom of the apparatus, respectively. All interstage flows are countercurrent. Two-section cascades are widely used in industry for multistage distillation.

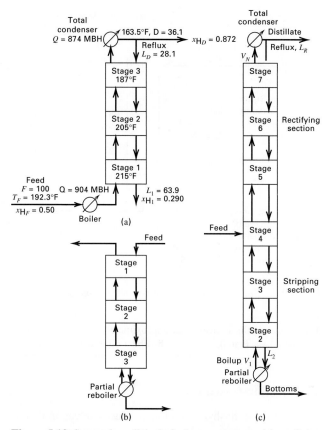

Figure 5.13 Successive adiabatic flash arrangements: (a) rectifying section; (b) stripping section; (c) multistage distillation.

The rectifying stages above the point of feed introduction purify the light product by contacting upward flowing vapor with successively richer liquid reflux. Stripping stages below the feed increase light-product recovery because vapor relatively low in volatile constituents strips light components out of the liquid. For the heavy product, the functions are reversed: The stripping section increases purity; the enriching section increases recovery.

Edmister [3] applied the Kremser group method for absorbers and strippers to distillation where two cascades are coupled to a condenser, a reboiler, and a feed stage. In Figure 5.14, five separation zones are shown: (1) partial condenser, C; (2) absorption or rectifying cascade (enriching section), E; (3) feed-flash stage, F; (4) stripping cascade (exhausting section), X; and (5) partial reboiler, B.

In Figure 5.14, N stages for the enricher are numbered from the top down and the overhead product is distillate; whereas for the exhauster, M stages are numbered from the bottom up. Component feeds to the enricher section are vapor, v_F, from the feed stage and liquid, l_C, from the condenser. Component feeds to the exhauster are liquid, l_F, from the feed stage and vapor, v_B, from the reboiler. Component flows leaving the enricher cascade are vapor, v_{TE}, from the top stage, 1, and liquid, l_{BE}, from the bottom stage, N. Component flows leaving the exhauster cascade are vapor, v_{TX}, from the top stage, M, and liquid, l_{BX}, from the bottom stage, 1. The recovery equations for the enricher are obtained from (5-54)

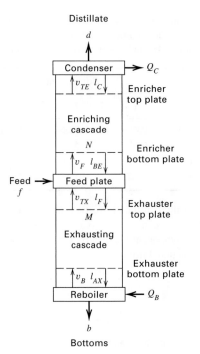

Distillate
d

Figure 5.14 Countercurrent distillation cascade.

by making the following substitutions, which are obtained from material balance and equilibrium considerations. For each component in the feed,

$$v_{TE} = l_C + d \tag{5-56}$$
$$v_F = l_{BE} + d \tag{5-57}$$

and

$$l_C = d A_C \tag{5-58}$$

where

$$A_C = \frac{L_C}{DK_C} \quad \text{(for a partial condenser)} \tag{5-59}$$

$$A_C = \frac{L_C}{D} \quad \text{(for a total condenser)} \tag{5-60}$$

The resulting enricher recovery equations for each species are

$$\frac{l_{BE}}{d} = \frac{A_C \phi_{SE} + 1}{\phi_{AE}} - 1 \tag{5-61}$$

or

$$\frac{v_F}{d} = \frac{A_C \phi_{SE} + 1}{\phi_{AE}} \tag{5-62}$$

where the additional subscript E on ϕ refers to the enricher.

The recovery equations for the exhauster are obtained in a similar manner, as

$$\frac{v_{TX}}{b} = \frac{S_B \phi_{AX} + 1}{\phi_{SX}} - 1 \tag{5-63}$$

$$\frac{l_F}{b} = \frac{S_B \phi_{AX} + 1}{\phi_{SX}} \tag{5-64}$$

where

$$S_B = K_B V_B / B$$

for a partial reboiler, and additional subscript X on ϕ denotes an exhauster.

For either an enricher or exhauster, ϕ_A and ϕ_E are given, from above, by (5-48) and (5-50), respectively, or from Figure 5.9.

To couple the enriching and exhausting cascades, a feed stage is employed for which the absorption factor is related to the streams leaving the feed stage by

$$A_F = \frac{L_F}{K_F V_F} = \frac{L_F}{(v_F L_F / V_F l_F) V_F} = \frac{l_F}{v_F} \tag{5-65}$$

For the distillation column of Figure 5.14, (5-62), (5-64), and (5-65) are combined to eliminate l_F and v_F. The result is

$$\frac{b}{d} = A_F \frac{[(A_C \phi_{SE} + 1)/\phi_{AE}]}{[(S_B \phi_{AX} + 1)/\phi_{SX}]} \tag{5-66}$$

To apply (5-66) for the calculation of component split ratios, b/d, it is necessary to establish values of absorption factors A_F and A_C, and the stripping factor S_B. Average values for factors A_E, A_X, S_E, and S_X for each component are also required for the two cascades to determine the corresponding ϕ values. To establish these values, it is necessary to estimate temperatures and molar vapor and liquid, V and L, flow rates. An approximate method for making these estimates is given in the following example.

EXAMPLE 5.4

The hydrocarbon gas of Example 5.3 is distilled at 400 psia (2.76 MPa), to separate ethane from propane, for the conditions shown in Figure 5.15. Estimate the distillate and bottoms compositions using (5-66). This example is best solved by using a spreadsheet computer program.

SOLUTION

Assume a feed stage temperature equal to the feed temperature, 105°F. To estimate the condenser and reboiler temperatures, assume a perfect split for the specified distillate rate of 530 lbmol/h, with all methane and ethane going to the distillate and all propane and heavier going to the bottoms. Thus the preliminary material balance is

Component	lbmol/h		
	Feed, f	Assumed Distillate, d	Assumed Bottoms, b
C_1	160	160	0
C_2	370	370	0
C_3	240	0	240
nC_4	25	0	25
nC_5	5	0	5
	800	530	270

For these assumed products, applying procedures in Section 4.4, a distillate temperature of 12°F is obtained from a dew-point calculation and a bottoms temperature of 165°F from a bubble-point calculation. Average temperatures of $(12 + 105)/2 = 59°F$ and $(105 + 165)/2 = 135°F$ are estimated for the enriching and exhausting cascades, respectively. Assuming that total molar flow rates are constant in each cascade, the following vapor and liquid flow-rate estimates are obtained by working down from the top of

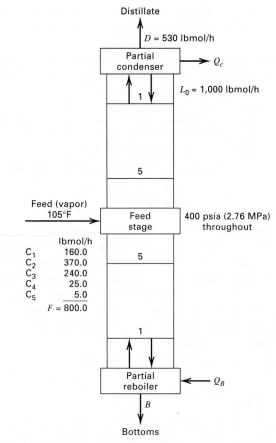

Figure 5.15 Specifications for fractionator of Example 5.4.

the column, where the liquid reflux is specified:

Stage or Section	Average Temperature, °F	Average Flow Rates, lbmol/h	
		Vapor	Liquid
Condenser	12	530	1,000
Enricher	59	1,530	1,000
Feed	105	1,530	1,000
Exhauster	135	730	1,000
Reboiler	165	730	270

From the column pressure and the estimated temperature values, K-values are read from Figure 2.8. These values are then used to estimate absorption and stripping factors for the five sections, with the following results:

	Component				
	C_1	C_2	C_3	nC_4	nC_5
A_C	0.356	2.45	10.48	43.9	172
A_E	0.131	0.511	1.72	5.78	18.7
$S_E = 1/A_E$	7.65	1.96	0.581	0.173	0.0536
A_F	0.111	0.363	1.108	3.19	9.08
A_X	0.205	0.652	1.73	4.72	12.5
$S_X = 1/A_X$	4.89	2.26	0.577	0.212	0.080
S_B	19.5	7.03	2.78	1.08	0.433

From the values of A_E, A_X, S_E, and S_X, and the numbers of theoretical stages specified in Figure 5.15, the following values of ϕ are computed from (5-48) and (5-50) or read from Figure 5.9:

	Component				
	C_1	C_2	C_3	nC_4	nC_5
ϕ_{AE}	0.869	0.498	0.0289	0.0001	0.000
ϕ_{AX}	0.796	0.377	0.028	0.00034	0.000
ϕ_{SE}	0.0000	0.0173	0.435	0.827	0.946
ϕ_{SX}	0.0003	0.0444	0.439	0.788	0.920

From the values in the above two tables, values of (b/d) are computed for each component from (5-66). Since an overall balance for each component is given by $f = d + b$, values of d and b can then be computed from

$$d = \frac{f}{1 + (b/d)} \qquad (1)$$

$$b = f - d \qquad (2)$$

The following results are obtained:

		lbmol/h	
	(b/d)	d	b
C_1	0.000002	160	0
C_2	0.00924	366.6	3.4
C_3	86.8	2.7	237.3
nC_4	937,000	0	25
nC_5	Very large	0	5
Totals		529.3	270.7

Total distillate rate is somewhat less than the 530.0 lbmol/h specified. Values of d_i and b_i can be corrected to force the total to 530.0 by the method of Lyster et al. [5], which involves finding the positive root of θ in the relation

$$D = \sum_i \frac{f_i}{1 + \theta(b_i/d_i)}$$

followed by recalculation of d_i from

$$d_i = \frac{f_i}{1 + \theta(b_i/d_i)}$$

and b_i from $f_i - d_i$. The resulting value of θ is 0.8973, which gives $d_{C_2} = 367$, $b_{C_2} = 3$, $d_{C_3} = 3$, and $b_{C_3} = 237$, with no changes for other components.

The separation achieved by distillation is considerably improved over the separation achieved by absorption in Example 5.3. Although overhead vapor flow rates are approximately the same (530 lbmol/h) in this example and in Example 5.3, a reasonably sharp split between ethane and propane occurs for distillation because of the two-section cascade, while the absorber, with only a one-section cascade, allows appreciable quantities of both ethane and propane to exit in the overhead vapor and bottoms liquid. Even if the absorbent rate in Example 5.3 is doubled so that the recovery of propane in the bottoms exit liquid approaches 100%, more than 50% of the ethane also appears in the bottoms.

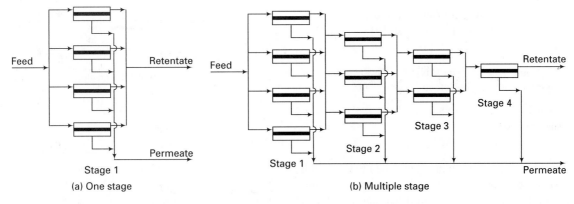

Figure 5.16 Parallel units of membrane separators.

5.5 MEMBRANE CASCADES

Membrane separation systems frequently consist of multiple-membrane units or modules. One reason for this is that a single module of the maximum size available may not be large enough to handle the required feed rate. In that case, it is necessary to use a number of modules of identical size in parallel as shown in Figure 5.16a, with retentates and permeates from each module combined, respectively, to obtain the final retentate and final permeate. For example, a membrane-separation system for separating hydrogen from methane might require a membrane area of 9,800 ft². If the largest membrane module available has 3,300 ft² of membrane surface, three modules in parallel are required. The parallel units in Figure 5.16a constitute a single stage of membrane separation. If, in addition, a large fraction of the feed is to become permeate, it may be necessary to carry out the membrane separation in two or more stages, as shown in Figure 5.16b for four stages, with the number of modules reduced for each successive stage as the flow rate on the feed-retentate side of the membrane decreases. The combined retentate from each stage becomes the feed for the next stage. The combined permeates for each stage differ in composition. They can be further combined to give an overall permeate, as shown in Figure 5.16b, or not, to give two or more permeate products of different composition.

A second reason for using multiple-membrane modules is that a single-membrane stage is often limited in the degree of separation achievable. In some cases, a high purity can be obtained, but only at the expense of a low recovery. In other cases, neither a high purity nor a high recovery can be obtained. The following table gives two examples of the degree of separation achieved for a single stage of gas permeation using a commercially available membrane.

Feed Molar Composition	More Permeable Component	Product Molar Composition	Percent Recovery
85% H₂ 15% CH₄	H₂	99% H₂ 1% N₂ in the permeate	60% of H₂ in the feed
80% CH₄ 20% N₂	N₂	97% CH₄ 3% N₂ in the retentate	57% of CH₄ in the feed

In the first example, the component of highest percentage in the feed is the most permeable component. The permeate purity is quite high, but the recovery is not. In the second example, the component of highest purity in the feed is not the most permeable component. The purity of the retentate is reasonably high, but, again, the recovery is not. To further increase the purity of one product and the recovery of the main component in that product, membrane stages are cascaded with recycle. Consider the separation of air to produce a high-purity nitrogen retentate and an oxygen-enriched permeate. Shown in Figure 5.17 are three membrane-separation systems, studied by Prasad et al. [6] for the production of high-purity nitrogen from air, using a membrane material that is more permeable to oxygen. The first system is just a single stage. The second system is a cascade of two stages, with recycle of permeate from the second stage to the first stage. The third system is a cascade of three stages with permeate recycles from stage 3 to stage 2 and stage 2 to stage 1. The two cascades are similar to the single-section, countercurrent stripping cascade shown in Figure 5.8b, with the membrane feed, permeate, and retentate corresponding, respectively, to the stripper entering liquid, exiting vapor, and exiting liquid. However, the membrane cascades do not include a stream corresponding to the stripper entering

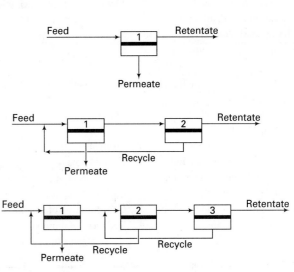

Figure 5.17 Membrane cascades.

vapor. Not shown in Figure 5.17 are recycle gas compressors. Typical calculations of Prasad et al. [6] give the following results:

Membrane System	Mol% N_2 in Retentate	% Recovery of N_2
Single stage	98	45
Two-stage cascade	99.5	48
Three-stage cascade	99.9	50

These results show that a high purity can be obtained with a single-section membrane cascade, but without major improvement in the recovery. To obtain both high purity and high recovery, a two-section membrane cascade is necessary, as discussed in Section 14.3.

5.6 HYBRID SYSTEMS

To reduce costs, particularly energy cost, make possible a difficult separation, and/or improve the degree of separation, *hybrid systems*, consisting of two or more separation operations of different types in series are used. Although combinations of membrane separators with other separation operations are the most common, other combinations have found favor. Table 5.1 is a partial list of hybrid systems that are used commercially or have received considerable attention. Examples of applications are included for some hybrid systems. Not included in Table 5.1 are hybrid systems consisting of distillation combined with extractive distillation, azeotropic distillation, and/or liquid–liquid extraction, which are very common and are considered in detail in Chapter 11.

The first example in Table 5.1 is a hybrid system that combines pressure–swing adsorption (PSA), to preferentially

Table 5.1 Hybrid Systems

Hybrid System	Separation Example
Adsorption—gas permeation	Nitrogen—Methane
Simulated moving bed adsorption—distillation	Metaxylene-paraxylene with ethylbenzene eluent
Chromatography—crystallization	—
Crystallization—distillation	—
Crystallization—pervaporation	—
Crystallization—liquid–liquid extraction	Sodium carbonate—water
Distillation—adsorption	Ethanol—water
Distillation—crystallization	—
Distillation—gas permeation	Propylene—propane
Distillation—pervaporation	Ethanol—water
Gas permeation—absorption	Dehydration of natural gas
Reverse osmosis—distillation	Carboxylic acids—water
Reverse osmosis—evaporation	Concentration of wastewater
Stripper—gas permeation	Recovery of ammonia and hydrogen sulfide from sour water

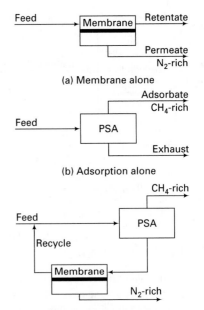

Figure 5.18 Separation of methane from nitrogen.

remove methane, with a gas-permeation membrane operation to preferentially remove nitrogen. The permeate is recycled to the adsorption step. Figure 5.18 shows this hybrid system compared to the use of just a single-stage gas-permeation membrane operation and a single-stage pressure-swing adsorption operation. Only the hybrid system is capable of making a relatively sharp separation between methane and nitrogen. Typical products obtained from these three processes are compared in Table 5.2 for 100,000 scfh of a feed containing 80 mol% methane and 20 mol% nitrogen. For all three processes, the methane-rich product contains 97 mol% methane. However, only the hybrid system gives a nitrogen-rich product containing a nitrogen composition greater than 90 mol%, and a high recovery of methane (98%). The methane recovery for a membrane alone is

Table 5.2 Typical Products for Processes in Figure 5.18

	Flow Rate, Mscfh	Mol% CH_4	Mol% N_2
Feed gas	100	80	20
Membrane only:			
Retentate	47.1	97	3
Permeate	52.9	65	35
PSA only:			
Adsorbate	70.6	97	3
Exhaust	29.4	39	61
Hybrid system:			
CH_4-rich	81.0	97	3
N_2-rich	19.0	8	92

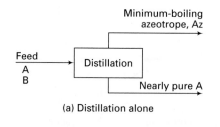

(a) Distillation alone

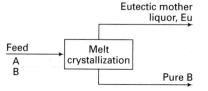

(b) Melt crystallization alone

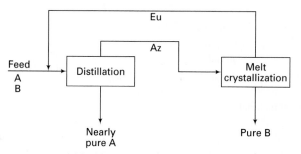

(c) Distillation–crystallization hybrid

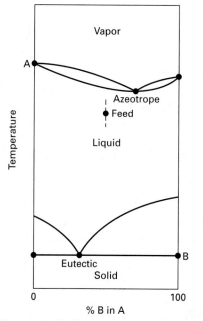

(d) Phase diagram for distillation–crystallization hybrid system.

Figure 5.19 Separation of an azeotropic- and eutectic-forming mixture.

only 57%, while the adsorber alone gives 86%. The hybrid system is clearly superior to a single membrane or adsorber.

No application is shown in Table 5.1 for hybrid systems of crystallization and distillation. However, there is much interest because Berry and Ng [7] show such systems can

overcome the limitations of eutectics in crystallization and azeotropes in distillation. Furthermore, although streams containing solids are more difficult to process than fluids, crystallization requires just a single stage to obtain high-purity crystals. Figure 5.19 includes one of the many distillation and crystallization hybrid configurations discussed by Berry and Ng [7]. The feed is a mixture of A and B, which, as shown in the accompanying phase diagram, form both an azeotrope in the vapor–liquid region and a eutectic in the liquid–solid region at a lower temperature. With respect to component B, the feed composition in Figure 5.19 lies between the eutectic and azeotropic compositions. If distillation alone is used with a sufficient number of stages, the distillate composition will approach that of the minimum-boiling azeotrope, Az, and the bottoms will approach pure A. If melt crystallization alone is used, the two products will be crystals of pure B and a mother liquor approaching the eutectic composition, Eu. The hybrid system in Figure 5.19 combines distillation with melt crystallization to produce both pure B and nearly pure A. The feed enters the distillation column, where the distillate of near-azeotropic composition is sent to the melt crystallizer. Here, the mother liquor of near-eutectic composition is recovered and recycled to the distillation column. The net result is a separation, with nearly pure A obtained as bottoms from the distillation column and pure B obtained from the crystallizer.

Another hybrid system receiving considerable attention is the combination of distillation and pervaporation for separation of azeotropic mixtures, particularly ethanol–water. As discussed in Section 14.7, distillation produces a bottoms of nearly pure water and a distillate of the azeotrope, which is sent to the pervaporation step, producing a nearly pure ethanol retentate and a water-rich permeate that is recycled to the distillation step.

5.7   DEGREES OF FREEDOM AND SPECIFICATIONS FOR COUNTERCURRENT CASCADES

The solution to a multicomponent, multiphase, multistage separation problem is found in the simultaneous solution of the material balance, energy balance, and phase equilibria equations. This implies that a sufficient number of design variables is specified so that the number of remaining unknown (output) variables exactly equals the number of independent equations. In this section, the degrees-of-freedom analysis discussed in Section 4.1 for a single equilibrium stage is extended to one- and multiple-section countercurrent cascades.

An intuitively simple, but operationally complex, method of finding N_D, the number of independent design variables, *degrees of freedom*, or *variance* in the process, is to enumerate all pertinent variables, N_V, and to subtract from these the total number of independent equations or relationships, N_E, relating the variables:

$$N_D = N_V - N_E \qquad (5\text{-}67)$$

This approach to separation process design was developed by Kwauk [8], and a modification of his methodology forms the basis for this discussion.

Typically, the variables in a separation process are intensive variables such as composition, temperature, and pressure; extensive variables such as flow rate or the heat-transfer rate; and equipment parameters such as the number of equilibrium stages. Physical properties such as enthalpy or K-values are not counted because they are functions of the intensive variables. The variables are relatively easy to enumerate, but to achieve an unambiguous count of N_E it is necessary to carefully seek out all independent relationships due to material and energy conservations, phase-equilibria restrictions, process specifications, and equipment configurations.

Separation equipment consists of physically identifiable elements (equilibrium stages, condensers, reboilers, etc.) as well as stream dividers and stream mixers. It is helpful to examine each element separately, before synthesizing the complete system.

Stream Variables

For each single-phase stream containing C components, a complete specification of intensive variables consists of C mole fractions (or other concentration variables) plus temperature and pressure, or $C + 2$ variables. However, only $C - 1$ of the feed mole fractions are independent, because the other mole fraction must satisfy the mole-fraction constraint:

$$\sum_{i=1}^{c} \text{mole fractions} = 1.0$$

Thus, only $C + 1$ intensive stream variables can be specified. This is in agreement with the phase rule, which states that, for a single-phase system, the intensive variables are specified by $C - \mathcal{P} + 2 = C + 1$ variables. To this number can be added the total flow rate of the stream, an extensive variable. Although the missing mole fraction is often treated implicitly, it is preferable for completeness to include the missing mole fraction in the list of stream variables and then to include in the list of equations the above mole-fraction constraint. Thus, associated with each stream are $C + 3$ variables. For example, for a liquid-phase stream, the variables are liquid mole fractions $x_1, x_2, \ldots, x_C$; total molar flow rate L; temperature T; and pressure P.

Adiabatic or Nonadiabatic Equilibrium Stage

For a single adiabatic or nonadiabatic equilibrium stage with two entering streams and two exit streams, as shown in Figure 5.20, the variables are those associated with the four streams plus the heat transfer rate to or from the stage. Thus:

$$N_V = 4(C + 3) + 1 = 4C + 13$$

The exiting streams V_{OUT} and L_{OUT} are in equilibrium, so there are equilibrium restrictions as well as component material balances, a total material balance, an energy balance, and mole fraction constraints. Thus, the equations relating

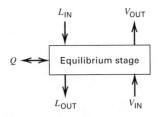

Figure 5.20 Equilibrium stage with heat addition.

these variables and N_E are

Equations	Number of Equations
Pressure equality $\quad P_{V_{OUT}} = P_{L_{OUT}}$	1
Temperature equality, $\quad T_{V_{OUT}} = T_{L_{OUT}}$	1
Phase equilibrium relationships, $\quad (y_i)_{V_{OUT}} = K_i(x_i)_{L_{OUT}}$	C
Component material balances, $\quad L_{IN}(x_i)_{L_{IN}} + V_{IN}(y_i)_{V_{IN}} = L_{OUT}(x_i)_{L_{OUT}} + V_{OUT}(y_i)_{V_{OUT}}$	$C - 1$
Total material balance, $\quad L_{IN} + V_{IN} = L_{OUT} + V_{OUT}$	1
Energy balance, $\quad Q + h_{L_{IN}} L_{IN} = h_{V_{IN}} V_{IN} = h_{L_{OUT}} L_{OUT} + h_{V_{OUT}} V_{OUT}$	1
Mole fraction constraints in entering and exiting streams $\quad$ e.g., $\sum_{i=1}^{C}(x_i)_{L_{IN}} = 1$	4
	$N_E = 2C + 7$

Alternatively, C, instead of $C - 1$, component material balances can be written. The total material balance is then a dependent equation obtained by summing the component material balances and applying the mole-fraction constraints to eliminate the mole fractions. From (5-67),

$$N_D = (4C + 13) - (2C + 7) = 2C + 6$$

Notice that the coefficient of C is equal to 2, the number of streams entering the stage.

Several different sets of design variables can be specified. A typical set includes complete specification of the two entering streams as well as the stage pressure and heat transfer rate.

Variable Specification	Number of Variables
Component mole fractions, $(x_i)_{L_{IN}}$	$C - 1$
Total flow rate, L_{IN}	1
Component mole fractions, $(y_i)_{V_{IN}}$	$C - 1$
Total flow rate, V_{IN}	1
Temperature and pressure of L_{IN}	2
Temperature and pressure of V_{IN}	2
Stage pressure, $(P_{V_{OUT}}$ or $P_{L_{OUT}})$	1
Heat transfer rate, Q	1
	$N_D = 2C + 6$

Specification of these $(2C + 6)$ variables permits calculation of the unknown variables L_{OUT}, V_{OUT}, $(x_C)_{L_{IN}}$, $(y_C)_{V_{IN}}$, all $(x_i)_{L_{OUT}}$, T_{OUT}, and all $(y_i)_{V_{OUT}}$, where C denotes the missing mole fractions in the two entering streams.

Single-Section, Countercurrent Cascade

Consider the N-stage, single-section, countercurrent cascade unit shown in Figure 5.21. This cascade consists of N adiabatic or nonadiabatic equilibrium-stage elements of the type shown in Figure 5.20. An algorithm is easily developed for enumerating variables, equations, and degrees of freedom for combinations of such elements to form a unit. The number of design variables for the unit is obtained by summing the variables associated with each element and then subtracting from the total variables the $C + 3$ variables for each of the N_R redundant interconnecting streams that arise when the output of one element becomes the input to another. Also, if an unspecified number of repetitions of any element occurs within the unit, an additional variable is added, one for each group of repetitions, giving a total of N_A additional variables. In a similar manner, the number of independent equations for the unit is obtained by summing the values of N_E for the units and then subtracting the N_R redundant mole-fraction constraints. The number of degrees of freedom is obtained as before, from (5-67). Thus,

$$(N_V)_{unit} = \sum_{\text{all elements, } e} (N_V)_e - N_R(C + 3) + N_A \quad (5\text{-}68)$$

$$(N_E)_{unit} = \sum_{\text{all elements, } e} (N_E)_e - N_R \quad (5\text{-}69)$$

Combining (5-67), (5-68), and (5-69), we have

$$(N_D)_{unit} = \sum_{\text{all elements, } e} (N_D)_e - N_R(C + 2) + N_A \quad (5\text{-}70)$$

or

$$(N_D)_{unit} = (N_V)_{unit} - (N_E)_{unit} \quad (5\text{-}71)$$

For the N-stage cascade unit of Figure 5.21, with reference to the above degrees-of-freedom analysis for the single

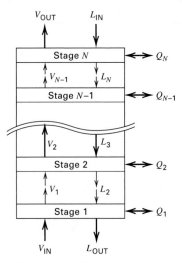

Figure 5.21 An N-stage cascade.

adiabatic or nonadiabatic equilibrium-stage element, the total number of variables from (5-68) is

$$(N_V)_{unit} = N(4C + 13) - [2(N - 1)](C + 3) + 1$$
$$= 7N + 2NC + 2C + 7$$

since $2(N - 1)$ interconnecting streams exist. The additional variable is the total number of stages (i.e., $N_A = 1$).

The number of independent relationships from (5-69) is

$$(N_E)_{unit} = N(2C + 7) - 2(N - 1) = 5N + 2NC + 2$$

since $2(N - 1)$ redundant mole-fraction constraints exist.

The number of degrees of freedom from (5-71) is

$$(N_D)_{unit} = N_V - N_E = 2N + 2C + 5$$

Note, again, that the coefficient of C is 2, the number of streams entering the cascade. For a cascade, the coefficient of N is always 2 (corresponding to stage P and Q).

One possible set of design variables is

Variable Specification	Number of Variables
Heat transfer rate for each stage (or adiabaticity)	N
Stage pressures	N
Stream V_{IN} variables	$C + 2$
Stream L_{IN} variables	$C + 2$
Number of stages	1
	$2N + 2C + 5$

Output variables for this specification include missing mole fractions for V_{IN} and L_{IN}, stage temperatures, and the variables associated with the V_{OUT} stream, L_{OUT} stream, and interstage streams. This N-stage cascade unit can represent simple absorbers, strippers, or liquid–liquid extractors.

Two-Section, Countercurrent Cascades

Two-section, countercurrent cascades can consist not only of adiabatic or nonadiabatic equilibrium-stage elements, but also of other elements of the type shown in Table 5.3, including total and partial reboilers; total and partial condensers; equilibrium stages with a feed, F, or a sidestream S; and stream mixers and dividers. These different elements can be combined into any of a number of complex cascades by applying to (5-68) to (5-71) the values of N_V, N_E, and N_D given in Table 5.3 for the different elements.

The design or simulation of multistage separation operations involves solving the variable relationships for output variables after selecting values of design variables to satisfy the degrees of freedom. Two cases are commonly encountered. In case I, the design case, recovery specifications are made for one or two key components and the number of required equilibrium stages is determined. In case II, the simulation case, the number of equilibrium stages is specified and component separations are computed. For rigorous calculations involving multicomponent feeds, the second case is more widely applied because less computational complexity is involved with the number of stages fixed. Table 5.4 is a

Table 5.3 Degrees of Freedom for Separation Operation Elements and Units

	Schematic	Element or Unit Name	N_V, Total Number of Variables	N_E, Independent Relationships	N_D, Degrees of Freedom
(a)		Total boiler (reboiler)	$(2C + 7)$	$(C + 3)$	$(C + 4)$
(b)		Total condenser	$(2C + 7)$	$(C + 3)$	$(C + 4)$
(c)		Partial (equilibrium) boiler (reboiler)	$(3C + 10)$	$(2C + 6)$	$(C + 4)$
(d)		Partial (equilibrium) condenser	$(3C + 10)$	$(2C + 6)$	$(C + 4)$
(e)		Adiabatic equilibrium stage	$(4C + 12)$	$(2C + 7)$	$(2C + 5)$
(f)		Equilibrium stage with heat transfer	$(4C + 13)$	$(2C + 7)$	$(2C + 6)$
(g)		Equilibrium feed stage with heat transfer and feed	$(5C + 16)$	$(2C + 8)$	$(3C + 8)$
(h)		Equilibrium stage with heat transfer and sidestream	$(5C + 16)$	$(3C + 9)$	$(2C + 7)$
(i)		N-connected equilibrium stages with heat transfer	$(7N + 2NC + 2C + 7)$	$(5N + 2NC + 2)$	$(2N + 2C + 5)$
(j)		Stream mixer	$(3C + 10)$	$(C + 4)$	$(2C + 6)$
(k)		Stream divider	$(3C + 10)$	$(2C + 5)$	$(C + 5)$

[a]Sidestream can be vapor or liquid.

[b]Alternatively, all streams can be vapor.

Table 5.4 Typical Variable Specifications for Design Cases

Unit Operation		N_D	Variable Specification[a]	
			Case I, Component Recoveries Specified	Case II, Number of Equilibrium Stages Specified
(a) Absorption (two inlet streams)		$2N + 2C + 5$	1. Recovery of one key component	1. Number of stages
(b) Distillation (one inlet stream, total condenser, partial reboiler)		$2N + C + 9$	1. Condensate at saturation temperature 2. Recovery of light key component 3. Recovery of heavy key component 4. Reflux ratio (> minimum) 5. Optimal feed stage[b]	1. Condensate at saturation temperature 2. Number of stages above feed stage 3. Number of stages below feed stage 4. Reflux ratio 5. Distillate flow rate
(c) Distillation (one inlet stream, partial condenser, partial reboiler, vapor distillate only)		$(2N + C + 6)$	1. Recovery of light key component 2. Recovery of heavy key component 3. Reflux ratio (> minimum) 4. Optimal feed stage[b]	1. Number of stages above feed stage 2. Number of stages below feed stage 3. Reflux ratio 4. Distillate flow rate
(d) Liquid–liquid extraction with two solvents (three inlet streams)		$2N + 3C + 8$	1. Recovery of key component 1 2. Recovery of key component 2	1. Number of stages above feed 2. Number of stages below feed
(e) Reboiled absorption (two inlet streams)		$2N + 2C + 6$	1. Recovery of light key component 2. Recovery of heavy key component 3. Optimal feed stage[b]	1. Number of stages above feed 2. Number of stages below feed 3. Bottoms flow rate

(continued)

Table 5.4 (*Continued*)

Unit Operation	N_D	Variable Specification[a]	
		Case I, Component Recoveries Specified	Case II, Number of Equilibrium Stages Specified
(f) Reboiled stripping (one inlet stream)	$2N + C + 3$	1. Recovery of one key component 2. Reboiler heat duty[d]	1. Number of stages 2. Bottoms flow rate
(g) Distillation (one inlet stream, partial condenser, partial reboiler, both liquid and vapor distillates)	$2N + C + 9$	1. Ratio of vapor distillate to liquid distillate 2. Recovery of light key component 3. Recovery of heavy key component 4. Reflux ratio (> minimum) 5. Optimal feed stage[b]	1. Ratio of vapor distillate to liquid distillate 2. Number of stages above feed stage 3. Number of stages below feed stage 4. Reflux ratio 5. Liquid distillate flow rate
(h) Extractive distillation (two inlet streams, total condenser, partial reboiler, single-phase condensate)	$2N + 2C + 12$	1. Condensate at saturation temperature 2. Recovery of light key component 3. Recovery of heavy key component 4. Reflux ratio (> minimum) 5. Optimal feed stage[b] 6. Optimal MSA stage[b]	1. Condensate at saturation temperature 2. Number of stages above MSA stage 3. Number of stages between MSA and feed stages 4. Number of stages below feed stage 5. Reflux ratio 6. Distillate flow rate
(i) Liquid–liquid extraction (two inlet streams)	$2N + 2C + 5$	1. Recovery of one key component	1. Number of stages
(j) Stripping (two inlet streams)	$2N + 2C + 5$	1. Recovery of one key component	1. Number of stages

[a]Does not include the following variables, which are also assumed specified: all inlet stream variables ($C + 2$ for each stream); all element and unit pressures; all element and unit heat transfer rates except for condensers and reboilers.

[b]Optimal stage for introduction of inlet stream corresponds to minimization of total stages.

[c]For case I variable specifications, MSA flow rates must be greater than minimum values for specified recoveries.

[d]For case I variable specifications, reboiler heat duty must be greater than minimum value for specified recovery.

summary of possible variable specifications for each of these two cases for a number of separator types discussed in Chapter 1 and shown in Table 1.1. For all separators in Table 5.4, it is assumed that all inlet streams are completely specified (i.e., $C - 1$ mole fractions, total flow rate, temperature, and pressure) and all element and unit pressures and heat transfer rates (except for condensers and reboilers) are specified. Thus, only variable specifications for satisfying the remaining degrees of freedom are listed.

EXAMPLE 5.5

Consider a multistage distillation column with one feed, one side-stream, a total condenser, a partial reboiler, and provisions for heat transfer to or from any stage. Determine the number of degrees of freedom and a reasonable set of specifications.

SOLUTION

This separator is assembled as shown in Figure 5.22, from the circled elements and units, which are all found in Table 5.3. The total variables are determined by summing the variables $(N_V)_e$ for each element from Table 5.3 and then subtracting the redundant variables due to interconnecting flows. As before, redundant mole-fraction constraints are subtracted from the summation of independent relationships for each element $(N_E)_e$. This problem was first treated by Gilliland and Reed [9] and more recently by Kwauk [8]. Differences in N_D obtained by various authors are due, in part, to their method of numbering stages. Here, the partial reboiler is the first equilibrium stage. From Table 5.3, element variables and relationships are as follows:

Element or Unit	$(N_V)_e$	$(N_E)_e$
Total condenser	$(2C + 7)$	$(C + 3)$
Reflux divider	$(3C + 10)$	$(2C + 5)$
$(N - S)$ stages	$[7(N - S) + 2(N - S)C + 2C + 7]$	$[5(N - S) + 2(N - S)C + 2]$
Sidestream stage	$(5C + 16)$	$(3C + 9)$
$(S - 1) - F$ stages	$[7(S - 1 - F) + 2(S - 1 - F)C + 2C + 7]$	$[5(S - 1 - F) + 2(S - 1 - F)C + 2]$
Feed stage	$(5C + 16)$	$(2C + 8)$
$(F - 1) - 1$ stages	$[7(F - 2) + 2(F - 2)C + 2C + 7]$	$[5(F - 2) + 2(F - 2)C + 2]$
Partial reboiler	$(3C + 10)$	$(2C + 6)$
	$\sum(N_V)_e = 7N + 2NC + 18C + 59$	$\sum(N_E)_e = 5N + 2NC + 4C + 22$

Subtracting $(C + 3)$ redundant variables for 13 interconnecting streams, according to (5-68), with $N_A = 0$ (no unspecified repetitions), gives

$$(N_V)_{\text{unit}} = \sum(N_V)_e - 13(C + 3) = 7N + 2NC + 5C + 20$$

Subtracting the corresponding 13 redundant mole-fraction constraints, according to (5-69),

$$(N_E)_{\text{unit}} = \sum(N_E)_e - 13 = 5N + 2NC + 4C + 9$$

Therefore, from (5-71),

$$N_D = (7N + 2NC + 5C + 20) - (5N + 2NC + 4C + 9)$$
$$= 2N + C + 11$$

Note that the coefficient of C is only 1, because there is only one feed, and, again, the coefficient of N is 2.

A set of feasible design variable specifications is

Variable Specification	Number of Variables
1. Pressure at each stage (including partial reboiler)	N
2. Pressure at reflux divider outlet	1
3. Pressure at total condenser outlet	1
4. Heat transfer rate for each stage (excluding partial reboiler)	$(N - 1)$
5. Heat transfer rate for divider	1
6. Feed mole fractions and total feed rate	C

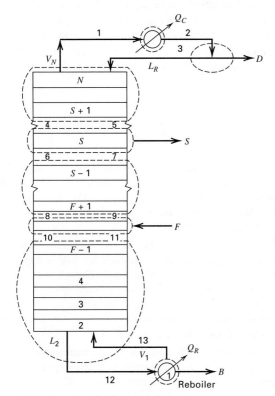

Figure 5.22 Complex distillation unit.

Variable Specification	Number of Variables
7. Feed temperature	1
8. Feed pressure	1
9. Condensate temperature (e.g., saturated liquid)	1
10. Total number of stages, N	1
11. Feed stage location	1
12. Sidestream stage location	1
13. Sidestream total flow rate, S	1
14. Total distillate flow rate, D or D/F	1
15. Reflux flow rate, L_R, or reflux ratio, L_R/D	1

$$N_D = (2N + C + 11)$$

In most separation operations, variables related to feed conditions, stage heat-transfer rates, and stage pressure are known or set. Remaining specifications have proxies, provided that the variables are mathematically independent of each other and of those already known. Thus, in the above list, the first nine entries are almost always known or specified. Variables 10 to 15, however, have surrogates. Some of these are

16. Condenser heat duty, Q_C
17. Reboiler heat duty, Q_R
18. Recovery or mole fraction of one component in bottoms
19. Recovery or mole fraction of one component in distillate

Heat duties Q_C and Q_R are not good design variables because they are difficult to specify. Condenser duty Q_C, for example, must be specified so that the condensate temperature lies between that corresponding to a saturated liquid and the freezing point of the condensate. Otherwise, a physically unrealizable (or no) solution to the problem is obtained. Similarly, it is much easier to calculate Q_R knowing the total flow rate and enthalpy of the bottom streams than vice versa. In general, Q_R and Q_C are so closely related that it is not advisable to specify both.

Other proxies are possible, such as a stage temperature, a flow rate leaving a stage, or any independent variable that characterizes the process. The problem of independence of variables requires careful consideration. Distillate product rate, Q_C, and L_R/D, for example, are not independent. It should also be noted that, for the design case, recoveries of no more than two species (items 18 and 19) are specified. These species are referred to as key components. Attempts to specify recoveries of three or four species will usually result in an unsuccessful solution of the equations.

The degrees of freedom for the complex distillation unit of Figure 5.22 can be determined quickly by modifying a similar unit operation in Table 5.4. The closest unit is (b), which differs from the unit in Figure 5.22 by only a sidestream. From Table 5.3, we see that an equilibrium stage with heat transfer but without a sidestream [element (f)] has $N_D = (2C + 6)$, while an equilibrium stage with heat transfer and with a sidestream [element (h)] has $N_D = (2C + 7)$ or one additional degree of freedom. In addition, when this sidestream stage is placed in a cascade, an additional degree of freedom is added for the location of the sidestream stage. Thus, two degrees of freedom are added to $N_D = 2N + C + 9$ for unit operation (b) in Table 5.4. The result is $N_D = 2N + C + 11$, which is identical to that determined in the above example.

In a similar manner, the above example can be readily modified to include a second feed stage. By comparing values of N_D for elements (f) and (g) in Table 5.3, it is seen that a feed adds $C + 2$ degrees of freedom. In addition, one more degree of freedom must be added for the location of this feed stage in a cascade. Thus, a total of $C + 3$ degrees of freedom are added, giving $N_D = 2N + 2C + 14$.

SUMMARY

1. A cascade is a collection of contacting stages arranged to: (a) accomplish a separation that cannot be achieved in a single stage, and/or (b) reduce the amount of mass- or energy-separating agent.

2. Cascades are single- or multiple-sectioned and may be configured in cocurrent, crosscurrent, or countercurrent arrangements. Cascades are readily computed when governing equations are linear in component split ratios.

3. Stage requirements for a countercurrent solid–liquid leaching and/or washing cascade, involving constant underflow and mass transfer of one component, are given by (5-10).

4. Stage requirements for a single-section, liquid–liquid extraction cascade assuming a constant distribution coefficient and immiscible solvent and carrier are given by (5-19), (5-22), and (5-29) for cocurrent, crosscurrent, and countercurrent flow arrangements, respectively. The countercurrent cascade is the most efficient.

5. Single-section stage requirements for a countercurrent cascade for absorption and stripping can be estimated with the Kremser equations, (5-48), (5-50), (5-54), and (5-55). A single-section, countercurrent cascade is limited in its ability to achieve a separation between two components.

6. The Kremser equations can be combined for a two-section cascade to give (5-66), which is suitable for making approximate calculations of component splits for distillation. A two-section, countercurrent cascade can achieve a sharp split between two key components. The rectifying section purifies the light components and increases recovery of heavy components. The stripping section provides the opposite function.

7. Equilibrium cascade equations involve parameters referred to as washing W, extraction E, absorption A, and stripping S, factors that involve distribution coefficients, such as K, K_D, and R, and phase flow ratios, such as S/F and L/V.

8. Single-section membrane cascades increase purity of one product and recovery of the main component in that product.

9. Hybrid systems of different types reduce energy expenditures, make possible separations that are otherwise difficult, and/or improve the degree of separation.

10. The number of degrees of freedom (number of specifications) for a mathematical model of a cascade is the difference between

the number of unique variables and the number of independent equations that relate the variables. For a single-section, countercurrent cascade, the recovery of one component can be specified. For a two-section countercurrent cascade, the recoveries of two components can be specified.

REFERENCES

1. Berdt, R.J., and C.C. Lynch, *J. Am. Chem. Soc.,* **66,** 282–284 (1944).

2. Kremser, A., *Natl. Petroleum News,* **22**(21), 43–49 (May 21, 1930).

3. Edmister, W.C., *AIChE J.,* **3,** 165–171 (1957).

4. Smith, B.D., and W.K. Brinkley, *AIChE J.,* **6,** 446–450 (1960).

5. Lyster, W.N., S.L. Sullivan, Jr., D.S. Billingsley, and C.D. Holland, *Petroleum Refiner,* **38**(6), 221–230 (1959).

6. Prasad, R., F. Notaro, and D.R. Thompson, *J. Membrane Science,* **94,** Issue 1, 225–248 (1994).

7. Berry, D.A., and K.M. Ng, *AIChE J.,* **43,** 1751–1762 (1997).

8. Kwauk, M., *AIChE J.,* **2,** 240–248 (1956).

9. Gilliland, E.R., and C.E. Reed, *Ind. Eng. Chem.,* **34,** 551–557 (1942).

EXERCISES

Section 5.1

5.1 Devise an interlinked cascade of the type shown in Figure 5.2e, but consisting of three columns for the separation of a four-component feed into four products.

5.2 A liquid–liquid extraction process is conducted batchwise as shown in Figure 5.23. The process begins in vessel 1 (original),

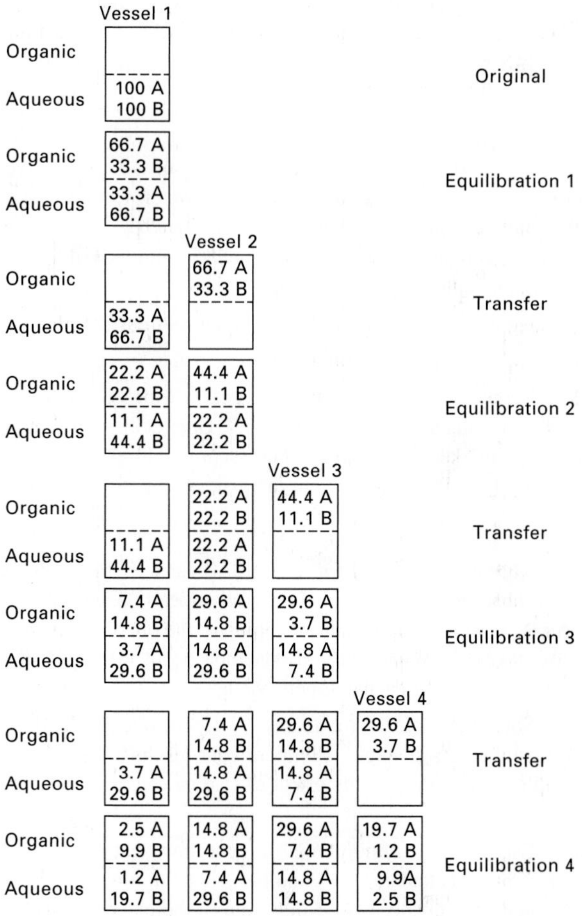

Figure 5.23 Liquid–liquid extraction process for Exercise 5.2.

where 100 mg each of solutes A and B are dissolved in 100 ml of water. After adding 100 ml of an organic solvent that is more selective for A than B, the distribution of A and B becomes that shown for equilibration 1 with vessel 1. The organic-rich phase is transferred to vessel 2 (transfer), leaving the water-rich phase in vessel 1 (transfer). Assume that water and the organic solvent are immiscible. Next, 100 ml of water is added to vessel 2, resulting in the phase distribution shown for vessel 2 (equilibration 2). Also, 100 ml of organic solvent is added to vessel 1 to give the phase distribution shown for vessel 1 (equilibration 2). The batch process is continued by adding vessel 3 and then 4 to obtain the results shown.

(a) Carefully study the process in Figure 5.23 and then draw a corresponding cascade diagram, labeled in a manner similar to Figure 5.2(b).

(b) Is the process of the cocurrent, countercurrent, or crosscurrent type?

(c) Compare the separation achieved with that for a single-batch equilibrium step.

(d) How could the process be modified to make it a countercurrent cascade [see O. Post and L.C. Craig, *Anal. Chem.,* **35,** 641 (1963)].

5.3 Nitrogen is to be removed from a gas mixture with methane by gas permeation (see Table 1.2) using a glassy polymer membrane that is selective for nitrogen. However, the desired degree of separation cannot be achieved in one stage. Draw sketches of two different two-stage membrane cascades that might be considered to perform the desired separation.

Section 5.2

5.4 In Example 4.9, 83.25% of the oil in soybeans is leached by benzene using a single equilibrium stage. Calculate the percent extraction of oil if:

(a) Two countercurrent equilibrium stages are used to process 5,000 kg/h of soybean meal with 5,000 kg/h of benzene.

(b) Three countercurrent equilibrium stages are used to process the same flows as in part (a).

(c) Also, determine the number of countercurrent equilibrium stages required to extract 98% of the oil if a solvent rate of twice the minimum value is used.

5.5 For Example 5.1, involving the separation of sodium carbonate from an insoluble oxide, compute the minimum solvent feed rate in pounds per hour. What is the ratio of actual solvent rate to the minimum solvent rate? Determine and plot the percent recovery of soluble solids with a cascade of five countercurrent equilibrium stages for solvent flow rates from 1.5 to 7.5 times the minimum value.

5.6 Aluminum sulfate, commonly called alum, is produced as a concentrated aqueous solution from bauxite ore by reaction with aqueous sulfuric acid, followed by a three-stage, countercurrent washing operation to separate soluble aluminum sulfate from the insoluble content of the bauxite ore, followed by evaporation. In a typical process, 40,000 kg/day of solid bauxite ore containing 50 wt% Al_2O_3 and 50% inert is crushed and fed together with the stoichiometric amount of 50 wt% aqueous sulfuric acid to a reactor, where the Al_2O_3 is reacted completely to alum by the reaction

$$Al_2O_3 + 3H_2SO_4 \rightarrow Al_2(SO_4)_3 + 3H_2O$$

The slurry effluent from the reactor (digester), consisting of solid inert material from the ore and an aqueous solution of aluminum sulfate is then fed to a three-stage, countercurrent washing unit to separate the aqueous aluminum sulfate from the inert material. If the solvent is 240,000 kg/day of water and the underflow from each washing stage is 50 wt% water on a solute-free basis, compute the flow rates in kilograms per day of aluminum sulfate, water, and inert solid in each of the two product streams leaving the cascade. What is the percent recovery of the aluminum sulfate? Would the addition of one more stage be worthwhile?

5.7 (a) When rinsing clothes with a given amount of water, would one find it more efficient to divide the water and rinse several times; or should one use all the water in one rinse? Explain.

(b) Devise a clothes-washing machine that gives the most efficient rinse cycle for a fixed amount of water.

Section 5.3

5.8 An aqueous acetic-acid solution containing 6.0 moles of acid per liter is to be extracted in the laboratory with chloroform at 25°C to recover the acid (B) from chloroform-insoluble impurities present in the water. The water (A) and chloroform (C) are essentially immiscible. If 10 liters of solution are to be extracted at 25°C, calculate the percent extraction of acid obtained with 10 liters of chloroform under the following conditions:

(a) Using the entire quantity of solvent in a single batch extraction

(b) Using three batch extractions with one-third of the total solvent used in each batch

(c) Using three batch extractions with 5 liters of solvent in the first, 3 liters in the second, and 2 liters in the third batch

Assume that the volumetric amounts of the feed and solvent do not change during extraction. Also, assume the distribution coefficient for the acid, $K''_{D_B} = (c_B)_C/(c_B)_A = 2.8$, where $(c_B)_C$ = concentration of acid in chloroform and $(c_B)_A$ = concentration of acid in water, both in moles per liter.

5.9 A 20 wt% solution of uranyl nitrate (UN) in water is to be treated with tributyl phosphate (TBP) to remove 90% of the uranyl nitrate. All operations are to be batchwise equilibrium contacts. Assuming that water and TBP are mutually insoluble, how much TBP is required for 100 g of solution if at equilibrium (g UN/g TBP) = 5.5(g UN/g H_2O) and:

(a) All the TBP is used at once in one stage?

(b) Half is used in each of two consecutive stages?

(c) Two countercurrent stages are used?

(d) An infinite number of crosscurrent stages is used?

(e) An infinite number of countercurrent stages is used?

5.10 The uranyl nitrate (UN) in 2 kg of a 20 wt% aqueous solution is to be extracted with 500 g of tributyl phosphate. Using the equilibrium data in Exercise 5.9, calculate and compare the percentage recoveries for the following alternative procedures:

(a) A single-stage batch extraction

(b) Three batch extractions with one-third of the total solvent used in each batch (the solvent is withdrawn after contacting the entire UN phase)

(c) A two-stage cocurrent extraction

(d) A three-stage countercurrent extraction

(e) An infinite-stage countercurrent extraction

(f) An infinite-stage crosscurrent extraction

5.11 One thousand kilograms of a 30 wt% dioxane in water solution is to be treated with benzene at 25°C to remove 95% of the dioxane. The benzene is dioxane-free, and the equilibrium data of Example 5.2 can be used. Calculate the solvent requirements for:

(a) A single batch extraction

(b) Two crosscurrent stages using equal amounts of benzene

(c) Two countercurrent stages

(d) An infinite number of crosscurrent stages

(e) An infinite number of countercurrent stages

5.12 Chloroform is to be used to extract benzoic acid from wastewater effluent. The benzoic acid is present at a concentration of 0.05 mol/liter in the effluent, which is discharged at a rate of 1,000 liter/h. The distribution coefficient for benzoic acid at process conditions is given by

$$c^I = K_D^{II} c^{II}$$

where $K_D^{II} = 4.2$, c^I = molar concentration of solute in solvent, and c^{II} = molar concentration of solute in water. Chloroform and water may be assumed immiscible. If 500 liters/h of chloroform is to be used, compare the fraction benzoic acid removed in

(a) A single equilibrium contact

(b) Three crosscurrent contacts with equal portions of chloroform

(c) Three countercurrent contacts

5.13 Repeat Example 5.2 with a solvent for which $E = 0.90$. Display your results in a plot like Figure 5.7. Does countercurrent flow still have a marked advantage over crosscurrent flow? Is it desirable to choose the solvent and solvent rate so that $E > 1$? Explain.

Section 5.4

5.14 Repeat Example 5.3 for $N = 1, 3, 10$, and 30 stages. Plot the percent absorption of each of the five hydrocarbons and the total feed gas, as well as the percent stripping of the oil versus the number of stages, N. What can you conclude about the effect of the number of stages on each component?

5.15 Solve Example 5.3 for an absorbent flow rate of 330 lbmol/h and three theoretical stages. Compare your results to the results of Example 5.3 and discuss the effect of trading stages for absorbent flow.

5.16 Estimate the minimum absorbent flow rate required for the separation calculated in Example 5.3 assuming that the key component is propane, whose flow rate in the exit vapor is to be 155.4 lbmol/h.

5.17 Solve Example 5.3 with the addition of a heat exchanger at each stage so as to maintain isothermal operation of the absorber at

(a) 125°F

(b) 150°F

What is the effect of temperature on absorption in the range of 100 to 150°F?

5.18 One million pound-moles per day of a gas of the following composition is to be absorbed by *n*-heptane at −30°F and 550 psia in an absorber having 10 theoretical stages so as to absorb 50% of the ethane. Calculate the required flow rate of absorbent and the distribution, in lbmol/h, of all the components between the exiting gas and liquid streams.

Component	Mole Percent in Feed Gas	K-value @ −30°F and 550 psia
C_1	94.9	2.85
C_2	4.2	0.36
C_3	0.7	0.066
nC_4	0.1	0.017
nC_5	0.1	0.004

5.19 A stripper operating at 50 psia with three equilibrium stages is used to strip 1,000 kmol/h of liquid at 300°F having the following molar composition: 0.03% C_1, 0.22% C_2, 1.82% C_3, 4.47% nC_4, 8.59% nC_5, 84.87% nC_{10}. The stripping agent is 1,000 kmol/h of superheated steam at 300°F and 50 psia. Use the Kremser equation to estimate the compositions and flow rates of the stripped liquid and exiting rich gas.

Assume a *K*-value for C_{10} of 0.20 and assume that no steam is absorbed. However, calculate the dew-point temperature of the exiting rich gas at 50 psia. If that temperature is above 300°F, what would you suggest be done?

5.20 In Figure 5.12, is anything gained by totally condensing the vapor leaving each stage? Alter the processes in Figure 5.12a and 5.12b so as to eliminate the addition of heat to stages 2 and 3 and still achieve the same separations.

5.21 Repeat Example 5.4 for external reflux flow rates L_0 of

(a) 1,500 lbmol/h

(b) 2,000 lbmol/h

(c) 2,500 lbmol/h

Plot d_{C_3}/b_{C_3} as a function of L_0 from 1,000 to 2,500 lbmol/h. In making the calculations, assume that stage temperatures do not change from the results of Example 5.4. Discuss the effect of reflux ratio on the separation.

5.22 Repeat Example 5.4 for the following numbers of equilibrium stages (see Figure 5.15):

(a) $M = 10, N = 10$

(b) $M = 15, N = 15$

Plot d_{C_3}/b_{C_3} as a function of $M + N$ from 10 to 30 stages. In making the calculations, assume that state temperatures and total flow rates do not change from the results of Example 5.4. Discuss the effect of the number of stages on the separation.

5.23 Use the Edmister group method to determine the compositions of the distillate and bottoms for the distillation operation shown in Figure 5.24. At column conditions, the feed is approximately 23 mol% vapor.

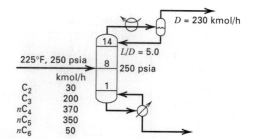

Figure 5.24 Conditions for Exercise 5.23.

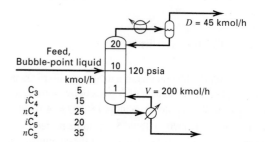

Figure 5.25 Conditions for Exercise 5.24.

5.24 A bubble-point liquid feed is to be distilled as shown in Figure 5.25. Use the Edmister group method to estimate the mole-fraction compositions of the distillate and bottoms. Assume initial overhead and bottoms temperatures are 150 and 250°F, respectively.

Section 5.7

5.25 Verify the values given in Table 5.3 for N_V, N_E, and N_D for a partial reboiler and a total condenser.

5.26 Verify the values given in Table 5.3 for N_V, N_E, and N_D for a stream mixer and a stream divider.

5.27 A mixture of maleic anhydride and benzoic acid containing 10 mol% acid is a product of the manufacture of phthalic anhydride. The mixture is to be distilled continuously in a column with a total condenser and a partial reboiler at a pressure of 13.2 kPa (100 torr) with a reflux ratio of 1.2 times the minimum value to give a product of 99.5 mol% maleic anhydride and a bottoms of 0.5 mol% anhydride. Is this problem completely specified?

5.28 Verify N_D for the following unit operations in Table 5.4: (*b*), (*c*), and (*g*). How would N_D change if two feeds were used instead of one?

5.29 Verify N_D for unit operations (*e*) and (*f*) in Table 5.4. How would N_D change if a vapor side stream was pulled off some stage located between the feed stage and the bottom stage?

5.30 Verify N_D for unit operation (*h*) in Table 5.4. How would N_D change if a liquid side stream was added to a stage that was located between the feed stage and stage 2?

5.31 The following are not listed as design variables for the distillation unit operations in Table 5.4:

(a) Condenser heat duty

(b) Stage temperature

(c) Intermediate-stage vapor rate

(d) Reboiler heat load

Under what conditions might these become design variables? If so, which variables listed in Table 5.4 would you eliminate?

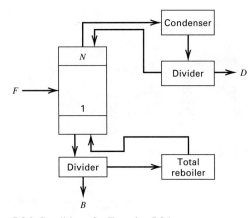

Figure 5.26 Conditions for Exercise 5.34.

5.32 Show for distillation that, if a total condenser is replaced by a partial condenser, the number of degrees of freedom is reduced by 3, provided that the distillate is removed solely as a vapor.

5.33 Unit operation (b) in Table 5.4 is to be heated by injecting live steam directly into the bottom plate of the column instead of by using a reboiler, for a separation involving ethanol and water. Assuming a fixed feed, an adiabatic operation, atmospheric pressure throughout, and a top alcohol concentration specification:

(a) What is the total number of design variables for the general configuration?

(b) How many design variables are needed to complete the design? Which variables do you recommend?

5.34 (a) For the distillation column shown in Figure 5.26, determine the number of independent design variables.

(b) It is suggested that a feed consisting of 30% A, 20% B, and 50% C, all in moles, at 37.8°C and 689 kPa, be processed in the unit of Figure 5.26, consisting of a 15-plate, 3-m-diameter column that is designed to operate at vapor velocities of 0.3 m/s and an L/V of 1.2. The pressure drop per plate is 373 Pa at these conditions, and the condenser is cooled by plant water at 15.6°C.

The product specifications in terms of the concentration of A in the distillate and C in the bottoms have been set by the process department, and the plant manager has asked you to specify a feed rate for the column. Write a memorandum to the plant manager pointing out why you can't do this, and suggest some alternatives.

5.35 Calculate the number of degrees of freedom for the mixed-feed, triple-effect evaporator system shown in Figure 5.27. Assume that the steam and all drain streams are at saturated conditions and the feed is an aqueous solution of a dissolved organic solid. Also,

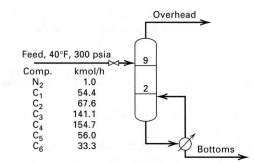

Figure 5.28 Conditions for Exercise 5.36.

assume that all overhead streams are pure water vapor, with no entrainment.

If this evaporator system is used to concentrate a feed containing 2 wt% dissolved organic to a product with 25 wt% dissolved organic, using 689-kPa saturated steam, calculate the number of unspecified design variables and suggest likely candidates. Assume perfect insulation against heat loss for each effect.

5.36 A reboiled stripper as shown in Figure 5.28 is to be designed for the task shown. Determine

(a) The number of variables.

(b) The number of equations relating the variables.

(c) The number of degrees of freedom and indicate.

(d) Which additional variables, if any, need to be specified.

5.37 The thermally coupled distillation system shown in Figure 5.29 is to be used to separate a mixture of three components into three products. Determine for the system

(a) The number of variables.

(b) The number of equations relating the variables.

(c) The number of degrees of freedom and propose.

(d) A reasonable set of design variables.

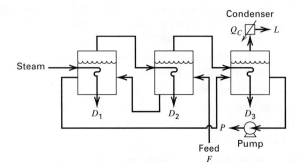

Figure 5.27 Conditions for Exercise 5.35.

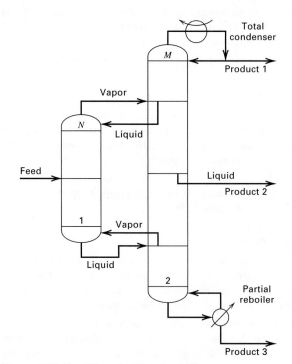

Figure 5.29 Conditions for Exercise 5.37.

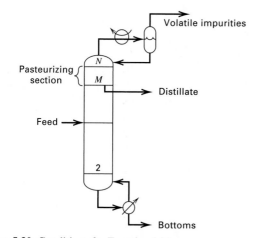

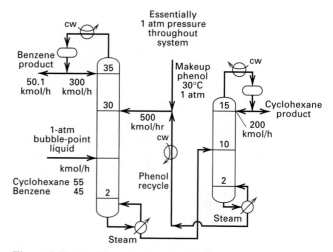

Figure 5.30 Conditions for Exercise 5.38.

Figure 5.32 Conditions for Exercise 5.40.

5.38 When the feed to a distillation column contains a small amount of impurities that are much more volatile than the desired distillate, it is possible to separate the volatile impurities from the distillate by removing the distillate as a liquid sidestream from a stage located several stages below the top stage. As shown in Figure 5.30, this additional top section of stages is referred to as a pasteurizing section.

(a) Determine the number of degrees of freedom for the unit

(b) Determine a reasonable set of design variables

5.39 A system for separating a mixture into three products is shown in Figure 5.31. For it, determine

(a) The number of variables.

(b) The number of equations relating the variables.

(c) The number of degrees of freedom and propose.

(d) A reasonable set of design variables.

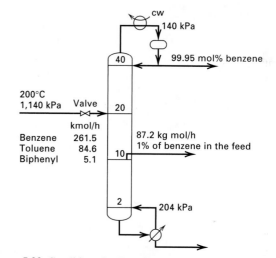

Figure 5.33 Conditions for Exercise 5.41.

5.40 A system for separating a binary mixture by extractive distillation, followed by ordinary distillation for recovery and recycle of the solvent, is shown in Figure 5.32. Are the design variables shown sufficient to specify the problem completely? If not, what additional design variables(s) would you select?

5.41 A single distillation column for separating a three-component mixture into three products is shown in Figure 5.33. Are the design variables shown sufficient to specify the problem completely? If not, what additional design variable(s) would you select?

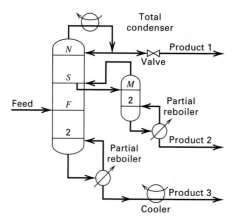

Figure 5.31 Conditions for Exercise 5.39.

Part 2

Separations by Phase Addition or Creation

Among the most widely used industrial separation operations are absorption, stripping, various types of distillation, and liquid–liquid extraction, all of which involve separations by selective mass transfer of components from one fluid phase to another. The other phase is created by thermal energy input (energy-separating agent) or by addition (mass-separating agent). In most cases, these operations are based on the use of countercurrent cascades of multiple stages. Detailed descriptions of, and design and analysis calculations for, these vapor–liquid and liquid–liquid operations are presented in Chapters 6 through 13. Two types of mathematical models are considered: (1) stages that attain thermodynamic phase equilibrium and (2) stages that do not reach phase equilibrium but are governed by rates of mass transfer. The less-complex equilibrium-stage models are more widely used, with a stage efficiency, but the availability of fast and inexpensive digital computations is encouraging an increase in the application of more tedious, but more accurate, mass-transfer models.

Absorption (vapor-phase feed) and stripping (liquid-phase feed) are covered in Chapter 6. These two operations usually rely on the addition of a mass-separating agent (liquid absorbent and vapor-stripping agent, respectively), but may also use heat transfer to produce the other phase. In general, these operations are not used to make a sharp separation, but can achieve a high recovery of a key component in the feed by its transfer to the other phase. Absorption and stripping equipment most often consists of columns containing trays or packing for contacting the two phases, with continuous flow of the fluid phases; calculation methods, graphical and algebraic, are presented for both types of contacting. Methods for estimating tray efficiency, column height, and diameter are also presented.

The continuous distillation of binary mixtures in multiple-stage, trayed or packed columns is covered in Chapter 7, with emphasis on the classical McCabe–Thiele graphical, equilibrium-stage model. Typically, this separation operation utilizes energy to achieve the separation and two sections (rectifying and stripping), which make possible, with non-azeotropic-forming mixtures, the separation of a binary mixture into two nearly pure products. Equipment-sizing methods of Chapter 6 generally apply to distillation in Chapter 7.

When the separation of a liquid binary mixture by distillation is infeasible or too expensive, liquid–liquid extraction using a selective solvent is considered, as presented in Chapter 8. Although many equipment configurations are available, columns or vessels with mechanically assisted agitation are the most useful when multiple stages are needed to achieve the desired recovery. This chapter emphasizes graphical, equilibrium-stage methods using triangular diagrams for treating ternary systems.

Equilibrium-stage models and calculations for multicomponent mixtures are considerably more complex than those for binary mixtures, as in distillation, and for ternary mixtures, as in absorption, stripping, and liquid–liquid extraction. Approximate algebraic methods are presented in Chapter 9, while rigorous algebraic methods are developed in Chapter 10. These methods are implemented in process simulators and widely used.

Chapter 11 considers equilibrium-stage calculation methods for so-called enhanced distillation of mixtures that are difficult to separate by conventional distillation or liquid–liquid extraction. An important aspect of enhanced distillation is the determination of feasible products, which uses residue-curve maps. Extractive, azeotropic, and salt distillation use mass-addition as well as thermal energy input to achieve separation.

Pressure-swing distillation involves use of two columns operating at different pressures. Reactive distillation strives to couple a chemical reaction with separation of the products. Included in Chapter 11 is supercritical-fluid extraction, which makes use of the favorable properties in the vicinity of the critical point to achieve a separation.

Mass-transfer models for multicomponent, multi-stage, vapor–liquid separation operations are available in several process simulators. These models are particularly useful in cases where stage efficiency is low or uncertain and are described in Chapter 12.

Batch distillation has become increasingly popular with the trend toward production of specialty products. Calculation methods, widely used in process simulators, for both binary and multicomponent mixtures are presented in Chapter 13, with an introduction to methods for determining an optimal set of operation steps.

Chapter 6

Absorption and Stripping of Dilute Mixtures

In *absorption* (also called *gas absorption, gas scrubbing,* and *gas washing*), a gas mixture is contacted with a liquid (the *absorbent* or *solvent*) to selectively dissolve one or more components by mass transfer from the gas to the liquid. The components transferred to the liquid are referred to as *solutes* or *absorbate*. Absorption is used to separate gas mixtures; remove impurities, contaminants, pollutants, or catalyst poisons from a gas; or recover valuable chemicals. Thus, the species of interest in the gas mixture may be all components, only the component(s) not transferred, or only the component(s) transferred.

The opposite of absorption is *stripping* (also called *desorption*), wherein a liquid mixture is contacted with a gas to selectively remove components by mass transfer from the liquid to the gas phase. As discussed in Chapter 5, absorbers are frequently coupled with strippers to permit regeneration (or recovery) and recycling of the absorbent. Because stripping is not perfect, absorbent recycled to the absorber contains species present in the vapor entering the absorber. When water is used as the absorbent, it is more common to separate the absorbent from the solute by distillation rather than stripping.

6.0 INSTRUCTIONAL OBJECTIVES

After completing this chapter, you should be able to:

- Explain the difference between absorption and stripping.
- Explain the difference between physical and chemical absorption.
- Explain why absorbers are best operated at high pressure and low temperature, while strippers are best operated at low pressure and high temperature.
- Enumerate different types of industrial equipment for absorption and stripping and explain which are most popular.
- Explain how vapor and liquid streams flow from one tray to another in a trayed tower.
- Compare three different types of trays.
- Explain the difference between random and structured packings and cite examples of each.
- Explain the importance of the liquid distributor and redistributors in a packed column with respect to liquid flow.
- Derive the "operating-line equation," used in graphical methods, starting with a component material balance.
- Calculate the minimum MSA flow rate to achieve a specified recovery of a key component in a single-section, countercurrent cascade.
- Determine graphically, by stepping off stages, or algebraically, the required number of equilibrium stages in a countercurrent cascade to achieve a specified recovery of a key component, given an MSA flow rate greater than the minimum value.
- Define the overall stage efficiency and explain why efficiency values are relatively low for absorbers and at a moderate level for strippers.
- Make preliminary estimates of overall stage efficiency of absorbers and strippers.
- Explain why multiple liquid-flow passes are necessary in trayed columns of moderate to large column diameter.
- Define Murphree point and tray vapor efficiencies and their relationship to overall stage efficiency.
- Explain how experimental stage-efficiency data from a small laboratory Oldershaw column can be scaled up to a large-diameter column.
- Explain two mechanisms by which a trayed column can flood.
- Enumerate the contributions to pressure drop in a trayed column.
- Estimate column diameter and tray pressure drop for a trayed column.
- Estimate tray efficiency from correlations of mass-transfer coefficients using two-film theory.
- Estimate weeping, entrainment, and downcomer backup in a trayed column.

- For a packed column, define the "height equivalent to a theoretical (equilibrium) stage (plate)," HETP, and explain how it and the number of equilibrium stages differ from "height of a transfer unit," HTU, and "number of transfer units," NTU, respectively.
- Explain differences between "loading point" and "flooding point" in a packed column.
- Estimate packed height, packed-column diameter, and pressure drop across the packing.
- Estimate HTU from correlations of mass-transfer coefficients.
- Explain how the number of theoretical stages is computed for concentrated solutions in which equilibrium and operating lines are curved.

Industrial Example

A typical absorption operation is shown in Figure 6.1. The feed, which contains air (21% O_2, 78% N_2, and 1% Ar), water vapor, and acetone vapor, is the gas leaving a dryer where solid cellulose acetate fibers, wet with water and acetone, are dried. The purpose of the 30-tray (equivalent to 10 equilibrium stages) absorber is to remove the acetone by contacting the gas with a suitable absorbent, water. By using countercurrent flow of gas and liquid in a multiple-stage device, the material balance, shown in Figure 6.1, indicates that 99.5% of the acetone is absorbed. The gas leaving the absorber contains only 143 ppm (parts per million) by weight of acetone vapor and can be recycled to the dryer or exhausted to the atmosphere. Although the major component transferred between phases is acetone, the material balance indicates that small amounts of oxygen and nitrogen are also absorbed by the water solvent. Because water is present in both the feed gas and the absorbent, it can be both absorbed and stripped. As seen in Figure 6.1, the net effect is that water is stripped because more water appears in the exit gas than in the feed gas. The exit gas is almost saturated with water vapor and the exit liquid is almost saturated with air. The temperature of the absorbent decreases by 3°C to supply the energy of vaporization needed to strip the water, which

in this example is greater than the energy of condensation liberated from the absorption of acetone.

As was shown in Figure 5.9, the fraction of a component absorbed in a countercurrent cascade depends on the number of equilibrium stages and the absorption factor, $A = L/(KV)$, for that component. For the conditions of Figure 6.1, using $L = 1943$ kmol/h and $V = 703$ kmol/h, estimated K-values and absorption factors, which range over many orders of magnitude, are

Component	$A = L/(KV)$	K-value
Water	89.2	0.031
Acetone	1.38	2.0
Oxygen	0.00006	45,000
Nitrogen	0.00003	90,000
Argon	0.00008	35,000

For acetone, the K-value is based on Eq. (4) of Table 2.3, the modified Raoult's law, $K = \gamma P^s/P$, with $\gamma = 6.7$ for a dilute solution of acetone in water at 25°C and 101.3 kPa. For oxygen and nitrogen, K-values are based on the use of Eq. (6) of Table 2.3, Henry's law, $K = H/P$, using constants from Figure 4.27 at 25°C. For water, the K-value is obtained from Eq. (3) of Table 2.3, Raoult's law, $K = P^s/P$, which applies because the mole fraction of water in the liquid phase is close to 1. For argon, the Henry's law constant at 25°C was obtained from the International Critical Tables [1].

Figure 5.9 shows that if the value of A is greater than 1, any degree of absorption can be achieved: the larger the value of A, the fewer the number of stages required to absorb a desired fraction of the solute. However, very large values of A can correspond to absorbent flow rates that are larger than necessary. From an economic standpoint, the value of A, for the main (key) species to be absorbed, should be in the range of 1.25 to 2.0, with 1.4 being a frequently recommended value. Thus, the above value of 1.38 for acetone is favorable.

For a given feed-gas flow rate and choice of absorbent, factors that influence the value of A are absorbent flow rate, temperature, and pressure. Because $A = L/(KV)$, the larger the absorbent flow rate is, the larger will be the value of A. The required absorbent flow rate can be reduced by reducing the K-value of the solute. Because the K-value for many solutes varies exponentially with temperature and is inversely proportional to pressure, this reduction can be achieved by

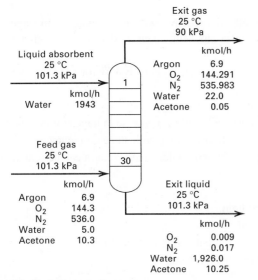

Figure 6.1 Typical absorption process.

reducing the temperature and/or increasing the pressure. Increasing the pressure also serves to reduce the diameter of the equipment for a given gas throughput. However, temperature adjustment by feed-gas refrigeration and/or absorbent refrigeration, and/or adjustment of the feed-gas pressure by gas compression can be expensive. For these reasons, the absorber in Figure 6.1 operates at near-ambient conditions.

For a stripper, the stripping factor, $S = 1/A = KV/L$, is crucial. To reduce the required flow rate of stripping agent, operation of the stripper at a high temperature and/or a low pressure is desirable, with an optimum stripping factor in the vicinity of 1.4.

Absorption and stripping are technically mature separation operations. Design procedures are well developed and commercial processes are common. Table 6.1 lists representative, commercial absorption applications. In most cases, the solutes are contained in gaseous effluents from chemical reactors. Passage of strict environmental standards with respect to pollution by emission of noxious gases has greatly increased the use of gas absorbers in the past decade.

When water and hydrocarbon oils are used as absorbents, no significant chemical reactions occur between the absorbent and the solute, and the process is commonly referred to as *physical absorption*. When aqueous sodium hydroxide (a strong base) is used as the absorbent to dissolve an acid gas, absorption is accompanied by a rapid and irreversible neutralization reaction in the liquid phase and the process is referred to as *chemical absorption* or *reactive absorption*. More complex examples of chemical absorption are processes for absorbing CO_2 and H_2S with aqueous solutions of monoethanolamine (MEA) and diethanolamine (DEA), where a reversible chemical reaction takes place in the liquid phase. Chemical reactions can increase the rate of absorption, increase the absorption capacity of the solvent, increase selectivity to preferentially dissolve only certain components of the gas, and convert a hazardous chemical to a safe compound.

In this chapter, trayed and packed-column equipment for conducting absorption and stripping operations is discussed and fundamental *equilibrium-based* and *rate-based* (mass-transfer) models and calculation procedures, both graphical and algebraic, are presented for physical absorption and stripping of mainly dilute mixtures. The methods also apply to reactive absorption with irreversible and complete chemical reactions of the solute in the liquid phase. Calculations for concentrated mixtures and reactive absorption with reversible chemical reactions are best handled by computer-aided calculations, which are discussed in Chapters 10 and 11. An introduction to calculations for concentrated mixtures in packed columns is given in the last section of this chapter.

Table 6.1 Representative, Commercial Applications of Absorption

Solute	Absorbent	Type of Absorption
Acetone	Water	Physical
Acrylonitrile	Water	Physical
Ammonia	Water	Physical
Ethanol	Water	Physical
Formaldehyde	Water	Physical
Hydrochloric acid	Water	Physical
Hydrofluoric acid	Water	Physical
Sulfur dioxide	Water	Physical
Sulfur trioxide	Water	Physical
Benzene and toluene	Hydrocarbon oil	Physical
Butadiene	Hydrocarbon oil	Physical
Butanes and propane	Hydrocarbon oil	Physical
Naphthalene	Hydrocarbon oil	Physical
Carbon dioxide	Aq. NaOH	Irreversible chemical
Hydrochloric acid	Aq. NaOH	Irreversible chemical
Hydrocyanic acid	Aq. NaOH	Irreversible chemical
Hydrofluoric acid	Aq. NaOH	Irreversible chemical
Hydrogen sulfide	Aq. NaOH	Irreversible chemical
Chlorine	Water	Reversible chemical
Carbon monoxide	Aq. cuprous ammonium salts	Reversible chemical
CO_2 and H_2S	Aq. monoethanolamine (MEA) or diethanolamine (DEA)	Reversible chemical
CO_2 and H_2S	Diethyleneglycol (DEG) or triethyleneglycol (TEG)	Reversible chemical
Nitrogen oxides	Water	Reversible chemical

6.1 EQUIPMENT

Absorption and stripping are conducted mainly in trayed towers (plate columns) and packed columns, and less often in spray towers, bubble columns, and centrifugal contactors, as shown schematically in Figure 6.2. A trayed tower is a vertical, cylindrical pressure vessel in which vapor and liquid, which flow countercurrently, are contacted on a series of trays or plates, an example of which is shown in Figure 6.3. Liquid flows across each tray, over an outlet weir, and into a downcomer, which takes the liquid by gravity to the tray below. Gas flows upward through openings in each tray, bubbling through the liquid on the tray. When the openings are holes, any of the five two-phase-flow regimes shown in

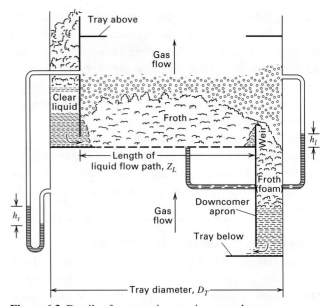

Figure 6.3 Details of a contacting tray in a trayed tower.
[Adapted from B.F. Smith, *Design of Equilibrium Stage Processes*, McGraw-Hill, New York (1963).]

Figure 6.4, and considered in detail by Lockett [2], may occur. The most common and favored regime is the *froth regime,* in which the liquid phase is continuous and the gas passes through in the form of jets or a series of bubbles. The *spray regime,* in which the gas phase is continuous, occurs for low weir heights (low liquid depths) at high gas rates. For low gas rates, the *bubble regime* can occur, in which the liquid is fairly quiescent and bubbles rise in swarms. At high liquid rates, small gas bubbles may be undesirably emulsified. If bubble coalescence is hindered, an undesirable foam forms. Ideally, the liquid carries no vapor bubbles (*occlusion*) to the tray below, the vapor carries no liquid droplets (*entrainment*) to the tray above, and there is no *weeping* of liquid through the holes in the tray. With good contacting, equilibrium between the exiting vapor and liquid phases is approached on each tray.

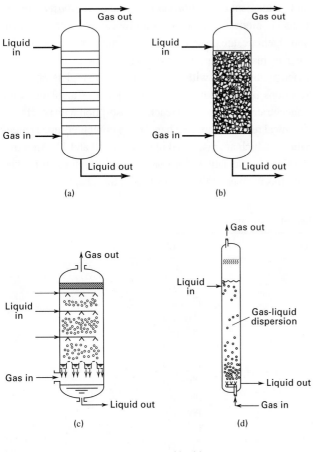

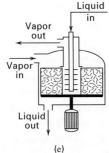

Figure 6.2 Industrial equipment for absorption and stripping: (a) trayed tower; (b) packed column; (c) spray tower; (d) bubble column; (e) centrifugal contactor.

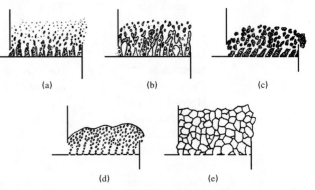

Figure 6.4 Possible vapor–liquid flow regimes for a contacting tray: (a) spray; (b) froth; (c) emulsion; (d) bubble; (e) cellular foam.

[Reproduced by permission from M.J. Lockett, *Distillation Tray Fundamentals,* Cambridge University Press, London (1986).]

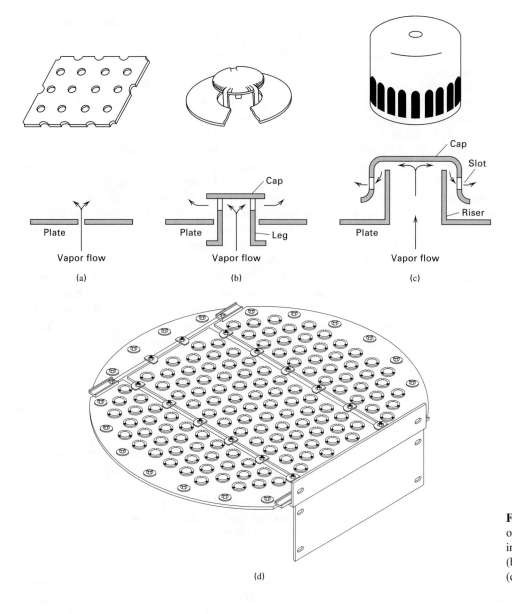

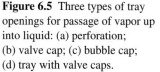

Figure 6.5 Three types of tray openings for passage of vapor up into liquid: (a) perforation; (b) valve cap; (c) bubble cap; (d) tray with valve caps.

As shown in Figure 6.5, openings in the tray for the passage of vapor are most commonly perforations, valves, and/or bubble caps. The simplest is perforations, usually $\frac{1}{8}$ to $\frac{1}{2}$ in. in diameter, used in a so-called *sieve tray* (also called a *perforated tray*). A *valve tray* has much larger openings, commonly from 1 to 2 in. in diameter. Each hole is fitted with a valve that consists of a cap, which overlaps the hole, with legs or a cage to limit the vertical rise while maintaining the horizontal location of the valve. With no vapor flow, each valve sits on the tray, over a hole. As the vapor rate is increased, the valve rises, providing a larger and larger peripheral opening for vapor to flow into the liquid to create a froth. A bubble-cap tray has bubble caps that consist of a fixed cap, 3 to 6 in. in diameter, mounted over and above a concentric riser of 2 to 3 in. in diameter. The cap has rectangular or triangular slots cut around its side. The vapor flows up through the tray opening into the riser, turns around, and passes out through the slots of the cap, into the liquid to form a froth. An 11-ft-diameter column might have trays with 50,000 $\frac{3}{16}$-in.-diameter perforations, or 1,000 2-in.-diameter valve caps, or 500 4-in.-diameter bubble caps.

As listed in Table 6.2, tray types are compared on the basis of cost, pressure drop, mass-transfer efficiency, vapor capacity, and flexibility in terms of turndown ratio (ratio of

Table 6.2 Comparison of Types of Trays

	Sieve Trays	Valve Trays	Bubble-Cap Trays
Relative cost	1.0	1.2	2.0
Pressure drop	Lowest	Intermediate	Highest
Efficiency	Lowest	Highest	Highest
Vapor capacity	Highest	Highest	Lowest
Typical turndown ratio	2	4	5

maximum to minimum vapor capacity). At the limiting vapor capacity, *flooding* of the column occurs because of excessive entrainment of liquid droplets in the vapor causing the liquid flow rate to exceed the capacity of the downcomer and, thus, go back up the column. At low vapor rates, weeping of liquid through the tray openings or vapor pulsation becomes excessive. Because of their low relative cost, sieve trays are preferred unless flexibility is required, in which case valve trays are best. Bubble-cap trays, which exist in many pre-1950 installations, are rarely specified for new installations, but may be preferred when the amount of liquid holdup on a tray must be controlled to provide adequate residence time for a chemical reaction or when weeping must be prevented.

A *packed column,* shown in detail in Figure 6.6, is a vertical, cylindrical pressure vessel containing one or more sections of a packing material over whose surface the liquid flows downward by gravity, as a film or as droplets between packing elements. Vapor flows upward through the wetted packing, contacting the liquid. The sections of packing are contained between a lower gas-injection support plate, which holds the packing, and an upper grid or mesh holddown plate, which prevents packing movement. A *liquid distributor,* placed above the hold-down plate, ensures uniform distribution of liquid over the cross-sectional area of the column as it enters the packed section. If the depth of packing is more than about 20 ft, liquid channeling may occur, causing the liquid to flow down the column mainly near the wall, and

gas to flow mainly up the center of the column, thus greatly reducing the extent of vapor–liquid contact. In that case, a liquid *redistributor* should be installed.

Commercial packing materials include *random* (dumped) packings, some of which are shown in Figure 6.7a, and *structured* (also called arranged, ordered, or stacked packings), some of which are shown in Figure 6.7b. Among the random packings, which are poured into the column, are the old (1895–1950) ceramic Raschig rings and Berl saddles, which are seldom specified for new installations. They have been largely replaced by metal and plastic Pall® rings, metal Bialecki® rings, and ceramic Intalox® saddles, which provide more surface area for mass transfer, a higher flow capacity, and a lower pressure drop. More recently, through-flow packings of a lattice-work design have been developed. These packings, which include metal Intalox® IMTP®; metal, plastic, and ceramic Cascade Mini-Rings®; metal Levapak®; metal, plastic, and ceramic Hiflow® rings; metal Tri-Packs®; and plastic Nor-Pac® rings, exhibit even lower pressure drop per unit height of packing and even higher mass-transfer rates per unit volume of packing. Accordingly, they are called "high-efficiency" random packings. Most random packings are available in nominal diameters, ranging from 1 in. to 3.5 in. As packing size increases, mass-transfer efficiency and pressure drop may decrease. Therefore, for a given column diameter an optimal packing size exists that represents a compromise between these two factors, since low pressure drop and high mass-transfer rates are both desirable. However, to minimize channeling of liquid, the nominal diameter of the packing should be less than one-eighth of the column diameter. Most recently, a "fourth generation" of random packings, including VSP® rings, Fleximax®, and Raschig super-rings, has been developed, which features a very open undulating geometry that promotes even wetting, but with recurrent turbulence promotion. The result is lower pressure drop, but sustained mass-transfer efficiency that may not decrease noticeably with increasing column diameter and may permit a larger depth of packing before a liquid redistributor is necessary. Metal packings are usually preferred because of their superior strength and good wettability. Ceramic packings, which have superior wettability but inferior strength, are used only to resist corrosion at elevated temperatures, where plastics would fail. Plastic packings, usually of polypropylene, are inexpensive and have sufficient strength, but may experience poor wettability, particularly at low liquid rates.

Representative structured packings include the older corrugated sheets of metal gauze, such as Sulzer® BX, Montz™ A, Gempak® 4BG, and Intalox® High-Performance Wire Gauze Packing. Newer and less-expensive structured packings, which are fabricated from sheet metal and plastics and may or may not be perforated, embossed, or surface roughened, include metal and plastic Mellapak™ 250Y, metal Flexipac,® metal and plastic Gempak® 4A, metal Montz™ B1, and metal Intalox® High-Performance Structured Packing.

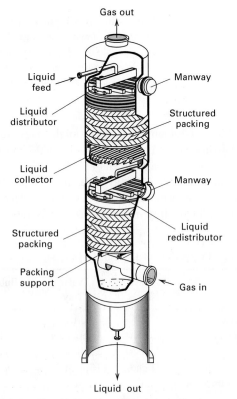

Figure 6.6 Details of internals used in a packed column.

Ceramic Raschig rings

Ceramic Berl saddle

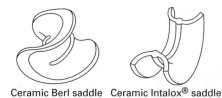
Ceramic Intalox® saddle

Plastic super
Intalox® saddle

Metal Intalox® IMTP

Metal Pall® ring

Plastic Flexiring®

Metal Bialecki® ring

Metal Fleximax®

Metal Cascade
Mini-ring® (CMR)

Metal Top-Pak®

Metal Raschig
Super-ring

Plastic Tellerette®

Plastic Hackett®

Plastic Hiflow® ring

Metal VSP® ring

(a)

Figure 6.7 Typical materials used in a packed column: (a) random packing materials; *(continued)*

Structured packings come with different size openings between adjacent corrugated layers and are stacked in the column. Although structured packings are considerably more expensive per unit volume than random packings, structured packings exhibit far less pressure drop per theoretical stage and have higher efficiency and capacity.

As shown in Table 6.3, packings are usually compared on the basis of the same factors used to compare tray types. However, the differences between random and structured packings are much greater than the differences among the three types of trays listed in Table 6.2.

Table 6.3 Comparison of Types of Packing

	Random		
	Raschig Rings and Saddles	"Through Flow"	Structured
Relative cost	Low	Moderate	High
Pressure drop	Moderate	Low	Very low
Efficiency	Moderate	High	Very high
Vapor capacity	Fairly high	High	High
Typical turndown ratio	2	2	2

If only one or two theoretical stages are required, only a very low pressure drop is allowed, and the solute is very soluble in the liquid phase, the use of a *spray tower* may be advantageous. As shown in Figure 6.2, a spray tower consists of a vertical, cylindrical vessel filled with gas into which liquid is sprayed. A *bubble column,* also shown in Figure 6.2, consists of a vertical, cylindrical vessel partially filled with liquid into which the vapor is bubbled. Vapor pressure drop is high, and only one or two theoretical stages can be achieved. Such a device has a low vapor throughput and should not be considered unless the solute has a very low solubility in the liquid and/or a slow chemical reaction takes place in the liquid phase, thus requiring an appreciable residence time. A novel device is the *centrifugal contactor,* one example of which, as shown in Figure 6.2, consists of a stationary, ringed housing, intermeshed with a ringed rotating section. The liquid phase is fed near the center of the packing, from which it is caused to flow outward by centrifugal force. The vapor phase flows inward by a pressure driving force. Very high mass-transfer rates can be achieved with only moderately high rotation rates. It is possible to obtain the equivalent of several equilibrium stages in a very compact unit. This type of contact is favored when headroom for a trayed tower or packed column is not available or when a short residence time is desired.

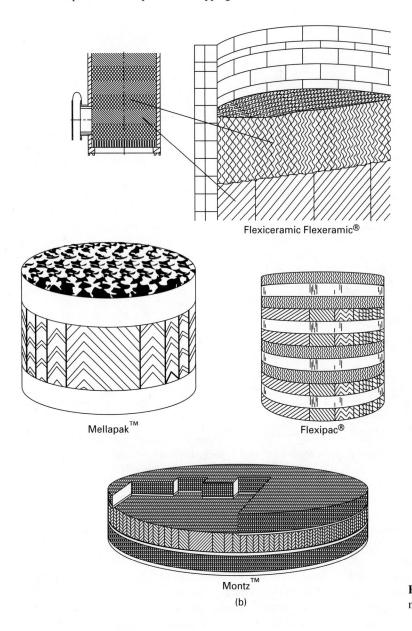

Flexiceramic Flexeramic®

Mellapak™

Flexipac®

Montz™

(b)

Figure 6.7 (*Continued*) (b) structured packing materials.

In most applications, the choice of contacting device is between a trayed tower and a packed column. The latter, using dumped packings, is almost always favored when a column diameter of less than 2 ft and a packed height of not more than 20 ft are sufficient. In addition, packed columns should be considered for corrosive services where ceramic or plastic materials are preferred over metals, in services where foaming may be severe if trays are used, and when pressure drop must be low, as in vacuum or near-ambient-pressure operations. Otherwise, trayed towers, which can be designed and scaled up more reliably, are preferred. Although structured packings are quite expensive, they may be the best choice for a new installation when pressure drop must be very low or for replacing existing trays (retrofitting) when a higher capacity or degree of separation is required in an existing column. Trayed towers are preferred when liquid velocities are low, while columns with random packings are best for high-liquid velocities. The use of structured

packings should be avoided at high-pressures (> 200 psia) and high-liquid flow rates (> 10 gpm/ft^2), Kister [33]. In general, a continuous, turbulent liquid flow is desirable if mass transfer is limiting in the liquid phase, while a continuous, turbulent gas flow is desirable if mass transfer is limiting in the gas phase.

6.2 GENERAL DESIGN CONSIDERATIONS

Design or analysis of an absorber (or stripper) requires consideration of a number of factors, including:

1. Entering gas (liquid) flow rate, composition, temperature, and pressure
2. Desired degree of recovery of one or more solutes
3. Choice of absorbent (stripping agent)
4. Operating pressure and temperature, and allowable gas pressure drop

5. Minimum absorbent (stripping agent) flow rate and actual absorbent (stripping agent) flow rate as a multiple of the minimum rate needed to make the separation

6. Number of equilibrium stages and stage efficiency

7. Heat effects and need for cooling (heating)

8. Type of absorber (stripper) equipment

9. Height of absorber (stripper)

10. Diameter of absorber (stripper)

The ideal absorbent should (a) have a high solubility for the solute(s) to minimize the need for absorbent, (b) have a low volatility to reduce the loss of absorbent and facilitate separation of absorbent from solute(s), (c) be stable to maximize absorbent life and reduce absorbent makeup requirement, (d) be noncorrosive to permit use of common materials of construction, (e) have a low viscosity to provide low pressure drop and high mass- and heat-transfer rates, (f) be nonfoaming when contacted with the gas so as to make it unnecessary to increase absorber dimensions, (g) be nontoxic and nonflammable to facilitate its safe use, and (h) be available, if possible, within the process, to make it unnecessary to provide an absorbent from external sources, or be inexpensive. As already indicated at the beginning of this chapter, the most widely used absorbents are water, hydrocarbon oils, and aqueous solutions of acids and bases. The most common stripping agents are steam, air, inert gases, and hydrocarbon gases.

In general, operating pressure should be high and temperature low for an absorber, to minimize stage requirements and/or absorbent flow rate and to lower the equipment volume required to accommodate the gas flow. Unfortunately, both compression and refrigeration of a gas are expensive. Therefore, most absorbers are operated at feed-gas pressure, which may be greater than ambient pressure, and ambient temperature, which can be achieved by cooling the feed gas and absorbent with cooling water, unless one or both streams already exist at a subambient temperature. Operating pressure should be low and temperature high for a stripper to minimize stage requirements or stripping agent flow rate. However, because maintenance of a vacuum is expensive, strippers are commonly operated at a pressure just above ambient. A high temperature can be used, but it should not be so high as to cause undesirable chemical reactions. Of course, operating temperature and pressure must be compatible with the necessary phase conditions of the streams being contacted. For example, an absorber should not be operated at a pressure and/or temperature that would condense the feed gas, and a stripper should not be operated at a pressure and/or temperature that would vaporize the feed liquid. The possibility of such conditions occurring can be checked by bubble-point and dew-point calculations, discussed in Chapter 4.

For given feed-gas (liquid) flow rate, extent of solute absorption (stripping), operating pressure and temperature, and absorbent (stripping agent) composition, a minimum

absorbent (stripping agent) flow rate exists that corresponds to an infinite number of countercurrent equilibrium contacts between the gas and liquid phases. In every design problem involving flow rates of the absorbent (stripping agent) and number of stages, a trade-off exists between the number of equilibrium stages and the absorbent (stripping agent) flow rate at rates greater than the minimum value. Graphical and analytical methods for computing the minimum flow rate and this trade-off are developed in the following sections for a mixture that is dilute in the solute(s). For this essentially isothermal case, the energy balance can be ignored. As discussed in Chapters 10 and 11, computer-aided methods are best used for concentrated mixtures, where multicomponent phase-equilibrium and mass-transfer effects can become complicated and it is necessary to consider the energy balance.

6.3 GRAPHICAL EQUILIBRIUM-STAGE METHOD FOR TRAYED TOWERS

Consider the countercurrent-flow, trayed tower for absorption (or stripping) operating under isobaric, isothermal, continuous, steady-state flow conditions shown in Figure 6.8. For convenience, the stages are numbered from top to bottom for the absorber and from bottom to top for the stripper. Phase equilibrium is assumed to be achieved at each of the N trays between the vapor and liquid streams leaving the tray. That is, each tray is treated as an equilibrium stage. Assume that the only component transferred from one phase to the

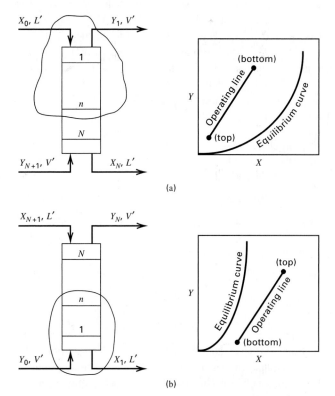

Figure 6.8 Continuous, steady-state operation in a countercurrent cascade with equilibrium stages: (a) absorber; (b) stripper.

other is the solute. For application to an absorber, let:

L' = molar flow rate of solute-free absorbent

V' = molar flow rate of solute-free gas (carrier gas)

X = mole ratio of solute to solute-free absorbent in the liquid

Y = mole ratio of solute to solute-free gas in the vapor

Note that with these definitions, values of L' and V' remain constant through the tower, assuming no vaporization of absorbent into carrier gas or absorption of carrier gas by liquid. For the solute at any equilibrium stage, n, the K-value is given in terms of X and Y as:

$$K_n = \frac{y_n}{x_n} = \frac{Y_n/(1 + Y_n)}{X_n/(1 + X_n)} \qquad (6\text{-}1)$$

where $Y = y/(1 - y)$ and $X = x/(1 - x)$.

For the fixed temperature and pressure and a series of values of x, equilibrium values of y in the presence of the solute-free absorbent and solute-free gas are estimated by methods discussed in Chapter 2. From these values, an equilibrium curve of Y as a function of X is calculated and plotted, as shown in Figure 6.8. In general, this curve will not be a straight line, but it will pass through the origin. If the solute undergoes, in the liquid phase, a complete irreversible conversion by chemical reaction, to a nonvolatile solute, the equilibrium curve will be a straight line of zero slope passing through the origin.

At either end of the towers shown in Figure 6.8, entering and leaving streams and solute mole ratios are paired. For the absorber, the pairs are $(X_0, L'$ and $Y_1, V')$ at the top and $(X_N, L'$ and $Y_{N+1}, V')$ at the bottom; for the stripper, $(X_{N+1}, L'$ and $Y_N, V')$ at the top and $(X_1, L'$ and $Y_0, V')$ at the bottom. These terminal pairs can be related to intermediate pairs of passing streams by the following solute material balances for the envelopes shown in Figure 6.8. The balances are written around one end of the tower and an arbitrary intermediate equilibrium stage, n.

For the absorber,

$$X_0 L' + Y_{n+1} V' = X_n L' + Y_1 V' \qquad (6\text{-}2)$$

or, solving for Y_{n+1},

$$Y_{n+1} = X_n (L'/V') + Y_1 - X_0 (L'/V') \qquad (6\text{-}3)$$

For the stripper,

$$X_{n+1} L' + Y_0 V' = X_1 L' + Y_n V' \qquad (6\text{-}4)$$

or, solving for Y_n,

$$Y_n = X_{n+1} (L'/V') + Y_0 - X_1 (L'/V') \qquad (6\text{-}5)$$

Equations (6-3) and (6-5), which are called *operating-line equations,* are plotted in Figure 6.8. The terminal points of these lines represent the conditions at the top and bottom of the towers. For the absorber, the operating line is above the equilibrium line because, for a given solute concentration in

the liquid, the solute concentration in the gas is always greater than the equilibrium value, thus providing the driving force for mass transfer of solute from the gas to the liquid. For the stripper, the operating line lies below the equilibrium line for the opposite reason. For the coordinate systems in Figure 6.8, the operating lines are straight with a slope of L'/V'.

For an absorber, the terminal point of the operating line at the top of the tower is fixed at X_0 by the amount of solute, if any, in the entering absorbent, and the specified degree of absorption of the solute, which fixes the value of Y_1 in the leaving gas. The terminal point of the operating line at the bottom of the tower depends on Y_{N+1} and the slope of the operating line and, thus, the flow rate, L', of solute-free absorbent.

Minimum Absorbent Flow Rate

Operating lines for four different absorbent flow rates are shown in Figure 6.9, where each operating line passes through the terminal point, (Y_1, X_0), at the top of the column, and corresponds to a different liquid absorbent rate and corresponding slope, L'/V'. To achieve the desired value of Y_1 for given Y_{N+1}, X_0, and V', the solute-free absorbent flow rate L', must lie in the range of ∞ (operating line 1) to $L'_{\min}$ (operating line 4). The value of the solute concentration in the outlet liquid, X_N, depends on L' by a material balance on

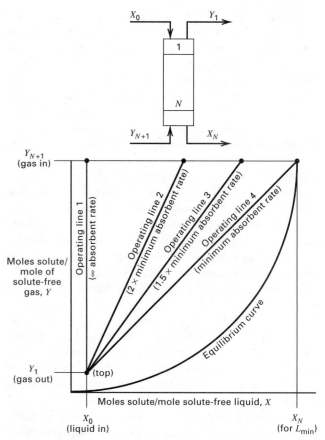

Figure 6.9 Operating lines for an absorber.

the solute for the entire absorber. From (6-2), for $n = N$,

$$X_0 L' + Y_{N+1} V' = X_N L' + Y_1 V' \qquad (6\text{-}6)$$

or

$$L' = \frac{V'(Y_{N+1} - Y_1)}{(X_N - X_0)} \qquad (6\text{-}7)$$

Note that the operating line can terminate at the equilibrium line, as for operating line 4, but cannot cross it because that would be a violation of the second law of thermodynamics.

The value of $L'_{\min}$ corresponds to a value of X_N (leaving the bottom of the tower) in equilibrium with Y_{N+1}, the solute concentration in the feed gas. It takes an infinite number of stages for this equilibrium to be achieved. An expression for $L'_{\min}$ of an absorber can be derived from (6-7) as follows.

For stage N, (6-1) becomes, for the minimum absorbent rate,

$$K_N = \frac{Y_{N+1}/(1 + Y_{N+1})}{X_N/(1 + X_N)} \qquad (6\text{-}8)$$

Solving (6-8) for X_N and substituting the result into (6-7) gives

$$L'_{\min} = \frac{V'(Y_{N+1} - Y_1)}{\{Y_{N+1}/[Y_{N+1}(K_N - 1) + K_N]\} - X_0} \qquad (6\text{-}9)$$

For dilute-solute conditions, where $Y \approx y$ and $X \approx x$, (6-9) approaches

$$L'_{\min} = V' \left(\frac{y_{N+1} - y_1}{\dfrac{y_{N+1}}{K_N} - x_0} \right) \qquad (6\text{-}10)$$

Furthermore, if the entering liquid contains no solute, that is, $X_0 \approx 0$, (6-10) approaches

$$L'_{\min} = V' K_N \text{ (fraction of solute absorbed)} \qquad (6\text{-}11)$$

This equation is reasonable because it would be expected that $L'_{\min}$ would increase with increasing V', K-value, and fraction of solute absorbed.

The selection of the actual operating absorbent flow rate is based on some multiple of $L'_{\min}$, typically from 1.1 to 2. A value of 1.5 corresponds closely to the value of 1.4 for the optimal absorption factor mentioned earlier. In Figure 6.9, operating lines 2 and 3 correspond to 2.0 and 1.5 times $L'_{\min}$, respectively. As the operating line moves from 1 to 4, the number of required equilibrium stages, N, increases from zero to infinity. Thus, a trade-off exists between L' and N, and an optimal value of L' exists.

A similar derivation of $V'_{\min}$, for the stripper of Figure 6.8, results in an expression analogous to (6-11):

$$V'_{\min} = \frac{L'}{K_N} \text{ (fraction of solute stripped)} \qquad (6\text{-}12)$$

Number of Equilibrium Stages

As shown in Figure 6.10a, the operating line relates the solute concentration in the vapor passing upward between

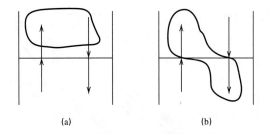

Figure 6.10 Vapor–liquid stream relationships: (a) operating line (passing streams); (b) equilibrium curve (leaving streams).

two stages to the solute concentration in the liquid passing downward between the same two stages. Figure 6.10b illustrates that the equilibrium curve relates the solute concentration in the vapor leaving an equilibrium stage to the solute concentration in the liquid leaving the same stage. This makes it possible, in the case of an absorber, to start from the top of the tower (at the bottom of the Y–X diagram) and move to the bottom of the tower (at the top of the Y–X diagram) by constructing a staircase alternating between the operating line and the equilibrium curve, as shown in Figure 6.11a. The number of equilibrium stages required for a particular absorbent flow rate corresponding to the slope of the operating line, which in Figure 6.11a is for $(L'/V') = 1.5(L'_{\min}/V')$, is stepped off by moving up the staircase, starting from the point (Y_1, X_0), on the operating line and moving horizontally to the right to the point (Y_1, X_1) on the equilibrium curve. From there, a vertical move is made to the point (Y_2, X_1) on the operating line. Proceeding in this manner, the staircase is climbed until the terminal point (Y_{N+1}, X_N) on the operating line is reached. As shown in Figure 6.11a, the stages are counted at the points of the staircase on the equilibrium curve. As the slope (L'/V') is increased, fewer equilibrium stages are required. As (L'/V') is decreased, more stages are required until $(L'_{\min}/V')$ is reached, at which the operating line and equilibrium curve intersect at a so-called *pinch point*, for which an infinite number of stages is required. Operating line 4 in Figure 6.9 has a pinch point at Y_{N+1}, X_N. If (L'/V') is reduced below $(L'_{\min}/V')$, the specified extent of absorption of the solute cannot be achieved.

The number of equilibrium stages required for stripping a solute is determined in a manner similar to that for absorption. An illustration is shown in Figure 6.11b, which refers to Figure 6.8b. For given specifications of Y_0, X_{N+1}, and the extent of stripping of the solute, which corresponds to a value of X_1, $V'_{\min}$ is determined from the slope of the operating line that passes through the points (Y_0, X_1), and (Y_N, X_{N+1}) on the equilibrium curve. The operating line in Figure 6.11b is for $V' = 1.5V'_{\min}$ or a slope of $(L'/V') = (L'/V'_{\min})/1.5$.

In Figure 6.11, the number of equilibrium stages for the absorber and stripper is exactly three each. These integer results are coincidental. Ordinarily, the result is some fraction above an integer number of stages, as is the case in the following example. In practice, the result is usually rounded to the next highest integer.

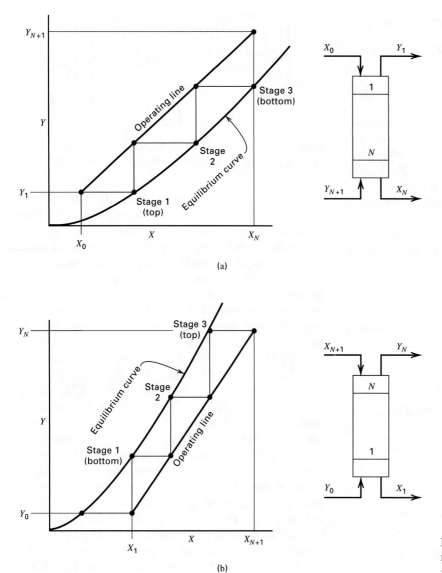

Figure 6.11 Graphical determination of the number of equilibrium stages for (a) absorber and (b) stripper.

EXAMPLE 6.1

When molasses is fermented to produce a liquor containing ethyl alcohol, a CO_2-rich vapor containing a small amount of ethyl alcohol is evolved. The alcohol can be recovered by absorption with water in a sieve-tray tower. For the following conditions, determine the number of equilibrium stages required for countercurrent flow of liquid and gas, assuming isothermal, isobaric conditions in the tower and neglecting mass transfer of all components except ethyl alcohol.

Entering gas:
  180 kmol/h; 98% CO_2, 2% ethyl alcohol; 30°C, 110 kPa

Entering liquid absorbent:
  100% water; 30°C, 110 kPa

Required recovery (absorption) of ethyl alcohol: 97%

SOLUTION

From Section 5.7 for a single-section, countercurrent cascade, the number of degrees of freedom is $2N + 2C + 5$. All stages operate adiabatically at a pressure of approximately 1 atm, taking $2N$ degrees of freedom. The entering gas is completely specified, tak-

ing $C + 2$ degrees of freedom. The entering liquid flow rate is not specified; thus, only $C + 1$ degrees of freedom are taken by the entering liquid. The recovery of ethyl alcohol takes one additional degree of freedom. Thus, the total number of degrees of freedom taken by the problem specification is $2N + 2C + 4$. This leaves one additional specification to be made, which in this example can be the entering liquid flow rate at, say, 1.5 times the minimum value.

The above application of the degrees of freedom analysis from Chapter 5 has assumed the use of an energy balance for each stage. The energy balances are assumed to result in the assumed isothermal operation at 30°C.

Assume that the exiting absorbent will be dilute in ethyl alcohol, whose K-value is determined from a modified Raoult's law, $K = \gamma P^s/P$. The vapor pressure of ethyl alcohol at 30°C is 10.5 kPa. At infinite dilution in water at 30°C, the liquid-phase activity coefficient of ethyl alcohol is taken as 6. Therefore, $K = (6)(10.5)/110 = 0.57$. The minimum solute-free absorbent rate is given by (6-11), where the solute-free gas rate, V', is $(0.98)(180) = 176.4$ kmol/h. Thus,

$$L'_{\min} = (176.4)(0.57)(0.97) = 97.5 \text{ kmol/h}$$

The actual solute-free absorbent rate, at 50% above the minimum rate, is

$$L' = 1.5(97.5) = 146.2 \text{ kmol/h}$$

The amount of ethyl alcohol transferred from the gas to the liquid is 97% of the amount of alcohol in the entering gas or

$$(0.97)(0.02)(180) = 3.49 \text{ kmol/h}$$

The amount of ethyl alcohol remaining in the exiting gas is

$$(1.00 - 0.97)(0.02)(180) = 0.11 \text{ kmol/h}$$

We now compute the alcohol mole ratios at both ends of the operating line as follows, referring to Figure 6.8a:

$$\text{top} \left\{ X_0 = 0, \qquad\qquad Y_1 = \frac{0.11}{176.4} = 0.0006 \right.$$

$$\text{bottom} \left\{ Y_{N+1} = \frac{0.11 + 3.49}{176.4} = 0.0204, \quad X_N = \frac{3.49}{146.2} = 0.0239 \right.$$

The equation for the operating line from (6-3) with $X_0 = 0$ is

$$Y_{N+1} = \left(\frac{146.2}{176.4} \right) X_N + 0.0006 = 0.829 X_N + 0.0006 \quad (1)$$

It is clear that we are dealing with a dilute system. The equilibrium curve for ethyl alcohol can be determined from (6-1) using the value of $K = 0.57$ computed above. From (6-1),

$$0.57 = \frac{Y/(1+Y)}{X/(1+X)}$$

Solving for Y, we obtain

$$Y = \frac{0.57X}{1 + 0.43X} \quad (2)$$

To cover the entire column, the necessary range of X for a plot of Y vs X is 0 to almost 0.025. From the Y–X equation, (2),

Y	X
0.00000	0.000
0.00284	0.005
0.00569	0.010
0.00850	0.015
0.01130	0.020
0.01410	0.025

For this dilute system in ethyl alcohol, the maximum error in Y is 1.0% if Y is taken simply as $Y = KX = 0.57X$.

The equilibrium curve, which is almost straight in this example, and a straight operating line drawn through the terminal points (Y_1, X_0) and (Y_{N+1}, X_N) is given in Figure 6.12. The determination of points for the operating line and the equilibrium curve, as well as the plot of the points, is conveniently done with a spreadsheet program on a computer using Eqs. (1) and (2). The theoretical stages are stepped off as shown starting from the top stage (Y_1, X_0) located near the lower left corner of Figure 6.12. The required number of theoretical stages for 97% absorption of ethyl alcohol is just slightly more than six. Accordingly, it is best to provide seven theoretical stages.

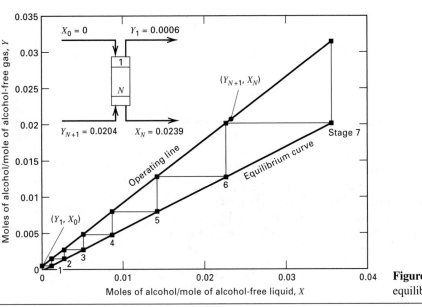

Figure 6.12 Graphical determination of number of equilibrium stages for an absorber.

6.4   ALGEBRAIC METHOD FOR DETERMINING THE NUMBER OF EQUILIBRIUM STAGES

Graphical methods for determining equilibrium stages have great educational value because a fairly complex multistage problem can be readily followed and understood. Furthermore, one can quickly gain visual insight into the phenomena involved. However, the application of a graphical method can become very tedious when (1) the problem specification fixes the number of stages rather than the percent recovery of solute, (2) when more than one solute is being absorbed or stripped, (3) when the best operating conditions of temperature and pressure are to be determined so that the location of the equilibrium curve is unknown, and/or (4) if very low or very high concentrations force the graphical construction to the corners of the diagram so that multiple y–x diagrams of varying sizes and dimensions are needed.

Then, the application of an algebraic method may be preferred.

The Kremser method for single-section cascades, as developed in Section 5.4, is ideal for absorption and stripping of dilute mixtures. For example, (5-48) and (5-50) can be written in terms of the fraction of solute absorbed or stripped as

$$\text{Fraction of a solute, } i, \text{ absorbed} = \frac{A_i^{N+1} - A_i}{A_i^{N+1} - 1} \quad (6\text{-}13)$$

and

$$\text{Fraction of a solute, } i, \text{ stripped} = \frac{S_i^{N+1} - S_i}{S_i^{N+1} - 1} \quad (6\text{-}14)$$

where the solute absorption and stripping factors are, respectively,

$$A_i = L/(K_i V) \quad (6\text{-}15)$$

$$S_i = K_i V/L \quad (6\text{-}16)$$

Values of L and V in moles per unit time may be taken as entering values. Values of K_i depend mainly on temperature, pressure, and liquid-phase composition. Methods for estimating K-values are discussed in detail in Chapter 2. At near-ambient pressure, for dilute mixtures, some common expressions are

$$K_i = P_i^s/P \qquad \text{(Raoult's law)} \qquad (6\text{-}17)$$

$$K_i = \gamma_{iL}^{\infty} P_i^s/P \qquad \text{(modified Raoult's law)} \quad (6\text{-}18)$$

$$K_i = H_i/P \qquad \text{(Henry's law)} \qquad (6\text{-}19)$$

$$K_i = P_i^s/x_i^* P \qquad \text{(solubility)} \qquad (6\text{-}20)$$

The first expression applies for ideal solutions involving solutes at subcritical temperatures. The second expression is useful for moderately nonideal solutions when activity coefficients are known at infinite dilution. For solutes at supercritical temperatures, the use of Henry's law may be preferable. For sparingly soluble solutes at subcritical temperatures, the fourth expression is preferred when solubility data in mole fractions, x_i^*, are available. This expression is derived by considering a three-phase system consisting of an ideal-vapor-containing solute, carrier vapor, and solvent; a pure or near-pure solute as liquid (1); and the solvent liquid (2) with dissolved solute. In that case, for solute, i, at equilibrium between the two liquid phases,

$$x_i^{(1)} \gamma_{iL}^{(1)} = x_i^{(2)} \gamma_{iL}^{(2)}$$

But,

$$x_i^{(1)} \approx 1, \quad \gamma_{iL}^{(1)} \approx 1, \quad x_i^{(2)} = x_i^*$$

Therefore,

$$\gamma_{iL}^{(2)} \approx 1/x_i^*$$

and from (6-18),

$$K_i^{(2)} = \gamma_{iL}^{(2)} P_i^s/P = P_i^s/(x_i^* P)$$

The advantage of (6-13) and (6-14) is that they can be solved directly for the percent absorption or stripping of a solute when the number of theoretical stages, N, and the absorption or stripping factor are known.

EXAMPLE 6.2

As discussed by Okoniewski [3], volatile organic compounds (VOCs) can be stripped from wastewater by air. Such compounds are to be stripped at 70°F and 15 psia from 500 gpm of wastewater with 3,400 scfm of air (standard conditions of 60°F and 1 atm) in an existing tower containing 20 plates. A chemical analysis of the wastewater shows three organic chemicals in the amounts shown in the following table. Included are necessary thermodynamic properties from the 1966 *Technical Data Book–Petroleum Refining* of the American Petroleum Institute. For all three organic compounds, the wastewater concentrations can be shown to be below the solubility values.

Organic Compound	Concentration in the Wastewater, mg/L	Solubility in Water at 70°F, mole fraction	Vapor Pressure at 70°F, psia
Benzene	150	0.00040	1.53
Toluene	50	0.00012	0.449
Ethylbenzene	20	0.000035	0.149

It is desirable that 99.9% of the total VOCs be stripped, but the plate efficiency of the tower is uncertain, with an estimated range of 5% to 20%, corresponding to one to four theoretical stages for the 20-plate tower. Calculate and plot the percent stripping of each of the three organic compounds for one, two, three, and four theoretical stages. Under what conditions can we expect to achieve the desired degree of stripping? What should be done with the exiting air?

SOLUTION

Because the wastewater is dilute in the VOCs, the Kremser equation may be applied independently to each of the three organic chemicals. We will ignore the absorption of air by the water and the stripping of water by the air. The stripping factor for each compound is given by $S_i = K_i V/L$, where V and L will be taken at entering conditions. The K-value may be computed from a modified Raoult's law, $K_i = \gamma_{iL} P_i^s/P$, where for a compound that is only slightly soluble, take $\gamma_{iL} = 1/x_i^*$, where x_i^* is the solubility in mole fraction. Thus, from (6-20), $K_i = P_i^s/x_i^* P$

$$V = 3,400(60)/(379 \text{ scf/lbmol}) \quad \text{or} \quad 538 \text{ lbmol/h}$$

$$L = 500(60)(8.33 \text{ lb/gal})/(18.02 \text{ lb/lbmol}) \quad \text{or} \quad 13,870 \text{ lbmol/h}$$

The corresponding K-values and stripping factors are

Component	K at 70°F, 15 psia	S
Benzene	255	9.89
Toluene	249	9.66
Ethylbenzene	284	11.02

From (6-14),

$$\text{Fraction stripped} = \frac{S^{N+1} - S}{S^{N+1} - 1}$$

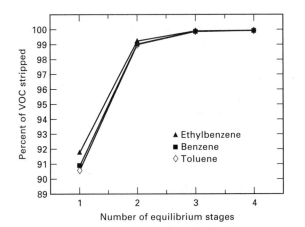

Figure 6.13 Results of Example 6.2 for stripping of VOCs from water with air.

The calculations when carried out with a spreadsheet computer program give the following results:

Component	Percent Stripped			
	1 Stage	2 Stages	3 Stages	4 Stages
Benzene	90.82	99.08	99.91	99.99
Toluene	90.62	99.04	99.90	99.99
Ethylbenzene	91.68	99.25	99.93	99.99

The results are quite sensitive to the number of theoretical stages as shown in Figure 6.13. To achieve 99.9% removal of the total VOCs, three theoretical stages are needed, corresponding to the necessity for a 15% stage efficiency in the existing 20-tray tower.

It is best to process the exiting air to remove or destroy the VOCs, particularly the benzene, which is a carcinogen [4]. The amount of benzene stripped is

(500 gpm)(60 min/h)(3.785 liters/gal)(150 mg/liters)
= 17,030,000 mg/h   or   37.5 lb/h

If benzene is valued at $0.30/lb, the annual value is approximately $100,000. It is doubtful that this would justify a recovery technique, such as carbon adsorption. It is perhaps preferable to destroy the VOCs by incineration. For example, the air can be sent to a utility boiler, a waste-heat boiler, or a catalytic incinerator. It is also to be noted that the amount of air was arbitrarily given as 3,400 scfm. To complete the design procedure, various air rates should be investigated. It will also be necessary to verify by methods given later in this chapter that, at the chosen air flow rates, no flooding or weeping will occur in the column.

6.5 STAGE EFFICIENCY

Graphical and algebraic methods for determining stage requirements for absorption and stripping assume equilibrium with respect to both heat and mass transfer at each stage. Thus, the number of *equilibrium stages* (*theoretical stages, ideal stages,* or *ideal plates*) is determined or specified when

using those methods. Except when temperature changes significantly from stage to stage, the assumption that vapor and liquid phases leaving a stage are at the same temperature is often reasonable. The assumption of equilibrium with respect to mass transfer, however, is not often reasonable and, for streams leaving a stage, vapor-phase mole fractions are not related to liquid-phase mole fractions simply by thermodynamic K-values. To determine the actual number of plates, the number of equilibrium stages must be adjusted with a *stage efficiency* (*plate efficiency* or *tray efficiency*).

Stage efficiency concepts are applicable to devices in which the phases are contacted and then separated, that is, when discrete stages can be identified. This is not the case for packed columns or continuous-contact devices. For these, the efficiency is imbedded into an equipment- and system-dependent parameter, an example of which is the HETP (height of packing equivalent to a theoretical plate).

The simplest approach for staged columns, in preliminary design studies and in the evaluation of the performance of an existing column, is to apply an overall stage (or column) efficiency, defined by Lewis [5] as

$$E_o = N_t/N_a \qquad (6-21)$$

where E_o is the fractional overall stage efficiency, usually less than 1.0; N_t is the calculated number of equilibrium (theoretical) stages; and N_a is the actual number of contacting trays or plates (usually greater than N_t) required. Based on the results of extensive research conducted over a period of more than 60 years, the overall stage efficiency has been found to be a complex function of the

1. Geometry and design of the contacting trays
2. Flow rates and flow paths of vapor and liquid streams
3. Compositions and properties of vapor and liquid streams

For well-designed trays and for flow rates near the capacity limit, E_o depends mainly on the physical properties of the vapor and liquid streams.

Values of E_o can be predicted by any of the following four methods:

1. Comparison with performance data from industrial columns for the same or similar systems
2. Use of empirical efficiency models derived from data on industrial columns
3. Use of semitheoretical models based on mass- and heat-transfer rates
4. Scale-up from data obtained with laboratory or pilot-plant columns

These methods, which are discussed in some detail in the following four subsections, are applied to other vapor–liquid separation operations, such as distillation, as well as to absorption and stripping. Suggested correlations of mass-transfer coefficients for trayed towers are deferred to Section 6.6, following the discussion of tray capacity.

Table 6.4 Performance Data for Absorbers and Strippers in Hydrocarbon Service

Service	Type of Tray	Column Diameter, ft	No. of Trays	Tray Spacing, in.	Average Pressure, psia	Average Temp., °F	Molar Average Liquid Viscosity, cP	Overall Stage Efficiency, %
Absorption of butane	Bubble cap	4	24	18	260	120	0.48	36
Absorption of butane	Bubble cap	5	16	30	254	132	0.31	50
Absorption of butane	Bubble cap	4	16	24	94	117	1.41	10.4
Steam stripping of kerosene	Bubble cap	5	4	30	68	448	0.205	57
Steam stripping of gas oil	Bubble cap	5	6	30	60	507	0.250	49

Source: H.G. Drickamer and J.R. Bradford [6].

Performance Data

Performance data obtained from industrial absorption and stripping columns equipped with trays generally include gas- and liquid-feed and product flow rates and compositions, average column pressure and temperature or pressures and temperatures at the bottom and top of the column, number of actual trays, N_a, column diameter, and type of tray with, perhaps, some details of the tray design. From these data, particularly if the system is dilute with respect to the solute(s), the graphical or algebraic methods, described in Sections 6.3 and 6.4, respectively, can be used to estimate the number of equilibrium stages, N_t, required. Then (6-21) can be applied to determine the overall stage efficiency, E_o. Values of E_o for absorbers and strippers are typically low, often less than 50%.

Table 6.4 presents performance data, from a study by Drickamer and Bradford [6], for five industrial hydrocarbon absorption and stripping operations using columns with bubble-cap trays. For the three absorbers, the stage efficiencies are based on the absorption of *n*-butane as the key component. For the two strippers, both of which use steam as the stripping agent, the key component is not given, but is probably *n*-heptane. Although the data cover a wide range of average pressure and temperature, the overall stage efficiencies, which cover a wide range of 10.4% to 57%, appear to depend primarily on the molar average liquid viscosity, a key factor for the rate of mass transfer in the liquid phase.

The gas feed to a hydrocarbon absorber contains a range of light hydrocarbons, each of which is absorbed to a different extent based on its *K*-value, as illustrated in Example 5.3. The data of Jackson and Sherwood [7] for a 9-ft-diameter hydrocarbon absorber equipped with 19 bubble-cap trays on 30-in. tray spacing and operating at 92 psia and 60°F, as analyzed by O'Connell [8] and summarized in Table 6.5, show that each component being absorbed has a different overall stage efficiency, which appears to increase with decreasing *K*-value (increasing solubility in the liquid absorbent). For

Table 6.5 Effect of Species on Overall Stage Efficiency in a 9-ft-Diameter Industrial Absorber Using Bubble-Cap Trays

Component	Overall Stage Efficiency, %
Ethylene	10.3
Ethane	14.9
Propylene	25.5
Propane	26.8
Butylene	33.8

Source: H.E. O'Connell [8].

the same molar-average liquid viscosity (1.90 cps), the overall stage efficiency is seen to vary from as low as 10.3% for ethylene, the most-volatile species considered, to 33.8% for butylene (presumably *n*-butene), the least-volatile species considered.

An even more dramatic effect of the species solubility in the absorbent on the overall stage efficiency is seen in Table 6.6, from a study by Walter and Sherwood [9] using small laboratory, bubble-cap tray columns ranging in size from 2 to 18 in. in diameter. Stage efficiencies vary over a very wide range from 0.65% to 69%. Comparing the data for the water absorption of ammonia (a very soluble gas) and carbon dioxide (a slightly soluble gas), it is clear that the solubility of the gas (i.e., the *K*-value) has a large effect on stage efficiency. Thus, low stage efficiency can occur when the liquid viscosity is high and/or the gas solubility is low (high *K*-value); high stage efficiency can occur when the liquid viscosity is low and the gas solubility is high (low *K*-value).

Empirical Correlations

Using 20 sets of performance data from industrial hydrocarbon absorbers and strippers, including the data in Table 6.4, Drickamer and Bradford [6] correlated the overall stage efficiency of the key component absorbed or stripped with just

Table 6.6 Performance Data for Absorption in Laboratory Bubble-Cap Tray Columns

Service	Column Diameter, in.	No. of Trays	Tray Spacing, in.	Average Pressure, psia	Average Temp., °F	Overall Stage Efficiency, %
Absorption of ammonia in water	18	1	—	14.7	57	69
Absorption of isobutylene in heavy naphtha	2	1	—	66	78.8	36.4
Absorption of propylene in gas oil	2	1	—	66	118.4	13.1
Absorption of propylene in gas lube oil	2	1	—	66	105.8	4.7
Absorption of carbon dioxide in water	18	1	—	14.7	50.4	2.0
Desorption of carbon dioxide from 43.7 wt% aqueous glycerol	5	4	11	14.7	77	0.65

Source: J.F. Walter and T.K. Sherwood [9].

the molar-average viscosity of the rich oil (liquid leaving an absorber or liquid entering a stripper) at the average tower temperature over a viscosity range of 0.19 to 1.58 cP. The empirical equation,

$$E_o = 19.2 - 57.8 \log \mu_L, \quad 0.2 < \mu_L < 1.6 \text{ cP} \quad (6-22)$$

where E_o is in percent and μ is in centipoise, fits the data with average- and maximum-percent deviations of 10.3% and 41%, respectively. A plot of the Drickamer and Bradford correlation, compared to performance data, is given in Figure 6.14. Equation (6-22) should not be used for absorption into nonhydrocarbon liquids and is restricted to the listed range of the liquid viscosity data used to develop the correlation.

Mass-transfer theory indicates that when the volatility of species being absorbed or stripped covers a wide range, the relative importance of liquid-phase and gas-phase mass-transfer resistances can shift. Thus, O'Connell [8] found that the Drickamer-Bradford correlation, (6-22), was inadequate for absorbers and strippers when applied to species covering a wide range of volatility or K-values. This additional effect is indicated clearly in the performance data of Tables 6.5 and 6.6, where liquid viscosity alone cannot correlate the data. O'Connell obtained a more general correlation by using a parameter that included not only the liquid viscosity but also the liquid density and the Henry's law constant of the species being absorbed or stripped. Edmister [10] and Lockhart and Leggett [11] suggested slight modifications to the O'Connell correlation to permit its use with K-values (instead of Henry's-law constants). An O'Connell-type plot of overall stage efficiency for absorption or stripping in bubble-cap tray columns is given in Figure 6.15.

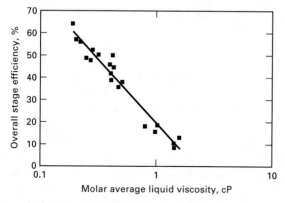

Figure 6.14 Drickamer and Bradford correlation for plate efficiency of hydrocarbon absorbers and strippers.

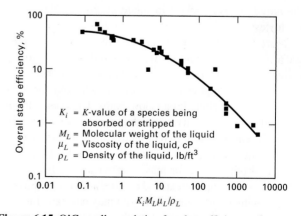

Figure 6.15 O'Connell correlation for plate efficiency of absorbers and strippers.

The correlating parameter, suggested by Edmister, is $K_i M_L \mu_L / \rho_L$, where:

K_i = K-value of species being absorbed or stripped

M_L = molecular weight of the liquid, lb/lbmol

μ_L = viscosity of the liquid, cP

ρ_L = density of the liquid, lb/ft^3

Thus, the correlating parameter has the units of cP-ft^3/lbmol. A reasonable fit to the 33 data points used by O'Connell is given by the empirical equation

$$\log\ E_o = 1.597 - 0.199 \log\left(\frac{K M_L \mu_L}{\rho_L}\right)$$
$$- 0.0896\left[\log\left(\frac{K M_L \mu_L}{\rho_L}\right)\right]^2 \quad (6\text{-}23)$$

The average and maximum deviations of (6-23) for the 33 data points of Figure 6.15 are 16.3% and 157%, respectively. More than 50% of the data points, including points for the highest- and lowest-observed efficiencies, are predicted to within 10%.

The 33 data points in Figure 6.15 cover a wide range of conditions:

Column diameter:	2 in. to 9 ft
Average pressure:	14.7 to 485 psia
Average temperature:	60 to 138°F
Liquid viscosity:	0.22 to 21.5 cP
Overall stage efficiency:	0.65 to 69%

Absorbents include both hydrocarbons and water. For the absorption or stripping of more than one species, because of the effect of species K-value, different stage efficiencies are predicted, as observed from performance data of the type shown in Table 6.5. The inclusion of the K-value also permits the correlation to be used for aqueous systems where the solute may exhibit a very wide range of solubility (e.g., ammonia versus carbon dioxide) as included in Table 6.6. In using Figure 6.15 or Eq. (6-23), the K-value and absorbent properties are best evaluated at the end of the tower where the liquid phase is richest in solute(s). Prudent designs use the lowest predicted efficiency.

Most of the data used to develop the correlation of Figure 6.15 are for columns having a liquid flow path across the active tray area of from 2 to 3 ft. Theory and experimental data show that higher efficiencies are achieved for longer flow paths. For short liquid flow paths, the liquid flowing across the tray is usually completely mixed. For longer flow paths, the equivalent of two or more completely mixed, successive liquid zones may be present. The result is a greater average driving force for mass transfer and, thus, a higher efficiency—perhaps greater than 100%. For example, a column with a 10-ft liquid flow path may have an efficiency as much as 25% greater than that predicted by (6-23). However, at high liquid rates, long liquid-path lengths are

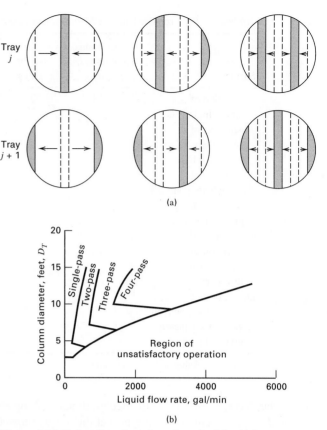

(a)

(b)

Figure 6.16 Estimation of number of required liquid flow passes. (a) Multipass trays: (1) two-pass; (2) three-pass; (3) four-pass. (b) Flow pass selection.
(Derived from *Koch Flexitray Design Manual, Bulletin 960,* Koch Engineering Co., Inc., Wichita, KA, 1960.)

undesirable because they lead to excessive hydraulic gradients. When the effective height of a liquid on a tray is appreciably higher on the inflow side than at the overflow weir, vapor may prefer to enter the tray in the latter region, leading to nonuniform bubbling action. Multipass trays, as shown in Figure 6.16a, are used to prevent excessive liquid gradients. Estimation of the desired number of flow paths can be made with Figure 6.16b, where, e.g., a 10-foot-diameter column with a liquid flow rate of 1000 gpm should use a three-pass tray.

Based on estimates of the number of actual trays and tray spacing, the height of a column between the top tray and the bottom tray is computed. By adding another 4 ft above the top tray for removal of entrained liquid and 10 ft below the bottom tray for bottoms surge capacity, the total column height is estimated. If the height is greater than 212 ft (equivalent to 100 trays on 24-in. spacing), two or more columns arranged in series may be preferable to a single column. Perhaps the tallest column in the world, located at the Shell Chemical Company complex in Deer Park, Texas, stands 338 ft tall [*Chem. Eng.,* **84** (26), 84 (1977)].

EXAMPLE 6.3

Performance data, given below, for a bubble-cap tray absorber located in a Texas petroleum refinery, were reported by Drickamer and Bradford [6]. Based on these data, back-calculate the overall stage efficiency for n-butane and compare the result with both the Drickamer–Bradford and O'Connell correlations. Lean oil and rich gas enter the tower; rich oil and lean gas leave the tower.

Performance Data

Number of plates	16
Plate spacing, in.	24
Tower diameter, ft	4
Tower pressure, psig	79
Lean oil temperature, °F	102
Rich oil temperature, °F	126
Rich gas temperature, °F	108
Lean gas temperature, °F	108
Lean oil rate, lbmol/h	368
Rich oil rate, lbmol/h	525.4
Rich gas rate, lbmol/h	946
Lean gas rate, lbmol/h	786.9
Lean oil molecular weight	250
Lean oil viscosity at 116°F, cP	1.4
Lean oil gravity, °API	21

Stream Compositions, Mol%

Component	Rich Gas	Lean Gas	Rich Oil	Lean Oil
C_1	47.30	55.90	1.33	
C_2	8.80	9.80	1.16	
$C_3^=$	5.20	5.14	1.66	
C_3	22.60	21.65	8.19	
$C_4^=$	3.80	2.34	3.33	
nC_4	7.40	4.45	6.66	
nC_5	3.00	0.72	4.01	
nC_6	1.90		3.42	
Oil absorbent			70.24	100
Totals	100.00	100.00	100.00	100

SOLUTION

Before computing the overall stage efficiency for n-butane, it is worthwhile to check the consistency of the plant data by examining the overall material balance and the material balance for each component. From the above stream compositions, it is apparent that the compositions have been normalized to total 100%.

The overall material balance is

Total flow into tower $= 368 + 946 = 1{,}314$ lbmol/h

Total flow from tower $= 525.4 + 786.9 = 1{,}312.3$ lbmol/h

These two totals agree to within 0.13%. This is excellent agreement.

The component material balance for the oil absorbent is

Total oil in $= 368$ lbmol/h

Total oil out $= (0.7024)(525.4) = 369$ lbmol/h

These two totals agree to within 0.3%. Again, this is excellent agreement.

Component material balances for other hydrocarbons from spreadsheet calculations are as follows.

	lbmol/h			
Component	Lean Gas	Rich Oil	Total Out	Total In
C_1	439.9	7.0	446.9	447.5
C_2	77.1	6.1	83.2	83.2
$C_3^=$	40.4	8.7	49.1	49.2
C_3	170.4	43.0	213.4	213.8
$C_4^=$	18.4	17.5	35.9	35.9
nC_4	35.0	35.0	70.0	70.0
nC_5	5.7	21.1	26.8	28.4
nC_6	0.0	18.0	18.0	18.0
	786.9	156.4	943.3	946.0

Again, we see excellent agreement. The largest difference is 6% for pentanes. Plant data are not always so consistent.

For the back-calculation of stage efficiency from the performance data, the Kremser equation is applied to compute the number of equilibrium stages required for the measured absorption of n-butane.

$$\text{Fraction of } nC_4 \text{ absorbed} = \frac{35}{70} = 0.50$$

From (6-13),
$$0.50 = \frac{A^{N+1} - A}{A^{N+1} - 1}$$

where $A = $ absorption factor $= \dfrac{L}{KV}$

Because L and V vary greatly through the column, let

$$L = \text{average liquid rate} = \frac{368 + 525.4}{2} = 446.7 \text{ lbmol/h}$$

and let

$$V = \text{average vapor rate} = \frac{946 + 786.9}{2} = 866.5 \text{ lbmol/h}$$

Assume average tower temperature $=$ the average of inlet and outlet temperatures $= (102 + 126 + 108 + 108)/4 = 111°F$. Also assume that the viscosity of the lean oil at 116°F equals the viscosity of the rich oil at 111°F. Therefore, $\mu = 1.4$ cP.

Assume the ambient pressure is 14.7 psia. Then

Tower pressure $= 79 + 14.7 = 93.7$ psia

From Figure 2.8, at 93.7 psia and 111°F, $K_{nC_4} = 0.7$. Thus,

$$A = \frac{446.7}{(0.7)(866.5)} = 0.736$$

Therefore,
$$0.50 = \frac{0.736^{N+1} - 0.736}{0.736^{N+1} - 1}$$

Solving, $\qquad N = N_t = 1.45$

From the performance data, $N_a = 16$

From (6-21), $\qquad E_o = \dfrac{1.45}{16} = 0.091 \quad \text{or} \quad 9.1\%$

Equation (6-22) is applicable to n-butane, because that component is absorbed to the extent of about 50% and thus can be considered one of the key components. Other possible key components are butenes and n-pentane.

From (6-22), $\qquad E_o = 19.2 - 57.8 \log(1.4) = 10.8\%$

To estimate the stage efficiency from the O'Connell correlation, use the following properties for the rich oil at 126°F, 93.7 psia, and 30 mol% light hydrocarbons/70 mol% of 250-MW oil, as obtained from a simulation program.

$K = 0.77$ for *n*-butane

$M_L = 195$

$\mu_L = 0.9$ cP

$\rho_L = 44.1$ lb/ft^3

Therefore, $\dfrac{KM_L\mu_L}{\rho_L} = \dfrac{0.77(195)(0.9)}{(44.1)} = 3.1$

From (6-23),

$\log E_o = 1.597 - 0.199 \log(3.1) - 0.0896[\log(3.1)]^2 = 1.48$

$E_o = 10^{1.48} = 30.2$

For this hydrocarbon absorber, the Drickamer and Bradford correlation (10.8%) gives better agreement than the O'Connell correlation (30.2%) with the plant performance data (9.1%).

Semitheoretical Models

A third method for predicting the overall stage efficiency involves the application of a semitheoretical tray model based on mass- and heat-transfer rates. With this model, the fractional approach to equilibrium, called the *plate* or *tray efficiency,* is estimated for each component in the mixture for each tray in the column. These efficiency values are then utilized to determine conditions for each tray, or averaged for the column to obtain the overall plate efficiency.

Tray efficiency models, in order of increasing complexity, have been proposed by Holland [12], Murphree [13], Hausen [14], and Standart [15]. All four models are based on the assumption that vapor and liquid streams entering each tray are of uniform compositions. The *Murphree vapor efficiency,* which is the oldest and most widely used, is derived with the additional assumptions of (1) complete mixing of the liquid flowing across the tray such that the liquid is of a uniform concentration, equal to the composition of the liquid leaving the tray and entering the next tray below, and (2) plug flow of the vapor passing up through the liquid, as indicated in Figure 6.17 for tray *n*. Considering species *i*, let

n = rate of mass transfer for absorption from the gas to the liquid

K_G = overall gas mass-transfer coefficient based on a partial-pressure driving force

a = vapor–liquid interfacial area per volume of combined gas and liquid holdup (froth or dispersion) on the tray,

A_b = active bubbling area of the tray (total cross-sectional area minus liquid down-comer areas)

Z_f = height of combined gas and liquid holdup on the tray

P = total absolute pressure

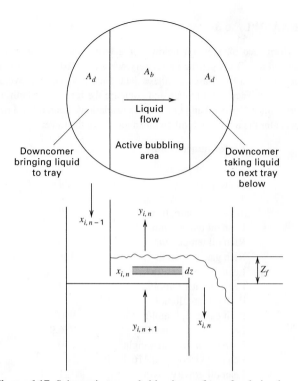

Figure 6.17 Schematic top and side views of tray for derivation of Murphree vapor-tray efficiency.

y_i = mole fraction of *i* in the vapor rising up through the liquid

y_i^* = vapor mole fraction of *i* in equilibrium with the completely mixed liquid on the tray

Then the differential rate of mass transfer for a differential height of holdup on tray *n*, numbered down from the top, is

$$dn_i = K_G a(y_i - y_i^*)PA_b\,dZ \qquad (6\text{-}24)$$

where K_G takes into account both gas- and liquid-phase resistances to mass transfer. By material balance, assuming a negligible change in V across the stage,

$$dn_i = -V\,dy_i \qquad (6\text{-}25)$$

where V = molar gas flow rate up through the liquid on the tray.

Combining (6-24) and (6-25) to eliminate dn_i, separating variables, and converting to integral form,

$$A_b \int_0^{Z_f} \frac{K_G a P}{V}\,dZ = \int_{y_{i,n+1}}^{y_{i,n}} \frac{dy_i}{y_{i,n}^* - y_i} = N_{OG} \qquad (6\text{-}26)$$

where a second subscript involving the tray number, *n*, has been added to the mole fraction of the vapor phase. The vapor enters tray *n* at $y_{i,n+1}$ and exits at $y_{i,n}$. This equation defines

N_{OG} = number of overall gas-phase mass-transfer units

Values of K_G, a, P, and V may vary somewhat as the gas flows up through the liquid on the tray, but if they as well as

y_i^* are taken to be constant, (6-26) can be integrated to give

$$N_{OG} = \frac{K_G a P Z_f}{(V/A_b)} = \ln\left(\frac{y_{i,n+1} - y_{i,n}^*}{y_{i,n} - y_{i,n}^*}\right) \quad (6\text{-}27)$$

A rearrangement of (6-27) in terms of the fractional approach of y_i to equilibrium defines the Murphree vapor efficiency as

$$E_{MV} = \frac{y_{i,n+1} - y_{i,n}}{y_{i,n+1} - y_{i,n}^*} = 1 - e^{-N_{OG}} \quad (6\text{-}28)$$

or

$$N_{OG} = -\ln(1 - E_{MV}) \quad (6\text{-}29)$$

Suppose that measurements give

$$y_i \text{ entering tray } n = y_{i,n+1} = 0.64$$

$$y_i \text{ leaving tray } n = y_{i,n} = 0.61$$

and, from thermodynamics or phase equilibrium data, y_i^* in equilibrium with x_i on and leaving tray $n = 0.60$.
Then, from (6-28),

$$E_{MV} = (0.64 - 0.61)/(0.64 - 0.60) = 0.75$$

or a 75% approach to equilibrium. From (6-29),

$$N_{OG} = -\ln(1 - 0.75) = 1.386$$

When $N_{OG} = 1$, $E_{MV} = 1 - e^{-1} = 0.632$.

The derivation of the Murphree vapor efficiency does not consider the exiting stream temperatures. However, it is implied that the completely mixed liquid phase is at its bubble-point temperature so that the equilibrium vapor phase mole fraction, $y_{i,n}^*$, can be computed.

For multicomponent mixtures, values of E_{MV} are component-dependent and can vary from tray to tray; but at each tray it can be shown that the number of independent values of E_{MV} is one less than the number of components. The dependent value of E_{MV} is determined by forcing $\sum y_i = 1$. It is thus possible that a negative value of E_{MV} can result for a component in a multicomponent mixture. Such negative efficiencies are possible because of mass-transfer coupling among concentration gradients in a multicomponent mixture, which is discussed in Chapter 12. However, for a binary mixture, values of E_{MV} are always positive and identical for the two components.

Only if liquid travel distance across a tray is small will the liquid on a tray approach the complete-mixing assumption used to derive (6-27). To handle the more general case of incomplete liquid mixing, a *Murphree vapor-point efficiency* is defined by assuming that liquid composition varies with distance of travel across a tray, but is uniform in the vertical direction. Thus, for species i on tray n, at any horizontal distance from the downcomer that directs liquid onto tray n, as shown in Figure 6.18,

$$E_{OV} = \frac{y_{i,n+1} - y_i}{y_{i,n+1} - y_i^*} \quad (6\text{-}30)$$

Because x_i varies across a tray, y_i^* and y_i also vary. However, the exiting vapor is then assumed to mix completely to give a uniform $y_{i,n}$ before entering the tray above. Because E_{OV} is

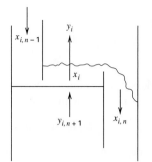

Figure 6.18 Schematic of tray for Murphree vapor-point efficiency.

a more fundamental quantity than E_{MV}, E_{OV} serves as the basis for semitheoretical estimates of tray efficiency and overall column efficiency.

Lewis [16] integrated E_{OV} over a tray for several cases. For complete mixing of liquid on a tray to give a uniform composition, $x_{i,n}$, it is obvious that

$$E_{OV} = E_{MV} \quad (6\text{-}31)$$

For plug flow of liquid across a tray with no longitudinal diffusion (no mixing of liquid in the horizontal direction), Lewis derived

$$E_{MV} = \frac{1}{\lambda}(e^{\lambda E_{OV}} - 1) \quad (6\text{-}32)$$

with

$$\lambda = mV/L \quad (6\text{-}33)$$

where V and L are gas and liquid molar flow rates, respectively, and $m = dy/dx = $ slope of the equilibrium line for a species, using the expression $y = mx + b$. If b is taken as zero, then m is the K-value, and for the key component, k, being absorbed,

$$\lambda = K_k V/L = 1/A_k$$

If A_k, the key-component absorption factor, is given the typical value of 1.4, $\lambda = 0.71$. Suppose the measured or predicted point efficiency is $E_{OV} = 0.25$. From (6-32),

$$E_{MV} = 1.4(e^{0.71(0.25)} - 1) = 0.27$$

which is only 9% higher than E_{OV}. However, if $E_{OV} = 0.9$, E_{MV} is 1.25, which is significantly higher and equivalent to more than a theoretical stage. This surprising result is due to the concentration gradient in the liquid across the length of travel on the tray, which allows the vapor to contact a liquid having an average concentration of species k that can be appreciably lower than that in the liquid leaving the tray.

Equations (6-31) and (6-32) represent extremes between complete mixing and no mixing of the liquid phase, respectively. A more realistic, but considerably more complex model that accounts for partial liquid mixing on the tray, as developed by Gerster et al. [17], is

$$\frac{E_{MV}}{E_{OV}} = \frac{1 - e^{-(\eta + N_{Pe})}}{(\eta + N_{Pe})\{1 + [(\eta + N_{Pe})/\eta]\}} + \frac{e^{\eta} - 1}{\eta\{1 + [\eta/(\eta + N_{Pe})]\}} \quad (6\text{-}34)$$

where

$$\eta = \frac{N_{\mathrm{Pe}}}{2}\left[\left(1+\frac{4\lambda E_{OV}}{N_{\mathrm{Pe}}}\right)^{1/2}-1\right] \qquad (6\text{-}35)$$

The dimensionless Peclet number, N_{Pe}, which serves as a partial-mixing parameter, is defined by

$$N_{\mathrm{Pe}}=Z_L^2/D_E\theta_L=Z_Lu/D_E \qquad (6\text{-}36)$$

where Z_L is the length of liquid flow path across the tray as shown in Figure 6.3, D_E is the eddy diffusion coefficient in the direction of liquid flow, θ_L is the average liquid residence time on the tray, and $u=Z_L/\theta_L$ is the mean liquid velocity across the tray. Equation (6-34) is plotted in Figure 6.19 for wide ranges of N_{Pe} and λE_{OV}. When $N_{\mathrm{Pe}}=0$, (6-31) holds; when $N_{\mathrm{Pe}}=\infty$, (6-32) holds.

From (6-36), the Peclet number can be viewed as the ratio of the mean liquid bulk velocity to the eddy-diffusion velocity. When N_{Pe} is small, eddy diffusion is important and the liquid approaches a well-mixed condition. When N_{Pe} is large, bulk flow predominates and the liquid approaches plug flow. Experimental measurements of D_E in bubble-cap and sieve-plate columns [18–21] cover a range of 0.02 to 0.20 ft²/s. Values of u/D_E typically range from 3 to 15 ft⁻¹. Based on the second form of (6-36), N_{Pe} increases directly with increasing Z_L and, therefore, column diameter. A typical value of N_{Pe} for a 2-ft-diameter column is 10; for a 6-ft-diameter column, N_{Pe} might be 30. For N_{Pe} values of this

magnitude, Figure 6.19 shows that values of E_{MV} can be significantly larger than E_{OV} for large values of λ.

Lewis [16] showed that when the equilibrium and operating lines are straight, but not necessarily parallel, the overall stage efficiency, defined by (6-21), is related to the Murphree vapor efficiency by

$$E_o=\frac{\log[1+E_{MV}(\lambda-1)]}{\log\lambda} \qquad (6\text{-}37)$$

When the two lines are not only straight but parallel, such that $\lambda=1$, (6-37) becomes $E_o=E_{MV}$. Also, when $E_{MV}=1$, then $E_o=1$ regardless of the value of λ.

Scale-up from Laboratory Data

When vapor–liquid equilibrium data for a system are unavailable or not well known, and particularly if the system forms a highly nonideal liquid solution with possible formation of azeotropes, tray requirements are best estimated, and the feasibility of achieving the desired degree of separation verified, by conducting laboratory tests. A particularly useful apparatus is a small glass or metal sieve-plate column with center-to-side downcomers developed by Oldershaw [22]

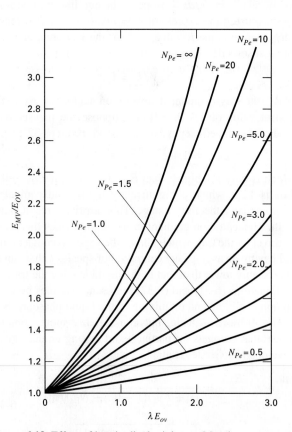
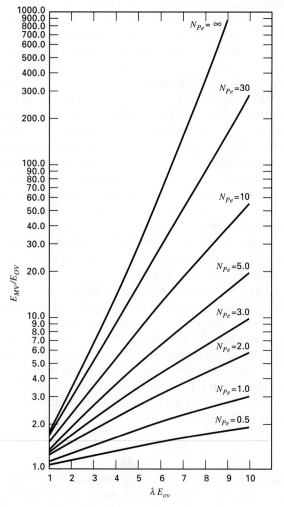

Figure 6.19 Effect of longitudinal mixing on Murphree vapor tray efficiency.

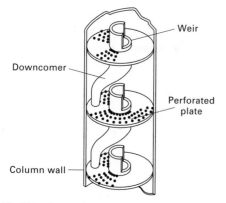

Figure 6.20 Oldershaw column.

and shown schematically in Figure 6.20. Oldershaw columns are typically 1 to 2 in. in diameter and can be assembled with almost any number of sieve plates, usually containing 0.035- to 0.043-in. holes with a hole area of approximately 10%. A detailed study by Fair, Null, and Bolles [23] showed that overall plate efficiencies of Oldershaw columns operated over a pressure range of 3 to 165 psia are in conservative agreement with distillation data obtained from sieve-tray, pilot-plant and industrial-size columns ranging in size from 18 in. to 4 ft in diameter when operated in the range of 40% to 90% of flooding. It may be assumed that similar agreement might be realized for absorption and stripping.

It is believed that the small-diameter Oldershaw column achieves essentially complete mixing of liquid on each tray, thus permitting the measurement of a point efficiency. As discussed above, somewhat larger efficiencies may be observed in much-larger-diameter columns due to incomplete liquid mixing, which results in a higher Murphree tray efficiency and, therefore, higher overall plate efficiency.

Fair et al. [23] recommend the following conservative scale-up procedure for the Oldershaw column:

1. Determine the flooding point.

2. Establish operation at about 60% of flooding (but 40 to 90% seems acceptable).

3. Run the system to find a combination of plates and flow rates that gives the desired degree of separation.

4. Assume that the commercial column will require the same number of plates for the same ratio of liquid to vapor molar flow rates.

If reliable vapor–liquid equilibrium data are available, they can be used with the Oldershaw data to determine the overall column efficiency, E_o. Then (6-37) and (6-34) can be used to estimate the average point efficiency. For the commercial-size column, the Murphree vapor efficiency can be determined from the Oldershaw column point efficiency using (6-34), which takes into account incomplete liquid mixing. In general, the tray efficiency of the commercial column, depending on the length of the liquid flow path, will be higher than for the Oldershaw column at the same percentage of flooding.

EXAMPLE 6.4

Assume that the column diameter for the absorption operation of Example 6.1 is 3 ft. If the overall stage efficiency, E_o, is 30% for the absorption of ethyl alcohol, estimate the average Murphree vapor efficiency, E_{MV}, and the possible range of the Murphree vapor-point efficiency, E_{OV}.

SOLUTION

For Example 6.1, the system is dilute in ethyl alcohol, the main component undergoing mass transfer. Therefore, the equilibrium and operating lines are essentially straight, and (6-37) can be applied. From the data of Example 6.1, $\lambda = KV/L = 0.57(180)/151.5 = 0.68$.

Solving (6-37) for E_{MV}, using $E_o = 0.30$,

$$E_{MV} = (\lambda^{E_o} - 1)/(\lambda - 1) = (0.68^{0.30} - 1)/(0.68 - 1) = 0.34$$

For a 3-ft-diameter column, the degree of liquid mixing probably lies intermediate between complete mixing and plug flow. From (6-31) for the former case, $E_{OV} = E_{MV} = 0.34$. From a rearrangement of (6-32) for the latter case, $E_{OV} = \ln(1 + \lambda E_{MV})/\lambda = \ln[1 + 0.68(0.34)]/0.68 = 0.31$. Therefore, E_{OV} lies in the range of 31% to 34%, probably closer to 34% for complete mixing. However, the differences between E_o, E_{MV}, and E_{OV} for this example are almost negligible.

6.6 TRAY DIAMETER, PRESSURE DROP, AND MASS TRANSFER

In the trayed tower shown in Figure 6.21, vapor flows vertically upward, contacting liquid in crossflow on each tray. When trays are designed properly, a stable operation is achieved wherein (1) vapor flows only through the perforations or open regions of the tray between the downcomers, (2) liquid flows from tray to tray only by means of the downcomers, (3) liquid neither weeps through the tray perforations nor is carried by the vapor as entrainment to the tray above, and (4) vapor is neither carried (occluded) down by the liquid in the downcomer to the tray below nor allowed to bubble up through the liquid in the downcomer. Tray design includes the determination of tray diameter and the division of the tray cross-sectional area, A, as shown in Figure 6.21, into active vapor bubbling area, A_a, and liquid downcomer area, A_d. With the tray diameter fixed, vapor pressure drop and mass-transfer coefficients can be estimated.

Tray Diameter

For a given liquid flow rate, as shown in Figure 6.22 for a sieve-tray column, a maximum vapor flow rate exists beyond which incipient column flooding occurs because of backup of liquid in the downcomer. This condition, if sustained, leads to carryout of liquid with the overhead vapor leaving the column. *Downcomer flooding* takes place when, in the absence of entrainment, liquid backup is caused by downcomers of inadequate cross-sectional area, A_d, to carry

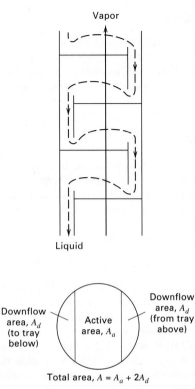

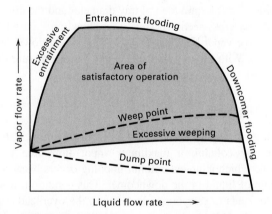

Figure 6.21 Vapor and liquid flow through a trayed tower.

the liquid flow. It rarely occurs if downcomer cross-sectional area is at least 10% of total column cross-sectional area and if tray spacing is at least 24 in. The usual design limit is *entrainment flooding,* which is caused by excessive carry-up of liquid, at the rate e, by vapor entrainment to the tray above. At incipient flooding, $(e + L) \gg L$ and downcomer cross-sectional area is inadequate for the excessive liquid load $(e + L)$. Tray diameter is determined as follows to avoid entrainment flooding.

Entrainment of liquid is due to carry-up of suspended droplets by rising vapor or to throw-up of liquid particles by vapor jets formed at tray perforations, valves, or bubble-cap slots. Souders and Brown [24] successfully correlated

Figure 6.22 Limits of stable operation in a trayed tower.

[Reproduced by permission from H.Z. Kister, *Distillation Design,* McGraw-Hill, New York (1992).]

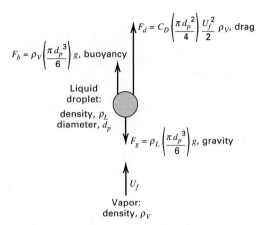

Figure 6.23 Forces acting on a suspended liquid droplet.

entrainment flooding data for 10 commercial trayed columns by assuming that carry-up of suspended droplets controls entrainment. At low vapor velocity, a droplet settles out; at high vapor velocity, it is entrained. At flooding or incipient entrainment velocity, U_f, the droplet is suspended such that the vector sum of the gravitational, buoyant, and drag forces acting on the droplet, as shown in Figure 6.23, are zero. Thus,

$$\sum F = 0 = F_g - F_b - F_d \qquad (6\text{-}38)$$

In terms of droplet diameter, d_p, terms on the right-hand side of (16-38) become, respectively,

$$\rho_L \left(\frac{\pi d_p^3}{6} \right) g - \rho_V \left(\frac{\pi d_p^3}{6} \right) g - C_D \left(\frac{\pi d_p^2}{4} \right) \frac{U_f^2}{2} \rho_V = 0 \qquad (6\text{-}39)$$

where C_D is the drag coefficient. Solving for flooding velocity,

$$U_f = C \left(\frac{\rho_L - \rho_V}{\rho_V} \right)^{1/2} \qquad (6\text{-}40)$$

where C = capacity parameter of Souders and Brown. According to the above theory,

$$C = \left(\frac{4 d_p g}{3 C_D} \right)^{1/2} \qquad (6\text{-}41)$$

Parameter C can be calculated from (6-41) if the droplet diameter d_p is known. In practice, however, d_p is distributed over a wide range and C is treated as an empirical parameter determined using experimental data obtained from operating equipment. Souders and Brown considered all the important variables that could influence the value of C and obtained a correlation for commercial-size columns with bubble-cap trays. Data covered column pressures from 10 mmHg to 465 psia, plate spacings from 12 to 30 in., and liquid surface tensions from 9 to 60 dyne/cm. In accordance with (6-41), the value of C increases with increasing surface tension, which increases d_p. Also, C increases with increasing tray spacing, since this allows more time for agglomeration to a larger d_p.

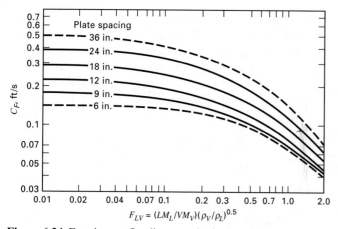

Figure 6.24 Entrainment flooding capacity in a trayed tower.

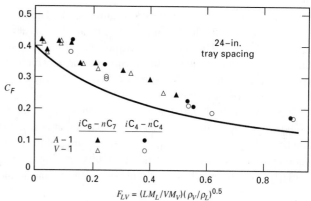

Figure 6.25 Comparison of flooding correlation with data for valve trays.

Using additional commercial operating data, Fair [25] produced the more general correlation of Figure 6.24, which is applicable to columns with bubble cap and sieve trays. Whereas Souders and Brown base the vapor velocity on the entire column cross-sectional area, Fair utilizes a net vapor flow area equal to the total inside column cross-sectional area minus the area blocked off by the downcomer, that is, $A - A_d$ in Figure 6.21. The value of C_F in Figure 6.24 depends on tray spacing and the abscissa ratio $F_{LV} = (LM_L/VM_V)(\rho_V/\rho_L)^{0.5}$ (where flow rates are in molar units), which is a kinetic energy ratio first used by Sherwood, Shipley, and Holloway [26] to correlate packed-column flooding data. The value of C in (6-41) is obtained from Figure 6.24 by correcting C_F for surface tension, foaming tendency, and the ratio of vapor hole area A_h to tray active area A_a, according to the empirical relationship

$$C = F_{ST}F_F F_{HA}C_F \qquad (6-42)$$

where

F_{ST} = surface tension factor = $(\sigma/20)^{0.2}$

F_F = foaming factor

F_{HA} = 1.0 for $A_h/A_a \geq 0.10$ and $5(A_h/A_a) + 0.5$ for $0.06 \leq A_h/A_a \leq 0.1$

σ = liquid surface tension, dyne/cm

For nonfoaming systems, $F_F = 1.0$; for many absorbers, F_F may be 0.75 or even less. The quantity A_h is the area open to the vapor as it penetrates into the liquid on a tray. It is the total cap slot area for bubble-cap trays and the perforated area for sieve trays.

Figure 6.24 appears to be conservative for valve trays. This is shown in Figure 6.25, where entrainment-flooding data of Fractionation Research, Inc. (FRI) [27,28], for a 4-ft-diameter column equipped with Glitsch type A-1 and V-1 valve trays on 24-in. spacing are compared to the correlation in Figure 6.24. For valve trays, the slot area A_h is taken as the full valve opening through which vapor enters the frothy liquid on the tray at a 90° angle with the axis of the column.

Column diameter D_T is based on a fraction, f, of flooding velocity U_f, which is calculated from (6-40), using C from (6-42), based on C_F from Figure 6.24. By the continuity equation from fluid mechanics (flow rate = velocity × flow area × density), the molar vapor flow rate is related to the flooding velocity by

$$V = (fU_f)(A - A_d)\frac{\rho_V}{M_V} \qquad (6-43)$$

where A = total column cross-sectional area = $\pi D_T^2/4$. Thus,

$$D_T = \left[\frac{4VM_V}{fU_f\pi(1 - A_d/A)\rho_V}\right]^{0.5} \qquad (6-44)$$

Typically, the fraction of flooding, f, is taken as 0.80.

Oliver [29] suggests that A_d/A be estimated from F_{LV} in Figure 6.24 by

$$\frac{A_d}{A} = \begin{cases} 0.1, & F_{LV} \leq 0.1 \\ 0.1 + \dfrac{(F_{LV} - 0.1)}{9}, & 0.1 \leq F_{LV} \leq 1.0 \\ 0.2, & F_{LV} \geq 1.0 \end{cases}$$

Column diameter is calculated at both the top and bottom of the column, with the larger of the two diameters used for the entire column unless the two diameters differ appreciably. Because of the need for internal access to columns with trays, a packed column, discussed later in this chapter, is generally used if the calculated diameter from (6-44) is less than 2 ft.

Tray spacing must be specified to compute column diameter using Figure 6.24. As spacing is increased, column height is increased but column diameter is reduced. A spacing of 24 in., which provides ease of maintenance, is optimal for a wide range of conditions; however, a smaller spacing may be desirable for small-diameter columns with a large number of stages; and larger spacing is frequently used for large-diameter columns with a small number of stages.

As shown in Figure 6.22, a minimum vapor rate exists below which liquid weeps (dumps) through tray perforations or risers instead of flowing completely across the active area and into the downcomer. Below this minimum, the degree of contacting of liquid with vapor is reduced, causing tray efficiency to decline. The ratio of the vapor rate at flooding to the minimum vapor rate is the *turndown ratio,* which is approximately 8 for bubble-cap trays, 5 for valve trays, but only about 2 for sieve trays.

When vapor and liquid flow rates change appreciably from tray to tray, column diameter, tray spacing, or hole area can be varied to reduce column cost and ensure stable operation at high tray efficiency. Variation of tray spacing is particularly applicable to columns with sieve trays because of their low turndown ratio.

High-Capacity Trays

Since the 1990s, a number of high-capacity trays have been introduced and installed in hundreds of columns. By various changes to the conventional tray design shown in Figure 6.3, capacity increases of more than 20% of that predicted by Figure 6.24 have been achieved with both perforated trays and valve trays. These changes, which are discussed in some detail by Sloley [71], have included:

1. Sloping or stepping of the downcomer to make the downcomer area smaller at the bottom than at the top so as to increase the active flow area.

2. Vapor flow through that portion of the tray located beneath the downcomer, in addition to the normal active flow area.

3. Use of staggered, louvered downcomer floor plates to impart a horizontal velocity to the liquid exiting the downcomer, thus enhancing the ability to allow vapor flow beneath the downcomer.

4. Elimination of vapor impingement from adjacent valves, in valve trays, by using bi-directional fixed valves.

5. Use of multiple-downcomer trays that provide very long outlet weirs leading to low crest heights and lower froth heights. The downcomers terminate in the active vapor space of the tray below.

6. Directional slotting of sieve trays to impart a horizontal component to the up-flowing vapor, enhance plug flow of liquid across the tray, and eliminate dead areas.

Regardless of the tray design, as shown by Stupin and Kister [72], an ultimate capacity, independent of tray spacing, exists for a countercurrent-flow, vapor–liquid contactor, in which the superficial vapor velocity in the column exceeds the settling velocity of large liquid droplets. Their correlation, based on FRI data, uses the following form of (6-40):

$$U_{S,ult} = C_{S,ult} \left(\frac{\rho_L - \rho_V}{\rho_V} \right)^{1/2} \quad (6-45)$$

where $U_{S,ult}$ is the superficial vapor velocity in m/s based on the column cross-sectional area. The ultimate capacity

parameter, $C_{S,ult}$ in m/s, is independent of the superficial liquid velocity, L_S in m/s, below a critical value; but above that value it decreases with increasing L_S. It is given by the smaller of C_1 and C_2, both in m/s, where

$$C_1 = 0.445(1 - F) \left(\frac{\sigma}{\rho_L - \rho_V} \right)^{0.25} - 1.4L_S \quad (6-46)$$

$$C_2 = 0.356(1 - F) \left(\frac{\sigma}{\rho_L - \rho_V} \right)^{0.25} \quad (6-47)$$

where

$$F = \frac{1}{\left[1 + 1.4 \left(\frac{\rho_L - \rho_V}{\rho_V} \right)^{1/2} \right]} \quad (6-48)$$

ρ is in kg/m^3 and σ is the surface tension in dynes/cm.

EXAMPLE 6.5

(a) Estimate the required tray diameter for the absorber of Example 6.1, assuming a tray spacing of 24 in., a foaming factor of $F_F = 0.90$, a fraction flooding of $f = 0.80$, and a surface tension of $\sigma = 70$ dynes/cm. (b) Estimate the ultimate superficial vapor velocity.

SOLUTION

Because tower conditions are almost the same at the top and bottom, the calculation of column diameter is made only at the bottom, where the gas rate is highest. From Example 6.1,

$$T = 30°C \quad P = 110 \text{ kPa}$$

$$V = 180 \text{ kmol/h}, \quad L = 151.5 + 3.5 = 155.0 \text{ kmol/h}$$

$$M_V = 0.98(44) + 0.02(46) = 44.0,$$

$$M_L = \frac{151.5(18) + 3.5(46)}{155} = 18.6$$

$$\rho_V = \frac{PM}{RT} = \frac{(110)(44)}{(8.314)(303)} = 1.92 \text{ kg/m}^3,$$

$$\rho_L = (0.986)(1,000) = 986 \text{ kg/m}^3$$

$$F_{LV} = \frac{(155)(18.6)}{(180)(44.0)} \left(\frac{1.92}{986} \right)^{0.5} = 0.016$$

(a) For tray spacing = 24 in., from Figure 6.24, $C_F = 0.39$ ft/s,

$$F_{ST} = \left(\frac{\sigma}{20} \right)^{0.2} = \left(\frac{70}{20} \right)^{0.2} = 1.285, \quad F_F = 0.90$$

Because $F_{LV} < 0.1$, $A_d/A = 0.1$ and $F_{HA} = 1.0$. Then, from (6-42),

$$C = 1.285(0.90)(1.0)(0.39) = 0.45 \text{ ft/s}$$

From (6-40),

$$U_f = 0.45 \left(\frac{986 - 1.92}{1.92} \right)^{0.5} = 10.2 \text{ ft/s}$$

From (6-44), using SI units and time in seconds,

$$D_T = \left[\frac{4(180/3{,}600)(44.0)}{(0.80)(10.2/3.28)(3.14)(1-0.1)(1.92)} \right]^{0.5}$$
$$= 0.81 \text{ m} = 2.65 \text{ ft}$$

(b) From (6-48),

$$F = \frac{1}{\left[1 + 1.4 \left(\dfrac{986 - 1.92}{1.92} \right)^{1/2} \right]} = 0.0306$$

From (6-47),

$$C_2 = 0.356(1 - 0.0306) \left(\frac{70}{986 - 1.92} \right)^{0.25}$$
$$= 0.178 \text{ m/s} = 0.584 \text{ ft/s}$$

If C_2 is the smaller value of C_1 and C_2, then from (6-45),

$$U_{S,\text{ult}} = 0.178 \left(\frac{986 - 1.92}{1.92} \right)^{1/2} = 4.03 \text{ m/s} = 13.22 \text{ ft/s}$$

To apply (6-46) to compute C_1, the value of L_S is required. This value is related as follows to the value of the superficial vapor velocity, U_S.

$$L_S = U_S \frac{\rho_V L M_L}{\rho_L V M_V} = U_S \frac{(1.92)(155)(18.6)}{(986)(180)(44.0)} = 0.000709 \, U_S$$

With this expression for L_S, (6-46) becomes

$$C_1 = 0.445(1 - 0.0306) \left(\frac{70}{986 - 1.92} \right)^{0.25} - 1.4(0.000709) \, U_S$$
$$= 0.223 - 0.000993 \, U_S, \text{ m/s}$$

If C_1 is the smaller, then, using (6-45),

$$U_{S,\text{ult}} = (0.223 - 0.000993 \, U_{S,\text{ult}}) \left(\frac{986 - 1.92}{1.92} \right)^{1/2}$$
$$= 5.05 - 0.0225 \, U_{S,\text{ult}}$$

Solving, $U_{S,\text{ult}} = 4.94$ m/s, which gives $C_1 = 0.223 - 0.000993(4.94) = 0.218$ m/s. Thus, C_2 is the smaller value and $U_{S,\text{ult}} = 4.03$ m/s $= 13.22$ ft/s. This ultimate velocity is 30% higher than the flooding velocity computed in part (a).

Tray Vapor Pressure Drop

Typical tray pressure drop for flow of vapor in a tower is from 0.05 to 0.15 psi/tray. Referring to Figure 6.3, pressure drop (head loss) for a sieve tray is due to friction for vapor flow through the tray perforations, holdup of the liquid on the tray, and a loss due to surface tension:

$$h_t = h_d + h_l + h_\sigma \tag{6-49}$$

where

h_t = total pressure drop/tray, in. of liquid

h_d = dry tray pressure drop, in. of liquid

h_l = equivalent head of clear liquid on tray, in. of liquid

h_σ = pressure drop due to surface tension, in. of liquid

The dry sieve-tray pressure drop is given by a modified orifice equation, applied to the holes in the tray,

$$h_d = 0.186 \left(\frac{u_o^2}{C_o^2} \right) \left(\frac{\rho_V}{\rho_L} \right) \tag{6-50}$$

where u_o = hole velocity (ft/s) and C_o depends on the percent hole area and the ratio of tray thickness to hole diameter. For a typical 0.078-in.-thick tray with $\frac{3}{16}$-in.-diameter holes and a percent hole area (based on the cross-sectional area of the tower) of 10%, C_o may be taken as 0.73. Otherwise, C_o lies between about 0.65 and 0.85.

The equivalent height of clear liquid holdup on a tray depends on weir height, liquid and vapor densities and flow rates, and downcomer weir length, as given by the following empirical expression developed from experimental data by Bennett, Agrawal, and Cook [30]:

$$h_l = \phi_e \left[h_w + C_l \left(\frac{q_L}{L_w \phi_e} \right)^{2/3} \right] \tag{6-51}$$

where

h_w = weir height, in.

ϕ_e = effective relative froth density (height of clear liquid/froth height)
$= \exp(-4.257 \, K_s^{0.91})$ (6-52)

K_s = capacity parameter, ft/s $= U_a \left(\dfrac{\rho_V}{\rho_L - \rho_V} \right)^{1/2}$ (6-53)

U_a = superficial vapor velocity based on active bubbling area,

$A_a = (A - 2A_d)$, of the tray, ft/s,

L_w = weir length, in.

q_L = liquid flow rate across tray, gal/min

$C_l = 0.362 + 0.317 \exp(-3.5 h_w)$ (6-54)

The second term in (6-51) is related to the Francis weir equation for a straight segmental weir, taking into account the froth nature of the liquid flow over the weir. For $A_d/A = 0.1$, $L_w = 73\%$ of the tower diameter.

As the gas emerges from the tray perforations, the bubbles must overcome surface tension. The pressure drop due to surface tension is given by the difference between the pressure inside the bubble and that of the liquid, according to the theoretical relation

$$h_\sigma = \frac{6\sigma}{g \rho_L D_{B(\max)}} \tag{6-55}$$

where, except for tray perforations much smaller than $\frac{3}{16}$-in. in diameter, $D_{B(\max)}$, the maximum bubble size, may be taken as the perforation diameter, D_H.

Methods for estimating vapor pressure drop for bubble-cap trays and valve trays are given by Smith [31] and Klein [32], respectively, and are discussed by Kister [33] and Lockett [34].

EXAMPLE 6.6

Estimate the tray vapor pressure drop for the absorber of Example 6.1, assuming use of sieve trays with a tray diameter of 1 m, a weir height of 2 in., and a hole diameter of $\frac{3}{16}$ in.

SOLUTION

From Example 6.5,

$$\rho_V = 1.92 \text{ kg/m}^3 \quad \rho_L = 986 \text{ kg/m}^3$$

At the bottom of the tower, vapor velocity based on the total cross-sectional area of the tower is

$$\frac{(180/3,600)(44)}{(1.92)[3.14(1)^2/4]} = 1.46 \text{ m/s}$$

For a 10% hole area, based on the total cross-sectional area of the tower,

$$u_o = \frac{1.46}{0.10} = 14.6 \text{ m/s} \quad \text{or} \quad 47.9 \text{ ft/s}$$

Using the above densities, (6-50) gives

$$h_d = 0.186 \left(\frac{47.9^2}{0.73^2}\right)\left(\frac{1.92}{986}\right) = 1.56 \text{ in. of liquid}$$

Take weir length as 73% of tower diameter, with $A_d/A = 0.10$. Then

$$L_w = 0.73(1) = 0.73 \text{ m} \quad \text{or} \quad 28.7 \text{ in.}$$

$$\text{Liquid flow rate in gpm} = \frac{(155/60)(18.6)}{986(0.003785)} = 12.9 \text{ gpm}$$

with $\quad A_d/A = 0.1 \quad A_a/A = (A - 2A_d)/A = 0.8$

Therefore, $U_a = 1.46/0.8 = 1.83 \text{ m/s} = 5.99 \text{ ft/s}$

From (6-53),

$$K_s = 5.99[1.92/(986 - 1.92)]^{0.5} = 0.265 \text{ ft/s}$$

From (6-52),

$$\phi_e = \exp[-4.257(0.265)^{0.91}] = 0.28$$

From (6-54),

$$C_l = 0.362 + 0.317 \exp[-3.5(2)] = 0.362$$

From (6-51),

$$h_l = 0.28[2 + 0.362(12.9/28.7/0.28)^{2/3}]$$
$$= 0.28(2 + 0.50) = 0.70 \text{ in.}$$

From (6-55), in metric units, using $D_{B(\max)} = D_H = \frac{3}{16}$ in. $= 0.00476$ m,

$$\sigma = 70 \text{ dynes/cm} = 0.07 \text{ N/m} = 0.07 \text{ kg/s}^2,$$
$$g = 9.8 \text{ m/s}^2, \text{ and } \rho_L = 986 \text{ kg/m}^3$$

$$h_\sigma = \frac{6(0.07)}{9.8(986)(0.00476)} = 0.00913 \text{ m} = 0.36 \text{ in.}$$

From (6-45), the total tray head loss is $h_t = 1.56 + 0.70 + 0.36 = 2.62$ in.

For $\rho_L = 986 \text{ kg/m}^3 = 0.0356 \text{ lb/in}^3$,

$$\text{tray vapor pressure drop} = h_t \rho_L = 2.62(0.0356) = 0.093 \text{ psi/tray}$$

Mass-Transfer Coefficients and Transfer Units

Following the determination of tower diameter and major details of the tray layout, an estimate of the Murphree vapor point efficiency, defined by (6-30), can be made using empirical correlations for mass-transfer coefficients, based on experimental data. For a vertical path for vapor flow up through the froth from a point on the bubbling area of the tray, (6-29) applies to the Murphree vapor-point efficiency:

$$N_{OG} = -\ln(1 - E_{OV}) \tag{6-56}$$

where

$$N_{OG} = \frac{K_G a P Z_f}{(V/A_b)} \tag{6-57}$$

The overall, volumetric mass-transfer coefficient, $K_G a$, is related to the individual volumetric mass-transfer coefficients by the sum of the mass-transfer resistances, which from equations in Section 3.7 can be shown to be

$$\frac{1}{K_G a} = \frac{1}{k_G a} + \frac{(K P M_L/\rho_L)}{k_L a} \tag{6-58}$$

where the two terms on the right-hand side are the gas- and liquid-phase resistances, respectively, and the symbols k_p for the gas and k_c for the liquid used in Chapter 3 have been replaced by k_G and k_L, respectively. In terms of individual transfer units, defined by

$$N_G = \frac{k_G a P Z_f}{(V/A_b)} \tag{6-59}$$

and

$$N_L = \frac{k_L a \rho_L Z_f}{M_L(L/A_b)} \tag{6-60}$$

we obtain from (6-57) and (6-58)

$$\frac{1}{N_{OG}} = \frac{1}{N_G} + \frac{(KV/L)}{N_L} \tag{6-61}$$

Important empirical mass-transfer correlations have been published by the AIChE [35] for bubble-cap trays, Chan and Fair [36, 37] for sieve trays, and Scheffe and Weiland [38] for one type of valve tray (Glitsch V-1). These correlations have been developed in terms of N_L, N_G, k_L, k_G, a, and N_{Sh} for either the gas or liquid phase. In this section, we present only correlations for sieve trays, as given for binary systems by Chan and Fair [36], who used a correlation for the liquid phase based on the work of Foss and Gerster [39] as reported by the AIChE [40], and who developed a correlation for the vapor phase from a fairly extensive experimental data bank of 143 points for towers 2.5 to 4.0 ft in diameter, operating at pressures from 100 mmHg to 400 psia.

Experimental data for sieve trays have validated the assumed direct dependence of mass transfer on the interfacial area between the gas and liquid phases, and on the residence times in the froth of the gas and liquid phases. Accordingly,

Chan and Fair give the following modifications of (6-59) and (6-60):

$$N_G = k_G \bar{a} \bar{t}_G \qquad (6\text{-}62)$$

$$N_L = k_L \bar{a} \bar{t}_L \qquad (6\text{-}63)$$

where $\bar{a}$ is the interfacial area per unit volume of equivalent clear liquid, $\bar{t}_G$ is the average residence time of the gas in the froth, and $\bar{t}_L$ is the average residence time of the liquid in the froth.

Average residence times are estimated from the following dimensionally consistent, theoretical continuity equations, using (6-51) for the equivalent head of clear liquid on the tray and (6-52) for the effective relative density of the froth:

$$\bar{t}_L = \frac{h_l A_a}{q_L} \qquad (6\text{-}64)$$

and

$$\bar{t}_G = \frac{(1 - \phi_e) h_l}{(\phi_e U_a)} \qquad (6\text{-}65)$$

where $(1 - \phi_e) h_l / \phi_e$ is the equivalent height of vapor holdup in the froth, and the residence times are usually computed in seconds.

Empirical expressions for $k_G \bar{a}$ and $k_L \bar{a}$ in units of s^{-1} are

$$k_G \bar{a} = \frac{1{,}030 D_V^{0.5}(f - 0.842 f^2)}{(h_l)^{0.5}} \qquad (6\text{-}66)$$

and

$$k_L \bar{a} = 78.8 D_L^{0.5}(F + 0.425) \qquad (6\text{-}67)$$

where the variables and their units are

D_V, D_L = diffusion coefficients, cm²/s

h_l = clear liquid height, cm

$f = U_a/U_f$, fractional approach to flooding

$F = F$-factor $= U_a \rho_V^{0.5}$, (kg/m)$^{0.5}$/s

From (6-66), it is seen that an important factor influencing the value of $k_G \bar{a}$ is the fractional approach to flooding, $f = U_a/U_f$. This effect is shown in Figure 6.26, where (6-66) is compared to experimental data. At gas rates corresponding to a fractional approach to flooding of greater than 0.60, the mass-transfer factor decreases with increasing value of f. This may be due to entrainment, which is discussed in the next sub-section. On an entrainment-free basis, the curve in Figure 6.26 might be expected to at least remain at its peak value for conditions above $f = 0.60$.

From (6-67), it is seen that the F-factor is an important consideration for liquid-phase mass transfer. Experimental data that support this are shown in Figure 6.27, where $k_L \bar{a}$ depends strongly on F but is almost independent of liquid flow rate and weir height. The Murphree vapor-point efficiency model of (6-56), (6-61), (6-64), (6-65), (6-66), and (6-67) correlates the 143 points of the Chan and Fair [36] data bank with an average absolute deviation of 6.27%. Lockett [34] pointed out that (6-67) implies that $k_L \bar{a}$ depends on tray spacing, which seems unreasonable.

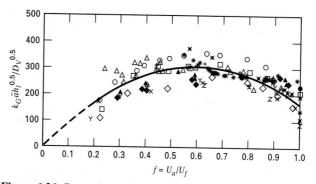

Figure 6.26 Comparison of experimental data to the correlation of Chan and Fair for gas-phase mass transfer.

[From H. Chan and J.R. Fair, *Ind. Eng. Chem. Process Des. Dev.*, **23**, 817 (1984) with permission.]

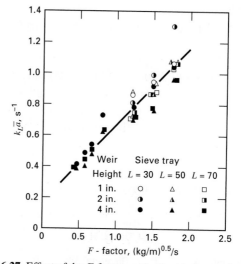

Figure 6.27 Effect of the F-factor on the liquid-phase volumetric mass-transfer coefficient for desorption of oxygen from water with air at 1 atm. and 25°C, where L = gal/(min)/(ft of average flow width).

However, the data bank did include data for tray spacings from 6 to 24 in.

EXAMPLE 6.7

Estimate the Murphree vapor-point efficiency for the absorber of Example 6.1, using results from Examples 6.5 and 6.6, for the tray of Example 6.6. In addition, determine the controlling resistance to mass transfer.

SOLUTION

Pertinent data for the two phases are as follows.

	Gas	Liquid
Molar flow rate, kmol/h	180.0	155.0
Molecular weight	44.0	18.6
Density, kg/m³	1.92	986
Ethanol diffusivity, cm²/s	7.86×10^{-2}	1.81×10^{-5}

Pertinent tray dimensions from Example 6.6 are $D_T = 1$ m, and $A = 0.785$ m^2; $A_a = 0.80$, $A = 0.628$ m$^2 = 6,280$ cm^2; $L_w = 28.7$ in. = 0.73 m.

From Example 6.6,

$$\phi_e = 0.28; \quad h_l = 0.70 \text{ in.} = 1.78 \text{ cm};$$
$$U_a = 5.99 \text{ ft/s} = 183 \text{ cm/s} = 1.83 \text{ m/s}$$

From Example 6.5,

$$U_f = 10.2 \text{ ft/s}; \quad f = U_a/U_f = 5.99/10.2 = 0.59$$
$$F = 1.83(1.92)^{0.5} = 2.54 \text{ (kg/m)}^{0.5}/\text{s}$$
$$q_L = \frac{(155.0)(18.6)}{986}\left(\frac{10^6}{3,600}\right) = 812 \text{ cm}^3/\text{s}$$

From (6-64),

$$\bar{t}_L = (1.78)(6,280)/812 = 13.8 \text{ s}$$

From (6-65),

$$\bar{t}_G = (1 - 0.28)(1.78)/[(0.28)(183)] = 0.025 \text{ s}$$

From (6-67),

$$k_L \bar{a} = 78.8(1.81 \times 10^{-5})^{0.5}(2.54 + 0.425) = 0.99 \text{ s}^{-1}$$

From (6-66),

$$k_G \bar{a} = 1,030(7.86 \times 10^{-2})^{0.5}[0.59 - 0.842(0.59)^2]/(1.78)^{0.5}$$
$$= 64.3 \text{ s}^{-1}$$

From (6-63),

$$N_L = (0.99)(13.8) = 13.7$$

From (6-62),

$$N_G = (64.3)(0.025) = 1.61$$

From Example 6.1, $K = 0.57$. Therefore, $KV/L = (0.57)(180)/155 = 0.662$.

From (6-61),

$$N_{OG} = \frac{1}{(1/1.61) + (0.662/13.7)} = \frac{1}{0.621 + 0.048} = 1.49$$

and the mass transfer of ethanol is seen to be controlled by the vapor-phase resistance. From (6-56), solving for E_{OV},

$$E_{OV} = 1 - \exp(-N_{OG}) = 1 - \exp(-1.49) = 0.77 = 77\%$$

Weeping, Entrainment, and Downcomer Backup

For a tray to operate at high efficiency, (1) weeping of liquid through the tray perforations must be small compared to flow over the outlet weir and into the downcomer, (2) entrainment of liquid by the gas must not be excessive, and (3) to prevent downcomer flooding, froth height in the downcomer must not approach tray spacing. The tray must operate in the stable region shown schematically in Figure 6.22. Weeping is associated with the lower limit of gas velocity, while entrainment flooding is associated with the upper limit.

Weeping occurs at low vapor velocities and/or high liquid rates when the clear liquid height on the tray exceeds the sum of the dry (no liquid flow) tray pressure drop, due to vapor flow, and the surface tension effect. Thus, to prevent weeping, it is necessary that

$$h_d + h_\sigma > h_l \qquad (6\text{-}68)$$

everywhere on the active area of the tray. If weeping occurs uniformly over the tray active area or mainly near the downcomer, a ratio of weep rate to downcomer liquid rate as high as 0.1 may not cause an unacceptable decrease in tray efficiency. Methods for estimating weep rates are discussed by Kister [33].

The prediction of fractional liquid entrainment by the vapor, defined as $\psi = e/(L + e)$, can be made by the correlation of Fair [41], given in Figure 6.28. As shown, entrainment becomes excessive at high values of fraction of flooding, $f = U_a/U_f$, particularly for small values of the kinetic-energy ratio, F_{LV}. The effect of entrainment on the Murphree vapor efficiency can be estimated by the following relation derived by Colburn [42], where E_{MV} is the usual "dry" efficiency and $E_{MV,\text{wet}}$ is the "wet" efficiency:

$$\frac{E_{MV,\text{wet}}}{E_{MV}} = \frac{1}{1 + eE_{MV}/L}$$
$$= \frac{1}{1 + E_{MV}[\psi/(1 - \psi)]} \qquad (6\text{-}69)$$

Equation (6-69) assumes that $\lambda = KV/L = 1$ and that the liquid is well mixed on the tray such that the composition of the entrained liquid is that of the liquid flowing to the tray below. For a given value of the entrainment ratio, ψ, the larger the value of E_{MV}, the greater is the effect of entrainment. For $E_{MV} = 1.0$ and $\psi = 0.10$, the "wet" efficiency is 0.90. An equation similar to (6-69) for the effect of weeping

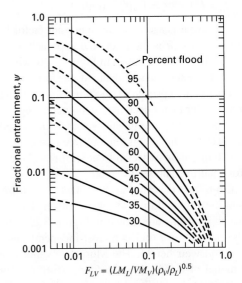

Figure 6.28 Correlation of Fair for fractional entrainment for sieve trays.

[Reproduced by permission from B.D. Smith, *Design of Equilibrium Stage Processes*, McGraw-Hill, New York (1963).]

is not available, because this effect depends greatly on the degree of liquid mixing on the tray and on the distribution of weeping over the active area of the tray. If weeping occurs only in the vicinity of the downcomer, no decrease in the value of E_{MV} is observed.

The height of clear liquid in the downcomer, h_{dc}, is always greater than the height of clear liquid on the tray because, by reference to Figure 6.3, the pressure difference across the froth in the downcomer is equal to the total pressure drop across the tray from which liquid enters the downcomer, plus the height of clear liquid on the tray below to which the liquid flows, and plus the head loss for liquid flow under the downcomer apron. Thus, the clear liquid head in the downcomer is

$$h_{dc} = h_t + h_l + h_{da} \qquad (6\text{-}70)$$

where h_t is given by (6-49) and h_l by (6-51), and the hydraulic gradient is assumed to be negligible. The head loss for liquid flow under the downcomer, h_{da}, in inches of liquid can be estimated from an empirical orifice-type equation:

$$h_{da} = 0.03 \left(\frac{q_L}{100\, A_{da}} \right)^2 \qquad (6\text{-}71)$$

where q_L is the liquid flow in gpm and A_{da} is the area in ft^2 for liquid flow under the downcomer apron. If the height of the opening under the apron (typically 0.5 in. less than h_w) is h_a, then $A_{da} = L_w h_a$. The height of the froth in the downcomer is

$$h_{df} = h_{dc}/\phi_{df} \qquad (6\text{-}72)$$

where the froth density, ϕ_{df}, can be taken conservatively as 0.5.

EXAMPLE 6.8

Using data from Examples 6.5, 6.6, and 6.7, estimate the entrainment rate, the froth height in the downcomer, and whether weeping occurs.

SOLUTION

Weeping criterion: From Example 6.6,

$$h_d = 1.56 \text{ in.} \quad h_\sigma = 0.36 \text{ in.} \quad h_l = 0.70 \text{ in.}$$

$$\text{From (6-68), } 1.56 + 0.36 > 0.70$$

Therefore, if the liquid level is uniform across the active area, no weeping occurs.

Entrainment: From Example 6.5,

$$F_{LV} = 0.016 \quad \text{From Example 6.7, } f = 0.59$$

From Figure 6.28, $\psi = 0.06$. Therefore, for $L = 155$ from Example 6.7, the entrainment rate is $0.06(155) = 9.3$ kmol/h. Assuming that (6-69) is reasonably accurate for $\lambda = 0.662$ from Example 6.7, and that $E_{MV} = 0.78$, the effect of ψ on E_M, is given by

$$\frac{E_{MV,\text{wet}}}{E_{MV}} = \frac{1}{1 + 0.78(0.06/0.94)} = 0.95$$

or

$$E_{MV} = 0.95(0.78) = 0.74$$

Downcomer backup:

From Example 6.6, $h_t = 2.62$ in.
From Example 6.7, $L_w = 28.7$ in.
From Example 6.6, $h_w = 2.0$ in.
Assume that $h_a = 2.0 - 0.5 = 1.5$ in. Then

$$A_{da} = L_w h_a = 28.7(1.5) = 43.1 \text{ in.}^2 = 0.299 \text{ ft}^2$$

From Example 6.6, $q_L = 12.9$ gpm

From (6-71), $h_{da} = 0.03 \left[\dfrac{12.9}{(100)(0.299)} \right]^2 = 0.006$ in.

From (6-70), $h_{dc} = 2.62 + 0.70 + 0.006 = 3.33$ in. of clear liquid backup

From (6-72), $h_{df} = \dfrac{3.33}{0.5} = 6.66$ in. of froth in the downcomer

Based on these results, neither weeping nor downcomer backup appear to be problems. An estimated 5% loss in tray efficiency occurs due to entrainment.

6.7 RATE-BASED METHOD FOR PACKED COLUMNS

Absorption and stripping are frequently conducted in packed columns, particularly when (1) the required column diameter is 2 ft or less; (2) the pressure drop must be low, as for a vacuum service; (3) corrosion considerations favor the use of ceramic or polymeric materials; and/or (4) low liquid holdup is desirable. Structured packing is often favored over random packing for revamps to overcome capacity limitations of trayed towers.

Packed columns are continuous, differential-contacting devices that do not have the physically distinguishable stages found in trayed towers. Thus, packed columns are best analyzed by mass-transfer considerations rather than by the equilibrium-stage concept described in earlier sections of this chapter for trayed towers. Nevertheless, in practice, packed-tower performance is often analyzed on the basis of equivalent equilibrium stages using a packed height equivalent to a theoretical (equilibrium) plate (stage), called the HETP or HETS and defined by the equation

$$\text{HETP} = \frac{\text{packed height}}{\text{number of equivalent equilibrium stages}} = \frac{l_T}{N_t} \qquad (6\text{-}73)$$

The HETP concept, unfortunately, has no theoretical basis. Accordingly, although HETP values can be related to mass-transfer coefficients, such values are best obtained by back-calculation from (6-73) using experimental data from laboratory or commercial-size columns. To illustrate the application of the HETP concept, consider Example 6.1, which involves the recovery of ethyl alcohol from a CO_2-rich vapor by absorption with water. The required number of equilibrium stages is found to be just slightly more than 6, say, 6.1. Suppose that experience shows that if 1.5-in. metal Pall rings are used in a packed tower, an average HETP of

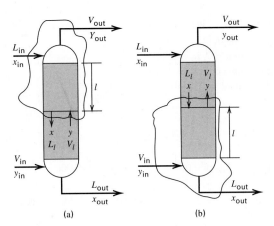

Figure 6.29 Packed columns with countercurrent flow:
(a) absorber; (b) stripper.

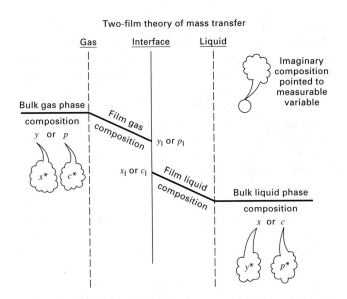

Figure 6.30 Interface properties in terms of bulk properties.

2.25 ft can be achieved. From (6-73), the required packed height, l_T, is $l_T = (HETP)N_t = 2.25(6.1) = 13.7$ ft. With metal Intalox IMTP #40 random packing, the HETP might be 2.0 ft, giving $l_T = 12.3$ ft. With Mellapak 250Y corrugated, sheet-metal structured packing, the HETP might be only 1.2 ft, giving $l_T = 7.3$ ft.

For packed columns, it is preferable to determine packed height from a more theoretically based method involving mass-transfer coefficients for the liquid and vapor phases. As with cascades of equilibrium stages, countercurrent flow of vapor and liquid is generally preferred over cocurrent flow. Consider the countercurrent-flow packed columns of packed height l_T, shown in Figure 6.29, which is analogous to Figure 6.8 for trayed towers. For packed absorbers and strippers, operating-line equations, that are analogous to those of Section 6.3 can be derived in terms of mole fractions and total molar flow rates. Thus, for the absorber in Figure 6.29a, a material balance around the upper envelope, for the solute, gives

$$x_{in}L_{in} + yV_l = xL_l + y_{out}V_{out} \qquad (6\text{-}74)$$

or solving for y, assuming dilute solutions such that $V_l = V_{in} = V_{out} = V$ and $L_l = L_{in} = L_{out} = L$

$$y = x\left(\frac{L}{V}\right) + y_{out} - x_{in}\left(\frac{L}{V}\right) \qquad (6\text{-}75)$$

Similarly for the stripper in Figure 6.29b,

$$y = x\left(\frac{L}{V}\right) + y_{in} - x_{out}\left(\frac{L}{V}\right) \qquad (6\text{-}76)$$

In Equations (6-74) to (6-76), mole fractions y and x represent, respectively, bulk compositions of the gas and liquid streams in contact with each other at any elevation of the packed part of the column. For the case of absorption, with mass transfer of the solute from the gas stream to the liquid stream, the two-film theory, developed in Section 3.7, can be applied as illustrated in Figure 6.30. A concentration gradient exists in each film. At the interface between the two phases, physical equilibrium is assumed to exist. Thus, as with trayed towers, an operating line and an equilibrium line are of great importance in a packed column. For a given

problem specification, the location of the two lines is independent of whether the tower is trayed or packed. Thus, the method for determining the minimum absorbent liquid or stripping vapor flow rates in a packed column is identical to the method for trayed towers, as presented in Section 6.3 and illustrated in Figure 6.9.

The rate of mass transfer for absorption or stripping in a packed column can be expressed in terms of mass-transfer coefficients for each phase. Coefficients, k, based on a unit area for mass transfer could be used, but the area for mass transfer in a packed bed is difficult to determine. Accordingly, as with mass transfer in the froth of a trayed tower, it is more common to use volumetric mass-transfer coefficients, ka, where the quantity a represents the area for mass transfer per unit volume of packed bed. Thus, ka is based on a unit volume of packed bed. At steady state in an absorber, in the absence of chemical reactions, and since species moles are conserved, the rate of solute mass transfer across the gas-phase film must equal the rate across the liquid-phase film. If the system is dilute with respect to the solute, unimolecular diffusion (UMD) may be approximated by the simpler equations for equimolar counterdiffusion (EMD) discussed in Chapter 3. The rate of mass transfer per unit volume of packed bed, r, may be written in terms of mole-fraction driving forces in each of the two phases or in terms of a partial-pressure driving force in the gas phase and a concentration driving force in the liquid phase, as indicated in Figure 6.30. Using the former, for absorption, with the subscript I to denote the interface:

$$r = k_y a(y - y_I) = k_x a(x_I - x) \qquad (6\text{-}77)$$

The composition at the interface depends on the ratio, $k_x a/k_y a$, of the volumetric mass-transfer coefficients, because (6-77) can be rearranged to

$$\frac{y - y_I}{x - x_I} = -\frac{k_x a}{k_y a} \qquad (6\text{-}78)$$

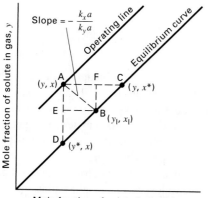

Figure 6.31 Interface composition in terms of the ratio of mass-transfer coefficients.

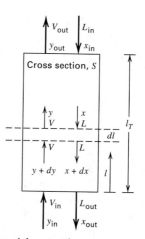

Figure 6.32 Differential contact in a countercurrent-flow, packed absorption column.

Thus, a straight line of slope $-k_x a / k_y a$ drawn from the operating line at point (y, x) intersects the equilibrium curve at (y_I, x_I). This result is shown graphically in Figure 6.31.

The slope $-k_x a / k_y a$ determines the relative resistances of the two phases to mass transfer. In Figure 6.31 the distance AE is the gas-phase driving force $(y - y_I)$, while AF is the liquid-phase driving force $(x_I - x)$. If the mass-transfer resistance in the gas phase is very low, y_I is approximately equal to y. Then, the resistance resides entirely in the liquid phase. This situation occurs in the absorption of a solute that is only slightly soluble in the liquid phase (i.e., a solute with a high K-value) and is referred to as a liquid-film resistance-controlling process. Alternatively, if the resistance in the liquid phase is very low, x_I is approximately equal to x. This situation occurs in the absorption of a solute that is very soluble in the liquid phase (i.e., a solute with a low K-value) and is referred to as a gas-film resistance-controlling process. It is important to know if one of the two resistances is controlling. If so, the rate of mass transfer can be increased by promoting turbulence in and/or increasing the dispersion of the controlling phase.

To avoid the need to determine the composition at the interface between the two phases, overall, volumetric mass-transfer coefficients can be defined in terms of overall driving forces for either the gas phase or the liquid phase. Thus, for mole-fraction driving forces,

$$r = K_y a(y - y^*) = K_x a(x^* - x) \qquad (6\text{-}79)$$

where, as shown in Figure 6.31, y^* is the fictitious vapor mole fraction that is in equilibrium with the mole fraction, x, in the bulk liquid; and x^* is the fictitious liquid mole fraction that is in equilibrium with the mole fraction, y, in the bulk vapor. By combining (6-77) to (6-79), the overall coefficients can be expressed in terms of the separate coefficients for the two phases. Thus,

$$\frac{1}{K_y a} = \frac{1}{k_y a} + \frac{1}{k_x a}\left(\frac{y_I - y^*}{x_I - x}\right) \qquad (6\text{-}80)$$

and

$$\frac{1}{K_x a} = \frac{1}{k_x a} + \frac{1}{k_y a}\left(\frac{x^* - x_I}{y - y_I}\right) \qquad (6\text{-}81)$$

However, from Figure 6.31, for dilute solutions when the equilibrium curve is approximately a straight line through the origin,

$$\frac{y_I - y^*}{x_I - x} = \frac{\overline{ED}}{\overline{BE}} = K \qquad (6\text{-}82)$$

and

$$\frac{x^* - x_I}{y - y_I} = \frac{\overline{CF}}{\overline{FB}} = \frac{1}{K} \qquad (6\text{-}83)$$

where K is the K-value for the solute. Combining (6-80) with (6-82) and (6-81) with (6-83),

$$\frac{1}{K_y a} = \frac{1}{k_y a} + \frac{K}{k_x a} \qquad (6\text{-}84)$$

and

$$\frac{1}{K_x a} = \frac{1}{k_x a} + \frac{1}{K k_y a} \qquad (6\text{-}85)$$

Determination of the packed height of a column most commonly involves the overall gas-phase coefficient, $K_y a$, because the liquid usually has a strong affinity for the solute so that resistance to mass transfer is mostly in the gas. This is analogous to a trayed tower, where the tray efficiency from mass transfer considerations is commonly based on $K_{OG}a$ or N_{OG}. Consider the countercurrent-flow absorption column in Figure 6.32. For a dilute system, a differential material balance for a solute being absorbed over a differential height of packing dl, gives:

$$-V\,dy = K_y a(y - y^*)S\,dl \qquad (6\text{-}86)$$

where S is the inside cross-sectional area of the tower. In integral form, with nearly constant terms placed outside the integral, (6-86) becomes

$$\frac{K_y a S}{V}\int_0^{l_T} dl = \frac{K_y a S l_T}{V} = \int_{y_{out}}^{y_{in}} \frac{dy}{y - y^*} \qquad (6\text{-}87)$$

Solving for the packed height gives

$$l_T = \frac{V}{K_y a S}\int_{y_{out}}^{y_{in}} \frac{dy}{y - y^*} \qquad (6\text{-}88)$$

Chilton and Colburn [43] suggested that the right-hand side of (6-88) be written as the product of two terms:

$$l_T = H_{OG} N_{OG} \qquad (6\text{-}89)$$

where

$$H_{OG} = \frac{V}{K_y aS} \qquad (6\text{-}90)$$

and

$$N_{OG} = \int_{y_{out}}^{y_{in}} \frac{dy}{y - y^*} \qquad (6\text{-}91)$$

If (6-89) is compared to (6-73), it is seen that H_{OG} is analogous to HETP, as is N_{OG} to N_t.

The term H_{OG} is called the *overall height of a transfer unit* (HTU) based on the gas phase. Experimental data show that the HTU varies less with V than $K_y a$. The smaller the HTU, the more efficient is the contacting. The term N_{OG} is called the *overall number of transfer units* (NTU) based on the gas phase. It represents the overall change in solute mole fraction divided by the average mole-fraction driving force. The larger the NTU, the greater is the extent of contacting required.

Equation (6-91) was first integrated by Colburn [44]. By using the linear equilibrium condition $y^* = Kx$ to eliminate y^* and using the linear, solute material-balance operating line, (6-75), to eliminate x, the result is

$$\int_{y_{out}}^{y_{in}} \frac{dy}{y - y^*} = \int_{y_{out}}^{y_{in}} \frac{dy}{(1 - KV/L)y + y_{out}(KV/L) - Kx_{in}} \qquad (6\text{-}92)$$

Letting $L/(KV) = A$, the absorption factor, and integrating (6-88), gives

$$N_{OG} = \frac{\ln\{[(A-1)/A][(y_{in} - Kx_{in})/(y_{out} - Kx_{in})] + (1/A)\}}{(A-1)/A} \qquad (6\text{-}93)$$

By applying (6-93) and (6-90), the required packed height, l_T, can be determined from (6-89). However, (6-93) is very sensitive when $A < 0.9$.

The NTU (e.g., N_{OG}) and the HTU (e.g., H_{OG}) should not be confused with the number of equilibrium (theoretical) stages, N_t, and the HETP, respectively. However, when the operating and equilibrium lines are not only straight but also parallel, NTU = N_t and HTU = HETP. Otherwise, the NTU is greater than or less than N_t as shown in Figure 6.33 for the case of absorption. When the operating and equilibrium lines are straight but not parallel, then

$$\text{HETP} = H_{OG}\frac{\ln(1/A)}{(1 - A)/A} \qquad (6\text{-}94)$$

and

$$N_{OG} = N_t \frac{\ln(1/A)}{(1 - A)/A} \qquad (6\text{-}95)$$

Although the most common applications of the HTU and NTU are based on (6-89) to (6-91) and (6-93), a number of alternative groupings have been used, depending on the selected driving force for mass transfer and whether the overall basis is the gas phase, as above, or the liquid phase, where H_{OL} and N_{OL} apply. These groupings are summarized in Table 6.7. Included are driving forces based on partial pressures, p; mole ratios, X, Y; and concentrations, c; as well as mole fractions, x, y. Also included in Table 6.7 for later reference in the last section of this chapter are groupings for unimolecular diffusion (UMD) when solute concentration is not dilute. It is frequently necessary to convert a mass-transfer coefficient based on one type of driving force to another coefficient based on a different type of driving force. Table 3.15 gives the relationships among the different mass-transfer coefficients. The relationships include coefficients based on a concentration and mole-fraction driving force. In addition, a partial-pressure driving force is included for the gas phase.

EXAMPLE 6.9

Repeat Example 6.1 for absorption in a tower packed with 1.5-in. metal Pall rings. If $H_{OG} = 2.0$ ft, compute the required packed height.

SOLUTION

From Example 6.1, $V = 180$ kmol/h, $L = 151.5$ kmol/h, $y_{in} = 0.020$, $x_{in} = 0.0$, and $K = 0.57$. For 97% recovery of ethyl alcohol, by material balance,

$$y_{out} = \frac{(0.03)(0.02)(180)}{180 - (0.97)(0.02)(180)} = 0.000612$$

$$A = \frac{L}{KV} = \frac{151.5}{(0.57)(180)} = 1.477$$

$$\frac{y_{in}}{y_{out}} = \frac{0.020}{0.000612} = 32.68$$

From (6-93),

$$N_{OG} = \frac{\ln\{[(1.477 - 1)/1.477](32.68) + (1/1.477)\}}{(1.477 - 1)/1.477}$$

$$= 7.5 \text{ transfer units}$$

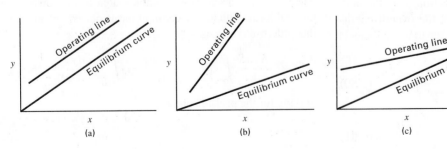

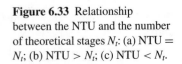

Figure 6.33 Relationship between the NTU and the number of theoretical stages N_t: (a) NTU = N_t; (b) NTU > N_t; (c) NTU < N_t.

Table 6.7 Alternative Mass-Transfer Coefficient Groupings

		Height of a Transfer Unit, HTU			Number of Transfer Units, NTU	
Driving Force	Symbol	EM Diffusion or Dilute UM Diffusion	UM Diffusion	Symbol	EM Diffusion[a] or Dilute UM Diffusion	UM Diffusion
1. $(y - y^*)$	H_{OG}	$\dfrac{V}{K_y a S}$	$\dfrac{V}{K_y' a (1-y)_{LM} S}$	N_{OG}	$\displaystyle\int \dfrac{dy}{(y-y^*)}$	$\displaystyle\int \dfrac{(1-y)_{LM} dy}{(1-y)(y-y^*)}$
2. $(p - p^*)$	H_{OG}	$\dfrac{V}{K_G a P S}$	$\dfrac{V}{K_G' a (1-y)_{LM} P S}$	N_{OG}	$\displaystyle\int \dfrac{dp}{(p-p^*)}$	$\displaystyle\int \dfrac{(P-p)_{LM} dp}{(P-p)(p-p^*)}$
3. $(Y - Y^*)$	H_{OG}	$\dfrac{V'}{K_Y a S}$	$\dfrac{V'}{K_Y a S}$	N_{OG}	$\displaystyle\int \dfrac{dY}{(Y-Y^*)}$	$\displaystyle\int \dfrac{dY}{(Y-Y^*)}$
4. $(y - y_I)$	H_G	$\dfrac{V}{k_y a S}$	$\dfrac{V}{k_y' a (1-y)_{LM} S}$	N_G	$\displaystyle\int \dfrac{dy}{(y-y_I)}$	$\displaystyle\int \dfrac{(1-y)_{LM} dy}{(1-y)(y-y_I)}$
5. $(p - p_I)$	H_G	$\dfrac{V}{k_p a P S}$	$\dfrac{V}{k_p' a (P-p)_{LM} S}$	N_G	$\displaystyle\int \dfrac{dp}{(p-p_I)}$	$\displaystyle\int \dfrac{(P-p)_{LM} dp}{(P-p)(p-p_I)}$
6. $(x^* - x)$	H_{OL}	$\dfrac{L}{K_x a S}$	$\dfrac{L}{K_x' a (1-x)_{LM} S}$	N_{OL}	$\displaystyle\int \dfrac{dx}{(x^*-x)}$	$\displaystyle\int \dfrac{(1-x)_{LM} dx}{(1-x)(x^*-x)}$
7. $(c^* - c)$	H_{OL}	$\dfrac{L}{K_L a (\rho_L/M_L) S}$	$\dfrac{L}{K_L' a (\rho_L/M_L - c)_{LM} S}$	N_{OL}	$\displaystyle\int \dfrac{dc}{(c^*-c)}$	$\displaystyle\int \dfrac{(\rho_L/M_L - c)_{LM} dx}{(\rho_L/M_L - c)(c^*-c)}$
8. $(X^* - X)$	H_{OL}	$\dfrac{L'}{K_X a S}$	$\dfrac{L'}{K_X a S}$	N_{OL}	$\displaystyle\int \dfrac{dX}{(X^*-X)}$	$\displaystyle\int \dfrac{dX}{(X^*-X)}$
9. $(x_I - x)$	H_L	$\dfrac{L}{k_x a S}$	$\dfrac{L}{k_x' a (1-x)_{LM} S}$	N_L	$\displaystyle\int \dfrac{dx}{(x_I-x)}$	$\displaystyle\int \dfrac{(1-x)_{LM} dx}{(1-x)(x_I-x)}$
10. $(c_I - c)$	H_L	$\dfrac{L}{k_L a (\rho_L/M_L) S}$	$\dfrac{L}{k_L' a (\rho_L/M_L - c)_{LM} S}$	N_L	$\displaystyle\int \dfrac{dc}{(c_I-c)}$	$\displaystyle\int \dfrac{(\rho_L/M_L - c)_{LM} dc}{(\rho_L/M_L - c)(c_I-c)}$

[a]The substitution $K_y = K_y' y_{B_{LM}}$ or its equivalent can be made.

The packed height, from (6-89), is

$$l_T = 2.0(7.5) = 15 \text{ ft}$$

Note that N_t for this example was determined in Example 6.1 to be about 6.1. The value of 7.5 for N_{OG} is greater than N_t because the slope of the operating line, L/G, is greater than the slope of the equilibrium line, K, so Figure 6.33b applies.

EXAMPLE 6.10

Experimental data have been obtained for air containing 1.6% by volume SO_2 being scrubbed with pure water in a packed column of 1.5 m^2 in cross-sectional area and 3.5 m in packed height. Entering gas and liquid flow rates are 0.062 and 2.2 kmol/s, respectively. If the outlet mole fraction of SO_2 in the gas is 0.004 and column temperature is near-ambient with $K_{SO_2} = 40$, calculate from the data:

(a) The N_{OG} for absorption of SO_2

(b) The H_{OG} in meters

(c) The volumetric, overall mass-transfer coefficient, $K_y a$ for SO_2 in kmol/m^3-s-(Δy).

SOLUTION

(a) Assume a straight operating line because the system is dilute in SO_2.

$$A = \frac{L}{KV} = \frac{2.2}{(40)(0.062)} = 0.89, \quad y_{in} = 0.016,$$
$$y_{out} = 0.004, \quad x_{in} = 0.0$$

From (6-93),

$$N_{OG} = \frac{\ln\{[(0.89 - 1)/0.89](0.016/0.004) + (1/0.89)\}}{(0.89 - 1)/0.89}$$
$$= 3.75$$

(b) $l_T = 3.5$ m. From (6-89), $H_{OG} = l_T/N_{OG} = 3.5/3.75 = 0.93$ m

(c) $V = 0.062$ kmol/s, $S = 1.5$ m^2.

From (6-90), $K_y a = V/H_{OG}S = 0.062/[(0.93)(1.5)] = 0.044$ kmol/m^3-s-(Δy)

EXAMPLE 6.11

A gaseous reactor effluent consisting of 2 mol% ethylene oxide in an inert gas is scrubbed with water at 30°C and 20 atm. The total gas feed rate is 2,500 lbmol/h, and the water rate entering the scrubber is 3,500 lbmol/h. The column, with a diameter of 4 ft, is packed in two 12-ft-high sections with 1.5-in. metal Pall rings. A liquid redistributor is located between the two packed sections. Under the operating conditions for the scrubber, the K-value for ethylene oxide is 0.85 and estimated values of $k_y a$ and $k_x a$ are 200 lbmol/h-ft^3-Δy and 165 lbmol/h-ft^3-Δx, respectively.

Calculate: (a) $K_y a$ and (b) H_{OG}.

SOLUTION

(a) From (6-84),

$$K_y a = \frac{1}{(1/k_y a) + (K/k_x a)} = \frac{1}{(1/200) + (0.85/165)}$$
$$= 98.5 \text{ lbmol/h-ft}^3\text{-}\Delta y$$

(b) $S = 3.14(4)^2/4 = 12.6 \text{ ft}^2$

From (6-90), $H_{OG} = V/K_y aS = 2,500/[(98.5)(12.6)] = 2.02 \text{ ft.}$

Note that in this example, both gas-phase and liquid-phase resistances are important.

The value of H_{OG} can also be computed from values of H_G and H_L using equations in Table 6.7:

$$H_G = V/k_y aS = 2,500/[(200)(12.6)] = 1.0 \text{ ft}$$
$$H_L = L/k_x aS = 3,500/[(165)(12.6)] = 1.68 \text{ ft}$$

Substituting these two expressions and (6-90) into (6-84) gives the following relationship for H_{OG} in terms of H_G and H_L:

$$H_{OG} = H_G + H_L/A$$
$$A = L/KV = 3,500/[(0.85)(2,500)] = 1.65 \quad (6\text{-}96)$$
$$H_{OG} = 1.0 + 1.68/1.65 = 2.02 \text{ ft}$$

6.8 PACKED-COLUMN EFFICIENCY, CAPACITY, AND PRESSURE DROP

Values of volumetric mass-transfer coefficients and corresponding HTUs depend on gas and/or liquid flow rates per unit inside cross-sectional area of the packed column. Therefore, column diameter must be estimated before determining required height of packing. The estimation of a suitable column diameter for a given system, packing, and operating conditions requires consideration of liquid holdup, flooding, and pressure drop.

Liquid Holdup

Typical experimental curves, taken from Billet [45] and shown also by Stichlmair, Bravo, and Fair [46], for specific pressure drop in meters of water head per meter of packed height, and specific liquid holdup in cubic meters per cubic meter of packed bed as a function of superficial gas velocity for different values of superficial water velocity are shown in Figures 6.34 and 6.35, respectively, for a 0.15-m-diameter column packed with 1-in. metal Bialecki® rings to a height of 1.5 m and operated at 25°C and 1 bar. In Figure 6.34, the

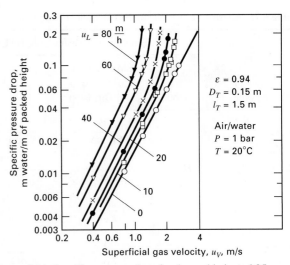

Figure 6.34 Specific pressure drop for dry and irrigated 25-mm metal Bialecki® rings.

[From R. Billet, *Packed Column Analysis and Design,* Ruhr-University Bochum (1989) with permission.]

lowest curve corresponds to zero liquid flow, that is, the dry pressure drop. Over an almost 10-fold range of superficial air velocity (the velocity the air would have in the absence of packing), the pressure drop for air flowing up through the packing is proportional to air velocity to the 1.86 power. As liquid flows down through the packing at an increasing rate, gas-phase pressure drop for a given gas velocity increases. However, below a certain limiting gas velocity, the curve for each liquid velocity is a straight line parallel to the dry-pressure-drop curve. In this region, the liquid holdup for each liquid velocity is constant, as shown in Figure 6.35. Thus, for a liquid velocity of 40 m/h, specific liquid holdup is 0.08 m^3/m^3 of packed bed until a superficial gas velocity of 1.0 m/h is reached. Instead of a packed-column void fraction, ϵ, of 0.94 for the gas to flow through (corresponding to zero liquid flow), the effective void fraction is reduced by

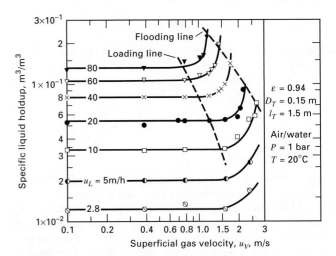

Figure 6.35 Specific liquid holdup for irrigated 25-mm metal Bialecki® rings.

[From R. Billet, *Packed Column Analysis and Design,* Ruhr-University Bochum (1989) with permission.]

the liquid holdup to $0.94 - 0.08 = 0.86$, causing an increased pressure drop. For a given liquid velocity, the upper limit to the gas velocity for a constant liquid holdup is termed the *loading point*. Below this point, the gas phase is the continuous phase. Above this point, liquid begins to accumulate or load the bed, replacing gas holdup and causing a sharp increase in pressure drop. Finally, a gas velocity is reached at which the liquid surface is continuous across the top of the packing and the column is flooded. At the *flooding point*, the drag force of the counterflowing gas is sufficient to entrain the entire liquid. Approximate loci of both loading and flooding points are included in Figure 6.35.

The region between the loading point and the flooding point is the *loading region*, where significant liquid entrainment is observed, liquid holdup increases sharply, mass-transfer efficiency decreases, and column operation is unstable. Typically, according to Billet [45], the superficial gas velocity at the loading point is approximately 70% of that at the flooding point. Although a packed column can operate in the loading region, most packed columns are designed to operate at or below the loading point, in the *preloading region*.

The specific liquid holdup in the preloading region has been found, from extensive experiments by Billet and Schultes [47, 69] for a wide variety of random and structured packings and, for a number of gas–liquid systems, to depend on packing characteristics, and the viscosity, density, and superficial velocity of the liquid, u_L, according to the dimensionless expression

$$h_L = \left(12 \frac{N_{Fr_L}}{N_{Re_L}}\right)^{1/3} \left(\frac{a_h}{a}\right)^{2/3} \qquad (6\text{-}97)$$

where

$$N_{Re_L} = \text{liquid Reynolds number} = \frac{\text{inertial force}}{\text{viscous force}}$$

$$= \frac{u_L \rho_L}{a \mu_L} = \frac{u_L}{a \nu_L} \qquad (6\text{-}98)$$

where ν_L is the kinematic viscosity.

$$N_{Fr_L} = \text{liquid Froude number} = \frac{\text{inertial force}}{\text{gravitational force}} \qquad (6\text{-}99)$$

$$= \frac{u_L^2 a}{g}$$

and the ratio of specific hydraulic area of packing, a_h, to specific surface area of packing, a, is given by

$$a_h/a = C_h N_{Re_L}^{0.15} N_{Fr_L}^{0.1} \quad \text{for } N_{Re_L} < 5 \qquad (6\text{-}100)$$

$$a_h/a = 0.85\, C_h N_{Re_L}^{0.25} N_{Fr_L}^{0.1} \quad \text{for } N_{Re_L} \geq 5 \qquad (6\text{-}101)$$

Values of $a_h/a > 1$ are reasonable because of the creation of droplets and jet flow beside the rivulets that cover the packing surface [70].

Values of a and C_h are characteristic of the particular type and size of packing, as listed, together with packing void fraction, ϵ, and other packing constants in Table 6.8. Because the specific liquid holdup is constant in the preloading region, as seen in Figure 6.35, (6-97) does not involve gas-phase properties or gas velocity.

At low liquid velocities, liquid holdup can become so small that the packing is no longer completely wetted. When this occurs, packing efficiency decreases dramatically, particularly for aqueous systems of high surface tension. To ensure adequate wetting of packing, proven liquid distributors and redistributors should be used and superficial liquid velocities should exceed the following values:

Type of Packing Material	$u_{L_{min}}$, m/s
Ceramic	0.00015
Oxidized or etched metal	0.0003
Bright metal	0.0009
Plastic	0.0012

EXAMPLE 6.12

An absorption column is to be designed using oil absorbent with a kinematic viscosity of three times that of water at 20°C. The superficial liquid velocity will be 0.01 m/s, which is safely above the minimum value for good wetting. The superficial gas velocity will be such that operation will be in the preloading region. Two packing materials are being considered: (1) randomly packed 50-mm metal Hiflow® rings and (2) metal Montz® B1-200 structured packing. Estimate the specific liquid holdup for each of these two packings.

SOLUTION

From Table 6.8,

Packing	a, m^2/m^3	ϵ	C_h
50-mm metal Hiflow® rings	92.3	0.977	0.876
Montz® metal™ B1-200	200.0	0.979	0.547

At 20°C for water, kinematic viscosity, $\nu = \mu/\rho = 1 \times 10^{-6}$ m^2/s. Therefore, for the oil, $\mu/\rho = 3 \times 10^{-6}$ m^2/s. From (6-98) and (6-99),

$$N_{Re_L} = \frac{0.01}{3 \times 10^{-6} a} \qquad N_{Fr_L} = \frac{(0.01)^2 a}{9.8}$$

Therefore,

Packing	N_{Re_L}	N_{Fr_L}
Hiflow®	36.1	0.000942
Montz™	16.67	0.00204

From (6-101), since $N_{Re_L} > 5$, for the Hiflow® packing, $a_h/a = (0.85)(0.876)(36.1)^{0.25}(0.000942)^{0.1} = 0.909$. For the Montz™ packing, $a_h/a = 0.85(0.547)(16.67)^{0.25}(0.00204)^{0.10} = 0.506$.

From (6-97), for the Hiflow® packing,

$$h_L = \left[\frac{12(0.000942)}{36.1}\right]^{1/3} (0.909)^{2/3} = 0.0637 \text{ m}^3/\text{m}^3$$

For the Montz™ packing,

$$h_L = \left[\frac{12(0.0204)}{16.67}\right]^{1/3} (0.506)^{2/3} = 0.0722 \text{ m}^3/\text{m}^3$$

Note that for the Hiflow® packing, the void fraction available for gas flow is reduced by the liquid flow from $\epsilon = 0.977$ (Table 6.8) to $0.977 - 0.064 = 0.913$ m^3/m^3. For the Montz™ packing, the reduction is from 0.979 to 0.907 m^3/m^3.

Table 6.8 Characteristics of Packings

Characteristics from Billet (columns C_h through C_{Fl})

Packing	Material	Size	F_P, ft²/ft³	a, m²/m³	ϵ, m³/m³	C_h	C_p	C_L	C_V	C_s	C_{Fl}
Random Packings											
Berl saddles	Ceramic	25 mm	110	260.0	0.680	0.620		1.246	0.387		1.896
Berl saddles	Ceramic	13 mm	240	545.0	0.650	0.833		1.364	0.232		1.885
Bialecki rings	Metal	50 mm		121.0	0.966	0.798	0.719	1.721	0.302	2.916	1.856
Bialecki rings	Metal	35 mm		155.0	0.967	0.787	1.011	1.412	0.390	2.753	1.912
Bialecki rings	Metal	25 mm		210.0	0.956	0.692	0.891	1.461	0.331	2.521	1.991
Dinpak®	Plastic	70 mm		110.7	0.938	0.991	0.378	1.527	0.326	2.970	1.522
Dinpak®	Plastic	47 mm		131.2	0.923	1.173	0.514	1.690	0.354	2.929	1.864
Envi Pac®	Plastic	80 mm, no. 3		60.0	0.955	0.641	0.358	1.603	0.257	2.846	2.012
Envi Pac®	Plastic	60 mm, no. 2		98.4	0.961	0.794	0.338	1.522	0.296	2.987	1.900
Envi Pac®	Plastic	32 mm, no. 1		138.9	0.936	1.039	0.549	1.517	0.459	2.944	1.760
Cascade Mini-Rings®	Metal	30 PMK		180.5	0.975	0.930	0.851	1.920	0.450	2.694	1.841
Cascade Mini-Rings®	Metal	30 P		164.0	0.959	0.851	1.056	1.577	0.398	2.564	1.870
Cascade Mini-Rings®	Metal	1.5″		174.9	0.974	0.935	0.632			2.697	1.996
Cascade Mini-Rings®	Metal	1.5″, T		188.0	0.972	0.870	0.627			2.790	2.178
Cascade Mini-Rings®	Metal	1.0″		232.5	0.971	1.040	0.641			2.703	
Cascade Mini-Rings®	Metal	0.5″		356.0	0.952		0.882	2.038	0.495	2.644	
Cascade Mini-Rings®	Metal	30 PMK		180.2	0.975	0.930					
Cascade Mini-Rings®	Metal	30 P		168.9	0.958	0.851					
Cascade Mini-Rings®	Metal	1.5″ CMR, T		188.0	0.972	0.870					
Cascade Mini-Rings®	Metal	1.5″ CMR		174.9	0.974	0.935					
Cascade Mini-Rings®	Metal	1.0″ CMR	29	232.5	0.971	1.040					
Cascade Mini-Rings®	Metal	0.5″ CMR	40	356.0	0.955	1.338					
Hackettes®	Plastic	45 mm	15	139.5	0.928	0.643	0.399	1.377	0.379	2.832	1.966
Hiflow® rings	Ceramic	75 mm	29	54.1	0.868		0.435	1.659	0.464	2.819	1.694
Hiflow® rings	Ceramic	50 mm	37	89.7	0.809		0.538	1.744	0.465	2.840	1.930
Hiflow® rings	Ceramic	38 mm		111.8	0.788		0.621				
Hiflow® rings	Ceramic	20 mm, 6 stg.		265.8	0.776	0.958	0.628				
Hiflow® rings	Ceramic	20 mm, 4 stg.		261.2	0.779	1.167					
Hiflow® rings	Metal	50 mm	16	92.3	0.977	0.876	0.421	1.168	0.408	2.702	1.626
Hiflow® rings	Metal	25 mm	42	202.9	0.962	0.799	0.689	1.641	0.402	2.918	2.177
Hiflow® rings	Plastic	90 mm	9	69.7	0.968		0.276				
Hiflow® rings	Plastic	50 mm, hydr.		118.4	0.925		0.311	1.553	0.369		
Hiflow® rings	Plastic	50 mm	20	117.1	0.924	1.038	0.327	1.487	0.345	2.894	1.871

Hiflow® rings	Plastic	25 mm		194.5	0.918	0.741		1.577	0.390	2.841	1.989
Hiflow® rings, super	Plastic	50 mm, S		82.0	0.942	0.414		1.219	0.342	2.866	1.702
Hiflow® saddles	Plastic	50 mm		86.4	0.938	0.454					
Intalox® saddles	Ceramic	50 mm	40	114.6	0.761	0.747					
Intalox® saddles	Plastic	50 mm	28	122.1	0.908	0.758					
Nor-Pac® rings	Plastic	50 mm	14	86.8	0.947	0.350	0.651	1.080	0.322	2.959	1.786
Nor-Pac® rings	Plastic	35 mm	21	141.8	0.944	0.371	0.587	0.756	0.425	3.179	2.242
Nor-Pac® rings	Plastic	25 mm, type B		202.0	0.953	0.397	0.601	0.883	0.366	3.277	2.472
Nor-Pac® rings	Plastic	25 mm, 10 stg.		197.9	0.920	0.383		0.976	0.410	2.865	2.083
Nor-Pac® rings	Plastic	25 mm	31	180.0	0.927		0.601				
Nor-Pac® rings	Plastic	22 mm		249.0	0.913	0.397					
Nor-Pac® rings	Plastic	15 mm		311.4	0.918	0.365	0.343				
Pall® rings	Ceramic	50 mm	43	155.2	0.754	0.233	1.066	1.278	0.333	3.793	3.024
Pall® rings	Metal	50 mm	27	112.6	0.951	0.763	0.784	1.192	0.410	2.725	1.580
Pall® rings	Metal	35 mm	40	139.4	0.965	0.967	0.644	1.012	0.341	2.629	1.679
Pall® rings	Metal	25 mm	56	223.5	0.954	0.957	0.719	1.440	0.336	2.627	2.083
Pall® rings	Metal	15 mm	70	368.4	0.933	0.990	0.590				
Pall® rings	Plastic	50 mm	26	111.1	0.919	0.698	0.593	1.239	0.368	2.816	1.757
Pall® rings	Plastic	35 mm	40	151.1	0.906	0.927	0.718	0.856	0.380	2.654	1.742
Pall® rings	Plastic	25 mm	55	225.0	0.887	0.865	0.528	0.905	0.446	2.696	2.064
Raflux® rings	Plastic	15 mm		307.9	0.894	0.595	0.491	1.913	0.370	2.825	2.400
Ralu® flow	Plastic	1		165	0.940	0.485	0.640	1.486	0.360	3.612	2.401
Ralu® flow	Plastic	2		100	0.945	0.350	0.640	1.270	0.320	3.412	2.174
Ralu® rings	Plastic	50 mm, hydr.		94.3	0.939		0.439	1.481	0.341		
Ralu® rings	Plastic	50 mm		95.2	0.983	0.468	0.640	1.520	0.303	2.843	1.812
Ralu® rings	Plastic	38 mm		150	0.930	0.672	0.640	1.320	0.333	2.843	1.812
Ralu® rings	Plastic	25 mm		190	0.940	0.800	0.719	1.320	0.333	2.841	1.989
Ralu® rings	Metal	50 mm		105	0.975	0.763	0.784	1.192	0.345	2.725	1.580
Ralu® rings	Metal	38 mm		135	0.965	1.003	0.644	1.277	0.341	2.629	1.679
Ralu® rings	Metal	25 mm		215	0.960	0.957	0.714	1.440	0.336	2.627	2.083
Raschig rings	Carbon	25 mm		202.2	0.720		0.623	1.379	0.471		
Raschig rings	Ceramic	25 mm	179	190.0	0.680		0.577	1.361	0.412		
Raschig rings	Ceramic	15 mm	380	312.0	0.690		0.648	1.276	0.401		
Raschig rings	Ceramic	10 mm	1,000	440.0	0.650		0.791	1.303	0.272		
Raschig rings	Ceramic	6 mm	1,600	771.9	0.620		1.094	1.130			
Raschig rings	Metal	15 mm	170	378.4	0.917		0.455				
Raschig rings	Ceramic	25		190.0	0.680	1.329	0.577	1.361	0.412	2.454	1.899
Raschig Super-rings	Metal	0.3		315	0.960	0.760	0.750	1.500	0.450	3.560	2.340

(Continued)

Table 6.8 (*Continued*)

Packing	Material	Size	F_P, ft²/ft³	a, m²/m³	ϵ, m³/m³	Characteristics from Billet					
						C_h	C_p	C_L	C_V	C_s	C_{Fl}
Raschig Super-rings	Metal	0.5		250	0.975	0.620	0.780	1.450	0.430	3.350	2.200
Raschig Super-rings	Metal	1		160	0.980	0.750	0.500	1.290	0.440	3.491	2.200
Raschig Super-rings	Metal	2		97.6	0.985	0.720	0.464	1.323	0.400	3.326	2.096
Raschig Super-rings	Metal	3		80	0.982	0.620	0.430	0.850	0.300	3.260	2.100
Raschig Super-rings	Plastic	2		100	0.960	0.720	0.377	1.250	0.337	3.326	2.096
Tellerettes®	Plastic	25 mm	40	190.0	0.930	0.588	0.538	0.899		2.913	2.132
Top-Pak® rings	Aluminum	50 mm		105.5	0.956	0.881	0.604	1.326	0.389	2.528	1.579
VSP® rings	Metal	50 mm, no. 2		104.6	0.980	1.135	0.773	1.222	0.420	2.806	1.689
VSP® rings	Metal	25 mm, no. 1		199.6	0.975	1.369	0.782	1.376	0.405	2.755	1.970
Structured Packings											
Euroform®	Plastic	PN-110		110.0	0.936	0.511	0.250	0.973	0.167	3.075	1.975
Gempak®	Metal	A2 T-304		202.0	0.977	0.678	0.344			2.986	2.099
Impulse®	Ceramic	100		91.4	0.838	1.900	0.417	1.317	0.327	2.664	1.655
Impulse®	Metal	250		250.0	0.975	0.431	0.262	0.983	0.270	2.610	1.996
Koch-Sulzer®	Metal	CY	70								
Koch-Sulzer®	Metal	BX	21								
Mellapak™	Plastic	250 Y	22	250.0	0.970	0.554	0.292			3.157	2.464
Montz™	Metal	B1-100		100.0	0.987	0.626					
Montz™	Metal	B1-200		200.0	0.979	0.547	0.355	0.971	0.390	3.116	2.339
Montz™	Metal	B1-300	33	300.0	0.930	0.482	0.295	1.165	0.422	3.098	2.464
Montz™	Plastic	C1-200		200.0	0.954		0.453	1.006	0.412		
Montz™	Plastic	C2-200		200.0	0.900		0.481	0.739		2.653	1.973
Ralu Pak®	Metal	YC-250		250.0	0.945	0.650	0.191	1.334	0.385	3.178	2.558

Column Diameter and Pressure Drop

Most packed columns consist of cylindrical vertical vessels. The column diameter is determined so as to safely avoid flooding and operate in the preloading region with a pressure drop of no greater than 1.5 in. of water head per foot of packed height (equivalent to 0.054 psi/ft of packing). In addition, for random packings, a nominal packing diameter not greater than one-eighth of the diameter of the column is selected; otherwise, poor distribution of liquid and vapor flow over the cross-sectional area of the column can occur, with liquid tending to migrate to the wall of the column.

Flooding data for packed columns with countercurrent flow of liquid and gas were first correlated successfully by Sherwood et al. [26], who used the same liquid-to-gas kinetic energy ratio, $F_{LV} = (LM_L/VM_V)(\rho_V/\rho_L)^{0.5}$, already discussed for the correlation of flooding and entrainment in trayed towers, as shown in Figures 6.24 and 6.28, respectively. The superficial gas velocity, u_V, was embedded in the dimensionless term $u_V^2 a/g\epsilon^3$, which was arrived at by considering the square of the actual gas velocity, u_V^2/ϵ^2, the hydraulic radius, $r_H = \epsilon/a$, which is the volume available for flow divided by the wetted surface area of the packing, and the gravitational acceleration, g, to give the dimensionless expression, $u_V^2 a/g\epsilon^3 = u_V^2 F_P/g$. The ratio, a/ϵ^3, is a function of the packing only, and is known as the packing factor, F_P. Values of a, ϵ, and F_P are included in Table 6.8. In some cases, F_P is a modified packing factor, treated as an empirical constant, backed out from experimental data so as to fit a generalized correlation. Additional factors were added by Sherwood et al. to account for liquid density and viscosity, and gas density.

In 1954, Leva [48] used experimental data on ring and saddle packings to extend the Sherwood et al. [26] flooding correlation to include lines of constant pressure drop, with the resulting chart becoming known as the generalized pressure-drop correlation (GPDC).

A modern version of the GPDC chart is that of Leva [49], as shown in Figure 6.36a. The abscissa is the same F_{LV} parameter, but the ordinate is given by

$$Y = \frac{u_V^2 F_P}{g}\left(\frac{\rho_V}{\rho_{H_2O_{(L)}}}\right) f\{\rho_L\}f\{\mu_L\} \qquad (6\text{-}102)$$

where the density of H_2O is taken as 62.4 lb/ft^3 with ρ_V in the same units. The functions $f\{\rho_L\}$ and $f\{\mu_L\}$ are corrections for liquid properties as given by Figures 6.36b and 6.36c, respectively.

For given fluid flow rates and properties, and a given packing material, the GPDC chart is used to compute u_{Vf}, the superficial gas velocity at flooding. Then a fraction of flooding, f, is selected (usually from 0.5 to 0.7), followed by calculation of the tower diameter from an equation similar to (6-44):

$$D_T = \left(\frac{4VM_V}{fu_{V,f}\pi\rho_V}\right)^{0.5} \qquad (6\text{-}103)$$

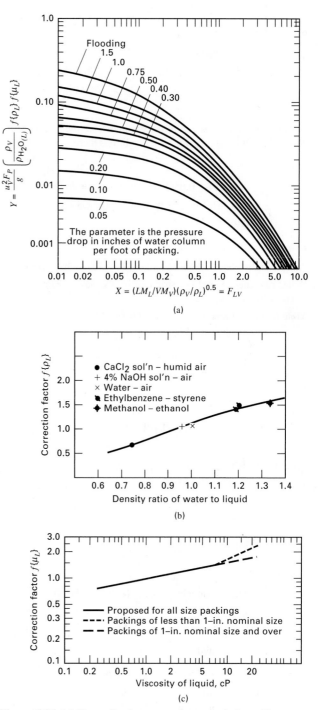

Figure 6.36 (a) Generalized pressure-drop correlation of Leva for packed columns. (b) Correction factor for liquid density. (c) Correction factor for liquid viscosity.
[From M. Leva, *Chem. Eng. Prog.*, **88** (1), 65–72 (1992) with permission.]

EXAMPLE 6.13

Air containing 5 mol% NH_3 at a total flow rate of 40 lbmol/h enters a packed column operating at 20°C and 1 atm, where 90% of the ammonia is scrubbed by a countercurrent flow of 3,000 lb/h of water. Use the GPDC chart of Figure 6.36 to estimate the superficial, gas-flooding velocity, the column inside diameter for operation at 70% of flooding, and the pressure drop per foot of packing for two packing materials:

(a) One-inch ceramic Raschig rings ($F_P = 179$ ft²/ft³)

(b) One-inch metal IMTP® packing ($F_P = 41$ ft²/ft³)

SOLUTION

Because the superficial gas velocity is highest at the bottom of the column, calculations are made for conditions there.

Inlet gas:

$$M_V = 0.95(29) + 0.05(17) = 28.4, \quad V = 40 \text{ lbmol/h}$$

$$\rho_V = PM_V/RT = (1)(28.4)/[(0.730)(293)(1.8)]$$
$$= 0.0738 \text{ lb/ft}^3$$

Exiting liquid:

Ammonia absorbed = $0.90(0.05)(40)(17) = 30.6$ lb/h or 1.8 lbmol/h

Water rate (neglecting any stripping by the gas) = 3,000 lb/h or 166.7 lbmol/h

Mole fraction of ammonia = $1.8/(166.7 + 1.8) = 0.0107$

$$M_L = 0.0107(17) + (0.9893)(18) = 17.9,$$
$$L = 1.8 + 166.7 = 168.5 \text{ lbmol/h}$$

Take: $\rho_L = 62.4$ lb/ft³ and $\mu_L = 1.0$ cP
Now,

$$X = F_{LV}(\text{abscissa in Figure 6.36a})$$
$$= \frac{(168.5)(17.9)}{(40)(28.4)}\left(\frac{0.0738}{62.4}\right)^{0.5} = 0.092$$

From Figure 6.36a, $Y = 0.125$ at flooding.
From Figure 6.36b, $f\{\rho_L\} = 1.14$.
From Figure 6.36c, $f\{\mu_L\} = 1.0$.
From (6-102),

$$u_V^2 = 0.125\left(\frac{g}{F_P}\right)\frac{62.4}{(0.0738)(1.14)(1.0)} = 92.7 \, g/F_P.$$

Using $g = 32.2$ ft/s²,

Packing Material	F_P, ft²/ft³	u_o, ft/s
Raschig rings	179	4.1
IMTP® packing	41	8.5

For $f = 0.70$, using (6-103),

Packing Material	$fu_{V,f}$, ft/s	D_T, in.
Raschig rings	2.87	16.5
IMTP® packing	5.95	11.5

From Figure 6.36a, for $F_{LV} = 0.092$ and $Y = 0.70^2(0.125) = 0.0613$ at 70% of flooding, the pressure drop is 0.88 in. of water head per foot of packed height for both packings.

Based on these results, the IMTP® packing has a much greater capacity than the Raschig rings, since the required column cross-sectional area is reduced by about 50%.

Experimental flooding-point data for a variety of packing materials are in reasonable agreement with the upper curve of the GPDC chart of Figure 6.36. Unfortunately, such good agreement is not always the case for pressure drop, particularly for operation at superficial vapor velocities

above 50% of flooding, where pressure drop is greater than 0.5 in. of water head per foot of packed height. Reasons for the difficulty of achieving a simple generalization of pressure drop measurements are discussed in detail by Kister [33]. As an example of the possible magnitude of the disparity, the predicted pressure drop of 0.88 in. of water per foot in Example 6.13 for operation with IMTP® packing at 70% of flooding is in poor agreement with the value of 0.63 in. of water head per foot determined from data supplied by the packing manufacturer.

If Figure 6.36a is crossplotted as pressure drop versus Y for constant values of F_{LV}, it is found that a pressure drop of from 2.5 to 3 in. of water head per foot is predicted at the flooding condition for all packings. However, studies by Kister and Gill [33, 50] for both random and structured packings show that the pressure drop at flooding is strongly dependent on the packing factor, F_P, by the empirical expression

$$\Delta P_{\text{flood}} = 0.115 F_P^{0.7} \tag{6-104}$$

where ΔP_{flood} has units of inches of water head per foot of packed height and F_P has units of ft²/ft³. As seen in Table 6.8, the range of F_P is from about 10 to 100. Thus, (6-103) predicts pressure drops at flooding from as low as 0.6 to as high as 3 in. of water head per foot of packed height. Kister and Gill also give an interpolation procedure for estimating pressure drop, which utilizes experimental data in conjunction with a GPDC-type plot.

Theoretically based models for predicting pressure drop in packed beds with countercurrent gas–liquid flows have been presented by Stichlmair et al. [46], who use a particle model, and Billet and Schultes [51, 69], who use a channel model. Both models extend well-accepted equations for dry-bed pressure drop to account for the effect of liquid holdup. Billet and Schultes [69] include a semitheoretical model for predicting the superficial vapor velocity at the loading point, $u_{V,l}$, which provides an alternative, perhaps more accurate, method for estimating column diameter. Their model, which is based on a liquid velocity of zero at the phase boundary at the loading point, gives

$$u_{V,l} = \left(\frac{g}{\Psi_l}\right)^{1/2}\left[\frac{\epsilon}{a^{1/6}} - a^{1/2}\xi_l^{1/3}\right]\xi_l^{1/6}\left(\frac{\rho_L}{\rho_V}\right)^{1/2} \tag{6-105}$$

where $u_{V,l}$ is in m/s,

$$g = \text{gravitational acceleration} = 9.807 \text{ m/s}^2$$

$$\Psi_l = \frac{g}{C^2}\left[F_{LV}\left(\frac{\mu_L}{\mu_V}\right)^{0.4}\right]^{-2n_s} \tag{6-106}$$

ϵ and a are obtained from Table 6.8,

$F_{LV} = $ kinetic energy ratio of Figures 6.24 and 6.36a,

$$\xi_l = \left(12\frac{\mu_L}{g\rho_L}u_{L,l}\right) \tag{6-107}$$

μ_L and μ_V are in kg/m-s

ρ_L and ρ_V are in kg/m³

$u_{L,l}$ = superficial liquid velocity at loading point

$$= u_{V,l} \frac{\rho_V L M_L}{\rho_L V M_V} \quad \text{in m/s}$$

The values for n_s and C depend on the value of the kinetic energy ratio as follows:

If $F_{LV} \leq 0.4$, the liquid trickles downward over the packing as a disperse phase and $n_s = -0.326$, while $C = C_s$ from Table 6.8.

If $F_{LV} > 0.4$, the column holdup reaches such a large value that the empty spaces within the bed close up and the liquid flows downward as a continuous phase while the gas rises in the form of bubbles, with $n_s = -0.723$ and

$$C = 0.695 \left(\frac{\mu_L}{\mu_V}\right)^{0.1588} C_s \quad \text{(from Table 6.8)} \quad (6\text{-}108)$$

Billet and Schultes [69] also present a model for predicting the superficial vapor velocity at the flooding point, $u_{V,f}$, that involves the flooding constant, C_{Fl}, in Table 6.8, but a suitable expression is

$$u_{V,f} = \frac{u_{V,l}}{0.7} \quad (6\text{-}109)$$

When a gas flows through a packed column under conditions of no liquid flow, a correlation for the pressure drop can be obtained in a manner similar to that for flow through an empty, straight pipe, by plotting a modified friction factor against a modified Reynolds number as shown in Figure 6.37 from the widely used study by Ergun [52]. In this plot, in which D_P is an effective packing material diameter, it can be seen that at low, superficial gas velocities (modified $N_{Re} < 10$), typical of laminar flow, the pressure drop per unit height is proportional to the superficial vapor velocity, u_V. At high gas velocities, typical of turbulent flow, the pressure drop per unit height approaches a dependency of the square of the gas velocity. Most packed columns used for separations operate in the turbulent region (modified $N_{Re} > 1,000$). Thus, dry pressure-drop data shown in Figure 6.34 for Bialecki rings show an exponential dependency on gas velocity of about 1.86. Also, as shown in Figure 6.34, when liquid flows countercurrent to the gas in the preloading region, this same dependency continues, but at a higher pressure drop because the volume for gas flow decreases due to liquid holdup.

Based on extensive experimental studies using more than 50 different packing materials, including structured packings, Billet and Schultes [51, 69] developed a correlation for dry-gas pressure drop, ΔP_o, similar in form to that of Figure 6.37. Their dimensionally consistent correlating equation is

$$\frac{\Delta P_o}{l_T} = \Psi_o \frac{a}{\epsilon^3} \frac{u_V^2 \rho_V}{2} \frac{1}{K_W} \quad (6\text{-}110)$$

where

l_T = height of packing

K_W = a wall factor

K_W can be important for columns with an inadequate ratio of effective packing diameter to inside column diameter, and is given by

$$\frac{1}{K_W} = 1 + \frac{2}{3}\left(\frac{1}{1-\epsilon}\right)\frac{D_P}{D_T} \quad (6\text{-}111)$$

where the effective packing diameter, D_P, is determined from

$$D_P = 6\left(\frac{1-\epsilon}{a}\right) \quad (6\text{-}112)$$

The dry-packing resistance coefficient (a modified friction factor), Ψ_o, is given by the empirical expression

$$\Psi_o = C_p \left(\frac{64}{N_{Re_V}} + \frac{1.8}{N_{Re_V}^{0.08}}\right) \quad (6\text{-}113)$$

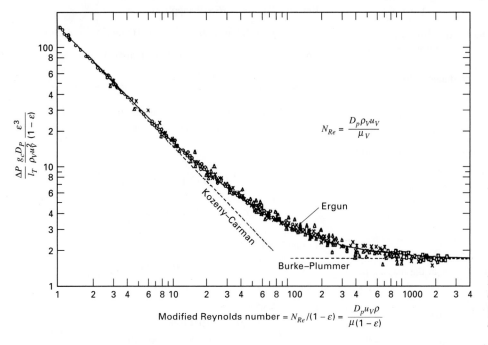

$$N_{Re} = \frac{D_p \rho_V u_V}{\mu_V}$$

Modified Reynolds number $= N_{Re}/(1-\varepsilon) = \dfrac{D_p u_V \rho}{\mu(1-\varepsilon)}$

Figure 6.37 Ergun correlation for dry-bed pressure drop.

[From S. Ergun, *Chem. Eng. Prog.* **48** (2), 89–94 (1952) with permission.]

where

$$N_{Re_V} = \frac{u_V D_P \rho_V}{(1-\epsilon)\mu_V} K_W \qquad (6\text{-}114)$$

and C_p is a packing constant, determined from experimental data, and tabulated for a number of packings in Table 6.8. In (6-113), the laminar-flow region is characterized by the term $64/N_{Re_V}$, while the next term characterizes the more common turbulent-flow regime.

When a packed tower is irrigated with a downward-flowing liquid, the cross-sectional area for gas flow is reduced by the liquid holdup and the surface structure exposed to the gas is changed as a result of the coating of the packing with a liquid film. The pressure drop now becomes dependent on the holdup and a two-phase flow resistance, and was found by Billet and Schultes [69] to depend on the liquid-flow Froude number as follows for flow rates up to the loading point:

$$\frac{\Delta P}{\Delta P_o} = \left(\frac{\epsilon}{\epsilon - h_L}\right)^{3/2} \exp\left[\frac{13300}{a^{3/2}}(N_{Fr_L})^{1/2}\right] \qquad (6\text{-}115)$$

where

h_L is given by (6-97) and is in m^2/m^3,

ϵ and a are given in Table 6.8, where a in (6-115) must be in m^2/m^3, and

N_{Fr_L} is given by (6-99).

EXAMPLE 6.14

A column packed with 25-mm metal Bialecki® rings is to be designed for the following vapor and liquid conditions:

	Vapor	Liquid
Mass flow rate, kg/h	515	1,361
Density, kg/m³	1.182	1,000
Viscosity, kg/m-s	1.78×10^{-5}	1.00×10^{-3}
Molecular weight	28.4	18.02
Surface tension, kg/s²		2.401×10^{-2}

Using the equations of Billet and Schultes, determine the vapor and liquid superficial velocities at the loading and flooding points, the specific liquid holdup at the loading point, the specific pressure drop at the loading point, and the column diameter for operation at the loading point.

SOLUTION

From Table 6.8, the following constants apply to the Bialecki rings:

$a = 210 \text{ m}^2/\text{m}^3$

$\epsilon = 0.956$

$C_h = 0.692$

$C_p = 0.891$

$C_s = 2.521$

First, compute the superficial vapor velocity at the loading point. From the abscissa label of Figure 6.36a,

$$F_{LV} = \frac{1,361}{515}\left(\frac{1.182}{1,000}\right)^{1/2} = 0.0908$$

Because $F_{LV} < 0.4$, $n_s = -0.326$ and C in (6-106) $= C_s = 2.521$.

From (6-106),

$$\Psi_l = \frac{9.807}{2.521^2}\left[0.0908\left(\frac{0.001}{0.0000178}\right)^{0.4}\right]^{-2(-0.326)} = 0.923$$

$$u_{L,l} = u_{V,l}\frac{\rho_V L M_L}{\rho_L V M_V} = u_{V,l}\frac{(1.182)(1,361)}{(1,000)(515)} = 0.00312\,u_{V,l}$$

From (6-107),

$$\xi_l = \left(12\frac{(0.001)}{(9.807)(1,000)}(0.00312)\,u_{V,l}\right) = 3.82 \times 10^{-9}u_{V,l}$$

From (6-105),

$$u_{V,l} = \left(\frac{9.807}{0.923}\right)^{1/2}\left[\frac{0.956}{210^{1/6}} - 210^{1/2}\left(3.82 \times 10^{-9}\,u_{V,l}\right)^{1/3}\right]$$

$$\times \left(3.82 \times 10^{-9}\,u_{V,l}\right)^{1/6}\left(\frac{1,000}{1.182}\right)^{1/2}$$

$$= 3.26\left[0.392 - 0.0227\,u_{V,l}^{1/3}\right]1.15\,u_{V,l}^{1/6}$$

$$= \left(1.47 - 0.0851\,u_{V,l}^{1/3}\right)u_{V,l}^{1/6}$$

Solving this nonlinear equation in $u_{V,l}$ gives $u_{V,l}$ = superficial vapor velocity at the loading point = 1.46 m/s.

The corresponding superficial liquid velocity = $u_{L,l}$ = 0.00312 $u_{V,l}$ = 0.00312(1.46) = 0.00457 m/s.

The superficial vapor flooding velocity = $u_{V,f} = \dfrac{u_{V,l}}{0.7} = \dfrac{1.46}{0.7} =$ 2.09 m/s.

The corresponding superficial liquid velocity =

$$u_{L,f} = \frac{0.00457}{0.7} = 0.00653 \text{ m/s}.$$

Next, compute the specific liquid holdup at the loading point. From (6-98) and (6-99),

$$N_{Re_L} = \frac{(0.00457)(1,000)}{(210)(0.001)} = 21.8$$

and

$$N_{Fr_L} = \frac{(0.00457)^2(210)}{9.807} = 0.000447$$

Because $N_{Re_L} > 5$, (6-101) applies:

$$a_h/a = 0.85(0.692)(21.8)^{0.25}(0.000447)^{0.1} = 0.588$$

From (6-97), the specific liquid holdup at the loading point is

$$h_L = \left(12\frac{0.000447}{21.8}\right)^{1/3}0.588^{2/3} = 0.0440 \text{ m}^3/\text{m}^3$$

Before computing the specific pressure drop at the loading point, we must compute the column diameter for operation at the loading point. Applying (6-103),

$$D_T = \left[\frac{4(515/3600)}{(1.46)(3.14)(1.182)}\right]^{1/2} = 0.325 \text{ m}$$

From (6-112),

$$D_P = 6\left(\frac{1 - 0.956}{210}\right) = 0.00126 \text{ m}$$

From (6-111),

$$\frac{1}{K_W} = 1 + \frac{2}{3}\left(\frac{1}{1 - 0.956}\right)\frac{0.00126}{0.325} = 1.059 \text{ and } K_W = 0.944$$

From (6-114),

$$N_{Re_V} = \frac{(1.46)(0.00126)(1.182)}{(1 - 0.956)(0.0000178)}(0.944) = 2,621$$

From (6-113),

$$\Psi_o = 0.891\left(\frac{64}{2,621} + \frac{1.8}{2,621^{0.08}}\right) = 0.876$$

From (6-110), the specific dry-gas pressure drop is

$$\frac{\Delta P_o}{l_T} = 0.876\frac{(210)(1.46)^2(1.182)}{(0.956)^3(2)}(1.059)$$

$$= 281 \text{ kg/m}^2\text{-s}^2 = \text{Pa/m}$$

From (6-115), the specific pressure drop at the loading point is

$$\frac{\Delta P}{l_T} = 281\left(\frac{0.956}{0.956 - 0.0440}\right)^{3/2}\exp\left[\frac{13300}{210^{3/2}}(0.000447)^{1/2}\right]$$

$$= 331 \text{ kg/m}^2\text{-s}^2$$

or 0.406 in. of water/ft

Mass-Transfer Efficiency

The mass-transfer efficiency of a packed column is incorporated in the HETP or the more theoretically based HTUs and volumetric mass-transfer coefficients. Although the HETP concept lacks a sound theoretical basis, its simplicity, coupled with the relative ease with which equilibrium-stage calculations can be made with computer-aided simulation programs, has made it a widely used method for estimating packing height. In the preloading region and where good distribution of vapor and liquid is initiated and maintained, values of the HETP depend mainly on packing type and size, liquid viscosity, and surface tension. For rough estimates the following relations, taken from Kister [33], can be used.

1. Pall rings and similar high-efficiency random packings with low-viscosity liquids:

   $$\text{HETP, ft} = 1.5D_P, \text{ in.} \qquad (6\text{-}116)$$

2. Structured packings at low-to-moderate pressure with low-viscosity liquids:

   $$\text{HETP, ft} = 100/a, \text{ ft}^2/\text{ft}^3 + 4/12 \qquad (6\text{-}117)$$

3. Absorption with viscous liquid:

   $$\text{HETP} = 5 \text{ to } 6 \text{ ft}$$

4. Vacuum service:

   $$\text{HETP, ft} = 1.5D_P, \text{ in.} + 0.5 \qquad (6\text{-}118)$$

5. High-pressure service (> 200 psia):

   HETP for structured packings may be greater than predicted by (6-117)

6. Small-diameter columns, $D_T < 2$ ft:

   $$\text{HETP, ft} = D_T, \text{ ft, but not less than 1 ft}$$

In general, lower values of HETP are achieved with smaller-size random packings, particularly in small-diameter columns, and with structured packings, particularly those with large values of a, the packing surface area per packed volume. The experimental data of Figure 6.38 for no. 2 (2-in.-diameter) Nutter rings from Kunesh [53] show that in the preloading

region, the HETP is relatively independent of the vapor-flow F-factor:

$$F = u_V(\rho_V)^{0.5} \qquad (6\text{-}119)$$

provided that the ratio L/V is maintained constant as the superficial gas velocity, u_V, is increased. Beyond the loading point, and as the flooding point is approached, the HETP can increase dramatically like the pressure drop and liquid holdup.

Experimental mass-transfer data for packed columns are usually correlated in terms of volumetric mass-transfer coefficients and/or HTUs, rather than in terms of HETPs. The data are obtained from experiments in which either the liquid-phase or the gas-phase mass-transfer resistance is negligible, so that the other resistance can be studied and correlated independently. For applications where both resistances may be important, the two resistances are added together according to the two-film theory of Whitman [54], as discussed in Chapter 3, to obtain the overall resistance. This theory assumes the absence of any mass-transfer resistance at the interface between the gas and liquid phases. Thus, the two phases are in equilibrium at the interface.

The two-film theory defines an overall coefficient in terms of the individual volumetric mass-transfer coefficients discussed in Section 6.7. Most commonly, reference is made to the overall gas-phase resistance, (6-84),

$$\frac{1}{K_ya} = \frac{1}{k_ya} + \frac{K}{k_xa}$$

for mass-transfer rates expressed in terms of mole-fraction driving forces by (6-77),

$$r = k_ya(y - y_I) = k_xa(x_I - x) = K_ya(y - y^*)$$

where K is the vapor–liquid equilibrium ratio.

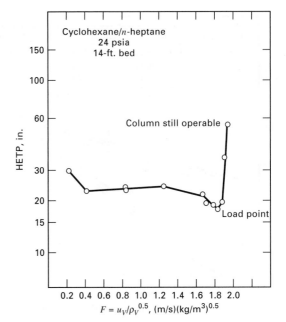

Figure 6.38 Effect of F-factor on HETP.

Alternatively, as summarized in Table 6.7, mass-transfer rates can be expressed in terms of liquid-phase concentrations and gas-phase partial pressure

$$r = k_p a(p - p_I) = k_L a(c_I - c) = K_G a(p - p^*) \quad (6\text{-}120)$$

If we define a Henry's-law constant at the equilibrium interface between the two phases by

$$p_I = H'c_I \quad (6\text{-}121)$$

and let

$$p^* = H'c \quad (6\text{-}122)$$

then

$$\frac{1}{K_G a} = \frac{1}{k_p a} + \frac{H'}{k_L a} \quad (6\text{-}123)$$

Alternatively, expressions can be derived for $K_x a$ and $K_L a$.

It should be noted that the units of various mass transfer coefficients differ:

	SI Units	American Engineering Units
r	mol/m^3-s	lbmol/ft^3-h
$k_y a, k_x a, K_x a, K_y a$	mol/m^3-s	lbmol/ft^3-h
$k_p a, K_G a$	mol/m^3-s-kPa	lbmol/ft^3-h-atm
$k_L a, k_G a, k_c a$	s^{-1}	h^{-1}
k_L, k_G, k_c	m/s	ft/h

Instead of using mass-transfer coefficients directly for column design, the transfer-unit concept of Chilton and Colburn [43,44] is often employed because HTUs: (1) have only one dimension (length), (2) generally vary with column conditions less than mass-transfer coefficients, and (3) are related to an easily understood geometrical quantity, namely, height per theoretical stage. Definitions of individual and overall HTUs are included in Table 6.7 for the dilute case. By substituting these definitions into (6-84),

$$H_{OG} = H_G + (KV/L)H_L \quad (6\text{-}124)$$

Alternatively, an expression can be derived for H_{OL}. In the absorption or stripping of very insoluble gases, the solute K-value or Henry's law constant, H' in (6-112), is very large, making the last terms in (6-84), (6-123), and (6-124) large such that the resistance of the gas phase is negligible and the rate of mass transfer is controlled by the liquid phase. Such data can then be used to study the effect of the variables on the volumetric, liquid-phase mass-transfer coefficient and HTU. Typical data are shown in Figure 6.39 for three different-size Berl-saddle packings for the stripping of oxygen from water by air, in a 20-in.-I.D. column operated at near-ambient temperature and pressure in the preloading region, as reported in an early study by Sherwood and Holloway [55]. The effect of liquid velocity on $k_L a$ is seen to be quite pronounced, with $k_L a$ increasing at about the 0.75 power of the liquid mass velocity. Gas velocity was observed to have no effect on $k_L a$ in the preloading region. Also included in Figure 6.39 are the data plotted in terms of H_L, where

$$H_L = \frac{M_L L}{\rho_L k_L a S} \quad (6\text{-}125)$$

As seen, H_L does not depend as strongly as $k_L a$ on liquid mass velocity, $M_L L/S$.

Another system for which the rate of mass transfer is controlled by the liquid phase is CO_2–air–H_2O, where CO_2 can be either absorbed or stripped. Measurements on this system for a variety of modern metal, ceramic, and plastic packings are reported by Billet [45]. Data on the effect of liquid loading on $k_L a$ in the preloading region for two different-size ceramic Hiflow® ring packings are shown in Figure 6.40. The effect of gas velocity on $k_L a$ in terms of the F-factor at a constant liquid rate is shown in Figure 6.41 for the same system, but with 50-mm plastic Pall® rings and Hiflow® rings. Up to an F-factor value of about 1.8 m$^{-1/2}$-s^{-1}-kg$^{1/2}$, which is in the preloading region, no effect of gas velocity is

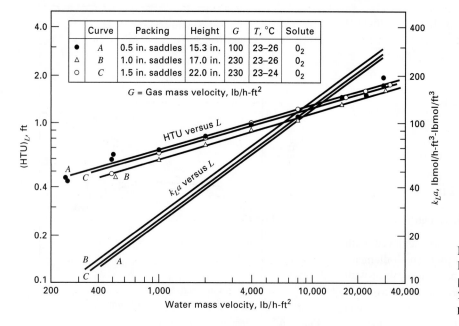

Figure 6.39 Effect of liquid rate on liquid-phase mass transfer of O_2.

[From T.K. Sherwood and F.A.L. Holloway, *Trans. AIChE.*, **36**, 39–70 (1940) with permission.]

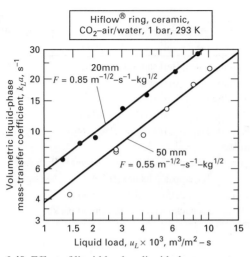

Figure 6.40 Effect of liquid load on liquid-phase mass transfer of CO_2.

[From R. Billet, *Packed Column Analysis and Design*, Ruhr-University Bochum (1989) with permission.]

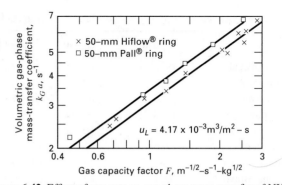

Figure 6.42 Effect of gas rate on gas-phase mass transfer of NH_3.

[From R. Billet, *Packed Column Analysis and Design*, Ruhr-University Bochum (1989) with permission.]

observed. Above the loading limit, $k_L a$ increases with increasing gas velocity because of increased liquid holdup, which increases interfacial surface area for mass transfer. Although it is not illustrated in Figures 6.39 to 6.41, another major factor that influences the rate of mass transfer in the liquid phase is the solute molecular diffusivity in the solvent. For a given packing, experimental data on different systems in the preloading region can usually be correlated satisfactorily by the following empirical expression, which includes only the liquid velocity and liquid diffusivity:

$$k_L a = C_1 D_L^{1/2} u_L^n \qquad (6\text{-}126)$$

where n has been observed by different investigators to vary from about 0.6 to 0.95, with 0.75 being a typical value. The exponent on the diffusivity is consistent with the penetration theory discussed in Chapter 3.

A convenient system for studying gas-phase mass transfer is NH_3–air–H_2O. The high solubility of NH_3 in H_2O corresponds to a relatively low K-value. Accordingly, the last terms in (6-84), (6-123), and (6-124) may be negligible so that the

gas-phase resistance controls the rate of mass transfer. The small effect of the liquid-phase resistance can be backed out using a correlation such as (6-126). The typical effect of superficial vapor velocity, expressed in terms of the F-factor of (6-119), on the volumetric, gas-phase mass-transfer coefficient in the preloading region is shown in Figure 6.42 for two different plastic packings at the same liquid velocity. The coefficients are proportional to about the 0.75 power of F. Figure 6.43 shows that the liquid velocity also affects the gas-phase mass-transfer coefficient, probably because as the liquid rate is increased, the holdup increases and more interfacial surface is created.

The volumetric, gas-phase mass-transfer coefficients, $k_G a$, plotted in Figures 6.42 and 6.43, are based on gas-phase molar concentrations. Thus, they have the same units as $k_L a$. For a given packing, experimental data on $k_p a$ or $k_G a$ for different systems in the preloading region can usually be correlated satisfactorily with empirical correlations of the form

$$k_p a = C_2 D_G^{0.67} F^{m'} u_L^{n'} \qquad (6\text{-}127)$$

where D_G is the gas diffusivity of the solute and m' and n' have been observed by different investigators to vary from 0.65 to 0.85 and from 0.25 to 0.5, respectively, a typical value for m' being 0.8.

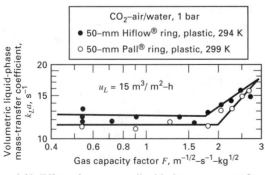

Figure 6.41 Effect of gas rate on liquid-phase mass transfer of CO_2.

[From R. Billet, *Packed Column Analysis and Design*, Ruhr-University Bochum (1989) with permission.]

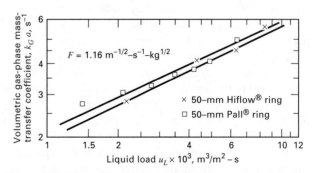

Figure 6.43 Effect of liquid rate on gas-phase mass transfer of NH_3.

[From R. Billet, *Packed Column Analysis and Design*, Ruhr-University Bochum (1989) with permission.]

Table 6.9 Generalized Correlations for Mass Transfer in Packed Columns

Investigator	Year	Ref. No.	Type of Correlations	Packings
Shulman et al.	1955	64	k_p, k_L, a	Raschig rings, Berl saddles
Cornell et al.	1960	56, 57	H_G, H_L	Raschig rings, Berl saddles
Onda et al.	1968	65	k_p, k_L, a	Raschig rings, Berl saddles
Bolles and Fair	1979, 1982	58, 59	H_G, H_L	Raschig rings, Berl saddles, Pall® rings
Bravo and Fair	1982	60	a	Raschig rings, Berl saddles, Pall® rings
Bravo et al.	1985	61	k_G, k_L	Sulzer®
Fair and Bravo	1987	62	k_G, k_L, a	Sulzer®, Gempak®, Mellapak®, Montz®, Ralu Pak®
Fair and Bravo	1991	63	k_G, k_L, a	Flexipac®, Gempak®, Intalox 2T®, Montz®, Mellapak®, Sulzer®
Billet and Schultes	1991	67	$k_G a, k_L a$	14 random packings and 4 structured packings
Billet and Schultes	1999	69	$k_G a, k_L a$	19 random packings and 6 structured packings

The development of separate generalized correlations for gas- and liquid-phase mass-transfer coefficients and/or HTUs, which began with the study of Sherwood and Holloway [55] on the liquid phase, has led to a significant number of empirical and semitheoretical equations, most of which are based on the application of the two-film theory by Fair and co-workers [56–63] and others [64, 65]. In some cases, values of k_G and k_L are correlated separately from a; in others, the combinations $k_G a$ and $k_L a$ are used. Important features of some of these correlations are summarized in Table 6.9. The development of such correlations from experimental data is difficult because, as shown by Billet [66] in a comprehensive study with metal Pall® rings, values of the mass-transfer coefficients are significantly affected by the technique used to pack the column and the number of liquid feed-distribution points per unit of column cross section, when this number is less than 10 points/ft². When 25 points/ft² are used and $D_T/D_P > 10$, column diameter has little, if any, effect on mass-transfer coefficients for packed heights up to 20 ft.

In an extensive investigation, Billet and Schultes [67] measured and correlated volumetric mass-transfer coefficients and HTUs for 31 different binary and ternary chemical systems with 67 different types and sizes of packings in columns of diameter ranging from 2.4 in. to 4.6 ft. Additional data are reported by Billet and Schultes [69], particularly for Hiflow® rings and Raschig Super-rings. The systems include some for which mass-transfer resistance resides mainly in the liquid phase and others for which resistance in the gas phase is predominant. They assume uniform distribution of gas and liquid over the cross-sectional area of the column and apply the two-film theory of mass transfer discussed in Chapter 3. For the liquid-phase resistance, they assume that the liquid flows in a thin film through the irregular channels of the packing, with continual remixing of the liquid at points of contact with the packing such that Higbie's penetration theory of diffusion [68], as developed in Chapter 3, can be applied. Thus, for the diffusing component, in terms of concentration units, the volumetric mass-transfer coefficient is defined by

$$r = (k_L a_{Ph})(c_{L_I} - c_L) \qquad (6\text{-}128)$$

From the penetration theory of Higbie, (3-194),

$$k_L = 2(D_L/\pi t_L)^{0.5} \qquad (6\text{-}129)$$

where t_L = time of exposure of the liquid film before remixing. Billet and Schultes assume that this time is governed by a length of travel equal to the hydraulic diameter of the packing:

$$t_L = h_L d_H/u_L \qquad (6\text{-}130)$$

where d_H, the hydraulic diameter, is equal to $4r_H$ or $4\epsilon/a$. Thus, in terms of the height of a liquid transfer unit, (6-129) and (6-130) give

$$H_L = \frac{u_L}{k_L a_{Ph}} = \frac{\sqrt{\pi}}{2}\left(\frac{4h_L\epsilon}{D_L a u_L}\right)^{1/2}\frac{u_L}{a_{Ph}} \qquad (6\text{-}131)$$

Equation (6-131) was modified to include an empirical constant, C_L, which is back-calculated for each packing to fit the experimental data. The final predictive equation given by Billet and Schultes is

$$H_L = \frac{1}{C_L}\left(\frac{1}{12}\right)^{1/6}\left(\frac{4h_L\epsilon}{D_L a u_L}\right)^{1/2}\frac{u_L}{a}\left(\frac{a}{a_{Ph}}\right) \qquad (6\text{-}132)$$

where values of C_L are included in Table 6.8.

A similar development was made by Billet and Schultes for the gas-phase resistance, except that the time of exposure of the gas between periods of mixing was determined empirically, to give

$$H_G = \frac{1}{C_V}(\epsilon - h_L)^{1/2}\left(\frac{4\epsilon}{a^4}\right)^{1/2}(N_{Re_V})^{-3/4}(N_{Sc_V})^{-1/3}\left(\frac{u_V a}{D_G a_{Ph}}\right) \qquad (6\text{-}133)$$

where C_V is included in Table 6.8 and

$$N_{Re_V} = \frac{u_V \rho_V}{a\mu_V} \qquad (6\text{-}134)$$

$$N_{Sc_V} = \frac{\mu_V}{\rho_V D_V} \qquad (6\text{-}135)$$

Equations (6-132) and (6-133) contain an area ratio, a_{Ph}/a, which is the ratio of the phase interface area to the packing surface area, which from Billet and Schultes [69] is not the same as the hydraulic area ratio, a_h/a, given by (6-100) and

(6-101). Instead, they give the following correlation:

$$\frac{a_{Ph}}{a} = 1.5(ad_h)^{-1/2}(N_{Re_L,h})^{-0.2}(N_{We_L,h})^{0.75}(N_{Fr_L,h})^{-0.45}$$

$$(6\text{-}136)$$

where

$$d_h = \text{packing hydraulic diameter} = 4\frac{\epsilon}{a} \quad (6\text{-}137)$$

and the following liquid-phase dimensionless groups use the packing hydraulic diameter as the characteristic length:

$$\text{Reynolds number} = N_{Re_L,h} = \frac{u_L d_h \rho_L}{\mu_L} \quad (6\text{-}138)$$

$$\text{Weber number} = N_{We_L,h} = \frac{u_L^2 \rho_L d_h}{\sigma} \quad (6\text{-}139)$$

$$\text{Froude number} = N_{Fr_L,h} = \frac{u_L^2}{g d_h} \quad (6\text{-}140)$$

Following the estimation of H_L and H_G from (6-132) and (6-133), respectively, the overall HTU value can be determined from (6-124), followed by the determination of packed height from

$$l_T = H_{OG}N_{OG} \quad (6\text{-}141)$$

where the determination of N_{OG} is discussed in Section 6.7.

EXAMPLE 6.15

For the absorption of ethyl alcohol from CO_2 with water, as considered in Example 6.1, a 2.5-ft-I.D. tower, packed with 1.5-in. metal Pall-like rings, is to be used. It is estimated that the tower will operate in the preloading region with a pressure drop of approximately 1.5 in. of water head per foot of packed height. From Example 6.9, the required number of overall transfer units based on the gas phase is 7.5. Estimate H_G, H_L, H_{OG}, HETP, and the required packed height in feet using the following estimates of flow conditions and physical properties at the bottom of the packing:

	Vapor	Liquid
Flow rate, lb/h	17,480	6,140
Molecular weight	44.05	18.7
Density, lb/ft³	0.121	61.5
Viscosity, cP	0.0145	0.63
Surface tension, dynes/cm	—	101
Diffusivity of ethanol, m²/s	7.75×10^{-6}	1.82×10^{-9}
Kinematic viscosity, m²/s	0.75×10^{-5}	0.64×10^{-6}

SOLUTION

Cross-sectional area of tower $= (3.14)(2.5)^2/4 = 4.91$ ft².
Volumetric liquid flow rate $= 6,140/61.5 = 99.8$ ft³/h.
$u_L = $ superficial liquid velocity $= 99.8/[(4.91)(3,600)] = 0.0056$ ft/s or 0.0017 m/s.
From Section 6.8, $u_L > u_{L,min}$, but the velocity is on the low side.
$u_V = $ superficial gas velocity $= 17,480/[(0.121)(4.91)(3,600)] = 8.17$ ft/s $= 2.49$ m/s.
Let the packing characteristics for the 1.5-inch metal Pall-like rings be as follows (somewhat different from values for Pall rings

in Table 6.8):

$$a = 149.6 \text{ m}^2/\text{m}^3, \quad \epsilon = 0.952$$
$$C_h = \text{approximately } 0.7, \quad C_L = 1.227, \quad C_V = 0.341$$

Estimation of specific liquid holdup, h_L:

From (6-98),

$$N_{Re_L} = \frac{0.0017}{(0.64 \times 10^{-6})(149.6)} = 17.8.$$

From (6-99),

$$N_{Fr_L} = \frac{(0.0017)^2(149.6)}{9.8} = 4.41 \times 10^{-5}$$

From (6-101),

$$\frac{a_h}{a} = 0.85(0.7)(17.8)^{0.25}(4.41 \times 10^{-5})^{0.10} = 0.045$$
$$a_h = 0.045(149.6) = 6.73 \text{ m}^2/\text{m}^3$$

From (6-97),

$$h_L = \left[\frac{12(4.41 \times 10^{-5})}{17.8}\right]^{1/3}(0.045)^{2/3} = 0.0128 \text{ m}^3/\text{m}^3$$

Estimation of H_L:

First compute a_{Ph}, the ratio of phase interface area to packing surface area.

From (6-137),

$$d_h = 4\frac{0.952}{149.6} = 0.0255 \text{ m}$$

From (6-138),

$$N_{Re_L,h} = \frac{(0.0017)(0.0255)}{(0.64 \times 10^{-6})} = 67.7$$

From (6-139),

$$N_{We_L,h} = \frac{(0.0017)^2[(61.5)(16.02)](0.0255)}{[(101)(0.001)]} = 0.000719$$

From (6-140),

$$N_{Fr_L,h} = \frac{(0.0017)^2}{(9.807)(0.0255)} = 1.156 \times 10^{-5}$$

From (6-136),

$$\frac{a_{Ph}}{a} = 1.5(149.6)^{-1/2}(0.0255)^{-1/2}(67.7)^{-0.2}$$
$$\times (0.000719)^{0.75}(1.156 \times 10^{-5})^{-0.45} = 0.242$$

Estimation of H_L:

From (6-132), using consistent SI units:

$$H_L = \frac{1}{1.227}\left(\frac{1}{12}\right)^{1/6}\left[\frac{(4)(0.0128)(0.952)}{(1.82 \times 10^{-9})(149.6)(0.0017)}\right]^{1/2}$$
$$\times \left(\frac{0.0017}{149.6}\right)\left(\frac{1}{0.242}\right) = 0.26 \text{ m} = 0.85 \text{ ft}$$

Estimation of H_G:

From (6-134),

$$N_{Re_V} = 2.49/[(149.6)(0.75 \times 10^{-5})] = 2,220$$

From (6-135),

$$N_{Sc_V} = 0.75 \times 10^{-5}/7.75 \times 10^{-6} = 0.968$$

From (6-133), using consistent SI units,

$$H_G = \frac{1}{0.341}(0.952 - 0.0128)^{1/2} \left[\frac{(4)(0.952)}{(149.6)^4} \right]^{1/2}$$

$$\times (2220)^{-3/4}(0.968)^{-1/3} \left[\frac{(2.49)}{7.75 \times 10^{-6}(0.242)} \right]$$

$$= 1.03 \text{ m} \quad \text{or} \quad 3.37 \text{ ft}$$

Estimation of H_{OG}:

From Example 6.1, the K-value for ethyl alcohol = 0.57,

$$V = 17{,}480/44.05 = 397 \text{ lbmol/h},$$

$$L = 6{,}140/18.7 = 328 \text{ lbmol/h},$$

and $1/A = KV/L = (0.57)(397)/328 = 0.69$

From (6-124),

$$H_{OG} = 3.37 + 0.69(0.85) = 3.96 \text{ ft}$$

The mass-transfer resistance in the gas phase is much larger than that in the liquid phase.

Estimation of Packed Height:

From (6-141),

$$l_T = 3.96(7.5) = 29.7 \text{ ft}$$

Estimation of HETP:

From (6-94), for straight operating and equilibrium lines, with $A = 1/0.69 = 1.45$,

$$\text{HETP} = 3.96 \left[\frac{\ln(0.69)}{(1 - 1.45)/1.45} \right] = 4.73 \text{ ft}$$

6.9 CONCENTRATED SOLUTIONS IN PACKED COLUMNS

When the solute concentration in the gas and/or liquid is concentrated so that the operating line and/or equilibrium line are noticeably curved, then the procedure given in Section 6.7 for determining N_{OG} and l_T cannot be used because (6-91) cannot be analytically integrated to give (6-93). Instead, alternative methods can be employed or the computer-aided methods discussed in Chapters 10 and 11 can be applied.

For concentrated solutions, the two columns in Table 6.7 labeled UM (unimolecular) diffusion apply. To obtain these columns from the two columns labeled EM (equimolar) diffusion, we let

$$L' = L(1 - x) \quad \text{and} \quad V' = V(1 - y)$$

where L' and V' are the constant flow rates of the inert (solvent) liquid and (carrier) gas, respectively on a solute-free basis. Then

$$d(Vy) = V'd\left(\frac{y}{1-y}\right) = V'\frac{dy}{(1-y)^2} = V\frac{dy}{(1-y)} \tag{6-141}$$

$$d(Lx) = L'd\left(\frac{x}{1-x}\right) = L'\frac{dx}{(1-x)^2} = L\frac{dx}{(1-x)} \tag{6-142}$$

Equation (6-88) now becomes

$$l_T = \int_{y_2}^{y_1} \left(\frac{V}{K_y'aS} \right) \frac{dy}{(1-y)(y-y^*)}$$

$$= \frac{V}{K_y'aS} \int_{y_2}^{y_1} \frac{dy}{(1-y)(y-y^*)} \tag{6-143}$$

where 1 refers to inlet and 2 refers to outlet conditions. Based on the liquid phase,

$$l_T = \int_{x_1}^{x_2} \left(\frac{L}{K_x'aS} \right) \frac{dx}{(1-x)(x^*-x)}$$

$$= \frac{L}{K_x'aS} \int_{x_1}^{x_2} \frac{dx}{(1-x)(x^*-x)} \tag{6-144}$$

where the overall mass-transfer coefficients are primed to signify UM diffusion.

If the numerators and denominators of (6-143) and (6-144) are multiplied by $(1 - y)_{LM}$ and $(1 - x)_{LM}$, respectively, where $(1 - y)_{LM}$ is the log mean of $(1 - y)$ and $(1 - y^*)$, and $(1 - x)_{LM}$ is the log mean of $(1 - x)$ and $(1 - x^*)$, we obtain the expressions in rows 1 and 6 of columns 4 and 7 in Table 6.7:

$$l_T = \int_{y_2}^{y_1} \left[\frac{V}{K_y'a(1-y)_{LM}S} \right] \frac{(1-y)_{LM}\,dy}{(1-y)(y-y^*)}$$

$$= \frac{V}{K_y'a(1-y)_{LM}S} \int_{y_2}^{y_1} \frac{(1-y)_{LM}\,dy}{(1-y)(y-y^*)} \tag{6-145}$$

$$l_T = \int_{x_1}^{x_2} \left[\frac{L}{K_x'a(1-x)_{LM}S} \right] \frac{(1-x)_{LM}\,dx}{(1-x)(x^*-x)}$$

$$= \frac{L}{K_x'a(1-x)_{LM}S} \int_{x_1}^{x_2} \frac{(1-x)_{LM}\,dx}{(1-x)(x^*-x)} \tag{6-146}$$

In these equations $K_y'(1 - y)_{LM}$ is equal to the concentration-independent K_y, and $K_x'(1 - x)_{LM}$ is equal to the concentration-independent K_x. If there is appreciable absorption, vapor flow rate V decreases from the bottom to the top of the absorber. However, the values of Ka are also a function of flow rate, such that the ratio V/Ka is approximately constant and HTU groupings, $[L/K_x'a(1 - x)_{LM}S]$ and $[V/K_y'a(1 - y)_{LM}S]$, can often be taken out of the integral sign without incurring errors larger than those inherent in experimental measurements of Ka. Usually, average values of V, L, and $(1 - y)_{LM}$ are used.

Another approach is to leave all of the terms in (6-145) or (6-146) under the integral sign and evaluate l_T by a stepwise or graphical integration. In either case, to obtain the terms $(y - y^*)$ or $(x^* - x)$, the equilibrium and operating lines must be established. The equilibrium curve is determined from appropriate thermodynamic data or correlations. To establish the operating line, which will not be straight if the solutions are concentrated, the appropriate material-balance equations must be developed. With reference to Figure 6.29, an overall balance around the upper part of the absorber gives

$$V + L_{\text{in}} = V_{\text{out}} + L \tag{6-147}$$

Similarly a balance around the upper part of the absorber for the component being absorbed, assuming a pure-liquid absorbent, gives:

$$V y = V_{out} y_{out} + L x \qquad (6\text{-}148)$$

An absorbent balance around the upper part of the absorber is:

$$L_{in} = L(1 - x) \qquad (6\text{-}149)$$

Combining (6-147) to (6-149) to eliminate V and L gives

$$y = \frac{V_{out} y_{out} + [L_{in} x/(1 - x)]}{V_{out} + [L_{in} x/(1 - x)]} \qquad (6\text{-}150)$$

Equation (6-150) allows the y–x operating line to be calculated from a knowledge of terminal conditions only.

A simpler approach to the problem of concentrated gas or liquid mixtures is to linearize the operating line by expressing all concentrations in mole ratios, and the gas and liquid flows on a solute-free basis, that is, $V' = (1 - y)V$, $L' = (1 - x)L$. Then, in place of (6-145) and (6-146), we have

$$l_T = \int_{Y_2}^{Y_1} \left(\frac{V'}{K_Y a S} \right) \frac{dY}{(Y - Y^*)} = \frac{V'}{K_Y a S} \int_{Y_2}^{Y_1} \frac{dY}{(Y - Y^*)} \qquad (6\text{-}151)$$

$$l_T = \int_{X_1}^{X_2} \left(\frac{L'}{K_X a S} \right) \frac{dX}{(X^* - X)} = \frac{L'}{K_X a S} \int_{X_1}^{X_2} \frac{dX}{(X^* - X)} \qquad (6\text{-}152)$$

This set of equations is listed in rows 3 and 8 of Table 6.7.

EXAMPLE 6.16

To remove 95% of the ammonia from an air stream containing 40% ammonia by volume, 488 lbmol/h of an absorbent per 100 lbmol/h of entering gas are to be used, which is greater than the minimum requirement.

Equilibrium data are given in Figure 6.44. Pressure is 1 atm and temperature is assumed constant at 298 K. Calculate the number of transfer units by:

(a) Equation (6-145) using a curved operating line determined from (6-150)

(b) Equation (6-151) using mole ratios.

SOLUTION

(a) Take as a basis $L_{in} = 488$ lbmol/h. Then $V_{out} = 100 - (40)(0.95) = 62$ lbmol/h, and $y_{out} = (0.05)(40)/62 = 0.0323$. From (6-150), it is possible to construct the curved operating line of Figure 6.44. For example, if $x = 0.04$,

$$y = \frac{(62)(0.0323) + [(488)(0.04)/(1 - 0.04)]}{62 + [(488)(0.04)/(1 - 0.04)]} = 0.27$$

It is now possible to calculate the following values of y, y^*, $(1 - y)_{LM} = [(1 - y) - (1 - y^*)]/\ln[(1 - y)/(1 - y^*)]$, and $(1 - y)_{LM}/[(1 - y)(y - y^*)]$ for use in (6-145). For example, in Figure 6.44, for $x = 0.044$, y (on the operating line) $= 0.30$, and y^*

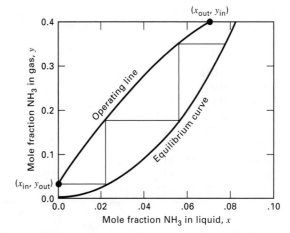

Figure 6.44 Determination of the number of theoretical stages for Example 6.15.

(on the equilibrium curve) $= 0.12$, from which the other four quantities in the following table follow.

y	y^*	$(y - y^*)$	$(1 - y)$	$(1 - y)_{LM}$	$\dfrac{(1 - y)_{LM}}{(1 - y)(y - y^*)}$
0.03	0.002	0.028	0.97	0.99	36.47
0.05	0.005	0.045	0.95	0.97	22.68
0.10	0.01	0.09	0.90	0.94	11.60
0.15	0.025	0.125	0.85	0.91	8.56
0.20	0.04	0.16	0.80	0.89	6.95
0.25	0.08	0.17	0.75	0.85	6.66
0.30	0.12	0.18	0.70	0.82	6.51
0.35	0.17	0.18	0.65	0.73	6.24
0.40	0.26	0.14	0.60	0.67	7.97

Note that since $(1 - y) \approx (1 - y)_{LM}$, these two terms frequently cancel out of the NTU equations, particularly when y is small.

Figure 6.45 is a plot of $(1 - y)_{LM}/[(1 - y)(y - y^*)]$ versus y to determine N_{OG}. The integral on the right-hand side of (6-145), between $y = 0.4$ and $y = 0.0322$, is $3.44 = N_{OG}$. This is approximately 1 more than the number of equilibrium stages of 2.6, as seen in the steps of Figure 6.44.

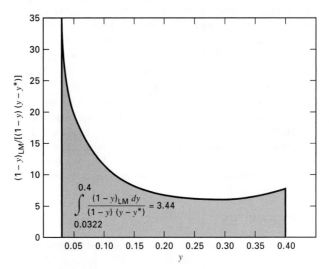

Figure 6.45 Determination of the number of transfer units for Examples 6.15.

(b) It is a simple matter to obtain the following values for $Y = y/(1 - y)$, $Y^* = y^*/(1 - y^*)$, $(Y - Y^*)$, and $(Y - Y^*)^{-1}$.

y	Y	y^*	Y^*	$(Y - Y^*)^{-1}$
0.03	0.031	0.002	0.002	34.48
0.05	0.053	0.005	0.005	20.83
0.1	0.111	0.01	0.010	9.9
0.15	0.176	0.025	0.026	6.66
0.20	0.250	0.04	0.042	4.8
0.25	0.333	0.08	0.087	4.06
0.30	0.43	0.12	0.136	3.40
0.35	0.54	0.17	0.205	2.98
0.40	0.67	0.26	0.310	2.78

Graphical integration of the right-hand-side integral of (6-151) is carried out by determining the area under the curve of Y versus $(Y - Y^*)^{-1}$ between $Y = 0.67$ and $Y = 0.033$. The result is $N_{OG} = 3.46$. Alternatively, the numerical integration can be performed on a computer with a spreadsheet.

It must be pointed out that for concentrated solutions, the assumption of constant temperature may not be valid and can result in a large error. If an overall energy balance indicates a temperature change that alters the equilibrium curve significantly, it is best to use a computer-aided method that includes the energy balance. Such methods are presented in Chapter 10.

SUMMARY

1. A liquid can be used to selectively absorb one or more components from a gas mixture. A gas can be used to selectively desorb or strip one or more components from a liquid mixture.

2. The fraction of a component that can be absorbed or stripped in a countercurrent cascade depends on the number of equilibrium stages and the absorption factor, $A = L/(KV)$, or the stripping factor, $S = KV/L$, respectively.

3. Absorption and stripping are most commonly conducted in trayed towers equipped with sieve or valve trays, or in towers packed with random or structured packings.

4. Absorbers are most effectively operated at high pressure and low temperature. The reverse is true for stripping. However, high costs of gas compression, refrigeration, and vacuum often preclude operation at the most thermodynamically favorable conditions.

5. For a given gas flow rate and composition, a desired degree of absorption of one or more components, a choice of absorbent, and an operating temperature and pressure, there is a minimum absorbent flow rate, given by (6-9) to (6-11), that corresponds to the use of an infinite number of equilibrium stages. For the use of a finite and reasonable number of stages, an absorbent rate of 1.5 times the minimum is typical. A similar criterion, (6-12), holds for a stripper.

6. The number of equilibrium stages required for a selected absorbent or stripping agent flow rate for the absorption or stripping of a dilute solution can be determined from the equilibrium line, (6-1), and an operating line, (6-3) or (6-5), using graphical, algebraic, or numerical methods. Graphical methods, such as Figure 6.11, offer considerable visual insight into stage-by-stage changes in compositions of the gas and liquid streams.

7. Rough estimates of overall stage efficiency, defined by (6-21), can be made with the correlations of Drickamer and Bradford (6-22), O'Connell (6-23), and Figure 6.15. More accurate and reliable procedures involve the use of a small Oldershaw column or semitheoretical equations, e.g., of Chan and Fair, based on mass-transfer considerations, to determine a Murphree vapor-point

efficiency, (6-30), from which a Murphree vapor tray efficiency can be estimated from (6-31) to (6-34), which can then be related to the overall efficiency using (6-37).

8. Tray diameter can be determined from (6-44) based on entrainment flooding considerations using Figure 6.24. Tray vapor pressure drop, the weeping constraint, entrainment, and downcomer backup can be estimated from (6-49), (6-68), (6-69), and (6-70), respectively.

9. Packed-column height can be estimated using the HETP, (6-73), or HTU/NTU, (6-89), concepts, with the latter having a more fundamental theoretical basis in the two-film theory of mass transfer. For straight equilibrium and operating lines, HETP is related to the HTU by (6-94), and the number of equilibrium stages is related to the NTU by (6-95).

10. Below a so-called loading point, in a preloading region, the liquid holdup in a packed column is independent of the vapor velocity. The loading point is typically about 70% of the flooding point, and most packed columns are designed to operate in the preloading region at from 50% to 70% of flooding. From the GPDC chart of Figure 6.36, the flooding point can be estimated, from which the column diameter can be determined with (6-102). The loading point can be estimated from (6-105).

11. One significant advantage of a packed column is its relatively low pressure drop per unit of packed height, as compared to a trayed tower. Packed-column pressure drop can be roughly estimated from Figure 6.36 or more accurately from (6-106) or (6-115).

12. Numerous rules of thumb are available for estimating the HETP of packed columns. However, the preferred approach is to estimate H_{OG} from separate semitheoretical mass-transfer correlations for the liquid and gas phases, such as those of (6-132) and (6-133) based on the extensive experimental work of Billet and Schultes.

13. Determination of theoretical stages for concentrated solutions involves numerical integration because of curved equilibrium and/or operating lines.

REFERENCES

1. WASHBURN, E.W., Ed.-in-Chief, *International Critical Tables,* McGraw-Hill, New York, Vol. **III**, p. 255 (1928).

2. LOCKETT, M., *Distillation Tray Fundamentals,* Cambridge University Press, Cambridge, UK, p. 13 (1986).

3. OKONIEWSKI, B.A., *Chem. Eng. Prog.,* **88** (2), 89–93 (1992).

4. SAX, N.I., *Dangerous Properties of Industrial Materials,* 4th ed., Van Nostrand Reinhold, New York, pp. 440–441 (1975).

5. LEWIS, W.K., *Ind. Eng. Chem.,* **14,** 492–497 (1922).

6. DRICKAMER, H.G., and J.R. BRADFORD, *Trans. AIChE,* **39,** 319–360 (1943).

7. JACKSON, R.M., and T.K. SHERWOOD, *Trans. AIChE,* **37,** 959–978 (1941).

8. O'CONNELL, H.E., *Trans. AIChE,* **42,** 741–755 (1946).

9. WALTER, J.F., and T.K. SHERWOOD, *Ind. Eng. Chem.,* **33,** 493–501 (1941).

10. EDMISTER, W.C., *The Petroleum Engineer,* C45–C54 (Jan. 1949).

11. LOCKHART, F.J., and C.W. LEGGETT, in K.A. KOBE and J.J. MCKETTA, Jr., Ed., *Advances in Petroleum Chemistry and Refining,* Vol. 1, Interscience, New York, Vol. **1,** pp. 323–326 (1958).

12. HOLLAND, C.D., *Multicomponent Distillation,* Prentice-Hall, Englewood Cliffs, NJ (1963).

13. MURPHREE, E.V., *Ind. Eng. Chem.,* **17,** 747 (1925).

14. HAUSEN, H., *Chem. Ing. Tech.,* **25,** 595 (1953).

15. STANDART, G., *Chem Eng. Sci.,* **20,** 611 (1965).

16. LEWIS, W.K., *Ind. Eng. Chem.,* **28,** 399 (1936).

17. GERSTER, J.A., A.B. HILL, N.H. HOCHGRAF, and D.G. ROBINSON, "Tray Efficiencies in Distillation Columns," Final Report from the University of Delaware, American Institute of Chemical Engineers, New York (1958).

18. *Bubble-Tray Design Manual,* AIChE, New York (1958).

19. GILBERT, T.J., *Chem. Eng. Sci.,* **10,** 243 (1959).

20. BARKER, P.E., and M.F. SELF, *Chem. Eng. Sci.,* **17,** 541 (1962).

21. BENNETT, D.L., and H.J. GRIMM, *AIChE J.,* **37,** 589 (1991).

22. OLDERSHAW, C.F., *Ind. Eng. Chem. Anal. Ed.,* **13,** 265 (1941).

23. FAIR, J.R., H.R. NULL, and W.L. BOLLES, *Ind. Eng. Chem. Process Des. Dev.,* **22,** 53–58 (1983).

24. SOUDERS, M., and G.G. BROWN, *Ind. Eng. Chem.,* **26,** 98–103 (1934).

25. FAIR, J.R., *Petro/Chem. Eng.,* **33,** 211–218 (Sept. 1961).

26. SHERWOOD, T.K., G.H. SHIPLEY, and F.A.L. HOLLOWAY, *Ind. Eng. Chem.,* **30,** 765–769 (1938).

27. *Glitsch Ballast Tray, Bulletin No. 159,* Fritz W. Glitsch and Sons, Dallas, TX (from FRI report of Sept. 3, 1958).

28. *Glitsch V-1 Ballast Tray, Bulletin No. 160,* Fritz W. Glitsch and Sons, Dallas, TX (from FRI report of Sept. 25, 1959).

29. OLIVER, E.D., *Diffusional Separation Processes. Theory, Design, and Evaluation,* John Wiley and Sons, New York, pp. 320–321 (1966).

30. BENNETT, D.L., R. AGRAWAL, and P.J. COOK, *AIChE J.,* **29,** 434–442 (1983).

31. SMITH, B.D., *Design of Equilibrium Stage Processes,* McGraw-Hill, New York (1963).

32. KLEIN, G.F., *Chem. Eng.,* **89** (9), 81–85 (1982).

33. KISTER, H.Z., *Distillation Design,* McGraw-Hill, New York (1992).

34. LOCKETT, M.J., *Distillation Tray Fundamentals,* Cambridge University Press, Cambridge, UK, p. 146 (1986).

35. American Institute of Chemical Engineers (AIChE), *Bubble-Tray Design Manual,* AIChE, New York, (1958).

36. CHAN, H., and J.R. FAIR, *Ind. Eng. Chem. Process Des. Dev.,* **23,** 814–819 (1984).

37. CHAN, H., and J.R. FAIR, *Ind. Eng. Chem. Process Des. Dev.,* **23,** 820–827 (1984).

38. SCHEFFE, R.D., and R.H. WEILAND, *Ind. Eng. Chem. Res.,* **26,** 228–236 (1987).

39. FOSS, A.S., and J.A. GERSTER, *Chem. Eng. Prog.,* **52,** 28-J to 34-J (Jan. 1956).

40. GERSTER, J.A., A.B. HILL, N.N. HOCHGRAF, and D.G. ROBINSON, "Tray Efficiencies in Distillation Columns," Final Report from University of Delaware, American Institute of Chemical Engineers (AIChE), New York (1958).

41. FAIR, J.R., *Petro./Chem. Eng.,* **33** (10), 45 (1961).

42. COLBURN, A.P., *Ind. Eng. Chem.,* **28,** 526 (1936).

43. CHILTON, T.H., and A.P. COLBURN, *Ind. Eng. Chem.,* **27,** 255–260, 904 (1935).

44. COLBURN, A.P., *Trans. AIChE,* **35,** 211–236, 587–591 (1939).

45. BILLET, R., *Packed Column Analysis and Design,* Ruhr-University Bochum (1989).

46. STICHLMAIR, J., J.L. BRAVO, and J.R. FAIR, *Gas Separation and Purification,* **3,** 19–28 (1989).

47. BILLET, R., and M. SCHULTES, *Packed Towers in Processing and Environmental Technology,* translated by J.W. Fullarton, VCH Publishers, New York (1995).

48. LEVA, M., *Chem. Eng. Prog. Symp. Ser.,* **50** (10), 51 (1954).

49. LEVA, M., *Chem. Eng. Prog.,* **88** (1), 65–72 (1992).

50. KISTER, H.Z., and D.R. GILL, *Chem. Eng. Prog.,* **87** (2), 32–42 (1991).

51. BILLET, R., and M. SCHULTES, *Chem. Eng. Technol.,* **14,** 89–95 (1991).

52. ERGUN, S., *Chem. Eng. Prog.,* **48** (2), 89–94 (1952).

53. KUNESH, J.G., *Can. J. Chem. Eng.,* **65,** 907–913 (1987).

54. WHITMAN, W.G., *Chem. and Met. Eng.,* **29,** 146–148 (1923).

55. SHERWOOD, T.K., and F.A.L. HOLLOWAY, *Trans. AIChE.,* **36,** 39–70 (1940).

56. CORNELL, D., W.G. KNAPP, and J.R. FAIR, *Chem. Eng. Prog.,* **56** (7) 68–74 (1960).

57. CORNELL, D., W.G. KNAPP, and J.R. FAIR, *Chem. Eng. Prog.,* **56** (8), 48–53 (1960).

58. BOLLES, W.L., and J.R. FAIR, *Inst. Chem. Eng. Symp. Ser.,* **56,** 3/35 (1979).

59. BOLLES, W.L., and J.R. FAIR, *Chem. Eng.,* **89** (14), 109–116 (1982).

60. BRAVO, J.L., and J.R. FAIR, *Ind. Eng. Chem. Process Des. Devel.,* **21,** 162–170 (1982).

61. BRAVO, J.L., J.A. ROCHA, and J.R. FAIR, *Hydrocarbon Processing,* **64** (1), 56–60 (1985).

62. FAIR, J.R., and J.L. BRAVO, *I. Chem. E. Symp. Ser.,* **104,** A183–A201 (1987).

63. FAIR, J.R., and J.L. BRAVO, *Chem. Eng. Prob.,* **86** (1), 19–29 (1990).

64. SHULMAN, H.L., C.F. ULLRICH, A.Z. PROULX, and J.O. ZIMMERMAN, *AIChE J.,* **1,** 253–258 (1955).

65. ONDA, K., H. TAKEUCHI, and Y.J. OKUMOTO, *J. Chem. Eng. Jpn.,* **1,** 56–62 (1968).

66. BILLET, R., *Chem. Eng. Prog.,* **63** (9), 53–65 (1967).

67. BILLET, R., and M. SCHULTES, *Beitrage zur Verfahrens-Und Umwelttechnik,* Ruhr-Universitat Bochum, pp. 88–106 (1991).

68. HIGBIE, R., *Trans. AIChE,* **31,** 365–389 (1935).

69. BILLET, R., and M. SCHULTES, *Chem. Eng. Res. Des., Trans. IChemE,* **77**A, 498–504 (1999).

70. M. SCHULTES, Private Communication (2004).

71. SLOLEY, A.W., *Chem. Eng. Prog.,* **95** (1), 23–35 (1999).

72. STUPIN, W.J., and H.Z. KISTER, *Trans. IChemE.,* **81**A, 136–146 (2003).

EXERCISES

Section 6.1

6.1 In any absorption operation, the absorbent is stripped to some extent depending on the K-value of the absorbent. In any stripping operation, the stripping agent is absorbed to some extent depending on its K-value. In Figure 6.1, it is seen that both absorption and stripping occur. Which occurs to the greatest extent in terms of kilomoles per hour? Should the operation be called an absorber or a stripper? Why?

6.2 Prior to 1950, only two types of commercial random packings were in common use: Raschig rings and Berl saddles. Starting in the 1950s, a wide variety of commercial random packings began to appear. What advantages do these newer packings have? By what advances in packing design and fabrication techniques were these advantages achieved? Why were structured packings introduced?

6.3 Bubble-cap trays were widely used in the design of trayed towers prior to the 1960s. Today sieve and valve trays are favored. However, bubble-cap trays are still occasionally specified, especially for operations that require very high turndown ratios or appreciable liquid residence time. What characteristics of bubble-cap trays make it possible for them to operate satisfactorily at low vapor and liquid rates?

Section 6.2

6.4 In Example 6.3, a lean oil of 250 MW is used as the absorbent. Consideration is being given to the selection of a new absorbent. Available streams are:

	Rate, gpm	Density, lb/gal	**MW**
C_5s	115	5.24	72
Light oil	36	6.0	130
Medium oil	215	6.2	180

Which stream would you choose? Why? Which streams, if any, are unacceptable?

6.5 Volatile organic compounds (VOCs) can be removed from water effluents by stripping in packed towers. Possible stripping agents are steam and air. Alternatively, the VOCs can be removed by carbon adsorption. The U.S. Environmental Protection Agency (EPA) has identified air stripping as the best available technology from an economic standpoint. What are the advantages and disadvantages of air compared to steam?

6.6 Prove by equations why, in general, absorbers should be operated at high pressure and low temperature, while strippers should be operated at low pressure and high temperature. Also prove, by equations, why a tradeoff exists between number of stages and flow rate of the separating agent.

Section 6.3

6.7 The exit gas from an alcohol fermenter consists of an air–CO_2 mixture containing 10 mol% CO_2 that is to be absorbed in a 5.0-N solution of triethanolamine, containing 0.04 mol of carbon dioxide per mole of amine solution. If the column operates isothermally at 25°C, if the exit liquid contains 78.4% of the CO_2 in the feed gas to the absorber, and if absorption is carried out in a six-theoretical-plate column, calculate:

(a) Moles of amine solution required per mole of feed gas.

(b) Exit gas composition.

Equilibrium Data

Y	0.003	0.008	0.015	0.023	0.032	0.043
X	0.01	0.02	0.03	0.04	0.05	0.06
Y	0.055	0.068	0.083	0.099	0.12	
X	0.07	0.08	0.09	0.10	0.11	

Y = moles CO_2/mole air; X = moles CO_2/mole amine solution

6.8 Ninety-five percent of the acetone vapor in an 85 vol% air stream is to be absorbed by countercurrent contact with pure water in a valve-tray column with an expected overall tray efficiency of 50%. The column will operate essentially at 20°C and 101 kPa pressure. Equilibrium data for acetone–water at these conditions are:

Mole percent acetone in water	3.30	7.20	11.7	17.1
Acetone partial pressure in air, torr	30.00	62.80	85.4	103.0

Calculate:

(a) The minimum value of L'/V', the ratio of moles of water per mole of air.

(b) The number of equilibrium stages required using a value of L'/V' of 1.25 times the minimum.

(c) The concentration of acetone in the exit water.

From Table 5.2 for N connected equilibrium stages, there are $2N + 2C + 5$ degrees of freedom. Specified in this problem are

Stage pressures (101 kPa)	N
Stage temperatures (20°C)	N
Feed stream composition	$C-1$
Water stream composition	$C-1$
Feed stream T, P	2
Water stream, T, P	2
Acetone recovery	1
L/V	1
	$2N + 2C + 4$

The remaining specification is the feed flow rate, which can be taken on a basis of 100 kmol/h.

6.9 A solvent-recovery plant consists of a plate-column absorber and a plate-column stripper. Ninety percent of the benzene (B) in the gas stream is recovered in the absorption column. Concentration of benzene in the inlet gas is 0.06 mol B/mol B-free gas. The oil entering the top of the absorber contains 0.01 mol B/mol pure oil. In the leaving liquid, $X = 0.19$ mol B/mol pure oil. Operating temperature is 77°F (25°C).

Open, superheated steam is used to strip benzene out of the benzene-rich oil at 110°C. Concentration of benzene in the oil = 0.19 and 0.01 (mole ratios) at inlet and outlet, respectively. Oil (pure)-to-steam (benzene-free) flow rate ratio = 2.0. Vapors are condensed, separated, and removed.

MW oil = 200   MW benzene = 78   MW gas = 32

Equilibrium Data at Column Pressures

X in Oil	Y in Gas, 25°C	Y in Steam, 110°C
0	0	0
0.04	0.011	0.1
0.08	0.0215	0.21
0.12	0.032	0.33
0.16	0.042	0.47
0.20	0.0515	0.62
0.24	0.060	0.795
0.28	0.068	1.05

Calculate:

(a) The molar flow rate ratio of B-free oil to B-free gas in the absorber; (b) The number of theoretical plates in the absorber; and (c) The minimum steam flow rate required to remove the benzene from 1 mol of oil under given terminal conditions, assuming an infinite-plates column.

6.10 A straw oil used to absorb benzene from coke-oven gas is to be steam-stripped in a sieve-plate column at atmospheric pressure to recover the dissolved benzene. Equilibrium conditions at the operating temperature are approximated by Henry's law such that, when the oil phase contains 10 mol% C_6H_6, the C_6H_6 partial pressure above the oil is 5.07 kPa. The oil may be considered non-volatile. The oil enters containing 8 mol% benzene, 75% of which is to be recovered. The steam leaving contains 3 mol% C_6H_6. (a) How many theoretical stages are required? (b) How many moles of steam are required per 100 mol of oil–benzene mixture? (c) If 85% of the benzene is to be recovered with the same oil and steam rates, how many theoretical stages are required?

Section 6.4

6.11 Groundwater at a flow rate of 1,500 gpm, containing three volatile organic compounds (VOCs), is to be stripped in a trayed tower with air to produce drinking water that will meet EPA standards. Relevant data are given below. Determine the maximum air flow rate in scfm (60F, 1 atm) and the number of equilibrium stages required if an air flow rate of twice the minimum is used and the tower operates at 25°C and 1 atm. Also determine the composition in parts per million for each VOC in the resulting drinking water.

		Concentration, ppm	
Component	**K-value**	**Ground water**	**Max. for Drinking water**
1,2-Dichloroethane (DCA)	60	85	0.005
Trichloroethylene (TCE)	650	120	0.005
1,1,1-Trichloroethane (TCA)	275	145	0.200

Note: ppm = parts per million by weight.

6.12 Sulfur dioxide and butadienes (B3 and B2) are to be stripped with nitrogen from the liquid stream as shown in Figure 6.46 so that

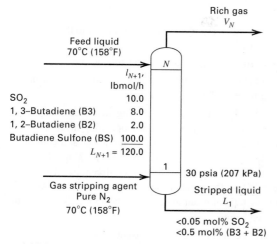

Figure 6.46 Data for Exercise 6.12.

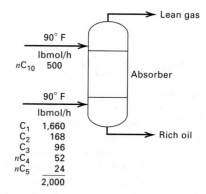

Figure 6.47 Data for Exercise 6.13.

butadiene sulfone (BS) product will contain less than 0.05 mol% SO_2 and less than 0.5 mol% butadienes. Estimate the flow rate of nitrogen, N_2, and the number of equilibrium stages required. At 70°C, K-values for SO_2, B2, B3, and BS are, respectively, 6.95, 3.01, 4.53, and 0.016.

6.13 Determine by the Kremser method the separation that can be achieved for the absorption operation indicated in Figure 6.47 for the following combinations of conditions: (a) Six equilibrium stages and 75 psia operating pressure, (b) Three equilibrium stages and 150 psia operating pressure, (c) Six equilibrium stages and 150 psia operating pressure. At 90°F and 75 psia, the K-value of $nC_{10} = 0.0011$.

6.14 One thousand kilomoles per hour of rich gas at 70°F with 25% C_1, 15% C_2, 25% C_3, 20% nC_4, and 15% nC_5 by moles is to be absorbed by 500 kmol/h of nC_{10} at 90°F in an absorber operating at 4 atm. Calculate by the Kremser method the percent absorption of each component for 4, 10, and 30 theoretical stages. What do you conclude from the results? (Note: The K-value of nC_{10} at 80°F and 4 atm is 0.0014.)

Section 6.5

6.15 Using the performance data of Example 6.3, back-calculate the overall stage efficiency for propane and compare the result with estimates from the Drickamer–Bradford and O'Connell correlations.

6.16 Several hydrogenation processes are being considered that will require hydrogen of 95% purity. A refinery stream of 800,000 scfm (at 32°F, 1 atm), currently being used for fuel and containing 72.5% H_2, 25% CH_4, and 2.5% C_2H_6 is available. To convert this gas to the required purity, oil absorption, activated charcoal adsorption, and membrane separation are being considered. For oil absorption, an available n-octane stream can be used as the absorbent. Because the 95% H_2 must be delivered to a hydrogenation process at not less than 375 psia, it is proposed to operate the absorber at 400 psia and 100°F. If at least 80% of the hydrogen fed to the absorber is to leave in the exit gas, determine:

(a) The minimum absorbent rate in gallons per minute.

(b) The actual absorbent rate if 1.5 times the minimum amount is used.

(c) The number of theoretical stages.

(d) The stage efficiency for each of the three species in the feed gas, using the O'Connell correlation.

(e) The number of trays actually required.

(f) The composition of the exit gas, taking into account the stripping of octane.

(g) If the octane lost to the exit gas is not recovered, estimate the annual cost of this lost oil if the process operates 7,900 h/year and the octane is valued at $1.00/gal.

6.17 The absorption operation of Examples 6.1 and 6.4 is being scaled up by a factor of 15, such that a column with an 11.5-ft diameter will be needed. In addition, because of the low efficiency of 30% for the original operation, a new tray design has been developed and tested in an Oldershaw-type column. The resulting Murphree vapor-point efficiency, E_{OV}, for the new tray design for the system of interest is estimated to be 55%. Estimate E_{MV} and E_o. (To estimate the length of the liquid flow path, Z_L, use Figure 6.16. Also, assume that $u/D_E = 6 \text{ ft}^{-1}$.)

Section 6.6

6.18 Conditions at the bottom tray of a reboiled stripper are as shown in Figure 6.48. If valve trays are used with a 24-in. tray spacing, estimate the required column diameter for operation at 80% of flooding.

6.19 Determine the flooding velocity and column diameter for the following conditions at the top tray of a hydrocarbon absorber equipped with valve trays:

Pressure	400 psia
Temperature	128°F
Vapor rate	530 lbmol/h
Vapor MW	26.6
Vapor density	1.924 lb/ft^3
Liquid rate	889 lbmol/h
Liquid MW	109
Liquid density	41.1 lb/ft^3
Liquid surface tension	18.4 dynes/cm
Foaming factor	0.75
Tray spacing	24 in.
Fraction flooding	0.85
Valve trays	

6.20 For Exercise 6.16, if a flow rate of 40,000 gpm of octane is used to carry out the absorption in a sieve-tray column using 24-in. tray spacing, a weir height of 2.5 in., and holes of $\frac{1}{4}$-in. diameter, determine for a foaming factor of 0.80 and a fraction flooding of 0.70:

(a) The column diameter based on conditions near the bottom of the column.

(b) The vapor pressure drop per tray.

(c) Whether weeping will occur.

(d) The entrainment rate.

(e) The fractional decrease in E_{MV} due to entrainment.

(f) The froth height in the downcomer.

6.21 Repeat the calculations of Examples 6.5, 6.6, and 6.7 for a column diameter corresponding to 40% of flooding.

6.22 For the acetone absorber of Figure 6.1, assuming the use of sieve trays with a 10% hole area and $\frac{3}{16}$-in. holes with an 18-in. tray spacing, estimate:

(a) The column diameter for a foaming factor of 0.85 and a fraction of flooding of 0.75.

(b) The vapor pressure drop per tray.

546.2 lbmol/h
6.192 cfs

	y, mol%
C_2	0.0006
C_3	0.4817
nC_4	60.2573
nC_5	32.5874
nC_6	6.6730

Bottom tray

230.5° F
150 psia

	x, mol%
C_2	0.0001
C_3	0.1448
nC_4	39.1389
nC_5	43.0599
nC_6	17.6563

621.3 lbmo/h
171.1 gpm

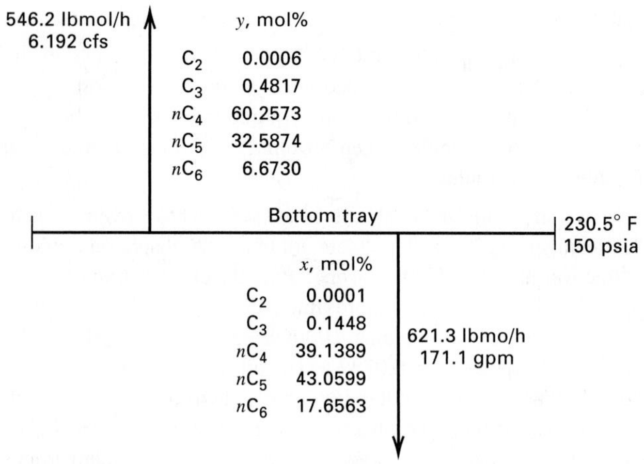

Figure 6.48 Data for Exercise 6.18.

(c) The number of transfer units, N_G and N_L, from (6-62) and (6-63), respectively.

(d) N_{OG} from (6-61).

(e) The controlling resistance to mass transfer.

(f) E_{OV} from (6-56).

From your results, determine if 30 actual trays are adequate.

6.23 Design a VOC stripper for the flow conditions and separation of Example 6.2 except that the wastewater and air flow rates are twice as high. To develop the design, determine:

(a) The number of equilibrium stages required.

(b) The column diameter for sieve trays.

(c) The vapor pressure drop per tray.

(d) Murphree vapor-point efficiencies using the Chan and Fair method.

(e) The number of trays actually required.

Section 6.7

6.24 Air containing 1.6 vol% sulfur dioxide is scrubbed with pure water in a packed column of 1.5-m^2 cross-sectional area and 3.5-m height packed with no. 2 plastic Super Intalox® saddles, at a pressure of 1 atm. Total gas flow rate is 0.062 kmol/s, the liquid flow rate is 2.2 kmol/s, and the outlet gas SO_2 concentration is $y = 0.004$. At the column temperature, the equilibrium relationship is given by $y^* = 40x$.

(a) What is L/L_{min}?

(b) Calculate N_{OG} and compare your answer to that for the number of theoretical stages required.

(c) Determine H_{OG} and the HETP from the operating data.

(d) Calculate $K_G a$ from the data, based on a partial-pressure driving force as in item 2 of Table 6.7.

6.25 An SO_2–air mixture is being scrubbed with water in a countercurrent-flow packed tower operating at 20°C and 1 atm. Solute-free water enters the top of the tower at a constant rate of 1,000 lb/h and is well distributed over the packing. The liquor leaving contains 0.6 lb SO_2/100 lb of solute-free water. The partial pressure of SO_2 in the spent gas leaving the top of the tower is 23 torr. The mole ratio of water to air is 25. The necessary equilibrium data are given below.

(a) What percent of the SO_2 in the entering gases is absorbed in the tower?

(b) In operating the tower it was found that the rate coefficients k_p and k_L remained substantially constant throughout the tower at the following values:

$$k_L = 1.3 \text{ ft/h}$$

$$k_p = 0.195 \text{ lbmol/h-ft}^2\text{-atm}$$

At a point in the tower where the liquid concentration is 0.001 lbmol SO_2 per lbmol of water, what is the liquid concentration at the gas–liquid interface in lbmol/ft^3? Assume that the solution has the same density as H_2O.

Solubility of SO_2 in H_2O at 20°C

lb SO_2 100 lb H_2O	Partial Pressure of SO_2 in Air, torr
0.02	0.5
0.05	1.2
0.10	3.2
0.15	5.8
0.20	8.5
0.30	14.1
0.50	26.0
0.70	39.0
1.0	59

6.26 A wastewater stream of 600 gpm, containing 10 ppm (by weight) of benzene, is to be stripped with air in a packed column operating at 25°C and 2 atm to produce water containing 0.005 ppm of benzene. The packing is 2-in. Flexirings® made of polypropylene. The vapor pressure of benzene at 25°C is 95.2 torr. The solubility of benzene in water at 25°C is 0.180 g/100 g. An expert in VOC stripping with air has suggested use of 1,000 scfm of air (60°F, 1 atm), at which condition one should achieve for the mass transfer of benzene:

$$k_L a = 0.067 \text{ s}^{-1} \quad \text{and} \quad k_G a = 0.80 \text{ s}^{-1}$$

Determine:

(a) The minimum air stripping rate in scfm. Is it less than the rate suggested by the expert? If not, use 1.4 times your minimum value.

(b) The stripping factor based on the air rate suggested by the expert.

(c) The number of transfer units, N_{OG}, required.

(d) The overall mass-transfer coefficient, $K_G a$, in units of mol/m^3-s-kPa and s^{-1}. Which phase controls mass transfer?

(e) The volume of packing in cubic meters.

Section 6.8

6.27 Germanium tetrachloride ($GeCl_4$) and silicon tetrachloride ($SiCl_4$) are used in the production of optical fibers. Both chlorides are oxidized at high temperature and converted to glasslike particles. However, the $GeCl_4$ oxidation is quite incomplete and it is necessary to scrub the unreacted $GeCl_4$ from its air carrier in a packed column operating at 25°C and 1 atm with a dilute caustic solution. At these conditions, the dissolved $GeCl_4$ has no vapor pressure and mass transfer is controlled by the gas phase. Thus, the equilibrium curve is a straight line of zero slope. Why? The entering gas is 23,850 kg/day of air containing 288 kg/day of $GeCl_4$. The air also contains 540 kg/day of Cl_2, which, when dissolved, also will have no vapor pressure. The two

liquid-phase reactions are

$$GeCl_4 + 5OH^- \rightarrow HGeO_3^- + 4Cl^- + 2H_2O$$

$$Cl_2 + 2OH^- \rightarrow ClO^- + Cl^- + H_2O$$

It is desired to absorb 99% of both $GeCl_4$ and Cl_2 in an existing 2-ft-diameter column that is packed to a height of 10 ft with $\frac{1}{2}$-in. ceramic Raschig rings. The liquid rate should be set so that the column operates at 75% of flooding. For the packing: $\epsilon = 0.63$, $F_P = 580 \text{ ft}^{-1}$, and $D_P = 0.01774$ m.

Gas-phase mass-transfer coefficients for $GeCl_4$ and Cl_2 can be estimated from the following empirical equations developed from experimental studies, where μ, ρ, and D_i are gas-phase properties:

$$K_y a = k_y a$$

$$\frac{k_y}{(V/S)} = 1.195 \left[\frac{D_p V'}{\mu (1 - \epsilon_o)} \right]^{-0.36} (N_{Sc})^{-2/3}$$

$$\epsilon_o = \epsilon - h_L$$

$$h_L = 0.03591 (L')^{0.331}$$

$$a = \frac{14.69 (808 \, V'/\rho^{1/2})^n}{(L')^{0.111}}$$

$$n = 0.01114 L' + 0.148$$

where

$S = $ column cross sectional area, m^2

$k_y = $ kmol/m^2-s

$V = $ molar gas rate, kmol/s

$D_p = $ equivalent packing diameter, m

$\mu = $ gas viscosity, kg/m-s

$\rho = $ gas density, kg/m^3

$N_{Sc} = $ Schmidt number $= \mu / \rho D_i$

$D_i = $ molecular diffusivity of component i in the gas, m^2/s

$a = $ interfacial area for mass transfer, m^2/m^3 of packing

$L' = $ liquid mass velocity, kg/m^2-s

$V' = $ gas mass velocity, kg/m^2-s

For the two diffusing species, take

$$D_{GeCl_4} = 0.000006 \text{ m}^2/\text{s}$$

$$D_{Cl_2} = 0.000013 \text{ m}^2/\text{s}$$

Determine:

(a) The dilute caustic flow rate in kilograms per second.

(b) The required packed height in feet based on the controlling species ($GeCl_4$ or Cl_2). Is the 10 ft of packing adequate?

(c) The percent absorption of $GeCl_4$ and Cl_2 based on the available 10 ft of packing. If the 10 ft of packing is not sufficient, select an alternative packing that is adequate.

6.28 For the VOC stripping task of Exercise 6.26, the expert has suggested that we use a tower diameter of 0.80 m for which we can expect a pressure drop of 500 N/m^2-m of packed height (0.612 in. H_2O/ft). Verify the information from the expert by estimating:

(a) The fraction of flooding using the GPDC chart of Figure 6.36 with $F_P = 24 \text{ ft}^2/\text{ft}^3$.

(b) The pressure drop at flooding.

(c) The pressure drop at the operating conditions of Exercise 6.26 using the GPDC chart.

(d) The pressure drop at operating conditions using the correlation of Billet and Schultes by assuming that 2-in. plastic Flexiring® packing has the same characteristics as 2-in. plastic Pall® rings.

6.29 For the VOC stripping task of Exercise 6.26, the expert suggested certain mass-transfer coefficients. Check this information by estimating the coefficients from the correlations of Billet and Schultes by assuming that 2-in. plastic Flexiring® packing has the same characteristics as 2-in. plastic Pall® rings.

6.30 A 2 mol% NH_3-in-air mixture at 68°F and 1 atm is to be scrubbed with water in a tower packed with 1.5-in. ceramic Berl saddles. The inlet water mass velocity will be 2400 lb/h-ft², and the inlet gas mass velocity 240 lb/h-ft². Assume that the tower temperature remains constant at 68°F, at which the gas solubility relationship follows Henry's law, $p = Hx$, where p is the partial pressure of ammonia over the solution, x is the mole fraction of ammonia in the liquid, and H is the Henry's law constant, equal to 2.7 atm/mole fraction.

(a) Calculate the required packed height for absorption of 90% of the NH_3.

(b) Calculate the minimum water mass velocity in lb/h-ft² for absorbing 98% of the NH_3.

(c) The use of 1.5-in. ceramic Hiflow® rings rather than the Berl saddles has been suggested. What changes would this cause in K_Ga, pressure drop, maximum liquid rate, K_La, column height, column diameter, H_{OG}, and N_{OG}?

6.31 You are to design a packed column to absorb CO_2 from air into fresh, dilute-caustic solution. The entering air contains 3 mol% CO_2, and a 97% recovery of CO_2 is desired. The gas flow rate is 5,000 ft³/min at 60°F, 1 atm. It may be assumed that in the range of operation, the equilibrium curve is $Y^* = 1.75X$, where Y and X are mole ratios of CO_2 to carrier gas and liquid, respectively. A column diameter of 30 in. with 2-in. Intalox® saddle packing can be assumed for the initial design estimates. Assume the caustic solution has the properties of water. Calculate:

(a) The minimum caustic solution-to-air molar flow rate ratio.

(b) The maximum possible concentration of CO_2 in the caustic solution.

(c) The number of theoretical stages at $L/V = 1.4$ times minimum.

(d) The caustic solution rate.

(e) The pressure drop per foot of column height. What does this result suggest?

(f) The overall number of gas transfer units, N_{OG}.

(g) The height of packing, using a K_Ga of 2.5 lbmol/h-ft³-atm.

Section 6.9

6.32 At a point in an ammonia absorber using water as the absorbent and operating at 101.3 kPa and 20°C, the bulk gas phase contains 10 vol% NH_3. At the interface, the partial pressure of NH_3 is 2.26 kPa. The concentration of the ammonia in the body of the liquid is 1 wt%. The rate of ammonia absorption at this point is 0.05 kmol/h-m².

(a) Given this information and the equilibrium curve in Figure 6.49, calculate X, Y, Y_I, X_I, X^*, Y^*, K_Y, K_X, k_Y, and k_X.

(b) What percent of the mass-transfer resistance is in each phase?

(c) Verify for these data that $1/K_Y = 1/k_Y + H'/k_X$.

6.33 One thousand cubic feet per hour of a 10 mol% NH_3 in air mixture is required to produce nitrogen oxides. This mixture is to be obtained by desorbing an aqueous 20 wt% NH_3 solution with air at 20°C. The spent solution should not contain more than 1 wt% NH_3.

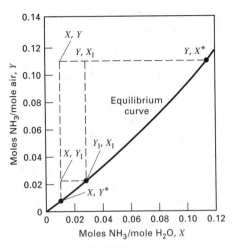

Figure 6.49 Data for Exercise 6.32.

Calculate the volume of packing required for the desorption column. Vapor–liquid equilibrium data for Exercise 6.32 can be used and $K_Ga = 4$ lbmol/h-ft³-atm partial pressure.

6.34 Ammonia, present at a partial pressure of 12 torr in an air stream saturated with water vapor at 68°F and 1 atm, must be removed to the extent of 99.6% by water absorption at the same temperature and pressure. Two thousand pounds of dry air per hour are to be handled.

(a) Calculate the minimum amount of water necessary using the equilibrium data for Exercise 6.32 in Figure 6.49.

(b) Assuming an operation at 2 times the minimum water flow and at one-half the flooding gas velocity, compute the dimensions of a column packed with 38-mm ceramic Berl Saddles.

(c) Repeat part (b) for 50-mm Pall® rings.

(d) Which of the two packings would you recommend?

6.35 Exit gas from a chlorinator consists of a mixture of 20 mol% chlorine in air. This concentration is to be reduced to 1% chlorine by water absorption in a packed column to operate isothermally at 20°C and atmospheric pressure. Using the following equilibrium x–y data, calculate for 100 kmol/h of feed gas:

(a) The minimum water rate in kilograms per hour.

(b) N_{OG} for twice the minimum water rate.

Data for x–y at 20°C (in chlorine mole fractions):

x	0.0001	0.00015	0.0002	0.00025	0.0003
y	0.006	0.012	0.024	0.04	0.06

6.36 Calculate the diameter and height for the column of Example 6.15 if the tower is packed with 1.5-in. metal Pall® rings. Assume that the absorbing solution has the properties of water and use conditions at the bottom of the tower, where flow rates are highest.

6.37 You are asked to design a packed column to recover acetone from air continuously, by absorption with water at 60°F. The air contains 3 mol% acetone, and a 97% recovery is desired. The gas flow rate is 50 ft³/min at 60°F, 1 atm. The maximum-allowable gas superficial velocity in the column is 2.4 ft/s. It may be assumed that in the range of operation, $Y^* = 1.75X$, where Y and X are mole ratios for acetone.

Calculate:

(a) The minimum water-to-air molar flow rate ratio.

(b) The maximum acetone concentration possible in the aqueous solution.

(c) The number of theoretical stages for a flow rate ratio of 1.4 times the minimum.

(d) The corresponding number of overall gas transfer units.

(e) The height of packing, assuming $K_ya = 12.0$ lbmol/h-ft^3-molar ratio difference.

(f) The height of packing as a function of the molar flow rate ratio, assuming that V and HTU remain constant.

6.38 Determine the diameter and packed height of a countercurrently operated packed tower required to recover 99% of the ammonia from a gas mixture that contains 6 mol% NH$_3$ in air. The tower, packed with 1-in. metal Pall® rings, must handle 2,000 ft^3/min of gas as measured at 68°F and 1 atm. The entering water-absorbent rate will be twice the theoretical minimum, and the gas velocity will be such that it is 50% of the flooding velocity. Assume isothermal operation at 68°F and 1 atm. Equilibrium data are given in Figure 6.49.

6.39 A tower, packed with Montz™ B1-200 metal structured packing, is to be designed to absorb SO$_2$ from air by scrubbing with water. The entering gas, at an SO$_2$-free flow rate of 6.90 lbmol/h-ft^2 of bed cross section, contains 80 mol% air and 20 mol% SO$_2$. Water enters at a flow rate of 364 lbmol/h-ft^2 of bed cross section. The exiting gas is to contain only 0.5 mol% SO$_2$. Assume that neither air nor water will be transferred between phases and that the tower operates at 2 atm and 30°C. Equilibrium data in mole fractions for SO$_2$ solubility in water at 30°C and 2 atm (*Perry's Chemical Engineers' Handbook,* 4th ed., Table 14.31, p. 14-6) have been fitted by a least-squares method to the equation

$$y = 12.697x + 3148.0x^2 - 4.724 \times 10^5 x^3 + 3.001 \times 10^7 x^4 - 6.524 \times 10^8 x^5$$

(a) Derive the following molar material balance operating line for SO$_2$ mole fractions:

$$x = 0.0189 \left(\frac{y}{1-y} \right) - 0.00010$$

(b) Write a computer program or use a spreadsheet program to calculate the number of required transfer units based on the overall gas-phase resistance.

Chapter 7

Distillation of Binary Mixtures

In *distillation* (*fractionation*), a feed mixture of two or more components is separated into two or more products, including, and often limited to, an overhead distillate and a bottoms, whose compositions differ from that of the feed. Most often, the feed is a liquid or a vapor–liquid mixture. The bottoms product is almost always a liquid, but the distillate may be a liquid or a vapor or both. The separation requires that (1) a second phase be formed so that both liquid and vapor phases are present and can contact each other on each stage within a separation column, (2) the components have different volatilities so that they will partition between the two phases to different extents, and (3) the two phases can be separated by gravity or other mechanical means. Distillation differs from absorption and stripping in that the second fluid phase is usually created by thermal means (vaporization and condensation) rather than by the introduction of a second phase that may contain an additional component or components not present in the feed mixture.

According to Forbes [1], the art of distillation dates back to at least the first century A.D. By the eleventh century, distillation was being used in Italy to produce alcoholic beverages. At that time, distillation was probably a batch process based on the use of just a single stage, the boiler. The feed to be separated, a liquid, was placed in a vessel to which heat was applied, causing part of the liquid to evaporate. The vapor passed out of the heating vessel and was cooled in another chamber by transfer of heat through the wall of the chamber to water, producing condensate that dripped into a product receiver. The word *distillation* is derived from the Latin word *destillare,* which means dripping or trickling down. By at least the sixteenth century, it was known that the extent of separation could be improved by providing multiple vapor–liquid contacts (stages) in a so-called Rectificatorium. The term *rectification* is derived from the Latin words *recte facere,* meaning to improve. Modern distillation derives its ability to produce almost pure products from the use of multistage contacting.

Throughout the twentieth century, multistage distillation was by far the most widely used industrial method for separating liquid mixtures of chemical components. Unfortunately, distillation is a very energy-intensive technique, especially when the relative volatility, α, of the components being separated is low (<1.50). Mix et al. [2] report that the energy consumption for distillation in the United States for 1976 totaled 2×10^{15} Btu (2 quads), which was nearly 3% of the entire national energy consumption. Approximately two-thirds of the distillation energy was consumed by petroleum refining, where distillation is widely used to separate crude oil into petroleum fractions, light hydrocarbons (C_2's to C_5's), and aromatic chemicals. The separation of other organic chemicals, often in the presence of water, is widely practiced in the chemical industry.

7.0 INSTRUCTIONAL OBJECTIVES

After completing this chapter, you should be able to:

- Explain the difference between distillation and absorption or stripping.
- Explain the need in distillation for a condenser to produce reflux and a reboiler to produce boilup
- Enumerate factors that influence design of a distillation column.
- Distinguish between required specifications and results that can be obtained from the McCabe–Thiele method for continuous binary distillation.
- Determine the five construction lines used in the McCabe–Thiele method using component material balances and vapor–liquid equilibrium relations.
- Explain the concept of "constant molar overflow," assumptions required for its validity, and why it eliminates the need for energy balances around stages.
- Distinguish among five possible phase conditions of the feed.
- Apply the McCabe–Thiele method for determining minimum reflux ratio, minimum number of equilibrium stages, number of equilibrium stages for a specified reflux ratio greater than minimum, and optimal feed-stage location, given the required split between the two feed components.

- Select an appropriate condenser and a suitable operating pressure for a distillation separation.
- Explain differences among the three most common types of reboilers.
- Calculate condenser and reboiler heat duties and consider use of a feed preheater.
- Determine the optimal reflux ratio.
- Use the Murphree vapor stage efficiency to determine actual number of stages (plates) from the number of equilibrium stages.
- Extend the McCabe–Thiele method to multiple feeds, side streams, and open steam (in place of a reboiler).
- Estimate overall stage efficiency for binary distillation from correlations and laboratory column data.
- Determine the diameter of a trayed tower and size of the reflux drum.
- Determine packed height and diameter of a packed column.
- Use an enthalpy-concentration diagram when assumption of "constant molar overflow" is not valid.

Industrial Example

The fundamentals of distillation are best understood by the study of binary distillation, the separation of a two-component mixture, which is the subject of this chapter. The more general and much more difficult case of a multicomponent mixture is covered in Chapters 10 and 11. A representative binary distillation operation is shown in Figure 7.1 for the separation of 620 lbmol/h (0.0781 kmol/s) of a binary mixture of 46 mol% benzene (the more volatile component) and 54 mol% toluene. The purpose of the 25-sieve-tray (equivalent to 20 theoretical stages plus a partial reboiler that acts as an additional theoretical stage)

distillation column is to separate the feed into a liquid distillate of 99 mol% benzene and a liquid bottoms product of 98 mol% toluene. The column operates at a pressure in the reflux drum of 18 psia (124 kPa), just slightly above ambient pressure. For a negligible pressure drop across the condenser and a vapor pressure drop of 0.1 psi/tray (0.69 kPa/tray), the pressure in the reboiler is $18 + 0.1(25) = 20.5$ psia (141 kPa). In this range of pressure, benzene and toluene form nearly ideal mixtures with a relative volatility of from 2.26 at conditions of the bottom tray to 2.52 at the top tray, as determined from Raoult's law by (2-44). The reflux ratio (reflux rate to distillate rate) is 2.215. If an infinite number of stages were used, the required reflux ratio would be a minimum value of 1.708. Thus, the ratio of reflux rate to minimum reflux rate for this example is 1.297. Most distillation columns are designed to operate with optimal-reflux-to-minimum-reflux ratios of 1.1 to 1.5. If an infinite ratio of reflux to minimum reflux were used, only 10.7 theoretical stages would be required. Thus, the ratio of theoretical stages to minimum theoretical stages for this example is $21/10.7 = 1.96$. For most distillation columns, this ratio is approximately 2. The stage efficiency is 20/25 or 80%. This is close to the average efficiency observed for distillation.

The feed to the separation operation of Figure 7.1 is a saturated liquid at 55 psia (379 kPa). A bubble-point calculation gives a temperature of 294°F (419 K). When this feed is flashed adiabatically across the feed valve to the feed tray pressure of 19.25 psia (133 kPa), the feed temperature drops to 220°F (378 K), causing 23.4 mol% of the feed to be vaporized. A total condenser is used to obtain saturated liquid reflux and liquid distillate at a bubble-point temperature of 189°F (360 K) at 18 psia (124 kPa). The duty of the condenser is 11,820,000 Btu/h (3.46 MW). At the bottom of the column, a partial reboiler is used to produce vapor boilup and a liquid bottoms product. Assuming that the boilup and bottoms are in physical equilibrium, the

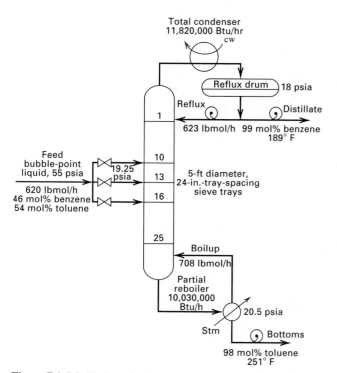

Figure 7.1 Distillation of a binary mixture of benzene and toluene.

partial reboiler functions as an additional theoretical stage, giving a total of 21 theoretical stages. Because the bottoms product is a saturated liquid, its temperature of 251°F (395 K) corresponds to the bubble point of the bottoms at 20.5 psia (141 kPa). The duty of the reboiler is 10,030,000 Btu/h (2.94 MW), which is within 15% of the condenser duty.

The inside diameter of the distillation column in Figure 7.1 is a constant 5 ft (1.53 m). At the top tray this diameter corresponds to 84% of flooding, while at the bottom tray the percent flooding is 81%. As shown, the column can be fed at any one of three trays. For the design conditions, the optimal feed entry is between trays 12 and 13. However, should the feed composition or product specifications change, one of the other two feed trays could become optimal.

Distillation columns similar to that of Figure 7.1 have been built for diameters up to at least 30 ft (9.14 m). With a 24-in. (0.61-m) tray spacing, the maximum number of trays included in a single column is usually no greater than 150. In general, for the sharp separation of a binary mixture with a relative volatility less than 1.05, distillation can require many hundreds of trays, so a more efficient separation

technique should be sought. Even when distillation is the most economical separation technique, its second-law efficiency, using the calculational procedure developed in Chapter 2, can be less than 10%.

Technically, distillation is the most mature separation operation. Design and operation procedures are well established; for example, see Kister [3, 4]. Only when vapor–liquid equilibrium or other data are uncertain is a laboratory and/or pilot-plant study necessary prior to the design of a commercial unit. Table 7.1, taken partially from the study of Mix et al. [2], lists just some of the more common commercial binary distillation operations in decreasing order of difficulty of separation. Included are representative nominal values of relative volatility, number of trays, column operating pressure, and reflux-to-minimum-reflux ratio. Although the data in Table 7.1 refer to trayed towers, distillation can also be carried out in packed columns. More and more frequently, additional distillation capacity is being achieved with existing trayed towers by replacing all or some of the trays with sections of random or structured packing.

Table 7.1 Representative Commercial Binary Distillation Operations [2]

Binary Mixture	Average Relative Volatility	Number of Trays	Typical Operating Pressure, psia	Reflux-to-Minimum-Reflux Ratio
1,3-Butadiene/vinyl acetylene	1.16	130	75	1.70
Vinyl acetate/ethyl acetate	1.16	90	15	1.15
o-Xylene/m-xylene	1.17	130	15	1.12
Isopentane/n-pentane	1.30	120	30	1.20
Isobutane/n-butane	1.35	100	100	1.15
Ethylbenzene/styrene	1.38	34	1	1.71
Propylene/propane	1.40	138	280	1.06
Methanol/ethanol	1.44	75	15	1.20
Water/acetic acid	1.83	40	15	1.35
Ethylene/ethane	1.87	73	230	1.07
Acetic acid/acetic anhydride	2.02	50	15	1.13
Toluene/ethylbenzene	2.15	28	15	1.20
Propane/1,3-butadiene	2.18	40	120	1.13
Ethanol azeotrope/water	2.21	60	15	1.35
Isopropanol/water	2.23	12	15	1.28
Benzene/toluene	3.09	34	15	1.15
Methanol/water	3.27	60	45	1.31
Cumene/phenol	3.76	38	1	1.21
Benzene/ethylbenzene	6.79	20	15	1.14
HCN/water	11.20	15	50	1.36
Ethylene oxide/water	12.68	50	50	1.19
Formaldehyde/methanol	16.70	23	50	1.17
Water/ethylene glycol	81.20	16	4	1.20

In this chapter, equipment for conducting distillation operations is discussed and fundamental equilibrium-based and rate-based calculational procedures are developed for binary mixtures. Trayed and packed distillation columns are identical in most respects to the absorption and stripping columns discussed in the previous chapter. Therefore, where appropriate, reference is made to Chapter 6 and only important differences are discussed in this chapter.

7.1 EQUIPMENT AND DESIGN CONSIDERATIONS

Industrial distillation operations are most commonly conducted in trayed towers, but packed columns are finding increasing use. Occasionally, distillation columns contain both trays and packing. Types of trays and packings are identical to those used for absorption and stripping, as described in Section 6.1, shown in Figures 6.2 to 6.7, and compared in Tables 6.2 and 6.3.

Factors that influence the design or analysis of a binary-distillation operation include:

1. Feed flow rate, composition, temperature, pressure, and phase condition
2. Desired degree of separation between two components
3. Operating pressure (which must be below the critical pressure of the mixture)
4. Vapor pressure drop, particularly for vacuum operation
5. Minimum reflux ratio and actual reflux ratio
6. Minimum number of equilibrium stages and actual number of equilibrium stages (stage efficiency)
7. Type of condenser (total, partial, or mixed)
8. Degrees of subcooling, if any, of the liquid reflux
9. Type of reboiler (partial or total)
10. Type of contacting (trays or packing or both)
11. Height of the column
12. Feed-entry stage
13. Diameter of the column
14. Column internals.

The phase condition (also called thermal condition) of the feed is determined at the feed-tray pressure by an adiabatic-flash calculation across the feed valve. As the molar fraction of vapor in the feed increases, the required reflux ratio (L/D) increases, but the corresponding boilup ratio (V/B) decreases. The column operating pressure in the reflux drum should correspond to a distillate temperature somewhat higher (e.g., 10 to 50°F or 6 to 28°C) than the supply temperature of the cooling water used as the coolant in the overhead condenser. However, if this pressure approaches the critical pressure of the more volatile component, then a lower operating pressure must be used and a refrigerant is required as coolant. For example, in Table 7.1, the separation of ethylene/ethane is conducted at 230 psia (1,585 kPa), giving a column top temperature of −40°F (233 K), which requires a refrigerant. Water at 80°F (300 K) cannot be used in the condenser because the critical temperature of ethylene is 48.6°F (282 K). If the estimated pressure is less than atmospheric pressure, the operating pressure at the top of the column is often set just above atmospheric pressure to avoid vacuum operation, unless the temperature at the bottom of the column is found to exceed a bottoms temperature limited by decomposition, polymerization, excessive corrosion, or other chemical reaction. In that case, vacuum operation is necessary. In Table 7.1, vacuum operation is required for the separation of ethyl-benzene from styrene to maintain a bottoms temperature sufficiently low to prevent polymerization of styrene.

For given (1) feed, (2) desired degree of separation, and (3) operating pressure, a minimum reflux ratio exists that corresponds to an infinite number of theoretical stages; and a minimum number of theoretical stages exists that corresponds to an infinite reflux ratio. A design trade-off is usually made between the number of stages and the reflux ratio. A graphical method for determining the data needed to make this trade-off and to determine the optimal feed-stage location is developed in the next section.

7.2 McCABE–THIELE GRAPHICAL EQUILIBRIUM-STAGE METHOD FOR TRAYED TOWERS

Consider the general countercurrent-flow, multistage, binary-distillation operation shown in Figure 7.2. The operation consists of a column containing the equivalent of N theoretical stages; a total condenser in which the overhead vapor leaving the top stage is totally condensed to give a liquid distillate product and liquid reflux that is returned to the top stage; a partial reboiler in which liquid from the bottom stage is partially vaporized to give a liquid bottoms product and vapor boilup that is returned to the bottom stage, and an intermediate feed stage. By means of multiple, countercurrent contacting stages arranged in a two-section cascade with reflux and boilup, as discussed in Section 5.4, it is possible to achieve a sharp separation between the two components in the feed unless an azeotrope is formed, in which case one of the two products will approach the azeotropic composition.

The feed, which contains a more-volatile (light) component (the *light key*, LK), and a less-volatile (heavy)

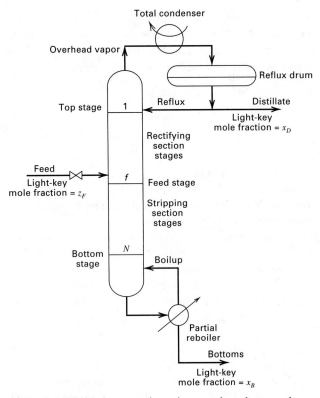

Figure 7.2 Distillation operation using a total condenser and partial reboiler.

component (the *heavy key,* HK), enters the column at a feed stage, *f*. At the feed-stage pressure, the feed may be liquid, vapor, or a mixture of liquid and vapor, with its overall mole-fraction composition with respect to the light component denoted by z_F. The mole fraction of the light key in the distillate is x_D, while the mole fraction of the light key in the bottoms product is x_B. Corresponding compositions with respect to the heavy key are $1 - z_F$, $1 - x_D$, and $1 - x_B$.

The goal of distillation is to produce from the feed a distillate, rich in the light key (i.e., x_D approaching 1.0), and a bottoms product, rich in the heavy key (i.e., x_B approaching 0.0). The ease or difficulty with which the separation can be achieved depends on the relative volatility, α, of the two components (LK = 1 and HK = 2), where

$$\alpha_{1,2} = K_1/K_2 \tag{7-1}$$

Methods for estimating K-values are discussed in Chapter 2.

If the two components form ideal solutions and follow the ideal-gas law in the vapor phase, Raoult's law applies to give

$$K_1 = P_1^s/P \quad \text{and} \quad K_2 = P_2^s/P$$

and from (7-1), the relative volatility is given simply by the ratio of vapor pressures, $\alpha_{1,2} = P_1^s/P_2^s$ and thus is a function only of temperature. As discussed in Section 4.2, as the temperature (and therefore the pressure) increases, $\alpha_{1,2}$

decreases. At the convergence pressure of the mixture, $\alpha_{1,2} = 1.0$ and a separation cannot be achieved at this or any higher pressure.

The relative volatility can be expressed in terms of equilibrium vapor and liquid compositions from the definition of the K-value as $K_i = y_i/x_i$. For a binary mixture,

$$\alpha_{1,2} = \frac{y_1/x_1}{y_2/x_2} = \frac{y_1(1 - x_1)}{x_1(1 - y_1)} \tag{7-2}$$

Solving (7-2) for y_1,

$$y_1 = \frac{\alpha_{1,2}x_1}{1 + x_1(\alpha_{1,2} - 1)} \tag{7-3}$$

For ideal binary mixtures of components with close boiling points, the temperature change over the column is small and $\alpha_{1,2}$ is almost constant. In any case, for a given pressure P and liquid-phase composition x_1, the Gibbs phase rule, discussed in Chapter 4, fixes the temperature and equilibrium-vapor composition. An equilibrium curve for the benzene–toluene system is shown in Figure 7.3, where y and x correspond to the light key, benzene, and the pressure is 1 atm, at which pure benzene and pure toluene boil at 176 and 231°F, respectively. Thus, these two components are not close-boiling. Using (7-3) with this curve, α varies from about 2.6 at the bottom of the curve to about 2.35 at the top of the curve. Representative equilibrium curves for some average values of α are shown in Figure 4.5. The higher the average value of α, the easier it is to achieve the desired separation. Average values of α for the distillation operations in Table 7.1 range from 1.16 to 81.2.

In 1925, McCabe and Thiele [5] published an approximate graphical method for combining the equilibrium curve of Figure 7.3 with operating-line curves to estimate, for a given binary-feed mixture and column operating pressure, the number of equilibrium stages and the amount of reflux required for a desired degree of separation of the feed.

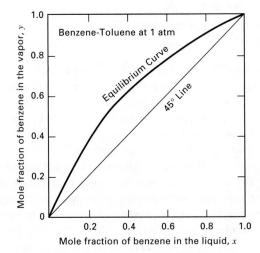

Figure 7.3 Equilibrium curve for benzene–toluene at 1 atm.

Although computer-aided methods, discussed later in Chapter 10, are more accurate and easier to apply, the graphical construction of the McCabe–Thiele method greatly facilitates the visualization of many of the important aspects of multistage distillation, and therefore the effort required to learn the method is well justified.

Typical problem specifications for and results from the McCabe–Thiele method are summarized in Table 7.2. This table applies to a simple, binary-distillation operation, like that in Figure 7.2, for a single feed and two products. The distillate can be a liquid from a total condenser, as shown in Figure 7.2, or a vapor from a partial condenser. The feed phase condition must be known at the column pressure, which is assumed to be uniform throughout the column for the McCabe–Thiele method. The type of condenser and reboiler must be specified, as well as the ratio of reflux to minimum reflux. From the specification of x_D and x_B for the light key, the distillate and bottoms flow rates, D and B, are fixed by material balance, since

$$F z_F = x_D D + x_B B$$

But, $B = F - D$ and therefore

$$F z_F = x_D D + x_B (F - D)$$

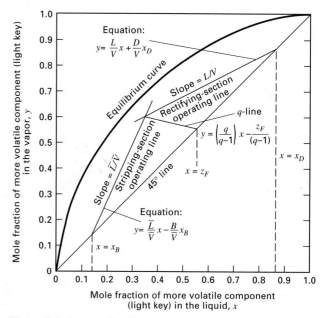

Figure 7.4 Construction lines for McCabe–Thiele method.

or

$$D = F \left(\frac{z_F - x_B}{x_D - x_B} \right)$$

This result requires $x_B < z_F < x_D$.

The McCabe–Thiele method determines not only N, the number of equilibrium stages, but also N_{min}, R_{min}, and the optimal stage for feed entry. Following the application of the McCabe–Thiele method, energy balances are applied to estimate condenser and reboiler heat duties.

Besides the equilibrium curve, the McCabe–Thiele method involves a 45° reference line, separate operating lines for the upper *rectifying* (enriching) section of the column and the lower *stripping* (exhausting) section of the column, and a fifth line (the *q-line* or feed line) for the phase or thermal condition of the feed. A typical set of these lines is shown in Figure 7.4. Equations for these lines are derived in the following subsection.

Rectifying Section

As shown in Figure 7.2, the rectifying section of equilibrium stages extends from the top stage, 1, to just above the feed stage, f. Consider a top portion of the rectifying stages, including the total condenser. A material balance for the light key over the envelope shown in Figure 7.5a for the total condenser and stages 1 to n is as follows, where y and x refer to vapor and liquid mole fractions, respectively, for the light key:

$$V_{n+1} y_{n+1} = L_n x_n + D x_D \qquad (7\text{-}4)$$

Table 7.2 Specifications for and Results from the McCabe–Thiele Method for Binary Distillation

Specifications	
F	Total feed rate
z_F	Mole-fraction composition of the feed
P	Column operating pressure (assumed uniform throughout the column)
	Phase condition of the feed at column pressure
	Vapor–liquid equilibrium curve for the binary mixture at column pressure
	Type of overhead condenser (total or partial)
	Type of reboiler (usually partial)
x_D	Mole-fraction composition of the distillate
x_B	Mole-fraction composition of the bottoms
R/R_{min}	Ratio of reflux to minimum reflux

Results	
D	Distillate flow rate
B	Bottoms flow rate
N_{min}	Minimum number of equilibrium stages
R_{min}	Minimum reflux ratio, L_{min}/D
R	Reflux ratio, L/D
V_B	Boilup ratio, $\bar{V}/B$
N	Number of equilibrium stages
	Optimal feed-stage location
	Stage vapor and liquid compositions

All mole fraction compositions are for the light key.

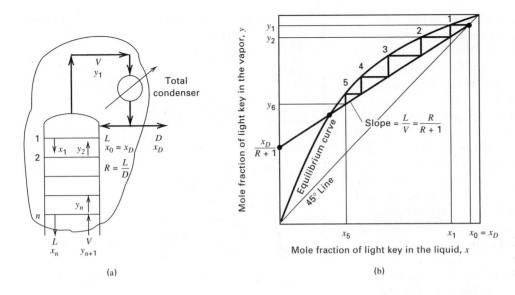

Figure 7.5 McCabe–Thiele operating line for the rectifying section.

Solving for y_{n+1}:

$$y_{n+1} = \frac{L_n}{V_{n+1}} x_n + \frac{D}{V_{n+1}} x_D \qquad (7\text{-}5)$$

Equation (7-5) relates the compositions, y_{n+1} and x_n, of two *passing* streams, V_{n+1} and L_n, respectively. For (7-5) to plot as a straight line of the form $y = mx + b$, which is the locus of compositions of all passing streams in the rectifying section, total molar flow rates L and V must not vary from stage to stage. This is the case if:

1. The two components have equal and constant molar enthalpies of vaporization (latent heats).

2. Component sensible-enthalpy changes ($C_P \Delta T$) and heat of mixing are negligible compared to latent heat changes.

3. The column is well insulated so that heat loss is negligible.

4. The pressure is uniform throughout the column (no pressure drop).

These assumptions are referred to as the *McCabe–Thiele assumptions* leading to the condition of *constant molar overflow* in the rectifying section, which refers to a molar liquid flow rate that remains constant as the liquid overflows each weir from one stage to the next. Since a total material balance for the rectifying-section envelope in Figure 7.5a gives $V_{n+1} = L_n + D$, if L is constant, then V is also constant for a particular value of D. Thus, (7-5) can be rewritten as

$$y = \frac{L}{V} x + \frac{D}{V} x_D \qquad (7\text{-}6)$$

as shown in Figure 7.4. Thus, the slope of the operating line is L/V, which is constant. Because $V > L$, $L/V < 1$ in the rectifying section, as seen in Figure 7.5b.

For constant molar overflow, it is not necessary to consider energy balances in either the rectifying or stripping sections; only material balances and a vapor–liquid equilibrium curve are required. However, energy balances are needed to determine condenser and reboiler duties, as discussed later.

The liquid entering the top stage is the external reflux rate, L_0, and its ratio to the distillate rate, L_0/D, is the reflux ratio, R. Because of the assumption of constant molar overflow, R is a constant in the rectifying section, equal to L/D. Since $V = L + D$, the slope of the operating line is readily related to the reflux ratio:

$$\frac{L}{V} = \frac{L}{L+D} = \frac{L/D}{L/D + D/D} = \frac{R}{R+1} \qquad (7\text{-}7)$$

Similarly,

$$\frac{D}{V} = \frac{D}{L+D} = \frac{1}{R+1} \qquad (7\text{-}8)$$

Combining (7-6), (7-7), and (7-8) produces the most useful form of the operating line for the rectifying section:

$$y = \left(\frac{R}{R+1}\right) x + \left(\frac{1}{R+1}\right) x_D \qquad (7\text{-}9)$$

If values of R and x_D are specified, (7-9) plots as a straight line with an intersection at $y = x_D$ on the 45° line, a slope of $L/V = R/(R+1)$, and an intersection at $y = x_D/(R+1)$ for $x = 0$, as shown in Figure 7.5b, which also contains a 45° line and an equilibrium curve. The equilibrium stages are stepped off in the manner described in Section 6.3 for absorption. Starting from the point ($y_1 = x_D$, $x_0 = x_D$) on the operating line and the 45° line, a horizontal line is drawn to the left until it intersects the equilibrium curve at (y_1, x_1), that is, the compositions of the *equilibrium phases* leaving the top equilibrium stage. A vertical line is now dropped until it intersects the operating line at the point (y_2, x_1), the compositions of the two *phases passing* each other between stages 1 and 2. The horizontal- and vertical-line constructions are

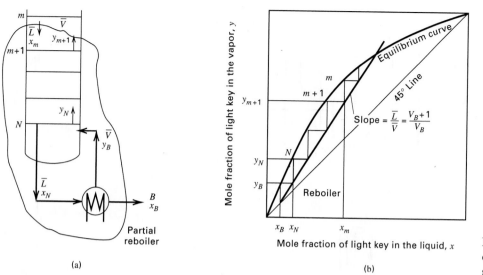

Figure 7.6 McCabe–Thiele operating line for the stripping section.

continued down the rectifying section in the manner shown in Figure 7.5b to give the staircase construction shown, which is arbitrarily terminated at stage 5. The optimal stage for termination is considered later.

Stripping Section

As shown in Figure 7.2, the stripping section of equilibrium stages extends from the feed to the bottom stage. In Figure 7.6a, consider a bottom portion of the stripping stages, including the partial reboiler and extending up from stage N to stage $m + 1$, located somewhere below the feed. A material balance for the light key over the envelope shown in Figure 7.6a results in

$$\bar{L}x_m = \bar{V}y_{m+1} + Bx_B \qquad (7\text{-}10)$$

Solving for y_{m+1}:

$$y_{m+1} = \frac{\bar{L}}{\bar{V}}x_m - \frac{B}{\bar{V}}x_B$$

or

$$y = \frac{\bar{L}}{\bar{V}}x - \frac{B}{\bar{V}}x_B \qquad (7\text{-}11)$$

where $\bar{L}$ and $\bar{V}$ are the total molar flows, which by the constant-molar-overflow assumption remain constant from stage to stage. The slope of this operating line for the compositions of passing steams in the stripping section is seen to be $\bar{L}/\bar{V}$. Because $\bar{L} > \bar{V}$, $\bar{L}/\bar{V} > 1$, as seen in Figure 7.6b. This is the inverse of conditions in the rectifying section.

The vapor leaving the partial reboiler is assumed to be in equilibrium with the liquid bottoms product. Thus, the partial reboiler acts as an additional equilibrium stage. The vapor rate leaving it is called the *boilup*, $\bar{V}_{N+1}$, and its ratio to the bottoms product rate, $V_B = \bar{V}_{N+1}/B$, is the *boilup ratio*. Because of the constant-molar-overflow assumption, V_B is constant in the stripping section. Since $\bar{L} = \bar{V} + B$,

$$\frac{\bar{L}}{\bar{V}} = \frac{\bar{V} + B}{\bar{V}} = \frac{V_B + 1}{V_B} \qquad (7\text{-}12)$$

Similarly,

$$\frac{B}{\bar{V}} = \frac{1}{V_B} \qquad (7\text{-}13)$$

Combining (7-11), (7-12), and (7-13), the operating-line equation for the stripping section becomes

$$y = \left(\frac{V_B + 1}{V_B}\right)x - \left(\frac{1}{V_B}\right)x_B \qquad (7\text{-}14)$$

If values of V_B and x_B are known, (7-14) can be plotted, together with the equilibrium curve and a 45° line, as a straight line with an intersection at $y = x_B$ on the 45° line and a slope of $\bar{L}/\bar{V} = (V_B + 1)/V_B$, as shown in Figure 7.6b. The equilibrium stages are stepped off, in a manner similar to that described for the rectifying section, starting from the point $(y = x_B,\ x = x_B)$ on the operating and 45° lines and moving upward on a vertical line until the equilibrium curve is intersected at $(y = y_B,\ x = x_B)$, which represents the equilibrium mole fractions in the vapor and liquid leaving the partial reboiler. From that point, the staircase is constructed by drawing horizontal and then vertical lines, moving back and forth between the operating line and equilibrium curve, as observed in Figure 7.6b, where the staircase is arbitrarily terminated at stage m. Next, we determine where to terminate the two operating lines.

Feed-Stage Considerations

Thus far, the McCabe–Thiele construction has not considered the feed to the column. In determining the operating lines for the rectifying and stripping sections, it is very important to note that although x_D and x_B can be selected independently, R and V_B are related by the feed phase condition.

Consider the five possible feed conditions shown in Figure 7.7, which assumes that the feed has been flashed adiabatically to the feed-stage pressure. If the feed is a bubble-point liquid, it adds to the reflux, L, coming from the stage above to give $\bar{L} = L + F$. If the feed is a dew-point

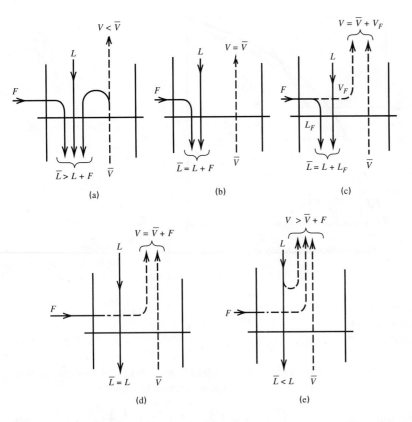

Figure 7.7 Possible feed conditions: (a) subcooled-liquid feed; (b) bubble-point liquid feed; (c) partially vaporized feed; (d) dew-point vapor feed; (e) superheated-vapor feed.

[Adapted from W.L. McCabe, J.C. Smith, and P. Harriott, *Unit Operations of Chemical Engineering,* 5th ed., McGraw-Hill, New York (1993).]

vapor, it adds to the boilup vapor, $\bar{V}$, coming from the stage below to give $V = \bar{V} + F$. For a partially vaporized feed, as shown in Figure 7.7c, $F = L_F + V_F$ and $\bar{L} = L + L_F$ and $V = \bar{V} + V_F$. If the feed is a subcooled liquid, it will cause a portion of the boilup, $\bar{V}$, to condense, giving $\bar{L} > L + F$ and $V < \bar{V}$. If the feed is a superheated vapor, it will cause a portion of the reflux, L, to vaporize, giving $\bar{L} < L$ and $V > \bar{V} + F$.

For cases (b), (c), and (d) of Figure 7.7, covering a range of feed conditions from a saturated liquid to a saturated vapor, the boilup $\bar{V}$ is related to the reflux L by the material balance:

$$\bar{V} = L + D - V_F \qquad (7\text{-}15)$$

and the boilup ratio, $V_B = \bar{V}/B$, is

$$V_B = \frac{L + D - V_F}{B} \qquad (7\text{-}16)$$

Alternatively, the reflux can be determined from the boilup by

$$L = \bar{V} + B - L_F \qquad (7\text{-}17)$$

Although distillation operations can be specified by either the reflux ratio R or the boilup ratio V_B, by tradition R or $R/R_{\min}$ is used because the distillate product is most often the more important product.

For the other two cases, (a) and (e) of Figure 7.7, V_B and R cannot be related by simple material balances alone. It is necessary to consider an energy balance to convert sensible enthalpy into latent enthalpy of phase change. This is most

conveniently done by defining a parameter, q, as the ratio of the increase in molar reflux rate across the feed stage to the molar feed rate,

$$q = \frac{\bar{L} - L}{F} \qquad (7\text{-}18)$$

or by material balance around the feed stage,

$$q = 1 + \frac{\bar{V} - V}{F} \qquad (7\text{-}19)$$

Values of q for the five feed conditions are

Feed condition	q
Subcooled liquid	>1
Bubble-point liquid	1
Partially vaporized	$L_F/F = 1 -$ molar fraction vaporized
Dew-point vapor	0
Superheated vapor	<0

To determine values of q for subcooled liquid and superheated vapor, a more general definition of q is applied:

$q =$ enthalpy change to bring the feed to a dew-point vapor divided by enthalpy of vaporization of the feed (dew-point vapor minus bubble-point liquid), that is,

$$q = \frac{(h_F)_{\text{sat'd vapor temperature}} - (h_F)_{\text{feed temperature}}}{(h_F)_{\text{sat'd vapor temperature}} - (h_F)_{\text{sat'd liquid temperature}}} \qquad (7\text{-}20)$$

For a subcooled liquid feed, (7-20) becomes

$$q = \frac{\Delta H^{\text{vap}} + C_{P_L}(T_b - T_F)}{\Delta H^{\text{vap}}} \qquad (7\text{-}21)$$

For a superheated vapor, (7-20) becomes

$$q = \frac{C_{P_V}(T_d - T_F)}{\Delta H^{\mathrm{vap}}} \qquad (7\text{-}22)$$

where C_{P_L} and C_{P_V} are the liquid and vapor molar heat capacities, respectively, ΔH^{vap} is the molar enthalpy change from the bubble point to the dew point, and T_F, T_d, and T_b are the feed, dew-point, and bubble-point temperatures, respectively, of the feed at the column operating pressure.

Instead of using (7-14) to locate the stripping operating line on the McCabe–Thiele diagram, it is more common to use an alternative method that involves a q-line (feed line), which is included in Figure 7.4. The q-line, one point of which is the intersection of the rectifying and stripping operating lines, is derived in the following manner. Combining (7-11) with (7-6) gives

$$y(V - \bar{V}) = (L - \bar{L})x + Dx_D + Bx_B \qquad (7\text{-}23)$$

But

$$Dx_D + Bx_B = Fz_F \qquad (7\text{-}24)$$

and a material balance around the feed stage gives

$$F + \bar{V} + L = V + \bar{L} \qquad (7\text{-}25)$$

Combining (7-23) to (7-25) with (7-18) gives

$$y = \left(\frac{q}{q-1}\right)x - \left(\frac{z_F}{q-1}\right) \qquad (7\text{-}26)$$

which is the equation for the q-line. This line is located on the McCabe–Thiele diagram by noting that when $x = z_F$, (7-26) reduces to the point $y = z_F = x$, which lies on the 45° line. From (7-26), the slope of the line is $q/(q-1)$. This construction is shown in Figure 7.4 for a partially vaporized feed, for which $0 < q < 1$ and $-\infty < [q/(q-1)] < 0$. Following the placement of the rectifying-section operating line and the q-line, the stripping-section operating line is located by drawing a straight line from the point $(y = x_B, x = x_B)$ on the 45° line to and through the point of intersection of the q-line and the rectifying-section operating line as shown in Figure 7.4. The point of intersection must lie somewhere between the equilibrium curve and the 45° line.

As q changes from a value greater than 1 (subcooled liquid) to a value less than 0 (superheated vapor), the slope of the q-line, $q/(q-1)$, changes from a positive value to a negative value and back to a positive value, as shown in Figure 7.8. For a saturated liquid feed, the q-line is vertical; for a saturated vapor, the q-line is horizontal.

Determination of Number of Equilibrium Stages and Feed-Stage Location

Following the construction of the five lines shown in Figure 7.4, the number of equilibrium stages required for the entire column, as well as the location of the feed stage, are determined by stepping off stages by any of several ways. The stages can be stepped off first from the top down and

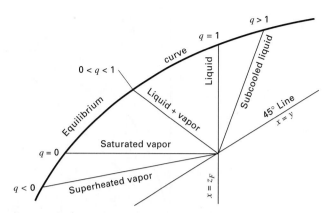

Figure 7.8 Effect of thermal condition of feed on slope of q-line.

then from the bottom up, as described above, until a point of merger is found for the feed stage. Alternatively, the stages can be stepped off from the bottom all the way to the top, or vice versa. Hardly ever will an integer number of stages result, but rather a fractional stage will appear near the middle, at the top, or at the bottom. Usually the staircase is stepped off from the top and continued all the way to the bottom, starting from the point $(y = x_D, x = x_D)$ on the 45° line, as shown in Figure 7.9 for the case of a partially vaporized feed. In that figure, point P is the intersection of the q-line with the two operating lines. The transfer point for stepping off stages between the rectifying-section operating line and the equilibrium curve to stepping off stages between the stripping-section operating line and the equilibrium curve occurs at the feed stage. In Figure 7.9a, the feed stage is stage 3 from the top and a fortuitous total of exactly five stages is required, where the last stage is the partial reboiler. In Figure 7.9b the feed stage is stage 5 and a total of about 6.4 stages is required. In Figure 7.9c, the feed stage is stage 2 and a total of about 5.9 stages is required. In Figure 7.9b, the stepping off of stages in the rectifying section can be continued indefinitely, finally approaching, but never reaching, point K. In Figure 7.9c, if the stepping off of stages had started from the partial reboiler at the point $(y = x_B, x = x_B)$ and proceeded upward, the staircase in the stripping section could have been continued indefinitely, finally approaching, but never reaching, point R. In Figure 7.9, it is seen that the smallest number of total stages occurs when the transfer is made at the first opportunity after a horizontal line of the staircase passes over point P, as in Figure 7.9a. This feed-stage location is optimal.

Limiting Conditions

For a given specification (Table 7.2), a reflux ratio can be selected anywhere from the minimum, $R_{\min}$, to an infinite value (total reflux) where all of the overhead vapor is condensed and returned to the top stage (thus, no distillate is withdrawn). As shown in Figure 7.10b, the minimum reflux corresponds to the need for an infinite number of stages, while in Figure 7.10a the infinite reflux ratio corresponds to

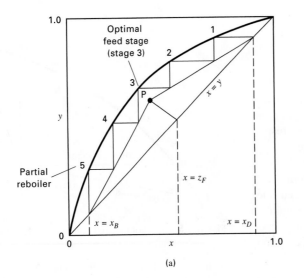

(a)

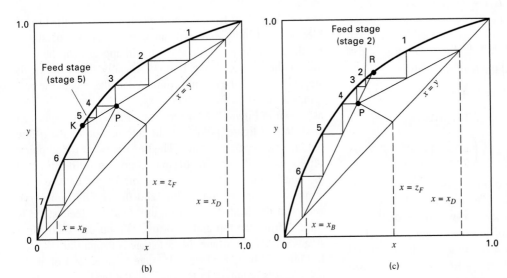

(b)   (c)

Figure 7.9 Optimal and nonoptimal locations of feed stage: (a) optimal feed-stage location; (b) feed-stage location below optimal stage; (c) feed-stage location above optimal stage.

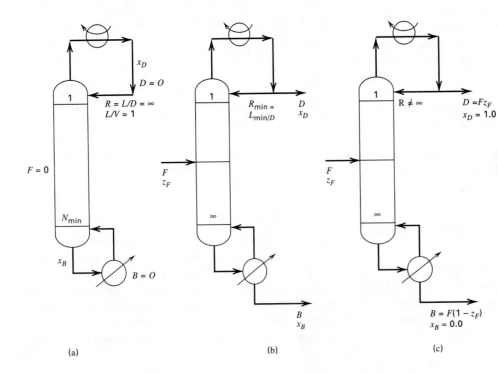

(a)   (b)   (c)

Figure 7.10 Limiting conditions for distillation: (a) total reflux, minimum stages; (b) minimum reflux, infinite stages; (c) perfect separation for nonazeotropic system.

the minimum number of equilibrium stages. The McCabe–Thiele graphical method can quickly determine the two limits, N_{min} and R_{min}. Then, for a practical operation, $N_{min} < N < \infty$ and $R_{min} < R < \infty$.

Minimum Number of Equilibrium Stages

As the reflux ratio is increased, the slope of the rectifying-section operating line, given by (7-7), increases from $L/V < 1$ to a limiting value of $L/V = 1$. Correspondingly, the boilup ratio increases and the slope of the stripping section operating line, given by (7-12), decreases from $\bar{L}/\bar{V} > 1$ to a limiting value of $\bar{L}/\bar{V} = 1$. Thus, at this limiting condition, both the rectifying and stripping operating lines coincide with the 45° line and neither the feed composition, z_F, nor the q-line influences the staircase construction. This is total reflux because when $L = V$, $D = B = 0$, and the total condensed overhead is returned to the column as reflux. Furthermore, all liquid leaving the bottom stage is vaporized and returned as boilup to the column. If both distillate and bottoms flow rates are zero, the feed to the column is also zero, which is consistent with the lack of influence of the feed condition. It is possible to operate a column at total reflux, and such an operation is convenient for the experimental measurement of tray efficiency because a steady-state operating condition is readily achieved.

A simple example of the McCabe–Thiele construction for this limiting condition is shown in Figure 7.11 for two equilibrium stages. Because the operating lines are located as far away as possible from the equilibrium curve, a minimum number of stages is required.

Minimum Reflux Ratio

As the reflux ratio decreases from the limiting case of infinity (i.e., total reflux), the intersection of the two operating lines and the q-line moves away from the 45° line toward the equilibrium curve. The number of equilibrium stages required increases because the operating lines move closer and closer to the equilibrium curve, thus requiring more and

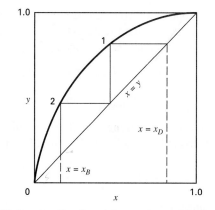

Figure 7.11 Construction for minimum stages at total reflux.

more stairs to move from the top of the column to the bottom. Finally a limiting condition is reached when the point of intersection is on the equilibrium curve, as shown in Figure 7.12. For binary mixtures that are not highly nonideal, the typical case is shown in Figure 7.12a, where the intersection, P, is at the feed stage. To reach that stage from either the rectifying section or the stripping section, an infinite number of stages is required. The point P is called a *pinch point* because the two operating lines each pinch the equilibrium curve.

For a highly nonideal binary system, the pinch point may occur at a stage above or below the feed stage. The former case is illustrated in Figure 7.12b, where the operating line for the rectifying section intersects the equilibrium curve before the feed stage is reached. The slope of this operating line cannot be reduced further because it would then cross over the equilibrium curve and thereby violate the second law of thermodynamics because of a reversal in the direction of mass transfer. This would require spontaneous mass transfer from a region of low concentration to a region of high concentration. This is similar to a second-law violation by a temperature crossover in a heat exchanger. Now, the pinch point occurs entirely in the rectifying section, where an infinite number of stages exists; the stripping section contains a finite number of stages.

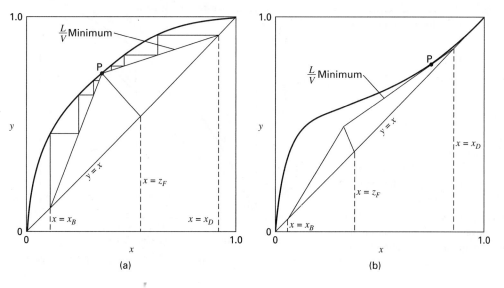

(a)   (b)

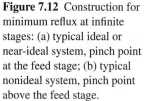

Figure 7.12 Construction for minimum reflux at infinite stages: (a) typical ideal or near-ideal system, pinch point at the feed stage; (b) typical nonideal system, pinch point above the feed stage.

From the slope of the limiting operating line for the rectifying section, the minimum reflux ratio can be determined. From (7-7), the minimum feasible slope is

$$(L/V)_{min} = R_{min}/(R_{min} + 1)$$

or

$$R_{min} = (L/V)_{min}/[1 - (L/V)_{min}] \qquad (7\text{-}27)$$

Alternatively, the limiting condition of infinite stages corresponds to a minimum boilup ratio for $(\bar{L}/\bar{V})_{max}$. From (7-12),

$$(V_B)_{min} = 1/[(\bar{L}/\bar{V})_{max} - 1] \qquad (7\text{-}28)$$

Perfect Separation

A third limiting condition of interest involves the degree of separation. As a perfect split $(x_D = 1, x_B = 0)$ is approached, for a reflux ratio at or greater than the minimum value, the number of stages required near the top and near the bottom of the column increases rapidly and without limit until pinches are encountered at $x_D = 1$ and $x_B = 0$. Thus, a perfect separation of a binary mixture that does not form an azeotrope requires an infinite number of stages in both sections of the column. However, this is not the case for the reflux ratio. In Figure 7.12a, as x_D is moved from, say, 0.90 toward 1.0, the slope of the operating line at first increases, but in the range of x_D from 0.99 to 1.0 the slope changes only slightly. Furthermore, the value of the slope, and therefore the value of R, is finite for a perfect separation. For example, if the feed is a saturated liquid, application of (7-4) and (7-7) gives the following equation for the minimum reflux of a perfect binary separation:

$$R_{min} = \frac{1}{z_F(\alpha - 1)} \qquad (7\text{-}29)$$

where the relative volatility, α, is evaluated at the feed condition.

EXAMPLE 7.1

A trayed tower is to be designed to continuously distill 450 lbmol/h (204 kmol/h) of a binary mixture of 60 mol% benzene and 40 mol% toluene. A liquid distillate and a liquid bottoms product of 95 mol% and 5 mol% benzene, respectively, are to be produced. The feed is preheated so that it enters the column with a molar percent vaporization equal to the distillate-to-feed ratio. Use the McCabe–Thiele method to compute the following, assuming a uniform pressure of 1 atm (101.3 kPa) throughout the column: (a) Minimum number of theoretical stages, N_{min}; (b) Minimum reflux ratio, R_{min}; and (c) Number of equilibrium stages N, for a reflux-to-minimum reflux ratio, R/R_{min}, of 1.3 and the optimal location of the feed stage.

SOLUTION

Calculate D and B. An overall material balance on benzene gives

$$0.60(450) = 0.95D + 0.05B \qquad (1)$$

A total balance gives $450 = D + B \qquad (2)$

Combining (1) and (2) to eliminate B, followed by solving the resulting equation for D and (2) for B gives $D = 275$ lbmol/h, $B = 175$ lbmol/h, and $D/F = 0.611$

Calculate the slope of the q-line:
$V_F/F = D/F$ for this example $= 0.611$ and q for a partially vaporized feed is

$$\frac{L_F}{F} = \frac{(F - V_F)}{F} = 1 - \frac{V_F}{F} = 0.389$$

From (7-26),

the slope of the q-line is $\dfrac{q}{q-1} = \dfrac{0.389}{0.389 - 1} = -0.637$

(a) In Figure 7.13, where y and x refer to benzene, the more-volatile component, with $x_D = 0.95$ and $x_B = 0.05$, the number of minimum equilibrium stages is stepped off from the top between the equilibrium curve and the 45° line, giving $N_{min} = 6.7$.

(b) In Figure 7.14, a q-line is drawn that has a slope of -0.637 and passes through the feed composition ($z_F = 0.60$) on the 45° line. For the minimum-reflux condition, an operating line for the rectifying section passes through the point $x = x_D = 0.95$ on the 45° line and through the point of intersection of the q-line and the equilibrium curve ($y = 0.684$, $x = 0.465$). The slope of this operating line is 0.55, which from (7-9) equals $R/(R + 1)$. Therefore, $R_{min} = 1.22$.

(c) The operating reflux ratio is $1.3R_{min} = 1.3(1.22) = 1.59$

From (7-9), the slope of the operating line for the rectifying section is

$$\frac{R}{R+1} = \frac{1.59}{1.59 + 1} = 0.614$$

The construction for the resulting two operating lines, together with the q-line, is shown in Figure 7.15, where the operating line for the stripping section is drawn to pass through the point $x = x_B = 0.05$ on the 45° line and the point of intersection of the q-line and the operating line for the stripping section. The number of equilibrium stages is stepped off between, first, the rectifying-section operating line and the equilibrium curve and then the stripping-section operating line and the equilibrium curve, starting from point A

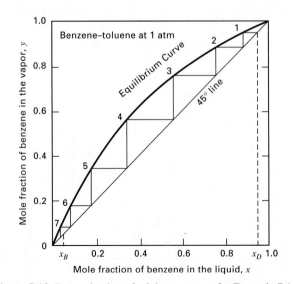

Figure 7.13 Determination of minimum stages for Example 7.1.

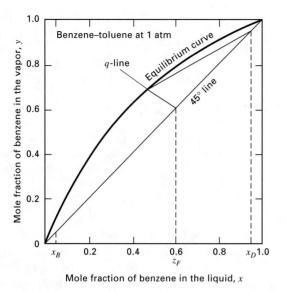

Figure 7.14 Determination of minimum reflux for Example 7.1.

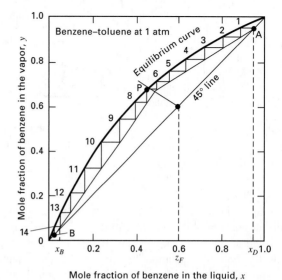

Figure 7.15 Determination of number of equilibrium stages and feed-stage location for Example 7.1.

(at $x = x_D = 0.95$) and finishing at point B (to the left of $x = x_B = 0.05$). For the optimal feed-stage location, the transfer from the rectifying-section operating line to the stripping-section operating line takes place at point P. The result is $N = 13.2$ equilibrium

stages, with stage 7 from the top being the feed stage. Thus, for this example, $N/N_{min} = 13.2/6.7 = 1.97$. The bottom stage is the partial reboiler, leaving 12.2 equilibrium stages contained in the column. If the plate efficiency were 0.8, 16 trays would be needed.

Column Operating Pressure and Condenser Type

For preliminary design, column operating pressure and condenser type are established by the procedure shown in Figure 7.16, which is formulated to achieve, if possible, a reflux-drum pressure, P_D, between 0 and 415 psia (2.86 MPa) at a minimum temperature of 120°F (49°C) (corresponding to the use of water as the coolant in the condenser). The pressure and temperature limits are representative only and depend on economic factors. Columns can operate at pressures higher than 415 psia if the critical or convergence

pressure of the mixture is not approached. A condenser pressure drop of 0 to 2 psi (0 to 14 kPa) and an overall, column pressure drop of 5 psi (35 kPa) may be assumed. However, when column tray requirements are known, more refined computations should result in approximately 0.1 psi/tray (0.7 kPa/tray) pressure drop for atmospheric and superatmospheric pressure operation and 0.05 psi/tray (0.35 kPa/tray) pressure drop for vacuum-column operation. Column bottom temperature must not result in bottoms decomposition or correspond to a near-critical condition. Therefore, after

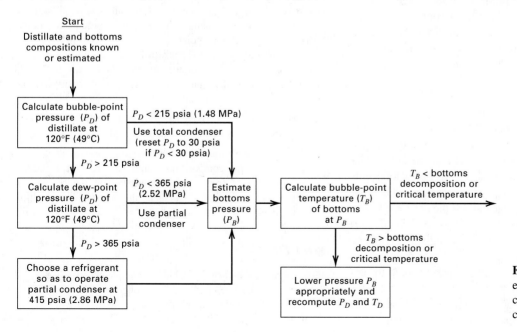

Figure 7.16 Algorithm for establishing distillation-column pressure and condenser type.

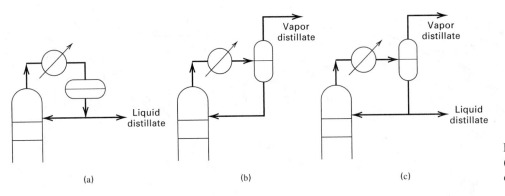

Figure 7.17 Condenser types: (a) total condenser; (b) partial condenser; (c) mixed condenser.

the bottoms pressure is estimated from the pressure in the reflux drum, a bubble-point temperature of the bottoms is computed at the bottoms pressure. If that temperature exceeds the bottoms decomposition or critical temperature, then the bottoms pressure is recomputed at or below the bubble-point decomposition or critical temperature. The pressure in the reflux drum will then be lower and must be recomputed, together with the distillate temperature, from the assumed column and condenser pressure drops. This will result often in vacuum operation. If the recomputed distillate temperature is less than 120°F (49°C), a refrigerant, rather than cooling water, is used for the condenser.

A total condenser is recommended for reflux drum pressures to 215 psia (1.48 MPa). A partial condenser is appropriate from 215 psia to 365 psia (2.52 MPa). However, a partial condenser can be used below 215 psia when a vapor distillate is desired. A mixed condenser can provide both vapor and liquid distillates. The three types of condenser configurations are shown in Figure 7.17. A refrigerant is often used as condenser coolant if pressure tends to exceed 365 psia.

When a partial condenser is specified, the McCabe–Thiele staircase construction for the case of a total condenser must be modified, as will be illustrated in Example 7.2, to account for the fact that the first equilibrium stage, counted down from the top, is now the partial condenser. This is based on the assumption that the liquid reflux leaving the reflux drum is in equilibrium with the vapor distillate.

Subcooled Reflux

Although most distillation columns are designed so that the reflux is a saturated (bubble-point) liquid, such is not always the case for operating columns. If the condenser type is partial or mixed, the reflux is a saturated liquid unless heat losses cause its temperature to decrease. For a total condenser, however, the operating reflux is often a subcooled liquid at column pressure, particularly if the condenser is not tightly designed and the distillate bubble-point temperature is significantly higher than the inlet cooling-water temperature. If the condenser outlet pressure is lower than the top-tray pressure of the column, the reflux is subcooled for any of the three types of condensers.

When subcooled reflux enters the top tray, its temperature rises and causes vapor entering the tray to condense. The latent enthalpy of condensation of the vapor provides the sensible enthalpy to heat the subcooled reflux to the bubble point. In that event, the internal reflux ratio within the rectifying section of the column is higher than the external reflux ratio from the reflux drum. The McCabe–Thiele construction should be based on the internal reflux ratio, which can be estimated by the following equation derived from an approximate energy balance around the top tray:

$$R_{\text{internal}} = R\left(1 + \frac{C_{P_L}\Delta T_{\text{subcooling}}}{\Delta H^{\text{vap}}}\right) \qquad (7\text{-}30)$$

where C_{P_L} and ΔH^{vap} are per mole and $\Delta T_{\text{subcooling}}$ is the degrees of subcooling. The internal reflux ratio replaces R, the external reflux ratio, in (7-9). If a correction is not made for subcooled reflux, the calculated number of equilibrium stages is somewhat more than required.

EXAMPLE 7.2

One thousand kmol/h of a feed containing 30 mol% n-hexane and 70% n-octane is to be distilled in a column consisting of a partial reboiler, one equilibrium (theoretical) plate, and a partial condenser, all operating at 1 atm (101.3 kPa). Thus, hexane is the light key and octane is the heavy key. The feed, a bubble-point liquid, is fed to the reboiler, from which a liquid bottoms product is continuously withdrawn. Bubble-point reflux is returned from the partial condenser to the plate. The vapor distillate, in equilibrium with the reflux, contains 80 mol% hexane, and the reflux ratio, L/D, is 2. Assume that the partial reboiler, plate, and partial condenser each function as equilibrium stages.

(a) Using the McCabe–Thiele method, calculate the bottoms composition and kmol/h of distillate produced.

(b) If the relative volatility α is assumed constant at a value of 5 over the composition range (the relative volatility actually varies from approximately 4.3 at the reboiler to 6.0 at the condenser), calculate the bottoms composition analytically.

SOLUTION

First determine whether the problem is completely specified. From Table 5.4c, we have $N_D = C + 2N + 6$ degrees of freedom, where N includes the partial reboiler and the stages in the column, but not the partial condenser. With $N = 2$ and $C = 2$, $N_D = 12$. Specified in

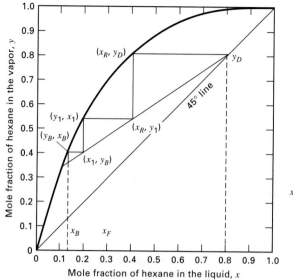

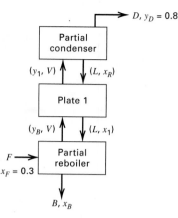

Figure 7.18 Solution to Example 7.2.

this problem are

Feed-stream variables	4
Plate and reboiler pressures	2
Condenser pressure	1
Q (=0) for plate	1
Number of stages	1
Feed-stage location	1
Reflux ratio, L/D	1
Distillate composition	1
Total	12

Thus, the problem is fully specified and can be solved.

(a) *Graphical solution.* A diagram of the separator is given in Figure 7.18 as is the McCabe–Thiele graphical solution, which is constructed in the following manner.

1. The point $y_D = 0.8$ at the partial condenser is located on the $x = y$ line.

2. Conditions in the condenser are fixed because x_R (reflux composition) is in equilibrium with y_D. Hence, the point (x_R, y_D) is located on the equilibrium curve.

3. Noting that $(L/V) = 1 - 1/[1 + (L/D)] = 2/3$, the operating line with slope $L/V = 2/3$ is drawn through the point $y_D = 0.8$ on the 45° line until it intersects the equilibrium curve. Because the feed is introduced into the partial reboiler, there is no stripping section.

4. Three theoretical stages (partial condenser, plate 1, and partial reboiler) are stepped off and the bottoms composition $x_B = 0.135$ is read.

The amount of distillate is determined from overall material balances. For hexane, $z_F F = y_D D + x_B B$. Therefore, $(0.3)(1,000) = (0.8)D + (0.135)B$. For the total flow, $B = 1,000 - D$. Solving these two equations simultaneously, $D = 248$ kmol/h.

(b) *Analytical solution.* For constant α, equilibrium liquid compositions for the light key, in terms of α and y are given by a rearrangement of (7-3):

$$x = \frac{y}{y + \alpha(1 - y)} \qquad (1)$$

where α is assumed constant at a value of 5.

The steps in the solution are as follows:

1. The liquid leaving the partial condenser at x_R is calculated from (1), for $y = y_D = 0.8$:

$$x_R = \frac{0.8}{0.8 + 5(1 - 0.8)} = 0.44$$

2. Then y_1 is determined by a material balance about the partial condenser:

$$V y_1 = D y_D + L x_R \quad \text{with} \quad D/V = 1/3 \quad \text{and} \quad L/V = 2/3$$
$$y_1 = (1/3)(0.8) + (2/3)(0.44) = 0.56$$

3. From (1), for plate 1, $x_1 = \dfrac{0.56}{0.56 + 5(1 - 0.56)} = 0.203$

4. By material balance around plate 1 and the partial condenser,

$$V y_B = D y_D + L x_1$$
and
$$y_B = (1/3)(0.8) + (2/3)(0.203) = 0.402$$

5. From (1), for the partial reboiler,

$$x_B = \frac{0.402}{0.402 + 5(1 - 0.402)} = 0.119.$$

By approximating the equilibrium curve with $\alpha = 5$, an answer of 0.119 is obtained rather than 0.135 for x_B obtained in part (a). Note that for a larger number of theoretical plates, part (b) can be readily computed with a spreadsheet program.

EXAMPLE 7.3

Consider Example 7.2. (a) Solve it graphically, assuming that the feed is introduced on plate 1, rather than into the reboiler. (b) Determine the minimum number of stages required to carry out the separation. (c) Determine the minimum reflux ratio.

SOLUTION

(a) The flowsheet and solution given in Figure 7.19 are obtained as follows.

1. The point x_R, y_D is located on the equilibrium line.

2. The operating line for the enriching section is drawn through the point $y = x = 0.8$, with a slope of $L/V = 2/3$.

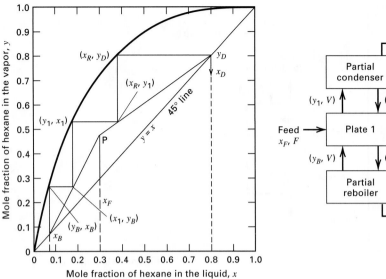

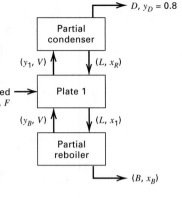

Figure 7.19 Solution to Example 7.3.

3. The intersection of the q-line, $x_F = 0.3$ (which, for a saturated liquid, is a vertical line), with the enriching-section operating line is located at point P. The stripping-section operating line must also pass through this point, but its slope and the point x_B are not known initially.

4. The slope of the stripping-section operating line is found by trial and error to give three equilibrium contacts in the column, with the middle stage involved in the switch from one operating line to the other. If the middle stage is the optimal feed-stage location, the result is $x_B = 0.07$, as shown in Figure 7.19. The amount of distillate is obtained from the combined total and hexane overall material balances to give $(0.3)(1,000) = (0.8D) + 0.07(1,000 - D)$. Solving, $D = 315$ kmol/h.

Comparing this result to that obtained in Example 7.2, we find that the bottoms purity and distillate yield are improved by introduction of the feed to plate 1, rather than to the reboiler. This improvement could have been anticipated if the q-line had been constructed in Figure 7.18. That is, the partial reboiler is not the optimal feed-stage location.

(b) The construction corresponding to total reflux ($L/V = 1$, no products, no feed, minimum equilibrium stages) is shown

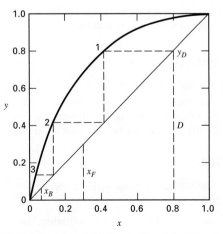

Figure 7.20 Solution for total reflux in Example 7.3.

in Figure 7.20. Slightly more than two stages are required for an x_B of 0.07, compared to the three stages previously required.

(c) To determine the minimum-reflux ratio, the vertical q-line in Figure 7.19 is extended from point P until the equilibrium curve is intersected, which is determined to be the point $(0.71, 0.3)$. The slope, $(L/V)_{min}$ of the operating line for the rectifying section, which connects this point to the point $(0.8, 0.8)$ on the $45°$ line is 0.18. Thus $(L/D)_{min} = (L/V_{min})/[1 - (L/V_{min})] = 0.22$. This is considerably less than the $L/D = 2$ specified.

Reboiler Type

Different types of reboilers are used to provide boilup vapor to the stripping section of a distillation column. For small laboratory and pilot-plant-size columns, the reboiler consists of a reservoir of liquid located just below the bottom plate to which heat is supplied from (1) a jacket or mantle that is heated by an electrical current or by condensing steam, or (2) tubes that pass through the liquid reservoir carrying condensing steam. Both of these types of reboilers have limited heat-transfer surface and are not suitable for industrial applications.

For plant-size distillation columns, the reboiler is usually an external heat exchanger, as shown in Figure 7.21, of either the kettle or vertical thermosyphon type. Both can provide the amount of heat-transfer surface required for large installations. In the former case, liquid leaving the sump (reservoir) at the bottom of the column enters the kettle, where it is partially vaporized by the transfer of heat from tubes carrying condensing steam or some other heating medium. The bottoms product liquid leaving the reboiler is assumed to be in equilibrium with the vapor returning to the bottom tray of the column. Thus the kettle reboiler is a partial reboiler equivalent to one equilibrium stage. The kettle

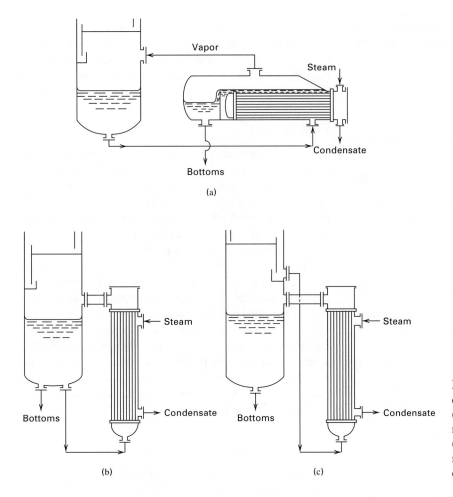

(a)

(b)   (c)

Figure 7.21 Reboilers for plant-size distillation columns: (a) kettle-type reboiler; (b) vertical thermosyphon-type reboiler, reboiler liquid withdrawn from bottom sump; (c) vertical thermosyphon-type reboiler, reboiler liquid withdrawn from bottom-tray downcomer.

reboiler is sometimes located in the bottom of the column to avoid piping.

The vertical thermosyphon reboiler may be of the type shown in Figure 7.21b or 7.21c. In the former, both the bottoms product and the reboiler feed are withdrawn from the column bottom sump. Circulation through the tubes of the reboiler occurs because of the difference in static heads of the supply liquid and the column of partially vaporized fluid flowing through the reboiler tubes. The partial vaporization provides enrichment of the exiting vapor in the more volatile component. However, the exiting liquid is then mixed with liquid leaving the bottom tray, which contains a higher percentage of the more volatile component. The result is that this type of reboiler arrangement provides only a fraction of an equilibrium stage and it is best to take no credit for it.

A more complex and less-common vertical thermosyphon reboiler is that of Figure 7.21c, where the reboiler liquid is withdrawn from the downcomer of the bottom tray. Partially vaporized liquid is returned to the column, where the bottoms product from the bottom sump is withdrawn. This type of reboiler does function as an equilibrium stage.

Kettle reboilers are common, but thermosyphon reboilers are favored when (1) the bottoms product contains thermally sensitive compounds, (2) bottoms pressure is high, (3) only a small ΔT is available for heat transfer, and (4) heavy

fouling occurs. Horizontal thermosyphon reboilers are sometimes used in place of the vertical types when only small static heads are needed for circulation, surface-area requirement is very large, and/or when frequent cleaning of the tubes is anticipated. A pump may be added for either thermosyphon type to improve circulation. Liquid residence time in the column bottom sump should be at least 1 minute and perhaps as much as 5 minutes or more.

Condenser and Reboiler Duties

Following the determination of the feed condition, reflux ratio, and number of theoretical stages by the McCabe–Thiele method, estimates of the heat duties of the condenser and reboiler are made. An energy balance for the entire column gives

$$Fh_F + Q_R = Dh_D + Bh_B + Q_C + Q_{\text{loss}} \quad (7\text{-}31)$$

Except for small and/or uninsulated distillation equipment, Q_{loss} is negligible and can be ignored. We can approximate the energy balance of (7-31) by applying the assumptions of the McCabe–Thiele method. An energy balance for a total condenser is

$$Q_C = D(R + 1)\,\Delta H^{\text{vap}} \quad (7\text{-}32)$$

where ΔH^{vap} = average molar heat of vaporization of the two components being separated. For a partial condenser,

$$Q_C = DR\,\Delta H^{\text{vap}} \tag{7-33}$$

For a partial reboiler,

$$Q_R = BV_B\Delta H^{\text{vap}} \tag{7-34}$$

When the feed is at the bubble point and a total condenser is used, (7-16) can be arranged to:

$$BV_B = L + D = D(R+1) \tag{7-35}$$

Comparing this to (7-34) and (7-32), note that $Q_R = Q_C$. When the feed is partially vaporized and a total condenser is used, the heat required by the reboiler is less than the condenser duty and is given by

$$Q_R = Q_C\left[1 - \frac{V_F}{D(R+1)}\right] \tag{7-36}$$

If saturated steam is the heating medium for the reboiler, the steam rate required is given by an energy balance:

$$m_s = \frac{M_s Q_R}{\Delta H_s^{\text{vap}}} \tag{7-37}$$

where

   m_s = mass flow rate of steam

   Q_R = reboiler duty (rate of heat transfer)

   M_s = molecular weight of steam

   ΔH_s^{vap} = molar enthalpy of vaporization of steam

The cooling water rate for the condenser is

$$m_{\text{cw}} = \frac{Q_C}{C_{P_{\text{H}_2\text{O}}}(T_{\text{out}} - T_{\text{in}})} \tag{7-38}$$

where

   m_{cw} = mass flow rate of cooling water

   Q_C = condenser duty (rate of heat transfer)

   $C_{P_{\text{H}_2\text{O}}}$ = specific heat of water

   $T_{\text{out}}, T_{\text{in}}$ = temperature of cooling water out of and into the condenser, respectively

Because the annual cost of reboiler steam can be an order of magnitude higher than the annual cost of cooling water, the feed to a distillation column is frequently preheated and partially vaporized to reduce Q_R, in comparison to Q_C, as indicated by (7-36).

Feed Preheat

The feed to a distillation column is usually a process feed, an effluent from a reactor, or a liquid product from another separator. The feed pressure must be greater than the pressure in the column at the feed-tray location. If so, any excess feed pressure is dropped across a valve, which may cause the feed to partially vaporize before entering the column; if not, additional pressure is added with a pump.

The temperature of the feed as it enters the column does not necessarily equal the temperature in the column at the feed-tray location. However, such equality will increase second-law efficiency. It is usually best to avoid a subcooled liquid or superheated vapor feed and supply a partially vaporized feed. This is achieved by preheating the feed in a heat exchanger with the bottoms product or some other process stream that possesses a suitably high temperature, to ensure a reasonable ΔT driving force for heat transfer, and a sufficient available enthalpy.

Optimal Reflux Ratio

An industrial distillation column must be operated between the two limiting conditions of minimum reflux and total reflux. As shown in Table 7.3, for a typical case adapted from Peters and Timmerhaus [6], as the reflux ratio is increased from the minimum value, the number of plates decreases, the column diameter increases, and the reboiler steam and condenser cooling-water requirements increase. When the annualized fixed investment costs for the column, condenser, reflux drum, reflux pump, and reboiler are added to the annual cost of steam and cooling water, an optimal reflux ratio is established, as shown, for the conditions of Table 7.3, in Figure 7.22. For this example the optimal R/R_{min} is 1.1.

Table 7.3   Effect of Reflux Ratio on Annualized Cost of a Distillation Operation

					Annualized Cost, $/yr			Total
R/R_{min}	Actual N	Diam., ft	Reboiler Duty, Btu/h	Condenser Duty, Btu/h	Equipment	Cooling Water	Steam	Annualized Cost, $/yr
1.00	Infinite	6.7	9,510,160	9,416,000	Infinite	17,340	132,900	Infinite
1.05	29	6.8	9,776,800	9,680,000	44,640	17,820	136,500	198,960
1.14	21	7.0	10,221,200	10,120,000	38,100	18,600	142,500	199,200
1.23	18	7.1	10,665,600	10,560,000	36,480	19,410	148,800	204,690
1.32	16	7.3	11,110,000	11,000,000	35,640	20,220	155,100	210,960
1.49	14	7.7	11,998,800	11,880,000	35,940	21,870	167,100	224,910
1.75	13	8.0	13,332,000	13,200,000	36,870	24,300	185,400	246,570

(Adapted from an example by Peters and Timmerhaus [6].)

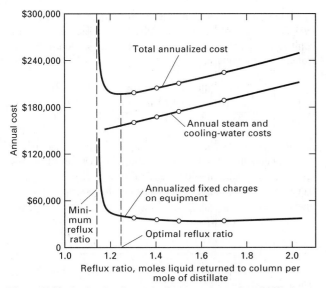

Figure 7.22 Optimal reflux ratio for a representative distillation operation.

[Adapted from M.S. Peters and K.D. Timmerhaus, *Plant Design and Economics for Chemical Engineers,* 4th ed., McGraw-Hill, New York (1991).]

The data in Table 7.3 show that although the condenser and reboiler duties are almost identical for a given reflux ratio, the annual cost of steam for the reboiler is almost eight times that of the cost of condenser cooling water. The total annual cost is dominated by the cost of steam except at the minimum-reflux condition. At the optimal reflux ratio, the cost of steam is 70% of the total annualized cost. Because the cost of steam is dominant, the optimal reflux ratio is sensitive to the steam cost. For example, at the extreme of zero cost for steam, the optimal R/R_{min} for this example is shifted from 1.1 to 1.32. This example assumes that the heat removed by cooling water in the condenser has no value.

The range of optimal ratio of reflux to minimum reflux often is from 1.05 to 1.50, with the lower value applying to a difficult separation (e.g., $\alpha = 1.2$) and the higher value applying to an easy separation (e.g., $\alpha = 5$). However, as seen in Figure 7.22, the optimal reflux ratio is not sharply defined. Accordingly, to achieve greater operating flexibility, columns are often designed for reflux ratios greater than the optimum.

Large Number of Stages

The McCabe–Thiele graphical construction is difficult to apply when conditions of relative volatility and/or product purities are such that a large number of stages must be stepped off. In that event, one of the following techniques can be used to determine the stage requirements.

1. Separate plots of expanded scales and/or larger dimensions are used for stepping off stages at the ends of the y–x diagram. For example, the additional plots might cover just the regions (1) 0.95 to 1.0 and (2) 0 to 0.05.

2. As described by Horvath and Schubert [7] and shown in Figure 7.23, a plot based on logarithmic coordinates is used for the low (bottoms) end of the y–x diagram, while for the high (distillate) end, the log–log graph is turned upside down and rotated 90°. Unfortunately, as seen in Figure 7.23, the operating lines become curved, but they can be plotted from a few points computed from (7-9) and (7-14). The 45° line remains straight and the normally curved equilibrium curve becomes nearly straight at the two ends.

3. The stages at the two ends are computed algebraically in the manner of part (b) of Example 7.2. This is readily done with a spreadsheet computer program.

4. If the equilibrium data are given in analytical form, commercially available McCabe–Thiele computer programs can be used.

5. The stages are determined by combining the McCabe–Thiele graphical construction, for a suitable region in the middle, with the Kremser equations of Section 5.4 for the low and/or high ends, where absorption and stripping factors are almost constant. This technique, which is often preferred, is illustrated in the following example.

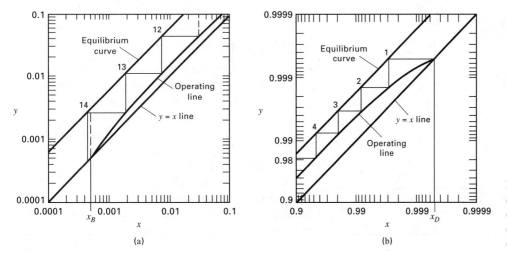

(a)

(b)

Figure 7.23 Use of log–log coordinates for McCabe–Thiele construction: (a) bottoms end of column; (b) distillate end of column.

EXAMPLE 7.4

Repeat part (c) of Example 7.1 for benzene distillate and bottoms purities of 99.9 and 0.1 mol%, respectively, using a reflux ratio of 1.88, which is about 30% higher than the minimum reflux of 1.44 for these new purities. At the top of the column, $\alpha = 2.52$; at the bottom, $\alpha = 2.26$.

SOLUTION

Figure 7.24 shows the McCabe–Thiele construction for the region of x from 0.028 to 0.956, where the stages have been stepped off in two directions, starting from the feed stage. In this middle region, seven stages are stepped off above the feed stage and eight below the feed stage, for a total of 16 stages, including the feed stage. The Kremser equations can now be applied to determine the remaining stages needed to achieve the desired high purities for the distillate and bottoms.

Additional stages for the rectifying section. With respect to Figure 5.8a, counting stages from the top down, from Figure 7.24:

From (7-3), for $(x_N)_{benzene} = 0.956$,

$$(y_{N+1})_{benzene} = 0.982 \quad \text{and} \quad (y_{N+1})_{toluene} = 0.018$$

Also, $(x_0)_{benzene} = (y_1)_{benzene} = 0.999$ and $(x_0)_{toluene} = (y_1)_{toluene} = 0.001$

Combining the Kremser equations (5.55), (5-34), (5-35), (5-48), and (5-50) and performing a number of algebraic manipulations:

$$N_R = \frac{\log\left[\frac{1}{A} + \left(1 - \frac{1}{A}\right)\left(\frac{y_{N+1} - x_0 K}{y_1 - x_0 K}\right)\right]}{\log A} \quad (7\text{-}39)$$

where N_R = additional equilibrium stages for the rectifying section. For that section, which is like an absorption section, it is best to apply (7-39) to toluene, the heavy key. Because $\alpha = 2.52$ at the top of the column, where $K_{benzene}$ is close to one, take $K_{toluene} = 1/2.52 = 0.397$. Since $R = 1.88$, $L/V = R/(R+1) = 0.653$.

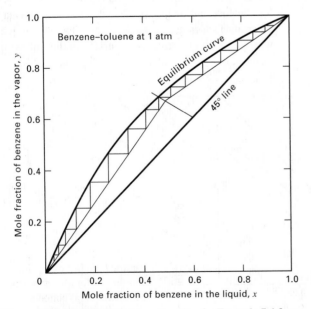

Figure 7.24 McCabe–Thiele construction for Example 7.4 from $x = 0.028$ to $x = 0.956$.

Therefore, the absorption factor for toluene is $A_{toluene} = L/(V K_{toluene}) = 0.653/0.397 = 1.64$ which is assumed to remain constant in the uppermost part of the rectifying section. Therefore, from (7-39) for toluene,

$$N_R = \frac{\log\left[\frac{1}{1.64} + \left(1 - \frac{1}{1.64}\right)\left(\frac{0.018 - 0.001(0.397)}{0.001 - 0.001(0.397)}\right)\right]}{\log 1.64} = 5.0$$

Additional stages for the stripping section. With respect to Figure 5.8b, counting stages from the bottom up, we have from Figure 7.24: $(x_{N+1})_{benzene} = 0.048$. Also, $(x_1)_{benzene} = (x_B)_{benzene} = 0.001$. Combining the Kremser equations for a stripping section gives

$$N_S = \frac{\log\left[\bar{A} + (1 - \bar{A})\left(\frac{x_{N+1} - x_1/K}{x_1 - x_1/K}\right)\right]}{\log(1/\bar{A})} \quad (7\text{-}40)$$

where

N_S = additional equilibrium stages for the stripping section

$\bar{A}$ = absorption factor in the stripping section = $\bar{L}/K\bar{V}$

Because benzene is being stripped in the stripping section, it is best to apply (7-40) to the benzene. At the bottom of the column, where $K_{toluene}$ is approximately 1.0, $\alpha = 2.26$, and therefore $K_{benzene} = 2.26$. By material balance, with flows in lbmol/h, $D = 270.1$. For $R = 1.88$, $L = 507.8$, and $V = 270.1 + 507.8 = 777.9$. From Example 7.1, $V_F = D = 270.1$ and $L_F = 450 - 270.1 = 179.9$. Therefore, $\bar{L} = L + L_F = 507.8 + 179.9 = 687.7$ lbmol/h and $\bar{V} = V - V_F = 777.9 - 270.1 = 507.8$ lbmol/h.

$$\bar{L}/\bar{V} = 687.7/507.8 = 1.354;$$
$$\bar{A}_{benzene} = \bar{L}/K\bar{V} = 1.354/2.26 = 0.599$$

Substitution into (7-40) gives

$$N_S = \frac{\log\left[0.599 + (1 - 0.599)\left(\frac{0.028 - 0.001/2.26}{0.001 - 0.001/2.26}\right)\right]}{\log(1/0.599)} = 5.9$$

This value includes the partial reboiler. Accordingly, the total number of equilibrium stages starting from the bottom is: partial reboiler + 5.9 + 8 + feed stage + 7 + 5.0 = 26.9.

Use of Murphree Efficiency

The McCabe–Thiele method assumes that the two phases leaving each stage are in thermodynamic equilibrium. In industrial, countercurrent, multistage equipment, it is not always practical to provide the combination of residence time and intimacy of contact required to approach equilibrium closely. Hence, concentration changes for a given stage are usually less than predicted by equilibrium.

As discussed in Section 6.5, a stage efficiency frequently used to describe individual tray performance for individual components is the Murphree plate efficiency. This efficiency can be defined on the basis of either phase and, for a given component, is equal to the change in actual composition in

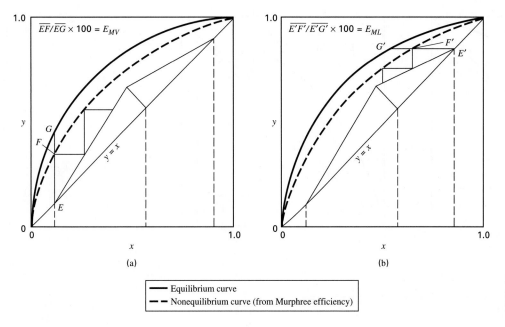

(a)

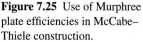

(b)

| Equilibrium curve |
| Nonequilibrium curve (from Murphree efficiency) |

Figure 7.25 Use of Murphree plate efficiencies in McCabe–Thiele construction.

the phase, divided by the change predicted by equilibrium. This definition applied to the vapor phase can be expressed in a manner similar to (6-28):

$$E_{MV} = \frac{y_n - y_{n+1}}{y_n^* - y_{n+1}} \qquad (7\text{-}41)$$

Where E_{MV} is the Murphree vapor efficiency for stage n, where $n + 1$ is the stage below and y_n^* is the composition in the hypothetical vapor phase in equilibrium with the liquid composition leaving stage n. Values of E_{MV} can be less than or somewhat more than 100%. The component subscript in (7-41) is dropped because values of E_{MV} are equal for the two components of a binary mixture.

In stepping off stages, the Murphree vapor efficiency, if known, can be used to dictate the percentage of the distance taken from the operating line to the equilibrium line; only E_{MV} of the total vertical path is traveled. This is shown in Figure 7.25a for the case of Murphree efficiencies based on the vapor phase. Figure 7.25b shows the case when the Murphree tray efficiency is based on the liquid. In effect, the dashed curve for actual exit-phase composition replaces the thermodynamic equilibrium curve for a particular set of operating lines. In Figure 7.25a, $E_{MV} = \overline{EF}/\overline{EG} = 0.7$ for the bottom stage.

Multiple Feeds, Side Streams, and Open Steam

The McCabe–Thiele method for a single feed and two products is readily extended to the case of multiple feeds and/or side streams by adding one additional operating line for each additional feed or side stream. A multiple-feed arrangement is shown in Figure 7.26. In the absence of side stream L_S, this arrangement has no effect on the material balance associated with the rectifying section of the column above the upper-feed point, F_1. The section of column between the upper-feed point

and the lower-feed point F_2 (in the absence of feed F) is represented by an operating line of slope L'/V', this line intersecting the rectifying-section operating line. A similar argument holds for the stripping section of the column. Hence it is possible to apply the McCabe–Thiele graphical construction shown in Figure 7.27a, where feed F_1 is a dew-point vapor, while feed F_2 is a bubble-point liquid. Feed F and side stream L_S of Figure 7.26 are not present. Thus, between the two feed points for this example, the molar vapor flow rate is $V' = V - F_1$ and $\bar{L} = L' + F_2 = L + F_2$. For given x_B, z_{F_2},

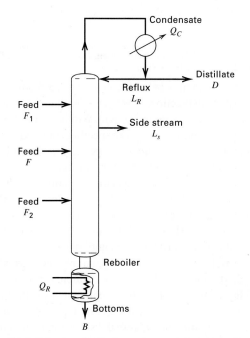

Figure 7.26 Complex distillation column with multiple feeds and side stream.

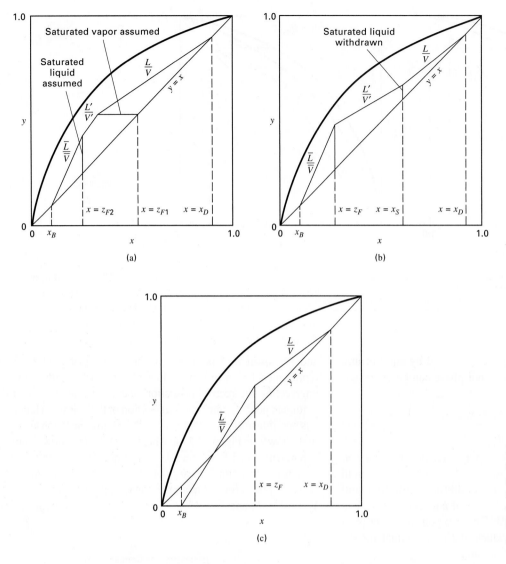

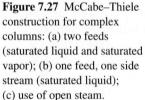

Figure 7.27 McCabe–Thiele construction for complex columns: (a) two feeds (saturated liquid and saturated vapor); (b) one feed, one side stream (saturated liquid); (c) use of open steam.

z_{F_1}, x_D, and L/D, the three operating lines in Figure 7.27a are readily constructed.

A side stream may be withdrawn from the rectifying section, the stripping section, or between multiple feed points, as a saturated vapor or saturated liquid. Within material-balance constraints, L_S and x_S can both be specified. In Figure 7.27b, a saturated-liquid side stream of composition x_S and molar flow rate L_S is withdrawn from the rectifying section above feed F. In the section of stages between the side stream-withdrawal stage and the feed stage, $L' = L - L_S$, while $V' = V$. The McCabe–Thiele constructions determine the location of the side stream stage. However, if it is not located directly above x_S, the reflux ratio must be varied until it does.

For certain types of distillation, an inert hot gas is introduced directly into the base of the column. Open steam, for example, can be used if one of the components in the mixture is water, or if water can form a second liquid phase, thereby reducing the boiling point, as in the steam distillation of fats, where heat is supplied by live, superheated steam and no reboiler is used. Most commonly, the feed contains water, which is removed as bottoms. In that application, Q_R of Figure 7.26 is replaced by a stream of composition $y = 0$ (pure steam) which, with $x = x_B$, becomes a point on the operating line, since the passing streams at this point actually exist at the end of the column. With open steam, the bottoms flow rate is increased by the flow rate of the open steam. The use of open steam rather than a reboiler for the operating condition $F_1 = F_2 = L_S = 0$ is represented graphically in Figure 7.27c.

EXAMPLE 7.5

A complex distillation column, equipped with a partial reboiler and total condenser, and operating at steady state with a saturated-liquid feed, has a liquid side stream draw-off in the enriching (rectifying) section. Making the usual simplifying assumptions of the McCabe–Thiele method: (a) Derive an equation for the two operating lines in the enriching section. (b) Find the point of intersection of these operating lines. (c) Find the intersection of the operating line between F and L_S with the diagonal. (d) Show the construction on a y–x diagram.

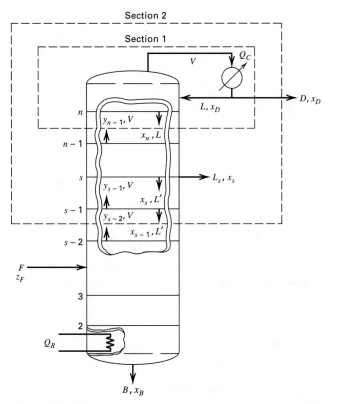

Figure 7.28 Distillation column with side stream for Example 7.5.

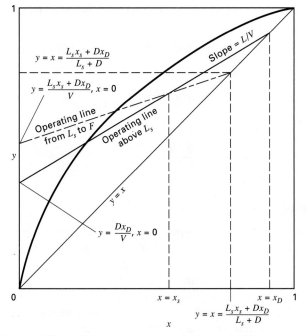

Figure 7.29 McCabe–Thiele diagram for Example 7.5.

SOLUTION

(a) By material balance over section 1 in Figure 7.28, $V_{n-1}y_{n-1} = L_n x_n + D x_D$. About section 2, $V_{s-2}y_{s-2} = L'_{s-1}x_{s-1} + L_s x_s + D x_D$. The two operating lines for conditions of

constant molar overflow become:

$$y = \frac{L}{V}x + \frac{D}{V}x_D \quad \text{and} \quad y = \frac{L'}{V}x + \frac{L_s x_s + D x_D}{V}$$

(b) Equating the two operating lines, the intersection occurs at $(L - L')x = L_s x_s$ and since $L - L' = L_s$, the point of intersection becomes $x = x_S$.

(c) The intersection of the lines

$$y = \frac{L'}{V}x + \frac{L_s x_s + D x_D}{V}$$

and $\quad y = x$ occurs at $x = \dfrac{L_s x_s + D x_D}{L_s + D}$

(d) The y–x diagram is shown in Figure 7.29.

7.3 ESTIMATION OF STAGE EFFICIENCY

Methods for estimating the stage efficiency for binary distillation are analogous to those for absorption and stripping, presented in Section 6.5. The efficiency is a complex function of tray design, fluid properties, and flow patterns. However, in hydrocarbon absorption and stripping, the liquid phase is often rich in heavy components so that liquid viscosity is high and mass-transfer rates are relatively low. This leads to low stage efficiencies, usually less than 50%. In contrast, for binary distillation, particularly of close-boiling mixtures, liquid viscosity is low, with the result that stage efficiencies, for well-designed trays and optimal operating conditions, are often higher than 70% and can be even higher than 100% for large-diameter columns where a cross-flow effect is present.

Performance Data

As discussed in *AIChE Equipment Testing Procedure* [8], performance data for an industrial distillation column are best obtained at conditions of total reflux (no feed or products) so as to avoid possible column-feed fluctuations, simplify location of the operating line, and avoid discrepancies between feed and feed-tray compositions. However, as shown by Williams, Stigger, and Nichols [9], efficiency measured at total reflux can differ markedly from that at design reflux ratio. Ideally, the column is operated in the range of 50% to 85% of flooding. If liquid samples are taken from the top and bottom of the column, the overall plate efficiency, E_o, can be determined from (6-21), where the number of theoretical stages required is determined by applying the McCabe–Thiele method at total reflux, as in Figure 7.11. If liquid samples are taken from the downcomers of intermediate trays, Murphree vapor efficiencies, E_{MV}, can be determined using (6-28). If liquid samples are withdrawn from different points on one tray, (6-30) can be applied to obtain point efficiencies, E_{OV}. Reliable values for these efficiencies require the availability of accurate vapor–liquid equilibrium data. For that reason, efficiency data for binary mixtures that form ideal solutions are preferred.

Table 7.4 Performance Data for the Distillation of a Mixture of Methylene Chloride and Ethylene Chloride

Company	Eastman Kodak
Location	Rochester, New York
Column diameter	5.5 ft (65.5 in. I.D.)
No. of trays	60
Tray spacing	18 in.
Type tray	10 rows of 3-in.-diameter bubble caps on 4-7/8-in. triangular centers.115 caps/tray
Bubbling area	20 ft^2
Length of liquid travel	49 in.
Outlet-weir height	2.25 in.
Downcomer clearance	1.5 in.
Liquid rate	24.5 gal/min-ft = 1,115.9 lb/min
Vapor F-factor	1.31 ft/s (lb/ft^3)$^{0.5}$
Percent of flooding	85
Pressure, top tray	33.8 psia
Pressure, bottom tray	42.0 psia

Liquid composition, mole % methylene chloride:

From tray 33	89.8
From tray 32	72.6
From tray 29	4.64

Source: J.A. Gerster, A.B. Hill, N.H. Hochgrof, and D.B. Robinson, *Tray Efficiencies in Distillation Columns, Final Report from the University of Delaware,* AIChE, New York (1958).

Table 7.4, from Gerster et al. [10], lists plant data, obtained from Eastman Kodak Company in Rochester, New York, for the distillation at total reflux of a methylene chloride (MC)–ethylene chloride (EC) mixture in a 5.5-ft-diameter column containing 60 bubble-cap trays on 18-in. tray spacing and operating at 85% of flooding at total reflux. MC is the light key.

EXAMPLE 7.6

Using the performance data of Table 7.4, estimate: (a) the overall tray efficiency for the section of trays from 35 to 29 and (b) the Murphree vapor efficiency for tray 32. Assume the following values for relative volatility:

x_{MC}	$\alpha_{MC,EC}$	y_{MC} from (7-3)
0.00	3.55	0.00
0.10	3.61	0.286
0.20	3.70	0.481
0.30	3.76	0.617
0.40	3.83	0.719
0.50	3.91	0.796
0.60	4.00	0.857
0.70	4.03	0.904
0.80	4.09	0.942
0.90	4.17	0.974
1.00	4.25	1.00

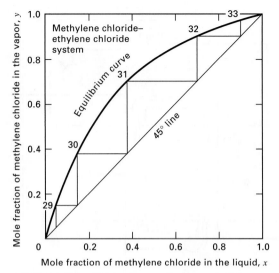

Figure 7.30 McCabe–Thiele diagram for Example 7.6.

SOLUTION

(a) The above x–α–y data are plotted in Figure 7.30. Four theoretical stages are stepped off from $x_{33} = 0.898$ to $x_{29} = 0.0464$ for total reflux. Since the actual number of stages is also 4, the overall stage efficiency from (6-21) is 100%.

(b) At total reflux conditions, passing vapor and liquid streams have the same composition. That is, the operating line is the 45° line. Using this together with the above performance data and the equilibrium curve in Figure 7.30, we obtain for methylene chloride, with trays counted from the bottom up:

$$y_{32} = x_{33} = 0.898 \quad \text{and} \quad y_{31} = x_{32} = 0.726$$

From (6-28),

$$(E_{MV})_{32} = \frac{y_{32} - y_{31}}{y_{32}^* - y_{31}}$$

From Figure 7.30, for $x_{32} = 0.726$, $y_{32}^* = 0.917$,

$$(E_{MV})_{32} = \frac{0.898 - 0.726}{0.917 - 0.726} = 0.90 \quad \text{or} \quad 90\%$$

Empirical Correlations

Based on 41 sets of performance data for bubble-cap-tray and sieve-tray columns, distilling mainly hydrocarbon mixtures and a few water and miscible organic mixtures, Drickamer and Bradford [11] correlated the overall stage efficiency for the separation of the two key components in terms of the molar-average liquid viscosity of the tower feed at the average tower temperature. The data covered average temperatures from 157 to 420°F, pressures from 14.7 to 366 psia, feed liquid viscosities from 0.066 to 0.355 cP, and overall tray efficiencies from 41% to 88%. The empirical equation

$$E_o = 13.3 - 66.8 \log \mu \tag{7-42}$$

where E_o is in percent and μ is in centipoise, fits the data with average and maximum percent deviations of 5.0% and 13.0%, respectively. A plot of the Drickamer and Bradford

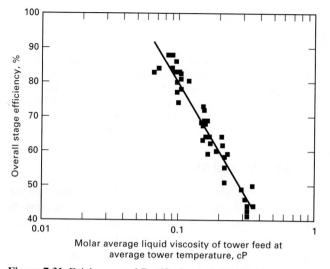

Figure 7.31 Drickamer and Bradford correlation for plate efficiency of distillation columns.

correlation, compared to performance data for distillation, is given in Figure 7.31. Equation (7-42) is restricted to the range of the data and is intended mainly for hydrocarbon mixtures.

Mass-transfer theory, discussed in Section 6.5, indicates that, when the relative volatility covers a wide range, the relative importance of liquid-phase and gas-phase mass-transfer resistances can shift. Thus, as might be expected, O'Connell [12] found that the Drickamer–Bradford correlation correlates data inadequately for fractionators operating on key components with large relative volatilities. Separate correlations in terms of a viscosity–volatility product were developed for fractionators and for absorbers and strippers by O'Connell. However, as shown in Figure 7.32, Lockhart and Leggett [13] were able to obtain a single correlation by using the product of liquid viscosity and an appropriate volatility as the correlating variable. For fractionators, the relative volatility of the key components was used; for hydrocarbon absorbers, the volatility was taken as 10 times the K-value of a selected key component, which must be one

that is reasonably distributed between top and bottom products. The data used by O'Connell cover a range of relative volatility from 1.16 to 20.5. A comprehensive study of the effect on E_o of the ratio of liquid-to-vapor molar flow rates, L/V, for eight different binary systems in a 10-in.-diameter column with bubble-cap trays was reported by Williams, et al. [9]. The systems included water, hydrocarbons, and other organic compounds. While L/V did have an effect, it could not be correlated. For fractionation with L/V nearly equal to 1.0 (i.e., total reflux), their distillation data, which are included in Figure 7.32, are in reasonable agreement with the O'Connell correlation. For the distillation of hydrocarbons in a column having a diameter of 0.45 m, Zuiderweg, Verburg, and Gilissen [14] found differences in E_o among bubble-cap, sieve, and valve trays to be insignificant at 85% of flooding. Accordingly, Figure 7.32 is assumed to be applicable to all three tray types, but may be somewhat conservative for well-designed trays. For example, data of Fractionation Research Incorporated (FRI) for valve trays operating with the cyclohexane/n-hexane and isobutane/n-butane systems are also included in Figure 7.32 and show efficiencies 10% to 20% higher than the correlation.

For the distillation data plotted in Figure 7.32, which cover a viscosity–relative volatility range for distillation of from 0.1 to 10 cP, the O'Connell correlation fits the empirical equation

$$E_o = 50.3(\alpha\mu)^{-0.226} \tag{7-43}$$

where E_o is in percent and μ is in centipoise. The relative volatility is determined for the two key components at average column conditions.

Most of the data for developing the correlation of Figure 7.32 are for columns having a liquid flow path across the active tray area of from 2 to 3 ft. Gautreaux and O'Connell [15], using theory and experimental data, showed that higher efficiencies are achieved for longer flow paths. For short liquid flow paths, the liquid flowing across the tray is usually mixed completely. For longer flow paths, the equivalent of two or more completely mixed, successive liquid zones may

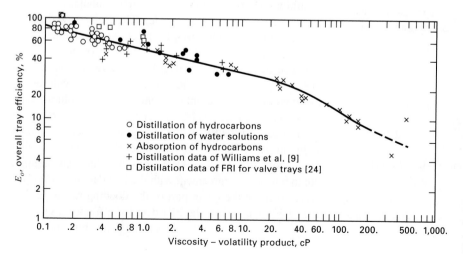

Figure 7.32 Lockhart and Leggett version of the O'Connell correlation for overall tray efficiency of fractionators, absorbers, and strippers.

[Adapted from F.J. Lockhart and C.W. Leggett, in *Advances in Petroleum Chemistry and Refining*, Vol. 1, Eds., K.A. Kobe and John J. McKetta, Jr., Interscience, New York, pp. 323–326 (1958).]

Table 7.5 Correction to Overall Tray Efficiency for Length of Liquid Flow Path ($0.1 \leq \mu\alpha \leq 1.0$)

Length of Liquid Flow Path, ft	Value to Be Added to E_o from Figure 7.32, %
3	0
4	10
5	15
6	20
8	23
10	25
15	27

Source: F.J. Lockhart and C.W. Leggett, in K.A. Kobe and J.J. McKetta, Jr., Eds., *Advances in Petroleum Chemistry and Refining,* Vol. 1, Interscience, New York, pp. 323–326 (1958).

be present. The result is a greater average driving force for mass transfer, and, thus, a higher efficiency—sometimes even greater than 100%. Provided that the viscosity–volatility product lies between 0.1 and 1.0, Lockhart and Leggett [13] recommend addition of the increments in Table 7.5 to the value of E_o from Figure 7.32 when the liquid flow path is greater than 3 ft. However, at large liquid rates, long liquid-path lengths are undesirable because they lead to excessive liquid gradients, causing maldistribution of vapor flow. The use of multipass trays, shown in Figure 6.16, to prevent excessive liquid gradients is discussed in Section 6.5.

EXAMPLE 7.7

For the benzene–toluene distillation of Figure 7.1, use the Drickamer–Bradford and O'Connell correlations to estimate the overall stage efficiency and number of actual plates required. Calculate the height of the tower assuming 24-in. tray spacing, with 4 ft above the top tray for removal of entrained liquid and 10 ft below the bottom tray for bottoms surge capacity. The separation requires 20 equilibrium stages plus a partial reboiler that acts as an equilibrium stage.

SOLUTION

For estimating overall stage efficiency, the liquid viscosity is determined at the feed-stage condition of 220°F, assuming a liquid composition of 50 mol% benzene.

$$\mu \text{ of benzene} = 0.10 \text{ cP}; \quad \mu \text{ of toluene} = 0.12 \text{ cP};$$
$$\text{Average } \mu = 0.11 \text{ cP}.$$

From Figure 7.3, take the average relative volatility as

$$\text{Average } \alpha = \frac{\alpha_{top} + \alpha_{bottom}}{2} = \frac{2.52 + 2.26}{2} = 2.39$$

From the Drickamer-Bradford correlation (7-42), $E_o = 13.3 - 66.8 \log(0.11) = 77\%$.

This is close to the value given in the description of this problem. Therefore, 26 actual trays are required and column height $= 4 + 2(26 - 1) + 10 = 64$ ft.

From the O'Connell correlation (7-43),

$$E_o = 50.3[(2.39)(0.11)]^{-0.226} = 68\%.$$

For a 5-ft-diameter column, the length of the liquid flow path is about 3 ft for a single-pass tray and even less for a two-pass tray. From Table 7.5, the efficiency correction is zero. Therefore, the actual number of trays required is $20/0.68 = 29.4$, or call it 30 trays. Column height $= 4 + 2(30 - 1) + 10 = 72$ ft.

Semi-Theoretical Models

In Section 6.5, semi-theoretical tray models based on the Murphree vapor efficiency and the Murphree vapor-point efficiency are applied to absorption and stripping. These same relationships are valid for distillation. However, because the equilibrium line is curved for distillation, λ must be taken as mV/L (not $KV/L = 1/A$), where m = local slope of the equilibrium curve = dy/dx. In Section 6.6, the method of Chan and Fair [16] is used for estimating the Murphree vapor-point efficiency from mass-transfer considerations. The Murphree vapor efficiency can then be estimated. The Chan and Fair correlation is specifically applicable to binary distillation because it was developed from experimental data that includes six different binary systems.

Scale-up from Laboratory Data

When binary mixtures form ideal or nearly ideal solutions, it is rarely necessary to obtain laboratory distillation data. Where nonideal solutions are formed and/or the possibility of azeotrope formation exists, use of a small laboratory Oldershaw column, of the type discussed in Section 6.5, should be used to verify the desired degree of separation and to obtain an estimate of the Murphree vapor-point efficiency. The ability to predict the efficiency of an industrial-size sieve-tray column from measurements with 1-in. glass and 2-in. metal diameter Oldershaw columns is shown in Figure 7.33, from the work of Fair, Null, and Bolles [17]. The measurements were made for the cyclohexane/n-heptane system at vacuum conditions (Figure 7.33a) and at near-atmospheric conditions (Figure 7.33b) and for the isobutane/n-butane system at 11.2 atm (Figure 7.33c). The Oldershaw data are correlated by the solid lines. Data for the 4-ft-diameter column with sieve trays of 8.3% and 13.7% open area were obtained by Sakata and Yanagi [18] and Yanagi and Sakata [19], respectively, of FRI. The Oldershaw column is assumed to measure point efficiency. The FRI column measured overall efficiency, but the relations of Section 6.5 were used to convert the FRI data to the point efficiencies shown in Figure 7.33. The data cover a percent of flooding ranging from about 10% to 95%. Data from the Oldershaw column are in reasonable agreement with the FRI data for 14% open area, except at the lower part of the flooding range. In Figures 7.33b and 7.33c, the FRI data for 8% open area show efficiencies as much as 10 percentage points higher.

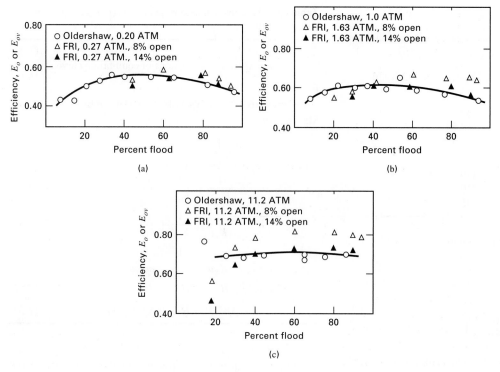

Figure 7.33 Comparison of Oldershaw column efficiency with point-efficiency in 4-ft-diameter FRI column with sieve trays: (a) cychlohexane/*n*-heptane system; (b) cyclohexane/*n*-heptane systems; (c) isobutane/*n*-butane system.

7.4 DIAMETER OF TRAYED TOWERS AND REFLUX DRUMS

In Section 6.6, methods for estimating tray diameter and pressure drop for absorbers and strippers are presented. These same methods apply to distillation columns. Calculations of column diameter are usually made for conditions at the top and bottom trays of the tower. If the diameters differ by 1 ft or less, the larger diameter is used for the entire column. If the diameters differ by more than 1 ft, it is often more economical to swage the column, using the different diameters computed for the sections above and below the feed.

Reflux Drums

Almost all commercial towers are provided with a cylindrical reflux drum, as shown in Figure 7.1. This drum is usually located near ground level, necessitating a pump to lift the reflux to the top of the column. If a partial condenser is used, the drum is often oriented vertically to facilitate the separation of vapor from liquid—in effect, acting as a flash drum. Vertical reflux and flash drums are sized by calculating a minimum drum diameter, D_T, to prevent liquid carryover by entrainment, using (6-44) in conjunction with the curve for 24-in. tray spacing in Figure 6.24 and a value of $F_{HA} = 1.0$ in (6-42). Also, $f = 0.85$ and $A_d = 0$ are used. To absorb process upsets and fluctuations, and otherwise facilitate control, vessel volume, V_V, is determined on the basis of liquid residence time, t, which should be at least 5 min, with the vessel half full of liquid [20]:

$$V_V = \frac{2LM_L t}{\rho_L} \qquad (7\text{-}44)$$

where L is the molar liquid flow rate leaving the vessel. Assuming a vertical, cylindrical vessel and neglecting the volume associated with the heads, the height H of the vessel is

$$H = \frac{4V_V}{\pi D_T^2} \qquad (7\text{-}45)$$

However, if $H > 4D_T$, it is generally preferable to increase D_T and decrease H to give $H = 4D$. Then

$$D_T = \frac{H}{4} = \left(\frac{V_V}{\pi}\right)^{1/3} \qquad (7\text{-}46)$$

A height above the liquid level of at least 4 ft is necessary for feed entry and disengagement of liquid droplets from the vapor. Within this space, it is common to install a wire mesh pad, which serves as a mist eliminator.

When vapor is totally condensed, a cylindrical, horizontal reflux drum is commonly employed to receive the condensate. Equations (7-44) and (7-46) permit estimates of the drum diameter, D_T, and length, H, by assuming a near-optimal value for H/D_T of 4, and the same liquid residence time suggested for a vertical drum. A horizontal drum is also used following a partial condenser when the liquid flow rate is appreciably greater than the vapor flow rate.

EXAMPLE 7.8

Equilibrium vapor and liquid streams leaving a flash drum, supplied by a partial condenser, are as follows:

Component	Vapor	Liquid
Pound-moles per hour:		
HCl	49.2	0.8
Benzene	118.5	81.4
Monochlorobenzene	71.5	178.5
Total	239.2	260.7
Pounds per hour	19,110	26,480
T, °F	270	270
P, psia	35	35
Density, lb/ft^3	0.371	57.08

Determine the dimensions of the flash drum.

SOLUTION

Using Figure 6.24,

$$F_{LV} = \frac{26,480}{19,110}\left(\frac{0.371}{57.08}\right)^{0.5} = 0.112$$

C_F at a 24-in. tray spacing is 0.34. Assume, in (6-24), that $C = C_F$. From (6-40),

$$U_f = 0.34\left(\frac{57.08 - 0.371}{0.371}\right)^{0.5} = 4.2 \text{ ft/s} = 15,120 \text{ ft/h}$$

From (6-44) with $A_d/A = 0$,

$$D_T = \left[\frac{(4)(19,110)}{(0.85)(15,120)(3.14)(1)(0.371)}\right]^{0.5} = 2.26 \text{ ft}$$

From (7-44), with $t = 5$ min $= 0.0833$ h,

$$V_V = \frac{(2)(26,480)(0.0833)}{(57.08)} = 77.3 \text{ ft}^3$$

From (7-43),

$$H = \frac{(4)(77.3)}{(3.14)(2.26)^2} = 19.3 \text{ ft}$$

However, $H/D_T = 19.3/2.26 = 8.54 > 4$. Therefore, redimension V_V for $H/D_T = 4$.

From (7-46),

$$D_T = \left(\frac{77.3}{3.14}\right)^{1/3} = 2.91 \text{ ft and } H = 4D_T = (4)(2.91) = 11.64 \text{ ft}$$

Height above the liquid level is $11.64/2 = 5.82$ ft, which is adequate.

Alternatively, with a height of twice the minimum disengagement height, $H = 8$ ft and $D_T = 3.5$ ft.

7.5 RATE-BASED METHOD FOR PACKED COLUMNS

With the availability of more efficient liquid distributors and economical and efficient packings, packed towers are finding increasing use in new distillation processes and for

Table 7.6   Modified Efficiency and Mass-Transfer Equations for Binary Distillation

$$\lambda = mV/L \qquad (7\text{-}47)$$
$$m = dy/dx = \text{local slope of equilibrium curve}$$

Efficiency:

Equations (6.31) to (6.37) hold if λ is defined by (7.47)

Mass transfer:

$$\frac{1}{N_{OG}} = \frac{1}{N_G} + \frac{\lambda}{N_L} \qquad (7\text{-}48)$$

$$\frac{1}{K_{OG}} = \frac{1}{k_G a} + \frac{mPM_L/\rho_L}{k_L a} \qquad (7\text{-}49)$$

$$\frac{1}{K_y a} = \frac{1}{k_y a} + \frac{m}{k_x a} \qquad (7\text{-}50)$$

$$\frac{1}{K_x a} = \frac{1}{k_x a} + \frac{1}{mk_y a} \qquad (7\text{-}51)$$

$$H_{OG} = H_G + \lambda H_L \qquad (7\text{-}52)$$

$$\text{HETP} = H_{OG} \ln\lambda/(\lambda - 1) \qquad (7\text{-}53)$$

retrofitting existing trayed towers. Methods in Section 6.8 for estimating packed-column efficiency, diameter, and pressure drop for absorbers are applicable to distillation. Methods for determining packed height are similar to those presented in Section 6.7 and are extended here for use in conjunction with the McCabe–Thiele diagram. Both the HETP and the HTU methods are discussed and illustrated. Unlike the case of absorption or stripping of dilute solutions, where values of HETP and HTU may be constant throughout the packed height, values of HETP and HTU can vary over the packed height of a distillation column, especially across the feed entry, where appreciable changes in vapor and liquid traffic occur. Also, because the equilibrium line for distillation is curved rather than straight, the mass-transfer equations of Section 6.8 must be modified by replacing $\lambda = KV/L = 1/A$ with

$$\lambda = \frac{mV}{L} = \frac{\text{slope of equilibrium curve}}{\text{slope of operating line}}$$

where $m = dy/dx$ varies with location in the tower. The modified efficiency and mass-transfer relationships are summarized in Table 7.6.

HETP Method

In the HETP method, the equilibrium stages are first stepped off on a McCabe–Thiele diagram. The case of equimolar counterdiffusion (EMD) applies to distillation. At each stage, the temperature, pressure, phase-flow ratio, and phase compositions are noted. A suitable packing material is selected and the column diameter is estimated for operation at, say, 70% of flooding by one of the methods of Section 6.8. Mass-transfer coefficients for the individual phases are estimated for the conditions at each stage from correlations also discussed in Section 6.8. From these coefficients, values of H_{OG} and HETP are estimated for each stage. The latter values are then summed to obtain the separate packed heights of the rectifying and stripping sections. If experimental values of HETP are available, they are used directly. In computing values of H_{OG} from H_G and H_L, or K_y from k_y and k_x, (6-92) and (6-80) must be modified because for binary distillation where the mole fraction of the light key may range from almost 0 at the bottom of the column to almost 1 at the top of the column, the ratio $(y_I - y^*)/(x_I - x)$ in (6-76) is no longer a constant equal to the K-value, but is dy/dx equal to the slope, m, of the equilibrium curve. The modified equations are included in Table 7.6.

EXAMPLE 7.9

For the benzene–toluene distillation of Example 7.1, determine packed heights of the rectifying and stripping sections based on a column diameter and packing material with the following values for the individual HTUs. Included are the L/V values for each section from Example 7.1.

	H_G, ft	H_L, ft	L/V
Rectifying section	1.16	0.48	0.62
Stripping section	0.90	0.53	1.40

SOLUTION

Slopes dy/dx of the equilibrium curve are obtained from Figure 7.15 and values of λ from (7-47). H_{OG} for each stage is determined from (7-52) in Table 7.6. HETP for each stage is determined from (7-53) in Table 7.6. The results are given in Table 7.7, where only 0.2 of stage 13 is needed and stage 14 is the partial reboiler.

Based on the results in Table 7.7, 10 ft of packing should be used in each of the two sections.

Table 7.7 Results for Example 7.9

Stage	m	$\lambda = \dfrac{mV}{L}$ or $m\dfrac{\bar{V}}{\bar{L}}$	H_{OG}, ft	HETP, ft
1	0.47	0.76	1.52	1.74
2	0.53	0.85	1.56	1.70
3	0.61	0.98	1.62	1.64
4	0.67	1.08	1.68	1.62
5	0.72	1.16	1.71	1.59
6	0.80	1.29	1.77	1.56
Total for rectifying section:				9.85
7	0.90	0.64	1.32	1.64
8	0.98	0.70	1.28	1.52
9	1.15	0.82	1.34	1.47
10	1.40	1.00	1.43	1.43
11	1.70	1.21	1.53	1.40
12	1.90	1.36	1.62	1.38
13	2.20	1.57	1.73	1.37(0.2) = 0.27
Total for stripping section:				9.11
Total packed height:				18.96

HTU Method

In the HTU method, equilibrium stages are not stepped off on a McCabe–Thiele diagram. Instead, the diagram provides data to perform an integration over the packed height of each section using either mass-transfer coefficients or transfer units.

Consider the schematic diagram of a packed distillation column and its accompanying McCabe–Thiele diagram in Figure 7.34. Assume that V, L, $\bar{V}$, and $\bar{L}$ are constant in their respective sections. For equimolar countercurrent diffusion (EMD), the rate of mass transfer of the light-key component from the liquid phase to the vapor phase is

$$n = k_x a(x - x_I) = k_y a(y_I - y) \qquad (7-54)$$

Rearranging:

$$-\frac{k_x a}{k_y a} = \frac{y_I - y}{x_I - x} \qquad (7-55)$$

Thus, as shown in Figure 7.34b, for any point (x, y) on the operating line, the corresponding interfacial point (x_I, y_I) on the equilibrium curve is obtained by drawing a line of slope $(-k_x a/k_y a)$ from the point (x, y) to the point where it intersects the equilibrium curve.

By material balance over an incremental section of packed height, assuming constant molar overflow,

$$V\, dy = k_y a(y_I - y)S\, dl \qquad (7-56)$$

$$L\, dx = k_x a(x - x_I)S\, dl \qquad (7-57)$$

where S is the cross-sectional area of the packed section. Integrating over the rectifying section,

$$(l_T)_R = \int_0^{(l_T)_R} dl = \int_{y_F}^{y_2} \frac{V\, dy}{k_y a S(y_I - y)} = \int_{x_F}^{x_D} \frac{L\, dx}{k_x a S(x - x_I)} \qquad (7-58)$$

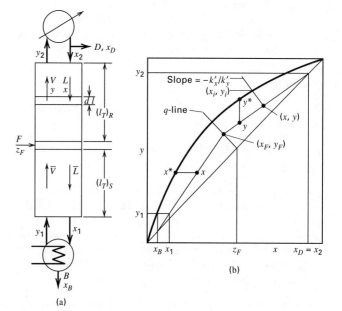

Figure 7.34 Distillation in a packed column.

or

$$(l_T)_R = \int_{y_F}^{y_2} \frac{H_G\,dy}{(y_I - y)} = \int_{x_F}^{x_D} \frac{H_L\,dx}{(x - x_I)} \quad (7\text{-}59)$$

Integrating over the stripping section,

$$(l_T)_S = \int_0^{(l_T)_S} dl = \int_{y_1}^{y_F} \frac{V\,dy}{k_y a S(y_I - y)} = \frac{L\,dx}{k_x a S(x - x_I)} \quad (7\text{-}60)$$

or

$$(l_T)_S = \int_{y_1}^{y_F} \frac{H_G\,dy}{(y_I - y)} = \int_{x_1}^{x_F} \frac{H_L\,dx}{(x - x_I)} \quad (7\text{-}61)$$

In general, values of k_y and k_x vary over the packed height, causing the slope $(-k_x a / k_y a)$ to vary. If $k_x a > k_y a$, the main resistance to mass transfer resides in the vapor and it is most accurate to evaluate the integrals in y. For $k_y a > k_x a$, the integrals in x are used. Usually, it is sufficient to evaluate k_y and k_x at just three points in each section, from which their variation with x can be determined. Then by computing and plotting their ratios from (7-55), a locus of points P can be found, from which values of $(y_I - y)$ for any value of y, or $(x - x_I)$ for any value of x can be read for use in integrals (7-58) to (7-61). These integrals can be evaluated either graphically or numerically to determine the packed heights.

EXAMPLE 7.10

Suppose that 250 kmol/h of a mixture of 40 mol% isopropyl ether in isopropanol is distilled in a packed column operating at 1 atm to obtain a distillate of 75 mol% isopropyl ether and a bottoms of 95 mol% isopropanol. At the feed entry, the mixture is a saturated liquid. A reflux ratio of 1.5 times minimum is used and the column is equipped with a total condenser and a partial reboiler. For the packing and column diameter, mass-transfer coefficients given below have been estimated from empirical correlations of the type discussed in Section 6.8. Compute the required packed heights of the rectifying and stripping sections.

SOLUTION

The distillate and bottoms rates are computed by an overall material balance on isopropyl ether:

$$0.40(250) = 0.75D + 0.05(250 - D)$$

Solving, $D = 125$ kmol/h and $B = 250 - 125 = 125$ kmol/h

The equilibrium curve for this mixture at 1 atm is shown in Figure 7.35, where it is noted that isopropyl ether is the light key and an azeotrope is formed at 78 mol% isopropyl ether. The distillate composition of 75 mol% is safely below the azeotropic composition. Also shown in Figure 7.35 are the q-line and the rectification-section operating line for the condition of minimum reflux. The slope of the latter line is measured to be $(L/V)_{min} = 0.39$. From (7-27),

$$R_{min} = 0.39/(1 - 0.39) = 0.64 \quad \text{and} \quad R = 1.5\,R_{min} = 0.96$$

$$L = RD = 0.96(125) = 120 \text{ kmol/h}$$

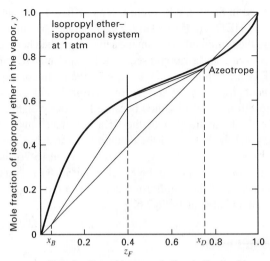

Figure 7.35 Operating lines and minimum-reflux line for Example 7.10.

and $V = L + D = 120 + 125 = 245$ kmol/h

$$\bar{L} = L + L_F = 120 + 250 = 370 \text{ kmol/h}$$

and $\bar{V} = V - V_F = 245 - 0 = 245$ kmol/h

Slope of rectification-section operating line $= L/V = 120/245$
$$= 0.49$$

This line and the stripping-section operating line are plotted in Figure 7.35. The partial reboiler, R, is stepped off in Figure 7.36 to give the following end points for determining the packed heights of the two sections, where the symbols refer to Figure 7.34a:

	Stripping Section	**Rectifying Section**
Top	$(x_F = 0.40, y_F = 0.577)$	$(x_2 = 0.75, y_2 = 0.75)$
Bottom	$(x_1 = 0.135, y_1 = 0.18)$	$(x_F = 0.40, y_F = 0.577)$

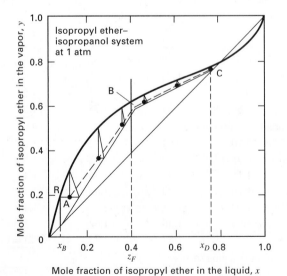

Figure 7.36 Mass-transfer driving forces for Example 7.10.

Mass-transfer coefficients at three values of x in each section are as follows:

x	k_ya kmol/m³-h-(mole fraction)	k_xa kmol/m³-h-(mole fraction)
Stripping section:		
0.15	305	1,680
0.25	300	1,760
0.35	335	1,960
Rectifying section:		
0.45	185	610
0.60	180	670
0.75	165	765

Slopes of the above mass-transfer coefficients are computed, for each point, x, on the operating line using (7-55), and drawn from the operating line to the equilibrium line, as shown in Figure 7.36. These lines are often referred to as tie lines because they tie the operating line to the equilibrium line. Using the tie lines as hypotenuses, right triangles are drawn, as shown in Figure 7.36. Dashed locus lines, AB and BC, are then drawn through the points at the 90° corners of the triangles. Using these locus lines, additional tie lines can quickly be added to the three plotted in each section, as needed, to give sufficient accuracy. From the tie lines, values of $(y_I - y)$ can be tabulated for values of y on the operating lines. Since the diameter of the column is not given, the packed volumes are determined from the following rearrangements of (7-58) and (7-60), where $V = Sl_T$:

$$V_R = \int_{y_F}^{y_2} \frac{V\,dy}{k_ya(y_I - y)} \tag{7-62}$$

$$V_S = \int_{y_1}^{y_F} \frac{V\,dy}{k_ya(y_I - y)} \tag{7-63}$$

Values of k_ya are interpolated as necessary. Results are given in the following table.

y	$(y_I - y)$	k_ya	$\dfrac{V(\text{or }\bar{V})}{k_ya(y_I - y)}$, m³
Stripping section:			
0.18	0.145	307	5.5
0.25	0.150	303	5.4
0.35	0.143	300	5.7
0.45	0.103	320	7.4
0.577	0.030	350	23.3
Rectifying section:			
0.577	0.030	187	43.7
0.60	0.033	185	40.1
0.65	0.027	182	49.9
0.70	0.017	175	82.3
0.75	0.010	165	148.5

By numerical integration, $V_S = 3.6$ m³ and $V_R = 12.3$ m³.

7.6 PONCHON–SAVARIT GRAPHICAL EQUILIBRIUM-STAGE METHOD FOR TRAYED TOWERS

The McCabe–Thiele method, in Section 7.2 for binary distillation, assumes that molar vapor and liquid flow rates are constant in each section of the column. This assumption

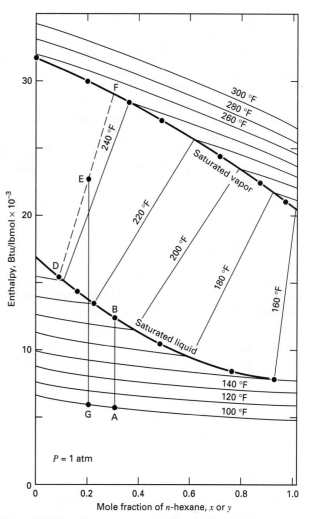

Figure 7.37 Enthalpy–concentration diagram for n-hexane/n-octane.

(constant molar overflow) eliminates the need to make an energy balance around each stage. For nonideal binary mixtures, such an assumption may not be valid and the McCabe–Thiele method may not be accurate. A graphical method that includes energy balances as well as material balances and phase equilibrium relations is the Ponchon–Savarit method [21, 22], which utilizes an enthalpy–composition diagram of the type shown in Figure 7.37 for the n-hexane/n-octane system at 1 atm. This diagram includes curves for the enthalpies of saturated vapor and liquid mixtures. Terminal points of tie lines connecting these two curves represent the equilibrium vapor and liquid compositions, together with vapor and liquid enthalpies, for the given temperature. Isotherms above the saturated vapor curve represent enthalpies of the superheated vapor, while isotherms below the saturated liquid curve represent the subcooled liquid. In Figure 7.37, a mixture of 30 mol% hexane and 70 mol% octane at 100°F (Point A) is a subcooled liquid. By heating it to Point B at 204°F, it becomes a liquid at its bubble point (Point B). When a mixture of 20 mol% hexane and 80 mol% octane at 100°F (Point G) is heated to 243°F (Point E), at equilibrium, it splits into a vapor phase at Point F and a liquid phase at Point D. The liquid phase contains 7 mol% hexane, while the vapor contains 29 mol% hexane.

The application of the enthalpy–concentration diagram to equilibrium-stage calculations may be illustrated by considering a single equilibrium stage, $n - 1$, where vapor from stage $n - 2$ below is mixed adiabatically with liquid from stage n above to give an overall mixture, denoted by mole-fraction z, and then brought to equilibrium. The process is represented schematically in two steps, mixing followed by equilibration, at the top of Figure 7.38. The energy-balance equations for stage $n - 1$ are

Mixing:

$$V_{n-2}H_{n-2} + L_n h_n = (V_{n-2} + L_n)h_z \qquad (7\text{-}64)$$

Equilibration:

$$(V_{n-2} + L_n)h_z = V_{n-1}H_{n-1} + L_{n-1}h_{n-1} \qquad (7\text{-}65)$$

where H and h are vapor and liquid molar enthalpies, respectively. The governing material-balance equations for the light component are

Mixing:

$$y_{n-2}V_{n-2} + x_n L_n = z(V_{n-2} + L_n) \qquad (7\text{-}66)$$

Equilibration:

$$z(V_{n-2} + L_n) = y_{n-1}V_{n-1} + x_{n-1}L_{n-1} \qquad (7\text{-}67)$$

Simultaneous solution of (7-64) and (7-66) gives

$$\frac{H_{n-2} - h_z}{y_{n-2} - z} = \frac{h_z - h_n}{z - x_n} \qquad (7\text{-}68)$$

which is the three-point form of a straight line plotted in Figure 7.38. Similarly, the simultaneous solution of (7-65) and (7-67) gives

$$\frac{H_{n-1} - h_z}{y_{n-1} - z} = \frac{h_z - h_{n-1}}{z - x_{n-1}} \qquad (7\text{-}69)$$

which is also the equation for a straight line. However, in this case y_{n-1} and x_{n-1} are in equilibrium and, therefore, the points (H_{n-1}, y_{n-1}) and (h_{n-1}, x_{n-1}) must lie on the

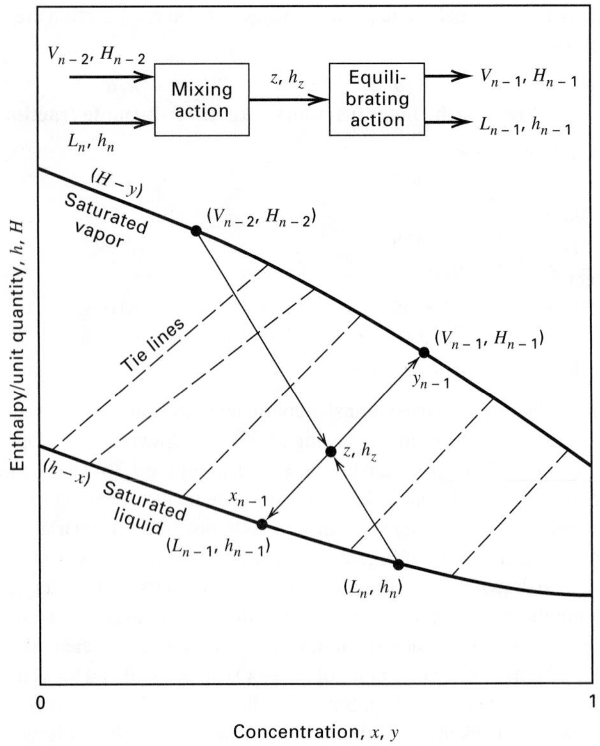

Figure 7.38 Two-phase mixing and equilibration on an enthalpy–concentration diagram.

opposite ends of the tie line that passes through the mixing point (h_z, z), as shown in Figure 7.38.

The Ponchon–Savarit method for binary distillation is an extension of the construction in Figure 7.38 to countercurrent cascades above and below the feed stage, with consideration of the condenser and reboiler. A detailed description of the method is not given here because the method has been largely superseded by the rigorous computer-aided calculation procedures, discussed in Chapter 10, which include energy balances and can be applied to multicomponent as well as binary mixtures. A detailed presentation of the Ponchon–Savarit method for binary distillation is given by Henley and Seader [23].

SUMMARY

1. A binary-liquid and/or binary-vapor mixture can be separated economically into two nearly pure products (distillate and bottoms) by distillation, provided that the value of the relative volatility of the two components is high enough, usually greater than 1.05.

2. Distillation is the most mature and widely used separation operation, with design procedures and operation practices well established.

3. The purities of the products from distillation depend on the number of equilibrium stages in the rectifying section above the feed entry and in the stripping section below the feed entry, and on the reflux ratio. Both the number of stages and the reflux ratio must be greater than the minimum values corresponding to total reflux and infinite stages, respectively. The optimal reflux-to-minimum-reflux ratio is usually in the range of 1.10 to 1.50.

4. Distillation is most commonly conducted in trayed towers equipped with sieve or valve trays, or in columns packed with random or structured packings. Many older towers are equipped with bubble-cap trays.

5. Most distillation towers are equipped with a condenser, cooled with cooling water, to provide reflux, and a reboiler, heated with steam, to provide boilup.

6. When the assumption of constant molar overflow is valid in each of the two sections of the distillation tower, the McCabe–Thiele graphical method is convenient for determining stage and reflux requirements. This method facilitates the visualization of many aspects of distillation and provides a procedure for locating the optimal feed-stage location.

7. Miscellaneous considerations involved in the design of a distillation tower include selection of operating pressure, type of condenser, degree of reflux subcooling, type of reboiler, and extent of feed preheat.

8. The McCabe–Thiele method can be extended to handle Murphree stage efficiency, multiple feeds, side streams, open steam, and use of interreboilers and intercondensers.

9. Rough estimates of overall stage efficiency, defined by (6-21), can be made with the Drickamer and Bradford, (7-42), or O'Connell, (7-43), correlations. More accurate and reliable procedures use data from a small Oldershaw column or the same semi-theoretical equations for mass transfer in Chapter 6 that are used for absorption and stripping.

10. Tray diameter, pressure drop, weeping, entrainment, and downcomer backup can all be estimated by the procedures in Chapter 6.

11. Reflux and flash drums are sized by a procedure based on avoidance of entrainment and provision for adequate liquid residence time.

12. Packed-column diameter and pressure drop are determined by the same procedures presented in Chapter 6 for absorption and stripping.

13. The height of a packed column may be determined by the HETP method, or preferably from the HTU method. Application of the latter method is similar to that of Chapter 6 for absorbers and strippers, but differs in the manner in which the curved equilibrium line must be handled, as given by (7-47).

14. The Ponchon–Savarit graphical method removes the assumption of constant molar overflow in the McCabe–Thiele method by employing energy balances with an enthalpy-concentration diagram. However, use of the Ponchon–Savarit method has largely been supplanted by rigorous computer-aided methods.

REFERENCES

1. FORBES, R.J., *Short History of the Art of Distillation*, E.J. Brill, Leiden (1948).

2. MIX, T.W., J.S. DWECK, M. WEINBERG, and R.C. ARMSTRONG, *Chem. Eng. Prog.*, **74** (4), 49–55 (1978).

3. KISTER, H.Z., *Distillation Design*, McGraw-Hill, New York (1992).

4. KISTER, H.Z., *Distillation Operation*, McGraw-Hill, New York (1990).

5. MCCABE, W.L., and E.W. THIELE, *Ind. Eng. Chem.*, **17**, 605–611 (1925).

6. PETERS, M.S., and K.D. TIMMERHAUS, *Plant Design and Economics for Chemical Engineers*, 4th ed., McGraw-Hill, New York (1991).

7. HORVATH, P.J., and R.F. SCHUBERT, *Chem. Eng.*, **65** (3), 129–132 (1958).

8. *AIChE Equipment Testing Procedure, Tray Distillation Columns*, 2nd ed., AIChE, New York (1987).

9. WILLIAMS, G.C., E.K. STIGGER, and J.H. NICHOLS, *Chem. Eng. Progr.*, **46** (1), 7–16 (1950).

10. GERSTER, J.A., A.B. HILL, N.H. HOCHGROF, and D.B. ROBINSON, *Tray Efficiencies in Distillation Columns, Final Report from the University of Delaware*, AIChE, New York (1958).

11. DRICKAMER, H.G., and J.R. BRADFORD, *Trans. AIChE*, **39**, 319–360 (1943).

12. O'CONNELL, H.E., *Trans. AIChE*, **42**, 741–755 (1946).

13. LOCKHART, F.J., and C.W. LEGGETT, in K.A. Kobe and John J. McKetta, Jr., Eds., *Advances in Petroleum Chemistry and Refining*, Vol. 1, Interscience, New York, pp. 323–326 (1958).

14. ZUIDERWEG, F.J., H. VERBURG, and F.A.H. GILISSEN, *Proc. International Symposium on Distillation*, Institution of Chem. Eng., London, 202–207 (1960).

15. GAUTREAUX, M.F., and H.E. O'CONNELL, *Chem. Eng. Prog.*, **51** (5) 232–237 (1955).

16. CHAN, H., and J.R. FAIR, *Ind. Eng. Chem. Process Des. Dev.*, **23**, 814–819 (1984).

17. FAIR, J.R., H.R. NULL, and W.L. BOLLES, *Ind. Eng. Chem. Process Des. Dev.*, **22**, 53–58 (1983).

18. SAKATA, M., and T. YANAGI, *I. Chem. E. Symp. Ser.*, **56**, 3.2/21 (1979).

19. YANAGI, T., and M. SAKATA, *Ind. Eng. Chem. Process Des. Devel.*, **21**, 712 (1982).

20. YOUNGER, A.H., *Chem. Eng.*, **62** (5), 201–202 (1955).

21. PONCHON, M., *Tech. Moderne*, **13**, 20, 55 (1921).

22. SAVARIT, R., *Arts et Metiers*, pp. 65, 142, 178, 241, 266, 307 (1922).

23. HENLEY, E.J., and J.D. SEADER, *Equilibrium-Stage Separation Operations in Chemical Engineering*, John Wiley and Sons, New York (1981).

24. *GLITSCH BALLAST TRAY*, Bulletin 159, Fritz W. Glitsch and Sons, Dallas (from FRI report of September 3, 1958).

EXERCISES

Unless otherwise stated, the usual simplifying assumptions of saturated-liquid reflux, optimal feed-stage location, no heat losses, steady state, and constant molar liquid and vapor flows apply to each of the following problems.

Section 7.1

7.1 List as many differences between absorption and distillation as you can. List as many differences between stripping and distillation as you can.

7.2 Prior to the 1980s, packed columns were rarely used for distillation unless column diameter was less than 2.5 ft. Explain why, in recent years, some existing trayed towers are being retrofitted with packing and some new large-diameter columns are being designed for packing rather than trays.

7.3 A mixture of methane and ethane is to be separated by distillation. Explain why water cannot be used as the coolant in the condenser. What would you choose as the coolant?

7.4 A mixture of ethylene and ethane is to be separated by distillation. Determine the maximum operating pressure of the column. What operating pressure would you suggest? Why?

7.5 Under what circumstances would it be advisable to conduct laboratory or pilot-plant tests of a proposed distillation separation?

7.6 Explain why an economic tradeoff exists between the number of trays and the reflux ratio.

Section 7.2

7.7 Following the development by Sorel in 1894 of a mathematical model for continuous, steady-state, equilibrium-stage

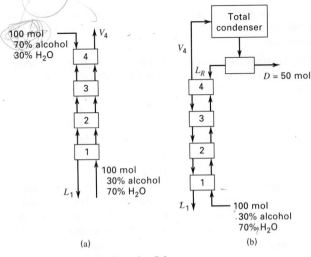

Figure 7.39 Data for Exercise 7.8.

distillation, a number of methods were proposed for solving the equations graphically or algebraically during an 18-year period from 1920 to 1938, prior to the availability of digital computers. Today, the only method from that era that remains in widespread use is the McCabe–Thiele method. What are the attributes of this method that are responsible for its continuing popularity?

7.8   (a) For the cascade shown in Figure 7.39a, calculate the compositions of streams V_4 and L_1. Assume atmospheric pressure, saturated liquid and vapor feeds, and the vapor–liquid equilibrium data given below. Compositions are in mole percent.

(b) Given the feed compositions in cascade (a), how many equilibrium stages are required to produce a V_4 containing 85 mol% alcohol?

(c) For the cascade configuration shown in Figure 7.39b, with $D = 50$ mol, what are the compositions of D and L_1?

(d) For the configuration of cascade (b), how many equilibrium stages are required to produce a D of 50 mol% alcohol?

EQUILIBRIUM DATA, MOLE-FRACTION ALCOHOL

x	0.1	0.3	0.5	0.7	0.9
y	0.2	0.5	0.68	0.82	0.94

7.9   Liquid air is fed to the top of a perforated-tray reboiled stripper operated at substantially atmospheric pressure. Sixty percent of the oxygen in the feed is to be drawn off in the bottoms vapor product from the still. This product is to contain 0.2 mol% nitrogen. Based on the assumptions and data given below, calculate:

(a) The mole percent of nitrogen in the vapor leaving the top plate.

(b) The moles of vapor generated in the still per 100 mol of feed.

(c) The number of theoretical plates required.

   Notes: To simplify the problem, assume constant molar overflow equal to the moles of feed. Liquid air contains 20.9 mol% of oxygen and 79.1 mol% of nitrogen. The equilibrium data [*Chem. Met. Eng.*, **35**, 622 (1928)] at atmospheric pressure are

Temperature, K	Mole-Percent N_2 in Liquid	Mole-Percent N_2 in Vapor
77.35	100.00	100.00
77.98	90.00	97.17
78.73	79.00	93.62

Temperature, K	Mole-Percent N_2 in Liquid	Mole-Percent N_2 in Vapor
79.44	70.00	90.31
80.33	60.00	85.91
81.35	50.00	80.46
82.54	40.00	73.50
83.94	30.00	64.05
85.62	20.00	50.81
87.67	10.00	31.00
90.17	0.00	0.00

7.10   A mixture of A (more volatile) and B is being separated in a plate distillation column. In two separate tests run with a saturated-liquid feed of 40 mol% A, the following compositions, in mol% A, were obtained for samples of liquid and vapor streams from three consecutive stages between the feed and total condenser at the top:

	Mol% A			
	Test 1		Test 2	
Stage	Vapor	Liquid	Vapor	Liquid
$M+2$	79.5	68.0	75.0	68.0
$M+1$	74.0	60.0	68.0	60.5
M	67.9	51.0	60.5	53.0

Determine the reflux ratio and overhead composition in each case, assuming that the column has more than three stages.

7.11   A saturated-liquid mixture containing 70 mol% benzene and 30 mol% toluene is to be distilled at atmospheric pressure to produce a distillate of 80 mol% benzene. Five procedures, described below, are under consideration. For each of the procedures, calculate and tabulate:

(a) Moles of distillate per 100 moles of feed,

(b) Moles of total vapor generated per mole of distillate,

(c) Mole percent benzene in the residue, and

(d) For each part, construct a y–x diagram. On this, indicate the compositions of the overhead product, the reflux, and the composition of the residue.

(e) If the objective is to maximize total benzene recovery, which, if any, of these procedures is preferred?

   Note: Assume that the relative volatility equals 2.5.

The procedures are as follows:

1. Continuous distillation followed by partial condensation. The feed is sent to the direct-heated still pot, from which the residue is continuously withdrawn. The vapors enter the top of a helically coiled partial condenser that discharges into a trap. The liquid is returned (refluxed) to the still, while the residual vapor is condensed as a product containing 80 mol% benzene. The molar ratio of reflux to product is 0.5.

2. Continuous distillation in a column containing one equilibrium plate. The feed is sent to the direct-heated still, from which residue is withdrawn continuously. The vapors from the plate enter the top of a helically coiled partial condenser that discharges into a trap. The liquid from the trap is returned to the plate, while the uncondensed vapor is condensed to form a distillate containing 80 mol% benzene. The molar ratio of reflux to product is 0.5.

3. Continuous distillation in a column containing the equivalent of two equilibrium plates. The feed is sent to the direct-heated still, from which residue is withdrawn continuously. The

vapors from the top plate enter the top of a helically coiled partial condenser that discharges into a trap. The liquid from the trap is returned to the top plate (refluxed) while the uncondensed vapor is condensed to form a distillate containing 80 mol% benzene. The molar ratio of reflux to product is 0.5.

4. The operation is the same as that described for Procedure 3 with the exception that the liquid from the trap is returned to the bottom plate.

5. Continuous distillation in a column containing the equivalent of one equilibrium plate. The feed at its boiling point is introduced on the plate. The residue is withdrawn continuously from the direct-heated still pot. The vapors from the plate enter the top of a helically coiled partial condenser that discharges into a trap. The liquid from the trap is returned to the plate while the uncondensed vapor is condensed to form a distillate containing 80 mol% benzene. The molar ratio of reflux to product is 0.5.

7.12 A saturated-liquid mixture of benzene and toluene containing 50 mol% benzene is distillated in an apparatus consisting of a still pot, one theoretical plate, and a total condenser. The still pot is equivalent to one equilibrium stage, and the pressure is 101 kPa.

The still is supposed to produce a distillate containing 75 mol% benzene. For each of the following procedures, calculate, if possible, the number of moles of distillate per 100 moles of feed. Assume a relative volatility of 2.5.

(a) No reflux with feed to the still pot.

(b) Feed to the still pot, reflux ratio $L/D = 3$.

(c) Feed to the plate with a reflux ratio of 3.

(d) Feed to the plate with a reflux ratio of 3. However, in this case, a partial condenser is employed.

(e) Part (b) using minimum reflux.

(f) Part (b) using total reflux.

7.13 A fractionation column operating at 101 kPa is to separate 30 kg/h of a solution of benzene and toluene containing 0.6 mass-fraction toluene into an overhead product containing 0.97 mass-fraction benzene and a bottoms product containing 0.98 mass-fraction toluene. A reflux ratio of 3.5 is to be used. The feed is liquid at its boiling point, feed is to the optimal tray, and the reflux is at saturation temperature.

(a) Determine the quantity of top and bottom products.

(b) Determine the number of stages required.

EQUILIBRIUM DATA IN MOLE-FRACTION BENZENE, 101 KPA

y	0.21	0.37	0.51	0.64	0.72	0.79	0.86	0.91	0.96	0.98
x	0.1	0.2	0.3	0.4	0.5	0.6	0.7	0.8	0.9	0.95

7.14 A mixture of 54.5 mol% benzene in chlorobenzene at its bubble point is fed continuously to the bottom plate of a column containing two theoretical plates. The column is equipped with a partial reboiler and a total condenser. Sufficient heat is supplied to the reboiler to give $\bar{V}/F = 0.855$, and the reflux ratio L/V in the top of the column is kept constant at 0.50. Under these conditions, what quality of product and bottoms (x_D, x_B) can be expected?

EQUILIBRIUM DATA AT COLUMN PRESSURE, MOLE-FRACTION BENZENE

x	0.100	0.200	0.300	0.400	0.500	0.600	0.700	0.800
y	0.314	0.508	0.640	0.734	0.806	0.862	0.905	0.943

7.15 A continuous distillation operation with a reflux ratio (L/D) of 3.5 yields a distillate containing 97 wt% B (benzene) and bottoms containing 98 wt% T (toluene). Due to weld failures, the 10 plates in the bottom section of the column are ruined, but the 14 upper plates are intact. It is suggested that the column still be used, with the feed (F) as saturated vapor at the dew point, with $F = 13,600$ kg/h containing 40 wt% B and 60 wt% T. Assuming that the plate efficiency remains unchanged at 50%: (a) Can this column still yield a distillate containing 97 wt% B, (b) How much distillate can we get, and (c) What will the composition of the residue be in mole percent?

For vapor–liquid equilibrium data, see Exercise 7.13.

7.16 A distillation column having eight theoretical stages (seven in the column + partial reboiler + total condenser) is being used to separate 100 kmol/h of a saturated-liquid feed containing 50 mol% A into a product stream containing 90 mol% A. The liquid-to-vapor molar ratio at the top plate is 0.75. The saturated-liquid feed is introduced on plate 5 from the top. Determine: (a) The composition of the bottoms, (b) The $\bar{L}/\bar{V}$ ratio in the stripping section, and (c) The moles of bottoms per hour.

Unbeknown to the operators, the bolts holding plates 5, 6, and 7 rust through, and the plates fall into the still pot. If no adjustments are made, what is the new bottoms composition?

It is suggested that, instead of returning reflux to the top plate, an equivalent amount of liquid product from another column be used as reflux. If this product contains 80 mol% A, what now is the composition of: (a) The distillate, and (b) The bottoms.

EQUILIBRIUM DATA, MOLE FRACTION OF A

y	0.19	0.37	0.5	0.62	0.71	0.78	0.84	0.9	0.96
x	0.1	0.2	0.3	0.4	0.5	0.6	0.7	0.8	0.9

7.17 A distillation unit consists of a partial reboiler, a column with seven equilibrium plates, and a total condenser. The feed consists of a 50 mol% mixture of benzene in toluene. It is desired to produce a distillate containing 96 mol% benzene, when operating at 101 kPa.

(a) With saturated-liquid feed fed to the fifth plate from the top, calculate: (1) Minimum reflux ratio $(L_R/D)_{min}$, (2) The bottoms composition, using a reflux ratio (L_R/D) of twice the minimum, and (3) Moles of product per 100 moles of feed.

(b) Repeat part (a) for a saturated-vapor feed fed to the fifth plate from the top.

(c) With saturated-vapor feed fed to the reboiler and a reflux ratio (L/V) of 0.9, calculate: (1) Bottoms composition, (2) Moles of product per 100 mole of feed.

Equilibrium data are given in Exercise 7.13.

7.18 A valve-tray fractionating column containing eight theoretical plates, a partial reboiler equivalent to one theoretical plate, and a total condenser is in operation separating a benzene–toluene mixture containing 36 mol% benzene at 101 kPa. Under normal operating conditions, the reboiler generates 100 kmol of vapor per hour. A request has been made for very pure toluene, and it is proposed to operate this column as a stripper, introducing the feed on the top plate as a saturated liquid, employing the same boilup at the still, and returning no reflux to the column. Equilibrium data are given in Exercise 7.13.

(a) What is the minimum feed rate under the proposed conditions, and what is the corresponding composition of the liquid in the reboiler at the minimum feed?

(b) At a feed rate 25% above the minimum, what is the rate of production of toluene, and what are the compositions in mole percent of the product and distillate?

7.19   A solution of methanol and water at 101 kPa containing 50 mol% methanol is continuously rectified in a seven-theoretical-plate, perforated-tray column, equipped with a total condenser and a partial reboiler heated by steam.

During normal operation, 100 kmol/h of feed is introduced on the third plate from the bottom. The overhead product contains 90 mol% methanol, and the bottoms product contains 5 mol% methanol. One mole of liquid reflux is returned to the column for each mole of overhead product.

Recently it has been impossible to maintain the product purity in spite of an increase in the reflux ratio. The following test data were obtained:

Stream	kmol/h	mol% alcohol
Feed	100	51
Waste	62	12
Product	53	80
Reflux	94	—

What is the most probable cause of this poor performance? What further tests would you make to establish definitely the reason for the trouble? Could some 90% product be obtained by further increasing the reflux ratio, while keeping the vapor rate constant?

Vapor–liquid equilibrium data at 1 atm [*Chem. Eng. Prog.* **48**, 192 (1952)] in mole-fraction methanol are

x	0.0321	0.0523	0.075	0.154	0.225	0.349	0.813	0.918
y	0.1900	0.2940	0.352	0.516	0.593	0.703	0.918	0.963

7.20   A fractionating column equipped with a partial reboiler heated with steam, as shown in Figure 7.40, and with a total condenser, is operated continuously to separate a mixture of 50 mol% A and 50 mol% B into an overhead product containing 90 mol% A and a bottoms product containing 20 mol% A. The column has three theoretical plates, and the reboiler is equivalent to one theoretical plate. When the system is operated at an $L/V = 0.75$ with

the feed as a saturated liquid to the bottom plate of the column, the desired products can be obtained. The system is instrumented as shown. The steam to the reboiler is controlled by a flow controller so that it remains constant. The reflux to the column is also on a flow controller so that the quantity of reflux is constant. The feed to the column is normally 100 kmol/h, but it was inadvertently cut back to 25 kmol/h. What would be the composition of the reflux, and what would be the composition of the vapor leaving the reboiler under these new conditions? Assume that the vapor leaving the reboiler is not superheated. Relative volatility for the system is 3.0.

7.21   A saturated-vapor mixture of maleic anhydride and benzoic acid containing 10 mol% acid is a by-product of the manufacture of phthalic anhydride. This mixture is distilled continuously at 13.3 kPa to give a product of 99.5 mol% maleic anhydride and a bottoms of 0.5 mol% anhydride. Using the data below, calculate the number of theoretical plates needed using an L/D of 1.6 times the minimum.

VAPOR PRESSURE, TORR:

Temperature, °C:	10	50	100	200	400
Maleic anhydride	78.7	116.8	135.8	155.9	179.5
Benzoic acid	131.6	167.8	185.0	205.8	227

7.22   A bubble-point binary mixture containing 5 mol% A in B is to be distilled to give a distillate containing 35 mol% A and a bottoms product containing 0.2 mol% A. If the relative volatility is constant at a value of 6, calculate the following algebraically, assuming that the column will be equipped with a partial reboiler and a partial condenser.

(a) The minimum number of equilibrium stages

(b) The minimum boilup ratio $\bar{V}/B$ leaving the reboiler

(c) The actual number of equilibrium stages for a boilup ratio equal to 1.2 times the minimum value

7.23   Methanol (M) is to be separated from water (W) by distillation as shown in Figure 7.41. The feed is subcooled such that $q = 1.12$. Determine the feed-stage location and the number of theoretical stages required. Vapor–liquid equilibrium data are given in Exercise 7.19.

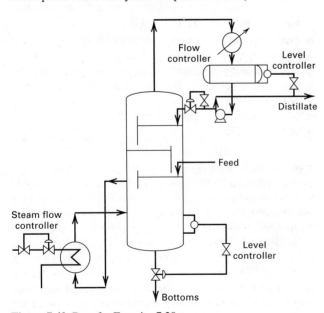

Figure 7.40 Data for Exercise 7.20.

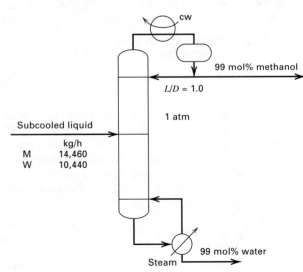

Figure 7.41 Data for Exercise 7.23.

7.24 A saturated-liquid mixture of 69.4 mol% benzene (B) in toluene (T) is to be continuously distilled at atmospheric pressure to produce a distillate containing 90 mol% benzene, with a yield of 25 moles of distillate per 100 moles of feed. The feed is sent to a steam-heated still (reboiler), where residue is to be withdrawn continuously. The vapors from the still pass directly to a partial condenser. From a liquid separator following the condenser, reflux is returned to the still. Vapors from the separator, which are in equilibrium with the liquid reflux, are sent to a total condenser and are continuously withdrawn as distillate. At equilibrium the mole ratio of B to T in the vapor may be taken as 2.5 times the mole ratio of B to T in the liquid. Calculate analytically and graphically the total moles of vapor generated in the still per 100 mol of feed.

7.25 A plant has a batch of 100 kmol of a liquid mixture containing 20 mol% benzene and 80 mol% chlorobenzene. It is desired to rectify this mixture at 1 atm to obtain bottoms containing only 0.1 mol% benzene. The relative volatility may be assumed constant at 4.13. There are available a suitable still to vaporize the feed, a column containing the equivalent of four theoretical plates, a total condenser, and a reflux drum to collect the condensed overhead. The run is to be made at total reflux. While the steady state is being approached, a finite amount of distillate is held in a reflux trap. When the steady state is reached, the bottoms contain 0.1 mol% benzene. With this apparatus, what yield of bottoms can be obtained? The liquid holdup in the column is negligible compared to that in the still and in the reflux drum.

7.26 A mixture of acetone and isopropanol containing 50 mol% acetone is to be distilled continuously to produce an overhead product containing 80 mol% acetone and a bottoms containing 25 mol% acetone. If a saturated-liquid feed is employed, if the column is operated with a reflux ratio of 0.5, and if the Murphree vapor efficiency is 50%, how many trays will be required? Assume a total condenser, partial reboiler, saturated-liquid reflux, and optimal feed stage. The vapor–liquid equilibrium data for this system are

EQUILIBRIUM DATA, MOLE-PERCENT ACETONE

Liquid	0	2.6	5.4	11.7	20.7	29.7	34.1	44.0	52.0
Vapor	0	8.9	17.4	31.5	45.6	55.7	60.1	68.7	74.3

Liquid	63.9	74.6	80.3	86.5	90.2	92.5	95.7	100.0
Vapor	81.5	87.0	89.4	92.3	94.2	95.5	97.4	100.0

7.27 A mixture of 40 mol% carbon disulfide (CS_2) in carbon tetrachloride (CCl_4) is continuously distilled. The feed is 50% vaporized ($q = 0.5$). The top product from a total condenser is 95 mol% CS_2, and the bottoms product from a partial reboiler is a liquid of 5 mol% CS_2.

The column operates with a reflux ratio, L/D, of 4 to 1. The Murphree vapor efficiency is 80%.

(a) Calculate graphically the minimum reflux, the minimum boilup ratio from the reboiler, $\bar{V}/B$, and the minimum number of stages (including reboiler).

(b) How many trays are required for the actual column at 80% efficiency by the McCabe–Thiele method.

The vapor–liquid equilibrium data at column pressure for this mixture in terms of CS_2 mole fraction are

x	0.05	0.1	0.2	0.3	0.4	0.5	0.6	0.7	0.8	0.9
y	0.135	0.245	0.42	0.545	0.64	0.725	0.79	0.85	0.905	0.955

7.28 A distillation unit consists of a partial reboiler, a bubble-cap column, and a total condenser. The overall plate efficiency is 65%.

The feed is a liquid mixture, at its bubble point, consisting of 50 mol% benzene in toluene. This liquid is fed to the optimal plate. The column is to produce a distillate containing 95 mol% benzene and a bottoms of 95 mol% toluene. Calculate for an operating pressure of 1 atm: (a) Minimum reflux ratio $(L/D)_{min}$, (b) Minimum number of actual plates to carry out the desired separation, (c) Using a reflux ratio (L/D) of 50% more than the minimum, the number of actual plates needed, (d) The kilograms per hour of product and residue, if the feed is 907.3 kg/h, (e) The saturated steam at 273.7 kPa required in kilograms per hour for heat to the reboiler using enthalpy data below and any assumptions necessary, and (f) A rigorous enthalpy balance on the reboiler, using the enthalpy data, tabulated below and assuming ideal solutions. Enthalpies in Btu/lbmol at reboiler temperature:

	h_L	h_V
benzene	4,900	18,130
toluene	8,080	21,830

Vapor–liquid equilibrium data are given in Exercise 7.13.

7.29 A continuous distillation unit, consisting of a perforated-tray column together with a partial reboiler and a total condenser, is to be designed to operate at atmospheric pressure to separate ethanol and water. The feed, which is introduced into the column as liquid at its bubble point, contains 20 mol% alcohol. The distillate is to contain 85 mol% alcohol, and the alcohol recovery is to be 97%.

(a) What is the molar concentration of the bottoms?

(b) What is the minimum value of:
    (1) The reflux ratio L/V?
    (2) The reflux ratio L/D?
    (3) The boilup ratio $\bar{V}/B$ from the reboiler?

(c) What is the minimum number of theoretical stages and the corresponding number of actual plates if the overall plate efficiency is 55%?

(d) If the reflux ratio L/V used is 0.80, how many actual plates will be required?

Vapor–liquid equilibrium for ethanol–water at 1 atm in terms of mole fractions of ethanol are [*Ind. Eng. Chem.,* **24,** 881 (1932)]:

x	y	T, °C	x	y	T, °C
0.0190	0.1700	95.50	0.3273	0.5826	81.50
0.0721	0.3891	89.00	0.3965	0.6122	80.70
0.0966	0.4375	86.70	0.5079	0.6564	79.80
0.1238	0.4704	85.30	0.5198	0.6599	79.70
0.1661	0.5089	84.10	0.5732	0.6841	79.30
0.2337	0.5445	82.70	0.6763	0.7385	78.74
0.2608	0.5580	82.30	0.7472	0.7815	78.41
			0.8943	0.8943	78.15

7.30 A solvent A is to be recovered by distillation from its water solution. It is necessary to produce an overhead product containing 95 mol% A and to recover 95% of the A in the feed. The feed is available at the plant site in two streams, one containing 40 mol% A and the other 60 mol% A. Each stream will provide 50 kmol/h of component A, and each will be fed into the column as saturated liquid. Since the less volatile component is water, it has been proposed to supply the necessary heat in the form of open steam. For the preliminary design, it has been suggested that the operating reflux ratio, L/D, be 1.33 times the minimum value. A total condenser will be employed. For this system, it is estimated that the overall plate efficiency will be 70%. How many plates will be required, and what will be the bottoms composition? The relative volatility may be assumed to be constant at 3.0. Determine analytically the points necessary to locate the operating lines. Each feed should enter the column at its optimal location.

7.31 A saturated-liquid feed stream containing 40 mol% *n*-hexane (H) and 60 mol% *n*-octane is fed to a plate column. A reflux ratio L/D equal to 0.5 is maintained at the top of the column. An overhead product of 0.95 mole fraction H is required, and the column bottoms is to be 0.05 mole fraction H. A cooling coil submerged in the liquid of the second plate from the top removes sufficient heat to condense 50 mol% of the vapor rising from the third plate down from the top.

(a) Derive the equations needed to locate the operating line.

(b) Locate the operating lines and determine the required number of theoretical plates if the optimal feed plate location is used.

7.32 One hundred kilogram-moles per hour of a saturated liquid mixture of 12 mol% ethyl alcohol in water is distilled continuously by direct steam at 1 atm introduced directly to the bottom plate. The distillate required is 85 mol% alcohol, representing 90% recovery of the alcohol in the feed. The reflux is saturated liquid with $L/D = 3$. Feed is on the optimal stage. Vapor–liquid equilibrium data are given in Exercise 7.29. Calculate:

(a) Steam requirement, kmol/h

(b) Number of theoretical stages

(c) The feed stage (optimal)

(d) Minimum reflux ratio, $(L/D)_{\min}$

7.33 A water–isopropanol mixture at its bubble point containing 10 mol% isopropanol is to be continuously rectified at atmospheric pressure to produce a distillate containing 67.5 mol% isopropanol. Ninety-eight percent of the isopropanol in the feed must be recovered. If a reflux ratio L/D of 1.5 times the minimum is used, how many theoretical stages will be required: (a) If a partial reboiler is used? (b) If no reboiler is used and saturated steam at 101 kPa is introduced below the bottom plate? (c) How many stages are required at total reflux?

Vapor–liquid equilibrium data in mole fraction of isopropanol at 101 kPa are

T, °C	93.00	84.02	82.12	81.25	80.62	80.16	80.28	81.51
y	0.2195	0.4620	0.5242	0.5686	0.5926	0.6821	0.7421	0.9160
x	0.0118	0.0841	0.1978	0.3496	0.4525	0.6794	0.7693	0.9442

Notes: Composition of the azeotrope is $x = y = 0.6854$. Boiling point of azeotrope = 80.22°C

7.34 An aqueous solution containing 10 mol% isopropanol is fed at its bubble point to the top of a continuous stripping column, operated at atmospheric pressure, to produce a vapor containing 40 mol% isopropanol. Two procedures are under consideration, both involving the same heat expenditure with V/F (moles of vapor generated/mole of feed) = 0.246 in each case. Scheme 1 uses a partial reboiler at the bottom of a plate-type stripping column, generating vapor by the use of steam condensing inside a closed coil. In Scheme 2, the reboiler is omitted and live steam is injected directly below the bottom plate. Determine the number of stages required in each case.

Equilibrium data for the system isopropanol–water are given in Exercise 7.33. The usual simplifying assumptions may be made.

7.35 Determine the optimal-stage location for each feed and the number of theoretical stages required for the distillation separation

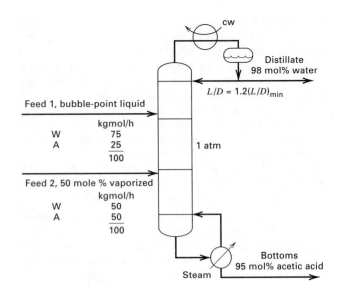

Figure 7.42 Data for Exercise 7.35.

shown in Figure 7.42 using the following equilibrium data in mole fractions.

WATER (W)/ACETIC ACID (A), 1 ATM

x_W	0.0055	0.053	0.125	0.206	0.297	0.510	0.649	0.803	0.9594
y_W	0.0112	0.133	0.240	0.338	0.437	0.630	0.751	0.866	0.9725

7.36 Determine the number of theoretical stages required and the optimal-stage locations for the feed and liquid side stream for the distillation process shown in Figure 7.43 assuming that methanol (M) and ethanol (E) form an ideal solution.

7.37 A mixture of *n*-heptane (H) and toluene (T) is separated by extractive distillation with phenol (P). Distillation is then used to recover the phenol for recycle as shown in Figure 7.44a, where the small amount of *n*-heptane in the feed is ignored. For the conditions shown in Figure 7.44a, determine the number of theoretical stages

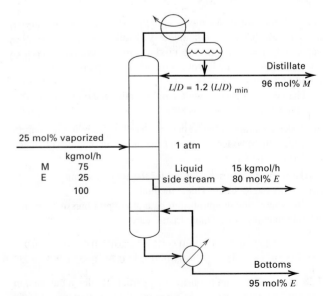

Figure 7.43 Data for Exercise 7.36.

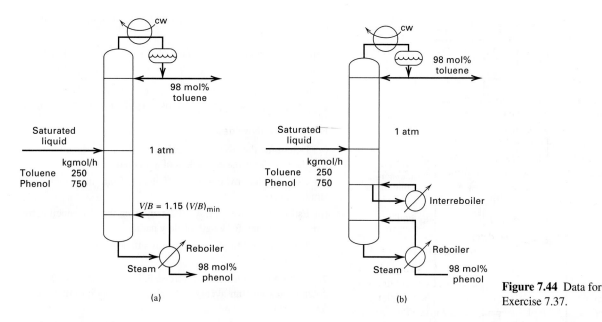

(a)

(b)

Figure 7.44 Data for Exercise 7.37.

required. Note that heat will have to be supplied to the reboiler at a high temperature because of the high boiling point of phenol.

Therefore, consider the alternative scheme in Figure 7.44b, where an interreboiler, located midway between the bottom plate and the feed stage, is used to provide 50% of the boilup used in Figure 7.44a. The remainder of the boilup is provided by the reboiler. Determine the number of theoretical stages required for the case with the interreboiler and the temperature of the interreboiler stage. Unsmoothed vapor–liquid equilibrium data at 1 atm are [*Trans. AIChE*, **41**, 555 (1945)]:

x_T	y_T	T, °C	x_T	y_T	T, °C
0.0435	0.3410	172.70	0.6512	0.9260	120.00
0.0872	0.5120	159.40	0.7400	0.9463	119.70
0.1186	0.6210	153.80	0.7730	0.9536	119.40
0.1248	0.6250	149.40	0.8012	0.9545	115.60
0.2190	0.7850	142.20	0.8840	0.9750	112.70
0.2750	0.8070	133.80	0.9108	0.9796	112.20
0.4080	0.8725	128.30	0.9394	0.9861	113.30
0.4800	0.8901	126.70	0.9770	0.9948	111.10
0.5898	0.9159	122.20	0.9910	0.9980	111.10
0.6348	0.9280	120.20	0.9939	0.9986	110.50
			0.9973	0.9993	110.50

7.38 A distillation column for the separation of *n*-butane from n-pentane was recently put into operation in a petroleum refinery. Apparently, an error was made in the design because the column fails to make the desired separation as shown in the following table [*Chem. Eng. Prog.*, **61** (8), 79 (1965)]:

	Design Specification	**Actual Operation**
Mol% nC_5 in distillate	0.26	13.49
Mol% nC_4 in bottoms	0.16	4.28

In order to correct the situation, it is proposed to add an intercondenser in the rectifying section to generate more reflux and an interreboiler in the stripping section to produce additional boilup. Show by use of a McCabe–Thiele diagram how such a proposed change can improve the operation.

7.39 In the production of chlorobenzenes by chlorination of benzene, two close-boiling isomers, *para*-dichlorobenzene (P) and *ortho*-dichlorobenzene (O), are separated by distillation. The feed to the column consists of 62 mol% of the para isomer and 38 mol% of the ortho isomer. Assume that the pressures at the bottom and top of the column are 20 psia (137.9 kPa) and 15 psia (103.4 kPa), respectively. The distillate is a liquid containing 98 mol% para isomer. The bottoms product is to contain 96 mol% ortho isomer. At column pressure, the feed is slightly vaporized with $q = 0.9$. Calculate the number of theoretical stages required for a reflux ratio equal to 1.15 times the minimum-reflux ratio. Base your calculations on a constant relative volatility obtained as the arithmetic average between the column top and column bottom using appropriate vapor pressure data and the assumption of Raoult's and Dalton's laws. The McCabe–Thiele construction should be supplemented at the two ends by use of the Kremser equations as illustrated in Example 7.4.

7.40 Relatively pure oxygen and nitrogen can be obtained by the distillation of air using the Linde double column, which, as shown in Figure 7.45, consists of a lower column operating at elevated pressure surmounted by an atmospheric-pressure column. The boiler of the upper column is at the same time the reflux condenser for both columns. Gaseous air plus enough liquid to take care of heat leak into the column (more liquid, of course, if liquid-oxygen product is withdrawn) enters the exchanger at the base of the lower column and condenses, giving up heat to the boiling liquid and thus supplying the vapor flow for this column. The liquid air enters an intermediate point in this column, as shown in Figure 7.45. The vapors rising in this column are partially condensed to form the reflux, and the uncondensed vapor passes to an outer row of tubes and is totally condensed, the liquid nitrogen collecting in an annulus, as shown. By operating this column at 4 to 5 atm, the liquid oxygen boiling at 1 atm is cold enough to condense pure nitrogen. The liquid that collects in the bottom of the lower column contains about 45 mol% O_2 and forms the feed for the upper column. Such a double column can produce very pure oxygen with high oxygen recovery, and relatively pure nitrogen. On a single McCabe–Thiele diagram—using equilibrium lines, operating lines, *q*-lines, 45°line, stepped-off stages, and other illustrative aids—show qualitatively how the stage requirements of the double column can be computed.

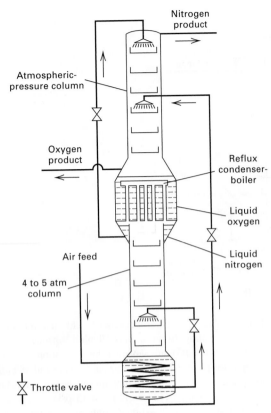

Figure 7.45 Data for Exercise 7.40.

Section 7.3

7.41 The following performance data have been obtained for a distillation tower separating a 50/50 by weight percent mixture of methanol and water:

Feed rate = 45,438 lb/h, Feed condition = bubble-point liquid at feed-tray pressure, Wt% methanol in distillate = 95.04, and Wt% methanol in bottoms = 1.00

Reflux ratio = 0.947; Reflux condition = saturated liquid

Boilup ratio = 1.138; Pressure in reflux drum = 14.7 psia

Type condenser = total; Type reboiler = partial

Condenser pressure drop = 0.0 psi; Tower pressure drop = 0.8 psi

Trays above feed tray = 5; Trays below feed tray = 6

Total trays = 12; Tray diameter = 6 ft

Type tray = single-pass sieve tray; Flow path length = 50.5 in.

Weir length = 42.5 in.; Hole area = 10%; Hole size = 3/16 in.

Weir height = 2 in.; Tray spacing = 24 in.

Viscosity of feed = 0.34 cP

Surface tension of distillate = 20 dyne/cm; Surface tension of bottoms = 58 dyne/cm

Temperature of top tray = 154°F; Temperature of bottom tray = 207°F

Vapor–liquid equilibrium data at column pressure in mole fraction of methanol are

y	0.0412	0.156	0.379	0.578	0.675	0.729	0.792	0.915
x	0.00565	0.0246	0.0854	0.205	0.315	0.398	0.518	0.793

Based on the above data:

(a) Determine the overall tray efficiency from the data, assuming that the reboiler is the equivalent of a theoretical stage.

(b) Estimate the overall tray efficiency from the Drickamer–Bradford correlation.

(c) Estimate the overall tray efficiency from the O'Connell correlation, accounting for length of flow path.

(d) Estimate the Murphree vapor tray efficiency by the method of Chan and Fair.

7.42 For the conditions of Exercise 7.41, a laboratory Oldershaw column measures an average Murphree vapor-point efficiency of 65%. Estimate E_{MV} and E_o.

Section 7.4

7.43 Conditions for the top tray of a distillation column are as shown in Figure 7.46. Determine the column diameter corresponding to 85% of flooding if a valve tray is used. Make whatever assumptions necessary.

7.44 A separation of propylene from propane is achieved by distillation as shown in Figure 7.47, where two columns in series are used because a single column would be too tall. The tray numbers refer to equilibrium stages. Determine the column diameters, tray efficiency using the O'Connell correlation, number of actual trays, and column heights if perforated trays are used.

7.45 Determine the height and diameter of a vertical flash drum for the conditions shown in Figure 7.48.

7.46 Determine the length and diameter of a horizontal reflux drum for the conditions shown in Figure 7.49.

7.47 Results of design calculations for a methanol–water distillation operation are given in Figure 7.50.

(a) Calculate the column diameter at the top tray and at the bottom tray for sieve trays. Should the column be swaged?

(b) Calculate the length and diameter of the horizontal reflux drum.

7.48 For the conditions given in Exercise 7.41, estimate for the top tray and the bottom tray: (a) Percent of flooding, (b) Tray pressure drop in psi, (c) Whether weeping will occur, (d) Entrainment rate, and (e) Froth height in the downcomer.

7.49 If the feed rate to the tower of Exercise 7.41 is increased by 30% with all other conditions except for tower pressure drop remaining the same, estimate for the top and bottom trays: (a) Per-

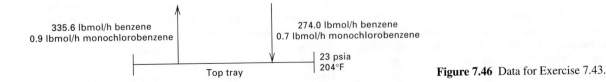

Figure 7.46 Data for Exercise 7.43.

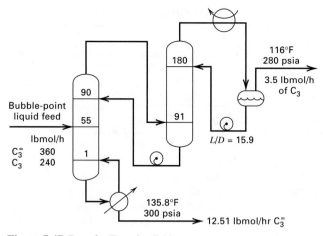

Figure 7.47 Data for Exercise 7.44.

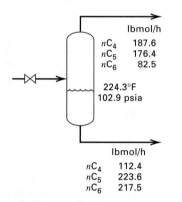

Figure 7.48 Data for Exercise 7.45.

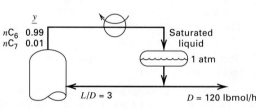

Figure 7.49 Data for Exercise 7.46.

cent of flooding, (b) Tray pressure drop in psi, (c) Entrainment rate, (d) Froth height in the downcomer. Will the new operation be acceptable? If not, should you consider a retrofit with packing? If so, should both sections of the column be packed or could just one section be packed to achieve an acceptable operation?

Section 7.5

7.50 A mixture of benzene and dichloroethane is used to test the efficiency of a packed column that contains 10 ft of packing and operates adiabatically at atmospheric pressure. The liquid is charged to the reboiler, and the column is operated at total reflux until equilibrium is established. At equilibrium, liquid samples from the distillate and reboiler, as analyzed by refractive index, give the following compositions for benzene: $x_D = 0.653$, $x_B = 0.298$.

Calculate the value of HETP in inches for this packing. What are the limitations on using this calculated value for design?

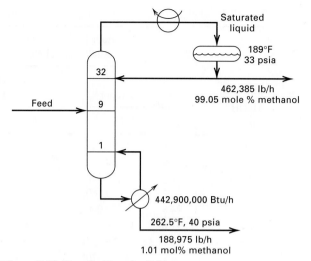

Figure 7.50 Data for Exercise 7.47.

Data for x–y at 1 atm (in benzene mole fractions) are

x	0.1	0.2	0.3	0.4	0.5	0.6	0.7	0.8	0.9
y	0.11	0.22	0.325	0.426	0.526	0.625	0.720	0.815	0.91

7.51 Consider a distillation column for separating ethanol from water at 1 atm. The following specifications are set:

Feed: 10 mol% ethanol (bubble-point liquid)

Bottoms: 1 mol% ethanol

Distillate: 80 mol% ethanol (saturated liquid)

Reflux ratio: 1.5 times the minimum

Constant molar overflow may be assumed and vapor–liquid equilibrium data are given in Exercise 7.29.

(a) How many theoretical plates are required above and below the feed if a plate column is used?

(b) How many transfer units are required above and below the feed if a packed column is used?

(c) Assuming that the plate efficiency is approximately 80% and the plate spacing is 18 in., how high is the plate column?

(d) Using an H_{OG} value of 1.2 ft., how high is the packed column?

(e) Assuming that you had HTU data available only on the benzene–toluene system, how would you go about applying the data to obtain the HTU for the ethanol–water system?

7.52 Plant capacity for the methanol–water distillation of Exercise 7.41 is to be doubled. Rather than installing a second trayed tower identical to the one in operation, a packed column is to be considered for the new installation. This column will have a feed location identical to the present trayed tower and will be expected to achieve the same product purities with the same top pressure and reflux ratio. Two packings are being considered:

1. 50-mm plastic NOR PAC rings (a random packing)
2. Montz metal B1-300 (a structured packing)

For each of these two packings, design a packed column to operate at 70% of flooding by calculating for each section: (a) Liquid holdup, (b) Column diameter, (c) H_{OG}, (d) Packed height, (e) Pressure drop.

What are the advantages, if any, of each of the packed-column designs over a second trayed tower? Which packing, if either, is preferable?

Table 7.8 Methanol–Water Vapor–Liquid Equilibrium and Enthalpy Data for 1 atm (MeOH = Methyl Alcohol)

| Mol% MeOH y or x | Enthalpy above 0°C, Btu/lbmol Solution | | | | Vapor Liquid Equilibrium Data | | |
| | Saturated Vapor | | Saturated Liquid | | Mol% MeOH in | | Temperature, °C |
	T, °C	h_V	T, °C	h_L	Liquid	Vapor	
0	100	20,720	100	3,240	0	0	100
5	98.9	20,520	92.8	3,070	2.0	13.4	96.4
10	97.7	20,340	87.7	2,950	4.0	23.0	93.5
15	96.2	20,160	84.4	2,850	6.0	30.4	91.2
20	94.8	20,000	81.7	2,760	8.0	36.5	89.3
30	91.6	19,640	78.0	2,620	10.0	41.8	87.7
40	88.2	19,310	75.3	2,540	15.0	51.7	84.4
50	84.9	18,970	73.1	2,470	20.0	57.9	81.7
60	80.9	18,650	71.2	2,410	30.0	66.5	78.0
70	76.6	18,310	69.3	2,370	40.0	72.9	75.3
80	72.2	17,980	67.6	2,330	50.0	77.9	73.1
90	68.1	17,680	66.0	2,290	60.0	82.5	71.2
100	64.5	17,390	64.5	2,250	70.0	87.0	69.3
					80.0	91.5	67.6
					90.0	95.8	66.0
					95.0	97.9	65.0
					100.0	100.0	64.5

Source: J.G. Dunlop, "Vapor–Liquid Equilibrium Data," M.S. thesis, Brooklyn Polytechnic Institute, Brooklyn, NY (1948).

7.53 For the specifications of Example 7.1, design a packed column using 50-mm metal Hiflow rings and operating at 70% of flooding by calculating for each section: (a) Liquid holdup, (b) Column diameter, (c) H_{OG}, (d) Packed height, and (e) Pressure drop.

What are the advantages and disadvantages of a packed column as compared to a trayed tower for this service?

Section 7.6

7.54 An enthalpy–concentration diagram is given in Figure 7.37 for a mixture of *n*-hexane (H), and *n*-octane (O) at 101 kPa. Using this diagram, determine the following:

(a) The mole-fraction composition of the vapor when a liquid containing 30 mol% H is heated from point A to the bubble-point temperature at point B.

(b) The energy required to vaporize 60 mol% of a mixture initially at 100°F and containing 20 mol% H (point G).

(c) The compositions of the equilibrium vapor and liquid resulting from part (b).

7.55 Using the enthalpy–concentration diagram of Figure 7.37, determine the following for a mixture of *n*-hexane (H) and *n*-octane (O) at 1 atm:

(a) The temperature and compositions of equilibrium liquid and vapor resulting from adiabatic mixing of 950 lb/h of a mixture of 30 mol% H in O at 180°F with 1,125 lb/h of a mixture of 80 mol% H in O at 240°F.

(b) The energy required to partially condense, by cooling, a mixture of 60 mol% H in O from an initial temperature of 260°F to 200°F. What are the compositions and amounts of the resulting vapor and liquid phases per pound-mole of original mixture?

(c) If the equilibrium vapor from part (b) is further cooled to 180°F, determine the compositions and relative amounts of the resulting vapor and liquid.

7.56 One hundred pound-moles per hour of a mixture of 60 mol% methanol in water at 30°C and 1 atm is to be separated by distillation at the same pressure into a liquid distillate containing 98 mol% methanol and a bottoms liquid product containing 96 mol% water. Enthalpy and equilibrium data for the mixture at 1 atm are given in Table 7.8. The enthalpy of the feed mixture is 765 Btu/lbmol.

(a) Using the given data, plot an enthalpy–concentration diagram.

(b) Devise a procedure to determine, from the diagram of part (a), the minimum number of equilibrium stages for the condition of total reflux and the required separation.

(c) From the procedure developed in part (b), determine N_{min}. Why is the value independent of the feed condition?

(d) What are the temperatures of the distillate and the bottoms?

Chapter **8**

Liquid–Liquid Extraction with Ternary Systems

In *liquid–liquid extraction,* a liquid feed of two or more components to be separated is contacted with a second liquid phase, called the *solvent,* which is immiscible or only partly miscible with one or more components of the liquid feed and completely or partially miscible with one or more of the other components of the liquid feed. Thus, the solvent, which is a single chemical species or a mixture, partially dissolves certain components of the liquid feed, effecting at least a partial separation of the feed. Liquid–liquid extraction is sometimes called *extraction, solvent extraction,* or *liquid extraction.* These, as well as the term *solid–liquid extraction,* are also applied to the recovery of substances from a solid by contact with a liquid solvent, such as the recovery of oil from seeds by an organic solvent. Solid–liquid extraction (leaching) is covered in Chapter 16.

According to Derry and Williams [1], liquid extraction has been practiced since at least the time of the Romans, who separated gold and silver from molten copper by extraction using molten lead as a solvent. This was followed by the discovery that sulfur could selectively dissolve silver from an alloy with gold. However, it was not until the early 1930s that the first large-scale liquid–liquid extraction process began operation. In that industrial process, named after its inventor L. Edeleanu, aromatic and sulfur compounds were selectively removed from liquid kerosene by liquid–liquid extraction with liquid sulfur dioxide at 10 to 20°F. Removal of aromatic compounds resulted in a cleaner-burning kerosene. Liquid–liquid extraction has grown in importance in recent years because of the growing demand for temperature-sensitive products, higher-purity requirements, more efficient equipment, and availability of solvents with higher selectivity.

The simplest liquid–liquid extraction involves only a ternary system. The feed consists of two miscible components, the *carrier, C,* and the *solute, A.* Solvent, *S,* is a pure compound. Components *C* and *S* are at most only partially soluble in each other. Solute *A* is soluble in *C* and completely or partially soluble in *S.* During the extraction process, mass transfer of *A* from the feed to the solvent occurs, with less transfer of *C* to the solvent, or *S* to the feed. However, complete or nearly complete transfer of *A* to the solvent is seldom achieved in just one stage, as discussed in Chapter 4. In practice, a number of stages are used in one- or two-section, countercurrent cascades, as discussed in Chapter 5.

8.0 INSTRUCTIONAL OBJECTIVES

After completing this chapter, you should be able to:

- Explain differences among liquid–liquid extraction, stripping, and distillation.
- List situations where liquid–liquid extraction might be preferred to distillation.
- Explain why so many different types of equipment are used for liquid–liquid extraction.
- List major types of equipment used for liquid–liquid extraction and compare their advantages and disadvantages.
- List major factors involved in the selection of extraction equipment.
- List factors that influence liquid–liquid extraction.
- List characteristics of an ideal solvent.
- Define the distribution coefficient and show its relationship to activity coefficients and relative selectivity of a solute between carrier and solvent.
- Make a preliminary selection of a solvent using group-interaction rules.
- Distinguish, for ternary mixtures, between Type I and Type II systems.
- For a specified recovery of a solute, calculate with the Hunter and Nash method, using a triangular diagram, minimum solvent requirement and number of equilibrium stages for ternary liquid–liquid extraction in a countercurrent cascade.

- Determine usefulness of extract reflux and carry out calculations with the Maloney and Schubert graphical method for a two-section extraction cascade that uses extract reflux.
- Design a cascade of mixer-settler units based on mass-transfer considerations.
- Determine the size of multicompartment extraction columns, including consideration of the effect of axial dispersion.

Industrial Example

Acetic acid is produced by methanol carbonylation or oxidation of acetaldehyde, or as a by-product of cellulose-acetate manufacture. In all three cases, a mixture of acetic acid (normal b.p. = 118.1°C) and water (normal b.p. = 100°C) must be separated to give glacial acetic acid (99.8 wt% min). When the mixture contains less than 50% acetic acid, separation by distillation is expensive because of the need to vaporize large amounts of the more volatile water, with its very high heat of vaporization. Accordingly, an alternative

liquid–liquid extraction process is often considered. A typical implementation is shown in Figure 8.1. In this process, it is important to note that two additional distillation separation steps are required to recover the solvent for recycle to the extractor. These additional separation steps are common to almost all extraction processes.

In the process of Figure 8.1, a feed of 30,260 lb/h, of 22 wt% acetic acid in water, is sent to a single-section extraction column, operating at near-ambient conditions, where the feed is countercurrently contacted with 71,100 lb/h of

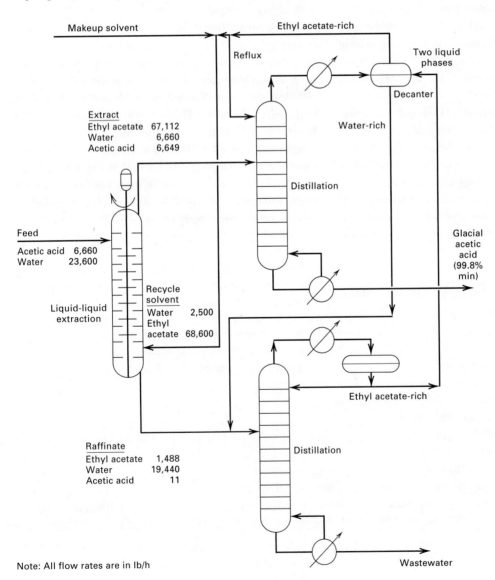

Note: All flow rates are in lb/h

Figure 8.1 Typical liquid–liquid extraction process.

ethyl-acetate solvent (normal b.p. = 77.1°C), saturated with water. The extract (solvent-rich product), being the low-density liquid phase, exits from the top of the extractor with 99.8% of the acetic acid originally contained in the feed. The raffinate (carrier-rich product), being the high-density liquid phase, exits from the bottom of the extractor and contains only 0.05 wt% acetic acid. The extract is sent to a distillation column, where glacial acetic acid is the bottoms product. The overhead vapor, which is rich in ethyl acetate but which also contains appreciable water vapor, splits into two liquid phases upon condensation. The two phases are separated by gravity in the decanter. The lighter ethyl-acetate-rich phase is divided into two streams. One is used for reflux for the distillation operation and the other is used for solvent recycle to the extractor.

The water-rich phase from the decanter is sent, together with the raffinate from the extractor, to a second distillation column, where wastewater is removed from the bottom and the ethyl-acetate-rich overhead distillate is recycled to the decanter. Makeup ethyl-acetate solvent is provided for solvent losses to the glacial acetic acid and wastewater products.

At an average extraction temperature of 100°F, six equilibrium stages are required to transfer 99.8% of the acetic acid from the feed to the extract using a solvent-to-feed ratio of 2.35 on a weight basis, where the recycled solvent is saturated with water. For six theoretical stages, a mechanically assisted extractor is preferred and a rotating-disk contactor (RDC), in a column configuration, is shown in Figure 8.1. The organic-rich phase is dispersed into droplets by rotating disks, while the water-rich phase is a continuous phase throughout the column. Dispersion and subsequent coalescence and settling takes place easily because at extractor operating conditions, liquid-phase viscosities are less than 1 cP, the phase-density difference is more than 0.08 g/cm^3, and the interfacial tension between the two phases is appreciable, at more than 30 dyne/cm.

The column has an inside diameter of 5.5 ft and a total height from the tangent of the top head to the tangent of the bottom head of 28 ft. The column is divided into 40 compartments, each 7.5 in. high and each containing a 40-in.-diameter rotor disk located between a pair of stator (donut) rings of 46-in. inside diameter. Above the top stator ring and below the bottom stator ring are settling zones. Because the light liquid phase is dispersed, the liquid-liquid interface is maintained near the top of the column. The rotors are mounted on a centrally located single shaft driven at a nominal 60 rpm by a 5-hp motor, equipped with a speed changer, the optimal disk speed being determined during plant operation. The HETP for the extractor is 50 in., equivalent to 6.67 compartments per theoretical stage. The HETP would be only 33 in. if axial (longitudinal) mixing did not occur.

Because of the corrosive nature of aqueous acetic acid solutions, the extractor is constructed of stainless steel. Since 1948, hundreds of extraction columns similar to that of Figure 8.1, with diameters ranging up to at least 25 ft, have been built. As discussed in Section 8.1, a number of other extraction devices are suitable for the process in Figure 8.1.

Liquid-liquid extraction is a reasonably mature separation operation, although not as mature or as widely applied as distillation, absorption, and stripping. Since the 1930s, more than 1,000 laboratory, pilot-plant, and industrial extractors have been installed. Procedures for determining the number of theoretical stages to achieve a desired solute recovery are well established. However, in the thermodynamics of liquid-liquid extraction, no simple limiting theory, such as that of ideal solutions for vapor-liquid equilibrium, exists. In many cases, experimental equilibrium data are preferred over predictions based on activity-coefficient correlations. However, such data can often be correlated well by semi-theoretical activity-coefficient equations such as the NRTL or UNIQUAC equations discussed in Chapter 2. Also, considerable laboratory effort may be required just to find an acceptable and efficient solvent. Furthermore, as will be discussed in the next section, a wide variety of industrial extraction equipment is available, making it necessary to consider many alternatives before making a final selection. Unfortunately, no generalized capacity and efficiency correlations are available for all equipment types. Often, equipment vendors must be relied upon to determine equipment size, or pilot-plant tests must be performed, followed by application of scale-up procedures recommended by the vendor or taken from sources such as this textbook.

Since the introduction of industrial liquid-liquid extraction processes, a large number of applications have been proposed and developed. The petroleum industry represents the largest-volume application for liquid-liquid extraction. By the late 1960s, more than 100,000 m^3/day of liquid feedstocks were being processed with physically selective solvents [2]. Extraction processes are well suited to the petroleum industry because of the need to separate heat-sensitive liquid feeds according to chemical type (e.g., aliphatic, aromatic, naphthenic) rather than by molecular weight or vapor pressure. Table 8.1 shows some representative, industrial extraction processes. Other major applications exist in the biochemical industry, where emphasis is on the separation of antibiotics and protein recovery from natural substrates; in the recovery of metals, such as copper from ammoniacal leach liquors, and in separations involving rare metals and radioactive isotopes from spent-fuel elements; and in the inorganic chemical industry, where high-boiling constituents

Table 8.1 Representative Industrial Liquid–Liquid Extraction Processes

Solute	Carrier	Solvent
Acetic acid	Water	Ethyl acetate
Acetic acid	Water	Isopropyl acetate
Aconitic acid	Molasses	Methyl ethyl ketone
Ammonia	Butenes	Water
Aromatics	Paraffins	Diethylene glycol
Aromatics	Paraffins	Furfural
Aromatics	Kerosene	Sulfur dioxide
Aromatics	Paraffins	Sulfur dioxide
Asphaltenes	Hydrocarbon oil	Furfural
Benzoic acid	Water	Benzene
Butadiene	1-Butene	*aq.* Cuprammonium acetate
Ethylene cyanohydrin	Methyl ethyl ketone	Brine liquor
Fatty acids	Oil	Propane
Formaldehyde	Water	Isopropyl ether
Formic acid	Water	Tetrahydrofuran
Glycerol	Water	High alcohols
Hydrogen peroxide	Anthrahydroquinone	Water
Methyl ethyl ketone	Water	Trichloroethane
Methyl borate	Methanol	Hydrocarbons
Naphthenes	Distillate oil	Nitrobenzene
Naphthenes/ aromatics	Distillate oil	Phenol
Phenol	Water	Benzene
Phenol	Water	Chlorobenzene
Penicillin	Broth	Butyl acetate
Sodium chloride	*aq.* Sodium hydroxide	Ammonia
Vanilla	Oxidized liquors	Toluene
Vitamin A	Fish-liver oil	Propane
Vitamin E	Vegetable oil	Propane
Water	Methyl ethyl ketone	*aq.* Calcium chloride

such as phosphoric acid, boric acid, and sodium hydroxide need to be recovered from aqueous solutions.

In general, extraction is preferred to distillation for the following applications:

1. In the case of dissolved or complexed inorganic substances in organic or aqueous solutions.

2. The removal of a component present in small concentrations, such as a color former in tallow or hormones in animal oil.

3. When a high-boiling component is present in relatively small quantities in an aqueous waste stream, as in the recovery of acetic acid from cellulose acetate. Extraction becomes competitive with distillation because of the expense of evaporating large quantities of water with its very high heat of vaporization.

4. The recovery of heat-sensitive materials, where extraction may be less expensive than vacuum distillation.

5. The separation of a mixture according to chemical type rather than relative volatility.

6. The separation of close-melting or close-boiling liquids, where solubility differences can be exploited.

7. Mixtures that form azeotropes.

The key to an effective extraction process is the discovery of a suitable solvent. In addition to being stable, nontoxic, inexpensive, and easily recoverable, a good solvent should be relatively immiscible with feed components(s) other than the solute and have a different density from the feed to facilitate phase separation. Also, it must have a very high affinity for the solute, from which it should be easily separated by distillation, crystallization, or other means. Ideally, the distribution coefficient for the solute between the two liquid phases should be greater than 1; otherwise a large solvent-to-feed ratio is required. When the degree of solute extraction is not particularly high and/or when a large extraction factor can be achieved, an extractor will not require many stages. This is fortunate because mass-transfer resistance in liquid–liquid systems is often high and stage efficiency is low in commercial contacting devices, unless mechanical agitation is provided.

In this chapter, equipment for conducting liquid–liquid extraction operations is discussed and fundamental equilibrium-based and rate-based calculation procedures are presented mainly for extraction in ternary systems. The use of graphical methods is emphasized. Except for systems dilute in solute(s), calculations for higher-order multicomponent systems are best conducted with computer-aided methods discussed in Chapter 10.

8.1 EQUIPMENT

Given the wide diversity of applications, one might expect a correspondingly large variety of liquid–liquid extraction devices. Indeed, such is the case. Equipment similar to that used for absorption, stripping, and distillation is sometimes used, but such devices are inefficient unless liquid viscosities are low and the difference in phase density is high. For that reason, centrifugal and mechanically agitated devices are often preferred. Regardless of the type of equipment, the

necessary number of theoretical stages is computed. Then the size of the device for a continuous, countercurrent process is obtained from experimental HETP or mass-transfer-performance-data characteristic of the particular piece of equipment. In extraction, some authors use the acronym HETS, height equivalent to a theoretical stage, rather than HETP. Also, the dispersed phase is sometimes referred to as the *discontinuous phase*, the other phase being the *continuous phase*.

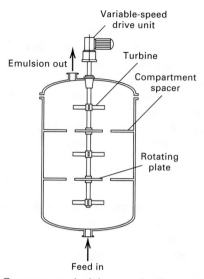

Figure 8.2 Compartmented mixing vessel with variable-speed turbine agitators.

[Adapted from R.E. Treybal, *Mass Transfer,* 3rd ed., McGraw-Hill, New York (1980).]

Mixer-Settlers

In mixer-settlers, the two liquid phases are first mixed (Figure 8.2) and then separated by settling (Figure 8.4). Any number of mixer-settler units may be connected together to form a multistage, countercurrent cascade. During mixing, one of the liquids is dispersed in the form of small droplets into the other liquid phase. The dispersed phase may be either the heavier or the lighter of the two phases. The mixing

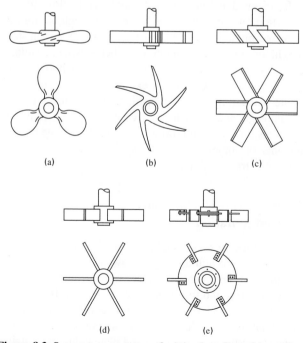

Figure 8.3 Some common types of mixing impellers: (a) marine-type propeller; (b) centrifugal turbine; (c) pitched-blade turbine; (d) flat-blade paddle; (e) flat-blade turbine.

[From R.E. Treybal, *Mass Transfer,* 3rd ed., McGraw-Hill, New York (1980) with permission.]

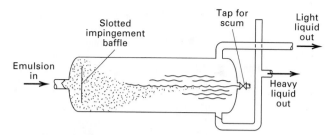

Figure 8.4 Horizontal gravity-settling vessel.

[Adapted from R.E. Treybal, *Liquid Extraction,* 2nd ed., McGraw-Hill, New York (1963) with permission.]

step is commonly conducted in an agitated vessel, with sufficient agitation and residence time so that a reasonable approach to equilibrium (e.g., 80% to 90% of a theoretical stage) is attained. The vessel may be compartmented as shown in Figure 8.2, and is usually agitated by means of impellers of the type shown in Figure 8.3. If dispersion is easily achieved and equilibrium is rapidly approached, as with liquids of low interfacial tension and viscosity, the mixing step can be carried out by impingement in a jet mixer; by turbulence in a nozzle mixer, orifice mixer, or other in-line mixing device; by shearing action if both phases are fed simultaneously into a centrifugal pump; or by injectors, wherein the flow of one liquid is induced by another.

The settling step is by gravity in a second vessel called a settler or decanter. In the configuration shown in Figure 8.4, a horizontal vessel, with an impingement baffle to prevent the jet of the entering two-phase dispersion (emulsion) from disturbing the gravity-settling process, is used. Vertical and inclined vessels are also used. A major problem in settlers is the emulsification in the mixing vessel, which may occur if the agitation is so intense that the dispersed droplet size falls below 1 to 1.5 micrometers. When this happens, coalescers, separator membranes, meshes, electrostatic forces, ultrasound, chemical treatment, or other ploys are required to speed settling. The rate of settling can also be increased by substituting centrifugal for gravitational force. This may be necessary if the phase-density difference is small.

A large number of commercial single- and multi-stage mixer-settler units are available, many of which are described by Bailes, Hanson, and Hughes [3] and by Lo, Baird, and Hanson [4]. Particularly worthy of mention is the Lurgi extraction tower [4], which was originally developed for extracting aromatics from hydrocarbon mixtures. In this device, the phases are mixed by centrifugal mixers stacked vertically outside the column and driven from a single shaft. Settling takes place in the column, with phases flowing interstagewise, guided by a complex baffle design located within the settling zones.

Spray Columns

The simplest and one of the oldest extraction devices is the spray column. Either the heavy phase or the light phase can be dispersed, as shown in Figure 8.5. The droplets of the

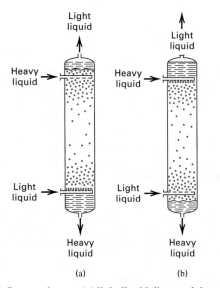

Figure 8.5 Spray columns: (a) light liquid dispersed, heavy liquid continuous; (b) heavy liquid dispersed, light liquid continuous.

dispersed phase are generated only at the inlet, usually by spray nozzles. Because of lack of column internals, throughputs are large, depending upon phase-density difference and phase viscosities. As in gas absorption, *axial dispersion* (*backmixing*) in the continuous phase limits these devices to applications where only one or two stages are required. Axial dispersion is so serious for columns with large diameter-to-length ratio that the continuous phase may be completely mixed. Therefore, spray columns are rarely used, despite their very low cost.

Packed Columns

Axial mixing in a spray column can be substantially reduced, but not eliminated, by packing the column. The packing also improves mass transfer by breaking up large drops to increase interfacial area and promotes mixing in drops by distorting droplet shape. With the exception of Raschig rings [5], the same packings used in distillation and absorption are

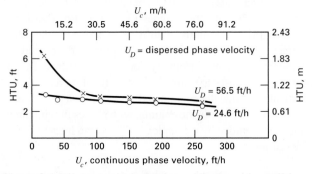

Figure 8.6 Efficiency of 1-in. Intalox saddles in a column 60 in. high with MEK–water–kerosene.

[From R.R. Neumatis, J.S. Eckert, E.H. Foote, and L.R. Rollinson, *Chem. Eng. Progr.*, **67**(**1**), 60 (1971) with permission.]

employed for liquid–liquid extraction. The choice of packing material, however, is somewhat more critical. A material preferentially wetted by the continuous phase is preferred. Figure 8.6 shows performance data, in terms of HTU, for Intalox saddles in an extraction service as a function of continuous, U_C, and discontinuous, U_D, phase superficial velocities. Because of backmixing, the HETP is generally larger than for staged devices. For that reason, packed columns are used only where few stages are needed.

Plate Columns

Sieve plates in a column also reduce axial mixing and achieve a more stagewise type of contact. The dispersed phase may be the light or the heavy phase. In the former case, the dispersed phase, analogous to vapor bubbles in distillation, flows vertically up the column, with redispersion at each tray. The heavy phase is the continuous phase, flowing at each stage through a *downcomer* and then across the tray the way a liquid does in a trayed distillation tower. If the heavy phase is dispersed, *upcomers* are used for the light phase. Columns have been built and successfully operated for diameters larger than 4.5 m. Holes from 0.64 to 0.32 cm in diameter and 1.25 to 1.91 cm apart are commonly used. Tray spacings are much closer than in distillation—10 to 15 cm in most applications involving low-interfacial-tension liquids. Plates are usually built without outlet weirs on the downspouts. A variation of the simple sieve column is the Koch Kascade Tower, where perforated plates are set in vertical arrays of complex designs.

If operated in the proper hydrodynamic flow regime, extraction rates in sieve-plate columns are high because the dispersed-phase droplets coalesce and re-form on each stage. This helps destroy concentration gradients, which develop if a droplet passes through the entire column without disturbance. Sieve-plate columns in extraction service are subject to the same limitations as distillation columns: flooding, entrainment, and, to a lesser extent, weeping. Additional problems, such as scum formation at interfaces due to small amounts of impurities, are frequently encountered in all types of extraction devices.

Columns with Mechanically Assisted Agitation

If the surface tension is high, and/or the density difference between the two liquid phases is low, and/or liquid viscosities are high, gravitational forces are inadequate for proper phase dispersal and the creation of turbulence. In that case, some type of mechanical agitation is necessary to increase interfacial area per unit volume and/or decrease mass-transfer resistance. For packed and plate columns, agitation is provided by an oscillating pulse to the liquid, either by mechanical or pneumatic means. Pulsed, perforated-plate columns found considerable application in the nuclear

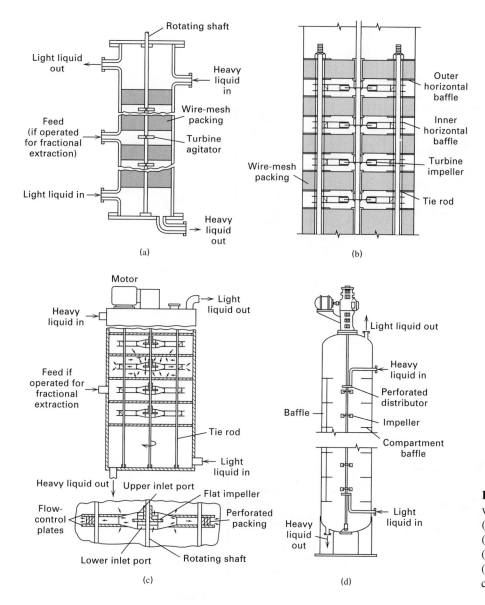

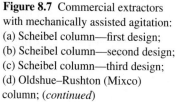

Figure 8.7 Commercial extractors with mechanically assisted agitation: (a) Scheibel column—first design; (b) Scheibel column—second design; (c) Scheibel column—third design; (d) Oldshue–Rushton (Mixco) column; (*continued*)

industry in the 1950s, but their popularity declined because of mechanical problems and the difficulty of propagating a pulse through a large volume [6]. The most important mechanically agitated columns are those that employ rotating agitators, driven by a shaft that extends axially through the column. The agitators create shear mixing zones, which alternate with settling zones in the column. Differences among the various agitated columns lie primarily in the mixers and settling chambers used. Nine of the more popular arrangements are shown in Figure 8.7. Agitation can also be induced in a column by moving the plates back and forth in a reciprocating motion (Figure 8.7j) or in a novel horizontal contactor (Figure 8.7k). Such devices are also included in Figure 8.7. These devices answer the plea of Fenske, Carlson, and Quiggle [7] in 1947 for equipment that can efficiently provide large numbers of equilibrium stages in a compact device without large numbers of pumps and

motors, and extensive piping. They stated, "Despite . . . advantages of liquid–liquid separational processes, the problems of accumulating twenty or more theoretical stages in a small compact and relatively simple countercurrent operation have not yet been fully solved." Indeed, in 1946 it was considered impractical to design for more than seven theoretical stages, which represented the number of mixer-settler units in the only large-scale, commercial, liquid-extraction process in use at that time.

Perhaps the first mechanically agitated column of importance was the Scheibel column [8] (Figure 8.7a), in which countercurrent liquid phases are contacted at fixed intervals by unbaffled, flat-bladed, turbine-type agitators (Figure 8.3) mounted on a vertical shaft. In the unbaffled separation or calming zones, located between the mixing zones, knitted wire-mesh packing is installed to prevent backmixing between mixing zones and to induce coalescence and

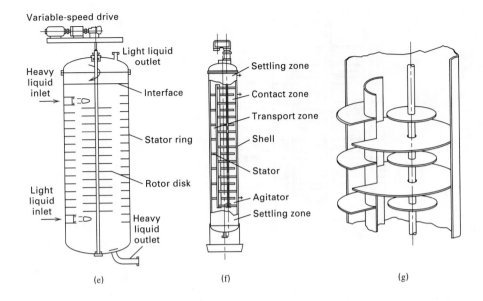

(e)   (f)   (g)

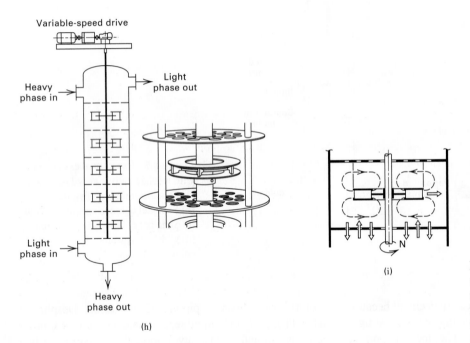

(h)

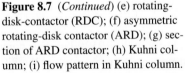

(i)

Figure 8.7 (*Continued*) (e) rotating-disk-contactor (RDC); (f) asymmetric rotating-disk contactor (ARD); (g) section of ARD contactor; (h) Kuhni column; (i) flow pattern in Kuhni column.

settling of drops. The mesh material must be wetted by the dispersed phase. For more economical designs for larger-diameter installations (>1 m), Scheibel [9] (Figure 8.7b) added outer and inner horizontal annular baffles to divert the vertical flow of the phases in the mixing zone and to ensure complete mixing. For systems with high interfacial surface tension and viscosities, the wire mesh is removed. The first two Scheibel designs did not permit removal of the agitator shaft for inspection and maintenance. Instead, the entire internal assembly (called the cartridge) had to be removed. To permit removal of just the agitator assembly shaft, especially for large-diameter columns (e.g., >1.5 m), and allow an access way through the column for any necessary

inspection, cleaning, and repair, Scheibel [10] offered a third design, shown in Figure 8.7c. Here the agitator assembly shaft can be removed because it has a smaller diameter than the opening in the inner baffle.

The Oldshue–Rushton extractor [11] (Figure 8.7d) consists of a column with a series of compartments separated by annular outer stator-ring baffles, each with four vertical baffles attached to the wall. The centrally mounted vertical shaft drives a flat-bladed turbine impeller in each compartment.

A third type of column with rotating agitators that appeared about the same time as the Scheibel and Oldshue–Rushton columns is the rotating-disk contactor (RDC)

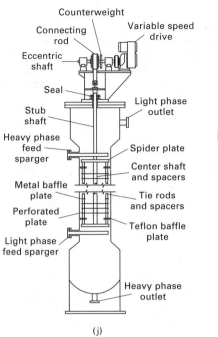

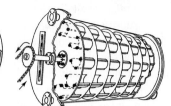

(k)

(j)

Figure 8.7 (*Continued*)
(j) Karr reciprocating-plate
column (RPC); (k) Graesser
raining-bucket (RTL)
extractor.

[12, 13] (Figure 8.7e), an example of which is described at the beginning of this chapter and shown in Figure 8.1. On a worldwide basis, it is probably the most extensively used liquid–liquid extraction device, with hundreds of units in use by 1983 [4]. Horizontal disks, mounted on a centrally located rotating shaft, are the agitation elements. Mounted at the column wall are annular stator rings with an opening larger than the agitator-disk diameter. Thus, the agitator assembly shaft is easily removed from the column. Because the rotational speed of the rotor controls the drop size, the rotor speed can be continuously varied over a wide range.

A modification of the RDC concept is the asymmetric rotating-disk contactor (ARD) [14], which has been in industrial use since 1965. As shown in Figure 8.7f, the contactor consists of a column, a baffled stator, and an offset multi-stage agitator fitted with disks. The asymmetric arrangement, shown in more detail in Figure 8.7g, provides contact and transport zones that are separated by a vertical baffle, to which is attached a series of horizontal baffles. Compared to the RDC, this design retains the efficient shearing action, but reduces backmixing because of the separate mixing and settling compartments.

Another extractor based on the Scheibel concept is the Kuhni extraction column [15]. As shown in Figure 8.7h, the column is compartmented by a series of stator disks made of perforated plates. On a centrally positioned shaft is mounted a series of double-entry, radial-flow, shrouded-turbine mixers, which promote, in each compartment, the circulation action shown in Figure 8.7i. For columns of diameter greater than 3 m, three turbine-mixer shafts on parallel axes are normally provided to preserve scale-up.

Three hundred of these extractors were in use, mainly in Europe, by 1983 [4].

Rather than provide agitation by rotating impellers on a vertical shaft, or by pulsing the liquid phases, Karr [16, 17] devised a reciprocating, perforated-plate extractor, also called the Karr column, in which the plates move up and down approximately two times per second with a stroke length of 0.75 inch. As shown in Figure 8.7j, annular baffle plates are provided periodically in the plate stack to minimize axial mixing. The perforated plates use large holes (typically 9/16-in. diameter) and a high hole area (typically 58%). The central shaft, which supports both sets of plates, is reciprocated by a drive mechanism located at the top of the column. A modification of the Karr column is the vibrating-plate extractor (VPE) of Prochazka et al. [18], which uses perforated plates of smaller hole size and smaller percent hole area than the Karr column. The small holes provide passage for the dispersed phase, while one or more large holes on each plate provide passage for the continuous phase. Some VPE columns operate like the Karr column with uniform motion of all plates; others are provided with two shafts to obtain countermotion of alternate plates.

Another novel device for providing agitation is the Graesser raining-bucket contactor (RTL), which was developed in the late 1950s [4], primarily for extraction processes involving liquids of small density difference, low interfacial tension, and a tendency to form emulsions As shown in Figure 8.7k, a series of disks is mounted inside a shell on a central, horizontal, rotating shaft, with a series of horizontal, C-shaped buckets fitted between and around the periphery of the disks. An annular gap between the disks and the inside perphery of the shell allows countercurrent, longitudinal flow of the phases. Dispersing action is very gentle, with each phase cascading through the other in opposite directions toward the two-phase interface, which is maintained close to the equatorial position.

A number of industrial centrifugal extractors have been available since 1944, when the Podbielniak (POD) extractor, with its short residence time, was successfully applied to penicillin extraction [19]. In the POD, several concentric sieve trays are arranged around a horizontal axis through which the two liquid phases flow countercurrently. Liquid inlet pressures of 4 to 7 atm are required to overcome pressure drop and centrifugal force. As many as five theoretical stages can be achieved in one unit.

Many of the commercial extractors described above have seen numerous industrial applications. Maximum loadings and sizes for column-type equipment, as given by Reissinger and Schroeter [5, 20] and Lo et al. [4], are listed in Table 8.2. As seen, the Lurgi tower, RDC, and Graesser extractors have been built in very large sizes. Throughputs per unit cross-sectional area are highest for the Karr extractor and lowest for the Graesser extractor.

The selection of an appropriate extractor is based on a large number of factors. Table 8.3 lists the advantages and disadvantages of the various types of extractors. Figure 8.8 shows a selection scheme for commercial extractors. For example, if only a small number of stages is required, a mixer-settler unit might be selected. If more than five theoretical stages, a high throughput, and a large load range (m^3/m^2-h) are needed, and floor space is limited, an RDC or ARD contactor should be considered.

Table 8.2 Maximum Size and Loading for Commercial Liquid–Liquid Extraction Columns

Column Type	Approximate Maximum Liquid Throughout, m^3/m^2-h	Maximum Column Diameter, m
Lurgi tower	30	8.0
Pulsed packed	40	3.0
Pulsed sieve tray	60	3.0
Scheibel	40	3.0
RDC	40	8.0
ARD	25	5.0
Kuhni	50	3.0
Karr	100	1.5
Graesser	<10	7.0

Above data apply to systems of:

1. High interfacial surface tension (30 to 40 dyne/cm).
2. Viscosity of approximately 1 cP.
3. Volumetric phase ratio of 1:1.
4. Phase-density difference of approximately 0.6 g/cm^3.

Table 8.3 Advantages and Disadvantages of Different Extraction Equipment

Class of Equipment	Advantages	Disadvantages
Mixer-settlers	Good contacting Handles wide flow ratio Low headroom High efficiency Many stages available Reliable scale-up	Large holdup High power costs High investment Large floor space Interstage pumping may be required
Continuous, counterflow contactors (no mechanical drive)	Low initial cost Low operating cost Simplest construction	Limited throughput with small density difference Cannot handle high flow ratio High headroom Sometimes low efficiency Difficult scale-up
Continuous, counterflow contactors (mechanical agitation)	Good dispersion Reasonable cost Many stages possible Relatively easy scale-up	Limited throughput with small density difference Cannot handle emulsifying systems Cannot handle high flow ratio
Centrifugal extractors	Handles low-density difference between phases Low holdup volume Short holdup time Low space requirements Small inventory of solvent	High initial costs High operating cost High maintenance cost Limited number of stages in single unit

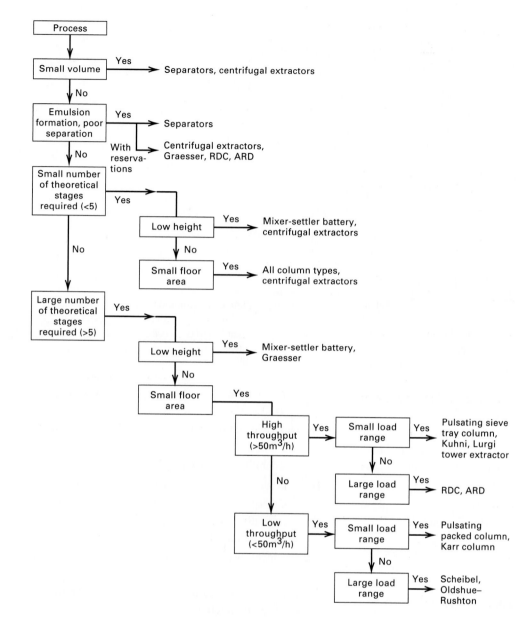

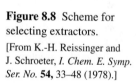

Figure 8.8 Scheme for selecting extractors.

[From K.-H. Reissinger and J. Schroeter, *I. Chem. E. Symp. Ser. No.* **54**, 33–48 (1978).]

8.2 GENERAL DESIGN CONSIDERATIONS

The design and analysis of a liquid–liquid extractor involves more factors than for vapor–liquid operations because of complications introduced by the two liquid phases. One of the three different cascade arrangements in Figure 8.9, or a more complex arrangement, must be selected. The single-section cascade of Figure 8.9a is similar to that used for absorption and stripping. It is designed to transfer to the solvent a certain percentage of the solute in the feed. The two-section cascade of Figure 8.9b is similar to distillation. Solvent enters at one end and reflux, derived from the extract, enters at the other end. The feed enters between the two sections. With two sections, depending on solubility considerations, it is sometimes possible to achieve a reasonably sharp separation between components of the feed; if not, a dual-solvent arrangement with two sections, as in Figure 8.9c, with or without reflux at the two ends, may be advantageous.

For the latter configuration, which involves a minimum of four components (two in the feed and two solvents), computer-aided calculations are preferred, as discussed in Chapter 10. Although the configurations in Figure 8.9 are shown with packed sections, any of the extractors discussed in Section 8.1 may be considered. The factors influencing extraction include:

1. Entering feed flow rate, composition, temperature, and pressure

2. Type of stage configuration (one-section or two-section)

3. Desired degree of recovery of one or more solutes for one-section cascades

4. Degree of separation of the feed for two-section cascades

5. Choice of liquid solvent(s)

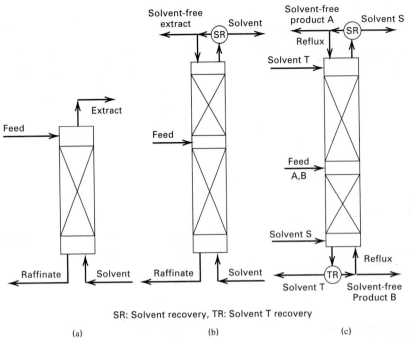

SR: Solvent recovery, TR: Solvent T recovery

(a)　　　　　　　(b)　　　　　(c)

Figure 8.9 Common liquid–liquid extraction cascade configurations: (a) single-section cascade; (b) two-section cascade; (c) dual solvent with two-section cascade.

6. Operating temperature

7. Operating pressure (greater than the bubble point of the system)

8. Minimum-solvent flow rate and actual solvent flow rate as a multiple of the minimum rate for one-section cascades or reflux rate and minimum reflux ratio for two-section cascades

9. Number of equilibrium stages

10. Emulsification and scum-formation tendency

11. Interfacial tension

12. Phase-density difference

13. Type of extractor

14. Extractor size and horsepower requirement

The ideal solvent has:

1. A high selectivity for the solute relative to the carrier, so as to minimize the need to recover carrier from the solvent

2. A high capacity for dissolving the solute, so as to minimize the solvent-to-feed ratio

3. A minimal solubility in the carrier

4. A volatility sufficiently different from the solute that recovery of the solvent can be achieved by distillation, but the vapor pressure should not be so high that a high extractor pressure is needed or so low that a high temperature is needed if the solvent is recovered by distillation

5. Stability to maximize the solvent life and minimize the solvent make-up requirement

6. Inertness to permit use of common materials of construction

7. A low viscosity to promote phase separation, minimize pressure drop, and provide a high solute mass-transfer rate

8. Nontoxic and nonflammable characteristics to facilitate its safe use

9. Availability at a relatively low cost

10. A moderate interfacial tension to balance the ease of dispersion and the promotion of phase separation

11. A large difference in density relative to the carrier to achieve a high capacity in the extractor

12. Compatibility with the solute and carrier to avoid contamination

13. A lack of tendency to form a stable rag or scum layer at the phase interface

14. Desirable wetting characteristics with respect to extractor internals

Solvent selection is frequently a compromise among all the properties listed above. However, initial consideration is usually given first to selectivity and environmental concerns, and second to capacity. From (2-20) in Chapter 2, the distribution coefficient for solute A between solvent S and carrier C is given by:

$$(K_A)_D = (x_A)^{II}/(x_A)^{I} = (\gamma_A)^{I}/(\gamma_A)^{II} \qquad (8\text{-}1)$$

where II is the extract phase, rich in S, and I is the raffinate phase, rich in C. Similarly, for the carrier and the solvent, respectively,

$$(K_C)_D = (x_C)^{II}/(x_C)^{I} = (\gamma_C)^{I}/(\gamma_C)^{II} \qquad (8\text{-}2)$$

$$(K_S)_D = (x_S)^{II}/(x_S)^{I} = (\gamma_S)^{I}/(\gamma_S)^{II} \qquad (8\text{-}3)$$

From (2-22), the relative selectivity of the solute with respect to the carrier is obtained by taking the ratio of (8-1)

Table 8.4 Group Interactions for Solvent Selection

Group	Solute	Solvent								
		1	2	3	4	5	6	7	8	9
1	Acid, aromatic OH (phenol)	0	−	−	−	−	0	+	+	+
2	Paraffinic OH (alcohol), water, imide or amide with active H	−	0	+	+	+	+	+	+	+
3	Ketone, aromatic nitrate, tertiary amine, pyridine, sulfone, trialkyl phosphate, or phosphine oxide	−	+	0	+	+	−	0	+	+
4	Ester, aldehyde, carbonate, phosphate, nitrite or nitrate, amide without active H; intramolecular bonding, e.g., *o*-nitrophenol	−	+	+	0	+	−	+	+	+
5	Ether, oxide, sulfide, sulfoxide, primary and secondary amine or imine	−	+	+	+	0	−	0	+	+
6	Multihaloparaffin with active H	0	+	−	−	−	0	0	+	0
7	Aromatic, halogenated aromatic, olefin	+	+	0	+	0	0	0	0	0
8	Paraffin	+	+	+	+	+	+	0	0	0
9	Monohaloparaffin or olefin	+	+	+	+	+	0	0	+	0

(+) Plus sign means that compounds in the column group tend to raise activity coefficients of compounds in the row group.

(−) Minus sign means a lowering of activity coefficients.

(0) Zero means no effect.

Choose a solvent that lowers the activity coefficient.

Source: Cusack, R.W., P. Fremeaux, and D. Glate, *Chem Eng.*, **98**(2), 66–76 (1991).

to (8-2), giving

$$\beta_{AC} = (K_A)_D/(K_C)_D = \frac{(x_A)^{\mathrm{II}}/(x_A)^{\mathrm{I}}}{(x_C)^{\mathrm{II}}/(x_C)^{\mathrm{I}}} = \frac{(\gamma_A)^{\mathrm{I}}/(\gamma_A)^{\mathrm{II}}}{(\gamma_C)^{\mathrm{I}}/(\gamma_C)^{\mathrm{II}}}$$

$$(8\text{-}4)$$

For high selectivity, the value of β_{AC} should be high, that is, at equilibrium there should be a high concentration of A and a low concentration of C in the solvent. A first estimate of β_{AC} is made from available values or predictions of the activity coefficients $(\gamma_A)^{\mathrm{I}}$, $(\gamma_A)^{\mathrm{II}}$, and $(\gamma_C)^{\mathrm{II}}$, at infinite dilution where $(\gamma_C)^{\mathrm{I}} = 1$, or by using liquid–liquid equilibrium data for the lowest tie line on a triangular diagram of the type discussed in Chapter 4. If A and C form a nearly ideal solution, the value of $(\gamma_A)^{\mathrm{I}}$ in (8-4) can also be taken as 1.

For high solvent capacity, the value of $(K_A)_D$ should be high. From (8-2) it is seen that this is difficult to achieve if A and C form nearly ideal solutions, such that $(\gamma_A)^{\mathrm{I}} = 1.0$, unless A and S have a great affinity for each other, which would result in a negative deviation from Raoult's law to give $(\gamma_A)^{\mathrm{II}} < 1$. Unfortunately, such systems are rare.

For ease in solvent recovery, $(K_S)_D$ should be as large as possible and $(K_C)_D$ as small as possible to minimize the presence of solvent in the raffinate and carrier in the extract. This will generally be the case if activity coefficients $(\gamma_S)^{\mathrm{I}}$ and $(\gamma_C)^{\mathrm{II}}$ at infinite dilution are large.

If a water-rich feed is to be separated, it is common to select an organic solvent; for an organic-rich feed, an aqueous solvent is often selected. In either case, it is desirable

to select a solvent that lowers the activity coefficient of the solute. Consideration of molecule group interactions can help narrow the search for such a solvent before activity coefficients are estimated or liquid–liquid equilibrium data are sought. A table of interactions for solvent-screening purposes, as given by Cusack et al. [21], based on a modification of the work of Robbins [22], is shown as Table 8.4, where the solute and solvent each belong to any of nine different chemical groups. In this table, a minus (−) sign for a given solute–solvent pair means that the solvent will desirably lower the value of the activity coefficient of the solute relative to its value in the feed solution. For example, suppose it is desired to extract acetone from water. Acetone, the solute, is a ketone. Thus, in Table 8.4, group 3 applies for the solute, and desirable solvents are of the type given in groups 1 and 6. In particular, trichloroethane, a group 6 compound, is known to be a highly selective solvent with high capacity for acetone over a wide range of feed compositions. However, if the compound is environmentally objectionable, it must be rejected. A more sophisticated solvent-selection method, based on the UNIFAC group-contribution method for estimating activity coefficients and utilizing a computer-aided constrained optimization approach, has been developed by Naser and Fournier [23].

In Chapter 4, ternary diagrams were introduced for representing liquid–liquid equilibrium data for three-component systems at constant temperature. Such diagrams are available for a large number of systems, as discussed by

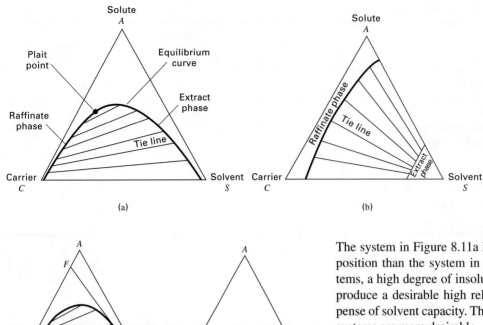

Figure 8.10 Most common classes of ternary systems: (a) type I, one immiscible pair; (b) type II, two immiscible pairs.

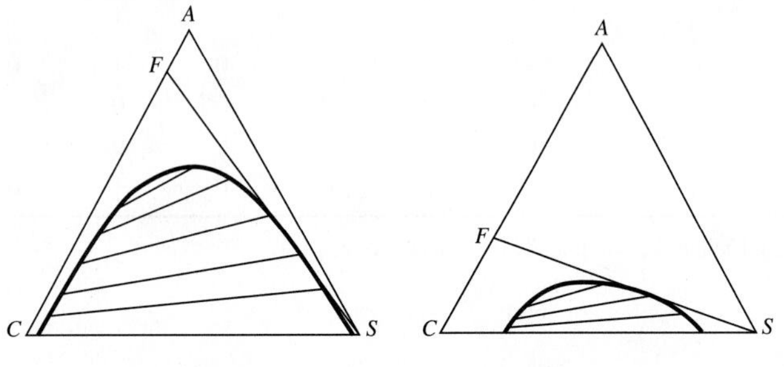

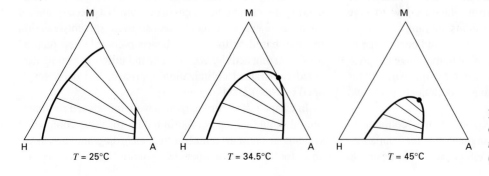

Figure 8.11 Effect of solubility on range of feed composition that can be extracted.

Humphrey et al. [6]. For liquid–liquid extraction with ternary systems, the most common diagram is Type I, shown in Figure 8.10a; much less common is Type II, shown in Figure 8.10b. Examples of Type II systems are (1) *n*-heptane/ aniline/methyl cyclohexane, (2) styrene/ethylbenzene/diethylene glycol, and (3) chlorobenzene/water/methylethyl ketone. For Type I, the solute and solvent are miscible in all proportions, while in Type II they are not. For Type I systems, the greater the two-phase region on line $\overline{CS}$, the greater will be the immiscibility of carrier and solvent. The closer the top of the two-phase region is to apex A, the greater will be the range of feed composition, along line $\overline{AC}$, that can be separated with solvent *S*. In Figure 8.11, it is possible to separate feed solutions only in the composition range from *C* to *F* because, regardless of the amount of solvent added, two liquid phases are not formed in the feed composition range of $\overline{FA}$ (i.e., $\overline{FS}$ does not pass through the two-phase region).

The system in Figure 8.11a has a wider range of feed composition than the system in Figure 8.11b. For Type II systems, a high degree of insolubility of *S* in *C* and *C* in *S* will produce a desirable high relative selectivity, but at the expense of solvent capacity. Thus, solvents that result in Type I systems are more desirable.

Whether a ternary system is of Type I or Type II often depends on the temperature. For example, data of Darwent and Winkler [24] for the ternary system *n*-hexane (H)/methylcyclopentane (M)/aniline (A) for temperatures of 25, 34.5, and 45°C are shown in Figure 8.12. At the lowest temperature, 25°C, we have a Type II system because both H and M are only partially miscible in the aniline solvent. As the temperature increases, the solubility of M in aniline increases more rapidly than the solubility of H in aniline until at 34.5°C, the critical solution temperature for M in aniline is reached. At this temperature, the system is at the border of Type II and Type I. At 45°C, the system is clearly of type I, with aniline more selective for M than H. Type I systems have a plait point (*P* in Figure 8.10a); type II systems do not.

Except in the near-critical region, pressure has little if any effect on liquid-phase activity coefficients and, therefore, on liquid–liquid equilibrium. It is only necessary to select an operating pressure of at least ambient, and greater than the bubble-point pressure of the two-liquid-phase mixture at any location in the extractor. Most extractors operate at near-ambient temperature. If feed and solvent enter the extractor at the same temperature, the operation will be nearly isothermal because the only thermal effect is the heat of mixing, which is usually small.

Figure 8.12 Effect of temperature on solubility for the system *n*-hexane (H)/methylcyclopentane (M)/aniline (A).

Laboratory or pilot-plant work, using actual or expected plant feed and solvent, is almost always necessary to ascertain dispersion and coalescence properties of the liquid–liquid system. Although rapid coalescence of drops is desirable, this reduces interfacial area, leading to reduced mass-transfer rates. Thus, compromises are necessary. Coalescence is enhanced when the solvent phase is continuous and mass transfer of solute is from the droplets. This phenomenon, called the *Marangoni effect,* is due to a lowering of interfacial tension by a significant presence of the solute in the interfacial film. When the solvent is the dispersed phase, the interfacial film is depleted of solute, causing an increase in interfacial tension and inhibition of coalescence.

For a given (1) feed liquid, (2) degree of solute extraction, (3) operating pressure and temperature, and (4) choice of solvent for a single-section cascade, a minimum solvent-to-feed flow-rate ratio exists that corresponds to an infinite number of countercurrent, equilibrium contacts. As with absorption and stripping, a trade-off then exists between the number of equilibrium stages and the solvent-to-feed flow-rate ratio. For a two-section cascade, as for distillation, the trade-off involves the reflux ratio and the number of stages. Algebraic methods, similar to those for absorption and stripping described in Chapter 6, for computing the minimum ratios and the trade-off are rapid, but are useful only for very dilute solutions, where values of the solute activity coefficients are essentially those at infinite dilution. When the carrier and the solvent are mutually insoluble, the algebraic method of Sections 5.3 and 5.4 can be used. For more general applications, use of the graphical methods described in this chapter is preferred for ternary systems. Computer-aided methods discussed in Chapter 10 are necessary for higher-order multicomponent systems.

8.3 HUNTER–NASH GRAPHICAL EQUILIBRIUM-STAGE METHOD

Stagewise extraction calculations for ternary systems of Type I and Type II (Figure 8.10) are most conveniently carried out with equilibrium diagrams [25]. In this section, procedures are developed and illustrated, using mainly triangular diagrams. The use of other diagrams is covered in the next section.

Consider a countercurrent-flow, N-equilibrium-stage contactor for liquid–liquid extraction of a ternary system operating under isothermal, continuous, steady-state flow conditions at a pressure sufficient to prevent vaporization, as shown in Figure 8.13. Stages are numbered from the feed end. Thus, the final extract is E_1 and the final raffinate is R_N. Equilibrium is assumed to be achieved at each stage, so that for any stage, n, the extract, E_n, and the raffinate, R_n, are in equilibrium with all three components. Mass transfer of all components occurs at each stage. The feed, F, contains the carrier, C, and the solute, A, and can also contain solvent, S, up to the solubility limit. The entering solvent, S, can contain C and A, but preferably contains little of either, if any. Because most liquid–liquid equilibrium data are given in mass rather than mole concentrations, let:

F = mass flow rate of feed to the cascade

S = mass flow rate of solvent to the cascade

E_n = mass flow rate of extract leaving stage n

R_n = mass flow rate of raffinate leaving stage n

$(y_i)_n$ = mass fraction of species i in extract leaving stage n

$(x_i)_n$ = mass fraction of species i in raffinate leaving stage n

Although Figure 8.13 might seem to imply that the extract is the light phase, either phase can be the light phase. Phase equilibrium may be represented, as discussed in Chapter 4, on an equilateral-triangle diagram, as proposed by Hunter and Nash [26], or on a right-triangle diagram as proposed by Kinney [27]. Assume, for illustration purposes, that the ternary system is A (solute), C (carrier), and S (solvent) at a particular temperature, T, such that the liquid–liquid equilibrium data are represented on the equilateral-triangle diagram of Figure 8.14, where the bold line is the equilibrium curve and the dashed lines are the tie lines that connect equilibrium phases of the equilibrium curve (also called the *binodal curve* because the plait point separates the curve into an extract and a raffinate). The equilibrium tie lines slope upward from the C side of the diagram toward the S side. Therefore, at equilibrium A has a concentration higher in S than in C. Thus, S is an effective solvent for extracting A from a mixture with C. On the other hand, because the tie lines slope downward from the S side toward the C side, C is not an effective solvent for extracting A from S.

Some ternary systems, such as isopropanol–water–benzene, exhibit a phenomenon called *solutropy,* wherein moving from the plait point, the tie lines first slope in one direction. However, the slope diminishes until an intermediate tie line becomes horizontal. Below that tie line, the remaining tie lines slope in the other direction. Sometimes, the solutropy phenomenon disappears if mole-fraction coordinates, rather than mass-fraction coordinates, are used.

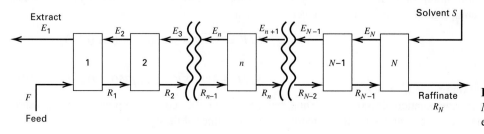

Figure 8.13 Countercurrent-flow, N-equilibrium-stage liquid–liquid extraction cascade.

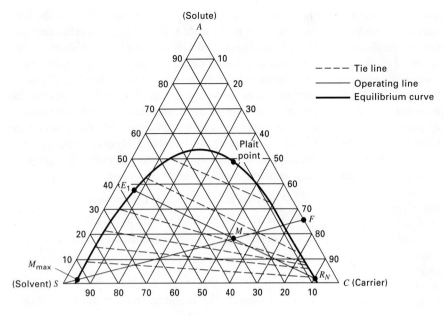

Figure 8.14 Construction 1: Location of product points.

Number of Equilibrium Stages

From the degrees-of-freedom discussion in Chapters 4 and 5, the following sets of specifications, for the cascade of Figure 8.13 with a ternary system, can be made, where all sets include the specification of F, $(x_i)_F$, $(y_i)_S$, and T:

Set 1. S and $(x_i)_{R_N}$	Set 4. N and $(x_i)_{R_N}$
Set 2. S and $(y_i)_{E_1}$	Set 5. N and $(y_i)_{E_1}$
Set 3. $(x_i)_{R_N}$ and $(y_i)_{E_1}$	Set 6. S and N

where values of $(x_i)_{R_N}$ and $(y_i)_{E_1}$ and all exiting phases must lie on the equilibrium curve.

Calculations for sets 1 to 3, which involve the determination of N, are made directly using the triangular diagram. Sets 4 to 6, which involve a specified N, require an iterative procedure. We first consider the calculational procedure for Set 1, with the procedures for Sets 2 and 3 being just minor modifications. The procedure, sometimes referred to as the *Hunter–Nash method* [26], involves three kinds of construction on the triangular diagram and is somewhat more difficult than the McCabe–Thiele staircase-type construction for distillation. Although the procedure is illustrated only for the Type I system, the principles are readily extended to a Type II system. The constructions are shown in Figure 8.14, where A is the solute, C is the carrier, and S is the solvent. On the binodal curve, all extract compositions lie on the equilibrium curve to the left of the plait point, while all raffinate compositions lie to the right. To determine the number of stages, given the flow rates and compositions of the feed and solvent, and the desired raffinate composition, the constructions are as follows:

Construction 1 (Product Points)

First, on Figure 8.14, we locate mixing point M, which represents the overall composition of the combination of feed, F, and entering solvent, S. Assume the following feed and solvent specifications, as plotted in Figure 8.14, where pure S is the solvent:

Feed	Solvent
$F = 250$ kg	$S = 100$ kg
$(x_A)_F = 0.24$	$(x_A)_S = 0.00$
$(x_C)_F = 0.76$	$(x_C)_S = 0.00$
$(x_S)_F = 0.00$	$(x_S)_S = 1.00$

The overall composition M for combined F and S is obtained from the following material balances:

$$M = F + S = 250 + 100 = 350 \text{ kg}$$
$$(x_A)_M M = (x_A)_F F + (x_A)_S S$$
$$= 0.24(250) + 0(100) = 60 \text{ kg}$$
$$(x_A)_M = 60/350 = 0.171$$

$$(x_C)_M M = (x_C)_F F + (x_C)_S S$$
$$= 0.76(250) + 0(100) = 190 \text{ kg}$$
$$(x_C)_M = 190/350 = 0.543$$

$$(x_S)_M M = (x_S)_F F + (x_S)_S S$$
$$= 0(250) + 1(100) = 100 \text{ kg}$$
$$(x_S)_M = 100/350 = 0.286$$

From any two of these $(x_i)_M$ values, point M is located, as shown, in Figure 8.14. Based on the properties of the triangular diagram, presented in Chapter 4, point M must be located somewhere on the straight line connecting F and S. Therefore, M can be located knowing just one value of $(x_i)_M$, say $(x_S)_M$. Also, the ratio S/F is given by the inverse-lever-arm rule as

$$S/F = \overline{MF}/\overline{MS} = 100/250 = 0.400$$

or

$$S/M = \overline{MF}/\overline{SF} = 100/350 = 0.286$$

Thus, point M can be located by two points or by measurement, employing either of these ratios.

With point M located, the composition of extract, E_1, exiting from a countercurrent, multistage extractor, can be determined from overall material balances:

$$M = R_N + E_1 = 350 \text{ kg}$$
$$(x_A)_M M = 60 = (x_A)_{R_N} R_N + (x_A)_{E_1} E_1$$
$$(x_C)_M M = 190 = (x_C)_{R_N} R_N + (x_C)_{E_1} E_1$$
$$(x_S)_M M = 100 = (x_S)_{R_N} R_N + (x_S)_{E_1} E_1$$

Because the raffinate, R_N, is assumed to be at equilibrium, its composition must lie on the equilibrium curve of Figure 8.14. Therefore, if we specify the value $(x_A)_{R_N} = 0.025$, we can locate the point R_N, and the values of $(x_C)_{R_N}$ and $(x_S)_{R_N}$ can be read from Figure 8.14. A straight line drawn from R_N through M will locate E_1 at the intersection of the equilibrium curve, from which, in Figure 8.14, the composition of E_1 can be read. Values of the flow rates R_N and E_1 can then be determined from the overall material balances above or from Figure 8.14 by the inverse-lever-arm rule:

$$E_1/M = \overline{MR_N}/\overline{E_1 R_N}$$
$$R_N/M = \overline{ME_1}/\overline{E_1 R_N}$$

with $M = 350$ kg for this illustration. By either method, we find:

Raffinate Product	Extract Product
$R_N = 198.6$ kg	$E_1 = 151.4$ kg
$(x_A)_{R_N} = 0.025$	$(x_A)_{E_1} = 0.364$
$(x_C)_{R_N} = 0.90$	$(x_C)_{E_1} = 0.075$
$(x_S)_{R_N} = 0.075$	$(x_S)_{E_1} = 0.561$

Also included in Figure 8.14 is the point M_{max}, which lies on the equilibrium curve along the straight line connecting F to S. M_{max} corresponds to the maximum possible solvent addition. If more solvent were added, two liquid phases could not exist.

Construction 2 (Operating Point and Lines)

In Chapters 6 and 7, we learned that an operating line is the locus of passing streams in a cascade. Referring to Figure 8.13, material balances around groups of stages from the feed end are

$$F - E_1 = \cdots = R_{n-1} - E_n = \cdots = R_N - S = P \quad (8\text{-}5)$$

Because the passing streams are differenced, P defines a *difference point* rather than a *mixing point*. From the same geometric considerations as apply to a mixing point, a difference point also lies on a straight line drawn through the points involved. However, while the mixing point always lies inside the triangular diagram and between the two end points, the difference point usually lies outside the triangular diagram along an extrapolation of the line through two points such as F and E_1, R_N and S, and so on.

To locate the difference point, two straight lines are drawn, respectively, through the point pairs (E_1, F) and (S, R_N), which are established by Construction 1 and shown in Figure 8.15. These lines are extrapolated until they intersect at difference point P. Figure 8.15 shows these lines and the difference point, P. From (8-5), straight lines drawn through points on the triangular diagram for any other pair of passing streams, such as (E_n, R_{n-1}), must also pass through point P. Thus, we refer to the difference point as an *operating point,* and the lines drawn through pairs of points for passing streams and extrapolated to point P as *operating lines*.

The difference point has properties similar to those of the mixing point. If $F - E_1 = P$ is rewritten as $F = E_1 + P$, we see that F can be interpreted as the mixing point for P and E_1. Therefore, by the inverse-lever-arm rule, the length of line $\overline{E_1 P}$ relative to the length of the line $\overline{FP}$ is given by

$$\frac{\overline{E_1 P}}{\overline{FP}} = \frac{E_1 + P}{E_1} = \frac{F}{E_1} \quad (8\text{-}6)$$

Thus, point P can be located, if desired, by measurement with a ruler using either pair of feed-product passing streams.

The operating point, P, lies on the feed or raffinate side of the triangular diagram in the illustration of Figure 8.15. Depending on the relative amounts of feed and solvent and the slope of the tie lines, point P may be located on the solvent or feed side of the diagram, and inside or outside the diagram.

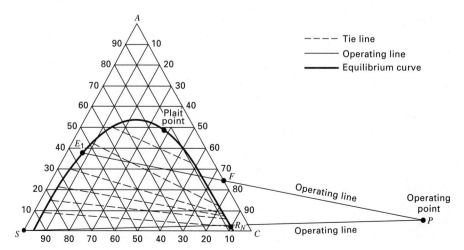

Figure 8.15 Construction 2: Location of operating point.

Construction 3 (Equilibrium Lines)

The third type of construction involves the dashed *tie lines* that connect opposite sides of the equilibrium curve, which is divided into the two sides by the plait point, which for type I diagrams is the point where the two equilibrium phases become one phase. A material balance around any stage n for any of the three components is

$$(x_i)_{n-1}R_{n-1} + (y_i)_{n+1}E_{n+1} = (x_i)_n R_n + (y_i)_n E_n \quad (8\text{-}7)$$

Because R_n and E_n are in equilibrium, their composition points are on the triangular diagram at the two ends of a tie line. Typically a diagram will not contain all the tie lines needed. Tie lines may be added by centering them between existing experimental or predicted tie lines, or by using either of two interpolation procedures illustrated in Figure 8.16. In Figure 8.16a, the conjugate line from the plait point to J is determined from four tie lines and the plait point. From tie line $\overline{DE}$, lines $\overline{DG}$ and $\overline{EF}$ are drawn parallel to triangle sides $\overline{CB}$ and $\overline{AC}$, respectively. The intersection at point H gives a second point on the conjugate curve. Subsequent intersections, using the other tie lines, establish additional points from which the conjugate curve is drawn. Then, using the curve, additional tie lines are drawn by reversing the procedure. If it is desired to keep the conjugate curve inside the two liquid-phase region of the triangular diagram, the procedure illustrated in Figure 8.16b is used, where lines are drawn parallel to triangle sides $\overline{AB}$ and $\overline{AC}$.

Stepping off Stages

Equilibrium stages are stepped off on the triangular diagram by alternating the use of tie lines and operating lines, as shown in Figure 8.17, where Constructions 1 and 2 have already been employed to locate the five points F, E, S, R_1, and P. We start at the feed end from point E_1. Referring to Figure 8.13, we see that R_1 is in equilibrium with E_1. Therefore, by Construction 3, R_1 in Figure 8.17 must be at the opposite end of a tie line (shown as a dashed line) connecting to E_1. From Figure 8.13, R_1 passes E_2. Therefore, by Construction 2, E_2 must lie at the

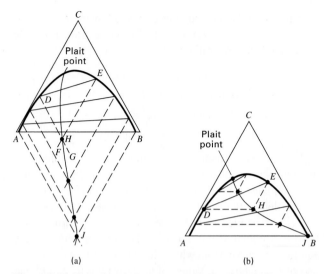

Figure 8.16 Use of conjugate curves to interpolate tie lines: (a) method of *International Critical Tables,* Vol. III, McGraw-Hill, New York, p. 393 (1928); (b) method of T.K. Sherwood, *Absorption and Extraction,* McGraw-Hill, New York, p. 242 (1937). [From R.E. Treybal, *Liquid Extraction,* 2nd ed., McGraw-Hill, New York (1963) with permission.]

intersection of the straight operating line, drawn through points R_1 and P, and back to the extract side of the equilibrium curve. From E_2, we locate R_2 with a tie line by Construction 3; from R_2, we locate E_3 by Construction 2. Continuing in this fashion by alternating between equilibrium tie lines and operating lines, we finally reach or pass the specified point R_N. If the latter, a fraction of the last stage is taken. In Figure 8.17 approximately 2.8 equilibrium stages are required, where stages are counted by the number of tie lines used.

Procedures for problem specification sets 2 and 3 are very similar to that for set 1. Sets 4 and 5 can be handled by iteration on assumed values for S and following the above procedure for set 1. Set 6 can also use the procedure of set 1 by iterating on E_1.

From (8-6), we see that if the ratio F/E_1 approaches a value of 1, the operating point, P, will be located at a large distance

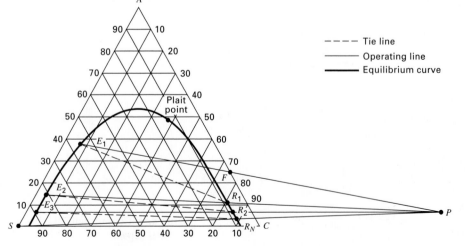

Figure 8.17 Determination of the number of equilibrium stages.

from the triangular diagram. In that case, using an arbitrary rectangular-coordinate system superimposed over the triangular diagram, the coordinates of P can be calculated from (8-6) using the equations for the two straight lines established in Construction 2. Operating lines for intermediate stages can then be located on the triangular diagram so as to pass through P. Details of this procedure are given by Treybal [25].

Minimum and Maximum Solvent-to-Feed Flow-Rate Ratios

The graphical procedure just described for determining the number of equilibrium stages to achieve a desired solute extraction for a given solvent-to-feed flow-rate ratio presupposes that this ratio is greater than the minimum ratio corresponding to the need for an infinite number of stages, but less than the maximum ratio that would prevent the formation of the required second liquid phase. In practice, one usually determines the minimum ratio before solving specification sets 1 or 2. In essence, we must solve set 4 with $N = \infty$, where, as in distillation, absorption, and stripping, the infinity of stages occurs at a pinch point of the equilibrium curve and the operating line(s). In ternary systems, the pinch point occurs when a tie line is coincident with an operating line. The calculation is somewhat involved because the location of the pinch point is not always at the feed end of the cascade. Consider the previous A–C–S system, as shown in Figure 8.18. The points F, S, and R_N are specified, but E_1 is not because the solvent rate has not yet been specified. The operating line $\overline{OL}$ is drawn through the points S and R_N and extended to the left and right of the diagram. This line is the locus of all possible material balances determined by adding S to R_N. Each tie line is then assumed to be a pinch point by extending each tie line until it intersects the line $\overline{OL}$. In this manner, a sequence of intersections, P_1, P_2, P_3, and so on, is found. If these points lie on the raffinate side of the diagram, as in Figure 8.18, the pinch point corresponds to the point P_{min} located at the greatest distance from R_N. If the triangular diagram does not have a sufficient number of tie lines to determine that point accurately, additional tie lines are introduced by a method described previously and illustrated in Figure 8.16. If we assume in Figure 8.18 that no other tie line gives a point P_i farther away from R_N than P_1, then $P_1 = P_{min}$.

With P_{min} known, an operating line can be drawn through point F and extended to E_1 at an intersection with the extract side of the equilibrium curve. From the compositions of the four points, S, R_N, F, and E_1, the mixing point M can be found and the following material balances can then be used to solve for S_{min}/F:

$$F + S_{min} = R_N + E_1 = M \qquad (8-8)$$

$$(x_A)_F F + (x_A)_S S_{min} = (x_A)_M M \qquad (8-9)$$

from which

$$\frac{S_{min}}{F} = \frac{(x_A)_F - (x_A)_M}{(x_A)_M - (x_A)_S} \qquad (8-10)$$

A solvent flow rate greater than S_{min} must be selected for the extraction to be conducted in a finite number of stages. In Figure 8.18, such a solvent rate results in an operating point P to the right of P_{min}, that is, at a location farther away from R_N. A reasonable value for S might be $1.5\,S_{min}$. From Figure 8.18, we find $(x_A)_M = 0.185$, from which, by (8-10), $S_{min}/F = 0.30$. In our example of Figure 8.17, we used $S/F = 0.40$, giving $S/S_{min} = 1.33$.

In Figure 8.18 the tie lines slope downward toward the raffinate side of the diagram. If the tie lines slope downward toward the extract side of the diagram, the above procedure for finding S_{min}/F must be modified. The sequence of points P_1, P_2, P_3, and so on, is now found on the other side of the diagram. However, the pinch point now corresponds to that point, P_{min}, that is closest to point S and an operating point, P, must be chosen between points P_{min} and S. For a system that exhibits solutropy, intersections P_1, P_2, and so on, will be found on both sides of the diagram. Those on the extract side will determine the minimum solvent-to-feed ratio.

In Figure 8.14, the mixing point M must lie in the two-phase region. As this point is moved along the line $\overline{SF}$ toward S, the ratio S/F increases according to the inverse-lever-arm rule. In the limit, a maximum S/F ratio is reached when $M = M_{max}$ arrives at the equilibrium curve on the

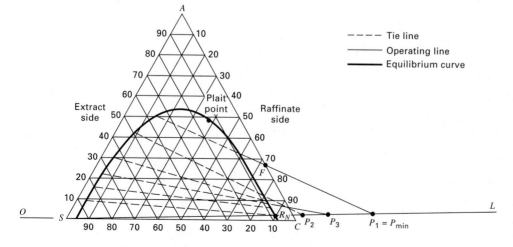

Figure 8.18 Determination of minimum solvent-to-feed ratio.

extract side. At this point, all of the feed is dissolved in the solvent, no raffinate is obtained, and only one stage is required. To avoid this impractical condition, as well as the other extreme of infinite stages, we must select a solvent ratio, S/F, such that $(S/F)_{min} < (S/F) < (S/F)_{max}$. In Figure 8.14, the mixing point M_{max} is located as shown, from which $(S/F)_{max}$ is determined to be about 16.

EXAMPLE 8.1

Acetone is to be extracted from a feed mixture of 30 wt% acetone (A) and 70 wt% ethyl acetate (C) at 30°C by using pure water (S) as the solvent by the cascade shown at the bottom of Figure 8.19. The final raffinate is to contain 5 wt% acetone on a water-free basis. Determine the minimum and maximum solvent-to-feed ratios and the number of equilibrium stages required for two intermediate S/F ratios. The equilibrium data, which are shown in Figure 8.19 and are taken from Venkataranam and Rao [28], correspond to a type I system, but with tie lines sloping downward toward the extract side of the diagram. Thus, although water is a convenient solvent, it does not have a high capacity, relative to ethyl acetate, for dissolving acetone. The flow diagram of the cascade in Figure 8.19 shows the nomenclature to be used for this example. Also, determine, for the feed, the maximum weight percent acetone that can enter the extractor. This example, as well as Example 8.2 later, are taken largely from an analysis by Sawistowski and Smith [29].

SOLUTION

Point B represents the solvent-free final raffinate. By drawing a straight line from B to S, the intersection with the equilibrium curve on the raffinate side, B', is the actual raffinate composition leaving stage N.

Minimum S/F. Because the tie lines slope downward toward the extract side of the diagram, we seek the extrapolated tie line that intersects the extrapolated line $\overline{SB}$ closest to S. This tie line, leading to P_{min}, is shown in Figure 8.20. The intersection is not shown because it occurs far to the left of the triangular diagram. Because this tie line is at the feed end of the extractor, the location of the extract composition, D'_{min}, is determined as shown in Figure 8.20. The mixing point, M_{min}, for $(S/F)_{min}$ is the intersection of lines $\overline{B'D'}$min and $\overline{SF}$. By the inverse-lever-arm rule, $(S/F)_{min} = \overline{FM}_{min}/\overline{SM}_{min} = 0.60$.

Maximum S/F. If M in Figure 8.20 is moved along line $\overline{FS}$ toward S, the intersection for $(S/F)_{max}$ occurs at the point shown on the extract side of the binodal curve. By the inverse-lever-arm rule, using line $\overline{FS}$, $(S/F)_{max} = \overline{FM}_{max}/\overline{SM}_{max} = 12$.

Equilibrium stages for other S/F ratios. First consider $S/F = 1.75$. In Figure 8.19, the composition of the saturated extract D' is obtained from a material balance about the extractor,

$$S + F = D' + B' = M$$

For $S/F = 1.75$, point M can be located such that $\overline{FM}/\overline{MS} = 1.75$. A straight line must also pass through D', B', and M. Therefore, D' can be located by extending $\overline{B'M}$ to the extract envelope.

The flow difference point P is located to the left of the triangular diagram. Therefore, $P = S - B' = D' - F$. It is located at the intersection of extensions of lines $\overline{FD'}$ and $\overline{B'S}$.

Stepping off stages poses no problem. Starting at D', we follow a tie line to L_1. Then V_2 is located by noting the intersection of the operating line $\overline{L_1P}$ with the phase envelope. Additional stages are stepped off in the same manner by alternating between the tie lines and operating lines. For the sake of clarity, only the first stage is shown; four are required.

For $S/F = 5(S/F)_{min} = 3.0$, M is determined and the stages are stepped off in a similar manner to give two equilibrium stages.

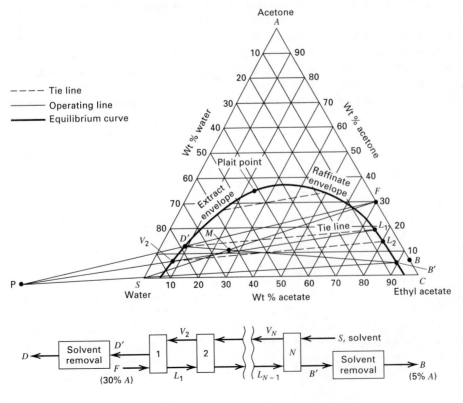

Figure 8.19 Determination of stages for Example 8.1 with $S/F = 1.75$.

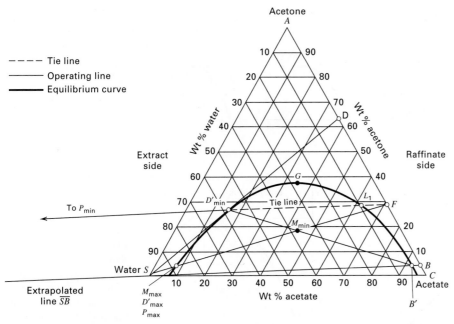
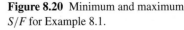

Figure 8.20 Minimum and maximum S/F for Example 8.1.

In summary, for the countercurrent cascade, we have.

S/F (solvent/feed ratio)	0.60	1.75	3	12
N (equilibrium stages)	∞	4	2	1
x_D (wt% acetone, solvent free)	64	62	50	30

If the wt% acetone in the feed mixture is increased from the base value of 30%, a feed composition will be reached that cannot be extracted because two liquid phases in equilibrium will not form (no phase splitting). This feed composition is determined by extending a line from S, tangent to the equilibrium curve, until it intersects $\overline{AC}$. This is shown as point D in Figure 8.20. The feed composition is 64 wt% acetone. Feed mixtures with a higher acetone content cannot be extracted with water.

Use of Right-Triangle Diagrams

As discussed in Chapter 4, diagrams other than the equilateral-triangle diagram are used for calculations involving ternary liquid–liquid systems. Ternary, countercurrent extraction calculations can also be made on a right-triangle diagram as shown by Kinney [27]; no new principles are involved. The disadvantage is that mass-percent compositions of only two of the components are plotted; the third being determined, when needed, by difference from 100%. The advantage of right-triangle diagrams is that ordinary rectangular-coordinates graph paper can be used and either one of the coordinates can be expanded, if necessary, to increase the accuracy of the constructions.

A right-triangle diagram can be developed from an equilateral-triangle diagram as shown in Figure 8.21a, where the coordinates in both diagrams are in mass fractions or in mole fractions. Point P on the equilibrium curve and tie line $\overline{RE}$ in the equilateral triangle become point P and tie line $\overline{RE}$ in the right-triangle diagram, which uses rectangular coordinates, x_A and x_C, where A is the solute and C is the carrier.

Consider the right-triangle diagram in Figure 8.22 for the A–C–S system of Figure 8.14. The compositions of S (the solvent) and A (the solute) are plotted in weight (mass) fractions, x_i. For example, point M represents a liquid mixture of overall composition ($x_A = 0.43$, $x_S = 0.28$). By difference, x_C, the carrier, which is not shown on the diagram, is $1 - 0.43 - 0.28 = 0.29$. Although lines of constant x_C

are included on the right triangle of Figure 8.22, such lines are usually omitted because they clutter the diagram. As with the equilateral-triangle diagram, Figure 8.22 for a right triangle includes the binodal curve, with extract and raffinate sides, tie lines connecting compositions of equilibrium phases, and the plait point, at $x_A = 0.49$.

Because point M falls within the phase envelope, the mixture separates into two liquid phases, whose compositions are given by points A' and A'' at the ends of the tie line that passes through point M. In this case, the extract at A'' is richer in the solute (A) and the solvent (S) than the raffinate at A'.

Point M might be the result of mixing a feed, point F, consisting of 26,100 kg/h of 60 wt% A in C ($x_A = 0.6, x_S = 0$), with 10,000 kg/h of pure furfural, point S. At equilibrium, the mixture splits into the phases represented by A' and A''. The location of point M and the amounts of extract and raffinate are given by the same mixing rule and inverse-lever-arm rule used for equilateral-triangle diagrams. The mixture separates spontaneously into 11,600 kg/h of raffinate ($x_S = 0.08, x_A = 0.32$) and 24,500 kg/h of extract ($x_S = 0.375, x_A = 0.48$).

Figure 8.23 represents the portion of an n-stage, countercurrent-flow cascade, where x and y are weight fractions of solute, A, in the raffinate and extract, respectively, and L and V are total amounts of raffinate and extract, respectively. The feed to stage N is $L_{N+1} = 180$ kg of 35 wt% A in a saturated mixture with C and S ($x_{N+1} = 0.35$), and

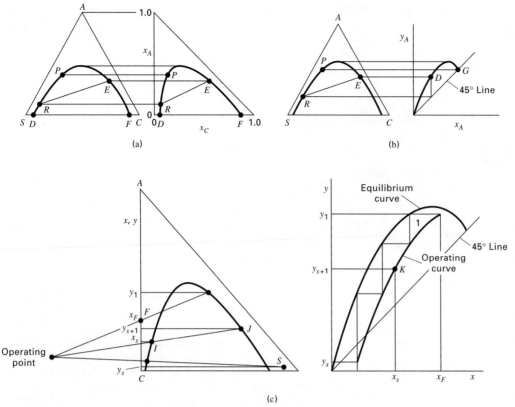

(a)

(b)

(c)

Figure 8.21 Development of other coordinate systems from the equilateral-triangle diagram: (a) to right-triangle diagram; (b) to auxiliary distribution curve. (c) Location of operating point on auxiliary McCabe–Thiele diagram.

[From R.E. Treybal, *Liquid Extraction,* 2nd ed., McGraw-Hill, New York (1963) with permission.]

the solvent to stage 1 is $V_W = 100$ kg of pure S ($y_W = 0.0$). Thus, the solvent-to-feed ratio is $100/180 = 0.556$. These two points are shown on the right-triangle diagram of Figure 8.24. The mixing point for L_{N+1} and V_W is shown as point M_1, as determined by the inverse lever-arm rule. Suppose

that the final raffinate, L_W, leaving stage 1 is to contain 0.05 weight-fraction glycol ($x_W = 0.05$). By an overall balance,

$$M_1 = V_W + L_{N+1} = V_N + L_W \qquad (8\text{-}11)$$

Applying the mixing rule, since V_W, L_{N+1}, and M_1 lie on a straight line, V_N, L_W, and M_1 must also lie on a straight line. Furthermore, because V_N leaves stage N at equilibrium and L_W leaves stage 1 at equilibrium, these two streams must lie on the extract and raffinate sides, respectively, of the equilibrium curve. The resulting points are shown in Figure 8.24, where it is seen that the weight fraction of glycol in the final extract is $y_N = 0.34$.

Figures 8.23b and 8.23c, and 8.24 include two additional cases of solvent-to-feed ratio, each with the same compositions for the solvent and the feed and the same

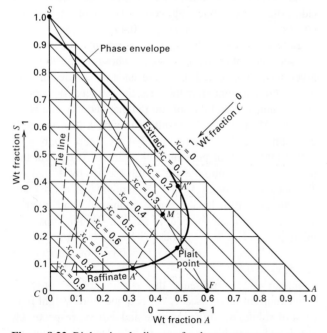

Figure 8.22 Right-triangle diagram for the ternary system of Figure 8.14.

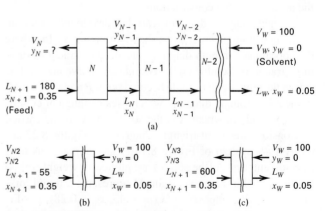

(a)

(b)

(c)

Figure 8.23 Multistage countercurrent contactors.

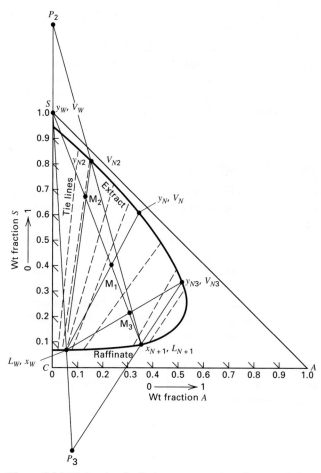

Figure 8.24 Right-triangle diagram constructions for cascade in Figure 8.23.

If we step off stages for Case 3, starting from y_{N3}, the tie line and the operating line coincide. That is, we have a pinch point. Thus, the solvent-to-feed ratio of 0.167 for this case is the minimum value corresponding to an infinite number of equilibrium stages.

For Case 1, where the solvent-to-feed ratio is in between that of Cases 2 and 3, the required number of equilibrium stages lies between 1 and ∞. Construction of the difference point and the steps for this case are not shown in Figure 8.14. The difference point is found to be located at a very large distance from the triangle because lines $\overline{L_W V_W}$ and $\overline{L_{N+1} V_N}$ are almost parallel. When the stages are stepped off, using operating lines parallel to $\overline{L_W V_W}$, it is found that between one and two stages are required.

Use of an Auxiliary Distribution Curve with a McCabe–Thiele Diagram

As the number of equilibrium stages to be stepped off on either one of the two types of triangular diagrams becomes more than a few, the diagram becomes cluttered. In that event, the use of a triangular diagram in conjunction with the McCabe–Thiele method becomes attractive. This is a method devised by Varteressian and Fenske [30]. The y–x diagram, discussed in Section 4.5 and illustrated in Figure 8.21b, is simply a plot of the tie-line data in terms of mass fractions (or mole fractions) of the solute in the extract (y_A) and equilibrium raffinate (x_A). The curve begins at the origin and terminates at the plait point where the curve intersects the 45° line. A tie line, such as $\overline{RE}$, in the triangular diagram becomes a point, in the equilibrium curve of Figure 8.25.

value for x_W:

Case	Feed L_{N+1}, kg	Solvent, V_W, kg	Solvent-to-Feed Ratio	Extract Designation	Mixing Point
1	180	100	0.556	V_N	M_1
2	55	100	1.818	V_{N2}	M_2
3	600	100	0.167	V_{N3}	M_3

For Case 2, a difference point, P_2, may be defined, as with the equilateral-triangle diagram, in terms of passing streams, as

$$P_2 = V_{N2} - L_{N+1} = V_W - L_W \qquad (8\text{-}12)$$

This point, shown in Figure 8.24, is located at the top of the diagram, where the lines $\overline{L_W V_W}$ and $\overline{L_{N+1} V_{N2}}$ intersect. For Case 3, the difference point, P_3, falls at the bottom of the diagram, where the lines $\overline{L_W V_W}$ and $\overline{L_{N+1} V_{N3}}$ intersect.

Equilibrium stages for Figure 8.24 are stepped off in a manner similar to that for an equilateral-triangle diagram by alternating use of equilibrium tie lines and operating lines passing through the difference point. For example, considering Case 2, with the high solvent-to-feed ratio of 1.818, and stepping off stages from stage N, a tie line from the point y_{N2} gives a value of $x_N = 0.04$. But this is less than the specified value of $x_w = 0.05$. Therefore, less than one equilibrium stage is required.

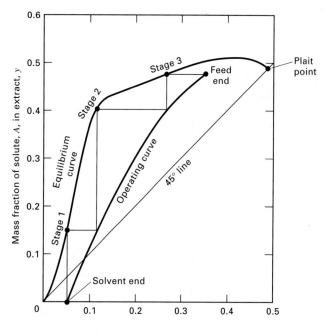

Figure 8.25 Stepping-off stages by the McCabe-Thiele method for the A–C–S system.

If an operating line is added to the equilibrium curve in Figure 8.21b, a staircase construction of the type used in the McCabe–Thiele method of Chapter 7 can rapidly determine the number of equilibrium stages. However, unlike distillation, where the operating line is straight because of the assumption of constant molar overflow, the operating line for liquid–liquid extraction in a ternary system will always be curved except in the low-solute-concentration region. Fortunately, the curved operating line is quite readily drawn using the following technique of Varteressian and Fenske [30]. In Figure 8.19 for the equilateral-triangle diagram, or in Figure 8.24 for the right-triangle diagram, the intersections of the equilibrium curve with a line drawn through a difference (operating) point represent the compositions of the passing streams. Thus, for each such operating line on the triangular diagram, one point of the operating line for the $y–x$ plot is determined. The operating lines passing through the difference point can be drawn at random; they need not coincide with passing streams of actual equilibrium-stage operating lines. Usually five or six such fictitious operating-line intersections, covering the expected range of compositions in the extraction cascade, are sufficient to establish the curved operating line in the $y–x$ plot. For example, in Figure 8.21c, the arbitrary operating line that intersects the equilibrium curve at I and J in the right-triangle diagram becomes a point K on

the operating line of the $y–x$ diagram. The $y–x$ plot of Figure 8.25 for the A–C–S system includes an operating line established in this manner, based on the data of Figure 8.24, but with a solvent-to-feed ratio of 0.208, that is, $V_W = 100$, $L_{N+1} = 480$ (25% greater than the minimum ratio of 0.167). The stages are stepped off in the McCabe–Thiele manner starting from the feed end. The result is seen to be almost exactly three equilibrium stages.

Extract and Raffinate Reflux

The simple, single-section, countercurrent, equilibrium-stage extraction cascade shown in Figure 8.13 can be refluxed, as in Figure 8.26a, to resemble distillation. In Figure 8.26a, L is used for raffinate flows, V is used for extract flows, and stages are numbered from the solvent end of the process. Extract reflux, L_R, is provided by sending the extract, V_N, to a solvent-recovery step, which removes most of the solvent, to give a solute-rich solution, $L_R + D$, which is divided into extract reflux, L_R, which is returned to stage N, and solute product, D. At the other end of the cascade, a portion, B, of the raffinate, L_1, is withdrawn in a stream divider and added as raffinate reflux, V_B, to fresh solvent, S. The remaining raffinate, B, is sent to a solvent-removal step (not shown) to produce a carrier-rich raffinate product. When using extract reflux,

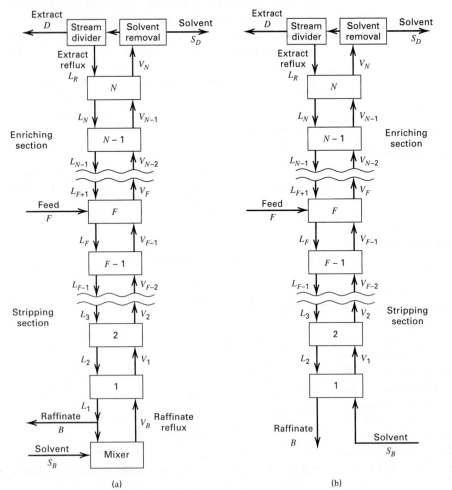

Figure 8.26 Liquid–liquid extraction with reflux: (a) with extract and raffinate reflux; (b) with extract reflux only.

Table 8.5 Analogy between Distillation and Extraction

Distillation	Extraction
Addition of heat	Addition of solvent
Reboiler	Solvent mixer
Removal of heat	Removal of solvent
Condenser	Solvent separator
Vapor at the boiling point	Solvent-rich solution saturated with solvent
Superheated vapor	Solvent-rich solution containing more solvent than that required to saturate it
Liquid below the boiling point	Solvent-lean solution, containing less solvent than that required to saturate it
Liquid at the boiling point	Solvent-lean solution saturated with solvent
Mixture of liquid and vapor	Two-phase liquid mixture
Relative volatility	Relative selectivity
Change of pressure	Change of temperature
D = distillate	D = extract product (solute on a solvent-free basis)
B = bottoms	B = raffinate (solvent-free basis)
L = saturated liquid	L = saturated raffinate (solvent-free)
V = saturated vapor	V = saturated extract (solvent-free)
A = more volatile component	A = solute to be recovered
C = less volatile component	C = carrier from which A is extracted
F = feed	F = feed
x = mole fraction A in liquid	X = mole or weight ratio of A (solvent-free), $A/(A+C)$
y = mole fraction A in vapor	$Y = S/(A+C)$

minimum- and total-reflux conditions, corresponding to infinite and minimum number of stages, bracket the optimal extract reflux ratio. Raffinate reflux is not processed through the solvent-removal unit because fresh solvent is added at this end of the cascade. It is necessary, however, to remove solvent from extract reflux at the enriching end of the cascade.

The analogy between a two-section liquid–liquid extractor with feed entering a middle stage, and distillation, is considered in some detail by Randall and Longtin [32]. Different aspects of the analogy are listed in Table 8.5. The most important analogy is that the solvent (a mass-separating agent) in extraction serves the same purpose as heat (an energy-separating agent) in distillation.

The use of raffinate reflux has been judged to be of little, if any, benefit by Skelland [31], who shows that the amount of raffinate reflux does not affect the number of stages required. Accordingly, we will consider a two-section, countercurrent cascade that includes only extract reflux, as shown in Figure 8.26b.

Analysis of a refluxed extractor, such as that of Figure 8.26b, involves relatively straightforward extensions of the procedures already developed. As will be shown, however, results for a type I system depend critically on the feed composition and the nature of the equilibrium-phase diagram, and it is very difficult to draw any general conclusions with respect to the effect (or even feasibility) of reflux.

For the two-section cascade with extract reflux shown in Figure 8.26b, a degrees-of-freedom analysis can be performed as described in Chapter 5. The result, using as elements two countercurrent cascades, a feed stage, a splitter, and a divider, is $N_D = 2N + 3C + 13$. All but four of the specifications will usually be

Variable Specification	Number of Variables
Pressure at each stage	N
Temperature for each stage	N
Feed-stream flow rate, composition, temperature, and pressure	$C+2$
Solvent composition, temperature, and pressure	$C+1$
Split of each component in the splitter (solvent removal step)	C
Temperature and pressure of the two streams leaving the splitter	4
Pressure and temperature of the divider	2
	$2N + 3C + 9$

The four additional specifications can be taken from one of the following sets:

Set 1	Set 2	Set 3
Solvent rate	Reflux ratio	Solvent rate
Solute concentration in extract (solvent free)	Solute concentration in extract (solvent-free)	Reflux ratio
Solute concentration in raffinate (solvent-free)	Solute concentration in raffinate (solvent-free)	Number of stages
Optimal feed-stage location	Optimal feed-stage location	Feed-stage location

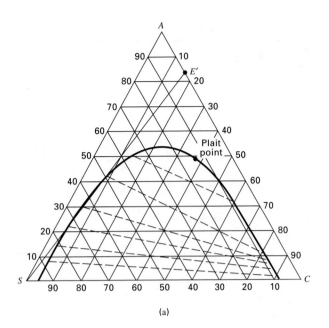

(a)

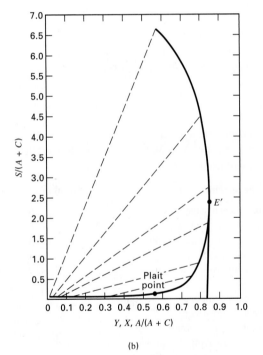

(b)

Figure 8.27 Limitation on product purity: (a) using an equilateral-triangle diagram; (b) using a Janecke diagram.

Sets 1 and 2 are of particular interest in the design of a new extractor because two of the specifications deal with the split of the feed into two products of designated purities, on a solvent-free basis. Set 2 is analogous to the design of a binary distillation column using the McCabe–Thiele graphical method, where the purities of the distillate and bottoms, the reflux ratio, and the optimal feed-stage location are specified. For a single-section cascade, it is not feasible to specify the split of the feed with respect to two key components. Instead, as in absorption and stripping, the recovery of just one component in the feed is specified.

It will be recalled for binary distillation that the purity of one of the products may be limited by the formation of an azeotrope. A similar limitation can occur for a type I system when using a two-section cascade with extract reflux, because of the plait point, which separates the two-liquid-phase region from the homogeneous, single-phase region. This limitation can be determined from a triangular diagram, but it is most readily observed on a Janecke diagram, of the type described in Chapter 4 and shown previously in Figure 4.14e. In Figure 8.27, liquid–liquid equilibrium data are repeated for the A–C–S system of Figure 8.14 where A is the solute and S is the solvent. In the triangular representation of Figure 8.27a, the maximum solvent-free solute concentration that can be achieved in the extract by a countercurrent cascade with extract reflux is determined by the intersection of line $\overline{SE'}$, drawn tangent to the binodal curve, from the pure solvent point S to the solvent-free composition line $\overline{AC}$, giving, in this case, 83 wt% solute. The same value is read from the binodal curve of the Janecke diagram of Figure 8.27b as the value of the abscissa for the point E' farthest to the right of the curve.

Without extract reflux, the maximum solvent-free solute concentration that can be achieved corresponds to an extract that is in equilibrium with the feed, when saturated with the extract. If this maximum value is close to the maximum value determined, as in Figure 8.27, then the use of extract reflux will be of little value. This is often the case for type I systems, as illustrated in the following example.

EXAMPLE 8.2

In Example 8.1, a feed mixture of 30 wt% acetone and 70 wt% ethyl acetate was extracted in a single-section, countercurrent cascade with pure water to obtain a raffinate of 5 wt% acetone on a water (solvent)-free basis. The maximum solvent-free solute concentration in the extract was found to be 64 wt%, as shown in Figure 8.20 at point D, corresponding to the condition of minimum $S/F = 0.60$ at infinite stages. For an actual $S/F = 1.75$ with four equilibrium stages, the extract contains 62 wt% acetone on a solvent-free basis. Thus, use of extract reflux for the purpose of producing a more pure (solvent-free) extract is not very attractive, given the particular phase-equilibrium diagram and feedstock composition. However, to demonstrate the technique, the calculation for extract reflux is carried out nevertheless. Also, the minimum number of equilibrium stages at total reflux and the minimum reflux ratio are determined.

SOLUTION

For the case of the single-section countercurrent cascade, the extract pinch point is at 57 wt% water, 27 wt% acetone, and 16 wt% acetate, as shown in Figure 8.20 at point $D'_{\min}$. If stages are added above the feed point, as in the two-section, refluxed cascade of Figure 8.26b, it is possible, theoretically, to reduce the water content of the extract to about 28 wt%, as shown by point G in Figure 8.20. However, the solvent (water)-free extract would not be as rich in acetone (51 wt%), which is determined from the line drawn through points S and G and extended to where it intersects the solvent-free line $\overline{AC}$.

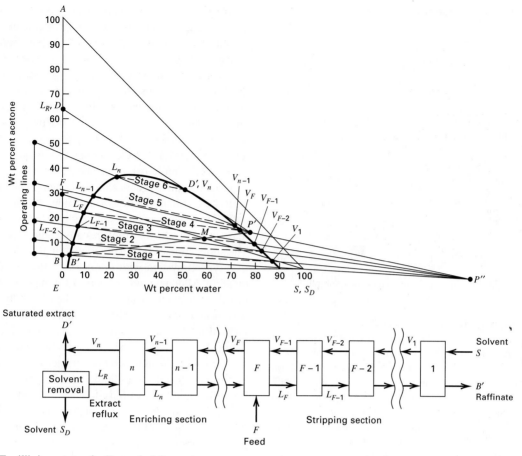

Figure 8.28 Equilibrium stages for Example 8.2.

To make this example more interesting, assume that a saturated extract containing 50 wt% water is required elsewhere in the process. Thus, the extraction cascade is that shown in Figure 8.28, rather than that of Figure 8.26b. The difference lies in the location of the solvent-removal step. This saturated-extract product is shown as point D' in Figure 8.28. Assume the ratio S/F to be 1.43, which is more than twice the minimum ratio found in Example 8.1. The desired raffinate composition is again 5 wt% acetone on a water-free basis (point B in Figure 8.28), which maps to point B' on the raffinate side of the binodal curve on a line connecting points B and S.

As with single-section cascades, the mixing point, M, in Figure 8.28, for the two streams entering the cascade, S and F, is determined by applying the inverse-lever-arm rule, using the S/F ratio, or by computing the overall composition of M, which in this case is 59 wt% water, 29 wt% acetate, and 12 wt% acetone.

The cascade in Figure 8.28 consists of an enriching section to the left of the feed point and a stripping section to the right, where extract is enriched in solute and raffinate is stripped of solute, respectively. A difference or operating point is needed for each section. We will let these be P' and P'', respectively, for the enriching and stripping sections. In the enriching section, referring to the cascade in Figure 8.28, $P' = V_n - L_R = V_{n-1} - L_n$. But, by material balance, $V_n - L_R = D' + S_D$. Therefore, $P' = D' + S_D$, that is, the total flow leaving the extract end of the cascade. Also, by overall material balance, $M = F + S = B' + D' + S_D = B' + P'$. Thus, P' must lie on a line drawn through points B' and M. To locate the position of P' on that line, we also note that V_n has the same composition as D', and L_R is simply D' with the solvent

removed (point D). From above, however, $P' = V_n - L_R$ or $V_n = P' + L_R$. Thus, point P' must also lie on a line drawn through points $V_n(D')$ and $L_R(D)$. Thus, point P' is located at the intersection of the extended lines $\overline{B'M}$ and $\overline{DD'}$ as shown in Figure 8.28.

The difference point, P'', for the stripping section is located in a similar manner, if we note that $P'' = B' - S = F - D' - S_D = F - P'$. Thus, P'' must be the intersection of extended lines $\overline{FP'}$ and $\overline{B'S}$ as shown to the right of the triangular diagram of Figure 8.28.

The stages can now be stepped off in the usual manner by starting from V_n and using the difference point P' until an operating line crosses the feed line $\overline{FP'}$. From there, the stages are stepped off using the difference point P'' until the raffinate composition is reached or exceeded. In Figure 8.28, it is seen that six equilibrium stages are required, with two in the enriching section and four in the stripping section. The feed enters the third stage from the left.

The reflux ratio, defined for this example as $(V_n - D')/D' = (L_R + S_D)/D'$, can be determined as follows. From above, $P' = D' + S_D$. Therefore, by the mixing rule, $S_D/D' = \overline{D'P'}/\overline{S_D P'}$. By material balance, $V_n - D' = L_R + S_D$. Therefore,

$$\frac{V_n - D'}{S_D} = \frac{L_R + S_D}{S_D} = \frac{\overline{L_R S_D}}{\overline{L_R D'}} \quad \text{and}$$

$$\frac{V_n - D'}{D'} = \left(\frac{S_D}{D'}\right)\left(\frac{V_n - D'}{S_D}\right) = \left(\frac{\overline{D'P'}}{\overline{S_D P'}}\right)\left(\frac{\overline{L_R S_D}}{\overline{L_R D'}}\right)$$

By measurement from Figure 8.28, $(V_n - D')/D' = (1.2)(2.0) = 2.4$. The reflux ratio is valid only for the selected solvent-to-feed ratio of 1.43.

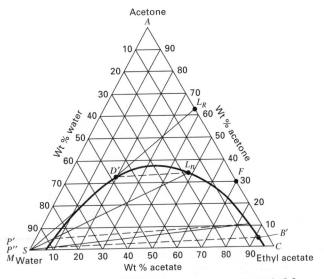

Figure 8.29 Total reflux and minimum stages for Example 8.2.

Next, we consider the case of total reflux, corresponding to the minimum number of stages. With reference to the equilateral-triangle diagram of Figure 8.29, compositions of existing streams are as previously specified or computed. With respect to acetone, we have 30 wt% in F, 4.9 wt%, in B', 33 wt%, in D', and 62 wt% in L_R. As in the case of the single-section cascade of Figure 8.20, as the solvent-to-feed ratio is increased, the mixing point $M = F + S$ moves toward the pure-solvent apex. At the maximum solvent addition, M lies at the intersection of the line through F and S with the extract side of the binodal curve. Difference points P' and P'' also move toward S because $P' = D' + S_D$ approaches S_D at total reflux and $P'' = F − P'$ approaches P', recalling that at total reflux $F = D' = 0$. As shown in Figure 8.29, the minimum number of equilibrium stages is three, as stepped off from the S apex.

Lastly, we consider the case of minimum reflux ratio at infinite stages, which also corresponds to the minimum solvent ratio. As the solvent ratio is reduced, point M moves toward the feed point, F, and point P'' moves away from the binodal curve. Also, point P' moves toward V_n. Ultimately, a value of the S/F ratio is reached where an operating line in either the enriching section or the stripping section coincides with a tie line, giving a pinch point and, therefore, an infinite number of stages. Often, this occurs for the extended tie line that passes through the feed point. Such is the case here, giving a minimum reflux ratio of about 0.6 and a corresponding minimum S/F ratio of about 0.75.

8.4 MALONEY–SCHUBERT GRAPHICAL EQUILIBRIUM-STAGE METHOD

For type II ternary systems, as shown in Figure 8.10b, use of a two-section cascade with extract reflux is particularly desirable. Without a plait point, the two-phase region extends all the way across the solute composition. Thus, while maximum solvent-free solute concentration in the extract is limited for a type I system as was shown in Figure 8.27, no such limit exists with a type II system. Accordingly, it is possible with extract reflux to achieve as sharp a separation as desired between the solute (A) and carrier (C).

When reflux is used, many stages may be required and the use of triangular diagrams is often not convenient. Instead, use can be made of a McCabe–Thiele-type diagram. Alternatively, a Janecke diagram, of the type shown earlier in Figure 4.14e, often in conjunction with a distribution diagram, has proved to be useful. In Janecke diagrams, which use convenient rectangular coordinates, solvent concentration on a solvent-free basis is plotted as the ordinate against solute concentration on a solvent-free basis as the abscissa, that is, $\%S/(\%A + \%C)$ against $\%A/(\%A + \%C)$, either with mass or mole percents. The Janecke diagram is analogous to an enthalpy–concentration diagram and is consistent with the distillation-extraction analogy of Table 8.5, where enthalpy is replaced by solvent concentration because a mass-separating agent replaces an energy-separating agent. The application of such a diagram to liquid–liquid extraction of a type II system with the use of reflux is considered in detail by Maloney and Schubert [33], who use an auxiliary distribution diagram of the McCabe–Thiele type, but on a solvent-free basis, to facilitate visualization of the stages. This method is also referred to as the *Ponchon–Savarit method for extraction*. Unlike the analogous method for distillation mentioned briefly at the end of Chapter 7 and which requires both enthalpy and vapor–liquid equilibrium data, the method for extraction requires only ternary liquid–liquid solubility data, which are far more common than combined vapor–liquid enthalpy and equilibrium data. Accordingly, despite the development of rigorous computer-aided methods, the Ponchon–Savarit method for extraction has remained useful, while the analogous method for distillation has rapidly declined in popularity. Although the Janecke diagram can also be applied to type I systems, it becomes difficult to use when the carrier and the solvent are highly immiscible, because the resulting values of the ordinate can become very large.

With the Janecke diagram, construction of tie lines, mixing points, operating points, and operating lines are all made in a manner similar to that for a triangular diagram [33]. Consider the case of extraction for a type II system with extract reflux, shown in Figure 8.26b. A representative Janecke diagram is shown in Figure 8.30, where all flow rates are on a solvent-free mass basis and the following solvent-free concentrations for both extract and raffinate phases are based on Janecke coordinates:

$$Y = \frac{\text{mass solvent}}{\text{mass of solvent-free liquid phase}}$$

$$X = \frac{\text{mass solute}}{\text{mass of solvent-free liquid phase}}$$

Values of the ordinate, Y, especially for the saturated-extract phase, can vary over a wide range depending on the solubility of the solute and carrier in the solvent. Values of the abscissa, X, vary from 0 to 1 (pure carrier to pure solute). Equilibrium tie lines relate concentrations in the saturated extract to the saturated raffinate. The Y location of the feed, F, is somewhere between zero and the saturated-raffinate curve. The extract, V_N, leaving stage N and prior to solvent

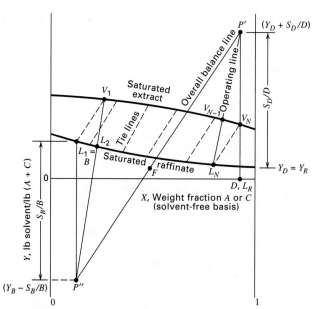

Figure 8.30 Construction of equilibrium stages on a Janecke diagram.

removal, is rich in solute and lies on the saturated extract curve. Upon solvent removal, the extract, D, and extract reflux, L_R, with identical compositions, lie on the $Y = 0$ horizontal line. The solvent removed from the cascade, S_D, is assumed to be pure. The raffinate, L_1, leaving stage 1 is rich in the carrier and lies on the saturated-raffinate line. Solvent S_B, fed to the cascade, is assumed to be pure.

Points P' and P'' are difference or operating points for the enriching and stripping sections, respectively, that are used to draw operating lines. The locations of P' and P'' are derived as follows: Referring to Figure 8.26b, solvent-free and solvent material balances around the solvent-recovery step, in terms of passing streams, are, respectively,

$$V_N - L_R = D \qquad (8\text{-}13)$$

$$Y_{V_N} V_N - Y_D L_R = S_D + Y_D D \qquad (8\text{-}14)$$

For a solvent difference balance around a section of top stages down to stage n, located above stage F in Figure 8.26b, we obtain

$$Y_{V_n} V_n - Y_{L_{n+1}} L_{n+1} = S_D + Y_D D \qquad (8\text{-}15)$$

Thus, any solvent flow difference between passing streams in the enriching section above the feed stage is given by $S_D + Y_D D$. If (8-13) and (8-14) are combined to eliminate V_N, we obtain

$$\frac{L_R}{D} = \frac{(Y_D + S_D/D) - Y_{V_N}}{Y_{V_N} - Y_D} \qquad (8\text{-}16)$$

In Figure 8.30, $(Y_{V_N} - Y_D)$ is the vertical distance between the points V_N and (D, L_R). The difference point, P', in Figure 8.30 becomes $(S_D + Y_D D)$ divided by D to give $(Y_D + S_D/D)$. That is,

$$P' = Y_D + S_D/D \qquad (8\text{-}17)$$

and $\qquad L_R/D = \overline{P'V_N}/\overline{V_N L_R} \qquad (8\text{-}18)$

Similarly, for the stripping section, it can be shown that,

$$P'' = Y_B - S_B/B \qquad (8\text{-}19)$$

$$D/B = \overline{FP''}/\overline{P'F} \qquad (8\text{-}20)$$

Stages are stepped off in a manner analogous to that for the triangular diagram, starting from either the extract, D, or the raffinate, B, alternating between operating lines and tie lines. For example, in Figure 8.30, we can start from the top of the cascade at the extract D and step off stages in the enriching section. The solute compositions of D, L_R, and V_N on a solvent-free basis are identical. Thus, the operating line for passing streams L_R and V_N is a vertical line passing through the difference point P'. From point V_N on the saturated extract curve, a tie line is followed down to the equilibrium raffinate phase, L_N. An operating line connecting points P' and L_N intersects the extract curve at the passing stream V_{N-1}. Subsequent stages in the enriching section are stepped off in a similar manner until the feed stage is reached. The optimal location of this stage is determined in a manner analogous to the intersection of the rectification and stripping operating lines in the McCabe–Thiele method. On the Janecke diagram, this intersection is the line $\overline{P'P''}$, which passes through the feed point, as shown in Figure 8.30. Thus, the transition from the enriching section (where the difference point P' is used) to the stripping section (where P'' is used) is made when an equilibrium tie line for a stage crosses the line $\overline{P'P''}$. Following the location of the feed stage, the remaining stripping-section stages are stepped off until the desired product raffinate solvent-free concentration is reached or crossed over.

The Janecke diagram can also be used to determine the two limiting conditions of total reflux (minimum stages) and minimum reflux (infinite stages). For total reflux, the difference points P' and P'' lie at $Y = +\infty$ and $-\infty$, respectively, because $F = B = D = 0$. Thus, all operating lines become vertical lines and the minimum number of stages are stepped off in the manner illustrated in Figure 8.31a.

For the condition of minimum reflux, a pinch condition is sought either at the feed stage or some other stage location. In Figure 8.31b, where the pinch is assumed to be at the feed stage, an operating line is drawn coincident with a tie line and the feed point, F, to determine points P' and P''. To determine if the pinch does occur at the feed stage, tie lines to the right of the feed-stage tie line are extended to an intersection with the vertical line through D. If a higher intersection occurs, then that P' may be the correct P'_{min} difference point for minimum reflux. In a similar manner, tie lines to the left of the feed-stage tie line are extended to an intersection with the vertical line through B. If a lower intersection occurs, then that P'' may be the factor that determines the minimum reflux. The former case is shown in Figure 8.31c, where P' is higher than P'_1. Thus, $P' = P'_{min}$. In any case, once the controlling P' or P'' is determined, a line through F determines the other difference point and the minimum reflux ratio is computed from (8-18) using P'_{min}.

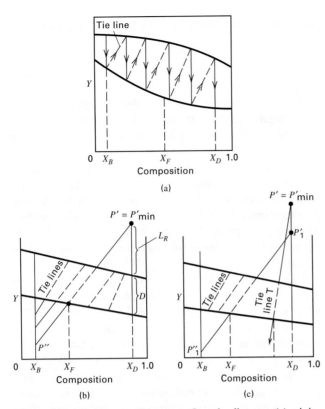

Figure 8.31 Limiting conditions on a Janecke diagram: (a) minimum number of stages at total reflux; (b) minimum reflux determined by a tie line through the feed point; (c) minimum reflux determined by a tie line to the right of the feed point.

EXAMPLE 8.3

As shown in Figure 8.32, a countercurrent, extraction cascade equipped with a perfect solvent separator to provide extract reflux is used to separate methylcyclopentane (A) and n-hexane (C) into a final extract and raffinate containing, on a solvent-free basis, 95 wt% and 5 wt% A, respectively, using aniline (S) as the solvent. The feed rate is 1,000 kg/h with 55 wt% A, and the mass ratio of solvent to feed is 4.0. The feed contains no aniline and the fresh solvent is pure. Recycle solvent is also assumed to be pure. Equilibrium curves and tie lines are given in Figure 8.33.

(a) Determine the reflux ratio and number of stages. Equilibrium data at extractor temperature and pressure are shown for mass units in the Janecke diagram of Figure 8.33. Feed is to enter at the optimal stage.

(b) Determine the minimum number of stages for the specified solvent-free extract and raffinate compositions.

(c) Determine the minimum reflux ratio for the specified feed and product compositions.

SOLUTION

(a) First, determine all product rates by material-balance calculations. An overall balance on solute plus carrier gives

$$D + B = 1,000 \text{ kg/h}$$

A solute balance gives $0.95D + 0.05B = (0.55)(1,000)$
$$= 550 \text{ kg/h}$$

Solving these two equations simultaneously gives $D = 556$ kg/h and $B = 444$ kg/h. Since $S_B = 4,000$ kg/h, $S_B/B = 9.0$. In Figure 8.30, point P'' is located at a distance of S_B/B below the raffinate composition, X_B, at point B. Since Y at point B, from Figure 8.33, is approximately 0.3, point P'' is located at $0.3 - 9.0 = -8.7$.

A line drawn through P'' and F, extended to the intersection with the vertical line through D, gives $P' = 6.7$. By measurement from Figure 8.33, using (8-16), $L_R/D = 3.7$.

In Figure 8.33, stages are stepped off starting from point D. At the third stage ($N - 2$), the tie line crosses line $\overline{P''FP'}$. Thus, this is the optimal feed stage. Three more stages are required to reach B, giving a total of six equilibrium stages.

(b) If the construction for minimum stages, shown in Figure 8.31a, is used in Figure 8.33, just less than five stages are determined.

(c) If the construction for minimum reflux, shown in Figure 8.31b for a pinch at the feed stage, is used in Figure 8.33, a value of $P' = 2.90$ is obtained. No other tie line in either section gives a larger value. Therefore, $P_{min} = 2.9$. By measurement, using (8-18), $(L_R/D)_{min} = 0.83$. Using the construction indicated in Figure 8.30, the corresponding $(S_B/B)_{min}$ is found to be 4.2. Thus, $(S_B)_{min} = 4.2(444) = 1,865$ kg/h or $(S_B/F)_{min} = 1,865/1,000 = 1.865$.

For this example, a relatively high reflux ratio and corresponding solvent-to-feed ratio is employed to keep the required number of equilibrium stages small. When the number of equilibrium stages is large, the Janecke diagram becomes cluttered with operating lines and tie lines. In that case, an auxiliary McCabe–Thiele-type plot of solute mass fraction in the extract layer versus solute mass fraction in the raffinate layer, both on a solvent-free basis, as in the Janecke diagram, can be drawn, with points on the enriching and stripping operating lines determined, as discussed above, from arbitrary operating lines on the Janecke diagram. Stages are

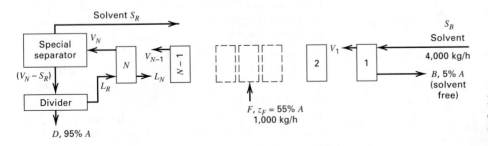

Figure 8.32 Countercurrent extraction cascade with extract reflux for Example 8.3.

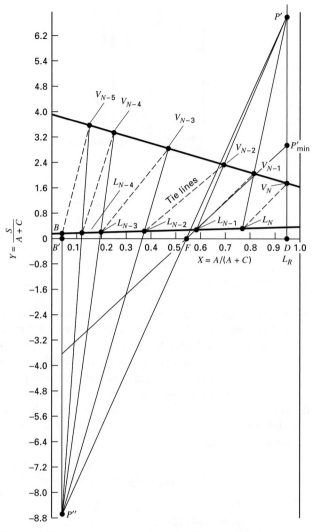

Figure 8.33 Maloney–Schubert constructions on Janecke diagram for Example 8.3.

then stepped off in the McCabe–Thiele manner. An example of the Janecke diagram with such an auxiliary McCabe–Thiele diagram is shown in Figure 8.34, taken from Maloney and Schubert [33].

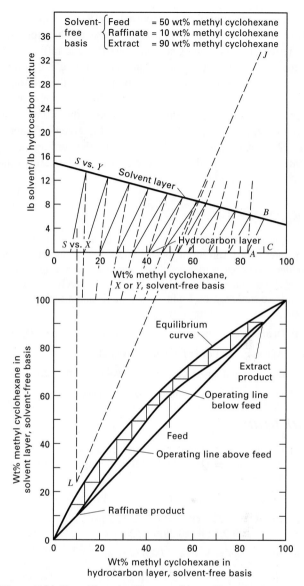

Figure 8.34 Use of Janecke diagram with auxiliary distribution diagram.

[From *Chemical Engineers' Handbook,* 5th ed., R.H. Perry and C.H. Chilton, Eds., McGraw-Hill, New York (1973).]

8.5 THEORY AND SCALE-UP OF EXTRACTOR PERFORMANCE

Following the estimation, by methods described in Sections 8.3 and 8.4, of the number of equilibrium stages, suitable extraction equipment can be selected using the scheme of Figure 8.8. Often, the choice is between a cascade of mixer-settler units or a multicompartment, column-type extractor with mechanical agitation, the main considerations being the number of stages required, and the floor space and head room available. Methods for estimating size and power requirements of these two general types of extractors are presented next. Column devices with no mechanical agitation are also considered.

Mixer-Settler Units

Sizing of mixer-settler units is done most accurately by scale-up from batch or continuous runs in laboratory or pilot-plant equipment. However, preliminary-sizing calculations can be made using available theory and empirical correlations. Experimental data of Flynn and Treybal [34] show that when liquid-phase viscosities are less than 5 cP and the specific-gravity difference between the two liquid phases is greater than about 0.10, the average residence time required of the two liquid phases in the mixing vessel to achieve at least 90% stage efficiency may be as low as 30 s and is usually not more than 5 min, when an agitator-power input per mixer volume of 1,000 ft-lbf/min-ft^3 (4 hp/1,000 gal) is used.

Based on experiments reported by Ryan, Daley, and Lowrie [35], the capacity of a settler vessel can be expressed in terms of C gal/min of combined extract and raffinate per square foot of phase-disengaging area. For a horizontal, cylindrical vessel of length L and diameter D_T, the economic ratio of L to D_T is approximately 4. Thus, if the phase interface is located at the middle of the vessel, the disengaging area is $D_T L$ or $4D_T^2$. A typical value of C given by Happel and Jordan [36] is about 5. Frequently, the settling vessel will be larger than the mixing vessel, as is the case in the following example.

EXAMPLE 8.4

Benzoic acid is to be continuously extracted from a dilute solution in water with a solvent of toluene in a series of discrete mixer-settler vessels operated in countercurrent flow. The flow rates of the feed and solvent are 500 and 750 gal/min, respectively. Assuming a residence time, t_{res}, of 2 min in each mixer and a settling vessel capacity of 5 gal/min-ft², estimate:

(a) Diameter and height of a mixing vessel, assuming $H/D_T = 1$

(b) Agitator horsepower for a mixing vessel

(c) Diameter and length of a settling vessel, assuming $L/D_T = 4$

(d) Residence time in a settling vessel in minutes

SOLUTION

(a) Q = total flow rate $= 500 + 750 = 1{,}250$ gal/min
V = volume $= Q t_{res} = 1{,}250(2) = 2{,}500$ gal or $2{,}500/7.48 =$
334 ft³

$$V = \pi D_T^2 H/4, \quad H = D_T, \quad \text{and} \quad V = \pi D_T^3/4$$

$$D_T = (4V/\pi)^{1/3} = [(4)(334)/3.14]^{1/3}$$
$$= 7.52 \text{ ft and } H = 7.52 \text{ ft}$$

(b) Horsepower $= 4(2{,}500/1{,}000) = 10$ hp

(c) $D_T L = 1{,}250/5 = 250$ ft²; $D_T^2 = 250/4 = 62.5$ ft²

$$D_T = 7.9 \text{ ft}; \quad L = 4D_T = 4(7.9) = 31.6 \text{ ft}$$

(d) Volume of settler $= \pi D_T^2 L/4 = 3.14(7.9)^2(31.6)/4 = 1{,}548$ ft³
or $1{,}548(7.48) = 11{,}580$ gal

$$t_{res} = V/Q = 11{,}580/1{,}250 = 9.3 \text{ min}$$

A typical single-compartment mixing tank for liquid–liquid extraction is shown in Figure 8.35. The vessel is closed with the two liquid phases entering at the bottom and the effluent, in the form of a two-phase emulsion, leaving at the top. Although flat tank heads are shown in Figure 8.35, rounded heads of the type in Figure 8.2 are preferred to eliminate stagnant fluid regions. Air or other gases must be evacuated from the vessel so no gas–liquid interface exists.

Mixing is accomplished by an appropriate, centrally located impeller selected from the many types available, some of which are shown in Figure 8.3. For example, a flat-blade turbine might be chosen as in Figure 8.35. A single turbine is adequate unless the vessel height is greater than the vessel diameter, in which case a compartmented vessel with two or

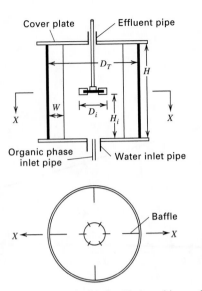

Figure 8.35 Agitated vessel with flat-blade turbine and baffles.

more impellers might be employed. When the vessel is open, vertical side baffles are mandatory to prevent vortex formation at the gas–liquid interface. For closed vessels that run full of liquid, vortexing will not occur. Nevertheless, it is common to install baffles, even in closed tanks, to minimize swirling and improve circulation patterns. Although no standards exist for vessel and turbine geometry, the following, with reference to Figure 8.35, give good dispersion performance in liquid–liquid agitation:

$$\text{Number of turbine blades} = 6;$$
$$\text{Number of vertical baffles} = 4$$

$$H/D_T = 1; \quad D_i/D_T = 1/3;$$
$$W/D_T = 1/12 \quad \text{and} \quad H_i/H = 1/2$$

To achieve a high stage efficiency for extraction in a mixing vessel—say, between 90 and 100%—it is necessary to provide fairly vigorous agitation. For a given type of impeller and vessel–impeller geometry, the agitator power, P, can be estimated from an empirical correlation in terms of a power number, N_{Po}, which depends on an impeller Reynolds number, N_{Re}, where

$$N_{Po} = \frac{P g_c}{N^3 D_i^5 \rho_M} \qquad (8\text{-}21)$$

$$N_{Re} = \frac{D_i^2 N \rho_M}{\mu_M} \qquad (8\text{-}22)$$

The impeller Reynolds number is the ratio of the inertial force to the viscous force:

$$\text{Inertial force} \propto (N D_i)^2 \rho_M D_i^2$$

$$\text{Viscous force} \propto \frac{\mu_M (N D_i) D_i^2}{D_i}$$

where N = rate of impeller rotation. Thus, the characteristic length in the impeller Reynolds number is the impeller diameter and the characteristic velocity is $N D_i$ = impeller peripheral velocity.

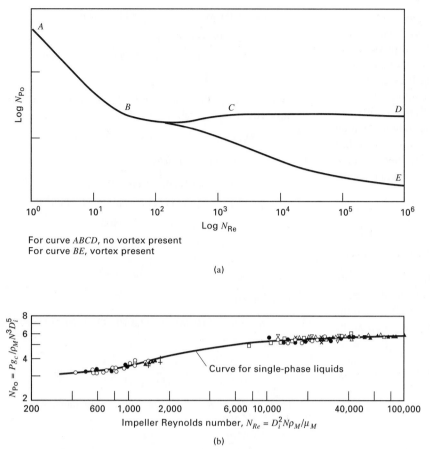

For curve *ABCD*, no vortex present
For curve *BE*, vortex present

(a)

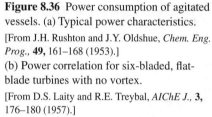

Curve for single-phase liquids

(b)

Figure 8.36 Power consumption of agitated vessels. (a) Typical power characteristics.
[From J.H. Rushton and J.Y. Oldshue, *Chem. Eng. Prog.*, **49**, 161–168 (1953).]
(b) Power correlation for six-bladed, flat-blade turbines with no vortex.
[From D.S. Laity and R.E. Treybal, *AIChE J.*, **3**, 176–180 (1957).]

The agitator power is proportional to the product of the volumetric liquid flow produced by the impeller and the applied kinetic energy per unit volume of fluid. The result is

$$P \propto (ND_i^3)[\rho_M(ND_i)^2/2g_c]$$

which can be rewritten as (8-21), where the constant of proportionality is $2N_{Po}$. Both the impeller Reynolds number and the power number (also called the Newton number) are dimensionless groups. Thus, any consistent set of units can be used. The power number for an agitated vessel serves the same purpose as the friction factor for the flow of a fluid through a pipe. This is illustrated, over a wide range of impeller Reynolds number, for a typical mixing impeller in Figure 8.36a, taken from the work of Rushton and Oldshue [37]. The upper curve, ABCD, pertains to a vessel with baffles, while the lower curve, ABE, pertains to the same tank with no baffles. In the low-Reynolds-number region, AB, viscous forces dominate and the impeller power is proportional to $\mu_M N^2 D_i^3$. Somewhere beyond a Reynolds number of about 200, a vortex appears if no baffles are present and the power-number relation is given by curve BE. In this region, the Froude number, $N_{Fr} = N^2 D_i/g$, which is the ratio of inertial to gravitational forces, also becomes a factor. With baffles present, and the Reynolds number greater than about 1,000, a region, CD, is reached where fully developed turbulent flow exists. Now, inertial forces dominate and the power is proportional to $\rho_M N^3 D_i^5$. It is clear that the addition of baffles greatly increases power requirements in the turbulent flow region.

A correlation of experimental data for liquid–liquid mixing in baffled vessels with six-bladed, flat-blade turbines is shown in Figure 8.36b, from a study by Laity and Treybal [38]. The range of impeller Reynolds number covers only the turbulent–flow region, where efficient liquid–liquid mixing is achieved. The solid line represents batch mixing of single-phase liquids. The data points represent liquid–liquid mixing, where agreement is achieved with the single-phase curve by computing two-phase mixture properties from

$$\rho_M = \rho_C \phi_C + \rho_D \phi_D \qquad (8\text{-}23)$$

$$\mu_M = \frac{\mu_C}{\phi_C}\left(1 + \frac{1.5\mu_D\phi_D}{\mu_C + \mu_D}\right) \qquad (8\text{-}24)$$

where ϕ is the volume fraction of holdup in the tank, with subscripts C for the continuous phase and D the dispersed phase, such that $\phi_D + \phi_C = 1$. When measurements were made for continuous flow from inlets at the bottom of the vessel to an outlet for the emulsion from the top of the vessel and with the impeller located at a position above the liquid–liquid interface when at rest, the data were correlated with the curve of Figure 8.36b.

With fully developed turbulent flow, the volume fraction of dispersed phase in the vessel closely approximates that in the feed to the vessel; otherwise the volume fraction may be different from that in the total feed to the vessel. That is, the residence times of the two phases in the vessel may not be the same. At best, spheres of uniform size can pack tightly to give a void fraction of 0.26. Therefore, $\phi_C > 0.26$ and $\phi_D < 0.74$ is sometimes quoted. However, some experiments have shown a 0.20–0.80 range. For continuous flow, the vessel is first filled with the phase to be continuous. Following initiation of agitation, the two-feed liquids are then introduced into the vessel in their desired volume ratio.

Based on the work of Skelland and Ramsay [39] and Skelland and Lee [40], a minimum impeller rate of rotation is required for complete and uniform dispersion of one liquid into another. For a flat-blade turbine in a baffled vessel of the type discussed above, this minimum rotation rate can be estimated from

$$\frac{N_{min}^2 \rho_M D_i}{g\,\Delta\rho} = 1.03 \left(\frac{D_T}{D_i}\right)^{2.76} \phi_D^{0.106} \left(\frac{\mu_M^2 \sigma}{D_i^5 \rho_M g^2 (\Delta\rho)^2}\right)^{0.084}$$

(8-25)

where $\Delta\rho$ is the absolute value of the difference in density and σ is the interfacial tension between the two liquid phases. The dimensionless group on the left-hand side of (8-25) is the two-phase Froude number. The dimensionless group at the far right of (8-25) is a ratio of forces:

$$\frac{(\text{viscous})^2 (\text{interfacial tension})}{(\text{inertial})(\text{gravitational})^2}$$

EXAMPLE 8.5

Furfural is to be continuously extracted from a dilute solution in water by toluene at 25°C in an agitated vessel of the type shown in Figure 8.35. The feed enters at a flow rate of 20,400 lb/h, while the solvent enters at 11,200 lb/h. For a residence time in the vessel of 2 min, estimate for either phase as the dispersed phase:

(a) The dimensions of the mixing vessel and the diameter of the flat-blade turbine impeller

(b) The minimum rate of rotation of the impeller for complete and uniform dispersion

(c) The power requirement of the agitator at the minimum rotation rate

SOLUTION

Mass flow rate of feed = 20,400 lb/h; feed density = 62.3 lb/ft^3

Volumetric flow rate of feed = Q_F = 20,400/62.3 = 327 ft^3/h

Mass flow rate of solvent = 11,200 lb/h

Solvent density = 54.2 lb/ft^3; volumetric flow rate of solvent = Q_S = 11,200/54.2 = 207 ft^3/h

Because of the dilute concentration of solute in the feed and sufficient agitation to achieve complete and uniform dispersion, assume fractional volumetric holdups of raffinate and extract in the vessel are equal to the corresponding volume fractions in the combined feed (raffinate, R) and solvent (extract, E) entering the mixer:

$$\phi_R = 327/(327 + 207) = 0.612; \quad \phi_E = 1 - 0.612 = 0.388$$

(a) Mixer volume = $(Q_F + Q_S)t_{res} = V = (327 + 207)(2/60) = 17.8$ ft^3. Assume a cylindrical vessel with $D_T = H$ and neglect the volume of the bottom and top heads and the volume

occupied by the agitator and the baffles. Then

$$V = \left(\pi D_T^2/4\right) H = \pi D_T^3/4$$
$$D_T = [(4/\pi)V]^{1/3} = [(4/3.14)17.8]^{1/3} = 2.83 \text{ ft}$$
$$H = D_T = 2.83 \text{ ft}$$

Make the vessel 3 ft in diameter by 3 ft high, giving a volume $V = 21.2$ ft^3 = 159 gal. Assume that

$$D_i/D_T = 1/3; \quad D_i = D_T/3 = 3/3 = 1 \text{ ft.}$$

(b) *Case 1—Raffinate phase dispersed:*

$$\phi_D = \phi_R = 0.612; \quad \phi_C = \phi_E = 0.388$$
$$\rho_D = \rho_R = 62.3 \text{ lb/ft}^3; \quad \rho_C = \rho_E = 54.2 \text{ lb/ft}^3$$
$$\mu_D = \mu_R = 0.89 \text{ cP} = 2.16 \text{ lb/h-ft};$$
$$\mu_C = \mu_E = 0.59 \text{ cP} = 1.43 \text{ lb/h-ft}$$
$$\Delta\rho = 62.3 - 54.2 = 8.1 \text{ lb/ft}^3;$$
$$\sigma = 25 \text{ dyne/cm} = 719,000 \text{ lb/h}^2$$

From (8-23),

$$\rho_M = (54.2)(0.388) + (62.3)(0.612) = 59.2 \text{ lb/ft}^3$$

From (8-24),

$$\mu_M = \frac{1.43}{0.388}\left[1 + \frac{1.5(2.16)(0.612)}{1.43 + 2.16}\right] = 5.72 \text{ lb/h-ft}$$

From (8-25), using American engineering units, with $g = 4.17 \times 10^8$ ft/h^2,

$$\frac{\mu_M^2 \sigma}{D_i^5 \rho_M g^2 (\Delta\rho)^2} = \frac{(5.72)^2 (719,000)}{(1)^5 (59.2)(4.17 \times 10^8)^2 (8.1)^2}$$
$$= 3.47 \times 10^{-14}$$

$$N_{min}^2 = 1.03\left(\frac{g\,\Delta\rho}{\rho_M D_i}\right)\left(\frac{D_T}{D_i}\right)^{2.76} \phi_D^{0.106}(3.47 \times 10^{-14})^{0.084}$$

$$= 1.03\left[\frac{(4.17 \times 10^8)(8.1)}{(59.2)(1)}\right]\left(\frac{3}{1}\right)^{2.76}(0.612)^{0.106}(0.0740)$$

$$= 8.56 \times 10^7 (\text{rph})^2$$

$N_{min} = 9,250$ rph = 155 rpm

Case 2—Extract phase dispersed: Calculations similar to case 1 result in $N_{min} = 8,820$ rph = 147 rpm

(c) *Case 1—Raffinate phase dispersed:*

From (8-22), $\quad N_{Re} = \frac{(1)^2 (9,250)(59.2)}{(5.72)} = 9.57 \times 10^4$

From Figure 8.36b, it is seen that a fully turbulent flow exists, with the power number given by its asymptotic value of $N_{Po} = 5.7$.

From (8-21),

$$P = N_{Po} N^3 D_i^5 \rho_M/g_c$$
$$= (5.7)(9,250)^3 (1)^5 (59.2)/(4.17 \times 10^8)$$
$$= 640,000 \text{ ft-lbf/h} = 0.323 \text{ hp}$$
$$P/V = 0.323(1000)/159 = 2.0 \text{ hp/1,000 gal}$$

Case 2—Extract phase dispersed:

Calculations similar to case 1 result in $P = 423,000$ ft-lbf/h = 0.214 hp.

$$P/V = 0.214(1000)/159 = 1.4 \text{ hp/1,000 gal}$$

Mass-Transfer Efficiency

When dispersion is complete and uniform, the contents of the vessel are perfectly mixed with respect to both phases. In that case, the concentration of the solute in each of the two phases in the mixing vessel is uniform and equal to the concentrations in the two-phase emulsion leaving the mixing vessel. This is the so-called ideal CFSTR or CSTR (continuous-flow stirred-tank reactor) model, sometimes called the completely back-mixed or perfectly mixed model, first discussed by MacMullin and Weber [41] and widely applied to reactor design. The Murphree dispersed-phase efficiency for liquid–liquid extraction, based on the raffinate as the dispersed phase, can be expressed as the fractional approach to equilibrium. In terms of bulk molar concentrations of the solute,

$$E_{MD} = \frac{c_{D,\text{in}} - c_{D,\text{out}}}{c_{D,\text{in}} - c_D^*} \qquad (8\text{-}26)$$

where c_D^* is the solute concentration in equilibrium with the bulk solute concentration in the exiting continuous phase, $c_{C,\text{out}}$. The molar rate of mass transfer of the solute, n, from the dispersed phase to the continuous phase can be expressed as

$$n = K_{OD}a(c_{D,\text{out}} - c_D^*)V \qquad (8\text{-}27)$$

where the concentration driving force for mass transfer is uniform throughout the well-mixed vessel and is equal to the driving force based on the exit concentrations, a is the interfacial area for mass transfer per unit volume of liquid phases, V is the total volume of liquid phases in the vessel, and K_{OD} is the overall mass-transfer coefficient based on the dispersed phase, which is given in terms of the separate resistances of the dispersed and continuous phases by

$$\frac{1}{K_{OD}} = \frac{1}{k_D} + \frac{1}{mk_C} \qquad (8\text{-}28)$$

where equilibrium is assumed at the interface between the two phases and m = the slope of the equilibrium curve for the solute plotted as c_C versus c_D:

$$m = dc_C/dc_D \qquad (8\text{-}29)$$

For dilute solutions, changes in volumetric flow rates of the raffinate and extract are small, and thus the rate of mass transfer based on the change in solute concentration in the dispersed phase is given by material balance:

$$n = Q_D(c_{D,\text{in}} - c_{D,\text{out}}) \qquad (8\text{-}30)$$

where Q_D is the volumetric flow rate of the dispersed phase.

To obtain an expression for E_{MD} in terms of $K_{OD}a$, (8-26), (8-27), and (8-30) are combined in the following manner. From (8-26),

$$\frac{E_{MD}}{1 - E_{MD}} = \frac{c_{D,\text{in}} - c_{D,\text{out}}}{c_{D,\text{out}} - c_D^*} \qquad (8\text{-}31)$$

Equating (8-27) and (8-30), and noting that the right-hand side of (8-31) is the number of dispersed-phase transfer units

for a perfectly mixed vessel with $c_D = c_{D,\text{out}}$,

$$N_{OD} = \int_{c_{D,\text{out}}}^{c_{D,\text{in}}} \frac{dc_D}{c_D - c_D^*} = \frac{c_{D,\text{in}} - c_{D,\text{out}}}{c_{D,\text{out}} - c_D^*} = \frac{K_{OD}aV}{Q_D} \qquad (8\text{-}32)$$

Combining (8-31) and (8-32) and solving for E_{MD},

$$E_{MD} = \frac{K_{OD}aV/Q_D}{1 + K_{OD}aV/Q_D} = \frac{N_{OD}}{1 + N_{OD}} \qquad (8\text{-}33)$$

When $N_{OD} = (K_{OD}aV/Q_D) \gg 1$, $E_{MD} = 1$.

Drop Size and Interfacial Area

From (8-33) and (8-28), it is seen that an estimate of E_{MD} requires generalized correlations of experimental data for the interfacial area for mass transfer, a, and the dispersed- and continuous-phase mass-transfer coefficients, k_D and k_C, respectively. The population of dispersed-phase droplets in an agitated vessel will cover a range of sizes and shapes. For each droplet, it is useful to define d_e, the equivalent diameter of a spherical drop, using the method of Lewis, Jones, and Pratt [42],

$$d_e = (d_1^2 d_2)^{1/3} \qquad (8\text{-}34)$$

where d_1 and d_2 are the major and minor axes, respectively, of an ellipsoidal-drop image. For a spherical drop, d_e is simply the diameter of the drop. For the population of drops, it is useful to define an average or mean drop diameter. A number of different definitions are available depending on whether weight-mean, mean-volume, surface-mean, mean-surface, length-mean, or mean-length diameter is appropriate [43]. For mass-transfer calculations, the surface-mean diameter, d_{vs} (also called the *Sauter mean diameter*), is most appropriate because it is the mean drop diameter that gives the same interfacial surface area as the entire population of drops for the same mass of drops. It is determined from experimental drop-size distribution data for N drops by the definition:

$$\frac{\pi d_{vs}^2}{(\pi/6)d_{vs}^3} = \frac{\pi \sum_N d_e^2}{(\pi/6)\sum_N d_e^3}$$

which, when solved for d_{vs}, gives

$$d_{vs} = \frac{\sum_N d_e^3}{\sum_N d_e^2} \qquad (8\text{-}35)$$

With this definition, the interfacial surface area per unit volume of a two-phase mixture is

$$a = \frac{\pi N d_{vs}^2 \phi_D}{\pi N d_{vs}^3/6} = \frac{6\phi_D}{d_{vs}} \qquad (8\text{-}36)$$

Equation (8-36) is used to estimate the interfacial area, a, from a measurement of d_{vs} or vice versa. Early experimental investigations, such as those of Vermeulen, Williams, and

Langlois [44], found that d_{vs} is dependent on a Weber number:

$$N_{We} = \frac{(\text{inertial force})}{(\text{interfacial tension force})} = \frac{D_i^3 N^2 \rho_C}{\sigma} \quad (8\text{-}37)$$

High Weber numbers give small droplets and high interfacial areas. Gnanasundram, Degaleesan, and Laddha [45] correlated d_{vs} over a wide range of N_{We}. Below a critical value of $N_{We} = 10{,}000$, d_{vs} is dependent on dispersed-phase holdup, ϕ_D, because of coalescence effects. For $N_{We} > 10{,}000$, inertial forces dominate so that coalescence effects are much less prominent and d_{vs} is almost independent of holdup up to $\phi_D = 0.5$. The recommended correlations are

$$\frac{d_{vs}}{D_i} = 0.052(N_{We})^{-0.6} e^{4\phi_D}, \; N_{We} < 10{,}000 \quad (8\text{-}38)$$

$$\frac{d_{vs}}{D_i} = 0.39(N_{We})^{-0.6}, \; N_{We} > 10{,}000 \quad (8\text{-}39)$$

Typical values of N_{We} for industrial extractors are less than 10,000, so (8-38) applies. Values of d_{vs}/D_i are frequently in the range of 0.0005 to 0.01.

Experimental studies, for example, those of Chen and Middleman [46] and Sprow [47], show that the dispersion produced in an agitated vessel is a dynamic phenomenon. Droplet breakup by turbulent pressure fluctuations dominates in the vicinity of the impeller blades, while for reasonable dispersed-phase holdup, coalescence of drops by collisions dominates away from the impeller. Thus, a distribution of drop sizes is found in the vessel, with smaller drops in the vicinity of the impeller blades and larger drops elsewhere. Typically, when both drop breakup and coalescence occur, the drop-size distribution is such that $d_{min} \approx d_{vs}/3$ and $d_{max} \approx 3d_{vs}$. Thus, the drop size varies over about a 10-fold range, and the distribution approximates a normal Gaussian distribution.

EXAMPLE 8.6

For the conditions and results of Example 8.5, with the extract phase as the dispersed phase, estimate the Sauter mean drop diameter, the range of drop sizes, and the interfacial area.

SOLUTION

$$D_i = 1 \text{ ft}; \quad N = 147 \text{ rpm} = 8{,}820 \text{ rph}$$
$$\rho_C = 62.3 \text{ lb/ft}^3; \quad \sigma = 718{,}800 \text{ lb/h}^2$$

From (8-37),

$$N_{We} = (1)^3(8{,}820)^2(62.3)/718{,}800 = 6{,}742; \quad \phi_D = 0.388$$

From (8-38),

$$d_{vs} = (1)(0.052)(6{,}742)^{-0.6} \exp[4(0.388)] = 0.00124 \text{ ft}$$

or

$$(0.00124)(12)(25.4) = 0.38 \text{ mm}$$

$$d_{min} = d_{vs}/3 = 0.126 \text{ mm}; \quad d_{max} = 3d_{vs} = 1.134 \text{ mm}$$

From (8-36), $\quad a = 6(0.388)/0.00124 = 1{,}880 \text{ ft}^2/\text{ft}^3$

Mass-Transfer Coefficients

Experimental studies, conducted since the early 1940s, show that mass transfer in mechanically agitated liquid–liquid systems is very complex. This is true for mass transfer in (1) the dispersed-phase droplets, (2) the continuous phase, and (3) at the interface. The reasons for this complexity are many. The magnitude of k_D depends on drop diameter, solute diffusivity, and fluid motion within the drop. When drop diameter is small (less than 1 mm according to Davies [48]), interfacial tension is high (say > 15 dyne/cm), and trace amounts of surface-active agents are present, droplets are rigid (internally stagnant), and they behave like solids. As droplets become larger, interfacial tension decreases, surface-active agents become relatively ineffective, and internal toroidal fluid circulation patterns, caused by viscous drag of the continuous phase, appear within the drops. For larger-diameter drops, the shape of the drop may oscillate between spheroid and ellipsoid or other shapes.

Mass-transfer coefficients, k_C, in the continuous phase depend on the relative motion between the droplets and the continuous phase, and whether the drops are forming or breaking, or are coalescing. Interfacial movements or turbulence, called *Marangoni effects*, occur due to interfacial-tension gradients. Such effects can induce substantial increases in mass-transfer rates.

A relatively conservative estimate of the overall mass-transfer coefficient, K_{OD}, in (8-28), can be made from estimates of k_D and k_C, by assuming rigid drops, the absence of Marangoni effects, and a stable drop size (i.e., no drop forming, breaking, or coalescing). For k_D, the asymptotic steady-state solution for mass transfer in a rigid sphere with negligible resistance of the surroundings is given by Treybal [25] as

$$(N_{Sh})_D = \frac{k_D d_{vs}}{D_D} = \frac{2}{3}\pi^2 = 6.6 \quad (8\text{-}40)$$

where D_D is the diffusivity of the solute in the droplet. N_{Sh} is the Sherwood number. Exercise 3.31 in Chapter 3 for diffusion from the surface of a sphere into an infinite, quiescent fluid gives the following result for the continuous-phase Sherwood number:

$$(N_{Sh})_C = \frac{k_C d_{vs}}{D_C} = 2 \quad (8\text{-}41)$$

where D_C is the diffusivity of the solute in the continuous phase. However, if other spheres of equal diameter are located near the sphere of interest, $(N_{Sh})_C$ may decrease to a value as low as 1.386, according to Cornish [49]. In an agitated vessel, the continuous-phase Sherwood number will usually be much greater than 1.386. A reasonable estimate can be made with the semi-theoretical correlation of Skelland and Moeti [50]. They fitted 180 data points for three different solutes, three different dispersed organic solvents, and water as the continuous phase. Mass transfer was from the dispersed phase to the continuous phase, but only for $\phi_D = 0.01$. Skelland and Moeti assumed an equation of the form

$$(N_{Sh})_C \propto (N_{Re})_C^y (N_{Sc})_C^x \qquad (8\text{-}42)$$

where

$$(N_{Sh})_C = k_C d_{vs}/D_C \qquad (8\text{-}43)$$

$$(N_{Sc})_C = \mu_C/\rho_C D_C \qquad (8\text{-}44)$$

For the Reynolds number, they assumed that the characteristic velocity is the square root of the mean-square, local fluctuating velocity in the vicinity of the droplet, based on the theory of local isotropic turbulence of Batchelor [51]:

$$\bar{u}^2 \propto \left(\frac{Pg_c}{V}\right)^{2/3} \left(\frac{d_{vs}}{\rho_C}\right)^{2/3} \qquad (8\text{-}45)$$

Thus,

$$(N_{Re})_C = \frac{(\bar{u}^2)^{1/2} d_{vs} \rho_C}{\mu_C} \qquad (8\text{-}46)$$

Combining (8-45) and (8-46), with omission of the proportionality constant:

$$(N_{Re})_C = \frac{d_{vs}^{4/3} \rho_C^{2/3} (Pg_c/V)^{1/3}}{\mu_C} \qquad (8\text{-}47)$$

As discussed previously in conjunction with Figure 8.36, in the turbulent-flow region,

$$Pg_c \propto \rho_M N^3 D_i^5 \quad \text{or} \quad \text{for low } \phi_D, \ Pg_c/V \propto \rho_C N^3 D_i^5/D_T^3$$

Thus,

$$(N_{Re})_C = \frac{d_{vs}^{4/3} \rho_C N D_i^{5/3}}{\mu_C D_T} \qquad (8\text{-}48)$$

Skelland and Moeti correlated their mass-transfer coefficient data with

$$k_C \propto D_C^{2/3} \mu_C^{-1/3} N^{3/2} d_{vs}^0$$

The exponents in this proportionality are used to determine the exponents y and x in (8-42) as $\frac{2}{3}$ and $\frac{1}{3}$, respectively. In addition, based on the work of previous investigators, a droplet Eotvos number,

$$N_{Eo} = \rho_D d_{vs}^2 g/\sigma \qquad (8\text{-}49)$$

where $N_{Eo} = $ (gravitational force)/(surface tension force) and the dispersed-phase holdup, ϕ_D, are incorporated into the following final correlation, which predicts 180 experimental data points to an average absolute deviation of 19.71%:

$$(N_{Sh})_C = \frac{k_C d_{vs}}{D_C} = 1.237 \times 10^{-5} \left(\frac{\mu_C}{\rho_C D_C}\right)^{1/3}$$

$$\times \left(\frac{D_i^2 N \rho_C}{\mu_C}\right)^{2/3} \phi_D^{-1/2} \left(\frac{D_i N^2}{g}\right)^{5/12} \qquad (8\text{-}50)$$

$$\times \left(\frac{D_i}{d_{vs}}\right)^2 \left(\frac{d_{vs}}{D_T}\right)^{1/2} \left(\frac{\rho_D d_{vs}^2 g}{\sigma}\right)^{5/4}$$

EXAMPLE 8.7

For the system, conditions, and results of Examples 8.5 and 8.6, with the extract as the dispersed phase, estimate:

(a) The dispersed-phase mass-transfer coefficient, k_D

(b) The continuous-phase mass-transfer coefficient, k_C

(c) The Murphree dispersed-phase efficiency, E_{MD}

(d) The fractional extraction of furfural

The molecular diffusivities of furfural in toluene (dispersed) and water (continuous) at dilute conditions are, respectively,

$$D_D = 8.32 \times 10^{-5} \text{ft}^2/\text{h} \quad \text{and} \quad D_C = 4.47 \times 10^{-5} \text{ ft}^2/\text{h}$$

The distribution coefficient for dilute conditions is $m = dc_C/dc_D = 0.0985$.

SOLUTION

(a) From (8-40), $k_D = 6.6(D_D)/d_{vs} = 6.6(8.32 \times 10^{-5})/0.00124 = 0.44$ ft/h

(b) To apply (8-50) to the estimation of k_C, first compute each of the dimensionless groups in that equation:

$$N_{Sc} = \mu_C/\rho_C D_C = 2.165/[(62.3)(4.47 \times 10^{-5})] = 777$$
$$N_{Re} = D_i^2 N \rho_C/\mu_C = (1)^2 (8,820)(62.3)/2.165 = 254,000$$
$$N_{Fr} = D_i N^2/g = (1)(8,820)^2/(4.17 \times 10^8) = 0.187$$
$$D_i/d_{vs} = 1/0.00124 = 806; \quad d_{vs}/D_T = 0.00124/3 = 0.000413$$
$$N_{Eo} = \rho_D d_{vs}^2 g/\sigma = (54.2)(0.00124)^2(4.17 \times 10^8)/718,800$$
$$= 0.0483$$

From (8-50),

$$N_{Sh} = 1.237 \times 10^{-5}(777)^{1/3}(254,000)^{2/3}(0.388)^{-1/2}(0.187)^{5/12}$$
$$\times (806)^2(0.000413)^{1/2}(0.0483)^{5/4} = 109$$

which is much greater than the value of 2 in a quiescent fluid.

$$k_C = N_{Sh} D_C/d_{vs} = (109)(4.47 \times 10^{-5})/0.00124 = 3.93 \text{ ft/h}$$

(c) From (8-28) and the results of Example 8.6,

$$K_{OD}a = \left\{ \frac{1}{1/0.44 + 1/[(0.0985)(3.93)]} \right\} 1,880 = 387 \text{ h}^{-1}$$

From (8-32), with $V = \pi D_T^2 H/4 = (3.14)(3)^2(3)/4 = 21.2 \text{ ft}^2$

$$N_{OD} = K_{OD}aV/Q_D = 387(212)/207 = 39.6$$

From (8-33),

$$E_{MD} = (N_{OD}/(1 + N_{OD}) = 39.6)/(1 + 39.6) = 0.975 = 97.5\%.$$

(d) By material balance,

$$Q_C(c_{C,\text{in}} - c_{C,\text{out}}) = Q_D c_{D,\text{out}} \qquad (1)$$

From (8-26),

$$E_{MD} = c_{D,\text{out}}/c_D^* = m c_{D,\text{out}}/c_{C,\text{out}} \qquad (2)$$

Combining (1) and (2) to eliminate $c_{D,\text{out}}$ gives

$$\frac{c_{C,\text{out}}}{c_{C,\text{in}}} = \frac{1}{1 + Q_D E_{MD}/(Q_C m)} \qquad (3)$$

and

$$f_{\text{Extracted}} = \frac{c_{C,\text{in}} - c_{C,\text{out}}}{c_{C,\text{in}}} = 1 - \frac{c_{C,\text{out}}}{c_{C,\text{in}}} = \frac{Q_D E_{MD}/(Q_C m)}{1 + Q_D E_{MD}/(Q_C m)}$$

$$\frac{Q_D}{Q_C}\frac{E_{MD}}{m} = \frac{(207)(0.975)}{(327)(0.0985)} = 6.27$$

Thus,

$$f_{\text{Extracted}} = \frac{6.27}{1 + 6.27} = 0.862 \quad \text{or} \quad 86.2\%$$

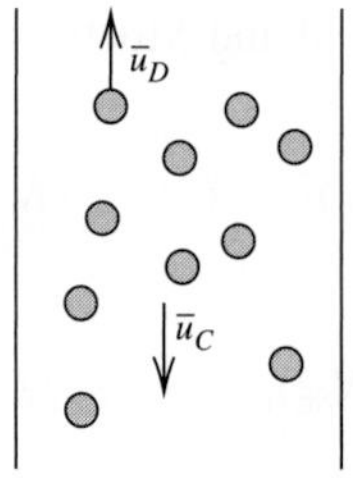

Multicompartment Columns

Sizing extraction columns, which may or may not include mechanical agitation, involves the determination of column diameter and column height. The diameter must be sufficiently large to permit the two phases to flow countercurrently through the column without flooding. The column height must be sufficient to achieve the number of equilibrium stages corresponding to the desired degree of extraction.

For small-diameter columns, rough estimates of the diameter and height can be made using the results of a study by Stichlmair [52] with the toluene–acetone–water system for $Q_D/Q_C = 1.5$. Typical ranges of l/HETS and the sum of the superficial phase velocities for a number of extractor types are given in Table 8.6.

Because of the large number of important variables, an accurate estimation of column diameter for liquid–liquid contacting devices is far more complex and more uncertain than for vapor–liquid contactors. These variables include individual phase flow rates, density difference between the two phases, interfacial tension, direction of mass-transfer, viscosity and density of the continuous phase, rotating or reciprocating speed, and geometry of internals. Column diameter is best determined by scale-up from tests run in standard laboratory or pilot-plant test units with a diameter of 1 in. or larger. The sum of the measured superficial velocities of the two liquid phases in the test unit can then be assumed to hold for larger commercial units. This sum is often expressed in total gallons per hour per square foot of empty column cross-section area.

In the absence of laboratory data, preliminary estimates of diameter for some columns can be made by a simplification of the theory of Logsdail, Thornton, and Pratt [53], which is

Table 8.6 Performance of Several Types of Column Extractors

Extractor Type	l/HETS, m^{-1}	$U_D + U_C$, m/h
Packed column	1.5–2.5	12–30
Pulsed packed column	3.5–6	17–23
Sieve-plate column	0.8–1.2	27–60
Pulsed-plate column	0.8–1.2	25–35
Scheibel column	5–9	10–14
RDC	2.5–3.5	15–30
Kuhni column	5–8	8–12
Karr column	3.5–7	30–40
RTL contactor	6–12	1–2

Source: J. Stichlmair, *Chemie-Ingenieur-Technik,* **52.** 253 (1980).

Figure 8.37 Countercurrent flows of dispersed and continuous liquid phases in a column.

compared to other procedures by Landau and Houlihan [54] in the case of the rotating-disk contactor. Because the relative motion between a dispersed droplet phase and a continuous phase is involved, this theory is based on a concept that is similar to that developed in Chapter 6 for liquid droplets dispersed in a vapor phase.

Consider the case of liquid droplets of the lower-density phase rising through the denser, downward-flowing, continuous liquid phase, as shown in Figure 8.37. If the average superficial velocities of the discontinuous (droplet) phase and the continuous phase are U_D in the upward direction and U_C in the downward direction (i.e., both of these velocities are positive), respectively, the corresponding average actual velocities relative to the column wall are

$$\bar{u}_D = \frac{U_D}{\phi_D} \tag{8-51}$$

and

$$\bar{u}_C = \frac{U_C}{1 - \phi_D} \tag{8-52}$$

The average droplet rise velocity relative to the continuous phase is the sum of (8-51) and (8-52):

$$\bar{u}_r = \frac{U_D}{\phi_D} + \frac{U_C}{1 - \phi_D} \tag{8-53}$$

This relative velocity (also called *slip velocity*) can be expressed in terms of a modified form of (6-40) where the continuous-phase density in the buoyancy term is replaced by the density of the two-phase mixture, ρ_M. Thus, after noting for the case here that the drag force, F_d, and gravitational force, F_g, act downward while buoyancy, F_b, acts upward, we obtain

$$\bar{u}_r = C\left(\frac{\rho_M - \rho_D}{\rho_C}\right)^{1/2} f\{1 - \phi_D\} \tag{8-54}$$

where C is the same parameter as in (6-41) and $f\{1 - \phi_D\}$ is a factor that allows for the hindered rising effect of neighboring droplets. The density ρ_M is a volumetric mean given by

$$\rho_M = \phi_D \rho_D + (1 - \phi_D)\rho_C \tag{8-55}$$

$$\rho_M - \rho_D = (1 - \phi_D)(\rho_C - \rho_D) \tag{8-56}$$

Substitution of (8-56) into (8-54) yields

$$\bar{u}_r = C\left(\frac{\rho_C - \rho_D}{\rho_C}\right)^{1/2}(1 - \phi_D)^{1/2} f\{1 - \phi_D\} \tag{8-57}$$

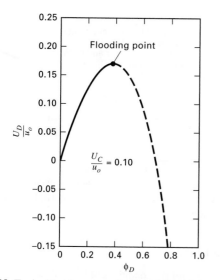

Figure 8.38 Typical holdup curve for liquid–liquid extraction column.

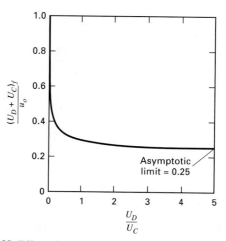

Figure 8.39 Effect of phase ratio on total capacity of liquid–liquid extraction column.

From experimental data, Gayler, Roberts, and Pratt [55] found that, for a given liquid–liquid system, the right-hand side of (8-57) can be expressed empirically as

$$\bar{u}_r = u_0(1 - \phi_D) \qquad (8-58)$$

where u_0 is a characteristic rise velocity for a single droplet, which depends on all the variables discussed above, except those on the right-hand side of (8-53). Thus, for a given liquid–liquid system, column design, and operating conditions, the combination of (8-53) and (8-58) gives

$$\frac{U_D}{\phi_D} + \frac{U_C}{1 - \phi_D} = u_0(1 - \phi_D) \qquad (8-59)$$

where u_0 is a constant. Equation (8-59) is cubic in ϕ_D, with a typical solution shown in Figure 8.38 for $U_C/u_0 = 0.1$. Thornton [56] argues that, with U_C fixed, an increase in U_D results in an increased value of the holdup ϕ_D, until the flooding point is reached, at which $(\partial U_D/\partial \phi_D)_{U_C} = 0$. Thus, in Figure 8.38, only that portion of the curve for $\phi_D = 0$ to $(\phi_D)_f$, the holdup at the flooding point, is realized in practice. Alternatively, with U_D fixed, $(\partial U_C/\partial \phi_D)_{U_D} = 0$ at the flooding point. If these two derivatives are applied to (8-59), we obtain, respectively,

$$U_C = u_0[1 - 2(\phi_D)_f][1 - (\phi_D)_f]^2 \qquad (8-60)$$
$$U_D = 2u_0[1 - (\phi_D)_f](\phi_D)_f^2 \qquad (8-61)$$

where the subscript f denotes flooding. Combining (8-60) and (8-61) to eliminate u_0 gives the following expression for $(\phi_D)_f$:

$$(\phi_D)_f = \frac{[1 + 8(U_C/U_D)]^{0.5} - 3}{4[(U_C/U_D) - 1]} \qquad (8-62)$$

This equation predicts values of $(\phi_D)_f$ ranging from zero at $U_D/U_C = 0$ to 0.5 at $U_C/U_D = 0$. At $U_D/U_C = 1$, $(\phi_D)_f = \frac{1}{3}$. The simultaneous solution of (8-59) and (8-62) results in Figure 8.39 for the variation of total capacity as a function of phase flow ratio. The largest total capacities are achieved, as might be expected, at the smallest ratios of dispersed-phase flow rate to continuous-phase flow rate.

For fixed values of column geometry and rotor speed, experimental data of Logsdail et al. [53] for a laboratory-scale RDC indicate that the dimensionless group $(u_0\mu_C\rho_C/\sigma \Delta\rho)$ is approximately constant. Data of Reman and Olney [57] and Strand, Olney, and Ackerman [58] for well-designed and efficiently operated commercial RDC columns ranging from 8 to 42 in. in diameter indicate that this dimensionless group has a value of roughly 0.01 for systems involving water as either the continuous or dispersed phase. This value is suitable for preliminary calculations of RDC and Karr column diameters, when the sum of the actual superficial phase velocities is taken as 50% of the estimated sum at flooding conditions.

EXAMPLE 8.8

Estimate the diameter of an RDC to extract acetone from a dilute toluene–acetone solution into water at 20°C. The flow rates for the dispersed organic and continuous aqueous phases are 27,000 and 25,000 lb/h, respectively.

SOLUTION

The necessary physical properties are

$\mu_C = 1.0$ cP (0.000021 lbf-s/ft^2) and $\rho_C = 1.0$ g/cm^3

$\Delta\rho = 0.14$ g/cm^3 and $\sigma = 32$ dyne/cm (0.00219 lbf/ft)

$$\frac{U_D}{U_C} = \left(\frac{27,000}{25,000}\right)\left(\frac{\rho_C}{\rho_D}\right) = \left(\frac{27,000}{25,000}\right)\left(\frac{1.0}{0.86}\right) = 1.26$$

From Figure 8.39, $(U_D + U_C)_f/u_0 = 0.29$.
Assume that $u_0\mu_C\rho_C/\sigma \Delta\rho = 0.01$.
Therefore,

$$u_0 = \frac{(0.01)(0.00219)(0.14)}{(0.000021)(1.0)} = 0.146 \text{ ft/s}$$

$$(U_D + U_C)_f = 0.29(0.146) = 0.0423 \text{ ft/s}$$

$$(U_D + U_C)_{50\% \text{ of flooding}} = \left(\frac{0.0423}{2}\right)(3{,}600) = 76.1 \text{ ft/h}$$

$$\text{Total ft}^3/\text{h} = \frac{27{,}000}{(0.86)(62.4)} + \frac{25{,}000}{(1.0)(62.4)} = 904 \text{ ft}^3/\text{h}$$

$$\text{Column cross-sectional area} = A_c = \frac{904}{76.1} = 11.88 \text{ ft}^2$$

$$\text{Column diameter} = D_T = \left(\frac{4A_c}{\pi}\right)^{0.5} = \left[\frac{(4)(11.88)}{3.14}\right]^{0.5} = 3.9 \text{ ft}$$

Note that from Table 8.6, a typical $(U_D + U_C)$ for an RDC is 15 to 30 m/h or 49 to 98.4 ft/h.

Despite their compartmentalization, mechanically assisted liquid–liquid extraction columns, such as the RDC and Karr columns, operate more nearly like differential contacting devices than like staged contactors. Therefore, it is more common to consider stage efficiency for such columns in terms of HETS (height equivalent to a theoretical stage) or as some function of mass-transfer parameters, such as HTU (height of a transfer unit). Although it is not on as sound a theoretical basis as the HTU, the HETS is preferred here because it can be applied directly to determine column height from the number of equilibrium stages.

Because of the great complexity of liquid–liquid systems and the large number of variables that influence contacting efficiency, general correlations for HETS have been difficult to develop. However, for well-designed and efficiently operated columns, the available experimental data indicate that the dominant physical properties influencing HETS are the interfacial tension, the phase viscosities, and the density difference between the phases. In addition, it has been observed by Reman [59] for RDC units and by Karr and Lo [60] for Karr columns that HETS increases with increasing column diameter because of axial mixing effects discussed in the next section.

It is preferred to obtain values of HETS by conducting small-scale laboratory experiments with systems of interest. These values are scaled to commercial-size columns by assuming that HETS varies with column diameter D_T, raised to an exponent, which may vary from 0.2 to 0.4 depending on the system.

In the absence of experimental data, the crude correlation of Figure 8.40 can be used for preliminary design if phase viscosities are no greater than 1 cP. The data points correspond to minimum reported HETS values for RDC and Karr units with the exponent on column diameter set arbitrarily to $\frac{1}{3}$. The points represent values of HETS that vary from as low as 6 in. for a 3-in.-diameter, laboratory-size column operating with a low-interfacial-tension/low-viscosity system such as methyl–isobutyl ketone/acetic acid/water, to as high as 25 in. for a 36-in.-diameter commercial column operating with a high-interfacial-tension/low-viscosity system such as xylenes–acetic acid–water. For systems having one phase of high viscosity, values of HETS can be 24 in. or more, even for a small, laboratory-size column.

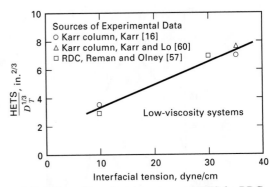

Figure 8.40 Effect of interfacial tension on HETS for RDC and Karr columns.

EXAMPLE 8.9

Estimate HETS for the conditions of Example 8.8.

SOLUTION

Because toluene has a viscosity of approximately 0.6 cP, this is a low-viscosity system. From Example 8.8, the interfacial tension is 32 dyne/cm. From Figure 8.40. $\text{HETS}/D_T^{1/3} = 6.9$. For $D_T = 3.9$ ft, $\text{HETS} = 6.9[(3.9)(12)]^{1/3} = 24.8$ in. Note that from Table 8.6, HETS for an RDC varies from 0.29 to 0.40 m or 11.4 to 15.7 in. for a small column.

More accurate estimates of flooding and HETS are discussed in detail by Lo et al. [4] and by Thornton [61]. Packed column design is considered by Strigle [62].

Axial Dispersion

In this and previous chapters covering liquid–liquid and vapor–liquid countercurrent-flow contactors, plug flow of each phase has been assumed. Each element of a phase is assumed to have the same residence time in the contactor, while each phase may have a different residence time. Because axial concentration gradients in the direction of bulk flow are established in each phase, diffusion of a species is superimposed on the bulk flow of the species in that phase. Axial diffusion degrades the efficiency of multistage separation equipment, and in the limit, a multistage separator behaves like a single well-mixed stage. In Figure 8.41, solute concentration profiles for the extract and raffinate phases of a liquid–liquid extraction column are shown for plug flow (dashed lines) and for flow with significant axial diffusion in each of the two phases (solid lines). The continuous phase is the feed/raffinate (x subscript), which enters the contactor at the top ($z = 0$). The dispersed phase is the solvent/extract (y subscript), which enters the contactor at the bottom ($z = H$). Solute transfer is from the continuous phase to the dispersed phase. Two effects of axial diffusion are seen: (1) The concentration curves in the presence of axial diffusion are closer together than for plug flow and (2) these close proximities are due partially to concentrations at the two ends, which are different from those in the original feed and

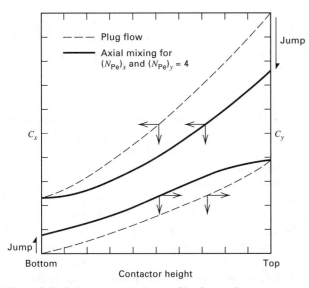

Figure 8.41 Solute concentration profiles for continuous, countercurrent extraction with and without axial mixing.

solvent. These differences are called *jumps* and are due to axial diffusion outside the region in the contactor where the two liquid phases are in contact. The jump at the top is caused by axial diffusion, superimposed on the bulk flow, in the feed liquid before it enters the contactor. This causes the concentration of solute in the feed just as it enters the contactor to be less than its concentration in the original feed liquid. Similarly, diffusion of solute into the incoming solvent causes the concentration of solute in the solvent just entering the bottom of the contactor to be greater than the concentration in the original solvent, which in Figure 8.41 is zero. The overall effect of axial diffusion is a reduction in the average driving force for mass transfer of the solute between the two phases, necessitating a taller column to accomplish the desired separation.

The effects shown in Figure 8.41 are actually due to a number of factors besides diffusion, which are lumped together into one overall effect, commonly referred to as *axial dispersion, axial mixing, longitudinal dispersion,* or *backmixing*. These factors include:

1. Molecular and turbulent diffusion of the continuous phase along concentration gradients
2. Circulatory motion of the continuous phase due to the droplets of the dispersed phase
3. Transport and shedding of the continuous phase in the wakes attached to the rear of droplets of the dispersed phase
4. Circulation of continuous and dispersed phases in mechanically agitated columns
5. Channeling and nonuniform velocity profiles leading to distributions of residence times in the two phases

In general, the effect of axial dispersion is most pronounced when (1) a high recovery of solute is necessary, (2) the contactor is short in height, (3) large circulation

patterns occur, (4) a wide range of droplet sizes is present, and/or (5) the feed-to-solvent flow ratio is very small or very large. Although axial-dispersion effects are generally negligible in extractors where phase separation occurs between stages, such as in mixer-settler cascades and sieve-plate columns with downcomers, axial dispersion can be significant in spray columns, packed columns, and RDCs. Although axial dispersion can occur in packed absorbers, packed strippers, and packed distillation columns, it is significant only when operating at very high liquid-to-gas ratios. However, axial dispersion can be significant in spray and bubble columns used for absorption.

Two types of models have been developed for predicting the extent and effect of axial mixing: (1) diffusion models for differential-type contactors, due to Sleicher [63] and Miyauchi and Vermeulen [64]; and (2) backflow models for staged extractors without complete phase separation between stages, due to Sleicher [65] and Miyauchi and Vermeulen [66]. Both types are discussed by Vermeulen et al. [67]. Diffusion models, which have received the most attention and have been most applied more frequently, are convenient for studying the complex nature of axial dispersion.

Consider a differential height, dz, of a differential contactor with countercurrent two-phase flow, as shown in Figure 8.42. Feed enters the top of the column at $z = 0$, while solvent enters the bottom of the column at $z = H$. Assume that: (1) axial dispersion in each phase is characterized by a constant turbulent-diffusion coefficient, E; (2) phase superficial velocities are each uniform over the cross section and

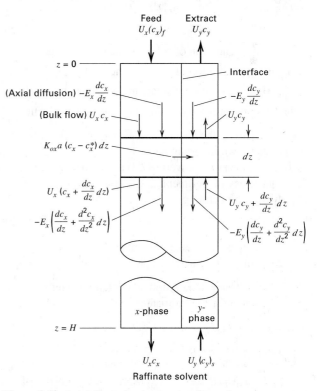

Figure 8.42 Axial dispersion in an extraction column.

[From J.D. Thornton, *Science and Practice of Liquid–Liquid Extraction*, Vol. 1, Clarendon Oxford, (1992) with permission.]

constant in the axial direction; (3) the volumetric, overall mass-transfer coefficients for the solute are constant; (4) only the solute undergoes mass transfer between the two phases; and (5) the phase equilibrium ratio for the solute is constant. Then the solute mass-balance equations for the feed/raffinate (x) and solvent/extract (y) phases, respectively, are

$$E_x \frac{d^2 c_x}{dz^2} - U_x \frac{dc_x}{dz} - K_{Ox} a (c_x - c_x^*) = 0 \quad (8\text{-}63)$$

$$E_y \frac{d^2 c_y}{dz^2} + U_y \frac{dc_y}{dz} + K_{Ox} a (c_x - c_x^*) = 0 \quad (8\text{-}64)$$

where c_x^* is the concentration of the solute in the raffinate that is equilibrium with the solute concentration in the bulk extract. For these two differential equations, the boundary conditions, which were first proposed by Danckwerts [68] and were further elucidated by Wehner and Wilhelm [69], are

at $z = 0$, $\quad U_x c_{x_f} - U_x c_{x_0} = -E_x \dfrac{dc_x}{dz} \quad (8\text{-}65)$

and

$$dc_y/dz = 0 \quad (8\text{-}66)$$

at $z = H$,

$$U_y c_{y_H} - U_y c_{y_s} = E_y \frac{dc_y}{dz} \quad (8\text{-}67)$$

and

$$dc_x/dz = 0 \quad (8\text{-}68)$$

where:

c_{x_f} = concentration of solute in the original feed

c_{x_0} = concentration of solute in the feed at $z = 0$

c_{y_H} = concentration of solute in the solvent at $z = H$

c_{y_s} = concentration of solute in the original solvent

The two terms on the left-hand sides of (8-65) and (8-67) are the jumps shown in Figure 8.41.

It is customary to convert (8-63) and (8-64) to alternative forms in terms of pertinent dimensionless groups. This is readily done by defining

$$Z = z/H \quad (8\text{-}69)$$

$N_{Pe_y} = U_y H/E_y$ = axial, turbulent column Peclet number for the extract phase $\quad (8\text{-}70)$

$N_{Pe_x} = U_x H/E_x$ = axial, turbulent column Peclet number for the raffinate phase $\quad (8\text{-}71)$

$N_{Ox} = K_{Ox} a H/U_x = K_{Ox} a V/Q_x \quad (8\text{-}72)$

Equations (8-63) and (8-64) then become

$$\frac{d^2 c_x}{dZ^2} - N_{Pe_x} \frac{dc_x}{dZ} - N_{Ox} N_{Pe_x} (c_x - c_x^*) = 0 \quad (8\text{-}73)$$

$$\frac{d^2 c_y}{dZ^2} - N_{Pe_y} \frac{dc_y}{dZ} - \left(\frac{U_x}{U_y} \right) N_{Ox} N_{Pe_y} (c_x - c_x^*) = 0 \quad (8\text{-}74)$$

The boundary conditions are transformed in a similar way. For a straight equilibrium curve, $c_x^* = m c_y$. Thus, we have a coupled set of ordinary differential equations, whose solutions for c_y and c_x are functions of c_{y_f}, c_{y_s}, m, N_{Pe_x}, N_{Pe_y}, N_{Ox}, U_x/U_y, and Z.

Further algebraic manipulations involving the substitution of dimensionless solute concentrations can reduce the number of variables from 10 to 7. In either case, the solution of the axial dispersion equations as obtained by Sleicher [65] and Miyauchi and Vermeulen [66] is very difficult to display in tabular or graphical form. However, the possible importance of axial dispersion is most commonly judged by the magnitudes of the Peclet numbers. A Peclet number of 0 corresponds to complete back-mixing, such that at most only one equilibrium stage is achieved; the entire column functions like a single mixer stage. A Peclet number of ∞ corresponds to an absence of axial dispersion. Experimental data on several different types of liquid-liquid extraction columns indicate that N_{Pe} for the dispersed phase is frequently greater than 50, while N_{Pe} for the continuous phase may be in the range of 5 to 30. Thus, as a first approximation, axial dispersion in the dispersed phase can be largely ignored. This effect was observed experimentally by Geankoplis and Hixson [70] in a spray extraction column and by Gier and Hougen [71] in spray and packed extraction columns. They reported end-concentration changes of significant magnitude at the continuous-phase entrance, but not at the dispersed-phase entrance.

A number of approximate solutions to the axial dispersion equations, (8-63) to (8-68), have been published, including one by Sleicher [63]. Alternatively, if the original solvent is free of solute, a rapid and somewhat conservative estimate of the effect of axial dispersion can be made by the method of Watson and Cochran [72] from an empirical relation for the column efficiency:

$$\frac{H_{\text{plug flow}}}{H_{\text{actual}}} = \frac{(\text{HTU}_{Ox})(\text{NTU}_{Ox})}{H}$$

$$= 1 - \frac{1}{1 + N_{Pe_x}(\text{HTU}_{Ox}/H) - E + (1/\text{NTU}_{Ox})}$$

$$- \frac{E}{N_{Pe_y}(\text{HTU}_{Ox}/H) - 1 + E + (1/\text{NTU}_{Ox})} \quad (8\text{-}75)$$

where

$H_{\text{actual}} = H$ = height of column taking into account axial dispersion

HTU_{Ox} = height of an overall transfer unit based on the raffinate phase for plug flow

NTU_{Ox} = number of overall transfer units based on the raffinate phase for plug flow

E = extraction factor = $m U_x / U_y$

$m = dc_x/dc_y$

The product of HTU_{Ox} and NTU_{Ox} is the column height for plug flow, which is $< H$. Thus, the ratio on the left-hand side of (8-75) is a column efficiency. The NTU_{Ox} is approximated by:

$$\text{NTU}_{Ox} = \ln\left(\frac{X}{XE + 1 - E}\right)\bigg/ (E-1) \quad (8\text{-}76)$$

where

$$X = (c_x)_{\text{out}}/(c_x)_{\text{in}} \quad (8\text{-}77)$$

The HTU_{Ox} is defined by:

$$\text{HTU}_{Ox} = U_x/K_{Ox}a \quad (8\text{-}78)$$

For given values of HTU_{Ox}, NTU_{Ox}, E, N_{Pe_x}, and N_{Pe_y}, (8-75) is solved for H. Caution must be exercised in using (8-75) because of its empirical nature. The equation is limited to $NTU_{Ox} \geq 2$, $E > 0.25$, $N_{\text{Pe}_x}(\text{HTU}_{Ox}/H) > 1.5$, and the calculated value of the column efficiency, $H_{\text{plug flow}}/H_{\text{actual}}$, must be ≥ 0.20. Within these restrictions, an extensive comparison by Watson and Cochran with the exact solution of (8-63) to (8-68) gives conservative efficiency values that deviate by no more than 0.07 (7%), with the highest accuracy for estimated efficiencies greater than 0.5 (50%).

EXAMPLE 8.10

Experiments conducted for a dilute system under laboratory conditions approximating plug flow give $\text{HTU}_{Ox} = 3$ ft. If a commercial column is to be designed for $\text{NTU}_{Ox} = 4$ and N_{Pe_x} and N_{Pe_y} are estimated to be 19 and 50, respectively, determine the necessary column height if $E = 0.5$.

SOLUTION

For plug flow, column height is $(\text{HTU}_{Ox})(\text{NTU}_{Ox})$. Substitution of the data into (8-75) gives

$$\frac{12}{H} = 1 - \frac{1}{(57/H) + 0.75} - \frac{0.5}{(180/H) - 0.25}$$

This is a nonlinear algebraic equation in H. Solving by an iterative method,

$$H = 17\,\text{ft}$$

Efficiency $= (\text{HTU}_{Ox})(\text{NTU}_{Ox})/H = 12/17 = 0.706 \quad (70.6\%)$

SUMMARY

1. A solvent can be used to selectively extract one or more components from a liquid mixture.

2. Although liquid–liquid extraction is a reasonably mature separation operation, considerable experimental effort is often needed to find a suitable solvent and to determine residence-time requirements or values of HETS, NTU, or mass-transfer coefficients.

3. Compared to vapor–liquid separation operations, extraction has a higher overall mass-transfer resistance. Stage efficiencies in columns are frequently low.

4. A wide variety of commercial extractors are available, as shown in Figures 8.2 to 8.7, ranging from simple columns with no mechanical agitation to centrifugal devices that may spin at several thousand revolutions per minute. A selection scheme, given in Table 8.3, is useful for choosing the most suitable extractors for a given separation.

5. Solvent selection is facilitated by consideration of a number of chemical factors given in Table 8.4 and physical factors discussed in Section 8.2.

6. For liquid–liquid extraction with ternary mixtures, phase equilibrium is conveniently represented on equilateral- or right-triangle diagrams for both type I (solute and solvent completely miscible) and the less common type II (solute and solvent not completely miscible) systems.

7. For determining equilibrium-stage requirements of single-section, countercurrent cascades for ternary systems, the graphical methods of Hunter and Nash (equilateral-triangle diagram), Kinney (right-triangle diagram), or Varteressian and Fenske (distribution diagram of McCabe–Thiele type) can be applied, as described in Section 8.3. These methods can also determine minimum and maximum solvent requirements.

8. A two-section, countercurrent cascade with extract reflux can be employed with a type II ternary system to enable a sharp separation of a binary feed mixture. The calculation of stage requirements of such a two-section cascade is conveniently carried out by the graphical method of Maloney and Schubert using a Janecke equilibrium diagram, as discussed in Section 8.4. The addition of raffinate reflux to such a cascade is of little value. The Maloney–Schubert method can also be applied to single-section cascades.

9. When only a few equilibrium stages are required, a cascade of mixer-settler units may be attractive because each mixer can be designed to closely approach an equilibrium stage. With many ternary and higher-order systems, the residence-time requirement may be only a few minutes for a 90% approach to equilibrium using an agitator input of approximately 4 hp/1,000 gal. Adequate phase-disengaging area for the settlers may be estimated from the rule of 5 gal of combined extract and raffinate per minute per square foot of disengaging area.

10. For mixers utilizing a six-flat-bladed turbine in a closed vessel with side vertical baffles, as shown in Figure 8.35, useful extractor design correlations are available for estimating, for a given extraction, the mixing-vessel dimensions, minimum impeller rotation rate for complete and uniform dispersion, impeller horsepower, mean droplet size, range of droplet sizes, interfacial area per unit volume, dispersed-phase and continuous-phase mass-transfer coefficients, and Murphree efficiency.

11. For column-type extractors, with and without mechanical agitation, correlations for determining column diameter, to avoid flooding, and column height are suitable only for very preliminary sizing calculations. For final extractor selection and design, recommendations of equipment vendors based on experimental data from pilot-size equipment are highly desirable.

12. Sizing of column-type extractors must consider axial dispersion, which can significantly reduce mass-transfer driving forces and thus increase the required column height. Axial dispersion effects are often most significant in the continuous phase.

REFERENCES

1. DERRY, T.K., and T.I. WILLIAMS, *A Short History of Technology,* Oxford University Press, New York (1961).

2. BAILES, P.J., and A. WINWARD, *Trans. Inst. Chem. Eng.,* **50,** 240–258 (1972).

3. BAILES, P.J., C. HANSON, and M.A. HUGHES, *Chem. Eng.,* **83** (2), 86–100 (1976).

4. LO, T.C., M.H.I. BAIRD, and C. HANSON, Eds., *Handbook of Solvent Extraction,* Wiley-Interscience, New York (1983).

5. REISSINGER, K.-H., and J. SCHROETER, "Alternatives to Distillation," *I. Chem. E. Symp. Ser. No. 54,* 33–48 (1978).

6. HUMPHREY, J.L., J.A. ROCHA, and J.R. FAIR, *Chem. Eng.,* **91** (19), 76–95 (1984).

7. FENSKE, M.R., C.S. CARLSON, and D. QUIGGLE, *Ind. Eng. Chem.,* **39,** 1932 (1947).

8. SCHEIBEL, E.G., *Chem. Eng. Prog.,* **44,** 681 (1948).

9. SCHEIBEL, E.G., *AIChE J.,* **2,** 74 (1956).

10. SCHEIBEL, E.G., U. S. Patent 3,389,970 (June 25, 1968).

11. OLDSHUE, J., and J. RUSHTON, *Chem. Eng. Prog.,* **48** (6), 297 (1952).

12. REMAN, G.H., *Proceedings of the 3rd World Petroleum Congress,* The Hague, Netherlands, Sec. III, 121 (1951).

13. REMAN, G.H., *Chem. Eng. Prog.,* **62** (9), 56 (1966).

14. MISEK, T., and J. MAREK, *Br. Chem. Eng.,* **15,** 202 (1970).

15. FISCHER, A., *Verfahrenstechnik,* **5,** 360 (1971).

16. KARR, A.E., *AIChE J.,* **5,** 446 (1959).

17. KARR, A.E., and T.C. LO, *Chem. Eng. Prog.,* **72** (11), 68 (1976).

18. PROCHAZKA, J., J. LANDAU, F. SOUHRADA, and A. Heyberger, *Br. Chem. Eng.,* **16,** 42 (1971).

19. BARSON, N., and G.H. BEYER, *Chem. Eng. Prog.,* **49** (5), 243–252 (1953).

20. REISSINGER, K.-H., and J. SCHROETER, "Liquid–Liquid Extraction, Equipment Choice," in J.J. McKetta and W.A. Cunningham, Eds., *Encyclopedia of Chemical Processing and Design,* Vol. 21, Marcel Dekker, New York (1984).

21. CUSACK, R.W., P. FREMEAUX, and D. GLATZ, *Chem. Eng.,* **98** (2), 66–76 (1991).

22. ROBBINS, L.A., *Chem. Eng. Prog.,* **76** (10), 58–61 (1980).

23. NASER, S.F., and R.L. FOURNIER, *Comput. Chem. Eng.,* **15,** 397–414 (1991).

24. DARWENT, B., and C.A. WINKLER, *J. Phys. Chem.,* **47,** 442–454 (1943).

25. TREYBAL, R.E., *Liquid Extraction,* 2nd ed., McGraw-Hill, New York (1963).

26. HUNTER, T.G., and A.W. NASH, *J. Soc. Chem. Ind.,* **53,** 95T–102T (1934).

27. KINNEY, G.F., *Ind. Eng. Chem.,* **34,** 1102–1104 (1942).

28. VENKATARANAM, A., and R.J. RAO, *Chem. Eng. Sci.,* **7,** 102–110 (1957).

29. SAWISTOWSKI, H., and W. SMITH, *Mass Transfer Process Calculations,* Interscience, New York (1963).

30. VARTERESSIAN, K.A., and M.R. FENSKE, *Ind. Eng. Chem.,* **28,** 1353–1360 (1936).

31. SKELLAND, A.H.P., *Ind. Eng. Chem.,* **53,** 799–800 (1961).

32. RANDALL, M., and B. LONGTIN, *Ind. Eng. Chem.,* **30,** 1063, 1188, 1311 (1938); **31,** 908, 1295 (1939); **32,** 125 (1940).

33. MALONEY, J.O., and A.E. SCHUBERT, *Trans. AIChE,* **36,** 741 (1940).

34. FLYNN, A.W. and R.E. TREYBAL, *AIChE J.,* **1,** 324–328 (1955).

35. RYON, A.D., F.L. DALEY, and R.S. LOWRIE, *Chem. Eng. Prog.,* **55** (10), 70–75 (1959).

36. HAPPEL, J., and D.G. JORDAN, *Chemical Process Economics,* 2nd ed., Marcel Dekker, New York (1975).

37. RUSHTON, J.H., and J.Y. OLDSHUE, *Chem Eng. Prog.,* **49,** 161–168 (1953).

38. LAITY, D.S., and R.E. TREYBAL, *AIChE J.,* **3,** 176–180 (1957).

39. SKELLAND, A.H.P., and G.G. RAMSEY, *Ind. Eng. Chem. Res.,* **26,** 77–81 (1987).

40. SKELLAND, A.H.P., and J.M. LEE, *Ind. Eng. Chem. Process Des. Dev.,* **17,** 473–478 (1978).

41. MacMULLIN, R.B., and M. WEBER, *Trans. AIChE,* **31,** 409–458 (1935).

42. LEWIS, J.B., I. JONES, and H.R.C. PRATT, *Trans. Inst. Chem. Eng.,* **29,** 126 (1951).

43. COULSON, J.M., and J.F. RICHARDSON, *Chemical Engineering,* Vol. 2, 4th ed., Pergamon, Oxford (1991).

44. VERMUELEN, T., G.M. WILLIAMS, and G.E. LANGLOIS, *Chem. Eng. Prog.,* **51,** 85F (1955).

45. GNANASUNDARAM, S., T.E. DEGALEESAN, and G.S. LADDHA, *Can. J. Chem. Eng.,* **57,** 141–144 (1979).

46. CHEN, H.T., and S. MIDDLEMAN, *AIChE J.,* **13,** 989–995 (1967).

47. SPROW, F.B., *AIChE J.,* **13,** 995–998 (1967).

48. DAVIES, J.T., *Turbulence Phenomena,* Academic Press, New York, p. 311 (1978).

49. CORNISH, A.R.H., *Trans. Inst. Chem. Eng.,* **43,** T332–T333 (1965).

50. SKELLAND, A.H.P., and L.T. MOETI, *Ind. Eng. Chem. Res.,* **29,** 2258–2267 (1990).

51. BATCHELOR, G.K., *Proc. Cambridge Phil. Soc.,* **47,** 359–374 (1951).

52. STICHLMAIR, J., *Chemie-Ingenieur-Technik,* **52,** 253 (1980).

53. LOGSDAIL, D.H., J.D. THORNTON, and H.R.C. PRATT, *Trans. Inst. Chem. Eng.,* **35,** 301–315 (1957).

54. LANDAU, J., and R. HOULIHAN, *Can. J. Chem. Eng.,* **52,** 338–344 (1974).

55. GAYLER, R., N.W. ROBERTS, and H.R.C. PRATT, *Trans. Inst. Chem. Eng.,* **31,** 57–68 (1953).

56. THORNTON, J.D., *Chem. Eng. Sci.,* **5,** 201–208 (1956).

57. REMAN, G.H., and R.B. OLNEY, *Chem. Eng. Prog.,* **52** (3), 141–146 (1955).

58. STRAND, C.P., R.B. OLNEY, and G.H. ACKERMAN, *AIChE J.,* **8,** 252–261 (1962).

59. REMAN, G.H., *Chem. Eng. Prog.,* **62** (9), 56–61 (1966).

60. KARR, A.E., and T.C. LO, "Performance of a 36-inch Diameter Reciprocating-Plate Extraction Column," paper presented at the 82nd National Meeting of AIChE, Atlantic City, NJ (Aug. 29–Sept. 1, 1976).

61. THORNTON, J.D., *Science and Practice of Liquid–Liquid Extraction,* Vol. 1, Clarendon Press, Oxford (1992).

62. STRIGLE, R.F., JR., *Random Packings and Packed Towers,* Gulf Publishing Company, Houston, TX (1987).

63. SLEICHER, C.A., JR., *AIChE J.,* **5,** 145–149 (1959).

64. MIYAUCHI, T., and T. VERMUELEN, *Ind. Eng. Chem. Fund.,* **2,** 113–126 (1963).

65. SLEICHER, C.A., JR., *AIChE J.,* **6,** 529–531 (1960).

66. MIYAUCHI, T., and T. VERMUELEN, *Ind. Eng. Chem. Fund.,* **2,** 304–310 (1963).

67. VERMEULEN, T., J.S. MOON, A. HENNICO, and T. MIYAUCHI, *Chem. Eng. Prog.,* **62** (9), 95–101 (1966).

68. DANCKWERTS, P.V., *Chem. Eng. Sci.,* **2,** 1–13 (1953).

69. WEHNER, J.F., and R.H. WILHELM, *Chem. Eng. Sci.,* **6,** 89–93 (1956).

70. GEANKOPLIS, C.J., and A.N. HIXSON, *Ind. Eng. Chem.,* **42,** 1141–1151 (1950).

71. GIER, T.E., and J.O. HOUGEN, *Ind. Eng. Chem.,* **45,** 1362–1370 (1953).

72. WATSON, J.S., and H.D. COCHRAN, Jr., *Ind. Eng. Chem. Process Des. Dev.,* **10,** 83–85 (1971).

EXERCISES

Section 8.1

8.1 Explain why it is preferable to separate a dilute mixture of benzoic acid in water by liquid–liquid extraction rather than distillation.

8.2 Why is liquid–liquid extraction preferred over distillation for the separation of a mixture of formic acid and water?

8.3 Based on the information in Table 8.3 and the selection scheme in Figure 8.8, is the choice of an RDC appropriate for the extraction of acetic acid from water by ethyl acetate in the process described in the introduction to this chapter and shown in Figure 8.1? What other types of extractors might be considered?

8.4 What is the major advantage of the ARD over the RDC? What is the disadvantage of the ARD compared to the RDC?

8.5 Under what conditions is a cascade of mixer-settler units probably the best choice of extraction equipment?

8.6 A petroleum reformate stream of 4,000 bbl/day is to be contacted with diethylene glycol to extract the aromatics from the paraffins. The ratio of solvent volume to reformate volume is 5. It is estimated that eight theoretical stages will be needed. Using Tables 8.2 and 8.3, and Figure 8.8, which types of extractors would be most suitable?

Section 8.2

8.7 Using Table 8.4, select possible liquid–liquid extraction solvents for separating the following mixtures: (a) water–ethyl alcohol, (b) water–aniline, and (c) water–acetic acid. For each case, indicate clearly which of the two components should be the solute.

8.8 Using Table 8.4, select possible liquid–liquid extraction solvents for removing the solute from the carrier in the following cases:

	Solute	Carrier
(a)	Acetone	Ethylene glyol
(b)	Toluene	*n*-Heptane
(c)	Ethyl alcohol	Glycerine

8.9 For the extraction of acetic acid (A) from a dilute solution in water (C) into ethyl acetate (S) at 25°C, estimate or obtain data for $(K_A)_D$, $(K_C)_D$, $(K_S)_D$, and β_{AC}. Does this system exhibit: (a) High selectivity, (b) High solvent capacity and (c) Ease in recovering the solvent? Can you select a solvent that would exhibit better factors than ethyl acetate?

8.10 Interfacial tension can be an important factor in liquid–liquid extraction. Very low values of interfacial tension result in stable emulsions that are difficult to separate, while very high values require large energy inputs to form the dispersed phase. It is best to measure the interfacial tension for the two-phase mixture of interest. However, in the absence of experimental data, propose a method for estimating the interfacial tension of a ternary system using only the compositions of the equilibrium phases and the values of surface tension in air for each of the three components.

Section 8.3

8.11 One thousand kilograms per hour of a 45 wt% acetone in-water solution is to be extracted at 25°C in a continuous, countercurrent system with pure 1,1,2-trichloroethane to obtain a raffinate containing 10 wt% acetone. Using the following equilibrium data, determine with an equilateral-triangle diagram:

(a) the minimum flow rate of solvent,

(b) the number of stages required for a solvent rate equal to 1.5 times the minimum, and

(c) the flow rate and composition of each stream leaving each stage.

	Acetone, Weight Fraction	Water, Weight Fraction	Trichloroethane, Weight Fraction
Extract	0.60	0.13	0.27
	0.50	0.04	0.46
	0.40	0.03	0.57
	0.30	0.02	0.68
	0.20	0.015	0.785
	0.10	0.01	0.89
Raffinate	0.55	0.35	0.10
	0.50	0.43	0.07
	0.40	0.57	0.03
	0.30	0.68	0.02
	0.20	0.79	0.01
	0.10	0.895	0.005

The tie-line data are:

Raffinate, Weight Fraction Acetone	Extract, Weight Fraction Acetone
0.44	0.56
0.29	0.40
0.12	0.18

8.12 Solve Exercise 8.11 with a right-triangle diagram.

8.13 A distillate containing 45 wt% isopropyl alcohol, 50 wt% diisopropyl ether, and 5 wt% water is obtained from the heads column of an isopropyl alcohol finishing unit. The company desires to recover the ether from this stream by liquid–liquid extraction in a column, with water, as the solvent, entering the top and the feed entering the bottom so as to produce an ether containing no more than 2.5 wt% alcohol and to obtain the extracted alcohol at a concentration of at least 20 wt%. The unit will operate at 25°C and 1 atm. Using the method of Varteressian and Fenske with a McCabe–Thiele diagram, find how many theoretical stages are required.

Is it possible to obtain an extracted alcohol composition of 25 wt%? Equilibrium data are given below.

PHASE EQUILIBRIUM DATA AT 25°C, 1 ATM

Ether Phase			Water Phase		
Wt% Alcohol	Wt% Ether	Wt% Water	Wt% Alcohol	Wt% Ether	Wt% Water
2.4	96.7	0.9	8.1	1.8	90.1
3.2	95.7	1.1	8.6	1.8	89.6
5.0	93.6	1.4	10.2	1.5	88.3
9.3	88.6	2.1	11.7	1.6	86.7
24.9	69.4	5.7	17.5	1.9	80.6
38.0	50.2	11.8	21.7	2.3	76.0
45.2	33.6	21.2	26.8	3.4	69.8

ADDITIONAL POINTS ON PHASE BOUNDARY

Wt% Alcohol	Wt% Ether	Wt% Water
45.37	29.70	24.93
44.55	22.45	33.00
39.57	13.42	47.01
36.23	9.66	54.11
24.74	2.74	72.52
21.33	2.06	76.61
0	0.6	99.4
0	99.5	0.5

8.14 Benzene and trimethylamine (TMA) are to be separated in a three-stage liquid–liquid extraction column using water as the solvent. If the solvent-free extract and raffinate products are to contain, respectively, 70 and 3 wt% TMA, find the original feed composition and the water-to-feed ratio with a right-triangle diagram. There is no reflux and the solvent is pure water. Equilibrium data are as follows:

TRIMETHYLAMINE–WATER–BENZENE COMPOSITIONS ON PHASE BOUNDARY

Extract, wt%			Raffinate, wt%		
TMA	H₂O	Benzene	TMA	H₂O	Benzene
5.0	94.6	0.4	5.0	0.0	95.0
10.0	89.4	0.6	10.0	0.0	90.0
15.0	84.0	1.0	15.0	1.0	84.0
20.0	78.0	2.0	20.0	2.0	78.0
25.0	72.0	3.0	25.0	4.0	71.0
30.0	66.4	3.6	30.0	7.0	63.0
35.0	58.0	7.0	35.0	15.0	50.0
40.0	47.0	13.0	40.0	34.0	26.0

The tie-line data are:

Extract, wt% TMA	Raffinate, wt% TMA
39.5	31.0
21.5	14.5
13.0	9.0
8.3	6.8
4.0	3.5

8.15 The system docosane–diphenylhexane (DPH)–furfural is representative of more complex systems encountered in the solvent

refining of lubricating oil. Five hundred kilograms per hour of a 40 wt% mixture of DPH in docosane are to be continuously extracted in a countercurrent system with 500 kg/h of a solvent containing 98 wt% furfural and 2 wt% DPH to produce a raffinate that contains only 5 wt% DPH. Calculate with a right-triangle diagram the number of theoretical stages required and the number of kilograms per hour of DPH in the extract at 45°C and at 80°C. Equilibrium data are as follows.

EQUILIBRIUM DATA: BINODAL CURVES IN DOCOSANE–DIPHENYLHEXANE–FURFURAL SYSTEM
[*IND. ENG. CHEM.*, 35, 711 (1943)]

Wt% at 45°C			Wt% at 80°C		
Docosane	DPH	Furfural	Docosane	DPH	Furfural
96.0	0.0	4.0	90.3	0.0	9.7
84.0	11.0	5.0	50.5	29.5	20.0
67.0	26.0	7.0	34.2	35.8	30.0
52.5	37.5	10.0	23.8	36.2	40.0
32.6	47.4	20.0	16.2	33.8	50.0
21.3	48.7	30.0	10.7	29.3	60.0
13.2	46.8	40.0	6.9	23.1	70.0
7.7	42.3	50.0	4.6	15.4	80.0
4.4	35.6	60.0	3.0	7.0	90.0
2.6	27.4	70.0	2.2	0.0	97.8
1.5	18.5	80.0			
1.0	9.0	90.0			
0.7	0.0	99.3			

The tie lines in the docosane–diphenylhexane–furfural system are:

Docosane Phase Composition, wt%			Furfural Phase Composition, wt%		
Docosane	DPH	Furfural	Docosane	DPH	Furfural
Temperature, 45°C:					
85.2	10.0	4.8	1.1	9.8	89.1
69.0	24.5	6.5	2.2	24.2	73.6
43.9	42.6	13.3	6.8	40.9	52.3
Temperature, 80°C:					
86.7	3.0	10.3	2.6	3.3	94.1
73.1	13.9	13.0	4.6	15.8	79.6
50.5	29.5	20.2	9.2	27.4	63.4

8.16 For each of the ternary systems shown in Figure 8.43, indicate whether: (a) simple, countercurrent extraction, or (b) countercurrent extraction with extract reflux, or (c) countercurrent extraction with raffinate reflux, or (d) countercurrent extraction with both extract and raffinate reflux would be expected to yield the most economical process.

8.17 Two solutions, feed F at the rate of 7,500 kg/h containing 50 wt% acetone and 50 wt% water, and feed F' at the rate of 7,500 kg/h containing 25 wt% acetone and 75 wt% water, are to be extracted in a countercurrent system with 5,000 kg/h of 1,1,2-trichloroethane at 25°C to give a raffinate containing 10 wt% acetone. Calculate the number of equilibrium stages required and the stage to which each feed should be introduced, using a right-triangle diagram. Equilibrium data are given in Exercise 8.11.

8.18 The three-stage extractor shown in Figure 8.44 is used to extract the amine from a fluid consisting of 40 wt% benzene (B) and 60 wt% trimethylamine (T). The solvent (water) flow to stage 3

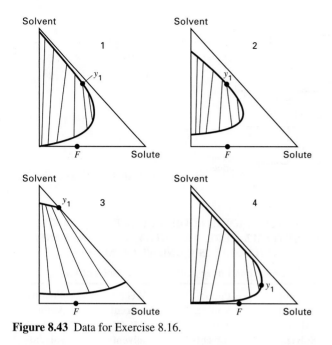

Figure 1 (Solvent/Solute triangle with y_1, F)

Figure 2 (Solvent/Solute triangle with y_1, F)

Figure 3 (Solvent/Solute triangle with y_1, F)

Figure 4 (Solvent/Solute triangle with y_1, F)

Figure 8.43 Data for Exercise 8.16.

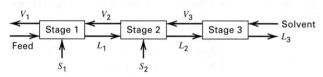

Figure 8.44 Data for Exercise 8.18.

is 5,185 kg/h and the feed flow rate is 10,000 kg/h. On a solvent-free basis V_1 is to contain 76 wt% T and L_3 is to contain 3 wt% T. Determine the required solvent flow rates S_1 and S_2 using an equilateral-triangle diagram. Equilibrium data are given in Exercise 8.14.

8.19 The extraction process shown Figure 8.45 is conducted in a multiple-feed, countercurrent unit without extract or raffinate reflux. Feed F' is composed of solvent and solute, and is an extract-phase feed. Feed F'' is composed of unextracted raffinate and solute and is a raffinate-phase feed. Derive the equations required to establish the three reference points needed to step off the theoretical stages in the extraction column. Show the graphical determination of these points on a right-triangle graph.

8.20 A mixture containing 50 wt% methylcyclohexane (MCH) in n-heptane is fed to a countercurrent, stage-type extractor at 25°C. Aniline is used as solvent. Reflux is used on both ends of the column.

An extract containing 95 wt% MCH and a raffinate containing 5 wt% MCH (both on solvent-free basis) are required. The minimum extract reflux ratio is 3.49. Using a right-triangle diagram with the equilibrium data of Exercise 8.22 below, calculate: (a) the raffinate

reflux ratio, (b) the amount of aniline that must be removed at the separator "on top" of the column, and (c) the amount of solvent that must be added to the solvent mixer at the bottom of the column.

8.21 In its natural state, zirconium, which is an important material of construction for nuclear reactors, is associated with hafnium, which has an abnormally high neutron-absorption cross section and must be removed before the zirconium can be used. Refer to Figure 8.46 for a proposed liquid–liquid extraction process wherein tributyl phosphate (TBP) is used as a solvent for the separation of hafnium from zirconium.

One liter per hour of 5.10-N HNO_3 containing 127 g of dissolved Hf and Zr oxides per liter is fed to stage 5 of the 14-stage extraction unit. The feed contains 22,000 g Hf per million g of Zr. Fresh TBP enters stage 14, while scrub water is fed to stage 1. Raffinate is removed at stage 14, while the organic extract phase that is removed at stage 1 goes to a stripping unit. The stripping operation consists of a single contact between fresh water and the organic phase. The following table gives experimental data. (a) Use these data to fashion a complete material balance for the process. (b) Check the data for consistency in as many ways as you can. (c) What is the advantage of running the extractor as shown? Would you recommend that all the stages be used?

STAGEWISE ANALYSES OF MIXER-SETTLER RUN

	Organic Phase			Aqueous Phase		
Stage	g oxide/ liter	N HNO_3	(Hf/Zr) × (100)	g oxide/ liter	N HNO_3	(Hf/Zn) × (100)
1	22.2	1.95	<0.010	17.5	5.21	<0.010
2	29.3	2.02	<0.010	27.5	5.30	<0.010
3	31.4	2.03	<0.010	33.5	5.46	<0.010
4	31.8	2.03	0.043	34.9	5.46	0.24
5	32.2	2.03	0.11	52.8	5.15	3.6
6	21.1	1.99	0.60	30.8	5.15	6.8
7	13.7	1.93	0.27	19.9	5.05	9.8
8	7.66	1.89	1.9	11.6	4.97	20
9	4.14	1.86	4.8	8.06	4.97	36
10	1.98	1.83	10	5.32	4.75	67
11	1.03	1.77	23	3.71	4.52	110
12	0.66	1.68	32	3.14	4.12	140
13	0.46	1.50	42	2.99	3.49	130
14	0.29	1.18	28	3.54	2.56	72
Stripper		0.65		76.4	3.96	<0.01

[Data From R.P. Cox, H.C. Peterson, and C.H. Beyer, *Ind. Eng. Chem.*, **50** (2), 141 (1958). Exercise adapted from E.J. Henley and H. Bieber, *Chemical Engineering Calculations*, McGraw-Hill, New York, p. 298 (1959).]

8.22 At 45°C, 5,000 kg/h of a mixture of 65 wt% docosane, 7 wt% furfural, and 28 wt% diphenylhexane is to be extracted with pure furfural to obtain a raffinate with 12 wt% diphenylhexane in a continuous, countercurrent, multistage liquid–liquid extraction system. Phase-equilibrium data for this ternary system are given in

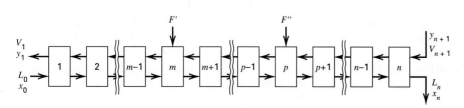

Figure 8.45 Data for Exercise 8.19.

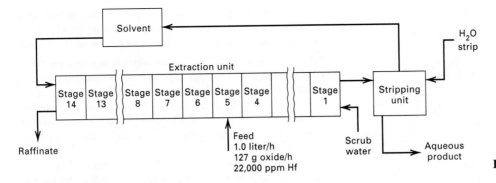

Figure 8.46 Data for Exercise 8.21.

Exercise 8.15. Determine:

(a) The minimum flow rate of solvent.

(b) The flow rate and composition of the extract at the minimum solvent flow rate.

(c) The number of equilibrium stages required if a solvent flow rate of 1.5 times the minimum is used.

8.23 At 45°C, 1,000 kg/h of a mixture of 0.80 mass fraction docosane and 0.20 mass fraction diphenylhexane is to be extracted with pure furfural to remove some of the diphenylhexane from the feed. Phase-equilibrium data for this ternary system are given in Exercise 8.15. Determine:

(a) The composition and flow rate of the extract and raffinate from a single equilibrium stage for solvent flow rates of 100, 1000, and 10000 kg/h.

(b) The minimum solvent flow rate to form two liquid phases.

(c) The maximum solvent flow rate to form two liquid phases.

(d) The composition and flow rate of the extract and raffinate if a solvent flow rate of 2000 kg/h and two equilibrium stages are used in a countercurrent flow system.

8.24 A liquid mixture of 27 wt% acetone and 73 wt% water is to be separated at 25°C into a raffinate and extract by multistage, steady-state, countercurrent liquid–liquid extraction with a solvent of pure 1,1,2-trichloroethane. Phase equilibrium data are given in Exercise 8.11. Determine:

(a) The minimum solvent-to-feed ratio to obtain a raffinate that is essentially free of acetone.

(b) The composition of extract at the minimum solvent-to-feed ratio.

(c) The composition of the extract stream leaving stage 2 (see Figure 8.13), if a large number of equilibrium stages is used with the minimum solvent rate.

Section 8.4

8.25 A feed mixture containing 50 wt% n-heptane and 50 wt% methylcyclohexane (MCH) is to be separated by liquid–liquid extraction into one product containing 92.5 wt% methylcyclohexane and another containing 7.5 wt% methylcyclohexane, both on a solvent-free basis. Aniline will be used as the solvent. Using the equilibrium data given below and the graphical method of Maloney and Schubert: (a) What is the minimum number of theoretical stages necessary to effect this separation? (b) What is the minimum extract reflux ratio? (c) If the reflux ratio is 7.0, how many theoretical contacts are required?

LIQUID–LIQUID EQUILIBRIUM DATA FOR THE SYSTEM n-HEPTANE/METHYLCYCLOHEXANE/ ANILINE AT 25°C AND AT 1 ATM (101 kPa)

Hydrocarbon Layer		Solvent Layer	
Weight Percent MCH, Solvent-Free Basis	Pounds Aniline/ Pound Solvent-Free Mixture	Weight Percent MCH, Solvent-Free Basis	Pounds Aniline/ Pound Solvent-Free Mixture
0.0	0.0799	0.0	15.12
9.9	0.0836	11.8	13.72
20.2	0.087	33.8	11.5
23.9	0.0894	37.0	11.34
36.9	0.094	50.6	9.98
44.5	0.0952	60.0	9.0
50.5	0.0989	67.3	8.09
66.0	0.1062	76.7	6.83
74.6	0.1111	84.3	6.45
79.7	0.1135	88.8	6.0
82.1	0.116	90.4	5.9
93.9	0.1272	96.2	5.17
100.0	0.135	100.0	4.92

8.26 Two liquids, A and B, which have nearly identical boiling points, are to be separated by liquid–liquid extraction with solvent C. The following data represent the equilibrium between the two liquid phases at 95°C.

EQUILIBRIUM DATA, WT%

Extract Layer			Raffinate Layer		
A, %	B, %	C, %	A, %	B, %	C, %
0	7.0	93.0	0	92.0	8.0
1.0	6.1	92.9	9.0	81.7	9.3
1.8	5.5	92.7	14.9	75.0	10.1
3.7	4.4	91.9	25.3	63.0	11.7
6.2	3.3	90.5	35.0	51.5	13.5
9.2	2.4	88.4	42.0	41.0	17.0
13.0	1.8	85.2	48.1	29.3	22.6
18.3	1.8	79.9	52.0	20.0	28.0
24.5	3.0	72.5	47.1	12.9	40.0
31.2	5.6	63.2		Plait point	

[Adapted from McCabe and Smith, *Unit Operations of Chemical Engineering*, 4th ed., McGraw-Hill, New York, p. 557 (1985).]

Determine the minimum amount of reflux that must be returned from the extract product to produce an extract containing 83% A and 17% B (on a solvent-free basis) and a raffinate product containing 10% A and 90% B (solvent-free basis). The feed contains 35% A and 65% B on a solvent-free basis and is a saturated raffinate. The raffinate is the heavy liquid. Determine the number of ideal stages on both sides of the feed required to produce the same end products from the same feed when the reflux ratio of the extract, expressed as pounds of extract reflux per pound of extract product (including solvent), is twice the minimum. Calculate the masses of the various streams per 1,000 lb of feed, all on a solvent-free basis. Solve the problem using equilateral-triangle coordinates, right-triangle coordinates, and solvent-free coordinates. Which method is best for this exercise?

8.27 Solve Exercise 8.20 by the graphical method of Maloney and Schubert.

Section 8.5

8.28 Acetic acid is continuously extracted from a 3 wt% dilute solution in water with a solvent of isopropyl ether in a mixer-settler unit. The flow rates of the feed and solvent are 12,400 and 24,000 lb/h, respectively. Assuming a residence time of 1.5 min in the mixer and a settling vessel capacity of 4 gal/min-ft^2, estimate: (a) Diameter and height of the mixing vessel, assuming $H/D_T = 1$, (b) Agitator horsepower for the mixing vessel, (c) Diameter and length of the settling vessel, assuming $L/D_T = 4$, and (d) Residence time in minutes in the settling vessel.

8.29 A cascade of six mixer-settler units is available, each unit consisting of a 10-ft-diameter by 10-ft-high mixing vessel equipped with a 20-hp agitator, and a 10-ft-diameter by 40-ft-long settling vessel. If this cascade is used for the acetic acid extraction described in the introduction to this chapter, estimate the pounds per hour of feed that could be processed.

8.30 Acetic acid is to be extracted from a dilute aqueous solution with isopropyl ether at 25°C in a countercurrent cascade of mixer-settler units. In one of the units, the following conditions apply:

	Raffinate	**Extract**
Flow rate, lb/h	21,000	52,000
Density, lb/ft^3	63.5	45.3
Viscosity, cP	3.0	1.0

Interfacial tension = 13.5 dyne/cm. If the raffinate is the dispersed phase and the mixer residence time is 2.5 minutes, estimate for the mixer: (a) The dimensions of a closed, baffled vessel, (b) The diameter of a flat-bladed impeller, (c) The minimum rate of rotation in revolutions per minute of the impeller for complete and uniform dispersion, and (d) The power requirement of the agitator at the minimum rate of rotation.

8.31 For the conditions of Exercise 8.30, estimate: (a) Sauter mean drop size, (b) Range of drop sizes, and (c) Interfacial area of the two-phase liquid–liquid emulsion.

8.32 For the conditions of Exercises 8.30 and 8.31, and the additional data given below, estimate: (a) The dispersed-phase mass-transfer coefficient, (b) The continuous-phase mass-transfer coefficient, (c) The Murphree dispersed-phase efficiency, and (d) The fraction of acetic acid extracted.

Additional data:

Diffusivity of acetic acid: in the raffinate, 1.3×10^{-9} m^2/s and in the extract, 2.0×10^{-9} m^2/s.

Distribution coefficient for acetic acid: $c_D/c_C = 2.7$

8.33 For the conditions and results of Example 8.4, involving the extraction of benzoic acid, from a dilute solution in water with toluene, determine the following when using a six-flat-blade turbine impeller in a closed vessel with baffles and with the extract phase dispersed, based on the physical properties given: (a) The minimum rate of rotation of the impeller for complete and uniform dispersion, (b) The power requirement of the agitator at the minimum rotation rate, (c) The Sauter mean droplet diameter, (d) The interfacial area, (e) The overall mass-transfer coefficient, K_{OD}, (f) The number of overall transfer units, N_{OD}, (g) The Murphree efficiency, E_{MD}, and (h) The fractional extraction of benzoic acid. Liquid properties are:

	Raffinate Phase	**Extract Phase**
Density, g/cm^3	0.995	0.860
Viscosity, cP	0.95	0.59
Diffusivity of benzoic acid, cm^2/s	2.2×10^{-5}	1.5×10^{-5}

Interfacial tension = 22 dyne/cm
Distribution coefficient for benzoic acid = $c_D/c_C = 21$

8.34 Estimate the diameter of an RDC column to extract acetic acid from water with isopropyl ether for the conditions and data of Exercises 8.28 and 8.30.

8.35 Estimate the diameter of a Karr column to extract benzoic acid from water with toluene for the conditions of Exercise 8.33.

8.36 Estimate the value of HETS for an RDC column operating under the conditions of Exercise 8.34.

8.37 Estimate the value of HETS for a Karr column operating under the conditions of Exercise 8.35.

Chapter 9

Approximate Methods for Multicomponent, Multistage Separations

$\mathbf{A}$lthough rigorous computer methods, discussed in Chapter 10, are available for solving multicomponent separation problems, approximate methods continue to be used in practice for various purposes, including preliminary design, parametric studies to establish optimal design conditions, process synthesis studies to determine optimal separation sequences, and for obtaining an initial approximation for a rigorous method.

In Section 5.4, the approximate methods of Kremser [1] for absorbers, and Edmister [2] for distillation are discussed. This chapter presents an additional approximate method

that is widely used for making preliminary designs and optimization of simple distillation. The method is commonly referred to as the *Fenske–Underwood–Gilliland* or FUG method. In addition, application of the Kremser method is extended to and illustrated for strippers and liquid–liquid extraction. Although these methods can be applied fairly readily by manual calculation if physical properties are independent of composition, computer calculations are preferred, and FUG models are included in most computer-aided process design programs.

9.0 INSTRUCTIONAL OBJECTIVES

After completing this chapter, you should be able to:

- For multicomponent distillation, select two key components, operating pressure, and type condenser.
- For the specified separation between two key components in a multicomponent distillation column, estimate minimum number of equilibrium stages and distribution of nonkey components by the Fenske equation, minimum reflux ratio by the Underwood method, number of equilibrium stages for a reflux ratio greater than minimum by the Gilliland correlation, and feed stage location.
- Estimate stage requirements for multicomponent absorption, stripping, and liquid–liquid extraction using the Kremser equation.

9.1 FENSKE–UNDERWOOD–GILLILAND METHOD

An algorithm for the empirical Fenske–Underwood–Gilliland method, named after the authors of the three important steps in the procedure, is shown in Figure 9.1 for a simple distillation column of the type shown in Figure 9.3. The column can be equipped with a partial or total condenser. From Table 5.4, the number of degrees of freedom with a total condenser is $2N + C + 9$. In this case, the following variables are generally specified with the partial reboiler counted as a theoretical stage:

	Number of Specifications
Feed flow rate	1
Feed mole fractions	$C - 1$
Feed temperature[1]	1
Feed pressure[1]	1
Adiabatic stages (excluding reboiler)	$N - 1$
Stage pressures (including reboiler)	N

	Number of Specifications
Split of light key component	1
Split of heavy key component	1
Feed-stage location	1
Reflux ratio (as multiple of minimum-reflux ratio)	1
Reflux temperature	1
Adiabatic reflux divider	1
Pressure of total condenser	1
Pressure at reflux divider	1
	$2N + C + 9$

Similar specifications can be written for columns with a partial condenser.

---

[1] Feed temperature and pressure may correspond to known stream conditions leaving the previous piece of equipment.

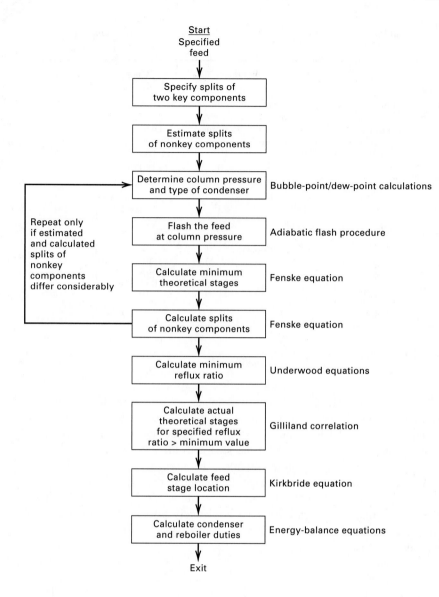

Figure 9.1 Algorithm for multicomponent distillation by FUG method.

Selection of Two Key Components

For multicomponent feeds, specification of two key components and their distribution between distillate and bottoms is accomplished in a variety of ways. Preliminary estimation of the distribution of nonkey components can be sufficiently difficult to require the iterative procedure indicated in Figure 9.1. However, generally only two and seldom more than three iterations are necessary.

Consider the multicomponent hydrocarbon feed in Figure 9.2. This mixture is typical of the feed to the recovery section of an alkylation plant [3]. Components are listed in order of decreasing volatility. A sequence of distillation columns including a deisobutanizer and a debutanizer is to be used to separate this mixture into the three products indicated. In Case 1 of Table 9.1, the deisobutanizer is selected as the first column in the sequence. Since the allowable quantities of n-butane in the isobutane recycle, and isobutane in the n-butane product, are specified, isobutane is the light key and

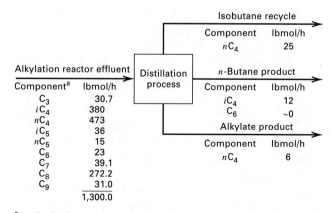

^{a}C$_6$, C$_7$, C$_8$, C$_9$ are taken as normal paraffins.

Figure 9.2 Separation specifications for alkylation-reactor effluent.

Table 9.1   Specifications of Key Component Splits and Preliminary Estimation of Nonkey Component Splits for Alkylation Reactor Effluent

Component	Feed, lbmol/h	Case 1, Deisobutanizer Column First, lbmol/h		Case 2, Debutanizer Column First (iC_5 is HK), lbmol/h		Case 3, Debutanizer Column First (C_6 is HK), lbmol/h	
		Distillate	Bottoms	Distillate	Bottoms	Distillate	Bottoms
C_3	30.7	(30.7)	(0)	(30.7)	(0)	(30.7)	(0)
iC_4	380	368[a]	12[b]	(380.0)	(0)	(380.0)	(0)
nC_4	473	25[b]	448[a]	467[a]	6[b]	467[a]	6[b]
iC_5	36	(0)	(36)	13[b]	23[a]	(13)	(23)
nC_5	15	(0)	(15)	(1)	(14)	(1)	(14)
C_6	23	(0)	(23)	(0)	(23)	0.01[b]	22.99[a]
C_7	39.1	(0)	(39.1)	(0)	(39.1)	(0)	(39.1)
C_8	272.2	(0)	(272.2)	(0)	(272.2)	(0)	(272.2)
C_9	31.0	(0)	(31.0)	(0)	(31.0)	(0)	(31.0)
	1,300.0	423.7	876.3	891.7	408.3	891.71	408.29

[a]By material balance.

[b]Specification.

(Preliminary estimate.)

n-butane is the heavy key. These two keys are adjacent in order of volatility. Because a fairly sharp separation between these two keys is indicated and the nonkey components are not close in volatility to the butanes, as a preliminary estimate we can assume the separation of the nonkey components to be perfect.

Alternatively, in Case 2, if the debutanizer is placed first in the sequence, specifications in Figure 9.2 require that *n*-butane be selected as the light key. However, selection of the heavy key is uncertain because no recovery or purity is specified for any component less volatile than *n*-butane. Possible heavy-key components for the debutanizer are iC_5, nC_5, or C_6. The simplest procedure is to select iC_5 so that the two keys are again adjacent.

For example, suppose we specify that 13 lbmol/h of iC_5 in the feed is allowed to appear in the distillate. Because the split of iC_5 is then not sharp and nC_5 is close in volatility to iC_5, it is probable that the quantity of nC_5 in the distillate will not be negligible. A preliminary estimate of the distributions of the nonkey components for Case 2 is given in Table 9.1. Although iC_4 may also distribute, a preliminary estimate of zero is made for it in the bottoms.

Finally, in Case 3, we select C_6 as the heavy key for the debutanizer at a specified rate of 0.01 lbmol/h in the distillate, as shown in Table 9.1. Now iC_5 and nC_5 will distribute between the distillate and bottoms in amounts to be determined; as a preliminary estimate, we assume the same distribution as in Case 2.

In practice, the deisobutanizer is usually placed first in the sequence. In Table 9.1, the bottoms for Case 1 then becomes the feed to the debutanizer, for which, if nC_4 and iC_5 are selected as the key components, component-separation specifications for the debutanizer are as indicated in Figure 9.3

with preliminary estimates of the separation of nonkey components shown in parentheses. This separation has been treated by Bachelor [4]. Because nC_4 and C_8 comprise 82.2 mol% of the feed and differ widely in volatility, the temperature difference between distillate and bottoms is likely to be large. Furthermore, the light key split is rather sharp, but the heavy key split is not. As will be shown later, this case provides a relatively severe test of the empirical design procedure discussed in this section.

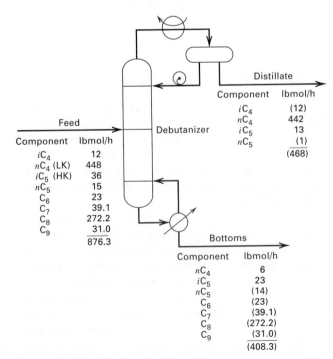

Figure 9.3 Specifications for debutanizer.

Column Operating Pressure

For preliminary design, column operating pressure and type of condenser can be established by the procedure discussed in Section 7.2 and shown in Figure 7.16, as illustrated in the following example. With column operating pressure established, the column feed can be flashed adiabatically at an estimated feed-tray pressure to determine feed-phase condition.

EXAMPLE 9.1

Determine column operating pressures and type of condenser for the debutanizer of Figure 9.3.

SOLUTION

Using the estimated distillate composition in Figure 9.3, we compute the distillate bubble-point pressure at 120°F (48.9°C) iteratively from (4-12) in a manner similar to Example 4.2. This procedure gives 79 psia as the reflux-drum pressure. Thus, a total condenser is indicated. Allowing a 2-psi condenser pressure drop, column top pressure is $(79 + 2) = 81$ psia; and allowing a 5-psi pressure drop through the column, the bottoms pressure is $(81 + 5) = 86$ psia. Assume a feed-tray pressure midway between the column top and bottom pressures or 83.5 psia.

Bachelor [4] sets column pressure at 80 psia throughout. He obtains a distillate temperature of 123°F. A bubble-point calculation for the bottoms composition at 80 psia gives 340°F. This temperature is sufficiently low to prevent decomposition.

Feed to the debutanizer is presumably bottoms from a deisobutanizer operating at a pressure of perhaps 100 psia or more. Results of an adiabatic flash of this feed, by the procedure of Section 4.4, to 80 psia are given by Bachelor [4] as follows.

Pound-Moles per Hour

Component	Vapor Feed	Liquid Feed
iC_4	3.3	8.7
nC_4	101.5	346.5
iC_5	4.6	31.4
nC_5	1.6	13.4
nC_6	1.3	21.7
nC_7	1.2	37.9
nC_8	3.2	269.0
nC_9	0.2	30.8
	116.9	759.4

The temperature of the flashed feed is 180°F (82.2°C). From above, the feed-mole-fraction vaporized is $(116.9/876.3) = 0.1334$.

Fenske Equation for Minimum Equilibrium Stages

For a specified separation between two key components of a multicomponent mixture, an exact expression is easily developed for the required minimum number of equilibrium stages, which corresponds to total reflux. This condition can be achieved in practice by charging the column with feedstock and operating it with no further input of feed and no withdrawal of distillate or bottoms, as illustrated in Figure 9.4. To

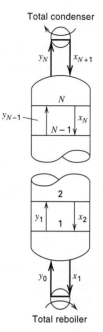

Figure 9.4 Distillation column operation at total reflux.

facilitate derivation of the Fenske equation, stages are numbered from the bottom up. All vapor leaving stage N is condensed and returned to stage N as reflux. All liquid leaving stage 1 is vaporized and returned to stage 1 as boilup. For steady-state operation within the column, heat input to the reboiler and heat output from the condenser are made equal (assuming no heat losses). Then, by a material balance, vapor and liquid streams passing between any pair of stages have equal flow rates and compositions, for example, $V_{N-1} = L_N$ and $y_{i,N-1} = x_{i,N}$. However, molar vapor and liquid flow rates will change from stage to stage unless the assumption of constant molar overflow is valid.

Derivation of an exact equation for the minimum number of equilibrium stages involves only the definition of the K-value and the mole-fraction equality between stages. For component i at stage 1 in Figure 9.4,

$$y_{i,1} = K_{i,1}x_{i,1} \tag{9-1}$$

But for passing streams

$$y_{i,1} = x_{i,2} \tag{9-2}$$

Combining these two equations,

$$x_{i,2} = K_{i,1}x_{i,1} \tag{9-3}$$

Similarly, for stage 2,

$$y_{i,2} = K_{i,2}x_{i,2} \tag{9-4}$$

Combining (9-3) and (9-4), we have

$$y_{i,2} = K_{i,2}K_{i,1}x_{i,1} \tag{9-5}$$

Equation (9-5) is readily extended in this fashion to give

$$y_{i,N} = K_{i,N}K_{i,N-1} \cdots K_{i,2}K_{i,1}x_{i,1} \tag{9-6}$$

Similarly, for component j,

$$y_{j,N} = K_{j,N} K_{j,N-1} \cdots K_{j,2} K_{j,1} x_{j,1} \quad (9\text{-}7)$$

Combining (9-6) and (9-7), we find that

$$\frac{y_{i,N}}{y_{j,N}} = \alpha_N \alpha_{N-1} \cdots \alpha_2 \alpha_1 \left(\frac{x_{i,1}}{x_{j,1}} \right) \quad (9\text{-}8)$$

or

$$\left(\frac{x_{i,N+1}}{x_{i,1}} \right) \left(\frac{x_{j,1}}{x_{j,N+1}} \right) = \prod_{k=1}^{N_{\min}} \alpha_k \quad (9\text{-}9)$$

where $\alpha_k = K_{i,k}/K_{j,k}$, the relative volatility between components i and j. Equation (9-9) relates the relative enrichments of any two components i and j over a cascade of N theoretical stages to the stage relative volatilities between the two components. Although (9-9) is exact, it is rarely used in practice because the conditions of each stage must be known to compute the set of relative volatilities. However, if the relative volatility is assumed constant, (9-9) simplifies to

$$\left(\frac{x_{i,N+1}}{x_{i,1}} \right) \left(\frac{x_{j,1}}{x_{j,N+1}} \right) = \alpha^N \quad (9\text{-}10)$$

or

$$N_{\min} = \frac{\log\{[(x_{i,N+1})/x_{i,1}][x_{j,1}/(x_{j,N+1})]\}}{\log \alpha_{i,j}} \quad (9\text{-}11)$$

Equation (9-11) is extremely useful. It is referred to as the *Fenske equation* [5]. When $i =$ the light key (LK) and $j =$ the heavy key (HK), the minimum number of equilibrium stages is influenced by the nonkey components only by their effect (if any) on the value of the relative volatility between the key components.

Equation (9-11) permits a rapid estimation of minimum equilibrium stages. A more convenient form of (9-11) is obtained by replacing the product of the mole-fraction ratios by the equivalent product of mole-distribution ratios in terms of component distillate and bottoms flow rates d and b, respectively,[2] and by replacing the relative volatility by a geometric mean of the top-stage and bottom-stage values. Thus,

$$N_{\min} = \frac{\log[(d_i/d_j)(b_j/b_i)]}{\log \alpha_m} \quad (9\text{-}12)$$

where the mean relative volatility is approximated by

$$\alpha_m = [(\alpha_{i,j})_N (\alpha_{i,j})_1]^{1/2} \quad (9\text{-}13)$$

Thus, the minimum number of equilibrium stages depends on the degree of separation of the two key components and their relative volatility, but is independent of feed-phase condition. Equation (9-12) in combination with (9-13) is exact for two minimum stages. For one stage, it is equivalent to the equilibrium-flash equation. In practice, distillation columns are designed for separations corresponding to as many as 150 minimum equilibrium stages.

The Fenske equation is quite reliable except when the relative volatility varies appreciably over the column, and/or when the mixture forms nonideal liquid solutions. In those cases, if the Fenske equation is applied with (9-13), it should be done with great caution, and should be followed by rigorous calculations of the type in Chapter 10.

EXAMPLE 9.2

For the debutanizer shown in Figure 9.3 and considered in Example 9.1, estimate the minimum equilibrium stages by the Fenske equation. Assume uniform operating pressure of 80 psia (552 kPa) throughout and utilize the ideal K-values given by Bachelor [4] as plotted in Figure 9.5.

SOLUTION

The two key components are n-butane and isopentane. Distillate and bottoms conditions based on the estimated product distributions for nonkey components in Figure 9.3 are

Component	$x_{N+1} = x_D$	$x_1 = x_B$
iC_4	0.0256	~0
nC_4 (LK)	0.9445	0.0147
iC_5 (HK)	0.0278	0.0563
nC_5	0.0021	0.0343
nC_6	~0	0.0563
nC_7	~0	0.0958
nC_8	~0	0.6667
nC_9	~0	0.0759
	1.0000	1.0000

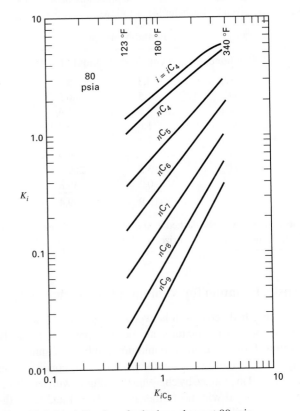

Figure 9.5 Ideal K-values for hydrocarbons at 80 psia.

---

[2] This substitution is valid even though no distillate or bottoms products are withdrawn at total reflux.

From Figure 9.5, at 123°F, the assumed top-stage temperature is

$$(\alpha_{nC_4, iC_5})_N = 1.03/0.495 = 2.08$$

At 340°F, the assumed bottom-stage temperature is

$$(\alpha_{nC_4, iC_5})_1 = 5.20/3.60 = 1.44$$

From (9-13),

$$\alpha_m = [(2.08)(1.44)]^{1/2} = 1.73$$

Noting that $(d_i/d_j) = (x_{D_i}/x_{D_j})$ and $(b_i/b_j) = (x_{B_i}/x_{B_j})$, (9-12) becomes

$$N_{\min} = \frac{\log[(0.9445/0.0278)(0.0563/0.0147)]}{\log 1.73} = 8.88 \text{ stages}$$

Distribution of Nonkey Components at Total Reflux

The Fenske equation is not restricted to the two key components. Once $N_{\min}$ is known, (9-12) can be used to calculate molar flow rates d and b for all nonkey components. These values provide a first approximation to the actual product distribution when more than the minimum number of stages is employed.

Let i = a nonkey component and j = the heavy key or reference component denoted by r. Then (9-12) becomes

$$\left(\frac{d_i}{b_i}\right) = \left(\frac{d_r}{b_r}\right)(\alpha_{i,r})_m^{N_{\min}} \qquad (9\text{-}14)$$

Substituting $f_i = d_i + b_i$ in (9-14) gives

$$b_i = \frac{f_i}{1 + (d_r/b_r)(\alpha_{i,r})_m^{N_{\min}}} \qquad (9\text{-}15)$$

or

$$d_i = \frac{f_i(d_r/b_r)(\alpha_{i,r})_m^{N_{\min}}}{1 + (d_r/b_r)(\alpha_{i,r})_m^{N_{\min}}} \qquad (9\text{-}16)$$

Equations (9-15) and (9-16) give the distribution of nonkey components at total reflux as predicted by the Fenske equation.

For accurate calculations, (9-15) and (9-16) should be used to compute the smaller of the two quantities b_i and d_i. The other quantity is best obtained by overall material balance.

EXAMPLE 9.3

Estimate the product distributions for nonkey components by the Fenske equation for the conditions of Example 9.2.

SOLUTION

All nonkey relative volatilities are calculated relative to isopentane using the K-values of Figure 9.5.

Component	α_{i, iC_5} 123°F	340°F	Geometric Mean
iC_4	2.81	1.60	2.12
nC_5	0.737	0.819	0.777
nC_6	0.303	0.500	0.389
nC_7	0.123	0.278	0.185
nC_8	0.0454	0.167	0.0870
nC_9	0.0198	0.108	0.0463

Based on $N_{\min} = 8.88$ stages from Example 9.2 and the above geometric-mean relative volatilities, values of $(\alpha_{i,r})_m^{N_{\min}}$ are computed relative to isopentane as tabulated below.

From (9-15), using the feed rate specifications in Figure 9.3 for f_i, the distribution of nonkey iC_4 is

$$b_{iC_4} = \frac{12}{1 + (13/23)790} = 0.0268 \text{ lbmol/h}$$

$$d_{iC_4} = f_{iC_4} - b_{iC_4} = 12 - 0.0268 = 11.9732 \text{ lbmol/h}$$

Results of similar calculations for the other nonkey components are included in the following table.

Component	$(\alpha_{i, iC_5})_m^{N_{\min}}$	d_i	b_i
iC_4	790	11.9732	0.0268
nC_4	130	442.0	6.0
iC_5	1.00	13.0	23.0
nC_5	0.106	0.851	14.149
nC_6	0.000228	0.00297	22.99703
nC_7	3.11×10^{-7}	6.87×10^{-6}	39.1
nC_8	3.83×10^{-10}	5.98×10^{-8}	272.2
nC_9	1.41×10^{-12}	2.48×10^{-11}	31.0
		467.8272	408.4728

Underwood Equations for Minimum Reflux

Minimum reflux is based on the specifications for the degree of separation between two key components. The minimum reflux is finite and feed product withdrawals are permitted. However, a column cannot operate under this condition because of the accompanying requirement of infinite stages. Nevertheless, minimum reflux is a useful limiting condition.

For binary distillation of an ideal mixture at minimum reflux, as shown in Figure 7.12a, most of the stages are crowded into a constant-composition zone that bridges the feed stage. In this zone, all vapor and liquid streams have compositions essentially identical to those of the flashed feed. This zone constitutes a single *pinch point* or *point of infinitude* as shown in Figure 9.6a. If nonideal phase conditions are such as to create a point of tangency between the equilibrium curve and the operating line in the rectifying

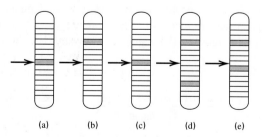

(a)   (b)   (c)   (d)   (e)

Figure 9.6 Location of pinch-point zones at minimum reflux: (a) binary system; (b) binary system, nonideal conditions giving point of tangency; (c) multicomponent system, all components distributed (Class 1); (d) multicomponent system, not all LLK and HHK distributing (Class 2); (e) multicomponent system, all LLK, if any, distributing, but not all HHK distributing (Class 2). (LLK = lighter than light key; HHK = heavier than heavy key.)

section, as shown in Figure 7.12b, the pinch point will occur in the rectifying section as in Figure 9.6b. Alternatively, the single pinch point can occur in the stripping section.

Shiras, Hanson, and Gibson [6] classified multicomponent systems as having one (Class 1) or two (Class 2) pinch points. For Class 1 separations, all components in the feed distribute to both the distillate and bottoms products. Then the single pinch point bridges the feed stage as shown in Figure 9.6c. Class 1 separations can occur when narrow-boiling-range mixtures are distilled or when the degree of separation between the key components is not sharp.

For Class 2 separations, one or more of the components appear in only one of the products. If neither the distillate nor the bottoms product contains all feed components, two pinch points occur away from the feed stage as shown in Figure 9.6d. Stages between the feed stage and the rectifying-section pinch point remove heavy components that do not appear in the distillate. Light components that do not appear in the bottoms are removed by the stages between the feed stage and the stripping-section pinch point. However, if all feed components appear in the bottoms, the stripping-section pinch point moves to the feed stage as shown in Figure 9.6e.

Consider the general case of a rectifying-section pinch point at or away from the feed stage as shown in Figure 9.7. A component material balance over all stages gives

$$y_{i,\infty} V_\infty = x_{i,\infty} L_\infty + x_{i,D} D \qquad (9\text{-}17)$$

A total balance over all stages is

$$V_\infty = L_\infty + D \qquad (9\text{-}18)$$

Since phase compositions do not change in the pinch zone, the phase equilibrium relation is

$$y_{i,\infty} = K_{i,\infty} x_{i,\infty} \qquad (9\text{-}19)$$

Combining (9-17) to (9-19) for components i and j to eliminate $y_{i,\infty}$, $y_{j,\infty}$, and V_∞; solving for the internal reflux ratio at the pinch point; and substituting $(\alpha_{i,j})_\infty = K_{i,\infty}/K_{j,\infty}$, we have

$$\frac{L_\infty}{D} = \frac{[(x_{i,D}/x_{i,\infty}) - (\alpha_{i,j})_\infty (x_{j,D}/x_{j,\infty})]}{(\alpha_{i,j})_\infty - 1} \qquad (9\text{-}20)$$

For Class 1 separations, flashed feed- and pinch-zone compositions are identical.[3] Therefore, $x_{i,\infty} = x_{i,F}$ and (9-20) for the light key (LK) and the heavy key (HK) becomes

$$\frac{(L_\infty)_{\min}}{F} =$$
$$\frac{(L_F/F)[(Dx_{\text{LK},D})/(L_F x_{\text{LK},F}) - (\alpha_{\text{LK},\text{HK}})_F (Dx_{\text{HK},D}/L_F x_{\text{HK},F})]}{(\alpha_{\text{LK},\text{HK}})_F - 1} \qquad (9\text{-}21)$$

This equation is attributed to Underwood [7] and can be applied to subcooled-liquid or superheated-vapor feeds by using fictitious values of L_F and $x_{i,F}$ computed by making a

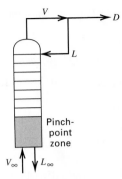

Figure 9.7 Rectifying-section pinch-point zone.

flash calculation outside the two-phase region. As with the Fenske equation, (9-21) applies to components other than the key components. Therefore, for a specified split of two key components, the distribution of nonkey components is obtained by combining (9-21) with the analogous equation for component i in place of the light key to give

$$\frac{Dx_{i,D}}{L_F x_{i,F}} = \left[\frac{(\alpha_{i,\text{HK}})_F - 1}{(\alpha_{\text{LK},\text{HK}})_F - 1}\right]\left(\frac{Dx_{\text{LK},D}}{L_F x_{\text{LK},F}}\right) \qquad (9\text{-}22)$$
$$+ \left[\frac{(\alpha_{\text{LK},\text{HK}})_F - (\alpha_{i,\text{HK}})_F}{(\alpha_{\text{LK},\text{HK}})_F - 1}\right]\left(\frac{Dx_{\text{HK},D}}{L_F x_{\text{HK},F}}\right)$$

For a Class 1 separation,

$$0 < \left(\frac{Dx_{i,D}}{F x_{i,F}}\right) < 1$$

for all nonkey components. If that is so, the external reflux ratio is obtained from the internal reflux by an enthalpy balance around the rectifying section in the form

$$\frac{(L_{\min})_{\text{external}}}{D} = (R_{\min})_{\text{external}}$$
$$= \frac{(L_\infty)_{\min}(h_{V_\infty} - h_{L_\infty}) + D(h_{V_\infty} - h_V)}{D(h_V - h_L)} \qquad (9\text{-}23)$$

where subscripts V and L refer to vapor leaving the top stage and external liquid reflux sent to the top stage, respectively. For conditions of constant molar overflow,

$$(R_{\min})_{\text{external}} = (L_\infty)_{\min}/D$$

Even when (9-21) is invalid, it is useful because, as shown by Gilliland [8], the minimum-reflux ratio computed by assuming a Class 1 separation is equal to or greater than the true minimum. This is because the presence of distributing nonkey components in the pinch-point zones increases the difficulty of the separation, thus increasing the reflux requirement.

EXAMPLE 9.4

Calculate the minimum internal reflux for the conditions of Example 9.2 assuming a Class 1 separation. Check the validity of this assumption.

---

[3] Assuming the feed is neither subcooled nor superheated.

SOLUTION

From Figure 9.5, the relative volatility between nC_4(LK) and iC_5(HK) at the feed temperature of 180°F is 1.93. Feed liquid and distillate quantities are given in Figure 9.3 and Example 9.1. From (9-21),

$$(L_\infty)_{min} = \frac{759.4[(442/346.5) - 1.93(13/31.4)]}{1.93 - 1} = 389 \text{ lbmol/h}$$

Distribution of nonkey components in the feed is determined by (9-22). The most likely nonkey component to distribute is nC_5 because its volatility is close to that of iC_5(HK), which does not undergo a sharp separation. For nC_5, using data for K-values from Figure 9.5, we have

$$\frac{Dx_{nC_5,D}}{L_F x_{nC_5,F}} = \left[\frac{0.765 - 1}{1.93 - 1}\right]\left(\frac{442}{346.5}\right) + \left[\frac{1.93 - 0.765}{1.93 - 1}\right]\left(\frac{13}{31.4}\right)$$

$$= 0.1963$$

Therefore, $Dx_{nC_5,D} = 0.1963(13.4) = 2.63$ lbmol/h of nC_5 in the distillate. This is less than the quantity of nC_5 in the total feed. Therefore, nC_5 distributes between the distillate and the bottoms. However, similar calculations for the other nonkey components give negative distillate flow rates for the other heavy components and, in the case of iC_4, a distillate flow rate greater than the feed rate. Thus, the computed reflux rate is not valid. However, as expected, it is greater than the true internal value of 298 lbmol/h reported by Bachelor [4].

For Class 2 separations, (9-17) to (9-20) still apply. However, (9-20) cannot be used directly to compute the internal minimum-reflux ratio because values of $x_{i,\infty}$ are not simply related to feed composition for Class 2 separations. Underwood [9] devised an ingenious algebraic procedure to overcome this difficulty. For the rectifying section, he defined a quantity Φ by

$$\sum_i \frac{(\alpha_{i,r})_\infty x_{i,D}}{(\alpha_{i,r})_\infty - \Phi} = 1 + (R_\infty)_{min} \qquad (9\text{-}24)$$

Similarly, for the stripping section, Underwood defined Φ' by

$$\sum \frac{(\alpha'_{i,r})_\infty x_{i,B}}{(\alpha'_{i,r})_\infty - \Phi'} = 1 - (R'_\infty)_{min} \qquad (9\text{-}25)$$

where $R'_\infty = L'_\infty/B$ and the prime refers to conditions in the stripping-section pinch-point zone. In his derivation, Underwood assumed that relative volatilities are constant in the region between the two pinch-point zones and that $(R_\infty)_{min}$ and $(R'_\infty)_{min}$ are related by the assumption of constant molar overflow in the region between the feed entry and the rectifying-section pinch point and in the region between the feed entry and the stripping-section pinch point. Hence,

$$(L'_\infty)_{min} - (L_\infty)_{min} = qF \qquad (9\text{-}26)$$

With these two critical assumptions, Underwood showed that at least one common root θ (where $\theta = \Phi = \Phi'$) exists between (9-24) and (9-25).

Equation (9-24) is analogous to the following equation derived from (9-19), and the relation $\alpha_{i,r} = K_i/K_r$,

$$\sum_i \frac{(\alpha_{i,r})_\infty x_{i,D}}{(\alpha_{i,r})_\infty - L_\infty/[V_\infty(K_r)_\infty]} = 1 + (R_\infty)_{min} \qquad (9\text{-}27)$$

where $L_\infty/[V_\infty(K_r)_\infty]$ is called the *absorption factor* for a reference component in the rectifying-section pinch-point zone. Although Φ is analogous to the absorption factor, a different root of Φ is used to solve for $(R_\infty)_{min}$, as discussed by Shiras et al. [6].

The common root θ may be determined by multiplying (9-24) and (9-25) by D and B, respectively, adding the two equations, substituting (9-25) to eliminate $(R'_\infty)_{min}$ and $(R_\infty)_{min}$, and utilizing the overall component balance $z_{i,F}F = x_{i,D}D + x_{i,B}B$ to obtain

$$\sum_i \frac{(\alpha_{i,r})_\infty z_{i,F}}{(\alpha_{i,r})_\infty - \theta} = 1 - q \qquad (9\text{-}28)$$

where q is the thermal condition of the feed from (7-20) and r is conveniently taken as the heavy key, HK. When only the two key components distribute, (9-28) is solved iteratively for a root of θ that satisfies $\alpha_{LK,HK} > \theta > 1$. The following modification of (9-24) is then solved for the internal reflux ratio $(R_\infty)_{min}$:

$$\sum_i \frac{(\alpha_{i,r})_\infty x_{i,D}}{(\alpha_{i,r})_\infty - \theta} = 1 + (R_\infty)_{min} \qquad (9\text{-}29)$$

If any nonkey components are suspected of distributing, estimated values of $x_{i,D}$ cannot be used directly in (9-29). This is particularly true when nonkey components are intermediate in volatility between the two key components. In this case, (9-28) is solved for m roots of θ, where m is one less than the number of distributing components. Furthermore, each root of θ lies between an adjacent pair of relative volatilities of distributing components. For instance, in Example 9.4, it was found the nC_5 distributes at minimum reflux, but nC_6 and heavier do not and iC_4 does not. Therefore, two roots of θ are necessary, where

$$\alpha_{nC_4,iC_5} > \theta_1 > 1.0 > \theta_2 > \alpha_{nC_5,iC_5}$$

With these two roots, (9-29) is written twice and solved simultaneously to yield $(R_\infty)_{min}$ and the unknown value of $x_{nC_5,D}$. The solution must, of course, satisfy the condition $\sum x_{i,D} = 1.0$.

With the internal reflux ratio $(R_\infty)_{min}$ known, the external reflux ratio is computed by enthalpy balance with (9-23). This requires a knowledge of the rectifying-section pinch-point compositions. Underwood [9] shows that

$$x_{i,\infty} = \frac{\theta x_{i,D}}{(R_\infty)_{min}[(\alpha_{i,r})_\infty - \theta]} \qquad (9\text{-}30)$$

with $y_{i,\infty}$ given by (9-17). The value of θ to be used in (9-30) is the root of (9-29) satisfying the inequality

$$(\alpha_{HNK,r})_\infty > \theta > 0$$

where HNK refers to the heaviest nonkey component in the distillate at minimum reflux. This root is equal to $L_\infty/[V_\infty(K_r)_\infty]$ in (9-27). With wide-boiling feeds, the external reflux can be significantly higher than the internal reflux. Bachelor [4] cites a case where the external reflux rate is 55% greater than the internal reflux.

For the stripping-section pinch-point composition, Underwood obtains

$$x'_{i,\infty} = \frac{\theta x_{i,B}}{[(R'_\infty)_{min}+1][(\alpha_{i,r})_\infty - \theta]} \qquad (9\text{-}31)$$

where, in this case, θ is the root of (9-29) satisfying the inequality

$$(\alpha_{HNK,r})_\infty > \theta > 0$$

where HNK refers to the heaviest nonkey in the bottoms product at minimum reflux.

Because of their relative simplicity, the Underwood minimum-reflux equations for Class 2 separations are widely used, but too often without examining the possibility of nonkey distribution. In addition, the assumption is frequently made that $(R_\infty)_{min}$ equals the external reflux ratio. When the assumptions of constant relative volatility and constant molar overflow in the regions between the two pinch-point zones are not valid, values of the minimum-reflux ratio computed from the Underwood equations for Class 2 separations can be appreciably in error because of the sensitivity of (9-28) to the value of q, as will be shown in Example 9.5. When the Underwood assumptions appear to be valid and a negative minimum-reflux ratio is computed, this may be interpreted to mean that a rectifying section is not required to obtain the specified separation. The Underwood equations show that the minimum reflux depends mainly on the feed condition and relative volatility and, to a lesser extent, on the degree of separation between the two key components. A finite minimum-reflux ratio exists even for a perfect separation.

An extension of the Underwood method for distillation columns with multiple feeds is given by Barnes, Hanson, and King [10]. Exact computer methods for determining minimum reflux are available [11]. For making rigorous distillation calculations at actual reflux conditions by the computer methods of Chapter 10, knowledge of the minimum reflux is not essential, but the minimum number of equilibrium stages is very useful.

EXAMPLE 9.5

Repeat Example 9.4 assuming a Class 2 separation and utilizing the corresponding Underwood equations. Check the validity of the Underwood assumptions. Also calculate the external reflux ratio.

SOLUTION

From the results of Example 9.4, assume that the only distributing nonkey component is n-pentane. Assuming that the feed temperature of 180°F is reasonable for computing relative volatilities in the pinch zone, the following quantities are obtained from Figures 9.3 and 9.5:

Species i	$z_{i,F}$	$(\alpha_{i,HK})_\infty$
iC_4	0.0137	2.43
nC_4 (LK)	0.5113	1.93
iC_5 (HK)	0.0411	1.00

Species i	$z_{i,F}$	$(\alpha_{i,HK})_\infty$
nC_5	0.0171	0.765
nC_6	0.0262	0.362
nC_7	0.0446	0.164
nC_8	0.3106	0.0720
nC_9	0.0354	0.0362
	1.0000	

The q for the feed is assumed to be the mole fraction of liquid in the flashed feed. From Example 9.1, $q = 1 - 0.1334 = 0.8666$. Applying (9-28), we have

$$\frac{2.43(0.0137)}{2.43-\theta} + \frac{1.93(0.5113)}{1.93-\theta} + \frac{1.00(0.0411)}{1.00-\theta}$$
$$+ \frac{0.765(0.0171)}{0.765-\theta} + \frac{0.362(0.0262)}{0.362-\theta} + \frac{0.164(0.0446)}{0.164-\theta}$$
$$+ \frac{0.072(0.3106)}{0.072-\theta} + \frac{0.0362(0.0354)}{0.0362-\theta} = 1 - 0.8666$$

Solving this equation by a bounded-Newton method for two roots of θ that satisfy

$$\alpha_{nC_4,iC_5} > \theta_1 > \alpha_{iC_5,iC_5} > \theta_2 > \alpha_{nC_5,iC_5}$$

or

$$1.93 > \theta_1 > 1.00 > \theta_2 > 0.765$$

$\theta_1 = 1.04504$ and $\theta_2 = 0.78014$. Because distillate rates for nC_4 and iC_5 are specified (442 and 13 lbmol/h, respectively), the following form of (9-29) is preferred:

$$\sum_i \frac{(\alpha_{i,r})_\infty (x_{i,D}D)}{(\alpha_{i,r})_\infty - \theta} = D + (L_\infty)_{min} \qquad (9\text{-}32)$$

with the restriction that

$$\sum_i (x_{i,D}D) = D \qquad (9\text{-}33)$$

Assuming that $x_{i,D}D$ equals 0.0 for components heavier than nC_5 and 12.0 lbmol/h for iC_4, we find that these two relations give the following three linear equations:

$$D + (L_\infty)_{min} = \frac{2.43(12)}{2.43-1.04504} + \frac{1.93(442)}{1.93-1.04504}$$
$$+ \frac{1.00(13)}{1.00-1.04504} + \frac{0.765(x_{nC_5,D}D)}{0.765-1.04504}$$

$$D + (L_\infty)_{min} = \frac{2.43(12)}{2.43-0.78014} + \frac{1.93(442)}{1.93-0.78014}$$
$$+ \frac{1.00(13)}{1.00-0.78014} + \frac{0.765(x_{nC_5,D}D)}{0.765-0.78014}$$

$$D = 12 + 442 + 13 + (x_{nC_5,D}D)$$

Solving these three equations gives

$$x_{nC_5,D}D = 2.56 \text{ lbmol/h}$$
$$D = 469.56 \text{ lbmol/h}$$
$$(L_\infty)_{min} = 219.8 \text{ lbmol/h}$$

The distillate rate for nC_5 is very close to the value of 2.63 computed in Example 9.4, if we assume a Class 1 separation. The internal minimum reflux ratio at the rectifying pinch point is considerably less than the value of 389 computed in Example 9.4 and is also much less than the true internal value of 298 reported by Bachelor [4]. The main reason for the discrepancy between the value of 219.8 and the

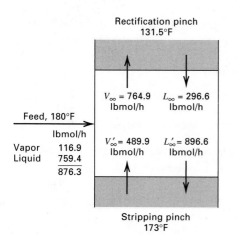

Rectification pinch
131.5°F

$V_\infty = 764.9$ lbmol/h $L_\infty = 296.6$ lbmol/h

Feed, 180°F

lbmol/h
Vapor   116.9
Liquid   759.4
         876.3

$V'_\infty = 489.9$ lbmol/h $L'_\infty = 896.6$ lbmol/h

Stripping pinch
173°F

Figure 9.8 Pinch-point region conditions for Example 9.5 from computations by Bachelor.

[From J.B. Bachelor, *Petroleum Refiner,* **36**(6), 161–170 (1957).]

true value of 298 is the invalidity of the assumption of constant molar overflow. Bachelor computed the pinch-point region flow rates and temperatures shown in Figure 9.8. The average temperature of the region between the two pinch regions is 152°F (66.7°C), which is appreciably lower than the flashed-feed temperature. The relatively hot feed causes additional vaporization across the feed zone. The effective value of q in the region between the pinch points is obtained from (7-18):

$$q_{eff} = \frac{L'_\infty - L_\infty}{F} = \frac{896.6 - 296.6}{876.3} = 0.685$$

This is considerably lower than the value of 0.8666 for q based on the flashed-feed condition. On the other hand, the value of $\alpha_{LK,HK}$ at 152°F (66.7°C) is not much different from the value at 180°F (82.2°C). If this example is repeated using q equal to 0.685, the resulting value of $(L_\infty)_{min}$ is 287.3 lbmol/h, which is only 3.6% lower than the true value of 298. Unfortunately, in practice, this corrected procedure cannot be applied because the true value of q cannot be readily determined.

To compute the external-reflux ratio from (9-23), rectifying pinch-point compositions must be calculated from (9-30) and (9-17). The root of θ to be used in (9-30) is obtained from the version of (9-29) used above. Thus,

$$\frac{2.43(12)}{2.43 - \theta} + \frac{1.93(442)}{1.93 - \theta} + \frac{1.00(13)}{1.00 - \theta}$$
$$+ \frac{0.765(2.56)}{0.765 - \theta} = 469.56 + 219.8$$

where $0.765 > \theta > 0$. Solving, $\theta = 0.5803$. Liquid pinch-point compositions are obtained from the following form of (9-30):

$$x_{i,\infty} = \frac{\theta(x_{i,D} D)}{(L_\infty)_{min}[(\alpha_{i,r})_\infty - \theta]}$$

with $(L_\infty)_{min} = 219.8$ lbmol/h.

For iC_4,

$$x_{iC_4,\infty} = \frac{0.5803(12)}{219.8(2.43 - 0.5803)} = 0.0171$$

From a combination of (9-17) and (9-18),

$$y_{i,\infty} = \frac{x_{i,\infty} L_\infty + x_{i,D} D}{L_\infty + D}$$

For iC_4,

$$y_{iC_4,\infty} = \frac{0.0171(219.8) + 12}{219.8 + 469.56} = 0.0229$$

Similarly, the mole fractions of the other components appearing in the distillate are

Component	$x_{i,\infty}$	$y_{i,\infty}$
iC_4	0.0171	0.0229
nC_4	0.8645	0.9168
iC_5	0.0818	0.0449
nC_5	0.0366	0.0154
	1.0000	1.0000

The temperature of the rectifying-section pinch point is obtained from either a bubble-point temperature calculation on $x_{i,\infty}$ or a dew-point temperature calculation on $y_{i,\infty}$. The result is 126°F. Similarly, the liquid-distillate temperature (bubble point) and the temperature of the vapor leaving the top stage (dew point) are both computed to be approximately 123°F. Because rectifying-section pinch-point temperature and distillate temperatures are very close, it is expected that $(R_\infty)_{min}$ and $(R_{min})_{external}$ will be almost identical. Bachelor [4] obtained a value of 292 lbmol/h for the external-reflux rate, compared to 298 lbmol/h for the internal reflux rate.

Gilliland Correlation for Actual Reflux Ratio and Theoretical Stages

To achieve a specified separation between two key components, the reflux ratio and the number of theoretical stages must be greater than their minimum values. The actual reflux ratio is generally established by economic considerations at some multiple of minimum reflux. The corresponding number of theoretical stages is then determined by suitable analytical or graphical methods or, as discussed in this section, by an empirical equation. However, there is no reason why the number of theoretical stages cannot be specified as a multiple of minimum stages and the corresponding actual reflux computed by the same empirical relationship. As shown in Figure 9.9, from studies by Fair and Bolles [12], the optimal value of R/R_{min} is approximately 1.05. However, near-optimal conditions extend over a relatively broad range of mainly larger values of R/R_{min}. In practice, superfractionators requiring a large number of stages are frequently designed for a value of R/R_{min} of approximately 1.10, while separations requiring a small number of stages are designed for a value of R/R_{min} of approximately 1.50. For intermediate cases, a commonly used rule of thumb is R/R_{min} equal to 1.30.

The number of equilibrium stages required for the separation of a binary mixture assuming constant relative volatility and constant molar overflow depends on $z_{i,F}$, $x_{i,D}$, $x_{i,B}$, q, R, and α. From (9-11), for a binary mixture, N_{min} depends on $x_{i,D}$, $x_{i,B}$, and α, while R_{min} depends on $z_{i,F}$, $x_{i,D}$, q, and α. Accordingly, a number of investigators have assumed empirical correlations of the form

$$N = N\left\{N_{min}\{x_{i,D}, x_{i,B}, \alpha\}, \ R_{min}\{z_{i,F}, x_{i,D}, q, \alpha\}, \ R\right\}$$

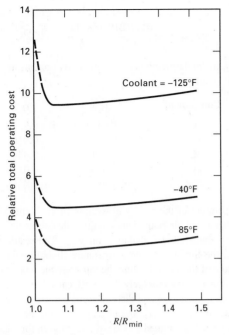

Figure 9.9 Effect of reflux ratio on cost.
[From J.R. Fair and W.L. Bolles, *Chem. Eng.*, **75**(9), 156–178 (1968).]

Furthermore, they have assumed that such a correlation might exist for nearly ideal multicomponent systems even though additional feed composition variables and nonkey relative volatilities also influence the value of R_{min}.

The most successful and simplest empirical correlation of this type is the one developed by Gilliland [13] and slightly modified in a later version by Robinson and Gilliland [14]. The correlation is shown in Figure 9.10, where the three sets of data points, which are based on accurate calculations, are

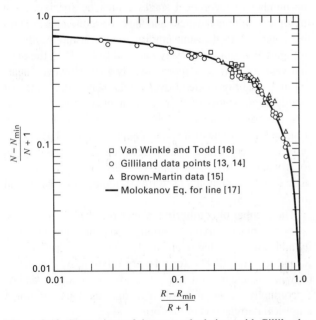

Figure 9.10 Comparison of rigorous calculations with Gilliland correlation.

the original points from Gilliland [13] and the multicomponent data points of Brown and Martin [15] and Van Winkle and Todd [16]. The 61 data points cover the following ranges of conditions:

1. Number of components: 2 to 11
2. q: 0.28 to 1.42
3. Pressure: vacuum to 600 psig
4. α: 1.11 to 4.05
5. R_{min}: 0.53 to 9.09
6. N_{min}: 3.4 to 60.3

The line drawn through the data represents the equation developed by Molokanov et al. [17]:

$$Y = \frac{N - N_{min}}{N + 1} = 1 - \exp\left[\left(\frac{1 + 54.4X}{11 + 117.2X}\right)\left(\frac{X - 1}{X^{0.5}}\right)\right]$$

(9-34)

where

$$X = \frac{R - R_{min}}{R + 1}$$

This equation satisfies the end points ($Y = 0$, $X = 1$) and ($Y = 1$, $X = 0$). At a value of R/R_{min} near the optimum of 1.3, Figure 9.10 predicts an optimal ratio for N/N_{min} of approximately 2. The value of N includes one stage for a partial reboiler and one stage for a partial condenser, if any.

The Gilliland correlation is very useful for preliminary exploration of design variables. Although it was never intended for final design, the Gilliland correlation was used, before the applicability of digital computers, to design many distillation columns for multicomponent separations without benefit of accurate stage-by-stage calculations. In Figure 9.11, a replot of the correlation in linear coordinates shows that a small initial increase in R above R_{min} causes a large decrease in N, but further changes in R have a much smaller effect on N. The knee in the curve of Figure 9.11 corresponds closely to the optimal value of R/R_{min} in Figure 9.9.

Robinson and Gilliland [14] state that a more accurate correlation should utilize a parameter involving the feed condition q. This effect is shown in Figure 9.12 using data points for the sharp separation of benzene–toluene mixtures from Guerreri [18]. The data, which cover feed conditions ranging

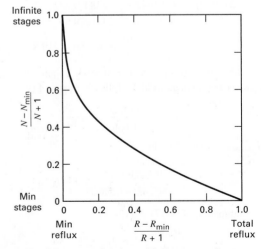

Figure 9.11 Gilliland correlation with linear coordinates.

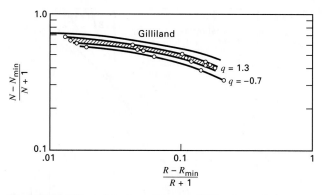

Figure 9.12 Effect of feed condition on Gilliland correlation.
[From G. Guerreri, *Hydrocarbon Processing*, **48**(8), 137–142 (1969).]

from subcooled liquid to superheated vapor (q equals 1.3 to -0.7), show a trend toward decreasing theoretical-stage requirements with increasing feed vaporization. The Gilliland correlation appears to be conservative for feeds having low values of q. Donnell and Cooper [19] state that this effect of q is important only when the α between the key components is high or when the feed is low in volatile components.

A serious problem with the Gilliland correlation can occur when stripping is much more important than rectification. For example, Oliver [20] considers a fictitious binary case with specifications of $z_F = 0.05$, $x_D = 0.40$, $x_B = 0.001$, $q = 1$, $\alpha = 5$, $R/R_{min} = 1.20$, and constant molar overflow. By exact calculations, $N = 15.7$. From the Fenske equation, $N_{min} = 4.04$. From the Underwood equation, $R_{min} = 1.21$. From (9-32) for the Gilliland correlation, $N = 10.3$. This is 34% lower than the exact value. This limitation, which is caused by ignoring boilup, is discussed further by Strangio and Treybal [21], who present a more accurate, but far more tedious, method for such cases.

EXAMPLE 9.6

Use the Gilliland correlation to estimate the theoretical-stage requirements for the debutanizer of Examples 9.1, 9.2, and 9.5 for an external reflux of 379.6 lbmol/h (30% greater than the exact value of the minimum-reflux rate from Bachelor).

SOLUTION

From the examples cited, values of R_{min} and $[(R - R_{min})/(R + 1)]$ are obtained using a distillate rate from Example 9.5 of 469.56 lbmol/h. Thus, $R = 379.6/469.56 = 0.808$. With $N_{min} = 8.88$,

$$R_{min} = 0.479, \quad \text{and} \quad \frac{R - R_{min}}{R + 1} = X = 0.182$$

From (9-34),

$$\frac{N - N_{min}}{N + 1} = 1 - \exp\left[\left(\frac{1 + 54.4(0.182)}{11 + 117.2(0.182)}\right)\left(\frac{0.182 - 1}{0.182^{0.5}}\right)\right]$$

$$= 0.476$$

$$N = \frac{8.88 + 0.476}{1 - 0.476} = 17.85$$

$$N - 1 = 16.85$$

where $N - 1$ corresponds to the equilibrium stages in the tower allowing one theoretical stage for the reboiler, but no stage for the total condenser.

It should be kept in mind that, had the exact value of R_{min} not been known and a value of R equal to 1.3 times R_{min} from the Underwood method been used, the value of R would have been 292 lbmol/h. But this, by coincidence, is only the true minimum reflux. Therefore, the desired separation would not be achieved.

Feed-Stage Location

Implicit in the application of the Gilliland correlation is the specification that the theoretical stages be distributed optimally between the rectifying and stripping sections. As suggested by Brown and Martin [15], the optimal feed stage can be located by assuming that the ratio of stages above the feed to stages below the feed is the same as the ratio determined by simply applying the Fenske equation to the separate sections at total reflux conditions to give

$$\frac{N_R}{N_S} \simeq \frac{(N_R)_{min}}{(N_S)_{min}}$$

$$= \frac{\log[(x_{LK,D}/z_{LK,F})(z_{HK,F}/x_{HK,D})]}{\log[(z_{LK,F}/x_{LK,B})(x_{HK,B}/z_{HK,F})]} \frac{\log[(\alpha_B \alpha_F)^{1/2}]}{\log[(\alpha_D \alpha_F)^{1/2}]} \tag{9-35}$$

Unfortunately, (9-35) is not reliable except for fairly symmetrical feeds and separations.

A reasonably good approximation of optimal feed-stage location can be made by employing the empirical equation of Kirkbride [22]:

$$\frac{N_R}{N_S} = \left[\left(\frac{z_{HK,F}}{z_{LK,F}}\right)\left(\frac{x_{LK,B}}{x_{HK,D}}\right)^2\left(\frac{B}{D}\right)\right]^{0.206} \tag{9-36}$$

An extreme test of both these equations is provided by the fictitious binary-mixture problem of Oliver [20] cited in the previous section. Exact calculations by Oliver and calculations using (9-35) and (9-36) give the following results:

Method	N_R/N_S
Exact	0.08276
Kirkbride (9-34)	0.1971
Fenske ratio (9-33)	0.6408

Although the result from the Kirkbride equation is not very satisfactory, the result from the Fenske ratio method is much worse.

EXAMPLE 9.7

Use the Kirkbride equation to determine the feed-stage location for the debutanizer of Example 9.1, assuming an equilibrium-stage requirement of 18.27.

SOLUTION

Assume that the product distribution computed in Example 9.3 for total-reflux conditions is a good approximation to the distillate and bottoms compositions at actual reflux conditions. Then

$$x_{nC_4,B} = \frac{6.0}{408.5} = 0.0147 \quad x_{iC_5,D} = \frac{13}{467.8} = 0.0278$$

$$D = 467.8 \text{ lbmol/h} \qquad B = 408.5 \text{ lbmol/h}$$

From Figure 9.3,

$$z_{nC_4,F} = 448/876.3 = 0.5112$$

and

$$z_{iC_5,F} = 36/876.3 = 0.0411$$

From (9-36),

$$\frac{N_R}{N_S} = \left[\left(\frac{0.0411}{0.5112} \right) \left(\frac{0.0147}{0.0278} \right)^2 \left(\frac{408.5}{467.8} \right) \right]^{0.206} = 0.445$$

Therefore, $N_R = (0.445/1.445)(18.27) = 5.63$ stages and $N_S = 18.27 - 5.63 = 12.64$ stages.

Rounding the estimated stage requirements leads to one stage as a partial reboiler, 12 stages below the feed, and six stages above the feed.

Distribution of Nonkey Components at Actual Reflux

For multicomponent mixtures, all components distribute to some extent between distillate and bottoms at total reflux conditions. However, at minimum-reflux conditions, none or only a few of the nonkey components distribute. Distribution ratios for these two limiting conditions are shown in Figure 9.13 for the debutanizer example. For total-reflux conditions, results from the Fenske equation in Example 9.3 plot as a straight line for the log–log coordinates. For minimum reflux, results from the Underwood equation in Example 9.5 are shown as a dashed line.

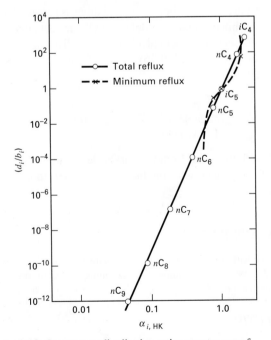

Figure 9.13 Component distribution ratios at extremes of distillation operating conditions.

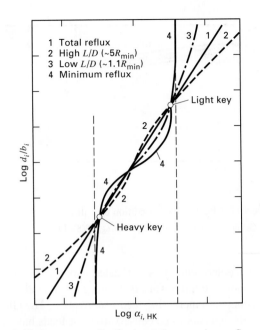

Figure 9.14 Component distribution ratios at various reflux ratios.

It might be expected that a product-distribution curve for actual reflux conditions would lie between the two limiting curves. However, as shown by Stupin and Lockhart [23], product distributions in distillation are complex. A typical result is shown in Figure 9.14. For a reflux ratio near minimum, the product distribution (curve 3) lies between the two limits (curves 1 and 4). However, for a high reflux ratio, the product distribution for a nonkey component (curve 2) may actually lie outside the limits, so that an inferior separation results.

For the behavior of the product distribution in Figure 9.14, Stupin and Lockhart provide an explanation that is consistent with the Gilliland correlation of Figure 9.10. As the reflux ratio is decreased from total reflux while maintaining the specified splits of the two key components, equilibrium-stage requirements increase only slowly at first, but then rapidly as minimum reflux is approached. Initially, large decreases in reflux cannot be adequately compensated for by increasing the number of stages. This causes inferior nonkey component distributions. However, as minimum reflux is approached, comparatively small decreases in reflux are more than compensated for by large increases in equilibrium stages; and the separation of nonkey components becomes superior to that at total reflux. It appears reasonable to assume that, at a near-optimal reflux ratio of 1.3, nonkey-component distribution is close to that estimated by the Fenske equation for total-reflux conditions.

9.2 KREMSER GROUP METHOD

Many multicomponent separators are cascades of stages where the two contacting phases flow countercurrently. Approximate calculation procedures have been developed to relate compositions of streams entering and exiting cascades to the number of equilibrium stages required. These approximate procedures are called *group methods* because they provide only an overall treatment of the stages in the cascade without considering detailed changes in temperature, flow rates, and composition in the individual stages. In this section, single cascades used for absorption, stripping, and liquid–liquid extraction are considered.

Kremser [1] originated the group method. He derived overall species material balances for a multistage countercurrent absorber. Subsequent articles by Souders and Brown [24] Horton and Franklin [25] and Edmister [26] improved the method. The Kremser equations are derived and applied to absorption in Section 5.4. These equations are illustrated for strippers and extractors here. Another treatment by Smith and Brinkley [27] emphasizes liquid–liquid separations.

Strippers

The vapor entering a stripper is often steam or an inert gas. When the stripping agent contains none of the species in the feed liquid, is not present in the entering liquid, and is not absorbed or condensed in the stripper, the only direction of mass transfer is from the liquid to the gas phase. Then, only values of the effective stripping factor, S_e, as defined by (5-51), are needed to apply the group method via (5-49) and (5-50). The equations for strippers are analogous to those for absorbers.

For optimal stripping, temperatures should be high and pressures low. However, temperatures should not be so high as to cause decomposition, and vacuum should be used only if necessary. The minimum stripping-agent flow rate, for a specified value of ϕ_S for a key component K corresponding to an infinite number of stages, can be estimated from an equation obtained from (5-50) with $N = \infty$,

$$(V_0)_{\min} = \frac{L_{N+1}}{K_K}(1 - \phi_{S_K}) \qquad (9\text{-}37)$$

This equation assumes that $A_K < 1$ and the fraction of liquid feed stripped is small.

EXAMPLE 9.8

Sulfur dioxide and butadienes (B3 and B2) are to be stripped with nitrogen from the liquid stream given in Figure 9.15 so that butadiene sulfone (BS) product will contain less than 0.05 mol% SO_2 and less than 0.5 mol% butadienes. Estimate the flow rate of nitrogen, N_2, and the number of equilibrium stages required.

SOLUTION

Neglecting stripping of BS, the stripped liquid must have the following component flow rates, and corresponding values for ϕ_S:

Species	l_1, lbmol/h	$\phi_S = \dfrac{l_1}{l_{N+1}}$
SO_2	<0.0503	<0.00503
B3 + B2	<0.503	<0.0503
BS	100.0	—

Thermodynamic properties can be computed based on ideal solutions at low pressures. For butadiene sulfone, the vapor pressure is

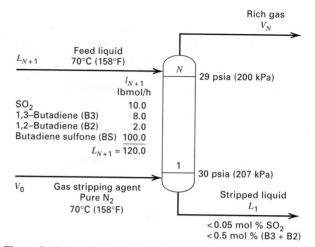

Figure 9.15 Specifications for stripper of Example 9.8.

given by

$$P_{BS}^s = \exp\left(17.30 - \frac{11{,}142}{T + 459.67}\right)$$

where P_{BS}^s is in pounds force per square inch and T is in degrees Fahrenheit. The liquid enthalpy of BS is

$$(h_L)_{BS} = 50T$$

where $(h_L)_{BS}$ is in British thermal units per pound-mole and T is in degrees Fahrenheit.

The entering flow rate of the stripping agent V_0 is not specified. The minimum rate at infinite stages can be computed from (9-37), provided that a key component is selected. Suppose we choose B2, which is the heaviest component to be stripped to a specified extent. At 70°C, the vapor pressure of B2 is 90.4 psia. From Raoult's law at 30 psia total pressure,

$$K_{B2} = \frac{90.4}{30} = 3.01$$

From (9-37), using $(\phi_S)_{B2} = 0.0503$, we have

$$(V_0)_{\min} = \frac{120}{3.01}(1 - 0.0503) = 37.9 \text{ lbmol/h}$$

For this value of $(V_0)_{\min}$, (9-42) can now be used to determine ϕ_S for B3 and SO_2. The K-values for these two species are 4.53 and 6.95, respectively. From (9-37), at infinite stages with $V_0 = 37.9$ lbmol/h,

$$(\phi_S)_{B3} = 1 - \frac{4.53(37.9)}{120} = -0.43$$

and

$$(\phi_S)_{SO_2} = 1 - \frac{6.95(37.9)}{120} = -1.19$$

These negative values indicate complete stripping of B3 and SO_2. Therefore, the total butadienes in the stripped liquid would be only $(0.0503)(2.0) = 0.1006$, compared to the specified value of 0.503. We can obtain a better estimate of $(V_0)_{\min}$ by assuming that all of the butadiene content of the stripped liquid is due to B2. Then $(\phi_S)_{B2} = 0.503/2 = 0.2515$, and $(V_0)_{\min}$ from (9-37) is 29.9 lbmol/h. Values of $(\phi_S)_{B3}$ and $(\phi_S)_{SO_2}$ are still negative.

The actual entering flow rate for the stripping vapor must be greater than the minimum value. To estimate the effect of V_0 on the theoretical stage requirements and ϕ_S values for the nonkey components, the Kremser approximation is used with K-values at 70°C and 30 psia, $L = L_{N+1} = 120$ lbmol/h, and $V = V_0$ equal to a series of multiples of 29.9 lbmol/h. The calculations are greatly facilitated if values of N are selected and values of V are determined from (5-51), where S is obtained from Figure 5.9. Because B3 will be found to some extent in the stripped liquid, $(\phi_S)_{B2}$ will be held below 0.2515. By making iterative calculations, one can choose $(\phi_S)_{B2}$ so that $(\phi_S)_{B2+B3}$ satisfies the specification of, say, 0.05. For 10 theoretical stages, assuming essentially complete stripping of B3 such that $(\phi_S)_{B2} \approx 0.25$, $S_{B2} \approx 0.76$ from Figure 5.9. From (5-51),

$$V = V_0 = \frac{(120)(0.76)}{3.01} = 30.3 \text{ lbmol/h}$$

For B3, from (5-51),

$$S_{B3} = \frac{(4.53)(30.3)}{120} = 1.143$$

From Figure 5.9,

$$(\phi_S)_{B3} = 0.04$$

Thus,

$$(\phi_S)_{B2+B3} = \frac{0.25(2) + 0.04(8)}{10} = 0.082$$

This is considerably above the specification of 0.05. Therefore, repeat the calculations with, say, $(\phi_S)_{B2} = 0.09$ and continue to repeat until the specified value of $(\phi_S)_{B2+B3}$ is obtained. In this manner, calculations for various numbers of theoretical stages are carried out with converged results as shown.

N	V_0, lbmol/h	$V_0/(V_0)_{min}$	Fraction Not Stripped				
			ϕ_{SO_2}	ϕ_{B3}	ϕ_{B2}	ϕ_{B2+B3}	ϕ_{BS}
∞	29.9	1.00	0.0	0.0	0.2515	0.0503	0.9960
10	33.9	1.134	0.0005	0.017	0.18	0.050	0.9955
5	45.4	1.518	0.0050	0.029	0.117	0.040	0.9940
3	94.3	3.154	0.0050	0.016	0.045	0.022	0.9874

These results show that the specification on SO_2 is also met for all four values of N.

Liquid–Liquid Extraction

A schematic representation of a countercurrent extraction cascade is shown in Figure 9.16, with stages numbered from the top down and solvent V_{N+1} entering at the bottom.[4] The group method of calculation can be applied, with the equations written by analogy to absorbers. In place of the K-value, the distribution coefficient is used:

$$K_{D_i} = \frac{y_i}{x_i} = \frac{v_i/V}{l_i/L} \qquad (9\text{-}38)$$

Here, y_i is the mole fraction of i in the solvent or extract phase and x_i is the mole fraction in the feed or raffinate phase. Also, in place of the stripping factor, an *extraction*

---

[4] In a vertical extractor, solvent would have to enter at the top if of greater density than the feed.

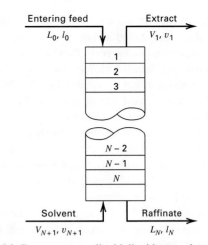

Figure 9.16 Countercurrent, liquid–liquid extraction cascade.

factor, E, is used, where

$$E_i = \frac{K_{D_i} V}{L} \qquad (9\text{-}39)$$

The reciprocal of E is

$$U_i = \frac{1}{E_i} = \frac{L}{K_{D_i} V} \qquad (9\text{-}40)$$

The working equations for each component are

$$v_1 = v_{N+1}\phi_U + l_0(1 - \phi_E) \qquad (9\text{-}41)$$

$$l_N = l_0 + v_{N+1} - v_1 \qquad (9\text{-}42)$$

where

$$\phi_U = \frac{U_e - 1}{U_e^{N+1} - 1} \qquad (9\text{-}43)$$

$$\phi_E = \frac{E_e - 1}{E_e^{N+1} - 1} \qquad (9\text{-}44)$$

For the Kremser approximation, values of E_i and U_i in (9-39) and (9-40) are based on the feed and solvent at entering conditions. However, in liquid–liquid extraction, values of V, L, and K_D may change considerably from stage to stage. Therefore a better approximation is desirable. This is achieved, with reference to Figure 9.16, by the following relations due to Horton and Franklin [25] and Edmister [26], which use average values of E_i and U_i based on estimates of values of V, L, and K_D at each end of the cascade. These equations are applied in Example 9.9.

$$V_2 = V_1\left(\frac{V_{N+1}}{V_1}\right)^{1/N} \qquad (9\text{-}45)$$

$$L_1 = L_0 + V_2 - V_1 \qquad (9\text{-}46)$$

$$V_N = V_{N+1}\left(\frac{V_1}{V_{N+1}}\right)^{1/N} \qquad (9\text{-}47)$$

$$E_e = [E_1(E_N + 1) + 0.25]^{1/2} - 0.5 \qquad (9\text{-}48)$$

$$U_e = [U_N(U_1 + 1) + 0.25]^{1/2} - 0.5 \qquad (9\text{-}49)$$

If desired, (9-38) through (9-49) can be applied using mass units rather than mole units. No enthalpy-balance equations are required because ordinarily temperature changes in an adiabatic extractor are not great unless the feed and solvent enter at appreciably different temperatures or the heat of mixing is large. Unfortunately, the group method is not always reliable for liquid–liquid extraction cascades because the distribution coefficient, as discussed in Chapter 2, is a ratio of activity coefficients, which can vary drastically with composition. Accordingly, rigorous methods of Chapter 10 are preferred.

EXAMPLE 9.9

Countercurrent, liquid–liquid extraction with methylene chloride is to be used at 25°C to recover dimethylformamide from an aqueous stream as shown in Figure 9.17. Estimate flow rates and compositions of extract and raffinate streams by the group method using mass units. Distribution coefficients for all components except DMF are essentially constant over the expected composition range and on a mass-fraction basis are

Component	K_{D_i}
MC	40.2
FA	0.005
DMA	2.2
W	0.003

The distribution coefficient for DMF depends on concentration in the water-rich phase as shown in Figure 9.18.

SOLUTION

Although the Kremser approximation could be applied for the first trial calculation, the following values will be assumed from guesses based on the magnitudes of the K_D-values.

Pounds per Hour

Component	Feed, l_0	Solvent, v_{11}	Raffinate, l_{10}	Extract, v_1
FA	20	0	20	0
DMA	20	0	0	20
DMF	400	2	2	400
W	3,560	25	3,560	25
MC	0	9,973	88	9,885
	4,000	10,000	3,670	10,330

From (9-45) through (9-49), we have

$$V_2 = 10,330 \left(\frac{10,000}{10,330} \right)^{1/10} = 10,297 \text{ lb/h};$$

$$L_1 = 4,000 + 10,297 - 10,330 = 3,967 \text{ lb/h};$$

$$V_{10} = 10,000 \left(\frac{10,330}{10,000} \right)^{1/10} = 10,033 \text{ lb/h}$$

From (9-39), (9-40), (9-48), and (9-49), assuming a mass fraction of 0.09 for DMF in L_1 in order to obtain $(K_D)_{DMF}$ for stage 1, we have

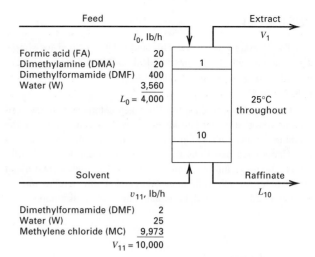

Feed — Extract

l_0, lb/h — V_1

Formic acid (FA) 20
Dimethylamine (DMA) 20
Dimethylformamide (DMF) 400
Water (W) 3,560
$L_0 = 4,000$

1

25°C throughout

10

Solvent — Raffinate

v_{11}, lb/h — L_{10}

Dimethylformamide (DMF) 2
Water (W) 25
Methylene chloride (MC) 9,973
$V_{11} = 10,000$

Figure 9.17 Specifications for extractor of Example 9.9.

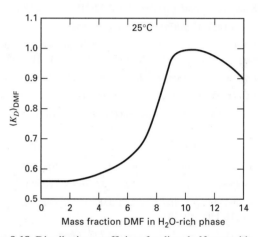

Figure 9.18 Distribution coefficient for dimethylformamide between water and methylene chloride.

Component	E_1	E_{10}	U_1	U_{10}	E_e	U_e
FA	0.013	0.014	—	—	0.013	—
DMA	5.73	6.01	—	—	5.86	—
DMF	2.50	1.53	0.400	0.653	2.06	0.579
W	0.0078	0.0082	128	122	0.0078	125
MC	—	—	0.0096	0.0091	—	0.0091

From (9-44), (9-43), (9-41), and (9-42), we have

Pounds per Hour

Component	ϕ_E	ϕ_U	Raffinate, l_{10}	Extract, v_1
FA	0.9870	—	19.7	0.3
DMA	0.0	—	0.0	20.0
DMF	0.000374	0.422	1.3	400.7
W	0.9922	0.0	3,557.2	37.8
MC	—	0.9909	90.8	9,882.2
			3,669.0	10,331.0

The calculated total flow rates L_{10} and V_1 are almost exactly equal to the assumed rates. Therefore, an additional iteration is not necessary. The degree of extraction of DMF is very high. It would be worthwhile to calculate additional cases with less solvent and/or fewer equilibrium stages.

SUMMARY

1. The Fenske–Underwood–Gilliland (FUG) method for simple distillation of ideal and nearly ideal multicomponent mixtures is useful for making preliminary estimates of stage and reflux requirements.

2. Based on a specified split of two key components in the feed mixture, the theoretical Fenske equation is used to determine the minimum number of equilibrium stages at total reflux. The theoretical Underwood equations are used to determine the minimum-reflux ratio for an infinite number of stages. The empirical Gilliland correlation relates the minimum stages and minimum reflux ratio to the actual reflux ratio and the actual number of equilibrium stages.

3. Estimates of the distribution of nonkey components and the feed-stage location can be made with the Fenske and Kirkbride equations, respectively.

4. The Underwood equations are more restrictive than the Fenske equation and must be used with care and caution.

5. The Kremser group method can be applied to simple strippers and liquid–liquid extractors to make approximate estimates of component recoveries for specified values of entering flow rates and number of equilibrium stages.

REFERENCES

1. KREMSER, A., *Natl. Petroleum News,* **22**(21), 43–49 (1930).

2. EDMISTER, W.C., *AIChE J.,* **3**, 165–171 (1957).

3. KOBE, K.A., and J.J. MCKETTA, JR., Eds, *Advances in Petroleum Chemistry and Refining,* Vol. 2, Interscience, New York, 315–355 (1959).

4. BACHELOR, J.B., *Petroleum Refiner,* **36**(6), 161–170 (1957).

5. FENSKE, M.R., *Ind. Eng. Chem.,* **24**, 482–485 (1932).

6. SHIRAS, R.N., D.N. HANSON, and C.H. GIBSON, *Ind. Eng. Chem.,* **42**, 871–876 (1950).

7. UNDERWOOD, A.J.V., *Trans. Inst. Chem. Eng.,* **10**, 112–158 (1932).

8. GILLILAND, E.R., *Ind. Eng. Chem.,* **32**, 1101–1106 (1940).

9. UNDERWOOD, A.J.V., *J. Inst. Petrol.,* **32**, 614–626 (1946).

10. BARNES, F.J., D.N. HANSON, and C.J. KING, *Ind. Eng. Chem., Process Des. Dev.,* **11**, 136–140 (1972).

11. TAVANA, M., and D.N. HANSON, *Ind. Eng. Chem., Process Des. Dev.,* **18**, 154–156 (1979).

12. FAIR, J.R., and W.L. BOLLES, *Chem. Eng.,* **75**(9), 156–178 (1968).

13. Gilliland, E.R., *Ind. Eng. Chem.,* **32**, 1220–1223 (1940).

14. ROBINSON, C.S., and E.R. GILLILAND, *Elements of Fractional Distillation,* 4th ed., McGraw-Hill, New York, pp. 347–350 (1950).

15. BROWN, G.G., and H.Z. MARTIN, *Trans. AIChE,* **35**, 679–708 (1939).

16. VAN WINKLE, M., and W.G. TODD, *Chem. Eng.,* **78**(21), 136–148 (1971).

17. MOLOKANOV, Y.K., T.P. KORABLINA, N.I. MAZURINA, and G.A. NIKIFOROV, *Int. Chem. Eng.,* **12**(2), 209–212 (1972).

18. GUERRERI, G., *Hydrocarbon Processing,* **48**(8), 137–142 (1969).

19. DONNELL, J.W., and C.M. COOPER, *Chem. Eng.,* **57**, 121–124 (1950).

20. OLIVER, E.D., *Diffusional Separation Processes: Theory, Design, and Evaluation,* John Wiley and Sons, New York, pp. 104–105 (1966).

21. STRANGIO, V.A., and R.E. TREYBAL, *Ind. Eng. Chem., Process Des. Dev.,* **13**, 279–285 (1974).

22. KIRKBRIDE, C.G., *Petroleum Refiner,* **23**(9), 87–102 (1944).

23. STUPIN, W.J., and F.J. LOCKHART, "The Distribution of Non-Key Components in Multicomponent Distillation," presented at the 61st Annual Meeting of the AIChE, Los Angeles, CA, December 1–5, 1968.

24. SOUDERS, M., and G.G. BROWN, *Ind. Eng. Chem.,* **24**, 519–522 (1932).

25. HORTON, G., and W.B. FRANKLIN, *Ind. Eng. Chem.,* **32**, 1384–1388 (1940).

26. EDMISTER, W.C., *Ind. Eng. Chem.,* **35**, 837–839 (1943).

27. SMITH, B.D., and W.K. BRINKLEY, *AIChE J.,* **6**, 446–450 (1960).

EXERCISES

Section 9.1

9.1 A mixture of propionic and *n*-butyric acids, which can be assumed to form ideal solutions, is to be separated by distillation into a distillate containing 95 mol% propionic acid and a bottoms product containing 98 mol% *n*-butyric acid. Determine the type of condenser to be used and estimate the distillation-column operating pressure.

9.2 A sequence of two distillation columns is to be used to produce the products indicated in Figure 9.19. Establish the type of condenser and an operating pressure for each column for: (a) The direct sequence (C_2/C_3 separation first) and (b) The indirect sequence (C_3/nC_4 separation first). Use K-values from Figures 2.8 and 2.9.

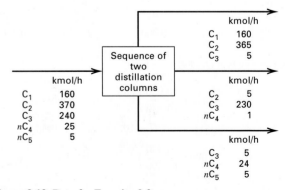

Figure 9.19 Data for Exercise 9.2.

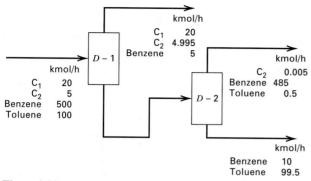

Figure 9.20 Data for Exercise 9.3.

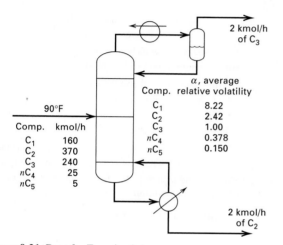

Figure 9.21 Data for Exercise 9.4.

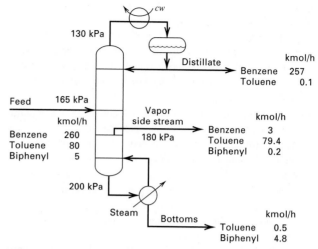

Figure 9.22 Data for Exercise 9.5.

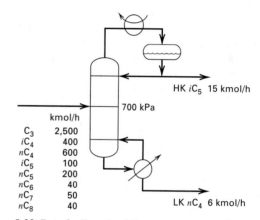

Figure 9.23 Data for Exercise 9.7.

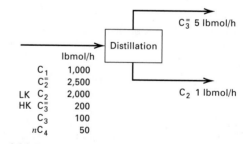

Figure 9.24 Data for Exercise 9.8.

9.3 For each of the two distillation separations (D-1 and D-2) indicated in Figure 9.20, establish the type of condenser and an operating pressure.

9.4 A deethanizer is to be designed for the separation indicated in Figure 9.21. Estimate the number of equilibrium stages required, assuming it is equal to 2.5 times the minimum number of equilibrium stages at total reflux.

9.5 For the complex distillation operation shown in Figure 9.22, use the Fenske equation to determine the minimum number of stages required between: (a) The distillate and feed, (b) The feed and the side stream, and (c) The side stream and bottoms. The K-values can be obtained from Raoult's law.

9.6 A 25 mol% mixture of acetone (A) in water (W) is to be separated by distillation at an average pressure of 130 kPa into a distillate containing 95 mol% acetone and a bottoms containing 2 mol% acetone. The infinite-dilution activity coefficients are

$$\gamma_A^\infty = 8.12 \quad \gamma_W^\infty = 4.13$$

Calculate by the Fenske equation the number of equilibrium stages required. Compare the result to that calculated from the McCabe–Thiele method. Is the Fenske equation reliable for this separation?

9.7 For the distillation operation indicated in Figure 9.23, calculate the minimum number of equilibrium stages and the distribution of the nonkey components by the Fenske equation, using Figures 2.8 and 2.9 for K-values.

9.8 For the distillation operation shown in Figure 9.24, establish the type of condenser and an operating pressure, calculate the minimum number of equilibrium stages, and estimate the distribution of the nonkey components. Obtain K-values from Figures 2.8 and 2.9.

9.9 For 15 minimum equilibrium stages at 250 psia, calculate and plot the percent recovery of C_3 in the distillate as a function of distillate flow rate for the distillation of 1,000 lbmol/h of a feed containing 3% C_2, 20% C_3, 37% nC_4, 35% nC_5, and 5% nC_6 by moles. Obtain K-values from Figures 2.8 and 2.9.

9.10 Use the Underwood equations to estimate the minimum external-reflux ratio for the separation by distillation of 30 mol% propane in propylene to obtain 99 mol% propylene and 98 mol% propane, if the feed condition at a column operating pressure of 300 psia is: (a) Bubble-point liquid, (b) Fifty mole percent vaporized, and (c) Dew-point vapor. Use K-values from Figures 2.8 and 2.9.

9.11 For the conditions of Exercise 9.7, compute the minimum-external-reflux rate and the distribution of the nonkey components at minimum reflux by the Underwood equation if the feed is a bubble-point liquid at column pressure.

9.12 Calculate and plot the minimum-external-reflux ratio and the minimum number of equilibrium stages against percent product purity for the separation by distillation of an equimolar bubble-point liquid feed of isobutane/n-butane at 100 psia. The distillate is to have the same iC_4 purity as the bottoms is to have nC_4 purity. Consider percent purities from 90% to 99.99%. Discuss the significance of the results.

9.13 Use the Fenske–Underwood–Gilliland shortcut method to determine the reflux ratio required to conduct the distillation operation indicated in Figure 9.25 if $N/N_{min} = 2.0$, the average relative volatility $= 1.11$, and the feed is at the bubble-point temperature at column feed-stage pressure. Assume that external reflux equals internal reflux at the upper pinch zone. Assume a total condenser and a partial reboiler.

9.14 A feed consisting of 62 mol% *para*-dichlorobenzene in *ortho*-dichlorobenzene is to be separated by distillation at near atmospheric pressure into a distillate containing 98 mol% para isomer and bottoms containing 96 mol% ortho isomer.

If a total condenser and partial reboiler are used, $q = 0.9$, average relative volatility $= 1.154$, and reflux/minimum reflux $= 1.15$, use the Fenske–Underwood–Gilliland procedure to estimate the number of theoretical stages required.

9.15 Explain why the Gilliland correlation can give erroneous results for cases where the ratio of rectifying to stripping stages is small.

9.16 The hydrocarbon feed to a distillation column is a bubble-point liquid at 300 psia with the mole fraction composition, $C_2 = 0.08$, $C_3 = 0.15$, $nC_4 = 0.20$, $nC_5 = 0.27$, $nC_6 = 0.20$, and $nC_7 = 0.10$.

(a) For a sharp separation between nC_4 and nC_5, determine the column pressure and type of condenser if condenser outlet temperature is 120°F.

(b) At total reflux, determine the separation for eight theoretical stages overall, specifying 0.01 mole fraction nC_4 in the bottoms product.

(c) Determine the minimum-reflux ratio for the separation in part (b).

(d) Determine the number of theoretical stages at $L/D = 1.5$ times minimum using the Gilliland correlation.

9.17 The following feed mixture is to be separated by ordinary distillation at 120 psia so as to obtain 92.5 mol% of the nC_4 in the liquid distillate and 82.0 mol% of the iC_5 in the bottoms.

Component	lbmol/h
C_3	5
iC_4	15
nC_4	25
iC_5	20
nC_5	35
	100

(a) Estimate the minimum number of equilibrium stages required by applying the Fenske equation. Obtain K-values from Figures 2.8 and 2.9.

(b) Use the Fenske equation to determine the distribution of non-key components between distillate and bottoms.

(c) Assuming that the feed is at its bubble point, use the Underwood method to estimate the minimum-reflux ratio.

(d) Determine the number of theoretical stages required by the Gilliland correlation assuming $L/D = 1.2(L/D)_{min}$, a partial reboiler, and a total condenser.

(e) Estimate the feed-stage location.

9.18 Consider the separation by distillation of a chlorination effluent to recover C_2H_5Cl. The feed is a bubble-point liquid at the column pressure of 240 psia with the following composition and K-values for the column conditions:

Component	Mole Fraction	K
C_2H_4	0.05	5.1
HCl	0.05	3.8
C_2H_6	0.10	3.4
C_2H_5Cl	0.80	0.15

Specifications are: (x_D/x_B) for $C_2H_5Cl = 0.01$
$\qquad\qquad\quad (x_D/x_B)$ for $C_2H_6 = 75$

Calculate the product distribution, the minimum theoretical stages, the minimum reflux, and the theoretical stages at 1.5 times minimum L/D and locate the feed stage. The column is to have a partial condenser and a partial reboiler.

9.19 One hundred kilogram-moles per hour of a three-component bubble-point mixture to be separated by distillation has the following composition:

Component	Mole Fraction	Relative Volatility
A	0.4	5
B	0.2	3
C	0.4	1

(a) For a distillate rate of 60 kmol/h, five theoretical stages, and total reflux, calculate the distillate and bottoms compositions by the Fenske equation.

(b) Using the separation in part (a) for components B and C, determine the minimum reflux and minimum boilup ratio by the Underwood equation.

(c) For an operating reflux ratio of 1.2 times the minimum, determine the number of theoretical stages and the feed-stage location.

9.20 For the conditions of Exercise 9.6, determine the ratio of rectifying to stripping equilibrium stages by: (a) Fenske equation, (b) Kirkbride equation, and (c) McCabe–Thiele diagram. Discuss your results.

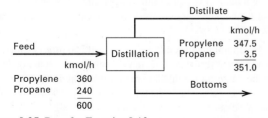

Figure 9.25 Data for Exercise 9.13.

Section 9.2

9.21 Starting with equations like (5-46) and (5-47), show that for two stages, $S_e = \sqrt{0.25 + S_2(S_1 + 1)} - 0.5$.

9.22 Determine by the Kremser group method the separation that can be achieved for the absorption operation indicated in Figure 9.26 for the following combinations of conditions: (a) Six equilibrium stages and 75-psia operating pressure, (b) Three equilibrium stages and 150-psia operating pressure, and (c) Six equilibrium stages and 150-psia operating pressure.

9.23 One thousand kilogram-moles per hour of rich gas at 70°F with 25% C_1, 15% C_2, 25% C_3, 20% nC_4, and 15% nC_5 by moles is to be absorbed by 500 kmol/h of nC_{10} at 90°F in an absorber operating at 4 atm. Calculate by the Kremser group method the percent absorption of each component for: (a) Four theoretical stages, (b) Ten theoretical stages, and (c) Thirty theoretical stages. Use Figures 2.8 and 2.9 for K-values.

9.24 For the flashing and stripping operation indicated in Figure 9.27, determine by the Kremser group method the kilogram-moles per hour of steam if the stripper is operated at 2 atm and has five theoretical stages.

9.25 A stripper operating at 50 psia with three equilibrium stages is used to strip 1,000 kmol/h of liquid at 250°F having the following molar composition: 0.03% C_1, 0.22% C_2, 1.82% C_3, 4.47% nC_4, 8.59% nC_5, 84.87% nC_{10}. The stripping agent is 100 kmol/h of superheated steam at 300°F and 50 psia. Use the group method to estimate the compositions and flow rates of the stripped liquid and rich gas.

9.26 One hundred kilogram-moles per hour of an equimolar mixture of benzene (B), toluene (T), n-hexane (C_6), and n-heptane (C_7)

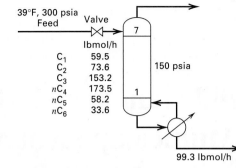

	lbmol/h
C_1	59.5
C_2	73.6
C_3	153.2
nC_4	173.5
nC_5	58.2
nC_6	33.6

Figure 9.28 Data for Exercise 9.27.

is to be extracted at 150°C by 300 kmol/h of diethylene glycol (DEG) in a countercurrent, liquid–liquid extractor having five equilibrium stages. Estimate the flow rates and compositions of the extract and raffinate streams by the group method. In mole-fraction units, the distribution coefficients for the hydrocarbon can be assumed essentially constant at the following values:

Component	$K_{D_i} = y$(solvent phase)$/x$(raffinate phase)
B	0.33
T	0.29
C_6	0.050
C_7	0.043

For diethylene glycol, assume $K_D = 30$. [E.D. Oliver, *Diffusional Separation Processes,* John Wiley and Sons, New York, p. 432 (1966).]

9.27 A reboiled stripper in a natural-gas plant is to be used to remove mainly propane and lighter components from the feed shown in Figure 9.28. Determine by the group method the compositions of the vapor and liquid products.

9.28 A mixture of ethylbenzene and xylenes is to be distilled as shown in Figure 9.29. Assuming the applicability of Raoult's and Dalton's laws:

(a) Use the Fenske–Underwood–Gilliland method to estimate the number of stages required for a reflux-to-minimum reflux ratio of 1.10. Estimate the feed stage location by the Kirkbride equation.

(b) From the results of part (a) for reflux, stages, and distillate rate, use the Edmister group method of Section 5.4 to predict the compositions of the distillate and bottoms. Compare the results with the specifications.

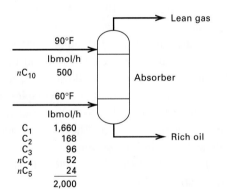

	lbmol/h
nC_{10}	500

	lbmol/h
C_1	1,660
C_2	168
C_3	96
nC_4	52
nC_5	24
	2,000

Figure 9.26 Data for Exercise 9.22.

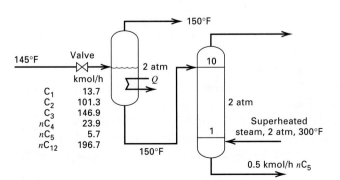

	kmol/h
C_1	13.7
C_2	101.3
C_3	146.9
nC_4	23.9
nC_5	5.7
nC_{12}	196.7

Figure 9.27 Data for Exercise 9.24.

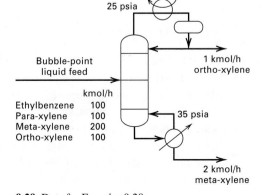

	kmol/h
Ethylbenzene	100
Para-xylene	100
Meta-xylene	200
Ortho-xylene	100

Figure 9.29 Data for Exercise 9.28.

Chapter 10

Equilibrium-Based Methods for Multicomponent Absorption, Stripping, Distillation, and Extraction

Previous chapters have considered graphical, empirical, and approximate group methods for the solution of multistage separation problems involving equilibrium stages. Except for simple cases, such as binary distillation, these methods are suitable only for preliminary-design studies. Final design of multistage equipment for conducting multicomponent separations requires rigorous determination of temperatures, pressures, stream flow rates, stream compositions, and heat-transfer rates at each stage. (However, rigorous calculational procedures may not be justified when multicomponent physical properties or stage efficiencies are not well known.) This determination is made by solving material balance, energy balance, and equilibrium relations for each stage. Unfortunately, these relations are nonlinear algebraic equations that interact strongly. Consequently, solution procedures are relatively difficult and tedious. However, once the procedures are programmed for a high-speed digital computer, solutions are achieved fairly rapidly and almost routinely. Such programs are readily available and widely used. This chapter discusses the solution methods used by such programs, with applications to absorption, stripping, distillation, and liquid–liquid extraction. Applications to extractive, azeotropic, and reactive distillation are covered in Chapter 11.

This chapter begins in Section 10.1 with the development of a mathematical model for a general equilibrium stage for vapor–liquid contacting. The resulting equations, when collected together for a countercurrent cascade of stages, are often referred to as the MESH equations. A number of strategies for solving these equations have been proposed, as summarized in Section 10.2, with those most important considered in detail here. All of these methods utilize an algorithm for solving a tridiagonal-matrix equation, described in Section 10.3. When the feed(s) to the cascade contains components of a narrow boiling-point range, the bubble-point (BP) method is very efficient. When the components cover a wide range of volatilities, the sum-rates (SR) method is a better choice. The BP and SR methods are relatively simple, but are restricted to ideal and nearly ideal mixtures, and are limited in allowable specifications. Sections 10.4 and 10.5 present more complex methods, Newton–Raphson (NR) and Inside-Out, respectively, which are required for nonideal systems. These two methods, which are widely available in process simulators such as ASPEN PLUS, CHEMCAD, HYSYS, and PRO/II, also provide many specification options.

10.0 INSTRUCTIONAL OBJECTIVES

After completing this chapter, you should be able to:

- Write MESH equations for an equilibrium stage in a multicomponent vapor–liquid cascade.
- Explain how equilibrium stages can be combined to form a countercurrent cascade of N equilibrium stages that can be applied to absorption, stripping, distillation, and extraction.
- Discuss different methods to solve the MESH equations and the use of the tridiagonal-matrix algorithm.
- Solve, rigorously, with a simulation program, countercurrent-flow, multi-equilibrium stage, multicomponent separation problems by the bubble-point, sum rates, Newton–Raphson, and inside-out methods. Select the best method to use for a given problem.

10.1 THEORETICAL MODEL FOR AN EQUILIBRIUM STAGE

Consider a general, continuous, steady-state vapor–liquid or liquid–liquid separator consisting of a number of stages arranged in a countercurrent cascade. Assume that: (1) phase equilibrium is achieved at each stage, (2) no chemical reactions occur, and (3) entrainment of liquid drops in vapor and occlusion of vapor bubbles in liquid are negligible. A general schematic representation of an equilibrium stage j is shown in Figure 10.1 for a vapor–liquid separator, where the stages are numbered down from the top. The same representation applies to a liquid–liquid separator if the higher-density liquid phases are represented by liquid streams and the lower-density liquid phases are represented by vapor streams.

Entering stage j can be one single- or two-phase feed of molar flow rate F_j, with overall composition in mole fractions $z_{i,j}$ of component i, temperature T_{F_j}, pressure P_{F_j}, and corresponding overall molar enthalpy h_{F_j}. Feed pressure is assumed equal to or greater than stage pressure P_j. Any excess feed pressure $(P_F - P_j)$ is reduced to zero adiabatically across valve F.

Also entering stage j can be interstage liquid from stage $j - 1$ above, if any, of molar flow rate L_{j-1}, with composition in mole fractions $x_{i,j-1}$, enthalpy $h_{L_{j-1}}$, temperature T_{j-1}, and pressure P_{j-1}, which is equal to or less than the pressure of stage j. Pressure of liquid from stage $j - 1$ is increased adiabatically by hydrostatic head change across head L.

Similarly, from stage $j + 1$ below, interstage vapor of molar flow rate V_{j+1}, with composition in mole fractions $y_{i,j+1}$, enthalpy $h_{V_{j+1}}$, temperature T_{j+1}, and pressure P_{j+1} can enter stage j. Any excess pressure $(P_{j+1} - P_j)$ is reduced to zero adiabatically across valve V.

Leaving stage j is vapor of intensive properties $y_{i,j}$, h_{V_j}, T_j, and P_j. This stream can be divided into a vapor side stream of molar flow rate W_j and an interstage stream of molar flow rate V_j to be sent to stage $j - 1$ or, if $j = 1$, to leave the separator as a product. Also leaving stage j is liquid of intensive properties $x_{i,j}$, h_{L_j}, T_j, and P_j, which is in equilibrium with vapor $(V_j + W_j)$. This liquid can be divided also into a liquid side stream of molar flow rate U_j and an interstage or product stream of molar flow rate L_j to be sent to stage $j + 1$ or, if $j = N$, to leave the multistage separator as a product.

Heat can be transferred at a rate Q_j from $(+)$ or to $(-)$ stage j to simulate stage intercoolers, interheaters, intercondensers, interreboilers, condensers, or reboilers as shown in Figure 1.8. The model in Figure 10.1 does not allow for pumparounds of the type shown in Figure 10.2. Such pumparounds are often used in columns having side streams in order to conserve energy and balance column vapor loads. Some simulator models can handle pumparounds.

Associated with each general theoretical stage are the following indexed equations expressed in terms of the variable set in Figure 10.1. However, variables other than those shown in Figure 10.1 can be used. For example, component flow rates can replace mole fractions, and side-stream flow rates can be expressed as fractions of interstage flow rates.

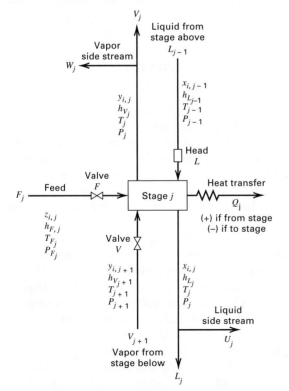

Figure 10.1 General equilibrium stage.

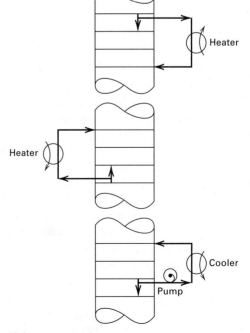

Figure 10.2 Pumparounds.

The equations are similar to those of Section 5.7[1] and are often referred to as the MESH equations after Wang and Henke [1].

1. *M* equations—*M*aterial balance for each component (*C* equations for each stage).

$$M_{i,j} = L_{j-1}x_{i,j-1} + V_{j+1}y_{i,j+1} + F_j z_{i,j} \\ - (L_j + U_j)x_{i,j} - (V_j + W_j)y_{i,j} = 0 \tag{10-1}$$

2. *E* equations—phase-*E*quilibrium relation for each component (*C* equations for each stage), from (2-19).

$$E_{i,j} = y_{i,j} - K_{i,j}x_{i,j} = 0 \tag{10-2}$$

where $K_{i,j}$ is the phase equilibrium ratio.

3. *S* equations—mole-fraction *S*ummations (one for each stage),

$$(S_y)_j = \sum_{i=1}^{C} y_{i,j} - 1.0 = 0 \tag{10-3}$$

$$(S_x)_j = \sum_{i=1}^{C} x_{i,j} - 1.0 = 0 \tag{10-4}$$

4. *H* equation—energy balance (one for each stage).

$$H_j = L_{j-1}h_{L_{j-1}} + V_{j+1}h_{V_{j+1}} + F_j h_{F_j} \\ - (L_j + U_j)h_{L_j} - (V_j + W_j)h_{V_j} - Q_j = 0 \tag{10-5}$$

where kinetic- and potential-energy changes are ignored.

A total material balance equation can be used in place of (10-3) or (10-4). It is derived by combining these two equations and $\sum_j z_{i,j} = 1.0$ with (10-1) summed over the *C* components and over stages 1 through *j* to give

$$L_j = V_{j+1} + \sum_{m=1}^{j} (F_m - U_m - W_m) - V_1 \tag{10-6}$$

In general, $K_{i,j} = K_{i,j}\{T_j, P_j, \mathbf{x}_j, \mathbf{y}_j\}, h_{V_j} = h_{V_j}\{T_j, P_j, \mathbf{y}_j\}$, and $h_{L_j} = h_{L_j}\{T_j, P_j, \mathbf{x}_j\}$, where $\mathbf{x}_j$ and $\mathbf{y}_j$ are vectors of component mole fractions in streams leaving stage *j*. If these relations are not counted as equations and the three properties are not counted as variables, each equilibrium stage is defined only by the $2C + 3$ MESH equations. A countercurrent cascade of *N* such stages, as shown in Figure 10.3, is represented by $N(2C + 3)$ such equations in $[N(3C + 10) + 1]$ variables. If *N* and all $F_j, z_{i,j}, T_{F_j}, P_{F_j}, P_j, U_j, W_j,$ and Q_j are specified, the model is represented by $N(2C + 3)$ simultaneous algebraic equations in $N(2C + 3)$ unknown (output) variables comprising all $x_{i,j}, y_{i,j}, L_j, V_j,$ and T_j, where the *M*, *E*, and *H* equations are nonlinear. If other variables are

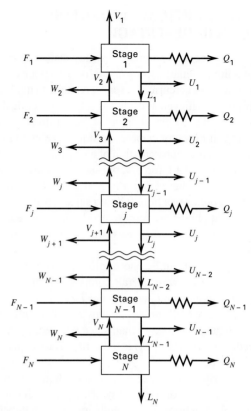

Figure 10.3 General countercurrent cascade of *N* stages.

specified, as they often are, corresponding substitutions are made to the list of output variables. Regardless of the specifications, the result is a set containing nonlinear equations that must be solved by iterative techniques.

10.2 GENERAL STRATEGY OF MATHEMATICAL SOLUTION

A wide variety of iterative solution procedures for solving nonlinear, algebraic equations has appeared in the literature. In general, these procedures make use of equation partitioning in conjunction with equation tearing and/or linearization by Newton–Raphson techniques, which are described in detail by Myers and Seider [2]. The equation-tearing method was applied in Section 4.4 for computing an adiabatic flash.

Early, pre-computer attempts to solve (10-1) to (10-5) or equivalent forms of these equations resulted in the classical *stage-by-stage, equation-by-equation* calculational procedures of Lewis–Matheson [3] in 1932 and Thiele–Geddes [4] in 1933 based on equation tearing for solving simple fractionators with one feed and two products. Composition-independent *K*-values and component enthalpies were generally employed. The Thiele–Geddes method was formulated to handle the Case II variable specification in Table 5.4 wherein the number of equilibrium stages above and below the feed, the reflux ratio, and the distillate flow rate are specified, and stage temperatures and interstage vapor (or liquid) flow rates are the iteration (tear) variables. Although widely

---

[1] Unlike the treatment in Section 5.7, all *C* component material balances are included here, and the total material balance is omitted. Also, the separate but equal temperature and pressure of the equilibrium phases are replaced by the stage temperature and pressure.

used for hand calculations in the years immediately following its appearance in the literature, the Thiele–Geddes method was found often to be numerically unstable when attempts were made to program it for a digital computer. However, Holland [5] developed an improved Thiele–Geddes procedure called the *theta method,* which in various versions has been applied with considerable success.

The Lewis–Matheson method is also an equation-tearing procedure. It was formulated according to the Case I variable specification in Table 5.4 to determine stage requirements for specifications of the separation of two key components, a reflux ratio, and a feed-stage location criterion. Both outer and inner iterations are required. The outer-loop tear variables are the mole fractions or flow rates of nonkey components in the products. The inner-loop tear variables are the interstage vapor (or liquid) flow rates. The Lewis–Matheson method was widely used for hand calculations, but it also proved often to be numerically unstable when implemented on a digital computer.

Rather than using an equation-by-equation solution procedure, Amundson and Pontinen [6] in a significant development in 1958, showed that (10-1), (10-2), and (10-6) of the MESH equations for a Case II specification could be combined and solved component-by-component from simultaneous-linear-equation sets for all N stages by an equation-tearing procedure using the same tear variables as the Thiele–Geddes method. Although too tedious for hand calculations, such equation sets are readily solved with a digital computer.

In a classic study in 1964, Friday and Smith [7] systematically analyzed a number of tearing techniques for solving the MESH equations. They carefully considered the choice of output variable for each equation. They showed that no one tearing technique could solve all types of problems. For separators where the feed(s) contains only components of similar volatility (narrow-boiling case), a modified Amundson–Pontinen approach termed the bubble-point (BP) method was recommended. For a feed(s) containing components of widely different volatility (wide-boiling case) or solubility, the BP method was shown to be subject to failure and a so-called sum-rates (SR) method was suggested. For intermediate cases, the equation-tearing technique may fail to converge; in that case, Friday and Smith indicated that either a Newton–Raphson method or a combined tearing and Newton–Raphson technique was necessary. Boston and Sullivan [8] in 1974 presented an alternative, robust approach to obtaining a solution to the MESH equations. They defined energy and volatility parameters, which are used as the primary successive-approximation variables. A third parameter, which is a combination of the phase flow rates and temperature at each stage, is employed to iterate on the primary variables; thus the name *inside-out method.* Current practice is based mainly on the BP, SR, Newton–Raphson, and inside-out methods, all of which are treated in this chapter. The latter two methods are the most widely used because they permit considerable flexibility in the choice of specified variables and generally are capable of solving most problems. However, the first iteration of the BP or SR method is frequently used to initiate the Newton–Raphson (NR) or inside-out method.

10.3 EQUATION-TEARING PROCEDURES

In general, the modern equation-tearing procedures are readily programmed, are rapid, and require a minimum of computer storage. Although they can be applied to a wider variety of problems than the classical Thiele–Geddes tearing procedure, they are usually limited to the same choice of specified variables. Thus, neither product purities, species recoveries, interstage flow rates, nor stage temperatures can be specified.

Tridiagonal Matrix Algorithm

The key to the success of the BP and SR tearing procedures is the tridiagonal matrix that results from a modified form of the M equations (10-1) when they are torn from the other equations by selecting T_j and V_j as the tear variables, which leaves the modified M equations linear in the unknown liquid mole fractions. This set of equations, one for each component, is solved by a highly efficient and reliable modified Gaussian elimination algorithm due to Thomas as applied by Wang and Henke [1]. The modified M equations are obtained by substituting (10-2) into (10-1) to eliminate y and by substituting (10-6) into (10-1) to eliminate L. Thus, equations for calculating y and L are partitioned from the other equations. The result for each component and each stage is as follows, where the i subscripts have been dropped from the B, C, and D terms.

$$A_j x_{i,j-1} + B_j x_{i,j} + C_j x_{i,j+1} = D_j \qquad (10\text{-}7)$$

where

$$A_j = V_j + \sum_{m=1}^{j-1}(F_m - W_m - U_m) - V_1, \quad 2 \le j \le N \tag{10-8}$$

$$B_j = -\left[V_{j+1} + \sum_{m=1}^{j}(F_m - W_m - U_m) \right. \\ \left. -V_1 + U_j + (V_j + W_j)K_{i,j} \right], \quad 1 \le j \le N \tag{10-9}$$

$$C_j = V_{j+1}K_{i,j+1}, \quad 1 \le j \le N-1 \tag{10-10}$$

$$D_j = -F_j z_{i,j}, \quad 1 \le j \le N \tag{10-11}$$

with $x_{i,0} = 0$, $V_{N+1} = 0$, $W_1 = 0$, and $U_N = 0$, as indicated in Figure 10.3. If the modified M equations are grouped by component, they can be partitioned by writing them as a series of separate tridiagonal matrix equations, one for each of the C components, where the output variable for each matrix equation is x_i over the entire countercurrent cascade of N stages.

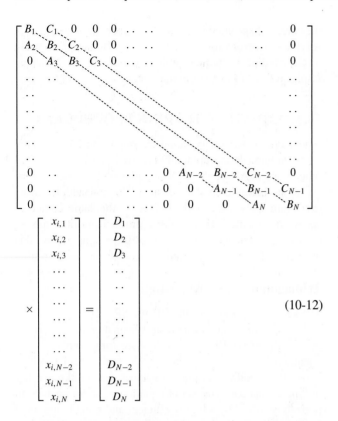

$$\begin{bmatrix} B_1 & C_1 & 0 & 0 & 0 & \cdots & & & & 0 \\ A_2 & B_2 & C_2 & 0 & 0 & \cdots & & & & 0 \\ 0 & A_3 & B_3 & C_3 & 0 & \cdots & & & & 0 \\ \cdots & & & & & & & & & \cdots \\ \cdots & & & & & & & & & \cdots \\ \cdots & & & & & & & & & \cdots \\ \cdots & & & & & & & & & \cdots \\ \cdots & & & & & & & & & \cdots \\ \cdots & & & & & & & & & \cdots \\ 0 & & & & \cdots & 0 & A_{N-2} & B_{N-2} & C_{N-2} & 0 \\ 0 & & & & \cdots & 0 & 0 & A_{N-1} & B_{N-1} & C_{N-1} \\ 0 & & & & \cdots & 0 & 0 & 0 & A_N & B_N \end{bmatrix}$$

$$\times \begin{bmatrix} x_{i,1} \\ x_{i,2} \\ x_{i,3} \\ \cdots \\ \cdots \\ \cdots \\ \cdots \\ \cdots \\ \cdots \\ x_{i,N-2} \\ x_{i,N-1} \\ x_{i,N} \end{bmatrix} = \begin{bmatrix} D_1 \\ D_2 \\ D_3 \\ \cdots \\ \cdots \\ \cdots \\ \cdots \\ \cdots \\ \cdots \\ D_{N-2} \\ D_{N-1} \\ D_N \end{bmatrix} \qquad (10\text{-}12)$$

$$\begin{bmatrix} B_1 & C_1 & 0 & 0 & 0 \\ A_2 & B_2 & C_2 & 0 & 0 \\ 0 & A_3 & B_3 & C_3 & 0 \\ 0 & 0 & A_4 & B_4 & C_4 \\ 0 & 0 & 0 & A_5 & B_5 \end{bmatrix} \cdot \begin{bmatrix} x_1 \\ x_2 \\ x_3 \\ x_4 \\ x_5 \end{bmatrix} = \begin{bmatrix} D_1 \\ D_2 \\ D_3 \\ D_4 \\ D_5 \end{bmatrix}$$

(a)

$$\begin{bmatrix} 1 & p_1 & 0 & 0 & 0 \\ 0 & 1 & p_2 & 0 & 0 \\ 0 & 0 & 1 & p_3 & 0 \\ 0 & 0 & 0 & 1 & p_4 \\ 0 & 0 & 0 & 0 & 1 \end{bmatrix} \cdot \begin{bmatrix} x_1 \\ x_2 \\ x_3 \\ x_4 \\ x_5 \end{bmatrix} = \begin{bmatrix} q_1 \\ q_2 \\ q_3 \\ q_4 \\ q_5 \end{bmatrix}$$

(b)

$$\begin{bmatrix} 1 & 0 & 0 & 0 & 0 \\ 0 & 1 & 0 & 0 & 0 \\ 0 & 0 & 1 & 0 & 0 \\ 0 & 0 & 0 & 1 & 0 \\ 0 & 0 & 0 & 0 & 1 \end{bmatrix} \cdot \begin{bmatrix} x_1 \\ x_2 \\ x_3 \\ x_4 \\ x_5 \end{bmatrix} = \begin{bmatrix} r_1 \\ r_2 \\ r_3 \\ r_4 \\ r_5 \end{bmatrix}$$

(c)

Figure 10.4 The coefficient matrix for the modified M-equations of a given component at various steps in the Thomas algorithm for five equilibrium stages. (Note that the i subscript is deleted from x.) (a) Initial matrix. (b) Matrix after forward elimination. (c) Matrix after backward substitution.

Constants B_j and C_j for each component depend only on tear variables T and V provided that K-values are composition independent. If not, compositions from the previous iteration may be used to estimate the K-values.

The Thomas algorithm for solving the linearized equation set (10-12) is a Gaussian–elimination procedure that involves forward elimination starting from stage 1 and working toward stage N to finally isolate $x_{i,N}$. Other values of $x_{i,j}$ are then obtained starting with $x_{i,N-1}$ by backward substitution. For five stages, the matrix equations at the beginning, middle, and end of the procedure are as shown in Figure 10.4.

The equations used in the Thomas algorithm are as follows:

For stage 1, (10-7) is $B_1 x_{i,1} + C_1 x_{i,2} = D_1$, which can be solved for $x_{i,1}$ in terms of unknown $x_{i,2}$ to give

$$x_{i,1} = \frac{D_1 - C_1 x_{i,2}}{B_1}$$

Let

$$p_1 = \frac{C_1}{B_1} \quad \text{and} \quad q_1 = \frac{D_1}{B_1}$$

Then

$$x_{i,1} = q_1 - p_1 x_{i,2} \qquad (10\text{-}13)$$

Thus, the coefficients in the matrix become $B_1 \leftarrow 1$, $C_1 \leftarrow p_1$, and $D_1 \leftarrow q_1$, where $\leftarrow$ means "is replaced by." Only values for p_1 and q_1 need be stored.

For stage 2, (10-7) can be combined with (10-13) and solved for $x_{i,2}$ to give

$$x_{i,2} = \frac{D_2 - A_2 q_1}{B_2 - A_2 p_1} - \left(\frac{C_2}{B_2 - A_2 p_1} \right) x_{i,3}$$

Let

$$q_2 = \frac{D_2 - A_2 q_1}{B_2 - A_2 p_1} \quad \text{and} \quad p_2 = \frac{C_2}{B_2 - A_2 p_1}$$

Then

$$x_{i,2} = q_2 - p_2 x_{i,3}$$

Thus, $A_2 \leftarrow 0$, $B_2 \leftarrow 1$, $C_2 \leftarrow p_2$, and $D_2 \leftarrow q_2$. Only values for p_2 and q_2 need be stored.

In general, we can define

$$p_j = \frac{C_j}{B_j - A_j p_{j-1}} \qquad (10\text{-}14)$$

$$q_j = \frac{D_j - A_j q_{j-1}}{B_j - A_j p_{j-1}} \qquad (10\text{-}15)$$

Then

$$x_{i,j} = q_j - p_j x_{i,j+1} \qquad (10\text{-}16)$$

with $A_j \leftarrow 0$, $B_j \leftarrow 1$, $C_j \leftarrow p_j$, and $D_j \leftarrow q_j$. Only values of p_j and q_j need be stored. Thus, starting with stage 1, values of p_j and q_j are computed recursively in the order $p_1, q_1, p_2, q_2, \ldots, p_{N-1}, q_{N-1}, q_N$. For stage N, (10-16) isolates $x_{i,N}$ as

$$x_{i,N} = q_N \qquad (10\text{-}17)$$

Successive values of x_i are computed recursively by backward substitution from (10-16) in the form

$$x_{i,j-1} = q_{j-1} - p_{j-1} x_{i,j} = r_{j-1} \qquad (10\text{-}18)$$

Equation (10-18) corresponds to the identity coefficient matrix.

The Thomas algorithm, when applied in this fashion, generally avoids buildup of computer truncation errors because usually none of the steps involves subtraction of nearly equal quantities. Furthermore, computed values of $x_{i,j}$ are almost always positive. The algorithm is highly efficient, requires a minimum of computer storage as noted previously, and is superior to alternative matrix-inversion routines. A modified Thomas algorithm for difficult cases is given by Boston and Sullivan [9]. Such cases can occur for columns having large numbers of equilibrium stages and with components whose absorption factors [see (5-38)] are less than unity in one section of stages and greater than unity in another section.

Bubble-Point (BP) Method for Distillation

Frequently, distillation involves species that cover a relatively narrow range of vapor–liquid equilibrium ratios (*K*-values). A particularly effective solution procedure for this case was suggested by Friday and Smith [7] and developed in detail by Wang and Henke [1]. It is referred to as the bubble-point (BP) method because a new set of stage temperatures is computed during each iteration from bubble-point equations of the type discussed in Section 4.4. In the method, all equations are partitioned and solved sequentially except for the modified *M* equations, which are solved separately for each component by the tridiagonal matrix technique.

The algorithm for the Wang–Henke BP method is shown in Figure 10.5. A FORTRAN computer program for the method is available [10]. Problem specifications consist of conditions and stage location of all feeds, pressure at each stage, total flow rates of all side streams (note that liquid distillate flow rate, if any, is designated as U_1), heat transfer rates to or from all stages except stage 1 (condenser) and stage *N* (reboiler), total number of stages, external bubble-point reflux flow rate, and vapor distillate flow rate. A sample problem specification is shown in Figure 10.6.

To initiate the calculations, values for the tear variables are assumed. For most problems, it is sufficient to establish an initial set of V_j values based on the assumption of constant molar interstage flows using the specified reflux, distillate, feed, and side stream flow rates. A generally adequate initial set of T_j values can be provided by computing or assuming both the bubble-point temperature of an estimated bottoms product and the dew-point temperature of an assumed vapor distillate product; or computing or assuming bubble-point temperature if distillate is liquid or a temperature in-between the dew-point and bubble-point temperatures if distillate is mixed (both vapor and liquid); and then determining the other stage temperatures by assuming a linear variation of temperature with stage location.

To solve (10-12) for x_i by the Thomas method, $K_{i,j}$ values are required. When they are composition dependent, initial

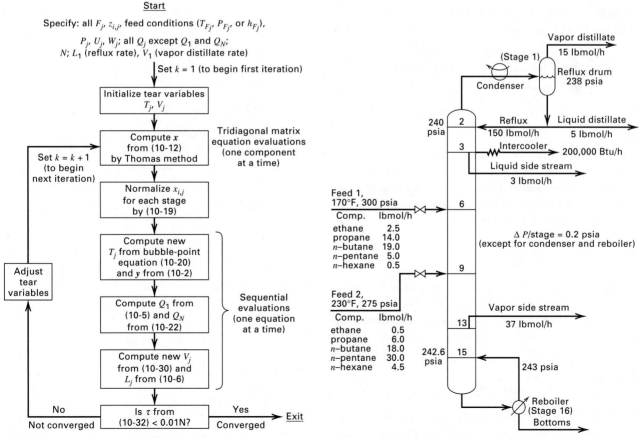

Figure 10.5 Algorithm for Wang–Henke BP method for distillation.

Figure 10.6 Sample specification for application of Wang–Henke BP method to distillation.

assumptions for all $x_{i,j}$ and $y_{i,j}$ values are also needed unless ideal K-values are employed for the first iteration. For each iteration, the computed set of $x_{i,j}$ values for each stage will, in general, not satisfy the summation constraint given by (10-4). Although not mentioned by Wang and Henke, it is advisable to normalize the set of computed $x_{i,j}$ values by the relation

$$(x_{i,j})_{\text{normalized}} = \frac{x_{i,j}}{\sum\limits_{i=1}^{C} x_{i,j}} \qquad (10\text{-}19)$$

These normalized values are used for all subsequent calculations involving $x_{i,j}$ during the iteration.

A new set of temperatures T_j is computed stage by stage by computing bubble-point temperatures from the normalized $x_{i,j}$ values. Friday and Smith [7] showed that bubble-point calculations for stage temperatures are particularly effective for mixtures having a narrow range of K-values because temperatures are not then sensitive to composition. For example, in the limiting case where all components have identical K-values, the temperature corresponds to the conditions of $K_{i,j} = 1$ and is not dependent on $x_{i,j}$ values. At the other extreme, however, bubble-point calculations to establish stage temperatures can be very sensitive to composition. For example, consider a binary mixture containing one component with a high K-value that changes little with temperature. The second component has a low K-value that changes very rapidly with temperature. Such a mixture is methane and n-butane at 400 psia. The effect on the bubble-point temperature of small quantities of methane dissolved in liquid n-butane is very large, as indicated by the following results.

Liquid Mole Fraction of Methane	Bubble-Point Temperature, °F
0.000	275
0.018	250
0.054	200
0.093	150

Thus, the BP method is best when components have a relatively narrow range of K-values.

The necessary bubble-point equation is obtained in the manner described in Chapter 4 by combining (10-2) and (10-3) to eliminate $y_{i,j}$, giving

$$\sum_{i=1}^{C} K_{i,j} x_{i,j} - 1.0 = 0 \qquad (10\text{-}20)$$

which is nonlinear in T_j and must be solved iteratively. Wang and Henke prefer to use Muller's iterative method [11] because it is reliable and does not require the calculation of derivatives. Muller's method requires three initial assumptions of T_j. For each assumption, the value of S_j is computed from

$$S_j = \sum_{i=1}^{C} K_{i,j} x_{i,j} - 1.0 = 0 \qquad (10\text{-}21)$$

The three sets of (T_j, S_j) are fitted to a quadratic equation for S_j in terms of T_j. The quadratic equation is then employed to predict T_j for $S_j = 0$, as required by (10-20). The validity of

this value of T_j is checked by using it to compute S_j in (10-21). The quadratic fit and S_j check are repeated with the three best sets of (T_j, S_j) until some convergence tolerance is achieved, say $\left| T_j^{(n)} - T_j^{(n-1)} \right| / T_j^{(n)} \leq 0.0001$, with T in absolute degrees, where n is the iteration number for the temperature loop in the bubble-point calculation, or one can use $S_j \leq 0.0001\, C$, which is preferred.

Values of $y_{i,j}$ are determined along with the calculation of stage temperatures using the E equations, (10-2). With a consistent set of values for $x_{i,j}$, T_j, and $y_{i,j}$, molar enthalpies are computed for each liquid and vapor stream leaving a stage. Since F_1, V_1, U_1, W_1, and L_1 are specified, V_2 is readily obtained from (10-6), and the condenser duty, a $(+)$ quantity, is obtained from (10-5). Reboiler duty, a $(-)$ quantity, is determined by summing (10-5) for all stages to give

$$Q_N = \sum_{j=1}^{N} (F_j h_{F_j} - U_j h_{L_j} - W_j h_{V_j}) \qquad (10\text{-}22)$$

$$- \sum_{j=1}^{N-1} Q_j - V_1 h_{V_1} - L_N h_{L_N}$$

A new set of V_j tear variables is computed by applying the following modified energy balance, which is obtained by combining (10-5) and (10-6) twice to eliminate L_{j-1} and L_j. After rearrangement,

$$\alpha_j V_j + \beta_j V_{j+1} = \gamma_j \qquad (10\text{-}23)$$

where

$$\alpha_j = h_{L_{j-1}} - h_{V_j} \qquad (10\text{-}24)$$

$$\beta_j = h_{V_{j+1}} - h_{L_j} \qquad (10\text{-}25)$$

$$\gamma_j = \left[\sum_{m=1}^{j-1} (F_m - W_m - U_m) - V_1 \right] \left(h_{L_j} - h_{L_{j-1}} \right) \qquad (10\text{-}26)$$
$$+ F_j \left(h_{L_j} - h_{F_j} \right) + W_j \left(h_{V_j} - h_{L_j} \right) + Q_j$$

and enthalpies are evaluated at the stage temperatures last computed rather than at those used to initiate the iteration. Written in didiagonal matrix form (10-23) applied over stages 2 to $N - 1$ is:

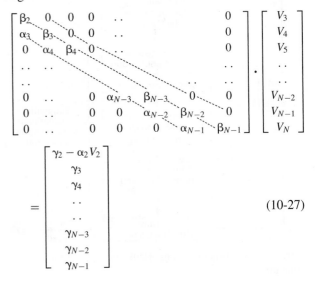

$$(10\text{-}27)$$

Matrix equation (10-27) is readily solved one equation at a time by starting at the top where V_2 is known and working down recursively. Thus,

$$V_3 = \frac{\gamma_2 - \alpha_2 V_2}{\beta_2} \qquad (10\text{-}28)$$

$$V_4 = \frac{\gamma_3 - \alpha_3 V_3}{\beta_3} \qquad (10\text{-}29)$$

or, in general

$$V_j = \frac{\gamma_{j-1} - \alpha_{j-1} V_{j-1}}{\beta_{j-1}} \qquad (10\text{-}30)$$

and so on. Corresponding liquid flow rates are obtained from (10-6).

The solution procedure is considered to be converged when sets of $T_j^{(k)}$ and $V_j^{(k)}$ values are within some prescribed tolerance of corresponding sets of $T_j^{(k-1)}$ and $V_j^{(k-1)}$ values, where k is the iteration index. One possible convergence criterion is

$$\sum_{j=1}^{N}\left[\frac{T_j^{(k)} - T_j^{(k-1)}}{T_j^{(k)}}\right]^2 + \sum_{j=1}^{N}\left[\frac{V_j^{(k)} - V_j^{(k-1)}}{V_j^{(k)}}\right]^2 \le \epsilon \quad (10\text{-}31)$$

where T is the absolute temperature and ϵ is some prescribed tolerance. However, Wang and Henke suggest that the following simpler criterion, which is based on successive sets of T_j values only, is adequate.

$$\tau = \sum_{j=1}^{N}\left[T_j^{(k)} - T_j^{(k-1)}\right]^2 \le 0.01 N \qquad (10\text{-}32)$$

Successive substitution is often employed for iterating the tear variables; that is, values of T_j and V_j generated from (10-20) and (10-30), respectively, during an iteration are used directly to initiate the next iteration. However, experience indicates that it is desirable frequently to adjust the values of the generated tear variables prior to beginning the next iteration. For example, upper and lower bounds should be placed on stage temperatures, and any negative values of interstage flow rates should be changed to near-zero positive values. Also, to prevent oscillation of the iterations, damping can be employed to limit changes in the values of V_j and absolute T_j from one iteration to the next to, say, 10%.

EXAMPLE 10.1

For the distillation column shown in Figure 10.7, do one iteration of the BP method up to and including the calculation of a new set of T_j values from (10-20). Use composition-independent K-values.

SOLUTION

By overall total material balance

Liquid distillate $= U_1 = F_3 - L_5 = 100 - 50 = 50$ lbmol/h

Then

$$L_1 = (L_1/U_1)U_1 = (2)(50) = 100 \text{ lbmol/h}$$

By total material balance around the total condenser

$$V_2 = L_1 + U_1 = 100 + 50 = 150 \text{ lbmol/h}$$

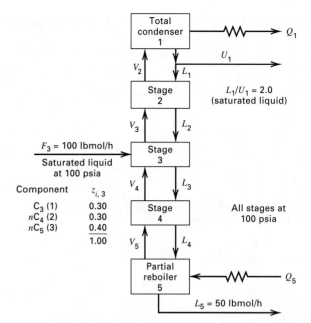

Figure 10.7 Specifications for distillation column of Example 10.1.

Initial guesses of tear variables are

Stage j	V_j, lbmol/h	T_j, °F
1	(Fixed at 0 by specifications)	65
2	(Fixed at 150 by specifications)	90
3	150	115
4	150	140
5	150	165

At 100 psia, the estimated K-values at the assumed stage temperatures are

	$K_{i,j}$				
Stage	1	2	3	4	5
$C_3(1)$	1.23	1.63	2.17	2.70	3.33
$nC_4(2)$	0.33	0.50	0.71	0.95	1.25
$nC_5(3)$	0.103	0.166	0.255	0.36	0.49

The matrix equation (10-12) for the first component C_3 is developed as follows. From (10-8) with $V_1 = 0$, $W = 0$,

$$A_j = V_j + \sum_{m=1}^{j-1}(F_m - U_m)$$

Thus, $A_5 = V_5 + F_3 - U_1 = 150 + 100 - 50 = 200$ lbmol/h.
Similarly, $A_4 = 200$, $A_3 = 100$, and $A_2 = 100$ in the same units.
From (10-9) with $V_1 = 0$, $W = 0$,

$$B_j = -\left[V_{j+1} + \sum_{m=1}^{j}(F_m - U_m) + U_j + V_j K_{i,j}\right]$$

Thus, $B_5 = -\left[F_3 - U_1 + V_5 K_{1,5}\right] = -[100 - 50 + (150)3.33] = -549.5$ lbmol/h.
Similarly, $B_4 = -605$, $B_3 = -525.5$, $B_2 = -344.5$, and $B_1 = -150$ in the same units.
From (10-10), $C_j = V_{j+1} K_{1,j+1}$. Thus, $C_1 = V_2 K_{1,2} = 150(1.63) = 244.5$ lbmol/h.

Similarly, $C_2 = 325.5$, $C_3 = 405$, and $C_4 = 499.5$ in the same units.

From (10-11), $D_j = -F_j z_{1,j}$. Thus, $D_3 = -100(0.30) = -30$ lbmol/h.

Similarly, $D_1 = D_2 = D_4 = D_5 = 0$.

Substitution of the above values in (10-7) gives

$$
\begin{bmatrix}
-150 & 244.5 & 0 & 0 & 0 \\
100 & -344.5 & 325.5 & 0 & 0 \\
0 & 100 & -525.5 & 405 & 0 \\
0 & 0 & 200 & -605 & 499.5 \\
0 & 0 & 0 & 200 & -549.5
\end{bmatrix}
\begin{bmatrix}
x_{1,1} \\
x_{1,2} \\
x_{1,3} \\
x_{1,4} \\
x_{1,5}
\end{bmatrix}
=
\begin{bmatrix}
0 \\
0 \\
-30 \\
0 \\
0
\end{bmatrix}
$$

Using (10-14) and (10-15), we apply the forward step of the Thomas algorithm as follows.

$$p_1 = \frac{C_1}{B_1} = 244.5/(-150) = -1.630$$

$$q_1 = \frac{D_1}{B_1} = 0/(-150) = 0$$

$$p_2 = \frac{C_2}{B_2 - A_2 p_1} = \frac{325.5}{-344.5 - 100(-1.630)} = -1.793$$

By similar calculations, the matrix equation after the forward-elimination procedure is

$$
\begin{bmatrix}
1 & -1.630 & 0 & 0 & 0 \\
0 & 1 & -1.793 & 0 & 0 \\
0 & 0 & 1 & -1.170 & 0 \\
0 & 0 & 0 & 1 & -1.346 \\
0 & 0 & 0 & 0 & 1
\end{bmatrix}
\begin{bmatrix}
x_{1,1} \\
x_{1,2} \\
x_{1,3} \\
x_{1,4} \\
x_{1,5}
\end{bmatrix}
=
\begin{bmatrix}
0 \\
0 \\
0.0867 \\
0.0467 \\
0.0333
\end{bmatrix}
$$

Applying the backward steps of (10-17) and (10-18) gives

$$x_{1,5} = q_5 = 0.0333$$
$$x_{1,4} = q_4 - p_4 x_{1,5} = 0.0467 - (-1.346)(0.0333) = 0.0915$$

Similarly,

$$x_{1,3} = 0.1938, \ x_{1,2} = 0.3475, \ x_{1,1} = 0.5664$$

The matrix equations for nC_4 and nC_5 are solved in a similar manner to give

Stage	$x_{i,j}$				
	1	2	3	4	5
C_3	0.5664	0.3475	0.1938	0.0915	0.0333
nC_4	0.1910	0.3820	0.4483	0.4857	0.4090
nC_5	0.0191	0.1149	0.3253	0.4820	0.7806
$\sum_i x_{i,j}$	0.7765	0.8444	0.9674	1.0592	1.2229

After these compositions are normalized, bubble-point temperatures at 100 psia are computed iteratively from (10-20) and compared to the initially assumed values,

Stage	$T^{(2)}$, °F	$T^{(1)}$, °F
1	66	65
2	94	90
3	131	115
4	154	140
5	184	165

The rate of convergence of the BP method is unpredictable, and, as shown in Example 10.2, it can depend drastically on the assumed initial set of T_j values. In addition, cases with high reflux ratios can be more difficult to converge than cases with low reflux ratios. Orbach and Crowe [12] describe a generalized extrapolation method for accelerating convergence based on periodic adjustment of the tear variables when their values form geometric progressions during at least four successive iterations.

EXAMPLE 10.2

Calculate stage temperatures, interstage vapor and liquid flow rates and compositions, reboiler duty, and condenser duty by the BP method for the distillation column specifications given in Example 5.4.

SOLUTION

The computer program of Johansen and Seader [10] based on the Wang–Henke procedure was used. In this program, no adjustments to the tear variables are made prior to the start of each iteration, and the convergence criterion is (10-32). The K-values and enthalpies are computed from correlations for hydrocarbons. The only initial assumptions required are distillate and bottoms temperatures shown previously for four cases.

The significant effect of initially assumed distillate and bottoms temperatures on the number of iterations required to satisfy (10-32) is indicated by the following results.

Case	Assumed Temperatures, °F		Number of Iterations for Convergence
	Distillate	Bottoms	
1	11.5	164.9	29
2	0.0	200.0	5
3	20.0	180.0	12
4	50.0	150.0	19

The terminal temperatures of Case 1 were within a few degrees of the exact values and were much closer estimates than those of the other three cases. Nevertheless, Case 1 required the largest number of iterations. Figure 10.8 is a plot of τ from (10-32) as a function of the number of iterations for each of the four cases. Case 2 converged rapidly to the criterion of $\tau < 0.13$. Cases 1, 3, and 4 converged rapidly for the first three or four iterations, but then moved only slowly toward the criterion. This was particularly true of Case 1, for which application of a convergence-acceleration method would be particularly desirable. In none of the four cases did oscillations of values of the tear variables occur; rather the values approached the converged results in a monotonic fashion.

The overall results of the converged calculations, as taken from Case 2, are shown in Figure 10.9. Product-component flow rates were not quite in material balance with the feed. Therefore, adjusted values that do satisfy overall material-balance equations were determined by averaging the calculated values and are included in Figure 10.9. A smaller value of τ would have improved the overall material balance. Figures 10.10 to 10.13 are plots of converged values for stage temperatures, interstage flow rates, and mole-fraction compositions from the results of Case 2. Results from the other

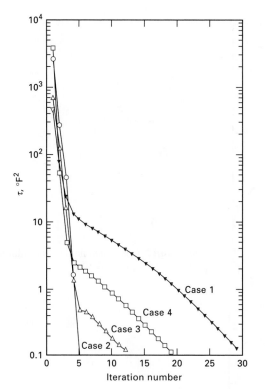

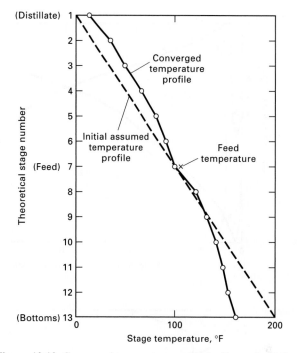

Figure 10.10 Converged temperature profile for Example 10.2.

In Figure 10.11, it is seen that the assumption of constant interstage molar flow rates does not hold in the rectifying section. Both liquid and vapor flow rates decrease in moving down from the top stage toward the feed stage. Because the feed is vapor near the dew point, the liquid rate changes only slightly across the feed stage. Correspondingly, the vapor rate decreases across the feed stage by an amount equal to the feed rate. For this problem, the interstage molar flow rates are almost constant in the stripping section. However, the assumed vapor flow rate in this section based on adjusting the rectifying-section rate across the feed zone is approximately 33% higher than the average converged vapor rate. A much better initial estimate of the vapor rate in the stripping section can be made by first computing the reboiler duty from the condenser duty based on the specified reflux rate and then determining the corresponding vapor rate leaving the partial reboiler.

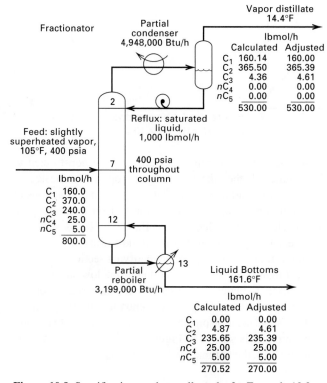

Figure 10.8 Convergence patterns for Example 10.2.

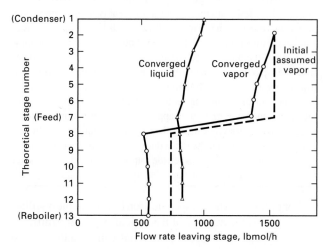

Figure 10.11 Converged interstage flow rate profiles for Example 10.2.

Figure 10.9 Specifications and overall results for Example 10.2.

three cases were almost identical to those of Case 2. Included in Figure 10.10 is the initially assumed linear temperature profile. Except for the bottom stages, it does not deviate significantly from the converged profile. A jog in the profile is seen at the feed stage. This is a common occurrence.

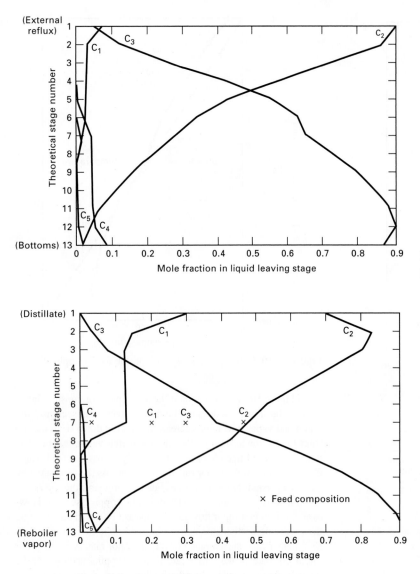

Figure 10.12 Converged liquid composition profiles for Example 10.2.

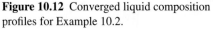

Figure 10.13 Converged vapor composition profiles for Example 10.2.

For this problem, the separation is between C_2 and C_3. Thus, these two components can be designated as the light key (LK) and heavy key (HK), respectively. Thus C_1 is a lighter-than-light key (LLK), and nC_4 and nC_5 are heavier than the heavy key (HHK). Each of these four designations exhibits a different type of composition-profile curve as shown in Figures 10.12 and 10.13. Except at the feed zone and at each end of the column, both liquid and vapor mole fractions of the light key (C_2) decrease smoothly and continuously from the top of the column to the bottom. The inverse occurs for the heavy key (C_3). Mole fractions of methane (LLK) are almost constant over the rectifying section except near the top. Below the feed zone, methane rapidly disappears from both vapor and liquid streams. The inverse is true for the two HHK components. In Figure 10.13, it is seen that the feed composition is somewhat different from the composition of either the vapor entering the feed stage from the stage below or the vapor leaving the feed stage.

For problems where a specification is made of the distillate flow rate and the number of theoretical stages, it is difficult to specify the feed-stage location that will give the highest degree of separation. However, once the results of a rigorous calculation are available, a modified McCabe–Thiele plot

based on the key components [13] can be constructed to determine whether the feed stage is optimally located or whether it should be moved. For this plot, mole fractions of the light-key component are computed on a nonkey-free basis. The resulting diagram for Example 10.2 is shown in Figure 10.14. It is seen that the trend toward a pinched-in region is more noticeable in the rectifying section just above stage 7 than in the stripping section just below stage 7. This suggests that a better separation between the key components might be made by shifting the feed entry to stage 6. The effect of feed-stage location on the percent loss of ethane to the bottoms product is shown in Figure 10.15. As predicted from Figure 10.14, the optimal feed stage is stage 6.

Sum-Rates Method for Absorption and Stripping

The chemical components present in most absorbers and strippers cover a relatively wide range of volatility. Hence, the BP method of solving the MESH equations will fail because calculation of stage temperature by bubble-point

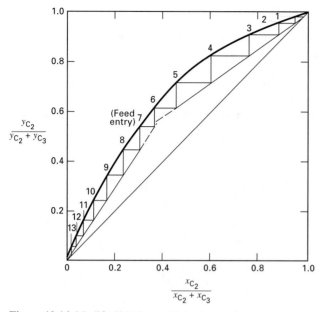

$\dfrac{y_{C_2}}{y_{C_2}+y_{C_3}}$ (Feed entry)

$\dfrac{x_{C_2}}{x_{C_2}+x_{C_3}}$

Figure 10.14 Modified McCabe–Thiele diagram for Example 10.2.

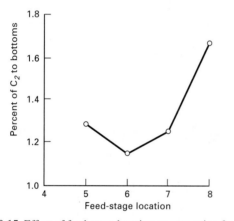

Figure 10.15 Effect of feed-stage location on separation for Example 10.2.

determination (10-20) is too sensitive to liquid-phase composition, and the stage energy balance (10-5) is much more sensitive to stage temperatures than to interstage flow rates. In this case, Friday and Smith [7] showed that an alternative procedure devised by Sujata [14] could be successfully applied. This procedure, termed the *sum-rates* (SR) *method,* was further developed in conjunction with the tridiagonal-matrix formulation for the modified *M* equations by Burningham and Otto [15].

Figure 10.16 shows the algorithm for the Burningham–Otto SR method. A FORTRAN computer program for the method is available [16]. Problem specifications consist of conditions and stage locations for all feeds, pressure at each stage, total flow rates of any side streams, heat-transfer rates to or from any stages, and total number of stages.

An initial set of tear variables T_j and V_j is assumed to initiate the calculations. For most problems it is sufficient to

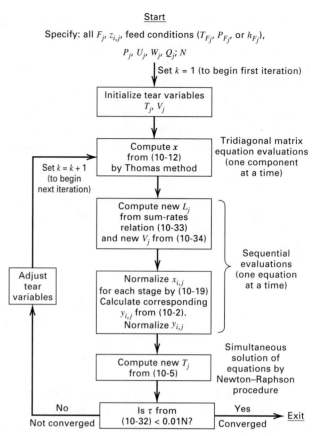

Figure 10.16 Algorithm for Burningham–Otto SR method for absorption/stripping.

assume a set of V_j values based on the assumption of constant-molar interstage flows, working up from the bottom of the absorber using specified vapor feeds and any vapor side stream flows. Generally, an adequate initial set of T_j values can be derived from assumed top-stage and bottom-stage values and a linear variation with stages in-between.

Values of $x_{i,j}$ are obtained by solving (10-12) by the Thomas algorithm. However, the values obtained are not normalized at this step but are utilized directly to produce new values of L_j by applying (10-4) in the form referred to as the *sum-rates equation:*

$$L_j^{(k+1)} = L_j^{(k)} \sum_{i=1}^{C} x_{i,j} \qquad (10\text{-}33)$$

where values of $L_j^{(k)}$ are obtained from values of $V_j^{(k)}$ by (10-6). Corresponding values of $V_j^{(k+1)}$ are obtained from a total material balance, which is derived by summing (10-1) over the C components, combining the result with (10-3) and (10-4), and summing that result over stages j through N to give

$$V_j = L_{j-1} - L_N + \sum_{m=j}^{N} (F_m - W_m - U_m) \qquad (10\text{-}34)$$

Normalized values of $x_{i,j}$ are next calculated from (10-19). Corresponding values of $y_{i,j}$ are computed from (10-2).

A new set of values for stage temperatures T_j is obtained by solving the simultaneous set of energy-balance relations for the N stages given by (10-5). The temperatures are embedded in the specific enthalpies corresponding to the unspecified vapor and liquid flow rates. In general, these enthalpies are nonlinear in temperature. Therefore, an iterative solution procedure is required, such as the Newton–Raphson method [10].

In the Newton–Raphson method, the simultaneous, nonlinear equations in terms of variables x_i are written in zero form:

$$f_i\{x_1, x_2, \ldots, x_n\} = 0, \quad i = 1, 2, \ldots, n \quad (10\text{-}35)$$

Initial guesses, marked by asterisks, are provided for the n variables and each function is expanded about these guesses in a Taylor's series that is terminated after the first derivatives to give

$$0 = f_i\{x_1, x_2, \ldots, x_n\} \quad (10\text{-}36)$$

$$\approx f_i\{x_1^*, x_2^*, \ldots, x_n^*\} + \left.\frac{\partial f_i}{\partial x_1}\right|^* \Delta x_1 + \left.\frac{\partial f_i}{\partial x_2}\right|^* \Delta x_2$$
$$+ \cdots + \left.\frac{\partial f_i}{\partial x_n}\right|^* \Delta x_n \quad (10\text{-}37)$$

where $\Delta x_j = x_j - x_j^*$.

Equations (10-36) are linear and can be solved directly for the corrections Δx_i. If the corrections are all found to be zero, the guesses are correct and equations (10-35) have been solved; if not, the corrections are added to the guesses to provide a new set of guesses that are applied to (10-36). The procedure is repeated, in a series of r iterations, until all the corrections, and thus the functions, become zero to within some tolerance. In recursion form (10-36) and (10-37) are

$$\sum_{j=1}^{n} \left[\left(\frac{\partial f_i}{\partial x_j} \right)^{(r)} \Delta x_j^{(r)} \right] = -f_i^{(r)}, \quad i = 1, 2, \ldots, n \quad (10\text{-}38)$$

$$x_j^{(r+1)} = x_j^{(r)} + \Delta x_j^{(r)}, \quad j = 1, 2, \ldots, n \quad (10\text{-}39)$$

EXAMPLE 10.3

Solve the simultaneous, nonlinear equations

$$x_1 \ln x_2 + x_2 \exp(x_1) = \exp(1)$$
$$x_2 \ln x_1 + 2x_1 \exp(x_2) = 2 \exp(1)$$

for x_1 and x_2 to within ± 0.001, by the Newton–Raphson method.

SOLUTION

In the form of (10-35), the two equations are

$$f_1\{x_1, x_2\} = x_1 \ln x_2 + x_2 \exp(x_1) - \exp(1) = 0$$
$$f_2\{x_1, x_2\} = x_2 \ln x_1 + 2x_1 \exp(x_2) - 2 \exp(1) = 0$$

From (10-38), the linearized recursive form of these equations is

$$\left(\frac{\partial f_1}{\partial x_1} \right)^{(r)} \Delta x_1^{(r)} + \left(\frac{\partial f_1}{\partial x_2} \right)^{(r)} \Delta x_2^{(r)} = -f_1^{(r)}$$

$$\left(\frac{\partial f_2}{\partial x_1} \right)^{(r)} \Delta x_1^{(r)} + \left(\frac{\partial f_2}{\partial x_2} \right)^{(r)} \Delta x_2^{(r)} = -f_2^{(r)}$$

The solution of these two equations is readily obtained by the method of determinants to give

$$\Delta x_1^{(r)} = \frac{\left[f_2^{(r)} \left(\frac{\partial f_1}{\partial x_2} \right)^{(r)} - f_1^{(r)} \left(\frac{\partial f_2}{\partial x_2} \right)^{(r)} \right]}{D}$$

and

$$\Delta x_2^{(r)} = \frac{\left[f_1^{(r)} \left(\frac{\partial f_2}{\partial x_1} \right)^{(r)} - f_2^{(r)} \left(\frac{\partial f_1}{\partial x_1} \right)^{(r)} \right]}{D}$$

where

$$D = \left(\frac{\partial f_1}{\partial x_1} \right)^{(r)} \left(\frac{\partial f_2}{\partial x_2} \right)^{(r)} - \left(\frac{\partial f_1}{\partial x_2} \right)^{(r)} \left(\frac{\partial f_2}{\partial x_1} \right)^{(r)}$$

and the derivatives as obtained from the equations are

$$\left(\frac{\partial f_1}{\partial x_1} \right)^{(r)} = \ln \left(x_2^{(r)} \right) + x_2^{(r)} \exp \left(x_1^{(r)} \right),$$

$$\left(\frac{\partial f_2}{\partial x_1} \right)^{(r)} = \frac{x_2^{(r)}}{x_1^{(r)}} + 2 \exp \left(x_2^{(r)} \right)$$

$$\left(\frac{\partial f_1}{\partial x_2} \right)^{(r)} = \frac{x_1^{(r)}}{x_2^{(r)}} + \exp \left(x_1^{(r)} \right),$$

$$\left(\frac{\partial f_2}{\partial x_2} \right)^{(r)} = \ln \left(x_1^{(r)} \right) + 2x_1^{(r)} \exp \left(x_2^{(r)} \right)$$

As initial guesses, take $x_1^{(1)} = 2$, $x_2^{(1)} = 2$. Applying the Newton–Raphson procedure, one obtains the following results where at the sixth iteration, values of $x_1 = 1.0000$ and $x_2 = 1.0000$ correspond closely to the required values of zero for f_1 and f_2.

r	$x_1^{(r)}$	$x_2^{(r)}$	$f_1^{(r)}$	$f_2^{(r)}$	$(\partial f_1/\partial x_1)^{(r)}$	$(\partial f_1/\partial x_2)^{(r)}$	$(\partial f_2/\partial x_1)^{(r)}$	$(\partial f_2/\partial x_2)^{(r)}$	$\Delta x_1^{(r)}$	$\Delta x_2^{(r)}$
1	2.0000	2.0000	13.4461	25.5060	15.4731	8.3891	15.7781	30.2494	−0.5743	−0.5436
2	1.4247	1.4564	3.8772	7.3133	6.4354	5.1395	9.6024	12.5880	−0.3544	−0.3106
3	1.0713	1.1457	0.7720	1.3802	3.4806	3.8541	7.3591	6.8067	−0.0138	−0.1878
4	1.0575	0.9579	−0.0059	0.1290	2.7149	3.9830	6.1183	5.5679	−0.0591	0.0417
5	0.9984	0.9996	−0.0057	−0.0122	2.7126	3.7127	6.4358	5.4244	0.00159	0.000368
6	1.0000	1.0000	5.51×10^{-6}	2.86×10^{-6}	2.7183	3.7183	6.4366	5.4366	12.1×10^{-6}	-3.0×10^{-6}
7	1.0000	1.0000	0.0	-2×10^{-9}	2.7183	3.7183	6.4366	5.4366	—	—

As applied to the solution of a new set of T_j values from the energy equation (10-5), the recursion equation for the Newton–Raphson method is

$$\left(\frac{\partial H_j}{\partial T_{j-1}}\right)^{(r)} \Delta T_{j-1}^{(r)} + \left(\frac{\partial H_j}{\partial T_j}\right)^{(r)} \Delta T_j^{(r)} + \left(\frac{\partial H_j}{\partial T_{j+1}}\right)^{(r)} \Delta T_{j+1}^{(r)} = -H_j^{(r)}$$

(10-40)

where

$$\Delta T_j^{(r)} = T_j^{(r+1)} - T_j^{(r)} \tag{10-41}$$

$$\frac{\partial H_j}{\partial T_{j-1}} = L_{j-1}\frac{\partial h_{L_{j-1}}}{\partial T_{j-1}} \tag{10-42}$$

$$\frac{\partial H_j}{\partial T_j} = -(L_j + U_j)\frac{\partial h_{L_j}}{\partial T_j} - (V_j + W_j)\frac{\partial h_{V_j}}{\partial T_j} \tag{10-43}$$

$$\frac{\partial H_j}{\partial T_{j+1}} = V_{j+1}\frac{\partial h_{V_{j+1}}}{\partial T_{j+1}} \tag{10-44}$$

The partial derivatives depend upon the enthalpy correlations that are utilized. For example, if composition-independent polynomial equations in temperature are used, then

$$h_{V_j} = \sum_{i=1}^{C} y_{i,j}(A_i + B_i T + C_i T^2) \tag{10-45}$$

$$h_{L_j} = \sum_{i=1}^{C} x_{i,j}(a_i + b_i T + c_i T^2) \tag{10-46}$$

and the partial derivatives are

$$\frac{\partial h_{V_j}}{\partial T_j} = \sum_{i=1}^{C} y_{i,j}(B_i + 2C_i T) \tag{10-47}$$

$$\frac{\partial h_{L_j}}{\partial T_j} = \sum_{i=1}^{C} x_{i,j}(b_i + 2c_i T) \tag{10-48}$$

The N relations given by (10-40) form a tridiagonal-matrix equation that is linear in $\Delta T_j^{(r)}$. The form of the matrix equation is identical to (10-12) where, for example, $A_2 = (\partial H_2/\partial T_1)^{(r)}$, $B_2 = (\partial H_2/\partial T_2)^{(r)}$, $C_2 = (\partial H_2/\partial T_3)^{(r)}$, $x_{i,2} \leftarrow \Delta T_2^{(r)}$, and $D_2 = -H_2^{(r)}$. The matrix of partial derivatives is called the *Jacobian correction matrix*. The Thomas algorithm can be employed to solve for the set of corrections $\Delta T_j^{(r)}$. New guesses of T_j are then determined from

$$T_j^{(r+1)} = T_j^{(r)} + t\Delta T_j^{(r)} \tag{10-49}$$

where t is a scalar attenuation factor that is useful when initial guesses are not reasonably close to the true values. Generally, as in (10-39), t is taken as 1, but an optimal value can be determined at each iteration to minimize the sum of the squares of the functions,

$$\sum_{j=1}^{N} \left[H_j^{(r+1)}\right]^2$$

When all the corrections $\Delta T_j^{(r)}$ have approached zero, the resulting values of T_j are used with criteria such as (10-31) or (10-32) to determine whether convergence has been achieved. If not, before beginning a new k iteration, one can adjust values of V_j and T_j as indicated in Figure 10.16 and

previously discussed for the BP method. Rapid convergence is generally observed for the sum-rates method.

EXAMPLE 10.4

Calculate stage temperatures and interstage vapor and liquid flow rates and compositions by the rigorous SR method for the absorber column specifications given in Figure 5.11.

SOLUTION

The digital computer program of Shinohara et al. [16], based on the Burningham–Otto solution procedure, was used. Initial assumptions for the top-stage and bottom-stage temperatures were 90°F (32.2°C) (entering liquid temperature) and 105°F (40.6°C) (entering gas temperature), respectively. The corresponding number of iterations to satisfy the convergence criterion of (10-32) was seven. Values of τ were as follows.

Iteration Number	τ, (°F)^2
1	9,948
2	2,556
3	46.0
4	8.65
5	0.856
6	0.124
7	0.0217

The overall results of the converged calculations are shown in Figure 10.17. Adjusted values of product-component flow rates that satisfy overall material-balance equations are included. Figures 10.18 to 10.20 are plots of converged values for stage temperatures, interstage total flow rates, and interstage component vapor flow rates, respectively. Figure 10.18 shows that the initial assumed linear-temperature profile is grossly in error. Because of the substantial degree of absorption and accompanying high heat released by

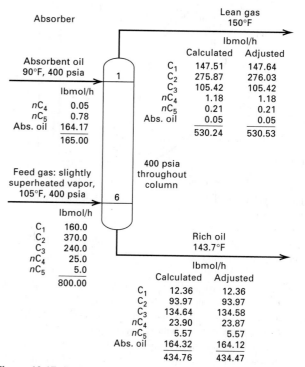

Figure 10.17 Specifications and overall results for Example 10.4.

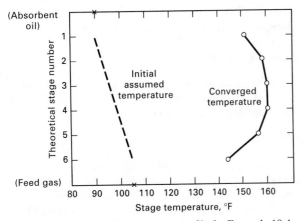

Figure 10.18 Converged temperature profile for Example 10.4.

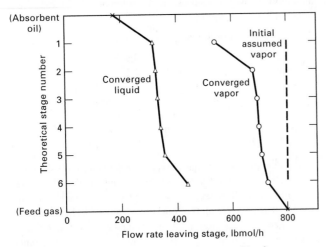

Figure 10.19 Converged interstage flow rate profiles for Example 10.4.

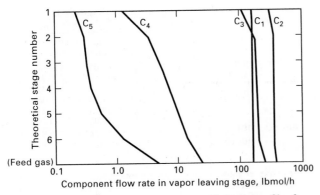

Figure 10.20 Converged component vapor flow rate profiles for Example 10.4.

absorption, stage temperatures are considerably greater than the two entering stream temperatures. The heat is absorbed by both the vapor and liquid streams. The peak stage temperature is essentially at the midpoint of the column. Figure 10.19 shows that the bulk of the overall absorption occurs at the two terminal stages. In Figure 10.20, it is seen that absorption of C_1 and C_2 occurs almost exclusively at the top and bottom stages. Absorption of C_3 occurs throughout the column, but mainly at the two terminal stages. Absorption of C_4 and C_5 also occurs throughout the column, but for C_5 mainly at the bottom where vapor first contacts absorption oil.

Isothermal Sum-Rates Method for Liquid–Liquid Extraction

Multistage liquid–liquid extraction equipment is operated frequently in an adiabatic manner. When entering streams are at the same temperature and heat of mixing is negligible, the operation is also isothermal. For this condition, or when stage temperatures are specified, as indicated by Friday and Smith [7] and shown in detail by Tsuboka and Katayama [17], a simplified isothermal version of the sum-rates method (ISR) can be applied. It is based on the same equilibrium-stage model presented in Section 10.1. However, with all stage temperatures specified, values of Q_j can be computed from stage energy balances, which can be partitioned from the other equations and solved in a separate step following the calculations discussed here. In the ISR method, particular attention is paid to the possibility that phase compositions may strongly influence K_{ij} values.

Figure 10.21 shows the algorithm for the Tsuboka–Katayama ISR method. Liquid-phase and vapor-phase symbols correspond to raffinate and extract, respectively. Problem specifications consist of flow rates, compositions, and stage locations for all feeds; stage temperatures (frequently all equal); total flow rates of any side streams; and total number of stages. Stage pressures need not be specified but are understood to be greater than corresponding stage bubble-point pressures to prevent vaporization.

With stage temperatures specified, the only tear variables are V_j (extract flow rates) values. An initial set is obtained by assuming a perfect separation among the components of the feed and neglecting mass transfer of the solvent to the raffinate phase. This gives approximate values for the flow rates of the exiting raffinate and extract phases. Intermediate values of V_j are obtained by linear interpolation over the N stages. Modifications to this procedure are necessary for side streams or intermediate feeds. As shown in Figure 10.21, the tear variables are reset in an outer iterative loop.

The effect of phase compositions is often considerable on K_D-values (distribution coefficients) for liquid–liquid extraction. Therefore, it is best also to provide initial estimates of $x_{i,j}$ and $y_{i,j}$ from which initial values of $K_{i,j}$ are computed. Initial values of $x_{i,j}$ are obtained by linear interpolation, with stage, of the compositions of the known entering and assumed exit streams. Corresponding values of $y_{i,j}$ are computed by material balance from (10-1). Values of $\gamma_{iL,j}$ and $\gamma_{iV,j}$ are determined from an appropriate correlation—for example, the van Laar, NRTL, UNIQUAC, or UNIFAC equations discussed in Chapter 2. Corresponding K_D-values are obtained from the following equation, which is equivalent to (2-30).

$$K_{i,j} = \frac{\gamma_{iL,j}}{\gamma_{iV,j}} \qquad (10\text{-}50)$$

A new set of $x_{i,j}$ values is obtained by solving (10-12) by the Thomas algorithm of Section 10.3. These values are compared to the assumed values by computing

$$\tau_1 = \sum_{j=1}^{N} \sum_{i=1}^{C} \left| x_{i,j}^{(r-1)} - x_{i,j}^{(r)} \right| \qquad (10\text{-}51)$$

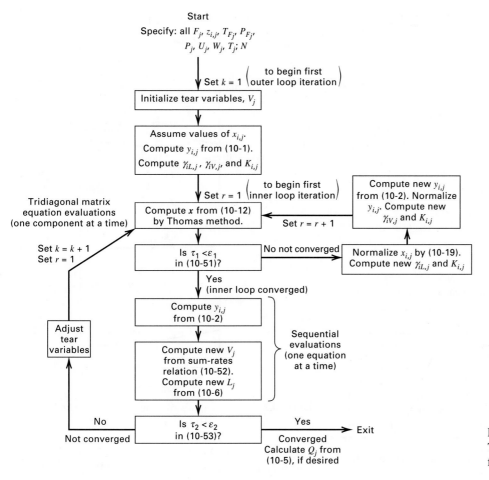

Figure 10.21 Algorithm for Tsuboka–Katayama ISR method for liquid–liquid extraction.

where r is an inner loop index. If $\tau_1 > \epsilon_1$, where, for example, the convergence criterion ϵ_1 might be taken as 0.01 NC, the inner loop is used to improve values of $K_{i,j}$ by using normalized values of $x_{i,j}$ and $y_{i,j}$ to compute new values of $\gamma_{iL,j}$ and $\gamma_{iV,j}$.

When the inner loop is converged, values of $x_{i,j}$ are used to calculate new values of $y_{i,j}$ from (10-2). A new set of tear variables V_j is then computed from the sum-rates relation

$$V_j^{(k+1)} = V_j^{(k)} \sum_{i=1}^{C} y_{i,j} \qquad (10\text{-}52)$$

where k is an outer-loop index. Corresponding values of $L_j^{(k+1)}$ are obtained from (10-6).

The outer loop is converged when

$$\tau_2 = \sum_{j=1}^{N} \left(\frac{V_j^{(k)} - V_j^{(k-1)}}{V_j^{(k)}} \right)^2 \leq \epsilon_2 \qquad (10\text{-}53)$$

where, for example, the convergence criterion ϵ_2 may be taken as 0.01 N.

Before beginning a new k iteration, we can adjust values of V_j as previously discussed for the BP method. Convergence of the ISR method is generally rapid but is subject to the extent to which $K_{i,j}$ depends upon composition.

EXAMPLE 10.5

The separation of benzene (B) from n-heptane (H) by ordinary distillation is difficult. At atmospheric pressure, the boiling points differ by 18.3°C. However, because of liquid-phase nonideality, the relative volatility decreases to a value less than 1.15 at high benzene concentrations [18]. An alternative method of separation is liquid–liquid extraction with a mixture of dimethylformamide (DMF) and water [19]. The solvent is much more selective for benzene than for n-heptane at 20°C. For two different solvent compositions, calculate interstage flow rates and compositions by the rigorous ISR method for the countercurrent, liquid–liquid extraction cascade, which contains five equilibrium stages and is shown schematically in Figure 10.22.

SOLUTION

Experimental phase-equilibrium data for the quaternary system [19] were fitted to the NRTL equation by Cohen and Renon [20]. The resulting binary-pair constants in (2-92) and (2-93) are

Binary Pair, ij	τ_{ij}	τ_{ji}	α_{ji}
DMF, H	2.036	1.910	0.25
Water, H	7.038	4.806	0.15
B, H	1.196	−0.355	0.30
Water, DMF	2.506	−2.128	0.253
B, DMF	−0.240	0.676	0.425
B, Water	3.639	5.750	0.203

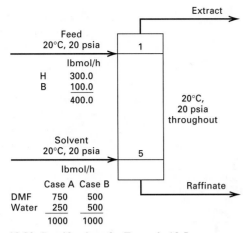

Figure 10.22 Specifications for Example 10.5.

For Case A, estimates of V_j (the extract phase), $x_{i,j}$, and $y_{i,j}$ are as follows, based on a perfect separation and linear interpolation by stage.

Stage		$y_{i,j}$				$x_{i,j}$			
j	V_j	H	B	DMF	Water	H	B	DMF	Water
1	1100	0.0	0.0909	0.6818	0.2273	0.7895	0.2105	0.0	0.0
2	1080	0.0	0.0741	0.6944	0.2315	0.8333	0.1667	0.0	0.0
3	1060	0.0	0.0566	0.7076	0.2359	0.8824	0.1176	0.0	0.0
4	1040	0.0	0.0385	0.7211	0.2404	0.9375	0.0625	0.0	0.0
5	1020	0.0	0.0196	0.7353	0.2451	1.0000	0.0	0.0	0.0

The converged solution is obtained by the ISR method with the following corresponding stage flow rates and compositions:

Stage		$y_{i,j}$				$x_{i,j}$			
j	V_j	H	B	DMF	Water	H	B	DMF	Water
1	1113.1	0.0263	0.0866	0.6626	0.2245	0.7586	0.1628	0.0777	0.0009
2	1104.7	0.0238	0.0545	0.6952	0.2265	0.8326	0.1035	0.0633	0.0006
3	1065.6	0.0213	0.0309	0.7131	0.2347	0.8858	0.0606	0.0532	0.0004
4	1042.1	0.0198	0.0157	0.7246	0.2399	0.9211	0.0315	0.0471	0.0003
5	1028.2	0.0190	0.0062	0.7316	0.2432	0.9438	0.0125	0.0434	0.0003

Computed products for the two cases are:

	Extract, lbmol/h		Raffinate, lbmol/h	
	Case A	Case B	Case A	Case B
H	29.3	5.6	270.7	294.4
B	96.4	43.0	3.6	57.0
DMF	737.5	485.8	12.5	14.2
Water	249.9	499.7	0.1	5.0
	1113.1	1034.1	286.9	365.9

On a percentage extraction basis, the results are:

	Case A	Case B
Percent of benzene feed extracted	96.4	43.0
Percent of n-heptane feed extracted	9.8	1.87
Percent of solvent transferred to raffinate	1.26	1.45

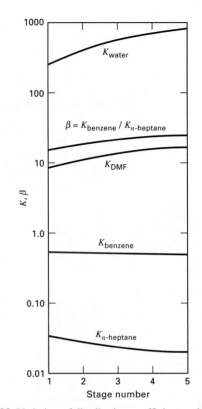

Figure 10.23 Variation of distribution coefficient and relative selectivity for Example 10.5, Case A.

Thus, the solvent with 75% DMF extracts a much larger percentage of the benzene, but the solvent with 50% DMF is more selective between benzene and n-heptane.

For Case A, the variations with stage of K-values and the relative selectivity are shown in Figure 10.23, where the relative selectivity is $\beta_{B,H} = K_B/K_H$. The distribution coefficient for n-heptane varies by a factor of almost 1.75 from stage 5 to stage 1, while the coefficient for benzene is almost constant. The relative selectivity varies by a factor of almost 2.

10.4 NEWTON–RAPHSON METHOD

The BP and SR methods for vapor–liquid contacting converge only with difficulty or not at all for separations involving very nonideal liquid mixtures or for cases where the separator is like an absorber or stripper in one section and a fractionator in another section (e.g., a reboiled absorber). Furthermore, BP and SR methods are generally restricted to the very limited specifications stated previously. More general procedures capable of solving all types of multicomponent, multistage, separation problems are based on the simultaneous solution of all the MESH equations, or combinations thereof, by simultaneous-correction (SC) techniques, often using the Newton–Raphson method.

In order to develop an SC procedure that uses the Newton–Raphson method, one must select and order the unknown variables and the corresponding functions (MESH equations) that contain them. As discussed by Goldstein and Stanfield [21], grouping of the functions by type is computationally most efficient for problems involving a large number of components, but few stages. Alternatively, it is most efficient to group the functions according to stage location for problems involving many stages, but relatively few components. The latter grouping, presented here, is described by Naphtali [22] and was implemented by Naphtali and Sandholm [23].

The SC procedure of Naphtali and Sandholm is developed in detail because it utilizes many of the mathematical techniques presented in Section 10.3 on tearing methods. A computer program for their method is given by Fredenslund, Gmehling, and Rasmussen [24]. However, that program does not have the flexibility of specifications in Newton–Raphson implementations found in commercial simulators for computer-aided process design.

The equilibrium-stage model of Figures 10.1 and 10.3 is again employed. However, rather than solving the $N(2C + 3)$ MESH equations simultaneously, we combine (10-3) and (10-4) with the other MESH equations to eliminate $2N$ variables and thus reduce the problem to the simultaneous solution of $N(2C + 1)$ equations. This is done by first multiplying (10-3) and (10-4) by V_j and L_j, respectively, to give

$$V_j = \sum_{i=1}^{C} v_{i,j} \qquad (10\text{-}54)$$

$$L_j = \sum_{i=1}^{C} l_{i,j} \qquad (10\text{-}55)$$

where we have used the mole-fraction definitions

$$y_{i,j} = \frac{v_{i,j}}{V_j} \qquad (10\text{-}56)$$

$$x_{i,j} = \frac{l_{i,j}}{L_j} \qquad (10\text{-}57)$$

Equations (10-54) to (10-57) are now substituted into (10-1), (10-2), and (10-5) to eliminate V_j, L_j, $y_{i,j}$, and $x_{i,j}$ and introduce component flow rates $v_{i,j}$ and $l_{i,j}$. As a result, the following $N(2C + 1)$ equations are obtained, where $s_j = U_j/L_j$ and $S_j = W_j/V_j$ are dimensionless side stream flow rates:

Material Balance

$$M_{i,j} = l_{i,j}(1 + s_j) + v_{i,j}(1 + S_j) - l_{i,j-1} - v_{i,j+1} - f_{i,j} = 0 \qquad (10\text{-}58)$$

Phase Equilibria

$$E_{i,j} = K_{i,j} l_{i,j} \frac{\sum\limits_{\kappa=1}^{C} v_{\kappa,j}}{\sum\limits_{\kappa=1}^{C} l_{\kappa,j}} - v_{i,j} = 0 \qquad (10\text{-}59)$$

Energy Balance

$$\begin{aligned}
H_j = {}& h_{L_j}(1 + s_j) \sum_{i=1}^{C} l_{i,j} + h_{V_j}(1 + S_j) \sum_{i=1}^{C} v_{i,j} \\
& - h_{L_{j-1}} \sum_{i=1}^{C} l_{i,j-1} - h_{V_{j+1}} \sum_{i=1}^{C} v_{i,j+1} \\
& - h_{F_j} \sum_{i=1}^{C} f_{i,j} - Q_j = 0
\end{aligned} \qquad (10\text{-}60)$$

where $f_{i,j} = F_j z_{i,j}$.

If N and all $f_{i,j}$, T_{F_j}, P_{F_j}, P_j, s_j, S_j, and Q_j are specified, the M, E, and H functions are nonlinear in the $N(2C + 1)$ unknown (output) variables $v_{i,j}$, $l_{i,j}$, and T_j for $i = 1$ to C and $j = 1$ to N. Although other sets of specified and unknown variables are possible, we consider these sets first.

Equations (10-58), (10-59), and (10-60) are solved simultaneously by the Newton–Raphson iterative method in which successive sets of the output variables are completed until the values of the M, E, and H functions are driven to within some tolerance of zero. During the iterations, nonzero values of the functions are called *discrepancies* or *errors*. Let the functions and output variables be grouped by stage in order from top to bottom. As will be shown, this is done to produce a block-tridiagonal structure for the Jacobian matrix of partial derivatives so that a matrix form of the Thomas algorithm can be applied. Let

$$\mathbf{X} = [\mathbf{X}_1, \mathbf{X}_2, \ldots, \mathbf{X}_j, \ldots, \mathbf{X}_N]^T \qquad (10\text{-}61)$$

and

$$\mathbf{F} = [\mathbf{F}_1, \mathbf{F}_2, \ldots, \mathbf{F}_j, \ldots, \mathbf{F}_N]^T \qquad (10\text{-}62)$$

where $\mathbf{X}_j$ is the vector of output variables for stage j arranged in the order

$$\begin{aligned}
\mathbf{X}_j = [{}& v_{1,j}, v_{2,j}, \ldots, v_{i,j}, \ldots, v_{C,j}, T_j, l_{1,j}, \\
& l_{2,j}, \ldots, l_{i,j}, \ldots, l_{C,j}]^T
\end{aligned} \qquad (10\text{-}63)$$

and $\mathbf{F}_j$ is the vector of functions for stage j arranged in the order

$$\begin{aligned}
\mathbf{F}_j = [{}& H_j, M_{1,j}, M_{2,j}, \ldots, M_{i,j}, \ldots, M_{C,j}, E_{1,j}, \\
& E_{2,j}, \ldots, E_{i,j}, \ldots, E_{C,j}]^T
\end{aligned} \qquad (10\text{-}64)$$

The Newton–Raphson iteration is performed by solving for the corrections $\Delta \mathbf{X}$ to the output variables from (10-38), which in matrix form becomes

$$\Delta \mathbf{X}^{(k)} = - \left[\left(\frac{\partial \mathbf{F}}{\partial \mathbf{X}} \right)^{-1} \right]^{(k)} \mathbf{F}^{(k)} \qquad (10\text{-}65)$$

These corrections are used to compute the next approximation to the set of output variables from

$$\mathbf{X}^{(k+1)} = \mathbf{X}^{(k)} + t \, \Delta \mathbf{X}^{(k)} \qquad (10\text{-}66)$$

The quantity $(\partial \mathbf{F}/\partial \mathbf{X})$ is the following Jacobian or $(N \times N)$ matrix of blocks of partial derivatives of all the functions with respect to all the output variables.

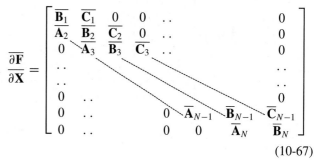

$$\frac{\partial \mathbf{F}}{\partial \mathbf{X}} = \begin{bmatrix}
\overline{\mathbf{B}}_1 & \overline{\mathbf{C}}_1 & 0 & 0 & .. & & 0 \\
\overline{\mathbf{A}}_2 & \overline{\mathbf{B}}_2 & \overline{\mathbf{C}}_2 & 0 & .. & & 0 \\
0 & \overline{\mathbf{A}}_3 & \overline{\mathbf{B}}_3 & \overline{\mathbf{C}}_3 & .. & & 0 \\
.. & & & & & & .. \\
.. & & & & & & .. \\
0 & .. & & & & & 0 \\
0 & .. & & & 0 & \overline{\mathbf{A}}_{N-1} & \overline{\mathbf{B}}_{N-1} & \overline{\mathbf{C}}_{N-1} \\
0 & .. & & & 0 & 0 & \overline{\mathbf{A}}_N & \overline{\mathbf{B}}_N
\end{bmatrix}$$

$$(10\text{-}67)$$

This Jacobian is of a block-tridiagonal form, like (10-12), because functions for stage j are only dependent on output variables for stages $j - 1$, j, and $j + 1$. Each $\bar{\mathbf{A}}$, $\bar{\mathbf{B}}$, or $\bar{\mathbf{C}}$ block in (15-67) represents a $(2C + 1)$ by $(2C + 1)$ submatrix of partial derivatives, where the arrangements of output variables and functions are given by (10-63) and (10-64), respectively. Blocks $\bar{\mathbf{A}}_j$, $\bar{\mathbf{B}}_j$, and $\bar{\mathbf{C}}_j$ correspond to submatrices of partial derivatives of the functions on stage j with respect to the output variables on stages $j - 1$, j, and $j + 1$, respectively. Thus, using (10-58), (10-59), and (10-60), and denoting only the nonzero partial derivatives by $+$, or by row or diagonal strings of $+ \cdots +$, or by the following square or rectangular blocks enclosed by connected strings,

$$
\begin{matrix}
+ \cdots + \\
\vdots \qquad \vdots \\
+ \cdots +
\end{matrix}
$$

we find that the blocks have the following form, where $+$ is replaced by a numerical value (-1 or 1) in the event that the partial derivative has only that value.

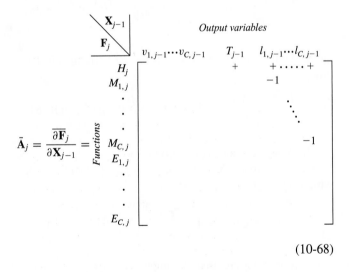

$$\bar{\mathbf{A}}_j = \frac{\partial \overline{\mathbf{F}}_j}{\partial \mathbf{X}_{j-1}} = \qquad (10\text{-}68)$$

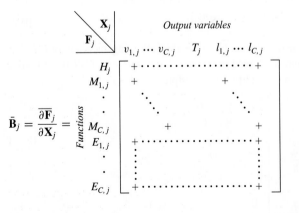

$$\bar{\mathbf{B}}_j = \frac{\partial \overline{\mathbf{F}}_j}{\partial \mathbf{X}_j} = \qquad (10\text{-}69)$$

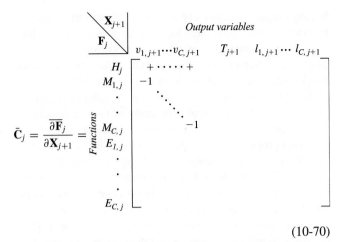

$$\bar{\mathbf{C}}_j = \frac{\partial \overline{\mathbf{F}}_j}{\partial \mathbf{X}_{j+1}} = \qquad (10\text{-}70)$$

Thus, (10-65) consists of a set of $N(2C + 1)$ simultaneous, linear equations in the $N(2C + 1)$ corrections $\Delta \mathbf{X}$. For example, the $2C + 2$ equation in the set is obtained by expanding function H_2 (10-60) into a Taylor's series like (10-36) around the $N(2C + 1)$ output variables. The result is as follows after the usual truncation of terms involving derivatives of order greater than one:

$$0(\Delta v_{1,1} + \cdots + \Delta v_{C,1}) - \frac{\partial h_{L_1}}{\partial T_1} \sum_{i=1}^{C} l_{i,1}(\Delta T_1)$$

$$- \left(\frac{\partial h_{L_1}}{\partial l_{1,1}} \sum_{i=1}^{C} l_{i,1} + h_{L_1} \right) \Delta l_{1,1} - \cdots$$

$$- \left(\frac{\partial h_{L_1}}{\partial l_{C,1}} \sum_{i=1}^{C} l_{i,1} + h_{L_1} \right) \Delta l_{C,1}$$

$$+ \left[\left(\frac{\partial h_{V_2}}{\partial v_{1,2}} \right) (1 + S_2) \sum_{i=1}^{C} v_{i,2} + h_{V_2}(1 + S_2) \right] \Delta v_{1,2} + \cdots$$

$$+ \left[\left(\frac{\partial h_{V_2}}{\partial v_{C,2}} \right) (1 + S_2) \sum_{i=1}^{C} v_{i,2} + h_{V_2}(1 + S_2) \right] \Delta v_{C,2}$$

$$+ \left[\left(\frac{\partial h_{L_2}}{\partial T_2} \right) (1 + s_2) \sum_{i=1}^{C} l_{i,2} + \left(\frac{\partial h_{V_2}}{\partial T_2} \right) (1 + S_2) \sum_{i=1}^{C} v_{i,2} \right] \Delta T_2$$

$$+ \left[\left(\frac{\partial h_{L_2}}{\partial l_{1,2}} \right) (1 + s_2) \sum_{i=1}^{C} l_{i,2} + h_{L_2}(1 + s_2) \right] \Delta l_{1,2} + \cdots$$

$$+ \left[\left(\frac{\partial h_{L_2}}{\partial l_{C,2}} \right) (1 + s_2) \sum_{i=1}^{C} l_{i,2} + h_{L_2}(1 + s_2) \right] \Delta l_{C,2}$$

$$- \left(\frac{\partial h_{V_3}}{\partial v_{1,3}} \sum_{i=1}^{C} v_{i,3} + h_{V_3} \right) \Delta v_{1,3} - \cdots$$

$$- \left(\frac{\partial h_{V_3}}{\partial v_{C,3}} \sum_{i=1}^{C} v_{i,3} + h_{V_3} \right) \Delta v_{C,3}$$

$$- \frac{\partial h_{V_3}}{\partial T_3} \sum_{i=1}^{C} v_{i,3} \Delta T_3 + 0(\Delta l_{1,3} + \cdots + \Delta l_{C,N}) = -H_2$$

$$(10\text{-}71)$$

Although lengthy, equations such as (10-71) are handled readily in computer programs.

As a further example, the entry in the Jacobian matrix for row $(2C + 2)$ and column $(C + 3)$ is obtained from (10-71) as

$$\frac{\partial H_2}{\partial l_{2,1}} = -\frac{\partial h_{L_1}}{\partial l_{2,1}} \sum_{i=1}^{C} l_{i,1} + h_{L_1} \qquad (10\text{-}72)$$

All partial derivatives are stated by Naphtali and Sandholm [23].

Partial derivatives of enthalpies and K-values depend upon the particular correlation utilized for these properties and are sometimes simplified by including only the dominant terms. For example, suppose that the Chao–Seader correlation is to be used for K-values. In general,

$$K_{i,j} = K_{i,j} \left\{ P_j, T_j, \frac{l_{i,j}}{\sum\limits_{\kappa=1}^{C} l_{\kappa,j}}, \frac{v_{i,j}}{\sum\limits_{\kappa=1}^{C} v_{\kappa,j}} \right\}$$

In terms of the output variables, the partial derivatives $\partial K_{i,j}/\partial T_j$; $\partial K_{i,j}/\partial l_{i,j}$; and $\partial K_{i,j}/\partial v_{i,j}$ all exist and can be expressed analytically or evaluated numerically if desired. However, for some problems, the terms that include the first and second of these three groups of derivatives may be the dominant terms so that the third group may be taken as zero.

EXAMPLE 10.6

Derive an expression for $(\partial h_V/\partial T)$ from the Redlich–Kwong equation of state.

SOLUTION

From (2-53),

$$h_V = \sum_{i=1}^{C} (y_i h_{iV}^\circ) + RT \left[Z_V - 1 - \frac{3A}{2B} \ln\left(1 + \frac{B}{Z_V}\right) \right]$$

where h_{iV}°, Z_V, A, and B all depend on T, as determined from (2-36) and (2-46) to (2-50). Thus,

$$\frac{\partial h_V}{\partial T} = \sum_{i=1}^{C} \left[y_i \left(\frac{\partial h_{iV}^\circ}{\partial T} \right) \right] + R \left[Z_V - 1 - \frac{3A}{2B} \ln\left(1 + \frac{B}{Z_V}\right) \right]$$
$$+ RT \left\{ \left(\frac{\partial Z_V}{\partial T} \right) - \frac{3}{2} \left(\frac{\partial(A/B)}{\partial T} \right) \ln\left(1 + \frac{B}{Z_V}\right) \right.$$
$$\left. - \frac{3A}{2B} \left[\frac{1}{Z_V} \left(\frac{\partial B}{\partial T} \right) - \frac{B}{Z_V^2} \left(\frac{\partial Z_V}{\partial T} \right) \right] \right\}$$

From (2-36) and (2-35),

$$\left(\frac{\partial h_{iV}^\circ}{\partial T} \right) = \sum_{k=0}^{4} (a_k)_i T^k = (C_{P_V}^\circ)_i$$

From (2-48) and Table 2.5,

$$B = \frac{bP}{RT} \quad \text{and} \quad b = \frac{0.08664 RT_c}{P_c}$$

Thus,

$$\frac{\partial B}{\partial T} = -\frac{B}{T}$$

From (2-47) to (2-50) and Table 2.5,

$$\frac{A}{B} = \frac{a}{bRT} \quad \text{and} \quad a = \frac{0.42748 R^2 T_c^{2.5}}{P_c T^{0.5}}$$

Thus,

$$\frac{\partial(A/B)}{\partial T} = -1.5 \frac{A}{BT}$$

From (2-46),

$$Z_V^3 - Z_V^2 + (A - B - B^2)Z_V - AB = 0$$

By implicit differentiation,

$$3Z_V^2 \frac{\partial Z_V}{\partial T} - 2Z_V \frac{\partial Z_V}{\partial T} + (A - B - B^2)\frac{\partial Z_V}{\partial T}$$
$$+ \frac{Z_V}{T}(-2.5A + B + 2B^2) + \frac{3.5AB}{T} = 0$$

which reduces to

$$\frac{\partial Z_V}{\partial T} = \frac{(Z_V/T)(2.5A - B - 2B^2) - 3.5AB/T}{3Z_V^2 - 2Z_V + (A - B - B^2)}$$

Because the Thomas algorithm can be applied to the block-tridiagonal structure of (10-67), submatrices of partial derivatives are computed only as needed. The solution of (10-65) follows the scheme in Section 10.3, given by (10-13) to (10-18) and represented in Figure 10-4, where matrices and vectors $\bar{\mathbf{A}}_j$, $\bar{\mathbf{B}}_j$, $\bar{\mathbf{C}}_j$, $-\bar{\mathbf{F}}_j$, and $\Delta \mathbf{X}_j$ correspond to variables A_j, B_j, C_j, D_j, and x_j, respectively. However, the simple multiplication and division operations in Section 10.3 are changed to matrix multiplication and inversion, respectively. The steps are as follows:

Starting at stage 1, $\bar{\mathbf{C}}_1 \leftarrow (\bar{\mathbf{B}}_1)^{-1}\bar{\mathbf{C}}_1$, $\mathbf{F}_1 \leftarrow (\bar{\mathbf{B}}_1)^{-1}\mathbf{F}_1$, and $\bar{\mathbf{B}}_1 \leftarrow \mathbf{I}$ (the identity submatrix). Only $\bar{\mathbf{C}}_1$ and $\mathbf{F}_1$ are saved. For stages j from 2 to $(N - 1)$, $\bar{\mathbf{C}}_j \leftarrow (\bar{\mathbf{B}}_j - \bar{\mathbf{A}}_j \bar{\mathbf{C}}_{j-1})^{-1}\bar{\mathbf{C}}_j$, $\mathbf{F}_j \leftarrow (\bar{\mathbf{B}}_j - \bar{\mathbf{A}}_j \bar{\mathbf{C}}_{j-1})^{-1}(\mathbf{F}_j - \bar{\mathbf{A}}_j \mathbf{F}_{j-1})$. Then $\bar{\mathbf{A}}_j \leftarrow 0$, and $\bar{\mathbf{B}}_j \leftarrow \mathbf{I}$. Save $\bar{\mathbf{C}}_j$ and $\mathbf{F}_j$ for each stage. For the last stage, $\mathbf{F}_N \leftarrow (\mathbf{B}_N - \mathbf{A}_N \mathbf{C}_{N-1})^{-1}(\mathbf{F}_N - \mathbf{A}_N \mathbf{F}_{N-1})$, $\bar{\mathbf{A}}_N \leftarrow 0$, $\bar{\mathbf{B}}_N \leftarrow \mathbf{I}$, and therefore $\Delta \mathbf{X}_N = -\mathbf{F}_N$. This completes the forward steps. Remaining values of $\Delta \mathbf{X}$ are obtained by successive, backward substitution from $\Delta \mathbf{X}_j = -\mathbf{F}_j \leftarrow -(\mathbf{F}_j - \bar{\mathbf{C}}_j \mathbf{F}_{j+1})$.

For the last stage, $\mathbf{F}_N \leftarrow (\mathbf{B}_N - \mathbf{A}_N \mathbf{C}_{N-1})^{-1}(\mathbf{F}_N - \mathbf{A}_N \mathbf{F}_{N-1})$, $\bar{\mathbf{A}}_N \leftarrow 0$, $\bar{\mathbf{B}}_N \leftarrow \mathbf{I}$, and therefore $\Delta \mathbf{X}_N = -\mathbf{F}_N$. This completes the forward steps. Remaining values of $\Delta \mathbf{X}$ are obtained by successive, backward substitution from $\Delta \mathbf{X}_j = -\mathbf{F}_j \leftarrow -(\mathbf{F}_j - \bar{\mathbf{C}}_j \mathbf{F}_{j+1})$. This procedure is illustrated by the following example.

EXAMPLE 10.7

Solve the following matrix equation, which has a block-tridiagonal structure, by the matrix form of the Thomas algorithm.

$$
\begin{bmatrix}
1 & 2 & 1 & 2 & 2 & 1 & 0 & 0 & 0 \\
2 & 1 & 1 & 2 & 1 & 0 & 0 & 0 & 0 \\
1 & 2 & 2 & 1 & 2 & 0 & 0 & 0 & 0 \\
\hline
0 & 1 & 3 & 1 & 2 & 1 & 1 & 2 & 1 \\
0 & 0 & 1 & 2 & 2 & 0 & 1 & 2 & 0 \\
0 & 0 & 2 & 2 & 1 & 1 & 1 & 1 & 0 \\
\hline
0 & 0 & 0 & 0 & 1 & 2 & 2 & 1 & 1 \\
0 & 0 & 0 & 0 & 0 & 2 & 1 & 1 & 1 \\
0 & 0 & 0 & 0 & 0 & 1 & 2 & 1 & 2
\end{bmatrix}
\cdot
\begin{bmatrix}
\Delta x_1 \\ \Delta x_2 \\ \Delta x_3 \\ \hline \Delta x_4 \\ \Delta x_5 \\ \Delta x_6 \\ \hline \Delta x_7 \\ \Delta x_8 \\ \Delta x_9
\end{bmatrix}
=
\begin{bmatrix}
9 \\ 7 \\ 8 \\ \hline 12 \\ 8 \\ 8 \\ \hline 7 \\ 5 \\ 6
\end{bmatrix}
$$

SOLUTION

The matrix equation is in the form

$$
\begin{bmatrix}
\bar{\mathbf{B}}_1 & \bar{\mathbf{C}}_1 & 0 \\
\bar{\mathbf{A}}_2 & \bar{\mathbf{B}}_2 & \bar{\mathbf{C}}_2 \\
0 & \bar{\mathbf{A}}_3 & \bar{\mathbf{B}}_3
\end{bmatrix}
\cdot
\begin{bmatrix}
\Delta\mathbf{X}_1 \\ \Delta\mathbf{X}_2 \\ \Delta\mathbf{X}_3
\end{bmatrix}
= -
\begin{bmatrix}
\mathbf{F}_1 \\ \mathbf{F}_2 \\ \mathbf{F}_3
\end{bmatrix}
$$

Following the procedure just given, starting at the first block row,

$$
\bar{\mathbf{B}}_1 = \begin{bmatrix} 1 & 2 & 1 \\ 2 & 1 & 1 \\ 1 & 2 & 2 \end{bmatrix}, \quad
\bar{\mathbf{C}}_1 = \begin{bmatrix} 2 & 2 & 1 \\ 2 & 1 & 0 \\ 1 & 2 & 0 \end{bmatrix}, \quad
\mathbf{F}_1 = \begin{bmatrix} -9 \\ -7 \\ -8 \end{bmatrix}
$$

By standard matrix inversion

$$
(\bar{\mathbf{B}}_1)^{-1} = \begin{bmatrix} 0 & 2/3 & -1/3 \\ 1 & -1/3 & -1/3 \\ -1 & 0 & 1 \end{bmatrix}
$$

By standard matrix multiplication

$$
(\bar{\mathbf{B}}_1)^{-1}(\bar{\mathbf{C}}_1) = \begin{bmatrix} 1 & 0 & 0 \\ 1 & 1 & 1 \\ -1 & 0 & -1 \end{bmatrix}
$$

which replaces $\bar{\mathbf{C}}_1$, and

$$
(\bar{\mathbf{B}}_1)^{-1}(\mathbf{F}_1) = \begin{bmatrix} -2 \\ -4 \\ 1 \end{bmatrix}
$$

which replaces $\mathbf{F}_1$. Also

$$
\mathbf{I} = \begin{bmatrix} 1 & 0 & 0 \\ 0 & 1 & 0 \\ 0 & 0 & 1 \end{bmatrix} \text{ replaces } \bar{\mathbf{B}}_1
$$

For the second block row

$$
\bar{\mathbf{A}}_2 = \begin{bmatrix} 0 & 1 & 3 \\ 0 & 0 & 1 \\ 0 & 0 & 2 \end{bmatrix}, \quad
\bar{\mathbf{B}}_2 = \begin{bmatrix} 1 & 2 & 1 \\ 2 & 2 & 0 \\ 2 & 1 & 1 \end{bmatrix},
$$

$$
\bar{\mathbf{C}}_2 = \begin{bmatrix} 1 & 2 & 1 \\ 1 & 2 & 0 \\ 1 & 1 & 0 \end{bmatrix}, \quad
\bar{\mathbf{F}}_2 = \begin{bmatrix} -12 \\ -8 \\ -8 \end{bmatrix}
$$

By matrix multiplication and subtraction

$$
(\bar{\mathbf{B}}_2 - \bar{\mathbf{A}}_2\bar{\mathbf{C}}_1) = \begin{bmatrix} 3 & 1 & 3 \\ 3 & 2 & 1 \\ 4 & 1 & 3 \end{bmatrix}
$$

which upon inversion becomes

$$
(\bar{\mathbf{B}}_2 - \bar{\mathbf{A}}_2\bar{\mathbf{C}}_1)^{-1} = \begin{bmatrix} -1 & 0 & 1 \\ 1 & 3/5 & -6/5 \\ 1 & -1/5 & -3/5 \end{bmatrix}
$$

By multiplication

$$
(\bar{\mathbf{B}}_2 - \bar{\mathbf{A}}_2\bar{\mathbf{C}}_1)^{-1}\bar{\mathbf{C}}_2 = \begin{bmatrix} 0 & -1 & -1 \\ 2/5 & 2 & 1 \\ 1/5 & 1 & 1 \end{bmatrix}
$$

which replaces $\bar{\mathbf{C}}_2$. In a similar manner, the remaining steps for this and the third block row are carried out to give

$$
\begin{bmatrix}
\begin{bmatrix} 1 & 0 & 0 \\ 0 & 1 & 0 \\ 0 & 0 & 1 \end{bmatrix} & \begin{bmatrix} 1 & 0 & 0 \\ 1 & 1 & 1 \\ -1 & 0 & -1 \end{bmatrix} & \begin{bmatrix} 0 & 0 & 0 \\ 0 & 0 & 0 \\ 0 & 0 & 0 \end{bmatrix} \\
\begin{bmatrix} 0 & 0 & 0 \\ 0 & 0 & 0 \\ 0 & 0 & 0 \end{bmatrix} & \begin{bmatrix} 1 & 0 & 0 \\ 0 & 1 & 0 \\ 0 & 0 & 1 \end{bmatrix} & \begin{bmatrix} 0 & -1 & -1 \\ 2/5 & 2 & 1 \\ 1/5 & 1 & 1 \end{bmatrix} \\
\begin{bmatrix} 0 & 0 & 0 \\ 0 & 0 & 0 \\ 0 & 0 & 0 \end{bmatrix} & \begin{bmatrix} 0 & 0 & 0 \\ 0 & 0 & 0 \\ 0 & 0 & 0 \end{bmatrix} & \begin{bmatrix} 1 & 0 & 0 \\ 0 & 1 & 0 \\ 0 & 0 & 1 \end{bmatrix}
\end{bmatrix}
\cdot
\begin{bmatrix}
\Delta X_1 \\ \Delta X_2 \\ \Delta X_3 \\ \Delta X_4 \\ \Delta X_5 \\ \Delta X_6 \\ \Delta X_7 \\ \Delta X_8 \\ \Delta X_9
\end{bmatrix}
= -
\begin{bmatrix}
-2 \\ -4 \\ +1 \\ +1 \\ -22/5 \\ -16/5 \\ -1 \\ -1 \\ -1
\end{bmatrix}
$$

Thus, $\Delta X_7 = \Delta X_8 = \Delta X_9 = 1$.

The remaining backward steps begin with the second block row where

$$
\bar{\mathbf{C}}_2 = \begin{bmatrix} 0 & -1 & -1 \\ 2/5 & 2 & 1 \\ 1/5 & 1 & 1 \end{bmatrix}, \quad
\bar{\mathbf{F}}_2 = \begin{bmatrix} 1 \\ -22/5 \\ -16/5 \end{bmatrix}
$$

$$
(\mathbf{F}_2 - \bar{\mathbf{C}}_2\bar{\mathbf{F}}_3) = \begin{bmatrix} -1 \\ -1 \\ -1 \end{bmatrix}
$$

Thus, $\Delta X_4 = \Delta X_5 = \Delta X_6 = 1$. Similarly, for the first block row, the result is

$$
\Delta X_1 = \Delta X_2 = \Delta X_3 = 1
$$

Usually, it is desirable to specify certain top- and bottom-stage variables other than the condenser duty and/or reboiler duty. (In fact, the condenser and reboiler duties are usually so interdependent that specification of both values is not recommended.) Specifying other variables is readily accomplished by removing heat balance functions H_1 and/or H_N from the simultaneous equation set and replacing them with discrepancy functions depending upon the desired specification(s). Functions for alternative specifications for a column with a partial condenser are listed in Table 10.1.

If desired, (10-54) can be modified to permit real rather than theoretical stages to be computed. Values of the Murphree vapor-phase plate efficiency must then be specified. These values are related to phase compositions by

Table 10.1   Alternative Functions for H_1 and H_N

Specification	Replacement for H_1	Replacement for H_N
Reflux or reboil (boilup) ratio, (L/D) or (V/B)	$\sum l_{i,1} - (L/D)\sum v_{i,1} = 0$	$\sum v_{i,N} - (V/B)\sum l_{i,N} = 0$
Stage temperature, T_D or T_B	$T_1 - T_D = 0$	$T_N - T_B = 0$
Product flow rate, D or B	$\sum v_{i,1} - D = 0$	$\sum l_{i,N} - B = 0$
Component flow rate in product, d_i or b_i	$v_{i,1} - d_i = 0$	$l_{i,N} - b_i = 0$
Component mole fraction in product, y_{iD} or x_{iB}	$v_{i,1} - (\sum v_{i,1})y_{iD} = 0$	$l_{i,N} - (\sum l_{i,N})x_{iB} = 0$

the definition

$$\eta_j = \frac{y_{i,j} - y_{i,j+1}}{K_{i,j}x_{i,j} - y_{i,j+1}} \qquad (10\text{-}73)$$

In terms of component flow rates, (10-73) becomes the following discrepancy function, which replaces (10-59).

$$E_{i,j} = \frac{\eta_j K_{i,j} l_{i,j} \sum\limits_{\kappa=1}^{C} v_{\kappa,j}}{\sum\limits_{\kappa=1}^{C} l_{\kappa,j}} - v_{i,j} + \frac{(1 - \eta_j)v_{i,j+1} \sum\limits_{\kappa=1}^{C} v_{\kappa,j}}{\sum\limits_{\kappa=1}^{C} v_{\kappa,j+1}} = 0$$

$$(10\text{-}74)$$

If a total condenser with subcooling is desired, it is necessary to specify the degrees of subcooling, if any, and to replace (10-59) or (10-74) with functions that express identity of reflux and distillate compositions as discussed by Naphtali and Sandholm [23].

The algorithm for the Naphtali–Sandholm implementation of the Newton–Raphson method is shown in Figure 10.24. Problem specifications are quite flexible. Pressure, compositions, flow rates, and stage locations are necessary specifications for all feeds. The thermal condition of each feed can be given in terms of enthalpy, temperature, or molar fraction vaporized. If a feed is found to consist of two phases, the phases can be sent to the same stage or the vapor can be directed to the stage above the designated feed stage. Stage pressures and stage efficiencies can be designated by specifying top- and bottom-stage values. Remaining values are obtained by linear interpolation. By default, intermediate stages are assumed to be adiabatic unless Q_j or T_j values are specified. Vapor and/or liquid side streams can be designated in terms of total flow rate or flow rate of a specified component, or by the ratio of the side stream flow rate to the flow rate remaining and passing to the next stage. The top- and bottom-stage specifications are selected from Q_1 or Q_N, and/or more generally from the other specifications listed in Table 10.1.

In order to achieve convergence, the Newton–Raphson procedure requires that reasonable guesses be provided for the values of all output variables. Rather than provide all these guesses a priori, we can generate them if T, V, and L are guessed for the bottom and top stages and, perhaps, for one or more intermediate stages. Remaining guessed values of T_j, V_j, and L_j are readily obtained by linear interpolation of the given T_j values and computed (V_j/L_j) values. Initial values for $v_{i,j}$ and $l_{i,j}$ are then obtained by either of two

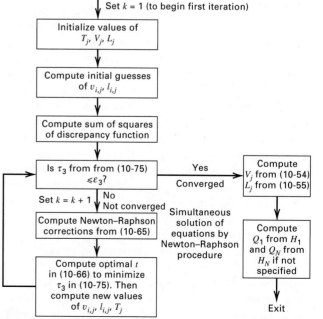

Figure 10.24 Algorithm for the Newton–Raphson method of Naphtali–Sandholm for all vapor–liquid separators.

techniques. If K-values are composition independent or can be approximated as such, one technique is to compute $x_{i,j}$ values and corresponding $y_{i,j}$ values from (10-12) and (10-2) as in the first iteration of the BP or SR method. A much cruder estimate is obtained by flashing the combined feeds at some average column pressure and a V/L ratio that approximates the ratio of overheads to bottoms products. The resulting mole-fraction compositions of the equilibrium vapor and liquid phases are assumed to hold for each stage. The second technique works surprisingly well, but the first technique is preferred for difficult cases. For either technique, the initial component flow rates are computed by using the $x_{i,j}$ and $y_{i,j}$ values to solve (10-56) and (10-57) for $l_{i,j}$ and $v_{i,j}$, respectively.

Based on initial guesses for all output variables, the sum of the squares of the discrepancy functions is computed and compared to the convergence criterion

$$\tau_3 = \sum_{j=1}^{N}\left\{(H_j)^2 + \sum_{i=1}^{C}[(M_{i,j})^2 + (E_{i,j})^2]\right\} \leq \epsilon_3 \quad (10\text{-}75)$$

In order that the values of all discrepancies be of the same order of magnitude, it is necessary to divide energy-balance functions H_j by a scale factor approximating the latent heat of vaporization (e.g., 1,000 Btu/lbmol). If the convergence criterion is computed from

$$\epsilon_3 = N(2C+1)\left(\sum_{j=1}^{N} F_j^2\right)10^{-10} \quad (10\text{-}76)$$

resulting converged values of the output variables will generally be accurate, on the average, to four or more significant figures. When employing (10-76), most problems are converged in 10 iterations or less.

Generally, the convergence criterion is far from satisfied during the first iteration when guessed values are assumed for the output variables. For each subsequent iteration, the Newton–Raphson corrections are computed from (10-65). These corrections can be added directly to the present values of the output variables to obtain a new set of values for the output variables. Alternatively, (10-66) can be employed where t is a nonnegative, scalar step factor. At each iteration, a single value of t is applied to all output variables. By permitting t to vary from, say, slightly greater than zero up to 2, it can serve to dampen or accelerate convergence, as appropriate. For each iteration, an optimal value of t is sought to minimize the sum of the squares given by (10-75). Generally, optimal values of t proceed from an initial value for the second iteration at between 0 and 1 to a value nearly equal to or slightly greater than 1 when the convergence criterion is almost satisfied. An efficient optimization procedure for finding t at each iteration is the Fibonacci search [25]. If no optimal value of t can be found within the designated range, t can be set to 1, or some smaller value, and the sum of squares can be allowed to increase. Generally, after several iterations, the sum of squares will decrease for every iteration.

If the application of (10-66) results in a negative component flow rate, Naphtali and Sandholm recommend the following mapping equation, which reduces the value of the unknown variable to a near-zero, but nonnegative, quantity.

$$X^{(k+1)} = X^{(k)} \exp\left[\frac{t\,\Delta X^{(k)}}{X^{(k)}}\right] \quad (10\text{-}77)$$

In addition, it is advisable to limit temperature corrections at each iteration.

The Naphtali–Sandholm SC method is readily extended to staged separators involving two liquid phases (e.g., extraction) and three coexisting phases (e.g., three-phase distillation), as shown by Block and Hegner [26], and to interlinked separators as shown by Hofeling and Seader [27].

EXAMPLE 10.8

A reboiled absorber is to be designed to separate the hydrocarbon vapor feed of Examples 10.2 and 10.4. Absorbent oil of the same composition as that of Example 10.4 will enter the top stage. Complete specifications are given in Figure 10.25. The 770 lbmol/h (349 kmol/h) of bottoms product corresponds to the amount of C_3 and heavier in the two feeds. Thus, the column is to be designed as a deethanizer. Calculate stage temperatures, interstage vapor and liquid flow rates and compositions, and reboiler duty by the Newton–Raphson method. Assume all stage efficiencies are 100%. Compare the degree of separation of the feed to that achieved by ordinary distillation in Example 10.2.

SOLUTION

A digital computer program for the method of Naphtali and Sandholm was used. The K-values and enthalpies were assumed independent of composition and were computed by linear interpolation

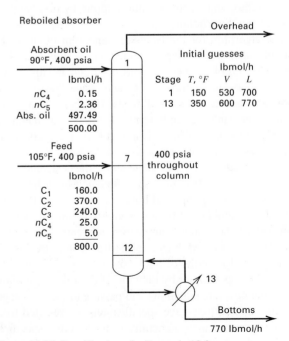

Figure 10.25 Specifications for Example 10.8.

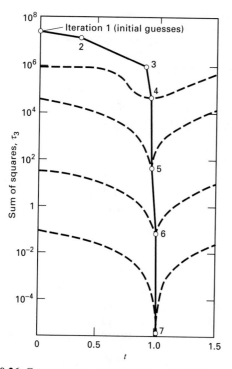

Figure 10.26 Convergence pattern for Example 10.8.

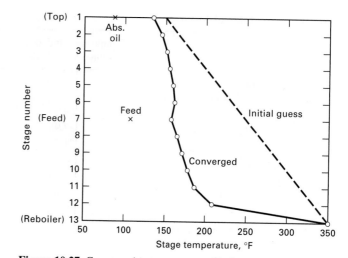

Figure 10.27 Converged temperature profile for Example 10.8.

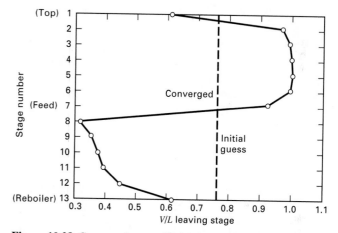

Figure 10.28 Converged vapor–liquid ratio profile for Example 10.8.

between tabular values given at $100°F$ increments from 0 to $400°F$ (-17.8 to $204.4°C$).

From (10-76), the convergence criterion is

$$\varepsilon_3 = 13[2(6) + 1](500 + 800)^2 10^{-10} = 2.856 \times 10^{-2}$$

Figure 10.26 shows the reduction in the sum of the squares of the 169 discrepancy functions from iteration to iteration. Seven iterations were required to satisfy the convergence criterion. The initial iteration was based on values of the unknown variables computed from interpolation of the initial guesses shown in Figure 10.25 together with a flash of the combined feeds at 400 psia (2.76 MPa) and a V/L ratio of 0.688 (530/770). Thus, for the first iteration, the following mole-fraction compositions were computed and were assumed to apply to every stage.

Species	y	x
C_1	0.2603	0.0286
C_2	0.4858	0.1462
C_3	0.2358	0.1494
nC_4	0.0153	0.0221
nC_5	0.0025	0.0078
Abs. Oil	0.0003	0.6459
	1.0000	1.0000

The corresponding sum of squares of the discrepancy functions, τ_3, of 2.865×10^7 was very large. Subsequent iterations employed the Newton–Raphson method. For iteration 2, the optimal value of t was found to be 0.34. However, this caused only a moderate reduction in the sum of squares. The optimal value of t increased to 0.904 for iteration 3, and the sum of squares was reduced by an order of magnitude. For the fourth and subsequent iterations, the effect of t on the sum of squares is included in Figure 10.26. Following iteration 4, the sum of squares was reduced by at least two orders of magnitude for each iteration. Also, the optimal value of t was rather

sharply defined and corresponded closely to a value of 1. An improvement of τ_3 was obtained for every iteration.

In Figures 10.27 and 10.28, converged temperature and V/L profiles are compared to the initially guessed profiles. In Figure 10.27, the converged temperatures are far from linear with respect to stage number. Above the feed stage, the temperature profile increases from the top down in a gradual and declining manner. The relatively cold feed causes a small temperature drop from stage 6 to stage 7. Temperature also increases from stage 7 to stage 13. A particularly dramatic increase occurs in moving from the bottom stage in the column to the reboiler, where heat is added. In Figure 10.28, the V/L profile is also far from linear with respect to stage number. Dramatic changes in this ratio occur at the top, middle, and bottom of the column.

Component flow-rate profiles for the two key components (ethane vapor and propane liquid) are shown in Figure 10.29. The initial guessed values are in very poor agreement with the converged values. The propane-liquid profile is quite regular except at the bottom, where a large decrease occurs because of vaporization in the reboiler. The ethane-vapor profile has large changes at the top, where entering oil absorbs appreciable ethane, and at the feed stage, where substantial ethane vapor is introduced.

Table 10.2 Product Compositions and Reboiler Duty for Example 10.8

	Composition-Independent Tabular Properties	Chao–Seader Correlation	Soave–Redlich–Kwong Equation
Overhead component flow rates, lbmol/h			
C_1	159.99	159.98	159.99
C_2	337.96	333.52	341.57
C_3	31.79	36.08	28.12
nC_4	0.04	0.06	0.04
nC_5	0.17	0.21	0.18
Abs. oil	0.05	0.15	0.10
	530.00	530.00	530.00
Bottoms component flow rates, lbmol/h			
C_1	0.01	0.02	0.01
C_2	32.04	36.4	28.43
C_3	208.21	203.92	211.88
nC_4	25.11	25.09	25.11
nC_5	7.19	7.15	7.18
Abs. oil	497.44	497.34	497.39
	770.00	770.00	770.00
Reboiler duty, Btu/h	11,350,000	10,980,000	15,640,000
Bottoms temperature, °F	346.4	338.5	380.8

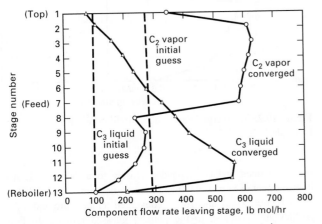

Figure 10.29 Converged flow rates for key components in Example 10.8.

Converged values for the reboiler duty and overhead and bottoms compositions are given in Table 10.2. Also included are converged results for two additional solutions that used the Chao–Seader and Soave–Redlich–Kwong equations for K-values and enthalpies in place of interpolation of composition-independent tabular properties. With the Soave–Redlich–Kwong equation, a somewhat sharper separation between the two key components is predicted. In addition, the Soave–Redlich–Kwong equation predicts a substantially higher bottoms temperature and a much larger reboiler duty. As discussed in Chapter 4, the effect of physical properties on equilibrium-stage calculations can be significant.

It is interesting to compare the separation achieved with the reboiled absorber of this example to the separation achieved by ordinary distillation of the same feed in Example 10.2 as shown in Figure 10.9. The latter separation technique results in a much sharper

separation and a much lower bottoms temperature and reboiler duty for the same number of stages. However, refrigeration is necessary for the overhead condenser, and the reflux flow rate is twice the absorbent oil flow rate. If the absorbent-oil flow rate for the reboiled absorber is made equal to the reflux flow rate, calculations give a separation almost as sharp as for ordinary distillation. However, the bottoms temperature and reboiler duty are increased to almost 600°F (315.6°C) and 60,000,000 Btu/h (63.3 GJ/h), respectively.

10.5 INSIDE-OUT METHOD

In the bubble-point (BP) and sum-rates (SR) methods described in Section 10.3 and the Newton–Raphson method described in Section 10.4, a large percentage of the computational effort is expended in calculating K-values, vapor-phase enthalpies, and liquid-phase enthalpies, particularly when rigorous thermodynamic-property models (e.g., Soave–Redlich–Kwong, Peng–Robinson, Wilson, NRTL, UNIQUAC) are utilized. As seen in Figures 10.30a and 10.30b, these property calculations are made at each iteration. Furthermore, at each iteration, derivatives are required of: (1) all three thermodynamic properties with respect to temperature and compositions of both phases, for the Newton–Raphson method; (2) K-values with respect to temperature for the BP method, unless Mullers method is used to compute bubble points; and (3) vapor and liquid enthalpies with respect to temperature for the SR method.

In 1974, Boston and Sullivan [28] presented an algorithm designed to significantly reduce the time spent in computing thermodynamic properties when designing steady-state,

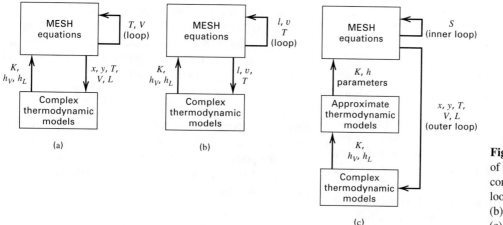

Figure 10.30 Incorporation of thermodynamic property correlations into interactive loops. (a) BP and SR methods. (b) Newton–Raphson method. (c) Inside-out method.

multicomponent separation operations. As shown in Figure 10.30c, two sets of thermodynamic-property models are employed: (1) a simple, approximate empirical set used frequently to converge inner loop calculations, and (2) the rigorous and complex set used less often in the outer loop. The MESH equations are always solved in the inner loop with the approximate set. The parameters in the empirical equations for the approximate set are updated in the outer loop by the rigorous equations, but only at infrequent intervals. A distinguishing feature of the Boston–Sullivan method is these inner and outer loops; hence the name *inside-out* for this class of methods. Another name, less frequently used, is *two-tier* methods.

Another difference that distinguishes the inside-out method, as shown in Figure 10.30, is the choice of iteration variables. For the Newton–Raphson method, the iteration variables are $l_{i,j}$, v_{ij}, T_j. For the BP and SR methods, the choice is $x_{i,j}$, $y_{i,j}$, T_j, L_j, and V_j. For the inside-out method, the iteration variables for the outer loop are the parameters in the approximate equations for the thermodynamic properties. The iteration variables for the inner loop are related to the stripping factors, $S_{i,j} = K_{i,j} V_j / L_j$.

In the original presentation of the inside-out method in 1974, the development and application of the method was restricted to hydrocarbon distillation (moderately nonideal systems) for the Case II variable specification in Table 5.4, but with multiple feeds, side streams, and intermediate heat exchangers. For these applications, the inside-side method was shown to be rapid and robust. Since 1974, the method has been extended and improved in a number of published articles [29, 30, 31, 32, 33, 34] and proprietary implementations in simulation computer programs. These extensions permit the inside-out method to be applied to almost any type of steady-state, multicomponent, multistage vapor–liquid separation operation. In the extensive implementation of the inside-out method by ASPEN technology in ASPEN PLUS, in computer programs called RADFRAC and MULTIFRAC, these applications include:

1. Absorption, stripping, reboiled absorption, reboiled stripping, extractive distillation, and azeotropic distillation

2. Three-phase (vapor-liquid-liquid) systems

3. Reactive systems

4. Highly nonideal systems requiring activity-coefficient models

5. Interlinked systems of separation units, including pumparounds, bypasses, and external heat exchangers

6. Narrow-boiling, wide-boiling, and dumbbell (mostly heavy and light components with little in between) feeds

7. Presence of free water

8. Wide variety of specifications other than Case II of Table 5.2 for the reflux ratio and product rates (e.g. product purities)

9. Use of Murphree-stage efficiencies

The inside-out method takes advantage of the following characteristics of the iterative calculations:

1. Component relative volatilities vary much less than component K-values.

2. Enthalpy of vaporization varies less than phase enthalpies.

3. Component stripping factors combine effects of temperature and liquid and vapor flows at each stage.

The inner loop of the inside-out method uses relative volatility, energy, and stripping factors to improve stability and reduce computing time. A widely used implementation of the inside-out method is that of Russell [31], which is described here together with further refinements suggested and tested by Jelinek [33].

MESH Equations

As with the BP, SR, and Newton–Raphson methods, the equilibrium-stage model of Figures 10.1 and 10.3 is again employed. The form of the equations is similar to the Newton–Raphson method in that component flow rates are utilized. However, in addition, the following inner-loop variables are defined:

$$\alpha_{i,j} = K_{i,j}/K_{b,j} \tag{10-78}$$

$$S_{b,j} = K_{b,j}V_j/L_j \tag{10-79}$$

$$R_{Lj} = 1 + U_j/L_j \tag{10-80}$$

$$R_{Vj} = 1 + W_j/V_j \tag{10-81}$$

where K_b is the K-value for a base or hypothetical reference component, $S_{b,j}$ is the stripping factor for the base component, R_{Lj} is a liquid-phase withdrawal factor, and R_{Vj} is a vapor-phase withdrawal factor. For stages without side streams, R_{Lj} and R_{Vj} reduce to 1. With the defined variables of (10-78) to (10-81), (10-54) to (10-57) still apply, but the MESH equations, (10-58) to (10-60), become as follows, where (10-83) results from the use of (10-80) to (10-82) to eliminate the variables in V and the side stream ratios s and S:

Phase Equilibria:

$$v_{i,j} = \alpha_{i,j}S_{b,j}l_{i,j}, \quad i = 1 \text{ to } C, \quad j = 1 \text{ to } N \tag{10-82}$$

Component Material Balance:

$$l_{i,j-1} - (R_{Lj} + \alpha_{i,j}S_{b,j}R_{Vj})l_{i,j} + (\alpha_{i,j+1}S_{b,j+1})l_{i,j+1}$$
$$= -f_{i,j}, \quad i = 1 \text{ to } C, \quad j = 1 \text{ to } N \tag{10-83}$$

Energy Balance:

$$H_j = h_{Lj}R_{Lj}L_j + h_{Vj}R_{Vj}V_j - h_{Lj-1}L_{j-1} - h_{Vj+1}V_{j+1}$$
$$- h_{Fj}F_j - Q_j = 0, \quad j = 1 \text{ to } N \tag{10-84}$$

where $S_{i,j} = \alpha_{i,j}S_{b,j}$.

In addition, discrepancy functions of the type shown in Table 10.1 for the Newton–Raphson method can be added to the MESH equations to permit any reasonable set of product specifications.

Rigorous and Complex Thermodynamic Property Models

The complex thermodynamic models referred to in Figure 10.30 can include any of the types of models discussed in Chapter 2, including those based on P–v–T equations of state (e.g., Soave–Redlich–Kwong and Peng–Robinson) and those based on free-energy models for predicting liquid-phase activity coefficients (e.g. Wilson, NRTL, and UNIQUAC). These models are used to generate parameters in the approximate thermodynamic-property models. In general, the rigorous property models are of the form:

$$K_{i,j} = K_{i,j}\{P_j, T_j, \mathbf{x}_j, \mathbf{y}_j\} \tag{10-85}$$

$$h_{Vj} = h_{Vj}\{P_j, T_j, \mathbf{y}_j\} \tag{10-86}$$

$$h_{Lj} = h_{Lj}\{P_j, T_j, \mathbf{x}_j\} \tag{10-87}$$

Approximate Thermodynamic Property Models

K-Values

The approximate models used in the inside-out method are designed to facilitate the calculation of stage temperatures and stripping factors. The approximate K-value model of Russell [31] and Jelinek [33], which differs only slightly from the model of Boston and Sullivan [28] and originates from a proposal in the classic textbook by Robinson and Gilliland [35], is (10-78) combined with

$$K_{b,j} = \exp(A_j - B_j/T_j) \tag{10-88}$$

Either a component in the feed or a hypothetical reference component can be selected as the base, b, component, with the latter preferred. For that case, the base component is determined from a vapor-composition weighting using the following relations:

$$K_{b,j} = \exp\left(\sum_i w_{i,j} \ln K_{i,j}\right) \tag{10-89}$$

where $w_{i,j}$ are weighting functions given by

$$w_{i,j} = \frac{y_{i,j}[\partial \ln K_{i,j}/\partial(1/T)]}{\sum_i y_{i,j}[\partial \ln K_{i,j}/\partial(1/T)]} \tag{10-90}$$

A unique K_b model and values of $\alpha_{i,j}$ in (10-78) are derived for each stage j from values of $K_{i,j}$ determined from the rigorous model. At the top stage, the base component will be close to one of the light components, while at the bottom stage, the base component will be close to a heavy component. The derivatives in (10-90) are obtained numerically or analytically from the rigorous model. To determine the values of A_j and B_j in (10-88), two temperatures must be selected for each stage. For example, the estimated or current temperatures of the two adjacent stages, $j - 1$ and $j + 1$, might be selected. Calling these two temperatures T_1 and T_2 and using (10-88) at each stage, b:

$$B = \frac{\ln(K_{b_{T_1}}/K_{b_{T_2}})}{\left(\dfrac{1}{T_2} - \dfrac{1}{T_1}\right)} \tag{10-91}$$

and

$$A = \ln K_{b_{T_1}} + B/T_1 \tag{10-92}$$

If highly nonideal-liquid solutions are involved, it is advisable to separate the rigorous K-value into two parts, as in (2-27). Thus,

$$K_i = \gamma_{iL}(\phi_{iL}/\bar{\phi}_{iV}) \tag{10-93}$$

Then, $(\phi_{iL}/\bar{\phi}_{iV})$ is used to determine K_b and, as proposed by Boston [30], values of γ_{iL} at each stage are fitted at a reference temperature T^*, to the liquid-phase mole fraction by the linear function

$$\gamma_{iL}^* = a_i + b_i x_i \tag{10-94}$$

to obtain the approximate estimates, γ_{iL}^*. Equation (10-83) is then modified by replacing $\alpha_{i,j}$ with $\alpha_{i,j}\gamma_{iL}^*$, where

$$\alpha_{i,j} = \frac{(\phi_{iL}/\bar{\phi}_{iV})_j}{K_{b,j}} \tag{10-95}$$

rather than the $\alpha_{i,j}$ given by (10-78).

Enthalpies

Boston and Sullivan [28] and Russell [31] employ the same approximate enthalpy models. Jelinek [33] does not use approximate enthalpy models, because the additional complexity involved in the use of two enthalpy models may not always be justified to the extent that the use of both approximate and rigorous K-value models is justified.

The basis for the enthalpy calculations is the same as for the rigorous equations discussed in Chapter 2. Thus, for either phase, from Table 2.6,

$$h = h_V^\circ + (h - h_V^\circ) = h_V^\circ + \Delta H \qquad (10\text{-}96)$$

where h_V° is the ideal-gas mixture enthalpy, as given by the polynomial equations, (2-35) and (2-36), based on the vapor-phase composition for h_V and the liquid-phase composition for h_L. The ΔH term is the enthalpy departure, $\Delta H_V = (h_V - h_V^\circ)$ for the vapor phase, which accounts for the effect of pressure, and $\Delta H_L = (h_L - h_V^\circ)$ for the liquid phase, which accounts for the enthalpy of vaporization and the effect of pressure on both liquid and vapor phases, as indicated in (2-57). Of particular importance is the enthalpy of vaporization, which dominates the ΔH_L term. The time-consuming parts of the enthalpy calculations are the two enthalpy-departure terms, which are complex when an equation of state is used. Therefore, in the approximate enthalpy equations, the rigorous enthalpy departures are replaced by the simple linear functions,

$$\Delta H_{Vj} = c_j - d_j(T_j - T^*) \qquad (10\text{-}97)$$

and

$$\Delta H_{Lj} = e_j - f_j(T_j - T^*) \qquad (10\text{-}98)$$

where the departures are modeled in terms of enthalpy per unit mass instead of per unit mole, and T^* is a reference temperature. The parameters c, d, e, and f are evaluated from the rigorous models at each iteration of the outer loop.

Inside-Out Algorithm

The inside-out algorithm of Russell [31] involves an initialization procedure, inner-loop iterations, and outer-loop iterations.

Initialization Procedure

Before inner- or outer-loop calculations can begin, it is necessary to provide reasonably good estimates of all stage values of $x_{i,j}$, $y_{i,j}$, T_j, V_j, and L_j. Boston and Sullivan [28] suggest the following procedure:

1. Specify the number of theoretical stages, conditions of all feeds, feed-stage locations, and column pressure profile.
2. Specify stage locations for each product withdrawal (including side streams) and for each heat exchanger.
3. Provide an additional specification for each product and each intermediate heat exchanger.

4. If not specified, estimate each product withdrawal rate, and estimate each value of V_j. Estimate values of L_j from the total material-balance equation, (10-6).
5. Estimate an initial temperature profile, T_j, by combining all feed streams (composite feed) and determining the bubble- and dew-point temperatures at the average column pressure. The dew-point temperature is taken as the top-stage temperature, T_1, whereas the bubble-point temperature is taken as the bottom-stage temperature, T_N. Intermediate-stage temperatures are estimated by linear interpolation. Reference temperatures T^* for use with (10-94), (10-97), and (10-98) are set equal to T_j.
6. Flash the composite feed isothermally at the average column pressure and average column temperature. The resulting vapor and liquid compositions, y_i and x_i, are the estimated compositions for each stage.
7. Using the initial estimates from Steps 1 through 7, use the selected complex thermodynamic-property correlation to determine values of the stagewise outside-loop K and h parameters A_j, B_j, $a_{i,j}$, $b_{i,j}$, c_j, d_j, e_j, f_j, $K_{b,j}$, and $\alpha_{i,j}$ of the approximate models.
8. Compute initial values of $S_{b,j}$, R_{Lj}, and R_{Vj} from (10-79), (10-80), and (10-81).

Inner-Loop Calculation Sequence

An iterative sequence of inner-loop calculations begins with a set of values for the outside-loop parameters listed in Step 7, obtained initially from the initialization procedure and later from outer-loop calculations, using results from the inner loop, as shown in Figure 10.30c.

9. Compute component liquid flow rates, $l_{i,j}$, from the set of N equations (10-83) for each of the C components by the tridiagonal-matrix algorithm.
10. Compute component vapor flow rates, v_{ij}, from (10-82).
11. Compute a revised set of total flow rates, V_j and L_j, from the component flow rates by (10-54) and (10-55), respectively.
12. To calculate a revised set of stage temperatures, T_j, as follows, compute a set of x_i values for each stage from (10-57), then a revised set of $K_{b,j}$ values from a combination of the bubble-point equation, (4-12), $\sum_i K_i x_i = 1$, with (10-78), which gives

$$K_{b,j} = 1 \Big/ \sum_{i=1}^{C} (\alpha_{i,j} x_{i,j}) \qquad (10\text{-}99)$$

From this new set of $K_{b,j}$ values, compute a new set of stage temperatures from the following rearrangement of (10-88):

$$T_j = \frac{B_j}{A_j - \ln K_{b,j}} \qquad (10\text{-}100)$$

At this point in the inner-loop iterative sequence, we have a revised set of values for $v_{i,j}$, $l_{i,j}$, and T_j, which satisfy the component material-balance and phase-equilibria equations for the estimated thermodynamic properties. However, these values do not satisfy the energy balance and specification equations unless the estimated base-component stripping factors and product-withdrawal rates are correct.

13. Select inner-loop iteration variables as

$$\ln S_{b,j} = \ln(K_{b,j}V_j/L_j) \qquad (10\text{-}101)$$

together with any other iteration variables. For a simple distillation column of the type shown in Figure 10.9, no other inner-loop iteration variables would be needed if the condenser and reboiler duties were specified. If the reflux ratio (L/D) and bottoms flow rate (B) are specified in place of the two duties, which is the more common situation, one adds, in place of the two (10-84) equations for H_1 and H_N, the following two specification equations from Table 10.1 in the form of discrepancy functions, D_1 and D_2:

$$D_1 = L_1 - (L/D)V_1 = 0 \qquad (10\text{-}102)$$
$$D_2 = L_N - B = 0 \qquad (10\text{-}103)$$

For each side stream, a side-stream-withdrawal factor is added as an inner-loop iteration variable, e.g., $\ln(U_j/L_j)$ and $\ln(W_j/V_j)$, together with a specification equation on purity or some other variable.

14. Compute enthalpies of all streams from (10-96) to (10-98).

15. Compute normalized discrepancies of H_j, D_1, D_2, etc., from the energy balances (10-84) and (10-102), (10-103), etc., except compute Q_1 from H_1 and Q_N from H_N where appropriate. A typical normalization is discussed in Section 10.4 for the Newton–Raphson method.

16. Compute the Jacobian of partial derivatives of H_j, D_1, D_2, etc., with respect to the iteration variables of (10-101), etc. This is done by successive perturbation of each iteration variable and recalculation of the discrepancies through Steps 9 to 15, numerically or by differentiation.

17. Compute corrections to the inner-loop iteration variables by a Newton–Raphson iteration of the type discussed for the SR method in Section 10.3 and the Newton–Raphson method in Section 10.4.

18. Compute new values of the iteration variables from the sum of the previous values and the corrections with (10-66), using damping if necessary to reduce the sum of the squares of the normalized discrepancies.

19. Check whether the sum of the squares is sufficiently small. If so, proceed to the outer-loop calculation procedure given next. If not, repeat Steps 15 to 18

using the latest values of the iteration variables. For any subsequent cycles through Steps 15 to 18, Russell [31] uses Broyden [36] updates to avoid reestimation of the Jacobian partial derivatives, whereas Jelinek [33] recommends the standard Newton–Raphson method of recalculating the partial derivatives for each inner-loop iteration.

20. Upon convergence of Steps 15 to 19, Steps 9 through 12 will have produced an improved set of primitive variables $x_{i,j}$, $v_{i,j}$, $l_{i,j}$, T_j, V_j, and L_j. From (10-56), corresponding values of $y_{i,j}$ can be computed. The values of these variables are not correct until the approximate thermodynamic properties are in agreement with the properties from the rigorous models. The primitive variables are input to the outer-loop calculations to bring the approximate and complex models into successively better agreement.

Outer-Loop Calculation Sequence

21. Using the values of the primitive variables from Step 20, compute relative volatilities and stream enthalpies from the complex thermodynamic models. If they are in close agreement with the previous values used to initiate a set of inner-loop iterations, both the outer-loop and inner-loop iterations are converged and the problem is solved. If not, proceed to Step 22.

22. Determine values of the stagewise outside-loop K and h parameters of the approximate models from the complex models as in initialization Step 7.

23. Compute values of $S_{b,j}$, R_{Lj}, and R_{Vj}, as in initialization Step 8.

24. Repeat the inner-loop calculation sequence of Steps 9 through 20.

Convergence of the inside-out method is not guaranteed. However, for most problems, the method is robust and rapid. Convergence can encounter difficulty because of poor initial estimates, resulting in negative or zero flow rates at certain locations in the column. To counteract this tendency, all component stripping factors are scaled with a scalar multiplier, S_b, sometimes called the base stripping factor, to give

$$S_{i,j} = S_b \alpha_{i,j} S_{b,j} \qquad (10\text{-}104)$$

The value of S_b is initially chosen to force the results of the initialization procedure to give a reasonable distribution of component flows throughout the column. Russell recommends that S_b be chosen only once, whereas Boston and Sullivan compute a new value S_b for each new set of $S_{b,j}$ values.

For highly nonideal-liquid mixtures, use of the inside-out method may become quite difficult. When that occurs, the Newton–Raphson method may be preferred. If the Newton–Raphson method also fails to converge, relaxation or continuation methods, described by Kister [37], are usually successful, but computing time may be an order of magnitude longer than that for similar problems converged successfully with the inside-out method.

EXAMPLE 10.9

For the conditions of the distillation column shown in Figure 10.7, obtain a converged solution by the inside-out method, using the SRK equation-of-state for thermodynamic properties.

SOLUTION

A computer solution was obtained with the equipment module TOWR (an inside-out method) of the CHEMCAD process simulation program of Chemstations, Inc. The only initial assumptions are a condenser outlet temperature of 65°F and a bottoms-product temperature of 165°F. The bubble-point temperature of the feed is computed as 123.5°F. In the initialization procedure, the constants A and B in (10-88), with T in °R, are determined from the SRK equation, with the following results:

Stage	T, °F	A	B	K_b
1	65	6.870	3708	0.8219
2	95	6.962	4031	0.7374
3	118	7.080	4356	0.6341
4	142	7.039	4466	0.6785
5	165	6.998	4576	0.7205

Values of the enthalpy coefficients c, d, e, and f in (10-97) and (10-98) are not tabulated here but are also computed for each stage, based on the initial temperature distribution.

In the inner-loop calculation sequence, component flow rates are computed from (10-83) by the tridiagonal-matrix method. The resulting bottoms-product flow rate deviates somewhat from the specified value of 50 lbmol/h. However, by modifying the component stripping factors with a base stripping factor, S_b, in (10-104) of 1.1863, the error in the bottoms flow rate is reduced to 0.73%.

The initial inside-loop error from the solution of the normalized energy-balance equations, (10-84), is found to be only 0.04624. This is reduced to 0.000401 after two iterations through the inner loop.

At this point in the inside-out method, the revised column profiles of temperature and phase compositions are used in the outer loop with the complex SRK thermodynamic models to compute updates of the approximate K and h constants. Only one inner-loop iteration is required to obtain satisfactory convergence of the energy equations. The K and h constants are again updated in the outer loop. After one inner-loop iteration, the approximate K and h

values are found to be sufficiently close to the SRK values that overall convergence is achieved. Thus, a total of only three outer-loop iterations and four inner-loop iterations are required.

To illustrate the efficiency of the inside-out method to converge this example, the results from each of the three outer-loop iterations are summarized in the following table:

Outer-Loop Iteration	Stage Temperatures, °F				
	T_1	T_2	T_3	T_4	T_5
Initial guess	65	—	—	—	165
1	82.36	118.14	146.79	172.66	193.20
2	83.58	119.50	147.98	172.57	192.53
3	83.67	119.54	147.95	172.43	192.43

Outer-Loop Iteration	Total Liquid Flows, lbmol/h				
	L_1	L_2	L_3	L_4	L_5
Specification	100	—	—	—	—
1	100.00	89.68	187.22	189.39	50.00
2	100.03	89.83	188.84	190.59	49.99
3	100.0	89.87	188.96	190.56	50.00

Outer-Loop Iteration	Component Flows in Bottoms Product, lbmol/h			
	C_3	nC_4	nC_5	L_5
1	0.687	12.045	37.268	50.000
2	0.947	12.341	36.697	49.985
3	0.955	12.363	36.683	50.001

From this table it is seen that the stage temperatures and total liquid flows are already close to the converged solution after only one outer-loop iteration. However, the composition of the bottoms product, specifically with respect to the lightest component, C_3, is not close to the converged solution until after two iterations. The inside-out method does not always converge so dramatically but is usually quite efficient, as shown in the following table.

Problem	Total Number of Inner Loops	Number of Outer-Loop Iterations
Exercise 10.11	7	6
Exercise 10.25	6	3
Exercise 10.37	17	9
Exercise 10.41	16	5

Computing times for each of these four exercises was less than 1 second on a PC with a Pentium 4 processor at 2.4 GHz.

SUMMARY

1. Rigorous methods are readily available for computer-solution of equilibrium-based models for multicomponent, multistage absorption, stripping, distillation, and liquid–liquid extraction.

2. The equilibrium-based model for a countercurrent-flow cascade provides for multiple feeds, vapor side streams, liquid side streams, and intermediate heat exchangers. Thus, the model can handle almost any type of column configuration.

3. The model equations include component material balances, total material balances, phase equilibria relations, and energy balances.

4. Some or all of the model equations can usually be grouped so as to obtain tridiagonal-matrix equations, for which an efficient solution algorithm is available.

5. Widely used methods for iteratively solving all of the model equations are the bubble-point (BP) method, the sum-rates (SR) method, the Newton–Raphson method, and the inside-out method.

6. The BP method is generally restricted to distillation problems involving narrow-boiling feed mixtures.

7. The SR method is generally restricted to absorption and stripping problems involving wide-boiling feed mixtures or in the ISR form to extraction problems.

8. The Newton–Raphson and inside-out methods are designed to solve any type of column configuration for any type of feed mixture. Because of its computational efficiency, the inside-out method is often the method of choice; however, it may fail to converge when highly nonideal-liquid mixtures are involved, in which case the slower Newton–Raphson method should be tried. Both methods permit considerable flexibility in specifications.

9. When both the Newton–Raphson and inside-out methods fail, resort can be made to much slower relaxation and continuation methods.

REFERENCES

1. WANG, J.C., and G.E. HENKE, *Hydrocarbon Processing* **45**(8), 155–163 (1966).

2. MYERS, A.L., and W.D. SEIDER, *Introduction to Chemical Engineering and Computer Calculations,* Prentice-Hall, Englewood Cliffs, NJ, 484–507 (1976).

3. LEWIS, W.K., and G.L. MATHESON, *Ind. Eng. Chem.* **24**, 496–498 (1932).

4. THIELE, E.W., and R.L. GEDDES, *Ind. Eng. Chem.* **25**, 290 (1933).

5. HOLLAND, C.D., *Multicomponent Distillation.* Prentice-Hall, Englewood Cliffs, NJ (1963).

6. AMUNDSON, N.R., and A.J. PONTINEN, *Ind. Eng. Chem.* **50**, 730–736 (1958).

7. FRIDAY, J.R., and B.D. SMITH, *AIChE J.* **10**, 698–707 (1964).

8. BOSTON, J.F., and S.L. SULLIVAN, JR., *Can. J. Chem. Eng.* **52**, 52–63 (1974).

9. BOSTON, J.F., and S.L. SULLIVAN, JR., *Can. J. Chem. Eng.* **50**, 663–669 (1972).

10. JOHANSON, P.J., and J.D. SEADER, *Stagewise Computations—Computer Programs for Chemical Engineering Education* (ed. by J. Christensen), Aztec Publishing, Austin, TX, pp. 349–389, A-16 (1972).

11. LAPIDUS, L., *Digital Computation for Chemical Engineers,* McGraw-Hill, New York, pp. 308–309 (1962).

12. ORBACH, O., and C.M. CROWE, *Can. J. Chem. Eng.* **49**, 509–513 (1971).

13. SCHEIBEL, E.G., *Ind. Eng. Chem.* **38**, 397–399 (1946).

14. SUJATA, A.D., *Hydrocarbon Processing* **40**(12), 137–140 (1961).

15. BURNINGHAM, D.W., and F.D. OTTO, *Hydrocarbon Processing* **46**(10), 163–170 (1967).

16. SHINOHARA, T., P.J. JOHANSEN, and J.D. SEADER, *Stagewise Computations—Computer Programs for Chemical Engineering Education,* J. Christensen, Ed., Aztec Publishing, Austin, TX, pp. 390–428, A-17 (1972).

17. TSUBOKA, T., and T. KATAYAMA, *J. Chem. Eng. Japan* **9**, 40–45 (1976).

18. HÁLA, E., I. WICHTERLE, J. POLAK, and T. BOUBLIK, *Vapor–Liquid Equilibrium Data at Normal Pressures,* Pergamon, Oxford, p. 308 (1968).

19. STEIB, V.H., *J. Prakt. Chem.* **4**, Reihe, Bd. 28, 252–280 (1965).

20. COHEN, G., and H. RENON, *Can. J. Chem. Eng.* **48**, 291–296 (1970).

21. GOLDSTEIN, R.P., and R.B. STANFIELD, *Ind. Eng. Chem., Process Des. Develop.* **9**, 78–84 (1970).

22. NAPHTALI, L.M., "The distillation column as a large system," paper presented at the AIChE 56th National Meeting, San Francisco, May 16–19, 1965.

23. NAPHTALI, L.M., and D.P. SANDHOLM, *AIChE J.* **17**, 148–153 (1971).

24. FREDENSLUND, A., J. GMEHLING, and P. RASMUSSEN, *Vapor–Liquid Equilibria Using UNIFAC, A Group Contribution Method.* Elsevier, Amsterdam (1977).

25. BEVERIDGE, G.S.G., and R.S. SCHECHTER, *Optimization: Theory and Practice,* McGraw-Hill, New York, pp. 180–189 (1970).

26. BLOCK, U., and B. HEGNER, *AIChE J.* **22**, 582–589 (1976).

27. HOFELING, B., and J.D. SEADER, *AIChE J.* **24**, 1131–1134 (1978).

28. BOSTON, J.F., and S.L. SULLIVAN, JR., *Can. J. Chem. Engr.* **52**, 52–63 (1974).

29. BOSTON, J.F., and H.I. BRITT, *Comput. Chem. Engng.* **2**, 109–122 (1978).

30. BOSTON, J.F., *ACS Symp. Ser.* No. 124, 135–151 (1980).

31. RUSSELL, R.A., *Chem. Eng.* **90**(20), 53–59 (1983).

32. TREVINO-LOZANO, R.A., T.P. KISALA, and J.F. BOSTON, *Comput. Chem. Engng.* **8**, 105–115 (1984).

33. JELINEK, J., *Comput. Chem. Engng.* **12**, 195–198 (1988).

34. VENKATARAMAN, S., W.K. CHAN, and J.F. BOSTON, *Chem. Eng. Prog.* **86**(8), 45–54 (1990).

35. ROBINSON, C.S., and E.R. GILLILAND, *Elements of Fractional Distillation,* 4th edition, pp. 232–236. McGraw-Hill, New York (1950).

36. BROYDEN, C.G., *Math Comp.* **19**, 577–593 (1965).

37. KISTER, H. Z., *Distillation Design,* McGraw-Hill, Inc., New York (1992).

EXERCISES

The exercises for this chapter are most conveniently divided into two groups: (1) those that can be solved manually, and (2) those that are best solved with computer implementation of the methods discussed in this chapter. The first group is referenced to section numbers of this chapter. The second group of problems follows the first group and is referenced to the type of separator. Computer implementations for use with the second group are found in the following widely available programs and simulators:

ASPEN PLUS of Aspen Technology

CHEMCAD of Chemstations

HYSYS of Aspen Technology

PRO/II of SimSci-Esscor

Section 10.1

10.1 Show mathematically that (10-6) is not independent of (10-1), (10-3), and (10-4).

10.2 Revise the MESH equations to account for entrainment, occlusion, and chemical reaction.

Section 10.2

10.3 Revise the MESH equations (10-1) to (10-6) to allow for pumparounds of the type shown in Figure 10.2 and discussed by Bannon and Marple [*Chem. Eng. Prog.* **74**(7), 41–45 (1978)] and Huber [*Hydrocarbon Processing* **56**(8), 121–125 (1977)]. Combine the equations to obtain modified M equations similar to (10.7). Can these equations still be partitioned in a series of C tridiagonal-matrix equations?

10.4 Use the Thomas algorithm to solve the following matrix equation for x_1, x_2, and x_3.

$$\begin{bmatrix} -160 & 200 & 0 \\ 50 & -350 & 180 \\ 0 & 150 & -230 \end{bmatrix} \cdot \begin{bmatrix} x_1 \\ x_2 \\ x_3 \end{bmatrix} = \begin{bmatrix} 0 \\ -50 \\ 0 \end{bmatrix}$$

10.5 Use the Thomas algorithm to solve the following tridiagonal matrix equation for the **x** vector.

$$\begin{bmatrix} -6 & 3 & 0 & 0 & 0 \\ 3 & -4.5 & 3 & 0 & 0 \\ 0 & 1.5 & -7.5 & 3 & 0 \\ 0 & 0 & 4.5 & -7.5 & 3 \\ 0 & 0 & 0 & 4.5 & -4.5 \end{bmatrix} \cdot \begin{bmatrix} x_1 \\ x_2 \\ x_3 \\ x_4 \\ x_5 \end{bmatrix} = \begin{bmatrix} 0 \\ 0 \\ 100 \\ 0 \\ 0 \end{bmatrix}$$

Section 10.3

10.6 On page 162 of their article, Wang and Henke [1] claim that their method of solving the tridiagonal matrix for the liquid-phase mole fractions does not involve subtraction of nearly equal quantities. Prove or disprove their statement.

10.7 Derive an equation similar to (10-7), but with $v_{i,j} = y_{i,j} V_j$ as the variables instead of the liquid-phase mole fractions. Can the resulting equations still be partitioned into a series of C tridiagonal-matrix equations?

10.8 In a computer program for the Wang–Henke bubble-point method, 10,100 storage locations are wastefully set aside for the four indexed coefficients of the tridiagonal-matrix solution of the component material balances for a 100-stage distillation column.

$$A_j x_{i,j-1} + B_j x_{i,j} + C_j x_{i,j+1} - D_j = 0$$

Determine the minimum number of storage locations required if the calculations are conducted in the most efficient manner.

10.9 Solve by the Newton–Raphson method the simultaneous, nonlinear equations

$$x_1^2 + x_2^2 = 17$$
$$(8x_1)^{1/3} + x_2^{1/2} = 4$$

for x_1 and x_2 to within ± 0.001. As initial guesses, assume

(a) $x_1 = 2, x_2 = 5$.
(b) $x_1 = 4, x_2 = 5$.
(c) $x_1 = 1, x_2 = 1$.
(d) $x_1 = 8, x_2 = 1$.

10.10 Solve by the Newton–Raphson method the simultaneous, nonlinear equations

$$\sin(\pi x_1 x_2) - \frac{x_2}{2} - x_1 = 0$$
$$\exp(2x_1)\left[1 - \frac{1}{4\pi}\right] + \exp(1)\left[\frac{1}{4\pi} - 1 - 2x_1 + x_2\right] = 0$$

for x_1 and x_2 to within ± 0.001. As initial guesses, assume

(a) $x_1 = 0.4, x_2 = 0.9$.
(b) $x_1 = 0.6, x_2 = 0.9$.
(c) $x_1 = 1.0, x_2 = 1.0$.

10.11 One thousand kilogram-moles per hour of a saturated liquid mixture of 60 mol% methanol, 20 mol% ethanol, and 20 mol% *n*-propanol is fed to the middle stage of a distillation column having three equilibrium stages, a total condenser, a partial reboiler, and an operating pressure of 1 atm. The distillate rate is 600 kmol/h, and the external reflux rate is 2,000 kmol/h of saturated liquid. Assuming that ideal solutions are formed such that K-values can be obtained from vapor pressures and assuming constant molar overflow such that the vapor rate leaving the reboiler and each stage is 2,600 kmol/h, calculate one iteration of the BP method up to and including a new set of T_j values. To initiate the iteration, assume a linear-temperature profile based on a distillate temperature equal to the normal boiling point of methanol and a bottoms temperature equal to the arithmetic average of the normal boiling points of the other two alcohols.

Section 10.4

10.12 Solve the following nine simultaneous linear equations, which have a block tridiagonal matrix structure, by the Thomas algorithm.

$$x_2 + 2x_3 + 2x_4 + x_6 = 7$$
$$x_1 + x_3 + x_4 + 3x_5 = 6$$
$$x_1 + x_2 + x_3 + x_5 + x_6 = 6$$
$$x_4 + 2x_5 + x_6 + 2x_7 + 2x_8 + x_9 = 11$$
$$x_4 + x_5 + 2x_6 + 3x_7 + x_9 = 8$$
$$x_5 + x_6 + x_7 + 2x_8 + x_9 = 8$$
$$x_1 + 2x_2 + x_3 + x_4 + x_5 + 2x_6 + 3x_7 + x_8 = 13$$
$$x_2 + 2x_3 + 2x_4 + x_5 + x_6 + x_7 + x_8 + 3x_9 = 14$$
$$x_3 + x_4 + 2x_5 + x_6 + 2x_7 + x_8 + x_9 = 10$$

10.13 Naphtali and Sandholm group the $N(2C + 1)$ equations by stage. Instead, group the equations by type (i.e., enthalpy balances, component balances, and equilibrium relations). Using a three-component, three-stage example, show whether the resulting matrix structure is still block tridiagonal.

10.14 Derivatives of properties are needed in the Naphtali–Sandholm SC method. For the Chao–Seader correlation, determine analytical derivatives for

$$\frac{\partial K_{i,j}}{\partial T_j}, \quad \frac{\partial K_{i,j}}{\partial v_{i,k}}, \quad \frac{\partial K_{i,j}}{\partial l_{i,k}}$$

10.15 A rigorous partial NR method for multicomponent, multistage vapor–liquid separations can be devised that is midway between the complexity of the BP/SR methods on the one hand and the NR methods on the other hand. The first major step in the procedure is to solve the modified M equations for the liquid-phase mole fractions by the usual tridiagonal-matrix algorithm. Then, in the second major step, new sets of stage temperatures and total vapor flow rates leaving a stage are computed simultaneously by a Newton–Raphson method. These two major steps are repeated until a sum-of-squares criterion is satisfied. For this partial NR method:

(a) Write the two indexed equations you would use to simultaneously solve for a new set of T_j and V_j.

(b) Write the truncated Taylor series expansions for the two indexed equations in the T_j and V_j unknowns, and derive complete expressions for all partial derivatives, except that derivatives of physical properties with respect to temperature can be left as such. These derivatives are subject to the choice of physical property correlations.

(c) Order the resulting linear equations and the new variables ΔT_j and ΔV_j into a Jacobian matrix that will permit a rapid and efficient solution.

10.16   Revise equations (10-58) to (10-60) to allow two interlinked columns of the type shown in Figure 10.31 to be solved simultaneously by the NR method. Does the matrix equation that results from the Newton–Raphson procedure still have a block tridiagonal structure?

10.17   In Equation (10-63), why is the variable order selected as v, T, l? What would be the consequence of changing the order to l, v, T? In Equation (10-64), why is the function order selected as H, M, E? What would be the consequence of changing the order to E, M, H?

Section 10.5

10.18   Suggest in detail a method for determining the scalar multiplier, S_b, in (10-104).

10.19   Suggest in detail an error function, similar to (10-75), that could be used to determine convergence of the inner-loop calculations for the inside-out method.

Distillation Problems

10.20   Calculate product compositions, stage temperatures, interstage vapor and liquid flow rates and compositions, reboiler duty, and condenser duty for the following distillation-column specifications.

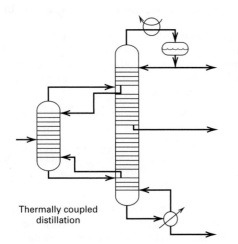

Thermally coupled distillation

Figure 10.31   Data for Exercise 10.15.

Feed (bubble-point liquid at 250 psia and 213.9°F):

Component	Lbmol/h
Ethane	3.0
Propane	20.0
n-Butane	37.0
n-Pentane	35.0
n-Hexane	5.0

Column pressure = 250 psia

Partial condenser and partial reboiler

Distillate rate = 23.0 lbmol/h

Reflux rate = 150.0 lbmol/hr

Number of equilibrium stages (exclusive of condenser and reboiler) = 15

Feed is sent to middle stage

For this system at 250 psia, K-values and enthalpies may be computed by the Soave–Redlich–Kwong equations.

10.21   Determine the optimal feed stage location for Exercise 10.20.

10.22   Revise Exercise 10.20 so as to withdraw a vapor side stream at a rate of 37.0 lbmol/h from the fourth stage from the bottom.

10.23   Revise Exercise 10.20 so as to provide an intercondenser on the fourth stage from the top with a duty of 200,000 Btu/h and an interreboiler on the fourth stage from the bottom with a duty of 300,000 Btu/h.

10.24   Using the Peng–Robinson equations for thermodynamic properties, calculate the product compositions, stage temperatures, interstage vapor and liquid flow rates and compositions, reboiler duty, and condenser duty for the following multiple-feed distillation column, which has 30 equilibrium stages exclusive of a partial condenser and a partial reboiler and operates at 250 psia.

Feeds (both bubble-point liquids at 250 psia):

	Pound-moles per Hour	
Component	**Feed 1 to Stage 15 from the Bottom**	**Feed 2 to Stage 6 from the Bottom**
Ethane	1.5	0.5
Propane	24.0	10.0
n-Butane	16.5	22.0
n-Pentane	7.5	14.5
n-Hexane	0.5	3.0

Distillate rate = 36.0 lbmol/hr.
Reflux rate = 150.0 lbmol/hr.

Determine whether the feed locations are optimal.

10.25   Use the Chao–Seader or Grayson–Streed correlation for thermodynamic properties to calculate product compositions, stage temperatures, interstage flow rates and compositions, reboiler duty, and condenser duty for the distillation specifications in Figure 10.32.

Compare your results with those given in the *Chemical Engineers' Handbook*, Sixth Edition, pp. 13–42 to 13–45. Why do the two solutions differ?

10.26   Solve Exercise 10.11 using the UNIFAC method for K-values and obtain the converged solution.

10.27   Calculate with the Peng–Robinson equations for thermodynamic properties, the product compositions, stage temperatures,

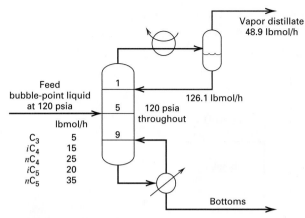

Figure 10.32 Data for Exercise 10.25.

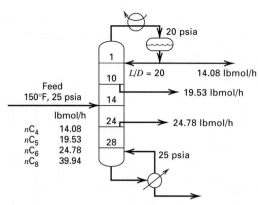

Figure 10.33 Data for Exercise 10.27.

interstage flow rates and compositions, reboiler duty, and condenser duty for the distillation specifications in Figure 10.33, which represent an attempt to obtain four nearly pure products from a single distillation operation. Reflux is a saturated liquid. Why is such a high reflux ratio required?

10.28 Repeat Exercise 10.25, but substitute the following specifications for the specifications of vapor distillate rate and reflux rate:

Recovery of nC_4 in distillate = 98%

Recovery of iC_5 in bottoms = 98%

If the calculations fail to converge, the number of stages may be less than the minimum value. If so, increase the number of stages, revise the feed location, and repeat until convergence is achieved.

10.29 A saturated liquid feed at 125 psia contains 200 lbmol/h of 5 mol% iC_4, 20 mol% nC_4, 35 mol% iC_5, and 40 mol% nC_5. This feed is to be distilled at 125 psia with a column equipped with a total condenser and partial reboiler. The distillate is to contain 95% of the nC_4 in the feed, and the bottoms is to contain 95% of the iC_5 in the feed. Use the SRK equation for thermodynamic properties to determine a suitable design. Twice the minimum number of equilibrium stages, as estimated by the Fenske equation in Chapter 9, should provide a reasonable number of equilibrium stages.

10.30 A depropanizer distillation column is designed to operate at an average total pressure of 315 psia for separating a feed into distillate and bottoms with the flow rates shown next:

	lbmol/h		
	Feed	**Distillate**	**Bottoms**
Methane (C_1)	26	26	
Ethane (C_2)	9	9	
Propane (C_3)	25	24.6	0.4
n-Butane (C_4)	17	0.3	16.7
n-Pentane (C_5)	11		11
n-Hexane (C_6)	12		12
Totals	100	59.9	40.1

The thermal condition of the feed is such that it is 66 mol% vapor at tower pressure. Steam at 315 psia and cooling water at 65°F are available for the reboiler and condenser. The total pressure drop across the column may be taken to be 2 psi as a first approximation.

(a) Should a total condenser be used for this column?

(b) What are the feed temperature, K-values, and relative volatilities (with reference to C_3) at the feed temperature and pressure?

(c) If the reflux ratio is 1.3 times the minimum reflux, what is the actual reflux ratio? How many theoretical plates are needed in the rectifying and stripping sections?

(d) Compute the separation of species. How will the separation differ, if a reflux ratio of 1.5, 15 theoretical plates, and feed at the 9th plate are chosen.

(e) For part (c), compute the temperature and concentrations on each stage. What is the effect of feed plate location? How will the results differ if a reflux ratio of 1.5 and 15 theoretical plates are used?

10.31 Toluene is to be separated from biphenyl by ordinary distillation. The specifications for the separation are as follows:

	lbmol/h		
	Feed	**Distillate**	**Bottoms**
Benzene	3.4		
Toluene	84.6		2.1
Biphenyl	5.1	1.0	

Temperature = 264°F; Pressure = 37.1 psia for the feed
Reflux ratio = 1.3 times minimum reflux with total condenser
Top pressure = 36 psia; bottom pressure = 38.2 psia

(a) Determine the actual reflux ratio and the number of theoretical trays in the rectifying and stripping sections.

(b) For a D/F ratio of $(3.4 + 82.5 + 1.0)/93.1$, compute the separation of species. Compare the results to the preceding specifications.

(c) If the separation of species computed in part (b) is not sufficiently close to the specified split, adjust the reflux ratio to achieve the specified toluene flow in the bottoms.

10.32 The following stream at 100°F and 480 psia is to be separated by two ordinary distillation columns into the indicated products.

	lbmol/h			
Species	**Feed**	**Product 1**	**Product 2**	**Product 3**
H_2	1.5	1.5		
CH_4	19.3	19.2	0.1	
C_6H_6 (benzene)	262.8	1.3	258.1	3.4
C_7H_8 (toluene)	84.7		0.1	84.6
$C_{12}H_{10}$ (biphenyl)	5.1			5.1

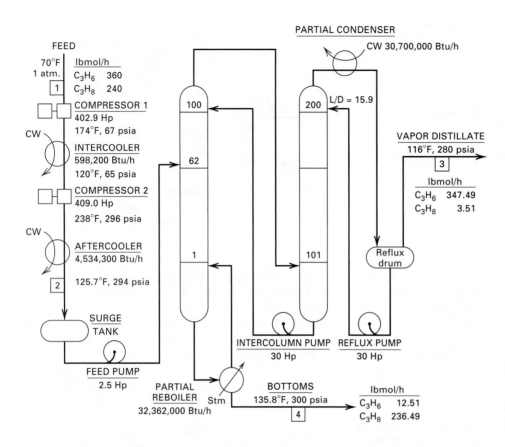

PARTIAL CONDENSER

CW 30,700,000 Btu/h

FEED

70°F
1 atm.

	lbmol/h
C_3H_6	360
C_3H_8	240

1

COMPRESSOR 1
402.9 Hp
174°F, 67 psia

CW

INTERCOOLER
598,200 Btu/h
120°F, 65 psia

COMPRESSOR 2
409.0 Hp
238°F, 296 psia

CW

AFTERCOOLER
4,534,300 Btu/h

125.7°F, 294 psia

2

SURGE
TANK

FEED PUMP
2.5 Hp

PARTIAL
REBOILER   Stm
32,362,000 Btu/h

100

62

1

200   L/D = 15.9

101

INTERCOLUMN PUMP   REFLUX PUMP
30 Hp   30 Hp

BOTTOMS
135.8°F, 300 psia

4

Reflux
drum

VAPOR DISTILLATE
116°F, 280 psia

3

	lbmol/h
C_3H_6	347.49
C_3H_8	3.51

	lbmol/h
C_3H_6	12.51
C_3H_8	236.49

Figure 10.34 Data for
Exercise 10.33.

Two different distillation sequences are to be examined. In the first sequence, CH_4 is removed in the first column. In the second sequence, toluene is removed in the first column. Compute the two sequences in the following manner: Estimate the actual reflux ratio and theoretical-tray requirements for both sequences. Specify a reflux ratio equal to 1.3 times the minimum. Adjust isobaric column pressures so as to obtain distillate temperatures of about 130°F; however, no column pressure should be less than 20 psia. Specify total condensers, except that a partial condenser should be used when methane is taken overhead.

10.33 A process for the separation of a propylene–propane mixture to produce 99 mol% propylene and 95 mol% propane is shown in Figure 10.34. Because of the high product purities and the low relative volatility, 200 stages may be required. Assuming a tray efficiency of 100% and tray spacing of 24 inches, this will necessitate the two columns shown in series, because a single tower would be too tall. Assume a vapor distillate pressure of 280 psia, a pressure drop of 0.1 psi per tray, and a 2-psi drop through the condenser. The stage numbers and reflux ratio shown are only approximate. Determine the necessary reflux ratio for the stage numbers shown. Pay close attention to the determination of the proper feedstage location so as to avoid pinch or near-pinch conditions wherein several adjacent trays may not be accomplishing anything.

10.34 So-called stabilizers are distillation columns that are often used in the petroleum industry to perform relatively easy separations between light components and considerably heavier components when one or two single-stage flashes are inadequate. An example of a stabilizer is shown in Figure 10.35 for the separation of H_2, methane, and ethane from benzene, toluene, and xylenes. Such

columns can be difficult to calculate because a purity specification for the vapor distillate cannot be readily determined. Instead, it is more likely that the designer will be told to provide a column with 20 to 30 trays and a water-cooled partial condenser to provide 100°F reflux at a rate that will provide sufficient boilup at the bottom of the column to meet the purity specification there. It is desired to more accurately design the stabilizer column. The number of theoretical stages shown are just a first approximation and may

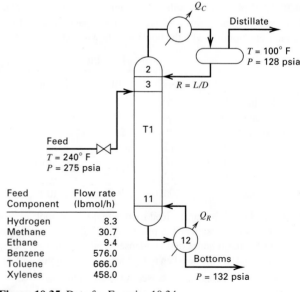

Q_C

Distillate

1

$T = 100°$ F
$P = 128$ psia

2
3    $R = L/D$

T1

Feed
$T = 240°$ F
$P = 275$ psia

11

Q_R

12

Bottoms
$P = 132$ psia

Feed Component	Flow rate (lbmol/h)
Hydrogen	8.3
Methane	30.7
Ethane	9.4
Benzene	576.0
Toluene	666.0
Xylenes	458.0

Figure 10.35 Data for Exercise 10.34.

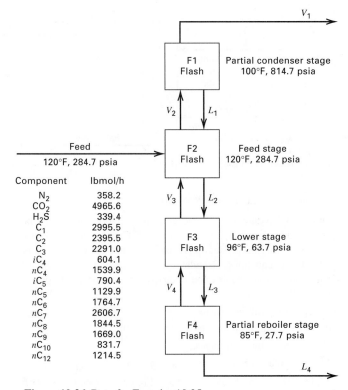

Figure 10.36 Data for Exercise 10.35.

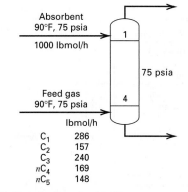

Figure 10.37 Data for Exercise 10.37.

be varied. Strive to achieve a bottoms product with no more than 0.05 mol% methane plus ethane and a vapor distillate temperature of about $100°F$.

These specifications may be achieved by varying the distillate rate and the reflux ratio. Reasonable initial estimates for these two quantities are 49.4 lbmol/h and 2. Assume a tray efficiency of 70%.

10.35 A multiple recycle-loop problem formulated by Cavett[1] and shown in Figure 10.36 has been used extensively to test tearing, sequencing, and convergence procedures. The flowsheet is the equivalent of a four-theoretical-stage, near-isothermal distillation (rather than the conventional near-isobaric type), for which a patent by Gunther[2] exists. The flowsheet does not include necessary mixers, compressors, pumps, valves, and heat exchangers to make it a practical system. For the specifications shown on the drawing, determine the component flow rates for all streams in the process.

Absorber and Stripper Problems

10.36 An absorber is to be designed for a pressure of 75 psia to handle 2,000 lbmol/h of gas at $60°F$ having the following composition.

Component	Mole Fraction
Methane	0.830
Ethane	0.084
Propane	0.048
n-Butane	0.026
n-Pentane	0.012

[1] R. H. Cavett, *Proc. Am. Petrol. Inst.* **43,** 57 (1963).

[2] A. Gunther, U.S. Patent 3,575,077 (April 13, 1971).

The absorbent is an oil, which can be treated as a pure component having a molecular weight of 161. Calculate product rates and compositions, stage temperatures, and interstage vapor and liquid flow rate and compositions for the following conditions.

	Number of Equilibrium Stages	Entering Absorbent Flow Rate lbmol/h	Entering Absorbent Temperature, °F
(a)	6	500	90
(b)	12	500	90
(c)	6	1,000	90
(d)	6	500	60

10.37 Calculate product rates and compositions, stage temperatures, and interstage vapor and liquid flow rates and compositions for an absorber having four equilibrium stages with the specifications in Figure 10.37. Assume the oil is nC_{10}.

10.38 In Example 10.4, temperatures of the gas and oil, as they pass through the absorber, increase substantially. This limits the extent of absorption. Repeat the calculations with a heat exchanger that removes 500,000 Btu/h from:

(a) Stage 2.

(b) Stage 3.

(c) Stage 4.

(d) Stage 5.

How effective is the intercooler? Which stage is the preferred location for the intercooler? Should the duty of the intercooler be increased or decreased assuming that the minimum-stage temperature is $100°F$ using cooling water? Assume the absorber oil is nC_{12}.

10.39 Calculate product rates and compositions, stage temperatures, and interstage vapor and liquid flow rates and compositions for the absorber shown in Figure 10.38.

10.40 Determine product compositions, stage temperatures, interstage flow rates and compositions, and reboiler duty for the reboiled absorber shown in Figure 10.39. Repeat the calculations without the interreboiler and compare both sets of results. Is the interreboiler worthwhile? Should an intercooler in the top section of the column be considered?

10.41 Calculate the product compositions, stage temperatures, interstage flow rates and compositions, and reboiler duty for the reboiled stripper shown in Figure 10.40.

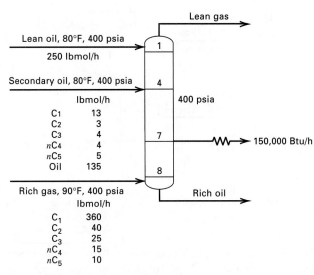

Figure 10.38 Data for Exercise 10.39.

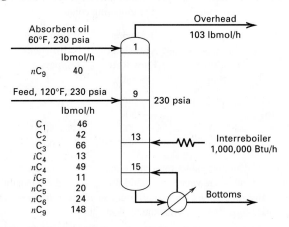

Figure 10.39 Data for Exercise 10.40.

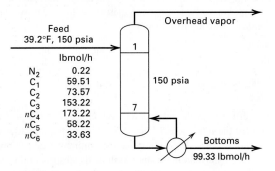

Figure 10.40 Data for Exercise 10.41.

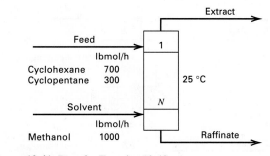

Figure 10.41 Data for Exercise 10.42.

Liquid–Liquid Extraction Problems

10.42 A mixture of cyclohexane and cyclopentane is to be separated by liquid–liquid extraction at 25°C with methanol. Phase equilibria for this system may be predicted by the NRTL or UNIQUAC equations. Calculate product rates and compositions and interstage flow rates and compositions for the conditions in Figure 10.41 with:

(a) $N = 1$ equilibrium stage.

(b) $N = 2$ equilibrium stages.

(c) $N = 5$ equilibrium stages.

(d) $N = 10$ equilibrium stages.

10.43 The liquid–liquid extractor in Figure 8.1 operates at 100°F and a nominal pressure of 15 psia. For the feed and solvent flows shown, determine the number of equilibrium stages to extract 99.5% of the acetic acid, using the NRTL equation for activity coefficients. The NRTL constants may be taken as follows:

1 = ethyl acetate

2 = water

3 = acetic acid

I	J	B_{IJ}	B_{JI}	α_{IJ}
1	2	166.36	1190.1	0.2
1	3	643.30	−702.57	0.2
2	3	−302.63	−1.683	0.2

Compare the computed compositions of the raffinate and extract products to those of Figure 8.1.

Chapter 11

Enhanced Distillation
and Supercritical Extraction

When two or more components differ in boiling point by less than approximately 50°C and form a nonideal-liquid solution, the relative volatility may be below 1.10. Then, separation by ordinary distillation may be uneconomical and if an azeotrope forms even impossible. In that event, the following separation techniques, referred to as *enhanced distillation* by Stichlmair, Fair, and Bravo [1], should be explored:

1. *Extractive Distillation:* A method that uses a large amount of a relatively high-boiling solvent to alter the liquid-phase activity coefficients of the mixture, so that the relative volatility of the key components becomes more favorable. The solvent enters the column above the feed entry and a few trays below the top, and exits from the bottom of the column without causing an azeotrope to be formed. If the feed to the column is an azeotrope, the solvent breaks it. Also, the solvent may reverse volatilities.

2. *Salt Distillation:* A variation of extractive distillation in which the relative volatility of the key components is altered by dissolving a soluble, ionic salt in the top reflux. Because the salt is nonvolatile, it stays in the liquid phase as it passes down the column.

3. *Pressure-Swing Distillation:* A method for separating a pressure-sensitive azeotrope that utilizes two columns operated in sequence at two different pressures.

4. *Homogeneous Azeotropic Distillation:* A method of separating a mixture by adding an entrainer that forms a homogeneous minimum- or maximum-boiling azeotrope with one or more feed components. The entrainer is added near the top of the column, to the feed, or near the bottom of the column, depending upon whether the azeotrope is removed from the top or bottom.

5. *Heterogeneous Azeotropic Distillation:* A more useful azeotropic-distillation method in which a minimum-boiling heterogeneous azeotrope is formed by the entrainer. The azeotrope splits into two liquid phases in the overhead condensing system. One liquid phase is sent back to the column as reflux, while the other liquid phase is sent to another separation step or is a product.

6. *Reactive Distillation:* A method that adds a separating agent to react selectively and reversibly with one or more of the constituents of the feed. The reaction product is subsequently distilled from the nonreacting components. The reaction is then reversed to recover the separating agent and the other reacting components. Reactive distillation also refers to the case where a chemical reaction and multistage distillation are conducted simultaneously in the same apparatus to produce other chemicals. This combined operation, sometimes referred to as *catalytic distillation* if a catalyst is used, is especially suited to chemical reactions limited by equilibrium constraints, since one (or more) of the products of the reaction is (are) continuously separated from the reactants.

For ordinary distillation of multicomponent mixtures, the determination of feasible distillation sequences, the design of the columns in the sequence by rigorous methods described in Chapters 10 and 12, and the optimization of the column operating conditions are tedious, but are relatively straightforward. In contrast, determining and optimizing feasible, enhanced-distillation sequences is a considerably more difficult task. In particular, rigorous calculations of enhanced distillation frequently fail because of liquid-solution nonidealities and/or the difficulty of specifying feasible separations. To significantly reduce the chances of failure, especially for ternary systems, graphical techniques, described by Partin [2] and developed largely by Doherty and co-workers, and by Stichlmair and co-workers, as referenced later, provide valuable guidance for the development of feasible, enhanced-distillation sequences prior to making rigorous calculations. This chapter presents an introduction to the principles of these graphical methods and applies them to enhanced distillation. Doherty and Malone [94], Stichlmair and Fair [95], and Siirola and Barnicki [96] give more detailed treatments of enhanced distillation.

Also included is a discussion of supercritical extraction, which differs considerably from conventional liquid–liquid extraction because of strong nonideal effects, and also requires considerable care in the development of an optimal system. The principles and techniques in this chapter are largely restricted to ternary systems; enhanced distillation and supercritical extraction are most commonly applied to

such systems because the expense of these operations often requires that a multicomponent mixture first be reduced, by distillation or other means, to a binary or ternary system.

11.0   INSTRUCTIONAL OBJECTIVES

After completing this chapter, you should be able to:

- Explain how enhanced-distillation methods work and how they differ from ordinary distillation.
- Explain how supercritical-fluid extraction differs from liquid–liquid extraction.
- Describe what residue-curve maps and distillation-curve maps represent on triangular diagrams for a ternary system.
- Explain how residue-curve maps and distillation-curve maps are constructed and under what conditions they are identical.
- Explain how residue-curve maps limit feasible product-composition regions in ordinary distillation and enhanced, ternary distillation.
- List requirements for an effective solvent in extractive distillation.
- Calculate, with a simulation program, a separation by extractive distillation.
- Explain how salt distillation differs from extractive distillation.
- Explain how pressure-swing distillation is used to separate a binary azeotropic mixture.
- Explain why it is difficult to find an entrainer that will permit use of homogeneous azeotropic distillation.
- Calculate, with a simulation program and a residue-curve map, a separation by homogeneous azeotropic distillation.
- List characteristics of an entrainer for heterogeneous azeotropic distillation.
- Calculate, with a simulation program, but using a residue-curve map and a bimodal curve, a separation by heterogeneous azeotropic distillation.
- List conditions necessary for carrying out reactive distillation.
- Calculate, with a simulation program, a separation by reactive distillation.
- Explain why enormous changes in properties can occur in the critical region.
- Calculate, with a simulation program, a separation by supercritical-fluid extraction.

11.1   USE OF TRIANGULAR GRAPHS

When a binary mixture at a given pressure is separated by continuous distillation in equilibrium stages, all possible equilibrium compositions are uniquely located on a vapor–liquid (y–x) equilibrium curve. Figure 11.1 shows typical isobaric vapor–liquid equilibrium curves in terms of the mole

fractions of the lowest-boiling component, A. In Figure 11.1a, possible compositions of the distillate and bottoms cover the entire range from pure B to pure A for a *zeotropic* (nonazeotropic) system. Temperatures, although not shown on the isobaric equilibrium curve, range from the boiling point of A to the boiling point of B. As the liquid and vapor

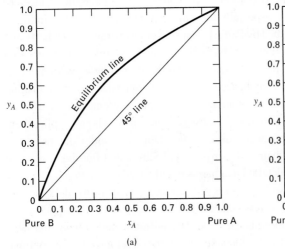

(a)

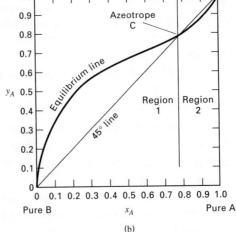

(b)

Figure 11.1 Vapor–liquid equilibria for binary systems. (a) Zeotropic system. (b) Azeotropic system.

compositions change from pure B to pure A, the temperature decreases.

In Figure 11.1b, a minimum-boiling azeotrope is formed at C and divides the plot into two regions. For distillation in Region 1, distillate and bottoms compositions can only vary from pure B to azeotrope C, and in Region 2, only from pure A to azeotrope C. For Region 1, as the composition changes from pure B to azeotrope C, the temperature decreases, as shown for example in Figure 4.6, where B is isopropyl alcohol, A is isopropyl ether, and the minimum-boiling azeotrope occurs at 78 mol% isopropyl ether and 66°C at 1 atm. In Region 2, the temperature also decreases as the composition changes from pure A to azeotrope C. Thus, a single distillation column operating at 1 atm cannot separate the mixture into two nearly pure products. Depending upon whether the feed composition lies in Region 1 or 2, the column, at best, can only produce a distillate of azeotrope C and a bottoms of either pure B or pure A. However, all possible equilibrium compositions still lie on the equilibrium curve. These results are consistent with the Gibbs phase rule, as discussed in Chapter 4. From (4-1), for two components and two phases, that rule gives two degrees of freedom. Thus, if the pressure and temperature are fixed, the equilibrium vapor and liquid compositions are fixed. However, as shown in Figure 11.2, for the case of an azeotrope-forming binary mixture, two feasible solutions exist within a certain temperature range. The particular solution observed depends on the overall composition of the two phases.

In the distillation of a ternary system, possible equilibrium compositions do not lie uniquely on a single, isobaric equilibrium curve because the Gibbs phase rule gives an additional degree of freedom. The other compositions are determined only if the temperature, pressure, and composition of one component in one phase are fixed.

As discussed in Chapters 4 and 8, the composition of a ternary mixture can be represented on a triangular diagram, either equilateral or right, where the three apexes of the triangle represent the pure components. Although Stichlmair [3] shows that vapor–liquid phase equilibria at a fixed pressure can be plotted by letting the triangular grid represent the liquid phase and superimposing lines of constant equilibrium-vapor composition for two of the three components, this representation is seldom used. It is more useful, when developing a feasible-separation process, to plot only equilibrium-liquid-phase compositions on the triangular diagram. Typical plots of this type at 1 atm, for three different ternary systems, are shown in Figure 11.3, where compositions are in mole fractions. Each curve in each diagram is the locus of possible equilibrium liquid-phase compositions that occur during distillation of a mixture, starting from any point on the curve. The boiling points of the three components and their binary and/or ternary azeotropes are included on the diagrams. The zeotropic alcohol system of Figure 11.3a does not form any azeotropes. If a mixture of these three alcohols is distilled, there is only one distillation region, similar to the binary system of Figure 11.1a. Accordingly, the distillate product can be nearly pure methanol (A) or the bottoms product can be nearly pure 1-propanol (C). However, nearly pure ethanol (B), the intermediate-boiling component, cannot be produced either as a distillate or bottoms. To separate this ternary mixture into the three components, a sequence of two ordinary-distillation columns is used, as shown in Figure 11.4, where the feed, distillate, and bottoms product compositions must lie on a straight (total material-balance) line within the triangular diagram. Thus, in the so-called *direct sequence* of Figure 11.4a, the feed, F, is first separated into distillate A and a bottoms of B and C; then B is separated from C in the second column. In the *indirect sequence* of Figure 11.4b, a distillate of A and B and a bottoms of C is produced in the first column, followed by the separation of A from B in the second column.

When a ternary mixture forms an azeotrope, the possible products from a single ordinary distillation column depend on the feed composition, as for a binary mixture. However, unlike the case of the binary mixture, where two distillation regions, shown in Figure 11.1b, are simple and well defined, the determination of possible distillation regions for azeotrope-forming ternary mixtures is complex. Consider first the example of Figure 11.3b, for a mixture of acetone (A), methanol (B), and ethanol (C), which are in the order of increasing boiling point. The only azeotrope formed at 1 atm is a minimum-boiling binary azeotrope, at 55.7°C, of the two lower-boiling components, acetone and methanol. The azeotrope contains 78.4 mol% acetone. For this type of system, as will be shown later, no distillation boundaries for the ternary mixture exist, even though an azeotrope is present. Thus, a feed composition located within the triangular diagram can be separated into two binary products, consistent with the straight (total material-balance) line. That is, ternary distillate or bottoms products can be avoided if the column split is properly selected. For example, the following five feed compositions can all produce, at a high reflux ratio and for a

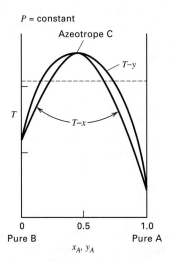

Figure 11.2 Multiple equilibrium solutions for an azeotropic system.

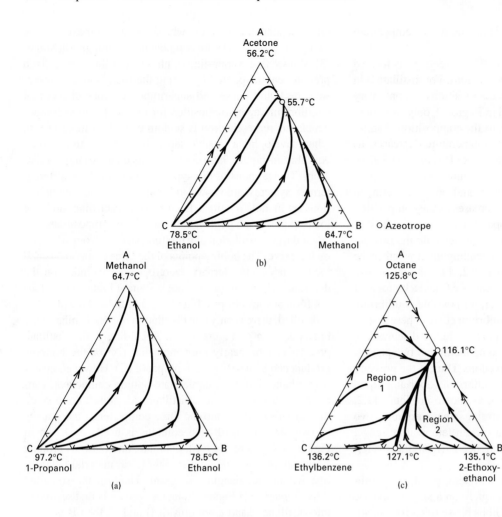

Figure 11.3 Distillation curves for liquid-phase compositions of ternary systems at 1 atm. (a) Mixture not forming an azeotrope. (b) Mixture forming one minimum-boiling azeotrope. (c) Mixture forming two minimum-boiling azeotropes.

large number of stages, a distillate of the minimum-boiling azeotrope of acetone and methanol, and a bottoms product containing methanol and ethanol. That is, little or no ethanol will be in the distillate and little or no acetone in the bottoms.

Case	Feed:		Distillate:		Bottoms:	
	$x_{acetone}$	$x_{methanol}$	$x_{acetone}$	$x_{methanol}$	$x_{acetone}$	$x_{methanol}$
1	0.1667	0.1667	0.7842	0.2158	0.0000	0.1534
2	0.1250	0.3750	0.7837	0.2163	0.0000	0.4051
3	0.2500	0.2500	0.7837	0.2163	0.0000	0.2658
4	0.3750	0.1250	0.7837	0.2163	0.0000	0.0412
5	0.3333	0.3333	0.7837	0.2163	0.0000	0.4200

Alternatively, the column split can be selected to obtain a bottoms of nearly pure ethanol and a distillate of acetone and methanol. For either split, the straight, total material-balance line passing through the feed point can extend to the sides of the triangle.

Next, consider the more complex case of the ternary mixture of *n*-octane (A), 2-ethoxyethanol (B), and ethylbenzene (C), shown in Figure 11.3c. A and B form a minimum-boiling binary azeotrope at 116.1°C, and B and C do the same at 127.1°C. A triangular diagram for this type of system is separated by a *distillation boundary* (shown as a bold curved line) into regions 1 and 2. A material-balance line connecting the feed to the distillate and bottoms cannot cross this distillation boundary, thus restricting the possible products of ordinary distillation of the ternary feed mixture. For example, a mixture with a feed composition inside Region 2 cannot produce a bottoms of ethylbenzene, the highest-boiling component in the mixture. It can be distilled to produce a distillate of the A–B azeotrope and a bottoms of a

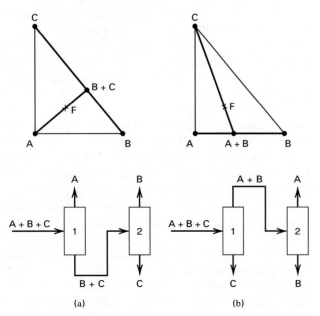

Figure 11.4 Distillation sequences for ternary zeotropic mixtures. (a) Direct sequence. (b) Indirect sequence.

mixture of B and C, or a bottoms of B and a distillate of all three components. If the feed composition lies in Region 1 of Figure 11.3c, ordinary distillation can produce the A–B azeotrope and a bottoms of a mixture of A and C, or a bottoms of C and a distillate of a mixture of A and B. Thus, each region produces unique products.

To further illustrate the restriction in product compositions caused by a distillation boundary, consider ordinary distillation with a feed mixture of 15 mol% A, 70 mol% B, and 15 mol% C. If the mixture is that in Figure 11.3a or b, a bottoms product of nearly pure C, the highest-boiling component, is obtained with a feed split corresponding to a distillate-to-bottoms ratio of 85/15. If, however, the mixture is that in Figure 11.3c, the same feed split ratio results in a bottoms of nearly pure B, the second-highest-boiling component.

Thus, products of a ternary mixture cannot be predicted merely from the boiling points of the components and azeotropes and a specified distillate-to-bottoms molar ratio, when distillation boundaries are present. These boundaries, as well as the mappings of distillation curves in the ternary plots of Figure 11.3, can be determined by either of the methods described in the next two sections.

Residue-Curve Maps

Consider the simple Rayleigh batch or differential distillation (no trays, packing, or reflux) shown schematically in Figure 13.1. For any component of a ternary mixture, a material balance for its vaporization from the liquid in the still, assuming that the liquid is perfectly mixed and at its bubble-point temperature, is given by (13-1), which can be written as

$$\frac{dx_i}{dt} = (y_i - x_i)\frac{dW}{W\,dt} \tag{11-1}$$

where

x_i = mole fraction of component i in W moles of perfectly mixed liquid residue in the still

y_i = mole fraction of component i in the vapor (instantaneous distillate) in equilibrium with x_i

Because W changes (decreases) with time, t, it is possible to combine W and t into a single variable. Following the treatment of Doherty and Perkins [4], let this variable be ξ, such that

$$\frac{dx_i}{d\xi} = x_i - y_i \tag{11-2}$$

Combining (11-1) and (11-2) to eliminate $dx_i/(x_i - y_i)$:

$$\frac{d\xi}{dt} = -\frac{1}{W}\frac{dW}{dt} \tag{11-3}$$

Let the initial condition be $x = 0$ and $W = W_0$ at $t = 0$. Then the solution to (11-3) for ξ at time t is

$$\xi\{t\} = \ln[W_0/W\{t\}] \tag{11-4}$$

Because $W\{t\}$ decreases monotonically with time, $\xi\{t\}$ must increase monotonically with time and is considered a dimensionless, warped time. Thus, for the ternary mixture, the simple distillation process can be modeled by the following set of differential-algebraic equations (DAEs), assuming that a second liquid phase does not form:

$$\frac{dx_i}{d\xi} = x_i - y_i, \quad i = 1,2 \tag{11-5}$$

$$\sum_{i=1}^{3} x_i = 1 \tag{11-6}$$

$$y_i = K_i x_i, \quad i = 1, 2, 3 \tag{11-7}$$

and the bubble-point-temperature equation:

$$\sum_{i=1}^{3} K_i x_i = 1 \tag{11-8}$$

where, in the general case, $K_i = K_i\{T, P, \boldsymbol{x}, \boldsymbol{y}\}$.

Thus, the system consists of seven equations in nine variables: P, T, x_1, x_2, x_3, y_1, y_2, y_3, and ξ. If the pressure is fixed, the next seven variables can be computed from (11-5) to (11-8) as a function of ξ, from a specified initial condition. The calculations can proceed in either the forward or backward direction of ξ. The results, when plotted on a triangular graph, are called a *residue curve* because the plot follows, with time, the liquid-residue composition in the still. A collection of residue curves, for a given ternary system at a fixed pressure, is a *residue-curve map*. A simple, but inefficient, procedure for calculating a residue curve is illustrated in the following example. Better, but more elaborate, procedures are given by Doherty and Perkins [4] and Bossen, Jørgensen, and Gani [5]. The last procedure is also applicable when two separate liquid phases form, as is a procedure by Pham and Doherty [6].

EXAMPLE 11.1

Calculate and plot a portion of a residue curve for the ternary system, *n*-propanol (1), isopropanol (2), and benzene (3) at 1 atm, starting from a bubble-point liquid with a composition of 20 mol% each of 1 and 2, and 60 mol% of component 3. For *K*-values, use the modified Raoult's law (see Table 2.3) with regular-solution theory [see (2-64)] for estimating the liquid-phase activity coefficient as a function of composition and temperature. The normal boiling points of the three components in °C are 97.3, 82.3, and 80.1, respectively. Minimum-boiling azeotropes are formed at 77.1°C for components 1,3 and at 71.7°C for 2,3.

SOLUTION

A bubble-point calculation, using (11-7) and (11-8), gives starting values of $\boldsymbol{y}$ of 0.1437, 0.2154, and 0.6409, respectively, and a value of 79.07°C for the starting temperature, from the ChemSep program of Taylor and Kooijman [7].

For a specified increment in the dimensionless time, ξ, the differential equations (11-5) can be solved for x_1 and x_2 using Euler's method with a spreadsheet. Then x_3 is obtained from (11-6). The corresponding values of $\boldsymbol{y}$ and T are then obtained from (11-7) and (11-8). This procedure is repeated for the next increment in ξ. Thus, from (11-5), for component 1:

$$x_1^{(1)} = x_1^{(0)} + \left(x_1^{(0)} - y_1^{(0)}\right)\Delta\xi$$
$$= 0.2000 + (0.2000 - 0.1437)0.1 = 0.2056$$

where superscripts (0) indicate starting values and a superscript of (1) indicates the value after the first increment in ξ. The value of

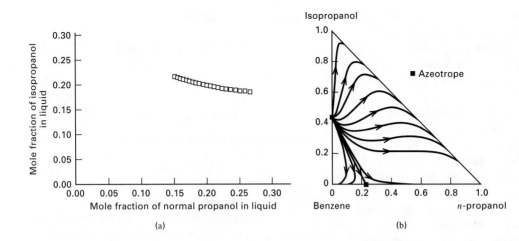

Figure 11.5 Residue curves for the normal propanol–isopropanol–benzene system at 1 atm for Example 11.1. (a) Calculated partial residue curve. (b) Residue-curve map.

0.1 for $\Delta\xi$ gives reasonable accuracy, since the change in x_1 is seen to be only 2.7%. Similarly:

$$x_2^{(1)} = 0.2000 + (0.2000 - 0.2154)0.1 = 0.1985$$

From (11-6):

$$x_3^{(1)} = 1 - x_1^{(1)} - x_2^{(1)} = 1 - 0.2056 - 0.1985 = 0.5959$$

From a bubble-point calculation using (11-7) and (11-8), from ChemSep,

$$y^{(1)} = [0.1474, 0.2134, 0.6392]^T \quad \text{and} \quad T^{(1)} = 79.14°C$$

The calculations are continued in the forward direction of ξ to $\xi = 1.0$. The calculations are also carried out in the backward direction back to $\xi = -1.0$. The results are in the table below, and the partial residue curve is plotted in Figure 11.5a. For comparison, the complete residue-curve map for this system, from Doherty [8], is given on a right-triangle diagram in Figure 11.5b.

ξ	x_1	x_2	y_1	y_2	$T, °C$
−1.0	0.1515	0.2173	0.1112	0.2367	78.67
−0.9	0.1557	0.2154	0.1141	0.2344	78.71
−0.8	0.1600	0.2135	0.1171	0.2322	78.75
−0.7	0.1644	0.2117	0.1201	0.2300	78.79
−0.6	0.1690	0.2099	0.1232	0.2278	78.83
−0.5	0.1737	0.2081	0.1264	0.2256	78.87
−0.4	0.1786	0.2064	0.1297	0.2235	78.91
−0.3	0.1837	0.2047	0.1331	0.2214	78.95
−0.2	0.1889	0.2031	0.1365	0.2194	79.00
−0.1	0.1944	0.2015	0.1401	0.2173	79.05
0.0	0.2000	0.2000	0.1437	0.2154	79.07
0.1	0.2056	0.1985	0.1474	0.2134	79.14
0.2	0.2115	0.1970	0.1512	0.2115	79.19
0.3	0.2175	0.1955	0.1550	0.2095	79.24
0.4	0.2237	0.1941	0.1589	0.2076	79.30
0.5	0.2302	0.1928	0.1629	0.2058	79.24
0.6	0.2369	0.1915	0.1671	0.2041	79.41
0.7	0.2439	0.1902	0.1714	0.2023	79.48
0.8	0.2512	0.1890	0.1758	0.2006	79.54
0.9	0.2587	0.1878	0.1804	0.1989	79.61
1.0	0.2665	0.1867	0.1850	0.1973	79.68

The residue-curve map in Figure 11.5b shows residue curves with arrows. The curves include the three border sides of the triangular diagram. The arrow on each curve points from a lower-boiling component or azeotrope to a higher-boiling component or azeotrope. In Figure 11.5b, all residue curves of the ternary mixture originate from the isopropanol–benzene azeotrope (lowest boiling point of 71.7°C). One of the curves terminates at the other azeotrope (n-propanol–benzene, which has a higher boiling point of 77.1°C). This is a special residue curve, called a *simple distillation boundary* because it divides the ternary region into two separate distillation regions. All residue curves lying above and to the right of the distillation boundary terminate at the n-propanol apex, which has the highest boiling point (97.3°C) for that region. All residue curves lying below and to the left of the distillation boundary are deflected to the benzene apex, whose boiling point of 80.1°C is the highest for this second region.

On the triangular diagram, all pure-component vertices and azeotropic points, whether binary azeotropes on the borders of the triangle, as in Figure 11.5b, or a ternary azeotrope within the triangle, are singular or fixed points of the residue curves because at these points, $d\mathbf{x}/d\xi = 0$. In the vicinity of these points, the behavior of a residue curve depends upon the two eigenvalues of (11-5). At each pure-component vertex, the two eigenvalues are identical. At each azeotropic point, the two eigenvalues are different. Three cases, illustrated by each of three pattern groups in Figure 11.6, are possible:

Case 1: Both eigenvalues are negative. This is the point reached as ξ tends to infinity. It is the point at which all residue curves in a given region terminate. Thus, it is the component or azeotrope with the highest boiling point in the region. This point is a *stable node* because it is like the low point of a valley, in which a rolling ball finds a stable position. In Figure 11.6b, the stable node is pure n-propanol.

Case 2: Both eigenvalues are positive. This is the point from which all residue curves in a given region originate. Thus, it is the component or azeotrope with the lowest boiling point in the region. This point is an *unstable node* because it is like the top of a peaked mountain from which a ball rolls toward a stable position. In Figure 11.6b, the unstable node is the isopropanol-benzene azeotrope.

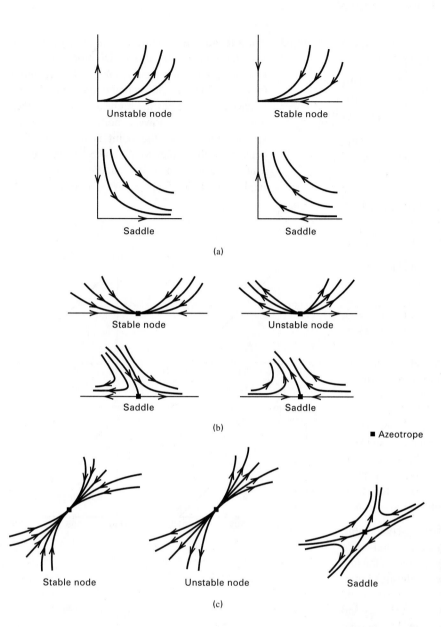

Figure 11.6 Residue-curve patterns (a) near pure-component vertices; (b) near binary azeotropes; (c) near ternary azeotropes.
[From M.F. Doherty and G.A. Caldarola, *IEC Fundam.*, **24**, 477 (1985) with permission.]

Case 3: One eigenvalue is positive and one is negative. Residue curves within the triangle move toward and then away from such points, which are *saddles*. For a given distillation region, all pure components and azeotropes intermediate in boiling point between the stable node and the unstable node are saddles. In Figure 11.5b, the upper distillation region has one saddle at the isopropanol vertex and another saddle at the normal propanol–benzene azeotrope.

From Example 11.1, it is clear that calculation of a residue-curve map requires a considerable effort. However, computer-aided simulation programs such as ASPEN PLUS [9] and CHEMCAD compute residue maps. Alternatively, as developed by Doherty and Perkins [10] and Doherty [8], the classification of singular points as stable nodes, unstable nodes, and saddles provides a rapid method for approximating a residue-curve map, including approximate distillation boundaries, from just the pure-component boiling points and

azeotrope boiling points and compositions. Boiling points of pure components are readily found in handbooks and component data banks of computer-aided simulation programs. Extensive listings of binary azeotropes are found in Horsley [11] and Gmehling et al. [12]. The former lists more than 1,000 binary azeotropes. The latter includes experimental data for more than 20,000 systems involving approximately 2,000 compounds, as well as material on selecting enhanced distillation systems. The listings of ternary azeotropes are undoubtedly quite incomplete. However, in lieu of experimental data, a homotopy-continuation method for estimating all homogeneous azeotropes of a multicomponent mixture from a thermodynamic model (e.g., Wilson, NRTL, UNIQUAC, UNIFAC) has been developed by Fidkowski, Malone, and Doherty [13]. Eckert and Kubicek [97] present an extension for computing heterogeneous azeotropes.

Based on experimental evidence, for ternary mixtures, with very few exceptions, there are at most three binary azeotropes and one ternary azeotrope. Accordingly, the following set of restrictions apply to a ternary system:

$$N_1 + S_1 = 3 \tag{11-9}$$

$$N_2 + S_2 = B \le 3 \tag{11-10}$$

$$N_3 + S_3 = 1 \quad \text{or} \quad 0 \tag{11-11}$$

where N is the number of stable and unstable nodes, S is the number of saddles, B is the number of binary azeotropes, and the subscript is the number of components at the node (stable or unstable) or saddle. Thus, S_2 is the number of binary azeotrope saddles. Doherty and Perkins [10] developed the following topological relationship among N and S:

$$2N_3 - 2S_3 + 2N_2 - B + N_1 = 2 \tag{11-12}$$

For the system of Figure 11.5b, which has no ternary azeotrope, we see that $N_1 = 2$, $N_2 = 1$, $N_3 = 0$, $S_1 = 1$, $S_2 = 1$, $S_3 = 0$, and $B = 2$. Applying (11-12) gives $0 - 0 + 2 - 2 + 2 = 2$. Equation (11-9) gives $2 + 1 = 3$, (11-10) gives $1 + 1 = 2$, and (11-11) gives $0 + 0 = 0$. Thus, all four relations are satisfied.

The topological relationships are especially useful for rapidly sketching, on a ternary diagram, an approximate residue-curve map, including distillation boundaries, as described in detail by Foucher, Doherty, and Malone [14]. Their procedure involves the following nine steps (0–8), which are partly illustrated by a hypothetical example taken from their article and shown in Figure 11.7. The procedure is summarized in Figure 11.8. Approximate maps are usually developed from data at 1 atm. In the description of the steps, the term *species* refers to pure components and azeotropes.

Step 0: Label the ternary diagram with the pure-component, normal-boiling-point temperatures. It is preferable to designate the top vertex of the triangle as the low boiler (L), the bottom-right vertex as the high boiler (H), and the bottom-left vertex as the intermediate boiler (I). Plot composition points for the binary and ternary azeotropes and add labels for their normal boiling points. This determines the value of B. See Figure 11.7, Step 0, where two minimum-boiling and one maximum-boiling binary azeotropes and one ternary azeotrope are designated by filled square markers. Thus, $B = 3$.

Step 1: Draw arrows on the edges of the triangle, in the direction of increasing temperature, for each pair of adjacent species. See Figure 11.7, Step 1, where there are six species on the edges of the triangle and six arrows have been added.

Step 2: Determine the type of singular point for each pure-component vertex, by using Figure 11.6 with the arrows drawn in Step 1. This determines the values for N_1 and S_1. If a ternary azeotrope exists, go to Step 3; if not, go to Step 5. In Figure 11.7, Step 2, L is a saddle because one arrow points toward L and one points away from L; H is a stable node because both arrows point toward H, and I is a saddle. Therefore, $N_1 = 1$ and $S_1 = 2$.

Step 3 **(for a ternary azeotrope):** Determine the type of singular point for the ternary azeotrope, if one exists. The point is a node if (a) $N_1 + B < 4$, and/or (b) excluding the pure-component saddles, the ternary azeotrope has the highest, second-highest, lowest, or second-lowest boiling point of all species. Otherwise, the point is a saddle. This determines the values for N_3 and S_3. If the point is a node, go to Step 5; if a saddle, go to Step 4. In Figure 11.7, Step 3, $N_1 + B = 1 + 3 = 4$.

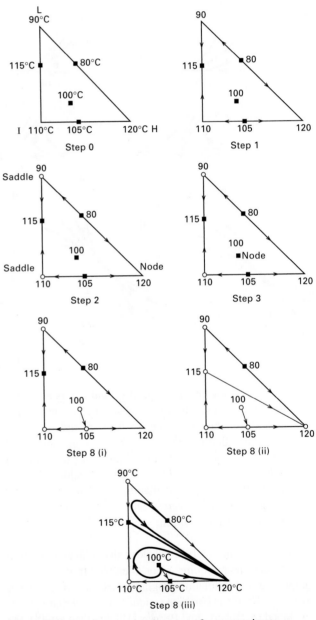

Figure 11.7 Step-by-step development of an approximate residue-curve map for a hypothetical system with two minimum-boiling binary azeotropes, one maximum-boiling binary azeotrope and one ternary azeotrope.

[From E.R. Foucher, M.F. Doherty, and M.F. Malone, *IEC Res.* **30**, 764 (1991) with permission.]

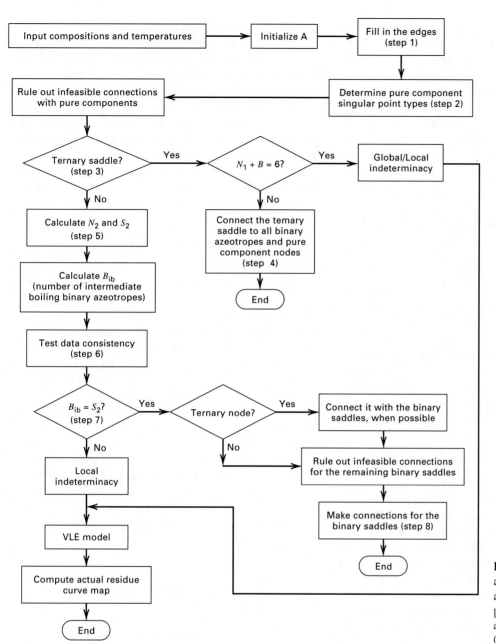

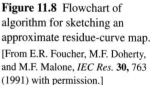

Figure 11.8 Flowchart of algorithm for sketching an approximate residue-curve map.

[From E.R. Foucher, M.F. Doherty, and M.F. Malone, *IEC Res.* **30,** 763 (1991) with permission.]

However, excluding L and I because they are saddles, the ternary azeotrope has the second-lowest boiling point. Therefore, the point is a node, and $N_3 = 1$ and $S_3 = 0$. The type of node, stable or unstable, is still to be determined.

Step 4 **(for a ternary saddle):** Connect the ternary saddle, by straight lines, to all binary azeotropes and to all pure-component nodes (but not to pure-component saddles) and draw arrows on the lines to indicate the direction of increasing temperature. Determine the type of singular point for each binary azeotrope, by using Figure 11.6 with the arrows drawn in this step. This determines the values for N_2 and S_2. These values should be consistent with (11-10) and (11-12). This completes

the development of the approximate residue-curve map, with no further steps needed. However, if $N_1 + B = 6$, then special checks must be made, as given in detail by Foucher, Doherty, and Malone [14]. This step does not apply to the example in Figure 11.7, because the ternary azeotrope is not a saddle.

Step 5 **(for a ternary node or no ternary azeotrope):** Determine the number of binary nodes, N_2, and binary saddles, S_2, from (11-10) and (11-12), where (11-12) can be solved for N_2 to give

$$N_2 = (2 - 2N_3 + 2S_3 + B - N_1)/2 \quad (11\text{-}13)$$

For the example of Figure 11.7, $N_2 = (2 - 2 + 0 + 3 - 1)/2 = 1$. From (11-10), $S_2 = 3 - 1 = 2$.

Step 6: Count the binary azeotropes that are intermediate boilers (i.e. that are not the highest- or the lowest-boiling species), (and call that number B_{ib}). Make the following two data consistency checks: (a) The number of binary azeotropes, B, less B_{ib} must equal N_2, and (b) S_2 must be $\leq B_{ib}$. For the system in Figure 11.7, both checks are satisfied because $B_{ib} = 2$, $B - B_{ib} = 1$, $N_2 = 1$, and $S_2 = 2$. If these two consistency checks are not satisfied, one or more of the species' boiling points may be in error.

Step 7: If $S_2 \neq B_{ib}$, this procedure cannot determine a unique residue-curve-map structure, which therefore must be computed from (11-5) to (11-8). If $S_2 = B_{ib}$, there is a unique structure, which is completed in Step 8. For the example in Figure 11.7, $S_2 = B_{ib} = 2$; therefore, there is a unique map.

Step 8: In this final step for a ternary node or no ternary azeotrope, the distillation boundaries (connections), if any, are determined and entered on the triangular diagram as straight lines, and, if desired, one or more representative residue curves are sketched as curved lines within each distillation region. This step applies to cases of $S_3 = 0$, $N_3 = 0$ or 1, and $S_2 = B_{ib}$. In all cases, the number of distillation boundaries equals the number of binary saddles, S_2. Each binary saddle must be connected to a node (pure component, binary, or ternary). A ternary node must be connected to at least one binary saddle. Thus, a pure-component node cannot be connected to a ternary node, and an unstable node cannot be connected to a stable node. The connections are made by determining a connection for each binary saddle such that (a) a minimum-boiling binary saddle connects to an unstable node that boils at a lower temperature and (b) a maximum-boiling binary saddle connects to a stable node that boils at a higher temperature. It is best to first consider connections with the ternary node and then examine the possible connections for the remaining binary saddles. In the example of Figure 11.7, $S_2 = 2$, with these saddles denoted as L-I, a maximum-boiling azeotrope at 115°C, and as I-H, a minimum-boiling azeotrope at 105°C. Therefore, we make two connections to establish two distillation boundaries. The ternary node at 100°C cannot connect to L-I because 100°C is not greater than 115°C. The ternary node can, however, connect, as shown in Step 8 (i), to I-H because 100°C is lower than 105°C. This marks the ternary node as unstable. The connection for L-I can only be to H, as shown in Step 8 (ii) because it is a node (stable), and 120°C is greater than 115°C. This completes the connections. Finally, as shown in Step 8 (iii) of Figure 11.7, three typical, but approximate, residue curves are added to the diagram. These curves originate from unstable nodes and terminate at stable nodes.

Residue-curve maps are used to determine feasible distillation sequences for nonideal ternary systems. Matsuyama and Nishimura [15] showed that the topological constraints just discussed limit the number of possible maps to about 113. However, Siirola and Barnicki [96] show 12 additional maps; all 125 maps are called distillation region diagrams (DRD). Doherty and Caldarola [16] provide sketches of 87 maps that contain at least one minimum-boiling binary azeotrope. These maps cover most of the cases found in industrial applications, since minimum-boiling azeotropes are much more common than maximum-boiling azeotropes.

Distillation-Curve Maps

A residue curve represents the liquid-residue composition with time as the result of a simple, one-stage batch distillation. The curve is pointed in the direction of increasing time, from a lower-boiling state to a higher-boiling state for the liquid residue. An alternative representation for distillation on a ternary diagram is a *distillation curve* for continuous, rather than batch, distillation. The curve is most readily determined for total reflux (infinite reflux ratio) at a constant pressure, usually 1 atm. The calculations are made down or up the column starting from any composition. Suppose we choose to make the calculations by moving up the column, starting from a stage designated as Stage 1, and numbering the stages upward. At any location between equilibrium stages j and $j + 1$, it will be recalled, from the McCabe–Thiele method for binary mixtures in Chapter 7 or the Fenske equation for multicomponent systems from Chapter 9, that passing vapor and liquid streams have the same composition. Thus,

$$x_{i,j+1} = y_{i,j} \qquad (11\text{-}14)$$

Also, liquid and vapor streams leaving the same stage are in equilibrium. Thus,

$$y_{i,j} = K_{i,j} x_{i,j} \qquad (11\text{-}15)$$

To calculate a distillation curve for a fixed pressure, an initial liquid-phase composition, $x_{i,1}$, is assumed. This liquid is at its bubble-point temperature, which is determined from (11-8), which also gives the equilibrium-vapor composition, $y_{i,1}$ in agreement with (11-15). The composition, $x_{i,2}$, of the passing liquid stream is equal to $y_{i,1}$ by (11-14). The process is then repeated to obtain $x_{i,3}$, then $x_{i,4}$, and so forth. The sequence of liquid-phase compositions, which corresponds to the operating line for the total-reflux condition, is plotted on the triangular diagram. The procedure is essentially that of Fenske for the determination of the minimum number of equilibrium stages for operation at total reflux to achieve a specified split of two key components, as discussed in Chapter 9. The distillation curve is analogous to the 45° line on a McCabe–Thiele diagram for a binary mixture. The calculation of a portion of a distillation curve is illustrated next.

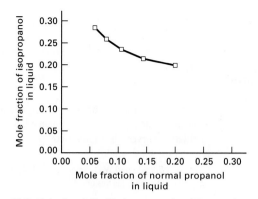

Figure 11.9 Calculated distillation curve for the normal propanol–isopropanol–benzene system at 1 atm for Example 11.2.

EXAMPLE 11.2

Calculate and plot a portion of a distillation curve for the same starting conditions as Example 11.1.

SOLUTION

The starting values, $x^{(1)}$, are 0.2000, 0.2000, and 0.6000 for components 1, 2, and 3, respectively. From Example 11.1, the bubble-point calculation gives a temperature of 79.07°C and the following values for $y^{(1)}$: 0.1437, 0.2154, and 0.6409. From (11-14), values of $x^{(2)}$ are 0.1437, 0.2154, and 0.6409. A bubble-point calculation for this composition gives $T^{(2)} = 78.62°C$ and $y^{(2)} = 0.1063, 0.2360$, and 0.6577. Subsequent calculations are summarized in the following table:

Equilibrium

Stage	x_1	x_2	y_1	y_2	T, °C
1	0.2000	0.2000	0.1437	0.2154	79.07
2	0.1437	0.2154	0.1063	0.2360	78.62
3	0.1063	0.2360	0.0794	0.2597	78.29
4	0.0794	0.2597	0.0592	0.2846	78.02
5	0.0592	0.2846	0.0437	0.3091	77.80

The resulting distillation curve is plotted in Figure 11.9, where points represent equilibrium stages and are connected by straight lines.

Distillation curves can be computed more rapidly than residue curves, and closely approximate them for reasons noted by Fidkowski, Doherty, and Malone [17]. If (11-5), which must be solved numerically as in Example 11.1, is written in a forward-finite-difference form, we obtain

$$(x_{i,j+1} - x_{i,j})/\Delta\xi = x_{i,j} - y_{i,j} \qquad (11\text{-}16)$$

In Example 11.1, $\Delta\xi$ was set to +0.1 for calculations that give increasing values of T and to −0.1 to give decreasing values of T. If we choose the latter direction to be consistent with the direction used in Example 11.2, but set $\Delta\xi$ equal to −1.0, (11-16) becomes identical to (11-14). Thus, residue curves (which are true, continuous curves) are equal to distillation curves (which are discrete points, through which a smooth curve is drawn), when the residue curves are approximated by a crude forward-finite-difference formulation, using $\Delta\xi = -1.0$.

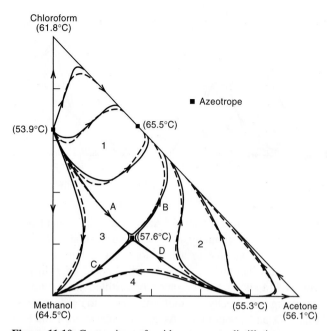

Figure 11.10 Comparison of residue curves to distillation curves. [From Z.T. Fidkowski, M.F. Malone, and M.F. Doherty, *AIChE J.*, **39**, 1303 (1993) with permission.]

A collection of distillation curves, including lines for distillation boundaries, is a *distillation-curve map,* an example of which from Fidkowski et al. [17] is shown in Figure 11.10 for the acetone–chloroform–methanol system at 1 atm. The Wilson equation was used to compute liquid-phase activity coefficients for the system. The dashed lines are the distillation curves; they approximate the residue curves, which are solid lines. This system has two minimum-boiling binary azeotropes, one maximum-boiling binary azeotrope, and a ternary saddle azeotrope. The map shows four distillation boundaries, designated by A, B, C, and D, consistent with Step 4 earlier. These computed boundaries, which define four distillation regions (1 to 4), are all curved lines rather than the approximate straight lines in the sketches of Figure 11.7.

Distillation-curve maps have been used extensively by Stichlmair and associates [1, 3, 18] for the development of feasible-distillation sequences. In their maps, arrows on the distillation curves are directed toward the lower-boiling species, rather than the higher-boiling species as in residue-curve maps.

Product-Composition Regions at Total Reflux (Bow-Tie Regions)

As mentioned above, the possible distillation regions for azeotrope-forming ternary mixtures are not obvious. Fortunately, residue-curve maps and distillation-curve maps can be used to make preliminary estimates of regions of *feasible-product compositions* for distillation of nonideal ternary mixtures. The product regions are determined by superimposing a column material-balance line on either curve-map diagram. Consider first the simpler zeotropic ternary system in Figure 11.11a, which shows a typical isobaric residue-curve

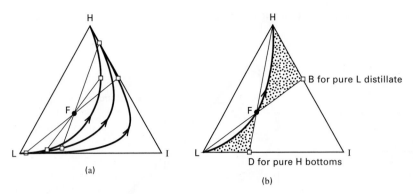

Figure 11.11 Product-composition regions for a zeotropic system. (a) Material-balance lines and distillation curves. (b) Product-composition regions shown shaded.

[From S. Widagdo and W.D. Seider, *AIChE J.*, **42**, 96–130 (1996) with permission.]

map with three residue curves. Assume that this map is identical to a corresponding distillation-curve map for total-reflux conditions and to a map for a finite, but very high reflux ratio. Suppose a ternary feed, denoted by *F* in Figure 11.11a, is to be continuously distilled in a column, operating isobarically at a high reflux ratio, to produce a distillate, *D*, and a bottoms, *B*. As shown in Chapters 4 and 8, if a straight line is drawn that connects distillate and bottoms compositions, the line must pass through the feed composition at some intermediate point to satisfy overall and component material-balance equations. This line is a *material-balance line,* three of which are included on Figure 11.11a. For a given material-balance line, a set of *D* and *B* composition points, designated by open squares, must lie on the same distillation curve. This causes the material-balance line to intersect the distillation curve at these two points and be a chord to the distillation curve.

The limiting distillate-composition point for this zeotropic system is pure low-boiling component, L. From the material-balance line passing through *F*, as shown in Figure 11.11b, the corresponding bottoms composition with the least amount of component L is point *B*. At the other extreme, the limiting bottoms-composition point is pure high-boiling component, H. A material-balance line from this point, through feed point *F*, ends at *D*. These two

lines and the distillation curve define the feasible product-composition regions, shown shaded. Note that because, for a given feed, both the distillate and bottoms compositions must lie on the same distillation curve, the shaded feasible regions lie on the convex side of the distillation curve that passes through the feed point. Because of its appearance, the feasible-product-composition region is referred to as a *bow-tie-region.*

For an azeotropic system, where distillation boundaries are present, a feasible-product-composition region can be found for each distillation region. Two examples are shown in Figure 11.12. The first, in Figure 11.12a, has two distillation regions caused by two minimum-boiling binary azeotropes. A curved distillation boundary connects the two minimum-boiling azeotropes. In the lower, right-hand distillation region (1), the lowest-boiling species is the *n*-octane–2-ethoxy-ethanol minimum-boiling azeotrope, while the highest-boiling species is 2-ethoxy-ethanol. Accordingly, for feed F_1, straight lines are drawn from the points for each of these two species, through the point F_1, and to a boundary (either a distillation boundary or a side of the triangle). Shaded, feasible-product-composition regions are then drawn on the outer side of the distillation curve that passes through the feed point. The result is that distillate compositions are confined to shaded region D_1 and bottoms compositions are

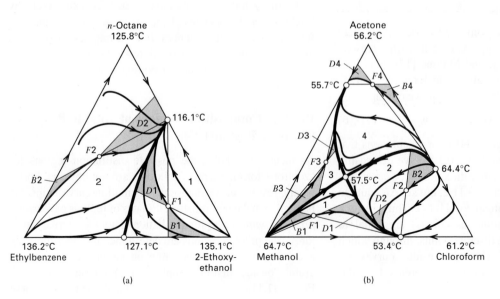

Figure 11.12 Product-composition regions for given feed compositions. (a) Ternary mixture with two minimum-boiling binary azeotropes at 1 atm. (b) Ternary mixture with three binary and one ternary azeotrope at 1 atm.

confined to shaded region B_1. For a given D_1, B_1 must lie on a straight line that passes through D_1 and F_1. At total reflux, D_1 and B_1 must also lie on the same distillation curve. A more complex distillation-curve map, with four distillation regions, is shown in Figure 11.12b for the acetone-methanol-chloroform system with two minimum-boiling binary azeotropes, one maximum-boiling binary azeotrope, and one ternary azeotrope. One shaded bow-tie region, determined in the same way as for region 1 in Figure 11.12a, is present for each distillation region. For this system, feasible-product-composition regions are highly restricted.

A complicated situation is observed in distillation region 1 on the left side of Figure 11.12a. In that region, the lowest-boiling species is the binary azeotrope of octane and 2-ethoxy-ethanol, while the highest-boiling species is the ethylbenzene. The complicating factor in distillation region 1 is that feed F_2 lies on or close to an inflection point of an S-shaped distillation curve. In this case, as discussed by Wahnschafft et al. [20], feasible-product-composition regions may lie on either side of the distillation curve that passes through the feed point. The feasible regions shown are similar to those determined by Stichlmair et al. [1], while other feasible regions are shown for this system by Wahnschafft et al. [20]. As they point out, for a situation such as this, mass-balance lines of the type drawn in Figure 11.12b do not limit the feasible regions. Hoffmaster and Hauan [98] provide a method for determining extended-product-feasibility regions in the presence of S-shaped distillation curves.

In Figures 11.11b, 11.12a, and 11.12b, each bow-tie region is confined to its distillation region, as defined by the distillation boundaries. In all cases, the feed, distillate, and bottoms points on the material-balance line lie within a distillation region, with the feed point between the distillate and bottoms points. The material-balance lines do not cross the distillation-boundary lines. Is this always so? The answer is no! Under conditions where the distillation-boundary line is highly curved, it can be crossed by material-balance lines to obtain feasible-product compositions. That is, a feed point can be on one side and the distillate and bottoms points on the other side of the distillation-boundary line. Consider the example in Figure 11.13, taken from Widagdo and Seider [19]. The distillation-boundary line, which is highly curved, extends from a minimum-boiling azeotrope K of H-I to the pure component L. This line divides the triangular diagram into two distillation regions, 1 and 2. Feed F_1 can be separated into products D_1 and B_1, which lie on distillation curve (a). In this case, the material-balance line and the distillation curve are both on the convex side of the distillation-boundary line. However, because the feed point F_1 lies close to the highly curved boundary line, F_1 can also be separated into D_2 and B_2 (or B_3), which lie on a distillation curve in region 2 on the concave side of the boundary. Thus, the material-balance line crosses the boundary from the convex to the concave side. Feed F_2 can be separated into D_4 and B_4, but not into D and B. In the latter case, the material-balance

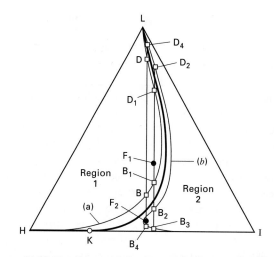

Figure 11.13 Feasible and infeasible crossings of distillation boundaries for an azeotropic system.

[From S. Widagdo and W.D. Seider, *AIChE J.*, **42**, 96–130 (1996) with permission.]

line cannot cross the boundary from the concave to the convex side, because the point F_2 does not lie between D and B on the material-balance line. The determination of the feasible-product-composition regions for Figure 11.13 is left for an exercise at the end of this chapter. A detailed treatment of product composition regions is given by Wahnschafft et al. [20].

11.2 EXTRACTIVE DISTILLATION

Extractive distillation is used to separate azeotropes and other mixtures that have key components with a relative volatility below about 1.1 over an appreciable range of concentration. If the feed is a minimum-boiling azeotrope, a solvent, with a lower volatility than the key components of the feed mixture, is added to a tray above the feed stage and a few trays below the top of the column so that (1) the solvent is present in the downflowing liquid phase to the bottom of the column, and (2) little solvent is stripped and lost to the overhead vapor. If the feed is a maximum-boiling azeotrope, the solvent enters the column with the feed. The components in the feed must have different affinities for the solvent so that the solvent causes an increase in the relative volatility of the key components, to the extent that separation becomes feasible and economical. The solvent should not form an azeotrope with any components in the feed. Generally, a molar ratio of solvent-to-feed on the order of 1 is required to achieve this goal. The bottoms from the extractive distillation column is processed further to recover the solvent for recycle and complete the feed separation. The name, extractive distillation, was introduced by Dunn et al. [21] in connection with the commercial separation of toluene from a paraffin–hydrocarbon mixture, using phenol as solvent.

Table 11.1 Some Industrial Applications of Extractive Distillation

Key Components in Feed Mixture	Solvent
Acetone–methanol	Aniline, ethylene glycol, water
Benzene–cyclohexane	Aniline
Butadienes–butanes	Acetone
Butadiene–butene-1	Furfural
Butanes–butenes	Acetone
Butenes–isoprene	Dimethylformamide
Cumene–phenol	Phosphates
Cyclohexane–heptanes	Aniline, phenol
Cyclohexanone–phenol	Adipic acid diester
Ethanol–water	Glycerine, ethylene glycol
Hydrochloric acid–water	Sulfuric acid
Isobutane–butene-1	Furfural
Isoprene–pentanes	Acetonitrile, furfural
Isoprene–pentenes	Acetone
Methanol–methylene bromide	Ethylene bromide
Nitric acid–water	Sulfuric acid
n-Butane–butene-2s	Furfural
Propane–propylene	Acrylonitrile
Pyridine–water	Bisphenol
Tetrahydrofuran–water	Dimethylformamide, propylene glycol
Toluene–heptanes	Aniline, phenol

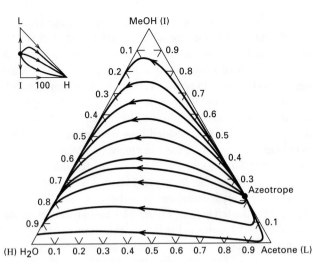

Figure 11.14 Residue-curve map for acetone–methanol–water system at 1 atm.

Mol% Water	Relative Volatility, $\alpha_{A,M}$			Liquid-Phase Activity Coefficient at Infinite Dilution	
	Methanol-rich	Acetone-rich	Equimolar	Acetone	Methanol
40	2.48	2.57	2.03	2.12	0.70
50	2.56	2.86	2.29	2.41	0.72

Table 11.1 lists a number of industrial applications of extractive distillation. Consider the case of the acetone–methanol system. At 1 atm, acetone (nbp = 56.2°C) and methanol (nbp = 64.7°C) form a minimum-boiling azeotrope of 80 mol% acetone at a temperature of 55.7°C. The UNIFAC program was used to predict the vapor–liquid equilibria for this system at 1 atm. The azeotrope was estimated to occur at 55.2°C with 77.1 mol% acetone. At infinite dilution with respect to methanol, the relative volatility of acetone (A) with respect to methanol (M), $\alpha_{A,M}$, is predicted to be 0.74, with a liquid-phase activity coefficient for methanol of 1.88. At infinite dilution with respect to acetone, $\alpha_{A,M}$ is 2.48; by coincidence, the liquid-phase activity coefficient for acetone is 1.88 also. Water is a possible solvent for the system because at 1 atm: (1) it does not form a binary or ternary azeotrope with acetone and/or methanol, and (2) it boils (100°C) at a higher temperature. The resulting residue-curve map with arrows directed from the azeotrope to pure water, computed by ASPEN PLUS using UNIFAC, is shown in Figure 11.14, where it is seen that no distillation boundaries exist. As discussed by Doherty and Caldarola [16], this is an ideal situation for the selection of an extractive distillation process. Their schematic residue-curve map for this type system (designated 100) is included as an insert in Figure 11.14.

Ternary mixtures of acetone, methanol, and water at 1 atm give the following separation factors, estimated from the UNIFAC equation, when appreciable solvent is present.

Thus, the presence of appreciable water increases the liquid-phase activity coefficient of acetone and decreases that of methanol, with the result that, over the entire concentration range of acetone and methanol, the relative volatility of acetone to methanol is at least 2.0. This makes it possible, with extractive distillation, to obtain a distillate of acetone and a bottoms of methanol and water. Furthermore, the relative volatilities of acetone to water and methanol to water average about 4.5 and 2.0, respectively. Thus, it is relatively easy to prevent an appreciable amount of water from reaching the distillate, and, in subsequent operations, to separate methanol from water by ordinary distillation.

EXAMPLE 11.3

Forty moles per second of a bubble-point mixture of 75 mol% acetone and 25 mol% methanol at 1 atm is separated by an extractive-distillation process, using water as the solvent, to produce an acetone product of not less than 95 mol% acetone, a methanol product of not less than 98 mol% methanol, and a water stream for recycle of at least 99.9 mol% purity. Prepare a preliminary process design, using the traditional sequence consisting of ordinary distillation followed by extractive distillation, and then ordinary distillation to recover the solvent, as shown for another system in Figure 11.15.

SOLUTION

In the first column, the feed mixture of acetone and methanol would be partially separated by ordinary distillation, where the distillate composition approaches that of the binary azeotrope. The bottoms

Table 11.2 Material and Energy Balances for Extractive Distillation Process of Example 11.3

	Material Balances					
	Flow rate, mol/s:					
Species	Column 2 Feed	Column 2 Solvent	Column 2 Distillate	Column 2 Bottoms	Column 3 Distillate	Column 3 Bottoms
Acetone	30	0	29.86	0.14	0.14	0.0
Methanol	10	0	0.016	9.984	9.926	0.058
Water	0	60	1.35	58.65	0.06	58.59
Total	40	60	31.226	68.774	10.126	58.648

	Energy Balances	
	Column 1	Column 2
Condenser duty, MW	4.71	1.07
Reboiler duty, MW	4.90	1.12

would be nearly pure acetone or nearly pure methanol depending upon whether the feed contains more or less than 80 mol% acetone, respectively. However, in this example, the feed composition is already close to the azeotrope composition; therefore, the first column is omitted.

Accordingly, the acetone–methanol feed is sent to the second column, an extractive-distillation column equipped with a total condenser and a partial reboiler to produce a distillate of at least 95 mol% acetone. The acetone recovery is better than approximately 99% to achieve methanol purity in the third column. The bottoms from the extractive-distillation column is pumped to that column, where methanol and water are separated by ordinary distillation to achieve the specified purities.

The ChemSep and ChemCAD programs were used to make the calculations, with the UNIFAC method for activity coefficients.

Equilibrium-stage, feed-stage, solvent entry-stage, solvent flow-rate, and reflux-ratio requirements were varied until a satisfactory design was achieved. The resulting material and energy balances are summarized in Table 11.2.

For the extractive-distillation column, a solvent flow rate of 60 mol/s of water is suitable. Using 28 theoretical trays, a 50°C solvent entry at Tray 6 from the top, a feed entry at Tray 12 from the top, and a reflux ratio of 4, a distillate composition of 95.6 mol% acetone is achieved. The impurity is mainly water. The acetone recovery is 99.5%. A 6-ft-diameter column with 60 sieve trays on 2-ft tray spacing is adequate for the operation. A liquid-phase composition profile is shown in Figure 11.16. The mole fraction of water (the solvent) in the liquid phase is appreciable, at least 0.35 for all of the stages below the solvent-entry stage. A distillation-curve map for the actual extractive distillation operation is given in Figure 11.17, where both the vapor and liquid curves are plotted. The arrows are directed from the bottom of the column to the top.

For the ordinary-distillation column, operation with 16 theoretical stages, a bubble-point feed-stage location of Stage 11 and a reflux ratio of 2 is adequate to achieve a methanol distillate of 98.1 mol% purity and a water bottoms, suitable for recycle, of 99.9 mol% purity. A McCabe–Thiele diagram in Figure 11.18

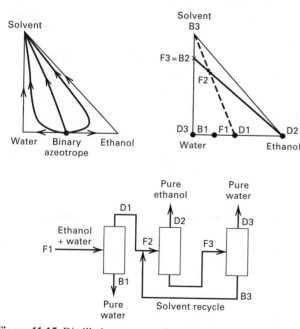

Figure 11.15 Distillation sequence for extractive distillation.
[From M.F. Doherty and G.A. Caldarola, *IEC Fundam.*, **24**, 479 (1985) with permission.]

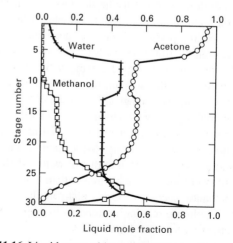

Figure 11.16 Liquid composition profile for extractive-distillation column of Example 11.3.

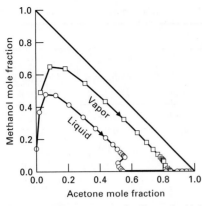

Figure 11.17 Distillation-curve map for Example 11.3. Data points are for theoretical stages.

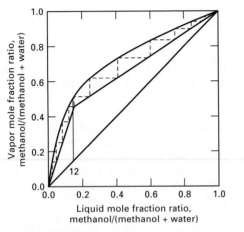

Figure 11.18 McCabe–Thiele diagram for methanol–water distillation in Example 11.3.

shows the locations of the theoretical stages. The feed stage is optimally located. Water makeup is less than 1.5 mol/s. A 2.5-ft-diameter column packed with 48 feet of 50-mm-diameter metal Pall rings is suitable for the separation.

One unfortunate aspect in the design of the extractive-distillation column in Example 11.3 is caused by the relatively low boiling point of water. With a solvent entry point of Tray 6 from the top, 1.35 mol/s (2.25% of the water solvent) is stripped from the liquid into the distillate. The use of two other higher-boiling solvents listed in Table 11.1, aniline (nbp = 184°C) or ethylene glycol (nbp = 198°C), results in far less stripping of solvent. Other possible solvents for the separation of acetone from methanol by extractive distillation include methylethylketone (MEK) and ethanol. MEK behaves in a fashion opposite to that of water: MEK causes the volatility of methanol to be greater than that of acetone. Thus, methanol becomes the distillate in the extractive-distillation column, leaving acetone to be separated from MEK in the subsequent column.

In selecting a solvent for extractive distillation, a number of factors are considered, including availability, cost,

corrosivity, vapor pressure, thermal stability, heat of vaporization, reactivity, toxicity, infinite-dilution activity coefficients in the solvent of the components to be separated, and ease of recovery for recycle. In addition, the solvent should not form azeotropes. Initial screening is based on the measurement or prediction of infinite-dilution activity coefficients. Berg [22] discusses, in detail, the selection of separation agents for both extractive and azeotropic distillation. He points out that all successful solvents for extractive distillation are highly hydrogen-bonded liquids, such as (1) water, amino alcohols, amides, and phenols that form three-dimensional networks of strong hydrogen bonds, and (2) alcohols, acids, phenols, and amines that are composed of molecules containing both active hydrogen atoms and donor atoms (oxygen, nitrogen, and fluorine). In general, it is very difficult or impossible to find a suitable solvent to economically separate components having the same functional groups.

Extractive distillation is also used to separate binary mixtures that form a maximum-boiling azeotrope, as shown in the following example.

EXAMPLE 11.4

Acetone (nbp = 56.16°C) and chloroform (nbp = 61.10°C) form a maximum-boiling homogeneous azeotrope at 1 atm and 64.43°C that contains 37.8 mol% acetone. Thus, they cannot be separated by ordinary distillation at 1 atm. Instead, it is proposed to separate them using extractive distillation in a two-column sequence, shown in Figure 11.19, with benzene (nbp = 80.24°C) as the solvent. Benzene does not form azeotropes with either of the feed components.

In the first column, the feed, blended with recycled solvent, is distilled to produce a distillate of 99 mol% acetone. The bottoms is sent to the second column, where 99 mol% chloroform leaves as distillate and the bottoms, which is rich in benzene, is recycled to the inlet of the first column with a small flow of makeup benzene. If the fresh feed is 21.8858 mol/s of 54.83 mol% acetone, with the balance chloroform, design a feasible two-column system using a ratio of 3.1667 moles of benzene per mole of acetone + chloroform in the combined feed to the first column. Both columns operate at a nominal pressure of 1 atm with total condensers, saturated liquid reflux, and partial reboilers. Use the UNIFAC method for

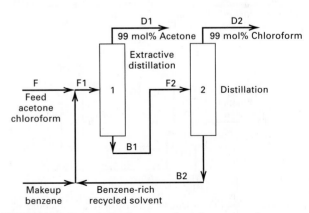

Figure 11.19 Process for the separation of acetone and chloroform in Example 11.4.

Table 11.3 Material and Energy Balances for Homogeneous Azeotropic Distillation of Example 11.4

Species	F	F_1	D_1	$B_1 = F_2$	D_2	B_2
Material Balances with Flows in mol/s						
Acetone	12.0000	12.0000	11.9948	0.0052	0.0052	0.0000
Chloroform	9.8858	12.0000	0.1046	11.8954	9.7812	2.1142
Benzene	0.0000	76.0000	0.0207	75.9793	0.0934	75.8859

Heat duty, kcal/h	Column 1	Column 2
Energy Balances		
Condenser	950,000	891,600
Reboiler	958,400	1,102,000

estimating activity coefficients. The combined feed to the first column is brought to the bubble point before entering the feed stage.

SOLUTION

The residue-curve map for the ternary system acetone–chloroform–benzene at 1 atm is shown in Figure 11.20. The only azeotrope is that formed by acetone and chloroform. A curved distillation boundary extending from that azeotrope to the pure benzene apex divides the diagram into two distillation regions. The first column, which produces nearly pure acetone, operates in Region 1, whereas the second column operates in upper Region 2.

This ternary system was studied in detail by Fidkowski, Doherty, and Malone [17]. A design, based on their studies and using the ChemCAD process simulator, is summarized in Table 11.3. The first column contains 65 theoretical stages with the combined feed entering Stage 30 from the top. With a reflux ratio of 10, the acetone distillate purity is achieved with an acetone recovery of better than 99.95%. In Column 2, which contains 50 theoretical stages with the feed entering at Stage 30 from the top, a reflux ratio of 11.783 gives the required chloroform purity in the distillate, but with a recovery of only 82.23%. However, this is not serious

because the remaining chloroform leaving in the bottoms is recycled with the benzene to Column 1, with the result being an overall recovery of chloroform of 98.9%. The benzene makeup rate is 0.1141 mol/s. Feed, distillate, and bottoms compositions are designated in Figure 11.20.

11.3 SALT DISTILLATION

The use of water as a solvent in the extractive distillation of acetone and methanol in Example 11.3 has the two disadvantages that a large amount of water is required to adequately alter the relative volatility and, even though the solvent is introduced into the column several trays below the top tray, enough water is stripped by vapor traffic into the distillate to reduce the acetone purity to 95.6 mol%. The water vapor pressure can be lowered, and thus the purity of acetone distillate increased, by use of an aqueous inorganic-salt solution as the solvent. For example, a 1927 patent application by Othmer [23] describes the use of a concentrated, calcium-chloride brine. Not only does calcium chloride, which is highly soluble in water, reduce the volatility of water, but it also has a strong affinity for methanol. Thus, the relative volatility of acetone with respect to methanol is further enhanced. The separation of the brine solution from methanol is easily accommodated in the subsequent distillation step, with the brine solution recycled to the extractive distillation column. The vapor pressure of the dissolved salt is so small that it never enters the vapor phase, provided that entrainment is avoided. An even earlier patent by Van Raymbeke [24] describes the extractive distillation of ethanol from water by using solutions of calcium chloride, zinc chloride, or potassium carbonate in glycerol.

Rather than using a solvent that contains a dissolved salt, the salt can be added as a solid or melt directly into the column by dissolving it in the liquid reflux before it enters the column. This technique was demonstrated experimentally by Cook and Furter [25] in a 4-inch-diameter, 12-tray rectifying column with bubble caps for the separation of ethanol from water using potassium acetate. At salt concentrations

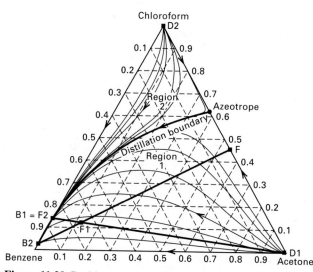

Figure 11.20 Residue-curve map for Example 11.4.

below saturation and between 5 and 10 mol%, an almost pure ethanol distillate was achieved. The salt, which must be soluble in the reflux, is recovered from the aqueous bottoms by evaporation and crystallization.

Salt distillation is accompanied by several potential problems. First and foremost is corrosion, particularly with aqueous chloride-salt solutions, which may require stainless steel or a more corrosion-resistant material. The feeding and dissolving of a salt into the reflux stream poses many problems, as described by Cook and Furter [25]. The solubility of the salt will be low in the reflux because it is rich in the more volatile component, while the salt will be most soluble in the less-volatile component. Consequently, the solid salt must be metered at a constant rate. The salt-feeding mechanism must avoid bridging and must prevent the entry of vapor, which could cause clogging upon condensation. The salt must be rapidly dissolved. The reflux must be maintained near the boiling point to avoid precipitation of already-dissolved salt in lines. In the column, the presence of the dissolved salt may increase the potential for foaming, which may require the addition of antifoaming agents and/or an

increase in column diameter. Some concern has been voiced for the possibility of crystallization of salt within the column. However, the concentration of the less-volatile component (e.g., water) increases down the column. Thus, the solubility of the salt increases down the column, while its concentration remains relatively constant. Therefore, the possibility of clogging and plugging due to solids formation in the column is highly unlikely.

In aqueous alcohol solutions, both *salting out* and *salting in* have been observed by Johnson and Furter [26], as shown in the vapor–liquid equilibrium data in Figure 11.21: in (a), sodium-nitrate salts out methanol, but in (b), mercuric chloride salts in methanol. Even low concentrations of potassium acetate can eliminate the ethanol–water azeotrope, as shown in Figure 11.21c. Mixed potassium- and sodium-acetate salts were used in Germany and Brazil from 1930 to 1965 for the separation of ethanol and water.

Surveys of the use of inorganic salts for extractive distillation, including effects on vapor–liquid equilibria, are given by Johnson and Furter [27], Furter and Cook [28], and Furter [29, 30]. A survey of methods for predicting the effect of

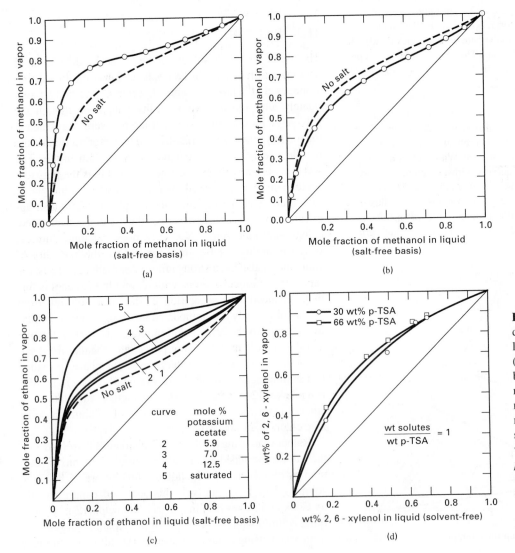

Figure 11.21 Effect of dissolved salts on vapor–liquid equilibria at 1 atm. (a) Salting-out of methanol by saturated aqueous sodium nitrate. (b) Salting-in of methanol by saturated aqueous mercuric chloride. (c) Effect of salt concentration on ethanol–water equilibria. (d) Effect of *p*-toluenesulfonic acid (*p*-TSA) on phase equilibria of 2,6 xylenol-*p*-cresol. [From A.I. Johnson and W.F. Furter, *Can. J. Chem. Eng.*, **43**, 356–358 (1965) with permission.]

inorganic salts on vapor–liquid equilibria is given by Kumar [31]. Column-simulation results, using the Newton–Raphson method, are presented by Llano-Restrepo and Aguilar-Arias [99] for the ethanol-water-calcium chloride system and by Fu [100], for the ethanol-water-ethanediol-potassium acetate system, who shows that his simulation results compare favorably with measurements on an industrial column.

Salt distillation can also be applied to the separation of organic compounds that have little capacity alone for dissolving inorganic salts by using a special class of organic salts called *hydrotropes*. Typical hydrotropic salts are alkali and alkaline-earth salts of the sulfonates of toluene, xylene, or cymene, and the alkali benzoates, thiocyanates, and salicylates. For example, Mahapatra, Gaikar, and Sharma [32] showed that the addition of aqueous solutions of 30 and 66 wt% *p*-toluenesulfonic acid to mixtures of 2,6-xylenol and *p*-cresol at 1 atm increased the relative volatility from approximately 1 to about 3, as shown in Figure 11.21d. Hydrotropes can also be used to enhance separations by liquid–liquid extraction, as shown by Agarwal and Gaikar [33].

11.4 PRESSURE-SWING DISTILLATION

When a binary azeotrope disappears at some pressure or changes composition by 5 mol% or more over a moderate range of pressure, consideration should be given to using, without a solvent, two distillation columns operating in series at different pressures. This process is referred to as *pressure-swing distillation* or *two-column distillation.* Knapp and Doherty [34] list 36 pressure-sensitive, binary azeotropes, taken mainly from the compilation of Horsley [11]. The effect of pressure on the temperature and composition of two minimum-boiling azeotropes is shown in Figure 11.22. The mole fraction of ethanol in the ethanol–water azeotrope increases from 0.8943 at 760 torr to more than 0.9835 at 90 torr. Although not shown in Figure 11.22b, the azeotrope finally disappears at below about 70 torr. A much more dramatic change in azeotropic composition with pressure is seen in Figure 11.22b for the ethanol–benzene system, which forms a minimum-boiling azeotrope at 44.8 mol%

ethanol and 1 atm. Applications of pressure-swing distillation, which was first noted by Lewis [35] in a 1928 patent, include the separations of the minimum-boiling azeotrope of tetrahydrofuran–water and the maximum-boiling azeotropes of hydrochloric acid–water and formic acid–water.

Consider the case, described by Van Winkle [36], of a minimum-boiling azeotrope for the mixture A–B, with T–y–x curves as shown in Figure 11.23a. As the pressure is decreased from P_2 to P_1, the azeotropic composition moves toward a smaller percentage of A. An operable pressure-swing sequence is shown in Figure 11.23b. The total feed, F_1, to Column 1, operating at the lower pressure, P_1, is the sum of the fresh feed, F, whose composition is richer in A than the azeotrope, and the recycled distillate, D_2, whose composition is close to that of the azeotrope at pressure, P_2. The compositions of D_2 and, consequently, F_1 are both richer in A than the azeotrope composition at P_1. The bottoms, B_1, leaving Column 1 is almost pure A. The distillate, D_1, which is slightly richer in A than the azeotrope, but less rich in A than the azeotrope at P_2, is fed to Column 2, where the bottoms, B_2, is almost pure B. Robinson and Gilliland [37] provide an example of the separation of ethanol and water, where the fresh-feed composition is less rich in ethanol than the azeotrope. For that case, the products are still removed as bottoms, but nearly pure B is taken from the first column and A from the second.

Pressure-swing distillation can also be applied to the separation of the less-common, maximum-boiling binary azeotropes. The sequence is shown in Figure 11.23c, where both products are withdrawn as distillates, rather than as bottoms. In this case, the composition of the azeotrope becomes richer in A as the pressure is decreased. The fresh feed, which is richer in A than the azeotrope at the higher pressure, is first distilled in Column 1 at the higher pressure, P_1, to produce a distillate of nearly pure A and a bottoms slightly richer in A than the azeotrope at the higher pressure. The bottoms is fed to Column 2, operating at the lower pressure, P_2, where the azeotrope composition is richer in A than the feed to that column. Accordingly, the distillate is nearly pure B, while the recycled bottoms from Column 2 is slightly less rich in A than the azeotrope at the lower pressure.

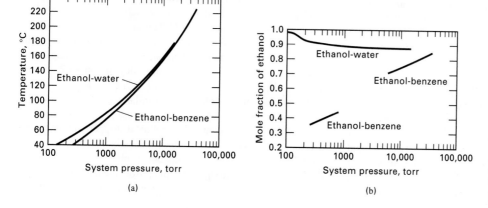

(a)   (b)

Figure 11.22 Effect of pressure on azeotrope conditions. (a) Temperature of azeotrope. (b) Composition of azeotrope.

[From *Perry's Chemical Engineers' Handbook* 6th ed., R.H. Perry and D.W. Green, Eds. McGraw-Hill (1984) with permission.]

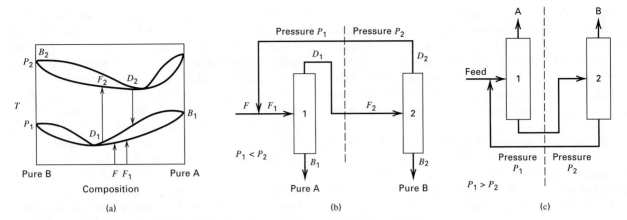

Figure 11.23 Pressure-swing distillation. (a) T–y–x curves at pressures P_1 and P_2 for minimum-boiling azeotrope. (b) Distillation sequence for minimum-boiling azeotrope. (c) Distillation sequence for maximum-boiling azeotrope.

For all pressure-swing-distillation sequences, the recycle ratio is an important factor in the design and depends on the difference in azeotropic composition for the column pressures. The following example illustrates the calculation and importance of the recycle stream.

EXAMPLE 11.5

Ninety mol/s of a mixture of two-thirds by moles ethanol and one-third benzene at the bubble point at 101.3 kPa is to be separated into 99 mol% ethanol and 99 mol% benzene. Ordinary distillation is not feasible because the mixture forms a minimum-boiling azeotrope at 760 torr with a composition of 44.8 mol% ethanol and a temperature of 68°C. If the pressure is reduced to 200 torr, as shown in Figure 11.22b, the azeotrope composition shifts to 36 mol% ethanol at 35°C. This magnitude of shift makes this a candidate for pressure-swing distillation.

Apply the sequence shown in Figure 11.23b. Let the first column operate with a top-tray pressure of 30 kPa (225 torr). Because the feed composition is greater than the azeotrope composition at the pressure of this column, the distillate composition approaches the minimum-boiling azeotrope at the top-tray pressure, and 99 mol% ethanol can be withdrawn as bottoms. The distillate is sent to the second column, which operates with a top-tray pressure of 106 kPa. The feed to this column has an ethanol content greater than that of the azeotrope at the pressure of the second column. Accordingly, the distillate composition approaches the azeotrope at the top-tray pressure, and 99 mol% benzene can be withdrawn as bottoms. The distillate is recycled to the first column. Design a pressure-swing distillation system for this separation.

SOLUTION

For the first column, which operates under vacuum, the reflux-drum and reboiler pressures are set at 26 and 40 kPa, respectively. For the second column, which operates just slightly above ambient pressure, the reflux-drum and reboiler pressures are set at 101.3 and 120 kPa, respectively. The bottoms compositions are specified at the required purities. The distillate composition for the first column is set at 37 mol% ethanol, slightly greater than the azeotrope

composition at 30 kPa. The distillate composition for the second column is set at 44 mol% ethanol, slightly less than the azeotrope composition at 106 kPa. With these composition specifications, material-balance calculations on ethanol and benzene give the following flow rates in moles per second.

Component	F	D_2	F_1	B_1	D_1	B_2
Ethanol	60.0	67.3	127.3	59.7	67.6	0.3
Benzene	30.0	85.6	115.6	0.6	115.0	29.4
Totals:	90.0	152.9	242.9	60.3	182.6	29.7

It is seen that the recycle molar flow rate, D_2, is about 10% greater than that of the fresh feed, F.

Equilibrium-stage calculations for the two columns were made with the ChemSep program, using total condensers and partial reboilers. For Column 1, a number of runs were made in an attempt to find optimal feed-tray locations for the fresh feed and the recycle, using a reasonable reflux rate that avoided any near-pinch conditions. The selected design uses seven theoretical trays (not counting the partial reboiler), with the recycle stream, at a temperature of 68°C, sent to Tray 3 from the top and the fresh feed to Tray 5 from the top. A reflux ratio of 0.5 is sufficient to achieve specifications. The resulting liquid-phase composition profile is shown in Figure 11.24a, where the desirable lack of composition pinch points is observed. The McCabe–Thiele diagram for Column 1 is given in Figure 11.24b, where the three operating lines are evident and optimal-feed locations are indicated. Because of the azeotrope, the operating lines and equilibrium curve all lie below the 45° line. The condenser duty is 9.88 MW, while the reboiler duty is 8.85 MW. The bottoms temperature is 56°C. This column was sized with the ChemCAD program for sieve trays on 24-inch tray spacing and a 1-inch weir height to minimize pressure drop. The resulting diameter is 3.2 meters (10.5 ft). A tray efficiency of about 47% is predicted, making the required number of trays equal to 15.

For the design of Column 2, a similar procedure was used to establish the optimal feed tray, total trays, and reflux ratio. The selected design turned out to be a refluxed stripper with only three theoretical stages (not counting the partial reboiler). A reflux rate of only 25.5 mol/s achieves the product specifications, with most of the liquid traffic in the stripper coming from the feed. The resulting liquid-phase composition profile is shown in Figure 11.25a, where,

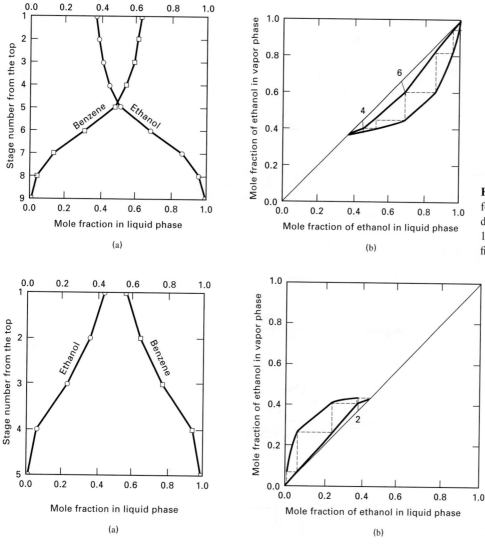

Figure 11.24 Computed results for Column 1 of pressure-swing distillation system in Example 11.5. (a) Liquid composition profiles. (b) McCabe–Thiele diagram.

Figure 11.25 Computed results for Column 2 of pressure-swing distillation system in Example 11.5. (a) Liquid composition profiles. (b) McCabe–Thiele diagram.

again, no composition pinches are evident. The McCabe–Thiele diagram for Column 2 is given in Figure 11.25b, where an optimal-feed location is indicated. The condenser duty is 6.12 MW, while the reboiler duty is 7.07 MW. The bottoms temperature is 84°C. This column was sized for the same conditions as Column 1, resulting in a column diameter of 2.44 meters (8 ft). A tray efficiency of 50% results in 6 actual trays.

11.5   HOMOGENEOUS AZEOTROPIC DISTILLATION

As discussed earlier, an azeotrope can be separated by extractive distillation, using a solvent that is higher boiling (less volatile) than the components in the feed, and that does not form an azeotrope with any of them. Alternatively, the separation can be made by *homogeneous azeotropic distillation,* using an *entrainer* that is not subject to such restrictions. Like extractive distillation, a sequence of two or three distillation columns is used. Alternatively, the sequence is a hybrid system that includes separation operations other than distillation, such as liquid–liquid extraction.

The conditions that a potential entrainer must satisfy for homogeneous azeotropic distillation to be feasible has been a subject of study by a number of investigators, including Doherty and Caldarola [16], Stichlmair, Fair, and Bravo [1], Foucher, Doherty, and Malone [14], Stichlmair and Herguijuela [18], Fidkowski, Malone, and Doherty [13], Wahnschafft and Westerberg [38], and Laroche, Bekiaris, Andersen, and Morari [39]. If it is assumed that a distillation boundary, if any, of a residue-curve map is straight or cannot be crossed, the conditions of Doherty and Caldarola apply. These conditions are based on the rule that for a potential entrainer, E, the two components, A and B, to be separated, or any product azeotrope, must lie in the same distillation region of the residue-curve map. Thus, a distillation boundary cannot be connected to the A-B azeotrope. Furthermore, A or B, but not both, must be a saddle. The maps suitable for a sequence that includes homogeneous azeotropic distillation together with ordinary distillation are classified into the five groups illustrated in Figure 11.26a, b, c, d, and e. The figure for each group includes the applicable residue-curve maps and the sequence of separation columns used to

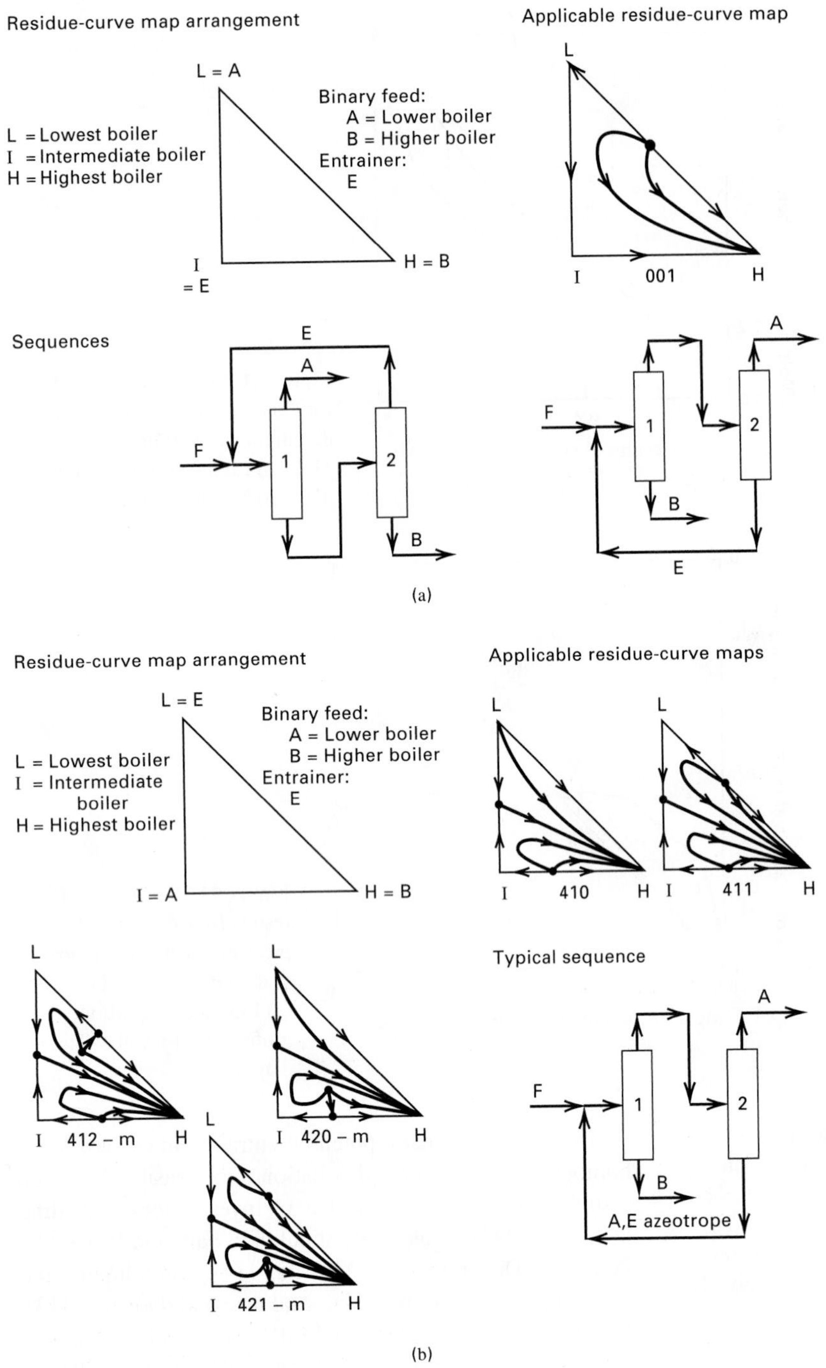

Residue-curve map arrangement

Applicable residue-curve map

Sequences

(a)

Residue-curve map arrangement

Applicable residue-curve maps

Typical sequence

Figure 11.26 Residue-curve maps and distillation sequences for homogeneous azeotropic distillation. (a) Group 1: A and B form a minimum-boiling azeotrope, I = E, E forms no azeotropes. (b) Group 2: A and B form a minimum-boiling azeotrope, L = E, E forms a maximum-boiling azeotrope with A. (*continued*)

(b)

separate A from B and recycle the entrainer. For all groups, the residue-curve map is drawn in the manner of Doherty and Caldarola [16], with the lowest-boiling component, L, at the top vertex; the intermediate-boiling component, I, at the bottom-left vertex; and the highest-boiling component, H, at the bottom-right vertex. Component A is the lower-boiling component of the binary mixture and B the higher-boiling. For the first three groups, A and B form a minimum-boiling azeotrope; for the other two groups, they form a maximum-boiling azeotrope.

In Group 1, the intermediate boiler, I, is E, which forms no azeotropes with A and/or B. As shown in Figure 11.26a,

this case, like extractive distillation, involves no distillation boundary. Two sequences are shown, both of which assume that the fresh feed, *F*, of A and B, as fed to Column 1, is close to the azeotropic composition. Thus, this feed may be the distillate from a previous column used to produce the azeotrope from the original mixture of A and B. Either the *direct sequence,* in which Column 2 is fed by the bottoms from Column 1, or the *indirect sequence,* in which Column 2 is fed by the distillate from Column 1, may be used. In the first sequence, the entrainer is recovered as distillate from Column 2 and recycled to Column 1. In the second sequence, the entrainer is recovered as bottoms from Column 2 and

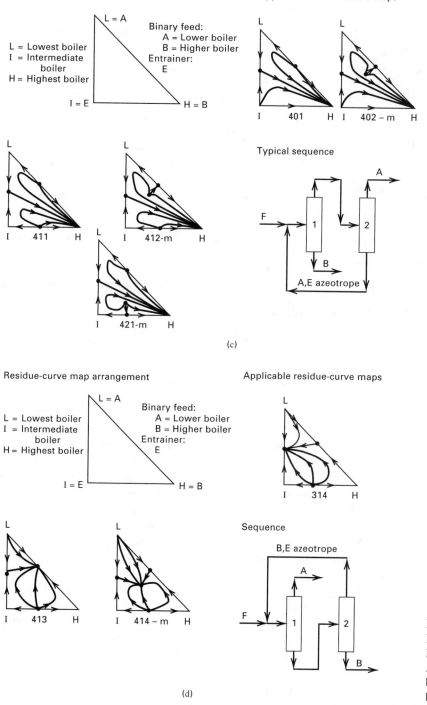

Residue-curve map arrangement

L = Lowest boiler
I = Intermediate boiler
H = Highest boiler

Binary feed:
A = Lower boiler
B = Higher boiler
Entrainer:
E

L = A

I = E   H = B

Applicable residue-curve maps

I   401   H   I   402 – m   H

I   411   H   I   412-m   H

I   421-m   H

Typical sequence

F

A

B

A,E azeotrope

(c)

Residue-curve map arrangement

L = Lowest boiler
I = Intermediate boiler
H = Highest boiler

Binary feed:
A = Lower boiler
B = Higher boiler
Entrainer:
E

L = A

I = E   H = B

Applicable residue-curve maps

I   314   H

I   413   H   I   414 – m   H

Sequence

B,E azeotrope

A

F

B

(d)

Figure 11.26 (*Continued*) (c) Group 3: A and B form a minimum-boiling azeotrope, I = E, E forms a maximum-boiling azeotrope with A. (d) Group 4: A and B form a maximum-boiling azeotrope, I = E, E forms a minimum-boiling azeotrope with B.

recycled to Column 1. Although both sequences show the entrainer being combined with the fresh feed before being fed to Column 1, the fresh feed and the recycled entrainer can be fed to different trays to enhance the separation.

In Group 2, the low boiler, L, is E, which forms a maximum-boiling azeotrope with A. Entrainer E may also form a minimum-boiling azeotrope with B, and/or a minimum-boiling (unstable node) ternary azeotrope. Thus, in Figure 11.26b, any of the five residue-curve maps shown may apply. In all five cases, a distillation boundary exists,

which is directed from the maximum-boiling azeotrope of A–E to pure B, the high boiler. A feasible indirect or direct sequence is restricted to the subtriangle bounded by the vertices of pure components A, B, and the binary azeotrope of A–E. An example of an indirect sequence is included in Figure 11.26b. In this case, the azeotrope of A–E is recycled to Column 1 from the bottoms of Column 2. Alternatively, as shown in Figure 11.26c for Group 3, A and E may be switched to make A the low boiler and E the intermediate boiler, which again forms a maximum-boiling azeotrope

Residue-curve map arrangement   Applicable residue-curve maps

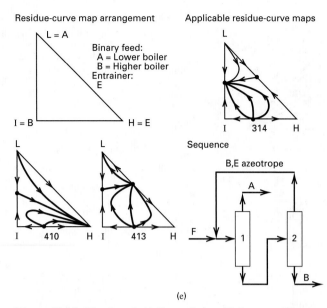

(e)

Figure 11.26 (*Continued*) (e) Group 5: A and B form a maximum-boiling azeotrope, H = E, E forms a minimum-boiling azeotrope with B.

with A. All sequences for Group 3 are confined to the same subtriangle as for Group 2.

Groups 4 and 5, shown in Figures 11.26d and e, respectively, are similar to Groups 2 and 3. However, A and B now form a maximum-boiling azeotrope. In Group 4, the entrainer is the intermediate boiler, which forms a minimum-boiling azeotrope with B. The entrainer may also form a maximum-boiling azeotrope with A, and/or a maximum-boiling (stable node) ternary azeotrope. A feasible sequence is restricted to the subtriangle formed by vertices A, B, and the B–E azeotrope. In the sequence shown, the distillate from Column 2, which is the minimum-boiling azeotrope of B and E, is mixed with the fresh feed to Column 1, which produces a distillate of pure A. The bottoms from Column 1 has a composition such that when fed to Column 2, a bottoms of pure B can be produced. Although just a direct sequence is shown, the indirect sequence can also be used. Alternatively, as shown in Figure 11.26e for Group 5, B and E may be switched to make E the high boiler. In the sequence shown, as in the sequence of Figure 11.26d, the bottoms from Column 1, again, has a composition such that when fed to Column 2, a bottoms of pure B can be produced. Otherwise, the other conditions and the sequences are the same as for Group 4.

The distillation boundaries for the hypothetical ternary systems in Figure 11.26 are shown as straight lines. When a distillation boundary is curved, it may be crossed, provided that both the distillate and bottoms products lie on the same side of the boundary.

It is often difficult to find an entrainer for a sequence involving homogeneous azeotropic distillation and ordinary distillation. However, azeotropic distillation can also be incorporated into a hybrid sequence involving separation operations other than distillation. In that case, some of the

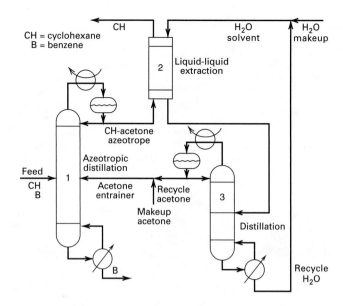

Figure 11.27 Separation sequence for separating cyclohexane and benzene using homogeneous azeotropic distillation with acetone entrainer.

[From *Perry's Chemical Engineers' Handbook,* 6th ed., R.H. Perry and D.W. Green, Eds., McGraw-Hill, New York (1984) with permission.]

restrictions for the entrainer and the resulting residue-curve map may not apply. For example, the separation of the very close-boiling and minimum-azeotrope-forming system of benzene and cyclohexane using acetone as the entrainer violates the restrictions for a distillation-only sequence because the ternary system involves only two minimum-boiling binary azeotropes. However, the separation can be achieved by the sequence shown in Figure 11.27, which involves: (1) homogeneous azeotropic distillation with acetone entrainer to produce a bottoms product of nearly pure benzene and a distillate close in composition to the minimum-boiling binary azeotrope of acetone and cyclohexane; (2) liquid–liquid extraction of the distillate with water to give a raffinate of nearly pure cyclohexane and an extract of acetone and water; and (3) ordinary distillation of the extract to recover the acetone for recycle. As shown in the following example, the azeotropic distillation column is still subject to product-composition-region restrictions.

EXAMPLE 11.6

Benzene (nbp = 80.13°C) and cyclohexane (nbp = 80.64°C) form a minimum-boiling homogeneous azeotrope at 1 atm and 77.4°C that contains 54.2 mol% benzene. Thus, they cannot be separated by ordinary distillation at 1 atm. Instead, it is proposed to separate them by using acetone as the entrainer in the separation sequence shown in Figure 11.27. The fresh feed to the azeotropic column consists of 100 kmol/h of 75 mol% benzene and 25 mol% cyclohexane. Determine a feasible acetone-addition rate to the feed so that nearly pure benzene can be obtained as the bottoms product. Acetone (nbp = 56.14°C) forms a minimum-boiling azeotrope with cyclohexane (but not benzene) at 53.4°C and 1 atm at 74.6 mol% acetone. The residue-curve map at 1 atm is given in Figure 11.28.

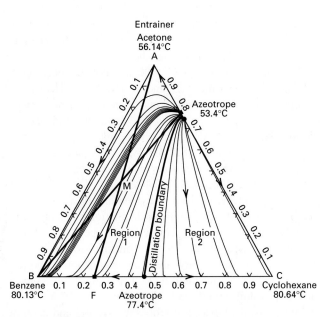

Figure 11.28 Residue-curve map for Example 11.6.

Table 11.4 Effect of Acetone-Entrainer Flow Rate on Benzene Purity for the Homogeneous-Azeotropic-Distillation Process of Example 11.6

Case	Acetone Flow Rate, kmol/h	Benzene Purity in Bottoms, %
1	50	88.69
2	75	94.21
3	100	99.781
4	125	99.779

SOLUTION

The residue-curve map shows a slightly curved distillation boundary connecting the two azeotropes and dividing the diagram into distillation regions, 1 and 2. The fresh-feed composition is designated in Figure 11.28 by a filled-in box labeled F. If a straight line is drawn from F to the pure acetone apex, A, the mixture of the fresh feed and the acetone entrainer must lie somewhere in Region 1 on this line. Suppose that the 100 kmol/h of fresh feed is combined with an equal flow rate of entrainer. The mixing point, M, is located at the midpoint of the line connecting F and A. If a line is drawn from the benzene apex, B, through M and to the side of the triangle that connects the acetone apex to the cyclohexane apex, it does not cross the distillation boundary separating the two regions, but lies completely in Region 1. Thus, the separation into a nearly pure benzene bottoms and a distillate mixture containing mainly acetone and cyclohexane is possible. This is confirmed by calculations with the ASPEN PLUS process simulator for a column operating at 1 atm with 38 theoretical stages, a total condenser, a partial reboiler, a reflux ratio of 4, a bottoms product flow rate of 75 kmol/h (equivalent to the benzene flow rate in the feed to the column), and a bubble-point combined feed sent to Stage 19 from the top. The resulting product flow rates are listed in Table 11.4 as Case 3, where it is seen that a bottoms of 99.8 mol% benzene is achieved with a benzene recovery of the same value. A higher entrainer flow rate of 125 kmol/h, included in Table 11.4 as Case 4, is also successful in achieving high benzene-bottoms-product purity and recovery. However, if only 75 kmol/h (Case 2) or 50 kmol/h (Case 1) of entrainer is used, a nearly pure benzene bottoms is not achieved because of the distillation boundary restriction.

11.6 HETEROGENEOUS AZEOTROPIC DISTILLATION

The requirement, for a distillation sequence based on homogeneous azeotropic distillation, that A and B must lie in the same distillation region of the residue-curve map with entrainer E, is so restrictive that it is usually difficult, if not impossible, to find a feasible entrainer. The Group 1 map in Figure 11.26a requires that the entrainer not form an azeotrope but yet be the intermediate-boiling component, while the other two components form a minimum-boiling azeotrope. Such systems are rare, because most intermediate-boiling entrainers form an azeotrope with one or both of the other two azeotrope-forming components. The other four groups in Figure 11.26 all require that at least one maximum-boiling azeotrope be formed. However, such azeotropes are far less common than minimum-boiling azeotropes. The result is that industrial applications of distillation sequences based on homogeneous azeotropic distillation are not common.

An alternative technique that does find wide industrial application is heterogeneous azeotropic distillation, which is used to separate close-boiling binary mixtures and minimum-boiling binary azeotropes by employing an entrainer that forms a binary and/or ternary heterogeneous azeotrope. As discussed in Section 4.3, a heterogeneous azeotrope is one involving more than one liquid phase. The overall composition of the liquid phases is equal to that of the vapor phase. Thus, all three phases have different compositions. The overhead vapor from the column is close to the composition of the heterogeneous azeotrope. When condensed, two liquid phases form in a decanter downstream of the condenser. After separation in the decanter, most or all of the entrainer-rich liquid phase is returned to the column as reflux, while most or all of the other liquid phase is sent to the next column for further separation. Because these two phases usually lie in different distillation regions of the residue-curve map, the restriction that usually dooms distillation sequences based on homogeneous azeotropic distillation is overcome. Thus, in heterogeneous azeotropic distillation, the components to be separated need not lie in the same distillation region.

Heterogeneous azeotropic distillation has been practiced for almost a century, first by batch and then by continuous processing. Two of the most widely used applications are (1) the use of benzene or one of a number of other entrainers to separate the minimum-boiling azeotrope of ethanol and water, and (2) the use of ethyl acetate or one of a number of other entrainers to separate the close-boiling mixture of acetic acid and water. Other applications, cited by Widagdo and Seider [19], include dehydrations of isopropanol with isopropylether, *sec*-butyl-alcohol with *disec*-butyl-ether, chloroform with mesityl oxide, formic acid with toluene, and acetic acid with toluene. Also, dehydration of tanker-transported feedstocks such as styrene and benzene is a major application.

Consider the separation of the azeotrope of ethanol and water by heterogeneous azeotropic distillation. The two most widely used entrainers are benzene and diethyl ether. A number of other entrainers are feasible, including *n*-pentane, illustrated later in Example 11.7, and cyclohexane. In 1902, Young [40] discussed the use of benzene as an entrainer for the batch dehydration of ethanol, in perhaps the first application of heterogeneous azeotropic distillation. In 1928, Keyes obtained a patent [41] on a continuous process, discussed in a 1929 article [42]. A residue-curve map, computed by Bekiaris, Meski, and Morari [43] for the ethanol (E)–water (W)–benzene (B) system at 1 atm, using the UNIQUAC equation (with parameters from ASPEN PLUS) for liquid-phase activity coefficients, is shown in Figure 11.29. Superimposed on the residue-curve map is a bold-dashed binodal curve for the boundary of the two liquid-phase region. The normal boiling points of E, W, and B are 78.4, 100, and 80.1°C, respectively. The UNIQUAC equation predicts that homogeneous minimum-boiling azeotropes AZ1 and AZ2 are formed by E and W at 78.2°C and 10.0 mol% W, and by E and B at 67.7°C and 44.6 mol% E, respectively. A heterogeneous minimum-boiling azeotrope AZ3 is predicted for W and B at 69.3°C, with a vapor composition of 29.8 mol% W. The overall composition of the two liquid phases is the same as that of the vapor, but each liquid phase is almost pure. The

B-rich liquid phase is predicted to contain 0.55 mol% W, while the W-rich liquid phase contains only 0.061 mol% B. A ternary minimum-boiling heterogeneous azeotropic AZ4 is predicted at 64.1°C, with a vapor composition of 27.5 mol% E, 53.1 mol% B, and 19.4 mol% W. The overall composition of the two liquid phases of the ternary azeotrope is the same as that of the vapor, but a thin, dashed tie line through the AZ4 point shows that the benzene-rich liquid phase contains 18.4 mol% E, 79.0 mol% B, and 2.6 mol% W, while the water-rich liquid phase contains 43.9 mol% E, 6.3 mol% B, and 49.8 mol% W.

In Figure 11.29, the map is divided into three distillation regions by three, thick, solid-line distillation boundaries that each extend from the ternary azeotrope to a binary azeotrope. Each distillation region contains one pure component. Because the ternary azeotrope is the lowest-boiling azeotrope, it is an unstable node. Because all three binary azeotropes boil below the boiling points of the three pure components, the binary azeotropes are saddles and the pure components are stable nodes. Accordingly, all residue curves begin at the ternary azeotrope and terminate at a pure component apex. Liquid–liquid solubility is shown as a thick, dashed, curved line. However, this curve is not like the usual ternary solubility curve, because it is for isobaric, rather than isothermal, conditions. Superimposed on the distillation boundary that separates distillation regions 2 and 3 are thick dashes that represent the vapor composition in equilibrium with two liquid phases. The compositions of the two equilibrium liquid phases for a particular vapor composition are obtained from the two ends of the straight tie line that passes through the vapor composition point and terminates at the liquid solubility curve. The only tie line shown in Figure 11.29 is a thin, dashed line that passes through the ternary azeotrope. Other tie lines, which would represent other temperatures, could be added; however, in most heterogeneous azeotropic distillation operations, an attempt is made to restrict the formation of two liquid phases to just the decanter downstream of the condenser where the composition approaches the ternary azeotrope.

Figure 11.29 clearly shows how a distillation boundary is crossed by the tie line through AZ4 to form two liquid phases in the decanter. This phase split is utilized in the following manner by a typical azeotropic-tower operation for the dehydration of ethanol by benzene. The tower is treated as a column with no condenser, a main feed that enters a few trays below the top of the column, and the reflux of benzene-rich liquid as a second feed. The composition of the combined two feeds lies in distillation region 1. Thus, from the directions of the residue curves, the products of the tower can be a bottoms of nearly pure ethanol and an overhead vapor approaching the composition of the ternary azeotrope. When that vapor is condensed, phase splitting occurs to give a water-rich phase that lies in distillation region 3 and an entrainer-rich phase in distillation region 2. Thus, if the water-rich phase is sent to a reboiled stripper, the residue curves indicate that a nearly pure-water bottoms can be

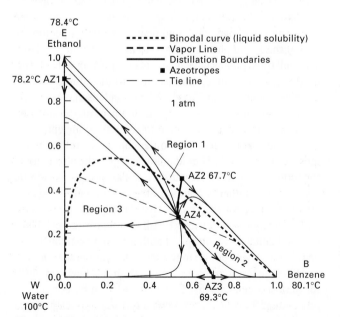

Figure 11.29 Residue-curve map for the ethanol–water–benzene system at 1 atm.

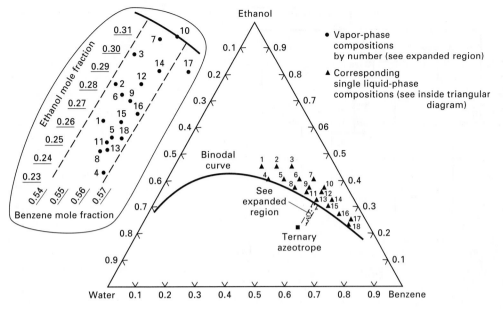

Figure 11.30 Overhead vapor compositions not in equilibrium with two liquid phases.

[From J. Prokopakis and W.D. Seider, *AIChE J.*, **29**, 49–60 (1983) with permission.]

produced, with the overhead vapor, rich in ethanol, recycled to the decanter. When the entrainer-rich phase in distillation region 2 is added to the main feed, which lies in distillation region 1, the overall composition lies in region 1.

It is preferable to restrict the formation of two liquid phases to the decanter. To avoid formation of two liquid phases on the top trays of the azeotropic tower, the composition of the vapor leaving the top tray must be such that the equilibrium liquid lies outside of the two-phase liquid region enclosed by the binodal curve and the base of the triangle in Figure 11.29. As shown in a comprehensive study by Prokopakis and Seider [44], vapor compositions that form two liquid phases when condensed, but are in equilibrium with only one liquid phase on the top tray, are restricted to a very small window, as shown in Figure 11.30. Furthermore, that window can only be achieved by adding to the reflux of entrainer-rich liquid phase a portion of the water-rich liquid phase or a portion of the condensed vapor prior to separation in the decanter.

A variety of column sequences for heterogeneous azeotropic distillation have been proposed. Three of these that utilize only distillation, taken from a study by Ryan and Doherty [45], are shown in Figure 11.31. Most common is the three-column sequence, in which an aqueous feed dilute in ethanol is first preconcentrated in Column 1 to obtain a nearly pure water-bottoms product and a distillate with composition approaching that of the binary azeotrope. The latter is fed to the azeotropic tower, Column 2, where nearly pure ethanol is recovered as the bottoms product and the tower is refluxed by most or all of the entrainer-rich liquid phase from the decanter. The water-rich phase, which contains ethanol and a small amount of entrainer, is sent to the entrainer-recovery column, which is a distillation column with both rectifying and stripping sections, or a stripper. The distillate from the recovery column is recycled to the azeotropic column. Alternatively, the distillate from

Column 3 could be recycled to the decanter. As shown in all three sequences of Figure 11.31, portions of either liquid phase from the decanter can be returned to the azeotropic tower or to the next column in the sequence to control phase splitting on the top trays of the azeotropic tower.

A four-column distillation sequence is shown in Figure 11.31b. The first column is identical to the first column of the three-column sequence of Figure 11.31a. The second column is the azeotropic column, which is fed by the near-azeotrope distillate of ethanol and water from Column 1 and by a recycle distillate of about the same composition from Column 4. The purpose of Column 3 is to remove, as distillate, the entrainer from the water-rich liquid phase leaving the decanter and recycle it back to the decanter. Ideally, the composition of this distillate is identical to that of the vapor distillate from Column 2. The bottoms from Column 3 is separated in Column 4 into a bottoms of nearly pure water, and a distillate that approaches the ethanol–water azeotrope and is therefore recycled to the feed to Column 2. A study by Pham and Doherty [46] found no advantage for the four-column sequence over the three-column sequence.

A novel two-column distillation sequence, due to Lynn and described by Ryan and Doherty [45], is shown in Figure 11.31c. The feed is sent to Column 2, which is a combined preconcentrator and entrainer recovery column. The distillate from this column is the feed to the azeotropic column. The bottoms from Column 1 is nearly pure ethanol, while Column 2 produces a bottoms of nearly pure water. For feeds that are very dilute in ethanol, Ryan and Doherty found that the two-column sequence has a lower investment cost, but a higher operating cost, than the three-column sequence. For feeds that are richer in ethanol, these two sequences are economically comparable.

The ethanol–benzene–water residue curve map of Figure 11.29 is only one of a number of different residue-curve maps that can lead to feasible distillation sequences that

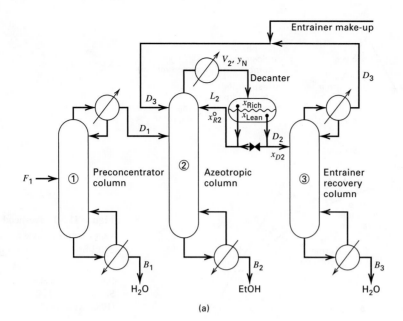

(a)

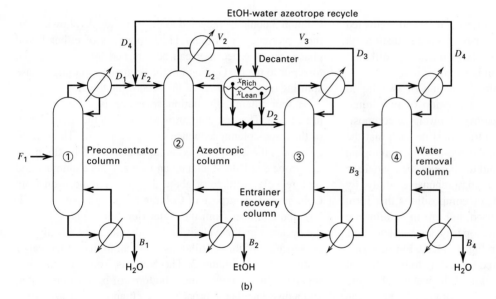

(b)

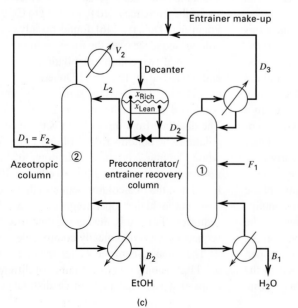

(c)

Figure 11.31 Distillation sequences for heterogeneous azeotropic distillation:
(a) Three-column sequence;
(b) four-column sequence;
(c) two-column sequence.

[From P.J. Ryan and M.F. Doherty, *AIChE J.,* **35,** 1592–1601 (1989) with permission.]

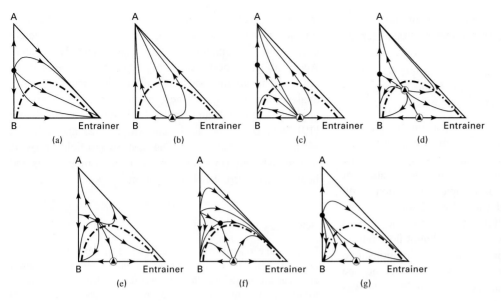

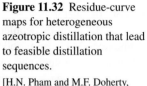

Figure 11.32 Residue-curve maps for heterogeneous azeotropic distillation that lead to feasible distillation sequences.

[H.N. Pham and M.F. Doherty, *Chem. Eng. Sci.,* **45,** 1845–1854 (1990) with permission.]

include heterogeneous azeotropic distillation. Pham and Doherty [46] note that a feasible entrainer is one that causes phase splitting over a portion of the three-component composition region, but does not cause the two components of the feed to be placed in different distillation regions. Figure 11.32 shows seven such maps, where the dash-dot lines are the liquid–liquid solubility (binodal) curves.

The convergence of rigorous calculations for heterogeneous azeotropic distillation columns by the methods described in Chapter 10 can be extremely difficult, especially when the convergence of the entire sequence is attempted. For calculation purposes, it is preferable to uncouple the

columns by using a residue-curve map to establish, by material-balance calculations, the flow rates and compositions of the feeds and products for each column. This procedure is illustrated for a three-column sequence in Figure 11.33, where the dash–dot lines separate the three distillation regions, the short-dash line is the liquid–liquid solubility curve, and the remaining lines are material-balance lines. Each column in the sequence can be computed separately. Even then, the calculations can be so sensitive, because of nonidealities in the liquid phase and possible phase splitting, that it may be necessary to use more robust methods. Among the most successful approaches for the most difficult cases are the boundary-value, tray-by-tray method of Ryan and Doherty [45], the homotopy-continuation method of Kovach and Seider [47], and the collocation method of Swartz and Stewart [48].

Multiplicity of Solutions

Solutions to mathematical models for operations of interest to chemical engineers are not always unique. The existence of multiple, steady-state solutions for the continuous, stirred-tank reactor (CSTR) has been known since at least 1922 and is described in detail in a number of textbooks on chemical reaction engineering. The existence of multiplicity in steady-state separation problems is a relatively new discovery. Gani and Jørgensen [49] define the following three types of multiplicity, all of which can, under certain conditions, occur in distillation simulations:

1. *Output multiplicity,* where all input variables are specified and more than one solution for the set of output variables is found. For example, for a distillation column, the feed condition, number of stages, feed-stage location, distillate flow rate, reflux ratio, type condenser and reboiler, and column-pressure profile might be specified and two or more sets of product compositions and column profiles found.

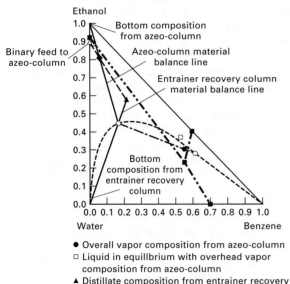

- ● Overall vapor composition from azeo-column
- □ Liquid in equilibrium with overhead vapor composition from azeo-column
- ▲ Distillate composition from entrainer recovery column
- ◆ Overall feed composition to azeo-column
- ■ Azeotrope

Figure 11.33 Material-balance lines for the three-column sequence of Figure 11.31a.

[From P.J. Ryan and M.F. Doherty, *AIChE J.,* **35,** 1592–1601 (1989) with permission.]

2. *Input multiplicity,* where one or more output variables are specified and multiple solutions are found for the unknown input variables.

3. *Internal-state multiplicity,* where multiple sets of internal conditions or profiles are found for the same values of the input and output variables.

Of particular interest here is output multiplicity for azeotropic distillation, which was first discovered by Shewchuk [50] in 1974. With different starting guesses, he found two steady-state solutions for the dehydration of ethanol by heterogeneous azeotropic distillation with benzene. In a more detailed study for the same system, Magnussen, Michelsen, and Fredenslund [51] found, with difficulty, for a rather narrow range of ethanol flow rate in the top feed to the column, three steady-state solutions, two of which are stable. The unstable solution can not be achieved in an operating plant. A similar multiplicity was found when pentane was used as the entrainer. One of the two stable solutions predicts a far purer ethanol bottoms product than the other stable solution. Thus, from a practical standpoint it is important to obtain all stable solutions when more than one exists. Subsequent studies, some contradictory, show that multiple solutions usually persist only over a narrow range of distillate- or bottoms-flow rate specifications, but may exist over a wide range of reflux rate provided that a sufficient number of stages are present. Composition profiles of five multiple solutions found by Kovach and Seider [47] for a 40-tray ethanol–water–benzene heterogeneous azeotropic distillation are shown in Figure 11.34. The variation in the profiles is extremely large. Again, when multiple solutions exist, it is important to locate them. Unfortunately, the use of current steady-state, computer-aided process design and simulation programs to find multiple solutions is fraught with a number of difficulties because: (1) azeotropic columns are difficult to converge to even one solution, (2) multiple solutions may exist only in a very restricted range, (3) the multiple solutions can only be found

in these programs by changing the initial guesses of the composition profiles, and (4) the choice of activity-coefficient correlation and interaction parameters can be crucial. Accordingly, the best results have been obtained when more advanced techniques such as continuation and bifurcation analysis are employed. These methods are described and applied by Kovach and Seider [47], Widagdo and Seider [19], Bekiaris, Meski, Radu, and Morari [52], and Bekiaris, Meski, and Morari [43]. The last two articles provide reasons why multiple solutions occur in homogeneous and heterogeneous azeotropic distillation.

EXAMPLE 11.7

Design and economic studies by Black and Ditsler [53] and Black, Golding, and Ditsler [54] show that *n*-pentane is a superior entrainer for the dehydration of ethanol. Like benzene, *n*-pentane forms a minimum-boiling heterogeneous ternary azeotrope with ethanol and water. Design a separation system for the dehydration of 16.8176 kmol/h of 80.937 mol% ethanol and 19.063 mol% water as a liquid at 344.3 K and 333 kPa, using *n*-pentane as an entrainer, to produce 99.5 mol% ethanol, and water with less than 100 ppm (by weight) of combined ethanol and *n*-pentane.

SOLUTION

A heterogeneous azeotropic distillation process for this ternary system has been studied extensively by Black [55], who proposed the two-column process flow diagram shown in Figure 11.35. The process consists of an 18-equilibrium-stage heterogeneous azeotropic distillation column (C-1) equipped with a total condenser and a partial reboiler, a decanter (D-1), a 4-equilibrium-stage reboiled stripper (C-2), and a condenser (E-1) to condense the overhead vapor from C-2. Each reboiler adds the equivalent of another equilibrium stage. Column C-1 operates at a bottoms pressure of 344.6 kPa with a column pressure drop of 13.1 kPa. Column C-2 operates at a top pressure of 308.9 kPa, with a column pressure drop of 3.0 kPa. These pressures permit the use of cooling water in the condensers. Purity specifications are placed on the bottoms products. The feed enters C-1 at Stage 3 from the top. The ethanol product is withdrawn from the bottom of C-1. A small *n*-pentane makeup stream, not shown in Figure 11.35, enters Stage 2 from the top. The overhead vapor from C-1 is condensed and sent to D-1, where a pentane-rich liquid phase and a water-rich liquid phase are formed. The pentane-rich phase is returned to C-1 as reflux, while the water-rich phase is sent to C-2, where the water is stripped of residual pentane and ethanol to produce a bottoms of the specified water purity. Twenty percent of the condensed vapor from C-2 is returned to D-1. To ensure that two liquid phases form in the decanter but not on the trays of C-1, the remaining 80% of the condensed vapor from C-2 is combined with the pentane-rich phase from D-1 for use as additional reflux to C-1. The specifications for the problem are included on Figure 11.35.

A very important step in the design of a heterogeneous azeotropic distillation column is the selection of a suitable method for predicting liquid-phase activity coefficients and the determination of the binary interaction parameters. The latter usually involves the regression of both vapor–liquid (VLE) and liquid–liquid (LLE) experimental equilibrium data for all binary pairs. If

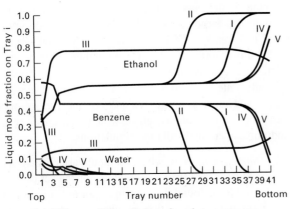

Figure 11.34 Five multiple solutions for a heterogeneous distillation operation.

[From J.W. Kovach III and W.D. Seider, *Computer Chem. Engng.,* **11,** 593 (1987) with permission.]

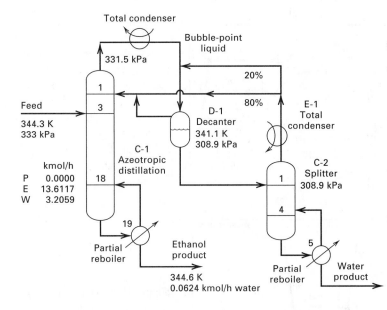

Figure 11.35 Process flow diagram for Example 11.7.
[From *Perry's Chemical Engineers' Handbook,* 6th ed., R.H. Perry and D.W. Green, Eds., McGraw-Hill, New York (1984) with permission.]

available, ternary data can also be included in the regression. Unfortunately, for most activity-coefficient prediction methods, it is difficult to simultaneously fit VLE and LLE data. Accordingly, often different binary-interaction parameters are used for the azeotropic column where VLE is important and for the decanter where LLE is important. This has been found to be particularly desirable for the ethanol–water–benzene system. For this example, however, the use of a single set of binary-interaction parameters with the modification by Black [56] of the Van Laar equation was deemed adequate. The binary-interaction parameters are listed by Black, Golding, and Ditsler [54].

The calculations were made with the Process simulation program of Simulation Sciences, Inc., using their rigorous distillation routine to model the columns and a three-phase-flash routine to model the decanter. Because the entrainer was internal to the system, except for a very small makeup rate, it was necessary to provide a reasonable initial guess for the component flow rates in the combined feed to the decanter. The guessed values in kilomoles per hour were 25.0 for *n*-pentane, 3.0 for ethanol, and 7.5 for water. The converged material balance is given in Table 11.5, where it is seen that the product specifications are met and approximately 22.6 kmol/h of *n*-pentane circulates through the top trays of the azeotropic distillation column. The computed condenser and

reboiler duties for Column C-1 are 1,116.5 and 1,135.0 MJ/h, respectively. The reboiler duty for Column C-2 is 486 MJ/h and the duty for Condenser E-1 is 438 MJ/h.

Because of the large effect of composition on liquid-phase activity coefficients, column profiles for azeotropic columns often show steep fronts. In Figure 11.36a to c, stage temperatures, total vapor and liquid flow rates, and liquid-phase compositions for Column C-1 vary only slightly from the reboiler (Stage 19) up to Stage 13. In this region, the liquid phase is greater than 99 mol% ethanol, whereas the *n*-pentane concentration slowly builds up from a negligible concentration in the bottoms to just less than 0.02 mol% at Stage 13. From Stage 13 to Stage 8, the *n*-pentane mole fraction in the liquid increases very rapidly to 53.8 mol%. In the same region, the temperature decreases sharply from 385.6 K to 348.4 K. Continuing up the column from Stage 8 to Stage 3, where the feed enters, the most significant change is the mole fraction of water in the liquid. Rather drastic changes in all variables take place about Stage 3. The large effects of *n*-pentane concentration on the relative volatility of water to ethanol, and of water concentration on the relative volatility of *n*-pentane to ethanol are shown in Figure 11.36d, where the variation over the column is about 10-fold for each pair.

No phase splitting occurs in either column, but two liquid phases of drastically different composition are formed and separated in the decanter. The light phase, which is almost twice the quantity of the heavy phase, is 95 mol% *n*-pentane, whereas the heavy phase is 90 mol% water. These extremely different compositions are due to the small amount of ethanol in the overhead vapor from C-1. Because of the high concentration of water in the feed to the stripper, C-2, the concentrations of ethanol and *n*-pentane in the liquid phase are quickly reduced to parts-per-million levels. Temperatures, vapor flow rates, and liquid flow rates in the stripper, C-2, are almost constant at 408 K, 15.6 kmol/h, and 12.4 kmol/h, respectively. Because of the large relative volatility of ethanol with respect to water (approximately 9) under the dilute ethanol conditions in C-2, the ethanol mole fraction decreases by almost an order of magnitude for each equilibrium stage. The extremely large relative volatility of *n*-pentane to water (more than 1,000) causes the *n*-pentane to be entirely stripped in just two stages.

Table 11.5 Converged Material Balance for Example 11.7

	Flow rate, kmol/h			
Stream	*n*-Pentane	Ethanol	Water	Total
C-1 feed	0.0000	13.6117	3.2059	16.8176
C-1 overhead	22.5565	2.1298	10.7269	35.4132
C-1 bottoms	0.0000	13.6117	0.0624	13.6741
C-1 reflux	22.5565	2.1298	7.5834	32.2697
D-1 *n*C5-rich	22.5500	1.0637	0.1129	23.7266
D-1 water-rich	0.0081	1.3326	12.4816	13.8223
C-2 overhead	0.0081	1.3326	9.3381	10.6788
C-2 bottoms	0.0000	0.0000	3.1435	3.1435

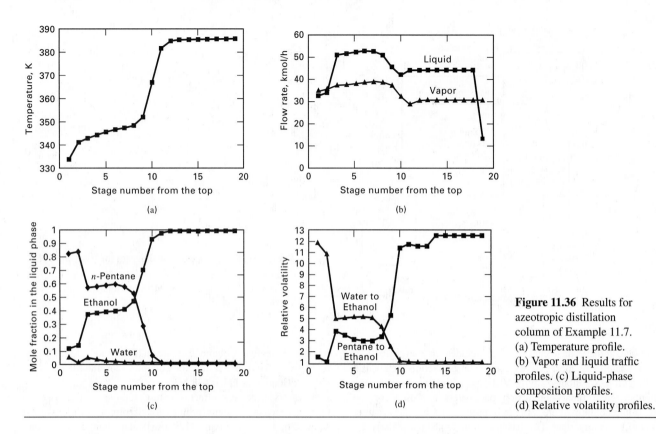

Figure 11.36 Results for azeotropic distillation column of Example 11.7. (a) Temperature profile. (b) Vapor and liquid traffic profiles. (c) Liquid-phase composition profiles. (d) Relative volatility profiles.

11.7 REACTIVE DISTILLATION

Reactive distillation involves simultaneous chemical reaction and distillation. The chemical reaction usually takes place in the liquid phase or at the surface of a solid catalyst in contact with the liquid phase. One general application of reactive distillation, described by Terrill, Sylvestre, and Doherty [57], is the separation of a close-boiling or azeotropic mixture of components A and B, where a reactive entrainer E is introduced into the distillation column. If A is the lower-boiling component, it is preferable that E be higher boiling than B and that it react selectively and reversibly with B to produce reaction product C, which also has a higher boiling point than component A and does not form an azeotrope with A, B, or E. Component A is removed from the distillation column as distillate, and components B and C, together with any excess E, are removed as bottoms. Components B and E are recovered from C in a separate distillation step, where the reaction is reversed to completely react C back to B and E; B is taken off as distillate, and E is taken off as bottoms and recycled to the first column. Terrill, Sylvestre, and Doherty [57] discuss the application of reactive entrainers to the separation of mixtures of *p*-xylene and *m*-xylene, whose normal boiling points differ by only 0.8°C, resulting in a relative volatility of only 1.029. Separation by ordinary distillation is impractical because, for example, to produce 99 mol% pure products from an equimolar feed, more than 500 theoretical stages are required. By reacting the *m*-xylene with a reactive entrainer such as *tert*-butylbenzene accompanied by a solid aluminum chloride catalyst, or

chelated sodium *m*-xylene dissolved in cumene, the stage requirements are drastically reduced.

Closely related to the use of reactive entrainers in distillation is the use of reactive absorbents in absorption, which finds wide application in industry. For example, sour natural gas is sweetened by the removal of hydrogen sulfide and carbon dioxide acid gases by absorption into aqueous alkaline solutions of mono- and di-ethanolamines. Fast and reversible reactions occur to form soluble-salt complexes such as carbonates, bicarbonates, sulfides, and mercaptans. The rich solution leaving the absorber is sent to a reboiled stripper where the reactions are reversed at higher temperatures to regenerate the amine solution as the bottoms and deliver the acid gases as overhead vapor.

A second application of reactive distillation involves taking into account undesirable chemical reactions that may occur during distillation. For example, Robinson and Gilliland [58] present an example involving the separation of cyclopentadiene from C_7 hydrocarbons. During distillation, cyclopentadiene dimerizes. The more volatile cyclopentadiene is taken overhead as distillate, but a small amount dimerizes in the lower section of the column and leaves in the bottoms with the C_7s. Alternatively, the cyclopentadiene can be dimerized to facilitate its separation by distillation from other constituents of a mixture. Then the dicyclopentadiene is removed as bottoms from the distillation column. However, during distillation, it is also necessary to account for possible depolymerization to produce cyclopentadiene, which would migrate to the distillate.

The most interesting application of reactive distillation, and the only one considered in detail in this section, involves combining chemical reaction(s) and separation by distillation in a single distillation apparatus. This concept appears to have been first pronounced by Backhaus, who, starting in 1921 [59], obtained a series of patents for esterification reactions in a distillation column. This concept of continuous and simultaneous chemical reaction and distillation in a single vessel was verified experimentally by Leyes and Othmer [60] for the esterification of acetic acid with an excess of *n*-butanol in the presence of sulfuric acid catalyst to produce butyl acetate and water. This type of reactive distillation should be considered as an alternative to the use of separate reactor and distillation vessels whenever the following hold:

1. The chemical reaction occurs in the liquid phase, in the presence or absence of a homogeneous catalyst, or at the interface of a liquid and a solid catalyst.

2. Feasible temperature and pressure for the reaction and distillation are the same. That is, reaction rates and distillation rates are of the same order of magnitude.

3. The reaction is equilibrium-limited such that if one or more of the products formed can be removed, the reaction can be driven to completion; thus, a large excess of a reactant is not necessary to achieve a high conversion. This is particularly advantageous when recovery of the excess reagent is difficult because of azeotrope formation. For reactions that are irreversible, it is more economical to take the reactions to completion in a reactor and then separate the products in a separate distillation column. In general, reactive distillation is not attractive for supercritical conditions, for gas-phase reactions, and for reactions that must take place at high temperatures and pressures, and/or that involve solid reactants or products.

Careful consideration must be given to the configuration of the distillation column when employing reactive distillation. Important factors are feed entry and product-removal stages, the possible need for intercoolers and interheaters when the heat of reaction is appreciable, and the method for obtaining required residence time for the liquid phase. In the following ideal cases, it is possible, as shown by Belck [61] and others for several two-, three-, and four-component systems, to obtain the desired products without the need for additional distillation.

Case 1: The reaction A ↔ R or A ↔ 2R, where R has a higher volatility than A. In this case, only a reboiled rectification section is needed. Pure A is sent to the column reboiler where all or most of the reaction takes place. As R is produced, it is vaporized, passing to the rectification column where it is purified. Overhead vapor from the column is condensed, with part of the condensate returned to the column as reflux. Chemical reaction may also take place in the column. If A and R form a maximum-boiling azeotrope, this configuration is still applicable if, under steady-state conditions, the mole fraction of R in the reboiler is greater than the azeotropic composition.

Case 2: The reaction A ↔ R or 2A ↔ R, where A has the lower boiling point or higher volatility. In this case, only a stripping section is needed. The feed of pure liquid A is sent to the top of the column, from which it flows down the column, reacting to produce R. The column is provided with a total condenser and a partial reboiler. No product is withdrawn from the top of the column. Product R is withdrawn from the reboiler. This configuration requires close examination because, at a certain location in the column, chemical equilibrium may be achieved, and if the reaction is allowed to proceed below that point, the reverse reaction can occur.

Case 3: The reactions 2A ↔ R + S or A + B ↔ R + S, where A and B are intermediate in volatility to R and S, and R has the highest volatility. In this case, the feed enters an ordinary distillation column somewhere near the middle, with R withdrawn as distillate and S withdrawn as bottoms. If B is less volatile than A, then B may enter the column separately and at a higher level than A.

Commercial applications of reactive distillation include the following

1. The esterification of acetic acid with ethanol to produce ethyl acetate and water

2. The reaction of formaldehyde and methanol to produce methylal and water, using a solid acid catalyst, as described by Masamoto and Matsuzaki [62]

3. The esterification of acetic acid with methanol to produce methyl acetate and water, using sulfuric acid catalyst, as patented by Agreda and Partin [63], and described by Agreda, Partin, and Heise [64]

4. The reaction of isobutene with methanol to produce methyl-*tert*-butyl ether (MTBE), using a solid, strong-acid ion-exchange resin catalyst, as patented by Smith [65–67] and further developed by DeGarmo, Parulekar, and Pinjala [68]

The first widely studied example of reactive distillation is the esterification of acetic acid (A) with ethanol (B) to produce ethyl acetate (R) and water (S). The respective normal boiling points in °C are 118.1, 78.4, 77.1, and 100. Also, minimum-boiling binary homogeneous azeotropes are formed by B–S at 78.2°C with 10.57 mol% B, and by B–R at 71.8°C with 46 mol% B. A minimum-boiling, binary heterogeneous azeotrope is formed by R–S at 70.4°C with 24 mol% S, and a ternary, minimum-boiling azeotrope is formed by B–R–S at 70.3°C with 12.4 mol% B and 60.1 mol% R. Thus, this system is exceedingly complex and nonideal. A number of studies, both experimental and computational, have been published, many of which are cited by Chang and Seader [69], who developed a robust computational procedure for reactive distillation based on a homotopy continuation method. More recently, other computational procedures, used in computer-aided process design programs, have been reported by Venkataraman, Chan, and Boston [70] and Simandl and Svrcek [71]. Kang, Lee, and Lee [72] obtained binary-interaction parameters for the UNIQUAC equation by fitting experimental data simultaneously for vapor–liquid equilibrium and liquid-phase chemical equilibrium.

In all of the computational procedures, a reaction-rate term must be added to the component material balance for a stage. For example, in the development of Chang and Seader [69], (10-58) for the Newton-Raphson procedure is modified to include a reaction-rate source term for the liquid phase, assuming that at each stage, the liquid phase is completely mixed:

$$M_{i,j} = l_{i,j}(1 + s_j) + v_{i,j}(1 + S_j) - l_{i,j-1} - v_{i,j+1} - f_{i,j}$$
$$- (V_{LH})_j \sum_{n=1}^{NRX} v_{i,n} r_{j,n}, \quad i = 1, \ldots C \quad (11\text{-}17)$$

where

$(V_{LH})_j$ = the volumetric liquid holdup at stage j

$v_{i,n}$ = stoichiometric coefficient for component i and reaction n using the customary convention of positive values for products and negative values for reactants

$r_{j,n}$ = reaction rate for reaction n on stage j, as the increase in moles of a reference reactant per unit time per unit volume of liquid phase

NRX = number of reversible and irreversible chemical reactions.

Typically, each reaction rate is expressed in a power-law form with liquid molar concentrations (where the n subscript is omitted in the following equation):

$$r_j = \sum_{p=1}^{2} k_p \prod_{q=1}^{NRC} c_{j,q}^m = \sum_{p=1}^{2} A_p \exp\left(-\frac{E_p}{RT_j}\right) \prod_{q=1}^{NRC} c_{j,q}^m \quad (11\text{-}18)$$

where

$c_{j,q}$ = concentration of component q on stage j

k_p = reaction rate constant for the pth term, where $p = 1$ indicates the forward reaction and $p = 2$ indicates the reverse reaction; k_1 is positive and k_2 is negative

m = the exponent on the concentration

NRC = number of components in the power-law expression

A_p = preexponential (frequency) factor

E_p = activation energy

With (11-17) and (11-18), a reaction may be treated as irreversible ($k_2 = 0$), reversible (k_2 negative and not equal to zero), or at equilibrium. The last can be achieved by using very large values for the volumetric liquid holdup at each stage in the case of a single, reversible reaction, or by multiplying each of the two frequency factors, A_1 and A_2, by the same large number, thus greatly increasing the forward and backward reactions, but maintaining the correct value for the chemical-

reaction equilibrium constant. For equilibrium reactions, it is important that the power-law expression for the backward reaction be derived from the power-law expression for the forward reaction and the reaction stoichiometry so as to be consistent with the expression for the chemical-reaction equilibrium constant. The volumetric liquid holdup for a stage, when using a trayed tower, depends on the active bubbling area of the tray, the height of the froth on the tray as influenced by the weir height, and the liquid-volume fraction of the froth. These factors are all considered in the section on pressure-drop calculations in Chapter 6. In general, the liquid backup in the downcomer is not included in the estimate of volumetric liquid holdup. When large holdups are necessary, bubble-cap trays are preferred because they do not allow weeping. When the chemical reaction is in the reboiler, a large liquid holdup can be provided. The following example illustrates the application of the computational procedure to the esterification of acetic acid with ethanol to produce ethyl acetate and water. In this example, the single, reversible chemical reaction is assumed to reach chemical equilibrium at each stage. Thus, no estimate of liquid holdup is needed. In a subsequent example, chemical equilibrium is not achieved and holdup estimates are made, which necessitates an estimate of tower diameter.

EXAMPLE 11.8

A reactive-distillation column containing the equivalent of 13 theoretical stages and equipped with a total condenser and partial reboiler is used to produce ethyl acetate (R) at 1 atm. A saturated liquid feed of 90 lbmol/h of acetic acid (A) enters Stage 2 from the top, while 100 lbmol/h of a saturated liquid of 90 mol% ethanol (B) and 10 mol% water (S) (close to the azeotropic composition) enters Stage 9 from the top. Thus, the acetic acid and ethanol are in stoichiometric ratio for esterification. The other specifications are a reflux ratio of 10 and a distillate rate of 90 lbmol/h in the hope that complete conversion to ethyl acetate (the low boiler) will occur. Kinetic data for the homogeneous reaction are given by Izarraraz, Bentzen, Anthony, and Holland [73], in terms of the rate law:

$$r = k_1 c_A c_B - k_2 c_R c_S$$

with $k_1 = 29,000 \exp(-14,300/RT)$ in L/(mol-min) with T in kelvins, and $k_2 = 7,380 \exp(-14,300/RT)$ in L/(mol-min) with T in kelvins. Because the activation energies for the forward and backward steps are the same, the chemical-equilibrium constant is independent of temperature and equal to $k_1/k_2 = 3.93$. Assume that chemical equilibrium is achieved at each theoretical stage. Thus, very large values of liquid holdup are specified for each stage. Binary-interaction parameters, for all six binary pairs, for predicting liquid-phase activity coefficients from the UNIQUAC equation are as follows, from Kang, Lee, and Lee [72]:

Components in Binary Pair, $i - j$	Binary Parameters	
	$u_{i,j}/R$, K	$u_{j,i}/R$, K
Acetic acid–ethanol	268.54	−225.62
Acetic acid–water	398.51	−255.84
Acetic acid–ethyl acetate	−112.33	219.41
Ethanol–water	−126.91	467.04
Ethanol–ethyl acetate	−173.91	500.68
Water–ethyl acetate	−36.18	638.60

Vapor-phase association of acetic acid is to be accounted for and the possible formation of two liquid phases is to be checked at each stage. Calculate the compositions of the distillate and bottoms products and determine the liquid-phase-composition and reaction-rate profiles.

SOLUTION

The calculations were made with the SCDS model (Newton–Raphson method) of the ChemCAD computer-aided process simulation program, where the total condenser is counted as the first stage. The only initial estimates provided were 163 and 198°F for the temperatures of the distillate and the bottoms, respectively. Convergence of the calculations required 17 iterations. A complete conversion to ethyl acetate was not achieved, as indicated by the following distillate and bottoms:

Component	Product Flow Rates, lbmol/h	
	Distillate	Bottoms
Ethyl acetate	49.52	6.39
Ethanol	31.02	3.07
Water	6.73	59.18
Acetic acid	2.73	31.36
Total	90.00	100.00

All four components appear in both products. The overall conversion to ethyl acetate is only 62.1%, with 88.6% of this going to the distillate. The distillate is 55 mol% acetate, while the bottoms is 59.2 mol% water. Only small changes in these compositions occur when the feed locations are varied. Two important factors in the failure to achieve a high conversion and nearly pure products are (1) the highly nonideal nature of the quaternary mixture, accompanied by the large number of azeotropes, and (2) the tendency of the reverse reaction to occur on certain stages. The former effect is shown in Figure 11.37a, where the relative volatilities between ethyl acetate and water and between ethanol and water in the top section of the column are no greater than 1.25, making the separations difficult. The liquid-phase mole-fraction distribution is shown in Figure 11.37b, where, in the section between the two feed points, compositions change slowly despite the esterification reaction. In Figure 11.37c, the reaction-rate profile is quite unusual. Above the upper feed stage (now Stage 3), the reverse reaction is dominant. From that feed point down to the second feed entry (now Stage 10), the forward reaction dominates, but mainly at the upper feed stage. The reverse reaction is dominant for Stages 11–13, whereas the forward reaction dominates at Stages 14 and 15 (the reboiler). The largest extents of forward reaction occur at Stages 3 and 15. Even when the number of stages is increased to 60, with the reaction confined to Stages 25 to 35, the distillate contains an appreciable fraction of ethanol and the bottoms contains a substantial fraction of acetic acid. For this example, the development of a reactive-distillation scheme for achieving a high conversion and nearly pure products represents a significant challenge.

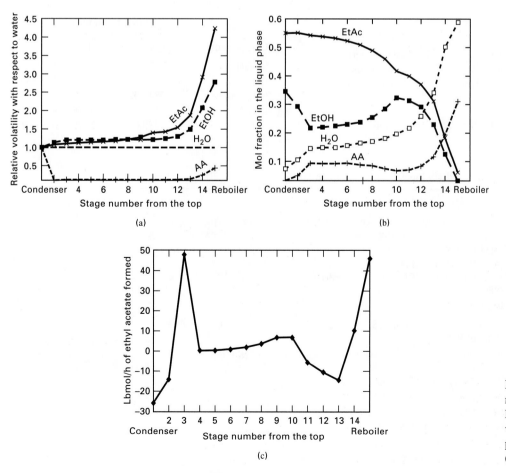

Figure 11.37 Profiles for reactive distillation in Example 11.8. (a) Relative volatility profile. (b) Liquid-phase mole-fraction profiles. (c) Reaction-rate profile.

EXAMPLE 11.9

Using thermodynamic and kinetic data from Rehfinger and Hoffmann [74] for the formation of MTBE from methanol (MeOH) and isobutene (IB), in the presence of n-butene (NB), both Jacobs and Krishna [75] and Nijhuis, Kerkhof, and Mak [76] computed, for catalyzed reactive distillation, with the ASPEN PLUS simulator, multiple solutions having drastically different isobutene conversions when the feed stage for methanol was varied. An explanation for these multiple solutions is given by Hauan, Hertzberg, and Lien [77].

Compute a converged solution for the following conditions, taking into account the kinetics of the reaction, but assuming vapor–liquid equilibrium at each stage. The distillation column has a total condenser, a partial reboiler, and 15 equilibrium stages in the column, which operates at 11 bar. Stages are counted down from the top, with the total condenser numbered stage 1 when using a process simulator, even though it is not an equilibrium stage. The mixed butenes feed, consisting of 195.44 mol/s of IB and 353.56 mol/s of NB enters stage 11 as a vapor at 350 K and 11 bar. The methanol at a flow rate of 215.5 mol/s enters stage 10 as a liquid at 320 K and 11 bar. The reflux ratio is 7 and the bottoms flow rate is set at 197 mol/s. The catalyst is provided only for Stages 4 through 11 (8 stages total), with 204.1 kg of catalyst per stage. The catalyst is a strong-acid ion-exchange resin with 4.9 equivalents of acid groups per kilogram of catalyst. Thus, the equivalents per stage are 1,000 or 8,000 for the eight stages. Compute the product compositions and column profiles using the RADFRAC model in ASPEN PLUS.

SOLUTION

The only chemical reaction considered is

$$\text{IB} + \text{MeOH} \leftrightarrow \text{MTBE}$$

with NB inert.

For the forward reaction, the rate law is formulated in terms of mole-fraction concentrations, instead of activities (products of activity coefficient and mole fraction) as in Rehfinger and Hoffmann [74]:

$$r_{\text{forward}} = 3.67 \times 10^{12} \exp(-92{,}440/RT)x_{\text{IB}}/x_{\text{MeOH}} \quad (1)$$

The corresponding backward rate law is

$$r_{\text{backward}} = 2.67 \times 10^{17} \exp(-134{,}454/RT)x_{\text{MTBE}}/x_{\text{MeOH}}^2 \quad (2)$$

where r is in moles per second per equivalent of acid groups, $R = 8.314$ J/mol-K, T is in kelvins, and x_i is liquid mole fraction.

The Redlich–Kwong equation of state is used to estimate vapor-phase fugacities with the UNIQUAC equation to estimate the liquid-phase activity coefficients. The UNIQUAC binary interaction parameters are as follows, where it is very important to include the inert NB in the system by assuming it has the same parameters as IB and that the two butenes form an ideal solution. The parameters are defined as follows, with all $a_{ij} = 0$.

$$T_{ij} = \exp\left(-\frac{u_{ij} - u_{jj}}{RT}\right) = \exp\left(a_{ij} + \frac{b_{ij}}{T}\right) \quad (3)$$

Components in Binary Pair, ij	Binary Parameters	
	b_{ij}, K	b_{ji}, K
MeOH–IB	35.38	−706.34
MeOH–MTBE	88.04	−468.76
IB–MTBE	−52.2	24.63
MeOH–NB	35.38	−706.34
NB–MTBE	−52.2	24.63

The only initial guesses provided are temperatures of 350 and 420 K, respectively, for Stages 1 and 17; liquid-phase mole fractions of 0.05 for MeOH and 0.95 for MTBE leaving Stage 17; and vapor-phase mole fractions of 0.125 for MeOH and 0.875 for MTBE leaving Stage 17. The ASPEN PLUS input data for release 10.1 are listed in Table 11.6. The converged temperatures for Stages 1 and 17, respectively, are 347 and 420 K. Converged product flow rates are as follows:

Component	Flow Rate, mol/s	
	Distillate	Bottoms
MeOH	28.32	0.31
IB	7.27	1.31
NB	344.92	8.64
MTBE	0.12	186.74
Total	380.63	197.00

Table 11.6   ASPEN PLUS Input Data for Example 11.9

```
TITLE 'mtbe'
IN-UNITS MET VOLUME-FLOW='CUM/HR' ENTHALPY-FLO='MMKCAL/HR' &
        HEAT-TRANS-C='KCAL/HR-SQM-K' PRESSURE=BAR TEMPERATURE=C &
        VOLUME=CUM DELTA-T=C HEAD=METER MOLE-DENSITY='KMOL/CUM' &
        MASS-DENSITY='KG/CUM' MOLE-ENTHALP='KCAL/MOL' &
        MASS-ENTHALP='KCAL/KG' HEAT=MMKCAL MOLE-CONC='MOL/L' &
        PDROP=BAR

DEF-STREAMS CONVEN ALL

DATABANKS PURECOMP / AQUEOUS / SOLIDS / INORGANIC / &
        NOASPENPCD

PROP-SOURCES PURECOMP / AQUEOUS / SOLIDS / INORGANIC

COMPONENTS
    MEOH CH4O MEOH /
    IB C4H8-5 IB /
    NB C4H8-1 NB /
    MTBE C5H12O-D2 MTBE

FLOWSHEET
    BLOCK B1 IN=1 2 OUT=3 4
```

Table 11.6   (*Continued*)

```
PROPERTIES SYSOP11
PROP-REPLACE SYSOP11 UNIQ-RK
    PROP PHILMX PHILMX11
    PROP HLMX HLMX11
    PROP GLMX GLMX11
    PROP SLMX SLMX11
    PROP MUVMX MUVMX02
    PROP MULMX MULMX02
    PROP KVMX KVMX02
    PROP DV DV01
PROP-DATA UNIQ-1
    IN-UNITS MET VOLUME-FLOW='CUM/HR' ENTHALPY-FLO='MMKCAL/HR'  &
        HEAT-TRANS-C='KCAL/HR-SQM-K' PRESSURE=BAR TEMPERATURE=K  &
        VOLUME=CUM DELTA-T=C HEAD=METER MOLE-DENSITY ='KMOL/CUM'  &
        MASS-DENSITY='KG/CUM' MOLE-ENTHALP='KCAL/MOL'  &
        MASS-ENTHALP='KCAL/KG' HEAT=MMKCAL MOLE-CONC='MOL/L'  &
        PDROP=BAR
    PROP-LIST UNIQ
    BPVAL MEOH IB 0.0 35.38 0.0 0.0 0.0 1000.000
        BPVAL IB MEOH 0.0 -706.34 0.0 0.0 0.0 1000.000
        BPVAL MEOH MTBE 0.0 88.04 0.0 0.0 0.0 1000.000
        BPVAL MTBE MEOH 0.0 -468.76 0.0 0.0 0.0 1000.000
        BPVAL IB MTBE 0.0 -52.2 0.0 0.0 0.0 1000.000
        BPVAL MTBE IB 0.0 24.63 0.0 0.0 0.0 1000.000
        BPVAL MEOH NB 0.0 35.38 0.0 0.0 0.0 1000.000
        BPVAL NB MEOH 0.0 -706.34 0.0 0.0 0.0 1000.000
        BPVAL NB MTBE 0.0 -52.2 0.0 0.0 0.0 1000.000
        BPVAL MTBE NB 0.0 24.63 0.0 0.0 0.0 1000.000

    PROP-SET VLE PHIMX GAMMA PL SUBSTREAM=MIXED PHASE=V L

    PROP-SET VLLE PHMIX GAMMA PL SUBSTREAM=MIXED PHASE=V L1 L2

    STREAM 1
        SUBSTREAM MIXED TEMP=320 <K> PRES=11 NPHASE=1 PHASE=L
        MOLE-FLOW MEOH 215.5 <MOL/SEC> / IB 0. <MOL/SEC> / NB &
            0. <MOL/SEC> / MTBE 0. <MOL/SEC>

    STREAM 2
        SUBSTREAM MIXED TEMP=350 <K> PRES=11 NPHASE=1
        MOLE-FLOW MEOH 0. <MOL/SEC> / IB 195.44 <MOL/SEC> / NB &
            353.56 <MOL/SEC> / MTBE 0. <MOL/SEC>

    BLOCK B1 RADFRAC
        PARAM NSTAGE=17 MAXOL=50 MAXIL=50 ILMETH=NEWTON
        FEEDS 1 10 / 2 11 ON-STAGE
        PRODUCTS 3 1 L / 4 17 L
        P-SPEC 1 11
        COL-SPECS DP-COL=0 MOLE-RDV=0 MOLE-B=197 <MOL/SEC> MOLE-RR=7
        REAC-STAGES 4 11 r-1
        HOLD-UP 4 11 MASS-LHLDP=8000
        T-EST 1 350 <K> / 17 420 <K>
        X-EST 17 MEOH .05 / 17 MTBE .95
        Y-EST 17 MEOH .125 / 17 MTBE .875
        TRAY-REPORT TRAY-OPTION=ALL-TRAYS PROPERTIES=VLE VLLE

    STREAM-REPOR PROPERTIES=VLE VLLE

    REACTIONS R-1 REAC-DIST
        REAC-DATA 1 KINETIC CBASIS=MOLEFRAC
        REAC-DATA 2 KINETIC CBASIS=MOLEFRAC
        RATE-CON 1 PRE-EXP=3.67E12 ACT-ENERGY=92400 <KJ/KMOL>
        RATE-CON 2 PRE-EXP=2.67E17 ACT-ENERGY=134454 <KJ/KMOL>
        STOIC 1 MEOH -1 / IB -1 / NB 0 / MTBE 1
        STOIC 2 MTBE -1 / NB 0 / MEOH 1 / IB 1
        POWLAW-EXP 1 MEOH -1 / IB 1 / NB 0 / MTBE 0
        POWLAW-EXP 2 MTBE 1 / NB 0. / MEOH -2 / IB 0
        ;
        ;
        ;
```

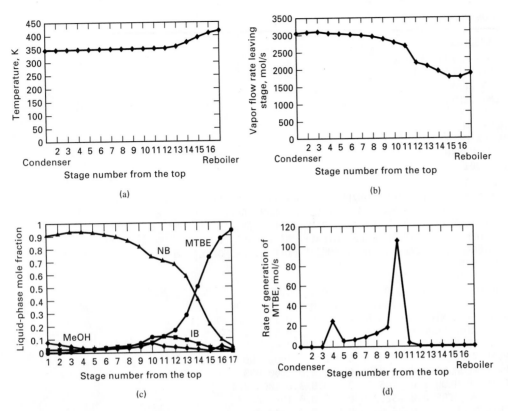

(a)

(b)

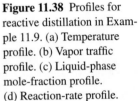

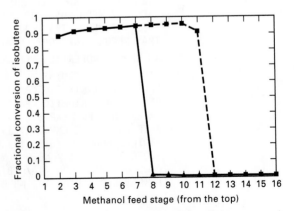

(c)

(d)

Figure 11.38 Profiles for reactive distillation in Example 11.9. (a) Temperature profile. (b) Vapor traffic profile. (c) Liquid-phase mole-fraction profile. (d) Reaction-rate profile.

The combined feeds to the reactive distillation contained a 10.3% mole excess of MeOH over IB. Therefore, IB was the limiting reactant and the preceding product distribution indicates that 95.6% of the IB, or 186.86 mol/s, reacted to form MTBE. The percent purity of the MTBE in the bottoms is 94.8%. Only 2.4% of the inert NB and 1.1% of the unreacted MeOH are found in the bottoms. The computed condenser and reboiler duties are, respectively, 53.2 and 40.4 MW.

Seven iterations were required to obtain a converged solution. The column profiles are in Figure 11.38. Figure 11.38a shows that most of the reaction occurs in a narrow temperature range of 348.6 to 353 K. The reaction temperature can be varied by adjusting the column pressure. Figure 11.38b shows that the vapor traffic in the column above the two feed entries changes by less than 11%, because of only small changes in temperature. As one moves down the column below the two feed entries, the temperature increases rapidly from 353 to 420 K, causing the vapor traffic to decrease by about 20%. In Figure 11.38c, the liquid-composition profiles show that the liquid is dominated by NB from the top stage down to Stage 13, thus drastically reducing the driving force for the reaction. Below Stage 11, the liquid quickly becomes richer in MTBE as the mole fractions of the other components decrease because of increasing temperature. In the section of the column above the reaction zone, the mole fraction of MTBE quickly decreases as one moves to the top stage. These changes are due mainly to the large differences between the K-values for MTBE and those for the other three components. The relative volatility of MTBE with respect to any of the other components ranges from about 0.24 at the top stage to about 0.35 at the bottom. Nonideality in the liquid phase influences mainly MeOH, whose liquid-phase activity coefficient varies from a high of 10 at Stage 5 to a low of 2.6 at Stage 17. This causes the unreacted MeOH to leave mainly with the NB in the distillate rather than with MTBE in the bottoms. The profile for the rate of reaction is shown in Figure 11.38d, where it is seen that the forward reaction dominates on every stage of the reaction section. However, 56% of the reaction occurs on Stage 10, which is the MeOH feed stage. The least amount of reaction occurs on Stage 11.

As mentioned earlier, the literature indicates that the percent conversion of IB to MTBE will vary depending upon the stage to which the MeOH is fed. Furthermore, in the range of MeOH feed stages from about 8 to 11, both a low-conversion and a high-conversion solution can be computed. This is shown in Figure 11.39, where the high-conversion solutions are mainly in the 90+ % range, while the low-conversion solutions are all less than 10%. However, if activities are used in the rate expressions, rather than mole fractions, the low-conversion solutions are higher because of the large values for the activity coefficient for MeOH. The results in Figure 11.39 were computed starting with the MeOH feed entering Stage 2. The resulting profiles for this run were used as the initial guesses for the run with MeOH entering Stage 3. Subsequent runs were performed in a similar manner, increasing the

Figure 11.39 Effect of MeOH feed stage location on conversion of IB to MTBE.

MeOH feed stage by 1 each time and initializing with the results of the previous run. High-conversion solutions were obtained for each run until the MeOH feed stage was lowered to Stage 12, at which point the conversion decreased dramatically. Further lowering of the MeOH feed stage to Stage 16 also resulted in a low-conversion solution. However, when the direction of change to the MeOH feed stage was reversed starting from Stage 12, a low-conversion was obtained until the feed stage was decreased to Stage 9, at which point the conversion jumped back to the high-conversion result.

Huss et al. [101] present a detailed study of reactive distillation for the acid-catalyzed reaction of acetic acid and methanol to produce methyl acetate and water, including the effect of the side reaction of methanol dehydration, using simulation models with a comparison to experimental measurements. They consider both finite reaction rates and chemical equilibrium, coupled with phase equilibrium. The results include consideration of reflux limits and multiplicity of solutions (multiple steady states).

11.8 SUPERCRITICAL-FLUID EXTRACTION

Solute extraction from a liquid or solid mixture is usually accomplished with a liquid solvent, as discussed in Chapters 8 and 16, respectively, at conditions of temperature and pressure that lie substantially below the critical temperature and pressure of the solvent. Following the extraction step, the solvent and dissolved solute are subjected to a subsequent separation step, such as distillation, to recover the solvent for recycle and purify the solute.

In 1879, Hannay and Hogarth [78] reported that solid potassium iodide could be dissolved in ethanol, as a dense gas, at supercritical conditions of $T > T_c = 516 \text{ K}$ and $P > P_c = 65 \text{ atm}$. The iodide could then be precipitated from the ethanol by reducing the pressure. This process was later referred to as *supercritical-fluid extraction, supercritical-gas extraction, supercritical extraction* (SCE), *dense-gas extraction,* or *destraction* (a combination of distillation and extraction). By the 1940s, as chronicled by Williams [79], proposed practical applications of SCE began to appear in the patent and technical literature. Figure 11.40 shows the

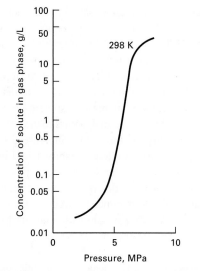

Figure 11.41 Effect of pressure on solubility of pICB in supercritical ethylene.

supercritical fluid region for CO_2, which has a critical point of 304.2 K and 73.83 bar.

The solvent power of a compressed gas can undergo an enormous change in the vicinity of its critical point. Consider, for example, the solubility of *p*-iodochlorobenzene (pICB) in ethylene, as shown in Figure 11.41, at 298 K for pressures from 2 to 8 MPa. This temperature is 1.05 times the critical temperature of ethylene (283 K) and the pressure range straddles the critical pressure of ethylene (5.1 MPa). At 298 K, pICB is a solid (melting point = 330 K) with a vapor pressure of the order of 0.1 torr. At 2 MPa, if pICB formed an ideal-gas solution with ethylene, the mole fraction of pICB in the gas in equilibrium with pure, solid pICB would be extremely small at about 6.7×10^{-6} or a concentration of 0.00146 g/L. The experimental concentration from Figure 11.41 is 0.015 g/L, which is an order of magnitude higher because of nonideal-gas effects. If the pressure is increased from 2 MPa to almost the critical pressure at 5 MPa (an increase by a factor of 2.5),

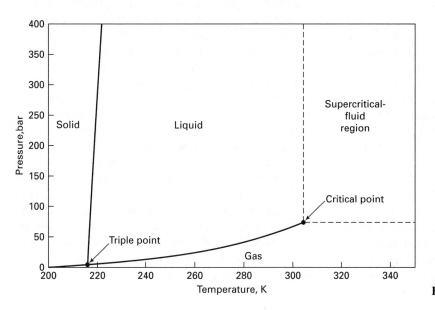

Figure 11.40 Supercritical fluid region for CO_2.

the equilibrium concentration of pICB is increased about 10-fold to 0.15 g/L. At 8 MPa, the concentration begins to level out at 40 g/L, which is 2,700 times higher than predicted from the vapor pressure for an ideal-gas solution. It is this dramatic increase in solubility of a solute at near-critical conditions of a solvent that makes SCE of interest.

Why such a dramatic increase in solvent power? The explanation lies in the change that occurs to the solvent density while the solubility of the solute increases. A pressure–enthalpy diagram for ethylene is shown in Figure 11.42, which includes the specific volume as a parameter, from which the density can be determined as the reciprocal. The range of variables and parameters straddles the critical point of ethylene. The density of ethylene compared to the solubility of pICB is as follows at 298 K:

Pressure, MPa	Ethylene Density, g/L	Solubility of pICB, g/L
2	25.8	0.015
5	95	0.15
8	267	40

Although there is far from a 1:1 correspondence in the increase of pICB solubility with density for ethylene over this range of pressure, there is a meaningful correlation. As the pressure is increased, closer packing of the solvent molecules allows them to surround and trap solute molecules. This phenomenon is most dramatic and useful at reduced temperatures from about 1.01 to 1.12.

Two other effects in the supercritical region are favorable for SCE. It will be recalled that the molecular diffusivity of a solute in an ambient-pressure gas is about four orders of magnitude higher than for a liquid. For a near-critical fluid, the diffusivity of solute molecules is usually one to two orders of magnitude higher than in a normal liquid solvent, thus resulting in a lower mass-transfer resistance in the solvent phase than might be expected. In addition, the viscosity of the supercritical fluid is about an order of magnitude less than that of a normal liquid solvent.

Many patents have been issued, proposals prepared, and experimental studies conducted on SCE as a possible alternative for distillation, enhanced distillation, and liquid–liquid extraction. However, in general, when these other techniques are feasible, SCE usually cannot compete economically because of high solvent-compression costs to reach near-critical pressure. SCE is most favorable for the extraction of small amounts of large, relatively nonvolatile solutes in solid or liquid mixtures. Such applications are cited by Williams [79] and McHugh and Krukonis [80].

Solvent selection depends on the composition of the feed mixture. If only the chemical(s) to be extracted is (are) soluble in a potential solvent, then high solubility is a key factor. However, if a chemical besides the desired solute is soluble in the potential solvent, then the selectivity of the solvent becomes as important as solubility. A number of light gases and other low-molecular-weight chemicals, including the following, have received attention as solvents for SCE:

Solvent	Critical Temperature, K	Critical Pressure, MPa	Critical Density, kg/m^3
Methane	192	4.60	162
Ethylene	283	5.03	218
Carbon dioxide	304	7.38	468
Ethane	305	4.88	203
Propylene	365	4.62	233
Propane	370	4.24	217
Ammonia	406	11.3	235
Water	647	22.0	322

Those solvents with a critical temperature below 373 K have been well studied. A particularly desirable solvent, particularly for the extraction of undesirable, valuable, or heat-sensitive chemicals from natural products such as foods, is carbon dioxide, which has a moderate critical pressure, a high critical density, and a critical temperature close to ambient temperature. Carbon dioxide is nonflammable, non-corrosive, nontoxic in low concentrations, readily available, inexpensive, and safe. Also, supercritical carbon dioxide has a relatively low viscosity and high molecular diffusivity. Separation of carbon dioxide from the solute is often possible by simply reducing the extract pressure. According to Williams [79], supercritical carbon dioxide has been used to extract caffeine from coffee, hops oil from beer, piperine from pepper, capsaicin from chilis, oil from nutmeg, and nicotine from tobacco.

Carbon dioxide is not a suitable solvent for all potential applications. McHugh and Krukonis [81] cite the energy crisis of the 1970s that led to substantial research on an energy-efficient separation of ethanol and water. The primary goal, which was to break the ethanol–water azeotrope, was not achieved by SCE with carbon dioxide because, although supercritical carbon dioxide has unlimited capacity to dissolve pure ethanol, water is also dissolved in significant amounts. A liquid–supercritical-fluid phase diagram for the ethanol–water–carbon dioxide ternary system at 308.2 K and 10.08 MPa, based on the experimental data of Takishima, Saiki, Arai, and Saito [82], is shown in Figure 11.43. These conditions correspond to $T_r = 1.014$ and $P_r = 1.366$ for carbon dioxide. For the binary mixture of water and carbon dioxide, two phases exist; a water-rich phase with about 2 mol% carbon dioxide and a carbon dioxide-rich phase with about 1 mol% water. Ethanol and carbon dioxide are completely soluble in each other. Ternary mixtures containing more than 40 mol% ethanol are completely miscible. If a near-azeotropic mixture of ethanol and water, say, 85 mol% ethanol and 15 mol% water, is extracted by carbon dioxide at the conditions of Figure 11.43, a mixing line drawn between this composition and a point for pure carbon dioxide does not appear to cross into the two-phase region. That is, regardless of the amount of solvent used, both water and ethanol are completely soluble in the carbon dioxide and no separation is possible at these temperatures and pressures. Alternatively, consider an ethanol–water broth from a

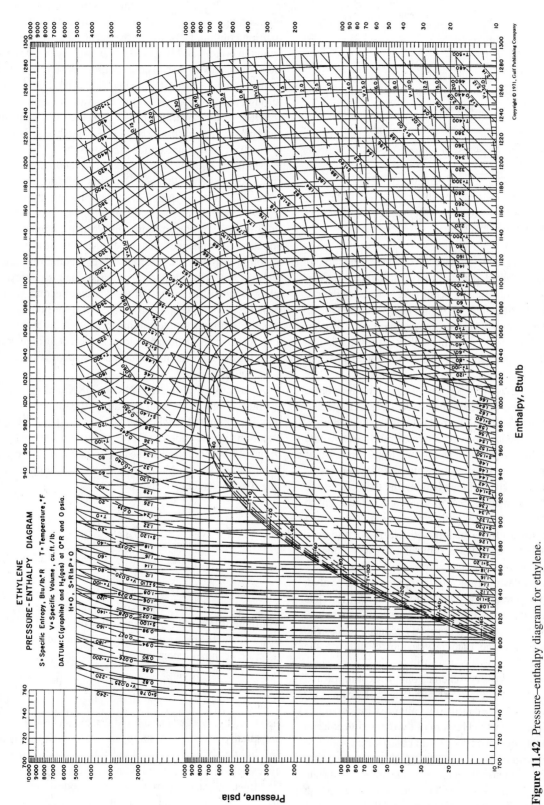

Figure 11.42 Pressure–enthalpy diagram for ethylene.

[From K.E. Starling, "Fluid Thermodynamic Properties for Light Petroleum Systems," Gulf Publishing, Houston (1973) reprinted with permission.]

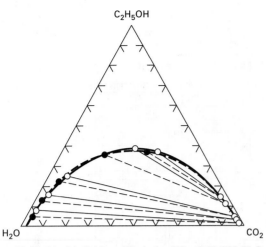

Figure 11.43 Liquid–fluid equilibria for CO_2–C_2H_5OH–H_2O at 308–313.2 K and 10.1–10.34 MPa.

fermentation reactor with 10 wt% (4.17 mol%) ethanol. If this mixture is extracted with supercritical carbon dioxide, complete dissolution will not occur and a modest degree of separation of ethanol from water can be achieved, as shown in the next example. The separation can be further enhanced by the use of a cosolvent, such as glycerol, that improves the selectivity, as shown by Inomata, Kondo, Ari, and Saito [83].

When CO_2 is used as a solvent, it must be recovered and recycled. Three schemes discussed by McHugh and Krukonis [81] are shown in Figure 11.44. In the first scheme, shown for the separation of ethanol and water, the ethanol–water feed is pumped as a liquid to the pressure of the extraction column, where it is contacted with supercritical carbon dioxide. The raffinate leaving the extractor at the bottom is enriched with respect to water and can be sent to another part of the plant for further processing. The extract stream, which leaves from the top of the extractor and contains most of the carbon dioxide, some ethanol, and a smaller amount of water, is expanded across a valve to a lower pressure. In a flash drum downstream of the valve, ethanol–water condensate is collected and the CO_2-rich gas is recycled through a gas compressor back to the extractor. However, unless the pressure is greatly reduced across the valve, resulting in large compression costs, little of the ethanol is condensed.

A second CO_2 recovery scheme, due to de Filippi and Vivian [84], is shown in Figure 11.44b. The flash drum is replaced by a high-pressure distillation column, which operates at a pressure just below the pressure of the extraction column to produce a CO_2-rich distillate and an ethanol-rich bottoms. The distillate is compressed and recycled through the reboiler and back to the extractor. Both the raffinate and the distillate are flashed to recover dissolved CO_2. This scheme, although more complicated than the first, is more versatile.

A third CO_2 recovery scheme, due to Katz et al. [85] for the decaffeination of coffee, is shown in Figure 11.44c. In the extractor, green, wet coffee beans are mixed with

supercritical CO_2 to extract caffeine. The extract is sent to a second extraction column, where the caffeine is extracted with water. The CO_2-rich raffinate from this column is recycled through a compressor (not shown) back to the first extraction column, from which the decaffeinated coffee leaves from the bottom and is sent to a roasting tower. The caffeine-rich water leaving the second column is sent to a reverse-osmosis unit, where the water is purified and recycled through a pump (not shown) to the water column. All three separation steps operate at high pressure. The concentrated caffeine–water mixture leaving the osmosis unit is sent to a crystallizer to produce caffeine crystals.

Multiple equilibrium stages in a countercurrent-flow contactor are generally needed to obtain the desired extent of extraction. A major problem in determining the number of stages required is the estimation of liquid–supercritical fluid phase-equilibrium constants. Most commonly, cubic-equation-of-state methods, such as the Soave–Redlich–Kwong (SRK) or Peng–Robinson (PR) equations, are used, but they have two shortcomings. First, their accuracy diminishes in the critical region of the solvent. Second, if the feed contains polar components that form a nonideal-liquid mixture, an appropriate mixing rule, such as that of Wong and Sandler [86], that provides a correct bridge between equation-of-state methods and activity-coefficient methods must be employed.

As discussed in Section 2.5, the SRK and PR equations for pure components both contain two parameters, a and b, that are computed from critical constants. The SRK and PR equations are extended to liquid or vapor mixtures by a *mixing rule* for computing values of a_m and b_m for the mixture from values for the pure components. The simplest mixing rule, due to van der Waals, is:

$$a_m = \sum_{i=1}^{C} \sum_{j=1}^{C} x_i x_j a_{ij} \qquad (11\text{-}19)$$

$$b_m = \sum_{i=1}^{C} \sum_{j=1}^{C} x_i x_j b_{ij} \qquad (11\text{-}20)$$

where x is a mole fraction in the vapor or liquid mixture. Although these two mixing-rule equations are identical in form, the following *combining rules* for a_{ij} and b_{ij} are quite different, with the former being a geometric mean and the latter an arithmetic mean:

$$a_{ij} = (a_i a_j)^{1/2} \qquad (11\text{-}21)$$
$$b_{ij} = (b_i + b_j)/2 \qquad (11\text{-}22)$$

As stated by Sandler, Orbey, and Lee [87], (11-19) to (11-22) are usually adequate for nonpolar mixtures of hydrocarbons and light gases when critical temperature and/or size differences between the molecules are not very large.

Molecular-size differences and/or modest degrees of polarity are handled by the following modified combining rules:

$$a_{ij} = (a_i a_j)^{1/2}(1\text{-}k_{ij}) \qquad (11\text{-}23)$$

$$b_{ij} = [(b_i + b_j)/2](1\text{-}l_{ij}) \qquad (11\text{-}24)$$

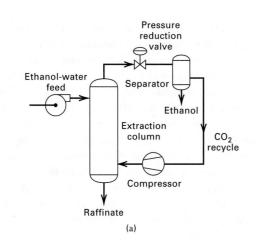

(a)

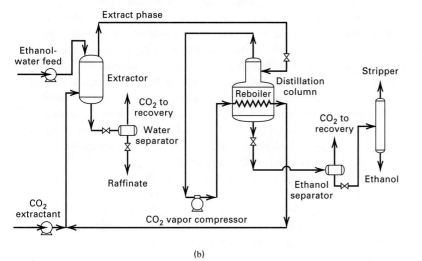

(b)

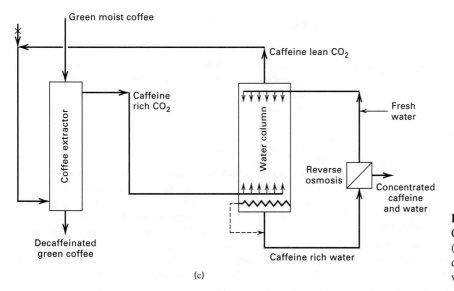

(c)

Figure 11.44 Techniques for recovery of CO_2 in supercritical extraction processes. (a) Pressure reduction. (b) High-pressure distillation. (c) High-pressure absorption with water.

where k_{ij} and l_{ij} are binary-interaction parameters back-calculated from experimental vapor–liquid equilibrium and/or density data. Often the latter parameter is set equal to zero. A tabulation of values of k_{ij}, suitable for use with the SRK and PR equations when the mixture contains hydrocarbons with CO_2, H_2S, N_2, and/or CO, is given by Knapp et al. [88]. In a study by Shibata and Sandler [89], using experimental phase-equilibria and phase-density data for the nonpolar binary system nitrogen–n-butane at 410.9 K over a pressure range of about 30 to 70 bar, reasonably good predictions, except in the critical region, were obtained using (11-19) and (11-20), with (11-23) and (11-24), and values of $k_{ij} = -0.164$ and $l_{ij} = -0.233$ in conjunction with the PR equation. Similar good agreement with experimental data was obtained for the systems nitrogen–cyclohexane, carbon dioxide–n-butane, and carbon dioxide–cyclohexane, and the ternary systems nitrogen–carbon dioxide–n-butane and nitrogen–carbon dioxide–cyclohexane.

For high pressures and mixtures containing one or more strongly polar components, the preceding rules are inadequate and it would seem desirable, in those cases, to combine the equation-of-state method with the activity-coefficient (free energy) method to handle strong nonidealities in the liquid phase. The following theoretically based mixing rule of Wong and Sandler [86] accomplishes such a bridge between a cubic equation of state and a free-energy or activity-coefficient equation. If, for example, the PR equation of state and the NRTL activity-coefficient equation are used, the Wong and Sandler mixing rule leads to the following expressions for computing a_m and b_m to be used in the PR equation:

$$a_m = RTQD/(1-D) \tag{11-25}$$

$$b_m = Q/(1-D) \tag{11-26}$$

where

$$Q = \sum_{i=1}^{C} \sum_{j=1}^{C} x_i x_j \left(b - \frac{a}{RT}\right)_{ij} \tag{11-27}$$

$$D = \sum_{i=1}^{C} x_i \frac{a_i}{b_i RT} + \frac{G^{ex}(x_i)}{\sigma RT} \tag{11-28}$$

$$\left(b - \frac{a}{RT}\right)_{ij} = \frac{1}{2}\left[\left(b_i - \frac{a_i}{RT}\right) + \left(b_j - \frac{a_j}{RT}\right)\right](1 - k_{ij}) \tag{11-29}$$

$$\sigma = \frac{1}{\sqrt{2}}[\ln(\sqrt{2} - 1)] \tag{11-30}$$

$$\frac{G^{ex}}{RT} = \sum_{i=1}^{C} x_i \left(\frac{\sum_{j=1}^{C} x_j \tau_{ji} g_{ji}}{\sum_{k=1}^{C} x_k g_{ki}}\right) \tag{11-31}$$

$$g_{ij} = \exp(-\alpha_{ij} \tau_{ij}) \tag{11-32}$$

with $a_{ij} = a_{ji}$.

From Equations (11-25) to (11-32), it is seen that for a binary system, using the NRTL equation, there are four adjustable binary-interaction parameters (BIPs): k_{ij}, α_{ij}, τ_{ij}, and τ_{ji}. These four parameters, for a temperature and pressure range of interest, are best obtained by regression of experimental binary-pair data for VLE, LLE, and/or VLLE. The parameters can then be used to predict phase equilibria for ternary and higher multicomponent mixtures. However, Wong, Orbey, and Sandler [90] show that when values of the latter three parameters are already available, even at just

near-ambient temperature and pressure conditions, from an experimental-data-compilation source such as that of Gmehling and Onken [91], those parameters can be assumed to be independent of temperature and used directly to make reasonably accurate predictions of phase equilibria, even at temperatures to at least 200°C and pressures to 200 bar. Furthermore, regression of experimental data to obtain a value of k_{ij} is not necessary either, because Wong, Orbey, and Sandler show that it can be determined from the other three parameters by choosing its value so that the excess Gibbs free energy computed from the equation of state matches that computed from the activity-coefficient mode. Thus, the application of the Wong–Sandler mixing rule to supercritical extraction conditions is facilitated.

Another phase-equilibrium prediction method applicable to wide ranges of pressure, temperature, molecular size, and polarity is the group-contribution equation of state (GC-EOS) of Skjold-Jørgensen [92]. This method, which combines features of the van der Waals equation of state, the Carnahan–Starling expression for hard spheres, the NRTL activity-coefficient equation, and the group-contribution principle, has been successfully applied to supercritical-extraction conditions. The GC-EOS method is particularly useful when all of the necessary binary data are not available to determine all binary interaction parameters.

When experimental K-values are available, or when the Wong–Sandler mixing rule or the GC-EOS can be applied, equilibrium-stage calculations for supercritical extraction can be made by conventional computer programs, as the following example illustrates.

EXAMPLE 11.10

One mol/s of 10 wt% ethanol in water is extracted by 3 mol/s of carbon dioxide at 305 K and 9.86 MPa in a countercurrent-flow extraction column with the equivalent of five equilibrium stages. Determine the flow rates and compositions of the exiting extract and raffinate.

SOLUTION

This problem, which was taken from Colussi, Fermeglia, Gallo, and Kikic [93], was solved with the Tower Plus model of the ChemCAD process simulator, under conditions of constant temperature and constant pressure, at which composition changes were small enough that K-values could be assumed constant. The following K-values, taken from Colussi et al., who used the GC-EOS method, and which are defined as the mole fraction in the extract divided by the mole fraction in the raffinate, are in reasonably good agreement with experimental data.

Component	K-Value
CO_2	34.5
Ethanol	0.115
Water	0.00575

Table 11.7 Calculated Flow and Composition Profiles for Example 11.10

Leaving Streams	Stage 1 Extract Mole Fraction	Stage 1 Raffinate Mole Fraction	Stage 2 Extract Mole Fraction	Stage 2 Raffinate Mole Fraction	Stage 3 Extract Mole Fraction	Stage 3 Raffinate Mole Fraction	Stage 4 Extract Mole Fraction	Stage 4 Raffinate Mole Fraction	Stage 5 Extract Mole Fraction	Stage 5 Raffinate Mole Fraction
Carbon dioxide	0.98999	0.02870	0.99002	0.02870	0.99012	0.02870	0.99043	0.02870	0.99138	0.02874
Ethanol	0.00466	0.04053	0.00463	0.04023	0.00452	0.03929	0.00419	0.03645	0.00319	0.02775
Water	0.00535	0.93077	0.00535	0.93107	0.00536	0.93201	0.00538	0.93485	0.00543	0.94351
Total flow, gmol/s	3.0013	1.0298	3.0311	1.0294	3.0308	1.0285	3.0298	1.0255	3.0268	0.9987

The percent extraction of ethyl alcohol was computed to be 33.6%, with an extract of 69 wt% pure ethanol (solvent-free basis) and a raffinate containing 93 wt% water (solvent-free basis). The calculated stage-wise flow rates and component mole fractions are listed in Table 11.7, where stages are numbered from the feed end of the cascade.

SUMMARY

1. Extractive distillation, salt distillation, pressure-swing distillation, homogeneous azeotropic distillation, heterogeneous azeotropic distillation, and reactive distillation are enhanced distillation techniques to be considered when separation by ordinary distillation is uneconomical or impossible. Reactive distillation can also be used to conduct, simultaneously and in the same apparatus, a chemical reaction and a separation by distillation.

2. For ternary systems, a composition plot on a triangular graph is very useful for finding feasible separations, especially when binary and ternary azeotropes form. With such a diagram, distillation paths, called residue curves or distillation curves, are readily tracked. The curves may be restricted to certain regions of the triangular diagram by distillation boundaries. Feasible product compositions at total reflux are readily determined.

3. Extractive distillation, using a low-volatility solvent that enters near the top of the column, is widely used to separate azeotropes and very close-boiling mixtures. Preferably, the solvent should not form an azeotrope with any component in the feed.

4. Certain salts, when added to a solvent, reduce the volatility of the solvent and increase the relative volatility between the two components to be separated. In this process, called salt distillation, the salt is dissolved in the solvent or added as a solid or melt to the reflux.

5. Pressure-swing distillation, utilizing two columns operating at different pressures, can be used to separate an azeotropic mixture when the azeotrope can be made to disappear at some pressure. If not, the technique may still be practical if the azeotropic composition changes by 5 mol% or more over a moderate range of pressure.

6. In homogeneous azeotropic distillation, an entrainer is added to a stage, usually above the feed stage. A minimum- or maximum-boiling azeotrope, formed by the entrainer with one or more feed components, is removed from the top or bottom of the column, respectively. Unfortunately, potential applications of this technique for difficult-to-separate mixtures are not common because of limitations due to distillation boundaries.

7. A more common and useful technique is heterogeneous azeotropic distillation, in which the entrainer forms, with one or more components of the feed, a minimum-boiling heterogeneous azeotrope. When condensed, the overhead vapor splits into organic-rich and water-rich phases. The azeotrope is broken by returning one liquid phase as reflux, with the other sent on as distillate for further processing.

8. A growing application of reactive or catalytic distillation is the combined operation of chemical reaction and distillation in one vessel. To be effective, it must be possible to carry out the reaction and phase separation at the same pressure and range of temperature, with reactants and products favoring different phases so that an equilibrium-limited reaction can go to completion.

9. Liquid–liquid or solid–liquid extraction can be carried out with a supercritical-fluid solvent at temperatures and pressures just above the critical because of favorable values for solvent density and viscosity, solute diffusivity, and solute solubility in the solvent. An attractive supercritical solvent is carbon dioxide, particularly for extraction of certain chemicals from natural products.

REFERENCES

1. STICHLMAIR, J., J.R. FAIR, and J.L. BRAVO, *Chem. Eng. Progress,* **85**(1), 63–69 (1989).

2. PARTIN, L.R., *Chem. Eng. Progress,* **89** (1), 43–48 (1993).

3. STICHLMAIR, J., "Distillation and Rectification," in *Ullmann's Encyclopedia of Industrial Chemistry,* 5th ed., VCH Verlagsgesellschaft Weinheim Vol. B3, pp. 4–1 to 4–94, (1988).

4. DOHERTY, M.F., and J.D. PERKINS, *Chem. Eng. Sci.,* **33,** 281–301 (1978).

5. BOSSEN, B.S., S.B. JØRGENSEN, and R. GANI, *Ind. Eng. Chem. Res.,* **32,** 620–633 (1993).

6. PHAM, H.N. and M.F. DOHERTY, *Chem. Eng. Sci.,* **45,** 1837–1843 (1990).

7. TAYLOR, R., and H.A. KOOIJMAN, *CACHE News,* No. 41, 13–19 (1995).

8. DOHERTY, M.F., *Chem. Eng. Sci.,* **40,** 1885–1889 (1985).

9. ASPEN PLUS, "What's New in Release 9," Aspen Technology, Cambridge, MA (1994).

10. DOHERTY, M.F., and J.D. PERKINS, *Chem. Eng. Sci.,* **34,** 1401–1414 (1979).

11. HORSLEY, L.H., "Azeotropic Data III," in *Advances in Chemistry Series,* American Chemical Society, Washington, D.C., Vol. 116 (1973).

12. GMEHLING, J., J. MENKE, J. KRAFCZYK, and K. FISCHER, *Azeotropic Data,* 2nd edition in 3 volumes, Wiley-VCH, Weinheim, Germany (2004).

13. FIDKOWSKI, Z.T., M.F. MALONE, and M.F. DOHERTY, *Computers Chem. Engng.,* **17,** 1141–1155 (1993).

14. FOUCHER, E.R., M.F. DOHERTY, and M.F. MALONE, *Ind. Eng. Chem. Res.,* **30,** 760–772 (1991) and **30,** 2364 (1991).

15. MATSUYAMA, H., and H.J. NISHIMURA, *J. Chem. Eng. Japan,* **10,** 181 (1977).

16. DOHERTY, M.F., and G.A. CALDAROLA, *Ind. Eng. Chem. Fundam.,* **24,** 474–485 (1985).

17. FIDKOWSKI, Z.T., M.F. DOHERTY, and M.F. MALONE, *AIChE J.,* **39,** 1303–1321 (1993).

18. STICHLMAIR, J.G., and J.-R. HERGUIJUELA, *AIChE J.,* **38,** 1523–1535 (1992).

19. WIDAGDO, S., and W.D. SEIDER, *AIChE J.,* **42,** 96–130 (1996).

20. WAHNSCHAFFT, O.M., J.W. KOEHLER, E. BLASS, and A.W. WESTERBERG, *Ind. Eng. Chem. Res.,* **31,** 2345–2362 (1992).

21. DUNN, C.L., R.W. MILLAR, G.J. PIEROTTI, R.N. SHIRAS, and M. SOUDERS, JR., *Trans. AIChE,* **41,** 631–644 (1945).

22. BERG, L., *Chem. Eng. Progress,* **65** (9), 52–57 (1969).

23. OTHMER, D.F., *AIChE Symp. Series,* **235** (79), 90–117 (1983).

24. VAN RAYMBEKE, U.S. Patent 1,474,216 (1922).

25. COOK, R.A., and W.F. FURTER, *Can. J. Chem. Eng.,* **46,** 119–123 (1968).

26. JOHNSON, A.I., and W.F. FURTER, *Can. J. Chem. Eng.,* **43,** 356–358 (1965).

27. JOHNSON, A.I., and W.F. FURTER, *Can. J. Chem. Eng.,* **38,** 78–87 (1960).

28. FURTER, W.F., and R.A. COOK, *Int. J. Heat Mass Transfer,* **10,** 23–36 (1967).

29. FURTER, W.F., *Can. J. Chem. Eng.,* **55,** 229–239 (1977).

30. FURTER, W.F., *Chem. Eng. Commun.,* **116,** 35 (1992).

31. KUMAR, A., *Sep. Sci. and Tech.,* **28,** 1799–1818 (1993).

32. MAHAPATRA, A., V.G. GAIKAR, and M.M. SHARMA, *Sep. Sci. and Tech.,* **23,** 429–436 (1988).

33. AGARWAL, M., and V.G. GAIKAR, *Sep. Technol.,* **2,** 79–84 (1992).

34. KNAPP, J.P., and M.F. DOHERTY, *Ind. Eng. Chem. Res.,* **31,** 346–357 (1992).

35. LEWIS, W.K., U.S. Patent 1,676,700 (1928).

36. VAN WINKLE, M., *Distillation,* McGraw-Hill, New York (1967).

37. ROBINSON, C.S., and E.R. GILLILAND, *Elements of Fractional Distillation,* 4th ed., McGraw-Hill, New York (1950).

38. WAHNSCHAFFT, O.M., and A.W. WESTERBERG, *Ind. Eng. Chem. Res.,* **32,** 1108 (1993).

39. LAROCHE, L., N. BEKIARIS, H.W. ANDERSEN, and M. MORARI, *AIChE J.,* **38,** 1309 (1992).

40. YOUNG, S., *J. Chem. Soc. Trans.,* **81,** 707–717 (1902).

41. KEYES, D.B., U.S. Patent 1,676,735 (1928).

42. KEYES, D.B., *Ind. Eng. Chem.,* **21,** 998–1001 (1929).

43. BEKIARIS, N., G.A. MESKI, and M. MORARI, *Ind. Eng. Chem. Res.,* **35,** 207–217 (1996).

44. PROKOPAKIS, G.J., and W.D. SEIDER, *AIChE J.,* **29,** 49–60 (1983).

45. RYAN, P.J., and M.F. DOHERTY, *AIChE J.,* **35,** 1592–1601 (1989).

46. PHAM, H.N., and M.F. DOHERTY, *Chem. Eng. Sci.,* **45,** 1845–1854 (1990).

47. KOVACH, III, J.W., and W.D. SEIDER, *Computers and Chem. Engng.,* **11,** 593 (1987).

48. SWARTZ, C.L.E., and W.E. STEWART, *AIChE J.,* **33,** 1977–1985 (1987).

49. GANI, R., and S.B. JØRGENSEN, *Computers Chem. Engng.,* **18,** Suppl., S55 (1994).

50. SHEWCHUK, C.F., "Computation of Multiple Distillation Towers," Ph.D. Thesis, University of Cambridge (1974).

51. MAGNUSSEN, T., M.L. MICHELSEN, and A. FREDENSLUND, *Inst. Chem. Eng. Symp. Series No. 56, Third International Symp. on Distillation,* Rugby, England (1979).

52. BEKIARIS, N., G.A. MESKI, C.M. RADU, and M. MORARI, *Ind. Eng. Chem. Res.,* **32,** 2023–2038 (1993).

53. BLACK, C., and D.E. DITSLER, *Advances in Chemistry Series,* ACS, Washington, D.C., Vol. 115, pp. 1–15 (1972).

54. BLACK, C., R.A. GOLDING, and D.E. DITSLER, *Advances in Chemistry Series,* ACS, Washington, D.C., Vol. 115, pp. 64–92 (1972).

55. BLACK, C., *Chem. Eng. Progress,* **76** (9), 78–85 (1980).

56. BLACK, C., *Ind. Eng. Chem.,* **50,** 403–412 (1958).

57. TERRILL, D.L., L.F. SYLVESTRE, and M.F. DOHERTY, *Ind. Eng. Chem. Proc. Des. Develop.,* **24,** 1062–1071 (1985).

58. ROBINSON, C.S., and E.R. GILLILAND, *Elements of Fractional Distillation,* 4th ed., McGraw-Hill, New York (1950).

59. BACKHAUS, A.A., U.S. Patent 1,400,849 (1921).

60. LEYES, C.E., and D.F. OTHMER, *Trans. AIChE,* **41,** 157–196 (1945).

61. BELCK, L.H., *AIChE J.,* **1,** 467–470 (1955).

62. MASAMOTO, J., and K. MATSUZAKI, *J. Chem. Eng. Japan,* **27,** 1–5 (1994).

63. AGREDA, V.H., and L.R. PARTIN, U.S. Patent 4,435,595 (March 6, 1984).

64. AGREDA, V.H., L.R. PARTIN, and W.H. HEISE, *Chem. Eng. Prog.,* **86** (2), 40–46 (1990).

65. SMITH, L.A., U.S. Patent 4,307,254 (Dec. 22, 1981).

66. SMITH, L.A., U.S. Patent 4,443,559 (April 17, 1984).

67. SMITH, L.A., U.S. Patent 4,978,807 (Dec. 18, 1990).

68. DEGARMO, J.L., V.N. PARULEKAR, and V. PINJALA, *Chem. Eng. Prog.,* **88** (3), 43–50 (1992).

69. CHANG, Y.A., and J.D. SEADER, *Computers Chem. Engng.,* **12,** 1243–1255 (1988).

70. VENKATARAMAN, S., W.K. CHAN, and J.F. BOSTON, *Chem. Eng. Progress,* **86** (8), 45–54 (1990).

71. SIMANDL, J., and W.Y. SVRCEK, *Computers Chem. Engng.,* **15,** 337–348 (1991).

72. KANG, Y.W., Y.Y. LEE, and W.K. LEE, *J. Chem. Eng. Japan,* **25,** 649–655 (1992).

73. IZARRARAZ, A., G.W. BENTZEN, R. G. ANTHONY, and C.D. HOLLAND, *Hydrocarbon Processing,* **59** (6), 195 (1980).

74. REHFINGER, A., and U. HOFFMANN, *Chem. Eng. Sci.*, **45**, 1605–1617 (1990).

75. JACOBS, R., and R. KRISHNA, *Ind. Eng. Chem. Res.*, **32**, 1706–1709 (1993).

76. NIJHUIS, S.A., F.P.J.M. KERKHOF, and N.S. MAK, *Ind. Eng. Chem. Res.*, **32**, 2767–2774 (1993).

77. HAUAN, S., T. HERTZBERG, and K.M. LIEN, *Ind. Eng. Chem. Res.*, **34**, 987–991 (1995).

78. HANNAY, J.B., and J. HOGARTH, *Proc. Roy. Soc. (London) Sec. A*, **29**, 324 (1879).

79. WILLIAMS, D.F., *Chem. Eng. Sci.*, **36**, 1769–1788 (1981).

80. MCHUGH, M., and V. KRUKONIS, *Supercritical Fluid Extraction—Principles and Practice*, Butterworths, Boston (1986).

81. MCHUGH, M., and V. KRUKONIS, *Supercritical Fluid Extraction—Principles and Practice*, 2nd ed., Butterworth-Heinemann, Boston (1994).

82. TAKISHIMA, S., A. SAIKI, K. ARAI, and S. SAITO, *J. Chem. Eng. Japan*, **19**, 48–56 (1986).

83. INOMATA, H., A. KONDO, K. ARAI, and S. SAITO, *J. Chem. Eng. Japan*, **23**, 199–207 (1990).

84. DE FILLIPI, R.P., and J.E. VIVIAN, U.S. Patent 4,349,415 (1982).

85. KATZ, S.N., J.E. SPENCE, M.J. O'BRIAN, R.H. SKIFF, G.J. VOGEL, and R. PRASAD, U.S. Patent 4,911,941 (1990).

86. WONG, D.S.H., and S.I. SANDLER, *AIChE J.*, **38**, 671–680 (1992).

87. SANDLER, S.I., H. ORBEY, and B-I. LEE, in *Models for Thermodynamic and Phase Equilibria Calculations*, S.I. Sandler, Ed., Marcel Dekker, New York, pp. 87–186 (1994).

88. KNAPP, H., R. DORING, L. OELLRICH, U. PLOCKER, and J.M. PRAUSNITZ, *Vapor–Liquid Equilibria for Mixtures of Low Boiling Substances*, Chem. Data Ser., Vol. VI, DECHEMA, pp. 771–793 (1982).

89. SHIBATA, S.K., and S.I. SANDLER, *Ind. Eng. Chem. Res.*, **28**, 1893–1898 (1989).

90. WONG, D.S.H., H. ORBEY, and S.I. SANDLER, *Ind. Eng. Chem. Res.*, **31**, 2033–2039 (1992).

91. GMEHLING, J., and U. ONKEN, *Vapor–Liquid Equilibrium Data Compilation*, DECHEMA Data Series, DECHEMA, Frankfurt (1977).

92. SKJOLD-JØRGENSEN, S., *Ind. Eng. Chem. Res.*, **27**, 110–118 (1988).

93. COLUSSI, I.E., M. FERMEGLIA, V. GALLO, and I. KIKIC, *Computers Chem. Engng.*, **16**, 211–224 (1992).

94. DOHERTY, M.F., and M.F. MALONE, *Conceptual Design of Distillation Systems*, McGraw-Hill, New York (2001).

95. STICHLMAIR, J.G., and J.R. FAIR, *Distillation Principles and Practices*, Wiley-VCH, New York (1998).

96. SIIROLA, J.J., and S.D. BARNICKI, in *Perry's Chemical Engineers' Handbook*, Seventh Edition, Perry, R.H., and D.W. Green, eds., McGraw-Hill, New York, pp. 13–54 to 13–85 (1997).

97. ECKERT, E., and M. KUBICEK, *Computers Chem. Eng.*, **21**, 347–350 (1997).

98. HOFFMASTER, W.R., and S. HAUAN, *AIChE J.*, **48**, 2545–2556 (2002).

99. LLANO-RESTREPO, M., and J. AGUILAR-ARIAS, *Computers Chem. Engng.*, **27**, 527–549 (2003).

100. FU, J., *AIChE J.*, **42**, 3364–3372 (1996).

EXERCISES

Section 11.1

11.1 For the ternary system, normal hexane–methanol–methyl acetate at 1 atm find, in suitable references, all the binary and ternary azeotropes, sketch an approximate residue-curve map on a right-triangular diagram, and indicate the distillation boundaries. Determine for each azeotrope and pure component whether it is a stable node, an unstable node, or a saddle.

11.2 For the same ternary system as in Exercise 11.1, use a process-simulation program with the UNIFAC equation to calculate a portion of a residue curve at 1 atm starting from a bubble-point liquid with a composition of 20 mol% normal hexane, 60 mol% methanol, and 20 mol% methyl acetate.

11.3 For the same conditions as Exercise 11.2, use a process-simulation program with the UNIFAC equation to calculate a portion of a distillation curve at 1 atm.

11.4 For the ternary system acetone, benzene, and *n*-heptane at 1 atm find, in suitable references, all the binary and ternary azeotropes, and sketch an approximate distillation-curve map on an equilateral-triangle diagram, and indicate the distillation boundaries. Determine for each azeotrope and pure component whether it is a stable node, an unstable node, or a saddle.

11.5 For the same ternary system as in Exercise 11.4, use a process-simulation program with the UNIFAC equation to calculate a portion of a residue curve at 1 atm starting from a bubble-point liquid with a composition of 20 mol% acetone, 60 mol% benzene, and 20 mol% *n*-heptane.

11.6 For the same conditions as Exercise 11.5, use a process-simulation program with the UNIFAC equation to calculate a portion of a distillation curve at 1 atm.

11.7 Develop the feasible product-composition regions for the system of Figure 11.13, using Feed F_1.

11.8 Develop the feasible product composition regions for the system of Figure 11.10 if the feed composition is 50 mol% chloroform, 25 mol% methanol, and 25 mol% acetone.

Section 11.2

11.9 Repeat Example 11.3, but with ethanol as the solvent.

11.10 Repeat Example 11.3, but with MEK as the solvent.

11.11 Repeat Example 11.4, but with toluene as the solvent.

11.12 An equimolar mixture of *n*-heptane and toluene at 200°F, 20 psia, and a flow rate of 400 lbmol/h is to be separated by extractive distillation at 20 psia, using phenol at 220°F as the solvent, at a flow rate of 1200 lbmol/h. Design a suitable two-column system, obtaining reasonable product purities, with only a small loss of solvent.

Section 11.4

11.13 Repeat Example 11.5, but with a feed of 100 mol/s of 55 mol% ethanol and 45 mol% benzene.

11.14 Determine the feasibility of separating 100 mol/s of a mixture of 20 mol% ethanol and 80 mol% benzene by pressure-swing distillation. If feasible, design such a system.

11.15 Design a pressure-swing distillation system to produce 99.8 mol% ethanol for 100 mol/s of an aqueous feed containing 30 mol% ethanol.

Section 11.5

11.16 In Example 11.6, a mixture of benzene and cyclohexane is separated in a separation sequence that begins with homogeneous azeotropic distillation using acetone as the entrainer. Can the same separation be achieved using methanol as the entrainer? If not, why not? [Ref.: Ratliff, R.A., and W.B. Strobel, *Petro. Refiner,* **33** (5), 151 (1954)].

11.17 Devise a separation sequence to separate 100 mol/s of an equimolar mixture of toluene and 2,5-dimethylhexane into nearly pure products. Include in the sequence a homogeneous azeotropic distillation column using methanol as the entrainer and determine a feasible design for that column. [Ref.: Benedict, M., and L.C. Rubin, *Trans. AIChE,* **41**, 353–392 (1945)].

11.18 A mixture of 55 wt% methyl acetate and 45 wt% methanol at a flow rate of 16,500 kg/h is to be separated into one product of 99.5 wt% methyl acetate and another product of 99 wt% methanol. It has been suggested that such a separation might be possible by using a sequence of one homogeneous azeotropic distillation column and one ordinary distillation column. Possible entrainers are *n*-hexane, cyclohexane, and toluene. Determine the feasibility of such a sequence. If feasible, prepare a process design. If not feasible, suggest an alternative process and prove its feasibility.

Section 11.6

11.19 Design a three-column distillation sequence to separate 150 mol/s of an azeotropic mixture of ethanol and water at 1 atm into nearly pure ethanol and nearly pure water using heterogeneous azeotropic distillation with benzene as the entrainer.

11.20 Design a three-column distillation sequence to separate 120 mol/s of an azeotropic mixture of isopropanol and water at 1 atm into nearly pure isopropanol and nearly pure water using heterogeneous azeotropic distillation with benzene as the entrainer. [Ref.: Pham, H.N., P.J. Ryan, and M.F. Doherty, *AIChE J.,* **35**, 1585–1591 (1989)].

11.21 Design a two-column distillation sequence to separate 1,000 kmol/h of 20 mol% aqueous acetic acid into nearly pure acetic acid and nearly pure water. The first column should use heterogeneous azeotropic distillation with *n*-propyl acetate as the entrainer.

Section 11.7

11.22 Repeat Example 11.9, with the entire range of methanol feed-stage locations. Compare your results for isobutene conversion with the values shown in Figure 11.39.

11.23 Repeat Exercise 11.22, but with activities, instead of mole fractions, in the reaction rate expressions. How do the results differ? Explain.

11.24 Repeat Exercise 11.22, but with the assumption of chemical equilibrium on stages where catalyst is employed. How do the results differ from Figure 11.39? Explain.

Section 11.8

11.25 Repeat Example 11.10, but with 10 equilibrium stages instead of 5. What is the effect of this change?

11.26 An important application of supercritical extraction is the removal of solutes from particles of porous natural materials. Such applications include the extraction of caffeine from coffee beans and the extraction of ginger oil from ginger root. When CO_2 is used as the solvent, the rate of extraction is found to be independent of the flow rate of CO_2 past the particles, but dependent upon the particle size. Develop a suitable mathematical model for the rate of extraction that is consistent with these observations. What parameter in the model would have to be determined by experiment?

11.27 Cygnarowicz and Seider [*Biotechnol. Prog.,* **6**, 82–91 (1990)] present a process design for the supercritical extraction of β-carotene from water with carbon dioxide using the GC-EOS method of Skjold-Jørgensen to estimate phase equilibria. Repeat the calculations for the conditions of their design using the Peng–Robinson EOS with the Wong–Sandler mixing rules. How do the two designs compare?

11.28 Cygnarowicz and Seider [*Ind. Eng. Chem. Res.,* **28**, 1497–1503 (1989)] present a process design for the supercritical extraction of acetone from water with carbon dioxide using the GC-EOS method of Skjold-Jørgensen to estimate phase equilibria. Repeat the calculations for the conditions of their design using the Peng–Robinson EOS with the Wong–Sandler mixing rules. How do the two designs compare?

Chapter 12

Rate-Based Models for Distillation

Chapter 10 contains rigorous, equilibrium-based models for continuous-flow, steady-state, multicomponent, multi-stage distillation, absorption, stripping, and liquid–liquid extraction based on component material balances, energy balances, and thermodynamic correlations and criteria for phase equilibria. These models are extended in Chapter 11 to supercritical extraction and enhanced distillation, including extractive distillation, azeotropic distillation, and reactive distillation. The fundamental equations for the equilibrium-based models were first published by Sorel [1] in 1893. His equations consisted of total and component material balances around top and bottom sections of equilibrium stages (theoretical plates), including a total condenser and a reboiler, and corresponding energy balances that included provision for heat losses, which are an important factor for small laboratory columns, but not for insulated, industrial columns. Sorel used graphs of phase-equilibrium data instead of equations. Because of the complexity of Sorel's model, it was not widely applied until 1921, when it was adapted to graphical-solution techniques for binary systems, first by Ponchon and then by Savarit, who used an enthalpy–concentration diagram. In 1925, a much simpler, but less-rigorous, graphical technique was developed by McCabe and Thiele, who eliminated the energy balances by assuming constant vapor and liquid molar flow rates from equilibrium stage to equilibrium stage except across feed or side-stream withdrawal stages. This is referred to as the constant-molar-overflow assumption. When applicable, the McCabe–Thiele graphical method, developed in detail in Chapter 7, is applied even today for binary distillation, because the method gives valuable insight into changes in phase compositions from stage to stage.

Because some of Sorel's equations are nonlinear, it is not possible to obtain algebraic solutions, unless simplifying assumptions are made. A notable achievement in this respect was made by Smoker [2] in 1938 for the distillation of a binary mixture by assuming not only constant molar overflow, but also constant relative volatility between the two components. Smoker's equation is still useful for superfractionators involving close-boiling binary mixtures, where that assumption is valid. Starting in 1932, two iterative, numerical methods were developed for obtaining a general solution to Sorel's model for the distillation of multicomponent mixtures. The Thiele–Geddes method [3] requires specification of the number of equilibrium stages, the feed stage, the reflux ratio, and the distillate flow rate, with the resulting distribution of the components between distillate and bottoms being calculated. The Lewis–Matheson method [4] computes the number of stages required and the location of the feed stage for a specified reflux ratio and split between two key components. These two methods were widely used for the simulation and design of single-feed, multicomponent distillation columns prior to the 1960s.

Attempts in the late 1950s and early 1960s to adapt the Thiele–Geddes and Lewis–Matheson methods to computations with a digital computer had limited success. The real breakthrough in computerization of equilibrium-stage calculations occurred when Amundson and co-workers, starting in 1958, applied techniques of matrix algebra. This led to a number of successful computer-aided methods, based on sparse-matrix algebra, for Sorel's equilibrium-based model. The most important of these models are presented in Chapter 10. Today, computer-aided design and simulation programs abound for the rigorous, iterative, numerical solution of Sorel's equilibrium-based model for a wide variety of column configurations and specifications. Although the iterative computations sometimes fail to converge, the methods are widely applied and have become more flexible and robust with each passing year.

The methods presented in Chapters 10 and 11 assume that equilibrium is achieved, at each stage, with respect to both heat and component mass transfer. Except when temperature changes significantly from stage to stage, the assumption of temperature equality for vapor and liquid phases leaving a stage is usually acceptable. However, in most industrial applications, the assumption of equilibrium with respect to exiting-phase compositions is not reasonable. In general, exiting vapor-phase mole fractions are not related to exiting liquid-phase mole fractions by thermodynamic K-values. To overcome this limitation of equilibrium-based models, Lewis [5], in 1922, proposed the use of an overall stage efficiency for converting theoretical stages to actual stages. Unfortunately, experimental data show that this efficiency varies, depending on the application, over a range of about

5 to 120%, where the high values are achieved for distillation in large-diameter, single-liquid-pass trays because of a cross-flow effect, whereas the lower values occur in absorption columns when a high-viscosity, high-molecular-weight absorbent is used.

A preferred procedure for accounting for nonequilibrium with respect to mass transfer, since its introduction by Murphree [6] in 1925, has been to incorporate the Murphree vapor-phase tray efficiency, $(E_{MV})_{i,j}$, directly into Sorel's model as a replacement for the equilibrium equation based on definition of the K-value. Thus, the equation

$$K_{i,j} = y_{i,j}/x_{i,j} \qquad (12\text{-}1)$$

where i refers to the component and j the stage, is replaced by

$$(E_{MV})_{i,j} = (y_{i,j} - y_{i,j+1})/(y_{i,j}^* - y_{i,j+1}) \qquad (12\text{-}2)$$

where stages are numbered from the top.

Thus, the efficiency is the ratio of the actual change in vapor-phase mole fraction to the change that would occur if equilibrium were achieved. The equilibrium value, $y_{i,j}^*$, is obtained from (12-1), with substitution into (12-2) giving

$$(E_{MV})_{i,j} = (y_{i,j} - y_{i,j+1})/(K_{i,j}x_{i,j} - y_{i,j+1}) \qquad (12\text{-}3)$$

Equations (12-2) and (12-3) assume the following:

1. Uniform concentrations of vapor and liquid streams entering into and exiting a tray
2. Complete mixing throughout the liquid flowing across the tray
3. Plug flow of the vapor up through the liquid
4. Negligible resistance to mass transfer in the liquid phase

Application of the Murphree efficiency using empirical correlations has proved to be adequate for binary and close-boiling, ideal, and near-ideal multicomponent vapor–liquid mixtures. However, deficiencies of the Murphree efficiency for general multicomponent vapor–liquid mixtures have long been recognized. Murphree himself stated clearly the deficiencies of his development for multicomponent mixtures and for cases where the efficiency is low. He even stated that the theoretical plate should not be the basis of calculation for ternary mixtures.

When the equilibrium-based model is applied to multicomponent mixtures, a number of problems arise. Values of E_{MV} differ from component to component and vary from stage to stage. But, at each stage, the number of independent values of E_{MV} must be determined so as to force the sum of the mole fractions in the vapor phase to sum to 1. This introduces the possibility that negative values of E_{MV} can result. This is in contrast to binary mixtures for which the values of E_{MV} are always positive and are identical for the two components. When using the Murphree vapor-phase efficiency, the temperatures of the exiting vapor and liquid phases are assumed to be the same and equal to the bubble-point temperature of the exiting liquid phase. Because the vapor phase is not in equilibrium with the liquid phase, the vapor temperature does not correspond to the dew-point temperature. It is even possible, algebraically, for the vapor temperature to correspond to a value below its dew-point temperature, which is physically impossible.

Values of E_{MV} can be obtained from experimental data or correlations. These values, however, are more likely to be Murphree vapor-point (rather than tray) efficiencies. Point efficiencies only apply to a particular location in the liquid on the tray. To convert these point efficiencies to tray efficiencies, vapor and liquid flow patterns must be assumed after the manner of Lewis [7], as discussed by Seader [8]. However, if the vapor and liquid phases are both completely mixed, the point efficiency equals the tray efficiency.

Walter and Sherwood [9] found that experimentally measured tray efficiencies covered an enormous range: 0.65 to 4.2% for absorption and stripping of carbon dioxide from water and glycerine solutions; 4.7 to 24% for absorption of olefins into oils; and 69 to 92% for absorption of ammonia, humidification of air, and rectification of alcohol.

In 1957, Toor [10] showed that diffusion in a ternary mixture is enormously more complex than in a binary mixture because of coupling among component concentration gradients, especially when components differ widely in size, shape, and polarity. Toor showed that, in addition to diffusion due to the conventional Fickian concentration driving force, the possible consequences of gradient coupling could result in: (1) diffusion against a driving force (reverse diffusion), (2) no diffusion even though a concentration driving force is present (diffusion barrier), and (3) diffusion with zero driving force (osmotic diffusion). Theoretical calculations by Toor and Burchard [11] predicted the possibility of negative values of E_{MV} in multicomponent systems, but values of E_{MV} for binary systems are restricted to the range from 0 to 100%.

In 1977, Krishna et al. [12] extended the theoretical work of Toor and Burchard and showed that when the vapor mole-fraction driving force of a component (call it A) is small compared to the other components in the mixture, the transport rate of A is controlled by the other components, with the result that E_{MV} for A is anywhere in the range from minus infinity to plus infinity. They confirmed this theoretical prediction by conducting experiments with the ethanol/tert-butanol/water system and obtained values of E_{MV} for tert-butanol ranging from −2,978% to +527%. In addition, values of E_{MV} for ethanol and water sometimes differed significantly.

Two other tray efficiencies are defined in the literature: the vaporization efficiency of Holland, which was first mentioned by McAdams, and the Hausen tray efficiency, which eliminates the assumption in E_{MV} that the exiting liquid is at its bubble point. The former cannot distinguish the Toor phenomena and can vary widely in a manner that is not ascribable to the particular component. The latter does appear to be superior to E_{MV}, but is considerably more complicated and difficult to use, and it has not found wide application.

Although the equilibrium-based model, modified to incorporate stage efficiency, is adequate for binary mixtures and for the major components in nearly ideal multicomponent mixtures, that model has serious deficiencies for more general cases and the development of a more realistic nonequilibrium, transport- or rate-based model has long been a desirable goal. In 1977, Waggoner and Loud [13] developed a rate-based, mass-transport model limited to nearly ideal, close-boiling, multicomponent systems. However, an energy-transport equation was not included (because thermal equilibrium would be closely approximated for a close-boiling mixture) and the coupling of component mass-transfer rates was ignored.

In 1979, Krishna and Standart [14] showed the possibility of applying rigorous multicomponent mass- and heat-transfer theory to calculations of simultaneous transport. The theory was further developed by Taylor and Krishna [15]. The availability of this theory led to the development in 1985 by Krishnamurthy and Taylor [16] of the first general,

rate-based, computer-aided model for application to trayed and packed columns for distillation and other continuous, countercurrent, vapor–liquid separation operations. This model applies the two-film theory of mass transfer discussed in Chapter 3, with phase equilibria assumed at the interface of the two phases, and provides options for vapor and liquid flow configurations in trayed columns, including plug flow and perfectly mixed flow, on each tray. Although the model does not require tray efficiencies or values of HETP, correlations of mass-transfer and heat-transfer coefficients are needed for the particular type of trays or packing employed. The model was extended in 1994 by Taylor, Kooijman, and Hung [17] to include: (1) effects of entrainment of liquid droplets in the vapor and occlusion of vapor bubbles in the liquid; (2) estimation of the column-pressure profile; (3) interlinking streams; and (4) axial dispersion in packed columns. In addition, unlike the 1985 model, which required the user to specify the column diameter and tray geometry or packing size, the 1994 version includes a design mode that estimates column diameter for a specified fraction of flooding or pressure drop. Rate-based models are implemented in several computer programs, including RATEFRAC [18] of Aspen Technology, ChemSep Release 3.1 [19], and CHEMCAD.

12.0 INSTRUCTIONAL OBJECTIVES

After completing this chapter, you should be able to:

- Write equations that model a nonequilibrium vapor–liquid stage, where the assumption of equilibrium is only applied at the interface between two phases.
- Explain component-coupling effects in multicomponent mass transfer.
- Explain the bootstrap problem and how it is handled for distillation.
- Cite available methods for estimating transport coefficients and interfacial areas required for rate-based calculations.
- Explain differences among ideal vapor–liquid flow patterns that are employed for rate-based calculations.
- Apply a simulation program to make a rate-based calculation for a multicomponent, multistage, vapor–liquid separation problem.

12.1 RATE-BASED MODEL

A schematic diagram of a nonequilibrium stage, consisting of a tray, a group of trays, or a segment of a packed section, is shown in Figure 12.1. Entering stage j, at pressure P_j, are liquid F_j^L and/or vapor F_j^V molar flow rates; component i molar flow rates, $f_{i,j}^L$ and $f_{i,j}^V$; and stream molar enthalpies, H_j^{LF} and H_j^{VF}. Also leaving from $(+)$ or entering to $(-)$ the liquid and/or vapor phases in the stage are heat transfer rates Q_j^V and Q_j^L, respectively. Also entering the stage from the stage above is liquid molar flow rate L_{j-1} at temperature T_{j-1}^L and pressure P_{j-1}, with molar enthalpy H_{j-1}^L and component mole fractions $x_{i,j-1}$; and entering the stage from the stage below is vapor molar flow rate V_{j+1} at temperature

T_{j+1}^V and pressure P_{j+1}, with molar enthalpy H_{j+1}^V and component mole fractions $y_{i,j+1}$. Within the stage, mass transfer of components occurs across the phase boundary at molar rates $N_{i,j}$ from the vapor phase to the liquid phase $(+)$ or vice versa $(-)$, and heat transfer occurs across the phase boundary at rates e_j from the vapor phase to the liquid phase $(+)$ or vice versa $(-)$. Leaving the stage is liquid at temperature T_j^L and pressure P_j, with molar enthalpy H_j^L; and vapor at temperature T_j^V and pressure P_j, with molar enthalpy H_j^V. A fraction, r_j^L, of the liquid exiting the stage may be withdrawn as a liquid side stream at molar flow rate U_j, leaving the molar flow rate L_j to enter the stage below or to exit the column. A fraction r_j^V, of the vapor exiting the stage may

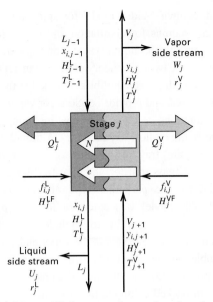

Figure 12.1 Nonequilibrium stage for rate-based method.

be withdrawn as a vapor side stream at molar flow rate W_j, leaving the molar flow rate V_j to enter the stage above or to exit the column. If desired, entrainment, occlusion, interlink flows, a second immiscible liquid phase, and chemical reaction(s) can be added to the model.

Recall that the equilibrium-stage model of Chapter 10 utilizes the $2C + 3$ MESH equations for each stage:

C mass balances for components

C phase equilibria relations

2 summations of mole fractions

1 energy balance

In the rate-based model, the mass and energy balances around each equilibrium stage are each replaced by separate balances for each phase around a stage, which can be a tray, a collection of trays, or a segment of a packed section. In residual form, the equations are as follows, where the residuals are on the left-hand sides and become zero when the computations are converged. When not converged, the residuals are used to determine the proximity to convergence.

Liquid-phase component material balance:

$$M_{i,j}^{L} \equiv \left(1 + r_j^{L}\right) L_j x_{i,j} - L_{j-1} x_{i,j-1} - f_{i,j}^{L} - N_{i,j}^{L} = 0,$$
$$i = 1, 2, \ldots, C \tag{12-4}$$

Vapor-phase component material balance:

$$M_{i,j}^{V} \equiv \left(1 + r_j^{V}\right) V_j y_{i,j} - V_{j+1} y_{i,j+1} - f_{i,j}^{V} + N_{i,j}^{V} = 0,$$
$$i = 1, 2, \ldots, C \tag{12-5}$$

Liquid-phase energy balance:

$$E_j^{L} \equiv \left(1 + r_j^{L}\right) L_j H_j^{L} - L_{j-1} H_{j-1}^{L}$$
$$- \left(\sum_{i=1}^{C} f_{i,j}^{L}\right) H_j^{LF} + Q_j^{L} - e_j^{L} = 0 \tag{12-6}$$

Vapor-phase energy balance:

$$E_j^{V} \equiv \left(1 + r_j^{V}\right) V_j H_j^{V} - V_{j+1} H_{j+1}^{V}$$
$$- \left(\sum_{i=1}^{C} f_{i,j}^{V}\right) H_j^{VF} + Q_j^{V} + e_j^{V} = 0 \tag{12-7}$$

where at the phase interface, I,

$$E_j^{I} = e_j^{V} - e_j^{L} = 0 \tag{12-8}$$

Equations (12-4) and (12-5) are coupled by the component mass-transfer rates:

$$R_{i,j}^{L} \equiv N_{i,j} - N_{i,j}^{L} = 0, \quad i = 1, 2, \ldots, C - 1 \tag{12-9}$$
$$R_{i,j}^{V} \equiv N_{i,j} - N_{i,j}^{V} = 0, \quad i = 1, 2, \ldots, C - 1 \tag{12-10}$$

The equations for the mole-fraction summation for each phase are applied at the vapor–liquid interface:

$$S_j^{LI} \equiv \sum_{i=1}^{C} x_{i,j}^{I} - 1 = 0 \tag{12-11}$$

$$S_j^{VI} \equiv \sum_{i=1}^{C} y_{i,j}^{I} - 1 = 0 \tag{12-12}$$

A hydraulic equation for stage pressure drop is given by

$$H_j \equiv P_{j+1} - P_j - (\Delta P_j) = 0, \quad j = 1, 2, 3, \ldots, N - 1 \tag{12-13}$$

where the stage is assumed to be at mechanical equilibrium such that

$$P_j^{L} = P_j^{V} = P_j \tag{12-14}$$

and ΔP_j is the gas-phase pressure drop from stage $j + 1$ to stage j. Equation (12-13) is optional. It is only included when it is desired to compute one or more stage pressures from hydraulics, as discussed in Chapter 6. Phase equilibrium for each component is assumed to exist only at the phase interphase:

$$Q_{i,j}^{I} \equiv K_{i,j} x_{i,j}^{I} - y_{i,j}^{I} = 0, \quad i = 1, 2, \ldots, C \tag{12-15}$$

Because only $C - 1$ equations are written for the component mass-transfer rates in (12-9) and (12-10), total phase material balances in terms of total mass-transfer rates, $N_{T,j}$, can be added to the system:

$$M_{T,j}^{L} \equiv \left(1 + r_j^{L}\right) L_j - L_{j-1} - \sum_{i=1}^{C} f_{i,j}^{L} - N_{T,j} = 0 \tag{12-16}$$

$$M_{T,j}^{V} \equiv \left(1 + r_j^{V}\right) V_j - V_{j+1} - \sum_{i=1}^{C} f_{i,j}^{V} + N_{T,j} = 0 \tag{12-17}$$

where

$$N_{T,j} = \sum_{i=1}^{C} N_{i,j} \tag{12-18}$$

Equations (12-4), (12-5), (12-9), (12-10), (12-16), (12-17), and (12-18) contain terms for component mass-transfer rates, estimated from diffusive and bulk-flow (convective) contributions. The former are based on interfacial area, average mole-fraction driving forces, and mass-transfer coefficients that account for component coupling effects through binary-pair coefficients. Empirical equations are used for the interfacial area and binary mass-transfer coefficients, based on correlations of experimental data for bubble-cap trays, sieve trays, valve trays, random packings, and structured packings. The average mole-fraction driving forces for diffusion depend upon the assumed vapor and liquid flow patterns. The simplest case is perfectly mixed flow for both the vapor and liquid phases, which simulates small-diameter, trayed columns. The case of countercurrent plug flow for vapor and liquid phases simulates a packed column with no axial dispersion.

Equations (12-6) to (12-8) contain terms for heat-transfer rates. These are estimated from convective and enthalpy-flow contributions, where the former are based on interfacial area, average temperature-driving forces, and convective heat-transfer coefficients, estimated from the Chilton–Colburn analogy for the vapor phase and the penetration theory for the liquid phase.

The K-values in (12-15) are estimated from the same equation-of-state or activity-coefficient models used with equilibrium-based models. Tray or packed-segment pressure drops are estimated from suitable correlations of the type discussed in Chapter 6.

The total number of independent equations, referred to as the MERSHQ equations, for each nonequilibrium stage, is $5C + 5$, as listed in Table 12.1. These equations apply for N stages, that is, $N_E = N(5C + 5)$ equations, in terms of $7NC + 14N + 1$ variables, listed in Table 12.2. The number of degrees of freedom is

$$N_D = N_V - N_E = (7NC + 14N + 1) - (5NC + 5N)$$
$$= 2NC + 9N + 1$$

Table 12.1 Summary of Independent Equations for Rate-Based Model

Equation	No. of Equations
$M_{i,j}^{L}$	C
$M_{i,j}^{V}$	C
$M_{T,j}^{L}$	1
$M_{T,j}^{V}$	1
E_{j}^{L}	1
E_{j}^{V}	1
E_{j}^{I}	1
$R_{i,j}^{L}$	$C - 1$
$R_{i,j}^{V}$	$C - 1$
S_{j}^{LI}	1
S_{j}^{VI}	1
H_{j}	(optional)
$Q_{i,j}^{I}$	C
	$5C + 5$

Table 12.2 List of Variables for Rate-Based Model

Variable Type No.	Variable	No. of Variables
1	No. of stages, N	1
2	$f_{i,j}^{L}$	NC
3	$f_{i,j}^{V}$	NC
4	T_{j}^{LF}	N
5	T_{j}^{VF}	N
6	P_{j}^{LF}	N
7	P_{j}^{VF}	N
8	L_{j}	N
9	$x_{i,j}$	NC
10	r_{j}^{L}	N
11	T_{j}^{L}	N
12	V_{j}	N
13	$y_{i,j}$	NC
14	r_{j}^{V}	N
15	T_{j}^{V}	N
16	P_{j}	N
17	Q_{j}^{L}	N
18	Q_{j}^{V}	N
19	$x_{i,j}^{I}$	NC
20	$y_{i,j}^{I}$	NC
21	T_{j}^{I}	N
22	$N_{i,j}$	NC
	$N_V = 7NC + 14N + 1$	

If variable types 1 to 7, 10, 14, and 16 to 18 in Table 12.2 are specified, a total of $2NC + 9N + 1$ variables are assigned values and the degrees of freedom are totally consumed. Then, the remaining $5C + 5$ independent variables in the $5C + 5$ equations are

$$x_{i,j}, \ y_{i,j}, \ x_{i,j}^{I}, \ y_{i,j}^{I}, \ N_{i,j}, \ T_{j}^{L}, \ T_{j}^{V}, \ T_{j}^{I}, \ L_{j}, \ \text{and} \ V_{j}$$

which are the variables to be computed from the equations. Properties $K_{i,j}^{I}$, H_{j}^{LF}, H_{j}^{VF}, H_{j}^{L}, and H_{j}^{V} are computed from thermodynamic correlations in terms of the remaining independent variables. Transport rates $N_{i,j}^{L}$, $N_{i,j}^{V}$, e_{j}^{L}, and e_{j}^{V} are computed from transport correlations and certain physical properties, in terms of the remaining independent variables. Stage pressures are computed from pressure drops, ΔP_j, stage geometry, fluid-mechanic equations, and certain physical properties, in terms of the remaining independent variables.

For a distillation column, it is preferable to specify $Q_1^{L} = 0$ and $Q_N^{V} = 0$. In that case, Q_1^{V} (heat-transfer rate from the vapor in the condenser) and Q_N^{L} (heat-transfer rate to the liquid in the reboiler) are not specified, but instead, as in the case of a column with a partial condenser, L_1 (reflux rate) and L_N (bottoms flow rate) are substitute specifications, which are sometimes referred to as standard specifications for ordinary distillation. For an adiabatic absorber or adiabatic stripper, however, all Q_j^{L} and Q_j^{V} are set equal to zero, with no substitution of specifications.

12.2 THERMODYNAMIC PROPERTIES AND TRANSPORT-RATE EXPRESSIONS

Rate-based models use the same K-value and enthalpy correlations as equilibrium-based models. However, the K-values apply only at the equilibrium interface between the vapor and liquid phases on trays or in packing. In general, the K-value correlation, whether based on an equation-of-state or an activity-coefficient model, is a function of phase-interface temperature and compositions, and tray pressure. Enthalpies are evaluated only at the conditions of the phases as they exit a tray. For the equilibrium-based model, the vapor is at the dew-point temperature and the liquid is at the bubble-point temperature, where both temperatures are equal and at the stage temperature. For the rate-based model, the liquid is subcooled and the vapor superheated.

The accuracy of enthalpies and, particularly, K-values is crucial to equilibrium-based models. For rate-based models, accurate predictions of heat-transfer rates and, particularly, mass-transfer rates are also required. These rates depend upon transport coefficients, interfacial area, and driving forces. It is important that mass-transfer rates account for component-coupling effects through binary-pair coefficients.

The general forms for component mass-transfer rates across the vapor and liquid films, respectively, on a tray or in a packed segment, are as follows, where both diffusive and convective (bulk-flow) contributions are included:

$$N_{i,j}^{V} = a_j^I J_{i,j}^{V} + y_{i,j} N_{T,j} \qquad (12\text{-}19)$$

and

$$N_{i,j}^{L} = a_j^I J_{i,j}^{L} + x_{i,j} N_{T,j} \qquad (12\text{-}20)$$

where a_j^I is the total interfacial area for the stage and $J_{i,j}^P$ is the molar diffusion flux relative to the molar-average velocity, where P stands for the phase (V or L). For a binary mixture, as discussed in Chapter 3, these fluxes, in terms of mass-transfer coefficients, are given by

$$J_i^{V} = c_t^{V} k_i^{V} \left(y_i^{V} - y_i^{I}\right)_{\text{avg}} \qquad (12\text{-}21)$$

and

$$J_i^{L} = c_t^{L} k_i^{L} \left(x_i^{I} - x_i^{L}\right)_{\text{avg}} \qquad (12\text{-}22)$$

where c_t^P is the total molar concentration, k_i^P is the mass-transfer coefficient for a binary mixture based on a mole-fraction driving force, and the last terms in (12-21) and (12-22) are the mean mole-fraction driving forces over the stage. The positive direction of mass transfer is assumed to be from the vapor phase to the liquid phase. From the definition of the molar diffusive flux:

$$\sum_{i=1}^{C} J_i = 0 \qquad (12\text{-}23)$$

Thus, for the binary system $(1, 2)$, $J_1 = -J_2$.

As discussed in detail by Taylor and Krishna [15], the general multicomponent case for mass transfer is considerably more complex than the binary case because of component-coupling effects. For example, for the ternary system $(1, 2, 3)$,

from Taylor and Krishna [15], the fluxes for the first two components are

$$J_1^{V} = c_t^{V} \kappa_{11}^{V} \left(y_1^{V} - y_1^{I}\right)_{\text{avg}} + c_t^{V} \kappa_{12}^{V} \left(y_2 - y_2^{I}\right)_{\text{avg}} \qquad (12\text{-}24)$$

$$J_2^{V} = c_t^{V} \kappa_{21}^{V} \left(y_1^{V} - y_1^{I}\right)_{\text{avg}} + c_t^{V} \kappa_{22}^{V} \left(y_2 - y_2^{I}\right)_{\text{avg}} \qquad (12\text{-}25)$$

The flux for the third component is not independent of the other two, but is obtained from (12-23):

$$J_3^{V} = -J_1^{V} - J_2^{V} \qquad (12\text{-}26)$$

In these equations, the binary-pair coefficients, κ^P, are complex functions related to inverse-rate functions described below and are called Maxwell–Stefan mass-transfer coefficients in binary mixtures.

For the general multicomponent system $(1, 2, \ldots, C)$, the independent fluxes for the first $C - 1$ components are given in matrix equation form as

$$\boldsymbol{J}^{V} = c_t^{V} [\boldsymbol{\kappa}^{V}](\boldsymbol{y}^{V} - \boldsymbol{y}^{I})_{\text{avg}} \qquad (12\text{-}27)$$

$$\boldsymbol{J}^{L} = c_t^{L} [\boldsymbol{\kappa}^{L}](\boldsymbol{x}^{I} - \boldsymbol{x}^{L})_{\text{avg}} \qquad (12\text{-}28)$$

where $\boldsymbol{J}^P$, $(\boldsymbol{y}^{V} - \boldsymbol{y}^{I})_{\text{avg}}$, and $(\boldsymbol{x}^{I} - \boldsymbol{x}^{L})_{\text{avg}}$ are column vectors of length $C - 1$ and $[\boldsymbol{\kappa}^P]$ is a $(C - 1) \times (C - 1)$ square matrix. The method for determining the average mole-fraction driving forces depends, as discussed in the next section, upon the flow patterns of the vapor and liquid phases.

The most fundamental theory for multicomponent diffusion is that of Maxwell and Stefan, who, in the period from 1866 to 1871, applied the kinetic theory of ideal gases. Their theory is presented most conveniently in terms of rate coefficients, $\boldsymbol{B}$, which are defined in reciprocal diffusivity terms [15]. Likewise, it is convenient to determine $[\boldsymbol{\kappa}^P]$ from a reciprocal mass-transfer coefficient function, $\boldsymbol{R}$, defined by Krishna and Standart [14]. For an ideal-gas solution:

$$[\boldsymbol{\kappa}^{V}] = [\boldsymbol{R}^{V}]^{-1} \qquad (12\text{-}29)$$

For a nonideal-liquid solution:

$$[\boldsymbol{\kappa}^{L}] = [\boldsymbol{R}^{L}]^{-1} [\boldsymbol{\Gamma}^{L}] \qquad (12\text{-}30)$$

where the elements of $\boldsymbol{R}^P$ in terms of general mole fractions, z_i, are:

$$R_{ii}^{P} = \frac{z_i}{k_{iC}^{P}} + \sum_{\substack{k=1 \\ k \neq i}}^{C} \frac{z_k}{k_{ik}^{P}} \qquad (12\text{-}31)$$

$$R_{ij}^{P} = -z_i \left(\frac{1}{k_{ij}^{P}} - \frac{1}{k_{iC}^{P}}\right) \qquad (12\text{-}32)$$

where here, j refers to the jth component and not the jth stage and the values of k are binary-pair mass-transfer coefficients obtained from correlations of experimental data.

For the four-component vapor-phase system, the combination of (12-27) and (12-29) gives

$$\begin{bmatrix} J_1^{V} \\ J_2^{V} \\ J_3^{V} \end{bmatrix} = c_t^{V} \begin{bmatrix} R_{11}^{V} & R_{12}^{V} & R_{13}^{V} \\ R_{21}^{V} & R_{22}^{V} & R_{23}^{V} \\ R_{31}^{V} & R_{32}^{V} & R_{33}^{V} \end{bmatrix}^{-1} \begin{bmatrix} (y_1^{V} - y_1^{I})_{\text{avg}} \\ (y_2^{V} - y_2^{I})_{\text{avg}} \\ (y_3^{V} - y_3^{I})_{\text{avg}} \end{bmatrix}$$

$$(12\text{-}33)$$

with

$$J_4^V = -\left(J_1^V + J_2^V + J_3^V\right) \tag{12-34}$$

and, for example, from (12-32) and (12-33), respectively:

$$R_{11}^V = \frac{y_1}{k_{14}^V} + \frac{y_2}{k_{12}^V} + \frac{y_3}{k_{13}^V} + \frac{y_4}{k_{14}^V} \tag{12-35}$$

$$R_{12}^V = -y_1\left(\frac{1}{k_{12}^V} - \frac{1}{k_{14}^V}\right) \tag{12-36}$$

The term $[\boldsymbol{\Gamma}^L]$ in (12-30) is a $(C-1) \times (C-1)$ matrix of thermodynamic factors that corrects for nonideality, which often is a necessary correction for the liquid phase. When an activity-coefficient model is used:

$$\Gamma_{ij}^L = \delta_{ij} + x_i\left(\frac{\partial \ln \gamma_i}{\partial x_j}\right)_{T,P,x_k,k \neq j=1,\dots,C-1} \tag{12-37}$$

For a nonideal vapor, a $[\boldsymbol{\Gamma}^V]$ term can be included in (12-29), but this is rarely necessary. For either phase, if an equation-of-state model is used, (12-37) can be rewritten by substituting $\bar{\phi}_i$, the mixture fugacity coefficient, for γ_i. The term δ_{ij} is the Kronecker delta, which is 1 if $i=j$ and 0 if not. The thermodynamic factor is required because it is generally accepted that the fundamental driving force for diffusion is the gradient of the chemical potential rather than the mole fraction or concentration gradient.

When mass-transfer fluxes are moderate to high, an additional correction term is needed in (12-29) and (12-30) to correct for distortion of the composition profiles. This correction, which can have a serious effect on the results, is discussed in detail by Taylor and Krishna [15]. The calculation of the low mass-transfer fluxes, according to (12-19) to (12-32), is illustrated by the following example.

EXAMPLE 12.1

This example is similar to Example 11.5.1 on page 283 of Taylor and Krishna [15]. The following results were obtained for tray n from a rate-based calculation of a ternary distillation at 14.7 psia, involving acetone (1), methanol (2), and water (3) in a 5.5-ft-diameter column using sieve trays with a 2-inch-high weir. Vapor and liquid phases are assumed to be completely mixed.

Component	y_n	y_{n+1}	y_n^I	K_n^I	x_n
1	0.2971	0.1700	0.3521	2.759	0.1459
2	0.4631	0.4290	0.4677	1.225	0.3865
3	0.2398	0.4010	0.1802	0.3673	0.4676
	1.0000	1.0000	1.0000		1.0000

The computed products of the gas-phase, binary mass-transfer coefficients and interfacial area, using the Chan–Fair correlations of Section 6.6, are as follows in lbmol/(h-unit mole fraction):

$$k_{12} = k_{21} = 1,955; \quad k_{13} = k_{31} = 2,407; \quad k_{23} = k_{32} = 2,797$$

(a) Compute the molar diffusion rates.

(b) Compute the mass-transfer rates.

(c) Calculate the Murphree vapor-tray efficiencies.

SOLUTION

Because rates instead of fluxes are given, the equations developed in this section are used with rates rather than fluxes.

(a) Compute the reciprocal rate functions, **R**, from (12-31) and (12-32), assuming linear mole-fraction gradients such that z_i can be replaced by $(y_i + y_i^I)/2$.

Thus:

$$z_1 = (0.2971 + 0.3521)/2 = 0.3246$$
$$z_2 = (0.4631 + 0.4677)/2 = 0.4654$$
$$z_3 = (0.2398 + 0.1802)/2 = 0.2100$$

$$R_{11}^V = \frac{z_1}{k_{13}} + \frac{z_2}{k_{12}} + \frac{z_3}{k_{13}} = \frac{0.3246}{2,407} + \frac{0.4654}{1,955} + \frac{0.2100}{2,407}$$
$$= 0.000460$$

$$R_{22}^V = \frac{z_2}{k_{23}} + \frac{z_1}{k_{21}} + \frac{z_3}{k_{23}} = \frac{0.4654}{2,797} + \frac{0.3246}{1,955} + \frac{0.2100}{2,797}$$
$$= 0.000408$$

$$R_{12}^V = -z_1\left(\frac{1}{k_{12}} - \frac{1}{k_{13}}\right) = -0.3246\left(\frac{1}{1,955} - \frac{1}{2,407}\right)$$
$$= -0.0000312$$

$$R_{21}^V = -z_2\left(\frac{1}{k_{21}} - \frac{1}{k_{23}}\right) = -0.4654\left(\frac{1}{1,955} - \frac{1}{2,797}\right)$$
$$= -0.0000717$$

Thus, in matrix form: $[\mathbf{R}^V] = \begin{bmatrix} 0.000460 & -0.0000312 \\ -0.0000717 & 0.000408 \end{bmatrix}$

From (12-29), by matrix inversion:

$$[\boldsymbol{\kappa}^V] = [\mathbf{R}^V]^{-1} = \begin{bmatrix} 2,200 & 168.2 \\ 386.6 & 2,480 \end{bmatrix}$$

Because the off-diagonal terms in the preceding 2×2 matrix are much smaller than the diagonal terms, the effect of coupling in this example is small.

From (12-27):

$$\begin{bmatrix} J_1^V \\ J_2^V \end{bmatrix} = \begin{bmatrix} \kappa_{11}^V & \kappa_{12}^V \\ \kappa_{21}^V & \kappa_{22}^V \end{bmatrix} \begin{bmatrix} (y_1 - y_1^I) \\ (y_2 - y_2^I) \end{bmatrix}$$

$$J_1^V = \kappa_{11}^V(y_1 - y_1^I) + \kappa_{12}^V(y_2 - y_2^I)$$
$$= 2,200(0.2971 - 0.3521) + 168.2(0.4631 - 0.4677)$$
$$= -121.8 \text{ lbmol/h}$$

$$J_2^V = \kappa_{21}^V(y_1 - y_1^I) + \kappa_{22}^V(y_2 - y_2^I)$$
$$= 386.6(0.2971 - 0.3521) + 2,480(0.4631 - 0.4677)$$
$$= -32.7 \text{ lbmol/h}$$

From (12-23):

$$J_3^V = -J_1^V - J_2^V = 121.8 + 32.7 = 154.5 \text{ lbmol/h}$$

(b) From (12-19), but with diffusion and mass-transfer rates instead of fluxes:

$$N_1^V = J_1^V + z_1 N_T^V = -121.8 + 0.3246 N_T^V \tag{1}$$

Similarly:

$$N_2^V = -32.7 + 0.4654 N_T^V \tag{2}$$

$$N_3^V = 154.5 + 0.2100 N_T^V \tag{3}$$

To determine the component mass-transfer rates, it is necessary to know the total mass-transfer rate for the tray, N_T^V. The problem of determining this quantity when the diffusion rates, J, are known is referred to as the *bootstrap problem* (p. 145 in Taylor and Krishna [15]). In chemical reaction with diffusion, N_T is determined by the stoichiometry. In distillation, N_T is determined by an energy balance, which gives the change in molar vapor rate across a tray. For the assumption of constant molar overflow, $N_T = 0$. In this example, that assumption is not valid, and the change is

$$N_T = V_{n+1} - V_n = -54 \text{ lbmol/h}$$

From (1), (2), (3):

$$N_1^V = -121.8 + 0.3246(-54) = -139.4 \text{ lbmol/h}$$
$$N_2^V = -32.7 + 0.4654(-54) = -57.8 \text{ lbmol/h}$$
$$N_3^V = -154.5 + 0.2100(-54) = 143.2 \text{ lbmol/h}$$

(c) Approximate values of the Murphree vapor-tray efficiency are obtained from (12-3), with K-values at phase interface conditions:

$$E_{MV_i} = (y_{i,n} - y_{i,n+1})/\left(K_{i,n}^I x_{i,n} - y_{i,n+1}\right) \quad (4)$$

From (4):

$$E_{MV_1} = \frac{(0.2971 - 0.1700)}{[(2.759)(0.1459) - 0.1700]} = 0.547$$
$$E_{MV_2} = \frac{(0.4631 - 0.4290)}{[(1.225)(0.3865) - 0.4290]} = 0.767$$
$$E_{MV_3} = \frac{(0.2398 - 0.4010)}{[(0.3673)(0.4676) - 0.4010]} = 0.703$$

The general forms for rates of heat transfer across the vapor and liquid films of a stage, respectively, are

$$e_j^V = a_j^I h^V(T^V - T^I) + \sum_{i=1}^C N_{i,j}^V \bar{H}_{i,j}^V \quad (12\text{-}38)$$

$$e_j^L = a_j^I h^L(T^I - T^L) + \sum_{i=1}^C N_{i,j}^L \bar{H}_{i,j}^L \quad (12\text{-}39)$$

where $\bar{H}_{i,j}^P$ are the partial molar enthalpies of component i for stage j and h^P are convective heat-transfer coefficients. The second terms on the right-hand sides of (12-38) and (12-39) account for the transfer of enthalpy by mass transfer. Temperatures T^V and T^L are the temperatures exiting the stage regardless of the assumed flow patterns for the vapor and liquid.

12.3 METHODS FOR ESTIMATING TRANSPORT COEFFICIENTS AND INTERFACIAL AREA

Equations (12-31) and (12-32) require binary-pair mass-transfer coefficients. In most rate-based model applications, the coefficients are estimated from empirical correlations of experimental data for different contacting devices.

As discussed in Section 6.6, for trayed columns, the most widely used correlations are the AIChE method [20] for bubble-cap trays, the correlations of Chan and Fair [24] for

sieve trays, and the correlations of Scheffe and Weiland [36] for Glitsch V-1 valve trays. Other important correlations are those of Harris [21] and Hughmark [22] for bubble-cap trays; and Zuiderweg [23], Chen and Chuang [25], Taylor and Krishna [15], and Young and Stewart [37, 38] for sieve trays.

Some mass-transfer correlations are presented in terms of the number of transfer units, N_V and N_L, where, by definition:

$$N_V \equiv k^V a h_f / u_s \quad (12\text{-}40)$$

$$N_L \equiv k^L a h_f z / (Q_L / W) \quad (12\text{-}41)$$

where

a = interfacial area/volume of froth on the tray

h_f = froth height

u_s = superficial vapor velocity based on the bubbling area of the tray

z = length of liquid-flow path across the bubbling area of the tray

Q_L = volumetric liquid flow rate

W = weir length

The interfacial area for a tray, a^I, is related to a by

$$a^I = a h_f A_b \quad (12\text{-}42)$$

where

A_b = bubbling area

Thus, k^P and a^I are obtained from correlations in terms of N_V and N_L.

For random (dumped) packings, empirical correlations for mass-transfer coefficients and interfacial-area density (area/packed volume) have been published by Onda, Takeuchi, and Okumoto [26] and Bravo and Fair [27]. For structured packings, the empirical correlations of Bravo, Rocha, and Fair for gauze packings [28] and for a wide variety of structured packings [29] are available. A semi-theoretical correlation by Billet and Schultes [30] based on over 3,500 data points for more than 50 test systems and more than 70 different types of packings, requires five packing parameters and is applicable to both random and structured packings. This correlation is discussed in Section 6.8.

Heat-transfer coefficients for the vapor film are usually estimated from the Chilton–Colburn analogy between heat and mass transfer, described in Chapter 3. Thus,

$$h^V = k^V \rho^V C_P^V (N_{Le})^{2/3} \quad (12\text{-}43)$$

where:

$$N_{Le} = \left(\frac{N_{Sc}}{N_{Pr}}\right) \quad (12\text{-}44)$$

For the liquid-phase film, a penetration model is preferred, where

$$h^L = k^L \rho^L C_P^L (N_{Le})^{1/2} \quad (12\text{-}45)$$

A more detailed heat-transfer model, specifically for sieve trays, is given by Spagnolo et al. [39].

12.4 VAPOR AND LIQUID FLOW PATTERNS

The simplest flow pattern for a stage corresponds to the assumption of a perfectly mixed vapor and a perfectly mixed liquid. Under these conditions, the mass-transfer driving forces in (12-27) and (12-28) are simplified to

$$(y^V - y^I)_{avg} = (y^V - y^I) \qquad (12\text{-}46)$$

$$(x^I - x^L)_{avg} = (x^I - x^L) \qquad (12\text{-}47)$$

where y^V and x^L are exiting stage mole fractions. These flow patterns are only valid for trayed towers with a short liquid flow path.

A plug-flow pattern for the vapor and/or liquid assumes that the phase moves through the froth without mixing. This pattern requires that the mass-transfer rates be integrated over the froth. An approximation of the integration is provided by Kooijman and Taylor [31], who assume constant mass-transfer coefficients and interface compositions. The resulting expressions for the average mole-fraction driving forces are the same as (12-46) and (12-47) except for a correction factor in terms of N^V or N^L, included on the right-hand side of each equation. Plug-flow patterns are generally more accurate for trayed towers than perfectly mixed flow patterns and are also applicable to packed towers.

The perfectly-mixed-flow and plug-flow patterns are the two patterns presented by Lewis [7] to convert Murphree vapor point efficiencies to Murphree vapor tray efficiencies, as discussed in Section 6.5. They represent the extreme situations. Fair, Null, and Bolles [32] recommend a more realistic partial mixing or dispersion model that utilizes a turbulent Peclet number, whose value can cover a wide range. This model provides a bridge between the two extremes.

For reactive distillation, a rate-based multicell (or mixed pool) model has proved useful. In this model, the liquid on the tray is assumed to flow horizontally across the tray through a series of perfectly mixed cells (perhaps 4 or 5). In the model of Higler, Krishna, and Taylor [40], which is available in v4.3 of the ChemSep program, the vapor phase is also assumed to be perfectly mixed in each cell. If desired, cells for each tray can also be stacked in the vertical direction. Thus, a tray model might consist of a 5×5 cell arrangement for a total of 25 perfectly mixed cells. Higler, Krishna, and Taylor assume that the vapor streams leaving the topmost cells on a tray are collected and mixed before being divided to enter the cells on the next tray. The rate-based multicell model of Pyhalahti and Jakobsson [41] allows only one set of cells in the horizontal direction, but the vapor streams leaving the cells on a tray may be mixed or not mixed before entering the cells on the next tray and the reversal of liquid flow direction from tray to tray, shown in Figure 6.21 for single-pass trays, is allowed.

12.5 METHOD OF CALCULATION

As indicated in Section 12.1, the number of equations to be solved for the single-cell per tray, rate-based model of Figure 12.1 is $N(5C + 5)$ when the pressure-drop equations are omitted, as summarized in Table 12.1. The equations contain the variables listed in Table 12.2. Other parameters in the equations are computed from these variables. When the number of equations is subtracted from the number of variables, the number of degrees of freedom is $2NC + 9N + 1$. If the total number of stages and all column feed conditions, including feed-stage locations ($2NC + 4N + 1$ variables) are specified, the number of remaining degrees of freedom, using the variable designations in Table 12.2, is $5N$. A computer program for the rate-based model would generally require the user to specify these $2NC + 4N + 1$ variables. The degree of flexibility provided to the user in the selection of the remaining $5N$ variables depends on the particular rate-based computer algorithm, three of which are widely available: (1) Chem-Sep Release v4.3 from R. Taylor and H. A. Kooijman, (2) RATEFRAC in Release 12 of ASPEN PLUS from Aspen Technology, Cambridge, Massachusetts, and (3) CHEMCAD v5.4. Both algorithms provide a wide variety of correlations for thermodynamic and transport properties. Both programs also provide considerable flexibility in the selection of the remaining $5N$ specifications. The basic $5N$ specifications are

$$r_j^L \text{ or } U_j, r_j^V \text{ or } W_j, P_j, Q_j^L, \text{ and } Q_j^V$$

However, substitutions can be made as discussed next.

ChemSep Program

The ChemSep program applies the transport equations to trays or short heights (called segments) of packing. The condenser and reboiler stages are treated as equilibrium stages. The specification options include:

1. r_j^L and r_j^V: From each stage, either a liquid or a vapor side stream can be specified as (a) a side-stream flow rate or (b) a ratio of the side-stream flow rate to the flow rate of the remaining fluid passing to the next stage, that is,

$$r_j^L = U_j/L_j \quad \text{or} \quad r_j^V = W_j/V_j \text{ in Figure 12.1.}$$

2. P_j: Four options are available:

   (a) Condenser pressure (if any) and a constant, but different, pressure for all stages in the tower and for the reboiler, if any.

   (b) Condenser pressure (if any), top tower pressure, and bottom pressure (bottom tower stage or reboiler, if any). Pressures of stages intermediate between top and bottom are obtained by linear interpolation.

   (c) Condenser pressure (if any), top tower pressure, and specified pressure drop per stage to obtain remaining stage pressures.

   (d) Condenser pressure (if any) and top tower pressure, with stage pressure drops estimated by ChemSep from hydraulic correlations.

3. Q_j^L and Q_j^V: The heat duty must be specified for all stage heaters and coolers except for the condenser and/or reboiler, if present. In addition, a heat loss for the tower can be specified that is divided equally over all stages. When a condenser (total without subcooling, total with subcooling, or partial) is present, one of the following specifications can replace the heat duty of the condenser: (a) molar reflux ratio, (b) condensate temperature, (c) distillate molar flow rate, (d) reflux molar flow rate, (e) component molar flow rate in distillate, (f) mole fraction of a component in distillate, (g) fractional recovery, from all feeds, of a component in the distillate, (h) molar fraction of all feeds to the distillate, and (i) molar ratio of two components in the distillate.

For distillation, an often-used specification is the molar reflux ratio.

When a reboiler (partial, total with a vapor product, or total with a superheated vapor product) is present, the following list of specification options, similar to those just given for a condenser, can replace the heat duty of the reboiler: (a) molar boilup ratio, (b) reboiler temperature, (c) bottoms molar flow rate, (d) reboiled-vapor (boilup) molar flow rate, (e) component molar flow rate in bottoms, (f) mole fraction of a component in bottoms, (g) fractional recovery, from all feeds, of a component in the bottoms, (h) molar fraction of all feeds to the bottoms, and (i) molar ratio of two components in the bottoms. For distillation, an often-used specification is the molar bottoms flow rate, which must be estimated if it is not specified.

The preceding number of optional specifications is considerable. In addition, ChemSep also provides "flexible" specifications that can substitute for the condenser and/or reboiler duties. These are advanced options supplied in the form of strings that contain values of certain allowable variables and/or combinations of these variables using the five common arithmetic operators (+, −, *, /, and exponentiation). The variables include stage variables (L, V, x, y, and T) and interface variables (x^I, y^I, and T^I) at any stage. Flow rates can be in mole or mass units.

Certain options and advanced options must be used with great care because values can be specified that cannot lead to a converged solution. For example, with a simple distillation column of a fixed number of stages, that number may be less than the minimum number to achieve specified distillate and bottoms purities. As always, it is generally wise to begin a simulation with a standard pair of top and bottom specifications, such as reflux ratio and a bottoms molar flow rate that corresponds to the desired distillate rate. These specifications are almost certain to converge unless interstage liquid or vapor flow rates tend to zero somewhere in the column. A study of the calculated results will provide valuable insight into possible limits in the use of other options.

The equations for the rate-based model, some linear and some nonlinear, are solved by Newton's method in a manner similar to that developed by Naphtali and Sandholm for the equilibrium-based model described in Chapter 10. Thus, the variables and equations are grouped by stage so that the Jacobian matrix is of the block-tridiagonal form. However, the equations to be solved number $5C + 6$ or $5C + 5$ per stage, depending on whether stage pressures are computed or specified, compared to just $2C + 1$ for the equilibrium-based method.

Calculations of transport coefficients and pressure drops require column diameter and dimensions of column internals. These may be specified (simulation mode) or computed (design mode). In the latter case, default dimensions are selected for the internals, with column diameter computed from a specified value for percent of flooding for a trayed or packed column, or a specified pressure drop per unit height for a packed column.

Computing time per iteration for the design mode is only approximately twice that for the simulation mode, which usually requires less than twice the time for the equilibrium-based model. The number of iterations required for the design mode can be two to three times that for the equilibrium-based model. Overall, the total computing time for the design mode is usually less than an order of magnitude greater than that for the equilibrium-based model. With today's fast workstations and PCs, computing times for the design mode of the rate-based model are usually less than one minute.

Like the Naphtali–Sandholm method, the rate-based model utilizes mainly analytical partial derivations in the Jacobian matrix, and requires initial estimates of all variables. These estimates are generated automatically by the ChemSep program using a method of Powers et al. [33]. In this method, the usual assumptions of constant molar overflow and a linear temperature profile are employed. The initialization of the stage mole fractions is made by performing several iterations of the bubble-point method using ideal K-values for the first iteration and nonideal K-values thereafter. Initial interface mole fractions are set equal to estimated bulk values and initial mass-transfer rates are arbitrarily set to values of $\pm 10^{-3}$ kmol/h with the sign dependent upon the component K-value.

To prevent oscillations and promote convergence of the iterations, corrections to certain variables from iteration to iteration can be limited. Defaults are 10 K for temperature and 50% for flows. When a correction to a mole fraction would result in a value outside of the feasible range of 0 to 1, the default correction is one-half of the step that would take the value to a limit. For very difficult problems, homotopy-continuation methods described by Powers et al. [33] can be applied to promote convergence.

Convergence of Newton's method is determined from values of the residuals of the functions, as in the Naphtali–Sandholm method, or from the corrections to the variables. ChemSep applies both criteria and terminates when either of the following are satisfied:

$$\left[\sum_{j=1}^{N} \sum_{k=1}^{N_j} f_{k,j}^2 \right]^{1/2} < \epsilon \qquad (12\text{-}48)$$

$$\sum_{j=1}^{N}\sum_{k=1}^{N_j}|\Delta X_{k,j}|/X_{k,j} < \epsilon \qquad (12\text{-}49)$$

where

$f_{k,j}$ = residuals in Table 12.1

N = number of stages

N_j = number of equations for the jth stage

$X_{k,j}$ = unknown variables from Table 12.2

ϵ = a small number with a default value of 10^{-4}

Unlike in the Naphtali–Sandholm method, the residuals are not scaled. Accordingly, the second criterion is usually satisfied first.

From the results of a converged solution, it is highly desirable to back-calculate Murphree vapor-tray efficiencies, component by component and tray by tray, from (12-3) for trayed columns, and HETP values for packed towers.

ChemSep can also perform rate-based calculations for liquid–liquid extraction.

EXAMPLE 12.2

A mixture of n-heptane and toluene cannot be separated at 1 atm by ordinary distillation. Accordingly, an enhanced-distillation scheme using methylethyl ketone as a solvent is used. As part of an initial design study, use the rate-based model of ChemSep with the specifications listed in Table 12.3 to calculate a sieve-tray column.

SOLUTION

The information in Table 12.3 was entered via the ChemSep menu and the program was executed. A converged solution was achieved in 8 iterations in 6 seconds on a PC with a Pentium 90 CPU, running the Windows 95 operating system. Initialization of all variables was done by the program.

The predicted separation is as follows:

Component	Distillate, lbmol/h	Bottoms, lbmol/h
n-Heptane	54.87	0.13
Toluene	0.45	44.55
Methylethyl ketone	199.68	0.32

Predicted-column profiles for pressure, liquid-phase temperature, total vapor and liquid flow rates, component vapor and liquid mole fractions, component mass-transfer rates, and Murphree vapor-tray efficiencies are shown in Figure 12.2, where stages are numbered from the top down and stages 2 to 21 are sieve trays. Back-calculated Murphree tray efficiencies are summarized as follows:

Component	Fractional Murphree Efficiencies Range	Median
n-Heptane	0.52 to 1.10	0.73
Toluene	0.70 to 0.79	0.79
Methylethyl ketone	−3.23 to 1.14	0.76

Table 12.3 Specifications for Example 12.2

Total condenser delivering saturated liquid
Partial reboiler
Pressure at condenser outlet = 14.7 psia
Pressure at condenser inlet = 15.0 psia
Reflux ratio = 1.5
Bottoms flow rate = 45 lbmol/h
Total number of trays = 20
Feed 1 to tray 10 from top:
  55 lbmol/h of n-heptane
  45 lbmol/h of toluene
  100 lbmol/h of methylethyl ketone (MEK)
  Saturated liquid at 20 psia
Feed 2 to tray 15 from top:
  100 lbmol/h of MEK
  Saturated liquid at 20 psia
UNIFAC for liquid-phase activity coefficients
Chan–Fair correlation for mass-transfer coefficients
Plug flow for vapor
Mixed flow for liquid
85% of flooding
Tray spacing = 0.5 m (19.7 inches)
Weir height = 2 inches

The median values, based on experience, seem reasonable and give confidence in the rate-based method. The 20 trays are equivalent to approximately 15 equilibrium stages.

For purposes of sizing, the column was divided into three sections: 9 trays above the top feed, 5 trays from the top feed to the bottom feed, and 6 trays below the bottom feed. Computed column diameters are, respectively, 1.75 m (5.74 ft), 1.74 m (5.71 ft), and 1.83 m (6.00 ft). Thus, a 1.83-m (6.00-ft)-diameter column is a reasonable choice. Average predicted pressure drop per tray is 0.06 psi. Computed heat-exchanger duties are as follows:

Condenser: 2.544 MW (8,680,000 Btu/h)

Reboiler: 2.482 MW (8,470,000 Btu/h)

EXAMPLE 12.3

Repeat Example 12.2 for a tower packed with Flexipac® 2 structured packing, operating at 75% of flooding. The packing heights are as follows:

Section	Packing Height, ft
Above top feed	13
Between top and bottom feeds	6.5
Below bottom feed	6.5

SOLUTION

Each 6.5 feet of packing was simulated by 50 segments. Because of the large number of segments, mixed flow is assumed for both vapor and liquid. Newton's method could not converge the calculations. Therefore, the homotopy-continuation option was selected.

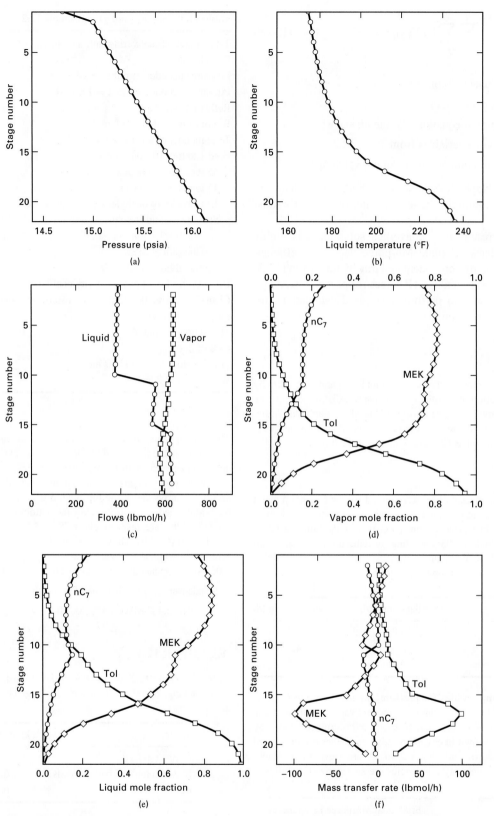

Figure 12.2 Column profiles for Example 12.2: (a) pressure profile; (b) liquid-phase temperature profile; (c) vapor and liquid flow rate profiles; (d) vapor mole-fraction profiles; (e) liquid mole-fraction profiles; (f) mass-transfer rate profiles. (*continued*)

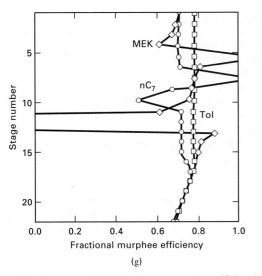

Figure 12.2 (*Continued*) (g) Murphree vapor-tray efficiencies.

Then convergence was achieved in 73 s after a total of 26 iterations. The predicted separation, which is just slightly better than that in Example 12.2, is as follows:

Component	Distillate, lbmol/h	Bottoms, lbmol/h
n-Heptane	54.88	0.12
Toluene	0.40	44.60
Methylethyl ketone	199.72	0.28

The HETP profile is plotted in Figure 12.3. Median values for *n*-heptane, toluene, and methylethyl ketone, respectively, are approximately 0.55 m (21.7 inches), 0.45 m (17.7 inches), and 0.5 m (19.7 inches). The HETP values for the ketone are seen to vary widely.

Predicted column diameters for the three sections, starting from the top, are 1.65, 1.75, and 1.85 m, which are very close to the predicted sieve-tray diameters.

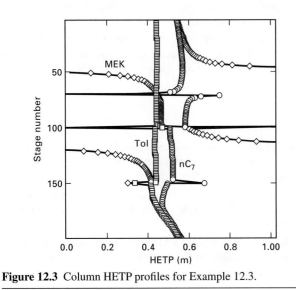

Figure 12.3 Column HETP profiles for Example 12.3.

RATEFRAC Program

The RATEFRAC program of Aspen Technology is designed to model columns used for reactive distillation. The latest version of ChemSep can also model reactive distillation. For RATEFRAC, the reactions can be equilibrium-based or kinetics-based, including reactions among electrolytes. For kinetically controlled reactions, built-in power-law expressions are selected, or the user supplies FORTRAN subroutines for the rate law(s). For equilibrium-based reactions, the user supplies a temperature-dependent equilibrium constant, or RATEFRAC computes reaction-equilibrium constants from free-energy values stored in the data bank. The user specifies the phase in which the reaction takes place. Flow rates of side streams and the column-pressure profile must be provided. The heat duty must be specified for each intercooler or interheater. The standard specifications for the rating mode are the reflux ratio and the bottoms flow rate. However, these specifications can be manipulated in the design mode to achieve any of the following substitute specifications:

(a) Purity of a product or internal stream with respect to one component or a group of components.

(b) Recovery of a component or group of components in a product stream.

(c) Flow rate of a component or group of components in a product or internal stream.

(d) Temperature of a product or internal vapor or liquid stream.

(e) Heat duty of condenser or reboiler.

(f) Value of a product or internal stream physical property.

(g) Ratio or difference of any pair of product or internal stream physical properties, where the two streams can be the same or different.

Mass-transfer correlations are built into RATEFRAC for bubble-cap trays, valve trays, sieve trays, and random packings. Users may provide their own FORTRAN subroutines for transport coefficients and interfacial area. Newton's method is used to converge the calculations.

EXAMPLE 12.4

Use RATEFRAC to predict the column profiles for a 3.5-ft-diameter, 20-bubble-cap tray absorber operating at the conditions listed in Table 12.4.

SOLUTION

No initial estimates of variables were provided. The program was run on a PC with a Pentium 90 CPU running under MS-DOS 6.22. Initial estimates of the values of the variables were provided by RATEFRAC. A total of five iterations were required following an initialization step. Total computing time, including translation, compiling, linking, and execution, was about 60 s. The following results were obtained for the product streams:

Component	Lean Vapor, lbmol/h	Rich Oil, lbmol/h
Hydrogen	216.6	1.4
Nitrogen	86.1	0.9
Methane	131.4	4.6
Ethane	120.0	19.0
Propane	73.4	44.6
Isobutane	1.1	4.9
n-Butane	0.0	2.0
Isopentane	0.0	43.0
n-Hexane	0.0	14.0
n-Heptane	0.0	4.0
n-Dodecane	0.0	165.0
n-Tridecane	0.0	165.0
	628.6	468.4

The back-calculated fractional Murphree vapor tray efficiencies are as follows:

Component	E_{MV} Range	E_{MV} Median Value
Hydrogen	−1.25 to 1.21	0.43
Nitrogen	−0.51 to 3.25	0.41
Methane	−0.13 to 0.87	0.41
Ethane	−0.03 to 1.02	0.30
Propane	−1.89 to 2.71	0.33
Isobutane	−6.51 to 1.16	0.42
n-Butane	−0.75 to 5.65	0.50
Isopentane	−3.17 to 1.15	0.57
n-Hexane	0.63 to 1.38	0.63
n-Heptane	0.64 to 0.88	0.64
n-Dodecane	−2.92 to 1.01	0.42
n-Tridecane	−5.04 to 0.97	0.42

Table 12.4 Specifications for Example 12.4

Column top pressure = 182 psia
Column bottom pressure = 185 psia
Weir height = 2 inches
Vapor completely mixed on each tray
Liquid completely mixed on each tray
AIChE correlations for binary mass-transfer coefficients and interfacial area
Chilton–Colburn analogy for heat transfer
Chao–Seader correlation for K-values

Vapor feed at 123°F and 184 psia:

Component	lbmol/h
Hydrogen	218
Nitrogen	87
Methane	136
Ethane	139
Propane	118
Isobutane	6
n-Butane	2
Isopentane	43
n-Hexane	14
n-Heptane	4
	767

Liquid absorbent feed at 100°F and 182 psia:

Component	lbmol/h
n-Dodecane	165
n-Tridecane	165

It is seen that the efficiencies vary widely from component to component and from tray to tray. For absorber simulation and design, a rate-based model is clearly superior to an equilibrium-based model.

As chemical engineers become more knowledgeable in the principles of mass and heat transfer, and improved correlations for mass-transfer and heat-transfer coefficients are developed for trays and packings, the use of rate-based models will accelerate. For best results, these models will also benefit from more realistic options for vapor and liquid flow patterns. More comparisons of rate-based models with industrial operating data are needed to gain confidence in the use of such models. Some recent comparisons are presented by Taylor, Kooijman, and Woodman [34], and Kooijman and Taylor [31]. Comparisons by Ovejero et al. [35], with distillation data obtained in a column packed with spheres and cylinders of known interfacial area, show very good agreement for three binary and two ternary systems.

SUMMARY

1. Rate-based models of multicomponent, multistage, vapor–liquid separation operations became available in the late 1980s. These models are potentially superior to equilibrium-based models for all but near-ideal systems.

2. Rate-based models incorporate rigorous procedures for treating component-coupling effects in multicomponent mass transfer.

3. The number of equations for a rate-based model is greater than for an equilibrium-based model because separate balances are needed for each of the two phases. In addition, rate-based models are influenced by the geometry of the column internals. Correlations are used to predict mass-transfer and heat-transfer rates. Tray or packing hydraulics are also incorporated into the rate-based model to enable prediction of column-pressure profile. Equilibrium is assumed at the phase interface.

4. Computing time for a rate-based model is not generally more than an order of magnitude greater than that for an equilibrium-based model.

5. Both the ChemSep and RATEFRAC rate-based computer programs offer considerable flexibility in user specifications, so much so that inexperienced users can easily specify impossible conditions. Therefore, it is best to begin simulation studies with standard specifications.

REFERENCES

1. SOREL, ERNEST, *La rectification de l' alcool,* Paris (1893).

2. SMOKER, E.H., *Trans. AIChE,* **34,** 165 (1938).

3. THIELE, E.W., and R.L. GEDDES, *Ind. Eng. Chem.,* **25,** 290 (1933).

4. LEWIS, W.K., and G.L. MATHESON, *Ind. Eng. Chem.,* **24,** 496–498 (1932).

5. LEWIS, W.K., *Ind. Eng. Chem.,* **14,** 492 (1922).

6. MURPHREE, E.V., *Ind. Eng. Chem.,* **17,** 747–750, 960–964 (1925).

7. LEWIS, W.K., *Ind. Eng. Chem.,* **28,** 399 (1936).

8. SEADER, J.D., *Chem. Eng. Prog.,* **85** (10), 41–49 (1989).

9. WALTER, J.F., and T.K. SHERWOOD, *Ind. Eng. Chem.,* **33,** 493–501 (1941).

10. TOOR, H.L., *AIChE J.,* **3,** 198 (1957).

11. TOOR, H.L., and J.K. BURCHARD, *AIChE J.,* **6,** 202 (1960).

12. KRISHNA, R., H.F. MARTINEZ, R. SREEDHAR, and G.L. STANDART, *Trans. I. Chem. E.,* **55,** 178 (1977).

13. WAGGONER, R.C., and G.D. LOUD, *Comput. Chem. Engng.,* **1,** 49 (1977).

14. KRISHNA, R., and G.L. STANDART, *Chem. Eng. Comm.,* **3,** 201 (1979).

15. TAYLOR, R., and R. KRISHNA, *Multicomponent Mass Transfer,* John Wiley and Sons, New York (1993).

16. KRISHNAMURTHY, R., and R. TAYLOR, *AIChE J.,* **31,** 449, 456 (1985).

17. TAYLOR, R., H.A. KOOIJMAN, and J.-S. HUNG, *Comput. Chem. Engng.,* **18,** 205–217 (1994).

18. *ASPEN PLUS Reference Manual—Volume 1,* Aspen Technology, Cambridge, MA (1994).

19. TAYLOR, R., and H.A. KOOIJMAN, *CACHE News,* No. 41, 13–19 (1995).

20. AIChE, *Bubble-Tray Design Manual,* New York (1958).

21. HARRIS, I.J., *British Chem. Engng.,* **10**(6), 377 (1965).

22. HUGHMARK, G.A., *Chem. Eng. Progress,* **61**(7), 97–100 (1965).

23. ZUIDERWEG, F.J., *Chem Eng. Sci.,* **37,** 1441 (1982).

24. CHAN, H., and J.R. FAIR, *Ind. Eng. Chem. Process Des. Dev.,* **23,** 814–827 (1984).

25. CHEN, G.X., and K.T. CHUANG, *Ind. Eng. Chem. Res.,* **32,** 701–708 (1993).

26. ONDA, K., H. TAKEUCHI, and Y.J. OKUMOTO, *J. Chem. Eng. Japan,* **1,** 56–62 (1968).

27. BRAVO, J.L., and J.R. FAIR, *Ind. Eng. Chem. Process Des. Devel.,* **21,** 162–170 (1982).

28. BRAVO, J.L., J.A. ROCHA, and J.R. FAIR, *Hydrocarbon Processing,* **64**(1), 56–60 (1985).

29. BRAVO, J.L., J.A. ROCHA, and J.R. FAIR, *I. Chem. E. Symp. Ser.,* No. 128, A489–A507 (1992).

30. BILLET, R., and M. SCHULTES, *I. Chem. E. Symp. Ser.,* No. 128, B129 (1992).

31. KOOIJMAN, H.A., and R. TAYLOR, *Chem. Eng. J.,* **57**(2), 177–188 (1995).

32. FAIR, J.R., H.R. NULL, and W.L. BOLLES, *Ind. Eng. Chem. Process Des. Dev.,* **22,** 53–58 (1983).

33. POWERS, M.F., D.J. VICKERY, A. AREHOLE, and R. TAYLOR, *Comput. Chem. Engng.,* **12,** 1229–1241 (1988).

34. TAYLOR, R., H.A. KOOIJMAN, and M.R. WOODMAN, *I. Chem. E. Symp. Ser.,* No. 128, A415–A427 (1992).

35. OVEJERO, G., R. VAN GRIEKEN, L. RODRIGUEZ, and J.L. VALVERDE, *Sep. Sci. Tech.,* **29,** 1805–1821 (1994).

36. SCHEFFE, R.D., and R.H. WEILAND, *Ind. Eng. Chem. Res.,* **26,** 228–236 (1987).

37. YOUNG, T.C., and W.E. STEWART, *AIChE J.,* **38,** 592–602 with errata on p. 1302 (1993).

38. YOUNG, T.C., and W.E. STEWART, *AIChE J.,* **41,** 1319–1320 (1995).

39. SPAGNOLO, D.A., E.L. PLAICE, H.J. NEUBURG, and K.T. CHUANG, *Can. J. Chem. Eng.,* **66,** 367–376 (1988).

40. HIGLER, A., R. KRISHNA, and R. TAYLOR, *AIChE J.,* **45,** 2357–2370 (1999).

41. PYHALAHTI, A., and K. JAKOBSSON, *Ind. Eng. Chem. Res.,* **42,** 6188–6195 (2003).

EXERCISES

Section 12.1

12.1 Modify the rate-based model of (12-4) to (12-18) to include entrainment and occlusion.

12.2 Modify the rate-based model of (12-4) to (12-18) to include a chemical reaction in the liquid phase under conditions of:

(a) Chemical equilibrium

(b) Kinetic rate law.

12.3 Explain how the number of rate-based modeling equations can be reduced. Would this be worthwhile?

Section 12.2

12.4 The following results were obtained at tray n from a rate-based calculation at 14.7 psia, for a ternary mixture of acetone (1), methanol (2), and water (3) in a sieve-tray column assuming that both phases are perfectly mixed.

Component	y_n	y_{n+1}	y_n^I	K_n^I	x_n
1	0.4913	0.4106	0.5291	1.507	0.3683
2	0.4203	0.4389	0.4070	0.900	0.4487
3	0.0884	0.1505	0.0639	0.3247	0.1830

The products of the computed gas-phase, binary mass-transfer coefficients and interfacial area from the Chan–Fair correlations are as follows in units of lbmol/(h-unit mole fractions).

$$k_{12} = k_{21} = 1,750$$
$$k_{13} = k_{31} = 2,154$$
$$k_{23} = k_{32} = 2,503$$

The computed vapor rates are $V_n = 1,200$ lbmol/h and $V_{n+1} = 1,164$ lbmol/h. Determine:

(a) The component molar diffusion rates.

(b) The mass-transfer rates.

(c) The Murphree vapor-tray efficiencies.

12.5 Write all the expanded equations (12-31) and (12-32) for $\mathbf{R}^P$ for a five-component system.

12.6 Repeat the calculations of Example 12.1, but using 1 = methanol, 2 = water, and 3 = acetone. Are the results any different? If not, why not? Prove your conclusion mathematically.

Section 12.3

12.7 Compare and discuss the advantages and disadvantages of the available correlations for estimating binary-pair mass-transfer coefficients for trayed columns.

12.8 Compare and discuss the advantages and disadvantages of the available correlations for estimating binary-pair mass-transfer coefficients for columns with random (dumped) and structured packings.

Section 12.4

12.9 Discuss how the method of Fair, Null, and Bolles [32] might be used to model the flow patterns in a rate-based model. How would the mole-fraction driving forces be computed?

Section 12.5

12.10 A bubble-point mixture of 100 kmol/h of methanol, 50 kmol/h of isopropanol, and 100 kmol/h of water at 1 atm is sent to the 25th tray from the top of a 40-sieve-tray column equipped with a total condenser and partial reboiler, operating at a nominal pressure of 1 atm. If the reflux ratio is 5 and the bottoms flow rate is 150 kmol/h, determine the separation achieved if the UNIFAC method is used to estimate K-values and the Chan–Fair correlations are used for mass transfer. Assume that both phases are perfectly mixed on each tray and that operation is at about 80% of flooding.

12.11 A sieve-tray column, operating at a nominal pressure of 1 atm, is used to separate a mixture of acetone and methanol by extractive distillation using water. The column has 40 trays with a total condenser and partial reboiler. The feed of 50 kmol/h of acetone and 150 kmol/h of methanol at 60°C and 1 atm enters tray 35 from the top, while 50 kmol/h of water at 65°C and 1 atm enters tray 5 from the top. Determine the separation for a reflux ratio of 10 and a bottoms flow rate of 200 kmol/h. Use the UNIFAC method for K-values and the AIChE method for mass transfer. Assume a perfectly mixed liquid and a vapor in plug flow on each tray, with operation at 80% of flooding. Also determine the number of equilibrium stages (to the nearest stage) to achieve the same separation.

12.12 Repeat Exercise 12.10, if a column packed with 2-in. stainless-steel Pall® rings is used with 25 ft of rings above the feed and 15 ft below. Be sure to use a sufficient number of segments for the calculations.

12.13 Repeat Exercise 12.10, if a column with structured packing is used with 25 ft above the feed and 15 ft below. Be sure to use a sufficient number of segments.

12.14 Solve Exercise 12.10 for combinations of the following values of percent flooding, weir height, and hole area, respectively:

40, 60 and 80%
1, 2, and 3 inches
6, 10, and 14%

12.15 The upper column of an air-separation system, of the type discussed and shown in Exercise 7.40, contains 48 sieve trays and

operates at a nominal pressure of 131.7 kPa. A feed at 80 K and 131.7 kPa enters the top plate at 1,349 lbmol/h with a composition of 97.868 mol% nitrogen, 0.365 mol% argon, and 1.767 mol% oxygen. A second feed enters tray 12 from the top at 83 K and 131.7 kPa at 1,832 lbmol/h with a composition of 59.7 mol% nitrogen, 1.47 mol% argon, and 38.83 mol% oxygen. The column has no condenser, but has a split reboiler. Vapor distillate leaves the top plate at 2,487 lbmol/h, with remaining products leaving the reboiler as 50 mol% vapor and 50 mol% liquid. Assume that ideal solutions are formed. Determine the effect of percent flooding on the separation and the median Murphree vapor-tray efficiency for oxygen.

12.16 The following bubble-point, organic-liquid mixture at 1.4 atm is distilled by extractive distillation with the following phenol-rich solvent at 1.4 atm and at the same temperature as the main feed:

Component	Feed, kmol/h	Solvent, kmol/h
Methanol	50	0
n-Hexane	20	0
n-Heptane	180	0
Toluene	150	10
Phenol	0	800

The column has 30 sieve trays, with a total condenser and a partial reboiler. The solvent enters the 5th tray and the feed enters tray 15, from the top. The pressure in the condenser is 1.1 atm; the pressure at the top tray is 1.2 atm, and the pressure at the bottom is 1.4 atm. The reflux ratio is 5 and the bottoms rate is 960 kmol/h. Thermodynamic properties can be estimated with the UNIFAC method for the liquid phase and the SRK equation for the vapor phase. The Antoine equation is suitable for vapor pressure. Use the nonequilibrium model of the ChemSep program to estimate the separation. Assume that the vapor and liquid are both well mixed and that the trays operate at 75% of flooding. Specify the Chan–Fair correlation for calculating mass-transfer coefficients. In addition, determine from the tray-by-tray results the average Murphree vapor-tray efficiency for each component (after discarding values that appear to be much different than the majority of values). Try to improve the sharpness of the split by changing the feed and solvent entry tray locations. How can you increase the sharpness of the separation? List as many ideas as you have.

12.17 A bubble-cap tray absorber is designed to absorb 40% of the propane from a rich gas at 4 atm. The specifications for the entering rich gas and absorbent oil are as follows:

	Absorbent Oil	Rich Gas
Flow rate, kmol/s	11.0	11.0
Temperature, °C	32	62
Pressure, atm	4	4
Mole fraction:		
Methane	0	0.286
Ethane	0	0.157
Propane	0	0.240
n-Butane	0.02	0.169
n-Pentane	0.05	0.148
n-Dodecane	0.93	0

(a) Determine the number of equilibrium stages required and the splits of all components.

(b) Determine the actual number of trays required and the splits and Murphree vapor-tray efficiencies of all components.

(c) Compare and discuss the equilibrium-based and rate-based results. What do you conclude?

12.18 A ternary mixture of methanol, ethanol, and water is distilled in a sieve-tray column to obtain a distillate with not more than 0.01 mol% water. The feed to the column is as follows:

Flow rate, kmol/h	142.46
Pressure, atm	1.3
Temperature, K	316
Mole fractions:	
Methanol	0.6536
Ethanol	0.0351
Water	0.3113

For a distillate rate of 93.10 kmol/h, a reflux ratio of 1.2, a condenser outlet pressure of 1.0 atm, and a top-tray pressure of 1.1 atm, determine using the UNIFAC method for activity coefficients:

(a) The number of equilibrium stages required and the corresponding split, if the feed enters at the optimal stage.

(b) The number of actual trays required if the column operates at about 85% of flooding and the feed is introduced to the optimal tray. Compare the split to that in part (a). In addition, compute the component Murphree vapor-tray efficiencies. What do you conclude about the two methods of calculations?

12.19 Repeat Exercise 12.18 for a column packed with 2-in. stainless-steel Pall® rings.

12.20 It is required to absorb 96% of the benzene from a gas stream with absorption oil in a sieve-tray column at a nominal pressure of 1 atm. The feed conditions are as follows:

	Vapor	Liquid
Flow rate, kmol/s	0.01487	0.005
Pressure, atm	1.0	1.0
Temperature, K	300	300
Composition, mol fraction:		
Nitrogen	0.7505	0
Oxygen	0.1995	0
Benzene	0.0500	0.005
n-Tridecane (C_{13})	0	0.995

Tray geometry is as follows:

Tray spacing, m	0.5
Weir height, m	0.05
Hole diameter, m	0.003
Sheet thickness, m	0.002

Determine column diameter for 80% of flooding, the number of actual trays required, and the Murphree vapor-tray efficiency profile for benzene for the possible combinations of vapor and liquid flow patterns on a tray. Could the equilibrium-based method be used to obtain a reliable solution to this problem?

Chapter 13

Batch Distillation

In batch-separation operations, a feed mixture is charged to the equipment and one or more products are withdrawn. A familiar example is laboratory distillation, shown in Figure 13.1, where a liquid mixture is charged to a still pot, retort, or flask and heated by a flame or electric mantle to boiling. The vapor formed is continuously removed and condensed to produce a distillate.

The composition of both the initial charge and distillate change with time; there is no steady state. The still temperature increases and the relative amount of lower-boiling components remaining in the still pot decreases as distillation proceeds.

Batch operations can be used to advantage under the following circumstances:

1. The capacity of a facility is too small to permit continuous operation at a practical rate.

2. It is necessary, because of seasonal demands, to distill with one unit different feedstocks to produce different products.

3. It is desired to produce several new products with one distillation unit for evaluation by potential buyers.

4. Upstream process operations are batchwise and the composition of feedstocks for distillation vary with time or from batch to batch.

5. The feed contains solids or materials that form solids, tars, or resin that can plug or foul a continuous distillation column.

13.0 INSTRUCTIONAL OBJECTIVES

After completing this chapter, you should be able to:

- List assumptions for and derive the Rayleigh equation for the simplest form of batch distillation (differential distillation).
- Calculate, by graphical and/or algebraic means, batch-still temperature, residue composition, instantaneous distillate composition, and average distillate composition for a binary mixture as a function of time for binary, differential distillation.
- Calculate, by modified McCabe–Thiele methods, residue and distillate compositions for binary, batch rectification under conditions of equilibrium stages, no liquid holdup, and for constant or variable-reflux ratio to achieve constant distillate composition.
- Explain the importance of taking into account liquid holdup.
- Calculate, using shortcut and rigorous equilibrium-stage methods with a simulator, multicomponent, multistage batch rectification that includes a sequence of operating steps to obtain specified products.
- Apply the principles of optimal control to optimize batch distillation.

13.1 DIFFERENTIAL DISTILLATION

The simplest case of batch distillation, as discussed by Lord Rayleigh [1], is *differential distillation*, which involves use of the apparatus shown in Figure 13.1. There is no reflux; at any instant, vapor leaving the still pot with composition y_D is assumed to be in equilibrium with perfectly mixed liquid in the still. For total condensation, $y_D = x_D$. Thus, there is only a single equilibrium stage, the still pot. This apparatus is useful for separating wide-boiling mixtures. The following nomenclature is used for variables that vary with time, t,

assuming that all compositions refer to a particular species in the multicomponent mixture.

D = instantaneous-distillate rate, mol/h

$y = y_D = x_D$ = instantaneous-distillate composition, mole fraction

W = moles of liquid left in still

$x = x_W$ = composition of liquid left in still, mole fraction

0 = subscript referring to $t = 0$

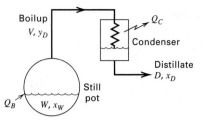

Figure 13.1 Differential (Rayleigh) distillation.

For any component in the mixture:

Instantaneous rate of output $= Dy_D$

$$\left.\begin{array}{c}\text{Instantaneous}\\\text{rate of depletion}\\\text{in the still}\end{array}\right\} = -\frac{d}{dt}(Wx_W) = -W\frac{dx_W}{dt} - x_W\frac{dW}{dt}$$

The distillate rate and, therefore, the rate of depletion of the liquid in the still depend on the rate of heat input to the still. By material balance at any instant:

$$\frac{d}{dt}(Wx_W) = W\frac{dx_W}{dt} + x_W\frac{dW}{dt} = -Dy_D \quad (13\text{-}1)$$

Multiplying by dt:

$$Wdx_W + x_W dW = y_D(-Ddt) = y_D dW$$

since by total balance $-Ddt = dW$. Separating variables and integrating from the initial charge condition of W_0 and x_{W_0}

$$\int_{x_{W_0}}^{x_W}\frac{dx_W}{y_D - x_W} = \int_{W_0}^{W}\frac{dW}{W} = \ln\left(\frac{W}{W_0}\right) \quad (13\text{-}2)$$

This is the well-known Rayleigh equation, which was first applied to the separation of wide-boiling mixtures such as HCl–H_2O, H_2SO_4–H_2O, and NH_3–H_2O. Without reflux, y_D and x_W are assumed to be in equilibrium and (13-2) simplifies to

$$\int_{x_0}^{x}\frac{dx}{y - x} = \ln\left(\frac{W}{W_0}\right) \quad (13\text{-}3)$$

where x_{W_0} is replaced by x_0.

Equation (13-3) is easily integrated only when pressure is constant, temperature change in the still pot is relatively small (close-boiling mixture), and K-values are composition independent. Then $y = Kx$, where K is approximately constant, and (13-3) becomes

$$\ln\left(\frac{W}{W_0}\right) = \frac{1}{K-1}\ln\left(\frac{x}{x_0}\right) \quad (13\text{-}4)$$

For a binary mixture, if instead, the relative volatility α is assumed constant, substitution of (4-8) into (13-3), followed by integration and simplification, gives

$$\ln\left(\frac{W_0}{W}\right) = \frac{1}{\alpha-1}\left[\ln\left(\frac{x_0}{x}\right) + \alpha\ln\left(\frac{1-x}{1-x_0}\right)\right] \quad (13\text{-}5)$$

If the equilibrium relationship $y = f\{x\}$ is in graphical or tabular form, integration of (13-3) can be performed graphically or numerically. The final liquid remaining in the still pot is often referred to as the *residue*.

The following three examples illustrate some of the applications of the Rayleigh equation to binary mixtures.

EXAMPLE 13.1

A batch still is loaded with 100 kmol of a liquid consisting of a binary mixture of 50 mol% benzene in toluene. As a function of time, make plots of (a) still temperature, (b) instantaneous vapor composition, (c) still-pot composition, and (d) average total-distillate composition. Assume a constant boilup rate of 10 kmol/h and a constant relative volatility of 2.41 at a pressure of 101.3 kPa (1 atm).

SOLUTION

Initially, $W_0 = 100$ kmol, $x_0 = 0.5$. Solving (13.5) for W at values of x from 0.5 in increments of 0.05, and determining corresponding values of time from $t = (W_0 - W)/10$, the following table is generated:

t, h	2.12	3.75	5.04	6.08	6.94	7.66	8.28	8.83	9.35
W, kmol	78.85	62.51	49.59	39.16	30.59	23.38	17.19	11.69	6.52
$x = x_W$	0.45	0.40	0.35	0.30	0.25	0.20	0.15	0.10	0.05

The instantaneous-vapor composition, y, is obtained from (4-8), which is $y = 2.41x/(1 + 1.41x)$, the equilibrium relationship for constant α. The average value of y_D or x_D over the time interval 0 to t is related to x and W at time t by combining overall component and total material balances to give

$$(x_D)_{\text{avg}} = (y_D)_{\text{avg}} = \frac{W_0 x_0 - Wx}{W_0 - W} \quad (13\text{-}6)$$

Equation (13-6) is much easier to apply than an equation that eventually integrates the distillate composition.

To obtain the temperature in the still, it is necessary to use experimental T–x–y data for benzene–toluene at 101.3 kPa as given in Table 13.1. The temperature and compositions as a function of time are shown in Figure 13.2.

Table 13.1 Vapor–Liquid Equilibrium Data for Benzene (B)–Toluene (T) at 101.3 kPa

x_B	y_B	T, °C
0.100	0.208	105.3
0.200	0.372	101.5
0.300	0.507	98.0
0.400	0.612	95.1
0.500	0.713	92.3
0.600	0.791	89.7
0.700	0.857	87.3
0.800	0.912	85.0
0.900	0.959	82.7
0.950	0.980	81.4

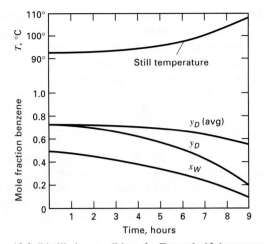

Figure 13.2 Distillation conditions for Example 13.1.

EXAMPLE 13.2

Repeat Example 13.1, except instead of using a constant value of 2.41 for the relative volatility, use the vapor-liquid equilibrium data for benzene-toluene at 101.3 kPa, given in Table 13.1, to solve the problem graphically or numerically with (13-3) rather than (13-5).

SOLUTION

Equation (13-3) can be solved by graphical integration by plotting $1/(y-x)$ versus x with a lower limit of $x_0 = 0.5$. Using the data of Table 13.1 for y as a function of x, points for the plot in terms of benzene are as follows:

x	0.5	0.4	0.3	0.2	0.1
$\dfrac{1}{y-x}$	4.695	4.717	4.831	5.814	9.259

The area under the plotted curve from $x_0 = 0.5$ to a given value of x is equated to $\ln(W/W_0)$, and W is computed for $W_0 = 100$ kmol. In the region from $x = 0.5$ to 0.3, the value of $1/(y-x)$ changes only slightly. Therefore, a numerical integration by the trapezoidal rule is readily made:

$x = 0.4$:

$$\ln\left(\frac{W}{W_0}\right) = \int_{0.5}^{0.4} \frac{dx}{y-x} \approx \Delta x \left[\frac{1}{y-x}\right]_{avg}$$

$$= (0.4 - 0.5)\left(\frac{4.695 + 4.717}{2}\right) = -0.4706$$

$$W/W_0 = 0.625, \ W = -0.625(100) = 62.5 \text{ kmol}$$

$x = 0.3$:

$$\ln\left(\frac{W}{W_0}\right) = \int_{0.5}^{0.3} \frac{dx}{y-x} \approx \Delta x \left[\frac{1}{y-x}\right]_{avg}$$

$$= (0.3 - 0.5)\left[\frac{4.695 + 4.717 + 4.717 + 4.831}{4}\right] = -0.948$$

$$W/W_0 = 0.388, \ W = 0.388(100) = 38.8 \text{ kmol}$$

These two values are in good agreement with those in Example 13.1. A graphical integration from $x_0 = 0.4$ to $x = 0.1$ gives $W = 10.7$, which is approximately 10% less than the result in Example 13.1, which uses a constant value of the relative volatility.

The Rayleigh equation (13-1) can be applied to any two components, i and j, of a multicomponent mixture. Thus, if we let M_i be the moles of i in the still pot,

$$\frac{dM_i}{dt} = \frac{d}{dt}(Wx_{W_i}) = -Dy_{D_i}$$

Then

$$dM_i/dM_j = y_{D_i}/y_{D_j} \tag{13-7}$$

For constant $\alpha_{i,j} = y_{D_i}x_{W_j}/y_{D_j}x_{W_i}$, (13-7) becomes

$$dM_i/dM_j = \alpha_{i,j}(x_{W_i}/x_{W_j}) \tag{13-8}$$

Substitution of $M_i = Wx_{W_i}$ for both i and j into (13-8) gives

$$dM_i/M_i = \alpha_{i,j} \, dM_j/M_j \tag{13-9}$$

Integration from the initial-charge condition gives

$$\ln(M_i/M_{i_0}) = \alpha_{i,j} \ln(M_j/M_{j_0}) \tag{13-10}$$

As shown in the following example, (13-10) is useful for determining the effect of relative volatility on the degree of separation achievable by Rayleigh distillation.

EXAMPLE 13.3

The charge to a simple batch still consists of an equimolar, binary mixture of A and B. For values of $\alpha_{A,B}$ of 2, 5, 10, 100, and 1,000, and 50% vaporization of A, determine the percent vaporization of B and the mole fraction of B in the total distillate.

SOLUTION

For $\alpha_{A,B} = 2$ and $M_A/M_{A_0} = 1 - 0.5 = 0.5$, (13-10) gives

$$M_B/M_{B_0} = (M_A/M_{A_0})^{1/\alpha_{A,B}} = (0.5)^{0.5} = 0.7071$$

Percent vaporization of B $= (1 - 0.7071)(100) = 29.29\%$.

For 200 moles of charge, the amounts of components in the distillate are $D_A = (0.5)(0.5)(200) = 50$ mol and $D_B = (0.2929)(0.5)(200) = 29.29$ mol

Mole fraction of B in the total distillate $= \dfrac{29.29}{50 + 29.29} = 0.3694$

Similar calculations for other values of $\alpha_{A,B}$ give the following results:

$\alpha_{A,B}$	% Vaporization of B	Mole Fraction of B in Total Distillate
2	29.29	0.3694
5	12.94	0.2057
10	6.70	0.1182
100	0.69	0.0136
1,000	0.07	0.0014

These results show that a sharp separation between A and B for 50% vaporization of A is only achieved if $\alpha_{A,B} \geq 100$. Furthermore, the purity achieved depends on the percent vaporization of A. For $\alpha_{A,B} = 100$, if 90% of A is vaporized, the mole fraction of B in the total distillate increases from 0.0136 to 0.0247. For this reason, as discussed in detail and illustrated by example in a later section of this chapter, it is common to conduct a binary, batch-distillation separation of light key (LK) and heavy key (HK) in the following manner:

1. Produce a distillate LK cut until the limit of impurity of HK in the total distillate is reached.

2. Continue the batch distillation to produce an intermediate cut of impure LK until the limit of impurity of LK in the liquid in the still is reached.

3. Empty the HK-rich cut from the still.

4. Recycle the intermediate cut to the next still charge.

For desired purities of the LK cut and the HK cut, the fraction of intermediate cut increases as the LK-HK relative volatility decreases.

13.2 BINARY BATCH RECTIFICATION WITH CONSTANT REFLUX AND VARIABLE DISTILLATE COMPOSITION

To achieve a sharp separation and/or reduce the intermediate-cut fraction, a trayed or packed column, located above the still, and a means of sending reflux to the column, is provided as shown for the batch rectifier of Figure 13.3. For a column of a given diameter, the molar vapor-boilup rate is usually fixed at a value safely below the column flooding point. Two modes of operation of batch rectification are cited most frequently because they are the most readily modeled. The first is operation at a constant reflux rate or ratio (same as a constant distillate rate), while the second is operation at a constant distillate composition. With the former, the distillate composition varies with time; with the latter, the reflux ratio or distillate rate varies with time. The first mode is most easily implemented because of the availability of rapidly responding flow sensors. For the second mode, rapidly responding composition sensors may not be readily available. In a third mode of operation, referred to here as the optimal control mode, both reflux ratio (or distillate rate) and distillate composition vary with time so as to maximize the amount of distillate, minimize the operation time, or maximize profit. The first mode of operation is now presented, followed by the second mode. A discussion of optimal control is deferred until the end of this chapter.

If the reflux ratio R or distillate rate D is fixed, instantaneous-distillate and still-bottoms compositions vary

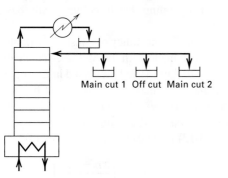

Figure 13.3 Batch rectification.

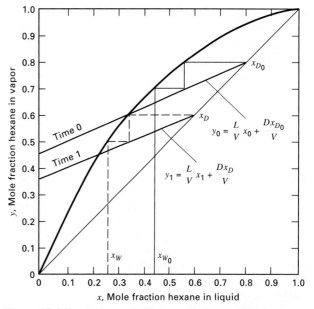

Figure 13.4 Batch binary distillation with fixed L/V and two theoretical stages.

with time. For a total condenser, negligible holdup of vapor and liquid in the condenser and the column, phase equilibrium at each stage, and constant molar overflow, (13-2) still applies with $y_D = x_D$. The analysis of such a batch rectification for a binary system is facilitated by the McCabe–Thiele diagram using the method of Smoker and Rose [2].

Initially, the composition of the light-key component in the liquid in the reboiler of the column in Figure 13.3 is the charge composition, x_{W_0}, which is given the value 0.43 in the McCabe–Thiele diagram of Figure 13.4. If there are two theoretical stages, the initial distillate composition x_{D_0} at time 0 can be found by constructing an operating line of slope $L/V = R/(R+1)$, such that exactly two stages are stepped off from x_{W_0} to the $y = x$ line in Figure 13.4. At an arbitrary later time, say time 1, at still-pot composition $x_W < x_{W_0}$, the instantaneous-distillate composition is x_D. A time-dependent series of points for x_D is thus established by trial and error, with L/V and the number of stages held constant.

Equation (13-2) cannot be integrated analytically because the relationship between y_D and x_W depends on the liquid-to-vapor ratio, the number of theoretical stages, and the phase-equilibrium relationship. However, it can be integrated graphically with pairs of values for x_W and $y_D = x_D$ obtained from the McCabe–Thiele diagram for a series of operating lines of the same slope.

The time t required for batch rectification at constant reflux ratio and negligible holdup in the column and condenser can be computed by a total material balance based on a constant boilup rate V, to give the following equation due to Block [3]:

$$t = \frac{W_0 - W_t}{V\left(1 - \dfrac{L}{V}\right)} = \frac{R+1}{V}(W_0 - W_t) \qquad (13\text{-}11)$$

With a constant-reflux policy, the instantaneous-distillate purity is above the specification at the beginning of distillation and below specification at the end of the run. By an overall material balance, the average mole fraction of the light-key component in the accumulated distillate at time t is given by

$$x_{D_{\text{avg}}} = \frac{W_0 x_0 - W_t x_{W_t}}{W_0 - W_t} \tag{13-12}$$

EXAMPLE 13.4

A three-theoretical-stage batch rectifier (first stage is the still pot) is charged with 100 kmol of a 20 mol% n-hexane in n-octane mixture. At a constant reflux ratio of 1 ($L/V = 0.5$), how many moles of charge must be distilled if an average product composition of 70 mol% nC_6 is required? The phase-equilibrium curve at column pressure is given in Figure 13.5. If the boilup rate is 10 kmol/h, calculate the distillation time.

SOLUTION

A series of operating lines and, hence, values of x_W are located by the trial-and-error procedure described earlier, as shown in Figure 13.5 for $x_{W_0} = 0.20$ and $x_W = 0.09$. It is then possible to construct the following table:

$y_D = x_D$	0.85	0.60	0.5	0.35	0.3
x_W	0.2	0.09	0.07	0.05	0.035
$\dfrac{1}{y_D - x_W}$	1.54	1.96	2.33	3.33	3.77

The graphical integration is shown in Figure 13.6. Assuming a final value of $x_W = 0.1$, for instance, integration of (13-2) gives

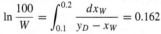

$$\ln \frac{100}{W} = \int_{0.1}^{0.2} \frac{dx_W}{y_D - x_W} = 0.162$$

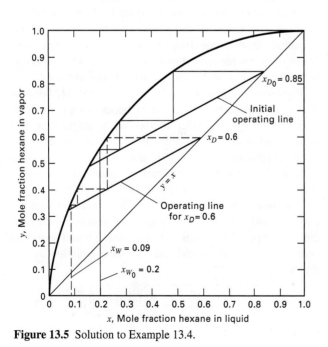

Figure 13.5 Solution to Example 13.4.

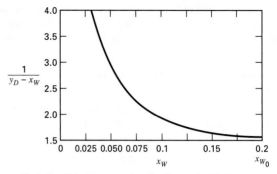

Figure 13.6 Graphical integration for Example 13.4.

Hence,

$$W = 85 \quad \text{and} \quad D = 15.$$

From (13-12):

$$(x_D)_{\text{avg}} = \frac{100(0.20) - 85(0.1)}{(100 - 85)} = 0.77$$

The $(x_D)_{\text{avg}}$ is higher than the desired value of 0.70; hence, another final x_W must be chosen. By trial, the correct answer is found to be $x_W = 0.06$, with $D = 22$, and $W = 78$, corresponding to a value of 0.25 for the integral.

From (13-11), the distillation time is $t = \frac{(1+1)}{10}(100 - 78) = 4.4$ h.

When Rayleigh differential distillation is used, Figure 13.5 shows that a 70 mol% hexane distillate is not achievable because the initial distillate is only 56 mol% hexane.

13.3 BINARY BATCH RECTIFICATION WITH CONSTANT DISTILLATE COMPOSITION AND VARIABLE REFLUX

The constant-reflux-ratio policy described in the previous section is simple and easy to implement. For small batch-rectification systems, it may be the least expensive policy. A more optimal operating policy, as discussed in Section 13.8, is to maintain a constant molar-vapor rate, but continuously vary the reflux ratio to achieve a constant distillate composition that meets the specified purity. This policy requires a more complex control system, including a composition sensor (or suitable substitute) on the distillate, which may be justified only for large batch rectification systems. Other methods of operating batch columns are described by Ellerbe [5].

Calculations for the policy of constant distillate composition can also be made with the McCabe–Thiele diagram, as described by Bogart [4] and illustrated in Example 13.5. The Bogart method assumes negligible liquid holdup and constant molar overflow. An overall material balance for the light-key component, at any time, t, is given by a rearrangement of (13-12) at constant x_D, for W as a function of x_W.

$$W = W_0 \left[\frac{x_D - x_{W_0}}{x_D - x_W} \right] \tag{13-13}$$

Differentiating (13-13) with respect to t for varying W and x_W gives

$$\frac{dW}{dt} = W_0 \frac{(x_D - x_{W_0})}{(x_D - x_W)^2} \frac{dx_W}{dt} \qquad (13\text{-}14)$$

For constant molar overflow, the rate of distillation is given by the rate of loss of charge, or

$$-\frac{dW}{dt} = (V - L) = \frac{dD}{dt} \qquad (13\text{-}15)$$

where D is now the amount of distillate, not the distillate rate. Substituting (13-15) into (13-14) and integrating:

$$t = \frac{W_0(x_D - x_{W_0})}{V} \int_{x_{W_t}}^{x_0} \frac{dx_W}{(1 - L/V)(x_D - x_W)^2} \qquad (13\text{-}16)$$

For fixed values of W_0, x_{W_0}, x_D, V, and the number of equilibrium stages, the McCabe–Thiele diagram is used to determine values of L/V for a series of values of still composition between x_{W_0} and the final value of x_W. These values are then used with (13-16) to determine, by graphical or numerical integration, the time for rectification or the time to reach any intermediate value of still composition. The required number of theoretical stages can be estimated by assuming total-reflux conditions for the final value of x_W. While rectification is proceeding, the instantaneous-distillate rate will vary according to (13-15), which can be expressed in terms of L/V as

$$\frac{dD}{dt} = V(1 - L/V) \qquad (13\text{-}17)$$

EXAMPLE 13.5

A three-stage batch still (boiler and the equivalent of two equilibrium plates) is loaded with 100 kmol of a liquid containing a mixture of 50 mol% n-hexane in n-octane. A liquid distillate of 0.9 mole fraction hexane is to be maintained by continuously adjusting the reflux ratio, while maintaining a distillate rate of 20 kmol/h. What should the reflux ratio be after 1 h when the accumulated distillate is 20 kmol? Theoretically, when must accumulation of the distillate cut be stopped? Assume negligible holdup on the plates and constant molar overflow.

SOLUTION

When the accumulated distillate = 20 kmol, $W = 80$ kmol, and the still residue composition with respect to the light-key is given by a rearrangement of (13-13):

$$x_W = \frac{W x_D - W_0(x_D - x_{W_0})}{W} = \frac{0.9(80) - 100(0.9 - 0.5)}{80} = 0.4$$

For $y_D = 0.9$, a series of operating lines of varying slope, $L/V = R/(R + 1)$, with three stages stepped off is used to determine the corresponding still residue composition. By trial and error, Line 1 in Figure 13.7 is found for $x_W = 0.4$, corresponding to an $L/V = 0.22$. The reflux ratio = $(L/V)/[1 - (L/V)] = 0.282$.

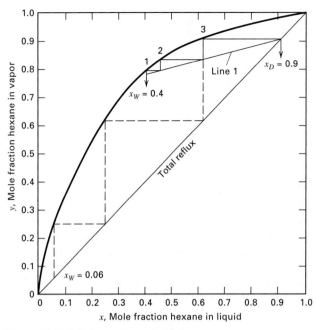

Figure 13.7 Solution to Example 13.5.

At the highest reflux rate possible, $L/V = 1$ (total reflux), and $x_W = 0.06$ according to the dashed-line construction shown in Figure 13.7. The corresponding time by material balance is given by $0.06(100 - 20t) = 50 - 20t(0.9)$. Solving, $t = 2.58$ h.

13.4 BATCH STRIPPING AND COMPLEX BATCH DISTILLATION

A batch stripper consisting of a large accumulator, a trayed or packed stripping column, and a reboiler is shown in Figure 13.8. The initial charge is placed in the accumulator rather than the reboiler. The mixture in the accumulator is fed to the top of the column and the bottoms cut is removed from the reboiler. A batch stripper is useful for removing small quantities of volatile impurities. For binary mixtures, the McCabe–Thiele construction applies, and the graphical methods described in Sections 13.2 and 13.3 can be modified to follow with time the change in composition in the

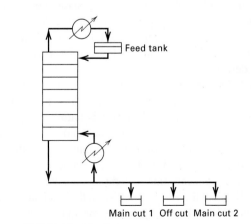

Figure 13.8 Batch stripping.

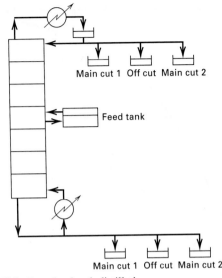

Figure 13.9 Complex batch distillation.

accumulator and the corresponding instantaneous and average composition of the bottoms cut.

A complex batch-distillation unit, of the type described by Hasebe et al. [6], that permits considerable operating flexibility is shown in Figure 13.9. The charge in the feed tank is fed to a suitable column location. Holdups in the reboiler and condenser are kept to a minimum. Products or intermediate cuts are withdrawn from the condenser, the reboiler, or both. In addition, the liquid in the column at the feed location can be recycled to the feed tank if it is desirable to make the composition in the feed tank close to the composition of the liquid at the feed location.

13.5 EFFECT OF LIQUID HOLDUP

Except at high pressure, vapor holdup in a rectifying column is negligible in batch distillation because of the small molar density of the vapor phase. However, the effect of liquid holdup on the trays and in the condensing and reflux system can be significant when the molar ratio of holdup to original charge is more than a few percent. This is especially true when a charge contains low concentrations of one or more of the components to be separated. In general, the effect of holdup in a trayed column is greater than in a packed column because of the lower amount of holdup in the latter. For either type of column, liquid holdup can be estimated by methods described in Chapter 6.

A batch rectifier is usually operated under total-reflux conditions for an initial period of time prior to the withdrawal of distillate product. During this initial time period, liquid holdup in the column increases and approaches a value that is reasonably constant for the remainder of the distillation cycle. Because of the total-reflux concentration profile, the initial concentration of light components in the remaining charge to the still is less than in the original charge. At high liquid holdups, this causes the initial purity and degree of separation to be reduced from estimates based on

methods that ignore liquid holdup. Liquid holdup can reduce the size of product cuts, increase the size of intermediate fractions that are recycled, increase the amount of residue, increase the batch-cycle time, and increase the total energy input. Although approximate methods for predicting the effect of liquid holdup were developed in the past, the complexity of the holdup effect is such that it is now considered best to use the rigorous computer-based, batch-distillation algorithms described in Section 13.7 to study the effect on a case-by-case basis.

13.6 SHORTCUT METHOD FOR MULTICOMPONENT BATCH RECTIFICATION WITH CONSTANT REFLUX

The batch rectification methods presented in Sections 13.2 and 13.3 are limited to binary mixtures, under the assumptions of constant molar overflow, and negligible vapor and liquid holdup. Shortcut methods for handling multicomponent mixtures under the same assumptions have been developed by Diwekar and Madhaven [7] for the two cases of constant distillate composition and constant reflux, and by Sundaram and Evans [8] for constant reflux. Both methods avoid tedious stage-by-stage calculations of vapor and liquid compositions by employing the Fenske–Underwood–Gilliland (FUG) shortcut procedure for continuous distillation, described in Chapter 9, at successive time steps. In essence, they treat batch rectification as a sequence of continuous, steady-state rectifications. As in the FUG method, no estimations of compositions or temperatures are made for intermediate stages.

Sundaram and Evans [8] apply their shortcut method to a column of the type shown in Figure 13.3. An overall mole balance for a constant distillate rate, D, gives

$$-\frac{dW}{dt} = D \qquad (13\text{-}18)$$

Therefore,

$$-\frac{dW}{dt} = \frac{V}{1 + R} \qquad (13\text{-}19)$$

For any component, i, an instantaneous mole balance around the column gives

$$\frac{d(x_{W_i} W)}{dt} = x_{D_i} \frac{dW}{dt} \qquad (13\text{-}20)$$

Expanding the LHS of (13-20) and solving for dx_{W_i}:

$$dx_{W_i} = (x_{D_i} - x_{W_i}) \frac{dW}{W} \qquad (13\text{-}21)$$

In finite-difference form, using Euler's method, (13-19) and (13-21) become, respectively:

$$W^{(k+1)} = W^{(k)} - \left(\frac{V}{1 + R}\right) \Delta t \qquad (13\text{-}22)$$

$$x_{W_i}^{(k+1)} = x_{W_i}^{(k)} + \left(x_{D_i}^{(k)} - x_{W_i}^{(k)}\right) \left[\frac{W^{(k+1)} - W^{(k)}}{W^{(k)}}\right] \qquad (13\text{-}23)$$

where k is the time-increment index. For a given Δt time increment, $W^{(k+1)}$ is computed from (13-22) and then $x_{W_i}^{(k+1)}$ is computed for each component from (13-23). However, (13-23) requires values for $x_{D_i}^{(k)}$.

Calculations are initiated at $k = 0$. The initial charge to the still is $W^{(0)}$. Values of $x_{W_i}^{(0)}$ are equal to the mole fractions of the initial charge. Corresponding values of $x_{D_i}^{(0)}$ depend on the method used to start up the batch rectification still. If total-reflux operation is employed as the start-up method, as mentioned in Section 13.5, the Fenske equation of Chapter 9 can be applied to compute values of $x_{D_i}^{(0)}$ for given values of $x_{W_i}^{(0)}$ if column and condenser holdups are negligible. For a given number of equilibrium stages, N:

$$N = \frac{\log\left[\left(\dfrac{x_{D_i}}{x_{W_i}}\right)\left(\dfrac{x_{W_r}}{x_{D_r}}\right)\right]}{\log \alpha_{i,r}} \tag{13-24}$$

Solving,

$$x_{D_i} = x_{W_i}\left(\frac{x_{D_r}}{x_{W_r}}\right)\alpha_{i,r}^{N} \tag{13-25}$$

where r is an arbitrary reference component of the mixture, such as the least volatile component. Since

$$\sum_{i=1}^{C} x_{D_i} = 1.0 \tag{13-26}$$

substitution of (13-25) in (13-26) gives

$$x_{D_r} = \frac{x_{W_r}}{\displaystyle\sum_{i=1}^{C} x_{W_i}\alpha_{i,r}^{N}} \tag{13-27}$$

The initial distillate composition, $x_{D_r}^{(0)}$, is computed from (13-27). The remaining values of $x_{D_i}^{(0)}$ are computed in turn from (13-25).

Using the initial set of values for $x_{D_i}^{(0)}$, values of $x_{W_i}^{(1)}$ are computed from (13-23) following the calculation of $W^{(1)}$ from (13-22). To compute each subsequent set of $x_{W_i}^{(k+1)}$ for $k > 0$, values of $x_{D_i}^{(k)}$ for $k > 0$ are needed. These are obtained by applying the FUG method. Equation (13-24) applies during batch rectification if N is replaced by $N_{min} < N$ with $i =$ LK and $r =$ HK. But N_{min} is related to N by the Gilliland correlation. A convenient, but approximate, equation for that correlation, due to Eduljee [9], is

$$\frac{N - N_{min}}{N + 1} = 0.75\left[1 - \left(\frac{R - R_{min}}{R + 1}\right)^{0.5668}\right] \tag{13-28}$$

An estimate of the minimum-reflux ratio, R_{min}, is provided by the Class I Underwood equation of Chapter 9, which assumes that all components in the charge distribute between the two products. Thus:

$$R_{min} = \frac{\left(\dfrac{x_{D_{LK}}}{x_{W_{LK}}}\right) - \alpha_{LK,HK}\left(\dfrac{x_{D_{HK}}}{x_{W_{HK}}}\right)}{\alpha_{LK,HK} - 1} \tag{13-29}$$

If one or more components fail to distribute, then Class II Underwood equations should be used. Sundaram and Evans use only (13-29) with LK and HK equal to the lightest component, 1, and the heaviest component, C, in the mixture, respectively.

If (13-25), with $i = 1$, $r = C$, and $N = N_{min}$, and (13-27) with $r = C$ are substituted into (13-29) with LK $= 1$ and HK $= C$, the result is

$$R_{min} = \frac{\alpha_{1,C}^{N_{min}} - \alpha_{1,C}}{(\alpha_{1,C} - 1)\displaystyle\sum_{i=1}^{C} x_{W_i}\alpha_{i,C}^{N_{min}}} \tag{13-30}$$

For specified values of N and R, (13-28) and (13-30) are solved for R_{min} and N_{min} simultaneously by an iterative method. The value of x_{D_C} is then computed from (13-27) with $N = N_{min}$, followed by the calculation of the other values of x_{D_i} from (13-25). Values of N_{min} and R_{min} change with time.

The procedure of Sundaram and Evans involves an inner loop for the calculation of $x_{D_i}^{(k)}$, and an outer loop for $W^{(k+1)}$ and $x_{W_i}^{(k+1)}$. The inner loop requires iterations because of the nonlinear nature of (13-28) and (13-30). Calculations of the outer loop are direct because (13-22) and (13-23) are linear. Application of the method is illustrated in the following example, where relative volatilities are assumed constant.

EXAMPLE 13.6

A charge of 100 kmol of a ternary mixture of A, B, and C with composition $x_{W_A}^{(0)} = 0.33$, $x_{W_B}^{(0)} = 0.33$, and $x_{W_C}^{(0)} = 0.34$ is distilled in a batch rectifier with $N = 3$ (including the reboiler), $R = 10$, and $V = 110$ kmol/h. Estimate the variation of the still, instantaneous distillate, and distillate-accumulator compositions as a function of time for 2 h of operation following an initial start-up period during which a steady-state operation at total reflux is achieved. Use $\alpha_{AC} = 2.0$ and $\alpha_{BC} = 1.5$, and neglect column holdup.

SOLUTION

The method of Sundaram and Evans is applied with $D = V/(1 + R) = 110/(1 + 10) = 10$ kmol/h. Therefore, 10 h would be required to distill the entire charge.

Start-up Period:

From (13-27), with C as the reference r,

$$x_{D_C}^{(0)} = \frac{0.34}{0.33(2)^3 + 0.33(1.5)^3 + 0.34(1)^3} = 0.0831$$

From (13-25),

$$x_{D_A}^{(0)} = 0.33\left(\frac{0.0831}{0.34}\right)2^3 = 0.6449$$

and

$$x_{D_B}^{(0)} = 0.33\left(\frac{0.0831}{0.34}\right)1.5^3 = 0.2720$$

Take time increments, Δt, of 0.5 h.

Table 13.2 Results for Example 13.6

Time, h	W, kmol	x_W			N_{min}	R_{min}	x_D			x of Accumulated Distillate		
		A	B	C			A	B	C	A	B	C
0.0	100	0.3300	0.3300	0.3400	—	—	0.6449	0.2720	0.0831	—	—	—
0.5	95	0.3143	0.3329	0.3528	2.6294	1.2829	0.5957	0.2962	0.1081	0.6283	0.2749	0.0968
1.0	90	0.2995	0.3348	0.3657	2.6249	1.3092	0.5803	0.3048	0.1149	0.6045	0.2868	0.1087
1.5	85	0.2839	0.3365	0.3796	2.6199	1.3385	0.5633	0.3142	0.1225	0.5912	0.2932	0.1156
2.0	80	0.2675	0.3378	0.3947	2.6143	1.3709	0.5446	0.3242	0.1312	0.5800	0.2988	0.1212

At t = 0.5 h for outer loop:

From (13-22), $W^{(1)} = 100 - \left(\dfrac{110}{1+10}\right)0.5 = 95$ kmol

From (13-23) with $k = 0$,

$$x_{W_A}^{(1)} = 0.33 + (0.6449 - 0.33)\left[\frac{95-100}{100}\right] = 0.3143$$

$$x_{W_B}^{(1)} = 0.33 + (0.2720 - 0.33)\left[\frac{95-100}{100}\right] = 0.3329$$

$$x_{W_C}^{(1)} = 0.34 + (0.0831 - 0.34)\left[\frac{95-100}{100}\right] = 0.3528$$

At t = 0.5 h for inner loop:

From (13-28)

$$\frac{3 - N_{min}}{3+1} = 0.75\left[1 - \left(\frac{10 - R_{min}}{10+1}\right)^{0.5668}\right]$$

Solving for R_{min},

$$R_{min} = 10 - 1.5835\, N_{min}^{1.7643} \qquad (1)$$

This equation holds for all values of time *t*.

From (13-30),

$$R_{min} = \frac{2^{N_{min}} - 2}{(2-1)[0.3143(2)^{N_{min}} + 0.3329(1.5)^{N_{min}} + 0.3528(1)^{N_{min}}]} \qquad (2)$$

Equations (1) and (2) are solved simultaneously for R_{min} and N_{min}. This can be done by numerical or graphical methods including successive substitution, Newton's method, or with a spreadsheet by plotting each equation as R_{min} versus N_{min} and determining the intersection. The result is $R_{min} = 1.2829$ and $N_{min} = 2.6294$.
From (13-27), with $N = 2.6294$,

$$x_{D_C}^{(1)} = 0.3528/[0.3143(2)^{2.6294} + 0.3329(1.5)^{2.6294} + 0.3528]$$
$$= 0.1081$$

From (13-25):

$$x_{D_A} = 0.3143\left(\frac{0.1081}{0.3528}\right)2^{2.6924} = 0.5959$$

$$x_{D_B} = 0.3329\left(\frac{0.1081}{0.3528}\right)1.5^{2.6294} = 0.2962$$

Subsequent, similar calculations give the results in Table 13.2.

13.7 STAGE-BY-STAGE METHODS FOR MULTICOMPONENT, BATCH RECTIFICATION

For final design studies or for the simulation of multicomponent, batch rectification, complete stage-by-stage temperature, flow, and composition profiles as a function of time are required. Such calculations are tedious, but can be carried out conveniently with either of two types of computer-based methods. Both methods are based on the same differential-algebraic equations for the distillation model, but differ in the way the equations are solved.

Rigorous Model

Meadows [10] developed the first rigorous, multicomponent, batch-distillation model, based on the assumptions of equilibrium stages, perfect mixing of liquid and vapor phases at each stage, negligible vapor holdup, constant-molar-liquid holdup, *M*, on a stage and in the condenser system, and adiabatic stages in the column. Distefano [11] extended the model and developed a computer-based method

for solving the set of equations. A more efficient method for solving the equations is presented by Boston et al. [12]. For more rapid calculations, Galindez and Fredenslund [13] developed a quasi-steady-state solution procedure that, at each step, utilizes the simultaneous-correction or inside-out methods for continuous distillation discussed in Chapter 10.

The Distefano model is based on the multicomponent, batch-rectification operation shown in Figure 13.10. The equipment consists of a partial reboiler, a column with N equilibrium stages or equivalent in packing, and a total condenser with a reflux drum. Also included, but not shown in Figure 13.10, are a number of accumulator or receiver drums equal to the desired number of overhead product and intermediate cuts. When product purity specifications cannot be made for successive distillate cuts, then intermediate (waste or slop) cuts are necessary. These cuts are usually recycled. To initiate operation, the feed is charged to the reboiler, to which heat is supplied. Vapor leaving Stage 1 at the top of the column is totally condensed and passes to the reflux drum. At first, a total-reflux condition is established for a steady-state, fixed-overhead vapor-flow rate. Depending upon the amount of liquid holdup in the column and in the condenser system,

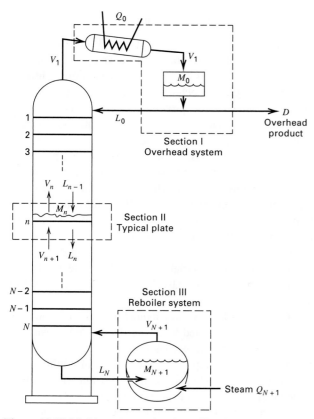

Figure 13.10 Multicomponent, batch-rectification operation.
[From G.P. Distefano, *AIChE J.,* **140,** 190 (1968) with permission.]

the amount and composition of the liquid in the reboiler at total reflux differs to some extent from the original feed.

Starting at time $t = 0$, distillate is removed from the reflux drum and sent to a receiver (accumulator) at a constant molar rate, and a reflux ratio is established. The heat-transfer rate to the reboiler is adjusted so as to maintain the overhead-vapor molar flow rate. Model equations are derived for the overhead condensing system, the column stages, and the reboiler, as illustrated for the overhead condensing system. For section I, in Figure 13.10, component material balances, a total material balance, and an energy balance are given, respectively, by

$$V_1 y_{i,1} - L_0 x_{i,0} - D x_{i,D} = \frac{d(M_0 x_{i,0})}{dt} \tag{13-31}$$

$$V_1 - L_0 - D = \frac{dM_0}{dt} \tag{13-32}$$

$$V_1 h_{V_1} - (L_0 + D) h_{L_0} = Q_0 + \frac{d(M_0 h_{L_0})}{dt} \tag{13-33}$$

where the derivative terms are accumulations due to holdup, which is assumed to be perfectly mixed. Also, for phase equilibrium at Stage 1 of the column:

$$y_{i,1} = K_{i,1} x_{i,1} \tag{13-34}$$

The working equations are obtained by combining (13-31) and (13-34) to obtain a revised component material balance

in terms of liquid-phase compositions, and by combining (13-22) and (13-33) to obtain a revised energy balance that does not include dM_0/dt. Equations for Sections II and III in Figure 13.10 are derived in a similar manner. The resulting working model equations for $t = 0+$ are as follows, where i refers to the component and j refers to the stage, and M is molar liquid holdup.

1. Component mole balances for the overhead-condensing system, column stages, and reboiler, respectively:

$$\frac{dx_{i,0}}{dt} = - \left[\frac{L_0 + D + \dfrac{dM_0}{dt}}{M_0} \right] x_{i,0} + \left[\frac{V_1 K_{i,1}}{M_0} \right] x_{i,1},$$
$$i = 1 \text{ to } C \tag{13-35}$$

$$\frac{dx_{i,j}}{dt} = \left[\frac{L_{j-1}}{M_j} \right] x_{i,j-1} - \left[\frac{L_j + K_{i,j} V_j + \dfrac{dM_j}{dt}}{M_j} \right] x_{i,j}$$
$$+ \left[\frac{K_{i,j+1} V_{j+1}}{M_j} \right] x_{i,j+1},$$
$$i = 1 \text{ to } C, \; j = 1 \text{ to } N \tag{13-36}$$

$$\frac{dx_{i,N+1}}{dt} = \left(\frac{L_N}{M_{N+1}} \right) x_{i,N}$$
$$- \left[\frac{V_{N+1} K_{i,N+1} + \dfrac{dM_{N+1}}{dt}}{M_{N+1}} \right] x_{i,N+1},$$
$$i = 1 \text{ to } C \tag{13-37}$$

where $L_0 = RD$.

2. Total mole balances for overhead-condensing system and column stages, respectively:

$$V_1 = D(R + 1) + \frac{dM_0}{dt} \tag{13-38}$$

$$L_j = V_{j+1} + L_{j-1} - V_j - \frac{dM_j}{dt}, \quad j = 1 \text{ to } N \tag{13-39}$$

3. Enthalpy balances around overhead-condensing system, adiabatic column stages, and reboiler, respectively:

$$Q_0 = V_1 (h_{V_1} - h_{L_0}) - M_0 \frac{dh_{L_0}}{dt} \tag{13-40}$$

$$V_{j+1} = \frac{1}{(h_{V_{j+1}} - h_{L_j})}$$
$$\times \left[V_j (h_{V_j} - h_{L_j}) - L_{j-1} (h_{L_{j-1}} - h_{L_j}) + M_j \frac{dh_{L_j}}{dt} \right],$$
$$j = 1 \text{ to } N \tag{13-41}$$

$$Q_{N+1} = V_{N+1} (h_{V_{N+1}} - h_{L_{N+1}}) - L_N (h_{L_N} - h_{L_{N+1}})$$
$$+ M_{N+1} \left(\frac{dh_{L_{N+1}}}{dt} \right) \tag{13-42}$$

4. Phase equilibrium on column stages and in the reboiler:

$$y_{i,j} = K_{i,j} x_{i,j}, \quad i = 1 \text{ to } C, \quad j = 1 \text{ to } N + 1 \tag{13-43}$$

5. Mole-fraction sums at column stages and in the reboiler:

$$\sum_{i=1}^{C} y_{i,j} = \sum_{i=1}^{C} K_{i,j} x_{i,j} = 1.0, \quad j = 0 \text{ to } N + 1 \tag{13-44}$$

6. Molar holdups in the condenser system and on the column stages based on constant-volume holdups, G_j, respectively:

$$M_0 = G_0 \rho_0 \tag{13-45}$$
$$M_j = G_j \rho_j, \quad j = 1 \text{ to } N \tag{13-46}$$

where ρ is liquid molar density.

7. Variation of molar holdup in the reboiler, where M_{N+1}^0 is the initial charge to the reboiler:

$$M_{N+1} = M_{N+1}^0 - \sum_{j=0}^{N} M_j - \int_0^t D \, dt \tag{13-47}$$

Equations (13-35) through (13-47) constitute an initial-value problem for a system of ordinary differential and algebraic equations (DAEs). The total number of equations is $(2CN + 3C + 4N + 7)$. If variables $N, D, R = L_0/D, M_{N+1}^0$, and all G_j are specified, and if correlations are available for computing liquid densities, vapor and liquid enthalpies, and K-values, the number of unknown variables, distributed as follows, is equal to the number of equations.

$x_{i,j}$	$CN + 2C$
$y_{i,j}$	$CN + C$
L_j	N
V_j	$N + 1$
T_j	$N + 2$
M_j	$N + 2$
Q_0	1
Q_{N+1}	1
	$2CN + 3C + 4N + 7$

Initial values at $t = 0$ for all these variables are obtained from the steady-state, total-reflux calculation, which depends only on values of $N, M_{N+1}^0, x_{N+1}^0, G_j$, and V_1.

Equations (13-35) through (13-42) include first derivatives of $x_{i,j}$, M_j, and h_{L_j}. Except for M_{N+1}, derivatives of the latter two variables can be approximated with sufficient accuracy by incremental changes over the previous time step. If the reflux ratio is high, as it often is, the derivative of M_{N+1} can also be approximated in the same manner. This leaves only the $C(N + 2)$ ordinary differential equations (ODEs) for the component material balances to be integrated in terms of the $x_{i,j}$ dependent variables.

Rigorous Integration Method

The nonlinear equations (13-35) to (13-37) cannot be integrated analytically. Distefano [11] developed a numerical method of solution based on an investigation of 11 different numerical integration techniques that step in time. Of particular concern were the problems of *truncation* error and *stability*, which make it difficult to integrate the equations rapidly and accurately. Such systems of ODEs or DAEs constitute so-called *stiff systems* as described further below.

Local truncation errors result from using approximations for the functions on the right-hand side of the ODEs at each time step. These errors may be small, but they can propagate through subsequent time steps, resulting in global truncation errors sufficiently large to be unacceptable. As truncation errors become large, the number of significant digits in the computed dependent variables gradually decrease. Truncation errors can be reduced by decreasing the size of the time step.

Stability problems are much more serious. When instability occurs, the computed values of the dependent variables become totally inaccurate, with no significant digits at all. Reducing the time step does not eliminate instability until a time-step criterion, which depends on the numerical method, is satisfied. Even then, a further reduction in the time step is required to prevent oscillations of dependent variables.

Problems of stability and truncation error are conveniently illustrated by comparing results obtained by using the explicit- and implicit-Euler methods, both of which are first-order in accuracy, as discussed by Davis [15] and Riggs [16].

Consider the nonlinear, first-order ODE:

$$\frac{dy}{dt} = f\{t, y\} = ay^2 t e^y \tag{13-48}$$

for $y\{t\}$, where initially $y\{t_0\} = y_0$. The explicit- (forward) Euler method approximates (13-48) with a sequence of discretizations of the form

$$\frac{y_{k+1} - y_k}{\Delta t} = ay_k^2 t_k e^{y_k} \tag{13-49}$$

where Δt is the time step and k the sequence index. The function $f\{t, y\}$ is evaluated at the beginning of the current time step. Solving for y_{k+1} gives the recursion equation:

$$y_{k+1} = y_k + ay_k^2 t_k e^{y_k} \Delta t \tag{13-50}$$

Regardless of the nature of $f\{t, y\}$ in (13-48), the recursion equation can be solved explicitly for y_{k+1} using results from the previous time step. However, as discussed later, this advantage is counterbalanced by a limitation on the magnitude of Δt to avoid instability and oscillations.

The implicit- (backward) Euler method also utilizes a sequence of discretizations of (13-48), but the function, $f\{t, y\}$, is evaluated at the end of the current time step. Thus:

$$\frac{y_{k+1} - y_k}{\Delta t} = ay_{k+1}^2 t_{k+1} e^{y_{k+1}} \tag{13-51}$$

Because the function $f\{t, y\}$ is nonlinear in y, (13-51) cannot be solved explicitly for y_{k+1}. This disadvantage is counterbalanced by unconditional stability with respect to selection of Δt. However, too large a value can result in unacceptable truncation errors.

When the explicit-Euler method is applied to (13-35) to (13-47) for batch rectification, as shown in the following example, the maximum value of Δt can be estimated from the maximum, absolute eigenvalue, $|\lambda|_{max}$, of the Jacobian matrix of (13-35) to (13-37). To prevent instability, $\Delta t_{max} \leq 2/|\lambda|_{max}$. To prevent oscillations, $\Delta t_{max} \leq 1/|\lambda|_{max}$. Applications of the explicit- and implicit-Euler methods are compared in the following batch-rectification example.

EXAMPLE 13.7

A charge of 100 kmol of an equimolar mixture of n-hexane (A) and n-heptane (B) is distilled at 15 psia in a batch rectifier consisting of a total condenser with a constant liquid holdup, M_0, of 0.10 kmol, a single equilibrium stage with a constant liquid holdup, M_1, of 0.01 kmol, and a reboiler. Initially the system is brought to the following total-reflux condition, with saturated liquid leaving the total condenser:

Stage	T, °F	x_A	K_A	K_B	M, kmol
Condenser	162.6	0.85935	—	—	0.1
Plate, 1	168.7	0.70930	1.212	0.4838	0.01
Reboiler, 2	178.6	0.49962	1.420	0.5810	99.89

Distillation begins ($t = 0$) with a reflux rate, L_0, of 10 kmol/h and a distillate rate, D, of 10 kmol/h. Calculate the mole fractions of n-hexane and n-heptane at $t = 0.05$ h (3 min), at each of the three rectifier locations, assuming constant molar overflow and constant K-values from above for this small period of elapsed time. Use both the explicit- and implicit-Euler methods to determine the influence of the choice of the time step, Δt.

SOLUTION

Based on the constant molar overflow assumption:

$$V_1 = V_2 = 20 \text{ kmol/h} \quad \text{and} \quad L_0 = L_1 = 10 \text{ kmol/h}$$

Using the K-values and liquid holdups given earlier, (13-35) to (13-37), with all $dM_j/dt = 0$, become as follows:

Condenser

$$\frac{dx_{A,0}}{dt} = -200x_{A,0} + 242.4x_{A,1} \tag{1}$$

$$\frac{dx_{B,0}}{dt} = -200x_{B,0} + 96.76x_{B,1} \tag{2}$$

Plate

$$\frac{dx_{A,1}}{dt} = 1,000x_{A,0} - 3,424x_{A,1} + 2,840x_{A,2} \tag{3}$$

$$\frac{dx_{B,1}}{dt} = 1,000x_{B,0} - 1,967x_{B,1} + 1,162x_{B,2} \tag{4}$$

Reboiler

$$\frac{dx_{A,2}}{dt} = \left(\frac{10}{M_2}\right)x_{A,1} - \left(\frac{28.40}{M_2}\right)x_{A,2} \tag{5}$$

$$\frac{dx_{B,2}}{dt} = \left(\frac{10}{M_2}\right)x_{B,1} - \left(\frac{11.62}{M_2}\right)x_{B,2} \tag{6}$$

where

$$M_2(t = t) = M_2(t = 0) - (V_2 - V_1)t$$
or $$M_2 = 99.89 - 10t \tag{7}$$

Equations (1) through (6) can be grouped by component into the following two matrix equations:

Component A:

$$\begin{bmatrix} -200 & 242.2 & 0 \\ 1,000 & -3,424 & 2,840 \\ 0 & 10/M_2 & -28.40/M_2 \end{bmatrix} \cdot \begin{bmatrix} x_{A,0} \\ x_{A,1} \\ x_{A,2} \end{bmatrix} = \begin{bmatrix} dx_{A,0}/dt \\ dx_{A,1}/dt \\ dx_{A,2}/dt \end{bmatrix} \tag{8}$$

Component B:

$$\begin{bmatrix} -200 & 96.76 & 0 \\ 1,000 & -1,967 & 1,160 \\ 0 & 10/M_2 & -11.62/M_2 \end{bmatrix} \cdot \begin{bmatrix} x_{B,0} \\ x_{B,1} \\ x_{B,2} \end{bmatrix} = \begin{bmatrix} dx_{B,0}/dt \\ dx_{B,1}/dt \\ dx_{B,2}/dt \end{bmatrix} \tag{9}$$

Although (8) and (9) do not appear to be coupled, they are because at each time step the sums $x_{A,j} + x_{B,j}$ do not equal 1. Accordingly, the mole fractions are normalized at each time step to force them to sum to 1. The initial eigenvalues of the Jacobian matrices (8) and (9) are computed from any of a number of computer programs, such as MathCad, Mathematica, MATLAB, or Maple, to be as follows, using $M_2 = 99.89$ kmol:

	Component A	Component B
λ_0	−126.54	−146.86
λ_1	−3,497.6	−2,020.2
λ_2	−0.15572	−0.03789

It is seen that $|\lambda|_{max} = 3,497.6$. Thus, for the explicit-Euler method, instability and oscillations can be prevented by choosing:

$$\Delta t \leq 1/3,497.6 = 0.000286$$

If we select $\Delta t = 0.00025$ h (just slightly smaller than the criterion), it takes $0.05/0.00025 = 200$ time steps to reach $t = 0.05$ h (3 min). No such restriction applies to the implicit-Euler method, but too large a Δt may result in an unacceptable truncation error.

Explicit-Euler Method

According to Distefano [11], the maximum step size for integration using an explicit method is nearly always limited by stability considerations, and usually the truncation error is small. Assuming this to be true for this example, the following results were obtained using $\Delta t = 0.00025$ h with a spreadsheet program by converting

(8) and (9) together with (7) for M_2, to the form of (13-50). Only the results for every 40 time steps are given.

Time, h	Normalized Mole Fractions in Liquid for *n*-Hexane			Normalized Mole Fractions in Liquid for *n*-Heptane		
	Distillate	Plate	Still	Distillate	Plate	Still
0.01	0.8183	0.6271	0.4993	0.1817	0.3729	0.5007
0.02	0.8073	0.6219	0.4991	0.1927	0.3781	0.5009
0.03	0.8044	0.6205	0.4988	0.1956	0.3795	0.5012
0.04	0.8036	0.6199	0.4985	0.1964	0.3801	0.5015
0.05	0.8032	0.6195	0.4982	0.1968	0.3805	0.5018

To show the instability effect, a time step of 0.001 h (four times the previous time step) gives the following unstable results during the first five time steps to an elapsed time of 0.005 h. Also included are values at 0.01 h for comparison to the preceding stable results.

Time, h	Normalized Mole Fractions in Liquid for *n*-Hexane			Normalized Mole Fractions in Liquid for *n*-Heptane		
	Distillate	Plate	Still	Distillate	Plate	Still
0.000	0.85935	0.7093	0.49962	0.14065	0.2907	0.50038
0.001	0.859361	0.559074	0.499599	0.140639	0.440926	0.500401
0.002	0.841368	0.75753	0.499563	0.158632	0.24247	0.500437
0.003	0.852426	0.00755	0.499552	0.147574	0.99245	0.500448
0.004	0.809963	0.884925	0.499488	0.190037	0.115075	0.500512
0.005	0.874086	1.154283	0.499546	0.125914	–0.15428	0.500454
0.01	1.006504	0.999254	0.493573	–0.0065	0.000746	0.506427

Much worse results are obtained if the time step is increased 10-fold to 0.01 h, as shown in the following table, where at $t = 0.01$ h, a very negative mole fraction has appeared.

Time, h	Normalized Mole Fractions in Liquid for *n*-Hexane			Normalized Mole Fractions in Liquid for *n*-Heptane		
	Distillate	Plate	Still	Distillate	Plate	Still
0.00	0.85935	0.7093	0.49962	0.14065	0.2907	0.50038
0.01	0.859456	–0.79651	0.49941	0.140544	1.796512	0.50059
0.02	2.335879	2.144666	0.497691	–1.33588	–1.14467	0.502309
0.03	1.284101	1.450481	0.534454	–0.2841	–0.45048	0.465546
0.04	1.145285	1.212662	8.95373	–0.14529	–0.21266	–7.95373
0.05	1.07721	1.11006	1.191919	–0.07721	–0.11006	–0.19192

Implicit-Euler Method

If (8) and (9) are converted to implicit equations like (13-51), they can be rearranged into a linear, tridiagonal set for each component. For example, the equation for component A on the plate becomes

$$(1,000 \, \Delta t)x_{A,0}^{(k+1)} - (1 + 3,424 \, \Delta t)x_{A,1}^{(k+1)} + (2,840 \, \Delta t)x_{A,2}^{(k+1)} = -x_{A,1}^{(k)}$$

The two tridiagonal equation sets can be solved by the tridiagonal-matrix algorithm or with a spreadsheet program using the iterative, circular-reference technique. For the implicit-Euler method, the selection of the time step, Δt, is not restricted by stability considerations. However, too large a Δt can lead to unacceptable truncation

errors. Normalized, liquid-mole-fraction results at $t = 0.05$ h for just component A are as follows for a number of different choices of Δt, all of which are greater than the 0.00025 h used earlier to obtain stable and oscillation-free results with the explicit-Euler method. Included for comparison is the explicit-Euler result for $\Delta t = 0.00025$ h.

Time = 0.05 h:

Δt, h	Normalized Mole Fractions in Liquid for *n*-Hexane		
	Distillate	Plate	Still
	Explicit-Euler		
0.00025	0.8032	0.6195	0.4982
	Implicit-Euler		
0.0005	0.8042	0.6210	0.4982
0.001	0.8042	0.6210	0.4982
0.005	0.8045	0.6211	0.4982
0.01	0.8049	0.6213	0.4982
0.05	0.8116	0.6248	0.4982

The preceding data show acceptable results with the implicit-Euler method using a time step about 200 times the $\Delta t_{\max}$ for the explicit-Euler method.

Another serious computational problem occurs when integrating the equations of batch distillation. Because the liquid holdups on the trays and in the condenser are small, the values of the corresponding liquid mole fractions, $x_{i,j}$, respond quickly to changes. The opposite holds for the reboiler (still) with its large liquid holdup. Hence, the required time step for accuracy is usually small, leading to a very slow response of the overall rectification system. Systems of ODEs having this characteristic constitute so-called stiff systems. For such a system, as discussed by Carnahan and Wilkes [17], an explicit method of solution must utilize a small time step for the entire period even though values of the dependent variables may all be changing slowly for a large portion of the time period. Accordingly, it is preferred to utilize a special implicit-integration technique developed by Gear [14] and others, as contained in the public-domain software package called ODEPACK. Gear-type methods strive for accuracy, stability, and computational efficiency by using multistep, variable order, and variable-step-size implicit techniques.

A commonly used measure of the degree of stiffness is the eigenvalue ratio $|\lambda|_{\max}/|\lambda|_{\min}$, where λ values are the eigenvalues of the Jacobian matrix of the set of ODEs. For the Jacobian matrix of (13-35) through (13-37), the Gerschgorin circle theorem, discussed by Varga [18], can be employed to estimate the eigenvalue ratio. The maximum absolute eigenvalue corresponds to the component with the largest K-value and the tray with the smallest liquid molar holdup. When the Gerschgorin theorem is applied to a row of the Jacobian matrix based on (13-36):

$$|\lambda|_{\max} \leq \left[\left(\frac{L_{j-1}}{M_j} \right) + \left(\frac{L_j + K_{i,j} V_j}{M_j} \right) + \left(\frac{K_{i,j+1} V_{j+1}}{M_j} \right) \right] \approx 2 \left[\frac{L_j + K_{i,j} V_j}{M_j} \right] \quad \text{(13-52)}$$

where i refers to the most-volatile component and j to the stage with the smallest liquid molar holdup. The minimum absolute eigenvalue almost always corresponds to a row of the Jacobian matrix for the reboiler. Thus, from (13-37):

$$|\lambda|_{\min} \leq \left[\left(\frac{L_N}{M_{N+1}} \right) + \left(\frac{V_{N+1} K_{i,N+1}}{M_{N+1}} \right) \right] \approx \left[\frac{L_N + K_{i,N+1} V_{N+1}}{M_{N+1}} \right] \quad \text{(13-53)}$$

where i now refers to the least-volatile component and $N + 1$ is the reboiler stage. The largest value of the reboiler holdup is M^0_{N+1}. The stiffness ratio, SR, is

$$\text{SR} = \frac{|\lambda|_{\max}}{|\lambda|_{\min}} \approx 2 \left(\frac{L + K_{\text{lightest}} V}{L + K_{\text{heaviest}} V} \right) \left(\frac{M^0_{N+1}}{M_{\text{tray}}} \right) \quad \text{(13-54)}$$

From (13-54), the stiffness ratio depends not only on the difference between tray and initial reboiler molar holdups, but also on the difference between K-values of the lightest and heaviest components in the charge to the still.

Davis [15] states SR = 20 is not stiff, SR = 1,000 is stiff, and SR = 1,000,000 is very stiff. For the conditions of Example 13.7, using (13-54):

$$SR \approx 2 \left[\frac{10 + (1.212)(20)}{10 + (0.581)(20)} \right] \left(\frac{100}{0.01} \right) = 31{,}700$$

which meets the criterion of a stiff problem. A modification of the computational procedure of Distefano [11], for solving (13-35) through (13-46), is as follows:

Initialization

1. Establish total-reflux conditions, based on vapor and liquid molar flow rates V_j^0 and $L_j^0 . V^0_{N+1}$ is the desired boilup rate or L_0^0 is based on the desired distillate rate and reflux ratio such that $L_0^0 = D(R + 1)$.
2. At $t = 0$, reduce L_0^0 to begin distillate withdrawal, but maintain the boilup rate established or specified for the total-reflux condition. This involves replacing all L_j^0 with $L_j^0 - D$. Otherwise, the initial values of all variables are those established for total reflux.

Time Step

3. In (13-35) to (13-37), replace liquid-holdup derivatives by total material balance equations:

$$\frac{dM_j}{dt} = V_{j+1} + L_{j-1} - V_j - L_j$$

Solve the resulting equations for the liquid mole fractions using an appropriate implicit-integration technique and a suitable time step. Normalize the mole fractions if they do not sum to one at each stage.
4. Compute a new set of stage temperatures and corresponding vapor-phase mole fractions from (13-44) and (13-43), respectively.
5. Compute liquid densities and liquid holdups from (13-45) and (13-46), and liquid and vapor enthalpies. Then determine derivatives of enthalpies and liquid holdups with respect to time by forward-finite-difference approximations.
6. Compute a new set of liquid and vapor molar flow rates from (13-38), (13-39), and (13-41).
7. Compute the new reboiler molar holdup from (13-47).
8. Compute condenser and reboiler heat-transfer rates from (13-40) and (13-42).

Iteration to Completion of Operation

9. Repeat Steps 3 through 8 for additional time steps until a specified operation is complete. The specified operation might be for a desired amount of distillate, desired mole fraction of a particular component in the accumulated distillate, etc.

New Operation

10. Dump the accumulated distillate into a receiver, change operating conditions, and repeat Steps 2 through 9. Terminate calculations following the final operation.

The foregoing procedure is limited to narrow-boiling feeds and the simple configuration shown in Figure 13.10. A more flexible and efficient method, specifically designed to cope with stiffness, is that of Boston et al. [12], which uses a modified inside-out algorithm of the type discussed for continuous distillation in Chapter 10, which can handle feeds ranging from narrow-boiling to wide-boiling, even for nonideal-liquid solutions. In addition, the Boston et al. method permits multiple feeds, side streams, tray heat transfer, vapor distillate, and considerable flexibility in operation specifications.

EXAMPLE 13.8

A charge of 100 kmol of 30 mol% acetone, 30 mol% methanol, and 40 mol% water at 60°C and 1 atm is to be distilled in a batch rectifier consisting of a reboiler, a column with five theoretical stages, a total condenser, a reflux drum, and three distillate accumulators. The molar liquid holdup of the condenser-reflux drum is 5 kmol, whereas the molar liquid holdup of each stage is 1 kmol. The pressure is assumed constant at 1 atm throughout the rectifier. The following four events are to occur, each with a reboiler duty of 1 million kcal h:

Event 1: Establishment of total-reflux conditions.

Event 2: Rectification with a reflux ratio of 3 until the purity of the accumulated distillate in the first accumulator drops to 73 mol%.

Event 3: Rectification with a reflux ratio of 3 and a second accumulator for 21 min.

Event 4: Rectification with a reflux ratio of 3 and a third accumulator for 27 min.

Determine accumulator and column conditions at the end of each event. Use the Wilson equation for computing K-values.

SOLUTION

The following results were obtained with the program DESIGN II/Batch of the ChemShare Corporation. The stiffness ratio, SR, may be computed from (13-54) based on total-reflux conditions at the end of Event 1. The conditions are as follows:

Event 1: Total Reflux Conditions:

			Mole Fraction in Liquid		
Stage	T, °C	L, kmol/h	Acetone	Methanol	Water
Condenser	55.6	138.9	0.770	0.223	0.007
1	55.6	138.6	0.761	0.227	0.012
2	55.7	138.0	0.747	0.235	0.018
3	55.9	137.0	0.722	0.247	0.031
4	56.2	134.8	0.673	0.269	0.058
5	57.3	128.7	0.560	0.306	0.134
Reboiler	62.2	—	0.252	0.307	0.441

The charge remaining in the still is $100 - 5 - 5(1) = 90$ kmol. The most-volatile component is acetone, with a K-value at the bottom stage of 1.203. The least-volatile component is water, with a corresponding K-value of 0.428. The stiffness ratio is

$$\text{SR} \approx 2\left[\frac{128.7 + (1.203)(134.8)}{128.7 + (0.428)(134.8)}\right]\left(\frac{90}{1}\right) \approx 281$$

Thus, this problem is not very stiff. A time step of 0.06 min is used.

Event 2

The time required to complete Event 2 is computed to be 57.5 minutes. The accumulated distillate in Tank 1 is 32.0 kmol with a composition as follows: 73.0 mol% acetone, 26.0 mol% methanol, and 1.0 mol% water. The liquid remaining in the reboiler is 58.0 kmol with the following composition: 2.8 mol% acetone, 30.0 mol% methanol, and 67.2 mol% water.

Event 3

The time required to complete this event is specified as 21 min. The accumulated distillate in Tank 2 is 11.3 kmol of the following composition: 47.2 mol% acetone, 51.8 mol% methanol, and 1.0 mol% water. This intermediate cut is recycled for addition to the next charge.

Event 4

At the end of the 27-min specification, the accumulated distillate in Tank 3 is 13.8 kmol of the following composition: 8.3 mol% acetone, 86.2 mol% methanol, and 5.5 mol% water. The composition of the remaining 32.9 kmol in the still is as follows: 0.0 mol% acetone, 0.4 mol% methanol, and 99.6 mol% water.

Rapid-Solution Method

As an alternative to integration of the stiff system of differential equations, the quasisteady-state procedure of Galindez and Fredenslund [13] can be used. With this method, the transient conditions are simulated as a succession of a finite number of continuous steady states of short duration, typically 0.05 h (3 min). Holdup is taken into account, but the stiffness of the problem is of no consequence. Results compare favorably with those from the rigorous integration method.

Consider an intermediate theoretical stage, j, with molar holdup, M_j, in the batch rectifier in Figure 13.11a. A material balance for component i, in terms of component flow rates, rather than mole fractions, is

$$l_{i,j} + v_{i,j} - l_{i,j-1} - v_{i,j+1} + \frac{d(M_j x_{i,j})}{dt} = 0 \quad (13\text{-}55)$$

Assume constant molar holdup. Also, assume that during a short time period, $dt = \Delta t = t_{k+1} - t_k$, the component flow rates given by the first four terms in (13-55) remain constant at values corresponding to time t_{k+1}. The component holdup term in (13-55) is

$$\frac{d(M_j x_{i,j})}{dt} = M_j\left[\frac{x_{i,j}\{t_{k+1}\} - x_{i,j}\{t_k\}}{\Delta t}\right] \quad (13\text{-}56)$$

But, $x_{i,j} = l_{i,j}/L_j$. Therefore, (13-56) can be rewritten as

$$\frac{d(M_j x_{i,j})}{dt} = \frac{M_j l_{i,j}}{L_j \Delta t} - \frac{M_j l_{i,j}\{t_k\}}{L_j\{t_k\}\Delta t} \quad (13\text{-}57)$$

If (13-57) is substituted into (13-55) and terms in the component flow rate $l_{i,j}$ are collected:

$$l_{i,j}\left(1 + \frac{M_j}{L_j \Delta t}\right) + v_{i,j} - l_{i,j-1} - v_{i,j+1} - \frac{M_j l_{i,j}\{t_k\}}{L_j\{t_k\}\Delta t} \quad (13\text{-}58)$$

If (13-58) for unsteady-state (batch) distillation is compared to (10-58) for steady-state (continuous) distillation, we see that the term $M_j/(L_j\,\Delta t)$ in (13-58) corresponds to the

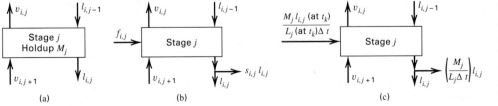

Figure 13.11 Simulation of holdup in a batch rectifier. (a) Stage in a batch rectifier with holdup. (b) Stage in a continuous fractionator. (c) Simulation of batch holdup in a continuous fractionator.

liquid side-stream ratio in (10-58) or that $M_j/\Delta t$ corresponds to a liquid side-stream flow rate. We also see that the term $M_j l_{i,j}\{t_k\}/(L_j\{t_k\}\Delta t)$ in (13-58) corresponds to a component feed rate in (10-58). The analogy is shown in parts (b) and (c) of Figure 13.11. Thus, the change in component liquid holdup per unit time, $d(M_j x_{i,j})/dt$ in (13-56), is interpreted for a small, finite-time difference as the difference between a component feed rate into the stage and a component flow rate in a liquid side stream leaving the stage. In a similar manner, the stage enthalpy holdup in the energy balance for the stage is interpreted as the difference over a small, finite-time interval between a heat input to the stage and an enthalpy output in a liquid side stream leaving the stage. The overall result is a system of steady-state equations, identical in form to the equations for the simultaneous-correction and inside-out methods of Chapter 10. Accordingly, as implemented in the Batch Column computer model of Chemstations, Inc., either of those two methods can be used to solve the system of component-material-balance, phase-equilibrium, and energy-balance equations at each time step. The initial guesses used to initiate each time step are the values at the end of the previous time step. Because the variables generally change by only a small amount for each time step, convergence of the Newton–Raphson or inside-out method is generally achieved in a small number of iterations.

EXAMPLE 13.9

A charge of 100 lbmol of a mixture of 25 mol% benzene (B), 50 mol% monochlorobenzene (MCB), and 25 mol% *ortho*-dichlorobenzene (DCB) is distilled in a batch rectifier consisting of a reboiler, 10 equilibrium stages, a reflux drum, and three distillate product accumulators. The condenser-reflux drum holdup is constant at 0.20 ft³, and each stage in the column has a liquid holdup of 0.02 ft³. Pressures are 17.5 psia in the reboiler and 14.7 psia in the reflux drum, with a linear pressure profile in the column from 15.6 psia at the top to 17 psia at the bottom. Following an initialization at total reflux, the batch is distilled in the following three operation steps, each with a vapor boilup rate of 200 lbmol/h and a reflux ratio of 3. Thus, the distillate rate is 50 lbmol/h.

Operation step 1: Terminate when the mole fraction of benzene in the instantaneous distillate drops below 0.100.

Operation step 2: Terminate when the mole fraction of MCB in the distillate drops below 0.40.

Operation step 3: Terminate when the mole fraction of DCB in the reboiler rises above 0.98.

Assume ideal solutions and the ideal-gas law.

SOLUTION

This problem is quite stiff, with a stiffness ratio of approximately 15,000. The quasi-steady-state procedure of Galindez and Fredenslund [13], as implemented in Batch Column, was used with a time increment of 0.005 h for each of the three operation steps. Although 0.05 h is normal for the Galindez and Fredenslund method, the high ratio of distillate rate to charge for this problem necessitated a smaller Δt. Computed results are given in

Table 13.3 Results at the End of Each Operation Step for Example 13.9

	Operation Step		
	1	2	3
Operation time, h	0.605	0.805	0.055
No. of time increments	121	161	11
Accumulated distillate:			
Total lbmol	33.65	41.96	2.73
Mole fractions:			
B	0.731	0.009	0.000
MCB	0.269	0.950	0.257
DCB	0.000	0.041	0.743
Reboiler holdup:			
Total lbmol	66.13	24.19	21.46
Mole fractions:			
B	0.006	0.000	0.000
MCB	0.616	0.044	0.018
DCB	0.378	0.956	0.982
Total heat duties, 10^6 Btu:			
Condenser	1.95	2.65	0.19
Reboiler	2.08	2.63	0.18

Table 13.3, where it is seen that the accumulated distillate cuts from operation steps 1 and 3 are quite impure with respect to benzene and DCB, respectively. The cut from Step 2 is 95 mol% pure MCB. The residual left in the reboiler after Step 3 is quite pure in DCB. A plot of the instantaneous-distillate composition as a function of total-distillate accumulation for all steps is shown in Figure 13.12.

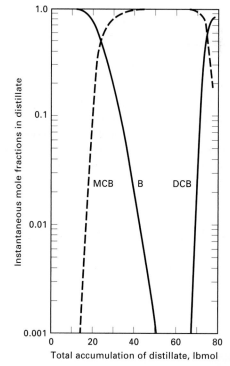

Figure 13.12 Instantaneous-distillate composition profile for Example 13.9.

[*Perry's Chemical Engineers' Handbook,* 6th ed., R.H. Perry and D.W. Green, Eds., McGraw-Hill, New York (1984) with permission.]

Table 13.4 Results of Alternative Operating Schedule for Example 13.9

Distillate Cut	Amount, lbmol	Composition, Mole Fractions		
		B	MCB	DCB
Benzene-rich	18	0.993	0.007	0.000
Intermediate 1	18	0.374	0.626	0.000
MCB-rich	34	0.006	0.994	0.000
Intermediate 2	8	0.000	0.536	0.464
DCB-rich residual	22	0.000	0.018	0.982
Total	100			

Changes in mole fractions occur very rapidly at certain times during the batch rectification, indicating that relatively pure cuts may be possible. This plot is useful in developing alternative schedules to obtain almost pure cuts. For example, suppose relatively rich distillate cuts of B, MCB, and DCB are desired. From Figure 13.12, an initial benzene-rich cut of, say, 18 lbmol might be taken, followed by an intermediate cut for recycle of, say, 18 lbmol. Then, an MCB-rich cut of 34 lbmol might be taken, followed by another intermediate cut of 8 lbmol, leaving a DCB-rich residual of 22 lbmol. For this series of operation steps, with the same vapor boilup rate of 200 lbmol/h and reflux ratio of 3, the computed results for each distillate accumulation (cut), using Batch Column with a time step of 0.005 h, are given in Table 13.4. As seen, all three product cuts are better than 98 mol% pure. However, (18 + 8) = 26 lbmol of intermediate cuts, or about one-fourth of the original charge, would have to be recycled. Further improvements in purities of the cuts or reduction in the amounts of intermediate cuts for recycle can be made by increasing the reflux ratio and/or the number of theoretical stages.

13.8 OPTIMAL CONTROL

As discussed by Luyben [19], design of a batch distillation process can be complex. Two aspects must be considered: (1) the products to be obtained and (2) the control method to be employed. Basic design parameters are the number of trays, the size of the initial charge to the still pot, the boilup ratio, and the reflux ratio as a function of time. Even if the feed is a binary, it may be necessary to take three products: a distillate rich in the most-volatile component, a residue rich in the least-volatile component, and an intermediate (slop or waste) cut containing both components. If the feed is a ternary system, one or two slop cuts may be necessary, in addition to two distillate cuts and the residue.

Slop Cuts

First consider the batch distillation of a binary mixture in the following example.

EXAMPLE 13.10

We wish to batch-distill 100 kmol of an equimolar mixture of *n*-hexane (C6) and *n*-heptane (C7) at 1 atm in a batch-rectification column with a total condenser. It is desired to produce two products, one with 95 mol% C6 and the other with 95 mol% C7. Neglect

Table 13.5 Batch Distillation of a C6-C7 Mixture

	Case 1	Case 2	Case 3	Case 4	Case 5
Reflux ratio	2	3	4	8	9.54
C6 product, kmol	15.1	36.0	42.4	49.2	50.0
C7 product, kmol	34.4	40.7	44.3	49.2	50.0
Slop cut, kmol	50.5	23.3	13.3	1.6	0.0
Mole fraction of C6 in slop cut	0.67	0.59	0.57	0.54	No slop cut
Total operation time, hours	1.97	2.37	2.78	4.57	5.27

holdup and assume a boilup rate of 100 kmol/h. Assume column operation at constant reflux ratio. Thus, the distillate composition will change with time. Determine a reasonable number of equilibrium stages and the effect of reflux ratio on the amount of slop cut.

SOLUTION

To determine the number of equilibrium stages, a McCabe–Thiele diagram, based on *K*-values from the SRK equation of state, is used in the manner of Figures 13.4 and 13.5. For the condition of total reflux ($y = x$, 45°line), the minimum number of stages to achieve a 95 mol% C6 from an initial feed of 50 mol% C6 is 3.1, where one stage is the boiler. For operation at a finite reflux ratio, assume twice the minimum, or five equilibrium plates plus the boiler, for a total of six equilibrium stages.

Now compute the products obtained at different reflux ratios. For each case, the first product is the accumulation of distillate of 95 mol% C6. At this point, if the residue contains less than 95 mol% C7, then, in a second step, a second accumulation of distillate (the slop cut) is made until the residue achieves the desired C7 composition. The reflux ratio is held constant throughout. The results (see Table 13.5) were obtained with the Batch Column model of Chemstations, which is included in the CHEMCAD simulation program. For a perfect separation (by material balance) the C6 and C7 products must each be 50 lbmol at 95 mol% purity. From Table 13.5, this is achieved at a constant reflux ratio of 9.54, with an operating time of 5.27 hours.

For lower reflux ratios, a slop cut whose amount increases as the reflux ratio decreases, is necessary. If the quantity of feed is much larger than the capacity of the still pot, the feed can be distilled in a sequence of charges. Then the slop cut for binary distillation of a batch can be recycled to the next batch. In this manner, each charge consists of fresh feed mixed with recycle slop cut. As discussed by Luyben [19], the composition of the slop cut is often not very different from the feed. This is largely confirmed in Table 13.5. If the number of stages in this example is increased from 6, the reflux ratio for eliminating the slop cut can be reduced. For example, if 10 equilibrium stages are used, the reflux ratio can be reduced from 9.54 to approximately 6.

Slop-cut strategy for batch distillation of a ternary mixture, as discussed by Luyben [19], is considerably more complex, as shown in the following example.

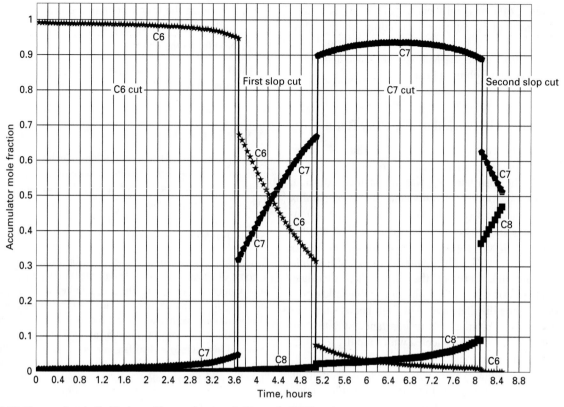

Figure 13.13 Ternary, batch distillation with two slop cuts in Example 13.11.

EXAMPLE 13.11

We wish to batch-distill 150 kmol of an equimolar ternary mixture of C6, C7, and normal octane (C8) at 1 atm in a batch-rectification column with a total condenser. We wish to produce three products: distillates of 95 mol% C6 and 90 mol% C7, and a residue of 95 mol% C8. Neglect holdup and assume a boilup rate of 100 kmol/h. Assume that column operation is controlled at constant reflux ratio. Thus, the distillate composition will change with time. Assume the column will contain five equilibrium stages, which together with the boiler, makes a total of six equilibrium stages. Determine the effect of reflux ratio on the slop cuts.

SOLUTION

The difficulty in this ternary example is to find operating conditions that will permit the purity specification of the intermediate component, C7, to be met. The difficulty lies in the specification for the termination of the second cut, which, unless the reflux ratio is high enough, is a slop cut. For example, suppose the reflux ratio is held constant at 4 and the intention is to terminate the second cut when the mole fraction of C7 in the instantaneous distillate reaches 90 mol% C7. Unfortunately, computer simulations show that only a value of 88 mol% C7 is reached. Therefore, try a higher reflux ratio, e.g., 8. In addition, terminate the third cut (the C7 product) when the mole fraction of C8 in that cut rises to 0.09 in the accumulator; and terminate the second slop cut when the mole fraction of C8 in the residue rises to 0.95, the desired purity. Note that no purity specification has been placed on the C7 product. Instead, it has been assumed that the desired purity of 90 mol% C7 will be achieved with impurities of 9 mol% C8 and 1 mol% C6. Acceptable

Table 13.6 Batch Distillation of a C6-C7-C8 Mixture

	Case 1	Case 2
Reflux ratio	4	8
C6 product, kmol	35.85	46.70
First slop cut:		
Amount, kmol	42.16	16.67
Mole fraction C6	0.373	0.316
Mole fraction C7	0.602	0.672
C7 product:		
Amount, kmol		35.43
Mole fraction C6		0.011
Mole fraction C7	0.877 max	0.898
Second slop cut:		
Amount, kmol		4.38
Mole fraction C6		0.000
Mole fraction C7		0.523
C8 product, kmol		46.82
Total operation time, hr		8.48

results are almost achieved for the reflux ratio of 8, as shown in Table 13.6 and Figure 13.13, where the desired purity of the C7 cut is 89.8 mol%. However, for a reflux ratio of 8, these results may not correspond to the optimal termination specification for the first slop cut. Furthermore, with a small adjustment in the reflux ratio, it may be possible to eliminate the second slop cut. These two considerations are the subject of Exercise 13.29.

Slop (waste, off) cuts and their recycle have been studied by a number of investigators, including Mayur, May, and Jackson [20], Luyben [19], Quintero–Marmol and Luyben [21], Farhat et al. [22], Mujtaba and Macchietto [23], Diehl et al. [24], and Robinson [25].

Optimal Control by Variation of Reflux Ratio

Two methods of controlling a batch distillation, under conditions of constant boilup rate, have been discussed. In Section 13.2, the distillate rate was maintained constant, which, at constant boilup, is equivalent to constant reflux ratio. This resulted in a variable composition for the instantaneous distillate and the accumulated distillate. This method of control is relatively simple and, accordingly, is widely practiced. In Section 13.3, the composition of the instantaneous distillate and, therefore, the accumulated distillate, was maintained constant. This requires a variable reflux ratio and accompanying distillate rate. Although not as simple as the constant reflux ratio method, the constant distillate composition method can be implemented with a rapidly responding composition (or surrogate) sensor.

What is the optimal way to control a batch distillation: (1) by constant reflux ratio, (2) by constant distillate composition, or (3) by some other means? With a simulation program, it is fairly straightforward to compare the first two methods. However, the results depend on the objective for the optimization. Diwekar [26] studied the following three objectives when the accumulated-distillate composition and/or the residual composition is specified:

1. Maximize the amount of accumulated distillate in a given time.

2. Minimize the time to obtain a given amount of accumulated distillate.

3. Maximize the profit.

In the following example, the first two methods of control are compared for the first two objectives.

EXAMPLE 13.12

Repeat Example 13.10 under conditions of constant distillate composition and compare the results to those of Example 13.10 for a constant reflux ratio of 4 with respect to both the amount of distillate and time of operation.

SOLUTION

For Example 13.10, from Table 13.5, for a reflux ratio of 4, the amount of accumulated distillate during the first operation step is 42.4 kmol of 95 mol% C6. The time required for this cut, which is not listed in Table 13.5, is 1.98 hours. Using a simulation program, the operation specifications for a constant composition operation are a boilup rate of 100 kmol/h, as in Example 13.10, with a constant instantaneous-distillate composition of 95 mol% C6. For the maximum distillate objective, the stop time for the first cut is 1.98 hours as in Example 13.10. The amount of distillate obtained is 43.5 kmol, which is 2.6% higher than for operation at constant reflux ratio. The variation of reflux ratio with time for constant composition control is shown in Figure 13.14, where the constant reflux ratio of 4 is also shown. The initial reflux ratio is 1.7; rising gradually at first and rapidly at the end. At 1 hour, the reflux ratio is 4, while at 1.98 hours, it is 15.4. For constant composition control, 42.4 kmol of accumulated distillate are obtained in 1.835 hours, compared to 1.98 hours for reflux-ratio control. Thus, again, constant composition control is more optimal, this time by almost 8%.

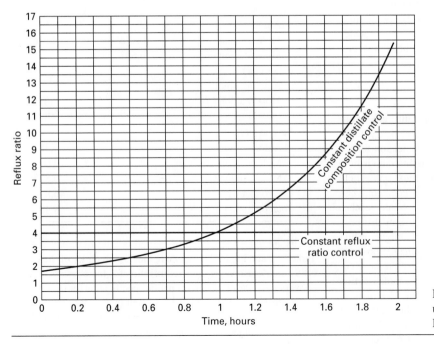

Figure 13.14 Binary, batch distillation under distillate-composition control in Example 13.12.

Studies by Converse and Gross [27], Coward [28, 29], and Robinson [30] for binary systems; by Robinson [30] and Mayur and Jackson [31] for ternary systems; and Diwekar et al. [32] for higher multicomponent systems show that further maximization of the amount of distillate or minimization of operation time, as well as maximization of profit, can be achieved by using an optimal-reflux-ratio policy. Often, this reflux-ratio policy is intermediate between the constant-reflux-ratio and constant-composition controls shown in Figure 13.14 for Example 13.12. Generally, the optimal-reflux-ratio control curve rises less sharply than that for the constant-distillate-composition control, with the result that savings in distillate, time, or money are highest for the more difficult separations. For relatively easy separations, savings for constant-distillate-composition control or optimal-reflux-ratio control may not be justified over the use of the simpler constant-reflux-ratio control.

Determination of optimal-reflux-ratio policy for complex operations requires a much different approach than that used in solving simpler optimization problems, which involve finding the optimal discrete value or set of values that will minimize or maximize some objective with respect to some algebraic function(s). For example, in Chapter 7, a single value of the optimal reflux ratio for a continuous distillation operation is found by plotting, as in Figure 7.22, the total annualized cost versus the reflux ratio, R, and locating the minimum in the curve. The optimal value is found to be approximately 1.24, which corresponds to $R/R_{\min} = 1.1$. In this section, we discuss establishing the optimal reflux ratio as a function of time, $R\{t\}$, for a batch distillation, which is modeled with differential or integral equations rather than algebraic equations. This type of problem requires optimal-control methods that include the calculus of variations, the maximum principle of Pontryagin, dynamic programming of Bellman, and nonlinear programming. Diwekar [33] describes these methods in detail. Their development by mathematicians in Russia and the United States were essential for the success of their respective space programs.

To illustrate one of the approaches to optimal control, consider the classic *Brachistochrone* (Greek for "shortest time") *problem* of Johann Bernoulli, one of the earliest variational problems, whose investigation by famous mathematicians, including Johann and Jakob Bernoulli, Gottfried Leibnitz, Guillaume de L'Hospital, and Isaac Newton, was the starting point for the development of the *calculus of variations,* a subject considered in detail by Weinstock [34]. A particle, e.g., a bead, is located in the x-y plane at (x_1, y_1), where the x-axis is horizontal to the right, while the y-axis is vertically downward. The problem is to find the frictionless path, $y = f\{x\}$, ending at the point (x_2, y_2), down which the particle will move, subject only to gravity, in the least time. Some possible paths from point 1 to point 2, shown in Figure 13.15, include a straight line, a circular arc, and a broken line consisting of two connected straight lines (one steep followed by one shallow). The shortest distance is the straight line, but Galileo Galilei proposed that the path of shortest

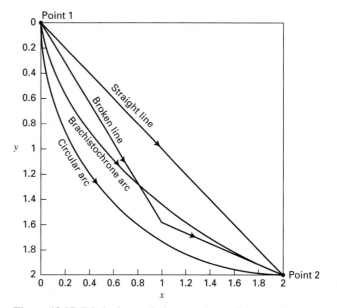

Figure 13.15 Frictionless paths between two points.

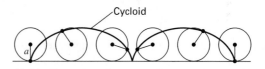

Figure 13.16 Generation of a cycloid from a circle of radius a.

time was the circular arc. However, the other aforementioned mathematicians proved that the solution is the arc of a cycloid, which, as included in Figure 13.15, is the locus of a point on the rim of a circle of radius a rolling along a straight line, as generated in Figure 13.16. The cycloid is given in parametric form as

$$x = a(\theta - \sin \theta) \quad \text{and} \quad y = a(1 - \cos \theta) \quad (13\text{-}59)$$

By eliminating θ, the Cartesian equation for the cycloid is

$$x = a \cos^{-1}(1 - y/a) - (2ay - y^2)^{0.5} \quad (13\text{-}60)$$

This optimal solution of the Brachistochrone problem is obtained by the calculus of variations as follows.

Let the arc length along the path be s. Then a differential length along the arc is the hypotenuse of a differential triangle, such that

$$ds = [(dy)^2 + (dx)^2]^{0.5} \quad (13\text{-}61)$$

or

$$ds = dx \sqrt{\left(\frac{dy}{dx}\right)^2 + 1} \quad (13\text{-}62)$$

The time, t_{12}, for the particle to travel from point 1 (P$_1$) to point 2 (P$_2$) is given by

$$t_{12} = \int_{P_1}^{P_2} \frac{ds}{v} \quad (13\text{-}63)$$

where v is the speed of the particle. By the conservation of energy, as the particle descends, its kinetic energy will increase as its potential energy decreases. Thus, if m is the mass of the particle and g is the acceleration due to gravity, where the velocity increases as the downward distance increases:

$$\frac{1}{2}mv^2 = mgy \qquad (13\text{-}64)$$

Solving for v,

$$v = \sqrt{2\,gy} \qquad (13\text{-}65)$$

Substituting (13-62) and (13-65) into (13-63) gives

$$t_{12} = \int_{P_1}^{P_2} \sqrt{\frac{\left(\dfrac{dy}{dx}\right)^2 + 1}{2\,gy}}\, dx = \int_{P_1}^{P_2} \sqrt{\frac{1 + (y')^2}{2\,gy}}\, dx \qquad (13\text{-}66)$$

where, $y' = dy/dx$. Thus, the function to be minimized is

$$f = \sqrt{\frac{1 + (y')^2}{2\,gy}} \qquad (13\text{-}67)$$

Equation (13.67) is of the following general form of a problem that can be solved by the calculus of variations:

Minimize the integral,

$$I = \int_1^2 f(x, y\{x\}, y'\{x\})\, dx \qquad (13\text{-}68)$$

Weinstock shows that a necessary condition for the solution of (13-68) is the following Euler-Lagrange equation:

$$\frac{\partial f}{\partial y} - \frac{d}{dx}\left(\frac{\partial f}{\partial y'}\right) = 0 \qquad (13\text{-}69)$$

There are two special cases of (13-69), resulting in the following simplifications:

1. If f is explicitly independent of y, then,

$$\frac{\partial f}{\partial y'} = C_1 \qquad (13\text{-}70)$$

2. If f is explicitly independent of x, then

$$y'\left(\frac{\partial f}{\partial y'}\right) - f = C_2 \qquad (13\text{-}71)$$

The Brachistochrone function of (13-67) is explicitly independent of x, so that (13-71) applies to give

$$(y')^2[1 + (y')^2]^{-\frac{1}{2}}(2\,gy)^{-\frac{1}{2}} \\ - [1 + (y')^2]^{\frac{1}{2}}(2\,gy)^{-\frac{1}{2}} = C_2 \qquad (13\text{-}72)$$

which simplifies to

$$\left[1 + \left(\frac{dy}{dx}\right)^2\right] y = \frac{1}{2\,gC_2^2} = 2a \qquad (13\text{-}73)$$

If point 1 is located at $x = 0$, $y = 0$, the solution of (13-73) is (13-60), which is the arc of a cycloid.

How much better is the cycloid-arc path compared to the other paths shown in Figure 13.15? If point 2 is taken at $x = 2$ ft and $y = 2$ ft, then with $g = 32.17$ ft/s^2, the calculated travel times, from the application of (13-66) to move from point 1 to point 2 are as follows, where the cycloid arc is just slightly better than a circular arc:

Path in Figure 13.15	Travel time, seconds
Straight line	0.498
Broken line	0.472
Circular arc of radius = 2 ft	0.460
Cycloid arc with $a = 1.145836$ ft	0.455

Application of the calculus of variations to the determination of the optimal-reflux-control strategy for batch distillation is carried out in a manner similar to the above for the Brachistochrone problem. For example, if it is desired to find the reflux-ratio, R, path that will minimize the time, t, required to obtain an accumulated distillate of given amount and composition for a fixed boilup rate, V, using a column with N equilibrium stages, the integral to be minimized is as follows, where the variables for the distillate are replaced by the variables for the residual, W, remaining in the still:

$$t = \int_{x_{W_0}}^{x_W} \frac{W}{V}\left(\frac{R+1}{y_N - x_W}\right) dx_W \qquad (13\text{-}74)$$

SUMMARY

1. The simplest case of batch distillation corresponds to the condensation of a vapor rising from a boiling liquid, called differential or Rayleigh distillation. The vapor leaving the liquid surface is assumed to be in equilibrium with the liquid. The compositions of the liquid and vapor continually change as distillation proceeds. The instantaneous vapor and liquid compositions can be computed as a function of time for a given vaporization rate.

2. A batch-rectifier system consists of a reboiler, a column with plates or packing that sits on top of the reboiler, a condenser, a reflux drum, and one or more distillate receivers.

3. For a binary system, a batch rectifier is usually operated at a constant reflux ratio or at a constant distillate composition. For either case, a McCabe–Thiele diagram can be used to follow the progress of the rectification, if the assumptions of constant molar overflow and negligible tray (or packing), condenser, and reflux drum liquid holdups are made.

4. A batch stripper is useful for removing small quantities of impurities from a charge. For complete flexibility, complex batch distillation involving both rectification and stripping can be employed.

5. Liquid holdup on the trays (or packing) and in the condenser and reflux drum can influence the course of batch rectification and the size and composition of the distillate cuts. The complexity of the holdup effect is such that it is best determined by rigorous calculations for specific cases.

6. For multicomponent, batch rectification, with negligible liquid holdup except in the reboiler, the shortcut method of Sundaram and Evans, based on successive applications of the Fenske–Underwood–Gilliland (FUG) method at a sequence of time intervals, can be used to obtain approximate distillate and charge compositions and amounts as a function of time.

7. For accurate and detailed multicomponent, batch-rectification compositions, the model of Distefano as implemented by Boston et al. should be used. It accounts for liquid holdup and permits a sequence of operation steps to produce multiple distillate cuts. The model consists of algebraic and ordinary differential equations, which, when stiff, are best solved by Gear-type implicit-integration methods. The Distefano model can also be solved by the method of Galindez and Fredenslund, which simulates the unsteady batch process by a succession of steady states of short duration, which are solved by either the Newton–Raphson or the inside-out methods of Chapter 10.

8. Two difficult aspects of batch distillation are (1) determination of the best set of operations for the production of the desired products and (2) the optimal control method to be used. The first, which involves the possibility that slop or waste cuts may be necessary, is solved by computational studies using a simulation program. The second, which requires consideration of the best reflux-rate policy, is solved by applying optimal-control methods.

REFERENCES

1. RAYLEIGH, J.W.S., *Phil. Mag. and J. Sci.*, Series 6, **4** (23), 521–537 (1902).

2. SMOKER, E.H., and A. ROSE, *Trans. AIChE*, **36**, 285–293 (1940).

3. BLOCK, B., *Chem. Eng.*, **68** (3), 87–98 (1961).

4. BOGART, M.J.P., *Trans. AIChE*, **33**, 139–152 (1937).

5. ELLERBE, R.W., *Chem. Eng.*, **80** (12), 110–116 (1973).

6. HASBE, S., B.B. ABDUL AZIZ, I. HASHIMOTO, and T. WATANABE, *Proc. IFAC Workshop, London, Sept. 7–8, 1992*, p. 177.

7. DIWEKAR, U.M., and K.P. MADHAVEN, *Ind. Eng. Chem. Res.*, **30**, 713–721 (1991).

8. SUNDARAM, S., and L.B. EVANS, *Ind. Eng. Chem. Res.*, **32**, 511–518 (1993).

9. EDULJEE, H.E., *Hydrocarbon Processing*, **56** (9), 120–122 (1975).

10. MEADOWS, E.L., *Chem. Eng. Progr. Symp. Ser. No. 46*, **59**, 48–55 (1963).

11. DISTEFANO, G.P., *AIChE J.*, **14**, 190–199 (1968).

12. BOSTON, J.F., H.I. BRITT, S. JIRAPONGPHAN, and V.B. SHAH, in *Foundations of Computer-Aided Chemical Process Design*, Vol. II, AIChE, R.H.S. Mah and W.D. Seider, Eds., pp. 203–237 (1981).

13. GALINDEZ, H., and A. FREDENSLUND, *Comput. Chem. Eng.*, **12**, 281–288 (1988).

14. GEAR, C.W., *Numerical Initial Value Problems in Ordinary Differential Equations*, Prentice-Hall, Englewood Cliffs, NJ (1971).

15. DAVIS, M.E., *Numerical Methods and Modeling for Chemical Engineers*, John Wiley and Sons, New York (1984).

16. RIGGS, J.B., *An Introduction to Numerical Methods for Chemical Engineers*, Texas Tech. Univ. Press, Lubbock, TX (1988).

17. CARNAHAN, B., and J.O. WILKES, "Numerical solution of differential equations—an overview," in *Foundations of Computer-Aided Chemical Process Design*, R.S.H. Mah and W.D. Seider, Eds., Engineering Foundation, New York, Vol. I, pp. 225–340 (1981).

18. VARGA, R.S., *Matrix Iterative Analysis*, Prentice-Hall, Englewood Cliffs, NJ (1962).

19. LUYBEN, W.L., *Ind. Eng. Chem. Res.*, **27**, 642–647 (1988).

20. MAYUR, D.N., R.A. MAY, and R. JACKSON, *Chem. Eng. Journal*, **1**, 15–21 (1970).

21. QUINTERO-MARMOL, E., and W.L. LUYBEN, *Ind. Eng. Chem. Res.*, **29**, 1915–1921 (1990).

22. FARHAT, S., M. CZERNICKI, L. PIBOULEAU, and S. DOMENECH, *AIChE J.*, **36**, 1349–1360 (1990).

23. MUJTABA, I.M., and S. MACCHIETTO, *Comput. Chem. Eng.*, **16**, S273–S280 (1992).

24. DIEHL, M., A. SCHAFER, H.G. BOCK, J.P. SCHLODER, and D.B. LEINEWEBER, *AIChE J.*, **48**, 2869–2874 (2002).

25. ROBINSON, E.R., *Chem. Eng. Journal*, **2**, 135–136 (1971).

26. DIWEKAR, U.M., *Batch Distillation—Simulation, Optimal Design and Control*, Taylor & Francis, Washington, D.C. (1995).

27. CONVERSE, A.O., and G.D. GROSS, *Ind. Eng. Chem. Fundamentals*, **2**, 217–221 (1963).

28. COWARD, I., *Chem. Eng. Science*, **22**, 503–516 (1967).

29. COWARD, I., *Chem. Eng. Science*, **22**, 1881–1884 (1967).

30. ROBINSON, E.R., I. COWARD, *Chem. Eng. Science*, **24**, 1661–1668 (1969).

31. MAYUR, D.N., and R. JACKSON, *Chem. Eng. Journal*, **2**, 150–163 (1971).

32. DIWEKAR, U.M., R.K. MALIK, and K.P. MADHAVAN, *Comput. Chem. Eng.*, **11**, 629–637 (1987).

33. DIWEKAR, U.M., *Introduction to Applied Optimization*, Kluwer Academic Publishers (2003).

34. WEINSTOCK, R., *Calculus of Variations*, McGraw-Hill Book Co., Inc., New York (1952).

EXERCISES

Section 13.1

13.1 (a) A bottle of pure *n*-heptane is accidentally poured into a drum of pure toluene in a commercial laboratory. One of the laboratory assistants, with almost no background in chemistry, suggests that, since heptane boils at a lower temperature than toluene, the following purification procedure can be used:

Pour the mixture (2 mol% *n*-heptane) into a simple still pot. Boil the mixture at 1 atm and condense the vapors until all heptane is boiled away. Obtain the pure toluene from the residue in the still pot.

You, being a chemical engineer, immediately realize that such a purification method will not work. Indicate this by a curve showing the composition of the material remaining in the pot after various quantities of the liquid have been distilled. What is the composition of the residue after 50 wt% of the original material has been distilled? What is the composition of the cumulative distillate?

(b) When one-half of the heptane has been distilled, what is the composition of the cumulative distillate and of the residue? What weight percent of the original material has been distilled?

Vapor–liquid equilibrium data at 1 atm [*Ind. Eng. Chem.*, **42**, 2912 (1949)] are as follows:

Mole Fraction *n*-Heptane			
Liquid	Vapor	Liquid	Vapor
0.025	0.048	0.448	0.541
0.062	0.107	0.455	0.540
0.129	0.205	0.497	0.577
0.185	0.275	0.568	0.637
0.235	0.333	0.580	0.647
0.250	0.349	0.692	0.742
0.286	0.396	0.843	0.864
0.354	0.454	0.950	0.948
0.412	0.504	0.975	0.976

13.2   A mixture of 40 mol% isopropanol in water is to be distilled at 1 atm by a simple batch distillation until 70 mol% of the charge has been vaporized (equilibrium data are given in Exercise 7.33). What will be the compositions of the liquid residue remaining in the still pot and of the collected distillate?

13.3   A 30 mol% feed of benzene in toluene is to be distilled in a batch operation. A product having an average composition of 45 mol% benzene is to be produced. Calculate the amount of residue, assuming $\alpha = 2.5$ and $W_0 = 100$.

13.4   A charge of 250 lb of 70 mol% benzene and 30 mol% toluene is subjected to batch, differential distillation at atmospheric pressure. Determine the compositions of the distillate and residue after one-third of the original mass is distilled off. Assume the mixture forms an ideal solution and apply Raoult's and Dalton's laws with vapor-pressure data.

13.5   A mixture containing 60 mol% benzene and 40 mol% toluene is subjected to batch, differential distillation at 1 atm, under three different conditions:

1. Until the distillate contains 70 mol% benzene
2. Until 40 mol% of the feed is evaporated
3. Until 60 mol% of the original benzene leaves in the vapor phase

Using $\alpha = 2.43$, determine for each of the three cases:

(a) The number of moles in the distillate for 100 mol of feed.

(b) The compositions of distillate and residue.

13.6   A mixture consisting of 15 mol% phenol in water is to be batch distilled at 260 torr. What fraction of the original batch remains in the still when the total distillate contains 98 mol% water? What is the residue concentration?

Vapor–liquid equilibrium data at 260 torr [*Ind. Eng. Chem.*, **17**, 199 (1925)]:

x, wt% (H_2O):

  1.54 4.95 6.87 7.73 19.63 28.44 39.73 82.99 89.95 93.38 95.74

y, wt% (H_2O):

  41.10 79.72 82.79 84.45 89.91 91.05 91.15 91.86 92.77 94.19 95.64

13.7   A still is charged with 25 mol of a mixture of benzene and toluene containing 0.35 mole fraction benzene. Feed of the same composition is supplied at a rate of 7 mol/h, and the heat rate is adjusted so that the liquid level in the still remains constant. No liquid leaves the still pot, and $\alpha = 2.5$. How long will it be before the distillate composition falls to 0.45 mole fraction benzene?

13.8   A distillation system consisting of a reboiler and a total condenser (no column) is to be used to separate A and B from a trace of nonvolatile material. The reboiler initially contains 20 lbmol of feed of 30 mol% A. Feed is to be supplied to the reboiler at the rate of 10 lbmol/h, and the heat input is so adjusted that the total moles of liquid in the reboiler remains constant at 20. No residue is withdrawn from the still. Calculate the time required for the composition of the overhead product to fall to 40 mol% A. The relative volatility may be assumed constant at 2.50.

Section 13.2

13.9   Repeat Exercise 13.2 for the case of a batch distillation carried out in a two-stage column with a reflux ratio of $L/V = 0.9$.

13.10   Repeat Exercise 13.3 assuming the operation is carried out in a three-stage column with $L/V = 0.6$.

13.11   One kilomole of an equimolar mixture of benzene and toluene is fed to a batch still containing three equivalent stages (including the boiler). The liquid reflux is at its bubble point, and $L/D = 4$. What is the average composition and amount of product at a time when the instantaneous product composition is 55 mol% benzene? Neglect holdup, and assume $\alpha = 2.5$.

13.12   The fermentation of corn produces a mixture of 3.3 mol% ethyl alcohol in water. If 20 mol% of this mixture is distilled at 1 atm by a simple, batch distillation, calculate and plot the instantaneous-vapor composition as a function of mole percent of batch distilled. If reflux with three theoretical stages (including the reboiler) is used, what is the maximum purity of ethyl alcohol that can be produced by batch distillation?

Equilibrium data are given in Exercise 7.29.

Section 13.3

13.13   An acetone–ethanol mixture of 0.5 mole fraction acetone is to be separated by batch distillation at 101 kPa.

Vapor–liquid equilibrium data at 101 kPa are as follows:

Mole Fraction Acetone									
y	0.16	0.25	0.42	0.51	0.60	0.67	0.72	0.79	0.87	0.93
x	0.05	0.10	0.20	0.30	0.40	0.50	0.60	0.70	0.80	0.90

(a) Assuming an L/D of 1.5 times the minimum, how many stages should this column have if we want the composition of the distillate to be 0.90 mole fraction acetone at a time when the residue contains 0.1 mole fraction acetone?

(b) Assume the column has eight stages and the reflux rate is varied continuously so that the top product is maintained constant at 0.9 mole fraction acetone. Make a plot of the reflux ratio versus the still-pot composition and the amount of liquid left in the still.

(c) Assume now that the same distillation is carried out at constant reflux ratio (and varying product composition). We wish to have a residue containing 0.1 and an (average) product containing 0.9 mole fraction acetone, respectively. Calculate the total vapor generated. Which method of operation is more energy intensive? Can you suggest operating policies other than constant reflux ratio and constant distillate compositions that might lead to equipment and/or operating cost savings?

13.14   A total of 2,000 gallons of 70 wt% ethanol in water, having a specific gravity of 0.871, is to be separated at 1 atm in a batch rectifier operating at constant distillate composition with a constant molar vapor boilup rate to obtain a distillate product of 85 mol% ethanol and a residual waste water containing 3 wt% ethanol. If the task is to be completed in 24 h, allowing 4 h for charging, start-up, shutdown, and cleaning, determine: (a) the number of theoretical plates required in the column, (b) the reflux ratio when the concentration of ethanol in the pot is 25 mol%, (c) the instantaneous distillate rate in lbmol/h when the concentration of ethanol in the pot is 15 mol%, (d) the lbmol of distillate product, and (e) the lbmol of residual wastewater.

Vapor–liquid equilibrium data are given in Exercise 7.29.

13.15   A charge of 1,000 kmol of a mixture of 20 mol% ethanol in water is to undergo batch rectification at 101.3 kPa at a vapor boilup rate of 100 kmol/h. If the column has the equivalent of six theoretical plates and the distillate composition is to be maintained at 80 mol% ethanol by varying the reflux ratio, determine: (a) the time in hours for the residue to reach an ethanol mole fraction of 0.05, (b) the kmol of distillate obtained when the condition of part (a) is achieved, (c) the minimum- and maximum-reflux ratios during the rectification period, and (d) the variation of the distillate rate in kmol/h during the rectification period. Assume constant molar overflow, neglect liquid holdup, and obtain vapor–liquid equilibrium data from Exercise 7.29.

13.16   A 500 lbmol mixture of 48.8 mol% A and 51.2 mol% B with a relative volatility $\alpha_{A,B}$ of 2.0 is to be separated in a batch rectifier consisting of a total condenser, a column with seven theoretical stages, and a partial reboiler. The reflux ratio is to be varied so as to maintain the distillation composition constant at 95 mol% A. The column can operate satisfactorily with a molar vapor boilup rate of 213.5 lbmol/h. The rectification is to be stopped when the mole fraction of A in the still drops to 0.192. Determine: (a) the time required for rectification, and (b) the total amount of distillate produced.

Section 13.4

13.17   Develop a procedure similar to that of Section 13.2 to calculate a binary batch stripping operation using the equipment arrangement of Figure 13.8.

13.18   A three-theoretical-stage batch stripper (one stage is the reboiler) is charged to the feed tank (see Figure 13.8) with 100 kmol of 10 mol% n-hexane in n-octane mix. The boilup rate is 30 kmol/h. If a constant boilup ratio (V/L) of 0.5 is used, determine the

instantaneous-bottoms composition and the composition of the accumulated bottoms product at the end of 2 h of operation.

13.19   Develop a procedure similar to that of Section 13.2 to calculate a complex, binary, batch-distillation operation using the equipment arrangement of Figure 13.9.

Section 13.5

13.20   For a batch rectifier with appreciable column holdup:

(a) Why is the composition of the charge to the still higher in the light component than the still composition at the start of rectification, assuming that total-reflux conditions are established before rectification begins?

(b) Why will separation be more difficult than with zero holdup?

13.21   For a batch rectifier with appreciable column holdup, why do tray compositions change less rapidly compared to a rectifier with negligible column holdup, and why is the degree of separation improved?

13.22   Based on the statements in Exercises 13.20 and 13.21, why is it difficult to predict the effect of holdup?

Section 13.6

13.23   Use the shortcut method of Sundaram and Evans to solve Example 13.7, but with zero condenser and stage holdups.

13.24   A charge of 100 kmol of an equimolar mixture of A, B, and C, with $\alpha_{A,B} = 2$ and $\alpha_{A,C} = 4$, is distilled in a batch rectifier containing the equivalent of four theoretical stages, including the reboiler. If holdup can be neglected, use the shortcut method with $R = 5$ and $V = 100$ kmol/h to estimate the variation of the still and instantaneous-distillate compositions as a function of time following a start-up period during which total reflux conditions are established.

13.25   A charge of 200 kmol of a mixture of 40 mol% A, 50 mol% B, and 10 mol% C with $\alpha_{A,C} = 2.0$ and $\alpha_{B,C} = 1.5$ is to be separated in a batch rectifier with a total of three theoretical stages and operating at a reflux ratio of 10, with a molar vapor boilup rate of 100 kmol/h. Holdup is negligible. Use the shortcut method to estimate instantaneous-distillate and bottoms compositions as a function of time for the first hour of operation following start-up to achieve total reflux conditions.

Section 13.7

13.26   A charge of 100 lbmol of 35 mol% n-hexane, 35 mol% n-heptane, and 30 mol% n-octane is to be distilled at 1 atm in a batch rectifier, consisting of a partial reboiler, a column, and a total condenser, at a constant boilup rate of 50 lbmol/h and a constant reflux ratio of 5. Before rectification begins, total reflux conditions are to be established. Then, the following three operation steps are to be carried out to obtain an n-hexane-rich cut, an intermediate cut for recycle, an n-heptane-rich cut, and an n-octane-rich residue:

Step 1:   Stop when the accumulated distillate purity drops below 95 mol% n-hexane.

Step 2:   Empty the n-hexane-rich cut produced in Step 1 into a receiver and resume rectification until the instantaneous distillate composition reaches 80 mol% n-heptane.

Step 3:   Empty the intermediate cut produced in Step 2 into a receiver and resume rectification until the accumulated distillate composition reaches 4 mol% n-octane.

For thermodynamic properties, assume ideal solutions and the ideal gas law.

Consider conducting the rectification in two different columns, each with the equivalent of 10 theoretical stages and a condenser-reflux drum liquid holdup of 1.0 lbmol. For each column, determine with a suitable batch-distillation computer program the compositions and amounts in lbmol of each of the four products.

Column 1: A plate column with a total liquid holdup of 8 lbmol.

Column 2: A packed column with a total liquid holdup of 2 lbmol.

Discuss the effect of liquid holdup for the two columns. Are the results what you expected?

13.27 A charge of 100 lbmol of a hydrocarbon mixture containing 10 mol% propane, 30 mol% n-butane, 10 mol% n-pentane, and the balance n-hexane is to be separated in a batch rectifier equipped with a partial reboiler, a total condenser with a liquid holdup of 1.0 ft^3, and a column with the equivalent of eight theoretical stages and a total holdup of 0.80 ft^3. The pressure in the condenser is 50.0 psia and the column pressure drop is 2.0 psi. The rectification campaign or operating policy, given as follows, is designed to produce cuts of 98 mol% propane, 99.8 mol% n-butane, and a residual cut of 99 mol% n-hexane, and two intermediate cuts, one of which may be a relatively rich cut of n-pentane. All five operating steps are conducted at a molar vapor boilup rate of 40 lbmol/h. Use a suitable batch-distillation computer program to determine the amounts and compositions of all cuts.

Step	Reflux Ratio	Stop Criterion
1	5	98% propane in accumulator
2	20	95% n-butane in instantaneous distillate
3	25	99.8% n-butane in accumulator
4	15	80% n-pentane in instantaneous distillate
5	25	99% n-hexane in the pot

Make suggestions as to how you might alter the operation steps so as to obtain larger amounts of the product cuts and smaller amounts of the intermediate cuts.

13.28 A charge of 100 lbmol of benzene (B), monochlorobenzene (MCB), and o-dichlorobenzene (DCB) is being distilled in a batch rectifier that consists of a total condenser, a column with 10 theoretical stages, and a partial reboiler. Following the establishment of total reflux, the first operation step begins for a boilup rate of 200 lbmol/h and a reflux ratio of about 3. At the end of 0.60 h, the following conditions exist for the top three stages in the column:

	Top Stage	Stage 2	Stage 3
Temperature, °F	267.7	271.2	272.5
V, lbmol/h	206.1	209.0	209.5
L, lbmol/h	157.5	158.0	158.1
M, lbmol	0.01092	0.01088	0.01087

Vapor Mole Fractions:

B	0.0994	0.0449	0.0331
MCB	0.9006	0.9551	0.9669
DCB	0.0000	0.0000	0.0000

Liquid Mole Fractions:

B	0.0276	0.0121	0.00884
MCB	0.9724	0.9879	0.99104
DCB	0.0000	0.0000	0.00012

In addition reboiler and condenser holdups at 0.6 h are 66.4 and 0.1113 lbmol, respectively.

For benzene, use the preceding data with (13-36) and (13-39) to estimate the liquid-phase mole fraction of benzene leaving Stage 2 at 0.61 h by using the explicit-Euler method with a Δt of 0.01 h. If the result is unreasonable, explain why with respect to stability and stiffness considerations.

13.29 A mixture of 100 kmoles of 30 mol% methanol, 30 mol% ethanol, and 40 mol% n-propanol is charged at a pressure of 120 kPa to a batch rectifier, consisting of a partial reboiler, a column containing the equivalent of 10 equilibrium stages, and a total condenser. After establishing a total-reflux condition, the column will begin a sequence of two operating steps, each for a duration of 15 hours at a distillate flow rate of 2 kmol/h and a reflux ratio of 10. Thus, the two accumulated distillates will equal in moles that of the methanol and that of the ethanol in the feed. Neglect the liquid holdup in the condenser and the column. The column pressure drop is 8 kPa, with a pressure drop of 2 kPa through the condenser. Using a process-simulation program with the UNIFAC method for liquid-phase activity coefficients, determine the mole-fraction composition and amount in kmoles of each of the three cuts.

13.30 Repeat Exercise 13.29 with the following modifications. Add a third operating step. For all three steps, use the same distillate rate and reflux rate as in Exercise 13.29. Use the following durations for the three steps: 13 hours for Step 1, 4 hours for Step 2, and 13 hours for Step 3. The distillate from Step 2 will be a slop cut. Determine the mole-fraction composition and amount in kmoles of each of the four cuts.

13.31 A mixture of 100 kmoles of 45 mol% acetone, 30 mol% chloroform, and 25 mol% benzene is charged at pressure of 101.3 kPa to a batch rectifier, consisting of a partial reboiler, a column containing the equivalent of 10 equilibrium stages, and a total condenser. After establishing a total-reflux condition, the column will begin a sequence of two operating steps, each at a distillate flow rate of 2 kmol/h and a reflux ratio of 10. The durations will be 13.3 hours for Step 1 and 24.2 hours for Step 2. Neglect pressure drops and the liquid holdup in the condenser and the column. Using a process-simulation program with the UNIFAC method for liquid-phase activity coefficients, determine the mole-fraction composition and amount in kmoles of each of the three cuts.

Section 13.8

13.32 Using a batch-distillation simulation program, make the following modifications to the C6-C7-C8 ternary example in Section 13.8:

(a) Increase the reflux above 8 to eliminate the second slop cut.

(b) Change the termination specification on the second step to reduce the amount of the first slop cut, without failing to meet all three product specifications.

Part 3

Separations by Barriers and Solid Agents

In recent years, the number of industrial applications of separations using barriers and solid agents have greatly increased because of progress in producing selective membranes and adsorbents. Chapter 14 presents a discussion of rates of mass transfer through membranes and calculation methods for the more widely used continuous membrane separations for gas and liquid feeds. These include gas permeation, reverse osmosis, dialysis, electrodialysis, pervaporation, ultrafiltration, and microfiltration.

Chapter 15 is concerned with separations of adsorption, ion exchange and chromatography that use solid agents. Discussions of equilibrium and rates of mass transfer in porous adsorbents are followed by calculation methods for batch and continuous equipment for liquid and gaseous feeds, including fixed-bed, pressure-swing, and simulated-moving-bed adsorption.

Chapter 14

Membrane Separations

In a membrane-separation process, a feed consisting of a mixture of two or more components is partially separated by means of a semipermeable barrier (the membrane) through which one or more species move faster than another or other species. The most general membrane process is shown in Figure 14.1 where the feed mixture is separated into a *retentate* (that part of the feed that does not pass through the membrane, i.e., is retained) and a *permeate* (that part of the feed that does pass through the membrane). Although the feed, retentate, and permeate are usually liquid or gas, they may also be solid. The barrier is most often a thin, nonporous, polymeric film, but may also be porous polymer, ceramic, or metal materials, or even a liquid or gas. The barrier must not dissolve, disintegrate, or break. The optional sweep, shown in Figure 14.1, is a liquid or gas, used to help remove the permeate. Many of the industrially important membrane-separation operations are listed in Tables 1.2 and 14.1.

In membrane separations: (1) the two products are usually miscible, (2) the separating agent is a semipermeable barrier, and (3) a sharp separation is often difficult to achieve. Thus, membrane separations differ in two or three of these respects from the more common separation operations of absorption, stripping, distillation, and liquid–liquid extraction.

Although membranes as separating agents have been known for more than 100 years [1], large-scale applications have only appeared in the past 50 years. In the 1940s, porous fluorocarbons were used to separate $^{235}UF_6$ from $^{238}UF_6$ [2]. In the mid-1960s, reverse osmosis with cellulose acetate was first used to desalinize seawater to produce potable water (drinkable water with less than 500 ppm by weight of dissolved solids) [3]. Commercial ultrafiltration membranes followed in the 1960s. In 1979, Monsanto Chemical Company introduced a hollow-fiber membrane of polysulfone to separate certain gas mixtures—for example, to enrich hydrogen- and carbon dioxide-containing streams [4]. Commercialization of alcohol dehydration by pervaporation began in the late 1980s, as did the large-scale application of emulsion liquid membranes for removal of metals and organics from wastewater.

The replacement of the more-common separation operations with membrane separations has the potential to save large amounts of energy. This replacement requires the production of high-mass-transfer-flux, defect-free, long-life membranes on a large scale and the fabrication of the membrane into compact, economical modules of high surface area per unit volume.

14.0 INSTRUCTIONAL OBJECTIVES

After completing this chapter, you should be able to:

- Explain membrane processes in terms of the membrane, feed, sweep, retentate, and permeate.
- List eight types of industrial membrane-separation processes.
- Discuss industrial, polymeric-membrane materials.
- Differentiate between the membrane mass-transfer parameters of permeability and permeance, and explain their relationship to mass-transfer coefficients.
- Differentiate between asymmetric and thin-layer composite membranes, and between dense and microporous membranes.
- Describe four membrane shapes.
- Describe six membrane modules.
- Explain mechanisms of mass transfer through membranes.
- Derive mass-transfer-rate equations for the solution-diffusion mechanism for liquid and gas mixtures.
- Explain four common idealized flow patterns in membrane modules.
- Explain use and advantages/disadvantages of recycle cascades of membrane modules.
- Explain concentration polarization.

- Calculate mass-transfer rates for dialysis and electrodialysis.
- Explain osmosis and how reverse osmosis can be achieved.
- Calculate mass-transfer rates for reverse osmosis.
- Calculate mass-transfer rates for gas permeation.
- Calculate mass-transfer rates for pervaporation.
- Calculate mass-transfer rates for ultrafiltration and microfiltration.

Industrial Example

A common large-scale application of membranes is to the separation of hydrogen from methane. Following World War II, during which large amounts of toluene were required to produce TNT (trinitrotoluene) explosives, petroleum refiners sought other markets for toluene. One potential market was the use of toluene as a feedstock for the manufacture of benzene, a precursor for nylon, and xylenes, precursors for a number of other chemicals, including polyesters. Toluene can be catalytically disproportionated to benzene and mixed xylenes in an adiabatic reactor with the feed entering at 950°F and a pressure greater than 500 psia. The main reaction is

$$2C_7H_8 \rightarrow C_6H_6 + C_8H_{10} \text{ isomers}$$

To suppress the formation of coke, which fouls the catalyst, the reactor feed must contain a substantial fraction of hydrogen at a partial pressure of at least 215 psia. Unfortunately, the hydrogen takes part in a side reaction for the hydrodealkylation of toluene to benzene and methane:

$$C_7H_8 + H_2 \rightarrow C_6H_6 + CH_4$$

Makeup hydrogen is usually not pure, but contains perhaps 15 mol% methane and 5 mol% ethane. Thus, typically, the reactor effluent contains H_2, CH_4, C_2H_6, C_6H_6, unreacted C_7H_8, and C_8H_{10} isomers. As shown in Figure 14.2a, for just the reaction section of the process, this effluent is cooled and partially condensed to 100°F at a pressure of 465 psia. At these conditions, a reasonably good separation is achieved between C_2H_6 and C_6H_6 in the flash drum. Thus, the vapor leaving the flash drum contains most of the H_2, CH_4, and C_2H_6, with most of the aromatic chemicals leaving in the

Table 14.1 Industrial Applications of Membrane Separation Processes

1. Reverse osmosis:
   - Desalinization of brackish water
   - Treatment of wastewater to remove a wide variety of impurities
   - Treatment of surface and ground water
   - Concentration of foodstuffs
   - Removal of alcohol from beer and wine
2. Dialysis:
   - Separation of nickel sulfate from sulfuric acid
   - Hemodialysis (removal of waste metabolites, excess body water, and restoration of electrolyte balance in human blood)
3. Electrodialysis:
   - Production of table salt from seawater
   - Concentration of brines from reverse osmosis
   - Treatment of wastewaters from electroplating
   - Demineralization of cheese whey
   - Production of ultrapure water for the semiconductor industry
4. Microfiltration:
   - Sterilization of drugs
   - Clarification and biological stabilization of beverages
   - Purification of antibiotics
   - Separation of mammalian cells from a liquid
5. Ultrafiltration:
   - Preconcentration of milk before making cheese
   - Clarification of fruit juice
   - Recovery of vaccines and antibiotics from fermentation broth
   - Color removal from Kraft black liquor in paper-making
6. Pervaporation:
   - Dehydration of ethanol–water azeotrope
   - Removal of water from organic solvents
   - Removal of organics from water
7. Gas permeation:
   - Separation of CO_2 or H_2 from methane and other hydrocarbons
   - Adjustment of the H_2/CO ratio in synthesis gas
   - Separation of air into nitrogen- and oxygen-enriched streams
   - Recovery of helium
   - Recovery of methane from biogas
8. Liquid membranes:
   - Recovery of zinc from wastewater in the viscose fiber industry
   - Recovery of nickel from electroplating solutions

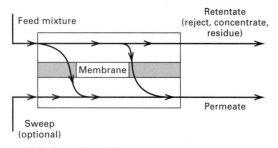

Figure 14.1 General membrane process.

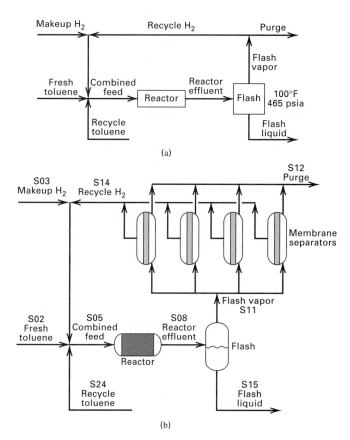

Figure 14.2 Reactor section of process to disproportionate toluene into benzene and xylene isomers. (a) Without a vapor separation step. (b) With a membrane separation step. Note: Heat exchangers, compressors, pump not shown.

liquid. Because of the large amount of hydrogen in the flash-drum vapor, it is important to recycle this stream to the reactor, rather than sending it to a flare or using it as a fuel. However, if all of the vapor were recycled, methane and ethane would build up in the recycle loop, since no other exit is provided. Before the development of acceptable membranes for the separation of H_2 from CH_4 by gas permeation, part of the vapor stream was customarily purged from the process, as shown in Figure 14.2a, to provide an exit for CH_4, and C_2H_6. With the introduction of a suitable

membrane in 1979, it became possible to apply membrane separators, as shown in Figure 14.2b.

Table 14.2 is the steady-state material balance of the reaction section of Figure 14.2b for a plant designed to process 7,750 barrels (42 gal/bbl) per operating day of fresh toluene feed. The gas permeation membrane system separates the flash vapor (stream S11) into an H_2-enriched permeate (S14, the recycled hydrogen), and a methane-enriched retentate (S12, the purge). The flash vapor to the membrane system contains 89.74 mol% H_2 and 9.26 mol% CH_4. No sweep fluid is necessary. The permeate is enriched to 94.46 mol% in H_2. The retentate is enriched in CH_4 to 31.18 mol%. The recovery of H_2 in the permeate is 90%. Thus, only 10% of the H_2 in the vapor leaving the flash drum is lost to the purge. Before entering the membrane separator system, the vapor is heated to a temperature of at least 200°F (the dew-point temperature of the retentate) at a pressure of 450 psia (heater not shown). Because the hydrogen content of the feed is reduced in passing through the membrane separator, the retentate becomes more concentrated in the heavier components. Without the heater, undesirable condensation would occur in the separator. The retentate leaves the separator at about the same temperature and pressure as that of heated flash vapor entering the separator. The permeate leaves at the much-lower pressure of 50 psia and a temperature somewhat lower than 200°F because of gas expansion.

The membrane is an aromatic polyamide polymer consisting of a 0.3-micron-thick, nonporous layer in contact with the feed, and a much-thicker porous support backing to give the membrane strength and ability to withstand the pressure differential of $450 - 50 = 400$ psi. This large pressure difference is needed to force the hydrogen through the nonporous membrane, which is in the form of a spiral-wound module made from flat membrane sheets. The average flux of hydrogen through the membrane is 40 scfh (standard ft^3/h at 60°F and 1 atm) per ft^2 of membrane surface area. From the material balance in Table 14.2, the total amount of H_2 transported through the membrane is

$$(1,685.1 \text{ lbmol/h})(379 \text{ scf/lbmol}) = 639,000 \text{ scfh}$$

Table 14.2 Material Balance for Membrane Separation Process in a Toluene Disproportionation Plant; Flow Rates in lbmol/h for Streams in Reactor Section of Figure 14.2b

Component	S02	S03	S24	S14	S05	S08	S15	S11	S12
Hydrogen		269.0		1,685.1	1,954.1	1,890.6	18.3	1,872.3	187.2
Methane		50.5		98.8	149.3	212.8	19.7	193.1	94.3
Ethane		16.8			16.8	16.8	5.4	11.4	11.4
Benzene			13.1		13.1	576.6	571.8	4.8	4.8
Toluene	1,069.4		1,333.0		2,402.4	1,338.9	1,334.7	4.2	4.2
p-Xylene			8.0		8.0	508.0	507.4	0.6	0.6
Total	1,069.4	336.3	1,354.1	1,783.9	4,543.7	4,543.7	2,457.4	2,086.3	302.4

Thus, the required membrane surface area is $639,000/40 = 16,000 \text{ ft}^2$. The membrane is packaged in pressure-vessel modules of 4,000 ft^2 each. Thus, four modules in parallel are used, as shown in Figure 14.2b. A disadvantage of the membrane separator in this application is the need to recompress the recycle hydrogen to the reactor inlet pressure. Unlike distillation, where the energy of separation is usually heat, the energy for gas permeation is the shaft work of gas compression.

Membrane separation is an emerging unit operation. Important progress is still being made in the development of efficient membrane materials and the packaging thereof for the processes listed in Table 14.1. Other novel methods for conducting separation with barriers for a wider variety of mixtures are being researched and developed. Applications covering wider ranges of temperature and types of membrane materials are being found. Already, membrane separation processes have found wide application in such diverse industries as the beverage, chemical, dairy, electronic, environmental, food, medical, paper, petrochemical, petroleum, pharmaceutical, and textile industries. Some of these applications are given in Table 1.2 and included in Table 14.1. Often, compared to other separation equipment, membrane separators are more compact, less capital intensive, and more easily operated, controlled, and maintained. However, membrane separators are usually modular in construction, with many parallel units required for large-scale applications, as contrasted with the more common separation techniques, where larger pieces of equipment are designed as plant size becomes larger.

The key to an efficient and economical membrane separation process is the membrane and the manner in which it is packaged and modularized. Desirable attributes of a membrane are (1) good permeability, (2) high selectivity, (3) chemical and mechanical compatibility with the processing environment, (4) stability, freedom from fouling, and reasonable useful life, (5) amenability to fabrication and packaging, and (6) ability to withstand large pressure differences across the membrane thickness. Research and development of membrane processes deals mainly with the discovery of suitable membrane materials and their fabrication.

This chapter discusses types of membrane materials, membrane modules, the theory of transport through membrane materials and modules, and the scale-up of membrane separators from experimental performance data. Emphasis is on dialysis, electrodialysis, reverse osmosis, gas permeation, pervaporation, and ultrafiltration, but many of the theoretical principles apply as well to emerging, but not-yet commercialized, membrane processes such as membrane distillation, membrane gas absorption, membrane stripping, membrane solvent extraction, perstraction, and facilitated transport, which are not covered here. The status of industrial membrane separation systems and directions in research to improve existing applications and make possible new applications are considered in detail by Baker et al. [5] in a study supported by the U.S. Department of Energy (DOE) and by a host of contributors in a recent handbook edited by Ho and Sirkar [6], which includes emerging membrane processes. The book, "Membrane Technology and Applications" by Baker [49], is a comprehensive treatment of theory and technology.

14.1 MEMBRANE MATERIALS

Almost all industrial membrane materials are made from natural or synthetic polymers (macromolecules). Natural polymers include wool, rubber, and cellulose. A wide variety of synthetic polymers has been developed and commercialized since 1930. Synthetic polymers are produced by polymerization of a monomer by condensation (step reactions) or addition (chain reactions), or by the copolymerization of two different monomers. The resulting polymer is categorized as having (1) a long linear chain, such as linear polyethylene; (2) a branched chain, such as polybutadiene; (3) a three-dimensional, highly cross-linked structure, such as phenol–formaldehyde; or (4) a moderately cross-linked structure, such as butyl rubber. The linear-chain polymers soften with an increase in temperature, are often soluble in organic solvents, and are referred to as *thermoplastic* polymers. At the other extreme, highly cross-linked polymers do not soften appreciably, are almost insoluble in most organic solvents, and are referred to as *thermosetting* polymers. Of more interest in the application of polymers to membranes is a classification based on the arrangement or conformation of the polymer molecules. At low temperatures, typically below 100°C, idealized polymers can be classified as *glassy* or *crystalline*. The former refers to a polymer that is brittle and glassy in appearance and lacks any crystalline structure (i.e., *amorphous*), whereas the latter refers to a polymer that is brittle, hard, and stiff, with a crystalline structure. If the temperature of a glassy polymer is increased, a point, called the *glass-transition temperature, T_g*, may be reached where the polymer becomes *rubbery*. If the temperature of a crystalline polymer is increased, a point, called the *melting temperature, T_m*, is reached where the polymer becomes a melt. However, a thermosetting polymer never melts. Many polymers have both amorphous and crystalline regions, that is, a certain degree of crystallinity that varies from 5 to 90%, making it possible for some polymers to have both a T_g and a T_m. Membranes made of glassy polymers can operate below or above T_g; membranes of crystalline polymers must

Table 14.3 Common Polymers Used in Membranes

Polymer	Type	Representative Repeat Unit	Glass Transition Temp., °C	Melting Temp., °C
Cellulose triacetate	Crystalline	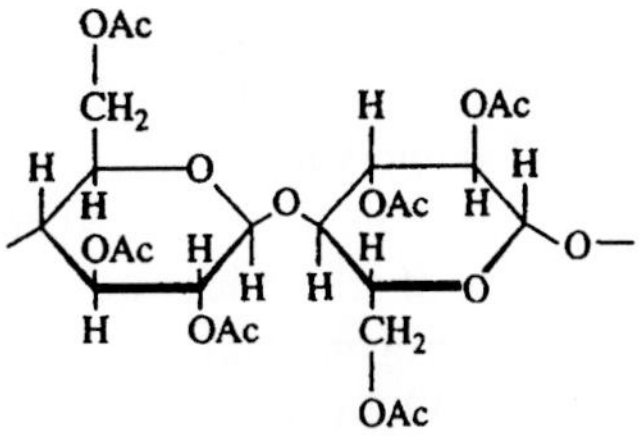		300
Polyisoprene (natural rubber)	Rubbery	$-[CH_2CH=CH_3\,|\,CCH_2]_n-$	−70	
Aromatic polyamide	Crystalline	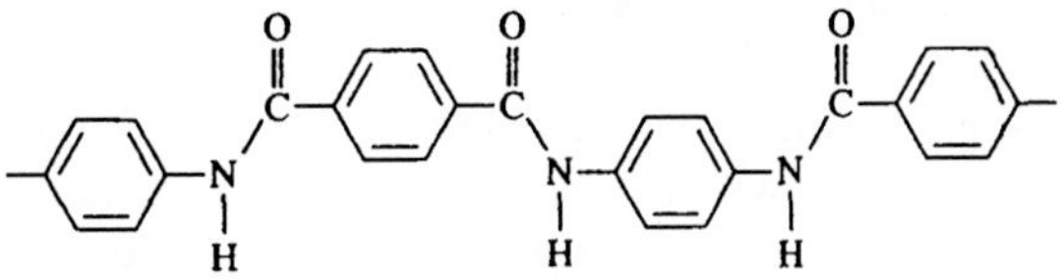		275
Polycarbonate	Glassy	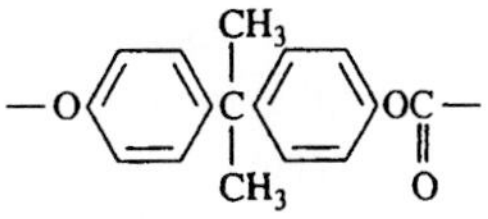	150	
Polyimide	Glassy	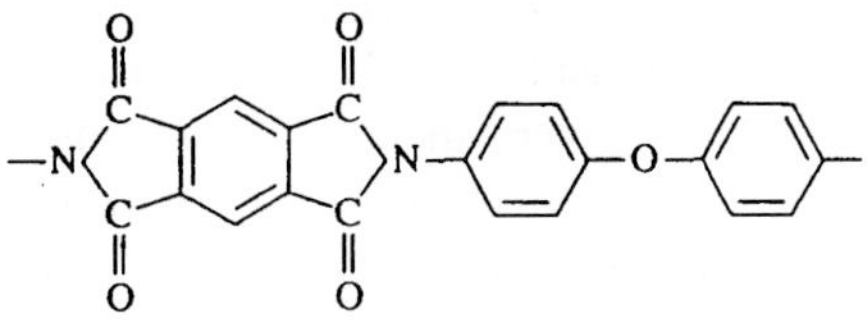	310–365	
Polystyrene	Glassy	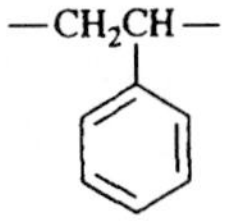 $-CH_2CH-$ (phenyl)	74–110	
Polysulfone	Glassy	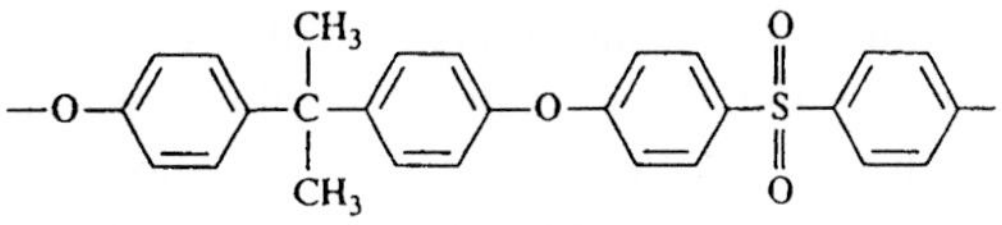	190	
Polytetrafluoroethylene (Teflon)	Crystalline	$-CF_2-CF_2-$		327

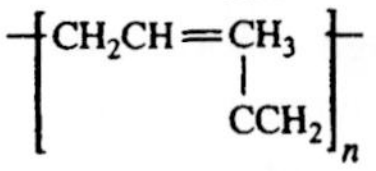

operate below T_m. Table 14.3 lists *repeat units* and values of T_g and/or T_m for several natural and synthetic polymers, from which membranes have been fabricated. Included are crystalline, glassy, and rubbery polymers. Cellulose triacetate is the reaction product of cellulose and acetic anhydride. Cellulose is the most readily available organic raw material in the world. The repeat unit of cellulose is identical to that shown for cellulose triacetate in Table 14.3, except that the acetyl, Ac (CH₃CO) groups are replaced by H. Typically, the number of repeat units (*degree of polymerization*) in cellulose is 1,000 to 1,500, whereas that in cellulose triacetate is about 300. Partially acetylated products are cellulose acetate and cellulose diacetate, with blends of two or three of the acetates being common. The triacetate is highly crystalline, of uniformly high quality, and hydrophobic.

Polyisoprene (natural rubber) is obtained from at least 200 different plants, with many of the rubber-producing countries being located in the Far East. Compared to the other polymers in Table 14.3, polyisoprene has a very low glass-transition temperature. Natural rubber has a degree of polymerization of from about 3,000 to 40,000 and is hard and rigid when cold, but soft, easily deformed, and sticky when hot. Depending on the temperature, it slowly crystallizes. To increase the strength, elasticity, and stability of rubber, it is vulcanized with sulfur, a process that introduces cross-links, but still allows unrestricted local motion of the polymer chain.

Aromatic polyamides (also called aramids) are high-melting crystalline polymers that have better long-term thermal stability and higher resistance to solvents than do aliphatic polyamides, such as nylon. Some aromatic polyamides are easily fabricated into fibers, films, and sheets. The polyamide structure shown in Table 14.3 is that of Kevlar, a trade name of DuPont.

Polycarbonates, which are characterized by the presence of the -OCOO- group in the chain, are mainly amorphous in structure. The polycarbonate shown in Table 14.3 is an aromatic form, but aliphatic forms also exist. Polycarbonates differ from most other amorphous polymers in that they possess ductility and toughness below T_g. Because polycarbonates are thermoplastic, they can be extruded into various shapes, including films and sheets.

Polyimides are characterized by the presence of aromatic rings and heterocyclic rings containing nitrogen and attached oxygen. The structure shown in Table 14.3 is only one of a number available. Polyimides are tough, amorphous polymers with high resistance to heat and excellent wear resistance. They can be fabricated into a wide variety of forms, including fibers, sheets, and films.

Polystyrene is a linear, amorphous, highly pure polymer of about 1,000 units of the structure shown in Table 14.3. Above a relatively low T_g, which depends on molecular weight, polystyrene becomes a viscous liquid that is easily fabricated by extrusion or injection molding. Like many other polymers, polystyrene can be annealed (heated and then cooled slowly) to convert it to a crystalline polymer with a melting point of 240°C. Styrene monomer can be copolymerized with a number of other organic monomers, including acrylonitrile and butadiene to form ABS copolymers.

Polysulfones are relatively new synthetic polymers, first introduced in 1966. The structure in Table 14.3 is just one of many, all of which contain the SO_2 group, which gives the polymers high strength. Polysulfones are easily spun into hollow fibers.

Polytetrafluoroethylene is a straight-chain, highly crystalline polymer with a very high degree of polymerization of the order of 100,000, which gives it considerable strength. It possesses exceptional thermal stability and can be formed into sheets, films, and tubing.

To be effective for separating a mixture of chemical components, a polymer membrane must possess high *permeance* and a high permeance ratio for the two species being separated by the membrane. The permeance for a given species diffusing through a membrane of given thickness is analogous to a mass-transfer coefficient, i.e., the flow rate of that species per unit cross-sectional area of membrane per unit driving force (concentration, partial pressure, etc.) across the membrane thickness. The molar transmembrane flux of species i is

$$N_i = \left(\frac{P_{M_i}}{l_M}\right)(\text{driving force}) = \bar{P}_{M_i}(\text{driving force}) \quad (14\text{-}1)$$

where $\bar{P}_{M_i}$ is the permeance, which is defined as the ratio of P_{M_i}, the *permeability*, to l_M, the membrane thickness.

Polymer membranes can be characterized as dense or microporous. For dense amorphous membranes, pores of microscopic dimensions may be present, but they are generally less than a few Angstroms in diameter, such that most, if not all, diffusing species must dissolve into the polymer and then diffuse through the polymer between the segments of the macromolecular chains. Diffusion can be difficult, but highly selective for glassy polymers. If the polymer is partly crystalline, diffusion will occur almost exclusively through the amorphous regions, with the crystalline regions decreasing the diffusion area and increasing the diffusion path.

A microporous membrane contains interconnected pores that are small (on the order of 0.001–10 μm; 10–100,000 Å), but large in comparison to the size of the molecules to be transferred. The pores are formed by a variety of proprietary techniques, some of which are described by Baker et al. [5]. Such techniques are especially valuable for producing symmetric, microporous, crystalline membranes. Permeability for microporous membranes is high, but selectivity is low, for small molecules. However, when molecules both smaller and larger than the pore size are in the feed to the membrane, the molecules may be separated almost perfectly by size.

Thus, for the separation of small molecules, we seem to be presented with a dilemma. We can have high permeability or a high separation factor, but not both. The beginning of the resolution of this dilemma occurred in 1963 with the fabrication by Loeb and Sourirajan [7] of an asymmetric membrane of cellulose acetate by a novel casting procedure. As shown in Figure 14.3a, the resulting membrane consists of a thin dense skin about 0.1–1.0 μm in. thick, called the *permselective* layer, formed over a much thicker microporous

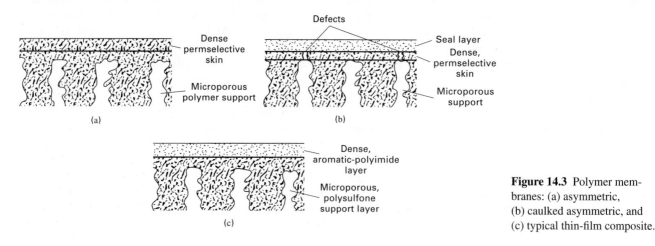

(a)

(b)

(c)

Figure 14.3 Polymer membranes: (a) asymmetric, (b) caulked asymmetric, and (c) typical thin-film composite.

layer that provides support for the skin. The flux rate of a species is controlled by the permeance of the very thin permselective skin. From (14-1), the permeance of species i can be high because of the very small value of l_M even though the permeability, P_{M_i}, is low because of the absence of pores. When large differences of P_{M_i} exist among molecules, both high permeance and high selectivity can be achieved with asymmetric membranes.

A very thin, asymmetric membrane is subject to minute defects or pinholes in the permselective skin, which can render the membrane useless for the separation of a gas mixture. A practical solution to the defect problem for an asymmetric polysulfone membrane was patented by Henis and Tripodi [8] of the Monsanto Company in 1980. They pulled silicone rubber, from a coating on the surface of the skin, into the defects by applying a vacuum. The resulting membrane is sometimes referred to as a *caulked membrane*, as shown in Figure 14.3b.

Another patent by Wrasidlo [9] in 1977, introduced the thin-film composite membrane as an alternative to the asymmetric membrane. In the first application, as shown in Figure 14.3c, a very thin, dense film of polyamide polymer, 250 to 500 Å in thickness, was formed on a thicker microporous polysulfone support. Today, both asymmetric and thin-film composite membranes are fabricated from a variety of polymers by a variety of techniques.

The application of polymer membranes is generally limited to temperatures below about 200°C and to the separation of mixtures that are chemically inert. When operation at high temperatures and/or with chemically active mixtures is necessary, membranes made of inorganic materials can be used. These include mainly microporous ceramics, metals, and carbon; and dense metals, such as palladium, that allow the selective diffusion of very small molecules such as hydrogen and helium.

Some examples of inorganic membranes are (1) asymmetric, microporous α-alumina tubes with 40–100 Å pores at the inside surface and 100,000 Å pores at the outside; (2) microporous glass tubes, the pores of which may or may not be filled with other oxides or the polymerization–pyrolysis product of trichloromethylsilane; (3) silica hollow fibers with extremely fine pores of 3–5 Å; (4) porous ceramic, glass, or polymer materials coated with a thin, dense film of palladium metal that is just a few microns thick; (5) sintered metal; (6) pyrolyzed carbon; and (7) zirconia on sintered carbon. Extremely fine pores (<10 Å) are necessary to separate gas mixtures. Larger pores (>50 Å) may be satisfactory for the separation of large molecules or solid particles from solutions containing small molecules.

EXAMPLE 14.1

A silica-glass membrane of 2 μm thickness and with very fine pores less than 10 Å in diameter has been developed for separating H_2 from CO at a temperature of 500°F. From laboratory data, the membrane permeabilities for hydrogen and carbon monoxide,

respectively, are 200,000 and 700 barrer, where the barrer, a commonly used unit for gas permeation, is defined by:

$$1 \text{ barrer} = 10^{-10} \text{ cm}^3 \text{ (STP)-cm/(cm}^2\text{-s-cmHg)}$$

where cm^3 (STP)/(cm^2-s) refers to the volumetric transmembrane flux of the diffusing species in terms of standard conditions of 0°C and 1 atm, cm refers to the membrane thickness, and cmHg refers to the transmembrane, partial-pressure driving force for the diffusing species.

The barrer unit is named for R. M. Barrer, who published an early article [10] on the nature of diffusion in a membrane, followed later by a widely referenced monograph on diffusion in and through solids [11].

If the transmembrane, partial-pressure driving forces for H_2 and CO, respectively, are 240 psi and 80 psi, calculate the transmembrane fluxes in kmol/(m^2-s). Compare the hydrogen flux to that for hydrogen in the commercial application discussed at the beginning of this chapter.

SOLUTION

At 0°C and 1 atm, 1 kmol of gas occupies 22.42×10^6 cm^3. Also, 2 μm thickness = 2×10^{-4} cm and 1 cmHg $\Delta P = 0.1934$ psi. Therefore, using (14-1):

$$N_{H_2} = \frac{(200,000)(10^{-10})(240/0.1934)(10^4)}{(22.42 \times 10^6)(2 \times 10^{-4})} = 0.0554 \frac{\text{kmol}}{\text{m}^2\text{-s}}$$

$$N_{CO} = \frac{(700)(10^{-10})(80/0.1934)(10^4)}{(22.42 \times 10^6)(2 \times 10^{-4})} = 0.000065 \frac{\text{kmol}}{\text{m}^2\text{-s}}$$

In the application discussed at the beginning of this chapter, the flux of H_2 for the polymer membrane is

$$\frac{(1685.1)(1/2.205)}{(16,000)(0.3048)^2(3600)} = 0.000143 \frac{\text{kmol}}{\text{m}^2\text{-s}}$$

Thus, the flux of H_2 through the ultramicroporous-glass membrane is more than 100 times higher than the flux through the dense-polymer membrane. Large differences in molar fluxes through different membranes are common.

The following are useful factors for converting barrers to SI and American Engineering units:

$$1.00 \text{ barrer} = 10^{-10} \frac{\text{cm}^3(\text{STP}) \cdot \text{cm}}{\text{cm}^2 \cdot \text{s} \cdot \text{cmHg}}$$

Multiply barrers by 3.348×10^{-19} to obtain units of (kmol $\cdot$ m)/(m$^2 \cdot$ s $\cdot$ Pa).

Multiply barrers by 5.584×10^{-12} to obtain units of (lbmol $\cdot$ ft)/(ft$^2 \cdot$ h $\cdot$ psi).

14.2 MEMBRANE MODULES

The asymmetric and thin-film, composite, polymer-membrane materials described in the previous section are available in one or more of the three shapes shown in Figure 14.4a, b, and c. Flat sheets have typical dimensions of 1 m by 1 m by 200 μm thick, with a dense skin or thin, dense layer 500 to 5,000 Å in thickness. Tubular membranes are

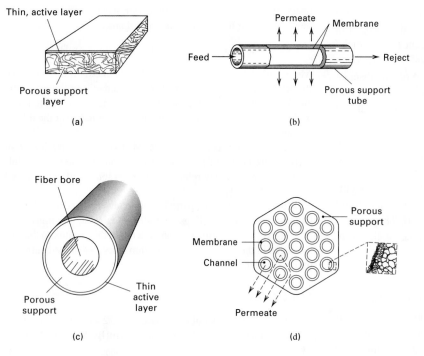

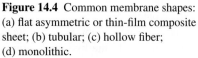

Figure 14.4 Common membrane shapes: (a) flat asymmetric or thin-film composite sheet; (b) tubular; (c) hollow fiber; (d) monolithic.

typically 0.5 to 5.0 cm in diameter and up to 6 m in length. The thin, dense layer is on either the inside, as shown in Figure 14.4b, or the outside surface of the tube. The porous supporting part of the tube is fiberglass, perforated metal, or other suitable porous material. Very small-diameter hollow fibers, first reported by Mahon [12, 13] in the 1960s, are typically 42 μm i.d. by 85 μm o.d. by 1.2 m long with a 0.1- to 1.0-μm-thick dense skin. Hollow fibers, shown in Figure 14.4c, provide a large membrane surface area per unit volume. A honeycomb, monolithic element for inorganic oxide membranes is included in Figure 14.4d. Elements of both hexagonal and circular cross-section are available [14]. The circular flow channels are typically 0.3 to 0.6 cm in diameter, with a 20- to 40-mm-thick membrane layer. The hexagonal element in Figure 14.4d has 19 channels and is 0.85 m long. Both the bulk support and the thin membrane layer are porous, but the pores of the latter can be very small, down to 40 Å. Still in the research stage are membranes based on nanotechnology.

The membrane shapes of Figure 14.4 are incorporated into compact, commercial modules and cartridges, some of which are shown in Figure 14.5. Flat sheets used in plate-and-frame modules are circular, square, or rectangular in cross-section. The sheets are separated by support plates that channel the permeate. In Figure 14.5a, a feed of brackish water flows across the surface of each membrane sheet in the stack. Pure water is the permeate product, whereas the retentate is a concentrated-brine solution.

Flat sheets are also fabricated into spiral-wound modules shown in Figure 14.5b. A laminate, consisting of two membrane sheets separated by spacers for the flow of the feed and permeate, is wound around a central, perforated, collection tube to form a module that is inserted into a pressure vessel.

The feed flows axially in the channels created between the membranes by the porous spacers. Permeate passes through the membrane, traveling inward in a spiral path to the central collection tube. From there, the permeate flows in either axial direction through and out of the central tube. A typical spiral-wound module is 0.1 to 0.3 m in diameter and 3 m long. Six such modules are often placed in series. The four-leaf modification in Figure 14.5c minimizes the pressure drop of the permeate because the permeate travel is less for the same membrane area.

The hollow-fiber module shown in Figure 14.5d, for a gas-permeation application, resembles a shell-and-tube heat exchanger. The pressurized feed enters the shell side at one end. While flowing over the fibers toward the other end, permeate passes through the fiber walls into the central fiber channels. Typically the fibers are sealed at one end and embedded into a tube sheet with epoxy resin at the other end. A commercial module might be 1 m long and 0.1 to 0.25 m in diameter and contain more than one million hollow fibers.

A tubular module is shown in Figure 14.5e. This module also resembles a shell-and-tube heat exchanger, but the feed flows through the tubes. Permeate passes through the wall of the tubes into the shell side of the module. Tubular modules contain up to 30 tubes.

The monolithic module in Figure 14.5f contains from 1 to 37 monolithic elements in a module housing. The feed flows through the circular channels and permeate passes through the membrane and porous support and into the open region between elements.

Table 14.4 is a comparison of the characteristics of four of the modules shown in Figure 14.5. The packing density is the membrane surface area per unit volume of module, for which the hollow-fiber membrane modules are clearly superior.

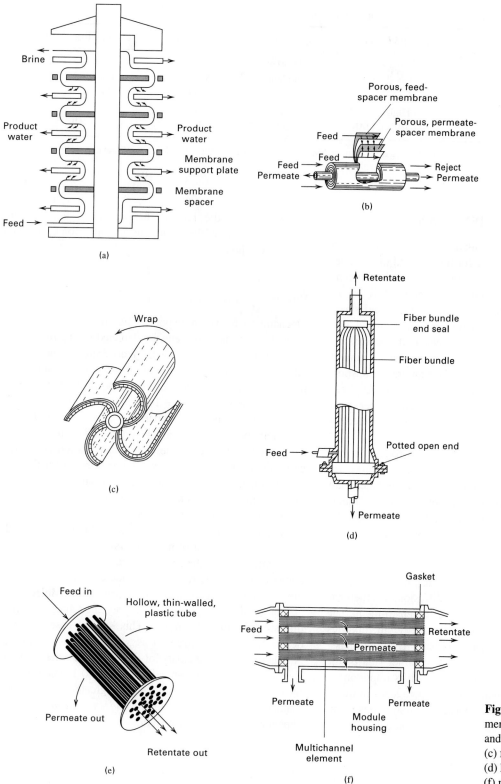

Brine

Product water

Product water

Membrane support plate

Membrane spacer

Feed →

(a)

Porous, feed-spacer membrane

Porous, permeate-spacer membrane

Feed

Feed

Feed
Permeate ←

→ Reject
→ Permeate

(b)

Wrap

(c)

↑ Retentate

Fiber bundle end seal

Fiber bundle

Feed →

Potted open end

↓ Permeate

(d)

Feed in

Hollow, thin-walled, plastic tube

Permeate out

Retentate out

(e)

Gasket

Feed

Retentate

Permeate

Permeate

Permeate

Module housing

Multichannel element

(f)

Figure 14.5 Common membrane modules: (a) plate-and frame, (b) spiral-wound, (c) four-leaf spiral-wound, (d) hollow-fiber, (e) tubular, (f) monolithic.

Although the plate-and-frame module has a high cost and a moderate packing density, it finds use in all membrane applications except gas permeation. It is the only module widely used for pervaporation. The spiral-wound module is very popular for most applications because of its low cost and reasonable resistance to fouling. Tubular modules are only used for small applications or when a high resistance to fouling and/or ease of cleaning are essential. Hollow-fiber modules, with a very high packing density and low cost, are popular where fouling does not occur and cleaning is not necessary.

Table 14.4 Typical Characteristics of Membrane Modules

	Plate and Frame	Spiral-Wound	Tubular	Hollow-Fiber
Packing density, m^2/m^3	30 to 500	200 to 800	30 to 200	500 to 9,000
Resistance to fouling	Good	Moderate	Very good	Poor
Ease of cleaning	Good	Fair	Excellent	Poor
Relative cost	High	Low	High	Low
Main applications	D, RO, PV, UF, MF	D, RO, GP, UF, MF	RO, UF	D, RO, GP, UF

Note: D, dialysis; RO, reverse osmosis; GP, gas permeation; PV, pervaporation; UF, ultrafiltration; MF, microfiltration.

14.3 TRANSPORT IN MEMBRANES

For a given application, the calculation of membrane surface area is based on laboratory data for the selected membrane. Although permeation can occur by one or more of the mechanisms discussed in this section, these mechanisms are all consistent with (14-1) in either its permeance form or its permeability form, with the latter being applied more widely. However, because both the driving force and the permeability or permeance depend markedly on the mechanism of transport, it is important to understand the nature of transport in membranes, which is the subject of this section. Applications to dialysis, reverse osmosis, gas permeation, pervaporation, ultrafiltration, and microfiltration are presented in subsequent sections.

Membranes can be macroporous, microporous, or dense (nonporous). Only microporous or dense membranes are permselective. However, macroporous membranes are widely used to support thin microporous and dense membranes when significant pressure differences across the membrane are necessary to achieve a reasonable throughput. The theoretical basis for transport through microporous membranes is more highly developed than that for dense membranes, so porous-membrane transport is discussed first, with respect to bulk flow, liquid diffusion, and then gas diffusion. This is followed by nonporous (dense)-membrane transport, including solution diffusion for liquid mixtures and solution diffusion for gas mixtures. External mass-transfer resistances in the fluid films on either side of the membrane are treated where appropriate. It is important to note that, because of the wide range of pore sizes in membranes, the distinction between porous and nonporous membranes is not always obvious. The distinction can be made based only on the relative permeabilities for diffusion through the pores of the membrane and diffusion through the solid, amorphous regions of the membrane.

Porous Membranes

Mechanisms for the transport of liquid and gas molecules through a porous membrane are depicted in Figure 14.6a, b, and c. If the pore diameter is large compared to the molecular diameter, and a pressure difference exists across the membrane, bulk or convective flow through the pores occurs, as shown in Figure 14.6a. Such a flow is generally undesirable because it is not permselective and, therefore, no separation between components of the feed occurs. If fugacity, activity, chemical potential, concentration, or partial pressure differences exist across the membrane for the various components, but the pressure is the same on both sides of the membrane so as not to cause a bulk flow, permselective diffusion of the components through the pores will take place, effecting a separation as shown in Figure 14.6b. If the pores are of the order of molecular size for at least some of the components in the feed mixture, the diffusion of those components will be restricted (hindered) as shown in Figure 14.6c, resulting in an enhanced separation. Molecules of size larger than the pores will be prevented altogether from diffusing through the pores. This special case is highly desirable and is referred to as *size exclusion* or *sieving*. Another special case exists for gas diffusion where the pore size and/or pressure (typically a vacuum) is such that the mean free path of the molecules is greater than the pore diameter, resulting in so-called *Knudsen diffusion*, which is dependent on molecular weight.

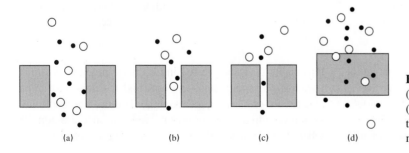

Figure 14.6 Mechanisms of transport in membranes. (Flow is downward.) (a) Bulk flow through pores; (b) diffusion through pores; (c) restricted diffusion through pores; (d) solution-diffusion through dense membranes.

Bulk Flow

Bulk flow is the principle mechanism of transfer through microporous membranes used for ultrafiltration and microfiltration, where separation is achieved mainly by sieving. Consider the bulk flow of a fluid, due to a pressure difference, through an idealized straight, cylindrical pore. If the flow is in the laminar regime ($N_{Re} = Dv\rho/\mu < 2,100$), which is almost always the case for flow in small-diameter pores, the flow velocity, v, is given by the Hagen–Poiseuille law [15] as being directly proportional to the transmembrane pressure drop:

$$v = \frac{D^2}{32\mu L}(P_0 - P_L) \qquad (14\text{-}2)$$

where D is the pore diameter, large enough to pass all molecules, μ is the viscosity of the fluid, and L is the length of the pore. This law assumes that a parabolic velocity profile exists across the pore radius for the entire length of the pore, that the fluid is Newtonian, and, if a gas, that the mean free path of the molecules is small compared to the pore diameter. If the membrane contains n such pores per unit cross-section of membrane surface area normal to flow, the porosity (void fraction) of the membrane is

$$\epsilon = n\pi D^2/4 \qquad (14\text{-}3)$$

Then the superficial fluid bulk-flow flux (mass velocity), N, through the membrane is

$$N = v\rho\epsilon = \frac{\epsilon\rho D^2}{32\mu l_M}(P_0 - P_L) = \frac{n\pi\rho D^4}{128\mu l_M}(P_0 - P_L) \qquad (14\text{-}4)$$

where l_M is the membrane thickness and ρ and μ are fluid properties.

In real porous membranes, pores may not be cylindrical and straight, making it necessary to modify (14-4). One procedure is that due to Carman and Kozeny, as extended by Ergun [16], where the pore diameter in (14-2) is replaced, as a rough approximation, by the hydraulic diameter:

$$d_H = 4\left(\frac{\text{Volume available for flow}}{\text{Total pore surface area}}\right)$$

$$= \frac{4\left(\dfrac{\text{Total pore volume}}{\text{Membrane volume}}\right)}{\left(\dfrac{\text{Total pore surface area}}{\text{Membrane volume}}\right)} = \frac{4\epsilon}{a} \qquad (14\text{-}5)$$

where the membrane volume includes the volume of the pores. The specific surface area, a_v, which is the pore surface area per unit volume of just the membrane material (not including the pores), is

$$a_v = a/(1 - \epsilon) \qquad (14\text{-}6)$$

Pore length is longer than the membrane thickness and can be represented by $l_M\tau$, where τ is a tortuosity factor >1. Substituting (14-5), (14-6), and the tortuosity factor into

(14-4) gives

$$N = \frac{\rho\epsilon^3(P_0 - P_L)}{2(1 - \epsilon)^2\tau a_v^2\mu l_M} \qquad (14\text{-}7)$$

In terms of a bulk-flow permeability, (14-7) becomes

$$N = \frac{P_M}{l_M}(P_0 - P_L) \qquad (14\text{-}8)$$

where

$$P_M = \frac{\rho\epsilon^3}{2(1 - \epsilon)^2\tau a_v^2\mu} \qquad (14\text{-}9)$$

Typically, τ is approximately 2.5, whereas a_v is inversely proportional to the average pore diameter, giving it values over a wide range.

Equation (14-7) may be compared to the following semi-theoretical Ergun equation [16], which represents the best fit of experimental data for flow of a fluid through a packed bed:

$$\frac{P_0 - P_L}{l_M} = \frac{150\mu v_0(1 - \epsilon)^2}{D_P^2\epsilon^3} + \frac{1.75\rho v_0^2(1 - \epsilon)}{D_P\epsilon^3} \qquad (14\text{-}10)$$

where D_P is the mean particle diameter, v_0 is the superficial fluid velocity through the membrane, and v_0/ϵ is the actual velocity in the pores. The first term on the right-hand side of (14-10) applies to the laminar flow region and is frequently referred to as Darcy's law. The second term applies to the turbulent region. For a spherical particle, the specific surface area is

$$a_v = \pi D_P^2/\left(\pi D_P^3/6\right) \quad \text{or} \qquad (14\text{-}11)$$
$$D_P = 6/a_v$$

Substitution of (14-11) into (14-10), for just the laminar-flow region, and rearrangement into the bulk-flow flux form gives

$$N = \frac{\rho\epsilon^3(P_0 - P_L)}{(150/36)(1 - \epsilon)^2 a_v^2\mu l_M} \qquad (14\text{-}12)$$

Comparing (14-12) to (14-7), we see that the term (150/36) in (14-12) corresponds to the term 2τ in (14-7). Thus, τ appears to have a value of 2.08, which seems reasonable. Accordingly, (14-12) can be used as a first approximation to the pressure drop for flow through a porous membrane when the pores are not straight cylinders. For gas flow, the density may be taken as the arithmetic average of the densities at the upstream and downstream faces of the membrane.

EXAMPLE 14.2

It is desired to pass water at 70°F through a supported, polypropylene membrane, with a skin of 0.003 cm thickness and 35% porosity, at the rate of 200 m^3/m^2-day. The pores can be considered as straight cylinders of uniform diameter equal to 0.2 micron. If the pressure on the downstream side of the membrane is 150 kPa, estimate the required pressure on the upstream side of the membrane. The pressure drop through the support is negligible.

SOLUTION

Equation (14-4) applies, where in SI units:

$$N/\rho = 200/(24)(3600) = 0.00232 \ \text{m}^3/\text{m}^2\text{-s}, \quad \epsilon = 0.35$$

$$D_P = 0.2 \times 10^{-6} \ \text{m}, \quad l_M = 0.00003 \ \text{m}$$

$$P_L = 150 \ \text{kPa} = 150,000 \ \text{Pa}, \quad \mu = 0.001 \ \text{Pa-s}$$

From (14-4),

$$P_0 = P_L + \frac{32\mu l_M (N/\rho)}{\epsilon D_P^2}$$

$$= 150,000 + \frac{(32)(0.001)(0.00003)(0.00232)}{(0.35)(0.2 \times 10^{-6})^2}$$

$$= 309,000 \ \text{Pa} \quad \text{or} \quad 309 \ \text{kPa}$$

Liquid Diffusion in Pores

Consider diffusion through the pores of a membrane from a fluid feed to a sweep fluid when identical total pressures but different component concentrations exist on both sides of the membrane. In that case, bulk flow through the membrane due to a pressure difference does not occur and if species diffuse at different rates, a separation can be achieved. If the feed mixture is a liquid of solvent and solutes i, the transmembrane flux for each solute is given by a modified form of Fick's law:

$$N_i = \frac{D_{e_i}}{l_M}(c_{i_0} - c_{i_L}) \qquad (14\text{-}13)$$

where D_{e_i} is the effective diffusivity, and c_i is the concentration of i in the liquid in the pores at the two faces of the membrane. In general the effective diffusivity is given by

$$D_{e_i} = \frac{\epsilon D_i}{\tau} K_{r_i} \qquad (14\text{-}14)$$

where D_i is the ordinary molecular diffusion coefficient (diffusivity) of the solute i in the solution, ϵ is the volume fraction of pores in the membrane, τ is the tortuosity, and K_r is a restrictive factor that accounts for the effect of pore diameter, d_p, in causing interfering collisions of the diffusing solutes with the pore wall, when the ratio of molecular diameter, d_m, to pore diameter exceeds about 0.01. The restrictive factor is approximated by Beck and Schultz [17] with:

$$K_r = \left[1 - \frac{d_m}{d_p}\right]^4, \quad (d_m/d_p) \le 1 \qquad (14\text{-}15)$$

From (14-15), when $(d_m/d_p) = 0.01$, $K_r = 0.96$, but when $(d_m/d_p) = 0.3$, $K_r = 0.24$. When $d_m > d_p$, $K_r = 0$, and the solute cannot diffuse through the pore. This is the sieving or size-exclusion effect illustrated in Figure 14.6c. In general, as illustrated in the following example, transmembrane fluxes for liquids through microporous membranes are very small because effective diffusivities are very low.

For solute molecules that are not subject to size exclusion, a useful selectivity ratio can be defined as

$$S_{ij} = \frac{D_i K_{r_i}}{D_j K_{r_j}} \qquad (14\text{-}16)$$

This ratio is greatly enhanced by the effect of restrictive diffusion when the solutes differ widely in molecular weight and one or more molecular diameters approach the pore diameter. This is shown in the next example.

EXAMPLE 14.3

Beck and Schultz [18] measured effective diffusivities of urea and several different sugars, in aqueous solutions, through microporous membranes of mica, which were especially prepared to give almost straight, elliptical pores of almost uniform size. Based on the following data for a membrane and two solutes, estimate transmembrane fluxes for the two solutes in g/cm^2-s at 25°C. Assume that the aqueous solutions on either side of the membrane are sufficiently dilute that no multicomponent diffusional effects are present.

Membrane:

Material	Microporous mica
Thickness, μm	4.24
Average pore diameter, Angstroms	88.8
Tortuosity, τ	1.1
Porosity, ϵ	0.0233

Solutes (in aqueous solution at 25°C):

Solute	MW	$D_i \times 10^6$ cm²/s	molecular diameter, d_m, Å	g/cm³ c_{i_0}	g/cm³ c_{i_L}
1 Urea	60	13.8	5.28	0.0005	0.0001
2 β-Dextrin	1135	3.22	17.96	0.0003	0.00001

SOLUTION

Calculate the restrictive factor and effective diffusivity from (14-15) and (14-14), respectively. For urea (1):

$$K_{r_1} = \left[1 - \left(\frac{5.28}{88.8}\right)\right]^4 = 0.783$$

$$D_{e_1} = \frac{(0.0233)(13.8 \times 10^{-6})(0.783)}{1.1} = 2.29 \times 10^{-7} \text{cm}^2/\text{s}$$

For β-dextrin (2):

$$K_{r_2} = \left[1 - \left(\frac{17.96}{88.8}\right)\right]^4 = 0.405$$

$$D_{e_2} = \frac{(0.0233)(3.22 \times 10^{-6})(0.405)}{1.1} = 2.78 \times 10^{-8} \text{cm}^2/\text{s}$$

Because of the large differences in molecular size, the two effective diffusivities differ by almost an order of magnitude. From (14-16), the selectivity is

$$S_{1,2} = \frac{(13.8 \times 10^{-6})(0.783)}{(3.22 \times 10^{-6})(0.405)} = 8.3$$

Calculate transmembrane fluxes from (14.13) noting the given concentrations are at the two faces of the membranes. Concentrations in the bulk solutions on either side of the membrane may differ from the concentrations at the faces depending upon the magnitudes of the external mass-transfer resistances in boundary layers or films adjacent to the two faces of the membrane.

For urea:

$$N_1 = \frac{(2.29 \times 10^{-7})(0.0005 - 0.0001)}{4.24 \times 10^{-4}} = 2.16 \times 10^{-7} \text{ g/cm}^2\text{-s}$$

For β-dextrin:

$$N_2 = \frac{(2.768 \times 10^{-8})(0.0003 - 0.00001)}{(4.24 \times 10^{-4})}$$

$$= 1.90 \times 10^{-8} \text{ g/cm}^2\text{-s}$$

Note that these fluxes are extremely low.

Gas Diffusion

When the mixture on either side of a microporous membrane is a gas, the rate of species diffusion can again be expressed in terms of Fick's law. If pressure and temperature on either side of the membrane are equal and the ideal-gas law holds, (14-13) can be written in terms of a partial-pressure driving force:

$$N_i = \frac{D_{e_i} c_M}{P l_M}(p_{i_0} - p_{i_L}) = \frac{D_{e_i}}{RT l_M}(p_{i_0} - p_{i_L}) \quad (14\text{-}17)$$

where c_M is the total concentration of the gas mixture given as P/RT by the ideal-gas law.

For a gas, diffusion through a pore may occur by ordinary diffusion, as with a liquid, and/or in series with Knudsen diffusion when pore diameter is very small and/or total pressure is low. In the Knudsen-flow regime, collisions occur primarily between gas molecules and the pore wall, rather than between gas molecules. Thus, in the absence of a bulk-flow effect or restrictive diffusion, (14-14) is modified for gas flow:

$$D_{e_i} = \frac{\epsilon}{\tau}\left[\frac{1}{(1/D_i) + (1/D_{K_i})}\right] \quad (14\text{-}18)$$

where D_{K_i} is the Knudsen diffusivity, which from the kinetic theory of gases applied to a straight, cylindrical pore of diameter d_p is given by

$$D_{K_i} = \frac{d_p \bar{v}_i}{3} \quad (14\text{-}19)$$

where $\bar{v}_i$ is the average molecule velocity given by:

$$\bar{v}_i = (8RT/\pi M_i)^{1/2} \quad (14\text{-}20)$$

where M is the molecular weight. Combining (14-19) and (14-20):

$$D_{K_i} = 4,850 d_p (T/M_i)^{1/2} \quad (14\text{-}21)$$

where D_K is cm^2/s, d_p is cm, and T is K. When Knudsen flow predominates, as it often does for the micropores in membranes, a selectivity based on the permeability ratio for species A and B is given from a combination of (14-1), (14-17), (14-18), and (14-21):

$$\frac{P_{M_A}}{P_{M_B}} = \left(\frac{M_B}{M_A}\right)^{1/2} \quad (14\text{-}22)$$

Except for gaseous species of widely differing molecular weight, the permeability ratio from (14-22) is not large, and the separation of gases by microporous membranes at low to moderate pressures that are equal on both sides of the membrane to minimize bulk flow is almost always impractical, as illustrated in the following example. However, it is important to note that the separation of the two isotopes of UF$_6$ by the United States government was accomplished by Knudsen diffusion, with a permeability ratio of only 1.0043, on a large scale at Oak Ridge, Tennessee, using thousands of stages and many acres of membrane surface.

EXAMPLE 14.4

A gas mixture of hydrogen (H) and ethane (E) is to be partially separated with a composite membrane having a 1-μm-thick porous skin with an average pore size of 20 Å and a porosity of 30%. The tortuosity can be assumed to be 1.5. The pressure on either side of the membrane is 10 atm and the temperature is 100°C. Estimate the permeabilities of the two components in Barrers.

SOLUTION

From (14-1), (14-17), and (14-18), the permeability can be expressed in gmol-cm/cm^2-s-atm:

$$P_{M_i} = \frac{\epsilon}{RT\tau}\left[\frac{1}{(1/D_i) + (1/D_{K_i})}\right]$$

where $\epsilon = 0.30$, $R = 82.06$ cm^3-atm/mol-K, $T = 373$ K, and $\tau = 1.5$.

At 100°C, the ordinary diffusivity is given by $D_H = D_E = D_{H,E} = 0.86/P$ in cm^2/s with total pressure P in atm. Thus, at 10 atm, $D_H = D_E = 0.086$ cm^2/s. Knudsen diffusivities are given by (14-21) with pore diameter, d_p, equal to 20×10^{-8} cm.

$$D_{K_H} = 4,850(20 \times 10^{-8})(373/2.016)^{1/2} = 0.0132 \text{ cm}^2/\text{s}$$

$$D_{K_E} = 4,850(20 \times 10^{-8})(373/30.07)^{1/2} = 0.00342 \text{ cm}^2/\text{s}$$

For both components, diffusion is controlled mainly by Knudsen diffusion.

For hydrogen: $\dfrac{1}{(1/D_H) + (1/D_{K_H})} = 0.0114$ cm^2/s.

For ethane: $\dfrac{1}{(1/D_E) + (1/D_{K_E})} = 0.00329$ cm^2/s.

$$P_{M_H} = \frac{0.30(0.0114)}{(82.06)(373)(1.5)} = 7.45 \times 10^{-8} \frac{\text{mol-cm}}{\text{cm}^2\text{-s-atm}}$$

$$P_{M_E} = \frac{0.30(0.00329)}{(82.06)(373)(1.5)} = 2.15 \times 10^{-8} \frac{\text{mol-cm}}{\text{cm}^2\text{-s-atm}}$$

To convert to Barrer as defined in Example 14.1, note that

$$76 \text{ cmHg} = 1 \text{ atm and } 22{,}400 \text{ cm}^3(\text{STP}) = 1 \text{ mol}$$

$$P_{M_H} = \frac{7.45 \times 10^{-8}(22{,}400)}{(10^{-10})(76)} = 220{,}000 \text{ Barrer}$$

$$P_{M_E} = \frac{2.15 \times 10^{-8}(22{,}400)}{(10^{-10})(76)} = 63{,}400 \text{ Barrer}$$

Nonporous Membranes

The transport of components through nonporous (dense) solid membranes is the predominant mechanism of membrane separators for reverse osmosis (liquid), gas permeation (gas), and pervaporation (liquid and vapor). As indicated in

Figure 14.6d, gas or liquid components absorb into the membrane at the upstream face, diffuse through the solid membrane, and desorb at the downstream face.

Liquid diffusivities are several orders of magnitude less than gas diffusivities, and diffusivities of solutes in solids are a few orders of magnitude less than diffusivities in liquids. Thus, differences between diffusivities in gases and solids are enormous. For example, at 1 atm and 25°C, diffusivities in cm²/s for water are as follows:

Water vapor in air	0.25
Water in ethanol liquid	1.2×10^{-5}
Dissolved water in cellulose acetate solid	1×10^{-8}

As might be expected, small molecules fare better than large molecules for diffusivities in solids. For example, from the *Polymer Handbook* [19], diffusivities in cm²/s for several components in low-density polyethylene at 25°C are

Helium	6.8×10^{-6}
Hydrogen	0.474×10^{-6}
Nitrogen	0.320×10^{-6}
Propane	0.0322×10^{-6}

Regardless of whether a nonporous membrane is used to separate a gas or liquid mixture, the *solution-diffusion model* of Lonsdale, Merten, and Riley [20] is most often applied to analyze experimental permeability data and design membrane separators. This model is based on Fick's law for diffusion through solid, nonporous membranes based on the driving force, $c_{i_0} - c_{i_L}$ shown in Figure 14.7b, where the concentrations are those for the solute dissolved in the

membrane. The concentrations in the membrane are related to the concentrations or partial pressures in the fluid adjacent to the membrane faces by assuming thermodynamic equilibrium for the solute between the fluid and membrane material at the fluid–membrane interfaces. This assumption has been validated experimentally by Motanedian et al. [21] for the case of permeation of light gases through dense cellulose acetate membranes at up to 90 atm.

Solution-Diffusion for Liquid Mixtures

Figures 14.7a and b show typical solute-concentration profiles for liquid mixtures with porous and nonporous membranes, respectively. Included in these diagrams is the drop in concentration across the membrane and, also, possible drops due to resistances in the fluid boundary layers or films on either side of the membrane. For porous membranes, of the type considered in the previous section, the concentration profile is continuous from the bulk-feed liquid to the bulk-permeate liquid because liquid is present continuously from one side to the other. The concentration c_{i_0} is the same in the liquid feed just adjacent to the membrane surface and in the liquid just within the entrance of the pore. This is not the case for the nonporous membrane in Figure 14.7b. Solute concentration c'_{i_0} is that in the feed liquid just adjacent to the upstream membrane surface, whereas c_{i_0} is that in the membrane just adjacent to the upstream membrane surface. In general, c_{i_0} is considerably smaller than c'_{i_0}, but the two are related by a thermodynamic equilibrium partition coefficient K_i, defined by

$$K_{i_0} = c_{i_0}/c'_{i_0} \qquad (14\text{-}23)$$

Similarly, at the other face:

$$K_{i_L} = c_{i_L}/c'_{i_L} \qquad (14\text{-}24)$$

Fick's law applied to the nonporous membrane of Figure 14.7b is:

$$N_i = \frac{D_i}{l_M}(c_{i_0} - c_{i_L}) \qquad (14\text{-}25)$$

where D_i is the diffusivity of the solute in the membrane material. If (14-23) and (14-24) are combined with (14-25), and the partition coefficient is assumed to be independent of concentration, such that $K_{i_0} = K_{i_L} = K_i$, we obtain for the flux

$$N_i = \frac{K_i D_i}{l_M}(c'_{i_0} - c'_{i_L}) \qquad (14\text{-}26)$$

If the mass-transfer resistances in the two fluid boundary layers or films are negligible:

$$N_i = \frac{K_i D_i}{l_M}(c_{i_F} - c_{i_P}) \qquad (14\text{-}27)$$

In (14-26) and (14-27), $K_i D_i$ is the permeability, P_{M_i}, for the solution-diffusion model, where K_i accounts for the solubility of the solute in the membrane and D_i accounts for diffusion through the membrane. Because D_i is generally

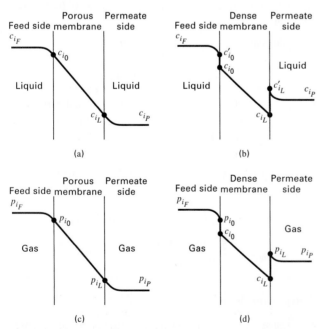

Figure 14.7 Concentration and partial pressure profiles for solute transport through membranes. Liquid mixture with (a) a porous and (b) a nonporous membrane; gas mixture with (c) a porous and (d) a nonporous membrane.

Table 14.5 Factors That Influence Permeability of Solutes in Dense Polymers

Factor	Value Favoring High Permeability
Polymer density	low
Degree of crystallinity	low
Degree of cross-linking	low
Degree of vulcanization	low
Amount of plasticizers	high
Amount of fillers	low
Chemical affinity of solute for polymer (solubility)	high

very small, it is important that the membrane material offers a large value for K_i and/or a small membrane thickness.

Both D_i and K_i, and therefore P_{M_i}, depend on the solute and the membrane. When solutes dissolve in a polymer membrane, it will swell, causing both D_i and K_i to increase. Other polymer membrane factors that can influence D_i, K_i, and P_{M_i} are listed in Table 14.5. However, the largest single factor is the chemical structure of the polymer. Because of the many factors involved, it is important to obtain experimental permeability data on the membrane and feed mixture of interest. The effect of external mass-transfer resistances is considered in a subsection near the end of this section.

Solution-Diffusion for Gas Mixtures

Figures 14.7c and d show typical solute profiles for gas mixtures with porous and nonporous membranes, respectively, including the effect of external-fluid boundary layer or film mass-transfer resistances. For the porous membrane, a continuous partial-pressure profile is shown. For the nonporous membrane, a concentration profile is shown within the membrane where the solute is dissolved in the membrane. Fick's law, given by (14-25), holds for transport through the membrane. Assuming that thermodynamic equilibrium exists at the two fluid–membrane interfaces, the concentrations in Fick's law can be related to the partial pressures adjacent to the membrane faces by Henry's law, which is a linear relation that is most conveniently written for membrane applications as

$$H_{i_0} = c_{i_0}/p_{i_0} \qquad (14\text{-}28)$$

and

$$H_{i_L} = c_{i_L}/p_{i_L} \qquad (14\text{-}29)$$

If we assume that H_i is independent of total pressure and that the temperature is the same at both membrane faces:

$$H_{i_0} = H_{i_L} = H_i \qquad (14\text{-}30)$$

Combining (14-25), (14-28), (14-29), and (14-30), the flux is

$$N_i = \frac{H_i D_i}{l_M}(p_{i_0} - p_{i_L}) \qquad (14\text{-}31)$$

If the external mass-transfer resistances are neglected, $p_{i_F} = p_{i_0}$ and $p_{i_L} = p_{i_P}$, giving

$$N_i = \frac{H_i D_i}{l_M}(p_{i_F} - p_{i_P}) = \frac{P_{M_i}}{l_M}(p_{i_F} - p_{i_P}) \qquad (14\text{-}32)$$

where

$$P_{M_i} = H_i D_i \qquad (14\text{-}33)$$

Thus, the permeability depends on both the solubility of the gas component in the membrane and the diffusivity of that component in the membrane material. An acceptable rate of transport through the membrane can be achieved only by using a very thin membrane and a high pressure on the feed side. The permeability of a gaseous component in a polymer membrane is subject to the factors listed in Table 14.5. Light gases do not interact with the polymer or cause it to swell. Thus, a light-gas–permeant-polymer combination is readily characterized experimentally. Often both solubility and diffusivity are measured. An extensive tabulation is given in the *Polymer Handbook* [19]. Representative data at 25°C are given in Table 14.6. In general, diffusivity decreases and solubility increases with increasing molecular weight of the gas species, making it difficult to achieve a high selectivity. The effect of temperature over a modest range of about 50°C can be represented for both solubility and diffusivity by Arrhenius equations. For example,

$$D = D_0 e^{-E_D/RT} \qquad (14\text{-}34)$$

In general, the modest effect of temperature on solubility may act in either direction. However, an increase in temperature can cause a substantial increase in diffusivity and, therefore, a corresponding increase in permeability. Typical activation energies of diffusion in polymers, E_D, range from 15 to 60 kJ/mol.

The application of Henry's law for rubbery polymers is well accepted, particularly for low-molecular-weight penetrants, but is less accurate for glassy polymers, for which alternative theories have been proposed. Foremost is the dual-mode model first proposed by Barrer and co-workers [22–24] as the result of a comprehensive study of sorption and diffusion in ethyl cellulose. In this model, sorption of the penetrant occurs by ordinary dissolution in the polymer chains, as described by Henry's law, and by Langmuir sorption into holes or sites between chains of glassy polymers. When the downstream pressure is negligible compared to the upstream pressure, the permeability for Fick's law is given by

$$P_{M_i} = H_i D_i + \frac{D_{L_i}ab}{1 + bP} \qquad (14\text{-}35)$$

where the second term refers to Langmuir sorption with D_{L_i} = diffusivity of Langmuir sorbed species, P = penetrant pressure, and a,b = Langmuir constants for sorption site capacity and site affinity, respectively.

Koros and Paul [25] found that the dual-mode theory accurately represents data for the sorption of CO_2 in

Table 14.6 Coefficients for Gas Permeation in Polymers

	Gas Species					
	H_2	O_2	N_2	CO	CO_2	CH_4
Low-Density Polyethylene:						
$D \times 10^6$	0.474	0.46	0.32	0.332	0.372	0.193
$H \times 10^6$	1.58	0.472	0.228	0.336	2.54	1.13
$P_M \times 10^{13}$	7.4	2.2	0.73	1.1	9.5	2.2
Polyethylmethacrylate:						
$D \times 10^6$	—	0.106	0.0301	—	0.0336	—
$H \times 10^6$	—	0.839	0.565	—	11.3	—
$P_M \times 10^{13}$	—	0.889	0.170	—	3.79	—
Polyvinylchloride:						
$D \times 10^6$	0.5	0.012	0.0038	—	0.0025	0.0013
$H \times 10^6$	0.26	0.29	0.23	—	4.7	1.7
$P_M \times 10^{13}$	1.3	0.034	0.0089	—	0.12	0.021
Butyl Rubber:						
$D \times 10^6$	1.52	0.081	0.045	—	0.0578	—
$H \times 10^6$	0.355	1.20	0.543	—	6.71	—
$P_M \times 10^{13}$	5.43	0.977	0.243	—	3.89	—

Note: Units: D in cm^2/s; H in cm^3 (STP)/cm^3-Pa; P_M in cm^3 (STP)-cm/cm^2-s-Pa.

polyethylene terephthalate below its glass-transition temperature of about 85°C. Above that temperature, the rubbery polymer obeys just Henry's law. Mechanisms of diffusion for the Langmuir mode have been suggested by Barrer [26].

The ideal, dense-polymer membrane has a high permeance, P_{M_i}/l_M, for the penetrant molecules and a high separation factor (selectivity) between the components to be separated. The separation factor is defined similarly to relative volatility in distillation:

$$\alpha_{A,B} = \frac{(y_A/x_A)}{(y_B/x_B)} \quad (14\text{-}36)$$

where y_i is the mole fraction in the permeate leaving the membrane, corresponding to the partial pressure p_{i_P} in Figure 14.7d, while x_i is the mole fraction in the retentate on the feed side of the membrane, corresponding to the partial pressure p_{i_F} in Figure 14.7d. Unlike the case of distillation, y_i and x_i are not in equilibrium.

For the separation of a binary gas mixture of species A and B in the absence of external boundary layer or film mass-transfer resistances, the transport fluxes are given by (14-32):

$$N_A = \frac{H_A D_A}{l_M}(p_{A_F} - p_{A_P}) = \frac{H_A D_A}{l_M}(x_A P_F - y_A P_P) \quad (14\text{-}37)$$

$$N_B = \frac{H_B D_B}{l_M}(p_{B_F} - p_{B_P}) = \frac{H_B D_B}{l_M}(x_B P_F - y_B P_P) \quad (14\text{-}38)$$

When no sweep gas is used, the ratio of N_A to N_B fixes the composition of the permeate so that it is simply the ratio of

y_A to y_B in the permeate gas. Thus,

$$\frac{N_A}{N_B} = \frac{y_A}{y_B} = \frac{H_A D_A(x_A P_F - y_A P_P)}{H_B D_B(x_B P_F - y_B P_P)} \quad (14\text{-}39)$$

If the downstream (permeate) pressure, P_P, is negligible compared to the upstream pressure, P_F, such that $y_A P_P \ll x_A P_F$ and $y_B P_P \ll x_B P_F$, (14-39) can be rearranged and combined with (14-36) to give an *ideal separation factor*:

$$\alpha_{A,B}^* = \frac{H_A D_A}{H_B D_B} = \frac{P_{M_A}}{P_{M_B}} \quad (14\text{-}40)$$

Thus, a high separation factor can be achieved from a high solubility ratio, a high diffusivity ratio, or both. The separation factor depends on both transport phenomena and thermodynamic equilibria.

When the downstream pressure is not negligible, (14-39) can be rearranged to obtain an expression for $\alpha_{A,B}$ in terms of the pressure ratio, $r = P_P/P_F$, and the mole fraction of A on the feed or retentate side of the membrane. Combining (14-36), (14-40), and the definition of r with (14-39):

$$\alpha_{A,B} = \alpha_{A,B}^* \left[\frac{(x_B/y_B) - r\alpha_{A,B}}{(x_B/y_B) - r} \right] \quad (14\text{-}41)$$

Because $y_A + y_B = 1$, we can substitute into (14-41) for x_B, the identity:

$$x_B = x_B y_A + x_B y_B$$

to give

$$\alpha_{A,B} = \alpha_{A,B}^* \left[\frac{x_B\left(\dfrac{y_A}{y_B}+1\right) - r\alpha_{A,B}}{x_B\left(\dfrac{y_A}{y_B}+1\right) - r} \right] \quad (14\text{-}42)$$

Table 14.7 Ideal Membrane Separation Factors of Binary Pairs for Two Membrane Materials

	PDMS, Silicon Rubbery Polymer Membrane	PC, Polycarbonate Glassy Polymer Membrane
$P_{M_{He}}$, Barrer	561	14
α^*_{He,CH_4}	0.41	50
α^*_{He,C_2H_4}	0.15	33.7
$P_{M_{CO_2}}$, Barrer	4,550	6.5
$\alpha^*_{CO_2,CH_4}$	3.37	23.2
$\alpha^*_{CO_2,C_2H_4}$	1.19	14.6
$P_{M_{O_2}}$, Barrer	933	1.48
$\alpha^*_{O_2,N_2}$	2.12	5.12

If we combine (14-36) and (14-42) and replace x_B with $1 - x_A$, we obtain for the separation factor:

$$\alpha_{A,B} = \alpha^*_{A,B} \left[\frac{x_A(\alpha_{A,B} - 1) + 1 - r\alpha_{A,B}}{x_A(\alpha_{A,B} - 1) + 1 - r} \right] \quad (14\text{-}43)$$

Equation (14-43) is an implicit equation for $\alpha_{A,B}$, in terms of the pressure ratio, r, and x_A, that is readily solved for $\alpha_{A,B}$ by rearranging the equation into the form of a quadratic equation. In the limit when $r = 0$, (14-43) reduces to (14-40), where $\alpha_{A,B} = \alpha^*_{A,B} = (P_{M_A}/P_{M_B})$. Many experimental investigators report values of $\alpha^*_{A,B}$. For example, Table 14.7, taken from the *Membrane Handbook* [6], gives data at 35°C for various binary pairs with polydimethyl siloxane (PDMS), a rubbery polymer, and bisphenol-A-polycarbonate (PC), a glassy polymer. For the rubbery polymer, permeabilities are high, but separation factors are low. The opposite is true for the glassy polymer. For a given feed composition, the separation factor places a definite limit on the degree of separation that can be achieved.

EXAMPLE 14.5

Air can be separated into nitrogen-enriched and oxygen-enriched streams by gas permeation with a number of different dense-polymer membranes. In all cases, the membrane is more permeable to oxygen. A total of 20,000 scfm of air is compressed, cooled, and treated to remove moisture and compressor oil prior to being sent to a membrane separator at 150 psia and 78°F. Assume the composition of the air is 79 mol% N_2 and 21 mol% O_2. A low-density polyethylene membrane in the form of a thin-film composite is being considered with solubilities and diffusivities given in Table 14.6. If the membrane skin is 0.2 μm thick, calculate the material balance and area in ft^2 for the membrane as a function of the cut, θ (fraction of feed permeated). Assume a pressure of 15 psia on the permeate side with perfect mixing on both sides of the membrane, such that compositions on both sides are each uniform and equal to exit compositions. Neglect pressure drop and mass-transfer resistances external to the membrane. Comment on the practicality of the membrane for making a reasonable separation.

SOLUTION

Assume standard conditions are 0°C and 1 atm (359 ft³/lbmol)

$$n_F = \text{Feed flow rate} = \frac{20,000}{359}(60) = 3,343 \text{ lbmol/h}$$

For the low-density polyethylene membrane, from Table 14.6, and applying (14-33), letting A = O_2 and B = N_2:

$$P_{M_B} = H_B D_B = (0.228 \times 10^{-6})(0.32 \times 10^{-6})$$
$$= 0.073 \times 10^{-12} \text{ cm}^3(\text{STP})\text{-cm/cm}^2\text{-s-Pa}$$

or, in American engineering units,

$$P_{M_B} = \frac{(0.073 \times 10^{-12})(2.54 \times 12)(3600)(101,300)}{(22,400)(454)(14.7)}$$
$$= 5.43 \times 10^{-12} \frac{\text{lbmol-ft}}{\text{ft}^2\text{-h-psia}}$$

Similarly, for oxygen: $P_{M_A} = 16.2 \times 10^{-12} \frac{\text{lbmol-ft}}{\text{ft}^2\text{-h-psia}}$

Permeance values are based on a 0.2-μm-thick membrane skin $(0.66 \times 10^{-6} \text{ ft})$

From (14-1), $\bar{P}_{M_i} = P_{M_i}/l_M$
$$\bar{P}_{M_B} = 5.43 \times 10^{-12}/0.66 \times 10^{-6}$$
$$= 8.23 \times 10^{-6} \text{ lbmol/ft}^2\text{-h-psia}$$
$$\bar{P}_{M_A} = 16.2 \times 10^{-12}/0.66 \times 10^{-6}$$
$$= 24.55 \times 10^{-6} \text{ lbmol/ft}^2\text{-h-psia}$$

Material balance equations

For N_2

$$x_{F_B} n_F = y_{P_B} n_P + x_{R_B} n_R \quad (1)$$

where n = flow rate in lbmol/h and subscripts F, P, and R refer, respectively, to the feed, permeate, and retentate. Let

$$\theta = \text{cut} = n_P/n_F, \quad \text{then}(1 - \theta) = n_R/n_F$$

Note that if all components of the feed have a finite permeability, the cut, θ, can vary from 0 to 1. For a cut of 1, all of the feed becomes permeate and no separation is achieved.
Substituting the definition of θ in (1) gives

$$x_{R_B} = \frac{x_{F_B} - y_{P_B}\theta}{1 - \theta} = \frac{0.79 - y_{P_B}\theta}{1 - \theta} \quad (2)$$

Similarly, for O_2,

$$x_{R_A} = \frac{0.21 - y_{P_A}\theta}{1 - \theta} \quad (3)$$

Separation factor

From the definition of the separation factor, (14-36), since both fluid sides are well mixed, such that compositions are those of the retentate and permeate,

$$\alpha_{A,B} = \frac{y_{P_A}/x_{R_A}}{(1 - y_{P_A})/(1 - x_{R_A})} \quad (4)$$

Transport equations

The transport of A and B through the membrane of area A_M, with partial pressures at exit conditions because of perfect mixing, can be written as

$$N_B = y_{P_B} n_P = A_M \bar{P}_{M_B}(x_{R_B} P_R - y_{P_B} P_P) \quad (5)$$
$$N_A = y_{P_A} n_P = A_M \bar{P}_{M_A}(x_{R_A} P_R - y_{P_A} P_P) \quad (6)$$

where A_M is the membrane area normal to flow, n_P, through the membrane. The ratio of (6) to (5) is y_{P_A}/y_{P_B}, and subsequent manipulations lead to (14-43), where

$$r = P_P/P_R = 15/150 = 0.1$$

and

$$\alpha_{A,B}^* = \alpha_{O_2,N_2} = \bar{P}_{M_{O_2}}/\bar{P}_{M_{N_2}}$$

$$= (24.55 \times 10^{-6})/(8.23 \times 10^{-6}) = 2.98$$

From (14-43):

$$\alpha_{A,B} = \alpha = 2.98 \left[\frac{x_{R_A}(\alpha - 1) + 1 - 0.1\alpha}{x_{R_A}(\alpha - 1) + 1 - 0.1} \right] \qquad (7)$$

Equations (3), (4), and (7) contain four unknowns: x_{R_A}, y_{P_A}, θ, and $\alpha_{A,B} = \alpha$. The variable θ is bounded between 0 and 1, so values of θ are selected in that range. The other three variables are computed in the following manner. Combine (3), (4), and (7) to eliminate α and x_{R_A}. Solve the resulting nonlinear equation for y_{P_A}. Then solve (3) for x_{R_A} and (4) for α. Solve (6) for the membrane area, A_M. Alternatively, the three equations can be solved simultaneously with a computer program such as Mathcad, Matlab, or Polymath. The following results are obtained:

θ	x_{R_A}	y_{P_A}	$\alpha_{A,B}$	A_M, ft^2
0.01	0.208	0.406	2.602	22,000
0.2	0.174	0.353	2.587	462,000
0.4	0.146	0.306	2.574	961,000
0.6	0.124	0.267	2.563	1,488,000
0.8	0.108	0.236	2.555	2,035,000
0.99	0.095	0.211	2.548	2,567,000

Note that the separation factor remains almost constant, varying by only 2% with a value of about 86% of the ideal value. The maximum oxygen content of the permeate (40.6 mol%) occurs with the smallest amount of permeate ($\theta = 0.01$). The maximum nitrogen content of the retentate (90.5 mol%) occurs with the largest amount of permeate ($\theta = 0.99$). With a retentate equal to 60 mol% of the feed ($\theta = 0.4$), the nitrogen content of the retentate has been increased only from 79 to 85.4 mol%. Furthermore, the membrane area requirements are very large. The low-density polyethylene membrane is not very practical. To achieve a more reasonable separation, say with $\theta = 0.6$ and a retentate of 95 mol% N$_2$, it is advisable to use a membrane with an ideal separation factor of 5 in a membrane module that approximates crossflow or countercurrent flow of permeate and retentate with no mixing and a much higher permeance for oxygen. For higher purities, a membrane cascade of two or more stages should be considered. These alternatives are developed in the next two subsections.

Module Flow Patterns

In Example 14.5, perfect mixing was assumed on both sides of the membrane. Three other idealized flow patterns, shown in Figure 14.8, have received considerable attention; all assume no mixing and are comparable to the idealized flow patterns used to design heat exchangers. These patterns are (b) countercurrent flow; (c) cocurrent flow; and (d) crossflow. For a given cut, θ, the flow pattern can significantly affect the degree of separation and the membrane area. For flow patterns (b) to (d), fluid on the feed or retentate side of the membrane flows along and parallel to the upstream surface of the membrane. For countercurrent and cocurrent flow, permeate fluid at a given location on the downstream side of the membrane consists of fluid that has just passed through the membrane at that location plus the permeate fluid flowing to that location. For the crossflow case, there is no flow of permeate fluid along the membrane surface. The permeate fluid that has just passed through the membrane at a given location is the only fluid there. For a given module geometry, it is not always obvious which idealized flow pattern to assume. This is particularly true for the spiral-wound module of Figure 14.5b. If the permeation rate is high, the fluid issuing from the downstream side of the membrane may continue to flow perpendicularly to the membrane surface until it finally mixes with the bulk permeate fluid flowing past the surface. In that case, the idealized crossflow pattern might be appropriate. Hollow-fiber modules may be designed to approximate idealized countercurrent, cocurrent, or crossflow patterns. The hollow-fiber module shown in Figure 14.5d is approximated by a countercurrent flow pattern.

Walawender and Stern [27] present solution methods for all four flow patterns of Figure 14.8 under the assumptions of a binary feed with constant pressure ratio, r, and constant ideal separation factor, $\alpha_{A,B}^*$. Exact analytical solutions are possible for the perfect mixing case (as shown in Example 14.5) and for the crossflow case, but numerical solutions are necessary for the countercurrent and cocurrent flow cases. A reasonably simple, but approximate, analytical solution for the crossflow case, derived by Naylor and Backer [28], is presented here.

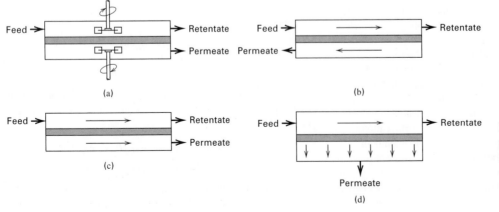

(a)

(b)

(c)

(d)

Figure 14.8 Idealized flow patterns in membrane modules: (a) perfect mixing; (b) countercurrent flow; (c) cocurrent flow; (d) crossflow.

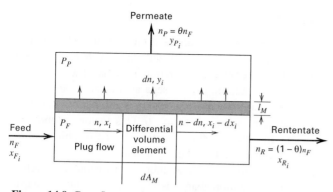

Figure 14.9 Crossflow model for membrane module.

Consider a membrane module with the crossflow pattern shown in Figure 14.9. The feed passes across the upstream membrane surface in plug flow with no longitudinal mixing. The pressure ratio, $r = P_P/P_F$, and the ideal separation factor, $\alpha_{A,B}^*$, are assumed constant. Boundary layer or film mass-transfer resistances external to the membrane are assumed negligible. At the differential element, the local mole fractions in the retentate and permeate, respectively, are x_i and y_i, and the penetrant molar flux is dn/dA_M. Also, the local separation factor is given by (14-43) in terms of the local x_A, r, and $\alpha_{A,B}^*$. An alternative expression for the local permeate composition in terms of y_A, x_A, and r, is obtained by combining (14-36) and (14-41).

$$\frac{y_A}{1 - y_A} = \alpha_{A,B}^* \left[\frac{x_A - r y_A}{(1 - x_A) - r(1 - y_A)} \right] \quad (14\text{-}44)$$

A material balance for A around the differential-volume element gives

$$y_A dn = d(nx_A) = x_A dn + n dx_A \quad \text{or}$$

$$\frac{dn}{n} = \frac{dx_A}{y_A - x_A} \quad (14\text{-}45)$$

which is identical to the Rayleigh equation for differential, batch distillation. If (14-36) is combined with (14-45) to eliminate y_A, we obtain

$$\frac{dn}{n} = \left[\frac{1 + (\alpha - 1)x_A}{x_A(\alpha - 1)(1 - x_A)} \right] dx_A \quad (14\text{-}46)$$

where $\alpha = \alpha_{A,B}$.

In the solution to Example 14.5, it was noted that $\alpha = \alpha_{A,B}$ is relatively constant over the entire range of cut, θ. Such is generally the case when the pressure ratio, r, is small. If the assumption of constant $\alpha = \alpha_{A,B}$ is made in (14-46) and integration is carried out from the intermediate location of the differential element to the final retentate, that is, from n to n_R and from x_A to x_{R_A}, the result is

$$n = n_R \left[\left(\frac{x_A}{x_{R_A}} \right)^{\left(\frac{1}{\alpha - 1} \right)} \left(\frac{1 - x_{R_A}}{1 - x_A} \right)^{\left(\frac{\alpha}{\alpha - 1} \right)} \right] \quad (14\text{-}47)$$

The mole fraction of A in the final permeate and the total membrane surface area are obtained by integrating the

values obtained from solving (14-44) to (14-46):

$$y_{P_A} = \int_{x_{F_A}}^{x_{R_A}} y_A dn / \theta n_F \quad (14\text{-}48)$$

By combining (14-48) with (14-46), (14-47), and the definition of α, the integral in n can be transformed to an integral in x_A, which when integrated gives

$$y_{P_A} = x_{R_A}^{\left(\frac{1}{1 - \alpha} \right)} \left(\frac{1 - \theta}{\theta} \right)$$

$$\times \left[(1 - x_{R_A})^{\left(\frac{\alpha}{\alpha - 1} \right)} \left(\frac{x_{F_A}}{1 - x_{F_A}} \right)^{\left(\frac{\alpha}{\alpha - 1} \right)} - x_{R_A}^{\left(\frac{\alpha}{\alpha - 1} \right)} \right]$$

$$(14\text{-}49)$$

where $\alpha = \alpha_{A,B}$ can be estimated from (14-43) by using $x_A = x_{F_A}$.

The differential rate of mass transfer of A across the membrane is given by

$$y_A dn = \frac{P_{M_A} dA_M}{l_M} [x_A P_F - y_A P_P] \quad (14\text{-}50)$$

from which the total membrane surface area can be obtained by integration:

$$A_M = \int_{x_{R_A}}^{x_{F_A}} \frac{l_M y_A dn}{P_{M_A}(x_A P_F - y_A P_P)} \quad (14\text{-}51)$$

The application of the crossflow model is illustrated in the next example.

EXAMPLE 14.6

For the conditions of Example 14.5, compute exit compositions for a spiral-wound module that approximates crossflow.

SOLUTION

From Example 14.5: $\alpha_{A,B}^* = 2.98$; $r = 0.1$; $x_{F_A} = 0.21$
  From (14-43), using $x_A = x_{F_A}$; $\alpha_{A,B} = 2.60$
An overall module material balance for O_2 (A) gives

$$x_{F_A} n_F = x_{R_A}(1 - \theta)n_F + y_{P_A}\theta n_F \quad \text{or} \quad x_{R_A} = \frac{(x_{F_A} - y_{P_A}\theta)}{(1 - \theta)} \quad (1)$$

Solving (1) and (14-49) simultaneously with a program such as Mathcad, Matlab, or Polymath gives the following results:

θ	x_{R_A}	x_{P_A}	Stage α_S
0.01	0.208	0.407	2.61
0.2	0.168	0.378	3.01
0.4	0.122	0.342	3.74
0.6	0.0733	0.301	5.44
0.8	0.0274	0.256	12.2
0.99	0.000241	0.212	1,120.

Comparing these results to those of Example 14.5, we see that for crossflow, the permeate is richer in oxygen and the retentate is richer in nitrogen. Thus, for a given cut, θ, crossflow is more efficient than perfect mixing, as might be expected.

Also included in the preceding table is the calculated degree of separation for the stage, α_S, defined on the basis of the mole fractions in the permeate and retentate exiting the stage by

$$(\alpha_{A,B})_S = \alpha_S = \frac{(y_{P_A}/x_{R_A})}{(1 - y_{P_A})/(1 - x_{R_A})} \quad (2)$$

Recall that the ideal separation factor, $\alpha_{A,B}^*$, for this example is 2.98. Also, if (2) is applied to the perfect mixing case of Example 14.5, it is found that α_S is 2.603 for $\theta = 0.01$ and decreases slowly with increasing θ until at $\theta = 0.99$, $\alpha_S = 2.548$. Thus, for perfect mixing, $\alpha_S < \alpha^*$ for all θ. Such is not the case for cross-flow. In the above table, $\alpha_S < \alpha^*$ for $\theta > 0.2$, and α_S increases with increasing θ. For $\theta = 0.6$, α_S is almost twice α^*.

Calculations of the degree of separation of a binary mixture in a membrane module utilizing cocurrent or countercurrent flow patterns involve the numerical solution of ordinary differential equations. Derivation of these equations and FORTRAN computer codes for their solution are given by Walawender and Stern [27]. A representative solution is shown in Figure 14.10 for the separation of air (20.9 mol% O_2) for conditions of $\alpha^* = 5$ and $r = 0.2$. For a given cut, θ, it is seen that the best separation is achieved with countercurrent flow. The curve for cocurrent flow lies between those for crossflow and perfect mixing. The readily computed crossflow case is considered to be a good, conservative estimate of membrane module performance. The perfect mixing case for binary mixtures is extended to multicomponent mixtures by Stern et al. [29], who present an iterative procedure. As with crossflow, countercurrent flow also offers the possibility of a separation factor for the stage, α_S, defined by (2) earlier, that can be considerably greater than α^*.

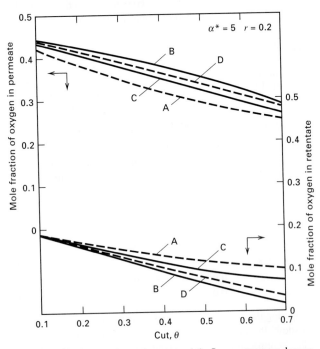

Figure 14.10 Effect of membrane module flow pattern on degree of separation of air. A, perfect mixing; B, countercurrent flow; C, cocurrent flow; D, crossflow.

Cascades

A single membrane module or a number of such modules arranged in parallel or in series without recycle constitutes a single-stage membrane separation process. The extent to which a feed mixture can be separated in a single stage is limited and, as shown in the previous subsection, is determined by the separation factor, α. This factor depends, in turn, on the module flow pattern, the permeability ratio (ideal separation factor), the cut, θ, and the driving force for mass transfer through the membrane. To achieve a higher degree of separation than possible with a single stage, a countercurrent cascade of membrane stages, such as are used in distillation, absorption, stripping, and liquid–liquid extraction, or a hybrid process that couples a membrane separator with another separation operation, such as distillation or adsorption, can be applied. Membrane cascades were presented briefly in Section 5.5. They are now discussed in more detail and illustrated with an example.

A countercurrent recycle cascade of membrane separators, similar to a distillation column, is shown in Figure 14.11a. The feed enters at stage F, somewhere near the middle of the column. Permeate is enriched in components of high permeability in an enriching section, while the retentate is enriched in components of low permeability in a stripping section. The final permeate is withdrawn from stage 1, while the final retentate is withdrawn from stage N. For a cascade, additional factors that affect the degree of separation of the feed are the number of stages and the recycle ratio (permeate recycle rate/permeate product rate). As discussed by Hwang and Kammermeyer [30], it is best to manipulate the cut and reflux rate at each stage so as to force the compositions of the two streams entering each stage to be identical. For example, the composition of retentate leaving Stage 1 and entering Stage 2 would be identical to the composition of permeate flowing from Stage 3 to Stage 2. This corresponds to the least amount of entropy production for the cascade and, thus, the highest second-law efficiency. Such a cascade is referred to as ideal. Calculation methods for cascades are discussed by Hwang and Kammermeyer [30] and utilize the single-stage methods that depend upon the module flow pattern, as discussed in the previous section. The calculations are best carried out with a computer program, but results for a binary mixture can be conveniently displayed on a McCabe–Thiele type diagram in terms of the mole fraction in the permeate leaving each stage, y_i, versus the mole fraction in the retentate leaving each stage, x_i. For a membrane cascade, the equilibrium curve becomes the selectivity curve in terms of the separation factor for the stage, α_S.

In Figure 14.11, it is assumed the pressure drop on the feed or upstream side of the membrane is negligible. Thus, only the permeate must be pumped, if a liquid, or compressed, if a gas, to be sent to the next stage. In the case of gas permeation, compression costs can be high. Thus, membrane cascades for gas permeation are often limited to just two or three stages, with the most common configurations shown in

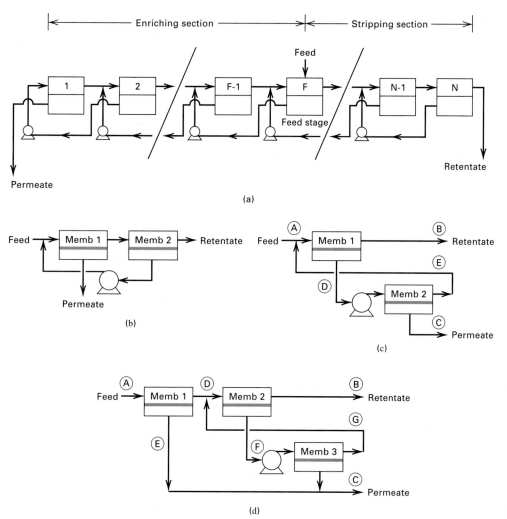

Figure 14.11 Countercurrent recycle cascades of membrane separators.
(a) Multiple-stage unit.
(b) Two-stage stripping cascade. (c) Two-stage enriching cascade.
(d) Two-stage enriching cascade with additional premembrane stage.

Figures 14.11b, c, and d. Compared to a single stage, the two-stage stripping cascade is designed to obtain a more-pure retentate, whereas a more-pure permeate is the goal of the two-stage enriching cascade. The addition of the premembrane stage, shown in Figure 14.11d, may be attractive when the feed concentration is low in the component to be passed preferentially through the membrane, the desired permeate purity is high, the separation factor is low, and/or a high recovery of the more permeable component is desired. An example of the application of the enrichment cascades is given by Spillman [31] for the removal of carbon dioxide from natural gas (assumed to be methane) using cellulose-acetate membranes in spiral-wound modules that approximate crossflow. The ideal separation factor, $\alpha^*_{CO_2, CH_4}$ is 21. Results of computer calculations are given in Table 14.8 for a single stage (not shown in Figure 14.11), a two-stage enriching cascade (Figure 14.11c), and a two-stage enriching cascade with an additional premembrane stage (Figure 14.11d). Carbon dioxide flows through the membrane faster than methane. In all three cases, the feed is 20 million (MM) scfd of 7 mol% CO_2 in methane at 850 psig (about 865 psia) and the retentate is 98 mol% in methane. For each stage, the

downstream (permeate side) membrane pressure is 10 psig (about 25 psia). In Table 14.8, for all three cases, stream A is the feed, stream B is the final retentate, and stream C is the final permeate. Case 1 achieves a 90.2% recovery of methane. Case 2 increases that recovery to 98.7%. Case 3 achieves an intermediate recovery of 94.6%. The following degrees of separation are computed from data given in Table 14.8:

Case	α_S for Membrane Stage		
	1	2	3
1	28	—	—
2	28	57	—
3	20	19	44

We can also compute overall degrees of separation for the cascades, α_C, for cases 2 and 3, giving values of 210 and 51, respectively.

External Mass-Transfer Resistances

Thus far, all of the resistance to mass transfer has been assumed to be associated with the membrane. Thus, concentrations in the fluid at the upstream and downstream faces of the

Table 14.8 Separation of N_2 and CH_4 with Membrane Cascades

Case 1: Single Membrane Stage:

	Stream		
	A Feed	B Retentate	C Permeate
Composition (mole%)			
CH_4	93.0	98.0	63.4
CO_2	7.0	2.0	36.6
Flow rate (MM SCFD)	20.0	17.11	2.89
Pressure (psig)	850	835	10

Case 2: Two-Stage Enriching Cascade (Figure 14.11c):

	Stream				
	A	B	C	D	E
Composition (mole%)					
CH_4	93.0	98.0	18.9	63.4	93.0
CO_2	7.0	2.0	81.1	36.6	7.0
Flow rate (MM SCFD)	20.00	18.74	1.26	3.16	1.90
Pressure (psig)	850	835	10	10	850

Case 3: Two-Stage Enriching Cascade with Premembrane Stage (Figure 14.11d):

	Stream						
	A	B	C	D	E	F	G
Composition (mole%)							
CH_4	93.0	98.0	49.2	96.1	56.1	72.1	93.0
CO_2	7.0	2.0	50.8	3.9	43.9	27.9	7.0
Flow rate (MM SCFD)	20.00	17.95	2.05	19.39	1.62	1.44	1.01
Pressure (psig)	850	835	10	840	10	10	850

Note: MM, million.

membrane have been assumed equal to the respective bulk-fluid concentrations on either side of the membrane. When mass-transfer resistances external to the membrane are not negligible, concentration or partial-pressure gradients exist in the boundary layers or films adjacent to the membrane surfaces, as is illustrated for four cases in Figure 14.7. For given bulk-fluid concentrations, the presence of these resistances reduces the driving force for mass transfer across the membrane and, therefore, the flux of penetrant.

In general, the solution-diffusion mechanism for gas permeation is quite slow compared to diffusion in the gas boundary layers or film adjacent to the membrane, so external mass-transfer resistances are negligible and $P_{i_F} = P_{i_0}$ and $P_{i_P} = P_{i_L}$ in Figure 14.7d. Because diffusion in liquid boundary layers and films can be slow, concentration polarization cannot be neglected in membrane processes that involve liquids, such as dialysis, reverse osmosis, and pervaporation. The need to consider the effect of concentration polarization is of particular importance in reverse osmosis,

where the effect can reduce the water flux and increase the salt flux, making it more difficult to obtain potable water.

Consider a membrane process of the type in Figure 14.7a, involving liquids with a porous membrane. At steady state, rates of mass transfer of a penetrating species, i, through the three resistances are as follows, assuming no change in the area for mass transfer across the thickness of the membrane,

$$N_i = k_{i_F}(c_{i_F} - c_{i_0}) = \frac{D_{e_i}}{l_M}(c_{i_0} - c_{i_L}) = k_{i_P}(c_{i_L} - c_{i_P})$$

where D_{e_i} is given by (14-14).

If these three equations are combined to eliminate the intermediate concentrations, c_{i_0} and c_{i_L}, we obtain

$$N_i = \frac{c_{i_F} - c_{i_P}}{\dfrac{1}{k_{i_F}} + \dfrac{l_M}{D_{e_i}} + \dfrac{1}{k_{i_P}}} \tag{14-52}$$

Now consider the membrane process in Figure 14.7b, involving liquids with a dense (nonporous) membrane, for which the solution-diffusion mechanism is used for mass transfer through the membrane, as given by (14-26). At steady state, for constant mass-transfer area, rates of species mass transfer through the three resistances are:

$$N_i = k_{i_F}(c_{i_F} - c'_{i_0}) = \frac{K_i D_i}{l_M}(c'_{i_0} - c'_{i_L}) = k_{i_P}(c'_{i_L} - c_{i_P})$$

If these three equations are combined to eliminate the intermediate concentrations, c'_{i_0} and c'_{i_L}, we obtain

$$N_i = \frac{c_{i_F} - c_{i_P}}{\dfrac{1}{k_{i_F}} + \dfrac{l_M}{K_i D_i} + \dfrac{1}{k_{i_P}}} \qquad (14\text{-}53)$$

in (14-52) and (14-53), where k_{i_F} and k_{i_P} are mass-transfer coefficients for the feed-side and permeate-side boundary layers or films. The three terms in the denominator of the right-hand side are the resistances to the mass flux. In general, the mass-transfer coefficients depend on fluid properties, flow-channel geometry, and flow regime (turbulent or laminar). In the laminar-flow regime, a long entry region may exist where the mass-transfer coefficient changes with the distance, L, from the entry of the membrane channel. Estimation of the coefficients is complicated by fluid velocities that change because of mass exchange between the two fluids. In (14-52) and (14-53), the membrane resistances, l_M/D_{e_i} and $l_M/K_i D_i$, respectively, can be replaced by l_M/P_{M_i} or $\bar{P}_{M_i}$.

Typical mass-transfer coefficients for channel flow can be obtained from the general empirical film-model correlation [32]:

$$N_{\text{Sh}} = k_i d_H / D_i = a N_{\text{Re}}^b N_{\text{Sc}}^{0.33} (d_H/L)^d \qquad (14\text{-}54)$$

where

$$N_{\text{Re}} = d_H v \rho / \mu$$
$$N_{\text{Sc}} = \mu / \rho D_i$$
$$d_H = \text{hydraulic diameter}$$
$$v = \text{velocity}$$

The constants a, b, and d are as follows:

Flow Regime	Flow Channel Geometry	d_H	a	b	d
Turbulent, ($N_{\text{Re}} > 10,000$)	Circular tube	D	0.023	0.8	0
	Rectangular channel	$2hw/(h+w)$	0.023	0.8	0
Laminar, ($N_{\text{Re}} < 2,100$)	Circular tube	D	1.86	0.33	0.33
	Rectangular channel	$2hw/(h+w)$	1.62	0.33	0.33

where

$w = $ width of channel

$h = $ height of channel

$L = $ length of channel

EXAMPLE 14.7

A dilute solution of solute A in solvent B is passed through a tubular-membrane separator, where the feed flows through the tubes. At a certain location, the solute concentrations are 5.0×10^{-2} kmol/m^3 and 1.5×10^{-2} kmol/m^3, respectively, on the feed and permeate sides. The permeance of the membrane for solute A is given by the membrane vendor as 7.3×10^{-5} m/s. If the tube-side Reynolds number is 15,000, the feed-side solute Schmidt number is 500, the diffusivity of the feed-side solute is 6.5×10^{-5} cm^2/s, and the inside diameter of the tube is 0.5 cm, estimate the flux of the solute through the membrane if the mass-transfer resistance on the permeate side of the membrane is negligible.

SOLUTION

The flux of the solute is given by the permeance form of (14-52) or (14-53):

$$N_A = \frac{c_{A_F} - c_{A_P}}{\dfrac{1}{k_{A_F}} + \dfrac{1}{\bar{P}_{M_A}} + 0}$$

$$c_{A_F} - c_{A_P} = 5 \times 10^{-2} - 1.5 \times 10^{-2} = 3.5 \times 10^{-2} \text{ kmol/m}^3 \qquad (1)$$

$$\bar{P}_{M_A} = 7.3 \times 10^{-5} \text{ m/s}$$

From (14-54), for turbulent flow in a tube, since $N_{\text{Re}} > 10,000$:

$$k_{A_F} = 0.023 \frac{D_A}{D} N_{\text{Re}}^{0.8} N_{\text{Sc}}^{0.33}$$

$$= 0.023 \left(\frac{6.5 \times 10^{-5}}{0.5}\right)(15,000)^{0.8}(500)^{0.33}$$

$$= 0.051 \text{ cm/s or } 5.1 \times 10^{-4} \text{ m/s}$$

From (1),

$$N_A = \frac{3.5 \times 10^{-2}}{\dfrac{1}{5.1 \times 10^{-4}} + \dfrac{1}{7.3 \times 10^{-5}}} = 2.24 \times 10^{-6} \text{ kmol/s} - \text{m}^2$$

The fraction of the total resistance due to the membrane is

$$\frac{\dfrac{1}{7.3 \times 10^{-5}}}{\dfrac{1}{5.1 \times 10^{-4}} + \dfrac{1}{7.3 \times 10^{-5}}} = 0.875 \text{ or } 87.5\%$$

Concentration Polarization and Fouling

When gases are produced during electrolysis, they accumulate on and around the electrodes of the electrolytic cell, reducing the flow of electric current. This phenomenon is referred to as polarization. A similar phenomenon, *concentration polarization*, occurs in membrane separators when the membrane is permeable to molecules of A, but relatively impermeable to molecules of B. Thus, molecules of B are carried by bulk flow to the upstream surface of the membrane where they accumulate, causing their concentration at the surface of the membrane to increase in a "polarization layer." The equilibrium concentration of B in this layer is reached when its back-diffusion to the bulk fluid on the feed-retentate side equals its bulk flow toward the membrane.

Concentration polarization is most common in pressure-driven membrane separations, such as reverse osmosis and ultrafiltration, where it can reduce the flux of molecules of A through the membrane. The polarization effect can be particularly serious if the concentration of B attains its solubility limit next to the membrane surface. A precipitate of gel may then form, the result being fouling on the membrane surface or within membrane pores, with a further reduction in the flux of A. In general, concentration polarization and fouling are most severe at high values of the flux of A. Examples of the effects of concentration polarization and fouling and simplified theoretical treatments are given in the sections on reverse osmosis and ultrafiltration.

14.4 DIALYSIS AND ELECTRODIALYSIS

In a dialysis membrane-separation process, shown in Figure 14.12, the feed is a liquid, at pressure P_1, containing solvent, solutes of type A, and solutes of type B and/or insoluble, but dispersed, colloidal matter. A sweep liquid or wash of the same solvent is fed at pressure P_2 to the other side of the membrane. The membrane is thin with micropores of a size such that solutes of type A can pass through by a concentration driving force. Solutes of type B are larger in molecular size than those of type A and pass through the membrane only with difficulty or not at all. This transport of solutes A and B through the membrane is called dialysis. Colloids do not pass through the membrane. With pressure $P_1 = P_2$, the solvent may also pass through the membrane, but by a concentration driving force acting in the opposite direction. The transport of the solvent is called osmosis. By elevating P_1 above P_2, solvent osmosis can be reduced or eliminated. The products of a dialysis unit (dialyzer) are a liquid *diffusate* (permeate) containing solvent, solutes of type A, and smaller amounts of solutes of type B; and a *dialysate* (retentate) of the solvent and remaining solutes of types A and B, and colloidal matter. Ideally, the dialysis unit would enable a perfect separation between solutes of type A and solutes of type B and any colloidal matter. However, at best only a fraction of the solutes of type A are recovered in the diffusate, even when solutes of type B do not pass through the membrane.

For example, when dialysis is used to recover sulfuric acid from an aqueous stream containing sulfate salts, the following results are obtained, as reported by Chamberlin and Vromen [33]:

	Streams in		Streams out	
	Feed	Wash	Dialysate	Diffusate
Flow rate, gph	400	400	420	380
H_2SO_4, g/L	350	0	125	235
$CuSO_4$, g/L as Cu	30	0	26	2
$NiSO_4$, g/L as Ni	45	0	43	0

Thus, about 64% of the H_2SO_4 is recovered in the diffusate, accompanied by only about 6% of the $CuSO_4$, and essentially no $NiSO_4$.

Dialysis is closely related to other membrane processes that use other driving forces for separating liquid mixtures, including (1) reverse osmosis, which depends upon a transmembrane pressure difference for solute and/or solvent transport; (2) electrodialysis and electro-osmosis, which depend upon a transmembrane electrical-potential difference for solute and solvent transport, respectively; and (3) thermal osmosis, which depends on a transmembrane temperature difference for solute and solvent transport.

Dialysis is attractive when the concentration differences for the main diffusing solutes are large and the permeability differences between those solutes and the other solute(s) and/or colloids is large. Although dialysis has been known since the work of Graham in 1861 [34], commercial applications of dialysis do not rival reverse osmosis and gas permeation. Nevertheless, dialysis has been applied to a number of separations, including (1) recovery of sodium hydroxide from a 17–20 wt% caustic viscose liquor contaminated with hemicellulose to produce a diffusate of 9–10 wt% caustic; (2) recovery of chromic, hydrochloric, and hydrofluoric acids from contaminating metal ions; (3) recovery of sulfuric acid from aqueous solutions containing nickel sulfate; (4) removal of alcohol from beer to produce a reduced-alcohol beer; (5) recovery of nitric and hydrofluoric acids from spent stainless-steel pickle liquor; (6) removal of mineral acids from organic compounds; (7) removal of low-molecular-weight contaminants from polymers; and (8) purification of pharmaceuticals. Also of great importance is hemodialysis, in which urea, creatine, uric acid, phosphates, and chlorides are removed from blood without removing essential higher-molecular-weight compounds and blood cells. This dialysis device is called an artificial kidney.

Typical microporous-membrane materials used in dialysis are hydrophilic, including cellulose, cellulose acetate, various acid-resistant polyvinyl copolymers, polysulfones, and polymethylmethacrylate, typically less than 50 μm thick and with pore diameters of 15 to 100 Å. The most common membrane modules are plate-and-frame and hollow-fiber. Compact hollow-fiber hemodialyzers, such as the one shown

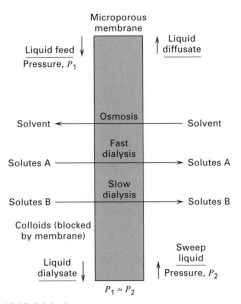

Figure 14.12 Dialysis.

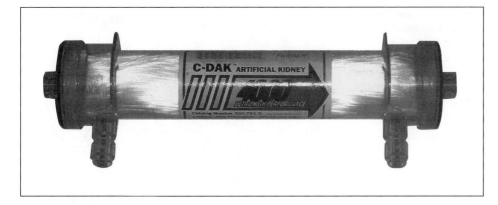

Figure 14.13 Artificial kidney.

in Figure 14.13, which are widely used, typically contain several thousand 200-μm-diameter fibers with a wall thickness of 20–30 μm and a length of 10–30 cm. Dialysis membranes can be thin because pressures on either side of the membrane are essentially equal.

At a differential location in a dialyzer, the rate of mass transfer of solute across the dialysis membrane is given by

$$dn_i = K_i(c_{i_F} - c_{i_P})dA_M \qquad (14\text{-}55)$$

where K_i is the overall mass-transfer coefficient, which is given in terms of the individual coefficients from the permeability form of (14-52):

$$\frac{1}{K_i} = \frac{1}{k_{i_F}} + \frac{l_M}{P_{M_i}} + \frac{1}{k_{i_P}} \qquad (14\text{-}56)$$

The determination of the membrane area is made by integrating (14-55) taking into account the module flow patterns, the bulk-concentration gradients, and the individual mass-transfer coefficients in (14-56).

One of the oldest membrane materials for use with aqueous solutions is porous cellophane, for which solute permeability is given by (14-14) with $P_{M_i} = D_{e_i}$ and $\bar{P}_{M_i} \cdot l_M$. In the presence of a solution, cellophane will swell to about twice its dry thickness. The wet thickness should be used for l_M. Typical values of parameters given in (14-13) to (14-15) for commercial cellophane are as follows:

> Wet thickness = $l_M = 0.004$ to 0.008 cm;
> porosity = $\epsilon = 0.45$ to 0.60
>
> Tortuosity = $\tau = 3$ to 5;
> pore diameter = $D = 30$ to 50 Å

If solute does not interact with the membrane material, the diffusivity, D_{e_i}, in (14-14) is the ordinary molecular diffusion coefficient, which depends only on solute and solvent properties. However, the membrane may have a profound effect on the solute diffusivity if any of a number of membrane–solute interactions occur, including covalent, ionic, and hydrogen bonding; physical adsorption and chemisorption; and membrane polymer flexibility. Thus, it is preferred to measure $\bar{P}_{M_i}$ experimentally using the actual process fluids.

Although the transport of solvents, such as water, which usually occurs in a direction opposite to the solute, could be formulated in terms of Fick's law, it is more common to measure the solvent flux and report the so-called *water-transport number,* which is the ratio of the water flux to the solute flux, with a negative value indicating transport of solvent in the same direction as the solute. The membrane can also interact with the solvent and even curtail solvent transport. Ideally, the water transport number should be a small value less than $+1.0$. The ideal experimental dialyzer is a batch cell with a variable-speed stirring mechanism on both sides of the membrane so that external mass transfer resistances, $1/k_{i_F}$ and $1/k_{i_P}$ in (14-56), are made negligible. Stirrer speeds greater than 2,000 rpm may be required.

A common dialyzer is the plate-and-frame type of Figure 14.5a. However, for dialysis applications, the frames are arranged vertically. A typical unit might contain 100 square frames, each 0.75×0.75 m on 0.6-cm spacing, equivalent to 56 m² of membrane surface. The dialysis rate for sulfuric acid might be 5 lb/day-ft². More recently developed dialysis units utilize hollow fibers of 200-μm inside diameter, 16-μm wall thickness, and 28-cm length packed into a heat-exchanger-type module to give 22.5 m² of membrane area in a volume that might be one-tenth of the volume of an equivalent plate-and-frame unit.

In a plate-and-frame dialyzer, the flow pattern is nearly countercurrent. Because total flow rates change little and solute concentrations are typically small, it is common to estimate the solute transport rate by assuming a constant overall mass-transfer coefficient with a log-mean concentration driving force. Thus, from (14-55):

$$n_i = K_i A_M (\Delta c_i)_{\text{LM}} \qquad (14\text{-}57)$$

where K_i is given by (14-56). This method is applied in the following example.

EXAMPLE 14.8

A countercurrent-flow, plate-and-frame dialyzer is to be sized to process 0.78 m³/h of an aqueous solution of 300 kg/m³ of H_2SO_4 and smaller amounts of copper and nickel sulfates. A wash water rate of 1.0 m³/h is to be used, and it is desired to recover 30% of the acid at

25°C. From batch laboratory experiments with an acid-resistant vinyl membrane, in the absence of external mass-transfer resistances, a permeance of 0.025 cm/min for the acid and a water transport number of +1.5 are measured. Membrane transport of copper and nickel sulfates is negligible. For these flow rates, experience with plate-and-frame dialyzers indicates that flow will be laminar and the combined external liquid-film mass-transfer coefficients will be 0.020 cm/min. Determine the membrane area required in m^2.

SOLUTION

$m_{H_2SO_4}$ in feed $= 0.78(300) = 234$ kg/h

$m_{H_2SO_4}$ transferred $= 0.3(234) = 70$ kg/h

m_{H_2O} transferred to dialysate $= 1.5(70) = 105$ kg/h

m_{H_2O} in entering wash $= 1.0(1,000) = 1,000$ kg/h

m_P leaving $= 1,000 - 105 + 70 = 965$ kg/h

For mixture densities, assume aqueous sulfuric acid solutions and use the appropriate table in *Perry's Chemical Engineers' Handbook:*

$\rho_F = 1,175$ kg/m^3 $\rho_R = 1,114$ kg/m^3 $\rho_P = 1,045$ kg/m^3

$m_F = 0.78(1,175) = 917$ kg/h

m_R leaving $= 917 + 105 - 70 = 952$ kg/h

Sulfuric acid concentrations:

$c_F = 300$ kg/m^3 $c_{wash} = 0$ kg/m^3

$$c_R = \frac{(234 - 70)}{952}(1,114) = 192 \text{ kg/m}^3$$

$$c_P = \frac{70}{965}(1,045) = 76 \text{ kg/m}^3$$

The log-mean driving force for H_2SO_4 with countercurrent flow of feed and wash:

$$(\Delta c)_{LM} = \frac{(c_F - c_P) - (c_R - c_{wash})}{\ln\left(\dfrac{c_F - c_P}{c_R - c_{wash}}\right)} = \frac{(300 - 76) - (192 - 0)}{\ln\left(\dfrac{300 - 76}{192 - 0}\right)}$$

$$= 208 \text{ kg/m}^3$$

The driving force is almost constant in the membrane module, varying only from 224 to 192 kg/m^3.

From (14-56),

$$K_{H_2SO_4} = \frac{1}{\dfrac{1}{\bar{P}_M} + \left(\dfrac{1}{k}\right)_{combined}} = \frac{1}{\dfrac{1}{0.025} + \dfrac{1}{0.020}}$$

$$= 0.0111 \text{ cm/min} \quad \text{or} \quad 0.0067 \text{ m/h}$$

From (14-57), using mass units instead of molar units:

$$A_M = \frac{m_{H_2SO_4}}{K_{H_2SO_4}(\Delta c_{H_2SO_4})_{LM}} = \frac{70}{0.0067(208)} = 50 \text{ m}^2$$

Electrodialysis

Electrodialysis began in the early 1900s as a modification to dialysis by the addition of electrodes and direct current to increase the rate of dialysis in electrolyte solutions. However, since the 1940s, electrodialysis has developed into a membrane-separation process that differs from dialysis in many ways. Today, electrodialysis refers to an electrolytic process for separating an aqueous, electrolyte feed solution into a concentrate or brine and a dilute or desalted water (diluate) by means of an electric field and ion-selective membranes. A typical electrodialysis process is shown in Figure 14.14, where the four ion-selective membranes shown are of two types arranged in an alternating-series pattern. The cation-selective membranes (C) carry a negative charge, and thus attract and pass positively charged ions (cations), while retarding negative ions. The anion-selective membranes (A) carry a positive charge that attracts and

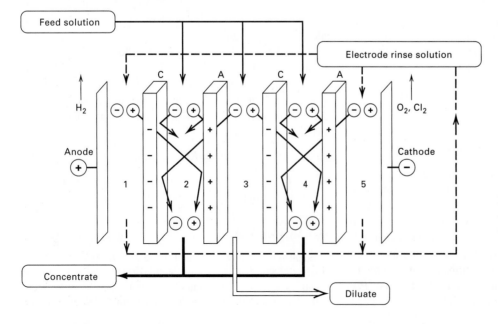

Figure 14.14 Schematic diagram of the electrodialysis process. C, cation transfer membrane; A, anion transfer membrane.

[Adapted from W. S. W. Ho and K. K. Sirkar, editors, "Membrane Handbook," Van Nostrand Reinhold, New York (1992).]

permits passage of negative ions (anions). Both types of membranes are impervious to water. The net result is that both anions and cations are concentrated in compartments 2 and 4, from which concentrate is withdrawn, and ions are depleted in compartment 3, from which the diluate is withdrawn. Compartment pressures are essentially equal. Compartments 1 and 5 are bounded on the far sides by the anode and cathode, respectively. A direct-current voltage is applied (e.g., with a battery or direct-current generator) across the anode and cathode, causing current to flow by metallic conduction of electrons through wiring from the anode to the cathode and then through the cell by ionic conduction from the cathode back to the anode. Both electrodes are chemically neutral metals, with the anode being typically stainless steel and the cathode typically platinum-coated tantalum, niobium, or titanium. Thus, the electrodes are neither oxidized nor reduced.

But half reactions must occur at the two electrodes. Typically, the most easily oxidized species is oxidized at the anode and the most easily reduced species is reduced at the cathode. With inert electrodes, the result at the cathode is the reduction of water by the half reaction

$$2H_2O + 2e^- \rightarrow 2OH^- + H_{2(g)}, \; E^0 = -0.828 \, \text{V}$$

The oxidation half reaction at the anode is

$$H_2O \rightarrow 2e^- + \tfrac{1}{2}O_{2(g)} + 2H^+, \; E^0 = -1.23 \, \text{V}$$

or, if chloride ions are present:

$$2Cl^- \rightarrow 2e^- + Cl_{2(g)}, \; E^0 = -1.360 \, \text{V}$$

where the electrode potentials are the standard values at 25°C for one molar solution of ions and partial pressures of one atmosphere for the gaseous products. Values of E^0 can be corrected for nonstandard conditions by the Nernst equation.

The corresponding overall cell reactions are:

$$3H_2O \rightarrow H_{2(g)} + \tfrac{1}{2}O_{2(g)} + 2H^+ + 2OH^-$$

or

$$2H_2O + 2Cl^- \rightarrow 2OH^- + H_{2(g)} + Cl_{2(g)}, \; E^0_{cell} = -2.058 \, \text{V}$$

The net reaction for the first case is

$$H_2O \rightarrow H_{2(g)} + \tfrac{1}{2}O_{2(g)}, \; E^0_{cell} = -2.188 \, \text{V}$$

The electrode rinse solution that circulates through compartments 1 and 5 is typically acidic to neutralize the OH ions formed in compartment 1 and prevent precipitation of compounds such as $CaCO_3$ and $Mg(OH)_2$.

The most widely used ion-exchange membranes for electrodialysis, first reported by Juda and McRae [35] in 1950, are (1) cation-selective membranes containing negatively charged groups fixed to a polymer matrix, and (2) anion-selective membranes containing positively charged groups fixed to a polymer matrix. The former, shown schematically in Figure 14.15, includes fixed anions, mobile cations (called counterions), and mobile anions (called co-ions). The latter are almost completely excluded from the polymer

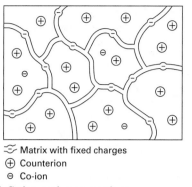

☞ Matrix with fixed charges
⊕ Counterion
⊖ Co-ion

Figure 14.15 Cation-exchange membrane.
[From H. Strathmann, *Sep. and Purif. Methods,* **14**(1), 41–66 (1985) with permission.]

matrix by electrical repulsion, called the Donnan effect. For perfect exclusion, only cations are transferred through the membrane. In practice, the exclusion is better than 90%.

A typical cation-selective membrane is made of polystyrene cross-linked with divinylbenzene and sulfonated to produce fixed sulfonate, $-SO_3^-$, anion groups. A typical anion-selective membrane of the same polymer contains quaternary ammonium groups such as $-NH_3^+$. Membranes are 0.2–0.5 mm in thickness and reinforced with a screen to provide mechanical stability. The membranes, which are made in flat sheets, contain 30 to 50% water and have a network of pores too small to permit water transport.

A cell pair or unit cell consists of one cation-selective membrane and one anion-selective membrane. Although Figure 14.14 shows an electrodialysis system with two cell pairs, a commercial electrodialysis system is a large stack of membranes patterned after a plate-and-frame configuration that, according to Applegate [2] and the *Membrane Handbook* [6], may contain 100 to 600 cell pairs. In a stack, membranes of from 0.4 to 1.5 m² surface area each are separated by from 0.5 to 2 mm with spacer gaskets. The total voltage or electrical potential applied across the cell includes (1) the electrode potentials discussed earlier, (2) overvoltages due to gas formation at the two electrodes, (3) the voltage required to overcome the ohmic resistance of the electrolyte in each compartment, (4) the voltage required to overcome the resistance in each membrane, and (5) the voltage required to overcome concentration-polarization effects caused by mass-transfer resistances in the electrolyte solutions adjacent to the membrane surface. For large stacks, the latter three voltage increments predominate and depend upon the current density (amps flowing through the stack per unit surface area of membranes). A typical voltage drop across a cell pair is 0.5–1.5 V. Current densities are in the range of 5–50 mA/cm². Thus, a stack of 400 membranes (200 unit cells) of 1 m² surface area each might require 200 V at 100 A. Typically 50 to 90% of brackish water is converted to potable water, depending on concentrate recycle.

As the current density is increased for a given membrane surface area, the concentration-polarization effect increases.

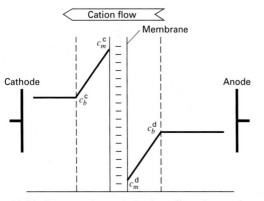

Figure 14.16 Concentration-polarization effects for a cation-exchange membrane.

[From H. Strathmann, *Sep. and Purif. Methods*, **14**(1), 41–66 (1985) with permission.]

A schematic diagram of this effect for a single cation-selective membrane is shown in Figure 14.16, where c_m refers to cation concentrations in the membrane, c_b refers to bulk electrolyte cation concentrations, and superscripts c and d refer to concentrate side and dilute side, respectively. The maximum or limiting current density is reached when c_m^d reaches zero. Typically, an electrodialysis cell is operated at 80% of the limiting current density, which is determined by experiment. The corresponding cell voltage or resistance is also determined experimentally.

The amounts of gases formed at the electrodes at the two ends of the stack are governed by *Faraday's law of electrolysis*. During electrolysis, one Faraday (96,520 coulombs) of electricity reduces at the cathode and oxidizes at the anode an equivalent of oxidizing and reducing agent corresponding to the transfer of 6.023×10^{23} (Avogadro's number) electrons through wiring from the anode to the cathode. In general, it takes a very large quantity of electricity to form appreciable quantities of gases in an electrodialysis process.

Of more importance in the design or operation of an electrodialysis process are the membrane area and electrical-energy requirements as discussed by Applegate [2] and Strathmann [36]. The membrane area is estimated from the current density, rather than from a permeability and mass-transfer resistances, by applying Faraday's law:

$$A_M = \frac{FQ\Delta c}{i\xi} \qquad (14\text{-}58)$$

where

A_M = total area of all cell pairs, m^2

F = Faraday's constant (96,520 amp-s/equivalent)

Q = volumetric flow rate of the diluate (potable water), m^3/s

Δc = difference between feed and diluate ion concentration in equivalents/m^3

i = current density, amps/m^2 of a cell pair, usually about 80% of $i_{\max}$

ξ = current efficiency < 1.00

The last variable accounts for the fact that not all of the current is effective in transporting the selected ions through the membranes. Inefficiencies are caused by a Donnan exclusion of less than 100%, some transfer of water through the membranes, current leakage through manifolds, etc.

Power consumption is given by

$$P = IE \qquad (14\text{-}59)$$

where

P = power, W, I = electric current flow through the stack, and E = voltage across the stack.

The electrical-current flow is given by a rearrangement of (14-58):

$$I = \frac{FQ\Delta c}{n\xi} \qquad (14\text{-}60)$$

where n is the number of cell pairs.

The main application of electrodialysis is to the desalinization of brackish water in the salt concentration range of 500 to 5,000 ppm (mg/L). Below this range, ion exchange is more economical, whereas above this range, to 50,000 ppm, reverse osmosis is preferred. However, electrodialysis cannot produce water with a very low dissolved-solids content because of the high electrical resistance of dilute solutions. Other applications include recovery of nickel and copper from electroplating rinse water; deionization of cheese whey, fruit juices, wine, milk, and sugar molasses; separation of salts, acids, and bases from organic compounds; and recovery of organic compounds from their salts. Bipolar membranes, prepared by laminating a cation-selective membrane and an anion-selective membrane back-to-back, can be used to produce sulfuric acid and sodium hydroxide from a sodium sulfate solution.

EXAMPLE 14.9

Estimate membrane area and electrical-energy requirements for an electrodialysis process to reduce the salt (NaCl) content of 24,000 m^3/day of brackish water from 1,500 mg/L to 300 mg/L with a 50% conversion. Assume each membrane has a surface area of 0.5 m^2 and each stack contains 300 cell pairs. A reasonable current density is 5 mA/cm^2 and the current efficiency is 0.8 (80%).

SOLUTION

Use (14-58) to estimate membrane area

$$F = 96{,}520 \text{ A/equiv,}$$
$$Q = (24{,}000)(0.5)/(24)(3{,}600) = 0.139 \text{ m}^3/\text{s}$$
$$\text{MW}_{\text{NaCl}} = 58.5, \ i = 5 \text{ mA/cm}^2 = 50 \text{ A/m}^2$$
$$\Delta c = (1{,}500 - 300)/58.5 = 20.5 \text{ mmol/L or } 20.5 \text{ mol/m}^3$$
$$= 20.5 \text{ equiv/m}^3$$
$$A_M = \frac{(1)(96{,}520)(0.139)(20.5)}{(50)(0.8)} = 6{,}876 \text{ m}^2$$

Each stack contains 300 cell pairs with a total area of $0.5(300) = 150 \, m^2$. Therefore, number of stacks $= 6,876/150 = 46$ in parallel.

From (14-60), electrical current flow is given by

$$I = \frac{(96,500)(0.139)(20.5)}{(300)(0.8)}$$
$$= 1,146 \, A \text{ or } I/\text{stack} = 1,146/46 = 25 \, A/\text{stack}$$

To obtain the electrical power, we need to know the average voltage drop across each cell pair. Assume a value of 1 V. From (14-59) for 300 cell pairs.

$$P = (1,146)(1)(300) = 344,000 \, W = 344 \, kW$$

Additional energy is required to pump feed, recycle concentrate, and electrode rinse.

It is also instructive to estimate the amount of feed that would be electrolyzed (say, as water to hydrogen and oxygen gases) at the electrodes. From the half-cell reactions presented earlier, half a molecule of H_2O is electrolyzed for each electron or, 0.5 mol H_2O is electrolyzed for each faraday of electricity.

1,146 amps $= 1,146$ coulombs/s or $(1,146)(3,600)(24) = 99,010,000$ coulombs/day or $99,010,000/96,520 = 1,026$ faradays/day. This electrolyzes $(0.5)(1,026) = 513$ mol/day of water. The feed rate is 12,000 m^3/day, or

$$\frac{(12,000)(10^6)}{18} = 6.7 \times 10^8 \, \text{mol/day}$$

Therefore, the amount of water electrolyzed is negligible.

14.5 REVERSE OSMOSIS

Osmosis, from the Greek word for "push," refers to the passage of a solvent, such as water, through a membrane that is much more permeable to the solvent (A) than to the solute(s) (B) (e.g., inorganic ions). The first recorded account of osmosis was given in 1748 by Nollet, whose experiments were conducted with water, an alcohol, and an animal-bladder membrane. The important aspects of osmosis are illustrated by example in Figure 14.17, where all solutions are at 25°C. In the initial condition (a), seawater of approximately 3.5 wt% dissolved salts and at 101.3 kPa is on the left side of the membrane, while pure water at the same pressure is on the right side. The dense membrane is permeable to water, but not to the dissolved salts. By osmosis, water passes from the right side to the seawater on the left side, causing dilution with respect to dissolved salts. At equilibrium, the condition of Figure 14.17b is reached, wherein some pure water still resides on the right side and seawater, less concentrated in salt, resides on the left side. The pressure, P_1, on the left side is now greater than the pressure, P_2, on the right side, with the difference, π, referred to as the *osmotic pressure*.

The process of osmosis is not useful as a separation process because the solvent is transferred in the wrong direction, resulting in mixing rather than separation. However, the direction of transfer of solvent through the membrane can be reversed, as shown in Figure 14.17c by applying a pressure, P_1, on the left side of the membrane, that is higher than the sum of the osmotic pressure and the pressure, P_2, on the right side: that is, $P_1 - P_2 > \pi$. Now water in the seawater is transferred to the pure water, and the seawater becomes more concentrated in dissolved salts. This phenomenon, called *reverse osmosis,* can be used to partially remove a solvent from a solute-solvent mixture. As discussed later, an important factor in developing a reverse osmosis separation process is the osmotic pressure, π, of the feed mixture. In general, as discussed in more detail later, π is proportional to the solute concentration. For pure water, $\pi = 0$.

In a reverse-osmosis (RO) membrane-separation process, as shown in Figure 14.18, the feed is a liquid at high pressure, P_1, containing solvent (e.g., water) and solubles (e.g., inorganic salts and, perhaps, colloidal matter). No sweep liquid is used, but the other side of the membrane is maintained at a much lower pressure, P_2. A dense membrane, such as an acetate or aromatic polyamide, is used that is permselective for the solvent. To withstand the large pressure differential, the membrane must be thick. Accordingly, asymmetric or thin-wall composite membranes, having a thin, dense skin or layer on a thick, porous support, are used. The products of reverse osmosis are a permeate of almost pure solvent and a retentate of solvent-depleted feed. However, a perfect separation between the solvent and solute is not achieved, since only a fraction of the solvent in the feed is transferred to the permeate.

Reverse osmosis is applied to the desalinization and purification of seawater, brackish water, and wastewater. Prior to 1980, multistage, flash distillation was the main process for the desalinization of water. By 1990, this situation was dramatically reversed, making RO the dominant process for new construction. The dramatic shift from a thermally

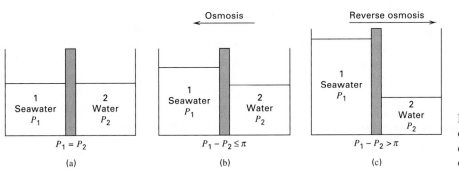

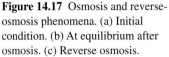

Figure 14.17 Osmosis and reverse-osmosis phenomena. (a) Initial condition. (b) At equilibrium after osmosis. (c) Reverse osmosis.

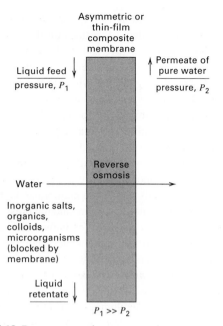

Figure 14.18 Reverse osmosis.

driven process to a more economical pressure-driven process was made possible through the development by Loeb and Sourirajan [7] of an asymmetric membrane that allows pressurized water to pass through at a high rate, while almost preventing transmembrane flows of dissolved salts, organic compounds, colloids, and microorganisms. Today more than 1,000 RO desalting plants are producing more than 750,000,000 gallons per day of potable water worldwide.

According to Baker et al. [5], the use of RO to desalinize water is accomplished mainly with spiral-wound and hollow-fiber membrane modules utilizing cellulose triacetate, cellulose diacetate, and aromatic polyamide membrane materials. Cellulose acetates are susceptible to biological attack, and acidic or basic hydrolysis back to cellulose, making it necessary to chlorinate the feed water and control the pH within the range of 4.5 to 7.5. Polyamides are not susceptible to biological attack and resist hydrolysis in the pH range of 4 to 11. However, polyamides are attacked by chlorine.

The preferred membrane for the desalinization of seawater, which contains about 3.5 wt% dissolved salts and has an osmotic pressure of 350 psia, is a spiral-wound, multileaf module of polyamide, thin-film composite operating at a feed pressure of 800 to 1,000 psia. With a transmembrane water flux of 9 gal/ft²-day (0.365 m³/m²-day), this module can recover 45% of the water at a purity of about 99.95 wt%. A typical cylindrical module is 8 inches in diameter by 40 inches long, containing 365 ft² (33.9 m²) of membrane surface. Such modules resist fouling by colloidal and particulate matter, but the seawater must be treated with sodium bisulfate to remove oxygen and/or chlorine.

For the desalinization of brackish water containing less than 0.5 wt% dissolved salts, hollow-fiber modules of high packing density, and containing fibers of cellulose acetates or aromatic polyamides, are used if fouling is not serious.

Because the osmotic pressure is much lower (<50 psi), feed pressures can be less than 250 psia. Transmembrane fluxes may be as high as 20 gal/ft²-day.

Other uses of reverse osmosis, usually on a smaller scale than the desalinization of water to produce potable water, include (1) the treatment of industrial wastewater to remove heavy-metal ions, nonbiodegradable substances, and other components of commercial value; (2) the treatment of rinse water from electroplating processes to obtain a metal-ion concentrate and a permeate that can be reused as a rinse; (3) the separation of sulfites and bisulfites from effluents in pulp and paper processes; (4) the treatment of wastewater in dyeing processes; (5) the recovery of constituents having food value from wastewaters in food-processing plants (e.g., lactose, lactic acid, sugars, and starches); (6) the treatment of municipal water to remove inorganic salts, low-molecular-weight organic compounds, viruses, and bacteria; (7) the dewatering of certain food products such as coffee, soups, tea, milk, orange juice, and tomato juice; and (8) the concentration of amino acids and alkaloids. In such applications, membranes must have chemical, mechanical, and thermal stability to be competitive with other processes.

As with all membrane processes where the feed being separated is a liquid, three resistances to mass transfer must be considered: the membrane resistance and the two fluid-film or boundary-layer resistances on either side of the membrane. If the permeate is pure solvent, then there is no film resistance on that side of the membrane.

Although the driving force for the transport of water through the dense membrane is the concentration or activity difference in and across the membrane, common practice is to use a driving force based on osmotic pressure. Consider the reverse-osmosis process of Figure 14.17c. At thermodynamic equilibrium, solvent chemical potentials of fugacities on the two sides of the membrane must be equal. Thus,

$$f_A^{(1)} = f_A^{(2)} \qquad (14\text{-}61)$$

From definitions in Table 2.2, rewrite (14-61) in terms of activities:

$$a_A^{(1)} f_A^0\{T, P_1\} = a_A^{(2)} f_A^0\{T, P_2\} \qquad (14\text{-}62)$$

For pure solvent, A, $a_A^{(2)} = 1$. For seawater, $a_A^{(1)} = x_A^{(1)} \gamma_A^{(1)}$. Substitution into (14-62) gives

$$f_A^0\{T, P_2\} = x_A^{(1)} \gamma_A^{(1)} f_A^0\{T, P_1\} \qquad (14\text{-}63)$$

Standard-state, pure-component fugacities f^0 increase with increasing pressure. Thus, if $x_A^{(1)} \gamma_A^{(1)} < 1$, then from (14-63), $P_1 > P_2$. The pressure difference $P_1 - P_2$ is shown as a hydrostatic-head difference in Figure 14.17b. This difference, which can be observed experimentally, is defined as the osmotic pressure, π.

To relate π to solvent or solute concentration, we apply the Poynting correction of (2-28), which for an incompressible liquid of specific volume, v_A, gives

$$f_A^0\{T, P_2\} = f_A^0\{T, P_1\} \exp\left[\frac{v_{A_L}(P_2 - P_1)}{RT}\right] \qquad (14\text{-}64)$$

Substitution of (14-63) into (14-64) gives

$$\pi = P_1 - P_2 = -\frac{RT}{v_{A_L}} \ln \left(x_A^{(1)} \gamma_A^{(1)} \right) \quad (14\text{-}65)$$

Thus, osmotic pressure is a thermodynamic quantity that replaces activity.

For a mixture, on the feed or retentate side of the membrane, that is dilute in the solute, $\gamma_A^{(1)} = 1$. Also, $x_A^{(1)} = 1 - x_B^{(1)}$ and $\ln(1 - x_B^{(1)}) \approx -x_B^{(1)}$. Substitution into (14-65) gives

$$\pi = P_1 - P_2 = RT \, x_B^{(1)} / v_{A_L} \quad (14\text{-}66)$$

Finally, since $x_B^{(1)} \approx n_B/n_A$, $n_A v_{A_L} = V$, and $n_B/V = c_B$, (14-66) becomes

$$\pi \approx RT c_B \quad (14\text{-}67)$$

which was cited in Exercise 1.8. For applications to the reverse osmosis of seawater, Applegate [2] suggests the approximate expression

$$\pi = 1.12T \sum \bar{m}_i \quad (14\text{-}68)$$

where π is in psia, T is in K, and $\sum \bar{m}_i$ is the summation of molarities of all dissolved ions and nonionic species in the solution in mol/L. More exact expressions for estimating π are those of Stoughton and Lietzke [38].

In the general case, when reverse osmosis takes place with solute on each side of the membrane, then at equilibrium, $(P_1 - \pi_1) = (P_2 - \pi_2)$. Accordingly, as discussed by Merten [37], the driving force for solvent transport through the membrane is $\Delta P - \Delta \pi$, and the rate of mass transport is

$$N_{H_2O} = \frac{P_{M_{H_2O}}}{l_M}(\Delta P - \Delta \pi) \quad (14\text{-}69)$$

where

ΔP = hydraulic pressure difference across the membrane
    = $P_{\text{feed}} - P_{\text{permeate}}$

$\Delta \pi$ = osmotic pressure difference across the membrane
    $\pi_{\text{feed}} - \pi_{\text{permeate}}$

Often, $\pi_{\text{permeate}} \approx 0$ because the permeate is almost pure solvent.

The flux of solute (e.g., salt) is given by (14-26) in terms of membrane concentrations, and thus is independent of the ΔP across the membrane. Accordingly, the higher the ΔP, the purer the permeate water. Alternatively, the flux of salt may be conveniently expressed in terms of *salt passage, SP,* defined by

$$SP = (c_{\text{salt}})_{\text{permeate}} / (c_{\text{salt}})_{\text{feed}} \quad (14\text{-}70)$$

Values of SP decrease with increasing ΔP. *Salt rejection* is given by $SR = 1 - SP$.

For brackish water of 1,500 mg/L as NaCl, at 25°C, (14-68) predicts $\pi = 17.1$ psia. For seawater of 35,000 mg/L as NaCl, at 25C, (14-68) predicts $\pi = 385$ psia, while Stoughton and Lietzke [38] give 368 psia. From (14-69),

ΔP must be greater than $\Delta \pi$ for reverse osmosis to occur. For the desalinization of brackish water by RO, ΔP is typically 400–600 psi, while for seawater, it is 800–1,000 psi.

The feed water to an RO unit may contain a variety of potential foulants, which are removed prior to passage of the feed through the membrane unit. Otherwise the foulants can adversely affect the performance and reduce the useful life of the membrane. Suspended solids and particulate matter can be removed by screening and filtration. Colloids can be flocculated and filtered. Scale-forming salts require acidification or water softening. Biological materials require chlorination or ozonation. Other organic foulants are removed by adsorption or oxidation.

Concentration polarization is particularly important on the feed side of reverse-osmosis membranes. This effect is illustrated in Figure 14.19, where typical concentrations are shown for water, c_w, and salt, c_s. Because of the high pressure, the activity of water on the feed side is somewhat higher than that of near-pure water on the permeate side, thus providing the necessary driving force for water transport through the membrane. The flux of water to the membrane carries with it salt by bulk flow. However, because the salt cannot readily penetrate the membrane, the concentration of the salt in the liquid adjacent to the surface of the membrane, c_{s_i}, is greater than that in bulk of the feed, c_{s_F}. This difference causes mass transfer of salt by diffusion from the membrane surface back to the bulk feed. The back rate of salt diffusion depends on the mass-transfer coefficient for the film or boundary layer on the feed side. The lower the mass-transfer coefficient, the higher the value of c_{s_i}. The value of c_{s_i} is important because it fixes the osmotic pressure, and influences the driving force for water transport according to (14-69).

Consider steady-state transport of water with back-diffusion of salt. A salt balance at the upstream membrane surface gives

$$N_{H_2O}c_{s_F}(SR) = k_s(c_{s_i} - c_{s_F}) \quad (14\text{-}71)$$

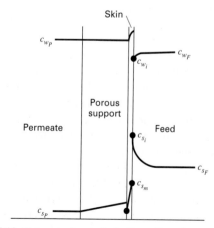

Figure 14.19 Concentration-polarization effects in reverse osmosis.

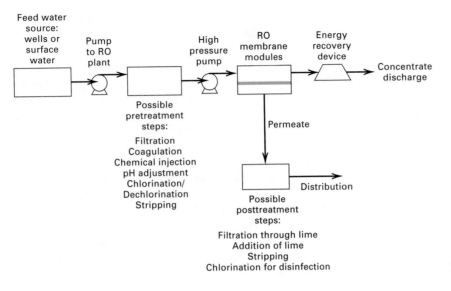

Figure 14.20 Reverse-osmosis process.

Solving for c_{s_i} gives

$$c_{s_i} = c_{s_F}\left(1 + \frac{N_{H_2O}(SR)}{k_s}\right)$$

Values of k_s are estimated from (14-54). As would be expected, the concentration-polarization effect is seen to be most significant for high water fluxes and low mass-transfer coefficients.

A quantitative estimate of the importance of concentration polarization can be derived by defining the following concentration polarization factor, Γ, in terms of (14-7):

$$\Gamma \equiv \frac{c_{s_i} - c_{s_F}}{c_{s_F}} = \frac{N_{H_2O}(SR)}{k_s} \qquad (14\text{-}72)$$

Values of SR are typically in the range of 0.97 to 0.995. If Γ is greater than, say, 0.2, the effect of concentration polarization may be significant, indicating that design changes to reduce the value of Γ should be considered.

Feed-side pressure drop is also important because, by (14-69), it causes a reduction in the driving force for water transport. Because of the complex geometries used for both spiral-wound and hollow-fiber modules, it is best to estimate pressure drops from experimental data. Feedside pressure drops for spiral-wound modules and hollow-fiber modules range from 43 to 85 and 1.4 to 4.3 psi, respectively [6].

A schematic diagram of a typical reverse-osmosis process for the desalinization of water is shown in Figure 14.20. The source of feed water may be a well or surface water, which is pumped through a series of pretreatment steps to ensure a long membrane life. Of particular importance is pH adjustment. The pretreated water is then fed by a high-pressure-discharge pump to an appropriate parallel-and-series network of reverse-osmosis modules of the spiral-wound or hollow-fiber type. The concentrate, which leaves the membrane system at a high pressure that is 10–15% lower than the inlet pressure, is then routed through a power-recovery turbine, which reduces the net power consumption of the process by 25 to 40% while reducing the pressure of the concentrate to

an appropriate low level. The permeate, which may be 99.95 wt% pure water and about 50% of the feed water, is sent to a series of posttreatment steps before it is ready to drink.

EXAMPLE 14.10

At a certain location in a spiral-wound membrane, the bulk conditions on the feed side are 1.8 wt% NaCl, 25°C, and 1,000 psia, while bulk conditions on the permeate side are 0.05 wt% NaCl, 25°C, and 50 psia. For the particular membrane being used, the permeance values are 1.1×10^{-5} g/cm²-s-atm for H_2O and 16×10^{-6} cm/s for the salt. If mass-transfer resistances are negligible on each side of the membrane, calculate the flux of water in gal/ft²-day and the flux of salt in g/ft²-day. If $k_s = 0.005$ cm/s, estimate the polarization factor.

SOLUTION

Bulk salt concentrations are approximately

$$c_{s_F} = \frac{1.8(1,000)}{58.5(98.2)} = 0.313 \text{ mol/L on feed side}$$

$$c_{s_P} = \frac{0.05(1,000)}{58.5(99.95)} = 0.00855 \text{ mol/L on permeate side}$$

For water transport, using (14-68) for osmotic pressure, noting that dissolved NaCl gives 2 ions per molecule:

$$\Delta P = (1,000 - 50)/14.7 = 64.6 \text{ atm}$$
$$\pi_{\text{feedside}} = 1.12(298)(2)(0.313) = 209 \text{ psia} = 14.2 \text{ atm}$$
$$\pi_{\text{permeate side}} = 1.12(298)(2)(0.00855) = 5.7 \text{ psia} = 0.4 \text{ atm}$$
$$\Delta P - \Delta \pi = 64.6 - (14.2 - 0.4) = 50.8 \text{ atm}$$
$$P_{M_{H_2O}}/l_M = 1.1 \times 10^{-5} \text{ g/cm}^2\text{-s-atm}$$

From (14-69),

$$N_{H_2O} = (1.1 \times 10^{-5})(50.8) = 0.000559 \text{ g/cm}^2\text{-s} \quad \text{or}$$

$$\frac{(0.000559)(3,600)(24)}{(454)(8.33)(1.076 \times 10^{-3})} = 11.9 \text{ gal/ft}^2\text{-day}$$

For salt transport:

$$\Delta c = 0.313 - 0.00855 = 0.304 \text{ mol/L} \quad \text{or} \quad 0.000304 \text{ mol/cm}^3$$
$$P_{M_{NaCl}}/l_M = 16 \times 10^{-6} \text{ cm/s}$$

From (14-26):

$$N_{NaCl} = 16 \times 10^{-6}(0.000304) = 4.86 \times 10^{-9} \text{ mol/cm}^2\text{-s}$$

or $\dfrac{(4.86 \times 10^{-9})(3,600)(24)(58.5)}{1.076 \times 10^{-3}} = 22.8 \text{ g/ft}^2\text{-day}$

We see that the flux of salt is very much smaller than the flux of water.

To estimate the concentration-polarization factor, first convert the water flux through the membrane into the same units as the salt mass-transfer coefficient, k_s, i.e., cm/s:

$$N_{H_2O} = \frac{0.000559}{1.00} = 0.000559 \text{ cm/s}$$

From (14-70), the salt passage is

$$SP = 0.00855/0.313 = 0.027$$

Therefore, the salt rejection $= SR = 1 - 0.027 = 0.973$
From (14-72), the concentration-polarization factor is

$$\Gamma = \frac{0.000559(0.972)}{0.005} = 0.11$$

Thus, here polarization is not particularly significant.

14.6 GAS PERMEATION

In gas permeation (GP), shown in Figure 14.21, the feed gas, at high pressure P_1, contains some low-molecular-weight species (MW < 50) to be separated from small amounts of higher-molecular-weight species. Usually a sweep gas is not used, but the other side of the membrane is maintained at a much lower pressure, P_2, often near-ambient pressure. The membrane, often dense but sometimes microporous, is permselective for certain of the low-molecular-weight species in the feed gas, shown in Figure 14.21 as the A species. If the membrane is dense, these species are absorbed at the surface and then transported through the membrane by one or more mechanisms. Thus, permselectivity depends on both membrane absorption and the membrane transport rate. Usually all mechanisms are formulated in terms of a partial-pressure or fugacity driving force using the solution-diffusion model of (14-32). The products are a permeate that is enriched in the A species and a retentate that is enriched in B. A near-perfect separation is generally not achievable. If the membrane is microporous, as for example in high-temperature applications, pore size is extremely important because it is usually necessary to block the passage of species B. Otherwise, unless molecular weights of A and B differ appreciably, only a very modest separation is achievable, as was discussed in connection with Knudsen diffusion, (14-22).

Since the early 1980s, applications of GP with dense, polymeric membranes have increased dramatically. Applications include (1) separation of hydrogen from methane; (2) adjustment of H_2-to-CO ratio in synthesis gas; (3) O_2 enrichment of air; (4) N_2 enrichment of air; (5) removal of CO_2; (6) drying of natural gas and air; (7) removal of helium; and (8) removal of organic solvents from air.

Gas permeation must compete with distillation at cryogenic conditions, absorption, and pressure-swing adsorption. Some of the advantages of gas permeation, as cited by Spillman and Sherwin [39], are low capital investment, ease of installation, ease of operation, absence of rotating parts, high process flexibility, low weight and space requirements, and low environmental impact. In addition, if the feed gas is already at so high a pressure that a gas compressor is not needed, then no utilities are required.

Since 1986, the most rapidly developing application for GP has been air separation, for which available membranes have separation factors for O_2 with respect to N_2 of 3 to 7. However, product purities are economically limited to a retentate of 95–99.9% N_2 and a permeate of 30–45% O_2. Thus, the largest application of GP for air separation is the production of nitrogen rather than oxygen.

Gas permeation also competes very favorably with other separation processes for hydrogen recovery because of the high separation factors achieved. For example, the rate of permeation of hydrogen through a typical dense polymer membrane is more than 30 times that for nitrogen. A typical GP process might achieve a 95% recovery of 90% pure hydrogen from a feed gas containing 60% hydrogen.

Early applications of GP used dense (nonporous) membranes of cellulose acetates and polysulfones, which are still predominant, although polyimides, polyamides, polycarbonates, polyetherimides, sulfonated polysulfones, Teflon, polystyrene, and silicone rubber are also finding applications for temperatures to at least 70°C. Although plate-and-frame and tubular modules can be used for gas permeation, almost all large-scale applications use spiral-wound or hollow-fiber modules because of their higher packing density. Commercial membrane modules for gas permeation are available from more than 20 suppliers. Feed-side pressure is typically 300 to 500 psia, but is as high as 1,650 psia. Typical refinery

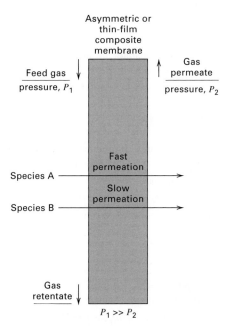

Figure 14.21 Gas permeation.

Table 14.9 Predicted Values of Diffusivity and Solubility of Light Gases in a Glassy and a Rubbery Polymer

Permeant	$D \times 10^{11}$, m²/s	$S \times 10^4$, gmol/m³-Pa	P_M, barrer
Polyvinyltrimethylsilane (Glassy Polymer)			
He	470	0.18	250
Ne	87	0.26	66
Ar	5.1	1.95	30
Kr	1.5	6.22	29
Xe	0.29	20.6	18
Rn	0.07	69.6	15
H_2	160	0.54	250
O_2	7.6	1.58	37
N_2	3.8	0.84	9
CO_2	4.0	13.6	160
CO	3.7	1.28	14
CH_4	1.9	3.93	22
C_2H_6	0.12	30.2	10
C_3H_8	0.01	98.1	2.8
C_4H_{10}	0.001	347	1.2
C_2H_4	0.23	17.8	12
C_3H_6	0.038	77.6	9
C_4H_8 (1)	0.0052	293	4.5
C_2H_2	0.58	16.8	32
C_3H_4 (m)	0.17	138.1	70
C_4H_6 (e)	0.053	318.5	50
C_3H_4 (a)	0.15	186.5	83
C_4H_6 (b)	0.03	226.1	20
Polyisoprene (Rubber-like Polymer)			
He	213	0.06	35
Ne	77.4	0.08	18
Ar	14.6	0.58	25
Kr	7.2	1.78	25
Xe	2.7	5.68	45
Rn	1.2	18.7	64
H_2	109	0.17	54
O_2	18.4	0.47	26
N_2	12.2	0.26	10
CO_2	12.6	3.80	140
CO	12.1	0.38	14
CH_4	8.0	1.14	27
C_2H_6	3.3	8.13	79
C_3H_8	1.6	25.4	123
C_4H_{10}	1.5	86.4	390
C_2H_4	4.3	4.84	62
C_3H_6	2.7	20.3	163
C_4H_8 (1)	1.5	73.3	333
C_2H_2	5.7	4.64	80
C_3H_4(m)	4.1	35.3	433
C_4H_6(e)	2.9	79.6	690
C_3H_4(a)	4.5	47.4	640
C_4H_6(b)	3.4	40.0	410

Note: m, methylacetylene; e, ethylacetylene; a, allene; b, butadiene.

applications involve feed-gas flow rates of 20 million scfd, but flow rates as large as 300 million scfd have been reported [40]. When the feed gas contains condensables, it may be necessary to preheat the gas prior to entry into the membrane system to prevent condensation on the membrane as the retentate becomes richer in the high-molecular-weight species. For high-temperature applications where polymers cannot be used, membranes of glass, carbon, and inorganic oxides are available, but are limited in their selectivity.

For dense membranes, external mass-transfer resistances or concentration-polarization effects are generally negligible, and (14-32) with a partial-pressure driving force can be used to compute the rate of species transport through the membrane. As discussed earlier in the subsection on module flow patterns, the appropriate partial-pressure driving force depends on the flow pattern. Cascades of the type discussed earlier are used to increase the degree of separation.

Progress is being made in the development of a method for the prediction of permeability of gases in glassy and rubbery homopolymers, random copolymers, and block copolymers. Teplyakov and Meares [41] present correlations at 25°C for the diffusion coefficient, D, and solubility, S, applied to 23 different gases for 30 different polymers. Predicted values for glassy polyvinyltrimethylsilane (PVTMS) and rubbery polyisoprene are listed in Table 14.9. Typically, D and S agree with experimental data to within ±20% and ±30%, respectively.

Gas permeation separators are claimed to be relatively insensitive to changes in feed flow rate, feed composition, and loss of membrane surface area [42]. This claim is tested in the following example.

EXAMPLE 14.11

The feed to a membrane separator consists of 500 lbmol/h of a mixture of 90% H_2 (H) and 10% CH_4 (M) at 500 psia. Permeance values based on a partial-pressure driving force are

$$\bar{P}_{M_H} = 3.43 \times 10^{-4} \text{ lbmol/h-ft}^2\text{-psi}$$

and $\quad \bar{P}_{M_M} = 5.55 \times 10^{-5} \text{ lbmol/h-ft}^2\text{-psi}$

The flow patterns in the separator are such that the permeate side is well mixed and the feed side is in plug flow. The pressure on the permeate side is constant at 20 psia and there is no pressure drop on the feed side.

(a) Compute the membrane area and permeate purity if 90% of the hydrogen is transferred to the permeate.

(b) For the membrane area determined in part (a), calculate the permeate purity and hydrogen recovery if

(1) the feed rate is increased by 10%.

(2) the feed composition is reduced to 85% H_2.

(3) 25% of the membrane area becomes inoperative.

SOLUTION

The following independent equations apply to all parts of this example. Component material balances:

$$n_{i_F} = n_{i_R} + n_{i_P}, \quad i = H, M \quad (1,2)$$

Dalton's law of partial pressures:

$$P_k = p_{H_k} + p_{M_k}, \quad k = F, R, P \qquad (3,4,5)$$

Partial pressure–mole relations:

$$p_{H_k} = P_k n_{H_k}/(n_{H_k} + n_{M_k}), \quad k = F, R, P \qquad (6,7,8)$$

Solution-diffusion transport rates are obtained using (14-32), assuming a log-mean partial-pressure driving force based on the exiting permeate partial pressures on the downstream side of the membrane because of the assumption of perfect mixing on that side:

$$n_{i_P} = \bar{P}_{M_i} A_M \left[\frac{p_{i_F} - p_{i_R}}{\ln\left(\dfrac{p_{i_F} - p_{i_P}}{p_{i_R} - p_{i_P}} \right)} \right], \quad i = H, M \quad (9,10)$$

Thus, we have a system of 10 equations in the following 18 variables:

$$
\begin{array}{cccccc}
A_M & n_{H_F} & n_{M_F} & P_F & P_R & P_P \\
\bar{P}_{M_H} & n_{H_R} & n_{M_R} & p_{H_F} & p_{H_R} & p_{H_P} \\
\bar{P}_{M_M} & n_{H_P} & n_{M_P} & p_{M_F} & p_{M_R} & p_{M_P}
\end{array}
$$

Thus, eight variables must be fixed. For all parts of this example, the following five variables are fixed:

$$\bar{P}_{M_H} \text{ and } \bar{P}_{M_M} \text{ given above}$$

$$P_F = 500 \text{ psia} \quad P_R = 500 \text{ psia} \quad P_P = 20 \text{ psia}$$

For each part, three additional variables must be fixed.

(a)

$$n_{H_F} = 0.9(500) = 450 \text{ lbmol/h}$$

$$n_{M_F} = 0.1(500) = 50 \text{ lbmol/h}$$

$$n_{H_P} = 0.9(450) = 405 \text{ lbmol/h}$$

Solving Equations (1)–(10) above, using a PC program such as MathCad, Matlab, or Polymath, we obtain

$$A_M = 3{,}370 \text{ ft}^2$$

$n_{M_P} = 20.0 \text{ lbmol/h} \quad n_{H_R} = 45.0 \text{ lbmol/h} \quad n_{M_R} = 30.0 \text{ lbmol/h}$

$p_{H_F} = 450 \text{ psia} \quad p_{M_F} = 50 \text{ psia} \quad p_{H_R} = 300 \text{ psia}$

$p_{M_R} = 200 \text{ psia} \quad p_{H_P} = 19.06 \text{ psia} \quad p_{M_P} = 0.94 \text{ psia}$

(b) Calculations are made in a similar manner using Equations (1)–(10). Results for Parts (1), (2), and (3) are:

	Part		
	(1)	(2)	(3)
Fixed:			
n_{H_F}, lbmol/h	495	425	450
n_{M_F}, lbmol/h	55	75	50
A_M, ft^2	3,370	3,370	2,528
Calculated, in lbmol/h:			
n_{H_P}	424.2	369.6	338.4
n_{M_P}	18.2	25.9	11.5
n_{H_R}	70.8	55.4	111.6
n_{M_R}	36.8	49.1	38.5
Calculated, in psia:			
p_{H_F}	450	425	450
p_{M_F}	50	75	50
p_{H_R}	329	265	372
p_{M_R}	171	235	128
p_{H_P}	19.18	18.69	19.34
p_{M_P}	0.82	1.31	0.66

From the above results, the following are computed:

	Part			
	(a)	(b1)	(b2)	(b3)
Mol% H$_2$ in permeate	95.3	95.9	93.5	96.7
% H$_2$ recovery in permeate	90	85.7	87.0	75.2

From these results, we see that when the feed rate is increased by 10% (Part b1), the hydrogen recovery drops about 5%, but the permeate purity is maintained. When the feed composition is reduced from 90% to 85% hydrogen (Part b2), the hydrogen recovery decreases by about 3% and the permeate purity decreases by about 2%. With 25% of the membrane area inoperative (Part b3), the hydrogen recovery decreases by about 15%, but the permeate purity is about 1% higher. Overall, percentage changes in hydrogen recovery and purity are less than the percentage changes in feed flow rate, feed composition, and membrane area, thus tending to confirm the insensitivity of gas permeation separators to changes in operating conditions.

14.7 PERVAPORATION

As shown in Figure 14.22, pervaporation (PV) differs from dialysis, reverse osmosis, and gas permeation in that the phase state on one side of the membrane is different from that on the other side. The feed to the membrane module is a liquid mixture (e.g., an alcohol–water azeotrope) at a pressure, P_1, that is usually ambient or elevated high enough to maintain a liquid phase as the feed is depleted of species A and B to produce the product retentate. A composite membrane is used that is selective for species A, but species B usually has some finite permeability. The dense, thin membrane film is in contact with the liquid side. The retentate is enriched in species B. Generally, a sweep fluid is not used

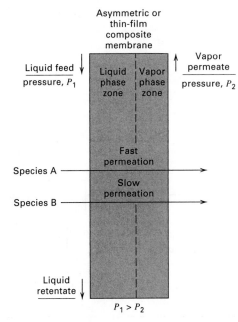

Figure 14.22 Pervaporation.

on the other side of the membrane, but a pressure, P_2, is maintained at or below the dew point of the permeate, making it vapor. Often, P_2 is a vacuum. Vaporization may occur near the downstream face of the membrane, such that the membrane can be considered to operate with two zones, a liquid-phase zone and a vapor-phase zone, as shown in Figure 14.22. Alternatively, the vapor phase may only exist on the permeate side of the membrane. The vapor permeate is enriched in species A. Overall permeabilities of species A and B depend upon their solubilities in and diffusion rates through the membrane. Generally, the solubilities cause the membrane to swell.

The term pervaporation is a combination of the two words, *perm*selective and *evaporation*. It was first reported in 1917 by Kober [43], who studied several experimental techniques for removing water from albumin/toluene solutions. Although the economic potential of PV was shown by Binning et al. [44] in 1961, commercial applications were delayed until the mid-1970s, when adequate membrane materials first became available. Major commercial applications now

include (1) dehydration of ethanol; (2) dehydration of other organic alcohols, ketones, and esters; and (3) removal of organics from water. The separation of organic mixtures, e.g., benzene–cyclohexane, is receiving much attention.

Pervaporation is best applied when the feed solution is dilute in the main permeant because sensible heat of the feed mixture provides the enthalpy of vaporization of the permeant. If the feed is rich in the main permeant, a number of membrane stages may be needed, with a small amount of permeant produced per stage and reheating of the retentate between stages. Even when only one membrane stage is sufficient, the feed may be heated before entering the membrane module.

Many pervaporation separation schemes have been proposed [6], with three of the more important ones shown in Figure 14.23. A hybrid process for integrating distillation with pervaporation to produce 99.5 wt% ethanol from a feed of 60 wt% ethanol is shown in Figure 14.23a. The feed is sent to a distillation column operating at near-ambient pressure, where a bottoms product of nearly pure water and an

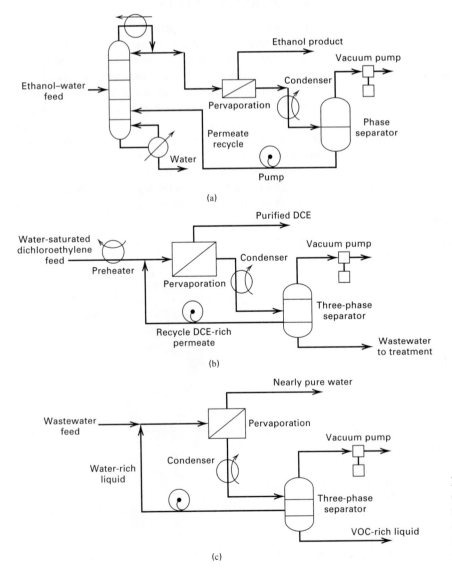

(a)

(b)

(c)

Figure 14.23 Pervaporation processes. (a) Hybrid process for removal of water from ethanol. (b) Dehydration of dichloroethylene. (c) Removal of volatile organic compounds (VOCs) from wastewater.

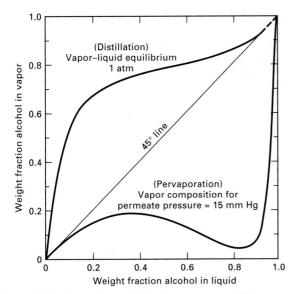

Figure 14.24 Comparison of ethanol–water separabilities.
[From M. Wesslein et al., *J. Membrane Sci.*, **51**, 169 (1990).]

ethanol-rich distillate of 95 wt% is produced. The distillate purity is limited because of the 95.6 wt% ethanol in water azeotrope. The distillate is sent to a pervaporation step where a permeate of 25 wt% alcohol and a retentate of 99.5 wt% ethanol is produced. The permeate vapor is condensed under vacuum and recycled to the distillation column. The vacuum is sustained with a vacuum pump. The dramatic difference in separability of the pervaporation membrane as compared to vapor–liquid equilibrium for distillation is shown with a 45° reference line in Figure 14.24 taken from Wesslein et al. [45]. For pervaporation, the compositions refer to a liquid feed (abscissa) and a vapor permeate (ordinate) at 60°C for a polyvinylalcohol (PVA) membrane and a vacuum of 15 torr. For this membrane, there is no limitation on ethanol purity and the separation index is very high for feeds containing more than 90 wt% ethanol.

A pervaporation process for dehydrating dichloroethylene (DCE) is shown in Figure 14.23b. The liquid feed, which is DCE saturated with water (0.2 wt%), is preheated to 90°C at 0.7 atm and sent to a PVA membrane system, which produces a retentate of almost pure DCE (<10 ppm H_2O) and a permeate vapor of 50 wt% DCE under vacuum. Following condensation, the two resulting liquid phases are separated, with the DCE-rich phase recycled back to the membrane system and the water-rich phase sent to an air stripper, steam stripper, adsorption unit, or hydrophobic, pervaporation, membrane system for residual DCE removal.

Pervaporation can be used for the removal of VOCs (e.g., toluene and trichloroethylene) from wastewater by pervaporation with hollow-fiber modules of silicone rubber, as shown in Figure 14.23c. The retentate is almost pure water (< 5 ppb of VOCs) and the permeate, after condensation, is

(1) a water-rich phase that is recycled to the membrane system and (2) a nearly pure VOC phase.

A pervaporation module may operate with heat transfer or adiabatically with the enthalpy of vaporization supplied by sensible enthalpy of the feed. Consider the adiabatic pervaporation of a binary liquid mixture of components A and B. Assume constant pure-component liquid specific heats, and ignore heat of mixing. For an enthalpy datum temperature of T_0, an enthalpy balance, in terms of mass flow rates, m, liquid sensible heats, C_P, and heats of vaporization, ΔH^{vap}, gives

$$(m_{A_F}C_{P_A} + m_{B_F}C_{P_B})(T_F - T_0)$$
$$= [(m_{A_F} - m_{A_P})C_{P_A} + (m_{B_F} - m_{B_P})C_{P_B}](T_R - T_0)$$
$$+ (m_{A_P}C_{P_A} + m_{B_P}C_{P_B})(T_P - T_0) + m_{A_P}\Delta H_A^{vap}$$
$$+ m_{B_P}\Delta H_B^{vap} \qquad (14\text{-}73)$$

where enthalpies of vaporization are evaluated at T_P. After collection of terms, (14-73) reduces to

$$(m_{A_F}C_{P_A} + m_{B_F}C_{P_B})(T_F - T_R) = (m_{A_P}C_{P_A} + m_{B_P}C_{P_B})$$
$$\times (T_P - T_R) + \left(m_{A_P}\Delta H_A^{vap} + m_{B_P}\Delta H_B^{vap}\right) \qquad (14\text{-}74)$$

The temperature of the permeate, T_P, is the permeate dew point at the permeate vacuum upstream of the condenser. The retentate temperature is computed from (14-74).

Membrane selection is critical in the commercial application of PV, when used in the presence of organic compounds. For water permeation, hydrophilic membrane materials are preferred. For example, a three-layer membrane is often used for the dehydration of ethanol, with water being the main permeating species. The support layer is porous polyester, which is cast on a microporous polyacrylonitrile or polysulfone membrane. The final layer, which provides the separation, is dense PVA of 0.1 μm in thickness. This composite combines chemical and thermal stability with adequate permeability. Hydrophobic membranes, such as silicone rubber and Teflon, are preferred when organics are the permeating species.

Commercial membrane modules for PV are almost exclusively of the plate-and-frame type because of the ease of using gasketing materials that are resistant to organic solvents and the ease of providing heat exchange for evaporation and high-temperature operation. However, considerable interest is evident in the use of hollow-fiber modules for the removal of VOCs from wastewater. Because feeds are generally clean and operation is at low pressure, membrane fouling and damage can be minimal, resulting in a useful membrane life of 2–4 years.

Various models for the transport of a permeant through a membrane by pervaporation have been proposed, based on the solution-diffusion model. They all assume equilibrium between the upstream liquid and the upstream membrane surface, and between the downstream vapor and the other side of the membrane. Transport through the membrane follows Fick's law with a concentration gradient of the permeant in the membrane as the driving force. However, because of the phase change and nonideal-solution effects in the liquid feed, simple equations like (14-55) for dialysis and (14-32) for gas permeation do not apply to pervaporation.

A particularly convenient PV model is that of Wijmans and Baker [46]. They express the driving force for permeation in terms of a partial-vapor-pressure difference. Because pressures on the both sides of the membrane are low, the gas phase follows the ideal-gas law. Therefore, at the upstream membrane surface (1), permeant activity for component i is expressed as

$$a_i^{(1)} = f_i^{(1)}/f_i^{(0)} = p_i^{(1)}/P_i^{s(1)} \qquad (14\text{-}75)$$

where P_i^s is the vapor pressure at the feed temperature. The liquid on the upstream side of the membrane is generally nonideal. Thus, from Table 2.2:

$$a_i^{(1)} = \gamma_i^{(1)} x_i^{(1)} \qquad (14\text{-}76)$$

Combining (14-75) and (14-76):

$$p_i^{(1)} = \gamma_i^{(1)} x_i^{(1)} P_i^{s(1)} \qquad (14\text{-}77)$$

On the downstream vapor side of the membrane (2), the partial pressure is

$$p_i^{(2)} = y_i^{(2)} P_P^{(2)} \qquad (14\text{-}78)$$

Thus, the driving force can be expressed as $(\gamma_i^{(1)} x_i^{(1)} P_i^{s(1)} - y_i^{(2)} P_P^{(2)})$.

The corresponding permeant flux, after dropping unnecessary superscripts, is

$$N_i = \frac{P_{M_i}}{l_M}\left(\gamma_i x_i P_i^s - y_i P_P\right) \qquad (14\text{-}79)$$

or

$$N_i = \bar{P}_{M_i}\left(\gamma_i x_i P_i^s - y_i P_P\right) \qquad (14\text{-}80)$$

where γ_i and x_i refer to the feed-side liquid, P_i^s is the vapor pressure at the feed-side temperature, y_i is the mole fraction in the permeant vapor, and P_P is the total permeant pressure.

Unlike gas permeation where P_{M_i} depends mainly on the permeant, the polymer, and temperature; the permeability for pervaporation depends additionally on the concentrations of permeants in the polymer, which can be large enough to cause polymer swelling and cross-diffusion effects. For a binary system it is best to back-calculate and correlate the permeant flux with feed composition at a given feed temperature and permeate pressure. Because of these nonideal effects, the selectivity can be a strong function of feed concentration and permeate pressure, causing inversion of selectivity in some cases, as illustrated in the following example.

EXAMPLE 14.12

Wesslein et al. [45] present the following experimental data for the pervaporation of liquid mixtures of ethanol (1) and water (2) at a feed temperature of 60°C for a permeate pressure of 76 mmHg, using a commercial polyvinylalcohol membrane:

wt% ethanol		Total Permeation Flux
Feed	Permeate	kg/m²-h
8.8	10.0	2.48
17.0	16.5	2.43
26.8	21.5	2.18
36.4	23.0	1.73
49.0	22.5	1.46
60.2	17.5	0.92
68.8	13.0	0.58
75.8	9.0	0.40

At 60°C, vapor pressures are 352 and 149 mmHg for ethanol and water, respectively.

Liquid-phase activity coefficients at 60°C for the ethanol (1)–water (2) system are given by the van Laar equations:

$$\ln \gamma_1 = 1.6276\left[\frac{0.9232 x_2}{1.6276 x_1 + 0.9232 x_2}\right]^2$$

$$\ln \gamma_2 = 0.9232\left[\frac{1.6276 x_1}{1.6276 x_1 + 0.9232 x_2}\right]^2$$

Calculate values of permeance for water and ethanol from (14-80).

SOLUTION

For the first row of data, the mole fractions in the feed mixture (x_i) and the permeate (y_i), using molecular weights of 46.07 and 18.02 for ethanol and water, respectively, are

$$x_1 = \frac{0.088/46.07}{\dfrac{0.088}{46.07} + \dfrac{(1.0 - 0.088)}{18.02}} = 0.0364$$

$$x_2 = 1.0 - 0.0364 = 0.9636$$

$$y_1 = \frac{0.10/46.07}{\dfrac{0.10}{46.07} + \dfrac{0.90}{18.02}} = 0.0416$$

$$y_2 = 1.0 - 0.0416 = 0.9584$$

The activity coefficients for the feed mixture are

$$\gamma_1 = \exp\left\{1.6276\left[\frac{0.9232(0.9636)}{1.6276(0.0364) + 0.9232(0.9636)}\right]^2\right\} = 4.182$$

$$\gamma_2 = \exp\left\{0.9232\left[\frac{1.6276(0.0364)}{1.6276(0.0364) + 0.9232(0.9636)}\right]^2\right\} = 1.004$$

From the total mass flux, the component molar fluxes are

$$N_1 = \frac{(2.48)(0.10)}{46.07} = 0.00538\,\frac{\text{kmol}}{\text{h} - \text{m}^2}$$

$$N_2 = \frac{(2.48)(0.90)}{18.02} = 0.1239\,\frac{\text{kmol}}{\text{h} - \text{m}^2}$$

From (14-80), permeance values are

$$\bar{P}_{M_1} = \frac{0.00538}{(4.182)(0.0364)(352) - (0.0416)(76)}$$

$$= 0.000107\,\frac{\text{kmol}}{\text{h} - \text{m}^2 - \text{mmHg}}$$

$$\bar{P}_{M_2} = \frac{0.1239}{(2.004)(1.0 - 0.0364)(149) - (1.0 - 0.0416)(76)}$$

$$= 0.001739 \frac{\text{kmol}}{\text{h} - \text{m}^2 - \text{mmHg}}$$

Results for other feed conditions are computed in a similar manner:

wt% Ethanol		Activity Coefficient in Feed		Permeance, kmol/h-m²-mmHg	
Feed	Permeate	Ethanol	Water	Ethanol	Water
8.8	10.0	4.182	1.004	1.07×10^{-4}	1.74×10^{-3}
17.0	16.5	3.489	1.014	1.02×10^{-4}	1.62×10^{-3}
26.8	21.5	2.823	1.038	8.69×10^{-5}	1.43×10^{-3}
36.4	23.0	2.309	1.077	6.14×10^{-5}	1.17×10^{-3}
49.0	22.5	1.802	1.158	4.31×10^{-5}	1.10×10^{-3}
60.2	17.5	1.477	1.272	1.87×10^{-5}	8.61×10^{-4}
68.8	13.0	1.292	1.399	7.93×10^{-6}	6.98×10^{-4}
75.8	9.0	1.177	1.539	3.47×10^{-6}	6.75×10^{-4}

The PVA membrane is hydrophilic. Thus, as the concentration of ethanol in the feed liquid increases, the sorption of feed liquid by the membrane decreases, resulting in a reduction of polymer swelling. The preceding results show that as swelling is reduced, the permeance of ethanol decreases more rapidly than that of water, thus increasing selectivity for water. For example, selectivity for water can be defined as

$$\alpha_{2,1} = \frac{(100 - w_1)_P/(w_1)_P}{(100 - w_1)_F/(w_1)_F}$$

where w_1 = weight fraction of ethanol. For cases of 8.8 and 75.8 wt% ethanol in the feed, the selectivities for water are, respectively, 0.868 (more selective for ethanol) and 31.7 (more selective for water).

14.8 ULTRAFILTRATION

Consider an aqueous stream that contains a number of possible dissolved and suspended solutes, including suspended particles like bacteria and blood cells, large, dissolved molecules (e.g., proteins and carbohydrates), and small, dissolved molecules (e.g., salts). It is desired to separate these solutes from the carrier solution. As discussed in an earlier subsection, reverse osmosis is widely used for purifying saltwater. Part of the water passes through the membrane, leaving a retentate (concentrate) that contains most of the salts and some of the water. For recovering the solutes listed above, ultrafiltration and microfiltration (covered in the next subsection) are more commonly used in the manner indicated in Table 14.10.

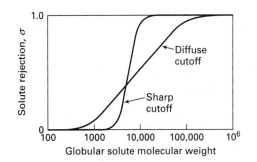

Figure 14.25 Molecular weight cutoff curves.

As with reverse osmosis, microfiltration and ultrafiltration are pressure driven, with the membrane permselective for the solvent (usually water). Microfiltration and ultrafiltration separate mainly by size exclusion of solute(s). Microfiltration retains particles of micron size. Ultrafiltration retains particles of submicron size by ultramicroporous membranes. Typically, ultrafiltration retains solutes in the 300–500,000 molecular weight range, including biomolecules, polymers, sugars, and colloidal particles. Ultrafiltration membranes are characterized by a nominal molecular weight cutoff (MWCO), defined as the smallest globular-solute molecular weight for which the membrane has at least a 90% rejection ($\sigma \geq 0.9$), with most commercial materials operating in the $1,000 \leq \text{MWCO} \leq 100,000$ range. The rejection, σ, is defined in terms of solute concentrations, c_{i_P} and c_{i_R} on either side of the membrane (i.e., retentate, R, and permeate, P):

$$\sigma_i = 1 - \frac{c_{i_P}}{c_{i_R}} \tag{14-81}$$

For an ultrafiltration membrane, a retention-cutoff curve is established experimentally. Two generic curves are shown in Figure 14.25. A sharp cutoff is desirable, but more typical is the diffuse cutoff curve, because of the difficulty in producing a membrane with a narrow pore-size distribution.

The flux of solvent through an ultrafiltration membrane depends on the pressure drop across the membrane, flow rates on either side of the membrane, and temperature. It is also a strong function of solute concentration, as shown in Figure 14.26 for experimental data with protein solutes. Often, over a wide range of solute concentration, flux is a function of the logarithm of the solute concentration. Alternatively, flux may be correlated for a given feed with

Table 14.10 Comparison of Reverse Osmosis to Ultrafiltration and Microfiltration

	Microfiltration	Ultrafiltration	Reverse Osmosis
Typical membrane materials	cellulose acetate, polysulfone, ceramics	cellulose acetate, polyamides, polysulfone	cellulose acetate, aromatic polyamides
Pore size (angstroms)	100,000–200	200–10	10–1
Pore size (microns)	10–0.02	0.02–0.001	0.001–0.0001
Membrane pressure drop (psi)	1–10	10–100	100–1,000
Permeate	water + dissolved molecules	water + small molecules	water
Retentate (concentrate)	water + large suspended particles	water + large molecules	water + solutes

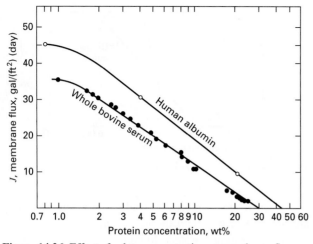

Figure 14.26 Effect of solute concentration on membrane flux.

concentration factor, CF, which is defined in terms of the volumetric flow rates of feed and retentate:

$$CF = \frac{Q_F}{Q_R} \qquad (14\text{-}82)$$

Concentration polarization, described in Subsection 14.3, can cause a decrease in the permeate flux for ultrafiltration by two mechanisms. In the first, increased rejected solute concentration on the upstream membrane surface increases the osmotic pressure on the retentate side. Because the driving force across membranes for ultrafiltration is the same as for reverse osmosis, i.e., (14-69), the driving force decreases with increasing osmotic pressure on the retentate side. In general, the effect of osmotic pressure in ultrafiltration is only important when concentration polarization occurs because, without polarization, large molecular sizes in the retentate contribute little in moles of solute and, therefore, little to the osmotic pressure as given, approximately, by (14-68). In the second, more generally accepted mechanism, the permeate flux becomes dependent, at least partially, on the mass-transfer resistances of the boundary layer or film and any gel layer on the retentate side of the membrane.

Concentration polarization effects in ultrafiltration are small at low pressures, low-solute concentrations, and high velocities on the retentate side of the membrane. Then, the permeate flux is controlled by transmembrane pressure, in response to which the flux increases linearly. However, as the transmembrane pressure or solute concentration increases or the velocity decreases, a point is reached where this linear relationship no longer holds and permeate flux reaches an asymptotic limiting value, controlled by mass transfer on the retentate side of the membrane. In the extreme case of a thick gel layer, it alone may control the permeate flux. Cheryan [50] and Zeman and Zydney [51] develop theories for predicting limiting permeate flux.

Fouling in ultrafiltration occurs by gel-layer consolidation, adsorption of solutes on the upstream membrane surface, and/or deposition of solutes within the membrane pores. Fouling, unlike concentration polarization, is a function of

time. While fouling can be reversed by periodic cleaning (e.g., back-flushing or chemical treatment), some of the fouling may be irreversible, resulting ultimately in the need to replace the membrane. Zeman and Zydney [51] give a treatment of membrane cleaning.

Process Configurations

An ultrafiltration process is commonly conducted in one of four configurations or combinations thereof: (1) batch ultrafiltration, (2) continuous bleed-and-feed ultrafiltration, (3) batch diafiltration, and (4) continuous bleed-and-feed diafiltration.

Batch Ultrafiltration

A batch configuration is shown in Figure 14.27, where the membrane consists of a number of membrane cartridges in parallel. Initially, the feed tank is filled with a batch, V_F, of feed solution of solute concentration, c_F. The solution is pumped through membrane cartridges, where permeate is continuously removed, but retentate (concentrate) is recycled, usually at a high volumetric flow rate, Q, to minimize fouling, back to the feed tank. Solute concentration in the retentate increases with time, as solvent (usually water) selectively passes through the membrane. As time passes, the retained volume of solution in the system, $V\{t\}$, decreases and its retentate solute concentration, $c_R\{t\}$, increases. Operation is terminated when the desired solute retentate concentration, c_R, is reached. At that point, the feed tank and associated equipment contains the final retentate, V_R, which can be drained to another tank. After cleaning, another batch is processed. The required time for batch processing depends on the membrane area, A, and the flux, J, of permeate through the membrane. The flux decreases with time because it depends strongly on the increasing solute concentration on the upstream side of the membrane as shown in Figure 14.26.

Assume that the batch feed contains completely rejected solutes and only partially rejected solutes. Assume that the flux is a linear function of the logarithm of the concentration

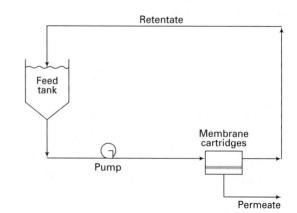

Figure 14.27 Batch ultrafiltration.

factor, CF, as discussed above, where for the batch process of Figure 14.27, the concentration factor is a function of time as defined by

$$CF = \frac{V_F}{\text{retained } V\{t\}} \quad (14\text{-}83)$$

Then it can shown that the average flux, J_{avg}, for the batch process is approximately

$$J_{avg} = J\{c_F\} - 0.33[J\{c_F\} - J\{c_R\}] \quad (14\text{-}84)$$

where values of $J\{c_F\}$ and $J\{c_R\}$ are obtained from experimental data like that in Figure 14.26. The required membrane area as a function of batch processing time, t, is given accordingly by

$$A = \frac{V_P}{t J_{avg}} = \frac{V_F - V_R}{t J_{avg}} \quad (14\text{-}85)$$

To obtain a material balance for a solute, we note that solute concentration in the retained volume is a function of both the reduction in retained volume and the amount of solute that passes through the membrane. A solute, i, material balance for a differential volume passing through the membrane is, by analogy to (14-45) for gas permeation or to (13-2) for batch distillation,

$$\frac{dV}{V} = \frac{dc_{i_R}}{c_{i_P} - c_{i_R}} \quad (14\text{-}86)$$

Combining with (14-81) for the definition of rejection, σ_i,

$$\frac{dV}{V} = -\frac{dc_{i_R}}{\sigma_i c_{i_R}} \quad (14\text{-}87)$$

Integrating this equation from initial feed to final retentate gives an equation for solute concentration in the retentate as a function of retained volume, where if the retained volume is the final retentate volume, then solute concentration is the concentration in the final retentate volume:

$$c_{i_R} = c_{i_F} \left(\frac{V_F}{V_R} \right)^{\sigma_i} = c_{i_F} (CF)^{\sigma_i} \quad (14\text{-}88)$$

The yield, Y_i, of solute, i, defined as the amount of feed solute that is retained in the retentate, is obtained from (14-88):

$$Y_i = \frac{c_{i_R} V_R}{c_{i_F} V_F} = \left(\frac{V_F}{V_R} \right)^{\sigma_i} \left(\frac{V_R}{V_F} \right) = \left(\frac{V_F}{V_R} \right)^{\sigma_i - 1} = CF^{\sigma_i - 1} \quad (14\text{-}89)$$

Application of (14-81) to (14-89) is illustrated in the following example of batch ultrafiltration.

EXAMPLE 14.13

An aqueous feed of 1,000 L is to undergo batch ultrafiltration with a polysulfone membrane. The solute concentrations and their measured rejection values are as follows:

Solute	Type molecule	MW	Concentration, c, g/L	Rejection, σ
Albumin	Globular	67,000	10	1.00
Cytochrome C	Globular	13,000	10	0.70
Polydextran	Linear	100,000	10	0.05

Note that polydextran has the highest molecular weight, but the lowest rejection because it is a linear molecule rather than a globular one. The volume of the final retentate is to be 200 L, which is achieved in a four-hour batch-processing time. Thus, from (14-82), $CF = 1,000/200 = 5$. From experimental measurements, the flux values are 30 L/m²-h at $CF = 1$ and 10 L/m²-h at $CF = 5$. Calculate the solute concentration in the final retentate, yield of each solute, and membrane area. Neglect changes in solution density.

SOLUTION

From (14-84), the average flux $= 30 - 0.33(30 - 10) = 23.4$ L/m²-h. The total permeate volume $= 1,000 - 200 = 800$ L. From (14-85), for $t = 4$ h,

$$A = \frac{800}{4(23.4)} = 8.55 \text{ m}^2$$

Using (14-89) and (14-88), the yield and concentration of each solute in the final retentate are

Solute	Concentration in final retentate, g/L	% Yield
Albumin	50.0	100.0
Cytochrome C	30.9	61.7
Polydextran	10.8	21.7

Note that although polydextran has a very low rejection, neither the final concentration in the retentate nor the % yield approach zero.

When processing biological materials, the batch configuration may be disadvantageous. Recycle of the retentate can cause damage to proteins and cells. Also, if the batch residence time is too long, unacceptable bacterial growth may occur. Continuous ultrafiltration, which is widely used for large-scale processes, is then preferred.

Continuous Feed-and-Bleed Ultrafiltration

Although, as shown in Figure 14.20, continuous reverse osmosis usually operates in a single-pass mode, continuous ultrafiltration rarely does; instead, it operates in a multipass mode, as shown in Figure 14.28, called single-stage *feed-and-bleed*. This is achieved by recycling, at steady state, a large fraction of the retentate. In effect, feed to the membrane is the sum of fresh feed and recycle retentate. The

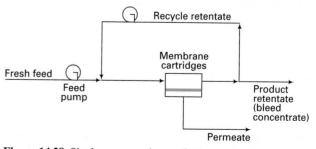

Figure 14.28 Single-stage continuous feed-and-bleed ultrafiltration.

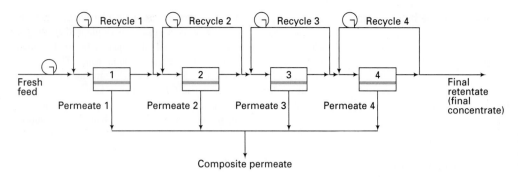

Figure 14.29 Multistage, continuous feed-and-bleed ultrafiltration.

bleed is that portion of the retentate that is not recycled, but is withdrawn as product retentate. At startup the entire retentate is recycled until the desired retentate concentration is achieved, at which time bleed is initiated. The advantages and disadvantages of feed-and-bleed operation, as compared to a single-pass mode (no recycle of retentate), are considered by Cheryan [50] and in detail by Zeman and Zydney [51]. The single-pass mode is usually unsuitable for ultrafiltration because the main product is the concentrate rather than the permeate (as in reverse osmosis), and high yields of permeate are required in order to adequately concentrate solutes in the retentate. Typically, the concentration factor, CF, defined by (14-82), has a value of 10. As a result, a single-pass ultrafiltration requires a very long membrane path or a very large membrane area. A disadvantage of the feed-and-bleed mode, however, is that with the high recycle ratio, the concentration of solutes on the retentate side is at its highest value, resulting, as shown in Figure 14.26, in the lowest flux, with a resulting membrane area larger than that required for the batch mode. To counter this, the feed-and-bleed mode is most often staged as shown for four stages in Figure 14.29, where the retentate (bleed) from each stage is sent to the next stage, while the permeates from the stages are collected into a final composite permeate. Solute concentrations increase incrementally as the retentates pass through the system. The final and highest concentration is only present in the final stage. As a result, retentate concentrations are lower and higher fluxes are achieved, compared to a single-stage, bleed-and-feed system, for all but the final stage. This leads to a smaller total membrane area. In practice, three to four feed-and-bleed stages are usually optimal.

For a single-stage, continuous, bleed-and-feed ultrafiltration, the following material-balance equations apply in terms of volumetric flow rates and concentrations:

Total balance:

$$Q_F = Q_R + Q_P \qquad (14\text{-}90)$$

Solute total balance:

$$c_{i_F} Q_F = c_{i_R} Q_R + c_{i_P} Q_P \qquad (14\text{-}91)$$

If the recycle rate is sufficiently high, concentration of the stream flowing on the upstream side of the membrane will be the retentate. Then, if (14-90) and (14-91) are combined with (14-81) and (14-82), rejection in the stage and CF are

constant and based on retentate, such that the following equation applies for computing the solute concentration in the retentate:

$$c_{i_R} = c_{i_F} \left[\frac{CF}{CF(1 - \sigma_i) + \sigma_i} \right] \qquad (14\text{-}92)$$

Membrane area is given by (14-85) in continuous-process form,

$$A = \frac{Q_P}{J \{\text{at } CF\}} \qquad (14\text{-}93)$$

Yield of solute is given in continuous-process form by combining the definition of yield in (14-89) with (14-82) and (14-92):

$$Y_i = \frac{c_{i_R} Q_R}{c_{i_F} Q_F} = \frac{1}{CF(1 - \sigma_i) + \sigma_i} \qquad (14\text{-}94)$$

For the four-stage, continuous, feed-and-bleed ultrafiltration system, shown in Figure 14.29, equations (14-90) to (14-94) are applied to each stage. It is usually assumed that the most desirable multistage system is one in which all stages have the same membrane area, which reduces cost of maintenance. The calculations, as described and illustrated in Example 14.14, are iterative in nature, using an outer loop in which membrane area per stage is assumed, and an inner loop in which an overall concentration parameter is assumed.

Diafiltration

As seen in Figure 14.26, when a high degree of solute separation is desired, the flux may drop to a low value. To overcome low flux levels, when it is necessary to continue removing permeable solutes from solutes of little or no permeability, it is common to employ diafiltration, which involves the addition of solvent (usually water) to the retentate, followed by filtration. Additional solvent dilutes the retentate so as to increase the flux. Thus ultrafiltration is employed to a certain limiting concentration of solutes, followed by diafiltration to further enhance solute separation. The final retentate may not be very concentrated in retained solutes, but it contains a smaller fraction of permeable solutes.

Diafiltration is conducted in the same modes as ultrafiltration, i.e., batch or continuous feed-and-bleed, including multistage systems. The added amount or flow rate of solvent is a variable, whose value, for preliminary calculations, may be set equal to the amount of permeate obtained during diafiltration.

Consider a batch diafiltration in which retentate from the previous step is added to the feed tank for diafiltration and recycled, without permeate withdrawal from the membrane unit during startup. Dilution solvent is then added at a continuous rate to the feed tank, under perfect-mixing conditions, with permeate withdrawal at a rate equal to the solvent-addition rate. This operation is sometimes referred to as fed-batch or semicontinuous. If the recycle rate is very high, the concentrations of solutes in the membrane unit will be uniform on each side of the membrane, so rejection in the membrane at any instant is given by (14-81), where both concentrations will change with time. Let c_i = the instantaneous concentration of solute in the recycle retentate. Initially, before solvent is added, its value is that of the feed to the diafiltration system, c_{i_F}. Let c_{i_P} = the instantaneous solute concentration in the permeate leaving the membrane unit. Then (14-81) becomes

$$\sigma_i = 1 - \frac{c_{i_P}}{c_i} \qquad (14\text{-}95)$$

With a constant volume, V_F, in the feed tank, before solvent is added, an instantaneous solute material balance equates the decrease in amount of solute in the feed tank to the amount of solute appearing in the permeate. But permeate flow rate, Q_P, is equal to the solvent addition rate, $Q_S = dV_S/dt$, giving for a solute material balance

$$-V_F \frac{dc_i}{dt} = c_{i_P} Q_P = c_{i_P} \frac{dV_S}{dt} \qquad (14\text{-}96)$$

Combining (14-95) and (14-96) to eliminate c_{i_P} gives, in integral form over time for diafiltration to the final retentate concentration,

$$\int_{c_{i_F}}^{c_{i_R}} \frac{dc_i}{c_i} = -\frac{(1-\sigma_i)}{V_F} \int_0^{V_{S_{total}}} dV_S \qquad (14\text{-}97)$$

Integration gives an equation for computing the final retentate concentration:

$$c_{i_R} = c_{i_F} \exp\left[-\frac{V_{S_{total}}}{V_F}(1-\sigma_i) \right] \qquad (14\text{-}98)$$

Calculations for continuous diafiltration are similar to those for continuous ultrafiltration, as illustrated in Example 14.14.

A major industrial application of ultrafiltration is in processes for manufacturing protein concentrates from skim milk. Skim milk is coagulated to render two products, (1) a thick precipitate called curd, rich in a phosphoprotein called casein, used to make cheese, plastics, paints, and adhesives, and (2) whey (or cheese whey), a watery, residual liquid. One hundred pounds of skim milk yields approximately 10 pounds of curd and 90 pounds of whey. Typically, whey consists, on a mass basis, of 93.35% water; 0.6% true protein (TP) of molecular weight ranging from about 10,000 to 200,000; 0.3% nonprotein nitrogen compounds (NPN); 4.9% lactose (a sugar of empirical formula $C_{12}H_{22}O_{11}$, and

molecular weight of 342, which has an ambient solubility in water of about 10 wt%); 0.2% lactic acid ($C_3H_6O_3$) of molecular weight 90, which is very soluble in water; 0.6% ash (inorganic salts of calcium, sodium, phosphorus, and potassium) of molecular weight ranging from about 20 to 100; and 0.05% butter fat.

Proteins are macromolecules consisting of sequences of amino acids, which contain both amino and carboxylic-acid functional groups. When digested, proteins become sources of amino acids, which are classified as nutritionally essential or nonessential. The *nonessential amino acids* are synthesized by a healthy body from metabolized food. *Essential amino acids* cannot be synthesized by the body, but must be ingested. Amino acids in proteins are the essential building blocks for health because they repair body cells, build and repair muscles and bones, regulate metabolic processes, and provide energy.

The proteins in whey, on a mass basis of total true protein, are betalactoglobulin (50–55%), alpha-lactalbumin (20–25%), immunoglobulins (10–15%), bovine serum albumin (5–10%), and smaller amounts of glycomacropeptide, lactoferrin, lactoperoxidase, and lysozyme. The first five of these eight proteins provide an excellent source of all eight essential amino acids: isoleucine, leucine, lysine, methionine, phenylalanine, threonine, tryptophan, and valine. Approximately 35 wt% of proteins in whey provide amino acids. The nonprotein nitrogen compounds include ammonia, creatine, creatinine, urea, and uric acid, with molecular weights ranging from 17 to 168.

To obtain dry protein concentrate from whey, a number of industrial processes are used. Most involve ultrafiltration, which separates by size exclusion that depends on molecular weight and molecular shape. For separation purposes, the compounds in whey form five groups: (1) true protein and butter fat, (2) nonprotein nitrogen, (3) lactose, (4) lactic acid and ash, and (5) water. A typical process is shown in Figure 14.30. Whey is pumped to ultrafiltration section I, where the exiting retentate (concentrate) contains all of the protein. The other whey-feed components leave mainly in the exiting permeate. The retentate is further concentrated in an evaporator and then spray dried to produce a *whey protein concentrate*. Permeate is pumped to ultrafiltration section II, where all remaining lactose is retained and sent to a second spray dryer to produce *lactose-rich concentrate*, while the permeate is sent to wastewater treatment. Whey-protein concentrate produced by this process contains too high a lactose content for the millions of individuals who are diagnosed as "lactose intolerant" because of susceptibility to digestive disorders, including gassy symptoms, bloating, and diarrhea. To produce so-called *whey protein isolate* of 90–97 wt% protein and almost no lactose or fat, the process of Figure 14.30 is modified by adding additional ultrafiltration. The following example, based on information provided in the 2001 AIChE National Student Design Competition, involves the process design of an ultrafiltration section for producing a protein concentrate.

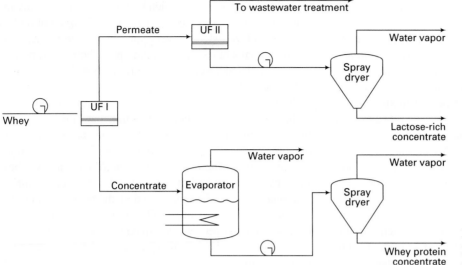

Figure 14.30 Whey process to produce protein and lactose concentrates.

EXAMPLE 14.14

A cheese plant produces a byproduct stream of 1,000,000 lb/day of whey. The whey is to be further processed to obtain a dry powder containing 85 wt% combined TP (true protein) and NPN (nonprotein nitrogen compounds). The process includes three sections: (1) four stages of continuous bleed-and-feed ultrafiltration to reach 55 wt% (dry basis), followed by (2) four stages of continuous diafiltration to reach 75 wt% (dry basis), followed by (3) one stage of batch diafiltration to reach the final 85 wt% (dry basis), with a batch-time limit of 4 hours. Diafiltration must be used above 55 wt% because the retentate from ultrafiltration becomes too viscous above that value. Each section will use PM 10 ultrafiltration hollow-fiber membrane cartridges from Koch Membrane Systems, which are 3 in. in diameter by 40 in. long, with 26.5 ft^2 of membrane area, at a purchase cost of $200.00 each. For each cartridge, the recirculation (recycle) rate is 23 gal/min. The number of cartridges is to be the same in each stage for Sections 1 and 2. The inlet pressure to each cartridge is 30 psig, with a crossflow pressure drop of 15 psi and a permeate pressure of 5 psig. For these conditions, the flux through the membrane has been experimentally measured for the whey and correlated as a function of the concentration factor, CF, by the following equation:

$$\text{membrane flux, gal/ft}^2 \cdot \text{day} = 27.9 - 5.3\ln(CF) \qquad (1)$$

where CF, for any stage n, is defined by reference to the fresh feed to Section 1, as $CF_n = F_{\text{Section }1}/R_n$.

The composition of the whey and measured membrane solute rejections, σ, are as shown in this table:

Component	Wt% in Whey	Flow rate in Whey, lb/day	Solute rejection, σ
Water	93.35	933,500	—
True protein, TP	0.6	6,000	0.970
Nonprotein nitrogen, NPN	0.3	3,000	0.320
Lactose	4.9	49,000	0.085
Ash	0.8	8,000	0.115
Butter fat	0.05	500	1.000

Based on the σ values, the membrane increases the concentration of TP while selectively removing low-molecular-weight solutes of lactose and ash. The density of the whey and all retentate and permeate streams can be assumed to be 8.5 lb/gal. The continuous sections of the process will operate 20 hr per day, leaving 4 hr per day for automatic cleaning to remove accumulated membrane foulants and kill any microorganisms that have attached to equipment surfaces during the 20-hr run period. For each section, calculate:

(1) Component material balances in lb/day of operation, including dilution water for diafiltration operations.

(2) Percent recovery from whey of TP and NPN in the intermediate and final 85 wt% concentrate.

(3) Number of membrane cartridges.

Also, for Section 1, make calculations for a single continuous stage, compare results to those for four stages, and discuss advantages and disadvantages of four stages versus one stage.

SOLUTION

The flow diagram for the specified ultrafiltration–diafiltration process is shown in Figure 14.31. Stream flow rates are those labeled in this diagram.

Single, Continuous, Ultrafiltration Stage for Section 1 to Reach 55 wt% TP + NPN

First compute results for Section 1 using a single stage of continuous, bleed-and-feed ultrafiltration, shown in Figure 14.28. To do this, assume 10 as the value for the concentration factor, CF, and compute by material balance from the whey feed rate, $F_1 = 1,000,000$ lb/day, flow rates of retentate (concentrate), R_1, and permeate, P_1. By definition of CF for this type of ultrafiltration, $R_1 = F_1/CF = 1,000,000/10 = 100,000$. Therefore, $P_1 = F_1 - R_1 = 1,000,000 - 100,000 = 900,000$ lb/day. Next, use a mass flow rate form of the yield equation, (14-94), to compute the flow rate of each solute in the concentrate. For example, for TP,

$$(m_{TP})_{R_1} = \frac{(m_{TP})_{F_1}}{[CF(1-\sigma)+\sigma]}$$

$$= \frac{6000}{[10(1-0.97)+0.97]} = 4724 \text{ lb/day}$$

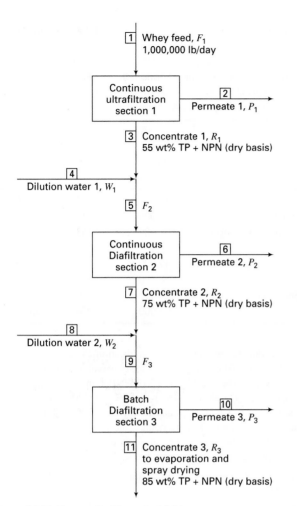

Figure 14.31 Process for Example 14.14.

In a similar manner, flow rates of other solutes in the concentrate, R_1, are computed with the following results, where the water rate is obtained by difference.

Concentrate for a Single Stage of Continuous Ultrafiltration in Section 1, for an Assumed $CF = 10$

Component	Wt% in Concentrate, C_1, lb/day	Flow Rate in Concentrate, C_1, lb/day
Water	88.157	88,157
True protein, TP	4.724	4,724
Nonprotein nitrogen, NPN	0.421	421
Lactose	5.306	5,306
Ash	0.892	892
Butter fat	0.500	500
Total	100.000	100,000

From this table, the wt% TP + NPN in the concentrate on a dry basis is $(4,724 + 421)/(100,000 - 88,157) = 0.4344$ or 43.44 wt% (dry basis). This is less than the 55-wt% target. Therefore, the assumed value of $CF = 10$ is too low. If calculations are made with a spreadsheet, the Solver function can be applied to find the value of CF that meets the 55-wt% target. The result is

$CF = 24.955$, giving the following concentrate:

Concentrate for a Single Stage of Continuous Ultrafiltration in Section 1, for the Correct $CF = 24.955$

Component	Wt% in Concentrate, C_1, lb/day	Flow Rate in Concentrate, C_1, lb/day
Water	83.372	33,409
True protein, TP	8.713	3,491
Nonprotein nitrogen, NPN	0.433	174
Lactose	5.335	2,138
Ash	0.899	360
Butter fat	1.248	500
Total	100.000	40,072

The wt% TP + NPN in the concentrate is now $(3,491 + 174)/(40,072 - 33,409) = 0.5500$ or 55.00%, the specified value.

It is also of interest to compute the % yield of TP + NPN, which is

$$(3491 + 174)/(6000 + 3000) = 0.4072 \text{ or } 40.72\%.$$

The membrane area for this single stage is computed from (14-93). The permeate rate is $P_1 = F_1 - R_1 = 1,000,000 - 40,072 = 959,928$ lb/day or $959,928/8.5 = 112,933$ gal/day. For 20-h/day operation, the volumetric permeate rate $= 112,933/20 = 5,647$ gal/h. From (1), for the computed CF,

$$\text{membrane flux} = 27.9 - 5.3\ln(CF) = 27.9 - 5.3\ln(24.955)$$
$$= 10.85 \text{ gal/ft}^2\cdot\text{day or } 10.85/24 = 0.452 \text{ gal/ft}^2\cdot\text{h}$$

Therefore, from (14-93), the membrane area $= 5,647/0.452 = 12,490$ ft². Each cartridge has an area of 26.5 ft². Therefore, $12,490/26.5 = 471$ parallel cartridges are needed. Total fresh feed rate based on 20 hr of operation is

$$\frac{1,000,000}{8.5(20)} = 5,882 \text{ gal/h}$$

Fresh feed rate to each cartridge is

$$\frac{5,882}{60(471)} = 0.208 \text{ gal/min/cartridge}$$

The combined flow rate (fresh plus recycle) to each cartridge is $0.208 + 23 = 23.208$ gal/min, which is a desirable high recycle ratio.

To increase the % yield and decrease the number of cartridges, a multistage section is needed. In the problem statement, four stages in series are specified. These are computed next.

Four, Continuous, Ultrafiltration Stages for Section 1 to Reach 55 wt% TP + NPN

The following procedure is based on an equal membrane area for each of the four stages in Figure 14.29. The calculations are accomplished by a double "trial-and-error" (nested iteration) procedure, best carried out using a spreadsheet with a Solver function.

Assume a membrane area per stage. Because the single-stage calculation resulted in an area of approximately 12,500 ft², and we are using four stages, the total area for the four stages will be less. First, assume a total area of 8,000 ft² or 2,000 ft² per stage $= A$. Next, find, by iteration, the overall concentration factor, $\overline{CF}$, that gives the fresh feed rate to the first Stage, as calculated above, of

5,882 gal/h. This is done with a spreadsheet starting from Stage 4 and working backward to Stage 1, using the following equations, where J_n = hourly membrane flux = $(1)/24$ of Eq. (1), based on a CF using F_1 and R_n. For Stage 4, CF_4 = the assumed $\overline{CF}$ and $R_4 = F_1/CF_4$. Then, for the calculations back to stage 1:

$$P_n = AJ_n, \quad R_{n-1} = P_n + R_n, \quad CF_{n-1} = F_1/R_{n-1}$$

When Stage 1 is reached by calculation of P_1, the fresh feed rate is computed from $F_1 = P_1 + R_1$. If F_1 is not 5,882 gal/h, new values of $\overline{CF}$ are assumed until the correct value of F_1 is obtained. This iteration can be done with the spreadsheet Solver function. Suppose $\overline{CF} = 20$ is assumed with the assumed $A = 2,000$ ft². The following results are obtained, where all flows are in gal/h:

CF	R_4	P_4	R_3	P_3	R_2	P_2	R_1	P_1	F_1
20	294	1002	1296	1657	2953	2021	4974	2250	7224
61.2	96.2	508	604	1320	1924	1831	3755	2127	5882

The tabulated results give $F_1 = 7,224$ gal/h, which is too high. Using the Solver function of the spreadsheet, a value of $\overline{CF} = 61.2$ gives the correct F_1, with the following corresponding computed values of CF_n and J_n:

$\overline{CF}$	CF_4	J_4	CF_3	J_3	CF_2	J_2	CF_1	J_1
61.2	61.2	0.254	9.735	0.660	3.057	0.916	1.566	1.063

However, the assumed value of membrane area per stage may not be correct. To check this, calculations, with a spreadsheet, similar to those above for a single stage are carried out, starting with Stage 1 and proceeding stage-by-stage to Stage 4. Pertinent results for $A = 2,000$ ft² per stage and $\overline{CF} = 61.2$ are:

Stage	1	2	3	4
Wt% TP + NPN in retentate from stage (dry basis)	17.47	25.46	44.96	74.16

Because the wt% TP + NPN (dry basis) in the retentate from Stage 4 is 74.16%, which is higher than the specified 55 wt%, the calculations must be repeated for other values of membrane area per stage. For each assumed membrane area, a new value of $\overline{CF}$ that gives the correct fresh feed rate must be found. The following spreadsheet results are obtained when iterating on A and $\overline{CF}$:

A, Membrane Area per Stage, ft²	$\overline{CF}$ for Correct Fresh Feed Rate	Wt% TP + NPN in Final Retentate (dry basis)
2,000	61.2	74.16
1,750	26.9	63.22
1,700	22.5	60.26
1,650	18.8	57.14
1,600	15.7	53.95
1,617	16.65	55.00

These results show that for a continuous, four-stage ultrafiltration system, with equal membrane area per stage, the desired value of 55 wt% (dry basis) for TP + NPN in the retentate (concentrate) from Stage 4 corresponds to $A = 1,617$ ft² per stage and an overall concentration factor, $\overline{CF}$, of 16.65. From these results, the material balance, which includes the combined permeate from four ultrafiltration stages and the computed retentate (concentrate) of 55 wt% that leaves Stage 4 and becomes the feed to the continuous

diafiltration section to increase the wt% TP + NPN to 75 (dry basis), is as presented in this table:

Concentrate and Combined Permeate from a Four-stage, Continuous Ultrafiltration in Section 1, for a $\overline{CF}$ of 16.65 and $A = 1,617$ ft²/Stage, which Meets the 55 wt% Specification

Component	Flow Rate of Whey, lb/day	Flow Rate of Concentrate, lb/day	Flow Rate of Combined Permeate, lb/day
Water	933,500	49,897	883,603
True protein, TP	6,000	5,245	755
Nonprotein nitrogen, NPN	3,000	353	2,647
Lactose	49,000	3,476	45,524
Ash	8,000	603	7,397
Butter fat	500	500	0
Total	1,000,000	60,074	939,926

For Section 1, the number of ultrafiltration cartridges required is $1,617/26.5 = 61$ cartridges per stage or a total of 244 cartridges for four stages. The % yield of TP + NPN in the concentrate is $(5,245 + 353)/(6,000 + 3,000) \times 100\% = 62.20\%$. These results compare to 471 cartridges and a % yield of 40.72% for a single stage in Section 1. On both counts, the four-stage system is preferred. A definitive economic analysis needs to include the additional piping and instrumentation costs for a four-module system.

Four, Continuous, Diafiltration Stages for Section 2 to Reach 75 wt% TP + NPN

The following procedure is based on an equal membrane area for each of the four stages, based on a flow diagram similar to Figure 14.29, differing only in the addition of water to the feed to each stage. The calculations for a continuous, multistage, diafiltration system require iteration on a single variable, the added water rate, to achieve the specified 75 wt% TP + NPN. This is conveniently done with a spreadsheet and its Solver function.

For each diafiltration stage, the added water rate, W_n, is the same and equal to W for each stage. Furthermore, for each stage, the permeate rate, P_n is set equal to the added water rate, W. Therefore, the feed rate to stage, F_1 for the first stage and R_{n-1} for the succeeding three stages (i.e., before the added water and the recycle), are all equal to the retentate (concentrate) rate, R_n, sent to the next stage. Thus, all retentate rates are the same, i.e., $F_1 = R_n$. These simplifications result in the same concentration factor, $\overline{CF}$ for every stage:

$$\overline{CF} = \frac{W + F_1}{F_1} \tag{2}$$

For a continuous diafiltration system of n stages, solute component, i, flow rates in the concentrate from the final stage are given by an equation, obtained by applying (14-94) successively to each stage, setting the solute flow rate in the feed to Stages 2, 3, and 4 equal to its flow rate in the retentate from the preceding stage:

$$(m_i)_{R_n} = (m_i)_{F_1} \left[\frac{1}{\overline{CF}(1 - \sigma_i) + \sigma_i} \right]^n \tag{3}$$

Using a spreadsheet, (2) and (3) are used where values of $(m_i)_{F_1}$ are the solute component flow rates in the concentrate leaving Section 1, as given in the above table. Values of the solute rejections, σ_i, are given in the problem statement. A value is assumed for the added water rate to each stage, W, and $\overline{CF}$ is computed from (2). From (3), the values of $(m_i)_{R_n}$ are computed for each solute. The wt% TP + NPN (dry basis) is then calculated. If it is not the specified 75 wt%, a new value of W is assumed. For example, assume a value for W equal to half of the feed rate, F_1, or 60,074/ 2 = 30,037 lb/day. Then, from (1), $\overline{CF} = (30,037 + 60,074)/ 60,074 = 1.50$. For TP, from (2),

$$(m_{TP})_{R_4} = 5,245 \left[\frac{1}{1.50(1 - 0.97) + 0.97} \right]^4 = 4,942 \text{ lb/day}$$

The calculations are repeated for the other components, and the water rate in the concentrate from stage 4 of Section 2 is determined so that the total concentrate flow rate equals that of the feed, 60,074 lb/day. The wt% TP + NPN in the concentrate is then calculated with a result of 78.2 wt%, which is higher than the specified 75 wt%. Using the Solver function, the correct water rate for each stage is found to be 23,332 lb/day or a total of 93,328 lb/day for the four stages, with a corresponding $\overline{CF} = 1.388$. The resulting material balance, which includes the combined permeate from the four diafiltration stages and the computed retentate (concentrate) of 75 wt% that leaves stage 4 and becomes the feed to the batch diafiltration section to increase the wt% TP + NPN to 85 (dry basis) is as presented in the next table.

Concentrate and Combined Permeate from a Four-stage, Continuous Diafiltration in Section 2, and an Added Water Rate of 23,332 lb/day per Stage, which Meets the 75 wt% Specification

Component	Flow Rate in Feed to Section 2, lb/day	Flow Rate in Concentrate, lb/day	Flow Rate in Combined Permeate, lb/day
Water	49,897	53,214	90,011
True protein, TP	5,245	5,007	238
Nonprotein nitrogen, NPN	353	138	215
Lactose	3,476	1,030	2,446
Ash	603	185	418
Butter fat	500	500	0
Total	60,074	60,074	93,328

From these results, the yield of TP + NPN from diafiltration is $(5,007 + 138)/(5,245 + 353) \times 100\% = 91.91\%$ for an overall yield to this point of $(0.9191)(0.6220) \times 100\% = 57.17\%$. The membrane flux for each stage is obtained from (1). However, the CF used in that equation is the ratio of the whey feed for the process to the retentate rate from the stage, which for the four stages of diafiltration is the same as that for the last stage of the ultrafiltration in Section 1. A value of $CF = 1,000,000/ 60,074 = 16.65$ applies, which results in a membrane flux of 0.5415 gal/h-ft². The volumetric permeate flow rate per stage = $93,328/[(20)(4)(8.5)] = 137$ gal/h. The membrane area required per stage = $137/0.5415 = 253$ ft². The number of cartridges

per stage = $253/26.5 = 9.5 \approx 10$ cartridges per diafiltration stage for a total of 40 diafiltration cartridges.

These results for four stages of continuous diafiltration may be compared to results obtained in a similar manner with just a single, continuous, diafiltration stage, which gives an added water rate of 167,200 gal/day (compared to 93,328 for four stages) and an overall TP + NPN yield of 55% (compared to 57% for four stages).

A Single, Batch, Diafiltration Stage for Section 3 to Reach 85 wt% TP + NPN

For the final membrane section, a single batch diafiltration is employed with maximum batch time of 4 hr. The common method for carrying out batch diafiltration is to initially fill the feed tank with the feed (concentrate from Section 2) and add water continuously over the 4-hr period to maintain the liquid level in the tank. With a high recycle ratio, the concentration of solutes in the retentate is maintained constant, with the flow rates of the solute component, i, in the daily feed of concentrate given by (14-98) in mass-flow form:

$$(m_i)_R = (m_i)_F \exp\left[-\frac{W}{F}(1 - \sigma_i) \right] \quad (4)$$

where W and F are total amounts of additional water and feed processed during the 4-hr period. To reach the target of 85 wt%, it is particularly important to remove the lactose from the feed. For example, suppose we choose a daily amount of added water equal to the daily amount of feed from Section 2. Then, $W/F = 1$. From the preceding table, that feed contains 1,030 lb/day of lactose, which has a rejection, $\sigma = 0.085$. Substitution into (4) gives

$$(m_{lactose})_R = 1,030 \exp[-1(1 - 0.085)] = 413 \text{ lb/day}$$

Flow rates of the other solutes in the concentrate from Section 3 are computed in a similar manner, from which the wt% TP + NPN (dry basis) is computed. If this is done with a spreadsheet, the Solver function can be used to determine the added water to achieve 85 wt%. The added water found in this manner is 80,520 lb/day.

The following material balance includes the permeate from the batch diafiltration stage and the computed retentate (concentrate) of 85 wt% (dry basis) from Section 3.

Concentrate and Permeate from a Single-stage Batch Diafiltration in Section 3, for an Added Water Rate of 80,520 lb/day, to Meet the 85 wt% Specification

Component	Flow Rate of Feed to Section 3, lb/day	Flow Rate of Concentrate, lb/day	Flow Rate of Permeate, lb/day
Water	53,214	54,351	79,383
True protein, TP	5,007	4,810	197
Nonprotein nitrogen, NPN	138	55	83
Lactose	1,030	302	728
Ash	185	56	129
Butter fat	500	500	0
Total	60,074	60,074	80,520

From this table the % yield of TP + NPN in Section 3 is $(4,810 + 55)/(5,007 + 138) \times 100\% = 94.56\%$. The overall yield of TP + NPN from the whey feed is $(4,810 + 55)/(9,000) \times 100\% = 54.06\%$.

The membrane flux is 0.5415 gal/h-ft^2, as in Section 2. The volumetric permeate flow rate over a 4-hour batch operation = $80,520/[(4)(8.5)] = 2,368$ gal/h. Therefore, the membrane area required = $2,368/0.5415 = 4,373$ ft^2. The number of cartridges needed in Section 3 is $4,373/26.5 = 165$.

14.9 MICROFILTRATION

As shown in Table 14.10, microfiltration is a pressure-driven, microporous membrane process used to retain matter as low as 0.02 micron in size, but more commonly of 0.1–10 microns. The matter may include large colloids, small and solid particles, blood cells, yeast, bacteria and other microbial cells, and very large and soluble macromolecules.

Membrane structures for microfiltration include *screen filters* that collect retained matter on the surface and *depth filters* that trap particles at constrictions within the membrane. As Porter discusses in Schweitzer [52], depth filters include: (1) relatively thick, high-porosity (80–85%) cast-cellulose-ester membranes having an open, tortuous, sponge-like structure; and (2) thin, low-porosity (nominal 10%) polyester or polycarbonate track-etch membranes of a sieve-like structure with narrow distribution of straight-through, cylindrical pores. The latter have a much sharper cutoff, resulting in enhanced separation factors. For example, a Nuclepore™ membrane of type 2 can separate a male-determining sperm from a female-determining sperm. As shown in Table 14.11, Nuclepore™ membranes come in pore sizes from 0.03 to 8.0 microns with water permeate flux

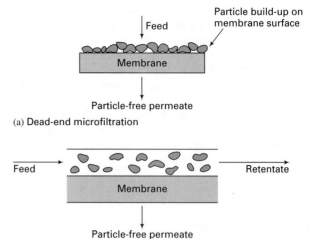

(a) Dead-end microfiltration

(b) Tangential-flow microfiltration

Figure 14.32 Common modes of microfiltration.

rates, at 70°F and a transmembrane pressure difference of 10 psi, ranging from 0.006 to 144 gal/min-ft^2.

Two different modes of microfiltration are employed, as shown in Figure 14.32. In *dead-end* (in-line) filtration (DEF), which is suitable for batch or semicontinuous filtration of dilute solutions, the entire solvent is forced under pressure through the membrane, in a manner similar to conventional filtration of solid–liquid slurries. With time, as retained material accumulates on or within the membrane, the pressure required to maintain a desired flow rate must increase or permeate flux will decrease. Thus, DEF may operate at constant pressure, constant flux, or a combination thereof. A

Table 14.11   Typical Specifications for Nuclepore™ Track-etch Microfiltration Membranes

Specified Pore Size, μm	Pore-size Range, μm	Nominal Pore Density, Pores/cm^2	Nominal Membrane Thickness, μm	Typical Flow Rates at 10 lb/in^2 ΔP, 70°F — Water, gal/(min)(ft^2)
8.0	6.9–8.0	1×10^5	8.0	144.0
5.0	4.3–5.0	4×10^5	8.6	148.0
3.0	2.5–3.0	2×10^6	11.0	121.0
1.0	0.8–1.0	2×10^7	11.5	67.5
0.8	0.64–0.80	3×10^7	11.6	48.3
0.6	0.48–0.60	3×10^7	11.6	16.3
0.4	0.32–0.40	1×10^8	11.6	17.0
0.2	0.16–0.20	3×10^8	12.0	3.1
0.1	0.08–0.10	3×10^8	5.3	1.9
0.08	0.064–0.080	3×10^8	5.4	0.37
0.05	0.040–0.050	6×10^8	5.4	1.12
0.03	0.024–0.030	6×10^8	5.4	0.006

combination, in which constant-flux operation is employed up to a limiting pressure, followed by constant-pressure operation until a minimum flux is reached, is a superior strategy to improve yield.

In *tangential-flow* (crossflow) *filtration* (TFF), which is more suitable for large-scale, continuous filtration, feed flows along the surface with only a fraction of the solvent passing through the membrane. Ideally, retained matter is carried out of the microfilter with retentate fluid. The tangential-flow mode is usually accompanied by a large retentate recycle in a manner similar to feed-and-bleed operations for ultrafiltration. Typical tangential flow velocities parallel to the membrane surface on the feed-retentate side are in the 3–25-ft/s range. The tangential-flow mode is also used, almost exclusively, for reverse osmosis and ultrafiltration, as discussed above in subsections 14.5 and 14.8. Significant attempts to improve tangential-flow filtration (TFF) have evolved into *high-performance tangential flow filtration* (HPTFF) for separation of macromolecular species. In a study by van Reis et al. [53], significant improvements in yield and purification factors were achieved in the separation of protein mixtures of close molecular weights with HPTFF, wherein the process is optimized with respect to membrane pore-size distribution, membrane chemistry (electrostatic interactions between the membrane material and the solutes), accompanying use of a solution buffer to control pH to obtain the proper electrostatic level, fluid dynamics (including crossflow velocity and trans-membrane pressure), and number of stages.

Membrane modules for microfiltration are mainly DEF plate-and-frame and TFF pleated cartridges. In the latter, the membrane is pleated and then folded around a permeate core. Many module types are inexpensive and disposable. A typical disposable cartridge is 2.5 in. in diameter and 10 in. long, with 3 ft^2 of membrane area. The cartridge may include a prefilter to extend filter life by removing larger particles, leaving the microporous membrane to make the required separation. Compared to TFF, DEF has a lower capital cost, a higher operating cost, and a simpler operation. DEF is most suitable for dilute solutions, while TFF is preferred for concentrated solutions.

Equations for computing TFF microfiltration are those developed for ultrafiltration. This includes batch, continuous feed-and-bleed, and diafiltration operation modes. Equations for DEF microfiltration are those for conventional, batch, solid–liquid, slurry filtration, frequently referred to as cake filtration. These equations are developed as follows, where the only resistances considered are those of the membrane and buildup of particles in a cake on the membrane surface. Any deposition of matter within the membrane is ignored, as is osmotic pressure, which is almost always negligible for microfiltration because of the high solute molecular weight.

At any instant of time, the volumetric permeate flux, J, is

$$J = \frac{1}{A_M}\frac{dV}{dt} = \frac{\Delta P}{\mu(R_m + R_c)} \quad (14\text{-}99)$$

where

$$\Delta P = \text{pressure drop across the cake and membrane}$$

$$\mu = \text{permeate viscosity}$$

$$R_m \text{ and } R_c = \text{respectively, resistances to permeate flow of membrane and cake}$$

$$V = \text{permeate volume}$$

$$A_M = \text{membrane area}$$

Assume that membrane resistance is constant and that the cake can be treated as a growing packed bed of particles, for which the Ergun correlation of Figure 6.37 can be applied. In the low-Reynolds-number region, which is typical of flow through a cake, pressure drop is governed by the Kozeny–Carman equation, which gives for cake resistance.

$$R_c = \frac{150 l_c (1 - \epsilon_c)^2}{D_P^2 \epsilon_c^3} = K_1 \frac{l_c (1 - \epsilon_c)^2}{\epsilon_c^3} \quad (14\text{-}100)$$

where

$l_c =$ cake thickness, which increases with time

$\epsilon_c =$ cake porosity

$D_P =$ effective diameter of the matter in the cake

$K_1 =$ experimental constant for a particular filtration system

The solid matter in the feed, m_c, that will be retained in the cake is

$$m_c = c_F V = \rho_c (1 - \epsilon_c) A_M l_c \quad (14\text{-}101)$$

where $c_F =$ the solid matter per unit volume of liquid in the feed.

Substitution of (14-101) into (14-100) to eliminate l_c and substitution of the resulting equation for R_c into (14-99) gives

$$J = \frac{1}{A_M}\frac{dV}{dt} = \frac{\Delta P}{\mu \left[R_m + K_1 \dfrac{c_F V (1 - \epsilon_c)}{\rho_c A_M \epsilon_c^3} \right]}$$

$$= \frac{\Delta P}{\mu \left[R_m + K_2 \dfrac{c_F V}{A_M} \right]} \quad (14\text{-}102)$$

where $K_2 =$ experimental constant for a particular filtration system.

Constant-Flux Operation

For operation with a constant permeate flux, J, and corresponding volumetric permeate flow rate, pressure drop will increase with time as cake thickness increases. From (14-102), $dV/dt = V/t$, with $V = 0$ at $t = 0$, and variation of pressure drop with time is given by

$$\Delta P = J\mu[R_m + K_2 c_F J t] \quad (14\text{-}103)$$

Initially, pressure depends only on membrane resistance, with $\Delta P = J\mu R_m$. Subsequently, pressure drop increases linearly with time as cake thickness increases.

Constant-Pressure Operation

For operation with a constant pressure drop, permeate flow rate and corresponding permeate flux will decrease with time as cake thickness increases. For this type of operation, the integral of (14-102) applies:

$$\int_0^V \left[R_m + K_2 \frac{c_F V}{A_M} \right] dV = \frac{A_M \Delta P}{\mu} \int_0^t dt \quad (14\text{-}104)$$

Integration of (14-104) gives

$$R_m V + \frac{K_2 c_F V^2}{2 A_M} = \frac{A_M \Delta P \, t}{\mu} \quad (14\text{-}105)$$

Equation (14-105) is a quadratic equation in V, which when solved gives the following positive root:

$$V = -\frac{R_m A_M}{K_2 c_F} + \frac{A_M}{K_2 c_F} \left[R_m^2 + \frac{2 K_2 c_F \Delta P \, t}{\mu} \right]^{\frac{1}{2}} \quad (14\text{-}106)$$

Differentiating (14-106) to obtain the permeate flux gives

$$\frac{1}{A_M} \frac{dV}{dt} = J = \frac{\Delta P}{\mu} \left[R_m^2 + \frac{2 K_2 c_F \Delta P \, t}{\mu} \right]^{-\frac{1}{2}} \quad (14\text{-}107)$$

Combined Operation

As mentioned above, improvements in yield can be achieved by a combined operation in which: (1) Constant-flux operation is employed in Stage 1 up to a limiting pressure drop, followed by (2) Constant-pressure operation in Stage 2 until a minimum flux is reached. Let V_{CF} = the volume of permeate obtained during Stage 1 for the time period t_{CF}. Because of the constant permeate flux,

$$V_{CF} = J A_M t_{CF} \quad (14\text{-}108)$$

Let ΔP_{UL} = the upper limit of the pressure drop across the cake and membrane. Then, for Stage 2 at constant pressure, (14-104) becomes

$$\int_{V_{CF}}^V \left[R_m + K_2 \frac{c_F V}{A_M} \right] dV = \frac{A_M \Delta P_{UL}}{\mu} \int_{t_{CF}}^t dt \quad (14\text{-}109)$$

Integrating (14-109) and solving the resulting quadratic equation for the positive root of V,

$$V = -\frac{R_m A_M}{K_2 c_F} + \left[\frac{A_M^2 R_m^2}{K_2^2 c_F^2} + \frac{2 A_M}{K_2 c_F} \right.$$
$$\left. \times \left(R_m V_{CF} + \frac{K_2 c_F V_{CF}^2}{2 A_M} + \frac{A_M \Delta P_{UL}(t - t_{CF})}{\mu} \right) \right]^{\frac{1}{2}} \quad (14\text{-}110)$$

where $V - V_{CF}$ = the volume of permeate obtained during Stage 2 for the time period $t - t_{CF}$ during Stage 2.

During Stage 2, the permeate flux decreases with time, according to the following equation, which is obtained by applying to (14-110) the definition of the permeate flux of the left-hand side and middle part of (14-102):

$$J = \frac{A_M \Delta P_{UL}}{K_2 c_F \mu} \left[\frac{A_M^2 R_m^2}{K_2^2 c_F^2} + \frac{2 A_M}{K_2 c_F} \right.$$
$$\left. \times \left(R_m V_{CF} + \frac{K_2 c_F V_{CF}^2}{2 A_M} + \frac{A_M \Delta P_{UL}(t - t_{CF})}{\mu} \right) \right]^{-\frac{1}{2}} \quad (14\text{-}111)$$

Application of the above equations for combined operation is illustrated in the following example.

EXAMPLE 14.15

Diluted skim milk with a protein concentration of 4.3 g/L is to undergo DEF microfiltration. Laboratory experiments have been performed using a cellulose-acetate membrane with an average pore diameter of 0.45 micron and a membrane area of 17.3 cm². For a Stage 1 operation at a constant permeate rate of 15 mL/min, pressure across the cake and membrane increases from 0.3 psi to an upper limit of 20 psi in 400 seconds. The permeate viscosity is 1 cP. If the operation is continued in a second stage at a constant pressure drop at the upper limit until the permeate rate drops to 5 mL/min, estimate the additional time of operation. Also, prepare plots of permeate volume in mL and permeate flux in mL/cm²-min as functions of time in seconds.

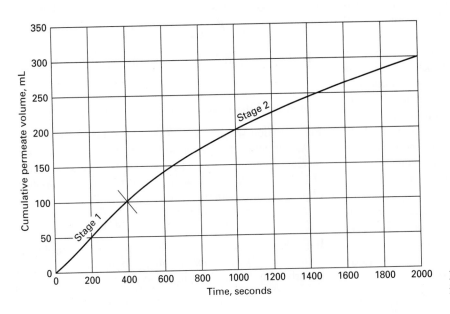

Figure 14.33 Cumulative permeate volume for Example 14.15.

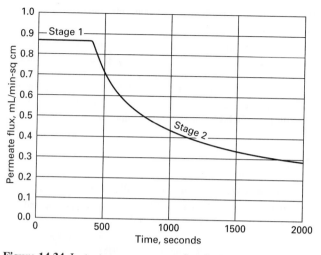

Figure 14.34 Instantaneous permeate flux for Example 14.15.

SOLUTION

In SI units, $c_F = 4.3$ kg/m^3, $A_M = 0.00173$ m^2, volumetric flow rate in Stage 1 $= 0.25 \times 10^{-6}$ m^3/s, P at zero time $= 2068$ Pa, P at 400 s $= 137900$ Pa, and $\mu = 0.001$ Pa-s. Based on the experimental results for Stage 1, calculate values of R_m and K_2, as follows. From (14-103),

$$R_m = \frac{(\Delta P \text{ at } t = 0)}{J\mu} = \frac{2068}{\left(\dfrac{0.25 \times 10^{-6}}{0.00173}\right)(0.001)}$$

$$= 1.43 \times 10^{10} \text{ m}^{-1}$$

For $t = 400$ s,

$$K_2 = \frac{\Delta P - J\mu R_m}{J^2 \mu c_F t} = \frac{\Delta P - (\Delta P \text{ at } t = 0)}{J^2 \mu c_F t}$$

$$= \frac{6895(20 - 0.3)}{\left(\dfrac{0.25 \times 10^{-6}}{0.00173}\right)^2 (0.001)(4.3)(400)} = 3.78 \times 10^{12} \text{ m/kg}$$

Assume that these values of R_m and K_2 are valid during Stage 2. At the end of Stage 2, the permeate flux $= J = (5 \text{ mL/min})/(17.3 \text{ cm}^2) = 0.289$ mL/min-cm$^2 = 4.82 \times 10^{-5}$ m/s.

Solving (14-111) for t gives 2,025 s. Plots of permeate volume and permeate flux versus time, shown in Figures 14.33 and 14.34, are obtained by solving, as a function of time, (14-110) and (14-111) with a spreadsheet during Stage 2.

SUMMARY

1. The separation of liquid and gas mixtures with membranes is an emerging separation operation. Applications began accelerating in the 1980s. The products of separation are retentate and permeate.

2. The key to an efficient and economical membrane separation process is the membrane. It must have good permeability, high selectivity, stability, freedom from fouling, and a long life (2 or more years).

3. Commercialized, membrane-separation processes include dialysis, electrodialysis, reverse osmosis, gas permeation, pervaporation, ultrafiltration, and microfiltration.

4. Most membranes for commercial separation processes are natural or synthetic, glassy or rubbery polymers. However, for high-temperature (>200°C) operations with chemically reactive mixtures, ceramics, metals, and carbon find applications.

5. To achieve high permeability and selectivity, dense, nonporous membranes are preferred. For mechanical integrity, membranes of 0.1 to 1.0 mm in thickness are incorporated as a surface layer or film onto or as part of a much thicker asymmetric or composite membrane.

6. To achieve a high surface area per unit volume, membranes are fabricated into spiral-wound or hollow-fiber modules. Less surface is available in plate-and-frame, tubular, and monolithic modules.

7. Permeation through a membrane can occur by a variety of mechanisms. For a microporous membrane, the mechanisms include bulk flow (with no selectivity), liquid diffusion, gas diffusion, Knudsen diffusion, restrictive diffusion (including sieving), and surface diffusion. For a nonporous membrane, a solution-diffusion mechanism, involving absorption, diffusion, and desorption, is commonly assumed.

8. Flow patterns in membrane modules have a profound effect on overall permeation rates. Idealized flow patterns for which theory has been developed include perfect mixing, countercurrent flow, cocurrent flow, and crossflow.

9. To overcome the limit of separation in a single membrane-module stage, modules can be arranged in series and/or parallel cascades.

10. In gas permeation, boundary-layer or film mass-transfer resistances on either side of the membrane are usually negligible compared to the membrane resistance. For membrane separation of liquid mixtures, however, external mass-transfer effects and concentration polarization can be significant.

11. For most membrane separators, the component mass-transfer fluxes through the membrane can be formulated as the product of two terms: concentration, partial pressure, fugacity, or activity driving force; and a permeance $\bar{P}_{M_i}$, which is the ratio of the permeability, P_{M_i}, to the membrane thickness, l_M.

12. In the dialysis of a liquid mixture, small solutes of type A are separated from the solvent and larger solutes of type B with a microporous membrane. The driving force is the concentration difference across the membrane. Transport of solvent can be minimized by adjusting pressure differences across the membrane to equal osmotic pressure.

13. In electrodialysis, a series of alternating cation- and anion-selective membranes are used with a direct-current voltage across an outer anode and an outer cathode to concentrate an electrolyte.

14. In reverse osmosis, the solvent of a liquid mixture is selectively transported through a dense membrane. By this means, seawater can be desalinized. The driving force for solvent transport through the membrane is fugacity difference, which is commonly expressed in terms of $\Delta P - \Delta \pi$, where π is the osmotic pressure.

15. In gas permeation, mixtures of gases are separated by differences in permeation rates through dense membranes. The driving force for each component is its partial pressure difference, Δp_i, across the membrane. Both permeance and permeability depend on membrane absorptivity for the particular gas species (usually as a Henry's law constant) and species diffusivity through the membrane. Thus, $P_{M_i} = H_i D_i$.

16. In pervaporation, a liquid mixture is separated with a dense membrane by pulling a vacuum on the permeate side of the membrane so as to evaporate the permeate. The driving force may be approximated as a fugacity difference expressed by $(\gamma_i x_i P_i^s - y_i P_P)$.

Permeability can vary greatly with concentration because of membrane swelling.

17. Microfiltration and ultrafiltration, like reverse osmosis, are pressure-driven operations that use microporous membranes to retain large molecules and particles, allowing water or other low-molecular-weight molecules to pass through. Typically, microfiltration retains matter of diameter between 0.02 and 10 microns, while ultrafiltration retains in the 0.001–0.02-micron range.

REFERENCES

1. LONSDALE, H.K., *J. Membrane Sci.*, **10**, 81 (1982).

2. APPLEGATE, L.E., *Chem. Eng.*, **91**(12), 64–89 (1984).

3. HAVENS, G.G., and D.B. GUY, *Chem. Eng. Progress Symp. Series*, **64**(90), 299 (1968).

4. BOLLINGER, W.A., D.L. MACLEAN, and R.S. NARAYAN, *Chem. Eng. Progress*, **78**(10), 27–32 (1982).

5. BAKER, R.W., E.L. CUSSLER, W. EYKAMP, W.J. KOROS, R.L. RILEY, and H. STRATHMANN, *Membrane Separation Systems—A Research and Development Needs Assessment*, Report DE 90-011770, Department of Commerce, NTIS, Springfield, VA (1990).

6. HO, W.S.W., and K.K. SIRKAR, Eds., *Membrane Handbook*, Van Nostrand Reinhold, New York (1992).

7. LOEB, S., and S. SOURIRAJAN, Advances in Chemistry Series, Vol. 38, *Saline Water Conversion II* (1963).

8. HENIS, J.M.S., and M.K. TRIPODI, U.S. Patent 4,230,463 (1980).

9. WRASIDLO, W.J., U.S. Patent 3,951,815 (1977).

10. BARRER, R.M., *J. Chem. Soc.*, 378–386 (1934).

11. BARRER, R.M., *Diffusion in and through Solids*, Cambridge Press, London (1951).

12. MAHON, H.I., U.S. Patent 3,228,876 (1966).

13. MAHON, H.I., U.S. Patent 3,228,877 (1966).

14. HSIEH, H.P., R.R. BHAVE, and H.L. FLEMING, *J. Membrane Sci.*, **39**, 221–241 (1988).

15. BIRD, R.B., W.E. STEWART, and E.N. LIGHTFOOT, *Transport Phenomena*, John Wiley and Sons, New York, pp. 42–47 (1960).

16. ERGUN, S., *Chem. Eng. Progress*, **48**, 89–94 (1952).

17. BECK, R.E., and J.S. SCHULTZ, *Science*, **170**, 1302–1305 (1970).

18. BECK, R.E., and J.S. SCHULTZ, *Biochim. Biophys. Acta*, **255**, 273 (1972).

19. BRANDRUP, J., and E.H. IMMERGUT, Eds., *Polymer Handbook*, 3rd ed., John Wiley and Sons, New York (1989).

20. LONSDALE, H.K., U. MERTEN, and R.L. RILEY, *J. Applied Polym. Sci.*, **9**, 1341–1362 (1965).

21. MOTAMEDIAN, S., W. PUSCH, G. SENDELBACH, T.-M. TAK, and T. TANIOKA, *Proceedings of the 1990 International Congress on Membranes and Membrane Processes*, Chicago, Vol. II, pp. 841–843.

22. BARRER, R.M., J.A. BARRIE, and J. SLATER, *J. Polym. Sci.*, **23**, 315–329 (1957).

23. BARRER, R.M., and J.A. BARRIE, *J. Polym. Sci.*, **23**, 331–344 (1957).

24. BARRER, R.M., J.A. BARRIE, and J. SLATER, *J. Polym. Sci.*, **27**, 177–197 (1958).

25. KOROS, W.J., and D.R. PAUL, *J. Polym. Sci., Polym. Physics Edition*, **16**, 1947–1963 (1978).

26. BARRER, R.M., *J. Membrane Sci.*, **18**, 25–35 (1984).

27. WALAWENDER, W.P., and S.A. STERN, *Separation Sci.*, **7**, 553–584 (1972).

28. NAYLOR, R.W., and P.O. BACKER, *AIChE J.*, **1**, 95–99 (1955).

29. STERN, S.A., T.F. SINCLAIR, P.J. GAREIS, N.P. VAHLDIECK, and P.H. MOHR, *Ind. Eng. Chem.*, **57**(2), 49–60 (1965).

30. HWANG, S.-T., and K.L. KAMMERMEYER, *Membranes in Separations*, Wiley-Interscience, New York, pp. 324–338 (1975).

31. SPILLMAN, R.W., *Chem. Eng. Progress*, **85**(1), 41–62 (1989).

32. STRATHMANN, H., "Membrane and Membrane Separation Processes," in Vol. A16, *Ullmann's Encyclopedia of Industrial Chemistry*, VCH, FRG, p. 237 (1990).

33. CHAMBERLIN, N.S., and B.H. VROMEN, *Chem. Engr.* **66**(9), 117–122 (1959).

34. GRAHAM, T., *Phil. Trans. Roy. Soc. London*, **151**, 183–224 (1861).

35. JUDA, W., and W.A. MCRAE, *J. Amer. Chem. Soc.*, **72**, 1044 (1950).

36. STRATHMANN, H., *Sep. and Purif. Methods*, **14**(1), 41–66 (1985).

37. MERTEN, U., *Ind. Eng. Chem. Fundamentals*, **2**, 229–232 (1963).

38. STOUGHTON, R.W., and M.H. LIETZKE, *J. Chem. Eng. Data*, **10**, 254–260 (1965).

39. SPILLMAN, R.W., and M.B. SHERWIN, *Chemtech*, 378–384 (June 1990).

40. SCHELL, W.J., and C.D. HOUSTON, *Chem. Eng. Progress*, **78**(10), 33–37 (1982).

41. TEPLYAKOV, V., and P. MEARES, *Gas Sep. and Purif.*, **4**, 66–74 (1990).

42. ROSENZWEIG, M.D., *Chem. Eng.*, **88**(24), 62–66 (1981).

43. KOBER, P.A., *J. Am. Chem. Soc.*, **39**, 944–948 (1917).

44. BINNING, R.C., R.J. LEE, J.F. JENNINGS, and E.C. MARTIN, *Ind. Eng. Chem.*, **53**, 45–50 (1961).

45. WESSLEIN, M., A. HEINTZ, and R.N. LICHTENTHALER, *J. Membrane Sci.*, **51**, 169 (1990).

46. WIJMANS, J.G., and R.W. BAKER, *J. Membrane Sci.*, **79**, 101–113 (1993).

47. RAUTENBACH, R., and R. ALBRECHT, *Membrane Processes*, John Wiley and Sons, New York (1989).

48. RAO, M.B., and S. SIRCAR, *J. Membrane Sci.*, **85**, 253–264 (1993).

49. BAKER, R., *Membrane Technology and Applications*, 2nd ed., John Wiley & Sons, New York (2004).

50. CHERYAN, M., *Ultrafiltration Handbook*, Technomic Publishing Co., Lancaster, PA (1986).

51. ZEMAN, L.J., and A.L. ZYDNEY, *Microfiltration and Ultrafiltration, Principles and Applications*, Marcel Dekker, Inc., New York (1996).

52. SCHWEITZER, P.A., *Handbook of Separation Techniques for Chemical Engineers*, 2nd ed., Section 2.1 by M.C. Porter, McGraw-Hill Book Co., New York (1988).

53. VAN REIS, R. et al., *Biotechnology and Bioengineering*, **56**(1), October 5, 1997.

EXERCISES

Section 14.1

14.1 Explain, as completely as you can, how membrane separations differ from:

(a) Absorption and stripping

(b) Distillation

(c) Liquid–liquid extraction

(d) Extractive distillation

14.2 For the commercial application of membrane separators discussed at the beginning of this chapter, calculate the permeabilities of hydrogen and methane in barrer.

14.3 A new asymmetric, polyimide, polymer membrane has been developed for the separation of N_2 from CH_4. At 30°C, permeance values are 50,000 and 10,000 barrer/cm for N_2 and CH_4, respectively. If this new membrane is used to perform the separation in Figure 14.35, determine the membrane surface area required in m^2, and the kmol/h of CH_4 in the permeate. Base the driving force for diffusion through the membrane on the arithmetic average of the partial pressures of the entering feed and the exiting retentate, with the permeate-side partial pressures at the exit condition.

Section 14.2

14.4 A hollow-fiber module has 4,000 ft^2 of membrane surface area based on the inside diameter of the fibers, which are 42 μm i.d. × 85 μm o.d. × 1.2 m long each. Determine:

(a) The number of hollow fibers in the module.

(b) The diameter of the module, assuming the fibers are on a square spacing of 120 μm center-to-center.

(c) The membrane surface area per unit volume of module (packing density) m^2/m^3. Compare your result with that in Table 14.4.

14.5 A typical spiral-wound module made from a flat sheet of membrane material is 0.3 m in diameter and 3 m long. If the packing density (membrane surface area/unit module volume) is 500 m^2/m^3, determine the center-to-center spacing of the membrane in the spiral, assuming a collection tube 1 cm in diameter.

14.6 A monolithic membrane element, of the type shown in Figure 14.4d, contains 19 flow channels that are 0.5 cm in inside diameter by 0.85 m long. If nine of these elements are placed into a cylindrical module of the type shown in Figure 14.5, determine reasonable values for:

(a) Module volume in m^3.

(b) Packing density in m^2/m^3. Compare your value with values for other membrane modules given in Table 14.4.

Section 14.3

14.7 Water at 70°C is to be passed through a porous polyethylene membrane of 25% porosity with an average pore diameter of 0.3 micron and an average tortuosity of 1.3. The pressures on the downstream and upstream sides of the membrane are 125 and 500 kPa, respectively. Estimate the flow rate of water through the membrane in m^3/m^2-day.

14.8 A porous-glass membrane, with an average pore diameter of 40 Å, is to be used to separate light gases at 25°C under conditions where Knudsen flow may be dominant. The downstream pressure is 15 psia, while the upstream pressure is not greater than 120 psia. The membrane has been calibrated with pure helium gas, giving a constant permeability of 117,000 barrer over the operating pressure range. Experiments with pure CO_2 over the pressure range give a permeability of 68,000 barrer.

Assuming that helium is in Knudsen flow, predict the permeability of CO_2. Is the value in agreement with the experimental value? If not, suggest an explanation. Reference: Kammermeyer, K., and L.O. Rutz, *C.E.P. Symp. Ser.,* **55**(24), 163–169 (1959).

14.9 Two mechanisms for the transport of gas components through a porous membrane that are not discussed in Section 14.3 or illustrated in Figure 14.6 are (1) partial condensation in the pores by some components of the gas mixture to the exclusion of other components and subsequent transport of the condensed molecules through the pore, and (2) selective adsorption on pore surfaces of certain components of the gas mixture and subsequent surface diffusion across the pores. In particular, Rao and Sircar [48] have found that the latter mechanism provides a potentially attractive means for separating hydrocarbons from hydrogen for low-pressure gas streams. In porous-carbon membranes with continuous pores 4–15 Å in diameter, little pore void space is available for the Knudsen diffusion of hydrogen when the hydrocarbons are selectively adsorbed.

Typically, the membranes are not more than 5 μm in thickness. Measurements at 295.1 K of permeabilities for five pure components and a mixture of the five components are as follows:

Component	**Permeability, barrer**		mol% in the
	As a Pure Gas	In the Mixture	Mixture
H_2	130	1.2	41.0
CH_4	660	1.3	20.2
C_2H_6	850	7.7	9.5
C_3H_8	290	25.4	9.4
nC_4H_{10}	155	112.3	19.9
			100.0

A refinery waste gas mixture of the preceding composition is to be processed through such a porous-carbon membrane. If the pressure of the gas is 1.2 atm and an inert sweep gas is used on the permeate side such that partial pressures of feed gas components on that side are close to zero, determine the permeate composition on a sweep-gas-free basis when the composition on the upstream-pressure side of the membrane is that of the feed gas. Explain why the component permeabilities differ so drastically between experiments with the pure gas and the gas mixture.

14.10 A mixture of 60 mol% propylene and 40 mol% propane at a flow rate of 100 lbmol/h and at 25°C and 300 psia is to be separated with a polyvinyltrimethylsilane polymer (see Table 14.9 for permeabilities). The membrane skin is 0.1 μm thick, and spiral-wound modules are used with a pressure of 15 psia on the permeate

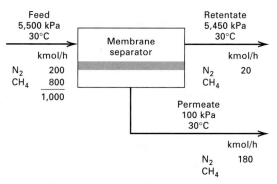

Figure 14.35 Data for Exercise 14.3.

side. Calculate the material balance and membrane area in m^2 as a function of the cut (fraction of feed permeated) for:

(a) Perfect-mixing flow pattern.

(b) Crossflow pattern.

14.11 Repeat part (a) of Exercise 14.10 for a two-stage stripping cascade and a two-stage enriching cascade, as shown in Figure 14.12. However, select just one set of reasonable cuts for the two stages of each case so as to produce 40 lbmol/h of final retentate.

14.12 Repeat Example 14.7 with the following changes:

Tube-side Reynolds number = 25,000

Tube inside diameter = 0.4 cm

Permeate-side mass-transfer coefficient = 0.06 cm/s

How important is concentration polarization?

Section 14.4

14.13 An aqueous process stream of 100 gal/h at 20°C contains 8 wt% Na$_2$SO$_4$ and 6 wt% of a high-molecular-weight substance (A). This stream is processed in a continuous, countercurrent-flow dialyzer using a pure water sweep of the same flow rate. The membrane is a microporous cellophane with pore volume = 50%, wet thickness = 0.0051 cm, tortuosity = 4.1, and pore diameter = 31Å. The molecules to be separated have the following properties:

	Na$_2$SO$_4$	A
Molecular weight	142	1,000
Molecular diameter, Å	5.5	15.0
Diffusivity, cm^2/s × 10^5	0.77	0.25

Calculate the membrane area in m^2 for only a 10% transfer of A through the membrane, assuming no transfer of water. What is the percent recovery of the Na$_2$SO$_4$ in the diffusate? Use log-mean concentration driving forces and assume that the mass-transfer resistances on each side of the membrane are each 25% of the total mass-transfer resistances for Na$_2$SO$_4$ and A.

14.14 A dialyzer is to be used to separate 300 L/h of an aqueous solution containing 0.1 M NaCl and 0.2 M HCl. Laboratory experiments with the microporous membrane to be used give the following values for the overall mass-transfer coefficient K_i in (14-57), for a log-mean concentration driving force:

	K_i, cm/min
Water	0.0025
NaCl	0.021
HCl	0.055

Determine the membrane area in m^2 for 90, 95, and 98% transfer of HCl to the diffusate. For each of the three cases, determine the complete material balance in kmol/h. A sweep of 300 L/h can be assumed.

14.15 A total of 86,000 gal/day of an aqueous solution of 3,000 ppm of NaCl is to be desalinized to 400 ppm by electrodialysis, with a 40% conversion. The process will be conducted in four stages, with three stacks of 150 cell pairs in each stage. The fractional desalinization will be the same in each stage and the expected current efficiency is 90%. The applied voltage for the first stage is 220 V. Each cell pair has an area of 1,160 cm^2. Calculate the current density in mA/cm^2, the current in A, and the power requirement in kW for the first stage. Reference: Mason, E.A., and T.A. Kirkham, *C.E.P. Symp. Ser.,* **55**(24), 173–189 (1959).

Section 14.5

14.16 A reverse-osmosis plant is used to treat 30,000,000 gal/day of seawater at 20°C containing 3.5 wt% dissolved solids to produce 10,000,000 gal/day of potable water with 500 ppm of dissolved solids, and the balance as brine containing 5.25 wt% dissolved solids. The feed-side pressure is 2,000 psia, while the permeate pressure is 50 psia. A single stage of spiral-wound membranes is used that approximates crossflow. If the total membrane area is 2,000,000 ft^2, estimate the permeance for water and the salt passage.

14.17 A reverse-osmosis process is to be designed to handle a feed flow rate of 100 gal/min. Three designs have been proposed, differing in the % recovery of potable water from the feed:

Design 1: A single stage consisting of four units in parallel to obtain a 50% recovery

Design 2: Two stages in series with respect to the retentate (four units in parallel followed by two units in parallel)

Design 3: Three stages in series with respect to the retentate (four units in parallel followed by two units in parallel followed by a single unit)

Draw the three designs and determine the percent recovery of potable water for Designs 2 and 3.

14.18 The production of paper involves a pulping step to break down wood chips into cellulose and lignin. In the Kraft process, an aqueous, pulping-feed solution, known as white liquor, is used that consists of dissolved inorganic chemicals such as sodium sulfide and sodium hydroxide. Following removal of the pulp (primarily cellulose), a solution known as weak (Kraft) black liquor (KBL) is left, which is regenerated to recover white liquor for recycle. In the conventional process, a typical 15 wt% (dissolved solids) KBL is concentrated to 45 to 70 wt% by multieffect evaporation. It has been suggested that reverse osmosis might be used to perform an initial concentration to perhaps 25 wt%. Higher concentrations may not be feasible because of the very high osmotic pressure, which at 180°F and 25 wt% solids is estimated to be 1,700 psia. The osmotic pressure for other conditions can be scaled with (14-68) using wt% instead of molality.

A two-stage RO process, shown in Figure 14.36, has been proposed to carry out this initial concentration for a feed rate of 1,000 lb/h at 180°F. A feed pressure of 1,756 psia is used for the first stage to yield a permeate of 0.4 wt% solids. The feed pressure to the second stage is 518 psia to produce water of 300 ppm dissolved solids and a retentate of 2.6 wt% solids. Permeate-side pressure for both stages is 15 psia. Equation (14-69) can be used to estimate membrane area, where the permeance for water can be taken as 0.0134 lb/ft^2-hr-psi in conjunction with an arithmetic-mean osmotic pressure for plug flow on the feed side. Complete the material balance for the process and estimate the required

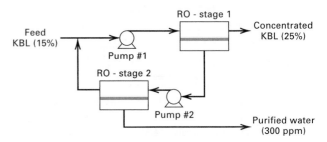

Figure 14.36 Data for Exercise 14.18.

membrane areas for each stage. Reference: Gottschlich, D.E., and D.L. Roberts. Final Report DE91004710, SRI International, Menlo Park, CA, Sept. 28, 1990.

Section 14.6

14.19 Gas permeation can be used to recover VOCs from air at low pressures using a membrane material that is highly selective for the VOCs. In a typical application, 1,500 scfm (0°C, 1 atm) of air containing 0.5 mol% acetone (A) is fed to a spiral-wound membrane module system at 40°C and 1.2 atm. A liquid-ring vacuum pump on the permeate side establishes a pressure of 4 cmHg. A silicone-rubber, thin-composite membrane with a 2-μm-thick skin gives permeabilities of 4 barrer for air and 20,000 barrer for acetone.

If the retentate is to contain 0.05 mol% acetone and the permeate is to contain 5 mol% acetone, determine the membrane area required in m^2, assuming crossflow. References: (1) Peinemann, K.-V., J.M. Mohr, and R.W. Baker, *C.E.P. Symp. Series,* **82**(250), 19–26 (1986); (2) Baker, R.W., N. Yoshioka, J.M. Mohr, and A.J. Khan, *J. Membrane Sci.,* **31**, 259–271 (1987).

14.20 The separation of air into nitrogen and oxygen is widely practiced. Cryogenic distillation is most economical for processing 100 to 5,000 tons of air per day, while pressure-swing-adsorption is favorable for 20 to 50 tons/day. For small-volume users requiring less than 10 tons/day, gas permeation finds applications where for a single stage, either an oxygen-enriched air (40 mol% oxygen) or 98 mol% nitrogen can be produced. It is desired to produce 5 tons/day (2,000 lb/ton) of 40 mol% oxygen and nitrogen, ideally of 90 mol% purity, by gas permeation. Assume pressures of 500 psia (feed side) and 20 psia (permeate). Two companies, who can supply the membrane modules, have provided the following data:

	Company A	**Company B**
Module type	Hollow-fiber	Spiral-wound
$\bar{P}_M$ for O_2, barrer/μm	15	35
$\bar{P}_{M_{O_2}}/\bar{P}_{M_{N_2}}$	3.5	1.9

Determine the required membrane area in m^2 for each company. Assume that both module types approximate crossflow.

14.21 A joint venture has been underway for several years to develop a membrane process to separate CO_2 and H_2S from high-pressure sour natural gas. Typical feed and product conditions are:

	Feed Gas	**Pipeline Gas**
Pressure, psia	1,000	980
Composition, mol%:		
CH_4	70	97.96
H_2S	10	0.04
CO_2	20	2.00

To meet these conditions, the following hollow-fiber membrane material targets have been established:

	Selectivity
CO_2–CH_4	50
H_2S–CH_4	50

where selectivity is the ratio of permeabilities.

$$P_{M_{CO_2}} = 13.3 \text{ barrer},$$

and membrane skin thickness is expected to be 0.5 μm.

Make calculations to show whether the targets can realistically meet the pipeline-gas conditions in a single stage with a reasonable

membrane area. Assume a feed-gas flow rate of 10×10^3 scfm (0°C, 1 atm) with crossflow.

Reference: Stam, H., in *Future Industrial Prospects of Membrane Processes,* L. Cecille and J.-C. Toussaint, Eds., Elsevier Applied Science, London, pp. 135–152 (1989).

Section 14.7

14.22 Pervaporation is to be used to separate ethyl acetate (EA) from water. The feed rate is 100,000 gal/day of water containing 2.0 wt% EA at 30°C and 20 psia. The membrane is dense polydimethylsiloxane with a 1-μm-thick skin in a spiral-wound module that approximates crossflow. The permeate pressure is 3 cmHg. The total measured membrane flux at these conditions is 1.0 L/m^2-h with a separation factor given by (14-36) of 100 for EA with respect to water. A retentate of 0.2 wt% EA is desired for a permeate of 45.7 wt% EA. Determine the required membrane area in m^2 and estimate the temperature drop of the feed. Reference: Blume, I., J.G. Wijans, and R.W. Baker, *J. Membrane Sci.,* **49**, 253–286 (1990).

14.23 For a temperature of 60°C and a permeate pressure of 15.2 mmHg, Wesslein et al. [45] measured a total permeation flux of 1.6 kg/m^2-h for a 17.0 wt% ethanol in water feed, giving a permeate of 12 wt% ethanol. Otherwise, conditions were those of Example 14.12. Calculate the permeances of ethyl alcohol and water for these conditions. Also, calculate the selectivity for water.

14.24 The separation of benzene (B) from cyclohexane (C) by distillation at 1 atm is impossible because of a minimum-boiling-point azeotrope at 54.5 mol% benzene. However, extractive distillation with furfural is feasible. For an equimolar feed, cyclohexane and benzene products of 98 and 99 mol%, respectively, can be produced. Alternatively, the use of a three-stage pervaporation process, with selectivity for benzene using a polyethylene membrane, has received attention, as discussed by Rautenbach and Albrecht [47]. Consider the second stage of this process where the feed is 9,905 kg/h of 57.5 wt% B at 75°C. The retentate is 16.4 wt% benzene at 67.5°C and the permeate is 88.2 wt% benzene at 27.5°C. The total permeate mass flux is 1.43 kg/m^2-h and the selectivity for benzene is 8. Calculate the flow rates of retentate and permeate in kg/h and the required membrane surface area in m^2.

Section 14.8

14.25 Based on the problem statement of Example 14.14, calculate for just Section 1 the component material balance in pounds per day of operation, the percent recovery (yield) from the whey of the TP and NPN in the final concentrate, and the number of cartridges required if only two stages are used instead of four.

14.26 Based on the problem statement of Example 14.14, design a four-stage diafiltration section to take the 55 wt% concentrate from Section 1 and achieve the desired 85 wt% concentrate, thus eliminating Section 3.

Section 14.9

14.27 Using the membrane and feed conditions of and values for R_m and K_2 determined in Example 14.15 for DEF microfiltration, compute and plot the permeate flux and cumulative permeate volume as a function of time. Assume a combined operation with Stage 1 at a constant permeate rate of 10 mL/min to an upper-limit pressure drop of 25 psi, followed by Stage 2 at this pressure drop until the permeate rate drops to a lower limit of 5 mL/min.

Chapter 15

Adsorption, Ion Exchange, and Chromatography

Adsorption, ion exchange, and chromatography are *sorption* operations, in which certain components of a fluid phase, called solutes, are selectively transferred to insoluble, rigid particles suspended in a vessel or packed in a column. Sorption, which is a general term introduced by J.W. McBain [*Phil. Mag.*, **18,** 916–935 (1909)], includes selective transfer to the surface and/or into the bulk of a solid or liquid. Thus, absorption of gas species into a liquid and penetration of fluid species into a nonporous membrane are also sorption operations. In a general sorption process, the sorbed solutes are referred to as *sorbate,* and the sorbing agent is the *sorbent.*

In an *adsorption* process, molecules, as shown in Figure 15.1a, or atoms or ions, in a gas or liquid diffuse to the surface of a solid, where they bond with the solid surface or are held there by weak intermolecular forces. The adsorbed solutes are referred to as *adsorbate,* whereas the solid material is the *adsorbent.* To achieve a very large surface area for adsorption per unit volume, highly porous solid particles with small-diameter interconnected pores are used, with the bulk of the adsorption occurring within the pores.

In an *ion-exchange* process, as shown in Figure 15.1b, ions of positive charge (*cations*) or negative charge (*anions*) in a liquid solution, usually aqueous, replace dissimilar and displaceable ions of the same charge contained in a solid *ion exchanger,* which also contains immobile, insoluble, and permanently bound co-ions of the opposite charge. Thus, ion exchange can be cation or anion exchange. Water softening by ion exchange involves a cation exchanger, in which the following reaction occurs to replace calcium ions with sodium ions.

$$Ca_{(aq)}^{2+} + 2NaR_{(s)} \leftrightarrow CaR_{2(s)} + 2Na_{(aq)}^{+}$$

where R is the residual material of the ion exchanger. The exchange of ions is reversible and does not cause any permanent change to the structure of the solid ion exchanger. Thus, it can be used and reused unless fouled by organic compounds in the liquid feeds that attach to exchange sites on and within the ion exchanger. The ion-exchange concept can be extended to the removal of essentially all inorganic salts from water by a two-step process called *demineralization* or *deionization.* In the first step, a cation resin exchanges hydrogen ions for cations such as calcium, magnesium, and sodium. In the second step, an anion resin exchanges hydroxyl ions for strongly and weakly ionized anions such as sulfate, nitrate, chloride, and bicarbonate. The hydrogen and hydroxyl ions that enter the water combine to form water. Regeneration of the cation and anion resins is usually accomplished with sulfuric acid and sodium hydroxide, respectively.

In *chromatography,* the sorbent may be a solid adsorbent, an insoluble, nonvolatile, liquid absorbent contained in the pores of a granular solid support, or an ion exchanger. In either case, the solutes to be separated move through the chromatographic separator, with an inert, eluting fluid, at different rates because of repeated sorption, desorption cycles.

During adsorption and ion exchange, the solid separating agent becomes saturated or nearly saturated with the molecules, atoms, or ions transferred from the fluid phase. To recover the sorbed substances and allow the adsorbent to be reused, it is regenerated by desorbing the sorbed substances. Accordingly, these two separation operations are carried out in a cyclic manner. In chromatography, regeneration occurs continuously, but at changing locations in the separator.

Adsorption processes may be classified as *purification* or *bulk separation,* depending on the concentration in the feed fluid of the components to be adsorbed. Although there is no sharp dividing concentration, Keller [1] has suggested 10 wt%. Early applications of adsorption involved only purification. For example, adsorption with charred wood to improve the taste of water has been known for centuries. The decolorization of liquid solutions by adsorption with bone char and other materials has been practiced for at least five centuries. Adsorption of gases by a solid (charcoal) was first described by C.W. Scheele in 1773. Commercial applications of bulk separation by gas adsorption began in the early 1920s, but did not escalate until the 1960s, following the inventions by Milton [2] of synthetic molecular-sieve zeolites, which provide high adsorptive selectivity, and by Skarstrom [3] of the *pressure-swing cycle,* which made possible the efficient operation of a fixed-bed, cyclic, gas-adsorption process. The commercial-scale bulk separation

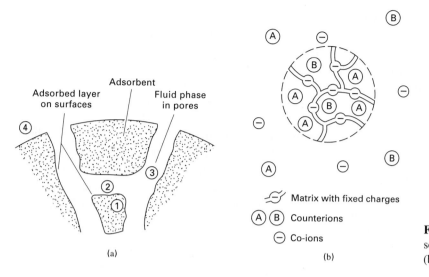

Figure 15.1 Sorption operations with solid-particle sorbents. (a) Adsorption. (b) Ion exchange.

of liquid mixtures also began in the 1960s, following the invention by Broughton and Gerhold [4] of the simulated moving bed for adsorptive separation.

The use of ion exchange dates back to at least the time of Moses, who, while leading his followers out of Egypt into the wilderness, sweetened the bitter waters of Marah with a tree [Exodus 15:23–26]. In ancient Greece, Aristotle observed that the salt content of water is reduced when it percolates through certain sands. Systematic studies of ion exchange were published in 1850 by both Thompson and Way, who experimented with cation exchange in soils before the discovery of the existence of ions.

The first major application of ion exchange, which occurred about 100 years ago, was for water treatment to remove the ions responsible for water hardness, such as calcium. Initially, the ion exchanger was a porous, natural, mineral zeolite containing silica. In 1935, synthetic, insoluble, polymeric-resin ion exchangers were introduced. Today they are dominant for water-softening and deionizing applications, but natural and synthetic zeolites still find some use.

Since the invention of chromatography by M.S. Tswett [5], a Russian botanist, in 1903, it has found widespread use as an analytical and preparative laboratory technique. Tswett separated a mixture of structurally similar yellow and green chloroplast pigments in leaf extracts by dissolving the extracts in carbon disulfide and passing the solution through a column packed with chalk particles. The pigments were separated by color. Hence, the name chromatography was coined by Tswett in 1906 from the Greek words *chroma,* meaning color, and *graphe,* meaning writing. Chromatography has revolutionized the laboratory chemical analysis of liquid and, particularly, gas mixtures. The large-scale, commercial applications described by Bonmati et al. [6] and Bernard et al. [7], however, did not begin until the 1980s.

15.0 INSTRUCTIONAL OBJECTIVES

After completing this chapter, you should be able to:

- List the major types of porous adsorbents and their most significant properties.
- Explain why a few grams of porous adsorbent can have an adsorption area as large as a football field.
- Differentiate between chemisorption and physical adsorption.
- Explain how ion-exchange resins work.
- List types of sorbents used in chromatography.
- Compare three major expressions (so-called isotherms) used for correlating single-component adsorption equilibria.
- List steps involved in adsorption of a solute, and which steps may control rate of adsorption.
- Estimate external and internal rates of adsorption.
- Describe major modes for contacting the adsorbent with a fluid containing solute(s) to be adsorbed.
- Describe major methods for regenerating adsorbent.
- Calculate vessel size or residence time for any of the major modes of slurry adsorption.
- List and explain assumptions for ideal fixed-bed adsorption and explain the concept of width of mass-transfer zone.

- Explain the concept of breakthrough in fixed-bed adsorption.
- Calculate bed height, bed diameter, and cycle time for fixed-bed adsorption.
- Compute separations for a simulated moving bed operation.
- Make design calculations for ion-exchange cycles.
- Calculate rectangular and Gaussian-distribution pulses in chromatography.

Industrial Example

The pressure-swing gas adsorption process is primarily used for the dehydration of air and for the separation of air into nitrogen and oxygen. A small unit for the dehydration of compressed air is described by White and Barkley [8] and shown in Figure 15.2. The unit consists of two fixed-bed adsorbers, each 12.06 cm in diameter and packed with 11.15 kg of 3.3-mm-diameter Alcoa F-200 activated-alumina beads to a height of 1.27 m. The external porosity (void fraction) of the bed is 0.442 and the alumina-bead bulk density is 769 kg/m^3.

The unit operates on a 10-min cycle, with 5 min for adsorption of water vapor from the air and 5 min for regeneration, which consists of depressurization, purging of the water vapor, and a 30-s repressurization. While one bed is adsorbing, the other bed is being regenerated. The adsorption (drying) step takes place with air entering at 21°C and 653.3 kPa (6.45 atm) with a flow rate of 1.327 kg/min, passing up through the bed with a pressure drop of 2.386 kPa. The dew-point temperature of the air at system pressure is reduced from 11.2 to –61°C by the adsorption process. During the 270-second period of purging, about a third of the dry air leaving one bed is directed to the other bed as a downward-flowing purge to regenerate the adsorbent. The purge is exhausted at a pressure of 141.3 kPa. By conducting the

purge flow countercurrent to the entering air flow, the highest degree of water-vapor desorption is achieved.

Other equipment, shown in Figure 15.2, includes an air compressor; an aftercooler; piping and valving to switch the beds from one step in the cycle to the other; a coalescing filter

Table 15.1   Industrial Applications of Sorption Operations

1. Adsorption
   Gas purifications:
      Removal of organics from vent streams
      Removal of SO$_2$ from vent streams
      Removal of sulfur compounds from gas streams
      Removal of water vapor from air and other gas streams
      Removal of solvents and odors from air
      Removal of NO$_x$ from N$_2$
      Removal of CO$_2$ from natural gas
   Gas bulk separations:
      N$_2$/O$_2$
      H$_2$O/ethanol
      Acetone/vent streams
      C$_2$H$_4$/vent streams
      Normal paraffins/isoparaffins, aromatics
      CO, CH$_4$, CO$_2$, N$_2$, A, NH$_3$/H$_2$
   Liquid purifications:
      Removal of H$_2$O from organic solutions
      Removal of organics from H$_2$O
      Removal of sulfur compounds from organic solutions
      Decolorization of solutions
   Liquid bulk separations:
      Normal paraffins/isoparaffins
      Normal paraffins/olefins
      p-xylene/other C$_8$ aromatics
      p- or m-cymene/other cymene isomers
      p- or m-cresol/other cresol isomers
      Fructose/dextrose, polysaccharides
2. Ion Exchange
   Water softening
   Water demineralization
   Water dealkalization
   Decolorization of sugar solutions
   Recovery of uranium from acid leach solutions
   Recovery of antibiotics from fermentation broths
   Recovery of vitamins from fermentation broths
3. Chromatography
   Separation of sugars
   Separation of perfume ingredients
   Separation of C$_4$–C$_{10}$ normal and isoparaffins

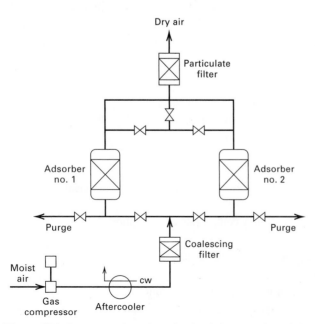

Figure 15.2 Pressure-swing adsorption for the dehydration of air.

to remove aerosols from the entering air; and a particulate filter for the exiting dry air to remove adsorbent fines. If the dry air is needed only at low-to-moderate pressures, an air turbine can be installed to recover energy while reducing the air pressure.

During the 5-min adsorption period of the cycle, the capacity of the adsorbent for water must not be exceeded. In this example, the water content of the air is reduced from 1.27×10^{-3} kg H_2O/kg air to the very low value of 9.95×10^{-7} kg H_2O/kg air. To achieve this exiting water vapor content, only a small fraction of the adsorbent capacity is utilized during the adsorption step, with most of the adsorption occurring in the first 0.2 m of the 1.27-m bed height.

The bulk separation of gas and liquid mixtures by adsorption is an emerging separation operation. Important progress is being made in the development of new and more-selective adsorbents and in more-efficient operation cycles. In addition, attention is being directed to hybrid systems that include membrane and other types of separation steps. Already, the three sorption operations addressed in this chapter have found numerous applications, some of which are listed in Table 15.1, compiled from listings in Rousseau [9]. The applications cover a very wide range of species molecular weight.

This chapter discusses sorbents, including their equilibrium, sieving, transport, and kinetic properties with respect to solutes being removed from solutions; techniques for conducting cyclic operations; and equipment configuration and design. Both equilibrium-stage and rate-based models are developed. Although emphasis is on adsorption, the basic principles of ion exchange and chromatography are also presented. Further descriptions of the three sorption operations are given by Rousseau [9] and Ruthven [10].

15.1 SORBENTS

To be suitable for commercial applications, a sorbent should have (1) high selectivity to enable sharp separations, (2) high capacity to minimize the amount of sorbent needed, (3) favorable kinetic and transport properties for rapid sorption, (4) chemical and thermal stability, including extremely low solubility in the contacting fluid, to preserve the amount of sorbent and its properties, (5) hardness and mechanical strength to prevent crushing and erosion, (6) a free-flowing tendency for ease of filling or emptying vessels, (7) high resistance to fouling for long life, (8) no tendency to promote undesirable chemical reactions, (9) the capability of being regenerated when used with commercial feedstocks that contain trace quantities of high-molecular-weight species that are strongly sorbed and difficult to desorb, and (10) relatively low cost.

Adsorbents

Most solids are able to adsorb species from gases and liquids. However, only a few have a sufficient selectivity and capacity to make them serious candidates for commercial adsorbents. Of considerable importance is a large specific surface area (area per unit volume), which is achieved by adsorbent manufacturing techniques that result in solids with a microporous structure. By the definition of the International Union of Pure and Applied Chemistry (IUPAC), a micropore is <20 Å, a mesopore is 20–500 Å, and a macropore is >500 Å (50 nm). Typical commercial adsorbents, which may be granules, spheres, cylindrical pellets, flakes, and/or powders of size ranging from 50 μm to 1.2 cm, have specific surface areas from 300 to 1,200 m²/g. Thus, just a few grams of adsorbent can have a surface area equal to that of a football field (120 × 53.3 yards or 5,350 m²)! Such a large area is made possible by a particle porosity from 30 to 85 vol% with average pore diameters from 10 to 200 Å. To quantify this, consider a cylindrical pore of diameter d_p and length L. The surface area-to-volume ratio is

$$S/V = \pi d_p L / (\pi d_p^2 L/4) = 4/d_p \qquad (15\text{-}1)$$

If the fractional particle porosity is ϵ_p and the particle density is ρ_p, the specific surface area, S_g, in area per unit mass of adsorbent is

$$S_g = 4\epsilon_p / \rho_p d_p \qquad (15\text{-}2)$$

Thus, if ϵ_p is 0.5, ρ_p is 1 g/cm³ $= 1 \times 10^6$ g/m³, and d_p is 20 Å (20×10^{-10} m), substitution into (15-2) gives $S_g = 1,000$ m²/g, a desirable value.

Depending upon the type of forces between the fluid molecules and the molecules of the solid, adsorption may be classified as *physical adsorption* (van der Waals adsorption) or *chemisorption* (activated adsorption). Physical adsorption from a gas occurs when the intermolecular attractive forces between molecules of a solid and the gas are greater than those between molecules of the gas itself. In effect, the resulting adsorption is like condensation, which is exothermic and thus is accompanied by a release of heat. The magnitude of the heat of adsorption can be less than or greater than the heat of vaporization, and changes with the extent of adsorption. Physical adsorption, which may be a monomolecular (unimolecular) layer, or may be two, three or more layers thick (multimolecular), occurs rapidly. If unimolecular, it is reversible; if multimolecular, such that capillary pores are filled, hysteresis may occur. The density of the adsorbate is of the order of magnitude of the liquid rather than the vapor state. As physical adsorption takes place, it begins as a monolayer, becomes multilayered, and then, if the pores are close to the size of the molecules, capillary condensation occurs, and the pores fill with adsorbate. Accordingly, the maximum capacity of a porous adsorbent can be more related to the pore volume than to the surface area. However, for gases at temperatures above their critical temperature, adsorption is confined to a monolayer.

In contrast, chemisorption involves the formation of chemical bonds between the adsorbent and adsorbate in a monolayer, often with a release of heat much larger than the heat of vaporization. Chemisorption from a gas generally takes place only at temperatures greater than 200°C and may be slow and irreversible. Commercial adsorbents rely on physical adsorption; catalysis relies on chemisorption.

Adsorption from a liquid is a more difficult phenomenon to measure experimentally or describe. When the fluid is a gas, experiments are conducted with pure gases or with mixtures. The amount of gas adsorbed in a confined space is determined from the measured decrease in total pressure. When the fluid is a liquid, no simple procedure for determining the extent of adsorption from a pure liquid exists; consequently, experiments are only conducted using liquid mixtures, including dilute solutions. When porous particles of adsorbent are immersed in a liquid mixture, the pores, if sufficiently larger in diameter than the molecules in the liquid, fill with liquid. At equilibrium, because of differences in the extent of physical adsorption among the different molecules of the liquid mixture, the composition of the liquid in the pores differs from that of the bulk liquid surrounding the adsorbent particles. The observed exothermic heat effect is referred to as the *heat of wetting*, which is much smaller than the heat of adsorption from the gas phase. As with gases, the extent of equilibrium adsorption of a given solute increases with concentration and decreases with temperature. Chemisorption can also occur with liquids.

Listed in Table 15.2 are six major types of solid adsorbents in use. Included are the nature of the adsorbent and representative values of the mean pore diameter, d_p; particle porosity (internal void fraction), ϵ_p; particle density, ρ_p; and specific surface area, S_g. In addition, for some adsorbents, the capacity for adsorbing water vapor at a partial pressure of 4.6 mmHg in air at 25°C is listed, as taken from Rousseau [9]. Not included is the specific pore volume, V_p, which can

be computed from the other properties by

$$V_p = \epsilon_p/\rho_p \qquad (15\text{-}3)$$

Also not included in Table 15.2, but of interest when the adsorbent is used in fixed beds, are the bulk density, ρ_b, and the bed porosity (external porosity), ϵ_b, which are related:

$$\epsilon_b = 1 - \frac{\rho_b}{\rho_p} \qquad (15\text{-}4)$$

In addition, the true solid density (also called the crystalline density), ρ_s, can be computed from a similar expression:

$$\epsilon_p = 1 - \frac{\rho_p}{\rho_s} \qquad (15\text{-}5)$$

The specific surface area of an adsorbent, S_g, is measured by adsorbing gaseous nitrogen, using the well-accepted BET method (Brunauer, Emmett, and Teller [11]). Typically, the BET apparatus operates at the normal boiling point of N_2 (–195.8°C) by measuring the equilibrium volume of pure N_2 physically adsorbed on several grams of the adsorbent at a number of different values of the total pressure in the vacuum range of 5 to at least 250 mmHg. Brunauer, Emmett, and Teller derived a theoretical equation to model the adsorption by allowing for the formation of multimolecular layers. Furthermore, they assumed that the heat of adsorption during monolayer formation (ΔH_{ads}) is constant and that the heat effect associated with subsequent layers is equal to the heat of condensation (ΔH_{cond}). The BET equation is

$$\frac{P}{v(P_0 - P)} = \frac{1}{v_m c} + \frac{(c-1)}{v_m c}\left(\frac{P}{P_0}\right) \qquad (15\text{-}6)$$

where

P = total pressure

P_0 = vapor pressure of adsorbate at test temperature

v = volume of gas adsorbed at STP (0°C, 760 mmHg)

Table 15.2 Representative Properties of Commercial Porous Adsorbents

Adsorbent	Nature	Pore Diameter d_p, Å	Particle Porosity, ϵ_p	Particle Density ρ_p, g/cm³	Surface Area S_g, m²/g	Capacity for H₂O Vapor at 25°C and 4.6 mmHg, wt% (Dry Basis)
Activated alumina	Hydrophilic, amorphous	10–75	0.50	1.25	320	7
Silica gel:	Hydrophilic/ hydrophobic,					
Small pore	amorphous	22–26	0.47	1.09	750–850	11
Large pore		100–150	0.71	0.62	300–350	—
Activated carbon:	Hydrophobic,					
Small pore	amorphous	10–25	0.4–0.6	0.5–0.9	400–1200	1
Large pore		>30	—	0.6–0.8	200–600	—
Molecular-sieve carbon	Hydrophobic	2–10	—	0.98	400	—
Molecular-sieve zeolites	Polar-hydrophilic, crystalline	3–10	0.2–0.5	1.4	600–700	20–25
Polymeric adsorbents	—	40–25	0.4–0.55	—	80–700	—

v_m = volume of monomolecular layer of gas adsorbed at STP

c = constant related to the heat of adsorption $\approx \exp[(\Delta H_{cond} - \Delta H_{ads})/RT]$

Experimental data for v as a function of P are plotted, according to (15-6), as $P/[v(P_0 - P)]$ versus P/P_0, from which v_m and c are determined from the slope and intercept of the best straight-line fit of the data. The value of S_g is then computed from

$$S_g = \frac{\alpha v_m N_A}{V} \qquad (15\text{-}7)$$

where

N_A = Avogadro's number = 6.023×10^{23} molecules/mol

V = Volume of gas per mole at STP conditions (0°C, 1 atm) = 22,400 cm³/mol

The quantity α is the surface area covered per adsorbed molecule. If we assume spherical molecules arranged in close two-dimensional packing, the projected surface area is:

$$\alpha = 1.091 \left(\frac{M}{N_A \rho_L} \right)^{2/3} \qquad (15\text{-}8)$$

where

M = molecular weight of the adsorbate

ρ_L = density of the adsorbate in g/cm³, taken as the liquid at the test temperature

Although the BET surface area may not always represent the surface area available for adsorption of a particular molecule, the BET test is reproducible and widely used in the characterization of adsorbents.

The specific pore volume, typically cm³ of pore volume/g of adsorbent, is determined for a small mass of adsorbent, m_p, by measuring the volumes of helium, V_{He}, and mercury, V_{Hg}, displaced by the adsorbent. The helium is not adsorbed, but fills the pores. At ambient pressure, the mercury cannot enter the pores because of unfavorable interfacial tension and contact angle. The specific pore volume, V_p, is then determined from

$$V_p = (V_{Hg} - V_{He})/m_p \qquad (15\text{-}9)$$

The particle density is obtained from

$$\rho_p = \frac{m_p}{V_{Hg}} \qquad (15\text{-}10)$$

The true solid density is obtained from

$$\rho_s = \frac{m_p}{V_{He}} \qquad (15\text{-}11)$$

The particle porosity is then obtained from (15-3) or (15-5).

The distribution of pore volume over the range of pore size, which is of great importance in adsorption, is measured by mercury porosimetry for large-diameter pores (>100 Å); by gaseous-nitrogen desorption for pores of 15–250 Å in diameter; and by molecular sieving, using molecules of different diameter, for pores <15 Å in diameter. In mercury porosimetry, the extent of mercury penetration into the pores

is measured as a function of applied hydrostatic pressure. A force balance along the axis of a straight pore of circular cross-section for the pressure and the interfacial tension between the mercury and the adsorbent surface gives

$$d_p = -\frac{4\sigma_I \cos \theta}{P} \qquad (15\text{-}12)$$

where for mercury: σ_I = interfacial tension = 0.48 N/m and θ = contact angle = 140°. With these values, (15-12) becomes

$$d_p(\text{Å}) = \frac{21.6 \times 10^5}{P \text{ (psia)}} \qquad (15\text{-}13)$$

Thus, forcing mercury into a 100-Å-diameter pore requires a very high pressure of 21,600 psia.

The nitrogen desorption method for determining pore-size distribution in the more important 15–250-Å-diameter range is an extension of the BET method described earlier for measuring specific surface area. By increasing the nitrogen pressure above 600 mmHg, the multilayer adsorbed films reach the point where they bridge the pore, resulting in capillary condensation. At $P/P_0 = 1$, the entire pore volume is filled with nitrogen. Then, by reducing the pressure in steps, nitrogen is desorbed selectively, starting with the larger pores. This selectivity occurs because of the effect of pore diameter on the vapor pressure of the condensed phase in the pore, as given by the Kelvin equation:

$$P_p^s = P^s \exp\left(-\frac{4\sigma v_L \cos \theta}{RT d_p} \right) \qquad (15\text{-}14)$$

where

P_p^s = vapor pressure of liquid in pore

P^s = the normal vapor pressure of liquid on a flat surface

σ = surface tension of liquid in pore

v_L = molar volume of liquid in pore

Thus, the vapor pressure of the condensed phase in the pore is less than its normal vapor pressure for a flat surface. The effect of d_p on P_p^s can be significant. For example, for liquid nitrogen at −195.8°C, P^s = 760 torr, σ = 0.00827 N/m, θ = 0, and v_L = 34.7 cm³/mol. Equation (15-14) then becomes

$$d_p(\text{Å}) = 17.9/\ln\left(P^s/P_p^s \right) \qquad (15\text{-}15)$$

From (15-15), for d_p = 30 Å, P_p^s = 418 torr, a reduction in vapor pressure of almost 50%. At 200 Å, the reduction is only about 10%. At 418 torr pressure, only pores less than 30 Å in diameter remain filled with liquid nitrogen. For greater accuracy in applying the Kelvin equation, a correction is needed for the thickness of the adsorbed layer. The use of this correction is discussed in detail by Satterfield [12]. For a monolayer, this thickness for nitrogen is about 0.354 nm, corresponding to a P/P_0 in (15-6) of between 0.05 and 0.10. At $P/P_0 = 0.60$ and 0.90, the adsorbed thicknesses are 0.75 and 1.22 nm, respectively. The correction is applied by subtracting twice the adsorbed thickness from d_p in (15-14) and (15-15).

EXAMPLE 15.1

Using data from Table 15.2, determine the volume fraction of pores in silica gel (small-pore type) filled with adsorbed water vapor when its partial pressure is 4.6 mmHg and the temperature is 25°C. At these conditions, the partial pressure is considerably below the vapor pressure of 23.75 mmHg. In addition, determine whether the amount of water adsorbed is equivalent to more than a monolayer, if the area of an adsorbed water molecule is given by (15-8) and the specific surface area of the silica gel is 830 m²/g.

SOLUTION

Take 1 g of silica gel particles as a basis. From (15-3) and data in Table 15.2, $V_p = 0.47/1.09 = 0.431$ cm³/g. Thus, for 1 g, pore volume is 0.431 cm³. From the capacity value in Table 15.2, amount of adsorbed water $= 0.11/(1 + 0.11) = 0.0991$ g. Assume density of adsorbed water is 1 g/cm³, volume of adsorbed water $= 0.0991$ cm³, fraction of pores filled with water $= 0.0991/0.431 = 0.230$, and surface area of 1 g $= 830$ m². From (15-8):

$$\alpha = 1.091 \left[\frac{18.02}{(6.023 \times 10^{23})(1.0)} \right]^{2/3} = 10.51 \times 10^{-16} \text{ cm}^2/\text{molecule}$$

Number of water molecules adsorbed $= \frac{(0.0991)(6.023 \times 10^{23})}{18.02} = 3.31 \times 10^{21}$ molecules

Number of water molecules in a monolayer for 830 m² $= \frac{830(100)^2}{10.51 \times 10^{-16}} = 7.90 \times 10^{21}$

Therefore, only $3.31/7.90$ or 42% of one monolayer is adsorbed.

The four most widely used adsorbents in decreasing order of commercial usage are carbon (activated and molecular-sieve), molecular-sieve zeolites, silica gel, and activated alumina. In Table 15.2, activated alumina, Al_2O_3, which includes activated bauxite, is made by removing water from hydrated colloidal alumina. Activated alumina has a moderately high specific surface area, with a capacity for adsorption of water sufficient to dry gases to less than 1 ppm moisture content. Because of its great affinity for water, activated alumina is widely used for the removal of water from gases and liquids.

Silica gel, SiO_2, which is made from colloidal silica, has a high surface area and high affinity for water and other polar compounds. Related silicate adsorbents include magnesium silicate, calcium silicate, various clays, Fuller's earth, and diatomaceous earth. Silica gel is also highly desirable for water removal. Both small-pore and large-pore types are available.

Activated carbon is made by processes that involve the partial oxidation of a number of materials, including coconut shells, fruit nuts, wood, coal, lignite, peat, petroleum residues, and bones. Because activated carbon is hydrophobic and has a high specific surface area, it is particularly useful for processes involving nonpolar and weakly polar organic molecules. Macropores within the carbon particles help transfer molecules to the micropores. Two commercial grades are produced, one with large pores for liquid applications and one with small pores for gas adsorption. As shown

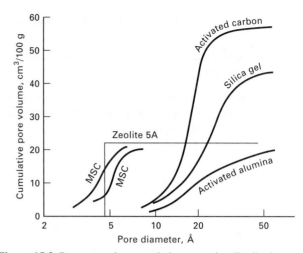

Figure 15.3 Representative cumulative pore-size distributions of adsorbents.

in Table 15.2, activated carbon is relatively hydrophobic and has a large surface area. Accordingly, it has found wide application for the purification and separation of gas and liquid mixtures containing nonpolar and weakly polar organic compounds, which adsorb much more strongly than water. In addition, the bonding strength of adsorption on activated carbon is low, resulting in a low heat of adsorption and ease of regeneration of the adsorbent.

Unlike activated carbon, which typically has pore diameters starting from 10 Å, molecular-sieve carbon (MSC) has much smaller pores ranging from 2 to 10 Å, making it possible to separate N_2 from air. The small pores, in one process, are made by depositing coke in the pore mouths of activated carbon.

Most commercial adsorbents have a range of pore sizes, as shown in Figure 15.3, where the cumulative pore volume is plotted against pore diameter. Exceptions are the molecular-sieve zeolites, which are crystalline, inorganic polymers of aluminosilicates and alkali or alkali-earth cation elements, such as Na, K, Mg, and Ca, with the general stoichiometric, unit-cell formula

$$M_{x/m}[(AlO_2)_x(SiO_2)_y]z \, H_2O$$

where M is the cation with valence m, z is the number of water molecules in each unit cell, and x and y are integers such that $y/x \geq 1$. The cations balance the charge of the AlO_2 groups, each having a net charge of -1. To activate the zeolite, the water molecules are removed by raising the temperature or pulling a vacuum. This leaves the remaining atoms spatially intact in interconnected, cagelike structures with six identical window apertures each, the size of which ranges from 3.8 to about 10 Å, depending on the cation and the crystal structure. These apertures act as sieves, which permit small molecules to enter the crystal cage, but exclude large molecules. Thus, compared to the other types of adsorbents, molecular-sieve zeolites are highly selective because all apertures have the same size. The properties and applications of five of the most commonly used molecular-sieve

Table 15.3 Properties and Applications of Some Molecular-Sieve Zeolites

Designation	Cation	Unit Cell Formula	Aperture Size, Å	Typical Applications
3A	K^+	$K_{12}[(AlO_2)_{12}(SiO_2)_{12}]$	2.9	Drying of reactive gases
4A	Na^+	$Na_{12}[(AlO_2)_{12}(SiO_2)_{12}]$	3.8	H_2O, CO_2 removal; air separation
5A	Ca^{2+}	$Ca_5Na_2[(AlO_2)_{12}(SiO_2)_{12}]$	4.4	Separation of air; separation of linear paraffins
10X	Ca^{2+}	$Ca_{43}[(AlO_2)_{86}(SiO_2)_{106}]$	8.0 ⎫	Separation of air;
13X	Na^+	$Na_{86}[(AlO_2)_{86}(SiO_2)_{86}]$	8.4 ⎭	removal of mercaptans

zeolites are given in Table 15.3, taken from Ruthven [13]. The zeolites separate not only by molecular size and shape, but also by polarity. Thus, they can also separate molecules of similar size. Some zeolites have circular apertures, whereas others have elliptical apertures. Adsorption in zeolites is actually a selective and reversible filling of crystal cages, so surface area is not a pertinent factor. Although naturally occurring zeolite minerals have been known for more than 200 years, molecular-sieve zeolites were first synthesized by Milton [2], who used very reactive materials at temperatures of 25–100°C.

The structure of the unit cell of a type A zeolite is shown in Figure 15.4a as a three-dimensional structure of silica and alumina tetrahedra, each formed by four oxygen atoms surrounding a silicon or aluminum atom. Oxygen and silicon atoms have two negative and four positive charges, respectively, causing the tetrahedra to build uniformly in four directions. Aluminum, with a valence of 3, causes the alumina tetrahedron to be negatively charged. The added cation provides the balance. In Figure 15.4a, an octahedron of tetrahedra is evident with six faces, with one near-circular window aperture at each face. A type X zeolite is shown in Figure 15.4b. This unit-cell structure results in a larger window aperture. Monographs by Barrer [14] and Breck [15] cover many important aspects of zeolites.

Of lesser commercial importance are polymeric adsorbents. Typically, they are spherical beads, 0.5 mm in diameter, made from microspheres about 10^{-4} mm in diameter. They are produced by polymerizing styrene and divinylbenzene for adsorbing nonpolar organics from aqueous solutions, and by polymerizing acrylic esters for adsorbing polar solutes. They are regenerated by leaching with organic solvents.

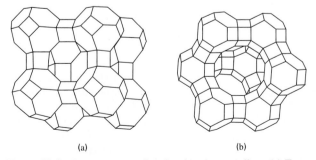

(a)   (b)

Figure 15.4 Line structures of molecular-sieve zeolites: (a) Type A unit cell. (b) Type X unit cell.

Ion Exchangers

The first ion exchangers were naturally occurring inorganic aluminosilicates (zeolites), used in experiments in the 1850s to exchange between ammonium ions in fertilizers and calcium ions in soils. Industrial water softeners using zeolites were introduced about 1910. However, the zeolites were unstable in the presence of mineral acids. The instability problem was solved by Adams and Holmes [16] in 1935, when they synthesized the first organic-polymer, ion-exchange resins by the polycondensation of phenol and aldehydes. Depending upon the nature of the phenolic group, the resin contains either sulfonic ($-SO_3^-$) or amine ($-NH_3^+$) groups, used for the reversible exchange of cations or anions. Today, the most widely used ion exchangers are synthetic, organic-polymer resins based on styrene- or acrylic-acid-type monomers, as described by D'Alelio in U.S. Patent 2,366,007 (Dec. 26, 1944).

Ion-exchange resins are generally solid gels in spherical or granular form, which consist of (1) a three-dimensional polymeric network, (2) ionic functional groups attached to the network, (3) counterions, and (4) a solvent. Strong-acid, cation-exchange resins and strong-base, anion-exchange resins that are fully ionized over the entire pH range are based on the copolymerization of styrene and a cross-linking agent, divinylbenzene, to produce the three-dimensional, cross-linked structure shown in Figure 15.5a. The degree of cross-linking is governed by the ratio of divinylbenzene to styrene. Weakly acid, cation exchangers are sometimes based on the copolymerization of acrylic acid and methacrylic acid, as shown in Figure 15.5b. These two cross-linked copolymers swell in the presence of organic solvents and have no ion-exchange properties.

To convert the copolymers to water-swellable gels with ion-exchange properties, ionic functional groups are added to the polymeric network by reacting the copolymers with various chemicals. For example, if the styrene–divinylbenzene copolymer is sulfonated, as shown in Figure 15.6a, a cation-exchange resin, as shown in Figure 15.6b, is obtained with ($-SO_3^-$) groups permanently attached or fixed to the polymeric network to give a negatively charged matrix and exchangeable, mobile, positive hydrogen ions (cations). The hydrogen ion can be exchanged on an equivalent basis with other cations, such as Na^+, Ca^{2+}, K^+, or Mg^{2+}, to maintain neutrality of the polymer. For example, two H^+ ions are

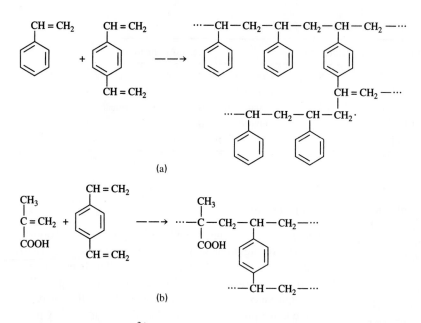

(a)

(b)

Figure 15.5 Ion-exchange resins: (a) Resin from styrene and divinylbenzene; (b) Resin from acrylic and methacrylic acid.

exchanged for one Ca^{2+} ion. The exchangeable ions are called *counterions*. The liquid whose ions are being exchanged also contains other ions of unlike charge, such as Cl^- for a solution of NaCl, where Na^+ is exchanged. These other ions are called *co-ions*. Often the liquid treated is water, which dissolves to some extent in the resin and causes it to swell. Other solvents, such as methanol, are also soluble in the resin. If the styrene–divinylbenzene copolymer is

chloromethylated and then aminated, a strong-base, anion-exchange resin is formed, as shown in Figure 15.6c, which can exchange Cl^- ions for other anions, such as OH^-, HCO_3^-, SO_4^{2-}, and NO_3^-.

Commercial ion exchangers in the hydrogen, sodium, and chloride form are available under the trade names of Amberlite, Duolite, Dowex, Ionac, and Purolite. Typically, they are in the form of spherical beads from about 40 μm

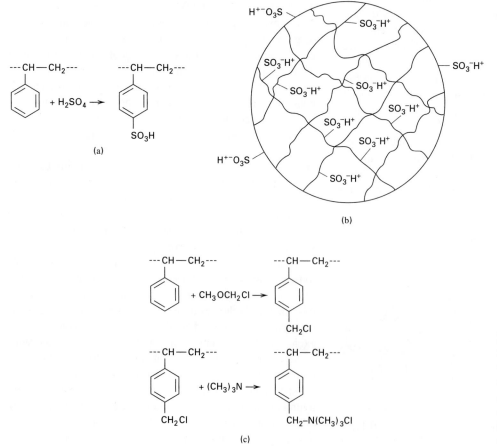

Figure 15.6 Introducing ionic functional groups into resins. (a) Sulfonation to a cation exchanger. (b) Fixed and mobile ions in a cation exchanger. (c) Chloromethylation and amination to an anion exchanger.

to 1.2 mm in diameter. When saturated with water, the beads have typical moisture contents from 40 to 65 wt%. When water-swollen, they have a particle density of 1.1–1.5 g/cm^3. When packed into a bed, they have bulk densities from 0.56 to 0.96 g/cm^3 with fractional bed porosities of 0.35–0.40.

When water is demineralized by ion exchange, potential organic foulants must be removed from the water feed. As discussed by McWilliams [17], this can be accomplished by coagulation, clarification, prechlorination, and the use of ion-exchanger traps.

The maximum ion-exchange capacity of a strong-acid cation or strong-base anion exchanger is stoichiometric, based on the number of equivalents of mobile charge in the resin. Thus, 1 mol H^+ is one equivalent, whereas 1 mol Ca^{2+} is two equivalents. The exchanger capacity is usually quoted as eq/kg of dry resin or eq/L of wet resin. The wet capacity depends on the water content and degree of swelling of a given resin, whereas the dry capacity is fixed. For the copolymer of styrene and divinylbenzene, the maximum capacity is determined on the assumption that each benzene ring in the resin contains one sulfonic-acid group.

EXAMPLE 15.2

A commercial, ion-exchange resin is made from 88 wt% styrene and 12 wt% divinylbenzene. Estimate the maximum ion-exchange capacity in eq/kg resin (same as meq/g resin).

SOLUTION

Basis: 100 g of resin before sulfonation.

	M	g	gmol
Styrene	104.14	88	0.845
Divinylbenzene	130.18	12	0.092
		100	0.937

Therefore, sulfonation at one location on each benzene ring requires 0.937 mol of H_2SO_4 to attach the sulfonic acid group ($M = 81.07$) and split out one water molecule. This is 0.937 equivalents with the addition in weight of 0.937(81.07) = 76 g. Total dry weight of sulfonated resin = 100 + 76 = 176 g maximum ion-exchange capacity, or

$$\frac{0.937}{(176/1,000)} = 5.3 \text{ eq/kg(dry)}$$

Depending on the extent of cross-linking, resins from copolymers of styrene and divinylbenzene are listed as having actual capacities of from 3.9 (high degree of cross-linking) to 5.5 (low degree of cross-linking). Although a low degree of cross-linking favors dry capacity, almost every other ion-exchanger property, including wet capacity and selectivity, is improved by cross-linking, as discussed by Dorfner [18].

Sorbents for Chromatography

Sorbents (called *stationary phases*) for chromatographic separations come in a wide variety of forms and chemical compositions because of the many ways in which chromatography is applied. Figure 15.7 shows a classification of analytical chromatographic systems, taken from Sewell and Clarke [19]. The mixture to be separated, after injection into the carrier fluid to form the *mobile phase,* may be a liquid (*liquid chromatography*) or a gas (*gas chromatography*). Often, the mixture is initially a liquid, but is vaporized without decomposition by the carrier gas, giving a gas mixture for the mobile phase. Gas carriers are inert and do not interact with the sorbent or components of the feed. Liquid carriers (solvents) can interact and must be selected carefully.

The stationary sorbent phase is a solid, a liquid supported on or bonded to a solid, or a gel. With a porous-solid adsorbent, the mechanism or mode of separation is adsorption. If an ion-exchange mechanism is desired, a synthetic, polymer-resin ion exchanger is used. With a polymer gel or a microporous solid, a separation based on sieving, called *exclusion,* can be applied. Unique to chromatography are the liquid-supported or liquid-bonded solids, where the mechanism is absorption into the liquid, also referred to as a *partition mode* of separation or *partition chromatography*. With mobile liquid phases, there is a tendency for the stationary liquid phase to be stripped or dissolved. Therefore, methods of chemically bonding the stationary liquid phase to the solid bonding support have been developed.

All sorbents can be used in columns. In packed columns >1 mm inside diameter, the sorbents are in the form of particles. In capillary columns <0.5 mm inside diameter, the sorbent is the inside wall or is a coating on the wall. If the inside wall of the capillary is liquid coated, the capillary column is referred to as a wall-coated, open-tubular (WCOT) column. If the coating is a layer of fine particulate support material to which a liquid adsorbent is added, the column is called a support-coated, open-tubular (SCOT) column. If the wall is coated with a porous adsorbent only, the column is referred to as a porous-layer, open-tubular (PLOT) column.

Each type of sorbent can be applied to sheets of glass, plastic, or aluminum for use in *thin-layer* (or planar) *chromatography* or to a sheet of cellulose material for use in *paper chromatography*. If a pump, rather than gravity, is used to pass a liquid mobile phase through a packed column, the name *high-performance liquid chromatography* (HPLC) is used.

The two most common adsorbents used in chromatography are porous alumina and porous silica gel. Of lesser importance are carbon, magnesium oxide, and various carbonates. Alumina is a polar adsorbent and is preferred for the separation of components that are weakly or moderately polar, with the more polar compounds retained more selectively by the adsorbent and, therefore, eluted from the column last. In addition, alumina is a basic adsorbent, preferentially retaining acidic compounds. Silica gel is less polar than alumina and is an acidic adsorbent, preferentially retaining basic compounds, such as amines. Carbon is a nonpolar (apolar) stationary phase with the highest attraction for larger nonpolar molecules.

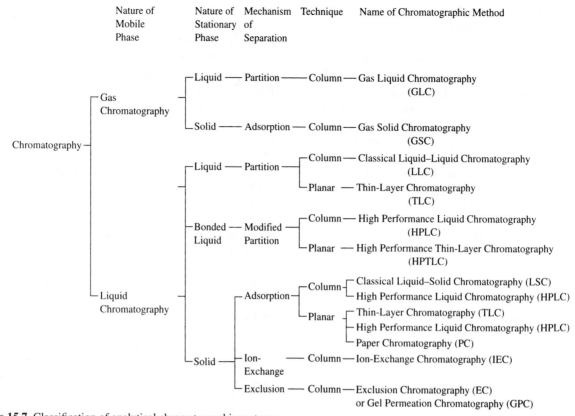

Figure 15.7 Classification of analytical chromatographic systems.
[From P.A. Sewell and B. Clarke, *Chromatographic Separations,* John Wiley and Sons, New York (1987) with permission.]

Adsorbent-type sorbents are better suited for the separation of a mixture on the basis of chemical type (e.g., olefins, esters, acids, aldehydes, alcohols) than for separation of individual members of a homologous series. For the latter, partition chromatography is preferred, wherein an inert-solid support, often silica gel, is coated with a liquid phase. For application to gas chromatography, the liquid must be nonvolatile. For liquid chromatography, the stationary liquid phase must be insoluble in the mobile phase. Since this is difficult to achieve, the stationary liquid phase is usually bonded to the solid support. An example of a bonded phase is the result of reacting silica with a chlorosilane. Both monofunctional and bifunctional silanes are used, as shown in Figure 15.8, where R is a methyl (CH_3) group and R' is a hydrocarbon chain (C_6, C_8, or C_{18}) where the terminal CH_3 group is replaced with a polar group, such as $-CN$ or $-NH_2$.

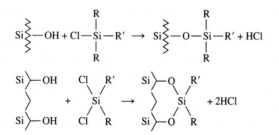

Figure 15.8 Bonded phases from the reaction of surface silanol groups with (a) Monofunctional and (b) Bifunctional chlorosilanes.

If the resulting stationary phase is more polar than the mobile phase, the technique is referred to as *normal-phase chromatography.* Otherwise, the name *reverse-phase chromatography* is used.

In liquid chromatography, the order of elution from the column of the solutes in the mobile phase can also be influenced by the solvent carrier of the mobile phase by matching the solvent polarity with the solutes and using more-polar adsorbents for less-polar solutes and less-polar adsorbents for more-polar solutes.

EXAMPLE 15.3

For the separation of each of the following mixtures, select an appropriate mode of chromatography from Figure 15.7: (a) gas mixture of O_2, CO, CO_2, and SO_2, (b) vaporized mixture of anthracene, phenanthrene, pyrene, and chrysene, and (c) aqueous solution containing Ca^{2+} and Ba^{2+}.

SOLUTION

(a) Use gas–solid chromatography, that is, with a gas mobile phase and a solid-adsorbent stationary phase.

(b) Use partition or gas–liquid chromatography, that is, with a gas mobile phase and a bonded liquid coating on a solid for the stationary phase.

(c) Use ion-exchange chromatography, that is, with a liquid as the mobile phase and polymer resin beads as the stationary phase.

15.2 EQUILIBRIUM CONSIDERATIONS

In adsorption, a dynamic phase equilibrium is established for the distribution of the solute between the fluid and the solid surface. This equilibrium is usually expressed in terms of (1) concentration (if the fluid is a liquid) or partial pressure (if the fluid is a gas) of the adsorbate in the fluid and (2) solute *loading* on the adsorbent, expressed as mass, moles, or volume of adsorbate per unit mass or per unit BET surface area of the adsorbent. Unlike vapor–liquid and liquid–liquid equilibria, where theory is often applied to estimate phase distributions, particularly in the form of K-values for the former type of equilibrium, no acceptable theory has been developed to estimate fluid–solid adsorption equilibria. Thus, it is necessary to obtain experimental equilibrium data for a particular solute, or mixture of solutes and/or solvent, and a sample of the actual solid-adsorbent material of interest. If the data are taken over a range of fluid concentrations at a constant temperature, a plot of solute loading on the adsorbent versus concentration or partial pressure in the fluid, called an *adsorption isotherm,* is made. This equilibrium isotherm places a limit on the extent to which a solute is adsorbed from a given fluid mixture on an adsorbent of given chemical composition and geometry for a given set of conditions. The rate at which the solute is adsorbed is also an important consideration and is discussed in Section 15.3.

Pure Gas Adsorption

For pure gases, experimental physical-adsorption isotherms have shapes, that are classified into five types by Brunauer et al. [20], as shown in Figure 15.9 and discussed in considerable detail by Brunauer [21]. The simplest isotherm is Type I, which corresponds to unimolecular adsorption, as characterized by a maximum limit in the amount adsorbed. This type applies often to gases at temperatures above their critical temperature. The more complex Type II isotherm is associated with multimolecular adsorption of the BET type and is observed for gases at temperatures below their critical temperature and for pressures below, but approaching, the saturation pressure (vapor pressure). The heat of adsorption for the first adsorbed layer is greater than that for the succeeding layers, each of which is assumed to have a heat of adsorption equal to the heat of condensation (vaporization). Both Types I and II are desirable isotherms, exhibiting strong adsorption.

The Type III isotherm in Figure 15.9, with its convex nature, is undesirable because the extent of adsorption is low except at high pressures. According to the BET theory, it corresponds to multimolecular adsorption where the heat of adsorption of the first layer is less than that of succeeding layers. Fortunately, this type of isotherm is rarely observed, an example being the adsorption of iodine vapor on silica gel. In the limit, as the heat of adsorption of the first layer approaches zero, adsorption is delayed until the saturation pressure is approached.

The derivation of the BET equation (15-6) assumes that an infinite number of molecular layers can be adsorbed.

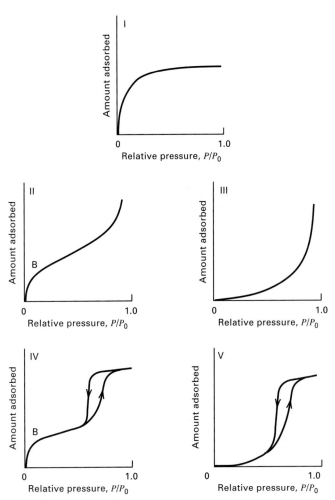

Figure 15.9 Brunauer's five types of adsorption isotherms. (P/P_0 = total pressure/vapor pressure.)

Thus, the equation precludes the possibility of capillary condensation. In a development by Brunauer et al. [20], subsequent to the BET equation, the number of layers is restricted by pore size, and capillary condensation is assumed to occur at a reduced vapor pressure in accordance with the Kelvin equation (15-14). The resulting equation is quite complex, but predicts adsorption isotherms of Types IV and V in Figure 15.9, where we see that the maximum extent of adsorption occurs before the saturation pressure is reached. Type IV is the capillary-condensation version of Type II; Type V is the capillary-condensation version of Type III.

As shown in Figure 15.9, a hysteresis phenomenon can occur in multimolecular adsorption regions for isotherms of types IV and V. The upward adsorption branch of the hysteresis loop is due to simultaneous, multimolecular adsorption and capillary condensation. Only capillary condensation occurs during the downward desorption branch of the loop. Hysteresis can also occur throughout any isotherm when strongly adsorbed impurities are present. Thus, measurements of pure-gas adsorption require adsorbents with clean pore surfaces, normally achieved by preevacuation.

Physical adsorption data of Titoff [22] for ammonia gas on charcoal, as discussed by Brunauer [21], are shown in Figure 15.10. The five adsorption isotherms of Figure 15.10a

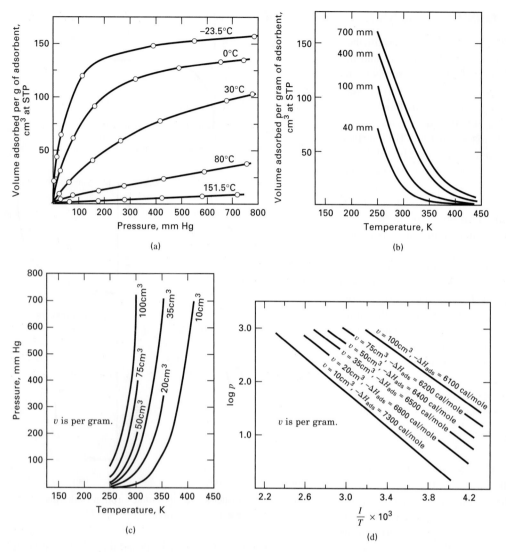

Figure 15.10 Different displays of adsorption equilibrium data for NH_3 on charcoal. (a) Adsorption isotherms. (b) Adsorption isobars. (c) Adsorption isosteres. (d) Isosteric heats of adsorption.

[From S. Brunauer, *The Adsorption of Gases and Vapors,* Vol. I, Princeton University Press (1943) with permission.]

cover pressures from vacuum to almost 800 mmHg and temperatures from –23.5 to 151.5°C. For ammonia, the normal boiling point is –33.3°C and the critical temperature is 132.4°C. For the lowest-temperature isotherm, up to 160 cm³ (STP) of ammonia per gram of charcoal is adsorbed, which is equivalent to 0.12 g NH_3/g charcoal. All five isotherms are of Type I. When the amount adsorbed is low (<25 cm³/g), the isotherms are almost linear and the following form of Henry's law, called the *linear isotherm,* is obeyed:

$$q = kp \qquad (15\text{-}16)$$

where q is equilibrium *loading* or amount adsorbed/unit mass of adsorbent (specific amount adsorbed), k is an empirical, temperature-dependent constant, and p is the partial pressure of the component in the gas. As the temperature increases, the amount adsorbed decreases because of Le Chatelier's principle for an exothermic process. This is shown more clearly for the same data in the crossplot of *adsorption isobars* in Figure 15.10b, where absolute temperature is employed. A third method of displaying the

experimental data is in the form of *adsorption isosteres,* also obtained by crossplotting, as shown in Figure 15.10c. These curves, representing constant amounts adsorbed, resemble vapor-pressure plots, for which the adsorption form of the Clausius–Clapeyron equation,

$$\frac{d \ln p}{dT} = \frac{-\Delta H_{ads}}{RT^2} \qquad (15\text{-}17)$$

or

$$\frac{d \log p}{d(1/T)} = \frac{-\Delta H_{ads}}{2.303RT} \qquad (15\text{-}18)$$

is applied to determine the heat of adsorption; which is negative because the effect is exothermic. The result is shown in Figure 15.10d, where it is seen that $-\Delta H_{ads}$ is initially 7,300 cal/mol, but decreases as the amount adsorbed increases, reaching 6,100 cal/mol at 100 cm³/g. These values can be compared to the heat of vaporization of NH_3, which at 30°C is 4,600 cal/mol.

Experimental, adsorption-isotherm data for 18 different pure gases and a variety of solid adsorbents are summarized and analyzed by Valenzuela and Myers [23]. The data show

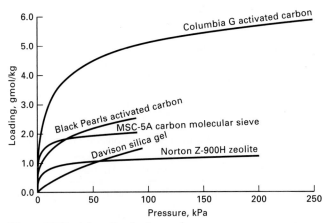

Figure 15.11 Adsorption isotherms for pure propane vapor at 298–303 K.

that adsorption isotherms for a given pure gas at a fixed temperature vary considerably with the adsorbent. This is shown for pure propane vapor in the narrow temperature range of 25–30°C in Figure 15.11 for pressures up to about 101.3 kPa. The highest specific adsorption is with Columbia G-grade activated carbon, while the lowest is with Norton Z-900H, a zeolite molecular sieve. Columbia G-grade activated carbon has about twice the adsorbate capacity of Cabot Black Pearls activated carbon.

The literature data compiled by Valenzuela and Myers [23] also show that for a given adsorbent, the loading depends strongly on the gas. This is illustrated in Table 15.4 for a temperature of 38°C and a narrow pressure range of 97.9 to 100 kPa from the data of Ray and Box [24] for Columbia L activated carbon. Included in the table are normal boiling points and critical temperatures. As might be expected, the species are adsorbed in approximately the inverse order of volatility.

The correlation of experimental adsorption isotherms for pure gases is the subject of a number of published articles and books. As summarized by Yang [25], approaches have ranged from empirical to theoretical. For practical applications, the classical equations of Freundlich and Langmuir

Table 15.4 Comparison of Equilibrium Adsorption of Pure Gases on 20–40 mesh Columbia L Activated Carbon Particles ($S_g = 1,152$ m^2/g) at 38°C and ~1 atm

Pure gas	q, mol/kg	T_b, °F	T_c, °F
H_2	0.0241	−423.0	−399.8
N_2	0.292	−320.4	−232.4
CO	0.374	−313.6	−220.0
CH_4	0.870	−258.7	−116.6
CO_2	1.64	−109.3	87.9
C_2H_2	2.67	−119	95.3
C_2H_4	2.88	−154.6	48.6
C_2H_6	3.41	−127.5	90.1
C_3H_6	4.54	−53.9	196.9
C_3H_8	4.34	−43.7	216.0

are still dominant because of their simplicity and ability to correlate isotherms of Type I in Figure 15.9.

Freundlich Isotherm

The equation attributed to Freundlich [26], but which was actually devised earlier by Boedecker and van Bemmelen, according to Mantell [27], is empirical and nonlinear in pressure:

$$q = kp^{1/n} \tag{15-19}$$

where k and n are temperature-dependent constants. Generally, n lies in the range of 1 to 5. With $n = 1$, (15-19) reduces to the Henry's law equation (15-16). Experimental q–p isothermal data can be fitted to (15-19) by a nonlinear, curve-fitting computer program or by converting (15-19) to a linear form as follows, and using a graphical method or a linear-regression program:

$$\log q = \log k + (1/n) \log p \tag{15-20}$$

If the graphical method is employed, the data are plotted as $\log q$ versus $\log p$. The best straight line through the data has a slope of $(1/n)$ and an intercept of $\log k$. In general, k decreases with increasing temperature, while n increases with increasing temperature and approaches a value of 1 at high temperatures. Equation (15-19) is derived by assuming a heterogeneous surface with a nonuniform distribution of the heat of adsorption over the surface, as discussed by Brunauer [21].

Langmuir Isotherm

The Langmuir equation [28], which is restricted to Type I isotherms, is derived from simple mass-action kinetics, assuming chemisorption. Assume that the surface of the pores of the adsorbent is homogeneous ($\Delta H_{ads} = $ constant) and that the forces of interaction between adsorbed molecules are negligible. Let θ be the fraction of the surface covered by adsorbed molecules. Therefore, $(1 - \theta)$ is the fraction of the bare surface. Then, the net rate of adsorption is the difference between the rates of adsorption on the bare surface and desorption on the covered surface:

$$dq/dt = k_a p(1 - \theta) - k_d \theta \tag{15-21}$$

At equilibrium, $dq/dt = 0$ and (15-21) reduces to

$$\theta = \frac{Kp}{1 + Kp} \tag{15-22}$$

where K is the adsorption-equilibrium constant ($= k_a/k_d$). Here,

$$\theta = q/q_m \tag{15-23}$$

where q_m is the maximum loading corresponding to complete coverage of the surface by the gas. Thus, the Langmuir adsorption isotherm is restricted to a monomolecular layer. Combining (15-23) with (15-22), we obtain the *Langmuir isotherm*:

$$q = \frac{Kq_m p}{1 + Kp} \tag{15-24}$$

At low pressures, if $Kp \ll 1$, (15-24) reduces to the linear Henry's law form (15-16), while at high pressures where $Kp \gg 1$, $q = q_m$. At intermediate pressures, (15-24) is non-linear in pressure. Although originally devised by Langmuir for chemisorption, (15-24) has been widely applied to physical adsorption data.

The quantities K and q_m in (15-24) are treated as empirical constants, obtained by fitting the nonlinear equation directly to experimental data or by employing the following linearized form, numerically or graphically:

$$\frac{p}{q} = \frac{1}{q_m K} + \frac{p}{q_m} \tag{15-25}$$

Using (15-25), the best straight line is drawn through a plot of points p/q versus p, giving a slope of $(1/q_m)$ and an intercept of $1/(q_m K)$. If the theory is reasonable, K should change rapidly with temperature, but q_m should not because it is related through v_m by (15-7) to the specific surface area of the adsorbent, S_g. It should be noted that the Langmuir isotherm predicts an asymptotic limit for q at high pressure, whereas the Freundlich isotherm does not.

Other Adsorption Isotherms

Valenzuela and Myers [23] fit pure gas adsorption-isotherm data to the more complex three-parameter isotherms of (1) Toth:

$$q = \frac{mp}{(b + p^t)^{1/t}} \tag{15-26}$$

where m, b, and t are constants for a given adsorbate–adsorbent system and temperature, and (2) Honig and Reyerson (called the UNILAN equation):

$$q = \frac{n}{2s} \ln \left[\frac{c + pe^s}{c + pe^{-s}} \right] \tag{15-27}$$

where n, s, and c are constants for a given adsorbate–adsorbent system and temperature. The Toth and UNILAN isotherms reduce to the Langmuir isotherm for $t = 1$ and $s = 0$, respectively.

EXAMPLE 15.4

The following experimental data for the equilibrium adsorption of pure methane gas on activated carbon (PCB from Calgon Corp.) at 296 K were obtained by Ritter and Yang [*Ind. Eng. Chem. Res.*, **26,** 1679–1686 (1987)]:

q, cm^3 (STP) of CH$_4$/g carbon	45.5	91.5	113	121	125	126	126
$P = p$, psia	40	165	350	545	760	910	970

Fit the data to: (a) the Freundlich isotherm, and (b) the Langmuir isotherm. Which isotherm provides a better fit to the data?

SOLUTION

By using the linearized forms of the isotherm equations, a spreadsheet or other computer program can be used to do a linear regression to obtain the constants.

(a) Using (15-20), we obtain $\log k = 1.213$, $k = 16.34$, $1/n = 0.3101$, and $n = 3.225$.

Thus, the Freundlich equation is $q = 16.34\,p^{0.3101}$.

(b) Using (15-25), we obtain $1/q_m = 0.007301$, $q_m = 137.0$, $1/(q_m K) = 0.5682$, and $K = 0.01285$.

Thus, the Langmuir equation is $q = \dfrac{1.760p}{1 + 0.01285p}$

The predicted values of q from the two isotherms are as follows:

	q, cm^3 (STP) of CH$_4$/g carbon		
p, psia	Experimental	Freundlich	Langmuir
40	45.5	51.3	46.5
165	91.5	79.6	93.1
350	113	101	112
545	121	115	120
760	125	128	124
910	126	135	126
970	126	138	127

For this example, the Langmuir isotherm fits the data significantly better than the Freundlich isotherm. Average percent deviations, in q, are computed to be 1.01% and 8.64%, respectively. One reason for the better fit of the Langmuir isotherm is the trend of the data to an asymptotic value for q at the highest pressures.

Gas Mixtures and Extended Isotherms

Commercial applications of physical adsorption involve mixtures rather than pure gases. If the adsorption of all components in the gas except one (A) is negligible, then the adsorption of A is estimated from its pure gas-adsorption isotherm using the partial pressure of A. If the adsorption of two or more components in the mixture is significant, the situation is quite complicated. Experimental data show that one component can increase, decrease, or have no influence on the adsorption of the other, depending on interactions of adsorbed molecules. A simple theoretical treatment is the extension of the Langmuir equation by Markham and Benton [29], who neglect interactions and assume that the only effect is the reduction of the vacant surface area for the adsorption of A because of the adsorption of other components. Consider a binary gas mixture of A and B. Let θ_A = fraction of the surface covered by A and θ_B = fraction of the surface covered by B. Then, $(1 - \theta_A - \theta_B)$ = fraction of vacant surface. At equilibrium:

$$(k_A)_a p_A (1 - \theta_A - \theta_B) = (k_A)_d \theta_A \tag{15-28}$$

$$(k_B)_a p_B (1 - \theta_A - \theta_B) = (k_B)_d \theta_B \tag{15-29}$$

Solving these equations simultaneously, and combining the results with (15-23) for each component, gives

$$q_A = \frac{(q_A)_m K_A p_A}{1 + K_A p_A + K_B p_B} \tag{15-30}$$

$$q_B = \frac{(q_B)_m K_B p_B}{1 + K_A p_A + K_B p_B} \tag{15-31}$$

where $(q_i)_m$ is the maximum amount of adsorption of species i for coverage of the entire surface. Equations (15-30) and (15-31) are readily extended to a multicomponent mixture of j components:

$$q_i = \frac{(q_i)_m K_i p_i}{1 + \sum_j K_j p_j} \qquad (15\text{-}32)$$

In a similar fashion, as shown by Yon and Turnock [30], the Freundlich equation can be combined with the Langmuir equation to give the following extended relation for gas mixtures:

$$q_i = \frac{(q_i)_0 k_i p_i^{1/n_i}}{1 + \sum_j k_j p_j^{1/n_j}} \qquad (15\text{-}33)$$

where $(q_i)_0$ is the maximum loading, which may differ from $(q_i)_m$ for a monolayer. Equation (15-33) represents data for nonpolar, multicomponent mixtures in molecular sieves reasonably well. Unfortunately, Broughton [31] has shown that the extended-Langmuir equation lacks thermodynamic consistency; such is also the case for the extended-Langmuir–Freundlich equation. Accordingly, both (15-32) and (15-33) are frequently referred to as nonstoichiometric isotherms. Nevertheless, for practical application, their simplicity often makes them the isotherms of choice. In particular, both the extended Langmuir and Freundlich adsorption isotherms of (15-32) and (15-33) are frequently referred to as *constant-selectivity-equilibrium* equations because they predict a separation factor (selectivity), $\alpha_{i,j}$, for each pair of components, i, j, in a multicomponent mixture that is constant for a given temperature and independent of mixture composition. For example, (15-32) gives

$$\alpha_{i,j} = \frac{q_i/q_j}{p_i/p_j} = \frac{(q_i)_m K_i}{(q_j)_m K_j}$$

As with multicomponent (three or more components) vapor–liquid and liquid–liquid phase equilibria, experimental data for binary and multicomponent gas–solid adsorbent equilibria are scarce and less accurate than corresponding pure gas data. Valenzuela and Myers [23] include experimental data on adsorption of gas mixtures from nine published studies on 29 binary systems, for which pure gas-adsorption isotherms were also obtained. They also describe procedures for applying the Toth and UNILAN equations to multicomponent mixtures based on the ideal-adsorbed-solution (IAS) theory of Myers and Prausnitz [32]. Unlike the extended-Langmuir equation (15-32), which is explicit in the amount adsorbed, the IAS theory, though more accurate, is not explicit and requires an iterative solution procedure. Additional experimental data for higher-order (ternary and/or higher) gas mixtures are given by Miller, Knaebel, and Ikels [33] for 5A molecular sieves and by Ritter and Yang [34] for activated carbon. Yang [25] presents a discussion of existing theories on adsorption of gas mixtures, together with comparisons of these theories with mixture data for activated carbon and zeolites. The data on zeolites are the

most difficult to correlate, with the simplified statistical thermodynamic model (SSTM) of Ruthven and Wong [35] giving best results.

EXAMPLE 15.5

The experimental work of Ritter and Yang, cited in Example 15.4, also includes adsorption isotherms for pure CO and CH_4, and a binary mixture of $CH_4(A)$ and $CO(B)$. For the pure gases, Ritter and Yang give relations over a temperature range of 296–480 K, for the two Langmuir constants. At 294 K, these constants are as follows:

	q_m, cm³(STP)/g	K, psi⁻¹
CH_4	133.4	0.01370
CO	126.1	0.00624

With these constants, use the extended-Langmuir equation to predict the specific adsorption volumes (STP) of CH_4 and CO for a vapor mixture of 69.6 mol% CH_4 and 30.4 mol% CO at 294 K and a total pressure of 364.3 psia. Compare the results with the following experimental data of Ritter and Yang:

Total volume adsorbed, cm³/(STP)/g	114.1
Mole fractions in adsorbate:	
CH_4	0.867
CO	0.133

SOLUTION

$$p_A = y_A P = 0.696(364.3) = 253.5 \text{ psia}$$
$$p_B = y_B P = 0.304(364.3) = 110.8 \text{ psia}$$

From (15-30):

$$q_A = \frac{133.4(0.0137)(253.5)}{1 + (0.0137)(253.5) + (0.00624)(110.8)} = 89.7 \text{ cm}^3(\text{STP})/\text{g}$$

$$q_B = \frac{126.1(0.00624)(110.8)}{1 + (0.0137)(253.5) + (0.00624)(110.8)} = 16.9 \text{ cm}^3(\text{STP})/\text{g}$$

The total amount adsorbed $= q = q_A + q_B = 89.7 + 16.9 = 106.6$ cm³(STP)/g, which is 6.6% lower than the experimental value. Estimated mole fractions in the adsorbate are $x_A = q_A/q = 89.7/106.6 = 0.841$ and $x_B = 1 - 0.841 = 0.159$. These adsorbate mole fractions deviate from the experimental values by a mole fraction of 0.026. For this example, the extended-Langmuir isotherm gives reasonable results.

Liquid Adsorption

When porous adsorbent particles are immersed in a pure gas, the pores fill with the gas, and the amount of adsorbed gas is determined by the decrease in total pressure. With a liquid, the pressure does not change, and no simple experimental procedure has been devised for determining the extent of adsorption of a pure liquid. If the liquid is a homogeneous binary mixture, it is customary to designate one component the solute (1) and the other the solvent (2). The assumption is then made that the change in composition of the bulk liquid in contact with the porous solid is due entirely to adsorption

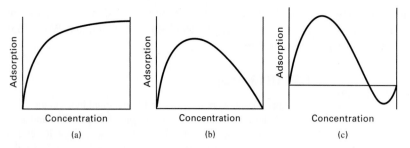

Figure 15.12 Representative isotherms of concentration change for liquid adsorption.

of the solute. That is, adsorption of the solvent is tacitly assumed not to occur. If the liquid mixture is dilute in the solute, the consequences are not serious. If, however, experimental data are obtained over the entire concentration range, the distinction between solute and solvent is arbitrary and the resulting adsorption isotherms, as discussed by Kipling [36], can exhibit curious shapes that are unlike those obtained for pure gases or gas mixtures. To demonstrate this, let

n^0 = total moles of binary liquid brought into contact with adsorbent

m = mass of adsorbent

x_1^0 = mole fraction of solute in the mixture before contact with adsorbent

x_1 = mole fraction of solute in the bulk solution after adsorption equilibrium is achieved

q_1^e = apparent moles of solute adsorbed per unit mass of adsorbent

A solute material balance, assuming no adsorption of solvent and a negligible change in the total moles of liquid mixture, gives

$$q_1^e = \frac{n^0(x_1^0 - x_1)}{m} \qquad (15\text{-}34)$$

If data are obtained at constant temperature over the entire concentration range and then processed with (15-34) and plotted as adsorption isotherms, the resulting curves are not of the expected type shown in Figure 15.12a. Instead, curves of the type shown in Figures 15.12b and c are obtained, where negative adsorption appears to occur in Figure 15.12c. Such isotherms are probably best referred to as *composite isotherms* or *isotherms of concentration change*, as suggested by Kipling [36]. Likewise, the adsorption loading, q_1^e, of (15-34) is more correctly referred to as the *surface excess*.

Under what conditions are composite isotherms of the form shown in Figures 15.12b and c obtained? This is shown by several examples in Figure 15.13, where various combinations of hypothetical adsorption isotherms for solute (A) and solvent (B) are shown together with the resulting composite isotherms. Thus, when the solvent is not adsorbed, as seen in Figure 15.13a, a composite curve without negative adsorption is obtained. In all other cases of Figure 15.13, negative values of the surface excess are obtained.

Valenzuela and Myers [23] tabulate literature values for the equilibrium adsorption of 25 different binary-liquid

mixtures. With one exception, all 25 mixtures give composite isotherms of the forms shown in Figures 15.12b and c. The one exception is a mixture of cyclohexane and n-heptane with silica gel, for which the surface excess is almost negligible $(0 \pm 0.05\,\text{mmol/g})$ over the composition range of $x_1 = 0.041$ to 0.911. They also include literature references to 354 sets of binary-mixture data, 25 sets of ternary-mixture data, and 3 sets of data for higher-order mixtures.

When data for the binary mixture are only available in the dilute region, the amount of adsorption, if any, of the solvent may be constant and all changes in the total amount adsorbed are due to just the solute. In that case, the adsorption isotherms are of the form of Figure 15.12a, which resembles the form obtained with pure gases. It is then common to fit the data with concentration forms of the Freundlich equation

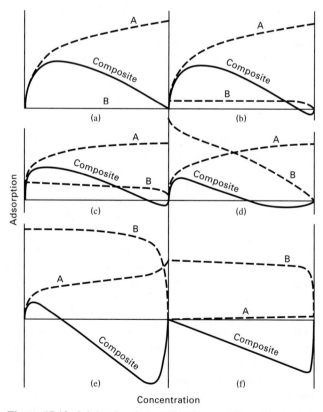

Figure 15.13 Origin of various types of composite isotherms for binary liquid adsorption.

[From J.J. Kipling, *Adsorption from Solutions of Non-electrolytes*, Academic Press, London (1965) with permission.]

(15-19) or the Langmuir equation (15-24):

$$q = kc^{1/n} \qquad (15\text{-}35)$$

$$q = \frac{K q_m c}{1 + Kc} \qquad (15\text{-}36)$$

Candidate systems for this case are small amounts of organic compounds dissolved in water and small amounts of water dissolved in hydrocarbons. For liquid mixtures that are dilute in two or more solutes, the multicomponent adsorption may be estimated from a concentration form of the extended Langmuir equation (15-32) based on the constants, q_m and K, obtained from experiments on the single solutes. However, when solute–solute interactions are suspected, it may be necessary to determine the constants from multicomponent data. As with gas mixtures, the concentration form of (15-32) also predicts a constant selectivity for each pair of components in a multicomponent mixture.

EXAMPLE 15.6

Small amounts of VOCs in water can be removed by adsorption. Generally, two or more VOCs are present. An aqueous stream containing small amounts of acetone (1) and propionitrile (2) is to be treated with activated carbon. Single-solute equilibrium data available from Radke and Prausnitz [37] have been fitted to the Freundlich and Langmuir isotherms, (15-35) and (15-36), with the average deviations indicated, for solute concentrations up to 50 mmol/L:

Acetone in Water (25°C):		Absolute Average Deviation of q, %
$q_1 = 0.141 c_1^{0.597}$	**(1)**	14.2
$q_1 = \dfrac{0.190 c_1}{1 + 0.146 c_1}$	**(2)**	27.3
Propionitrile in water (25°C):		
$q_2 = 0.138 c_2^{0.658}$	**(3)**	10.2
$q_2 = \dfrac{0.173 c_2}{1 + 0.0961 c_2}$	**(4)**	26.2

where

q_i = amount of solute adsorbed, mmol/g

c_i = solute concentration in aqueous solution, mmol/L

Use these single-solute results with an extended Langmuir-type isotherm to predict the equilibrium adsorption in a binary-solute aqueous system containing 40 and 34.4 mmol/L, respectively, of acetone and propionitrile at 25°C with the same adsorbent. Compare the results with the following experimental values from Radke and Prausnitz [37]:

$q_1 = 0.715$ mmol/g, $q_2 = 0.822$ mmol/g, and $q_{\text{total}} = 1.537$ mmol/g

SOLUTION

From (15-32), the extended Langmuir isotherm for the liquid phase is

$$q_i = \frac{(q_i)_m K_i c_i}{1 + \sum\limits_j K_j c_j} \qquad (5)$$

From (2), $(q_1)_m = 0.190/0.146 = 1.301$ mmol/g.
From (4), $(q_2)_m = 0.173/0.0961 = 1.800$ mmol/g.
From (5):

$$q_1 = \frac{1.301(0.146)(40)}{1 + (0.146)(40) + (0.0961)(34.4)} = 0.749 \text{ mmol/g}$$

$$q_2 = \frac{1.800(0.0961)(34.4)}{1 + (0.146)(40) + (0.0961)(34.4)} = 0.587 \text{ mmol/g}$$

$$q_{\text{total}} = 1.336 \text{ mmol/g}$$

Compared to experimental data, the percent deviations for q_1, q_2, and q_{total}, respectively, are 4.8%, −28.6%, and −13.1%. Better agreement is obtained by Radke and Prausnitz using an IAS theory. It is expected that a concentration form of (15-33) would also give better agreement, but that requires that the single-solute data be refitted for each solute to a Langmuir–Freundlich isotherm of the form

$$q = \frac{q_0 kc^{1/n}}{1 + kc^{1/n}} \qquad (6)$$

Ion Exchange Equilibria

Ion exchange differs from adsorption in that one sorbate (a counterion) is exchanged for a solute ion, and the exchange is governed by a reversible, stoichiometric, chemical-reaction equation. Thus, the selectivity of the ion exchanger for one counterion over another may be just as important as the capacity of the ion exchanger. Accordingly, for ion exchange, we apply the law of mass action to obtain an equilibrium ratio rather than fit data to a sorption isotherm such as the Langmuir or Freundlich equation.

As discussed by Anderson [38], two cases are important. In the first case, the counterion initially in the ion exchanger is exchanged with a counterion from an acid or base. For example:

$$\text{Na}^+_{(aq)} + \text{OH}^-_{(aq)} + \text{HR}_{(s)} \leftrightarrow \text{NaR}_{(s)} + \text{H}_2\text{O}_{(l)}$$

Note that the hydrogen ion leaving the ion exchanger immediately reacts with the hydroxyl ion in the aqueous solution to form water, leaving no counterion on the right-hand side of the reaction. Accordingly, the ion exchange will continue until the aqueous solution is depleted of sodium ions or the ion exchanger is depleted of hydrogen ions.

In the second case, which is more common than the first, the counterion being transferred from the ion exchanger to the fluid phase remains as an ion. For example, the exchange of counterions A and B is expressed by the reaction

$$\text{A}^{n\pm}_{(l)} + n\text{BR}_{(s)} \leftrightarrow \text{AR}_{n(s)} + n\text{B}^{\pm}_{(l)} \qquad (15\text{-}37)$$

where A and B must both be either cations (positive charge) or anions (negative charge). For this case, at equilibrium, we can define a conventional chemical-equilibrium constant according to the law of mass action:

$$K_{\text{A,B}} = \frac{q_{\text{AR}_n} c_{\text{B}\pm}^n}{q_{\text{BR}}^n c_{\text{A}^{n\pm}}} \qquad (15\text{-}38)$$

where molar concentrations c_i and q_i refer to the liquid and ion-exchanger phases, respectively. The constant, $K_{\text{A,B}}$ is not a rigorous thermodynamic equilibrium constant because (15-38) is written in terms of concentrations instead of activities. Although (15-38) could be corrected by including activity coefficients, it is usually applied in the form shown, with $K_{A,B}$ referred to as a *molar selectivity coefficient* for A entering the ion exchange resin and displacing B. For the resin phase, concentrations are in equivalents per unit mass or unit bed volume of ion exchanger. For the liquid solution, concentrations are in equivalents per unit volume of solution. For dilute liquid solutions, $K_{A,B}$ is reasonably constant for a given pair of counterions and a particular resin of a given degree of cross-linking.

When exchange is between two counterions of equal charge, (15-38) can be reduced to a simple equation in terms of just the equilibrium concentrations of A in the liquid solution and in the ion exchange resin. Because of (15-37), the total concentrations C and Q in equivalents of counterions in the liquid solution and the resin, respectively, remain constant during the exchange process. Accordingly:

$$c_i = Cx_i/z_i \tag{15-39}$$
$$q_i = Qy_i/z_i \tag{15-40}$$

where x_i and y_i are equivalent fractions, rather than mole fractions, of A and B, such that

$$x_A + x_B = 1 \tag{15-41}$$
$$y_A + y_B = 1 \tag{15-42}$$

and $z_i =$ valence of counterion i. Combining (15-38) to (15-42) results in

$$K_{A,B} = \frac{y_A(1 - x_A)}{x_A(1 - y_A)} \tag{15-43}$$

Thus, at equilibrium, x_A and y_A are independent of the total equivalent concentrations C and Q. Such is not the case when the two counterions are of unequal charge, as in the exchange of Ca^{2+} and Na^+. A derivation for this general case gives

$$K_{A,B} = \left(\frac{C}{Q}\right)^{n-1} \frac{y_A(1 - x_A)^n}{x_A(1 - y_A)^n} \tag{15-44}$$

For unequal counterion charges, we see that $K_{A,B}$ depends on the ratio C/Q and on the ratio of charges, n.

When experimental data for $K_{A,B}$ for a particular binary system of counterions with a particular ion exchanger are not available, the method of Bonner and Smith [39], as modified by Anderson [38], is used for screening purposes or preliminary calculations. In this method, the molar selectivity coefficient is

$$K_{ij} = K_i/K_j \tag{15-45}$$

where values for relative molar selectivities K_i and K_j are given in Table 15.5 for cations with an 8% cross-linked strong-acid resin and in Table 15.6 for anions with strong-base resins. For values of K in these tables, the units of C and Q are, respectively, eq/L of solution and eq/L of bulk bed volume of water-swelled resin.

A typical cation-exchange resin of the sulfonated styrene–divinylbenzene type, such as Dowex 50, as described by

Table 15.5   Relative Molar Selectivities for Cations with 8% Cross-linked Strong-Acid Resin

Li^+	1.0	Zn^{2+}	3.5
H^+	1.3	Co^{2+}	3.7
Na^+	2.0	Cu^{2+}	3.8
NH_4^+	2.6	Cd^{2+}	3.9
K^+	2.9	Be^{2+}	4.0
Rb^+	3.2	Mn^{2+}	4.1
Cs^+	3.3	Ni^+	3.9
Ag^+	8.5	Ca^{2+}	5.2
UO_2^{2+}	2.5	Sr^{2+}	6.5
Mg^{2+}	3.3	Pb^{2+}	9.9
		Ba^{2+}	11.5

Table 15.6   Approximate Relative Molar Selectivities for Anions with Strong-Base Resins

I^-	8	OH^- (Type II)	0.65
NO_3^-	4	HCO_3^-	0.4
Br^-	3	CH_3COO^-	0.2
HSO_4^-	1.6	F^-	0.1
NO_2^-	1.3	OH^- (Type I)	0.05–0.07
CN^-	1.3	SO_4^{2-}	0.15
Cl^-	1.0	CO_3^{2-}	0.03
BrO_3^-	1.0	HPO_4^{2-}	0.01

Bauman and Eichhorn [40] and Bauman, Skidmore, and Osmun [41], has an exchangeable ion capacity of 5 ± 0.1 meq/g of dry resin. As shipped, the water-wet resin might contain 41.4 wt% water. Thus, the wet capacity is $5(58.6/100) = 2.9$ meq/g of wet resin. If the bulk density of a drained bed of wet resin is 0.83 g/cm^3, the bed capacity is 2.4 eq/L of resin bed.

As with other separation processes, a separation factor, $S_{A,B}$, which ignores the valence of the exchanging ions, can be defined for an equilibrium stage. For the binary case, in terms of equivalent ionic fractions:

$$S_{A,B} = \frac{y_A(1 - x_A)}{x_A(1 - y_A)} \tag{15-46}$$

which is identical to (15-43). Experimental data for an exchange between Cu^{2+} (A) and Na^+ (B) (counterions of unequal charge) with Dowex 50 cation resin over a wide range of total-solution normality at ambient temperature are shown in terms of y_A and x_A in Figure 15.14, from Subba Rao and David [42]. At low, total-solution concentration, the resin is highly selective for copper ion, whereas at high, total-solution concentration, the selectivity is reversed to favor sodium ion slightly. A similar trend was observed by Selke and Bliss [43, 44] for exchange between Ca^{2+} and H^+ using a similar resin, Amberlite IR-120. This sensitivity of the selectivity is shown dramatically in Figure 15.15, from Myers and Byington [45], where the natural logarithm of the separation factor, S_{Cu^{2+},Na^+}, as computed from the data of Figure 15.14 with (15-46), is plotted as a function of

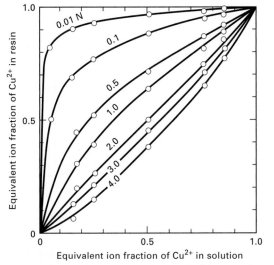

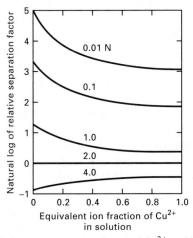

Figure 15.14 Isotherms for ion exchange of Cu^{2+} and Na^+ on Dowex 50-X8 as a function of total normality in the bulk solution.

[From A.L. Myers and S. Byington, *Ion Exchange Science and Technology*, M. Nijhoff, Boston (1986) with permission.]

Figure 15.15 Relative separation factor of Cu^{2+} and Na^+ for ion exchange on Dowex 50-X8 as a function of total normality in the bulk solution.

[From A.L. Myers and S. Byington, *Ion Exchange Science and Technology*, M. Nijhoff, Boston (1986) with permission.]

equivalent ionic fraction, $x_{Cu^{2+}}$. For dilute solutions of Cu^{2+}, S_{Cu^{2+},Na^+} ranges from about 0.5 at a total concentration of 4 N to 60 at 0.01 N. In terms of K_{Cu^{2+},Na^+} of (15-44), with $n = 2$, the corresponding variation is computed to be only from about 0.6 to 2.2.

EXAMPLE 15.7

An Amberlite IR-120 ion-exchange resin similar to that of Example 15.2, but with a maximum ion-exchange capacity of 4.90 meq/g of dry resin, is used to remove cupric ion from a waste stream containing 0.00975-M $CuSO_4$ (19.5 meq Cu^{2+}/L solution). The spherical resin particles range in diameter from 0.2 to just over 1.2 mm. The equilibrium ion-exchange reaction is of the divalent–monovalent

type:

$$Cu^{2+}_{(aq)} + 2HR_{(s)} \leftrightarrow CuR_{2(s)} + 2H^+_{(aq)}$$

As ion exchange takes place, the milliequivalents of cations in the aqueous solution and in the resin remain constant.

Experimental measurements by Selke and Bliss [43, 44] show an equilibrium curve of the type of Figure 15.14 at ambient temperature that is markedly dependent on the total equivalent concentration of the aqueous solution, with the following equilibrium data for the cupric ion with a 19.5 meq/liter solution:

c, meq Cu^{2+}/L Solution	0.022	0.786	4.49	10.3
q, meq Cu^{2+}/g Resin	0.66	3.26	4.55	4.65

These data follow a highly nonlinear isotherm.

(a) From the data, compute the molar selectivity coefficient, K, at each value of c for Cu^{2+} and compare it to the value estimated from (15-45) using Table 15.5.

(b) Predict the milliequivalents of Cu^{2+} exchanged at equilibrium from 10 L of 20 meq Cu^{2+}/L using 50 g of dry resin with 4.9 meq of H^+/g.

SOLUTION

(a) Selke and Bliss do not give a value for the resin capacity, Q, in eq/L of bed volume. Assume a value of 2.3. From (15-44):

$$K_{Cu^{2+},H^+} = \left(\frac{C}{Q}\right)\frac{y_{Cu^{2+}}(1 - x_{Cu^{2+}})^2}{x_{Cu^{2+}}(1 - y_{Cu^{2+}})^2}$$

where $(C/Q) = 0.0195/2.3 = 0.0085$

$$x_{Cu^{2+}} = c_{Cu^{2+}}/19.5 \quad \text{and} \quad y_{Cu^{2+}} = q_{Cu^{2+}}/4.9$$

Using the above values of c and q from Selke and Bliss:

q, meq Cu^{2+}/g	$x_{Cu^{2+}}$	$y_{Cu^{2+}}$	K_{Cu^{2+},H^+}
0.66	0.00113	0.135	1.35
3.26	0.0403	0.665	1.15
4.55	0.230	0.929	4.04
4.65	0.528	0.949	1.30

The average value of K is 2.0. The values in Table 15.5 when substituted into (15-45) predict

$$K_{Cu^{2+},H^+} = 3.8/1.3 = 2.9$$

which is somewhat higher.

(b) Assume a value of 2.0 for K_{Cu^{2+},H^+} with $Q = 2.3$ eq/L. The total solution concentration, C, is 0.02 eq/L. Equation (15-44) becomes

$$2.0 = \left(\frac{0.02}{2.3}\right)\frac{y_{Cu^{2+}}(1 - x_{Cu^{2+}})^2}{x_{Cu^{2+}}(1 - y_{Cu^{2+}})^2} \quad (1)$$

Initially, the solution contains $(0.02)(10) = 0.2$ equivalents of cupric ion with $x_{Cu^{2+}} = 1.0$. Let $a =$ equivalents of Cu exchanged. Then, at equilibrium, by material balance:

$$x_{Cu^{2+}} = \frac{0.02 - (a/10)}{0.02} \quad (2)$$

$$y_{Cu^{2+}} = \frac{(a/50)}{0.0049} \quad (3)$$

Substitution of (2) and (3) into (1) gives

$$2.0 = 0.0087 \frac{\left[\dfrac{(a/50)}{0.0049}\right]\left[1 - \dfrac{0.02 - (a/10)}{0.02}\right]^2}{\left[\dfrac{0.02 - (a/10)}{0.02}\right]\left[1 - \dfrac{(a/50)}{0.0049}\right]^2} \qquad (4)$$

Solving (4), a nonlinear equation, for a gives 0.1887 equivalents of Cu exchanged. Thus, $0.1887/[(0.020)(10)] = 0.944$ or 94.4% of the cupric ion is exchanged.

Equilibria in Chromatography

As discussed in Section 15.1, separation by chromatography involves sorption mechanisms of many types, including adsorption on porous solids, absorption or extraction (partitioning) in liquid-supported or bonded solids, and ion exchange in synthetic resins. Thus, at equilibrium, depending upon the sorption mechanism, equations such as (15-19), (15-24), (15-32), and (15-33) for gas adsorption; (15-35) and (15-36) for liquid adsorption; (6-17) to (6-20) for gas absorption; (8-1) for liquid extraction; and (15-38), (15-43), and (15-44) for ion exchange apply.

When the equilibrium (distribution or partition) constant is defined as

$$K_i = q_i/c_i \qquad (15\text{-}47)$$

where q is concentration in the stationary phase and c is concentration in the mobile phase, solutes with the highest equilibrium constants will elute from the chromatographic column at a slower rate than solutes with the smallest equilibrium constants.

15.3 KINETIC AND TRANSPORT CONSIDERATIONS

For the adsorption of a solute onto the porous surface of an adsorbent, the following steps are required:

1. External (interphase) mass transfer of the solute from the bulk fluid by convection, through a thin film or boundary layer, to the outer, solid surface of the adsorbent

2. Internal (intraphase) mass transfer of the solute by pore diffusion from the outer surface of the adsorbent to the inner surface of the internal porous structure

3. Surface diffusion along the porous surface

4. Adsorption of the solute onto the porous surface

For chemisorption, which involves bond formation, the rate of the fourth kinetic step may be slow and even controlling; for physical adsorption, however, step 4 is almost instantaneous because it depends only on the collision frequency and orientation of the molecules with the porous surface. Thus, only the first three steps need be considered here.

During regeneration of the adsorbent, the reverse of the four steps occurs, where the rate of physical desorption is instantaneous. Adsorption and desorption are accompanied

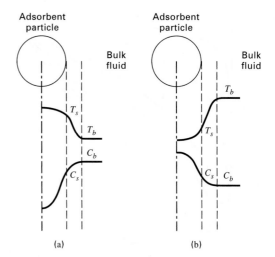

Figure 15.16 Solute concentration and temperature profiles for a porous adsorbent particle surrounded by a fluid: (a) Adsorption. (b) Desorption.

by heat transfer because of the exothermic heat of adsorption and the endothermic heat of desorption. However, although external mass transfer is limited to a convective mechanism, external heat transfer from the particle outer surface occurs not only by convection through the film or boundary layer surrounding each solid particle in the bed, but also by thermal radiation between particles when the fluid is a gas, and by conduction at points of contact by adjacent particles. Conduction and radiation mechanisms for heat transfer can also exist within the particle, in addition to convective heat transfer by the fluid within the pores.

In a fixed bed of adsorbent particles, solute concentration and temperature change continuously with time and location. For a given particle at a particular time, profiles of temperature and solute concentration in the fluid are as shown in Figures 15.16a and b for adsorption and desorption, respectively, where subscripts b and s refer to bulk fluid and particle outer surface, respectively. The fluid concentration gradient is usually steepest within the particle, whereas the temperature gradient is usually steepest in the fluid film or boundary layer surrounding the particle. Thus, although the major resistance to heat transfer is usually external to the adsorbent particle, the major resistance to mass transfer usually resides in the adsorbent particle. All four gradients in Figure 15.16 approach asymptotic values at the end points.

External Transport

Rates of convective mass and heat transfer between the outer surface of a particle and the surrounding bulk fluid during an adsorption process are given, respectively, by

$$n_i = \frac{d\mathcal{N}_i}{dt} = k_c A(c_{b_i} - c_{s_i}) \qquad (15\text{-}48)$$

$$q = \frac{dQ}{dt} = h A(T_s - T_b) \qquad (15\text{-}49)$$

For a spherical particle surrounded by an infinite, quiescent fluid, the mass- and heat-transfer coefficients are at their minimum values. Assume an insoluble, solid, spherical particle of radius R_p and diameter $D_p = 2R_p$, suspended in an infinite-fluid medium. The particle is heated so that, at steady state, its surface temperature is constant at T_s. The fluid medium is absolutely quiescent (no free convection) and radiation is ignored so that heat transfer through the fluid is by conduction only. The thermal conductivity k of the fluid is constant, and the temperature far from the particle is T_b. Fourier's second law of heat conduction in the fluid, for spherical coordinates, is

$$\frac{d}{dr}\left(kr^2\frac{dT}{dr}\right) = 0 \qquad (15\text{-}50)$$

for $r \geq R_p$, where r is the radial distance from the center of the particle. The boundary conditions are

$$T\{r = R_p\} = T_s \qquad (15\text{-}51)$$
$$T\{r = \infty\} = T_b \qquad (15\text{-}52)$$

If (15-50) is integrated twice with respect to r, we obtain:

$$T = -\frac{C_1}{r} + C_2 \qquad (15\text{-}53)$$

Substitution of the boundary conditions, (15-51) and (15-52), results in an expression for the temperature profile in the fluid:

$$\frac{T - T_b}{T_s - T_b} = \frac{R_p}{r}, \quad r \geq R_p \qquad (15\text{-}54)$$

The heat flux at the outer surface of the particle is given by Fourier's first law of heat conduction applied to the fluid adjacent to the particle:

$$\left.\frac{q}{A}\right|_{r=R_p} = -k\left.\frac{dT}{dr}\right|_{r=R_p} \qquad (15\text{-}55)$$

From (15-54):

$$\left.\frac{dT}{dr}\right|_{r=R_p} = -\frac{(T_s - T_b)}{R_p} \qquad (15\text{-}56)$$

We can also apply Newton's law of cooling for the heat flux at the outer surface of the particle:

$$\left.\frac{q}{A}\right|_{r=R_p} = h(T_s - T_b) \qquad (15\text{-}57)$$

Combining (15-55) to (15-57):

$$h = k/R_p \qquad (15\text{-}58)$$

which rearranges into a Nusselt number form:

$$N_{\text{Nu}} = hD_p/k = 2 \qquad (15\text{-}59)$$

A similar development for convective mass transfer using Fick's laws of diffusion gives

$$N_{\text{Sh}_i} = k_{c_i}D_p/D_i = 2 \qquad (15\text{-}60)$$

where D_i is the diffusivity of component i in the mixture.

When the fluid flows past a single particle, convection increases the convective mass- and heat-transfer coefficients above the values computed from (15-59) and (15-60).

Furthermore, the transport coefficients now vary around the periphery of the particle, with the largest value occurring where the fluid flow first impinges on the particle. Correlations of experimental transport data are usually developed for coefficients averaged over the surface of the particle. Typical correlations are those of Ranz and Marshall [46, 47] for Nusselt numbers as high as 30 and Sherwood numbers to 160:

$$N_{\text{Nu}} = 2 + 0.60\, N_{\text{Re}}^{1/2} N_{\text{Pr}}^{1/3} \qquad (15\text{-}61)$$

$$N_{\text{Sh}_i} = 2 + 0.60\, N_{\text{Re}}^{1/2} N_{\text{Sc}_i}^{1/3} \qquad (15\text{-}62)$$

where

$$N_{\text{Pr}} = \text{Prandtl number} = C_P\mu/k$$
$$N_{\text{Sc}_i} = \text{Schmidt number} = \mu/\rho D_i$$
$$N_{\text{Re}} = \text{Reynolds number} = D_pG/\mu$$

All fluid properties are evaluated at the average temperature of the film or boundary layer. Equations (15-61) and (15-62) reduce to (15-59) and (15-60), respectively, when the fluid mass velocity, G, is zero.

When particles are packed in a bed, the fluid-flow patterns are restricted, and the single-particle correlations of (15-61) and (15-62) cannot be used to estimate the average external-transport coefficients for the particles in the bed. However, Ranz [48] showed that equations of the same form as (15-61) and (15-62) correlate external-transport data for beds packed with spherical particles. Nevertheless, most early investigators, starting with Gamson, Thodos, and Hougen [49], developed correlations in the form of the Chilton and Colburn [50] j-factors:

$$j_D = (N_{\text{St}_M})(N_{\text{Sc}})^{2/3} = f\{N_{\text{Re}}\} \qquad (15\text{-}63)$$
$$j_H = (N_{\text{St}})(N_{\text{Pr}})^{2/3} = f\{N_{\text{Re}}\} \qquad (15\text{-}64)$$

with

$$N_{\text{St}_M} = k_c\rho/G \quad \text{and}\ N_{\text{St}} = h/C_PG$$

where different Reynolds-number functions apply to different regions. Various forms of the Reynolds number have been used, including D_pG/μ and $D_pG/\epsilon_b\mu$, in attempts to account for bed void fraction, ϵ_b, where G is the superficial mass velocity based on the empty-bed cross-sectional area, and G/ϵ_b (a larger value) is the effective mass velocity through the void region of the bed. Notable among correlations of this type are those of Sen Gupta and Thodos [51], Petrovic and Thodos [52], and Dwivedi and Upadhay [53].

A more-recent study by Wakao and Funazkri [54] reanalyzed 37 sets of previously published mass-transfer data, with Sherwood number corrections for axial dispersion. The resulting correlation, which represents a return to the form of (15-62), is

$$N_{\text{Sh}_i} = \frac{k_{c_i}D_p}{D_i} = 2 + 1.1\left(\frac{D_pG}{\mu}\right)^{0.6}\left(\frac{\mu}{\rho D_i}\right)^{1/3} \qquad (15\text{-}65)$$

The correlation is compared to 12 sets of gas-phase and 11 sets of liquid-phase data in Figure 15.17. The data cover

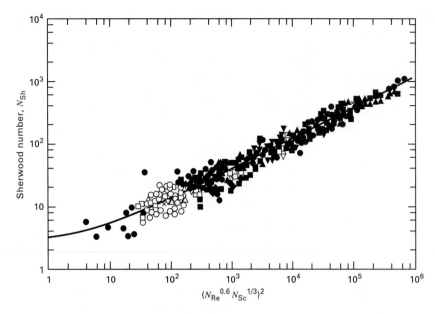

Figure 15.17 Correlation of experimental data for Sherwood number of external mass transfer in a packed bed.
[From N. Wakao and T. Funazkri, *Chem. Eng. Sci.*, **33**, 1375 (1978) with permission.]

a Schmidt number range of 0.6 to 70,600, a Reynolds number range of 3 to 10,000, and a particle diameter from 0.6 to 17.1 mm. Particle shapes include spheres, short cylinders, flakes, and granules. By analogy, the corresponding equation for fluid-particle convective heat transfer in packed beds is

$$N_{Nu} = \frac{hD_p}{k} = 2 + 1.1\left(\frac{D_p G}{\mu}\right)^{0.6}\left(\frac{C_P \mu}{k}\right)^{1/3} \quad (15\text{-}66)$$

When (15-65) and (15-66) are used with beds packed with nonspherical particles, D_p is the equivalent diameter of a spherical particle. The following suggestions have been proposed for computing the equivalent diameter from geometric properties of the particle. These suggestions may be compared by considering a short cylinder with diameter, D, equal to the length, L.

1. D_p = diameter of a sphere with the same external surface area:

$$\pi D_p^2 = \pi DL + \pi D^2/2$$
and $\quad D_p = (DL + D^2/2)^{0.5} = 1.225D$

2. D_p = diameter of a sphere with the same volume:

$$\pi D_p^3/6 = \pi D^2 L/4 \text{ and } D_p = (3D^2 L/2)^{1/3} = 1.145D$$

3. $D_p = 4$ times the hydraulic radius, r_H, where for a packed bed,

$$4r_H = 6/a_v$$
a_v = external particle surface area/volume of particle

Thus,

$$a_v = \frac{\pi DL + \pi D^2/2}{\pi D^2 L/4} = \frac{6}{D} \text{ and } D_p = 4r_H = \frac{6D}{6} = 1.0D$$

The use of the hydraulic radius concept is equivalent to replacing D_p in the Reynolds number by $\Psi D_p'$ where Ψ is the sphericity and D_p' is given by Suggestion 2: The sphericity is defined by:

$$\Psi = \frac{\text{Surface area of a sphere of same volume as particle}}{\text{Surface area of particle}}$$

For a cylinder of $D = L$,

$$\Psi = \frac{\pi D_p^2}{\pi DL + 2\left(\frac{\pi D^2}{4}\right)} = \frac{\pi(1.145D)^2}{\frac{3}{2}\pi D^2} = 0.874$$

and $\Psi D_p' = (0.874)(1.145D) = D$, the diameter of the cylinder.

Suggestions 2 and 3 are widely used. Suggestion 3 is conveniently applied to crushed particles of irregular surface, but with no obvious longer or shorter dimension, that is, isotropic in shape. In that case, D_p' is taken as the size of the particle and the sphericity is approximately 0.65, as discussed by Kunii and Levenspiel [55].

EXAMPLE 15.8

Acetone vapor in a nitrogen stream is removed by adsorption in a fixed bed of activated carbon. At a location in the bed where the pressure is 136 kPa, the bulk gas temperature is 297 K, and the bulk mole fraction of acetone is 0.05, estimate the external gas-to-particle mass-transfer coefficient for acetone and the external particle-to-gas heat-transfer coefficient. Additional data are as follows:

Average particle diameter = 0.0040 m and Gas superficial molar velocity
= 0.00352 kmol/m²-s

SOLUTION

Because the temperature and composition are known only for the bulk gas and not at the particle external surface, use gas properties at bulk gas conditions. From the ChemCAD process simulation program, relevant properties for use in (15-65) and (15-66) are as follows:

Viscosity $\mu = 0.0000165$ Pa-s (kg/m-s);   Density $\rho = 1.627$ kg/m³
Thermal conductivity $k = 0.0240$ W/m-K = 0.024×10^{-3} kJ/m-K-s
Heat capacity at constant pressure = 31.45 kJ/kmol-K
Molecular weight = $M = 29.52$
Thus, specific heat $C_P = 31.45/29.52 = 1.065$ kJ/kg-K

Other parameters are

Gas mass velocity $G = 0.00352(29.52) = 0.1039$ kg/m²-s
Assume a sphericity, ψ, of 0.65; therefore, $D_p = 0.65(0.004)$
$= 0.0026$ m

The diffusivity, D_i, of acetone in nitrogen at 297 K and 136 kPa is independent of the composition and is approximately 0.085×10^{-4} m²/s.

$N_{Re} = D_p G/\mu = 0.0026(0.1039)/(0.0000165) = 16.4$
$N_{Sc} = \mu/\rho D_i = 0.0000165/(1.627)(0.0000085) = 1.19$
$N_{Pr} = C_P \mu/k = (1.065)(0.0000165)/(0.000024) = 0.73$

From (15-65):

$$N_{Sh} = 2 + 1.1(16.4)^{0.6}(1.19)^{1/3} = 8.24$$

which from Figure 15.17 is well within the data range of the correlation. Thus, the mass-transfer coefficient for acetone is

$k_{c_i} = N_{Sh}(D_i/D_p) = 8.24(0.0000085/0.0026)$
$= 0.027$ m/s $= 0.088$ ft/s

From (15-66):

$N_{Nu} = 2 + 1.1(16.4)^{0.6}(0.73)^{1/3} = 7.31$
$h = N_{Nu}(k/D_p) = 7.31(0.0240/0.0026)$
$= 67.5$ W/m²-K $\quad$ or 11.9 Btu/h-ft²-°F

Internal Transport

Porous-adsorbent particles have a sufficiently high, effective thermal conductivity that temperature gradients within the particle are usually negligible. However, internal (intraphase) mass transfer in the particle must be considered. Mechanisms for mass transfer in the pores are those described for porous membranes in Section 14.3. However, in membranes, the transport is through the membrane. In sorption applications, transport is only into the interior of the particle during sorption and from the interior of the particle in desorption.

The mathematical model of internal transport in porous particles during adsorption or desorption is very similar to that for catalytic chemical reactions in porous catalyst pellets. The first pore model was that of Thiele [56], who

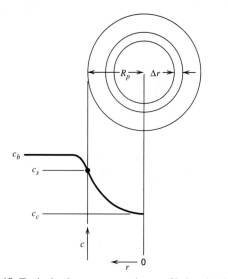

Figure 15.18 Typical solute concentration profile in adsorbent particle.

considered a first-order irreversible reaction taking place isothermally on the surface of a single, straight, cylindrical pore, closed at one end. Thiele's treatment was extended to a porous spherical pellet by Wheeler [57], who utilized an effective diffusivity applied to the case of sorption.

Consider the porous spherical pellet in Figure 15.18, where the fluid concentration, c, refers to the solute. A material balance in moles or mass per unit time over the spherical-shell volume of thickness Δr involves diffusion of the solute into the shell at radius $r + \Delta r$, adsorption within the shell, and diffusion out of the shell at radius r. Using Fick's first law:

$$4\pi(r + \Delta r)^2 D_e \frac{\partial c}{\partial r}\bigg|_{r+\Delta r} = 4\pi r^2 \Delta r \frac{\partial q}{\partial t} + 4\pi r^2 D_e \frac{\partial c}{\partial r}\bigg|_r$$
(15-67)

Dividing by $4\pi \Delta r$, taking the limit as $\Delta r \to 0$, and collecting terms gives

$$D_e \left(\frac{\partial^2 c}{\partial r^2} + \frac{2}{r} \frac{\partial c}{\partial r} \right) = \frac{\partial q}{\partial t}$$
(15-68)

The variable q is the amount adsorbed per unit volume of porous pellet. The effective diffusivity, D_e, applies to the entire surface area of the spherical shell, even though only about 50% of it is available as pores for diffusion. For liquid-phase diffusion in the pores, the effective diffusivity is given by (14-14), which involves the volume fraction of pores in the pellet, the solute molecular diffusivity in the fluid within the pore, the pore tortuosity, and a possible restrictive factor for relatively large solute molecules. For gas-phase diffusion in the pores, the effective diffusivity is given by (14-18), which accounts for the possibility of Knudsen diffusion, with diffusivity D_K for very small pore diameters and/or low total pressures. Although (14-14) and (14-18) strictly apply only to equimolar counterdiffusion, they can be used as an approximation for unimolecular diffusion for fluids dilute in the solute molecules. A diffusion mechanism not accounted for directly in (14-18) is that of surface diffusion along the pore wall due to the concentration gradient of the adsorbate (adsorbed solute) along the wall.

Fick's first law for molecular diffusion through a fluid in a pore can be written

$$n_i = -D_i A(dc_i/dx)$$
(15-69)

where n is the molar rate of ordinary diffusion of i through the fluid in the x-direction, perpendicular to the cross-sectional area, A, for diffusivity, D_i, and concentration, c_i, in moles/unit volume of fluid. A modified Fick's first law applies to surface diffusion, as suggested by Schneider and Smith [58]. Thus,

$$(n_i)_s = -(D_i)_s b d(c_i)_s/dx$$
(15-70)

where b is the perimeter of the surface, $(c_i)_s$ is the surface concentration of adsorbate in moles/unit surface area, and $(D_i)_s$ is the surface diffusivity as defined by (15-70).

For convenience, (15-70) is converted, as follows, to the flux form of (15-69) so that the two mechanisms of diffusion can be combined in a single transport rate equation. The flux form of (15-69) is

$$N_i = n_i/A = -D_i(dc_i/dx) \qquad (15\text{-}71)$$

The corresponding flux form of (15-70) is obtained by dividing both sides by the cross-sectional area of the pore and converting the surface concentration, $(c_i)_s$, in moles/unit surface area to the concentration, q, in mol/g of adsorbent, by using the product of the pore surface/pore volume times the reciprocal of the adsorbent particle density times the particle porosity. The result is

$$(N_i)_s = -(D_i)_s \frac{\rho_p}{\epsilon_p}\left(\frac{dq_i}{dx}\right) \qquad (15\text{-}72)$$

Assuming linear adsorption according to Henry's law:

$$q_i = K_i c_i \qquad (15\text{-}73)$$

Substituting (15-73) into (15-72) and adding the result to (15-71), the total flux is

$$N_i = -\left[D_i + (D_i)_s \frac{\rho_p K_i}{\epsilon_p}\right]\frac{dc_i}{dx} \qquad (15\text{-}74)$$

In terms of the effective diffusivity employed in (15-68):

$$D_e = \frac{\epsilon_p}{\tau}\left\{\left[\frac{1}{(1/D_i)+(1/D_K)}\right] + (D_i)_s \frac{\rho_p K_i}{\epsilon_p}\right\} \qquad (15\text{-}75)$$

Equation (15-75) needs to be used with caution, because, as discussed by Riekert [59], the tortuosity, τ, for pore-volume diffusion is not necessarily the same as that for surface diffusion.

Based on the study by Sladek, Gilliland, and Baddour [60], values of the surface diffusivity of light gases for physical adsorption are typically in the range of 5×10^{-3} to 10^{-6} cm²/s, with the larger values applying to cases of a low differential heat of adsorption. For nonpolar adsorbates, the surface diffusivity in cm²/s may be estimated from the correlation [60],

$$D_s = 1.6 \times 10^{-2}\ \exp[-0.45(-\Delta H_{ads})/mRT] \qquad (15\text{-}76)$$

where $m = 2$ for conducting adsorbents such as carbon and $m = 1$ for insulating adsorbents.

EXAMPLE 15.9

Porous silica gel of 1.0 mm particle diameter, with a particle density of 1.13 g/cm³, a porosity of 0.486, an average pore radius of 11 Å, and a tortuosity of 3.35 is to be used to adsorb propane from helium. At 100°C, diffusion in the pores is controlled by Knudsen and surface diffusion. Estimate the effective diffusivity. The differential heat of adsorption is $-5,900$ cal/mol. At 100°C, the adsorption constant (for a linear isotherm) is 19 cm³/g.

SOLUTION

From (14-21), the Knudsen diffusivity for propane is $D_K = 4,850$ $(22 \times 10^{-8})(373/44.06)^{1/2} = 3.7 \times 10^{-3}$ cm²/s. From (15-76), using $m = 1$, $D_s = 1.6 \times 10^{-2}\ \exp\{(-0.45)(5,900)/[(1)(1.987)(373)]\} = 4.45 \times 10^{-4}$ cm²/s.

Equation (15-75) reduces to $D_e = \frac{\epsilon_p}{\tau}D_K + \frac{\rho_p K}{\tau}D_s = \frac{0.486}{3.35}(3.17\times 10^{-3}) + \frac{(1.13)(19)}{3.35}(4.45 \times 10^{-4}) = 0.46 \times 10^{-3} + 2.85 \times 10^{-3} = 3.31 \times 10^{-3}$ cm²/s.

Experiments by Schneider and Smith [58] give a value of 1.22×10^{-3} cm²/s for D_e with a value of 0.88×10^{-3} for the contribution of surface diffusion. Thus, the estimated contribution from surface diffusion is high by a factor of about 3. In either case, the fractional contribution due to surface diffusion is large. A detailed review of surface diffusion is given by Kapoor, Yang, and Wong [61].

Mass Transfer in Ion Exchange and Chromatography

As discussed by Helfferich [62], two major mass-transfer resistances occur in ion exchange. The first is the external mass-transfer resistance due to the film or boundary layer surrounding the ion-exchange bead. The second is the internal diffusional resistance due to the resin bead. Either or both resistances can be rate-controlling; in either case, the diameter of the resin bead is an important factor. In general, external mass-transfer film diffusion is rate-controlling at very low exchange-ion concentrations, say below 0.01 N, whereas internal mass transfer (particle diffusion) controls at high concentrations (say above 1.0 N). It has also been observed that a large separation factor, as defined by (15-46), favors external mass-transfer control, and that divalent ions diffuse appreciably more slowly than monovalent ions through the resin bead. Usually, the rate-determining step is not the chemical reaction between the exchanging ions and the resin.

The external mass-transfer coefficient for flow of fluid through a fixed bed of ion-exchange resin beads is estimated from the same relation, (15-65), that is used for applications to adsorption in fixed beds. For internal mass transfer, it is customary to assume that the ion-exchange resin bead is a single quasi-homogeneous phase and that the diffusivity of the diffusing ion is constant at a given temperature. Under these conditions, (15-68) can be applied, where D_e is a diffusivity determined by experiment. In general, such diffusivities depend upon (1) the size and charge of the ion, with the smaller, monovalent ions diffusing the fastest; (2) the degree of cross-linking and resin swelling, with larger diffusivities favored by swelling and a small degree of cross-linking; and (3) temperature.

The most fundamental measurements of diffusivity in ion-exchange resins have been made with isotopes of the ions to obtain self-diffusion coefficients that are independent of ion concentration. Typical data are those of Soldano [63], shown in Figure 15.19 for Na^+, Zn^{2+}, and Y^{3+} in a sulfonated styrene–divinylbenzene cation exchanger at temperatures of 0.2 and 25°C. Recall that typical order-of-magnitude diffusivities for small molecules are as follows:

0.1 cm²/s in the gas phase

1×10^{-5} cm²/s in the liquid phase

1×10^{-7} cm²/s in polymers

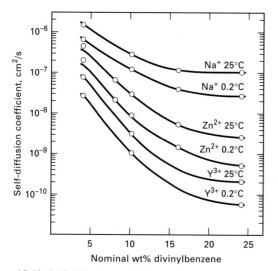

Figure 15.19 Self-diffusion coefficients for cations in a resin as a function of cross-linking with divinylbenzene.
[From B.A. Soldano, *Ann. NY Acad. Sci.,* **57**, 116 (1953) with permission.]

From Figure 15.19, it is seen that diffusivities depend strongly on the degree of cross-linking and the charge on the ion, with values much less than those found in liquids, especially for the divalent and trivalent ions, which have diffusivities even smaller than those observed for small molecules in polymers.

No new fundamental principles are required for formulating mass-transfer relations for chromatography. When packed beds are used, (15-65) and (15-66) are applied to determine external transport coefficients. If a coated flat plate or a tube with a coated inner wall is used, correlations of the type discussed in Chapter 3 are applicable. In some cases, an entry region of finite length exists, particularly for laminar flow, such that the transport coefficients vary with axial location, decrease in value with length, and eventually approach an asymptotic value. For internal diffusion in the sorbent, Fick's second law is applied where the effective diffusivity depends on factors discussed earlier in this section.

15.4 SORPTION SYSTEMS

A variety of equipment configurations and operating procedures are employed for commercial, sorption-separation operations. This variety is due mainly to the wide range of sorbent particle sizes used and the need, in most applications, to regenerate the solid sorbent.

Adsorption

For adsorption, the most widely used equipment configurations and operating procedures are listed in Table 15.7. For analysis purposes, the listed devices may be classified into the three modes of operations, shown schematically in Figure 15.20. An agitated vessel, shown in Figure 15.20a, is used with a batch of liquid to which is added a powdered adsorbent such as activated carbon, of particle diameter typically less than 1 mm, to form a slurry. With good agitation and small particles, the external resistance to mass transfer from the bulk liquid to the external surface of the adsorbent particles is small. For small adsorbent particles, the internal resistance to mass transfer within the pores of the particles is also small. Accordingly, the rate of adsorption is rapid. The required residence time of the slurry in a well-mixed agitated vessel is determined by how fast equilibrium is approached. The main application of this mode of operation is the removal of very small amounts of dissolved, and relatively large molecules, such as coloring agents, from water. Generally, the spent adsorbent, which is removed from the slurry by sedimentation or filtration, is discarded because of the difficulty of desorbing large molecules. The slurry adsorption system, also called *contact filtration,* is also operated continuously.

The cyclic-batch operating mode using a fixed bed, shown schematically in Figure 15.20b, is widely used with both liquid and gas feeds. Adsorbent particle size ranges from 0.05 to 1.2 cm. Bed pressure drop decreases with increasing particle size, but the solute transport rate increases with decreasing particle size. The optimal particle size is determined mainly from these two factors. To avoid jiggling

Table 15.7 Common Commercial Methods for Adsorption Separations

Phase Condition of Feed	Contacting Device	Adsorbent Regeneration Method	Main Application
Liquid	Slurry in an agitated vessel	Adsorbent discarded	Purification
Liquid	Fixed bed	Thermal reactivation	Purification
Liquid	Simulated moving bed	Displacement purge	Bulk separation
Gas	Fixed bed	Thermal swing (TSA)	Purification
Gas	Combined fluidized bed-moving bed	Thermal swing (TSA)	Purification
Gas	Fixed bed	Inert-purge swing	Purification
Gas	Fixed bed	Pressure swing (PSA)	Bulk separation
Gas	Fixed bed	Vacuum swing (VSA)	Bulk separation
Gas	Fixed bed	Displacement purge	Bulk separation

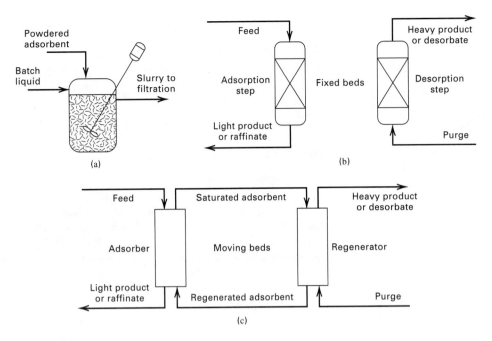

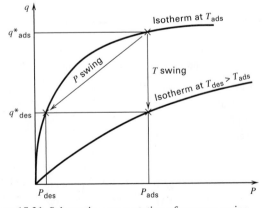

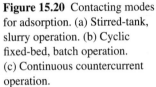

Figure 15.20 Contacting modes for adsorption. (a) Stirred-tank, slurry operation. (b) Cyclic fixed-bed, batch operation. (c) Continuous countercurrent operation.

or fluidizing the bed during adsorption, the flow of the liquid or gas feed is often downward. For removal of small amounts of dissolved hydrocarbons from water, the spent adsorbent is removed from the vessel and reactivated thermally at high temperature or it is discarded. Applications of fixed-bed adsorption, also called *percolation,* include the removal of dissolved organic compounds from water. For purification or bulk separation of gases, the adsorbent is almost always regenerated in place by one of the five methods listed in Table 15.7.

In the *thermal (temperature)-swing-adsorption* (TSA) method, the adsorbent is regenerated by desorption at a temperature higher than that used during the adsorption step of the cycle, as shown in Figure 15.21. The temperature of the bed is increased by (1) heat transfer from heating coils located in the bed followed by pulling a moderate vacuum or (2) more commonly, by heat transfer from an inert, nonadsorbing, hot purge gas, such as steam. Following desorption,

the bed is cooled before the adsorption step of the cycle is resumed. Because heating and cooling of the bed requires hours, a typical cycle time for TSA is hours to days. Therefore, if the quantity of adsorbent in the bed is to be reasonable, TSA is practical only for purification involving small rates of adsorption. Instead of using a fixed bed, a fluidized bed can be used for adsorption and a moving bed for desorption, as shown in Figure 15.22, provided that the adsorbent particles are attrition-resistant. In the adsorption section, sieve trays are used with the raw gas passing up through the perforations and fluidizing the adsorbent particles. The fluidized solids flow like a liquid across the tray, into the downcomer, and onto the tray below. From the adsorption section, the solids pass to the desorption section, where, as moving beds, they first flow down through preheating tubes and then through desorption tubes. Steam is used for indirect heating in both sets of tubes and for stripping in the desorption tubes. Moving beds, rather than fluidized beds on trays, are used in the desorption section because the stripping-steam flow rate is insufficient for fluidizing the solids. At the bottom of the unit, the regenerated solids are picked up by a carrier gas, which flows up through a gas-lift line to the top, where the solids settle out onto the top tray to repeat the adsorption part of the cycle. According to Keller [64], this configuration, which was announced in 1977, is used in more than 50 units worldwide to remove small amounts of solvents from air. Other applications of TSA include the removal of moisture, CO_2, and pollutants from gas streams.

In the *inert-purge-swing* method of regeneration, desorption is at the same temperature and pressure as the adsorption step, because the gas used for purging is nonadsorbing (inert) or only weakly adsorbing. This method is used only when the solute is weakly adsorbed, easily desorbed, and of little or no value. The purge gas must be inexpensive so that it does not have to be purified before recycle.

Figure 15.21 Schematic representation of pressure-swing and thermal-swing adsorption.

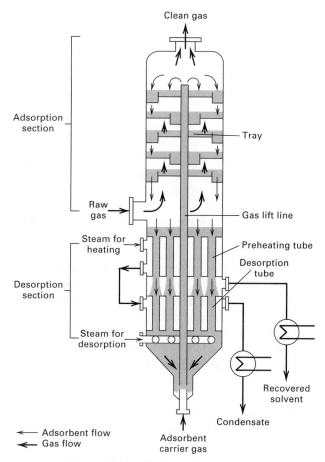

Clean gas

Adsorption section

Tray

Raw gas

Gas lift line

Steam for heating

Preheating tube

Desorption tube

Desorption section

Steam for desorption

Recovered solvent

Condensate

Adsorbent carrier gas

← Adsorbent flow
← Gas flow

Figure 15.22 Purasiv process with a fluidized bed for adsorption and moving bed for desorption.

[From G.E. Keller, "Separations: New Directions for an Old Field," *AIChE Monograph Series,* **83** (17) (1987) with permission.]

In the *pressure-swing-adsorption* (PSA) cycle, adsorption takes place at an elevated pressure, whereas desorption occurs at near-ambient pressure, as is shown in Figure 15.21. PSA is used for bulk separations because the bed can be depressurized and repressurized rapidly, making it possible to operate at cycle times of seconds to minutes. Because of these short times, the beds need not be large even when a substantial fraction of the feed gas is adsorbed. If adsorption takes place at near-ambient pressure and desorption under vacuum, the cycle is referred to as *vacuum-swing-adsorption* (VSA). PSA and VSA are widely used for the bulk separation of air. If a zeolite adsorbent is used, equilibrium is rapidly established and nitrogen is preferentially adsorbed. The nonadsorbed, high-pressure product gas is a mixture of oxygen and argon with a small amount of nitrogen. If a carbon molecular-sieve adsorbent is used, the particle diffusivity of oxygen is observed to be about 25 times that of nitrogen. As a result, the selectivity of adsorption is controlled by mass transfer, and oxygen is preferentially adsorbed. The resulting high-pressure product gas is nearly pure nitrogen. In both cases, the adsorbed gas, which is desorbed at low pressure, is quite impure. For the separation of air, large plants

use VSA because it is more energy-efficient than PSA. Small plants often use PSA because that cycle is simpler.

In the *displacement-purge (displacement desorption) cycle,* a strongly adsorbed purge gas is used in the desorption step to displace the adsorbed species. Another step is required to recover the purge gas. The displacement-purge cycle is considered only where TSA, PSA, and VSA cannot be used because of pressure or temperature limitations. One application is the separation of medium-molecular-weight linear paraffins (C_{10}-C_{18}) from mixtures of branched-chain and cyclic hydrocarbons by adsorption on 5A zeolite. Ammonia, which is easily separated from the paraffins by flash vaporization, is used as the purge.

Most commercial applications of adsorption involve fixed beds that cycle between adsorption and desorption. Thus, compositions, temperature, and/or pressure at a given location in the bed vary with time. Alternatively, a continuous, countercurrent operation, where such variations do not occur, can be envisaged, as shown in Figure 15.20c and discussed in detail by Ruthven and Ching [65]. The main difficulty with such a scheme is the need to circulate the solid adsorbent, as a moving bed, to achieve a steady-state operation. The first commercial application of countercurrent adsorption and desorption was the moving-bed Hypersorption process for the recovery, by adsorption on activated carbon, of light hydrocarbons from various gas streams in petroleum refineries, as discussed by Berg [66]. Only a few units were installed because of problems with attrition of the adsorbent, difficulties in regenerating the adsorbent when heavier hydrocarbons were present in the feed gas, and unfavorable economics compared to those of distillation. Newer adsorbents with a much higher resistance to attrition and possible applications to more difficult separations are reviving interest in moving-bed units.

A successful alternative countercurrent system for commercial application to the separation of liquid mixtures is the simulated moving-bed system, shown in a hybrid system with two distillation columns in Figure 15.23 and known generally as the UOP Sorbex process. As described by Broughton [67], the bed is held stationary in one column, which is equipped with a number (perhaps 12) of liquid feed entry and discharge locations. By shifting, with a rotary valve (RV), the locations of feed entry, desorbent entry, extract (adsorbed) removal, and raffinate (non-adsorbed) removal, a countercurrent movement of solids is simulated by a downward movement of liquid. For the valve positions shown in Figure 15.23, locations 2 (entering desorbent), 5 (exiting extract), 9 (entering feed), and 12 (exiting raffinate) are operational, with all other numbered lines closed. However, liquid is also circulated down through and back up (external to the column) to the top of the column by a pump. Ideally, an infinite number of entry and exit locations on the column would exist and the valve would continuously change the four operational locations. Since this is impractical, a finite number of locations are used and valve changes are made periodically. In Figure 15.23, when the valve is

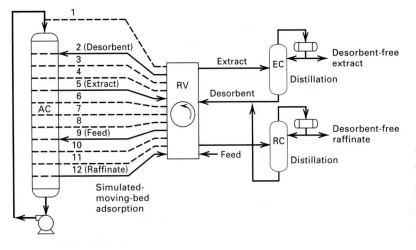

Figure 15.23 Sorbex hybrid simulated moving-bed process for bulk separation. AC, adsorbent chamber; RV, rotary valve; EC, extract column; RC, raffinate column.

[From D.B. Broughton, *Chem. Eng., Progress,* **64** (8), 60–65 (1968) with permission.]

moved to the next position, Lines 3, 6, 10, and 1 become operational. Thus, raffinate removal is relocated from the bottom of the bed to the top of the bed. The result is that the bed has no top or bottom. As discussed by Gembicki et al. [68], 78 Sorbex-type commercial units were installed during 1962–1989 for the bulk separation of p-xylene from C_8 aromatics; n-paraffins from branched and cyclic hydrocarbons; olefins from paraffins; p- or m-cymene (or cresol) from cymene (or cresol) isomers; and fructose from dextrose and polysaccharides. Humphrey and Keller [101] cite 100 commercial installations of Sorbex-type units and more than 50 different demonstrated separations.

Ion Exchange

Ion exchange employs the same modes of operation as shown for adsorption in Figure 15.20. Although the use of fixed beds in a cyclic operation is most common, stirred tanks are used for batch contacting, with an attached strainer or filter to separate the resin beads from the solution after

equilibrium conditions are approached. Agitation is mild to avoid resin attrition, but sufficient to achieve complete suspension of the resin. To increase resin utilization and achieve high ion-exchange reaction efficiency, much effort has been expended in the development of continuous, countercurrent contactors, two of which are shown in Figure 15.24. The Higgins contactor [69] operates as a moving, packed bed by using intermittent hydraulic pulses to move incremental portions of the bed from the contacting section, where ion exchange takes place, up, around, and down to the backwash region, down to the regenerating section, and back up through the rinse section to the contacting section to repeat the cycle. Liquid moves countercurrently to the resin. The Himsley contactor [70] has a series of trays, on each of which the resin beads are fluidized by the upward flow of liquid. Periodically, the flow is reversed to move incremental amounts of resin from one stage to the stage below. The batch of resin at the bottom is lifted to the wash column, then to the regeneration column, and then back to the top of the ion-exchange column for reuse.

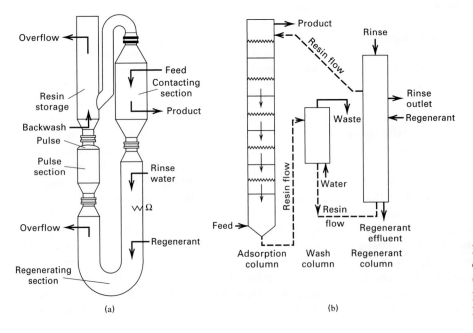

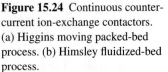

Figure 15.24 Continuous countercurrent ion-exchange contactors. (a) Higgins moving packed-bed process. (b) Himsley fluidized-bed process.

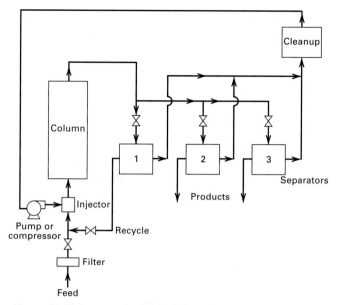

Figure 15.25 Large-scale, batch elution chromatography process.

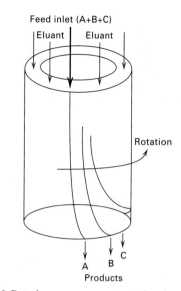

Figure 15.26 Rotating, cross-current, annular chromatograph.

Chromatography

Operation modes for large-scale, commercial application of chromatography are of two major types, as discussed in a book edited by Ganetsos and Barker [71]. The first, and the most common, is a transient mode that is a scaled-up version of an analytical chromatograph, referred to as large-scale, batch (or elution) chromatography. Packed columns of diameter up to 4.6 m and packed heights to 12 m have been reported. As shown in Figure 15.25 and discussed by Wankat in Chapter 14 of the *Handbook* edited by Rousseau [9], a recycled solvent or carrier gas is fed continuously into the sorbent-packed column. The feed mixture and recycle is pulsed into the column by an injector. A timer or detector (not shown) splits the effluent from the column, sending it to different separators (condensers, evaporators, distillation columns, etc.). Each separator is designed to remove a particular feed component from the carrier fluid. An additional cleanup step is required to purify the carrier fluid before it is recycled to the column. Separator one produces no product because it handles an effluent pulse that contains the carrier fluid and two or more of the feed components, which are recovered and recycled to the column. Thus, if properly designed and operated, the batch chromatograph operates somewhat like a batch-distillation column, producing a nearly pure cut for each component in the feed and slop cuts for recycle. The system shown in Figure 15.25 is designed to separate a binary system. If, say, three more separators are added, the system can separate a five-component feed into five nearly pure products.

The second major type of large-scale chromatograph is the countercurrent flow or simulated-moving-bed mode already discussed for adsorption. This mode is more efficient, but is more complicated and can only separate a mixture into two products. A third mode is the continuous, cross-current (or rotating) chromatograph, first conceived by Martin [72] and shown schematically in Figure 15.26. The packed annular bed rotates slowly about its axis, past the feed-inlet point. *Eluant* (solvent or carrier gas) enters the top of the bed uniformly over the entire cross-sectional area. Both feed and eluant are fed continuously and are carried downward and around by the rotation of the bed. Because of the different selectivities of the feed components for the sorbent, each component traces a different helical path since each spends a different amount of time in contact with the sorbent. Thus, each component is eluted from the bottom of the packed annulus at a different location. In principle, a multicomponent feed can be separated continuously into nearly pure components following separation of the carrier fluid from each eluted fraction. Units of up to 12 in. in diameter have successfully separated sugars, proteins, and metallic elements.

Slurry Adsorption (Contact Filtration)

Three modes of adsorption from a liquid in an agitated vessel are of interest. The first is the *batch mode* in which a batch of liquid is contacted with a batch of adsorbent for a period of time, followed by discharge of the slurry from the vessel, and filtration to separate the solids from the liquid. The second is the *continuous mode,* in which liquid and adsorbent are continuously added to and removed from the agitated vessel. In the third mode, called the *semibatch* or *semicontinuous mode,* the liquid is continuously fed to and removed from the agitated vessel, where it is contacted with the adsorbent, which is retained in a contacting zone of the vessel until it is nearly spent. Models for each of these three modes are developed next, followed by examples of their application. In all models, the slurry is assumed to be perfectly mixed by agitation in the turbulent regime to produce a fluidized bed of sorbent. Perfect mixing is approached by using a liquid depth of from one to two vessel diameters, four vertical wall baffles, and one or two marine propellers or pitched-blade turbines on a vertical shaft. With a proper impeller rotation rate, the axial flow achieves complete suspension. For semicontinuous operation, a clear liquid region is maintained above the suspension region for liquid withdrawal.

Because small particles are used in slurry adsorption and because the relative velocity between the particles and the liquid in an agitated slurry is low (small particles tend to move with the liquid), the rate of adsorption is assumed to be controlled by external, rather than internal, mass transfer.

Batch Mode

The rate of adsorption of solute, as controlled by external mass transfer, is

$$-\frac{dc}{dt} = k_L a(c - c^*) \tag{15-77}$$

where c is the concentration of solute in the bulk liquid; c^* is the concentration in equilibrium with the loading on the adsorbent, q; k_L is the external liquid-phase mass-transfer coefficient; and a is the external surface area of the adsorbent per unit volume of liquid. Starting from feed concentration, c_F, the instantaneous bulk concentration, c, at time t, is related to the instantaneous adsorbent loading, q, by material balance:

$$c_F Q = cQ + qS \tag{15-78}$$

where the adsorbent is assumed to be initially free of adsorbate, Q is the liquid volume (assumed to remain constant for dilute feeds), and S is the mass of adsorbent. The equilibrium concentration, c^*, is given by an appropriate adsorption isotherm: a linear isotherm, the Langmuir isotherm (15-36), or the Freundlich isotherm (15-35). For example, a rearrangement of the latter gives

$$c^* = (q/k)^n \tag{15-79}$$

To solve the system of equations for c and q as a function of time, starting from c_F at $t = 0$, (15-78) is combined with the equilibrium isotherm, for example, (15-79), to eliminate q. The resulting equation is combined with (15-77) to eliminate c^* to give an ODE for c in t, which is integrated analytically or numerically. Corresponding values of q are then obtained from (15-78).

If the equilibrium is represented by a linear isotherm,

$$c^* = q/k \tag{15-80}$$

an analytical integration gives

$$c = \frac{c_F}{\beta}[\exp(-k_L a\beta t) + \alpha] \tag{15-81}$$

where

$$\beta = 1 + \frac{Q}{Sk} \tag{15-82}$$

$$\alpha = \frac{Q}{Sk} \tag{15-83}$$

As the contact time approaches infinity, adsorption equilibrium is approached and for the linear isotherm, from (15-81) or combining (15-78), with $c = c^*$, and (15-80):

$$c\{t = \infty\} = c_F \alpha/\beta \tag{15-84}$$

Continuous Mode

When both liquid and solids flow continuously through a perfectly mixed vessel, (15-77) is converted to an algebraic equation because, as in a perfectly mixed reaction vessel (CSTR), the concentration, c, throughout the vessel, is equal to the exit (outlet) concentration, c_{out}. Thus, in terms of the residence time in the vessel, t_{res}:

$$\frac{c_F - c_{out}}{t_{res}} = k_L a(c_{out} - c^*) \tag{15-85}$$

or, rearranging:

$$c_{out} = \frac{c_F + k_L a t_{res} c^*}{1 + k_L a t_{res}} \tag{15-86}$$

Equation (15-78) becomes

$$c_F Q = c_{out} Q + q_{out} S \tag{15-87}$$

where Q and S are now flow rates. An appropriate adsorption isotherm relates c^* to q_{out}.

For a linear isotherm, (15-80) becomes $c^* = q_{out}/k$, which when combined with (15-87) and (15-86) to eliminate c^* and q_{out}, gives

$$c_{out} = c_F\left(\frac{1 + \gamma\alpha}{1 + \gamma + \gamma\alpha}\right) \tag{15-88}$$

where α is given by (15-83) and

$$\gamma = k_L a t_{res} \tag{15-89}$$

The corresponding q_{out} is given by a rearrangement of (15-87):

$$q_{out} = \frac{Q(c_F - c_{out})}{S} \tag{15-90}$$

For a nonlinear adsorption isotherm, such as (15-35) or (15-36), (15-85) and (15-87) are combined with the isotherm equation, but it may not be possible to express the result explicitly in q_{out}. In that event, a numerical solution is required, as illustrated below in Example 15.10.

Semicontinuous Mode

The most difficult mode to model is the semicontinuous mode, where the adsorbent is retained in the vessel, but the feed liquid enters and exits the vessel at a fixed, continuous flow rate. Both concentration, c, and loading, q, vary with time. With perfect mixing, the outlet concentration is given by (15-86), where t_{res} is the residence time of the liquid in the suspension, and c^* is related to q in the suspension by an appropriate adsorption isotherm. The variation of q in the batch of solids is given by (15-77), rewritten in terms of the change in q, rather than c:

$$S\frac{dq}{dt} = k_L a(c_{out} - c^*)t_{res}Q \qquad (15\text{-}91)$$

where, for this mode, S is the batch mass of adsorbent in the suspension and Q is the steady, volumetric-liquid flow rate.

Both (15-91) and (15-86) involve c^*, which can be replaced by a function of instantaneous q by selecting an appropriate isotherm. The resulting two equations are then combined to eliminate c_{out}. The resulting ODE is then integrated analytically or numerically to obtain q as a function of time, from which c_{out} as a function of time can be determined from (15-86) and the isotherm. The time-average value of c_{out} is then obtained by integration of c_{out} with respect to time. These steps are illustrated in the following example. For a linear isotherm, the derivation is left as an exercise.

EXAMPLE 15.10

An aqueous solution containing 0.010 mol phenol/L is to be treated at 20°C with activated carbon to reduce the concentration of phenol to 0.00057 mol/L. From Example 4.12, the adsorption equilibrium data are well fitted to the Freundlich equation:

$$q = 2.16c^{1/4.35} \qquad (1)$$

or

$$c^* = (q/2.16)^{4.35} \qquad (2)$$

where q and c are in mmol/g and mmol/L, respectively. In terms of kmol/kg and kmol/m^3, (2) becomes

$$c^* = (q/0.01057)^{4.35} \qquad (3)$$

All three modes of slurry adsorption are to be considered. From Example 4.12, the minimum amount of adsorbent is 5 g/L of solution. Laboratory experiments with adsorbent particles 1.5 mm in diameter in a well-agitated vessel have confirmed that the rate of adsorption is controlled by external mass transfer with $k_L = k_c = 5 \times 10^{-5}$ m/s. Particle surface area is 5 m^2/kg of particles.

(a) Using twice the minimum amount of adsorbent in an agitated vessel operated in the batch mode, determine the time in minutes to reduce the phenol content to the desired value.

(b) For operation in the continuous mode with twice the minimum amount of adsorbent, determine the required residence time in minutes in the agitated vessel. How does this compare to the batch time of part (a)?

(c) For operation in the semicontinuous mode with 1,000 kg of activated carbon, a liquid feed rate of 10 m^3/h, and a liquid residence time equal to 1.5 times the value computed in part (b), determine the run time to obtain a composite liquid product with the desired phenol concentration. Are the results reasonable, or should changes be made to the specifications?

SOLUTION

(a) Batch mode:

$$S/Q = 2(5) = 10 \text{ g/L} = 10 \text{ kg/m}^3$$
$$k_L a = 5 \times 10^{-5}(5)(10) = 2.5 \times 10^{-3}\text{s}^{-1}$$
$$c_F = 0.010 \text{ mol/L} = 0.010 \text{ kmol/m}^3$$

From (15-78),

$$q = \frac{c_F - c}{S/Q} = \frac{0.010 - c}{10} \qquad (4)$$

Substituting (4) into (3),

$$c^* = \left(\frac{0.10 - c}{0.1057}\right)^{4.35} \qquad (5)$$

Substituting (5) into (15-77),

$$-\frac{dc}{dt} = 2.5 \times 10^{-3}\left[c - \left(\frac{0.010 - c}{0.1057}\right)^{4.35}\right] \qquad (6)$$

where, $c = c_F = 0.010 \text{ kmol/m}^3$ at $t = 0$ and we want t for $c = 0.00057 \text{ kmol/m}^3$. By numerical integration of (6), $t = 1{,}140 \text{ s} = 19 \text{ min}$.

(b) Continuous mode:

Equation (15-85) applies, where all quantities are the same as those determined in part (a) and $c_{out} = 0.00057 \text{ kmol/m}^3$. Thus,

$$t_{res} = \frac{c_F - c_{out}}{k_L a(c_{out} - c^*)}$$

where c^* is given by (3) with $q = q_{out}$, and q_{out} is obtained from (15-87). Thus,

$$t_{res} = \frac{0.010 - 0.00057}{2.5 \times 10^{-3}\left[0.00057 - \left(\frac{0.010 - 0.00057}{0.1057}\right)^{4.35}\right]}$$

$$= 6{,}950 \text{ s or } 1.93 \text{ h}$$

This residence time is appreciably longer than the batch time of 1,140 s. In the batch mode, the concentration driving force for external mass transfer is initially $(c - c^*) = c_F = 0.010 \text{ kmol/m}^3$ and gradually declines to a much smaller final value, at 1,140 s, of

$$(c - c^*) = c_{final} - \left(\frac{0.010 - c_{final}}{0.1057}\right)^{4.35}$$

$$= 0.000543 \text{ kmol/m}^3$$

For the continuous mode with perfect mixing in the vessel, the concentration driving force for external mass transfer is always at the final batch value of 0.000543 kmol/m^3.

(c) Semicontinuous mode:

Equation (15-91) applies with

$$S = 1{,}000 \text{ kg}, \quad c_F = 0.010 \text{ kmol/m}^3$$
$$Q = 10 \text{ m}^3/\text{h}, \quad t_{res} = 10{,}425 \text{ s}, \quad k_L a = 2.5 \times 10^{-3} \text{ s}^{-1}$$

c^* is given in terms of q by (3) and c_{out} is given by (15-86).

Combining (15-91), (3), and (15-86) to eliminate c^* and c_{out} gives, after simplification,

$$\frac{dq}{dt} = \left(\frac{\gamma}{1+\gamma}\right)\frac{Q}{S}\left[c_F - \left(\frac{q}{0.01057}\right)^{4.35}\right] \qquad (7)$$

where γ is given by (15-89) and the time, t, is the time that the adsorbent remains in the vessel. For values of γ, Q/S, and c_F equal, respectively, to 26.06, 0.01 m^3/h-kg, and 0.010 kmol/m^3, (7) reduces to

$$\frac{dq}{dt} = 0.00963\left[0.010 - \left(\frac{q}{0.01057}\right)^{4.35}\right] \qquad (8)$$

where t is in hours and q is in kmol. By numerical integration of (8), starting from $q = 0$ at $t = 0$, we obtain q as a function of t as given in Table 15.8. Included are corresponding values of

Table 15.8 Results for Part (c), Semicontinuous Mode, of Example 15.10

Time t, h	q, kmol/kg	kmol/m^3	
		c_{out}	c_{cum}
0.0	0.0	0.000370	0.000370
5.0	0.000481	0.000371	0.000370
10.0	0.000962	0.000398	0.000375
15.0	0.001440	0.000535	0.000401
15.7	0.001506	0.000570	0.000407
20.0	0.001905	0.000928	0.000476
21.0	0.001995	0.001052	0.000501
22.0	0.002084	0.001195	0.000529
23.0	0.002172	0.001356	0.000561
23.2	0.002189	0.001390	0.000568
23.3	0.002197	0.001407	0.000572

c_{out} computed from (15-86) combined with (3) to eliminate c^*:

$$c_{out} = \frac{c_F + \gamma(q/0.01057)^{4.35}}{1+\gamma} = \frac{0.010 + 26.06(q/0.01057)^{4.35}}{27.06}$$

Also included in Table 15.8 are the cumulative values of c, for all of the liquid effluent that exits the vessel during the period from $t = 0$ to $t = t$, as obtained by integrating c_{out} with respect to time: $c_{cum} = \int_0^t c_{out}\, dt/t$.

From the results in Table 15.8, it is seen that the loading, q, increases almost linearly during the first 10 h, while the instantaneous phenol concentration c_{out} in the exiting liquid remains almost constant. At 15.7 h, c_{out} has increased to the specified value of 0.00057 kmol/m^3, but c_{cum} is only 0.000407 kmol/m^3. Therefore, the operation can continue. Finally, at between 23.2 and 23.3 h, c_{cum} reaches 0.00057 kmol/m^3 and the operation must be terminated. During operation, the vessel contains 1,000 kg or 2 m^3 of adsorbent particles. With a liquid residence time of almost 3 h, the vessel must contain 10(3) = 30 m^3. Thus, the vol% solids in the agitated vessel is 6.7. This is reasonable. If the mass of adsorbent in the vessel is increased to 2,000 kg, giving almost 12 vol% solids, the time of operation is doubled to 46.5 h.

Fixed-Bed Adsorption (Percolation)

In the continuous and semicontinuous modes of operation in slurry adsorption, the liquid exiting the vessel always contains unadsorbed solute. If a fixed bed is used, it is possible to obtain a nearly solute-free liquid or gas effluent until the adsorbent in the bed approaches saturation. A fixed bed is frequently used for gas purification and bulk separation.

Consider the flow, down through a fixed bed of adsorbent, of a fluid containing an adsorbable component (the solute). If (1) external and internal mass-transfer resistances are very small; (2) plug flow is achieved; (3) axial dispersion is negligible; (4) the adsorbent is initially free of adsorbate; and (5) the adsorption isotherm begins at the origin, then local equilibrium between the fluid and the adsorbent is achieved instantaneously, resulting, as shown in Figure 15.27, in a shock like wave, called a *stoichiometric front,* that moves as a sharp concentration front through the bed. This is *ideal (local equilibrium) fixed-bed adsorption.* Upstream of the front, the adsorbent is saturated with adsorbate and the concentration of solute in the fluid is that of the feed, c_F. The loading of adsorbate on the adsorbent is the q_F in equilibrium with c_F. The length (height) and weight of the bed section upstream of the front are LES and WES, respectively, where ES refers to the equilibrium section, called the *equilibrium zone.*

In the upstream region, the adsorbent is spent. Downstream of the stoichiometric front and in the exit fluid, the concentration of the solute in the fluid is zero, and the adsorbent is still adsorbate-free. In this section of the bed, the length and weight are LUB and WUB, respectively, where UB refers to unused bed.

After a period of time, called the *stoichiometric time,* the stoichiometric wave front reaches the end of the bed, the concentration of the solute in the fluid abruptly rises to the inlet value, c_F, no further adsorption is possible, and the adsorption step is terminated. This point is referred to as the *breakpoint* and the stoichiometric wave front becomes the ideal *breakthrough* curve.

For ideal fixed-bed adsorption, the location of the concentration wave front L, in Figure 15.27, as a function of time, is obtained solely by material balance and adsorption equilibrium considerations. Thus, at equilibrium, the loading in equilibrium with the feed is designated by $q_F = f\{c_F\}$, where $f\{c_F\}$ is given by an appropriate adsorption isotherm. By material balance on the adsorbate before breakthrough occurs: Solute in entering feed = adsorbate. Accordingly:

$$Q_F c_F t_{ideal} = q_F S L_{ideal}/L_B \qquad (15\text{-}92)$$

where Q_F is the volumetric flow rate of feed, c_F is the concentration of the solute in the feed, t_{ideal} is the time for an

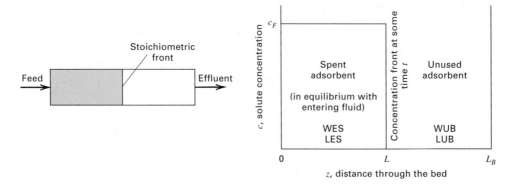

Figure 15.27 Stoichiometric (equilibrium) concentration front for ideal fixed-bed adsorption.

ideal front to reach $L_{\text{ideal}} < L_B$, q_F is the loading per unit mass of adsorbent that is in equilibrium with the feed concentration, S is the total mass of adsorbent in the bed, and L_B is the total bed length.

$$L_{\text{ideal}} = \text{LES} = \left(\frac{Q_F c_F t_{\text{ideal}}}{q_F S} \right) L_B \qquad (15\text{-}93)$$

$$\text{LUB} = L_B - \text{LES} \qquad (15\text{-}94)$$

$$\text{WES} = S \left(\frac{\text{LES}}{L_B} \right) \qquad (15\text{-}95)$$

$$\text{WUB} = S - \text{WES} \qquad (15\text{-}96)$$

In a real fixed-bed adsorber, the assumptions leading to (15-92) are not valid. Internal transport resistance and, in some cases, external transport resistance are finite. Axial dispersion can also be significant, particularly at low flow rates in shallow beds. These factors contribute to the development of broad concentration fronts like those in Figure 15.28. In Figure 15.28a, typical solute concentration profiles for the fluid are shown as a function of distance through the bed at increasing times t_1, t_2, and t_b from the start of flow through the bed. At t_1, no part of the bed is saturated. At t_2, the bed is almost saturated for a distance L_s. At L_f, the bed is almost clean. Beyond L_f, little mass transfer occurs at t_2 and the adsorbent is still unused. The region between L_s and L_f is called the mass-transfer zone, MTZ at t_2, where adsorption takes place. Because it is difficult to determine

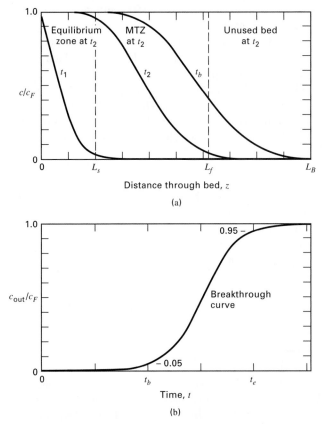

Figure 15.28 Solute wave fronts in a fixed-bed adsorber with mass-transfer effects. (a) Concentration–distance profiles. (b) Breakthrough curve.

where the MTZ zone begins and ends, L_f can be taken where $c/c_F = 0.05$, with L_s at $c/c_F = 0.95$. From time t_2 to time t_b, the S-shaped front moves through the bed. At t_b, the leading point of the MTZ just reaches the end of the bed. This is the breakthrough point. Rather than using $c/c_F = 0.05$, the breakthrough concentration can be taken as the minimum detectable or maximum allowable solute concentration in the effluent fluid.

Figure 15.28b is a typical plot of the ratio of the outlet-to-inlet solute concentration in the fluid as a function of time from the start of flow. The S-shaped curve is called the breakthrough curve. Prior to t_b, the outlet solute concentration is less than some maximum permissible value, say, $c_{\text{out}}/c_F = 0.05$. At t_b, this value is reached, the adsorption step is discontinued, and the regeneration part of the cycle is initiated or the spent adsorbent is discarded. If the adsorption step were to be continued for $t > t_b$, the outlet solute concentration would be observed to rise rapidly, eventually approaching the inlet concentration as the outlet end of the bed became saturated. The time to reach $c_{\text{out}}/c_F = 0.95$ is designated t_e.

The steepness of the breakthrough curve determines the extent to which the capacity of an adsorbent bed can be utilized. Thus, the shape of the curve is very important in determining the length of an adsorption bed. For the ideal case, with a stoichiometric wave front, (15-92) applies and all of the bed is utilized before breakthrough occurs. As the width of the breakthrough curve and the corresponding width of the MTZ for the concentration profiles increase, less and less of the bed capacity can be utilized. The situation is further complicated by the fact that the steepness of the concentration profiles shown in Figure 15.28a increases or decreases with time, depending on the shape of the adsorption isotherm, as shown by DeVault [73], in the following manner.

Assume: (1) plug flow of the fluid through the bed at a constant actual (interstitial) velocity, u; (2) instantaneous equilibrium of the solute in the bulk fluid with the adsorbate; (3) no axial dispersion; and (4) isothermal conditions. The bed is not initially free of adsorbate and/or the feed to the bed starting at time $t = 0$ is not at constant composition. The superficial fluid velocity is $\epsilon_b u$. A mass balance on the solute for the flow of fluid through a differential adsorption-bed length, dz, over a differential-time duration, dt, gives

$$\epsilon_b u A_b c|_z = \epsilon_b u A_b c|_{z+\Delta z} + \epsilon_b A_b \Delta z \frac{\partial c}{\partial t} + (1 - \epsilon_b) A_b \Delta z \frac{\partial q}{\partial t}$$
$$(15\text{-}97)$$

Dividing by Δz and taking the limit as $\Delta z \to 0$ gives

$$\frac{\partial c}{\partial t} + u \frac{\partial c}{\partial z} + \frac{(1 - \epsilon_b)}{\epsilon_b} \frac{\partial q}{\partial t} = 0 \qquad (15\text{-}98)$$

where q is the adsorption loading/unit volume of adsorbent particles, given by an appropriate adsorption isotherm. By the chain rule:

$$\frac{\partial q}{\partial t} = \frac{\partial q}{\partial c} \frac{\partial c}{\partial t} \qquad (15\text{-}99)$$

This hyperbolic PDE (15-98) gives $c = f\{z, t\}$. Therefore, by the rules of implicit partial differentiation:

$$u_c = \left(\frac{\partial z}{\partial t}\right)_c = -\frac{\left(\dfrac{\partial c}{\partial t}\right)}{\left(\dfrac{\partial c}{\partial z}\right)} \qquad (15\text{-}100)$$

where u_c is the velocity of the concentration wave front, $\partial z / \partial t$ at constant c. Combining (15-98) to (15-100):

$$u_c = \frac{u}{1 + \left(\dfrac{1 - \epsilon_b}{\epsilon_b}\right)\dfrac{dq}{dc}} \qquad (15\text{-}101)$$

This equation gives the velocity of the concentration wave front for the solute in terms of the interstitial fluid velocity and the slope, dq/dc, of the adsorption isotherm. If dq/dc is constant, the wave front moves at a constant value.

In general, the concentration wave front moves through the bed at a velocity, u_c, that is much less than the interstitial fluid velocity. For example, suppose that $\epsilon_b = 0.5$ and the equilibrium adsorption isotherm is given by $q = 5,000c$. Then $dq/dc = 5,000$. Then, from (15-101), $u_c/u = 0.0002$. If the interstitial velocity is 3 ft/s, the velocity of the concentration wave front is only 0.0006 ft/s. If the bed were 6 ft in height, it would take 2.78 h for the concentration wave front to pass through the bed. If the adsorption isotherm is curved, regions of the wave front at a higher concentration move at a velocity different from regions at a lower concentration. Thus, for a linear isotherm (curve A in Figure 15.29a), the width of the MTZ and the wave pattern remain constant. For a favorable isotherm of the Freundlich or Langmuir type (curve B in Figure 15.29a), high-concentration regions move faster than low-concentration regions, and the wavefront steepens with time until a constant pattern front (CPF) is developed, as shown in Figure 15.29b. For the much less common unfavorable type of isotherm (Curve C in Figure 15.29a), low-concentration regions travel faster and the wavefront broadens with time.

For the general case where external and internal mass-transfer resistances are finite and/or axial dispersion is not negligible, methods for predicting concentration profiles and breakthrough curves have been the subject of much study. As will be shown, when mass-transfer resistances are a factor, the concentration fronts develop quite differently from the equilibrium fronts just described. Solutions for a number

of simplified cases are discussed in detail by Ruthven [10]. The PDE for the governing dynamic behavior is a modification of (15-98):

$$-D_L\frac{\partial^2 c}{\partial z^2} + \frac{\partial(uc)}{\partial z} + \frac{\partial c}{\partial t} + \frac{(1 - \epsilon_b)}{\epsilon_b}\frac{\partial \bar{q}}{\partial t} = 0 \quad (15\text{-}102)$$

where the first term accounts for axial dispersion with eddy diffusivity D_L, the second term permits an axial variation in fluid velocity, and the fourth term is now based on $\bar{q}$, the volume-average adsorbate loading per unit mass. Thus, the latter term accounts for the variation of q throughout the adsorbent particle, due to internal mass-transfer resistance, by averaging the rate of adsorption over the adsorbent particle. The volume-average adsorbate loading for a spherical particle is given by

$$\bar{q} = \left(\frac{3}{R_p^3}\right)\int_0^{R_p} r^2 q\, dr \qquad (15\text{-}103)$$

where R_p is the radius of the adsorbent particle.

Equation (15-102) gives the concentration of solute in the bulk fluid as a function of time and location in the bed. Equation (15-68) gives the concentration of the solute in the fluid within the pores of an adsorbent particle. These two equations are coupled together by the continuity condition at the particle surface:

$$D_e\left(\frac{\partial c}{\partial r}\right)_{R_P} = k_c(c - c_{R_P}) \qquad (15\text{-}104)$$

where k_c is the external mass-transfer coefficient and D_e is the effective diffusivity in the particle, as discussed in Section 15.3. The simultaneous solution of (15-102), (15-103), (15-68), and (15-104) is a formidable task, which can be avoided by using the linear-driving-force (LDF) model formulated by Glueckauf [74, 75] and discussed in detail by Yang [25] and Ruthven [10]. This model, which is widely used to simulate and design fixed-bed adsorption, is based on the following relation, which replaces (15-68) and (15-104):

$$\frac{\partial \bar{q}}{\partial t} = k(q^* - \bar{q}) = kK(c - c^*) \qquad (15\text{-}105)$$

where q^* is the adsorbate loading in equilibrium with the solute concentration, c, in the bulk fluid; c^* is the concentration in equilibrium with average loading $\bar{q}$; k is the overall mass-transfer coefficient, which includes both external- and internal-transport resistances; and K is the adsorption-

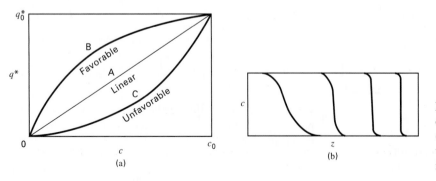

Figure 15.29 Effect of shape of isotherm on sharpness of concentration wavefront. (a) Isotherm shapes. (b) Self-sharpening wavefront caused by a favorable adsorption isotherm.

equilibrium constant for a linear adsorption isotherm of the form $q = Kc$.

A suitable relationship for the factor kK is

$$\frac{1}{kK} = \frac{R_p}{3k_c} + \frac{R_p^2}{15D_e} \qquad (15\text{-}106)$$

where the first term on the RHS represents the external mass-transfer resistance, $k_c a_v$, since for a sphere, the surface area/unit volume, a_v, is given by

$$4\pi R_p^2 / [(4/3)\pi R_p^3] = 3/R_p$$

The second term in (15-106) represents the internal resistance, which was first developed by Glueckauf [75], but can also be derived by assuming a parabolic adsorbate loading profile, in the particle, as shown by Liaw et al. [76]. Thus, let

$$q = a_0 + a_1 r + a_2 r^2 \qquad (15\text{-}107)$$

where the constants a_i depend on time and location in the bed, but are independent of r. Because $\partial q/\partial r = 0$ at $r = 0$ (symmetry condition), $a_1 = 0$. Equating Fick's first law for diffusion into the particle at the particle surface, to the rate of accumulation of adsorbate within the particle, assuming that effective diffusivity is independent of concentration, we obtain

$$\frac{\partial \bar{q}}{\partial t} = D_e a_v \frac{\partial q}{\partial r}\bigg|_{r=R_p} = \frac{3D_e}{R_p} \frac{\partial q}{\partial r}\bigg|_{r=R_p} \qquad (15\text{-}108)$$

At the particle surface, from (15-107):

$$q_{R_p} = a_0 + a_2 R_p^2 \qquad (15\text{-}109)$$

Substituting (15-107) with $a_1 = 0$ into (15-103) and integrating gives

$$\bar{q} = a_0 + \frac{3}{5} a_2 R_p^2 \qquad (15\text{-}110)$$

Combining (15-109) and (15-110) to eliminate a_0 gives

$$a_2 = \frac{5}{2R_p^2}(q_{R_p} - \bar{q}) \qquad (15\text{-}111)$$

From (15-107):

$$\frac{\partial q}{\partial r}\bigg|_{r=R_p} = 2a_2 R_p \qquad (15\text{-}112)$$

Combining (15-110), (15-111), and (15-108):

$$\frac{\partial \bar{q}}{\partial t} = \frac{15D_e}{R_p^2}(q_{R_p} - \bar{q}) \qquad (15\text{-}113)$$

Comparing (15-105) with (15-113), we see that the internal resistance is given by the second term in (15-106).

The analytical solution of a simplified form of (15-102), which assumes negligible axial dispersion, constant fluid velocity, u, and the LDF mass-transfer model, is summarized by Ruthven [10] and discussed in detail by Klinkenberg [77]. The solution was first obtained in terms of Bessel functions by Anzelius [78] for the analogous problem of heating or cooling a packed bed of depth z with a fluid. A useful approximate solution is that of Klinkenberg [79]:

$$\frac{c}{c_F} \approx \frac{1}{2}\left[1 + \text{erf}\left(\sqrt{\tau} - \sqrt{\xi} + \frac{1}{8\sqrt{\tau}} + \frac{1}{8\sqrt{\xi}}\right)\right] \qquad (15\text{-}114)$$

where

$$\xi = \frac{kKz}{u}\left(\frac{1 - \epsilon_b}{\epsilon_b}\right) = \text{Dimensionless distance coordinate} \qquad (15\text{-}115)$$

$$\tau = k\left(t - \frac{z}{u}\right) = \text{Dimensionless time coordinate corrected for displacement} \qquad (15\text{-}116)$$

$$\text{erf}(-x) = -\text{erf}(x) \qquad (15\text{-}117)$$

$$\text{erf}(x) = \frac{2}{\sqrt{\pi}}\int_0^x e^{-\eta^2} d\eta \qquad (15\text{-}118)$$

where ξ and τ are coordinate transformations for z and t, which convert the equations to a much simpler form. The approximation (15-114) is accurate to $<0.6\%$ error for $\xi > 2.0$. The erf(x), which is included as a function in most spreadsheet programs, is 0.0 at $x = 0$ and asymptotically approaches a value of 1.0 for $x > 2.0$, where x is a dummy variable.

Klinkenberg [79] also includes the following approximate solution for profiles of solute concentration in equilibrium with the average sorbent loading:

$$\frac{c^*}{c_F} = \frac{\bar{q}}{q_F^*} \approx \frac{1}{2}\left[1 + \text{erf}\left(\sqrt{\tau} - \sqrt{\xi} - \frac{1}{8\sqrt{\tau}} - \frac{1}{8\sqrt{\xi}}\right)\right] \qquad (15\text{-}119)$$

where $c^* = \bar{q}/K$ and $c^*/c_F = \bar{q}/q_F^*$, where q_F^* is the loading in equilibrium with c_F.

EXAMPLE 15.11

Air at 70°F and 1 atm, containing 0.9 mol% benzene, enters a fixed-bed adsorption tower at a flow rate of 23.6 lb/min. The tower is 2 ft in inside diameter and is packed to a height of 6 ft with 735 lb of 4 × 6 mesh silica gel (SG) particles having an effective diameter of 0.26 cm and an external void fraction of 0.5. The adsorption isotherm for benzene has been experimentally determined for the conditions of interest and found to be linear over the concentration range of interest, as given by

$$q = Kc^* = 5,120\, c^* \qquad (1)$$

where

$q = $ lb benzene adsorbed per ft^3 of silica gel particles

$c^* = $ equilibrium concentration of benzene in the gas, in lb benzene per ft^3 of gas

Mass-transfer experiments, simulating the conditions of the 2-foot-diameter bed, have been carried out and fitted to a linear-driving-force (LDF) model:

$$\frac{\partial \bar{q}}{\partial t} = 0.206K(c - c^*) \qquad (2)$$

where time is in minutes. The constant $k = 0.206$ min^{-1} includes resistances both in the gas film and in the adsorbent pores, with the latter resistance dominant.

Using the approximate concentration-profile equations of Klinkenberg [77], compute a set of breakthrough curves and determine the time when the concentration of benzene in the exiting air rises to 5% of the inlet concentration. Assume isothermal and isobaric operation. Compare the breakthrough time with the time predicted by the equilibrium model.

SOLUTION

For the equilibrium model, the bed becomes completely saturated with benzene at the inlet concentration.

MW of entering gas $= 0.009(78) + 0.991(29) = 29.44$
Density of entering gas $= (1)(29.44)/(0.730)(530) = 0.076/lb/ft^3$
Gas flow rate $= 23.6/0.0761 = 310\ ft^3/min$

$$\text{Benzene flow rate in entering gas} = \frac{(23.6)}{29.44}(0.009)(78)$$

$$= 0.562\ lb/min$$

or

$$c_F = \frac{0.562}{310} = 0.00181\ lb\ benzene/ft^3\ of\ gas$$

From (1),

$$q = 5{,}120(0.00181) = 9.27\frac{lb\ benzene}{ft^3\ SG}$$

The total adsorption of benzene at equilibrium

$$= \frac{9.27(3.14)(2)^2(6)(0.5)}{4} = 87.3\ lb$$

$$\text{Time of operation} = \frac{87.3}{0.562} = 155\ min$$

For the actual operation, taking into account external and internal mass-transfer resistances, from (15-115) and (15-116),

$$\xi = \frac{(0.206)(5{,}120)z}{u}\left(\frac{1-0.5}{0.5}\right) = 1{,}055\ z/u$$

$$u = \text{interstitial velocity} = \frac{310}{0.5\left(\frac{3.14 \times 2^2}{4}\right)} = 197\ ft/min \quad (3)$$

$$\xi = \frac{1{,}055}{197}z = 5.36z$$

where z is in feet.

When $z = $ bed height $= 6\ ft, \xi = 32.2$ and $\tau = 0.206\left(t - \frac{z}{197}\right)$ (4)

where t is in minutes. For $t = 155$ min (the ideal time), and $z = 6$ ft (the bed height), using (4), $\tau = 32$.

Thus, breakthrough curves should be computed from (15-114) for values of τ and ξ no greater than about 32. For example, when $\xi = 32.2$ (exit end of the bed), and $\tau = 30$, which corresponds to a time $t = 145.7$ minutes, the concentration of benzene in the exiting gas, from (15-114), is

$$\frac{c}{c_F} = \frac{1}{2}\left[1 + \text{erf}\left(30^{0.5} - 32.2^{0.5} + \frac{1}{8(30)^{0.5}} + \frac{1}{8(32.2)^{0.5}}\right)\right]$$

$$= \frac{1}{2}[1 + \text{erf}(-0.1524)] = \frac{1}{2}\text{erfc}(0.1524)$$

$$= 0.4147\ \text{or}\ 41.47\%$$

This far exceeds the specification of $c/c_F = 0.05$ or 5% at the exit. Thus, the time of operation of the bed is considerably less than the

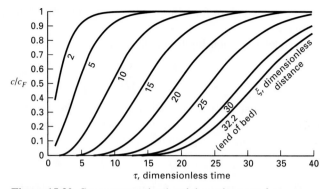

Figure 15.30 Gas concentration breakthrough curves for Example 15.11.

ideal time of 155 min. Figure 15.30 shows breakthrough curves computed from (15-114) over a range of the dimensionless time, τ, for values of the dimensionless distance, ξ, of 2, 5, 10, 15, 20, 25, 30, and 32.2, where the latter corresponds to the exit end of the bed. For $c/c_F = 0.05$ and $\xi = 32.2$, τ is seen to be about 20 (19.9 by calculation).

From (4), with $z = 6$ ft, the time to breakthrough is $t = \frac{20}{0.206} + \frac{6}{197} = 97.1$ min which is 62.3% of the ideal time.

Figure 15.30 or (15-114) can be used to compute the bulk concentration of benzene at various locations in the bed for $\tau = 20$. The results are as follows:

ξ	z, ft	c/c_F
2	0.373	1.00000
5	0.932	0.99948
10	1.863	0.97428
15	2.795	0.82446
20	3.727	0.53151
25	4.658	0.25091
30	5.590	0.08857
32.2	6.000	0.05158

We can also compute, at $\tau = 20$, the adsorbent loading, at various positions in the bed, from (15-119), using $q = 5{,}120c$. The maximum loading corresponds to c_F. Thus, $q_{max} = 9.28$ lb benzene/ft^3 of SG. Breakthrough curves for the solid loading are plotted in Figure 15.31. As expected, those curves are displaced to the right

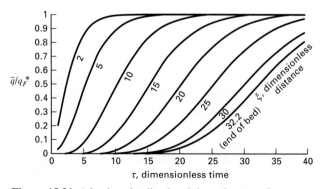

Figure 15.31 Adsorbent loading breakthrough curves for Example 15.11.

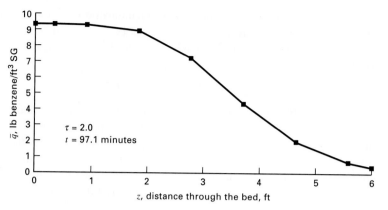

Figure 15.32 Adsorbent loading profile for Example 15.11.

from the curves of Figure 15.30. At $\tau = 20$:

ξ	z, ft	$\dfrac{c^*}{c_F} = \dfrac{\bar{q}}{q_F^*}$	$\bar{q}$, $\dfrac{\text{lb benzene}}{\text{ft}^3 \text{SG}}$
2	0.373	0.99998	9.28
5	0.932	0.99883	9.27
10	1.863	0.96054	8.91
15	2.795	0.77702	7.21
20	3.727	0.46849	4.35
25	4.658	0.20571	1.909
30	5.590	0.06769	0.628
32.2	6.000	0.03827	0.355

The values of $\bar{q}$ in the preceding table are plotted in Figure 15.32 and integrated over the 6-foot bed length to obtain the average bed loading:

$$\bar{q}_{\text{avg}} = \int_0^6 \bar{q}\, dz / 6$$

The result is 5.72 lb benzene/ft^3 of SG, which is 61.6% of the maximum loading based on the inlet benzene concentration.

If the bed were increased in height by a factor of 5, to 30 ft, $\xi = 161$. The ideal time of operation would be 780 min or 13 h. With mass-transfer effects taken into account, as before, the dimensionless operating time to breakthrough is computed to be $\tau = 132$, or breakthrough time from (4) is

$$t = \frac{132}{0.206} + \frac{30}{197} = 641 \text{ min}$$

which is 82.2% of the ideal time. This represents a substantial increase in bed utilization.

Scale-up for Constant-Pattern Front

In Example 15.11, the wavefront (of the type shown in Figure 15.28a), broadens as it moves through the bed. This is shown in Figure 15.33, where MTZ, the width of the mass-transfer zone, is plotted against the dimensionless time, τ, up to the value of 20 where the front breaks through the 6-foot-long bed. The MTZ in Figure 15.33 is based on a range of c/c_F from 0.95 to 0.05. As seen, MTZ increases from about 2 feet at $\tau = 6$ to about 4 feet at $\tau = 20$. As shown, with

increasing τ, the rate of broadening slows. However, for a deeper bed, it is found that even at $\tau = 100$, the wavefront is still slowly broadening.

The continual broadening of the wavefront determined in Example 15.11 is typical of that obtained with a linear adsorption isotherm (curve A in Figure 15.29a). The wavefront also continues to broaden with an unfavorable isotherm (curve C in Figure 15.29a). But, when the isotherm is of the favorable Langmuir or Freundlich type (curve B in Figure 15.29a), wavefront broadening rapidly diminishes and an asymptotic or *constant-pattern front* (CPF) is developed. For such a front, MTZ becomes constant and curves of c/c_F and $\bar{q}/q^*$ become coincident.

The bed depth at which the CPF is approached depends upon the nonlinearity of the adsorption isotherm and the importance of adsorption kinetics. The mathematical proof of the existence of an asymptotic wavefront solution is given by Cooney and Lightfoot [80], including the case of axial dispersion. Initially, the wavefront broadens because of mass-transfer resistance and/or axial dispersion. Eventually, the opposite influence of a favorable isotherm, as shown in Figure 15.29b, comes into play and an asymptotic wavefront pattern is approached. For a constant-pattern front, Sircar and Kumar [81] present some analytical solutions and Cooney [82] presents a rapid approximate method, illustrated with the Freundlich and Langmuir isotherms, to estimate concentration profiles and breakthrough curves when mass-transfer and equilibrium parameters are available.

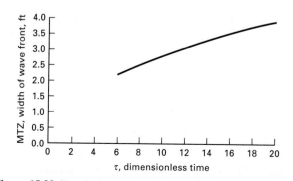

Figure 15.33 Broadening of wavefront in Example 15.11.

When the constant-pattern-front assumption is valid, it can be used to determine the length of a full-scale adsorbent bed from breakthrough curves obtained in small-scale laboratory experiments. This widely used technique is described by Collins [83] for purification applications. The adsorbent bed is considered to be the sum of two sections, analogous to those mentioned for ideal, fixed-bed adsorption. Thus, the total bed length is estimated to be the sum of the length, LES, of the ideal, fixed-bed adsorber plus an additional length, called the LUB, that depends on the observed width of the MTZ and the shape of the c/c_F profile within that zone. The total required bed length is

$$L_B = \text{LES} + \text{LUB} \qquad (15\text{-}120)$$

For the ideal, fixed-bed adsorber, with MTZ = 0, LUB is not necessary, but if $L_B > $ LES, then LUB is the length of unused bed. However, when an MTZ is present, then an LUB is necessary and is referred to as the equivalent length of unused bed. To determine LUB from an experimental breakthrough curve, for the same feed composition and superficial velocity to be used in the commercial adsorber, and for a CPF, the front is located such that in Figure 15.34, area A is equal to area B. Then:

$$\text{LUB} = \text{Ideal wavefront velocity} \times (t_s - t_b) = \frac{L_e}{t_s}(t_s - t_b) \qquad (15\text{-}121)$$

where L_e is the length of the experimental bed. For the ideal case, a solute mass balance for a cylindrical bed of diameter D gives

$$c_F Q_F t = q_F \rho_b \pi \frac{D^2}{4}(\text{LES}) \qquad (15\text{-}122)$$

where t is the time to breakthrough, from which the LES can be determined.

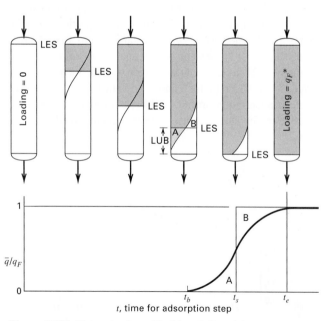

Figure 15.34 Determination of bed length from laboratory measurements.

Instead of positioning the stoichiometric front for equal areas in Figure 15.34, the LUB can be determined from the experimental breakthrough-curve data by computing t_s from

$$t_s = \int_0^{t_e} \left(1 - \frac{c}{c_F}\right) dt \qquad (15\text{-}123)$$

If, in Figure 15.34, t_s is located midway between t_b and t_e, such that the shape of the experimental breakthrough curve below area B is equivalent to the curve above area A, then LUB = MTZ/2, i.e., one-half of the width of the mass-transfer zone. In the absence of experimental breakthrough data, a conservative estimate of MTZ is 4 ft.

EXAMPLE 15.12

Collins [83] presents the following experimental data for the adsorption of water vapor from nitrogen in a fixed bed of 4A molecular sieves:

Bed depth = 0.88 ft, temperature = 83°F (negligible temperature change), pressure = 86 psia (negligible pressure drop), G = entering gas molar velocity = 29.6 lbmol/h-ft², entering water content = 1,440 ppm (by volume), initial adsorbent loading = 1 lb/100 lb sieves, and bulk density of bed = 44.5 lb/ft³. For the entering gas moisture content, c_F, the equilibrium loading, q_F, = 0.186 lb H$_2$O/lb solid.

c_{exit}, ppm (by volume)	Time, h	c_{exit}, ppm (by volume)	Time, h
<1	0–9.0	650	10.8
1	9.0	808	11.0
4	9.2	980	11.25
9	9.4	1,115	11.5
33	9.6	1,235	11.75
80	9.8	1,330	12.0
142	10.0	1,410	12.5
238	10.2	1,440	12.8
365	10.4	1,440	13.0
498	10.6		

Determine the bed height required for a commercial unit to be operated at the same temperature, pressure, and entering gas mass velocity and water content to obtain an exiting gas with no more than 9 ppm (by volume) of water vapor with a breakthrough time of 20 h.

SOLUTION

$$c_F = \frac{1,440(18)}{10^6} = 0.02592 \text{ lb H}_2\text{O/lbmol N}_2$$

$$G = \frac{Q_F}{\pi D^2/4} = 29.6 \text{ lbmol N}_2/\text{h-ft}^2 \text{of bed cross-section}$$

Initial moisture content of bed = 0.01 lb H$_2$O/lb solid

From (15-122),

$$\text{LES} = \frac{(0.02592)(29.6)(20)}{(0.186 - 0.01)(44.5)} = 1.96 \text{ ft}$$

Use the integration method to obtain LUB. From the data:

Take $\quad t_e = 12.8 \text{ h} (1,440 \text{ ppm}) \text{ and } t_b = 9.4 \text{ h} (9 \text{ ppm})$

By numerical integration of the breakthrough-curve data, using (15-123): $t_s = 10.93$ h

From (15-121),

$$LUB = \left(\frac{10.93 - 9.40}{10.93} \right) (0.88) = 0.12 \text{ ft}$$

From (15-120):

$L_B = 1.96 + 0.12 = 2.08$ ft or a bed utilization of $\frac{1.96}{2.08} \times 100\% = 94.2\%$.

Alternatively, the following approximate calculation can be made. Let t_b, the beginning of breakthrough, be 5% of the final ppm or $0.05(1440) = 72$ ppm. Using the experimental data, this corresponds to $t_b = 9.76$ h. Let t_e, the end of breakthrough, be 95% of the final ppm or $0.95(1440) = 1370$ ppm, corresponding to $t_e = 12.25$ h. Let $t_s =$ the midpoint or $(9.76 + 12.25)/2 = 11$ h. The ideal wavefront velocity $= L_e/t_s = 0.88/11 = 0.08$ ft/h. From (15-121), $LUB = 0.08(11 - 9.76) = 0.1$ ft. That is, the $MTZ = 0.2$ ft and $L_B = 1.96 + 0.1 = 2.06$ ft.

Thermal-Swing Adsorption

Thermal (temperature)-swing adsorption (TSA), in its simplest configuration, is carried out with two fixed beds in parallel, operating cyclically, as in Figure 15.20b. While one bed is adsorbing solute at near-ambient temperature, $T_1 = T_{ads}$, the other bed is regenerated by desorbing adsorbate at a higher temperature, $T_2 = T_{des}$, at which the equilibrium adsorbate loading is much less for a given concentration of solute in the fluid, as illustrated in Figure 15.21. Although the desorption step might be accomplished in the absence of a purge fluid by simply vaporizing the adsorbate, readsorption of some solute vapor would occur upon cooling the bed. Thus, it is best to remove the desorbed adsorbate with a purge. The desorption temperature is high, but not so high as to cause deterioration of the adsorbent. TSA is best applied to the removal of contaminants present at low concentrations in the feed fluid. In that case, nearly isothermal adsorption and desorption is achieved. An ideal cycle involves four steps: (1) adsorption at T_1 to breakthrough, (2) heating of the bed to T_2, (3) desorption at T_2 to a low adsorbate loading, and (4) cooling of the bed to T_1. Practical cycles do not operate with isothermal steps. Instead, Steps 2 and 3 are combined for the regeneration part of the cycle, with the bed being simultaneously heated and desorbed with preheated purge gas until the temperature of the effluent approaches that of the inlet purge. Steps 1 and 4 may also be combined because, as discussed in detail by Ruthven [10], the thermal wave precedes the MTZ front. Thus, adsorption takes place at essentially the feed-fluid temperature.

The heating and cooling steps cannot be accomplished instantaneously because of the relatively low bed thermal conductivity. Although heat transfer can be done indirectly from jackets surrounding the beds or from coils located within the beds, bed temperature changes are more readily achieved by preheating or precooling a purge fluid, as shown in Figure 15.35. The purge fluid can be a portion of the feed or effluent, or some other fluid. The purge fluid can also be used in the desorption step. When the adsorbate is valuable and easily condensed, the purge fluid might be a noncondensable gas. When the adsorbate is valuable, but not easily condensed, and is essentially insoluble in water, steam may be used as the purge fluid, followed by condensation of the steam to separate it from the desorbed adsorbate. When the adsorbate is not valuable, fuel and/or air can be used as the purge fluid, followed by incineration. Often the amount of purge used in the regeneration step is much less than the amount of feed sent to the bed in the adsorption step. In Figure 15.35, the feed fluid is a gas. The spent bed is heated and regenerated with preheated feed gas, which is then cooled to condense the desorbed adsorbate.

Because of the time to heat and cool a fixed bed, cycle times for TSA are long, usually extending over periods of hours or days. The longer the cycle time, the longer the required bed length, and the greater the percent utilization of the bed during adsorption. However, for a given cycle time,

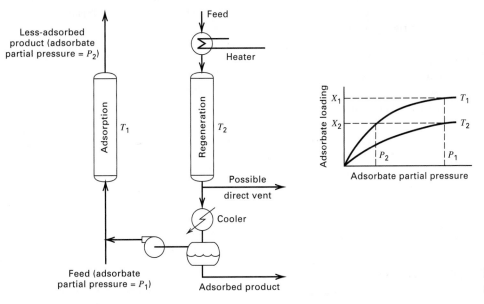

Figure 15.35 Temperature-swing adsorption cycle.

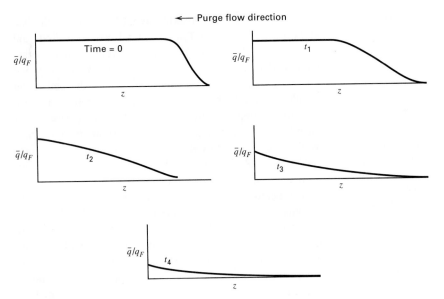

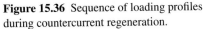

Figure 15.36 Sequence of loading profiles during countercurrent regeneration.

when the width of the MTZ is an appreciable fraction of the bed height, such that the capacity of the bed is poorly utilized, consideration should be given to a *lead-trim-bed* arrangement of two beds in series for the adsorption step. When the lead bed is spent, it is switched to regeneration. At this point in time, the trim bed has an MTZ occupying a considerable portion of the bed, and that bed becomes the lead bed, with a regenerated bed becoming the trim bed. In this manner only a fully spent bed is switched to regeneration. Thus, a total of three beds is used. If the flow rate of the feed stream is very high, beds in parallel may be required for both adsorption and desorption.

The adsorption step is usually conducted with the feed fluid flowing downward through the bed. The flow direction for desorption can be either downward or upward, but the upward, countercurrent direction is usually preferred because it is more efficient. Consider the sequence of loading fronts shown in Figure 15.36, for regeneration countercurrent to adsorption. Although the bed is shown in a horizontal position, it must be positioned vertically. The feed fluid flows downward, entering at the left and leaving at the right. At time $t = 0$, breakthrough has occurred, with a loading profile as shown at the top, where the MTZ is seen to be about 25% of the bed. If the purge fluid for regeneration also flows downward (entering at the left), all of the adsorbate will have to move through the unused portion of the bed. Thus, some

of the desorbed adsorbate will be readsorbed in the unused section and then desorbed a second time. If countercurrent regeneration is used, the unused portion of the bed is never contacted with desorbed adsorbate. During a countercurrent regeneration step, the loading profile changes progressively, as shown for a series of times in Figure 15.36. The right-side end of the bed, where the purge enters, is desorbed first. At the end of regeneration, the residual loading may be uniformly zero or, more likely, finite and nonuniform as shown at the bottom of Figure 15.36. If the latter, then the useful cyclic capacity, called the *delta loading,* is as shown in Figure 15.37.

Calculations of the concentration and loading profiles during desorption are only approximated by (15-114) and (15-119) because the loading is not uniform at the beginning of desorption. A numerical solution for the desorption step can be obtained in the following fashion using a procedure discussed by Wong and Niedzwiecki [84]. Although their method was developed for the adsorption step, it is readily applied to desorption. In the absence of axial dispersion and for constant fluid velocity, (15-102) and (15-105) are rewritten as

$$u\frac{\partial \phi}{\partial z} + \frac{\partial \phi}{\partial t} + \left(\frac{1 - \epsilon_b}{\epsilon_b}\right)kK(\phi - \psi) = 0 \qquad (15\text{-}123)$$

$$\frac{\partial \psi}{\partial t} = k(\phi - \psi) \qquad (15\text{-}124)$$

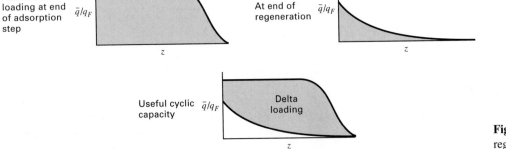

Figure 15.37 Delta loading for regeneration step.

where

$$\phi = c/c_F \qquad (15\text{-}125)$$

$$\psi = \bar{q}/q_F^* \qquad (15\text{-}126)$$

and c_F and q_F^* are the values at the beginning of the adsorption step. The boundary conditions are as follows:

At $t = 0$: $\phi = \phi\{z\}$ at the end of the adsorption step

          $\psi = \psi\{z\}$ at the end of the adsorption step

where, for countercurrent desorption, it is best to let z start from the bottom of the bed (called z') and increase in the direction of purge-gas flow. Thus, u in (15-123) is positive.

At $z' = 0$: $\phi = 0$ (no solute in the entering purge gas)

          $\psi = 0$

Partial differential equations (15-123) and (15-124) in independent variables z and t can be converted to a set of ordinary differential equations (ODEs) in independent variable t by the method of lines (MOL), which was first applied to parabolic PDEs by Rothe in 1930, as discussed by Liskovets [85], and subsequently to elliptic and hyperbolic PDEs. The MOL is developed in detail by Schiesser [86]. The lines refer to the z'-locations of the ODEs. To obtain the set of ODEs, the z'-coordinate is divided into N increments or $N + 1$ grid points that are usually evenly spaced. For many problems, 20 increments are sufficient. Letting i be the index for each grid point in z', starting from the end where the purge gas enters, and discretizing $\partial\phi/\partial z'$, (15-123) and (15-124) become

$$\frac{d\phi_i}{dt} = -u\left(\frac{\Delta\phi}{\Delta z'}\right)_i - \left(\frac{1-\epsilon_b}{\epsilon_b}\right) kK(\phi_i - \psi_i),$$

$$i = 1, N+1 \qquad (15\text{-}127)$$

$$\frac{d\psi_i}{dt} = k(\phi_i - \psi_i), \qquad i = 1, N+1 \qquad (15\text{-}128)$$

where the initial conditions ($t = 0$) for ϕ_i and ψ_i are as given above. Before we can integrate (15-127) and (15-128), we must provide a suitable approximation for $(\Delta\phi/\Delta z')_i$. In general, for a moving-front problem of the hyperbolic type here for adsorption and desorption, the simple central difference

$$\left(\frac{\Delta\phi}{\Delta z'}\right)_i \approx \frac{\phi_{i+1} - \phi_{i-1}}{2\Delta z'}$$

is not adequate. Instead, Wong and Niedzwiecki [84] found that a five-point, biased, upwind, finite-difference approximation, discussed by Schiesser [87], is very effective. This approximation, which is derived from a Taylor's series analysis, places emphasis on conditions upwind of the moving front. At an interior grid point:

$$\left(\frac{\Delta\phi}{\Delta z'}\right)_i \approx \frac{1}{12\Delta z'}[-\phi_{i-3} + 6\phi_{i-2} - 18\phi_{i-1} + 10\phi_i + 3\phi_{i+1}]$$

$$(15\text{-}129)$$

Note that the coefficients of the ϕ-factors, inside the square brackets, sum to 0. At the last grid point, $N + 1$, where the

purge gas exits, (15-129) is replaced by

$$\left(\frac{\Delta\phi}{\Delta z'}\right)_{N+1} \approx \frac{1}{12\Delta z'}[3\phi_{N-3} - 16\phi_{N-2} + 36\phi_{N-1} - 48\phi_N + 25\phi_{N+1}]$$

$$(15\text{-}130)$$

For the first three node points, the following approximations replace (15-129):

$$\left(\frac{\Delta\phi}{\Delta z'}\right)_1 \approx \frac{1}{12\Delta z'}[-25\phi_1 + 48\phi_2 - 36\phi_3 + 16\phi_4 - 3\phi_5]$$

$$(15\text{-}131)$$

$$\left(\frac{\Delta\phi}{\Delta z'}\right)_2 \approx \frac{1}{12\Delta z'}[-3\phi_1 - 10\phi_2 + 18\phi_3 - 6\phi_4 + \phi_5]$$

$$(15\text{-}132)$$

$$\left(\frac{\Delta\phi}{\Delta z'}\right)_3 \approx \frac{1}{12\Delta z'}[\phi_1 - 8\phi_2 + 0\phi_3 + 8\phi_4 - \phi_5]$$

$$(15\text{-}133)$$

However, because values of ϕ_1 (at $z' = 1$) are given as a boundary condition, (15-131) is not needed.

Equations (15-127) to (15-133) with boundary conditions for ϕ_1 and ψ_1 constitute a set of $2N$ ODEs as an initial-value problem, with time as the independent variable. However, the values of ϕ_i and ψ_i at the different axial locations can change with time at vastly different rates. For example, in Figure 15.36 for desorption fronts, if we divide the bed length, L, into 20 equal-width increments, starting from the right-hand side where the purge gas enters, we see that initially ψ_{21}, where the purge gas exits, is not changing at all, while ψ_5 is changing rapidly. Near the end of the desorption step, ψ_{21} is changing rapidly, while ψ_5 is not. The same observations hold for ϕ_i. This type of response is referred to as *stiffness*, as described by Schiesser [87] and in *Numerical Recipes* by Press et al. [88]. If we attempt to integrate the set of ODEs with simple Euler or Runge–Kutta methods, not only do we encounter truncation error, but, with time, the computed values of ϕ_i and ψ_i go through enormous instability, characterized by wild swings between large and impossible positive and negative values. Even if the length is divided into many more than 20 increments and very small time steps are used, instability is still often encountered.

The integration of a stiff set of ODEs is most efficiently carried out by variable-order/variable-step-size implicit methods of the type first developed by Gear [89]. These methods are included in a widely available software package called ODEPACK, described by Byrne and Hindmarsh [90]. The subject of stiffness is also discussed in Chapter 13.

EXAMPLE 15.13

In Example 15.11, benzene is adsorbed from air at $70°F$ and 1 atm onto silica gel in a fixed-bed adsorber, 6 ft in length. Breakthrough occurs at close to 97.1 min for $\phi = 0.05$. At that time, values of $\phi = c/c_F$ and $\psi = \bar{q}/q_F^*$ in the bed are distributed as follows, where z' is measured backwards from the exit of the bed for the adsorption step. These results were obtained by the numerical method just described, as applied to the adsorption step, and are in close agreement with the approximate, analytical Klinkenberg solution given in Example 15.11.

z', ft	$\phi = c/c_F$	$\psi = \bar{q}/q_F^*$
0	0.05227	0.03891
0.3	0.07785	0.05913
0.6	0.11314	0.08776
0.9	0.16008	0.12690
1.2	0.22017	0.17850
1.5	0.29394	0.24387
1.8	0.38042	0.32310
2.1	0.47678	0.41459
2.4	0.57825	0.51469
2.7	0.67861	0.61786
3.0	0.77108	0.71728
3.3	0.84969	0.80603
3.6	0.91057	0.87858
3.9	0.95281	0.93207
4.2	0.97848	0.96690
4.5	0.99172	0.98636
4.8	0.99731	0.99531
5.1	0.99921	0.99857
5.4	0.99987	0.99960
5.7	1.00000	1.00000
6.0	1.00000	1.00000

If the bed is regenerated isothermally with pure air at 1 atm and 145°F, and the desorption of benzene during the heat-up period is neglected, determine the loading, $\bar{q}$, profile at times of 15, 30, and 60 min for pure stripping air interstitial velocities of: (a) 197 ft/min, and (b) 98.5 ft/min. At 145°F and 1 atm, the adsorption isotherm, in the same units as in Example 15.11, is

$$q = 1{,}000c^* \qquad (1)$$

giving an equilibrium loading of about 20% of that at 70°F. Assume that k is unchanged from the value of 0.206 in Example 15.11.

SOLUTION

This problem is solved by the MOL with 20 increments in z', using the subroutine LSODE in ODEPACK to integrate the set of ODEs. The user supplies the FORTRAN MAIN program and the subroutine FEX, shown in Table 15.9, for the derivative functions given by (15-127) to (15-130) and (15-132) to (15-133). The program LSODE includes detailed instructions for writing these two routines. Note that the program in Table 15.9 actually includes both the adsorption and desorption steps for desorption conditions of 30 min at 197 ft/min.

The computed loading profiles for all conditions are plotted in Figures 15.38a and b, for desorption interstitial velocities of 197 and 98.5 ft/min, respectively, where z is the distance from the feed gas inlet end of the bed for the adsorption step. The curves are similar to those shown in Figure 15.30. For the 197 ft/min case, desorption is almost complete at 60 min with less than 1% of the bed still loaded with benzene. If this velocity were used, this would allow $97.1 - 60 = 37.1$ min for heating and cooling the bed before and

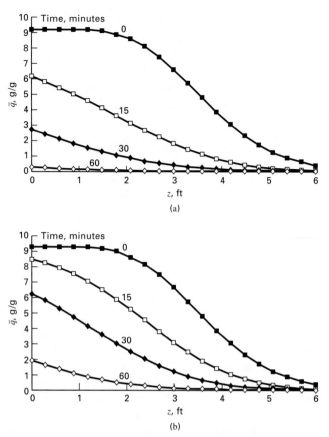

Figure 15.38 Regeneration loading profiles for Example 15.13.
(a) Regeneration air interstitial velocity = 197 ft/min.
(b) Regeneration air interstitial velocity = 98.5 ft/min.

after desorption. For the 98.5 ft/min case at 60 min, about 5% of the bed is still loaded with benzene. This may be acceptable, but the resulting adsorption step would take a little longer because initially the bed would not be clean. Several cycles are required to establish a cyclic steady state, whose development is considered in the next section on pressure-swing adsorption.

Pressure-Swing Adsorption

Pressure-swing adsorption (PSA) and vacuum-swing adsorption (VSA), in their simplest configurations, are carried out with two fixed beds in parallel, operating in a cycle, as in Figure 15.39. Unlike TSA, where thermal means is used to effect the separation, PSA and VSA use mechanical work to increase the pressure or create a vacuum. While one bed is adsorbing at one pressure, the other bed is desorbing at a lower pressure, as was illustrated in Figure 15.21. Unlike TSA, which can be used to purify gases or liquids, PSA and VSA are used only with gases, because a change in pressure has little or no effect on the equilibrium loading for liquid adsorption. PSA was originally used only for purification, as in the removal of moisture from air by the "heatless drier," which was invented by C.W. Skarstrom in 1960 to compete with TSA. However, by the early 1970s, PSA was being applied to bulk separations such as the partial separation of air

Table 15.9 FORTRAN Computer Program for Example 15.13

```
        PROGRAM tsa
        IMPLICIT DOUBLE PRECISION(A-H, O-Z)
        EXTERNAL FEX
        DIMENSION C(40),ATOL(60),RWORK(4162),IWORK(90),CH(40),DL(40)
        COMMON CF,VEL,AK,A(20)
        open (unit=3,file='n1.out')
        write(3,*)'desorption velocity=197ft/min, desorption time=30min'
        NEQ=60
        CF0=0.00181
        CF1=0.0
        TCA=97.1
        NUMCYCLE=1
        MXSTEP=2000
        DO 55 I=1,20
55      C(I)=0.0

        DO 56 I=21,40
        C(I)=0.0
56      CONTINUE

        T=0.D0
        TOUT=0.0
        ITOL=2
        RTOL=1.D-6
        DO 57 I=1, 60
        ATOL(I)=1.0 D-12
57      CONTINUE

        ITASK=1
        ISTATE=1
        IOPT=1
        IWORK(6)=2000
        LRW=4162
        LIW=90
        MF=22

        CF=CF0
        C0=CF
        AK=5120.
        Q0=AK*C0

        CALL LSODE(FEX,NEQ,C,T,TOUT,ITOL,RTOL,ATOL,ITASK,ISTATE,
1       IOPT,RWORK,LRW,IWORK,LIW,JEX,MF)
        WRITE(3,*)'CONDITIONS AT THE BEGINNING '
        WRITE(3,*)'TIME(SEC)=',TOUT
        WRITE(3,*)'CONC. GAS PHASE'
        WRITE(3,*)C0, (C(I),I=1,20)
        WRITE(3,*)'LODING gm/gm'
        WRITE(3,*)Q0,(C(I),I=21,40)
        write(3,128)
128     format(///)
        C0DL=1.0
        Q0DL=1.0
        DO 989 I=1,20
        DL(I)=C(I)/C0
        DL(I+20)=C(I+20)/Q0
989     CONTINUE
        WRITE(3,*)'DIMENSIONLESS CONDITIONS AT THE BEGINNING'
        WRITE(3,*)'TIME(SEC)=',TOUT
        WRITE(3,*)'DIMENSIONLESS GAS CONCENTRATION C/CF'
        WRITE(3,*)C0DL, (DL(I),I=1,20)
        WRITE(3,*)'DIMENSIONLESS LOADING Q/Q0'
        WRITE(3,*)Q0DL, (DL(I),I=21,40)
        write(3,129)
```

(continued)

Table 15.9   (*Continued*)

```
129       format(////////)
          DO 1000 KK=1,NUMCYCLE
C---------------------------------------------------------------------------
C-----------ADSORPTION STEP-------------------------------------------------
C---------------------------------------------------------------------------
          T=0.0

          CF=CF0
          C0=CF
          AK=5120.
          Q0=AK*C0

          VEL=197.0
          ISTATE=1
          TOUT= 97.1

          CALL LSODE(FEX,NEQ,C,T.TOUT,ITOL,RTOL,ATOL,ITASK,ISTATE,
     1    IOPT,RWORK,LRW,IWORK,LIW,JEX,MF)
          IF(KK.EQ.1)GOTO18
          IF((KK/25)*25.NE.KK)GOTO81
18        WRITE(3,*)'CONDITIONS AT THE END OF ADSORPTION STEP'
          WRITE(3,*)'STEP TIME(SEC)=',TOUT
          WRITE(3,*)'CONC. OF GAS PHASE'
          WRITE(3,*)C0,(C(I),I=1,20)
          WRITE(3,*)'LOADING gm/gm'
          WRITE(3,*)Q0,(C(I),I=21,40)
          WRITE(3,741)
741       FORMAT(///)
          C0DL=1.0
          Q0DL=1.0
          DO 990 I=1,20
          DL(I)=C(I)/C0
          DL(I+20)=C(I+20)/Q0
990       CONTINUE
          WRITE(3,*)'DIMENSIONLESS CONDITIONS AT THE END OF ADSORPTION'
          WRITE(3,*)'STEP TIME(SEC)=',TOUT
          WRITE(3,*)'DIMENSIONLESS GAS CONCENTRATION C/CF'
          WRITE(3,*)C0DL,(DL(I),I=1,20)
          WRITE(3,*'DIMENSIONLESS LOADING Q/Q0'
          WRITE(3,*)Q0DL,(DL(I),I=21,40)
          WRITE(3,238)
238       FORMAT(////////)
C---------------------------------------------------------------------------
C-------DESORPTION BY TEMPERATURE SWING-------------------------------------
C---------------------------------------------------------------------------
81        T=0.0
          VEL=197.0
          TOUT=30.0
          ISTATE=1

          CF=CF1
          C0=CF
          AK=1000.0
          Q0=AK*C0

          DO 91 I=1,40
          CH(I)=C(I)
91        CONTINUE

          DO 92 I=1,19
          C(I)=CH(20-I)
92        CONTINUE
          C(20)=CF0

          DO 95 I=1,19
```

Table 15.9 (*Continued*)

```
95      C(20+I)=CH(40-I)
        C(40)=CF0*5120.
        CALL LSODE (FEX,NEQ,C,T,TOUT,ITOL,RTOL,ATOL,ITASK,
     1  ISTATE,IOPT,RWORK,LRW,IWORK,LIW,JEX,MF)

        DO 93 I=1,40
        CH(I)=C(I)
93      CONTINUE

        DO 94 I=1,19
        C(I)=CH(20-I)
        C(20+I)=CH(40-I)
94      CONTINUE
        C0=C(20)
        Q0=C(40)
        C(20)=CF1
        C(40)=1000.*CF1
        IF(KK.EQ.1)GOTO38
        IF((KK/25)*25.NE.KK)GOTO1000
38      WRITE(3,*)'CONDITIONS AT THE END OF DESORPTION '
        WRITE(3,*)'STEP TIME(SEC)=',TOUT
        WRITE(3,*)'CONC. OF GAS PHASE '
        WRITE(3,*)C0,(C(I),I=1,20)
        WRITE(3,*)'LOADING gm/gm'
        WRITE(3,*)Q0,(C(I),I=21,40)
        WRITE(3,264)
264     FORMAT (///)
        C0DL=C0/CF0
        Q0DL=Q0/(CF0*5210.)
        DO 991 I=1,20
        DL(I)=C(I)/CF0
        DL(I+20)=C(I+20)/(CF0*5120.)
991     CONTINUE
        WRITE(3,*)'DIMENSIONLESS CONDITIONS AT THE END OF DESORPTION'
        WRITE(3,*)'STEP TIME(SEC)=',TOUT
        WRITE(3,*)'DIMENSIONLESS GAS CONCENTRATION C/CF'
        WRITE(3,*)C0DL,(DL(I),I=1,20)
        WRITE(3,*)'DIMENSIONLESS LOADING Q/Q0'
        WRITE(3,*)Q0DL,(DL(I),I=21,40)
        WRITE(3,365)
365     FORMAT(////////)
1000    CONTINUE

C-----------------------------------------------------------------------
C-----------------------------------------------------------------------
C-----------------------------------------------------------------------
        WRITE(3,60)IWORK(11),IWORK(12),IWORK(13)
60      FORMAT(/12H NO. STEPS =,I4,11H NO. F-S =,I4,11H NO. J-S =,I4)
        STOP
80      WRITE(3,90)ISTATE
90      FORMAT(///22H ERROR HALT.. ISTATE =,I3)
        close(unit=3)
        STOP
        END
C-----------------------------------------------------------------------
        SUBROUTINE FEX (NEQ,T,C,CDOT)
        IMPLICIT DOUBLE PRECISION(A-H,O-Z)
        DIMENSION C(40), CDOT(40)
        COMMON CF,VEL,AK,A(20)
        E=0.5
        C0=CF
        Q0=AK*C0
        DZ=6.0/20.0       !FT
```

(*continued*)

Table 15.9 (*Continued*)

```
          AA=-VEL
          BB=-(1.0-E)/E

          R4FDX=1./(12.*DZ)
          A(1)=R4FDX*
    1     (-3.*C0-10.*C(1)+18.*C(2)-6.*C(3)+1.*C(4))

          A(2)=R4FDX*
    1     (1.*C0-8.*C(1)+0.*C(2)+8.*C(3)-1.*C(4))

          A(3)=R4FDX*
    1     (-1.*C0+6.*C(1)-18.*C(2)+10.*C(3)+3.0*C(4))

          DO 455 I=4,19
          A(I)=R4FDX*
    1     (-1.*C(I-3)+6.*C(I-2)-18.*C(I-1)+10.*C(I)+3.*C(I+1))
  455     CONTINUE
          A(20)=R4FDX*
    1     (3.*C(16)-16.*C(17)+36.*C(18)-48.*C(19)+25.*C(20))

          DO 676 I=1,20
          CDOT(20+I)=0.206*(AK*C(I)-C(20+I))
          CDOT(I)=AA*A(I)+BB*CDOT(20+I)
  676     CONTINUE

          RETURN
          END
```

to produce either nitrogen or oxygen and to the removal of impurities and pollutants from other gas streams. PSA can also be used for vapor recovery, as discussed and illustrated by Ritter and Yang [91].

A typical sequence of steps in the Skarstrom cycle, operating with two beds, is shown in Figure 15.40. Each bed operates alternately in two half-cycles of equal duration: (1) pressurization followed by adsorption, and (2) depressurization (blowdown) followed by a purge. The feed gas is used for pressurization, while a portion of the effluent product gas is used for purge. Thus, in Figure 15.40, while adsorption is taking place in bed 1, part of the gas leaving bed 1

is routed to bed 2 to purge that bed in a direction countercurrent to the direction of flow of the feed gas during the adsorption step. When moisture is to be removed from air, the dry-air product is produced during the adsorption step in each of the two beds. In Figure 15.40, the adsorption and purge steps represent less than 50% of the total cycle time. In many commercial applications of PSA, these two steps consume a much greater fraction of the cycle time because pressurization and blowdown can be completed rapidly. Therefore, cycle times for PSA and VSA are short, typically seconds to minutes. Thus, small beds have relatively large throughputs. With the valving shown in Figure 15.39, the

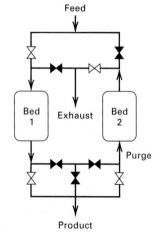

Figure 15.39 Pressure-swing-adsorption cycle.

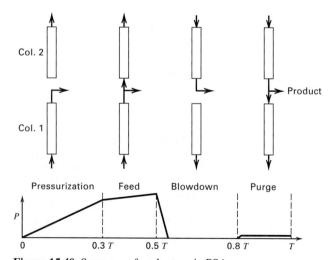

Figure 15.40 Sequence of cycle steps in PSA.

entire cyclic sequence can be programmed to operate automatically. With some valves open and others closed, as in Figure 15.39, adsorption takes place in Bed 1 and purge takes place in Bed 2. During the second half of the cycle, the valve openings and beds are switched.

Since the introduction of the Skarstrom cycle, numerous improvements have been made to increase product purity, product recovery, adsorbent productivity, and energy efficiency, as discussed by Yang [25] and by Ruthven, Farooq, and Knaebel [92]. Among these modifications are the use of (1) three, four, or more beds; (2) a pressure-equalization step in which both beds are equalized in pressure following purge of one bed and adsorption in the other; (3) pretreatment or guard beds to remove strongly adsorbed components that might interfere with the separation of other components; (4) purge with a strongly adsorbing gas; and (5) the use of an extremely short cycle time to approach isothermal operation, if a longer cycle causes an undesirable increase in temperature during adsorption and an undesirable decrease in temperature during desorption.

Separations by PSA and VSA are controlled by adsorption equilibrium or adsorption kinetics, where the latter refers to mass transfer external and/or internal to the adsorbent particle. Both types of control are important commercially. For the separation of air with zeolites, adsorption equilibrium is the controlling factor, with nitrogen more strongly adsorbed than oxygen and argon. For air with 21% oxygen and 1% argon, oxygen of about 96% purity can be produced. When carbon molecular sieves are used, oxygen and nitrogen have almost the same adsorption isotherms, but the effective diffusivity of oxygen is much larger than that of nitrogen. Consequently, a nitrogen product of very high purity (>99%) can be produced.

PSA and VSA cycles have been modeled successfully for both equilibrium and kinetic-controlled cases. The models and computational procedures are similar to those used for TSA. The models are particularly useful for optimizing cycles. Of particular importance in PSA and TSA is the determination of the cyclic steady state. In TSA, following the desorption step, the regenerated bed is usually clean. Thus, a cyclic steady state is closely approached in one cycle. In PSA and VSA, this is often not the case, and complete regeneration is seldom achieved or necessary. It is only required to attain a cyclic steady state whereby the product obtained during the adsorption step has the desired purity and at cyclic steady state, the difference between the loading profiles after adsorption and desorption is equal to the solute entering in the feed. Starting with a clean bed, the attainment of a cyclic steady state for a fixed cycle time may require tens or hundreds of cycles. Consider an example from a study by Mutasim and Bowen [93] on the removal of ethane and carbon dioxide from nitrogen with 5A zeolite, at ambient temperature with adsorption and desorption for 3 min each at 4 bar and 1 bar, respectively, in beds 0.25 m in length. Figures 15.41a and b show the computed development of the loading and gas concentration profiles at the end of each adsorption step for ethane, starting from a clean bed. At the end of the first cycle, the bed is still clean beyond about 0.11 m. By the end of the 10th cycle, a cyclic steady state has almost been attained, with a clean bed existing only near the very end of the bed. Experimental data points for ethane loading at the end of 10 cycles agree reasonably well with the computed profile from a mathematical model.

Modeling of PSA and VSA cycles is carried out with the same equations as for TSA. However, the assumptions of negligible axial diffusion and isothermal operation may be relaxed. For each cycle, the pressurization and blowdown steps are often ignored and the initial conditions for adsorption and desorption are the final conditions for the desorption and adsorption steps, respectively, of the previous cycle. This is illustrated in the following example. Calculations can also be made with Aspen Adsim of the Aspen Engineering Suite.

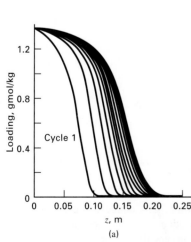

(a)

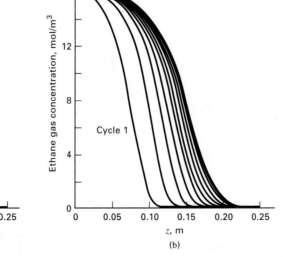

(b)

Figure 15.41 Development of cyclic steady-state profiles. (a) Loading profiles for first 11 cycles. (b) Ethane gas concentration profiles for first 16 cycles.

EXAMPLE 15.14

Ritter and Yang [91] conducted an experimental and theoretical study of the use of PSA to recover dimethyl methylphosphonate (DMMP) vapor from air. For the following data and operating conditions, starting with a clean bed, use the method of lines with a stiff integrator to estimate the concentration and loading profiles for the beds, the percent of the feed gas recovered as essentially pure air, and the average mole fraction of DMMP in the effluent gas leaving the desorption step during the third cycle.

Feed-Gas Conditions:
  236 ppm by volume of DMMP in dry air at 294 K and 3.06 atm

Adsorbent:
  BPL activated carbon, 5.25 g in each bed, 0.07 cm average particle diameter, and 0.43 bed porosity.

Bed dimensions: 1.1 cm i.d. by 12.8 cm each

Langmuir adsorption isotherm: $q = \dfrac{48,360\,p_{DMMP}}{1 + 98,700\,p_{DMMP}}$,

where q is in g/g and p is in atm.

Overall mass-transfer coefficient: $k = 5 \times 10^{-3}\,\text{s}^{-1}$

Cycle conditions (all at 294 K):

1. Pressurization with pure air from p_L to p_H in negligible time.

2. Adsorption at $p_H = 3.06$ atm with feed gas for 20 minutes. $u =$ interstitital velocity $= 10.465$ cm/s.

3. Blowdown from p_H to p_L with no loss of DMMP from the adsorbent or gas in the voids of the bed in negligible time.

4. Desorption at $p_L = 1.07$ atm with product gas (pure air) for 20 minutes. Interstitial velocity, u, corresponding to use of 41.6% of the product gas leaving the adsorption step.

SOLUTION

This example can be solved using the same equations and numerical techniques employed in Example 15.13, but noting that the units of q are different and a Langmuir isotherm replaces Henry's law. If the bed is not clean following the first desorption step, the results for the second and third cycles will differ from the first. The results are not presented here, but the calculations are required in Exercise 15.30.

Continuous, Countercurrent Adsorption Systems

In previous subsections, slurry and fixed-bed modes of adsorption, shown in Figures 15.20a and b, were considered. While these are traditional modes of adsorber operation, the third mode of operation in Figure 15.20c, continuous, countercurrent operation, has an important advantage because, as in a heat exchanger, an adsorber, and other separation cascades, countercurrent flow maximizes the average driving force for transport. In adsorption, this increases the efficiency of adsorbent use.

In Figure 15.20c, both liquid or gas mixtures undergoing separation and the solid adsorbent particles move through the system. However, as discussed in detail by Ruthven and Ching (65) and Wankat (97), the advantage of countercurrent operation can also be achieved by a simulated-moving-bed (SMB) operation, with one widely used implementation shown in Figure 15.23, wherein adsorbent particles remain fixed in a bed. In this subsection, the continuous, countercurrent system shown in Figure 15.20c is considered, while the next subsection covers the SMB. Both types of operation can be applied to purification or bulk separation.

McCabe–Thiele and Kremser Methods for Purification

Consider a binary mixture, dilute in a solute that is to be removed by adsorption in a continuous, countercurrent system of the type shown in Figure 15.42a. Only the solute is adsorbed. Feed F, with solute concentration c_F, enters the adsorption section, ADS, at Plane P_1, from which adsorbent S leaves with a solute loading q_F. Purified feed called the raffinate, with solute concentration c_R, leaves the adsorption section at Plane P_2, countercurrent to adsorbent of loading q_R, which enters at the top of the bed. At Plane P_3, a purge called the desorbent, D, with solute concentration c_D, enters at the bottom of the desorption section, DES, from which the adsorbent leaves to enter the adsorption section. We assume that the desorbent does not adsorb and exits from the desorption

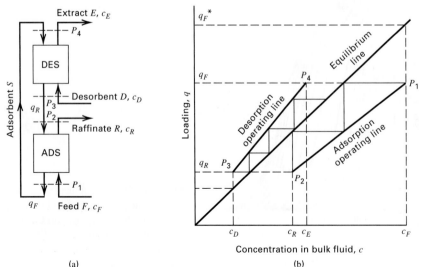

(a)   (b)

Figure 15.42 Continuous, countercurrent adsorption–desorption system. (a) System sections and flow conditions. (b) McCabe–Thiele diagram.

section as extract E, with solute concentration c_E, at Plane 4, where recycled adsorbent enters the desorption bed to complete the cycle.

If the system is dilute in the solute, if the solute adsorption isotherms for the feed solvent and the purge fluid are identical, and the system operates at constant temperature and pressure, the McCabe–Thiele diagram for the solute resembles that shown in Figure 15.42b, where the operating and equilibrium lines are straight because of the dilute condition. Note that the proper directions for mass transfer require that the adsorption and desorption operating lines lie below and above, respectively, the equilibrium line. These three lines are represented by the following equations:

Adsorption Operating Line:

$$q = \frac{F}{S}(c - c_F) + q_F \qquad (15\text{-}134)$$

Desorption Operating Line:

$$q = \frac{D}{S}(c - c_D) + q_R \qquad (15\text{-}135)$$

Equilibrium Line:

$$q = Kc \qquad (15\text{-}136)$$

where F, S, and D are solute-free mass flow rates, and all solute concentrations are per solute-free carrier.

In Figure 15.42b, as the concentration of solute in the entering desorbent (purge), c_D, approaches zero, and solute concentration in the exiting raffinate, c_R, approaches zero, in order to avoid a large number of stages, it is necessary to select the adsorbent and desorbent flow rates so that

$$\frac{F}{S} < K < \frac{D}{S}$$

Thus, because more purge, D, than feed, F, is required, this system is only economical when the purge fluid is inexpensive. From the equilibrium and operating lines in Figure 15.42b, 2 and 3.3 equilibrium stages are determined for the adsorption and desorption sections, respectively, by stepping off stages in the McCabe–Thiele diagram. When the equilibrium and operating lines are straight, as in Figure 15.42b,

the algebraic Kremser method, rather than the graphical McCabe–Thiele method, can be employed. The Kremser equation, discussed in Section 6.4, is written in the following end-point form for the adsorption or desorption section:

$$N_t = \frac{\ln\left[\dfrac{c_1 - q_1/K}{c_2 - q_2/K}\right]}{\ln\left[\dfrac{c_1 - c_2}{q_1/K - q_2/K}\right]} \qquad (15\text{-}137)$$

where 1 and 2 refer to opposite ends of the section, such as Planes 1 and 2 in Figure 15.42a, which are chosen so that $q_1 > q_2$.

If the operating conditions, e.g., temperature, for the two sections can be altered so as to place the equilibrium line for desorption below that for adsorption, it becomes possible to use a portion of the raffinate for desorption. This situation, shown in Figure 15.43, is achieved by desorbing at elevated temperature or, in the case of gas adsorption, at reduced pressure. Now, as shown in Figure 15.43, F/S can be greater than D/S. With a portion of raffinate used in Bed 2 (DES), the net raffinate product is $F - D$. Note that in this case, the two operating lines must intersect at the point (q_R, c_R). By adjusting D/F, this point can be moved closer and closer to the origin so as to achieve any raffinate purity, c_R, desired, but at the expense of more theoretical stages and, therefore, deeper beds.

For a number of theoretical stages, N_t, in either the adsorption or desorption sections, bed height L can be determined from

$$L = N_t(\text{HETP}) \qquad (15\text{-}138)$$

Values of HETP, which depend on mass-transfer resistances and axial dispersion, must be determined from experimental measurements. For large-diameter beds, typical values of HETP are in the range of 0.5–1.5 ft [97, 98].

McCabe–Thiele Method for Bulk Separation

Figure 15.44 shows a continuous, countercurrent adsorption–desorption process for bulk separation of a binary mixture.

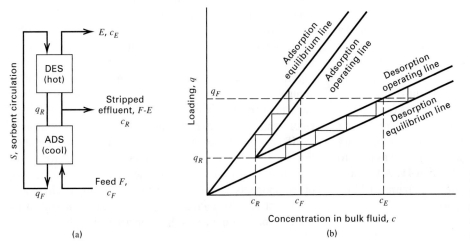

Figure 15.43 Continuous, countercurrent system with a temperature swing. (a) System sections and flow conditions. (b) McCabe–Thiele diagram.

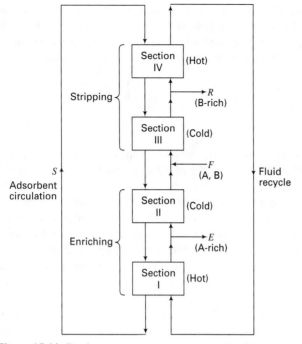

Figure 15.44 Continuous, countercurrent system for bulk separation.

The feed consists of component A, which is more strongly adsorbed, and B, which is less strongly adsorbed. The process consists of four sections (also called zones), numbered from the bottom up. Adsorbent S is circulated through the system, passing downward through the four sections, preferentially adsorbing A, leaving B to preferentially pass upward. To provide flexibility, a thermal swing is used, with Sections II and III operating at low or ambient temperature, while Sections I and IV operate at elevated temperature. The feed, F, enters between Sections II and III, passing up through Section III, where A is preferentially adsorbed at a relatively cold temperature. Product R, rich in B, is removed between Sections III and IV. At the higher temperature in Section IV, residual A and B is desorbed, with the fluid leaving from Section IV recycled to Section I. Adsorbent with mainly adsorbed A passes downward from Section III to Section II and then to Section I, where component A is desorbed to produce product E, rich in A, which is removed between Sections I and II. The system resembles an inverted distillation column, with the top two sections (III and IV) providing a stripping action to produce a product rich in the less strongly adsorbed component, while the bottom two sections provide an enriching action to produce a product rich in the more strongly adsorbed component. An equipment arrangement similar to that in Figure 15.44 was used in the Hypersorption moving bed process [66] for separating hydrogen and methane from ethane and heavier hydrocarbons, except that Section IV was a cooler, Section I was a steam stripper, and gas leaving Section IV was used to lift the adsorbent from Section I to Section IV. Additional flexibility can be achieved for the system in Figure 15.44 by providing separate adsorbent-circulation loops for the top two and bottom two sections.

EXAMPLE 15.15

One hundred pounds per minute (dry basis) of air at 80°F and 1 atm with 65% relative humidity is dehumidified isothermally and isobarically to 10% relative humidity in a continuous, countercurrent moving-bed adsorption unit. The adsorbent is dry silica gel (SG) having a particle-diameter range of 1.42 to 2.0 mm. Over the water-partial-pressure range of interest, the adsorption isotherm is given by measurements of Eagleton and Bliss [94] as

$$q_{H_2O} = 29c_{H_2O} \tag{1}$$

with concentration in lb H_2O/lb dry air and loading in lb H_2O/lb dry SG. If 1.5 times the minimum flow rate of silica gel is used, determine the number of equilibrium stages required.

SOLUTION

For relative humidities of 65% and 10%, the corresponding moisture contents are, from a humidity chart, 0.0143 and 0.0022 lb H_2O/lb dry air, respectively.

In this case, Figure 15.42b applies for the adsorption section. Using the nomenclature in that figure:

$$F = 100 \text{ lb/min}, \quad c_F = 0.0143 \text{ lb } H_2O/\text{lb dry air},$$
$$c_R = 0.0022 \text{ lb } H_2O/\text{lb dry air}, \quad \text{and } q_R = 0.$$

The value of q_F depends on adsorbent flow rate, S, which is 1.5 times the minimum value. At minimum-adsorbent rate, exiting adsorbent is in equilibrium with the entering gas. Therefore, from (1): $q_F^* = 29(0.0143) = 0.415$ lb H_2O/lb dry SG. The amount of water vapor adsorbed is $F(c_F - c_R) = 100(0.0143 - 0.0022) = 1.21$ lb/min. Therefore: $S_{min} = \frac{1.21}{0.415} = 2.92$ lb dry SG/min. If 1.5 times the minimum amount of silica gel is used: $S = 1.5 \, S_{min} = 1.5(2.92) = 4.38$ lb dry SG/min. By material balance: $q_F = \frac{1.21}{4.38} = 0.276$ lb H_2O/lb dry SG. From (15-137), with $K = 29$ from (1) and letting F be at plane 1 and R at plane 2:

$$N_t = \frac{\ln\left[\dfrac{0.0143 - 0.276/29}{0.0022 - 0}\right]}{\ln\left[\dfrac{0.0143 - 0.0022}{0.276/29 - 0}\right]} = 3.2 \text{ stages}$$

Simulated-Moving-Bed Systems

Continuous, countercurrent, moving-bed systems, often referred to as "true-moving-bed" (TMB) systems, encounter operating difficulties, including abrasion of adsorbent particles, failure to approach a plug flow of the particles as they move downward, and channeling of fluid through the moving bed. Alternatively, as shown in one implementation in Figure 15.23, a continuous countercurrent operation can be simulated by using a column containing a series of fixed beds and periodically moving the locations at which streams enter and leave the column. "Simulated-moving-bed" (SMB) systems have found widespread commercial application for liquid separations in the petrochemical, food, biochemical, pharmaceutical, and fine chemical industries, say of components A and B, when employing a circulating desorbent D (also called a diluent or eluent) to aid in the separation. In some cases the properties of D are such that, like A and B, it can be adsorbed. Then, D can displace A and/or B from the sorbent pores, while A and/or B can displace D. In that case, a hybrid process of SMB adsorption and distillation, as shown in Figure 15.23, is often utilized, where following the SMB, a D-free extract of A and a D-free raffinate of B are obtained by distillation, with recovered D recycled to the SMB. In other cases, D is a component of the feed and is not adsorbed, but simply acts as a carrier and stripping agent for separation of A from B. For example, an aqueous solution of glucose and fructose is often separated by an SMB into an extract of aqueous glucose and a raffinate of aqueous fructose, which may be the final products. In the literature, simulated-moving-bed operations are often referred to as chromatographic, rather than adsorptive, separations.

An SMB can be treated as a countercurrent cascade of sections (or zones), rather than stages, where stream entry or withdrawal points bound the sections. As discussed by Zang

and Wankat [99], two-, three-, and four-section systems for producing two products, and a nine-section system for three products are described in the literature, with the four-section system, shown in Figure 15.45a, being the most common commercial design. More recently, Kim and Wankat [100] proposed SMB designs with from 12 to 32 sections for separation of quaternary mixtures.

Operation of a simulated moving bed is best understood by studying the two representations of a four-section system and the accompanying fluid composition profile in Figure 15.45. The schematic representation in Figure 15.45a shows a TMB, with circulation of solid adsorbent S down through four dense-bed sections in a closed cycle, while Figure 15.45b represents an actual SMB system, comprised of four sections divided into 12 fixed-bed subsections, shown as rectangles, with periodic movement of fluid inlet and outlet ports, shown as circles. The sections in Figure 15.45a are sometimes

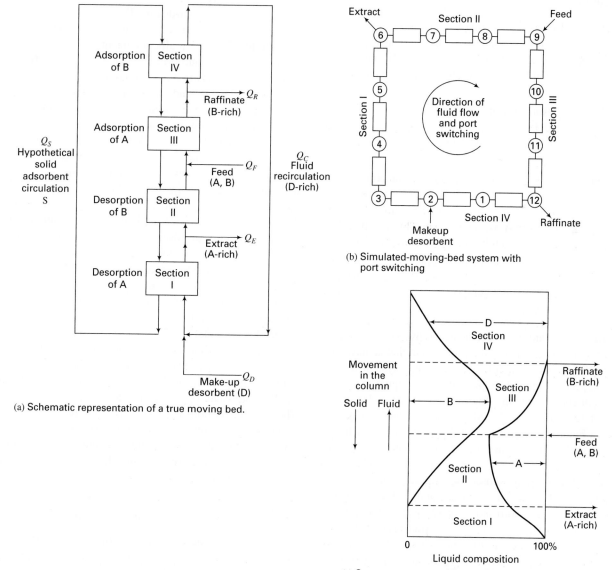

(a) Schematic representation of a true moving bed.

(b) Simulated-moving-bed system with port switching

(c) Component composition profile

Figure 15.45 Four-section system.

referred to as zones, and the fixed-bed subsections in Figure 15.45b are often referred to as beds and sometimes columns. In the equivalent TMB case of Figure 15.45a, fluid of changing composition with respect to feed components A and B, and desorbent D, flows upward through the downward-flowing adsorbent beds. From the top of Section IV, fluid rich in D is recirculated to Section I. Fluid feed is shown as a binary mixture of A and B, which enters between Sections II and III. Component A is more strongly adsorbed than D, which is more strongly adsorbed than B. The desired result is that A is almost completely separated from B. However, appreciable amounts of D may appear in both the B-rich raffinate and A-rich extract. Thus, makeup D is added to the recirculated fluid.

Each of the four sections in Figure 15.45a performs a different primary function, listed in Figure 15.45a. More detail follows for the case where D, as well as A and B, are adsorbed. A typical component composition profile is shown in Figure 15.45c.

Section I: Desorb A. Entering S contains adsorbed A and D. Ideally, entering fluid is nearly pure D. Exiting S contains adsorbed D. Exiting fluid is A and D, part of which is withdrawn as A-rich extract.

Section II: Desorb B. Entering S contains adsorbed A, B, and D. Entering fluid is A and D. Exiting S contains adsorbed A and D. Exiting fluid is A, B, and D.

Section III: Adsorb A. Entering S contains adsorbed B and D. Entering fluid is A, B, and D from section II and fresh feed of A and B. Exiting S contains adsorbed A, B, and D. Exiting fluid is B and D, part of which is withdrawn as B-rich raffinate.

Section IV: Adsorb B. Entering S contains adsorbed D. Entering fluid is B and D. Exiting S contains adsorbed B and D. Ideally, exiting fluid is nearly pure D.

The steady-state separation achieved by the TMB in Figure 15.45a can be a close approximation to that achieved by the SMB, shown for a commercial Sorbex system in Figure 15.23 and by a simpler representation in Figure 15.45b. In both figures, it is seen that four sections are provided with a total of 12 ports for fluid feeds to enter, or fluid products to exit. In Figure 15.45b, it is clear that ports divide each section into subsections, four for Section I, three for Section II, three for Section III, and two for Section IV. As each section is divided into more subsections (thereby adding more ports), the SMB system more closely approaches the separation achieved in the corresponding TMB. In Figure 15.45b, only ports 2, 6, 9, and 12 are open. After an increment of time (called the switching time or port switching interval, t^*), those ports are closed and 3, 7, 10, and 1 are opened. In this manner, the ports are closed and opened in sequence in the direction shown. By periodically shifting feed and

product positions by one port position in the direction of fluid flow, movement of solid adsorbent in the opposite direction within the sections is simulated. Because of stream additions and withdrawals between sections, flow rates in each of the four sections are different. Figure 15.23 shows a pump for controlling the fluid flow rate at the bottom of the SMB. Although sections are switched, the pump is not. Therefore, the pump must be programmed for four different flow rates depending on the section to which the pump is currently connected.

A number of models have been developed for designing and analyzing SMBs. These include: (1) TMB equilibrium-stage model using a McCabe–Thiele-type analysis, (2) TMB local adsorption-equilibrium model, (3) TMB rate-based model, and (4) SMB rate-based model. The first three assume steady-state conditions with continuous, countercurrent flows of fluid and solid adsorbent, approximating SMB operation with a TMB. The SMB rate-based model applies to transient operation for start-up, approach to a cyclic steady state, and shut-down. The simplest of the four approaches is the TMB equilibrium-stage model, but it is difficult to apply to multicomponent systems with non-linear adsorption-equilibrium isotherms. The TMB local adsorption-equilibrium model, although ignoring the effects of axial dispersion and fluid-particle mass transfer, has proved useful for establishing reasonable operating flow rates in multiple sections of an SMB because, for many applications, behavior of an SMB is determined largely by adsorption equilibria. For a linear adsorption isotherm, Wankat [102] has successfully applied this method to SMBs with up to 32 sections for feeds dilute in the solutes. Methods for solving the TMB local adsorption equilibrium model for multicomponent systems, including concentrated mixtures, with nonlinear adsorption isotherms, have been presented by a number of investigators, including Storti et al. [103], who extended the pioneering work of Rhee, Aris, and Amundson [104] for a single section to the commonly used four-section unit, and Mazzotti et al. [105] for multicomponent systems. For a final design, rate-based models are preferred. These models, which account for axial dispersion in the bed, particle-fluid mass-transfer resistances, and nonlinear adsorption isotherms, are available in the program Aspen Chromatography, of the Aspen Engineering Suite, for both TMB steady-state operation and SMB dynamic operation. The local adsorption equilibrium and rate-based models are described next, followed by illustrative examples, two of which are solved using Aspen Chromatography. Equations are presented for four-section units, but are readily extended to more sections.

Steady-state Local Adsorption Equilibrium TMB Model

The TMB model describes continuous, steady-state, multi-component adsorption with countercurrent flow of the fluid and solid adsorbent, as shown in Figure 15.46 for a single section of height Z of a multisection system, subject to these

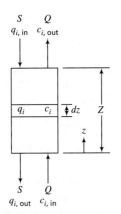

Figure 15.46 TMB local-adsorption-equilibrium model for a single section.

assumptions:

- One-dimensional plug flow of both phases with no channeling.
- Constant volumetric flow rates, of Q for the liquid and Q_S for the solid.
- Constant external void fraction, ϵ_b, of the solids bed.
- Negligible axial dispersion and particle-fluid mass-transfer resistances.
- Local adsorption equilibrium between solute concentrations, c_i, in the bulk liquid and adsorption loading, q_i, on the solid.
- Isothermal and isochoric conditions.

For a differential-bed thickness, dz, where component i undergoes mass transfer between the two phases, this mass balance applies:

$$Q\frac{dc_i}{dz} - S\frac{dq_i}{dz} = 0 \qquad (15\text{-}138)$$

Boundary conditions are

$$z = 0,\, c_i = c_{i,\text{in}} \quad \text{and} \quad z = Z,\, q_i = q_{i,\text{in}}$$

The solution to (15-138) depends on the equilibrium adsorption isotherm. Typically, when the fluid is a liquid dilute in

solutes, a linear (Henry's law) isotherm, $q_i = K_i c_i$, is used, where q_i is on a particle volume basis so that K_i is dimensionless. For the bulk separation of liquid mixtures, where concentrations of the feed components and desorbent are not small, a nonlinear, extended-Langmuir-equilibrium-adsorption isotherm of the constant-selectivity form, from Example (15.6) applies

$$q_i = \frac{(q_i)_m K_i c_i}{1 + \sum_j K_j c_j} \qquad (15\text{-}139)$$

In either case, the solution of Rhee, Aris, and Amundson [104], when extended to multiple (e.g., four) sections, as by Storti et al. [103], predicts constant component concentrations in each section, but with discontinuities at either one or both section boundaries. Typical concentration profiles are shown in Figure 15.47 for a four-solute system (1, 2, 3, and 4), where a set of stationary rectangular (shock-like) waves of constant concentration exists in the fluid phase in each section. The concentration profile for the desorbent (component 5) is not shown. Note that the concentrations of the four solutes for this local equilibrium assumption are negligible in Sections I and IV, where only desorbent is present.

The usefulness of local equilibrium theory is in approximate determinations of required solid adsorbent and fluid flow rates in each section of a TMB in order to achieve a perfect separation of two solutes. The description of the method, first developed by Ruthven and Ching [106] and extended by Zhong and Guiochon [107], is facilitated by applying local adsorption-equilibrium theory to the simple case of a feed dilute in binary solutes, A and B, that are to be completely separated. Assume diluent, D, does not adsorb and Henry's law governs adsorption equilibrium, with $K_A > K_B$ (i.e., A is more strongly adsorbed). First, we define a set of flow rate ratios, m_j, one for each section, j, as

$$m_j = \frac{Q_j}{Q_s} = \frac{\text{volumetric fluid phase flow rate}}{\text{volumetric solid particle phase flow rate}}$$

$$(15\text{-}140)$$

For conditions of local adsorption equilibrium, the following necessary and sufficient conditions apply to each section for

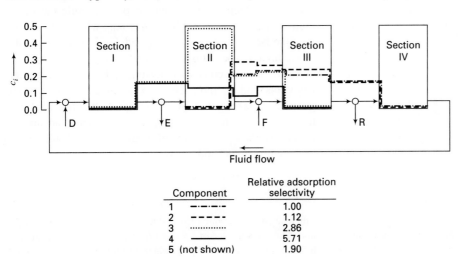

Component		Relative adsorption selectivity
1	—·—·—	1.00
2	– – – –	1.12
3	·············	2.86
4	————	5.71
5 (not shown)		1.90

Figure 15.47 Typical solute-concentration profiles for local adsorption equilibrium in a four-section unit.

complete separation:

$$K_A < m_I < \infty \qquad (15\text{-}141)$$

$$K_B < m_{II} < K_A \qquad (15\text{-}142)$$

$$K_B < m_{III} < K_A \qquad (15\text{-}143)$$

$$0 < m_{IV} < K_B \qquad (15\text{-}144)$$

Constraint (15-142) ensures that net flow rates of components A and B will be positive (upward) in section I. Constraint (15-144) ensures that the net flow rates of components A and B will be negative (downward) in Section IV. Constraints (15-142) and (15-143) are most important because they ensure sharpness of the separation. They cause net flow rates of A and B to be negative (downward) and positive (upward), respectively, in the two central sections II and III. Inequality constraints (15-141) to (15-144) may be converted to equality constraints, where β, the safety margin, is discussed shortly.

$$Q_I/Q_S = K_A \beta \qquad (15\text{-}145)$$

$$(Q_I - Q_E)/Q_S = K_B \beta \qquad (15\text{-}146)$$

$$(Q_I - Q_E + Q_F)/Q_S = K_A/\beta \qquad (15\text{-}147)$$

$$(Q_I - Q_E + Q_F - Q_R)/Q_S = K_B/\beta \qquad (15\text{-}148)$$

Solving (15-145) to (15-148) by eliminating Q_1 gives

$$Q_S = \frac{Q_F}{K_A/\beta - K_B \beta} \qquad (15\text{-}149)$$

$$Q_E = Q_S(K_A - K_B)\beta \qquad (15\text{-}150)$$

$$Q_R = Q_S(K_A - K_B)/\beta \qquad (15\text{-}151)$$

Then, using (15-145),

$$Q_I = Q_C + Q_D = Q_S K_A \beta \qquad (15\text{-}152)$$

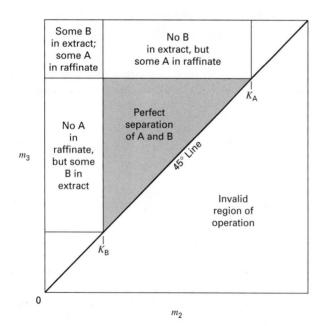

Figure 15.48 Triangle method for determining necessary values of flow rate ratios.

Therefore,

$$Q_C = Q_S K_A \beta - Q_D \qquad (15\text{-}153)$$

where Q_C = fluid recirculation rate before adding makeup desorbent. By an overall material balance,

$$Q_D = Q_E + Q_R - Q_F \qquad (15\text{-}154)$$

Restrictions on flow rate ratios, m_{II} and m_{III} in inequality constraints (15-142) and (15-143), are conveniently represented by the *triangle method* of Storti et al. [107], as shown in Figure 15.48. If values of m_{II} and m_{III} within the triangular region are selected, a perfect separation is possible. However, if $m_{II} < K_B$, some B will appear in the extract; if $m_{III} > K_A$, some A will appear in the raffinate. If $m_{II} < K_B$ and $m_{III} > K_A$, extract will contain some B and raffinate will contain some A.

The permissible range for safety margin, β, in (15-145) to (15-151) is determined from inequality constraints (15-142) and (15-143). Let

$$\gamma_{i,j} = \frac{m_j}{K_i} = \frac{Q_j}{Q_S K_i} \qquad (15\text{-}155)$$

In Section II, we require that $\gamma_{A,II} > 1$ and $\gamma_{B,II} < 1$. In terms of safety margin, β, we can apply (15-155) to give corresponding equalities, $Q_{II}/Q_S = K_A/\beta$ and $Q_{II}/Q_S = K_B \beta$, assuming equal β in all four sections. Equating these two equalities for the same safety margin gives $\beta = \sqrt{K_A/K_B}$, which is the maximum value of β for a perfect separation, the minimum value being 1.0. Above the maximum value of β, some sections will encounter negative fluid flow rates. Below a β value of 1.0, a perfect separation will not be achieved. As the value of β increases from minimum to maximum, fluid flow rates in the sections increase, often exponentially. Thus, estimation of operating flow rates is generally carried out using a value of β close to, but above, 1.0, e.g., 1.05 (unless it exceeds the maximum value of β). Note that as the separation factor, K_A/K_B, approaches 1.0, not only does the separation become more difficult, but also, the permissible range of β becomes smaller. In the triangle method, illustrated in Figure 15.48, the upper left corner of the triangle corresponds to $\beta = 1$, while the maximum value of β occurs when $m_{II} = m_{III}$, which falls on the 45° line between the values K_A and K_B. Extensions of the above binary procedures for estimating operating flow rates to cases of both constant selectivity Langmuir adsorption isotherms and to more complex nonlinear isotherms are given by Mazzotti et al. [109] and for multicomponent systems by Mazzotti et al. [105]. With nonlinear adsorption isotherms, the right triangle of Figure 15.48 is distorted to a shape with one or more curved sides.

EXAMPLE 15.16

Fructose (A) is separated from glucose (B) in a four-section SMB unit. The aqueous feed of 1.667 mL/min contains 0.467 g/min of A, 0.583 g/min of B, and 0.994 g/min of water. For the adsorbent and expected concentrations and temperature of the operation, Henry's law holds, with constants of $K_A = 0.610$ and $K_B = 0.351$ for fluid concentrations in g/mL and loadings in g/mL of adsorbent particles. Water is assumed not to adsorb. Estimate operating flow rates in mL/min to achieve a perfect separation of fructose from glucose for a TMB. Note that the extract will contain fructose, while raffinate will contain glucose. Conversion of the results to SMB operation will be made in Example 15.17.

SOLUTION

Equations (15-149) to (15-154) apply. The minimum value of β is 1.0, while the maximum value is $\sqrt{K_A/K_B} = \sqrt{0.610/0.351} = 1.32$. Calculations are most conveniently carried out with a spreadsheet. With reference to Figure 15.45 for the case of a TMB, the results for values of $\beta = 1.0, 1.05, 1.20$ are:

	Volumetric Flow Rates, mL/min		
	$\beta = 1.0$	$\beta = 1.05$	$\beta = 1.20$
Feed, Q_F	1.667	1.667	1.667
Solid particles, Q_S	6.436	7.848	19.132
Extract, Q_E	1.667	2.134	5.946
Raffinate, Q_R	1.667	1.936	4.129
Recirculation, Q_C	2.259	2.624	5.596
Make-up desorbent, Q_D	1.667	2.403	8.408
Q_I	3.926	5.027	14.004
Q_{II}	2.259	2.893	8.058
Q_{III}	3.926	4.560	9.725
Q_{IV}	2.259	2.624	5.596

Note that the lowest section fluid flow rates, Q_I–Q_{IV}, correspond to $\beta = 1.0$. At $\beta = 1.2$, section fluid flow rates, as well as the adsorbent particles flow rate, become significantly higher. The most concentrated products (extract and raffinate) and the smallest flow rate of make-up desorbent are also achieved with the lowest β value.

Steady-state TMB Model

This model, which assumes plug flow at isothermal, isobaric, and constant-fluid-velocity conditions in each section, j, ($j = 1$ to 4) requires for each component, i ($i = 1$ to C) the following equations, where each section begins at $z = 0$, where the fluid enters, and ends at $z = L_j$. Unlike the previous local adsorption-equilibrium model, axial dispersion and fluid-particle mass transfer are taken into account.

(1) Mass-balance equation for the bulk fluid phase, f, [similar to (15-102)]:

$$-D_{L_j}\frac{d^2c_{i,j}}{dz^2} + u_{f_j}\frac{dc_{i,j}}{dz} + \frac{(1-\epsilon_b)}{\epsilon_b}J_{i,j} = 0 \quad (15\text{-}156)$$

where the first term accounts for axial dispersion, J_i is the mass-transfer flux between the bulk fluid phase and the sorbate in the pores of the solid, and u_f is the interstitial fluid velocity, where for an adsorbent bed of cross-sectional area, A_b,

$$u_{f_j} = \frac{Q_j}{\epsilon_b A_b} \quad (15\text{-}157)$$

(2) Mass-balance equation for the sorbate, s, on the solid phase:

$$u_s\frac{d\bar{q}_{i,j}}{dz} - J_{i,j} = 0 \quad (15\text{-}158)$$

where u_s is the true moving-solid velocity, where

$$u_s = \frac{Q_S}{(1-\epsilon_b)A_b} \quad (15\text{-}159)$$

(3) Fluid-to-solid mass transfer [similar to (15-105)]:

$$J_{i,j} = k_{i,j}(q^*_{i,j} - \bar{q}_{i,j}) \quad (15\text{-}160)$$

(4) Adsorption isotherm [e.g., the multicomponent, extended-Langmuir equation of (15-139)]:

$$q^*_{i,j} = f\{\text{all } c_{i,j}\} \quad (15\text{-}161)$$

This system of $4C$ second-order ODEs and $4C$ first-order ODEs, together with the algebraic equations for mass-transfer rates and adsorption equilibria, requires $12C$ boundary conditions, i.e., $3C$ for each section.

At the entrance, $z = 0$, to each section, we require a boundary condition that accounts for axial dispersion. This has been discussed extensively in the literature, e.g., Danckwerts [110]. Most often used is

$$u_{f_j}(c_{i,j,0} - c_{i,j}) = -\epsilon_b D_{L_j}\frac{dc_{i,j}}{dz} \quad (15\text{-}162)$$

where $c_{i,j,0}$ is the concentration of component i entering ($z = 0$) section j.

For continuity of bulk fluid concentrations and sorbate loadings in moving from one section to another, the following boundary conditions apply at boundaries of adjacent sections:

At Sections I and II where extract is withdrawn:

$$c_{i,I,z=L_j} = c_{i,II,z=0} \quad (15\text{-}163)$$
$$q_{i,I,z=L_j} = q_{i,II,z=0} \quad (15\text{-}164)$$

At Sections III and IV where raffinate is withdrawn:

$$c_{i,III,z=L_j} = c_{i,IV,z=0} \quad (15\text{-}165)$$
$$q_{i,III,z=L_j} = q_{i,IV,z=0} \quad (15\text{-}166)$$

At Sections II and III where the feed enters:

$$c_{i,III,z=0} = \frac{Q_{II}}{Q_{III}}c_{i,II,z=L_{II}} + \frac{Q_F}{Q_{III}}c_{i,F} \quad (15\text{-}167)$$
$$q_{i,II,z=L_j} = q_{i,III,z=0} \quad (15\text{-}168)$$

At Sections IV and I where make-up desorbent enters:

$$c_{i,I,z=0} = \frac{Q_{IV}}{Q_I}c_{i,IV,z=L_{II}} + \frac{Q_D}{Q_I}c_{i,D} \quad (15\text{-}169)$$
$$q_{i,IV,z=L_j} = q_{i,I,z=0} \quad (15\text{-}170)$$

where the volumetric fluid flow rates, which change from section to section, are subject to

$$Q_I = Q_{IV} + Q_D \qquad (15\text{-}171)$$

$$Q_{II} = Q_I - Q_E \qquad (15\text{-}172)$$

$$Q_{III} = Q_{II} + Q_F \qquad (15\text{-}173)$$

$$Q_{IV} = Q_{III} - Q_E \qquad (15\text{-}174)$$

It is important to note that for an SMB, solid particles do not flow down through the unit, but are retained in stationary beds in each section. To obtain the same true velocity difference between the fluid and solid particle phase, the upward fluid velocity in the SMB must be the sum of the absolute true velocities in the upward-moving fluid and downward-moving solid particle phases in the TMB. Thus, using (15-157) and (15-159),

$$(Q_j)_{SMB} = (Q_j)_{TMB} + \left(\frac{\epsilon_b}{1 - \epsilon_b}\right)(Q_S)_{TMB} \quad (15\text{-}175)$$

The TMB model can be solved by any of a number of techniques, as discussed by Constantinides and Mostoufi [111], with the Newton shooting method being preferred. An example of the application of the steady-state TMB model is given after the next subsection that treats dynamic SMB models.

Dynamic SMB Model

The equations are subject to the same assumptions as the steady-state TMB model. Changes in the equations permit the model to take into account time of operation, t, and to use a fluid velocity relative to the stationary solid particles. In addition, equations now must be written for each bed subsection (also referred to as a column), k, between adjacent ports, as shown in Figure 15.45b. The revised equations are

(1) Mass-balance equation for the bulk fluid phase, f [similar to (15-102)]:

$$\frac{\partial c_{i,k}}{\partial t} - D_{L_j}\frac{\partial^2 c_{i,k}}{\partial z^2} + u_{f_k}\frac{\partial c_{i,k}}{\partial z} + \frac{(1 - \epsilon_b)}{\epsilon_b}J_{i,k} = 0$$

$$(15\text{-}176)$$

(2) Mass-balance equation for sorbate on the solid phase:

$$\frac{\partial \bar{q}_{i,k}}{\partial t} - J_{i,k} = 0 \qquad (15\text{-}177)$$

where the interstitial fluid velocity for SMB operation is related to that for TMB operation at a particular location by

$$(u_f)_{SMB} = (u_f)_{TMB} + |(u_s)|_{TMB} \qquad (15\text{-}178)$$

SMB and TMB models are further connected by an equation that relates solid velocity in the TMB model to a port-switching time, t^*, and bed height between adjacent ports, L_k, for use in this SMB model:

$$u_s = \frac{L_k}{t^*} \qquad (15\text{-}179)$$

The boundary conditions for the TMB model apply to SMB models. In addition, initial conditions are needed for fluid

concentrations, $c_{i,j}$, and sorbate loadings, $\bar{q}_{i,j}$, throughout the adsorbent beds; e.g., at $t = 0$, $c_{i,k} = 0$ and $\bar{q}_{i,k} = 0$.

The SMB model, which involves PDEs, rather than ODEs, is much more difficult to solve than the steady-state TMB, because it involves moving concentration fronts. In Aspen Chromatography, the dynamic SMB equations are solved by discretizing the first- and second-order spatial terms of the PDEs to obtain a large set of ODEs and algebraic equations, which constitute a DAE (differential algebraic equations) system. A number of discretization or differencing methods are provided. Each complete cycle of the SMB model provides a different result, which ultimately leads to a cyclic steady state. Studies have shown that if the number of bed subsections per section is at least four and the number of cycles is 10 or more, the steady-state TMB result closely approximates the SMB result. Therefore, if only steady-state results are of interest, the simpler steady-state TMB model is best employed.

All four models can be applied to a gas or liquid mixture, with the latter being the most widely applied to industrial separations. Regardless of the model used for design of an SMB (dynamic SMB or steady-state TMB), the basic information required is:

1. Flow rate and composition of the feed (binary of A and B, or multicomponent).

2. Selection of a suitable adsorbent, S, and desorbent, D.

3. Nominal bed operating temperature, T, and pressure, P.

4. A suitable adsorption isotherm for all components, with known constants at the bed operating conditions.

5. Desired separation, which may be purity (on a desorbent-free basis) and desired recovery of the most strongly adsorbed component in the extract.

Not initially known, but required before calculations can be made, are:

6. Total bed height and inside diameter of the adsorption column.

7. Amount of adsorbent in the column.

8. Desorbent recirculation rate.

9. Flow rates of extract and raffinate.

10. Overall mass-transfer coefficients for transport of solutes between bulk fluid and sorbate layer on the adsorbent.

11. Eddy diffusivity for axial dispersion.

12. Spacing of inlet and outlet ports.

Some guidance on initial values for items 6, 10, and 11 is sometimes provided in patents for similar separations. For example, for the separation of xylene mixtures using para diethylbenzene as desorbent, Minceva and Rodrigues [112] suggest:

- Molecular-sieve zeolite adsorbent with a spherical particle diameter, d_p, between 0.25 and 1.00 mm and a particle density, ρ_p, of 1.39 g/cm^3.

- Operating temperature between 140°C and 185°C with an operating pressure sufficient to maintain a liquid phase.
- Liquid interstitial velocity, u_f, between 0.4 and 1.2 cm/s.
- Four sections with eight to 24 subsections (beds).

For a commercial-size unit, the following are suggested:

- Bed height, L_k, in each subsection from 40 to 120 cm.
- Equation (15-106) for estimating overall mass-transfer coefficient, $k_{i,j}$, for solute transport between bulk fluid and sorbate layer on the adsorbent.
- An axial diffusivity, D_{L_j}, defined in terms of a Peclet number, where

$$N_{Pe} = \frac{u_f(\text{characteristic length})}{D_L} \quad (15\text{-}180)$$

Characteristic lengths equal to bed depth or particle diameter have been used. Most common for TMB and SMB is bed depth, with Peclet numbers in the 1000–2000 range.

EXAMPLE 15.17

Use the results of the fructose-glucose separation of Example 15.16, for β = 1.05, with the steady-state TMB model of Aspen Chromatography to estimate product compositions obtained with the following laboratory-size SMB unit:

Number of sections = 4

Number of subsections (beds) in each section (column) = 2

All bed diameters = 2.54 cm

All bed heights = 10 cm

Bed void fraction = 0.40

Particle diameter = 500 microns (0.5 mm)

Overall mass-transfer coefficient for A and B = 10 min^{-1}

Peclet number high enough that axial dispersion is negligible

SOLUTION

To use Aspen Chromatography, the recirculating liquid flow rate for a TMB must be converted to a SMB using (15-175), and solid-particle flow rate must be converted to a port switching time given by (15-179). From (15-175), using the results for β = 1.05 in Example 15.16,

$$(Q_C)_{SMB} = 2.624 + \left(\frac{0.40}{1 - 0.40}\right)7.848 = 7.856 \text{ mL/min}$$

The total liquid rate in Section I of the SMB is

$$(Q_I)_{SMB} = (Q_C)_{SMB} + Q_D = 7.856 + 2.403 = 10.259 \text{ mL/min}$$

This is the maximum volumetric flow rate in the SMB and it is of interest to calculate the corresponding interstitial fluid velocity. From (15-157),

$$(u_{f_i})_{SMB} = \frac{(Q_I)_{SMB}}{\epsilon_b A_b} = \frac{10.259}{0.40\left[\frac{3.14(2.54)^2}{4}\right]}$$

$$= 5.06 \text{ cm/min} = 0.0844 \text{ cm/s}$$

This fluid velocity is low, but it corresponds to a desirable bed diameter-to-particle diameter ratio of 2.54/0.05 = 49. To increase fluid velocity to, say, 0.4 cm/s, the bed diameter would be decreased to 1.17 cm, giving a bed diameter-to-particle diameter of 23, which would still be acceptable.

From (15-159), the true velocity of the solid particles in each bed is

$$u_s = \frac{Q_S}{(1 - \epsilon_b)A_b} = \frac{7.848}{(1 - 0.40)\left[\frac{3.14(2.54)^2}{4}\right]} = 2.58 \text{ cm/min}$$

From (15-179), port-switching time for subsection bed height, L, of 10 cm is,

$$t^* = \frac{L}{u_s} = \frac{10}{2.58} = 3.88 \text{ min}$$

The following results were obtained from Aspen Chromatography for a steady-state TMB:

	Feed	Desorbent	Extract	Raffinate
Flow rate, mL/min	1.667	2.403	2.134	1.936
Concentrations, g/L:				
Fructose	280.0	0.0	211.6	12.7
Glucose	350.0	0.0	8.4	295.3
Water	596.0	996.0	861.7	795.8
Mass fraction on water-free basis:				
Fructose	0.444		0.962	0.040
Glucose	0.556		0.038	0.960

As seen in the table, a reasonably sharp separation between fructose and glucose is achieved. In Exercise 15.39, modifications to the input data are studied in an attempt to improve separation sharpness.

EXAMPLE 15.18

Minceva and Rodrigues [112] consider the industrial-scale separation of paraxylene from a liquid mixture of C_8 aromatics in a four-section SMB. Feed to the unit is 1,450 L/min with the composition shown in a table below, which also contains results for this example. The adsorbent is a molecular-sieve zeolite with a particle density of 1.39 g/cm^3 and a particle diameter of 0.092 cm that packs a bed with an external void fraction of 0.39. The desorbent is paradiethylbenzene (PDEB). With reference to Figure 15.45, the number of subsections is 6, 9, 6, and 3, respectively, in Sections I to IV. The height of each bed subsection is 1.135 m, with a bed diameter of 4.117 m. The operation takes place at 180°C and a pressure above 12 bar, sufficient to prevent vaporization. At these conditions, the extended Langmuir adsorption isotherm (15-139) correlates adsorption equilibrium, yielding the following constants. Note that this is a constant-selectivity isotherm; therefore, the selectivity relative to paradiethylbenzene is tabulated.

Component	q_m, mg/g	K, cm^3/mg	Selectivity
Paraxylene	130.3	1.0658	0.9969
Paradiethylbenzene	107.7	1.2935	1.0000
Ethylbenzene	130.3	0.3067	0.2689
Metaxylene	130.3	0.2299	0.2150
Orthoxylene	130.3	0.1884	0.1762

Note that the desorbent does not have the most desirable equilibrium adsorption property because its selectivity does not lie between that of paraxylene and the C_8 components of the feed. Take the overall mass-transfer coefficient between sorbate and bulk fluid, in (15-160), as 2 min^{-1} for each component. For axial dispersion, assume a Peclet number of 700 in (15-180) with a characteristic length of bed height.

Using Aspen Chromatography with the TMB model as an approximation of the SMB, determine steady-state flow rates and compositions of extract and raffinate, together with the composition profiles in the four sections for the following operating conditions:

Extract flow rate = 1,650 L/min

Raffinate flow rate = 2,690 L/min

Circulation flow rate, $(Q_C)_{SMB}$, before adding makeup DPEB = 5,395 L/min

Port-switching interval, $t^* = 1.15$ min

SOLUTION

By an overall material balance, the DPEB makeup flow rate is

$$Q_D = Q_E + Q_R - Q_F = 1,650 + 2,690 - 1,450 = 2,890 \text{ L/min}$$

From the switching time, using (15-179), with a 1.135-m bed height,

$$u_s = 1.135/1.15 = 0.987 \text{ m/min} = 98.7 \text{ cm/min}$$

The adsorbent bed cross-sectional area, $A_b = 3.14(4.117)^2/4 = 13.31 \text{ m}^2$.

From (15-159), the volumetric flow rate of the solid particles in a TMB is

$$Q_S = u_s(1 - \epsilon)A_b = 0.987(1 - 0.39)(13.31)$$
$$= 8.014 \text{ m}^3/\text{min} = 8,014 \text{ L/min}$$

Liquid flow rates in the four sections are as follows, where both $(Q_j)_{SMB}$ and $(Q_j)_{TMB}$ flow rates are included, where the former are computed by material balance and the latter from (15-175). For example,

$$(Q_I)_{SMB} = (Q_C)_{SMB} + Q_D = 5,395 + 2,890 = 8,285 \text{ L/min}$$
$$(Q_I)_{TMB} = (Q_I)_{SMB} - [0.39/(1 - 0.39)]Q_S$$
$$= 8,285 - 0.639(8,014) = 3,164 \text{ L/min}$$

Section in Figure 15.45	$(Q_j)_{SMB}$, L/min	$(Q_j)_{TMB}$, L/min
I	8,285	3,164
II	6,635	1,514
III	8,085	2,964
IV	5,395	274

Results of the Aspen Chromatography calculations for the steady-state TMB model, but on an SMB basis are:

Wt% of component	Feed	Desorbent	Extract	Raffinate
Ethylbenzene	14.0	0.0	0.00	7.63
Metaxylene	49.7	0.0	0.00	27.09
Orthoxylene	12.7	0.0	0.00	6.92
PDEB	0.0	100.0	80.79	57.85
Paraxylene	23.6	0.0	19.21	0.51

Note that an excellent separation between paraxylene and the other feed components of the feed is achieved. However, both the extract and raffinate contain a substantial fraction of desorbent, PDEB. The desorbent in both products is recovered by the hybrid SMB-distillation process shown in Figure 15.23. Component concentration profiles in the four sections, as computed by Aspen Chromatography, are shown in Figure 15.49. In Sections I and III particularly, they differ considerably from the flat profile predictions of the simple, local-equilibrium TMB model. The circulating desorbent is predicted to be essentially pure PDEB.

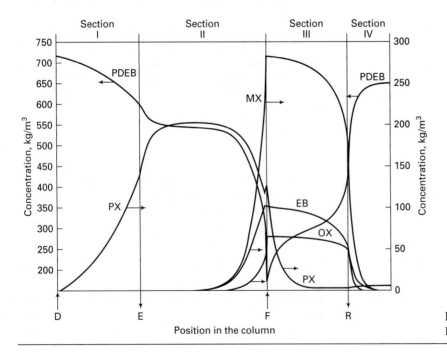

Figure 15.49 Concentration profiles in the liquid for SMB of Example 15.18.

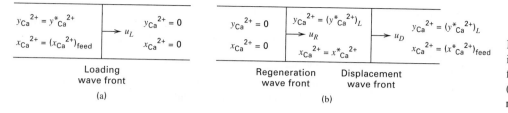

Figure 15.50 Ion exchange in a cyclic operation with a fixed bed. (a) Loading step. (b) Displacement and regeneration steps.

Ion-Exchange Cycle

Although ion exchange has a wide range of applications, water softening with gel resins continues to be the major one. Usually a fixed bed is used, which is operated in a cycle of four steps: (1) loading, (2) displacement, (3) regeneration, and (4) washing. The solute ions removed from water in the loading step are mainly Ca^{2+} and Mg^{2+}, which are absorbed by resin while an equivalent amount of Na^+ is transferred from resin to water as feed solution flows down through the bed. If mass transfer is rapid, the solution and resin are at equilibrium at all points in the bed. With a divalent ion (e.g., Ca^{2+}) replacing a monovalent ion (e.g., Na^+), the equilibrium expression is given by (15-44), where A is the divalent ion. If $(Q/C)^{n-1} K_{A,B} \gg 1$, equilibrium for the divalent ion is very favorable (see Figure 15.29a) and a self-sharpening front of the type shown in Figure 15.29b develops. In that case, which is common, ion exchange is well approximated using simple stoichiometric or shock-wave front theory for adsorption, assuming plug flow. As the front moves down through the bed, the resin behind or upstream of the front is in equilibrium with the feed composition. Ahead or downstream of the front, water is essentially free of the divalent ion(s). Breakthrough occurs when the front reaches the end of the bed.

Suppose the only cations in the feed are Na^+ and Ca^{2+}. Then, from (15-44):

$$K_{Ca^{2+},Na^+}\left(\frac{Q}{C}\right) = \frac{y_{Ca^{2+}}(1 - x_{Ca^{2+}})}{x_{Ca^{2+}}(1 - y_{Ca^{2+}})} \quad (15\text{-}181)$$

where Q is total concentration of the two cations in the resin, in eq/L of bed of wet resin, and C is total concentration of the two ions in the solution, in eq/L of solution. One mole of Na^+ is 1 equivalent, while 1 mole of Ca^{2+} is 2 equivalents. The quantities y_i and x_i are equivalent (rather than mole) fractions. From Table 15.5, using (15-45), the molar selectivity factor is

$$K_{Ca^{2+},Na^+} = 5.2/2.0 = 2.6$$

For a given loading step during water softening, values of Q and C remain constant. Thus, for a given equivalent fraction, $x_{Ca^{2+}}$ in the feed, (15-181) is solved for the equilibrium $y_{Ca^{2+}}$. By material balance, for a given bed volume, the time t_L for the loading step is computed. The loading wavefront velocity is $u_L = L/t_L$ where L is the height of the bed. Equivalent fractions ahead of and behind the loading front are shown in Figure 15.50a. Typically, feed-solution superficial mass velocities are about 15 gal/h-ft^2, but can be much higher at the expense of larger pressure drops.

At the end of the loading step, the bed voids are filled with feed solution, which must be displaced from the bed. This is best done with a regeneration solution, which is usually a concentrated salt solution that flows upwards through the bed. Thus, the displacement and regeneration steps are combined. Following displacement, mass transfer of Ca^{2+} from the resin beads to the regenerating solution takes place while an equivalent amount of Na^+ is transferred from the solution to the resin. In order for equilibrium to be favorable for regeneration with Na^+, it is necessary for $(Q/C)K_{Ca^{2+},Na^+} \ll 1$. In that case, which is just the opposite for loading, the wavefront during regeneration sharpens quickly into a shock-like wave. This criterion can be satisfied by using a saturated salt solution to give a large value for C.

During displacement and regeneration, two concentration waves move through the bed. The first is the displacement front; the second, the regeneration front. For plug flow and negligible mass-transfer resistance, the resin and solution are in equilibrium at all locations in the bed. Again (15-181) is used to solve for the equilibrium equivalent fractions, which are shown for the displacement and regeneration steps in Figure 15.50b. The displacement time, t_D, is determined from the interstitial velocity, u_D, of the fluid during displacement:

$$t_D = L/u_D \quad (15\text{-}182)$$

The regeneration time, t_R, is determined by material balance, from which the regeneration wavefront velocity is $u_R = L/t_R$. In general, the mass velocity of the regeneration solution is less than that of the feed solution. The cycle is completed by displacing, with water, the salt solution in the bed voids. The cycle calculations are illustrated by the following example.

EXAMPLE 15.19

Hard water, containing 500 ppm (by weight) of magnesium carbonate and 50 ppm of NaCl, is to be softened at 25°C in an existing fixed bed of gel resin of a cation capacity of 2.3 eq/L of bed volume. The bed is 8.5 ft in diameter and packed to a height of 10 ft, with a wetted-resin void fraction of 0.38. During the loading step, the recommended throughput is 15 gal/min-ft^2. During displacement, regeneration, and washing, the flow rate is reduced to 1.5 gal/min-ft^2. The displacement and regeneration solutions are water saturated with NaCl (26 wt%). Determine: (a) flow rate of feed solution, L/min, (b) loading time to breakthrough, h, (c) loading wavefront velocity, cm/min, (d) flow rate of regeneration solution, L/min, (e) displacement time, h, (f) additional time for regeneration, h, (g) regeneration wavefront velocity, cm/min, (h) amount of regeneration solution for one cycle, L, and (i) Washing time, h.

SOLUTION

Molecular weight, M, of $MgCO_3 = 83.43$

Concentration of $MgCO_3$ in feed $= \dfrac{500(1,000)}{83.43(1,000,000)}$

$= 0.006$ mol/L or 0.012 eq/L

M of $NaCl = 58.45$

Concentration of NaCl in feed $= \dfrac{50(1,000)}{58.45(1,000,000)}$

$= 0.000855$ mol/L or eq/L

(a) Bed cross-section area $= 3.14(8.5)^2/4 = 56.7$ ft^2.

Feed-solution flow rate $= 15(56.7)$

$= 851$ gpm or 3,219 L/min

(b) Behind the loading wavefront:

$$x_{Ca^{2+}} = \frac{0.012}{0.012 + 0.000855} = 0.9335$$

Since no NaCl in the feed is exchanged: $C = 0.012$ eq/L and $Q = 2.3$ eq/L

From Table 15.5,

$$K_{Mg^{2+},Na^+} = 3.3/2 = 1.65$$

From (15-181), for Mg^{2+} instead of Ca^{2+} as the exchanging ion, with $x_{Mg^{2+}} =$ that of the feed from Figure 15.50a:

$$1.65\left(\frac{2.3}{0.012}\right) = \frac{(y^*_{Mg^{2+}})(1 - 0.9335)}{(0.9335)(1 - y^*_{Mg^{2+}})}$$

Solving: $y^*_{Mg^{2+}} = 0.9998$. Thus, sodium ion is displaced from the resin almost completely.

Bed volume $= (56.7)(10) = 567$ ft^3 or 16,060 L

Total bed capacity $= 2.3(16,060) = 36,940$ eq

Mg^{2+} absorbed by resin $= 0.9998(36,940) = 36,930$ eq

Mg^{2+} entering bed in feed solution $= 0.012(3,219) = 38.63$ eq/min

$$t_L = \frac{36,930}{38.63} = 956 \text{ min or } 15.9 \text{ h}$$

(c) $u_L = L/t_L = 10/956 = 0.01046$ ft/min or 0.319 cm/min.

(d) Flow rate of regeneration solution $= (1.5/15)(3,219) = 321.9$ L/min.

(e) Displacement time $=$ time for 321.9 L/min to displace liquid in the voids.

Void volume $= 0.38(16,060) = 6,103$ L and $t_D = \dfrac{6103}{321.9} = 19$ min

(f) For a 26 wt% NaCl solution at 25°C, density from *Perry's Chemical Engineers' Handbook* $= 1.19443$ g/cm^3.

Flow rate of Na$^+$ in regeneration solution

$= \dfrac{321.9(1,000)(1.19443)(0.26)}{58.45} = 1,710$ eq/min

NaCl concentration in regenerating solution $= \dfrac{1,710}{321.9}$

$= 5.31$ eq/L $= c_R$

From (15-181), noting conditions in Figure 15.49:

$$\frac{Q}{c_R} K_{Mg^{2+},Na^+} = 1.65\left(\frac{2.3}{5.31}\right) = 0.715$$

This is less than 1, but not much less than 1. Therefore, the regeneration wavefront may not sharpen rapidly. Assume a shock-wave-like front anyway.

From (15-181),

$$0.715 = \frac{(0.09998)(1 - x^*_{Mg^{2+}})}{x^*_{Mg^{2+}}(1 - 0.9998)}$$

Solving: $x^*_{Mg^{2+}} = 0.9998$

So downstream of the regeneration wavefront, but upstream of the displacement wavefront, the liquid contains very few sodium ions.

Chromatographic Separations

The separation of multicomponent mixtures into more than two products usually requires more than one separation device. For example, if a four-component mixture (A, B, C, D) is to be separated by distillation into pure products, a sequence of three trayed columns is almost always used. If the order of decreasing volatility is A, B, C, and D, the first column might produce a distillate of nearly pure A; the second column a distillate of nearly pure B; and the third column a distillate of nearly pure C and a bottoms of nearly pure D. Four other sequences are possible, depending upon the selection of the split for each column.

Chromatography is one of the few separation techniques that can separate a multicomponent mixture into nearly pure components in a single device, generally a column packed with a suitable sorbent. The degree of separation depends upon column length and differences in component affinities for the sorbent.

As an example, consider a mixture of three components, A, B, and C, in order of decreasing affinity for the sorbent, S. If the separation is achieved by adsorption, then A is the most strongly adsorbed. A feed mixture, insufficient to load the sorbent, is introduced as a pulse into one end (feed end) of the packed chromatographic column. The resulting initial concentrations for the three components are shown in Figure 15.51a, where most of the bed remains clean of adsorbates. An *elutant,* such as a carrier gas or solvent that has little or no affinity for the sorbent, is now introduced continuously into the feed end of the bed, causing the three components to desorb, with C desorbing most readily. However, as the desorbed components are carried down the bed by the elutant into cleaner regions of the bed, the components are successively readsorbed and then redesorbed to produce three waves, as shown in Figure 15.51b. Because of differences in affinities for the sorbent, the three waves, which initially overlap considerably, gradually overlap less (Figure 15.51c), and finally, if the column is long enough, become completely separated, as in Figure 15.51d and e. In that case, the components are eluted from the column, one at a time. In Figure 15.51e, all components but A have been eluted. As the separated waves elute, the area under each component wave is proportional to the mass of the component moving through the packed column.

Chromatography, as discussed here, is limited to batch processes in which feed is introduced as pulses into an elutant carrier gas or solvent, which then contacts the sorbent. All feed components have affinities for the sorbent, but the elutant does not. This type of chromatography is sometimes

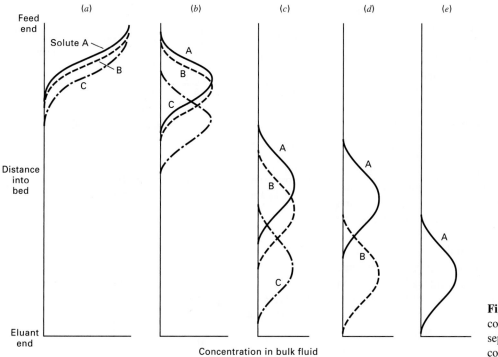

Figure 15.51 Movement of concentration waves during separation in a chromatographic column.

referred to as batch or elution chromatography. Chromatography in the broader sense, as mentioned by Ruthven [10], refers to any separation process involving partitioning of components between a flowing fluid and a solid adsorbent (or a solid-supported liquid absorbent). The previously presented simulated-moving-bed (SMB) system, which uses a circulating desorbent that may also partition, may be viewed as a chromatographic process.

Equilibrium Wave Pulse Theory for Linear Isotherm

A simple and useful wave theory for chromatography is based on isothermal, plug flow, negligible axial dispersion, and local equilibrium everywhere. This theory, when developed for adsorption, results in the stoichiometric wavefront that was shown in Figure 15.27. For chromatography, where solutes are pulsed into the column, a wave pulse rather than a wavefront results, as shown in Figure 15.52a. For a stoichiometric (equilibrium) wave, the pulse is a square wave rather than a Gaussian-distribution-like wave of the type shown in Figure 15.51c. The latter type of wave results when axial dispersion occurs, mass-transfer resistances are important, radial variations of the fluid velocity occur, and/or the pulse of solutes is not a square wave.

If the sorbent is nonporous, such as a gel, and the adsorption isotherm is linear for each solute i ($q = Kc$), then (15-101) applies and each solute wave velocity, u_i, in terms of the interstitial fluid velocity is given by

$$u_i = \frac{u}{1 + \dfrac{1 - \epsilon_b}{\epsilon_b} K_i} \qquad (15\text{-}183)$$

This equation applies to both the leading and trailing edges of the feed pulse to produce solute movement diagrams, used extensively by Wankat [95]. When the elutant is dilute

in the solutes, such that the sorption equilibrium constants do not depend on composition, but only on temperature, (15-183) applies independently to each solute in a multicomponent mixture. For a strongly sorbed solute, such as B in Figure 15.52b, K is large, and the second term in the

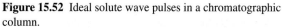

Figure 15.52 Ideal solute wave pulses in a chromatographic column.

denominator dominates, so that solute moves much slower through the bed than does the elutant. For a weakly sorbed solute, such as A in Figure 15.52b, K is small and the denominator is not much greater than 1. For such a solute, its velocity may not be significantly smaller than the elutant velocity.

In Figure 15.52b, the wave velocities of both the leading and trailing edges of the wave pulses are constant, with $u_A > u_B$. For each solute, the pulse time to move through a column of length L is $t_i = L/u_i$. The wave pulse of A reaches the end of the column in less time than the B wave pulse. In Figure 15.52c, the product-concentration ratios, c_A/c_{A_F} and c_B/c_{B_F}, are shown at the end of the bed as a function of time. The widths of these product waves are identical to the widths of the feed pulses, as illustrated in the following example. This simple, equilibrium-wave pulse theory for linear isotherms can be used to obtain an approximate estimate of the separation achievable in a chromatographic column. Unfortunately, the estimate is not conservative when computing the necessary column length, because the pulses broaden, as shown after the following example.

EXAMPLE 15.20

An aqueous solution of 3 g/cm³ each of glucose (G), sucrose (S), and fructose (F) is to be separated in a chromatographic column, packed with an ion-exchange resin of the calcium form. In the range of expected solute concentrations, the sorption isotherms are linear and independent, with $q_i = K_i c_i$, where q_i is in grams sorbent per 100 cm³ resin and c_i is in grams solute per 100 cm³ solution. From experiment:

Solute	K
Glucose	0.26
Sucrose	0.40
Fructose	0.66

The superficial solution velocity is 0.031 cm/s and bed void fraction is 0.39. If a 500-second pulse, t_P, of feed is followed by elution with pure water, what length of column packing is needed to separate the three solutes if sorption equilibrium is assumed? How soon after the first pulse begins can a second 500-second pulse begin?

SOLUTION

Interstitial solution velocity = 0.031/0.39 = 0.0795 cm/s.
Wave velocity for glucose from (15-183):

$$u_G = \frac{0.0795}{1 + \left(\frac{1 - 0.39}{0.39}\right)(0.26)} = 0.0565 \text{ cm/s}$$

Similarly,

$$u_S = 0.0489 \text{ cm/s} \quad \text{and} \quad u_F = 0.0391 \text{ cm/s}$$

The smallest difference in wave velocities is between glucose and sucrose. Therefore, the separation between these two waves determines the column length. The minimum column length, assuming equilibrium, corresponds to the time at which the trailing edge of the glucose wave pulse, together with the leading edge of the

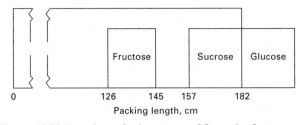

Figure 15.53 Locations of solute waves of first pulse for Example 15.20 at 3,718 s.

sucrose wave pulse, leaves the column. Thus, if t_P is the duration of the first pulse and L is the length of the packing:

$$t_P + \frac{L}{u_G} = \frac{L}{u_S} \tag{1}$$

Thus,

$$500 + \frac{L}{0.0565} = \frac{L}{0.0489}$$

Solving, length of packing, $L = 182$ cm. The glucose just leaves the column at

$$500 + \frac{182}{0.0565} = 3,718 \text{ s}$$

The locations of the three wave fronts in the column at 3,718 s are shown in Figure 15.53.

The time at which the second pulse begins is determined so that the trailing edge of the first fructose wave pulse just leaves the column as the second pulse of glucose begins to leave the column. This time, based on the fructose, is $500 + 182/0.0391 = 5,155$ s. It takes the leading edge of a glucose wave $182/0.0565 = 3,220$ s to pass though the column. Therefore, the second pulse can begin at $5,155 - 3,220 = 1,935$ s. This establishes the following ideal cycle: pulse: 500 s, elute: 1,435 s, pulse: 500 s, elute: 1,435 s, etc. In the real case, where we account for mass-transfer resistance, as shown in the next example, the column will have to be longer.

Analytical Solution for Rate-based Chromatography with Linear Isotherm

A plot of solute concentrations in the elutant as a function of time is a chromatogram. When mass-transfer resistances, axial dispersion, and other non-ideal phenomena are not negligible, the solute concentrations in a chromatogram will not appear as square waves, but will exhibit the wave shapes in Figure 15.51. Carta [96] developed analytical solutions for chromatographic response to periodic injections of rectangular feed pulses, taking into account mass-transfer resistances for solute mixtures having linear, independent adsorption isotherms. Carta's solution for the LDF approximation is readily applied to the determination of the necessary length of packing and frequency of feed pulses for the chromatographic separation of a feed mixture.

For each solute in the feed, (15-102) is simplified by neglecting axial dispersion and assuming a constant interstitial fluid velocity, u, through the packing:

$$\frac{\partial c_i}{\partial t} + u\frac{\partial c_i}{\partial z} + \frac{(1 - \epsilon_b)}{\epsilon_b}\frac{\partial \bar{q}_i}{\partial t} = 0 \tag{15-184}$$

The linear driving force approximation (15-105), the linear isotherm, and (15-106) for the overall mass-transfer resistance are assumed to apply for each solute. For periodic, rectangular feed pulses, the boundary conditions for feed pulses of duration, t_F, each followed by an elution period of duration, t_E, are for each solute concentration, $c_i\{z, t\}$:

Initial condition:

$$\text{At } t = 0, c_i\{z, 0\} = 0 \qquad (15\text{-}185)$$

Feed pulse:

$$\text{At } z = 0, c_i\{0, t\} = (c_i)_F \qquad (15\text{-}186)$$
$$\text{for } (j - 1)(t_F + t_E) < t < j(t_F + t_E) - t_E$$

Elution period:

$$\text{At } z = 0, c_i\{0, t\} = 0 \qquad (15\text{-}187)$$
$$\text{for } j(t_F + t_E) - t_E < t < j(t_F + t_E)$$

where $j = 1, 2, 3, \ldots$ is an index that accounts for the periodic nature of the feed and elution pulses. Thus, with $j = 1$, the feed pulse takes place from $t = 0$ to $t = t_F$ and the elution pulse is from t_F to $t_F + t_E$.

Carta solved the linear system of (15-184), (15-105), and (15-106) for conditions (15-185) to (15-187) by the Laplace transform method to obtain the following series solution, in terms of dimensionless parameters, which is applied to each solute in the feed pulse:

$$X = \frac{r_F}{2r} + \frac{2}{\pi} \sum_{m=1}^{\infty} \left[\frac{1}{m} \exp\left(-\frac{m^2 n_f}{m^2 + r^2} \right) \sin\left(\frac{m\pi r_F}{2r} \right) \right.$$

$$\left. \times \cos\left(\frac{m\theta_f}{r} - \frac{m\pi r_F}{2r} - \frac{m\beta n_f}{r} - \frac{mr n_f}{m^2 + r^2} \right) \right] \qquad (15\text{-}188)$$

where

$$X = c/c_F \qquad (15\text{-}189)$$

$$r = \frac{k}{2\pi K}(t_F + t_E) \qquad (15\text{-}190)$$

$$r_F = \frac{k}{\pi K} t_F \qquad (15\text{-}191)$$

$$n_f = \frac{(1 - \epsilon_b)kz}{\epsilon_b u} \qquad (15\text{-}192)$$

$$\theta_f = \frac{kt}{K} \qquad (15\text{-}193)$$

$$\beta = \frac{\epsilon_b}{(1 - \epsilon_b)K} \qquad (15\text{-}194)$$

$$K = q/c \qquad (15\text{-}195)$$

and where

$$k = \frac{1}{\dfrac{R_p}{3k_c} + \dfrac{R_p^2}{15D_e}} \qquad (15\text{-}196)$$

When nonlinear adsorption isotherms such as the Freundlich equation (15-35), the Langmuir equation (15-36), or extensions thereof to multicomponent mixtures [e.g., concentration forms of (15-32) or (15-33)] are necessary, the analytical solution of Carta is not applicable. However, the method of lines, using five-point, biased, upwind, finite-difference approximations, as described earlier in the section on thermal-swing adsorption, can be applied to obtain a numerical solution.

EXAMPLE 15.21

Use Carta's equation with the following properties to compute the chromatogram for the conditions of Example 15.20 with a packing length of 182 cm. Does a significant overlap of peaks result?

Property	Glucose	Sucrose	Fructose
K	0.26	0.40	0.66
D_e, cm^2/s	1.1×10^{-8}	1.8×10^{-8}	2.8×10^{-8}
k_c, cm/s	5.0×10^{-3}	5.0×10^{-3}	5.0×10^{-3}

$$\epsilon_b = 0.39, \ R_p = 0.0025 \text{ cm}, \ u = 0.0795 \text{ cm/s},$$
$$z = 182 \text{ cm}, \ t_E = 2{,}000 \text{ s}, \text{ and } t_F = 500 \text{ s}$$

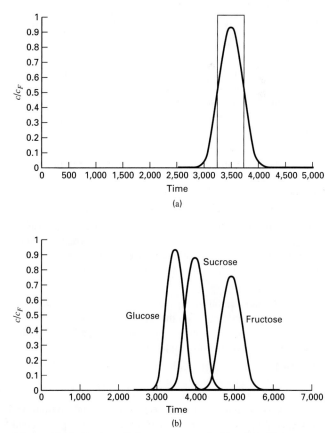

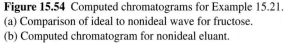

Figure 15.54 Computed chromatograms for Example 15.21.
(a) Comparison of ideal to nonideal wave for fructose.
(b) Computed chromatogram for nonideal eluant.

SOLUTION

Values of k and the computed dimensionless parameters from (15-148) to (15-152) are as follows:

	Glucose	Sucrose	Fructose
r	40.22	42.66	40.06
r_F	16.09	17.07	16.03
n_f	94.13	153.6	238.0
θ_f	$0.1011\,t$	$0.1072\,t$	$0.1007\,t$
β	2.459	1.598	0.9687
k, s^{-1}	0.0263	0.0429	0.0665

where t is in seconds.

Values of $X = c/c_F$ are computed with these parameters using (15-188) for values of time, t, in the neighborhood of times for the equilibrium-based waves. The resulting chromatogram for glucose is shown in Figure 15.54a, compared to the equilibrium rectangular wave (shown as a dashed line) determined in Example 15.20 using (15-183). The areas under the two curves should be identical. The equilibrium-based wave appears to be centered in time within the mass-transfer-based wave.

In Figure 15.54b, the complete computed chromatogram is plotted for the three carbohydrates. It is seen that the effect of mass transfer is to cause the peaks to overlap significantly. To obtain a sharp separation, it is necessary to lengthen the column or reduce the feed pulse time, t_F.

SUMMARY

1. Sorption is a generic term for the selective transfer of a solute from the bulk of a liquid or gas to the surface and/or into the bulk of a solid or liquid. Thus, sorption includes adsorption and absorption. The sorbed solute is commonly called the sorbate.

2. For commercial applications, a sorbent should have high selectivity, high capacity, rapid solute transport rates, stability, strength, and ability to be regenerated. An adsorbent should have small pores so as to give a large surface area per unit volume.

3. Physical adsorption of pure gases and gas mixtures is easily measured. Adsorption of pure liquids and liquid mixtures is not easily measured.

4. The most widely used commercial adsorbents are carbon (activated and molecular-sieve), molecular-sieve zeolites, silica gel, and activated alumina.

5. The most widely used ion exchangers are water-swellable, solid gel resins based on the copolymerization of styrene and a cross-linking agent, such as divinylbenzene. They can be cation or anion exchangers. Ions are exchanged stoichiometrically on an equivalent basis. Thus, Ca^{2+} is exchanged for $2\,Na^+$.

6. Sorbents for chromatographic separations are typically solid adsorbents, liquid absorbents supported on or bonded to an inert solid, or a gel.

7. The most commonly used adsorption isotherms for gases and liquids are Henry's law (linear isotherm), the Freundlich isotherm, and the Langmuir isotherm. The latter asymptotically approaches the linear isotherm at low concentrations and an asymptotic value, representing maximum surface coverage, at high concentrations. For mixtures, extended versions of the Freundlich and Langmuir isotherms are often used.

8. Ion-exchange equilibrium is most commonly represented by an equilibrium constant based on the law of mass action. Because of the dilute conditions common in chromatography, a linear equilibrium isotherm is commonly employed.

9. For physical adsorption, the rate of adsorption is almost instantaneous after the solute reaches the sorbing surface. Thus, only external and internal mass-transfer resistances need be considered. External mass-transfer coefficients are generally obtained from empirical correlations of the Chilton–Colburn j-factor type. Internal mass transfer is generally based on a modified Fick's first law using an effective diffusivity that depends on particle porosity, pore tortuosity, bulk molecular diffusivity, and surface diffusivity. Diffusivities in ion-exchange resin gels depend strongly on the degree of cross-linking.

10. A wide variety of sorption systems are used, including slurry adsorption in various modes of operation, fixed-bed adsorption, and simulated, continuous, countercurrent adsorption. When sorbent regeneration is necessary, the system must be operated on a cycle. For fixed beds, the most common cycles are temperature-swing adsorption (TSA) and pressure-swing adsorption (PSA). Ion exchange almost always includes a regeneration step, using a displacement fluid. In a chromatographic separation, adsorption and regeneration take place in the same column.

11. For the design and operation of all sorption systems, the adsorption isotherm is of great importance because it relates, at equilibrium, the concentration of the solute in the fluid to its loading as a sorbate in and/or on the sorbent. Most commonly, the overall rate of adsorption is expressed in the form of a linear driving force (LDF) model, where the driving force is the difference between bulk concentration and concentration in equilibrium with the loading. The coefficient in the LDF equation is a combined overall mass-transfer coefficient and area for sorption.

12. In ideal, fixed-bed operation, solute–sorbate equilibrium between the flowing fluid and the static bed is assumed everywhere. For plug flow and negligible axial dispersion, the result is a sharp concentration front that moves like a shock wave (stoichiometric front) through the bed. Upstream of the front, the sorbent is spent and in equilibrium with the feed mixture. Downstream of the front, the sorbent is clean of sorbate. Typically, the stoichiometric front travels through the bed at a much slower velocity than the interstitial velocity of the fluid feed. The time for the concentration front to reach the end of the bed is the breakthrough time.

13. When mass-transfer effects are taken into account, the concentration front broadens into an S-shaped curve such that at breakthrough only a portion of the sorbent is fully loaded. When mass-transfer coefficients and sorption isotherms are known, these curves can be readily computed with the Klinkenberg equations. Alternatively, when the shapes of experimental concentration fronts appear to exhibit a constant pattern, because of favorable adsorption equilibrium, commercial-size adsorption beds can be scaled-up directly from experimental breakthrough data by the method of Collins.

14. Thermal-swing adsorption (TSA) can be used to remove small concentrations of solutes from gas and liquid mixtures. Typically, adsorption is carried out at ambient temperature and desorption at an elevated temperature. Because bed heating and cooling between the adsorption and desorption steps is not instantaneous, TSA cycles are long, typically hours or days. The desorption step, starting with a partially loaded bed, can be computed numerically by the method of lines, using a stiff integrator.

15. Pressure-swing adsorption (PSA) is used to separate air and enrich hydrogen-containing streams. Adsorption is carried out at an elevated or ambient pressure, whereas desorption occurs at ambient pressure or in a vacuum; the latter is often referred to as vacuum-swing adsorption (VSA). Because pressure swings can be made rapidly, PSA cycles are short, typically seconds or minutes. Usually, it is not necessary to regenerate the bed completely. When that circumstance exists and the cycle is fixed, a number of cycles may be needed to approach a cyclic steady-state operation.

16. Although continuous, countercurrent adsorption with a moving bed is difficult to achieve successfully in practice, an SMB system is becoming popular, particularly for separation of solutes in dilute aqueous solutions and for bulk-liquid separations. Design procedures for SMB systems, which require solution of systems of differential-algebraic equations, are highly developed.

17. Design calculations for ion-exchange operations are based on the equilibrium assumption for both the loading and regeneration steps.

18. In chromatography, the feed is periodically pulsed into a column packed with sorbent. Between feed pulses, an elutant is passed through the column, causing the less strongly sorbed solutes to move through the column more rapidly than other solutes. If the column is long enough, a multicomponent feed can be completely separated, with solutes eluted one by one from the column. In the absence of mass-transfer resistances, a rectangular feed pulse is separated into individual solute rectangular pulses, whose position–time curves are readily established. When mass-transfer effects are important, the rectangular pulses take on a Gaussian distribution that can be predicted by the analytical solution of Carta, provided that a linear adsorption isotherm applies and axial dispersion is negligible.

REFERENCES

1. KELLER, G.E. II, in *Industrial Gas Separations,* T.E. Whyte, Jr., C.M. Yon, and E.H. Wagner, eds., ACS Symposium Series No. 223, American Chemical Society, Washington, D.C., p. 145 (1983).

2. MILTON, R.M., U.S. Patents 2,882,243 and 2,882,244 (1959).

3. SKARSTROM, C.W., U.S. Patent 2,944,627 (1960).

4. BROUGHTON, D.B., and C.G. GERHOLD, U.S. Patent 2,985,589 (May 23, 1961).

5. ETTRE, L.S., and A. ZLATKIS, Eds., *75 Years of Chromatography—A Historical Dialog,* Elsevier, Amsterdam (1979).

6. BONMATI, R.G., G. CHAPELET-LETOURNEUX, and J.R. MARGULIS, *Chem. Engr.,* **87** (6), 70–72 (1980).

7. BERNARD, J.R., J.P. GOURLIA, and M.J. GUTTIERREZ, *Chem. Engr.,* **88** (10), 92–95 (1981).

8. WHITE, D.H., JR., and P.G. BARKLEY, *Chem. Eng. Progress,* **85** (1) 25–33 (1989).

9. ROUSSEAU, R.W., Ed., *Handbook of Separation Process Technology,* Wiley-Interscience, New York (1987).

10. RUTHVEN, D.M., *Principles of Adsorption and Adsorption Processes,* John Wiley and Sons, New York (1984).

11. BRUNAUER, S., P.H. EMMETT, and E. TELLER, *J. Am. Chem. Soc.,* **60,** 309 (1938).

12. SATTERFIELD, C.N., *Heterogeneous Catalysis in Practice,* McGraw-Hill, New York (1980).

13. RUTHVEN, D.M., in *Kirk-Othmer Encyclopedia of Chemical Technology,* Vol. 1, 4th ed., Wiley-Interscience, New York (1991).

14. BARRER, R.M., *Zeolites and Clay Minerals as Sorbents and Molecular Sieves,* Academic Press, New York (1978).

15. BRECK, D.W., *Zeolite Molecular Sieves,* John Wiley and Sons, New York (1974).

16. ADAMS, B.A., and E.L. HOLMES, *J. Soc. Chem. Ind.,* **54,** 1–6T (1935).

17. MCWILLIAMS, J.D., *Chem. Engr.,* **85** (12), 80–84 (1978).

18. DORFNER, K., *Ion Exchangers, Properties and Applications,* 3rd ed., Ann Arbor Science, Ann Arbor, MI (1971).

19. SEWELL, P.A., and B. CLARKE, *Chromatographic Separations,* John Wiley and Sons, New York (1987).

20. BRUNAUER, S., L.S. DEMING, W.E. DEMING, and E. TELLER, *J. Am. Chem. Soc.,* **62,** 1723–1732 (1940).

21. BRUNAUER, S., *The Adsorption of Gases and Vapors,* Vol. I, *Physical Adsorption,* Princeton University Press (1943).

22. TITOFF, A., *Z. Phys. Chem.,* **74,** 641 (1910).

23. VALENZUELA, D.P., and A.L. MYERS, *Adsorption Equilibrium Data Handbook,* Prentice-Hall, Englewood Cliffs, NJ (1989).

24. RAY, G.C., and E.O. BOX, JR., *Ind. Eng. Chem.,* **42,** 1315–1318 (1950).

25. YANG, R.T., *Gas Separation by Adsorption Processes,* Butterworths, Boston (1987).

26. FREUNDLICH, H., *Z. Phys. Chem.,* **73,** 385–423 (1910).

27. MANTELL, C.L., *Adsorption,* 2nd ed., McGraw-Hill, New York, p. 25 (1951).

28. LANGMUIR, J., *J. Am. Chem. Soc.,* **37,** 1139–1167 (1915).

29. MARKHAM, E.C., and A.F. BENTON, *J. Am. Chem. Soc.,* **53,** 497–507 (1931).

30. YON, C.M., and P.H. TURNOCK, *AIChE Symp. Series,* **67** (117), 75–83 (1971).

31. BROUGHTON, D.B., *Ind. Eng. Chem.,* **40,** 1506–1508 (1948).

32. MYERS, A.L., and J.M. PRAUSNITZ, *AIChE J.,* **11,** 121–127 (1965).

33. MILLER, G.W., K.S. KNAEBEL, and K.G. IKELS, *AIChE J.,* **33,** 194–201 (1987).

34. RITTER, J.A., and R.T. YANG, *Ind. Eng. Chem. Res.,* **26,** 1679–1686 (1987).

35. RUTHVEN, D.M., and F. WONG, *Ind. Eng. Chem. Fundam.,* **24,** 27–32 (1985).

36. KIPLING, J.J., *Adsorption from Solutions of Nonelectrolytes,* Academic Press, London (1965).

37. RADKE, C.J., and J.M. PRAUSNITZ, *AIChE J.,* **18,** 761–768 (1972).

38. ANDERSON, R.E., "Ion-Exchange Separations," in *Handbook of Separation Techniques for Chemical Engineers,* 2nd ed., P.A. Schweitzer, Ed., McGraw-Hill, New York (1988).

39. BONNER, O.D., and L.L. SMITH, *J. Phys. Chem.* **61,** 326–329 (1957).

40. BAUMAN, W.C., and J. EICHHORN, *J. Am. Chem. Soc.,* **69,** 2830–2836 (1947).

41. BAUMAN, W.C., J.R. SKIDMORE, and R.H. OSMUN, *Ind. Eng. Chem.,* **40,** 1350–1355 (1948).

42. SUBBA RAO, H.C., and M.M. DAVID, *AIChE Journal,* **3,** 187–190 (1957).

43. SELKE, W.A., and H. BLISS, *Chem. Eng. Prog.,* **46,** 509–516 (1950).

44. SELKE, W.A., and H. BLISS, *Chem. Eng. Prog.,* **47,** 529–533 (1951).

45. MYERS, A.L., and S. BYINGTON, in *Ion Exchange: Science and Technology,* A.E. Rodrigues, Ed., Martinus Nijhoff, Boston, pp. 119–145.

46. RANZ, W.E., and W.R. MARSHALL, JR., *Chem. Eng. Prog.,* **48,** 141–146 (1952).

47. RANZ, W.E., and W.R. MARSHALL, JR., *Chem. Eng. Prog.,* **48,** 173–180 (1952).

48. RANZ, W.E., *Chem. Eng. Prog.,* **48,** 247–253 (1952).

49. GAMSON, B.W., G. THODOS, and O.A. HOUGEN, *Trans. AIChE,* **39,** 1–35 (1943).

50. CHILTON, T.H., and A.P. COLBURN, *Ind. Eng. Chem.,* **26,** 1183–1187 (1934).

51. SEN GUPTA, A., and G. THODOS, *AIChE J.,* **9,** 751–754 (1963).

52. PETROVIC, L.J., and G. THODOS, *Ind. Eng. Chem. Fundamentals,* **7,** 274–280 (1968).

53. DWIVEDI, P.N., and S.N. UPADHAY, *Ind. Eng. Chem. Process Des. Dev.,* **16,** 157–165 (1977).

54. WAKAO, N., and T. FUNAZKRI, *Chem. Eng. Sci.,* **33,** 1375–1384 (1978).

55. KUNII, D., and O. LEVENSPIEL, *Fluidization Engineering,* 2nd ed., Butterworth-Heinemann, Boston, Chap. 3 (1991).

56. THIELE, E.W., *Ind. Eng. Chem.,* **31,** 916–920 (1939).

57. WHEELER, A., *Advances in Catalysis,* Vol. 3, Academic Press, New York, pp. 249–327 (1951).

58. SCHNEIDER, P., and J.M. SMITH, *AIChE J.,* **14,** 886–895 (1968).

59. RIEKERT, L., *AIChE J.,* **31,** 863–864 (1985).

60. SLADEK, K.J., E.R. GILLILAND, and R.F. BADDOUR, *Ind. Eng. Chem. Fundam.,* **13,** 100–105 (1974).

61. KAPOOR, A., R.T. YANG, and C. WONG, "Surface Diffusion," *Catalyst Reviews,* **31,** 129–214 (1989).

62. HELFFERICH, F., *Ion Exchange,* McGraw-Hill, New York (1962).

63. SOLDANO, B.A., *Ann. NY Acad. Sci.,* **57,** 116–124 (1953).

64. KELLER, G.E., "Separations: New Directions for an Old Field," *AIChE Monograph Series,* **83** (17) (1987).

65. RUTHVEN, D.M., and C.B. CHING, *Chem. Eng. Sci.,* **44,** 1011–1038 (1989).

66. BERG, C., *Trans. AIChE,* **42,** 665–680 (1946).

67. BROUGHTON, D.B., *Chem. Eng. Progress,* **64** (8), 60–65 (1968).

68. GEMBICKI, S.A., A.R. OROSKAR, and J.A. JOHNSON, in *Encyclopedia of Chemical Technology,* 4th edition, Vol. 1, John Wiley, New York, pp. 573–600 (1991).

69. HIGGINS, I.R., and J.T. ROBERTS, *Chem. Engr. Prog. Symp. Ser.,* **50** (14), 87–92 (1954).

70. HIMSLEY, A., Canadian Patent 980,467 (Dec. 23, 1975).

71. GANETSOS, G., and P.E. BARKER, Ed., *Preparative and Production Scale Chromatography,* Marcel Dekker, New York (1993).

72. MARTIN, A.J.P., *Disc. Faraday Soc.,* **7,** 332 (1949).

73. DEVAULT, D., *J. Am. Chem Soc.,* **65,** 532–540 (1943).

74. GLUECKAUF, E., *Trans. Faraday Soc.,* **51,** 1540–1551 (1955).

75. GLUECKAUF, E., and J.E. COATES, *J. Chem. Soc.,* 1315–1321 (1947).

76. LIAW, C.H., J.S.P. WANG, R.A. GREENKORN, and K.C. CHAO, *AIChE J.,* **25,** 376–381 (1979).

77. KLINKENBERG, A., *Ind. Eng. Chem.,* **46,** 2285–2289 (1954).

78. ANZELIUS, A., Z., *Angew. Math u. Mech.,* **6,** 291–294 (1926).

79. KLINKENBERG, A., *Ind. Eng. Chem.,* **40,** 1992–1994 (1948).

80. COONEY, D.O., and E.N. LIGHTFOOT, *IEC Fundamentals,* **4,** 233–236 (1965).

81. SIRCAR, S., and K. KUMAR, *Ind. Eng. Chem. Process Des. Dev.,* **22,** 271–280 (1983).

82. COONEY, D.O., *Chem. Eng. Comm.,* **91,** 1–9 (1990).

83. COLLINS, J.J., *Chem. Eng. Prog. Symp. Ser.* **63** (74), 31–35 (1967).

84. WONG, Y.W., and J.L. NIEDZWIECKI, *AIChE Symposium Series,* **78** (219), 120–127 (1982).

85. LISKOVETS, O.A., *Differential Equations* (a translation of *Differentsial'nye Uravneniya*) **1,** 1308–1323 (1965).

86. SCHIESSER, W.E., *The Numerical Method of Lines Integration of Partial Differential Equations,* Academic Press, San Diego (1991).

87. SCHIESSER, W.E., *Computational Mathematics in Engineering and Applied Science,* CRC Press, Boca Raton, FL (1994).

88. PRESS, W.H., S.A. TEUKOLSKY, W.T. VETTERLING, and B.P. FLANNERY, *Numerical Recipes in FORTRAN,* 2nd ed., Cambridge University Press, Cambridge (1992).

89. GEAR, C.W., *Numerical Initial Value Problems in Ordinary Differential Equations,* Prentice-Hall, Englewood Cliffs, NJ (1971).

90. BYRNE, G.D., and A.C. HINDMARSH, *J. Comput. Phys.,* **70,** 1–62 (1987).

91. RITTER, J.A., and R.T. YANG, *Ind. Eng. Chem. Res.,* **30,** 1023–1032 (1991).

92. RUTHVEN, D.M., S. FAROOQ, and K.S. KNAEBEL, *Pressure-Swing Adsorption,* VCH, New York (1994).

93. MUTASIM, Z.Z., and J.H. BOWEN, *Trans. I. Chem. E.,* **69,** Part A, 108–118 (March 1991).

94. EAGLETON, L.C., and H. BLISS, *Chem. Eng. Progress,* **49,** 543–548 (1953).

95. WANKAT, P.C., *Rate-Controlled Separations,* Elsevier Applied Science, New York (1990).

96. CARTA, G., *Chem. Eng. Sci.,* **43,** 2877–2883 (1988).

97. WANKAT, P.C., *Large-Scale Adsorption and Chromatography,* Vols. I and II, CRC Press, Inc., Boca Raton (1986).

98. BROUGHTON, D.B., R.W. NEUZIL, J.M. PHARIS, and C.S. BREARBY, *Chem. Eng. Prog.,* **66** (9), 70–75 (1970).

99. ZANG, Y., and P.C. WANKAT, *Ind. Eng. Chem. Res.,* **41,** 5283–5289 (2002).

100. KIM, J.K., and P.C. WANKAT, *Ind. Eng. Chem. Res.,* **43,** 1071–1080 (2004).

101. HUMPHREY, J.L., and G.E. Keller II, *Separation Process Technology,* McGraw-Hill, New York (1997).

102. WANKAT, P.C., *Ind. Eng. Chem. Res.,* **40,** 6185–6193 (2001).

103. STORTI, G., M. MASI, S. CARRA, and M. MORBIDELLI, *Chem. Eng. Sci.,* **44,** 1329 (1989).

104. RHEE, H.-K., R. ARIS, and N.R. AMUNDSON, *Phil. Trans. Royal Soc. London, Series A,* **269** (No. 1194), 187–215 (Feb. 5, 1971).

105. MAZZOTTI, M., G. STORTI, and M. MORBIDELLI, *AIChE Journal,* **40,** 1825–1842 (1994).

106. RUTHVEN, D.M., and C.B. CHING, *Chem. Eng. Sci.,* **44,** 1011–1038 (1989).

107. ZHONG, G., and G. GUIOCHON, *Chem. Eng. Sci.,* **51,** 4307–4319 (1996).

108. STORTI, G., M. MAZZOTTI, M. MORBIDELLI, and S. CARRA, *AIChE Journal,* **39,** 471–492 (1993).

109. MAZZOTTI, M., G. STORTI, and M. MORBIDELLI, *J. Chromatography A,* **769,** 3–24 (1997).

110. DANCKWERTS, P.V., *Chem. Eng. Sci.,* **2,** 1 (1953).

111. CONSTANTINIDES, A., and N. MOSTOUFI, *Numerical Methods for Chemical Engineers with MATLAB Applications,* Prentice Hall PTR, Upper Saddle River, NJ (1999).

112. MINCEVA, M. and A. E. RODRIGUES, *Ind. Eng. Chem. Res.,* **41,** 3454–3461 (2002).

EXERCISES

Section 15.1

15.1 Porous particles of activated alumina have a BET surface area of 310 m^2/g, a particle porosity of 0.48, and a particle density of 1.30 g/cm^3. Determine: (a) specific pore volume in cm^3/g, (b) true solid density, g/cm^3, and (c) approximate pore diameter in angstroms from (15-2).

15.2 Carbon molecular sieves are available in two forms from a Japanese manufacturer:

	Form A	Form B
Pore volume, cm^3/g	0.18	0.38
Average pore diameter	5 Å	2.0 μm

Estimate the surface area of each form.

15.3 Representative properties of small-pore silica gel are as follows: pore diameter = 24 Å; particle porosity = 0.47; particle density = 1.09 g/cm^3; and specific surface area = 800 m^2/g

(a) Are these values reasonably consistent? (b) If the adsorption capacity for water vapor at 25°C and 6 mmHg partial pressure is 18% by weight, what fraction of a monolayer is adsorbed?

15.4 The following data were obtained in a BET apparatus for adsorption equilibrium of nitrogen on silica gel (SG) at –195.8°C. Estimate the specific surface area in m^2/g of silica gel. How does your value compare with that in Table 15.2?

N_2 Partial Pressure, torr	Volume of N_2 Adsorbed in cm^3 (0°C, 1 atm) per gram SG
6.0	6.1
24.8	12.7
140.3	17.0
230.3	19.7
285.1	21.5
320.3	23.0
430	27.7
505	33.5

15.5 Estimate the maximum ion-exchange capacity in meq/g resin for an ion-exchange resin made from 8 wt% divinylbenzene and 92 wt% styrene.

Section 15.2

15.6 Shen and Smith [*Ind. Eng. Chem. Fundam.,* **7,** 100–105 (1968)] measured equilibrium-adsorption isotherms at four different temperatures for pure benzene vapor on silica gel, having the following properties: surface area = 832 m^2/g, pore volume = 0.43 cm^3/g, particle density = 1.13 g/cm^3, and average pore diameter = 22 Å.

The adsorption data are as follows:

Partial Pressure of Benzene, atm	Moles Adsorbed/g Gel $\times 10^5$			
	70°C	90°C	110°C	130°C
5.0×10^{-4}	14.0	6.7	2.6	1.13
1.0×10^{-3}	22.0	11.2	4.5	2.0
2.0×10^{-3}	34.0	18.0	7.8	3.9
5.0×10^{-3}	68.0	33.0	17.0	8.6
1.0×10^{-2}	88.0	51.0	27.0	16.0
2.0×10^{-2}	—	78.0	42.0	26.0

(a) For each temperature, obtain a best fit of the data to (1) linear, (2) Freundlich, and (3) Langmuir isotherms. Which isotherm(s), if any, fit the data reasonably well?

(b) Do the data represent less than a monolayer of adsorption?

(c) From the data, estimate the heat of adsorption. How does this value compare to the heat of vaporization (condensation) of benzene?

15.7 The separation of propane and propylene is accomplished by distillation, but at the expense of more than 100 trays and a reflux ratio of greater than 10. Consequently, the use of adsorption has been investigated in a number of studies. Jarvelin and Fair [*Ind. Eng. Chem. Research,* **32,** 2201–2207 (1993)] measured adsorption-equilibrium data at 25°C for three different zeolite molecular sieves (ZMSs) and activated carbon. The data were fitted to the Langmuir isotherm with the following results:

Adsorbent	Sorbate	q_m	K
ZMS 4A	C_3	0.226	9.770
	$C_3^=$	2.092	95.096
ZMS 5A	C_3	1.919	100.223
	$C_3^=$	2.436	147.260
ZMS 13X	C_3	2.130	55.412
	$C_3^=$	2.680	100.000
Activated carbon	C_3	4.239	58.458
	$C_3^=$	4.889	34.915

where q and q_m are in mmol/g and p is in bar.

(a) Which component is most strongly adsorbed by each of the adsorbents? (b) Which adsorbent has the greatest adsorption capacity? (c) Which adsorbent has the greatest selectivity? (d) Based on equilibrium considerations, which adsorbent is best for the separation?

15.8 Ruthven and Kaul [*Ind. Eng. Chem. Res.,* **32,** 2047–2052 (1993)] measured adsorption isotherms for a series of gaseous aromatic hydrocarbons on well-defined crystals of NaX zeolite over ranges of temperature and pressure. For 1,2,3,5-tetramethylbenzene at 547 K, the following equilibrium data were obtained with a vacuum microbalance:

q, wt%	7.0	9.1	10.3	10.8	11.1	11.5
p, torr	0.012	0.027	0.043	0.070	0.094	0.147

Obtain a best fit of the data to the linear, Freundlich, and Langmuir isotherms, with q in mol/g and pressure in atm. Which isotherm gives the best fit?

15.9 Lewis, Gilliland, Chertow, and Hoffman [*J. Am. Chem. Soc.,* **72**, 1153–1157 (1950)] measured adsorption equilibria for pure propane, pure propylene, and binary mixtures thereof, on activated carbon and silica gel. Adsorbate capacity was high on carbon, but selectivity was poor. Selectivity was high on silica gel, but capacity was low. For silica gel (751 m²/g), the following pure component data were obtained at 25°C:

Propane		Propylene	
P, torr	q, mmol/g	P, torr	q, mmol/g
11.1	0.0564	34.2	0.3738
25.0	0.1252	71.4	0.7227
43.5	0.1980	91.6	0.7472
71.4	0.2986	194.3	1.129
100.0	0.3850	198.3	1.168
158.9	0.5441	271.5	1.401
227.5	0.7020	353.2	1.562
304.2	0.843	550.7	1.918
387.0	1.010	555.2	1.928
468.0	1.138	760.6	2.184
569.0	1.288		
677.8	1.434		
775.0	1.562		

The following mixture data were measured at 25°C, over a pressure range of 752–773 torr:

Total Pressure, torr	Millimoles of Mixture Adsorbed/g	y_{C_3}, Mole Fraction in Gas Phase	x_{C_3}, Mole Fraction in Adsorbate
769.2	2.197	0.2445	0.1078
760.9	2.013	0.299	0.2576
767.8	2.052	0.4040	0.2956
761.0	2.041	0.530	0.2816
753.6	1.963	0.5333	0.3655
766.3	1.967	0.5356	0.3120
754.0	1.974	0.6140	0.3591
753.6	1.851	0.6220	0.5550
754.0	1.701	0.6252	0.7007
760.0	1.686	0.7480	0.723
—	2.180	0.671	0.096
760.0	1.993	0.8964	0.253
760.0	1.426	0.921	0.401

(a) Fit the pure component data to Freundlich and Langmuir isotherms. Which gives the best fit? Which component is most strongly adsorbed?

(b) Use the results of the Langmuir fits in part (a) to predict binary-mixture adsorption using the extended Langmuir equation, (15-32). Are the predictions adequate?

(c) Ignoring the pure-component data, fit the binary-mixture data to the extended Langmuir equation, (15-32). Is the fit better than that obtained in part (b)?

(d) Ignoring the pure-component data, fit the binary mixture data to the extended Langmuir–Freundlich equation, (15-33). Is the fit adequate? Is the fit better than that in part (c)?

(e) For the binary-mixture data, compute the relative selectivity,

$$\alpha_{C_3, C_3^=} = y_{C_3}(1 - x_{C_3})/[x_{C_3}(1 - y_{C_3})]$$

for each condition. Does α vary widely or is the assumption of constant α reasonable?

15.10 In Example 15.6, pure-component, liquid-phase adsorption data are used with the extended-Langmuir isotherm to predict a binary-solute data point. Use the following mixture data to obtain the best fit to an extended Langmuir–Freundlich isotherm of the form

$$q_i = \frac{(q_0)_i k_i c_i^{1/n_i}}{1 + \sum_j k_j c_j^{1/n_j}} \tag{1}$$

Data for binary-mixture adsorption on activated carbon (1000 m²/g) at 25°C for acetone (1) and propionitrile (2) are as follows:

Solution Concentration, mol/L		Loading, mmol/g	
c_1	c_2	q_1	q_2
5.52E − 5	7.46E − 5	0.0192	0.0199
6.14E − 5	7.71E − 5	0.0191	0.0198
1.06E − 4	1.35E − 4	0.0308	0.0320
1.12E − 4	1.46E − 4	0.0307	0.0319
3.03E − 4	2.32E − 3	0.0378	0.263
3.17E − 4	2.34E − 3	0.0378	0.264
3.25E − 4	3.89E − 4	0.0644	0.0672
1.42E − 3	1.58E − 3	0.161	0.169
1.42E − 3	1.61E − 3	0.161	0.169
1.43E − 3	1.60E − 3	0.161	0.169
2.09E − 3	3.84E − 4	0.250	0.0390
2.17E − 3	3.85E − 4	0.251	0.0392
4.99E − 3	5.24E − 3	0.291	0.307
5.06E − 3	5.31E − 3	0.288	0.305
7.41E − 3	2.42E − 2	0.237	0.900
7.52E − 3	2.47E − 2	0.236	0.896
2.79E − 2	7.59E − 3	0.802	0.251
4.00E − 2	3.44E − 2	0.715	0.822
4.02E − 2	3.42E − 2	0.717	0.834

15.11 Sircar and Myers [*J. Phys. Chem.,* **74**, 2828–2835 (1970)] measured liquid-phase adsorption at 30°C for a binary mixture of cyclohexane (1) and ethyl alcohol (2) on activated carbon. Assuming no adsorption of ethyl alcohol, they used (15-34) to obtain the following results:

x_1	q_1^e, mmol/g	x_1	q_1^e, mmol/g
0.042	0.295	0.440	0.065
0.051	0.485	0.470	0.000
0.072	0.517	0.521	−0.129
0.148	0.586	0.537	−0.362
0.160	0.669	0.610	−0.643
0.213	0.661	0.756	−1.230
0.216	0.583	0.848	−1.310
0.249	0.595	0.893	−1.180
0.286	0.532	0.920	−1.230
0.341	0.383	0.953	−0.996
0.391	0.192	0.974	−0.470

(a) Plot the data as q_1^e against x_1. Explain the shape of the curve.

(b) In what regions of concentration could the Freundlich isotherm be fitted to the data? Make the fits.

15.12 Both the adsorptive removal of small amounts of toluene from water and small amounts of water from toluene are important in the process industries. Activated carbon is particularly effective for removing soluble organic compounds (SOCs) from water. Activated alumina is effective for removing soluble water from toluene. Fit each of the following two sets of equilibrium data for 25°C to both the Langmuir and Freundlich isotherms. For each case, which isotherm provides the better fit? Could a linear isotherm be used?

Toluene (in Water) Activated Carbon		Water (in Toluene) Activated Alumina	
c, mg/L	q, mg/g	c, ppm (by Weight)	q, g/100g
0.01	12.5	25	1.9
0.02	17.1	50	3.1
0.05	23.5	75	4.2
0.1	30.3	100	5.1
0.2	39.2	150	6.5
0.5	54.5	200	8.2
1	90.1	250	9.5
2	70.2	300	10.9
5	125.5	350	12.1
10	165	400	13.3

15.13 Derive (15-44). Use this equation to solve the following problem. Sulfate ion is to be removed from 60 L of water by exchanging it with chloride ion on 1 L of a strong-base resin with relative molar selectivities as listed in Table 15.6 and an ion-exchange capacity of 1.2 eq/L of resin. The water to be treated has a sulfate-ion concentration of 0.018 eq/L and a chloride-ion concentration of 0.002 eq/L. Following the attainment of equilibrium ion exchange, the treated water will be removed and the resin will be regenerated with 30 L of 10 wt% aqueous NaCl.

(a) Write the ion-exchange reaction.

(b) Determine the value of $K_{SO_4^{2-}, Cl^-}$.

(c) Calculate equilibrium concentrations $c_{SO_4^{2-}}$, c_{Cl^-}, $q_{SO_4^{2-}}$, and q_{Cl^-} in eq/L for the initial ion-exchange step.

(d) Calculate the concentration of Cl^- in eq/L for the regenerating solution.

(e) Calculate $c_{SO_4^{2-}}$, c_{Cl^-}, $q_{SO_4^{2-}}$, and q_{Cl^-} upon reaching equilibrium in the regeneration step.

(f) Are the separations sufficiently selective?

15.14 Silver ion in methanol was exchanged with sodium ion using Dowex 50 cross-linked with 8% divinylbenzene by Gable and Stroebel [*J. Phys. Chem.*, **60**, 513–517 (1956)]. The molar selectivity coefficient was found to vary somewhat with the equivalent fraction of Na^+ in the resin as follows:

x_{Na^+}	0.1	0.3	0.5	0.7	0.9
K_{Ag^+, Na^+}	11.2	11.9	12.3	14.1	17.0

If the wet capacity of the resin is 2.5 eq/L and the resin is initially saturated with Na^+, calculate the equilibrium equivalent fractions if 50 L of 0.05-M Ag^+ in methanol is treated with 1 L of wet resin.

15.15 Ion exclusion is a process that uses ion-exchange resins to separate nonionic organic compounds from ionic species contained in a polar solvent, usually water. The resin is presaturated with the same ions as in the solution, thus eliminating ion exchange. However, in the presence of the polar solvent, resins undergo considerable swelling by absorbing the solvent. Experiments have shown that a nonionic solute will distribute between the solution outside the resin and the solution within the resin, while the ions can only exchange.

A feed solution of 1,000 kg contains 6 wt% NaCl, 35 wt% glycerol, and 47 wt% water. This solution is to be treated with Dowex-50 ion-exchange resin in the sodium form, after prewetting with water, to recover 75% of the glycerol. The following data for the glycerol distribution coefficient,

$$K_d = \frac{\text{mass fraction in solution inside resin}}{\text{mass fraction in solution outside resin}}$$

were reported by Asher and Simpson [*J. Phys. Chem.*, **60**, 518–521 (1956)]:

Mass Fraction Glycerol in Solution Outside Resin	K_d	
	6 wt% NaCl	12 wt% NaCl
0.10	0.75	0.91
0.20	0.80	0.93
0.30	0.83	0.95
0.40	0.85	0.97

If the prewetted resin contains 40 wt% water, determine the kilograms of resin (dry basis) required.

Section 15.3

15.16 Benzene vapor in an air stream is adsorbed in a fixed bed of 4×6 mesh silica gel packed to an external void fraction of 0.5. The bed is 2 feet in inside diameter and the air flow rate is 25 lb/min (benzene-free basis). At a location in the bed where the pressure is 1 atm, the temperature is 70°F, and the bulk mole fraction of benzene is 0.005, estimate the external, gas-to-particle mass-transfer and heat-transfer coefficients.

15.17 Water vapor in an air stream is to be adsorbed in a 12.06-cm-inside-diameter column packed with 3.3-mm-diameter Alcoa F-200 activated alumina beads with an external porosity of 0.442. At a location in the bed where the pressure is 653.3 kPa, the temperature is 21°C, the gas flow rate is 1.327 kg/min, and the dew-point temperature is $11.2°C$, estimate the external, gas-to-particle mass-transfer and heat-transfer coefficients.

15.18 For the conditions of Example 15.8 estimate the effective diffusivity of acetone vapor in the pores of activated carbon with the following properties: particle density $= 0.85$ g/cm³, particle porosity $= 0.48$, average pore diameter $= 25$ Å, and tortuosity $= 2.75$.

Consider both bulk and Knudsen diffusion, but ignore surface diffusion.

15.19 For the conditions of Exercise 15.16, estimate the effective diffusivity of benzene vapor in the pores of silica gel with the following properties: particle density $= 1.15$ g/cm³, particle porosity $= 0.48$, average pore diameter $= 30$ Å, and tortuosity $= 3.2$.

Consider all mechanisms of diffusion. The adsorption equilibrium constant is given in Example 15.11, and the differential heat of adsorption is $-11,000$ cal/mol.

15.20 For the conditions of Exercise 15.17, estimate the effective diffusivity of water vapor in the pores of activated alumina with the following properties: particle density $= 1.38$ g/cm³, particle porosity $= 0.52$, average pore diameter $= 60$ Å, and tortuosity $= 2.3$.

Consider all mechanisms of diffusion except surface diffusion.

Section 15.4

15.21 Adsorption with activated carbon, made from bituminous coal, of soluble organic compounds (SOCs) to purify surface and ground water is a proven technology, as discussed by Stenzel [*Chem. Eng. Prog.*, **89** (4), 36–43 (1993)]. The less-soluble organic compounds, such as chlorinated organic solvents and aromatic solvents, are the more strongly adsorbed. Water containing 3.3 mg/L of trichloroethylene (TCE) is to be treated with activated carbon to obtain an effluent with only 0.01 mg TCE/L. At $25°C$, adsorption equilibrium data for TCE on activated carbon are correlated with the following Freundlich equation:

$$q = 67\, c^{0.564} \qquad (1)$$

where

$q = $ mg TCE/g carbon and $c = $ mg TCE/L solution

The TCE is to be removed by slurry adsorption using a powdered form of the activated carbon, with an average particle diameter of 1.5 mm. In the absence of any laboratory data on mass-transfer rates, assume that the rate of adsorption for the small particles is controlled by external mass transfer with a Sherwood number of 30. Particle surface area is 5 m²/kg. The molecular diffusivity of TCE in low concentrations in water at $25°C$ may be determined from the Wilke–Chang equation.

(a) Determine the minimum amount of adsorbent needed.

(b) For operation in the batch mode with twice the minimum amount of adsorbent, determine the time to reduce the TCE content to the desired value.

(c) For operation in the continuous mode using twice the minimum amount of adsorbent, determine the required residence time.

(d) For operation in the semicontinuous mode at a feed rate of 50 gpm and for a liquid residence time equal to 1.5 times that computed in part (c), determine the amount of activated carbon to give a reasonable vol% solids in the tank and a run time of not less than 10 times the liquid residence time.

15.22 Repeat Exercise 15.21 for water containing 0.324 mg/L of benzene (B) and 0.630 mg/L of m-xylene (X).

Adsorption isotherms at $25°C$ for these low concentrations are essentially independent and are given by

$$q_B = 32\, c_B^{0.428} \qquad (1)$$
$$q_X = 125\, c_X^{0.333} \qquad (2)$$

The feed concentrations of the SOCs in the feed are to be reduced to 0.002 mg/L each.

15.23 Repeat Exercise 15.21 for water containing 0.223 mg/L chloroform, whose concentration is to be reduced to 0.010 mg/L. The adsorption isotherm at $25°C$ is given by

$$q = 10\, c^{0.564} \qquad (1)$$

15.24 Three fixed-bed adsorbers containing 10,000 lb of granules of activated carbon ($\rho_b = 30$ lb/ft³) each are to be used to treat 250 gpm of water containing 4.6 mg/L of 1,2-dichloroethane (D) to reduce the concentration to less than 0.001 mg/L. Each carbon bed has a height equal to twice the diameter. Two beds are to be placed in series so that when Bed 1 (the lead bed) becomes saturated with D at the feed concentration, that bed is removed. Bed 2 (the trailing bed), which is partially saturated at this point, depending upon the width of the MTZ, becomes the lead bed, and previously idle Bed 3 takes the place of Bed 2. While Bed 1 is off-line, its spent carbon is removed and replaced with fresh carbon. The spent carbon is incinerated. The equilibrium adsorption isotherm for D is given by $q = 8\, c^{0.57}$, where q is in mg/g and c is in mg/L. Once the cycle is established, how often must the carbon in a bed be replaced? What is the maximum width of the MTZ that will allow saturated loading of the lead bed?

15.25 The fixed-bed adsorber series arrangement of Exercise 15.24 is to be used to treat 250 gpm of water containing 0.185 mg/L of benzene (B) and 0.583 mg/L of m-xylene (X). However, because the two solutes may have considerably different breakthrough times, more than two operating beds in series may be needed. The adsorption isotherms are given in Exercise 15.22, where q is in mg/g and c is in mg/L. From laboratory measurements, the widths of the mass-transfer zones are estimated to be $MTZ_B = 2.5$ ft and $MTZ_X = 4.8$ ft. Once the cycle is established, how often must the carbon in the bed be replaced?

15.26 Air at $80°F$, 1 atm, 80% relative humidity, and a superficial velocity of 100 ft/min passes through a 5-ft-high bed of 2.8-mm-diameter spherical particles of silica gel ($\rho_b = 39$ lb/ft³). The adsorption equilibrium isotherm at $80°F$ is given by

$$q_{H_2O} = 15.9\, p_{H_2O} \qquad (1)$$

where q is in lb H_2O/lb gel and p is in atm. The overall mass-transfer coefficient can be estimated from (15-106), using an effective diffusivity of 0.05 cm²/s and with k_c estimated from (15-65). Using the approximate concentration-profile equations of Klinkenberg, compute a set of breakthrough curves and determine the time when the humidity of the exiting air reaches 0.0009 lb H_2O/lb dry air. Assume isothermal and isobaric operation. Compare the time to breakthrough with the time for the equilibrium model. At breakthrough, what is the approximate width of the mass-transfer zone. What is the average loading of the bed at breakthrough?

15.27 A train of four 55-gallon cannisters of activated carbon is to be used to reduce the nitroglycerine (NG) content of 400 gph of wastewater from 2,000 ppm by weight to less than 1 ppm. Each cannister has a diameter of 2 ft and holds 200 lb activated carbon ($\rho_b = 32$ lb/ft^3). Each cannister is equipped with a liquid-flow distributor to promote plug flow through the bed of carbon. The effluent from the first cannister is monitored so that when a 1 ppm threshold of NG is reached, that cannister is removed from the train and a fresh cannister is added to the end of the train. The spent carbon is mixed with coal for use as a fuel in a coal-fired power plant at the process site. Using the following pilot-plant data, estimate how many cannisters are needed each month and the monthly cannister cost at $700 per cannister.

Pilot-plant data:

Tests with the same 55-gallon cannister to be used in the commercial process; water flow rate = 10 gpm; NG content in feed = 1,020 ppm by weight.

Breakthrough correlation:

$t_B = 3.90\,L - 2.05$, where t_B = time, h, at breakthrough of the 1 ppm threshold and L = bed depth of carbon in feet.

15.28 Air at a flow rate of 12,000 scfm (60°F, 1 atm) and containing 0.5 mol% ethyl acetate (EA) and no water vapor is to be treated with activated carbon (C) ($\rho_b = 30$ lb/ft^3) with an equivalent particle diameter of 0.011 ft in a fixed-bed adsorber to remove the ethyl acetate, which will be subsequently stripped from the carbon by steam at 230°F. Based on the following data, determine the diameter and height of the carbon bed, assuming adsorption at 100°F and 1 atm and a time-to-breakthrough of 8 h with a superficial gas velocity of 60 ft/min. If the bed height-to-diameter is unreasonable, what change in design basis would you suggest?

Adsorption isotherm data (100°F) for EA:

p^{EA}, atm	q, lb EA/lb C	p^{EA}, atm	q, lb EA/lb C
0.0002	0.125	0.0020	0.227
0.0005	0.164	0.0050	0.270
0.0010	0.195	0.0100	0.304

Breakthrough data at 100°F and 1 atm for EA in air at a gas superficial velocity of 60 ft/min in a 2-ft dry bed:

Mole Fraction EA in Effluent	Time, Min	Mole Fraction EA in Effluent	Time, Min
0.00005	60	0.00100	95
0.00010	66	0.00250	120
0.00025	75	0.00475	160
0.00050	84		

15.29 In Examples 15.11 and 15.13, benzene is adsorbed from air at 70°F in a 6-ft-high bed of silica gel and then stripped with air at 145°F. If the bed height is changed to 30 ft, the following data are obtained for breakthrough at 641 minutes for the adsorption step:

z, ft	$\phi = c/c_F$	$\psi = \bar{q}/q_F^*$	z, ft	$\phi = c/c_F$	$\psi = \bar{q}/q_F^*$
0–12	1.000	1.000	14	1.000	1.000
13	1.000	1.000	15	1.000	1.000

z, ft	$\phi = c/c_F$	$\psi = \bar{q}/q_F^*$	z, ft	$\phi = c/c_F$	$\psi = \bar{q}/q_F^*$
16	0.999	0.999	24	0.599	0.575
17	0.997	0.997	25	0.468	0.444
18	0.992	0.990	26	0.343	0.321
19	0.978	0.975	27	0.235	0.217
20	0.951	0.944	28	0.150	0.137
21	0.901	0.890	29	0.090	0.081
22	0.825	0.808	30	0.050	0.044
23	0.722	0.701			

If the bed is regenerated isothermally with pure air at 1 atm and 145°F, and the desorption of benzene during the heatup period is neglected, determine the loading, $\bar{q}$, profile at a time sufficient to remove 90% of the benzene from the bed if an interstitial pure air velocity of 98.5 ft/min is used. Values of k and K at 145°F are given in Example 15.13.

15.30 Use the method of lines with a five-point, biased, upwind finite-difference approximation and a stiff integrator to perform PSA cycle calculations that approach the cyclic steady state for the data and design basis in Example 15.14, starting from: (a) a clean bed, and (b) a bed saturated with the feed. Are the two cyclic steady states essentially the same?

15.31 Solve Example 15.14 for $P_L = 0.12$ atm and an interstitial velocity during desorption that corresponds to the use of 44.5% of the product gas from the adsorption step.

15.32 For the separation of air by PSA, adsorption of both O_2 and N_2 must be considered. Develop a model for this case taking into account two species mass balances, overall mass balance, two species mass-transfer rates, and two extended-Langmuir isotherms. Each of the two main steps can be isothermal and isobaric. Can your PDE equations still be solved by the method of lines with a stiff integrator? If so, outline a procedure for doing it.

15.33 Two adsorption-based separation processes not considered in this chapter because of lack of significant commercial application are (1) parametric pumping, first conceived by R.H. Wilhelm in the early 1960s, and (2) cycling-zone adsorption, invented by R.L. Pigford and co-workers in the late 1960s. The status of and future for these two processes was assessed by Sweed in 1984 [*AIChE Symp. Series*, **80** (233), 44–53 (1984)]. Describe in detail each of these processes. Can either be used for both gas-phase and liquid-phase adsorption?

15.34 A gas mixture containing 55 mol% propane and 45 mol% propylene is to be separated into products containing 10 and 90 mol% propane by adsorption in a continuous, countercurrent adsorption system operating at 25°C and 1 atm. The adsorbent is silica gel, for which equilibrium data are given in Exercise 15.9. Determine by the McCabe–Thiele method: (a) the adsorbent flow rate per 1,000 m^3 of feed gas at 25°C and 1 atm if 1.2 times the minimum rate is used, and (b) the number of theoretical stages required.

15.35 Repeat Example 15.19, except for a feed containing 400 ppm (by weight) of $CaCl_2$ and 50 ppm of $NaCl$.

15.36 An aqueous solution, buffered to a pH of 3.4 by sodium citrate and containing 20 mol/m^3 each of glutamic acid, glycine, and valine, is separated in a chromatographic column, packed with Dowex 50W-X8 in the sodium form to a depth of 470 mm. The resin is 0.07 mm in diameter and packs to a bed void fraction of 0.374. Equilibrium data follow Henry's law, as in Example 15.20, with the following dimensionless constants, determined by Takahashi and Goto [*J. Chem. Eng. Japan*, **24**, 121–123 (1991)]:

Solute	K
Glutamic acid	1.18
Glycine	1.74
Valine	2.64

The superficial solution velocity is 0.025 cm/s. Using equilibrium theory, what pulse duration can be used to achieve complete separation? How long must the elution step be before the second pulse can begin?

15.37 Repeat Exercise 15.36, but using Carta's equation to account for mass transfer with the following effective diffusivities:

Solute	D_e, cm^2/s
Glutamic acid	1.94×10^{-7}
Glycine	4.07×10^{-7}
Valine	3.58×10^{-7}

Assume $k_c = 1.5 \times 10^{-3}$ cm/s. Establish a cycle of feed pulses and elution periods that will give the desired separation.

15.38 A dilute feed of 3-phenyl-1-propanol (A) and 2-phenyl ethanol (B) in a 60/40 wt% ratio, methanol–water mixture is to be fed to a four-section laboratory SMB to separate A from B. The feed rate is 0.16 mL/min with 0.091 g/L of A and 0.115 g/L of B. The desorbent is a 60/40 wt% methanol–water mixture. For this dilute mixture of A and B and the adsorbent to be used, Henry's law applies with $K_A = 2.36$ and $K_B = 1.40$. Assume that neither the methanol nor the water adsorb. The external void fraction of the adsorbent beds, ϵ_b, is 0.572. A switching time of 10 minutes is to be used. Using the steady-state, local-composition TMB model for a perfect separation of A from B with $\beta = 1.15$, estimate initial values for the volumetric flow rates of the extract, raffinate, desorbent, and the solid particles. Convert the value of the recirculation rate for the TMB to that for the SMB and compute the resulting volumetric-liquid flow rates in each of the four sections.

15.39 In the steady-state TMB run of Example 15.17, results for Example 15.16, from the application of the steady-state, local-equilibrium TMB model, were used with a β of 1.05 to establish the flow rates of the raffinate, extract, makeup desorbent, recirculation, and solid particles so as to approach a perfect separation between fructose and glucose. While the separation was not perfect, it was reasonably good. What results are achieved in Example 15.17 if the overall mass-transfer coefficients approach infinity?

Part 4

Separations That Involve a Solid Phase

Chapters 16, 17, and 18 cover separation operations in which one or more components in a solid phase undergo mass transfer to or from a fluid phase. Chapter 16 covers selective leaching of material from a solid to a liquid solvent. This operation is widely used in the food industry. Crystallization from a liquid and desublimation from a vapor are discussed in Chapter 17, where evaporation, which often precedes crystallization, is included. Both solution crystallization to produce inorganic crystals and melt crystallization to produce organic crystals are considered. Chapter 18 is concerned with the drying of solids and the myriad types of equipment used industrially. A section on psychrometry is included.

Chapter 16

Leaching and Washing

Leaching, sometimes called solid–liquid (or liquid–solid) extraction, involves the removal of a soluble fraction (the solute or *leachant*) of a solid material by a liquid solvent. The solute diffuses from the solid into the surrounding solvent. Either the extracted solid fraction or the insoluble solids, or both may be the valuable product. Leaching is widely used in the metallurgical, natural product, and food industries. In the former, leaching may involve oxidation or reduction reactions of the solid with the solvent. Equipment is available to conduct leaching under batch, semicontinuous, or continuous operating conditions. Effluents from a leaching stage are essentially solids-free liquid, called the *overflow*, and wet solids, called the *underflow*. To reduce the concentration of solute in the liquid portion of the underflow, leaching is often accompanied by two or more countercurrent-flow washing stages. The combined process produces a final overflow, called the *extract*, which contains some of the solvent and most of the solute; and a final underflow, called the *extracted* or *leached solids*, which are wet with almost-pure, remaining solvent. Ideally, the soluble solids are perfectly separated from the insoluble solids, but solvent is distributed to both products. Additional processing of the extract and the leached solids is necessary to recover the solvent for recycle.

Some industrial applications of leaching include: (1) removal of copper from ore with sulfuric acid, (2) recovery of gold from ore with sodium-cyanide solution, (3) extraction of sugar from sugar beets with hot water, (4) extraction of tannin from tree bark with water, and (5) the removal of caffeine from green coffee beans with supercritical CO_2.

16.0 INSTRUCTIONAL OBJECTIVES

After completing this chapter, you should be able to:

- Describe equipment used for batch and continuous leaching.
- Explain differences between leaching and washing.
- List assumptions for an ideal, equilibrium leaching or washing stage.
- Calculate recovery of a solute for a continuous, countercurrent leaching and washing system for a constant ratio of liquid to solids in the underflow or for a variable-underflow ratio.
- Apply rate-based leaching to food-processing applications.
- Develop and apply the rate-based, shrinking-core model to reactive mineral processing.

Industrial Example

As an example of leaching, consider the extraction of vegetable oil from soybeans with a commercial hexane solvent in a pilot-plant size countercurrent-flow, multistage leaching unit, as described by Othmer and Agarwal [1]. Although edible oils can be extracted from a number of different field and tree crops, including coconuts, cottonseeds, palm, peanuts, grapeseed, soybeans, and sunflower seeds, the highest percentage of edible oil is from soybeans. In 1979–1980, world production of soybean oil was 14 million metric tons. Oil from soybeans is high in polyunsaturated fats and, thus, is less threatening from a cholesterol standpoint. When the oil content of seeds and beans is high, some of the oil can be removed by compressing the solids in a process known as *expression*, as discussed in Perry and Green [2]. For soybeans, whose oil content is typically less than 0.30 lb per lb of dry and oil-free solids, leaching is a more desirable technique than expression because a higher yield of oil can be achieved by leaching.

The soybeans are dehulled, cleaned, cracked, and flaked before being fed to the extractor. Typically, the cleaned soybeans contain 8 wt% moisture and 20 wt% oil. The dry and oil-free soybeans have a particle specific gravity of about 1.425 and the oil has a specific gravity of 0.907 with a viscosity at 25°C of 50 cP. Approximately 50% of the flake volume is taken up by oil, moisture, and air. It might be expected that whole soybeans, rather than flakes, might be fed to the extractor, with leaching taking place by molecular diffusion of the

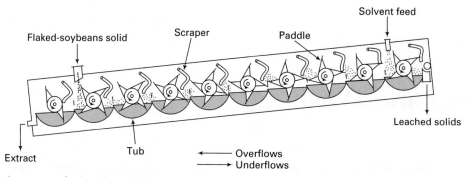

Figure 16.1 Kennedy extractor for leaching of oil from soybeans.

solvent uniformly into the seed, followed by diffusion of the oil through the solvent and out of the seed. If so, the mass-transfer process within the seed could be modeled with Fick's second law. However, experiments by King, Katz, and Brier [3] found that although extraction of oil with solvent in uniformly porous inorganic solids, like porous clay plates, obeyed Fick's law of molecular diffusion, extraction of oil from soybeans did not, presumably because of the complex internal structure of soybeans. Furthermore, Othmer and Agarwal [1], using whole and carefully cut half soybeans, found that diffusion was extremely slow. After 168 hours in contact with hexane, less than 0.08% of the original oil in the whole beans and less than 0.19% of the oil in the half beans was extracted. Such a slow diffusion rate for particles that are approximately 5 mm in diameter is probably due to the location of the oil within the insoluble cell walls, requiring that the oil pass through the walls by slow osmotic pressure differences.

The extent and rate of extraction of the oil is greatly enhanced by flaking the soybeans to thicknesses in the 0.005–0.02-in. range. The flaking process ruptures the cell walls, greatly facilitating contacting of the oil with the solvent. Using trichloroethylene [3] or *n*-hexane [1] as the solvent, with flakes of diameters ranging from 0.04 to 0.24 in., approximately 90% of the oil can be extracted in 100 minutes. The ideal solvent for commercial applications of the leaching of soybeans should have a high solubility of the oil to minimize the amount of solvent needed, a high volatility to facilitate the recovery of solvent from the oil by evaporation or distillation, nonflammability to eliminate the chance of fire and explosion, low cost, ready availability, chemical stability, low toxicity, and compatibility with inexpensive materials of construction

to minimize or eliminate corrosion. Although in a number of respects, especially nonflammability, trichloroethylene is an ideal solvent, it is now considered to be a very hazardous chemical because of its high toxicity. The favored solvent is commercial hexane (mostly *n*-hexane), which presents a fire hazard but has a low toxicity.

The pilot-plant leaching unit used by Othmer and Agarwal, and known as the Kennedy extractor, is shown in Figure 16.1. The soybeans enter continuously at the low end and are leached in a countercurrent cascade of tubs by hexane solvent, which enters at the upper end. The flakes and solvent are agitated and the underflows are pushed uphill from one tub to the next by slowly rotating paddles and scrapers, while the overflows move downhill from tub to tub. The paddles are perforated to drain the solids when they are lifted above the liquid level in the tub by the paddle. The cascade can contain as many tubs as required; Othmer and Agarwal used 15.

Soybean flakes of 0.012 in. average thickness and containing, on the average, 10.67 wt% moisture and 0.2675 g oil/g dry oil-free flakes were fed to the Kennedy extractor at a continuous flow rate of 6.375 lb/h. The solvent flow rate was 10.844 lb/h. Leaching took place at ambient conditions. After 11 hours of operation, a steady state was achieved that delivered an extract, called a *miscella,* of 7.313 lb/hr, containing 15.35 wt% oil. The leached solids contained 0.0151 g oil/g dry oil-free flakes. Thus, 94.4% of the oil was extracted. The residence time in each tub was 3 minutes, giving a total residence time of 45 minutes. From these data, a mass-balance check can be made for oil and solvent, and the liquid-to-solids ratio in the leached solids can be estimated. These calculations are left as an exercise.

16.1 EQUIPMENT FOR LEACHING

Often, the solids to be leached must undergo pretreatment before being fed to the extraction equipment, in order to obtain reasonable leaching times. For example, seeds and beans are dehulled, cracked, and flaked, as described above

for soybeans. When vegetable and animal material cannot be flaked, it may be possible to cut it into thin slices, as is done with sugar beets prior to the leaching of the sugar with water. In this case, the cell walls are left largely intact to minimize the leaching of undesirable material, such as colloids and albumens. Metallurgical ores are crushed and ground to small particles because the small regions of leachable material may be surrounded by relatively impermeable insoluble material. When the insoluble material is quartzite, leaching may be extremely slow. Van Arsdale [4] cites the very important effect of particle size on the time required for

effective leaching by aqueous sulfuric acid of a copper ore. The times for particle diameters of 150 mm, 6 mm, and less than 0.25 mm are approximately 5 years, 5 days, and 5 hours, respectively. When the solid to be leached contains a high percentage of solute, pretreatment may not be necessary because disintegration of the remaining skeleton of insoluble material may take place at the surface of the particles as leaching takes place. When the entire solid is soluble, leaching may be rapid, such that only one stage of extraction, called *dissolution,* may be required.

Industrial equipment for solid–liquid extraction is designed for either batchwise or continuous processing. The method of contacting of the solid with the solvent is either by *percolation* of the solvent through a bed of solids or by *immersion* of the solid in the solvent with agitation of the solid–liquid mixture. When immersion is used, countercurrent, multistage operation is common. With percolation, either a stagewise or differential contacting device can be used. An extractor must be efficient for its particular application so as to minimize the need for solvent because of the high cost of solvent recovery.

Batch Extractors

When the solids to be leached are in the form of fine particles, perhaps smaller than 0.1 mm in diameter, batch leaching is conveniently conducted in an agitated vessel. A simple configuration is the Pachuca tank [5], shown in Figure 16.2a and used extensively in the metallurgical industry. The tank is a tall, cylindrical vessel constructed of wood, concrete, or a metal-like steel, which can be lined with an inert,

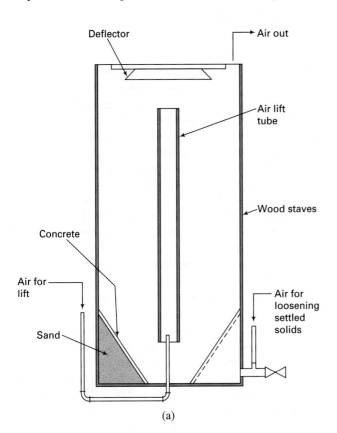

(a)

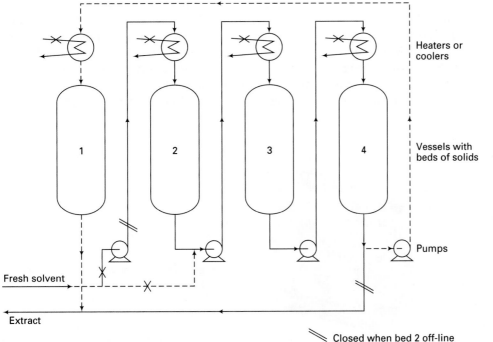

(b)

Figure 16.2 (a) Pachuca tank for batch leaching of small particles.

[From *Handbook of Separation Techniques for Chemical Engineers,* 2nd edition, P.A. Schweitzer, Editor-in-chief, McGraw-Hill, New York (1988) with permission.]

(b) Shanks countercurrent multibatch battery system for leaching of large particles by percolation.

[From C.J. King, *Separation Processes,* 2nd edition, McGraw-Hill, New York (1980) with permission.]

noncorrosive material such as lead. The solvent and solids are placed in the tank and agitation is achieved by an air lift, whereby air bubbles entering at the bottom of a circular tube, concentric with the tank, cause upward flow and subsequent circulation of the solid–liquid suspension. During agitation, air continuously enters and leaves the vessel. When the desired degree of leaching has been accomplished, agitation is stopped and the solids are allowed to settle into a sludge at the bottom, from where the sludge is removed with the assistance of air. The supernatant extract is removed by siphoning from the top of the tank. Agitation can also be achieved by a paddle stirrer or by the use of a propeller mounted in a draft tube in a manner to provide upward flow and subsequent circulation of the solid–liquid suspension, much like that in the Pachuca tank.

When the solids are too coarse to be easily suspended by immersion in the solvent with agitation, the percolation technique can be used. Again a tall, cylindrical vessel is employed. The solids to be leached are dumped into the vessel, followed by percolation of the solvent down through the bed of solids, much like fixed-bed adsorption. To achieve a high concentration of the solute in the solvent, a series of vessels is arranged in a multibatch, countercurrent-leaching technique developed in 1841 by James Shanks and often called an extraction battery using the Shanks system. This technique can be used for such applications as the batch removal of tannin from wood or bark, sugar from sugar beets, and water-soluble substances from coffee, tea, and spices. A typical arrangement of vessels is shown in Figure 16.2b, where Vessel 1 is shown off-line for emptying and refilling of solids. The solvent enters and percolates down through the solids in Vessel 2, and then percolates through Vessels 3 and 4, leaving as final extract from Vessel 4. The extraction of solids in Vessel 2 is completed first. When that occurs, Vessel 2 is taken off-line for emptying and refilling of solids and Vessel 1 is placed on-line. Now the fresh solvent first enters Vessel 3, followed by Vessels 4 and 1. In this manner, fresh solvent always contacts the solids that have been leached for the longest time, thus realizing the benefits of countercurrent contacting. Heat exchangers are provided between vessels to adjust the temperature of the liquid phase, and pumps can be used to move the liquid from vessel to vessel. Any number of vessels can be included in the battery. Although the Shanks system is batchwise with respect to the solids, the system is continuous with respect to the solvent and extract.

Espresso Machine

A widely used batch-leaching machine is the espresso coffee maker. Although a beverage from coffee beans was first made about 1100 BC on the Arabian peninsula, it was not until many centuries later that a method and a machine was devised to produce a high-quality coffee, called espresso (a term that connotes a cup of coffee expressly for you, made quickly and individually, and intended to be drunk right away). The prototype of the espresso machine was created in France in 1822, and the first commercial espresso machine was manufactured in Italy in 1905. By the 1990s, espresso machines were in common use in many countries of the world, producing billions of cups annually. Today, coffee, and, particularly, espresso, is a world commodity that is second only to oil.

A photograph of a typical consumer espresso machine is shown in Figure 16.3a, while a simpler diagram to help understand its operation is presented in Figure 16.3b. In the machine, 7 to 9 grams of coffee beans are ground to a powder of particle size ranging from 250 to 750 microns by a special burr grinder that minimizes the temperature increase of the grounds. The bed of powder, contained as a thin layer in a filter housing, is tamped to increase its uniformity. Water is pumped to a pressure of 9–15 atmospheres and heated rapidly to from 88 to 92°C. The high pressure is required for pressure infusion of hot water into the particles so that extraction can proceed rapidly. During a period of from 20 to 30 sec, the hot water is percolated through the bed of coffee powder to produce a 45-ml shot. The shot has a viscosity of warm honey, and is topped by a thick, dark, golden-cream foam ("crema"). A typical machine produces two shots, which can be added to water, milk, or other liquids or blends to produce various beverages, including

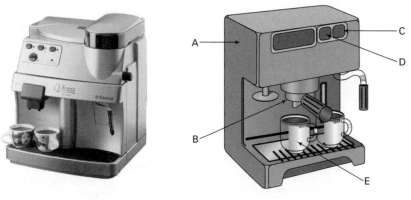

(a)

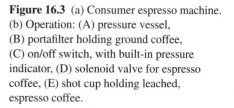

(b)

Figure 16.3 (a) Consumer espresso machine. (b) Operation: (A) pressure vessel, (B) portafilter holding ground coffee, (C) on/off switch, with built-in pressure indicator, (D) solenoid valve for espresso coffee, (E) shot cup holding leached, espresso coffee.

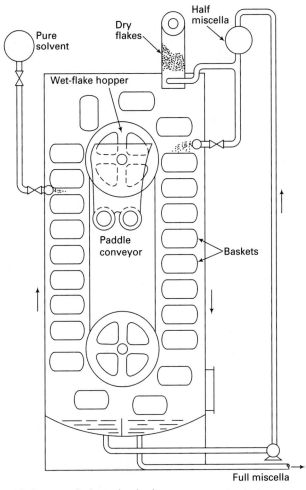

(a) Bollman vertical, moving-basket, conveyor extractor

Americano, Breve, Cappuccino, Latte, Lungo, Macchiato, Mocha, or Ristretto.

The short-time interval between grinding and leaching, the short residence time of the leaching process, the very small particle size of the coffee powder, and the controlled temperature and pressure of the water during leaching combine to maximize the extraction of the favorable flavor-and-aroma chemicals and minimize the extraction of chemicals associated with bitterness, such as quinine and caffeine. Furthermore, the creama traps the aroma, preventing its escape into the surrounding air. At the pull of a lever or the push of a button, espresso coffee is produced that is concentrated, full in body, and rich in flavor and aroma. A number of experimental studies of espresso production are reported in the literature, e.g., Andueza et al. [16, 17, 18].

Continuous Extractors

When leaching is to be carried out on a large scale, it is preferable to use an extraction device that can be operated with continuous flow of both solids and liquid. Many such patented devices are commercially available, especially for applications in the food industry, as discussed by Schwartzberg [6]. Some of the more widely discussed extractors are shown schematically in Figure 16.4. These devices differ mainly with respect to the manner in which the solids are transported through and the degree to which agitation of the solid–liquid mixture is provided in the equipment.

According to Schwartzberg, several extractors that have been described in the literature are now either obsolete or infrequently used, because of various limitations, including ineffective contacting of solid and liquid phases, bypassing,

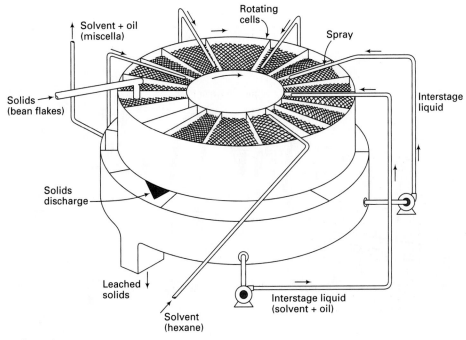

(b) Rotocel extractor

Figure 16.4 Equipment for continuous leaching.

[From *Handbook of Separation Techniques for Chemical Engineers,* 2nd edition, P.A. Schweitzer, Editor-in-chief, McGraw-Hill, New York (1988) with permission.]

[From R.E. Treybal, *Mass-Transfer Operations,* 3rd edition, McGraw-Hill, New York (1980) with permission.]

(continued)

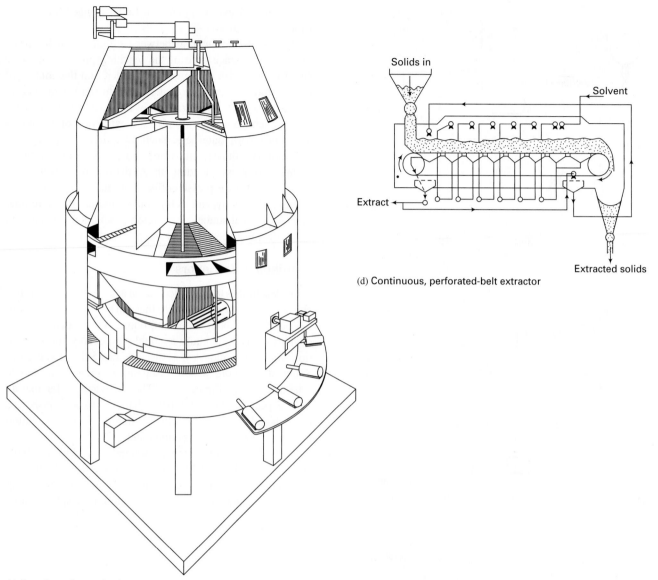

(c) French stationary-basket extractor

(d) Continuous, perforated-belt extractor

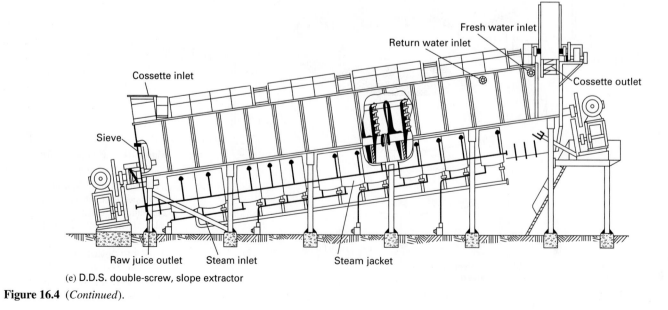

(e) D.D.S. double-screw, slope extractor

Figure 16.4 (*Continued*).

and fines entrainment. These include the Hildebrand, Detrex, Anderson, Allis Chalmers, and Bonotto extractors. The Kennedy extractor described above for application to oil extraction from soybeans may also have a low efficiency in some applications, but is still used and still available.

The Bollman vertical moving-basket conveyor extractor, shown in Figure 16.4a, has been widely used to extract oil from flaked seeds and beans. Baskets with perforated bottoms are moved around a vertical loop by a motor-driven chain drive. The solvent percolates down, from basket to basket, through the solids. When a basket reaches the top of the extractor, the basket is inverted to dump the extracted solids and then filled with fresh solids. The flow of liquid is countercurrent to the solids contained in ascending baskets and cocurrent in descending baskets. Fresh solvent enters near the top of the ascending leg and collects as "half miscella" in the left-hand part of the sump at the bottom. From there, the half miscella is pumped to the top of the descending leg, from which it flows down to the right-hand part of the sump and is withdrawn as final extract, called "full miscella." A typical Bollman extractor is 14 m high, with each basket filled with solids to a depth of about 0.5 m. According to Coulson et al. [7], the baskets are rotated very slowly at about one revolution per hour so as to give solids residence times of about 60 minutes. Each basket contains about 350 kg of flaked solids. Thus, for the 23 baskets shown in Figure 16.4a, almost 200,000 kg of solids can be extracted per day. About equal mass flows of flaked solids and solvent are fed to the extractor, and the full miscella is essentially solids-free with about 25 wt% oil.

Another widely used continuous extractor for flaked seeds and beans is the Rotocel extractor, shown in Figure 16.4b. In this device, which resembles a carousel and conveniently simulates the Shanks system, walled, annular sectors, called cells, on a horizontal plane, are slowly rotated by a motor. The cells, which hold the solids and are perforated on the bottom for solvent drainage, successively pass a solids feed area, a series of solvent sprays, a final spray and drainage area, and a solids-discharge area. Fresh solvent is supplied to the cell located just below the final spray and drainage area, from where the drained liquid is collected and pumped to the preceding cell location. The drainage from that cell is collected and pumped to the cell preceding that cell and so on. In this manner, a countercurrent flow of solids and liquid is achieved. The extracted solids typically contain 25–30 wt% liquid. Rotocel extractors are typically 3.4–11.3 m in diameter, 6.4–7.3 m in height, with bed depths of 1.8–3.0 m. They have processed up to three-million kg/day of flaked soybeans. The number of cells can be varied and residence time can be varied by varying the rate of rotation. A widely used variation of the Rotocel extractor is the French stationary-basket extractor, which is shown in Figure 16.4c and has about the same size and capacity as the Rotocel extractor. The sectored cells do not move. Instead, the solids feed spout and solids discharge zone rotate, with periodic switching of the solvent feed and discharge connections. Thus, the weight of moving parts is reduced.

Continuous, perforated-belt extractors of the type shown in Figure 16.4d are widely used to process sugar cane, sugar beets, oilseeds, and apples (for apple juice). The feed solids are fed from a hopper to a slow-moving, continuous and non-partitioned, perforated belt driven by motorized sprockets at either end. The height of the solids on the belt can be controlled by a damper at the outlet of the feed hopper. The belt speed is automatically adjusted to maintain the desired depth of solids. Extracted solids are discharged into an outlet hopper at the end of the belt by a scraper. Side walls prevent the solids from falling off the sides of the belt. Below the belt are compartments for collecting solvent. Fresh solvent is sprayed over the solids and above the compartments in a countercurrent fashion, starting from the discharge-end of the belt, in as many as 17 passes. Bed depths may range from 0.8 to 2.6 m. Units from 7 to 37 m long and with belts from 0.5 to 9.5 m wide have processed as much as 7,000,000 kg/day of sugar cane or sugar beets.

The D.D.S. (De Danske Sukker–fabriker) double-screw, slope extractor, shown in Figure 16.4e, is a very versatile unit. Although used mainly for extraction of sugar beets, the device has been applied successfully to a very wide range of other feed materials, including sugar cane, flaked seeds and beans, apples, pears, grapes, cherries, ginger, licorice, red beets, carrots, fishmeal, coffee, and tea. The opposite-turning screws of the metal ribbons are pitched so that they both move the solids uphill in parallel, cylindrical troughs. Extract flows through the screw surface downhill to achieve a differential, countercurrent flow with the solids. A novel feature is the ability to turn one screw slightly faster and then slightly slower than the other screw, causing the solids to be periodically squeezed. Units range in size from 2 to 3.7 m in diameter and 21–27 m in length and have been used to process as much as 3,000,000 kg/day of sugar beets in the form of cossettes (long, thin strips).

Continuous, Countercurrent Washing

When leaching is very rapid, as with small particles containing very soluble solutes, or when leaching has already been completed, or when solids are formed by chemical reactions in a solution, it is common to countercurrently wash the solids to reduce the concentration of the solute in the liquid adhering to the solids. This can be accomplished in a series of *gravity thickeners* or centrifugal thickeners, called *hydroclones,* arranged for countercurrent flow of the underflows and overflows as shown in Figure 16.5, and sometimes called a continuous, countercurrent decantation system. A typical continuous gravity thickener is shown in detail in Figure 16.6a. The combined feed to the thickener consists of feed solids or underflow from an adjacent thickener, together with fresh solvent or overflow from an adjacent thickener. The thickener has a threefold function. First, it must thoroughly mix the liquid and solids to obtain a uniform concentration of solute in the liquid. The thickener must also produce an overflow free of solids and an underflow with as

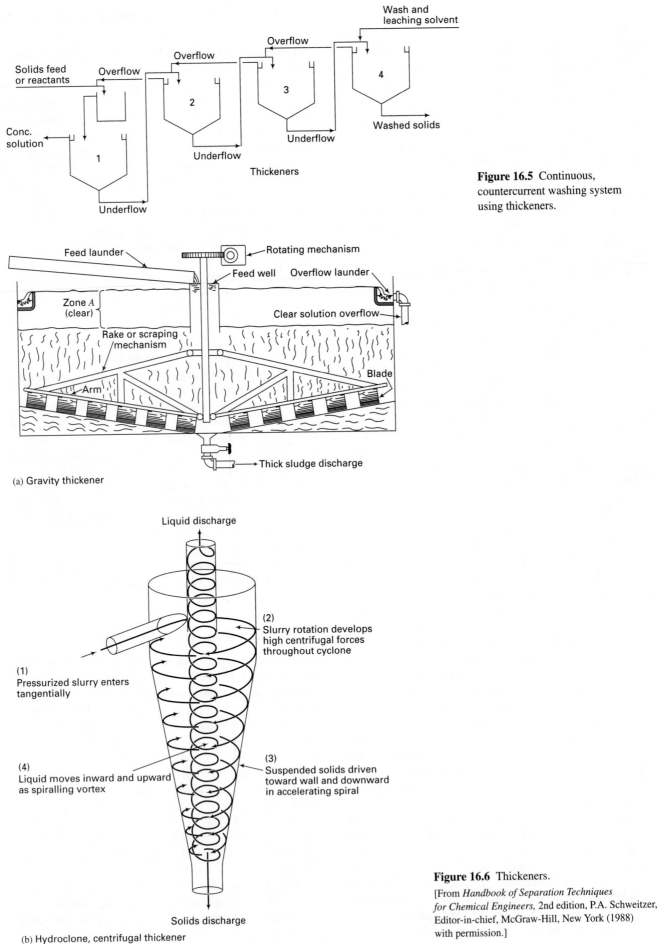

Figure 16.5 Continuous, countercurrent washing system using thickeners.

(a) Gravity thickener

(b) Hydroclone, centrifugal thickener

Figure 16.6 Thickeners.

[From *Handbook of Separation Techniques for Chemical Engineers,* 2nd edition, P.A. Schweitzer, Editor-in-chief, McGraw-Hill, New York (1988) with permission.]

high a fraction of solids as possible. The thickener consists of a large-diameter shallow tank with a flat or slightly conical bottom. The combined feed enters the tank near the center by a means of a feed launder that discharges into a feed well. Settling and sedimentation of the solid particles occur by gravity due to a solid particle density that is greater than the liquid density. In essence, solids flow downward and liquid flows upward. Around the upper, inner periphery of the tank is an overflow launder or weir for continuously removing clarified liquid. Solids settling to the bottom of the tank are moved inward toward a thick sludge discharge by a slowly rotating motor-driven rake. Thickeners as large as 100 m in diameter and 3.5 m high have been constructed. In large thickeners, the rake revolves at about 2 rpm.

The residence times of solids and liquids in a gravity thickener are often large (minutes or hours) and, as such, are sufficient to provide adequate residence time for mass transfer and mixing when small particles are involved. When long residence times are not needed and the overflow need not be perfectly clear of solids, the hydroclone, shown in Figure 16.6b, should be considered. The pressurized feed slurry enters the device tangentially to create, by centrifugal force, a downward spiraling motion. The higher-density, suspended solids are, by preference, driven to the wall, which becomes conical as it extends downward, and discharged as a thickened slurry at the bottom of the hydroclone. The liquid, which is forced to move inward and upward as a spiraling vortex, exits from a vortex-finder pipe that extends downward from the closed top of the hydroclone to a location just below the feed entry.

16.2 EQUILIBRIUM-STAGE MODEL FOR LEACHING AND WASHING

The simplest model for a continuous, countercurrent leaching and washing system, as shown in Figure 16.7, assumes that the solid feed consists of a solute, which is completely soluble in the solvent, and an inert substance or carrier that is not soluble in the solvent. Furthermore, the rate of leaching is rapid such that it is completed in a single leaching stage, which is followed by a series of one or more washing stages to reduce the concentration of solute in the liquid adhering to the solids in the underflow streams. All overflow

streams are assumed to be free of solids. In Figure 16.7:

S = mass flow rate of inert solids, which is constant from stage to stage.

V = mass flow rate of entering solvent or overflow liquid (solvent plus solute), which varies from stage to stage.

L = mass flow rate of underflow liquid (solvent plus solute), which varies from stage to stage.

y = mass fraction of solute in the overflow liquid.

x = mass fraction of solute in the underflow liquid.

Alternatively, V and L can refer to mass flow rates of solvent on a solute-free basis and the symbols Y and X can be used to refer to mass ratios of solute to solvent in the overflow liquid and underflow liquid, respectively. Mole or volume flow rates can also be used.

An ideal leaching or washing stage is defined by Baker [8] as one where:

1. Any entering solid solute is completely dissolved into the liquid in the stage (assuming that the liquid contains sufficient solvent).

2. The composition of the liquid in the stage is uniform throughout, including any liquid within pores of the inert solid.

3. Solute is not adsorbed on the surfaces of the inert solid.

4. The inert solids leaving in the underflow from each stage are wet with liquid, such that mass ratio of solvent in that liquid (or the total liquid) to inert solids is constant from stage to stage.

5. Because of (2), the concentration of solute in the overflow is equal to that in the liquid portion of the underflow. This is equivalent to an equilibrium assumption.

6. Overflows contain no solids.

7. Solvent is not vaporized, adsorbed, or crystallized in a stage.

For the continuous, countercurrent system of ideal leaching stages in Figure 16.7, solute and total-liquid material balances can be used to solve various types of problems, including (1) the determination of the number of ideal stages required to achieve a specified degree of washing and (2) the

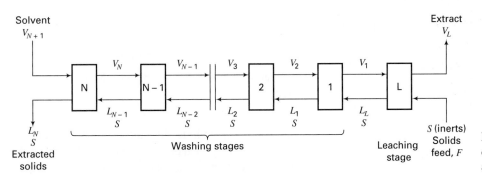

Figure 16.7 Continuous, countercurrent, ideal-stage leaching and washing system.

determination of the effect of washing for a specified degree of washing for a specified number of ideal stages. Depending on the nature of the problem, either an algebraic or a graphical method may be preferred. For most problems, it is usually best to consider the leaching stage separate from the washing stages, as illustrated in the following examples.

EXAMPLE 16.1

A finely divided solids feed, F, of 150 kg/h, containing 1/3 water-soluble Na_2CO_3 and 2/3 insoluble ash is to be leached and washed at 30°C in a two-stage, countercurrent system with 400 kg/h of water. The leaching stage consists of an agitated vessel that discharges the slurry into a thickener. The washing stage consists of a second thickener. Experiments show that the sludge underflow from each thickener will contain 2 kg of liquid (water and carbonate) per kg of insoluble ash. Assume ideal stages.

(a) Calculate the percent (%) recovery of carbonate in the final extract.

(b) If a third stage is added, calculate the amount of additional carbonate that will be recovered.

SOLUTION

At 30°C, the solubility of the carbonate in water is 38.8 kg/100 kg of water. Therefore, there is sufficient water to dissolve all the carbonate. Referring to Figure 16.7:

(a) $N = 1$ (L is the leaching stage)

$$S = \frac{2}{3}(150) = 100 \text{ kg/h of insoluble ash}$$

$$Na_2CO_3 \text{ in entering solids} = \frac{1}{3}(150) = 50 \text{ kg/h}$$

$$V_2 = \text{entering solvent} = 400 \text{ kg/h}$$

$$L_L = L_1 = 2S = 200 \text{ kg/h}$$

By total liquid material balances on Stage 1 and Stage L,

$$V_1 = V_2 + L_L - L_1 = 400 + 200 - 200 = 400 \text{ kg/h}$$
$$V_L = V_1 + 50 - L_L = 400 + 50 - 200 = 250 \text{ kg/h}$$

Na_2CO_3 *material balance around washing Stage 1:*

$$x_L L_L = y_1 V_1 + x_1 L_1 \tag{1}$$
$$200 x_L = 400 y_1 + 200 x_1 \tag{2}$$

But, $y_1 = x_1$ from item (5) above for an ideal stage.

Combining (1) and (2),

$$x_L = 3x_1 \tag{3}$$

Na_2CO_3 *material balance around leaching Stage L:*

$$y_1 V_1 + 50 = x_L L_L + y_L V_L \tag{4}$$
$$400 y_1 + 50 = 200 x_L + 250 y_L \tag{5}$$

But, again $y_1 = x_1$ and $y_L = x_L$ $\qquad$ (6)

Combining (4), (5), and (6),

$$x_1 = 1.125 x_L - 0.125 \tag{7}$$

Combining (3) and (7) and solving,

$$x_L = 0.158$$

Therefore, $y_L = 0.158$

From (7),

$$x_1 = 0.0526$$

$$\text{Recovery of } Na_2CO_3 = \frac{y_L V_L}{50} = \frac{(0.158)(250)}{50}$$

$$= 0.79 \text{ or } 79\%$$

(b) $N = 2$ washing stages

$$V_3 = 400 \text{ kg/h}$$

$$L_L = L_1 = L_2 = 2S = 200 \text{ kg/h}$$

$$V_2 = V_1 = 400 \text{ kg/h}$$

$$V_L = 250 \text{ kg/h}$$

Na_2CO_3 *material balance around Stage 2:*

$$x_1 L_1 = y_2 V_2 + x_2 L_2$$
$$200 x_1 = 400 y_2 + 200 x_2 \tag{8}$$

But, $\qquad\qquad y_2 = x_2$ $\qquad\qquad\qquad$ (9)

Combining (8) and (9),

$$x_1 = 3x_2 \tag{10}$$

Na_2CO_3 *material balance around Stage 1:*

$$y_2 V_2 + x_L L_L = x_1 L_1 + y_1 V_1$$
$$400 y_2 + 200 x_L = 200 x_1 + 400 y_1 \tag{11}$$

But, $\qquad\qquad y_2 = x_2$ $\qquad\qquad\qquad$ (12)
$\qquad\qquad\qquad y_1 = x_1$ $\qquad\qquad\qquad$ (13)

Combining (11), (12), and (13),

$$x_L = 3x_1 - 2x_2 \tag{14}$$

Na_2CO_3 *material balance around Stage L:*

This is the same as (7) in Part (a).

Solving (10), (14), and (7), which are all linear,

$$x_L = 0.1795$$

$$y_L = 0.1795$$

$$x_1 = 0.0769$$

$$x_2 = 0.0256$$

Recovery of $Na_2CO_3 = \dfrac{y_L V_L}{50} = \dfrac{(0.1795)(250)}{50} = 0.898$ or 89.8%

From Part (a), for two stages, recovery of Na_2CO_3 is $0.158(250) = 39.5$ kg/h.

For three stages, recovery is $0.1795(250) = 44.9$ kg/h

Recover $44.9 - 39.5 = 5.4$ kg/h more Na_2CO_3 with three stages.

For this example, it is difficult to apply a graphical McCabe–Thiele-type method because only the slope of the operating line is known and not either end point.

EXAMPLE 16.2

Baker [8] presents the following problem, for which a McCabe–Thiele graphical method can be applied. Two tons (4,000 lb) per day of waxed paper containing 25 wt% soluble wax and 75 wt% insoluble pulp are to be dewaxed by leaching with kerosene in a continuous, countercurrent contacting system of the type shown in Figure 16.7. The wax will be completely dissolved by the kerosene in the leaching stage, L. Subsequent washing stages will be used to reduce the wax content in the liquid adhering to the pulp leaving the last stage, N, to 0.2 lb wax/100 lb pulp. The kerosene entering the system is recycled from a solvent-recovery system and contains 0.05 lb wax/100 lb kerosene. The final extract is to contain 5 lb wax/100 lb kerosene. Experiments show that the underflow from each stage will contain 2 lb kerosene/lb insoluble pulp. Determine the number of washing stages required, exclusive of the leaching stage.

SOLUTION

Referring to Figure 16.7 and using the nomenclature defined at the beginning of this section for the case of concentrations on a mass-ratio basis and flow rates on a solute-free basis, the following material balance equations apply:

Overall mass balance on solute (wax):

$$0.25(4,000) + \frac{0.05}{100}V_{N+1} = \frac{5}{100}V_L + \frac{0.2}{100}(0.75)(4,000)$$

or

$$0.05V_L - 0.0005V_{N+1} = 994 \tag{1}$$

Overall mass balance on solvent (kerosene):

$$V_{N+1} = V_L + 2(0.75)(4,000)$$

or

$$V_{N+1} = V_L + 6,000 \tag{2}$$

Solving (1) and (2)

$$V_{N+1} = \text{kerosene in entering solvent} = 26,140 \text{ lb/day}$$

$$V_L = \text{kerosene in exiting extract} = 20,140 \text{ lb/day}$$

Thus, the final underflow contains $26,140 - 20,140 = 6,000$ lb/day of kerosene. Also,

$$X_N = \left[\frac{0.2}{100}(0.75)(4,000)\right]\Big/6,000$$

$$= 0.001 \text{ lb wax/lb kerosene in the final underflow.}$$

Material balances can now be made around the leaching stage.

Mass balance on kerosene:

$$V_1 = V_L + 2(0.75)(4,000)$$

Thus,

$$V_1 = 20,140 + 6,000 = 26,140 \text{ lb/day}$$

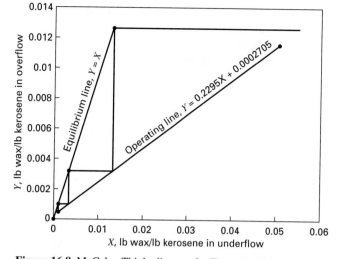

Figure 16.8 McCabe–Thiele diagram for Example 16.2.

Mass balance on wax:

$$26,140\,Y_1 + (0.25)(4,000) = \frac{5}{100}(20,140)$$
$$+ X_L(2)(0.75)(4,000) \tag{3}$$

But

$$X_L = Y_L = 0.05 \text{ lb wax/lb kerosene} \tag{4}$$

Substituting (4) into (3) and solving,

$$Y_1 = 0.01174 \text{ lb wax/lb kerosene}$$

We now have the end points (solute compositions) at the two ends of the washing cascade. Referring to Figure 16.7,

$$Y_{N+1} = 0.0005 \quad \text{and} \quad X_N = 0.001 \text{ in lb wax/lb kerosene}$$
$$Y_1 = 0.01174 \quad \text{and} \quad X_L = 0.05 \text{ in lb wax/lb kerosene}$$

Furthermore, for the washing section, the mass ratio of kerosene in the underflow to kerosene in the overflow is constant at a value of

$$\frac{L_{n-1}}{V_n} = \frac{2(0.75)(4,000)}{26,140} = 0.2295$$

Therefore, the operating line on a McCabe–Thiele plot of Y versus X will be a straight line through the two end points, with a slope of 0.2295. This corresponds to a line given by the equation, $Y = 0.2295X + 0.0002705$. The equilibrium line is simply $Y = X$. The operating and equilibrium lines are plotted in Figure 16.8. The ideal washing stages, which can be stepped off starting from either end of the operating line, are stepped off here from stage N. It is seen that slightly less than three ideal washing stages are needed.

McCabe–Smith Algebraic Method

It is seen in Figure 16.8 that it can be difficult to accurately step off the number of washing stages. When the above ideal-stage model of Baker [8] applies, a more accurate algebraic method, developed by McCabe and Smith [9] from the Kremser equation of Chapter 5, can be applied. The method is developed here using solute concentrations in mass fractions, but the final equations can also be applied with mass ratios.

Combining (5-48), (5-50), (5-51), and (5-54),

$$y_{N+1} = y_1 \left(\frac{1 - A^{N+1}}{1 - A} \right) - y_0^* A \left(\frac{1 - A^N}{1 - A} \right) \quad (16\text{-}1)$$

where N = number of ideal washing stages and

$$y_0^* = K x_0 = K x_L = y_L \quad (16\text{-}2)$$

For washing,

$$K = y/x = 1 \text{ and, therefore, } A = L/KV = L/V$$

Equation (16-1) can be written as follows by collecting terms in A^{N+1} and A:

$$A^{N+1}(y_1 - y_L) = A(y_{N+1} - y_L) + (y_1 - y_{N+1}) \quad (16\text{-}3)$$

This equation can be simplified by writing an overall solute balance around all washing stages:

$$y_{N+1} V_{N+1} + x_L L_L = y_1 V_1 + x_N L_N \quad (16\text{-}4)$$

But, for ideal washing stages, since

$$V_{N+1} = V_1, \quad L_L = L_N, \quad \text{and} \quad A = L/V,$$

(16-4) can be simplified to

$$y_{N+1} = y_1 + A x_N - A x_L \quad (16\text{-}5)$$

But $y_L = x_L$. Therefore, (16-5) can be written:

$$(y_1 - y_{N+1}) = A(y_L - x_N) \quad (16\text{-}6)$$

Combining (16-3) and (16-6) and rearranging gives

$$A^N = \left(\frac{y_{N+1} - x_N}{y_1 - y_L} \right) \quad (16\text{-}7)$$

Solving (16-7) for N with $A = L/V$, gives

$$N = \frac{\log \left(\dfrac{x_N - y_{N+1}}{y_L - y_1} \right)}{\log(L/V)} \quad (16\text{-}8)$$

The argument of the log term of the denominator can be written in terms of the end points to give

$$\frac{L}{V} = \left(\frac{y_1 - y_{N+1}}{x_L - x_N} \right) = \left(\frac{y_1 - y_{N+1}}{y_L - x_N} \right) \quad (16\text{-}9)$$

Combining (16-8) and (16-9),

$$N = \frac{\log \left(\dfrac{x_N - y_{N+1}}{y_L - y_1} \right)}{\log \left(\dfrac{y_1 - y_{N+1}}{y_L - x_N} \right)} \quad (16\text{-}10)$$

When the constant underflow liquid is given in terms of total (solvent plus solute) liquid, (16-8) or (16-9) is used directly, where L and V are total liquid flow rates. When the underflow liquid is given in terms of just the solvent, the y and x solute mass fractions in (16-8) and (16-9) are replaced by Y and X solute mass ratios, and V and L are liquid flow rates of solute-free solvent.

EXAMPLE 16.3

Solve for the number of ideal, continuous, countercurrent washing stages for the conditions of Example 16.2 using the McCabe–Smith equations.

SOLUTION

From the problem statement and the overall material balance and leaching stage calculations for Example 16.2,

$$Y_{N+1} = 0.0005$$
$$X_N = 0.001$$
$$Y_2 = 0.01174$$
$$Y_L = X_L = 0.05$$
$$L/V = 0.2295$$

From the mass ratio form of (16-8), the number of ideal leaching stages is:

$$N = \frac{\log \left(\dfrac{0.001 - 0.0005}{0.05 - 0.01174} \right)}{\log(0.2295)} = 2.95$$

The same result is obtained if (16-10) is used.

When the rate of leaching is slow, several countercurrent stages may be required, during which the effect of washing will be diminished. This is illustrated in the following example.

EXAMPLE 16.4

In Example 16.1, part (b), leaching was assumed to be completed in one stage, with two additional stages provided for washing. The recovery of the solute, Na_2CO_3, in the extract was 89.8%. Recalculate this example, assuming that $1/2$ of the carbonate is leached in the first stage and the remaining $1/2$ in the second stage, leaving only the last stage as a true washing stage. For convenience, number the stages as in Figure 16.9, which includes flow rates that are given or easily computed.

SOLUTION

Na_2CO_3 material balance around Stage 3:

$$x_2 L_2 = y_3 V_3 + x_3 L_3$$
$$200 x_2 = 400 y_3 + 200 x_3 \quad (1)$$

But, $$y_3 = x_3 \quad (2)$$

Combining (1) and (2),

$$x_2 = 3 x_3 \quad (3)$$

Na_2CO_3 material balance around Stage 2:

$$y_3 V_3 + x_1 L_1 = x_2 L_2 + y_2 V_2$$
$$400 y_3 + 200 x_1 + 25 = 200 x_2 + 425 y_2 \quad (4)$$

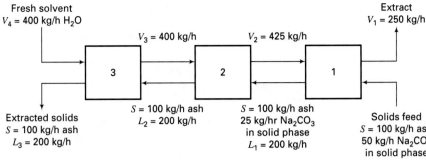

Extracted solids
$S = 100$ kg/h ash
$L_3 = 200$ kg/h

$S = 100$ kg/h ash
$L_2 = 200$ kg/h

$S = 100$ kg/h ash
25 kg/hr Na_2CO_3
in solid phase
$L_1 = 200$ kg/h

Solids feed
$S = 100$ kg/h ash
50 kg/h Na_2CO_3
in solid phase

Figure 16.9 Leaching and washing system for Example 16.4.

But, $$y_2 = x_2 \qquad (5)$$

Combining (2), (4), and (5),

$$x_1 = 3.125x_2 - 2x_3 - 0.125 \qquad (6)$$

Na_2CO_3 material balance around Stage 1:

$$y_2 V_2 + 50 = x_1 L_1 + y_1 V_1 + 25$$
$$425 y_2 + 50 = 200 x_1 + 250 y_1 + 25 \qquad (7)$$

But, $$y_1 = x_1 \qquad (8)$$

Combining (5), (7), and (8)

$$x_2 = 1.059 x_1 - 0.0588 \qquad (9)$$

Solving (3), (6), and (9),

$$x_1 = y_1 = 0.1681$$
$$x_2 = 0.1192$$
$$x_3 = 0.0397$$

Recovery of

$$Na_2CO_3 = \frac{y_1 V_1}{50} = \frac{(0.1681)(250)}{50} = 0.841 \text{ or } 84.1\%$$

which is almost 6% less than in Part (b) of Example 16.1, where leaching was completed in one stage.

Variable Underflow

In previous Examples 16.1–16.4, the ratio of liquid to solids in the underflow from stage to stage was assumed to be constant. Experiments by Ravenscroft [10] on the extraction of oil from granulated halibut livers by diethylether showed that the ratio of liquid to solids in the underflow increased significantly with increasing concentration of oil in the liquid in the range of 0.04–0.64 gal oil/gal liquid. In the leaching experiments, equilibrium was achieved in 2–3 minutes of agitation, but 10 minutes was used. Thirty minutes was allowed for settling of the extracted livers, after which the free solution (overflow) was decanted, leaving the underflow. Ravenscroft ascribed the variable-underflow effect to an appreciable increase in viscosity and density of the liquid as the concentration of the oil was increased. When neither density nor viscosity vary significantly, experimental data of Othmer and Agarwal [1] show that the main variable affecting the liquid-to-solid ratio in the underflow is surface area-to-volume ratio of the solids. When the flow rate of underflow varies, the operating line on a McCabe–Thiele diagram will be curved instead of straight. The curvature can be established by computing two arbitrary, intermediate points, as illustrated in the following example.

EXAMPLE 16.5

Oil is to be extracted from 10,000 lb/h of granulated halibut livers, based on oil-free livers, which contain 0.043 gal of extractable oil per pound of oil-free livers. It is desired to extract 95% of the oil in a countercurrent extraction system using oil-free diethylether as the solvent. The final extract is to contain 0.65 gal oil per gal of extract. Assume that the volumes of oil and ether are additive, and that

leaching will be completed in one stage. Although the experimental underflow data of Ravenscroft show some scatter, use the following smoothed data to predict underflow rates:

Gal oil per gal liquid	Gal liquid retained per lb oil-free livers
0.00	0.035
0.10	0.042
0.20	0.049
0.30	0.058
0.40	0.069
0.50	0.083
0.60	0.100
0.70	0.132

Determine the number of ideal stages required.

SOLUTION

First, determine the flow rate of solvent.

Oil in liver feed = 10,000(0.043) = 430 gal liquid/h

Oil in extract = 430(0.95) = 408.5 gal/h

Oil in underflow of extracted livers = 430 − 408.5 = 21.5 gal/h

By an iterative procedure, determine the gal/h of ether in the final underflow of extracted livers:

Assume 0.10 gal oil/gal liquid:

From the above data table, have 0.042 gal liquid/lb oil-free livers

Therefore, have 0.042(10,000) = 420 gal liquid/h and 0.10(420) = 42 gal/h oil, which is higher than the required value of 21.5 gal/h from above.

Assume 0.05 gal oil/gal liquid:

Linear interpolation of the above data table gives 0.0385 gal liquid/lb oil-free livers.

Therefore, have $0.0385(10,000) = 385$ gal liquid/h

$0.05(385) = 19.3$ gal oil/h, which is too low

By interpolation, have 0.055 gal oil/gal liquid and 0.0389 gal liquid/lb oil-free livers

Therefore, final underflow contains $0.0389(10,000) - 21.5 = 367.5$ gal ether/h

The final extract contains 408.5 gal oil/h, with given value of 65 vol % oil.

Therefore, final extract contains

$$\frac{0.35}{0.65}(408.5) = 220 \text{ gal ether/h}$$

Ether in solvent feed $= 220 + 367.5 = 587.5$ gal/h.

Overall material balance is

	Solvent feed	Livers feed	Final extract	Final underflow
Oil-free livers, lb/h	0	10,000	0	10,000
ether, gal/h	587.5	0	220	367.5
oil, gal/h	0	430	408.5	21.5

Ideal Leaching Stage:

Referring to Figure 16.7, the underflow leaving the leaching stage, L, will have the same concentration of oil as in the final extract. That concentration is the specification of 0.65 gal oil/gal liquid. From the above data table, by linear interpolation, gal liquid retained/lb oil-free livers $= 0.116$.

Therefore, letting x_j and $y_j =$ volume fractions,

$$L_L = 0.116(10,000) = 1,160 \text{ gal liquid}$$

$$x_L = 0.65 \text{ gal oil/gal liquid}$$

Oil material balance around leaching stage:

$$y_1 V_1 + 430 = x_L L_L + 408.5$$

or $\quad y_1 V_1 = 0.65(1,160) + 408.5 - 430 = 732.5$ gal oil

Ether material balance around leaching stage

$$(1 - y_1)V_1 = 0.35(1,160) + 220 = 626 \text{ gal ether/h}$$

$$y_1 = \frac{732.5}{732.5 + 626} = 0.539 \text{ gal oil/gal liquid}$$

We now have the end points of the operating line for the washing stages:

Stage N	Stage 1
$y_{N+1} = 0$	$y_1 = 0.539$
$x_N = 0.055$	$x_L = 0.65$

These two pairs of points together with the straight equilibrium line, $y = x$ are plotted on the McCabe–Thiele diagram of Figure 16.10. Because the underflow is variable, a straight line does not connect the end points of the operating line.

Calculation of Intermediate Points on the Operating Line

Determine by oil and total liquid material balances, values of y_{n+1} for $x_n = 0.30$ and 0.50. The material balance can be made from the solvent feed end to an arbitrary stage n.

For $x_n = 0.3$, with above data table giving 0.058 gal liquid/lb oil-free livers

Oil mass balance:

$$0.3(10,000)(0.058) = y_{n+1} V_{n+1} + 21.5 \tag{1}$$

Total liquid mass balance:

$$10,000(0.058) + 587.5 = V_{n+1} + (367.5 + 21.5) \tag{2}$$

Solving (1) and (2),

$$V_{n+1} = 778.5 \text{ gal/h}$$

$$y_{n+1} = 0.196 \text{ gal oil/gal liquid}$$

For $x_n = 0.50$, with 0.083 gal liquid/lb oil-free livers,

Oil mass balance:

$$0.5(10,000)(0.083) = y_{n+1} V_{n+1} + 21.5 \tag{3}$$

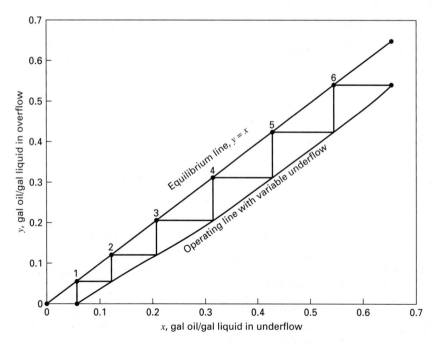

Figure 16.10 McCabe-Thiele diagram for Example 16.5.

Total liquid mass balance:

$$10,000(0.083) + 587.5 = V_{n+1} + (367.5 + 21.5) \quad (4)$$

Solving (3) and (4),

$$V_{n+1} = 1,028.5 \text{ gal/h}$$

$$y_{n+1} = 0.383 \text{ gal oil/gal liquid}$$

When these two sets of x_n, y_{n+1} points are plotted, a curved operating line is obtained as shown in Figure 16.10. A total of almost exactly six washing stages can be stepped off. Adding the leaching stage to this gives a total of seven ideal stages.

16.3 RATE-BASED MODEL FOR LEACHING

Leaching involves the transfer of a solute from the interior of a solid into the bulk of a liquid solvent or extract. The process can be modeled by considering two steps (in series): (1) molecular diffusion of the solute through the solid, and (2) convection and eddy diffusion of the solute through the solvent or extract that is exterior to the solid. Practical rates of molecular diffusion through the solid are only achieved after the solvent penetrates the solid to become occluded liquid, unless the solvent is initially present in the solid. The solute then dissolves into that liquid and diffuses at a reasonable rate to the surface of the solid, leaving behind the insoluble solids and any sparingly soluble materials in the form of a framework. If relative motion exists between the solids and the exterior solvent solution, the resistance to mass transfer in the fluid phase may be negligible compared to that in the solid, and the entire leaching process can be modeled by diffusion through the solid.

Food Processing

Schwartzberg and Chao [11] present a summary of published experimental and theoretical studies of solute diffusion in the leaching of food materials in the form of slices, near-cylinders, and nearly spherical particles, including a compilation of effective diffusivities. As with diffusion of liquids and gases in porous-solid adsorbents, the diffusivity can be expressed as a true molecular diffusivity in the occluded fluid phase, or as an effective diffusivity through the entire solid, including the insoluble-solid framework, sometimes called the *marc*, and the occluded liquid. When an effective diffusivity, D_e, is used, and Fick's laws are applied, the concentration driving force is taken to be the concentration of solute, X_i in mass per unit volume of solid particle. Thus, if r is the direction of diffusion, Fick's first law for the solute, i, is

$$n_i = -D_e A \left(\frac{\partial X_i}{\partial r} \right) \quad (16-11)$$

Fick's second law for constant effective diffusivity in the direction r is

$$\frac{\partial X_i}{\partial t} = \frac{D_e}{r^{v-1}} \frac{\partial}{\partial r} \left[r^{v-1} D_e \frac{\partial X_i}{\partial r} \right] \quad (16-12)$$

where by comparison to (3-69), (3-70), and (3-71), $v = 1, 2,$ and 3 for rectangular, cylindrical, and spherical coordinates, respectively.

The rate of mass transfer through the solvent external to the solid can be written in terms of a mass-transfer coefficient and the concentration of the solute in the solvent or extract Y_i, as in (3-105), where subscripts s and b refer to the solid–liquid interface and the bulk liquid, respectively,

$$n_i = k_c A[(Y_i)_s - (Y_i)_b] \quad (16-13)$$

At the solid–liquid interface at the exterior surface of the solid, (16-11) and (16-13) can be equated:

$$-D_e \left(\frac{\partial X_i}{\partial r} \right)_s = k_c[(Y_i)_s - (Y_i)_b] \quad (16-14)$$

Let

$$m = Y_i / X_i$$
$$a = \text{characteristic dimension of the solid,}$$
$$\text{e.g., the radius of a cylinder or spherical} \quad (16-15)$$
$$\text{particle, or the half-thickness of a slice.}$$

Combining (16-14) and (16-15) and expressing the result in the form of dimensionless groups,

$$-\left[\frac{\partial \left(\dfrac{Y_i}{(Y_i)_s - (Y)_b} \right)}{\partial (r/a)} \right]_s = \left(\frac{m k_c a}{D_e} \right) \quad (16-16)$$

The dimensionless group on the right-hand side of (16-16) is called the Biot number for mass transfer:

$$(N_{Bi})_M = \frac{m k_c a}{D_e} \quad (16-17)$$

which is analogous to the more common Biot number for heat transfer,

$$(N_{Bi}) = \frac{ha}{k} \quad (16-18)$$

Biot numbers are quantitative measures of the ratio of internal (solid) resistance to external (fluid) resistance to transport. In Section 3.3, transient solutions are given to (16-12) for different geometries for the case of an initial uniform concentration, $X_o = c_o$, of the solute in the solid. At time $t = 0$, the solute concentration in the solid phase at the solid-fluid interface is suddenly brought to and then held at $X_s = c_s$. The solutions given are

Geometry	Concentration Profile	Rate of Mass Transfer at Interface	Average Concentration in Solid
Semi-infinite	(3-75)	(3-78)	—
Slab of finite thickness	(3-80), (3-81), and Figure 3.8	(3-82)	(3-85) and Figure 3.9
Infinite cylinder	Figure 3.10	—	Figure 3.9
Sphere	Figure 3.11	—	Figure 3.9

These solutions apply to the case in which the Biot number for mass transfer is infinite, such that the resistance in the fluid phase is negligible and $(Y_i)_b = (Y_i)_s$.

For an infinite Biot number, as indicated above, the solute concentration profile as a function of time is given in Figure 3.8 while the average solute concentration in the solid is given in Figure 3.9. In these plots of the solutions, a dimensionless time, the Fourier number for mass transfer, is used, where

$$(N_{Fo})_M = \frac{D_e t}{a^2} \qquad (16\text{-}19)$$

and a = flake or slice half thickness. When the internal resistance to mass transfer is negligible, which is almost never the case in the leaching of foods, the solution for the uniform concentration of solute in the solid is given by the following equation, whose derivation is left as an exercise:

$$\frac{X_i - \dfrac{(Y_i)_b}{m}}{(X_i)_o - \dfrac{(Y_i)_b}{m}} = \exp\left(\frac{-k_c t m}{a}\right) \qquad (16\text{-}20)$$

When $(N_{Bi})_M > 200$, the external (fluid) mass-transfer resistance is negligible and Figures 3.8 and 3.9 can be applied. When $(N_{Bi})_M < 0.001$, the internal (solid) mass-transfer resistance is negligible and (16-20) applies. When $(N_{Bi})_M$ lies between these two extremes, both resistances must be taken into account. Solutions for this general case are given by Schwartzberg and Chao [11].

Effective diffusivities for solutes in solids are complicated because they depend on the volume fraction of and concentration of solute in the occluded solvent, temperature, tortuosity of the diffusion path, and extent of adsorption of the solute by the marc. Values of solute effective diffusivities in a variety of foods, with water as the solvent, are tabulated by Schwartzberg and Chao [11]. Typical values for sucrose, when the cell walls are hard (e.g., sugar cane and coffee), range from 0.5 to 1.0×10^{-6} cm^2/s. When cells walls are soft (e.g., sugar beets, potatoes, apples, celery, and onions), values are higher, ranging from 1.5 to 4.5×10^{-6} cm^2/s. When the solute is a salt (e.g., NaCl and KCl), effective diffusivities are about four times higher. As mentioned above, the diffusion of oil from flaked oil seeds does not follow Fick's law. If, nevertheless, Fick's law is applied to determine the effective diffusivity, the values are found to decrease significantly with time. For example, data of Karnofsky [12], who leached oil from soybeans, cottonseeds, and flaxseeds, with hexane, give values of effective diffusivity that decrease over the course of extraction by about one order-of-magnitude. Other foods exhibit the same trend under certain conditions. Frequently, the diffusivity is not a constant, but varies with flake or slice thickness and solute concentration. Schwartzberg [11] discusses possible reasons for these effects.

A thin slice or flake of solid can be treated as a slab of finite thickness, with mass transfer from the thin edges ignored. For this case, (3-85) or Figure 3.9 can be used to determine the effective diffusivity from experimental leaching data or predict the rate of leaching. In (3-85), $E_{\text{avg}_{\text{slab}}}$

is the fractional unaccomplished approach to equilibrium for extraction, which decreases with time. As seen in Figure 3.9 and as can be demonstrated with (3-85), when $(N_{Fo})_M > 0.10$, the series solution is converged to less than a 2% error with only one term of the infinite series, given by

$$E_{\text{avg}_{\text{slab}}} = \frac{8}{\pi^2} \exp\left(-\frac{(N_{Fo})_M \pi^2}{4}\right)$$

or

$$\ln E_{\text{avg}_{\text{slab}}} = \ln(8/\pi^2) - \frac{\pi^2}{4}(N_{Fo})_M \qquad (16\text{-}21)$$

Thus, if Fick's law holds, and the diffusivity is constant, a plot of experimental data as log $E_{\text{avg}_{\text{slab}}}$ against time should yield a straight line with a negative slope from which the effective diffusivity can be determined, as illustrated in the following example.

EXAMPLE 16.6

In the commercial extraction of sugar (sucrose) from sugar beets with water, the process is controlled by diffusion through the sugar beet. Yang and Brier [13] conducted diffusion experiments with beets that were sliced into cossettes that were 0.0383 in. thick × 0.25 in. wide and 0.5–1.0 in. long. Typically, the cossettes contained 16 wt% sucrose, 74 wt% water, and 10 wt% insoluble fiber. Experiments were conducted at temperatures ranging from 65 to 80°C, with solvent water rates from 1.0 to 1.2 lb/lb fresh cossettes. For a temperature of 80°C and a solvent water rate of 1.2 lb/lb fresh cossettes, the following smoothed data were obtained:

E_{avg}	t, min.
1.0	0
0.39	10
0.19	20
0.10	30
0.050	40
0.025	50
0.0135	60

These data are plotted in Figure 16.11, where it is seen that a straight line can be passed through the data in the range of time from 10 to 60 minutes. From the slope of this line, using (16-19) and (16-21),

$$\frac{\pi^2}{4}\left(\frac{D_e}{a^2}\right) = 0.00113 \text{ sec}^{-1}$$

Since a = half thickness = $\dfrac{0.0383}{2}(2.54) = 4.86 \times 10^{-2}$ cm

Therefore,

$$D_e = \frac{0.00113(4.86 \times 10^{-2})^2(4)}{(3.14)^2} = 1.1 \times 10^{-6} \text{ cm}^2/\text{s}$$

For a continuous, countercurrent extractor, (16-21) can be used to determine the approximate time required for leaching the solids. The time is given in terms of $E = E_{\text{avg}}$ by

$$t = \int_{E_{\text{in}}}^{E_{\text{out}}} \frac{dE}{\left(\dfrac{dE}{dt}\right)} \qquad (16\text{-}22)$$

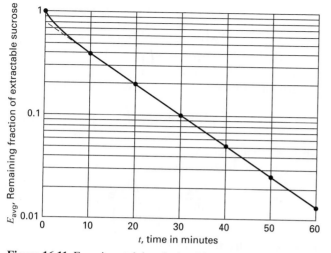

Figure 16.11 Experimental data for leaching of sucrose from sugar beets with water for Example 16.6.

If the solute diffusivity is constant, (dE/dt), except for small values of time, can be obtained by differentiating (16-21) and combining the result to eliminate time, t, to give

$$\frac{dE}{dt} = -\frac{\pi^2 D_e E}{4a^2}$$ (16-23)

Substitution of (16-23) into (16-22), followed by integration, gives

$$t = \frac{4a^2}{\pi^2 D_e} \ln\left(\frac{E_{\text{in}}}{E_{\text{out}}}\right)$$ (16-24)

When the solute diffusivity is not constant, which is more common, experimental plots of E as a function of time can be used directly to obtain values of (dE/dt) for use in (16-22), which can be graphically or numerically integrated, as shown by Yang and Brier [13].

EXAMPLE 16.7

The sucrose in 10,000 lb/h of sugar beets containing 16 wt% sucrose, 74 wt% water, and 10 wt% insoluble fiber is extracted in a continuous, countercurrent extractor at 80°C with 12,000 lb/h of water. If 98% of the sucrose is extracted and no net mass transfer of water occurs, determine the residence time in minutes for the beets. Assume the beets are sliced to 1 mm in thickness and that the effective sucrose diffusivity is that computed in Example 16.6.

SOLUTION

By material balance, the extracted beets contain

$$0.02(0.16)(10,000) = 32 \text{ lb/h sucrose}$$
$$0.74(10,000) = 7,400 \text{ lb/h water}$$
$$0.10(10,000) = 1,000 \text{ lb/h insoluble fiber}$$
$$\text{Total} = 8,432 \text{ lb/h}$$

Thus,

$$X_{\text{out}} = 32/8,432 = 0.0038 \text{ lb/lb}$$
$$X_{\text{in}} = 1,600/10,000 = 0.160 \text{ lb/lb}$$

where X is expressed on a weight fraction basis.

At the beet inlet end,

$$E_{\text{in}} = \frac{0.16 - (Y/m)_{\text{extract out}}}{0.16 - (Y/m)_{\text{extract out}}} = 1.0$$

At the beet outlet end, $(Y/m)_{\text{solvent in}} = 0$

Therefore,

$$E_{\text{out}} = \frac{0.0038}{1.0} = 0.0038$$

From (16-24),

$$t = \frac{4\left(\frac{0.1}{2}\right)^2}{(3.14)^2(1.1 \times 10^{-6})} \ln\left(\frac{1.0}{0.0038}\right) = 5,140 \text{ s} = 85.6 \text{ min}$$

Mineral Processing

Leaching can be used to recover valuable metals from low-grade ores. The leaching process is accomplished by reacting part of the ore with a constituent of the leach liquor, to produce ions of the metal, which are soluble in the liquid. In general, the reaction can be written as

$$A_{(l)} + bB_{(s)} \rightarrow \text{Products}$$ (16-25)

The removal of reactant B from the ore leaves pores in the solid particle for reactant A to diffuse through to reach reactant B in the interior of the particle.

Figure 16.12 shows a spherical mineral particle undergoing leaching. As the process proceeds, an outer porous leached shell develops, leaving an unleached core. The steps involved are:

1. Mass transfer of reactant A from the bulk liquid to the outer surface of the particle.

2. Pore diffusion of reactant A through the leached shell.

3. Chemical reaction at the interface between the leached shell and the unleached core.

4. Pore diffusion of the reaction products back through the leached shell.

5. Mass transfer of the reaction products back into the bulk liquid surrounding the particle.

Because the diameter of the unleached core shrinks with time, a mathematical model for the process, first conceived for application to gas–solid combustion reactions by Yagi and Kunii [14] in 1955 and extended to liquid–solid leaching by Roman, Benner, and Becker [15] in 1974 is referred to as the *shrinking-core model*. Although any one or more of the above five steps can control the process, the rate of leaching is often controlled by Step 2. Therefore, although the general model has been developed for all possibilities, the leaching model presented here is derived on the assumption that Step 2 is controlling.

Referring to Figure 16.12, assume that dr_c/dt, the rate of movement of the reaction interface at r_c, is small with respect to the diffusion velocity of reactant A, in (16-25), through the porous, leached layer. This is referred to as the

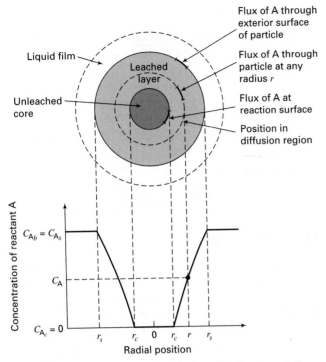

Figure 16.12 Shrinking-core model when diffusion through the leached shell is controlling.

pseudo-steady-state assumption. Although it is valid for the gas–solid case, it is less satisfactory for the liquid–solid case here. The importance of this assumption is that it allows us to neglect the accumulation of reactant A as a function of time in the leached layer as that layer increases in thickness, with the result that the model can be formulated as an ordinary differential equation rather than as a partial differential equation. Thus, the rate of diffusion of reactant A through the porous, leached layer is given by Fick's second law, (3-71), ignoring the term on the left-hand side and replacing the molecular diffusivity with an effective diffusivity:

$$\frac{D_e}{r^2}\frac{d}{dr}\left(r^2\frac{dc_A}{dr}\right) = 0 \tag{16-26}$$

with boundary conditions:

$$c_A = c_{A_s} = c_{A_b} \quad \text{at} \quad r = r_s$$
$$c_A = 0 \quad \text{at} \quad r = r_c$$

These boundary conditions hold because the mass-transfer resistance in the liquid film or boundary layer is assumed negligible and the interface reaction is assumed to be instantaneous and complete, respectively.

If (16-26) is integrated twice and the boundary conditions are applied, the result after simplification is

$$c_A = c_{A_b}\left[\frac{1 - \dfrac{r_c}{r}}{1 - \dfrac{r_c}{r_s}}\right] \tag{16-27}$$

To obtain a relationship between r_c and time, t, differentiate (16-27) with respect to r and evaluate the differential at

$r = r_c$:

$$\left.\frac{dc_A}{dr}\right|_{r=r_c} = \frac{c_{A_b}}{r_c\left(1 - \dfrac{r_c}{r_s}\right)} \tag{16-28}$$

The rate of diffusion at $r = r_c$ is given by Fick's first law:

$$n_A = \frac{d\mathcal{N}_A}{dt} = -4\pi r_c^2 D_e\left(\frac{dc_A}{dr}\right)_{r=r_s} \tag{16-29}$$

where $\mathcal{N}_A$ = moles of A

Combining (16-28) and (16-29),

$$-\frac{d\mathcal{N}_A}{dt} = \frac{4\pi r_c D_e c_{A_b}}{\left(1 - \dfrac{r_c}{r_s}\right)} \tag{16-30}$$

By stoichiometry, from (16-25),

$$\frac{d\mathcal{N}_A}{dt} = \frac{1}{b}\frac{d\mathcal{N}_B}{dt} \tag{16-31}$$

By material balance,

$$\frac{d\mathcal{N}_B}{dt} = \frac{\rho_B}{M_B}\frac{d}{dt}\left(\frac{4}{3}\pi r_c^3\right) = \frac{4\pi r_c^2 \rho_B}{M_B}\frac{dr_c}{dt} \tag{16-32}$$

where

ρ_B = initial mass of reactant B per unit volume of solid particle

M_B = molecular weight of B

Combining (16-30) with (16-32),

$$\frac{-\rho_B}{M_B}\left(\frac{1}{r_c} - \frac{1}{r_s}\right)r_c^2 dr_c = bD_e c_{A_b} dt \tag{16-33}$$

Integration of (16-33) and application of the boundary condition,

$$r_c = r_s \quad \text{at} \quad t = 0$$

gives

$$t = \frac{\rho_B r_s^2}{6D_e b M_B c_{A_b}}\left[1 - 3\left(\frac{r_c}{r_s}\right)^2 + 2\left(\frac{r_c}{r_s}\right)^3\right] \tag{16-34}$$

For complete leaching, $r_s = 0$, and (16-34) becomes

$$t = \frac{\rho_B r_s^2}{6D_e b M_B c_{A_b}} \tag{16-35}$$

EXAMPLE 16.8

A copper ore containing 2 wt% CuO is to be leached with 0.5-M H_2SO_4. The reaction is

$$\frac{1}{2}CuO_{(s)} + H^+_{(aq)} \rightarrow \frac{1}{2}Cu^{2+}_{(aq)} + \frac{1}{2}H_2O_{(aq)} \tag{1}$$

The leaching process is controlled by the diffusion of hydrogen ions through the leached layer. The effective diffusivity, D_e, of the hydrogen ion has been determined by laboratory tests to be 0.6×10^{-6} cm²/s. The specific gravity of the ore is 2.7. For ore particles of diameter equal to 10 mm, estimate the time required to leach 98% of the copper, assuming that the CuO is uniformly distributed throughout the particles. Also, check the validity of the pseudo-steady-state assumption by comparing the amount of hydrogen ions held up in the liquid in the pores with the amount reacted with CuO.

SOLUTION

If 98% of the cupric oxide is leached, then r_c corresponds to 2% of the particle volume. Thus,

$$\frac{4}{3}\pi r_c^3 = (0.02)\frac{4}{3}\pi r_s^3$$

or

$$r_c = (0.02)^{1/3} r_s = (0.02)^{1/3}(0.5) = 0.136 \text{ cm}$$

$$\rho_B = 0.02(2.7) = 0.054 \text{ g/cm}^3 = \text{density of CuO in the ore}$$

$$M_B = 79.6 = \text{molecular weight of CuO}$$

From (16-25) and (1), $b = 0.5$.

For 0.5-M H_2SO_4, $c_{H_b}^+ = \frac{2(0.5)}{1000} = 0.001 \text{ mol/cm}^3$.

From (16-34), with $r_c/r_s = 0.136/0.500 = 0.272$,

$$t = \frac{(0.054)(0.5)^2}{6(0.6 \times 10^{-6})(0.5)(79.6)(0.001)}[1 - 3(0.272)^2 + 2(0.272)^3]$$
$$= 77,000 \text{ sec} = 21.4 \text{ h}$$

Now, check the validity of the pseudo-steady-state assumption.

The specific gravity of CuO is 6.4 g/cm^3.

100 g of ore occupies $100/2.7 = 37.0$ cm^3.

The CuO in this amount of ore occupies

$$0.02(100)/6.4 = 0.313 \text{ cm}^3 \quad \text{or} \quad 0.845\% \text{ of the particle volume}$$

$$\text{Volume of one particle} = \frac{4}{3}\pi r_s^3 = \frac{4}{3}(3.14)(0.5)^3 = 0.523 \text{ cm}^3.$$

Volume of CuO as pores in one particle $= 0.00845(0.523) = 0.0044$ cm^3.

Mols of H^+ in pores, based on the bulk concentration to be conservative:

$$0.001(0.0044) = 4.4 \times 10^{-6} \text{ mol}$$

98% of CuO leached in a particle, in mol units:

$$\frac{0.98(0.02)(2.7)(0.523)}{79.6} = 3.5 \times 10^{-4} \text{ mol}$$

which requires 7.0×10^{-4} mol H^+ for reaction.

Because this value is approximately two orders or magnitude larger than the conservative estimate of H^+ in the pores, the pseudo-steady-state assumption is reasonable.

SUMMARY

1. Leaching is similar to liquid–liquid extraction, except that the solute initially resides in a solid. Leaching is widely used to remove solutes from foods and minerals.

2. When leaching is rapid, it can be accomplished in one stage. However, the leached solid will retain surface liquid that contains the solute. To recover most of the solute in the extract, it is desirable to add one or more washing stages in a countercurrent arrangement.

3. Leaching of large solids can be very slow because of very small values of diffusivities in solids. Therefore, it is common to reduce the size of the solids by crushing, grinding, flaking, slicing, etc.

4. Industrial leaching equipment is available for batch or continuous processing. The solids are contacted with the solvent by either percolation or immersion. Large, continuous, countercurrent extractors can process up to 7,000,000 kg/day of food solids.

5. Washing of large flow rates of leached solids is commonly carried out in thickeners that can be designed to produce a clear liquid overflow and a concentrated solids underflow. When a clear overflow is not critical, hydroclones can replace thickeners.

6. An equilibrium-stage model is widely used for continuous, countercurrent systems when leaching is rapid and washing is desirable for high solute recovery. The model assumes that the concentration of the solute in the overflow leaving a stage equals that in the liquid retained on the solid leaving the stage in the underflow.

7. When the ratio of liquid to solids in the underflow is constant form stage to stage, the equilibrium-stage model can be applied algebraically by a modified Kremser method or graphically by a modified McCabe-Thiele method. If the underflow is variable, the graphical method with a curved operating line is applied.

8. When leaching is slow, as with food solids or low-grade ores, leaching calculations must be done on a rate basis. In some cases, the diffusion of solutes in food solids does not obey Fick's law, because of complex membrane and fiber structures.

9. The leaching of low-grade ores by reactive-leaching is conveniently carried out with a shrinking-core diffusion model, using a pseudo-steady-state assumption.

REFERENCES

1. OTHMER, D.F., and J.C. AGARWAL, *Chem. Eng. Progress*, **51**, 372–373 (1955).

2. PERRY, R.H., and D.W. GREEN, Ed., *Perry's Chemical Engineers' Handbook*, 6th ed., McGraw-Hill, New York (1984), Section 19.

3. KING, C.O., D.J. KATZ, and J.C. BRIER, *Trans. AIChE*, **40**, 533–537 (1944).

4. VAN ARSDALE, G.D., *Hydrometallurgy of Base Metals*, McGraw-Hill, New York (1953).

5. LAMONT, A.G.W., *Can. J. Chem. Eng.*, **36**, 153 (1958).

6. SCHWARTZBERG, H.G., *Chem. Eng., Progress*, **76** (4), 67–85 (1980).

7. COULSON, J.M., J.F. RICHARDSON, J.R. BACKHURST, and J.H. HARKER, *Chemical Engineering*, Vol. 2, 4th ed., Pergamon Press, Oxford (1991).

8. BAKER, E.M., *Trans. AIChE.*, **32**, 62–72 (1936).

9. McCABE, W.L., and J.C. SMITH, *Unit Operations of Chemical Engineering*, 604–608, McGraw-Hill, New York (1956).

10. RAVENSCROFT, E.A., *Ind. Eng. Chem.*, **28**, 851–855 (1936).

11. SCHWARTZBERG, H.G., and R.Y. CHAO, *Food Tech.*, **36** (2), 73–86 (1982).

12. KARNOFSKY, G., *J. Am. Oil Chem. Soc.,* **26,** 564–569 (1949).

13. YANG, H.H., and J.C. BRIER, *AIChE J.,* **4,** 453–459 (1958).

14. YAGI, S., and D. KUNII, "Fifth Symposium (International) on Combustion," Reinhold, New York (1955), pp. 231–244.

15. ROMAN, R.J., B.R. BENNER, and G.W. BECKER, *Trans. Soc. Mining Engineering of AIME,* **256,** 247–256 (1974).

16. ANDUEZA, S., L. MAEZTU, B. DEAN, M.P. DE PENA, J. PELLO, and C. CID, *J. Agric. Food Chem.,* **50,** 7426–7431 (2002).

17. ANDUEZA, S., L. MAEZTU, L. PASCUAL, C. IBANEZ, M.P. DE PENA, and C. CID, *J. Sci. Food Agric.,* **83,** 240–248 (2003).

18. ANDUEZA, S., M.P. DE PENA, and C. CID, *J. Agric. Food Chem.,* **51,** 7034–7039 (2003).

EXERCISES

Section 16.1

16.1 Using experimental data from pilot-plant tests of soybean extraction by Othmer and Agarwal, summarized at the beginning of this chapter, check the mass balances for oil and hexane around the extractor, assuming the moisture is retained in the flakes, and compute the mass ratio of liquid oil to flakes in the leached solids leaving the extractor.

Section 16.2

16.2 Barium carbonate, which is essentially water insoluble, is to be made by precipitation from an aqueous solution containing 120,000 kg/day of water and 40,000 kg/day of barium sulfide, with the stoichiometric amount of solid sodium carbonate. The reaction also produces a by-product of water-soluble sodium sulfide. The process will be carried out in a continuous, countercurrent system of five thickeners. The reaction will take place completely in the first thickener to which will be fed the solid sodium carbonate, the aqueous solution of barium sulfide, and the overflow from the second thickener. Sufficient fresh water will enter the last thickener so that the overflow from the first thickener will be 10 wt% sodium sulfide, assuming that the underflow from each thickener contains two parts of water per one part of barium carbonate by weight.

(a) Draw a schematic diagram of the process and label it with all the given information.

(b) Determine the kg/day of sodium carbonate required and the kg/day of barium carbonate and sodium sulfide produced by the reaction.

(c) Determine the kg/day of fresh water needed, the wt% of sodium sulfide in the liquid portion of the underflow that leaves each thickener, and the kg/day of sodium sulfide that will remain with the barium carbonate product after it is dried.

16.3 Calcium-carbonate precipitate can be produced by the reaction of an aqueous solution of sodium carbonate and calcium oxide. The by-product is aqueous sodium hydroxide. Following decantation, the slurry leaving the precipitation tank is 5 wt% calcium carbonate, 0.1 wt% sodium hydroxide, and the balance water. One hundred thousand lb/h of this slurry is fed to a two-stage, continuous, countercurrent washing system to be washed with 20,000 lb/h of fresh water. The underflow from each thickener will contain 20 wt% solids. Determine the percent recovery of sodium hydroxide in the extract and wt% sodium hydroxide in the dried, calcium-carbonate product. Based on calculations, is it worthwhile to add a third stage?

16.4 Zinc is to be recovered from an ore containing zinc sulfide. The ore is first roasted with oxygen to produce zinc oxide, which is then leached with aqueous sulfuric acid to produce water-soluble zinc sulfate and an insoluble, worthless residue called gangue. The decanted sludge of 20,000 kg/h contains 5 wt% water, 10 wt% zinc sulfate, and the balance as gangue. This sludge is to be washed with water in a continuous, countercurrent washing system to produce an extract, called a strong solution, of 10 wt% zinc sulfate in water, with a 98% recovery of the zinc sulfate. Assume that the underflow from each washing stage contains, by weight, two parts of water (sulfate-free basis) per part of gangue. Determine the number of stages required.

16.5 Fifty-thousand kg/h of flaked soybeans, containing 20 wt% oil, is to be leached of the oil with the same flow rate of *n*-hexane in a countercurrent-flow system consisting of an ideal leaching stage and three ideal washing stages. Experiments show that the underflow from each stage will contain 0.8 kg liquid/kg soybeans (oil-free basis).

(a) Determine the % recovery of oil in the final extract.

(b) If leaching requires three of the four stages, such that one-third of the leaching occurs in each of these three stages, followed by just one true washing stage, determine the % recovery of oil in the final extract.

16.6 One hundred tons per hour of a feed containing 20 wt% Na_2CO_3 and the balance insoluble solids is to be leached and washed with water in a continuous, countercurrent system. Assume that leaching will be completed in one ideal stage. It is desired to obtain a final extract containing 15 wt% solute, with a 98% recovery of solute. The underflow from each stage will contain 0.5 lb solution/lb insoluble solids. Determine the number of ideal washing stages required.

16.7 Titanium dioxide, which is the most common white pigment in paint, can be produced from the titanium mineral, rutile, by chlorination to $TiCl_4$, followed by oxidation to TiO_2. To purify the insoluble titanium dioxide, it is washed free of soluble impurities in a continuous, countercurrent system of thickeners with water. Two hundred thousand kg/h of 99.9 wt% titanium dioxide pigment is to be produced by washing, followed by filtering and drying. The feed contains 50 wt% TiO_2, 20 wt% soluble salts, and 30 wt% water. The wash liquid is pure water at a flow rate equal to that of the feed on a mass-flow basis.

(a) Determine the number of washing stages required if the underflow from each stage is 0.4 kg solution/kg TiO_2.

(b) Determine the number of washing stages required if the underflow is variable as follows:

Concentration of solute, kg/solute/kg solution	Retention of solution, kg solution/kg TiO_2
0.0	0.30
0.2	0.34
0.4	0.38
0.6	0.42

Section 16.3

16.8 Derive (16-20), assuming that $(Y_i)_b$, k_c, m, and a are constants and that $(X_i)_o$ is uniform through the solid.

16.9 Derive (16-24).

16.10 Data of Othmer and Agarwal [1] for the batch extraction of oil from soybeans by oil-free *n*-hexane at 80°F are as follows:

Time, min	Oil content of Soybeans, g/g Dry, Oil-free Soybeans
0	0.203
0.5	0.1559
1	0.1359
2	0.1190
4	0.0981
7	0.0775
12	0.0591
20	0.04197
35	0.03055
60	0.02388
120	0.02107

Determine whether these data are consistent with a constant effective diffusivity of oil in soybeans.

16.11 Estimate the molecular diffusivity of sucrose in water at infinite dilution at 80°C, noting that the value is 0.54×10^{-5} cm²/s at 25°C. Give reasons for the difference between the value you obtain and the value for effective diffusivity in Example 16.6.

16.12 The sucrose in ground coffee particles of an average diameter of 2 mm is to be extracted with water in a continuous, countercurrent extractor at 25°C. The diffusivity of the sucrose in the particles has been determined to be about 1.0×10^{-6} cm²/s. Estimate the time in minutes to leach 95% of the sucrose. For a sphere, with $N_{\mathrm{Fo_M}} > 0.10$,

$$E_{\mathrm{ave}} = \frac{6}{\pi^2} \exp\left(\frac{-\pi^2 D_e t}{a^2}\right)$$

16.13 For the conditions of Example 16.8, determine the effect on leaching time of particle size over the range of 0.5 mm to 50 mm.

16.14 For the conditions of Example 16.8, determine the effect of % recovery of copper over the range of 50–100%.

16.15 Repeat Example 16.8, except that the ore contains 3 wt% Cu_2O.

16.16 For the shrinking-core model, if the rate of leaching is controlled by an interface chemical reaction that is first order in the concentration of reactant A, derive the expression,

$$t = \frac{\rho_B r_s}{b M_B k C_{A_b}} \left(1 - \frac{r_c}{r_s}\right)$$

where k = first-order rate constant.

Chapter 17

Crystallization, Desublimation, and Evaporation

Crystallization is a solid–fluid separation operation in which crystalline particles are formed from a homogeneous fluid phase. Ideally, the crystals are a pure chemical, obtained in a high yield with a desirable shape and a reasonably uniform and desirable size. Crystallization is one of the oldest known separation operations, with the recovery of sodium chloride as salt crystals from water by evaporation dating back to antiquity. Even today, the most common applications are the crystallization from aqueous solution of various inorganic salts, a short list of which is given in Table 17.1. All these cases are referred to as *solution crystallization* because the inorganic salt is clearly the solute, which is crystallized, and water is the solvent, which remains in the liquid phase. The phase diagram for systems suitable for solution crystallization is a solubility curve, such as shown in Figure 17.1a and described earlier in Chapter 4.

For the formation of organic crystals, organic solvents such as acetic acid, ethyl acetate, methanol, ethanol, acetone, ethyl ether, chlorinated hydrocarbons, benzene, and petroleum fractions may be preferred choices, but they must be used with great care when they are toxic or flammable with a low flash point and a wide range of explosive limits.

For either aqueous or organic solutions, crystallization is effected by cooling the solution, evaporating the solvent, or a combination of the two. In some cases, a mixture of two or more solvents may be best, examples of which include water with the lower alcohols, and normal paraffins with chlorinated solvents. Also, the addition of a second solvent is sometimes used to reduce the solubility of the solute. When water is the additional solvent, the process is called *watering-out;* when an organic solvent is added to an aqueous salt solution, the process is called *salting-out.* For both of these cases of solvent addition, fast crystallization called *precipitation* can occur, resulting in large numbers of very small crystals. Precipitation also occurs when one product of two reacting solutions is a solid with low solubility. For example, when aqueous solutions of silver nitrate and sodium chloride are mixed together, insoluble silver chloride is precipitated leaving an aqueous solution of mainly soluble sodium nitrate.

When both components of a homogeneous, binary solution have melting (freezing) points not far removed from each other, the solution is referred to as a *melt.* If, as in Figure 17.1b, the phase diagram for the melt exhibits a eutectic point, it is possible to obtain, in one step called *melt crystallization,* pure crystals of one component or the other, depending on whether the composition of the melt is to the left or right of the eutectic composition. If, however, solid solutions form, as shown in Figure 17.1c, a process of repeated melting and freezing steps, called *fractional melt crystallization,* is required to obtain nearly pure crystalline products. A higher degree of purity can be achieved by a technique called *zone melting* or *refining.* Examples of binary organic systems that form eutectics include metaxylene-paraxylene and benzene-naphthalene. Binary systems of naphthalene-beta naphthol and naphthalene-β naphthylamine, which form solid solutions, are not as common.

Crystallization can also occur from a vapor mixture by a process more properly called *desublimation.* A number of pure compounds, including phthalic anhydride and benzoic acid, are produced in this manner. When two or more compounds tend to desublime, a fractional desublimation process can be employed to obtain near-pure products.

Crystallization of a compound from a dilute, aqueous solution is often preceded by *evaporation* in one or more vessels, called *effects,* to concentrate the solution, and followed by partial separation and washing of the crystals from the resulting slurry, called the *magma,* by centrifugation or filtration. The process is completed by drying the crystals to a specified moisture content.

17.0 INSTRUCTIONAL OBJECTIVES

After completing this chapter, you should be able to:

- Describe different types of crystallization.
- Explain how crystals grow.
- Explain how crystal-size distribution can be measured, tabulated, and plotted.

- Explain the importance of supersaturation in crystallization.
- Differentiate between primary and secondary nucleation of crystals.
- Use mass-transfer theory to determine rate of crystal growth.
- Describe major types of batch and continuous solution-crystallization equipment.
- Apply the MSMPR model to design of a continuous, vacuum, evaporating crystallizer of the draft-tube baffled (DTB) type.
- Understand precipitation.
- Describe equipment for melt crystallization.
- Apply mass-transfer theory to a falling-film melt crystallizer.
- Apply the ideal zone-melting model.
- Differentiate between crystallization and desublimation.
- Describe evaporation equipment.
- Derive and apply the ideal evaporator model.
- Design multiple-effect evaporation systems.

Industrial Example

Consider the crystallization of $MgSO_4 \cdot 7H_2O$ (Epsom salt) from an aqueous solution. The solid–liquid phase diagram for the $MgSO_4 \cdot H_2O$ system at 1 atm is shown in Figure 17.2. Depending on the temperature, four different hydrated forms of $MgSO_4$ are possible: $MgSO_4 \cdot H_2O$, $MgSO_4 \cdot 6H_2O$, $MgSO_4 \cdot 7H_2O$, and $MgSO_4 \cdot 12H_2O$. Furthermore, a eutectic of the latter hydrate with ice is possible. To obtain the usually desired heptahydrate, crystallization must occur in the temperature range from $36°F$ to $118°F$ (Point b to Point c). Within this range, the solubility of $MgSO_4$ (anhydrous or hydrate-free basis) increases almost linearly from about 21 to 33 wt%.

A representative commercial process for producing 4,205 lb/hr (dry basis) of $MgSO_4 \cdot 7H_2O$ crystals from a 10 wt% aqueous solution at 1 atm and $70°F$ is shown in Figure 17.3. This solution is first concentrated in a double-effect evaporation system with forward feed and then mixed with recycled mother liquors from the hydroclone and centrifuge. The combined feed of 14,326 lb/h containing 31.0 wt% $MgSO_4$ at $120°F$ and 1 atm enters an evaporative, vacuum crystallizer constructed of 316 stainless steel and shown in more detail in Figure 17.4.

The crystallizer utilizes internal circulation of 6,000 gpm of magma up through a draft tube equipped with a 3-Hp marine-propeller agitator to obtain near-perfect mixing of

Table 17.1 Some Inorganic Salts Recovered from Aqueous Solutions

Chemical Name	Formula	Common Name	Crystal System
Ammonium chloride	NH_4Cl	sal-ammoniac	cubic
Ammonium sulfate	$(NH_4)_2SO_4$	mascagnite	orthorhombic
Barium chloride	$BaCl_2 \cdot 2H_2O$		monoclinic
Calcium carbonate	$CaCO_3$	calcite	rhombohedral
Copper sulfate	$CuSO_4 \cdot 5H_2O$	blue vitriol	triclinic
Magnesium sulfate	$MgSO_4 \cdot 7H_2O$	Epsom salt	orthorhombic
Magnesium chloride	$MgCl_2 \cdot 6H_2O$	bischofite	monoclinic
Nickel sulfate	$NiSO_4 \cdot 6H_2O$	single nickel salt	tetragonal
Potassium chloride	KCl	muriate of potash	cubic
Potassium nitrate	KNO_3	nitre	hexagonal
Potassium sulfate	K_2SO_4	arcanite	orthorhombic
Silver nitrate	$AgNO_3$	lunar caustic	orthorhombic
Sodium chlorate	$NaClO_3$		cubic
Sodium chloride	$NaCl$	salt, halite	cubic
Sodium nitrate	$NaNO_3$	chile salt petre	rhombohedral
Sodium sulfate	$Na_2SO_4 \cdot 10H_2O$	glauber's salt	monoclinic
Sodium thiosulfate	$Na_2S_2O_3 \cdot 5H_2O$	hypo	monoclinic
Zinc sulfate	$ZnSO_4 \cdot 7H_2O$	white vitriol	orthorhombic

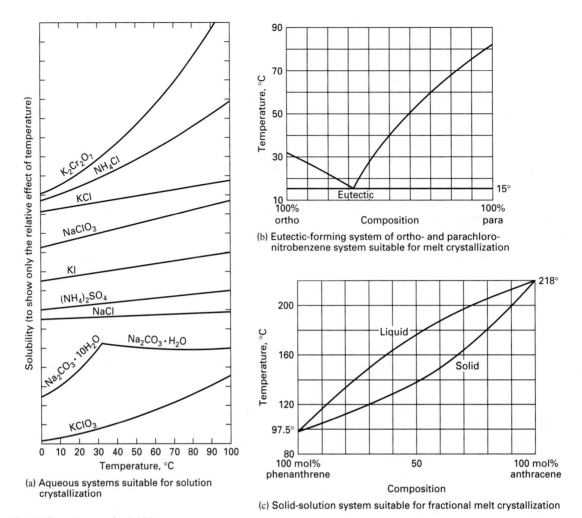

Figure 17.1 Different types of solubility curves.
[From *Handbook of Separation Techniques for Chemical Engineers,* 2nd ed., P.A. Schweitzer, Editor-in-chief, McGraw-Hill, New York (1988) with permission.]

the magma. Mother liquor, which is separated from crystals during upward flow outside of the skirt baffle, is circulated externally at the rate of 625 gpm, by a 10-Hp stainless-steel pump, through a 300-ft^2 stainless-steel, plate-and-frame heat exchanger, where 2,052,000 Btu/hr of heat is transferred to the solution from 2,185 lb/h of condensing 20 psig steam to provide supersaturation and energy to evaporate 2,311 lb/h of water in the crystallizer.

The vapor leaving the top of the crystallizer is condensed by direct contact with cooling water in a barometric condenser, attached to which are ejectors to pull a vacuum of 0.867 psia in the vapor space of the crystallizer. The product magma, at a temperature of 105°F, consists of 7,810 lb/h of mother liquor saturated with 30.6 wt% MgSO$_4$ and 4,205 lb/h of crystals. This corresponds to a magma containing 35% crystals by weight or 30.2% crystals by volume, based on a crystal density of 1.68 g/cm^3 and a mother liquor density of 1.35 g/cm^3. The boiling-point elevation of the saturated

mother liquor at 105°F is 8°F. Thus, the vapor leaving the crystallizer is superheated by the same 8°F. The magma residence time in the crystallizer is 4 hours, which is sufficient to produce the following crystal-size distribution:

> 35 wt% on 20 mesh U.S. screen
>
> 80 wt% on 40 mesh U.S. screen
>
> 99 wt% on 100 mesh U.S. screen

The crystallizer for the representative process is 30 ft high with a vapor-space diameter of 5-1/2 ft and a magma-space diameter of 10 ft. The magma is thickened to 50 wt% crystals in a hydroclone, from which the mother-liquor overflow is recycled to the crystallizer and the underflow slurry is sent to a continuous centrifuge, where the slurry is further thickened to 65 wt% crystals and washed. Filtrate mother liquor from the centrifuge is also recycled to the crystallizer. The centrifuge cake is fed to a continuous direct-heat rotary dryer to reduce the moisture content of the crystals to 1.5 wt%.

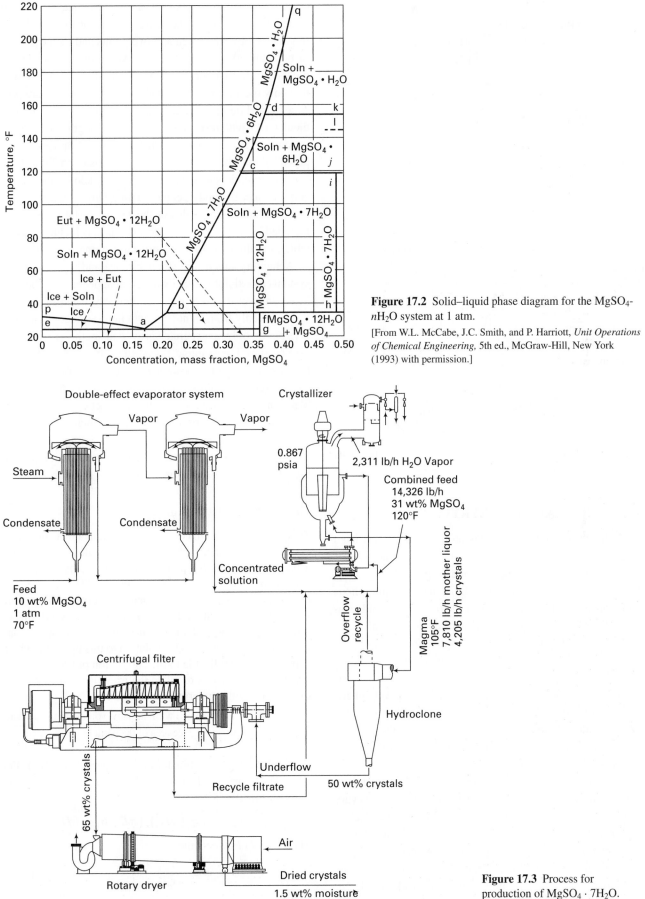

Figure 17.2 Solid–liquid phase diagram for the MgSO₄-*n*H₂O system at 1 atm.

[From W.L. McCabe, J.C. Smith, and P. Harriott, *Unit Operations of Chemical Engineering,* 5th ed., McGraw-Hill, New York (1993) with permission.]

Figure 17.3 Process for production of MgSO₄ · 7H₂O.

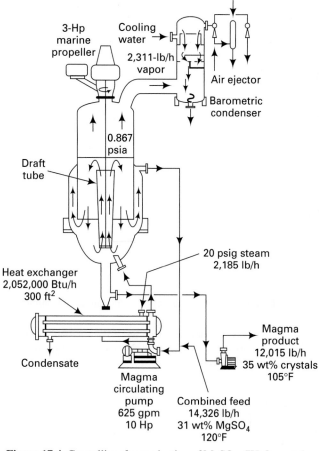

Figure 17.4 Crystallizer for production of $MgSO_4 \cdot 7H_2O$ crystals.

crystal. This led to the concept of a space lattice as a regular arrangement of points (molecules, atoms, or ions) such that if a line is drawn between any two points and then extended in both directions, the line will pass through other lattice points with an identical spacing. In 1848, Bravais showed that only the 14 space lattices shown in Figure 17.5 are possible. Based on the symmetry of the three mutually perpendicular axes with respect to their relative lengths (a, b, c) and the angles (α, β, γ) between the axes, the 14 lattices can be classified into the seven crystal systems listed in Table 17.2. For example, the cubic (regular) system includes the simple cubic lattice, the body-centered cubic lattice, and the face-centered lattice. Examples of the seven crystal systems are included in Table 17.1. The five sodium salts included in that table form three of the seven crystal systems.

Actual crystals of a given substance and a given crystal system can exhibit markedly different appearances when the faces grow at different rates, particularly when these rates vary markedly from stunted growth in one direction, so as to give plates, or by exaggerated growth in another direction, to give needles. For example, potassium sulfate, which belongs to the orthorhombic-crystal system, can take on any of the shapes (crystal habits) shown in Figure 17.6, including plates, needles, and prisms. When product crystals of a particular crystal habit are desired, experimental research may be required to find the necessary processing conditions. Modifications of crystal habit are most often accomplished by deliberate addition of impurities to the solution.

17.1 CRYSTAL GEOMETRY

In a solid, the motion of molecules, atoms, or ions is restricted largely to oscillations about fixed positions. If the solid is amorphous, these positions are not arranged in a regular or lattice pattern; if the solid is crystalline, they are. Amorphous solids are isotropic, such that physical properties are independent of the direction of measurement; crystalline solids are anisotropic, unless the crystals are cubic in structure.

When crystals grow, unhindered by other surfaces such as container walls and other crystals, they form polyhedrons with flat sides and sharp corners. Crystals are never spherical in shape. Although two crystals of a given chemical may appear quite different in size and shape, they always have something in common, known as the Law of Constant Interfacial Angles, proposed by Hauy in 1784. This law states that the angles between corresponding faces of all crystals of a given substance are constant, even though the crystals vary in size and in the development of the various faces (called the *crystal habit*). The interfacial angles and lattice dimensions can be measured accurately by x-ray crystallography.

As discussed by Mullin [1], early investigators found that crystals consist of many units, each shaped like the larger

Crystal-Size Distributions

Typical magmas from a crystallizer contain a distribution of crystal sizes and shapes. It is highly desirable to characterize a batch of crystals (or particles in general) by an average crystal size and a crystal-size distribution. This is often accomplished by defining a characteristic crystal dimension. However, as shown in Figure 17.6, some crystal shapes might require two characteristic dimensions, while one might suffice for others. One solution to this problem, which is particularly applicable to the correlation of transport rates involving particles, is to relate the irregular-shaped particle to a sphere by the sphericity, ψ, defined as

$$\psi = \frac{\text{surface area of a sphere with the same volume as the particle}}{\text{surface area of the particle}}$$

(17-1)

For a sphere, $\psi = 1$, while for all other particles, $\psi < 1$. For a spherical particle of diameter, D_p, the surface area, s_p, to volume, v_p, ratio is

$$(s_p/v_p)_{\text{sphere}} = \left(\pi D_p^2\right)/\left(\pi D_p^3/6\right) = 6/D_p$$

Therefore, (17-1) becomes

$$\psi = \frac{6}{D_p}\left(\frac{v_p}{s_p}\right)_{\text{particle}}$$

(17-2)

Table 17.2 The Seven Crystal Systems

Crystal System	Space Lattices	Length of Axes	Angles between Axes
Cubic (regular)	Simple cubic Body-centered cubic Face-centered cubic	$a = b = c$	$\alpha = \beta = \gamma = 90°$
Tetragonal	Square prism Body-centered square prism	$a = b < c$	$\alpha = \beta = \gamma = 90°$
Orthorhombic	Simple orthorhombic Body-centered orthorhombic Base-centered orthorhombic Face-centered orthorhombic	$a \neq b \neq c$	$\alpha = \beta = \gamma = 90°$
Monoclinic	Simple monoclinic Base-centered monoclinic	$a \neq b \neq c$	$\alpha = \beta = 90°$ $\gamma \neq 90°$
Rhombohedral (trigonal)	Rhombohedral	$a = b = c$	$\alpha = \beta = \gamma \neq 90°$
Hexagonal	Hexagonal	$a = b \neq c$	$\alpha = \beta = 90°$ $\gamma \neq 120°$
Triclinic	Triclinic	$a \neq b \neq c$	$\alpha \neq \beta \neq \gamma \neq 90°$

Simple cubic

Body-centered cubic

Face-centered cubic

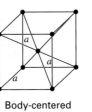

Simple tetragonal

Body-centered tetragonal

Simple orthorhombic

Body-centered orthorhombic

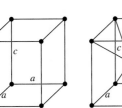

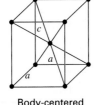

Base-centered orthorhombic

Face-centered orthorhombic

Rhombohedral

Hexagonal

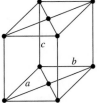

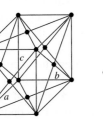

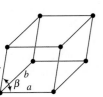

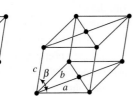

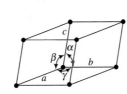

Simple monoclinic

Base-centered monoclinic

Triclinic

Figure 17.5 The 14 space lattices of Bravais.

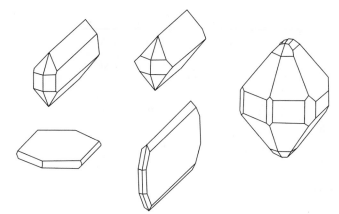

Figure 17.6 Some crystal habits of orthorhombic, potassium-sulfate crystals.

EXAMPLE 17.1

Estimate the sphericity of a cube of dimension a on each side.

SOLUTION

$$v_{cube} = a^3$$

$$s_{cube} = 6a^2$$

From (17-2),

$$\psi = \frac{6}{D_p}\left(\frac{a^3}{6a^2}\right) = \frac{a}{D_p}$$

Because the volumes of the sphere and the cube must be equal,

$$\pi D_p^3/6 = a^3$$

Solving,

$$D_p = 1.241\,a$$

Then,

$$\psi = a/(1.241\,a) = 0.806$$

The most common methods for measuring particle size are listed in Table 17.3 together with their useful particle-size ranges. Because of the irregular shapes of crystals, it should not be surprising that the different methods can give results that may differ by as much as 50%. Crystal-size distributions are most often determined with U.S. (or British)

Table 17.3 Methods of Measuring Particle Size

Method	Size Range, Microns
Woven-wire screen	32–5600
Coulter electrical sensor	1–200
Gravity sedimentation	1–50
Optical microscopy	0.5–150
Laser-light scattering	0.04–2000
Centrifugal sedimentation	0.01–5
Electron microscopy	0.001–5

Table 17.4 U.S. Standard Screens ASTM EII

Mesh Number	Opening of Square Aperture		
	in.	mm	μm
3-1/2	0.220	5.60	5600
4	0.187	4.75	4750
5	0.157	4.00	4000
6	0.132	3.35	3350
7	0.110	2.80	2800
8	0.0929	2.36	2360
10	0.0787	2.00	2000
12	0.0669	1.70	1700
14	0.0551	1.40	1400
16	0.0465	1.18	1180
18	0.0394	1.000	1000
20	0.0335	0.850	850
25	0.0280	0.710	710
30	0.0236	0.600	600
35	0.0197	0.500	500
40	0.0167	0.425	425
45	0.0140	0.355	355
50	0.0118	0.300	300
60	0.00984	0.250	250
70	0.00835	0.212	212
80	0.00709	0.180	180
100	0.00591	0.150	150
120	0.00492	0.125	125
140	0.00417	0.106	106
170	0.00354	0.090	90
200	0.00295	0.075	75
230	0.00248	0.063	63
270	0.00209	0.053	53
325	0.00177	0.045	45
400	0.00150	0.038	38
450	0.00126	0.032	32

standard wire-mesh screens [ASTM Ell (1989)] derived from the earlier Tyler standard screens. The U.S. standard is based on a 1-mm (1000-μm)-square *aperture-opening* screen called Mesh No. 18 because there are 18 apertures per inch. The standard Mesh numbers are listed in Table 17.4, where each successively smaller aperture differs from the preceding aperture by a factor of approximately $(2)^{1/4}$. Mechanical shaking is applied to conduct the sieving operation, using a stack of ordered screens.

When wire-mesh screens are used to determine crystal-size distribution, the crystal size is taken to correspond to the screen aperture through which the crystal just passes. However, because of the irregularity of particle shape, this should be considered as a nominal value only. This is particularly true for plates and needles, as illustrated in Figure 17.7.

Particle-size-distribution data, called a *screen analysis,* are presented in the form of a table, from which differential and cumulative plots can be made, usually on a mass-fraction

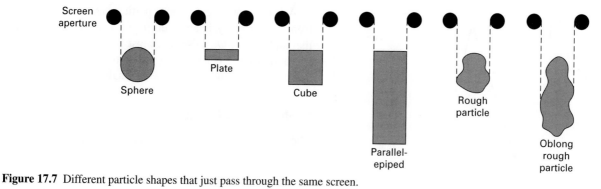

Figure 17.7 Different particle shapes that just pass through the same screen.

basis. Consider the following laboratory screen-analysis data presented by Graber and Taboada [2] for crystals of $Na_2SO_4 \cdot 10H_2O$ (Glauber's salt) grown at about 18°C during an average residence time of 37.2 min in a well-mixed, laboratory, cooling crystallizer. The smallest screen used was 140 mesh, with particles passing through that screen being retained on a pan.

Mesh Number	Aperture, D_p, mm	Mass Retained on Screen, Grams	% Mass Retained
14	1.400	0.00	0.00
16	1.180	9.12	1.86
18	1.000	32.12	6.54
20	0.850	39.82	8.11
30	0.600	235.42	47.95
40	0.425	89.14	18.15
50	0.300	54.42	11.08
70	0.212	22.02	4.48
100	0.150	7.22	1.47
140	0.106	1.22	0.25
Pan	—	0.50	0.11
		491.00	100.00

Differential Screen Analysis

A *differential screen analysis* is made by determining the arithmetic-average aperture for each mass fraction that passes through one screen but not the next screen. Thus, from the above table, a mass fraction of 0.0186 passes through a screen of 1.400-mm aperture, but does not pass through a screen of 1.180-mm aperture. The average of these two apertures is 1.290 mm, which is taken to be the nominal particle size for that mass fraction. The following differential analysis is computed in this manner, where the designation $-14 + 16$ refers to those particles passing through a 14-mesh screen and retained on a 16-mesh screen.

Mesh Range	$\bar{D}_p$, Average Particle Size, mm	Mass Fraction, x_i
$-14 + 16$	1.290	0.0186
$-16 + 18$	1.090	0.0654
$-18 + 20$	0.925	0.0811
$-20 + 30$	0.725	0.4795
$-30 + 40$	0.513	0.1815
$-40 + 50$	0.363	0.1108

Mesh Range	$\bar{D}_p$, Average Particle Size, mm	Mass Fraction, x_i
$-50 + 70$	0.256	0.0448
$-70 + 100$	0.181	0.0147
$-100 + 140$	0.128	0.0025
$-140 + (170)$	0.098	0.0011
		1.0000

A plot of the differential screen analysis is shown in Figure 17.8 both as (a) an *x-y* plot and as (b) a histogram. If a wide range of screen aperture is covered, it is best to use a log scale for that variable.

Cumulative Screen Analysis

Screen analysis data can also be plotted as cumulative-weight-percent oversize or (which is more common) undersize as a function of screen aperture. For the above data of Graber and Taboada [2], the two types of *cumulative screen analysis* are as follows:

Aperture, D_p, mm	Cumulative wt% Undersize	Cumulative wt% Oversize
1.400	100.00	0.00
1.180	98.14	1.86
1.000	91.60	8.40
0.850	83.49	16.51
0.600	35.54	64.46
0.425	17.39	82.61
0.300	6.31	93.69
0.212	1.83	98.17
0.150	0.36	99.64
0.106	0.11	99.89

Because 0.11 wt% passed through a 0.106-mm aperture but was retained on a pan with no indication of just how small these retained particles were, the cumulative wt% undersize and oversize cannot be taken to 0 and 100%, respectively.

The above cumulative screen analyses are plotted in Figure 17.8c. The two curves, which are mirror images of each other, cross at a median size where 50 wt% is larger in size and 50 wt% is smaller. As with differential plots, a log scale is preferred if a large range of screen aperture is covered. A log scale for the cumulative wt% may also be preferred if an appreciable fraction of the data points lie below 10%.

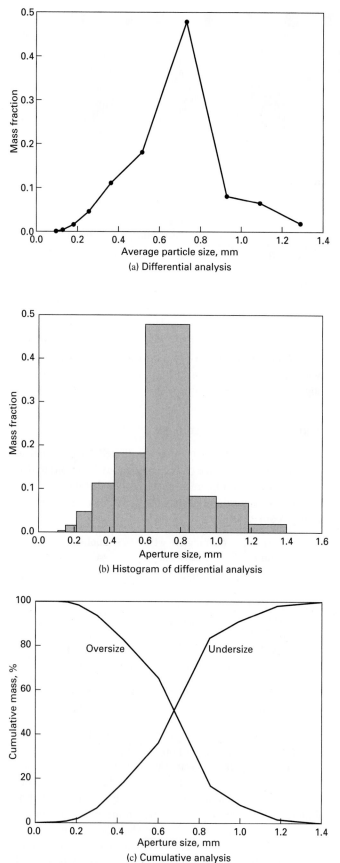

(a) Differential analysis

(b) Histogram of differential analysis

(c) Cumulative analysis

Figure 17.8 Screen analyses for data of Graber and Taboada [2].

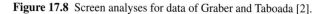

A number of different mean particle sizes that are derived from screen analysis are used in practice, depending upon the application. Of these, the most useful are: (1) surface-mean diameter, (2) mass-mean diameter, (3) arithmetic-mean diameter, and (4) volume-mean diameter.

Surface-Mean Diameter

The specific surface area (area/mass) of a particle of spherical or other shape is defined by

$$A_w = s_p/m_p = s_p/v_p \rho_p \qquad (17\text{-}3)$$

Combining (17-2) and (17-3)

$$A_w = 6/\psi \rho_p D_p \qquad (17\text{-}4)$$

For n mass fractions, x_i, each of average aperture $\bar{D}_{p_i}$, from a screen analysis, the overall specific surface area is given by

$$A_w = \sum_{i=1}^{n} \frac{6x_i}{\psi \rho_p \bar{D}_{p_i}} = \frac{6}{\psi \rho_p} \sum \frac{x_i}{\bar{D}_{p_i}} \qquad (17\text{-}5)$$

The surface-mean diameter is defined by

$$A_w = \frac{6}{\psi \rho_p \bar{D}_S} \qquad (17\text{-}6)$$

Combining (17-5) and (17-6),

$$\bar{D}_S = \frac{1}{\sum_{i=1}^{n} \dfrac{x_i}{\bar{D}_{p_i}}} \qquad (17\text{-}7)$$

which can be used to determine $\bar{D}_S$ from a screen analysis. This mean diameter is sometimes referred to as the Sauter mean diameter and as the volume-surface-mean diameter. It is often used for skin friction, heat-transfer, and mass-transfer calculations involving particles.

Mass-Mean Diameter

The mass-mean diameter is defined by

$$\bar{D}_W = \sum_{i=1}^{n} x_i \bar{D}_{p_i} \qquad (17\text{-}8)$$

Arithmetic-Mean Diameter

The arithmetic-mean diameter is defined in terms of the number of particles, N_i, in each size range:

$$\bar{D}_N = \frac{\sum_{i=1}^{n} N_i \bar{D}_{p_i}}{\sum N_i} \qquad (17\text{-}9)$$

The number of particles is related to the mass fraction of particles by

$$x_i = \frac{\text{mass of particles of average size } \bar{D}_{p_i}}{\text{total mass}}$$

$$= \frac{N_i f_v \left(\bar{D}_{p_i}\right)^3 \rho_p}{M_t} \qquad (17\text{-}10)$$

where

f_v = volume shape factor defined by $v_p = f_v \bar{D}_{p_i}^3$

M_t = total mass   (17-11)

For spherical particles, $f_v = \pi/6$.

If (17-10) is solved for N_i, substituted into (17-9), and simplified, we obtain

$$\bar{D}_N = \frac{\sum\limits_{i=1}^{n} \left(\dfrac{x_i}{\bar{D}_{p_i}^2} \right)}{\sum\limits_{i=1}^{n} \left(\dfrac{x_i}{\bar{D}_{p_i}^3} \right)} \qquad (17\text{-}12)$$

Volume-Mean Diameter

The volume-mean diameter, $\bar{D}_V$, is defined by

$$\left(f_v \bar{D}_V^3 \right) \sum_{i=1}^{n} N_i = \sum_{i=1}^{n} \left(f_v \bar{D}_{p_i}^3 \right) N_i \qquad (17\text{-}13)$$

Solving (17-13) for $\bar{D}_V$ for a constant value of f_v gives

$$\bar{D}_V = \left(\frac{\sum\limits_{i=1}^{n} N_i \bar{D}_{p_i}^3}{\sum\limits_{i=1}^{n} N_i} \right) \qquad (17\text{-}14)$$

The corresponding relation in terms of x_i rather than N_i is obtained by combining (17-14) with (17-10), giving

$$\bar{D}_V = \left(\frac{1}{\sum \dfrac{x_i}{\bar{D}_{p_i}^3}} \right)^{1/3} \qquad (17\text{-}15)$$

EXAMPLE 17.2

Using the screen analysis data of Graber and Taboada given above, compute all four mean diameters.

SOLUTION

Since the data are given in weight (mass) fractions, use (17-7), (17-8), (17-12), and (17-15).

$\bar{D}_p$, mm	x	$x/\bar{D}_p$	$x\bar{D}_p$	$x/\bar{D}_p^2$	$x/\bar{D}_p^3$
1.290	0.0186	0.0144	0.0240	0.0112	0.0087
1.090	0.0654	0.0600	0.0713	0.0550	0.0505
0.925	0.0811	0.0877	0.0750	0.0948	0.1025
0.725	0.4795	0.6614	0.3476	0.9122	1.2583
0.513	0.1815	0.3538	0.0931	0.6897	1.3444
0.363	0.1108	0.3052	0.0402	0.8409	2.3164
0.256	0.0448	0.1750	0.0115	0.6836	2.6703
0.181	0.0147	0.0812	0.0027	0.4487	2.4790
0.128	0.0025	0.0195	0.0003	0.1526	1.1921
0.098	0.0011	0.0112	0.0001	0.1145	1.1687
	1.0000	1.7695	0.6658	4.0032	12.5909

From (17-7),

$$\bar{D}_S = \frac{1}{1.7695} = 0.565 \text{ mm}$$

From (17-8),

$$\bar{D}_W = 0.666 \text{ mm}$$

From (17-12),

$$\bar{D}_N = \frac{4.0032}{12.5909} = 0.318 \text{ mm}$$

From (17-15),

$$\bar{D}_V = \left(\frac{1}{12.5909} \right)^{1/3} = 0.430 \text{ mm}$$

Thus, the mean diameters vary significantly.

17.2 THERMODYNAMIC CONSIDERATIONS

Solubility and Material Balances

Important thermodynamic properties for crystallization operations include melting point, heat of fusion, solubility, heat of crystallization, heat of solution, heat of transition, and supersaturation. For binary systems of water and soluble inorganic and organic chemicals, Mullin [1] presents extensive tables of solubility, as a function of temperature, and heat of solution at infinite dilution and room temperature (approximately 18–25°C). Data in water are listed in Table 17.5 for the inorganic salts of Table 17.1, where solubility data are given on a hydrate-free basis.

Solubilities are seen to vary widely from as low as 4.8 g/100 g of water for Na_2SO_4 (as the decahydrate) at 0°C to 952 g/100 g of water for $AgNO_3$ at 100°C. For KNO_3, the solubility increases by a factor of 18.6 for the same temperature increase.

The solubility of an inorganic compound can be even much lower than that shown for Na_2SO_4. Such compounds are generally considered to be just slightly or sparingly soluble or almost insoluble. The solubility of such compounds is usually expressed as a solubility product, K_c, in terms of ion concentrations. Data for several compounds are given in Table 17.6. For example, consider $Al(OH)_3$, which is sparingly soluble with a solubility product of $K_c = 1.1 \times 10^{-15}$ at 18°C and dissolves according to the equation

$$Al(OH)_{3_{(s)}} \Leftrightarrow Al^{3+}_{(aq)} + 3OH^-_{(aq)}$$

By the law of mass action, the equilibrium constant, called the solubility product for dissolution, is given by

$$K_c = \frac{(c_{Al^{3+}})(c_{OH^-})^3}{a_{Al(OH)_3}} = (c_{Al^{3+}})(c_{OH^-})^3 = 1.1 \times 10^{-15}$$

where the activity of $Al(OH)_3$ solid is taken as 1.0. Since, by stoichiometry,

$$(c_{OH^-}) = 3(c_{Al^{3+}}) \quad \text{and} \quad K_c = (c_{Al^{3+}})^4 (3)^3 = 1.1 \times 10^{-15}$$

then,

$$(c_{Al^{3+}}) = c_{\text{dissolved } Al(OH)_3} = 8 \times 10^{-5} \text{ gmoles/L}$$

which is a very small concentration.

Table 17.5 Solubility and Heat of Solution at Infinite Dilution of Some Inorganic Compounds in Water (A Positive Heat of Solution Is Endothermic)

Compound	Heat of Solution of Stable Hydrate (at Room Temperature) kcal/mole Compound	Solubility (Hydrate-free Basis) g/100 g H_2O at T, °C								Stable Hydrate at Room Temperature
		0	10	20	30	40	60	80	100	
NH_4Cl	+3.8	29.7	33.4	37.2	41.4	45.8	55.2	65.6	77.3	0
$(NH_4)_2SO_4$	+1.5	71.0	73.0	75.4	78.0	81.0	88.0	95.3	103.3	0
$BaCl_2$	+4.5	31.6	33.2	35.7	38.2	40.7	46.4	52.4	58.3	2
$CuSO_4$	+2.86	14.3	17.4	20.7	25.0	28.5	40.0	55.0	75.4	5
$MgSO_4$	+3.18	22.3	27.8	33.5	39.6	44.8	55.3	56.0	50.0	7
$MgCl_2$	−3.1	52.8	53.5	54.5	56.0	57.5	61.0	66.0	73.0	6
$NiSO_4$	+4.2	26	32	37	43	47	55	63	—	7
KCl	+4.4	27.6	31.0	34.0	37.0	40.0	45.5	51.1	56.7	0
KNO_3	+8.6	13.3	20.9	31.6	45.8	63.9	110	169	247	0
K_2SO_4	+6.3	7.4	9.3	11.1	13.1	14.9	18.3	21.4	24.2	0
$AgNO_3$	+5.4	122	170	222	300	376	525	669	952	0
$NaClO_3$	+5.4	80	89	101	113	126	155	189	233	0
$NaCl$	+0.93	35.6	35.7	35.8	36.1	36.4	37.1	38.1	39.8	0
$NaNO_3$	+5.0	72	78	85	92	98	—	133	163	0
Na_2SO_4	+18.7	4.8	9.0	19.4	40.8	48.8	45.3	43.7	42.5	10
$Na_2S_2O_3$	+11.4	52	61	70	84	103	207	250	266	5
Na_3PO_4	+15.0	1.5	4	11	20	31	55	81	108	12

Table 17.6 Concentration Solubility Products of Some Sparingly Soluble Inorganic Compounds

Compound	T, °C	K_c
Ag_2CO_3	25	6.15×10^{-12}
$AgCl$	25	1.56×10^{-10}
$Al(OH)_3$	15	4×10^{-13}
$Al(OH)_3$	18	1.1×10^{-15}
$BaSO_4$	18	0.87×10^{-10}
$CaCO_3$	15	0.99×10^{-8}
$CaSO_4$	10	1.95×10^{-4}
$CuSO_4$	16–18	2×10^{-47}
$Fe(OH)_3$	18	1.1×10^{-36}
$MgCO_3$	12	2.6×10^{-5}
ZnS	18	1.2×10^{-23}

For less sparingly soluble compounds, the equilibrium constant, called K_a, is more rigorously expressed in terms of ionic activities or activity coefficients:

$$K_a = \frac{(a_{Al^{3+}})(a_{OH^{-1}})^3}{a_{Al(OH)_3}} = (\gamma_{Al^{3+}})(c_{Al^{3+}})(\gamma_{OH^{-}})^3(c_{OH^{-}})^3$$

In general, $\gamma \approx 1.0$ for $c < 1 \times 10^{-3}$ gmoles/L. As c rises above 1×10^{-3} gmoles/L, γ decreases, but may pass through a minimum and then increase. Mullin [1] presents activity-coefficient data at 25°C for soluble inorganic compounds over a wide range of concentration.

Although the solubility of most inorganic compounds increases with increasing temperature, a few common compounds exhibit a so-called negative or inverted solubility, in certain ranges of temperature, where solubility decreases with increasing temperature. These compounds are the so-called hard salts, which include anhydrous Na_2SO_4 and $CaSO_4$.

A considerable change in the solubility curve can occur when a phase transition from one stable hydrate to another takes place. For example, in Table 17.5, $Na_2SO_4 \cdot 10H_2O$ is the stable form from 0°C to about 32.4°C. In that temperature range, the solubility increases rapidly from 4.8 to 49.5 g (hydrate-free basis)/100 g H_2O. From 32.4°C to 100°C, the stable form is Na_2SO_4, whose solubility decreases slowly from 49.5 to 42.5 g/100 g H_2O. In the phase diagram of Figure 17.2 for the $MgSO_4$-water system, the solubility-temperature curves of each of the four hydrated forms has a distinctive slope.

The solubility characteristic of a solute in a particular solvent is, by far, the most important property for determining: (1) the best method for causing crystallization, and (2) the ease or difficulty in growing crystals. Crystallization by cooling is only attractive for compounds having a solubility that decreases rapidly with decreasing temperature above ambient temperature. Such is not the case for most of the compounds in Table 17.5. For NaCl, crystallization by cooling would be undesirable because the solubility decreases only by about 10% when the temperature decreases from 100 to 0°C. For most soluble inorganic compounds, cooling by evaporation is the preferred technique.

Solid compounds with a very low solubility can be produced by reacting two soluble compounds. For example, in Table 17.6, solid $Al(OH)_3$ can be formed by the reaction

$$AlCl_{3(aq)} + 3NaOH_{(aq)} \Leftrightarrow Al(OH)_{3(ppt)} + 3NaCl_{(aq)}$$

However, the reaction is so fast that only very fine crystals, called a precipitate, are produced, with no simple method to cause them to grow to large crystals.

EXAMPLE 17.3

The concentrate from an evaporation system is 4,466 lb/h of 37.75 wt% $MgSO_4$ at 170°F and 20 psia. It is mixed with 9,860 lb/h of a saturated aqueous recycle filtrate of $MgSO_4$ at 85°F and 20 psia and sent to a vacuum crystallizer, operating at 85°F and 0.58 psia in the vapor space, to produce water vapor and a magma of 20.8 wt% crystals and 79.2 wt% saturated solution. The magma is sent to a filter, from which filtrate is recycled as mentioned above. Determine the lb/h of water evaporated and the maximum production rate of crystals in tons/day (dry basis for 2000 lb/ton).

SOLUTION

For the saturated filtrate at 85°F, the weight fraction of $MgSO_4$, from Figure 17.2, is 28 wt%. Therefore, $MgSO_4$ in the recycle filtrate is 9,860(0.28) = 2,760 lb/h. By material balance around the mixing step,

	lb/h		
Component	Feed	Recycle Filtrate	Crystallizer Feed
$MgSO_4$	1,686	2,760	4,446
H_2O	2,780	7,100	9,880
	4,466	9,860	14,326

The material balance around the crystallizer is conveniently made by a balance on $MgSO_4$. At 85°F, from Figure 17.2, the magma is 20.8 wt% $MgSO_4 \cdot 7H_2O$ crystals and 79.2 wt% of 28 wt% aqueous $MgSO_4$ liquid. Because the MW of $MgSO_4$ and $MgSO_4 \cdot 7H_2O$ are 120.4 and 246.4, respectively, the crystals are $120.4/246.4 = 0.4886$ mass fraction $MgSO_4$. Therefore, by a $MgSO_4$

balance,

$$4,446 = 0.28 L + 0.4886 S \tag{1}$$

where

L = lb/h of liquid

S = lb/h of crystals

Also,

$$S = 0.208(S + L) \tag{2}$$

Solving (1) and (2) simultaneously,

$$S = 2,856 \, \text{lb/h}$$
$$L = 10,876 \, \text{lb/h}$$

By a total material balance around the crystallizer,

$$F = V + L + S$$

where F = total feed rate and V = evaporation rate. Therefore,

$$14,326 = V + 10,876 + 2,856 \tag{3}$$

Solving,

$$V = 594 \, \text{lb/h}$$

The results in tabular form are:

	lb/h for crystallizer			
Component	Feed	Vapor	Liquid	Crystals
$MgSO_{4(aq)}$	4,446	0	3,045	0
$MgSO_4 \cdot 7H_2O_{(s)}$	0	0	0	2,856
H_2O	9,880	594	7,831	0
	14,326	594	10,876	2,856

The maximum production rate of crystals is

$$\frac{2,856}{2,000}(24) = 34.3 \, \text{tons/day}$$

A large number of organic compounds, particularly organic acids with relatively moderate melting points (125–225°C), are also soluble in water. Some data are given in Table 17.7, where it is seen that the solubility often increases significantly with increasing temperature. For example, the solubility of o-phthalic acid increases from a very low value of 0.56 to 18.0 g/100 g H_2O when the temperature increases from 20 to 100°C.

Table 17.7 Solubility and Melting Point of Some Organic Compounds in Water

Compound	Melting Point, °C	Solubility, g/100 g H_2O at T, °C							
		0	10	20	30	40	60	80	100
Adipic acid	153	0.8	1.0	1.9	3.0	5.0	18	70	160
Benzoic acid	122	0.17	0.20	0.29	0.40	0.56	1.16	2.72	5.88
Fumaric acid (trans)	287	0.23	0.35	0.50	0.72	1.1	2.3	5.2	9.8
Maleic acid	130	39.3	50	70	90	115	178	283	—
Oxalic acid	189	3.5	6.0	9.5	14.5	21.6	44.3	84.4	—
o-phthalic acid	208	0.23	0.36	0.56	0.8	1.2	2.8	6.3	18.0
Succinic acid	183	2.8	4.4	6.9	10.5	16.2	35.8	70.8	127
Sucrose	d	179	190	204	219	238	287	362	487
Urea	133	67	85	105	135	165	250	400	730
Uric acid	d	0.002	0.004	0.006	0.009	0.012	0.023	0.039	0.062

EXAMPLE 17.4

Oxalic acid is to be crystallized from a saturated aqueous solution initially at 100°C. To what temperature does the solution have to be cooled to crystallize 95% of the acid as the dihydrate?

SOLUTION

Assume a basis of 100 g of water. From Table 17.7, the amount of dissolved oxalic acid at 100°C is 84.4 g.

Amount to be crystallized = 0.95(84.4) = 80.2 g.

Amount of oxalic acid left in solution = 84.4 − 80.2 = 4.2 g.

MW of oxalic acid = 90.0.

MW of water = 18.0.

Water of hydration for $2H_2O = \dfrac{2(18.0)}{90.0} = 0.4 \dfrac{g\,H_2O}{g\,oxalic\,acid}$.

Therefore water of crystallization = 0.4(80.2) = 32.1 g H_2O.

Liquid water remaining = 100 − 32.1 = 67.9 g.

Final solubility must be $\dfrac{4.2}{67.9} \times 100 = 6.19 \dfrac{g}{100\,g\,H_2O}$.

From Table 17.7, by linear interpolation, temperature = 10.6°C.

Enthalpy Balances

When an anhydrous solid compound, whose solubility increases with increasing temperature, dissolves isothermally in water or some other solvent, heat must be absorbed by the solution. This amount of heat per mole of compound in an infinite amount of solvent varies with temperature and is referred to as the *heat of solution at infinite dilution* (ΔH_{sol}^{∞}). For example, in Table 17.5, the solubility of anhydrous NaCl is seen to increase slowly with increasing temperature from 10 to 100°C. Correspondingly, the heat of solution at infinite dilution in Table 17.5 is modestly endothermic (+) at room temperature. In contrast, the solubility of anhydrous KNO_3 increases more rapidly with increasing temperature, resulting in a higher endothermic heat of solution. For compounds that form hydrates, the heat of solution at infinite dilution may be negative (exothermic) for the anhydrous form, but becomes less negative and often positive as higher hydrates are formed by the reaction:

$$A \cdot nH_2O_{(s)} \rightarrow A_{(aq)} + nH_2O$$

For example, the following heats of solution at infinite dilution in kJ/mol of compound at 18°C for four hydrates of $MgSO_4$ clearly show this effect:

$MgSO_4$	−88.3
$MgSO_4 \cdot H_2O$	−58.6
$MgSO_4 \cdot 6H_2O$	−2.3
$MgSO_4 \cdot 7H_2O$	+13.3

Heats of a solution for a number of hydrated and anhydrous compounds are listed in Table 17.5.

As a solid compound continues to dissolve in a solvent, the heat of solution, which is now referred to as the integral

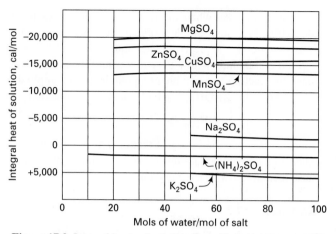

Figure 17.9 Integral heats of solution for sulfates in water at 25°C.

heat of solution, varies somewhat, as shown in Figure 17.9 for several compounds as a function of concentration. The integral heat of solution at saturation is numerically equal, but opposite in sign, to the heat of crystallization. The difference between the integral heat of solution at saturation and the heat of solution at infinite dilution is the heat of dilution:

$$\Delta H_{sol}^{sat} - \Delta H_{sol}^{\infty} = \Delta H_{dil}$$

with

$$\Delta H_{sol}^{sat} = -\Delta H_{crys}$$

As indicated in Figure 17.9, the heat of dilution is relatively small; therefore, it is common to use:

$$\Delta H_{crys} \approx -\Delta H_{sol}^{\infty}$$

An energy-balance calculation around a crystallizer is complex because it can involve not only the integral heat of solution and/or heat of crystallization, but also the specific heats of the solute and solvent and the heat of vaporization of the solvent. The calculation is readily made if an enthalpy-mass fraction diagram is available for the system, including solubility and phase-equilibria data. Mullin [1] lists 11 aqueous binary systems for which such a diagram has been constructed. A diagram for the $MgSO_4$-H_2O system is shown in Figure 17.10. The enthalpy datum is pure liquid water at 32°F (consistent with steam tables in American Engineering Units) at Point p and solid $MgSO_4$ at 32°F (not shown in Figure 17.10).

Points a to l, n, p, and q in the enthalpy-mass fraction diagram of Figure 17.10 correspond to the same points in the phase diagram of Figure 17.2. In Figure 17.10, the isotherms in the region above Curve pabcdq pertain to enthalpies of unsaturated solutions of $MgSO_4$. The straightness of these isotherms indicates that the heat of dilution is almost negligible. In this solid-free region, a 30 wt% aqueous solution of $MgSO_4$ has a specific enthalpy at 110°F of −31 Btu/lb solution.

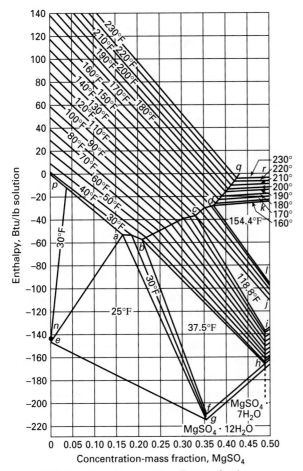

Figure 17.10 Enthalpy-concentration diagram for the $MgSO_4$-H_2O system at 1 atm.

In the region below the solubility curve *pabcdq*, in both Figures 17.2 and 17.10, the following phases exist at equilibrium:

Region	Temperature Range, °F	Phases
pae	25–32	ice and aqueous solution of $MgSO_4$
ea	25	ice and eutectic mixture
ag	25	eutectic and $MgSO_4 \cdot 12H_2O$
abfg	25–37.5	saturated solution and $MgSO_4 \cdot 12H_2O$
bcih	37.5–118.8	saturated solution and $MgSO_4 \cdot 7H_2O$
cdlj	118.8–154.4	saturated solution and $MgSO_4 \cdot 6H_2O$
dqrk	154.4–	saturated solution and $MgSO_4 \cdot H_2O$

Pure ice exists at Point *e*, where in Figure 17.10 the specific enthalpy is –147 Btu/lb, which is the heat of crystallization of water at 32°F. If a 30 wt% aqueous solution of $MgSO_4$ is cooled from 110°F to 70°F, the equilibrium magma will consist of a saturated solution of 26 wt% $MgSO_4$ and crystals of $MgSO_4 \cdot 7H_2O$ (49 wt% $MgSO_4$) as determined from the ends of the 70°F tie line that extends

from solubility curve *bc* to $MgSO_4 \cdot 7H_2O$ solid line *ih*. The relative amounts of the two equilibrium phases can be computed from a $MgSO_4$ balance. For a basis of 100 lb of mixture, let *S* be the pounds of crystals and *A* be the pounds of saturated aqueous solution. Thus, the $MgSO_4$ balance is

$$0.30(100) = 0.49\,S + 0.26\,A$$

where

$$100 = S + A$$

Solving, $S = 17.4$ lb and $A = 82.6$ lb. The enthalpy of the mixture at 70°F is –65 Btu/lb, which is equivalent to enthalpies of –46 and –155 Btu/lb, respectively, for the solution and crystals.

EXAMPLE 17.5

For the conditions of Example 17.3, calculate the Btu/h of heat addition for the crystallizer.

SOLUTION

An overall energy balance around the crystallizer gives

$$m_{feed}H_{feed} + Q_{in} = m_{vapor}H_{vapor} + m_{liquid}H_{liquid} + m_{crystals}H_{crystals} \quad (1)$$

where liquid refers to the saturated-liquid portion of the magma.

From the solution to Example 17.3, the feed consists of two streams:

$$m_{feed1} = 4{,}466 \text{ lb/hr of } 37.75 \text{ wt\%} \quad MgSO_4 \text{ at } 170°F$$
$$m_{feed2} = 9{,}860 \text{ lb/hr of } 28.0 \text{ wt\%} \quad MgSO_4 \text{ at } 85°F$$

From Figure 17.10,

$$H_{feed1} = -20 \text{ Btu/lb}$$
$$H_{feed2} = -43 \text{ Btu/lb}$$

Therefore,

$$m_{feed}\,H_{feed} = 4{,}466(-20) + 9{,}860(-43) = -513{,}000 \text{ Btu/h}$$
$$m_{vapor} = 594 \text{ lb/h at } 85°F \text{ and } 0.58 \text{ psia}$$

The vapor enthalpy does not appear on Figure 17.10, but enthalpy tables for steam can be used since they are based on the same datum (i.e., liquid water at 32°F).

Therefore, $H_{vapor} = 1099$ Btu/lb from steam tables and

$$m_{vapor}\,H_{vapor} = 594(1099) = 653{,}000 \text{ Btu/h}$$

The liquid plus crystals can be treated together as the magma. From the solution to Example 17.3,

$$m_{liquid} + m_{crystals} = m_{magma} = 10{,}876 + 2{,}856$$
$$= 13{,}732 \text{ lb/h of } 32.4 \text{ wt\%} \; MgSO_4 \text{ at } 85°F$$

From Figure 17.10,

$$H_{magma} = -67 \text{ Btu/lb}$$

and

$$m_{magma}H_{magma} = 13{,}732(-67) = -920{,}000 \text{ Btu/lb}$$

From (1),

$$Q_{in} = 653{,}000 - 920{,}000 - (-513{,}000) = 246{,}000 \text{ Btu/h}$$

In the absence of an enthalpy-mass fraction diagram, a reasonably accurate energy balance can be made if data for heat of crystallization and specific heats of the solutions are available or can be estimated and the heat of dilution is neglected, as shown in the next example.

EXAMPLE 17.6

The feed to a cooling crystallizer is 1,000 lb/h of 32.5 wt% $MgSO_4$ in water at 120°F. This solution is cooled to 70°F to form crystals of the heptahydrate. Estimate the heat removal rate in Btu/h.

SOLUTION

Material balance

From Figure 17.2, the feed at 120°F contains no crystals, but the magma at 70°F consists of crystals of the heptahydrate and a mother liquor of 26 wt% $MgSO_4$. By material balance in the manner of Example 17.3, the following results are obtained:

	lb/hr		
	Feed	Mother Liquor	Crystals
H_2O	675	530	0
$MgSO_4$	325	186	0
$Mg_2SO_4 \cdot 7H_2O$	0	0	284
	1,000	716	284

Take a thermodynamic path consisting of cooling the feed from 120°F to 70°F followed by crystallization at 70°F. From Hougen, Watson, and Ragatz [3], the specific heat of the feed is approximately constant over the temperature range at 0.72 Btu/lb-°F. Therefore, the heat that must be removed to cool the feed to 70°F is

$$1,000(0.72)(120 - 70) = 36,000 \, \text{Btu/h}$$

For data presented earlier in this section, the heat of crystallization can be taken as the negative of the heat of solution at infinite dilution:

$$-13.3 \, \text{kJ/mole of heptahydrate}$$

or

$$-23.2 \, \text{Btu/lb of heptahydrate}$$

Therefore, the heat that must be removed during crystallization of the heptahydrate is

$$284(23.2) = 6,600 \, \text{Btu/h}$$

The total heat removal is $36,000 + 6,600 = 42,600$ Btu/h.

If this example is solved with Figure 17.10, in the manner of Example 17.5, the result is 44,900 Btu/h, which is 5.4% higher.

17.3   KINETIC AND TRANSPORT CONSIDERATIONS

Crystallization is a complex phenomenon that involves three steps: *nucleation*, mass transfer of the solute to the crystal surface, and incorporation of the solute into the crystal lattice structure. Collectively, these phenomena are referred to as *crystallization kinetics*. Experimental data show that the driving force for all three steps is *supersaturation*.

Supersaturation

The solubility property discussed in the previous section refers to relatively large crystals of the size that can be seen by the naked eye, i.e., larger than 20 μm in diameter. As crystal size decreases below this diameter, solubility noticeably increases, making it possible to supersaturate a solution if it is cooled slowly without agitation. This phenomenon, based on the work of Miers and Isaac [4] in 1907, is represented in Figure 17.11, where the normal solubility curve, c_s is represented as a solid line. The solubility of very small crystals can fall in the metastable region which is shown to have a metastable limiting solubility, c_m, given by the dashed line.

Consider a solution at a temperature, T_1, given by the vertical line in Figure 17.11. If the concentration is given by Point a, the solution is undersaturated and crystals of all sizes dissolve. At Point b, equilibrium exists between a saturated solution and crystals that can be seen by the naked eye. In the metastable region at Point c, crystals can grow but cannot nucleate. If no crystals are present, none can form. For that concentration, the difference between the temperature at Point e on the solubility curve and Point c in the metastable region is the supersaturation temperature difference, which may be about 2°F. The supersaturation, $\Delta c = c - c_s$, is the difference in concentration between Points c and b.

At Point d, spontaneous nucleation of very small crystals, invisible to the naked eye, occurs. The difference in temperature between Points f and d is the limit of the supersaturation temperature difference. The limiting supersaturation is $\Delta c_{\text{limit}} = c_m - c_s$.

The relationship between solubility and crystal size is given quantitatively by the Kelvin equation (also known as

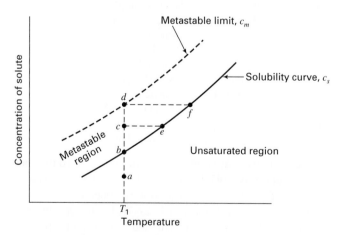

Figure 17.11 Representative solubility-supersolubility diagram.

the Gibbs–Thompson and Ostwald equations):

$$\ln\left(\frac{c}{c_s}\right) = \left(\frac{4v_s\sigma_{s,L}}{vRTD_p}\right) \qquad (17\text{-}16)$$

where

v_s = molar volume of the crystals

$\sigma_{s,L}$ = interfacial tension

v = number of ions/molecule of solute

c/c_s = supersaturation ratio = S

Measured values of interfacial tension (also called surface energy) range from as low as 0.001 J/m^2 for very soluble compounds to 0.170 for compounds of low solubility.

As might be expected, (17-16), in a more general form, applies to the effect of droplet diameter on vapor pressure and solubility in another liquid phase.

It is common to define a relative supersaturation, s, by

$$s = \frac{c - c_s}{c_s} = \frac{c}{c_s} - 1 = S - 1 \qquad (17\text{-}17)$$

In practice, s is usually less than 0.02 or 2%. For such small values, $\ln(c/c_s)$ can be approximated by s with no more than a 1% error.

EXAMPLE 17.7

Determine the effect of crystal diameter on the solubility of KCl in water at 25°C.

SOLUTION

From Table 17.5, by interpolation, $c_s = 35.5$ g/100 g H$_2$O.
Because KCl dissociates into K$^+$ and Cl$^-$, $v = 2$.
MW of KCl = 74.6.
Density of KCl crystals = 1980 kg/m^3.

$$v_s = \frac{74.6}{1980} = 0.0376 \,\text{m}^3/\text{kmol}$$

$$T = 298\,\text{K}$$

$$R = 8314\,\text{J/kmol-K}$$

From Mullin [1], page 200,

$$\sigma_{s,L} = 0.028\,\text{J/m}^2$$

From (17-16),

$$c = c_s \exp\left(\frac{4v_s\sigma_{s,L}}{vRTD_p}\right)$$

$$= 35.5 \exp\left[\frac{4(0.0376)(0.028)}{2(8315)(298)D_p}\right] \text{ for } D_p \text{ in m, or} \qquad (1)$$

$$= 35.5 \exp(0.00085/D_p, \mu\text{m})$$

D_p, μm	c/c_s	c, g KCl/100 g H$_2$O
0.01	1.0887	38.65
0.10	1.0085	35.80
1.00	1.00085	35.53
10.00	1.000085	35.50
100.00	1.0000085	35.50

Nucleation

To determine the volume or residence time of the magma in a crystallizer, the rate of *nucleation* (birth) of crystals and their rate of growth must be known or estimated. The relative rates of nucleation and growth are very important because they determine crystal size and size distribution. Nucleation may be *primary* or *secondary* depending on whether the supersaturated solution is free of crystalline surfaces or contains crystals, respectively. Primary nucleation requires a high degree of supersaturation and is the principal mechanism occurring in precipitation. The theory of primary nucleation is well developed and applies as well to the condensation of liquid droplets from a supersaturated vapor and the formation of droplets of a second liquid phase from an initial liquid phase. However, secondary nucleation is the principal mechanism in commercial crystallizers, where crystalline surfaces are present and large crystals are desirable.

Primary Nucleation

Primary nucleation can be homogeneous or heterogeneous. The former occurs within the supersaturated solution in the absence of any foreign matter, such as dust. Molecules in the solution first associate to form a *cluster*, which may dissociate or grow. If a cluster gets large enough to take on the appearance of a lattice structure, it becomes an *embryo*. Further growth can result in a stable crystalline *nucleus* whose size exceeds that given by (17-16) for the prevailing degree of supersaturation.

The rate of homogeneous nucleation is given by classical chemical kinetics in conjunction with (17-16), as discussed by Nielsen [5]. The resulting expression is

$$B^o = A \exp\left[\frac{-16\pi v_s^2\sigma_{s,L}^3 N_a}{3v^2(RT)^3\left[\ln\left(\frac{c}{c_s}\right)\right]^2}\right] \qquad (17\text{-}18)$$

where

B^o = rate of homogeneous primary nucleation, number of nuclei/cm^3-s

A = frequency factor

N_a = Avogadro's number = 6.022×10^{26} molecules/kmol

Theoretically, $A = 10^{30}$ nuclei/cm³-s; however, observed values are generally different due to the unavoidable presence of foreign matter. Thus, (17-18) can also be applied to heterogeneous primary nucleation, where A is determined experimentally and may be many orders of magnitude different from the theoretical value. A value of 10^{25} is often quoted.

The rate of primary nucleation is extremely sensitive to the supersaturation ratio, S, defined by (17-17), as illustrated in the following example.

EXAMPLE 17.8

Using the data in Example 17.7, estimate the effects of relative supersaturation on the primary homogeneous nucleation of KCl from an aqueous solution at 25°C. Use values of s corresponding to values of c/c_s of 2.0, 1.5, and 1.1.

SOLUTION

For $c/c_s = 2.0$, $\ln(c/c_s) = 0.693$. From (17-18), using data in Example 17.7,

$$B^o = 10^{30} \exp\left[\frac{-16(3.14)(0.0376)^2(0.028)^3(6.022 \times 10^{26})}{3(2)^2(8315)^3(298)^3(0.693)^2}\right]$$

$$= 2.23 \times 10^{25} \text{ nuclei/cm}^3\text{-s}$$

Calculations for the other values of c/c_s are obtained in the same manner with the following results:

c/c_s	B^o, nuclei/cm³-s
2.0	2.23×10^{25}
1.5	2.60×10^{16}
1.1	0

Since large values of the supersaturation ratio ($c/c_s > 1.02$) are essentially impossible for crystallization of solutes of moderate to high solubility (e.g., solutes listed in Tables 17.5 and 17.7), primary nucleation for these solutes never occurs. However, for relatively insoluble solutes (e.g., solutes listed in Table 17.6), large values of c/c_s can be generated rapidly from ionic reactions causing rapid precipitation of very fine particles. If $A = 10^{25}$ is used, B^o is divided by 10^5.

Secondary Nucleation

Nucleation in industrial crystallizers occurs mainly by secondary nucleation caused by the presence of existing crystals in the supersaturated solution. Secondary nucleation can occur by: (1) fluid shear past crystal surfaces that sweeps away nuclei, (2) collisions of crystals with each other, and (3) collisions of crystals with metal surfaces such as the crystallizer vessel wall or agitator blades. The latter two mechanisms, which are referred to as *contact nucleation* are the most common types since they can occur at the low values of relative supersaturation that are typically encountered in industrial applications.

In the absence of a theory for the complex phenomena of secondary nucleation, the following empirical power-law function, which correlates much of the experimental data, is widely used:

$$B^o = k_N s^b M_T^j N^r \qquad (17\text{-}19)$$

where M_T = mass of crystals per volume of magma and N = agitation rate (e.g., rpm of an impeller). The constants, $k_N, b, j,$ and r, are determined from experimental data, on the system of interest, as discussed below in the section on a crystallizer model.

Crystal Growth

In 1897, Noyes and Whitney [6] presented a mass-transfer theory of crystal growth based on equilibrium at the crystal-solution interface. Thus, they wrote

$$dm/dt = k_c A(c - c_s) \qquad (17\text{-}20)$$

where dm/dt = rate of mass deposited on the crystal surface, A = surface area of the crystal, k_c = mass-transfer coefficient, c = mass solute concentration in the bulk supersaturated solution, and c_s = mass solute concentration in the solution at saturation. Nernst [7] proposed the existence of a thin stagnant film of solution adjacent to the crystal face through which molecular diffusion of the solute took place. Thus, $k_c = \mathcal{D}/\delta$, where $\mathcal{D}$ = diffusivity and δ = film thickness, where the latter was assumed to depend on velocity of the solution past the crystal as determined by the degree of agitation.

The theory of Noyes and Whitney was challenged by Miers [8], who showed experimentally that an aqueous solution in contact with crystals of sodium chlorate was not saturated at the crystal–solution interface, but was supersaturated. This finding led to a two-step theory of crystal growth, referred to as the diffusion-reaction theory, as described by Valeton [9]. Mass transfer of solute from the bulk of the solution to the crystal–solution interface occurs in the first step, as given by a modification of (17-20):

$$dm/dt = k_c A(c - c_i) \qquad (17\text{-}21)$$

where c_i is the supersaturated concentration at the interface. In the second step, a first-order reaction is assumed to occur at the crystal–solution interface, in which solute molecules are integrated into the crystal-lattice structure. Thus, for this kinetic step,

$$dm/dt = k_i A(c_i - c_s) \qquad (17\text{-}22)$$

If (17-21) and (17-22) are combined to eliminate c_i, we obtain

$$dm/dt = \frac{A(c - c_s)}{1/k_c + 1/k_i} \qquad (17\text{-}23a)$$

Typically, k_c will depend on the velocity of the solution as shown in Figure 17.12. At low velocities, the growth rate

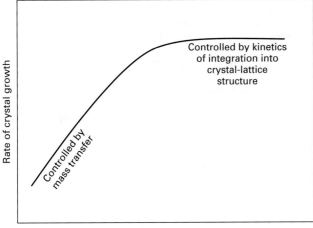

Figure 17.12 Effect of solution velocity past crystal on the rate of crystal growth.

may be controlled by the first step. The second step can be important, especially when the solution velocity past the crystal surface is high, such that k_c is large compared to k_i. In adsorption, the kinetic step is rarely important. It is also unimportant in dissolution, the reverse of crystallization.

The mass-transfer coefficient, k_c, for the first step, is independent of the crystallization process and can be estimated from general fluid-solid particle mass-transfer-coefficient correlations described in Chapters 3 and 15. The kinetic coefficient, k_i, is peculiar to the crystallization process. A number of theories have been advanced for the kinetic step, as discussed in Myerson [10]. One prominent theory is that of Burton, Cabrera, and Frank [11], which is based on a growth spiral starting from a screw dislocation, as shown in Figure 17.13 and verified in some experimental studies using scanning-electron microscopy. A dislocation is an imperfection in the crystal structure. The screw-dislocation theory predicts a growth rate proportional to: $(c_i - c_s)^2$ at low supersaturation and to $(c_i - c_s)$ at high supersaturation. Unfortunately, the theory does not provide a means to predict k_i. Accordingly, (17-23a) is generally applied with k_c estimated from available correlations and k_i back-calculated from experimental data.

Although crystals do not grow as spheres, let us develop an equation for the rate of increase of the diameter of a

spherical crystal. Rewriting (17-23a) in terms of an overall coefficient,

$$dm/dt = K_c A(c - c_s) \qquad (17\text{-}23\text{b})$$

Since
$$A = \pi D_p^2 \quad \text{and} \quad m = \frac{\pi D_p^3}{6}\rho$$

Equation (17-23b) becomes

$$\frac{dD_p}{dt} = \frac{2K_c(c - c_s)}{\rho} = \frac{2K_c(\Delta c)}{\rho} \qquad (17\text{-}24)$$

If the rate of growth is controlled by k_i, which is assumed to be independent of D_p, then

$$\frac{\Delta D_p}{\Delta t} = \frac{2k_i \Delta c}{\rho} \qquad (17\text{-}25)$$

and the rate of increase of crystal size is linear in time for a constant supersaturation. If the rate of growth is controlled by k_c at a low velocity, then, from (15-60),

$$K_c = k_c = 2\mathcal{D}/D_p \qquad (17\text{-}26)$$

where $\mathcal{D}$ is solute diffusivity.

Substitution of (17-26) into (17-24) gives

$$\frac{dD_p}{dt} = \frac{4\mathcal{D}(\Delta c)}{D_p \rho} \qquad (17\text{-}27)$$

Integration from D_{p_0} to D_p gives

$$\frac{D_p^2 - D_{p_0}^2}{2} = \frac{4\mathcal{D}(\Delta c)}{\rho} t \qquad (17\text{-}28)$$

If $D_{p_0} \ll D_p$, (17-28) reduces to

$$D_p = \left(\frac{8\mathcal{D}(\Delta c)t}{\rho}\right)^{1/2} \qquad (17\text{-}29)$$

In this case, the increase in crystal diameter slows with time.

At higher solution velocities where k_c still controls, the use of (15-62) results in

$$K_c = k_c = C_1/D_p^{1/2} \qquad (17\text{-}30)$$

For this case, the increase in crystal diameter also slows with time, but not as rapidly as predicted by (17-29). It is

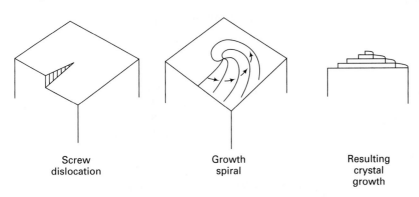

Screw dislocation   Growth spiral   Resulting crystal growth

Figure 17.13 Screw-dislocation mechanism of crystal growth.

common to assume that the rate of crystal growth is controlled by k_i and, thus, is not dependent on crystal size and is, therefore, invariant with time. This assumption will be utilized in a latter section of this chapter to develop a useful crystallizer model.

EXAMPLE 17.9

The heptahydrate of $MgSO_4$ is to be crystallized batchwise from a seeded aqueous solution. A low supersaturation is to be used to avoid primary nucleation and a mild agitation is to be applied to minimize secondary nucleation. The temperature in the crystallizer will be maintained at 35°C, at which the solubility of $MgSO_4$ in water is 30 wt%. The crystallizer will be charged with 3,000 lb of a saturated solution at that temperature. To this solution will be added 2 lb of heptahydrate seed crystals of 50 μm in diameter. A supersaturation of 0.01 gm of heptahydrate per gm of solution at 35°C will be maintained during crystallization by operating the crystallizer at vacuum and using heat exchange and the heat of crystallization to gradually evaporate water from the solution. Based on the assumptions and data listed below, determine the following if the final crystal size is to be 400 μm.

(a) Yield of crystals of heptahydrate in pounds.

(b) Number of crystals.

(c) Amount of water in pounds that will have to be evaporated.

(d) Product magma density in pounds of crystals per pound of solution.

(e) Crystallizer volume in gallons if the volume occupied by the magma during operation is at the most 50% of the crystallizer volume.

(f) Crystallizer pressure in psia and the boiling-point elevation in °F.

(g) Time in minutes to grow the crystals to the final size.

(h) Amount of heat transfer in Btu.

(i) Also, determine whether crystal growth is controlled by mass transfer, by surface reaction (incorporation into the lattice), or by both.

Assumptions and Data:

1. No primary or secondary nucleation.

2. Properties of aqueous 30 wt% $MgSO_4$ at 35°C:

   density = 1.34 g/cm^3

   viscosity = 8 cP

   diffusivity of $MgSO_4$ = 1.10×10^{-5} cm^2/s

3. Density of the crystals = 1.68 g/cm^3.

4. Crystal shape can be approximated as a sphere.

5. Average crystal-face growth rate, including effects of both mass transfer and surface reaction = 0.005 mm/min.

6. Solution velocity past crystal face = 5 cm/s.

SOLUTION

Molecular weights:
$MgSO_4$	120.5
$MgSO_4 \cdot 7H_2O$	246.6
H_2O	18.02

Initial charge:

$$\frac{\begin{array}{l} 900 \text{ lb aq } MgSO_4 \\ 2,100 \text{ lb } H_2O \end{array}}{3,000 \text{ lb}}$$

$+2$ lb $MgSO_4 \cdot 7H_2O$ crystal seeds

(a) Crystals grow from 50 μm to 400 μm in diameter:

$$\text{yield} = 2\left(\frac{400}{50}\right)^3 = 1,024 \text{ lb} \quad MgSO_4 \cdot 7H_2O$$

(b) Number of crystals, based on the crystal seeds,

$$= \frac{\text{mass of seeds}}{\text{mass/seed}}$$

$$= \frac{2(454)}{1.68\left(\dfrac{3.14}{6}\right)(50 \times 10^{-4})^3}$$

$$= 8.26 \times 10^9, \text{ where mass/seed} = \rho_p V_p = \rho \frac{\pi D_p^3}{6}$$

(c) Pounds of heptahydrate crystallized = $1,024 - 2 = 1,022$ lb

$$H_2O \text{ of hydration} = 1,022\left(\frac{246.6 - 120.5}{246.6}\right) = 523 \text{ lb}$$

$MgSO_4$ in crystals = $1,022 - 523 = 499$ lb

$MgSO_4$ left in solution = $900 - 499 = 401$ lb

$$H_2O \text{ left in solution} = 401\left(\frac{70}{30}\right) = 936 \text{ lb}$$

H_2O evaporated = $2,100 - 523 - 936 = 641$ lb

(d) Final mother liquor = $936 + 401 = 1,337$ lb

$$\text{Magma density} = \left(\frac{1,024}{1,337}\right) = 0.766 \left(\frac{\text{lb crystals}}{\text{lb mother liquor}}\right)$$

(e) Initially, using the factor of 8.33 lb/gal for 1.0 g/cm^3,

$$\text{Solution volume} = \frac{3,000}{8.33(1.34)} = 269 \text{ gal}$$

Finally,

$$\text{Solution volume} = \frac{1,337}{8.33(1.34)} = 120 \text{ gal}$$

$$\text{Crystal volume} = \frac{1,024}{8.33(1.68)} = 73 \text{ gal}$$

$$\text{Total volume} = 193 \text{ gal}$$

Therefore, initial conditions before crystallization will control the volume.

$$\text{Make crystallizer volume} = \frac{269}{0.5} = 538 \text{ gal}$$

(f) Calculate H_2O partial pressure by Raoult's law applied to mother liquor:

$$\text{lbmol } MgSO_4 = \frac{401}{120.5} = 3.33$$

$$\text{lbmol } H_2O = \frac{936}{18.02} = 51.95$$

$$\text{Total} = 55.28 \text{ lbmol}$$

$$x_{H_2O} = \frac{51.95}{55.28} = 0.94$$

At $35°C = 95°F$, $P_{H_2O}^s = 0.8153$ psia.

Therefore, $p_{H_2O} = x_{H_2O}P^s_{H_2O} = 0.94(0.8153) = 0.766$ psia.

This corresponds to a saturation temperature of 93°F.

Therefore, boiling-point elevation = 2°F.

(g) Growth rate = 0.005 mm/min.

Therefore, diameter grows at 0.01 mm/min.

Must grow from 50 μm = 0.05-mm diameter to 0.40-mm diameter.

$$\text{Time} = \frac{0.40 - 0.05}{0.01} = 35 \text{ min}$$

(h) Enthalpy balance:

$$FH_F + Q = VH_V + MH_M \tag{1}$$

Assume charge is at 35°C = 95°F; then, from Figure 17.10, $H_F = -40$ Btu/lb.

From the steam tables, $H_V = 1,102.2 + 0.9$ (for the boiling-point elevation) = 1,103.1 Btu/lb.

$$\text{Wt\% MgSO4 in magma} = \frac{(499 + 401 + 1)}{(3,000 + 2 - 641)} = \frac{901}{2,361}$$
$$= 0.38.$$

From Figure 17.10, $H_M = -90$ Btu/lb.

From (1),

$$Q = 641(1,103.1) + 2,361(-90) - 3,002(-40)$$
$$= 615,000 \text{ Btu} = \text{heat transferred to crystallizer}$$

(i) Assume mass-transfer controlled, using molar amount of crystals, n, and molar concentrations, c.

The molar form of (17-21) is

$$\frac{dn}{dt} = 4\pi r^2 k_c (\Delta c), \tag{2}$$

where r is the crystal radius

$$n = \frac{4}{3}\pi r^3 \rho_M, \text{ using a molar density.}$$

Therefore,

$$\frac{dn}{dt} = 4\pi r^2 \rho_M \frac{dr}{dt} \tag{3}$$

Equating (2) and (3),

$$4\pi r^2 k_c(\Delta c) = 4\pi r^2 \rho_M \frac{dr}{dt}$$

$$\frac{dr}{dt} = \frac{k_c \Delta c}{\rho_M} \tag{4}$$

From (15-62),

$$N_{Sh} = \frac{k_c D_p}{\mathcal{D}} = \frac{2k_c r}{\mathcal{D}} = 2 + 0.6(N_{Re})^{1/2}(N_{Sc})^{1/3}$$

For $r = \frac{50}{2} = 25$ μm,

$$N_{Re} = \frac{D_p v \rho}{\mu} = \frac{(50 \times 10^{-6})(100)(5)(1.34)}{0.08} = 0.42$$

$$N_{Sc} = \frac{\mu}{\rho \mathcal{D}} = \frac{0.08}{1.34(1.10 \times 10^{-5})} = 5,430$$

$$N_{Sh} = 2 + 0.6(0.42)^{1/2}(5,430)^{1/3}$$

$$= 2 + 6.8 = 8.8 = \frac{2k_c r}{\mathcal{D}}$$

$$k_c = \frac{(8.8)(1.1 \times 10^{-5})}{2(25 \times 10^{-4})} = 0.019 \text{ cm/s}$$

For $r = \frac{400}{2} = 200$ μm,

$$N_{Re} = 0.42\left(\frac{400}{50}\right) = 3.36$$

$$N_{Sh} = 2 + 0.6(3.36)^{1/2}(5,430)^{1/3}$$
$$= 2 + 19.3 = 21.3$$

$$k_c = \frac{(21.3)(1.1 \times 10^{-5})}{2(200 \times 10^{-4})} = 0.006 \text{ cm/s}$$

Thus, k_c changes by a factor of 3 as crystal size changes from 25 to 200 μm.

Assume a Δc based on the given supersaturation of 0.01 g crystal per g solution.

$$\Delta c = \frac{0.01(1.34)}{246.6} = 54.3 \times 10^{-6} \frac{\text{mol}}{\text{cm}^3}$$

$$\rho_M \text{ of crystals} = \frac{1.68}{246.6} = 6.81 \times 10^{-3} \frac{\text{mol}}{\text{cm}^3}$$

From (4),

$$\frac{dr}{dt} = k_c \frac{54.3 \times 10^{-6}}{6.81 \times 10^{-3}} = 0.008 \, k_c$$

for the lowest value of $k_c = 0.006$ cm/s,

$$\frac{dr}{dt} = 0.008(0.006) = 48 \times 10^{-6} \text{ cm/s}$$

or

$$480 \times 10^{-6}(60) = 0.029 \text{ mm/min.}$$

But this is six times faster than the growth rate. Therefore, crystal growth is largely controlled by surface reaction.

17.4 EQUIPMENT FOR SOLUTION CRYSTALLIZATION

Before developing a crystallizer model in Section 17.5, it is useful to describe the most widely used equipment for solution crystallization. Such equipment may be classified according to the three schemes shown in Table 17.8.

Table 17.8 Classification of Equipment for Solution Crystallization

Operation Modes	Methods for Achieving Supersaturation	Crystallizer Features for Achieving Desired Crystal Growth
Batch Continuous	Cooling Evaporation	Agitated or nonagitated Baffled or unbaffled Circulating liquor or circulating magma Classifying or nonclassifying Controlled or uncontrolled Cooling jacket or cooling coils

Although industrial equipment is available for batch or continuous operation, the latter is generally preferred. The choice of method for achieving supersaturation depends strongly on the effect of temperature on solubility. From the data in Table 17.5, it is seen that for many inorganic compounds in the near-ambient temperature range, e.g., $10-40°C$, the change in solubility is small and may be insufficient to utilize the cooling method. This is certainly the case for $MgCl_2$ and $NaCl$. For KNO_3 and Na_2SO_4, crystallization by cooling may be feasible. The majority of industrial crystallizers use the evaporation method or a combination of cooling and evaporation. Direct-contact cooling with agitation and evaporation can be achieved by bubbling air through the magma.

To produce crystals of a desired size distribution, considerable attention must be paid to the selection of features of the design of the crystallizer. The use of mechanical agitation can result in smaller and more uniformly sized crystals of a higher purity that are produced in less time. The use of vertical baffles can promote more uniform mixing. Supersaturation and uniformity can be controlled by circulation between a crystallizing zone and a supersaturation zone. In a circulating-liquor design, only the mother liquor is circulated, while in the circulating-magma design the mother liquor and crystals are circulated together. The circulation may be limited to the crystallizer vessel or may include pumping through an external heat exchanger. In a classifying crystallizer, the smaller crystals are separated from the larger and retained in the crystallizing zone for further growth or removed from the zone and redissolved. In a controlled design, one or more techniques are used to control the degree or supersaturation to avoid undesirable nucleation.

Cooling crystallizers may use a vessel jacket or internal coils, with the latter preferred because of the ease of wiping the crystals off the cooling surface.

A large number of patented commercial crystallizer designs have been developed. Many of them are described in Myerson [10] and by Mullin [1]. Only four of the more common solution crystallizers are described and illustrated here. The designs suffice to illustrate most of the features found in the many other designs.

Circulating, Batch Crystallizers

Although batch crystallizers can be operated without agitation or circulation by simply charging a hot solution to an open vessel and allowing the solution to stand as it cools by natural convection, the resulting crystals may be undesirably large, interlocked, impure because of entrapment of mother liquor, and difficult to remove from the vessel. Therefore, if the desired product of the crystallization operation is the crystals, it is preferable to use a more elaborate crystallizer configuration similar to either of the two batch crystallizers shown in Figure 17.14. In the design with external circulation, a high magma velocity is used for the flow of magma through the tubes of the heat exchanger to obtain a reasonable heat-transfer rate with a small temperature-driving force and minimization of crystal formation on the heat exchanger tube surfaces. This design can also be used for continuous crystallization, when the solubility-temperature curve dictates cooling crystallization.

In Figure 17.14b, crystallization is accomplished by evaporation under a vacuum pulled by a two-stage, steam-jet-ejector system though a water-cooled condenser. The

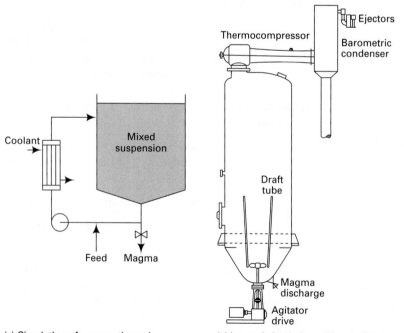

(a) Circulation of magma through an external, cooling heat exchanger

(b) Internal circulation with a draft tube

Figure 17.14 Circulating, batch cooling crystallizers.

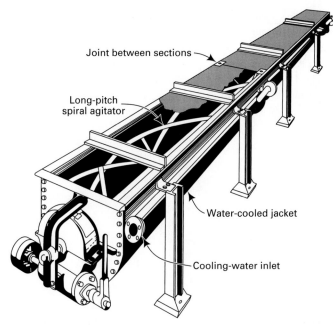

Figure 17.15 Swenson–Walker continuous, cooling crystallizer.

magma is circulated internally through a draft tube by a propeller. The energy for evaporation is supplied from the hot feed. A typical cycle, which includes charging the feed to the vessel, crystallization, and removal of the magma, may range from 2 to 8 h.

Continuous, Cooling Crystallizers

Figure 17.15 is a schematic diagram of a typical scraped-surface crystallizer known as the Swenson–Walker continuous, cooling crystallizer, described in detail by Seavoy and Caldwell [12]. The feed flows through a semicylindrical trough, typically 1 m wide × 3–12 m long. The trough has a water-cooled jacket and is provided with a low-speed (3–10 rpm) helical agitator-conveyor that scrapes the wall and prevents growth of crystals on the trough wall and promotes crystal growth by gentle agitation. Standard-size units can be linked together. The crystallization process may be controlled by the rather slow rate of heat transfer, with the major resistance due to the magma on the inside. Overall heat-transfer coefficients of only 10–25 Btu/h-ft²-°F (57–142 W/m²-K) are typically observed, based on a log-mean temperature difference between the magma and the coolant. Production rates of up to 20 tons per day of crystals of $Na_3PO_4 \cdot 12 H_2O$ and $Na_2SO_4 \cdot 10 H_2O$ of moderate size and uniformity have been achieved. Both salts show a significant decrease in solubility with decreasing temperature making the cooling crystallizer a viable choice.

EXAMPLE 17.10

The cooling crystallizer of Example 17.6 is to be a scraped-surface unit with 3 ft² of cooling surface per foot of running length of crystallizer. Cooling will be provided by a countercurrent flow of chilled water entering the cooling jacket at 60°F and leaving at 85°F. The overall heat-transfer coefficient, U, is expected to be 20 Btu/hr-ft²-°F. What length of crystallizer is needed?

SOLUTION

From Example 17.6, using Figure 17.10, the required rate of heat transfer is 44,900 Btu/h. The log mean temperature driving force for heat transfer is

$$\Delta T_{LM} = \frac{(120 - 85) - (70 - 60)}{\ln\left(\dfrac{120 - 85}{70 - 60}\right)} = 20°F$$

The area for heat transfer is

$$A = \frac{Q}{U \Delta T_{LM}} = \frac{44,900}{20(20)} = 112 \text{ ft}^2$$

The crystallizer length $= 112/3 = 37$ ft.

Continuous, Vacuum, Evaporating Crystallizers

A large number of designs for continuous, vacuum, evaporating crystallizers have been developed. A particularly successful and widely used design of this type is the Swenson draft-tube, baffled (DTB) crystallizer, described by Newman and Bennett [13] and shown in one of several variations in Figure 17.16. In the main body of the crystallizer, evaporation occurs, under vacuum, at the boiling surface, which is located several inches above a draft tube that extends down to within several inches of the bottom of the main body of the crystallizer vessel. Near the bottom and inside of the draft tube is a low-rpm propeller that directs the magma upward through the draft tube toward the boiling surface under conditions of a small degree of supercooling and in the absence of any violent flashing action. Thus, nucleation and buildup of crystals on the walls are minimized. Surrounding the draft tube is an annular space where the magma flows back downward for re-entry into the draft tube. The outer wall of the annular space is a skirt baffle, surrounded by an annular settling zone, whose outer wall is the wall of the crystallizer. A portion (perhaps 10%) of the magma flowing downward through the first annular space turns around and flows outward and upward through the settling zone where larger crystals can settle, leaving a mother liquor containing fine crystals, which flows to a circulating pipe where it is joined by the feed and flows upward through a pump and then a heat exchanger. The circulating solution is heated several degrees to provide the energy for feed preheat and subsequent evaporation and to dissolve the finer crystals. The circulating magma re-enters the main body of the crystallizer just below the bottom of the draft tube. Further classification of crystals by size can be accomplished by providing an elutriation leg, as shown at the bottom of the main body of the crystallizer in Figure 17.16. In that case, product magma is withdrawn through a pipe from a nozzle located near the bottom of the elutriation leg where the largest crystals are present. Otherwise, the product magma may be withdrawn from the lower part of the annular region surrounding the draft tube.

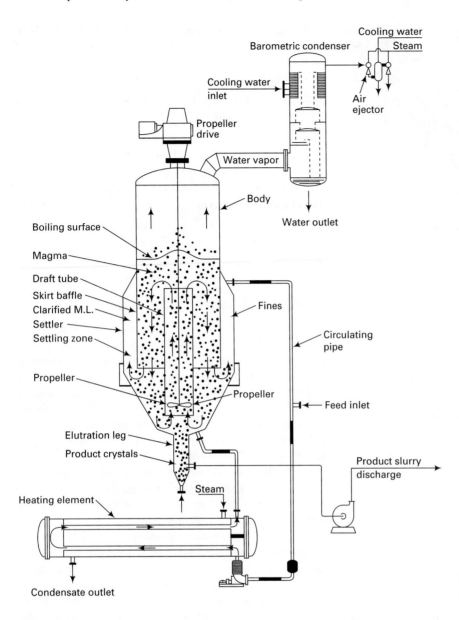

Figure 17.16 Swenson draft-tube, baffled (DTB) crystallizer.

17.5   THE MSMPR CRYSTALLIZATION MODEL

Because of the popularity of the DTB crystallizer, a mathematical model, due to Randolph [14], for its design and analysis is quite useful and is now found in the process equipment model libraries of a few continuous, steady-state, computer-aided simulation programs. This model is referred to as the *Mixed-Suspension, Mixed-Product-Removal* (*MSMPR*) model and is based on the following assumptions:

1. Continuous, steady-flow, steady-state operation.
2. Perfect mixing of the magma.
3. No classification of crystals.
4. Uniform degree of supersaturation throughout the magma.
5. Crystal growth rate independent of crystal size.
6. No crystals in the feed, but seeds are added initially.

7. No crystal breakage.
8. Uniform temperature.
9. Mother liquor in product magma in equilibrium with the crystals.
10. Nucleation rate constant and uniform and due to secondary nucleation by crystal contact.
11. Crystal-size distribution (CSD) uniform in the crystallizer and equal to that in the magma.
12. All crystals have the same shape.

Modifications to the model to account for: (1) classification of crystals due to settling, elutriation, and fines dissolving, and (2) variable growth rate are discussed by Randolph and Larson [15]. The central part of the MSMPR model is the estimation, by a crystal-population balance, of the crystal-size distribution (CSD), which is determined in practice by the rpm of the propeller in the draft tube and the external circulation rate. It is relatively easy to conduct experiments in a small laboratory crystallizer that approximates the MSMPR model and can provide some of the necessary crystal nucleation rate and growth rate data to design a large-scale, industrial crystallizer.

Crystal-Population Balance

The crystal-population balance accounts for all crystals in the magma and, together with the mass balance, makes possible the determination of the CSD. Let L = a characteristic crystal size (e.g., from a screen analysis), N = cumulative number of crystals of size L and smaller in the magma in the crystallizer, and V_{ML} = volume of mother liquor in the magma in the crystallizer. A typical cumulative-numbers undersize plot based on these variables is shown in Figure 17.17, where the slope of the curve, n, at a given value of L is the number of crystals per unit size per unit volume, given by

$$n = \frac{d(N/V_{ML})}{dL} = \frac{1}{V_{ML}}\frac{dN}{dL} \qquad (17\text{-}31)$$

The limits of n, as shown in Figure 17.17, vary from n° at $L = 0$ to 0 at $L = L_T$, the largest crystal size. In the MSMPR model, the cumulative plot of Figure 17.17 is independent of time and location in the magma. The plot is, in fact, the numbers-cumulative CSD for the crystals in the product magma.

For a constant, crystal-size growth rate, independent of crystal size, $G = dL/dt$, or

$$\Delta L = G\Delta t \qquad (17\text{-}32)$$

and

$$L = Gt_L \qquad (17\text{-}33)$$

where t_L = residence time in the magma in the crystallizer for crystals of size L. Equation (17-32) is referred to as the ΔL *law of McCabe* [16], who found that the law correlated experimental data on the growth of crystals of KCl and $CuSO_4 \cdot 5H_2O$ surprisingly well. Although McCabe's experiments were conducted in a small, laboratory crystallizer under ideal conditions, his resulting ΔL law is applied to commercial crystallization even though conditions may be far from ideal.

From (17-31), $dN = nV_{ML}\,dL$ = number of crystals in the size range dL. Now withdraw ($\Delta n\,dL$) crystals per unit volume in time increment Δt. Because of the perfect mixing assumption for the magma,

$$\frac{\dfrac{\text{number of crystals withdrawn}}{\text{mother-liquor volume withdrawn}}}{\dfrac{\text{number of crystals}}{\text{mother-liquor volume in the crystallizer}}}$$

Therefore,

$$\frac{\dfrac{\text{number of crystals withdrawn}}{\text{number of crystals in crystallizer}}}{\dfrac{\text{mother-liquor volume withdrawn}}{\text{mother-liquor volume in the crystallizer}}}$$

or

$$\frac{\Delta n\,dL}{n\,dL} = -\frac{\Delta n}{n} = \frac{Q_{ML}\,\Delta t}{V_{ML}} \qquad (17\text{-}34)$$

where Q_{ML} = volumetric flow rate of mother liquor in the withdrawn product magma and V_{ML} = volume of mother liquor in the crystallizer. Combining (17-32) and (17-34) and taking the limit,

$$-\frac{dn}{dL} = \frac{Q_{ML}\,n}{GV_{ML}} \qquad (17\text{-}35)$$

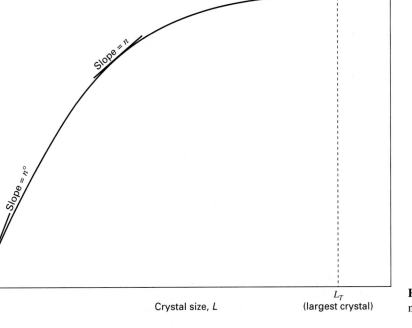

Figure 17.17 Typical cumulative-numbers-undersize distribution.

which is a simplified version of the following more-general, transient-population balance equation that allows for crystals in the feed and a nonuniform growth rate, but does assume a constant volume of mother liquor in the crystallizer:

$$\frac{\partial n}{\partial t} + \frac{\partial (nG)}{\partial L} + \frac{(Q_{ML})_{\text{out}}\, n}{V_{ML}} - \frac{(Q_{ML})_{\text{in}}\, n_{\text{in}}}{V_{ML}} = 0 \quad (17\text{-}36)$$

The retention time of mother liquor in the crystallizer is $\tau = V_{ML}/Q_{ML}$. Therefore, (17-35) can be rewritten as

$$-\frac{dn}{n} = \frac{dL}{G\tau} \quad (17\text{-}37)$$

If (17-37) is integrated for a constant growth rate and residence time, we obtain

$$n = n^{\circ} \exp\left(-L/G\tau\right) \quad (17\text{-}38)$$

This equation is the starting point for determining distribution curves for crystal population, crystal size or length, crystal surface area, and crystal volume or mass.

For example, to obtain crystal population, the number of crystals per unit volume of mother liquor below size L is

$$N/V_{ML} = \int_0^L n\, dL \quad (17\text{-}39)$$

The total number of crystals per unit volume of mother liquor is

$$N_T/V_{ML} = \int_0^{\infty} n\, dL \quad (17\text{-}40)$$

Combining (17-38) to (17-40), the cumulative number of crystals of size smaller than L, as a fraction of the total is

$$x_n = \frac{\int_0^L n^{\circ} e^{-L/G\tau}\, dL}{\int_0^{\infty} n^{\circ} e^{-L/G\tau}\, dL} = 1 - \exp\left(-L/G\tau\right) \quad (17\text{-}41)$$

Or if we define a dimensionless crystal size as

$$z = L/G\tau \quad (17\text{-}42)$$

then,

$$x_n = 1 - e^{-z} \quad (17\text{-}43)$$

A plot of (17-43), shown in Figure 17.18a, is referred to as the cumulative distribution or cumulative crystal population. For a given value of z, x_n is the fraction of crystals having a smaller value of z. Also of interest is the corresponding differential plot of dx_n/dz, shown in Figure 17.18b, where from (17-43),

$$\frac{dx_n}{dz} = e^{-z} \quad (17\text{-}44)$$

The differential plot gives the fraction of the crystals in a given interval of z. At small values of z, the fraction is seen to be large, while at large values of z, the fraction is small.

In Figures 17.18a and b, four different distribution plots are shown. From statistics, (17-41) is one of a number of *moment equations*, which for a relation $n = f(z)$, are given by

$$x_k = \frac{\int_0^z n z^k\, dz}{\int_0^{\infty} n z^k\, dz} \quad (17\text{-}45)$$

where k is the order of the moment. Thus, (17-43) is obtained by setting $k = 0$ and thus, is the zeroth moment of the distribution. Results for this moment and the corresponding first (length or size), second (area), and third (volume or mass) moments are summarized in Table 17.9. Corresponding cumulative and differential plots are included in Figure 17.18.

Of particular interest in the design and operation of a crystallizer is the predominant crystal size, L_{pd}, in terms of the volume or mass distribution. This size corresponds to the peak of the differential-mass distribution and is derived as follows: From Table 17.9,

$$dx_m/dz = (z^3/6)e^{-z} \quad (17\text{-}46)$$

At the peak,

$$\frac{d\left(\dfrac{dx_m}{dz}\right)}{dz} = 0 = \frac{3z^2\, e^{-z}}{6} - \frac{z^3\, e^{-z}}{6} \quad (17\text{-}47)$$

Solving (17-47) for z,

$$z = 3 = \frac{L}{G\tau} \quad (17\text{-}48)$$

Therefore,

$$L_{pd} = 3G\tau \quad (17\text{-}49)$$

Similar developments using the differential expressions in Table 17.9 show that the most populous sizes in terms of the number of crystals, the size of crystals, and the surface area of crystals are 0, $G\tau$, and $2G\tau$, respectively. If L_{pd} is selected, (17-41) and (17-43) can be used to estimate cumulative and differential screen analyses.

To utilize the distributions of Table 17.9, values of G and τ are needed. The growth rate depends on the supersaturation and the degree of agitation. The residence time depends on the design and operation of the crystallizer. It is also useful to know B° and n°, which are related as follows:

$$\frac{1}{V_{ML}}\frac{dN}{dt} = \frac{1}{V_{ML}}\frac{dN}{dL}\left(\frac{dL}{dt}\right) \quad (17\text{-}50)$$

$$\lim_{L\to 0} \frac{1}{V_{ML}}\frac{dN}{dt} = B^{\circ}$$

$$\frac{dL}{dt} = G$$

$$\lim_{L\to 0} \frac{1}{V_{ML}}\frac{dN}{dL} = n^{\circ}$$

Therefore,

$$B^{\circ} = Gn^{\circ} \quad (17\text{-}51)$$

Combining (17-51) with (17-38),

$$n = \frac{B^{\circ}}{G} \exp\left(-L/G\tau\right) \quad (17\text{-}52)$$

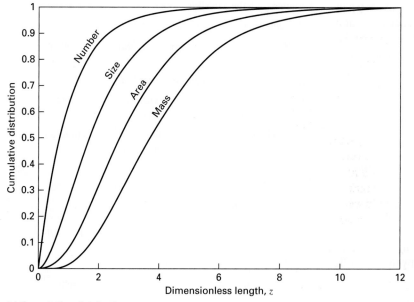

(a) Cumulative distributions

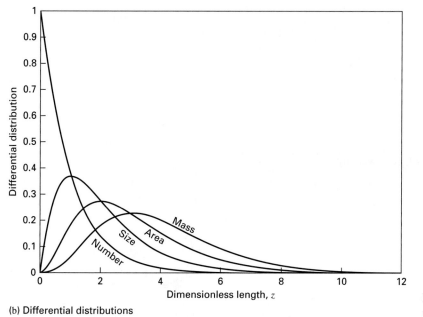

(b) Differential distributions

Figure 17.18 Crystal-size distributions predicted by the MSMPR model.

Table 17.9   Cumulative and Differential Plots for Moments of Crystal Distribution for Constant Growth Rate

Moment	Distribution Basis	Cumulative	Differential
Zeroth	Number	$x_n = 1 - e^{-z}$	$dx_n/dz = e^{-z}$
First	Size or length	$x_L = 1 - (1 + z)e^{-z}$	$dx_L/dz = ze^{-z}$
Second	Area	$x_a = 1 - \left(1 + z + \dfrac{z^2}{2}\right)e^{-z}$	$dx_a/dz = \dfrac{z^2}{2}e^{-z}$
Third	Volume or mass	$x_m = 1 - \left(1 + z + \dfrac{z^2}{2} + \dfrac{z^3}{6}\right)e^{-z}$	$dx_m/dz = \dfrac{z^3}{6}e^{-z}$

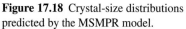

$z = L/G\tau$.

This equation can be used with experimental data, as shown in the following example, to obtain values of nucleation and growth rates for a given set of operating conditions when the assumptions of the MSMPR model hold. The effect of operating conditions on B° can be expressed as an empirical, power-law function of the form given by (17-19). However, since the growth rate can be proportional to the relative supersaturation, s, raised to an exponent, (17-19) can be rewritten as

$$B^{\circ} = k'_N \, G^i \, M_T^j \, N^r \tag{17-53}$$

Unfortunately, k'_N can be sensitive to the size of the equipment and is, therefore, best determined from data for a commercial crystallizer as discussed by Zumstein and Rousseau [17]. The exponents i, j, and r can be determined from small-scale experiments.

The necessary nucleation rate for a crystallization operation is related to the predominant crystal size by the MSMPR model. From (17-40) and (17-38),

$$n_c = N_T / V_{ML} = \int_0^\infty n \, dL = n^\circ \tau G \int_0^\infty e^{-z} \, dz = n^\circ \tau G \tag{17-54}$$

The mass of crystals per unit volume of mother liquor is

$$m_c = \int_0^\infty m_p n \, dL \tag{17-55}$$

where $m_p = $ mass of particle, given by

$$m_p = f_v L^3 \rho_p \tag{17-56}$$

where f_v is defined in (17-11).

Combining (17-55), (17-56), and (17-38), followed by integration gives

$$m_c = 6 f_v \, \rho_p \, n^\circ (G\tau)^4 \tag{17-57}$$

Combining (17-54) and (17-57), the number of crystals per unit mass of crystals is

$$\frac{n_c}{m_c} = \frac{1}{6 f_v \, \rho_p \, (G\tau)^3} \tag{17-58}$$

Combining (17-48) with (17-58),

$$\frac{n_c}{m_c} = \frac{9}{2 f_v \, \rho_p \, L_{pd}^3} \tag{17-59}$$

or the corresponding required nucleation rate is

$$B^{\circ} = \frac{n_c C}{m_c V_{ML}} = \frac{9C}{2 f_v \rho_p V_{ML} L_{pd}^3} \tag{17-60}$$

where $C = $ mass rate of production of crystals.

EXAMPLE 17.11

A continuous, vacuum-evaporating crystallizer of the DTB type is to be used to produce 2,000 lb/hr of $Al_2(SO_4)_3 \cdot 18H_2O$ ($\rho_p = 105$ lb/ft^3). The magma contains 0.15 ft^3 crystals/ft^3 magma and the residence time of the magma in the crystallizer in 2 h. The desired predominant crystal size on a mass basis is 0.417 mm.

Estimate with the MSMPR model:

(a) Required crystal growth rate in ft/h

(b) Necessary nucleation rate in nuclei/h-ft^3 of mother liquor in the crystallizer

(c) Number of crystals produced per hour

(d) Tables and plots of estimated cumulative and differential screen analyses of the product crystals on a mass or volume basis

Also explain how the required growth and nucleation rates for the operating crystallizer might be achieved.

SOLUTION

(a) From (17-48),

$$G = \frac{L_{pd}}{3\tau} = \frac{(0.417/304.8)}{3(2)} = 2.28 \times 10^{-4} \text{ ft/h}$$

(b) Need volume of mother liquor in the crystallizer.

$$\text{Volume of crystals produced} = \frac{2,000}{105} = 19.1 \text{ ft}^3\text{h}.$$

Volume of crystals in crystallizer $= 19.1(2) = 38.2$ ft^3.

$$\text{Volume of magma in crystallizer} = \frac{38.2}{0.15} = 255 \text{ ft}^3.$$

Volume of mother liquor in crystallizer $= V_{ML}$
$$= 255 - 38.2 = 217 \text{ ft}^3.$$

From (17-60), assuming $f_v = 0.5$,

$$B^{\circ} = \frac{9(2,000)}{2(0.5)(105)(217)(0.417/304.8)^3} = 3.1 \times 10^8 \frac{\text{nuclei}}{\text{h-ft}^3}$$

(c) Number of crystals produced $= 3.1 \times 10^8(217)$
$$= 6.7 \times 10^{10}/\text{h}.$$

(d)
$$z = \frac{L}{G\tau} = \frac{L}{(2.28 \times 10^{-4})(2)}$$
$$= 2.2 \times 10^3 \, L, \text{ ft} \quad \text{or} \quad 7.19 \, L, \text{ mm}$$

From Table 17.9,

$$x_m = 1 - \left(1 + z + \frac{z^2}{2} + \frac{z^3}{6}\right) e^{-z} \tag{1}$$

$$\frac{dx_m}{dz} = \frac{z^3}{6} e^{-z} \tag{2}$$

Using (1) and (2) with a spreadsheet for the older Tyler mesh sizes, the following results are obtained for the cumulative and differential screen analyses:

Tyler Mesh	Opening, mm	Dimensionless Length, z	Cumulative Screen Analysis, %	Differential Screen Analysis %
8	2.357	16.96	100.00	0.00
10	1.667	11.99	99.77	0.18
14	1.179	8.48	96.95	2.11
20	0.833	5.99	84.82	8.95
28	0.589	4.24	61.16	18.31
35	0.417	3.00	35.20	22.40
48	0.295	2.12	16.50	19.05
65	0.208	1.50	6.54	12.53
100	0.147	1.06	2.29	6.87
150	0.104	0.75	0.73	3.31
200	0.074	0.53	0.22	1.46

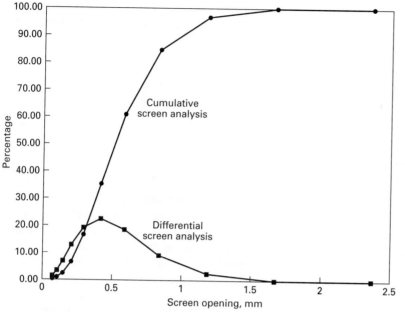

Figure 17.19 Predicted cumulative and differential screen analyses for Example 17.11.

A plot of the cumulative and differential screen analyses is shown in Figure 17.19.

Both growth and nucleation rates depend on the supersaturation. The growth may also depend on the relative velocity between the crystals and the mother liquor. The nucleation rate depends upon the degree of agitation. With a DTB crystallizer, the agitator rpm and the magma circulation rate through the external heat exchanger can be adjusted to achieve the required growth and nucleation rates.

17.6   PRECIPITATION

Solution crystallization, as discussed above, occurs when a solution containing a moderately to strongly soluble solute is cooled or partially evaporated, causing the solute to exceed its solubility sufficiently to partially crystallize. The degree of supersaturation is small, primary nucleation is avoided, and crystal growth is slow. If the process is carried out under controlled conditions, reasonably large and desirable crystals can be grown. In many respects, the process of precipitation is just the opposite of solution crystallization. Precipitation, as discussed by Nielsen [18], involves solutes that are only sparingly soluble. The precipitate is formed by changing solution pH, solvent concentration, solution temperature, or most commonly by the addition of a reagent to a solution, resulting in a reaction to produce another chemical that is almost insoluble in the resulting solution. The latter is commonly referred to as *reactive crystallization*. The degree of supersaturation produced by the reaction is very large, causing a high degree of primary nucleation. Although some growth occurs as the supersaturation is depleted, precipitates generally consist of very small particles that may be crystalline in nature and are frequently *aggregates* and *agglomerates*. Aggregates are masses of crystallites that are weakly bonded together. Agglomeration can follow, cementing aggregates together. Typical chemicals that are produced by precipitation are the sparingly soluble compounds listed in Table 17.6. Precipitation may be thought of as *fast crystallization*.

In precipitation, particle size is related to the solubility by (17-16), the Kelvin equation. However, for precipitates formed from ionic reactions in solution, the supersaturation ratio, $S = c/c_s$, is replaced by $(\pi/K_c)^{1/v}$, where $\pi =$ the ionic concentration product for the reaction, $K_c =$ (equilibrium) solubility product, and $v =$ sum of the number of cations and anions that form the precipitated compound. Thus, for aluminum hydroxide at 15°C,

$$\pi = \left(c_{Al}^{+3}\right)\left(c_{OH}^{-}\right)^3$$

with K_c given in Table 17.6 and $v = 1 + 3 = 4$. The Kelvin relation is of great importance in precipitation because nucleation can be due to homogeneous primary nucleation.

The extent of supersaturation is a very important factor in precipitation. When a reagent is added to a solution to form a sparingly soluble compound, a very high supersaturation can be developed, depending upon the ionic concentrations in the reagent and solution before mixing. For example, consider the formation of a precipitate of $BaSO_4$ from an aqueous reagent solution containing Ba^{++} ions and an aqueous solution of, e.g., sulfuric acid.

From Table 17.6, the solubility product of $BaSO_4$ at 18°C is 0.87×10^{-10}. Figure 17.20a, taken from Nielsen [19], is a plot of sulfate ion concentration versus barium ion concentration, both in the solution just after mixing and before precipitation, with contours of constant supersaturation ratio, S. The dashed lines refer to an ideal situation where activity

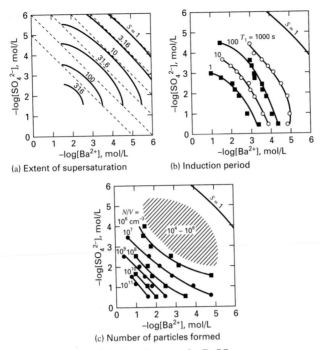

(a) Extent of supersaturation
(b) Induction period
(c) Number of particles formed

Figure 17.20 Precipitation diagrams for BaSO$_4$.

coefficients are 1.0 (concentrations = activities) and no complexes of Ba^{++} and SO$_4^=$, are formed. The solid, curved contours take these deviations from the ideal situation into account. Note the very large supersaturation ratios that can occur. For example, if the SO$_4^=$ concentration in the solution is 0.01 mol/L and the Ba^{++} concentration is 4.7×10^{-4} mol/L, the supersaturation ratio = 100.

Typically, precipitation does not take place immediately upon development of a large supersaturation because of the slow growth of very small particles. However, after a certain period of time, called the induction period, T_1 visible precipitation begins. As shown in Figure 17.20b, this period is highly dependent on the initial-ion concentrations and, by superposition of Figure 17.20a, on the supersaturation ratio. For example, at a supersaturation ratio of about 300, the induction period is only one second, while at a supersaturation ratio of about 10, the induction period is more than one minute.

As shown in Figure 17.20c, the number of particles formed per unit volume of solution, N/V, also depends strongly on the initial concentrations of the anion and cation and, therefore, on the supersaturation at high ion concentrations. A minimum number of particles per unit volume or a corresponding maximum particle size is frequently observed. For BaSO$_4$, this maximum size occurs for a supersaturation ratio of about 300. The number of particles formed, also depends on particle growth rate, which may be controlled by mass transfer of the ions to the particle surface and/or integration of ions into the particle crystalline structure (surface reaction). Particle growth rates during

precipitation generally follow one of the following laws, which exhibit different dependencies on the relative supersaturation, $s = S - 1$:

Linear

$$G = k_1 s \qquad (17\text{-}61)$$

Parabolic

$$G = k_2 s^2 \qquad (17\text{-}62)$$

Exponential

$$G = k_3 f\{s\} \qquad (17\text{-}63)$$

The linear rate law, which often applies for $G > 10$ nm/s, may be due to mass-transfer control or surface-adsorption control, where the latter depends more strongly on temperature. The parabolic rate law, for which G may be < 10 nm/s, applies to screw-dislocation-controlled growth. The exponential law, where $f\{s\}$ can involve a complex exponential, log, and/or power-law dependency corresponds to growth control by surface nucleation. The latter two mechanisms can occur in parallel. Rate constants for many electrolytes for the three different rate laws are given by Nielsen [20]. When growth is rapid, coprecipitation of soluble electrolytes may occur by entrapment, thus making it difficult to obtain a pure precipitate product.

Because a precipitate is formed at considerable supersaturation, the resulting particle shapes may be far from the shape corresponding to a minimum Gibbs energy, which depends on the particle surface area and interfacial tension. However, if the precipitate and mother liquor are allowed to age, then, as discussed by Nielsen, [20], the precipitate particle sizes and shapes can tend toward equilibrium values by: (1) flocculation and sintering of fine particles (2) transport of ions over the surface, and (3) ripening by dissolution and redeposition. Ripening can result in the release of coprecipitates, thus increasing precipitate purity.

The small particles produced in abundance during precipitation have a tendency to cluster together by interparticle-collision phenomena, variously referred to as agglomeration, aggregation, and flocculation. Such clusters, often called agglomerates, are common when the number of particles/cm^3 of solution exceeds 10^7. For BaSO$_4$, Figures 17.20a and c show that agglomeration may require a very large initial supersaturation ratio. Agglomeration is important mainly when particles sizes are between 1 μm and 50 μm.

EXAMPLE 17.12

Fitchett and Tarbell [21] studied the effect of impeller rpm on the crystal-size distribution obtained for the continuous precipitation of barium sulfate when mixing solutions of sodium sulfate and barium chloride. The contents of the 1.8-L crystallizer were assumed to be perfectly mixed. In their Run 15, they used feeds with dissolved salts in stoichiometric ratio to give a sodium-chloride concentration of 0.15 mol/L and an average residence time of 38 s.

Results for two different impeller speeds were as follows:

Size, μm	ln n, Number Density of Crystals, ln $(\mu m^{-1} L^{-1})$	
	950 rpm, 0.361 J/s · kg Feeds	400 rpm, 0.028 J/s · kg Feeds
7	22.07	23.45
9	21.66	22.75
11	21.35	22.08
13	20.97	21.36
15	20.77	20.75
17	20.41	20.14
19	20.04	19.57
21	19.77	18.94
23	19.48	18.43
25	19.09	17.75
27	18.85	
29	18.49	
31	18.11	
33	17.87	

Using the MSMPR model, determine from the two sets of data:

(a) n^o, number density of nuclei, nuclei/μm-L

(b) G, linear growth rate, μm/s

(c) B^o, nucleation rate, nuclei/L-s

(d) mean crystal length, μm

(e) n_c, number of crystals/volume of mother liquor, crystals/m^3

SOLUTION

(a) From (17-38), if each set of data is fitted to

$$\ln n = \ln n^o - L/G\tau \qquad (1)$$

the best straight line yields an intercept of $\ln n^o$ and a slope of $-1/G\tau$, from which n^o and G can be determined for $\tau = 38$ s.

Using a spreadsheet, the results for 950 rpm are

intercept = 23.13, slope = −0.1601

Therefore, $n^o = \exp(23.13) = 1.11 \times 10^{10}$ crystals/μm · L

(b) $G = 1/[(38)(0.1601)] = 0.164$ μm/s

(c) The nucleation rate is given by (17-51):

$B^o = Gn^o = 0.164(1.11 \times 10^{10}) = 1.82 \times 10^9$ nuclei/L · s

(d) The mean particle length is determined from the value of z for the maximum value of dx_L/dz, which, from Table 17.9 is ze^{-z}.

$$d\frac{(ze^{-z})}{dz} = e^{-z} - ze^{-z} = 0$$

Therefore, $z = 1 = L_{mean}/G\tau$ from (17-42)

$$L_{mean} = G\tau = 0.164(38) = 6.23 \,\mu m$$

(e) The number of crystals per unit volume of mother liquor is given by (17-54).

$n_c = n^o \tau G = B^o \tau = 1.82 \times 10^9 (38)$
$= 6.92 \times 10^{10}$ crystals/L $= 6.92 \times 10^7$ crystals/m^3

(f) In a similar manner, the following results are obtained for 400 rpm:

Intercept = 25.53, slope = −0.313

$n^o = 1.22 \times 10^{11}$ crystals/μm · L

$G = 0.0841$ μm/s

$B^o = 1.03 \times 10^{10}$ nuclei/L · s

$L_{mean} = 3.20 \,\mu m$

$n_c = 3.91 \times 10^8$ crystals/m^3

Comparing the two sets of results, we see that a higher agitator speed gives a larger mean crystal size, a larger growth rate, and a lower nucleation rate.

17.7   MELT CRYSTALLIZATION

Solution crystallization refers to crystallization of a solute from a solvent. It is most commonly conducted with aqueous solutions of dissolved inorganic salts. The phase equilibrium diagram for a water-salt system, e.g., Figure 17.2, typically includes temperatures ranging from the eutectic temperature (below 0°C) to a temperature exceeding the melting point of ice, but not greater than 200°C. Since the melting points of inorganic salts greatly exceed 200°C, the pure-salt melting point is, accordingly, not included on the diagram.

For mixtures of organic compounds, the situation is quite different. An analysis by Matsuoka et al. [22] found that more than 70% of the common organic compounds had melting points between 0 and 200°C. For binary mixtures of such compounds, the phase-equilibrium diagrams will typically include the melting points of both compounds. Typical diagrams are shown in Figure 17.1. In (b), crystals of ortho-chloronitrobenzene can be formed if the feed composition is less than the eutectic composition based on para-chloronitrobenzene; otherwise, pure para-chloronitrobenzene can be formed. The exception is the eutectic composition. *Eutectic-forming* systems consist of compounds that can not substitute for each other in the crystal lattice. Thus, the eutectic mixture consists of two different solid phases. The two solubility curves separating the liquid-phase region from the two solid–liquid regions are sometimes referred to as freezing-point curves, and mixtures at conditions in the liquid-phase region are referred to as *melts*.

Much less common are *solid-solution-forming* systems of the type shown in Figure 17.1c for the phenanthrene-anthracene system. These systems consist of compounds so nearly alike in structure that they can substitute for each other in the lattice structure to form a single solid phase of the two compounds over a wide range of composition. With this type of system, a mixture is crystallized. The liquid–solid phase diagram resembles that for vapor–liquid equilibrium, where the freezing-point and melting-point curves replace the dew-point and bubble-point curves, respectively. Mixtures in the liquid-phase region above the freezing-point curve are also referred to as melts. A mixture in the region between the two curves separates into a liquid phase and a solid phase, neither of which is pure.

Crystallization of melts from either eutectic-forming or solid-solution-forming mixtures is called melt crystallization. Although theoretically, melt crystallization of eutectic-forming systems, like solution crystallization of such systems, can produce pure crystals, the crystalline product from commercial single-stage crystallizers may not meet purity specifications for reasons discussed in detail by Wilcox [23]. In that case, repeated stages of melting and crystallization may be necessary. With eutectic-forming systems, multiple stages are mandatory to produce crystals of high purity.

The separation of organic mixtures is most commonly achieved by distillation. However, if distillation: (1) requires more than 100 theoretical stages, (2) cannot produce products that meet specifications (e.g., purity and color), (3) causes decomposition of feed components, or (4) requires extreme conditions of temperature or pressure (e.g., vacuum), other separation methods should be considered. According to Wynn [24], if the compounds to be separated are: (1) disubstituted benzenes, diphenyl alkyls, phenones, secondary or tertiary aromatic or aliphatic amines, isocyanates, fused-ring compounds, heterocyclic compounds, and carboxylic acids of MW < 150; (2) have a melting-temperature range from 0 to 160°C; (3) required to be very high-purity products; or if (4) a laboratory test produces a clearly defined solid phase from which the liquid phase drains freely, then melt crystallization should be considered as an alternative or supplement, as in a hybrid process, to distillation.

Equipment for Melt Crystallization

As with solution crystallization, a large number of crystallizer designs have been proposed for melt crystallization.

Only some of the more widely used commercial units are discussed here. Myerson [10] gives a detailed discussion. In all cases, crystallization is caused by cooling the mixture.

The two major methods used in melt crystallizers are *suspension crystallization* and *layer crystallization* by progressive freezing. In the former, crystals of a desired size distribution are grown slowly in a suspension by subcooling a seeded-feed melt. In the latter, crystals of uncontrolled size are grown rapidly on a cooled surface, wherein subcooling is supplied through the crystallized layer. In suspension crystallization, the remaining melt must be separated from the crystals by centrifugation, filtration, and/or settling. In layer crystallization, the remaining melt or residual liquid is drained from the solid layer, followed by melting of that layer.

Figure 17.21 shows a two-stage, scraped-wall-crystallizer system used for suspension crystallization. A cooling medium is used to control the surface temperatures of the two scraped-wall units, causing crystals to grow, which are subsequently scraped off by screws. The melt-crystal mixture is circulated through a ripening vessel. The two scraped-wall units of a typical system are 3.6 m long with 3.85 m² of total heat-transfer surface area. The screws are driven by a 10-kW motor.

Of more importance in commercial applications of melt crystallization is the falling-film crystallizer developed by Sulzer Brothers Limited and shown in Figure 17.22. This equipment is particularly useful for producing high-purity crystals (>99.9%) at high capacity (>10,000 tons/y). A large pair of units, each 4 m in diameter and containing 1,100 12-m-high tubes can produce 100,000 tons/yr of very pure crystals, with typical layer growth rates of 1 in./h. The feed melt flows as a film down the inside of the tubes, over a crystal layer that forms and grows by progressive freezing because the wall of the tube is cooled from the outside by a coolant. When a predetermined crystal layer thickness, typically 5–20 mm, is

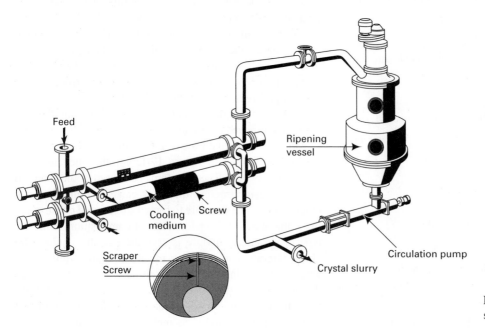

Figure 17.21 Two-stage, scraped-wall, melt crystallizer.

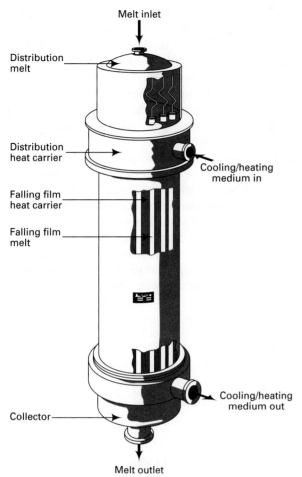

Figure 17.22 Sulzer falling-film melt crystallizer.

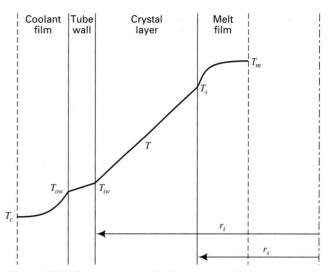

Figure 17.23 Temperature profile for melt crystallization from a falling film at a vertical location along the tube.

reached, the feed melt flow rate is stopped. The tubes are now slightly warmed to cause partial melting, called sweating, to remove impurities that may be stuck or bonded to the crystal layer. This is followed by complete melting of the remaining layer, which is of high purity. During the initial crystallization phase, melt is circulated at a high rate compared to the crystallization rate so that a uniform temperature and melt composition is approached down the length of the tube. The coolant also flows as a film down along the outside surface of the tubes.

Consider the freezing step in a falling-film crystallizer. A typical temperature profile is shown in Figure 17.23. The melt enters at the top of the tube and flows as a film down the inside wall. A coolant, at a temperature below the freezing point of the melt, also enters at the top of the tube and flows as a film down the outside wall. Heat is transferred from the melt to the coolant, causing the melt to form a crystal layer at the inside wall of the tube. As the melt and coolant flow down the tube, their temperatures decrease and increase, respectively. The thickness of the crystal layer increases with time, with the latent heat of fusion transferred from the crystal-melt interface to the coolant.

For a eutectic-forming, binary melt, only one component will crystallize, although small amounts of the other component may be trapped in the crystal layer, particularly if the rate of crystal formation is too fast. The temperature at the interface of the crystal layer and the melt can be assumed to be the equilibrium temperature corresponding to the saturated melt composition on the solubility curve. If mass transfer of the crystallizing component from the melt film to the phase interface is rapid, the interface temperature will correspond to the melt-film composition at that vertical location. Furthermore, if the thermal resistances of the coolant film, tube wall, and melt film are negligible compared to that of the crystal layer, and if the heat capacity of the crystal layer and metal wall are negligible, then the following simple model for the rate of increase of thickness of the crystal layer with time can be constructed.

At a particular vertical location, $T_s \approx T_m$, the melting point, and $T_{iw} \approx T_c$, the coolant bulk temperature. The rate of heat released by freezing, ΔH_f, is equal to the rate of heat conduction through the crystal layer. Thus, for a planar wall, referring to Figure 17.23, the heat evolved from fusion is equated to the rate of heat conduction through the crystal layer to give

$$\Delta H_f \frac{dm}{dt} = -A\rho_c(\Delta H_f)\frac{dr_s}{dt} = \frac{k_c A(T_m - T_c)}{r_i - r_s} \quad (17\text{-}64)$$

with an initial condition $r_s = r_i$ at $t = 0$. Integration of (17-64) gives

$$\frac{(r_i - r_s)^2}{2} = \frac{k_c(T_m - T_c)t}{\rho_c \,\Delta H_f} \quad (17\text{-}65)$$

or crystal layer thickness is

$$(r_i - r_s) = \sqrt{\frac{2k_c(T_m - T_c)t}{\rho_c \, \Delta H_f}} \qquad (17\text{-}66)$$

A similar derivation for a cylindrical tube wall, where $r_i =$ inside radius of the tube, gives

$$\frac{1}{4}\left(r_i^2 - r_s^2\right) - \frac{r_s^2}{2}\ln\left(\frac{r_i}{r_s}\right) = \frac{k_c(T_m - T_c)t}{\rho_c \, \Delta H_f} \qquad (17\text{-}67)$$

As would be expected, the value of the left-hand side of (17-67) approaches the value of the left-hand side of (17-65) as the value of r_s approaches the value of r_i. For the planar wall, (17-66) shows that the crystal layer grows as the square root of time. Thus, for a given period of time, the growth during the first half of the time period is $(1/\sqrt{2}) = 70.7\%$ of the total growth. For the cylindrical-tube wall, if the growth is from the wall to half of the radius, 67.9% of that growth occurs during the first half of the time period to produce

75.2% of the crystal layer. The time required for this thickness of growth for the cylindrical tube is only 80% of the time for the planar wall. Thus, a conservative result is obtained by using the much simpler planar wall case of (17-66).

During operation of the falling-film crystallizer, the temperature of the melt film decreases as it flows down the tube because of heat transfer from the melt to the colder crystal layer. Based on the earlier assumptions, the melt temperature at any elevation will be the temperature corresponding to the solubility curve for the melt composition at that elevation. If we: (1) assume that the sensible heat from the cooling of the melt layer is negligible compared to the latent heat of fusion, (2) neglect any sensible-heat storage in the crystal layer and tube wall, (3) neglect vertical conduction in the tube wall, crystal layer, and melt layer, and (4) assume a constant coolant temperature, we can couple (17-66) with a material balance for the depletion of the crystallizing component as it flows down the inside tube wall. The result, which is left as an exercise, gives the crystal-layer thickness as a function of time and vertical location.

EXAMPLE 17.13

A melt of 80 wt% naphthalene and 20 wt% benzene at the saturation temperature is fed to a falling-film crystallizer, where coolant enters at 15°C. Estimate the time required for the crystal-layer thickness near the top of the 8-cm i.d. tubes to reach 2 cm.

SOLUTION

From Figure 17.24, by extrapolation the saturation temperature of the melt is 62°C. Therefore, naphthalene will crystallize, with

$$T_m - T_c = (62 - 15)1.8 = 84.6°F$$

$$r_i = 8/2 = 4\,\text{cm} = 0.131\,\text{ft}$$

$$r_s = 4 - 2 = 2\,\text{cm} = 0.0655\,\text{ft}$$

The estimate can be made with (17-67), based on the following properties for naphthalene:

$$\rho_c = 71.4\,\text{lb/ft}^3$$

$$k_c = 0.17\,\text{Btu/h-ft-°F}$$

$$\Delta H_f = 63.9\,\text{Btu/lb}$$

From (17-67), solving for time, t,

$$t = \frac{(71.4)(63.9)}{(0.17)(84.6)}$$

$$\times\left[\frac{1}{4}(0.131^2 - 0.0655^2) - \frac{0.0655^2}{2}\ln\left(\frac{0.131}{0.0655}\right)\right] = 0.549\,\text{h}$$

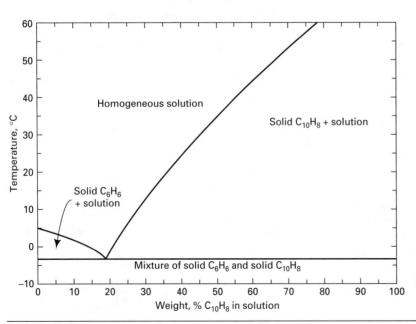

Figure 17.24 Solid–liquid phase diagram for the naphthalene-benzene system.

17.8 ZONE MELTING

When the melt consists of two components that form a solid solution, such as shown in Figure 17.1c, the liquid and solid phases at equilibrium contain amounts of both components, similar to the behavior previously discussed for vapor-liquid equilibrium. Accordingly, multiple stages of crystallization are required to obtain products of high purity. A particularly useful technique for doing this, especially when impurities are to be removed to achieve a very high-purity product, is zone melting, as developed by Pfann [25, 26] and discussed further by Zief and Wilcox [27].

The zone-melting technique, when carried out batchwise, is illustrated in Figure 17.25. Starting with an impure crystal slab, a melt zone, which is thin compared to the length of the slab, is passed slowly (typically at 1 cm/h for organic mixtures) through the slab from one end to the other by fixing the slab and using a moving heat source or, less commonly, by moving the slab through a fixed heat source. Radiofrequency (RF) induction heating is particularly convenient and causes mixing in the melt. The slab can be arranged horizontally or vertically, with the latter orientation preferred because it can take advantage of any density difference be-

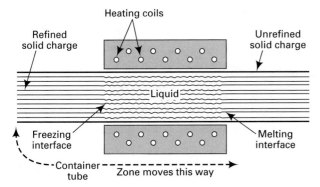

Figure 17.25 Zone melting.
[From *Perry's Chemical Engineers' Handbook,* 6th ed., R.H. Perry, D.W. Green, and J.O. Maloney, eds., McGraw-Hill, New York (1984) with permission.]

tween the crystals and the melt. Zone melting can be applied to the composition region near either end of the phase-equilibrium diagram. For example, for the binary mixture of phenanthrene (P) and anthracene (A) in Figure 17.1c, zone melting could be used to remove small amounts of A from P or small amounts of P from A.

A solid–liquid equilibrium distribution coefficient, K, can be defined in a manner analogous to that for liquid–liquid equilibrium. Thus,

$$K_i = \frac{\text{concentration of impurity, } i, \text{ in the solid phase}}{\text{concentration of impurity, } i, \text{ in the melt phase}}$$

(17-68)

For example, if the concentration is expressed in mole fractions and the impurity is anthracene in the system of Figure 17.1c, then at 120°C, $K_i = 0.30/0.12 = 2.50$. At 200°C, with phenanthrene as the impurity $K_i = 0.11/0.28 = 0.393$. Thus, K_i can be greater than or less than one. When greater than one, the impurity raises the melting point and the impurity concentrates in the solid phase; when less than one, the impurity lowers the melting point and concentrates in the melt phase. However, when K_i approaches a value of 1, purification by zone melting becomes very difficult.

Near either end of the composition range, the equilibrium curves for the solid and liquid phases approach straight lines, and therefore, the values of K_i become constants. In these composition ranges, an equation is readily developed to predict concentration of the impurity in the solid phase upstream of the moving melt zone as a function of distance down the crystal layer in the direction of movement of the melt zone. Figure 17.26 shows the position of the melt zone during zone melting. The melt zone is shown to move a distance dz. Assume that: (1) the melt zone of width, ℓ, is perfectly mixed with impurity concentration, w_L, in weight fraction; (2) no diffusion of impurity in the solid phases occurs; (3) the initial concentration of the impurity is uniform at w_o; and (4) the concentration of the impurity in the melt zone is in equilibrium with that in the solid phase upstream of the melt zone. A mass balance on the impurity for a dz movement of the melt

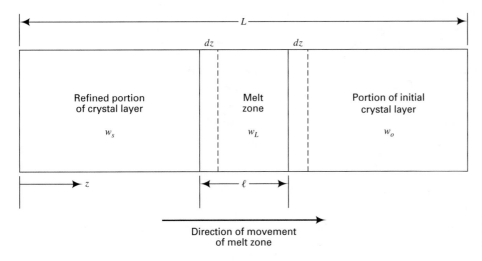

Figure 17.26 Ideal, zone-melting model.

zone is given by

(Mass of impurity added to melt zone)
− (mass of impurity removed from melt zone)
= (increase in the mass of impurity in the melt zone)

Thus,

$$w_o \, \rho_c \, A_c \, dz - w_s \, \rho_c \, A_c \, dz = \rho_L \, A_c \ell \, dw_L \quad (17\text{-}69)$$

From (17-68), let

$$K = w_s/w_L \quad (17\text{-}70)$$

Therefore,

$$dw_L = dw_s/K \quad (17\text{-}71)$$

Combining (17-69) to (17-71) to eliminate w_L and assuming $\rho_L = \rho_c$, we obtain

$$\frac{K}{\ell} \int_o^z dz = \int_{Kw_o}^{w_s} \frac{dw_s}{w_o - w_s} \quad (17\text{-}72)$$

Integration gives

$$\frac{w_s}{w_o} = 1 - (1 - K)\exp(-Kz/\ell) \quad (17\text{-}73)$$

for

$$z/\ell = 0 \quad \text{to} \quad z/\ell = \frac{L}{\ell} - 1$$

EXAMPLE 17.14

A crystal layer of 1 wt% phenanthrene and 99 wt% anthracene is subjected to zone melting with a melt-zone width equal to 0.1 of the length of the crystal layer ($\ell/L = 0.1$). The distribution coefficient for phenanthrene in the dilute composition region is 0.36. Determine the phenanthrene concentration profile in weight fractions at the conclusion of zone melting when the melt reaches the last 10% of the crystal layer length.

SOLUTION

From Figure 17.1c and the value of K, the phenanthrene favors distribution to the melt phase. Therefore, as the melt zone moves down the crystal layer, the phenanthrene will migrate from the upstream portion of the crystal layer into the melt zone. When the leading edge of the melt zone reaches the end of the crystal layer, all of the phenanthrene that has migrated will be in the melt layer at a uniform concentration equal to $w_s\{z/\ell = 0.9\}/K$. Assume instant freezing of this melt zone so as to maintain the uniform composition. From (17-73), with $K = 0.36$ and $w_o = 0.01$,

$$w_s = 0.01[1 - 0.64 \exp(-0.36 \, z/\ell)] \quad (1)$$

Solving (1) for values of $z/\ell = 0$ to 9 gives

z/ℓ	w_s
0	0.0036
1	0.0055
2	0.0069
3	0.0078

z/ℓ	w_s
4	0.0085
5	0.0089
6	0.0093
7	0.0095
8	0.0096
9	0.0097

In the melt zone, $w_L = w_s/K = 0.0097/0.36 = 0.0269$. A plot of the predicted profile is shown in Figure 17.27. To further refine the anthracene, additional passes of zone melting can be made to move more of the impurity into the melt zone. However, for each pass after the first, (17-73) is not valid because at the beginning of each of the additional passes, w_o is not uniform. For example, at the beginning of the second pass, w_o becomes w_s in Figure 17.27. For the additional passes, it is necessary to numerically solve the following differential-equation form of (17-72).

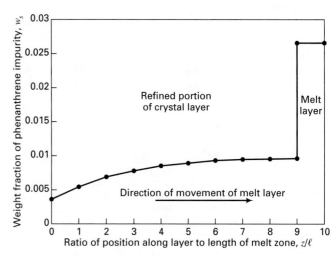

Figure 17.27 Predicted impurity profile for Example 17.14.

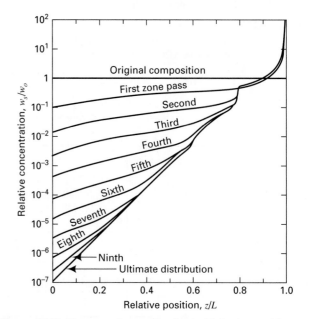

Figure 17.28 Concentration profiles for multiple zone-melting passes.

$$\frac{dw_s}{dz} = \frac{K}{\ell}(w_o - w_s) \qquad (17\text{-}74)$$

where $w_o = w_o\{z = z\}$ from the results for w_s for the results from the previous pass, and $w_s\{z = 0\} = Kw_o\{z = 0\}$.

Typical impurity-concentration profiles for multiple melt-zone passes are shown in Figure 17.28 from calculations by Burris, Stockman, and Dillon [28] for the case of $K = 0.1$ and a melt-zone width of 20% of the crystal-layer length ($\ell/L = 0.2$). However, in their calculations, the melt zone was allowed to diminish to zero as the heater moved away from the crystal layer, resulting in a steep gradient from $z/L = 0.8$ to 1.0. It is seen that an ultimate distribution of the impurity is approached after 9 passes. The number of passes required to approach the ultimate distribution is given approximately by $2(L/\ell) + 1$, as observed by Herington [29]. In this example, a highly pure crystalline product can be achieved if a portion of the more pure end of the final crystal layer is taken.

17.9 DESUBLIMATION

When crystallization occurs from a vapor phase, rather than from a liquid solution or melt, the operation is referred to as *desublimation*. The reverse operation, i.e., the vaporization of a solid directly to a vapor, is *sublimation*. To understand how such a phase change can occur without going through the liquid phase, consider the phase diagram for naphthalene shown in Figure 17.29. As with most chemicals that have a vapor pressure of much less than 1 atm at the melting temperature, that temperature coincides, within a fraction of a degree Celsius, with the triple point. Below the triple-point temperature of 80.2°C, corresponding to a vapor pressure of 7.8 torr, naphthalene cannot exist as a liquid, regardless of the pressure. Below 80.2°C, naphthalene exists as a solid provided that the pressure is greater than the vapor pressure of naphthalene at the prevailing temperature. However, if the pressure falls below the vapor pressure, solid naphthalene will sublime directly to a vapor.

If naphthalene is present in an ideal gas mixture with a noncondensible inert gas, sometimes called an entrainer, at a temperature below the triple point of 80.2°C, desublimation of the naphthalene to a crystalline solid will occur if the partial pressure of naphthalene in the gas is increased to a value that exceeds its vapor pressure. For example, consider a vapor mixture of 5 mol% naphthalene in nitrogen at 70°C and a total pressure of 40 torr. The partial pressure of naphthalene is 0.05(40) = 2 torr. From Figure 17.29, this partial pressure is less than the vapor pressure of 3.8 torr. If the total pressure is increased, desublimation will begin when the total pressure is increased to 3.8/0.05 = 76 torr. Alternatively, if the pressure is maintained at 40 torr, but the temperature is reduced, desublimation will begin to take place at 61°C, where the vapor pressure of naphthalene is 2 torr. Unless nitrogen is trapped, the naphthalene crystals will be pure.

The most common occurrence of desublimation is the formation of snow crystals by condensation of water vapor from air at temperatures below 0°C. At temperatures above −40°C, the snow crystals form by heterogeneous nucleation on fine mineral particles of perhaps 10^{-5}–10^{-2} mm in diameter. At temperatures below −40°C, homogeneous nucleation can take place. As the snow crystals fall through the atmosphere, they cluster together into snowflakes. The difference between the melting point of ice and the triple point of water is less than 0.01°C, with a corresponding vapor pressure of only 4.6 torr.

As discussed by Nord [30] and Kudela and Sampson [31], a number of chemicals are amenable to purification by desublimation, preceded perhaps by sublimation. In general, they are chemicals that are solids at ambient temperature and pressure and have solid vapor pressures greater than 5 μm Hg at moderate operating temperatures. These chemicals are listed in Table 17.10. Major applications of desublimation, to obtain near-pure solid chemicals or pure-chemical solid films on substrates, include crystallization of water-insoluble

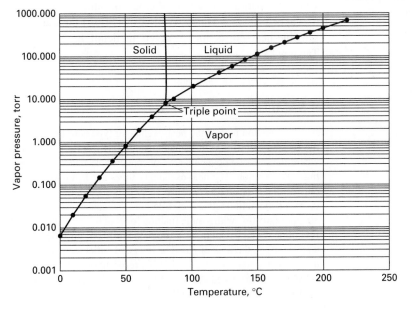

Figure 17.29 Vapor pressure of naphthalene.

Table 17.10 Chemicals Amenable to Purification by Desublimation

Aluminum chloride	Molybdenum Trioxide
Anthracene	Naphthalene
Anthranilic acid	β-Naphthol
Anthraquinone	Phthalic anhydride
Benzanthrone	o-Phthalimide
Benzoic acid	Pyrogallol
Calcium	Salicylic acid
Camphor	Sulfur
Chromium chloride	Terephthalic acid
Cyanuric chloride	Titanium tetrachloride
Ferric chloride	Thymol
Hafnium tetrachloride	Uranium hexafluoride
Iodine	Zirconium tetrachloride
Magnesium	

organic chemicals from mixtures with inert, nonvolatile gases and vapor-deposition of metals.

Desublimation is almost always effected by cooling the gas mixture at nearly constant pressure. As discussed by Holden and Bryant [32], cooling may be achieved by any of the following techniques:

1. Heat transfer from the gas through a solid surface, on which the crystals form.

2. Quenching by addition of a vaporizable liquid.

3. Quenching by addition of a cold, noncondensable gas.

4. Expansion of the gas mixture through a nozzle.

The first technique, which is widely used, has the disadvantage that the crystals must be removed by scraping of or by melting from the heat-transfer surface. The second technique, which is less common, must use a liquid in which the sublimate is not soluble. If excess water is used, this technique will produce a slurry of the crystals and nonvaporized liquid. The third technique, which is also common, produces dry crystals, sometimes called a snow. The fourth technique is much less common, because the degree of cooling necessary may require either a high pressure upstream of the nozzle or a vacuum downstream.

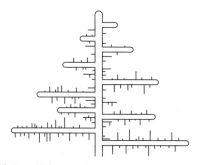

Figure 17.30 Dendritic growth of a crystal.

The number, size, and shape of the crystals produced in the snow of the third technique depend on the relative rates of nucleation, crystal growth, and crystal agglomeration. Frequently, as commonly observed with snow crystals, dendritic growth occurs, producing undesirable crystals of the shape shown in Figure 17.30. The main stem grows rapidly, followed by slower rates of growth of primary branches, and much slower growth of secondary branches. If the dendritic crystals can be suspended in the vapor long enough, the spaces between the branches can fill in to produce a more-desirable dense shape.

Desublimation in a Heat Exchanger

This technique is used in industry to recover, from a gas mixture, a number of organic chemicals, including anthracene, maleic anhydride, naphthalene, phthalic anhydride, and salicylic acid. The crystals are deposited on the outside of the tubes while a coolant flows through the inside of the tubes. If the rate of desublimation is controlled by the rate of conduction of heat through the deposited crystal layer, then a relationship for the time to deposit a crystal layer of given thickness is derived in a manner similar to that for the falling-film melt crystallizer in Section 17.8. The result, which is similar to (17-67), except that the crystal layer grows outward from the outside radius of the cylindrical tube, r_o, instead of inward from radius, r_i in (17-67), is

$$ t = \frac{\rho_c \Delta H_s}{k_c(T_g - T_c)} \left[\frac{r_s^2}{2} \ln\left(\frac{r_s}{r_o}\right) - \frac{1}{4}\left(r_s^2 - r_a^2\right) \right] \quad (17\text{-}75) $$

where r_s is the radius to the crystal layer-gas interface. Equation (17-75) ignores any sensible heat associated with cooling of the gas. If the sensible heat is not negligible, then it must be added in (17-75) to the heat of sublimation, ΔH_s. If the heat-transfer resistance in the gas phase is negligible, the temperature, T_g, in (17-75) corresponds to the temperature where the partial pressure of the desubliming solute equals its vapor pressure. The derivation of (17-75) including the sensible-heat correction is given in detail by Singh and Tawney [33].

EXAMPLE 17.15

A desublimation unit of the heat-exchanger type is to be sized for the recovery of 100 kg/h of naphthalene (N) from a gas stream, where the other components are noncondensable under the conditions of operation. The heat-exchanger tubes are 1 m long with an outside diameter of 2.5 cm. Tube spacing is such that N can build up to a maximum thickness of 1.25 cm. The gas enters the unit at 800 torr and 80°C with a mole fraction for N of 0.0095. The coolant is cooling water, which flows through the inside of the tubes, countercurrently to the flow of gas. The inlet and outlet temperatures of the cooling water are 25°C and 45°C, respectively. Determine the number of tubes needed and the time required to reach the maximum thickness, if the gas leaves at 60°C.

SOLUTION

Assume the properties

C_P of gas = 0.26 cal/g-°C

MW of naphthalene = 128.2

k_c of solid naphthalene = 1.5 cal/h-cm-°C

ΔH_s of naphthalene = 115 cal/g

ρ_c of solid naphthalene = 1.025 g/cm^3

When the maximum thickness of N is achieved, the amount of sublimate per tube, if uniform, is

$$m = \pi\left(r_s^2 - r_o^2\right)L\rho_c$$
$$= 3.14[(1.25 + 1.25)^2 - 1.25^2]100(1.025)$$
$$= 1,510 \text{ g} = 1.51 \text{ kg}$$

The entering gas has a partial pressure for N of 0.0095(800) = 7.6 torr. From Figure 17.29, the saturation temperature for this partial pressure is 79.7°C. This is just slightly below the entering-gas temperature of 80°C, which is less than the triple-point temperature. Therefore, N will condense as a solid. At the exit temperature of the gas, 60°C, the vapor pressure of N is 1.8 torr. Assuming saturation at the exit and no pressure drop, the mole fraction of N in the exit gas will be 1.8/800 = 0.00225. Thus, per mole of entering gas, 0.0073 moles of N will be condensed. If the gas is assumed to be nitrogen and N, then M_r, the ratio of the mass of gas mixture to the mass of N condensed, will be

$$M_r = \frac{28(0.9905) + 128.2(0.0095)}{(128.2)(0.0073)} = 30.94$$

The sensible heat plus the heat of fusion is

$$M_r C_{P_g}(T_{\text{in}} - T_{\text{out}})_g + \Delta H_s = 30.94(0.26)(80 - 60) + 115$$
$$= 276 \text{ cal/g}$$

Thus, for this example, with a very low mole fraction of the chemical undergoing desublimation, the sensible heat effect is large. From (17-75), the time to reach the maximum thickness of N, using the temperature driving force across the solid N of 35°C, is

$$t = \frac{(1.025)(276)}{(1.5)(35)}\left[\frac{(2.5)^2}{2}\ln\left(\frac{2.5}{1.25}\right) - \frac{1}{4}(2.5^2 - 1.25^2)\right]$$
$$= 5.37 \text{ h}$$

The number of tubes required is

$$\frac{(100)(5.37)}{1.51} = 356$$

17.10 EVAPORATION

Before crystallization of a solute from an aqueous solution takes place, it is customary to bring the concentration of the solute close to the solubility curve. This is accomplished by evaporation of the solvent in an evaporator. Such a device is also used to concentrate solutions even when the solute is not subsequently crystallized, e.g., solutions of sodium hydroxide. When the vapor formed is essentially pure, there is no mass-transfer resistance in the vapor phase. When the liquid phase is agitated, mass-transfer in the liquid phase is sufficiently rapid that the rate of evaporation of solvent can be determined by the rate of heat transfer from the heating medium, usually condensing steam, to the solution.

Evaporators differ in configuration and the degree of agitation of the liquid phase. The five most widely used continuous-flow evaporators are shown schematically in Figure 17.31. Their main characteristics are as follows:

(a) **Horizontal-tube evaporator**. This unit consists of a horizontal cylindrical vessel, equipped in the lower section with a horizontal bundle of tubes, inside of which steam condenses and outside of which the solution to be concentrated boils. The vapor produced leaves the surface of the liquid solution. Agitation is provided only by the movement of the bubbles formed. Therefore, this type of unit is only suitable for low-viscosity solutions that do not deposit scale on the heat-transfer surfaces.

(b) **Short-vertical-tube evaporator**. This unit differs significantly from the horizontal-tube evaporator. The tube bundle is arranged vertically, with the solution

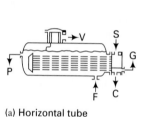

(a) Horizontal tube

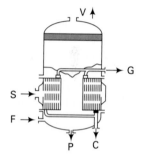

(b) Short vertical tube

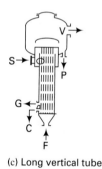

(c) Long vertical tube

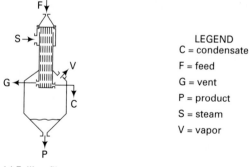

(d) Forced circulation

(e) Falling film

LEGEND
C = condensate
F = feed
G = vent
P = product
S = steam
V = vapor

Figure 17.31 Common types of evaporators.

[From *Perry's Chemical Engineers' Handbook*, 6th edition, R.H. Perry, D.W. Green, and J.O. Maloney, eds., McGraw-Hill, New York (1984) with permission.]

inside the tubes and steam condensing outside. Boiling inside the tube causes the solution to circulate, thus providing additional agitation and, therefore, higher heat-transfer coefficients. Nevertheless, this type of evaporator is not suitable for very viscous solutions.

(c) *Long-vertical-tube evaporator*. By lengthening the vertical tubes and providing a separate vapor-liquid disengagement chamber, as shown in Figure 17.31c, a higher tube-entering liquid velocity can be achieved and, thus, an even higher heat-transfer coefficient.

(d) *Forced-circulation evaporator*. To handle very viscous solutions, a pump is used to force the solution upward through relatively short tubes.

(e) *Falling-film evaporator*. This unit is widely used to concentrate heat-sensitive solutions such as fruit juices. The solution enters at the top and flows as a film down the inside walls of the tubes. The concentrate and the vapor produced are separated at the bottom.

For a given pressure in the vapor space of an evaporator, the boiling temperature of an aqueous solution will be equal to that of pure water if the solute is not dissolved in the water, but rather consists of small, insoluble, colloidal material. If the solute is soluble, the boiling temperature will be greater than that of pure water by an amount known as the

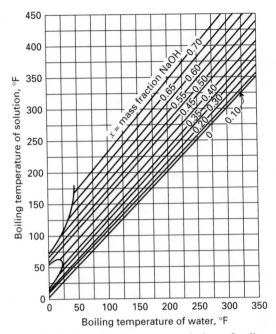

Figure 17.32 Dühring chart for aqueous solutions of sodium hydroxide.

[From W.L. McCabe, J.C. Smith, and P. Harriott, *Unit Operations of Chemical Engineering,* 5th ed., McGraw-Hill, New York (1993) with permission.]

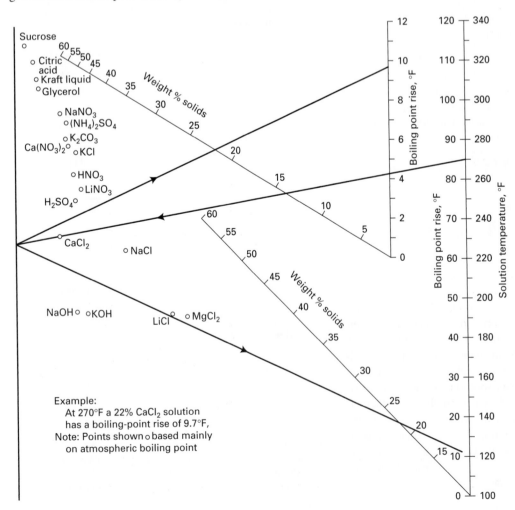

Example:
At 270°F a 22% CaCl₂ solution has a boiling-point rise of 9.7°F,
Note: Points shown ○ based mainly on atmospheric boiling point

Figure 17.33 Nomograph for boiling-point elevation of aqueous solutions.

[From *Perry's Chemical Engineers' Handbook*, 6th ed., R.H. Perry, D.W. Green, and J.O. Maloney, eds., McGraw-Hill, New York (1984) with permission.]

boiling-point elevation of the solution. If, as is usually the case, the solute has little or no vapor pressure, the evaporator pressure is equal to the partial pressure of the water in the solution. Then, by a modified Raoult's law,

$$P = p_{H_2O} = \gamma_{H_2O} x_{H_2O} P^s_{H_2O} \qquad (17\text{-}76)$$

Thus, for a given P and x_{H_2O}, a temperature can be determined to give the required $P^s_{H_2O}$ provided that γ_{H_2O} can be estimated. For solutions dilute in the solute, γ_{H_2O} approaches a value of 1.0. For concentrated solutions, the value of γ_{H_2O} can be estimated from correlations for electrolyte solutions as discussed by Reid, Prausnitz, and Poling [34].

Alternatively, the boiling temperature of the solution can be estimated by using a *Dühring-line chart* if it is available for the particular solute. Such a chart is shown in Figure 17.32 for the sodium hydroxide-water solution. The straight lines on this chart for different mass fractions of NaOH obey *Dühring's rule,* which states that as the pressure is increased, the boiling temperature of the solution increases linearly with the boiling temperature of the pure solvent. A nomograph for other solutes in water is given in Figure 17.33. The use of Figures 17.32 and 17.33, and (17-76) is illustrated in the following example.

EXAMPLE 17.16

An aqueous, NaOH solution is being evaporated at 6 psia. If the solution is 35-wt% NaOH, determine:

(a) The boiling temperature of the solution.

(b) The boiling-point elevation.

(c) The activity coefficient for water from (17-76).

SOLUTION

From steam tables, pure water has a vapor pressure of 6 psia at 170°F.

(a) From the Dühring-line chart of Figure 17.32, the boiling temperature of the solution is 207°F.

(b) The boiling-point elevation is:

$$207 - 170 = 37°F$$

Alternatively, the nomograph of Figure 17.33 may be used by drawing a straight line through the point for NaOH and a solution (boiling-point) temperature of 207°F. That line is extended to the left to the intersection with the leftmost vertical line. A straight line is then drawn from that intersection point through the lower, inclined line labeled weight percent solids at 35 (wt% NaOH). This second line is extended so as to intersect the right, vertical line. The value of the boiling-point elevation at this point of interaction is read as 36°F, which is close to the value of 37°F from the Dühring chart.

(c) The mole fraction of water in the solution is, for complete ionization of NaOH:

$$x_{H_2O} = \frac{0.65/18}{0.65/18 + 2(0.35)/40} = 0.674$$

Vapor pressure of pure water at 207°F = 13.3 psia. From (17-76),

$$\gamma_{H_2O} = \frac{6}{0.674(13.3)} = 0.67$$

Evaporator Model

The following mathematical model is widely used to make material balance, energy balance, and heat-transfer rate calculations to size evaporators operating under continuous-flow, steady-state conditions. The model is based on the schematic diagram in Figure 17.34. A so-called thin liquor at temperature, T_f, with weight-fraction solute, w_f, is fed to the evaporator at mass flow rate, m_f. A heating medium, e.g., saturated steam, is fed to the heat-exchanger tubes at temperature and pressure, T_s and P_s, and mass flow rate, m_s. Saturated condensate leaves the heat exchanger at the same temperature and pressure. Heat-transfer rate, Q, to the solution in the evaporator causes the solution in the evaporator at temperature, T_e, to partially evaporate to produce a vapor at temperature, T_v, with flow rate, m_v. The so-called thick-liquor concentrate leaves at temperature, T_p, with weight-fraction solute, w_p, at mass flow rate, m_p. The heat exchanger has a heat-transfer area, A, and overall heat-transfer coefficient, U, based on that area.

Key assumptions in formulating the mathematical model are:

1. The thin-liquor feed has only one volatile component, e.g., water.

2. Only the latent heat of the heating steam at T_s is available for heating and vaporizing the solution in the evaporator.

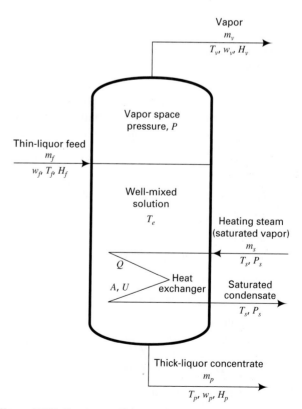

Figure 17.34 Continuous-flow, steady-state model for an evaporator.

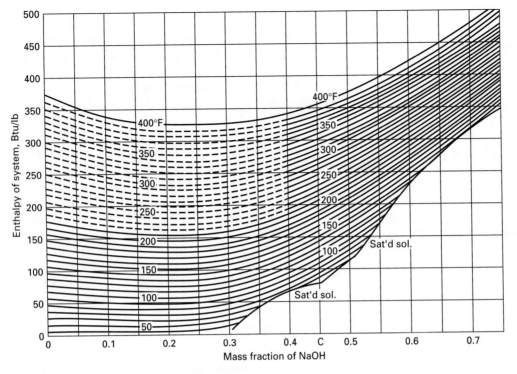

Figure 17.35 Enthalpy-concentration diagram for sodium hydroxide–water system.

[From W.L. McCabe, J.C. Smith, and P. Harriott, *Unit Operations of Chemical Engineering*, 5th ed., McGraw-Hill, New York (1993) with permission.]

3. The boiling action on the heat-exchanger surfaces agitates the solution, in the evaporator, sufficiently to achieve perfect mixing. Therefore, the temperature of the solution in the evaporator equals the exiting temperature of the thick-liquor concentrate. Thus, $T_e = T_p$. Also, $T_v = T_p$.

4. Because of Assumptions 2 and 3, the overall temperature driving force for heat transfer $= \Delta T = T_s - T_p$.

5. The ΔT is high enough to achieve nucleate boiling and not so high as to cause film boiling.

6. The exiting vapor temperature, $T_v = T_p = T_e$, corresponds to evaporator vapor-space pressure, P, taking into account the boiling-point elevation of the solution unless the solute is small, insoluble particles, such as colloidal matter.

7. No heat loss from the evaporator.

Based on the above assumptions, the mathematical model is as follows:

Total mass balance:

$$m_f = m_p + m_v \qquad (17\text{-}77)$$

Mass balance on the solute:

$$w_f m_f = w_p m_p \qquad (17\text{-}78)$$

Energy (enthalpy) balance on the solution:

$$Q = m_v H_v + m_p H_p - m_f H_f \qquad (17\text{-}79)$$

where H = enthalpy per unit mass

Energy (enthalpy) balance on the heating steam:

$$Q = m_s \Delta H^{\text{vap}} \qquad (17\text{-}80)$$

Heat-transfer rate:

$$Q = UA(T_s - T_p) \qquad (17\text{-}81)$$

The procedure used to solve these five equations depends upon the specifications of the problem. The following example is just one illustration. The solution of the energy-balance equations is greatly facilitated if an enthalpy-concentration diagram is available for the solute–solvent system. A diagram of this type for the NaOH–water system is given in Figure 17.35. In this diagram, the enthalpy datum for water is the pure liquid at 32°F, so that the diagram has the same datum as the steam tables. For NaOH, the datum is NaOH at infinite dilution in water at 20°C (68°F). Figure 17.35, together with the Dühring chart of Figure 17.32, were first prepared by McCabe [35].

EXAMPLE 17.17

An existing forced-circulation evaporator is to be used to concentrate 44 wt% NaOH to 65 wt% NaOH using steam at 3 atm gage (barometer reads 1 atm). The feed temperature will be 40°C and the pressure in the vapor space of the evaporator will be 2.0 psia. The evaporator has a heat-transfer area of 232 m². The overall heat-transfer coefficient for the given conditions is estimated to be 2,000 W/m²-°C. The density of the feed solution is 1,450 kg/m³. Neglect heat losses from the evaporator. Determine:

(a) The temperature of the solution in the evaporator in °C.

(b) The heating steam requirement in kg/h.

(c) The m³/h of feed solution that can be sent to the evaporator.

(d) The kg/h of concentrated NaOH solution leaving the evaporator.

(e) The rate of evaporation in kg/h.

SOLUTION

(a) At 2.0 psia, the boiling temperature of pure water is 126°F. From Figure 17.32, for 65 wt% NaOH, the solution boiling point is 240°F or 116°C. This is a considerable boiling-point elevation of 114°F.

(b) In American engineering units,

$$A = 232/0.0929 = 2,497 \text{ ft}^2$$
$$U = 2,000/5.674 = 352.5 \text{ Btu/h-ft}^2\text{-°F}$$
$$T_s = 291°F \text{ for 4 atm saturated steam}$$

The driving force for heat transfer is

$$\Delta T = T_s - T_p = 291 - 240 = 51°F$$

From (17-81),

$$Q = UA\,\Delta T = 352.5(2,497)(51) = 44,900,000 \text{ Btu/h}$$

The heat of vaporization of steam at 291°F is 917 Btu/lb. From (17-80),

$$m_s = Q/\Delta H^{\text{vap}} = 44,900,000/917 = 48,950 \text{ lb/h}$$

or

$$Q = 48,950(0.4536) = 22,200 \text{ kg/h}$$

It should be noted that the heat flux is $44,900,000/2,497 = 18,000 \text{ Btu/h-ft}^2$. This heat flux is safely in the nucleate-boiling region.

(c) From (17-79), using Figure 17.35 for NaOH solutions and the steam tables for water vapor, the energy balance on the solution is as follows, where the evaporated water is superheated steam at 240°F and 2.0 psia., and $m_v = m_f - m_p$:

$$44,900,000 = (m_f - m_p)(1168) + m_p(340) - m_f(115) \quad (1)$$

From (17-78),

$$0.44m_f = 0.65m_p \quad (2)$$

Substituting (2) into (1) to eliminate m_p, we obtain

$$m_f = \frac{44,900,000}{\left(1 - \dfrac{0.44}{0.65}\right)(1,168) + \left[\dfrac{0.44}{0.65}(340) - 115\right]}$$

$$= 91,170 \text{ lb/h}$$

or

$$m_f = (91,170)(0.4536) = 41,350 \text{ kg/h}$$

(d) From (2),

$$m_p = \frac{0.44}{0.65}(41,350) = 27,990 \text{ kg/h}$$

(e)
$$m_v = m_f - m_p = 41,350 - 27,990 = 13,360 \text{ kg/h}$$

Multiple-Effect Evaporator Systems

When condensing steam is used to evaporate water from an aqueous solution, the heat of condensation of the higher-temperature condensing steam is less than the heat of vaporization of the lower-temperature boiling water. Consequently, less than 1 kilogram of vapor is produced per kilogram condensation of heating steam. This ratio is called the *economy*. In Example 17.17, the economy is 13,360/22,200 = 0.602 or 60.2%. To reduce the amount of steam required and, thereby, increase the economy, a series of evaporators, called *effects*, can be used as shown in Figure 17.36. The increased economy is achieved by operating the effects

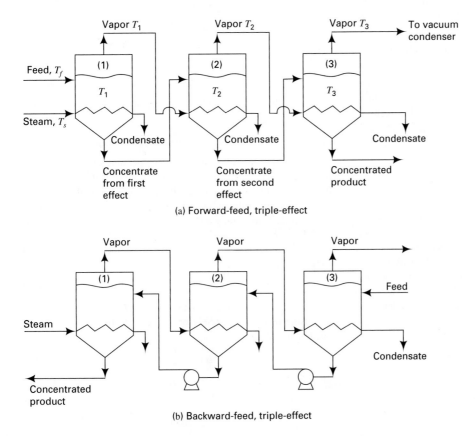

(a) Forward-feed, triple-effect

(b) Backward-feed, triple-effect

Figure 17.36 Multiple-effect evaporator systems.

at different pressures, and thus at different boiling temperatures, so that vapor produced in one effect can be condensed to supply the heat in another effect.

In Figure 17.36a, referred to as a *forward-feed, triple-effect-evaporator* system, approximately one-third of the total evaporation occurs in each effect. The fresh feed solution and steam both enter the first effect, which operates at pressure P_1. The concentrate from the first effect is sent to the second effect. The vapor produced in the first effect is also sent to the second effect where it condenses, giving up its heat of condensation to cause additional evaporation of the solution. To achieve a temperature-driving force for heat transfer in the second effect, the pressure of the second effect, P_2, is lower than that of the first effect. This procedure is repeated in the third effect. For three effects, the flow rate of steam entering the first is only about one-third of the amount of steam that would be required if only one effect were used. However, the temperature-driving force in each of the three effects is only about one-third of that in a single effect. Therefore, the heat transfer area of each of the three evaporators in a triple-effect system is approximately the same as for the one evaporator in a single-effect unit. Therefore, the savings in annual heating-steam cost must offset the additional capital investment for equipment.

When the temperature of the fresh feed is significantly below its saturation temperature corresponding to the pressure in the first effect, *backward-feed* operation is desirable, as shown in Figure 17.36b. The cold fresh feed is sent to the third effect, which operates at the lowest pressure and, therefore, the lowest temperature. Unlike the forward-feed system, pumps are required to move the concentrate from one effect to the next because $P_1 > P_2 > P_3$. However, unlike gas compressors, liquid pumps are not high-cost items.

Calculations for multieffect evaporator systems involve the same types of mass-balance, energy-balance, and heat-transfer equations as for a single-effect system. These equations are usually solved by an iterative method, especially when boiling-point elevations occur. The particular iterative procedure depends on the problem specifications. The following example illustrates a typical procedure.

EXAMPLE 17.18

A feed of 44,090 lb/h of an aqueous solution containing 8 wt% colloids is to be concentrated to 45 wt% colloids in a triple-effect-evaporator system using forward feed. The feed enters the first effect at 125°F, and the third effect operates at 1.94 psia in the vapor space. Fresh saturated steam at 29.3 psia is used for heating in the first effect. The specific heat of the colloids can be assumed constant at 0.48 Btu/lb-°F. Overall heat-transfer coefficients are estimated to be:

Effect	U, Btu/h-ft²-°F
1	350
2	420
3	490

If the heat-transfer areas of each of the three effects are to be equal, determine:

Evaporation temperatures, T_1 and T_2, in the first two effects.

Heating steam flow rate, m_s.

Solution flow rates, m_1, m_2, and m_3 leaving the three effects.

SOLUTION

The unknowns, which number seven, are $A(= A_1 = A_2 = A_3)$, T_1, T_2, m_s, m_1, m_2, and m_3. Therefore, seven independent equations are needed. Because the solute is colloids (insolubles), boiling-point elevations do not occur.

It is convenient to add three additional unknowns and, therefore, three additional equations, making a total of 10 equations. The added unknowns are the heat-transfer rates, Q_1, Q_2, and Q_3 in the three effects. The 10 equations, which are similar to (17-78) to (17-81), are

Overall colloid mass balance

$$w_f m_f = w_3 m_3 \tag{1}$$

Energy balances on the solutions

$$Q_1 = (m_f - m_1)H_{v_1} + m_1 H_1 - m_f H_f \tag{2}$$
$$Q_2 = (m_1 - m_2)H_{v_2} + m_2 H_2 - m_1 H_1 \tag{3}$$
$$Q_3 = (m_2 - m_3)H_{v_3} + m_3 H_3 - m_2 H_2 \tag{4}$$

Energy balances on steam and water vapors

$$Q_1 = m_s \Delta H_s^{vap} \tag{5}$$
$$Q_2 = (m_f - m_1)\Delta H_2^{vap} \tag{6}$$
$$Q_3 = (m_1 - m_2)\Delta H_3^{vap} \tag{7}$$

Heat-transfer rates

$$Q_1 = U_1 A_1 (T_s - T_1) = U_1 A_1 \Delta T_1 \tag{8}$$
$$Q_2 = U_2 A_2 (T_1 - T_2) = U_2 A_2 \Delta T_2 \tag{9}$$
$$Q_3 = U_3 A_3 (T_2 - T_3) = U_3 A_3 \Delta T_3 \tag{10}$$

From (1),

$$m_3 = (w_f/w_3)m_f = (0.08/0.45)(44,090) = 7,838 \text{ lb/h}$$

Also, the flow rate of colloids in the feed is $(0.08)(44,090) = 3,527$ lb/h.

Initial estimates of solution temperature in each effect:

With no boiling-point elevation, the temperature of the solution in the third effect is the saturation temperature of water at the specified pressure of 1.94 psia or 125°F. The temperature of the heating steam entering the first effect is the saturation temperature of 249°F at 29.3 psia. Thus, if only one effect were used, the temperature driving force for heat transfer, ΔT, would be $249 - 125 = 124$°F. With no boiling-point elevations, the ΔTs for the three effects must sum to the value for one effect. Thus,

$$\Delta T_1 + \Delta T_2 + \Delta T_3 = 124°F \tag{11}$$

As a first approximation, assume that the ΔTs for the three effects are, using (8)–(10), inversely proportional to the given values of U_1, U_2, and U_3. Thus,

$$\Delta T_1 = \frac{U_3}{U_1}\Delta T_3 \tag{12}$$

$$\Delta T_2 = \frac{U_3}{U_2}\Delta T_3 \tag{13}$$

Solving (11), (12), and (13), we obtain

$$\Delta T_1 = 48.6°F$$
$$\Delta T_2 = 40.6°F$$
$$\Delta T_3 = 34.8°F$$
$$T_1 = T_s - \Delta T_1 = 249 - 48.6 = 200.4°F$$
$$T_2 = T_1 - \Delta T_2 = 200.4 - 40.6 = 159.8°F$$
$$T_3 = T_2 - \Delta T_3 = 159.8 - 34.8 = 125°F$$

Initial Estimates of m_1 and m_2:

The total evaporation rate for the three effects is $m_f - m_3 = 44,090 - 7,838 = 36,252$ lb/h. Assume, as a first approximation, equal amounts of vapor produced in each effect. Then,

$$m_f - m_1 = 36,252/3 = 12,084 \text{ lb/h}$$
$$m_1 = 44,090 - 12,084 = 32,006 \text{ lb/h}$$
$$m_2 = 32,006 - 12,084 = 19,922 \text{ lb/h}$$
$$m_3 = 19,922 - 12,084 = 7,838 \text{ lb/h}$$

Corresponding estimates of the mass fractions of colloids are

$$w_1 = 3,527/32,006 = 0.110$$
$$w_2 = 3,527/19,922 = 0.177$$
$$w_3 = 3,527/7,838 = 0.450 \text{ (given)}$$

The remaining calculations are iterative in nature to obtain corrected values of T_1, T_2, m_1, and m_2, as well as values of A, m_s, Q_1, Q_2, and Q_3. These calculations are best carried out on a spreadsheet. Each iteration consists of the following steps:

Step 1

Combine (2) through (7) to eliminate Q_1, Q_2, and Q_3. Using the approximations for T_1, T_2, T_3, w_1, and w_2, the specific enthalpies for the resulting equations are calculated and the equations are solved for new approximations of m_s, m_1, and m_2. Corresponding approximations for w_1 and w_2 are computed.

For the first iteration, the enthalpy values are

$$\Delta H_s^{vap} = 946.2 \text{ Btu/lb}$$
$$\Delta H_2^{vap} = 977.6 \text{ Btu/lb}$$
$$\Delta H_3^{vap} = 1,002.6 \text{ Btu/lb}$$
$$H_{v_1} = 1,146 \text{ Btu/lb}$$
$$H_{v_2} = 1,130 \text{ Btu/lb}$$
$$H_{v_3} = 1,116 \text{ Btu/lb}$$
$$H_f = 0.92(92.9) + 0.08(0.48)(125 - 32) = 89.0 \text{ Btu/lb}$$
$$H_1 = 0.89(168.4) + 0.110(0.48)(200.4 - 32) = 158.8 \text{ Btu/lb}$$
$$H_2 = 0.823(127.7) + 0.177(0.48)(159.8 - 32) = 116.0 \text{ Btu/lb}$$
$$H_3 = 0.55(92.9) + 0.45(0.48)(125 - 32) = 71.2 \text{ Btu/lb}$$

When these enthalpy values are substituted into the combined energy balances, the following equations are obtained:

$$m_s = 49,250 - 1.043 m_1 \tag{14}$$
$$44,090 = 1.994 m_1 - 1.037 m_2 \tag{15}$$
$$8,168 = 1.997 m_2 - m_1 \tag{16}$$

Solving (14), (15), and (16),

$$m_s = 15,070 \text{ lb/h}$$
$$m_1 = 32,770 \text{ lb/h}$$
$$m_2 = 20,500 \text{ lb/h}$$

Corresponding values of colloid mass fractions are

$$w_1 = 0.108$$
$$w_2 = 0.172$$

It may be noted that these values of m_1, m_2, w_1, and w_2 are close to the first approximations. This is often the case.

Step 2

Using the values computed in Step 1, values of Q are determined from (5), (6), and (7); and values of A are determined from (8), (9), and (10).

$$Q_1 = 15,070(946.2) = 14,260,000 \text{ Btu/h}$$
$$Q_2 = (44,090 - 32,770)(977.6) = 11,070,000 \text{ Btu/h}$$
$$Q_3 = (32,770 - 20,500)(1,002.6) = 12,400,000 \text{ Btu/h}$$

$$A_1 = \frac{14,260,000}{(350)(48.6)} = 838 \text{ ft}^2$$

$$A_2 = \frac{11,070,000}{(420)(40.6)} = 649 \text{ ft}^2$$

$$A_3 = \frac{12,400,000}{(490)(34.8)} = 727 \text{ ft}^2$$

Step 3

Because the three areas are not equal, calculate the arithmetic-average, heat-transfer area and a new set of ΔT driving forces from (8), (9), and (10). Normalize these values so they sum to the overall ΔT (124°F in this example). From the ΔT values, compute T_1 and T_2.

$$A_{avg} = \frac{838 + 649 + 727}{3} = 738 \text{ ft}^2$$

$$\Delta T_1 = \frac{14,260,000}{(350)(738)} = 55.2°F$$

$$\Delta T_2 = \frac{11,070,000}{(420)(738)} = 35.7°F$$

$$\Delta T_3 = \frac{12,400,000}{(490)(738)} = 34.3°F$$

These values sum to 125.2°F. Therefore, they are normalized to

$$\Delta T_1 = 55.2(124/125.2) = 54.7°F$$
$$\Delta T_2 = 35.7(124/125.2) = 35.3°F$$
$$\Delta T_3 = 34.3(124/125.2) = 34.0°F$$
$$T_1 = 249 - 54.7 = 194.3°F$$
$$T_2 = 194.3 - 35.3 = 159.0°F$$

Steps 1 through 3 are now repeated using the new values of T_1 and T_2 from Step 3 and the new values of w_1 and w_2 from Step 1. The iterations are continued until the values of the unknowns no longer change significantly and $A_1 = A_2 = A_3$. The subsequent iterations for this example are left as an exercise. Based on the results of the first iteration, the economy of the three-effect system is

$$\frac{44,090 - 7,838}{15,070} = 2.41 \quad \text{or} \quad 241\%$$

Overall Heat-Transfer Coefficients in Evaporators

In an evaporator, the overall heat-transfer coefficient, U, depends mainly on the steamside, condensening coefficient, the solution-side coefficient, and a scale or fouling resistance on the solution side. The conduction resistance of the metal wall of the heat-exchanger tubes is usually negligible. Steam condensation is generally of the film, rather than dropwise, type. When boiling occurs on the surfaces of the heat-exchanger tubes, it is of the nucleate-boiling, rather than film-boiling, regime. In the absence of boiling on the tube surfaces, heat transfer is by forced convection to the solution. Local coefficients for film condensation, nucleate boiling, and forced convection of aqueous solutions are all relatively large, of the order of 1,000 Btu/h-ft^2-°F (5,700 W/m^2-K). Thus, the overall coefficient would be about one-half of this. However, when fouling occurs, the overall coefficient can be substantially less. Table 17.11, taken from Geankoplis [36], lists ranges of overall heat-transfer coefficients for different types of evaporators. The higher coefficients in forced-circulation evaporators are mainly a consequence of the greatly reduced fouling due to the high liquid velocity in the tubes.

Table 17.11   Typical Heat-Transfer Coefficients in Evaporators

Type Evaporator	U	
	Btu/h-ft^2-°F	W/m^2-K
Horizontal-tube	200–500	1,100–2,800
Short-tube vertical	200–500	1,100–2,800
Long-tube vertical	200–700	1,100–3,900
Forced circulation	400–2,000	2,300–11,300

SUMMARY

1. Crystallization involves the formation of solid crystalline particles from a homogeneous fluid phase. However, if the fluid is a gas, the process is usually referred to as desublimation.

2. In crystalline solids, as opposed to amorphous solids, the molecules, atoms, and/or ions are arranged in a regular lattice pattern. When crystals grow unhindered, they form polyhedrons with flat sides and sharp corners. Although the faces of a crystal may grow at different rates, referred to as crystal habit, the Law of Constant Interfacial Angles restricts the angles between corresponding faces to be constant. Crystals can form only seven different crystal systems, which include 14 different space lattices. Because of crystal habit, a given crystal system can take on different shapes, e.g., plates, needles, and prisms, but not spheres.

3. Crystal-size distributions can be determined or formulated in terms of differential or cumulative analyses, which are convertible, one from the other. A number of different, mean-particle sizes can be derived from crystal-size distribution data.

4. The most important thermodynamic properties for crystallization calculations are melting point, solubility, heat of fusion, heat of crystallization, heat of solution, heat of transition, and supersaturation. Solubilities of inorganic salts in water can vary widely from a negligible value to a concentration of greater than 50 wt%. Many salts crystallize in hydrated forms, with the number of waters of crystallization of the stable hydrate depending upon temperature. The solubility of sparingly soluble compounds is usually expressed in terms of a solubility product. When available, phase diagrams and enthalpy-concentration diagrams are extremely useful for making material- and energy-balance calculations.

5. Crystals smaller in size than can be seen by the naked eye (<20 mm) are more soluble than the normally listed solubility. Supersaturation ratio for a given crystal size is the ratio of the actual solubility of a small-size crystal to the solubility of larger crystals that can be seen by the naked eye. The driving force for nucleation and growth of crystals is supersaturation.

6. Primary nucleation, which requires a high degree of supersaturation, occurs in systems free of crystalline surfaces, and can be homogeneous or heterogeneous. Secondary nucleation occurs when crystalline surfaces are present. Crystal growth involves the mass transfer of the solute up to the crystal surface followed by incorporation of the solute into the crystal-lattice structure.

7. Equipment for solution crystallization can be classified according to operation mode (batch or continuous), method for achieving supersaturation (cooling or evaporation), and features for achieving desired crystal growth (e.g., agitation, baffles, circulation, and classification). Of primary importance is the effect of temperature on solubility. Three of the most widely used types of equipment for solution crystallization are: (1) batch crystallizer with external or internal circulation, (2) continuous, cooling crystallizer, and (3) continuous, vacuum, evaporating crystallizer.

8. The MSMPR crystallization model is widely used to simulate the often used continuous, vacuum, evaporating draft-tube, baffled crystallizer. Some of the assumptions of this model are perfect mixing of the magma, no classification of crystals, uniform degree of supersaturation throughout the magma, crystal growth rate independent of crystal size, no crystals in the feed, no crystal breakage, uniform temperature, equilibrium in product magma between mother liquor and crystals, constant and uniform nucleation rate due to secondary nucleation by crystal contact; uniform crystal-size distribution, and all crystals with the same shape.

9. For a specified crystallizer feed, magma density, magma residence time, and predominant crystal size, the MSMPR model can predict the required nucleation rate and crystal-growth rate, number of crystals produced per unit time, and crystal-size distribution.

10. Precipitation, leading to very small crystals, occurs with solutes that are only sparingly soluble. The precipitate is often produced by reactive crystallization from the addition of two soluble salt solutions, producing one soluble and one insoluble salt. Unlike solution crystallization, which takes place at a low degree of supersaturation, precipitation occurs at a very high supersaturation that results in very small crystals.

11. When both components of a mixture can be melted at reasonable temperatures (e.g., certain mixtures of organic compounds), melt crystallization can be used to separate the components. If the components form a eutectic mixture, pure crystals of one of the components can be formed. If the components form a solid solution, repeated stages of melting and crystallization are required to achieve high purity.

12. A large number of crystallizer designs have been proposed for melt crystallization. The two major methods are suspension crystallization and layer crystallization. Of particular importance is the falling-film crystallizer, which can be designed and operated for high production rates when the components form eutectic mixtures. For components that form solid solutions, the zone-melting technique developed by Pfann can be employed to produce nearly pure compounds.

13. A number of chemicals are amenable to purification by desublimation, preceded perhaps by sublimation. Desublimation is almost always achieved by cooling the gas mixture at constant pressure. The cooling can be accomplished by heat transfer, quenching with a vaporizable liquid, or quenching with a cold, noncondensible gas.

14. Evaporation can be used to concentrate a solute prior to solution crystallization. Common evaporators include the horizontal-tube unit, short-vertical-tube unit, long-vertical-tube unit, forced-circulation unit, and the falling-film unit. For a given evaporation pressure, the presence of the solution can cause a boiling-point elevation.

15. The most widely used evaporator model assumes that the liquor being evaporated is well-mixed such that the temperature and solute concentration are uniform and at exiting conditions.

16. The economy of an evaporator is defined as the mass ratio of water evaporated to heating steam required. The economy can be increased by using multiple evaporator effects that operate at different pressures such that vapor produced in one effect can be used as heating steam in a subsequent effect. The solution being evaporated can progress through the effects in forward, backward, or mixed directions.

17. Evaporators typically operate so that solutions are in the nucleate-boiling regime. Overall, heat-transfer coefficients are generally high because boiling occurs on one side and condensation on the other side.

REFERENCES

1. MULLIN, J.W., *Crystallization,* 3rd ed., Butterworth-Heinemann, Boston (1993).

2. GRABER, S., T.A., and M.E., TABOADA, M., *Chem. Eng. Ed.,* **25,** 102–105 (1991).

3. HOUGEN, O.A., K.M. WATSON, and R.H., RAGATZ, *Chemical Process Principles, Part I, Material and Energy Balances,* 2nd ed., John Wiley & Sons (1954).

4. MIERS, H.A., and F. ISAAC, *Proc. Roy. Soc.,* **A79,** 322–351 (1907).

5. NIELSEN, A.E., *Kinetics of Precipitation,* Pergamon Press, New York (1964).

6. NOYES, A.A., and W.R. WHITNEY, *J. Am. Chem. Soc.,* **19,** 930–934 (1897).

7. NERNST, W., *Zeit. für Physik. Chem.,* **47,** 52–55 (1904).

8. MIERS, H.A., *Phil. Trans.,* **A202,** 492–515 (1904).

9. VALETON, J.J.P., *Zeit. für Kristallographie,* **59,** 483 (1924).

10. MYERSON, A.S., Ed., *Handbook of Industrial Crystallization,* Butterworth-Heinemann, Boston (1993).

11. BURTON, W.K., N. CABRERA, and F.C. FRANK, *Phil. Trans.,* **A243,** 299–358 (1951).

12. SEAVOY, G.E., and H.B. CALDWELL, *Ind. Eng. Chem.,* **32,** 627–636 (1940).

13. NEWMAN, H.H., and R.C. BENNETT, *Chem. Eng. Prog.,* **55** (3), 65–70 (1959).

14. RANDOLPH, A.D., *AIChE Journal,* **11,** 424–430 (1965).

15. RANDOLPH, A.D., and M.A. LARSON, *Theory of Particulate Processes,* 2nd ed., Academic Press, New York (1988).

16. McCABE, W.L., *Ind. Eng. Chem.* **21,** 30–33 and 112–19 (1929).

17. ZUMSTEIN, R.C., and R.W. ROUSSEAU, *AIChE Symp. Ser.,* **83** (253), 130 (1987).

18. NIELSEN, A.E., *Kinetics of Precipitation,* Pergamon Press, Oxford, England (1964).

19. NIELSEN, A.E., Chapter 27 in *Treatise on Analytical Chemistry, Part 1,* Volume 3, 2nd ed., Editors I.M. Kolthoff and P.J. Elving, John Wiley & Sons, New York (1983).

20. NIELSEN, A.E., *J. Crys. Gr.,* **67,** 289–310 (1984).

21. FITCHETT, D.E., and J.M. TARBELL, *AIChE J.,* **36,** 511–522 (1990).

22. MATSUOKA, M., M. OHISHI, A. SUMITANI, and K. OHORI, World Congress III of Chemical Engineers, Tokyo, Sept. 21-325, 1986, pp. 980–983.

23. WILCOX, W.R., *Ind. Eng. Chem.,* **60** (3), 13–23 (1968).

24. WYNN, N., *Chemical Engineering,* **98** (7) 149–154 (1991).

25. PFANN, W.G., *Trans. AIME.,* **194,** 747 (1952).

26. PFANN, W.G., *Zone Melting,* 2nd ed., John Wiley and Sons, New York (1966).

27. ZIEF, M., and W.R. WILCOX, *Fractional Solidification,* Marcel Dekker, New York (1967).

28. BURRIS, JR., L., C.H. STOCKMAN, and I.G. DILLON, *Trans. AIME,* **203,** 1017 (1955).

29. HERINGTON, E.F.G., *Zone Melting of Organic Compounds,* John Wiley & Sons, New York (1963).

30. NORD, M., *Chem. Eng.,* **58** (9), 157–166 (1951).

31. KUDELA, L., and M.J. SAMPSON, *Chem. Eng.,* **93** (12), 93–98 (1986).

32. HOLDEN, C.A., and H.S. BRYANT, *Sep. Sci.,* **4** (1), 1–13 (1969).

33. SINGH, N.M., and R.K. TAWNEY, *Ind. J. Tech.,* **9,** 445–447 (1971).

34. POLING, B.E., J.M. PRAUSNITZ, and J.P. O'CONNELL, *The Properties of Gases and Liquids,* 5th ed., McGraw-Hill Book Co., New York (2001), p. 8.191.

35. McCABE, W.L., *Trans. AIChE,* **31,** 129–164 (1935).

36. GEANKOPLIS, C.J., *Transport Processes and Unit Operations,* 3rd ed., Prentice Hall, Englewood Cliffs, NJ (1993).

EXERCISES

Section 17.1

17.1 Estimate the sphericities of the following simple particle shapes:

(a) a cylindrical needle with a height, H, equal to 5 times the diameter, D

(b) a rectangular prism of sides a, $2a$, and $3a$

17.2 A certain circular plate of diameter, D, and thickness, t, has a sphericity of 0.594. What is the ratio of t to D?

17.3 A laboratory screen analysis for a batch of crystals of hypo (sodium thiosulfate) is as follows. Prepare both differential and cumulative-undersize plots of the data, using a spreadsheet.

U.S. Screen	Mass Retained, gm
6	0.0
8	8.8
12	21.3
16	138.2
20	211.6
30	161.7
40	81.6
50	44.1
70	28.7
100	13.2
140	9.6
170	8.8
230	7.4
	735.0

In preparing your plots, determine whether arithmetic, semilog, or log-log plots are preferred.

17.4 Derive expressions for the surface-mean and mass-mean diameter from a particle-size analysis based on counting, rather than weighing, particles in given size ranges, letting N_i be the number of particles in a given size range of average diameter, $\bar{D}_{p_i}$.

17.5 Using the screen analysis of Exercise 17.3, calculate, with a spreadsheet, the surface-mean, mass-mean, arithmetic-mean, and volume-mean crystal diameters, assuming that all particles have the same sphericity and volume shape factor.

17.6 A precipitation process for producing perfect spheres of silica has been developed. The individual particles are so small that most cannot be discerned by the naked eye. Using optical microscopy, the particle size distribution has been measured, with results given in the table below. Using these data on a spreadsheet program:

(a) Produce plots of the differential and cumulative particle-size analyses

(b) Determine:

(1) surface-mean diameter

(2) arithmetic-mean diameter

(3) mass-mean diameter

(4) volume-mean diameter

PSD of Silica Spheres

Particle-Size Interval, μm	Number of Particles
1.0–1.4	2
1.4–2.0	5
2.0–2.8	14

PSD of Silica Spheres

Particle-Size Interval, μm	Number of Particles
2.8–4.0	60
4.0–5.6	100
5.6–8.0	190
8.0–12.0	250
12.0–16.0	160
16.0–22.0	110
22.0–30.0	70
30.0–42.0	28
42.0–60.0	10
60.0–84.0	1
Total	1,000

17.7 A screen analysis for a sample of glauber's salt from a commercial crystallizer is as follows, where the crystals can be assumed to have a uniform sphericity and volume shape factor.

U.S. Screen	Mass Retained, gm
14	0.0
16	0.9
18	25.4
20	111.2
25	113.9
30	225.9
35	171.7
40	116.5
45	55.1
50	31.5
60	8.7
70	10.5
80	4.4
	875.7

Use a spreadsheet to determine in microns:

(a) a plot of the differential analysis

(b) a plot of the cumulative oversize analysis

(c) a plot of the cumulative undersize analysis

(d) the surface-mean diameter

(e) the mass-mean diameter

(f) the arithmetic-mean diameter

(g) the volume-mean diameter

Section 17.2

17.8 1,000 grams of water is mixed with 50 grams of Ag_2CO_3 and 100 grams of AgCl. At equilibrium at 25°C, calculate the concentrations in moles/liter of Ag^+, Cl^-, and CO_3^- ions and the grams of Ag_2CO_3 and AgCl in the solid phases.

17.9 5,000 lb/h of a saturated aqueous solution of $(NH_4)_2SO_4$ at 80°C is cooled to 30°C. At equilibrium, what is the amount of crystals formed in lb/h. If during the cooling process, 50% of the water is evaporated, what is the amount of crystals formed in lb/h?

17.10 7,500 lb/h of a 50 wt% aqueous solution of $FeCl_3$ at 100°C is cooled to 20°C. At 100°C, the solubility of the $FeCl_3$ is 540 g/100 g of water. At 20°C, the solubility is 91.8 g/100 g water and crystals of $FeCl_3$ are the hexahydrate. At equilibrium at 20°C, determine the lb/h of crystals formed.

17.11 The concentrate from an evaporation system is 5,870 lb/h of 35 wt% $MgSO_4$ at 180°F and 25 psia. It is mixed with 10,500 lb/h of saturated aqueous recycle filtrate of $MgSO_4$ at 80°F and 25 psia. The mixture is sent to a vacuum crystallizer, operating at 85°F and 0.58 psia in the vapor space, to produce steam and a magma of 25 wt% crystals and 75 wt% saturated solution. Determine the lb/h of water evaporated and the maximum production rate of crystals in tons/day (dry basis for 2000 lb/ton).

17.12 Urea is to be crystallized from an aqueous solution that is 90% saturated at 100°C. If 90% of the urea is to be crystallized in the anhydrous form and the final solution temperature is to be 30°C, what fraction of the water must be evaporated?

17.13 In Examples 17.3 and 17.5, heat addition to the crystallizer is by an external heat exchanger through which magma is circulated, as shown in Figure 17.16. If instead the heat is added to the feed, determine the new feed temperature. Which is the preferable way to add the heat?

17.14 For the conditions of Exercise 17.11, determine the rate at which heat must be added to the system.

17.15 For the conditions of Example 17.4, calculate the amount of heat in calories/100 grams of water that must be removed to cool the solution from 100 to 10.6°C.

Section 17.3

17.16 Based on the following data, compare the effect of crystal size on solubility in water at 25°C for (1) KCl (see Example 17.7), a soluble inorganic salt, with that for (2) $BaSO_4$, an almost insoluble inorganic salt, and (3) sucrose, a very soluble organic compound.

$$\sigma_{s,L} \text{ for barium sulfate} = 0.13 \text{ J/m}^2$$

$$\sigma_{s,L} \text{ for sucrose} = 0.01 \text{ J/m}^2$$

What conclusions can you draw from the results?

17.17 Determine the supersaturation ratio, S, required to permit 0.5-μm-diameter crystals of sucrose (MW = 342 and $\rho_c = 1,590$ kg/m³) to grow if $\sigma_{s,L} = 0.01$ J/m².

17.18 The Kelvin equation, (17-16), predicts that solubility increases to infinity as the crystal diameter decreases to zero. However, measurements by L. Harbury [*J. Phys. Chem.*, **50**, 190–199 (1946)] for several inorganic salts in water show a maximum in the solubility curve and a solubility that approaches zero as crystal size is reduced to zero. Harbury's explanation is that the surface energy of the crystals depends not only on interfacial tension, but also on surface electrical charge, given by

$$2q^2 v_s / \pi \kappa RT D_p^4$$

where

q = electrical charge on the crystal

κ = dielectric constant

Modify (17-16) to take into account electrical charge. Make sure your equation predicts a maximum.

17.19 Using the following data, compare the effect of supersaturation ratio over the range of 1.005 to 1.02 on the primary homogeneous nucleation of $AgNO_3$, $NaNO_3$, and KNO_3 from aqueous solutions at 25°C:

	AgNO₃	NaNO₃	KNO₃
Crystal density, g/cm³	4.35	2.26	2.11
Interfacial tension, J/m²	0.0025	0.0015	0.0030

17.20 Estimate the effect of relative supersaturation on the primary, homogeneous nucleation of $BaSO_4$ from an aqueous solution at 25°C, if

$$\text{Crystal density} = 4.50 \text{ g/cm}^3$$

$$\text{Interfacial tension} = 0.12 \text{ J/m}^2$$

17.21 Repeat parts (g) and (i) of Example 17.9 if the solution velocity past the crystal face is reduced from 5 cm/s to 1 cm/s.

Section 17.4

17.22 The feed to a cooling crystallizer is 2,000 kg/h of 30 wt% Na_2SO_4 in water at 40°C. This solution is to be cooled to a temperature at which 50% of the solute will be crystallized as the decahydrate. Estimate the required heat-transfer area in m² if an overall heat-transfer coefficient of 15 Btu/h-ft²-°F can be achieved. Assume a constant specific heat for the aqueous solution of 0.80 cal/g-°C. Chilled cooling water will flow countercurrently to the crystallizing solution, entering the crystallizer at 10°C, and exiting at a temperature sufficient to give a log-mean driving force of at least 10°C.

17.23 Two tons per hour of the dodecahydrate of sodium phosphate ($Na_3PO_4 \cdot 12H_2O$) is to be crystallized by cooling, in a cooling crystallizer, an aqueous solution that enters saturated at 40°C and leaves at 20°C. Chilled cooling water flows countercurrently, entering at 10°C and exiting at 25°C. The expected overall heat-transfer coefficient is 20 Btu/h-ft²-°F. The average specific heat of the solution is 0.80 cal/g-°C. Estimate:

(a) The tons (2,000 lb) per hour of feed solution.

(b) The heat-transfer area in ft².

(c) The number of crystallizer units required if each 10-ft-long unit contains 30 ft² of heat-transfer surface.

Section 17.5

17.24 An aqueous feed of 10,000 kg/h, saturated with $BaCl_2$ at 100°C, enters a crystallizer that can be simulated with the MSMPR model. However, crystallization is achieved with negligible evaporation. The magma leaves the crystallizer at 20°C with crystals of the dihydrate. The crystallizer has a volume (vapor-space free basis) of 2.0 m³. From laboratory experiments, the crystal growth rate is essentially constant at 4.0×10^{-7} m/s. Using the data below, determine:

(a) The kg/h of crystals in the magma product.

(b) The predominant crystal size in mm.

(c) The mass fraction of crystals in the size range from U.S. Standard 20 mesh to 25 mesh.

Data:

Density of the dihydrate crystals = 3.097 g/cm³

Density of an aqueous, saturated solution of barium chloride at 20°C = 1.29 g/cm³

17.25 The feed to a continuous crystallizer that can be simulated with the MSMPR model is 5,000 kg/h of 40 wt% sodium acetate in water. Monoclinic crystals of the trihydrate will be formed. The pressure in the crystallizer and the heat-transfer rate in the associated heat exchanger are such that 20% of the water in the feed will be evaporated at a crystallizer temperature of 40°C. The crystal growth rate, G, is 0.0002 m/h and a predominant crystal size, L_{pd}, of 20 mesh is desired.

Determine:

(a) The kg/h of crystals in the exiting magma.

(b) The kg/h of mother liquor in the exiting magma.

(c) The volume in m³ of magma in the crystallizer if

density of the crystals = 1.45 g/cm³

density of the mother liquor = 1.20 g/cm³

Solubility data:

T, °C	Solubility, g sodium acetate/100 g H$_2$O
30	54.5
40	65.5
60	139

17.26 An MSMPR-type crystallizer is to be designed to produce 2,000 lb/h of crystals of the heptahydrate of magnesium sulfate with a predominant crystal size of 35 mesh. The magma will be 15 vol% crystals. The temperature in the crystallizer will be 50°C and the residence time will be 2 h. The densities of the crystals and mother liquor are 1.68 and 1.32 g/cm³, respectively. Determine:

(a) The exiting flow rates in cubic feet per hour of

Crystals

Mother liquor

Magma

(b) The crystallizer volume in gallons, if the vapor space equals the magma space.

(c) The approximate dimensions in feet of the crystallizer, if the body is cylindrical with a height equal to twice the diameter.

(d) The required crystal growth rate in feet per hour.

(e) The necessary nucleation rate in nuclei per hour per cubic feet of mother liquor in the crystallizer.

(f) The number of crystals produced per hour.

(g) A screen analysis table covering a U.S. mesh range of 3-1/2 to 200, giving the predicted % cumulative and % differential screen analyses of the product crystals.

(h) Plots of the screen analyses predicted in part (g).

Section 17.6

17.27 Refer to Example 17.12. In Run 15, Fitchett and Tarbell also made measurements of number density of crystals at 200 rpm, for which the data can be fitted well by the equation

$$\ln n = 26.3 - 0.407 \, L \qquad (1)$$

where

n = number density of crystals

L = crystal size, μm

Using the MSMPR model, determine in the same units as for Example 17.12:

(a) n^o

(b) G

(c) B^o

(d) mean crystal length

(e) n_c

(f) Are your results consistent with the trends found in Example 17.12?

(g) Using your results and those in Example 17.12, predict the growth rate and mean-crystal length if no agitation is used.

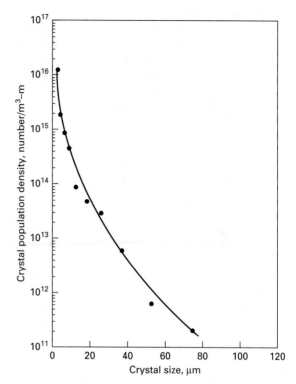

Figure 17.37 Population density of CaCO$_3$ for Exercise 17.28.

17.28 Tai and Chen [*AIChE J.*, **41**, 68–77 (1995)] studied the precipitation of calcium carbonate by mixing aqueous solutions of sodium carbonate and calcium chloride in an MSMPR crystallizer with pH control, such that the form of CaCO$_3$ was calcite rather than aragonite or vaterite. In Run S-2, which was conducted at 30°C, a pH of 8.65, and 800 rpm, with a residence time of 100 min, the crystal population density data were as shown in Figure 17.37.

Because the data do not plot as a straight line, they do not fit (17-38).

(a) Develop an empirical equation that will fit the data and determine, by regression, the constants.

(b) Can nucleation rate and growth rate be determined from the data? If so, how?

17.29 Tsuge and Matsuo ["Crystallization as a Separation Process," *ACS Symposium Series 438*, edited by Myerson and Toyokura, ACS, Washington, DC (1990), pp. 344–354] studied the precipitation of Mg(OH)$_2$ by reacting aqueous solutions of MgCl$_2$

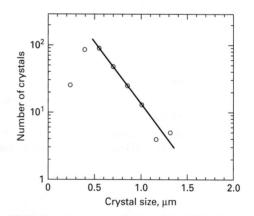

Figure 17.38 Crystal-size distribution of Mg(OH)$_2$ for Exercise 17.29.

and $Ca(OH)_2$ in a 1-liter MSMPR crystallizer operating at 450 rpm and 25°C. Crystal sizes were measured by a scanning electron microscope (SEM) and analyzed by a digitizer. Crystal size was taken to be the maximum length. A typical plot of the crystal-size distribution is given in Figure 17.38 for an assumed residence time of 5 min. Assuming that the number of crystals is proportional to $\exp(-L/G\tau)$, as in (17-38), determine:

(a) Growth rate

(b) Nucleation rate

(c) Predominant crystal size

Section 17.7

17.30 The feed to the top of a falling-film crystallizer is a melt of 60 wt% naphthalene and 40 wt% benzene at saturation conditions. If the coolant enters at the top at 10°C, determine the crystal-layer thickness for up to 2 cm, as a function of time. Necessary physical-property data are given in Example 17.13.

17.31 Paradichlorobenzene melts at 53°C, while orthodichlorobenzene melts at −17.6°C. They form a eutectic of 87.5 wt% of the ortho isomer at −23°C. The normal boiling points of these two isomers differ by about 5°C. A mixture of 80 wt% of the para isomer at the saturation temperature of 43°C is fed to the top of a falling-film crystallizer, where coolant enters at 15°C. If 8-cm i.d. tubes are used, determine the time for the crystal-layer thickness at the top of the tube to reach 2 cm. Which isomer will crystallize? Necessary physical properties are given in Perry's Chemical Engineers' Handbook, except for crystal thermal conductivity, for which we assume a value of 0.15 Btu/h-ft-°F for either isomer.

17.32 Derive (17-67).

Section 17.8

17.33 Derive the following expression for the average impurity concentration over a particular length of crystal layer, $z_2 - z_1$, after one pass or partial pass of zone melting.

$$w_{avg} = w_0 \left\{ \frac{\ell(1-K)}{K(z_2-z_1)}[\exp(-z_2 K/\ell) - \exp(-z_1 K/\ell)] + 1 \right\}$$
(1)

Using the results of Example 17.14, calculate w_{avg} for $z_1 = 0$ and $z_2/\ell = 9$.

17.34 In Example 17.14, let the last 20% of the crystal layer be removed, following the first pass, to $z/\ell = 9$. Calculate from (1), in Exercise 17.33, the average impurity concentration in the remaining crystal layer.

17.35 A bar of 98 wt% Al with 2 wt% of Fe impurity is to be subjected to one pass of zone refining. The solid–liquid equilibrium distribution coefficient for the impurity is 0.29. If $z/\ell = 10$ and the resulting bar is cut off at $z_2 = 0.75z$, calculate the concentration profile for Fe and the average concentration from (1) in Exercise 17.33.

Section 17.9

17.36 A desublimation unit of the heat-exchanger type is to be sized for the recovery of 200 kg/h of benzoic acid (BA) from a gas stream containing 0.8 mol% BA and 99.2 mol% N_2. The gas enters the unit at 780 torr at 130°C and leaves without pressure drop at 80°C. The coolant is pressurized cooling water that enters at 40°C

and leaves at 90°C, in countercurrent flow to the gas. The heat-exchanger tubes are of the type in Example 17.15.

Some properties of benzoic acid are given in Exercise 17.37. In addition,

$$k_c \text{ of solid benzoic acid} = 1.4 \text{ cal/h-cm-°C}$$

$$\rho_c \text{ of solid benzoic acid} = 1.316 \text{ g/cm}^3$$

Determine the number of tubes needed and the time required to reach the maximum thickness of benzoic acid of 1.25 cm.

17.37 Benzoic acid is to be crystallized by bulk-phase desublimation from N_2 using a novel method described by Vitovec, Smolik, and Kugler [*Coll. Czech. Chem. Commun.*, **42**, 1108–1117 (1977)]. The gas, containing 6.4 mol% benzoic acid and the balance N_2, flows at 3 m^3/h at 1 atm and a temperature of 10°C above the dew point. The gas is directly cooled by the vaporization of 150 cm^3/h of a water spray at 25°C. The gas is further cooled in two steps by nitrogen quench gas at 1 atm as follows:

Step	Quench Gas Flow Rate, m^3/h	Quench Gas Temp., °C
1	1.5	105
2	2.0	25

The quench gases enter through porous walls of the vessel so as to prevent crystallization on the vessel wall. Based on the following data for benzoic acid, determine the final gas temperature and the fractional yield of benzoic-acid crystals, assuming equilibrium in the exiting gas.

Melting point = 122.4°C

Specific heat of solid and vapor = 0.32 cal/g-°C

Heat of sublimation = 134 cal/g

Vapor pressure:

T, °C	Vapor Pressure, torr
96	1
105	1.7
119.5	5
132.1	10
146.7	20
162.6	40
172.8	60

The vapor pressure data can be extrapolated to lower temperatures by the Antoine equation.

17.38 Derive (17-75).

Section 17.10

17.39 Fifty-thousand pounds per hour of a 20 wt% aqueous solution of NaOH at 120°F is to be fed to an evaporator operating at 3.7 psia, where the solution is concentrated to 40 wt% NaOH. The heating medium is saturated steam at a temperature 40°F higher than the exiting temperature of the caustic solution. Determine:

(a) Boiling-point elevation of the solution

(b) Saturated-heating-steam temperature and pressure

(c) Evaporation rate

(d) Heat-transfer rate

(e) Heating-steam flow rate

(f) Economy

(g) Heat-transfer area if $U = 300$ Btu/h-ft²-°F

17.40 A 10 wt% aqueous solution of NaOH at 100°F and a flow rate of 30,000 lb/h is to be concentrated to 50 wt% by evaporation using saturated steam at 115 psia.

(a) If a single-effect evaporator is used with $U = 400$ Btu/h-ft^2-°F and a vapor-space pressure of 4 in. Hg, determine the heat-transfer area and the economy.

(b) If a double-effect evaporator system is used with forward feed and $U_1 = 450$ Btu/h-ft^2-°F and $U_2 = 350$ Btu/h-ft^2-°F, and a vapor-space pressure of 4 in. Hg in the second effect, determine the heat-transfer area of each effect, assuming equal areas, and the overall economy.

17.41 A 10 wt% aqueous solution of MgSO$_4$ at 14.7 psia and 70°F is sent to a double-effect evaporator system with forward feed at a flow rate of 16,860 lb/h, to be concentrated to 30 wt% MgSO$_4$. The pressure in the second effect is 2.20 psia. The heating medium is saturated steam at 230°F. Estimated heat-transfer coefficients in Btu/h-ft^2-°F are 400 for the first effect and 350 for the second effect. If the heat-transfer areas of the two effects are to be the same, and boiling-point elevations are neglected, determine:

(a) The pressure in the first effect

(b) The percent of the total evaporation occurring in the first effect.

(c) The heat-transfer area of each effect.

(d) The economy.

Chapter 18

Drying of Solids

Drying involves the removal of moisture (either water or other volatile compounds) from solids, solutions, slurries, and pastes to give solid products, which often, after drying, are final products ready to be packaged. In the feed to a dryer, the moisture may be a liquid, a solute in a solution, or a solid. In the first two cases, the moisture is evaporated; in the latter case, the moisture is sublimed. The term drying is also applied to a gas mixture in which a condensable vapor is removed from a noncondensable gas by cooling, as discussed in Chapter 4; and to the removal of moisture from a liquid or gas by sorption, as discussed in Chapters 6 and 16. This chapter deals only with drying operations that give solid products in various sizes and shapes.

Drying is widely used in industrial processes. Applications include the removal of moisture from: (1) crystalline particles of inorganic salts and organic compounds to cause them to be free-flowing; (2) biological materials, including foods, to prevent spoilage and decay from micro-organisms that cannot live without water; (3) pharmaceuticals; (4) detergents; (5) lumber, paper, and fiber products; (6) dyestuffs; (7) solid catalysts; (8) milk; and (9) films and coatings.

Drying can be expensive, especially when large amounts of water, with its high heat of vaporization, must be evaporated. Therefore, it is important, before drying, to remove as much moisture as possible by mechanical means such as expression; gravity, vacuum, or pressure filtration; settling; and by centrifugal means.

Because drying involves vaporization or sublimation of the moisture, heat must be transferred to the material being dried. The most commonly employed modes of heat transfer for drying are: (1) convection from a hot gas in contact with the material, (2) conduction from a hot, solid surface in contact with the material, (3) radiation from a hot gas or hot surface in view of the material, and (4) heat generation within the material by dielectric, or microwave heating. These different modes can sometimes be used to advantage, depending on whether the moisture to be removed is on the surface of the solid and/or inside the solid.

Of importance in the drying of solids is the temperature at which the moisture evaporates. When the first mode is employed and the moisture is a continuous liquid film or is rapidly supplied to the surface from the interior of the solid, the rate of evaporation is independent of the properties of the solid and can be determined by the rate of convective heat transfer from the gas to the surface. Then, the temperature of the evaporating surface is the wet-bulb temperature of the gas provided that the dryer operates adiabatically. If the convective heat transfer is supplemented by radiation, the temperature of the evaporating surface will be higher than the wet-bulb temperature of the gas. In the absence of contact with a convective-heating gas, as in the latter three modes, and when a sweep gas is not present, such that the dryer operates nonadiabatically, the temperature of the evaporating moisture is its boiling-point temperature at the pressure in the dryer. In evaporators, if the moisture contains dissolved, nonvolatile substances, the boiling-point temperature will be elevated.

18.0 INSTRUCTIONAL OBJECTIVES

After completing this chapter, you should be able to:

- Describe two common modes of drying.
- Discuss industrial drying equipment.
- Use a psychrometric chart to determine drying temperature.
- Differentiate between the adiabatic-saturation and wet-bulb temperatures.
- Explain equilibrium-moisture content of solids.
- Explain types of moisture content used in making dryer calculations.
- Describe the four different periods in direct-heat drying.
- Calculate drying rates for different periods.
- Apply models for a few common types of dryers.

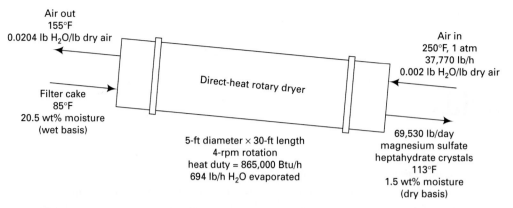

Air out
155°F
0.0204 lb H$_2$O/lb dry air

Air in
250°F, 1 atm
37,770 lb/h
0.002 lb H$_2$O/lb dry air

Direct-heat rotary dryer

Filter cake
85°F
20.5 wt% moisture
(wet basis)

5-ft diameter × 30-ft length
4-rpm rotation
heat duty = 865,000 Btu/h
694 lb/h H$_2$O evaporated

69,530 lb/day
magnesium sulfate
heptahydrate crystals
113°F
1.5 wt% moisture
(dry basis)

Figure 18.1 Process for drying magnesium-sulfate-heptahydrate filter cake.

Industrial Example

As an example of an industrial application of drying, consider the continuous production of 69,530 lb/day of MgSO$_4 \cdot$ 7H$_2$O crystalline solids containing 0.015 lb H$_2$O/lb dry solid. The feed to the dryer, shown in Figure 18.1, consists of a filter cake from a continuous, rotary-drum, vacuum filter. The cake is at 85°F and contains 20.5 wt% moisture on the wet basis. Because the crystals are relatively coarse and free-flowing when partially dry, and nonsticking, a direct-heat rotary dryer consisting of a slightly inclined, rotating, cylindrical shell is used. The filter-cake feed enters the high end of the dryer from an inclined, vibrated chute. Heated air at 250°F and atmospheric pressure, with an absolute humidity of 0.002 lb H$_2$O/lb dry air enters at the other end of the dryer, at a flow rate of 37,770 lb/h. In order to obtain good contact between the wet crystals and the hot air, the dryer is provided with internal, longitudinal flights that extend the entire length of the shell. As the shell rotates, the flights lift the solids until they reach their angle of repose and then shower down through the hot air, which is in countercurrent flow to the net longitudinal direction of flow of the solids. The dry solids discharge at 113°F through a rotating valve into a screw conveyor. The air, which has been cooled to 155°F and humidified to 0.0204 lb H$_2$O/lb dry air, by contact with the wet solids, exits from the opposite end of the dryer, where the moist air is drawn through a fan and exhausted to the surrounding atmosphere.

The hot air causes the evaporation of 694 lb/h of water. Most of this evaporation takes place at a temperature of approximately 94.5°F, which is the average of the entering and exiting gas wet-bulb temperatures of 95.5 and 93.5°F, respectively. In addition, the hot air must heat the solids from 85°F to 113°F and the evaporated moisture to 155°F. The total rate of convective heat transfer, Q, from the gas to the solids is 865,000 Btu/h. This value ignores heat loss from the shell to the surroundings and thermal radiation to the solids from the hot gas or the inside surfaces of the flighted shell.

Of the total heat load, approximately 83% is required to evaporate the moisture, with the balance of only 17% supplying sensible heat. Therefore, a reasonably accurate log-mean, temperature-driving force can be based on the assumption of a constant temperature at the gas–wet solids interface equal to the average air wet-bulb temperature of 94.5°F. Thus,

$$\Delta T_{\mathrm{LM}} = \frac{(250 - 94.5) - (155 - 94.5)}{\ln\left(\dfrac{250 - 94.5}{155 - 94.5}\right)} = 100.6°\text{F}$$

For a direct-heat, rotary dryer, it is convenient to characterize the convective heat transfer by an overall, volumetric heat-transfer coefficient, Ua, which for this example is 14.6 Btu/h-ft^3-°F. The required cylindrical shell volume, V, from $Q = UaV\,\Delta T_{\mathrm{LM}}$, is 590 ft^3. The dryer diameter is 5 ft, which gives an entering, superficial hot-air velocity of 9.56 ft/s, which is sufficiently low to prevent entrainment of solid particles in the exiting air. The length of the cylindrical shell is 30 ft. While moving through the dryer, the bulk solids, with a bulk density of 62 lb dry solids/ft^3, occupy 8 vol% of the dryer, and have a residence time of one hour. The shell rotates at 4 rpm.

18.1 DRYING EQUIPMENT

Many different forms of materials are sent to drying equipment, including granular solids, pastes, slabs, films, slurries, and liquids. No one device can handle efficiently such a wide variety of materials. Accordingly, a large number of different types of commercial dryers have been developed. These dryers can be classified in a number of ways. Perhaps most important is the mode of operation with respect to the material being dried. Batch operation is generally indicated when the production rate is less than 500 lb/h of dried solid, while continuous operation is preferred for a production rate of more than 2,000 lb/h. In the example above, the production rate is 2,900 lb/h and a continuous drying operation was selected.

A second method of classification is the mode used to supply heat to evaporate the moisture. As mentioned above, direct-heat (also called adiabatic or convection) dryers contact material with a hot gas, which not only provides the required energy to heat the material and evaporate the moisture, but also sweeps away the moisture. When the continuous mode of operation is used, the hot gas can flow countercurrently, cocurrently, or in crossflow to the material being dried. Countercurrent flow is the most efficient configuration, but cocurrent flow may be required if the material being dried is temperature-sensitive.

Indirect-heat (also called nonadiabatic) dryers provide the heat to the material indirectly by conduction and/or radiation from a hot surface. The energy may also be generated within the material by dielectric, radio frequency, or microwave heating. Indirect-heat dryers may also be operated under vacuum to reduce the temperature at which the moisture is evaporated. A sweep gas is not necessary, but can be provided to help remove the moisture. In general, indirect-heat dryers are more expensive than direct-heat dryers. Therefore, the former type is generally used only when the material is either temperature-sensitive or subject to breakage of crystals, with dust or fines formation.

A third method for classifying dryers is the degree to which the material to be dried is agitated. In some dryers, the material is stationary. At the opposite extreme is the fluidized-bed dryer. Agitation increases the rate of heat transfer to the material but, if too severe, can cause crystal breakage and dust formation. Agitation in a continuous dryer may be necessary if the material is sticky.

In the following, only some of the more widely used commercial dryers are described. A more complete coverage is given in the *Handbook of Industrial Drying* [1]. Extensive performance data for many of the types of dryers are given in *Perry's Chemical Engineers' Handbook* [2] and by Walas [3]. Batch dryers are discussed first, followed by continuous dryers.

Batch Operation

Equipment for drying batches includes: (1) tray (also called cabinet, compartment, or shelf) dryers; and (2) agitated dryers. Together, these two types cover many of the modes of heat transfer and agitation discussed above and can handle a wide variety of wet-solid feeds, such as slurries, filter cake, and particulate materials.

Tray Dryers

The oldest and simplest batch dryer is the tray dryer, which is shown schematically in Figure 18.2 and is particularly useful when low production rates of multiple products are involved and where drying times vary from hours to days. The material to be dried is loaded to a depth of typically 0.5–4 in. in removable trays that may measure 30 × 30 × 3 in. and are stacked on shelves about 3 in. apart in a cabinet or on a truck that is wheeled into a chamber. If the wet solids are

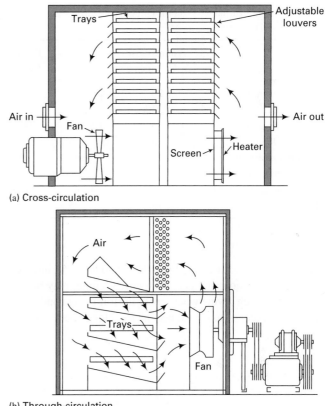

(a) Cross-circulation

(b) Through-circulation

Figure 18.2 Tray dryers.

granular or shaped into briquettes, noodles, or pellets, with appreciable voids, the tray bottom can be perforated or can be a screen so that the heating gas can be passed down through the material (called through-circulation) as shown in Figure 18.2b. Otherwise, the tray bottom is solid and the hot gas is passed at velocities of typically 3-30 ft/s over the open surface of the tray (called cross-circulation), as shown in Figure 18.2a.

Although fresh hot gas might be used for each pass through or across the material, it is almost always more economical to recirculate the gas, providing venting and makeup fresh gas at rates of 5–50% of the circulation rate to maintain the humidity at an acceptable level. Typically the gas is heated with an annular, finned-tube heat exchanger by steam condensing inside the tubes. If the moisture being evaporated is water, steam requirements can range from 1.5 to 7.5 lb steam/lb water evaporated. It is important to provide baffles in tray dryers to promote uniform distribution of hot gas across or through the trays and thus, achieve uniform drying.

Tray dryers are also available for operation under vacuum and with indirect heating. In one configuration, the trays are placed on hollow shelves that carry condensing steam and, thus, act as heat exchangers. Heat is transferred by conduction to a tray from the top of the shelf supporting it and by radiation from the bottom of the shelf located directly above the tray. Typical performance data for direct-heat, crossflow-circulation tray dryers are given in Table 18.1.

Table 18.1 Performance Data for Direct-Heat, Crossflow-Circulation Tray Dryers

Material	Aspirin-Base Granules	Chalk	Filter Cake
Number of trays	20	72	80
Area/tray, ft^2	3.5	15.7	3.5
Total loading, lb wet	56	1,800	2,800
Depth of loading, in.	0.5	2.0	1.0
% Initial moisture	15	46	70
% Final moisture	0.5	2	1
Maximum air temp., °F	122	180	200
Drying time, h	14	4.5	45

Agitated Dryers

As discussed by van't Land [4] and Uhl and Root [5], indirect heat with agitation and, perhaps, under vacuum, may be desirable for batch drying when any of the following conditions exist: (1) material oxidizes, becomes explosive, or becomes dusty during drying; (2) moisture is valuable, toxic, flammable, or explosive; (3) material tends to agglomerate or set up if not agitated; and (4) maximum product temperature is less than about 30°C. In most applications, the rate of heat transfer is controlled by contact resistance at the inner wall of the jacketed vessel and conduction into the material being dried. A wide variety of heating fluids can be used, including hot water, steam, Dowtherm, hot oil, and molten salt.

When only Condition 3 applies, the atmospheric, agitated-pan dryer, shown in Figure 18.3a, may be employed, particularly when the feed is a liquid, slurry, or paste. This dryer consists of a shallow (2–3-ft high), jacketed, flat-bottomed vessel, equipped with a paddle agitator that rotates at 2–20 rpm and scrapes the inner wall to help prevent cake buildup. Typical units range in size from 3 to 10 ft in diameter, with a capacity of up to 1,000 gallons and from 15 to 300 ft^2 of heat-transfer surface. When using steam as the heating medium in the jacket, overall heat-transfer coefficients may vary from 5 to 75 Btu/h-ft^2-°F. The material to be dried occupies about two-thirds of the volume of the vessel. The

(a) Atmospheric pan dryer

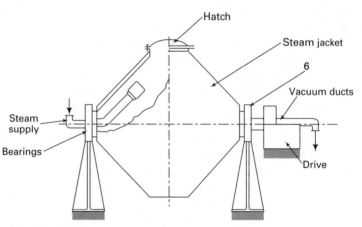

(b) Rotating, double-cone vacuum dryer

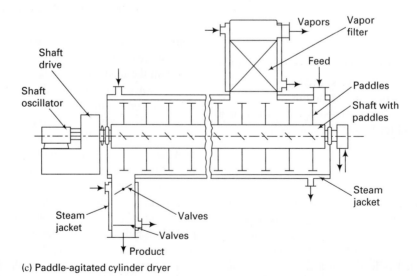

(c) Paddle-agitated cylinder dryer

Figure 18.3 Agitated dryers.

[From *Perry's Chemical Engineers' Handbook,* 6th ed., R.H. Perry, D.W. Green, and J.O. Maloney, Eds., McGraw-Hill, New York (1984) with permission.]

degree of agitation can be varied during the drying cycle. For example, with a thin-liquid feed, agitation may vary from very low initially to very high if a sticky paste forms, followed by a moderate degree of agitation when the final product of a granular solid begins to form. Typically, several hours are required for drying. Units that can be operated under vacuum are also available.

When any or all of the above four conditions apply, but only a mild degree of agitation is required, the jacketed, rotating, double-cone (also called tumbler) vacuum dryer, shown schematically in Figure 18.3b, can be used. V-shaped tumblers are also available. The conical shape facilitates discharge of the dried product, but no means is provided to prevent cake buildup on the inner walls. Double-cone volumes range from 0.13 to 16 m^3, with heat-transfer surface areas of 1 to 56 m^2. Additional heat-transfer surface can be provided by internal tubes or plates. Up to 70% of the volume can be occupied by the feed. A typical H_2O evaporation rate when operating at 10 torr with heating steam at 2 atm is 1 lb/h-ft^2 of heat-transfer surface.

A more widely used agitated dryer that is applicable when any or all of the above four conditions are relevant is the ribbon- or paddle-agitated, horizontal-cylinder dryer, shown in the paddle form in Figure 18.3c. The cylinder is jacketed and stationary. The ribbons or paddles provide agitation and scrape the inner walls to prevent solids buildup. As discussed by Uhl and Root [5], cylinder dimensions range up to diameters of 6 ft and lengths up to 40 ft. The agitator can be rotated over a wide range of rates, with values of 4–140 rpm having been reported, resulting in overall heat-transfer coefficients of 5–35 Btu/h-ft^2-°F. Typically from 20 to 70% of the cylinder volume is filled with feed and drying times vary from 4 to 16 hours. In more advanced versions of this type of dryer, as discussed by McCormick [6], one or two parallel rotating shafts can be provided that intermesh with stationary, lump-breaking bars to increase the range of application. The paddles can also be hollow to provide additional heat-transfer

surface. This type of dryer can also be operated in a continuous mode.

Continuous Operation

A wide variety of industrial drying equipment for continuous operation is available. The following descriptions cover most of the widely used types, organized by the nature of the wet feed: (1) granular, crystalline, and fibrous solids, cakes, extrusions, and pastes; (2) liquids and slurries; and (3) sheets and films. In addition, infrared, microwave, and freeze drying is described.

Tunnel Dryers

The simplest, most widely applicable, and perhaps oldest continuous dryers are the tunnel dryers, which are suitable for any material that can be placed into trays and is not subject to dust formation. The trays are stacked onto wheeled trucks, which are conveyed progressively in series through a tunnel where the material in the trays is contacted by cross-circulation of hot gases. As shown in Figure 18.4, the hot gases can flow countercurrently or cocurrently to the movement of the trucks. More complex flow configurations are also possible. As a truck of dried material is removed from the discharge end of the tunnel, a truck of wet material enters at the feed end. The overall drying operation is not truly continuous because wet material must be loaded into the trays and dried material removed from the trays manually outside the tunnel. Tray spacings and dimensions, as well as hot-gas velocities, are the same as for batch tray dryers. A typical tunnel might be 100 ft long and able to house 15 trucks.

Belt or Band Dryers

A truly continuous operation can be achieved by carrying the solids as a layer on a belt conveyor, with hot gases passing over the material. The endless belt is constructed of hinged,

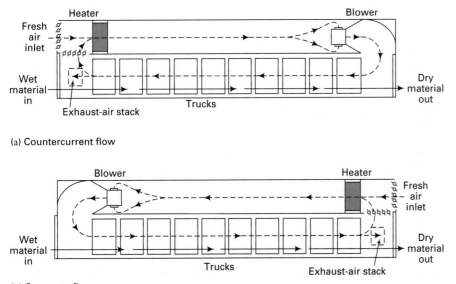

(a) Countercurrent flow

(b) Cocurrent flow

Figure 18.4 Tunnel dryer.

slotted-metal plates, or, preferably a thin metal band, which is ideal for slurries, pastes, and sticky materials. The bands are up to 1.5 m wide × 1 mm thick.

Much more common are screen or perforated-belt or band-conveyor dryers, which, as shown in Figure 18.5a, use circulation of heated gases upward and/or downward through a moving, permeable, layered, bed of wet material from 1 to 6 in. in depth. Multiple sections shown in Figure 18.5b, each with a fan and set of gas-heating coils, can be arranged in series to provide a dryer, with a single belt, as long as 150 ft with a 6-ft width, giving drying times up to 2 h with a belt speed of about 1 ft/min. To be permeable, the wet material must be of a granular-like form. If not, the material can be preformed by scoring, granulation, extrusion, pelletization, flaking, or briquetting. Particle sizes typically range from 30 mesh to 2 in. Hot-gas superficial velocities through the bed typically range from 0.5 to 1.5 m/s, with maximum bed pressure drops of 50-mm head of water. Heating gases are usually provided by heat transfer from condensing steam in finned-tube heat exchangers to temperatures in the range of 50–180°C, but temperatures up to 325°C are feasible by other means. Continuous, through-circulation conveyor dryers have been used to remove moisture from a wide variety of materials, some of which are listed in Table 18.2, which includes, in parenthesis, the method of preforming, if necessary.

In a typical application with a perforated-band-conveyor dryer, 50-ft long × 75-in. wide, 1,800 lb/h of calcium carbonate with a moisture content of 0.005 lb H_2O/lb carbonate is produced in a residence time of 40 minutes from 6-mm-diameter carbonate extrusions with a moisture content of 1.5 lb H_2O/lb carbonate, using air heated to 320°F by 160 psig steam and passing through the extrusions at a superficial

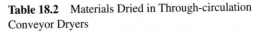

Table 18.2 Materials Dried in Through-circulation Conveyor Dryers

Aluminum hydrate (scored on filter)
Aluminum stearate (extruded)
Asbestos fiber
Breakfast food
Calcium carbonate (extruded)
Cellulose acetate (granulated)
Charcoal (briquetted)
Cornstarch
Cotton linters
Cryolite (granulated)
Dye intermediates (granulated)
Fluorspar
Gelatin (extruded)
Kaolin (granulated)
Lead arsenate (granulated)
Lithopone (extruded)
Magnesium carbonate (extruded)
Mercuric oxide (extruded)
Nickel hydroxide (extruded)
Polyacrylic nitrile (extruded)
Rayon staple and waste
Sawdust
Scoured wool
Silica gel
Soap flakes
Soda ash
Starch (scored on filter)
Sulfur (extruded)
Synthetic rubber (briquetted)
Tapioca
Titanium dioxide (extruded)
Zinc stearate (extruded)

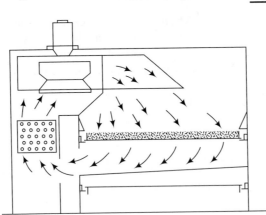

(a) Single downflow section

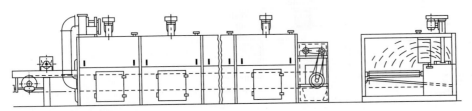

(b) Multiple sections

Figure 18.5 Perforated-belt or band-conveyor dryer.

velocity of 2.7 ft/s. The heating steam consumption is 1.75 lb/lb H_2O evaporated.

Turbo-Tray Tower Dryers

When floor space is limited, but head-room is available, the turbo-tray or rotating-shelf dryer, shown in Figure 18.6, may be a good choice for rapid drying of free-flowing, nondusting, granular solids. Annular shelves, mounted one above the other, are slowly rotated at up to one rpm by a central shaft. Wet feed enters through the roof onto the top shelf as it rotates under the feed opening. At the end of one revolution, a stationary wiper causes the material to fall through a radial slot onto the shelf below, where it is spread into a pile of uniform thickness by a stationary leveler. This action is repeated on each shelf until the dried material is discharged from the bottom of the unit. Also mounted on the central

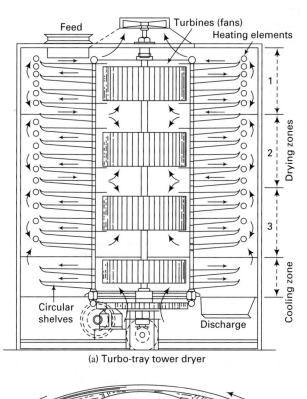

(a) Turbo-tray tower dryer

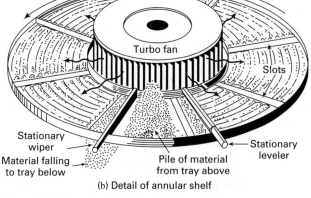

(b) Detail of annular shelf

Figure 18.6 Rotating-shelf dryer.

shaft are turbo-type fans that provide cross-circulation of hot gases at velocities of 2 to 8 ft/s across the shelves, and heating elements located at the outer periphery of the unit. The bottom section of shelves can be used as a solids-cooling zone. Because the solids are showered through the hot gases and redistributed from shelf to shelf, drying time is reduced from that for a cross-circulation, stationary-tray dryer. Typical turbo-tray dryers range in size from 2 to 20 m in height and 2 to 11 m in diameter with as many as 58 shelves, giving shelf areas from 5.5 to 1,675 m^2.

Overall heat-transfer coefficients, based on shelf area, of 30–120 J/m^2-s-K have been observed, giving, in some cases, moisture-evaporation rates comparable to through-circulation, belt- or band-conveyor dryers. Materials successfully handled in turbo-tray dryers include calcium hypochlorite, urea, calcium chloride, sodium chloride, antibiotics, antioxidants, and water-soluble polymers. The unit is particularly useful when product contamination must be avoided and the wet solids contain volatiles besides water. Capacities of up to 24,000 lb/h of dried product are quoted.

Direct-Heat Rotary Dryers

A widely used dryer for the evaporation of water from free-flowing granular, crystalline, and flaked solids of relatively small size, especially when breakage of the solids can be tolerated, is the direct-heat rotary dryer. As shown in Figure 18.7a, the dryer consists of a rotating, cylindrical shell that is slightly inclined from the horizontal with a slope of less than 8 cm/m. Wet solids enter the cylinder through a chute at the high end and dry solids discharge from the low end. Hot gases (heated air, flue gas, or superheated steam) generally flow countercurrent to the solids, but cocurrent flow can also be employed for temperature-sensitive solids. With cocurrent flow, the cylinder may not need to be inclined because the gas will help move the solids to the discharge end. To enhance the rate of heat transfer from the gas to the solids, longitudinal lifting flights, available in several different designs, two of which are shown in Figure 18.7b, are mounted on the inside of the rotating shell, causing the solids to be lifted and then showered through the hot gas during each revolution of the cylinder. Typically the bulk solids occupy 10–18% of the cylinder volume, with residence times from 5 min to 2 h. Resulting water-evaporation rates are 5–50 kg/h-m^3 of dryer volume. The gas blower can be located so as to push or pull the gas through the dryer, with the latter location favored if the material tends to form appreciable dust. Knockers, located on the outside wall of the shell, are sometimes used to prevent solids from sticking to the inside wall of the shell. Rotary dryers are available in a wide range of sizes, typically from 1 to 20 ft in diameter and 4-150 ft long. Superficial gas velocities may be limited by dust entrainment and are usually in the range of 0.5–10 ft/s. The peripheral velocity of the shell is typically 1 ft/s. A wide variety of materials, some of which are listed in Table 18.3, are dried in direct-heat rotary dryers.

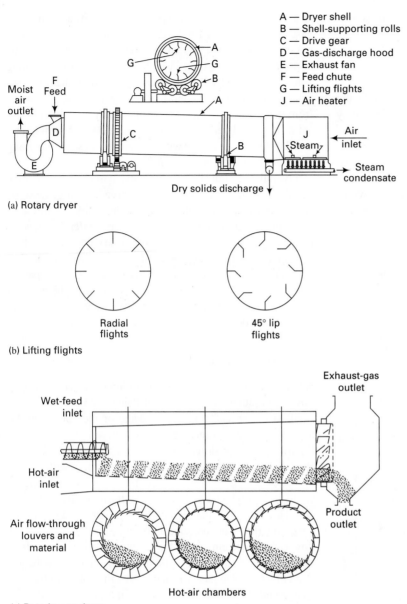

A — Dryer shell
B — Shell-supporting rolls
C — Drive gear
D — Gas-discharge hood
E — Exhaust fan
F — Feed chute
G — Lifting flights
J — Air heater

(a) Rotary dryer

Radial flights

45° lip flights

(b) Lifting flights

(c) Roto-louvre dryer

Figure 18.7 Direct-heat rotary dryer.

[From W.L. McCabe, J.C. Smith, and P. Harriott, *Unit Operations of Chemical Engineering,* 5th ed., McGraw-Hill, New York (1993) with permission.] [From *Perry's Chemical Engineers' Handbook,* 6th edition, R.H. Perry, D.W. Green, and J.O. Maloney, Eds., McGraw-Hill, New York (1984) with permission.]

Table 18.3 Materials Dried in Direct-heat Rotary Dryers

Ammonium nitrate prills	Sand
Ammonium sulfate	Sodium chloride
Blast furnace slag	Sodium sulfate
Calcium carbonate	Stone
Cast-iron borings	Styrene
Cellulose acetate	Sugar beet pulp
Coppers	Urea crystals
Fluorspar	Urea prills
Illmenite ore	Vinyl resins
Oxalic acid	Zinc concentrate

Roto-Louvre Dryer

A further improvement in the rate of heat transfer from hot gas to solids in a rotating cylinder is the through-circulation

action achieved in the Roto-Louvre dryer, shown in Figure 18.7c. A double wall provides an annular passage for hot gas, which passes through louvers and then through the rotating bed of solids. Because the gas pressure drop through the bed may be significant, both inlet and outlet gas blowers are often provided to maintain an internal pressure close to atmospheric. These dryers range in size from 3 to 12 ft in diameter and 9–36 ft long, with water-evaporation rates reported as high as 12,300 lb/hr. Roto-Louvre dryers are useful for processing coarse, free-flowing, dust-free granular solids.

Indirect-Heat, Steam-Tube, Rotary Dryers

When materials are: (1) free-flowing and granular, crystalline, or flaked; (2) wet with water or organic solvents;

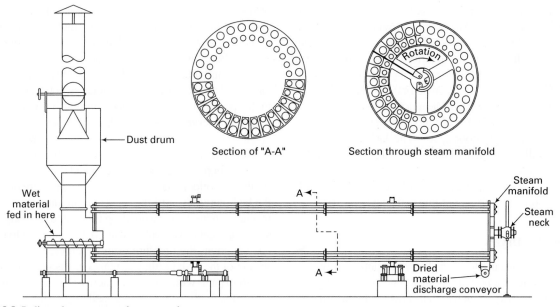

Figure 18.8 Indirect-heat, steam-tube, rotary dryer.

[From *Perry's Chemical Engineers' Handbook,* 6th ed., R.H. Perry, D.W. Green, and J.O. Maloney, Eds., McGraw-Hill, New York (1984) with permission.]

and/or (3) subject to undesirable breakage, dust formation, or contamination by air or flue gases, an indirect-heat, steam-tube, rotary dryer is often selected. A version of this dryer, shown in Figure 18.8, consists of a rotating cylinder that houses two concentric rows of longitudinal finned or unfinned tubes that carry condensing steam and rotate with the cylinder. Wet solids are fed into one end of the cylinder through a chute or by a screw conveyor. A gentle solids-lifting action is provided by the tubes. The dried product discharges from the other end after suitable contact with the hot-tube surfaces. The moisture (water or solvent) evaporates at about the boiling temperature, but can be swept out by a small purge of inert gas. Steam enters the tubes through a central revolving inlet manifold. Condensate is discharged into a collection ring. With unfinned tubes, overall heat-transfer coefficients based on the surface area of the tubes

range from 30 to 85 J/m^2-s-K, when the solids occupy 10–20% of the dryer volume. Steam-tube rotary dryers range in size from 3 to 8 ft in diameter by 15–80 ft long, with one or two rows of 14–90 tubes each, of 2.5–4.5 in. in diameter. The largest-size dryers contain a single row of 90 tubes. Rotation rates range from 3 to 6 rpm. Materials successfully dried include inorganic crystals, silica, mica, flotation concentrates, pigment filter cakes, precipitated calcium carbonate, distillers' grains, brewers' grains, citrus pulp, cellulose acetate, starch, and high-moisture organic compounds.

Screw-Conveyor Dryers

Used less often than rotary dryers is the screw-conveyor dryer, as shown in Figure 18.9, consisting of a trough or cylinder that carries a hollow screw inside of which steam

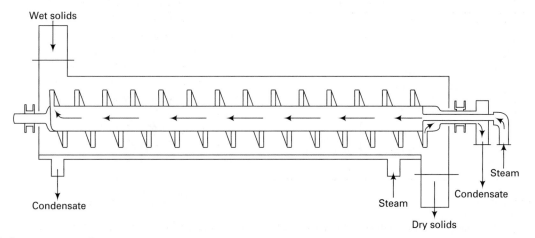

Figure 18.9 Screw-conveyor dryer.

condenses to provide heat for drying the material being conveyed. Additional heat transfer can be provided by jacketing the trough or cylindrical shell. A wide range of materials can be dried including slurries, solutions, and solvent-laden solids. The boiling moisture can be purged with a small amount of inert gas. Standard conveyor dryers are as large as 3 ft in diameter by 20 ft long. More drying time can be provided by arranging a number of units in series, with one unit above another to save floor space. The last unit can be a cooler. Overall heat-transfer coefficients are comparable to, but somewhat less than, those for indirect-heat steam-tube rotary dryers. Major applications include removal of solvents from solids and drying of fine and sticky materials.

Fluidized-Bed Dryers

Free-flowing, moist particles can be dried continuously with a residence time of a few minutes by contact with hot gases in a fluidized-bed dryer, such as that shown in Figure 18.10a. This dryer consists of a cylindrical or rectangular fluidizing chamber to which wet particles are fed from a bin through a star valve or by a screw conveyor and fluidized by hot gases that are blown through a heater and into a plenum chamber

below the bed, from where they pass into the fluidizing chamber through a perforated distributor plate. The hot gases pass up through the bed, transferring heat for evaporation of the moisture, and pass out the top of the fluidizing chamber and into a cyclone for dust removal. The solids are circulated by the action of the hot gases in the bed, but eventually pass out of the chamber through an overflow duct, which also serves to establish the height of the fluidized bed.

At low gas velocities, solids are not fluidized, but form a fixed bed through which the gas flows upward with a decrease in pressure due to friction and drag of the particles. As the gas velocity is increased, the gas pressure drop across the bed increases until the minimum fluidization velocity is reached where the pressure drop is equal to the weight of the solids per unit cross-sectional area of bed normal to gas flow. At this point, the pressure drop is sufficient to support the weight of the bed. Further increases in gas velocity cause the bed to expand with little or no increase in gas pressure drop. Typically, fluidized-bed dryers are designed for gas velocities of about twice the minimum required for fluidization. That value depends mainly on the particle size and density, and gas density and viscosity. Superficial gas velocities in fluidized-bed dryers generally are in the range of 0.5–5.0 ft/s, which

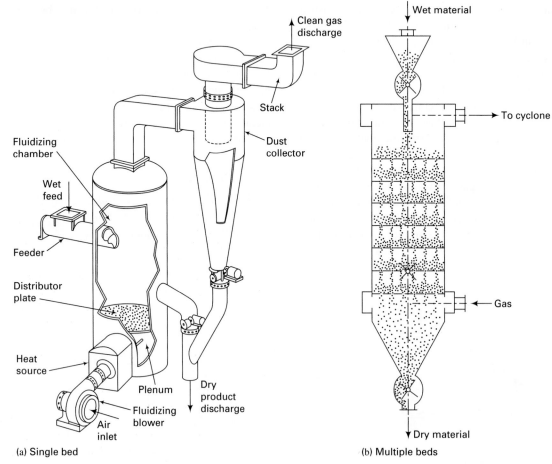

(a) Single bed   (b) Multiple beds

Figure 18.10 Fluidized-bed dryers.

[From W.L. McCabe, J.C. Smith, and P. Harriott, *Unit Operations of Chemical Engineering*, 5th ed., McGraw-Hill, New York (1993) with permission.]

provides stable, bubbling fluidization. Higher velocities can lead to undesirable slugging of large gas bubbles through the bed.

The capital and operating cost of a blower to provide sufficient gas pressure for the pressure drops across the distributor plate and the bed can be substantial. Therefore, the required solids residence time for drying is achieved by a shallow bed height and a large chamber cross-sectional area. Typical fluidized-bed heights range from 0.5 to 2.0 ft, with chamber diameters from 3 to 10 ft. However, chamber heights are much greater than fluidized-bed heights, because it is desirable to provide at least 6 ft of free-board height above the top surface of the fluidized bed so that the larger dust particles can settle back into the bed rather than being carried by the gas into the cyclone.

Because of intense mixing, the temperatures of the gas and solids in a fluidized bed are equal and uniform at the temperature of the discharged gas and solids. However, there is a substantial residence-time distribution for the particles in the bed. A fraction of the particles short-circuit from the feed inlet to the discharge duct with little residence time and opportunity to dry. Another fraction of the particles spends much more than the necessary residence time for complete drying. Thus, the nonuniform moisture content of the product solids may not meet specifications. When the final moisture content is critical, it may be advisable to smooth out the residence-time distribution by using a more elaborate multiple fluidized-bed dryer of the type shown in Figure 18.10b. Materials that are successfully dried in fluidized-bed dryers include coal, sand, limestone, iron ore, clay granules, granular fertilizer, granular desiccant, salt, sodium perborate, and polyvinylchloride (PVC). Large fluidized-bed dryers for coal and iron ore produce more than 0.5 million lb/h of dried material.

Spouted-Bed Dryers

When wet, free-flowing particles are larger than 1 mm in diameter, but are uniform in size and of low density, as in the case of various grains, a spouted-bed dryer, shown in Figure 18.11, is a good choice, particularly when the required drying time is more than just a few minutes. A high-velocity, hot gas enters the bottom of the drying chamber, entrains particles, and flows upward through a draft tube, above which the cross-sectional area for gas flow is significantly increased in a conical section, causing the gas velocity to decrease such that the particles are released to an annular-downcomer region. A fraction of the circulated solids is discharged from a duct in the conical section. The gas exits from the top of the vessel, following passage through a free-board region above the bed. For the drying of grains, entrainment of dust in the exiting gas is minimal.

Pneumatic-Conveyor (Flash) Dryers

When only surface moisture must be evaporated from materials that can be reduced to particles by a pulverizer, disinte-

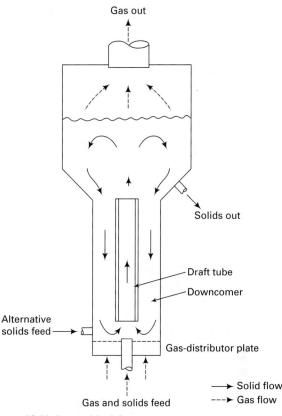

Figure 18.11 Spouted-bed dryer.

gration mill, or other deagglomeration device, a pneumatic-conveyor (gas-lift or flash) dryer is widely used and is particularly desirable when the material is temperature-sensitive, oxidizable, explosive, flammable, and/or non-erosive. A common flash dryer configuration is shown in Figure 18.12. Wet solids are fed into a paddle-conveyor mixer and dropped into a hammer mill where the solids are disintegrated. Air is pulled, by an exhaust fan, through a furnace into the hammer mill, where the disintegrated solids are picked up and further deagglomerated into discrete particles while being pneumatically conveyed at upward high velocity in a duct where much of the drying takes place. Typically the holdup of solids in the gas is less than 2% by volume. The particle-gas mixture is separated in a cyclone separator, from which the solids are discharged. Because the particles travel upward in the drying duct at a velocity almost equal to that of the gas, a residence time of less than 5 s is provided. If additional time is needed, up to 30 s can be achieved by partial recycle of the solids leaving the cyclone separator. Recycle of solids is also useful for the disintegration of materials that are sticky or pasty. The deagglomerated particle sizes typically range from −30 to −300 mesh. If the particles are crystalline or friable, they may be subject to excessive breakage. Pneumatic conveying velocities range from 10 to 30 m/s, usually about 3 m/s greater than the terminal (free-fall) velocity of the largest particle to be conveyed out of the disintegrator. The distribution of remaining moisture in the product particles can be wide because of the

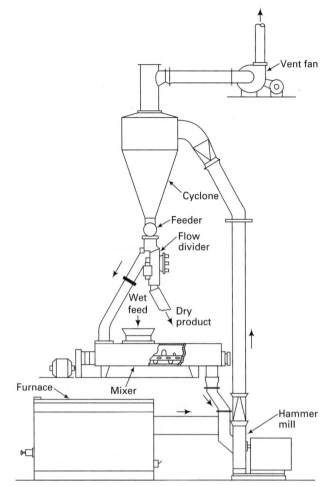

Figure 18.12 Pneumatic-conveyor (flash) dryer.
[From W.L. McCabe, J.C. Smith, and P. Harriott, *Unit Operations of Chemical Engineering*, 5th ed., McGraw-Hill, New York (1993) with permission.]

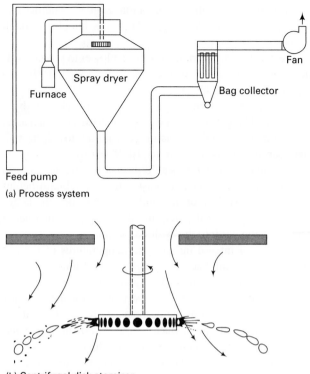

(a) Process system

(b) Centrifugal disk atomizer

Figure 18.13 Spray dryer.

distribution of particle residence times. However, surface drying is rapid because inlet gas temperatures as high as 1500°F can be employed. Nevertheless, because: (1) the particles flow cocurrently with the gas, (2) the gas temperature decreases significantly, and (3) the particle residence time is short and evaporation of moisture is incomplete, the particles do not attain temperatures greater than about 200°F.

Large flash dryers are provided with pneumatic-conveying dryer ducts 1 m in diameter and 12 m high, with evaporation capacities of 20,000 lb/h of water. Compared to many other dryers, they have small floor-area requirements and are used for drying filter cakes, centrifuge cakes and slurries, yeast cake, whey, starch, sewage sludge, gypsum, fruit pulp, copper sulfate, clay, coal, chicken droppings, adipic acid, polystyrene beads, ammonium sulfate, and hexamethylene tetramine.

Spray Dryers

When solutions, slurries, or pumpable pastes, containing more than 50 wt% moisture at feed rates greater than

1,000 lb/h are to be dried, a spray dryer should be considered. One of many spray-dryer configurations is shown in Figure 18.13a. The drying chamber frequently has a conical-shaped bottom section with a diameter at the top that may be nearly equal to the chamber height. The feed is pumped to the top center of the chamber where it is dispersed into small droplets or particles, ranging in size from 2 to 2,000 μm, by any of the following three types of atomizers: (1) single-fluid pressure nozzles, (2) pneumatic nozzles, and (3) centrifugal disks or spray wheels. Hot gas enters the chamber, after being heated in a furnace or heat exchanger, causing the moisture in the atomized feed to rapidly evaporate. The gas flows mainly cocurrently to the solids. The dried solids and gas are either partially separated in the chamber, followed by removal of dust from the gas by a cyclone separator, or, as shown in Figure 18.13a, are sent together to a cyclone separator, bag filter, or other gas-solid separator. The gas may be pushed or pulled through the unit by a fan.

In many respects, the spray dryer is similar in operating conditions to a pneumatic-conveyor dryer because the particles are small, the entering gas temperature can be very high, the residence time of the particles is short, mainly surface moisture is removed, and temperature-sensitive materials can be handled. However, a rather unique characteristic of a spray dryer is its ability, with some materials such as dyes, foods, and detergents, to produce, from a solution, rounded porous particles of fairly uniform size that can be rapidly dissolved or reacted in subsequent industrial or domestic applications. Although particle residence times are less than

5 s if only surface moisture is removed, residence times of up to 30 s can be provided for additional evaporation of internal moisture. Spray drying is also unique in that it combines, into one compact operation, the equivalent of some or all of the following sequence of operations: evaporation, crystallization or precipitation, filtration or centrifugation, size reduction, classification, and drying.

The most influential part of a spray dryer is the atomizer. Each of the three types has advantages and disadvantages and, therefore, the selection of the atomizer for a particular application is not straightforward. Pneumatic (two-fluid) nozzles impinge air, steam, or another gas on the feed at relatively low pressures of up to 100 psig, but are not efficient at high capacities. Consequently, they find applications only in pilot plants and commercial plants of low capacity. Exceptions are the dispersion of stringy and fibrous materials, thick pastes, certain filter cakes, and polymer solutions because with high atomizing gas-to-feed ratios, the smallest particles can be produced.

Pressure (single-fluid) nozzles, with orifice diameters of 0.012–0.15 in., require solution inlet pressures of 300–4,000 psig to achieve break-up of the feed stream. These nozzles can deliver the narrowest range of droplet sizes, but the droplet sizes are the largest delivered by the three types of atomizers, and multiple nozzles are required in large-diameter spray dryers. Also, orifice wear and plugging can be problems with some feeds. Because the spray is largely downward, chambers are relatively slender and tall, with height-to-diameter ratios of 4–5.

Centrifugal disks (spray wheels), of the type shown in Figure 18.13b, handle solutions or slurries, delivering thin sheets of feed that break up into small droplets in a nearly radial direction at high capacities. However, disks have the largest-diameter spray pattern and therefore require the largest-diameter drying chambers to prevent particles from striking the chamber wall while in a sticky state. Centrifugal disks range in diameter from a few inches in small units to up to 32 in. in very large units. The disks spin at 3,000–50,000 rpm, but a given disk can operate over a wide range of feed rate and rotation rate, without significantly affecting the particle-size distribution that occurs with variation of the feed pressure or by enlargement of or other damage to the orifice of a pressure nozzle.

Spray-dryer diameters for industrial applications are large, with 8–30 ft being common. Evaporation rates of up to 2,600 lb/h have been achieved in an 18-ft-diameter by 18-ft-high spray dryer equipped with a centrifugal-disk atomizer and supplied with an aqueous feed solution of 7 wt% dissolved solids and 11,000 cfm of air at 600°F. Larger spray dryers can evaporate up to 15,000 lb/h. Solutions, slurries, and pastes of the following materials are spray dried: detergents, blood, milk, eggs, starch, yeast, zinc sulfate, lignin, aluminum hydroxide, silica gel, magnesium chloride, manganese sulfate, aluminum sulfate, urea resin, sodium sulfide, coffee extract, tanning extract, color pigments, tea, tomato juice, polymer resins, and ceramics.

Drum Dryers

Approximately 100 years ago, and before the 1920s when spray dryers were introduced with applications for the drying of milk and detergent solutions, drum dryers were introduced to process solutions, slurries, and pastes with indirect heat. The first such dryer was the double-drum dryer, shown in Figure 18.14a, which is still the most versatile and widely applied drum dryer. It consists of two metal, cylindrical drums of identical size (1–5 ft diameter by 1.5–12 ft long), mounted side-by-side. One drum is movable so that the smallest distance between the two drums (nip) can be adjusted. The drums are heated on the inside by condensing steam at pressures as high as 12 atm. The feed to be dried enters at the top from a perforated pipe that runs the length of the two drums or from a pipe that swings like a pendulum from end-to-end of the drums. The drums are rotated in opposite directions toward each other at the top, as shown, causing the feed to form, on the hot surface of the drums, a clinging film or coating whose thickness is controlled by the nip. As the drums rotate, heat is transferred from the drum surfaces to the coating, causing it to dry. If the moisture is water, it exits as steam through a vapor hood mounted above the drums. If the moisture is a solvent or if the solid is dustable, a dryer enclosure can be provided. When the coating has made about three-quarters of a complete rotation, it is scraped off the drum surfaces by doctor blades that run the length of the drums. The dried material falls off as surface powder, chips, or more commonly as flakes, which typically are 1–3 mm thick, into conveyors. By adjusting (1) drum rotation rate, usually from 1 to 30 rpm, (2) drum-surface temperature, which is usually just a few degrees below the inside, condensing-steam temperature; (3) feed temperature; and (4) coating thickness, the moisture content of the dried material can be controlled. Drying times are in the range of 3–20 s. Performance data given by Walas [3] show capacities of double-drum dryers to be in the range of 1–60-kg dried product/h-m^2 of drum surface for feed moisture contents ranging from 10 to 90 wt%.

For drum dryers to be effective, the coating must adhere to the drum surface, which is often chrome plated. When necessary, other drum and feeding arrangements can be employed. For example, the twin-drum dryer with top feed shown in Figure 18.14b is not influenced by drum spacing because the drums rotate away from each other at the top. Much thicker coatings can then be formed and materials, like inorganic crystals, that might score or cause other damage to closely spaced drums can be processed. To improve the likelihood of the feed adhering to the drums, the feed may be splashed onto the surfaces of the twin-drum dryer as shown in Figure 18.14c.

For very viscous solutions or pastes that might cause undue pressure to be put on the surfaces of a double-drum dryer, a single-drum dryer can be used. The coating can be applied by using a top feed with applicator rolls, as shown in Figure 18.14d. If a porous product (e.g., malted milk) is desired or if the material is temperature-sensitive, a

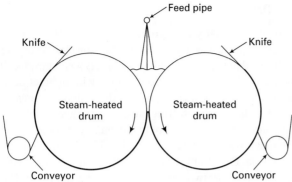

(a) Double-drum dryer

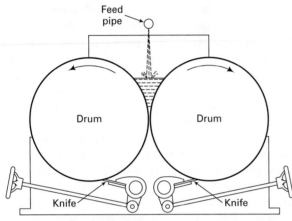

(b) Twin-drum dryer with top feed

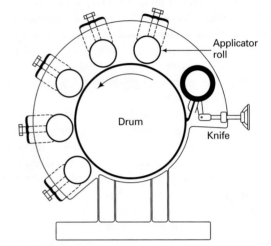

(c) Twin-drum dryer with splash feed

(d) Single-drum dryer with applicator feed

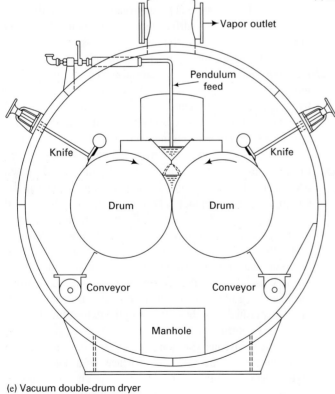

(e) Vacuum double-drum dryer

Figure 18.14 Drum dryers.

single-drum dryer or double-drum dryer, as shown in Figure 18.14e, can be enclosed so that a vacuum can be pulled to reduce the boiling point of the moisture.

Drum dryers are used to dry a wide variety of materials, including starch solutions, brewers yeast, potatoes, skim milk, malted milk, coffee, tanning extract, and vegetable glue; slurries of $Mg(OH)_2$, $Fe(OH)_2$, and $CaCO_3$; and solutions of sodium acetate, Na_2SO_4, Na_2HPO_4, $CrSO_4$, and various organic compounds. However, many of the applications of drum dryers are now handled with spray dryers. Drum dryers belong to a class of hot-cylinder dryers. Units with large numbers of cylinders in series and parallel are used to dry continuous sheets of woven fabrics and paper pulp at evaporation rates of about 10 kg/h-m^2.

Infrared Drying

In direct-heat dryers, the transfer of heat by convection from hot gases to the wet material is often supplemented by thermal radiation from hot surfaces that surround the material. However, this radiant-heat contribution is usually a minor one, such that it is often ignored. For the drying of certain films, sheets, and coatings, however, the use of thermal radiation as the major source of heat offers distinct advantages.

Radiant energy is released from all matter as a result of oscillations and transitions of its electrons. For gases and transparent solids and liquids, the radiation can be emitted from throughout the volume of the matter. For opaque solids and liquids, this internal radiation is quickly absorbed by adjoining molecules so that the net transfer of energy by radiation is only from the surface. Of greatest importance in radiation heat transfer for the drying of solids is the transfer of radiation from a hot, opaque surface through a nonabsorbing gas or vacuum to the material being dried. This transfer can be viewed as the propagation of discrete photons (quanta) and/or as the propagation of electromagnetic waves, consisting, as shown in Figure 18.15a, of electric and magnetic fields that oscillate at right angles to each other and to the direction in which the radiation travels. As shown in the electromagnetic spectrum of Figure 18.15b, the wavelength, λ, of the radiation, which depends on the manner in which it is generated, covers an exceedingly wide range from gamma rays of 10^{-8} μm to long radio waves of 10^{10} μm. Regardless of the wavelength, all radiation waves travel at the speed of light, c, in the particular medium, which for a vacuum is 2.998×10^8 m/s. Accordingly, the following relationship exists between the frequency of the wave, v, and the wavelength, λ:

$$v = c/\lambda \qquad (18\text{-}1)$$

The frequency is usually expressed in Hz, which is one cycle/s. The energy transmitted by the wave depends on its frequency and is expressed in terms of the energy, E, of a photon by

$$E = hv \qquad (18\text{-}2)$$

where h = Planck's constant = 6.62608×10^{-34} J-s.

A solid, opaque surface can emit infrared radiation by virtue of its temperature. This type of radiation is invisible

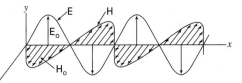

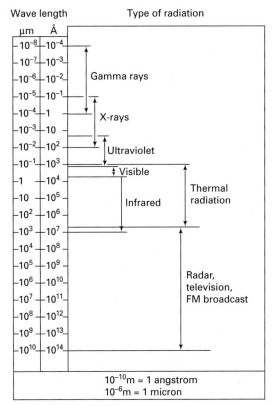

z E is electric component   H is magnetic component
E_o and H_o are amplitudes

(a) Electromagnetic wave

10^{-10}m = 1 angstrom
10^{-6}m = 1 micron

(b) Electromagnetic spectrum

Figure 18.15 Radiation.

and has a wavelength, as shown in Figure 18.15b, in the range of 0.75–100 μm. If the surface emitting the radiation is a so-called blackbody, b, such that the maximum amount of radiation will be emitted, that amount will be distributed over a range of wavelengths, depending on the temperature, as governed by the Planck distribution, which, in terms of radiant heat leaving diffusely from a unit area of surface, is

$$E_{\lambda,b} = C_1/\{\lambda^5[\exp(C_2/\lambda T) - 1]\} \qquad (18\text{-}3)$$

where the units of $E_{\lambda,b}$ are W/m^2-μm, $C_1 = 3.742 \times 10^8$, $C_2 = 1.439 \times 10^4$, T is in K, and λ is in μm. When (18-3) is integrated over the entire range of wavelengths, the result is the Stefan-Boltzmann equation

$$E_b = \sigma T^4 \qquad (18\text{-}4)$$

where $\sigma = 5.67051 \times 10^{-8}$ W/m^2-K^4 and T is in K. Thus, as the temperature of the infrared heat source is increased, the rate of heat transfer increases exponentially.

Sources of infrared radiant heat at surface temperatures in the range of 600–2,500 K are electrically heated metal-sheath rods, quartz tubes, and quartz lamps; and ceramic-enclosed gas burners. When the radiant energy reaches the material being dried, it is usually absorbed at the surface, from where it is transferred into the interior by conduction. In this respect, infrared radiant heat transfer to the surface is much like convective heat transfer to the surface. However, if the effective thermal conductivity of the material is low, the surface temperature may rise to an undesirable value, particularly if high-temperature, infrared-radiation sources are used. For that reason, infrared drying must be used with caution. Applications of growing interest include the drying of paper, paints, enamels, inks, glue-on flaps, and textiles. Continuous infrared dryers are more common than batch infrared dryers.

Dielectric and Microwave Drying

In contrast to infrared drying, dielectric and microwave drying involves the low-frequency, long-wavelength end of the electromagnetic spectrum of Figure 18.15b, where radio waves and microwaves reside. With nonelectrically conducting materials, heat is not absorbed at the surface but is generated throughout the material, reducing the need for heat conduction within the material and, thus, making this type of drying unique. Furthermore, the rate of energy dissipation in the material can be controlled over a very wide range. Other advantages over the more conventional drying methods include: (1) efficiency of energy usage because the energy dissipation occurs mainly in the moisture rather than in the solid substrate of other parts of the dryer; (2) operation at low temperatures, thus avoiding high material surface temperatures; and (3) more rapid drying. Dielectric and microwave drying is particularly useful for preheating materials and for removing the final traces of internal moisture. Dielectric drying is confined to frequencies between 1 and 150 MHz, while microwave drying utilizes frequencies from 300 MHz to 300 GHz. However, by international agreement, most microwave drying is done at only 915 and 2,450 MHz, as discussed by Mujumdar [1].

Equipment for dielectric and microwave drying consists of an energy generator and an applicator. The primary source of energy is the common alternating current of 50–60 Hz. A generator is used to boost this frequency to the much higher values quoted above. A negative-grid triode tube is used with dielectric systems, while magnetron or klystron tubes are used with microwave systems. Dielectric energy is usually applied by electrodes of various types and shapes, between which is placed the material to be dried. Microwave systems often use hollow, rectangular, metallic waveguides.

Dielectric systems are used to dry bulky materials such as lumber, ceramic monoliths, foam rubber, breakfast cereals, dog biscuits, crackers, biscuits, and cookies, as well as films, coatings, and other materials, such as paper, inks, adhesives, textiles, and penicillin, where high surface temperatures must be avoided. Microwave systems are used to dry pasta, onions, seaweed, baseball bats, potato chips, pharmaceuticals, ceramic filters, and sand casting molds.

Freeze Drying

In freeze drying (also called lyophilization), the moisture in the feed material is first frozen, by cooling the material, and then sublimed by conductive, convective, and/or radiant heating. Because the structure and properties of the material to be dried are hardly altered by freeze drying, it has been widely applied industrially to the drying of biological materials, pharmaceuticals, and foodstuffs. Products of freeze drying are porous and nonshrunken. When foodstuffs are dehydrated by freeze drying and then stored under a dry, inert gas, they evade deterioration almost indefinitely and can be rehydrated almost perfectly to their original state for later consumption. The first major application of freeze drying was for the preservation of blood plasma during World War II.

When the moisture is water, the material must be cooled to at least $0°C$ to freeze the water if it is free, and even lower if the water is dissolved in the material. Most freeze drying is conducted with the material at $-10°C$ or lower. At this temperature, ice has a vapor pressure of only 2 torr. Therefore, freeze drying must be conducted under a high vacuum. The feed material is usually in the form of a slab or particles. In both cases, heat for sublimation must be transferred from the heat source to the material under controlled conditions so that the moisture does not reach the melting point. In some cases, an even lower temperature, called the scorch point, must not be exceeded to avoid degradation of the material. During the drying period, which may take up to 20 hours, the resistance to heat transfer increases because an interface develops, between the porous freeze-dried layer and the frozen material, and gradually recedes into the material. The thermal resistance of the porous structure is greater than that of the frozen material.

For small quantities of biological and pharmaceutical materials, freeze drying is generally conducted batchwise on trays in vacuum cabinets, where the drying step follows the freezing step. The sublimed ice is desublimed on a cold metal plate that resides either inside the cabinet and adjacent to the trays or in a separate adjoining vessel. During the sublimation step, heat transfer is usually by conduction from the bottom and side surfaces of the tray, which contains coils of passages through which a heating fluid, e.g., vacuum steam, passes. For large quantities of foodstuffs, continuous freeze drying can be employed, as shown in Figure 18.16, using trays of prefrozen materials that are transported through a tunnel, past fixed heating platens, with vacuum locks at either end. With granular materials, drying times of less than 1 h can be achieved. Continuous freeze drying of small particles can also be accomplished rapidly in a fluidized bed, where heat transfer for sublimation is by convection and radiation. However, the resulting agitation of

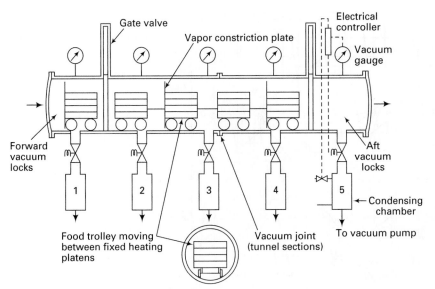

Figure 18.16 Tunnel freeze dryer.

the particles can cause breakage and dusting. In some cases, freeze drying can utilize infrared and microwave heating. Freeze drying is used for the sublimation of moisture from seafood, meat, vegetables, fruits, coffee, concentrated beverages, pharmaceuticals, blood plasma, and biological materials.

18.2 PSYCHROMETRY

For moisture to be evaporated from a wet solid, it must be heated to a temperature at which the vapor pressure of the moisture exceeds the partial pressure of the moisture in the gas in contact with the wet solid. In an indirect-heat dryer,

Table 18.4 Definitions of quantities useful in psychrometry: A = moisture, B = moisture-free gas, ideal-gas conditions

Quantity	Definition	Relationship	
Absolute, mass humidity	Moisture content of a gas by mass	$\mathcal{H} = \dfrac{M_A p_A}{M_B(P - p_A)}$	(18-5)
Molal humidity	Moisture content of a gas by mols	$\mathcal{H}_m = \dfrac{p_A}{P - p_A}$	(18-6)
Saturation humidity	Humidity at saturation	$\mathcal{H}_s = \dfrac{M_A P_A^s}{M_B(P - P_A^s)}$	(18-7)
Relative humidity (relative saturation as a percent)	Ratio of partial pressure of moisture to partial pressure of moisture at saturation	$\mathcal{H}_R = 100\% \times \dfrac{p_A}{P_A^s}$	(18-8)
Percentage humidity (percent saturation)	Ratio of humidity to humidity at saturation	$\mathcal{H}_P = 100\% \times \dfrac{\mathcal{H}}{\mathcal{H}_s}$	(18-9)
Humid volume	Volume of moisture–gas mixture per unit mass of moisture-free gas	$v_H = \dfrac{RT}{P}\left(\dfrac{1}{M_B} + \dfrac{\mathcal{H}}{M_A}\right)$	(18-10)
Humid heat	Specific heat of moisture–gas mixture per unit mass of moisture-free gas	$C_s = (C_P)_B + (C_P)_A \mathcal{H}$	(18-11)
Total enthalpy	Enthalpy of moisture–gas mixture per unit mass of moisture-free gas referred to temperature, T_o	$H = C_s(T - T_o) + \Delta H_o^{\text{vap}} \mathcal{H}$	(18-12)
Dew-point temperature	Temperature at which moisture begins to condense when mixture is cooled	T_{dew}	
Dry-bulb temperature	Temperature of mixture	T_d	
Wet-bulb temperature	Steady-state temperature attained by a wet-bulb thermometer	T_w	
Adiabatic-saturation temperature	Temperature attained when a gas is saturated with moisture in an adiabatic process	T_s	

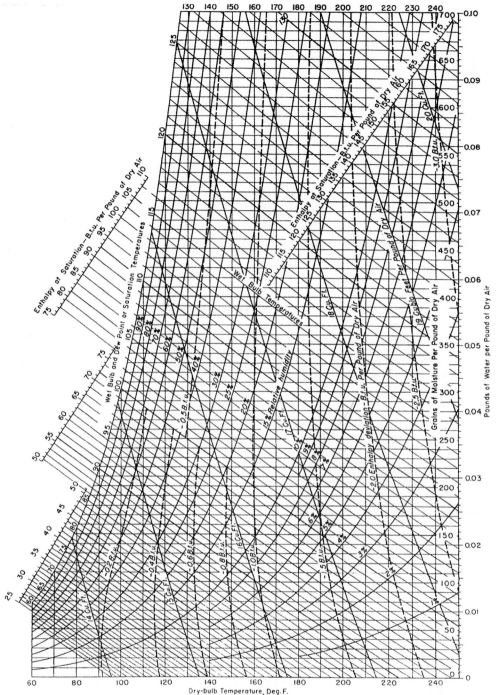

Figure 18.17 Psychrometric (humidity) chart for air–water at 1 atm.

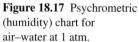

where little or no gas is used to carry away the moisture as vapor, the partial pressure of the moisture approaches the total pressure and the temperature of evaporation approaches the boiling point of the moisture at the prevailing pressure as long as the moisture is free liquid at the surface of the solid. If the moisture interface recedes into the solid, a temperature higher than the boiling point is necessary at the solid–gas interface to transfer the heat for evaporation from that interface to the liquid–gas interface. In addition, if the moisture level drops to a point where it is all sorbed by the solid, the vapor pressure of this sorbed moisture is less than the

pure vapor pressure and an even higher temperature is required to evaporate the moisture. In a direct-heat dryer, similar situations can occur except that the temperature at which moisture evaporation occurs is dependent on the moisture content of the gas. Calculations involving the properties of moisture–gas mixtures for application to drying are most conveniently carried out with psychrometric charts.

A typical chart, given in Figure 18.17, is that for air–water vapor mixtures at 1-atm total pressure. Included in this chart are properties that are included in the list of definitions in Table 18.4, which applies to general moisture–gas mixtures that obey the ideal-gas law. The definitions given by (18-5)–(18-9) for humidity and by (18-10)–(18-12) for humid volume, humid heat, and enthalpy are in terms of a unit mass or mole of moisture-free gas. In the case of a water vapor–air mixture, the term dry air is used. The use of Figure 18.17 and its basis in the relationships of Table 18.4 is illustrated in the following example.

EXAMPLE 18.1

Air at 131°F and 1 atm enters a direct-heat dryer with a humidity, $\mathcal{H}$, of 0.03 lb H_2O/lb H_2O-free air. Determine the following from the psychrometric chart of Figure 18.17 and/or relationships of Table 18.4:

(a) Molal humidity, $\mathcal{H}_m$

(b) Saturation humidity, $\mathcal{H}_s$

(c) Relative humidity, $\mathcal{H}_R$

(d) Percentage humidity, $\mathcal{H}_P$

(e) Humid volume, v_H

(f) Humid (specific) heat, C_s

(g) Enthalpy, H

SOLUTION

At 131°F, the vapor pressure of water is 118 torr = 0.155 atm. A = H_2O, B = air, and $\mathcal{H}$ = 0.03 lb H_2O/lb H_2O-free air.

(a) Combining (18-5) and (18-6),

$$\mathcal{H}_m = \frac{M_B}{M_A}\mathcal{H} = \frac{M_{air}}{M_{H_2O}}\mathcal{H} = \frac{28.97}{18.02}(0.03) = 0.048\frac{\text{lb mol } H_2O}{\text{lb mol dry air}}$$

(b) From (18-7),

$$\mathcal{H}_s = \frac{18.02}{28.97}\left(\frac{0.155}{1-0.155}\right) = 0.114\frac{\text{lb } H_2O}{\text{lb dry air}}$$

(c) From a rearrangement of (18-6),

$$p_{H_2O} = \frac{P\mathcal{H}_m}{1+\mathcal{H}_m} = \frac{(1)(0.048)}{1+0.048} = 0.0458 \text{ atm}$$

From (18-8),

$$\mathcal{H}_R = 100\left(\frac{0.0458}{0.155}\right) = 29.5\%$$

The same result is obtained from Figure 18.17.

(d) From (18-9),

$$\mathcal{H}_P = 100\left(\frac{0.03}{0.114}\right) = 26.3\%$$

(e) From (18-10), for $R = 0.730\frac{\text{atm-ft}^3}{\text{lbmol-}°R}$, $T = 131 + 460$

$$= 591°R,$$

$$v_H = \frac{0.730(591)}{1}\left(\frac{1}{28.97} + \frac{0.03}{18.02}\right) = 15.6 \text{ ft}^3/\text{lb dry air}$$

which agrees with Figure 18.17.

(f) From (18-11), using

$$(C_P)_{air} = 0.24 \text{ Btu/lb-}°F$$

$$(C_P)_{steam} = 0.45 \text{ Btu/lb-}°F$$

$$C_s = 0.24 + (0.45)(0.03) = 0.254\frac{\text{Btu}}{\text{lb dry air}}$$

(g) Equation (18-12) assumes that the enthalpy datum refers to air as a gas and water as a liquid. Taking $T_o = 32°F$ and $\Delta H_o^{vap} = 1075$ Btu/lb, (18-12) gives

$$H = 0.254(131 - 32) + 1,075(0.03) = 57.4 \text{ Btu/lb dry air}$$

EXAMPLE 18.2

At a certain location in a dryer where benzene is being evaporated from a solid, nitrogen gas at 50°F and 1.2 atm has a relative humidity for benzene of 35%.

Determine:

(a) Partial pressure of the benzene if the vapor pressure of benzene at 50°F = 45.6 torr.

(b) Humidity of the nitrogen-benzene mixture.

(c) Saturation humidity of the mixture.

(d) Percentage humidity of the mixture.

SOLUTION

$$\mathcal{H}_R = 35\%, \ P = 1.2\text{ atm} = 912\text{ torr}$$

$$A = \text{benzene}, \ B = N_2$$

$$M_A = 78.1, \ M_B = 28, \ p_{benzene}^s = 45.6\text{ torr}$$

(a) From (18-8),

$$p_{benzene} = \frac{p_{benzene}^s \mathcal{H}_R}{100} = \frac{(45.6)(35)}{100} = 16\text{ torr}$$

(b) From (18-5),

$$\mathcal{H} = \left(\frac{78.1}{28}\right)\left(\frac{16}{912-16}\right) = 0.050\frac{\text{lb benzene}}{\text{lb dry nitrogen}}$$

(c) From (18-7),

$$\mathcal{H}_s = \left(\frac{78.1}{28}\right)\left(\frac{45.6}{912-45.6}\right) = 0.147\frac{\text{lb benzene}}{\text{lb dry nitrogen}}$$

(d) From (18-9),

$$\mathcal{H}_P = 100\left(\frac{0.050}{0.147}\right) = 34\%$$

Wet-bulb Temperature

In general, the temperature at which moisture evaporates in a direct-heat dryer is complex to determine and can vary from the inlet to the outlet of the dryer. However, when the dryer operates isobarically and adiabatically, with all of the energy for moisture evaporation supplied from the hot gas by convective heat transfer and with no energy required for heating the wet solid to the evaporation temperature, application of the following simplified heat- and mass-transfer equations leads to an expression for the temperature of evaporation at a particular location in a dryer operating under continuous, steady-state conditions, or at a particular point in time for a dryer operating batchwise.

Because it is further assumed that the moisture being evaporated is free liquid, exerting its full vapor pressure, at the surface of the solid, this temperature of evaporation is referred to as the wet-bulb temperature, T_w, because it can be measured by covering the bulb of a thermometer with a wick saturated with the moisture to be evaporated, and passing a large quantity of partially saturated gas past the wick, as indicated in Figure 18.18a.

Refer to Figure 18.18b, where the wetted wick is replaced by an incremental amount of wet solid. Assume that the area for heat transfer = area for mass transfer = A. At steady-state conditions, the rate of convective heat transfer from the gas to the wet solid is given by Newton's law of cooling:

$$Q = h(T - T_w)A \qquad (18\text{-}13)$$

The molar rate of mass transfer of the evaporated moisture from the wet surface of the solid, A, is

$$n_A = \frac{k_y(y_{A_w} - y_A)A}{(1 - y_A)_{LM}} \qquad (18\text{-}14)$$

An enthalpy balance on the moisture being evaporated and heated to the gas temperature couples the heat and mass transfer to give

$$Q = n_A M_A \left[\Delta H_w^{vap} + (C_P)_A (T - T_w) \right] \quad (18\text{-}15)$$

To obtain a simplified relationship for the coupling in terms of T and $\mathcal{H}$, assume that the mole fraction of moisture in the gas is small in the bulk gas and at the wet solid–gas interface. Then, the bulk-flow effect in (18-14) becomes $(1 - y_A)_{LM} \sim 1.0$. Also, from (18-5), replacing p_A with $y_A P$

and solving for y_A, we obtain

$$y_A = \frac{\mathcal{H}M_A}{\dfrac{1}{M_B} + \dfrac{\mathcal{H}}{M_A}} \approx \frac{\mathcal{H}M_B}{M_A} \qquad (18\text{-}16)$$

In addition, assume that the latent-heat effect in (18-15) is much greater than the sensible-heat effect. Then

$$\Delta H_w^{vap} + (C_P)_A(T - T_w) \approx \Delta H_w^{vap} \qquad (18\text{-}17)$$

Combining (18-13) to (18-16) after making all these simplifications, we obtain

$$T_w = T - \frac{k_y M_B \Delta H_w^{vap}}{h}(\mathcal{H}_w - \mathcal{H}) \qquad (18\text{-}18)$$

Assume that the Chilton–Colburn analogy for mass and heat transfer applies. Then, from (3-165) and (3-228),

$$\frac{k_y M}{G} N_{Sc}^{2/3} = \frac{h}{C_P G} N_{Pr}^{2/3} \qquad (18\text{-}19)$$

If $M \approx M_B$, (18-19) simplifies to

$$\frac{k_y M_B}{h} = \frac{1}{(C_P)_B}\left(\frac{1}{N_{Le}}\right)^{2/3} \qquad (18\text{-}20)$$

where

$$N_{Le} = \text{Lewis number} = N_{Sc}/N_{Pr} \qquad (18\text{-}21)$$

The reciprocal of the Lewis number is sometimes referred to as the Luikov number, N_{Lu}. Substitution of (18-20) into (18-18) gives

$$T_w = T - \frac{\Delta H_w^{vap}}{C_P}\left(\frac{1}{N_{Le}}\right)^{2/3}(\mathcal{H}_w - \mathcal{H}) \quad (18\text{-}22)$$

In Equation (18-18), which defines the wet-bulb temperature, or (18-22), it is important to note that $\mathcal{H}_w$ refers to the saturation humidity at temperature T_w.

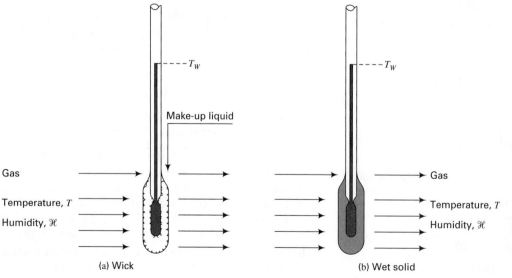

(a) Wick

(b) Wet solid

Figure 18.18 Wet-bulb temperature.

[From W.L. McCabe, J.C. Smith, and P. Harriott, *Unit Operations of Chemical Engineering*, 5th ed., McGraw-Hill, New York (1993) with permission.]

Table 18.5   Lewis Number for Liquids Evaporating into Air at 25°C

Liquid	N_{Le}	$(N_{Le})^{2/3} = $ Psychrometric Ratio
Benzene	2.44	1.812
Carbon tetrachloride	2.67	1.923
Chloroform	3.08	2.114
Ethyl acetate	2.58	1.880
Ethylene tetrachloride	3.05	2.101
Metaxylene	3.18	2.165
Methanol	1.37	1.233
Propanol	1.85	1.506
Toluene	2.64	1.908
Water	0.855	0.901

By coincidence, $(N_{Le})^{2/3}$ for the air–water vapor system at 25°C is close to 1.0 (actually 0.901). For air–organic vapor systems, as listed in Table 18.5, which is taken from Keey [7], $(N_{Le})^{2/3}$ is much less than 1.0. As will be shown in the next section, the value of $(N_{Le})^{2/3}$ has a great impact on the variation with location or time of the temperature of moisture evaporation in a direct-heat dryer.

Adiabatic-Saturation Temperature

In order to determine the possible change in the wet-bulb temperature of the wet solid, as it dries with location or time in the dryer, it is necessary to consider changes in the temperature and humidity of the gas as it cools because of the transfer of heat to the wet solid, and the transfer of moisture from the wet solid. A simplified relationship between gas temperature and humidity can be derived by considering the adiabatic saturation of a gas with an excess of moisture as pure liquid. Referring to the system of Figure 18.19, partially saturated gas at temperature T, humidity $\mathcal{H}$, and mass flow rate m_B (dry basis), together with liquid entering at temperature T_L and mass flow rate m_L, enters an isobaric and adiabatic chamber where a fraction of the liquid, (E/m_L), is

vaporized to saturate the gas. The gas and excess liquid leave the chamber in equilibrium at temperature T_s. An enthalpy balance using (18-12) for the enthalpy of the gas phase, but with the reference temperature $T_o = T_s$ (so as to simplify the balance), gives

$$m_L(C_{P_A})_L(T_L - T_s) + m_B[C_{s_{in}}(T - T_s) + \Delta H_s^{vap}\mathcal{H}]$$
$$= m_B[C_{s_{out}}(T_s - T_s) + \Delta H_s^{vap}\mathcal{H}_s] \qquad (18\text{-}23)$$
$$+ (m_L - E)(C_{P_A})_L(T_s - T_s)$$

Assume that the sensible heat required to heat the liquid from T_L to T_s is negligible. Then (18-23) simplifies to an equation for the adiabatic saturation temperature:

$$T_s = T - \frac{\Delta H_s^{vap}}{(C_s)_{in}}(\mathcal{H}_s - \mathcal{H}) \qquad (18\text{-}24)$$

Equation (18-24) also applies for the determination of a gas temperature and humidity intermediate between T and T_s and $\mathcal{H}$ and $\mathcal{H}_s$, respectively, if sensible heat for the liquid is ignored. If (18-24) is compared to (18-18), it is seen that the wet-bulb and adiabatic-saturation temperatures would be equal if the $h/k_y M_B$ were equal to $(C_s)_{in}$. For the air–water system, as shown by Lewis [8] and referred to as the Lewis relation, this is almost the case, with

$$\frac{h}{k_y M_B} \approx 0.216 \, \text{Btu/lb-°F}$$

compared to $C_s \approx 0.24$ Btu/lb-°F. However, typically, a small amount of radiation heat transfer to the wet solid supplements the convective heat transfer. If h in the psychrometric ratio is replaced by $(h_c + h_r)$, the corrected ratio is almost identical to the humid heat at low humidities. Accordingly, for the air–water system, the wet-bulb temperature is taken to be equal to the adiabatic-saturation temperature and only one family of lines is shown for these two temperatures on the psychrometric chart. From (18-24), these lines have a negative slope of $\Delta H_s^{vap}/(C_s)_m$. The following example illustrates the use of the chart in Figure 18.17 to determine the relationship between gas (dry-bulb) temperature, wet-bulb temperature, and humidity. The example also illustrates the use of (18-24).

EXAMPLE 18.3

For the conditions of Example 18.1, determine the wet-bulb temperature, assuming it is equal to the adiabatic-saturation temperature, by using (a) the psychrometric chart and (b) (18-24).

SOLUTION

(a) In Figure 18.17, the point $T = T_d = 131$°F and $\mathcal{H} = 0.03$ lb H_2O/lb dry air is plotted. This point lies just above the adiabatic-saturation-temperature line of 95°F at about 96.5°F.

(b) Equation (18-24) involves an iterative calculation to determine $T_w = T_s$ because ΔH_s^{vap} and $\mathcal{H}_s$ are unknown.

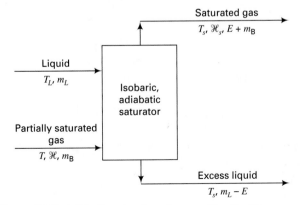

Figure 18.19 Adiabatic saturation of a gas with a liquid.

Assume

$$T_s = 95°F$$

$$\Delta H_s^{\text{vap}} = 1039.8 \text{ Btu/lb from steam tables}$$

$$(C_s)_{\text{in}} = 0.254 \text{ Btu/lb from Example 18.1}$$

$$P_A^s = 42.2 \text{ torr}$$

$$= \text{vapor pressure of water from Example 18.1}$$

From (18-7),

$$\mathcal{H}_s = \frac{18.02}{28.97}\left(\frac{42.2}{760 - 42.2}\right) = 0.0366 \frac{\text{lb H}_2\text{O}}{\text{lb dry air}}$$

From (18-24),

$$T_s = 131 - \frac{1039.8}{0.254}(0.0366 - 0.03) = 104°F$$

Assume

$$T_s = 97°F$$

$$\Delta H_s^{\text{vap}} = 1038.7 \text{ Btu/lb}$$

$$P_A^s = 45 \text{ torr}$$

$$\mathcal{H}_s = \frac{18.02}{28.97}\left(\frac{45}{760 - 45}\right) = 0.0391 \frac{\text{lb H}_2\text{O}}{\text{lb dry air}}$$

$$T_s = 131 - \frac{1038.7}{0.254}(0.0391 - 0.03) = 94°F$$

As can be seen, this iterative calculation is very sensitive. By interpolation, T_s is almost identical to the value of 96.5°F read from Figure 18.17.

For systems other than air–water, the Lewis relation does not hold and it is common to indicate this by defining a psychrometric ratio as $h/(k_y C_s M_B)$. Calculated values of this ratio, using the Chilton–Colburn analogy, are included in Table 18.5 for several systems. Comparing (18-18) for the wet-bulb temperature to (18-24) for the adiabatic-saturation temperature, it can be noted that since the inverse of the psychrometric ratio for all air–organic vapor systems in Table 18.5 is less than one, such systems will have slopes of adiabatic-saturation lines that are less than slopes of wet-bulb lines. An example of this is the psychrometric chart for the air–toluene system at 1 atm shown in Figure 18.20. The use of this chart is illustrated in the next example.

EXAMPLE 18.4

Air is used to dry a solid wet with toluene at 1 atm. At a location where the air has a temperature of 140°F and a relative humidity of 10%, determine the humidity, the adiabatic-saturation temperature, and the wet-bulb temperature.

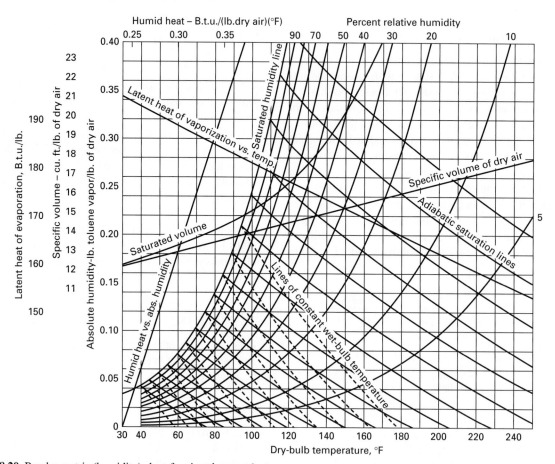

Figure 18.20 Psychrometric (humidity) chart for air–toluene at 1 atm.
[From *Perry's Chemical Engineers' Handbook*, 6th ed., R.H. Perry, D.W. Green, and J.O. Maloney, Eds., McGraw-Hill, New York (1984) with permission.]

SOLUTION

From Figure 18.20, the humidity is 0.062 lb toluene/lb dry air at the intersection point of a vertical temperature line and a curved percent relative-humidity line.

By following an interpolated adiabatic-saturation line from that intersection point to the saturated-humidity line, $T_s = 83°F$.

By following an interpolated wet-bulb line from the intersection point to the saturated-humidity line, $T_w = 92°F$.

Thus, the wet-bulb temperature is higher than the adiabatic-saturation temperature. In the next section, this causes a reduction in the driving force available for heat transfer during the drying of solids wet with organic moisture.

Moisture-Evaporation Temperature

The combination of the concepts of the adiabatic-saturation temperature and the wet-bulb temperature can be used to track the gas temperature and the moisture-evaporation (solid) temperature, T_v, with respect to location or time when removing surface moisture from a wet solid. The accuracy of the tracking is subject to the degree to which the assumptions of the two concepts are valid. If the moisture is water, T_v will be constant and equal to the constant T_w of the gas. If the moisture is an organic compound with properties similar to those in Table 18.5, T_v will still be equal to T_w, but, as shown next, that temperature will not be constant, but will decrease as the gas temperature decreases.

Let T_o and $\mathcal{H}_o$ be hypothetical entering conditions of a gas being used to dry a wet solid in an adiabatic, direct-heat dryer. At any point in the dryer, the conditions of the gas are given from (18-24) as

$$T_o - T_g = \frac{\Delta H_s^{vap}}{(C_s)_o}(\mathcal{H}_g - \mathcal{H}_o) \qquad (18\text{-}25)$$

Assuming a quasi-steady-state transport condition at any point, the gas–liquid moisture interface conditions of T_v and $\mathcal{H}_v$ are related by the following form of (18-18):

$$T_v - T_g = \frac{k_y M_B \Delta H_s^{vap}}{h}(\mathcal{H}_g - \mathcal{H}_v) \qquad (18\text{-}26)$$

For a continuous dryer, (18-26) holds regardless of whether the flows of gas and wet solid are countercurrent or cocurrent. Equations (18-25) and (18-26) can be combined to give

$$\frac{T_v - T_g}{T_o - T_g} = \left(\frac{k_y M_B (C_s)_o}{h}\right)\left(\frac{\mathcal{H}_g - \mathcal{H}_v}{\mathcal{H}_g - \mathcal{H}_o}\right) \qquad (18\text{-}27)$$

The coefficient of the right-hand side of (18-27) is the inverse of the psychrometric ratio. For the air–water system, that ratio can be assumed to be 1.0, giving

$$\frac{T_v - T_g}{T_o - T_g} = \frac{\mathcal{H}_g - \mathcal{H}_v}{\mathcal{H}_g - \mathcal{H}_o} \qquad (18\text{-}28)$$

Now assume that the wet solid in contact with the initial or entering gas is at the wet-bulb temperature, T_w, of the gas.

Then, at that point,

$$\frac{T_w - T_g}{T_o - T_g} = \frac{\mathcal{H}_g - \mathcal{H}_w}{\mathcal{H}_g - \mathcal{H}_o} \qquad (18\text{-}29)$$

But, since T_o is fixed, T_v must remain at the value T_w regardless of the value of T_g. The result, which is independent of the direction of flow of the wet solid, is shown for the outcome of Example 18.5 below and in Figure 18.21 for air–water and air–toluene systems.

In the general case, (18-27) applies. Assume that the wet solid is dried batchwise, or flows in cocurrent flow to the gas in a continuous dryer, with an initial temperature equal to the wet-bulb temperature of the gas. Because the inverse of the psychrometric ratio is less than 1.0, T_w will now decrease as T_g decreases, as shown in the following example.

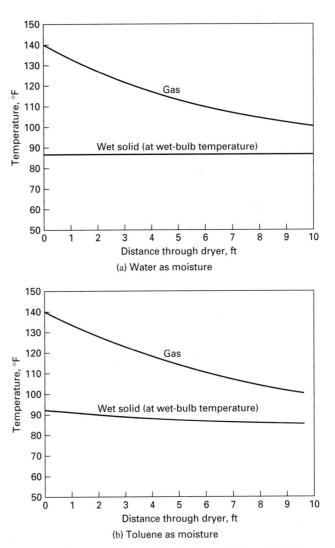

Figure 18.21 Temperature profiles for dryer of Example 18.5.

EXAMPLE 18.5

Air enters a continuous, adiabatic, direct-heat dryer at 140°F and 1 atm with the relative humidity below, and exits at 100°F. The wet solid enters in cocurrent flow at the wet-bulb temperature of the entering gas. Determine and plot the variation of the moisture-evaporation temperature, T_v, as a function of the distance through the dryer, for an exponential decrease of T_g according to the relation

$$T_g - T_s = (T_o - T_s) \exp(-0.1377z) \quad (1)$$

where z is distance through the dryer in feet and temperatures are in °F for:

(a) water moisture with entering air of 12.5% $\mathcal{H}_R$

(b) toluene moisture with entering air of 10% $\mathcal{H}_R$

SOLUTION

$$T_o = 140°F, \quad \mathcal{H}_R = 12.5\%$$

(a) From Figure 18.17, $\mathcal{H}_o = 0.015$ lb H$_2$O/lb dry air. $T_w = 86.5°F = T_s$. As the gas cools, its humidity will follow the adiabatic-saturation line.

Using (1) and Figure 18.17,

z, ft	T_g, °F	$\mathcal{H}_g$, $\dfrac{\text{lb H}_2\text{O}}{\text{lb dry air}}$
0	140.0	0.015
2	127.1	0.018
4	117.3	0.020
6	109.9	0.022
8	104.3	0.0235
10	100.0	0.0245

Equation (18-28) is only satisfied for

$$T_v = T_w = 86.5°F$$

$$\mathcal{H}_v = \mathcal{H}_w = 0.0275 \text{ lb H}_2\text{O/lb dry air}$$

For example, take $T_g = 109.9°F$, $\mathcal{H}_g = 0.022$ lb H$_2$O/lb dry air. Using (18-28), take values of $T_v = 80°F$ and 90°F, and compute the temperature and humidity ratios:

T_v, °F	$\mathcal{H}_v$, $\dfrac{\text{lb H}_2\text{O}}{\text{lb dry air}}$	$\dfrac{T_v - T_g}{T_o - T_g}$	$\dfrac{\mathcal{H}_g - \mathcal{H}_v}{\mathcal{H}_g - \mathcal{H}_o}$
86.5	0.0275	−0.777	−0.786
80.0	0.0223	−0.993	−0.043
90.0	0.0310	−0.661	−1.300

(b) From Figure 18-20, for $T_o = 140°F$, $\mathcal{H}_R = 10\%$,

$$\mathcal{H}_o = 0.062 \text{ lb toluene/lb dry air}$$

$$T_w \text{ for entering air} = 92°F$$

$$T_s = 83°F$$

From Table 18.4, 1/(psychrometric ratio) = 1/1.908 = 0.524. From (18-27),

$$\frac{T_v - T_g}{140 - T_g} = 0.524 \left(\frac{\mathcal{H}_g - \mathcal{H}_v}{\mathcal{H}_g - 0.062} \right) \quad (2)$$

From (1),

$$T_g = T_v + (140 - T_v) \exp(-0.1377\, z) \quad (3)$$

Equation (2) is solved iteratively for T_v for each value of T_g, where $\mathcal{H}_v$ is the saturation humidity at T_v, as determined from Figure 18.20. For example, for $T_g = 115.9°F$, with the humidity following the adiabatic saturation line, $\mathcal{H}_v = 0.095$, and (2) becomes:

$$\frac{T_v - 115.9}{140 - 115.9} = 0.524 \left(\frac{0.095 - \mathcal{H}_v}{0.095 - 0.062} \right)$$

or

$$T_v = 115.9 + 382.7(0.095 - \mathcal{H}_v) \quad (4)$$

This equation is solved by assuming T_v, determining $\mathcal{H}_v$ and then computing T_v.

Assumed T_v, °F	$\mathcal{H}_v$ from Fig. 18.20, lb toluene/lb dry air	T_v, °F from (4)
90	0.180	83.4
88	0.173	86.0
87	0.165	89.0

By interpolation, $T_v = 87.5°F$.

This result can also be obtained graphically from Figure 18.20 using the construction shown in Figure 18.22, with an essentially identical result. Calculations for the other values of T_g give

T_g, °F	T_v, °F
140.0	92.0
126.3	90.0
115.9	87.5
107.9	86.5
100.0	85.5

From (3), the following values of z are computed:

T_g, °F	z, ft
140.0	0.0
126.3	2.3
115.9	4.5
107.9	6.7
100.0	9.6

Thus, the moisture-evaporation temperature, which is plotted in Figure 18.21b, decreases with decreasing gas temperature as discussed above.

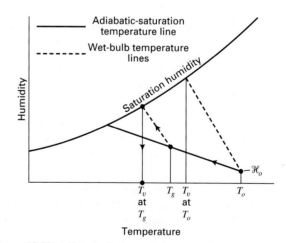

Figure 18.22 Adiabatic drying path for general vapor–moisture mixtures.

18.3 EQUILIBRIUM-MOISTURE CONTENT OF SOLIDS

As discussed by Faust et al. [9], wet solids can be grouped into the two following categories according to their drying behavior:

1. *Granular or crystalline solids that hold moisture in open pores between particles.* These are mainly inorganic materials, examples of which are crushed rocks, sand, catalysts, titanium dioxide, zinc sulfate, and sodium phosphates. During the drying process, the solid is unaffected by the removal of the moisture such that the selection of drying conditions and rate of drying are not critical to the final properties and appearance of the dried product. Materials in this category can be dried rapidly to very low moisture contents.

2. *Fibrous, amorphous, and gel-like materials that dissolve moisture or trap moisture in fibers or very fine pores.* These are mainly organic solids, including tree, plant, vegetable, and animal materials, such as wood, leather, soap, wood, eggs, glues, cereals, starch, cotton, and wool. These materials are significantly affected by moisture removal, often shrinking when dried and swelling when wetted. With this class of materials, drying in the later stages can be slow. If the surface is caused to dry too rapidly, moisture and temperature gradients can cause checking, warping, case hardening, and/or cracking. Therefore, the selection of drying conditions is a critical factor. Drying to low moisture contents is only possible when using a gas of low humidity.

In a direct-heat drying process, the extent to which moisture can be removed from a solid is limited, particularly for the second category of materials, by the *equilibrium-moisture content* of the solid, which depends on several factors including temperature, pressure, and moisture content of the gas. Even if the drying conditions produce a completely dry solid, subsequent exposure of the solid to an atmosphere of different humidity conditions can result in resorption of moisture.

A number of terms have been defined to describe and utilize the equilibrium-moisture content. These are shown in Figure 18.23 with reference to a hypothetical equilibrium isotherm. The moisture content, X, is expressed as mass of moisture per 100 mass units of bone-dry solid. This is the most commonly used way to express moisture content of a solid and is equivalent to weight % moisture on a dry-solid basis. This is analogous to expressions for humidity and is most convenient in drying calculations where the mass of bone-dry solid and bone-dry gas remain constant while moisture is transferred from the solid to the gas. Less common is weight % moisture on a wet-solid basis, W. The two moisture contents are related by the expression

$$X = \frac{100W}{100 - W} \qquad (18\text{-}30)$$

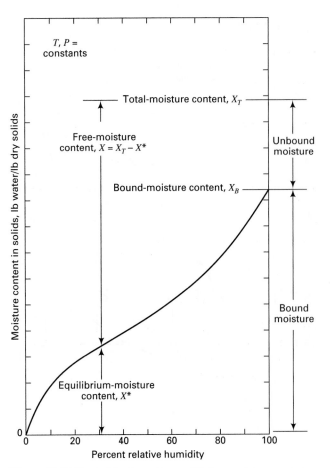

Figure 18.23 Typical isotherm for equilibrium moisture content.

or

$$W = \frac{100X}{100 + X} \qquad (18\text{-}31)$$

Rarely used is moisture content on a volume basis because the volume occupied by wet solids of the second category shrinks during drying. Also, moisture content is almost never expressed on a mole basis because the molecular weight of the dry solid may not be known.

In Figure 18.23, the equilibrium-moisture content, X^*1, is plotted for a typical solid of the second category, for a given temperature and pressure against the relative humidity, $\mathcal{H}_R$, where the limit is 100%. In some cases, humidity, $\mathcal{H}$, is used with a limit of the saturation humidity, $\mathcal{H}_s1$. At $\mathcal{H}_R = 100\%$, the equilibrium-moisture content is called *bound moisture*, X_B. If the wet solid has a total moisture content, X_T, greater than X_B, the excess, $X_T - X_B$, is called *unbound moisture*. At an intermediate relative humidity less than 100%, the excess of X_T over the equilibrium-moisture content, i.e., $X_T - X^*$, is called the *free-moisture* content.

In the presence of a saturated gas, only the unbound moisture can be removed during the drying process. For a partially saturated gas, only the free moisture can be removed. But if $\mathcal{H}_R = 0$, all solids, given enough time, may be dried to a bone-dry state. Solid materials that can contain bound moisture are *hygroscopic*. Bound moisture exhibits a vapor

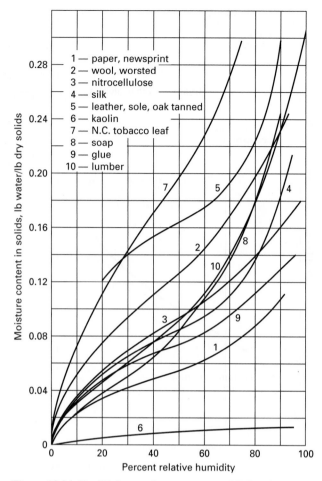

Figure 18.24 Equilibrium-moisture content at 25°C and 1 atm.

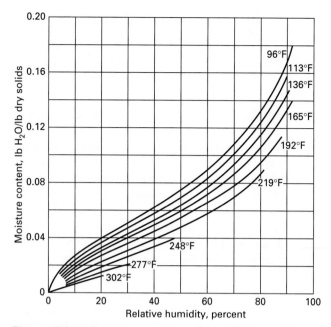

Figure 18.25 Effect of temperature on the equilibrium-moisture content of raw cotton at 1 atm.

pressure less than the normal vapor pressure. The bound-moisture content of cellular materials such as wood is often referred to as the fiber-saturation point.

Equilibrium-moisture isotherms at 25°C and 1 atm are shown for several materials of the second category in Figure 18.24. Such curves must be determined experimentally. At low values of $\mathcal{H}_R$, e.g., < 10%, moisture is bound to the solid on its surfaces as an adsorbed monomolecular layer. Such bound moisture can also be present on solids of the first category. At intermediate values of $\mathcal{H}_R$, e.g., 20–60%, multi-molecular layers may be built up successively on the mono-layer. At large values of $\mathcal{H}_R$, e.g., > 60%, moisture is held in micropores so small (e.g., < 1 μm is radius) that a lowering of the vapor pressure occurs according to the Kelvin equation (15-14). In cellular materials, such as plant and tree matter, some of the moisture is held osmotically in fibers behind semipermeable membranes of cell walls.

Temperature has a significant effect on the equilibrium-moisture content, an example of which is shown in Figure 18.25 for cotton over a temperature range of 96–302°F. At an $\mathcal{H}_R$ of 20%, the equilibrium-moisture content decreases from 0.037 to 0.012 lb H_2O/lb dry cotton. The experimental determination of equilibrium-moisture isotherms is complicated by an apparent hysteresis effect, as shown in

Figure 18.26 for sulfite pulp. The sorption and desorption curves were obtained by wetting and drying the solid, respectively. The equilibrium-moisture content measured in drying experiments is always somewhat higher, particularly in the percent relative-humidity range of 30% to 80%. According to Luikov [10] the hysteresis effect may be due to either: (1) a failure to achieve true equilibrium or (2) irreversibility of evaporation and condensation in capillaries. For the latter, a possible explanation is based on the representations of moisture in necked capillaries, as shown in Figure 18.26. For drying (desorption), the capillary contains more moisture than for wetting (sorption). Thus, for a given relative humidity, the equilibrium moisture content for drying is less than for wetting.

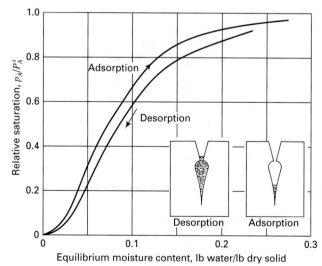

Figure 18.26 Effect of hysteresis on equilibrium-moisture content of sulfite pulp.

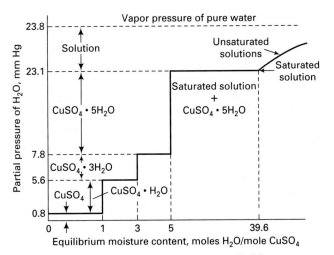

Figure 18.27 Equilibrium-moisture content for $CuSO_4$ at 25°C.

Bound moisture can also be defined as moisture that is held chemically as, e.g., water of hydration of inorganic crystals. This is one example of bound moisture dissolved in the solid. The vapor pressure of such moisture is lowered significantly below the true vapor pressure. For inorganic salts that form one or more hydrates, the hydrated form of the product will depend not only on the temperature, but also on the relative humidity of the gas in contact with the crystals. The effect of the latter for $CuSO_4$, in terms of the partial pressure of water, is shown in Figure 18.27. At 25°C, the stable hydrate is $CuSO_4 \cdot 5H_2O$. However, if the partial pressure of H_2O is between 5.6 and 7.8 torr, the trihydrate tends to form. From 5.6 to 0.8 torr, the monohydrate is favored. Below 0.8 torr, $CuSO_4$ crystals are completely free of water.

EXAMPLE 18.6

One-kilogram blocks of wet Borax laundry soap with an initial water content of 20.2 wt% on the dry basis are dried with air in a tunnel dryer at 1 atm. In the limit, if the soap were brought to equilibrium with the air at 25°C and a relative humidity of 20%, determine the kg of moisture evaporated from each block.

SOLUTION

The initial moisture content of the soap on a wet basis is obtained from a rearrangement of (18-30):

$$W = \frac{100X}{100 + X} = \frac{100(20.2)}{100 + 20.2} = 16.8\,\text{wt\%}$$

Initial weight of moisture $= 0.168(1.0) = 0.168$ kg H_2O

Initial weight of dry soap $= 1 - 0.168 = 0.832$ kg dry soap

From Figure 18.24, for soap at $\mathcal{H}_R = 0.20$, $X^* = 0.037$

Final weight of moisture $= 0.037(0.832) = 0.031$ kg

Moisture evaporated $= 0.168 - 0.031 = 0.137$ kg H_2O/kg soap block

---

18.4   DRYING PERIODS

During drying of either category of wet solids, the decrease in average moisture content, X, as a function of time, t, for fixed gas conditions in a direct-heat dryer was observed experimentally by Sherwood [11, 12] to exhibit generally the type of relationship shown in Figure 18.28a provided that the exposed surface of the solid is initially covered with observable moisture. If that curve is differentiated with respect to time and multiplied by the ratio of the mass of dry solid to the interfacial area of contact between the mass of wet solid and the gas, a plot can be made of drying-rate flux, R,

$$R = \frac{dm_v}{A\,dt} = -\frac{m_s}{A}\frac{dX}{dt} \qquad (18\text{-}32)$$

where

$m_v =$ mass of moisture evaporated

$m_s =$ mass of bone-dry solid

as a function of moisture content, as shown in Figure 18.28b. In both plots, the final equilibrium-moisture content is X^*. Although both plots can exhibit four drying periods, the periods are more distinct in the drying-rate curve. For some wet materials and/or some hot gas conditions, less than four

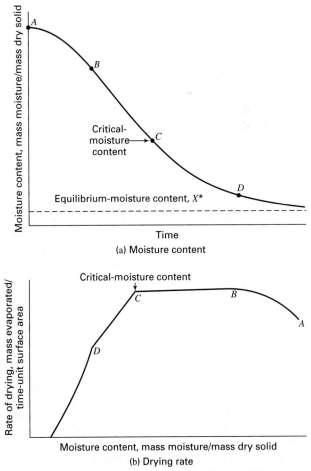

Figure 18.28 Drying curves for constant drying conditions.

drying periods may be observed. In the period from A to B, the wet solid is being preheated to an exposed-surface temperature equal to the wet-bulb temperature of the gas. Some moisture is evaporated in this preheat period, at an increasing rate, as the surface temperature increases.

At the end of the preheat period, if the wet solid is of the granular character of the first category, a cross section has the appearance of Figure 18.29a, where the exposed surface is still covered by a film of moisture. A wet solid of the second category is covered on the exposed surface by free moisture. The drying rate now becomes constant during the period from B to C, which prevails as long as free moisture still covers the exposed surface.

This surface moisture may be part of the original moisture that covered the surface or it may be moisture brought to the surface by capillary action in the case of wet solids of the first category or by liquid diffusion in the case of wet solids of the second category. In either case, the rate of drying is controlled by external mass and heat transfer between the exposed surface of the wet solid and the bulk

Drying-gas flow →

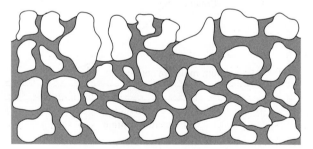

(a) Constant-rate period

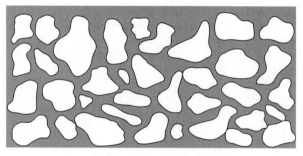

(b) First falling-rate period

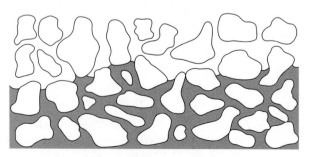

(c) Second falling-rate period

Figure 18.29 Drying stages for granular solids.

gas. The migration of moisture from the interior of the wet solid to the exposed surface is not a rate-affecting factor. This period is referred to as the *constant-rate drying period*. It terminates at point C, referred to as the *critical moisture content*. When drying wet solids of the first category under agitated conditions, as in a direct-heat rotary dryer, fluidized-bed dryer, flash dryer, or agitated batch dryer, such that all particle surfaces are in direct contact with the gas, the constant-rate drying period may extend all the way to X^*.

At the beginning of the period from C to D, the moisture just barely covers the exposed surface. From then until point D is reached, as shown in Figure 18.29b, the surface tends to a dry state because the rate of liquid travel by diffusion or capillary action to the exposed surface is not sufficiently fast. In this period, the exposed-surface temperature remains at the wet-bulb temperature if heat conduction is adequate, but the wetted exposed area for mass transfer decreases. Consequently, the rate of drying decreases linearly with decreasing average moisture content. This period is referred to as the *first falling-rate drying period*. It is not always observed with wet solids of the second category.

During the period from C to D, the liquid in the pores of wet solids of the first category begins to recede from the exposed surface. In the final period from D to E, as shown in Figure 18.29c, evaporation occurs from liquid surfaces in the pores, where the wet-bulb temperature prevails. However, the temperature of the exposed surface in the solid rises to approach the dry-bulb temperature of the gas. During this period, called the *second falling-rate drying period,* the rate of drying may be controlled by vapor diffusion for wet solids of the first category and by liquid diffusion for wet solids of the second category. The rate falls exponentially with decreasing moisture content.

Constant-Rate Drying Period

In direct-heat equipment, drying involves the transfer of heat from the gas to the surface and interior of the wet solid, and mass transfer of moisture from the interior and surface of the solid to the gas. During the constant-rate period, the rate of mass transfer is determined by the resistance of the boundary layer or film of the gas phase in contact with the exterior wet surface of the solid. The wet solid is assumed to be at a uniform temperature so that the only resistance to convective heat transfer is also in the gas phase. The rate of moisture evaporation can then be based on convective heat transfer or mass transfer, according to the following conventional, but simplified, transport relationships, where thermal radiation, the bulk-flow effect for mass transfer, and the sensible-heat effect for the evaporated moisture are ignored:

$$\frac{dm_v}{dt} = \frac{h(T_g - T_i)A}{\Delta H_i^{\mathrm{vap}}} = M_A k_y (y_i - y_g)A \quad (18\text{-}33)$$

where the subscript, i, refers to the gas at the interface with the solid.

As discussed in the previous section, the interface at these conditions is at the wet-bulb temperature, T_w. Although drying-rate calculations could be based on mass transfer using (18-33), it is more common to use the heat-transfer relation of (18-33) when air is the gas and water is the moisture because of the wide availability of the psychrometric chart for that system, the equality of the wet-bulb and adiabatic-saturation temperatures, and a wider availability of correlations for convective heat transfer than for mass transfer, although transport analogies can often be used to derive one from the other. Combining (18-32) and (18-33), the drying-rate flux for the constant-rate drying period, R_c, becomes

$$R_c = \frac{h(T_g - T_w)}{\Delta H_w^{vap}} \qquad (18\text{-}34)$$

while a less-useful equivalent mass-transfer form is obtained by combining (18-16), (18-32), and (18-33):

$$R_c = M_B k_y (\mathcal{H}_w - \mathcal{H}_d) \qquad (18\text{-}35)$$

where d and w refer to gas dry-bulb and wet-bulb conditions, respectively.

For some dryers, it is preferable to use a volumetric heat-transfer coefficient, (ha), defined by

$$\frac{dm_v}{dt} = \frac{(ha)(T_g - T_i)V}{\Delta H_w^{vap}} \qquad (18\text{-}36)$$

where

a = external surface area of wet solids per unit volume of dryer

V = volume of dryer

Then, the drying rate per unit dryer volume during the constant-rate drying period is

$$(R_c)_V = \frac{(ha)(T_d - T_w)}{\Delta H_w^{vap}} \qquad (18\text{-}37)$$

Equations for estimating interphase heat-transfer coefficients were discussed for several geometries in Chapter 3. Empirical equations that are particularly useful for drying-rate calculations are summarized in Mujumbar [1]. Representative equations, when the gas is air, are listed here in Table 18.6. There, in (1), G is the mass velocity of air in the flow channel that passes over the wet surface. In (2), G is the mass velocity of the air impinging on the wet surface. In (3) to (8), d_p is the particle diameter and G is the superficial mass velocity.

The dramatic effect of exposed surface area of wet solids in drying operations was shown by Marshall and Hougen [13] and is illustrated in the following two examples that deal with batch drying. In Example 18.7, cross-circulation, batch tray drying is used to dry slabs of filter cake. In Example 18.8, the filter cake is extruded and then dried by through-circulation. The difference in the two drying times for the constant-rate drying period is very significant.

Table 18.6 Empirical Equations for Interphase Heat-transfer Coefficients for Application to Dryers (h in W/m^2-K, G in kg/hr-m^2, d_p in m)

Geometry	Equation
Flat-plate, parallel flow	$h = 0.0204\, G^{0.8}$
	($T_d = 45 - 150°C$,
	$G = 2{,}450 - 29{,}300$) (1)
Flat-plate, perpendicular,	$h = 1.17\, G^{0.37}$
impingement flow	($G = 3{,}900 - 19{,}500$) (2)
Packed beds,	$h = 0.151\, G^{0.59}/d_p^{0.41}$,
through-circulation	($N_{Re} > 350$) (3)
	$h = 0.214\, G^{0.49}/d_p^{0.51}$
	($N_{Re} < 350$) (4)
Fluidized beds	$N_{Nu} = 0.0133\, N_{Re}^{1.6}$
	($0 < N_{Re} < 80$) (5)
Pneumatic conveyors	$N_{Nu} = 0.316\, N_{Re}^{0.8}$
	($8 < N_{Re} < 500$) (6)
Droplets in spray dryers	$N_{Nu} = 2 + 1.05\, N_{Re}^{0.5}\, N_{Pr}^{1/3}\, N_{Gu}^{0.175}$
	($N_{Re} < 1000$) (7)
Spouted beds	$N_{Nu} = 0.0005\, N_{Re_s}^{1.46}(u/u_s)^{1/3}$ (8)

$N_{Re} = d_p G / \mu$

$N_{Nu} = h d_p / k$

$N_{Pr} = C_P \mu / k$

$N_{Re_s} = d_p G_s / \mu$

  G_s = mass velocity for incipient spouting

  u = velocity

  u_s = incipient spouting velocity

  $N_{Gu} = (T_d - T_w)/T_d$ in absolute temperature

  d_P = particle site

  C_P = specific heat of gas

  μ = viscosity of gas

  k = thermal conductivity of gas

EXAMPLE 18.7

A filter cake of calcium carbonate contained in a tray is to be dried by cross-circulation from the top surface. Each tray is 2.5 cm high with an area of 1.5 m^2 and is completely filled with 73 kg of wet filter cake having a water content of 30% on the dry basis. The heating medium is air at 1 atm and 170°F with a relative humidity of 10%. The average velocity of the air passing across the wet solid is 4 m/s. Estimate the time in hours to reach the experimentally determined, critical-moisture content (end of the constant-rate period) of 10% on the dry basis, if the preheat period is neglected.

SOLUTION

$$H_2O \text{ in wet cake} = \left(\frac{30}{130}\right)(73) = 16.8\,\text{kg}$$

$$H_2O \text{ in cake at } X_c = 0.10(73 - 16.8) = 5.6\,\text{kg}$$

$$m_v = H_2O \text{ evaporated} = 16.8 - 45.6 = 11.2\,\text{kg}$$

For the constant-rate drying period, the heat-transfer form of (18-33) applies, which upon integration gives

$$t_c = \frac{m_v \Delta H_w^{vap}}{h(T_d - T_w)A}$$

where t_c is the time to reach the critical moisture content. From the humidity chart of Figure 18.17,

$$T_w = 100°F \text{ and } \mathcal{H} = 0.026 \frac{\text{lb } H_2O}{\text{lb dry air}}$$

$$T_d = 170°F = 76.7°C$$

$$T_d - T_w = 170 - 100 = 70°F = 38.9 \text{ K}$$

$$\text{At } T_w = 100°F, \Delta H_w^{\text{vap}} = 1037.2 \text{ Btu/lb} = 2,413 \text{ kJ/kg}$$

$$G = u_{\text{avg}}\rho$$

From (18-10), Table 18.4,

$$v_H = 0.730(170 + 460)\left(\frac{1}{28.97} + \frac{0.026}{18.02}\right) = 16.5 \text{ ft}^3/\text{lb dry air}$$

or

$$16.5/(1 + 0.026) = 16.1 \text{ ft}^3/\text{lb moist air}$$
$$= 1.004 \text{ m}^3/\text{kg moist air} = 1/\rho$$

$$G = 4/1.004 = 3.98 \text{ kg/m}^2\text{-s} = 14,300 \text{ kg/m}^2\text{-h}$$

$$A = 1.5 \text{ m}^2$$

From Table 18.6, (1) applies for turbulent flow with T_d and G within the allowable range.

$$h = 0.0204(14,300)^{0.8} = 43 \text{ W/m}^2\text{-K} = 43 \text{ J/s-m}^2\text{-K}$$

From (1), using SI units

$$t_c = \frac{(11.2)[(2,413)(1,000)]}{(43)(38.9)(1.5)} = 10,800 \text{ s} = 2.99 \text{ h}$$

EXAMPLE 18.8

The filter cake of Example 18.7 is extruded into cylindrical-shaped pieces of 1/4-in. diameter and 1/2-in. length to form a bed that is 1.5 m^2 in cross-sectional area and 5 cm high, with an external porosity of 50%. Air at 170°F and 10% relative humidity passes through the bed at a superficial velocity of 2 m/s (average interstitial velocity of 4 m/s). Estimate the time in hours to reach the critical-moisture content, if the preheat period is neglected. Compare this time to that estimated in Example 18.7.

SOLUTION

Compared to the tray of Example 18.7, the bed is twice as high with the same cross-sectional area. Therefore, for a porosity of 50%, the bed contains the same amount of wet solids. Thus, as in Example 18.7,

$$m_{\text{wet cake}} = 73 \text{ kg}$$

$$m_v = 11.2 \text{ kg } H_2O \text{ evaporated}$$

Also,

$$\Delta H_w^{\text{vap}} = 2,413 \text{ kJ/kg}$$

$$T_d - T_w = 38.9 \text{ K}$$

Assume that the density of the extrusions equals the density of the filter cake.

$$\rho_{\text{filter cake}} = \frac{73}{1.5\left(\frac{2.5}{100}\right)} = 1,950 \text{ kg/m}^3.$$

$$\text{Volume of one extrusion} = \pi D^2 L/4 = \frac{3.14(0.25)^2(0.5)}{4}$$
$$= 0.0245 \text{ in}^3 = 4.01 \times 10^{-7} \text{ m}^3.$$

$$\text{Number of extrusions} = 1.5\left(\frac{2.5}{100}\right)(0.5)/4.01 \times 10^{-7}$$
$$= 46,800.$$

$$\text{Surface area/extrusion} = \pi DL + \pi D^2/2$$

$$= 3.14\left[(0.25)(0.50) + \frac{(0.25)^2}{2}\right]$$

$$= 0.49 \text{ in}^2 = 0.000316 \text{ m}^2$$

$$A = 46,800(0.000316) = 14.8 \text{ m}^2$$

Thus, the transport area is $14.8/1.5 = 9.9$ times that for Example 18.7. From Table 18.6, (3) or (4) applies for estimating h, depending on N_{Re}. From Example 18.7, but with a superficial bed velocity of 50% of the crossflow velocity,

$$G = 3.98/2 = 1.96 \text{ kg/m}^2\text{-s}$$

Equations (3) and (4) refer to the work of Gamson, Thodos, and Hougen [14] for $N_{\text{Re}} > 350$ and Wilke and Hougen [15] for $N_{\text{Re}} < 350$, respectively. For both correlations, d_p is taken as the diameter of a sphere of the same surface area as the particle. For the extrusions of this example with $L = 2D$,

$$\pi d_p^2 = \frac{\pi D^2}{2} + 2\pi D^2 = 2.5\pi D^2$$

Solving (1),

$$d_p = D\sqrt{2.5} = 0.25\sqrt{2.5} = 0.395 \text{ in.} = 0.010 \text{ m}$$

$$\mu \approx 0.02 \text{ cP} = 2 \times 10^{-5} \text{ kg/m-s}$$

$$N_{\text{Re}} = \frac{d_p G}{\mu} = \frac{(0.010)(1.96)}{2 \times 10^{-5}} = 980$$

Therefore, (3) applies and

$$h = 0.151(14,300/2)^{0.59}/(0.010)^{0.41} = 188 \frac{\text{J}}{\text{s-m}^2\text{-K}}$$

The h is $188/43 = 4.4$ times greater than in Example 18.7. From (1) in that example,

$$t_c = \frac{(11.2)[(2,413)(1,000)]}{(188)(38.9)(14.8)} = 250 \text{ s} = 4.16 \text{ min}.$$

This example and the preceding example show that cross-circulation drying can take hours, while through-circulation drying may require only minutes.

Falling-rate Drying Period

When the rate of drying in the constant-rate period is high and/or the distance that interior moisture must travel to reach the surface is large, the moisture may eventually fail to reach the surface fast enough to maintain a constant rate of drying. Then, a transition to the falling-rate period will occur. In Examples 18.7 and 18.8, the constant rates of drying are, respectively, from (18-34):

$$R_c = \frac{43(38.9)(3,600)}{(2,413)(1,000)} = 2,50 \text{ kg/h-m}^2$$

and

$$R_c = \frac{(188)(38.9)(3,600)}{(2,413)(1,000)} = 10.9 \text{ kg/h-m}^2$$

However, in Example 18.7, the moisture may have to travel from as far away as 25 mm to reach the exposed surface, while in Example 18.8, the distance is only 3.2 mm. Therefore, as a first approximation, it might be expected that the critical-moisture contents for the two examples might not be the same. The value of 10% on the dry basis was taken from through-circulation drying experiments.

During drying, when moisture travels from the interior of a wet solid to the surface, a moisture profile develops in the wet solid. The shape of this profile depends on the nature of the moisture movement, as discussed by Hougen, McCauley, and Marshall [16]. If the wet solid is of the first category, where the moisture is not held in solution or in fibers, but is held as free moisture in the interstices of powders and granular solids such as paint pigments, minerals, clays, soil, and sand, or is moisture above the fiber-saturation point in textiles, paper, wood, and leather, then moisture movement occurs by capillary action. For wet solids of the second category, the internal moisture is bound moisture, as in the last stages of the drying of clay, starch, flour, textiles, paper and wood, or soluble moisture, as in soap, glue, gelatin, and paste. This type of moisture migrates to the surface by liquid diffusion. Moisture can also migrate by gravity, external pressure, and by vaporization-condensation sequences in the presence of a temperature gradient. In addition, vapor diffusion through the solid can occur in indirect-heat dryers when heating and vaporization occur at opposed surfaces.

A typical moisture profile for capillary flow is shown in Figure 18.30a. The profile is concave upward near the exposed surface, concave downward near the opposed surface, and with a point of inflection in between. For flow of moisture by diffusion, as shown in Figure 18.30b, the profile is concave downward throughout. If the diffusivity is independent of the moisture content, the solid curve applies. If, as is often the case, the diffusivity decreases with decreasing moisture content, due mainly to shrinkage, the dashed profile applies.

During the falling-rate period of drying, idealized theories for capillary flow and diffusion can be applied to estimate drying rates. Alternatively, the estimate could be made by a strictly empirical approach that ignores the mechanism of moisture movement, but instead relies on the experimental determination of drying rate as a function of average moisture content for a particular set of drying conditions. The empirical approach is examined first.

Empirical Approach

The empirical approach ignores the preheat period and is usually applied to experimental data in the form of Figure 18.31a (Case 1), but can be modified to be applied to data of Figures 18.31b (Case 2) and 18.31c (Case 3). In these plots, the abscissa is the free-moisture content, $X = X_T - X$, as shown in Figure 18.23, for the particular drying conditions. This allows all three plots to be extended to the origin. However, for a given application, all free moisture may not be removed.

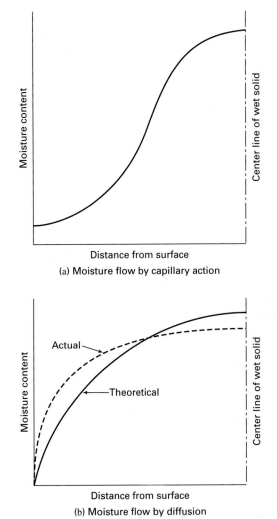

Figure 18.30 Moisture distribution in wet solids during drying. [From W.L. McCabe, J.C. Smith, and P. Harriott, *Unit Operations of Chemical Engineering*, 5th ed., McGraw-Hill, New York (1993) with permission.]

From (18-32),

$$\int dt = -\frac{m_s}{A}\int \frac{dX}{R} \tag{18-38}$$

For the constant-rate period, and ignoring the preheat period, $R = R_c = $ constant. Starting from an initial free-moisture content of X_o at time $t = 0$, the time to reach the critical free-moisture content, X_c, at time $t = t_c$ is obtained by integrating (18-38):

$$t_c = \frac{m_s(X_o - X_c)}{AR_c} \tag{18-39}$$

For Case 1 (Figure 18.31a) of the falling-rate period, the rate of drying is linear with X and terminates at the origin, according to

$$R = R_cX/X_c \tag{18-40}$$

Substituting (18-40) into (18-38) and integrating t from t_c to $t > 0$ and X from X_c to $X > 0$ gives the following

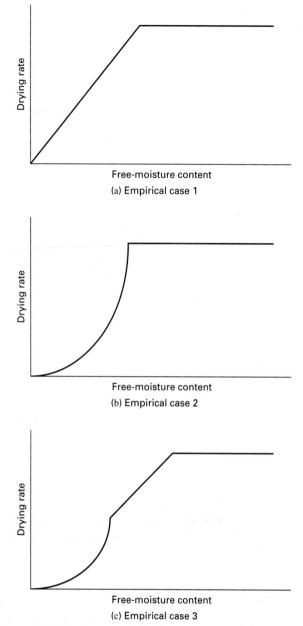

Figure 18.31 Drying-rate curves.

(a) Empirical case 1

(b) Empirical case 2

(c) Empirical case 3

expression for the drying time in the falling-rate period, t_f:

$$t_f = t - t_c = \frac{m_s X_c}{A R_c} \ln\left(\frac{X_c}{X}\right) = \frac{m_s X_c}{A R_c} \ln\left(\frac{R_c}{R}\right) \qquad (18\text{-}41)$$

The total drying time, t_T, is the sum of (18-39) and (18-41):

$$t_T = t_c + t_f = \frac{m_s}{A R_c}\left[(X_o - X_c) + X_c \ln\left(\frac{X_c}{X}\right)\right] \qquad (18\text{-}42)$$

EXAMPLE 18.9

Marshall and Hougen [13] present experimental data for the through-circulation drying of 5/16-in. extrusions of ZnO in a bed of 1 ft^2 cross section by 1-in. high, using air of $T_d = 158°F$ and

$T_w = 100°F$ at a flow rate of 340 ft^3/min. The data show a constant-rate period from $X_o = 33\%$ to $X_c = 13\%$, with $R_c = 1.42$ lb H$_2$O/h-lb bone-dry solid, followed by a falling-rate period that approximates Case 1 in Figure 18.31a. Calculate the drying time for the constant-rate period and the additional time in the falling-rate period to reach a free moisture content, X, of 1%.

SOLUTION

In (18-38), the drying rate, R, corresponds to mass of moisture evaporated per unit time per unit of exposed area of the wet material. In this example, the drying rate is not given per unit area, but per mass of bone-dry solid, with some associated exposed area. Equations (18-38) and (18-42) can be rewritten in terms of $R' = RA/m_s$, respectively, as

$$\int dt = -\int \frac{dX}{R'} \qquad (1)$$

and

$$t_T = t_c + t_f = \frac{1}{R'_c}\left[(X_o - X_c) + X_c \ln\left(\frac{X_c}{X}\right)\right] \qquad (2)$$

From (2), for just the constant-rate period,

$$t_c = \frac{1}{1.42}[0.33 - 0.13] = 0.141\,\text{h} = 8.45\,\text{min}$$

From (2), for just the falling-rate period,

$$t_f = \frac{1}{1.42}\left[0.13 \ln\left(\frac{0.13}{0.01}\right)\right] = 0.235\,\text{h} = 14.09\,\text{min}$$

The total drying time, ignoring the preheat period, is

$$t_T = t_c + t_f = 8.45 + 14.09 = 22.5\,\text{min}$$

For Case 2 (Figure 18.31b), R in the falling-rate period can be expressed as a parabolic function.

$$R = aX + bX^2 \qquad (18\text{-}43)$$

The values of the parameters a and b are obtained by fitting (18-43) to the experimental drying-rate plot, subject to the constraint that $R = R_c$ at $X = X_c$. If (18-43) is substituted into (18-38) and the result is integrated for the falling-rate period from $X = X_c$ down to some final value of X_f and corresponding R_f, the time for the falling-rate period is found to be

$$t_f = t - t_c = \frac{m_s}{aA} \ln\left[\frac{X_c(a + bX_f)}{X_f(a + bX_c)}\right] = \frac{m_s}{aA} \ln\left[\frac{X_c^2 R_f}{X_f^2 R_c}\right] \qquad (18\text{-}44)$$

EXAMPLE 18.10

Experimental data for through-circulation drying of 1/4-in.-diameter spherical pellets of a nonhygroscopic carburizing compound exhibit constant-rate drying of 1.9 lb H$_2$O/h-lb dry solid from $X_o = 30\%$ to $X_c = 21\%$, followed by a falling-rate period to $X_f = 4\%$ that fits (18-43) with $a = 3.23$ and $b = 27.7$ (both in lb H$_2$O/h-lb dry solid) for X as a fraction and R replaced by R' in lb H$_2$O/h-lb dry solid. Calculate the time for drying in the falling-rate period. Note that the values of a and b satisfy the constraint of $R'_c = 1.9$ at $X_c = 0.21$.

SOLUTION

For R' in the given units, (18-44) becomes

$$t_f = \frac{1}{a} \ln \left[\frac{X_c(a + bX_f)}{X_f(a + bX_c)} \right] \qquad (1)$$

Thus,

$$t_f = \frac{1}{3.23} \ln \left\{ \frac{0.21}{0.04} \left[\frac{3.23 + 27.7(0.04)}{3.23 + 27.7(0.21)} \right] \right\} = 0.286\,\text{h} = 17.1\,\text{min}$$

For Case 3 (Figure 18.31c), the falling-rate period consists of two subregions. In the first subregion, which is linear,

$$R = \alpha X + \beta \qquad (18\text{-}45)$$

with the constraints that $R = R_{c_1}$ at X_{c_1}, and $R = R_{c_2}$ at X_{c_2}. In the second subregion, (18-44) applies, but with the constraint that $R = R_{c_2}$ at X_{c_2}.

EXAMPLE 18.11

Experimental data of Sherwood [12] for the surface drying of a 3.18-cm-thick $\times$ 6.6-cm^2 cross-sectional area slab of a thick paste of CaCO$_3$ (whiting) from both sides by air at $T_d = 39.8°C$ and $T_w = 23.5°C$ and a cross-circulation velocity of 1 m/s exhibit the complex type of drying-rate curve shown in Figure 18.31c, with the following constants:

Constant-rate period:

$$X_o = 10.8\%$$
$$X_{c_1} = 8.3\%$$
$$R_{c_1} = 0.053\,\text{g H}_2\text{O/h-cm}^2$$

First falling-rate period:

$$X_{c_2} = 3.7\%$$
$$R_{c_2} = 0.038\,\text{g H}_2\text{O/h-cm}^2$$

Second falling-rate period to $X = 2.2\%$:

$$R = 29.03\,X^2 - 0.048\,X \qquad (1)$$

Determine the time to dry a slab of the same dimensions at the same drying conditions, but form $X_o = 0.14$ to $X = 0.01$, ignoring the preheat period. Assume that the initial weight of the slab is 46.4 g. Drying is from both sides.

SOLUTION

Constant-rate period:

$$X_o = 0.14, \ X_{c_1} = 0.083, \ R_{c_1} = 0.053\,\text{g/h-cm}^2$$

$$m_s = 46.4 \left(\frac{1}{1.14} \right) = 40.7\,\text{g of moisture-free solid}$$

$$A = 2(6.6) = 13.2\,\text{cm}^2$$

From (18-39),

$$t_c = \frac{40.7(0.14 - 0.083)}{13.2(0.053)} = 3.32\,\text{h}$$

First falling-rate period:

In this period, R is linear with end points of $(R_{c_1} = 0.053, X_{c_1} = 0.083)$ and $(R_{c_2} = 0.038, X_{c_2} = 0.037)$. This gives for (18-45),

$$R = 0.0259 + 0.326X \qquad (2)$$

Substitution of (2) into (18-38) and integration gives

$$t_{f_1} = -\frac{m_s}{A} \int_{X_{c_1}}^{X_{c_2}} \frac{dX}{0.0259 + 0.326X}$$

$$= \frac{m_s}{A} \frac{1}{0.326} \ln \left(\frac{0.0259 + 0.326X_{c_1}}{0.0259 + 0.326X_{c_2}} \right)$$

$$= \frac{m_s}{A} \frac{1}{0.326} \ln \left(\frac{R_{c_1}}{R_{c_2}} \right)$$

$$= \frac{40.7}{13.2(0.326)} \ln \left(\frac{0.053}{0.038} \right) = 3.15\,\text{h}$$

Second falling-rate period:

This period extends from $X_{c_2} = 0.037$ to $X = 0.022$, with R given by (1) for (18-43), with $a = -0.048$ and $b = 29.03$.

From (18-44),

$$t_{f_2} = \frac{40.7}{(-0.048)(13.2)} \ln \left\{ \frac{0.037[-0.048 + 29.03(0.022)]}{0.022[-0.048 + 29.03(0.037)]} \right\}$$

$$= 2.08\,\text{h}$$

the total drying time is

$$t_T = 3.32 + 3.15 + 2.08 = 8.6\,\text{h}$$

For drying-rate curves of shapes other than those of Figure 18.31, the time for drying from any X_o to any X can be determined by numerical or graphical integration of (18-38) or (1) in Example 18.9, as illustrated in the following example.

EXAMPLE 18.12

Marshall and Hougen [13] present the following experimental data for the through-circulation drying of rayon waste. Determine the drying time if $X_o = 100\%$ and the final X is 10%. Assume that all moisture is free moisture.

X, lb H$_2$O/lb dry solid	R', $\dfrac{\text{lb H}_2\text{O}}{\text{h-lb dry solid}}$
1.40	24
1.00	24
0.75	24
0.73	21
0.70	18
0.65	15.3
0.55	13
0.475	12.3
0.44	12.2
0.40	11
0.20	5.5
0	0

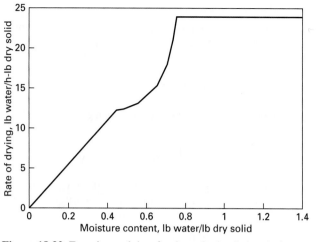

Figure 18.32 Experimental data for through-circulation drying of rayon waste.

SOLUTION

The data are plotted in Figure 18.32, where three distinct drying-rate periods are seen, but the two falling-rate periods are in the reverse order of Figure 18.31c. By numerical integration of (1) in Example 18.9 with a spreadsheet, the following drying times are obtained, noting that $R' = 2.75$ lb H_2O/h-lb dry solid at $X = 0.10$, $R_{c_1} = 24$ at $X_{c_1} = 0.75$, and $R_{c_2} = 12.2$ at $X_{c_2} = 0.44$.

$$t_c = 0.027\,\text{h} = 1.63\,\text{min}$$

$$t_{f_1} = 1.28\,\text{min}$$

$$t_{f_2} = 3.21\,\text{min}$$

$$t_T = 1.63 + 1.28 + 3.21 = 6.12\,\text{min}$$

Liquid-Diffusion Theory

The application of the empirical approach, in the preceding section, for determining drying time in the falling-rate period is limited to the particular conditions for which the experimental drying-rate curve is established. A more general approach, particularly for nonporous wet solids of the second category, discussed in Section 18.3, is the use of Fick's laws of diffusion. Once the moisture diffusion coefficient is established from experimental data for the particular wet solid, Fick's laws can be used to predict drying rates and moisture profiles for wet solids of other sizes and shapes and drying conditions during the falling-rate period.

Mathematical formulations of liquid diffusion in solids are readily obtained by analogy to the many solutions available for transient heat conduction in solids, as summarized, for example, by Carslaw and Jaeger [17] and discussed in Section 3.3. The following two solutions are of particular interest for the drying of slabs in the falling-rate period, where the area of the edges is small compared to the area of the two faces, or the edges are sealed to prevent escape of moisture. As in heat-conduction calculations, the equations also apply when one of the two faces is sealed so that only the other face is exposed to gas:

Case 1: Initially uniform moisture profile in the wet solid with negligible resistance to mass transfer of moisture in the gas phase.

Case 2: Initially parabolic moisture profile in the wet solid with negligible resistance to mass transfer in the gas phase.

Although the equations for these two cases are developed here only for a slab with sealed edges, solutions are available in Carslaw and Jaeger [17] for a sphere and a cylinder with sealed edges. When edges of slabs and cylinders are not sealed, the method of Newman [18] can be applied as discussed in Section 3.3.

Case 1 This case applies to slow-drying materials for which the rate of drying is controlled by the internal diffusion of moisture to the exposed surface. This can occur if initially the wet solid has no surface liquid film and the external resistance to mass transfer is negligible, such that there is no constant-rate drying period. Alternatively, the wet solid can have a surface liquid film, but during the evaporation of that film in a constant-rate drying period controlled by gas-phase mass transfer, no moisture diffuses to the surface and following the completion of evaporation of that film, the resistance to mass transfer is entirely due to internal diffusion in a falling-rate period.

The slab, of thickness $2a$, is pictured in Figure 3.7a, where the edges at $x = \pm c$ and $y = \pm b$ are sealed to mass transfer. Internal diffusion of moisture is in the z direction only toward exposed faces at $z = \pm a$. Alternatively, the slab may be of thickness a with the face at $z = 0$ sealed to mass transfer. Initially, the moisture content throughout the slab, not counting any surface liquid film, is uniform at X_o. At the beginning of the falling-rate period, $t = 0$, the exposed faces are (face is) brought to the equilibrium moisture content, X^*. For constant moisture diffusivity, D_{AB}, Fick's second law, as discussed in Section 3.3, applies:

$$\frac{\partial X}{\partial t} = D_{AB}\frac{\partial^2 X}{\partial z^2} \qquad (18\text{-}46)$$

for $t \geq 0$ in the region $-a \leq z \leq a$, where the boundary conditions are

$$X = X_o \quad \text{at} \quad t = 0 \quad \text{for} \quad -a < z < a$$

and

$$X = X^* \quad \text{at} \quad z = \pm a \quad \text{for} \quad t \geq 0$$

The solution to (18-46) for the moisture profile as a function of time under these boundary conditions, as discussed in Section 3.3, and first proposed for drying applications by Sherwood [11], is in terms of the unaccomplished free-moisture change, the following modification of (3-80),

$$E = \frac{X - X^*}{X_o - X^*} = \frac{4}{\pi} \sum_{n=0}^{\infty} \frac{(-1)^n}{(2n+1)}$$

$$\times \exp\left[-\frac{\pi^2(2n+1)^2}{4}\left(\frac{D_{AB}t}{a^2}\right)\right]\cos\left[\frac{\pi(2n+1)}{2}\left(\frac{z}{a}\right)\right]$$

$$(18\text{-}47)$$

Thus, E is a function of two dimensionless groups, the Fourier number for diffusion, $D_{AB}t/a^2$, and the position ratio, z/a. This solution is plotted as $(1 - E)$ in terms of these two groups in Figure 3.8.

The rate of mass transfer from one face is also given in Chapter 3 as (3-82), which in terms of R, the rate of drying expressed as the mass of moisture evaporated per unit time per unit area, is

$$R = \frac{2D_{AB}(X_o - X^*)\rho_s}{a}$$

$$\times \sum_{n=0}^{\infty} \exp\left[-\frac{\pi^2(2n+1)^2}{4}\left(\frac{D_{AB}t}{a^2}\right)\right]$$

$$(18\text{-}48)$$

where $\rho_s = $ mass of dry solid/volume of slab.

Also of interest is the average moisture content of the slab during drying. From (3-85), which is derived in Chapter 3,

$$E_{avg} = \frac{X_{avg} - X^*}{X_o - X^*}$$

$$= \frac{8}{\pi^2} \sum_{n=0}^{\infty} \frac{1}{(2n+1)^2} \exp\left[-\frac{\pi^2(2n+1)^2}{4}\left(\frac{D_{AB}t}{a^2}\right)\right]$$

$$(18\text{-}49)$$

Equations (18-47)–(18-49) can be used to determine the moisture diffusivity, D_{AB}, from experimental data, and then that value is used to estimate drying rates for other conditions as illustrated in the following example. However, such calculations must be made with caution because often the diffusivity is not constant, as shown by Sherwood [11] for the drying of slabs of soap, but decreases with decreasing moisture content because of shrinkage and/or case-hardening. In that case, numerical solutions of (18-46) are necessary.

EXAMPLE 18.13

A piece of poplar wood 15.2 cm long × 15.2 cm wide × 1.9 cm thick, with the edges sealed with a waterproofing cement was dried from both faces in a tunnel dryer using cross-circulation of air at 1 m/s. The initial moisture content was 39.7% on the dry basis. The total initial weight of the wet piece was 264 g, and no shrinkage occurred during drying. The direction of moisture diffusion was perpendicular to the grain. The equilibrium moisture content was 5% on the dry basis. The following experimental data were obtained for the average moisture content as a function of time. Included are values of E_{avg}, computed from its definition, given in (18-49), and values of t/a^2, where $a = 0.5(1.90) = 0.95$ cm.

t, h	X_{avg}, g H_2O/g dry wood	t/a^2, h/cm²	E_{avg}
0.36	0.362	0.40	0.900
0.90	0.328	1.00	0.800
1.53	0.303	1.70	0.730
1.94	0.291	2.15	0.694
2.89	0.267	3.20	0.626
3.47	0.255	3.85	0.591
4.02	0.245	4.45	0.562
4.92	0.230	5.45	0.520
5.82	0.218	6.45	0.483
6.95	0.204	7.70	0.443
8.03	0.192	8.90	0.409
8.98	0.183	9.95	0.382

Using the data, determine the average value of the diffusivity by nonlinear regression of (18-49) and use that value to determine the drying time from $X_o = 45\%$ to $X = 10\%$ with $X^* = 6\%$ for a piece of poplar measuring 72 in. long × 12 in. wide × 1 in. thick, neglecting mass transfer from the edges and assuming that all drying takes place in the falling-rate period with negligible resistance in the gas phase.

SOLUTION

$$m_s = 264\left(\frac{1}{1 + 0.397}\right) = 189 \text{ g dry wood}$$

$$A \text{ for two faces} = 2(15.2)^2 = 462 \text{ cm}^2$$

At any instant, from (18-38),

$$R = -\frac{m_s}{A}\frac{dX_{avg}}{dt} \tag{1}$$

From a plot of the data, approximate values of R as a function of X_{avg} are computed to be

R, g H_2O/h-cm²	X_{avg}, g H_2O/g dry solid
0.02622	0.345
0.01573	0.315
0.01258	0.297
0.01019	0.279
0.00847	0.261
0.00760	0.250
0.00661	0.238
0.00582	0.224
0.00503	0.211
0.00446	0.198
0.00404	0.187

These results are plotted in Figure 18.33, where it appears that all of the drying takes place in the falling-rate period. Thus, the data may be consistent with the Case 1 diffusion theory.

To determine the average moisture diffusivity, a spreadsheet is used to prepare a semi-log plot of the data points as E_{avg} against t/a^2, as shown in Figure 18.34. Equation (18-49) is then evaluated on the spreadsheet for different values of the moisture diffusivity until the best fit of the data is obtained, based on minimizing the error sum of squares (ESS) of the differences between E_{avg} of the data points and corresponding E_{avg} values calculated from (18-49).

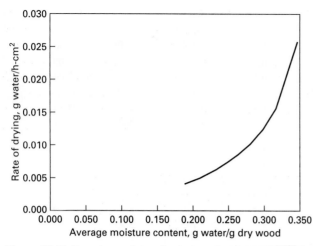

Figure 18.33 Experimental data for drying of poplar wood.

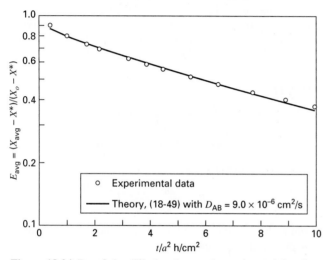

Figure 18.34 Best fit by diffusion theory of experimental data for drying of poplar wood.

The best fit is found for $D_{AB} = 9.0 \times 10^{-6}$ cm²/s with an ESS = 0.001669. The best fit of (18-49) is included as a line in Figure 18.34. It may be noted that for values of $N_{Fo_M} > 0.1$, only the first term in the infinite series of (18-49) is significant and therefore (18-49) approaches a straight line on a semi-log plot, as can be observed for the theoretical line in Figure 18.34, when $t/a^2 > 3.2$ h/cm².

To determine the drying time for the 72-in. × 12-in. × 1-in. piece of poplar, assume that all drying takes place in the diffusion-controlled, falling-rate period and that mass transfer from the edges is negligible. Further assume that the drying time will be long enough that $N_{Fo_M} > 0.1$. Then, as just mentioned, (18-49) reduces to

$$\ln\left(\frac{X_{avg} - X^*}{X_o - X^*}\right) = \ln\left(\frac{8}{\pi^2}\right) - \frac{\pi^2}{4}\left(\frac{D_{AB}t}{a^2}\right) \tag{2}$$

Solving (2) for $(D_{AB}t/a^2)$,

$$N_{Fo_M} = \frac{D_{AB}t}{a^2} = \frac{4}{\pi^2}\ln\left[\frac{8}{\pi^2}\left(\frac{X_o - X^*}{X_{avg} - X^*}\right)\right] \tag{3}$$

$X_{avg} = 0.10$, $X_o = 0.45$, $X^* = 0.06$

From (3),

$$N_{Fo_M} = \frac{4}{(3.14)^2}\ln\left[\frac{8}{(3.14)^2}\left(\frac{0.45 - 0.06}{0.10 - 0.06}\right)\right] = 0.839$$

Since $N_{Fo_M} > 0.1$, (2) and (3) are valid,

$$a = 0.5\,\text{in.} = 1.27\,\text{cm}$$

$D_{AB} = 9.0 \times 10^{-6}$ cm²/s from the above experiments

$$t = \frac{a^2 N_{Fo_M}}{D_{AB}} = \frac{(1.27)^2(0.839)}{(9.0 \times 10^{-6})(3600)} = 41.8\,\text{h}$$

Case 2　When a liquid-diffusion-controlled, falling-rate drying period is preceded by a constant-rate period, that rate of drying is determined by external mass transfer in the gas phase, as discussed earlier, but diffusional resistance to the flow of moisture in the solid causes a moisture profile to be established in the solid. This moisture profile approaches a parabolic distribution, as discussed by Sherwood [19] and Gilliland and Sherwood [20].

For the slab of Figure 3.7a, Fick's second law, as given by (18-46), still applies, with $X = X_o$ at $t = 0$ for $-a < z < a$. However, during the constant-rate drying period, the slab–gas interface boundary conditions are changed from Case 1 to the conditions $\partial X/\partial z = 0$ at $z = 0$ for $t \geq 0$ and $R_c = -D_{AB}\rho_s(\partial X/\partial z)$ at $z = \pm a$ for $t \geq 0$. This latter boundary condition is more conveniently expressed in the form

$$\frac{\partial X}{\partial z} = -\frac{R_c}{\rho_s D_{AB}} \tag{18-50}$$

where the term on the right-hand side is a constant during the constant-rate period. This is analogous to a constant-heat-flux boundary condition in heat transfer. The solution to this case for the moisture profile as a function of time during the constant-rate drying period is given from Walker, et al. [21] as

$$X = X_o - \frac{R_c a}{D_{AB}\rho_s}\left\{\frac{1}{2}\left(\frac{z}{a}\right)^2 - \frac{1}{6} + \frac{D_{AB}t}{a^2}\right. $$
$$\left. - \frac{2}{\pi^2}\sum_{m=1}^{\infty}\frac{(-1)^m}{m^2}\exp\left[-m^2\pi^2\left(\frac{D_{AB}t}{a^2}\right)\right]\cos\left(\frac{\pi m z}{a}\right)\right\} \tag{18-51}$$

where for small values of $D_{AB}t/z^2$, the infinite series term is significant and converges very slowly.

The average moisture content in the slab at any time during the constant-rate period is defined by

$$X_{avg} = \frac{1}{a}\int_o^a X\,dz \tag{18-52}$$

If (18-52) is integrated after substitution of X from (18-51), the result is

$$(X_o - X_{avg})\frac{D_{AB}\rho_s}{R_c a} = \frac{D_{AB}t}{a^2} = N_{Fo_M} \tag{18-53}$$

From (18-51), it is seen that the generalized moisture profile during the constant-rate drying period, $(X_o-X)D_{AB}\rho_s/(R_c a)$, is a function of the dimensionless position ratio, z/a, and N_{Fo_M}, where the latter is equal to the generalized, average moisture content, as given by (18-53). A plot of (18-51) for six different position ratios, is given in Figure 18.35a.

Equation (18-51) is based on the assumption that during the constant-rate drying period, moisture will be supplied to the surface by liquid diffusion at a rate sufficient to maintain the constant moisture-evaporation rate. As discussed above, the average moisture content at which the constant-rate period ends and the falling-rate period begins is called the critical moisture content, X_c. In the empirical approach to the falling-rate period, X_c must be known from experiment for the particular conditions being evaluated because X_c is not a constant for a given material, but also depends on

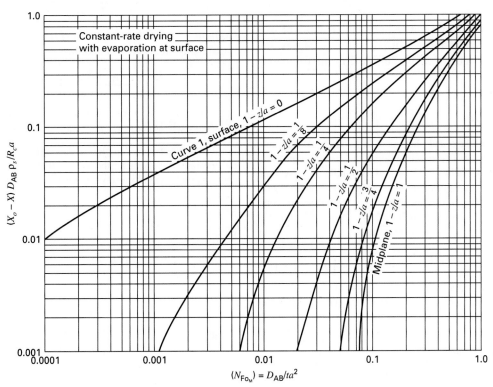

(a) Moisture profile change

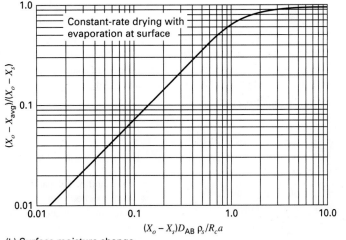

(b) Surface moisture change

Figure 18.35 Changes in moisture concentration during constant-rate period while diffusion in the solid occurs.

[From W.H. Walker, W.K. Lewis, W.H. McAdams, and E.R. Gilliland, *Principles of Chemical Engineering*, 3rd ed., McGraw-Hill, New York (1937) with permission.]

a number of other factors, including moisture diffusivity, slab thickness, initial and equilibrium moisture contents, and all factors that influence the moisture-evaporation flux in the constant-rate drying period. A very useful aspect of (18-51) is that it can be used to predict X_c. The basis for the prediction is the assumption that the falling-rate period will begin when the moisture content at the surface reaches the equilibrium-moisture content corresponding to the conditions of the surrounding gas. This prediction is facilitated, as described by Walker, et al. [21], by replotting an extension of Curve 1 in Figure 18.35a for the moisture content at the surface, X_s, in the form shown in Figure 18.35b. The use of Figure 18.35b and the predicted influence of several variables on the value of X_c is illustrated in the following example.

EXAMPLE 18.14

Experiments by Gilliland and Sherwood [20] with brick clay mix show that for certain drying conditions, the moisture profiles conform reasonably well to the Case 2 diffusion theory. Use Figure 18.35b to predict the critical-moisture content for the drying of clay slabs from the two faces only under three different sets of conditions. For all three sets, $X_o = 0.30$, $X^* = 0.05$, $\rho_s = 1.6\,\mathrm{g/cm^3}$, and $D_{AB} = 0.3\,\mathrm{cm^2/h}$. The other conditions are

	Set 1	Set 2	Set 3
a, slab half-thickness, cm	0.5	0.5	1.0
R_c, drying rate in constant-rate drying period, g/cm²-h	0.2	0.4	0.2

SOLUTION

For Set 1, using $X_s = X^* = 0.05$,

$$(X_o - X_s)\frac{D_{AB}\rho_s}{R_c a} = (0.30 - 0.05)\frac{(0.3)(1.6)}{(0.2)(0.5)} = 1.20$$

From Figure 18.35b,

$$\frac{X_o - X_{avg}}{X_o - X_s} = 0.7$$

Solving,

$$X_{avg} = X_c = 0.25 - 0.7(0.25 - 0.05) = 0.11$$

In a similar manner,

$$X_c \text{ for Set 2} = 0.16$$

$$X_c \text{ for Set 3} = 0.16$$

These results show that doubling the rate of drying in the constant-rate period or doubling the slab thickness substantially increases the critical-moisture content.

For sufficiently large values of time, corresponding to a Fourier number for diffusion $N_{FoM} = \frac{D_{AB}t}{a^2} > 0.5$, the term for the infinite series in (18-51) approaches a value of zero, and, at all locations in the slab, X becomes a parabolic function of z.

A simple equation for the parabolic distribution can be formulated as follows from (18-51) in terms of the moisture contents at the surface and midplane of the slab. At the surface $z = \pm a$, the long-time form is

$$X_o - X_s = \frac{R_c a}{D_{AB}\rho_s}\left[\frac{1}{3} + N_{FoM}\right] \qquad (18\text{-}54)$$

Similarly, at the midplane, $z = 0$, where $X = X_m$,

$$X_o - X_m = \frac{R_c a}{D_{AB}\rho_s}\left[-\frac{1}{6} + N_{FoM}\right] \qquad (18\text{-}55)$$

Combining (18-51), (18-54), and (18-55), the dimensionless moisture-content profile becomes

$$\frac{X_m - X}{X_m - X_s} = \left(\frac{z}{a}\right)^2 \qquad (18\text{-}56)$$

EXAMPLE 18.15

For the conditions of Example 18.14, determine the drying time for the constant-rate drying period and whether the parabolic moisture-content profile will be closely approached by the end of that period.

SOLUTION

From (18-39),

$$t_c = \frac{m_s(X_o - X_c)}{A R_c} \qquad (1)$$

For a half-slab of thickness a,

$$m_s = \rho_s a A \qquad (2)$$

Combining (1) and (2),

$$t_c = \frac{\rho_s a}{R_c}(X_o - X_c) \qquad (3)$$

For Set 1 of Example 18.14,

$$t_c = \frac{(1.6)(0.5)}{(0.2)}(0.30 - 0.11) = 0.76\,\mathrm{h}$$

$$N_{FoM} = \frac{D_{AB}t_c}{a^2} = \frac{(0.3)(0.76)}{(0.5)^2} = 0.91$$

Because $N_{FoM} > 0.5$, the parabolic profile will be closely approached. In a similar manner, the following results are obtained for Sets 2 and 3:

	Set 2	Set 3
t_c, h	0.28	1.12
N_{FoM}	0.34	0.34
Parabolic profile closely approached	no	no

For Sets 2 and 3, the parabolic moisture-content profiles are not closely approached. However, the absolute errors in $X_o - X$ at the surface and midplane are determined from (18-51) to be only 1.1% and 4.3%, respectively.

An approximate theoretical estimate of the additional drying time required for the falling-rate period is derived as follows from the development by Walker, et al. [21]. At the end of the constant-rate period, the rate of flow of moisture by Fickian diffusion to the surface of the slab, where it is then evaporated, may be equated to the reduction in average moisture content of the slab. Thus,

$$R = -\frac{\rho_s a A}{A} \frac{dX_{avg}}{dt} = -D_{AB}\rho_s \frac{dX}{dz}\bigg|_s \quad (18\text{-}57)$$

From the parabolic moisture profile of (18-56), at the surface, $z = +a$,

$$-\frac{dX}{dz}\bigg|_{z=+a} = \frac{2}{a}(X_m - X_s) \quad (18\text{-}58)$$

However, it is more desirable to convert this expression from one in terms of X_m to one in terms of X_{avg}. To do this, (18-56) can be substituted into (18-52) for the definition of X_{avg}, followed by integration to give

$$X_{avg} = \frac{2}{3}X_m + \frac{1}{3}X_s \quad (18\text{-}59)$$

which can be rewritten as

$$X_m - X_s = \frac{3}{2}(X_{avg} - X_s) \quad (18\text{-}60)$$

Substitution of (18-60) in (18-58), followed by substitution of the result into (18-57), gives

$$R = -a\rho_s \frac{dX_{avg}}{dt} = \frac{3D_{AB}\rho_s}{a}(X_{avg} - X_s) \quad (18\text{-}61)$$

The falling-rate period is assumed to begin, as discussed above, with $X_s = X^*$. If the parabolic moisture profile is maintained during the falling-rate period and if $X_s = X^*$ remains constant, then (18-61) applies during that period and a straight-line falling-rate period of the type shown in Figure 18.31a will be obtained. Integration of (18-61) from the start of the falling-rate period when $X_{avg} = X_c$, gives

$$t_f = \frac{a^2}{3D_{AB}} \ln\left[\frac{X_c - X^*}{X_{avg} - X^*}\right] \quad (18\text{-}62)$$

Thus, the duration of the falling-rate period is predicted to be directly proportional to the square of the slab half-thickness and inversely proportional to the moisture liquid diffusivity. Equation (18-62) gives reasonable predictions for nonporous slabs of materials such as wood, clay, and soap when the slabs are thick and D_{AB} is low. However, serious deviations can occur when D_{AB} depends strongly on X and/or temperature. In that case, an average value of D_{AB} can be used to obtain an approximate result. A summary of experimental average moisture liquid diffusivities for a wide range of water-wet solids is tabulated in Chapter 4 of Mujumdar [1].

EXAMPLE 18.16

Gilliland and Sherwood [20] obtained data of the drying of water-wet $7 \times 7 \times 2.54$-cm slabs of 193.9 g (bone-dry) brick clay mix for direct-heat convective air drying from the two faces in both the constant-rate and falling-rate periods. For $X_o = 0.273$, $X^* = 0.03$, the rate of drying in the constant-rate period to $X_c = 0.165$ was

0.157 g/h-cm^2. The air velocity past the two faces was 15.2 m/s with $T_d = 25°C$ and $T_w = 17°C$. During the falling-rate period, experimental average slab moisture contents were as follows:

Time from start of the constant-drying rate period, minutes	X_{avg}
67	0.165 (critical value)
87	0.145
102	0.134
119	0.124
138	0.114
162	0.106
183	0.099
205	0.095
216	0.090

At the end of the constant-drying-rate period, the moisture profile was shown to be very nearly parabolic. From other experiments, $D_{AB} = 0.72 \times 10^{-4}$ cm^2/s.

Use (18-62) to predict values of X_{avg} during the falling-rate period and compare the predicted values to the experimental values.

SOLUTION

Solving (18-62) for X_{avg} gives

$$X_{avg} = X^* + (X_c - X^*)\exp(-3D_{AB}t_f/a^2) \quad (1)$$

where t_f is the time from the start of the falling-rate period.
  For $t_f = 87 - 67 = 20$ min, from (1),

$$X_{avg} = 0.03 + (0.165 - 0.03)$$
$$\times \exp[-3(0.72 \times 10^{-4})(20)(60)/(1.27)^2] = 0.145 \text{ cm}^2/\text{s}$$

Calculations for other values of time give the following results:

t_f, Time from start of falling-rate period, min	Experimental X_{avg}	Predicted X_{avg}
0	0.165	0.165
20	0.145	0.145
35	0.134	0.132
52	0.124	0.119
71	0.114	0.106
95	0.106	0.093
116	0.099	0.083
138	0.095	0.075
149	0.090	0.071

Comparing the predicted values of X_{avg} with the experimental values, it is seen that the deviation increases with increasing time. If the value of D_{AB} is reduced to 0.53×10^{-4} cm^2/s, much better agreement is obtained with the ESS decreasing from 0.0013 to 0.000154 cm^4/s^2.

Capillary-Flow Theory

For wet solids of the first category as discussed in Section 18.3, moisture is held, e.g., as free moisture in the interstices of powders and granular solids. Movement of moisture from the interior to the surface can occur by capillary action in the interstices, but may be opposed by gravity.

Forces holding liquid molecules together are cohesive forces. Additionally, liquid molecules may be attracted to a solid surface by adhesive forces. Thus, water held in a glass tube will tend to creep up the side of the tube until the adhesive forces are balanced by the weight of the liquid. For an ideal case of a capillary tube of small diameter partially immersed in a vertical orientation in a liquid reservoir, the liquid will rise in the tube to a height above the surface of the liquid in the reservoir. At equilibrium the height, h, will be

$$h = 2\sigma/\rho_L g r \qquad (18\text{-}63)$$

where σ is the surface tension of the liquid and r is the radius of the capillary. This equation shows that the smaller the radius of the capillary, the larger the capillary effect. Unlike mass transfer by diffusion, which causes moisture to move from a region of higher concentration to one of lower concentration, moisture in interstices flows to regions of highest capillary regardless of concentration.

For capillary flow in granular beds of wet solids, the variable size and shape of the particles make it extremely difficult to develop a usable theory for predicting the rate of drying in the falling-rate period in terms of permeability and capillarity. However, interesting discussions and idealized theories are presented by Keey [7, 23] and Ceaglske and Kiesling [22]. For practical calculations, it appears that, despite pleas to the contrary, it is common to apply diffusion theory with effective diffusivities determined from experiment. In general, these diffusivities are lower than those for true diffusion of moisture in nonporous materials. Some values are included in a tabulation in Chapter 4 of Mujumdar [1]. For example, values for the effective diffusivity of water in beds of sand particles are given that cover a range of 1.0×10^{-2} to 8.0×10^{-4} cm^2/s.

18.5 DRYER MODELS

In previous sections, general mathematical models for estimating drying rates and moisture profiles have been developed and discussed, with applications to batch tray dryers of the cross-circulation and through-circulation type. More specific models have been developed over the years for a number of continuous dryers. In this section, models are presented for three such dryers: (1) belt dryer with through-circulation, (2) direct-heat rotary dryer, and (3) fluidized-bed dryer, all of which are direct-heat dryers. Other models are considered in a special issue of "Drying Technology," edited by Genskow [24], and in Mujumdar [1].

Material and Energy Balances for Direct-heat Dryers

Consider the general case of a continuous, direct-heat dryer, as shown in Figure 18.36. Although countercurrent flow is shown, the following development of energy balances applies equally well to other flow configurations such as cocurrent flow and crossflow. Assume that the dryer is perfectly insulated such that the operation is adiabatic. As the solid is dried, moisture is transferred to the gas. Further assume that no solid is entrained into the gas and that changes in kinetic energy and potential energy are negligible. The flow rates of dry solid, m_s, and dry gas, m_g, do not change as drying proceeds. Therefore, at steady state, a material balance on the moisture is given by

$$X_{ws}m_s + \mathcal{H}_{gi}m_g = X_{ds}m_s + \mathcal{H}_{go}m_g \qquad (18\text{-}64)$$

The rate of moisture evaporation, m_v, is given by a rearrangement of (18-64):

$$m_v = m_s(X_{ws} - X_{ds}) = m_g(H_{go} - H_{gi}) \qquad (18\text{-}65)$$

where the subscripts are ws (wet solid), ds (dry solid), gi (gas in), and go (gas out).

A steady-state energy balance can be written in terms of enthalpies or in terms of specific heat and heat of vaporization. In either case, it is convenient to treat the dry gas, dry solid, and moisture (liquid and vapor) separately, and assume ideal mixtures. In terms of enthalpies, the steady-state energy balance is as follows, where s, g, and m refer, respectively, to dry solid, dry gas, and moisture:

$$\begin{aligned}
m_s(H_s)_{ws} &+ X_{ws}m_s(H_m)_{ws} + m_g(H_g)_{gi} + \mathcal{H}_{gi}m_g(H_m)_{gi} \\
&= m_s(H_s)_{ds} + X_{ds}m_s(H_m)_{ds} \\
&\quad + m_g(H_g)_{go} + \mathcal{H}_{go}m_g(H_m)_{go}
\end{aligned} \qquad (18\text{-}66)$$

A factored rearrangement of (18-66) is

$$\begin{aligned}
m_s[(H_s)_{ds} &- (H_s)_{ws} + X_{ds}(H_m)_{ds} - X_{ws}(H_m)_{ws}] \\
&= m_g[(H_g)_{gi} - (H_g)_{go} + \mathcal{H}_{gi}(H_m)_{gi} \\
&\quad - \mathcal{H}_{go}(H_m)_{go}]
\end{aligned} \qquad (18\text{-}67)$$

where any convenient reference temperatures can be used to determine the enthalpies.

When the system is air, water, and a solid, a more convenient form of (18-67) can be obtained by evaluating the enthalpies of the solid and the air from specific heats, and obtaining moisture enthalpies from the steam tables. Often, the specific heat of the solid is almost constant over the temperature range of interest, and in the range from 25°C (78°F) to 400°C (752°F), the specific heat of dry air increases by

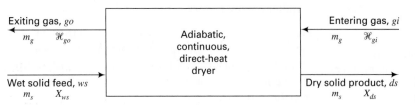

Figure 18.36 General configuration for a continuous, direct-heat dryer.

less than 3%, so that the use of a constant value of 0.242 Btu/lb-°F introduces little error. If the enthalpy reference temperature of the water is taken as T_o (usually 0°C or 32°F for liquid water when using the steam tables), (18-67) can be rewritten as

$$m_s[(C_P)_s(T_{ds} - T_{ws}) + X_{ds}(H_{H_2O})_{ds} - X_{ws}(H_{H_2O})_{ws}]$$
$$= m_g[(C_P)_{air}(T_{gi} - T_{go}) + \mathcal{H}_{gi}(H_{H_2O})_{gi} \qquad (18\text{-}68)$$
$$- \mathcal{H}_{go}(H_{H_2O})_{go}]$$

A further simplification in the energy balance for the air–water–solid system can be made by replacing enthalpies for water by their equivalents in terms of specific heats for liquid water and steam, and the heat of vaporization. In the range from 25°C (78°F) to 100°C (212°F), the specific heat of liquid water and steam are almost constant at 1 Btu/lb-°F and 0.447 Btu/lb-°F, respectively. The heat of vaporization of water over this same range decreases from 1049.8 to 970.3 Btu/lb, a change of almost 8%. Combining (18-65) with (18-68), and taking a thermodynamic path of water evaporation at the moisture-evaporation temperature, denoted T_v, the simplified energy balance is

$$Q = m_s\{(C_P)_s(T_{ds} - T_{ws}) + X_{ws}(C_P)_{H_2O(\ell)}(T_v - T_{ws})$$
$$+ X_{ds}(C_P)_{H_2O_{(\ell)}}(T_{ds} - T_v)$$
$$+ (X_{ws} - X_{ds})[\Delta H_v^{vap} + (C_P)_{H_2O(g)}(T_{go} - T_v)]\}$$
$$= m_g\{[(C_P)_{air} + \mathcal{H}_{gi}(C_P)_{H_2O(v)}](T_{gi} - T_{go})\} \qquad (18\text{-}69)$$

Equations (18-64) to (18-69) are useful for determining the required gas flow rate for drying a given flow rate of wet solids, as illustrated in the example below. Also of interest for sizing the dryer is the required heat-transfer rate, Q from the gas to the solid. For the air–water–solid system, this rate is equal to either the left-hand side or right-hand side of (18-69), as indicated. In the general case,

$$Q = m_g\{(H_g)_{gi} - (H_g)_{go} + \mathcal{H}_{gi}[(H_m)_{gi} - (H_m)_{go}]\}$$
$$= m_s\{(H_s)_{ds} - (H_s)_{ws} + X_{ds}(H_m)_{ds} - X_{ws}(H_m)_{ws}$$
$$+ m_g[(H_m)_{go}(\mathcal{H}_{go} - \mathcal{H}_{gi})]\}$$

$$(18\text{-}70)$$

EXAMPLE 18.17

A continuous, cocurrent-flow, direct-heat dryer is to be used to dry crystals of Epsom salt (magnesium sulfate heptahydrate). The feed to the dryer, a filter cake from a rotary, vacuum filter, consists of 2,854 lb/h of crystals (dry basis) with a moisture content of 25.8 wt% (dry basis) at 85°F and 14.7 psia. Air enters the dryer at 14.7 psia, with dry-bulb and wet-bulb temperatures of 250°F and 117°F, respectively. The final moisture content of the dried crystals is to be 1.5 wt% (dry basis), at a temperature of no more than 118°F to prevent decomposition to the hexahydrate (see Figure 17.2). Determine:

(a) The rate of moisture evaporation.

(b) The outlet temperature of the air.

(c) The rate of heat transfer.

(d) The entering air flow rate.

The average specific heat of Epsom salt is 0.361 Btu/lb-°F in the temperature range of 70–120°F.

SOLUTION

$$m_s = 2854 \text{ lb/h}, X_{ws} = 0.258, \ X_{ds} = 0.015$$
$$T_{ws} = 85°F, \ T_{ds} = 118°F, \ T_{gi} = 250°F, \ T_v = 117°F$$

From Figure 18.17, for $T_{db} = 250°F$ and $T_{wb} = 117°F$, $\mathcal{H}_{gi} = 0.0405$

(a) From (18-65),

$$m_v = 2,854(0.258 - 0.015) = 694 \text{ lb/h}$$

(b) Because the dryer operates cocurrently, the outlet temperature of the gas must be greater than the outlet temperature of the dry solid, which is taken as 118°F. The best value for T_{go} is obtained by optimizing the cost of the drying operation. A reasonable value for T_{go} can be estimated by using the concept of the number of heat-transfer units, which is analogous to the number of transfer units for mass transfer, as developed in Chapter 6. For heat transfer in a dryer, where the solids temperature throughout most of the dryer will be at T_v, the number of heat-transfer units is defined by

$$N_T = \ln\left[\frac{T_{gi} - T_v}{T_{go} - T_v}\right] \qquad (1)$$

where economical values of N_T are usually in the range of 1.0–2.5. Assume a value here of 2.0. From (1),

$$2 = \ln\left[\frac{250 - 117}{T_{go} - 117}\right]$$

from which $T_{go} = 135°F$

(c) The rate of heat transfer is obtained from (18-69) using the conditions for the solid flow.

$$Q = 2,854\{0.361(118 - 85) + 0.258(1)(117 - 85)$$
$$+ 0.015(1)(118 - 117) + (0.258 - 0.015)$$
$$\times [1,027.5 + (0.447)(135 - 117)]\}$$
$$= 2,854[11.9 + 8.3 + 0.02 + 249.7 + 2.0]$$
$$= 2,854(271.9) = 776,000 \text{ Btu/h}$$

It should be noted that the heat required to vaporize the 694 lb/h of moisture at 117°F is $(249.7/271.9) \times 100\% = 91.8\%$ of the total heat load.

(d) The entering air flow rate is obtained from (18-69) using the far right-hand side of that equation with the above value of Q.

$$m_g = \frac{776,000}{[(0.242) + (0.0405)(0.447)](250 - 135)} = 25,940 \text{ lb/h}$$

The total entering air, including the humidity, is

$$25,940(1 + 0.0405) = 27,000 \text{ lb/h}$$

Belt Dryer with Through-Circulation

Consider the continuous, two-zone, through-circulation belt dryer shown in Figure 18.37a. A bed of wet-solid particles is conveyed continuously into Zone 1, where contact is made

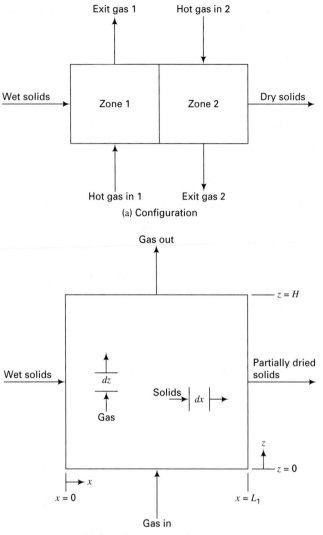

Figure 18.37 Continuous, two-zone, through-circulation belt dryer.

with hot gas passing upward through the bed. Because the temperature of the gas decreases as it passes through the bed, the temperature-driving force decreases so that the moisture content of solids near the bottom of the moving bed decreases more rapidly than for solids near the top. To obtain a dried solid of more uniform moisture content, the gas flow direction through the bed is reversed in Zone 2.

Based on the work of Thygeson and Grossmann [25], a mathematical model for Zone 1 can be developed as follows, using the coordinate system shown in Figure 18.37b. The model is based on the following assumptions:

1. Wet solids enter Zone 1 with a uniform moisture content of X_o on the dry basis.

2. Gas passes up through the moving bed in plug flow with no mass transfer in the vertical direction (i.e., no axial dispersion).

3. Drying takes place in the constant-rate period, controlled by the rate of heat transfer from the gas to the

surfaces of the solid particles, where the temperature is the adiabatic-saturation temperature.

4. Sensible-heat effects are negligible compared to latent-heat effects.

5. The void fraction of the bed is uniform and constant, and no mixing of solid particles occurs.

6. The solids are conveyed at a uniform linear speed, S.

Based on these assumptions, the temperature of the gas decreases with increasing distance z from the bottom of the bed, and is independent of the distance, x, in the direction in which the solids are conveyed, i.e., $T = T\{z\}$. The moisture content of the solids, however, varies in both the z and x directions, decreasing more rapidly near the bottom of the bed where the gas temperature is higher, i.e., $X = X\{z, x\}$.

Zone 1 With no mixing of solids, a material balance on the moisture in the solids at any vertical location, z, is given, for Zone 1, by

$$\frac{dX_1}{dt} = S\frac{dX_1}{dx} = -\frac{(ha)(T_1 - T_v)}{\Delta H_v^{\text{vap}}(\rho_b)_{ds}} \qquad (18\text{-}71)$$

where $a =$ surface area of solid particles per unit volume of bed, $T_1 =$ bulk temperature of the gas in Zone 1, which depends on z, and $(\rho_b)_{ds}$ is the bulk density of solids when dry. The initial condition is $X_1 = X_o$ for $x = 0$.

An energy balance for the gas phase at any location, x, is given by

$$\rho_g(C_P)_g u_s \frac{dT_1}{dz} = -(ha)(T_1 - T_v) \qquad (18\text{-}72)$$

where $\rho_g =$ density of the gas and $u_s =$ superficial velocity of gas through the bed. The initial condition is $T_1 = T_{gi}$ for $z = 0$.

Equation (18-71) is coupled to (18-72), which is independent of (18-71). Therefore, we can separate variables and integrate (18-72) to obtain

$$T_1 = T_v + (T_{gi} - T_v)\exp\left(-\frac{haz}{\rho_g(C_P)_g u_s}\right) \qquad (18\text{-}73)$$

At $z = H$ at the top of the bed,

$$T_{go_1} = T_v + (T_{gi} - T_v)\exp\left(-\frac{haH}{\rho_g(C_P)_g u_s}\right) \qquad (18\text{-}74)$$

Equation (18-71) can now be solved by combining it with (18-73) to eliminate T_1, followed by separation of variables and integration. The result is

$$\frac{X_1}{X_o} = 1 - \frac{xha(T_{gi} - T_v)\exp\left(-\dfrac{haz}{\rho_g(C_P)_g u_s}\right)}{S\Delta H_v^{\text{vap}}(\rho_b)_{ds}} \qquad (18\text{-}75)$$

The moisture content $(X_1)_{L_1}$ at $x = L_1$ is obtained by replacing x with L_1.

If desired, X_{avg} at $x = L_1$ can be determined from

$$X_{avg} = \int_o^H (X_1)_{L_1} dz \qquad (18\text{-}76)$$

Zone 2 In Zone 2, (18-71) still applies with X_1 and T_1 replaced by X_2 and T_2, but the initial condition for X_o is $(X_1)_{L_1}$ from (18-75) for $x = L_1$, which depends on z. Equation (18-72) also applies with T_1 replaced by T_2, but the initial condition is $T_2 = T_{gi}$ at $z = H$. The resulting integrated equations are

$$T_2 = T_v + (T_{gi} - T_v)\exp\left[-\frac{h\,a(H-z)}{\rho_g(C_P)_g u_s}\right] \qquad (18\text{-}77)$$

with T_{go_2} given by (18-74), where T_{go_1} is replaced by T_{go_2}, and

$$\frac{X_2}{(X_1)_{L_1}} = 1 - \frac{x h\,a(T_{gi} - T_v)\exp\left(-\dfrac{h\,a(H-z)}{\rho_g(C_P)_g u_s}\right)}{S\Delta H_v^{vap}(\rho_b)_{ds}} \qquad (18\text{-}78)$$

where $(X_1)_{L_1}$ is the value from (18-75) for $x = L_1$ and the value of z in (18-78). The value of x in (18-78) is the distance from the start of Zone 2. Values of $(X_2)_{L_2}$ at any z are obtained from (18-78) for $x = L_2$. The average moisture content over the height of the moving bed leaving Zone 2 is then obtained from (18-76) with $(X_1)_{L_1}$ replaced by $(X_2)_{L_2}$. The application of the above relationships is illustrated in the following example.

EXAMPLE 18.18

The extruded filter cake of calcium carbonate in Example 18.8 is to be dried continuously with a belt dryer using through-circulation. The dryer is 6 ft wide and has a belt speed of 1 ft/min. The dryer consists of two drying zones, each 12 ft long. Air at 170°F and 10% relative humidity enters both zones, passing upward through the moving bed in the first zone, and downward through the second zone, at a superficial velocity of 2 m/s. The bed height on the belt is constant at 2 in. Predict the moisture-content distribution with height at the end of each zone, and the average moisture content at the end of Zone 2. Assume all drying is in the constant-rate period and neglect the preheat period.

SOLUTION

From data in Examples 18.7 and 18.8,

$$X_o = 0.30$$

$$(\rho_b)_{ds} = \frac{1.00}{1.30}(1,950) = 1,500\,\text{kg/m}^3$$

$$\epsilon_b = 0.50$$

$$T_v = 37.8°C = 311\,\text{K}, \quad T_{gi} = 76.7°C = 350\,\text{K}$$

$$\Delta H_v^{vap} = 2,413\,\text{kJ/kg}$$

From extrusion area and volume in Example 18.8,

$$a = \frac{(3.16\times10^{-4})(0.5)}{(4.01\times10^{-7})} = 395\,\text{m}^2/\text{m}^3\text{ bed}$$

For $u_s = 2$ m/s, $h = 0.188$(kJ/s-m²-K²) from Example 18.8.

$$(C_P)_g = 1.09\,\text{kJ/kg-K}$$
$$\rho_g \text{ at 1 atm} = 0.942\,\text{kg/m}^3$$
$$S = 1\,\text{ft/min} = 0.00508\,\text{m/s}$$

The cross-sectional area of the moving bed normal to the conveying direction is $6(2/12) = 1\,\text{ft}^2 = 0.0929\,\text{m}^2$. For a belt speed of 1 ft/min = 0.305 m/min, the volumetric flow of solids is $(0.0929)(0.305) = 0.0283\,\text{m}^3/\text{min}$. The mass rate of flow is $0.0283(1,500) = 42.5$ kg/min (dry basis).

Zone 1

$$H = 0.167\,\text{ft} = 0.0508\,\text{m}$$
$$L_1 = 12\,\text{ft} = 3.66\,\text{m}$$

From (18-74), the gas temperature leaving the bed is

$$T_{go_1} = 37.8 + (76.7 - 37.8)\exp\left[-\frac{(0.188)(395)(0.0508)}{(0.942)(1.09)(2)}\right]$$
$$= 44°C = 317\,\text{K}$$

The moisture-content distribution at $x = L_1$ is obtained from (18-75). For $z = 0$,

$$X_1 = 0.30\left[1 - \frac{(3.66)(0.188)(395)(76.7 - 37.8)}{(0.00508)(2,413)(1,500)}\right] = 0.127$$

For other values of z, using a spreadsheet, the following values are obtained:

z, m	$(X_1)_{L_1}$
0	0.127
0.0127	0.191
0.0254	0.231
0.0381	0.257
0.0508	0.273

Because of the considerable decrease in bulk-gas temperature as it passes upward through the bed, the moisture content varies considerably over the bed depth.

Zone 2 The flow of air is now reversed to further the drying and smooth out the moisture-content distribution. The value of X_o is replaced by values of $(X_1)_{L_1}$ above for corresponding values of z. Using (18-78) with a spreadsheet, the following distribution is obtained at $L_2 = 3.66$ m for a total length of both zones = 24 ft = 7.32 m:

z, m	$(X_2)_{L_2}$
0	0.116
0.0127	0.163
0.0254	0.178
0.0381	0.163
0.0508	0.116

A much more uniform moisture distribution is achieved. From (18-76) for Zone 2, using numerical integration with a spreadsheet, $(X_2)_{avg} = 0.155$.

Direct-Heat Rotary Dryer

As discussed by Kelly in Mujumdar [1], the design of a direct-heat rotary dryer, of the type shown in Figure 18.7, for drying a specified solids feed rate of initial moisture content X_{ws} to a final moisture content X_{ds}, involves the selection and determination of a number of factors, including heating gas inlet and outlet conditions, dryer-cylinder slope and rotation rate, number and type of lifting flights, heating-gas velocity and flow direction, and dryer cylinder diameter and length. Also of interest are the solids holdup, as a percent of dryer-cylinder volume, and solids residence time.

Ideally, a commercial-size direct-heat rotary dryer should be scaled up from pilot-plant data obtained in a laboratory unit. However, if a representative sample of the wet solid is not available for testing, the following procedure and model, based on test results with several materials in both pilot-plant-size and commercial-size dryers, is useful for a preliminary design.

The hot gas can flow countercurrently or cocurrently to the flow of the solids. In general, cocurrent flow is used for very wet, sticky solids, with high inlet-gas temperatures, and for nonhygroscopic solids. Countercurrent flow is preferred for low-to-moderate inlet-gas temperatures, where thermal efficiency becomes a factor. When solids are not subject to thermal degradation, melting, or sublimation, a high-inlet gas temperature up to approximately 1000°F can be used. The exit-gas temperature is determined largely from economics, as discussed in Example 18.17, where (1) can be used with N_T in the range of 1.5–2.5. The allowable gas velocity is determined from the dusting characteristics of the dried solid particles, and can vary widely with particle-size distribution and particle density. Some typical values for allowable gas velocity are as follows:

Material	Particle Density, ρ_p, lb/ft³	Average Particle Size, d_p, μm	Allowable Gas Velocity, u_{all}, ft/s
Plastic granules	69	920	3.5
Ammonium nitrate	104	900	4.5
Sand	164	110	1.0
Sand	164	215	2.0
Sand	164	510	5.0
Sawdust	27.5	640	1.0

Using an appropriate allowable gas velocity, u_{all}, the mass flow rate of gas at the gas discharge end, $(m_g)_{exit}$, of the dryer, and the gas density $(\rho_g)_{exit}$ at that end, the dryer diameter, D, can be estimated by the continuity equation

$$D = \left[\frac{4(m_g)_{exit}}{\pi u_{all}(\rho_g)_{exit}}\right]^{0.5} \qquad (18-79)$$

The residence time of the solids in the dryer, θ, is related to the fractional volume holdup of solids, V_H, by the relation

$$\theta = \frac{L V_H}{F_V} \qquad (18-80)$$

where L = length of dryer cylinder and F_V = solids volumetric velocity in volume/unit cross-sectional area-unit time. A conservative estimate of the holdup, including the effect of gas velocity, is obtained by combining (18-80) with a relation in [2]:

$$V_H = \frac{0.23 F_V}{S N^{0.9} D} \pm 0.6 \frac{G(5/d_p^{0.5})}{\rho_p} \qquad (18-81)$$

where

F_V = ft³ solids/(ft² cross section)-h

S = dryer cylinder slope, ft/ft

N = dryer cylinder rate of rotation, rpm

D = dryer diameter, ft

G = gas superficial mass velocity, lb/h-ft²

d_p = mass-average particle size, μm

The plus (+) sign on the second term corresponds to countercurrent flow that tends to increase the holdup, while the minus (−) sign corresponds to cocurrent flow. Equation (18-81) holds for dryers with lifting flights that have lips, but is limited to gas velocities less than 3.5 ft/s. An improved, but more complex, model by Matchett and Sheikh [26] is valid for gas velocities to 10 ft/s. Optimal solids holdup is in the range of 10–18% of dryer volume so that flights run full and all or most of the solids are showered during each revolution.

When drying is in the constant-rate period such that the rate can be determined from the rate of heat transfer from the gas to the wet surface of the solids at the wet-bulb temperature, a volumetric heat-transfer coefficient, ha, can be used, which is defined by

$$Q = (ha)V \Delta T_{LM} \qquad (18-82)$$

where V = volume of dryer cylinder = $\pi D^2 L/4$

$$\Delta T_{LM} = \frac{(T_g)_{in} - (T_g)_{out}}{\ln\left[\dfrac{(T_g)_{in} - T_v}{(T_g)_{out} - T_v}\right]} \qquad (18-83)$$

and ha = volumetric heat-transfer coefficient based on dryer cylinder volume as given by the empirical correlation of McCormick [27], when the heating gas is air:

$$ha = K G^{0.67}/D \qquad (18-84)$$

where

ha is in Btu/h-ft³-°F

G is in lb/h-ft²

D is in ft

A value of $K = 0.5$ is recommended in [2] for dryers operating at a peripheral cylinder speed of 1.0–1.25 ft/s and with a flight count of 2.4D–3.0D per circle. When pilot-plant data are available, (18-84) can be used for scale-up to a larger diameter and a different value of G, where K is determined from the pilot-plant data.

It might be expected that a correlation for the volumetric heat-transfer coefficient, ha, would take into account the particle diameter because the solids are lifted and showered through the gas. However, the solids shower as curtains of some thickness, with the gas passing between curtains. Thus, particles inside the curtains do not receive significant exposure to the gas and the effective heat transfer area is more likely determined by the areas of the curtains. Nevertheless, (18-84) accounts for only two of the many possible variables and the inverse relation with dryer diameter is not well-supported by experimental data. A more satisfactory, but more complex, model for heat transfer is that of Schofield and Glikin [28], as modified by Langrish, Bahu, and Reay [29]. That model treats h and a separately.

EXAMPLE 18.19

Ammonium nitrate, with a moisture content of 15 wt% (dry basis), is to be fed into a direct-heat rotary dryer at a feed rate of 700 lb/min (dry basis) and a temperature of 70°F. Air at 250°F and 1 atm, with a humidity of 0.02 lb H_2O/lb dry air enters the dryer at the feed end and passes cocurrently with the solid through the dryer. The final moisture content of the solid is to be 1 wt% (dry basis). It may be assumed that all drying will take place in the constant-rate period. Make a preliminary estimate of the dryer diameter and length, assuming that such dryers are available in: (1) diameters from 1 to 5 ft in increments of 0.5 ft and from 5 to 20 ft in increments of 1.0 ft, and (2) lengths from 5 to 150 ft in increments of 5 ft.

SOLUTION

From the psychrometric chart (Figure 18.17), $T_{wb} = 107$°F. Assume that all drying takes place at this temperature of the solid. A reasonable outlet temperature for the air can be estimated from (1) of Example 18.17, assuming $N_T = 1.5$. From that equation,

$$1.5 = \ln\left[\frac{250 - 107}{T_{go} - 107}\right]$$

Solving,

$$T_{go} = 140°F$$

Assume solids outlet temperature $= T_{ds} = 135$°F

Heat-transfer rate

$m_s = 700(60) = 42,000$ lb/h of solids (dry basis)

$(C_P)_s = 0.4$ Btu/lb-°F

$T_{ws} = 70$°F

$X_{ws} = 0.15$

$T_v = T_{wb} = 107$°F

$X_{ds} = 0.01$

$\Delta H_v^{vap} = 1,033$ Btu/lb

From (18-65),

$m_v = 42,000(0.15 - 0.01) = 5,880$ lb/h H_2O evaporated

From (18-69),

$$Q = 42,000\{(0.4)(135 - 70) + (0.15)(1)(107 - 70) \\ + (0.01)(1)(135 - 70) + (0.15 - 0.01)[1,033 \\ + (0.447)(140 - 107)]\} = 7,510,000\,\text{Btu/h}$$

Air flow rate:

$$m_g = \frac{7,510,000}{[(0.242) + (0.02)(0.447)](250 - 135)} \\ = 260,000\,\text{lb/h entering dry air}$$

Dryer diameter:

Assume an allowable gas velocity at the exit end of the dryer of 4.5 ft/s.

$(m_g)_{exit} = 260,000(1 + 0.02) + 5,880 = 271,000$ lb/h total gas

$$(\rho_g)_{exit} = \frac{PM}{RT_{go}}$$

$$M = \frac{271,000}{\dfrac{260,000}{29} + \dfrac{11,000}{18}} = 28.3$$

$$(\rho_g)_{exit} = \frac{(1)(28.3)}{(0.730)(600)} = 0.0646\,\text{lb/ft}^3$$

From (18-79),

$$D = \left[\frac{4(271,000)}{(3.14)(4.5)(3,600)(0.0646)}\right]^{0.5} = 18\,\text{ft}$$

Dryer length:

$$G = G_{exit} = \frac{(271,000)(4)}{(3.14)(18)^2} = 1,070\,\text{lb/h-ft}^2$$

From (18-84),

$$ha = 0.5(1,070)^{0.67}/18 = 3\,\text{Btu/h-ft}^3\text{-°F}$$

From (18-83), neglecting the periods of wet solids heating up to 107°F and the dry solids heating up to 135°F,

$$\Delta T_{LM} = \frac{250 - 140}{\ln\left[\dfrac{250 - 107}{140 - 107}\right]} = 75°F$$

From (18-82)

$$V = \frac{7,510,000}{(3)(75)} = 33,400\,\text{ft}^3$$

Cross-sectional area $= (3.14)(18)^2/4 = 254$ ft^2

$$L = \frac{33,400}{254} = 130\,\text{ft}$$

Fluidized-Bed Dryer

Consider the behavior of a bed of solid particles when a gas is passed up through the bed, as shown in Figure 18.38. At a very low gas velocity, the bed remains fixed. At a very high gas velocity, the bed disappears because the particles are pneumatically transported away by the gas when its local velocity exceeds the terminal settling velocity of the particles. At an

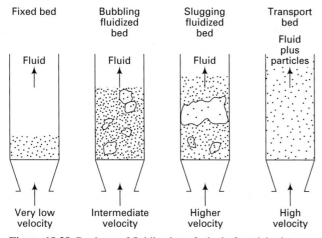

Figure 18.38 Regimes of fluidization of a bed of particles by a gas.

intermediate gas velocity, the bed is expanded, but particles are not carried out by the gas. Such a bed is said to be fluidized, because the bed of solids takes on some of the properties of a fluid. *Fluidization* is initiated when the gas velocity reaches the point where all the particles are just suspended by the gas. As the gas velocity is increased further, the bed expands and bubbles of gas are observed to pass up through the bed. This regime of fluidization is referred to as *bubbling fluidization* and is the most desirable regime for most fluidized-bed operations, including drying. If the gas velocity is increased further, a transition to *slugging fluidization* eventually occurs where bubbles coalesce and spread to a size that approximates the diameter of the vessel containing the bed.

Before fluidization occurs, when the bed of solids is fixed, the pressure drop across the bed for gas flow, ΔP_b, is predicted well by the Ergun [30] equation, discussed in Chapters 6 and 14:

$$\frac{\Delta P_b}{L_b} = 150 \frac{(1 - \epsilon_b)^2}{\epsilon_b^3} \frac{\mu u_s}{(\phi_s d_p)^2} + 1.75 \frac{(1 - \epsilon_b)}{\epsilon_b^3} \frac{\rho_g u_s^2}{\phi_s d_p}$$

$$(18\text{-}85)$$

where L_b = bed height, u_s = superficial gas velocity, and ϕ_s = particle sphericity. The first term on the right-hand side is dominant at low-particle Reynolds numbers where streamline flow exists and the second term dominates at high-particle Reynolds numbers where turbulent flow exists.

The onset of fluidization occurs when the drag force on the particles by the upward-flowing gas becomes equal to the weight of the particles (accounting for displaced gas):

$$\begin{pmatrix} \Delta P \\ \text{across} \\ \text{bed} \end{pmatrix} \begin{pmatrix} \text{cross-sectional} \\ \text{area} \\ \text{of bed} \end{pmatrix} = \begin{pmatrix} \text{Volume} \\ \text{of} \\ \text{bed} \end{pmatrix}$$

$$\times \begin{pmatrix} \text{Volume} \\ \text{fraction} \\ \text{of solid} \\ \text{particles} \end{pmatrix} \left[\begin{pmatrix} \text{Density} \\ \text{of} \\ \text{solid} \\ \text{particles} \end{pmatrix} - \begin{pmatrix} \text{Density} \\ \text{of} \\ \text{displaced} \\ \text{gas} \end{pmatrix} \right]$$

Thus,

$$\Delta P_b A_b = A_b L_b (1 - \epsilon_b)[(\rho_p - \rho_g)g] \quad (18\text{-}86)$$

The minimum gas-fluidization superficial velocity, u_{mf}, is obtained by solving (18-85) and (18-86) simultaneously for $u = u_{mf}$. For $N_{Re,p} = d_p u_{mf} \rho_g / \mu < 20$, the turbulent-flow contribution to (18-85) is negligible and the result is

$$u_{mf} = \frac{d_p^2 (\rho_p - \rho_g) g}{150 \mu} \left(\frac{\epsilon_b^3 \phi_s^2}{1 - \epsilon_b} \right) \quad (18\text{-}87)$$

For desirable operation in the bubbling fluidization regime, a superficial gas velocity of $u_s = 2u_{mf}$ is a reasonable choice. At this velocity, the bed will be expanded by about 10%, with no further increase in pressure drop across the bed. In this regime, the solid particles are well mixed and the bed temperature is uniform. Consequently, if the fluidized bed is operated continuously at steady-state conditions rather than batchwise with respect to the particles, the particles will have a residence-time distribution like that of a fluid in a continuous-stirred-tank reactor (CSTR). Some particles will be in the dryer for only a very short period of time and will experience almost no decrease in moisture content. Other particles will be in the dryer for a long time and may come to equilibrium before that time has elapsed. Thus, the dried solids will have a distribution of moisture content. This is in contrast to a batch-fluidization process where all particles have the same residence time and, therefore, a uniform final moisture content. This is an important distinction because continuous, fluidized-bed dryers are usually scaled up from data obtained in small-batch, fluidized-bed dryers. Therefore, it is important to have a relationship between batch drying time and continuous drying time.

The distribution of residence times for the effluent from a perfectly mixed vessel operating at continuous, steady-state conditions is given by Fogler [31] as

$$E\{t\} = \exp(-t/\tau)/\tau \quad (18\text{-}88)$$

where τ is the average residence time and $E\{t\}$ is defined such that $E\{t\}dt$ = the fraction of effluent with a residence time between t and $t + dt$. Thus, $\int_0^{t_1} E\{t\}dt$ = fraction of the effluent with a residence time less than t_1. For example, if the average particle residence time is 10 min, 63.2% of the particles will have a residence time of less than 10 min. If the particles are small and porous such that all drying takes place in the constant-rate period and θ is the time for complete drying, then

$$\frac{X}{X_o} = 1 - \frac{t}{\theta}, \qquad t \leq \theta \quad (18\text{-}89)$$

The average moisture content of the dried solids leaving the fluidized-bed reactor is obtained by integrating the following expression from 0 to only θ because $X = 0$ for $t > \theta$.

$$\bar{X}_{ds} = \int_0^{\theta} X E\{t\}dt = \int_0^{\theta} X_o \left(1 - \frac{t}{\theta}\right) E\{t\}dt \quad (18\text{-}90)$$

Combining (18-88) and (18-90) and integrating gives

$$\bar{X}_{ds} = X_o \left[1 - \frac{1 - \exp(-\theta/\tau)}{\theta/\tau} \right] \quad (18\text{-}91)$$

If the particles are nonporous and without surface moisture, such that all drying takes place in the falling-rate period, diffusion theory may apply such that the following empirical exponential expression may be used for the moisture content as a function of time:

$$\frac{X}{X_o} = \exp(-Bt) \quad (18\text{-}92)$$

In this case, the combination of (18-92) with (18-90), followed by integration from $t = 0$ to $t = \infty$ gives

$$\bar{X}_{ds} = 1/(1 + B\tau) \quad (18\text{-}93)$$

Values of θ and B must be determined from experiments with laboratory batch fluidized-bed dryers for scale-up to large commercial dryers operating under the same conditions.

EXAMPLE 18.20

10,000 lb/h of wet sand at 70°F with a moisture content of 20% (dry basis) is to be dried to a moisture content of 5% (dry basis) in a continuous, fluidized-bed dryer operating at a pressure of 1 atm in the freeboard region above the bed. The sand has a narrow size range with an average particle size of 500 μm and a sphericity, ϕ_s, of 0.67. The particle density is 2.6 g/cm³. When the sand is dry, the void fraction, ϵ_b, of a bed of the sand is 0.55. Fluidizing air will enter the bed at a temperature of 1,000°F with a humidity of 0.01 lb H₂O/lb dry air. The adiabatic-saturation temperature is estimated to be 145°F. Batch pilot-plant tests with a fluidization velocity of twice the minimum value show that drying takes place in the constant-rate period and all moisture can be removed in 8 minutes using air at the same conditions and with a bed temperature of 145°F. Determine the bed height and diameter for the large, continuous unit.

SOLUTION

$$d_p = 500 \ \mu m = 0.0500 \ cm$$

Heat-transfer rate:

$$(C_P)_s = 0.20 \ \text{Btu/lb-°F}$$
$$m_s = 10{,}000/(1 + 0.2) = 8{,}330 \ \text{lb/h dry sand}$$
$$T_v = 145°F = T_{go} = T_{ds}$$
$$\Delta H_v^{\text{vap}} = 1{,}011 \ \text{Btu/lb}$$
$$T_{ws} = 70°F$$

From (18-69),

$$Q = 8{,}330\{0.20(145 - 70) + 0.20(1)(145 - 70)$$
$$+ (0.20 - 0.05)(1{,}011)\} = 1{,}510{,}000 \ \text{Btu/h}$$

Air rate:

$$m_g = \frac{1{,}510{,}000}{[(0.242) + (0.01)(0.447)](1{,}000 - 145)}$$
$$= 7{,}170 \ \text{lb/h dry air}$$

From (18-65),

$$m_v = 8{,}330(0.20 - 0.05) = 1{,}250 \ \text{lb/h evaporated moisture}$$
$$\mathcal{H}_{go} = \frac{(7{,}170)(0.01) + 1{,}250}{7{,}170} = 0.184 \ \text{lb H}_2\text{O/lb dry air}$$

Total exiting gas flow rate $= 7{,}170(1 + 0.184) = 8{,}500 \ \text{lb/h}$

Minimum fluidization velocity:

$$\text{M of exiting gas} = \frac{8{,}500}{\dfrac{7{,}170}{29} + \dfrac{1{,}330}{18}} = 26.5$$

$$(\rho_g)_{\text{exit}} = \frac{PM}{RT_g} = \frac{(1)(26.5)}{(0.730)(605)} = 0.060 \ \text{lb/ft}^3$$
$$= 0.00096 \ \text{g/cm}^3$$
$$\mu = 0.048 \ \text{lb/ft-h} = 0.00020 \ \text{g/cm-s}$$

For small particles, assume streamline flow at u_{mf} so that (18-87) applies, but check to see if $N_{\text{Re},p} < 20$, using cgs units,

$$u_{mf} = \frac{(0.0500)^2(2.6 - 0.00096)(980)(0.55)^3(0.67)^2}{150(0.00020)(1 - 0.55)}$$
$$= 35.3 \ \text{cm/s}$$

$$N_{\text{Re},p} = \frac{d_p u_{mf} \rho_g}{\mu} = \frac{(0.0500)(35.3)(0.00096)}{0.00020} = 8.5$$

Since $N_{\text{Re},p} < 20$, (18-87) does apply.

Use an actual superficial-gas velocity of $2(35.3) = 70.6 \ \text{cm/s} = 8{,}340 \ \text{ft/h}$

Bed diameter:

Equation (18-79) applies:

$$D = \left[\frac{4(8{,}500)}{3.14(8{,}340)(0.060)} \right]^{0.5} = 4.7 \ \text{ft}$$

Bed density:

Fixed-bed density $= \rho_s(1 - \epsilon_b) = 2.6(1 - 0.55)(62.4) = 73.0 \ \text{lb/ft}^3$. Assume bed expands by 10% upon fluidization to $u = 2u_{mf}$:

$$\rho_b = \frac{73.0}{1.10} = 66 \ \text{lb/ft}^3 \ (\text{dry basis})$$

Average particle residence time:

For constant-rate drying in a batch dryer, all particles have the same residence time. From pilot-plant data, $\theta = 8$ min for complete drying.

For the large, continuous operation, (18-91) applies with $\bar{X}_{ds} = 0.05$, $X_o = 0.20$. Thus,

$$0.05 = 0.20 \left[1 - \frac{1 - \exp\left(-\dfrac{8}{\tau}\right)}{(8/\tau)} \right]$$

Solving this nonlinear equation, $\tau = 13.2$ minutes average residence time for particles. Only

$$\frac{(0.20 - 0.05)}{(0.20 - 0.0)}(8) = 6 \text{ min}$$

residence time would be required in a batch dryer to dry to 5% moisture. Therefore, more than double the residence time is needed in the continuous unit.

Bed height:

To achieve the average residence time of $13.2 \text{ min} = 0.22 \text{ h}$, the expanded-bed volume must be

$$V_b = \frac{m_s \tau}{\rho_b} = \frac{8,330(0.22)}{66} = 27.8 \text{ ft}^3$$

$$H_b = \frac{V_b}{\pi D^2/4} = \frac{27.8(4)}{3.14(4.7)^2} = 1.6 \text{ ft}$$

SUMMARY

1. Drying, in this chapter refers, to the removal of moisture from wet solids, solutions, slurries, and pastes. The moisture may be water or some other volatile liquid.

2. The two most common modes of drying are direct by heat transfer from a hot gas and indirect by heat transfer from a hot wall. The hot gas is frequently air, but can be a combustion gas, steam, nitrogen, or any other nonreactive gas.

3. Industrial drying equipment can be classified by operation (batch or continuous), mode (direct or indirect), and the degree to which the material being dried is agitated. Batch dryers include tray dryers and agitated dryers. Continuous dryers include tunnel dryers, belt or band dryers, turbo-tray tower dryers, rotary dryers, screw-conveyor dryers, fluidized-bed dryers, spouted-bed dryers, pneumatic-conveyor dryers, spray dryers, and drum dryers. Drying can also be accomplished with infrared radiation, dielectric and microwave radiation, and also from the frozen state by freeze drying.

4. Psychrometry, which deals with the properties of air–water mixtures and other gas–moisture mixtures, is useful in making drying calculations. Psychrometric (humidity) charts are particularly useful for determining the temperature at which surface moisture will be evaporated.

5. For the air–water system, the adiabatic-saturation temperature and the wet-bulb temperature are, by coincidence, almost identical. Thus, surface moisture is evaporated at the wet-bulb temperature. This greatly simplifies drying calculations.

6. Most solids can be grouped into one of two categories. Granular or crystalline solids that hold moisture in open pores between particles can be dried to very low moisture contents. Fibrous, amorphous, and gel-like materials that dissolve moisture or trap moisture in fibers or very fine pores can be dried to low moisture contents only with a gas of low humidity. The second category of materials can exhibit a significant equilibrium-moisture content that depends on the temperature, pressure, and humidity of the gas in contact with the material.

7. For drying calculations, the moisture content of a solid and a gas is usually based on the bone-dry solid and bone-dry gas, respectively. The bound-moisture content of a material in contact with a gas is the equilibrium-moisture content when the gas is saturated with the moisture. The excess-moisture content is called the unbound-moisture content. When the gas is not saturated, the excess-moisture content above the equilibrium-moisture content is called the free-moisture content. Solid materials that can contain bound moisture are hygroscopic. Bound moisture can be held chemically as water of hydration.

8. Drying by direct heat often takes place in four periods. The first is a preheat period accompanied by a rise in temperature but with little moisture removal. This is followed by a constant-rate period during which surface moisture is evaporated at the wet-bulb temperature. This moisture may be originally on the surface or moisture brought rapidly to the surface by diffusion or capillary action. The third period is a falling-rate period during which the rate of drying decreases linearly with time with little change in temperature. A fourth period may occur when the rate of drying falls off exponentially with time and the temperature rises.

9. The drying rate in the constant-rate period is governed by the rate of heat transfer from the gas to the surface of the solid. Empirical expressions for the heat-transfer coefficient have been formulated for a number of different types of direct-heat dryers.

10. The drying rate in the falling-rate period can be determined by using empirical expressions with experimental data. Diffusion theory can also be applied in some cases when moisture diffusivity is available or can be measured.

11. For direct-heat dryer models, material and energy balances can be applied to determine the rate of heat transfer from the gas to the wet solid and the gas flow rate.

12. A useful model for a two-zone belt dryer with through-circulation takes into account the changes in solids moisture content both vertically through the bed and in the direction of travel of the belt.

13. A model for preliminary sizing of a direct-heat rotary dryer is based on the use of a volumetric heat-transfer coefficient, assuming that the gas flows through curtains of cascading solids.

14. A model for sizing a large fluidized-bed dryer is based on the assumption of perfect mixing of the solids in the dryer when operating in the bubbling-fluidization regime. The procedure involves taking drying-time data from batch operation of a laboratory fluidized-bed dryer and correcting it for the expected solid particle residence-time distribution in the large dryer.

REFERENCES

1. *Handbook of Industrial Drying,* 2nd ed., A.S. Mujumdar, ed., Marcel Dekker, New York (1995).

2. *Perry's Chemical Engineers' Handbook,* 7th ed., R.H. Perry, D.W. Green, and J.O. Maloney, eds., McGraw-Hill, New York (1997).

3. WALAS, S.M., *Chemical Process Equipment,* Butterworths, Boston (1988).

4. VAN'T LAND, C.M., *Indusrial Drying Equipment,* Marcel Dekker, New York (1991).

5. UHL, V.W., and W.L. ROOT, *Chem. Eng. Progress,* **58,** 37–44 (1962).

6. McCORMICK, P.Y., in *Encyclopedia of Chemical Technology,* 4th ed., Vol. 8, John Wiley & Sons, New York (1993), pp. 475–519.

7. KEEY, R.B., *Introduction to Industrial Drying Operations,* Pergamon Press, Oxford (1978).

8. LEWIS, W.K., *Mech. Eng.,* **44,** 445–446 (1922).

9. FAUST, A.S., L.A. WENZEL, C.W. CLUMP, L. MAUS, and L.B. ANDERSON, *Principles of Unit Operations,* John Wiley & Sons, New York (1960).

10. LUIKOV, A.V., *Heat and Mass Transfer in Capillary-Porous Bodies,* Pergamon Press, London (1966).

11. SHERWOOD, T.K., *Ind. Eng. Chem.,* **21,** 12–16 (1929).

12. SHERWOOD, T.K., *Ind. Eng. Chem.,* **21,** 976–980 (1929).

13. MARSHALL, W.R., JR., and O.A. HOUGEN, *Trans. AIChE,* **38,** 91–121 (1942).

14. GAMSON, B.W., G. THODOS, and O.A. HOUGEN, *Trans. AIChE,* **39,** 1–35 (1943).

15. WILKE, C.R., and O.A. HOUGEN, *Trans. AIChE,* **41,** 445–451 (1945).

16. HOUGEN, O.A., H.J. McCAULEY, and W.R. MARSHALL, JR., *Trans. AIChE,* **36,** 183–209 (1940).

17. CARSLAW, H.S., and J.C. JAEGER, *Heat Conduction in Solids,* 2nd ed., Oxford University Press, London (1959).

18. NEWMAN, A.B., *Trans. AIChE,* **27,** 310–333 (1931).

19. SHERWOOD, T.K., *Ind. Eng. Chem.,* **24,** 307–310 (1932).

20. GILLILAND, E.R., and T.K. SHERWOOD, *Ind. Eng. Chem.,* **25,** 1134–1136 (1933).

21. WALKER, W.H., W.K. LEWIS, W.H. McADAMS, and E.R. GILLILAND, *Principles of Chemical Engineering,* 3rd ed., McGraw-Hill, New York (1937).

22. CEAGLSKE, N.H., and F.C. KIESLING, *Trans. AIChE,* **36,** 211–225 (1940).

23. KEEY, R.B., *Drying Principles and Practice,* Pergamon Press, Oxford (1972).

24. GENSKOW, L.R., ed., *Scale-Up of Dryers* in *Drying Technology,* **12** (1, 2), 1–416 (1994).

25. THYGESON, J.R. JR., and E.D. GROSSMANN, *AIChE Journal,* **16,** 749–754 (1970).

26. MATCHETT, A.J., and M.S. SHEIKH, *Trans. Inst. Chem. Engrs.,* **68,** Part A, 139–148 (1990).

27. McCORMICK, P.Y., *Chem. Eng. Progress,* **58** (6), 57–61 (1962).

28. SCHOFIELD, F.R., and P.G. GLIKIN, *Trans. Inst. Chem. Engrs.,* **40,** 183–190 (1962).

29. LANGRISH, T.A.G., R.E. BAHU, and D. REAY, *Trans. Inst. Chem. Engrs.,* **69,** Part A, 417–424 (1991).

30. ERGUN, S., *Chem. Eng. Progr.,* **48,** (2), 89–94 (1952).

31. FOGLER, H.S., *Elements of Chemical Reaction Engineering,* 3rd ed., Prentice-Hall, Upper Saddle River, NJ (1999).

EXERCISES

Section 18.1 (Use of the Internet is encouraged for the exercises of this section)

18.1 The surface moisture of crystals of NaCl of 0.5-mm average particle size is to be removed in a continuous, direct-heat dryer without a significant change to the particle size. What types of dryers would be suitable? How high could the gas feed temperature be?

18.2 A batch dryer is to be selected to dry 100 kg/h of a toxic, temperature-sensitive material (maximum of 50°C) of an average particle size of 350 μm. What dryers are suitable?

18.3 A thin, milk-like liquid is to be dried to produce a fine powder. What types of continuous, direct-heat dryers would be suitable? The material should not be heated above 200°C.

18.4 The selection of a batch or continuous dryer is determined largely by the condition of the feed, the temperature-sensitivity of the dried material, and the form of the dried product. What types of batch and continuous dryers would be suitable for the following cases:

(a) A temperature-insensitive paste that must be maintained in slab form.

(b) A temperature-insensitive paste that can be extruded.

(c) A temperature-insensitive slurry.

(d) A thin liquid from which flakes are to be produced.

(e) Pieces of lumber.

(f) Pieces of pottery.

(g) Temperature-insensitive inorganic crystals where particle size is to be maintained and only surface moisture is to be removed.

(h) Orange juice to produce a powder.

18.5 Solar drying has been used for centuries to dry and, thus, preserve fish, fruit, meat, plants, seeds, and wood. What are the advantages and disadvantages of this type of drying? What other types of dryers can be used to dry such materials? What type of dryer would you select to continuously dry beans?

18.6 Fluidized-bed dryers are used to dry a variety of vegetables, including potato granules, peas, diced carrots, and onion flakes. What are the advantages of this type of dryer for these types of materials?

18.7 Powdered milk can be produced from liquid milk in a three-stage process: (1) vacuum evaporation in a falling-film evaporator to a high-viscosity liquid of less than 50 wt% water; (2) spray drying to 7 wt% moisture; and (3) fluidized–bed drying to 3.6 wt% moisture. Give reasons why this three-stage process is preferable to a single-stage process involving just spray drying.

18.8 Deterioration must be strictly avoided when drying pharmaceutical products. Furthermore, such products are often produced from a nonaqueous solvent such as ethanol, methanol, acetone, etc. Explain why a closed-cycle spray dryer using nitrogen is frequently a good choice of dryer.

18.9 Paper is made from a suspension of fibers in water. The process begins by draining the fibers to a water-to-fiber ratio of 6:1, followed by pressing to a 2:1 ratio. What type of dryer could then be used to dry a continuous sheet to an equilibrium-moisture content of 8 wt% (dry basis)?

18.10 Green wood contains from 40 to 110 wt% moisture (dry basis) and must be dried before use to just under its equilibrium-moisture content when in the final environment. This moisture content is usually in the range from 6 to 15 wt% (dry basis). Why is it important to dry the wood, and what is the best way to do it so as to avoid distortion, cracks, splits, and checks?

18.11   Wet coal is usually dried to a moisture content of less than 20 wt% (dry basis) before being transported, briquetted, coked, gasified, carbonized, or burned. What types of direct-heat dryers are suitable for drying coal? Can a spouted-bed dryer be used? If air is used as the heating medium, is there a fire and explosion hazard? Could superheated steam be used as the heating medium?

18.12   Drying is widely used to remove solvents from coated webs, which include coated paper and cardboard, coated plastic films and tapes (e.g., photographic films and magnetic tapes), and coated metallic sheets. The coatings may be water-based or other solvent-based. Solid coatings are also used. Typical coatings are 0.1 mm in wet thickness. Much of the drying time is usually spent in the falling-rate period where the rate of drying decreases in an exponentially decaying fashion with time. What types of dryers can be used with coated webs? Are infrared dryers a possibility? Why?

18.13   A number of polymers, including polyvinylchloride, polystyrene, and polymethylmethacrylate are made by suspension or emulsion polymerization, in which the product of polymerization reaction is finely divided solvent- or water-wet beads. For large production rates, direct-heat dryers are commonly used with air, inert gas, or superheated steam as the heating medium. Why are rotary dryers, fluidized-bed dryers, and spouted-bed dryers popular choices for the drying operation?

18.14   What are the advantages and disadvantages of superheated steam compared to air as the heating medium? Why might superheated steam be superior to air for the drying of lumber?

Section 18.2

18.15.   A direct-heat dryer is to operate with air entering at 250°F and 1 atm with a wet-bulb temperature of 105°F. Determine from the psychrometric chart and/or relationships of Table 18.3 the following:

(a) Humidity.
(b) Molal humidity.
(c) Percentage humidity.
(d) Relative humidity.
(e) Saturation humidity.
(f) Humid volume.
(g) Humid heat.
(h) Enthalpy.
(i) Adiabatic-saturation temperature.
(j) Mole fraction of water in the air.

18.16   Air at 1 atm, 200°F, and a relative humidity of 15% enters a direct-heat dryer. Determine the following from the psychrometric chart and/or relationships of Table 18.3.

(a) Wet-bulb temperature.
(b) Adiabatic-saturation temperature.
(c) Humidity.
(d) Percentage humidity.
(e) Saturation humidity.
(f) Humid volume.
(g) Humid heat.
(h) Enthalpy.
(i) Partial pressure of water in the air.

18.17   Repeat Example 18.1 if the air is at 1.5 atm instead of 1.0 atm.

18.18   n-Hexane is being evaporated from a solid with nitrogen gas. At a point in the dryer where the gas is at 70°F and 1.1 atm, with a relative humidity for hexane of 25%, determine:

(a) Partial pressure of hexane at that point.
(b) Humidity of the nitrogen-hexane mixture.
(c) Percentage humidity of the nitrogen-hexane mixture.
(d) Mole fraction of hexane in the gas.

18.19   At a location in a dryer for evaporating toluene from a solid with air, the air is at 180°F, 1 atm, and a relative humidity of 15%. Determine the humidity, the adiabatic-saturation temperature, the wet-bulb temperature, and the psychrometric ratio.

18.20   Repeat Example 18.5 for water only, with air entering at 180°F and 1 atm, with a relative humidity of 15%, for an exit temperature of 120°F. In addition, plot temperature through the dryer.

18.21   Air enters a dryer at 1,000°F with a humidity of 0.01 kg H_2O/kg dry air. Determine the wet-bulb temperature if the air pressure is

(a) 1 atm.
(b) 0.8 atm.
(c) 1.2 atm.

18.22   Paper is being dried with recirculating air in a two-stage drying system operating at 1 atm. The air enters the first dryer at 180°F, where the air is adiabatically saturated with moisture. The air is then reheated in a heat exchanger to 174°F before entering the second dryer, where the air is adiabatically humidified to 80% relative humidity. The air is then cooled to 60°F in a second heat exchanger, causing some of the moisture to be condensed. This is followed by a third heater to heat the air to 180°F before it returns to the first dryer.

(a) Draw a process-flow diagram of the system and enter all of the given data.
(b) Determine the lb H_2O evaporated in each dryer per lb of dry air being circulated.
(c) Determine the lb H_2O condensed in the second heat exchanger per lb of dry air circulated.

18.23   Before being recirculated to a dryer, air at 96°F, 1 atm, and 70% relative humidity is to be dehumidified to 10% relative humidity. Cooling water is available at 50°F. Determine a method for carrying out the dehumidification, draw a labeled flow diagram of your process, and calculate the cooling-water requirement in lb H_2O per lb of dry air being circulated.

Section 18.3

18.24   Nitrocellulose fibers with an initial total water content of 40 wt% (dry basis) are dried in trays in a tunnel dryer operating at 1 atm. If the fibers are brought to equilibrium with air at 25°C and a relative humidity of 30%, determine the kg of moisture evaporated per kg of bone-dry fibers. The equilibrium-moisture content of the fibers is given in Figure 18.24.

18.25   Wet lumber of the type in Figure 18.24 is slowly dried from an initial total moisture content of 50 wt% to a moisture content in equilibrium with atmospheric air at 25°C and 40% relative humidity. Determine:

(a) The unbound moisture in the wet lumber before drying in lb water/lb bone-dry lumber.

(b) The bound moisture in the wet lumber before drying in lb water/lb bone-dry lumber.

(c) The free moisture in the wet lumber before drying, referred to as the final dried lumber, in lb water/lb bone-dry lumber.

(d) The lb of moisture evaporated per lb of bone-dry wood.

18.26 Fifty pounds of cotton cloth containing 20% total moisture content (dry basis) are hung in a closed room containing 4,000 ft^3 of air at 1 atm. Initially, the air is at 100°F at a wet-bulb temperature of 69°F. If the air is kept at 100°F, without admitting new air or venting the air, and the air is brought to equilibrium with the cotton cloth, determine the moisture content of the cotton cloth and the relative humidity of the air. Assume that the equilibrium-moisture content for cotton cloth at 100°F is the same as that of glue at 25°C as shown in Figure 18.24. Neglect the effect of the increase in air pressure, but calculate the final air pressure.

Section 18.4

18.27 Raw cotton having an initial total moisture content of 95% (dry basis) and a dry density 43.7 lb/ft^3 is to be dried batchwise to a final moisture content of 10% (dry basis) in a cross-circulation tray dryer. The trays, which are insulated on the bottom, are each 3 cm high with an area of 1.5 m^2 and are completely filled. The heating medium, which is air at 160°F and 1 atm with a relative humidity of 30%, flows across the top surface of the tray at a mass velocity of 500 lb/h-ft^2. Equilibrium-moisture content isotherms for the cotton are given in Figure 18.25. Experiments have shown that under the given conditions, the critical-moisture content will be 0.4 lb water/lb bone-dry cotton and that the falling-rate drying period will be like that of Figure 18.31a, based on the free-moisture content. Determine:

(a) The amount of raw cotton in pounds (wet basis) that can be dried in one batch if the dryer contains 16 trays.

(b) The drying time for the constant-rate period.

(c) The drying time for the falling-rate period.

(d) The total drying time if the preheat period is 1 h.

18.28 Slabs of filter cake with a bone-dry density of 1,600 kg/m^3 are to be dried from an initial free-moisture content of 110% (dry basis) to a final free-moisture content of 5% (dry basis) batchwise in trays that are 1 m long by 0.5 m wide with a depth of 3 cm. Drying will take place only from the top surface. The drying air conditions are 1 atm, 160°C, and a 60°C wet-bulb temperature. The air velocity across the trays is 3.5 m/s. Experiments under these drying conditions show a critical free-moisture content of 70% (dry basis), with a falling-rate period like that of Figure 18.31a, based on free-moisture content. Determine:

(a) The drying time for the constant-rate period.

(b) The drying time for the falling-rate period.

18.29 The filter cake of Exercise 18.28 is extruded into cylindrical-shaped pieces measuring 1/4 in. in diameter by 3/8 in. long that are placed in trays that are 6 cm high × 1 m long × 0.5 m wide and through which the air passes. The external porosity is 50%. If the superficial velocity of the air, which is at the same conditions as in Exercise 18.28, is 1.75 m/s, determine:

(a) The drying time for the constant-rate period.

(b) The drying time for the falling-rate period.

18.30 It takes 5 h to dry a wet solid, contained in a tray, from 36 to 8% moisture content, using air at constant conditions. Additional experiments give critical- and equilibrium-moisture contents of

15% and 5%, respectively. If the length of the preheat period is negligible and the falling-rate period is like that of Figure 18.31a, determine, for the same conditions, the drying time if the initial moisture content is 40% and a final moisture content of 7% is desired. All moisture contents are on the dry basis.

18.31 A tunnel dryer is to be designed to dry, by crossflow with air, a wet solid that will be placed in trays measuring 1.5 m long × 1.2 m wide × 25 cm deep. Drying will be from both sides. The initial total moisture content is 116% (dry basis) and the desired final average total moisture content is 10% (dry basis). The air conditions are 90°F and 1 atm with a relative humidity of 15%. The following laboratory data were obtained under the same conditions:

Equilibrium Moisture Content

% relative humidity	10	20	30	40	50	60	70	90
Moisture content, % (dry basis)	3.0	3.2	4.1	4.8	5.4	6.1	7.2	10.7

Drying Test		Drying Test	
Time, min	Moisture content, % (dry basis)	Time, min	Moisture content, % (dry basis)
0	116	362	31.4
36	106	415	28.6
125	81	465	24.8
194	61.8	506	22.8
211	57.4	601	15.4
242	49.6	635	13.5
277	42.8	785	11.4
313	37.1	822	10.2

Determine the total drying time to dry the same material from 110% to 10% moisture content if the air conditions are changed to 125°F and 20% relative humidity. Assume that the critical moisture content will not change and that the drying rate is proportional to the difference between the dry-bulb and wet-bulb temperatures of the air. All moisture contents are on the dry basis.

18.32 A piece of hemlock wood measuring 15.15 × 14.8 × 0.75 cm is to be dried from the two large faces from an initial total moisture content of 90% to a final average total moisture content of 10%, with drying taking place in the falling-rate period with liquid-diffusion controlling. The moisture diffusivity has been experimentally determined as 1.7×10^{-6} cm^2/s. Estimate the drying time if bone-dry air is used. All moisture contents are on a dry basis.

18.33 Gilliland and Sherwood [20] obtained data for the drying of a water-wet piece of Hemlock wood measuring 15.15 × 14.8 × 0.75 cm, where only the two largest faces were exposed to the drying air, which was at a temperature of 25°C and passed over the faces at 3.7 m/s. The wet-bulb temperature of the air and the pressure may be assumed to be 17°C and 1 atm, respectively. The data below were obtained for the average moisture content (dry basis) of the wood as a function of time. From these data, determine whether Case 1 or Case 2 for the diffusion of moisture in solids applies. If Case 1 applies, determine the effective diffusivity. If Case 2 applies, determine:

(a) The drying rate in g/h-cm^2 for the constant-rate period, assuming a wood density of 0.5 g/cm^3 (dry basis) and no shrinkage upon drying.

(b) The critical moisture content.

(c) The predicted parabolic moisture content profile at the beginning of the falling-rate period.

(d) The effective diffusivity during the falling-rate period.

In addition, for either case, describe what else could be determined from the data and explain how it could be determined.

Time, h	Avg. Moisture Content, % (dry basis)	Time, h	Avg. Moisture Content, % (dry basis)
0	127	9	41.8
1	112	10	38.5
2	96.8	12	30.8
3	83.5	14	26.4
4	73.6	16	20.9
5	64.9	18	16.5
6	57.2	20	14.3
7	51.7	22	12.1
8	46.1	∞	6.6

18.34 When Case 1 of liquid diffusion is controlling during the falling-rate period, the time for drying can be determined from (3) under Example 18.13. Using that equation, derive an equation for the rate of drying to show that it varies inversely with the square of the thickness of the solid. If capillary movement controls the falling-rate period, an equation for the rate of drying can be derived by assuming the laminar flow of moisture takes place from the interior of the solid to the surface such that the rate of drying varies linearly with the average free-moisture content. If so, derive equations for the rate of drying and the time for drying in the falling-rate period to show that the rate of drying is inversely proportional to just the thickness of the solid. Outline an experimental procedure that could be used to determine whether diffusion or capillary flow governed in a given material.

18.35 In a cross-circulation tray dryer, the equations for the constant-rate period neglect radiation and assume that the bottoms of the trays are insulated so that heat transfer takes place only by convection from the gas to the surface of the solid where evaporation takes place. Under these conditions, evaporation occurs at the wet-bulb temperature of the gas when the moisture is water. In actual tray dryers, the bottoms of the trays are not insulated and heat transfer to the evaporating surface can also take place by convection from the gas to the tray bottom and thence by conduction through the tray and then through the wet solid. Derive an equation similar to (18-34) for the case where heat transfer by convection and conduction from the bottom side is taken into account. However, the conduction resistance of the tray bottom can be neglected. Show by combining your equation with the mass-transfer equation (18-35) that evaporation will now take place at a temperature higher than the wet-bulb temperature of the gas. What effect would heat transfer by radiation from the bottom surface of a tray to the tray below have on the temperature of evaporation?

Section 18.5

18.36 A tunnel dryer is to be used to dry 30 lb/h of raw cotton (dry basis) with a countercurrent flow of 1,800 lb/h of air (dry basis). The cotton enters at 70°F with a moisture content of 100% (dry basis) and exits at 150°F with 10% moisture (dry basis). The

air enters at 200°F and 1 atm with a relative humidity of 10%. The specific heat of dry cotton can be taken constant at 0.35 Btu/lb-°F. Calculate:

(a) The rate of evaporation of moisture.

(b) The outlet temperature of the air.

(c) The rate of heat transfer.

18.37 A 25 wt% solution of coffee in water at 70°F is spray dried to a final moisture content of 5% (dry basis) with air that enters at 450°F and 1 atm with a humidity of 0.01 lb/lb (dry basis) and exits at 200°F. Assuming a reasonable value for the specific heat of coffee, calculate:

(a) The air rate in lb dry air/lb coffee solution.

(b) The temperature of evaporation.

(c) The heat-transfer rate in Btu/lb coffee solution.

18.38 7,000 lb/h of wet, pulverized, clay particles with 27% moisture (dry basis) at 15°C and 1 atm enter a flash dryer where they are dried to a moisture content of 5% (dry basis) with a cocurrent flow of air that enters at 525°C. The dried solids exit at the air wet-bulb temperature of 50°C, while the air exits at 75°C. Assuming a reasonable value for the specific heat of clay, calculate:

(a) The flow rate of air in lb/h (dry basis).

(b) The rate of evaporation of moisture.

(c) The heat-transfer rate in Btu/h.

18.39 5,000 lb/h of wet isophthalic acid crystals with 30 wt% moisture (wet basis) at 30°C and 1 atm enter an indirect-heat, steam-tube rotary dryer, where they are dried to a moisture content of 2 wt% (wet basis) by 25 psig steam (14 psia barometer) condensing inside the tubes. Evaporation takes place at 100°C, which is also the exit temperature of the crystals. The specific heat of isophthalic acid can be taken as 0.2 cal/g-°C. Determine:

(a) The rate of evaporation of moisture.

(b) The rate of heat transfer.

(c) The quantity of steam required in lb/h.

18.40 The extruded filter cake of Examples 18.8 and 18.18 is to be dried under the same conditions as in Example 18.18 except that three drying zones 8 ft long each will be used, with flow upward in the first and third zones and downward in the second zone. Predict the moisture-content distribution with height at the end of each zone and the final average moisture content.

18.41 Repeat the calculations of Example 18.18 if the extrusions are 3/8 in. in diameter × 1/2 in. long. Compare your results with those of Example 18.18 and comment.

18.42 A direct-heat, countercurrent-flow, rotary dryer with a 6-ft diameter and 60-ft length is available to dry titanium dioxide particles at 70°F and 1 atm with a moisture content of 30% (dry basis) to a moisture content of 2% (dry basis). Hot air is available at 400°F with a humidity of 0.015 lb/lb dry air. Experiments show that an air-mass velocity of 500 lb/h-ft^2 will not cause serious dusting. The specific heat of solid titanium dioxide is 0.165 Btu/lb-°F, and its true density is 240 lb/ft^3. Determine:

(a) A reasonable production rate in lb/h of dry titanium dioxide (dry basis).

(b) The heat-transfer rate in Btu/h.

(c) A reasonable air rate in lb/h (dry basis).

(d) Reasonable values for the exit air and exit solids temperatures.

18.43 A fluidized-bed dryer is to be sized to dry 5,000 kg/h (dry basis) of spherical polymer beads that are closely sized to 1 mm in diameter. The beads will enter the dryer at 25°C with a moisture content of 80% (dry basis). The drying medium will be superheated steam, which will enter the bed at 250°C. The pressure in the vapor space above the bed will be 1 atm. A fluidization velocity of twice the minimum will be used to obtain bubbling fluidization. The bed temperature, exit solids temperature, and exit vapor temperature can all be assumed to be 100°C. The beads are to be dried to a moisture content of 10% (dry basis). The void fraction of the bed before fluidization is 0.47. The specific heat of the dry polymer is 1.15 kJ/kg-K, while the density is 1,500 kg/m^3. Batch fluidization experiments show that drying will all take place in the falling-rate period, as governed by diffusion according to (18-92), where 50% of the moisture is evaporated in 150 s. Bed expansion is expected to be about 20%. Determine the dryer diameter, average bead residence time, and expanded bed height. Is the dryer size reasonable? If not, what changes in operation could be made to make the size reasonable? In addition, calculate the entering superheated-steam flow rate and the necessary heat-transfer rate.

Index

Physical Constants

Universal (ideal) gas law constant, R
  1.987 cal/mol $\cdot$ K or Btu/lbmol $\cdot$ °F
  8315 J/kmol $\cdot$ K or Pa $\cdot$ m³/kmol $\cdot$ K
  8.315 kPa $\cdot$ m³/kmol $\cdot$ K
  0.08315 bar $\cdot$ L/mol $\cdot$ K
  82.06 atm $\cdot$ cm³/mol $\cdot$ K
  0.08206 atm $\cdot$ L/mol $\cdot$ K
  0.7302 atm $\cdot$ ft³/lbmol $\cdot$ °R
  10.73 psia $\cdot$ ft³/lbmol $\cdot$ °R
  1544 ft $\cdot$ lbf/lbmol $\cdot$ °R
  62.36 mmHg $\cdot$ L/mol $\cdot$ K
  21.9 in. Hg $\cdot$ ft³/lbmol $\cdot$ °R

Atmospheric pressure (sea level)
  101.3 kPa = 101,300 Pa = 1.013 bar
  760 torr = 29.92 in. Hg
  1 atm = 14.696 psia

Avogadro's number
  6.022×10^{23} molecules/mol

Boltzmann constant
  1.381×10^{-23} J/K $\cdot$ molecule

Faraday's constant
  96490 charge/g-equivalent

Gravitational acceleration (sea level)
  9.807 m/s² = 32.174 ft/s²

Joule's constant (mechanical equivalent of heat)
  4.184 J/cal
  778.2 ft $\cdot$ lb$_f$/Btu

Planck's constant
  6.626×10^{-34} J $\cdot$ s/molecule

Speed of light in vacuum
  2.998×10^{8} m/s

Stefan-Boltzmann constant
  5.671×10^{-8} W/m² $\cdot$ K⁴
  0.1712×10^{-8} Btu/h $\cdot$ ft² $\cdot$ °R⁴